PRENTICE HALL MATHEMATICS

PRENTICE HALL THE FINE ART OF TEACHING

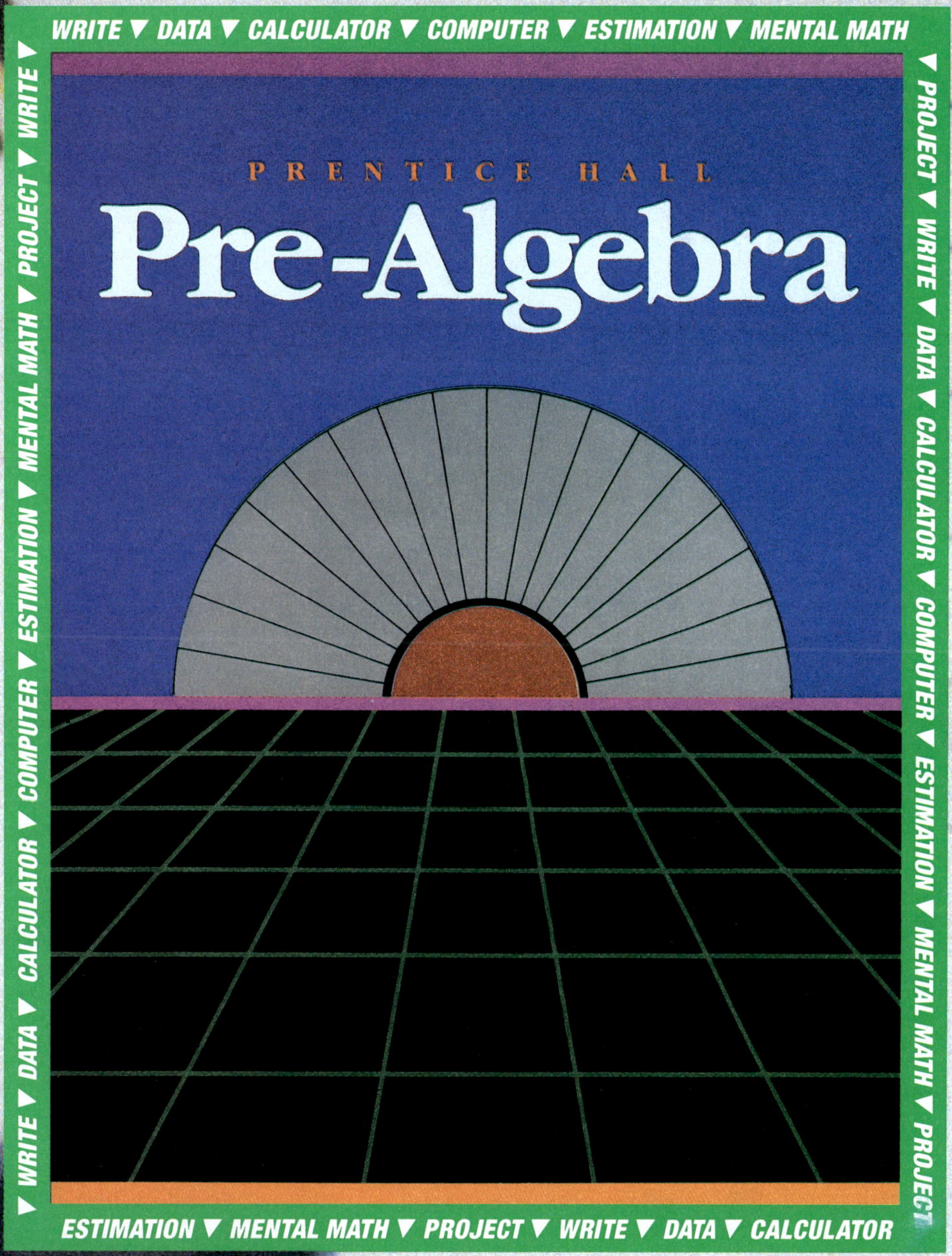

PRENTICE HALL

THE TOOLS TO PROVIDE THE

The most complete

Teacher's Edition

- Chapter Overviews (preceding each chapter)
- Professional Section "Putting Ideas to Work"
- Focus - Instruction - Guided Practice - Closure Lesson Plan (alongside student text pages)
- Answers in place on reduced student pages

Teacher's Resource Book

- Practice (one page per lesson)
- Enrichment (one page per lesson)
- Exploring (alternative activities for each chapter)
- Tests
- Overhead Manipulatives Kit
- Transparencies

MENTAL MATH ▼ PROJECT ▼ WRITE ▼ D

MEANS AND THE MOTIVATION

teaching package available.

▼ Uniquely developed to help you implement the Professional Standards for Teaching Mathematics.

▼ Complete with in-text correlations to the Curriculum and Evaluation Standards.

The complete teaching package also includes. . .

- Computer Test Bank
- Teaching Transparencies
- Prentice Hall Mathematics File Box

RITE ▼ DATA ▼ CALCULATOR ▼ COMPUTER ▼

With

PRENTICE HALL PRE-ALGEBRA © 1992

CONTINUE TO. . .

- build student **confidence** in mathematics
- create an **active** learning environment
- expand the **student's role**
- empower the **teacher** to communicate
- foster the **analysis** of teaching and assessment

PRE-ALGEBRA

TABLE OF CONTENTS

Chapter 1	Integers and Expressions
Chapter 2	Solving Equations
Chapter 3	Decimals and Equations
Chapter 4	Number Theory
Chapter 5	Rational Numbers and Expressions
Chapter 6	Ratios, Proportions, and Percent
Chapter 7	Equations and Inequalities
Chapter 8	Graphing in the Coordinate Plane
Chapter 9	Algebra in Geometry and Measurement
Chapter 10	Area and Volume Formulas
Chapter 11	Right Triangles in Algebra
Chapter 12	Statistics and Probability
Chapter 13	Polynomials

Prentice Hall Mathematics
We open the door to the future.

For more information, please write to:
Prentice Hall
A School Division of Simon & Schuster
4350 Equity Drive
Columbus, OH 43216-2649
1-800-848-9500

A Paramount Communications Company

CALCULATOR ▼ COMPUTER ▼ ESTIMATION ▼ MENTAL MATH ▼ PROJECT ▼ WRITE ▼ DATA ▼ CALCULATOR ▼ COMPUTER ▼ ESTIMATION ▼ PROJEC

TEXAS TEACHER'S EDITION

PRENTICE HALL

Pre-Algebra

With Lesson Plans to Support
Texas Teacher Appraisal System

DAVID M. DAVISON
MARSHA LANDAU
LEAH McCRACKEN
LINDA THOMPSON

Prentice Hall
Needham, Massachusetts
Englewood Cliffs, New Jersey

AUTHORS

David M. Davison, *Eastern Montana College, Billings, Montana*

Marsha S. Landau, *National-Louis University, Evanston, Illinois*

Leah McCracken, *Lockwood Junior High School, Billings, Montana*

Linda Thompson, *Warrenton, Oregon*

Staff Credits

PUBLISHER	Stephanie Rogalin
EDITORIAL DIRECTOR	Mimi Jigarjian
MARKETING DIRECTOR	Bridget A. Hadley
PROJECT EDITOR	Edward De Leon
EDITORS	Judith D. Buice Renee S. Heffernan Jo Ann M. Webber
BUYER	Roger Powers
MANUFACTURING	Bill Wood
ART DIRECTOR AND COVER ART DIRECTION	L. Christopher Valente
BOOK DESIGN/PRODUCTION COORDINATOR	Jonathan Pollard
PHOTO RESEARCHER	Russell Lappa/Karen McDonough, Assistant

Outside Credits

NAVA EDIT, INC., MONTCLAIR, N.J.
PUBLICOM, INC., LEXINGTON, MA.
DESIGN CONTINUUM, BOSTON, MA.

The section *Putting Ideas to Work* was produced entirely on a Macintosh® computer using desktop publishing software.

ISBN 0-13-721903-2
Printed in the United States of America.
4 5 95 94

Contents

PUTTING IDEAS TO WORK		T9
TEXAS CORRELATION		T29
CHAPTER 1	OVERVIEW	2A
CHAPTER 2	OVERVIEW	52A
CHAPTER 3	OVERVIEW	96A
CHAPTER 4	OVERVIEW	138A
CHAPTER 5	OVERVIEW	180A
CHAPTER 6	OVERVIEW	230A
CHAPTER 7	OVERVIEW	272A
CHAPTER 8	OVERVIEW	312A
CHAPTER 9	OVERVIEW	360A
CHAPTER 10	OVERVIEW	404A
CHAPTER 11	OVERVIEW	450A
CHAPTER 12	OVERVIEW	486A
CHAPTER 13	OVERVIEW	534A
INDEX		614

PUTTING ideas TO WORK

THE DYNAMIC ART OF TEACHING MATHEMATICS

A • SPECIAL • REPORT • FROM • PRENTICE • HALL

GLOBAL DATA

Mathematics is "an activity inherent to the human being's existence, spontaneously practiced and determined by the socio-cultural context and the material reality in which the individual lives."

Brazilian Ubiratan D'Ambrosio

Math is changing. Since the National Council of Teachers of Mathematics published its *Curriculum and Evaluation Standards for School Mathematics* in 1989, the art of teaching secondary mathematics has become especially dynamic. Everywhere, teachers are looking for practical ways to put new ideas to work. They're exploring new approaches and planning new activities. As teachers work, they gain fresh insights.

To pass along some of these insights, and to share research-based suggestions, Prentice Hall has created this special report. As you put ideas to work in your own classroom, you'll develop your own insights. We hope you'll share them with us. Please write to the Prentice Hall Mathematics Department, Prentice Hall, 160 Gould St., Needham Heights, MA 02194-2310.

Contents

Problem solving in the real world
It requires a sense of adventure as much as a set of strategies

Handy tools
Manipulatives keep kids in touch with math concepts

Sometimes it takes teamwork
Cooperative learning's secrets revealed

Why write?
Teachers tell how something so simple can make a real difference

United we stand
What to do when students are diverse, distracted, or disheartened

Whole assessment catalog
Fresh approaches tell you what you really want to know

With a little help from your friends . . .
Working with colleagues, families, businesses, and local groups

Editor: Mary Harbaugh. Design by Design Continuum, Boston. Photography by Richard Haynes. Published by Prentice-Hall, 160 Gould Street, Needham Heights, MA 02194-2310.

PROBLEM SOLVING in the real world

Developing kids' aptitude for solving unfamiliar problems is among education's most important goals for this decade and beyond. We've learned it's not a tidy process that can be neatly delineated and handed over to students. Instead, it's gloriously messy. It requires a sense of adventure as much as a set of strategies. Teaching it requires both our creativity and our good sense.

GLOBAL DATA

Unschooled children of the Ivory Coast's Dioula people perform mathematics as well as their schooled counterparts.

H.P. Ginsberg, J.K. Posner, R.L. Russell

Here's a true story: A teacher gave his students several questions that led them to explore actual census data in their search for solutions. After kids had worked with the data for a few days, the teacher asked them to predict the population in 20 years. Students came up with predictions using a number of strategies, considering a number of variables—and then wanted to know which one of their answers was *correct.* The teacher was chagrined to realize that the traditional mindset—*any question asked in math class must have a single correct answer*—had overpowered their common sense.

"Kids come into classrooms with a lot of beliefs that are counterproductive," acknowledges Marsha Landau of National-Louis University in Evanston, Illinois. "Many kids don't have a very sophisticated notion of problem solving. They think it is rule-bound, key-word, one-step, one-answer, one-way-to-get-the-answer, and you should get the answer instantly.

"Teachers need to deal with those notions directly, to have a discussion and get them out in the open, to be explicit about what problem solving is," Landau says. "Even kids who have been successful in math in a traditional way—kids who are good symbol-pushers—can find it scary when confronted by a situation without rules. Students need to understand that problem solving is what you do when you don't know what to do."

The Right Stuff

"Becoming a good mathematical problem solver," observes University of Pittsburgh researcher Lauren Resnick, "may be as much a matter of acquiring the habits and dispositions of interpretation and sense-making as acquiring any particular set of skills, strategies, or knowledge."

How can I make sense of this? What's going on here? These are key questions for students approaching an unfamiliar problem. How can you lead students to approach a problem with an open mind and a can-do attitude? Here's one suggestion: Pose problems in a meaningful context.

It sounds obvious now, but for years, many experts underrated the importance of context in attempts to help students develop into able mathematical problem solvers. The goal then was to identify some pure linear process, but what happened was that many kids couldn't make any sense out of the process without some context clues, nor did they have any authentic reason to try. When they did try to apply that pure linear process to messy-but-motivating real-world problems, it didn't always fit. Now we know that "context plays an active role in supporting the reasoning toward the solution," according to Claude Janvier of the University of Quebec.

It's also clear now that attitude and context are interrelated. We need to motivate students to pursue questions and persist in trying to find solutions. One way to do that is to offer questions that someone would really ask, questions with answers that matter, questions in contexts that kids can envision. By bringing real-world situations into instruction, we can tap students' sense-making contextual knowledge and their motivation to wrestle with nonroutine problems.

Real-World Data

A good textbook can provide interesting problems with real data for students to explore. But to make the most of the power of real-world contexts, and to bridge the gap between in-class and outside-class experience, it's worth looking for additional examples that touch kids' lives.

TECHNOLOGY

Do you still have any doubts about teaching with calculators?

Listen to James Fey of the University of Maryland at College Park: "A recent meta-analysis of over 70 research studies concluded that wise use of calculators can enhance student conceptual understanding, problem solving, and attitudes toward mathematics without apparent harm to acquisition of traditional skills."

Fey points out that students using calculators and computers still need to

"Sometimes teachers are worried that real data can be too complex to work with," says Linda L. Thompson, an Oregon education writer, consultant, and former teacher. "But with calculators, there's no reason to avoid real data. It's fun for the kids, because they can see how real it is. When you ask students to come up with some problems and gather data themselves, they're really motivated to find answers. I think the problems are more interesting for teachers too, who can say, 'Let's see what we get.' "

Estimation and Mental Math

"If you can use estimation and mental math in working with real data, it gives you another tool," Thompson points out. "I really like to encourage kids to use estimation whenever they solve a problem, as a way of seeing how sensible an answer is.

figure out what they want to know and to interpret what they find out.

"Calculators and computers can help kids see a problem in its totality," observes Montana teacher Leah McCracken. "Real-world data isn't neat and tidy, but you can grind it out using a calculator. The use of technology allows kids to conjecture—'this is what I think is going to happen'—and it gives them a quick sense of whether their idea is reasonable," she adds.

Calculators and computers help in forming generalizations about number patterns and functions. They can save time. They support use of real-world data. And they can motivate students to find solutions.

I remind them that they use estimation in everyday situations—buying groceries, shopping at the record store. I remind them that they have a way of thinking: How much am I spending? Is the cashier charging me the right amount? They already know that having a good sense of what to expect with numbers can save you from being cheated. I try to bring that thinking into the classroom.

"One-right-answer is a hard mentality to get around," admits Thompson. "We need to talk about problems that might not have one correct answer, or an estimate that is an adequate answer, or where differing estimates are OK."

Exploring Strategies

Kids need opportunities to scrutinize and try out each problem-solving strategy in realistic contexts. Researchers tell us that specific strategies should be taught with an exemplar problem, with think-aloud modeling by the teacher. After students apply the strategy to different kinds of problems, we need to help them learn whether to use it with an unfamiliar problem. This kind of learning takes time, because in searching for a solution, kids may try out several strategies. They may also want to represent the problem several ways, such as with manipulatives or in a graph. This exploring is just what you want students to do.

"If at first a kid's solution doesn't look like what you were expecting, take another look," advises Landau. "It may be that the student has taken another path. We want to encourage divergent thinking. We need to explicitly emphasize different ways to solve problems. We need to give kids more of the richness of what problem solving is all about, more of the sense of adventure. Problem solving should be a way of life in a classroom. It should be an attitude, a *modus operandi.*"

GLOBAL DATA

In many other countries, rather than having a separate course in algebra and a separate course in geometry, they teach a little of each all along. The student has an algebra strand each semester, a geometry strand each semester.

Graham Porter

Handy Tools

Manipulatives keep kids in touch with math concepts

"I'm trying for a visual approach to math concepts. I give the kids tiles to use for factoring, decimals, fractions, and percentages. It seems to work. Sometimes the kids think they're too old for it, so I start with something that's not too trivial, something that gives them a challenge. You have to approach it so the kids figure: 'This is the way this math class is going to be.'. . . I stayed away from geoboards for the longest time because I thought they'd shoot the rubber bands. I said, 'Are you crazy?!' when I heard about geoboards. But the kids don't shoot the rubberbands. You just tell them they can do the problem with the geoboard or they can do it on paper. They choose the geoboard. . . .I'm not always successful, but it's fun to try."

– Teacher Anne Marie Suk, Nampa, ID

GLOBAL DATA

In Australia, math is much less textbook-dominated. Textbooks are a source of problems or examples. Teachers select the ones that will do the job they want.

David Davison

Kids take a concept in hand, with the help of a geoboard. By using colored rubber bands on a pegboard-like surface, students create and transform figures and explore linear and spatial relationships.

Pattern blocks help students see fractional relationships

Students use algebra tiles to combine like terms (below) and a Mira to investigate symmetry, congruence and transformations (below right).

Sometimes it takes TEAMWORK

COOPERATIVE LEARNING HAS PROVEN POWER TO MOTIVATE KIDS AND STIMULATE LEARNING—IF YOU KNOW ITS SECRETS. HOW CAN YOU GET THE MOST OUT OF A COOPERATIVE APPROACH? TO FIND OUT, PRENTICE HALL SPOKE WITH ROBERT SLAVIN OF JOHNS HOPKINS.

RESEARCH ON THE BENEFITS OF COOPERATIVE LEARNING IS COMPELLING. TEACHERS ARE EXCITED BY ITS POTENTIAL, AND MANY USE IT SUCCESSFULLY. BUT THE PHRASE MEANS DIFFERENT THINGS TO DIFFERENT PEOPLE. IS THERE A RISK THAT SOME COOPERATIVE EFFORTS AREN'T SUPPORTED BY RESEARCH?

Some approaches people are using may not produce the benefits they expect. There are two things I'm hearing about. Some people are having kids work together without any group purpose. That undermines the whole idea, which is to build psychological support for learning through team dynamics. There has to be some kind of group goal or group reward, which motivates kids to care about helping one another learn.

The other thing I'm hearing is that students are working together without individual accountability. That allows one student to solve all the problems, while others don't learn much. That's not fair to anyone. Group success must depend on individual learning of every student in the group.

SO JUST ASKING STUDENTS TO WORK IN SMALL GROUPS WON'T NECESSARILY GET YOU ANYWHERE?

Simply telling people to work together and solve problems is not likely to lead to behavior that improves achievement. There's just no evidence that those simpler forms of group work are going to pay off. It won't hurt—my guess is that it would be a lot more fun than working in rows—but there's no evidence for it.

YOU MENTIONED THAT GROUP SUCCESS SHOULD DEPEND ON INDIVIDUAL LEARNING. HOW DOES THAT WORK?

Group success should depend on all members of the group. A way to make sure it does is to total individual assessments to create team scores. We [at Johns Hopkins] use a team scoring method based on improving over your own former achievement, so improving low-scorers still help their teams. Grades should be tied to individual assessments, not team scores. Teams are rewarded with other forms of recognition.

HOW OFTEN SHOULD YOU USE COOPERATIVE LEARNING?

To get a benefit, you need to use it frequently. It's part of a cycle of activities: The teacher explores a

Robert Slavin is principal researcher and a director of the Johns Hopkins University Center for Social Organization of Schools. The Johns Hopkins Team Learning Project has been recognized in effective schools research as significantly improving student achievement.

concept with the whole class, then cooperative groups meet to reconsider the concept and solve problems, then some individual assessment is made. You can look at it as replacing individual seat work.

PEOPLE FAMILIAR WITH SOME OF YOUR MODELS MAY THINK COOPERATIVE LEARNING ISN'T SUITED TO PROBLEM SOLVING AND CONCEPTUAL UNDERSTANDING. WHAT ABOUT THAT?

In some places, people use cooperative learning just for learning basic algorithms; in other places, people use it just for complex problem solving. It's fine for both.

For complex problem solving, one thing that's essential is thinking out loud: It helps the thinker form thoughts, and it helps the students listening who don't have a clue. Cooperative learning is ideal for getting people to share perspectives. Each student has more opportunity to think aloud and to respond to another's thoughts. Cooperative learning groups can also encourage self-assessment. Kids ask one another, "Do you understand this?"

ARE THERE ANY DRAWBACKS?

The biggest source of opposition to cooperative learning is from people concerned about high achievers. Will they be learning themselves? The answer is, of course, because they're getting the same fresh material everyone else is getting, and they're not teaching—they're discussing and trying out ideas. By giving individual improvement scores, you retain the challenge.

GLOBAL DATA

From age 10, most German students are sorted into tracks, with the highest achievers and the upper class going on to the Gymnasium; middle-class kids going to the *Realschule*, and lower-class youth going to the *Hauptschule* until they join the work force at age 15 or 16. Germany also has a network of heterogeneous high schools, some experiment by keeping small groups of students with the same team of teachers for six years.

A. Ratzki, A. Fischer

Cooperative Learning Tips

Research "supports the importance not only of a cooperative task structure, but also of group rewards, of individual accountability, and probably of group competition as well," say researchers F. M. Newmann and J. A. Thompson. How do you do it? Here are tips.

❶ Create heterogeneous teams of four or five students. (Identify the six highest achievers and six lowest achievers in class; assign one of each to a group. Assign others to create a cross-section of gender, past performance levels, and ethnicity.) Keep groups long enough for kids to feel like a team—six to eight weeks. ❷ Take time to explain what you are doing and why. Emphasize that the primary goal is to learn a concept. Use groups for both one-day and multi-day discussions and problem solving. ❸ Give teams a get-acquainted activity, such as planning an imaginary party with limited resources. Allow time for kids to organize and pick team names. ❹ Explain when you will assist. Monitor by walking around and listening, but try not to intervene too quickly if kids encounter difficulties. The goal is for groups to work through ideas themselves. Ask how they chose strategies, what difficulties they encountered, how they decided if answers were reasonable. ❺ Individual accountability can take many forms, such as quizzes or written explanations. Base group scores on individual improvement over previous performance. Group scores accumulate toward reward levels. ❻ Establish a signal for quiet, but expect a more lively atmosphere. Your goal is to generate the kind of peer support, excitement, and camaraderie of sports teams. With that, you've got some of the most powerful motivating qualities known in social psychology.

WHY WRITE?

Can something so simple make a real difference?

Just ask these teachers—

"Writing in math is one of my professional goals this year. At first, kids feel that they ought not to have to write. But when they are pushed to write their conclusions, they learn the concept in a very different way. Lots of students can do mathematics as a set of algorithmic processes, but they haven't thought about generalizing from that. Having students put a concept into words helps them understand it so they can transfer it and apply it to other areas. I ask them to write what they've learned or what conclusions they can draw from a discovery activity. Sometimes they say, 'This is hard. I can do the problem, but I can't explain it.' They seem not to be able to draw on their own history of mathematics learning. Writing is one way to help them make that learning their own."

– Susan Eddins, Glen Ellyn, IL

'This is hard. I can do the problem, but I can't explain it.'

"I'm sort of a rookie with writing in math, but I try to have a question on each of my tests that requires writing, such as 'Explain the concept of . . .' With new approaches like writing, you just start out slow. You can't be afraid to try."

– Patricia Baltzley, Union Bridge, MD

"I require students to keep math journals. They see what they learned and what they didn't understand. I'm trying to get them to be more responsible for their own learning. I say, 'You are in a good position to tell me what you don't know. I don't want to wait till test time to learn what you don't understand.' I look at the journals at least every six weeks—usually more often. It seems like a lot of work, but it's worth it. Of course, at first students aren't used to it, but later they write that they appreciate the innovative ways I'm using to help them understand. It's sort of touching. The students say, 'This works.' They like it."

– Larry DeVan Williams, Eutaw, AL

"I'm trying to get the kids to spend the first few minutes of each class writing about math they've used at home or wherever. I also use an activity for kids who think they don't need to write. I'll ask the class to write down the concept or an explanation of the rule and pass their papers to a classmate. I put a problem on the board, and each student must use what's on the paper—no more and no less—to try to solve the problem. Kids find out that they don't

communicate their math ideas very well. I'll also sometimes take what a student has written and work it out on the board. The student will say, 'You know what I meant!' But I point out that they need to think through the idea so they can express it more clearly."

– *Margaret Faircloth, Macon, GA*

"I have students do a lot of reading, then come back and talk about it, and then write a short comment on what they learned. Or I'll get them to do an investigation in partners and write a paragraph about it together. Or I'll ask my cooperative groups to write a comment about a concept we've just learned, then one student from each group reads it before the whole class. I'm trying to get students to understand and think—and writing focuses that."

– *Ian DeGroot, North Vancouver, BC*

"I try to get students to write their own problems. Students today like things to be relevant, up-to-date. It's hard for teachers to come up with new problems all the time relating to issues kids care about: troops in Saudi Arabia or the latest music group, for example. But kids can create their own. And you know what? I get a big extension of the concept. I also find it meets the needs of both the kid who's not very strong in the concept and the one who is. In addition to writing the problem, they also present it to the class using the board or the overhead. I believe students have to verbalize mathematics. They have to explain, so they dig deeper."

– *Larry Thomason, Spencer, NC*

The students say 'This works.' They like it.

"I use writing for diagnostic purposes. Sometimes I say, 'I want you to write down the key concepts we discussed today.' Kids' writing is very revealing about what they do and don't understand. I also use writing to get at attitudes about math. Early in the year, I ask students to write a short mathematical autobiography. They write about what they liked and what they didn't, and why. It's been helpful, because you have to deal with all sorts of preconceived notions that kids bring with them. I also have the kids share their writing—such as an explanation of their solution to a problem—with someone outside the class. That person reads the problem and the explanation and signs a paper if it's understandable. If it's not clear, the student rewrites the explanation and tries again."

– *Leslie Paoletti, Wallingford, CT*

"A few times a week I'll give a starter sentence and they'll write a paragraph. It helps them internalize the ideas they've been learning over time and make connections."

– *Beverly Van Camp, Bethel, AK*

GLOBAL DATA

China and Taiwan each have national curricula, with a single math textbook series used for the whole country.

K. Fuson, J. Stigler, K. Bartsch

CRITICAL THINKING

How can you encourage your students to really think? One clue comes from cognitive research. Studies indicate that thinking, speaking, reading, listening, and writing are interrelated: One reinforces another as students construct knowledge. Each is also an opportunity for reflection.

Notes Harvard's David Perkins: "There's broad agreement that students need to become reflective about their subject-matter learning. They need to become strategic learners."

Reflection—prompted by thoughtful questions—can lead students to consider problem-solving patterns. The more aware students are that such patterns exist, the more they are likely to try them. By writing, speaking, listening, and reading about mathematics, students learn patterns from one another.

United We Stand

Trust. *Willingness to say you don't understand. Interest in the strategy of someone different. Inclination to tackle an unfamiliar problem. Respect for intellectual courage. Understanding that mistakes are natural. Good humor. Confidence that you can learn and succeed . . . These are attitudes we naturally want for all students, because they seem to support learning and make life in classrooms a pleasure.*

When your students are diverse, distracted, or disheartened, a supportive environment holds learning together.

Camaraderie in Portsmouth, New Hampshire

Now we're learning that they offer special promise for helping students at risk. These attitudes are at the heart of what we know about the social context of learning and the power of peer support. They also make a lot of sense in classrooms where critical thinking and problem solving are valued.

"In all my classes, we spend the first three weeks of the year building a classroom community, developing a sense of unity," says teacher Patricia Baltzley of Union Bridge, Maryland. "It's really important for the kids. They have to break through the social barriers before they can get to the math part."

Until now, there's been relatively little emphasis on building a social scaffold for learning. But in the face of new challenges and new understanding of why kids fail, math teachers are considering how to strengthen this kind of support.

Among the key issues in creating a sense of trust and community are the linguistic and cultural diversity of students.

Linguistic and Cultural Diversity

Researchers talk about the significance of cultural dissonance in understanding causes of low achievement. Cultural dissonance is what happens when the culture of a school or classroom—the ways things are done, the values implied, the behaviors considered appropriate, the language used—are out of harmony with a student's culture outside of school. It's a tear in the fabric of learning. Cultural dissonance can occur any time a student's life experiences have been markedly different from what he or she experiences in a classroom. One aspect of dissonance concerns the representation and context of mathematical ideas, another the flavor of classroom life.

"In math, it's not the ethnic or racial factor per se that's important, it's that some kids are not masters of white, middle-class, mainstream American English, the language of textbooks," explains David Davison of Eastern Montana College, a researcher on math learning among American Indians. "A textbook can't possibly be all things to all people. It's a good place to start, but the teacher is the critical link."

GLOBAL DATA

Female eighth graders in Thailand generally outperform their male counterparts in math.

Corinna Ethington

In culturally diverse classrooms, activities involving listening, writing, and speaking take on special importance in developing mathematical vocabulary and in supporting acquisition of English skills. Manipulatives and other visuals are doubly helpful in conveying concepts when language is not clear to the student.

Davison also points out that some students may find the mathematics curriculum lacking in relevance to their lives outside of school. Teachers can address this aspect of cultural dissonance in two important ways: first, by tapping local events and situations when posing mathematical problems and discussing applications, and second, by creating a new in-class relevance, a sense of community that provides its own motivation and meaning. Cooperative learning groups, for example, can lead students of different backgrounds to care about one another's learning.

Education policy analysts Michael Knapp, Brenda Turnbull, and Patrick Shields note that students are more likely to do well when teachers respect their ethnic and linguistic backgrounds and express that appreciation; when kids are both encouraged to draw on their own experiences and exposed to new experiences and ways of thinking; and when the assumptions, expectations, and ways of doing things in class are made explicit to students by teachers.

ELEVEN IDEAS for empowering students to learn

ONE
Consider the underlying messages your teaching methods convey. Do they convey your respect for students' dignity and your faith in their desire and ability to learn?

TWO
Let kids know explicitly that you have high expectations of them. Tell them you expect errors. Express your belief that they are capable of solving most problems you pose.

THREE
Explain what students will learn and why. Be specific about goals and criteria for success.

FOUR
Draw on students' experiences in posing problems or discussing applications.

FIVE
Emphasize ongoing learning and progress, rather than absolute mastery of a given concept at a given time.

SIX
Remind students that people learn more when they're willing to risk being wrong. Help students see errors as useful clues to refining strategies.

SEVEN
Tell students it's intelligent to ask questions if they're confused.

EIGHT
Examine why a student thinks he or she has failed. Point out relationships between the student's use of specific strategies and specific accomplishments, so the student can see how he or she contributed to the successful result.

NINE
Ask students to explain and justify their strategies. This not only helps kids clarify their thinking for themselves and for classmates, it also makes them aware that they are making choices.

TEN
Look for opportunities to strengthen attitudes that generate self-motivation for learning.

ELEVEN
Offer kids choices and encourage them to take moderate risks.

GLOBAL DATA

Thanks to *perestroika*, most Soviet teachers may now choose the texts they use and have more flexibility overall in planning instruction. Grading is also less rigid than in the past.

Zoya Zarubina

Challenge and Choice

Another factor in creating social scaffolding for mathematical learning is establishing a shared understanding about trying new ideas and taking risks. We must allow all students "the privilege of learning from mistakes," insists Margaret Clifford of the University of Iowa. "Moderate challenge—implying considerable error-making—is essential for maximizing learning and optimizing motivation," she notes. "Moderate risk-taking increases performance, persistence, perceived competence, self-knowledge, pride, and satisfaction."

By giving students some choice in selecting or defining a problem and a strategy for solving it, you not only enhance their motivation, you imply both respect for their judgment and faith that they will learn something worthwhile, even if they make errors.

"From every risk-taking endeavor—whether it ends in failure or success, risk takers learn something about their skill and choice of strategy, and what they learn usually prompts them to seek another risk-taking opportunity," explains Clifford. Studies show that "autonomous behavior—that which is self-determined, freely chosen, and personally controlled—elicits high task interest, creativity, cognitive flexibility, positive emotion, and persistence," she adds.

This encouragement of intellectual risk-taking, while often used with higher-achievers, is unfortunately less common with average or low-achieving students. "We've tended to assume that you move into problem-solving and higher order thinking and risk-taking only after you've mastered 'basic skills,'" says Davison. "But that assumption is being pretty widely challenged now. The challenging assignment should not be denied to the kid whose abilities are untapped and unmeasured by traditional mechanisms."

It's important to follow up when offering choices and encouraging students to accept challenges. Teachers can encourage students to share strategies as they work. When students are unable to solve a problem they've chosen, you can explore what can be learned from the experience. When students succeed, teachers can help them understand why. Notes M. Kay Alderman of the University of Akron: "It is not enough that the student achieve success; in order to acquire a high degree of motivation, the student must know how he or she personally contributed to this success."

WHOLE ASSESSMENT CATALOG

Do you know the dictum *form follows function*? If you want to find out how well kids reason mathematically, solve problems, and apply learning in real contexts, you need assessment methods in which students do just that. It makes perfect sense.

"Yes, tests also need to change."

NCTM Standards

Thank goodness, then, that math educators are now developing a whole catalog of options for all your testing purposes. By using a number of these methods, you get a look at different kinds of mathematical thinking, an understanding how kids deal with the same concept from different approaches, and a much more complete view of student progress. This holistic notion of assessment reflects those teacher "instincts" you've been using all along, only now you're acquiring some acknowledged formats to use. When you need to make instructional decisions, you have a well-balanced bank of information to draw on. When you need to demonstrate your accountability, you've got appropriate tools for doing so.

Alternative assessment choices include innovative multiple-choice, open-ended questions, interviews, homework, projects, journals, demonstrations, class presentations, and more. Because these assessment techniques are based in ongoing instruction, they also offer direct instructional benefits, unlike traditional tests.

Let's take a closer look at some of the most popular alternative options: performance and portfolio assessment.

Holistic, informal, alternative—here's a collection of fresh approaches to help you find out what you really want to know.

Signs of the Times

The National Assessment of Educational Progress now includes estimation and innovative open-ended situational problems that require students to write about their answers.

QUICK CONVERSATION:

Researcher Lauren Resnick

How can we evaluate a student's conceptual understanding and problem-solving abilities?

The basic idea is extremely simple: What you do for instruction should be what you do for assessment. Assessment is not separate from instruction. If you believe it's a good idea for kids to work on a single complicated problem, then that's your assessment. You've got to work out what to expect—what you're hoping to see and what you don't want to see. You define what you want, then set up chances for students to show that kind of skill or concept. Grades and scores can be based on your judgment of the whole performance.

What about standardized tests?

Teachers worry: "How am I going to do this new stuff when I'm expected to prepare kids for these other tests?" Teachers want to know that new assessment ideas aren't pie-in-the-sky. They aren't. A lot of things are happening. There's massive work going on in improving tests.

In the meantime, teachers need to ally with others who are concerned with revisions in the curriculum. They need to press for change on state and local assessments. Teachers have to have faith that they can make it happen. Teachers can say, "We want to be held to high standards. We need the right kind of assessments to be measured against." The time is right for teachers and supervisors to press for these changes.

Lauren B. Resnick is a professor of psycholology and director of the Learning Research and Development Center at the University of Pittsburgh.

In 1993, students will be permitted to use calculators for portions of the PSAT. In 1994, calculators will be permitted for parts of the SAT.

Performance assessment can take many forms. The goal is to evaluate kids doing math in significant contexts. In one form of performance assessment, teachers observe and question students as they work through a nonroutine problem. Observations or interviews can focus on key areas: How well does the student define the problem? How well does the student identify strategies for solving it? How well does the student perform the appropriate separate operations? To what extent does the student recognize connections among the operations and draw conclusions? How persistent is the student in finding and checking a solution?

The 1993 edition of the Iowa Test of Basic Skills is expected to involve more data analysis, use of calculators, and estimation.

GLOBAL DATA

Great Britain recently introduced a national assessment that blurs the distinction between testing and teaching. The assessment contains scaffolded exercises and allows teachers to prompt students during the test, and if necessary, teach new concepts.

Jessica Richter

Portfolios, one form of performance assessment, are kept by each student and include samples of their best thinking and work. In Vermont, where statewide testing includes evaluation of portfolios, collections may include: ◆ Entries from a math journal. ◆ Solutions to a problem given as homework or on a quiz. ◆ Student-created problems, with or without a solution. ◆ A paper done for another subject that contains math. ◆ A report of a group project with individual comments. ◆ A sketch of a student's work with objects as a solution to a problem or as a description of a math concept. ◆ A teacher description of a student activity that displays conceptual understanding. ◆ A videotape of a student or group giving a presentation. ◆ A report on an application of a math concept.

How do you judge a portfolio? Consider how consistently and thoroughly the samples reveal evidence of understanding of a problem or concept, development of problem-solving skills, use of strategies, and use of appropriate tools and technology. Look also for consideration of reasonableness, self-reflection on mathematical thinking, use of math language and notation, and math confidence and flexibility.

Of course there's much more in the alternative-assessment catalog that you may want to try out. You'll find methods that tell you what you want to know, and you'll convey two powerful ideas to students: You care about how well they reason and you want to figure out why they have difficulties when they do.

The centerpiece of Vermont's new approach to state-wide testing is performance assessment through student math portfolios.

With a little help from your friends . . . you can extend math learning beyond the classroom

YOUR COLLEAGUES, KIDS' FAMILIES, LOCAL BUSINESSFOLK, AND COMMUNITY GROUPS ALL HAVE A STAKE IN STUDENTS' MATHEMATICAL GROWTH. SO WHY NOT MAKE USE OF THESE RESOURCES?

YOUR COLLEAGUES

Does the math in math class have anything to do with the math in science? Or the math in social studies? Or the math in art? Or in band? Or in shop? Or in home ec? Or in sports? Students don't always make the connections, and that's a real waste, because math encountered and applied in other subject areas can give a powerful boost to learning.

Making cross-curricular connections isn't always quick and easy, but all it really takes is a few teachers willing to sit down together. "We've set up a process whereby the faculty will be communicating more," says Susan Eddins of her colleagues in Aurora, Illinois. "We will be forming multidisciplinary teams so we know where we can help make those connections for kids. My hope is that as we begin to identify those concepts that are really important in mathematics, it's also making us more able to explain to people in other disciplines what it is that we do."

Teacher Charlene Moore Kincaid, of Gulf Breeze, Florida, works with science teachers. "We toss ideas back and forth, and teach each other ideas we can use in

GLOBAL DATA

Papua New Guinea has at least 225 different counting systems, one involving the counting of 27 points defined on body parts.

D.F. Lancy, G.B. Saxe

our own subjects," she explains. One of their most popular collaborations involves the space program. "I lead kids through an exercise involving the earth's surface and altitudes, the shuttle's velocity, the shape of orbits. After you've talked about it, the kids are surprised at the results. It's another way to show them that the concepts do exist in the real world."

KIDS' FAMILIES

Students' families can play an enormous role in their mathematical learning. The trick is to figure out just the right form for family involvement to take, and then to get on the phone or keyboard and spell out your request.

With some students, technology is the key: Home computers, calculators, and other devices hold a special cachet for parents and kids, so they provide a natural context for discussing mathematical issues. You'll want to offer parents some specific suggestions or questions to explore with their kids.

With other students, the best approach may be a workshop for parents. The workshop could help parents connect concrete examples involving everyday activities with abstract representations that students are expected to become literate in. Or guest speakers could emphasize the importance of math in 21st century life and work.

Sometimes just an encouraging note or call can make a big difference. Here's bit of advice from David Davison of Eastern Montana College: "When you talk with parents who are uncomfortable or unfamiliar with secondary math, you need to be careful not to exacerbate their anxieties or to reinforce the notion that they can't help their kids with math. At the secondary level, kids are dealing with math many parents have forgotten, and the parents are embarrassed."

LOCAL BUSINESSFOLK

One of the most important favors local businesspeople can do for your students is to broaden their understanding of how math is used in a wide range of careers. Often, all you have to do is ask, and you'll have a diverse panel of highly credible math advocates. College students can also convey excitement about the mathematics learning in your students' futures.

"We've brought people in to talk to the kids—business people, people in the community," notes teacher Raymond Fisher of Topsham, Maine. "We've also taken kids over to the ironworks to see math on site. We've also taken students to building sites to look at the construction of houses. It's a touch of reality that strikes a chord with the kids."

COMMUNITY GROUPS

"If you want kids to be thinking about math outside the classroom, look at common-sense math that friends, parents, neighbors can relate to, interact with," suggests Davison. "Include ideas that kids can chew on away from class—ideas that aren't symbolically sophisticated."

Even in tiny Bethel, Alaska, teacher Beverly Van Camp recognized an opportunity. "We have a dog sled race here every January, and mushers come in from everywhere. We also have a dance festival, and dancers come in from all over." she says. "Kids use these events to come up with their own problems to solve. It really gets them involved, because it shows respect for their culture. I choose the best ones—with the authors' names attached—to use from year to year. Kids see the names of older sudents they admire and say, 'Wow!' "

By getting students directly involved in community activities, you can give related lessons extra impact. Community service, in particular, promotes learning not only because it provides significant opportunities for applying mathematical concepts, but also because it can raise the kids' self-esteem.

In Union Bridge, New Jersey, for example, Patricia Baltzey encourages her students to design placemats for a local nursing home, collect money for natural-disaster victims, and make hundreds of peanut-butter-and-jelly sandwiches for a homeless shelter. The math they learn along the way is not soon forgotten.

Resources

ASSESSMENT

- *Curriculum and Evaluation Standards for School Mathematics,* National Council of Teachers of Mathematics (NCTM), Reston, VA, 1989.
- Designing District Evaluation Instruments for Math and Science Process Skills, an educators' network established in 1987 by the Association for Supervision and Curriculum Development. Contact Shelley Lipowich, 6321 N. Canon Del Pajaro, Tucson, AZ 85715.
- Mathematics Portfolio Project, Vermont Department of Education, 120 State St., Montpelier, VT 05602.
- "Tests as Standards of Achievement in Schools" by Lauren Resnick and Daniel Resnick, *The Uses of Standardized Tests in American Education,* Educational Testing Service, Princeton, NJ, 1989.

AT-RISK STUDENTS

- "Assessing and Improving the Psychosocial Environment of Mathematics Classrooms" by Barry Fraser and others, *Journal for Research in Mathematics Education,* NCTM, 1989.
- *Linguistic and Cultural Influences on Learning Mathematics,* edited by Rodney Cocking and Jose Mestre, Lawrence Erlbaum Associates, Hillsdale, NJ, 1988.
- "Students Need Challenge, Not Easy Success" by Margaret Clifford, *Educational Leadership,* September 1990.

COOPERATIVE LEARNING

- Team Learning Project, Center for Research on Elementary and Middle Schools, The Johns Hopkins University, 3505 N. Charles St., Baltimore, MD 21218.
- "Using Work-Groups in Mathematics Instruction" by Thomas L. Good and others, *Educational Leadership,* December 1989/January 1990.
- *Effects of Cooperative Learning on Achievement in Secondary Schools: A Summary of Research,* by F. M. Newmann and J. A. Thompson, National Center for Effective Secondary Schools, Madison, WI, 1987.

MANIPULATIVES

- *Junior High Cooperative Problem Solving with Pattern Blocks* by Ann Roper and *Junior High Cooperative Problem Solving with Geoboards* by Judy Goodnow, Creative Publications, Oak Lawn, IL, 1990.
- *Mathematics Through Paper Folding* by Alton T. Olson, NCTM, 1975.
- *Mira Math Activities for High School,* Mira Math Co., Willowdale, Ontario, 1973.

PARENTS, FAMILIES, AND COMMUNITY

- "Strengthening Partnerships with Parents and Community," a theme issue of *Educational Leadership,* October 1989.
- *Family Math* by Jean Kerr Stenmark, Virginia Thompson, and Ruth Cossey, Lawrence Hall of Science, University of California, Berkeley, CA, 1986.

PROBLEM SOLVING

- "Contextualization and Mathematics for All" by Claude Janvier and "Teaching Mathematics with a Vision: Integrating Computers into Instruction" by Marc Swadener and William Blubaugh, *Teaching and Learning Mathematics in the 1990s,* NCTM, 1990.
- *The Teaching and Assessing of Mathematical Problem Solving,* edited by Randall Charles and Edward A. Silver, NCTM, 1989.

WRITING IN MATH

- *"Writing as a Tool for Teaching Mathematics: The Silent Revolution"* by Aggie Azzolino, *Teaching and Learning Mathematics in the 1990s,* NCTM, 1990.

P R E N T I C E   H A L L

A • SPECIAL • REPORT • FROM • PRENTICE • HALL

Putting Ideas to Work with *Prentice Hall Pre-Algebra* ©1992

PROBLEM SOLVING IN THE REAL WORLD
- Problem Solving Strategy Lessons
- Problem Solving Application Lessons
- Think
- Think and Discuss
- Data Files
- Mixed Review
- Exercise Sets
- Critical Thinking
- Interdisciplinary
- Careers
- Projects
- Calculator
- Computer

HANDY TOOLS/ SOMETIMES IT TAKES TEAMWORK
- Interdisciplinary
- Projects
- Data Files
- Exploring Lessons
- Exercise Sets
- Teaching Tips
- Math Minutes
- Calculator
- Computer

WHY WRITE?
- Writing in Math
- Exercise Sets
- Critical Thinking
- Exploring Lessons

UNITED WE STAND
- Meeting Individual Needs
- Practice, Enrichment, Exploring
- Assignment Guide
- Exploring Lessons
- Interdisciplinary

WHOLE ASSESSMENT CATALOG
- Chapter Test
- Alternative Assessments
- Test Yourself
- Mixed Review
- Flashbacks
- Problem Solving Hints
- Think and Discuss
- Chapter Review

CRITICAL THINKING
- Decision Making
- Think
- Think and Discuss
- Writing
- Exploring Critical Thinking
- Interdisciplinary

For more information, call or write:

PRENTICE HALL

A School Division of Simon & Schuster

4350 Equity Drive
Columbus, OH 43216-2649
1-800-848-9500

A Paramount Communications Company

Texas Essential Elements

ESSENTIAL ELEMENT 1: Problem solving—experience in solving problems by selecting and matching strategies to given situations.

ESSENTIAL ELEMENT	STUDENT TEXT (pages)		TRB (pages)
1A how to communicate an understanding of a problem by describing and discussing the problem and recording the relevant information	**Ch. 1:**	**Lesson 1-2** (10-13)	A-3, A-4
		Lesson 1-3 (14-17)	A-5, A-6
		Lesson 1-4 (19-21)	A-7, A-8
		Lesson 1-5 (22-25)	A-9, A-10
		Lesson 1-6 (26-29)	A-11, A-12
		Lesson 1-7 (31-33)	A-13
		Lesson 1-8 (34-37)	A-14, A-15
		Lesson 1-9 (40-43)	A-16, A-17
		Lesson 1-10 (44-46)	A-18, A-19
	Ch. 2:	**Lesson 2-7** (80-82)	B-13, B-14
		Lesson 2-8 (85-86)	B-15, B-16
		Lesson 2-9 (87-89)	B-17
	Ch. 3:	**Lesson 3-2** (104-107)	C-3, C-4
		Lesson 3-3 (108-110)	C-5, C-6
		Lesson 3-6 (125-127)	C-11
		Lesson 3-7 (128-131)	C-12, C-13
		Lesson 3-8 (132-133)	C-14, C-15
	Ch. 4:	**Lesson 4-2** (145-148)	D-3, D-4
		Lesson 4-4 (153-155)	D-7
		Lesson 4-5 (159-162)	D-8, D-9
		Lesson 4-8 (172-174)	D-14, D-15
	Ch. 5:	**Lesson 5-6** (203-204)	E-11, E-12
		Lesson 5-10 (218-221)	E-19, E-20
		Lesson 5-11 (222-224)	E-21
	Ch. 6:	**Lesson 6-1** (233-236)	F-1, F-2
		Lesson 6-2 (237-240)	F-3, F-4
		Lesson 6-3 (241-243)	F-5
		Lesson 6-4 (245-248)	F-6, F-7
		Lesson 6-5 (252-255)	F-8, F-9
		Lesson 6-8 (264-266)	F-14, F-15
	Ch. 7:	**Lesson 7-3** (284-286)	G-5, G-6
		Lesson 7-6 (296-299)	G-11, G-12
		Lesson 7-7 (300-303)	G-13, G-14
		Lesson 7-8 (305-307)	G-15
	Ch. 8:	**Lesson 8-4** (330-333)	H-7, H-8
		Lesson 8-5 (334-336)	H-9, H-10
		Lesson 8-8 (348-351)	H-15, H-16
		Lesson 8-9 (353-355)	H-17
	Ch. 9:	**Lesson 9-9** (394-395)	I-17, I-18
		Lesson 9-10 (397-399)	I-19
	Ch. 10:	**Lesson 10-5** (425-426)	J-9, J-10
		Lesson 10-6 (427-429)	J-11, J-12
		Lesson 10-7 (430-432)	J-13, J-14
		Lesson 10-8 (433-435)	J-15
		Lesson 10-9 (437-440)	J-16, J-17
		Lesson 10-10 (441-444)	J-18, J-19
	Ch. 11:	**Lesson 11-3** (461-464)	K-5, K-6
		Lesson 11-4 (465-468)	K-7, K-8
		Lesson 11-5 (469-472)	K-9, K-10
		Lesson 11-6 (473-476)	K-11, K-12
	Ch. 12:	**Lesson 12-1** (488-491)	L-1, L-2
		Lesson 12-2 (492-495)	L-3, L-4
		Lesson 12-3 (498-501)	L-5, L-6
		Lesson 12-4 (502-505)	L-7, L-8
		Lesson 12-5 (508-511)	L-9, L-10
		Lesson 12-6 (512-515)	L-11, L-12
		Lesson 12-7 (518-520)	L-13, L-14
		Lesson 12-8 (522-525)	L-15, L-16
		Lesson 12-9 (527-529)	L-17
	Ch. 13:	**Lesson 13-3** (546-548)	M-5, M-6
		Lesson 13-4 (550-553)	M-7, M-8
		Lesson 13-5 (555-557)	M-9
		Lesson 13-6 (558-560)	M-10, M-11
1B how to select appropriate strategies from a variety of approaches	**Ch. 1:**	**Lesson 1-4** (19-21)	A-7, A-8
		Lesson 1-7 (31-33)	A-13
	Ch. 2:	**Lesson 2-8** (85-86)	B-15, B-16
	Ch. 3:	**Lesson 3-6** (125-127)	C-11
		Lesson 3-7 (128-131)	C-12, C-13
		Lesson 3-8 (132-133)	C-14, C-15
	Ch. 4:	**Lesson 4-8** (172-174)	D-14, D-15
	Ch. 5:	**Lesson 5-6** (203-204)	E-11, E-12
		Lesson 5-11 (222-224)	E-21
	Ch. 6:	**Lesson 6-2** (237-240)	F-3, F-4
		Lesson 6-3 (241-243)	F-5

1B continued	**Ch. 6:** **Lesson 6-5** (252-255) **Lesson 6-8** (264-266) **Ch. 7:** **Lesson 7-1** (275-278) **Lesson 7-3** (284-286) **Lesson 7-7** (300-303) **Lesson 7-8** (305-307) **Ch. 8:** **Lesson 8-5** (334-336) **Lesson 8-9** (353-355) **Ch. 9:** **Lesson 9-9** (394-395) **Lesson 9-10** (397-399) **Ch. 10:** **Lesson 10-2** (410-413) **Lesson 10-5** (425-426) **Lesson 10-8** (433-435) **Lesson 10-10** (441-444) **Ch. 11:** **Lesson 11-2** (458-459) **Lesson 11-4** (465-468) **Lesson 11-5** (469-472) **Lesson 11-6** (473-476) **Lesson 11-7** (477-479) **Ch. 12:** **Lesson 12-5** (508-511) **Lesson 12-7** (518-520) **Lesson 12-9** (527-529) **Ch. 13:** **Lesson 13-3** (546-548) **Lesson 13-4** (550-553) **Lesson 13-5** (555-557) **Lesson 13-6** (558-560)	F-8, F-9 F-14, F-15 G-1, G-2 G-5, G-6 G-13, G-14 G-15 H-9, H-10 H-17 I-17, I-18 I-19 J-3, J-4 J-9, J-10 J-15 J-18, J-19 K-3, K-4 K-7, K-8 K-9, K-10 K-11, K-12 K-13 L-9, L-10 L-13, L-14 L-17 M-5, M-6 M-7, M-8 M-9 M-10, M-11
1C how to select appropriate methods and materials for solving problems including concrete manipulatives, models, mental computation, paper and pencil, calculators and/or computers	**Ch. 1:** **Lesson 1-1** (4-7) **Lesson 1-2** (10-13) **Lesson 1-3** (14-17) **Lesson 1-5** (22-25) **Lesson 1-8** (34-37) **Lesson 1-9** (40-43) **Lesson 1-10** (44-46) **Ch. 2:** **Lesson 2-5** (72-75) **Lesson 2-6** (76-79) **Lesson 2-7** (80-82) **Lesson 2-8** (85-86) **Lesson 2-9** (87-89) **Ch. 3:** **Lesson 3-1** (100-103) **Ch. 4:** **Lesson 4-1** (141-144) **Lesson 4-4** (153-155) **Lesson 4-8** (172-174) **Ch. 5:** **Lesson 5-5** (199-202) **Lesson 5-6** (203-204) **Lesson 5-11** (222-224) **Ch. 6:** **Lesson 6-5** (252-255) **Ch. 7:** **Lesson 7-1** (275-278) **Lesson 7-4** (287-290) **Lesson 7-6** (296-299) **Lesson 7-7** (300-303) **Ch. 9:** **Lesson 9-9** (394-395) **Lesson 9-10** (397-399) **Ch. 10:** **Lesson 10-1** (406-409) **Lesson 10-2** (410-413) **Lesson 10-3** (414-417) **Lesson 10-5** (425-426) **Lesson 10-6** (427-429) **Lesson 10-7** (430-432) **Lesson 10-8** (433-435) **Lesson 10-9** (437-440) **Lesson 10-10** (441-444) **Ch. 11:** **Lesson 11-1** (454-457) **Lesson 11-2** (458-459) **Lesson 11-3** (461-464) **Lesson 11-4** (465-468) **Lesson 11-5** (469-472) **Lesson 11-6** (473-476) **Lesson 11-7** (477-479) **Ch. 12:** **Lesson 12-1** (488-491) **Lesson 12-3** (498-501) **Lesson 12-4** (502-505) **Lesson 12-5** (508-511) **Lesson 12-6** (512-515) **Lesson 12-7** (518-520) **Lesson 12-8** (522-525) **Ch. 13:** **Lesson 13-1** (538-541) **Lesson 13-2** (542-545) **Lesson 13-3** (546-548) **Lesson 13-4** (550-553) **Lesson 13-5** (555-557) **Lesson 13-6** (558-560)	A-1, A-2 A-3, A-4 A-5, A-6 A-9, A-10 A-14, A-15 A-16, A-17 A-18, A-19 B-9, B-10 B-11, B-12 B-13, B-14 B-15, B-16 B-17 C-1, C-2 D-1, D-2 D-7 D-14, D-15 E-9, E-10 E-11, E-12 E-21 F-8, F-9 G-1, G-2 G-7, G-8 G-11, G-12 G-13, G-14 I-17, I-18 I-19 J-1, J-2 J-3, J-4 J-5, J-6 J-9, J-10 J-11, J-12 J-13, J-14 J-15 J-16, J-17 J-18, J-19 K-1, K-2 K-3, K-4 K-5, K-6 K-7, K-8 K-9, K-10 K-11, K-12 K-13 L-1, L-2 L-5, L-6 L-7, L-8 L-9, L-10 L-11, L-12 L-13, L-14 L-15, L-16 M-1, M-2 M-3, M-4 M-5, M-6 M-7, M-8 M-9 M-10, M-11
1D how to solve a problem, evaluate the outcome for reasonableness (including appropriateness of units), make revisions as needed, describe and discuss the process and solution, and make a decision based on the solution	**Ch. 2:** **Lesson 2-7** (80-82) **Lesson 2-8** (85-86) **Lesson 2-9** (87-89) **Ch. 3:** **Lesson 3-2** (104-107) **Lesson 3-3** (108-110) **Lesson 3-4** (111-114) **Lesson 3-5** (115-118) **Lesson 3-6** (125-127)	B-13, B-14 B-15, B-16 B-17 C-3, C-4 C-5, C-6 C-7, C-8 C-9, C-10 C-11

1D continued	**Ch. 3:** **Lesson 3-7** (128-131)	C-12, C-13
	Lesson 3-8 (132-133)	C-14, C-15
	Ch. 4: **Lesson 4-4** (153-155)	D-7
	Lesson 4-6 (163-166)	D-10, D-11
	Lesson 4-8 (172-174)	D-14, D-15
	Ch. 5: **Lesson 5-6** (203-204)	E-11, E-12
	Lesson 5-10 (218-221)	E-19, E-20
	Lesson 5-11 (222-224)	E-21
	Ch. 6: **Lesson 6-2** (237-240)	F-3, F-4
	Lesson 6-3 (241-243)	F-5
	Lesson 6-7 (260-263)	F-12, F-13
	Lesson 6-8 (264-266)	F-14, F-15
	Ch. 7: **Lesson 7-3** (284-286)	G-5, G-6
	Lesson 7-8 (305-307)	G-15
	Ch. 8: **Lesson 8-4** (330-333)	H-7, H-8
	Lesson 8-5 (334-336)	H-9, H-10
	Ch. 9: **Lesson 9-9** (394-395)	I-17, I-18
	Lesson 9-10 (397-399)	I-19
	Ch. 10: **Lesson 10-5** (425-426)	J-9, J-10
	Lesson 10-7 (430-432)	J-13, J-14
	Lesson 10-8 (433-435)	J-15
	Lesson 10-9 (437-440)	J-16, J-17
	Lesson 10-10 (441-444)	J-18, J-19
	Ch. 11: **Lesson 11-3** (461-464)	K-5, K-6
	Ch. 12: **Lesson 12-3** (498-501)	L-5, L-6
	Lesson 12-7 (518-520)	L-13, L-14
	Lesson 12-8 (522-525)	L-15, L-16
	Lesson 12-9 (527-529)	L-17
1E how to generate, extend, and generalize problems and solutions	**Ch. 1:** **Lesson 1-3** (14-17)	A-5, A-6
	Lesson 1-4 (19-21)	A-7, A-8
	Lesson 1-5 (22-25)	A-9, A-10
	Lesson 1-6 (26-29)	A-11, A-12
	Lesson 1-7 (31-33)	A-13
	Lesson 1-8 (34-37)	A-14, A-15
	Lesson 1-9 (40-43)	A-16, A-17
	Ch. 2: **Lesson 2-9** (87-89)	B-17
	Ch. 4: **Lesson 4-3** (149-152)	D-5, D-6
	Lesson 4-4 (153-155)	D-7
	Lesson 4-5 (159-162)	D-8, D-9
	Lesson 4-6 (163-166)	D-10, D-11
	Lesson 4-8 (172-174)	D-14, D-15
	Ch. 6: **Lesson 6-1** (233-236)	F-1, F-2
	Lesson 6-2 (237-240)	F-3, F-4
	Ch. 8: **Lesson 8-9** (353-355)	H-17
	Ch. 9: **Lesson 9-9** (394-395)	I-17, I-18
	Lesson 9-10 (397-399)	I-19
	Ch. 10: **Lesson 10-5** (425-426)	J-9, J-10
	Lesson 10-8 (433-435)	J-15
	Lesson 10-10 (441-444)	J-18, J-19
	Ch. 12: **Lesson 12-4** (502-505)	L-7, L-8
	Lesson 12-5 (508-511)	L-9, L-10
	Lesson 12-8 (522-525)	L-15, L-16
	Lesson 12-9 (527-529)	L-17
	Ch. 13: **Lesson 13-5** (555-557)	M-9
	Lesson 13-6 (558-560)	M-10, M-11

ESSENTIAL ELEMENT 2: Number and numeration concepts—concepts and skills associated with the understanding of numbers and the place value system.

ESSENTIAL ELEMENT	STUDENT TEXT (pages)	TRB (pages)
2A scientific notation and its use with calculators	**Ch. 4:** **Lesson 4-3** (149-152)	D-5, D-6
	Lesson 4-4 (153-155)	D-7
	Ch. 5: **Lesson 5-8** (211-214)	E-15, E-16
2B how to find factors, prime factors, and multiples using divisibility rules	**Ch. 4:** **Lesson 4-5** (159-162)	D-8, D-9
	Lesson 4-6 (163-166)	D-10, D-11
2C concepts involving least common multiple and greatest common factor as applied to fractions and algebraic monomials	**Ch. 4:** **Lesson 4-7** (167-170)	D-12, D-13
	Ch. 5: **Lesson 5-1** (184-187)	E-1, E-2
	Lesson 5-4 (196-198)	E-7, E-8
	Lesson 5-5 (199-202)	E-9, E-10
	Lesson 5-9 (215-217)	E-17, E-18
	Ch. 7: **Lesson 7-4** (287-290)	G-7, G-8
2D properties and characteristics of subsystems of the real number system	**Ch. 1:** **Lesson 1-1** (4-7)	A-1, A-2
	Lesson 1-2 (10-13)	A-3, A-4
	Lesson 1-3 (14-17)	A-5, A-6
	Ch. 3: **Lesson 3-1** (100-103)	C-1, C-2
	Lesson 3-2 (104-107)	C-3, C-4
	Lesson 3-4 (111-114)	C-7, C-8
	Lesson 3-5 (115-118)	C-9, C-10
	Ch. 4: **Lesson 4-2** (145-148)	D-3, D-4
	Lesson 4-3 (149-152)	D-5, D-6
	Lesson 4-6 (163-166)	D-10, D-11
	Lesson 4-7 (167-170)	D-12, D-13

2D continued	**Ch. 5:** **Lesson 5-1** (184-187)	E-1, E-2
	Lesson 5-2 (188-191)	E-3, E-4
	Lesson 5-3 (192-195)	E-5, E-6
	Lesson 5-4 (196-198)	E-7, E-8
	Lesson 5-7 (206-208)	E-13, E-14
	Ch. 6: **Lesson 6-2** (237-240)	F-3, F-4
	Lesson 6-7 (260-263)	F-12, F-13
	Ch. 13: **Lesson 13-2** (542-545)	M-3, M-4
	Lesson 13-3 (546-548)	M-5, M-6
	Lesson 13-4 (550-553)	M-7, M-8
2E how to simplify polynomials by combining like terms and using the distributive property	**Ch. 2:** **Lesson 2-4** (66-69)	B-7, B-8
	Ch. 3: **Lesson 3-3** (108-110)	C-5, C-6
	Lesson 3-5 (115-118)	C-9, C-10
	Ch. 7: **Lesson 7-2** (279-282)	G-3, G-4
	Lesson 7-3 (284-286)	G-5, G-6
	Lesson 7-4 (287-290)	G-7, G-8
	Lesson 7-7 (300-303)	G-13, G-14
	Lesson 7-8 (305-307)	G-15
	Ch. 8: **Lesson 8-2** (320-323)	H-3, H-4
	Ch. 13: **Lesson 13-2** (542-545)	M-3, M-4
	Lesson 13-3 (546-548)	M-5, M-6
	Lesson 13-4 (550-553)	M-7, M-8
2F how to write symbolic expressions for word phrases	**Ch. 1:** **Lesson 1-1** (4-7)	A-1, A-2
	Lesson 1-2 (10-13)	A-3, A-4
	Lesson 1-3 (14-17)	A-5, A-6
	Lesson 1-5 (22-25)	A-9, A-10
	Lesson 1-6 (26-29)	A-11, A-12
	Lesson 1-8 (34-37)	A-14, A-15
	Lesson 1-9 (40-43)	A-16, A-17
	Lesson 1-10 (44-46)	A-18, A-19
	Ch. 3: **Lesson 3-1** (100-103)	C-1, C-2
	Lesson 3-2 (104-107)	C-3, C-4
	Lesson 3-3 (108-110)	C-5, C-6
	Lesson 3-4 (111-114)	C-7, C-8
	Lesson 3-5 (115-118)	C-9, C-10
	Ch. 4: **Lesson 4-3** (149-152)	D-5, D-6
	Lesson 4-7 (167-170)	D-12, D-13
	Ch. 5: **Lesson 5-9** (215-217)	E-17, E-18
	Ch. 6: **Lesson 6-1** (233-236)	F-1, F-2
	Lesson 6-2 (237-240)	F-3, F-4
	Lesson 6-6 (256-259)	F-10, F-11
	Ch. 7: **Lesson 7-1** (275-278)	G-1, G-2
	Lesson 7-2 (279-282)	G-3, G-4
	Lesson 7-3 (284-286)	G-5, G-6
	Lesson 7-4 (287-290)	G-7, G-8
	Lesson 7-5 (292-295)	G-9, G-10
	Lesson 7-6 (296-299)	G-11, G-12
	Lesson 7-7 (300-303)	G-13, G-14
	Lesson 7-8 (305-307)	G-15
	Ch. 8: **Lesson 8-5** (334-336)	H-9, H-10
	Lesson 8-8 (348-351)	H-15, H-16
	Lesson 8-9 (353-355)	H-17
	Ch. 11: **Lesson 11-2** (458-459)	K-3, K-4
	Lesson 11-3 (461-464)	K-5, K-6
	Lesson 11-5 (469-472)	K-9, K-10
	Lesson 11-6 (473-476)	K-11, K-12
	Lesson 11-7 (477-479)	K-13
	Ch. 12: **Lesson 12-9** (527-529)	L-17
	Ch. 13: **Lesson 13-2** (542-545)	M-3, M-4
	Lesson 13-3 (546-548)	M-5, M-6
	Lesson 13-4 (550-553)	M-7, M-8
	Lesson 13-5 (555-557)	M-9
	Lesson 13-6 (558-560)	M-10, M-11

ESSENTIAL ELEMENT 3: Operations and computation—basic operations on numbers, their properties, and their uses.

ESSENTIAL ELEMENT	STUDENT TEXT (pages)	TRB (pages)
3A how to use basic operations and properties of rational numbers to solve relevant problems, using a calculator, mental techniques, and pencil/paper, as appropriate	**Ch. 2:** **Lesson 2-1** (54-57)	B-1, B-2
	Lesson 2-2 (58-61)	B-3, B-4
	Lesson 2-3 (62-65)	B-5, B-6
	Lesson 2-4 (66-69)	B-7, B-8
	Lesson 2-5 (72-75)	B-9, B-10
	Lesson 2-6 (76-79)	B-11, B-12
	Lesson 2-7 (80-82)	B-13, B-14
	Lesson 2-8 (85-86)	B-15, B-16
	Lesson 2-9 (87-89)	B-17
	Ch. 3: **Lesson 3-2** (104-107)	C-3, C-4
	Lesson 3-3 (108-110)	C-5, C-6
	Lesson 3-4 (111-114)	C-7, C-8
	Lesson 3-5 (115-118)	C-9, C-10
	Lesson 3-6 (125-127)	C-11
	Lesson 3-7 (128-131)	C-12, C-13
	Lesson 3-8 (132-133)	C-14, C-15
	Ch. 4: **Lesson 4-3** (149-152)	D-5, D-6
	Lesson 4-4 (153-155)	D-7

3A	continued		**Lesson 4-5** (159-162)	D-8, D-9
			Lesson 4-6 (163-166)	D-10, D-11
			Lesson 4-7 (167-170)	D-12, D-13
		Ch. 5:	**Lesson 5-5** (199-202)	E-9, E-10
			Lesson 5-7 (206-208)	E-13, E-14
			Lesson 5-8 (211-214)	E-15, E-16
			Lesson 5-9 (215-217)	E-17, E-18
			Lesson 5-10 (218-221)	E-19, E-20
		Ch. 6:	**Lesson 6-1** (233-236)	F-1, F-2
			Lesson 6-3 (241-243)	F-5
			Lesson 6-4 (245-248)	F-6, F-7
			Lesson 6-6 (256-259)	F-10, F-11
			Lesson 6-8 (264-266)	F-14, F-15
		Ch. 7:	**Lesson 7-1** (275-278)	G-1, G-2
			Lesson 7-3 (284-286)	G-5, G-6
			Lesson 7-4 (287-290)	G-7, G-8
			Lesson 7-6 (296-299)	G-11, G-12
			Lesson 7-7 (300-303)	G-13, G-14
			Lesson 7-8 (305-307)	G-15
		Ch. 8:	**Lesson 8-5** (334-336)	H-9, H-10
			Lesson 8-9 (353-355)	H-17
		Ch. 9:	**Lesson 9-4** (374-376)	I-7, I-8
			Lesson 9-5 (377-379)	I-9, I-10
			Lesson 9-8 (390-393)	I-15, I-16
			Lesson 9-9 (394-395)	I-17, I-18
			Lesson 9-10 (397-399)	I-19
		Ch. 11:	**Lesson 11-1** (454-457)	K-1, K-2
			Lesson 11-2 (458-459)	K-3, K-4
			Lesson 11-3 (461-464)	K-5, K-6
			Lesson 11-4 (465-468)	K-7, K-8
			Lesson 11-5 (469-472)	K-9, K-10
			Lesson 11-6 (473-476)	K-11, K-12
			Lesson 11-7 (477-479)	K-13
		Ch. 12:	**Lesson 12-1** (488-491)	L-1, L-2
			Lesson 12-4 (502-505)	L-7, L-8
			Lesson 12-5 (508-511)	L-9, L-10
			Lesson 12-6 (512-515)	L-11, L-12
			Lesson 12-7 (527-529)	L-13, L-14
			Lesson 12-8 (522-525)	L-15, L-16
			Lesson 12-9 (527-529)	L-17
		Ch. 13:	**Lesson 13-1** (538-541)	M-1, M-2
			Lesson 13-2 (542-545)	M-3, M-4
			Lesson 13-3 (546-548)	M-5, M-6
			Lesson 13-4 (550-553)	M-7, M-8
			Lesson 13-5 (555–557)	M-9
			Lesson 13-6 (558-560)	M-10, M-11
3B	how to use the order of operations to simplify expressions and solve multistep problems	**Ch. 1:**	**Lesson 1-9** (40-43)	A-16, A-17
			Lesson 1-10 (44-46)	A-18, A-19
		Ch. 2:	**Lesson 2-1** (54-57)	B-1, B-2
			Lesson 2-4 (66-69)	B-7, B-8
			Lesson 2-9 (87-89)	B-17
		Ch. 3:	**Lesson 3-3** (108-110)	C-5, C-6
			Lesson 3-5 (115-118)	C-9, C-10
		Ch. 5:	**Lesson 5-3** (192-195)	E-5, E-6
			Lesson 5-7 (206-208)	E-13, E-14
		Ch. 7:	**Lesson 7-1** (275-278)	G-1, G-2
			Lesson 7-3 (284-286)	G-5, G-6
			Lesson 7-4 (287-290)	G-7, G-8
			Lesson 7-7 (300-303)	G-13, G-14
			Lesson 7-8 (305-307)	G-15
		Ch. 8:	**Lesson 8-9** (353-355)	H-17
		Ch. 11:	**Lesson 11-2** (458-459)	K-3, K-4
			Lesson 11-3 (461-464)	K-5, K-6
			Lesson 11-4 (465-468)	K-7, K-8
			Lesson 11-5 (469-472)	K-9, K-10
			Lesson 11-7 (477-479)	K-13
		Ch. 12:	**Lesson 12-1** (488-491)	L-1, L-2
			Lesson 12-7 (518-520)	L-13, L-14
			Lesson 12-9 (527-529)	L-17
		Ch. 13:	**Lesson 13-2** (542-545)	M-3, M-4
			Lesson 13-3 (546-548)	M-5, M-6
			Lesson 13-4 (550-553)	M-7, M-8
			Lesson 13-5 (555-557)	M-9
			Lesson 13-6 (558-560)	M-10, M-11
3C	how to evaluate an expression that contains absolute values	**Ch. 1:**	**Lesson 1-1** (4-7)	A-1, A-2
			Lesson 1-2 (10-13)	A-3, A-4
			Lesson 1-3 (14-17)	A-5, A-6
			Lesson 1-5 (22-25)	A-9, A-10
			Lesson 1-6 (26-29)	A-11, A-12
			Lesson 1-9 (40-43)	A-16, A-17
			Lesson 1-10 (44-46)	A-18, A-19
		Ch. 5:	**Lesson 5-3** (192-195)	E-5, E-6
			Lesson 5-5 (199-202)	E-9, E-10
			Lesson 5-10 (218-221)	E-19, E-20
		Ch. 7:	**Lesson 7-5** (292-295)	G-9, G-10
		Ch. 8:	**Lesson 8-7** (344-347)	H-13, H-14

3D laws of exponents and their applications including zero and negative integral exponents	**Ch. 4:**	**Lesson 4-1** (141-144)	D-1, D-2
		Lesson 4-2 (145-148)	D-3, D-4
		Lesson 4-3 (149-152)	D-5, D-6
		Lesson 4-4 (153-155)	D-7
	Ch. 5:	**Lesson 5-8** (211-214)	E-15, E-16
	Ch. 7:	**Lesson 7-8** (305-307)	G-15
	Ch. 11:	**Lesson 11-5** (469-472)	K-9, K-10
	Ch. 12:	**Lesson 12-5** (508-511)	L-9, L-10
	Ch. 13:	**Lesson 13-3** (546-548)	M-5, M-6
		Lesson 13-4 (550-553)	M-7, M-8
3E how to simplify and estimate numeric radical expressions involving square roots using a calculator when appropriate	**Ch. 10:**	**Lesson 10-3** (414-417)	J-5, J-6
	Ch. 11:	**Lesson 11-1** (454-457)	K-1, K-2
		Lesson 11-3 (461-464)	K-5, K-6
		Lesson 11-4 (465-468)	K-7, K-8
		Lesson 11-5 (469-472)	K-9, K-10
		Lesson 11-6 (473-476)	K-11, K-12
	Ch. 13:	**Lesson 13-1** (538-541)	M-1, M-2
3F how to solve simple linear equations and inequalities involving rational numbers	**Ch. 2:**	**Lesson 2-1** (54-57)	B-1, B-2
		Lesson 2-5 (72-75)	B-9, B-10
		Lesson 2-6 (76-79)	B-11, B-12
		Lesson 2-7 (80-82)	B-13, B-14
		Lesson 2-8 (85-86)	B-15, B-16
		Lesson 2-9 (87-89)	B-17
	Ch. 5:	**Lesson 5-9** (215-217)	E-17, E-18
		Lesson 5-10 (218-221)	E-19, E-20
	Ch. 7:	**Lesson 7-1** (275-278)	G-1, G-2
		Lesson 7-2 (279-282)	G-3, G-4
		Lesson 7-4 (287-290)	G-7, G-8
		Lesson 7-6 (296-299)	G-11, G-12
		Lesson 7-7 (300-303)	G-13, G-14
	Ch. 8:	**Lesson 8-7** (344-347)	H-13, H-14
		Lesson 8-8 (348-351)	H-15, H-16
	Ch. 9:	**Lesson 9-4** (374-376)	I-7, I-8
	Ch. 13:	**Lesson 13-2** (542-545)	M-3, M-4
		Lesson 13-3 (546-548)	M-5, M-6
		Lesson 13-4 (550-553)	M-7, M-8
3G how to estimate and solve application problems by writing and solving simple two-step equations	**Ch. 2:**	**Lesson 2-8** (85-86)	B-15, B-16
		Lesson 2-9 (87-89)	B-17
	Ch. 6:	**Lesson 6-2** (237-240)	F-3, F-4
	Ch. 7:	**Lesson 7-1** (275-278)	G-1, G-2
		Lesson 7-2 (279-282)	G-3, G-4
		Lesson 7-3 (284-286)	G-5, G-6
		Lesson 7-4 (287-290)	G-7, G-8
		Lesson 7-8 (305-307)	G-15
	Ch. 11:	**Lesson 11-4** (465-468)	K-7, K-8
	Ch. 13:	**Lesson 13-2** (542-545)	M-3, M-4
		Lesson 13-3 (546-548)	M-5, M-6
		Lesson 13-4 (550-553)	M-7, M-8
		Lesson 13-5 (555-557)	M-9
		Lesson 13-6 (558-560)	M-10, M-11
3H evaluation of expressions and formulas containing variables, using a calculator when appropriate	**Ch. 1:**	**Lesson 1-8** (34-37)	A-14, A-15
		Lesson 1-10 (44-46)	A-18, A-19
	Ch. 2:	**Lesson 2-1** (54-57)	B-1, B-2
		Lesson 2-4 (66-69)	B-7, B-8
		Lesson 2-5 (72-75)	B-9, B-10
		Lesson 2-6 (76-79)	B-11, B-12
		Lesson 2-7 (80-82)	B-13, B-14
		Lesson 2-8 (85-86)	B-15, B-16
	Ch. 3:	**Lesson 3-7** (128-131)	C-12, C-13
	Ch. 4:	**Lesson 4-1** (141-144)	D-1, D-2
		Lesson 4-2 (145-148)	D-3, D-4
	Ch. 5:	**Lesson 5-5** (199-202)	E-9, E-10
		Lesson 5-7 (206-208)	E-13, E-14
		Lesson 5-8 (211-214)	E-15, E-16
	Ch. 6:	**Lesson 6-5** (252-255)	F-8, F-9
		Lesson 6-6 (256-259)	F-10, F11
	Ch. 7:	**Lesson 7-1** (275-278)	G-1, G-2
		Lesson 7-3 (284-286)	G-5, G-6
		Lesson 7-4 (287-290)	G-7, G-8
		Lesson 7-5 (292-295)	G-9, G-10
		Lesson 7-6 (296-299)	G-11, G-12
		Lesson 7-7 (300-303)	G-13, G-14
		Lesson 7-8 (305-307)	G-15
	Ch. 10:	**Lesson 10-1** (406-409)	J-1, J-2
		Lesson 10-2 (410-413)	J-3, J-4
		Lesson 10-3 (414-417)	J-5, J-6
	Ch. 11:	**Lesson 11-3** (461-464)	K-5, K-6
		Lesson 11-4 (465-468)	K-7, K-8
		Lesson 11-6 (473-476)	K-11, K-12
	Ch. 12:	**Lesson 12-7** (518-520)	L-13, L-14
		Lesson 12-9 (527-529)	L-17
	Ch. 13:	**Lesson 13-1** (538-541)	M-1, M-2
		Lesson 13-2 (542-545)	M-3, M-4
		Lesson 13-3 (546-548)	M-5, M-6
		Lesson 13-4 (550-553)	M-7, M-8
		Lesson 13-5 (555-557)	M-9
		Lesson 13-6 (558-560)	M-10, M-11

3I how to estimate and solve problems using ratios, proportions, and percent and check the reasonableness of the result	**Ch. 6:** **Lesson 6-2** (237-240)	F-3, F-4
	Lesson 6-3 (241-243)	F-5
	Lesson 6-5 (252-255)	F-8, F-9
	Lesson 6-6 (256-259)	F-10, F-11
	Lesson 6-7 (260-263)	F-12, F-13
	Lesson 6-8 (264-266)	F-14, F-15
	Ch. 7: **Lesson 7-3** (284-286)	G-5, G-6
	Lesson 7-6 (296-299)	G-11, G-12
	Ch. 8: **Lesson 8-5** (334-336)	H-9, H-10
	Lesson 8-9 (353-355)	H-17
	Ch. 9: **Lesson 9-5** (377-379)	I-9, I-10
	Lesson 9-7 (384-387)	I-13, I-14
	Ch. 11: **Lesson 11-4** (465-468)	K-7, K-8
	Lesson 11-5 (469-472)	K-9, K-10
	Lesson 11-6 (473-476)	K-11, K-12
	Ch. 12: **Lesson 12-5** (508-511)	L-9, L-10
	Lesson 12-7 (518-520)	L-13, L-14
	Lesson 12-8 (522-525)	L-15, L-16
	Lesson 12-9 (527-529)	L-17

ESSENTIAL ELEMENT 4: Geometry and measurement—concepts and skills using metric and customary units.

ESSENTIAL ELEMENT	STUDENT TEXT (pages)	TRB (pages)
4A geometric figures and their characteristics	**Ch. 9:** **Lesson 9-1** (362-364)	I-1, I-2
	Lesson 9-2 (365-368)	I-3, I-4
	Lesson 9-3 (370-372)	I-5, I-6
	Lesson 9-4 (374-376)	I-7, I-8
	Lesson 9-5 (377-379)	I-9, I-10
	Lesson 9-6 (380-383)	I-11, I-12
	Lesson 9-7 (384-387)	I-13, I-14
	Lesson 9-9 (394-395)	I-17, I-18
	Lesson 9-10 (397-399)	I-19
	Ch. 10: **Lesson 10-2** (410-413)	J-3, J-4
	Lesson 10-4 (420-423)	J-7, J-8
	Lesson 10-6 (427-429)	J-11, J-12
	Lesson 10-7 (430-432)	J-13, J-14
	Lesson 10-9 (437-440)	J-16, J-17
	Lesson 10-10 (441-444)	J-18, J-19
	Ch. 11: **Lesson 11-4** (465-468)	K-7, K-8
	Lesson 11-5 (469-472)	K-9, K-10
	Lesson 11-6 (473-476)	K-11, K-12
4B distance, perimeter, circumference, area, surface area, and volume problems using metric and customary units in relevant problem situations	**Ch. 3:** **Lesson 3-6** (125-127)	C-11
	Lesson 3-8 (132-133)	C-14, C-15
	Ch. 7: **Lesson 7-3** (284-286)	G-5, G-6
	Lesson 7-8 (305-307)	G-15
	Ch. 9: **Lesson 9-7** (384-387)	I-13, I-14
	Lesson 9-8 (390-393)	I-15, I-16
	Lesson 9-9 (394-395)	I-17, I-18
	Lesson 9-10 (397-399)	I-19
	Ch. 10: **Lesson 10-1** (406-409)	J-1, J-2
	Lesson 10-2 (410-413)	J-3, J-4
	Lesson 10-3 (414-417)	J-5, J-6
	Lesson 10-6 (427-429)	J-11, J-12
	Lesson 10-7 (430-432)	J-13, J-14
	Lesson 10-9 (437-440)	J-16, J-17
	Lesson 10-10 (441-444)	J-18, J-19
	Ch. 11: **Lesson 11-5** (469-472)	K-9, K-10
	Ch. 12: **Lesson 12-7** (518-520)	L-13, L-14
	Lesson 12-9 (527-529)	L-17
	Ch. 13: **Lesson 13-2** (542-545)	M-3, M-4
	Lesson 13-3 (546-548)	M-5, M-6
	Lesson 13-4 (550-553)	M-7, M-8
	Lesson 13-5 (555-557)	M-9
	Lesson 13-6 (558-560)	M-10, M-11
4C indirect measurement with similar figures	**Ch. 9:** **Lesson 9-2** (365-368)	I-3, I-4
	Lesson 9-7 (384-387)	I-13, I-14
	Ch. 11: **Lesson 11-4** (465-468)	K-7, K-8
4D right triangle problems using the Pythagorean theorem and its converse	**Ch. 11:** **Lesson 11-3** (461-464)	K-5, K-6
	Lesson 11-4 (465-468)	K-7, K-8
	Lesson 11-5 (469-472)	K-9, K-10
	Lesson 11-6 (473-476)	K-11, K-12
	Ch. 13: **Lesson 13-5** (555-557)	M-9
	Lesson 13-6 (558-560)	M-10, M-11
4E application of algebra to geometric situations	**Ch. 9:** **Lesson 9-4** (374-376)	I-7, I-8
	Ch. 10: **Lesson 10-1** (406-409)	J-1, J-2
	Lesson 10-2 (410-413)	J-3, J-4
	Lesson 10-3 (414-417)	J-5, J-6
	Lesson 10-9 (437-440)	J-16, J-17
	Ch. 11: **Lesson 11-3** (461-464)	K-5, K-6
	Lesson 11-4 (465-468)	K-7, K-8
	Lesson 11-5 (469-472)	K-9, K-10
	Lesson 11-6 (473-476)	K-11, K-12

4E continued	**Ch. 13:** **Lesson 13-2** (542-545) **Lesson 13-3** (546-548) **Lesson 13-4** (550-553)	M-3, M-4 M-5, M-6 M-7, M-8

ESSENTIAL ELEMENT 5: Probability, statistics, and graphing—use of probability and statistics to collect and interpret data.

ESSENTIAL ELEMENT	STUDENT TEXT (pages)	TRB (pages)
5A sample spaces constructed by listing, tree diagram, etc.	**Ch. 5:** **Lesson 5-6** (203-204) **Lesson 5-11** (222-224) **Ch. 12:** **Lesson 12-5** (508-511) **Lesson 12-6** (512-515) **Lesson 12-7** (518-520) **Lesson 12-8** (522-525) **Lesson 12-9** (527-529)	E-11, E-12 E-21 L-9, L-10 L-11, L-12 L-13, L-14 L-15, L-16 L-17
5B how to determine the number of possible outcomes	**Ch. 12:** **Lesson 12-5** (508-511) **Lesson 12-6** (512-515) **Lesson 12-7** (518-520) **Lesson 12-8** (522-525) **Lesson 12-9** (527-529) **Ch. 13:** **Lesson 13-5** (555-557) **Lesson 13-6** (558-560)	L-9, L-10 L-11, L-12 L-13, L-14 L-15, L-16 L-17 M-9 M-10, M-11
5C the probability of independent and dependent events	**Ch. 12:** **Lesson 12-8** (522-525)	L-15, L-16
5D how to collect, organize, and graph data relevant to pre-algebra problems	**Ch. 2:** **Lesson 2-9** (87-89) **Ch. 5:** **Lesson 5-10** (218-221) **Ch. 8:** **Lesson 8-7** (344-347) **Ch. 9:** **Lesson 9-5** (377-379) **Ch. 12:** **Lesson 12-1** (488-491) **Lesson 12-3** (498-501) **Lesson 12-4** (502-505) **Lesson 12-8** (522-525)	B-17 E-19, E-20 H-13, H-14 I-9, I-10 L-1, L-2 L-5, L-6 L-7, L-8 L-15, L-16
5E how to interpret graphs and analyze data to solve relevant problems	**Ch. 1:** **Lesson 1-1** (4-7) **Ch. 2:** **Lesson 2-9** (87-89) **Ch. 5:** **Lesson 5-2** (188-191) **Lesson 5-10** (218-221) **Ch. 8:** **Lesson 8-5** (334-336) **Lesson 8-7** (344-347) **Ch. 12:** **Lesson 12-2** (492-495) **Lesson 12-3** (498-501) **Lesson 12-4** (502-505) **Lesson 12-8** (522-525)	A-1, A-2 B-17 E-3, E-4 E-19, E-20 H-9, H-10 H-13, H-14 L-3, L-4 L-5, L-6 L-7, L-8 L-15, L-16
5F measures of central tendency and their uses in relevant problem situations	**Ch. 12:** **Lesson 12-1** (488-491) **Lesson 12-2** (492-495) **Lesson 12-3** (498-501) **Lesson 12-4** (502-505) **Lesson 12-7** (518-520) **Lesson 12-9** (527-529)	L-1, L-2 L-3, L-4 L-5, L-6 L-7, L-8 L-13, L-14 L-17

ESSENTIAL ELEMENT 6: Concepts and skills involved in the analysis and graphing of relations and functions.

ESSENTIAL ELEMENT	STUDENT TEXT (pages)	TRB (pages)
6A the graphs of solution sets of equations and inequalities on a number line, using computer graphics when appropriate	**Ch. 7:** **Lesson 7-5** (292-295) **Lesson 7-6** (296-299) **Ch. 8:** **Lesson 8-6** (340-343) **Lesson 8-7** (344-347) **Lesson 8-8** (348-351)	G-9, G-10 G-11, G-12 H-11, H-12 H-13, H-14 H-15, H-16
6B the graphs of linear equations and inequalities on a coordinate plane, and exploration of characteristics of these graphs with computer graphics	**Ch. 8:** **Lesson 8-1** (316-319) **Lesson 8-2** (320-323) **Lesson 8-3** (324-327) **Lesson 8-6** (340-343) **Lesson 8-7** (344-347) **Lesson 8-8** (348-351)	H-1, H-2 H-3, H-4 H-5, H-6 H-11, H-12 H-13, H-14 H-15, H-16
6C the slope and intercepts of the graph of a linear equation	**Ch. 8:** **Lesson 8-3** (324-327) **Lesson 8-4** (330-333) **Lesson 8-7** (344-347)	H-5, H-6 H-7, H-8 H-13, H-14
6D the solution of systems of linear equations by graphing	**Ch. 8:** **Lesson 8-3** (324-327) **Lesson 8-6** (340-343) **Lesson 8-7** (344-347) **Lesson 8-8** (348-351)	H-5, H-6 H-11, H-12 H-13, H-14 H-15, H-16

Pre-Algebra

DAVID M. DAVISON
MARSHA LANDAU
LEAH McCRACKEN
LINDA THOMPSON

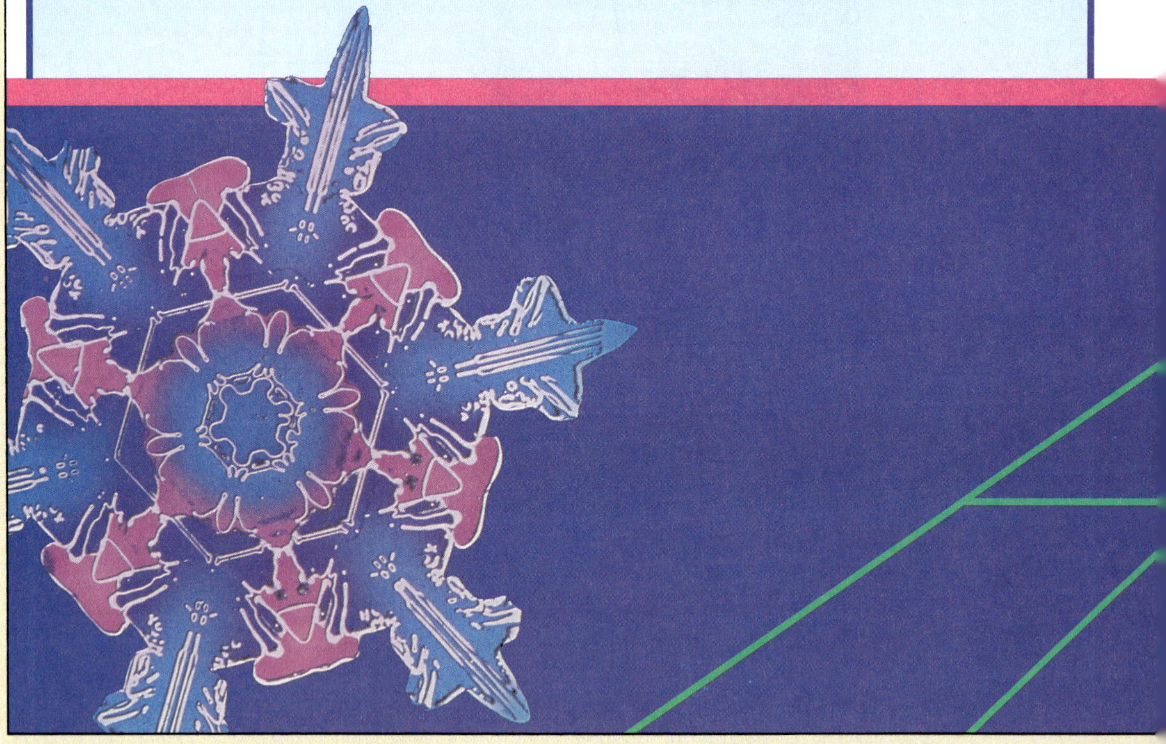

PRENTICE HALL

Pre-Algebra

NEEDHAM, MASSACHUSETTS
ENGLEWOOD CLIFFS, NEW JERSEY

AUTHORS

David M. Davison, *Eastern Montana College, Billings, Montana*

Marsha S. Landau, *National-Louis University, Evanston, Illinois*

Leah McCracken, *Lockwood Junior High School, Billings, Montana*

Linda Thompson, *Warrenton, Oregon*

REVIEWERS

Bettye C. Hall, *Director of Mathematics (retired), Houston Independent School District, Houston, Texas*

Joanne Martin, *Mathematics Teacher, New Prague, Minnesota*

Jeffrey S. McIntire, *Mathematics Teacher, Maumee High School, Maumee, Ohio*

Elizabeth C. McNair, *Mathematics Teacher, Burns Junior High School, Brandon, Florida*

Staff Credits

PUBLISHER	Stephanie Rogalin
EDITORIAL DIRECTOR	Mimi Jigarjian
MARKETING DIRECTOR	Bridget A. Hadley
PROJECT EDITOR	Edward De Leon
EDITORS	Judith D. Buice Renee S. Heffernan Jo Ann M. Webber
BUYER	Roger Powers
MANUFACTURING	Bill Wood
BOOK PRODUCTION EDITORS	Jo-Ann Connolly Anne Benaquist
ASST. PRODUCTION MANAGER	Pauline Wright
ART DIRECTOR AND COVER ART DIRECTION	L. Christopher Valente
BOOK DESIGN/PRODUCTION COORDINATOR	Jonathan Pollard/Richard Dalton
PHOTO RESEARCHER	Russell Lappa/Karen McDonough, Assistant

Outside Credits

EDITORIAL	Barbara J. Huffman
ART AND DESIGN ASSISTANTS	Eve Melnechuk—Project Leader; Sarah Aubry, James Brown, Helen Byers, Meral Dabcovich, Andrea Grassi, Fran Jarvis, Kerry McGuire, Kevin Spotts, and Sarah Prince
COVER DESIGN	Hannus Design Assoc.
BOOK DESIGN	The Chestnut House Group, Inc.
FRONT/END MATTER/ *DATA FILE* DESIGNS	Maureen Lauran

ISBN 0-13-693912-0
Printed in the United States of America.
3 4 5 95 94 93 92

PRENTICE HALL
dedicates
this mathematics program
to
all mathematics educators
and
their students.

Contents

1 *Integers and Expressions*

DATA FILE 1 — 2

1-1	Integers	*4*
	Exploring Integers	*8*
1-2	Adding Integers	*10*
	CRITICAL THINKING: Exploring Classification	
1-3	Subtracting Integers	*14*
	Exploring Number Patterns	*18*
1-4	***Problem Solving Strategy:*** Look for a Pattern	*19*
	CRITICAL THINKING: Exploring Venn Diagrams	
1-5	Multiplying Integers	*22*
1-6	Dividing Integers	*26*
	TEST YOURSELF	
	Practice	*30*
1-7	***Problem Solving Application:*** Fitness and Health	*31*
	Decision Making	
1-8	Expressions and Variables	*34*
	Exploring Spreadsheets	*38*
1-9	Order of Operations	*40*
1-10	Evaluating Expressions	*44*
	Problem Solving Practice	*47*

REVIEW • *48* TEST • *50*
CUMULATIVE REVIEW • *51*

How do fish live in a frozen pond? see p. 2

What is a proof set? see p. 97

2 *Solving Equations*

DATA FILE 2 — 52

2-1	Variables and Equations	*54*
2-2	Properties of Operations	*58*
2-3	The Distributive Property	*62*
	CRITICAL THINKING: Exploring Logical Thinking	
2-4	Simplifying Variable Expressions	*66*
	TEST YOURSELF	
	Exploring Equations	*70*
2-5	Addition and Subtraction Equations	*72*
2-6	Multiplication and Division Equations	*76*
	CRITICAL THINKING: Exploring Visual Thinking	
2-7	Writing Equations	*80*
	TEST YOURSELF	
	Exploring Graphing	*83*
2-8	***Problem Solving Strategy:*** Guess and Test	*85*
2-9	***Problem Solving Application:*** Ecology	*87*
	Decision Making	
	Practice	*90*
	Problem Solving Practice	*91*

REVIEW • *92* TEST • *94*
CUMULATIVE REVIEW • *95*

IN EVERY CHAPTER, LOOK FOR:

ROJECT ▼ WRITE ▼ DATA ▼ CALCULATOR ▼ COMPUTER ▼ ESTIMATION

3 Decimals and Equations

DATA FILE 3 — 96

Exploring Decimals — 98
3-1 Decimals — 100
3-2 Estimating with Decimals — 104
3-3 Expressions with Decimals — 108
CRITICAL THINKING: Exploring Patterns in Division
3-4 Addition and Subtraction Equations — 111
TEST YOURSELF
3-5 Multiplication and Division Equations — 115
CRITICAL THINKING: Exploring Logical Families
Practice — 119
Exploring Data in Tables — 120
Exploring Data Bases — 122
Problem Solving Practice — 124
3-6 *Problem Solving Application:* Using the Metric System — 125
Decision Making
3-7 Using Formulas — 128
TEST YOURSELF
3-8 *Problem Solving Strategy:* Simplify the Problem — 132
REVIEW • 134 TEST • 136
CUMULATIVE REVIEW • 137

4 Number Theory

DATA FILE 4 — 138

Exploring Square Numbers — 140
4-1 Exponents — 141
CRITICAL THINKING: Exploring Number Patterns
4-2 Rules About Exponents — 145
CRITICAL THINKING: Exploring Pascal's Triangle
4-3 Scientific Notation — 149
TEST YOURSELF
4-4 *Problem Solving Application:* Water Resources — 153
Decision Making
Exploring Factors and Multiples — 156
Practice — 158
4-5 Factors, Multiples, and Divisibility — 159
TEST YOURSELF
4-6 Prime Factorization — 163
CRITICAL THINKING: Exploring Classification
4-7 GCF and LCM — 167
Exploring Counting Problems — 171
4-8 *Problem Solving Strategy:* Account for All Possibilities — 172
Problem Solving Practice — 175
REVIEW • 176 TEST • 178
CUMULATIVE REVIEW • 179

What makes up population change? see p. 139

5 Rational Numbers and Expressions

DATA FILE 5 — 180

Exploring Fractions — 182
5-1 Equivalent Fractions and Lowest Terms — 184
5-2 Fractions and Decimals — 188
5-3 Rational Numbers — 192
TEST YOURSELF
5-4 Comparing and Ordering Rational Numbers — 196
5-5 Adding and Subtracting Rational Numbers — 199
CRITICAL THINKING: Exploring Visualization
5-6 *Problem Solving Strategy:* Working Backwards — 203
Practice — 205
5-7 Multiplying and Dividing Rational Numbers — 206
Exploring Infinity — 209
5-8 Rational Numbers with Exponents — 211
CRITICAL THINKING: Exploring Patterns in Scientific Notation
5-9 Addition and Subtraction Equations — 215

▼ DATA ▼ CALCULATOR ▼ COMPUTER ▼ ESTIMATION ▼ MENTAL MATH

5-10 Multiplication Equations 218
TEST YOURSELF
5-11 *Problem Solving Application:*
The Stock Market 222
Decision Making
Problem Solving Practice 225
REVIEW • *226* TEST • *228*
CUMULATIVE REVIEW • *229*

How long do most teens study per day? see p. 230

6 Ratios, Proportions, and Percent

DATA FILE 6 230
Exploring Ratios 232
6-1 Ratios, Proportions, and Rates 233
6-2 Solving Proportions 237
CRITICAL THINKING:
Exploring Venn Diagrams
6-3 *Problem Solving Application:*
Scale Drawing 241
Decision Making
Exploring Percents 244
6-4 Percent 245
Practice 249
Exploring Number Relationships 250
6-5 Using Proportions to Find Percent 252
TEST YOURSELF
6-6 Percents and Equations 256
CRITICAL THINKING:
Exploring Visual Perception
6-7 Percent of Change 260
TEST YOURSELF
6-8 *Problem Solving Strategy:*
Draw a Diagram 264
Problem Solving Practice 267
REVIEW • *268* TEST • *270*
CUMULATIVE REVIEW • *271*

7 Equations and Inequalities

DATA FILE 7 272
Exploring Two-step Equations 274
7-1 Two-step Equations 275
7-2 Simplifying and Solving Equations 279
TEST YOURSELF
Problem Solving Practice 283
7-3 *Problem Solving Strategy:*
Writing Equations 284
CRITICAL THINKING:
Exploring Clock Arithmetic
7-4 Equations with Variables on Both Sides 287
Exploring Inequalities 291
7-5 Inequalities and Their Graphs 292
7-6 Solving One-step Inequalities 296
TEST YOURSELF
7-7 Solving Two-step Inequalities 300
CRITICAL THINKING:
Exploring Quantitative Comparisons
Practice 304
7-8 *Problem Solving Application:*
Buying a Car 305
Decision Making
REVIEW • *308* TEST • *310*
CUMULATIVE REVIEW • *311*

8 Graphing in the Coordinate Plane

DATA FILE 8 312
Exploring Verbal Communication 314
8-1 The Coordinate Plane 316
CRITICAL THINKING:
Exploring Graphing
8-2 Solving Equations 320
TEST YOURSELF
8-3 Graphing Linear Equations 324
Exploring Slope 328
8-4 Slope and y-intercept 330
8-5 *Problem Solving Strategy:*
Solve by Graphing 334
Problem Solving Practice 337
Exploring Simulations 338
8-6 Solving Systems of Linear Equations 340
TEST YOURSELF

IN EVERY CHAPTER, LOOK FOR:

ROJECT ▼ WRITE ▼ DATA ▼ CALCULATOR ▼ COMPUTER ▼ ESTIMATION

8-7 Solving Linear Inequalities 344
CRITICAL THINKING:
Exploring Graphing
8-8 Graphing Linear Inequalities 348
Practice 352
8-9 *Problem Solving Application:*
Direct and Indirect Variation 353
Decision Making
REVIEW • *356* TEST • *358*
CUMULATIVE REVIEW • *359*

What is the loft of a golf club? see p. 360

9 Algebra in Geometry and Measurement

DATA FILE 9 360

9-1 Introduction to Geometry 362
9-2 Angles 365
Exploring Segments and Angles 369
9-3 Polygons and Quadrilaterals 370
Exploring with LOGO 373
9-4 Triangles 374
CRITICAL THINKING:
Exploring Inductive Reasoning
9-5 Circles 377
9-6 Congruence and Symmetry 380
TEST YOURSELF
9-7 Similar Figures 384
CRITICAL THINKING:
Exploring Deductive Reasoning
Practice 388
Exploring Pi 389
9-8 Perimeter and Circumference 390
TEST YOURSELF
9-9 *Problem Solving Strategy:*
Draw a Diagram 394
Problem Solving Practice 396
9-10 *Problem Solving Application:*
Tessellations 397
Decision Making
REVIEW • *400* TEST • *402*
CUMULATIVE REVIEW • *403*

10 Area and Volume Formulas

DATA FILE 10 404

10-1 Area of Rectangles and Parallelograms 406
10-2 Area of Triangles and Trapezoids 410
10-3 Area of Circles 414
TEST YOURSELF
Exploring Pick's Theorem 418
10-4 Space Figures 420
CRITICAL THINKING:
Exploring Visualization
Exploring Patterns in a Cube 424
10-5 *Problem Solving Strategy:*
Make a Model 425
10-6 Surface Area: Prisms and Cylinders 427
10-7 Surface Area: Pyramids, Cones, Spheres 430
10-8 *Problem Solving Application:*
Parade Floats 433
Decision Making
Practice 436
10-9 Volume: Prisms and Cylinders 437
TEST YOURSELF
10-10 Volume: Pyramids, Cones, Spheres 441
CRITICAL THINKING:
Exploring a Hypothesis
Problem Solving Practice 445
REVIEW • *446* TEST • *448*
CUMULATIVE REVIEW • *449*

What is an animal's home range? see p. 405

▼ DATA ▼ CALCULATOR ▼ COMPUTER ▼ ESTIMATION ▼ MENTAL MATH

11 Right Triangles in Algebra

DATA FILE 11 450

Exploring Square Roots 452
11–1 Finding Square Roots 454
CRITICAL THINKING: Exploring Patterns
11–2 *Problem Solving Strategy:* Simulating a Problem 458
Exploring Right Triangles 460
11–3 Pythagorean Theorem 461
11–4 Similar Right Triangles 465
TEST YOURSELF
11–5 Special Right Triangles 469
CRITICAL THINKING: Exploring Logic
11–6 Trigonometric Ratios 473
TEST YOURSELF
11–7 *Problem Solving Application:* Precision Drawing 477
Decision Making
Practice 480
Problem Solving Practice 481
REVIEW • 482 TEST • 484
CUMULATIVE REVIEW • 485

12 Statistics and Probability

DATA FILE 12 486

12–1 Mean, Median, and Mode 488
12–2 Line Plots and Frequency Tables 492
Exploring Misuse of Statistics 496
12–3 Stem and Leaf Plots 498
CRITICAL THINKING: Exploring Histograms
12–4 Box and Whisker Plots 502
TEST YOURSELF
Exploring Fair and Unfair Games 506
12–5 Counting Principle 508
CRITICAL THINKING: Exploring Visual Thinking
12–6 Probability 512
Exploring Pascal's Triangle 516
12–7 *Problem Solving Strategy:* Simulate the Problem 518
Practice 521
12–8 Independent and Dependent Events 522
TEST YOURSELF
Problem Solving Practice 526
12–9 *Problem Solving Application:* Making Predictions 527
Decision Making
REVIEW • 530 TEST • 532
CUMULATIVE REVIEW • 533

Who designed the 1495 parachute? see p. 450

13 Polynomials

DATA FILE 13 534

Exploring Polynomials 536
13–1 Polynomials 538
CRITICAL THINKING: Exploring Conclusions
13–2 Adding and Subtracting Polynomials 542
13–3 Multiplying a Polynomial by a Monomial 546
Exploring Mind-reading Tricks 549
13–4 Multiplying Binomials 550
TEST YOURSELF
Practice 554
13–5 *Problem Solving Application:* Using Matrices 555
Decision Making
13–6 *Problem Solving Strategy:* Using Multiple Strategies 558
Problem Solving Practice 561
REVIEW • 562 TEST • 564
CUMULATIVE REVIEW • 565

Extra Practice 566
Tables 579
Glossary 581
Selected Answers 589
Index 605
Acknowledgments 612

IN EVERY CHAPTER, LOOK FOR:

ROJECT ▼ WRITE ▼ DATA ▼ CALCULATOR ▼ COMPUTER ▼ ESTIMATION

IN EVERY CHAPTER LOOK FOR:

PROJECT ▼ DATA ▼ WRITE ▼ CALCULATOR ▼ COMPUTER ▼ ESTIMATION ▼ MENTAL MATH

ESTIMATION

17, 88, 89, 104, 107, 110, 130, 147, 183, 201, 221, 247, 259, 339, 453

EXPLORING

Counting Problems 171
Data Bases 122
Data in Tables 120
Decimals 98
Equations 70
Factors and Multiples 156
Fair and Unfair Games 506
Fractions 182
Graphing 83
Inequalities 291
Infinity 209
Integers 8
LOGO 373
Mind-reading Tricks 549
Misuse of Statistics 496
Number Patterns 18
Number Relationships 250
Pascal's Triangle 516
Patterns in a Cube 424
Percents 244
Pi 389
Pick's Theorem 418
Polynomials 536
Ratios 232
Right Triangles 460
Segments and Angles 369
Simulations 338
Slope 328
Spreadsheets 38
Square Numbers 140
Square Roots 452
Two-step Equations 274
Verbal Communication 314

COMPUTER

38, 43, 91, 118, 144, 147, 209, 295, 299, 338, 373, 417, 457

CRITICAL THINKING

EXPLORING 13, 21, 65, 79, 110, 118, 144, 148, 166, 202, 214, 240, 259, 286, 303, 319, 347, 376, 387, 423, 444, 457, 472, 501, 511, 541

THINK ABOUT IT . . . 3, 53, 97, 139, 181, 231, 273, 313, 361, 405, 451, 487, 535

DECISION MAKING 32, 88, 126, 154, 222, 242, 306, 354, 398, 434, 478, 528, 556

PROJECT

6, 7, 12, 29, 30, 39, 75, 82, 84, 89, 90, 113, 119, 123, 131, 144, 151, 152, 155, 158, 187, 191, 205, 214, 220, 232, 235, 239, 240, 249, 263, 299, 304, 319, 323, 333, 336, 351, 352, 364, 383, 387, 388, 409, 423, 429, 436, 439, 480, 497, 504, 505, 507, 515, 521, 525, 554

WRITE

7, 8, 13, 15, 18, 25, 43, 46, 57, 61, 65, 82, 103, 107, 110, 114, 117, 140, 143, 144, 152, 157, 191, 195, 209, 217, 224, 232, 244, 248, 251, 254, 255, 257, 262, 263, 290, 315, 336, 351, 363, 364, 386, 387, 419, 444, 452, 491, 497, 499, 504, 511, 515, 517, 525, 541, 545, 547, 553

MENTAL MATH

12, 16, 24, 28, 42, 45, 56, 60, 61, 64, 75, 77, 78, 109, 113, 117, 143, 187, 198, 201, 202, 208, 217, 254, 258, 262, 277, 281, 289, 303, 318, 319, 322, 327, 332, 378, 456, 463, 467, 471, 475, 490, 510, 541, 552

CALCULATOR

7, 12, 16, 24, 28, 42, 45, 46, 56, 75, 78, 109, 113, 117, 130, 131, 143, 147, 151, 157, 165, 170, 187, 189, 190, 191, 198, 201, 204, 247, 254, 258, 262, 277, 281, 289, 294, 295, 322, 346, 379, 392, 408, 409, 413, 429, 432, 443, 463, 464, 467, 472, 481, 490, 510, 541, 547

DATA

7, 12, 17, 20, 25, 29, 37, 43, 46, 47, 50, 57, 65, 75, 78, 82, 87, 91, 103, 107, 124, 126, 136, 147, 151, 153, 154, 165, 170, 175, 191, 195, 198, 224, 225, 236, 239, 240, 248, 254, 258, 259, 262, 263, 266, 267, 278, 282, 286, 290, 294, 295, 299, 303, 327, 343, 387, 393, 396, 416, 445, 459, 472, 476, 481, 491, 495, 500, 511, 515, 525, 553, 560, 561

CAREERS

Acoustical Physicist 205
Actuary 521
Airline Pilot 30
Chef 249
Civil Engineer 352
Demographer 158
Director of Marketing 304
Finance Manager 554
Landscape Architect 480
Medical Record Technician 119
Operations Research Analyst 436
Optician 388
Urban and Regional Planner 90

1

Chapter At a Glance

Chapter 1 introduces the students to the concept of integers. The relationship between integers and whole numbers and the natural numbers is established early in the chapter.

Reading the Chapter

This chapter contains words and concepts that can be difficult for students who are not familiar with reading mathematics. Words such as *positive, negative,* and *variable* have different meanings in a mathematical context than within the context of nonmathematical English. Review new terms because understanding the vocabulary is essential to understanding the ideas in the chapter.

Chapter Resources

Practice and Enrichment TRB A-1 to A-12, A-14 to A-19
Alternate Application TRB A-13
Chapter Test Form A, A-21
Chapter Test Form B, A-23
Computer Test Bank

References

STUDENT BIBLIOGRAPHY

Benson, William H., and Oswald Jacoby. *Recreations with Magic Squares.* (Dover, New York 1976).

TEACHER BIBLIOGRAPHY

Kohn, Judith B. "A Physical Model for the Operations with Integers." *Mathematics Teacher.* December 1978, pp. 734-736.

SOFTWARE

Appleworks (Apple Computer), Apple II family.
Lotus 1-2-3 (Lotus Development Corp.), IBM computers.
Microsoft Works (Microsoft Corp.), IBM or Macintosh.

FILMS, VIDEOTAPES, AND OTHER MEDIA

Classroom Management: A Proactive Approach to Creating an Effective Learning Environment (ASCD), VHS, or Beta.

Vocabulary

absolute value
cell
cursor
evaluate
integer
intersection
mean
negative
numerical expression
opposite
order of operations
positive
quantity
spreadsheet
variable
variable expression
Venn Diagram

Materials

algebra tiles (M*)
calculator
computer (optional)
ruler
paper
math journal
blank transparency
index cards
blue and white squares of paper
red paper rectangles
number cubes in two colors
half-sheets of paper with integers written on each

M*—Manipulative Kit

Resource Bank

The items listed below can be used as supplementary material to the chapter.

Class Activities

FIELD TRIPS

▼ Aquarium—to see how sea life has adapted to different altitudes below sea level.
▼ Airport—to observe how the tower keeps track of the altitude of airplanes.

CLASS SPEAKERS

▼ Oceanographer—to describe conditions and animal life at depths below sea level.
▼ Airplane pilot—to discuss the changes in temperature and pressure at different altitudes.

Bulletin Board

Have students research and record the temperatures of the hottest and coldest places in the world and in the United States. Have students locate the places on a map of the world. To indicate the temperature, ask students to stretch a piece of string from the location on the map to the temperature on the thermometer.

What is the hottest recorded temperature? the coldest? (136°F—El Azizia, Libya; -129°F—Soviet Antarctica)

Find the difference between the coldest and hottest recorded temperatures in the world and in the U.S. (265°F; 214°F)

What is the difference between the coldest recorded temperatures in the world and in the U.S.? (49°F)

Project: Find the hottest and coldest recorded temperatures in your city.

FIND THE HOTTEST AND COLDEST
Recorded Temperatures

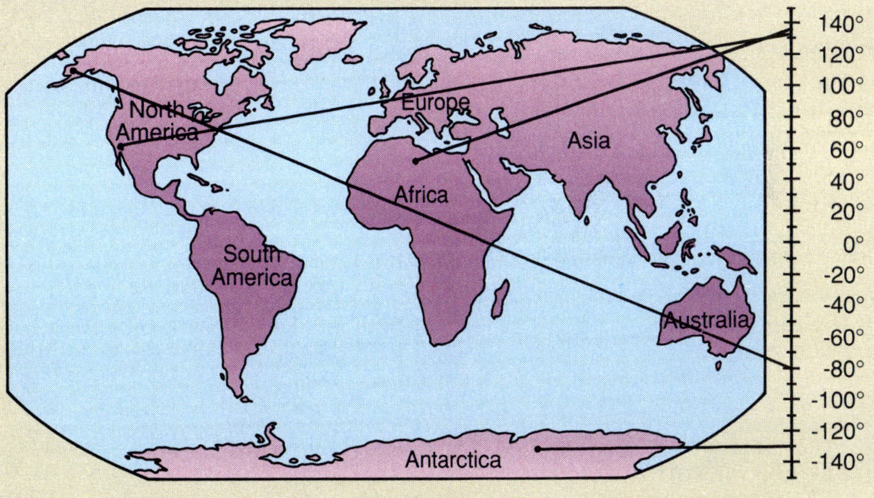

Extra-Credit Problems

BASIC

▼ List the integers between -10 and 10. Then find the average of the integers. Does it matter if the numbers -10 and 10 are included in the average? (0, no)

AVERAGE

▼ A set of numbers contains 10 integers. If 5 is added to each integer, would the average of the new set of integers be 50 more than the average of the original set? Justify your answer. (No, 5 more)

ENRICHED

▼ The average of three consecutive integers is zero. What are the three integers? (-1, 0, 1)
▼ What if the average is 1? Can the same problem be solved? Why or why not? (0, 1, 2; yes)

Chapter Planning Guide

OBJECTIVES	ASSIGNMENT GUIDE			ASSESSMENT	
	Basic	Average	Enriched	Review/ Tests	TRB*
1-1 Integers To represent integers, opposites, and absolute value.	1-2, 7-8, 13-18, 21-23, 28-29, 32-35, 40-44a, 45, 49-50, 53-54, MR* All	3-4, 9-10, 17-21, 24-25, 30-31, 34-37, 40-44a, 46-47, 50-51, 53-54, MR All	5-6, 11-12, 17-21, 26-27, 36-44b, 47-48, 51-54, MR All	Extra Practice p. 566	Practice A-1 Enrichment A-2
Exploring Integers Use models to discover rules about integers.	**Manipulative Connection** All	All	All		
1-2 Adding Integers To add integers using models, patterns, and rules.	1-11, 14-22, 29-30, 34-37, 44, 47, MR, CT* All	2, 6, 9-16, 23-31, 34-36, 38-42, 45-47, MR, CT All	3, 10-16, 29-36, 38-43, 45-49, MR, CT All	Extra Practice p. 566	Practice A-3 Enrichment A-4
1-3 Subtracting Integers To subtract integers using models, patterns, and rules.	1-8, 16-18, 25-36, 43-44, 47-50, 56-60, MR All	3-6, 10-13, 19-21, 29-38, 43, 45, 47-51, 59-64, MR All	12-15, 22-24, 33-42, 46-49, 52-55, 59-64, MR All	Extra Practice p. 566	Practice A-5 Enrichment A6
Exploring Number Patterns To explore and describe number patterns.	All	All	All		
1-4 Problem Solving Strategy: Look for a Pattern To solve problems by looking for a pattern.	**Problem Solving Connection** All	All	All	Problem Solving Practice p. 47	Practice A-7 Enrichment A-8 TY A-20
1-5 Multiplying Integers To multiply integers using repeated addition, patterns, and rules.	1-2, 5-6, 9-10, 13-15, 17-37 odd, 38-41, 46-47, 52-53, MR All	3-4, 7-12, 15-16, 18-36 even, 40-44, 47, 49-50, 52-53, MR All	9-12, 15-16, 18-36 even, 40-45, 48-53, MR All	Extra Practice p. 566	Practice A-9 Enrichment A-10
1-6 Dividing Integers To divide integers.	1-6, 9, 11, 13, 15-29, 38-43, 54-55, 57-60, MR, TY* All	3-4, 6-8, 10, 12, 14, 21-35, 38-41, 43-44, 46-47, 50-51, 54-56, 59-60, MR, TY All	6-8, 12-14, 27-41, 44-49, 52-56, 59-60, MR, TY All	Extra Practice p. 566 TY p. 29	Practice A-11 Enrichment A-12
1-7 Problem Solving Application: Fitness and Health To apply integer operations to problems involving fitness and health.	All	All	All		Alternate Application A-13
1-8 Expressions and Variables To assign variables and write variable expressions.	1-5, 8-21, 25-27, 33-37, 39, 41a, 42a, MR All	6-11, 16-24, 27-30, 34-41, 42a, MR All	8-11, 16-24, 29-32, 34, 38-46, MR All	Extra Practice p. 566	Practice A-14 Enrichment A-15 TY A-20
Exploring Spreadsheets To explore the uses of a computer spreadsheet.	All	All	All		
1-9 Order of Operations To study order of operations using pencil and paper and a calculator.	1-11, 16-24, 37-39, 45-49, 51, MR All	2-9, 12-13, 17-23 odd, 24-26, 31-36, 38-44, 48-52 55-56, MR All	4-9, 14-15, 17-23 odd, 24-28, 31-36 42-44, 48-50 53-61, MR All	Extra Practice p. 566	Practice A-16 Enrichment A-17
1-10 Evaluating Expressions To evaluate variable expressions.	1-18, 21-28, 38-39, 42, MR All	7-26, 28-31, 38-40, 43, MR All,	7-20, 31-36, 40-41, 44-47, MR All	Extra Practice p. 566	Practice A-18 Enrichment A-19
Chapter Test	All	All	All	pp. 48-51	Tests A-21–A-27

 *TRB—Teacher's Resource Book MR—Mixed Review TY—Test Yourself CT—Critical Thinking

CONNECTIONS							NCTM CORRELATION
Algebra	Critical Thinking	Problem Solving	Estimation/ Mental Math	Technology	Manipulatives/ Resources	Writing/Reading in Math	
		W*: 7-12, 21, 53		W: 44	W: 21, 53-54	C*: TD* 3 W: 21, 40-43, 54	Communication Connections Structure
	6e, 11b				1-12	1-12	Communication Connections
W: 47-49	CT*: 1-6	C: 3-5 W: 34-36, 44-46	W: 10-13	W: 14-16	C: 6-12 W: 4-9, 34-36	C: 3-5 W: 37-40, 46	Communication Connections
W: 43, 50-55	C: TD 1-3	W: 43, 47-49	W: 7-15, 56-64	W: 16-24	C: 9-10 W: 1-4, 47	C: 11 W: 44-49	Communication Connections Statistics
	3-6	1-6	1-6			1b, 2a, 2f, 3b	Communication Problem Solving
	CT: All	C: 1-2 W: 1-7 CT: All	W: 3	W: 1-7	W: 5	CT: All	Problem-Solving Communication Discrete Mathematics
	C: 7-8, TD 2-3	C: 7-8 W: 46-48, 52-53	C: 1-4 W: 9-12	W: 13-16	W: 52-53	C: TD 3 W: 38-42, 53c	Communication Statistics Connections
		W: 54-60	C: 1-4 W: 5-8	W: 9-14	W: 58-59	W: 60	Communication Statistics
		C: 2 W: 9-11		W: 1-11	C: 1-2 W: 1-11	W: 9-11	Problem Solving Connections Statistics
C: 1-16 W: 1-7, 12-24, 33-42	C: TD 3	W: 36-42			C: 5-7 W: 1-7, 42-46	C: 12-16, TD 3 W: 25-32, 39-40	Communication Algebra
5-9	6, 9	9, 11		1-11	1-11	11	Communication Connections
	C: TD 1-4 W: 61	W: 37-44, 50	W: 10-15	W: 16-26, 55-61	W: 50, 55-61	W: 44-49, 51-54	Communication Problem Solving
C: 1-14 W: 1-47	W: 46-47	C: 13-14 W: 37-41	W: 15-20	W: 21-26, 37-38	W: 39	W: 46	Communication Algebra

*C—Class Exercises W—Written Exercises TD—Think and Discuss CT—Critical Thinking

Exploring

In this manipulative activity, students use algebra tiles to model and solve subtraction problems.
See TRB Exploring O-1.

Exploring Subtracting Integers

To subtract integers using algebra tiles, it may be necessary to add zero pairs so that there are enough tiles to "take away."

Let white tiles represent positive integers.
Let blue or colored tiles represent negative integers.

Remember □□ is a zero pair.

1. Model the integer 5. □□□□□ □□
   Add 1 zero pair. What integer is represented? **5**
   Add another zero pair. What integer is represented? **5**
2. Model -6. □□□□□□ □□□□□□
   Add 3 zero pairs. What integer is represented? **-6**
   If you added 100 zero pairs, what integer would be represented? **-6**
3. What can you say about adding zero pairs to a model of an integer?
   Adding zero pairs does not change what the model represents.
4. You can use models to represent subtraction.
   To find 7 − 4, model the integer 7. □□□☒☒☒☒
   Remove 4 tiles. How many tiles are left? **3**
5. Use tiles to represent the 7 in the expression 7 − 9. ☒☒☒☒☒☒☒
   Do you have enough white tiles to remove 9? **no**
   Add 1 zero pair.
   Now do you have enough white tiles to remove 9? **no**
   Add as many zero pairs as you need to remove 9 positive tiles.
   Write the integer represented by the tiles that are left. **-2**
   What is 7 − 9? **7 − 9 = -2**
6. Model 14 in the expression 14 − (-8) □□□□□□□□□□□□□□
   Do you have enough negative tiles to remove 8? **no**
   Add as many zero pairs as you need to remove 8 negative tiles.
   Write the integer represented by the tiles that are left. **22**
   What is 14 − (-8)? **14 − (-8) = 22**
7. Model 11 in the expression 11 − (-2) − (-5). □□□□□□□□□□□
   How many zero pairs must you add so you can remove 2 negative tiles? **2**
   How many zero pairs must you add so you can remove 5 negative tiles? **5**
   What is 11 − (-2) − (-5)? **11 − (-2) − (-5) = 18**

O-1

Exploring Multiplying Integers

Models that show repeated addition can also represent multiplication.

The models at the right can represent 2 + 2 + 2. They can also represent 3 groups of 2 or 3 • 2.

Write a multiplication sentence for the model.

1. **4 • 2 = 8**
2. **2 • 3 = 6**
3. **2 • 4 = 8**
4. **4 • 3 = 12**

Use tiles to model each multiplication. Write the product. **Check students' models.**

5. 2 • 6 **12**
6. 3 • 5 **15**
7. 5 • 2 **10**
8. 6 • 1 **6**
9. What is the sign of the product when two positive integers are multiplied? **positive**

You can use models to multiply with negative integers.

The models at the right can represent -2 + (-2) + (-2). They can also represent 3 groups of -2, or 3(-2).

Make the model. Then write a multiplication sentence for the model.

10. **2(-2) = -4**
11. **3(-4) = -12**
12. **4(-3) = -12**
13. **8(-2) = -16**

Use tiles to model each multiplication. Write the product. **Check students' models.**

14. 4(-2) **-8**
15. 3(-6) **-18**
16. 5(-1) **-5**
17. 5(-4) **-20**
18. When you multiply a positive integer and a negative integer, what is the sign of the product? **negative**

O-2

In this manipulative activity, students make models to represent repeated addition or multiplication.
See TRB Exploring O-2.

In this computer activity, students use a spreadsheet and formulas to answer questions that involve critical thinking.
See TRB Exploring O-3.

Exploring Integers Using a Spreadsheet

If the values in one or more cells of a spreadsheet change, the values in all related cells change.

Set up the computer spreadsheet, inputting the labels shown below.

	A	B	C	D	E	F	G	H
1	a	b	c	ab	bc	(-ab)	(-bc)	abc
2								
3								

Then input the following values and formulas in the cells of row 3.

	A	B	C	D	E	F	G	H
3	2	3	4	=A3*B3	=B3*C3	=-A3*B3	=-B3*C3	=A3*B3*C3
4								
5								

1. What value is stored in each cell? D3 **6** E3 **12** F3 **-6** G3 **-12** H3 **24**

Now input these values and formulas in the cells of row 4.

	A	B	C	D	E	F	G	H
4	-2	-3	-4	=A4*B4	=B4*C4	=-A4*B4	=-B4*C4	=A4*B4*C4
5								
6								

2. What value is stored in each cell? D4 **6** E4 **12** F4 **-6** G4 **-12** H4 **-24**

Use the values at the right for columns A, B, and C in rows 5 through 10. Find all the possible products. Then analyze the results you see on the computer screen to answer the following questions.

	A	B	C
5	−2	3	4
6	−2	−3	4
7	−2	3	−4
8	2	−3	4
9	2	3	−4
10	2	−3	−4

3. If $a < 0$ and $b < 0$, what can be said about ab? **ab > 0**
4. If $b > 0$ and $c < 0$, what can be said about $-bc$? **-bc > 0**
5. If $-ab > 0$, what is true about a and b? **a < 0 or b < 0**
6. If $abc < 0$ and $a < 0$, what can be said about b and c?
   b > 0 and c > 0 or b < 0 and c < 0
7. If $abc > 0$ and $a < 0$, what can be said about b and c? **b < 0 or c < 0**

O-3

Exploring Venn Diagrams

The cook at the Taco Palace kept a tally of how many times tomatoes, cheese, and peppers were used as taco toppings in one day.

Topping	Times Used
Tomatoes	114
Cheese	69
Peppers	84

Although some tacos had only one kind of topping, most orders had several toppings. These were the tacos ordered for one day:

- 5 tacos with all three toppings
- 16 tacos with tomatoes and cheese only
- 41 tacos with tomatoes and peppers only
- 23 tacos with peppers and cheese only

Taco Toppings For One Day

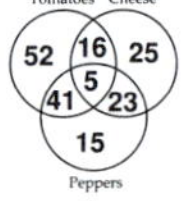

Use the given information to fill in the appropriate parts of the Venn diagram. Then complete the diagram by finding how many of the tacos had each of the following toppings.

1. tomatoes only **52**
2. peppers only **15**
3. cheese only **25**
4. tomatoes and cheese only **16**
5. tomatoes and peppers only **41**

Use the Venn diagram to answer the questions.

6. How many tacos had some tomatoes on them? **114**
7. How many tacos had both cheese and peppers on them? **28**
8. How many tacos had peppers but not cheese? **56**
9. How many tacos had tomatoes but not peppers? **68**
10. How many tacos had toppings that did not include peppers? **93**
11. How many tacos with at least one of the three toppings were ordered today? **177**
12. Can you assume that tomatoes, cheese, and peppers are the only toppings served at the Taco Palace? Explain.
    No; there could be more. The beginning statement does not say these are the only toppings.
13. Suppose you needed to draw another circle to represent 45 tacos with none of the above toppings. Would the circles intersect? **no**

O-4

In this problem solving activity, students use Venn Diagrams to solve problems.
See TRB Exploring O-4.

Meeting Individual Needs

GIFTED STUDENTS

Mathematically gifted students present a welcome learning opportunity for the entire class. The skills and energies of the gifted student can benefit all the students in the class.

Cooperative learning groups of 4-6 students, with at least one gifted student, have proven to be an excellent means of integrating mathematically gifted students with students of more average mathematical abilities. In groups, students work together on a particular task or problem and are encouraged to share knowledge and abilities. This setting allows the gifted student to reinforce his or her skills while helping others to improve their own.

To expand the topics of this chapter, ask groups of students to explore working with a finite number base system (the traditional system is base 10). Suggest that students use base 8 or base 12. Have students create addition, subtraction, multiplication, and division tables for the number base system of their choice.

AT-RISK STUDENTS

At-risk students often need some help perceiving the relevance of mathematics to their daily lives. For such students, it is important to emphasize the practical aspects of mathematics and the ways in which good math skills are increasingly useful in today's job market. To connect the material, use math skills in everyday situations. For example, in this chapter, sections on dieting, weather, and sports show the everyday uses of integers.

Another technique for motivating at-risk students is the cooperative learning group, in which their contribution to the group is needed and rewarded. The cooperative learning setting emphasizes the task and how well the group performs, rather than focusing on the individual grade. Many activities in this chapter can be completed by pairs or small groups of students.

To increase students' awareness of the ways math is needed in the job market, ask them to research one of the careers mentioned in the chapter. Have students find out the educational level required, the salary range, and future market for this career. Then ask students to discuss what math skills might be needed to do the job.

ENGLISH AS A SECOND LANGUAGE

Although mathematics is considered a universal language, it holds special challenges for ESL students. In the classroom, the ESL student must adapt to a new culture and language while trying to assimilate mathematical language and ideas.

To help the ESL student, use clear, concise, unidiomatic speech. Keep sentences short and avoid colloquialisms. Allow students time to form responses. This will encourage the ESL student to interact orally within the classroom.

To help students understand the material in the chapter and improve their listening skills, have students work in pairs to answer questions that require them to write numerical expressions. One student reads the phrase or expression slowly, while the other student listens. The first student reads the phrase or expression again, and the second student writes the phrase or expression. After two correct phrases, the students change roles.

See TRB Teaching Resources p. 51.

Problem Solving Situation

For the "Think About It" question, tell students that a rapid change in temperature will affect the structure of a material. Cold makes most substances contract while heat makes them expand. A cold glass will expand so rapidly in hot water that the glass will break. The question also asks how engineers design products to withstand rapid changes in temperature. To help students understand the dynamics of such changes, ask them to research ways in which the knowledge of extremes of temperature is used in scientific applications. Students will then share the information with the class.

DATA COLLECTION

Using encyclopedias, science books, and input from science teachers, students can research one of the following topics: cryogenics, thermodynamics, heat, superconductivity, freezing and melting points, and the Kelvin scale. For each topic, students will find a definition and identify current and potential uses. Students may choose to present the material using charts, graphs, temperature scales for comparison, pictures, or other forms of visual aids.

DATA ANALYSIS

Ask students to use the visual aids to speculate about the limitations and possibilities of the technology they researched. Because the topics may be difficult for the students to comprehend, it may be helpful to invite the science or physics teacher to clarify the concepts involved.

DATA FILE

Just how cold is it?

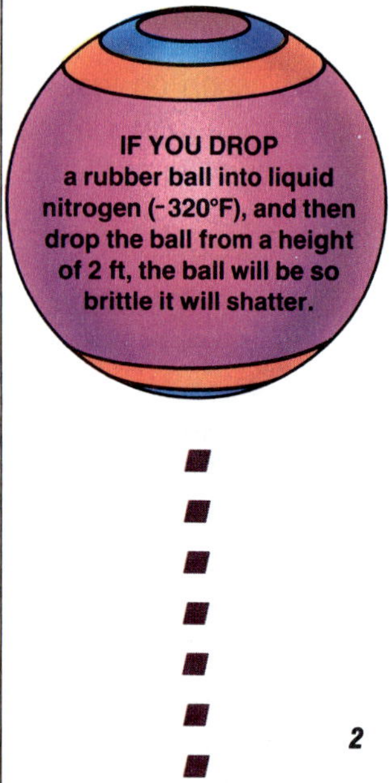

2

CHAPTER 1

Integers and Expressions

Think about it...

If you plunge a very cold glass into hot water, the glass will break. Why does this happen? How do engineers design products to withstand rapid changes in temperature?

See Problem Solving Situation

FREEZING POINTS of common substances

- Water 32°F
- Vinegar 2°F
- Gasoline −70°F
- Sugar 300°F
- Salt 1,474°F

WIND CHILL

IN 1939, wind chill measurements were developed from experiments done in Antarctica. In 1943, Siple and Passell of the United States Army Climatic Research Unit used the concept of wind chill to help determine a soldier's clothing needs in very cold climates.

AIR TEMP (°F)	WIND SPEED IN MILES PER HOUR									
	0	5	10	15	20	25	30	35	40	
	Equivalent wind chill temperatures									
35	35	32	22	16	12	8	6	4	3	Little danger of frostbite
30	30	27	16	9	4	1	-2	-4	-5	
25	25	22	10	2	-3	-7	-10	-12	-13	
20	20	16	3	-5	-10	-15	-18	-20	-21	Increased danger of frostbite
15	15	11	-3	-11	-17	-22	-25	-27	-29	
10	10	6	-9	-18	-24	-29	-33	-35	-37	
5	5	0	-15	-25	-31	-36	-41	-43	-45	
0	0	-5	-22	-31	-39	-44	-49	-52	-53	
-5	-5	-10	-27	-38	-46	-51	-56	-58	-60	
-10	-10	-15	-34	-45	-53	-59	-64	-67	-69	
-15	-15	-21	-40	-51	-60	-66	-71	-74	-76	High danger of frostbite
-20	-20	-26	-46	-58	-67	-74	-79	-82	-84	
-25	-25	-31	-52	-65	-75	-81	-86	-89	-92	

3

Cooperative Learning

This research project lends itself well to a cooperative learning format in which the class is divided into heterogeneous groups of 3 to 5 students. Each group can collect and analyze the data together.

To encourage students' participation, each group member is assigned a role. Possible roles include Writer, Scientist, Illustrator, Encourager, Presenter, and Coordinator. After collecting the data, the group will discuss the question raised in "Think About It." Explain to students that each group member must understand the conclusions reached by the group and be able to explain the group's research to the class. Comment that each member's contributions are important to the entire process.

Questions related to this Data File can be found on pages 7, 47, 165, 286, 409, 463, and 566.

Lesson Focus

Materials

Ruler, paper, pencil, and TRB Teaching Resources

Vocabulary/Symbols

absolute value
integer
negative
opposite
positive

MOTIVATING THE LESSON

Begin the lesson by discussing how numbers are used to measure or describe temperatures, distances above and below sea level, and debits and credits in a bank account. Explain that zero and numbers above and below zero are needed in all these situations.

Instruction

1 Ask students to list pairs of opposite integers that might be used to describe other familiar situations. Record pairs of integers as they are suggested.

2 Explain that in the set of integers, zero is the reference point. We use a zero reference point to separate negative and positive numbers. Note that zero is neither positive nor negative.

3 Use the number line transparency master to demonstrate graphing integers. Ask volunteers to graph pairs of points from the list of pairs made at the beginning of the lesson.

Teaching Tip

Use the number line master to make copies for the students. Have the students use these number lines in working the examples or graphing solutions to exercises.

OBJECTIVE: *To represent integers, opposites, and absolute value.*

1-1 Integers

▼ Suppose you earn $25 babysitting. The next week you spend $25 on sports equipment. You know that earning $25 is the *opposite* of spending $25. How can we represent this with numbers?

To represent earning $25, use a *positive integer:* +25, or 25.
To represent spending $25, use a *negative integer:* -25.

READ -25 as *negative 25* or *the opposite of 25*.

1

THINK What integer is neither negative nor positive? 0

2

Integers The whole numbers and their opposites form the set of integers.

..., -4, -3, -2, -1, 0, 1, 2, 3, 4, ...

negative zero positive

Example 1 **Write an integer to represent each situation.**

a. earning $50 **b.** a debt of $30

Solution **a.** +50 or 50 **b.** -30

▼ You can represent integers on a number line.

Example 2 **Write the integer represented by each point on the number line.**

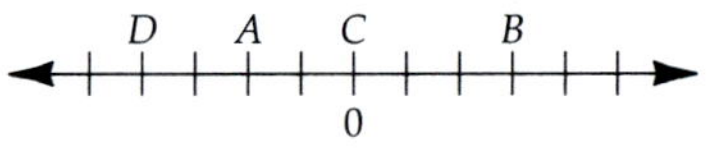

a. *A* **b.** *B* **c.** *C* **d.** *D*

Solution **a.** -2 **b.** 3 **c.** 0 **d.** -4

Opposites Opposites are two integers the same distance from zero on a number line, but in opposite directions.

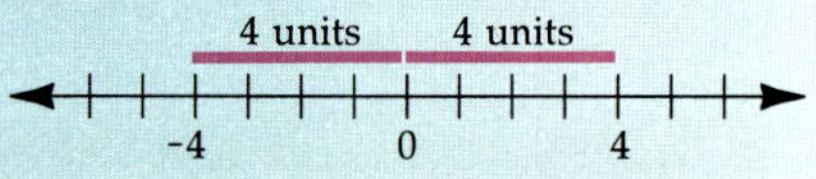

Positive integers relate to high tide and negative integers to low tide.
How can you use integers to describe water levels at high tide and low tide?

Example 3 **Graph -6 and its opposite on a number line.**

Solution Draw a number line. Make sure the tic marks are evenly spaced. Mark points on the number line at -6 and 6.

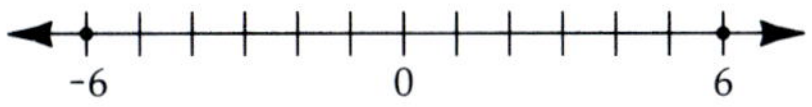

Mark zero on the number line first.

3

▼ Opposite integers are the same distance from zero on a number line. This means they have the same *absolute value*.

Absolute Value	The absolute value of an integer is its distance from zero on a number line.

Example 4 **a.** $|-5| = 5$ **b.** $|5| = 5$

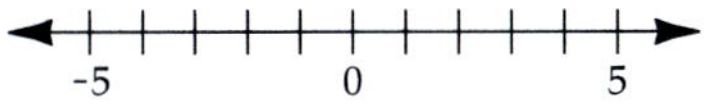

-5 is 5 units from 0.
5 is 5 units from 0.

▼ You can also use situations and number lines to compare integers.

Example 5 On Monday the temperature was -10°. On Tuesday it was -15°. It was colder on Tuesday since 15° below zero is colder than 10° below zero. Write: $-15 < -10$.

Example 6 To compare -3 and -6, think of a number line. On a horizontal number line, the integer farther to the right is the greater integer.

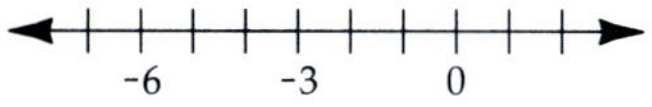

$-3 > -6$ and $-6 < -3$

-3 is to the right of -6, so -3 is greater than -6 and -6 is less than -3.

THINK Why can't the absolute value of a number be negative? Absolute value measures distance and distance is always nonnegative.

FLASHBACK

= *is equal to*
< *is less than*
> *is greater than*

CLASS EXERCISES

Give an integer to represent each situation.

1. a profit of \$250 250
2. 18° below 0 -18
3. 45 s before launch -45

Give the integer represented by each point on the number line.

4. *A* 2
5. *B* 5
6. *C* -4

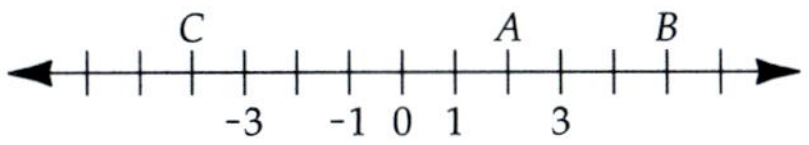

THINK AND DISCUSS See below.

1. Think of a number that is not an integer.
2. Explain how a thermometer is like a vertical number line.
3. Compare the set of positive integers with the set of nonnegative integers. Are they the same set? Explain.

1. Possible answers: $\frac{1}{4}$, 2.7
2. Positive and negative temperatures are above and below zero.
3. No. The nonnegative integers include 0. The positive integers do not.

ADDITIONAL EXAMPLES

1. **a.** a descent of 2 miles (-2)
   b. an ascent of 2 miles (2)
2. What integers are between points *A* and *B* on the number line? (-1, 0, 1, 2)
3. Graph 7 and its opposite on the number line.

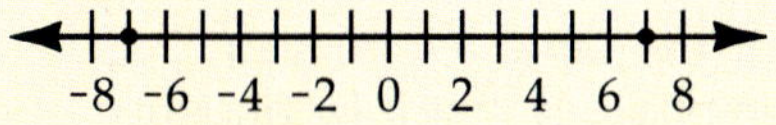

4. Find the absolute value.
   a. 18 (18) **b.** -18 (18)
5. Compare -20 and -21. ($-20 > -21$ and $-21 < -20$)

Guided Practice

Class Exercises Have students work these exercises orally. Ask students to explain how they arrived at their answers for questions 7-14.

Think and Discuss If students have difficulty with question 1, have them review the definition of *integers*. Then help the students rephrase the question as: *Think of a number between 1 and 2.* (1.1, 1.5, etc.) For question 3, students need to recognize that the set of *nonnegative* integers includes zero. The set of positive integers does not include zero.

MATH MINUTES

Make a two-column list or prepare cards for each set of ideas below. Have students match each idea with its opposite.
Set 1: cold, dry, positive, -24, loss, zero, |24|
Set 2: 24, negative, zero, gain, hot, -24, wet

A ***common error*** is to assume that zero is a positive integer and that its opposite is -0. Help students to see that zero separates the positive and negative integers. Use a number line with zero highlighted or with a longer tic mark for zero than for the other integers to emphasize that it belongs to neither group of integers.

Closure

After ensuring that students understand the Class Exercises, ask them to explain in their own words the concepts of *opposites* and *absolute value*.

Independent Practice

For questions 17-20, encourage students to translate the words into numerals and symbols before they attempt to find the answers.

Writing in Math For question 54, discuss the three suggestions for journal entries to ensure that students understand what each entails. Students can use the last suggestion to remind themselves of concepts or skills that they find confusing. Encourage students to use what they wrote to seek assistance. (Students should always have a math journal available for class.)

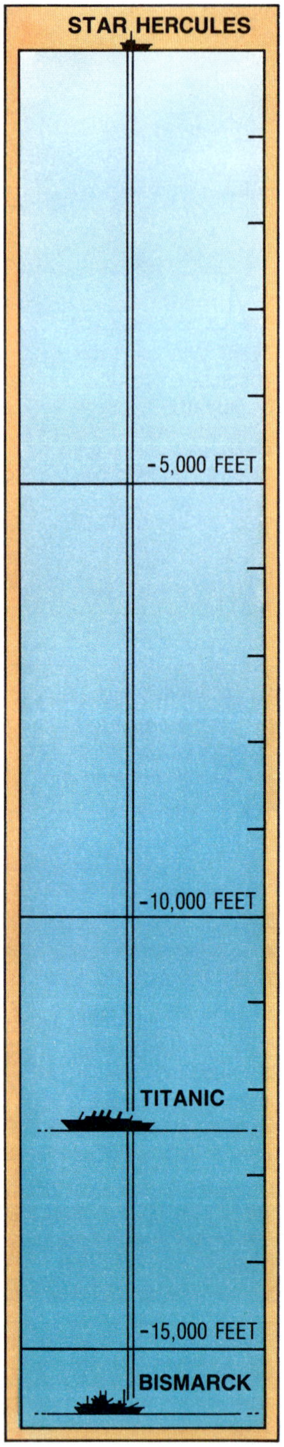

Give each number.

7. the opposite of 4 -4 **8.** the opposite of -9 9 **9.** |18| 18 **10.** |-3| 3

Name the greater integer.

11. -2, -4 -2 **12.** 8, -9 8 **13.** -12, -9 -9 **14.** 0, -6 0

WRITTEN EXERCISES

Write an integer to represent each situation. 1. 110 2. -50 3. -300

1. a deposit of $110 **2.** a debt of 50 **3.** 300 ft below sea level

4. win by 7 points 7 **5.** a loss of 8 yd -8 **6.** an elevation of 3,400 ft 3,400

Describe a situation that each integer could represent.

▼▼ *SAMPLE* -8: 8 min before liftoff Answers will vary.

7. -5 **8.** -1,000 **9.** 28 **10.** 7 **11.** 0 **12.** -126

Write each integer.

13. the opposite of 6 -6 **14.** the opposite of -2 2 **15.** |-3| 3 **16.** |13| 13

17. the absolute value of negative four 4

18. the absolute value of the opposite of sixteen 16

19. the opposite of the absolute value of negative nine -9

20. the opposite of the opposite of eight 8

Use the article below and the graph at the left.

21. a. Write integers that represent the positions of the *Titanic*, the *Bismarck*, and the *Star Hercules*. -12,500; -15,617; 0

b. ***PROJECT*** Research the *Titanic* or the *Bismarck*. Find out more about how and when each ship sank and was found.

Search Ends, Ships Found

In 1990 two historic ships were found deep in the North Atlantic Ocean. Both state-of-the-art ships survived less than one week.

The luxury passenger liner *Titanic* struck an iceberg. It came to rest 12,500 ft below sea level. The *Titanic* was 882 ft long, 92 ft wide, and displaced 66,000 t of water. The mighty warship *Bismarck* sank in battle. The *Bismarck* was 823 ft long, 118 ft wide, and displaced 50,000 t of water.

The *Star Hercules*, only 269 ft long, towed the underwater camera sled *Argo* that found the *Bismarck* under 15,617 ft of water.

Write the integer for each point on the number line,

22. *A* 6 **23.** *B* -2

24. *C* 2 **25.** *D* -8

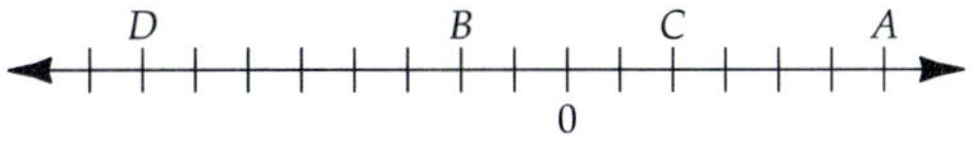

26. 8 units to the left of -6 -14 **27.** 10 units to the right of -2 8

Check students' graphs.

Graph each integer and its opposite on a number line.

28. 1, -3 **29.** -2, -8 **30.** 5, 0 **31.** 4, -1

Compare. Use >, <, or =.

32. -8 < 0 **33.** 4 > -25 **34.** |3| < |8| **35.** |-1| < |50|

36. -9 < -2 **37.** |-6| < |-12| **38.** -1 > -5 **39.** |10| = |-10|

Complete with a word that makes each statement true.

40. All ▒ integers are less than zero. negative

41. An integer is negative, positive, or ▒. zero

42. The opposite of a ▒ number is negative. positive

43. The absolute value of an integer is never ▒. negative

44. *CALCULATOR* 44. a. The opposite of the original integer; 25 (+/-)

a. Enter any positive integer on your calculator. Then enter (+/-). What appears in the display? What do you enter to display -25?

b. What integer will appear in the display after you enter these keys? 6 (+/-)(+/-)(+/-)(+/-). Check by entering the keys on your calculator. 6

For 45–52, answers will vary. One example is given.

Complete with an integer that makes each statement true.

45. -5 > ▒ -6 **46.** ▒ < 6 1 **47.** |-1| > ▒ 0 **48.** |▒| < 8 3

Write an integer between the given integers.

49. -6, 2 1 **50.** 0, -4 -2 **51.** 5, 1 4 **52.** -8, -12 -10

53. *DATA FILE 1 (pp. 2–3)*

a. Find the wind chill temperature when the temperature is -5°F and the wind speed is 10 mi/h. -27°

b. Starting at what wind chill temperature is there an increased danger of frostbite? a high danger of frostbite? -20°; -71°

c. ***PROJECT*** Find the freezing points of two substances that are not listed. Answers will vary.

54. *WRITE* Start a math journal. Use your journal to explain how to do an exercise you understand, to define new words in your own language, or to describe something you don't understand completely. Check students' journals.

MIXED REVIEW

Find each answer.

1. 812 + 95 907

2. 1,061 − 247 814

3. 136 × 8 1,088

4. 378 ÷ 9 42

5. 7,920 − 48 7,872

6. How many triangles are in the figure shown? 13

Lesson Follow-up

Reteaching Activity Prepare some slips of paper that each contain one integer from -7 through 7. Include absolute values and opposites on some of the slips. Have students draw a number line and place each slip of paper in its correct position on the number line.

LESSON RESOURCES

TRB Practice A-1
TRB Enrichment A-2

Practice

For use after 1-1 (pp. 4–7)

Integers

Give a numerical expression to represent each situation.

1. a loss of $70 **-70**
2. a temperature of 42° above 0 **42**
3. an elevator goes down 16 floors **-16**
4. an increase of 213 **213**

Write each integer.

5. the opposite of 17 **-17**
6. |-9| **9**
7. |21| **21**
8. the opposite of -8 **8**
9. |0| **0**
10. |-1| **1**
11. the absolute value of the opposite of negative 5 **5**
12. the absolute value of the opposite of 11 **11**

Write the integer for each point on the number line.

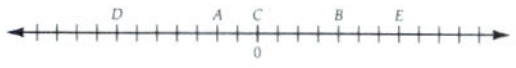

13. A **-2** 14. B **4** 15. C **0** 16. D **-7** 17. E **7**

Graph each integer on the number line.

18. -6 19. 2 20. |-4| 21. |7| 22. -|-3|

Compare. Use >, <, =.

23. -3 (<) 4 24. 5 (>) 1 25. -2 (>) -6 26. 7 (<) |8|
27. |-2| (=) |2| 28. |-1| (>) -6 29. |4| (<) |-5| 30. 0 (<) |-7|

True or false?

31. All integers are negative. **false**
32. The absolute value of every integer is its opposite. **false**
33. Two integers that are opposites have the same absolute value. **true**
34. The absolute value of an integer is always positive. **true**
35. The opposite of an integer is always negative. **false**
36. The absolute value of a postive integer is the integer. **true**
37. The absolute value of an integer is always greater than the integer. **false**
38. A positive and negative integer will always add to zero. **false**

Chapter 1 A-1

Enrichment

For use after 1-1 (pp. 4–7)

Absolute Value Equations

The absolute value of an integer is its distance from zero on a number line. You can use that fact to solve absolute value equations.

Example 1 Solve $|n| = 2$.

Solution Both -2 and 2 are at a distance of 2 units from 0 on the number line. Therefore, $n = -2$ or $n = 2$.

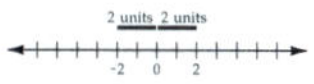

Example 2 Solve $|n - 2| = 3$.

Solution Both -1 and 5 are at a distance of 3 units from 2 on the number line. Therefore, $n = -1$ or $n = 5$.

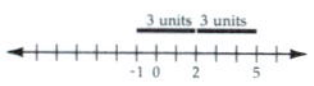

Solve by naming the possible values of *n*. Use the number line.

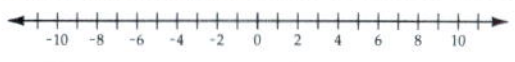

1. $|n| = 4$ $n =$ **-4** or **4**
2. $|n| = 1$ $n =$ **-1** or **1**
3. $|n| = 8$ $n =$ **-8** or **8**
4. $|n - 3| = 1$ $n =$ **2** or **4**
5. $|n - 2| = 5$ $n =$ **-3** or **7**
6. $|n - (-4)| = 6$ $n =$ **-10** or **2**
7. $|n - (-6)| = 3$ $n =$ **-9** or **-3**
8. $|n - 0| = 5$ $n =$ **-5** or **5**
9. $|n - 7| = 3$ $n =$ **4** or **10**
10. $|5 - n| = 1$ $n =$ **4** or **6**
11. $|2 - n| = 3$ $n =$ **-1** or **5**
12. $|-6 - n| = 2$ $n =$ **-8** or **-4**
13. $|n| = 0$ $n =$ **0**
14. $|n - 5| = 0$ $n =$ **5**
15. $|-3 - n| = 0$ $n =$ **-3**
16. $|-n| = 10$ $n =$ **-10** or **10**
17. $|-n + 5| = 17$ $n =$ **22** or **-12**
18. $|6 + (-n)| = 6$ $n =$ **0** or **12**
19. $|n - 11| = 2$ $n =$ **9** or **13**
20. $|0 - n| = 4$ $n =$ **4** or **-4**
21. $|3 - n| = 0$ $n =$ **3**
22. How do you know that the equation $|n| = -3$ has no solution?

Distance is always positive.

A-2 Chapter 1

LESSON QUIZ

Write the opposite and the absolute value of each integer.

1. 3 (-3, 3) **2.** -5 (5, 5)
3. 0 (0, 0) **4.** 14 (-14, 14)

Compare. Use >, <, or =.

5. 6 ▒ -5 (>)
6. -4 ▒ -9 (>)

Assignment Guide

Basic 1-2, 7-8, 13-18, 21-23, 28-29, 32-35, 40-44a, 45, 49-50, 53-54, MR All

Average 3-4, 9-10, 17-21, 24-25, 30-31, 34-37, 40-44a, 46-47, 50-51, 53-54, MR All

Enriched 5-6, 11-12, 17-21, 26-27, 36-44b, 47-48, 51-54, MR All

FOR THE NEXT LESSON

algebra tiles, checkers, or blue and white squares of paper, math journal

Lesson Focus

Materials

Algebra tiles or squares of blue and white paper, math journal

Vocabulary/Symbols

negative
positive

MOTIVATING THE LESSON

Discuss with students the different ways negative numbers are represented. For example, a negative sign, writing parentheses around negative amounts in checkbooks and financial statements, and writing negative amounts in red. Remind students that they will use tiles in two colors to represent positive and negative integers in this lesson.

Instruction

Have students form groups of three or four. Use the tiles to demonstrate how to model a positive number and a negative number. Ask students to record their work in their math journals.

(1) When groups have reached this point, call the class together to discuss their observations. Ask students to write the results in their journals. Have the students continue with their group work.

As students proceed through the lesson, monitor the groups, intervening to clarify a procedure if necessary.

OBJECTIVE: *Use models to discover rules about integers.*

MATERIALS

- Algebra tiles, checkers, or squares of paper in two different colors
- Math journal to record work

Exploring Integers

▼ You can use models, such as colored tiles, to represent integers.

Let white tiles represent positive integers. □□□ 3

Let blue tiles represent negative integers. ■■ -2

1. Write the integer represented by each model.

a. □□ 2 **b.** ■■■■■ -5

c. ■ -1 **d.** □□□□□□□□ 8

Check students' work.

2. Use models to represent each integer and its opposite.

a. -3 **b.** 5 **c.** -8 **d.** 7

▼ You can use models to represent number sentences.

□□□ + □ = □□□□ and ■■ + ■■■ = ■■■■■

3 + 1 = 4 and -2 + (-3) = -5

3. ***Model*** the sum 2 + 4 = 6. Model the sum -2 + (-4) = -6. How are the sums the same? How are they different? The absolute values are the same. The signs are different.

4. ***Write*** a number sentence for each model.

a. □□□□□ + □□ = □□□□□□□ 5 + 2 = 7

b. □□□ + □□ = □□□□□ 3 + 2 = 5

c. ■■ + ■■■■ = ■■■■■■ -2 + (-4) = -6

d. ■■■■■■ + ■■■ = ■■■■■■■■■ -6 + (-3) = -9

5. Use your answers from Exercise 4. ***Discuss*** the following. (1)

a. the sum of two *positive* integers The sum is positive.

b. the sum of two *negative* integers The sum is negative.

▼ Suppose you earn $5 and then spend $5. The result is a zero change in money. You can use models to represent zero. An equal number of white tiles and blue tiles combine to make zero.

■□ represents zero *or* ■ + □ = 0

6. a. Use models to represent a different sum of zero. How many of each color tile did you use? ***Compare*** with another group.

6. a. Use the same number of each color tile to represent a sum of zero.

b. If you use 12 white tiles, you will need ▒ blue tiles to make zero. 12

c. If you use 31 blue tiles, you will need ▒ white tiles to make zero. 31

d. What is the fewest number of blue and white tiles you can combine to make zero? 2, one of each

e. ***Summarize*** The sum of an integer and its opposite is ▒. zero

Teaching Tip

Place the algebra tiles or colored pieces of paper in envelopes. Have enough packets of tiles ready for all the groups in the class.

7. Combine tiles to make zeros. Write the integer for the remaining tiles.

8. ***Explore*** Use a different model to represent each integer from Exercise 7. Use both positive and negative tiles for each model. Compare your models with those of another group. In how many different ways can you represent a given integer?

8. Check students' work. There are an infinite number of ways to represent any integer.

9. Complete the model. Write a number sentence to show the sum.

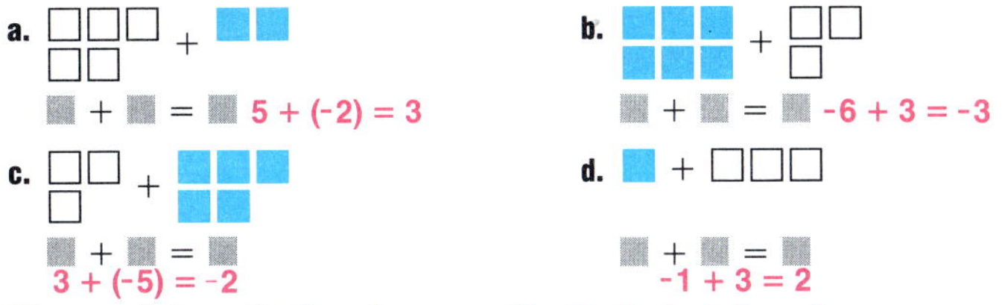

10. Use models to find each sum. Check students' work.

a. -7 + 3 **b.** 10 + (-4) **c.** -5 + (-4) **d.** 8 + (-11)

11. ***Analyze*** your answers from Exercises 9 and 10.

a. When is the sum *positive?* When is the sum *negative?*

b. ***Summarize*** Write a rule for the sign of the sum when you add two integers with different signs. *Hint:* Use absolute value.

11. a. The sum is positive when there are more white tiles. The sum is negative when there are more blue tiles.
11. b. The sign of the sum is the same as that of the integer with the greater absolute value.

12. Complete the model. Then write a number sentence for each sum.

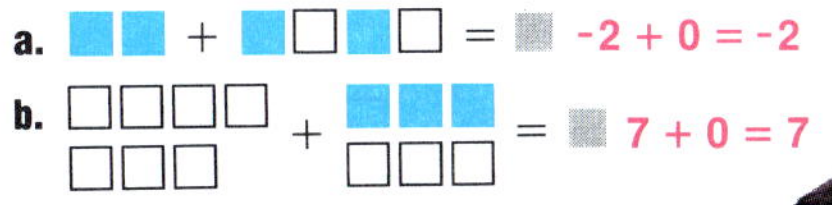

Closure

Summarize by asking groups to discuss their conclusions for questions 6e, 8, 11, and 12. If any group had different observations from the rest of the class, ask those students to explain how they arrived at their conclusions. If necessary, have them adjust their entries in their math journals.

Assignment Guide

Basic All
Average All
Enriched All

FOR THE NEXT LESSON

algebra tiles or blue and white squares of paper, ruler, paper, pencil, math journal, number cubes in two different colors

Lesson Focus

Materials

Algebra tiles or blue and white squares of paper, ruler, paper, pencil, math journal, number cubes in two different colors

MOTIVATING THE LESSON

Ask students to think of integers that express the answers for these situations: **If you climb 6 ft and then climb another 5 ft, how far up are you?** (11 ft; the integer is 11.) **If you dive 6 ft under water and then go 5 ft deeper, how deep are you?** (-11 ft; the integer is -11.) Draw a diagram on the board for each situation.

Instruction

Have pairs of students use tiles or colored squares of paper to model the examples in *Motivating the Lesson*. Then ask the students to write the number sentence for each model. (6 + 5 = 11, -6 + (-5) = -11)

1 As you discuss the introductory situation, ask students to model the number sentence using tiles. Have students look at Example 1, and ask how the addition in this number sentence differs from the others they have modeled. (The addends have different signs.) Remind students to match positive and negative squares in order to model the sum.

2 Ask students to describe in their own words how the number line model works for adding integers. Then ask the students to draw a number line to show -9 + 8. (-1)

OBJECTIVE:
To add integers using models, patterns, and rules.

1-2 Adding Integers

1

▼ Adding integers can be easier if you think of a familiar situation. Suppose you borrow \$5 from your friend. The next day you borrow \$2 more from your friend. You owe your friend \$7, so you are \$7 in debt. Here is your situation written as a sum of integers.

-5	+	-2	=	-7
borrow \$5		borrow \$2 more		\$7 in debt

▼ You can use models to add integers.

Example 1 6 + (-4)

Solution

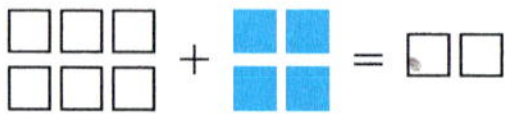

6 + (-4) = 2

Four positive tiles and four negative tiles make a zero pair.

Two positive tiles are left.

2

▼ Another useful model for adding integers is the number line.

Example 2 -8 + 9

Solution

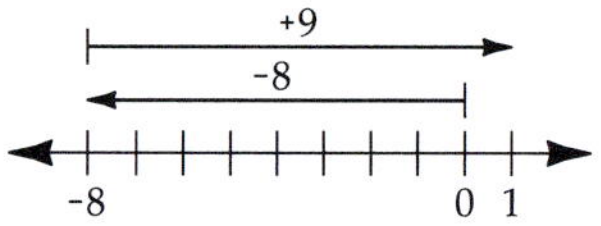

Notice that the sum falls on the same side of zero as the addend with the greater absolute value.

1. Begin at 0. Move 8 units to the *left* to *negative* eight.
2. Begin at -8. Move 9 units to the *right* to *positive* one.

-8 + 9 = 1

▼ It's not always convenient to use number lines or models. So, you can use these rules to add integers.

Adding Two Integers with the Same Sign	To add two integers with the same sign, add the absolute values of the integers. The sum has the same sign as the addends.
Adding Two Integers with Different Signs	To add two integers with different signs, find the *difference* of the absolute values of the addends. The sum has the sign of the integer with the greater absolute value.

NOTES & QUOTES

The early Egyptians drew pairs of legs walking in different directions to stand for addition and subtraction.

- *How is this model like the tiles and the number line?*

The legs are like the tiles since they model and distinguish between positive and negative. They are like a number line since they point in opposite directions.

Teaching Tip

Make copies of the graph paper from the TRB Teaching Resources. Students can use the graph paper to model the algebra tiles (outlining and filling in squares). They can also use graph paper to draw number lines.

Example 3 $-20 + 7$

Solution
1. Find absolute values. $|-20| = 20, |7| = 7$
2. Subtract. $20 - 7 = 13$
3. The sum is negative. $-20 + 7 = -13$

▼ You can use mental math or a calculator to find a sum.

Example 4 a. $-4 + (-1) + 4$ Use mental math.

opposites
$-4 + (-1) + 4 = -1$

Look for numbers to add in your head.

b. $-865 + 77 + 240$ Use a calculator.
865 (+/-) (+) 77 (+) 240 (=) -548

▼ Sometimes you will need to add integers to solve problems.

Example 5 To win a computer game, a player must have a positive number of points after ten rounds. The points for ten rounds are shown. Did the player win the game?

Solution Make a table to organize the data into positive and negative points. Use a calculator to add.

The player did not win the game since the final score was negative.

FLASHBACK
The sum of an integer and its opposite is zero.

Positive	Negative
6	-10
24	-55
19	-33
21	-18
12	
30	
112	+ (-116) = -4

1. Yes. You can add in any order.
2. Combine opposites.

CLASS EXERCISES

Give a numerical expression for each situation. Find the sum.

1. Susan deposited $120, then wrote a check for $25. 120 + (-25) = 95
2. A submarine at 35 ft below sea level moved up 10 ft. -35 + 10 = -25

Give a situation that describes each sum. Find the sum. Check students' work.

3. $235 + (-420)$ -185
4. $100 + (-100)$ 0
5. $-9 + (-1) + (-4)$ -14

Give a numerical expression for the model. Find the sum.

6. ■■■■ + □□□□□□□ -4 + 7 = 3
7. □□□□□ + □■ 5 + 0 = 5

Represent each sum on a number line. Find the sum. Check students' work.

8. $5 + (-7)$ -2
9. $-7 + (-6)$ -13
10. $4 + (-3) + 2$ 3

THINK AND DISCUSS

1. Is the sum $6 + (-5)$ the same as $-5 + 6$? Explain.
2. Explain how you could find each sum mentally.
$3 + (-3) + (-2) + 5 + 2$
$50 + 100 + 25 + (-50)$
3. Without adding, tell whether each sum is positive, negative, or zero. Explain your reasoning.
a. $-4 + (-10)$ b. $11 + (-3)$
c. $-8 + 5$ d. $6 + (-6)$

3. a. neg. (sum of 2 negs.)
b. pos. (11 has greater abs. val.)
c. neg. (-8 has greater abs. val.)
d. zero (opposites)

3 Ask students why the sum is negative. (The absolute value of -20 is greater than the absolute value of 7.) **Which addition rule is used in this example?** (second rule)

4 Ask students to explain why it is helpful to add opposites when finding the sum mentally. (The sum of opposites is zero, and zero plus a number is the number.)

ADDITIONAL EXAMPLES

1. Use tiles to find $-8 + 3$. (-5)
2. Use a number line to find $6 + (-2)$. [4]
3. Use the rules to find $-4 + (-8)$. [-12]
4. a. Use mental math: $7 + (-4) + (-7)$. [-4]
b. Use a calculator: $234 + (-723)$. [-489]
5. Can a player win with these scores: -8, -12, 6, 15, -23, 14, -3, -2, -5, 7? (No)

Guided Practice

Think and Discuss In question 1, students may recognize that the commutative property of addition is being illustrated.

For question 3, have students explain in their own words the rules for adding two integers. Ask if the rules can be applied directly to an expression such as $-14 + 32 + (-17)$ to determine the sign of the sum mentally. Ask students to explain their answers. (No. The absolute values of the positive and negative numbers cannot be compared unless -14 and -17 are added first.)

MATH MINUTES

Students, in groups, use two different color number cubes. One cube represents positive numbers, the other represents negative numbers. Students toss the cubes and determine if the sum is positive, zero, or negative.

Class Exercises Remind students that in questions 11-12 the first integer in the number sentence should correspond with the arrow that starts at zero.

A ***common error*** is to confuse the sign of the operation with the sign of an integer. Remind the students to read an addition sign as plus and to distinguish the symbol for a negative integer from a minus sign. Have students practice reading exercises. For example, read 4 + (-3) as *four plus negative three.*

Closure

After discussing the Class Exercises, ask students to write a description in their math journals of two integers whose sum is zero. (The two integers are opposites or both zero.) Repeat the process for the following situations: **1.** When the sum of two integers is positive (both integers are positive, one integer is positive and one is zero, or the integers have different signs and the absolute value of the positive integer is greater than the absolute value of the negative integer); **2.** When the sum of two integers is negative (both integers are negative, one integer is negative and one is zero, or the integers have different signs and the absolute value of the negative integer is greater than the absolute value of the positive integer).

Independent Practice

Students will need to use an almanac or other resource containing local weather data for question 36.

Point out that questions 47-49 have many correct answers.

11. -4 + (-2) = -6
The sum of the abs. val. is 6. Both integers are neg. so the sum is neg.

12. 3 + (-8) = -5
The difference of the abs. val. is 5. The integer with the greater abs. val. is neg. so the sum is neg.

Give the number sentence for the number line. In your own words, describe how each problem fits the rules.

11. 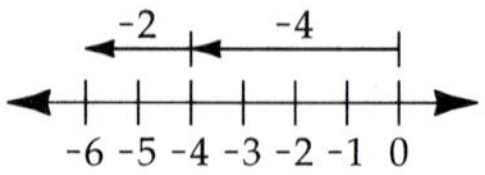

12. 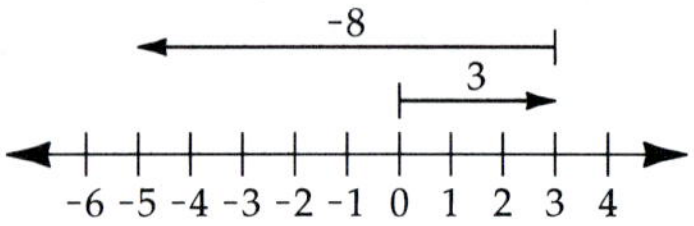

WRITTEN EXERCISES

1. -20 + 18 = -2; still owe $2 2. 200 + (-75) = 125; $125 left

Write a sum for each situation. Explain the result.

1. borrow $20, then pay back $18 **2.** save $200, then spend $75

3. temperature: -10°F drops 2°, rises 8°, drops 5°, drops 13°, rises 1° -10 + (-2) + 8 + (-5) + (-13) + 1 = -21; temp. is 21° below.

Use models to represent each sum. Find the sum. Check students' models.

4. -5 + 1 -4 **5.** -3 + (-6) -9 **6.** -4 + 4 + (-5) + 0 + 8 3

Represent each sum on a number line. Find the sum. Check students' number lines.

7. -10 + 14 4 **8.** 5 + (-8) -3 **9.** -1 + (-6) + 12 + (-7) + 8 + 3 9

MENTAL MATH **Find each sum mentally.**

10. -5 + 5 + 16 16 **11.** -4 + (-2) + (-2) -8

12. -1 + (-1) + (-1) + 1 + (-1) + 1 -2 **13.** -120 + 100 + (-20) -40

CALCULATOR **Find each sum.**

14. 145 + (-88) 57 **15.** -355 + (-492) -847 **16.** -192 + 825 + (-862) + 69 -160

Find each sum. Choose a method to use.

17. -2 + (-3) -5 **18.** -1 + 10 9 **19.** -9 + 9 0 **20.** 8 + (-12) -4

21. 14 + (-11) 3 **22.** 0 + (-9) -9 **23.** -6 + (-7) -13 **24.** -18 + 4 -14

25. -5 + 20 15 **26.** 6 + (-6) 0 **27.** -10 + (-3) -13 **28.** -94 + 68 -26

29. -8 + 7 + 5 4 **30.** 3 + (-2) + (-4) -3 **31.** -1 + (-9) + 4 -6

32. -3 + 2 + (-7) + 7 + 13 12 **33.** -20 + (-89) + 112 + 9 + (-3) 9

Temperatures in Sioux Week of January 22

Day	High (F°)	Low (F°)
Mon	8	–15
Tues	0	–24
Wed	4	–11
Thurs	–6	–21
Fri	1	–15
Sat	4	0
Sun	26	2

Use the *DATA* at the left. 34. -24, -21, -15, -15, -11, 0, 2

34. Write the low temperatures in order from lowest to highest.

35. Compare the lowest high temperature with the highest low temperature for the week. Use > or <. -6 < 2

36. ***PROJECT*** Find the lowest and highest recorded temperatures in your city and state. Answers will vary.

Write a numerical expression for each phrase. Find the sum.

37. negative two plus negative seven **-2 + (-7) = -9**

38. twelve plus the absolute value of nine **12 + |9| = 21**

39. positive three plus the opposite of eight **3 + (-8) = -5**

40. one hundred added to negative nineteen **-19 + 100 = 81**

Compare. Write >, <, or =.

41. -6 + 1 **<** 5 + 1 **42.** 0 + 3 **>** -2 + 0

43. -20 + (-7) **<** -11 + (-11) **<** 16 + 6

Solve.

44. Maria had $123. She spent $35, loaned $20 to a friend, and received her $90 paycheck. How much does she have now? **$158**

45. A football team gained 14 yd, lost 22 yd, gained 15 yd, lost 8 yd, and then lost 9 yd. Find the net gain or loss in yards. **10 yd loss**

46. ***WRITE*** and solve a word problem that uses the integers -10, 3, 5, -6, and -7. **Check students' work.**

Complete with integers. **Answers will vary. Examples are given.**

47. **3** + **4** = 7 **48.** **2** + **-10** = -8 **49.** **-15** + **3** = -12

PROBLEM SOLVING HINT

Using a number line or model may be helpful.

Critical Thinking

EXPLORING CLASSIFICATION

You can organize or *classify* numbers into categories by choosing a common *attribute* or characteristic. **Answers will vary. Examples are given.**

-7	-198	0
	-19 -47	
525	-50	-78

Look at the numbers in the box.

1. Classify three integers as odd. **-7, 525, -47**
2. Classify three integers as less than -10. **-19, -50, -198**
3. Classify three integers as having a 7 as a digit. **-7, -78, -47**
4. Classify -7, -19, and -50 by one common attribute. **all negative**
5. Classify three integers as between -49 and 0. Write three more integers that fit in this classification. **-47, -19, -7; -5, -25, -40**
6. Now choose some classifications of your own. Design at least three more ways to classify the numbers. List your numbers, exchange with a partner, and analyze each other's lists to figure out the classification. **Check students' work.**

MIXED REVIEW

Find each number.

1. the opposite of 8 **-8**

2. the opposite of -12 **12**

3. |-10| **10** **4.** |16| **16**

Complete with > or <.

5. -9 **<** -6 **6.** -2 **>** -7

7. In how many ways can you make change for a quarter? **12**

8. Name the integer five units to the left of negative eight. **-13**

Lesson Follow-up

Reteaching Activity To help students use the rules for adding integers, make a large number line on the floor or draw one on the chalkboard. Have students move along the line to illustrate the addition of two integers. Each time an example is acted out, ask students to explain which rule is demonstrated.

LESSON RESOURCES

TRB Practice A-3
TRB Enrichment A-4

Practice

For use after 1-2 (pp. 10–13)

Adding Integers

Write a numerical expression for each situation. Find the sum.

1. climb up 26 steps, then climb down 9 steps **26 + (-9); 17**
2. earn $100, spend $62, earn $35, spend $72 **100 + (-62) + 35 + (-72); $1**

Draw a model to represent each number sentence.

3. -7 + 6 = -1 4. 3 + (-6) + 4 = 1

Write an equation to represent each sum on the number line.

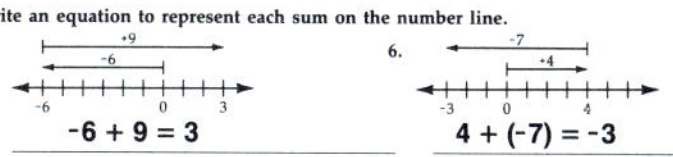

5. **-6 + 9 = 3** 6. **4 + (-7) = -3**

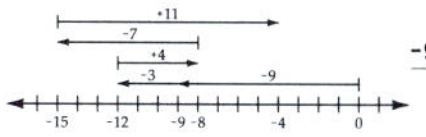

7. **-9 + (-3) + 4 + (-7) + 11 = -4**

MENTAL MATH **Find each sum mentally.**

8. -8 + 8 + -11 **-11** 9. -15 + 7 + 15 **7**
10. 12 + (-7) + 3 + (-8) **0** 11. -5 + (-16) + 5 + 8 + 16 **8**

CALCULATOR **Find each sum.**

12. 192 + (-129) **63** 13. -417 + (-296) **-713** 14. -87 + 175 **88**

Find each sum. Choose a method to use.

15. -8 + (-3) **-11** 16. 6 + (-6) **0** 17. -12 + (-17) **-29**
18. 9 + (-11) **-2** 19. -4 + (-6) **-10** 20. 18 + (-17) **1**

Compare. Write >, <, or =.

21. -7 + 5 (>) 3 + (-6) 22. 4 + (-9) (=) 6 + (-7) + (-4)
23. An elevator went up 15 floors, down 9 floors, up 11 floors, and down 19 floors. Find the net change. **down 2 floors.**

Chapter 1 A-3

Enrichment

For use after 1-2 (pp. 10–13)

Clock Numbers

A *finite number system* is one that contains a limited number of numbers. The finite number system on a clock face consists of the numbers from 1 to 12. The clock system is called Modulo-twelve, which is abbreviated *mod 12*.

Since "14" o'clock equals 2 o'clock, we can write 14 = 2(mod 12).

The integer 14 is equivalent to the number 2 in the mod 12 system. Every integer has an equivalent in mod 12. To find the equivalent of an integer, add or subtract a multiple of 12 to obtain a number from 1 to 12.

Examples 55 = 55 − 4(12) = 55 − 48 = 7(mod 12)
-13 = -13 + 2(12) = -13 + 24 = 11(mod 12)

Find the mod 12 equivalent. Each answer must be a number from 1 to 12.

1. 18 **6** 2. 85 **1** 3. -5 **7**
4. -64 **8** 5. 149 **5** 6. -97 **11**

The numbers 13, 25, and 37 are all equivalent to 1 (mod 12). When integers have the same equivalent, they are said to be *congruent*. The numbers 13, 25, and 37 are congruent in mod 12.

Write four integers, two positive and two negative, that are congruent in mod 12 to the given number. **Answers may vary; examples are given.**

7. 3 **15, 27, -9, -21** 8. 8 **20, 32, -4, -16** 9. 12 **24, 36,-12, -24**

To add in mod 12, find the sum in the usual manner. Then write the mod 12 equivalent.

Find the sum in mod 12.

10. 6 + 11 **5** 11. 9 + 5 + 12 **2** 12. 4 + (-7) + (-8) **1**
13. 3 + 11 + (-5) + (-16) **5** 14. -35 + (-47) + 28 + (-77) **1**
15. 29 + (-33) + (-2) + 14 **8** 16. -22 + (-11) + (-5) + (-19) **3**

A-4 Chapter 1

LESSON QUIZ

Find each sum.

1. -6 + 3 [-3]
2. -8 + (-9) [-17]
3. 11 + (-4) [7]
4. -6 + 6 [0]
5. 19 + 102 [121]
6. 24 + (-21) [3]
7. -6 + 8 + (-4) [-2]
8. 3 + (-7) + (-3) + (-9) [-16]

Assignment Guide

Basic 1-11, 14-22, 29-30, 34-37, 44, 47, MR, CT All

Average 2, 6, 9-16, 23-31, 34-36, 38-42, 45-47, MR, CT All

Enriched 3, 10-16, 29-36, 38-43, 45-49, MR, CT All

FOR THE NEXT LESSON

algebra tiles or blue and white squares of paper

Lesson Focus

Materials

Algebra tiles or blue and white squares of paper

MOTIVATING THE LESSON

Draw a vertical number line that ranges from 7 to -2 on the chalkboard. Label the zero point *lobby*. Explain that the number line represents an elevator. Say that the elevator is on the seventh floor and then moves down 2 floors. Ask students, **Where is it now?** (fifth floor) Continue by telling students that the elevator moves down 6 more floors. Ask, **Where is it now?** (-1, or one floor below the lobby) Ask students, **What two subtraction sentences represent the elevator's trip?** (7 − 2 = 5, 5 − 6 = -1)

Instruction

1 Relate the above example to the thermometer example. Point out that when using a number line to show subtraction, always start at zero and move to the first number (in this example, 44). This movement on the vertical number line is shown by an arrow pointing up. The arrow down shows the drop of 100°.

2 ***Number Sense*** Have pairs of students use models to show the subtraction. When they have finished, ask them to look at Example 1a, and compare the first number to the difference. Ask students, **What is the result when a positive number is subtracted?** (The difference is less than the first number.) Have students look at Example 1b. Ask students, **What seems to happen when a negative number is subtracted?** (The difference is greater than the first number.) Display the numerical sentence: -1 > -3.

1-3 Subtracting Integers

OBJECTIVE: *To subtract integers using models, patterns, and rules.*

In one day in January 1916 the temperature change in Browning, Montana set a record. The temperature fell 100° from 44°F to -56°F. You can write the situation as subtraction of integers.

44	−	100	=	-56
44° above zero		drop of 100°		new temperature

1 You can think of a thermometer as a number line. The thermometer at the left shows the subtraction as a move along a number line.

You can also use models to show subtraction situations.

2 ***Example 1*** **a.** You have $5. You give away $1. You now have $4.

b. You owe $3. You pay back $2. You now owe $1.

Solution **a.** 5 − 1 = 4

b. -3 − (-2) = -1

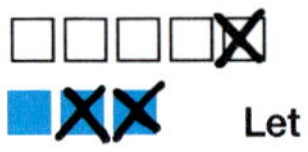

Let 3 negative tiles stand for the money you owe. Take away 2 tiles.

3 ***Example 2*** **a.** 4 − 5 **b.** 5 − (-6)

Solution **a.** You don't have 5 positive tiles. So, add a zero pair.

Now take away 5 positives.

4 − 5 = -1

b. You don't have 6 negative tiles. Add some zero pairs.

Now take away 6 negatives.

5 − (-6) = 11

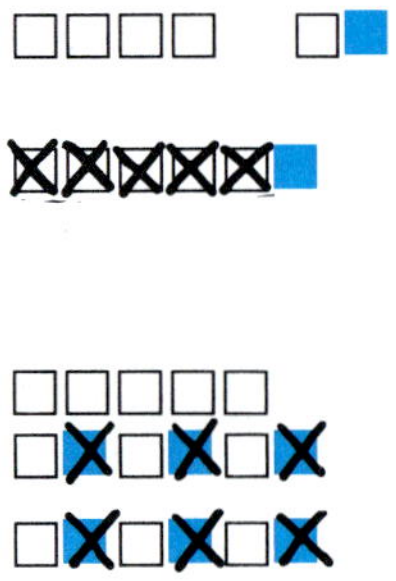

THINK How does adding zeros make it easier to subtract negative integers? **Adding zeros doesn't change the value. It does give you enough negatives to take away as needed.**

You know how to subtract positive integers and how to add any integers. Look at the pattern for the sums and differences you know.

5 − 4 = 1	7 − 3 = 4	12 − 5 = 7	14 − 6 = 8
5 + (-4) = 1	7 + (-3) = 4	12 + (-5) = 7	14 + (-6) = 8

4 You can see the same pattern when adding or subtracting any integers.

-10 − (-6) = -4	7 − 8 = -1	2 − (-9) = 11
-10 + 6 = -4	7 + (-8) = -1	2 + 9 = 11

THINK What is the same for the number sentences in each pair? What is different? **The answers are the same. The operations and the signs of the second addend are different.**

Teaching Tip

Have students work in pairs or small groups when working with algebra tiles. The students' understanding of the concept will be enhanced by group discussion.

▼ The models and patterns suggest the following rule for subtracting integers.

Subtracting Integers	To subtract an integer, add its opposite.

Example 3 **a.** $-8 - 2$ **b.** $-12 - (-4)$ **c.** $|-3| - |-7|$

Solution

a. $-8 - 2 = -8 + (-2)$ Add the opposite of 2.
$= -10$

b. $-12 - (-4) = -12 + 4$ Add the opposite of -4.
$= -8$

c. $|-3| - |-7| = 3 - 7$ Find absolute values.
$= 3 + (-7)$ Add the opposite of 7.
$= -4$

5

CLASS EXERCISES

Write each subtraction as an equivalent addition.

1. $2 - 3$ $2 + (-3)$
2. $-5 - (-6)$ $-5 + 6$
3. $-9 - (-7)$ $-9 + 7$
4. $1 - 8$ $1 + (-8)$
5. $33 - (-18)$ $33 + 18$
6. $0 - 75$ $0 + (-75)$
7. $-12 - (-12)$ $-12 + 12$
8. $|-10| - |7|$ $|-10| + (-|7|)$

Model each situation and give the result. Check students' models.

9. You are $3 in debt. You get a $10 gift. You have $7.
10. You pay a bill for $15. You earn $7. You have -$8.
11. ***WRITE*** In your own words, summarize the rules for adding and subtracting integers. Use examples if necessary. Check students' journals.

WRITTEN EXERCISES

Write a number sentence for each model or number line.

1. 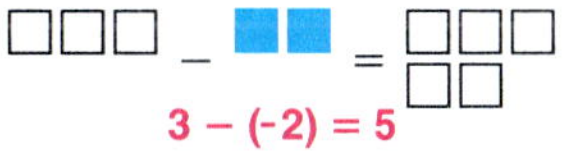
$3 - (-2) = 5$

2. 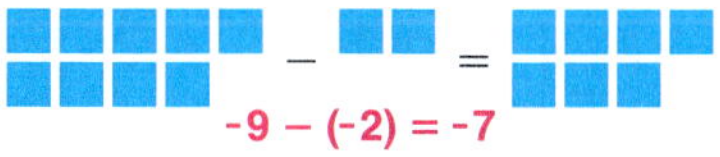
$-9 - (-2) = -7$

3. 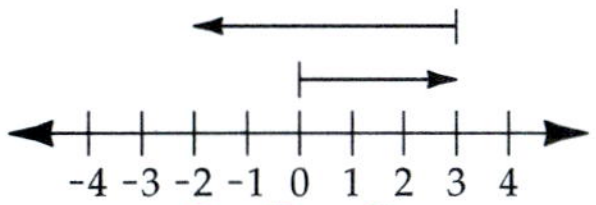

$3 - 5 = -2$

4. 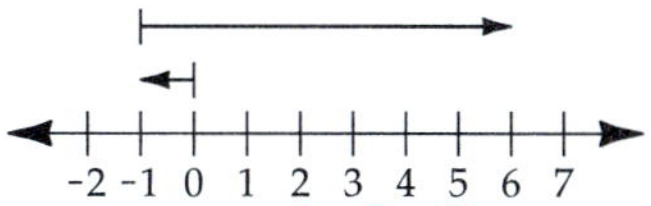

$-1 + 7 = 6$

1. When the integers have the same sign and the first integer has the greater abs. val.
2. all other cases
3. never

THINK AND DISCUSS See above

1. When is the absolute value of a difference the *same* as the difference of the absolute values?
2. When is the absolute value of a difference *greater* than the difference of the absolute values?
3. When is the absolute value of a difference *less* than the difference of the absolute values?

3 Examples 2a and 2b give students a strategy for removing tiles when there does not appear to be enough to remove. After the students have displayed the subtraction models, ask what happens when a positive number is subtracted from a number and when a negative number is subtracted from a number. (The difference is less than the first number; the difference is greater than the first number.)

4 As you discuss these questions, invite students to offer a generalization. (Subtracting a number or adding its opposite give the same result.)

5 Have students use models for $-8 - 2$ and $-8 + (-2)$ to verify the rule.

ADDITIONAL EXAMPLES

1. Use a model to show: You owe $10. You pay back $6. You still owe $4.
[$-10 - (-6) = -4$]
2. Use a model to solve.
   a. $4 - (-12)$ [16]
   b. $4 - (-8)$ [12]
3. Use the rule to subtract.
   a. $11 - 15$ (-4)
   b. $-6 - 3$ (-9)
   c. $|6| - |-8|$ (-2)

Guided Practice

Class Exercises Have students write a response for question 11 in their math journals. Ask for volunteers to read their entries to the class.

Think and Discuss To help the students interpret the questions use examples, such as:
$|6 - 4|$ ▪ $|6| - |4|$ and
$|6 - (-4)|$ ▪ $|6| - |-4|$
Next ask students to classify the pairs by determining whether they are equal, or whether the first value is greater or less than the second value.

MATH MINUTES

Using the numbers -6, 5, -2, 0, 3, and -7, have students find two numbers that, when subtracted, give the greatest difference and least difference.
$(5 - (-7) = 12)$
$(-7 - 5 = -12)$

A ***common error*** in subtracting integers is to change the operation to addition and to add the given number instead of its opposite. To remediate this error, give students subtraction problems, and have them use the rule to write the corresponding addition expressions. For example, -8 − 5 = -8 + (-5). Students write the addition expressions and read aloud negative eight plus the opposite of five.

Closure

Ask the students to explain what to do to find 3 − 7 and 3 − (-7). (-4, 10) Then ask the students to compare what happens when either 7 or -7 are subtracted from the same number. (Since subtracting 7 is the same as adding its opposite, -7, the difference is less than the number. Subtracting -7 is the same as adding 7, so the difference is greater than the number.)

Independent Practice

Tell students that there are many correct answers for questions 50-55.

For questions 56-64, point out that rounding integers is similar to rounding whole numbers.

NOTES & QUOTES

The magic square below uses the integers from 1–16 once. This magic square appeared in an engraving by Albrecht Dürer (1471–1528) in 1514, a number that appears in the square.

16	3	2	13
5	10	11	8
9	6	7	12
4	15	14	1

- *Find the magic sum.* 34
- *Find the sum of the four corner numbers.* 34
- *Find the sum of the numbers in the shaded squares.* 34
- *Find two other sets of four numbers that have the magic sum.* Ans. will vary. 1. The four middle squares. 2. 5, 9, 8, 12

Write a numerical expression for each phrase.

5. A plane climbs 3,000 ft and then descends 600 ft. 3,000 − 600

6. The temperature increases to 15° and then drops 25°. 15 − 25

MENTAL MATH **Find each difference.**

7. -2 − 3 -5 **8.** -7 − (-9) 2 **9.** -14 − 2 -16

10. -6 − (-8) 2 **11.** -45 − 15 -60 **12.** -7 − (-7) + (-7) -7

13. 100 − (-50) 150 **14.** 20 − (-10) − 20 10 **15.** -11 + 22 − (-55) 66

CALCULATOR **Find each difference.**

16. 88 − 97 -9 **17.** -235 − (-39) -196 **18.** -49 − 75 -124

19. 121 − (-57) 178 **20.** -81 − (-13) -68 **21.** 989 − 76 913

22. -59 − (-17) -42 **23.** -91 − (-79) − 19 -31 **24.** 815 + 35 − (-79) 929

Find each sum or difference. Use any method you wish.

25. 16 − (-9) 25 **26.** 11 − 5 6 **27.** 802 + (-977) -175 **28.** 75 + (-25) 50

29. -144 − 278 -422 **30.** 87 − (-9) 96 **31.** 22 + (-7) 15 **32.** 35 + (-15) 20

33. 100 − (-91) 191 **34.** -45 − 15 -60 **35.** -92 + (-9) -101 **36.** 167 + (-3) 164

37. |68| − |-12| 56 **38.** |-80| + |-28| 108

39. |-555| − |199| 356 **40.** |217| + |-317| + |0| 534

41. |-12| + |36| − |-10| 38 **42.** |-3| − |-2| − |-1| 0

43. Copy and complete. The first one is done for you.

8 − (-4) = 12
12 − (-4) = ■ 16
16 − (-4) = ■ 20
20 − (-4) = ■ 24
24 − (-4) = ■ 28

If you begin at 8 and subtract -4 five times, the result is ■. 28

In a magic square, each row, column, and diagonal has the same sum. Copy and complete each magic square.

44.

5	-9	■ 1
■ -5	-1	■ 3
-3	■ 7	-7

Sum = ■ -3

45.

-2	■ -7	■ -6
-9	-5	■ -1
-4	■ -3	■ -8

Sum = ■ -15

46.

■ 9	-5	■ -4	6
■ -2	4	3	■ 1
2	0	■ -1	5
-3	■ 7	■ 8	-6

Sum = ■ 6

ADDITIONAL ANSWERS

50. 2 − 2 = 0
51. 22 − 12 = 10; 8 − (-2) = 10
52. 4 − 10 = -6; -8 − (-2) = -6
53. 2 − 17 = -15; -20 − (-5) = -15
54. |-6| − |-3| = |-3|; -3 + 6 = |-3|
55. 15 − |-4| = |11|; -5 − (-16) = |11|

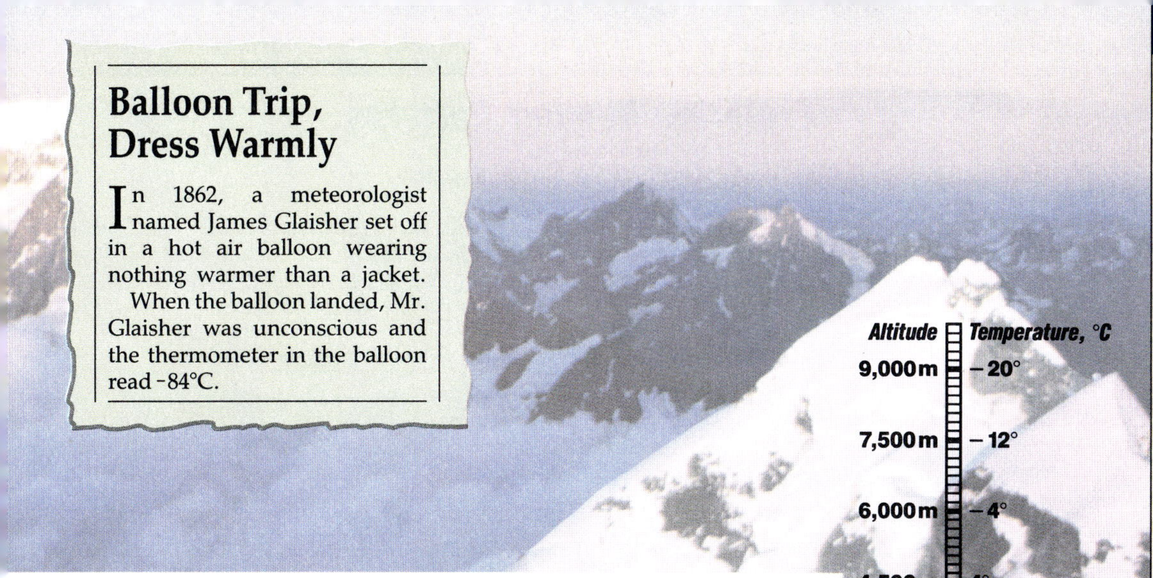

Balloon Trip, Dress Warmly

In 1862, a meteorologist named James Glaisher set off in a hot air balloon wearing nothing warmer than a jacket. When the balloon landed, Mr. Glaisher was unconscious and the thermometer in the balloon read -84°C.

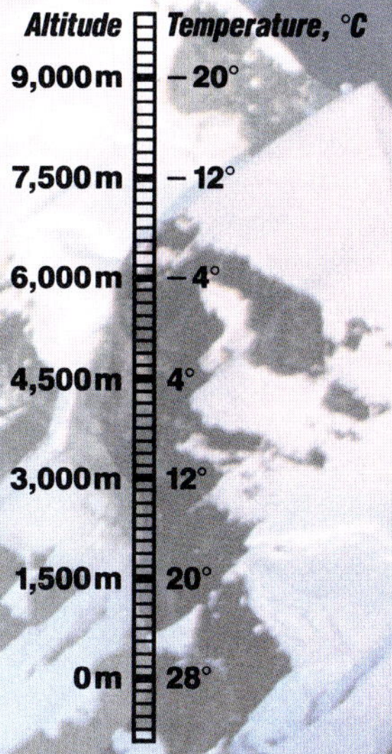

47. ***DATA*** Use the article above and the graph at the right.

- **a.** As the altitude increases, what happens to the temperature? The temperature decreases.
- **b.** By how much does the temperature change from 1,500 m to 6,000 m? 24°
- **c.** For every 1,500 m increase in altitude there is an 8° (increase, decrease) in temperature. decrease
- **d.** Use the given thermometer reading to estimate the height Mr. Glaisher's balloon reached. 21,000 m

48. How much warmer is it when the temperature is 20° than when the temperature is -7°? 27°

49. Suppose you had a score of 35 in a game. You then get a 50 point penalty. What is your new score? -15

Answers will vary. Examples are given. See Additional Answers.

Use positive and negative integers to write two different subtraction number sentences for each difference.

▼▼ *SAMPLE* ▒ − ▒ = -5 → -20 − (-15) = -5   17 − 22 = -5

50. ▒ − ▒ = 0   **51.** ▒ − ▒ = 10   **52.** ▒ − ▒ = -6

53. ▒ − ▒ = -15   **54.** ▒ − ▒ = |-3| **55.** ▒ − ▒ = |11|

ESTIMATION **Round each number to a convenient place. Estimate each sum or difference.** Accept reasonable estimates. Examples are given.

▼▼ *SAMPLE* Estimate -2,216 + 488.
-2,200 + 500 = -1,700

56. -45 + (-86) -140   **57.** 227 − 49 180   **58.** 398 − 67 330

59. -186 + 122 -70   **60.** 88 + 521 610   **61.** 3,321 − 924 2,400

62. 5,436 − (-4,725) 10,000   **63.** -864 + (-2,735) -3,600   **64.** 4,599 − 3,099 1,500

MIXED REVIEW

Add or subtract.

1. -17 − 12 -29
2. |60 − (-5)| 65
3. -8 + 15 7
4. -9 + (-4) + 7 -6
5. Name an integer between -5 and -10.
6. 6 + ▒ = 0 -6

Write a number sentence and solve.

7. A submarine at the surface dives 800 ft and then another 125 ft. Find the final depth. -925 ft

5. -6, -7, -8, or -9

Lesson Follow-up

Reteaching Activity For students having difficulty with the concept of subtracting integers, assign *Exploring Subtracting Integers*, TRB O-1. See also page 2E of the Chapter Overview.

LESSON RESOURCES

TRB Practice A-5
TRB Enrichment A-6

Practice

For use after 1-3 (pp. 14–17)

Subtracting Integers

Write a number sentence for each number line.

1. -6 + 5 = -1   2. -2 − 5 = -7

Draw an algebra model to represent each number sentence.

3. 4 − (-3) = 7   4. -6 − (-5) = -1

Write a numerical expression for each phrase.

5. A balloon goes up 2,300 ft, then goes down 600 ft. 2,300 − 600
6. You lose $50, then spend $35. -50 − 35

MENTAL MATH **Find each difference.**

7. 6 − 9 -3   8. 14 − 8 6   9. -15 − 3 -18
10. -25 − 25 -50   11. -16 − (-16) 0   12. 32 − (-17) − 32 17

CALCULATOR **Find each difference.**

13. 51 − 89 -38   14. -222 − (-117) -105   15. 843 − 677 166
16. -98 − 183 -281   17. 366 − (-429) 795   18. -83 − (-48) − 65 -100

Find each difference. Use any method you wish.

19. 8 − 12 -4   20. 13 − 6 7   21. 9 − (-12) 21
22. 57 − 39 18   23. -173 − 162 -335   24. 71 − (123) -52
25. |-71| − |83| -12   26. 48 − |-116| -68   27. -|275| − |-175| -450

ESTIMATION **Round each number to a convenient place. Estimate the answer.**

28. -57 + (-98) -160   29. 448 − 52 400   30. -191 + (-511) -700
31. -361 − (-58) -300   32. 888 + 1,177 2,100   33. -484 − 1,695 -2,200

Solve.

34. The Glasers had $317 in their checking account. They wrote checks for $74, $132, and $148. By how much was their account overdrawn? $37

Chapter 1 A-5

Enrichment

For use after 1-3 (pp. 14–17)

Time Lines

A time line is a number line marked off in dates rather than in integers. On the History of Mathematics time line below, dates labeled B.C. fall where the negative integers normally lie. Dates labeled A.D. replace the positive integers. Years given are dates of birth.

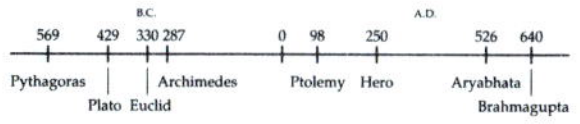

Find the number of years between the given events.

1. the births of Euclid and Hero 580
2. the births of Pythagoras and Archimedes 282
3. the births of Brahmagupta and Ptolemy 542
4. the births of Archimedes and Hero 537
5. Legend has it that Rome was founded in 753 B.C. How many years after the founding of Rome was Plato born? 324
6. One mathematician was born as many years before Ptolemy as Aryabhata was born after Ptolemy. Which one? Euclid
7. Which mathematician was born 1,069 years before Brahmagupta? Plato

Use the number line below to construct a time line. Write the letters of the given events below the appropriate tic mark. Above the line, write dates. Then, choose five other events relating to you, your family or friends, and include these on the time line. Also, include the year of your birth.

8.

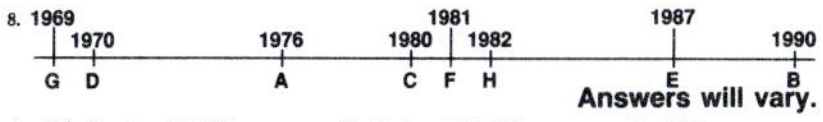

Answers will vary.

A. U.S. Bicentennial, 1976   B. East and West Germany reunite, 1990
C. Mount St. Helens erupts, 1980   D. First Earth Day, 1970
E. Halley's Comet returns, 1987   F. First shuttle flight, 1981
G. First moon landing, 1969   H. Compact disks introduced, 1982

A-6 Chapter 1

LESSON QUIZ

1. -9 − 4 (-13)
2. 8 − 19 (-11)
3. 5 − (-9) [14]
4. -12 − (-12) [0]
5. 388 − 284 (104)
6. -46 − 18 (-64)
7. -8 − (-22) [14]
8. 13 − 19 (-6)

Assignment Guide

Basic 1-8, 16-18, 25-36, 43-44, 47-50, 56-60, MR All
Average 3-6, 10-13, 19-21, 29-38, 43, 45, 47-51, 59-64, MR All
Enriched 12-15, 22-24, 33-42, 46-49, 52-55, 59-64, MR All

FOR THE NEXT LESSON

math journal

Lesson Focus

Materials

Math journal

MOTIVATING THE LESSON

Ask students to give examples of familiar patterns, including color, shape, and number patterns.

Instruction

Have students work in groups of two or three and record their answers in their math journals.

(1) Explain that it is necessary to find 2 − 4 for the second row, 6 − 8 + 10 for the third row, and so on.

(2) Help students understand the multiplication pattern: the fifth term is the number of the term (or 5) times the next counting number (6), or 30; the 100th term is 100 times the next counting number, 101, or 10,100.

Closure

Ask groups to share their results and explain what they have learned about number patterns.

ADDITIONAL ANSWERS

2b. Row 6: 1, 2, 3, 4, 5, 6, 5, 4, 3, 2, 1; Row 7: 1, 2, 3, 4, 5, 6, 7, 6, 5, 4, 3, 2, 1; Row 8: 1, 2, 3, 4, 5, 6, 7, 8, 7, 6, 5, 4, 3, 2, 1; Row 9: 1, 2, 3, 4, 5, 6, 7, 8, 9, 8 , 7, 6, 5, 4, 3, 2, 1.

3a. Row 6: 30, 32, 34, 36, 38, 40; Row 7: 42, 44, 46, 48, 50, 52, 54; Row 8: 56, 58, 60, 62, 64, 66, 68, 70; Row 9: 72, 74, 76, 78, 80, 82, 84, 86, 88; Row 10: 90, 92, 94, 96, 98, 100, 102, 104, 106, 108; Row 11: 110, 112, 114, 116, 118, 120, 122, 124, 126, 128, 130.

OBJECTIVE:
To explore and describe number patterns.

Exploring Number Patterns

Some number patterns are familiar.

2, 4, 6, 8, ■, ■, ■ 10, 12, 14; add 2
30, 25, 20, 15, ■, ■, ■ 10, 5, 0; subtract 5
2, −2, 2, −2, 2, ■, ■, ■ -2, 2, -2; 2 and its opposite
1, 3, 4, 12, 13, 39, 40, ■, ■, ■ 120, 121, 363; multiply by 3 then add 1.

1. a. ***Write*** the next three numbers in each pattern.
   b. ***Write*** a rule to describe each pattern.
   c. ***Discuss***—How are the patterns alike? How are they different? Answers will vary.

2. ***Analyze*** the integer triangle.

1
1 2 1
1 2 3 2 1
1 2 3 4 3 2 1
1 2 3 4 5 4 3 2 1

   a. ***Describe*** a pattern for each row and column.
   b. Copy the triangle and add four more rows. See additional answers.
   c. What is the middle number in each row?
   d. How many numbers appear in each row?
   e. What is the sum of the numbers in each row?
   f. ***Describe*** each pattern you found. Answers will vary.

2. a. The numbers in each row increase then decrease. The numbers in each column increase by one.
c. the greatest number in the row
d. double the middle number minus one
e. the middle number times itself

3. ***Analyze*** the triangle of even numbers.

0
2 4
6 8 10
12 14 16 18
20 ■ ■ ■ ■

   a. Copy the pattern and add six more rows. See additional answers.
   b. (1) ***Describe*** what happens when you alternate subtracting and adding the numbers in each row. If *n* is an even row, the result equals −*n*. If *n* is an odd row, the result equals the middle number in the row.
   Copy and complete the table. Use the table to search for patterns. ***Compare*** with another group.

Row	1	2	3	4	5	6	7	8	9	10
Result	0	−2	8	−4	■ 24	■ -6	■ 48	■ -8	■ 80	■ -10

4. Make up a pattern of your own. Exchange with a partner. Figure out each other's pattern. Check students' work.

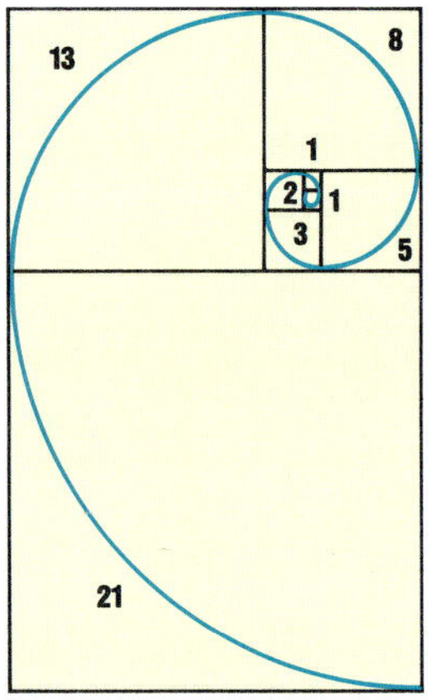

The spiral of a nautilus shell is an example of the Fibonacci numbers 1, 1, 2, 3, 5, 8, 13, . . .
- ***Explain and extend the pattern.***
- ***RESEARCH Find out how Fibonacci numbers occur in nature, music, and art.*** Ans. will vary.

You can think about some sets of numbers in more than one way. Look at the following pattern in two different ways.

2, 6, 12, 20, ... (differences +4, +6, +8)   2, 6, 12, 20, ... ($1 \cdot 2$, $2 \cdot 3$, $3 \cdot 4$, $4 \cdot 5$)

5. a. Extend the above patterns. Do they result in the same set of numbers? Is this what you expected? yes
   b. (2) Which method of extending the pattern would you use to find the 10th number in the pattern? the 100th number? Answers will vary.
6. Make up a number pattern that you can extend in two different ways. Exchange with a partner and solve. Answers will vary.

Teaching Tip

To encourage participation within groups, require that all students agree on their answers. As you monitor progress, ask individual students to explain how the group arrived at a particular answer.

Assignment Guide

Basic All
Average All
Enriched All

OBJECTIVE: *To solve problems by looking for a pattern.*

1-4 Look for a Pattern

You can solve many types of problems using patterns.

PROBLEM

1 News spreads quickly at River Dell High School. Each student who hears a story repeats that story to two other students in 15 min, and then tells no one else. A student hears some news at 8:00 A.M. How many students will know the news at 10:00 A.M.?

SOLUTION

READ Answer these questions to understand the given information.
How many students does each student tell? 2
How long does it take the news to reach two students? 15 min

PLAN Make a table to organize the given information. Then look for a pattern.

SOLVE Answer these questions to complete the table.
How many *new* students hear the news every 15 min? twice as many
How many students know the news after 15 min? $1 + 2 = 3$
after 30 min? $3 + 4 = 7$
after 45 min? $7 + 8 = 15$
Continue the pattern until you reach 10:00 A.M.

Time	8:00	8:15	8:30	8:45	9:00	9:15	9:30	9:45	10:00
Number of new students told	1	2	4	8	16	32	64	128	256
Number of students who know	1	3	7	15	31	63	127	255	511

By 10:00 A.M., 511 students know the story.

LOOK BACK One way to check a problem is to solve it another way. A tree diagram is a visual means to solving the problem.

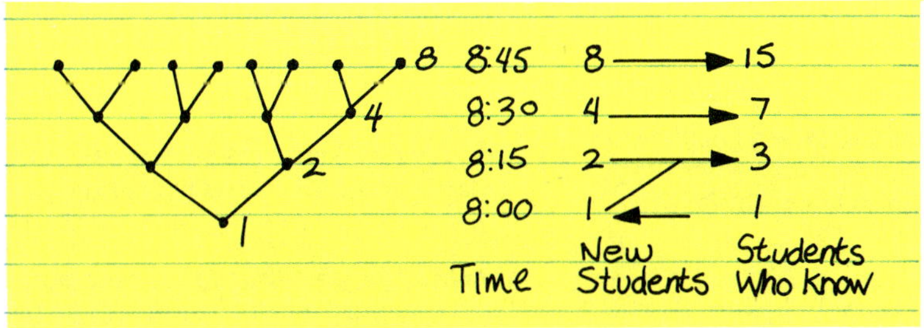

Lesson Focus

Materials
Math journal

Vocabulary/Symbols
intersection
Venn Diagram

MOTIVATING THE LESSON

Explain that you are going to ask two students to stand up. Then those two students each will tell two students to stand up, and so on. Ask how many students will be standing after three rounds of choosing.

Students may realize that there is no way to get an immediate answer. Make the point that a strategy for obtaining an answer is needed. Suggest that the situation be acted out.

Use a tree diagram to record the names of the students who are standing. (i.e., round 1: 2 students; round 2: 4 more students; round 3: 8 more students; 14 students are standing.)

Instruction

1 Display the problem and have students act it out. Have them suggest what is needed to keep a record of how the story spreads. (time, number of new students told, number of students who know) Record the results in a table as the problem is acted out. When all the students in the class "know the story," have them suggest how to fill in the table to 10:00 A.M.

Teaching Tip

As you assign the exercises, have small groups of students work together. Ask them to act out a problem when possible, or to model the problem, such as using pieces of paper for pennies in question 4.

Guided Practice

Class Exercises Ask students to predict by what time an entire town of 15,000 people would know the story. (11:15 A.M.) Then have them verify their prediction using the pattern.

A ***common error*** is for students to generalize too early with a number pattern. For example, show them this pattern: 2, 4, 6, 12. Students may hastily assume that the pattern is the even numbers, instead of alternately adding and multiplying by 2.

Closure

Writing in Math Summarize the lesson by asking students to recall all the different strategies they used to solve problems in this lesson. (For example, act out a problem, draw a diagram, make a table.) Ask students to describe these strategies in their math journals.

Independent Practice

For question 3, ask students for a generalization about the pattern. [Example: Let n be a whole number. The product $n \cdot n$ is one more than the product $(n - 1) \cdot (n + 1)$.]

CLASS EXERCISES

Refer to the problem on page 19.

1. Describe two ways to use the pattern to tell the number of students who know at 10:15 A.M.
   1. Double the number of students who know at 10:00 and add 1; continue the tree diagram.
2. Suppose the news continues to spread through the school. There are 1,735 students at River Dell High. By what time will every student know the story? **10:30 A.M.**

NOTES & QUOTES

When the German mathematician Karl Friedrich Gauss (1777–1855) was about ten years old, his teacher became annoyed with the class. As punishment, the teacher asked the class to compute the sum of the first 100 counting numbers. Gauss thought for a moment and then wrote the correct answer on his slate.

RESEARCH Find out how Gauss used patterns to solve this problem.

WRITTEN EXERCISES

Use a CALCULATOR where appropriate.

Solve by looking for a pattern.

1. The students in the town of Brighton are going to march in a parade. There will be one first-grader, two second-graders, three third-graders, and so on through the twelfth grade. How many students will march in the parade? **78 students**
2. Caroline is training for a swim meet. The first week she swims 1 lap per day. The second week she swims 3 laps per day. The third week she swims 6 laps per day. The fourth week she swims 10 laps per day. If she keeps to this training pattern, how many laps per day will Caroline swim in the eighth week? **36 laps**
3. Find each product and look for a pattern.

$2 \times 2 =$ ■ **4**	$3 \times 3 =$ ■ **9**
$1 \times 3 =$ ■ **3**	$2 \times 4 =$ ■ **8**
Difference = ■ **1**	Difference = ■ **1**
$4 \times 4 =$ ■ **16**	$5 \times 5 =$ ■ **25**
$3 \times 5 =$ ■ **15**	$4 \times 6 =$ ■ **24**
Difference = ■ **1**	Difference = ■ **1**

   a. Which is greater, 10×12 or 11×11? how much greater? **1**
   b. Suppose you know that $47 \times 47 = 2{,}209$. How can you find 46×48? **Subtract 1; 2,208.**
   c. Suppose you know that $64 \times 66 = 4{,}224$. How can you find 65×65? **Add 1; 4,225.**
4. Every day Maria saves twice as many pennies as she saved the day before. She starts by saving one penny on January 1. How much money will she have by January 10? **$10.23**
5. ***DATA FILE 8 (pp. 312–313)*** Did the population grow more between 1800 and 1810 or 1810 and 1820? How much more? **1810 and 1820; 467,174**

MATH MINUTES

Outline a large Venn Diagram on the floor. Distribute slips of paper with one or two school subjects on them. Call out two subjects. Students holding papers with those subjects arrange themselves in the Venn Diagrams.

Solve. Use any strategy you wish.

6. You can cut a pizza into two pieces with one straight cut. With two cuts you get four pieces. Three cuts will result in a maximum of seven pieces. What is the maximum number of pieces you can get with four cuts? with five cuts? **11; 16**

7. A restaurant offers special prices for groups dining together. For a buffet dinner, the restaurant charges $10 for one person, $20 for two people, $29 for three, $37 for four, and so on.
   a. How much does a buffet dinner for 8 cost? How much does the group save by eating together rather than alone? **$59; $21**
   b. The buffet costs the restaurant $6 per person. What size group can the restaurant serve without losing money? **10 people**

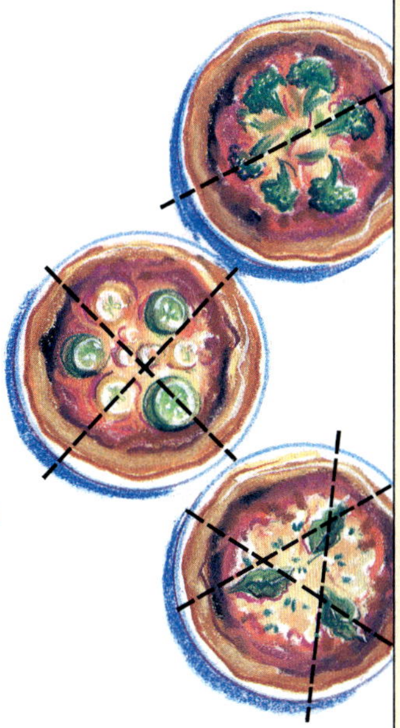

Critical Thinking

EXPLORING VENN DIAGRAMS

The principal wants to send invitations to a Science Fair to the homes of all students enrolled in biology and chemistry. Use the enrollment figures at the right to tell how many invitations are needed. A *Venn diagram* will help.

Course	Enrollment
Biology	127
Chemistry	124
Biology and Chemistry	17

- Draw intersecting circles. Label the circles with the given information.

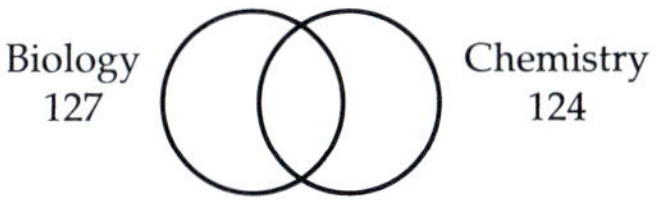

- The *intersection* (overlap) tells the number of students enrolled in both courses. Write the number in the intersection.

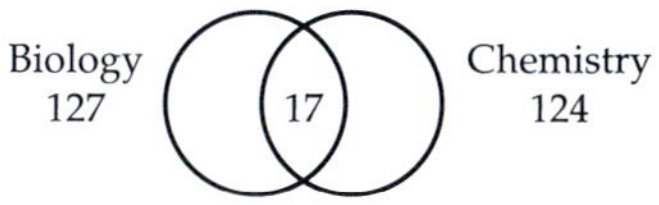

- The total in the biology circle is 110 plus the 17 in the intersection. What number goes in the chemistry circle? **107**

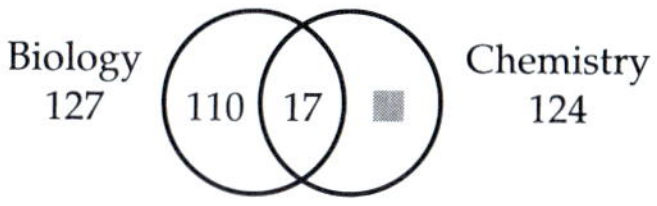

- Add the numbers in the three sections. How many invitations does the principal need? **234**

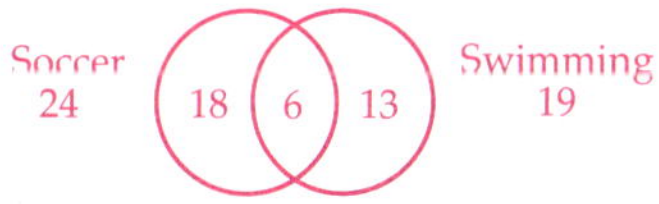

Use a Venn diagram to solve.

A coach needs to notify soccer players and swimmers of a revised schedule. Use the data at the right and a Venn diagram to tell how many notices the coach must send. **37**

Sport	Players
Soccer	24
Swimming	19
Soccer and swimming	6

Critical Thinking Point out that a Venn Diagram is a model to help visualize data in certain types of problems. The overlapping circles represent the number of students in *both* biology and chemistry. The other parts of the circles represent students *only* in biology and *only* in chemistry.

Lesson Follow-up

LESSON RESOURCES

TRB Practice A-7
TRB Enrichment A-8

Practice For use after 1-4 (pp. 19–21)

Look for a Pattern

Use the given numbers to find a pattern. Then write the next three numbers in the pattern.

1. 3, 6, 9, 12, 15, **18**, **21**, **24**
2. 1, 2, 4, 8, 16, **32**, **64**, **128**
3. 6, 7, 14, 15, 30, 31, **62**, **63**, **126**
4. 27, 33, 39, 45, 51, **57**, **63**, **69**
5. 1, 4, 16, 64, **256**, **1,024**, **4,096**
6. 34, 27, 20, 13, 6, **−1**, **−8**, **−15**
7. Each row in a window display of floppy disk cartons contains two more boxes than the row above.

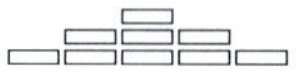

   a. Complete the table.

Row number	1	2	3	4	5	6
Total boxes in the display	**1**	**4**	**9**	**16**	**25**	**36**

   b. Describe the pattern in the numbers you wrote. **The total in the display is the row number multiplied by itself (that is, the number squared).**
   c. Find the number of rows in a display containing the given number of boxes.
   81 **9** 144 **12** 400 **20**
   d. Describe how you can use the number of boxes in the display to calculate the number of rows. **Find the number that, when multiplied by itself, equals the given number.**
8. A computer multiplied 9 by itself 100 times. You can use patterns to find the ones' digit of the product. $9 \times \underbrace{9 \times 9 \times 9 \times 9 \times \cdots \times 9}_{100 \text{ times}}$
   a. Find the ones' digit when 9 is multiplied by itself:
   1 time **1** 2 times **9** 3 times **1** 4 times **9**
   b. Describe the pattern. **When 9 is multiplied by itself an even number of times, the ones' digit is 9. When 9 is multiplied by itself an odd number of times, the ones' digit is 1.**
   c. What is the ones' digit of the computer product? **1**
9. Use the method of Exercise 8 to find the ones' digit of the product when 4 is multiplied by itself 100 times. **4**

Chapter 1 A-7

Enrichment For use after 1-4 (pp. 19–21)

Bode's Pattern

Occasionally, a scientist discovers a pattern of numbers which seems to suggest a natural law. The scientist must prove that the pattern is not simply accidental but that there is a reason behind it.

The table lists the planets known in 1772, and their relative distances from the sun, taking the earth's distance as 10. In that year the astronomer Johann Bode discovered an amazing pattern of numbers that closely matched the planetary distances.

Planet	Relative Distance (Earth = 10)
Mercury	5
Venus	7
Earth	10
Mars	16
Jupiter	52
Saturn	98

1. To find Bode's pattern, start with 1.5 and double each term.
   1.5, 3, **6**, **12**, **24**, **48**, **96**
2. Now add 4 to each term: 5.5, **7**, **10**, **16**, **28**, **52**, **100**
3. With one exception, note the close correlation between the pattern and the relative distances in the table.
   a. Two planetary distances are off slightly. Which two?
   Mercury and Saturn
   b. What is the exception?
   There is no planet corresponding to the pattern term 28.
4. In 1781, Uranus was discovered at a relative distance of 196 from the Sun. Calculate the next number in Bode's pattern in Exercise 2. Does the pattern correctly predict the discovery of Uranus? **yes**
5. In 1801, Ceres, the first and largest of the asteroids or "minor" planets, was discovered at a relative distance of 28.
   Does the pattern correctly predict the discovery of Ceres? **yes**
6. In 1846, the planet Neptune was discovered at a relative distance of 301. Had Bode discovered a law of planetary distance? Explain.
   No; the next term after 196 (Uranus) is 388. To fit the pattern, Neptune would have to be much farther away.

A-8 Chapter 1

Assignment Guide

Basic All
Average All
Enriched All

FOR THE NEXT LESSON

algebra tiles or blue squares of paper, math journal

Lesson Focus

Materials

Algebra tiles or blue squares of paper, math journal

MOTIVATING THE LESSON

Ask pairs or groups of students to use algebra tiles to model this situation: You spend $4 a day for 3 days. (three groups of four blue tiles) Ask what the model represents. (the total amount spent in 3 days, or $12) Ask students to write an addition sentence for the model. (-4 + (-4) + (-4) = -12) Remind students that repeated addition and multiplication are related operations. Ask students to suggest a multiplication sentence for the model. (3 × (-4) = -12)

Instruction

Illustrate the pattern below to show the product of a positive and a negative number.

3 × 2 = 6	3 × (-2) = -6
3 × 1 = 3	3 × (-3) = -9
3 × 0 = 0	3 × (-4) = -12
3 × (-1) = -3	

Ask students what happens to the product as one factor decreases. (the product decreases by 3) Point out that using tiles and patterns gives the same result for 3 × (-4). Ask students what sign the product of a positive and a negative number will have. (negative)

1 Continue the patterning so students can discover that the product of two negative numbers is positive.

OBJECTIVE: *To multiply integers using repeated addition, patterns, and rules.*

1-5 Multiplying Integers

▼ Suppose a football team loses 2 yd on each of 4 plays. At the end of the 4 plays the team loses 8 yd.

4	×	-2	=	-8
4 plays		lose 2 yd/play		lose 8 yd

▼ You also can think of multiplication as repeated addition.

Example 1 3 × (-7)

Solution 3 × (-7) = (-7) + (-7) + (-7)
= -21

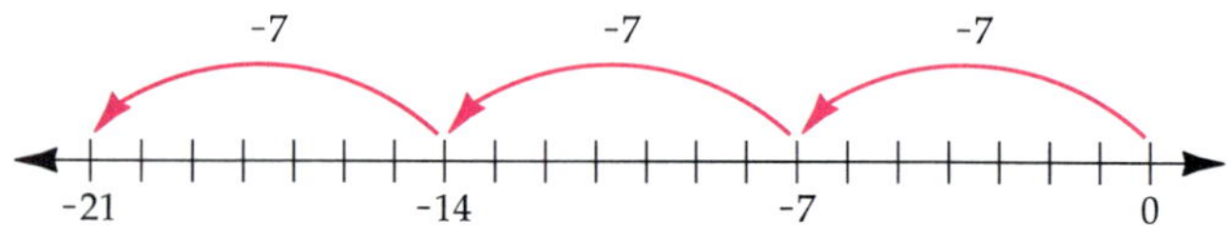

FLASHBACK

Symbols for multiplication

×	-3 × 5
( )	-3(5)
·	-3 · 5
*	-3 * 5

From now on, we will use · or ( ) for multiplication.

▼ Notice that the product of a negative number and a positive number is negative. You can use patterns to help find the product of two negative numbers.

Example 2 **Use patterns to find -3(-6) and -4(-5).**

Solution

2(-6) = -12	3(-5) = -15	Start with products you know.
1(-6) = -6	2(-5) = -10	
0(-6) = 0	1(-5) = -5	
	0(-5) = 0	
-1(-6) = 6	-1(-5) = 5	Continue the pattern for the product of two negative integers.
-2(-6) = 12	-2(-5) = 10	
-3(-6) = 18	-3(-5) = 15	
	-4(-5) = 20	

1

THINK Is the product increasing or decreasing? by how much? increasing; by 6, by 5

▼ From what you know about multiplying whole numbers and from the examples, you can write these rules for multiplying integers.

Multiplying Integers

To multiply two integers, find the product of the absolute values of the integers. Then use these rules.

1. The product of two integers with the same sign is positive.
   (+)(+) = +   (−)(−) = +
2. The product of two integers with different signs is negative.
   (+)(−) = −   (−)(+) = −

Teaching Tip

Encourage students to use a two-step method. First, find the sign of the product using the rules for multiplying integers. Once they have found the sign for the product, students can then multiply. Following a consistent procedure such as this can help students avoid errors.

▼ The rules are useful for finding the product of more than two integers.

Example 3 **Use the rules and mental math to find the product of $-2 \cdot 8(-5)$.**

Solution $-2 \cdot 8(-5)$

$10 \cdot 8 = 80$

$-2 \cdot 8(-5) = 80$

You can multiply integers in any order. Choose factors that are easy to multiply in your head.

▼ You can use integers to solve problems.

Example 4 Your average time for a 10-km race is 54 min. You would like to take 2 min off your average time each month for 3 mo. What will be your new average time?

Solution

1. By how much do you want to reduce your average time? $2 \cdot 3$
2. Compute your new average time. $54 - (2 \cdot 3) = 48$

Your new average time will be 48 min.

2

CLASS EXERCISES

Without computing, tell whether the product is positive, negative, or zero.

1. $-3(8)(-24)$ positive
2. $8(-83)$ negative
3. $2(-4) \cdot 29$ negative
4. $3(-21)(-12)$ positive

Write each sum as a product. Find the product.

5. $-8 + (-8) + (-8) + (-8)$ $4(-8) = -32$
6. $-2 + (-2) + (-2) + (-2) + (-2)$ $5(-2) = -10$

Complete the pattern. Is the product increasing or decreasing? by how much? 7. increasing by 3 8. increasing by 9

7. $3(-3) = ■$ -9
$2(-3) = ■$ -6
$1(-3) = ■$ -3
$0(-3) = ■$ 0
$-1(-3) = ■$ 3
$-2(-3) = ■$ 6
$-3(-3) = ■$ 9

8. $2(-9) = ■$ -18
$1(-9) = ■$ -9
$0(-9) = ■$ 0
$-1(-9) = ■$ 9
$-2(-9) = ■$ 18
$-3(-9) = ■$ 27
$-4(-9) = ■$ 36

9. Write a number sentence for the product shown on the number line. $5(-2) = -10$

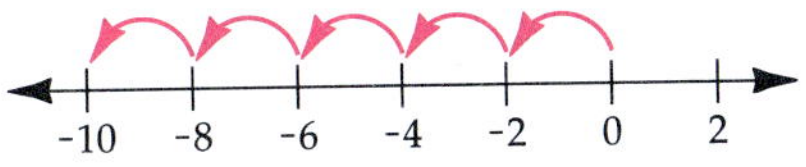

1. 6; opposite; opposite
2. negative; positive; negative
3. When an odd number of signs are all negative, the product is negative.

THINK AND DISCUSS See above

1. What is the product of -6 and -1? How are the product and -6 related? Complete: The product of any integer and -1 is the ■ of the integer.
2. Will the product of three negative integers be positive or negative? what about four negative integers? five negative integers?
3. Write a rule to use in deciding the sign for the product of more than two integers.

2 Emphasize that it is important to translate the information in a problem into an appropriate number sentence. Once the number sentence is identified, then compute the answer.

ADDITIONAL EXAMPLES

1. Use repeated addition to solve $4 \times (-4)$. [-16]
2. Use a pattern to show that $-3(-5) = 15$. [$2(-5) = -10$, $1(-5) = -5$, $0(-5) = 0$, $-1(-5) = 5$, $-2(-5) = 10$, $-3(-5) = 15$]
3. Use the rules for multiplying integers to find $-5(6)(4)$. [-120]
4. Your monthly electric bill was $78. You reduced it by $2 each month for 5 months. What is your monthly bill now? ($68)

Guided Practice

Class Exercises All questions may be done orally. Ask students to explain how they arrived at their answers for questions 1-4.

Think and Discuss Have students form small groups. For question 1, suggest that they try other examples such as $4(-1)$ and $-8(-1)$ before they make a conclusion.

Encourage students to write examples for each part of question 2 to arrive at the answers. The examples students generate will help them generalize the rule asked for in question 3. Have groups test each other's rules and explain why the rule works.

MATH MINUTES

Ask students to find all the pairs of factors that can have these products: 20 and -20. [20: 1(20), 2(10), 4(5), -1(-20), -2(-10), -4(-5); -20: -1(20), 1(-20), -2(10), 2(-10), -4(5), 4(-5)]

A ***common error*** is to write the incorrect sign for a product when three or more factors are involved. If students have difficulty with the rule generated in *Think and Discuss*, have them use the following procedure to find the product, -5(2)(-1)(-3). (-30) Work from left to right to determine the sign of the product.

-5(2)(-1)(-3)

(-)(-)(-)

(+)(-)

(-)

Then multiply the absolute values of the numbers.

Closure

Writing in Math Have students write in their math journals, a description of the rules for multiplying two integers. Then ask students, **When will a product be positive regardless of the number of factors?** (when there is an even number of negative factors or when all factors are positive) **When will the product be negative, regardless of the number of factors?** (when there is an odd number of negative factors)

Independent Practice

If necessary, remind students how to enter negative numbers when using a calculator for questions 13-16.

Questions 38-42 reinforce the terminology used when multiplying integers. Encourage students to read each question with care. Then ask them to write the numerical expression, read it to themselves to check that it matches the words, and finally find the product.

WRITTEN EXERCISES

Use repeated addition to find each product.

1. 10(-6) -60 **2.** -12 · 4 -48 **3.** 5(-6) -30 **4.** 4(-11) -44

Use patterns to find each product.

5. -7(-3) 21 **6.** -3(-6) 18 **7.** -5(-4) 20 **8.** -9(-12) 108

MENTAL MATH **Find the point on the number line that shows each product.**

9. -2 · 0 *A* **10.** 4(-2) *D* **11.** |-2| · |-2| *B* **12.** 2(-2) *C*

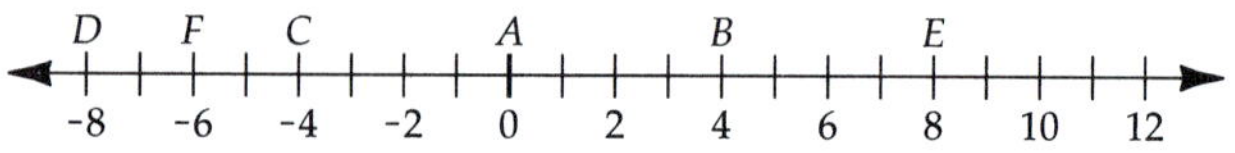

CALCULATOR **Find each product.**

13. -59(-79) 4,661 **14.** 243(-88) -21,384 **15.** -1,078(-43) 46,354 **16.** 23(-54) · 42(-39) 2,034,396

Find each answer. Choose a method to use.

17. -5(-3) 15 **18.** -6 · 10 -60 **19.** 8 · 3(-4) -96

20. -18(-12) 216 **21.** -11 · 20 -220 **22.** -8 · 0 0

23. 24(-16)(-32) 12,288 **24.** -9(-8)(-5) -360 **25.** -8 · 25 -200

26. |-9| + |-8| 17 **27.** |-2| · (-7) · 4 -56 **28.** |-2| · (-7) -14

29. -9 − (-2) -7 **30.** -8 + 6 + (-6) -8 **31.** 0(-12) · 4 0

32. 14 · 9 126 **33.** -20 + (-6) -26 **34.** 17 − (-3) 20

35. 9(-9) -81 **36.** 38(-2) -76 **37.** -15 + (-4) -19

Write a product for each word phrase. Then find the product.

38. negative eleven times negative five -11(-5) = 55

39. eight times the opposite of five 8(-5) = -40

40. the product of fourteen and negative seven 14(-7) = -98

41. the product of six and negative nine 6(-9) = -54

42. the absolute value of the product of negative twelve and ten |(-12) · 10| = 120

Compare. Use >, <, or = to make a true statement.

43. (-9)(-6) ■ 8(-10) > **44.** 5(-2) ■ (-6)(-1) < **45.** |-6||-2| ■ |-6(-2)| =

Solve.

46. The temperature dropped 5° each hour for 7 h. Use an integer to represent the total change in temperature. -35

MIXED REVIEW

Complete with <, >, or =.

1. -3 + (-8) ■ 12 − (-6) <

2. -9 + 13 ■ 24 − 30 >

3. |-6| − |12| ■ -8 + |-12| <

Write each number.

4. negative eleven -11

5. three thousand three 3,003

6. the absolute value of negative twenty |-20|

7. the opposite of -6 6

8. How many degrees difference is there between the freezing point of water (32°F) and the boiling point of water (212°F)? 180°

47. The price of a stock fell $3 each day over a 12-day period.

a. What was the total change in price? -$36

b. The original stock price was $76 per share. What was the price after the drop? $40

48. A car loan requires equal payments of $378 per month for four years.

a. What is the total amount paid for the four years? $18,144

b. Suppose a down payment of $2,500 was made. What was the total cost of the car? $20,644

PROBLEM SOLVING HINT
How many months are in four years? 48

Find two integers that fit the given description. Answers will vary. Examples are given.

49. sum: -7
product: 12
-3, -4

50. sum: 0
product: -9
3, -3

51. sum: 4
product: -5
5, -1

52. ***DATA FILE 9 (pp. 360–361)*** Suppose your score for the first nine holes at the St. Andrew's golf course is 4 under par. What would be your score? 32

53. ***DATA*** Use the chart below.

a. Can a balloon carry more weight at 5,000 ft at 40° or at 4,000 ft at 50°? 5,000 ft at 40°

b. Suppose people who weigh 150 lb, 112 lb, 129 lb, 183 lb, 108 lb, 75 lb, and 56 lb are planning a balloon trip. The balloon weighs 620 lb. Can the group rise to 4,000 ft if the temperature is 70°? no

c. ***WRITE*** Describe two patterns in the chart. Compare with a classmate. Did you find different patterns? Answers will vary.

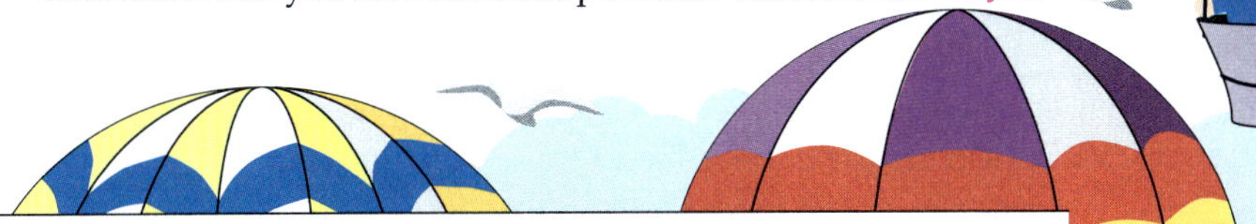

MAXIMUM WEIGHT A HOT AIR BALLOON CAN CARRY* (in pounds)

Altitude	Temperature (°F) 30°	40°	50°	60°	70°	80°	90°
2,000 ft	1,415	1,315	1,215	1,100	1,000	885	800
3,000 ft	1,370	1,270	1,170	1,070	970	850	770
4,000 ft	1,330	1,230	1,130	1,030	930	820	750
5,000 ft	1,285	1,185	1,085	985	900	800	715
10,000 ft	1,060	985	900	815	750	660	600
15,000 ft	885	815	750	670	615	550	485

*Limits include weight of the balloon

Hot air balloon trips are usually planned for early morning. Why do you think this is true? Balloons rise higher when it's cooler. Morning is usually the coolest time of day.

Suggest that students use a guess and test strategy for questions 49-51 and that they check that each answer fits both conditions.

Lesson Follow-up

Reteaching Activity Assign *Exploring Multiplying Integers*, TRB O-2. See also page 2E of the Chapter Overview.

LESSON RESOURCES

TRB Practice A-9
TRB Enrichment A-10

Practice

For use after 1-5 (pp. 22–25)

Multiplying Integers

Use repeated addition to find each product.

1. -3(5) -15
2. 4(-5) -20
3. 6(-2) -12
4. -15 • 3 -45

Use patterns to find each product.

5. -5(-2) 10
6. -7(-4) 28
7. -3(-8) 24
8. -9(-6) 54

CALCULATOR **Find each product.**

9. -36(29) -1,044
10. -117(-89) 10,413
11. -7(-56)(-13) -5,096

Find each answer. Choose a method to use.

12. -9(-2) 18
13. 0(-7) 0
14. 6(-8) -48
15. -12(-10) 120
16. -29(-1) 29
17. -2(-2)(7) 28
18. -17 • 3 -51
19. 8 • 7(-6) -336
20. -14|-3| -42
21. -7 • 50 -350
22. |-18| • |18| 324
23. 23 • 16 368

Write a numerical expression for each phrase. Then find the product.

24. the opposite of nine times negative seven -9(-7); 63
25. the product of the absolute value of negative twelve and the opposite of five |-12|(-5); -60
26. six times the product of negative eight and the opposite of three 6(-8)(-3); 144
27. the absolute value of negative nine times the sum of five and negative two |-9|[5 + (-2)]; 27
28. the opposite of the product of thirty and negative four -[30 • (-4)]; 120

Compare. Use >, <, or = to make a true statement.

29. -3(3) < 4(-2)
30. |-5| • |-7| > 6(-6)
31. -1(2) < -3(0)
32. -3(3) > -13
33. 49 • 2 = (-49)(-2)
34. 6(2 • 3) > 2 • 3

Solve.

35. The Red Sox fell 3 games farther back in the league standings each week for 7 consecutive weeks. Use an integer to represent the team's total drop in the standings. -21

Chapter 1 A-9

Enrichment

For use after 1-5 (pp. 22–25)

Curious Methods of Multiplication

A. Finger Multiplication
Adding on your fingers is easy. Here is how you can *multiply* numbers from 6 through 10 using your fingers.

Example Multiply: 9 • 8

Solution Imagine the fingers of both your hands are numbered from 6 (thumb) through 10 (little finger). Touch finger 9 on one hand to finger 8 on the other hand. Bend any fingers beyond those fingers that are touching.

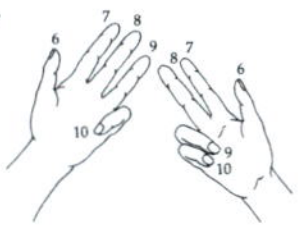

To find tens digit of product: Add the upright fingers. 4 + 3 = 7

To find ones digit of product: Multiply the number of bent fingers on one hand times the number of bent fingers on the other hand. 1 • 2 = 2

Product: 72

Use your fingers to multiply.

1. 7 • 8 56
2. 9 • 9 81
3. 9 • 7 63
4. 6 • 8 48

B. Binary Multiplication
Using only multiplication and division by 2, you can find the product of any two numbers.

Example Multiply: 39(-13)

Solution Write the factors side by side. 39 -13

On the left side, divide by 2, dropping any remainder. On the right side multiply by 2. Continue until you reach 1 on the left.

39	-13
19	-26
9	-52
4	-104
2	-208
1	-416

List those numbers from the right side that are positioned across from odd numbers on the left side. Add them.

-13
-26
-52
-416
Product: -507

Find each product using binary multiplication.

5. 22 • 17 374
6. 45(-25) -1,125
7. 68(-33) -2,244
8. 75(-41) -3,075

A-10 Chapter 1

LESSON QUIZ

1. 9(-3) [-27]
2. 8 · 2 (16)
3. -4(-9) [36]
4. -8 · 7 [-56]
5. -6(2)(-4) [48]
6. 4(-3)(0)(-11) [0]
7. (-3)(-8)(-1)(5) [-120]

Assignment Guide

Basic 1-2, 5-6, 9-10, 13-15, 17-37 odd, 38-41, 46-47, 52-53, MR All

Average 3-4, 7-12, 15-16, 18-36 even, 40-44, 47, 49-50, 52-53, MR All

Enriched 9-12, 15-16, 18-36 even, 40-45, 48-53, MR All

FOR THE NEXT LESSON

algebra tiles or squares of blue paper, blank transparency, math journal

Lesson Focus

Materials

Algebra tiles or squares of blue paper, blank transparency, math journal

Vocabulary/Symbols

mean

MOTIVATING THE LESSON

Have students model the following situation using the blue or negative tiles: You owe your friend \$8. You plan to pay back the same amount each week for 4 weeks. (four groups of two tiles) Ask students what the result of the grouping represents. (You pay back \$2 each week for 4 weeks.) **What operation does this situation call for?** (division) **What number sentence can you suggest for this situation?** ($-8 \div 4 = -2$)

Instruction

1 Have students use tiles to model the change in temperature. Then ask them to explain what an average change of $-7°$ per hour means. (a drop of 7° each hour)

2 As you discuss the *Flashback*, point out that another division sentence related to $8 \cdot 0 = 0$ cannot be written because zero cannot be used as a divisor. Also mention that attempting to divide by zero on a calculator results in an error message.

3 ***Writing in Math*** Finding the mean is introduced here as an application of division. Ask students for a practical example where a mean would be used and to write the example in their math journals. An example could be the average morning temperature for a certain month. Have students include a description of how to find the mean.

OBJECTIVE: *To divide integers.*

1-6 Dividing Integers

▼ A storm system moved through Minneapolis one day and the temperature dropped 28° in 4 h. To find the average change per hour, you can divide.

1

$$-28 \div 4 = -7$$

You can use a related multiplication sentence to see why the quotient is negative.

$$\text{Since } -7 \cdot 4 = -28,\ -28 \div 4 = -7$$

FLASHBACK

Symbols for division

$\div$	$48 \div 8$
$\overline{)\ }$	$8\overline{)48}$
—	$\frac{48}{8}$
/	48/8

▼ Multiplication and division are inverse operations. One *undoes* the other. Because $3 \cdot 2 = 6$, we can write $6 \div 3 = 2$ and $6 \div 2 = 3$. You can use this relationship to find quotients of integers. The table shows related multiplication and division sentences.

Multiplication Number Sentence	Division Number Sentence
$8(6) = 48$	$48 \div 6 = 8$
$(-8)(6) = -48$	$-48 \div 6 = -8$
$8(-6) = -48$	$-48 \div (-6) = 8$
$(-8)(-6) = 48$	$48 \div (-6) = -8$

▼ The examples demonstrate the following rules for dividing integers.

Dividing Integers

To divide two integers, find the quotient of the absolute values of the integers. Then use the following rules.

1. The quotient of two integers with the same sign is positive.
   $(+) \div (+) = +$ $\quad (-) \div (-) = +$
2. The quotient of two integers with different signs is negative.
   $(+) \div (-) = -$ $\quad (-) \div (+) = -$

2

FLASHBACK

$8 \cdot 0 = 0$
$0 \div 8 = 0$
Zero divided by any nonzero number is zero. You can't divide by zero.

Example 1 **a.** $27 \div 9$ **b.** $-18 \div (-3)$ **c.** $100 \div (-10)$ **d.** $-56 \div 7$

Solution

a. $27 \div 9 = 3$ $\quad (+) \div (+) = +$
b. $-18 \div (-3) = 6$ $\quad (-) \div (-) = +$
c. $100 \div (-10) = -10$ $\quad (+) \div (-) = -$
d. $-56 \div 7 = -8$ $\quad (-) \div (+) = -$

Teaching Tip

Encourage students to follow the same procedure each time they divide. For example, determine the sign of the quotient first; then divide the numbers without their signs.

▼ Many of the numbers you see every day are averages. The *mean* is the average you see most often. Many of the problems you encounter will ask you to compute the mean.

3

Finding the Mean	To find the mean of a group of numbers: **1.** Find the sum of the numbers. **2.** Divide the sum by the number of items.

Example 2 A student's scores on five math tests were 98, 90, 87, 95, and 90. Find the mean score.

Solution

1. Find the sum.
98 [+] 90 [+] 87 [+] 95 [+] 90 = 460

2. Divide by 5, the number of items.
460 [÷] 5 [=] 92

The mean score is 92.

CLASS EXERCISES

Without computing, state whether the quotient is positive, negative or zero. Explain your reasoning.

1. $-25 \div 5$ neg. $(-) \div (+) = (-)$
2. $39 \div (-3)$ neg. $(+) \div (-) = (-)$
3. $35 \div (-7)$ neg. $(+) \div (-) = (-)$
4. $8 \div 3$ pos. $(+) \div (+) = (+)$

For each product, write a related division sentence and solve.

5. $-9 \cdot 4$ $-36 \div 4 = -9$
6. $9(-3)$ $-27 \div (-3) = 9$
7. $-11 \cdot 11$ $-121 \div 11 = (-11)$
8. $8 \cdot 3$ $24 \div 3 = 8$

Find the quotient.

9. $144 \div 12$ 12
10. $55 \div 11$ 5
11. $34 \div (-17)$ −2
12. $-210 \div (-30)$ 7

Find the mean.

13. temperatures: −9°, −12°, 9°, 4°, −2° −2

14. feet above and below sea level: 135, −56, 92, −29, −88, −60 −1 ft

15. test scores: 80, 75, 90, 88, 87 84

16. golf scores: 3, 5, 8, 6, 4, 5, 3, 6, 5 5

17. bank balance: $200, −$85, $120, $200, $280 $143

WRITTEN EXERCISES

For each quotient, write a related multiplication sentence and solve.

1. $-90 \div (-9)$ $-9(10) = -90$
2. $35 \div (-7)$ $(-7)(-5) = 35$
3. $56 \div 8$ $8(7) = 56$
4. $88 \div 11$ $11(8) = 88$

THINK AND DISCUSS See below

1. Use a related multiplication sentence to explain why you can't divide by zero.

2. Compare the signs of quotients of integers with the signs of products of integers. What do you discover? They're the same.

3. Must the mean of a group of a numbers be one of the numbers in the group? no

1. For $n \div 0 = ?$ the related multiplication is $? \cdot 0 = n$. Since 0 times any number is 0, $n \div 0$ is not defined.

ADDITIONAL EXAMPLES

1. a. $480 \div 60$ (8)
b. $-24 \div (-8)$ [3]
c. $36 \div (-3)$ [−12]
d. $-54 \div 9$ (−6)

2. Find the mean: −48, 32, 57, −29, −24, −18 (−5)

Guided Practice

Class Exercises To reinforce the rules for dividing integers, ask students to first tell the sign of the quotient and then find the quotient in questions 9-12.

Think and Discuss For question 1, give this example: *Suppose 2 ÷ 0 is some number. Then some number times zero must equal two.* See if students can demonstrate why 0 ÷ 0 is not possible. Help students realize that the division is undefined and the quotient can not be determined. When attempting to write a related multiplication sentence, students will find that any number times zero equals zero.

For question 3, have students demonstrate their answers by using an example such as, *the mean of 48 and 50 is 49*, which is not one of the numbers in the group.

A ***common error*** when finding the mean of a set of numbers is to divide by an incorrect number or to err in adding the numbers. Although the mean need not be a number in the set, it cannot be greater than the greatest number in the set or less than the least number in the set. Students can use this approach as a guideline to determine if the mean they find is reasonable.

MATH MINUTES

Give each team member a paper with: ÷(−3), ×(−1), ×4, or ÷(−2). To start a relay race, call out any integer that is a multiple of 3. One team member divides the integer by −3 and passes the result to the next team member.

Closure

Ask students to summarize the rules for dividing two integers. Also ask students to compare the rules to those for multiplying two integers. (The rules are the same for the two numbers being multiplied and the two numbers being divided.)

Independent Practice

Students may use the guess and test strategy for question 60, or they may work backwards by first determining what the sum of the scores is and the average of the scores, and then comparing Sam Adam's average to the averages needed to score a B before determining the score he needed on the first test.

Lesson Follow-Up

Reteaching Activity Give students a set of division exercises. For each exercise, ask them to write the sign of the integer under each number as shown below. Have students read aloud the signs and the sign of the quotient as: *Positive divided by negative is negative.*

$$45 \div (-5) = ?\ [-9]$$
$$+ \quad \div \quad - \quad = \quad -$$

Then students may finish the division.

MENTAL MATH **Find each quotient.**

5. $-63 \div 9$ **-7**
6. $66 \div 6$ **11**
7. $250 \div (-50)$ **-5**
8. $1{,}200 \div (-40)$ **-30**

CALCULATOR **Find each quotient.**

9. $-432 \div 48$ **-9**
10. $693 \div 21$ **33**
11. $-10{,}584 \div (-84)$ **126**
12. $50{,}840 \div (-328)$ **-155**
13. $13{,}272 \div 237$ **56**
14. $-62{,}937 \div (-111)$ **567**

Find each answer. Choose a method to use.

15. $48 \div 12$ **4**
16. $1{,}000 \div (-50)$ **-20**
17. $-38 \div (-2)$ **19**
18. $-3{,}132 \div 36$ **-87**
19. $24 \div (-24)$ **-1**
20. $0 \div (-56)$ **0**
21. $225 \div (-15)$ **-15**
22. $18 \div (-1)$ **-18**
23. $-64 \cdot 6$ **-384**
24. $-33 + 11$ **-22**
25. $5{,}959 \div (-101)$ **-59**
26. $-58 \div (-1)$ **58**
27. $-200 - 25$ **-225**
28. $736 \div (-23)$ **-32**
29. $-72 + (-8)$ **-80**
30. $204 \div (-12)$ **-17**
31. $-1{,}225 \div 35$ **-35**
32. $0 \div (-8)$ **0**
33. $128 + (-64)$ **64**
34. $150 - (-15)$ **165**
35. $225 \cdot 15$ **3,375**
36. $|-56 \cdot 12| \div (-24)$ **-28**
37. $(-|-24(9)| \div |3(-8)|)$ **-9**

Write a numerical expression for each word phrase. Then evaluate the expression.

38. negative twenty-four divided by negative eight **-24 ÷ (-8); 3**
39. negative forty-two multiplied by three **-42(3); -126**
40. zero divided by negative seven **0 ÷ (-7); 0**
41. two hundred subtracted from negative twenty-five **-25 − 200; -225**

Find the mean.

42. temperature: -12°, -8°, -24°, 32°, 0°, -6° **-3°**
43. weekly allowance: $3, $2, $5, $2, $3, $2, $2, $5 **$3**
44. salary: $24,000; $18,000; $52,000; $27,000; $15,000 **$27,200**
45. score: -203, 813, -446, -231, 466, -155, -329, -228, 312, 1 **0**

Write >, <, or = to make a true statement.

46. $-10 \div (-2)$ **>** $25 \div (-5)$
47. $-(-15 \div 5)$ **<** $-100 \div (-20)$
48. $|-25| \div |-5|$ **=** $|-25 \div (-5)|$
49. $-|-28| \div 7$ **<** $-28 \div (-7)$

Write an integer between the given integers. **Answers will vary. Examples are given.**

50. $-2 \cdot (-2)$ and $2 \cdot 3$ **5**
51. $10 + (-7)$ and $10 \div (-5)$ **-1**
52. $121 \div (-11)$ and $|-7| - |7|$ **-1**
53. $50 + (-48)$ and $80 \div (-20)$ **-1**

Solve.

54. An integer multiplied by -8 equals -96. What is the integer? **12**

NOTES & QUOTES

The universe is written in the language of mathematics.

—Galileo Galilei (1564–1642)

MIXED REVIEW

Find each answer.

1. $5(-9)$ **-45**
2. $-8(-3)$ **24**
3. $|-3| \cdot 8 \cdot (-2)$ **-48**
4. $|-5 \cdot (-2) \cdot 3|$ **30**

Write the next three numbers in each pattern.

5. -7, -2, 3, 8, ■, ■, ■
6. 1, 4, 9, 16, ■, ■, ■
7. Make up your own pattern. Write a rule to describe it. **Answers will vary.**

5. 13 18 23
6. 25 36 49

55. An integer multiplied by 9 equals -135. What is the integer? -15

56. Find two integers with a sum of -10 and a product of -75. -15, 5

57. A scuba diver descended to a depth of 50 ft in 25 s. How many feet per second did she dive? 2 ft/s

58. ***DATA FILE 9 (pp. 360–361)*** Find the average length in yards for the first 9 holes at St. Andrew's golf course. 389 yd

59. ***PROJECT*** Find the shoe sizes of ten classmates who are the same gender as you. Use the information to predict the average shoe size of students of your gender and age. Answers will vary.

60. The grade book shows students' math scores.

Name	Test 1	Test 2	Test 3	Test 4	Test 5
Abrams, Joel	88	87	74	69	92
Adams, Sam	66	72	88	81	88
Barcos, Elena	99	91	90	95	90
Cuomo, Terri	67	70	72	71	80

LETTER GRADES

A	90 - 100
B	80 - 89
C	70 - 79
D	60 - 69
F	0 - 59

a. Find each student's average grade. Use the chart at the right to assign a letter grade. 82—B; 79—C; 93—A; 72—C

b. What would Sam Adams have needed to score on his first test to raise his grade to a B? 71

PROBLEM SOLVING HINT

Try guess and test.

TEST YOURSELF

Compare. Use >, <, or =.

1. 3 > -8
2. -10 < -6
3. -4 + 3 = 3 + (-4)

Find each number.

4. |-8| 8
5. -|-85| -85
6. the opposite of 12 -12

Evaluate.

7. 3 + (-11) -8
8. 12 − (-8) 20
9. -9 · 5 -45
10. -64 ÷ (-8) 8

Find the mean.

11. -4, 7, 0, -3, -2, 20 3
12. 20, 40, 25, 35, 100 44

LESSON RESOURCES

TRB Practice A-11
TRB Enrichment A-12

Practice

For use after 1-6 (pp. 26–29)

Dividing Integers

Find each quotient. Write a related multiplication sentence.

1. -36 ÷ (-4) 9; 9(-4) = -36
2. 30 ÷ (-5) -6; -6(-5) = 30

MENTAL MATH Find each quotient.

3. 16 ÷ (-2) -8
4. -35 ÷ (-7) 5
5. -80 ÷ 10 -8

CALCULATOR Find each quotient.

6. -299 ÷ 13 -23
7. 255 ÷ 15 17
8. -779 ÷ (-19) 41

Find each answer. Choose a method to use.

9. 54 ÷ (-9) -6
10. -37 ÷ (-1) 37
11. 91 ÷ 7 13
12. 77 ÷ (-77) -1
13. -416 ÷ 52 -8
14. 0 ÷ (-95) 0
15. -2,500 ÷ 25 -100
16. -363 ÷ (-33) 11
17. 21 ÷ (-7) -3
18. -200 ÷ 5 -40
19. 5,984 ÷ (-68) -88
20. 72 ÷ 9 8

Write a numerical expression for each word phrase. Then evaluate the expression.

21. the opposite of forty-nine divided by the opposite of seven -49 ÷ (-7); 7
22. negative eighty-eight divided by the absolute value of negative forty-four -88 ÷ |-44|; -2

Find the mean.

23. stock price changes: +\$3, +\$7, -\$5, +\$4, -\$12, \$9 +\$1
24. scores: 377, 161, -418, -529, 485, -771, 654, -463 -63
25. weight changes: +2, -10, -3, +14, -1, +5, -9, -6 -1

Write >, <, or = to make a true statement.

26. -18 ÷ 9 > 36 ÷ (-12)
27. -|35 ÷ (-7)| < -60 ÷ (-12)
28. 108 ÷ 9 = 24 ÷ 2
29. |33 ÷ (-3)| > 50 ÷ 10

Solve.

30. An integer multiplied by 18 equals -90. What is the integer? -5
31. At 1 A.M. the temperature was -17°F. At 5 A.M. it had dropped to -53°F. Find the average change in temperature per hour. -9°F

Chapter 1 A-11

Enrichment

For use after 1-6 (pp. 26–29)

Profit and Loss

In the following problems, write income and profit figures as positive numbers. Write expense and loss figures as negative numbers. Income and expense figures for the Widget Company for January through April are shown here.

Period	Income	Expenses
January	\$32,518	-\$41,728
February	35,229	- 40,600
March	40,760	- 38,211
April	41,905	- 44,165

1. Find the total income for the period. \$150,412
2. Find the total expenses for the period. -\$164,704
3. Calculate the profit or loss for the period by finding the difference between income and expenses. -\$14,292
4. Assuming that the average profit or loss remains unchanged, what will be the company's yearly profit or loss? -\$42,876
5. Find the average monthly profit or loss, January–April. -\$3,573

During the next two months, business improved. By the end of June, the average monthly loss, January–June, had dropped to -\$2,815.

6. Find the total loss during May and June. -\$2,598
7. Find the company's total loss, January–June. -\$16,890
8. Total income during May and June was \$83,109. Find the total expenses. -\$85,707
9. Find the total income, January–June. \$233,521
10. Find the total expenses, January–June. -\$250,411
11. Assuming that the January–June rate of loss remains unchanged, what will be the company's yearly profit or loss? -\$33,780

A-12 Chapter 1

LESSON QUIZ

1. 25 ÷ (-5) [-5]
2. 64 ÷ 4 (16)
3. 0 ÷ (-3) [0]
4. 72 ÷ (-8) [-9]
5. (-132) ÷ 11 (-12)
6. 34 ÷ (-34) [-1]
7. Find the mean: -32, -46, -29, -13 (-30)

Assignment Guide

Basic 1-6, 9, 11, 13, 15-29, 38-43, 54-55, 57-60, MR, TY All

Average 3-4, 6-8, 10, 12, 14, 21-35, 38-41, 43-44, 46-47, 50-51, 54-56, 59-60, MR, TY All

Enriched 6-8, 12-14, 27-41, 44-49, 52-56, 59-60, MR, TY All

Career Information

Plan to invite a flight instructor or pilot from a local airport to talk to the group. Encourage students to ask questions about the skills necessary for a career in the airline industry.

Ask the speaker to cover such math-related topics as:

- the training and testing necessary to obtain a commercial pilot's license or certificate
- the role of computers and new technology in today's transportation industry
- the skills necessary to calculate and coordinate the data for altitude, temperature, and weight during take off and landing
- other responsibilities—record keeping, scheduling, and instructing

CAREER

Help Wanted: Airline pilot

Minimum of two years of college required. Good health and flying aptitude a must.

For more information, write to Future Aviation Professionals of America, 4291 J. Memorial Drive, Atlanta, GA 30032.

An airline pilot makes many calculations before, during, and after a flight. The pilot studies weather conditions to determine the safest altitude, route, and speed for the trip. To calculate takeoff speed, the pilot considers temperature, wind direction, and the weight of the plane. Support staff and equipment aid the pilot and crew.

PROJECT

Interview an airline pilot or other transportation worker. Find out how math is used in the job.

Practice

Find each sum.

1. 27 + 28 **55**
2. 12 + (-4) **8**
3. -15 + (-8) **-23**
4. -25 + 38 **13**
5. 0 + (-19) **-19**
6. 59 + (-62) **-3**
7. -125 + 258 **133**
8. 278 + 179 **457**
9. -187 + (-147) **-334**
10. -26 + 38 + (-28) **-16**
11. 99 + (-127) + 268 + (-99) **141**
12. -712 + 0 + (-88) **-800**
13. 999 + (-90) + (-9) + (-1,000) **-100**

Find each difference.

14. 36 − 17 **19**
15. 42 − (-21) **63**
16. -44 − (-35) **-9**
17. -57 − 32 **-89**
18. -48 − (-44) **-4**
19. 0 − 62 **-62**
20. 125 − 63 **62**
21. -167 − 71 **-238**
22. -214 − (-158) **-56**
23. -89 − (-12) − 147 **-224**
24. 268 − 188 − (-12) **92**
25. -439 − 0 − 255 − 11 **-705**
26. 856 − (-327) − (-144) **1,327**

Find each product.

27. 9 · 7 **63**
28. 10 · (-2) **-20**
29. -6 · 14 **-84**
30. -15(-4) **60**
31. -18(9) **-162**
32. 21(14) **294**
33. -25 · (-15) **375**
34. -27 · 0 **0**
35. -32 · 28 **-896**
36. -1 · (-1) · (-1) · (-1) **1**
37. -12 · 4 · 2(-3) **288**
38. 33 · 17 · 0 · (-199) **0**
39. -248 · 4 · (-2) · (-250) **-496,000**

Find each quotient.

40. 28 ÷ 4 **7**
41. 27 ÷ (-3) **-9**
42. -35 ÷ (-7) **5**
43. -42 ÷ 7 **-6**
44. -51 ÷ (-17) **3**
45. 65 ÷ (-13) **-5**
46. -1 ÷ (-1) ÷ 1 ÷ (-1) **-1**
47. -333 ÷ 3 ÷ (-3) **37**
48. 0 ÷ 23 ÷ (-34) ÷ (-13) **0**
49. 444 ÷ 2 ÷ 2 ÷ (-3) **-37**

Find each answer.

50. -75 + 24 **-51**
51. 132 − (-21) **153**
52. -12 · 13 **-156**
53. -96 ÷ (-12) **8**
54. -162 − 83 **-245**
55. 316 + (-174) **142**
56. 19(-24) **-456**
57. 340 ÷ (-17) **-20**
58. 418 − (-319) **737**
59. 483 ÷ (-161) **-3**
60. -163 · (-83) **13,529**
61. -512 + 512 **0**
62. -48 + (-13) − 12 + (-3) **-76**
63. 127 + (-23) − (-14) − 63 **55**
64. -12 · 3 · (-8) ÷ 6 **48**
65. 24 · (-5) · (-4) ÷ 48 **10**
66. [12 + (-3)] · (14 − 2) **108**
67. (-14 + 29) ÷ (-12 − 3) **-1**

PROBLEM SOLVING APPLICATION

OBJECTIVE:
To apply integer operations to problems involving fitness and health.

1-7 Fitness and Health

Calories (C) measure the energy provided by the food or drink you consume. You burn calories during any activity.

When you eat a banana, you *take in* 100 C: +100.
When you walk for 20 min, you *burn* 100 C: -100.

Food	Calories
apple	70
banana	100
wheat bread, 1 slice	55
corn cereal, 1 c	210
muffin	70
cooked oatmeal, 1 c	130
scrambled egg	110
cheese, 1 slice	45
salad dressing, 1 c	80
mayonnaise, 1T	100
carrot	20
potato chips, 10	115
grape juice, 1 c	165
skim milk, 1 c	90
lemonade, 1 c	110
hamburger, 3 oz	245
canned tuna, 3 oz	170
chicken, 3 oz	115
frankfurter	170

Example 1 Use the graph below and the chart at the left. Find the net calories if you swim for 40 min and then drink 2 c lemonade.

1

Solution

$40 \cdot (-10) = -400$ Swimming burns 10 C/min.
$2 \cdot 110 = 220$ Lemonade has 110 C/c.
$-400 + 220 = -180$ There is a net loss of 180 C.

For a weight loss or gain program, doctors recommend a change of diet combined with an exercise program.

To *lose* a pound, *burn* 3,500 extra calories.
To *gain* a pound, *take in* 3,500 extra calories.

Example 2 Jo wants to gain about 1 lb per week for 10 weeks. About how many extra calories should she take in each day?

2

Solution $3{,}500 \div 7 = 500$

Jo should take in about 500 extra calories per day.

Example 3 Find the net calories if you swim for 30 min and then drink 1 c grape juice.

Solution

$30 \cdot (-10) = -300$ Swim for 30 min at 10 C/min.
$+165$ C Drink 1 c grape juice.
$-300 + 165 = -135$ net calories

Calories used during activities

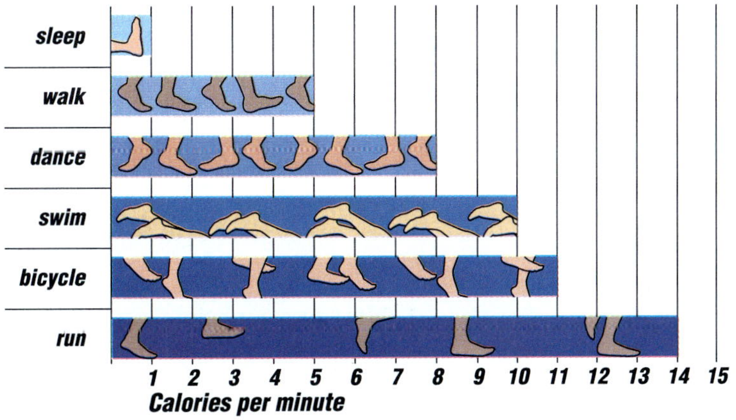

Teaching Tip

Display the calorie chart shown on page 31 with enough space for students to make additional entries. Ask students to find out the number of calories in a favorite food item and add it to the chart.

Lesson Focus

Materials
Math journal

MOTIVATING THE LESSON

Ask students for their views on why their bodies need food, and what may happen if they eat too much or too little food. Also, discuss the important role that exercise plays in keeping fit and healthy.

Instruction

Discuss the foods listed in the chart and the calories they provide. Then have students read the graph at the bottom of the page. Explain that the data is based on averages. Ask students how the rate at which they run or cycle might affect the number of calories used. (The faster they run or cycle, the more calories they use.)

1 Point out that using calories can be thought of as a negative situation, so -400 represents calories used.

2 Ask students why 3,500 is divided by 7 to find the answer. (Jo wants to gain 1 lb in a week, or 7 days; dividing by 7 gives the average number of calories she should take in per day.)

Guided Practice

Class Exercises Ask students how the net calories George took in per day were found. (The total number of calories he took in and the total number of calories he used were found. Then the net gain or loss was calculated for the day.)

Closure

Writing in Math Ask students to write a short essay in their math journals on how energy provided by food is measured, how exercise affects calorie intake, and how it is possible to monitor one's net calorie intake.

Independent Practice

The *Decision Making* feature allows students to apply the content in this lesson to their own lives.

For students to use the method described in 1a, they will need reference materials that list more foods than are given in the chart on page 31.

For questions 7-8, remind students that a net loss should be written as a negative number.

CLASS EXERCISES

Use the line graph below.

1. On which days did George

a. take in more calories than he used?

b. use more calories than he took in?

c. burn and take in the same number of calories?
a. Tues., Wed., Fri. b. Thurs., Sun. c. Mon., Sat.

2. Find George's net calories for the week. 200 C

a. Did he have a net gain or a net loss of calories? net gain

b. If this pattern continues, how will George's weight change in a month? in a year? He will gain weight.

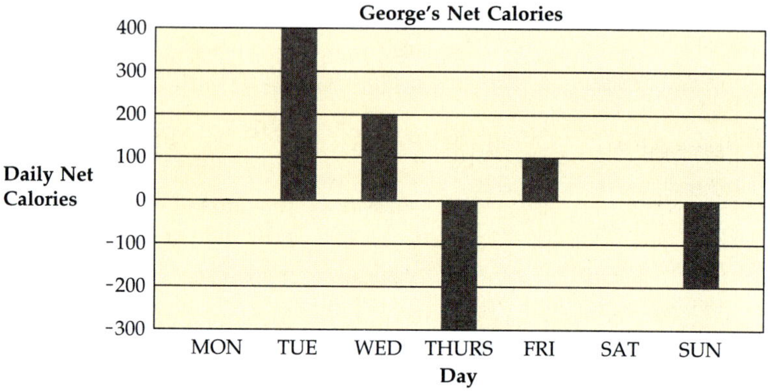

Decision Making **DECISION MAKING** Decision Making Decision Making Decision Making

FITNESS AND HEALTH

To be healthy, you must make good decisions about food and exercise. A nutritionally balanced diet and regular exercise will help you to look and feel your best.

COLLECT DATA Check students' work.

1. Find your normal daily intake of calories by one of the methods below.

a. Record your daily calorie intake for a week, then find the average.

b. Use an estimate of 16 C/lb of your weight.

SAMPLE 135 lb: $135 \cdot 16 = 2{,}160$ C per day

2. Measure your height and weight. Look in a science book or an encyclopedia to find the ideal weight for a person your height and age.

MATH MINUTES

Provide today's lunch menu from the school cafeteria. Ask the students what they plan to eat for lunch, and have them find the total number of calories they would take in. (Most almanacs have a list of foods and calories.)

WRITTEN EXERCISES

Use a CALCULATOR where appropriate.

Use the graph and table on page 31.

How many calories does each food provide?

1. 2 slices of cheese 90 C
2. 2 c skim milk 180 C
3. 20 potato chips 230 C

How many calories does each activity burn?

4. bicycle 40 min 440 C
5. dance 15 min 120 C
6. sleep 8 h 480 C

Find the net calories.

7. dance 2 h, eat a tuna sandwich -580 C
8. run 20 min, eat a muffin -210 C

Solve.

9. Suppose you bicycle for an hour. About how many calories would you burn? About how many hours would it take to burn 3,500 C? 660 C; about 5 h
10. Juan began running 30 min each day. In order to maintain his weight, he also increased his calorie intake. How many calories should he add to his daily diet? 420 C
11. Elizabeth joined the swim team. She swims 90 min every week day. How many calories will she burn swimming each day? How many in four weeks (20 days of swimming)? 900 C; 18,000 C

Decision Making ■ Decision Making ■ Decision Making ■ Decision Making ■ Decision Making ■ Decision Making

ANALYZE DATA

3. Analyze your eating habits.
   a. List the foods you eat often and the number of calories each food provides.
   b. List other healthy foods that would help balance your diet.
4. Analyze your activities.
   a. List the activities you do and the calories each activity burns.
   b. List other activities you might like to do.

MAKE DECISIONS

5. Decide on a plan to maintain or improve your health.
   a. Set a reasonable goal to gain, maintain, or lose weight.
   b. Decide how to change your eating habits and activities.

Lesson Follow-up

LESSON RESOURCES

TRB Alternate Application A-13

Alternate Application

For use after 1-7 (pp. 31–33)

Changes in Temperature

Two of the most important factors influencing temperature are elevation and latitude. (Latitude is position on the earth's surface measured in degrees north or south of the equator, from 0° to 90°.)

Elevation Rule
For every 300-ft gain in altitude, subtract 1°F.

Latitude Rule (January)
For every 2 degrees of latitude north or south of the equator, subtract 3°F.

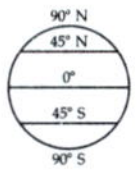

In Exercises 1-10, elevations and latitudes are approximate.
Compare Albuquerque and Portland in January.

Location	Latitude	Altitude (ft)
Albuquerque, NM	35°N	5,100
Chicago, IL	42°N	0
Mount Massive, CO	40°N	14,400
Portland, ME	43°N	0

1. How much warmer is Albuquerque due to latitude? 12°
2. How much colder is Albuquerque due to altitude? 17°
3. Find the net difference. 5°
4. Which city is colder in January? By how much? Albuquerque; 5°F

Compare Chicago and Mount Massive in January.

5. How much warmer is Mount Massive due to latitude? 3°
6. How much colder is Mount Massive due to altitude? 48°
7. Find the net difference. 45°
8. Which location is colder? By how much? Mount Massive; 45°F

Solve.

9. Moscow, USSR, has a latitude of 56°N and an altitude of 400 ft. Mexico City, Mexico, has a latitude of 20°N and an altitude of 7,300 ft. Which location is colder? By how much? Moscow, USSR; 31°F
10. Peking, China, has a latitude of 40°N and an altitude of 150 ft. St. Louis, Missouri, has a latitude of 38°N and an altitude of 450 ft. Which location is colder? By how much? Peking, China; 2°F

Chapter 1 A-13

Assignment Guide

Basic All
Average All
Enriched All

FOR THE NEXT LESSON

algebra tiles or blue and white squares of paper and red paper rectangles, math journal

Lesson Focus

Materials

Algebra tiles or blue and white paper squares and red paper rectangles, math journal

Vocabulary/Symbols

numerical expression
variable
variable expression

MOTIVATING THE LESSON

Ask students to write a numerical expression for these situations: Phil has $6. Robin has $2 more than Phil. (6 + 2) Joe has $1 less than Phil. (6 − 1) Tracey has twice as much money as Phil. (2 · 6)

Instruction

Ask students to suggest ways to write an expression for this situation: Megan has $5 more than Ted. To represent Ted's money, students may suggest a blank line, a box, a question mark, or a letter. Record all suggestions and point out that a standard way of representing an unknown quantity is to use a letter, or variable. Comment that in the variable expression $5 + t$, t represents Ted's money. The expression represents Megan's money.

1 As you discuss Example 1, ask students to explain the difference between a numerical expression and a variable expression.

2 Have pairs or small groups of students make the pictured models. Ask students to suggest reasons why a different shape is used for a variable. (Answers may vary; using a different shape may make it easier to remember that the shape stands for an unknown number.)

Work with students to model $x + (-4)$ (1 rectangle plus 4 blue tiles) and $5 - 2a$, (5 white tiles minus 2 rectangles).

OBJECTIVE: *To assign variables and write variable expressions.*

1-8 Expressions and Variables

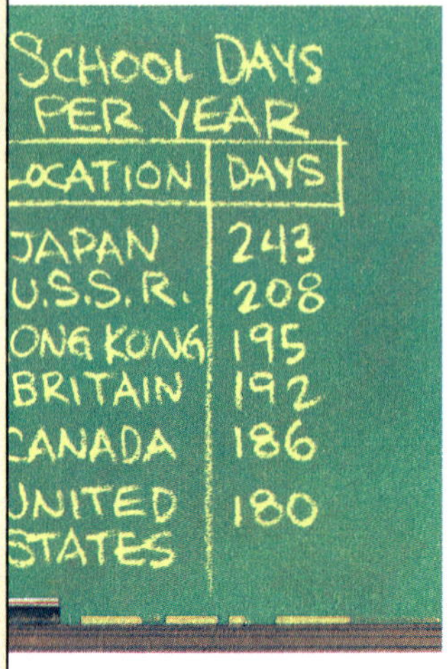

▼ You can write a *numerical expression* to show the number of weeks in each school year if you assume a 5-day school week.

	Days Per Year		Days Per Week
Japan	243	÷	5
U.S.S.R.	208	÷	5
Hong Kong	195	÷	5
Britain	192	÷	5
Canada	186	÷	5
United States	180	÷	5

If you don't know the number of days in the school year, you can use a letter or other symbol, called a *variable,* to stand for the number. You can write a *variable expression* to show the number of weeks in the school year.

Days Per Year		Days Per Week
d	÷	5

Variable A variable is a symbol (usually a letter) that stands for a number.

Variable Expression A variable expression is an expression that contains at least one variable.

FLASHBACK
We usually write products involving variables without multiplication symbols. Write $60h$ not $60 \cdot h$ or $h60$.

Example 1 Write an expression for the number of minutes in:
a. 1 hour **b.** 5 hours **c.** h hours

Solution **a.** $60 \cdot 1$ **b.** $60 \cdot 5$ (numerical expressions) **c.** $60h$ (variable expression)

▼ Just as we use models to stand for integers, we can use models for variable expressions. Use red rectangles for variables.

FLASHBACK
Use white tiles for positive integers and blue tiles for negative integers.

Example 2

Expression	Model
$2x$	
$3y + 3$	

Teaching Tip

To help students become comfortable with using rectangles (in algebra tiles) to represent variables, use a piece of tape or a label to write variables on the rectangles.

▼ You can use any letter or symbol as a variable. Mathematicians often use the first letter of a word, such as t for time, w for week, or a for age. The most commonly used variables are x, y, z, n, a, and b.

Example 3

Word Phrase	Variable Expression
a number plus nine	$y + 9$
three times a quantity decreased by four	$3z - 4$
negative one divided by a number	$-1 \div a$
the calories in two slices of toast with c calories per slice	$2c$
the cost of b books at \$5 per book	$5b$

③

▼ You can write variable expressions to describe situations.

Example 4 A student has several pencils in his desk. He takes out 5 pencils. Write an expression for the number of pencils left in the desk.

Solution
1. Choose a variable, say p, for the original number of pencils.
2. Then $p - 5$ is an expression for the number of pencils left in the desk.

Example 5 Suppose you study history for a different length of time each day for 3 days. Write an expression for the average amount of time you studied history each day.

Solution
1. Choose a variable, say t, for the total time spent studying for the three days.
2. Then $t \div 3$ is the average time spent each day.

CLASS EXERCISES

Tell whether each is a numerical expression or a variable expression. If it is a variable expression, name the variable.

1. $b + 6$ variable; b
2. $9x$ variable; x
3. $80 \div 8$ numerical
4. $14 - n$ variable; n

Choose a variable and write a variable expression for each model.

5.  $2x + 2$
6.  $-3 - y$
7. 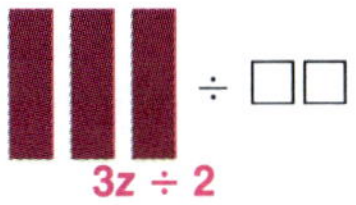 $3z \div 2$

Write a variable expression for each word phrase.

8. 16 more than m — $m + 16$
9. y decreased by -4 — $y - (-4)$
10. the quotient of 6 and z — $6 \div z$
11. the product of c and 3 — $3c$

NOTES & QUOTES

René Descartes (1596–1650) was the first to use x, y, and z as variables.

THINK AND DISCUSS See below

1. How does the word *vary* relate to *variable*?
2. Why do you think letters of the alphabet are used as variables?
3. Describe how the expressions for each phrase differ.
   - some number added to 12 $12 + n$
   - 12 equals some number $12 = n$
   - the sum of some numbers is 12 $a + b = 12$

1. The value of a variable varies.
2. Letters are less likely to be confused with numbers and are convenient.

③ Stress the importance of reading a word phrase carefully as well as more than once. Ask students to offer another word phrase for: $y + 9$ (a number increased by nine), $3z - 4$ (four less than the product of three and a number), and $2c$ (two times a number).

ADDITIONAL EXAMPLES

1. Write an expression for the number of hours in:
   a. 2 days $(24 \cdot 2)$
   b. d days $(24d)$
2. Model the expression $4y - 2$. (4 rectangles minus 2 white tiles)
3. Write a variable expression for, "three more than two times a number." $(2x + 3)$
4. Several people form groups with three people in each group. Write the expression for the number of groups there are. $(p \div 3)$

Guided Practice

Class Exercises Have students share their word phrases with the class for questions 12-15.

Think and Discuss Encourage multiple answers for question 2. Comment that letters sometimes help in remembering special expressions, or formulas, because often the variable used is the first letter of the word it represents.

For question 3, remind students that mathematical language is precise, and it is important to examine *every* word in a phrase.

MATH MINUTES

To strengthen listening skills, prepare a set of word phrases and read them aloud, one at a time. Ask students to write a variable expression for each word phrase.

A ***common error*** is to reverse the variable expression when the word phrase denotes a subtraction situation. For example, students often write *eight less than a number* as $8 - n$ instead of $n - 8$. Tell students to substitute a number for the unknown value. For example, try 5 for the unknown number and think, *eight less than five. I subtract eight from five. I write* $5 - 8$.

Closure

Writing in Math Write $6 + a$, $4 - n$, $3x$, and $b - (-2)$ on the chalkboard. Ask pairs of students or small groups to write in their math journals as many word phrases as they can for each expression.

Independent Practice

For questions 12-24, to avoid errors, ask students to reread each word phrase after they have written the variable expression to check that the variable expression matches the word phrase.

Write a word phrase for each variable expression. Answers will vary.

12. $m + 3$ **13.** $8 - t$ **14.** $-6k$ **15.** $t \div 12$

Write an expression for the situation.

16. The number of eggs in:

a. 1 dozen $1 \cdot 12$ **b.** 5 dozen $5 \cdot 12$ **c.** d dozen $12d$

WRITTEN EXERCISES

Choose a variable and write a variable expression for each model.

1.  $3x - 3$

2.  $5y$

3. 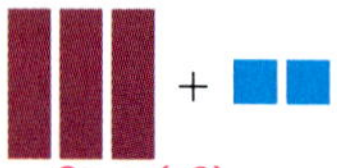 $3z + (-2)$

Use a model to represent each variable expression. Check students' models.

4. $6x$ **5.** $y + 5$ **6.** $-2 + 2a$ **7.** $m - (-4)$

Write a numerical expression for each word phrase.

8. the quotient of fourteen and negative seven $14 \div (-7)$

9. the product of twenty-three and negative nine $23(-9)$

10. four more than one thousand $1{,}000 + 4$

11. eight less than the opposite of six $-6 - 8$

Write a variable expression for each word phrase.

12. six subtracted from k $k - 6$

13. m more than nineteen $19 + m$

14. eight less than z $z - 8$

15. twelve times x $12x$

16. the sum of a and b $a + b$

17. n divided by negative one $n \div (-1)$

18. one more than a number $3p$ $3p + 1$

19. the product of g and four times r $g \cdot 4r$

20. the difference of a number and three $n - 3$

21. the product of ten and a number $10a$

22. sixty-four decreased by a number $64 - k$

23. a number increased by two hundred $t + 200$

24. twice a number plus the absolute value of negative seven $2a + |-7|$

Answers will vary.

Write two different word phrases for each variable expression.

25. $x + 2$ **26.** $12 - y$ **27.** $15 \div s$ **28.** $-20 + q$

29. $-5y$ **30.** $w - (-4)$ **31.** $100 + (-y)$ **32.** $|n| + 1$

Write an expression for each situation.

33. The number of days in:

a. 1 week $7 \cdot 1$ **b.** 4 weeks $7 \cdot 4$ **c.** w weeks $7w$

34. The value, in cents, of:

a. 10 pennies $10 \cdot 1$ **b.** 7 nickels $7 \cdot 5$ **c.** q quarters $25q$

35. Pam is 15 years old. Write an expression for Pam's age:

a. 3 years ago $15 - 3$ **b.** p years ago $15 - p$

c. 10 years from now $15 + 10$ **d.** f years from now $15 + f$

36. Peter has c cousins. Paul has 4 more cousins than Peter. How many cousins does Paul have? $c + 4$

37. Susan has $20 less than Charlotte. Charlotte has d dollars. How many dollars does Susan have? $d - 20$

38. There are twice as many sophomores as freshmen.

a. If there are f freshmen, how many sophomores are there? $2f$

b. If there are s sophomores, how many freshmen are there? $s \div 2$

39. Richard ran b miles. Write a related situation for each expression.

a. $b + 3$ Answers will vary. **b.** $2b$ **c.** $10 - b$

40. A hot air balloon is at an altitude of m meters. Write a related situation for each expression. Answers will vary.

a. $m + 34$ **b.** $m - 2{,}000$ **c.** $3m$

41. Jeans sell for $25 and T-shirts sell for $12.

a. Write a numerical expression for the selling price of 2 pairs of jeans and 4 T-shirts. $(2 \cdot 25) + (4 \cdot 12)$

b. Write a variable expression for the selling price of j pairs of jeans and t T-shirts. $25j + 12t$

42. ***DATA*** Use the calorie chart on page 31.

a. Write a numerical expression for the number of calories in 3 eggs and a slice of wheat bread. $3(110) + 55$

b. Write a variable expression for the number of calories in e eggs and s slices of wheat bread. $110e + 55s$

Match each variable expression with a model.

43. $4x$ b

44. $4 + x$ a

45. $x \div 4$ d

46. $x - 4$ c

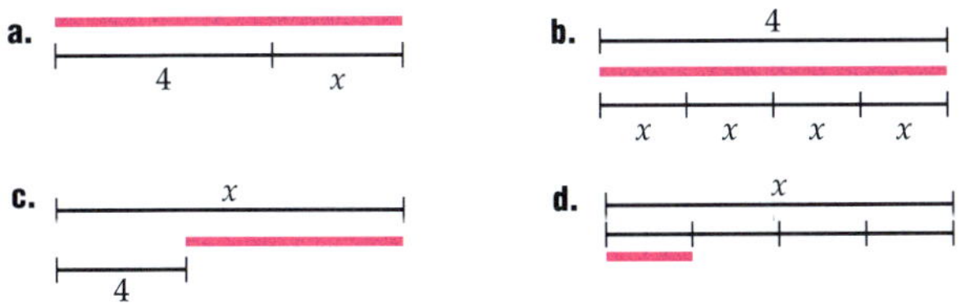

MIXED REVIEW

Find each answer.

1. $-10 \div 2$ -5
2. $-120 \div (-6)$ 20
3. $0 \div (-12)$ 0
4. $-6 + (-17)$ -23
5. $-20(-1)(5)$ 100
6. $4 + (-12) + 16$ 8
7. How much change from a five-dollar bill must the cashier give a customer who buys orange juice for $.95 and two apples for $.55 each? $2.95

For questions 33-42, have students discuss how they arrived at the expressions.

Lesson Follow-up

Reteaching Activity Write or draw related sets of variable expressions, word phrases, and pictures of algebra tile models on index cards. Mix the cards. Have students find related sets.

LESSON RESOURCES

TRB Practice A-14
TRB Enrichment A-15

Practice For use after 1-8 (pp. 34–37)

Expressions and Variables

Write a variable expression for each model.

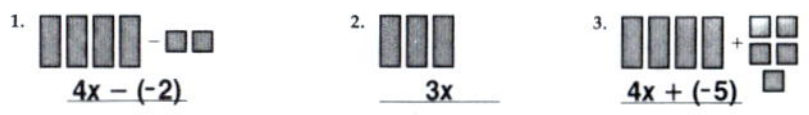

1. $4x - (-2)$ 2. $3x$ 3. $4x + (-5)$

Draw a model to represent each variable expression.

4. $3y + 4$ 5. $-3 - n$

Write a numerical expression for each word phrase.

6. five less than the opposite of nine $-9 - 5$
7. the product of eight and negative six $8(-6)$

Write a variable expression for each word phrase.

8. the sum of x and y $x + y$
9. p divided by k $\frac{p}{k}$ or $p \div k$
10. h less than negative two $-2 - h$
11. x times g xg
12. eleven more than z $z + 11$
13. three increased by twice h $3 + 2h$

Write two different word phrases for each variable expression. Answers may vary.

14. $g + 7$ 7 more than g; the sum of g and 7
15. $n \div 3$ the quotient of n and 3; n divided by 3

Write a variable expression for each situation.

16. the number of minutes in h hours $60h$
17. the value, in cents, of d dimes $10d$
18. the number of inches in f feet $12f$
19. the number of hours in d days $24d$

A book contains p pages. Write an expression for the number of pages in a book:

20. with 35 fewer pages $p - 35$
21. longer by 82 pages $p + 82$
22. with three times as many pages $3p$
23. with 26 more than double the number of pages $2p + 26$
24. with 32 less than double the number of pages $2p - 32$
25. with one third as many pages $\frac{1}{3}p$ or $\frac{p}{3}$

A-14 Chapter 1

Enrichment For use after 1-8 (pp. 34–37)

A Variable Maze

To find your way through the maze, look for the word phrase that corresponds to each numbered junction. Choose the direction of the letter beside the variable expression which correctly represents the phrase.

1.	the product of m and n	(A) mn	B. $m + n$
2.	two less than k	A. $2 - k$	(B) $k - 2$
3.	the sum of a and b	(A) $a + b$	B. ab
4.	h increased by ten	(A) $h + 10$	B. $h(10)$
5.	the quotient of x and 4	(A) $x \div 4$	B. $x(4)$
6.	eight more than p	A. $8(p)$	(B) $p + 8$
7.	the difference of y and 3	(A) $y - 3$	B. $y \div 3$
8.	twenty decreased by c	(A) $20 - c$	B. $20 + c$
9.	the absolute value of negative five	A. $-\lvert 5 \rvert$	(B) $\lvert -5 \rvert$
10.	t less than the opposite of two	A. $t - (-2)$	(B) $-2 - t$
11.	k increased by twice x	(A) $k + 2x$	B. $k + 2 + x$
12.	s divided by n	A. $n \div s$	(B) $s \div n$
13.	y times the sum of 2 and e	(A) $y(2 + e)$	B. $(y + 2)e$

Chapter 1 A-15

LESSON QUIZ

Write a numerical or variable expression for each.

1. eight more than negative twelve $(-12 + 8)$

2. the difference between a number d and fourteen $(d - 14)$

3. four more than the absolute value of three $(\lvert 3 \rvert + 4)$

4. the product of negative two and a number a $(-2a)$

Assignment Guide

Basic 1-5, 8-21, 25-27, 33-37, 39, 41a, 42a, MR All

Average 6-11, 16-24, 27-30, 34-41, 42a, MR All

Enriched 8-11, 16-24, 29-32, 34, 38-46, MR All

FOR THE NEXT LESSON

TRB Teaching Resources (blank spreadsheet form), math journal

Lesson Focus

Materials

TRB Teaching Resources (blank spreadsheet form), math journal

Vocabulary/Symbols

cell
cursor
spreadsheet

MOTIVATING THE LESSON

Display the blank computer spreadsheet on the overhead projector. Discuss the uses of electronic spreadsheets, such as keeping financial data for businesses, industries, and family budgets. Mention that spreadsheet software stores data in tabular form and has the built-in capability to perform calculations.

Instruction

Appleworks, *Lotus 1-2-3*, or *Excel* are excellent programs that can be used with this lesson.

1 Use the transparency to point out cells and show how they are named. Mention that the cursor is in cell B6. Copy the labels in column A and in row 1. Tell the class that these labels are not used in operations.

If students are not using a computer, have them work in pairs or groups of three or four. Have them record their answers in their math journals.

2 Remind students that the computer recalculates the totals when data in a cell are changed. This may be done automatically or on command.

OBJECTIVE:
To explore the uses of a computer spreadsheet.

MATERIALS

- paper and pencil
- computer (optional)
- Math journal to record work

DATAPOINT

Cells can store numbers, letters, or a combination of letters and numbers.

Exploring Spreadsheets

■ A spreadsheet is a tool for organizing and analyzing data. The data are arranged in rows and columns. The spreadsheet below shows data for a school walkathon.

1

	A	B	C	D
1	Class	No. Students	Mi/Student	Tot. Distance
2	9th Grade	250	10	2,500
3	10th Grade	234	15	3,510
4	11th Grade	199	20	3,980
5	12th Grade	176	20	3,520
6			TOTAL	13,510
7				

■ We call each section of a spreadsheet a *cell*. Cell B2 stores the value for the number of ninth-grade students who participated in the walkathon. **1. the distance walked by each grade; total distance**

1. What amounts are stored in cells D2 through D5? in cell D6?
2. How are the amounts in column D found? **Multiply amounts in B by amounts in C.**
3. Copy the spreadsheet shown above. Use a calculator or paper and pencil to complete column D. **2,500; 3,510; 3,980; 3,520; 13,510**

■ You enter data in a spreadsheet one cell at a time. The *cursor* shows you where the next character will be entered.

4. Look at the spreadsheet above. In which cell is the cursor? **B6**

■ For a spreadsheet to perform a calculation, you need to type a formula in a cell. Look at the spreadsheet below.

	A	B	C	D
1	Type of Seat	Tickets Sold	Ticket Price	Total
2	Balcony	25	\$7.50	
3	Mezzanine	34	\$8.50	=B3*C3
4	Front row	38	\$15.00	=B4*C4
5	Standing room	42	\$4.00	=B5*C5
6			TOTAL	=D2 + D3 + D4 + D5
7				

Cell D2 stores the formula B2*C2. The computer multiplies the values in cells B2 and C2 and displays the product in cell D2.

5. Write formulas for cells D3 to D5.
6. What happens if you change the value in cell B4? Which other cells will change? **D4 and D6**

2

Teaching Tip

If students are not using a computer, make copies of the blank spreadsheet for this lesson for students to use. Students who do use a computer can still use the form in designing their spreadsheets.

■ The spreadsheet below shows test results for five students.

	A	B	C	D	E	F	G	
1	Student	Test 1	Test 2	Test 3	Test 4	Test 5	Average	
2	Jane C.	76	87	88	85	92		
3	Art G.	87	84	75	83	94		=sum (B3:F3)/5
4	Jose F.	95	84	78	93	92		=sum (B4:F4)/5
5	Kim C.	76	77	83	84	100		=sum (B5:F5)/5
6	Dan H.	85	91	79	78	93		=sum (B6:F6)/5
7								

One formula for G2 is (B2 + C2 + D2 + E2 + F2)/5. A shorter formula using a range of cells is Sum(B2 : F2)/5.

7. Write formulas for cells G3 to G6.

8. A student wrote the formula Sum(B6 : G6)/5 for G6. What is wrong with this formula? What do you think would happen if you used this formula in cell G6? **It includes G6; you would get an error message.**

9. Suppose you drop the lowest score for each student.

a. How will the formulas in column G change? Write the new formulas. **=sum (B2:F2)/4; =sum (B3:F3)/4; =sum (B4:F4)/4; =sum (B5:F5)/4; =sum (B6:F6)/4**

b. What value should you enter in the cell that has the lowest score for each student? **zero**

c. Use a spreadsheet to find the average when the lowest score is dropped. If you do not have a computer, use a calculator or paper and pencil. **Jane 88, Art 87, José 91, Kim 86, Dan 87** (4)

10. ***Discuss*** ideas for different types of spreadsheets with the class. Make a list of your ideas.

Check students' work.

11. ***PROJECT*** Create a spreadsheet using one of the ideas from Exercise 10. Use these guidelines to plan your spreadsheet.

- Include at least five rows and four columns of data.
- At least one row or column should involve computation using the data in the other cells.
- The formulas in the spreadsheet should use at least two of the four basic operations.

a. Write a list of all the formulas used in your spreadsheet. Use these operation symbols: +, −, *, /.

b. Draw your spreadsheet on a sheet of lined paper. Show it to your class and explain how the spreadsheet is organized.

c. Use a spreadsheet program to create your spreadsheet. Change the data for various cells to check that your formulas are correct.

DATAPOINT

Check your spreadsheet program manual to learn how to show the formula for adding across a range of cells. (3)

(3) If students are using a computer, have a set of simplified directions for the software available for students to use.

(4) Encourage students to share their ideas for creating spreadsheets to complete this question, as well as questions 10 and 11. Help students realize the variety of uses a spreadsheet can have.

Closure

Writing in Math Ask students to think of one additional use of a spreadsheet and write it in their math journals. Then have them look at the spreadsheet at the top of page 38 and describe what happens if they change one entry in either column B or C. (The total distance in that row and the overall total in cell D6 would change.)

As a follow up, refer to the computer activity on page 2E of the Chapter Overview. See also TRB O-3.

Assignment Guide

Basic All
Average All
Enriched All

FOR THE NEXT LESSON

calculators, math journal, index cards

Lesson Focus

Materials

Calculators, math journal, index cards

Vocabulary/Symbols

evaluate
order of operations
quantity

MOTIVATING THE LESSON

Write the expression $4 + 3 \cdot 8$ on the chalkboard and ask students to find its value. (28) Most students should get either 28 or 56 as an answer. Have students explain how they arrived at their answers.

Instruction

Have students discuss why it might be desirable to have a single, agreed-upon value for an expression. Explain that, to avoid confusion, a set of rules, called the *order of operations*, is used to determine the value of an expression.

1 As you discuss order of operations, ask students to use the rules to verify that the value of $4 + 3 \cdot 8$ is 28.

2 Ask students to explain why it is important to work from left to right in Step 2. (Adding and subtracting out of order will give a different value (-20).)

3 Point out that the symbol for absolute value can be regarded in the same way as parentheses; i.e., the computation inside the symbol should be done first.

OBJECTIVE: *To study order of operations using paper and pencil and a calculator.*

1-9 Order of Operations

Suppose you spend \$4 and then earn \$5/h mowing lawns for 3 h. The numerical expression $-4 + 5 \cdot 3$ shows how much money you have. How do you *evaluate* this numerical expression?

$$(-4 + 5) \cdot 3 = 3 \qquad \text{or} \qquad -4 + (5 \cdot 3) = 11$$

Since you earned \$15 and spent \$4, you would have \$11 left. The order in which you compute changes your answer.

1

To avoid confusion and have a standard way to compute expressions with several operations, mathematicians agree on an *order of operations*.

FLASHBACK

Grouping symbols:
( ) parentheses
[ ] brackets

Order of Operations	
	1. Do all operations within grouping symbols first.
	2. Multiply or divide in order from left to right.
	3. Add or subtract in order from left to right.

You can use order of operations to evaluate a numerical expression.

Example 1 **Evaluate $-3 \cdot 5 - 8 \div 4 + 3$.**

2

Solution

1. There are no grouping symbols. So, multiply and divide from left to right. $-3 \cdot 5 - 8 \div 4 + 3$
2. Add and subtract from left to right. $-15 - 2 + 3$; $-17 + 3 = -14$

Example 2 **Evaluate $3(-8 + 5) - 12$.**

Solution

1. Work in parentheses. $3(-8 + 5) - 12$
2. Multiply. $3\ (-3) - 12$
3. Subtract. $-9 - 12 = -21$

When there are two or more sets of grouping symbols, start at the inside and work out.

Example 3 **Evaluate $-2[(-6 + 4) \div (3 - 5)] + 6$.**

Solution

1. Parentheses first. $-2[(-6 + 4) \div (3 - 5)] + 6$
2. Work inside brackets. $-2\ [-2 \div (-2)] + 6$
3. Multiply. $-2\ (1) + 6$
4. Add. $-2 + 6 = 4$

Teaching Tip

As you discuss the order of operations, introduce the expression *Please, My Dear Aunt Sally* as a device to help students remember that the order is *P*arentheses, *M*ultiplication, *D*ivision, *A*ddition, and *S*ubtraction.

▼ Absolute value symbols are a kind of grouping symbol.

Example 4 **Evaluate -5 − |8 − (-2)|.**

Solution
1. Work inside the absolute value symbols. -5 − |8 − (-2)|
2. Find the absolute value. -5 − |10| = -5 − 10
3. Subtract. = -15

▼ A scientific calculator follows order of operations. A standard calculator does not.

Example 5 **Use a calculator to evaluate 3 · 5 − 4 ÷ 2 + (5 + 4).**

Solution **Scientific calculator**

3 [×] 5 [−] 4 [÷] 2 [+] [(] 5 [+] 4 [)] [=] 22

Standard calculator

5 [+] 4 [=] [M+]	grouping symbols
3 [×] 5 [=] [M+]	multiplication
4 [÷] 2 [=] [M−]	division
[MRC] 22	

THINK Why must you use the memory key with the standard calculator? to insure correct order of operations

CLASS EXERCISES

Which operation would you perform first? Explain.

1. 35 · 98 − 50 multiplication
2. -29 − (87 + 115) addition
3. 4(67 ÷ 6) division

Evaluate.

4. 2 − 6 ÷ 3 0
5. 14(-6) − 12 -96
6. |13 − 21| + 5 13
7. 2(1 − 9) · 9 -144
8. 7 + 3(8 ÷ 4) 13
9. [2 + (6 · 8)] − 1 49

THINK AND DISCUSS

Is the answer positive or negative? Explain.

1. 586 − 25 · 30 neg.
2. (387 − 521) · (-86) pos.
3. -3|-5 · 4| neg.
4. Why do we need to agree on an order of operations?

4. so that anyone simplifying an expression will get the same result

WRITTEN EXERCISES

Which operation would you perform first? Explain.

1. 14 + 15 − 10 addition
2. -11 ÷ 4 + 99 division
3. 75 · 398 + |15 − 16| subtraction inside the absolute value symbols

Evaluate.

4. 15 · 3 − 2 43
5. -12 ÷ 4 − (-2) -1
6. 2 + (-3) · 24 -70
7. |56 − 5| ÷ 17 3
8. (-21 + 15) ÷ (-3) 2
9. 2 · 2 + 0 · (-4) 4

④ Have students evaluate the expression using their calculators to determine whether they have standard or scientific calculators.

Ask students why 5 + 4 was stored in memory first. (5 + 4 is in parentheses.) Next, ask students to explain why the sum is stored using the M+ key instead of the M− key. (A plus sign precedes the parentheses.) Finally ask why the multiplication is done before the division in this example. (It occurs first when working from left to right.) Point out that M− is used to store 4 ÷ 2 because the division is preceded by a minus sign.

Students may also note that their calculators may have separate MR and MC keys.

ADDITIONAL EXAMPLES

1. Evaluate 3 + 5 · 2 − 22. (-9)
2. Evaluate 4 + (-9)(-2 + 6). [-32]
3. Evaluate. -5 [(-4 + 8) − 14 ÷ 2] + (-5). [10]
4. Evaluate. 2|-8 − (-4)| + 12. (20)
5. Use a calculator to evaluate 23 (-7 + 8) + (-255). [-232]

Guided Practice

Class Exercises Ask students to explain how they arrived at their answers for questions 4-9.

Think and Discuss In questions 2 and 3 students do not need to estimate, as they did in question 1. Determining the signs of the factors will allow students to determine the sign of the answer.

MATH MINUTES

Give students an expression such as 4 + 3 · (-8)-2. Ask them to insert one pair of parentheses in different places to see how the value of the expression changes. The numbers must remain in the given order.

A ***common error*** is to perform operations in the order in which they appear. To help students remember to follow the rules for the order of operations, encourage them to write parentheses around expressions that should be evaluated first. For example, for 3 − 6 · 5 + 6, students may write 3 − (6 · 5) + 6 and then find the value. (-21)

Closure

Write the following expressions on the chalkboard.

(-4 − 5) ÷ 3 + 2 (-1)
-2 [5 − (-3 + 1)] [-14]

Ask students to tell when they might add first in an expression. (when the only operations are addition and subtraction, and addition is the leftmost operation; when addition is in parentheses)

Independent Practice

For question 43, point out that the digits of the year must remain in order.

Lesson Follow-up

Reteaching Activity Ask students to write this short version of the order of operations on an index card or a piece of paper.

1. Parentheses
2. Multiply and divide, left to right.
3. Add and subtract, left to right.

Give students a set of numerical expressions and ask them to write the number of the rule above the operation.

  3 2 3 1
-7 + 15 ÷ (-5) + (4 − 8)

Then have students evaluate the expression following their numbers. (-14)

MENTAL MATH Evaluate.

10. 12 − 8 ÷ 2 **8** **11.** 3(-8) + 4 **-20** **12.** 6 ÷ (2)(-9) **-27**
13. 3(-4) − 18 **-30** **14.** -21 ÷ 7 − (-15) **12** **15.** 3(-6) + 15 ÷ 3 **-13**

CALCULATOR Evaluate.

16. 538 + 18 · 24 − 677 **293** **17.** 450 ÷ 2 + 18 **243**
18. -8 − 3 · 2 − (-8) **-6** **19.** 4 ÷ (-4) · (-4) − (-4) + (-4) **4**
20. 2[8 + (3 − 5)] − 8 **4** **21.** 25(6 + 2)(-8) ÷ 4 + 6 **-394**
22. 6 ÷ 3 − 9 · 4 **-34** **23.** -11 − 27 ÷ 9 ÷ (-1) **-8**

Write an expression to match these keys on a standard calculator.

24. 5 [×] 4 [÷] 10 [+] 25 [÷] 9 [+/−] [=] -3 **[(5 × 4) ÷ 10 + 25] ÷ (-9) = -3**
25. 7 [+] 8 [÷] 5 [+] 5 [×] 2 [+/−] [=] -16 **-2[(7 + 8) ÷ 5 + 5] = -16**
26. How will the expressions for Exercises 24 and 25 change if you use a scientific calculator? **The parentheses will be in different places.**

Compare. Use >, <, or =.

27. 8 + 12 ÷ (-4) **>** (8 + 12) ÷ (-4)
28. (18 − (-15)) ÷ (5 + 6) **<** 18 − (-15) ÷ 5 + 6

Insert grouping symbols to make each number sentence true.

29. (7 + 4) · 6 = 66 **30.** 7 · (8 − 6) + 3 = 17
31. 3 · (8 − 2 + 5 − 12) = -3 **32.** 2 · (3 − 5) − 8 · (2 + 1) = -28

Which of the following equal 18? **34; 35**

33. 3 · 2 + 4 **no** **34.** (10 − 18) ÷ (-4) + (15 − (-17)) ÷ 2 **yes**
35. 27 − 13 · 2 − 17(5 − 6) **yes** **36.** 16 · 3 + 5 ÷ 5 − 18(-13) **no**

Solve.

37. Carmen worked 4 h on Monday and 7 h/day for the next 3 days. How many hours did she work in all? **25 h**

38. Sam bought 8 CDs at $12 each and 4 tapes at $6 each. How much did he spend? **$120**

39. Alice's bowling score is 15 less than Ray's. Together, they scored 221. What did each score? **Alice 103; Ray 118**

40. A cup of tomato juice has half the calories of a cup of skim milk. Together, they have 135 C. How many calories are in each? **tomato juice 45 C; skim milk 90 C**

41. Use the numbers -6, -8, 2, 4, and 6 exactly once to write a numerical expression with a value less than -100. **[-6 + (-8)](6 + 4) − 2**

42. Use the digits 1–9 in order. Insert addition and subtraction signs, brackets, and parentheses to get an answer of 100. **Answers will vary. An example is given. 1(2 + 3)4(5 + 6 − 7 − 8 + 9)**

43. Use the digits in the number of the year you were born, in order, plus operation symbols and parentheses. Write the greatest possible number and the least possible number that uses each digit exactly once. **Answers will vary.**

▼▼ *SAMPLE* born 1966

greatest $(1 + 9) \cdot 6 \cdot 6 = (10)(36) = 360$
least $(1 + 9)(-6 \cdot 6) = -360$

44. ***WRITE*** a word problem to fit the numerical expression $3(4 + 3) + 2 \cdot 6$ and then solve. **Check students' problems.**

46. 21 − [15 + (-5)]; 11 47. 17 − (25 ÷ 5); 12

Write and evaluate the numerical expression for each phrase.

45. five added to the product of four and nine **5 + (4)(9); 41**
46. twenty-one minus the sum of fifteen and negative five
47. seventeen minus the quotient of twenty-five and five
48. one hundred divided by twenty plus the product of negative six and three
49. one hundred thirty added to the difference of one hundred sixteen and eight
50. ***DATA FILE 3 (pp. 96–97)*** Write a numerical expression for the value of the quarters minted each day. **3,500,000(0.25)**

48. (100 ÷ 20) + (-6)(3); -13 49. 130 + (116 − 8) = 238

Write a description of each numerical expression.

▼▼ *SAMPLE* The word *quantity* is a description for a grouping symbol.

The product of two and the quantity three plus four describes the numerical expression $2(3 + 4)$. **Answers will vary.**

51. $2(3 + 5)$
52. $16 \div [3 - (-1)]$
53. $|4 + (-2)| \cdot (-3)$
54. $3(6 - 3) \div 9$

COMPUTER **Write using computer symbols, then evaluate.**

▼▼ *SAMPLE* The symbols for addition and subtraction are + and −. The multiplication symbol is an asterisk (*). The division symbol is a slash (/).

55. $74 + 5 \cdot 9 + (-7)$
56. $123 + (-5) \div (-1) + 18$
57. $70 + (8)(-9)$
58. $255 \div 5 + 117$
59. $2{,}087 \cdot 37 - 1{,}951$
60. $876 \div 12 + 13 \cdot 89$

61. ***COMPUTER*** Do computers follow order of operations? Use a computer to evaluate the expressions in Exercises 55–60. *Hint:* In BASIC, you use a PRINT statement such as PRINT 3*4/6 to evaluate the expression $(3 \cdot 4) \div 6$.
Yes, computers follow order of operations.

55. 74 + 5*9 + -7 = 112 **56. 123 + -5/-1 + 18 = 146**
57. 70 + 8*-9 = -2 **58. 255/5 + 117 = 168**
59. 2,087*37 − 1,951 = 75,268 **60. 876/12 + 13*89 = 1,230**

MIXED REVIEW

Write a variable expression.

1. the product of a number and 6 ***6n***
2. six less than a number **2. x − 6**
3. the sum of a number and the absolute value of -7 ***a* + |-7|**

Complete with <, >, or =.

4. -16 **<** -12
5. 8(-6) **<** 48
6. 11 + (-15) **<** -11 + 15
7. How many whole numbers between 10 and 200 have exactly two identical digits? **36**

LESSON RESOURCES

TRB Practice A-16
TRB Enrichment A-17

Practice

For use after 1-9 (pp. 40–43)

Order of Operations

Which operation would you perform first? Explain.

1. 25 − 16 ÷ 4 × 2 + 6 **Division; no grouping symbols, so multiply and divide left to right.**
2. 16 ÷ (10 − 2) • 5 **Subtraction; do work in parentheses first.**

Evaluate.

3. 24 − 16 + 5 **13**
4. 20 ÷ 2 • 5 **50**
5. 9 + 6 ÷ 3 **11**
6. 4 + (-3) • 5 **-11**
7. -19 + 20 ÷ (-2) **-29**
8. -3 − |4 − 10| **-9**

MENTAL MATH Evaluate.

9. 14 − 6 − 3 **5**
10. 14 − (6 − 3) **11**
11. -6 • 4 ÷ (-2) **12**
12. -2 • 5 + 3 • (-6) **-28**
13. 5 − 42 ÷ (-7) **11**
14. 25 + 5(-2) **15**

CALCULATOR Evaluate.

15. 77 + 83 − 196 **-36**
16. 67 − 13 − 12 **42**
17. -6 • 13 − 91 ÷ 7 **-91**
18. -3(-3) + 12 ÷ (-3) **5**
19. 49 − [37 − (29 − 84)] − 61 **-104**
20. 18 − 9(-4) + 105 ÷ (-5) **33**

Compare. Use >, <, or =.

21. 9 + 27 ÷ (-9) **(>)** -8 − 5(-2)
22. |12 − 14 − 16| **(=)** 13 − (7 − 12)

Insert grouping symbols to make each sentence true.

23. 8 • **(**3 − 5**)** = -16
24. -30 ÷ **(**5 × 2**)** = -3
25. 8 − 7 − **(**6 − 5**)** = 0
26. -4 • **(**8 − 3**)** • 2 = -40
27. **(**5 − 3**)** ÷ (-2) = -1
28. **(**9 + 15**)** ÷ 8 − 20 = -17

Solve.

29. Mark purchased four pens at $0.69 per pen. The sales tax on his purchase was $0.14. He paid for his purchase with a $10 bill. How much change did he get back? **$7.10**
30. Find the total cost of 5 apples costing $0.23 apiece, 3 pears costing $0.32 apiece, and 6 peaches costing $0.15 apiece. **$3.01**
31. Use the numbers 2, 4, and 6 exactly once each to write an expression equal to -2. **6 − 4 • 2 Answers may vary. An example is given.**

A-16 Chapter 1

Enrichment

For use after 1-9 (pp. 40–43)

Binary Operations

A *binary operation* is an operation performed on two numbers. Addition, subtraction, multiplication, and division are all binary operations. Once you know how to use a binary operation, you can perform it on any two numbers.

Here is a new binary operation. # means "multiply the numbers, then add the second number to the product."

Example 5 # 4 = 5 × 4 + 4 = 24

Use the operation # to solve.

1. 3 # 2 **8**
2. 8 # 5 **45**
3. 1 # 7 **14**
4. 10 # 9 **99**
5. 4 # (2 # 7) **105**
6. (3 # 4) # 5 **85**
7. Evaluate 3 # 5 + 2 doing the operation # first. **22**
8. Evaluate 3 # 5 + 2 doing the operation + first. **28**
9. Complete the following order of operation rule, guaranteeing that the value of 3 # 5 + 2 will be 28: When evaluating an expression involving # and +, **Do the operation + first.**
10. Use your rule to evaluate 8 # 6 + 3. **81**

Discover how to use each binary operation by studying the examples. Then perform the operation on the given numbers.

Example 2 * 5 = 20 4 * 3 = 24 7 * 5 = 70 10 * 10 = 200

11. 3 * 6 **36**
12. 5 * 5 **50**
13. 8 * 2 **32**
14. 12 * 12 **288**
15. (2 * 2) * 4 **64**
16. 2 * (2 * 4) **64**

Example 4 $ 1 = 17 6 $ 3 = 39 9 $ 2 = 83 10 $ 1 = 101

17. 5 $ 3 **28**
18. 2 $ 7 **11**
19. 7 $ 13 **62**
20. 12 $ 10 **154**
21. (2 $ 3) $ 4 **53**
22. 2 $ (3 $ 4) **17**

Now use the operations * and $ together.

23. (2 * 6) $ 5 **581**
24. (8 $ 3) * 9 **1,206**
25. 7 $ (1 * 3) **55**
26. 2 $ (12 * (-3)) **-68**
27. (0 * 99) $ (3 * 9) **54**
28. (19 * 1) * (3 $ 7) **1,216**
29. (8 − 2) $ (2 * 2) **44**
30. [-5 * (-13)] * (10 $ 3) **26,780**

Chapter 1 A-17

LESSON QUIZ

Evaluate.

1. 5 + 6 · (-2) [-7]
2. (15 − 18) · 7 (-21)
3. -32 − 5 ÷ 5 (-33)
4. 3 + [4 − (-8)(-6 + 2)] (-25)
5. -4 · |3 − 9| ÷ (-6) [4]
6. two minus the product of negative three and seven (23)

Assignment Guide

Basic 1-11, 16-24, 37-39, 45-49, 51, MR All
Average 2-9, 12-13, 17-23 odd, 24-26, 31-36, 38-44, 48-52, 55-56, MR All
Enriched 4-9, 14-15, 17-23 odd, 24-28, 31-36, 42-44, 48-50, 53-61, MR All

FOR THE NEXT LESSON

four half-sheets of paper or index cards with an integer written on each, math journal

Lesson Focus

Materials

Four half-sheets of paper or index cards with an integer written on each, math journal

Vocabulary/Symbols

evaluate

MOTIVATING THE LESSON

Write $4 - 9 \div 3$ and ask students to find the value. Next, write a variable expression such as $m - 3$ and ask what the value is. (It cannot be determined without the value of m.)

Instruction

Ask students to look at the expression on the chalkboard. Demonstrate how to replace the variable with a number by covering m with a sheet of paper containing an integer. Have students evaluate the expression for each number you place over m. Ask on what the value of the expression depends. (the value that is substituted for the variable)

1 Remind students to follow the order of operations as they evaluate the resulting numerical expression.

2 Help students understand that in order to write an expression for a situation, they must define what the variable represents. Ask what quantity is unknown, or varies, in this situation. (the number of subscriptions) The variable s will stand for the number of subscriptions.

OBJECTIVE: *To evaluate variable expressions.*

1-10 Evaluating Expressions

▼ Major league baseball teams use an average of 42 baseballs per game. The expression $42g$ represents the number of baseballs used in g games. You can *evaluate* $42g$ by replacing g with a number.

Suppose a team plays 50 games. Replace g with 50 to evaluate the expression.

$$42g = 42 \cdot 50$$
$$= 2{,}100$$

Evaluate an Expression — To evaluate an expression, replace each variable with a number. Then compute, following order of operations.

Example 1 **Evaluate each expression for the given value of the variable.**

a. $12b$ for $b = 3$

b. $4y - 15$ for $y = -30$

Solution **1**

a. $12b = 12 \cdot 3 = 36$ — Use mental math.

b. $4y - 15 = 4(-30) - 15 = -120 - 15 = -135$ — Multiply first, then subtract.

▼ A variable expression can have more than one variable.

Example 2 **Evaluate $3ab - 2c$ for $a = -2$, $b = -8$, and $c = -10$.**

Solution $3ab - 2c = 3(-2)(-8) - 2(-10)$ — Replace each variable with a number.

$= 48 - (-20)$

$= 68$ — Follow order of operations.

▼ You can write and evaluate expressions to solve problems.

Example 3 Bob sells magazine subscriptions. He earns \$20 per week plus \$2 for each subscription he sells.

a. Write an expression for his weekly earnings.

b. Find Bob's weekly earnings if he sells 14 subscriptions. **2**

Solution

a. Let s stand for the number of subscriptions. Then $20 + 2s$ is an expression for Bob's weekly earnings.

b. $20 + 2(14) = 20 + 28 = 48$ — Replace s with 14.

Bob earns \$48 by selling 14 subscriptions in a week.

Teaching Tip

To help students remember that they need to use the order of operations when evaluating, display a poster-size paper that lists the rules.

CLASS EXERCISES

Evaluate each expression for $x = 2$, $y = -3$, and $z = 10$.

1. $x + 5$ 7	**2.** $16 - z$ 6	**3.** $4y$ -12	**4.** $-8 \div x$ -4
5. $2z - 4$ 16	**6.** $x - z$ -8	**7.** $y + 5y$ -18	**8.** $x \div 2 + (-9)$ -8
9. xyz -60	**10.** $8y \div x$ -12	**11.** $3z - \|x\|$ 28	**12.** $(z + x) \div y$ -4

Solve.

13. A stenographer types 55 words/min. How many words does the stenographer type in m min? in 20 min? 55m, 1,100

14. An appliance repair center charges a $25 flat fee plus a fee of $10/h for labor. Find the cost of an oven repair that takes 3 h. $55

THINK AND DISCUSS See below

1. When is $6x$ positive? negative?
2. When is $-a + 12$ equal to zero?
3. José has 10 more books than Barbara. Barbara has 18 books. Does José have 8 books or 28 books? Explain.
4. Karen has 4 fewer tropical fish than Aaron. Aaron has 8 fish. Does Karen have 4 fish or 12 fish? Explain.

1. $6x$ is positive when x is positive and negative when x is negative.
2. when $a = 12$
3. José has 28, 10 more than Barbara.
4. Karen has 4; that's 4 fewer than 8.

WRITTEN EXERCISES

Evaluate each expression for the given values of the variables.

1. $-12a$ for $a = 2$ -24	**2.** $x - 6$ for $x = -16$ -22
3. $2a + 5$ for $a = -5$ -5	**4.** $-z$ for $z = 7$ -7
5. $\|a\| + (-17)$ for $a = 5$ -12	**6.** $\|n - 10\|$ for $n = -4$ 14
7. $-6 \div a + 8$ for $a = -2$ 11	**8.** $19 - (a - 4)$ for $a = 8$ 15
9. $-3ab$ for $a = 1, b = -7$ 21	**10.** $16 - 4mn$ for $m = 0, n = -3$ 16
11. $4a - b$ for $a = 3, b = 5$ 7	**12.** $3(a + b)$ for $a = 7, b = 9$ 48
13. $2\|a - b\|$ for $a = 9, b = 19$ 20	**14.** $(x - y) \div (-4)$ for $x = 52, y = 12$ -10

MENTAL MATH Evaluate each expression for the given values of the variables.

15. $-7b$ for $b = 5$ -35	**16.** $5 - b$ for $b = 4$ 1
17. $-3b$ for $b = -7$ 21	**18.** $x - 8$ for $x = 10$ 2
19. $41 - 4b$ for $b = 10$ 1	**20.** $5a + 7$ for $a = 20$ 107

CALCULATOR Evaluate each expression for the given values of the variables.

21. $5m$ for $m = 85$ 425	**22.** $-48 + n$ for $n = 933$ 885
23. $-288 \div c$ for $c = -16$ 18	**24.** $6ab$ for $a = 17, b = -21$ -2,142
25. $7a - 13b$ for $a = 0, b = -9$ 117	**26.** $5y - 5$ for $y = -178$ -895

Find a value for each variable that makes the statement true. Answers will vary.

27. $n > 1$ 2	**28.** $n < -7$ -8	**29.** $\|n\| = 8$ 8, -8	**30.** $3n = 12$ 4
31. $n + 6 = 0$ -6	**32.** $-4n = -4$ 1	**33.** $3 + n = 3$ 0	**34.** $n + 5 < 9$ 3

ADDITIONAL EXAMPLES

1. a. Evaluate $y \div 4$ for $y = -32$ (-8)

b. Evaluate $2 - 5a$ for $a = 6$. (-28)

2. Evaluate $-ac + 6b$ for $a = -3$, $b = 4$, and $c = 3$. (33)

3. It costs $85 plus $12 per hour to rent a limousine for the prom. Write an expression for the cost of the car and find how much it costs for 5 hours. (h is the number of hours; $85 + 12h$; $145)

Guided Practice

Class Exercises Have students work in small groups. To verify that students understand the material, require that all members agree on one answer.

Think and Discuss For questions 1 and 2, encourage students to think about the rules for multiplying integers and for adding integers. As a follow-up to question 1, ask when $6x$ is zero. (when $x = 0$)

For questions 3 and 4, have students write the numerical expression that represents the situation.

A ***common error*** is to drop the sign of the substituted value in a subtraction expression. For example, $3 - a$ when $a = -2$. Students write the expression as $3 - 2$. To help students avoid this error, have them rewrite the expression with the value in parentheses before they perform any operations.

Closure

Writing in Math Ask students to describe the procedure for evaluating a variable expression in their math journals.

MATH MINUTES

Divide the class into groups of four. Give each group a spinner labeled with integers. Write a variable expression on the chalkboard. Groups spin to get a number, then pairs compete to correctly evaluate the expression first.

Independent Practice

For question 46, ask what the pattern would be if $n = 3$.

Lesson Follow-up

Reteaching Activity Prepare 4-8 cards with an integer on each. Have students write $y - 5$ on their paper. Ask them to evaluate the expression by actually laying the card on the variable.

LESSON RESOURCES

TRB Practice A-18
TRB Enrichment A-19

Practice

For use after 1-10 (pp. 44–46)

Evaluating Expressions

Evaluate each expression for the given values of the variables.

1. $6x$ for $x = -3$ **-18**
2. $9 - k$ for $k = -2$ **11**
3. $17 \div p$ for $p = -1$ **-17**
4. $|y - 8|$ for $y = 5$ **3**
5. $-2 + n$ for $n = -3$ **-5**
6. $-m$ for $m = -1$ **1**
7. $10 - r - 5$ for $r = 9$ **-4**
8. xy for $x = -3$, $y = -5$ **15**
9. $24 - p \cdot 5$ for $p = -4$ **44**
10. $5a + b$ for $a = 6$, $b = -3$ **27**
11. $c + |5d|$ for $c = -10$, $d = -2$ **0**
12. $m + n \div 6$ for $m = 12$, $n = -18$ **9**

MENTAL MATH Evaluate each expression for the given values of the variables.

13. $9k$ for $k = -3$ **-27**
14. $n - 9$ for $n = -2$ **-11**
15. $-7a$ for $a = 5$ **-35**
16. $10 - x$ for $x = -3$ **13**
17. $4m + 3$ for $m = 5$ **23**
18. $35 - 3x$ for $x = 10$ **5**

CALCULATOR Evaluate each expression for the given values of the variables.

19. $-178 - g$ for $g = 263$ **-441**
20. $1{,}221 \div (-x)$ for $x = -37$ **33**
21. $p + 851$ for $p = -215$ **636**
22. $7y + z$ for $y = 77$, $z = -691$ **-152**
23. $18a - 9b$ for $a = 12$, $b = -15$ **351**
24. $-129h + k$ for $h = -17$, $k = 973$ **3,166**

Find a value for each variable that makes the statement true. **Answers may vary.**

25. $k < -7$ **-8**
26. $k > 17$ **18**
27. $|k| = 9$ **-9 or 9**
28. $8k = 40$ **5**
29. $-k = 15$ **-15**
30. $k - 9 = 12$ **21**
31. $18 - k = -3$ **21**
32. $2k + 5 = 19$ **7**

Solve.

33. Elliot is 58 years old.
    a. Write an expression for the number of years by which Elliot's age exceeds that of his daughter, who is y years old. **58 – y**
    b. If his daughter is 25, how much older is Elliot? **33 years**
34. A tree grew 27 in. each year.
    a. Write an expression for the tree's height after x years. **27x**
    b. When the tree is 36 years old, how tall will it be? **972 in. or 81 ft**
    c. A 135-in. tree is how many years old? **5**

A-18 *Chapter 1*

Enrichment

For use after 1-10 (pp. 44–46)

Equal Expressions

The value of a variable expression depends upon the value of the variable. By choosing the right variable, you can cause two expressions to have the same value. The expressions $x - 6$ and $2x - 11$, for example, both have the same value when $x = 5$:

$$x - 6 = 5 - 6 = -1 \qquad 2x - 11 = 2(5) - 11 = 10 - 11 = -1$$

Complete the tables for the given values of the variables. Then in the space to the right, name the value of the variable for which the two expressions are equal.

1.

k	10	11	12	13	14	15	16	17
$5 + k$	15	16	17	18	19	20	21	22
$2k - 9$	11	13	15	17	19	21	23	25

$k =$ **14**

2.

x	-4	-3	-2	-1	0	1	2	3
$10 - x$	14	13	12	11	10	9	8	7
$3x + 18$	6	9	12	15	18	21	24	27

$x =$ **-2**

3.

n	1	2	3	4	5	6	7	8
$2n - 3$	-1	1	3	5	7	9	11	13
$27 - 3n$	24	21	18	15	12	9	6	3

$n =$ **6**

4.

h	-2	-1	0	1	2	3	4	5
$5h + 7$	-3	2	7	12	17	22	27	32
$-2 - 4h$	6	2	-2	-6	-10	-14	-18	-22

$h =$ **-1**

5. Explain why there is no value of x for which the expressions $x + 3$ and $x + 4$ are equal.
   For every value of x, $x + 4 > x + 3$
6. Find all the values of x that make the expressions $|x + 1|$ and $|x - 1|$ equal.
   true for $x = 0$

Chapter 1 A-19

35. Find a value of x for which $4x$ and $x + 9$ are equal. **3**
36. Find values of a and b such that $a + b$ is 12 and $a - b$ is 16. **a = 14, b = -2**

Solve.

37. Every minute about 265 babies are born in the world.
    a. Write an expression for the number of babies born in m min. **265m**
    b. *CALCULATOR* About how many babies are born in 6 min? **1,590**
    c. *CALCULATOR* About how many babies are born in one day? **381,600**
38. *CALCULATOR* The fastest speed of a sailfish is 68 mi/h.
    a. Write an expression for the number of miles a sailfish travels in h hours swimming at 68 mi/h. **68h**
    b. How many miles would a sailfish travel swimming for 3 h at 68 mi/h? **204 mi**
39. *DATA* Use the chart on page 31 to find how many calories are used per minute in running.
    a. Write an expression for the number of calories used in running m min. **14m**
    b. How many calories are used in running 25 min? **350 C**
40. A carnival charges a \$3 admission fee plus \$1 per ride.
    a. Write an expression for the cost of riding r rides. **3 + r**
    b. Find the cost of riding 6 rides. **\$9**
    c. How many rides can you afford if you have \$10 to spend? **7 rides**
41. A club requires a \$100 initiation fee and \$25 each month. Find the cost of a one-year membership in the club. **\$400**

Copy and complete.

42.

x	$x + 5$
0	5
1	**6**
2	**7**
5	**10**
15	20

43.

n	$6n$
1	**6**
2	**12**
-2	**-12**
6	**36**
4	24

44.

a	$10 - a$
0	**10**
5	**5**
10	**0**
15	**-5**
20	-10

45. Evaluate $1 + 2 + 3 + 4 + \ldots + n$ for $n = 10$. **55**
46. Evaluate each number in the pattern $1n, 2n, 3n, 4n, \ldots 10n$ for $n = 2$. *WRITE* a description of the pattern. **2, 4, 6, 8, 10, 12, 14, 16, 18, 20; even numbers from 2 to 20**
47. Find the length of each red segment for $x = 8$.

a. **13** (segment: x, 5)

b. **24** (segment: x, x, x)

MIXED REVIEW

Evaluate.

1. $3(-9) - 27$ **-54**
2. $3 \cdot 4 \div (-2) \cdot 8$ **-48**
3. $3 \cdot (-7) + 2 \cdot 8$ **-5**
4. $4[2 \cdot (3 - 6)]$ **-24**

Write a word phrase for each expression.

5. $6 + x$ **the sum of 6 and a number**
6. $2(n - 2)$
7. $-12y$
8. Valerie has test grades of 97, 82, 78, and 75. What is her test average? **83**

6. twice the quantity of a number minus 2
7. the opposite of 12 times a number

LESSON QUIZ

Evaluate for the given variables.

1. $5n$, for $n = 3$ (15)
2. $2a - 7$, for $a = -4$ (-15)
3. $4ab - 3c$, for $a = 0$, $b = -4$, and $c = 9$ (-27)
4. $-5m + 2n$ for $m = -5$ and $n = -2$ (21)
5. $-7(a + b)$ for $a = -6$ and $b = -8$ (98)

Assignment Guide

Basic 1-18, 21-28, 38-39, 42, MR All
Average 7-26, 28-31, 38-40, 43, MR All
Enriched 7-20, 31-36, 40-41, 44-47, MR All

FOR THE NEXT LESSON

math journal, transparency, overhead projector, index cards

Problem Solving Practice

PROBLEM SOLVING STRATEGIES

Look for a Pattern
Guess and Test
Simplify the Problem
Account for all Possibilities
Make an Organized List
Work Backwards
Make a Table
Write an Equation
Solve by Graphing
Draw a Diagram
Make a Model
Simulate the Problem

1. a. Each month the interest increases by a penny more than the previous month.
b. May $1.04; $105.10; June $1.05; $106.15; July $1.06; $107.21; August $1.07; $108.28

The aurora borealis, or northern lights, occurs above the thermosphere.

Solve. The list at the left shows some possible strategies you can use.

1. Marjorie opened a savings account with $100 at the beginning of January. The table shows the interest earned each month for four months.

Month	Interest	Balance
January	$1.00	$101.00
February	$1.01	$102.01
March	$1.02	$103.03
April	$1.03	$104.06

a. Describe the pattern for the values under the interest column.

b. Use the pattern to extend the table for the next four months.

2. Particles have a positive charge, a negative charge, or no charge at all.

⊕ ⊕ ⊕ ⊕ ⊕

a. How much charge do these particles have? **+5**

b. How would you show a charge of −7? ⊖⊖⊖⊖⊖⊖⊖

c. Opposite charges cancel each other out. What is the total amount of charge when 12 negative charges combine with 24 positive charges? **+12**

3. The stratosphere is higher in altitude than the troposphere, but not as high as the thermosphere. The mesosphere is just below the thermosphere. Arrange these regions in order from the highest altitude to the lowest. **thermosphere, mesosphere, stratosphere, troposphere**

4. ***DATA FILE 9 (pp. 360–361)*** The chart below shows a golfer's score at St. Andrew's golf course. Complete the third row.

Hole	1	2	3	4	5	6	7	8	9
Score	5	5	3	3	2	3	5	5	6
Par	4	4	4	4	5	4	4	3	4

Is the total score above or below par? by how much? **above; 1 stroke**

5. ***DATA FILE 1 (pp. 2–3)*** Suppose the temperature was −15°F. The wind speed changed from 5 mi/h to 20 mi/h. Find the change in wind chill temperature. **−39°F**

6. Jeff earns $1,200 more than his brother. Together they earn $65,200. How much does each earn? **Jeff $33,200; brother $32,000**

Instruction

This page provides a variety of problems that can be used to reinforce and enhance the students' problem solving skills. Encourage students to read each problem carefully. Then have them refer to the list of problem solving strategies to help them decide how to solve the problem.

Point out, however, that not all questions require a strategy for solving, nor are all the strategies in the list used in this lesson.

Study Skills

A well-organized, comfortable environment enhances any student's ability to study. Proper lighting and seating, and the quiet necessary for a reflective exploration of ideas, are a few of the elements fundamental to making study pleasurable and fruitful.

Encourage students to discuss what they think they need in a study space, and how they might organize such a space at home. Reassure them that if there is no place at home to study, they can use the school or public libraries. You might also devote part of the class time to study.

Help students understand that their frame of mind, ability to concentrate, and study habits are as important as proper surroundings. If a student is confused about something being studied, clarifying it with a teacher, friend, or parent can be useful.

Finally, share the following study hints:

- Clean and organize desks before studying.
- Use a three-ring binder in which to store and organize notes and handouts for several subjects.
- Keep necessary materials such as sharpened pencils, rulers, calculators, and reference books available.

Chapter 1 Review

Complete each statement. Use the vocabulary words given.

VOCABULARY

integers
variable
opposite
variable expression
absolute value
order of operations
mean
grouping symbols

1. Parentheses, brackets, and the absolute value symbol are ■ that help you determine ■. grouping symbols; order of operations
2. The ■ of an integer is the same distance from zero on a number line as the integer, but in the opposite direction. opposite
3. The whole numbers, their opposites, and zero form the ■. integers
4. The ■ of an integer is its distance from zero on a number line. absolute value
5. A ■ is an expression that contains at least one ■. variable expression; variable
6. The ■ is an average. mean

Comparing Integers — 1-1

To compare integers, think of the number line. The integer farther to the right is the greater integer.

Compare. Use <, >, or =.

7. -7 ■ -9 > **8.** 0 ■ -3 > **9.** |-5| ■ |5| = **10.** -6 ■ 2 < **11.** -4 ■ -(-5) < **12.** |-3| ■ 1 >

Adding Integers and Subtracting Integers — 1-2, 1-3

To add integers with the *same* sign, *add* the absolute values of the integers. The sum has the same sign as the addends. To add integers with *different* signs, *subtract* the absolute values of the integers. The sum has the sign of the integer with the greater absolute value. To subtract an integer, add its opposite.

Add or subtract.

13. 8 + (-15) -7 **14.** -9 + 21 12 **15.** -15 − (-6) -9 **16.** 9 − (-5) 14 **17.** -8 − 4 -12

18. 32 − 48 -16 **19.** -62 − (-59) − 24 -27 **20.** 14 + (-9) + (-20) -15 **21.** -4 + 12 + (-3) + (-6) -1

Multiplying and Dividing Integers — 1-5, 1-6

To multiply or divide integers, multiply or divide the absolute values of the integers. If the integers have the *same* sign, the product or quotient is *positive*. If the integers have *different* signs, the product or quotient is *negative*.

Multiply or divide.

22. 7(-6) -42 **23.** 250 ÷ (-50) -5 **24.** (-9)(-8) 72 **25.** -56 ÷ (-8) 7 **26.** -120 ÷ 40 -3 **27.** -15(11) -165

Translating Word Phrases 1-8

To translate word phrases to algebraic expressions, look for key words that indicate operations.

addition (more than, sum) subtraction (difference, less than)
multiplication (times, product) division (quotient, ratio)

Write an algebraic expression for each phrase.

28. twenty-five less than x **$x - 25$**

29. the product of n and $3r$ **$3rn$**

30. two more than y **$y + 2$**

Using Order of Operations 1-9

To evaluate numerical expressions:

1. Do all operations within grouping symbols first.
2. Multiply and divide from left to right.
3. Add and subtract from left to right.

Evaluate.

31. $7 + 2 \cdot 28 - 3 \cdot 9$ **36**

32. $9 \cdot 5 - 4(18 \div 6)$ **33**

33. $3 \cdot 8 - 6 + 49 \div 7$ **25**

34. $3 \cdot 7 + 6 \div 2$ **24**

35. $|-5 + 6| \cdot |3|$ **3**

36. $4 + 8 \div 2 \cdot 0$ **4**

Evaluating Algebraic Expressions 1-10

To evaluate an algebraic expression, substitute a number for the variable(s), and simplify. Follow order of operations.

Evaluate.

37. $3x + 4$ for $x = 5$ **19**

38. $10 - n$ for $n = 4$ **6**

39. $|y - 6| + 8$ for $y = -2$ **16**

40. $|m - 7|$ for $m = -7$ **14**

41. $15t \cdot 10$ for $t = -3$ **-450**

42. $z + 3z$ for $z = 4$ **16**

Problem Solving 1-4, 1-7

To solve a problem, use a pattern.

Use a pattern to solve.

43. To run a classified ad for 7 days costs \$28 for 4 lines. Each additional line costs \$10.50. What is the cost of a 12-line ad? **\$112**

44. The graph at the right gives Sue's net calories for this week. If Sue continues to eat as she did this week, will she gain or lose weight in a month? **She will gain weight.**

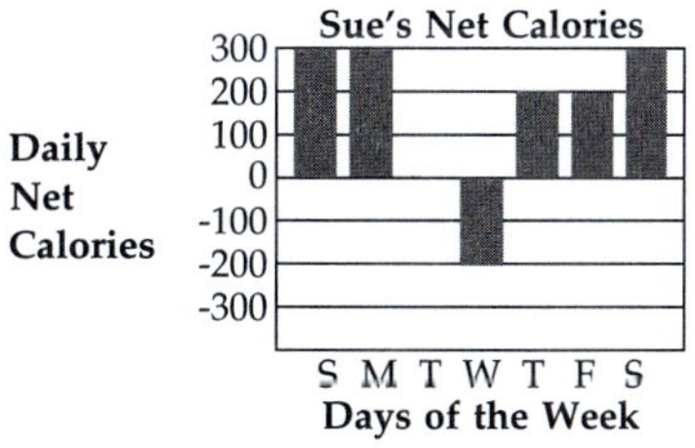

49

RESOURCES

TRB A-21–A-24

Scoring Chart

No. Correct	Score
37	100
36	97
35	95
34	92
33	89
32	86
31	84
30	81
29	78
28	76
27	73
26	70
25	68
24	65
23	62
22	57
21	57
20	51
19	51
18	49
17	46
16	43
15	41
14	38
13	35
12	32
11	30
10	27
9	24
8	22
7	19
6	16
5	14
4	11
3	8
2	5
1	3

Chapter 1 Test

Write an integer.

1. opposite of 7 -7
2. opposite of -9 9
3. $|-5|$ 5
4. $|12|$ 12

Compare. Use <, >, or =.

5. -6 ■ -5 <
6. 8 ■ -10 >
7. -3 ■ 3 <
8. 0 ■ -7 >

Find each answer.

9. $15 + (-7)$ 8
10. $-8 - (-12)$ 4
11. $-9(-7)$ 63
12. $54 \div (-6)$ -9
13. $-6 \cdot 48$ -288
14. $-56 \div (-7)$ 8
15. $119 - (-24)$ 143
16. $-47 + (-21)$ -68
17. $-83 + (-17) + 13$ -87
18. $5 \cdot (-12) \cdot (-3) \cdot (-1)$ -180
19. $420 \div (-6) \div 7 \div (-2)$ 5
20. $8 \cdot 6 \div (2 + 1)$ 16
21. $4 + 7 \cdot 2 + 8$ 26
22. $4(11 + 7) - 9 \cdot 8$ 0
23. $29 - 2 \cdot 3(9 - 4) \div 6$ 24
24. $16 - 2 \cdot 5 + 3 - 6$ 3
25. $|14 - (-9)|2$ 46

Write an expression for each word phrase.

26. ten less than the absolute value of negative three $|-3| - 10$
27. the product of x and negative five $-5x$
28. a number increased by nineteen $n + 19$
29. the opposite of the quantity five more than y $-(5 + y)$

Evaluate each expression for the given values of the variables.

30. $3a + 5$ for $a = -5$ -10
31. $5m + 9 + 7n$ for $m = 8, n = 1$ 56
32. $3|x - y| + x$ for $x = 1, y = 8$ 22
33. $20 - 2(a - b)$ for $a = 3, b = 2$ 18

Solve.

34. A submarine was at a depth of 250 m below sea level. It rose 75 m. Use an integer to describe the new depth of the submarine. -175 m
35. ***DATA*** Use the chart on page 31 to solve. John went swimming for $\frac{1}{2}$ h and cycling for an hour. He then ate two 3-oz hamburgers on wheat bread, and an apple, and drank 1 c of grape juice. Find his net calories. He lost 15 C
36. You are in an elevator on the seventh floor. Go down 4 floors. Go up 8 floors. Go down 3 floors. Go up 9 floors. Following this pattern, what floor will you be on if the elevator goes down again and you get off? 15th floor
37. A shirt costs $15 and jeans cost $25.
   a. Write an expression for the cost of j jeans and s shirts. $25j + 15s$
   b. Find the cost of 3 pairs of jeans and 5 shirts. $150
   c. How many pairs of jeans can you buy for $60? 2

Chapter 1 Cumulative Review

Choose the correct answer. Write A, B, C, or D.

1. What makes -1 > ▒ true? A

A. -2 **B.** 1
C. 0 **D.** not given

2. 7 + (-12) C

A. 5 **B.** -19
C. -5 **D.** not given

3. $|-10|$ C

A. -10 **B.** $\frac{1}{10}$
C. 10 **D.** not given

4. What is the variable expression for five less than $3n$? C

A. $5 - 3n$ **B.** $3(n - 5)$
C. $3n - 5$ **D.** not given

5. 1, 2, 3, 5, 8, 13, . . . D

A. 13 **B.** 20
C. 16 **D.** not given

6. -9(-7) A

A. 63 **B.** -16
C. -63 **D.** not given

7. ***DATA*** Use the chart on page 31. How many calories are burned walking 2 h? B

A. 10 calories **B.** 600 calories
C. 300 calories **D.** not given

8. Name the integer represented by the point on the number line. B

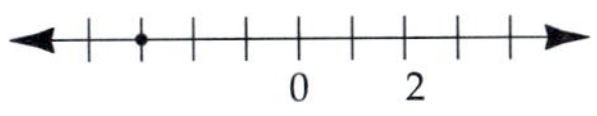

A. 3 **B.** -3
C. -4 **D.** not given

9. 210 ÷ (-3) B

A. -7 **B.** -70
C. -63 **D.** not given

10. -8 − (-6) B

A. -14 **B.** -2
C. 2 **D.** not given

11. Evaluate $6x - 9$ for $x = 2$. B

A. -21 **B.** 3
C. 12 **D.** not given

12. Which operation would you perform first? 12 ÷ 4(3 − 5) B

A. division **B.** subtraction
C. multiplication **D.** not given

13. What is the opposite of 8? A

A. -8 **B.** $\frac{1}{8}$
C. 8 **D.** not given

14. Evaluate $-10 \div |-2 + 3|$. D

A. 8 **B.** 10
C. -2 **D.** not given

15. -15 + (-3) D

A. -5 **B.** -20
C. 5 **D.** not given

16. Give the number sentence for the number line. C

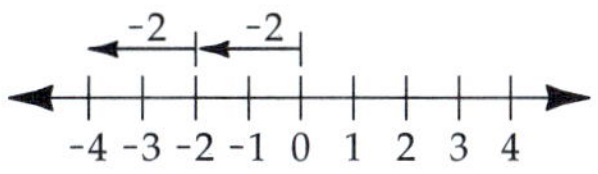

A. -2 + 2 = 0 **B.** 2 − (-2) = 4
C. -2 + (-2) = -4 **D.** not given

Study Skills

For many students, taking a test is so stressful that performance is impeded, and the test fails to provide a true measure of a student's skills and understanding.

Encourage students to discuss their feelings about tests, and ask them to share whatever ways they can think of to make test-taking less threatening. For example, students can think of tests as an opportunity to learn where their strengths and weaknesses are. This way, they can continue to build on their strengths as they work on the areas that need improvements.

Some of the following ideas can help foster a more relaxed attitude about tests:

- Provide an outline early in the term of scheduled tests and the topics each test will cover.
- Teach reviewing and test-taking techniques.
- Provide alternative evaluation methods when possible for students who have difficulty with written tests.
- Be available before the test for students who may need help.

Offer students these suggestions:

- Ask for help with problems prior to the test.
- Avoid cramming by reviewing material daily.
- Develop good study skills.

2 SOLVING EQUATIONS

Chapter At a Glance

In Chapter 2, students learn to read, write, and solve equations. Teaching emphasis is on the properties of operations.

Reading the Chapter

In this chapter, students are re-introduced to the commutative, associative, and distributive properties. This may be the first time that students will see the properties in an algebraic context. It may help students if they visualize the relationships the terms indicate. For instance, the word "commutative" stems from the word "commute," which means to interchange. To visualize the commutative property, students might try to think of an express commuter train. All the cars on the train are interchangeable, and they are all headed for one destination. After students have had a chance to envision this, you can explain that in the commutative property, the order in which numbers are added or multiplied does not affect the result. The numbers are interchangeable, just as the cars are on the imaginary commuter train. After discussing this, have students suggest relationships or situations that can help them remember the other two properties.

Chapter Resources

Practice and Enrichment TRB B-1 to B-16
Alternate Application TRB B-17
Chapter Test Form A, B-19
Chapter Test Form B, B-21
Computer Test Bank

References

STUDENT BIBLIOGRAPHY

Murphy, Elaine C. *Developing Skills with Tables and Graphs.* (Dale Seymour Publications, Palo Alto 1981).

TEACHER BIBLIOGRAPHY

Thornburg, David D. *Picture This! An Introduction to Computer Graphics for Kids of All Ages.* (Addison Wesley, Menlo Park 1982).

SOFTWARE

Excel (Microsoft Corp.), for IBM or Macintosh computers.
Graph (MECC), for Apple II family of computers.

FILMS, VIDEOTAPES, AND OTHER MEDIA

Visualizing Algebra (Sunburst Communications. VHS or Beta).

Vocabulary

additive identity
associative property
commutative property
distributive property
like terms
multiplicative identity
numerical coefficient
open equation
property of equality
solution

Materials

math journals
algebra tiles (M*)
graph paper (T*)
computer (optional)
graphing software (optional)

*M—Manipulative Kit
*T—Teacher's Resource Book

Resource Bank

Class Activities

FIELD TRIPS

▼ Radio station—to find out how sound is transmitted via radio waves and what "tuning in" a station means.

▼ City or town planning board office—to inquire about zoning laws and what data and/or formulas are used in establishing and changing the laws.

CLASS SPEAKERS

▼ Ham radio operator—To discuss transmitting messages by voice and Morse code.

▼ Urban and regional planner—to discuss how variable factors, such as predicted population, affect plans for urban expansion and revitalization.

Bulletin Board

Prepare a time line, leaving blank the boxed entries shown below, and make cards using the given clues. Ask students to use the clues to find the date of the event, and then write the event on the time line.

It took 6 years from his first transmission for Marconi to send radio waves across the Atlantic Ocean. (1901)

Marconi transmitted radio waves through the air 30 years after their discovery. (1895)

The first radio network was established in the U.S. 43 years before the astronauts landed on the moon. (1926)

Project: Have students research other such significant events involving the radio, and display them on the time line.

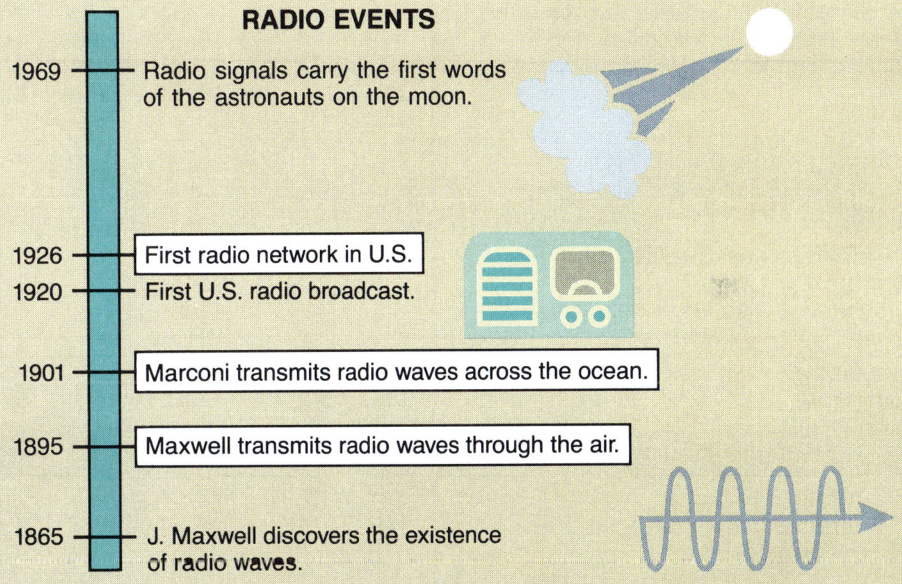

Extra-Credit Problems

BASIC

▼ Bill ate 100 raisins in five days. Each day he ate 6 more than on the day before. How many did he eat on each of the five days? (8, 14, 20, 26, 32 raisins)

AVERAGE

▼ The equations $3x + 9 = 18$ and $6x + 18 = 36$ have the same solution and are related to each other. Find the solution and describe the relationship. ($x = 3$; second is double the first)

ENRICHED

▼ A block and 2 marbles on one side of a scale are balanced by 6 marbles on the other side of the scale. How many marbles does it take to balance 2 blocks and 3 marbles? (11 marbles)

Chapter Planning Guide

OBJECTIVES	ASSIGNMENT GUIDE			ASSESSMENT	
	Basic	Average	Enriched	Review/ Tests	TRB*
2-1 Variables and Equations To determine if an equation is true or false and whether a given number is a solution of an open equation.	1-8, 11-16, 21-26, 31-32, 37-38, 43, 46-53, MR* All	2-3, 7-13, 17-18, 22-27, 33-34, 39-40, 44, 48-55, MR All	3-4, 8-14, 19-24, 28-30, 35-36, 41-43, 45, 48-57, MR All	Extra Practice p. 567	Practice B-1 Enrichment B-2
2-2 Properties of Operations To use the commutative, associative, and identity properties of addition and multiplication.	1-9 odd, 11-12, 15-19, 24-25, 29-30, 34-36, MR All	1-9 odd, 12-13, 15-20, 26-27, 31-32, 34, 36, MR All	2-10 even, 13-14, 16-17, 21-23, 27-28, 32-36, MR All	Extra Practice p. 567	Practice B-3 Enrichment B-4
2-3 The Distributive Property To use the distributive property.	1-6, 10-12, 16-17, 22-23, 26-28, 36, 38, MR, CT* All	2-3, 6-8, 14-15, 18-19, 23-24, 29-32, 36-38, MR, CT All	3, 7-9, 13, 15, 19-21, 24-25, 32-38, MR, CT All	Extra Practice p. 567	Practice B-5 Enrichment B-6
2-4 Simplifying Variable Expressions To use properties of addition and multiplication to simplify expressions.	1-6, 9-10, 12-13, 18-19, 21-24, 29, 36-38, 41, MR, TY* All	5-7, 10-11, 14-15, 19-21, 25-31, 38-39, 41-42, MR, TY All	7-8, 11, 16-17, 19-21, 28-35, 38-40, 41-42, MR, TY All	Extra Practice p. 567 TY p. 69	Practice B-7 Enrichment B-8 TY B-18
Exploring Equations Use models to solve addition and subtraction equations	**Manipulative Connection** All	All	All		
2-5 Addition and Subtraction Equations To solve one-step equations involving addition and subtraction.	1-7, 12-13, 18-26, 30-37, 45-50, MR All	2-10, 14-21, 26-32, 38-46, 49-52, 55 MR All	3-5, 10-17, 22-33, 40-46, 51-55, MR All	Extra Practice p. 567	Practice B-9 Enrichment B-10
2-6 Multiplication and Division Equations To solve one-step equations involving multiplication and division.	1-12, 19-21, 28-41, 52, 54-58, MR, CT All	2-15, 22-24, 28-35, 42-47, 52-59, MR, CT All	3, 7-9, 16-18, 25-35 odd, 48-50, 52-60, MR, CT All	Extra Practice p. 567	Practice B-11 Enrichment B-12
2-7 Writing Equations To write an equation for a word problem or model.	1-4, 7-10, MR, TY All	1-2, 4-5, 7-10, MR, TY All	2, 4-10, MR, TY All	Extra Practice p. 567 TY p. 82	Practice B-13 Enrichment B-14 TY B-18
Exploring Graphing To explore the uses of a computer graphing program.	All	All	All		
2-8 Problem Solving Strategy: Guess and Test To use guess and test to solve mathematical problems.	**Problem Solving Connection** All	All	All	Problem Solving Practice p. 91	Practice B-15 Enrichment B-16
2-9 Problem Solving Application: Ecology To apply solving equations to ecology problems.	All	All	All		Alternate Application B-17
Chapter Test	All	All	All	pp. 92-95	Tests B-19–B-25

 *TRB—Teacher's Resource Book MR—Mixed Review CT—Critical Thinking TY—Test Yourself

CONNECTIONS							NCTM CORRELATION
Algebra	**Critical Thinking**	**Problem Solving**	**Estimation/ Mental Math**	**Technology**	**Manipulatives/ Resources**	**Writing/Reading in Math**	
C*: 10-15 W*: 11-14 43-57	C: TD* 1-3 W: 43-45	W: 43-54	C: 7-9 W: 1-10, 37-42	W: 31-36	W: 53-54	C: 1-6 W: 46-55	Algebra Communication Connections
C: 1-6 W: 1-23	C: TD 1-2 W: 34-35	W: 36	C: 7-15 W: 24-33			W: 34-36	Algebra Communication Problem Solving
C: 4-18 W: 4-21, 26-35	C: TD 1-3 CT*: 1-5	C: 1-3 W: 36-38 CT: 1-4	W: 22-25		C: 1-3 W: 1-3, 38	CT: 5	Algebra Problem Solving Reasoning
C: 1-4, 9-16 W: 1-8, 12-20, 22-44 TY*: 4-15	C: TD 1-2 W: 21 TY 1-3	W: 41-44	W: 18-20, 22-30		C: 5-8 W: 9-11	W: 41-44	Algebra Reasoning Problem Solving
1-11	1-11	3, 5, 8, 10			3, 5, 9, 10	2, 4a-b, 8	Functions Reasoning Communication
C: 7-10 W: 6-23, 34-44	C: 3-6, TD 1-3 W: 45-46	W: 47-55	C: 7-10 W: 24-29	W: 30-33	C: 1-2 W: 1-5, 49, 51, 53, 55	W: 4-5	Functions Connections Reasoning
C: 4-9 W: 10-27, 36-50, 54-60	C: 1-3, TD: 1-3 W: 54-56 CT: 1-5	W: 51-53 CT: 1-5	W: 4-9, 28-31	W: 32-35	W: 1-3, 8-9, 52		Functions Reasoning Connections
C: 1-9 W: 1-10 TY: 1-8	C: 1-2 W: 1-2	W: 8-9, 10a TY: 7-8			W: 7, 10a-b	C: 3-9 W: 3-9, 10b	Reasoning Problem Solving Communication
	3-5, 6b-c	4, 5a, 6-7	6c	1-7	3, 7	1-2, 6a, 7	Functions Statistics Reasoning
C: 5 W: 1-5, 8-11	C: 5	C: 1-4 W: 1-11	C: 1-4 W: 9, 11	W: 1-11	C: 1-2 W: 6-7		Problem Solving Algebra Reasoning
C: 1-3 W: 1-4		C: 1-3 W: 1-7	W: 3-4, 6b	C: 1-3 W: 1-6	C: 1-3 W: 1-3, 7		Problem Solving Connections Functions

*C—Class Exercises W—Written Exercises TD—Think and Discuss CT—Critical Thinking TY—Test Yourself

Exploring

With this manipulative activity, students use algebra tiles to model the associative and commutative properties of addition and multiplication. **See TRB Exploring O-5.**

Exploring Properties of Operations

Write a number sentence for the sums.

1. **-2 + 3 = 1** / **3 + (-2) = 1**
2. **-3 + (-1) = -4** / **-1 + (-3) = -4**
3. Look at the first pair of models, then at the second pair.
   a. What is the same about each pair? **the addends, the sum**
   b. What is different about each pair? **the order of the addends**
4. Write a rule about adding numbers in any order. **You can add numbers in any order without changing the sum.**
5. Combine these models as shown. Write a number sentence to show each sum.
   a. combine first **3 + [-2 + (-4)] = -3**
   b. combine first **[3 + (-2)] + (-4) = -3**
6. What is the same about the above models? **the addends, the sums.**
7. What is different about them? **the grouping.**
8. Write a rule about changing the grouping when you add. **You can change the grouping and then add without changing the sum.**
9. Write a multiplication sentence for each model.
   a. **2 • 4 = 8**
   b. **4 • 2 = 8**
10. What do you notice about changing the order when you multiply? **You can multiply in any order without changing the product.**
11. For the model at the right, a row represents 3 • 2, and two rows represent 2 • (3 • 2). Make a model to show a change in grouping, or 3 • (2 • 2). Start with the parentheses. What is 3 • (2 • 2)? **12** **Check students' models.**
12. What do you conclude about changing the grouping of factors when you multiply? **You can change the grouping of factors without changing the product.**

O-5

In this problem solving activity, students explore logical thinking using clock addition with finite number systems. **See TRB Exploring O-6.**

Exploring Finite Number Systems

Our number system has an endless, or infinite set of numbers. We can do calculations with numbers containing many digits. In a finite number system, the set of numbers is limited.

Think of a clock as a finite number system. It has only twelve numbers. You can add with a 12-number system by counting in a clockwise direction. The number you stop at is the sum.

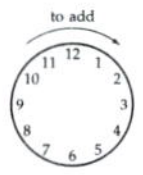

	Find	Add	Sum	Equation
1.	9	3	**12**	**9 + 3 = 12**
2.	11	5	**4**	**11 + 5 = 4**
3.	1	13	**2**	**1 + 13 = 2**
4.	6	7	**1**	**6 + 7 = 1**

5. What can you say about adding with twelve numbers? **Answers may vary. Possible answers: The greatest possible sum is 12; the sums will repeat.**
6. Suppose a finite number system has only four numbers, 0, 1, 2, and 3. Use the clock to help you complete the table. Find all the possible sums. The symbol, ⊕, represents the operation.

Hint: For 2 ⊕ 3 = x, find the first addend, 2, in the left column. Find the second addend, 3, in the top row. Write the sum where the row and column meet.

⊕	0	1	2	3
0	**0**	**1**	**2**	**3**
1	**1**	**2**	**3**	**0**
2	**2**	**3**	**0**	**1**
3	**3**	**0**	**1**	**2**

7. What is the greatest sum possible? **3**
8. Look at the table. What patterns do you see? **Answers may vary. Possible answers: The sums repeat after 3; a diagonal fold from ⊕ to bottom right 2 will cause equal sums to touch.**
9. What do you expect to see in a table for a five-number system, 0, 1, 2, 3, 4? Then, on a separate piece of paper, draw a table to verify your answer. **Answers may vary. Possible answers: The greatest possible sum is 4; the numbers repeat after 4; a diagonal fold will cause equal sums to touch.**

O-6

In this calculator activity, students rewrite equations, using the associative property of addition and distributive property of multiplication. **See TRB Exploring O-7.**

Exploring Number Sense

Suppose you want to use your calculator, but the [7] key is broken. You cannot enter the digit 7 into the calculator.

1. How can you find the sum of 279 and 685 without using the [7] key? **Enter 279 as the sum of two numbers with no digit that is 7, (240 + 39) + 685.**
2. Suppose you enter 250 [+] 29 [+] 685 to find the sum. Does it work? **yes** What algebraic property can you use to rewrite the expression? **associative**
3. Suggest three other ways to find the sum with the calculator. **Answers will vary. (300 − 21) + 685; (260 + 19) + 685; (265 + 14) + 685**
4. How can you use the calculator to find 297 [−] 176? **Answers will vary. Enter 297 as (295 + 2), 176 as (150 + 26).**
5. Suppose you enter 295 [+] 2 [−] 150 [+] 26. Does it work? **no** Verify the result on a calculator.
6. How can you change the expression to get the correct difference of 121? **Enter 295 + 2 − (150 + 26) or use the memory keys to store the sum of 295 + 2 and then subtract from memory the sum of 150 + 26.**
7. How can you use the calculator to find 47 [×] 29? **(45 + 2)29** Name an algebraic property that you might use. **distributive**
8. Suggest two other ways to find the product with the calculator. **Answers may vary. (43 + 4)29; (35 + 12)29**
9. Suppose you can only use the [+], [1], and [0] keys on a calculator. How can you show 374 on the display, using the fewest addition steps? **Enter 111 + 111 + 111 + 11 + 10 + 10 + 10**
10. Suppose you can use the [+], [−], and [1] keys on a calculator. How can you show 299 on the display, using the fewest steps? **Enter 111 + 111 + 111 − 11 − 11 − 11 − 1**
11. Suppose you can only use the [+], [−], and [7] keys on a calculator. How can you show 104 on the display, using the fewest steps? **Enter 777 ÷ 7 − 7**

O-7

In this computer activity, students explore computer graphing by constructing bar, circle, and line graphs and then analyzing the data. **See TRB Exploring O-8.**

Exploring Computer Graphing

The 20 members of the Student Council sold T-shirts as a school fund raiser. Four teams of 5 students sold 1,000 T-shirts at $6.50 each.

Use a computer to graph and analyze the data.

Teams	Number Sold
blue	250
red	375
gold	125
white	250

Construct a bar graph that shows the number of shirts sold by each team. **Check students' graphs.**

1. Which team sold the greatest number of shirts? **red**
2. About how many times as great as the gold team was the number of shirts sold by the white team? **2 times**
3. About how many times as great as the gold team was the number of shirts sold by the red team? **3 times**

Construct a circle graph to show the part of the total number of shirts sold by each team. **Check students' graphs.**

4. Estimate the fraction of the total number of shirts the red team sold. **about $\frac{1}{3}$ (actual $\frac{3}{8}$)**
5. What fraction of the total number of shirts did the blue and white team sell? **$\frac{1}{2}$**
6. What does a circle graph show about the data that the bar graph does not show? **fraction of the total distributed by each part**

Construct a line graph that compares the number of shirts sold and the amount of money raised. Show data for 100, 200, 300, . . . 1,000 shirts sold. **Check students' graphs.**

7. Use the graph to estimate the amount of money raised for 350 T-shirts; 525 T-shirts. **Accept reasonable estimates. about $2,000; about $3,500**
8. About how many shirts need to be sold to raise $4,500? **about 700**
9. Would you use a bar graph, line graph, or circle graph for each of the following situations?
   a. to show how a whole number amount is distributed **circle graph**
   b. to show something changes over time **line graph**
   c. to compare similar data **bar graph**

O-8

Meeting Individual Needs

GIFTED STUDENTS

When the mathematically gifted student is able to share his or her unique insights into mathematical relationships, the classroom experience can be enriched for all students. For example, many students are intimidated by mathematical puzzles, but the gifted student is often fascinated by them and can convey that enthusiasm to others.

In a cooperative learning framework, a gifted student can help the entire group solve a puzzle and understand its mathematical basis. In turn, the other students in the group can share with the gifted student *their* particular strengths (e.g. social or organizational skills).

Present the following puzzle to groups of four to six students, with at least one gifted child per group. Ask groups to repeat the steps starting with different numbers to discover that the final answer is always 5. Then ask the groups to demonstrate in algebraic terms that five is always the result.

Choose any whole number.	8
Add the next consecutive number.	$8 + 9 = 17$
Add 9 to the sum.	$17 + 9 = 26$
Divide the sum by 2.	$26 \div 2 = 13$
Subtract the original number.	$13 - 8 = 5$

AT-RISK STUDENTS

At-risk students can be motivated to work with such abstract ideas as variables and equations in cooperative groups. These groups foster personal responsibility and provide the supportive atmosphere in which at-risk students can feel comfortable practicing the skills needed to solve the problems in this chapter.

Pairs Check, is a group structure involving 4 students, separated into pairs. Partners alternate the roles of Problem Solver and Coach. After partners solve two problems, they compare their answers to those of the other pair. If the two pairs find that they have different answers to either problem, all four students work together until they agree on an answer. After all the exercises in a set have been completed, students use an answer key to check their work. If groups meet a set of criteria, they go to another set of exercises or a new activity. If they do not meet the criteria, groups work on a set of similar exercises and/or consult with the teacher.

Working in this way, at-risk students are helped to see connections between mathematical thinking and the solution of everyday problems, and begin to see themselves as competent mathematical thinkers.

ENGLISH AS A SECOND LANGUAGE

Some of the terms in this chapter, such as "commutative property" or "associative property," might be particularly difficult for the ESL student. It can be helpful if a more advanced student who speaks the same language can translate and explain these terms to the less advanced ESL student. Inability to complete a written assignment can sometimes indicate that the ESL student is having trouble understanding a task. For example, if the student cannot write an equation for the sentence, "Four times the opposite of five is negative twenty," (see p. 57, question 46), try to determine whether the problem is related to math skills or language skills by modeling the sentence using algebra tiles. If the student still cannot write the equation, the problem is probably math related. If you think that language is the stumbling block, have the ESL student underline the words or phrases in the problem that represent operations. Refer the student to bilingual dictionaries or math texts from earlier grades to help with the vocabulary.

See TRB Teaching Resources p. 52.

Problem Solving Situation

The question posed in the "Think About It" box asks students to think about what factors determine the speed of sound in a given substance. To help students generate their own ideas about this, encourage them to make a list of possible factors such as size, shape, and density.

Encourage the students to experiment by sitting quietly and listening to the sounds around them. Then have each student press his or her ear to the wall and note what sounds they hear. (Students should be able to hear more sounds.) Ask the students if they hear as many sounds after they move away from the wall. (no)

DATA COLLECTION

Students should make a list of what sounds they heard when they pressed their ears against the wall.

DATA ANALYSIS

Have students interpret their results based on what they know about the speed of sound in objects. How does the speed of sound in wood or metal account for being able to hear more sounds? (Sound travels faster and better through these materials.)

DATA FILE

RING THOSE BELS

The *bel*, named after Alexander Graham Bell, is a unit for measuring sound intensity. Ten bels equal one *decibel* (db).

	db	
db above 192 cause shock waves	190	
	180	
	170	some motorized toys
	160	
prolonged noise causes permanent deafness	150	rocket launch
	140	threshold of pain
sound levels above 90 are banned in factories	130	
	120	race cars, amplified rock band
	110	jet planes
	100	
	90	
	80	loud music, subway train
	70	snoring, telephone ring
	60	
safe range of sound intensity	50	loud conversation
	40	typewriter
	30	human speech
	20	whisper
	10	
	0	
below the threshold of human hearing	−10	

SONAR (sound navigation and ranging) is a method that uses sound to locate underwater objects. A sonar device emits a sharp pulse of sound, which is reflected back when it hits an object. You can find the distance to the object by measuring the time it takes the sound to return. You can use a formula to find the distance.

Let s = the speed of sound
t = the time for the sound to strike the object and return

Then $\frac{ts}{2} = d$ is a formula to find the distance to an object.

Commercial fishing boats use sonar to detect schools of fish.

52

CHAPTER 2

Solving Equations

THUNDER is a shock wave produced by a lightning flash. Because light (299,460 km/s) travels faster than sound (346 m/s), you see lightning before you hear thunder. To determine how far away lightning is, you can count the seconds between the lightning flash and the thunder, then multiply by 346.

Think about it...

Look at the data about the speed of sound. What factors do you think determine the speed of sound in a given substance?

See Problem Solving Situation

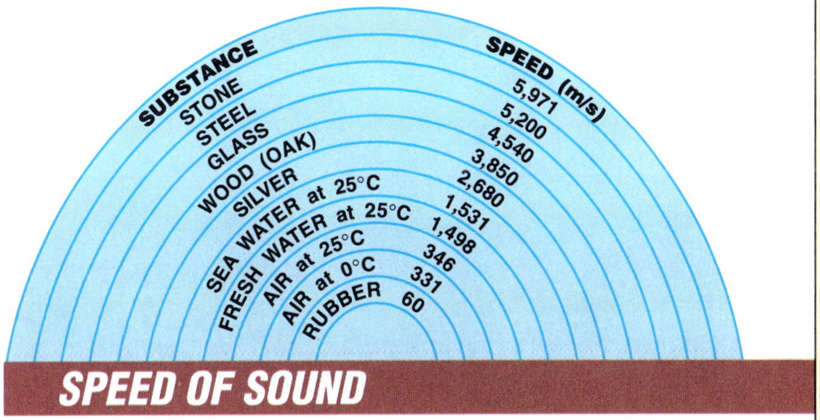

Cooperative Learning

Have students work together in small groups, and ensure that there is a good mix of abilities and personalities within the group. For example, each group will need members who can locate data, interpret it, display the group's finding, and present it to the entire class.

Observe students while they are working in their groups. Encourage passive group members to take a more active role by reminding them that everyone should be prepared to explain the group's results. After work sessions, have students discuss ways in which their group worked well and ideas for improving the group.

Questions related to this Data File can be found on pages 75, 124, 426, 429, and 567.

Lesson Focus

Materials

Math journal, transparency, overhead projector, index cards

Vocabulary/Symbols

equation
open equation
solution

MOTIVATING THE LESSON

Present the following mathematical sentences to the class.

$$a + 1 = 1 + a$$
$$2x = 5$$
$$3 + 4 = 7$$

Discuss what the sentences have in common and what is different about them. (All are equations; $a + 1 = 1 + a$ is a true equation for all numbers; $2x = 5$ is a true equation for the number 2.5 only; and $3 + 4 = 7$ is a true equation.)

Instruction

1 To help students remember the definition, point out that the only requirement needed to make a mathematical sentence an equation is that it contain an equal sign.

2 Have students explain the difference between an expression and an equation. (An expression does not have an equal sign. An expression is a mathematical phrase. Equations consist of expressions with an equal sign.)

3 Ask students to give examples of true, false and open equations to be sure they understand the distinction. As suggestions are offered, list them in a chart on the chalkboard or on a transparency for the overhead projector.

OBJECTIVE: *To determine if an equation is true or false and whether a given number is a solution of an open equation.*

2-1 Variables and Equations

A deep sea diver weighs 135 lb. When she puts on her diving equipment she weighs 165 lb. Let the variable w represent the weight of the equipment. You can write an *equation* to show this relationship.

$$135 + w = 165$$

1

Equation	An equation is a mathematical sentence with an equal sign.

An equation has an equal sign, an expression does not. **2**

THINK What is the difference between an expression and an equation?

There are many types of equations.

Example 1
a. $9 + 2 = 11$ — a numerical expression equal to a numerical expression
b. $x + 7 = 37$ — a variable expression equal to a numerical expression
c. $a + (-3) = 2a + 5$ — a variable expression equal to a variable expression

3

An equation can be true or false.

Example 2 This is a true equation. $6 + 12 = 18$ This is a false equation. $6 = 4 + 3$

Some equations are neither true nor false. They are called *open equations.*

Open Equation	An open equation is an equation that contains one or more variables.

Example 3 These are open equations.
a. $c + 9 = 24$ b. $6x = -3 + 5x$

Teaching Tip

To facilitate reteaching and minimize preparation for future classes, keep a diary that notes concepts and skills that students find difficult.

▼ You can replace a variable with a number to determine whether the number is a *solution* of the open equation.

Solution A solution is a number that replaces a variable to make an open equation true.

4

Example 4 Is 30 a solution of the open equation $170 + x = 200$?

Solution
$170 + x = 200$
$170 + 30 = 200$ Replace x with 30.
$200 = 200$ True, so 30 is a solution.

Example 5 Is 17 a solution of the open equation $9 = y - 10$?

Solution
$9 = y - 10$
$9 = 17 - 10$ Replace y with 17.
$9 = 7$ False, so 17 is not a solution.

5

THINK Find a solution to the equation $9 = y - 10$ using mental math. 19

▼ You can write a sentence as an equation.

6

Example 6 **Write an equation for the sentence. Identify the equation as true, false, or open.**
Nine times the opposite of five is forty-five.

Solution $9 \cdot (-5) = 45$ This is a false equation.

FLASHBACK
The phrases *is equal to*, *equals*, and *is* denote the equal sign.

CLASS EXERCISES

Write true or false.

1. An equation can be false. true
2. An open equation is true. false
3. $3w - 7$ is an open equation. false
4. An open equation must contain a variable. true
5. $4 + 2x = 12$ is an equation. true
6. An expression contains an equal sign. false

State whether each equation is true, false, or open.

7. $15 = 3 \cdot 5$ true
8. $4x - 8 = 25$ open
9. $3(-9) = -36 + 6$ false

Replace c with -2. State whether the equation is true or false.

10. $c + 5 = 3$ true
11. $24 = 2c + 29$ false
12. $c \div 2 - 8 = 3(-3)$ true

Is 7 a solution of each equation?

13. $d + 4 = 12$ no
14. $-12 = -2d + 2$ yes
15. $y - 2 = y - 2$ yes

THINK AND DISCUSS See below.

1. Name one similarity between a sentence and an equation.
2. Can an expression be true or false? Why or why not?
3. Can zero be a solution to the equation $1 \div x = 2$? Why or why not?

1. They each contain a verb (is, =).
2. No; an expression does not have a relation symbol.
3. No; division by zero is undefined.

4 To illustrate the concept of a solution, have students substitute different values for the variable in the open equation $c + 9 = 24$. Ask students to describe the result in each case. (For all values except 15, the result will be a false statement. For $c = 15$, the result is a true statement, so 15 is the solution.

5 Ask students to describe how to use mental math to find a solution to $9 = y - 10$. (Since $19 - 10 = 9$ and $y - 10 = 9$, then $y = 19$.)

6 Students might also write $9(-5) = 45$, which is equivalent to the solution given.

ADDITIONAL EXAMPLES

Identify each equation as true, false, or open.

1. $4(-5) = 20$ (false)
2. $2m = 50$ (open)
3. $-3(-8) = -(-20) + 4$ (true)
4. Is 3 a solution of $6(a) - 6 = 12$? (yes)
5. Is -5 a solution of $13 - x = 8$? (no)
6. Write an equation and identify as true, false, or open: The sum of twenty three and negative four is twenty seven. $(23 + (-4) = 27$, false)

Guided Practice

Class Exercises To check for understanding, ask students to restate each sentence they labeled as false as a true statement for questions 1-6.

Think and Discuss If students have difficulty with question 2, ask if they can tell whether 3 + 4 is true or false and to explain why. (No; an equal sign followed by a number is needed before you can decide.)

MATH MINUTES

Make two sets of cards, one with equations, one with solutions. Have students match the cards.

Equations	*Solutions*
$-4 + x = 0$	4
$2(m - 3) = -8$	-1
$-5c = 18 - 3$	-3
$2d + 6 = 2(20 - 7)$	10

For question 14, be sure that students understand that the same value must be used for *y* each time it appears in the equation.

A ***common error*** occurs when students think that a false equation is not an equation. To remediate, show students these equations: $3 = 3$, $3 + 2 = 3$, and $3 + 3 = 6$. Have students decide if any are not equations and determine which are true and which are false. (All are equations; $3 + 2 = 3$ is false.)

Closure

Have students write an explanation in their math journals of how to determine if a value is a solution for a given equation.

Independent Practice

Encourage students to review the order of operations shown in the Flashback before they attempt questions 11-30.

For questions 31-36, ask students to explain how they might use the calculator's memory key. For example, 114 might be stored in memory for question 33.

For questions 37-42, comment that it is not necessary to find solutions. For example, in question 37 the product of a positive number and a negative number is negative, so *x* must be positive.

Writing in Math For question 55, students' answers do not have to be about presidents or baseball.

NOTES & QUOTES

Robert Recorde (ca. 1510–1558) first used the equal sign as we know it in *The Whetstone of Witte,* which was published in 1557. He chose two parallel line segments of equal length because he believed that no two things could be more equal.

FLASHBACK

Order of operations:

1. Do all operations within grouping symbols.

2. Multiply and divide from left to right.

3. Add and subtract from left to right.

WRITTEN EXERCISES

State whether each is an equation.

1. $10 = 5 + x$ yes
2. $2c + 6$ no
3. $3(-4) = 12$ yes
4. $-3 + 6b$ no

State whether each equation is true, false, or open.

5. $4c - 12 = 20$ open
6. $18 = -3(-6)$ true
7. $36 \div 6 + 1 = 5 + 3$ false
8. $6[-3 - (-5)] = 2(-4 + 10)$ true
9. $-24(-3 + 3) = 18(4 - 2)$ false
10. $-9 + x = 50 \div 10 + 3$ open

Replace *y* with 5. State whether the equation is true or false.

11. $-5 + y = 0$ true
12. $-2y - 3 = 7$ false
13. $-11 = 4 - 3y$ true
14. $2(y - 5) = 5 - y$ true

Replace each variable with the given number. State whether the equation is true or false.

15. $20 - c = 12; c = 8$ true
16. $8 = 2a + 3; a = 0$ false
17. $-x - 5 = -(-6); x = 1$ false
18. $3 - w = 2w + 12; w = -3$ true
19. $3b \div 18 = 2; b = 12$ true
20. $2(-4g) = 10 \cdot 5 - 2; g = -6$ true

Is 2 a solution of each equation?

21. $4 + d = 6$ yes
22. $12 = 26 \div x$ no
23. $4b - 9 = 8b - 17$ yes
24. $4(-8 + t) = 5t - 32$ no

Is the given number a solution of the equation?

25. $3a = 12 + a; a = 6$ yes
26. $-6w = -2w + 32; w = -11$ no
27. $2m = m + 6; m = 4$ no
28. $2x \div 5 = 2(10 - 5); x = 25$ yes
29. $-4c = 6 - c; c = -2$ yes
30. $9 - 3y = 2y + 24; y = -3$ yes

CALCULATOR **Which of the numbers -2, 0, 2, 4, 6 is a solution?**

31. $253a = 0$ 0
32. $259 = 261 - a$ 2
33. $20 - 114z = 248$ -2
34. $53m + 106 = 53(m + 2)$ all
35. $10(-2 + x) = 10x \div 2$ 4
36. $-21x + 129 = 3(-100 + 101)$ 6

MENTAL MATH **Is the solution greater than or less than zero?**

37. $-3x = -15$ >
38. $-5 + y = -35$ <
39. $-27 \div a = 9$ <
40. $-3 = y + 17$ <
41. $x + 25 = 50$ >
42. $-2 \cdot (-6) = -3a$ <

State which of the numbers listed are sensible replacements for the variable. Answers are underlined.

43. Let p represent the number of passengers on a fifty-passenger school bus. Can p be 30? $27\frac{1}{2}$? -5? 48?

44. Let c represent the amount of change equal to one dollar. Can c be 5 quarters? 10 dimes? 100 pennies? 17 nickels?

45. Let d represent the day of the month. Can d be 15? 56? 28? 0?

Write an equation for the sentence. Identify the equation as true, false, or open.

46. Four times the opposite of five equals negative twenty. $4 \cdot (-5) = -20$; true

47. Zero times negative seven is negative seven. $0 \cdot (-7) = -7$; false

48. Negative twelve divided by negative four is equal to three.

49. The sum of fifteen and a number n is fifty. $15 + n = 50$; open

50. The product of negative twenty and nine is negative eleven.

51. The sum of negative seven and twelve equals negative five. $-7 + 12 = -5$; false

52. The difference of twenty-five and negative fifteen equals the sum of negative fifteen and twenty-five. $25 - (-15) = -15 + 25$; false

53. ***DATA FILE 3 (pp. 96–97)*** Three sheets of paper times the number of bills per sheet is equal to ninety-six. $3 \cdot 32 = 96$; true

54. ***DATA FILE 3 (pp. 96–97)*** The value of 1940 proof set in 1965 equals the value of a 1955 proof set in 1990, minus twenty. $50 = 77 - 20$; false

55. ***WRITE*** Sentences are similar to equations. Answers will vary.

a. The sentence *Abraham Lincoln was an American president* is a true sentence. Write two true sentences.

b. The sentence *Eleanor Roosevelt was an American president* is a false sentence. Write two false sentences.

c. The sentence *He is a professional baseball player* is an open sentence. It is not clear to whom the word *he* refers. Write two open sentences.

56. Write an equation relating the number of tickets (t) to the number of dollars (d). $5t = d$

tickets	1	2	3	t
dollars	5	10	15	d

57. Write an equation relating the number of flowers (f) to the number of bouquets (b). $13b = f$

flowers	13	26	39	f
bouquets	1	2	3	b

MIXED REVIEW

Evaluate when $a = 3$ and $b = -2$.

1. $a + 2b$ -1

2. $b - a + 15$ 10

3. $(3b - 2a) \div 4$ -3

4. $|5b - 9 \div 3|$ 13

Write the next three numbers in each pattern.

5. 2, 3, 5, 8 . . . 13, 21, 34 or 12, 17, 23

6. 63, 48, 35, 24 . . . 15, 8, 3

Solve.

7. Kim worked 4 h a day for 3 days to build a suspension bridge model. How many hours were spent on the project? 12 h

48. $-12 \div (-4) = 3$; true
50. $-20(9) = -11$; false

Lesson Follow-up

Reteaching Activity Prepare a set of index cards to form this equation.

[2] [(] [x] [+] [3] [)] [=] [1] [0]

Also prepare cards that show substitution values for x. Place a card over x and ask students to decide if that value is a solution.

LESSON RESOURCES

TRB Practice B-1
TRB Enrichment B-2

Practice

For use after 2-1 (pp. 54–57)

Variables and Equations

True or false?

1. $7 + 5$ is an equation. false
2. $14 = x - 9$ is an open equation. true
3. $8 + 7 = 10$ is an equation. true
4. An equation is either true or false. false
5. $-7(5 - 9) = 19 - (3)(-3)$ is a true equation. true

If n is replaced by 7, is the equation true or false?

6. $3 - n = -4$ true
7. $-n + 16 = 9$ true
8. $9 = 12 - 3n$ false
9. $3(4 + n) = 5n - 2$ true

If the variable is replaced by the given number, is the equation true or false?

10. $-k + 15 = 6$; $k = 21$ false
11. $3c + 5 = 14$; $c = 3$ true
12. $-m - 4 = m$; $m = -2$ true
13. $30 \div z = 11 + z$, $z = -5$ false
14. $24 - 6h = h \div 4$; $h = 4$ false
15. $9(4 + n) = 3n$; $n = -6$ true

Is 5 a solution of each equation?

16. $x + 13 = 18$ yes
17. $8p \div (-4) = 10$ no
18. $4 - 2y = 3(y - 7)$ yes
19. $-m - 3m = 15 - m$ no

Is the given number a solution of the equation?

20. $9k = 10 - k$, $k = -1$ no
21. $-7r - 15 = -2r$; $r = -3$ yes
22. $3g \div (-6) = 5 - g$; $g = -10$ no
23. $(4 - 7)p = 4p + 35$; $p = -5$ yes
24. $8 - e = 2e - 16$; $e = 8$ yes
25. $5(1 - 3s) = 8 - 16s$; $s = 3$ yes

CALCULATOR **Is the given number a solution of the equation?**

26. $-19x = -741$; $x = 39$ yes
27. $38 - 12n = -130$; $n = -14$ no
28. $816 \div (31 - p) = 12$; $p = -37$ yes
29. $429 + 2x = 357 - x$; $x = -24$ yes

MENTAL MATH **Does the equation have a solution greater than zero?**

30. $-6 = x + 12$ no
31. $-24 \div x = -3$ yes
32. $t + 17 = 44$ yes
33. $7x = -35$ no
34. $-12x + 10 = 5$ yes
35. $x \div (-1) + 1 = 0$ yes

Chapter 2 B-1

Enrichment

For use after 2-1 (pp. 54–57)

Keeping Time

The world's most accurate clocks are those at the U.S. Naval Research Laboratory in Washington, D.C. The clocks, called *masers,* are accurate to 1 second in 1,700,000 years. The masers are keyed to the activity of hydrogen electrons.

To find out how many times a hydrogen electron changes position every second, choose the number that is a solution of each equation. Write the *absolute value* of the solution above the number of the exercise at the bottom of the page.

Example $8 + k = 6$ $k = 3, -2, -4$

Solution $k = -2$
$|-2| = 2$
You would write 2 in the correct space.

Solve each equation. Write the absolute value of each solution below.

1. $m + 3 = 2$ $m = -1, -2, -3$
2. $-6k \div 8 = 3$ $k = -2, -4, -6$
3. $7 - 3p = 4p - 7$ $p = -1, -2, -3$
4. $-5r + 6 = 6$ $r = -1, 0, 2$
5. $-4x + 2 = -2(x + 3)$ $x = -2, 3, 4$
6. $2n - 13 = -n + 2$ $n = -2, 4, 5$
7. $-12k \div 3 = k$ $k = 0, 2, 6$
8. $6t + 24 = 3t + 3$ $t = -7, -2, 5$
9. $5(h - 3) = 35 - 5h$ $h = -3, 2, 5$
10. $-m - 1 = 0$ $m = -1, 0, 2$
11. $8(9 - 2y) = -4y$ $y = -4, 3, 6$
12. $5e + 36 = e$ $e = -12, -9, -4$
13. $17 - 9c = -3(c + 3) + 2$ $c = 2, 3, 4$

Answer: 1 (1) , 4 (2) 2 (3) 0 (4) , 4 (5) 5 (6) 0 (7) , 7 (8) 5 (9) 1 (10) , 6 (11) 9 (12) 4 (13) times per second

Write a summary of the results of the exercise above.

14. A hydrogen electron changes position 1,420,450,751,694 times every second.

B-2 Chapter 2

LESSON QUIZ

Is each equation true, false, or open?

1. $3x + 1 = -15$ (open)

2. $4(-5) - (-3) = 17$ (false)

3. $2 - (-2) = -4(9 - 10)$ (true)

Is the given number a solution of the equation?

4. $3z - 12 = -3$, $z = 3$ (yes)

5. $-4m = -2m + 23$, $m = -12$ (no)

Assignment Guide

Basic 1-8, 11-16, 21-26, 31-32, 37-38, 43, 46-53, MR All

Average 2-3, 7-13, 17-18, 22-27, 33-34, 39-40, 44, 48-55, MR All

Enriched 3-4, 8-14, 19-24, 28-30, 35-36, 41-43, 45, 48-57, MR All

FOR THE NEXT LESSON

algebra tiles or blue and white squares of paper, math journal

Lesson Focus

Materials

Algebra tiles or blue and white squares of paper, math journal

Vocabulary/Symbols

Additive Identity
Associative Property of Addition
Associative Property of Multiplication
Commutative Property of Addition
Commutative Property of Multiplication
Multiplicative Identity

MOTIVATING THE LESSON

Ask students what would be the easiest way to find a sum such as ⁻6 + 2 + ⁻4 + 8. (Add the negatives, ⁻10; add the positives, 10; then combine the results, 0.) Then have students add the numbers in order and see if they get the same result. (yes)

Instruction

(1) If students have difficulty with the algebraic notation, have them substitute values for *a* and *b*.

(2) Point out that grouping numbers differently sometimes simplifies computation and will be very useful when students evaluate more complex algebraic expressions and solve equations.

OBJECTIVE: *To use the commutative, associative, and identity properties of addition and multiplication.*

2-2 Properties of Operations

▼ Stephanie won 6 tennis games in the first set and 4 games in the second set. Marcia won 4 games in the first set and 6 in the second set. Each won 10 games.

You can add 6 and 4 in any order and still get the same sum. This suggests the following property.

(1)

Commutative Property of Addition	You can add in any order without changing the sum.	
	Arithmetic	**Algebra**
	6 + 4 = 4 + 6	$a + b = b + a$

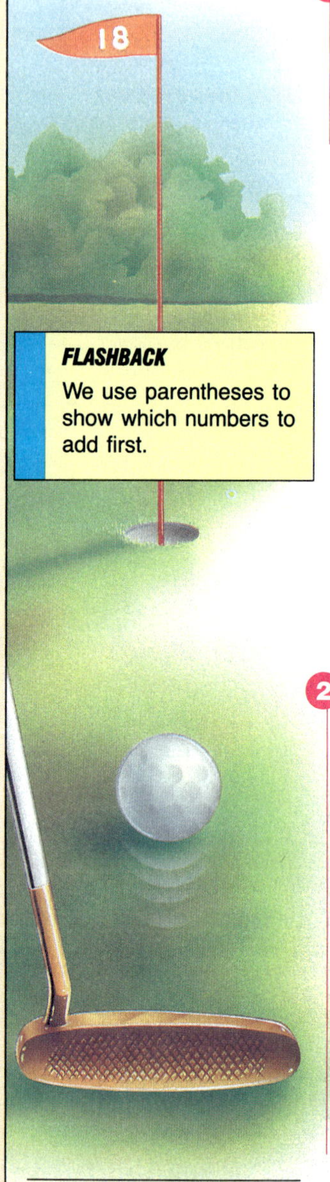

FLASHBACK

We use parentheses to show which numbers to add first.

▼ You can also change the grouping of numbers before you add them. You may want to regroup numbers to add mentally.

Example 1 Carlos rented golf clubs for $7 and a golf cart for $8. He paid a greens fee of $12. Here are two ways he can calculate how much money he spent.

a. $(7 + 8) + 12 = 15 + 12$
$= 27$

b. $7 + (8 + 12) = 7 + 20$
$= 27$

No matter how you group the numbers, the sum is 27.

▼ The mathematical name for the grouping property is the associative property.

(2)

Associative Property of Addition	You can change the grouping and then add without changing the sum.	
	Arithmetic	**Algebra**
	(3 + 7) + 2 = 3 + (7 + 2)	$(a + b) + c = a + (b + c)$

▼ You can also use the commutative and associative properties to multiply.

Commutative Property of Multiplication	You can multiply in any order without changing the product.	
	Arithmetic	**Algebra**
	8 · 4 = 4 · 8	$a \cdot b = b \cdot a$

Teaching Tip

Have posters available that list and define the properties of operations. Change the posters often to refresh students' memories.

Associative Property of Multiplication	You can change the grouping and then multiply without changing the product.	
	Arithmetic	**Algebra**
	$(7 \cdot 3)2 = 7(3 \cdot 2)$	$(ab)c = a(bc)$

▼ You can use the commutative and associative properties to write equivalent expressions.

Example 2 **a.** $5 \cdot 7 = 7 \cdot 5$ — commutative property

b. $(5 + 4) + 9 = 5 + (4 + 9)$ — associative property

c. $7 + a = a + 7$ — commutative property

d. $5(xy) = (5x)y$ — associative property

▼ Properties are helpful when adding or multiplying mentally.

Example 3 **Use mental math to evaluate 81 + 6 + 9.**

Solution

$$\begin{aligned} 81 + 6 + 9 &= 81 + 9 + 6 \\ &= (81 + 9) + 6 && \text{associative property} \\ &= (90) + 6 && \text{Add.} \\ &= 96 \end{aligned}$$

▼ There are *identity* elements for addition and multiplication. You can compute with an identity element without changing the value of a number.

Additive Identity	The additive identity is zero.	
	Arithmetic	**Algebra**
	$12 + 0 = 12$	$a + 0 = a$

Multiplicative Identity	The multiplicative identity is one.	
	Arithmetic	**Algebra**
	$10 \cdot 1 = 10$	$a \cdot 1 = a$

Example 4 **Use mental math to find the missing value.**

a. $5 + ■ = 5$ **b.** $9 \cdot ■ = 9$

Solution

a. $5 + 0 = 5$ When you add zero, the value does not change.

b. $9 \cdot 1 = 9$ When you multiply by one, the value does not change.

THINK How do reordering and regrouping help you to add mentally? **They allow you to group numbers by tens.**

THINK How could you model the additive identity using algebra tiles?

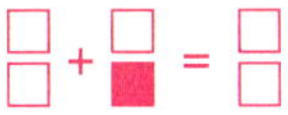

3 Ask students what the sum of a number and zero will be. (the number itself) Then ask what the product of a number and one will be. (the number itself) Point out that zero and one are called *identities* and that the result of adding zero to a number and the result of multiplying one and a number leave the number unchanged.

ADDITIONAL EXAMPLES

Use the given property to write an equivalent expression.

1. $4a + 3$, commutative $(3 + 4a)$

2. $8(xy)$, associative $[(8x)y]$

Use mental math to evaluate each expression.

3. $2(-13)(50)$ $(-1{,}300)$

4. $30 + 0$ (30)

5. $1(34x)$ $(34x)$

MATH MINUTES

Divide the class into two teams. Ask one team to state an expression and the other team to identify the property that is illustrated. After each team member has given an illustration, the teams reverse roles.

Guided Practice

Class Exercises To help students identify the property that was used, ask them to state what has changed in each question.

Think and Discuss Point out that $4 + (-3) = (-3) + 4$ because the negative sign is the sign of the number 3. However, $3 - 4 \neq 4 - 3$ because the minus sign is the sign of the operation. Some students may observe that subtraction can be *rewritten* as addition (and division as multiplication), after which the commutative and associative properties would apply.

A ***common error*** occurs when students assume that the associative property has been used, or must be used, because parentheses are involved. Have the students look at the following equation and explain whether the associative property applies: $3 - (2 + 4)$. (No. The property works for addition and multiplication, not subtraction.)

Closure

Ask students to write a description of each property in their math journals.

Independent Practice

For questions 1-10, point out that some properties may be used more than once.

In question 13, students may use the commutative property of addition or multiplication to change the given expression.

Writing in Math Discuss the answers in questions 34-35 with students to ensure that they understand that the rules for the order of operations must be followed.

THINK AND DISCUSS See below.

1. Does the commutative property apply to subtraction? to division? Use examples to support your answers.
2. Does the associative property apply to subtraction? to division? Use examples to support your answer.

1. no; $5 - 3 \neq 3 - 5$ no; $10 \div 2 \neq 2 \div 10$
2. no; $(3 - 2) - 1 \neq 3 - (2 - 1)$ no; $(20 \div 4) \div 2 \neq 20 \div (4 \div 2)$

MIXED REVIEW

Is each equation true, false, or open?

1. $5 = 2 - (-3)$ true
2. $7 = a + 4$ open

For $w = -5$, is each equation true?

3. $9w = 45$ no
4. $|w - 8| = 13$ yes
5. $2(w + 7) = -4$ no
6. $w + 0 = w$ yes

Use ( ), +, −, · to write a true equation.

7. 3 4 7 6 = -3 $3 \cdot (4 - 7) + 6 = -3$

Solve.

8. Mrs. Laurel has two sons. The product of their ages is 36 and the sum is 13. How old is each son? 4 and 9

CLASS EXERCISES

C: commutative A: associative MI: multiplicative identity AI: Additive identity

State which property is shown.

1. $\square + 6 = 6 + \square$ C
2. $0 + 8 = 8$ AI
3. $\triangle + s = s + \triangle$ C
4. $(5x)y = 5(xy)$ A
5. $999 \cdot 1 = 999$ MI
6. $\blacksquare \cdot \triangle = \triangle \cdot \blacksquare$ C

What numbers would you combine first to evaluate mentally?

7. $5 + 36 + 95$ 95 and 5
8. $5 \cdot 17 \cdot 2$ 5 and 2
9. $50 \cdot 2 \cdot 43$ 50 and 2

MENTAL MATH Evaluate.

10. $10 \cdot 13 \cdot 10$ 1,300
11. $23 + 15 + 85$ 123
12. $25 + 157 + 75$ 257
13. $5 \cdot 20 \cdot 66$ 6,600
14. $140 + 17 + 60$ 217
15. $30 \cdot 30 \cdot 6$ 5,400

WRITTEN EXERCISES

Match each equation with the property illustrated.

1. $a \cdot 1 = 1 \cdot a$ c
2. $\triangle + \blacktriangle = \blacktriangle + \triangle$ a
3. $(6x)y = 6(xy)$ d
4. $(6 + 5) + x = 6 + (5 + x)$ b
5. $6 \cdot 1 = 6$ f
6. $5 + 8 = 8 + 5$ a
7. $(3 \cdot 4)\blacksquare = 3(4 \cdot \blacksquare)$ d
8. $(3 + 2)(4 + 5) = (4 + 5)(3 + 2)$ c
9. $ab = ba$ c
10. $999 + 0 = 999$ e

a. commutative property of addition
b. associative property of addition
c. commutative property of multiplication
d. associative property of multiplication
e. additive identity
f. multiplicative identity

Use the commutative property to write an equivalent expression.

11. $25z$ $z \cdot 25$
12. $n + 2$ $2 + n$
13. $(a + b)5$ $5(a + b)$
14. $(a + b)(c + d)$ $(c + d)(a + b)$

Use the associative property to write an equivalent expression.

15. $(3 \cdot 25) \cdot 4$ $3 \cdot (25 \cdot 4)$
16. $34 + (16 + 35)$ $(34 + 16) + 35$
17. $(4a)b$ $4(ab)$

Use the commutative and associative properties to evaluate.

18. $725 + 563 + 275$ 1,563
19. $250 \cdot 47 \cdot 4$ 47,000
20. $200 + 423 + 800$ 1,423
21. $5 \cdot 20 \cdot 28$ 2,800
22. $5 \cdot 11 \cdot 20 \cdot 3$ 3,300
23. $79 + 17 + 1 + 3$ 100

What numbers would you combine first to evaluate mentally?

24. $\underline{5} \cdot 79 \cdot \underline{20}$ **25.** $\underline{3} + \underline{7} + 67$ **26.** $\underline{10} \cdot 37 \cdot \underline{10}$

27. $\underline{730} + 693 + \underline{270}$ **28.** $\underline{5} \cdot \underline{50} \cdot \underline{20} \cdot \underline{2}$

Answers are underlined.

MENTAL MATH **Evaluate.**

29. $35 + 15 + 8$ **58** **30.** $25 \cdot 4 \cdot 8$ **800** **31.** $42 + 17 - 2 + 3$ **60**

32. $125 + 18 + 75 + 162$ **380** **33.** $4 \cdot 6 \cdot 25 \cdot 50 \cdot 2$ **60,000**

WRITE **Explain your answer to each question below.**

34. a. Can you use the commutative property of addition with $4 + 2$ to evaluate the expression $3 \cdot 4 + 2 \div (-2)$?

b. Can you use the commutative property of multiplication? **Yes, the numbers 4 and 3 would still be multiplied first.**

35. a. Can you use the associative property of addition with $-4 + 20 + 30$ to evaluate $6 \cdot 5 \cdot (-4) + 20 + 30 \div 5$? **No, it would change the order of operations.**

b. Can you use the associative property of multiplication? **Yes, it does not matter in what order you multiply 6, 5, and -4.**

Use the article below to answer each question.

A Fair Fare in Alaska

Railroads are a popular means of transportation in Alaska. The *Anchorage-Fairbanks Express* is a major line that travels 356 mi from Anchorage to Fairbanks.

A one-way coach fare for the $11\frac{1}{2}$-h trip is \$98. A first-class fare is \$140. Because of long hours of daylight in summer, passengers are sure to get their money's worth and not miss out on any of Alaska's beautiful wilderness and wildlife.

36. a. How much would it cost a family of four to travel first class when traveling round trip from Anchorage to Fairbanks? How much money could they save if they went coach? **\$1,120; \$336**

b. The *Anchorage-Fairbanks Express* departed Fairbanks at 11:26 A.M. At what time will it arrive in Anchorage? **10:56 P.M.**

34. a. No, it would change the order of operations.

Lesson Follow-up

Reteaching Activity For students who have difficulty with this lesson, refer to the manipulative activity on p. 52E of the Chapter Overview. See also TRB Exploring O-5.

LESSON RESOURCES

TRB Practice B-3

TRB Enrichment B-4

Practice

For use after 2-2 (pp. 58–61)

Properties of Operations

Write the letter of the property illustrated.

1. $14(mn) = (14m)n$ **d**
2. $19 + 11 = 11 + 19$ **a**
3. $k \cdot 1 = k$ **f**
4. $(x + y) + z = x + (y + z)$ **b**
5. $65t = t(65)$ **c**
6. $p = 0 + p$ **e**
7. $n = 1 \cdot n$ **f**
8. $(x + p) + (r + t) = (r + t) + (x + p)$ **a**
9. $(h + 0) + 4 = h + 4$ **e**
10. $x + yz = x + zy$ **c**

a. commutative property of addition
b. associative property of addition
c. commutative property of multiplication
d. associative property of multiplication
e. additive identity
f. multiplicative identity

Use the commutative property to write an equivalent expression.

11. $x + 7$ **7 + x** 12. $9n$ **n(9)** 13. $h + 4$ **4 + h**

14. $-3y$ **y(-3)** 15. xy **yx** 16. $q + r$ **r + q**

Use the associative property to write an equivalent expression. **Answers may vary. Possible answers given.**

17. $2(9 \cdot 8)$ **(2 • 9)8** 18. $(7 + p) + 3$ **7 + (p + 3)**

19. $2a(bc)$ **(2a)bc** 20. $x + (3 + y)$ **(x + 3) + y**

Use the multiplicative or additive identity to evaluate.

21. $8 + 0$ **8** 22. $456 \cdot 1$ **456** 23. $1(abc)$ **abc**

24. $xyz \cdot 1$ **xyz** 25. $298q \cdot 1$ **298q** 26. $60mn \cdot 1$ **60mn**

Use the commutative and associative properties to evaluate.

27. $4 \cdot 13 \cdot 25$ **1,300** 28. $700 + 127 + 300$ **1,127**

29. $68 + 85 + 32$ **185** 30. $2 \cdot 3 \cdot 4 \cdot 5$ **120**

31. $14 + 71 + 29 + 86$ **200** 32. $125 \cdot 9 \cdot 8$ **9,000**

MENTAL MATH Evaluate.

33. $20 \cdot 7 \cdot 5$ **700** 34. $217 + 545 - 17$ **745**

35. $39 + 27 + 11$ **77** 36. $4 \cdot 12 \cdot 250$ **12,000**

Chapter 2 B-3

Enrichment

For use after 2-2 (pp. 58-61)

Reflexive, Symmetric, and Transitive Properties

Suppose ⊙ is a relationship between two quantities. The following properties may hold for ⊙.

Property	Statement
Reflexive	a ⊙ a.
Symmetric	If a ⊙ b, then b ⊙ a.
Transitive	If a ⊙ b and b ⊙ c, then a⊙c.

Example Which of the properties hold for the relationship ">?"

Solution Substitute > in each of the above statements. Substitute values for a and b. If you can find a single example for which the statement is not true, then the relationship does not have the given property. If every example you try produces a true statement, the relationship appears to have the given property.

Reflexive $a > a$

Test: Is $5 > 5$? No. The relationship ">" is not reflexive.

Symmetric If $a > b$, then $b > a$.

Test: $8 > 6$. Is $6 > 8$? No. The relationship ">" is not symmetric.

Transitive If $a > b$ and $b > c$, then $a > c$.

Test: $10 > 7$ and $7 > 3$. Is $10 > 3$? Yes.
$-1 > -2$ and $-2 > -5$. Is $-1 > -5$? Yes.
$12 > -3$ and $-3 > -6$. Is $12 > -6$? Yes.
The relationship ">" appears to be transitive.

Test the relationship for the reflexive, symmetric, and transitive properties. Tell which properties, if any, appear to hold. Write *R*, *S*, or *T*.

1. = **R, S, T** 2. ≥ **R, T**
3. "is older than" **T** 4. "is the father of" **none**
5. "is a sister of" **S, T** 6. "has the same zip code as" **R, S, T**
7. "is a friend of" **S** 8. "has the same color hair as" **R, S, T**
9. "is a teacher to" **none** 10. "is as smart as" **R, S, T**

B-4 Chapter 2

LESSON QUIZ

Identify the property that is illustrated.

1. $-21 \cdot 1 = -21$ (mult. iden.)
2. $c + 0 = c$ (Add. Iden.)
3. Use the commutative property to write an equivalent expression for $3(a + 4)$. [$3(4 + a)$ or $(a + 4)3$]

Assignment Guide

Basic 1-9 odd, 11-12, 15-19, 24-25, 29-30, 34-36, MR All

Average 1-9 odd, 12-13, 15-20, 26-27, 31-32, 34, 36, MR All

Enriched 2-10 even, 13-14, 16-17, 21-23, 27-28, 32-36, MR All

FOR THE NEXT LESSON

rectangle models or graph paper, index cards, math journal

Lesson Focus

Materials

rectangle models or graph paper, index cards, math journal

Vocabulary/Symbols

Distributive Property of Multiplication over Addition
Distributive Property of Multiplication over Subtraction

MOTIVATING THE LESSON

Begin the lesson by asking students to describe how they would evaluate an expression such as $4[(-2) + 8]$ using the order of operations. (Add -2 and 8, then multiply the result by 4.)

Instruction

1 Have students use models of rectangles or graph paper to illustrate the two methods of finding total area.

2 Stress the importance of multiplying both numbers inside a set of parentheses by the number outside the parentheses. Ask students why the two forms of the distributive property are equivalent. (Commutative Property of Multiplication)

3 Remind students that subtraction can be rewritten as addition; that is, $a - b = a + (-b)$. Therefore, the distributive property of multiplication can be used for subtraction as well as addition. For example:

$$a(b - c) = a[b + (-c)] = ab + a(-c) = ab - ac$$

2-3 The Distributive Property

OBJECTIVE: *To use the distributive property.*

FLASHBACK
Area = length · width
$A = l \cdot w$

▼ Two rectangles, each having the same width but different lengths, are placed end to end. Find the total area.

Method 1 Placing the rectangles end to end forms one large rectangle. The length is $(6 + 8)$ and the width is 4.

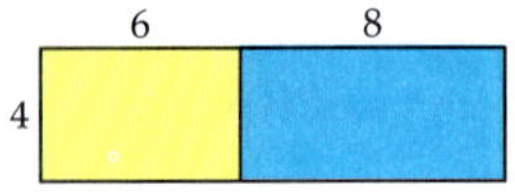

total area $= 4(6 + 8)$
$= 4(14)$
$= 56$

Method 2 You can find the area of each individual rectangle and then add the areas together.

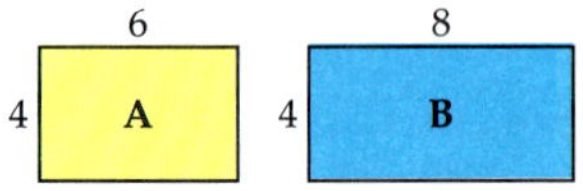

Area **A** $= 4 \cdot 6$ Area **B** $= 4 \cdot 8$

total area = area A + area B
$= 24 + 32$
$= 56$

The total area is the same no matter how you do the calculation. Therefore, the expression $4(6 + 8)$ has the same value as the expression $24 + 32$. This illustrates the *distributive property*.

THINK Why does $a(b + c) = (b + c)a$? commutative property of multiplication

Distributive Property of Multiplication over Addition	You can distribute a factor to each term inside a set of parentheses.	
	Arithmetic	$3(2 + 6) = 3 \cdot 2 + 3 \cdot 6$ $(2 + 6)3 = 2 \cdot 3 + 6 \cdot 3$
	Algebra	$a(b + c) = ab + ac$ $(b + c)a = ba + ca$

Example 1 **Evaluate $9(5 + 2)$ using the distributive property.**

Solution

$9(5 + 2) = 9 \cdot 5 + 9 \cdot 2$	Distribute 9.
$= 45 + 18$	Multiply and add.
$= 63$	

▼ You can also use the distributive property with subtraction.

THINK Why does $a(b - c) = (b - c)a$? commutative property of multiplication

Distributive Property of Multiplication over Subtraction	You can distribute a factor to each term inside a set of parentheses.	
	Arithmetic	$6(7 - 4) = 6 \cdot 7 - 6 \cdot 4$ $(7 - 4)6 = 7 \cdot 6 - 4 \cdot 6$
	Algebra	$a(b - c) = ab - ac$ $(b - c)a = ba - ca$

Teaching Tip

Display the algebraic form of an equation on the bulletin board. Assign values for each variable and place them under the equation. Leave the display up until you are satisfied that students understand the algebraic form.

Example 2 **Evaluate (3 − 1)6 using the distributive property.**

Solution $(3 - 1)6 = 3 \cdot 6 - 1 \cdot 6$ Distribute 6.
$= 18 - 6$ Multiply and subtract.
$= 12$

▼ You can use the distributive property in reverse.

Example 3 **Write 5 · 3 + 5 · 7 using the distributive property.**

Solution $5 \cdot 3 + 5 \cdot 7$ 5 multiplies 3 and 5 multiplies 7.
$5(3 + 7)$ 5 multiplies (3 + 7).

▼ You can use the distributive property to multiply mentally and solve word problems.

Example 4 **Use the distributive property to evaluate 20(102) mentally.**

Solution $20(102) = 20(100 + 2)$ Think of 102 as (100 + 2)
$= 20 \cdot 100 + 20 \cdot 2$ Distribute 20.
$= 2{,}000 + 40$ Multiply and add.
$= 2{,}040$

Example 5 The PTA sold 397 tickets for their annual pancake breakfast. Each patron will receive four pancakes. How many pancakes will the PTA members make?

Solution $4(397) = 4(400 - 3)$ Think of 397 as (400 − 3).
$= 4 \cdot 400 - 4 \cdot 3$ Distribute 4.
$= 1{,}600 - 12$ Multiply and subtract.
$= 1{,}588$

CLASS EXERCISES

Write an expression to describe the total area.

1. 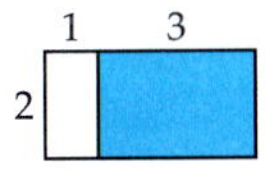

2(1 + 3)

2. 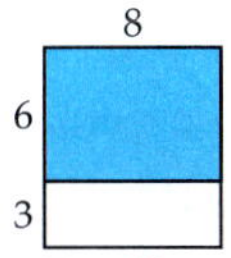

8(6 + 3)

3. 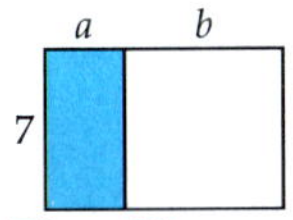

7(a + b)

State the number or variable that can be distributed.

4. $9(5 - 3)$ 9 **5.** $(5 + 7 + a)2$ 2 **6.** $z(x - y)$ z

4 Ask students to explain why the distributive property can be used. Then ask them to evaluate both expressions to check their explanation. [5 · 3 + 5 · 7 = 15 + 35 = 50 and 5(3 + 7) = 5(10) = 50]

5 Remind students that to *evaluate mentally* means to do the calculations without writing anything on paper. The calculations written in the text are for illustration.

ADDITIONAL EXAMPLES

Evaluate using the distributive property.

1. 8(6 + 3) (72)
2. (7 − 4)2 (6)
3. 14 · 8 + 14 · 9 (238)

Evaluate mentally.

4. 30(204) (6,120)
5. 14(299) (4,186)

Guided Practice

Class Exercises Have students use rectangle models to illustrate questions 1-3.

MATH MINUTES

For use with Critical Thinking p. 65.

◇	0	1	2	3
0	0	0	0	0
1	0	1	2	3
2	0	2	0	2
3	0	3	2	1

Think and Discuss In discussing question 3, have students distribute 6 to the quantity 3 + 11 and 4, 6[(3 + 11) + 4]; then have them distribute 6 to 3 and the quantity 11 + 4, 6[3 + (11 + 4)], to get the final result.

A ***common error*** is to multiply only the first term inside a set of parentheses by the factor outside the parentheses. Have students evaluate several numerical exercises, such as 7(9 − 3), using both the distributive property and order of operations to be sure the results are equivalent. (42) It may also help if students use arrows to indicate which number or variable is being distributed.

$(x - 5)3 = (x \cdot 3) - (5 \cdot 3)$

Closure

Writing in Math Ask students to write a procedure for using the distributive property.

Independent Practice

For questions 10-15, remind students that if the number being distributed is on the right, use the second form of the distributive property. That is, $a \cdot b + 3 \cdot b$ is written $(a + 3)b$ rather than $b(a + 3)$.

For questions 22-25, ask students to write each number using the distributive property of multiplication over addition to check their answers.

Critical Thinking For question 5, students investigate properties of addition for a modular system. Students may discover that the table is symmetric about its diagonal, illustrating commutativity. See the problem solving activity on p. 52E of the Chapter Overview (TRB O-6).

THINK AND DISCUSS See below.

1. What does the word *distribute* mean? How can you relate it to the way the distributive property is used?
2. How does the distributive property affect the rules for the order of operations?
3. How would you evaluate 6(3 + 11 + 4) using the distributive property?

1. To share or give to each one; you distribute a factor to each factor inside the parentheses.
2. Instead of doing what is inside the grouping symbols first, you multiply using the dist. property.
3. Distribute 6, then add.

State the number or variable that has been distributed. Rewrite using the distributive property in reverse.

7. $y \cdot 4 + y \cdot 6$ y; y(4 + 6)
8. $a \cdot (-3) - b \cdot (-3)$ -3; (a − b)(-3)

Complete with the appropriate number or variable.

9. $9(5 + 4) = 9 \cdot$ ■ $+ 9 \cdot 4$ 5
10. $(y - 6)z = y \cdot z - 6 \cdot$ ■ z
11. $12(3 + 5) = 12 \cdot 3 +$ ■ $\cdot$ ■ 12, 5
12. $a(3 - b) = 3$■ $-$ ■b a a

Use the distributive property to evaluate.

13. 6(4 + 8) 72
14. (14 − 9)2 10
15. -4(7 + 3 + 2) -48
16. 7(3 + 5) 56
17. 12(10 − 2) 96
18. 3(2 + 2) 12

WRITTEN EXERCISES

Write an expression to describe the total area.

1. 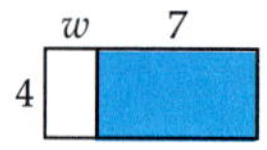

4(w + 7)

2. 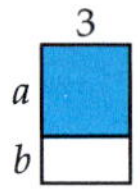

3(a + b)

3. 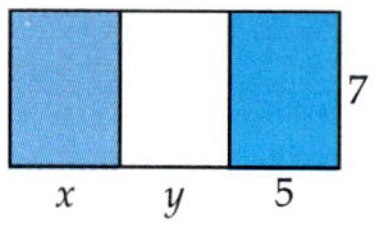

7(x + y + 5)

State the number or variable which can be distributed.

4. (10 − 2)7 7
5. -4(6 + 7 + 9) -4
6. [3 + (-1)]x x
7. (x − 4)c c
8. r(s − t) r
9. -w[9 + (-x)] -w

State the number or variable that has been distributed. Rewrite using the distributive property in reverse.

10. $9 \cdot 3 + 9 \cdot 4$ 9; 9(3 + 4)
11. $-9 \cdot x + (-9) \cdot y$ -9; -9(x + y)
12. $a \cdot b + 3 \cdot b$ b; (a + 3)b
13. $(b \cdot e) - (c \cdot e) - (d \cdot e)$ 13. e; (b − c − d)e
14. $-6 \cdot 4 + 9 \cdot 4$ 4; (-6 + 9)4
15. $-1 \cdot a + (-1) \cdot b$ -1; -1(a + b)

Complete with the appropriate number or variable.

16. $6(3 + 7) = 6 \cdot 3 +$ ■ $\cdot 7$ 6
17. $(w - x - z)y = ($■ $\cdot y) - (x \cdot$ ■$) - (z \cdot y)$ w, y
18. $6 \cdot b + 12 \cdot b = ($■ $+ 12)$■ 6, b
19. $[-2 + (-4)](-6) =$ ■ $\cdot (-6) +$ ■ $\cdot (-6)$ -2, -4
20. $-7 \cdot 12 - (-7) \cdot 17 =$ ■ $($■ $- 17)$ -7, 12
21. $[10 - (-2)]5 = 10 \cdot$ ■ $-$ ■ $\cdot 5$ 5, -2

MENTAL MATH Use the distributive property to evaluate.

22. 2(122) 244
23. (280)4 1,120
24. 2(670) 1,340
25. 5(1,015) 5,075

Use the distributive property to evaluate.

26. $4(5 + 11)$ 64 **27.** $(3 - 6)4$ -12 **28.** $-5(6 - 7)$ 5

29. $[9 - (-1)](-3)$ -30 **30.** $-3[2 + (-9)]$ 21 **31.** $-7[2 + (-3)]$ 7

32. $6(8 + 2 + 12)$ 132 **33.** $[26 + (-4) + 35]5$ 285

34. $2(5 + 3) - 3(4 - 2)$ 10 **35.** $(20 - 12)6 - 4[5 - (-3)]$ 16

Solve using the distributive property.

36. A theater was filled to capacity, 294, three nights in a row. How many people were at the show in these three nights? 882

37. It is 1,549 mi from Boston to Dallas. How many miles would you travel if you drove round trip? 3,098 mi

38. ***DATA FILE 3 (pp. 96–97)*** Find the total number of pennies produced in 5 days by the United States Mint. 195,000,000

MIXED REVIEW

State the property shown.

1. $3(6 \cdot 2) = 3(2 \cdot 6)$ C
2. $8 = 8 + 0$ AI
3. $4(8 \cdot 3) = (4 \cdot 8)3$ A
4. $1 \cdot 3 = 3$ MI

Compare. Use <, >, or =.

5. $(-8 \cdot 3) + 9$ ▪ $8(3 + 9)$ <
6. $3(4 + 5)$ ▪ $3 \cdot 4 + 3 \cdot 5$ =
7. $-5 \cdot 2 - |-9|$ ▪ $2(-2)3$ <

Solve.

8. ***DATA (p. 31)*** Find the total calories in a 3-oz hamburger, 20 potato chips, a carrot, and 1 c of lemonade. 605

Critical Thinking

EXPLORING LOGICAL THINKING See Math Minutes for the corresponding multiplication table.

The table shows a mathematical operation, symbolized by ✦. In the table you can see that 2 ✦ 3 = 1.

✦	0	1	2	3
0	0	1	2	3
1	1	2	3	0
2	2	3	0	1
3	3	0	1	2

1. Use the table to find the following.
   a. 1 ✦ 2 3 **b.** 1 ✦ 3 0 **c.** 3 ✦ 2 1
2. To investigate the properties, let us determine if ✦ is associative.

   ▼▼ *SAMPLE* 1 ✦ (3 ✦ 2) ▪ (1 ✦ 3) ✦ 2
   1 ✦ (1) ▪ (0) ✦ 2
   2 = 2

   a. Compare 3 ✦ (2 ✦ 2) ▪ (3 ✦ 2) ✦ 2. Think of other combinations. Check them. =
   b. Is the operation associative? yes
3. Is ✦ commutative? Does 2 ✦ 3 = 3 ✦ 2? Check other combinations. yes
4. Is there an identity? If so, what is it? yes; zero
5. ***WRITE*** a description of the operation ✦. It is an operation which is associative and commutative. It has an identity of zero.

Lesson Follow-up

Reteaching Activity Write expressions such as $3(n + 2)$ on index cards, one expression to each card. Have a student choose a card and use the distributive property to rewrite the expression on the chalkboard. Then have another student check the expression by substituting 2 for the variable.

LESSON RESOURCES

TRB Practice B-5
TRB Enrichment B-6

Practice

For use after 2-3 (pp. 62-65)

The Distributive Property

Write an expression for the total area.

1. 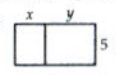

5(x + y)

2. 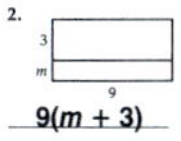

9(m + 3)

3. 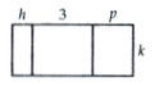

k(h + 3 + p)

Write the number or variable which can be distributed.

4. $8(14 - 4)$ 8 5. $(2 + k)9$ 9 6. $(x + 3)(-5)$ -5
7. $(n - 12)h$ h 8. $-c(m - p)$ -c 9. $(12 + 5 \div r)s$ 1s

Write the number or variable which has been distributed. Rewrite using the distributive property in reverse.

10. $8 \cdot 7 + 9 \cdot 7$ 7 (8 + 9)7 11. $3 \cdot 12 - 5 \cdot 12$ 12 (3 − 5)(12)
12. $k \cdot 15 + k(-11)$ k k(15 − 11) 13. $x \cdot m - x \cdot n$ x x(m − n)

Complete with the appropriate number or variable.

14. $12(5 + 9) = 12 \cdot 5 +$ 12 $\cdot 9$
15. $3 \cdot 7 - 8 \cdot 7 = ($ 3 $- 8)$ 7
16. $z(a - b - c) =$ z $\cdot a - z \cdot$ b − z $\cdot$ c
17. $[14 + (-3)]7 = 14 \cdot$ 7 + (-3) $\cdot 7$
18. $p[(-3) + n] = p \cdot$ (-3) + p $\cdot$ n

MENTAL MATH **Use the distributive property to evaluate.**

19. $98 \cdot 7$ 686 20. $290 \cdot 4$ 1,160 21. $30 \cdot 105$ 3,150
22. $78 \cdot 8$ 624 23. $195 \cdot 6$ 1,170 24. $598 \cdot 3$ 1,794
25. $899 \cdot 5$ 4,495 26. $7 \cdot 2{,}009$ 14,063 27. $9 \cdot 28$ 252

Use the distributive property to evaluate.

28. $3(9 + 5)$ 42 29. $(8 - 12)5$ -20 30. $-4(3 + 5)$ -32
31. $3(4 + 5 + 6)$ 45 32. $6[7 + (-2)]$ 30 33. $[10 + (-3)]7$ 49

Solve using the distributive property.

34. A shipping container holds 144 boxes. How many boxes can be shipped in 4 containers? 576

Chapter 2 B-5

Enrichment

For use after 2-3 (pp. 62–65)

Multiplying Binomials

A *term* is part of an expression. Terms are separated by addition and subtraction symbols. A *binomial* is an expression with two or more terms.

Binomials	Not binomials
2 + 7	12n (only one term)
3x + 4y	6 + 5k − 3py (three terms)
-7k − 9	-8(2g) (terms not separated by + or -)

You can multiply binomials using the distributive property.

Complete the following to find the product $(a + b)(c + d)$. (*Hint:* Think of $a + b$ as a single quantity. Distribute it to both c and d: $\boxed{a + b}$ $(c + d)$)

1. $(a + b)(c + d) = (a + b)($ c $) + (a + b)($ d $)$
$= ac +$ bc $+ ad +$ bd

To multiply two binomials, multiply each of the terms in the first binomial by each of the terms in the second binomial.

Example Find the product: $(3 + k)(m - 4)$

Solution $(3 + k)(m - 4) = 3 \cdot m + k \cdot m - 3 \cdot 4 - k \cdot 4$
$= 3m + km - 12 - 4k$

Find the product.

2. $(h + w)(c + d)$ hc + wc + hd + wd
3. $(5 + p)(9 + t)$ 45 + 9p + 5t + pt
4. $(x + y)(7 - m)$ 7x + 7y − xm − ym
5. $(n - 5)(y - 6)$ ny − 5y − 6n + 30

You can use binomials to find the product of two numbers.

$12 \times 8 = (10 + 2)(10 - 2)$
$= 10 \cdot 10 + 2 \cdot 10 - 10 \cdot 2 - 4$
$= 100 + 20 - 20 - 4$
$= 96$

Find the product by multiplying two binomials.

6. $23 \times 17 = (20 + 3)(20 - 3) =$ 391
7. 55×45 2,475 8. 76×84 6,384

B-6 Chapter 2

LESSON QUIZ

Rewrite using the distributive property in reverse.

1. $a \cdot 4 + b \cdot 4$ [$(a + b)4$]
2. $2 \cdot m + 2 \cdot n$ [$2(m + n)$]
3. $3(n - c + 2)$
($3 \cdot n - 3 \cdot c + 3 \cdot 2$)
4. Use the distributive property to evaluate 5(199). (995)

Assignment Guide

Basic 1-6, 10-12, 16-17, 22-23, 26-28, 36, 38, MR, CT All

Average 2-3, 6-8, 14-15, 18-19, 23-24, 29-32, 36-38, MR, CT All

Enriched 3, 7-9, 13, 15, 19-21, 24-25, 32-38, MR, CT All

FOR THE NEXT LESSON

algebra tiles or graph paper, scissors, index cards, math journal

Lesson Focus

Materials

Algebra tiles or graph paper, scissors, index cards, math journal

Vocabulary/Symbols

constant
like terms
numerical coefficient
simplify an expression
term

MOTIVATING THE LESSON

Ask students to state how 3*x*, 2*x*, and 4*y* are alike and how they are different. Record the similarities and differences on the chalkboard. (All contain variables; the first two quantities are alike because they both contain the same variable; the last quantity is different because it contains the variable *y*.)

Instruction

1 Comment that the rectangles and stars can be thought of as algebra tiles representing *r* and *s* and that *r* and *s* should not be considered as abbreviations for rectangle and star.

2 Stress that even though *x* and *xy* contain one variable that is the same, they are not like terms. You may need to demonstrate that they represent different quantities by substituting values for the variables.

3 Point out that the numerical coefficient for *x* is understood to be 1 because $x = 1x$ (by the identity property for multiplication). Have students use graph paper or tiles to model the sum.

OBJECTIVE: *To use properties of addition and multiplication to simplify expressions.*

2-4 Simplifying Variable Expressions

▼ The Texas state flag is made up of several rectangles and a star. You can represent the rectangles with the variable *r* and the star with the variable *s*. You can describe the parts of the flag by the following expression.

1

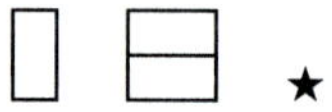

$1r \; + \; 2r \; + \; 1s$

This expression has three *terms*: $1r$, $2r$, and $1s$. The terms are separated by addition symbols.

Term	A term is a part of an expression. Terms are separated by addition and subtraction symbols.

THINK Are the terms *r* and 1*r* equivalent? Why or why not? **Yes, one is the multiplicative identity.**

▼ The term $2r$ consists of two parts. The variable is *r* and the *numerical coefficient* is 2.

Numerical Coefficient	A numerical coefficient is a number that is multiplied by a variable.

▼ $1r$ and $2r$ have the same variable. They are *like terms*.

Like Terms	Like terms have the same variable(s).

Example 1 **State the number of terms in each expression. Name the numerical coefficients and the like terms.**

a. $a + 5b - 3b$ **b.** $10xy + 5y + xy - 20$

THINK Why is 10*xy* considered to be one term? **There are no addition or subtraction signs.**

2

Solution

a. three terms: a, $5b$, and $3b$
numerical coefficients: 1, 5, and 3
like terms: $5b$ and $3b$

b. four terms: $10xy$, $5y$, xy, and 20
numerical coefficients: 10, 5, and 1
like terms: $10xy$ and xy

▼ Sometimes a term consists of a number without a variable. We call a term with no variable a *constant*.

Example 2 In the expression $5x + 2$, the constant is 2. No matter what value is substituted for x, the constant 2 remains the same.

Teaching Tip

Have index cards available that contain the derivation of difficult vocabulary words.

▼ You can use models to represent an expression before you combine like terms to simplify the expression.

Example 3 **Use algebra tiles or colored paper to simplify the expression $x + 1 + 2x$.**

Solution

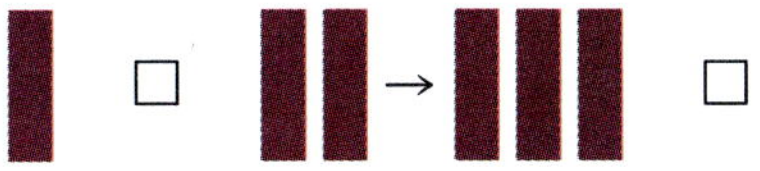

$x + 1 + 2x \rightarrow 3x + 1$

3 **THINK** What is the value of the expression $3x + 1$ if $x = 3$? if $x = -2$? 10; -5

▼ You can combine the numerical coefficients of like terms by using the distributive property.

Example 4 **Simplify the expression $7x - 2x$ using the distributive property.**

Solution

$7x - 2x = 7 \cdot x - 2 \cdot x$	x is distributed to 7 − 2.
$= (7 - 2)x$	Subtract coefficients.
$= 5 \cdot x$	Multiply.
$= 5x$	

4

▼ Combining like terms and using algebraic properties are helpful in simplifying variable expressions.

5

Simplify an Expression	To simplify an expression, replace it with an equivalent expression that contains no like terms or parentheses.

Example 5 **Simplify $4(5b)$.**

Solution

$4(5b) = (4 \cdot 5)b$	associative property
$= 20 \cdot b$	Multiply.
$= 20b$	

▼ Sometimes you will need to use more than one property to simplify a variable expression.

Example 6 **Simplify $4x + 3(3 + x)$.**

Solution

$4x + 3(3 + x) = 4x + 9 + 3x$	distributive property
$= 4x + 3x + 9$	commutative property
$= (4x + 3x) + 9$	associative property
$= (4 \cdot x + 3 \cdot x) + 9$	x is distributed.
$= (4 + 3)x + 9$	Add coefficients.
$= 7 \cdot x + 9$	Multiply.
$= 7x + 9$	

NOTES & QUOTES

Francois Viète (1540–1603) introduced the use of vowels to represent unknown quantities.

4 Explain to the class that one way they can make use of the distributive property is in combining like terms. Ask students to use models to verify the solution. Then ask if there is another way the solution can be verified. (Evaluate the expression by substituting a value for x, such as, $x = 2$, $7x - 2x = 7 \cdot 2 - 2 \cdot 2 = 10$ and $5x = 5 \cdot 2 = 10$)

5 To make sure students understand the concept, write several expressions on the chalkboard and ask the class to identify expressions that are *not* in simplest form.

ADDITIONAL EXAMPLES

State the number of terms, the numerical coefficients, and the like terms.

1. $3a + 2b + 4a$ (3; 3, 2, 4; $3a$ and $4a$)

2. $3x - 4x - 5 + 9x$ (4; 3, 4, 9; $3x$, $4x$, and $9x$)

Simplify.

3. $38x - 24x$ ($14x$)

4. $-2(a + 3b) - 5b$ ($-2a - 11b$)

5. $6(12x)$ ($72x$)

6. $5x - 3(4 - x)$ ($8x - 12$)

MATH MINUTES

Display pairs of unlike terms such as the following: $3x$ and 3, $6a$ and $7b$, and $8xy$ and $3y$. Ask students to work in groups to write a description of why the terms in each pair are not like terms.

Guided Practice

Class Exercises To model question 7, students will need to combine the tiles for $3x$ and $4x$. Only then can they remove tiles representing $7x$.

For questions 13-16, ask students how they will simplify each expression.

Think and Discuss If students have difficulty with question 2, suggest that they substitute x and y for □ and △.

A ***common error*** when students are identifying like terms such as $10xy$ and xy, is to omit the numerical coefficient 1 and to decide that the terms are unlike. Show students the following list of terms and have them identify the like terms: a, $10a$, $a \cdot 5$, $10ab$, $7b$, ba. (a, $10a$, $a \cdot 5$; $10ab$, ba; $7b$)

Closure

Have students discuss the procedure for simplifying variable expressions. Ask them to illustrate with Class Exercise 16 and to name the properties involved. (distributive, associative)

Independent Practice

Have students evaluate the original expression in questions 36-40, using the same values of x, y, and z to check that their results are correct.

THINK AND DISCUSS See below.

1. What are some advantages of simplifying expressions?
2. □ and △ each stand for a different number. Which expression is equivalent to □ + 4△ + 7 + 3△ − 2□?

a. 15□△
b. □ + 7△ + 7
c. □ + 14△
d. □ + 7△ + 4 + 3
e. 7△ − □ + 7

1. easier to understand, easier to work with
2. e

Class Exercises
1. 4; 3, 5, 6; 3x, 6x; 3
2. 2; 2; none; 7
3. 5; 1, 3, 6, 5; 1a, 5a; 1
4. 3; 2, 1, 6; 2r, 6r; none

Written Exercises
1. 2; 5, 8; 5a, 8a; none
2. 3; 6, -2, 1; -2b, b; none
3. 2; 2; none; -7
4. 2; 6, 5; 6xy, 5xy; none
5. 2; -7; none; 3
6. 3; 1, 1, 1; x, x, x; none
7. 4; 6, 4, 1; 6ab, 4ba, ab; 8
8. 5; 4, 7, 9, 1; 4x, 9x and 7w, w; 12

CLASS EXERCISES

Give the number of terms in each expression. Name the numerical coefficients, the like terms, and the constants, if any. Answers in side column.

1. $3x + 5y - 6x - 3$ **2.** $2x - 7$
3. $1a + 3b - 6c + 5a + 1$ **4.** $2r - 1s + 6r$

Combine the like terms using a model. Check students' work.

5. $12a + 7a$ 19a **6.** $5 + 2b$ 5 + 2b
7. $4x - 7x + 3x$ 0 **8.** $a + 2a + 3a - 4a$ 2a

D: distributive property

Give the property that you can use to simplify each expression.

9. $-3(5x)$ A **10.** $(n + 7) + 3$ A **11.** $9x + 5x$ D **12.** $(-5y)(-9)$ C then A

Simplify each expression.

13. $6(5b)$ 30b **14.** $-6x + (9x - 3)$ 3x − 3
15. $-4(a + 3) + 7a$ 3a − 12 **16.** $4 + 3(a + b) - 6b$ 3a − 3b + 4

WRITTEN EXERCISES

Give the number of terms in each expression. Name the numerical coefficients, the like terms, and the constants, if any. Answers in side column.

1. $5a + 8a$ **2.** $6a + (-2b) + b$ **3.** $2x - (-7)$
4. $6xy - 5xy$ **5.** $-7c + 3$ **6.** $x + x + x$
7. $6ab - 4ba + 8 + ab$ **8.** $12 - 4x + 7w - 9x - w$

Combine the like terms using a model. Check students' work.

9. $4a + 5a$ 9a **10.** $w - 3w + 2w$ 0 **11.** $7b + b - 3b$ 5b

State the property that you can use to simplify each expression.

12. $6x + (-2x)$ D **13.** $3(8k)$ A **14.** $(4n + 3) + 8$ A
15. $8z + (-15z)$ D **16.** $-2 + (7 + 8y)$ A **17.** $-6(-5s)$ A

Combine the like terms using the distributive property.

18. $16z + 24z$ 40z **19.** $52a - 47a$ 5a **20.** $r + 6r - 3r$ 4r

Complete with the appropriate number or variable.

21. $6a + 4a + 7 = (\blacksquare a + 4a) + 7$ 6
$= (6 + \blacksquare)a + 7$ 4
$= \blacksquare a + 7$ 10

Simplify each expression. See below.

22. $5 + 2x + 8$ **23.** $5m + (-4m)$ **24.** $4(-3y) + 7 - 3$

25. $3a + 2a + a$ **26.** $9(4t) + 8$ **27.** $18 + 6(9k) - 13$

28. $8z + 8y + 3z$ **29.** $3(g + 5) + 2g$ **30.** $6(3k + 2k)$

31. $4(w + 2x) + 9(-4w)$ **32.** $-7(2f + 5e) + 8(6 + 4e)$

33. $-12(5x) + 3(-7x) - 2x$ **34.** $(2t + 4)3 + 6(-5t) - (-8)$

35. $5(4a + b - c) + 8(3a - 8b + 3c)$

Simplify. Evaluate when $x = 3$, $y = -5$, and $z = 7$.

36. $2x + x + y$ 3x + y; 4 **37.** $-3(4y)$ -12y; 60 **38.** $z(2x + 3x)$ 5xz; 105

39. $2y + 2z + y - 16$ 3y + 2z − 16; -17 **40.** $6(2x + y) + 2(x + 2y)$ 14x + 10y; -8

Write an expression for each situation. Simplify.

41. Six bus loads containing x students each came to band day from one school, and 7 bus loads containing x students came from another school. Fourteen students came by car. 6x + 7x + 14; 13x + 14

42. Arleen unloaded 4 boxes each containing v videotapes and 3 boxes each containing $y + 2$ videotapes. 4v + 3(y + 2); 4v + 3y + 6

43. Janet bought three folders costing x cents each and two report covers costing x cents each. She also purchased a binder for $1.89. 3x + 2x + 189; 5x + 189

44. Mr. Unruh purchased five movie tickets for x dollars each. He also purchased a soda for $1.25 and popcorn for $2.25. 5x + 1.25 + 2.25; 5x + 3.50

22. 2x + 13 23. 1m 24. -12y + 4 25. 6a 26. 36t + 8 27. 54k + 5
28. 8y + 11z 29. 5g + 15 30. 30k 31. 8x − 32w 32. -3e − 14f + 48
33. -83x 34. -24t + 20 35. 44a − 59b + 19c

TEST YOURSELF

Is -4 a solution of each equation?

1. $2x - 6 = -14$ yes **2.** $16 \div 2y = 2$ no **3.** $-21 = -2(3 - x)$ no

State which property is shown.

4. $3 \cdot (-6) = -6 \cdot 3$ C **5.** $(3a)b = 3(ab)$ A

6. $17 \cdot 1 = 17$ MI **7.** $6 + 0 = 0 + 6$ C

8. $(3 + 2)(4 - 7) = (4 - 7)(3 + 2)$ C **9.** $4(3 - 12) = 4 \cdot 3 - 4 \cdot 12$ D

Simplify.

10. $5(10b)$ 50b **11.** $96 + 73 + 4$ 173 **12.** $9y - 3y + 12y$ 18y

13. $3(a + 2a)$ 9a **14.** $-14 + 2(9 - 3)$ -2 **15.** $7(2w) + 2(w - 3)$ 16w − 6

MIXED REVIEW

Complete.

1. $7(2 - 5) = 7 \cdot 2 - ■ \cdot 5$ 7

2. $3 \cdot 5 + ■ \cdot 8 = 3(5 + 8)$ 3

3. $(2 + 5) + 7 = 2 + (■ + 7)$ 5

4. $(5 \cdot 3) = (■ \cdot 5)$ 3

Find the mean.

5. -7, -8, 2, -3 -4

6. 88, 93, 76, 82, 91 86

Use a Venn diagram to solve.

7. There are 25 students in the French class and 29 in the Spanish class. Eight students are taking both classes. What is the total number of students enrolled? 46

7.

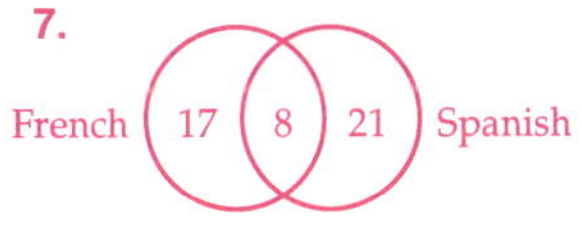

Lesson Follow-up

Reteaching Activity Display index cards containing one term each (for example, 8x, -3x, 2x). Ask students to pick two like terms, model them with tiles, then "simplify" the models.

LESSON RESOURCES

TRB Practice B-7
TRB Enrichment B-8

Practice

For use after 2-4 (pp. 66–69)

Simplifying Variable Expressions

Complete.

	Number of Terms	Numerical Coefficients	Like Terms
1. $3x + 7$	2	3	None
2. $4m + (-3n) + n$	3	4, -3, 1	-3n, n
3. $6kp + 9h + kp - 14$	4	6, 9, 1	6kp, kp
4. $-8y + 6ab + 7 - 3ba$	4	-8, 6, -3	6ab, -3ba
5. $c + 2c + c - 5c + 1$	5	1, 2, 1, -5	c, 2c, c, -5c

Name the property that you can use to simplify each expression.

6. $7k + (-3k)$ distributive
7. $5x + (3x + 7)$ distributive or associative (add.)
8. $11(12p)$ associative (mult.)
9. $-9y + 2y + 7$ distributive
10. $2m - 3m - 8n + n$ distributive
11. $2x + (x + y)$ associative (add.) or distributive

Combine the like terms by using the distributive property.

12. $12x - 3x$ 9x
13. $19a + 11a$ 30a
14. $85k + 36h + 13 - 9h$ 85k + 27h + 13
15. $7p + 6p + 12p - 5p$ 20p
16. $-5a + 2a - 10 + 2$ -3a − 8
17. $-33 - x + 33x + 1$ -32 + 32x

Complete with the appropriate term, number, or variable.

18. $5(n + 4) + 9n = 5 \cdot$ n + 5 $\cdot 4 + 9n$
= 5n + 20 $+ 9n$
$= 5n + 9n + 20$
= (5 $+ 9)n + 20$
= 14 $n + 20$

Simplify each expression.

19. $16 + 7y - 8$ 8 + 7y
20. $18m - 7 + 12m$ 30m − 7
21. $5(3t) - 7(2t)$ t
22. $2x - 9y + 7x + 20y$ 9x + 11y
23. $3(9k - 4) - 4(5n - 3)$ 27k − 20n
24. $6(g - h) - 6(g - h)$ 0
25. $-21(a + 2b) + 14a - 9b$ -7a − 51b
26. $-7a + 3(a - c) + 5c$ -4a + 2c
27. $-2(-5)q - (-72)(-q)$ -62q
28. $3(x + y) - 5x$ 3y − 2x

Chapter 2 B-7

Enrichment

For use after 2-4 (pp. 66–69)

Find the Speed

In 1988, Captain Elden Joersz set the world aircraft speed record. On July 28, 1976, he flew a Lockheed SR-71 at Beale Air Force Base in California. How fast did Captain Joersz fly?

To find out, simplify each expression. Find the answer in the digital read-out below and shade the region containing the answer.

1. $9x + (-5x)$ 4x
2. $-7y + 12y$ 5y
3. $-17p + 12p - 11p$ -16p
4. $3(2a + 5a)$ 21a
5. $4(3f) - 8f$ 4f
6. $n(8 + 6 - 19)$ -5n
7. $-2(4x) + 7x$ -x
8. $7(3f - 2g) + 14(g - f)$ 7f
9. $-15y + 11y - 9y$ -13y
10. $-3[12p + (-7p)]$ -15p
11. $-2n(5 - 7) + 3(3n)$ 13n
12. $6(b - a) + 2(4a - 3b)$ 2a
13. $-9(4f - 6f) + (-15f)$ 3f
14. $8(5 - n) - 19 + 17n - 21$ 9n
15. $8(3x - 2y) + 4[(-3x) + 4y]$ 12x
16. $-5p + 3p - 2p - 9p + 7p - 3p$ -9p
17. $-9(7y - 8y) + 12(6y - 7y)$ -3y
18. $8(5a - 4n + 3p) - 4(6p - 8n + 11a)$ -4a

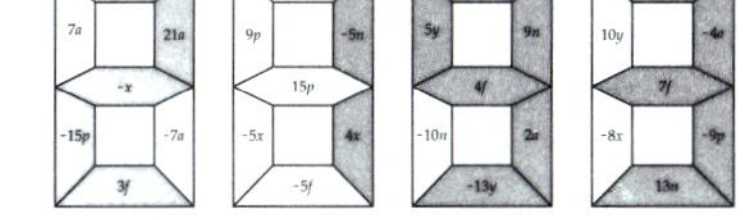

Answer: 2 1 9 3 mi/h

Write a summary of the results from the activity above.

19. The Lockheed SR-71 flew at a speed of 2,193 mi/h.

B-8 Chapter 2

LESSON QUIZ

State the number of terms. Name the numerical coefficients and the like terms.

1. $4a + 3ab - b - 2a$ (4; 4, 3, 1, 2; 4a and 2a)

Simplify and evaluate when $x = -2$ and $y = 4$.

2. $6(x + 2y) - 2(-x + y)$ (8x + 10y; 24)

Assignment Guide

Basic 1-6, 9-10, 12-13, 18-19, 21-24, 29, 36-38, 41, MR, TY All

Average 5-7, 10-11, 14-15, 19-21, 25-31, 38-39, 41-42, MR, TY All

Enriched 7-8, 11, 16-17, 19-21, 28-35, 38-40, 41-42, MR, TY All

FOR THE NEXT LESSON

algebra tiles or white and blue paper squares, red paper rectangles, math journal

Lesson Focus

Materials

Algebra tiles or white and blue paper squares and red paper rectangles, math journal

Vocabulary/Symbols

isolate the variable

MOTIVATING THE LESSON

Begin the lesson by having students explain the difference between an expression and an equation, and tell what a variable represents.

Instruction

1 Ask students how they can tell whether an algebra tile represents a positive or negative number (by the color). Then ask how they can tell that the models represent open equations. (The models contain a red tile, which indicates a variable, and a vertical bar, which indicates an equal sign.)

Suggest that students use their math journals to record notes of the discussions and answers to the questions.

2 Ask students why they think the goal in solving an equation is to isolate the variable. (The variable represents an unknown value, so when the variable is isolated, as in the equation $x = 4$, the unknown value is identified and the solution becomes apparent.) Then ask why two negative tiles were put on the left side of the model. (to make the numerical term zero)

OBJECTIVE:
Use models to solve addition and subtraction equations.

Exploring Equations

MATERIALS

- Algebra tiles or two different color squares of paper and rectangles of a third color
- Math journal to record work

▼ You can model an addition equation using algebra tiles. Use rectangles for variables and squares for positive and negative integers.

Equation 1: Equation 2:

1

1. ***Discuss*** What does the vertical bar represent? =
2. ***Write*** an equation for each model. $x + 3 = 5$, $x + 3 = -3$
3. ***Model*** each equation. Check students' work.

   a. $x + 3 = 5$ b. $z + 2 = -6$
   c. $y + 1 = 4$ d. $-3 = a - 4$
   e. $2b + 2 = 8$ f. $3 + 3x = -6$

▼ One way to find the solution of an equation is to *isolate* the variable. To isolate the variable, you use operations and properties to get the variable alone on one side of the equal sign.

2

4. Model Equation / Isolate Variable / Solution

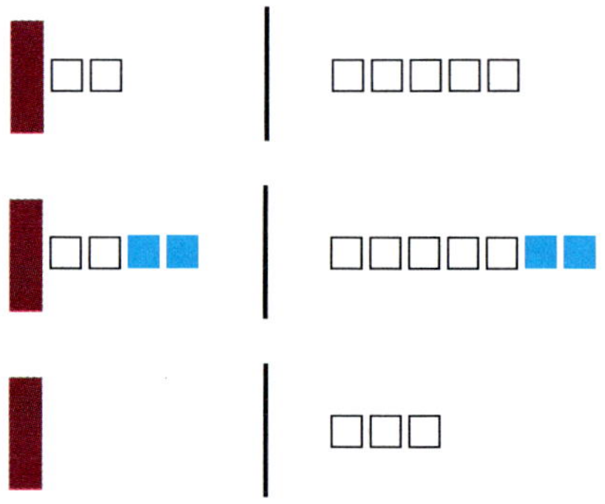

   a. ***Write*** What is the equation modeled in Step 1? $x + 2 = 5$
   b. ***Analyze*** What was done to isolate the variable? -2 was added to each side.
   c. What property of integers was used to isolate the variable? property of opposites
   d. What is the solution to the equation? 3
   e. Show mathematically what is represented in Steps 1–3.

4. e. $x + 2 = 5$
$x + 2 + (-2) = 5 + (-2)$
$x = 3$

5. ***Model*** and solve each equation. Check students' work.

   a. $x + 3 = 6$ 3 b. $4 + y = -7$ -11 c. $-3 = w + 2$ -5

6. ***Describe*** another way you could isolate a variable in an addition equation. Subtract the same amount from each side of the equation.

Teaching Tip

The overhead projector is useful for demonstrating lessons that use small materials. Allow students to use the overhead projector to share their use of models with the entire class.

▼ You can also model subtraction equations with algebra tiles.

$y - 3 = 4$ $\qquad$ $-3 = y - 2$

WRITE $y + (-3) = 4$ $\qquad$ **WRITE** $-3 = y + (-2)$

3

7. Why were the subtraction equations written as addition equations? Explain why you can write $y - 3 = 4$ as $y + (-3) = 4$.

8. ***Analyze*** How would you model each equation?
   a. $x - (-5) = 2$ $x + 5 = 2$ $\qquad$ b. $x - (-3) = 5$ $x + 3 = 5$

7. Addition equations can always be modeled, subtraction equations cannot; adding a negative is the same as subtracting.
8. Change each subtraction equation to an addition equation.

▼ You can also solve subtraction equations using algebra tiles.

$$x - 3 = 5$$

9. How would you model this equation? Check students' models. $x + (-3) = 5$
   a. What do you need to add to both sides of the equation to isolate the variable? 3
   b. What is the solution to the equation? Compare your solution with those of other students in the class. 8

10. Model and solve each equation. Check students' work.
   a. $x - 4 = 2$ 6 $\qquad$ b. $-5 = y - 3$ -2
   c. $-6 = z - 3$ -3

11. Is the equality of an equation affected when you add or subtract the same value on both sides? no

3 To help students understand how to model a subtraction equation, have students try to model $y - (-3) = 4$. They should see that it is not possible to model "$-(-3)$." As a result, the subtraction equation has to be written as an addition equation.

Closure

Ask students to describe how they could find the solution of an equation such as $x + 3 = 8$. Have them choose whether to explain using models or using a verbal description.

Once students understand how to model equations and what isolating the variable entails, they will be ready to explore solving equations.

Assignment Guide
Basic All
Average All
Enriched All

FOR THE NEXT LESSON
algebra tiles or blue and white paper squares and red paper rectangles, overhead projector, math journal

Lesson Focus

Materials

Algebra tiles or blue and white paper squares, red paper rectangles, overhead projector, math journal

Vocabulary/Symbols

Addition Property of Equality
equivalent equations
inverse operations
Subtraction Property of Equality

MOTIVATING THE LESSON

Give students the *true equation* $2 + 3 = 7 - 2$. Ask them to subtract the quantities 4, 3, 2, and 1, one at a time, from both sides of the equation and determine whether the equation is true in each case. Then have students add each of the quantities to both sides of the equation and determine if the equation is true in each case.

Instruction

1 Have students use tiles to model the equations. Ask what happens when 6 tiles are removed from each side. (The value of the variable a is known.) Then ask what operation represents removing 6 tiles (subtracting 6 tiles).

2 Point out that the Subtraction Property of Equality was used when the solution to the equation $p + 6 = 9$ was modeled.

OBJECTIVE: *To solve one-step equations involving addition and subtraction.*

2-5 Addition and Subtraction Equations

In the sixteenth century, only six planets were known to exist. By the twentieth century, all nine planets had been discovered. How many planets were discovered between the sixteenth and twentieth centuries?

If we let p represent the number of planets, we can describe the situation by an equation.

$$p + 6 = 9$$

THINK Is there another way to isolate the variable? Add 6 negative tiles to each side.

You can solve the equation $p + 6 = 9$ using a model.

1

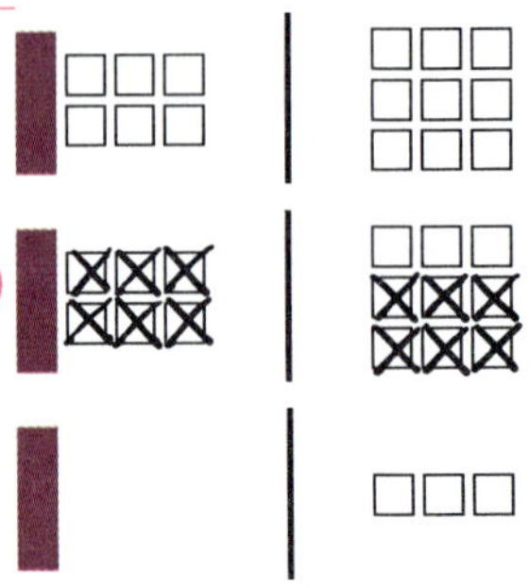

Represent the equation.

Remove 6 tiles from each side to isolate the variable.

p has a value of 3.

Three planets were discovered.

Addition and subtraction are *inverse operations*. You can use subtraction to undo addition. When you subtract the same quantity from both sides of an equation, the result is an *equivalent equation*.

2

Subtraction Property of Equality	You can subtract the same value from both sides of an equation.	
	Arithmetic	**Algebra**
	$9 = 9$	If $a = b$,
	$9 - 4 = 9 - 4$	then $a - c = b - c$.

Teaching Tip

Have students make an algebra tile workmat that contains a vertical bar (which represents the equal sign). Students can use the workmat to model equations.

Example 1 Solve.

$a + 22 = 28$

Solution

$a + 22 = 28$

$a + 22 - 22 = 28 - 22$ Subtract 22 from each side.

$a + 0 = 6$

$a = 6$

Check $a + 22 = 28$

$6 + 22 = 28$ Replace a with 6.

$28 = 28$ ✓ True, so 6 is the solution.

FLASHBACK
A solution makes an open equation true.

▼ To solve an equation involving subtraction, add the same value to both sides.

Addition Property of Equality

You can add the same value to both sides of an equation.

Arithmetic	**Algebra**
$5 = 5$	If, $a = b$,
$5 + 3 = 5 + 3$	then $a + c = b + c$.

Example 2 Solve.

$b - 12 = 59$

Solution

$b - 12 = 59$

$b - 12 + 12 = 59 + 12$ Add 12 to each side.

$b + 0 = 71$

$b = 71$

Check $b - 12 = 59$

$71 - 12 = 59$ Replace b with 71.

$59 = 59$ ✓ True, so 71 is the solution.

THINK How would you graph a solution of 71 on the number line?

▼ If you remember that the sum of a number and its opposite is zero, you can solve equations in a different way.

Example 3 Solve $x - 3 = -2$ using a model.

Solution

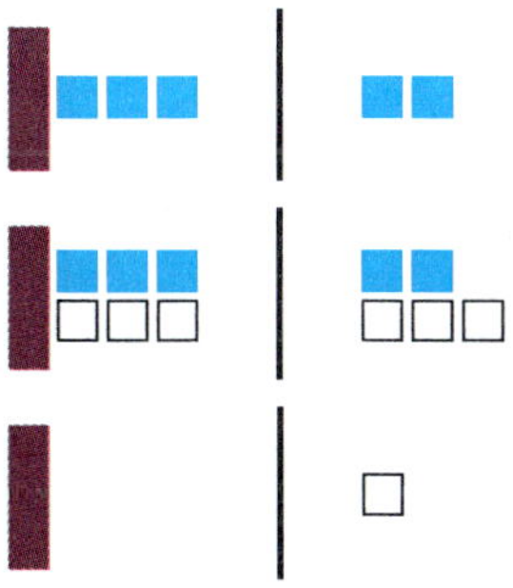

Represent the equation.

Add positive three to each side.

x has a value of 1.
The check is left for you.

FLASHBACK
$a + (-a) = 0$ and
$-a + a = 0$.

FLASHBACK
□■ represents zero.

3 Ask students to explain why 22 was subtracted from both sides of the equation. (Subtracting 22 undoes adding 22 and isolates the variable.) Help students understand the importance of checking solutions to equations.

4 Write the equation $a - 2 = 9$ on the chalkboard or overhead projector and ask students what will undo subtracting 2. (Adding 2.) Summarize both equality properties by pointing out that for a given equation, the same number can be added to (or subtracted from) both sides of the equation. The result will be an equivalent equation.

ADDITIONAL EXAMPLES

Solve each equation using the subtraction property.

1. $y + 28 = 93$ (65)

2. $217 + n = 201$ (-16)

Solve each equation using the addition property.

3. $y - 9 = -2$ (7)

4. $x - 310 = 288$ (598)

MATH MINUTES

Ask students to write five equivalent equations for each of these equations: $15 = 8 + n$ and $c - 19 = -4$. Have students compare results to discover how many different equivalent equations were found.

Guided Practice

Class Exercises It may be helpful for students to use models for questions 7, 9, and 10.

Think and Discuss In question 2, students should understand that either method (addition or subtraction) can be used to solve for *x*.

A ***common error*** in solving equations is to forget to add the same quantity to (or subtract the same quantity from) both sides of an equation. An excellent form of remediation is to encourage students to check solutions by substitution.

Closure

Writing in Math Have students write a description in their math journals of how they would solve the following equations:
$4 + v = 9 \quad -3 = c - 8$

Independent Practice

For questions 6-11, ask students who are having difficulty what they must do to isolate the variable in each equation.

For students who have difficulty solving equations, refer them to the calculator activity on p. 52E of the Chapter Overview. See also TRB Exploring O-7.

1. Adding a negative is the same as subtracting.
2. Answers will vary.
3. commutative, associative, and additive identity

Example 4 Solve $x + 4 = -5$ using opposites.

Solution

$$x + 4 = -5$$
$$x + 4 + (-4) = -5 + (-4) \quad \text{Add -4 to each side.}$$
$$x + 0 = -9$$
$$x = -9 \quad \text{The check is left for you.}$$

THINK AND DISCUSS See above.

1. Why is subtracting 8 from both sides of the equation $x + 8 = 17$ equivalent to adding -8 to both sides?
2. Which method do you prefer to use in solving an equation of the form $x + a = b$? Explain.
3. What properties do you need to use to show that $117 + n - 117$ is equal to n?

CLASS EXERCISES

Use a model to solve each equation. Graph each solution. Check students' work.

1. $6 + b = 9$ $b = 3$ **2.** $-3 = n - 4$ $n = 1$

State the first step in solving each equation.

3. $a + 8 = 12$ Subtract 8 from both sides. **4.** $54 + x = 98$ Subtract 54 from both sides. **5.** $34 = x - 19$ Add 19 to both sides. **6.** $-900 = 365 + x$ Subtract 365 from both sides.

Solve each addition equation.

7. $x + 35 = 15$ -20 **8.** $450 = x + 325$ 125

Solve each subtraction equation.

9. $x - 34 = 20$ 54 **10.** $-25 = b - 10$ -15

WRITTEN EXERCISES

Use a model to solve each equation. Check students' work.

1. $x + 5 = 7$ $x = 2$ **2.** $b - 4 = 3$ $b = 7$ **3.** $-6 = w - 4$ $w = -2$

Write and solve the equation represented by each model.

4. 
$x + 3 = -2$; -5

5. 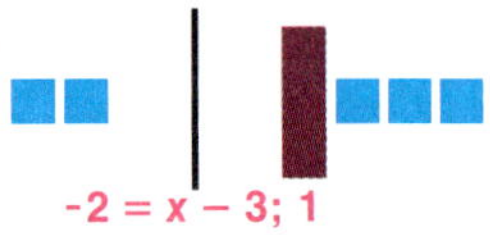
$-2 = x - 3$; 1

Solve each equation using the subtraction property.

6. $c + 9 = 37$ 28 **7.** $b + 24 = 19$ -5 **8.** $65 = n + 34$ 31

9. $-47 = 7 + y$ -54 **10.** $-45 = x + (-3)$ -42 **11.** $298 + n = 924$ 626

Solve each equation using the addition property.

12. $b - 15 = -9$ 6 **13.** $43 = g - 39$ 82 **14.** $x - 366 = -415$ -49

15. $-27 = w - 14$ -13 **16.** $-34 = c - 12$ -22 **17.** $8{,}923 = r - 1{,}298$ 10,221

Solve each equation using opposites.

18. $x - 19 = 34$ **53** **19.** $13 + c = 54$ **41** **20.** $432 = m - 391$ **823**

21. $48 = x + 9$ **39** **22.** $c - 42 = 12$ **54** **23.** $w + 3 = -8$ **-11**

MENTAL MATH Solve each equation.

24. $130 = 30 + s$ **100** **25.** $x + 800 = 500$ **-300** **26.** $95 = x - 15$ **110**

27. $-45 = b - 45$ **0** **28.** $25 = x + 425$ **-400** **29.** $r - 316 = -8$ **308**

CALCULATOR Solve each equation.

30. $x + 49{,}023 = 15{,}911$ **-33,112** **31.** $265{,}970 = b - 1{,}098{,}645$ **1,364,615**

32. $398{,}452 = x + 799{,}376$ **-400,924** **33.** $c - 36{,}000 = 41{,}098$ **77,098**

Solve each equation.

34. $v - 493 = 513$ **1,006** **35.** $400 + x = 900$ **500** **36.** $c + (-90) = -58$ **32**

37. $56 = c - 9$ **65** **38.** $-5 = -5 + n$ **0** **39.** $-25 = -5 + n$ **-20**

40. $32 + a = -32$ **-64** **41.** $2{,}314 = k + 716$ **1,598** **42.** $e + (-43) = -45$ **-2**

43. $34 + n + 12 = 78 - 7$ **25** **44.** $n - 29 + (-16) = -24$ **21**

Complete.

45. If $x + a = b$, then $x = ■$. **b − a** **46.** If $x - a = b$, then $x = ■$. **b + a**

Write an equation for each sentence. Solve for the variable.

47. A number d plus five is equal to seventeen. **d + 5 = 17; d = 12**

48. Negative five is the same as x minus eight. **-5 = x − 8; x = 3**

49. ***DATA FILE 2 (pp. 52–53)*** The number of decibels in a whisper is d decibels less than the number of decibels in a rocket launch. **20 + d = 150; d = 130**

50. Three hundred twenty-three is negative one hundred fifty-five plus y. **323 = -155 + y; y = 478**

51. ***DATA FILE 2 (pp. 52–53)*** The speed of sound travels through steel h m/s faster than it travels through silver. **5,200 = 2,680 + h; h = 2,520**

52. Thirty-three more than m is the same as negative seventeen. **m + 33 = -17; m = -50**

53. ***DATA FILE 4 (pp. 138–139)*** The number of people who speak Chinese is p people more than the number who speak German. **700 = 119 + p; p = 581 million**

54. Fifty-four less than a number k is equal to negative twenty-nine. **k − 54 = -29; k = 25**

55. ***PROJECT*** Visit your town hall. Research the population of your town or city. Has the population increased or decreased? by how many people? **Check students' work.**

MIXED REVIEW

Find the number of terms.

1. $5w - 3z + 2w$ **3**
2. $8 + 9a + 3(2a)$ **3**

Simplify.

3. $9x - 5x + 4$ **4x + 4**
4. $5q + 8 - 3q + (-2)$ **2q + 6**
5. $27 \div (-3) + 4(3)$ **3**
6. $|3(-8) + 6 \cdot 2|$ **12**

Solve.

7. George studied math for 1 min the first week of school, 2 min the second, and 4 min the third. He continued to double his study time each week. How many minutes did he study the tenth week? **512 min**

LESSON QUIZ

Solve each equation.

1. $n - 9 = -13$ (-4)
2. $2 + a = -3$ (-5)
3. $-92 = b - 92$ (0)
4. $-23 = a + 21$ (-44)
5. $200 = s - 150$ (350)

Assignment Guide

Basic 1-7, 12-13, 18-26, 30-37, 45-50, MR All

Average 2-10, 14-21, 26-32, 38-46, 49-52, 55, MR All

Enriched 3-5, 10-17, 22-33, 40-46, 51-55, MR All

FOR THE NEXT LESSON

algebra tiles or white and blue squares of paper and red paper rectangles, math journal

Lesson Follow-up

Reteaching Activity Make several drawings of pan balances on the chalkboard or overhead projector. Write a series of equations for each pan balance drawing. Ask different students to write above each pan what they would do to solve the equation (or balance the pan). Have other students check that the pans "balance" and complete the solution.

LESSON RESOURCES

TRB Practice B-9
TRB Enrichment B-10

Practice

For use after 2-5 (pp. 72–75)

Addition and Subtraction Equations

Solve each equation using the subtraction property of equality.

1. $x + 14 = 21$ **x = 7** 2. $31 = p + 17$ **p = 14**
3. $-19 = k + 9$ **k = -28** 4. $87 + y = 19$ **y = -68**
5. $36 + n = 75$ **n = 39** 6. $-176 = h + (-219)$ **h = 43**

Solve each equation using the addition property of equality.

7. $m - 17 = -8$ **m = 9** 8. $k - 55 = 67$ **k = 122**
9. $-44 + n = 36$ **n = 80** 10. $-36 = p - 91$ **p = 55**
11. $x - 255 = 671$ **x = 926** 12. $19 = c - (-12)$ **c = 7**

Solve each equation using opposites.

13. $84 = z + 37$ **z = 47** 14. $k - 55 = 23$ **k = 78**
15. $r + 7 = -16$ **r = -23** 16. $68 = p - 41$ **p = 109**
17. $144 + g = 78$ **g = -66** 18. $311 = y - 281$ **y = 592**

MENTAL MATH Solve each equation.

19. $-52 = -52 + k$ **k = 0** 20. $837 = p + 37$ **p = 800**
21. $x - 155 = 15$ **x = 170** 22. $180 = 80 + n$ **n = 100**
23. $2{,}000 + y = 9{,}500$ **y = 7,500** 24. $81 = x - 19$ **x = 100**
25. $111 + f = 100$ **f = -11** 26. $w - 6 = -16$ **w = -10**

CALCULATOR Solve each equation.

27. $e + 3{,}777 = 2{,}431$ **e = -1,346** 28. $45{,}709 = v - 78{,}666$ **v = 124,375**
29. $145{,}719 = y + 247{,}623$ **y = -101,904** 30. $m - 807{,}543 = -623{,}744$ **m = 183,799**
31. $x + 558{,}399 = 845{,}919$ **x = 287,520** 32. $3{,}725{,}444 = c + 4{,}210{,}500$ **c = -485,056**

Solve each equation.

33. $17 + m = 21$ **m = 4** 34. $y - 34 = 43$ **y = 77**
35. $t + 9 = -9$ **t = -18** 36. $15 = z + 6$ **z = 9**
37. $41 + k = 7$ **k = -34** 38. $1{,}523 + c = 2{,}766$ **c = 1,243**
39. $-88 + z = 0$ **z = 88** 40. $-33 + (-7) = 29 + m$ **m = -69**
41. $t + (-2) = -66$ **t = -64** 42. $-390 + x = 11 - 67$ **x = 334**

Chapter 2 B-9

Enrichment

For use after 2-5 (pp. 72–75)

Two Equations, One Solution

The equation $x + y = 16$ has many solutions.

$x = 9,\ y = 7 \longrightarrow x + y = 16$
$x = 22,\ y = -6 \longrightarrow x + y = 16$

The equation $x - y = 6$ also has many solutions.

$x = 13,\ y = 7 \longrightarrow x - y = 6$
$x = 24,\ y = 18 \longrightarrow x - y = 6$

Only one pair of numbers solves *both* equations: $x = 11$, $y = 5$

$x + y \stackrel{?}{=} 16$ $x - y \stackrel{?}{=} 6$
$11 + 5 = 16$ ✓ $11 - 5 = 6$ ✓

Answer these questions to find the pair of numbers that solves both of the following equations.

Equation 1: $x + y = 8$ Equation 2: $x - y = 20$

1. Find y in Equation 1 if $x = 12$. **-4**
   Find y in Equation 2 if $x = 12$. **-8**
2. Find y in Equation 1 if $x = 13$. **-5**
   Find y in Equation 2 if $x = 13$. **-7**
3. Find y in Equation 1 if $x = 14$. **-6**
   Find y in Equation 2 if $x = 14$. **-6**
4. Write the pair of numbers that solves both equations. $x =$ **14**, $y =$ **-6**

Complete the table to find the pair of numbers that solves both equations.

5. Equation 1: $x + y = -6$ Equation 2: $x - y = -12$

x	-7	-8	-9	-10	-11
y (Equation 1)	1	2	3	4	5
y (Equation 2)	5	4	3	2	1

Solution: $x =$ **-9** $y =$ **3**

6. Equation 1: $x + y = 12$ Equation 2: $x - y = 38$

x	21	22	23	24	25
y (Equation 1)	-9	-10	-11	-12	-13
y (Equation 2)	-17	-16	-15	-14	-13

Solution: $x =$ **25** $y =$ **-13**

B-10 Chapter 2

Lesson Focus

Materials

Algebra tiles or white and blue squares of paper and red paper rectangles, math journal

Vocabulary/Symbols

Division Property of Equality
Multiplication Property of Equality

MOTIVATING THE LESSON

On the chalkboard, write the equation 45 = 45. Ask students to divide both sides of the equation by 9 and to state their observations. (A true equation results.) Provide other examples of similar equations and have students either multiply or divide both sides by the same number. Ask if a true equation results each time. (yes.)

Instruction

1 Ask students to explain why each side was divided into two equal groups. (So that each group would contain a model to represent one m.) Have students verify that the value, 3, satisfies the equation.

2 Point out that if divisors were not restricted to nonzero values, the resulting equation would have no meaning since division by zero is not defined.

3 Discuss the importance of checking the solution by substituting the value of r in the original equation.

OBJECTIVE: *To solve one-step equations involving multiplication and division.*

2-6 Multiplication and Division Equations

▼ Mark and Sara each have the same amount of money to spend at the basketball game. If they have a total of \$6, how much does each have? Let m represent this amount. You can describe the situation by the following equation.

$$2m = 6$$

▼ You can solve a multiplication equation using a model.

Example 1 **Solve the equation $2m = 6$ using a model.**

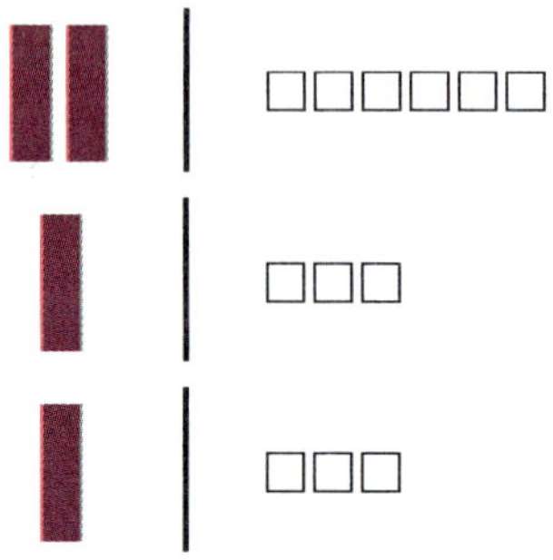

Represent the equation with a model.

1 Divide each side of the equation into two equal parts so each m has a value of 3.

Each person has \$3 to spend.

THINK Why must you divide by a nonzero value? **Division by zero is undefined.**

▼ Division undoes multiplication. When you divide both sides of an equation by the same nonzero number, the result is an equivalent equation.

2

Division Property of Equality	You can divide both sides of an equation by the same nonzero value.	
	Arithmetic	**Algebra**
	$9 = 9$	If $a = b$,
	$9 \div 3 = 9 \div 3$	then $a \div c = b \div c$.
	$\frac{9}{3} = \frac{9}{3}$	$\frac{a}{c} = \frac{b}{c}$, $c \neq 0$

THINK How do you know that $\frac{5r}{5}$ is the same as $5r \div 5$? **The division bar and the division sign mean the same thing.**

Example 2 **Solve $5r = -20$ using the division property of equality.**

Solution

$5r = -20$

$\frac{5r}{5} = \frac{-20}{5}$ Divide both sides by 5.

$r = -4$

3 **Check** $5r = -20$ Replace r with -4.

$5 \cdot (-4) = -20$ ✓

Teaching Tip

To develop reasoning skills, have students defend their answers. This is particularly important in working with equations, since some students apply rules by rote.

▼ To solve an equation involving division, multiply both sides by the same value.

Multiplication Property of Equality	You can multiply both sides of an equation by the same value.
	Arithmetic $12 = 12$, $12 \cdot 2 = 12 \cdot 2$ — **Algebra** If $a = b$, then $ac = bc$.

4

Example 3 Solve $\frac{x}{-9} = -3$ using the multiplication property of equality.

Solution

$$\frac{x}{-9} = -3$$

$$-9\left(\frac{x}{-9}\right) = -3 \cdot (-9) \quad \text{Multiply both sides by -9.}$$

$$x = 27$$

5

Check

$$\frac{x}{-9} = -3 \quad \text{Replace } x \text{ with 27.}$$

$$\frac{27}{-9} = -3$$

$$-3 = -3 ✓$$

CLASS EXERCISES

State the first step in solving each equation.

1. $6x = 96$ Divide both sides by 6.
2. $32 = c \div 3$ Multiply both sides by 3.
3. $\frac{r}{-5} = -4$ Multiply both sides by -5.

Solve each multiplication equation.

4. $8x = -48$ -6
5. $-2x = 12$ -6
6. $108 = 9x$ 12

Solve each division equation.

7. $\frac{v}{3} = 14$ 42
8. $-6 = n \div 4$ -24
9. $\frac{m}{-2} = -20$ 40

THINK AND DISCUSS See below.

1. Why is $1y = y$?
2. How would you solve the equation $\frac{25}{x} = 1$?
3. How are the procedures used to solve $3x = 9$ and $x + 3 = 12$ alike? How are they different?

1. $1 \cdot y = y$ by the multiplicative identity.
2. Multiply both sides by x.
3. Inverse operations are used; one is divided, the other subtracted.

WRITTEN EXERCISES

Use a model to solve each equation. Graph each solution. Check students' work.

1. $2g = 8$ g = 4
2. $10 = 2m$ m = 5
3. $4h = -12$ h = -3

MENTAL MATH Is -3 a solution of each equation?

4. $-6 = 2m$ yes
5. $\frac{b}{-3} = 1$ yes
6. $45p = 145$ no
7. $\frac{-18}{k} = -6$ no

MATH MINUTES

Ask groups of students to write as many equivalent equations as they can for the equations below.
$7x = -63$ $\frac{x}{-3} = 6$
Have the class compare and analyze the results.

4 Ask students what happens when they multiply both sides of an equation by zero. (The result is the true equation 0 = 0.) Point out that the Multiplication Property of Equality does not restrict factors to nonzero numbers since a true equation results when zero is the factor.

5 Ask students why -9 was chosen for the factor. (To undo division by -9.)

ADDITIONAL EXAMPLES

Solve each equation using the division property of equality.
1. $-2v = -24$ (12)
2. $102 = 3m$ (34)

Solve each equation using the multiplication property of equality.
3. $\frac{x}{8} = -5$ (-40)
4. $21 = \frac{m}{-3}$ (-63)

Guided Practice

Class Exercises It may be helpful if students state the first step in solving each of the equations in questions 1-9 before they solve them.

Think and Discuss Ask students what they can assume about x in question 2 and why. (x is a nonzero value; if x were zero it could not be used as a divisor.)

A ***common error*** is to divide one side of the equation by a number while multiplying the other side by the same number. Remind students that the same operation and number must be used on both sides of the equation. Encourage students to check their work for this error.

Closure

Writing in Math Ask students to describe the multiplication and division properties of equality in their math journals.

Independent Practice

Encourage students to check their solutions in questions 10-18 to be sure the solutions satisfy the original equations.

For questions 54-56, remind students that the absolute value of a number is never negative.

If students have difficulty with questions 57-60, encourage them to substitute values for variables *a* and *b* and use those values to help solve the equations.

Critical Thinking Questions 1-3 are designed to help students develop a plan for deciding what the characteristics are of the missing piece of the puzzle.

MIXED REVIEW

Solve.

1. $-4 = a + 7$ **-11**
2. $n - 5 = 12$ **17**
3. $t - (-4) = -15$ **-19**

Write the expression.

4. Subtract 3 from *a*, then add *b*. **$a - 3 + b$**
5. Multiply 7 times the difference of 9 and *w*. **$7(9 - w)$**
6. The absolute value of the sum of -8 and *q*. **$|-8 + q|$**

Solve.

7. The number of seconds it takes the sound of thunder to reach you is about five times the number of miles between you and lightning. How long does it take the sound to reach you if lightning is 12 mi away? **60 s or 1 min**

Write and solve each equation represented by the model.

8. 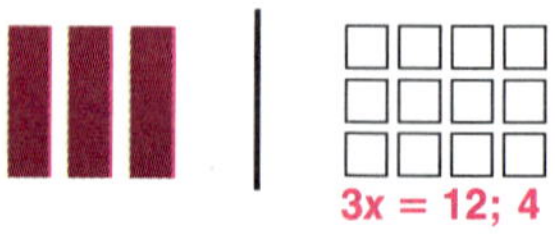
$3x = 12$; 4

9. 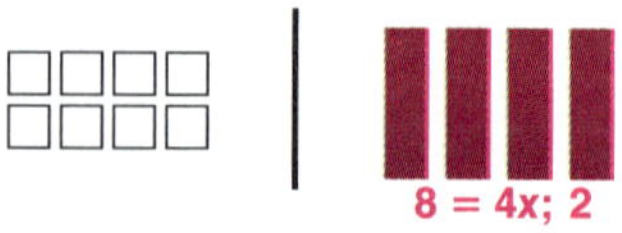
$8 = 4x$; 2

Solve each equation using the division property.

10. $4a = 28$ **7**
11. $-2b = 30$ **-15**
12. $-45 = 9a$ **-5**
13. $15c = 90$ **6**
14. $5w = 95$ **19**
15. $-28 = 7m$ **-4**
16. $-10d = 100$ **-10**
17. $125 = 25d$ **5**
18. $-35 = 5n$ **-7**

Solve each equation using the multiplication property.

19. $\frac{m}{4} = 13$ **52**
20. $\frac{b}{-6} = 20$ **-120**
21. $-2 = d \div 8$ **-16**
22. $f \div 3 = -4$ **-12**
23. $-50 = \frac{n}{-6}$ **300**
24. $9 = \frac{n}{8}$ **72**
25. $\frac{w}{12} = -2$ **-24**
26. $13 = n \div -4$ **-52**
27. $7 = \frac{y}{6}$ **42**

MENTAL MATH **Solve each equation.**

28. $20b = 2{,}000$ **100**
29. $\frac{v}{-50} = 300$ **-15,000**
30. $75m = -7{,}500$ **-100**
31. $3{,}823 = \frac{s}{100}$ **382,300**

CALCULATOR **Solve each equation.**

32. $358c = 80{,}550$ **225**
33. $x \div (-392) = 108$ **-42,336**
34. $4{,}523 = \frac{n}{-921}$ **-4,165,683**
35. $-48z = 76{,}128$ **-1,586**

Solve each equation using any method.

36. $4c = -36$ **-9**
37. $\frac{s}{-32} = 24$ **-768**
38. $-84t = 0$ **0**
39. $-88 = 11w$ **-8**
40. $6c = -96$ **-16**
41. $56 = f \div 9$ **504**
42. $-34c = 34$ **-1**
43. $15n = 225$ **15**
44. $-364 = \frac{c}{-3}$ **1,092**
45. $16s = 496$ **31**
46. $f \div 31 = 27$ **837**
47. $352 = 32v$ **11**
48. $25y = 500$ **20**
49. $0 = \frac{u}{254}$ **0**
50. $43x = 4{,}257$ **99**

51. A honeybee hive contains 35,000 cells. How many cells are there in 25 honeybee hives? **875,000**
52. ***DATA FILE 10 (pp. 404–405)*** How many acres does one gray wolf require to survive? **76,194 acres**
53. Write an expression for the number of eggs in *d* dozen. **12*d***

For what values of x is each equation true?

54. $|x| = 7$ **-7, 7** **55.** $\frac{|x|}{3} = 2$ **-6, 6** **56.** $2|x| = 8$ **-4, 4**

Solve each equation for x.

57. $ax = b$ $\frac{b}{a}$ **58.** $\frac{x}{a} = b$ **ba**

59. $x - a = b$ **b + a** **60.** $x + a = b$ **b − a**

Critical Thinking

EXPLORING VISUAL THINKING

Find the missing piece.

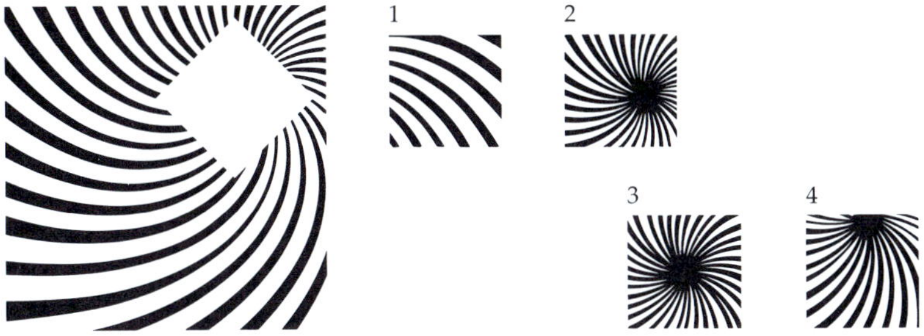

To solve this problem, you must develop a plan. What characteristics does the missing piece have? **Check students' work.**

1. What happens to the lines of the pattern as they swirl inward? **they get thinner**
2. At some point the lines will form a dot. Where will the dot be? **where all of the lines meet**
3. Is all of the dot on the missing piece or is part of it on the design? **All of the dot is on the missing piece.**
4. Based on the characteristics, which piece is the best choice? **2**
5. Here is another puzzle. Which figure will connect with part A to form a circle? Describe how you know. **3**

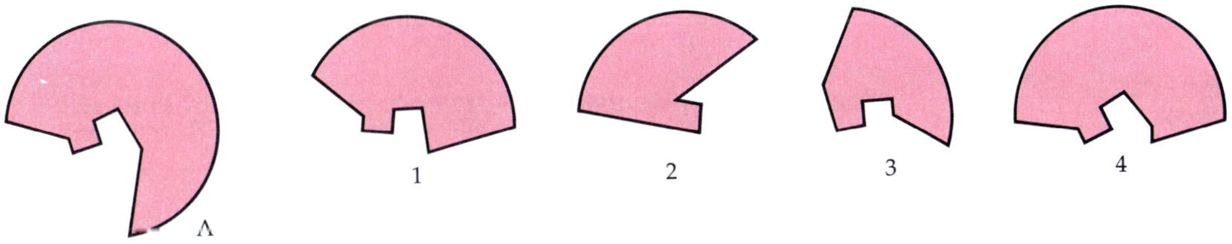

Lesson Follow-up

Reteaching Activity Provide a worksheet containing several equations in one column and the solutions in another column. Have students match each equation with its solution. Ask them to tell what they must do to find the solution. If you prefer, assign the Exploring Activity described on page 52E of the Chapter Overview.

LESSON RESOURCES

TRB Practice B-11
TRB Enrichment B-12

Practice

For use after 2-6 (pp. 76–79)

Multiplication and Division Equations

Solve each equation using the division property of equality.

1. $7m = 35$ **m = 5** 2. $-3x = 18$ **x = -6** 3. $-56 = 8y$ **y = -7**
4. $90 = 10k$ **k = 9** 5. $8p = -8$ **p = -1** 6. $-4s = -32$ **s = 8**
7. $100 = -20n$ **n = -5** 8. $14h = 42$ **h = 3** 9. $-175 = 25g$ **g = -7**
10. $-87{,}654y = 0$ **y = 0** 11. $-42 = 6m$ **m = -7** 12. $-2x = 34$ **x = -17**

Solve each equation using the multiplication property of equality.

13. $\frac{b}{8} = -3$ **b = -24** 14. $\frac{k}{-5} = -5$ **k = 25** 15. $-3 = \frac{n}{7}$ **n = -21**
16. $1 = \frac{n}{14}$ **n = 14** 17. $\frac{x}{12} = 0$ **x = 0** 18. $-6 = \frac{m}{-2}$ **m = 12**
19. $\frac{p}{15} = 5$ **p = 75** 20. $\frac{y}{-4} = -12$ **y = 48** 21. $\frac{s}{30} = 6$ **s = 180**
22. $\frac{1}{4}m = -12$ **m = -48** 23. $\frac{1}{9}z = 0$ **z = 0** 24. $-\frac{m}{55} = 1$ **m = -55**

MENTAL MATH Solve each equation.

25. $\frac{w}{100} = -24$ **w = -2,400** 26. $50k = 500$ **k = 10**
27. $-30m = 30{,}000$ **m = -1,000** 28. $\frac{p}{500} = -5$ **p = -2,500**
29. $-40m = 4{,}000$ **m = -100** 30. $\frac{n}{60} = -6$ **n = -360**

CALCULATOR Solve each equation.

31. $511 = \frac{x}{-23}$ **x = -11,753** 32. $-144k = -9{,}360$ **k = 65**
33. $\frac{n}{357} = -266$ **n = -94,962** 34. $915y = 70{,}455$ **y = 77**
35. $58y = 12{,}006$ **y = 207** 36. $\frac{b}{473} = 68$ **b = 32,164**

Solve each equation using any method.

37. $54 = -18v$ **v = -3** 38. $\frac{x}{-9} = -11$ **x = 99** 39. $216 = 9w$ **w = 24**
40. $15 = \frac{k}{4}$ **k = 60** 41. $-17v = -17$ **v = 1** 42. $-161 = 23t$ **t = -7**
43. $\frac{f}{71} = 0$ **f = 0** 44. $56h = 3{,}136$ **h = 56** 45. $20 = \frac{e}{-25}$ **e = -500**
46. $-12 = \frac{j}{-3}$ **j = 36** 47. $4{,}200 = 30x$ **x = 140** 48. $\frac{y}{-21} = -21$ **y = 441**
49. $382 = 2x$ **x = 191** 50. $\frac{m}{-3} = 21$ **m = -63** 51. $4{,}000 = \frac{-x}{40}$ **x = -160,000**

Chapter 2 B-11

Enrichment

For use after 2-6 (pp. 76–79)

Free Fall

Because of gravity, a free falling object falls faster and faster, or *accelerates*, as it descends. The distance (d) in feet that an object falls in t seconds is given by the equation $d = 16t^2$. (Recall that t^2 means $t \cdot t$.)

1. A penny thrown into a wishing well takes 3 s to reach the bottom. How deep is the well? **144 ft**
2. A rivet that falls off an airplane falls for 42 s before hitting the ground. How high is the airplane? **28,224 ft**
3. Complete the table showing how far an object falls each of the first 8 s.

t (sec)	1	2	3	4	5	6	7	8
d (ft)	16	64	144	256	400	576	784	1,024

4. a. How far does an object fall during the fifth second (from $t = 4$ to $t = 5$)? **144 ft**
   During the sixth second? **176 ft**
   During the seventh second? **208 ft**
   b. How much farther does an object fall during the eighth second than it falls during the seventh second? **32 ft**
   During the fifth second than during the fourth second? **32 ft**
   During the third second than during the second second? **32 ft**

You can use the division property to write a new free-fall equation.

5. Complete: If $d = 16t^2$, then $t^2 =$ $\frac{d}{16}$

Use your new equation to solve. You may wish to use a calculator.

6. A rock fell from the top of a 2,304-ft cliff to the base of the cliff. How long did the rock fall? **12 s**

B-12 Chapter 2

LESSON QUIZ

Solve each equation.

1. $-8x = 64$ (-8)
2. $92 = \frac{m}{3}$ (276)
3. $-356 = 4t$ (-89)
4. $\frac{t}{-8} = -301$ (2,408)

Assignment Guide

Basic 1-12, 19-21, 28-41, 52, 54-58, MR, CT All

Average 2-15, 22-24, 28-35, 42-47, 52-59, MR, CT All

Enriched 3, 7-9, 16-18, 25-35 odd, 48-50, 52-60, MR, CT All

FOR THE NEXT LESSON

math journal, index cards

Lesson Focus

Materials

Math journal, index cards

MOTIVATING THE LESSON

Ask students to write variable expressions for the following phrases: 4 more than a number m $(m + 4)$; 3 less than a number w $(w - 3)$; the product of -4 and the quantity t plus 6 $[-4(t + 6)]$.

Instruction

1 Remind students that they are using a known formula (for finding area) to write this equation. Point out that formulas are not available for all word problems. In most cases, students translate information in the problem to write an equation.

2 Comment on the importance of clearly identifying the variable before writing an equation to fit a word problem. Remind students that the word *is* often means *equals*.

3 Ask students to make up another possible word problem to fit the equation.

ADDITIONAL EXAMPLES

Write an equation.

1. Three times the quantity x plus five is negative ten. $[3(x + 5) = -10]$

2. The area of a rectangle with length 34 m and width w is 272 m^2. $(34w = 272)$

3. Write a word problem for the equation $x - 342 = 108$.

OBJECTIVE: *To write an equation for a word problem or model.*

2-7 Writing Equations

▼ A group of artists is painting a large outdoor mural. The length of the mural is 410 ft. The mural has an area of 28,700 ft². How could you find the width of the painting?

To find the width, you can write an equation.

1

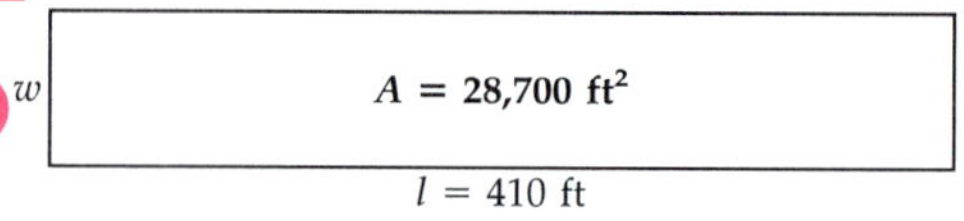

Sketch a diagram when appropriate.

Area = *length* · *width*

↕ ↕

$410 \cdot w = 28{,}700$ — Let w equal the width.

▼ To solve a word problem, you need to translate a sentence into an equation.

Example 1 **Write an equation.**

One more than three times the number of students in the class is equal to sixty-four.

Solution In this situation, we need to find the number of students.

2

Let s = *number of students.*	Identify the variable.
$3s$ = *three times the number of students.*	Include the coefficient.
$3s + 1$ = *one more than three times the number of students.*	Write the variable expression.
$3s + 1 = 64$	Write the equation.

▼ Before you write an equation decide on a variable to represent one unknown. Then write the other unknowns in terms of that variable.

THINK What if you let m = Joan's money? How would you write the expression for Eve's money? $m \div 2$

Example 2 **Write an equation.**

Joan collected twice as much money as Eve for the walkathon. Together they collected \$120.

Solution In this situation, we need to find the money that each collected.

Let m = Eve's money	Identify the variable.
$2m$ = Joan's money	Write any like terms.
$m + 2m = 120$	Write the equation.

Teaching Tip

Encourage students to consider alternative ways of answering questions. Have them rewrite their own equations or those of fellow students. Display all possibilities on the chalkboard so that students can see that there is often more than one way of finding a solution.

▼ Equations can represent real situations.

Example 3 **Write a word problem for the equation $12x = 496$.**

Solution One possible word problem is the following:

Janet makes 12 monthly payments for her automobile insurance. The yearly premium is \$496. How much will she pay per month?

3

CLASS EXERCISES

Choose the best equation for each problem. Do not solve.

1. Kendra uses 14 C/min while running. If she burned 154 C, how many minutes (m) did she run? b

   a. $m - 14 = 154$ b. $14m = 154$
   c. $m \div 154 = 14$ d. $m + 14 = 154$

2. Three less than the quantity $y - 7$ is equal to -6. c

   a. $y - 7 + 3 = -6$ b. $3 - (y - 7) = -6$
   c. $-6 = (y - 7) - 3$ d. $-6 = 3(y - 7)$

Write an equation for each problem. Do not solve.

3. The product of a number and 40 is equal to 360. $40n = 360$
4. The sum of 45 and some number is -30. $45 + n = -30$
5. A number decreased by -5 is 18. $n - (-5) = 18$
6. Three more than six times the number of books on the shelf is 63. $6n + 3 = 63$
7. The length of a rectangle is twice the width. The perimeter is 120. $120 = 2w + w + 2w + w$

Write a word problem for each equation. Answers will vary.

8. $500 = 10t$
9. $120 = 150 - x$

WRITTEN EXERCISES

Choose the best equation for each problem. Do not solve.

1. Suppose you travel 55 mi/h. How many hours (h) would it take you to go 275 mi? c

   a. $55 + h = 275$ b. $h \div 55 = 275$
   c. $55h = 275$ d. $275 = h - 55$

THINK AND DISCUSS See below.

1. Is there always just one correct equation that can be written for a given problem? Use an example to support your answer.
2. Are the equations $x + 5 = 10$ and $x = 10 - 5$ equivalent equations? Why or why not?

1. No; $x - (-2) = 6$ can also be written as $2 + x = 6$.
2. Yes; when 5 is subtracted from each side of the first equation it results in the second equation.

Guided Practice

Class Exercises For question 1, help students find the correct equation by asking them how to find the number of calories Kendra uses in 2 min, 3 min, and so on.

Think and Discuss Often one student will interpret a given situation as multiplication and another will interpret it as division. If the equations are equivalent, both are correct.

A ***common error*** is that students misinterpret the statement in a given word problem and write an incorrect equation. Have students write expressions for these statements: three more than a number; three times a number; three plus three times a number. ($x + 3$; $3x$; $3x + 3$)

Closure

Writing in Math Ask students to use their math journals to describe how they would write an equation to fit a word problem.

Independent Practice

For question 9, remind students that the perimeter of a triangle is the sum of its sides.

MATH MINUTES

Have students work in pairs. One student makes up a word problem and the other writes an equation for the problem. After they check their work, students reverse roles and repeat the process.

Lesson Follow-up

Reteaching Activity Write word problems on cards, one to a card. On a second set of cards, write equations for the problems. Have students match each problem with an appropriate equation.

LESSON RESOURCES

TRB Practice B-13
TRB Enrichment B-14

Practice

For use after 2-7 (pp. 80–82)

Writing Equations

Choose the best equation for each problem. Do not solve.

1. If you were to earn $6/h, how many hours ($h$) would it take you to earn $84? **c**
   a. $h + 6 = 84$ b. $\frac{h}{6} = 84$ c. $6h = 84$ d. $h = 84 - 6$
2. The quantity $k - 5$ is 4 more than -8. **b**
   a. $(k - 5) + 4 = -8$ b. $(k - 5) - 4 = -8$
   c. $4(k - 5) = -8$ d. $4 - (k - 5) = -8$

Write an equation using the given variable. Do not solve. **Equations may vary.**

3. Ernie scored 8 more points than Mike scored. Mike scored 16 points. How many points (p) did Ernie score? **$p - 8 = 16$**
4. One-fifth of a number (n) is equal to -7. **$\frac{n}{5} = -7$**
5. A bamboo tree grew 3 in. per day. How many days (d) did it take the tree to grow 144 in.? **$3d = 144$**
6. A truck driver drove 468 miles on Tuesday. That was 132 miles farther than she drove on Monday. How far (d) did she drive on Monday? **$d + 132 = 468$**

Write an equation for each problem. Then solve.

7. The product of a number and -7 is -91.
   Equation **$-7n = -91$** Solution **$n = 13$**
8. Carl's car averages 33 miles per gallon of gas. How much gas will he use driving 561 miles?
   Equation **$33g = 561$** Solution **17 gal**
9. The combined enrollment in the three grades at Jefferson Middle School is 977. There are 356 students in the seventh grade and 365 in the eighth grade. How many students are there in the ninth grade?
   Equation **$n + 356 + 365 = 977$** Solution **256 students**

Chapter 2 B-13

Enrichment

For use after 2-7 (pp. 80–82)

Discovering Variable Expressions

1. Complete the table by evaluating the variable expression $5x + 7$ for each of the given values of x.

x	1	2	3	4	5	6
$5x + 7$	12	17	22	27	32	37

2. By how much does the value of $5x + 7$ appear to increase each time the value of x increases by 1? **5**
3. Complete the table for the variable expression $8x - 3$.

x	1	2	3	4	5	6
$8x - 3$	5	13	21	29	37	45

4. By how much does the value of $8x - 3$ appear to increase each time the value of x increases by 1? **8**
5. Compare your answers to Exercises 2 and 4 with the variable expressions that produced them. What part of the variable expression could you have used to predict the answers? **the coefficient of x**
6. Suppose you were to complete a table like the ones above for the variable expression $6x + 11$. By how much would you expect the value of the expression to increase when the value of x increased by 1? **6**

The table gives the values of a variable expression for four values of x.

x	1	2	3	4
?	6	19	32	45

7. What is the numerical coefficient of x in the variable expression? **13**
8. According to the table, when $x = 1$, the value of the variable expression is 6. If you multiply your answer to Exercise 7 by 1, how much must you then subtract to obtain a value of 6? **7**
9. What is the variable expression used to create the table? **13x − 7**
10. Find the variable expression used to create the table at the right. **9x − 2**

x	1	2	3	4
?	7	16	25	34

B-14 Chapter 2

2. The quantity $x - 9$ times -3 is equal to 21. **b**
   a. $x - 9 \cdot (-3) = 21$ b. $(x - 9)(-3) = 21$
   c. $21 = 3 + (x - 9)$ d. $(x - 9) - 3 = 21$

Write an equation for each problem. Do not solve.

3. A number decreased by 24 is equal to -9. **$n - 24 = -9$**
4. Ten less than seven times the number of guests is sixty. **$7g - 10 = 60$**
5. Sean bought 15 notebooks. Each cost the same amount. He spent a total of $30. How much was each notebook? **$15c = 30$**
6. On Tuesday, 80 students were absent. The remaining 478 students were in school. How many students attend the school? **$s - 80 = 478$**
7. ***PROJECT*** Write an equation using one variable to represent the number of hours you study and the number of hours you watch television every week. **Answers will vary.**

Write an equation for each problem. Then solve.

8. Kirsten sent out invitations for a surprise party. She then decided to invite eight more people. She sent out 52 invitations in all. How many invitations did she originally send? **$I + 8 = 52$; 44**
9. Two sides of a triangle have lengths 46 mm and 54 mm. The perimeter is 150 mm. What is the length of the third side? **$x + 46 + 54 = 150$; 50 mm**
10. ***DATA FILE 3 (pp. 96–97)*** **10. a. $2p + p = 10{,}500$; 3,500; 1937**
    a. Jean has twice as many proof sets as Jim. They have the same type of proof sets. The 1990 value of their sets totals $10,500. What year were the proof sets made?
    b. ***WRITE*** a word problem using any of the data in the file. Exchange problems with a student and solve. **Answers will vary.**

MIXED REVIEW

Evaluate using mental math.

1. 5(103) **515** 2. (180)3 **540**

State the first step in solving.

3. $5x = 35$ **Divide by 5.**
4. $5 + x = 35$ **Subtract 5.**
5. $35 = x - 5$ **Add 5.**
6. $\frac{x}{5} = 35$ **Multiply by 5.**

Solve.

7. Badwater, California, (-282 ft alt.) is the lowest point in the Western Hemisphere. The Dead Sea (-1,310 ft alt.) is the lowest surface point on Earth. How much lower is the Dead Sea than Badwater? **1,028 ft**

TEST YOURSELF

Solve.

1. $a + 92 = 112$ **20** 2. $-17 = y \div 4$ **-68** 3. $-46 = -12 + x$ **-34**
4. $96 = 3r$ **32** 5. $b - 16 = -39$ **-23** 6. $-12w = 156$ **-13**

Write an equation for each problem. Solve.

7. Twelve more than some number is twenty. **$n + 12 = 20$; 8**
8. Three less than the quantity $c - 6$ is equal to negative ten. **$(c - 6) - 3 = -10$; -1**

LESSON QUIZ

Write an equation for each problem. Do not solve.

1. A number decreased by 15 is equal to -6 ($n - 15 = -6$)
2. Eight times the quantity x minus 3 is equal to -64. [$8(x - 3) = -64$]
3. Carla spent n dollars on 5 notebooks. Each one cost $2. ($\frac{n}{5} = 2$ or $n = 2 \cdot 5$)

Assignment Guide

Basic 1-4, 7-10, MR, TY All
Average 1-2, 4-5, 7-10, MR, TY All
Enriched 2, 4-10, MR, TY All

FOR THE NEXT LESSON

graph paper, pencil, computer and graphing software (optional), protractor, compass, math journal

OBJECTIVE:
To explore the uses of a computer graphing program.

MATERIALS

- Graph paper
- Computer and graphing software (optional)
- Math journal to record work

Exploring Graphing

■ Sometimes looking at a graph is an easier way to analyze data. You can use the data from a spreadsheet to create a graph.

1

	A	B
1	Month	Income
2	Jan	$500.00
3	Feb	$600.00
4	Mar	$550.00
5	Apr	$700.00
6	May	$850.00
7	Jun	$925.00

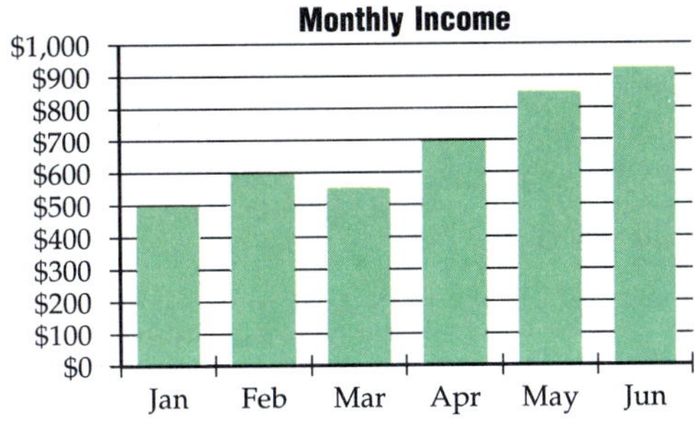

DATAPOINT

Many spreadsheet programs allow you to graph the spreadsheet data.

1. What part of the spreadsheet does the horizontal axis stand for in the bar graph? What does the vertical axis represent? **month; income**

2. Does the spreadsheet or bar graph let you see exact values? **spreadsheet**

3. On a sheet of graph paper, draw a bar graph using twice the value of the spreadsheet data. How is your graph similar to the one shown above? How is it different? **The shape is the same; different values.**

■ You can also use a computer to draw a line graph.

2

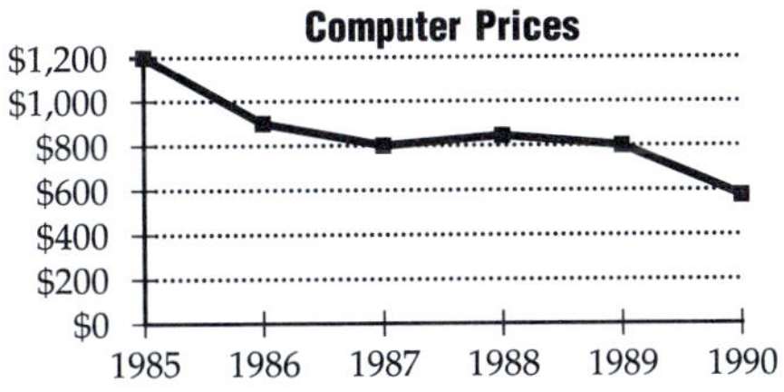

4. a. Could you use the data in the line graph to construct a bar graph? Why or why not?

b. ***Analyze*** Why would a line graph be inappropriate for the data from the spreadsheet shown earlier? Explain.

4. a. No; the data are continuous.
4. b. A line graph assumes continuity of data. A line graph of the monthly income would be misleading. A line graph is for continuous data; a bar graph is for discrete data.

Lesson Focus

Materials

Graph paper, pencil, computer and graphing software (optional), protractor, compass, math journal

Vocabulary/Symbols

bar graph
circle graph
horizontal axis
line graph
vertical axis

MOTIVATING THE LESSON

Begin the lesson by asking students to think of situations in which looking at a chart made it easier for them to understand data. (Answers will vary. Examples may include lists of heights/weights and running times.)

Instruction

1 Help students read the graph. Ask how the data in the bar graph was obtained. (from the spreadsheet) Ask why the spreadsheet might be preferred over the graph. (Exact amounts may be difficult to read on a graph.)

2 Have students discuss when a line graph is helpful for displaying data. (to indicate change or trends over time)

Teaching Tip

Have available several copies of the manual for the software you are using. Then groups of students can research any technical problems on their own.

3 If you are having students draw the circle graph, ask them to explain how to change from dollars to percent of the total amount (divide dollars by total amount) and then from percent to the number of degrees in a circle. (percent times 360) Demonstrate or ask a student to demonstrate how to draw an angle using a protractor.

Closure

4 Have students display their projects for others to view.

Writing in Math Ask students to use a page in their math journals to state in their own words when it would be most appropriate to use a circle graph (to compare parts of the whole), a bar graph (to show comparisons), and a line graph (to show changes, or trends over time).

As a follow-up activity, have students work the computer activity on p. 52E of the Chapter Overview. See also TRB Exploring O-8.

DATAPOINT

Here are some other types of graphs that you can draw with a graphing program:

- Stem and leaf plots
- Box and whisker plots
- Scattergrams
- Three-dimensional graphs

■ You can also use a computer to construct double bar graphs. The spreadsheet data below were used to construct the graph.

	A	B	C
1	Month	Income	Expenses
2	Jan	$500.00	$150.00
3	Feb	$600.00	$175.00
4	Mar	$550.00	$275.00
5	Apr	$700.00	$800.00

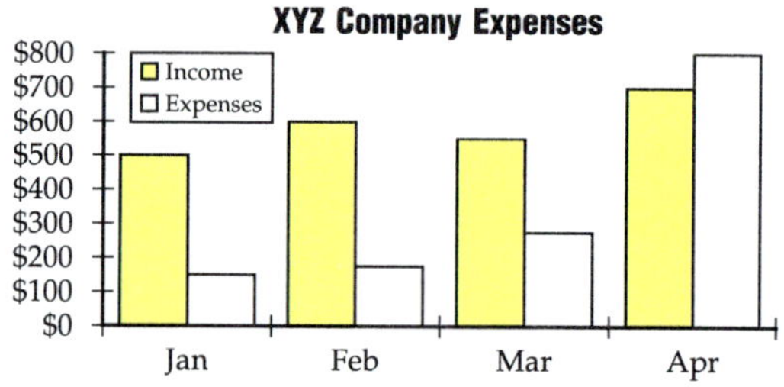

6. a. food stand; paper drive; wedges representing a percent

5. a. ***Analyze*** Based on the graph, is this business doing well financially? Why or why not? Yes; it only showed a loss in April.

b. ***Explore*** How does a graph help in analyzing data? You can visually see differences between sets of data and you can see patterns over a period of time.

■ You can also draw a circle graph with your data. The spreadsheet below shows the results of a class fund-raising drive.

3

	A	B
1	Car wash	$150.00
2	Paper drive	$75.00
3	Book sale	$225.00
4	Food stand	$378.00

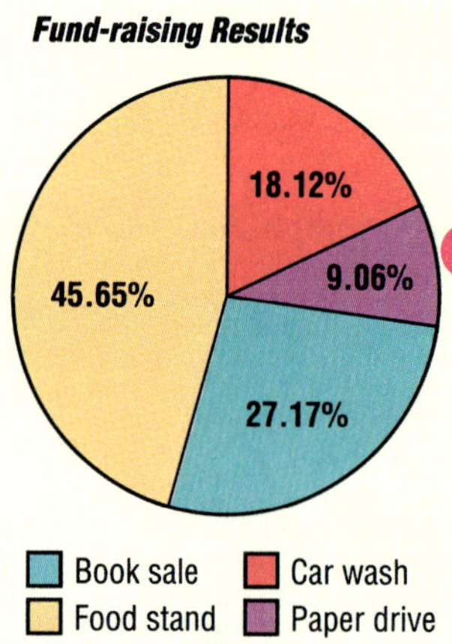

6. The data are shown on the circle graph at the left.

a. Which activity raised the most money? the least money? How are these two amounts shown on the graph?

b. ***Discuss*** Would a bar graph of the data be more useful? Which type of graph would you use for the data? Explain. Answers will vary.

c. How could the circle graph be used in planning next year's fund-raiser? Which activities should the class emphasize? See below.

4

7. ***PROJECT*** Draw a graph using data that you collect. Choose the two most appropriate types of graphs to display your data. If you have a computer graphing program, draw the graphs. Check students' work.

6. c. You can determine which fund-raisers were most profitable; food stand and book sale.

Assignment Guide

Basic All
Average All
Enriched All

FOR THE NEXT LESSON

math journal

PROBLEM SOLVING STRATEGY

OBJECTIVE:
To use guess and test to solve mathematical problems.

NOTES & QUOTES

Certainly, let us learn proving, but also let us learn guessing.
George Polya

2-8 Guess and Test

READ PLAN SOLVE LOOK BACK

It is sometimes useful when solving a mathematics problem to guess what the answer will be. You can use the guessing strategy to solve the following problem.

PROBLEM

Ronald Reagan is the oldest man to be elected president of the United States. John F. Kennedy was the youngest. The sum of their ages at the time of their election is 112. The difference is 26. How old was each man when elected president?

SOLUTION

READ What do you want to find? — Reagan's and Kennedy's ages when they were elected president

What is the sum of their ages when each was elected? — 112

What is the difference? — 26

PLAN 1 Find two numbers with a sum of 112.
Test whether their difference is 26.
Use each incorrect guess to make a better estimate.
Keep a record of your work in a table.

SOLVE 2

Guess	Test	Outcome
50 and 62	$50 + 62 = 112$ $62 - 50 = 12$	The difference is too small, so the numbers have to be farther apart.
40 and 72	$40 + 72 = 112$ $72 - 40 = 32$	The difference is too great, so the numbers have to be closer together.
43 and 69	$43 + 69 = 112$ $69 - 43 = 26$	Correct.

LOOK BACK Ronald Reagan was 69 years old when elected president.
John F. Kennedy was 43 years old when elected president.

CLASS EXERCISES

Answers will vary.

Make a reasonable guess for each question. Test your answer.

1. What is the length of your classroom? the width? the area?
2. How many times does your heart beat in a minute?

Teaching Tip

Ask questions that motivate students to analyze their problem solving strategies. If students get stuck when solving a problem, ask them to describe their solution process.

Lesson Focus

Materials
Math journal

MOTIVATING THE LESSON

Have students discuss strategies for solving this problem: Suppose that there are some people and dogs in a room. There are twice as many people as dogs and the total number of legs of dogs and people is 40. (One strategy is to test different numbers to see if they fit the conditions of the problem.)

Instruction

1 Tell the students that guesses should be reasonable numbers. For example, the minimum age of a president is 35, so a guess of 25 for John F. Kennedy's age would not be reasonable. Remind students that each pair of numbers must have a sum of 112.

2 Ask students if they think the first guess is reasonable and why. (Yes, 50 and 62 are good starting points since each number is more than 35.)

Guided Practice

Class Exercises Ask students why only one reasonable guess is necessary for questions 1 and 2. (Once the measurements are made, the answer is evident.)

A ***common error*** occurs when students do not organize their guesses. One solution is to have students make a table such as the one on page 85.

Closure

Writing in Math Ask students to describe in their math journals how to solve a problem using guess and test.

EE: 1A, 1B, 1C, 1D, 3A, 3F, 3G, 3H; TAAS: 2, 6, 7, 8, 9, 11, 12, 13

Independent Practice

Encourage students to organize their guesses in table form so they do not overlook the correct answer. Ask the students to identify what information in questions 2-4 they would use to guide their choice of guesses.

Lesson Follow-up

LESSON RESOURCES

TRB Practice B-15
TRB Enrichment B-16

Practice

For use after 2-8 (pp. 85–86)

Guess and Test

Use guess and test to solve each problem.

1. The length of a rectangle is 9 in. greater than the width. The area is 36 in.^{2} Find the dimensions. **3 in. by 12 in.**
2. Michael Jordan scored 30 points on 2-point and 3-point goals. He hit 5 more 2-pointers than 3-pointers. How many of each did he score? **4 3-pointers; 9 2-pointers**
3. The sums and products of pairs of integers are given. Find each pair of integers.
   a. sum = -12, product = 36 **-6, -6**
   b. sum = -12, product = 35 **-7, -5**
   c. sum = -12, product = 32 **-8, -4**
   d. sum = -12, product = 11 **-11, -1**
   e. sum = -12, product = 0 **-12, 0**
4. Jess had 3 more nickels than dimes for a total of $1.50. How many of each coin did he have? **9 dimes; 12 nickels**
5. A brush cost $2 more than a comb. The brush and a comb together cost $3.78. Find the cost of each. **brush: $2.89; comb: $.89**
6. The hard-cover edition of a book costs 3 times as much as the paperback edition. Both editions together cost $26.60. Find the cost of each. **paperback: $6.65; hardcover: $19.95**
7. The Wolverines scored 42 points in a football game. They scored 2 more field goals (3 points each) than touchdowns (6 points each). How many field goals and touchdowns did they score? **4 touchdowns; 6 field goals**
8. The volume of a box is equal to the product of the length, width, and height ($V = lwh$). The height and width of a shoe box are equal. The height is 6 in. less than the length. The volume of the box is 896 in.^{3} Find the length, width, and height of the box. **length = 14 in.; width = 8 in.; height = 8 in.**

Chapter 2 B-15

Enrichment

For use after 2-8 (pp. 85–86)

Super Guess and Test

In many problems, you have used the guessing strategy to find two numbers. Here is a classic problem which involves finding three numbers. Finding three numbers is more complicated than finding two, but you can still do it by guessing and testing.

A farmer sold cows for $100 apiece, sheep for $20, and rabbits for $1. The farmer sold 100 animals in all and earned exactly $2,000.

Complete to find the number of each type of animal sold.

1. Let c = the number of cows sold, s = the number of sheep sold, and r = the number of rabbits sold. Write an equation using c, s, and r. **$c + s + r = 100$**
2. Write a variable expression for:
   a. the amount of money earned on the sale of cows **100c**
   b. the amount of money earned on the sale of sheep **20s**
   c. the amount of money earned on the sale of rabbits **r**
3. Use the above expressions to write an equation expressing the total amount of money the farmer earned. **$100c + 20s + r = 2{,}000$**
4. How do you know that the number of cows is less than 20? **20 cows would bring in $2,000. 100 animals were sold.**
5. Complete the table.

No. cows sold	14	15	16	17	18	19
No. sheep sold	3	5	**6**	**8**	9	**2**
No. rabbits sold	**83**	**80**	78	75	**73**	79
Amount earned	**$1,543**	**$1,680**	**$1,798**	**$1,935**	**$2,053**	**$2,019**

6. Find the values in the table which produce earnings closest to $2,000. Adjust them using guess and test until you find the correct answer.
   Number of cows **19** Number of sheep **1** Number of rabbits **80**

B-16 Chapter 2

3. What are two whole numbers whose sum is 20 and whose difference is 2? **9 and 11**
4. What are two whole numbers whose product is 50 and whose quotient is 2? **5 and 10**

Which of the numbers 1, 2, 3, 4, or 5 is a solution?

5. a. $4y - 2 + 3y = 19$ **3** b. $3 + \frac{b}{2} = 5$ **4**

WRITTEN EXERCISES

Use a CALCULATOR where appropriate.

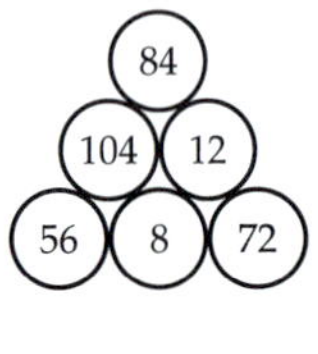

FLASHBACK
Consecutive integers have a difference of 1.
. . . -2, -1, 0, 1, 2 . . .

Use guess and test to solve each problem.

1. Find two pairs of numbers in the diagram whose quotient is 7. **12, 84 and 8, 56**
2. The teller's drawer has some $5 bills, $10 bills, and some $20 bills. There are 15 bills worth a total of $185. How many $5 bills, $10 bills, and $20 bills are there? **7, 1, 7; 5, 4, 6; 3, 7, 5; or 1, 10, 4**
3. The Smiths have two children. The sum of their ages is 23 and the product is 132. How old are the children? **11, 12**
4. The average of three consecutive integers is 10. Their sum is 30. What is the middle number? **10**

Use any strategy to solve each problem.

5. Three consecutive integers have a sum of -9 and a product of -24. What are the three integers? **-2, -3, -4**
6. A vegetable garden has a length of 5 ft and a width of 8 ft. Two feet are added to the length. By how much will this increase the area? **16 ft^2**
7. Trains leave New York for Boston every 40 min. The first train leaves at 5:20 A.M. What is the departure time closest to 12:55 P.M.? **12:40 P.M.**
8. Jean's age of 16 is the same as Rafi's age divided by three. How old is Rafi? **48**
9. A number that when multiplied by itself and then by itself again gives -1,000. What is the number? **-10**
10. In a collection of quarters and nickels, there are four more nickels than quarters. How many nickels and quarters are there if the collection is worth $2.30? **11, 7**
11. A triangle has sides of lengths $3x$, $4x$, and $5x$. The perimeter is 120 ft. What is the length of each side? **30 ft, 40 ft, 50 ft**

MATH MINUTES

Have students work in pairs. One student makes up a word problem and the other makes three reasonable guesses to test the conditions of the problem. Then the students reverse roles.

Assignment Guide

Basic All
Average All
Enriched All

FOR THE NEXT LESSON

math journal

PROBLEM SOLVING APPLICATION

READ PLAN SOLVE LOOK BACK

OBJECTIVE:
To apply solving equations to ecology problems.

2-9 Ecology

Americans throw away an average of four to five pounds of waste per person each day. As landfills across the United States fill up and the number of landfills decreases, disposing of solid waste is becoming a serious problem. One way to reduce the problem is to reduce the amount of solid waste.

FLASHBACK
1 ton (t) = 2,000 lb

Example 1 One aluminum can weighs about $\frac{1}{28}$ lb. Suppose your school starts a recycling drive. How many cans will your school need to recycle to reduce the trash by 1 t?

Solution Write an equation.

1

$c \div 28 = 2{,}000$ — Let c = number of cans.

$c \div 28 \cdot 28 = 2{,}000 \cdot 28$ — Multiply both sides by 28.

$c = 56{,}000$

Your school will need to recycle 56,000 cans.

Some companies pay for materials to recycle. The prices paid vary due to market conditions.

Example 2 Suppose recyclers pay \$5/t for newspapers. How many tons of newspaper would your school have to recycle to earn \$65?

Solution Write an equation.

2

$5n = 65$ — Let n = number of tons of newspaper.

$\frac{5n}{5} = \frac{65}{5}$ — Divide both sides by 5.

$n = 13$

Your school will have to recycle 13 t of newspaper.

CLASS EXERCISES

United States Solid Waste (lb/person/day)

Year	Generated	Recycled
1965	2.77	0.17
1970	3.16	0.21
1975	3.11	0.23
1980	3.35	0.32

Use the ***DATA*** at the left to solve.

1. In 1975, how many pounds of solid waste did the average person generate in seven days? 21.77 lb
2. In 1980, what was the net amount of solid waste that one person generated? The net amount of waste is the difference between the waste generated and the waste recycled. 3.03 lb
3. In 1970, the United States population was approximately 203,000,000 people. About how much solid waste did the entire population generate each day? 641,480,000 lb

Lesson Focus

Materials
Math journal

MOTIVATING THE LESSON

Begin the lesson by asking students what types of solid waste they must separate from the trash that they discard. (e.g. newspapers, aluminum cans, tin cans, cardboard, glass, plastic containers)

Instruction

1 To help students understand how the equation was obtained, ask them how many cans there would be in one pound of cans. (28)

2 Ask students to explain why a multiplication equation was written to solve the problem. (\$5 times the variable *n* equals \$65.)

Guided Practice

Class Exercises For question 3, ask students if 600 million pounds is a reasonable estimate for the answer and to tell why. (Yes; 203,000,000 is approximately 200 million, 3.16 is approximately 3, and 200 million times 3 is 600 million.)

Closure

Writing in Math Ask students to use their math journals to describe how they estimate the total amount of waste generated in one day by all the students in the school. (Have each student in the class keep a record of the waste he or she generates in one day. Find the class average and multiply the average by the total number of students.)

Teaching Tip

When students need help solving problems, ask them questions that provide hints. For example, asking *What is the total?* implies that the operation of addition or multiplication is necessary.

EE: 1A, 1C, 1D, 1E, 3A, 3B, 3F, 3G, 5D, 5E; TAAS: 2, 3, 4, 6, 7, 8, 9, 10, 11, 12

Independent Practice

For question 6a, ask students what they will substitute for the phrase *weight of waste* in the area formula. (63,000 · 1,800)

For question 7, have students work in small groups on the Project. Then have the groups share their findings.

Decision Making This feature allows students to relate the content of the lesson to their school and community. In order to complete the chart under question 1, students will need the cooperation of the other classes in the school or the maintenance crew.

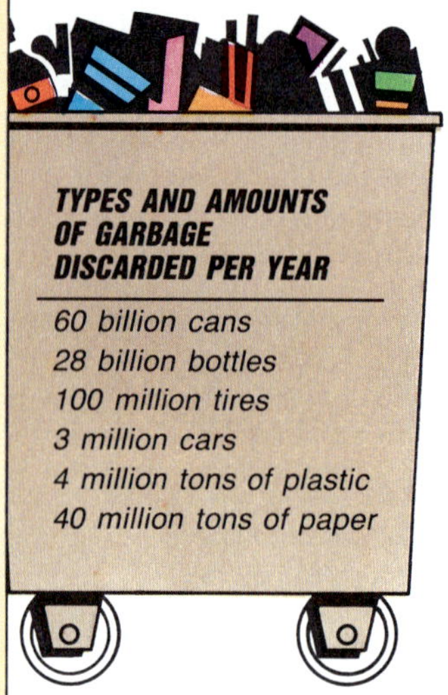

WRITTEN EXERCISES

Use a CALCULATOR where appropriate.

Use the *DATA* at the left to solve each problem.

1. It costs \$65/t to pick up garbage. How much will it cost per year to pick up just the paper and plastic discarded? \$2,860,000,000

2. Tires weigh an average of 9 lb. It costs \$72/t to dispose of tires. How much does it cost to dispose of one year's discarded tires? \$32,400,000

3. ***ESTIMATION*** The weight of twenty-eight aluminum cans is approximately 1 lb.

a. Estimate the weight of the cans discarded per year.

b. Recycling one pound of cans saves 8 kW · h of electricity. About how many kilowatt hours of electricity would be saved if all the discarded cans were recycled?

3. a. about 2 billion lb   3. b. about 16 billion kW · h

4. A town with a population of 38,000 people discards an average of 4 lb of waste per person each day. About how many tons of waste do they discard in a week? about 500 t

Decision Making **DECISION MAKING** *Decision Making* *Decision Making* *Decision Making*

ECOLOGY

COLLECT DATA Answers will vary.

1. How much waste does your school produce each week? Keep track of what you throw away. Make a chart like the one below to tally your results.

Type of Trash	Amount of Trash (in pounds)							
	Mon	Tues	Wed	Thur	Fri	Sat	Sun	Total
Newspaper								
Other paper								
Metal								
Plastic								
Glass								
Other								

2. Find out what kinds of recycling services are available in your community.

MATH MINUTES

Have students estimate the weight of one sheet of notebook paper. Then have them estimate the amount of paper they use in a week. Finally, have them estimate the weight of a year's worth of notebook paper.

5. Suppose a state with a population of 5 million reduces the average waste per person from 5 lb to 3 lb. How much could they save on disposal costs if the waste pickup costs \$65/t? **\$325,000**

6. A town of 63,000 collects 1,800 lb of waste per person in a year. The area of landfill (in square yards) needed for this waste is found using the formula $A = \frac{\text{weight of waste}}{1{,}000 \cdot \text{depth in yards}}$.

   a. Find the area in square yards of the town's landfill if the landfill's depth will be 3 yd. **37,800 yd^2**

   b. ***ESTIMATION*** There are 4,840 yd^2 in an acre. Estimate the number of acres needed per year for the town's landfill. **about 8 acres**

7. ***PROJECT*** **Check students' work.**

   a. Technically, solid waste, trash, and garbage are not synonyms. Find out which materials are in each category.

   b. Determine ways, other than reducing the landfill problem, in which recycling benefits the environment.

   c. Research the environmental damage that can result from an improperly managed landfill. Find out what we can do to minimize these dangers.

Decision Making ■ Decision Making ■ Decision Making ■ Decision Making ■ Decision Making ■ Decision Making

ANALYZE DATA

3. Analyze the garbage your school discards.

   a. What materials can be recycled?

   b. Could you use the garbage in other ways?

4. Other than recycling or reusing items, what ways can you think of to reduce the amount of your school's garbage?

MAKE DECISIONS

5. Plan to reduce the amount of waste your school discards.

   a. Set a reasonable goal for reducing your amount of trash.

   b. Decide on methods you can use to meet your goal.

6. Many fast-food companies use plastic wrappers or foam containers instead of coated paper to wrap their foods.

   a. Compare the ease of recycling these materials.

   b. Which wrapping do you think is better? Explain.

   c. How do you think the companies could reduce their trash?

Lesson Follow-up

LESSON RESOURCES

TRB Alternate Application B-17

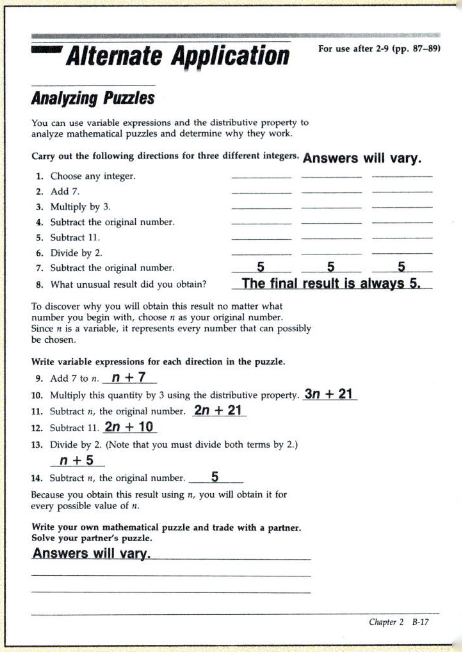

Alternate Application For use after 2-9 (pp. 87–89)

Analyzing Puzzles

You can use variable expressions and the distributive property to analyze mathematical puzzles and determine why they work.

Carry out the following directions for three different integers. **Answers will vary.**

1. Choose any integer. ______ ______ ______
2. Add 7. ______ ______ ______
3. Multiply by 3. ______ ______ ______
4. Subtract the original number. ______ ______ ______
5. Subtract 11. ______ ______ ______
6. Divide by 2. ______ ______ ______
7. Subtract the original number. **5** **5** **5**
8. What unusual result did you obtain? **The final result is always 5.**

To discover why you will obtain this result no matter what number you begin with, choose n as your original number. Since n is a variable, it represents every number that can possibly be chosen.

Write variable expressions for each direction in the puzzle.

9. Add 7 to n. **$n + 7$**
10. Multiply this quantity by 3 using the distributive property. **$3n + 21$**
11. Subtract n, the original number. **$2n + 21$**
12. Subtract 11. **$2n + 10$**
13. Divide by 2. (Note that you must divide both terms by 2.) **$n + 5$**
14. Subtract n, the original number. **5**

Because you obtain this result using n, you will obtain it for every possible value of n.

Write your own mathematical puzzle and trade with a partner. Solve your partner's puzzle.

Answers will vary.

Chapter 2 B-17

Assignment Guide

Basic All
Average All
Enriched All

FOR THE NEXT LESSON

decimal squares or graph paper, overhead transparencies, math journal

Career Information

If possible invite the regional or city planner to talk to the class. Encourage students to ask questions about the skills necessary for a career in urban planning.

Ask the speaker to cover such topics as:

- the course work necessary to work effectively as an urban planner
- the role of demographics and the use of statistical data in preparing for the community's growth
- the communication skills that are needed for interviewing potential users of services and for advising them of the limitations of those services
- the type of computer software that is used to assist the planner in making predictions and decisions

CAREER

Help Wanted: Urban and Regional Planner

Bachelor's degree in city planning, architecture, or engineering required. Master's degree in urban or regional planning preferable.

For more information, write to American Planning Association, 1776 Massachusetts Ave. NW, Washington, DC 20036.

Urban planners prepare for a community's growth and revitalization. Planners consider social, environmental, and economic concerns in long-range community development. Planners may examine medical facilities and schools to insure they can meet the population's needs.

PROJECT

Find out if your area has a planner and, if so, what his or her responsibilities are.

Practice

State whether each equation is true, false, or open.

1. $-17 = x + 5$ open
2. $\frac{-325}{-5} = 65$ true
3. $29 + (-3) = 26$ true
4. $25 \cdot 6 = 150$ true
5. $2x + 5 = 3x - 7$ open
6. $6 = -3[4 + (-2)]$ false

State whether the given number is a solution of the equation.

7. $12 = 2x + (-4); x = 4$ no
8. $-3[2y - (-5)] = 9; y = -4$ yes
9. $2r - 4 + r = 89; r = 31$ yes
10. $-2 = s - (-4) + 2s; s = 6$ no
11. $\frac{x}{5} - (-2) = -12; x = -50$ no
12. $2m + 6 = 2(m + 3); m = -2$ yes

State which property is shown.

13. $(3r)s = 3(rs)$ A
14. $(a - b)(r + s) = (r + s)(a - b)$ C
15. $653 + 0 = 0 + 653$ C
16. $6(t - 5) = 6 \cdot t - 6 \cdot 5$ D
17. $17 \cdot 1 = 17$ MI
18. $(3 + 9) + 0 = 3 + (9 + 0)$ A

Evaluate.

19. $-135 + (-341)$ -476
20. $(8 - 5)6 + 37$ 55
21. $120 + 16 + 80$ 216
22. $-15 - (-56)$ 41
23. $-550 \div 50 \cdot 2$ -22
24. $-6[4 + (-12)] + 9$ 57
25. $4\left(\frac{-12}{-6}\right) + (-3) \cdot 5 - 3(-11)$ 26
26. $-17 + [5 - (-7) + 2]3$ 25
27. $-3 + (-6) - [5 - (-2)]$ -16
28. $7 \cdot (-4) + [8 - (-5)]3$ 11

Simplify each expression.

29. $-5(3k)$ $-15k$
30. $2x + 5 - x$ $x + 5$
31. $(g + 5)3 + 2g - 7$ $5g + 8$
32. $8 + (3c + 3)4$ $12c + 20$
33. $4(x + y - z)$ $4x + 4y - 4z$
34. $8(a - b) + 2b - 2a$ $6a - 6b$
35. $\frac{-9}{3} + 12w - (-7) - 3(5w)$ $-3w + 4$
36. $23 + 2[b + (-15) + 3b]$ $8b - 7$

Solve each equation.

37. $c + 7 = 34$ 27
38. $-550 = 10w$ -55
39. $\frac{k}{-4} = -3$ 12
40. $78 = t - (-47)$ 31
41. $6d = 54$ 9
42. $17 = z - 3$ 20
43. $27 + v = -12$ -39
44. $h + (-3) = 53$ 56
45. $-192 = \frac{w}{-16}$ 3,072
46. Find the sum of the magic square. Write and solve equations to find a, b, and c.
 $a + 7 + 4 = 21$; 10
 $b + 7 + 5 = 21$; 9
 $c + 7 + 8 = 21$; 6

a	3	8
5	7	b
c	11	4

Write an equation for each problem. Solve.

47. The sum of a number and 5 is equal to -123. $n + 5 = -123$; -128
48. Six less than a is equal to 47. $a - 6 = 47$; 53

Problem Solving Practice

PROBLEM SOLVING STRATEGIES

Look for a Pattern
Guess and Test
Simplify the Problem
Make an Organized List
Work Backwards
Account for All Possibilities
Make a Table
Write an Equation
Solve by Graphing
Draw a Diagram
Make a Model
Solve Another Way
Simulate the Problem

Solve each problem. The list at the left shows some possible strategies you can use.

1. ***DATA FILE 5 (pp. 180–181)*** Look at the following two equations that are expressed with musical notes.

   a. How many quarter notes are equal to a whole note? 4
   b. How many eighth notes are equal to a whole note? 8
   c. Is there a pattern? Can you use the pattern to find the number of sixteenth and half notes in a whole note? Yes; 16; 2

2. The average age of five students is 15. Two of the students are 12 years old. Three students are not 12, but are the same age. How old are the other three students? 17

3. ***COMPUTER*** Look at the spreadsheet below.

	A	B	C	D
1	Employee	Hours	Wage	
2	Jones, C.	4	4.50	= B2×C2
3	Smith, G.	5	4.25	= B3×C3
4	Garcia, H.	12	5.15	= B4×C4
5				
6				
7				

   a. What values do the formulas in cells D2–D4 represent? total wages for each employee
   b. Suppose you wanted to find the total number of hours and the total amount of wages. What formulas would you use? In which cells would you enter the formulas? = SUM (B2:B4); = SUM (D2:D4); place in B5 and D5, respectively

4. You are given these directions to get to a friend's house:

   Drive east on State Street. Take a right on Main. Continue on Main and take a right on Broadway. Continue on Broadway and take a left on Center Street. The house is on the corner of Center and High streets.

   Assume the streets are arranged in a grid and there are no one-way streets. Is there a shorter route? If so, describe it. Answers may vary.

5. ***DATA FILE 3 (pp. 96–97)***
   a. Tom bought a 1937 proof set in 1965 and sold it in 1968. How much did he have to add to his profit in order to purchase three 1940 proof sets the next day? $260
   b. How long would it take to produce $1,920,000 in $5 bills? $1\frac{1}{2}$ h

Instruction

This page provides a variety of problems that can be used to reinforce and enhance the students' problem solving skills. Encourage students to read each problem carefully. Then have them refer to the list of problem solving strategies to help them decide how to solve the problem.

Point out, however, that not all questions require a strategy for solving, nor are all the strategies in the list used in this lesson.

Study Skills

Time-management is essential to the education process. Learning to schedule time wisely can help prevent procrastination and other behaviors that can lead to underachievement and low self-esteem. Time-management can easily be taught in the classroom.

Have students discuss how they spend their time. Emphasize the feelings they have when they put off assignments until the last moment. Then discuss strategies that can help students use their time more efficiently.

Here are a few ways to encourage good time-management techniques:

- Provide blank schedule sheets weekly.
- Teach students to prioritize and set study goals.
- Remind students of assignment due dates.

Offer students the following suggestions:

- Plan ahead for the week, month, or term.
- Budget time for schoolwork and other activities.
- Use a To-Do list, cross off each task as it is completed.
- Prioritize assignments and do important tasks first.
- Plan a reward after studying.

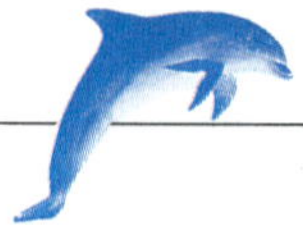

Chapter 2 Review

Match each word with the example that illustrates its meaning.

1. Commutative property f
2. Associative property a
3. Solution e
4. Identity i
5. Distributive property k
6. Term j
7. Numerical coefficient b
8. Constant c
9. Simplify an expression d
10. Property of equality g
11. False equation h

a. $a \cdot (b \cdot c) = (a \cdot b) \cdot c$
b. 4 in $4x + 5 = 9$
c. 5 in $4x + 5 = 9$
d. $3x + 4 + 5 = 3x + 9$
e. $x = 2$ for $3 + x = 5$
f. $a + b = b + a$
g. $4x + 5 - 5 = 9 - 5$
h. $-4 - 7 = 3$
i. $a + 0 = a$ and $a \cdot 1 = a$
j. $4x$ in $4x + 5 = 9$
k. $a(b + c) = (a \cdot b) + (a \cdot c)$

Using the Properties of Operations 2-1, 2-2

To evaluate an expression, use the commutative property to change the order. Use the associative property to change the grouping.

Use the commutative and the associative properties to evaluate.

12. $125 + 347 + 75$ 547
13. $58 + 16 + 2 + 4$ 80
14. $4 \cdot 7 \cdot 25 \cdot 1$ 700
15. $(20 \cdot 65) \cdot 5$ 6,500

Using the Distributive Property 2-3

To evaluate an expression with parentheses, use the distributive property to distribute a factor to each term inside the parentheses.

Use the distributive property to evaluate.

16. $5(20 + 3)$ 115
17. $4(50 - 2)$ 192
18. $2(25 + 8) + 2(15 - 8)$ 80
19. $6(40 - 21) - 6(20 - 1)$ 0

Simplifying Variable Expressions 2-4

To simplify a variable expression, combine like terms and eliminate parentheses using the distributive property.

Simplify each expression.

20. $5x + 3y + 3x + 2y$ 8x + 5y
21. $4 + 6(a + 2) + 3a$ 9a + 16

Addition and Subtraction Equations 2-5

To solve an addition or subtraction equation, add or subtract the same value from both sides of the equation.

Solve each equation.

22. 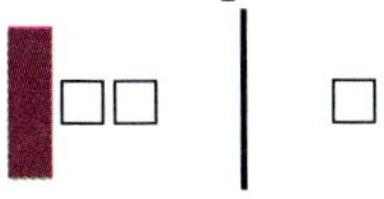 $x = -1$

23.  $x = 3$

24. $a - 7 = 28$ $a = 35$

25. $x + 19 = 30$ $x = 11$

26. $38 + y - 18 = 500$ $y = 480$

27. $n + (-13) = 7$ $n = 20$

Multiplication and Division Equations 2-6

To solve a multiplication or division equation, multiply or divide both sides of the equation by the same nonzero value.

Solve each equation.

28. 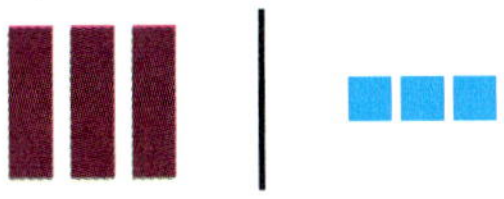 $x = -1$

29. 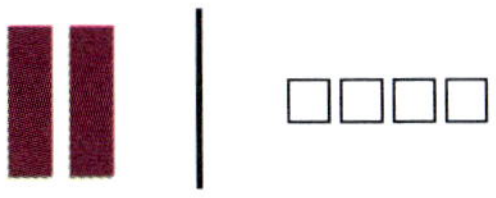 $x = 2$

30. $\frac{m}{8} = -9$ $m = -72$

31. $8b = 96$ $b = 12$

32. $\frac{c}{12} = 24$ $c = 288$

33. $-3k = -54$ $k = 18$

Writing Equations 2-7

To write an equation for a word problem, you need to recognize words that imply the variable(s), the operation(s), and the equality.

Write an equation for each problem.

34. Twice a number increased by 28 is 54. $2x + 28 = 54$

35. Seventeen less than a number is 12. $x - 17 = 12$

Write a word problem to describe the equation. Answers will vary. Check students' work.

36. $x + 7 = 95$

37. $4x = 17$

Problem Solving 2-8, 2-9

To solve some problems, guess at the solution. Then use each incorrect guess to make a better estimate of the correct answer.

Use guess and test to solve.

38. Paper plates come in packages of 15 or 20. Helene bought 9 packages and had 155 plates. How many packages of 15 and how many packages of 20 did she buy? 5 pkgs of 15 plates, 4 pkgs of 20 plates

Use the data on page 88.

39. How many pounds of paper are discarded per year? 80 billion lb

RESOURCES

TRB B-19–B-25

Scoring Chart

No. Correct	Score
30	100
29	97
28	94
27	91
26	88
25	85
24	82
23	78
22	74
21	70
20	66
19	62
18	59
17	56
16	53
15	50
14	46
13	42
12	38
11	34
10	30
9	27
8	25
7	22
6	19
5	16
4	13
3	10
2	7
1	4

Chapter 2 Test

State whether each equation is true, false, or open.

1. $24 = 3(-8)$ **false**

2. $5x + 28 = 153$ **open**

3. $18(-7 \div 7) = (-2)(9)$ **true**

4. $-6 + 15 = (120 \div 20) - (5 - 8)$ **true**

Use the commutative and the associative properties to evaluate.

5. $250 \cdot 38 \cdot 2$ **19,000**

6. $675 + (-8) - (75 - 8)$ **600**

Complete with the appropriate number or variable.

7. $9(8 + 5) = 9 \cdot 8 + \square \cdot 5$ **9**

8. $(x + y)\square = xz + yz$ **z**

9. $-3 \cdot 4 + \square \cdot 11 = \square(4 + 11)$ **-3**

10. $4x + 6\square = \square(4 + 6)$ **x**

Simplify each expression.

11. $6y + 4(y + 1)$ **10y + 4**

12. $5a + 2b + 3a - 7b$ **8a − 5b**

13. $3(m + 2n) - 2n$ **3m + 4n**

14. $3(2r - 5) + 8(r + 2)$ **14r + 1**

15. $2(x + y) - 2y$ **2x**

16. $(-2c + 3d)(-5) + 3(-2c) - (-8d)$ **4c − 7d**

Solve each equation.

17. 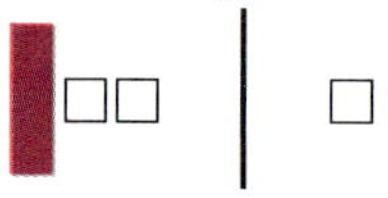 **x = -1**

18. 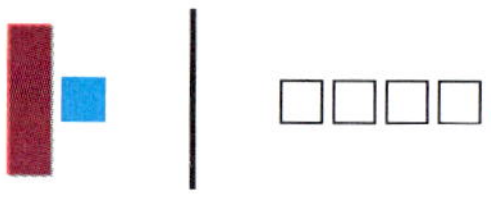 **x = 5**

19.  **x = -2**

20. $k - 23 = 17$ **k = 40**

21. $\frac{t}{-5} = 15$ **t = -75**

22. $-3f = -42$ **f = 14**

23. $120 = 38 + p$ **p = 82**

24. $7w = -217$ **w = -31**

25. $\frac{h}{12} = 12$ **h = 144**

Write an equation for each problem. Do not solve.

26. Five less than 8 times the number of students is 163. **8x − 5 = 163**

27. Three times the quantity $t + 9$ is equal to -18. **3(t + 9) = -18**

Solve.

28. The length of a room is 4 m longer than the width. The perimeter of the room is 28 m. Find the width of the room. **5 m**

29. Brian bought a used bike for $25 less than its original price. He paid a total of $88 for the bike. What was the original price of the bike? **$113**

30. Write a word problem to describe the equation $4x = 2$.
Answers will vary. Check students' work.

Chapters 1–2 Cumulative Review

Choose the correct answer. Write A, B, C, or D.

1. 0, 1, 4, 9, 16, 25, . . . B

A. 26 **B.** 36
C. 35 **D.** not given

2. Which property is used? C

$(ab)c = c(ab)$

A. associative **B.** identity
C. commutative **D.** not given

3. $12 - (-15)$ C

A. -3 **B.** 3
C. 27 **D.** not given

4. Simplify $(3x + 4)2 + 3(-2x)$. D

A. $-3x + 8$ **B.** $-3x + 4$
C. $12x + 8$ **D.** not given

5. Solve $x - (-3) = 12$. A

A. 9 **B.** 4
C. 15 **D.** not given

6. What is the variable expression for *six less than the absolute value of a number*? A

A. $|n| - 6$ **B.** $6 - |n|$
C. $|n - 6|$ **D.** not given

7. Evaluate $3xy - 2x$ for $x = -1$, $y = 2$. A

A. -4 **B.** 8
C. 4 **D.** not given

8. Solve $x + (-9) = 36$. C

A. 27 **B.** -4
C. 45 **D.** not given

9. What is the opposite of $|-3|$? C

A. 3 **B.** 0
C. -3 **D.** not given

10. Which equation has the solution -2? B

A. $x + 7 = 9$ **B.** $\frac{x}{-2} = 1$
C. $(-3)x = -6$ **D.** not given

11. Complete $4(9 + 7) = (4 \cdot \blacksquare) + (4 \cdot 7)$. B

A. 7 **B.** 9
C. 4 **D.** not given

12. Find two numbers whose sum is 10 and whose product is -24. D

A. -6 and 4 **B.** -12 and 2
C. -8 and -3 **D.** not given

13. Solve $\frac{y}{4} = -12$. C

A. 3 **B.** -3
C. -48 **D.** not given

14. Evaluate $98 + 2 \cdot 7 + 3$. C

A. 110 **B.** 118
C. 115 **D.** not given

15. Solve $-8y = 72$. B

A. -7 **B.** -9
C. 9 **D.** not given

16. Write an equation for the model. B

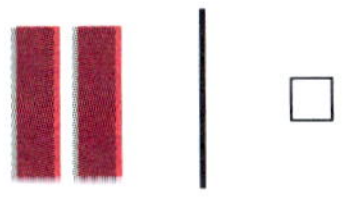

A. $x + 2 = 1$ **B.** $2x = 1$
C. $x - 2 = 1$ **D.** $2x = -1$

Study Skills

Reviewing for tests in an organized way can help to reduce test anxiety and increase the likelihood that students will do well.

Encourage students to begin the review process at least a week before a test. Several short periods of study can be more productive than one or two long sessions.

Suggest that students begin by assembling notes, handouts, and preview tests. Students can outline the major topics covered in class by using the table of contents at the front of the textbook. Then they can use the outline to systematically review each type.

Specific suggestions may include having students:

- Rewrite and summarize class notes.
- Pretest themselves using the questions at the end of the chapter.
- Use a tape recorder to record and replay important facts.
- Review together in cooperative learning groups.

Chapter At a Glance

In Chapter 3 students learn to use decimals in estimation, addition, subtraction, multiplication, and division equations.

Reading the Chapter

This chapter helps students refine reading skills by introducing a new element: reading and writing decimals. Students are asked to alternate between reading decimals as words, such as 70 hundredths, and writing them as numbers, such as 0.70. Encourage students to read each problem carefully.

Explain that when students are reading word problems with decimals, it can be helpful to read once for a general idea of the problem, and then reread to understand the numbers. If the decimals are written as words, ask students to change them to numbers. When they find the answer, have them change it back to words. Then have students reread the problem to see if their answer makes sense.

Chapter Resources

Practice and Enrichment TRB C-1 to C-10, C-12 to C-15
Alternate Application TRB C-11
Chapter Test Form A, C-17
Chapter Test Form B, C-19
Computer Test Bank

References

STUDENT BIBLIOGRAPHY

Seymour, Dale. *Developing Skills in Estimation, Books A and B.* (Dale Seymour Publications, Palo Alto, Calif. 1981).

TEACHER BIBLIOGRAPHY

Hess, Adrien L. *Mathematics Projects Handbook,* 2nd ed. (NCTM, Reston, Va. 1982).

SOFTWARE

Bank Street School Filer (Sunburst Communications), for Apple II family of computers.
PFS: File (Software Publishing), for IBM computers.

FILMS, VIDEOTAPES, AND OTHER MEDIA

Using Databases in the Classroom (Sunburst Communications, VHS or Beta).

Vocabulary

compatible numbers
data base
decimal point
field
formula
front-end estimation
record
rounding

Materials

decimal squares (T*)
data base software (optional)
graph paper (T)
math journal
computer (optional)

*T—Teacher's Resource Book

Resource Bank

Class Activities

FIELD TRIPS

▼ Bank—to observe how deposits and withdrawals are made and recorded.
▼ Small retail business—to find out what reports a cash register can generate.

CLASS SPEAKERS

▼ Coin collector—to discuss various grades of coins and how their values are established.
▼ Medical record technician—to ask how math is involved in compiling commonly used reports.

Bulletin Board

Display the following bar graph showing only the number of pennies produced at the Denver Mint for the given years. Have students research the total number of pennies made for the given years, and then extend the bars to show the data in tenths of billions.

In which years did the Denver Mint produce at least half of all the pennies made? (1985, 1986, 1987)

What is the total value of all the pennies produced in 1988? (about $113 million)

Why do you think the number of pennies made in more recent years is less than the number made in earlier years? (Answers will vary.)

Project: Have students research the metals in a penny and express these amounts as decimals less than one.

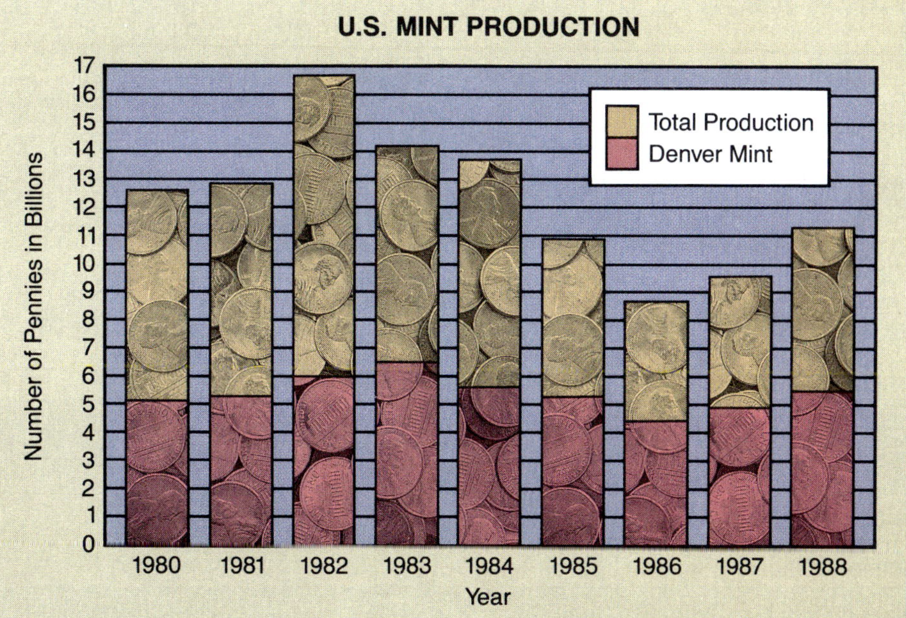

Extra-Credit Problems

BASIC

▼ Between any two numbers there is another number. Name a number between 0.001 and 0.002. (Answers will vary.)

▼ Name nine numbers between 0.001 and 0.002. Arrange these numbers in order from least to greatest. (Answers will vary.)

AVERAGE

▼ What is the least number expressed in hundredths that rounds to 6? (5.50) What is the greatest number? (6.49)

ENRICHED

▼ Write two equivalent integer equations for each decimal equation.
a. $0.14x + 4.5 = 28.4$
b. $3.45x - 23.76 = 0.001$
(Possible answers: $14x + 450 = 2{,}840$; $3{,}450x - 23{,}760 = 1$)

Chapter Planning Guide

OBJECTIVES	ASSIGNMENT GUIDE			ASSESSMENT	
	Basic	Average	Enriched	Review/ Tests	TRB*
Exploring Decimals To explore decimals using decimal square models.	**Manipulative Connection** All	All	All		
3-1 Decimals To compare, round, and order decimals.	1-2, 5-8, 11-14, 19-21, 23-26, 29-31, 35-37, MR* All	2-3, 6, 8-9, 13-16, 19-21, 24-27, 30-32, 35-37, MR All	3-4, 6, 9-10, 15-18, 20-22, 26-28, 32-35, 37-39, MR All	Extra Practice p. 568	Practice C-1 Enrichment C-2
3-2 Estimating with Decimals To estimate sums, differences, products, and quotients of decimals.	1-6, 11-12, 15-17, 21-24, 29-32, 37, 40-44, MR All	1-2, 5-8, 12-13, 16-18, 23-26, 32-34, 37-38, 40-45, MR All	2, 7-10, 13-14, 18-20, 25-28, 32, 36, 39-40, 43-46, MR All	Extra Practice p. 568	Practice C-3 Enrichment C-4
3-3 Expressions with Decimals To evaluate and simplify expressions involving decimals.	1-15 odd, 18-24 even, 27-31, 36-37, 40-41, MR, CT* All	2-16 even, 19-25, 29-33, 37-40, 42, MR, CT All	3-4, 7-8, 13-21, 24-26, 31-40, 43, MR, CT All	Extra Practice p. 568	Practice C-5 Enrichment C-6
3-4 Addition and Subtraction Equations To solve addition and subtraction equations involving decimals.	1-10, 17-18, 21-22, 25-30, 37, 38-40, 43-44, MR, TY* All	2-3, 10-15, 18-19, 22-23, 28-33, 37, 39-41, 44-45, MR, TY All	3-4, 11-16, 19-20, 23-24, 31-37, 40-42, 45-47, MR, TY All	Extra Practice p. 568	Practice C-7 Enrichment C-8 TY C-16
3-5 Multiplication and Division Equations To solve multiplication and division equations involving decimals	1-4, 9-13, 18-20, 23-29 odd, 31-33, 37-38, 40-43, 45, MR, CT All	3-17 odd, 19-21, 24-30 even, 32-34, 37-38, 42-45, MR, CT All	6-16 even, 20-22, 25-26, 29-30, 34-39, 44-47c, MR, CT All	Extra Practice p. 568	Practice C-9 Enrichment C-10
Exploring Data in Tables To explore using data in tables.	All	All	All		
Exploring Data Bases To explore the uses of a computer data base.	All	All	All		
3-6 Problem Solving Application: Using the Metric System To solve density problems using metric units.	All	All	All		Alternate Application C-11
3-7 Using Formulas To use formulas.	1-2, 5-7, 11-14, 18-19, 22, 24-25, 27-28, MR, TY All	2-3, 7-9, 13-16, 19-20, 22, 24-25, 28-30, MR, TY All	3-4, 8-17, 20-21, 23, 25-26, 29-31, MR, TY All	Extra Practice p. 568 TY p. 131	Practice C-12 Enrichment C-13 TY C-16
3-8 Problem Solving Strategy: Simplify the Problem To solve problems by using a simpler problem.	**Problem Solving Connection** All	All	All	Problem Solving Practice p. 124	Practice C-14 Enrichment C-15
Chapter Test	All	All	All	pp. 134-137	Tests C-17–C-23

*TRB—Teacher's Resource Book. MR—Mixed Review CT—Critical Thinking. TY—Test Yourself

CONNECTIONS							NCTM CORRELATION
Algebra	Critical Thinking	Problem Solving	Estimation/ Mental Math	Technology	Manipulatives/ Resources	Writing/Reading in Math	
	5c, 10				1-10	1-10	Communication Connections
W*: 38	C*: TD* 1-3 W: 37	W: 35-38			W: 7-10, 35-36	C: 3-8 W: 1-6, 39	Algebra Communication Problem Solving Connections
	C: TD 1-3 W: 46	W: 37-45	C: 1-8 W: 1-46		W: 37-40, 44-45	W: 46	Algebra Communication Problem Solving Statistics
C: 1-11, TD 1-2 W: 1-39, 41-43	C: TD 1-3 CT: 1-4	W: 41-43	W: 18-21, 36-39	W: 22-26		C: TD 1, 3 W: 40 CT: 4	Algebra Communication Problem Solving
C: 1-8, TD 1-2 W: 1-36, 38-47	C: TD 1-2 W: 42a, 43-47	W: 37, 43-47	W: 17-20 TY*: 7-9	W: 21-24	W: 25-36, 37a-c	W: 37, 42b, 43-47	Algebra Communication Problem Solving
C: 1-12, TD 3 W: 1-44, 46-47	C: 1-4, TD 1-3 CT: 1-4	W: 40-47	W: 18-22	W: 23-30, 47	CT: 3, 4 W: 47	C: 1-4 W: 45 CT: 1	Algebra Communication Problem Solving Statistics
	1-3, 6, 8, 10-14	1-15			1-5, 13-14	1b-c, 2, 3, 4b-c, 6, 8, 10, 12, 15	Communication Statistics Reasoning
	10	9		1-11	1-11	1-11	Communication Connections Statistics
	W: 7, 10	C: 1-3 W: 1-10	C: 1-3 W:1-2, 8	W: 1-10		W: 1-10	Communication Connections Problem Solving
C: 1-4, TD 2-3 W: 1-31 TY: 1-10	C: TD 1-3	C: 1-4 W: 22-31 TY: 10c	W: 9, 10	W: 22-23 TY: 1-6	W: 11-21, 24-26, 27a-b, 28-29	W: 29	Communication Problem Solving Statistics
		C: 1-3 W: 1-10		W: 1-10		C: 1-3 W: 1-10	Communication Problem Solving

*C—Class Exercises W—Written Exercises CT—Critical Thinking TY—Test Yourself TD—Think and Discuss

Exploring

In this manipulative activity, students use decimal square models to add tenths and hundredths. **See TRB Exploring O-9.**

Exploring Adding Decimals

You can use decimal squares to model sums of decimal numbers that have the same place values.

1. What number is represented by each model? Write the number under the model.

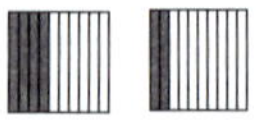

0.4 0.2

2. What equation can you write to express the sum represented by the models? **0.4 + 0.2 = 0.6**

You can also use decimal squares to model the sums of decimal numbers that have different place values.

3. What number is represented by each model? Write the number under the model.

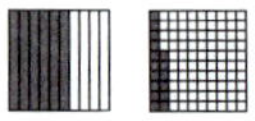

0.6 0.16

4. What equation can you write to express the sum represented by the models? **0.6 + 0.16 = 0.76**

Use decimal squares or graph paper to model each number. Then write an equation expressing the sum of the numbers.

5. six tenths; five tenths **0.6 + 0.5 = 1.1**
6. thirty-eight hundredths; fifty-five hundredths **0.38 + 0.55 = 0.93**
7. forty-five hundredths; two tenths **0.45 + 0.2 = 0.65**
8. three tenths; twenty-seven hundredths **0.3 + 0.27 = 0.57**
9. What different ways can you think of to model the sum 1.82? A possible example is 10 tenths and 82 hundredths.
Answers will vary.
Possible answers: 0.9 + 0.92; 1.8 + 0.02.

O-9

In this problem solving activity, students use thinking skills, and the strategy *guess and test*, to break a code. **See TRB Exploring O-10.**

Exploring Breaking a Code

Thinking skills and problem solving strategies are needed to break codes. In the codes below each symbol represents a number in an equation. Break the code by using the strategy guess and test, along with what you know about operations and properties.

Operation Symbols
◇ means +
▼ means −
☆ means ×
/ means ÷
| means =

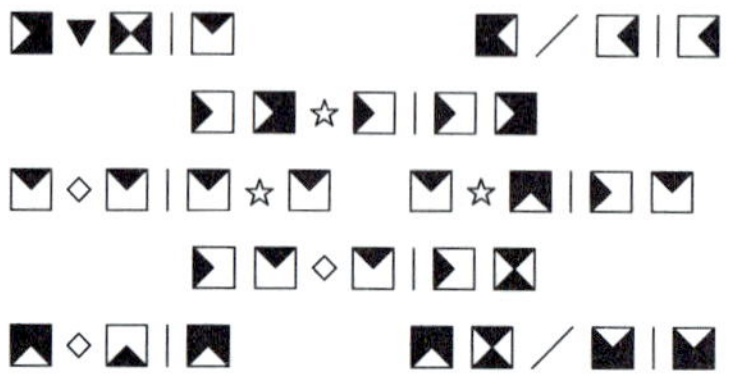

Decide what number each symbol represents. Draw the symbol here.
(Hint: Did you find the additive identity and the multiplicative identity?)

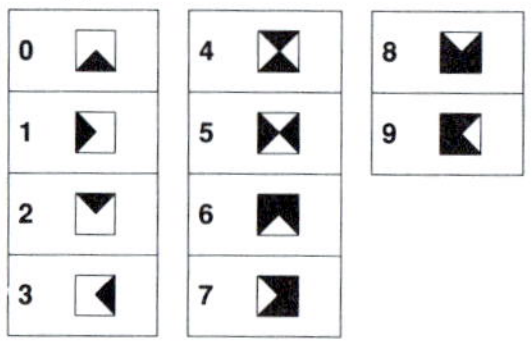

O-10

In this calculator activity, students use a calculator to solve problems involving formulas. **See TRB Exploring O-11.**

Exploring Formulas

You can use a calculator to help you with computations involving formulas.

Use the formula $d = rt$ to express distance traveled as a product of the rate of speed and the time traveled.

1. A car travels at the rate of 35 mi/h for 3.5 h. What distance did the car travel? **122.5 mi**
2. The same car travels at one and a half times the original rate for 3.5 h. What distance did the car travel?
rate **52.5 mi/h** distance **183.75 mi**
What happens to the distance when the rate increases? **Distance increases.**
How does the new distance compare with the original distance? **$1\frac{1}{2}$ times greater**
3. Suppose the same car travels at one and a half times the original rate and time. What are the rate, time, and distance?
rate **52.5 mi/h** time **5.25 h** distance **275.625 mi**
4. When the rate and time are both increased one and a half times, how does the distance compare with the original distance?
The distance is $2\frac{1}{4}$ times greater.
5. A car travels 135 mi in 4.5 h. What is the rate of travel? **30 mi/h**
6. A car travels 250 mi at the rate of 40 mi/h. How long did the car travel? **6 h and 15 min**

Use the formula $d = 16t^2$ to express distance in feet that an object will fall in t seconds. (Hint: $t^2 = t \cdot t$.)

7. How far will an object fall in 30 s? **14,400 ft**
8. If the time is doubled, how far will the object fall? **57,600 ft**
How does this distance compare with the original distance? **It is 4 times greater.**
9. If the time is halved, how far will the object fall? **3,600 ft**
How does this distance compare with the original? **It is $\frac{1}{4}$ the original distance.**
10. An object falls 200 ft. How long did it take to fall? Round your answer to the nearest tenth. **about 3.5 s**

O-11

In this computer activity, students learn about computer data bases and explore different ways to sort data. **See TRB Exploring O-12.**

Exploring Computer Data Bases

Suppose you are a devoted baseball fan. You use a computer data base to keep track of league, team, and player statistics. Here is part of that data base. It contains data about players who had batting averages over 0.320 for one season.

Name	League	Team	Average
Smith	National	Chicago Cubs	.324
Boggs	American	Boston Red Sox	.330
Larkin	National	Cincinnati Reds	.342
Gwynn	National	San Diego Padres	.336
Puckett	American	Minnesota Twins	.339
Clark	National	San Francisco Giants	.333
Harper	American	Minnesota Twins	.325
Lansford	American	Oakland Athletics	.336

Enter the above information into a data base.

1. How many fields does the data base contain? **4**
   a. What are the fields? **Name, League, Team, Average**
   b. How many records are there? **8**
2. Have the computer arrange the names in alphabetical order. Record the new order.
Boggs, Clark, Gwynn, Harper, Lansford, Larkin, Puckett, Smith
3. Have the computer sort the records by average. Which player had the highest batting average? Which player had the lowest?
Larkin, 0.342; Smith, 0.324
4. Have the computer sort the records to find the players whose average is greater than or equal to 0.330 and less than or equal to 0.340. Write the results.
Puckett, Lansford, Gwynn, Clark, Boggs
5. How else could you have the computer sort the records? **Answers may vary. Possible answers: by league; by team; by other ranges of batting averages.**
6. Research other baseball statistics. What additional fields might be useful to add to the data base?
Answers may vary. Possible answers: position; salary; number of years playing.

O-12

Meeting Individual Needs

GIFTED STUDENTS

Mathematically gifted students are often challenged by activities that allow them to refine their problem-solving skills. The pleasure such students take in mathematical thinking, and their perseverance, can also inspire other students to enjoy the trial and error nature of problem solving.

For the following activity, the class can be separated into small groups, with at least one gifted student per group. Explain that the task is for the groups to discover the formula for finding Fahrenheit temperature when all they know is that to find degrees Fahrenheit, you multiply a number by the Celsius temperature and either add or subtract 32. Their clues are that 212°F = 100°C, 32°F = 0°C, and 98.6°F = 37°C. Comment that students can use only this information to find the correct formula. (F = 1.8°C + 32) Then students can write the formula to find Celsius temperature when they know the Fahrenheit temperature.

Encourage students to brainstorm by writing and testing formulas. If a group is having trouble, point out that substituting 32°F = 0°C will tell them that 32 is added, not subtracted, in the formula. ($32 = n \cdot 0^0C + 32$) Students can then substitute 212°F = 100°C. ($212 = n \cdot 100 + 32$) Solving for n will give them the factor 1.8.

AT-RISK STUDENTS

To motivate the at-risk student, it is important to keep the classroom lessons relevant and to be supportive of the student's efforts and positive interactions.

In the following activity, students reinforce their ability to estimate with decimals. Note that the activity seeks to engage at-risk students in a meaningful activity. It also tries to provide an occasion for supportive social interaction.

Have students work in groups of four. Ask that each group bring in a cashier's receipt showing charges for ten or more items. The group will choose an estimation method and estimate the total, writing the problem and estimation strategy used on an index card. For example: Use front-end estimation to add these receipt items—$45.66, $109.72, $67.08, etc. Then ask groups to exchange receipts, using their method of estimation to find the new totals, until all groups have estimated all receipts. When every group has finished, invite the class to discuss solutions. Call on one member from each group to explain the group's estimation strategy.

ENGLISH AS A SECOND LANGUAGE

In this chapter, students are asked to recognize and verbalize numbers in which the place values extend to hundredths and thousandths. This can be challenging for English-language and ESL students alike. To help them, distribute copies of the place-value chart on page 100 and work with students as they practice recognizing and pronouncing names of the place values on the chart.

Student understanding can be further enhanced with the following activity: Ask the class to separate into pairs. (ESL students can be paired with English-proficient students who are best able to understand and communicate mathematical ideas.) Give one member of each pair a list of decimal numbers. Ask that student to recite the numbers while the other student writes them in their numerical form. Then have students switch roles. Monitor students to be sure that they are reading, verbalizing, and writing correctly.

See TRB Teaching Resources p. 53.

Problem Solving Situation

For the "Think About It" question, encourage students to think of factors that may affect the value of a coin. Make a list on the board. Lead students to suggest that the greater the number of proof sets produced, the less valuable the set becomes. Among the factors that affect the value of currency is the exchange rate. Ask students to collect and analyze data on the exchange rate between the dollar and another currency over a several week period. Have students choose a common currency, such as the lira or the pound.

DATA COLLECTION

Students can find the necessary data from a local bank. Have students keep their data in a table.

DATA ANALYSIS

Students can display the data in graph form. If possible, they should use a computer and a graphing program to construct their graph. A bar graph can help students visualize and compare data.

When students have organized and analyzed their data, ask them to look for trends and speculate on their causes.

A PRINTING PRESS can produce 8,000 sheets of 32 bills every hour.

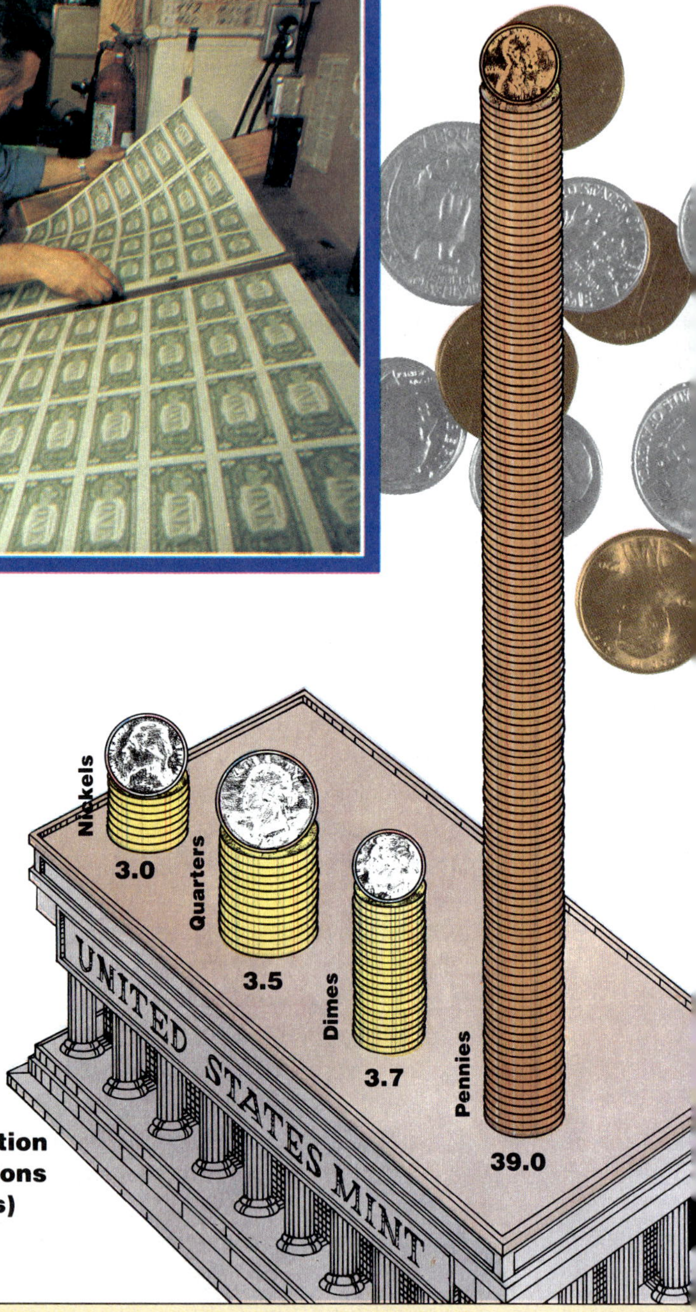

Daily Coin Production (in millions of coins)

CHAPTER 3

Decimals and Equations

Proof sets are special editions of a particular year's currency issued for collectors. A proof set comprises a penny, a nickel, a dime, a quarter, a half dollar, and, from 1973 to 1981, a silver dollar.

PROOF SETS

Year	Number Produced	Value			
		At time of Issue	in 1965	in 1968	in 1990
1936	3,837	$1.81	$400.00	$5,100.00	$5,050.00
1937	5,542	$1.81	$160.00	$3,500.00	$3,500.00
1940	11,246	$1.81	$50.00	$1,200.00	$1,450.00
1955	378,200	$2.10	$11.00	$60.00	$77.00
1961	3,028,244	$2.10	$3.00	$20.00	$16.00

Think about it...

Look at the information about proof sets. What factors do you think would increase or decrease the value of a coin?

See Problem Solving Situation

A dollar bill has a 16-mo. life expectancy

Amount of payment for a partially destroyed bill

FULL VALUE if $\frac{3}{5}$ or more is left

HALF VALUE if between $\frac{2}{5}$ and $\frac{3}{5}$ is left

NOTHING if less than $\frac{2}{5}$ is left

Questions related to this Data File can be found on pages 43, 57, 65, 82, 91, 147, 224, 236, 263, 278, 282, 327, 393, 387, and 568.

Cooperative Learning

To enable students to share ideas and support each other's efforts, have them work in groups of 4 or 6 to complete this project. In each group, pairs of students can collect the data. Then the groups can reconvene to compare, coordinate, and analyze the data collected. A role can be assigned to each group member to encourage individual participation. The roles might include: Recorder, Prober, Encourager, and Summarizer.

Explain to students that only one set of data will be collected from each group, and therefore everyone is responsible for the results. To foster individual responsibility, remark that you will randomly pick one student from each group to report the group's results to the class.

When the activity is completed, ask students to discuss the ways in which they functioned well as a group. Encourage students to examine the ways in which their group's interaction could be improved.

Lesson Focus

Materials

Decimal squares or graph paper, overhead transparencies, math journal

MOTIVATING THE LESSON

Show students a number line with two consecutive numbers marked. Explain that the types of numbers learned so far (integers) describe only the points on the number line, not the numbers between the points.

Instruction

1 If students are using decimal squares, have them select the model that shows one tenth. If students are using graph paper, have them draw a model similar to the one shown. It may be necessary to tell students to ignore the horizontal lines on the paper and have them make the vertical lines darker.

2 To help students understand decimals, have them explain the phrases *all but three tenths* and *all but seven tenths.* (three tenths are not shaded, seven tenths are not shaded)

OBJECTIVE:
To explore decimals using decimal square models.

MATERIALS

- Decimal squares or graph paper
- Math journal to record work

Exploring Decimals

▼ You can use decimal squares to model numbers less than 1.

1

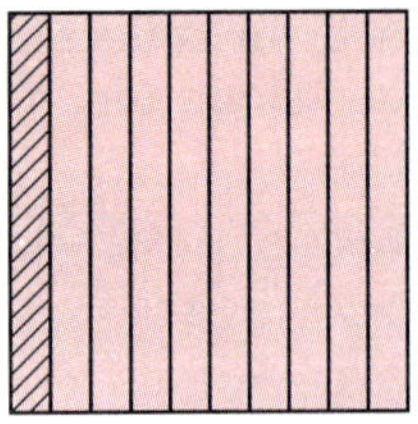

The figure above is divided into ten regions. Each region is called a *tenth.*

1. How many tenths are shaded? How many are not shaded? 9 1

2

2. ***Model*** each number using tenths' squares or graph paper. Check students' work.
   a. two tenths
   b. four tenths
   c. all but three tenths
   d. all but seven tenths

▼ In the figure below, each tenth has been divided into ten squares. Each small square is called a *hundredth.*

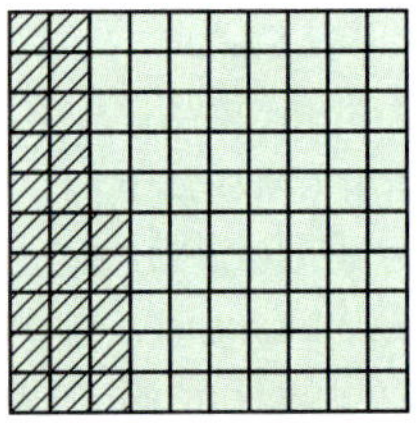

3. How many hundredths' are shaded? How many are not shaded? 25; 75
4. ***Model*** each number using hundredths' squares or graph paper. Check students' work.
   a. five hundredths
   b. twenty-six hundredths
   c. all but twenty hundredths
   d. all but ninety hundredths

Teaching Tip

Prepare two overhead transparencies: one with decimal and hundredth squares (no regions shaded), and one blank. Place the two transparencies on the overhead, with the blank one on top. Shade in the area of the decimal squares on the blank transparency.

▼ You can use decimal squares to find equivalent decimals.

5. a. *Model* three tenths using tenths' squares. Check students' work.

b. *Model* thirty hundredths using hundredths' squares. Check students' work.

c. *Describe* and discuss how your models are different. How are they alike?

d. Is three tenths equivalent to thirty hundredths? Explain.

▼ You can write each phrase as a decimal.

three tenths = 0.3 thirty hundredths = 0.30

Because three tenths and thirty hundredths are less than one, we write a zero as a place holder to the left of the *decimal point*. (3)

6. *Write* each phrase as a decimal.

a. four tenths 0.4 **b.** nine tenths 0.9 **c.** six tenths 0.6

d. eighty-nine hundredths 0.89 **e.** fifteen hundredths 0.15

7. *Determine* which of the following pairs of decimals are equivalent. ***Model*** the decimals to justify your answer. Check students' models.

a. three hundredths, three tenths not equivalent

b. seventy hundredths, seven tenths equivalent

c. 0.5, 0.57 not equivalent **d.** 0.4, 0.40 equivalent

5. c. One tenth is a larger unit than one hundredth. Three tenths and thirty hundredths cover the same amount of area.
5. d. Yes; you can cut out thirty hundredths and place it over the model of three tenths to show they are equivalent.

▼ You can compare decimals using models.

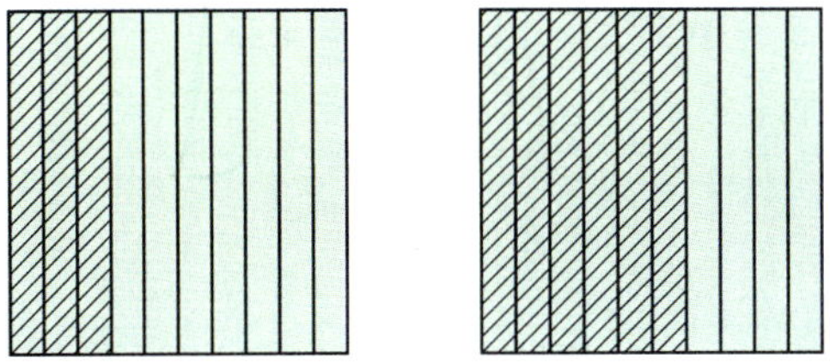

8. *Write* the decimals represented by the two models above. 0.3 and 0.6

a. Which decimal is greater? How do you know? 0.6; It covers more area.

b. Write an inequality using the decimals. $0.3 < 0.6$

9. *Determine* which decimal is greater. ***Model*** the decimals to justify your answer. Check students' models.

a. 0.60 or 0.65 0.65 **b.** 0.5 or 0.47 0.5

10. How can you compare the two decimals without using a model? (4)
Express each decimal to the hundredths' place. Read the numbers after the decimal point. Since $65 > 50$, then $0.65 > 0.50$.

(3) Comment that although it is not necessary to write a zero to the left of the decimal point for numbers less than 1, it is helpful to do so because it makes the decimal point much easier to spot.

(4) Some students may need help writing an answer. Remind them that the procedure for comparing decimals is the same as the one for comparing whole numbers. Have students write the answer in their math journals.

Closure

Review the concepts of equivalent and nonequivalent decimals with the students. Using decimal squares, model several numbers. Ask one student to select two models, name the decimals, write the decimals on the chalkboard, and compare them. The student may call on other students for assistance. Repeat this activity several times with different decimals.

As a follow-up, refer students to the manipulative activity found on p. 96E of the Chapter Overview. See also TRB Exploring O-9.

Assignment Guide

Basic All
Average All
Enriched All

FOR THE NEXT LESSON

decimal squares or graph paper, math journal

Lesson Focus

Materials

Decimal squares or graph paper, math journal

MOTIVATING THE LESSON

Ask students to consider the weight of their math textbooks. Tell them it weighs approximately 3 or 4 pounds. A more precise value for the weight means using a number between 3 and 4. Point out that other quantities, such as the distances between cities or Olympic record times, require precise values.

Instruction

1 Students should be able to read decimals by now. However, review with the class that when a decimal number is read, the word *and* is used only to represent the decimal point. The number 25.35 is read twenty-five *and* thirty-five hundredths. Ask students which place value is used to name the decimal. (the rightmost)

2 Review the number line. Explain that there is equal distance between each interval. When the interval between 2 and 3 is divided into small intervals, the distance between each interval will remain equal. When there are no marks to represent subintervals, estimation is used to position the decimal.

OBJECTIVE: *To compare, round, and order decimals.*

3-1 Decimals

▼ The smallest flowering plant is a water plant called duckweed. Its length is 0.02 in. and its width is 0.008 in. The length is read as *two hundredths* and the width is read as *eight thousandths.* The place value chart below shows how to read and write decimals.

1

hundred thousands	ten thousands	thousands	hundreds	tens	ones	and	tenths	hundredths	thousandths	ten-thousandths	hundred-thousandths
				7	2	.	9				
					0	.	0	0	2		

You read and write these decimals as *seventy-two and nine tenths* and *two thousandths*.

FLASHBACK

Zeros added to the right of a decimal do not change its value.
0.4 = 0.40 = 0.400

▼ You can graph decimals on a number line.

Example 1 **Give the decimal name for each point.**

2

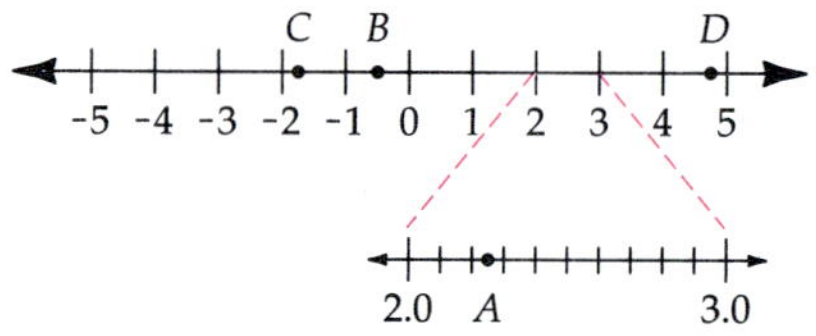

Solution $A = 2.25$ $B = -0.5$ $C = -1.75$ $D = 4.75$

Teaching Tip

To start an informal system of sharing creative ideas with fellow teachers, post an "idea" list in the teacher's lounge and invite everyone to add to it.

▼ You can compare decimal numbers using a model.

Example 2 **Compare the decimals 0.57 and 0.69 using a model.**

Solution

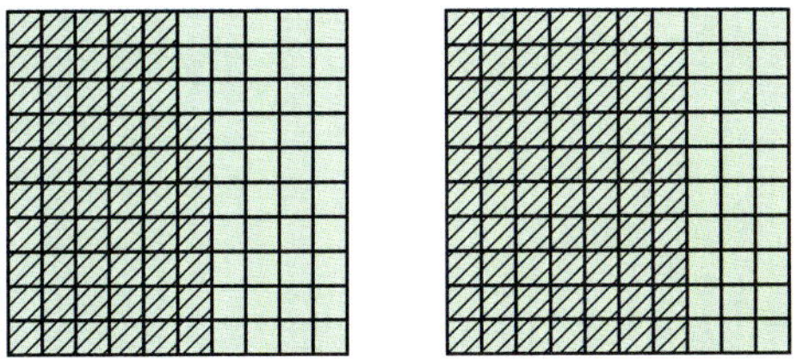

The area covered by 0.69 is greater than the area covered by 0.57. Therefore, 0.57 < 0.69.

▼ You can compare decimals using a number line.

Example 3 **Compare the decimals -0.5 and -1.25.**

Solution On a horizontal number line, numbers are greater as you move to the right.

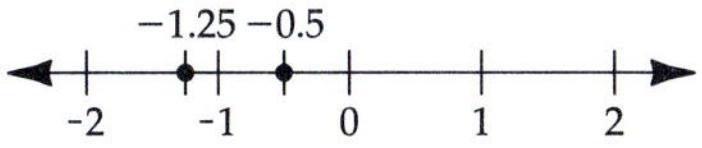

The decimal -0.5 is to the right of -1.25. So, -1.25 < -0.5.

▼ You can compare decimals by comparing corresponding digits.

Example 4 Does Wilmington or Philadelphia receive more rain? Use the data at the right.

Solution Compare the decimals 41.38 and 41.42.

The digits in the tens' and ones' places are the same. Compare the tenths' digits: 4 > 3. So, 41.42 > 41.38. Therefore, Philadelphia receives more rain.

▼ You can arrange decimals in order.

Example 5 **Order the cities according to their level of rainfall. Use the data at the right.**

Solution Compare the decimals 41.84, 41.42, 41.76, and 41.38.

The digits in the tens' and ones' places are the same. Compare the tenths' digits: 8 > 7 > 4 > 3. Therefore, 41.84 > 41.76 > 41.42 > 41.38. So, the cities ranked from greatest rainfall to least rainfall are: Baltimore, Raleigh, Philadelphia, and Wilmington.

THINK How could you compare the decimals 0.53 and 0.64 using the > symbol? 0.64 > 0.53

Average Yearly Rainfall

City	Amount (in.)
Baltimore, MD	41.84
Philadelphia, PA	41.42
Raleigh, NC	41.70
Wilmington, DE	41.38

3 Students should observe that the model with fewer squares shaded represents the lesser decimal. Help students understand that the inequality is read as fifty-seven hundredths *is less than* sixty-nine hundredths.

4 Draw a horizontal number line on the chalkboard with intervals from -3 through 3. Approximate the positions of 0.75, 1.25, and 2.5 and label them on the line. Ask students how they can tell which number is greater. (The number to the right on a number line is the greater number.) Now approximate the position of the negative numbers and several decimals. Ask students if the same rule applies. (yes)

5 Some students may find it helpful to write the numbers one above the other, with the decimal points lined up. Students can easily compare the value of the digits in each place. This works only for positive decimals.

ADDITIONAL EXAMPLES

1. Write each decimal.
   a. five and twelve thousandths (5.012)
   b. sixteen and four hundredths (16.04)
   c. the point midway between -3 and -4 on the number line. (-3.5)

2. Compare using >, <, or = .
   a. 4.04 ■ 4.2 (<)
   b. -0.08 ■ -0.08000 (=)
   c. 4.98 ■ -35.98 (>)

3. Write a decimal between 0.013 and 0.018. (Answers will vary. 0.0131, 0.014, 0.0179)

MATH MINUTES

Present the following decimals: 0.76, 1.114, and 9.99. For each decimal, have students give a decimal one tenth less and another decimal one tenth greater. (0.66, 0.86; 1.014, 1.214; 9.89, 10.09)

Guided Practice

Class Exercises Since the locations on the number line are approximate, answers to questions 9-12 will be estimates. Accept all answers close to those given.

Think and Discuss For question 1, have students identify several different decimals, all with the same two digits in the tenths place. Point out that the word *always* indicates that they have to find only one exception.

For question 3, have students replace *a* with several different positive and negative numbers.

A ***common error*** occurs when students count the number of decimal places to determine which one of two or more decimals is greatest. Show students the following decimals: 3.125, 3.124579, 3.2. Have them find the decimal with the smallest value. (3.124579)

Closure

Writing in Math Ask students to describe the procedure they use to compare two positive decimals and two negative decimals. Students should write the procedure in their math journals.

Independent Practice

For question 26, ask students why increasing the tenths digit will result in a wrong answer. It may help students to check their answers by graphing them on a number line.

To help students with questions 29-34, review the rules for rounding whole numbers. Then have students apply the same rules to round decimal numbers.

THINK AND DISCUSS See below.

1. The digit in the tenths' place of one decimal is less than the digit in the tenths' place of another decimal. Is the first decimal always less than the second? Explain.

2. Is 0.2 equivalent to 0.02? Why or why not?

3. For all nonzero decimals, is $a > -a$, where $-a$ is the opposite of a? Why or why not?

NOTES & QUOTES

Simon Stevin (1548–1620), a Flemish mathematician, extended decimal places to the right in 1585. Up until the sixteenth century, fractions were used instead of decimals.

1. No; you must compare the digits with the greatest place value first.
2. No; 0.2 is two tenths and 0.02 is two hundredths. Two tenths is greater than two hundredths.
3. No; $a > -a$ when a is positive and $a < -a$ when a is negative.

CLASS EXERCISES

Write the decimal represented by the shaded region.

1. 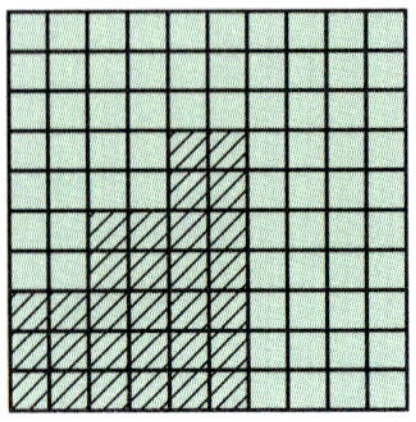 0.30

2. 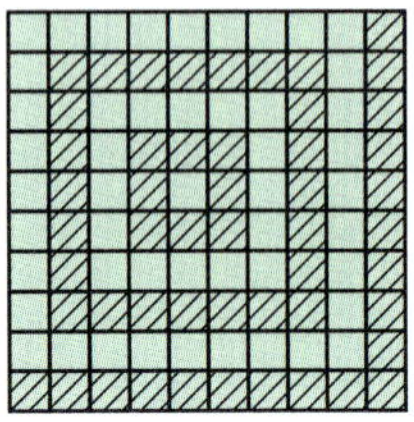 0.51

Read each decimal. Check students' work.

3. 0.42 **4.** 0.006 **5.** 4.801 **6.** 28.036

Write each decimal.

7. One kilometer is equivalent to *six hundred twenty-one thousandths* miles. 0.621

8. One fathom is *one and eight thousand two hundred eighty-eight ten-thousandths* meters. 1.8288

Give the decimal name for each point.

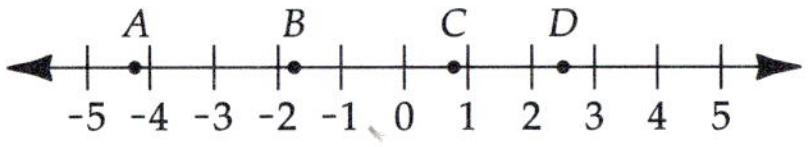

9. A -4.25 **10.** B -1.75 **11.** C 0.75 **12.** D 2.5

Compare using >, <, or =.

13. 0.6 ■ 0.06 > **14.** 0.84 ■ 0.840 = **15.** -3.862 ■ -3.859 <

WRITTEN EXERCISES

Write each decimal in words. Check students' work.

1. 0.83 **2.** 2.006 **3.** 392.9075 **4.** 0.00003

Write each decimal.

5. The world speed record for a motorcycle is *five hundred twelve and seventy-three hundredths* kilometers per hour. 512.73 km/h

6. The smallest book published has a length and width of *one and four tenths* millimeters. 1.4 mm

Graph each decimal on a number line. Check students' work.

7. 3.6 **8.** -2.9 **9.** |-9.65| **10.** -|-2.75|

Compare. Use >, <, or =.

11. 3.8 ■ 3.08 > **12.** -5.6 ■ -5.60 = **13.** -3.9 ■ -3.9000 =

14. -0.05 ■ 0.005 < **15.** -1.01 ■ -1.101 > **16.** 4.721 ■ 4.712 >

17. |-0.6| ■ |-0.09| > **18.** 24.3333 ■ 24.33333 <

Order from greatest to least.

19. 4.05, 4.5, 4.049 4.5, 4.05, 4.049

20. -4.98, -4.908, -4.098 -4.098, -4.908, -4.98

21. 0.03, 0.030008, 3.003, 0.30, 0.3002 3.003, 0.3002, 0.30, 0.030008, 0.03

22. 27.618, 27.681, -54.091, 27.6801, 54.0900 54.0900, 27.681, 27.6801, 27.618, -54.091

Write a decimal between the given decimals. Answers will vary. Example given.

23. 1.5 and 2.5 2.2 **24.** 0.6 and 0.9 0.7 **25.** 23.5 and 23.6 23.55

26. -0.5 and 0 -0.1 **27.** -0.678 and -0.679 -0.6785 **28.** 3.57 and 3.58 3.578

Round each decimal to the indicated place.

▼▼ *SAMPLE* Round 45.68 to the nearest tenth.

The number to the right of the tenths' place is eight. Since eight is greater than five, we increase the number in the tenths' place by one. So, 45.68 rounded to the nearest tenth is 45.7.

29. 0.76, nearest tenth 0.8 **30.** -9.095, nearest hundredth -9.10

31. 0.3632, nearest hundredth 0.36 **32.** 4.9677, nearest thousandth 4.968

33. 365,987.092, nearest integer 365,987 **34.** 5.9999, nearest thousandth 6.000

35. ***DATA FILE 10 (pp. 404–405)*** Order from least to greatest the home range in acres required by the species listed. Check students' work.

36. Order the data in the chart at the right from fastest winning speed to slowest winning speed of the Indianapolis 500.

37. Use the decimal 47.8364 to answer the following questions.

a. Will interchanging the tenths' and hundredths' digits produce a greater or lesser decimal? lesser

b. Will interchanging the hundredths' and thousandths' digits produce a greater or lesser decimal? greater

c. When will interchanging digits produce a greater decimal? when the digit to the right is greater than the digit to the left

38. Which decimals make each equation true?

a. $|x| = 0.03$ -0.03 or 0.03 **b.** $|a| = 80.123$ -80.123 or 80.123 **c.** $|x| = |-3.86|$ -3.86 or 3.86

39. ***WRITE*** Use absolute value to describe a method for deciding when one negative number is greater than another. If the absolute value of a negative number is less than the absolute value of another negative number, the first number is greater than the second number.

MIXED REVIEW

Solve each equation.

1. $-3 + x = -8$ -5

2. $y - 12 = -9$ 3

3. $-9y = 81$ -9

4. $\frac{a}{-3} = 15$ -45

5. What are two whole numbers that when added give you 10 and when multiplied give you 21? 3 and 7

6. Mario had d dollars in his bank account and wrote a check for $40. He then had $182 in his account. Write an equation for this problem and solve for d. $d - 40 = 182$; $222

Winning Speeds for Indianapolis 500

Year	Speed (mi/h)	
1979	158.899	8
1980	142.862	11
1981	139.029	12
1982	162.029	7
1983	162.117	6
1984	162.962	4
1985	152.982	9
1986	170.722	2
1987	162.175	5
1988	144.809	10
1989	167.581	3
1990	185.984	1

Lesson Follow-up

Reteaching Activity Students having trouble should work with the models in the Exploring Decimals lesson on page 98.

LESSON RESOURCES

TRB Practice C-1
TRB Enrichment C-2

Practice

For use after 3-1 (pp. 100–103)

Decimals

Write each decimal in words.

1. 7.316 **seven and three hundred sixteen thousandths**
2. 504.07 **five hundred four and seven hundredths**
3. 0.20008 **twenty thousand eight hundred-thousandths**
4. Graph each decimal on the number line. Write the letter above the point.

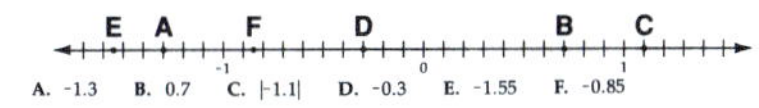

A. -1.3 B. 0.7 C. |-1.1| D. -0.3 E. -1.55 F. -0.85

Compare. Write >, <, or =.

5. -2.1 < 1.4 6. 5.3 < 5.33 7. -3 = -3.0
8. 9.80 > 9.08 9. -6.1 < -6.01 10. -8.336 > -8.363
11. 17.5 = 17.50 12. -0.03 < -0.003 13. 7.001 > 7.0001

Order from greatest to least.

14. 3.7, 3.07, 3.077 **3.7, 3.077, 3.07** 15. -1.9, -1.99, -1.09 **-1.09, -1.9, -1.99**
16. 0.06, 0.006, 0.60, 0.66 **0.66, 0.60, 0.06, 0.006**
17. -8.014, -8.0041, -8.401, -8.104 **-8.0041, -8.014, -8.104, -8.401**

Write a decimal between the given decimals. **Possible answers given.**

18. 0 and 0.1 **0.05** 19. 8.7 and 8.8 **8.75**
20. -0.3 and -0.2 **-0.25** 21. -1.44 and -1.43 **-1.435**

Round each decimal to the indicated place.

22. 0.5339, nearest hundredth **0.53** 23. -2.665, nearest tenth **-2.7**
24. 2.4051, nearest hundredth **2.41** 25. -0.9996, nearest thousandth **-1.000**
26. Order from fastest to slowest the winning speeds of the Olympic men's downhill skiing event. **1:45.50; 1:45.59; 1:45.73; 1:51.43; 1:59.85; 2:18.16**

Year	Time
1964	2 min 18.16 s
1968	1 min 59.85 s
1972	1 min 51.43 s
1976	1 min 45.73 s
1980	1 min 45.50 s
1984	1 min 45.59 s

Chapter 3 C-1

Enrichment

For use after 3-1 (pp. 100–103)

Hexadecimal Numbers

Numbers are written in the decimal number system using ten digits, 0 to 9. Numbers are written in the *hexadecimal* system using sixteen digits, 0 to 9, and the six letters, A through F.

Decimal	0	1	2	3	4	5	6	7	8	9	10	11	12	13	14	15
Hexadecimal	0	1	2	3	4	5	6	7	8	9	A	B	C	D	E	F

Hexadecimal numbers are based on the number 16. The value of each place in a hexadecimal number is 16 times that of the place to its right.

4,096's (16 × 256)	256's (16 × 16)	16's (16 × 1)	1's
C	8	E	5

Example Write $C8E5_{16}$ in base 10.

Solution
$$\begin{aligned} C8E5 &= (C \times 4{,}096) + (8 \times 256) + (E \times 16) + (5 \times 1) \\ &= (12 \times 4{,}096) + (8 \times 256) + (14 \times 16) + (5 \times 1) \\ &= 49{,}152 + 2{,}048 + 224 + 5 \\ &= 51{,}429 \\ C8E5_{16} &= 51{,}429_{10} \end{aligned}$$

Write in base 10.

1. D_{16} = **13**$_{10}$ 2. $5C_{16}$ = **92**$_{10}$ 3. $A9_{16}$ = **169**$_{10}$ 4. 27_{16} = **39**$_{10}$
5. $3F_{16}$ = **63**$_{10}$ 6. CD_{16} = **205**$_{10}$ 7. $F38_{16}$ = **3,896**$_{10}$ 8. $B46_{16}$ = **2,886**$_{10}$

Example Write 952_{10} in base 16.

Solution Divide by successively smaller base-16 place values. Begin with the largest place value smaller than the base-10 number. Divide remainders by the next smaller place values.

256)952 = ③ R 184
16)184 = ⑪ R 8
1)8 = ⑧

$952_{10} = 3B8_{16}$

Write in base 16.

9. 9_{10} = **9**$_{16}$ 10. 34_{10} = **22**$_{16}$ 11. 121_{10} = **79**$_{16}$
12. 200_{10} = **C8**$_{16}$ 13. 191_{10} = **BF**$_{16}$ 14. $1{,}701_{10}$ = **6A5**$_{16}$

C-2 *Chapter 3*

LESSON QUIZ

Write the decimal in words.
1. 0.231 (two hundred thirty-one thousandths)
Write the decimal.
2. one and six ten-thousandths (1.0006)
Order from greatest to least.
3. -9.03, -9.3, -9.33, -9.033 (-9.03, -9.033, -9.3, -9.33)
Round to the nearest tenth.
4. 3.0485 (3.0)

Assignment Guide

Basic 1-2, 5-8, 11-14, 19-21, 23-26, 29-31, 35-37, MR All
Average 2-3, 6, 8-9, 13-16, 19-21, 24-27, 30-32, 35-37, MR All
Enriched 3-4, 6, 9-10, 15-18, 20-22, 26-28, 32-35, 37-39, MR All

FOR THE NEXT LESSON

math journal, overhead projector, TRB Teaching Resources

Lesson Focus

Materials
math journal, overhead projector, TRB Teaching Resources

Vocabulary/Symbols
compatible numbers
front-end estimation
rounding

MOTIVATING THE LESSON
Identify several situations that involve decimals and do not require an exact answer, such as the cost of items on a shopping list or a tip in a restaurant. Ask students to identify other situations and describe how they would find the approximate answer.

Instruction

1 Ask several students to describe the procedure they would use to estimate the shaded area of each square. Reassure students that it is all right for their procedures and answer to be different because the answer will be an estimate. On an overhead projector, place a tenths square over each shaded square. Then have students estimate the shaded areas of the squares again and compare the estimates.

2 Remind students that in these examples the goal of rounding is to make it easier to estimate. For addition and subtraction, students may find it more convenient to round all numbers to the same place value. For multiplication, it is usually more practical to round all numbers to the leftmost place.

OBJECTIVE:
To estimate sums, differences, products, and quotients of decimals.

3-2 Estimating with Decimals

▼ You can estimate using decimals.

Example 1 **Estimate how much of each square is shaded. Write each estimate using decimal numbers.**

a. **b.**

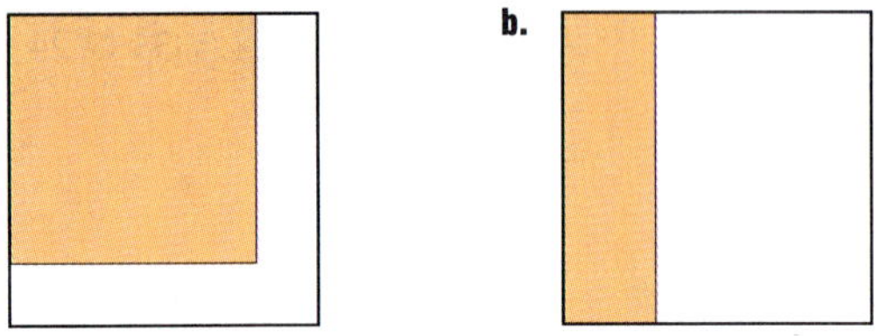

1

Solution **a.** 0.7 **b.** 0.3

▼ You can use *rounding* to estimate the sum, difference, product, or quotient of decimals.

FLASHBACK
The symbol ≈ means is approximately equal to.

Example 2 **Estimate the sum of 4.75, 2.2, and 9.86.**

Solution
$4.75 + 2.2 + 9.86$ Round to the nearest integer.
$\approx 5 + 2 + 10$ Add.
≈ 17
The sum is approximately 17.

2 **Example 3** **One orange contains 1.4 g of protein. Estimate how many grams of protein 2.5 oranges contain.**

Solution
$1.4 \cdot 2.5$ Round to the nearest integer.
$\approx 1 \cdot 3$ Multiply.
≈ 3
There are approximately 3 g of protein in 2.5 oranges.

Teaching Tip

Prepare a transparency with decimal and hundredth squares to use as a grid for estimating the areas of the shaded regions. The shaded regions should be on a separate transparency, and they should be the same size as the decimal and hundredth squares.

▼ You can use *front-end estimation* when adding decimals. This method is especially helpful when estimating dollar amounts.

Front-end Estimation	To use front-end estimation: 1. Add the front-end digits. 2. Adjust by estimating the sum of the remaining digits. 3. Add the two values.

Example 4 The junior class held three events to raise money for their prom. Estimate their profit.

Solution
1. Add front-end digits. $100 + 500 + 300$ is 900.
2. Adjust. $56.35 + 42.75 \approx 100$. $100 + 72.70 \approx 170$. ③
3. Add the two values. $900 + 170 = 1{,}070$

The class earned a profit of about $1,070.

Junior Class Fund Raisers

Event	Profit
Bake Sale	$156.35
Car Wash	$542.75
Raffle	$372.70

▼ You can estimate quotients using *compatible numbers*.

Compatible Numbers	Compatible numbers are two numbers that are easy to divide mentally.

Example 5 Taylor earns $13.75 per hour. Last week he earned $385. About how many hours did he work?

Solution
$385 \div 13.75$
$\approx 390 \div 13$ Use compatible numbers 390 and 13.
≈ 30

Taylor worked approximately 30 h. ④

1. grocery shopping, deciding how much money to take on a trip
2. 60
3. by rounding up and estimating, then rounding down and estimating

CLASS EXERCISES

Estimate using rounding. Accept reasonable estimates.

1. $-86.5 + 45.99 + (-91.21)$ -130
2. $858.32 - 281.319$ 600
3. $-92.81 \cdot (-48.33)$ 4,500
4. $318.09 \div 48.33$ 6

Estimate using front-end estimation. Accept reasonable estimates.

5. $3.75 + $14.10 + $23.30 $41
6. $88.50 − $29.60 $59

Estimate using compatible numbers. Accept reasonable estimates.

7. $0.8622 \div (-4)$ -0.2
8. $-43.08 \div 5.21$ -8

THINK AND DISCUSS See above.

1. In what situations would an estimate be preferred over an exact amount?
2. What is the quotient of $4{,}702 \div 81$ using rounding?
3. How could you get an estimate of the range within which an answer will fall?

MATH MINUTES

Students work in small groups to decide if each answer is reasonable.
1. $9.05 + 13.631 + 17.9$
Estimate: 42 (yes)
2. $89.672 - 1.9561$
Estimate: 70 (no)
3. $25.25 \cdot 0.5$
Estimate: 125 (no)

③ Remind students that when adding front-end digits, they must account for the place value of these digits.

④ Students have been taught to round 13.75 to 14. Explain that rounding is often not helpful in estimating a quotient because rounding may not yield a divisor that divides the dividend evenly. To be effective, students need to round the divisor and then round the dividend to a compatible number so that the remainder is zero.

ADDITIONAL EXAMPLES

1. Join the two squares in Example 1 to make a rectangle, then estimate the shaded region. (0.5)

Estimate using rounding.
2. $56.98 + 82.188$ (139)
3. $0.88 - 0.582$ (0.3)

Use front-end estimation.
4. $5.99 + 3.28 + 7.55$ (17)

Use compatible numbers
5. $7.028 \div 0.39$ (20)

Guided Practice

Class Exercises Have students discuss how they round numbers for questions 1-4.

For questions 7-8, ask students how they choose the compatible numbers.

Think and Discuss For question 3, ask students to describe situations when the estimate of a range is useful (to determine reasonable answers, to determine the amount of food needed for a party.)

A ***common error*** when using front end estimation with numbers containing a different number of digits is to add the front end digits as if they all represent the same place value. For example, in 398 + 27 + 116, students might add 300, 200 and 100 to get 600 instead of adding 300 and 100 to get 400. To help students avoid this error, have them identify the place value of the front-end digits and then add the digits.

Closure

Writing in Math Ask students to describe in their math journals the steps they use for each estimating technique: rounding; front-end estimation; and using compatible numbers. Have students work in pairs to compare the steps described and revise their descriptions if there are incorrect statements or omissions.

Independent Practice

For questions 21-28, ask students to discuss which estimation technique is appropriate for each question.

Ask students to tell whether they used rounding or front-end estimation to place each decimal point in questions 29-36 and to explain why they chose that technique.

WRITTEN EXERCISES

Estimate the shaded region using decimal numbers. Accept reasonable estimates.

1.  0.25

2.  0.30

Estimate using rounding. Accept reasonable estimates.

3. 34.99 + 27.302 60
4. 416.98 − 28.301 390
5. 0.08 · 400 40
6. 16.092 · 9.21 160
7. 329.08 · (-56) -18,000
8. $378.90 ÷ 42 10
9. 45.87 + 35.912 + 126.08 + 83.234 300
10. 0.043 + 0.0591 + 0.088 + 0.0241 + 0.0473 0.26

Estimate using front-end estimation. Accept reasonable estimates.

11. 3.57 + 2.95 + 1.681 8
12. $7.25 + $9.08 + $6.88 + $3.69 $27
13. 9.033 + 2.82 + 6.18 + 8.953 27
14. $9.01 + $8.94 + $5.63 + $6.48 + $8.23 $38

Estimate using compatible numbers. Accept reasonable estimates.

15. 9.392 ÷ 2.9 3
16. -483.09 ÷ 72.3 -7
17. -7.75 ÷ -1.98 4
18. $32.43 ÷ $4.68 $8
19. 0.5863 ÷ 26.2 0.02
20. $78.92 ÷ $8.55 $10

Estimate using the technique which seems best. Accept reasonable estimates.

21. $43.92 · 54 $2,000
22. −0.98 + (−0.34) + 0.66 -0.6
23. 0.083 + 0.149 0.23
24. 416.98 − 28.301 390
25. -18.9 · (-12.02) 200
26. 293.7 ÷ 42.03 7
27. -2.843 + (-5.022) + (-8.45) + (-3.991) -20
28. 21.88 + (-9.88) + 35.901 + 28.03 + (-13.99) + 26.92 100

Use estimation to place the decimal point in each answer.

29. 7.008 · 3.2 = 224256 22.4256
30. 98.003 (-1.8) = -1764054 -176.4054
31. 106.88 ÷ 0.5 = 21376 213.76
32. 14.39 + 6.132 + 0.684 = 21206 21.206
33. 94.02 + 9.011 + 18.34 + (-11.8) = 109571 109.571
34. 115.67 + 88.09 + (-113.6) = 9016 90.16
35. 46.872 · 0.05 + 65 = 673436 67.3436
36. 0.5 · 200.8 ÷ 2 = 502 50.2

MIXED REVIEW

Round to the nearest hundredth.

1. -0.883 -0.88
2. 4.1253 4.13

Find each product or quotient.

3. -3 · (-8) 24
4. 12 ÷ (-6) -2
5. -9 · 12 -108
6. -88 ÷ (-8) 11

Solve.

7. Jon baked three pans of muffins. Each pan holds 12 muffins. How many muffins did Jon bake? 36

ESTIMATION Use the table at the right to solve.

37. Can biscuits, puppy food, and a collar be purchased for $10? yes

38. Amy has $20 to buy a collar and a leash. With the remaining money, what is the greatest number of toys she can buy? 4

39. Todd is in the check-out line with all six items and only $17. What is the least expensive item he can put back and still pay for the other five items? food

Items to Buy for the New Puppy

Item	Cost
Leash	$6.37
Biscuits	$1.79
Food	$3.29
Collar	$4.37
Toy	$2.19
Shampoo	$1.97

Use the article below to answer each question. Accept reasonable estimates.

English, Anyone?

In 1989, the *Oxford English Dictionary* was revised for the second time. It was dedicated to Queen Elizabeth II. The dictionary consists of 20 volumes. It contains 21,728 pages and defines 616,500 words. The longest word defined is *pneumonoultramicroscopicsilicovolcanoconiosis*, a disease of the lungs. The 20 volumes weigh almost 138 lb and take up 45 in. of shelf space. Printing the first 10,000 sets of the dictionary required 6,243 lb of ink. The *Oxford English Dictionary* can be purchased for $2,500.

40. a. About how many pages are contained in each volume? 1,100

b. Estimate the weight of each volume. 7 lb

c. About how many inches of shelf would 6 volumes require? 14 in.

d. What is the approximate value of each volume? $120

Solve. Accept reasonable estimates.

41. Extra-large eggs cost $1.19 per dozen. Medium eggs cost $.98 per dozen. Estimate the savings on 10 dozen eggs if you buy medium instead of extra-large. $2

42. Grapes cost $1.14/lb. Estimate the cost of three bunches weighing 1.3 lb, 2.6 lb, and 1.9 lb. Explain your method. $6

43. The cost of sending a package is $23.80. Estimate the cost of sending 156 such packages. $3,200

44. ***DATA FILE 4 (pp. 138–139)*** About how many children are born in a minute? an hour? a day? a week? a month? a year? 300; 18,000; 432,000; 3,024,000; 12,096,000; 145,152,000

45. ***DATA FILE 9 (pp. 360–361)*** Estimate the total number of yards at St. Andrew's golf club for holes one through nine. about 3,500 yd

46. **WRITE** Explain how you might estimate the total cost of your purchases at the grocery store. Round each item to the nearest dollar.

Lesson Follow-up

Reteaching Activity Supply students with several copies of grocery tapes. Have students estimate the totals using front-end estimation and rounding. Students should compare their answers with a partner's and use a calculator to find the exact amount.

LESSON RESOURCES

TRB Practice C-3
TRB Enrichment C-4

Practice

For use after 3-2 (pp. 104–107)

Estimating with Decimals

Estimate how much of each region is shaded. Use decimal numbers. 1–34: Estimates may vary.

1. 0.75
2. 0.5
3. 0.3

Estimate using rounding.

4. 14.66 + 25.19 40
5. 8.7 + 3.21 + 3.899 16
6. 194.78 − 12.31 183
7. 1.88(8.13) 16
8. 80.4 ÷ 19.72 4
9. 36.19 + 18.4 + 23.85 + 11.9 90
10. 29.571 • 20.44 600
11. 507.639 ÷ 9.84 50.8

Estimate using front-end estimation.

12. 6.3 + 8.55 + 4.09 19
13. 34 + 68 + 85 190
14. 460 + 553 + 789 1,800
15. $6.14 + $9.38 + $2.53 $18
16. $39.65 + $25.84 + $11.45 $77
17. 19.71 + 3.94 + 8.4 32

Estimate using compatible numbers.

18. 11.57 ÷ 3.09 4
19. 43.68 ÷ 8.7 5
20. 29.5 ÷ 5.1 6
21. $41.09 ÷ $6.88 $6
22. 148.8 ÷ 9.8 15
23. $76.77 ÷ $24.19 $3

Estimate using the technique which seems best.

24. 6.3 + 8.7 + 24.8 + 10.2 50
25. 3.8(7.9) 32
26. 85.7 + 19.66 106
27. 15.831 ÷ 7.87 2
28. 521.063 − 25.19 495
29. $49.65 • 19.7 $1,000
30. 18.75 + (-14.33) + 36.5 + 21.31 + (-5.81) 57

Use estimation to place the decimal point in each answer.

31. 62.77(29.8) = 1 8 7 0.5 4 6
32. 16.132 ÷ 2.96 = 5.4 5
33. 85.2 + 39.6 + 14.97 = 1 3 9.7 7
34. Apples cost $.89 per lb. Estimate the cost of three 5-lb bags. $13.50

Chapter 3 C-3

Enrichment

For use after 3-2 (pp. 104–107)

Scrambled Digits

The answer to each problem is correct but the digits in each problem are scrambled. Unscramble the digits and write them in the correct order in the empty boxes.

1. 137 + 825 = 629 → 371 + 258 = 629
2. 8765 − 4329 = 2384 → 5678 − 3294 = 2384
3. 2273 + 8365 = 9405 → 3722 + 5683 = 9405
4. 91.4 × 63.8 = 748.84 → 19.4 × 38.6 = 748.84
5. 1.52 × 6.39 = 35.4816 → 5.12 × 6.93 = 35.4816
6. 67.027 ÷ 3.71 = 2.9 → 20.677 ÷ 7.13 = 2.9

C-4 Chapter 3

LESSON QUIZ

Estimate using the technique that seems best.

1. 23.87 · 92 (2,400)
2. 0.88243 + 0.074 (1.0)
3. $45.87 + $56.99 + $10.23 + $15.60 ($129)
4. 35.92 − 8.007 (28)
5. -3.021(-523.91) (1,500)

Assignment Guide

Basic 1-6, 11-12, 15-17, 21-24, 29-32, 37, 40-44, MR All

Average 1-2, 5-8, 12-13, 16-18, 23-26, 32-34, 37-38, 40-45, MR All

Enriched 2, 7-10, 13-14, 18-20, 25-28, 32, 36, 39-40, 43-46, MR All

FOR THE NEXT LESSON

math journal, calculator

Lesson Focus

Materials

Math journal, calculator

MOTIVATING THE LESSON

Ask students to name the property of addition or multiplication exhibited by these statements:

1. $3 + 2 = 2 + 3$
2. $(9 + 8) + 7 = 9 + (8 + 7)$
3. $(3 \cdot 5) + (3 \cdot 1) = 3(5 + 1)$
4. $x + (y + z) = (x + y) + z$
5. $6(a + b) = (6a + 6b)$
6. $a \cdot b = b \cdot a$

(1. commutative property,
2. associative property,
3. distributive property,
4. associative property,
5. distributive property,
6. commutative property)

Instruction

1 Reassure students that the process for evaluating an expression that contains decimals is similar to the process for evaluating expressions involving integers.

2 Ask students to identify the number of terms and the terms that are alike in the expression. (3; $3.1x$ and $8.4x$) Help students understand that if they use the distributive property to combine like terms, there will be fewer terms and the expression will be simplified.

ADDITIONAL EXAMPLES

1. Evaluate each expression for $x = 2.5$ and $y = -1.8$.
   a. $5.1 - y$ (6.9)
   b. $2x + 1.4y - 2.91$ (-0.43)
2. Simplify each expression.
   a. $4.8a - 9.01a$ (-4.21a)
   b. $3.5(2.1c) - 0.6(a - c)$

[$7.95c - 0.6a$]

OBJECTIVE: *To evaluate and simplify expressions involving decimals.*

3-3 Expressions with Decimals

▼ The number of hours a growing child should sleep each night depends on the child's age. You can find the recommended number of hours by evaluating the expression $17 - 0.5a$, where a is the child's age.

To find the number of hours a 7-year-old child should sleep in a night, evaluate the expression $17 - 0.5a$ for $a = 7$.

1 $17 - 0.5a = 17 - 0.5(7)$ — Replace a with 7.

$= 17$ [−] 0.5 [×] 7 [=] 13.5

A 7-year-old child should sleep 13.5 h.

▼ Expressions may contain more than one variable.

Example 1 **Use a calculator to evaluate $3.7a - 4b$ for $a = -3.2$ and $b = 6.1$.**

Solution $3.7a - 4b = 3.7(-3.2) - 4(6.1)$ — Replace a with -3.2 and b with 6.1.

3.7 [×] 3.2 [+/−] [−] 4 [×] 6.1 [=] -36.24

▼ You can simplify expressions involving decimals by using properties of addition and multiplication.

Example 2 **Simplify $3.1x + 2.3y + 8.4x$.**

Solution

2

$3.1x + 2.3y + 8.4x$	
$3.1x + 8.4x + 2.3y$	commutative property
$(3.1x + 8.4x) + 2.3y$	associative property
$(3.1 + 8.4)x + 2.3y$	distributive property
$11.5x + 2.3y$	Add coefficients.

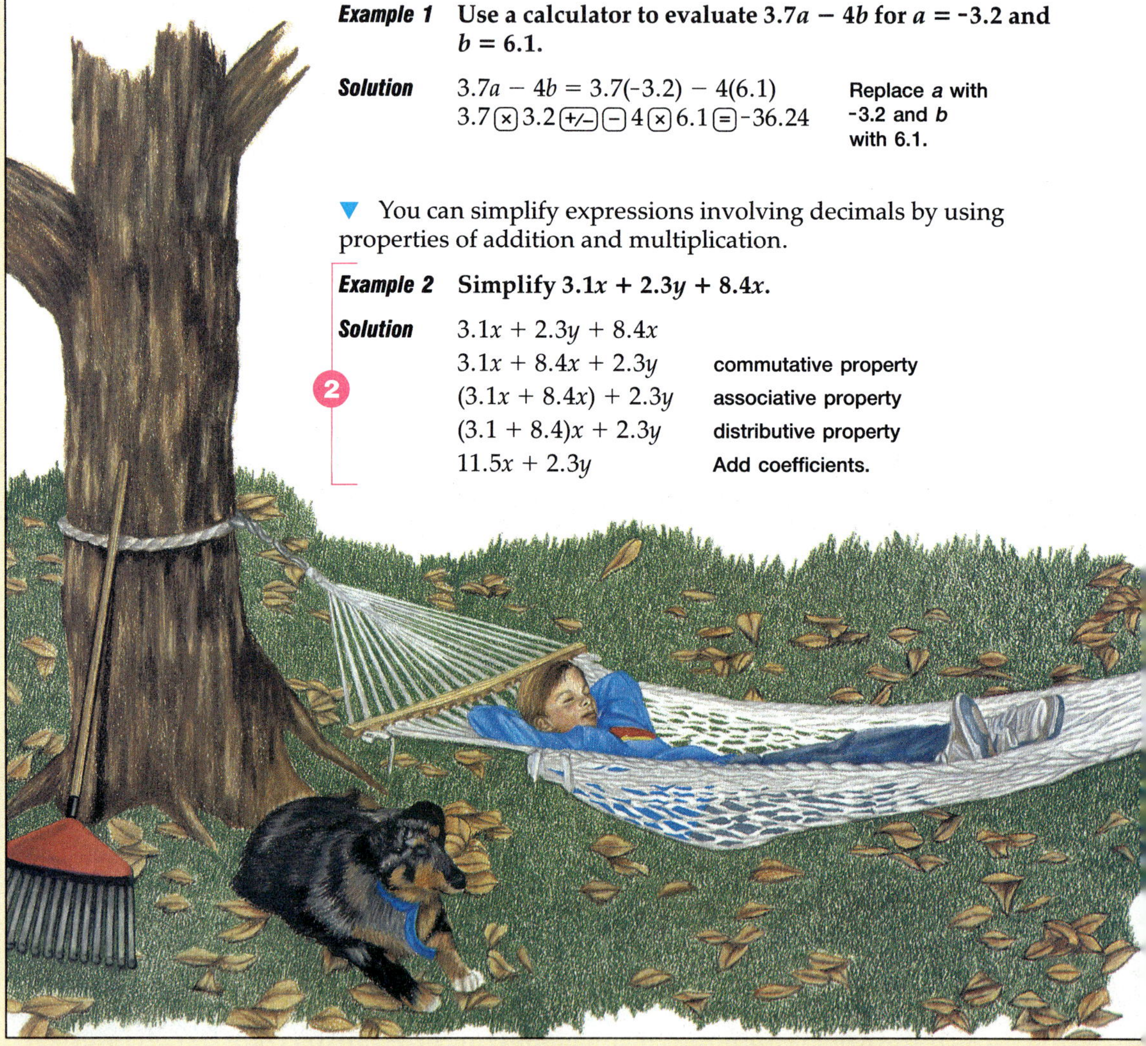

Teaching Tip

In your plan book or journal, keep track of specific problems students have in differentiating like terms. Address these problem areas when you review the chapter with the students. These notes will be valuable references to use in the future when teaching this concept.

CLASS EXERCISES

Evaluate each expression for $x = -1.9$ and $y = 2.4$.

1. $2.5x$ -4.75
2. $3 - y$ 0.6
3. $x - y$ -4.3
4. $x + y$ 0.5
5. $-x + 2y$ 6.7
6. $3x - 8 + 2y$ -8.9

Simplify each expression.

7. $2.1x + 3.4x$ 5.5x
8. $-1.2(7.9b)$ -9.48b
9. $-2(4.3a - 2.2a)$ -4.2a
10. $-9.4 + 3a + 16.25$ 3a + 6.85
11. $2.03b + 0.08a - 4.211b$ 0.08a − 2.181b

WRITTEN EXERCISES

Use the expression $17 - 0.5a$ to find the number of hours a child of each age should sleep each day.

1. $a = 11$ 11.5 h
2. $a = 1.5$ 16.25 h
3. $a = 5.25$ 14.375 h
4. $a = 0.5$ 16.75 h

Evaluate each expression for $x = 3.9$.

5. $-2x$ -7.8
6. $28.07 - x$ 24.17
7. $58.89 \div x$ 15.1
8. $x + (-4.03)$ -0.13

Evaluate each expression for $m = -7.06$ and $n = 13.2$.

9. $m + n$ 6.14
10. $m - n$ -20.26
11. $2m - 6.5$ -20.62
12. $-4m + 18.234$ 46.474
13. $-2m - 3n$ -25.48
14. $1.5(m + n)$ 9.21
15. $\frac{m - 4n}{4}$ -14.965
16. $\frac{3m - 2n}{-118.95}$ 0.4
17. $\frac{-n - m}{4 \div (3 - 1)}$ -3.07

MENTAL MATH Evaluate each expression.

18. $-100x$ for $x = -3.882$ 388.2
19. $50x$ for $x = -0.5$ -25
20. $x + y$ for $x = -8.22$ and $y = 8.22$ 0
21. $2x - y$ for $x = -4.22$ and $y = 12$ -20.44

CALCULATOR Evaluate each expression.

22. $3.98x$ for $x = -42.91$ -170.7818
23. $\frac{x}{9.8}$ for $x = 29.4098$ 3.001
24. $22.8x - 15.4y$ for $x = -0.092$ and $y = 21.3$. -330.1176
25. $(x - a) - (y - b)$ for $a = -4.96$, $b = 12.03$, $x = -2.3$, and $y = 7$. 7.69
26. $x(2.703y - 5.6701)$ for $x = 0.051$ and $y = -3.682$. -0.7967498

Simplify.

27. $2.4(-16.84w)$ -40.416w
28. $-5.23x \cdot 14.1$ -73.743x
29. $8.24m \div (-10.3)$ -0.8m
30. $3.78(4.01m) \div 0.02$ 757.89m
31. $-5.6x + 13.2x$ 7.6x

THINK AND DISCUSS See below.

1. Would the expression given at the beginning of this lesson work for a 36-year-old person? Explain.
2. For what numbers will the expression $-3.6w$ be positive? negative?
3. What does it mean to evaluate an expression?

1. No; the answer would be negative.
2. when w is negative; when w is positive
3. to find the value of the expression

Accept reasonable estimates.

MIXED REVIEW

Estimate each answer.

1. \$35.98 + \$155.23 \$200
2. 56,000.3 − 38,412.9 20,000
3. $-5.33 \cdot 0.992$ -5
4. $0.9341 \div 8.1$ 0.1

Solve each equation.

5. $x + (-23) = -19$ 4
6. $20 = a - 15$ 35

Solve.

7. Ricky bought 9 pencils at \$.23 per pencil and 5 pens at \$.37 per pen. How much more did he spend on pencils than pens? \$.22

Guided Practice

Class Exercises For questions 7-11, have students identify like terms in each expression before they try to simplify the expression.

Think and Discuss For question 1, ask students to determine the range of ages that is reasonable for this formula.

For question 2, remind students that they are not finding the exact value of w, but the kind of numbers the answer will be.

A ***common error*** is to think that unlike terms—such as $2a$ and ab—that contain the same variable are like terms. Ask students to describe the procedures they use for identifying like terms. Help students develop a method to correctly identify like terms.

Closure

Writing in Math Ask students to write a description in their math journals of the procedure for evaluating an expression such as $4.5y - x$ for $y = -3.8$ and $x = -2.1$.

Independent Practice

Review the use of the memory features on the calculator for questions 22-26.

MATH MINUTES

Students work in small groups to write four expressions, and evaluate the expressions on a separate page. Groups switch expressions, evaluate them, and compare answers.

Lesson Follow-up

Reteaching Activity Provide students with several expressions and a list of values. Have students substitute values in each expression and evaluate.

LESSON RESOURCES

TRB Practice C-5
TRB Enrichment C-6

Practice For use after 3-3 (pp. 108–110)

Expressions with Decimals

Find the value of $38 - 0.6n$ for the given value of n.

1. $n = 5$ **35** 2. $n = 4.2$ **35.48** 3. $n = 52$ **6.8**

Evaluate each expression for $h = 8.5$.

4. $-7.4h$ **-62.9** 5. $31.97 + h$ **40.47**
6. $-9.4 - h$ **-17.9** 7. $84.15 \div h$ **9.9**

Evaluate each expression for $x = 4.37$ and $y = -2.9$.

8. $x + y$ **1.47** 9. $x - y$ **7.27**
10. $-x + 8.55$ **4.18** 11. $2y + 3x$ **7.31**
12. $-4y - 17.9$ **-6.3** 13. $2.2(y - x)$ **-15.994**
14. $\frac{2x + 0.36}{5}$ **1.82** 15. $\frac{2x + y}{-0.73}$ **-8**
16. $\frac{4 - y}{x + 18.63}$ **0.3** 17. $x(1 - 8y)$ **105.754**

MENTAL MATH **Evaluate each expression.**

18. $-10m$ for $m = 7.381$ **-73.81** 19. $e + f$ for $e = 8.7$ and $f = -3.7$ **5**
20. $0.01k$ for $k = 2{,}371.6$ **23.716** 21. $5p$ for $p = -3.07$ **-15.35**

CALCULATOR **Evaluate each expression.**

22. $\frac{45.549}{x}$ for $x = -7.23$ **-6.3** 23. $-0.84x$ for $x = -1.23$ **1.0332**
24. $6.2x - 3.9y$ for $x = -0.07$ and $y = 1.4$ **-5.894**
25. $x - y - x(32 - 5y)$ for $x = 3.4$ and $y = 2.52$ **-65.08**

Simplify.

26. $19.4 \cdot 7.7x$ **149.38x** 27. $6.154x \div 1.81$ **3.4x**
28. $9.85(4.71x) \div 1.97$ **23.55x** 29. $44.73x - 72.19x$ **-27.46x**

ESTIMATION **Estimate the value if $x = 17.82$ and $y = 3.17$.** **Estimates may vary.**

30. $x + y$ **21** 31. $x - 3y$ **9**
32. $x \div 2y$ **3** 33. $40 - 2x$ **4**

Solve.

34. The total cost, including tax, of an item selling for x dollars is $1.05x$. Find the total cost of a chair that sells for $240. **$252**

Chapter 3 C-5

Enrichment For use after 3-3 (pp. 108–110)

Using Decimal Expressions

Solve.

1. $K = C + 273.15$ K = Kelvin temperature (°K); C = Celsius temperature (°C)
   a. Find the Kelvin temperature corresponding to a temperature of 538.98°C. **812.13°K**
   b. Find the Celsius temperature corresponding to a temperature of 481.16°K. **208.01°C**
2. $F = 1.8C + 32$ F = Fahrenheit Temperature (°F); C = Celsius Temperature (°C)
   Find the Fahrenheit temperature corresponding to a temperature of 55.5°C. **131.9°F**
3. $R = 39.95d + 0.14m$ R = cost of rental car ($); d = number of days rented; m = number of miles driven
   Find the rental charge.
   a. 5 days, 856 miles **$319.59** b. 2 days, 137 miles **$99.08**
4. $T = c + 0.055c + 0.0106c$ T = total cost of item including 5.5% sales tax and 1.06% federal excise tax ($); c = marked price of item ($)
   Find the total cost of an item marked $960. **$1,022.98**
5. $P = 0.444f$ (ocean water) $P = 0.433f$ (fresh water)
   P = water pressure $\left(\frac{\text{lb}}{\text{in.}^2}\right)$ f = depth (f)
   a. Find the water pressure at a depth of 35,820 ft, the deepest ocean descent by humans. **15,904.08 lb/in.²**
   b. Find the difference between ocean pressure and fresh-water pressure at a depth of 200 ft. **2.2 lb/in.²**

C-6 Chapter 3

32. $0.007m - 0.04m$ **-0.033m**

33. $9.0578a - 4(6.057 - 2.0473a)$ **17.247a − 24.228**

34. $\frac{-4.79x + 1.79x}{0.003}$ **-1,000x**

35. $\frac{(0.2x)(4.98) - 37.2x}{20}$ **-1.8102x**

ESTIMATION **Estimate the value if $x = 38.953$ and $y = 127.06$.** **Accept reasonable estimates.**

36. $x + y$ **170** **37.** $x - y$ **-90** **38.** $3x - 10$ **110** **39.** $\frac{y}{x}$ **3**

40. ***WRITE*** a paragraph explaining how to find the value of $3.8 + (-7.1)x$ if x is -4.25. Tell how to find the sign as well as the numerical value of the answer. **Multiply (-7.1) by (-4.25). The product is positive. Then add to 3.8.**

Solve.

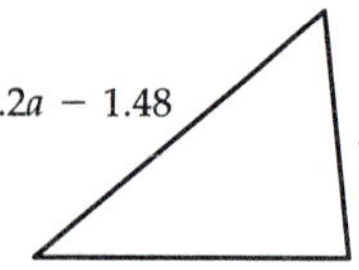

41. a. Express the perimeter of the triangle in simplest form. **12.3a − 1.08**

b. What is the perimeter if $a = 8.401$? **102.2523**

42. The interest earned on a bank account is $0.087x$, where x is the amount in the account. Find the interest earned on $10,000. **$870**

43. The selling price of a videotape is $c + 3.24$, where c is the cost of the tape. Find the selling price of a tape costing $18.43. **$21.67**

Critical Thinking

EXPLORING PATTERNS IN DIVISION

1. Follow the steps in the division problem below. Continue the pattern and describe what happens. **See left.**

0.2 and 1.6	Begin with two numbers.
1.6 ÷ 0.2 = 8	Divide the second by the first.
8 ÷ 1.6 = 5	Divide the quotient by the number above it.
5 ÷ 8 = 0.625	Continue dividing each quotient by the number above it for four more divisions.
0.625 ÷ 5 = 0.125	

2. Follow the same pattern with the numbers 2.5 and 2. Describe your results. Did it matter that you began with a decimal number and a whole number? **See left. By following the pattern, you arrive at the numbers you began with; no.**

3. Follow the pattern with the numbers 2 and -0.2. Describe your results. Did it matter that you began with a positive and a negative number? **See left. By following the pattern, you arrive at the numbers you began with; no.**

4. ***Write*** a paragraph describing this pattern. **By following this pattern, you arrive at the numbers you began with. It does not matter whether you use decimals, whole numbers, positive numbers, or negative numbers.**

1. 0.125 ÷ 0.625 = 0.2
0.2 ÷ 0.125 = 1.6
1.6 ÷ 0.2 = 8

2. 2.5 and 2
2 ÷ 2.5 = 0.8
0.8 ÷ 2 = 0.4
0.4 ÷ 0.8 = 0.5
0.5 ÷ 0.4 = 1.25
1.25 ÷ 0.5 = 2.5
2.5 ÷ 1.25 = 2
2 ÷ 2.5 = 0.8

3. 2 and -0.2
-0.2 ÷ 2 = -0.1
-0.1 ÷ (-0.2) = 0.5
0.5 ÷ (-0.1) = -5
-5 ÷ 0.5 = -10
-10 ÷ (-5) = 2
2 ÷ (-10) = -0.2
-0.2 ÷ 2 = -0.1

LESSON QUIZ

Evaluate each expression for $x = -4.2$ and $y = 0.8$.

1. $4x - y$ (-17.6)
2. $-2.1x + 3y$ (11.22)
3. $100(x - y)$ [-500]

Simplify.

4. $-9.2(4.2xy)$ [-38.64xy]
5. $3.072a - 5.8(22.07 - 2.088a)$ [15.1824a − 128.006]

Assignment Guide

Basic 1-15 odd, 18-24 even, 27-31, 36-37, 40-41, MR, CT All

Average 2-16 even, 19-25, 29-33, 37-40, 42, MR, CT All

Enriched 3-4, 7-8, 13-21, 24-26, 31-40, 43, MR, CT All

FOR THE NEXT LESSON

math journal, calculator, number cubes

OBJECTIVE:
To solve addition and subtraction equations involving decimals.

3-4 Addition and Subtraction Equations

▼ Earth is 93 million miles from the sun. Mars is 141.71 million miles from the sun. What is the minimum number of miles from Earth to Mars? Let m represent the minimum number of millions of miles from Earth to Mars. You can describe the situation by the following equation. (1)

$$93 + m = 141.71$$

You can solve this equation by using the subtraction property of equality.

$$93 + m = 141.71$$
$$93 + m - 93 = 141.71 - 93$$ Subtract 93 from each side.
$$m = 48.71$$

(2) **Check** $93 + m = 141.71$
$93 \boxed{+} 48.71 \boxed{=} 141.71$ ✓ Replace m with 48.71.

The minimum distance from Earth to Mars is 48.71 million miles.

▼ You can solve an equation involving subtraction by using the addition property of equality.

(3) **Example 1** **Solve $n - 29.1 = -30.85$ and graph the solution.**

Solution
$$n - 29.1 = -30.85$$
$$n - 29.1 + 29.1 = -30.85 + 29.1$$ Add 29.1 to each side.
$$n = -1.75$$

Check $n - 29.1 = -30.85$
$-1.75 \boxed{-} 29.1 \boxed{=} -30.85$ ✓ Replace n with -1.75.

Graph -1.75 on the number line.

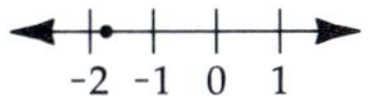

THINK How could you use estimation to decide if the solution is reasonable? Round each number to the nearest integer. Solve the equation. $-2 \approx -1.75$.

We know more about Mars than any other planet except Earth. Most of our information is from Mariner 9.

RESEARCH Find out when the Mariner 9 orbited Mars and what some of its findings were.

Teaching Tip

To encourage more student-to-student contact during class discussions, ask students if they agree with another student's answer or explanation. If there is a disagreement, help students keep the discussion positive.

Lesson Focus

Materials

Math journal, calculator, number cubes

MOTIVATING THE LESSON

Write the following equations on the chalkboard and then ask students which property of equality they would use to solve each equation: $a + 7 = 28$ (subtraction, $a = 21$), $b - 13 = 32$ (addition, $b = 45$). Ask students to describe how they solved each equation. Then write these equations and ask the same questions: $c + 5.3 = 9.2$ (subtraction, $c = 3.9$) $d - 10.6 = 4.2$ (addition, $d = 14.8$)

Help students understand that they use the same properties for equations with decimals as they do for equations with integers.

Instruction

(1) Point out to students that drawing a diagram is often useful in solving word problems. Ask them to label all the known and unknown distances.

(2) Point out that to check an answer, it is important to substitute the value for the variable in the equation.

(3) Ask students why the addition property of equality is used in an equation involving subtraction. (Addition is the opposite of subtraction and undoes it.) Ask students how they know which number to add to each side of the equation. (Since 29.1 is the value that is subtracted, it is also the value to add.)

4 Ask students why the opposite of 2.02 and not the opposite of 5 is added to both sides of the equation. (Because 2.02 is on the same side as the variable and the goal is to get the variable by itself.) Then ask which property is used to solve $2.98 = a + 0$. (additive identity property)

ADDITIONAL EXAMPLES

1. Solve each equation using the addition or subtraction property of equality.

a. $m - 3.8 = -10.22$ $(m = -6.42)$

b. $4.9 = 0.08 + x$ $(x = 4.82)$

2. Solve each equation using opposites.

a. $w + 4.02 = -8.41$ $(w = -12.43)$

b. $b - 0.003 = 0.0211$ $(b = 0.0241)$

3. Solve and graph: $p - 38.4 = -29.9$. $(p = 8.5)$

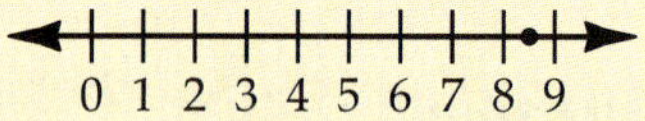

Guided Practice

Class Exercises Ask students to describe two ways to do the first step for questions 1-4.

For questions 5-8, ask students to describe step-by-step how they will solve each equation.

Think and Discuss Remind students not to solve the equation in question 1. They can answer the question by comparing 38.1 and 14.07.

To help students understand question 2, have them write $a - 15.23$ as $a + (-15.23)$.

▼ You can also use opposites to solve addition and subtraction equations.

Example 2 **Solve $5 = a + 2.02$ by using opposites.**

4 ***Solution***

$$5 = a + 2.02$$
$$5 + (-2.02) = a + 2.02 + (-2.02) \quad \text{Add -2.02 to each side.}$$
$$2.98 = a + 0$$
$$2.98 = a$$

The check is left for you.

THINK AND DISCUSS See below.

1. Will the solution to $38.1 + x = 14.07$ be greater than or less than zero? Explain.

2. What properties would you need to use to show $a - 15.23 + 15.23$ is equivalent to a?

1. Less than zero; a greater number is being subtracted from a lesser number.
2. associative property, additive identity

CLASS EXERCISES

State the first step in solving each equation.

1. $x + 4.9 = 18.8$ Sub. 4.9.

2. $a - 19.2 = 24$ Add 19.2.

3. $12.703 = n - 16.51$ Add 16.51.

4. $3.78 + m = 0$ Sub. 3.78.

Solve each addition equation.

5. $a + 0.98 = 0.24$ -0.74

6. $-4.26 + c = 22.991$ 27.251

Solve each subtraction equation.

7. $m - 43.23 = 80.9$ 124.13

8. $-0.09 = a - 0.224$ 0.134

WRITTEN EXERCISES

Solve each equation using the subtraction property.

1. $x + 4.38 = -9.011$ -13.391

2. $0 = y + 39.4$ -39.4

3. $1.77 + c = 3.4$ 1.63

4. $-8.32 = y + 3.211$ -11.531

Solve each equation using opposites.

5. $0.402 + c = 0.0322$ -0.3698

6. $s + 12.85 = 4.9$ -7.95

7. $-0.021 = x + 0.0023$ -0.0233

8. $309.462 + y = 500$ 190.538

Solve each equation using the addition property.

9. $11.03 = w - 1.55$ 12.58

10. $8.9 = b - 8.88$ 17.78

11. $-3.98 = m - 5.012$ 1.032

12. $u - 400.32 = -912.268$ -511.948

Solve each equation using opposites.

13. $w + (-5.07) = 8.24$ 13.31

14. $0 = n - 29.335$ 29.335

15. $-45.02 = m - 21.9$ -23.12

16. $t + (-0.66) = 0.66$ 1.32

MATH MINUTES

Write the following equation on the board: $x + \blacksquare.\blacksquare = 9.3$. Have students roll a number cube twice to fill in the digits of the decimal. Then solve the equation.

MENTAL MATH Solve each equation.

17. $a - 2.33 = 2.33$ 4.66

18. $t + 45.023 = 45.023$ 0

19. $9.45 = w + 7.45$ 2

20. $m - 0.003 = 18.29$ 18.293

CALCULATOR Solve each equation.

21. $60{,}000 = w - 392.0034$ 60,392.0034

22. $-0.34264 + x = 5.920154$ 6.262794

23. $m + 912.87 = 920.001$ 7.131

24. $t - 982.0012 = -893.20876$ 88.79244

Solve each equation using any method. Graph each solution. Check students' work.

25. $y + 0.05 = 3.95$ 3.9

26. $t - 79.4 = -46.7$ 32.7

27. $m - (-0.88) = 0.88$ 0

28. $-23.9 = x + 14.1$ -38

29. $p - 18.8 = -24.2$ -5.4

30. $k + (-35.9) = 24.8$ 60.7

31. $48.003 + r = 50.903$ 2.9

32. $b + 6.7 = 9.90$ 3.2

33. $w - (-0.34) = 0.74$ 0.4

34. $t - 43.8 = 0$ 43.8

35. $b + 16 = -43.9$ -59.9

36. $-1.78 = v - 0.98$ -0.8

Use the article below and the table at the right to answer each question.

pHase Out Acid Rain

The pH scale measures the acidity of a substance. The scale ranges from most acidic, 0, to least acidic, 14. Many plants and animals can only survive in a narrow range of acidity. Acid rain changes the acidity level of the water and soil where organisms live. As a result, many plants and animals can be severely damaged or even die. Studies have shown that industry and motor vehicle emissions are among the major sources of acid rain.

Average PH Levels

Item	PH
Unpolluted rain	5.6
Acid rain	4.6
Grapes	4.1
Grapefruit	3.2
Orange	3.5
Apple	3.1
Sea water	7.36–8.21
Milk	6.5

37. a. Order the fruits from most acidic to least acidic. apple, grapefruit, orange, grape

b. Hydrangeas have blue flowers when grown in soil with a pH level less than 7. The flowers are pink when grown in soil with a pH level greater than 7. What color would the flowers be if grown in soil having the same pH level as grapes? blue

c. ***PROJECT*** Research smog, another form of air pollution. Write a short paragraph. Be sure to include the pH level. Check students' work.

Solve each equation for *a*.

38. $a - b = c$ $a = c + b$

39. $b = a + c$ $b - c = a$

40. $-c + a = b$ $a = b + c$

41. Find values for x and y that solve both equations: $x + y = 0.03$ and $x - y = 0.13$. $x = 0.08$ and $y = -0.05$

A ***common error*** occurs when students do not perform the same operation on both sides of the equation. For example, in the equation $x + 3.6 = 15.6$, students may subtract 3.6 from one side, and arrive at the answer $x = 15.6$; or, they may add 3.6 to the other side. Remind students that the equation must remain balanced, i.e. whatever operation is done to one side must be done to the other side. Also, have students check their answer in the equation to help catch errors.

Closure

Writing in Math Have students write a description in their math journals of the concepts of *equality, the addition property of equality,* and *the subtraction property of equality.*

Independent Practice

For questions 25-36, it may be helpful to have students work in pairs.

For question 37c, ask students to work in small groups. If your city has a smog problem, have students find out what is being done to reduce smog. Remind students to list their sources.

Lesson Follow-up

Reteaching Activity Provide a worksheet of problems in this format:

$$w - 4.8 = 2.11$$

$$w - 4.8 + \_\_\_\_ = 2.11 + \_\_\_\_$$

Have students find each solution and explain what they did.

LESSON RESOURCES

TRB Practice C-7
TRB Enrichment C-8

Practice

For use after 3-4 (pp. 111–114)

Addition and Subtraction Equations

Solve each equation using the subtraction property.

1. $9.36 + k = 14.8$ **k = 5.44**
2. $-22 = p + 13.7$ **p = -35.7**
3. $y + 3.85 = 2.46$ **y = -1.39**
4. $-13.8 = h + 15.603$ **h = -29.403**

Solve each equation using opposites.

5. $x + 82.7 = 63.5$ **x = -19.2**
6. $-0.08 = f + 0.07$ **f = -0.15**
7. $0 = a + 27.98$ **a = -27.98**
8. $117.345 + m = 200$ **m = 82.655**

Solve each equation using the addition property.

9. $3.8 = n - 3.62$ **n = 7.42**
10. $x - 19.7 = -17.48$ **x = 2.22**
11. $12.5 = t - 3.55$ **t = 16.05**
12. $k - 263.48 = -381.09$ **k = -117.61**

Solve each equation using opposites.

13. $y - 48.763 = 0$ **y = 48.763**
14. $6.21 = e + (-3.48)$ **e = 9.69**
15. $x + (-0.0025) = 0.0024$ **x = 0.0049**
16. $-58.109 = v - 47.736$ **v = -10.373**

MENTAL MATH Solve each equation.

17. $k + 23.7 = 23.7$ **k = 0**
18. $5.63 = n + 1.63$ **n = 4**
19. $x - 3.2 = 4.1$ **x = 7.3**
20. $p - 0.7 = 9.3$ **p = 10**

CALCULATOR Solve each equation.

21. $-8.557 + y = 17.498$ **y = 26.055**
22. $265.89 = n - 188.775$ **n = 454.665**
23. $k - 0.88425 = 3.7429$ **k = 4.62715**
24. $298{,}416.7 + p = -377{,}598.86$ **p = -676,015.56**

Solve each equation using any method.

25. $n - 17.9 = -31.05$ **n = -13.15**
26. $4{,}365.77 = c + 7{,}935.49$ **c = -3,569.72**
27. $4.4 = p + 1.1$ **p = 3.3**
28. $k + (-8.5) = -0.6$ **k = 7.9**
29. $t + 18.5 = -41$ **t = -59.5**
30. $y - 33.4 = 81.9$ **y = 115.3**
31. $-15.7 = k + 20.4$ **k = -36.1**
32. $h - 215.876 = -114.015$ **h = 101.861**
33. $z - 81.6 = -81.6$ **z = 0**
34. $5.4 = t + (-6.1)$ **t = 11.5**
35. $-4.095 + b = 18.665$ **b = 22.76**
36. $4.87 = n + 0.87$ **n = 4**

Chapter 3 C-7

Enrichment

For use after 3-4 (pp. 111–114)

Postal Rates

The cost of mailing a 1-oz letter is \$0.25. Each additional ounce or part of an ounce costs \$0.20.

Solve.

1. Find the cost of mailing a 6-oz letter. **\$1.25**
2. Find the cost of mailing a 9.4-oz letter. **\$2.05**
3. A letter weighed n oz. It cost \$1.45 to mail. Which of the following equations can you solve to find the value of n? **b**
   a. $0.25n = 1.44$
   b. $0.25 + 0.2(n - 1) = 1.45$
   c. $0.25 + 0.2n = 1.45$
   d. $0.25n + 0.2n = 1.45$
   e. Solve the equation and find the weight of a letter costing \$1.45 to mail. **7 oz**

Suppose postage rates increased to \$0.35 for the first oz and \$0.28 for each additional oz or part of an oz.

4. Write and solve equation to find n, the weight of a letter costing \$2.31 to mail.
   $0.35 + 0.28(n - 1) = 2.31$; $n = 8$; 8 oz

For mail weighing more than 11 oz, Priority Mail rates apply.

Find the total charge.

5. 7.4-lb package mailed to Zone 4 **\$6.23**
6. 15-oz letter **\$2.40**
7. 5-lb package mailed to Zone 1 **\$3.61**
8. Two 3-lb packages mailed to Zone 4 and three 3.7-lb packages mailed to Zone 6 **\$19.91**
9. How much can you save by wrapping three 3-lb packages together and sending a single package to Zone 8, rather than sending them separately? **\$2.27**

Priority Mail Rates

Weight, up to but not exceeding— pound(s)	Zones: Local, 1, 2 and 3	4	5	6	7	8
2	\$2.40	\$2.40	\$2.40	\$2.40	\$2.40	\$2.40
3	2.74	3.16	3.45	3.74	3.96	4.32
4	3.18	3.75	4.13	4.53	4.92	5.33
5	3.61	4.32	4.86	5.27	5.81	6.37
6	4.15	5.08	5.71	6.31	6.91	7.66
7	4.58	5.66	6.39	7.09	7.80	8.67
8	5.00	6.23	7.07	7.87	8.68	9.68
9	5.43	6.81	7.76	8.66	9.57	10.69
10	5.85	7.39	8.44	9.44	10.45	11.70

C-8 Chapter 3

MIXED REVIEW

Find each value.

1. $|-4.53| + 3.099$ 7.629
2. $-0.993 - |-0.6712|$ -1.6642

Evaluate.

3. $3x$ when $x = -0.82$ -2.46
4. $x + 2y$ when $x = 3.266$ and $y = -4.01$ -4.754

Simplify

5. $4.3(-9.2x)$ -39.56x
6. $2.9y - 4.88y + 1.7y$ -0.28y

Solve.

7. A large juice is \$.83 and a small juice is \$.57. Ann buys one juice per day. What is her weekly savings if she buys small juices instead of large ones? \$1.82

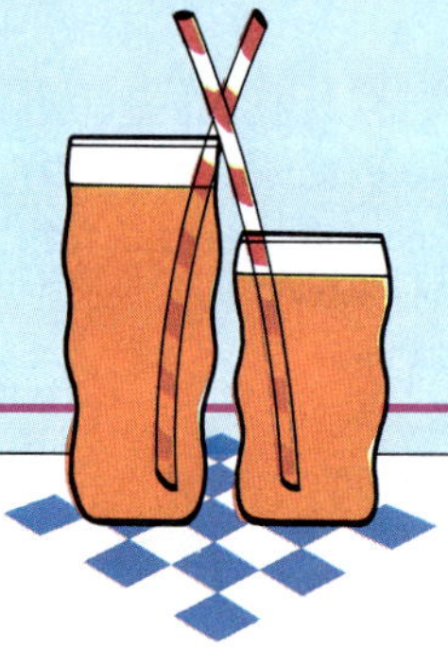

42. a. Are the equations $1.7x + 2.4 = 1.5$ and $17x + 24 = 15$ equivalent? How do you know? 42. a. Yes; both sides were multiplied by 10.

b. ***WRITE*** a procedure for changing a decimal equation into one with integers. Show how your method would work for any of the equations in this lesson. Multiply both sides of the equation by a power of ten so that the decimals are eliminated.

Write an equation for each sentence. Solve. Check students' work.

43. Fifty-eight thousandths less than some number is equal to fifty-eight hundredths. Find the number. $n - 0.058 = 0.58$; 0.638

44. Seven and three hundred thirty-nine thousandths equals five and one hundred seventy-eight thousandths more than some number. Find the number. $7.339 = n + 5.178$; 2.161

45. One and ninety-nine ten-thousandths equals two more than some number. Find the number. $1.0099 = n + 2$; $n = -0.9901$

46. Eight hundred ninety-two and thirty-two hundredths less a number is equal to one thousand, four hundred eleven and twelve thousandths. Find the number. $892.32 - n = 1{,}411.012$; $n = -518.692$

47. Three plus some number equals eight and sixteen hundredths. Find the number. $3 + n = 8.16$; $n = 5.16$

TEST YOURSELF

Compare. Write $>$, $<$, or $=$.

1. 3.088 ■ 3.808 $<$
2. 2.3 ■ 2.300 $=$
3. -4.23 ■ -4.991 $>$

Round to the nearest tenth.

4. -9.65 -9.7
5. 4.3088 4.3
6. 17.952 18.0

Estimate the value of each expression. Accept reasonable estimates.

7. $0.8823 \div 3$ 0.3
8. $5.9 \cdot 3.88$ 24
9. $437.02 - 188.54$ 250

Evaluate if $x = 5.8$ and $y = -2.3$.

10. xy -13.34
11. $x - 2$ 3.8
12. $-2y - x$ -1.2

Simplify.

13. $2.99 + x + (-3.08)$ $x - 0.09$
14. $-3.55y - 9.01y$ $-12.56y$

Solve each equation.

15. $x - 9.09 = -15.8$ -6.71
16. $24.011 = y + 23.9$ 0.111

LESSON QUIZ

Solve each equation using the addition or subtraction property of equality.

1. $-7.33 + t = 4.86$ (12.19)
2. $4.02 = m - 3.067$ (7.087)
3. Solve using opposites. $u - 372.01 = -1{,}000.92$ ($u = -628.91$)

Assignment Guide

Basic 1-10, 17-18, 21-22, 25-30, 37, 38-40, 43-44, MR, TY All

Average 2-3, 10-15, 18-19, 22-23, 28-33, 37, 39-41, 44-45, MR, TY All

Enriched 3-4, 11-16, 19-20, 23-24, 31-37, 40-42, 45-47, MR, TY All

FOR THE NEXT LESSON

decimal squares or graph paper, math journal, calculator

OBJECTIVE:
To solve multiplication and division equations involving decimals.

3-5 Multiplication and Division Equations

▼ On average, an oil well produces 16.8 barrels each day. In how many days will it produce 184.8 barrels? If we represent the number of days by the variable d, we can write an equation.

$$16.8d = 184.8$$

1 You can solve this equation using the division property of equality.

$$16.8d = 184.8$$

$$\frac{16.8d}{16.8} = \frac{184.8}{16.8}$$ Divide both sides by 16.8, the coefficient of d.

$$d = 11$$

Check $16.8d = 184.8$

$$16.8(11) = 184.8$$ Replace d with 11.

$$184.8 = 184.8 \checkmark$$

It will take 11 days to produce 184.8 barrels of oil.

▼ You can solve equations involving division by using the multiplication property of equality.

Example 1 **Solve $\frac{x}{2.1} = -0.9$ and graph the solution.**

Solution

2 $$\frac{x}{2.1} = -0.9$$

$$2.1\left(\frac{x}{2.1}\right) = -0.9(2.1)$$ Multiply both sides by 2.1.

$$x = -1.89$$

Check Use estimation to see if the solution is reasonable.

$$\frac{x}{2.1} = -0.9$$ Round each decimal to the nearest integer.

$$\approx \frac{-2}{2} = -1$$ Since -1 is close to -0.9, the solution is reasonable.

3 Graph -1.89 on the number line.

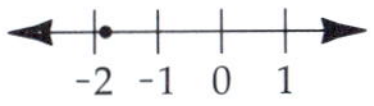

Teaching Tip

Prepare a number line transparency with tenths marked. Use it to graph solutions to decimal equations. It can also be used to estimate the graphing of hundredths.

Lesson Focus

Materials

Decimal squares or graph paper, math journal, calculator

MOTIVATING THE LESSON

Have students describe how they would solve an equation such as $3y = 9$. Ask if they think the same method could be used to solve an equation involving decimals. Provide students with a simple equation involving decimals, such as $4w = 0.32$, and ask them to apply the method they used to solve the first equation. Have them verify their solution by substitution. Repeat the entire process with an equation such as $\frac{y}{0.8} = 2$.

Instruction

1 If students are unsure how to solve this equation, ask how they would solve a similar but simpler equation, such as $15d = 150$. (Use the division property of equality to divide each side by 15.) Then ask how they would solve $1.5d = 15$. (Divide both sides by 1.5.)

2 Students may also need a simpler equation to help them understand this example. Display this equation: $\frac{x}{21} = 7$. Ask students what property they could use to solve it. (multiplication property of equality) Then ask them to solve $\frac{x}{2.1} = 0.7$.

3 Mention to students that the point for -1.89 must be between -1 and -2 on the number line. Have students decide whether -1.89 is closer to -1 or -2 to help them determine its approximate location. (closer to -2)

4 Ask students to write the first step for solving this equation.

$\frac{3.98x}{3.98} = \frac{470}{3.98}$

Have them identify the order in which they would enter the numbers in the calculator. ($470 \div 3.98$)

ADDITIONAL EXAMPLES

1. Solve each equation using the division property.
   - a. $-2.9y = 14.79$ ($y = -5.1$)
   - b. $0.15 = 0.003x$ ($x = 50$)
2. Solve each equation using the multiplication property.
   - a. $\frac{n}{-2.3} = 0.08$ ($n = -0.184$)
   - b. $-24.8 = \frac{t}{-2.05}$ ($t = 50.84$)
3. Solve and graph: $1{,}000t = 4{,}500$. ($t = 4.5$)

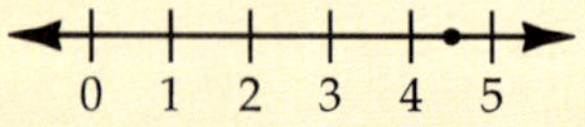

Guided Practice

Class Exercises For questions 5-12, continue to have students identify the first step in solving each equation before they find the solution.

Think and Discuss To provide further discussion for question 1, ask students to use compatible numbers to estimate the solution for Class Exercise 8.

▼ You may need to round an answer when solving a problem.

Example 2 **4** **Use a calculator to solve $3.98x = 470$. Round to the nearest tenth.**

Solution 470 (÷) 3.98 (=) 118.09045
≈ 118.1 Round to the nearest tenth.

THINK AND DISCUSS See below.

1. What numbers would you choose to check the solution in Example 2 using compatible numbers?
2. State a situation in everyday life where it is necessary to round a decimal to the nearest hundredth.
3. Are the equations $1.2x = 2.5$ and $12x = 25$ equivalent? Why or why not?

1. 480 and 4
2. sales tax
3. Yes; both sides of the first equation were multiplied by 10.

CLASS EXERCISES

State the first step in solving each equation.

1. $0.7x = -0.63$ **Divide by 0.7.**
2. $\frac{y}{0.6} = 1.2$ **Mult. by 0.6.**
3. $1.5 = d \div 15$ **Mult. by 15.**
4. $-1.2 = -0.4m$ **Divide by -0.4.**

Solve each multiplication equation.

5. $-0.5y = -0.73$ **1.46**
6. $0.8x = 0.448$ **0.56**
7. $-540 = -1.8t$ **300**
8. $13.133 = -2.3w$ **-5.71**

Solve each division equation.

9. $\frac{y}{2.3} = -4.8$ **-11.04**
10. $0.97 = \frac{c}{-2}$ **-1.94**
11. $2{,}390 = z \div 0.033$ **78.87**
12. $\frac{m}{0.19} = -492.05$ **-93.4895**

WRITTEN EXERCISES

Solve using the division property of equality.

1. $2x = -4.88$ **-2.44**
2. $-0.3y = 7.53$ **-25.1**
3. $6.4x = 0.2816$ **0.044**
4. $-0.00051z = -2.026791$ **3,974.1**
5. $1.92 = 1.6s$ **1.2**
6. $0.004m = 0.12$ **30**
7. $3.17n = 135.042$ **42.6**
8. $2.21 = 1.7w$ **1.3**

Solve using the multiplication property of equality.

9. $\frac{n}{1.7} = 0.22$ **0.374**
10. $\frac{k}{2.01} = 0.04$ **0.0804**
11. $4.5 = m \div (-3.3)$ **-14.85**
12. $-33.04 = \frac{z}{-0.03}$ **0.9912**
13. $-0.45 = x \div 12$ **-5.4**
14. $\frac{m}{0.89} = 3{,}488$ **3,104.32**
15. $\frac{w}{-3.4} = -25.5$ **86.7**
16. $12{,}088.25 = \frac{v}{3.8}$ **45,935.35**
17. $\frac{c}{12.56} = 0.245$ **3.0772**

MATH MINUTES

Groups of four trade and solve equations: give each group 4 index cards. Ask them to write on each card one decimal equation for each of the four properties, and write the property and solution on the back of each card.

MENTAL MATH Solve each equation.

18. $0.7x = 2.8$ 4 **19.** $\frac{m}{7.08} = -100$ -708 **20.** $6 = a \div 1.5$ 9

21. $10{,}000r = 483.08$ 0.048308 **22.** $0.55t = 0.0055$ 0.01

CALCULATOR Solve each equation. Round each answer to the nearest tenth.

23. $0.46x = 89.23$ 194.0 **24.** $45.08t = -2{,}917.335$ -64.7

25. $-0.93 = 0.0221z$ -42.1 **26.** $-9.03m = 499{,}812.4$ -55,350.2

CALCULATOR Solve each equation. Round each answer to the nearest hundredth.

27. $4.55x = 43.225$ 9.50 **28.** $\frac{m}{7.08} = -35.992$ -254.82

29. $y \div 84.6 = 2.79$ 236.03 **30.** $90.43n = -298.0113$ -3.30

Check students' work.

Solve each equation using any method. Graph each solution.

31. $0.9 = \frac{m}{41}$ 36.9 **32.** $100t = -45$ -0.45

33. $\frac{z}{-0.4} = 0.5$ -0.2 **34.** $3.94z = -21.67$ -5.5

35. $t \div 0.4 = -15$ -6 **36.** $-99.252 = -8.271r$ 12

Solve each equation for x.

37. $x \div y = z$ $x = zy$ **38.** $xz = y$ $x = \frac{y}{z}$ **39.** $y = \frac{x}{z}$ $x = yz$

Write an equation for each sentence. Solve.

40. The quotient of some number t divided by -4.5 equals 200.6. Find the number. $\frac{t}{-4.5} = 200.6$; -902.7

41. Four thousandths times some number is equal to eighty-eight hundredths. Find the number. $0.004n = 0.88$; 220

42. The cost of an adult ticket is c. The cost of eight adult tickets is \$71.60. What is the price of one ticket? $8c = 71.60$; \$8.95

43. A number divided by -2.35 is equal to 400.9. Find the number. 43. $\frac{n}{-2.35} = 400.9$; -942.115

44. Nineteen and five thousand five hundred twenty-five ten thousandths is the same as five and five tenths times some number. Find the number. $19.5525 = 5.5n$; 3.555

45. ***WRITE*** Write a paragraph explaining how to use estimation to place the decimal point in a multiplication or division computation. Check students' work.

46. Find values for x and y that satisfy both equations: $xy = 0.42$ and $x + y = 1.3$. 0.6 and 0.7

A ***common error*** occurs when students divide on one side of the equation and multiply on the other side. For example, to solve $15x = 150$, students may divide the left side by 15, but multiply the right side by 15. To help students remember which operation to use, have them write a small division sign (or multiplication sign if appropriate) on each side of the equation.

Closure

Ask students to write a paragraph in their math journals that describes how they decide which of the four properties of equality to use to solve a decimal equation.

Independent Practice

For questions 40-44, suggest that students copy the verbal statements and then write the symbols they will use over the phrases in the statement.

Writing in Math For question 45, ask students to supply a numerical example to support their verbal explanation.

For question 46, students may solve for x or y or use the guess-and-test strategy.

For question 47, students can use *Appleworks* for an Apple computer or *Microsoft Works* for an IBM.

Lesson Follow-up

Reteaching Activity Provide students with worksheets that have room for the students to write each step needed for both sides of the equation.

LESSON RESOURCES

TRB Practice C-9
TRB Enrichment C-10

Practice

For use after 3-5 (pp. 115–118)

Multiplication and Division Equations

Solve using the division property of equality.

1. $-9k = 2.34$ **k = -0.26**
2. $-12.42 = 0.03p$ **p = -414**
3. $-7.2y = 61.2$ **y = -8.5**
4. $-0.1035 = 0.23n$ **n = -0.45**
5. $1.5m = 3.03$ **m = 2.02**
6. $-0.007h = 0.2002$ **h = -28.6**
7. $8.13t = -100.812$ **t = -12.4**
8. $0.546 = 0.42y$ **y = 1.3**

Solve using the multiplication property of equality.

9. $\frac{p}{2.9} = 0.55$ **p = 1.595**
10. $9.1 = \frac{x}{-0.7}$ **x = -6.37**
11. $-6.4 = \frac{y}{8.5}$ **y = -54.4**
12. $\frac{k}{-1.2} = -0.07$ **k = 0.084**
13. $277.4 = \frac{n}{3.5}$ **n = 970.9**
14. $\frac{e}{-0.76} = 2{,}809$ **e = -2,134.84**
15. $\frac{a}{27} = -32.3$ **a = -872.1**
16. $\frac{p}{-1.52} = -3{,}600$ **p = 5,472**

MENTAL MATH Solve each equation.

17. $0.7h = 4.2$ **h = 6**
18. $\frac{x}{2.5} = -3$ **x = -7.5**
19. $38.7 = -100k$ **k = -0.387**
20. $-45.6e = -4.56$ **e = 0.1**

CALCULATOR Solve each equation. Round each answer to the nearest hundredth.

21. $-3.77p = 19.84$ **p = -5.26**
22. $\frac{k}{0.852} = -91.76$ **k = -78.18**
23. $0.0046 = 0.0041x$ **x = 1.12**
24. $417.92c = 316.55$ **c = 0.76**
25. $-2{,}885.9 = 0.32y$ **y = -9,018.44**
26. $-55.7 = \frac{z}{-83.6}$ **z = 4,656.52**
27. $\frac{x}{3.04} = 3.04$ **x = 9.24**
28. $\frac{-x}{-27.3} = 0.98$ **x = 26.75**

Write an equation for each sentence. Solve. **Letter for variable may vary.**

29. The opposite of seventy-five hundredths times some number equals twenty-four thousandths.
$-0.75n = 0.024$; $n = -0.032$
30. A number divided by -3.88 equals negative two thousand.
$-\frac{n}{3.88} = -2{,}000$; $n = 7{,}760$
31. Four hundredths times some number equals thirty-three and four tenths.
$0.04x = 33.4$; $x = 835$

Chapter 3 C-9

Enrichment

For use after 3-5 (pp. 115–118)

Foreign Currencies

The table lists the values of nine foreign currencies you could buy with 1 U.S. dollar on a recent day.

Value of 1 U.S. dollar	
Country	**Value**
Egypt	2.17 pounds
Finland	4.32 markkaas
Ghana	12 cedis
Italy	1,291 liras
Morocco	8.4 dirhams
New Zealand	1.71 dollars
Paraguay	240 guaranis
Tunisia	0.82 dinars
United Kingdom	0.59 pounds

1. Find the number of currency units you could buy with the given number of U.S. dollars.
   a. Italian lira, \$400 **516,400**
   b. Tunisian dinars, \$650 **533**
   c. Egyptian pounds, \$42.40 **92.008**
2. A banker used the equation $4.32d = 327.24$ to find the number of U.S. dollars (d) it would take to purchase 327.24 Finnish markkaas. Solve the equation.
$d = 75.75$
3. Write and solve an equation to find the number of dollars (d) it would take to purchase the given amount.
   a. 3,528 Moroccan dirhams **$8.4d = 3{,}528$; $d = 420$**
   b. 501.5 United Kingdom pounds **$0.59d = 501.5$; $d = 850$**
   c. 4.428 Tunisian dinars **$0.82d = 4.428$; $d = 5.4$**
4. A vacationer received 581.4 of a certain country's currency for \$340 U.S. Which country was it?
New Zealand
5. You can use the equation $12c = 1$ to find the value of 1 Ghanaian cedi in U.S. dollars. Solve the equation, expressing your answer as a decimal rounded to the nearest thousandth.
$c = 0.083$
6. Write and solve an equation to find the value in U.S. dollars. Round to the nearest thousandth.
   a. 1 Egyptian pound (p) **$2.17p = 1$; $p = 0.461$**
   b. 1 Paraguayan guarani (g) **$240g = 1$; $g = 0.004$**
   c. 1 United Kingdom pound (p) **$0.59p = 1$; $p = 1.695$**

C-10 Chapter 3

MIXED REVIEW

Solve each equation.

1. $m - 0.88 = 0.566$ 1.446
2. $45.92 = -38.8 + t$ 84.72
3. $z + 400.23 = 400.23$ 0
4. $0.332 = z - (-0.221)$ 0.111

Simplify. 5. -1,056 6. -360

5. $8 \cdot (-2) \cdot (-6)(-4 - 7)$
6. $45 \cdot (-4) + (-9 \cdot 4)5$
7. How many squares are in the figure shown? 14

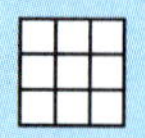

47. ***COMPUTER*** A school needs to buy a minimum of 2,000 pencils, 1,000 pens, 500 notebooks, and 150 reams of paper. The budget cannot exceed \$3,000.

	A	B	C	D
1	Item	Quantity	Unit Price	Total
2	Pencils		0.10	=B2*C2
3	Pens		0.50	=B3*C3
4	Notebooks		2.25	=B4*C4
5	Paper (ream)		4.75	=B5*C5
6			TOTAL	=SUM(D2:D5)
7				

a. What is the least amount the school can spend? \$2,537.50
b. Suppose the remaining money is spent on notebooks. How many more notebooks can be bought? 205
c. Suppose the price of pencils increases to \$.15 and the price of a ream of paper drops to \$4.55. Can the school afford its minimum amount of supplies? yes

Critical Thinking

EXPLORING LOGICAL FAMILIES

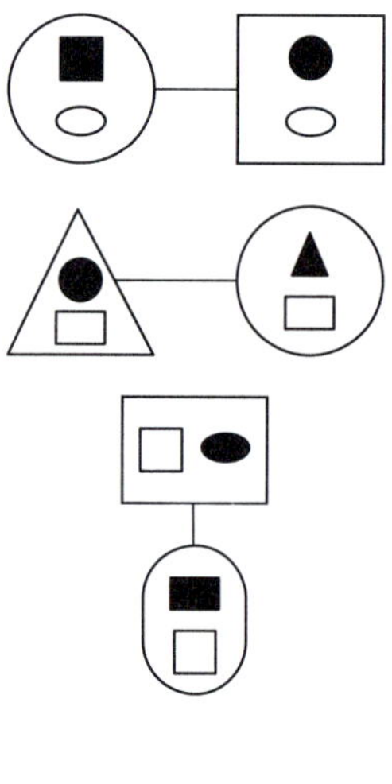

The figures at the left are called Quipps.

1. What are the distinguishing characteristics of a Quipp?
2. Which of the following are Quipps? c and d

a. b. c. d.

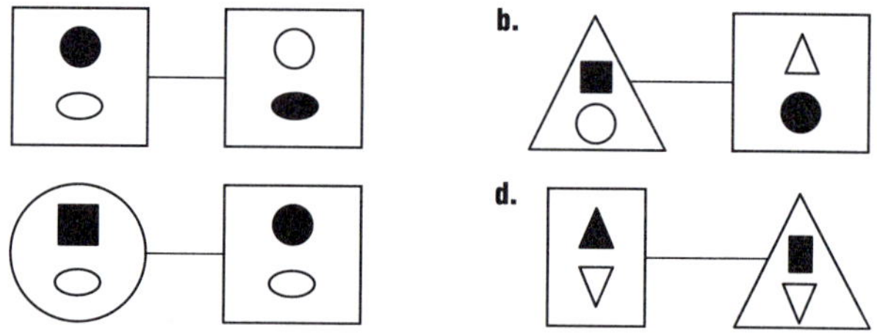

3. Create two of your own Quipps. Check students' work.
4. Create a logical family consisting of three members. Trade with a classmate to see if they can determine the distinguishing characteristic. Check students' work.

1. The outside shape becomes shaded inside; the shaded shape inside becomes the outside shape.

LESSON QUIZ

Solve each equation using the multiplication or division property.

1. $\frac{m}{3.7} = -0.03$ (-0.111)
2. $4.09t = 15.542$ (3.8)
3. $-0.008x = -2.8$ (350)
4. $0.84 = \frac{c}{-3.05}$ (-2.562)
5. Solve and graph: $-3.9w = 16.38$ (-4.2)

Assignment Guide

Basic 1-4, 9-13, 18-20, 23-29 odd, 31-33, 37-38, 40-43, 45, MR, CT All
Average 3-17 odd, 19-21, 24-30 even, 32-34, 37-38, 42-45, MR, CT All
Enriched 6-16 even, 20-22, 25-26, 29-30, 34-39, 44-47c, MR, CT All

FOR THE NEXT LESSON

math journal

Practice

Write each decimal in words.

1. 0.67 sixty-seven hundredths
2. 2.90 two and ninety hundredths
3. 637.0004 six hundred thirty-seven and four ten thousandths
4. 0.00007 seven hundred thousandths

Write each decimal.

5. two hundred fifteen and seventy-four hundredths 215.74
6. six and eight tenths 6.8
7. forty-two and seven hundredths 42.07
8. five hundred four thousandths 0.504

Compare using >, <, or =.

9. 9.9 ■ 9.09 >
10. -4.2 ■ -4.20 =
11. -8.600 ■ 8.6 <
12. -0.06 ■ 0.006 <
13. -3.47 ■ -3.547 >
14. 85.706 ■ 85.7 >
15. $|-0.2|$ ■ $|0.2|$ =
16. $|8.9|$ ■ $|-10.7|$ <
17. 0.36 ■ 0.3600 =

Estimate using rounding, front-end estimation, or compatible numbers. Answers may vary. Possible answer given.

18. $57.96 · 45 $3,000
19. 95.27 ÷ 5.2 20
20. 0.029 + 0.999 1
21. 0.98 ÷ 1.03 1
22. -490.6 − 25.302 -525
23. $10.25 + $36.32 + $9.05 $56

Evaluate each expression for $x = -2.3$ and $y = 8.92$.

24. $-x - y$ -6.62
25. $y - (-x)$ 6.62
26. $4x + 7.39$ -1.81
27. $-10y - 4.8$ -94
28. $x - 4y$ -37.98
29. $-6.3(x + y)$ -41.706
30. $\frac{x - y + y}{-x}$ -1
31. $\frac{-x - y}{x + y}$ -1
32. $-x + y$ 11.22

Solve each equation.

33. $x + 0.25 = 8$ $x = 7.75$
34. $\frac{w}{18} = -2.7$ $w = -48.6$
35. $a - (-42.4) = 42.4$ $a = 0$
36. $n \div 8 = 3.02$ $n = 24.16$
37. $0.008z = 0.24$ $z = 30$
38. $m - (-9.4) = 0$ $m = -9.4$
39. $-0.96 = 0.8t$ $t = -1.2$
40. $0.59 + s = -1.0$ $s = -1.59$
41. $y - 42.76 = -0.05$ $y = 42.71$
42. $-132 = 66i$ $i = -2$
43. $25j = -100.9$ $j = -4.036$
44. $d - (-0.04) = 0.74$ $d = 0.7$
45. $8x = -15.52$ $x = -1.94$
46. $-100.05 = c + 5$ $-105.05 = c$
47. $-18.07 - r = 0.5$ $r = -18.57$

Solve each equation for m.

48. $m + n = p$ $m = p - n$
49. $n = \frac{m}{p}$ $m = np$
50. $p + m = n$ $m = n - p$
51. $p = \frac{n}{m}$ $m = \frac{n}{p}$
52. $n - m = p$ $m = n - p$
53. $mn = p$ $m = \frac{p}{n}$

CAREER

Help Wanted: Medical Record Technician

Associate degree required.
For more information, write to American Medical Record Association, 875 N. Michigan Ave., Suite 1850, Chicago, IL 60611.

Medical record technicians keep patient information on written charts or in computer files. They assign codes to each diagnosis for insurance purposes. Technicians also compile statistics. Medical record technicians prepare reports for the government, law firms, and public planners. Medical records administrators use the records to monitor hospital spending, testify in court, and compile statistics.

PROJECT

Find out what other types of statistics medical record technicians gather.

Career Information

Plan to invite a medical record technician or medical record administrator to visit the class. Encourage students to ask questions about the skills necessary for these careers.

Ask the speaker to cover such topics as:

- the type of training or education received and the entrance requirements for this career
- how algebra, geometry, statistics, and calculus are used in the job
- how computers and software programs can help
- the types of reports that are prepared and how statistics are gathered and compiled
- how the compiled information is used in monitoring hospital and patient costs

Lesson Focus

Materials

Math journal

MOTIVATING THE LESSON

Have students look through their textbooks, some newspapers, or business magazines to find examples of different types of tables. Ask students why they think tables are used so often in this medium. (Tables provide a clear, concise way to organize data.)

Instruction

Have students work in groups to answer the questions and record the results in their math journals.

1 Mention that individuals with identical scores have the same rank.

2 Remind students that for all the teams to be balanced, the fastest player should be on the same team as the slowest player.

3 Comment that there are many different correct answers for making up the teams.

Closure

Have the class discuss their answers for questions 10-15 and explain their reasoning.

As a follow-up, refer students to the problem solving activity on p. 96E of the Chapter Overview. See also TRB Exploring O-10.

OBJECTIVE: *To explore using data in tables.*

MATERIALS

- Math journal to record work

Exploring Data in Tables

■ A sports camp holds track and field trials on the first day of camp. The table below shows the performance of the campers on several different measures.

	Running		Jumping (cm)		Shooting
Name	**50 m**	**1,500 m**	**high jump**	**long jump**	**basketball throw (out of 20 shots)**
Aimee	8.6 s	12 min 25 s	127.0	325.1	10
Barbara	10.3 s	15 min 49 s	106.7	299.7	8
Dwayne	8.5 s	12 min 4 s	139.7	365.8	15
Floyd	8.7 s	10 min 54 s	124.5	327.7	14
Hannah	7.9 s	11 min 0 s	129.6	294.6	17
Isaiah	8.1 s	10 min 37 s	116.8	342.9	15
Jane	8.4 s	10 min 37 s	132.1	332.7	15
Manuel	8.7 s	13 min 9 s	121.9	309.9	13
Nancy	7.4 s	9 min 4 s	139.2	364.8	18
Sydney	8.6 s	12 min 2 s	127.1	327.7	11
Tran	8.4 s	12 min 46 s	137.1	340.4	16
Warren	7.7 s	8 min 18 s	134.6	368.4	18

1. a. Which event best demonstrates the campers' endurance? 1,500-m run
   b. Which best demonstrates the campers' speed? Which best demonstrates accuracy?
   c. Does the camper who runs the 50 m fastest also run the 1,500 m fastest? Is this true for the second-fastest runner? the third fastest? Is there a relationship between the time run in the 50 m and the time run in the 1,500 m? Explain.
   d. Who would you consider a better runner, the camper who runs the 50 m faster or the 1,500 m faster? Why?
2. a. Who is the highest jumper? Who is the longest jumper? Dwayne; Warren
   b. Is the highest jumper the same person as the longest jumper? Is this always true? Explain.
   c. Is there a relationship between the best runners and the best jumpers? Explain. Sometimes; often good athletes are good at many different events.
3. Who is the best shooter? Is there a relationship between the campers' running, jumping, and shooting skills? Nancy and Warren; not necessarily, good athletes may be good at different events.
4. Who is the fastest runner in the 50-m run? Nancy
   a. List the campers in order from fastest to slowest.
   b. 1 Could you list the campers in order from slowest to fastest? Explain. Yes; begin with Barbara and write the list in the reverse order.

1. b. 50-m run; basketball throw
1. c. No; no; no; not necessarily, campers who run the 50 m fast may not have the endurance for the 1,500 m. Often, though, the best athlete is best at many events and the weakest athlete is weak at many events.
1. d. Either; the winner of the 50 m is fast for short distances; the winner of the 1,500 m demonstrates endurance.
2. b. No; not necessarily; often good athletes are good at many different events, although different skills are needed for the high jump and the long jump.
4. a. Nancy, Warren, Hannah, Isaiah, Jane-Tran, Dwayne, Aimee-Sydney, Floyd-Manuel, Barbara

Teaching Tip

Analyze how the members of each group interact to determine if the group members complement each others' strengths and weaknesses. Rearrange groups as needed.

c. Tran and Jane both ran the 50 m in 8.4 s. In what order will you list them? Explain. Tran then Jane; Tran slower in 1,500 m

d. In what order will you list Aimee, Sydney, Floyd, and Manuel? Manuel, Floyd, Aimee, Sydney

5. Continue to list the campers in order in the 1,500-m run, the high jump, the long jump, and the basketball throw. See additional answers.

6. In the jumping events, could you order the campers from shortest distance jumped to longest distance jumped? Explain.

■ The camp uses the results to assign campers to teams. The first event is a 2,000-m relay race between three evenly matched teams.

7. How far must each camper run in the relay race? 500 m

8. Who do you think would run best in the relay race: the campers who ran the fastest in the 50 m or the 1,500 m? Why? 1,500-m race; they have better endurance.

a. Would you take the campers' jumping and basketball throw scores into consideration? Explain.

b. What other factors will help you decide who should be on the three evenly matched teams?

9. Make a list of the three teams. ***Compare*** with another group. Answers may vary. One example is given. (2)

■ The second sporting event is a basketball game. There will be two evenly matched teams, with five on a team. The two extra campers will rotate positions in the game.

10. The two guards need to be the fastest runners, and they also must have good endurance and shooting skills. Who would you pick? Why? Answers may vary. Possible answer: Nancy and Warren; they are the best runners in both races and the top shooters.

11. The center doesn't need to be as fast as the guards. It's important for the center to be a good shooter and jumper, and have good endurance. Who would you pick? Why? Answers may vary. Possible answer: Tran is a good shooter and good at the high jump.

12. The two forwards need to have excellent endurance and strength. They need to be good jumpers and they must be fast. Who would you pick? Why? Answers may vary. Possible answer: Hannah and Isaiah; Hannah is a fair jumper and a good shooter. (3)

13. Make a list of the two teams. ***Compare*** with another group.

14. Did you choose the same teams the other groups chose? ***Describe*** and discuss why you made your choices. Is there one right answer?

15. ***Summarize*** For which event was it easiest to pick the teams? Why?

ADDITIONAL ANSWERS

5.

1,500-m run	High Jump
Warren	Dwayne
Nancy	Nancy
Jane-Isaiah	Tran
Floyd	Warren
Hannah	Jane
Sydney	Hannah
Dwayne	Sydney
Aimee	Aimee
Tran	Floyd
Manuel	Manuel
Barbara	Isaiah
	Barbara

Long Jump	Basketball
Warren	Nancy-Warren
Dwayne	Hannah
Nancy	Tran
Isaiah	Jane-Dwayne-Isaiah
Tran	Floyd
Jane	Manuel
Floyd-Sydney	Sydney
Aimee	Aimee
Manuel	Barbara
Barbara	
Hannah	

6. Yes; but it is helpful to order campers the same way for each event. Your order should be best scores to worst scores, or worst scores to best scores, for all events, to be consistent.

8. a. no; only if the relay requires jumping and throwing.
   b. Answers may vary. Possible answers; leadership, cooperation.

9.

Team 1	Team 2	Team 3
Warren	Nancy	Jane
Isaiah	Floyd	Hannah
Dwayne	Sydney	Aimee
Barbara	Manuel	Tran

13.

Team 1	Team 2
Nancy	Warren
Tran	Isaiah
Hannah	Jane
Dwayne	Aimee
Sydney	Floyd

Assignment Guide
Basic All
Average All
Enriched All

FOR THE NEXT LESSON
computer and data base software (optional), math journal

Lesson Focus

Materials

Computer and data base software (optional), math journal

Vocabulary/Symbols

data base
field
file
record

MOTIVATING THE LESSON

Ask students to discuss the different ways their school might keep track of their grades and other pertinent data. Then ask them how they think the records might be organized (e.g., by last name, by class). Mention that an efficient method is to use a computer data base.

Instruction

1 Review the term *data base* with students and differentiate between rows and columns in the chart.

2 Help students realize that the data are arranged alphabetically by state. Encourage them to think of arrangements—other than last name—that are possible, such as zip code or city.

3 Explain to students that data bases are valuable and useful because new information can be added easily and information that is no longer needed can be deleted.

4 Review the terms *field* and *record*. In the data base in question 4, the field is "Student" and one record is "Adams, Karen."

OBJECTIVE: *To explore the uses of a computer data base.*

Exploring Data Bases

1 ■ A *data base* is a collection of information. The data is usually arranged in rows and columns and can be rearranged in any order. Below is a data base of names and addresses.

Name	Street	City	State	Zip Code
Smith, Greg	123 Main St.	Tucson	Arizona	85726
Cruz, Maria	97 South St.	Burbank	California	91505
Chien, Janice	876 Water St.	Honolulu	Hawaii	96820
Jones, Carl	8 Division St.	Albany	New York	12266

2 **1.** In what order are the data arranged? How else can they be arranged? **alphabetically by state; alphabetically by last name**

2. Suppose you wanted to arrange the data in alphabetical order. Show the results on a sheet of paper (by last name). **Chien, Cruz, Jones, Smith**

3 **3.** Suppose you need to update the data base with this information.

Name	Street	City	State	Zip Code
Ales, George	24 Broadway	Dallas	Texas	74356
Costa, Anna	568 Beach St.	Miami	Florida	48576

Write the new data base in alphabetical order by name. **Ales, Chien, Costa, Cruz, Jones, Smith**

MATERIALS

- Paper
- Computer and data base software (optional)
- Math journal to record work

4 ■ In a data base, each column is known as a *field*. Each row is known as a *record*.

4. Look at the data base shown below.

Student	Grade	Grade Point Average
Adams, Karen	8	3.2
Eng, Charles	9	3.5
Garcia, Fran	8	3.4
Mitchell, Dennis	9	3.2

a. How many fields are there? What are they? **3; student, grade, grade point average**

b. How many records are there? **4**

5. The first field showing the student names could have been made up of two separate fields. What are they? **first name, last name**

6. Arrange the data base by grade point average. Show the results of the new data base on a sheet of paper. **See margin.**

7. How did you sort the two records with the same grade point average in Question 6? In what other way could you have sorted them? **alphabetically; by grade**

DATAPOINT

Notice that the first field contains words and the other two contain numbers.

6.

Student	Grade	Grade Point Average
Charles Eng	9	3.5
Fran Garcia	8	3.4
Dennis Mitchell	9	3.2
Karen Adams	8	3.2

DATAPOINT

When you sort a data base, you rearrange the data in a particular order.

Teaching Tip

If students are working as partners or in small groups on the computer, be sure that every student has an opportunity to actually work on the computer. If necessary rotate partners so that two students who may seem shy or reluctant are paired.

■ You enter data in a data base one record at a time. For example, the record below is from a data base that is used to keep track of inventory at a shoe store.

```
ITEM : Boots
QUANTITY (PAIRS) : 150
NEXT SHIPMENT DUE : Dec 10
```

8. How many field names are there? What are they? 3; item, quantity, next shipment due

■ After entering several records, you can view the data in rows and columns. This is part of the shoe store data base in column form.

Item	Quantity (on hand)	Next Shipment Due
Boots	150	Dec 10
Running shoes	34	Dec 5
Sandals	25	Dec 17
Loafers	16	Dec 28

9. Suppose it is the morning of December 1. On average, the store sells nine pairs of running shoes per day. Will there be enough before the next shipment arrives, which is scheduled for an evening delivery? no

10. ***Explore*** What other fields could you add to this data base to keep an accurate count of the inventory? Share your results with your classmates. Answers may vary. Possible answers are size and color.

■ A data base can help you keep track of information.

Check students' work.

11. ***PROJECT*** Make up your own data base. Choose any data that you like. Use the following questions as a guide.

a. What kind of information will your data base contain?

b. How many fields will the data base have? Have you included every field needed to make a useful data base?

c. How will you gather the data? Is the data simple to obtain?

d. How many records will your data base have? Is there a limit to the number of records it can have?

e. How often do you need to update the data? How will you know when you need to change the data for a particular field?

f. How will you arrange the data?

g. If you have a computer and data base software, create a data base.

5

DATAPOINT

When you receive a form letter with your name on it, you know your name is part of a data base.

DATAPOINT

Here are some suggestions for data bases:

- Books that you own
- Any of the data found in the chapter Data Files
- CDs and records
- Address book

5 Tell students that a useful data base requires careful planning before any data is entered. In addition to the questions in the text, students need to think about who will use the data, how the data will be used, how often the data will be accessed, and how the data may be sorted.

Students can use *Appleworks* for the Apple II computer, *Microsoft Works* for the IBM, or *Excel* for the Macintosh to create their data bases. They should save their data bases on disk and print them out, if possible.

Closure

Ask students what uses there are for sorting data by different fields. (same data are used by different people for different purposes) Ask the class to think of different uses of data bases that they may encounter. Possible uses include inventory and price lists in a grocery store, directory of books at the library, directory of phone numbers, and airline reservations.

As a follow-up, refer students to the computer activity on p. 96E of the Chapter Overview. See also TRB Exploring O-12.

Assignment Guide

Basic All
Average All
Enriched All

FOR THE NEXT LESSON

math journal, bowls of water, variety of objects to test if they sink or float

Instruction

This page provides a variety of problems that can be used to reinforce and enhance the students' problem solving skills. Encourage students to read each problem carefully. Then have them refer to the list of problem solving strategies to help them decide how to solve the problem.

Point out, however, that not all questions require a strategy for solving, nor are all the strategies in the list used in this lesson.

ADDITIONAL ANSWERS

2.

$.05	$.10	$.15
9		
7	1	
6		1
5	2	
4	1	1
3	3	
3		2
1	4	
	3	1
		3
2	2	1
1	1	2

Problem Solving Practice

PROBLEM SOLVING STRATEGIES

Look for a Pattern
Guess and Test
Simplify the Problem
Make an Organized List
Work Backwards
Account for All Possibilities
Make a Table
Write an Equation
Solve by Graphing
Draw a Diagram
Make a Model
Solve Another Way
Simulate the Problem

Solve each problem. The list at the left shows some possible strategies you can use.

1. Lisa is slower than Christine, but faster than Nicole. Nicole is slower than Lisa, but faster than Jo Ann. Order the girls from fastest to slowest. Christine, Lisa, Nicole, JoAnn
2. The convenience store sells pens for $0.05, $0.10, and $0.15. List all the ways that Joseph can spend exactly $0.45 on pens. See margin.
3. Miguel lives 2.75 mi from school. His friend lives 1.35 mi from school in the opposite direction. If Miguel rode his bicycle to school, visited his friend, and then returned home, how many miles would he ride in all? 8.2 mi
4. Janet purchased three spools of thread at $.89 each, five yards of material at $2.29 per yard, one sewing pattern at $1.89, and five yards of ribbon at $.45 per yard. She gave the cashier a $20 bill. How much change should she receive? How many more spools of thread can she buy with her change? $1.74; 1
5. ***DATA FILE 2 (pp. 52–53)*** Maria hears thunder 1.4 s after she sees a flash of lightning. Paul hears thunder 2.2 s after he sees a flash of lightning. How many meters farther from the lightning is Paul than Maria? 276.8 m
6. An art teacher is purchasing sketch paper for her students. She can buy a 12-package box with 300 sheets per package for $72.00, or a 6-package box with 200 sheets of paper per package for $30.00. Which is the better purchase? the 12-package box
7. The alarm on a clock rang at 6:00 A.M. It continued ringing at regular intervals. At 6:08 the buzzer was on, at 6:10 the buzzer was off, at 6:16 it was on, and at 6:48 it was on. Will the buzzer be on or off at 7:52? at 9:18? on; off
8. Five women ran the 100-m dash. Their times were 11.6 s, 10.2 s, 9.9 s, 10.6 s, and 11.9 s. What was the average time? Is 10.787 s above or below this average? 10.84 s; below
9. It is 398 mi from Buffalo to Boston. Nan leaves in her car and drives at an average of 47 mi/h. Barb leaves in her car and drives at an average rate of 55 mi/h. If they both drive straight through, how many hours to the nearest tenth will Barb have to wait for Nan to arrive? 1.2 h
10. June earns d dollars per hour for the first 40 h of work each week and $1.5d$ for each hour over 40. June worked 46 hours and 15 minutes last week. How much did she earn if d equals 5? $246.88

PROBLEM SOLVING APPLICATION

OBJECTIVE: *To solve density problems using metric units.*

3-6 Using the Metric System

The units of length, mass, and capacity in the metric system are related by water. A cube measuring 1 cm on each side has a volume of 1 cm^3. This is equivalent to 1 mL. A centimeter cube filled with water has a mass of 1 g.

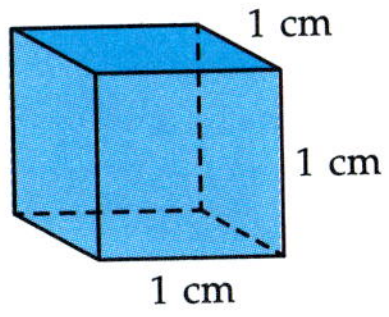

The *density* of a substance is its mass per cubic centimeter. The density of water is 1 g/cm^3. Substances with densities greater than 1 g/cm^3 sink in water. Substances with densities less than 1 g/cm^3 float in water.

You can use the following formula to find density.

$$\text{density } (d) = \frac{\text{mass}}{\text{volume}}$$

Example 1 A block of wood has a volume of 20.52 cm^3. Its mass is 17 g. Find the density of the block in grams per cubic centimeter. Tell whether the block will float or sink in water.

1

Solution

$\text{density} = \frac{\text{mass}}{\text{volume}}$ — Use the formula for density.

$d = 17 \div 20.52$ — Substitute values.

$\approx 0.83 \text{ g/cm}^3$

Since the density of the block is less than that of water, the block will float in water.

You can use the formula to find mass when density and volume are known or volume when density and mass are known.

Example 2 A diamond has a density of 3.5 g/cm^3 and a volume of 0.5 cm^3. Find the mass of the diamond.

Solution

2

$\text{density} = \frac{\text{mass}}{\text{volume}}$ — Use the formula for density.

$3.5 = \frac{m}{0.5}$ — Substitute values.

$0.5 \cdot 3.5 = \frac{m}{0.5} \cdot 0.5$ — Multiply each side by 0.5.

$m = 1.75 \text{ g}$

FLASHBACK

Prefix	Meaning
milli-	0.001
centi-	0.01
deci-	0.10
kilo-	1,000

Symbol	
cm	centimeter
cm^3	cubic centimeter
g	gram
L	liter
mL	milliliter

Lesson Focus

Materials

Math journal, bowls of water, variety of objects to test if they sink or float

Vocabulary/Symbols

density

MOTIVATING THE LESSON

Ask students why two objects with the same dimensions, such as a brick and a piece of foam rubber, do not have the same weight. (Density is different.)

Instruction

1 Have students write the units for the problem for mass and volume to help them understand the units in the answer.

2 Make sure that students understand the units for each value in the formula and in the answer.

Teaching Tip

Activities that involve a lot of discussion in groups may require that you remind students to use low voices and to raise their hands when they need information from you. All students should be encouraged to contribute to the group's discussion.

EE: 1A, 1B, 1D, 3A, 4B; TAAS: 4, 6, 8, 9, 11, 12, 13

Guided Practice

Class Exercises Have students solve question 1 at the chalkboard, labeling each value.

Closure

Writing in Math Have students write in their math journals the meaning of density. Then have them explain why objects with the same dimensions might have different weights.

Independent Practice

For questions 3-4, remind students that for a substance to float, it must have a lower density than the substance it is floating in.

For question 10, have students work in groups to research density and buoyancy.

Decision Making Have students work in groups to complete the questions in this activity and record their work in their math journals.

For question 1, be sure that groups have a variety of objects to test.

For questions 3-6, remind students that they do not need to find the answer, only the method they would use. Mention that members of the group must agree on the method.

Masses of Metal Blocks ($V = 8$ cm³)

aluminum	22 g
gold	154.4 g
silver	84 g

CLASS EXERCISES

Use the *DATA* at the left. Round to the nearest tenth.

1. Find the density of aluminum, gold, and silver. aluminum 2.8 g/cm³, gold 19.3 g/cm³, silver 10.5 g/cm³
2. How much greater is the mass of a cubic centimeter of gold than a cubic centimeter of silver? 8.8 g
3. How many milliliters of water equal the mass of a block of gold having a volume of 125 cm³? 2,412.5 mL

Densities of Selected Substances (g/cm³)

gasoline	0.68
ethyl alcohol	0.79
rubber	1.34
iron	7.9
copper	8.9
mercury	13.5

WRITTEN EXERCISES

Use a CALCULATOR where appropriate.

Use the *DATA* at the left to solve each problem. Express answers to the nearest tenth.

1. What is the mass in grams of a lump of copper that has a volume of 7.5 cm³? 66.8 g
2. What is the volume in cubic centimeters of a block of iron that has a mass of 100 g? 12.7 cm³
3. Which of the substances listed will float in water? in mercury? water: ethyl alcohol, gasoline; mercury: copper, gasoline, ethyl alcohol, rubber, and iron

Decision Making ■ **DECISION MAKING** ■ Decision Making ■ Decision Making ■ Decision Making

USING THE METRIC SYSTEM

COLLECT DATA

1. Test various objects in the classroom to find out which ones sink and which float in water. First guess, then check your guess by putting the object in water to see whether it sinks or floats. Keep a record of your results. Check students' work.

Object	Substance	Sink	Float

2. Why would it not be good to have objects such as pencils or sponges on your list to use for Exercise 1?

ANALYZE DATA

Suppose you have a ruler, a clear drinking glass with vertical sides and flat bottom, a scale that measures in grams, and several marbles of the same size.

MATH MINUTES

Provide several objects, giving the class *two* of the three measures: density, volume, mass. Have students find the unknown measure. Then change one known measure and have them explain how that affects the other measures.

4. Will copper float in mercury? in ethyl alcohol? yes; no

5. How much greater is the mass of 1,000 mL of ethyl alcohol than 1,000 mL of gasoline? 110 g

6. What is the mass in kilograms of 20.75 L of gasoline? 14.11 kg

Solve.

7. Water expands when it freezes. If you freeze 11 cm^3 of water, you will get about 12 cm^3 of ice. Use this fact to find the density of ice. 0.917 g/cm^3

8. ***ESTIMATION*** A gasoline truck is going to take on 1,289 L more gasoline. The density of gasoline is 0.68 g/cm^3. About how many kilograms more is the truck going to take on? 875 kg

9. A silver necklace has a volume of 100 cm^3. Use the chart on page 126 to find the density. Then find the mass of the necklace. 1,050 g

10. ***PROJECT*** Find out what the relationship is between density and buoyancy. Explain why a steel ocean liner floats in water. Buoyancy is upward force caused by fluid pressure. The density of a substance determines its buoyancy. A ship is a steel shell that's hollow inside. The combined density of the steel and air is less dense than the water, hence the ship floats.

Ice floats since it is less dense than water. How does this relate to the fact that 0.9 of an iceberg is below the surface of the water? Ice is only slightly less dense than water so part of the iceberg is above the water while most is below the water.

Decision Making ■ Decision Making ■ Decision Making ■ Decision Making ■ Decision Making ■ Decision Making

3. If you pour some water in the glass, how can you use the materials to find out how many cubic centimeters of space the water occupies? See Additional Answers.

4. If you take several marbles, how can you use the materials to find how many cubic centimeters of space they occupy?

■ ***MAKE DECISIONS***

5. Decide how you could use the materials listed in the *Analyze Data* section to find the density of the glass from which the marbles are made. Describe your plan in a paragraph. Use diagrams as needed to explain your plan.

6. Decide how you could use the materials to find the density of something that will not sink in water. Describe your plan in a paragraph.

Lesson Follow-up

LESSON RESOURCES

TRB Alternate Application C-11

ADDITIONAL ANSWERS

3. Compute the volume or weigh the empty glass and the full glass and compute the difference; then use the mass and density of water to compute the volume.
4. Place a full glass of water and the marbles on a scale. Add the marbles to the glass allowing the water to overflow into a second container. The volume of the displaced water is the volume of the marbles.
5. Use the mass of the marbles and the volume calculated in part (4).

Assignment Guide

Basic All
Average All
Enriched All

FOR THE NEXT LESSON

calculator

Alternate Application For use after 3-6 (pp. 125–127)

Diving Scores

Each judge in a diving competition grades every dive from 0 to 10, with 10 representing a perfect dive. High and low scores are thrown out. (If there are duplications of high and low scores, only one of the duplicated scores is thrown out.) The remaining scores are added and their sum is multiplied by the degree of difficulty.

Degrees of Difficulty			
Position:	Tuck	Pike	Layout
Forward $1\frac{1}{2}$ somersault	1.4	1.7	2.1
Back double somersault	1.9	2.2	2.5
Reverse dive	1.5	1.7	1.8
Inward $2\frac{1}{2}$ somersault	2.8	——	——
Back $1\frac{1}{2}$ somersault $2\frac{1}{2}$ twist	——	3.1	——

1. A back double somersault in the tuck position is awarded these scores: 6, 5, 7, 7, 8, 6, 7, 6
   a. Which scores should be thrown out? **5, 8**
   b. Find the total score for the dive. **74.1**

Complete.

	Dive	Position	Judge's scores	Diver's score
2.	Reverse	Layout	6, 8, 6, 8, 7, 7, 6, 6	**72**
3.	Forward $1\frac{1}{2}$ somersault	Pike	3, 4, 6, 4, 5, 5, 3, 7	**45.9**
4.	Inward $2\frac{1}{2}$ somersault	Tuck	9, 9, 8, 7, 8, 7, 9, 9	**140**

5. A diver received scores of 6, 9, 7, 9, 7, 7, 8, and 5 and earned a total score of 136.4. What dive did the diver perform?
   Back $1\frac{1}{2}$ somersault $2\frac{1}{2}$ twist

6. To win, a diver needs to score 110 points on her final dive. She thinks she could average 7.5 points from each of the 8 judges. What degree of difficulty should she attempt?
   2.5 or greater

Chapter 3 C-11

Lesson Focus

Materials

Calculator

Vocabulary/Symbols

formula

MOTIVATING THE LESSON

Ask students to think of formulas they already know, such as $A = lw$ and $A = \frac{1}{2}bh$. Explore the steps students normally use to solve problems using formulas.

Instruction

1 Often formulas have been written about relationships that are discovered on the basis of experimentation. Point out that when a formula is written, the variables must be defined.

2 Ask students to explain why n is replaced by 32. (n stands for the number of chirps per minute, which is 32.)

3 Ask students to substitute values in the distance formula for a situation in which Sara drove at the same rate but for only 30 min. ($d = 49.7 \cdot 0.5$) Have students explain why 0.5 is used, instead of 30, to represent t. (Time must be given in hours, not minutes, because the rate is given as miles per hour.)

OBJECTIVE: *To use formulas.*

3-7 Using Formulas

▼ You can use the number of chirps a cricket makes in a minute to estimate the temperature in degrees Fahrenheit.

Chirps per min (n)	Temperature (°F)
12	40°
16	41°
20	42°
24	43°
28	44°

▼ A science class conducted an experiment. For five weekdays the class counted the number of chirps their pet cricket made in a minute. They recorded this information in a table, along with the actual temperature. The class observed that if they divided the number of chirps per minute (n) by 4, and added 37, the result was the temperature in degrees Fahrenheit (F). A student wrote the following *formula* on the board. **1**

$$F = \frac{n}{4} + 37$$

Formula	A formula is an equation that shows the relationship between two or more variables.

THINK Will the formula $F = \frac{n}{4} + 37$ ever produce negative temperatures? Why or why not? **No; you cannot have fewer than zero chirps.**

Example 1 If the class counted 32 cricket chirps in 1 min, what is the Fahrenheit temperature?

Solution **2**

1. Write the formula. $F = \frac{n}{4} + 37$
2. Substitute values. $F = \frac{32}{4} + 37$

   $= 45$

The Fahrenheit temperature is 45°.

▼ The *distance formula* is an important and useful formula. The distance formula is $d = rt$, where d is the distance, r is the rate at which you travel, and t is the time you spend traveling.

Example 2 **Find the distance Sara traveled if she drove 8.5 h at an average speed (rate) of 49.7 mi/h.**

Solution **3**

1. Write the formula. $d = rt$
2. Substitute values. $d = (49.7)(8.5)$

   $d = 422.45$

Sara traveled 422.45 mi.

▼ You can write a formula in more than one way.

Example 3 State whether each formula is equivalent to $d = rt$. If so, state what was done to each side of the equation.

a. $\frac{d}{r} = t$ **b.** $d + r = t$

Solution **a.** yes; both sides divided by r **b.** no

Teaching Tip

Be aware of students who avoid answering questions or limit their responses as much as possible. These students may be having difficulty understanding the concepts or the vocabulary.

Batting average is determined by the formula $a = \frac{h}{n}$, where a is the batting average, h is the number of hits made, and n is the number of times up at bat.

Example 4 Use a calculator to solve. How many hits did Babe Ruth make if he was up at bat 8,399 times and had a batting average of 0.342?

Solution

4

1. Write the formula.	$a = \frac{h}{n}$
2. Substitute values.	$0.342 = \frac{h}{8{,}399}$
3. Multiply.	0.342 ⊠ 8,399 ⊜ 2,872.458
4. Round to nearest integer.	2,872

Babe Ruth made 2,872 hits.

CLASS EXERCISES

Solve. Use the appropriate formula.

1. Find the temperature if a cricket chirps 44 times per minute. 48°F
2. A snail travels about 0.14 cm/s. To find how far the snail travels, in what unit should time be given? How far will the snail travel in one hour? seconds; 504 cm
3. The spine-tailed swift is the fastest creature alive. It has been clocked traveling at 106.25 mi/h. At that speed, how far could the swift travel in 2.5 h? 265.625 mi
4. Florence Griffith-Joyner holds the American record in the 100-m run. She ran in 10.49 s. If she could continue at this rate, how long would it take her to run 1,000 m? 104.9 s

WRITTEN EXERCISES

Use the formula $F = \frac{n}{4} + 37$, where n is the number of cricket chirps per minute, to find the temperature in degrees Fahrenheit.

1. $n = 200$ 87°F **2.** $n = 288$ 109°F **3.** $n = 60$ 52°F **4.** $n = 104$ 63°F

Use the formula $d = rt$ to find the distance traveled.

5. $r = 38.5$ mi/h, $t = 12.1$ h 465.85 mi **6.** $r = 280$ mi/h, $t = 9.75$ h 2,730 mi

7. $r = 213$ cm/s, $t = 8$ s 1,704 cm **8.** $r = 0.08$ ft/s, $t = 2.5$ h 720 ft

THINK AND DISCUSS See below.

1. What would you expect to happen to the number of times a cricket would chirp in 1 min if the temperature increased?
2. How many variables are there in the distance formula? To use the formula, how many variables must equal a number?
3. How are formulas and equations alike? How are they different?

1. It would increase.
2. 3; 2
3. They both contain equal signs and variables; the variables in a formula stand for specific items.

4 In step 2, ask students to identify what each number and variable represents. (0.342 is the batting average, h is the number of hits, and 8,399 is the number of times up at bat)

ADDITIONAL EXAMPLES

1. Find the Fahrenheit temperature if a cricket chirps 120 times per minute. (67°)
2. Use the formula $d = rt$ to find the distance traveled when $r = 110$ cm/s and $t = 3.88$ s. (426.8 cm)
3. Use the formula $C = \pi d$ to find the diameter (d) of a circle if its circumference (C) is 9.734 m. Use 3.14 for π. (3.1 m)

Guided Practice

Class Exercises For questions 2-4, students will need to use the distance formula.

In question 4, have students identify which parts of the formula should be substituted with information given in the question, and which variable represents the unknown. (distance and time; r) Ask students to describe the steps they will follow to find the rate.

Think and Discuss Have students work in small groups to answer question 3 and then have groups compare answers. (A formula is a type of equation. A formula shows relationships between variables that are true in almost all cases. Formulas are intended to provide plans to find solutions.)

MATH MINUTES

Have students work in small groups to create their own formulas. Tell them that each variable must be defined. For example: $x = q + d + n$ where x = total amount of money, q = quarter, d = dime, n = nickel.

A ***common error*** occurs when students substitute the wrong values for variables in a formula. For example, a student may substitute the value for n in place of the variable h in the formula $a = \frac{h}{n}$. Encourage students to copy the formula before substituting values for variables.

Closure

To be sure students understand the concept of a formula, have them write formulas for different situations you describe, such as: The total floor space in a building is the product of the length and width of one floor times the number of floors. Be sure students identify the variables they use in the formulas.

Independent Practice

For question 10, after students have solved for t, have them determine how many days the journey would take.

Number Sense For questions 22 and 23, ask students if their answers seem reasonable. Then have them describe the steps they used to find the Fahrenheit temperature in question 23.

For question 29, have students work in small groups.

For questions 30 and 31, if students are having difficulty solving for a specific variable after they have copied the formula, suggest that they first substitute numbers for some of the variables.

9. ***ESTIMATE*** The first plane to fly faster than the speed of sound was piloted by Chuck Yeager. He flew at 670 mi/h. About how far could he fly in 9.8 h? Is this estimate high or low? 6,700 mi; high

10. The solar probe *Helios B* reached a speed of 149,125 mi/h. Suppose the probe traveled at this speed for its entire journey. About how long would it take to reach the sun, which is 93,000,000 mi away? Round to the nearest hundredth. 623.64 h

Major League Lifetime Leading Batters

Player	At bats (n)	Hits (h)
Browning	4,795	1,664
Hornsby	8,173	2,930
Cobb	11,436	4,190
Delahanty	7,493	2,593
Keeler	8,570	2,955
Jackson	4,981	1,774

Use the formula $a = \frac{h}{n}$ and the table at the left to determine each batting average. Round answers to the nearest thousandth.

11. Browning 0.347
12. Hornsby 0.358
13. Cobb 0.366
14. Delahanty 0.346
15. Keeler 0.345
16. Jackson 0.356
17. Order each player from highest batting average to lowest. Cobb, Hornsby, Jackson, Browning, Delahanty, Keeler

Use the formula $w = rt + 1.5r \cdot o$ to find a person's wage. Let w be the wage, r the hourly rate of pay, t the number of hours worked at the hourly rate, and o the number of hours worked overtime. Round to the nearest cent where necessary.

	Worker	r	t	o	
18.	Jeremy	6.20	40	8	$322.40
19.	Elaine	9.75	40	10	$536.25
20.	Eric	4.35	38	3	$184.88
21.	Serena	12.55	41	9	$683.98

CALCULATOR **Use the formula $F = 1.8C + 32$ to find the temperature in degrees Fahrenheit, where C is the temperature in degrees Celsius.**

22. Find the Fahrenheit temperature corresponding to 58°C, the highest recorded temperature in the world. 136.4°F

23. Find the Fahrenheit temperature corresponding to -89°C, the lowest recorded temperature in the world. -128.2°F

Bread and Milk

Years	Loaf of bread	Half gal of milk
1890s	$.03	$.14
1930s	$.09	$.28
1950s	$.18	$.48
1970s	$.24	$.66
1980s	$1.39	$1.09

Use the formula $I = N - O$ to find the price increase. Let I be the price increase, N the new price, and O the old price. Use the chart at the left to answer the following questions.

24. What was the price increase of a half gallon of milk from the 1930s to the 1970s? $.38

25. What was the price increase of a half gallon of milk from the 1970s to the 1980s? $.43

26. How much more was spent in the 1970s than in the 1890s for 5 loaves of bread and 6 half gallons of milk? $4.17

Use the formula $k = \frac{t \cdot w}{1{,}000}$ to find the number of kilowatt hours used. Let k = kilowatt hours, t = hours, and w = watts. Refer to the data at the right.

Appliance	Watts of electricity used by appliance
Television	145
Computer	155
Hair dryer	1,000
Microwave	1,500
Conventional oven	12,200
Stereo	109
VCR	45

27. a. How many kilowatt hours are used by cooking in the microwave for 0.75 h per evening for one week? How many are used by cooking in a conventional oven? **7.875; 64.05**

b. How much would each cost if the electric company charges $.04 per kilowatt hour? **$.32 and $2.56**

c. What is the savings over a year by using a microwave instead of a conventional oven? **$116.48**

28. Are more kilowatt hours used by working on the computer for 2.25 h or by drying your hair for 0.25 h? **working on computer**

29. ***PROJECT*** Find the local rate for electricity. Develop a plan for cutting back on your use of electricity. Determine how much money you would save in a year. **Check students' work.**

30. Gross earnings g is equal to net pay p, plus deductions d. Write a formula to find gross earnings. Solve the formula for d. **$g = p + d$; $d = g - p$**

31. The voltage, V, across any part of a circuit is the product of the current I and the resistance R. Write a formula to find the voltage. Solve the formula for R. **$V = IR$; $R = \frac{V}{I}$**

TEST YOURSELF

CALCULATOR **Solve each equation.**

1. $16.9376 = 6.32s$ **2.68**

2. $-8.125 = \frac{b}{0.023}$ **−0.186875**

3. $n \div (-7.34) = -2.758$ **20.24372**

4. $\frac{w}{105.1} = 0.7352$ **77.26952**

5. $0.0812g = 3.248$ **40**

6. $10.20807 = 3.369c$ **3.03**

Solve for variable a.

7. $c = ab$ **$a = \frac{c}{b}$**

8. $a + c = b$ **$a = b - c$**

9. $c = \frac{a}{b}$ **$a = cb$**

Use the formula $d = rt$.

10. a. Find d when $r = 48$ and $t = 3.5$. **168**

b. Find t when $d = 693$ and $r = 63$. **11**

c. Hal drove a distance of 562 mi at an average rate of 58 mi/h. How many hours, to the nearest tenth, did Hal drive? **9.7 h**

MIXED REVIEW

Solve each equation.

1. $-3n = 92.01$ **−30.67**

2. $\frac{x}{0.43} = -5.2$ **−2.236**

3. $10{,}000 \cdot 0.0034 = k$ **34**

Evaluate when $x = -3.1$ and $y = 10.7$.

4. $4x - 2y$ **−33.8**

5. $x(-12 + y)$ **4.03**

Solve.

6. A stockholder owns 1,000 shares of stock. She receives a dividend of $.65 per share four times per year. How much is the yearly dividend? **$2,600**

Lesson Follow-up

Reteaching Activity For students who still have difficulty with formulas, refer them to the calculator activity on p. 96E of the Chapter Overview. See also TRB Exploring O-11.

LESSON RESOURCES

TRB Practice C-12
TRB Enrichment C-13

Practice

For use after 3-7 (pp. 128–131)

Using Formulas

Use the formula $C = 3.14d$ to approximate the circumference of the circle where C is the circumference and d is the diameter.

1. $d = 18$ in. **56.52 in.**
2. $d = 12.4$ cm **38.936 cm**
3. $d = 224.5$ cm **704.93 cm**
4. $d = 0.06$ km **0.1884 km**

Use the formula $A = 180 - (B + C)$ to find the measure of angle A where A, B, and C are the measures of angles A, B, and C of $\triangle ABC$.

5. $B = 27°$, $C = 95°$ **58°**
6. $B = 110.6°$, $C = 29.8°$ **39.6°**
7. $B = 0.85°$, $C = 3.44°$ **175.71°**
8. $B = 144.32°$, $C = 16.99°$ **18.69°**

Use the formula $V = lwh$ to find the volume of the rectangular prism where V is the volume, l is the length, w is the width, and h is the height.

9. $l = 38$ in., $w = 14.25$ in., $h = 19.5$ in. **10,559.25 in.^{3}**
10. $l = 8.4$ cm, $w = 6.4$ cm, $h = 3.2$ cm **172.032 cm^3**

Use the formula $S = 6.28r(r + h)$ to approximate the surface area of the cylinder where S is the surface area, r is the radius, and h is the height.

11. $r = 3.8$ cm, $h = 10.2$ cm **334.096 cm^2**
12. $r = 0.46$ m, $h = 0.85$ m **3.784328 m^2**

Use the formula $E = \frac{9r}{I}$ and the table to find each pitcher's earned run average where E is the earned run average, r is the number of runs, and I is the number of innings pitched. Round answers to the nearest hundredth.

Pitcher	Innings	Runs
Marichal	300	70
Seaver	286	56
Ryan	149	28
Blue	312	63
Hunter	318	88
Clemens	254	70

13. Marichal **2.10**
14. Seaver **1.76**
15. Ryan **1.69**
16. Blue **1.82**
17. Hunter **2.49**
18. Clemens **2.48**
19. Order the pitchers from lowest earned run average to highest. **Ryan, Seaver, Blue, Marichal, Clemens, Hunter**

Use the formula $I = Prt$ to find the amount of interest where I is interest, P is principal, r is rate, and t is time.

20. $P = \$800$, $r = 0.06$, $t = 3$ **$144**
21. $P = \$2{,}400$, $r = 0.085$, $t = 2.5$ **$510**

C-12 Chapter 3

Enrichment

For use after 3-7 (pp. 128–131)

Car Rally

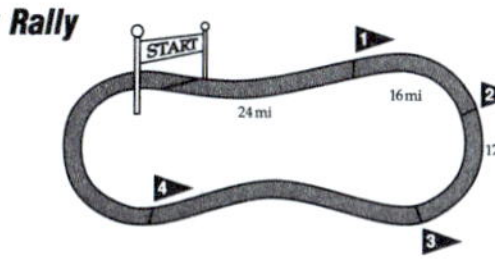

Solve these problems about Maria's car-rally performance.

1. Maria left the starting line at 8:30 A.M. and drove at an average rate of 40 mi/h. What time did she arrive at Checkpoint 1? **9:06 A.M.**
2. She continued to Checkpoint 2, taking 20 min to make the trip at an average rate of 48 mi/h. How far is it between Checkpoints 1 and 2? What time did she arrive at Checkpoint 2? **16 mi; 9:26 A.M.**
3. Maria needed 25 min to drive the leg between Checkpoints 2 and 3. What was her average rate of speed? What time did she arrive at Checkpoint 3? **42 mi/h; 9:51 A.M.**
4. Between Checkpoints 3 and 4, Maria used 1.44 gal of gas and averaged 25 mi/gal. Her average speed was 54 mi/h. Find the distance between Checkpoints 3 and 4. What time did she arrive at Checkpoint 4? **36 mi; 10:31 A.M.**
5. Maria arrived back at the starting line at 11:16 A.M., traveling the final leg of the rally course at an average rate of 46 mi/h. Find the length of the final leg. **34.5 mi**
6. Find the length of the entire course, the length of time it took Maria to finish, and her average overall speed. If necessary, round answers to the nearest hundredth. **128 mi; 2.77 h; 46.21 mi/h**

Chapter 3 C-13

LESSON QUIZ

Use the formula $d = rt$ to find the distance traveled.

1. $r = 40$ ft/s, $t = 2.5$ min ($d = 6000$ ft)

Use $F = 1.8C + 32$ to find the Fahrenheit temperature for each value of C.

2. 56° (132.8° F)

3. 80.5° (176.9° F)

Assignment Guide

Basic 1-2, 5-7, 11-14, 18-19, 22, 24-25, 27-28, MR, TY All

Average 2-3, 7-9, 13-16, 19-20, 22, 24-25, 28-30, MR, TY All

Enriched 3-4, 8-17, 20-21, 23, 25-26, 29-31, MR, TY All

FOR THE NEXT LESSON

math journal

Lesson Focus

Materials
Math journal

MOTIVATING THE LESSON

Mention that in past chapters the class has learned the problem solving strategies *look for a pattern* and *guess and test*. Comment that in this lesson students will learn the strategy *simplifying the problem*.

Instruction

1 Write the number groups 1-9, 10-99, and 100-476 on the chalkboard so students can see how many 1-, 2-, and 3-digit numbers are involved. Then write the number of digits under each number group.

Guided Practice

Class Exercises For question 3, remind students that the numbers need to be consecutive because they are page numbers.

A ***common error*** occurs when students forget to include the first numbers in a range. For example, in Written Exercise 1 students might consider the numbers 2-231 instead of 1-232. It may help students to use a table to account for all the numbers and digits.

Closure

Ask students to describe what they found difficult about solving the problems and how the strategy *simplifying the problem* was useful.

PROBLEM SOLVING STRATEGY

OBJECTIVE:
To solve problems by using a simpler problem.

3-8 Simplify the Problem

Sometimes when you solve a problem, it helps to first solve one or more simpler problems that have similar conditions.

PROBLEM

A typesetter needs one piece of type for each digit in the page numbers of the book. How many pieces of type will the printer need to number pages 1–476?

SOLUTION

READ What do you want to find?

number of pieces of type required to number pages 1–476

PLAN Simplify the problem. **1**

How many one-digit page numbers are there?	pages 1–9 9 pieces of type
How many two-digit page numbers are there?	pages 10–99 90 two-digit page numbers $90 \cdot 2 = 180$ pieces of type
How many three-digit page numbers are there?	pages 100–476 377 three-digit pages $377 \cdot 3 = 1{,}131$ pieces of type

SOLVE Add the numbers for the pieces of type required for the one-, two-, and three-digit numbers.

$$9 + 180 + 1{,}131 = 1{,}320$$

LOOK BACK Interpret your answer.
The printer will need 1,320 pieces of type to number pages 1–476.

CLASS EXERCISES

1. Janetta numbered the pages in her diary from 1 to 58. How many digits did she write? **107**
2. A printer used 330 pieces of type to number the pages of a book. The first page is numbered one. How many numbered pages are in the book? **146**
3. A book is opened. The product of the two page numbers that appear is 272. What are the two page numbers? **16 and 17**

Teaching Tip

As a motivational device, let students rewrite the problems in their textbooks to reflect their own experiences. Students can change a scenario or theme of a problem (or the name of one of the characters in a word problem) to one they prefer.

WRITTEN EXERCISES

Use a CALCULATOR where appropriate.

Solve by using simpler problems.

1. A printer is typesetting a book. He needs one piece of type for each digit in the page numbers of the book. How many twos will he need to number pages 1–232? 87
2. One pastry chef can decorate 12 cupcakes in 14 min. The bakery receives an order for 672 cupcakes. To the nearest hour, how long will it take four pastry chefs to decorate the cupcakes? 3 h

Solve using any strategy.

3. A city has a population of 586,785. The area is 25 mi^2. Find the population per square mile. 23,471.4
4. To accommodate a wheelchair, Tom installed counter tops that were 0.72 ft lower than the original ones in his house. The new counter tops are 2.5 ft high. How high were the original counter tops? 3.22 ft
5. The houses on Wheeler Avenue are numbered 1 to 138. How many house numbers contain at least one digit 6? 23
6. On their way to a concert, Betsy, Meg, and Mary each took turns driving. Mary drove seven miles more than Meg. Meg drove three times as far as Betsy. Betsy drove nine miles. How many total miles did they drive? 70 mi
7. The carnival has two types of rides for children. Each airplane seats 4 children and each spaceship seats 6 children. Altogether there are 24 airplanes and spaceships that seat a total of 128 children. How many of each are there? 8 airplanes; 16 spaceships
8. Jim has $8 in his savings account. Jo has $12 in her savings account. Jim will add $1 to his account each week and Jo will add $3 to her account each week. After how many weeks will Jo's account have twice as much money as Jim's? 4 wk
9. Julie had to number 275 dance tickets by hand. How many digits did she have to write? 717
10. Aaron can work 21.5 h per week at the gas station for $5.75/h. He could also work 27.75 h per week at the convenience store for $4.80/h. At which job will he earn more per week? the convenience store

MATH MINUTES

Have partners solve: Mia paints numbers on uniforms. She can use numbers from 5 through 150. Mia is paid either $1 a digit or a lump sum of $300. Find how many digits she paints and which is the better deal. (338 digits, $1 a digit.)

Assignment Guide

Basic All
Average All
Enriched All

FOR THE NEXT LESSON

graph paper or algebra tiles, math journal

Independent Practice

In question 2, students need to simplify the situation. Suggest that they determine how long it would take one chef, then two, and finally four. Then students need to decide how to find the answer in hours instead of minutes.

Lesson Follow-up

LESSON RESOURCES

TRB Practice C-14
TRB Enrichment C-15

Practice

For use after 3-8 (pp. 132–133)

Simplify the Problem

Solve by using simpler problems.

1. At the inauguration, the President was honored with a 21-gun salute. The report from each gunshot lasted 1 s. Four seconds elapsed between shots. How long did the salute last? 101 s
2. Bernie began building a model airplane on day 7 of his summer vacation and finished building it on day 65. He worked on the plane each day. How many days did it take? 59 days
3. A house-number manufacturer sold numbers to retail stores for $.09 per digit. A hardware store bought enough digits for two of every house number from 1 to 999. How many digits did the store purchase for house numbers:
   a. 1–9 18 b. 10–99 360 c. 100–999 5,400
   d. Find the total cost of the house numbers. $520.02
4. A tic-tac-toe diagram uses 2 vertical lines and 2 horizontal lines to create 9 spaces. How many spaces can you create using:
   a. 1 vertical line and 1 horizontal line 4
   b. 2 vertical lines and 1 horizontal line 6
   c. 3 vertical lines and 3 horizontal lines 16
   d. 4 vertical lines and 5 horizontal lines 30
   e. 17 vertical lines and 29 horizontal lines 540

5. Each side of each triangle in the figures has length 1 cm. The perimeter (the distance around) the first triangle is 3 cm. Find the perimeter of the figure formed by connecting:
   a. 2 triangles 4 cm b. 3 triangles 5 cm
   c. 4 triangles 6 cm d. 50 triangles 52 cm

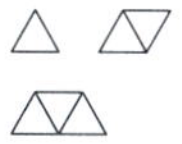

C-14 Chapter 3

Enrichment

For use after 3-8 (pp. 132–133)

Stranded in the Desert

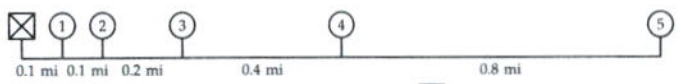

A party of explorers is stranded in the desert at ☒. Water supplies, 10 in all, have previously been buried at the increasingly greater distances from the party's camp shown on the map. In order to gauge how long the party can survive, the leader must calculate the difficulty of retrieving water from the supplies.

1. Write a rule that the leader can use to calculate distances between water supplies.
   Each distance is twice the previous distance.
2. Find the round-trip distances from camp to:
   supply 1 0.2 mi supply 2 0.4 mi supply 3 0.8 mi
   supply 4 1.6 mi supply 5 3.2 mi supply 6 6.4 mi
3. Because of their weakened condition, water-retrieval parties can travel at a rate of only 0.2 mi/h. Find the round-trip times from camp to:
   supply 1 1 h supply 2 2 h supply 3 4 h
   supply 4 8 h supply 5 16 h supply 6 32 h
4. The leader wished to calculate the total amount of time the party would spend retrieving water from all 10 supplies. Find the time needed to retrieve the indicated supplies.
   the first 1 1 h the first 2 3 h the first 3 7 h
   the first 4 15 h the first 5 31 h the first 6 63 h
5. The leader stumbled onto a simple method for calculating total water-retrieval times.
   Complete: $(2) - 1 =$ 1
   $(2 \times 2) - 1 =$ 3
   $(2 \times 2 \times 2) - 1 =$ 7
   $(2 \times 2 \times 2 \times 2) - 1 =$ 15
6. Use the leader's method to calculate the total amount of time the party will need to retrieve all 10 water supplies.
   1,023 h

Chapter 3 C-15

Study Skills

Developing good *listing skills* can help students maximize retention of what they learned during class and minimize the need to spend extra time reviewing lessons at home. Students can learn to screen out distractions and focus their attention on the lesson being taught.

Here are a few ways to develop listening skills:

- ▼ Plan activities and games that sharpen listening skills.
- ▼ Group desks for listening comfort.
- ▼ Minimize distractions.
- ▼ Move about the classroom as you teach to maximize eye contact with students.

Offer students the following suggestions:

- ▼ Read the lesson before class.
- ▼ Participate in class discussions by answering and asking questions.
- ▼ Ask questions when something is unclear.
- ▼ Eat, sleep, and exercise appropriately to maintain a high energy level.

Chapter 3 Review

True or false? If false, change the underlined word(s) to make the statement true.

1. You read and write the decimal 0.05 as five <u>tenths</u>. F; hundredths
2. The steps for front-end estimation are:
   - Add the front-end digits.
   - Adjust by <u>finding</u> the sum of the remaining digits. F; estimating
   - Add these two values.
3. <u>Compatible numbers</u> are used to make estimation easier. T
4. To rank a set of data, write the data <u>in order</u>. T
5. A formula is an equation that shows the relationship between two or more <u>numbers</u>. F; variables

Comparing Decimals — 3-1

To compare decimals, think of a number line. The decimal farther to the right is the greater decimal.

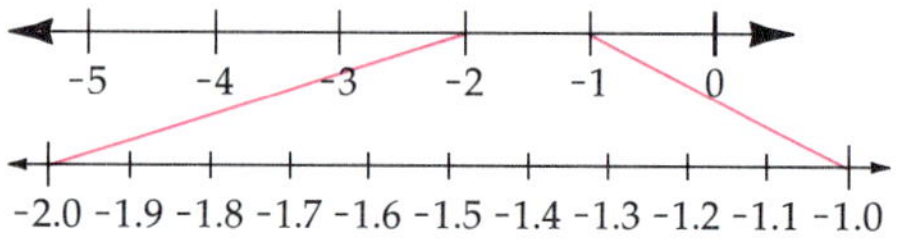

Compare using <, >, or =.

6. −1.9 ▒ −1.3 <
7. −1.0 ▒ −1 =
8. −2.0 ▒ −2.3 >
9. −1.25 ▒ −1.2 <
10. −1.4 ▒ 1.4 <

Estimating Decimals — 3-2

To estimate decimals, use rounding, front-end estimation, or compatible numbers.

ESTIMATION **Estimate each shaded region using decimals.**

11. 0.25

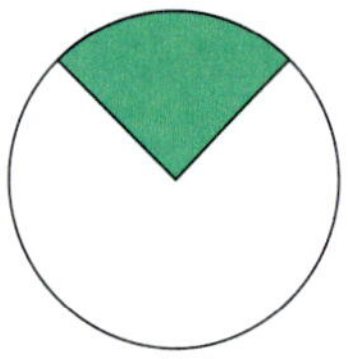

12. 0.5

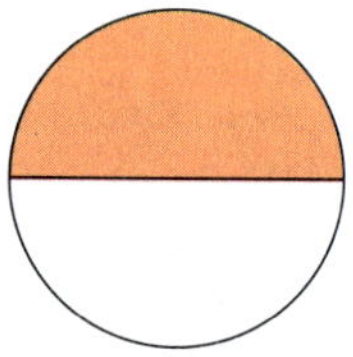

13. 0.25

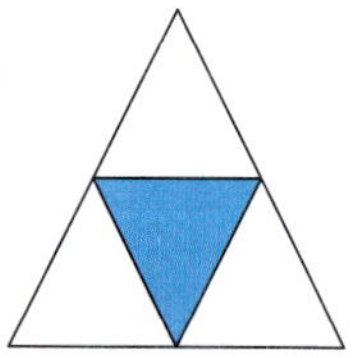

ESTIMATION **Write the technique that seems best. Then estimate.** Accept reasonable estimates.

14. 9.21 + 28.301 + 16.092 53
15. 2.531 ÷ 3.915 0.75
16. \$6.15 + \$9.28 + \$3.69 + \$5.90 \$24
17. 0.7845 ÷ 4.3 0.2
18. 12,909.3 − 3.899 12,905
19. \$48.75 + \$22.95 + \$7.50 \$79

Evaluating Expressions with Decimals — 3-3

To evaluate an algebraic expression, substitute a number for the variable(s), and simplify. Follow order of operations.

Evaluate each expression for $x = -0.5$ and $y = 0.3$.

20. $2x + 3y - 1$ -1.1 **21.** $4(x + 2y)$ 0.4 **22.** $|2x| - 3y + 4$ 4.1 **23.** $5x - 5y + |-15|$ 11

Simplifying Expressions with Decimals — 3-3

To simplify a variable expression, combine like terms and eliminate parentheses using the distributive property.

Simplify each expression.

24. $7.5a - 3(a - 0.2) + 2.5$ $4.5a + 3.1$ **25.** $\frac{(0.8x)(2.5) + 3x}{10}$ $\frac{x}{2}$ **26.** $5a + a(3 + b) + 12ab$ $8a + 13ab$

Solving Equations — 3-4, 3-5

To solve an equation, use the properties of equality.

Solve.

27. $-3.8x = 19$ -5 **28.** $m + 2.45 = 3$ 0.55 **29.** $\frac{k}{3.5} = 2.1$ 7.35 **30.** $y - 2.9 = 8.1$ 11

31. $5.3t = 53$ 10 **32.** $a - (-5.6) = 12.9$ 7.3 **33.** $14w = 42$ 3 **34.** $r + 7 = 10.8$ 3.8

Using Formulas — 3-7

To use a formula, substitute the known value(s) of the variable(s), and solve for the unknown value.

Use the formula $d = rt$. Solve.

35. $r = 55$ mi/h, $t = 3.5$ h 192.5 mi **36.** $r = 30$ ft/s, $t = 1.2$ s 36 ft **37.** $d = 365$ km, $r = 40$ km/h 9.125 h

Problem Solving — 3-6, 3-8

To solve a problem that has difficult numbers, use simpler numbers to see *how* to solve the problem. Then use the real data.

Solve.

38. An auditorium was filled to capacity with 3,500 people. An usher estimated there were 3 adults for every 2 children. How many children were in the auditorium? 1,400 children

39. Use the table on page 126. Find the density of silver. Use the formula density $= \frac{\text{mass}}{\text{volume}}$. 10.5 g/cm^3

RESOURCES

TRB C-17–C-23

Scoring Chart	
No. Correct	**Score**
37	100
36	97
35	95
34	92
33	89
32	86
31	84
30	81
29	78
28	76
27	73
26	70
25	68
24	65
23	62
22	57
21	57
20	51
19	51
18	49
17	46
16	43
15	41
14	38
13	35
12	32
11	30
10	27
9	24
8	22
7	19
6	16
5	14
4	11
3	8
2	5
1	3

Chapter 3 Test

Write each decimal.

1. four hundred fifty-three and fifty-nine hundredths **453.59**
2. three and seven thousand eight hundred fifty-three ten thousandths **3.7853**

Compare using >, <, or =.

3. 0.125 ■ 0.333 $<$
4. -7.656 ■ -0.777 $<$
5. 0.1001 ■ 0.10010 $=$
6. 0.05 ■ -0.05 $>$

Round each decimal to the indicated place.

7. 2.547 nearest tenth **2.5**
8. 8.029 nearest hundredth **8.03**
9. 159.809 nearest unit **160**
10. -0.352 nearest tenth **-0.4**
11. 0.295 nearest tenth **0.3**
12. 14.953 nearest tenth **15.0**

ESTIMATION **Estimate the shaded region using decimal numbers.** **Accept reasonable estimates.**

13. **0.75**

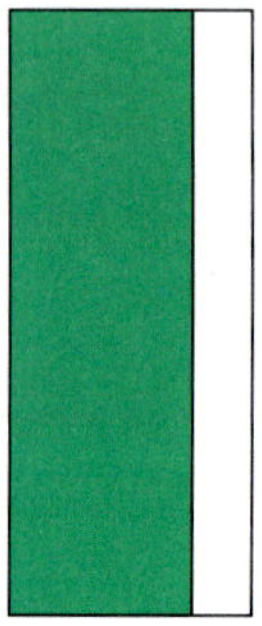

14. **0.5**

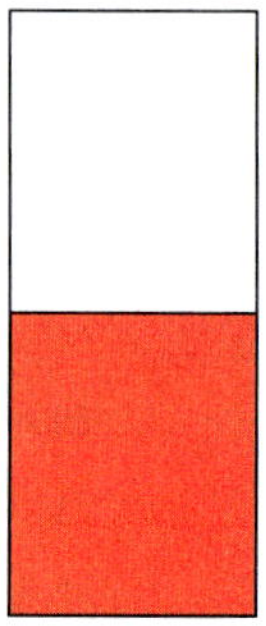

15. **0.25**

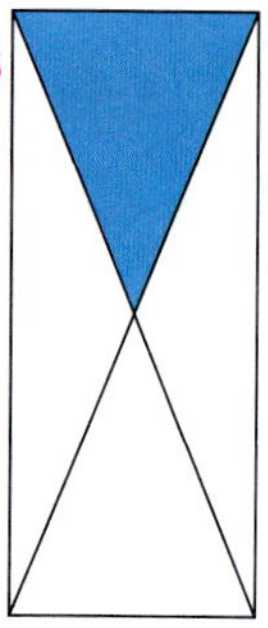

ESTIMATION **Use the technique that seems best.**

16. \$24.79 ÷ 62 **\$.40**
17. -45.167 ÷ 13.92 **-3**
18. 9.057 − 4.01 **5**
19. 31.597 ÷ 19 **1.5**
20. 300.5 + 98.2 **400**
21. 17.63 − 3.58 **14**
22. 1.78 + 2.12 + 18.49 + 7.23 + (-5.54) **24**

Solve each equation.

23. $m + 7.8 = 5.2$ **-2.6**
24. $z - (-8.9) = -2.1$ **-11**
25. $-4r = -2.8$ **0.7**
26. $\frac{h}{11} = -0.3$ **-3.3**
27. $a + (-3.24) = 5.8$ **9.04**
28. $\frac{x}{-0.2} = 0.6$ **-0.12**
29. $0.9k = 2.7$ **3**
30. $b - 9.4 = 0.6$ **10**

Use the formula $C = \frac{(F - 32)}{1.8}$ to find the Celsius temperature where F is the temperature in degrees Fahrenheit. Round to the nearest tenth.

31. 86°F **30°C**
32. 50°F **10°C**
33. 4°F **-15.6°C**
34. 65°F **18.3°C**
35. -10°F **-23.3°C**

Solve.

36. A city has a population of 8,276,386 people. The area is 359 mi^2. What is the population per square mile? **23,054 people/mi^2**
37. ***DATA*** Use the table on page 120. Dwayne is the top high jumper. How does Dwayne rank in the long jump? Who ranks second in the high jump? **2nd; Nancy**

Chapters 1–3 Cumulative Review

Choose the correct answer. Write A, B, C, or D.

1. Name the decimal represented by the point on the number line. B

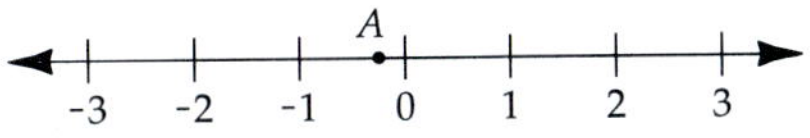

A. 0.25 **B.** −0.25
C. −1.75 **D.** not given

2. two hundred five and six hundredths C

A. 205.600 **B.** 200.56
C. 205.06 **D.** not given

3. What integer represents 15 s before launch? A

A. −15 **B.** +15
C. $|15|$ **D.** not given

4. $|-3| \cdot (-6) \cdot 2$ B

A. 36 **B.** −36
C. −18 **D.** not given

5. Find the mean temperature: 6°, −5°, 2°, 0°, −8°. C

A. −5° **B.** 1°
C. −1° **D.** not given

6. What replacement for a will make the equation $4a - 7 = 25$ true? D

A. 5 **B.** 7
C. −8 **D.** not given

7. $3 \cdot [2 \cdot (-4)] = (3 \cdot 2) \cdot (\blacksquare)$. A

A. −4 **B.** 2
C. 3 **D.** not given

8. Write an expression for the total area of the figure. C

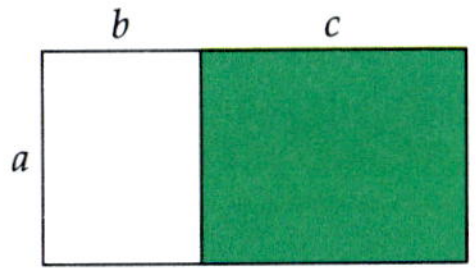

A. $a \cdot (b \cdot c)$ **B.** $a + (b + c)$
C. $a(b + c)$ **D.** not given

9. Round 6.54901 to the nearest tenth. A

A. 6.5 **B.** 6.55
C. 7 **D.** not given

10. Solve $x - 2.5 = -5$. A

A. −2.5 **B.** 7.5
C. −7.5 **D.** not given

11. Simplify $3x + (5 - 2x)0.5$. A

A. $2x + 2.5$ **B.** $x + 2.5$
C. $x - 2.5$ **D.** not given

12. Solve $\frac{y}{1.2} = -3.6$. B

A. −0.3 **B.** −4.32
C. 0.3 **D.** not given

13. Find the distance for $r = 50.5$ mi/h, and $t = 3$ h. Use $d = rt$. C

A. 16.5 mi **B.** 350 mi
C. 151.5 mi **D.** not given

14. Solve $5.6 = x + 3.5$. B

A. 9.1 **B.** 2.1
C. 8.5 **D.** not given

Study Skills

Since most tests have a time limit, *time-management* is an essential part of test-taking skills. One way to encourage mastery of time-management strategies is to schedule practice tests whose results will not affect grades.

Have a classroom clock available or encourage students to be prepared with watches. Suggest that they start by reading through the test and underlining words and phrases that are important, such as: *round to the nearest, estimate, in lowest terms, define,* and *compare.* Then suggest that they mark the more difficult questions. After making an estimate of how much time to allow for each question, students can begin by answering those questions that are easiest to answer. When students feel they have spent more than the allotted time on a question, they should go on to another question.

When the test is finished, have students discuss their feelings about the test, including which techniques helped them do well and which didn't help. Subsequent practice tests will allow students the opportunity to improve.

Some additional ideas:

- Be sure students know how long the test will be.
- Encourage a positive "I can do it" attitude.
- Suggest students use extra time to check their answers.
- Remind students when time will soon be up.

4 Chapter At a Glance

Chapter 4 develops the concepts of exponents, scientific notation, and factoring.

Reading the Chapter

Explain the new terms introduced in the chapter, and encourage students to look up the terms if they need to refresh their understanding. Also review with the class the terms *Greatest Common Factor* and *Least Common Multiple*.

Since students are learning to work with equations involving exponents, remind them to analyze each problem carefully. A hasty evaluation of the expression 2^{12}, for instance, may mean that students multiply 2 and 12. Students should be especially careful in evaluating exponential expressions with negative integers: -3^2, $(-3)^2$, x^2 for $x = -3$. (Only the first expression is negative.)

Chapter Resources

Practice and Enrichment TRB D-1 to D-6, D-8 to D-15
Alternate Application TRB D-7

Chapter Test Form A, D-17
Chapter Test Form B, D-19

Computer Test Bank

References

STUDENT BIBLIOGRAPHY

Stern, David P. *Math Squared: Graph Paper Activities for Fun* (Teachers College Press, New York 1981).

TEACHER BIBLIOGRAPHY

Jacobs, Russel F. *Problem Solving with the Calculator* (Jacobs Publishing Co, Phoenix, Ariz. 1977).

SOFTWARE

Data Insights (Sunburst Communications), for the Apple II family of computers.

FILMS, VIDEOTAPES, AND OTHER MEDIA

Bears, Monsters, and Frogs (Sunburst Communications, VHS or Beta).

Vocabulary

algebraic factorization
base
bit
byte
common factor
common multiple
composite number
exponent
factor
Greatest Common Factor (GCF)
kilobyte
Least Common Multiple (LCM)
megabyte
multiples
prime factorization
prime number
scientific notation
square number
square root
standard notation
systematically

Materials

graph paper (T*)
algebra tiles (M*)
math journal
overhead projector
overhead transparencies

*M—Manipulative Kit
*T—Teacher's Resource Book

Resource Bank

Class Activities

FIELD TRIPS

▼ Planetarium—to use a telescope, if available, and to learn how distances to planets and stars are measured.

▼ Local school board office—to look at maps to see how boundaries are drawn to determine which neighborhoods fall into a particular school district.

CLASS SPEAKERS

▼ Demographer—to inquire how predictions are made regarding local population growth or decline.

▼ Insurance representative—to discuss the use of demographics in establishing automobile insurance rates.

Bulletin Board

Display a road map of your state (a map of western Pennsylvania is shown), and have colored circles of paper available (see key). Have students find the population of villages, towns, etc., and use the appropriate color circle to record the rounded data as shown. Students then pin the circle to the map.

What areas of your state are densely populated? (Answers will vary.)

What areas are sparsely populated? (Answers will vary.)

Choose a sparsely populated area. Name one reason why it has fewer people than other areas. (Answers will vary.)

Project: Find the area in square miles of your city, town, or county. Then divide the population by the area to find the population density (people per square mile).

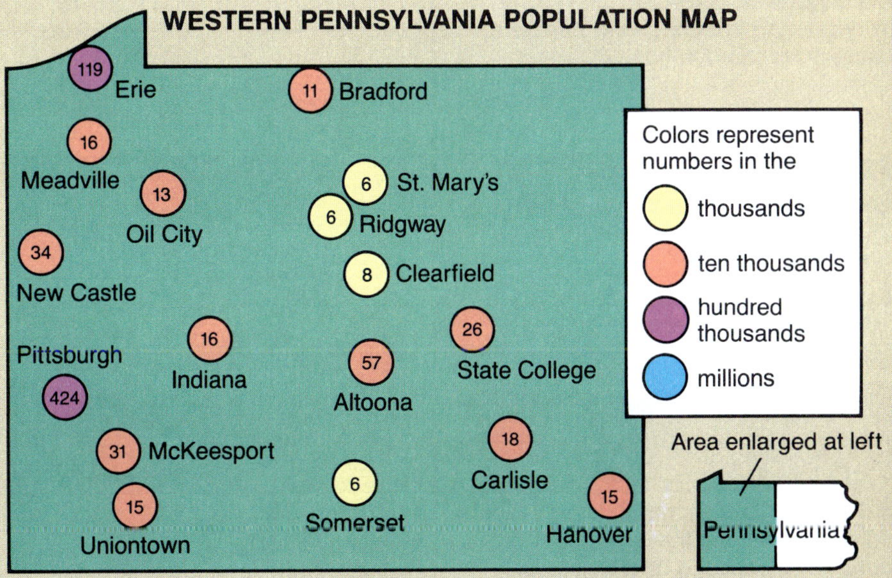

Extra-Credit Problems

BASIC

▼ The prime factorization of 12 is $2 \cdot 2 \cdot 3$. The sum of the factors is $2 + 2 + 3 = 7$. Find the three numbers that have prime factorizations with factors that total 8. (15, 16, 18)

AVERAGE

▼ The digits of a two-digit prime number are reversed and the result is subtracted from the prime number itself. The difference is a positive multiple of 4. What is the prime number? (73)

ENRICHED

▼ If you know the GCF and the LCM of a pair of numbers, can you determine what the numbers are? (Sometimes: if GCF is 5 and LCM is 10, then the two numbers are 5 and 10.)

Chapter Planning Guide

OBJECTIVES	ASSIGNMENT GUIDE			ASSESSMENT	
	Basic	Average	Enriched	Review/ Tests	TRB*
Exploring Square Numbers To explore square numbers using patterns.	**Manipulative Connection** All	All	All		
4-1 Exponents To evaluate and write powers.	1-5, 7-10, 12-15, 22-38, 42-45, MR*, CT* All	1-18, 22-40 even, 43-45, 49, MR, CT All	7-11, 16-34 even, 36-50, MR, CT All	Extra Practice p. 569	Practice D-1 Enrichment D-2
4-2 Rules About Exponents To multiply expressions with powers.	1-8, 19-21, 28-30, 43-45, 50, MR, CT All	1-17 odd, 19-27 odd, 28-31, 35-42, 43-49 odd, 50, 52, MR, CT All	5-27, 32-38, 46-49, 51-54, MR, CT All	Extra Practice p. 569	Practice D-3 Enrichment D-4
4-3 Scientific Notation To write and compute with numbers using scientific notation.	1-5, 6-13 odd, 14, 17-20, 23-24, 28-30, 35-38, 41-42, MR, TY* All	1-11, 15-18, 21-24, 27-29, 31-35, 39, 44, MR, TY All	3-4, 9-14, 16-20, 25-35, 38-41, 43-44, MR, TY All	Extra Practice p. 569 TY p.152	Practice D-5 Enrichment D-6 TY D-16
4-4 Application: Water Resources To apply scientific notation to solve problems concerning water resources and usage.	All	All	All		Alternate Application D-7
Exploring Factors and Multiples To explore factors and multiples.	All	All	All		
4-5 Factors, Multiples, and Divisibility To identify whole number factors and multiples of a number and to determine divisibility by 2, 3, 4, 5, 6, 9, or 10.	2-24 even, 30-31, 35-36, MR, TY All	1-12, 13-29 odd, 30-37, MR, TY All	7-12, 22-29, 31-38, MR, TY All	Extra Practice p. 569 TY p.162	Practice D-8 Enrichment D-9 TY D-16
4-6 Prime Factorization To determine whether a number is prime or composite and to find its prime factorization.	1-15 odd, 23-24, 27, MR, CT All	2-20 even, 23-24, 26-28, MR, CT All	12-30 even, 29, MR, CT All	Extra Practice p. 569	Practice D-10 Enrichment D-11
4-7 GCF and LCM To find the GCF or LCM by listing or using prime factorization.	1-4, 10-12, 19, 22, 24-27, 30-31, MR All	4-7, 13-28, 30-31, MR All	5-9, 13-23, 26-31, MR All	Extra Practice p. 569	Practice D-12 Enrichment D-13
Exploring Counting Problems To explore a problem using a systematic approach.	All	All	All		
4-8 Problem Solving Strategy: Account for All Possibilities To solve problems by accounting for all possibilities.	**Problem Solving Connection** All	All	All	Problem Solving Practice p. 175	Practice D-14 Enrichment D-15
Chapter Test	All	All	All	pp. 176-179	Tests D-17–D-23

*TRB—Teacher's Resource Book MR—Mixed Review TY—Test Yourself CT—Critical Thinking

			CONNECTIONS				NCTM CORRELATION
Algebra	Critical Thinking	Problem Solving	Estimation/ Mental Math	Technology	Manipulatives/ Resources	Writing/Reading in Math	
	2-4, 9	7-9			1-5, 8	6	Communication Problem Solving Connections
C*: 10, 13-14, TD* 3-4 W*: 2-11, 16-21, 37, 40-50	C: TD 1-4 W: 35-37, 39, 45-48, 49a CT*: 1-4	W: 39-41, 48-50	C: 1-4, 5-10 W: 1-9, 11, 36-37	W: 22-26, 41, 49-50	W: 49-50	W: 38, 49b	Communication Problem Solving Functions Discrete Math
C: 6-18, TD 1-3 W: 5-33, 50-51	C: TD 1-3 W: 28-42, 49-53 CT: 1-4	W: 49-54	C: 1-4, 6-10 W: 1-8, 28-32, 43-45, 53b	W: 52-53	W: 52-53		Communication Problem Solving
TY 1-2	C: TD 1-2 W: 40	C: 8-11 W: 3-5, 33-44	C: 1-4, 10-11	W: 27-32	W: 27-36, 41-44	W: 36-44	Communication Connections
C: 1-2 W: 1-8		C: 1-2 W: 1-8	W: 3, 5-7	C: 1-2 W: 1-8	W: 8	W: 8	Problem Solving Connections
4, 14	6, 8, 11, 15-17	1-18		18	1-18	17	Algebra Probability
W: 35-37 TY*: 16-19	C: 1-9, TD 1-4 W: 30f, 35-38	W: 30f, 35-38	W: 22-29,			C: 13 W: 31b	Communication Functions
W. 20-23, 30	C: TD 1-3 W: 20-22, 29 CT: 1-7	W: 23 30 CT: 1-7	W: 30	W: 13-15, 29d	W: 13-15, 27 CT: 1-7		Communication Problem Solving Reasoning
C: 12 W: 5-8, 14-16, 22-23	C: 12, TD 1-3	C: 12 W: 23-31	W: 30-31	W: 19-21	W: 19-21, 30-31		Communication Problem Solving Reasoning
7-10	1-13	1-13			1-13	9, 11	Communication Reasoning Algebra
W: 7, 10-11		C: 1-5 W: 1-14		W: 1-14			Communication Problem Solving Reasoning

*C—Class Exercises W—Written Exercises CT—Critical Thinking TY—Test Yourself TD—Think and Discuss

Exploring

In this manipulative activity, students explore prime numbers through modeling and factorization. **See TRB Exploring O-13.**

Exploring Prime Numbers

You can use models to discover relationships among numbers.

Use graph paper to model 19 and 20. Make as many different rectangular regions as you can for each number. **Check students' work.**

1. How many different rectangles did you make for 19? Describe the rectangle(s).
   1; 1 row of 19 squares
2. How many different rectangles did you make for 20? Describe the rectangle(s).
   3; 1 row of 20 squares, 2 rows of 10 squares, 4 rows of 5 squares
3. Look at the model for 19. How many factors of 19 does the model show? What are they?
   2; 1 and 19
4. Now look at the models for 20. How many factors of 20 do the models show? What are they?
   6; 1, 2, 4, 5, 10, 20

Use graph paper to model the following numbers. Find as many different rectangular regions as you can. Write the factors.

5. 12 **1, 2, 3, 4, 6, 12**
6. 3 **1, 3**
7. 24 **1, 2, 3, 4, 6, 8, 12, 24**
8. 17 **1, 17**
9. 5 **1, 5**
10. 6 **1, 2, 3, 6**
11. Which of the numbers you have modeled have only two factors, 1 and itself?
    3, 5, 17, 19

A number with only two factors, 1 and the number itself, is called a *prime number*.

12. Describe how you can use models to find prime numbers.
    Draw as many different rectangular regions as you can for each number. If a number can be modeled by only one rectangle, then the number is prime.
13. Using what you have learned about prime numbers, do you think that the number 1 is a prime number? Explain why or why not.
    No. A prime number has only two factors, 1 and itself. The number 1 does not have a second factor.

O-13

In this problem-solving activity, students use Pascal's Triangle to explore number patterns and square numbers. **See TRB Exploring O-14.**

Exploring Pascal's Triangle

The triangular pattern at the right is called Pascal's triangle, after the seventeenth-century mathematician and philosopher who studied it and discovered many of its properties.

```
            1
          1   1
        1   2   1
      1   3   3   1
    1   4   6   4   1
  1   5  10  10   5   1
1   6  15  20  15   6   1
```

1. What number is at the beginning and the end of each row?
   1
2. Find the sum of the numbers in each row.
   1, 2, 4, 8, 16, 32, 64
3. Analyze the results. What pattern do you see?
   The numbers are powers of 2.
4. What do you think would be the sum of each of the next three rows?
   128, 256, 512
5. Look at the diagonals. What pattern can you see in the second diagonal? the third diagonal? **Possible answers: 2nd diagonal: counting numbers; differences increase by one: 3rd diagonal: add 2, 3, 4, 5**
6. Look at the numbers other than 1. Then look at the numbers directly above, to the left and right of each number. Describe the relationship.
   Each number is the sum of the two numbers directly above it to the left and right.
7. Copy the triangle onto a separate piece of paper. Use what you learned about Pascal's triangle to extend the triangle up to and including row 10.
8. Explain the method you used to extend the triangle.
   Possible answers: extended the 2nd and 3rd diagonals from the left and right sides using patterns; added the top left and right numbers to get the sum below.
9. Look at your extended triangle again. Write the sum of the numbers in each of the odd rows. What do you call these numbers?
   1, 4, 16, 64, 256; Square numbers or powers of 4
10. What would be the sum of the numbers in row 11? Complete row 11 to test your prediction.
    1,024

O-14

In this paper-and-pencil activity students explore number patterns to find rules for dividing by 2, 3, 5, 9, and 10. **See TRB Exploring O-15.**

Exploring Divisibility

Look at the numbers at the right. Each is divisible by one or more of the numbers below. One number is *divisible* by another if it divides that number with no remainder.

297 | 655 | 1,010 | 42,933 | 42 | 9,065 | 828 | 710 | 414 | 5,886 | 160

Using a calculator, write the numbers at the right that are divisible by each of the following numbers.

1. 2 **42 160 414 710 828 1,010 5,886**
2. 3 **42 297 414 828 5,886 42,933**
3. 5 **160 655 710 1,010 9,065**
4. 9 **297 414 828 5,886**
5. 10 **160 710 1,010**
6. Look at the last digit of the numbers that are divisible by 2. Write a divisibility test that can help you decide if a number is divisible by 2 without using calculations.
   A number is divisible by 2 if the last digit is 0, 2, 4, 6 or 8.
7. Repeat the procedure for the numbers divisible by 5 and those divisible by 10. Write a divisibility test for each.
   a. Test for 5: **A number is divisible by 5 if the last digit ends in 0 or 5.**
   b. Test for 10: **A number is divisible by 10 if the last digit ends in 0.**
8. Look at the numbers divisible by 3. There is a relationship between the sum of the digits of each number and 3. Write the sums.
   6 18 9 18 27 21
9. What is the relationship? **The sums are all divisible by 3.**
10. Write a divisibility test for 3. **A number is divisible by 3 if the sum of the digits is divisible by 3.**
11. Analyze the numbers divisible by 9. Can the same relationship you discovered in question 9 be seen in these numbers?
    The sums are all divisible by 9.
12. Write a divisibility test for 9. **A number is divisible by 9 if the sum of the digits is divisible by 9.**

O-15

In this calculator activity, students explore numbers written in scientific notation using a scientific calculator. **See TRB Exploring O-16.**

Exploring Scientific Notation on a Calculator

You can use a scientific calculator to help you with calculations that use numbers written in scientific notation. **Answers may vary.**

1. What do you think is the greatest number your calculator can display in standard notation?
   99,999,999

Enter the number on the calculator and add 1. If the number is the greatest, the result will be in scientific notation.

2. Write the number you see in the display. Why do you think the calculator changed notation?
   1E8; the display cannot show numbers with more than 8 digits.
3. Without using a calculator write the product of 20,000 and 30,000 in scientific notation.
   6×10^8
4. Now use the calculator to find the product. Write the number you see. On a calculator 6E8 means 6×10^8.
   6E8

Using pencil and paper, write each product in scientific notation. Then use the calculator. Write what the display shows.

	Pencil and Paper	Calculator
5. $10{,}000 \times 200{,}000$	**2×10^9**	**2E9**
6. $23 \times 10{,}000{,}000$	**2.3×10^8**	**2.3E8**
7. $475 \times 1{,}000{,}000$	**4.75×10^8**	**4.75E8**

The (EXP) key will help you enter numbers in scientific notation into the calculator.

8. To enter 1.3×10^2, press 1.3 (EXP) 2. Write what the display shows.
   1.3E2
9. What keys do you press to find the product of (2.5×10^5) and (3.7×10^8)?
   2.5 (EXP) 5 (×) 3.7 (EXP) 8 (=)
10. Write the product. Explain the relationship between the exponent in the product and the exponents in the factors.
    9.25×10^{13}; It is the sum of the powers.

O-16

Meeting Individual Needs

GIFTED STUDENTS

One of the most gratifying outlets for the energies and talents of the mathematically gifted student is the cooperative learning group. Such groups allow gifted students to explore creatively and to share their explorations and discoveries with other students.

To amplify the lesson on prime factorization in Chapter 4, separate the class into small groups that include at least one gifted student per group. Explain that the task is to find a way to express 1,000,000 as a product of two prime numbers, each raised to a power. ($1,000,000 = 10^6 = (2 \cdot 5)^6 = 2^6 \cdot 5^6$)

As students brainstorm, monitor the groups. If a group is having difficulty solving the problem, the teacher can suggest that if students are testing two- or three-digit prime numbers, they might try testing smaller numbers. When groups have discovered the solution, ask them to write a formula to express any power of 10 (greater than 1) as a product of primes. ($10^n = (2 \cdot 5)^n = (2^n \cdot 5^n)$ when $n > 0$)

AT-RISK STUDENTS

At-risk students need encouragement as they work to develop problem solving skills. They also need frequent opportunities for reviewing and modeling problem-solving techniques.

This chapter can be useful to at-risk students, since it discusses how to approach problem solving systematically and introduces the strategy "Account for All Possibilities." Review the definition of *systematically* and ask students to discuss reasons for using a systematic approach to solve a problem.

Students may benefit from seeing the strategy "Account for All Possibilities" modeled several times. Teachers can make up new problems or use the written exercises on page 173. For question 2, model the use of an organized list to identify all the different amounts of postage that are possible with four 25-cent stamps and three 30-cent stamps. First draw four 25-cent stamps and then three 30-cent stamps on the board. Start at the left and list the first stamp. Continue by listing the amounts of postage found by adding the cost of the first and second stamps, the first, second, and third stamps, and so on. Then list the amounts found by adding the cost of the second stamp in combination with all the stamps to its right, and so on until all combinations have been found. Cross out any duplicate amounts. There are 19 different amounts of postage. It may help students if you explain each step of the process as you go along.

ENGLISH AS A SECOND LANGUAGE

ESL students might find it helpful for new terms to be reviewed carefully before each lesson. This can enable the ESL student to focus on the mathematical skills being taught rather than the manner in which the skill is worded. An expression such as *power raised to a power* can be confusing to the ESL student. Other new terms that can be reviewed and explained include *scientific notation*, *square numbers, factors, multiples,* and *prime* and *composite numbers.* In addition to discussing these terms in class, students can be encouraged to define the terms in their own words.

As part of a chapter review, the teacher can introduce a game that allows students to match a term with an appropriate example of that term. For instance, a student would match 4^2 to the terms *base* and *exponent.* This game can be played by students either individually or in small groups. In either case, the teacher would distribute worksheets with both the terms and the examples. Students would draw a line linking the term to the correct example.

See TRB Teaching Resources p. 54.

Problem Solving Situation

The "Think About It" question on this page offers an opportunity for students to develop and apply their own problem solving strategies. Ask students to work out a prediction; then call on students to explain the strategies they used. Have students discuss the data and any patterns that were relevant.

To practice the strategies that they've developed, ask students to research population statistics on two or three types of birds or other wildlife that live in the area. Students can trace the populations over several decades, noting and offering explanations for whether they've increased or decreased. Students can also make predictions about future populations.

DATA COLLECTION

Students can contact local bird-watching clubs, a university ornithologist or zoologist, the Audubon Society, or veterinarians to gather data on the types of birds or wildlife they are researching.

DATA ANALYSIS

Have students analyze and then make a line graph of their data. Groups should use their research and graphs to make predictions about future populations.

DATA FILE

TOP 10 MOST SPOKEN LANGUAGES
(millions of speakers)

CHINESE	700
ENGLISH	400
RUSSIAN	265
SPANISH	240
HINDI	230
ARABIC	146
PORTUGUESE	145
BENGALI	144
GERMAN	119
JAPANESE	116

Population Facts

Most crowded:

Monaco with 15,000 people/km^2

Least crowded:

Western Sahara with 0.5 people/km^2

Fraction of population under 15 years of age:

Africa: about $\frac{1}{2}$ Europe: about $\frac{1}{5}$

DEMOGRAPHERS look for patterns in the size, movement, density, and other characteristics of human populations. Demographers use graphs, like the population pyramid, to look for trends and determine future growth patterns.

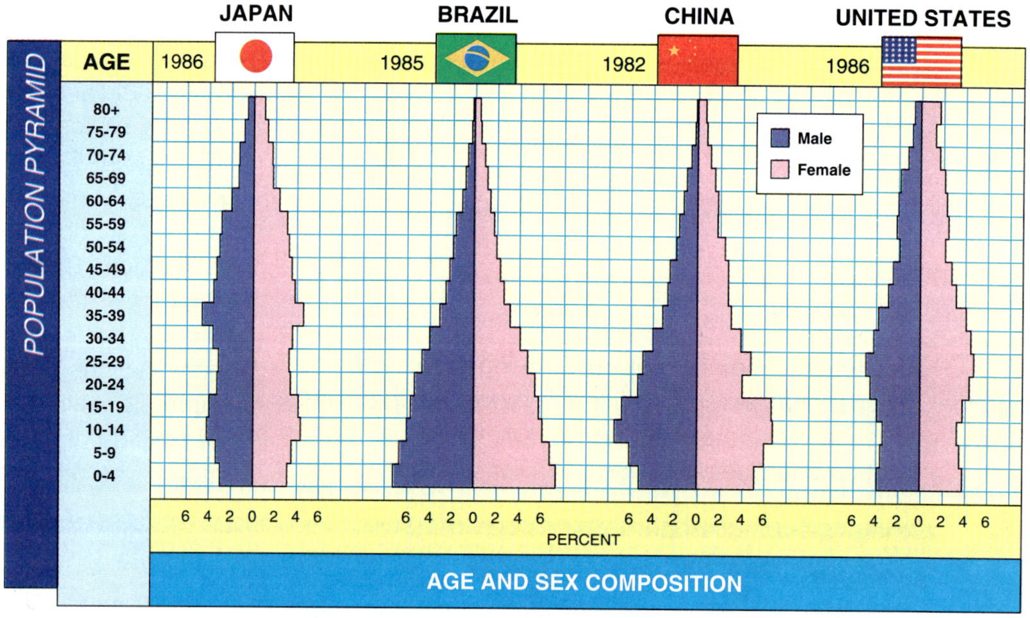

CHAPTER 4

Number Theory

WORLD POPULATION GROWTH ■ From 1450 to 1750 the world population doubled. It doubled again from 1750 to 1855, from 1855 to 1950, and from 1950 to 1989.

Think about it...

Look at the population growth graph. In 1989 the world population was about 5 billion. When do you think the population will double again?

See Problem Solving Situation

BIRTHS − DEATHS + IMMIGRANTS − EMIGRANTS = POPULATION CHANGE

Cooperative Learning

This activity can easily be done in cooperative work groups of 3-4 students. Group members can work together to decide what information they need and then research that data. Then they can discuss the best way to graph the data, and work out a strategy for making their predictions. They can also work together on a report that one group member can present to the rest of the class. Possible roles include Chart Maker, Statistician, Writer, and Fact Checker.

When groups have completed the activity, call on students from each group to describe and discuss their group's findings. Remind students that everyone in the group must agree upon and be able to explain the group's decisions about analyzing the data and making predictions.

Questions related to this Data File can be found on pages 75, 107, 151, 165, 175, 254, 267, 294, 569, and 570.

Lesson Focus

Materials

Overhead projector, unit squares for the overhead projector, math journal

Vocabulary/Symbols

square numbers

MOTIVATING THE LESSON

Use the overhead projector and 10 unit squares to show a rectangular region containing 2 rows of 5 squares. Have volunteers demonstrate other figures that can be made with the squares using all or some of the tiles. (Possible answers: line segment, triangle, square.)

Instruction

1 Use the overhead projector and unit squares to help students visualize the problem. Show the number of airplanes in each new group, then in the total number as the students extend the table.

2 Help students observe the pattern of odd numbers in the middle column, and explain that the number in the last column is a multiple of the group number in the first column.

3 Demonstrate that each total number of airplanes when modeled, form a square region. Help students recall that any number that can be modeled as a square is called a square number. Some students may make the connection that the product of a number multiplied by itself is a square number.

Closure

Writing in Math Ask students to describe in their math journals how to find any square number using only odd numbers. (In general, k^2 = the sum of the first k odd numbers.)

OBJECTIVE: *To explore square numbers using patterns.*

MATERIALS

- Graph paper or algebra tiles
- Math journal to record work

Exploring Square Numbers

▼ Each year the aviation club sponsors an air show. As part of the routine, the planes fly in formation. First one plane flies up, then three planes, then five, then seven.

1. Look at the table below. What was the total number of planes in the air after the third takeoff? after the fourth? **9; 16**

Group	Number of New Planes	Total Number in Air
1	1	1
2	3	4
3	5	9
4	7	16

2. If the flight pattern continued, how many planes would be in the fifth group? in the sixth group? Extend the table to organize your data. **9; 11**

3. What is the total number of planes in the air after the fifth group? after the sixth group? **25; 36**

4. ***Describe*** and discuss the relationship between the group number and the total number of planes in the air. **See left.**

4. The total number of planes is obtained by multiplying the group number times itself.

1 4 9

5. How do the models at the left show the relationship between the group number and the total number of planes in the air? **The number of squares used to make the figure equals the number of planes in the air.**

▼ The numbers in the last column of the table are called *square numbers* or *squares*.

6. ***Write*** two characteristics that describe square numbers. **See left.** **Answers will vary. Examples given.**

6. The two factors of a square number are identical; tiles used to represent square numbers are perfect squares.

7. Without using the table, figure out how many planes would be in the air if there could be a group 20. **400 planes**

▼ The models below show the relationship between the number of new planes and square numbers.

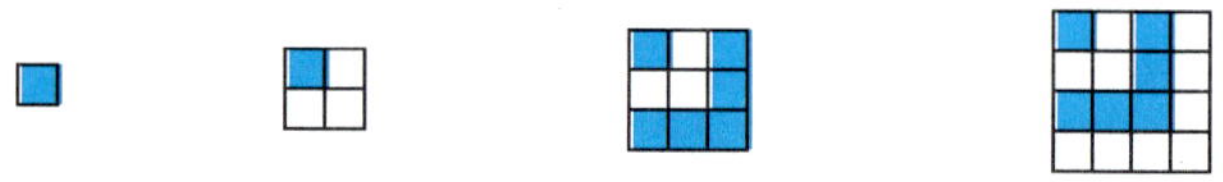

$1 = 1 \cdot 1 = 1$ $\quad$ $1 + 3 = 2 \cdot 2 = 4$ $\quad$ $1 + 3 + 5 = 3 \cdot 3 = 9$ $\quad$ $1 + 3 + 5 + 7 = 4 \cdot 4 = 16$

8. Use tiles or graph paper to model this relationship for the next three square numbers. ***Compare*** with another group. **Check students' work.**

9. ***Summarize***—How can you use an odd number of tiles to form a square number? **The sum of consecutive odd numbers form square numbers.**

Teaching Tip

Keep a supply of colored paper squares handy to be used in place of algebra tiles or unit squares, as well as for representing number patterns and visual patterns.

Assignment Guide

Basic All
Average All
Enriched All

FOR THE NEXT LESSON

calculator, math journal

OBJECTIVE: *To evaluate and write powers.*

4-1 Exponents

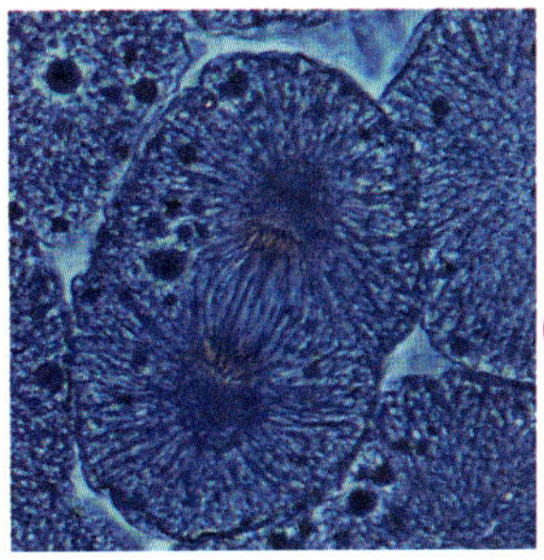

▼ Suppose a cell splits into two cells every hour. How many cells will there be after 12 hours?

The number of cells doubles every hour. You find the total by multiplying 2 twelve times.

$$2 \cdot 2 \cdot 2 \cdot 2 \cdot 2 \cdot 2 \cdot 2 \cdot 2 \cdot 2 \cdot 2 \cdot 2 \cdot 2 = 4{,}096$$

1 You can also express the product using an *exponent*.

$$\text{base} \rightarrow 2^{12} \leftarrow \text{exponent}$$

READ 2^{12} as *2 to the twelfth power.*

Base	The base is the number used as a factor.
Exponent	The exponent shows the number of times the base is used as a factor.

THINK Why does a^1 equal a? a^1 means a is used as a factor 1 time.

▼ You can refer to expressions with exponents as *powers*.

Write	Read	Evaluate
2 2^0	*two to the zero power*	1
12^1	*twelve to the first power*	12
5^2	*five to the second power, or five squared*	$5 \cdot 5 = 25$
6^3	*six to the third power, or six cubed*	$6 \cdot 6 \cdot 6 = 216$
3 -3^4	*the opposite of the fourth power of three*	$-(3 \cdot 3 \cdot 3 \cdot 3) = -81$

THINK What is the base in -3^4? in $(-3)^4$? 3; -3

Example 1 **Write using exponents.**

a. $5 \cdot 5 \cdot 5$ **b.** $-2 \cdot a \cdot 7 \cdot b \cdot b$

Solution **a.** 5^3

b. $-2 \cdot 7 \cdot a \cdot b \cdot b = -14ab^2$ Use the commutative and associative properties of multiplication.

FLASHBACK
Grouping symbols include parentheses ( ), brackets [ ], and absolute value | | symbols.

▼ You can extend the order of operations to include powers.

1. Do all operations within grouping symbols first.
2. Evaluate powers.
3. Multiply and divide from left to right.
4. Add and subtract from left to right.

Lesson Focus

Materials
Calculator, math journal

Vocabulary/Symbols
base
exponent
power
square root (√)

MOTIVATING THE LESSON

Ask students to name a product for which only one factor is repeated. For example: $5 \cdot 5 \cdot 5 \cdot 5 = 625$. Explain that it is convenient to express products of this type in a compact way.

Instruction

1 Explain that an exponent can be used any time a number is multiplied by itself a given number of times.

2 To demonstrate that a zero power is equal to 1 write the following table on the chalkboard.

2^4	$2 \times 2 \times 2 \times 2$	16
2^3	$2 \times 2 \times 2$	8
2^2	2×2	4
2^1	2	2
2^0	1	1

Help students observe the descending pattern of the powers of 2. Each power is the result of dividing the power above by a factor of 2. So, 2^0 equals 1. Chapter 5 Lesson 8 expands this topic.

3 You may need to emphasize that -3^4 is read *the opposite of* 3^4. Explain that 3^4 equals 81; so its opposite is -81. Students may find it easier to understand if you write the expression with parentheses: -(3)(3)(3)(3).

Teaching Tip

When using calculators that do not have an x^y key the value of expressions with exponents can be found by repeated multiplication with the [=] key. For example: 5^8 can be found by pressing [5] [×] [=] [=] [=] [=] [=] [=] [=].

4 Help students understand that in the key sequence needed to evaluate 8^5 the base (or 8) is pressed first; then the x^y key is pressed; next the exponent (or 5) is pressed; and finally the = key is pressed.

ADDITIONAL EXAMPLES

1. Write using exponents.
   a. (-2)(-2)(-2)(-2) $[(-2)^4]$
   b. $-3 \cdot m \cdot m \cdot m \cdot n \cdot 5 \cdot m$ $(-15m^4n)$
2. Evaluate $3x^2 - y^3$ for $x = 4$, $y = -2$. (56)
3. Evaluate 7^6 using a calculator. (117,649)

Guided Practice

Class Exercises You may wish to have students work in pairs to do these exercises.

Think and Discuss If students are having difficulty with question 2, remind them that for $(-6)^2$, the parentheses enclose a negative number, which is squared: (-6)(-6) = 36. Without the parentheses we read -6^2 as the opposite of 6^2 or -36.

For question 3, students may see that 1 raised to any power will still be 1, so the opposite of 1^n, or -1^n will be -1.

To explain the answer for question 4 have students consider the following pattern where in each case, the right side is divided by a:

$$a^4 = a \cdot a \cdot a \cdot a$$
$$a^3 = a \cdot a \cdot a$$
$$a^2 = a \cdot a$$
$$a^1 = a$$
so $a^0 = a \div a = 1$

NOTES & QUOTES

Edward Kasner (1878–1955), an American mathematician, asked his nine-year-old nephew what he would call the number 1 followed by 100 zeros or 10^{100}. The child responded with the word *googol*.

THINK AND DISCUSS

1. How and when are exponents useful? See below.
2. Compare the values of -6^2 and $(-6)^2$. Why are they different? See below.
3. Let n be any positive integer. Find the value of 1^n, -1^n. $1^n = 1$; $-1^n = -1$
4. In the expression a^0, how many times is a used as a factor? What is the value of a^0? 0; 1

1. Exponents are useful to express very large numbers.
2. -6^2 means the opposite of 6^2, or -36. $(-6)^2$ means the square of the quantity (-6), or 36.

Example 2 Evaluate.

a. $5(3 + 2)^2$

b. $-2x^3 + 4y$ for $x = 2$, $y = -3$

Solution

a. $5(5)^2 = 5 \cdot 25$
$= 125$

b. $-2(2)^3 + 4(-3) = -2 \cdot 8 + (-12)$
$= -16 + (-12)$
$= -28$

▼ A calculator is useful for evaluating expressions with exponents.

4 **Example 3** Evaluate 8^5 using a calculator.

Solution Use the x^y key.
8 x^y 5 = 32,768

CLASS EXERCISES

Name the base and the exponent.

1. 7^4 7; 4
2. $(-10)^2$ -10; 2
3. x^y x; y
4. -2^5 2; 5

Read aloud and evaluate.

5. 4^4 256
6. $(-9)^0$ 1
7. -4^3 -64
8. -2^1 -2

Write using exponents.

9. $6 \cdot 6 \cdot 6 \cdot 6 \cdot 6$ 6^5
10. $4 \cdot y \cdot y \cdot y \cdot 3 \cdot x$ $12xy^3$

Evaluate.

11. $10(5 + 4)^2$ 810
12. $[15 + (-18)]^2 + (-2)^3$ 1
13. $x^2 + y^3$; $x = 3$ and $y = -1$ 8
14. $-x^2 - 3 \cdot 2x$; $x = 3$ -27

WRITTEN EXERCISES

Write using exponents.

1. $8 \cdot 8 \cdot 8$ 8^3
2. $p \cdot p \cdot p \cdot p$ p^4
3. $2 \cdot r \cdot r \cdot r \cdot r \cdot s \cdot s$ $2r^4s^2$
4. $5 \cdot 5 \cdot y \cdot y$ 5^2y^2
5. $x \cdot x \cdot y \cdot y \cdot z$ x^2y^2z
6. $-9 \cdot m \cdot m \cdot (-4) \cdot n$ $36m^2n$
7. $\underbrace{n \cdot n \cdot n \cdot \ldots \cdot n}_{30 \text{ factors}}$ n^{30}
8. $\underbrace{m \cdot m \cdot m \cdot \ldots \cdot m}_{y \text{ factors}}$ m^y
9. $\underbrace{a \cdot a \cdot a \cdot \ldots \cdot a}_{a \text{ factors}}$ a^a
10. $(-4n)(-4n)(-4n)(-4n)(-4n)$ $(-4n)^5$
11. $(a + 1)(a + 1)(a + 1)$ $(a + 1)^3$

MATH MINUTES

Provide two sets of cards, one numbered 1 through 11 and one numbered 0 through 10. The first set are base numbers and the second are exponents. Students choose a card from each set and evaluate.

Evaluate.

12. **a.** 5^3 125 **b.** 3^5 243

13. **a.** 10^0 1 **b.** 10^6 1,000,000

14. **a.** -1^8 -1 **b.** $(-1)^9$ -1

15. **a.** -2^4 -16 **b.** $(-2)^0$ 1

16. x^5 for $x = 2$ 32

17. $(3a)^2$ for $a = -2$ 36

18. $-(4y)^2$ for $y = 3$ -144

19. $-(x)^2$ for $x = -2$ -4

20. $-2x^5$ for $x = -1$ 2

21. $(-x)^5$ for $x = -1$ 1

CALCULATOR **You can use a calculator to find the positive *square root* of a number.**

▼▼ *SAMPLE* 25 ⊘ ⊜ 5. So, $\sqrt{25} = 5$ since $5^2 = 25$

Find the square root.

22. $\sqrt{36}$ 6 **23.** $\sqrt{49}$ 7 **24.** $\sqrt{64}$ 8 **25.** $\sqrt{81}$ 9 **26.** $\sqrt{100}$ 10

Compare. Use <, >, or =.

27. 3^3 > 5^2 **28.** 4^3 = 8^2 **29.** $64 \div 8$ = 2^3 **30.** $4 \cdot 10$ < 10^4

31. -5^3 < -3^0 **32.** 10^2 < 2^{10} **33.** 3^4 > 4^3 **34.** -10^3 = $(-10)^3$

Solve.

35. Which of the following is equal to 1? d

a. -1^2 **b.** $(-1)^3$ **c.** $-(-1)^2$ **d.** $|-1|^3$

36. ***MENTAL MATH*** Given that $2^{10} = 1{,}024$, find 2^{11} mentally. Multiply by 2; 2,048.

37. ***MENTAL MATH*** Explain how to find $(-1)^{100}$ mentally.

38. ***WRITE*** a sentence that explains the roles of the base and the exponent of 8^4. The exponent, 4, tells you to multiply the base, 8, four times.

39. Make a table of the powers of 10: 10^0, 10^1, 10^2, 10^3, 10^4. Describe the pattern you see. Use the pattern to find 10^6 and 10^{10}. The power tells you the number of zeros in the product. $10^6 = 1{,}000{,}000$; $10^{10} = 10{,}000{,}000{,}000$

40. The formula for the area of a square is $A = s^2$, where A is area and s is the length of a side.

a. Find the area of a square with a side of length 3.5 cm. 12.25 cm²

b. Find the length of a side of a square with area 144 m². 12 m

41. ***CALCULATOR***

a. Copy and complete the table.

n	$4n$	4^n	n^4
0	0	1	0
1	4	4	1
2	8	16	16
3	12	64	81
4	16	256	256

b. For what value(s) of n is each true?

$4^n = n^4$ $n = 2, 4$

$4^n < n^4$ $n = 3$

$4^n > n^4$ $n = 0, n = 1, n > 4$

MIXED REVIEW

Solve.

1. $m - 7.5 = 12$ 19.5

2. $2.8h = -53.2$ -19

Use the formula $d = rt$ to find distance traveled.

3. $r = 55$ mi/h, $t = 2.25$ h

4. $r = 80$ km/h, $t = 3.5$ h

5. $r = 635$ mi/h, $t = 0.45$ h

3. 123.75 mi 4. 280 km 5. 285.75 mi

Simplify.

6. $3x - 2y + x$ $4x - 2y$

7. $w + 8 - 4w - 15$ $-3w - 7$

8. Sara wished to average her grades which were 79, 83, 74, 86, and 93. What is the average? 83

37. (-1) raised to an even power is positive; 1

PROBLEM SOLVING HINT

Draw a diagram.

A ***common error*** is to multiply the base by the exponent. For example, students may write $4^3 = 12$ instead of $4^3 = 64$. Remind students that the exponent indicates the number of times the base is used as a factor.

Closure

Writing in Math Have students write x^4 in their math journals. Ask them to name the base, the exponent, and the name of the whole expression. Then have them write the meaning of the expression. (base: x; exponent: 4; x to the fourth power)

Independent Practice

Remind students to be aware of the order of operations as they complete questions 17-21.

For questions 22-26, point out that although $(-5)^2 = 25$, the square root sign yields a positive square root.

For questions 49-50, students can use *Appleworks* for the Apple computer or *Microsoft Works* for the IBM to construct their bar graphs.

Critical Thinking The sum of the numbers in each row is the cube of the row number.

Lesson Follow-up

Reteaching Activity Write the following expressions: 8^3, 4^5, 7^2, 2^6, a^5, b^4. Then ask students to name the base and exponent and give the meaning of each expression.

LESSON RESOURCES

TRB Practice D-1
TRB Enrichment D-2

Practice — For use after 4-1 (pp. 141–144)

Exponents

Write using exponents.

1. $3 \cdot 3 \cdot 3 \cdot 3$ **3^4**
2. $k \cdot k \cdot k \cdot k \cdot k$ **k^5**
3. $(-9)(-9)(-9)m \cdot m \cdot m$ **$(-9)^3m^3$**
4. $g \cdot g \cdot g \cdot g \cdot h$ **g^4h**
5. $7 \cdot a \cdot a \cdot b \cdot b \cdot b$ **$7a^2b^3$**
6. $2 \cdot 2 \cdot 2 \cdot 2 \cdot 2 \cdot 2 \cdot 2$ **2^7**
7. $\underbrace{15 \cdot 15 \cdot 15 \cdot \ldots \cdot 15}_{25 \text{ factors}}$ **15^{25}**
8. $\underbrace{x \cdot x \cdot x \cdot x \cdot \ldots \cdot x}_{e \text{ factors}}$ **x^e**
9. $(5m)(5m)(5m)(5m)(5m)$ **$(5m)^5$**
10. $(y - 3z)(y - 3z)(y - 3z)(y - 3z)$ **$(y - 3z)^4$**

Evaluate.

11. 2^5 **32**
12. 6^3 **216**
13. 3^4 **81**
14. 1^{10} **1**
15. 10^5 **100,000**
16. 0^{12} **0**
17. $(-1)^9$ **-1**
18. $(-2)^3$ **-8**
19. -2^3 **-8**
20. m^4 for $m = 4$ **256**
21. $(5a)^3$ for $a = -1$ **-125**
22. $-(2p)^2$ for $p = 7$ **-196**
23. $-n^6$ for $n = 2$ **-64**
24. b^6 for $b = -1$ **1**
25. $(e - 2)^3$ for $e = 11$ **729**

CALCULATOR Use a calculator to find the square root.

26. $\sqrt{256}$ **16**
27. $\sqrt{400}$ **20**
28. $\sqrt{121}$ **11**
29. $\sqrt{225}$ **15**
30. $\sqrt{144}$ **12**
31. $\sqrt{576}$ **24**
32. $\sqrt{900}$ **30**
33. $\sqrt{324}$ **18**
34. $\sqrt{1,600}$ **40**
35. $\sqrt{529}$ **23**

Compare. Use <, >, or =.

36. 3^2 **>** 2^3
37. 4^3 **>** 5^2
38. $2 \cdot 3^2$ **>** $3 \cdot 2^2$
39. 4^4 **<** 5^5
40. -2^3 **<** 2^3
41. $(-3)^2$ **>** -3^2
42. 10^2 **<** 2^{10}
43. 2543^0 **=** 1^{310}
44. 8^1 **=** 2^3
45. -8^1 **=** -2^3
46. $(-1)^5$ **<** 1^5
47. -1^5 **<** 1^5

48. ***MENTAL MATH*** Given that $3^5 = 243$, find 3^6 mentally. **729**

Chapter 4 D-1

Enrichment — For use after 4-1 (pp. 141–144)

Exponent Patterns

Complete the patterns. Then use your results to make the indicated predictions.

1. $5^1 =$ **5**; $5^2 =$ **25**; $5^3 =$ **125**; $5^4 =$ **625**
   Predict:
   the ones' digit of 5^{43} **5**
   the tens' digit of 5^{74} **2**
2. $6^1 =$ **6**; $6^2 =$ **36**; $6^3 =$ **216**; $6^4 =$ **1,296**
   Predict:
   the ones' digit of 6^{113} **6**
3. $4^1 =$ **4**; $4^2 =$ **16**; $4^3 =$ **64**; $4^4 =$ **256**
   Predict:
   the ones' digit of 4^{29} **4**
   the ones' digit of 4^{82} **6**
4. $9^1 =$ **9**; $9^2 =$ **81**; $9^3 =$ **729**; $9^4 =$ **6,561**
   Predict:
   the ones' digit of 9^{114} **1**
   the ones' digit of 9^{223} **9**
5. $2^1 =$ **2**; $2^2 =$ **4**; $2^3 =$ **8**; $2^4 =$ **16**; $2^5 =$ **32**; $2^6 =$ **64**; $2^7 =$ **128**; $2^8 =$ **256**
   Predict:
   the ones' digit of 2^{84} **6**
   the ones' digit of 2^{43} **8**
6. $1^2 =$ **1**; $11^2 =$ **121**; $111^2 =$ **12,321**; $1,111^2 =$ **1,234,321**
   Predict:
   $11,111^2$ **123,454,321**
   $111,111^2$ **12,345,654,321**
7. Study the pattern of the ones' digits in the powers of 3 and the powers of 7. Then predict the ones' digit of each number.
   a. 3^{50} **9**
   b. 7^{101} **7**
   c. 3^{81} **3**
   d. 7^{44} **1**

D-2 Chapter 4

PROBLEM SOLVING HINT
Try guess and test.

Find the value of n.

42. $5^n = 125$ $n = 3$ **43.** $3^n = 1$ $n = 0$ **44.** $n^3 = 8$ $n = 2$

For what positive value(s) of x is each true?

45. $1x = x^1$ x = any positive value **46.** $2x = x^2$ $x = 2$ **47.** $9x = x^3$ 3

48. Evaluate $(-1)^m$ for $m = 2$, 4, and 6. Now let $m = 1$, 3, and 5. ***WRITE*** a rule for raising a negative number to an even power or an odd power. A negative number raised to an even power is positive; to an odd power it is negative.

49. ***COMPUTER*** You can use a spreadsheet to construct a bar graph for the powers of 2. Note that the caret (^) is used to indicate exponentiation. Use the spreadsheet data to construct a graph.

	A	B
1	Number	Power
2	2	=A2^2
3	2	=A3^3
4	2	=A4^4
5	2	=A5^5
6	2	=A6^6
7		

a. What do you notice about the bar for cell B2 compared to the bar for cell B3? Is the same true for cells B3 and B4? It is half as tall; yes.

b. ***WRITE*** a description for the bar graph of the powers of 2. Each bar is twice as tall as the previous bar.

50. ***PROJECT*** Use a spreadsheet format similar to Exercise 49 to construct a bar graph for the powers of 3. Answer Exercise 49 (a) and (b) for this bar graph. Each bar is three times as tall as the previous bar.

Critical Thinking

EXPLORING NUMBER PATTERNS

You know that the sum of consecutive odd integers is a square number. Now consider the odd numbers shown at the left.

1
3 5
7 9 11

1. Copy the odd number formation and add two more rows.
2. The sum of the first row is 1. What is the sum of the second row? the third row? the fourth row? 8; 27; 64
3. What is 1^3? 2^3? 3^3? 4^3? 1; 8; 27; 64
4. Predict the sum of the tenth row. 1,000

LESSON QUIZ

Name the base and the exponent.

1. $(-5)^3$ [base: (-5); exponent: 3]

Write using exponents.

2. $(-4)(-4)(-4) \cdot a \cdot a$ $[(-4)^3a^2]$
3. $2 \cdot x \cdot 2 \cdot x \cdot y \cdot y \cdot 3$ $(12x^2y^2)$

Evaluate.

4. $2x^4$ for $x = -3$ (162)
5. $(3x)^3$ for $x = -2$ (-216)

FOR THE NEXT LESSON

calculator, math journal

Assignment Guide

Basic 1-5, 7-10, 12-15, 22-38, 42-45, MR, CT All
Average 1-18, 22-40 even, 43-45, 49, MR, CT All
Enriched 7-11, 16-34 even, 36-50, MR, CT All

OBJECTIVE: *To multiply expressions with powers.*

4-2 Rules About Exponents

▼ Without evaluating the expressions, compare $7^2 \cdot 7^3$ and $7^4 \cdot 7^1$. Use what you've learned about exponents.

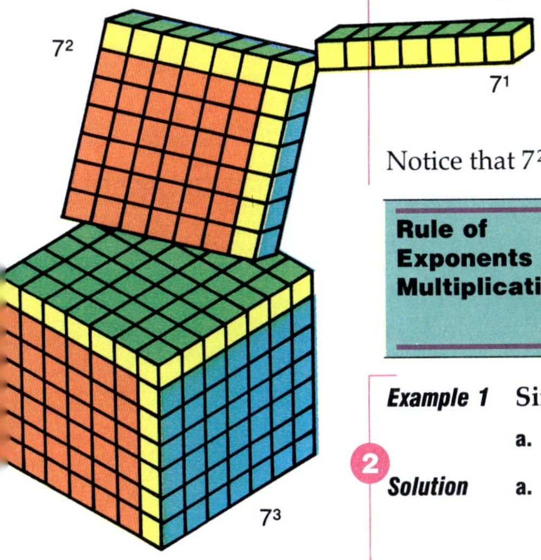

$$7^2 \cdot 7^3 = (7 \cdot 7) \cdot (7 \cdot 7 \cdot 7) = 7^5$$
$$7^4 \cdot 7^1 = (7 \cdot 7 \cdot 7 \cdot 7) \cdot (7) = 7^5$$
$$7^2 \cdot 7^3 = 7^4 \cdot 7^1 \text{ since } 7^5 = 7^5$$

Notice that $7^2 \cdot 7^3 = 7^{2+3} = 7^5$. This suggests the following rule.

Rule of Exponents for Multiplication	To multiply numbers or variables with the *same* base, add exponents. **Arithmetic** $2^3 \cdot 2^4 = 2^{3+4} = 2^7$ **Algebra** $a^m \cdot a^n = a^{m+n}$

Example 1 **Simplify using exponents.**

a. $a \cdot a^3 \cdot a^4$ **b.** $xy^2 \cdot x^3$

Solution **a.** $a \cdot a^3 \cdot a^4 = a^{1+3+4} = a^8$

b. $xy^2 \cdot x^3 = x \cdot x^3 \cdot y^2 = x^{1+3} \cdot y^2 = x^4y^2$

How do the blocks show 7^1, 7^2, and 7^3? $7^1 = 7$, $7^2 = 49$, $7^3 = 343$

THINK How would you simplify $(5^0)^3$? What is the result? Multiply $5^{0 \cdot 3} = 5^0$; $5^0 = 1$.

▼ You can raise a power to a power.

Example 2 **Simplify $(5^2)^3$.**

Solution **a.** Use the meaning of exponents.
$$(5^2)^3 = 5^2 \cdot 5^2 \cdot 5^2 = (5 \cdot 5)(5 \cdot 5)(5 \cdot 5) = 5^6$$

b. Use the rule for multiplication.
$$(5^2)^3 = 5^2 \cdot 5^2 \cdot 5^2 = 5^{2+2+2} = 5^6$$

▼ Notice that $(5^2)^3 = 5^6$. This suggests the following rule.

Rule of a Power Raised to a Power	To raise a power to a power, multiply the exponents. **Arithmetic** $(2^3)^4 = 2^{3 \cdot 4} = 2^{12}$ **Algebra** $(a^m)^n = a^{m \cdot n}$

Example 3 **Simplify.**

a. $(3^2)^3$ **b.** $(a^6)^2$

Solution **a.** $(3^2)^3 = 3^{2 \cdot 3} = 3^6 = 729$

b. $(a^6)^2 = a^{6 \cdot 2} = a^{12}$

Lesson Focus

Materials

Calculator, math journal

MOTIVATING THE LESSON

Remind students that exponents are a convenient way of expressing a product. Instead of writing $3 \cdot 3 \cdot 3 \cdot 3$, the number is written 3^4. Ask students how they would multiply two numbers with exponents. To help students, point out that multiplying $3 \cdot 3$ is helpful in evaluating $3^2 \cdot 3^2$.

Instruction

1 Have students observe that the exponents and not the bases are added.

2 You may need to remind students that the exponent 1 is understood in the term a, so $a = a^1$.

3 As students discuss the possibilities, help them realize that any number raised to the zero power is equal to 1. They may be able to reason that $(5^0)^3 = 1$, since $1 \cdot 1 \cdot 1 = 1$.

Teaching Tip

Have class "fact-finding" sessions. Pose a problem that students are not to solve. Ask the class what they need to know about the problem, including all assumptions or inferences. List all ideas on the chalkboard.

ADDITIONAL EXAMPLES

Simplify using exponents.

1. $x^3 \cdot x$ (x^4)
2. $x^2y^3 \cdot x^3y$ (x^5y^4)

Simplify.

3. $(x^2)^5$ $[x^{10}]$
4. $(-2x^2)^3$ $[-8x^6]$
5. $-(3m)^2$ $[-9m^2]$
6. $(7yz^4)^3$ $[343y^3z^{12}]$

Guided Practice

Class Exercises For questions 1-18 have students work in small groups. Then have a member of each group use the chalkboard to show how selected answers were found.

Think and Discuss Some students may need assistance understanding question 2. It might help to say that just as -6 can be expressed as $-1 \cdot 6$, then $-(2^3)^2$ can be expressed as $-1 \cdot (2^3)^2$ and $(-2^3)^2$ can be expressed as $(-1 \cdot 2^3)^2$. Now evaluate both expressions:

$$-1 \cdot (2^3)^2 = -1 \cdot 2^6 = -1 \cdot 64 = -64$$

and

$$(-1 \cdot 2^3)^2 = (-1)^2 \cdot (2^3)^2 = 1 \cdot 64 = 64$$

The concept underlying question 3 is developed further in Chapter 5, Lesson 8.

▼ You can also raise a product to a power.

Example 4 **Simplify $(xy^2)^3$.**

Solution Use the meaning of exponents.

$$\begin{aligned}(xy^2)^3 &= xy \cdot y \cdot xy \cdot y \cdot xy \cdot y \\ &= (x \cdot x \cdot x) \cdot (y \cdot y \cdot y \cdot y \cdot y \cdot y) \\ &= x^3y^6\end{aligned}$$

▼ Notice that $(xy^2)^3 = x^3y^6$. This suggests the following rule.

Rule of a Product Raised to a Power

To raise a product to a power, raise each factor to the power and then use the rule of exponents for multiplication.

Arithmetic	Algebra
$(10^3 \cdot 10^2)^4 = 10^{12} \cdot 10^8 = 10^{20}$	$(ab)^m = a^mb^m$

Example 5 **Simplify $(6x^4)^3$.**

Solution $(6x^4)^3 = 6^3 \cdot (x^4)^3 = 216x^{12}$

1. No, the bases are different.
2. $-(2^3)^2$ means take the opposite of the quantity $(2^3)^2$, or -64. $(-2^3)^2$ means square the quantity (-2^3), or 64.

THINK AND DISCUSS

1. Can you simplify $x^6 \cdot y^7$? Explain. See above.
2. Evaluate $-(2^3)^2$ and $(-2^3)^2$. Compare your answers. See above.
3. How many times is the expression xy^2 used as a factor in $(xy^2)^0$? What is the value of $(xy^2)^0$? 0; 1

CLASS EXERCISES

Evaluate or simplify. No variable has a value of zero.

1. $(-4)^3$ **-64**
2. -4^2 **-16**
3. $(-2 + 3)^7$ **1**
4. $2^3 \cdot 2^0$ **8**
5. $4 \cdot 4^3$ **256**
6. $x \cdot x^2 \cdot x^5$ x^8
7. $4^7 \cdot 4^{10}$ 4^{17}
8. $w^2(w^6)$ w^8
9. $(y^3)^5$ y^{15}
10. $(4x^5)^0$ **1**

Evaluate for $a = -1$, $b = -3$, $c = 2$.

11. a^3b **3**
12. abc^0 **3**
13. $(abc)^0$ **1**
14. $2a^5c^1$ **-4**
15. $(a^3)^6$ **1**
16. $(a^2b^3)^5$ **-14,348,907**
17. $(c^2 \cdot c)^8$ **16,777,216**
18. $(4c^2 \cdot 4c^3)^2$ **262,144**

WRITTEN EXERCISES

Evaluate. No variable has a value of zero.

1. $(-2)^3$ **-8**
2. $(2)^3$ **8**
3. $-(-2)^3$ **8**
4. $10^2 \cdot 10^5$ **10,000,000**
5. $(x^6)(x^3)$ x^9
6. $(x^2)(y^5)(x)$ x^3y^5
7. $5x^3 \cdot 2x^6$ $10x^9$
8. $a^{10} \cdot a^2$ a^{12}
9. $(2^3)(3^2)$ **72**
10. $(4^5)(4^3)$ **65,536**
11. $(-x^4)(-x)^4$ $-x^8$
12. $(z^6)(-z^3)(z^2)$ $-z^{11}$
13. $(-2a)^3 \cdot (-2a^3)$ $16a^6$
14. $(a^4b^4)^4$ $a^{16}b^{16}$
15. $(6y^3)^4$ $1{,}296y^{12}$
16. $-(3xy)^2$ $-9x^2y^2$
17. $(-3y^4)^3$ $-27y^{12}$
18. $(x^5)^7$ x^{35}

MATH MINUTES

Write this matrix. Have students draw a line through one item in each column, and evaluate the resulting expression as shown.

x^2	y^4	2
x^3	y^3	3
x^4	y^2	4

$[(x^4 \cdot y^3)^3 = x^{12} \cdot y^9]$

Evaluate each expression for $a = -3$, $b = 2$, $c = -1$.

19. $(2c)^3$ **-8** **20.** $5abc^0$ **-30** **21.** $(a^2)^3 \cdot (a^3)^2$ **531,441**

22. bc **-2** **23.** $(ac)^b$ **9** **24.** ab^3c^{100} **-24**

25. $a^2 + b^3$ **17** **26.** $(a + b)^3$ **-1** **27.** $(ab)^3$ **-216**

True or False? Explain your answer. No variable has a value of zero.

28. $x^8 \cdot x^2 = x^5 \cdot x^5$ **T; both equal x^{10}** **29.** $x^5 \cdot x^3 = x^{15}$

30. $x^3 \cdot y^4 = (xy)^7$ **31.** $5^0 = 7^0$

32. $(-r^3)^2 > 0$ **33.** $-(r^2)^3 < 0$ **34.** $1^8 = 1^{23}$

Which of the following is equal to 2^{13}? **36**

35. $(2^3)^{10}$ **36.** $2^5 \cdot 2^8$ **37.** $2^1 \cdot 2^0 \cdot 2^{13}$ **38.** 8,190

Which of the following is twice the value of 2^{15}? **40; 41; 42**

39. 2^{30} **40.** 2^{16} **41.** $2 \cdot 2^{15}$ **42.** 65,536

Compare. Use >, <, or =.

43. 5^2 **<** $(5^3)^2$ **44.** $7^2 \cdot 7^5$ **<** $(7^6)^2$ **45.** $(-2^2)^3$ **<** 2^5

46. 25^2 **=** $(5^2)^2$ **47.** $(2^7)^7$ **<** $(2^{25})^2$ **48.** $(4^5 \cdot 4^2)^3$ **>** $(4^4)^0$

49. Without computing, determine which is greater: 2^{75} or 3^{50}. *Hint:* write each as an expression with the same exponent. **3^{50}; $2^{75} = (2^3)^{25}$ and $3^{50} = (3^2)^{25}$.**

Solve. **50. Yes; multiplication of exponents is commutative.**

50. Is $(a^3)^4 = (a^4)^3$ a true equation? Explain your answer.

51. If $(3^2 + 3^2 + 3^2) \cdot 3^x = 243$ is a true equation, find x. **$x = 2$**

52. ***CALCULATOR*** Find a pattern in the last digits of the powers of 7.

a. What is the last digit of 7^{10}? What is the last digit of 7^{11}? **9; 3**

b. Predict the last digit of 7^{21}. What method did you use to find the answer? **7; the last digits form a pattern: 7, 9, 3, 1.**

53.** ***COMPUTER In computer science, information is measured and stored in *bits*. Eight bits are equal to one *byte*.

a. One *kilobyte* is defined as 2^{10} bytes. How many bytes are in a kilobyte? **1,024**

b. ***ESTIMATION*** What is an approximate value for a kilobyte using a power of ten? **10^3**

c. Write the number of bits in one byte as a power of 2. **2^3**

d. A *megabyte* is 2^{20} bytes. Write the number of bytes in a megabyte as the product of exponents. **$2^{10} \cdot 2^{10}$**

e. Write the number of bits there are in one megabyte as the product of exponents. **$2^{20} \cdot 2^3$**

MIXED REVIEW

Estimate. **4. $-14c^2d^3$**

1. 3.8 + 4.62 − 5.3 **4**

2. 42.7 · 8.5 **360**

Write using exponents.

3. $-5 \cdot a \cdot a \cdot a \cdot b \cdot b$ **$-5a^3b^2$**

4. $2 \cdot c \cdot c \cdot (-7) \cdot d \cdot d \cdot d$

Evaluate for $x = 3$, $y = -2$.

5. $4(x + 3y)$ **-12**

6. $\sqrt{x^3 + 3x}$ **6**

7. $(1 + y)^x$ **-1**

8. ***DATA FILE 3 (pp. 96–97)*** Approximately how many times did the value of a 1961 proof set increase by 1990? **8**

29. F; add exponents; $x^5 \cdot x^3 = x^8$
30. F; bases are not the same; $x^3 \cdot y^4 = x^3y^4$
31. T: 1 = 1
32. T; a negative number squared is positive.
33. T; $(r^2)^3$ is positive, so the opposite is less than 0.
34. T; 1 = 1

A ***common error*** is to use the rule of exponents for multiplication when simplifying expressions involving addition. For example, students may mistakenly write $a^2 + a^3 = a^5$. Remind students that exponents are added only when multiplication of the same bases is indicated.

Closure

Writing in Math Have students state the rules for multiplying expressions with exponents, and for raising a power to a power. Have students demonstrate how to apply the rules to examples they create and record them in their math journals.

Independent Practice

For questions 39-42, point out to students that doubling the base is not the same as doubling the exponent.

Critical Thinking For a more detailed analysis of Pascal's Triangle, refer students to *Exploring Pascal's Triangle,* found on p. 138E of the Chapter Overview. See also TRB Exploring O-14.

Lesson Follow-up

Reteaching Activity Review the rules of exponents for multiplication and raising a power to a power and present examples using the expanded form.

$$2^3 \cdot 2^4 = (2 \cdot 2 \cdot 2)(2 \cdot 2 \cdot 2 \cdot 2) = 2^{3+4} = 2^7$$

$$(2^3)^4 = (2^3)(2^3)(2^3)(2^3) = 2^{3+3+3+3} = 2^{12}$$

LESSON RESOURCES

TRB Practice D-3
TRB Enrichment D-4

Practice

For use after 4-2 (pp. 145–148)

Rules About Exponents

Evaluate. No variable has a value of zero.

1. $(-5)^2$ **25**
2. -5^2 **-25**
3. $-(5^2)$ **-25**
4. $(x^3)(x^4)$ **x^7**
5. $(x^4)(y^2)(x^2)$ **x^6y^2**
6. $3x^2 \cdot 2x^3$ **$6x^5$**
7. $y^4 \cdot y^5$ **y^9**
8. $(2^4)(4^2)$ **256**
9. $(4^3)(3^4)$ **5,184**
10. $(-y^5)(y^2)$ **$-y^7$**
11. $(x^3)(-x^2)$ **$-x^5$**
12. $(z^3)^5$ **z^{15}**
13. $(3y^2)(2y^3)$ **$6y^5$**
14. $-(m^4)^3$ **$-m^{12}$**
15. $(-m^4)^3$ **$-m^{12}$**

Evaluate each expression for $a = 2$, $b = -1$, and $c = 4$.

16. $(ab)^2$ **4**
17. $(c^2)^2$ **256**
18. $(2a)^3$ **64**
19. a^5b^{10} **32**
20. $(b + c)^3$ **27**
21. a^3c^2 **128**
22. a^4b^4 **16**
23. $a^3 + c^2$ **24**
24. $a^4b^2c^0$ **16**

True or False?

25. $y^2 \cdot x^3 = (xy)^5$ **false**
26. $4^6 = 6^4$ **false**
27. $z^2 \cdot z^6 = z^3 \cdot z^4$ **false**

Is the given expression equal to 3^{12}? Write yes or no.

28. $(3^3)^4$ **yes**
29. $3^6 + 3^6$ **no**
30. $3^4 \cdot 3^3$ **no**
31. $3^9 \cdot 3^3$ **yes**

Is the value of the given expression three times the value of 3^8? Write yes or no.

32. 3^{24} **no**
33. $3 \cdot 3^8$ **yes**
34. 19,683 **yes**
35. 3^{11} **no**

Compare. Use >, <, or =.

36. $(4^3)^2$ **=** $(4^2)^3$
37. $5^3 \cdot 5^4$ **<** 5^{10}
38. $(3^5)^4$ **>** 3^{10}
39. 25^0 **<** 5^2
40. $(3^{12})^0$ **=** $(12^3)^0$
41. $4^2 \cdot 4^3$ **=** 4^5
42. $(6^2)^0$ **=** 6^0
43. $5^0 \cdot 5^6$ **<** 5^7
44. $(8^3)^2$ **=** $(8^2)^3$

Chapter 4 D-3

Enrichment

For use after 4-2 (pp. 145–148)

Amazing Exponents

Solve.

1. 9^{99} is a very great number, but you can write an even greater one using three nines. What is the greatest number that can be written using three nines? (Written out, this number contains over 300 million digits.)
   $9^{(9^9)}$
2. Every person has 2 parents, 2^2 grandparents, 2^3 great-grandparents, and so on. There are, on average, 25 years in each generation.
   a. Use exponents to write the number of ancestors you had in the generation in which Columbus discovered America.
   2^{20}
   b. Circle the number closest to your answer to Part a.
   40   100   1,000   (1,000,000)

Circle the letter of the best estimate.

3. the height of a stack of 10^6 pennies
   a. 1 yd   b. 100 yd   (c.) 1 mi   d. 1 million mi
4. the weight of a stack of 10^6 dollar bills
   a. 1 lb   b. 10 lb   c. 100 lb   (d.) 1 ton
5. the length of time in 10^9 minutes
   a. 1 day   b. 1 yr   c. 100 yr   (d.) 2,000 years
6. the number of years in 10^7 days
   a. 1 yr   b. 1 million years   c. 10 years   (d.) 30,000 years

A sheet of paper is about 0.003 in. thick. Suppose you folded a sheet of paper in half the given number of times. What factor would you multiply 0.003 by to find the thickness of the folded sheet? Write the factor in exponential form.

7. once **$2^1 = 2$**
8. twice **$2^2 = 4$**
9. 50 times **2^{50}**
10. You could not actually fold a piece of paper in half 50 times. However, if you were able to do so, circle the best estimate of the thickness of the folded sheet.
    a. 1 ft   b. 1 mi   c. 1,000 mi   (d.) 50,000,000 mi

D-4 Chapter 4

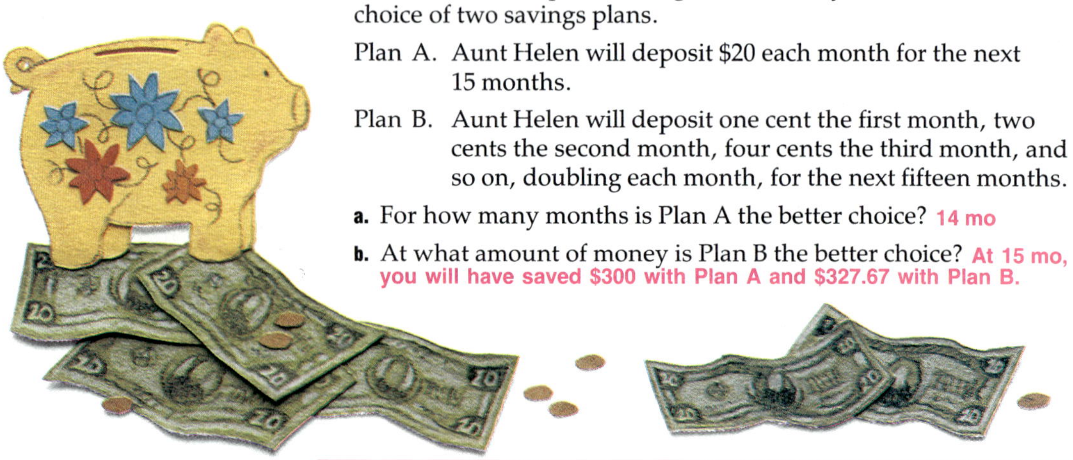

54. Aunt Helen will open a savings account for you. You have a choice of two savings plans.

Plan A. Aunt Helen will deposit $20 each month for the next 15 months.

Plan B. Aunt Helen will deposit one cent the first month, two cents the second month, four cents the third month, and so on, doubling each month, for the next fifteen months.

a. For how many months is Plan A the better choice? **14 mo**

b. At what amount of money is Plan B the better choice? **At 15 mo, you will have saved $300 with Plan A and $327.67 with Plan B.**

Critical Thinking

EXPLORING PASCAL'S TRIANGLE

The number triangle below was published in China about 1300 A.D. It is usually called *Pascal's triangle,* after the French mathematician Blaise Pascal, who wrote about many of its patterns in a 1653 paper.

1
1 1
1 2 1
1 3 3 1
1 4 6 4 1
1 5 10 10 5 1
1 6 15 . . .

1. Analyze the pattern.
Complete: Each entry in a row is the sum of ▇. **the two numbers diagonally above it in the preceding row**

If we call the row in which the number 2 first appears Row 2, the top row is Row 0.

2. What numbers appear in Row 6? **1, 6, 15, 20, 15, 6, 1**

3. What is the sum of the numbers in Row 3? **8**

4. Find the sum of the numbers in the twentieth row of Pascal's triangle. You may wish to use a calculator. *Hint:* Starting with Row 0, make a table that shows each row number and the sum of the numbers in that row. Look for a pattern. **1,048,576**

LESSON QUIZ

Simplify.

1. $(-5)^3$ [-125]
2. $2^2 \cdot 2^2$ (16)
3. $3 \cdot 3^3$ (81)
4. $(5x^2)^3$ [$125x^6$]
5. $a^{11} \cdot a^3 \cdot a$ (a^{15})
6. $(2^4)(-3^2)$ [-144]
7. $(2a^2b)^3$ [$8a^6b^3$]

Evaluate.

8. ab^2c^0 for $a = -2$, $b = -3$, $c = 48$ (-18)

FOR THE NEXT LESSON

scientific calculator, math journal

Assignment Guide

Basic 1-8, 19-21, 28-30, 43-45, 50, MR, CT All
Average 1-17 odd, 19-27 odd, 28-31, 35-42, 43-49 odd, 50, 52, MR, CT All
Enriched 5-27, 32-38, 46-49, 51-54, MR, CT All

OBJECTIVE: *To write and compute with numbers using scientific notation.*

4-3 Scientific Notation

▼ You can write a number in different ways without changing its value. Here are some ways to write 75.

$60 + 15$ $2(37.5)$ $150 \div 2$ $100 - 25$

FLASHBACK

The powers of 10:
$10^0 = 1$
$10^1 = 10$
$10^2 = 100$
⋮ ⋮

▼ You can also write numbers using powers of 10. Here are some ways you can write 4,000 using powers of 10.

$$4 \cdot 1{,}000 = 4 \cdot 10^3$$
$$40 \cdot 100 = 40 \cdot 10^2$$
$$400 \cdot 10 = 400 \cdot 10^1$$

▼ *Scientific notation* is a way to write numbers using powers of 10.

Example 1 The numbers below on the left are written in scientific notation. The numbers below on the right are not.

1

3.45×10^3	34.5×10^2
9.5×10^2	0.95×10^1
2.0×10^4	200×10^2

Scientific Notation A number is in scientific notation when it is written as the product of a number greater than or equal to 1 and less than 10, and a power of 10.

THINK How can you use mental math to find the exponent when writing a number in scientific notation? **You need to know that the exponent is the power of 10, or you can count how many places you moved the decimal point.**

▼ You can write large numbers in scientific notation.

Example 2 **Write 5,460,000 in scientific notation.**

2

Solution
1. Express as a number between 1 and 10. — 5.460000
2. To keep the value the same, multiply by a power of 10. — $5.46 \cdot 1{,}000{,}000$
3. Write the power of 10 with exponents. — 5.46×10^6

THINK How can you use mental math to multiply by a power of 10? **Move the decimal point to the right for each power of 10.**

▼ By multiplying, you can write any number that is in scientific notation in *standard notation.*

Example 3 **Write each number in standard notation.**
a. 9.3×10^7 **b.** 4.235×10^2

Solution **a.** $9.3 \cdot 10{,}000{,}000 = 93{,}000{,}000$ **b.** $4.235 \cdot 100 = 423.5$

Lesson Focus

Materials
Scientific calculator, math journal

Vocabulary/Symbols
scientific notation
standard notation

MOTIVATING THE LESSON

Ask students to give examples of data that are often expressed in very large numbers. (the distance of planets and stars, the number of cells in the human body, and so forth) Point out that it is often difficult to read, write, and compute with very large numbers. Explain that mathematicians have developed a method for writing very large numbers. It is called scientific notation.

Instruction

1 Have students describe what the numbers in the first column have in common. Help them to observe that each number has a decimal point after the first digit, and that the numbers are between 1 and 9 inclusive.

2 Point out that when writing a number in scientific notation, the power of ten will be the same as the number of places the decimal point should be moved to the left.

Teaching Tip

Try a different approach by giving the students an answer and asking them to write an appropriate question.

3 You may wish to introduce multiplication of numbers in scientific notation with the example below. It involves four steps to obtain the simplified product.

$(2.34 \times 10^4)(3.1 \times 10^2)$
$[7.254 \times 10^6]$

4 To help the students understand how to simplify the expression in Example 4, write the following on the chalkboard:

$(3.46 \times 10^5)(9.2 \times 10^3) =$
$(3.46 \times 9.2)(10^5 \times 10^3)$

Ask what properties make this equation true. (commutative and associative)

ADDITIONAL EXAMPLES

1. Is each number in scientific notation?
   a. 43.92×10^4 (no)
   b. 2.96×10^2 (yes)
2. Write in scientific notation.
   a. 43,710,000 (4.371×10^7)
   b. 821.9 (8.219×10^2)
3. Write in standard notation.
   a. 1.27×10^0 (1.27)
   b. 9.872×10^4 (98,720)
   c. 7.6×10^5 (760,000)
4. Write the product in scientific notation.
   a. $(2.34 \times 10^4)(3.1 \times 10^2)$ $[7.254 \times 10^6]$
   b. $(6.805 \times 10^5)(1.9 \times 10^1)$ $[1.29295 \times 10^7]$

Guided Practice

Class Exercises Have volunteers work the exercises at the chalkboard. If students are having difficulty with questions 4-7, refer them to the definition for scientific notation on p. 149. Have them verify the need to meet *two* conditions.

▼ You can multiply numbers in scientific notation.

3 ***Example 4*** **Write the product in scientific notation.**
$(3.46 \times 10^5)(9.2 \times 10^3)$

4 ***Solution***

1. Multiply the decimals. $3.46 \times 9.2 = 31.832$
2. Multiply the powers of 10. $10^5 \cdot 10^3 = 10^8$
3. Write the product. 31.832×10^8
4. Write the first factor in scientific notation. $(3.1832 \times 10) \times 10^8$
5. Simplify. 3.1832×10^9

1. It is easier to keep track of the zeros.
2. 100,000 + 10,000 = 110,000

THINK AND DISCUSS See above.

1. How does scientific notation prevent careless error when working with very large numbers?
2. How would you add 1×10^5 and 1×10^4?

4. No; not a product of a number from 1 to 10.
5. No; not a number between 1 and 10.
6. Yes
7. No; 3^4 is not a power of 10.

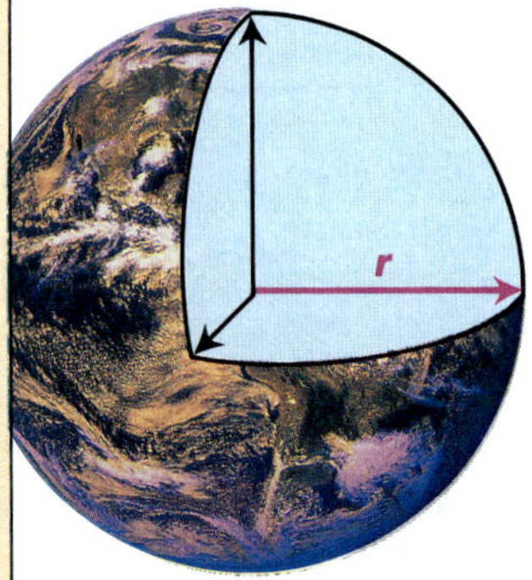

CLASS EXERCISES

Write each number in standard notation.

1. 10^6 1,000,000
2. 10^7 10,000,000
3. 0.93×10^4 9,300

Is each number written in scientific notation? Explain. See left.

4. 10^4
5. 0.12×10^1
6. 5.24×10^1
7. 7.2×3^4

Write each number in scientific notation.

8. Some computers can process 3.5 million instructions per second. 3.5×10^6 instructions per second
9. Light travels 299,790,000 m/s. 2.9979×10^8 m/s

Write each number in standard notation.

10. The radius of the earth is 5×10^6 m. 5,000,000 m
11. The estimated age of Earth is 4.7×10^9 years. 4,700,000,000 y

Write the result in scientific notation.

12. $(3.45 \times 10^6)(1.84 \times 10^2)$ 6.348×10^8
13. $(4.32 \times 10^3) \cdot (2.4 \times 10^1)$ 1.0368×10^5

WRITTEN EXERCISES

For 1 and 2, answers may vary. Some examples are given.

1. Write 625,000,000 three different ways using powers of 10.
2. Write 4.97×10^7 three different ways using powers of 10.

1. 6.25×10^8; 62.5×10^7; 625×10^6 2. 497×10^5; $4{,}970 \times 10^4$; $49{,}700 \times 10^3$

Write each number in scientific notation.

3. The average adult human male, weighing about 160 lb, consists of about 60,000,000,000,000 cells. 6×10^{13}

MATH MINUTES

Have students work in pairs. Each student should write five numbers in standard notation, exchange them with their partner and then write the numbers in scientific notation. Have students check their answers.

4. The temperature inside the sun is greater than 16,000,000°C. 1.6×10^7

5. ***DATA FILE 10 (pp. 404–405)*** Write the area of the Pacific Ocean in scientific notation. 6.382×10^7

Write each number in standard notation.

6. 10^8 100,000,000
7. 10^5 100,000
8. 59×10 590
9. 7.654×10^3 7,654
10. 1.45×10^{10} 14,500,000,000
11. 6.0032×10^2 600.32
12. 9.84×10^5 984,000
13. 4.06×10^3 4,060
14. 1.7×10^2 170

Simplify. Write each answer in scientific and standard notation. Round to the nearest tenth.

15. $(3 \times 10^2)(5 \times 10^2)$ 1.5×10^5; 150,000
16. $(4.2 \times 10^3)(3.84 \times 10^1)$ 1.6×10^5; 160,000
17. $389{,}000 \cdot 25{,}475{,}000$ 9.9×10^{12}; 9,900,000,000,000
18. $0.00125 \times 50{,}000$ 6.3×10^1; 63
19. $(9.087 \times 10^6)0.52$ 4.7×10^6; 4,700,000
20. $4(2.3 \times 10^5)$ 9.2×10^5; 920,000

Complete.

21. $8{,}450 = 8.45 \times 10^{■}$ 3
22. $■ \times 10^2 = 8.45 \times 10^3$ 84.5
23. $84.5 = ■ \times 10^1$ 8.45
24. $8.45 = 8.45 \times 10^{■}$ 0
25. $0.845 = ■ \times 10^3$ 0.000845
26. $■ \times 10^4 = 8.45 \times 10^3$ 0.845

CALCULATOR **Solve using a scientific calculator. Round each answer to the nearest hundredth.**

▼▼ *SAMPLE* $(6.28 \times 10^{24}) \times (5.3 \times 10^7)$

Enter: 6.28 (EXP) 24 (×) 5.3 (EXP) 7 (=)

Display: 3.3284 E 32

27. $(7.892 \times 10^{17})(3.16 \times 10^3)$ 2.49×10^{21}
28. $(7.892 \times 10^{17})^2$ 6.23×10^{35}
29. $(7.892 \times 10^{17}) \cdot 10^3$ 7.89×10^{20}
30. $(4.32 \times 10^5) \cdot 673$ 2.91×10^8
31. $905{,}200{,}000{,}000{,}000{,}000 \cdot 3{,}560{,}090{,}000{,}000$ 3.22×10^{30}
32. How many seconds are there in 1,000 years? 3.16×10^{10} s

DATA FILE 4 (pp. 138–139) **Solve.**

33. Write the number of speakers of Chinese in scientific notation. 7×10^8

34. Find the increase in population from 1900 to 1980. Express the number in scientific notation. 2.8×10^9

35. ***PROJECT*** Use an encyclopedia, almanac, or other reference book. Find the answers to the following questions and write in scientific notation. Check students' work.
   a. What is the diameter of a water molecule? 0.0000000276 cm
   b. How far from Earth is Mars? between 4.87×10^7 and 2.34×10^8 miles
   c. What is the shutter speed of the fastest still camera? 0.008 s

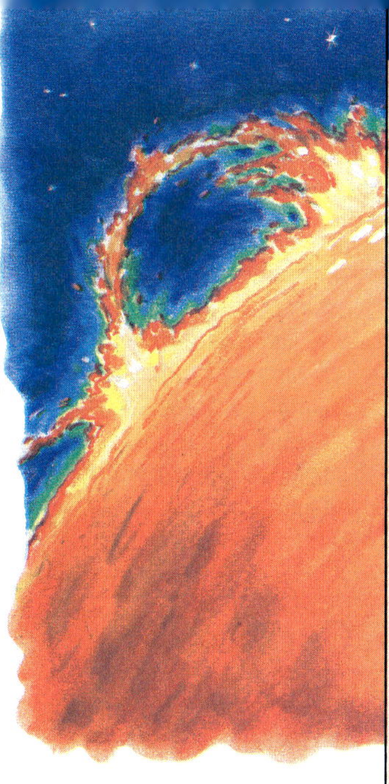

MIXED REVIEW

State the property shown. See below.

1. $a \cdot 1 = a$
2. $a + b = b + a$
3. $a + 0 = a$
4. $a(b + c) = ab + ac$

Simplify. No variable equals zero.

5. $3m^3 \cdot m^4$ $3m^7$
6. $n^3 \cdot n^3$ n^6
7. $5 \cdot 5 \cdot y \cdot y \cdot y$ $25y^3$
8. You have 2^1 parents, and 2^2 grandparents. You also had 2^3 great-grandparents. Show how many great-great-great-grandparents you had. 2^5 or 32

1. multiplicative identity
2. commutative prop. for addition
3. additive identity
4. distributive property

Think and Discuss For question 2, you might want to show students the alternate way to do the addition.

$1 \times 10^5 + 1 \times 10^4 =$
$1 \times 10^5 + 0.1 \times 10^5 = 1.1 \times 10^5$

A ***common error*** is to place the decimal point incorrectly when writing a number in scientific notation. For example, students may write 42.3×10^2 for 4,230 instead of 4.23×10^3. Emphasize that when writing a number in scientific notation, the decimal point must be moved to a position that shows the first factor as a number greater than 1 and less than 10.

Closure

Have students explain three things they learned in today's lesson. Answers may include the definition of scientific notation, why it is used, and how to multiply numbers in scientific notation.

Writing in Math Then ask students to record their comments about today's lesson in their math journals.

Independent Practice

For the calculator sample exercise, you may wish to check that the students understand the key sequence before they begin working independently.

For additional practice in using a scientific calculator with numbers written in scientific notation, refer students to the second calculator activity on p. 138E of the Chapter Overview. See also TRB Exploring O-16.

Lesson Follow-up

Reteaching Activity Discuss the procedures for moving the decimal point in scientific notation. Write the equations and ask students to identify the missing exponent.

$4{,}000 = 4.0 \times 10^{\blacksquare}$ (3)

$65{,}000 = 6.5 \times 10^{\blacksquare}$ (4)

$287{,}000 = 2.87 \times 10^{\blacksquare}$ (5)

$27{,}500{,}000 = 2.75 \times 10^{\blacksquare}$ (7)

LESSON RESOURCES

TRB Practice D-5

TRB Enrichment D-6

Practice

For use after 4-3 (pp. 149–152)

Scientific Notation

Write each number in scientific notation.

1. Pluto is about 3,653,000,000 mi from the sun. 3.653×10^9
2. There are 63,360 in. in a mile. 6.336×10^4
3. At its closest, Mercury is about 46,000,000 km from the sun. 4.6×10^7
4. 77,250,000 7.725×10^7
5. 526,000 5.26×10^5
6. 8 billion 8×10^9
7. 26 2.6×10^1
8. 745 million 7.45×10^8
9. 8,100,000,000 8.1×10^9
10. 888,200,000 8.882×10^8
11. 5,700 5.7×10^3

Write each number in standard notation.

12. 10^4 10,000
13. 10^6 1,000,000
14. 3.77×10^4 37,700
15. 8.5×10^3 8,500
16. 9.002×10^5 900,200
17. 1.91×10^5 191,000
18. 5.32×10^7 53,200,000
19. 3.21×10^9 3,210,000,000
20. 8.002×10^2 800.2
21. 4.6×10^3 4,600

Simplify. Write each number in scientific notation. Round to the nearest tenth. Then write the rounded number in standard notation.

	Scientific	Standard
22. $(2 \times 10^5)(3 \times 10^2)$	6×10^7	60,000,000
23. $(1.665 \times 10^6) \div (3.7 \times 10^4)$	4.5×10^1	45
24. $72{,}000 \times 143{,}000$	1.0×10^{10}	10,000,000,000
25. $(1.5 \times 10^5)(4 \times 10^9)$	6×10^{14}	600,000,000,000,000
26. $(3.2 \times 10^6)(5.1 \times 10^3)$	1.6×10^{10}	16,000,000,000
27. $9{,}400(2.5 \times 10^5)$	2.4×10^9	2,400,000,000

Complete.

28. $62.4 = \underline{6.24} \times 10^1$
29. $6.24 = 6.24 \times 10^{\underline{0}}$
30. $62{,}400 = 6.24 \times 10^{\underline{4}}$
31. $\underline{624} = 6.24 \times 10^2$

Chapter 4 D-5

Enrichment

For use after 4-3 (pp. 149–152)

Star Travel

Solve. Write your answers in scientific notation unless otherwise directed.

1. An unmanned space craft sets out to explore the moon, Jupiter, and Alpha Centauri, the closest star in our galaxy. A typical rocket travels about 20,000 mi/h. Write this number in scientific notation.
   2.0×10^4
2. The trip to the moon will take about 12 h. Use your answer to Exercise 1 and the formula $d = rt$ to find the distance to the moon.
   2.4×10^5 mi
3. The trip from the moon to Jupiter will take about 24,000 h.
   a. Write the number of hours in scientific notation. 2.4×10^4 h
   b. Find the distance to Jupiter.
   4.8×10^8 mi
   c. Write the number of days the journey will take in standard notation. (1 da = 24 h)
   1,000 da
4. From Earth, the trip to Alpha Centauri will take about 1.25×10^9 h.
   a. Find the distance to Alpha Centauri.
   2.5×10^{13} mi
   b. Find the number of hours in 1 year. (1 y = 365.25 da). Write the number in standard notation.
   8,766 h
   c. Find the length of the trip in years. Write the number in standard notation and round to the nearest ten thousand.
   140,000 y
5. The most distant star in the Milky Way is about 2.5×10^4 times as far from Earth as Alpha Centauri. In standard notation, how many years would that journey take?
   3,500,000,000 y
   a. Find the number of days the journey would take. 1.278375×10^{12}
   b. Write the number of days in standard notation. 1,278,375,000,000

D-6 Chapter 4

The Average Distance to the Sun (*km*)

Planet	Distance
Neptune	4.497×10^9
Mercury	5.79×10^7
Jupiter	7.783×10^8
Venus	1.082×10^8
Saturn	1.427×10^9
Earth	1.496×10^8
Uranus	2.869×10^9
Mars	2.279×10^8
Pluto	5.9×10^9

Our Nearest Neighbor

The moon is Earth's only satellite. Although it is not a planet, the moon is the closest object to Earth in our solar system. The moon is about 380,000 km from Earth.

On July 20, 1969, American astronauts from the *Apollo* mission were the first men to set foot on the moon. They returned to Earth with 382 kg of moon rocks and dust. The footsteps the astronauts left on the moon will probably be visible for at least 10 million years.

Use the table at the left to solve.

36. Which is closer to the sun, Saturn or Jupiter? Earth or Venus? Jupiter; Venus
37. Order the distances from the least to greatest. See below.
38. ***WRITE*** a procedure for ordering numbers in scientific notation. See below.
39. Alpha Centauri, the star closest to the sun, is about 40,600,000,000,000 km away. Write this in scientific notation. 4.06×10^{13} km
40. Why do you think astronomers prefer scientific notation? Scientific notation is shorter and fewer mistakes with zeros are made.
41. ***PROJECT*** Astronomers measure very great distances in *light years*. Find out what a light year is and the distance in light years from one side of our galaxy to the other. The distance light travels in one year.; about 100,000 light years

Use the article at the left to solve.

42. How would you express the distance from Earth to the moon in meters? Express this distance using scientific notation. 380,000,000 m; 3.8×10^8 m
43. How many 0.5-meter footsteps are there from here to the moon? Express using scientific notation. 7.6×10^8
44. ***PROJECT*** Find out why the footprints on the moon will be visible for 10 million years. There are no erosive forces such as wind or rain on the moon.

37. 5.79×10^7, 1.082×10^8, 1.496×10^8, 2.279×10^8, 7.783×10^8, 1.427×10^9, 2.869×10^9, 4.497×10^9, 5.9×10^9 38. First arrange the powers of 10 in order. Then order the factors that multiply the powers of 10.

TEST YOURSELF

Write using exponents.

1. $a \cdot b \cdot a \cdot b \cdot b$ a^2b^3
2. $-2 \cdot x \cdot (-2) \cdot y \cdot x \cdot x$ $4x^3y$

Evaluate.

3. 4^3 64
4. 12^2 144
5. $(-3)^5$ -243
6. $(3^2)^0$ 1
7. -5^4 -625
8. $2^2 \cdot 2^3 \cdot 2^4$ 512

Write each number in scientific notation.

9. 10,000 1×10^4
10. 100,000 1×10^5
11. 10,000,000 1×10^7
12. 75,000 7.5×10^4
13. 854,000 8.54×10^5
14. 1,645,123 1.645123×10^6

LESSON QUIZ

1. Write in scientific notation.
   a. 52.56 (5.256×10^1)
   b. 83,050,000 (8.305×10^7)
2. Write in standard notation.
   a. 2.87×10^6 (2,870,000)
   b. 3.05×10^5 (305,000)
3. Write the product in scientific notation.
   $(4.02 \times 10)(9.2 \times 10)$ $[3.6984 \times 10^3]$

FOR THE NEXT LESSON

math journal

Assignment Guide

Basic 1-5, 6-13 odd, 14, 17-20, 23-24, 28-30, 35-38, 41-42, MR, TY All

Average 1-11, 15-18, 21-24, 27-29, 31-35, 39, 44, MR, TY All

Enriched 3-4, 9-14, 16-20, 25-35, 38-41, 43-44, MR, TY All

PROBLEM SOLVING APPLICATION

OBJECTIVE: *To apply scientific notation to solve problems concerning water resources and usage.*

4-4 Water Resources

Water is critical for life. Our bodies are 0.65 water. About 0.95 of the total weight of a tomato plant is water. We need water to grow the food we eat and to manufacture the goods we use. Without water, we could not exist.

Earth's Water Supply (for every gallon)

Source	Amount
ocean	0.97
icecaps and glaciers	0.023
lakes and rivers	0.0001
underground	0.0059
atmosphere	0.001

2

Example 1 The total amount of water on Earth is about 3.26×10^8 mi^3. A cubic mile of water contains about 1.1×10^{12} gal of water. In gallons, what is the total amount of water on Earth?

1 **Solution** Multiply the number of cubic miles of water by the number of gallons in each cubic mile.

$$(3.26 \times 10^8) \times (1.1 \times 10^{12}) = (3.26 \times 1.1) \times (10^8 \times 10^{12})$$
$$= 3.586 \times 10^{20}$$

There are about 3.586×10^{20} gal of water on Earth.

Example 2 Find the total number of gallons of ocean water.

Solution Multiply 0.97 by the result from Example 1.

$$(0.97)(3.586 \times 10^{20}) = (0.97 \times 3.586) \times (10^{20})$$
$$= 3.47842 \times 10^{20}$$

There are 3.47842×10^{20} gal of ocean water.

You can find the portion of water given an amount.

Example 3 For every 50,000 gal of Earth's water supply, how much is found in icecaps and glaciers?

Solution Multiply the amount by 0.023 and express the product in scientific notation.

$$(0.023)(50{,}000) = 1{,}150$$
$$= 1.15 \times 10^3$$

The amount of water is 1.15×10^3 gal.

CLASS EXERCISES

Use the *DATA* at the left to solve each problem. Give your answers in scientific notation.

1. There are 358,600,000,000,000,000,000 gal of water on Earth. What is the total amount from underground sources? 2.11574×10^{18} gal
2. For every 75,000 gal of water, how much comes from lakes and rivers? 7.5 gal

Teaching Tip

When teaching problem solving, try to include a generalization of the problem. Generalizations will help the students apply the problem solving method to other problems and may generate a new solution or method.

Lesson Focus

Materials

Math journal

MOTIVATING THE LESSON

Mention that most of the Earth's surface is comprised of water and that the Pacific Ocean alone is larger than all the Earth's land surfaces put together. Comment that in this lesson the students will learn some other interesting facts about water, as they use what they have learned about scientific notation to solve problems.

Instruction

1 Point out to students that a cubic mile can be pictured as a cube with length, width, and height all measuring 1 mi.

Comment that if the number of gallons in one cubic mile is 1.1×10^{12}, then the amount in 3.26×10^8 mi^3 must be a number so great that it makes sense to write it in scientific notation.

2 Mention that to find the total number of gallons of ocean water, students must first look at the chart at the left and find the part of the Earth's water supply that is ocean water. Ask a volunteer what percent of the total water supply the decimal 0.97 represents. (97%) Examine the chart further, and help the students realize that when all the decimal amounts are added together the total should be 1.0, or 100% of the water supply.

Guided Practice

Class Exercises Remind students to use the chart to solve these problems. Ask how students would solve question 2. (Look at the chart for the amount of water in lakes and rivers; multiply 0.0001 and 75,000; write the product in scientific notation.)

Closure

Writing in Math Ask students to write a problem in their math journals that involves the chart on page 153 and scientific notation. Have the students explain how their problem can be solved.

Independent Practice

For questions 3-4, encourage students to check their answers at each step. An incorrect answer to question 3.a. will produce incorrect answers to questions 3.b. and 4, although the method may be correct.

For question 8, have students work in small groups. Be sure students understand the concepts of evaporation and condensation.

Decision Making For question 1, have students research standard estimates of water use at a local library. They can compare the standard estimates to their own estimates.

To help students analyze the data in question 2, have them make a bar graph.

For question 7, have students compare their results.

Average Water Usage in United States Cities (per person/day)

home	70 gal
factories and businesses	70 gal
city services	10 gal

WRITTEN EXERCISES

Use a CALCULATOR where appropriate.

Use the *DATA* at the left to solve each problem. Express answers in scientific notation.

1. There are about 187,500,000 people that live in cities in the United States. About how many gallons of water will these people use each day in their homes? 1.3125×10^{10}

2. In all, about how many gallons of water do the 187,500,000 people in cities use in a day for all purposes? 2.8125×10^{10}

3. The population of the city of Chicago is about 4,000,000 people.

a. About how many gallons of water do the people in Chicago use each day in their homes? 2.8×10^{8}

b. Chicago gets its drinking water from Lake Michigan, the largest body of fresh water in the United States. If you took the top inch of water from Lake Michigan, you would get about 3.9×10^{11} gal of water. Suppose this water is used exclusively by Chicago's homes. About how many days of water would it supply? about 1,393 da

Decision Making **DECISION MAKING** *Decision Making* *Decision Making* *Decision Making*

WATER RESOURCES

COLLECT DATA

1. Survey how much water your family uses in one week. You can estimate the amounts. For example, to measure the amount of water it takes to fill a bathtub, estimate how many quart or one-gallon milk cartons of water it would take to cover the bottom of the bathtub. Record your estimates in a chart like the one below. **Check students' work.**

Amount of Water Used Each Day (gal)							
Type of Use	**Mon**	**Tues**	**Wed**	**Thur**	**Fri**	**Sat**	**Sun**
bath/shower							
dish washing							
drinking							
brushing teeth							
washing hands/ laundry							

MATH MINUTES

Have the class work in pairs. One student writes an expression with exponents and reads the expression to his or her partner. The other student writes the expression and evaluates it. Repeat this several times, exchanging roles.

4. Chicago's main water purification plant can produce nearly 1.75 billion gal of water each day. The plant serves about 4.5 million people in the city and suburbs. About how many times greater is the plant capacity than the daily water needed for homes, factories, businesses, and city services? **about 2.5 times**

Solve.

5. Many places in the world do not have access to fresh water from lakes and rivers. They must purify ocean water. A desalting plant in Key West, Florida, can produce about 2 million gal of fresh water per day. Assume the plant operates every day. About how many gallons of pure water can the plant produce in a year? **7.3×10^8**
6. The Great Salt Lake in Utah holds about 5.7 mi^3 of water and about 4.0×10^9 t of salt. About how many tons of salt is this per cubic mile? **about 7×10^8**
7. The total amount of water on the planet is about 326,000,000 mi^3. About how many cubic miles are in glaciers and icecaps? **7.498×10^6**
8. ***PROJECT*** Water is naturally recycled by the processes of *evaporation* and *condensation*. Research how ocean water and fresh water are recycled through these processes.

Decision Making ■ Decision Making ■ Decision Making ■ Decision Making ■ Decision Making ■ Decision Making

ANALYZE DATA

2. What is the total amount of water your family uses in one week? Which activities use more water?
3. Estimate the amount of water used for brushing your teeth if the water is left running. How much water would be used in a year?
4. What ways can you think of to reduce the amount of water you and your family use?

MAKE DECISIONS

5. Discuss ways your classmates and families can conserve water. Make a list of ideas that you can use at home.
6. Try several water-saving ideas. Keep track of your family's water consumption for another week. Record your data.
7. How much water did you save in one week? Compare your results with those of your classmates.
8. Make a poster with at least three suggestions and the amount of water they would save in a year.

Lesson Follow-up

LESSON RESOURCES

TRB Alternate Application D-7

Assignment Guide

Basic All
Average All
Enriched All

FOR THE NEXT LESSON

graph paper, lined paper, math journal

Alternate Application

For use after 4-4 (pp. 153–155)

Geometric Sequences

A *geometric sequence* is a sequence of numbers in which each term is multiplied by a constant ratio to obtain the next term.

Example Find the next term in the geometric sequence 6, 12, 24, **?** .

Solution ratio = $\frac{12}{6} = \frac{24}{12} = 2$

next term = 24 • 2 = 48

Find the next three terms in each geometric sequence.

1. 5, 15, 45, **135**, **405**, **1,215**
2. 4, 24, 144, **864**, **5,184**, **31,104**
3. 640, 320, 160, **80**, **40**, **20**
4. $7^5, 7^7, 7^9$, **7^{11}**, **7^{13}**, **7^{15}**
5. 23 and 207 are adjacent terms in a geometric sequence. What is the next term? **1,863**
6. What term is found between 5 and 80 in a geometric sequence? **20**

You can use exponents to write the terms of a geometric sequence.

Example The ratio between terms of a geometric sequence is 5. The first term is 38. Write the next three terms.

Solution

1st	2nd	3rd	4th
38	$38 \cdot 5^1$	$38 \cdot 5 \cdot 5 = 38 \cdot 5^2$	$38 \cdot 5 \cdot 5 \cdot 5 = 38 \cdot 5^3$

The next three terms are 190, 950, and 4,750.

Write the next three terms in the geometric sequence.

7. first term = 2, ratio = 13 **$2 \cdot 13 = 26,\ 2 \cdot 13^2 = 338,\ 2 \cdot 13^3 = 4,394$**
8. first term = 20, ratio = 3 **$20 \cdot 3 = 60,\ 20 \cdot 3^2 = 180,\ 20 \cdot 3^3 = 540$**
9. first term = A, ratio = r **Ar, Ar^2, Ar^3**

Study the exponents above. Write an expression involving exponents for the indicated term of the geometric sequence.

10. forty-sixth term (first term = 11, ratio = 2) **$11 \cdot 2^{45}$**
11. ninety-fifth term (first term = 8, ratio = h) **$8 \cdot h^{94}$**
12. nth term (first term = A, ratio = r) **Ar^{n-1}**

Chapter 4 D-7

Lesson Focus

Materials

Graph paper, lined paper, math journal

Vocabulary/Symbols

factor
multiple

MOTIVATING THE LESSON

Write the set of multiples of 5 on the chalkboard: {5, 10, 15, 20, . . . }. Ask students to describe the set. Then ask, **What is true of every number or element in the set?** (Each is a multiple of 5, and 5 is a factor of each member of the set.)

Instruction

After students read the problem, remind them that two strategies for solving a complicated problem is to organize the data and break the problem into a series of simpler problems.

1 Mention that the students who changed locker 20 are students 1, 2, 4, 5, 10, and 20. All these numbers that represent the students are factors of 20. Help students make this generalization: a locker will be changed by those students whose position number is a factor of the locker's number.

OBJECTIVE: *To explore factors and multiples.*

MATERIALS

- Graph paper or lined paper
- Math journal to record work

Exploring Factors and Multiples

▼ A high school has 1,000 lockers, numbered 1 to 1,000, and 1,000 students. The students enter the building one at a time.

- The first student opens all the lockers.
- The second student starts with locker 2 and closes every second locker.
- The third student starts with locker 3 and moves to every third locker. The student opens the closed lockers and closes the open ones.
- The fourth student changes every fourth locker starting with locker 4, and so on for the remaining students.

Answer the questions below to find out which lockers are open and which lockers are closed after all the students enter.

1. Which of the first nine lockers will be open after the first student passes through? the second student? **all of the lockers; the odd numbered lockers**
2. Using a table like the one below, find out which lockers will be open after the third student passes through. **1, 5, 6, 7**

Student Number	Locker Number 1	2	3	4	5	6	7	8	9
1	O	O	O	O	O	O	O	O	O
2	O	C	O	C	O	C	O	C	O
3	O	C	C	C	O	O	O	C	C
4	O	C	C	O	O	O	O	O	C

KEY
O Open
C Closed

3. Which students change the condition of locker 1? 2? 3? **1; 1, 2; 1, 3**
4. ***Extend*** and complete the table for 12 students and 12 lockers. **Check students' work.**
5. Which students changed the condition of locker 12? Will any of the other 1,000 students change locker 12? Explain. **1, 2, 3, 4, 6, 12; No; other students start with higher numbers**
6. ***Describe*** and discuss the relationship between the numbers of the students who changed locker 12 and the number 12. **The student numbers 1, 2, 3, 4, 6, 12 all divide into 12 with no remainder.**

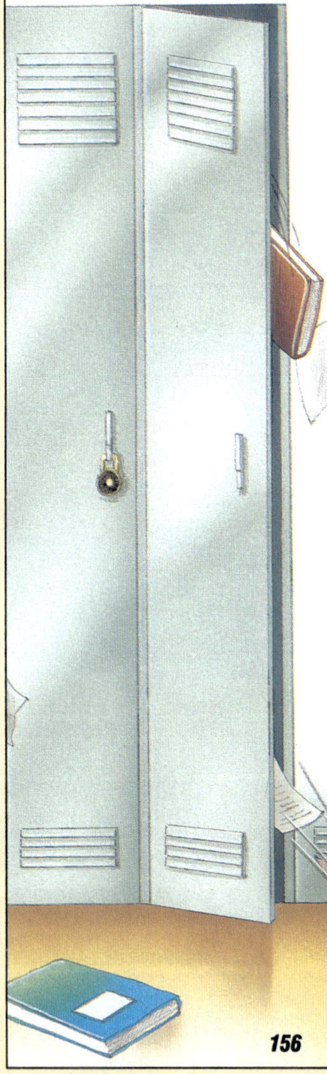

Teaching Tip

Remind students that it is all right to make mistakes. Emphasize that a mistake can be an opportunity to learn. You may wish to have students research discoveries and inventions that developed through error.

▼ The numbers that correspond to the students who changed locker 12 are *factors* of the number 12.

7. Use the relationship between the student number and the locker number to predict which students will change locker 20. 1, 2, 4, 5, 10, 20
8. ***Test*** your prediction by extending and completing the table for 20 students and 20 lockers. Check students' work. (1)
9. List all the factors of each number.
   a. 1 1   b. 4 1, 2, 4   c. 6 1, 2, 3, 6   d. 9 1, 3, 9

▼ Student 3 changed six of the first 20 lockers. List them. 3, 6, 9, 12, 15, 18

10. Will student 3 change any other of the 1,000 lockers? Explain. yes; every third locker starting with locker 3
11. ***Describe*** and discuss the relationship between the student number and the number of lockers the student changes. Every locker number that a student changes can be divided by the student number with no remainder.

▼ The numbers of the lockers that student 3 changed are *multiples* of the number 3.

12. Use the relationship between the student and the locker number to predict which lockers student 25 will change. 25, 50, 75, . . .
13. List the multiples found in your table for each number.
    a. 1 1, 2, 3, . . . 20   b. 4 4, 8, 12, 16, 20   c. 5 5, 10, 15, 20   d. 6 6, 12, 18
14. Refer to your table.
    a. Which of the 20 lockers will be open after 20 students pass through? 1, 4, 9, 16
    b. ***Predict*** the next two lockers that will be open. 25, 36
    c. What do you call these numbers? Do they have an odd or even number of factors? Is this true for all numbers of this type? square numbers; odd; yes (2)

▼ After the 1,000 students pass through, each locker with an open door corresponds to a number with an odd number of factors.

15. Why do you think an odd number of factors would correspond to a locker with an open door? An odd number of students must change the locker so that it stays open.
16. Does an even number of factors correspond to a locker with a closed door? Explain. Yes; every second student closes the locker.
17. ***Write*** a rule that explains which lockers are open and which are closed after the 1,000 students pass through. All square numbered lockers are open; all others are closed.
18. ***Calculator*** Make a list of the numbers that correspond to an open locker once all the students have passed through. Check students' work. All square numbers between 1 and 1,000, inclusive. (3)

(2) If necessary, review the properties of square numbers with students.

(3) Ask students to use the table to find which lockers were changed by exactly two students. Have the class make a list. Then ask, **What can be said about each of the numbers in your list?** (Each number has exactly two factors, the number and 1.) Remind students that numbers that fit this definition are called prime numbers.

Closure

Writing in Math Ask students to write a description of the problem they solved in their math journals, and ask how they were able to predict the number of lockers that remained opened after all 1,000 students had passed through the school. Have students describe how understanding the properties of factors and multiples helped them solve the problem.

Assignment Guide

Basic All
Average All
Enriched All

FOR THE NEXT LESSON

graph paper, math journal

Career Information

Invite a demographer to address the class on the topic of a career in demographics. To locate a speaker, contact the sociology department of a local college or university, government agencies, such as the U.S. Bureau of the Census, or insurance companies.

Ask the speaker to discuss topics such as:

- the kinds of information demographers gather
- what industries and institutions use demographic information, e.g., insurance, advertising, and political consulting companies, and how the information is used
- how demographic information is gathered, analyzed, and displayed
- the occupational outlook for demographers, including job availability and salaries
- the education and skills requirements for a career in demography

Practice

CAREER

Help Wanted: Demographer

Bachelor's degree in sociology required. A masters degree is suggested. For more information, write to Population Association of America, 806 15th St. NW, Suite 640, Washington, DC 20036.

Demographers study size, characteristics, and movement of populations to forecast future trends. Lawmakers, administrators, and marketers use demographers' findings to help improve intergroup relations, to solve social problems, and formulate better public policy.

PROJECT

Find out how the population in your town or city changed in the past ten years. Make a graph to show the change in population. **Answers will vary.**

Write using exponents.

1. $5 \cdot 3 \cdot 3 \cdot 5 \cdot 5$ **$3^2 \cdot 5^3$**
2. $-3 \cdot t \cdot s \cdot s \cdot 4 \cdot s$ **$-12s^3t$**
3. $(-3a)(-3a)(-3a)(-3a)(-3a)$ **$(-3a)^5$**
4. $(2n - 1)(2n - 1)(2n - 1)$ **$(2n - 1)^3$**

Evaluate.

5. 8^3 **512**
6. $(-1)^{17}$ **-1**
7. $(-3)^5$ **-243**
8. -4^3 **-64**
9. 4^3 **64**
10. -2^5 **-32**
11. $(-4)^2$ **16**
12. $\sqrt{36}$ **6**
13. $7^3 \cdot 7^6$ **40,353,607**
14. $6^5 \cdot 6^3$ **1,679,616**
15. $16 \cdot 4^2$ **256**
16. $(4^5 \cdot 4^9)^0$ **1**
17. $-3[5 - (-2)]^2$ **-147**
18. $[7 + (-2)]^4$ **625**
19. $(3^2)(3^6)$ **6,561**
20. $(2^4)(4^2) \div 4$ **64**
21. $4^4 \cdot 4^6 \cdot 2^3$ **8,388,608**
22. $5^3 - 2^3 \cdot 2^1$ **109**

Simplify.

23. $b \cdot b^2$ **b^3**
24. $g^{12} \cdot g^{10}$ **g^{22}**
25. $k^5 \cdot k^8$ **k^{13}**
26. $6x^4y^3$ **$6x^4y^3$**
27. $c \cdot c^6 \cdot c^4$ **c^{11}**
28. $a^3b^2c^6 \cdot a^0b^4c^5$ **$a^3b^6c^{11}$**
29. $(-y^3)(-y)^3$ **y^6**
30. $ab^6c^3 \cdot a^5bc^8$ **$a^6b^7c^{11}$**
31. $(3r^3)^5$ **$243r^{15}$**
32. $(-3s)^4 \cdot 9^1$ **$729s^4$**
33. $(3m^2)^4 \cdot 5m^2$ **$405m^{10}$**
34. $(-4m^3y^7)^2$ **$16m^6y^{14}$**

Evaluate for $a = 4$, $b = -1$, and $c = -5$.

35. $-4b^3 + ac^2$ **104**
36. $a^3 + 6cb^0$ **34**
37. $abc \cdot \sqrt{100}$ **200**
38. $(c^2 - a^2) \cdot (-3b^8c^0)$ **-27**
39. $a^2c + b$ **-81**
40. $-(c)^2 - (b)^2 \cdot (a^2 + 3b)$ **-38**
41. $(ab^2)^4 \cdot (4a^2b)$ **-16,384**

Write each number in standard notation.

42. 5×10^5 **500,000**
43. 24×10^7 **240,000,000**
44. 35.6×10^6 **35,600,000**
45. 10^7 **10,000,000**
46. 63.57×10^5 **6,357,000**
47. 9.83675×10^4 **98,367.5**

Write each number in scientific notation.

48. The height of Mt. Everest is 10,000 m. **1×10^4 m**

49. The sun's diameter is 3,392,000 km. **3.392×10^6 km**

50. In an average lifetime a heart beats about 2,500,000,000,000 times. **2.5×10^{12}**

51. The approximate weight of Earth is 5,880,000,000,000,000,000,000 t. **5.88×10^{21}**

52. The radius of our solar system is 100,000,000,000 m. **1×10^{11} m**

53. The radius of the Milky Way galaxy is 10,000,000,000,000,000,000 m. **1×10^{19} m**

54. In the 1864 presidential election, 2,218,388 votes were cast for Abraham Lincoln. **2.218388×10^6**

OBJECTIVE: *To identify whole number factors and multiples of a number and to determine divisibility by 2, 3, 4, 5, 6, 9, or 10.*

4-5 Factors, Multiples, and Divisibility

▼ The diagram at the right shows all the different rectangles that you can make with 12 squares. The dimensions are 1 by 12, 2 by 6, and 3 by 4. The numbers 1, 2, 3, 4, 6, and 12 are the *factors* of 12. The number 12 is *divisible* by its factors.

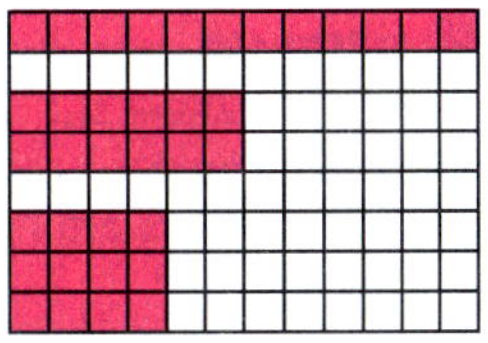

Divisible	A number is divisible by a second number if the second number divides the first with no remainder.

Factor	One number is a factor of another if it divides that number with no remainder.

Example 1 **Find the factors of 15.**

Solution $1 \cdot 15$ $3 \cdot 5$

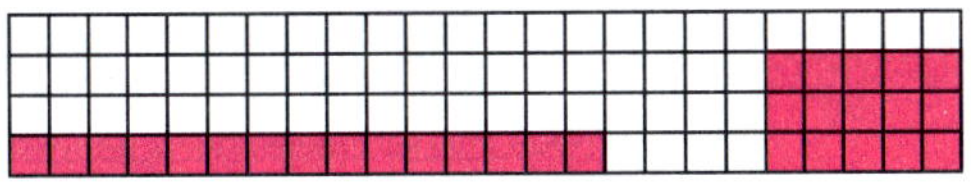

The factors of 15 are 1, 3, 5, and 15.

THINK Why is 8 not a factor of 45? **8 does not divide 45 with no remainder.**

Example 2 Are 5 and 8 factors of 45?

Solution 45 ÷ 5 = 9 ✓ Yes, 5 is a factor of 45.

45 ÷ 8 = 5.625 ✗ No, 8 is not a factor of 45.

▼ You can find the *multiples* of any number. The multiples of 12 are 0, 12, 24, 36, . . .

1

Multiple	A multiple of a number is the product of that number and any other whole number.

Example 3 **Find the first three multiples of each number.**

a. 2 **b.** 10

Solution

a. $2 \cdot 0 = 0$
$2 \cdot 1 = 2$
$2 \cdot 2 = 4$

b. $10 \cdot 0 = 0$
$10 \cdot 1 = 10$
$10 \cdot 2 = 20$

Example 4 Is 45 a multiple of 15?

Solution The multiples of 15 are 0, 15, 30, 45, . . .

Yes, 45 is a multiple of 15 because $3 \cdot 15 = 45$.

Lesson Focus

Materials

Graph paper, math journal

Vocabulary/Symbols

divisible

MOTIVATING THE LESSON

Have students recall what they have already learned about factors and multiples. Ask students to name all the factors and some of the multiples of 8. (Factors: 1, 2, 4, 8; Multiples: 0, 8, 16, 24, . . .) Explain that this lesson involves these concepts, as well as divisibility.

Instruction

1 Ask students why zero is a multiple of every whole number. Remind students that zero is a whole number, so the least multiple of a number is the product of the number and zero, which is zero.

Teaching Tip

Prepare an overhead transparency with a graph paper background. Use the clear side for demonstrating ideas and concepts on graph paper.

EE: 1A, 1E, 2B, 3A; TAAS: 2, 9, 11, 13

2 After discussing the table with the class, ask students to guess whether 337 and 111 are divisible by 3. (no; yes)

3 Have students guess whether 5,967 and 11,112 are divisible by 9. (yes; no)

4 Students may answer that the multiples of 9 are divisible by 3 because 3 is a factor of 9.

ADDITIONAL EXAMPLES

1. Find all the factors of 30. (1, 2, 3, 5, 6, 10, 15, 30)

2. Tell whether the first number is a factor of the second number.

- **a.** 5; 8,764,395 (yes)
- **b.** 9; 7,145,011 (no)
- **c.** 4; 24,568 (yes)

3. Write the first seven multiples of 13. (0, 13, 26, 39, 52, 65, 78)

4. Determine if the following numbers are divisible by 2, 3, 5, or 9.

- **a.** 814,536 (2, 3, 9)
- **b.** 721,302 (2, 3)

▼ You can use divisibility tests to find out if one number is divisible by another.

FLASHBACK
Even numbers are divisible by 2.
Odd numbers are not divisible by 2.

Divisibility Tests

A number is divisible by 2 if the ones' digit is 0, 2, 4, 6, or 8.
A number is divisible by 5 if the ones' digit is 0 or 5.
A number is divisible by 10 if the ones' digit is 0.

▼ To discover a divisibility test for 3, analyze the following table.

2

Number	Sum of digits	Is the sum divisible by 3?	Is the number divisible by 3?
136	1 + 3 + 6 = 10	no	no
462	4 + 6 + 2 = 12	yes	yes
216	2 + 1 + 6 = 9	yes	yes
1,017	1 + 0 + 1 + 7 = 9	yes	yes

Divisible by 3

A number is divisible by 3 if the sum of its digits is divisible by 3.

▼ To discover a divisibility test for 9, analyze the following table.

3

Number	Sum of digits	Is the sum divisible by 9?	Is the number divisible by 9?
136	1 + 3 + 6 = 10	no	no
462	4 + 6 + 2 = 12	no	no
216	2 + 1 + 6 = 9	yes	yes
1,017	1 + 0 + 1 + 7 = 9	yes	yes

THINK If a number is divisible by 3, is it also divisible by 9? Is the reverse true? Explain. **No, 12 is divisible by 3, but not by 9; yes, because 9 is divisible by 3.**

4

Divisible by 9

A number is divisible by 9 if the sum of its digits is divisible by 9.

Example 5 **Is the first number divisible by the second?**

a. 567; 2 **b.** 567; 3 **c.** 567; 5 **d.** 567; 9

Solution

a. No, the ones' digit is not 0, 2, 4, 6, or 8.

b. Yes, the sum of the digits is 18, which is divisible by 3. Write: 3|567 (3 divides 567 with no remainder).

c. No, the ones' digit is not 0 or 5.

d. Yes, the sum of the digits is divisible by 9. Write: 9|567.

MATH MINUTES

Write several money amounts on the chalkboard. (For example, $2.25, $45.75, $123.36.) Have students use the divisibility rules to decide which amounts can be shared by 2, 3, 5, or 9 students.

CLASS EXERCISES

1. True; the sum of the digits is divisible by 3.
2. True; the sum of the digits is divisible by 9.

True or *false*? Explain your answer.

1. 3|555
2. 981 is divisible by 9.
3. 9 is a multiple of 18.
4. 5 is a factor of 435.
5. 6 divides 56.
6. 2 divides every even number.
7. If a number is divisible by both 2 and 5, it is divisible by 10. See below.
8. If a number is divisible by 6, then it is divisible by 2. True, because 6 is divisible by 2.
9. If a number is divisible by 2, then it is divisible by 6. False; 10 is divisible by 2, 10 is not divisible by 6.
10. List the first ten nonzero multiples of 11. 11, 22, 33, 44, 55, 66, 77, 88, 99, 110
11. Write a missing digit so that the resulting number is divisible by 3. 4,826,■51 Possible answers: 1, 4, 7.
12. Write a missing digit so that the resulting number is divisible by 9. 4,826,■51 1
13. State a rule for divisibility by 100. A number is divisible by 100 if the last two digits are 0.

7. True, to be divisible by both 2 and 5, 2 and 5 must both be factors. Since $2 \cdot 5 = 10$, the number is divisible by 10.

3. False; 18 times any whole number does not equal 9.
4. True; the last digit is 5.
5. False; $56 \div 6 = 9.3$
6. True, the ones' digits of even numbers are 0, 2, 4, 6, or 8.

THINK AND DISCUSS

1. What number is a factor of all numbers? 1
2. How many multiples does any number have?
3. What number is a multiple of all numbers? 0
4. What number can never be a factor of a number? 0

2. an infinite number

WRITTEN EXERCISES

Decide whether the first number is a factor of the second.

1. 8; 72 yes
2. 12; 54 no
3. 7; 91 yes
4. 6; 68 no
5. 9; 621 yes
6. 3; 101 no
7. 5; 46,582 no
8. 4; 128 yes
9. 10; 75,020 yes
10. 9; 74,520 yes
11. 3; 876 yes
12. 15; 120 yes

List all the factors of each number.

13. 30 1, 2, 3, 5, 6, 10, 15, 30
14. 1 1
15. 55 1, 5, 11, 55
16. 126 1, 2, 3, 6, 7, 9, 14, 18, 21, 42, 63, 126
17. 29 1, 29

List the first five multiples of each number.

18. 1 0, 1, 2, 3, 4
19. 12 0, 12, 24, 36, 48
20. 16 0, 16, 32, 48, 64
21. 25 0, 25, 50, 75, 100

State whether each number is divisible by 2, 3, 5, or 9.

22. 213 3
23. 630 2, 3, 5, 9
24. 138 2, 3
25. 204 2, 3
26. 131 none
27. 4,805 5
28. 288 2, 3, 9
29. 4,719 3

30. Which numbers are divisible by both 2 and 3?
    a. 10 no
    b. 66 yes
    c. 898 no
    d. 47,820 yes
    e. 975 no
    f. ***WRITE*** a divisibility rule for 6. If the number is divisible by both 2 and 3, then it is divisible by 6.

NOTES & QUOTES

A perfect number is a number that is the sum of all its possible factors, excluding itself. Six is the first perfect number ($1 + 2 + 3 = 6$). At present, only 30 perfect numbers have been calculated.

RESEARCH Can you find the second perfect number? $1 + 2 + 4 + 7 + 14 = 28$

Guided Practice

Class Exercises Have students work in pairs to complete the questions. Remind them to support their answers with explanations.

Think and Discuss For question 1, explain that 1 is a factor of every number since every whole number can be divided by 1.

For question 2, point out that since there are an infinite number of whole numbers, there are an infinite number of multiples for any given number.

For question 4, tell students that division by zero is undefined. An alternative explanation to mention is that zero does not divide any number such that the remainder is zero.

A ***common error*** is to assume that the "sum of the digits" rule can be extended to test for divisibility by any number. Show that it does not work for all numbers. For example, 17 is not divisible by 4 even though the sum of the digits is divisible by 4.

Closure

Writing in Math Ask students to write the rules for divisibility in their math journals. Have them complete the following sentence: 12 is a ■ of 72, and 72 is a ■ of 12. (factor, multiple)

Independent Practice

Research Comment that the second perfect number is 28.

Encourage students to work carefully in finding the rules for divisibility by 6 and 4.

Lesson Follow-up

Reteaching Activity For students who have difficulty with the concept of divisibility, refer to *Exploring Divisibility*, on p. 138E of the Chapter Overview. See also, TRB Exploring O-15.

LESSON RESOURCES

TRB Practice D-8
TRB Enrichment D-9

Practice

For use after 4-5 (pp. 159–162)

Factors, Multiples, and Divisibility

Decide whether the first number is a factor of the second. Write yes or no.

1. 6; 54 yes 2. 4; 76 yes 3. 3; 107 no
4. 5; 424 no 5. 10; 6,270 yes 6. 9; 711 yes
7. 3; 555 yes 8. 6; 566 no 9. 12; 504 yes
10. 9; 251,613 yes 11. 5; 13,360 yes 12. 4; 428 yes

List all the factors of each number.

13. 12 1, 2, 3, 4, 6, 12 14. 35 1, 5, 7, 35
15. 41 1, 41 16. 54 1, 2, 3, 6, 9, 18, 27, 54

List the first five multiples of each number.

17. 3 0, 3, 6, 9, 12 18. 7 0, 7, 14, 21, 28
19. 18 0, 18, 36, 54, 72 20. 35 0, 35, 70, 105, 140

Decide whether each number is divisible by 2, 3, 5, or 9. Write yes or no for each number.

21. 215 no, no, yes, no 22. 432 yes, yes, no, yes
23. 770 yes, no, yes, no 24. 1,011 no, yes, no, no
25. 975 no, yes, yes, no 26. 2,070 yes, yes, yes, yes
27. 3,707 no, no, no, no 28. 5,715 no, yes, yes, yes

Write the missing digit to make each number divisible by 9.

29. 7[1]1 30. 2,2[3]2 31. 88[8]12

32. There are four different digits which, when inserted in the blank space in the number 4 ▒ 5, make the number divisible by 3. Write them. 0, 3, 6, 9

33. There are two different digits which, when inserted in the blank space in the number 7,16 ▒ , make the number divisible by 5. Write them. 0, 5

34. There are five different digits which, when inserted in the blank space in the number 99,99 ▒ , make the number divisible by 2. Write them. 0, 2, 4, 68

D-8 Chapter 4

Enrichment

For use after 4-5 (pp. 159–162)

More on Divisibility

You can test numbers for divisibility by 7 and by 11.

Example Test 959 for divisibility by 7.

Solution
1. Drop the ones' digit. 95~~9~~
2. Subtract twice the ones' digit from the number that remains. 95 − 18 = 77 (18 = 2 × 9)
3. The result, 77, is divisible by 7. Therefore, the original number, 959, is divisible by 7.

Test each number for divisibility by 7. Write yes or no.

1. 133 yes 2. 189 yes 3. 267 no 4. 385 yes
5. 553 yes 6. 689 no 7. 784 yes 8. 987 yes

Example Test 4,378,396 for divisibility by 11.

Solution
1. Add alternate digits, beginning with the first. **4**, 3**7**8, **3**9**6** $4 + 7 + 3 + 6 = 20$
2. Add alternate digits, beginning with the second. 4, **3**7**8**, 3**9**6 $3 + 8 + 9 = 20$
3. The sums are equal. Therefore, 4,378,396 is divisible by 11.

Test each number for divisibility by 11. Write yes or no.

9. 5,265 no 10. 32,175 yes 11. 222,222 yes
12. 805,969 no 13. 496,485 yes 14. 75,886,849 no

Divisibility by Multiples	If a number is divisible by both m and n, and m and n have no common factor other than 1, then the number is divisible by $m \times n$. If m and n have common factors other than 7, the number may or may not be divisible by $m \times n$.

Use the rule to test for divisibility. Write yes or can't tell.

15. 810 is divisible by both 5 and 9. Is 810 divisible by 45? yes
16. 3,654 is divisible by both 7 and 9. Is 3,654 divisible by 63? yes
17. 3,720 is divisible by both 6 and 8. Is 3,720 divisible by 48? can't tell
18. 189 is divisible by both 3 and 9. Is 189 divisible by 27? can't tell

Chapter 4 D-9

MIXED REVIEW

Write an equation for each. Do not solve.

1. A number decreased by 5 equals -24. $n - 5 = -24$
2. Ten less than 3 times a number is 57. $3n - 10 = 57$

Evaluate for $p = -3$, $r = 4$.

3. $5p - 2r$ -23
4. $4p^2$ 36
5. $-r^2 + p$ -19

Write in scientific notation.

6. 3,480,000 3.48×10^6
7. 250 2.5×10^2
8. On a hot summer day in Germany, the temperature is 35°C. Using the formula F = 1.8C + 32, find the temperature in °F. 95°F

31. a. Copy and complete the table.

Number	Last two digits	Last 2 digits divisible by 4?	Is the number divisible by 4?
136	36	yes	yes
1,268	68	yes	yes
314	14	no	no
1,078	78	no	no
696	96	yes	yes

b. ***WRITE*** a divisibility rule for 4. If the last two digits are divisible by 4, the number is divisible by 4.

Write the missing digit to make each number divisible by 9.

32. 29▒,634 3 **33.** 4▒,817 7 **34.** 8,03▒,373 3

Solve.

35. If a is divisible by 2, is $a + 1$ even or odd? odd

36. If a is divisible by 9, is $2a$ divisible by 9? yes

37. If a^2 is divisible by both 2 and 5, can $a = 10$? yes

38. Find a number less than 100 that is divisible by the first six positive integers. 60

TEST YOURSELF

Write in standard notation.

1. 9.604×10^3 9,604 **2.** 1.23×10^4 12,300 **3.** $(42 \times 10^9)(0.68 \times 10^6)$ 28,560,000,000,000,000

Write in scientific notation.

4. 9,650,000 9.65×10^6 **5.** 548 5.48×10^2 **6.** 30×10^5 3×10^6

List all the factors of each number.

7. 27 1, 3, 9, 27 **8.** 45 1, 3, 5, 9, 15, 45 **9.** 60 1, 2, 3, 4, 5, 6, 10, 12, 15, 20, 30, 60

State whether each number is divisible by 2, 3, 5, or 9.

10. 45 3, 5, 9 **11.** 300 2, 3, 5 **12.** 369 3, 9

Write using exponents.

13. $4 \cdot 4 \cdot 4 \cdot 4$ 4^4 **14.** $17 \cdot 17 \cdot 17$ 17^3 **15.** $z \cdot z \cdot z \cdot z$ z^4

Evaluate for $x = 2$ and $y = 5$.

16. x^2y 20 **17.** y^2x 50 **18.** $x^2 + y^2$ 29 **19.** $x^4 - y^2$ -9

LESSON QUIZ

Write True or False.

1. If a number is divisible by 2 and 6, it is also divisible by 12. (F)
2. Every factor of 12 is also a factor of 24. (T)
3. Every multiple of 9 is also a multiple of 18. (F)

Assignment Guide

Basic 2-24 even, 30-31, 35-36, MR, TY All
Average 1-12, 13-29 odd, 30-37, MR, TY All
Enriched 7-12, 22-29, 31-38, MR, TY All

FOR THE NEXT LESSON

graph paper, math journal

OBJECTIVE:
To determine whether a number is prime or composite and to find its prime factorization.

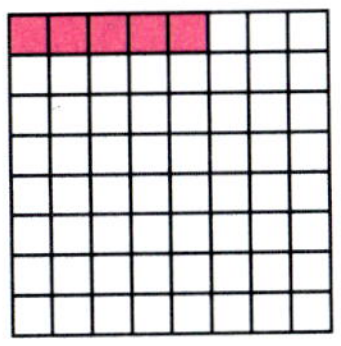

4-6 Prime Factorization

▼ The graph paper diagram shows all the possible factors of 5. Notice there is only one rectangle. The factors of 5 are 1 and 5. We call the number 5 a *prime number*.

Prime Number	A prime number is a whole number greater than one with exactly two factors, 1 and the number itself.

1

▼ Numbers greater than 1 that are not prime are *composite numbers*.

Example 1 **Prime numbers** 2, 3, 5, 7, 11, 13 . . . **Composite numbers** 4, 6, 8, 9, 10, 12, 14, 15 . . .

THINK Why is 2 the only even prime number? All other even numbers have 2 as a factor.

Composite Number	A composite number is a whole number greater than one with more than two factors.

▼ A composite number is divisible by prime factors.

Example 2 **Tell whether each number is prime or composite.**

a. 129 **b.** 23

Solution To test for divisibility, start with the smallest prime. Stop when you reach a prime whose square is greater than the number you are testing.

a. Is 129 divisible by 2? No, it is odd.

Is 129 divisible by 3? Yes, the sum of the digits is 12, a multiple of 3.

129 is composite.

b. Is 23 divisible by 2? No, it is odd.

Is 23 divisible by 3? No, the sum of the digits is 5.

Is 23 divisible by 5? No, the ones' digit is not 0 or 5.

Since $5^2 > 23$, 23 is prime.

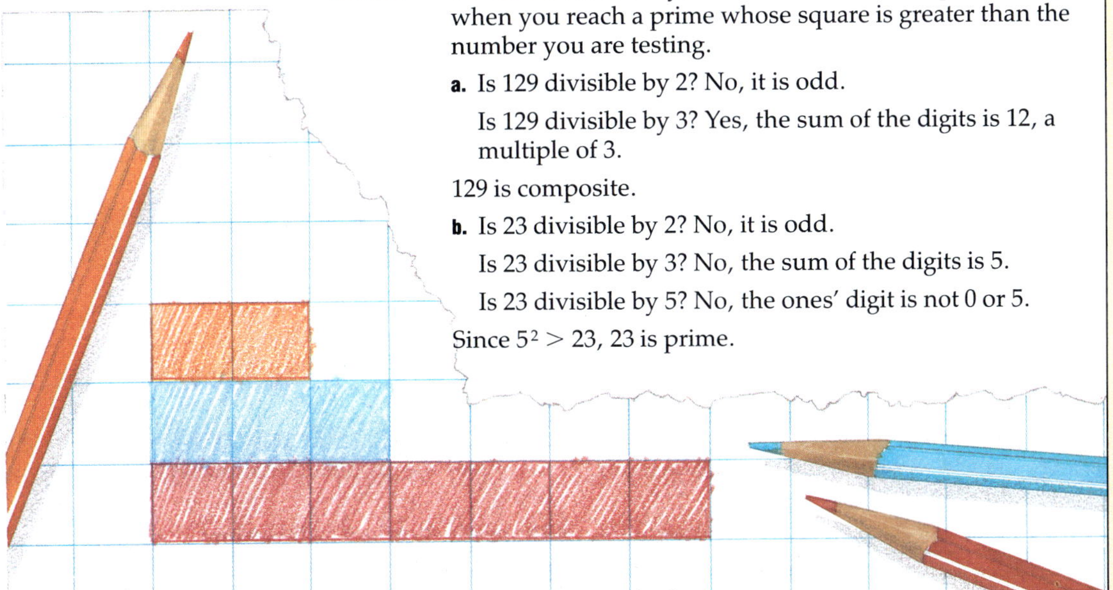

What prime numbers are shown? How do you know they are prime?
2, 3, 7; no other rectangles can be formed

Lesson Focus

Materials
Graph paper, math journal

Vocabulary/Symbols
composite number
prime number

MOTIVATING THE LESSON
Have students use graph paper to draw as many rectangular regions as possible with 7, 11, and 13 squares. Ask students to discuss what they observe. (Only one rectangle can be made with each number. There are only two factors for each number, 1 and the number itself.) Explain that numbers with only two factors are called prime numbers.

Instruction

1 Note that the number 1 is neither prime nor composite. It is the only number with exactly one factor.

Teaching Tip
Ask volunteers to be "math historians." Have them research history facts for current math topics and share them with the class. Use a bulletin board or scrapbook to display the reports, diagrams, and pictures.

(2) The factor tree may start with any two numbers whose product is 825. One pair is 25×33. Some other pairs are: 3×275, 11×75, and 15×55. You may wish to have groups of students complete factor trees for each of the above factor pairs. Have them observe that the same prime factorization results. Emphasize that, except for the order of the factors, there is only one prime factorization of a number.

ADDITIONAL EXAMPLES

1. Tell whether each number is prime or composite.
a. 51 **b.** 91 **c.** 101
(**a.** c **b.** c **c.** p)

2. Write the prime factorization of 540 using division. ($2^2 \cdot 3^3 \cdot 5$)

3. Write the prime factorization of 450 using a factor tree. ($2 \cdot 3^2 \cdot 5^2$)

Guided Practice

Class Exercises You may wish to have students work in pairs, then have volunteers show their work at the chalkboard.

Think and Discuss For question 3, explain that there is no largest prime number. Since there are an infinite number of numbers, there must be an infinite number of prime numbers.

A ***common error*** is to complete a factor tree with numbers that are not prime numbers. For 450, students may show the prime factorization as $2 \cdot 5^2 \cdot 9$. Remind them that the prime factorization of a number is the product of all its *prime factors*.

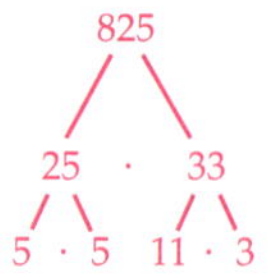

THINK What other ways can you write the prime factorization of 825 using a factor tree? (2)

1. 1 is not prime because it has only one factor, itself. 1 is not composite because it does not have more than 2 factors.
2. Zero is not prime because it has an infinite number of factors. Zero is not composite because it is less than 1.
3. No, because there is no largest number.

THINK AND DISCUSS See above.

1. Why is the number 1 neither prime nor composite?
2. Why is zero neither prime nor composite?
3. Do you think there is a largest prime number? Explain.

▼ You can write a composite number as a product of its prime factors, called the *prime factorization*.

Example 3 **Write the prime factorization of 60 using division.**

Solution Divide by prime numbers until the quotient is 1.

1. Is 60 divisible by 2? Yes. $60 \div 2 = 30$
2. Is 30 divisible by 2? Yes. $30 \div 2 = 15$
3. Is 15 divisible by 2? No.
4. Is 15 divisible by 3? Yes. $15 \div 3 = 5$
5. Is 5 divisible by 3? No.
6. Since $5 \div 5 = 1$, you are done. $5 \div 5 = 1$

$60 = 2 \cdot 2 \cdot 3 \cdot 5$, or $2^2 \cdot 3 \cdot 5$

Example 4 **Write the prime factorization of 825 using a factor tree.**

Solution

1. Write the composite number as the product of two factors.
2. Continue Step 1 with any remaining composite factors.
3. Stop when all factors are prime.
4. Write the prime factorization.

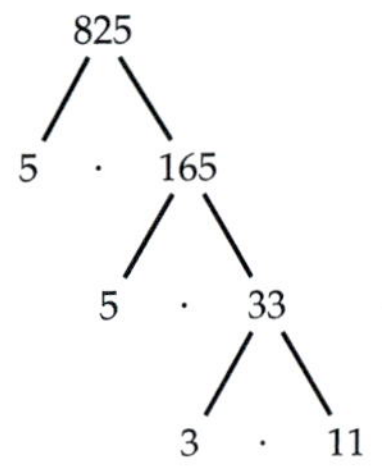

$5 \cdot 5 \cdot 3 \cdot 11 = 5^2 \cdot 3 \cdot 11$

CLASS EXERCISES

Tell whether each number is prime or composite.

1. 102 composite
2. 197 prime
3. 253 composite
4. 367 prime
5. 221 composite
6. 209 composite

Write the prime factorization using division. Check students' work.

7. 150 $2 \cdot 3 \cdot 5^2$
8. 280 $2^3 \cdot 5 \cdot 7$
9. 225 $3^2 \cdot 5^2$

Write the prime factorization using a factor tree. Check students' work.

10. 236 $2^2 \cdot 59$
11. 294 $2 \cdot 3 \cdot 7^2$
12. 275 $5^2 \cdot 11$

Find the number with the given prime factorization.

13. $2 \cdot 3^2 \cdot 5^2$ 450
14. $3 \cdot 5 \cdot 7^2$ 735

MATH MINUTES

Goldbach's Conjecture states that every even number greater than 2 can be written as the sum of two prime numbers. For example, $24 = 5 + 19$. Have students test Goldbach's Conjecture.

WRITTEN EXERCISES

Tell whether each number is prime or composite.

1. 45 composite **2.** 87 composite **3.** 97 prime

4. 109 prime **5.** 301 composite **6.** 1,001 composite

Write the prime factorization using division. Check students' work.

7. 425 $5^2 \cdot 17$ **8.** 240 $2^4 \cdot 3 \cdot 5$ **9.** 186 $2 \cdot 3 \cdot 31$

Write the prime factorization using a factor tree. Check students' work.

10. 650 $2 \cdot 5^2 \cdot 13$ **11.** 1,575 $3^2 \cdot 5^2 \cdot 7$ **12.** 1,617 $3 \cdot 7^2 \cdot 11$

CALCULATOR Find the number with the given prime factorization.

13. $2^5 \cdot 3 \cdot 11$ 1,056 **14.** $2 \cdot 5 \cdot 17^2$ 2,890 **15.** $2^5 \cdot 5 \cdot 7^3 \cdot 13^2$ 9,274,720

Use 5, 11, and 23 to find the prime factors of each number.

16. 115 $5 \cdot 23$ **17.** 621 $3^3 \cdot 23$ **18.** 3,105 $3^3 \cdot 5 \cdot 23$ **19.** 253 $11 \cdot 23$

Will each integer expression be even or odd?

20. $2ab^2$ even **21.** $2(a + b)^2$ even **22.** $2a^2b + 1$ odd

Solve.

23. The numbers 2, 3, and 7 are factors of x. Find four other factors of x. 6, 14, 21, 42

24. Find a number between 50 and 100 whose prime factorization has two factors. Answers may vary. One answer is 51.

25. Ms. Schwartz wrote a number on the chalkboard and said, "I know that to be sure this number is prime, I must check each prime divisor from 2 to 29." What is the least number Ms. Schwartz could have written? What is the greatest number she could have written? 841; 960

26. Kim and her grandmother have the same birthday and they have a family party together every year. Kim's age was a divisor of her grandmother's age for six birthdays in a row. What were their ages at each of those birthdays? 1, 61; 2, 62; 3, 63, 4, 64; 5, 65; 6, 66

27. ***DATA FILE 4 (pp. 138–139)***

a. Which country has the greatest percent of its population between the ages of 10 and 14? China

b. Which country has the greatest percent of its population between the ages of 0 and 4? Brazil

NOTES & QUOTES

The largest prime number yet calculated (on a CRAY supercomputer in Houston, Texas, in 1985) has 65,050 digits. It is mathematically written as $2^{216091} - 1$. The computer worked at a rate of 400 million calculations per second for 3 h to insure that this number was in fact prime.

MIXED REVIEW

1. List the factors of 8. 1, 2, 4, 8

2. List the first four multiples of 8. 0, 8, 16, 24

3. List the factors of 36. 1, 2, 3, 4, 6, 9, 12, 18, 36

Evaluate.

4. $54 \div (9 - 15)$ -9

5. $5(-4 + 7)^2$ 45

6. $36 \div [2 - (-1)]^2$ 4

7. ***DATA FILE 1 (pp. 2–3)*** Predict the equivalent wind chill temperature for air temperature -30°F and wind speed 10 mi/h. -57 to -59°F

PROBLEM SOLVING HINT

Use guess and test.

Closure

Ask students to define prime and composite numbers, and give three examples of each type of number.

Writing in Math Have students write the definitions and the examples in their math journals.

Independent Practice

If students are having difficulty solving question 26, give them a hint by telling them that Kim was 1 year old when they shared their first birthday. Since Kim's age is a factor of her grandmother's age, and the birthdays are consecutive, Kim's grandmother's age must be divisible by, 1 the first year, 2 the second year, 3 the third year, and so forth.

Lesson Follow-up

Reteaching Activity For students having difficulty with prime numbers, refer to the manipulative activity on p. 138E of the Chapter Overview. See also, TRB Exploring O-13.

LESSON RESOURCES

TRB Practice D-10
TRB Enrichment D-11

NOTES & QUOTES

Eratosthenes of Cyrene (c. 276–195 B.C.) was a Greek mathematician. He established a procedure for determining all prime numbers less than a given value. This procedure is called the *Sieve of Eratosthenes.*

RESEARCH Find out how Eratosthenes used his sieve to determine prime numbers. Use his method to determine all prime numbers less than 150.

28. Find two prime numbers whose product is 221. 13, 17

29. Twin primes are prime numbers whose difference is 2. For example, 11 and 13 are twin primes. 3, 5; 5, 7; 11, 13; 17, 19; 29, 31; 41, 43; 59, 61

a. List the first seven pairs of twin primes.

b. What do these numbers have in common? They are all odd.

c. Predict whether this will be true for all twin primes. yes

d. Use a computer or sieve (see below) to generate primes less than 100. Check students' work.

e. Test your prediction. Check students' work.

30. Find a pattern in the chart. Write an equation relating c and t.

chairs	5	10	15	c
tables	1	2	3	t

$c = 5t$; $\frac{c}{t} = 5$; or $t = \frac{c}{5}$

Critical Thinking

EXPLORING CLASSIFICATION

▼ You can use a *sieve* to find prime numbers. The first step is to list the numbers from 1 to 100 in rows of six on paper.

Mark out 1, since it's not prime. Circle 2, since it is prime. Mark out every multiple of 2.

1	2	3	4	5	6
7	8	9	10	11	12
91	92	93	94	95	96
97	98	99	100		

1. What pattern do you notice for the multiples of 2? They are all located in columns 2, 4, and 6.

2. Circle the first number after 2 that is unmarked. This is the next prime number. Mark out all of its multiples. What pattern do you notice for these multiples? They are all located in columns 3 and 6.

3. The next prime is 5. Circle it and mark out all of its multiples. Describe the pattern formed by the multiples of 5. The diagonals from right to left, starting with 5, 30, 60, and 90.

4. What is the next prime number? Mark out its multiples. 7

5. Eleven is prime. Circle it. Why have you already marked out all of the multiples of 11? The multiples of 11 are multiples of 2, 3, 5, or 7.

6. What do you notice about the rest of the unmarked numbers? Why is this true? all primes; All other numbers are crossed out because they are multiples of 2, 3, 5, and 7.

7. Find all the primes less than 200. When is it no longer necessary to mark out multiples? 2, 3, 5, 7, 11, 13, 17, 19, 23, 29, 31, 37, 41, 43, 47, 53, 59, 61, 67, 71, 73, 79, 83, 89, 97, 101, 103, 107, 109, 113, 127, 131, 137, 139, 149, 151, 157, 163, 167, 173, 179, 181, 191, 193, 197, 199; stop at 13 on the sieve.

LESSON QUIZ

1. Tell whether each number is prime or composite.
a. 47 (p) **b.** 169 (c)

2. Write the prime factorization of 315 using division. $(3^2 \cdot 5 \cdot 7)$

3. Write the prime factorization of 675 using a factor tree. $(3^3 \cdot 5^2)$

4. Find the number with the given prime factorization:
a. $3^2 \cdot 7 \cdot 13$ (819)
b. $2^3 \cdot 5 \cdot 17$ (680)

Assignment Guide

Basic 1-15 odd, 23-24, 27, MR, CT All
Average 2-20 even, 23-24, 26-28, MR, CT All
Enriched 12-30 even, 29, MR, CT All

FOR THE NEXT LESSON

math journal

Practice

For use after 4-6 (pp. 163–166)

Prime Factorization

Tell whether each number is prime or composite. Write p or c.

1. 19 p 2. 38 c 3. 57 c 4. 83 p
5. 171 c 6. 365 c 7. 137 p 8. 543 c

Write the prime factorization using division.

9. 75 $3 \cdot 5^2$ 10. 152 $2^3 \cdot 19$ 11. 143 $11 \cdot 13$
12. 432 $2^4 \cdot 3^3$ 13. 588 $2^2 \cdot 3 \cdot 7^2$ 14. 369 $3^2 \cdot 41$

Write the prime factorization using a factor tree.

15. 160 $2^5 \cdot 5$ 16. 108 $2^2 \cdot 3^3$ 17. 531 $3^2 \cdot 59$

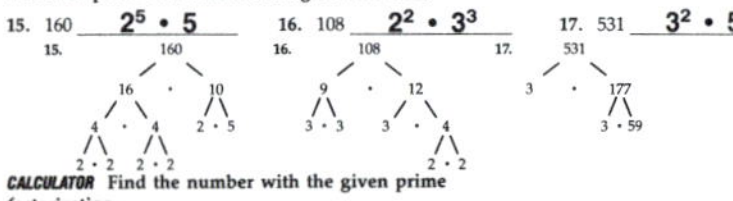

CALCULATOR Find the number with the given prime factorization.

18. $3^2 \cdot 5^3 \cdot 7$ 7,875 19. 11^3 1,331 20. $7^2 \cdot 13^2$ 8,281
21. $2^5 \cdot 3 \cdot 13$ 1,248 22. $13 \cdot 29 \cdot 43$ 16,211 23. $2^3 \cdot 7 \cdot 11 \cdot 17$ 10,472
24. $2^5 \cdot 3^4 \cdot 5^1$ 12,960 25. 7^4 2,401 26. $7 \cdot 11^2$ 847

ESTIMATION Use the numbers 7, 11, 17, and 23 to find the prime factors.

27. 187 $11 \cdot 17$ 28. 161 $7 \cdot 23$ 29. 539 $7^2 \cdot 11$
30. 391 $17 \cdot 23$ 31. 3,703 $7 \cdot 23^2$ 32. 2,401 7^4

Solve.

33. The numbers 3, 5, and 7 are factors of n. Find four other factors of n.
15, 21, 35, 105

34. Both a and b are odd whole numbers. Is $2ab - 1$ even or odd? Explain.
Odd; $2ab$ is even, so $2ab - 1$ is odd.

35. Both m and n are prime numbers. Is $m \times n$ prime? Explain.
No; $m \times n$ has factors m and n.

D-10 Chapter 4

Enrichment

For use after 4-6 (pp. 163–166)

Number of Factors

1. Find the factors of 23. 1, 23
2. Find the factors of 24. 1, 2, 3, 4, 6, 8, 12, 24

As the above examples show, two numbers may have greatly different numbers of factors, even though the numbers are nearly equal.

3. How many factors does 23 have? 2
4. How many factors does 24 have? 8

You can use prime factorizations to find out how many factors a number has.

Example How many factors does 40 have?

Solution
1. Write the prime factorization. $40 = 2^3 \cdot 5^1$
2. Take each exponent and add 1. $3 + 1 = 4$ $1 + 1 = 2$
3. Multiply the sums. $4 \times 2 = 8$

40 has 8 factors.

Find the number of factors.

5. 48 10 6. 36 9 7. 135 8 8. 224 12
9. 55 4 10. 63 6 11. 64 7 12. 1,000 16

Complete the table. The first one is done for you.

	Number	Prime Factorization	Number of Factors	Factors
	6	$2 \cdot 3$	4	1, 2, 3, 6
13.	44	$2^2 \cdot 11$	6	1, 2, 4, 11, 22, 44
14.	91	$7 \cdot 13$	4	1, 7, 13, 91
15.	125	5^3	4	1, 5, 25, 125
16.	54	$2 \cdot 3^3$	8	1, 2, 3, 6, 9, 18, 27, 54
17.	664	$2^3 \cdot 83$	8	1, 2, 4, 8, 83, 166, 332, 664
18.	369	$3^2 \cdot 41$	6	1, 3, 9, 41, 123, 369
19.	475	$5^2 \cdot 19$	6	1, 5, 19, 25, 95, 475
20.	222	$2 \cdot 3 \cdot 37$	8	1, 2, 3, 6, 37, 74, 111, 222

Chapter 4 D-11

OBJECTIVE:
To find the GCF or LCM by listing or using prime factorization.

4-7 GCF and LCM

▼ At a school fund raiser, the math class raised $150 and the science class raised $120. Each class will divide the money it raised equally, in whole dollar amounts, and give to the same number of charities. What method could you use to find the possible number of charities? the greatest possible number of charities?

List the factors of 150 and 120. Find the factors that are the same.

150: **1, 2, 3, 5, 6, 10, 15,** 25, **30,** 50, 75, 150

120: **1, 2, 3,** 4, **5, 6,** 8, **10,** 12, **15,** 20, 24, **30,** 40, 60, 120

The *common factors* are 1, 2, 3, 5, 6, 10, 15, and 30. The *greatest common factor (GCF)* is 30. The classes can distribute whole dollar amounts of money to 1, 2, 3, 5, 6, 10, 15, or, at most, 30 charities.

Common Factor	The factors that are the same for a given set of numbers are their common factors.

Greatest Common Factor (GCF)	The greatest common factor of a set of numbers is the greatest number that is a factor of the given numbers.

▼ Another way to find the GCF is to use prime factorization.

Example 1 **Find the GCF of 40 and 140.**

Solution

1. Write the prime factorization for each number. $40 = 2 \cdot 2 \cdot 2 \cdot 5$, $140 = 2 \cdot 2 \cdot 5 \cdot 7$
2. Circle each pair of common factors.
3. Multiply common factors. $2 \cdot 2 \cdot 5$

The GCF is 20.

▼ When the GCF of two numbers is 1, the numbers are *relatively prime.*

Example 2 **Find the GCF of 28 and 33.**

Solution

1. Write the prime factorization. $28 = 2 \cdot 2 \cdot 7$
2. Circle common factors. $33 = 3 \cdot 11$

There are no common prime factors. The GCF is 1.

THINK Do relatively prime numbers have any common factors besides 1? **No, by definition of relatively prime numbers.**

1

Lesson Focus

Materials

Math journal

Vocabulary/Symbols

algebraic factorization
common factor
common multiple
greatest common factor (GCF)
least common multiple (LCM)
relatively prime

MOTIVATING THE LESSON

Tell students that you have a certain number of baskets to fill with 12 oranges and 18 apples. You would like the baskets to have the same number of oranges and the same number of apples. How many different ways can the baskets be filled and how many baskets could be filled? Have students list the factors for 12 and 18. (1, 2, 3, 4, 6, 12; 1, 2, 3, 6, 9, 18) Then ask which factors are common to both 12 and 18. (1, 2, 3, 6). Finally, ask students which is the greatest common factor. (6) Students should be able to see that they could fill 1, 2, 3, or 6 baskets with fruit.

Instruction

1 Comment that if the GCF is 1, it must also be the only common factor.

Teaching Tip

Students often cannot see the relevancy of certain math concepts and skills. In some cases, it helps to tell the class how they will use the skill in future lessons. In other cases, students may appreciate the historical background and development for the concept.

2 Help students understand that to find the LCM using the prime factorization method, they must choose the set of factors that appears the greatest number of times in either factorization.

ADDITIONAL EXAMPLES

Find the GCF and the LCM of each set of numbers or expressions.

1. 45, 60 (15, 180)
2. 12, 18, 24 (6, 72)
3. $9a^2b$, $15ab^4$ ($3ab$, $45a^2b^4$)
4. 15, 16 (1, 240)
5. $8x$, $21y^2$ (1, $168xy^2$)
6. Which set of numbers shown above are relatively prime? (15 and 16)

Solve.

7. A group of friends had a collection of 133 baseball cards and 247 football cards, which they were able to share equally. How many friends were there? (19 friends)

Guided Practice

Class Exercises For question 10, draw a table on the chalkboard to help students keep track of all the values that they found and to see the pattern that develops.

Think and Discuss For question 2, help students realize that this is always true.

In question 3, have students find the LCM of several pairs of prime numbers. Help them understand that the LCM of two different primes is always the product of the primes.

THINK How many common multiples do any two numbers have? Any two numbers have an infinite number of common multiples.

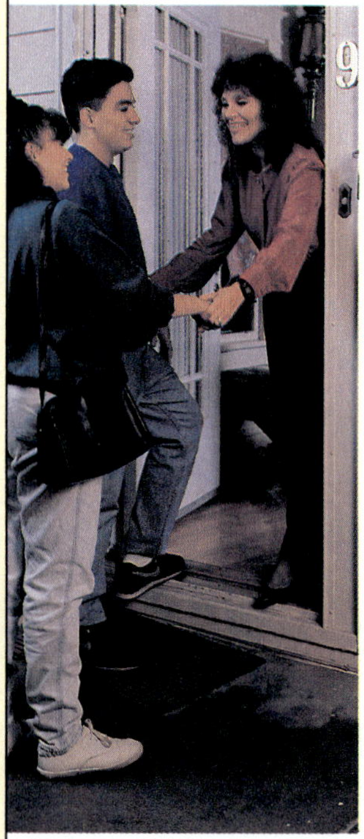

▼ You can find the GCF of variable expressions by writing each expression as the product of its factors. This is called *algebraic factorization*.

Example 3 **Find the GCF of $12a^3b$ and $15a^2b^2$.**

Solution

1. Write the algebraic factorizations.
2. Circle the common factors.
3. Multiply.

$12a^3b = 2 \cdot 2 \cdot 3 \cdot a \cdot a \cdot a \cdot b$
$15a^2b^2 = 3 \cdot 5 \cdot a \cdot a \cdot b \cdot b$

$3 \cdot a \cdot a \cdot b$

The GCF is $3a^2b$.

▼ Some problems require that you work with *common multiples* and the *least common multiple (LCM)*.

Common Multiples	The multiples that are the same for a given set of whole numbers are the common multiples.
Least Common Multiple	The least common multiple is the least number that is a common multiple of two or more given numbers.

Example 4 Aisha and Tom visited their aunt today. Aisha visits every 4 days; Tom visits every 3 days. When will Aisha and Tom visit their aunt again on the same day?

Solution

1. List the multiples of each number.
2. Circle common multiples.

4: 4, 8, 12, 16, 20, 24 . . .
3: 3, 6, 9, 12, 15, 18, 21, 24 . . .

3. Write the least common multiple. LCM = 12

Aisha and Tom will visit their aunt together in 12 days.

▼ You can find the LCM using prime factorization.

2

Example 5 **Find the LCM of 72 and 60.**

Solution

1. Write the prime factorizations.
2. Circle each factor the greatest number of times it appears.
3. Multiply.

$72 = 2 \cdot 2 \cdot 2 \cdot 3 \cdot 3$
$60 = 2 \cdot 2 \cdot 3 \cdot 5$

$2 \cdot 2 \cdot 2 \cdot 3 \cdot 3 \cdot 5$

The LCM is 360.

MATH MINUTES

Working in pairs, have one student write a number, then have the other student write two numbers that have the given number as the GCF. For example, 7 (21 and 35). Have students alternate roles.

You can find the LCM of variable expressions using algebraic factorization.

Example 6 Find the LCM of $4a^2$ and $6ab$.

Solution
1. Write the algebraic factorizations. $4a^2 = 2 \cdot 2 \cdot a \cdot a$; $6ab = 2 \cdot 3 \cdot a \cdot b$
2. Circle each factor the greatest number of times it appears.
3. Multiply. $2 \cdot 2 \cdot 3 \cdot a \cdot a \cdot b$

The LCM is $12a^2b$.

CLASS EXERCISES

Find the GCF of each set of numbers or expressions.

1. 10, 45 5
2. 6, 8, 12 2
3. $12r^3$, $8r$ $4r$

Find the LCM of each set of numbers or expressions.

4. 10, 45 90
5. 6, 8, 12 24
6. $12r^3$, $8r$ $24r^3$

Is the pair of numbers relatively prime? Write *yes* or *no*.

7. 51, 17 no
8. 9, 10 yes
9. 13, 23 yes

10. Find the GCF and the LCM for each pair of numbers. Find the product of the GCF and the LCM. Find the product of the two original numbers. What do you notice? Will this always be true? Is it true that the LCM of two numbers is a multiple of their GCF? See right.
    a. 6, 8 2; 24; 48; 48
    b. 15, 18 3; 90; 270; 270
    c. 20, 30 10; 60; 600; 600
    d. 20, 25 5; 100; 500; 500
    e. 12, 30 6; 60; 360; 360
    f. 84, 120 12; 840; 10,080; 10,080
11. Name four numbers that have both 6 and 10 as factors. What is the least number that has 6 and 10 as factors? 30, 60, 90, 120; 30
12. *True* or *false*? Explain.
    a. If a and b are positive integers and $a|b$, then the LCM of a and b is b.
    b. The set of all common factors of two given positive integers is finite. True, all integers have a finite number of factors.

WRITTEN EXERCISES

Find the GCF of each set of numbers or expressions.

1. 14, 21 7
2. 54, 144 18
3. 52, 65 13

THINK AND DISCUSS

1. Can any two numbers have a common multiple? Explain. See below.
2. Can two consecutive positive integers be relatively prime?
3. Is the LCM of two different prime numbers equal to their product? yes

1. Yes, multiply the two numbers. Their product is the common multiple.
2. yes, 9 and 10, for example

10. The product of the GCF and LCM equals the product of the two original numbers; yes; yes, the LCM is the product of the prime factors that appear the greatest number of times in both numbers. The GCF is the product of the common factors.
12. a. True, if b is divisible by a, then b is a multiple of a. b is the LCM of a and b.

A ***common error*** is to confuse the meaning of GCF and LCM. For example, to find the GCF of 40 and 140 ($2 \cdot 2 \cdot 2 \cdot 5$; $2 \cdot 2 \cdot 5 \cdot 7$) they will find the product $2 \cdot 2 \cdot 2 \cdot 5$ (40) instead of $2 \cdot 2 \cdot 5$ (20). To help students avoid this error, point out that the "F" in GCF refers to factors, and the "M" in LCM refers to multiples.

Closure

Writing in Math Ask students to explain in their math journals how to find the GCF and LCM for a given pair of numbers. Then have students compare and discuss their methods to be sure they are equivalent and then to make any necessary corrections.

Independent Practice

For question 24, remind students that the solution is one more than the LCM of 2, 3, 4, and 5.

Lesson Follow-up

Reteaching Activity Provide pairs of numbers. Have students find all the factors, circle the common factors, and identify the GCF.

LESSON RESOURCES

TRB Practice D-12
TRB Enrichment D-13

Practice

For use after 4-7 (pp. 167–170)

GCF and LCM

Find the GCF of each set.

1. 8, 12 **4**
2. 36, 54 **18**
3. 63, 81 **9**
4. 69, 92 **23**
5. 15, 28 **1**
6. 21, 35 **7**
7. $30m$, $36n$ **6**
8. $75x^3y^2$, $100xy$ **$25xy$**
9. 15, 24, 30 **3**
10. 48, 80, 128 **16**
11. $36hk^3$, $60k^2m$, $84k^4n$ **$12k^2$**
12. $2mn$, $4m^2n^2$ **$2mn$**

Find the LCM of each set.

13. 6, 8 **24**
14. 10, 12 **60**
15. 13, 15 **195**
16. 7, 21 **21**
17. 24, 32 **96**
18. 15, 50 **150**
19. $9a^3b$, $18abc$ **$18a^3bc$**
20. $28xy^2$, $42x^2y$ **$84x^2y^2$**
21. 3, 4, 5 **60**
22. 9, 12, 16 **144**
23. $6m^3$, $14mp^4$, $21mp^2$ **$42m^3p^4$**
24. $10x^2y$, $100x^5y^7z$, $5z$ **$100x^5y^7z$**
25. $2mn$, $26m^7n^3$, $8mn^{10}$ **$104m^7n^{10}$**

CALCULATOR Find the GCF and LCM for each set of numbers.

26. 192, 288 GCF **96** LCM **576**
27. 133, 551 GCF **19** LCM **3,857**
28. 138, 368, 828 GCF **46** LCM **3,312**

Solve.

29. A quality control inspector in an egg factory checks every forty-eighth egg for cracks and every fifty-fourth egg for weight. What is the number of the first egg each day that the inspector checks for both qualities?
432
30. The numbers m and n are both prime numbers.
   a. What is the GCF of m and n? **1**
   b. What is the LCM of m and n? **mn**
31. The LCM of 24 and x is 888. What is the LCM of 8, 24, and x?
888
32. The GCF of a and b is 15. What is the GCF of a, b, and 5?
5

D-12 Chapter 4

Enrichment

For use after 4-7 (pp. 167–170)

Euclid's Algorithm

More than 2,000 years ago, the mathemadician Euclid of Alexandria invented a simple method for finding the GCF of two numbers. The method is called *Euclid's algorithm*.

Example Find the GCF of 187 and 407.

Solution
1. Divide the greater number by the lesser. Write the remainder as a whole number. 187)407 = 2 R 33
2. Divide the previous divisor by the previous remainder. 33)187 = 5 R 22
3. Repeat Step 2 until finding a remainder of 0. 22)33 = 1 R 11
4. The last divisor is the GCF of the original numbers. 11)22 = 2 R 0

The GCF of 187 and 407 is 11.

Find the GCF using Euclid's algorithm.

1. 111 and 629 **37**
2. 272 and 935 **17**
3. 141 and 893 **47**
4. 208 and 464 **16**
5. 207 and 529 **23**
6. 527 and 1,023 **31**
7. 266 and 437 **19**
8. 490 and 1,183 **7**

Solve using Euclid's algorithm.

9. Two pieces of lumber measure 817 cm and 989 cm. Each is to be cut into equal-length boards. If all boards cut from both pieces must be the same length, what is the longest that each can be?
43 cm
10. Two groups of trees are each to be planted in a rectangular array of rows and columns like the one shown at the right. One group includes 767 trees. The other includes 1,121 trees. Both arrays must have the same number of rows. What is the maximum number of rows the arrays can have?
59 rows

Chapter 4 D-13

4. 18, 30 **6**
5. $27x^2y^4$, $46x^2yz$ **x^2y**
6. 8, 15, 20 **1**
7. $180a^2$, $210ab$ **$30a$**
8. $6a^3b$, $8ab^2$ **$2ab$**
9. 12, 15, 18 **3**

Find the LCM of each set of numbers or expressions.

10. 10, 55 **110**
11. 12, 20 **60**
12. 54, 36 **108**
13. 180, 210 **1,260**
14. $8x$, $25y$ **$200xy$**
15. $6a^3b$, $8ab^2$ **$24a^3b^2$**
16. $6cd^3$, $8c$, $12d^2$ **$24cd^3$**
17. 12, 15, 18 **180**
18. 14, 18, 21 **126**

CALCULATOR Find the GCF and LCM for each set of numbers.

19. 32, 12 **4; 96**
20. 119, 391 **17; 2,737**
21. 135; 280; 300 **5; 37,800**

22. The GCF of 30 and x is 6. Could x be each of the following?
   a. 15 **no**
   b. 24 **yes**
   c. 60 **no**
   d. 84 **yes**

*23. The LCM of 8 and x is a. Explain why a is divisible by 2^3. **If a is the LCM of 8 and x, then a is divisible by both 8 and x. Since $8 = 2^3$, a is divisible by 2^3.**

Solve.

24. When Jim sorts his stamps into piles of 2, 3, 4, or 5, there is always one stamp left over. What is the smallest number of stamps Jim can have? **61 stamps**
25. Two neon signs turn on at the same time. One blinks on every 10 s, the other blinks on every 6 s. How many times per minute do the signs blink on together? **2 times per minute**
26. The numbers of students attending a conference from three schools are 42, 48, and 60. The students will form discussion groups with an equal number of students from each school in each group. What is the greatest number of discussion groups that can be formed? **6 discussion groups**
27. There are two sizes of tables in a banquet hall. One size seats exactly 5 people and the other size seats exactly 8 people. Last night, 66 people were seated at fewer than 10 tables with no empty seats. How many tables of each size were there? **2 tables that seated 5 people; 7 tables that seated 8 people.**
28. A band of pirates shared 187 pieces of silver and 136 gold coins. Each pirate received a fair share. How many pirates were in the band? **17**
29. A farmer has three pieces of timber with lengths of 63 ft, 84 ft, and 105 ft. What is the length of the longest logs of equal length the farmer can cut from the timber? **21 ft**

DATA FILE 10 (pp. 404–405) Solve.

30. About how many acres should 22 grizzly bears have for a home range? **20,652,060 acres**
31. About how many striped skunks can live within 15,000 acres? **59**

MIXED REVIEW

Write the prime factorization using a factor tree. **Check students' work.**

1. 12 **$2^2 \cdot 3$**
2. 123 **$3 \cdot 41$**
3. What are all the factors of 12? **1, 2, 3, 4, 6, 12**
4. State the first four multiples of 12. **0, 12, 24, 36**

Solve.

5. $3.5n = 14$ **4**
6. $-7.3 = p + 4.1$ **-11.4**
7. $\frac{n}{3.2} = -5$ **-16**

Use guess and test.

8. Find two numbers that have a sum of 11 and product of 24. **8, 3**

LESSON QUIZ

Find the GCF and LCM.
1. 14, 35 **2.** $3a^4$, $4b$
Solve.
3. Carol visits the zoo every six days, and Mike visits the zoo every 8 days. If they both visit the zoo on August 1, what is the earliest date on which they will both visit again?

QUIZ ANSWERS

1. (7; 70)
2. (1; $12a^4b$)
3. (August 25)

Assignment Guide

Basic 1-4, 10-12, 19, 22, 24-27, 30-31, MR All
Average 4-7, 13-28, 30-31, MR All
Enriched 5-9, 13-23, 26-31, MR All

FOR THE NEXT LESSON

graph paper, math journal

OBJECTIVE:
To explore a problem using a systematic approach.

MATERIALS

- Graph paper
- Math journal to record work

Exploring Counting Problems

■ Counting figures can be confusing.

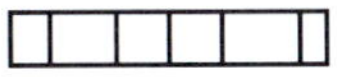

1. Count the rectangles. Are there more than six? Did you count the large rectangle bordering the figure? **Yes**
2. ❶ Can a rectangle be made up of smaller rectangles? Can the rectangles overlap? How would this affect your total count? **Yes; yes; this would increase the total count.**
3. How can you count the rectangles *systematically,* so you are sure you counted them all? ***Discuss*** with a partner. **Answer may vary. Possible answer: make a table.**

■ One way to count the rectangles is to trace around the different rectangles, starting at the left and working across.

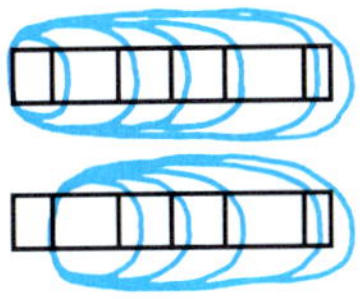

4. Start at the left vertical segment and count all the rectangles that use that segment as a left side. How many are there? **6**
5. Continue with the next vertical segment.
6. Continue counting. When do you know you are done? **when you only have the last rectangle on the right to count**

■ Another systematic way to count the rectangles is to account for the different combinations of smaller rectangles. To keep track, label the smaller rectangles. Refer to the figure at the left.

Number of Small Rectangles	Labeled Rectangles
1	A, B, C, D, E, F, G, H
2	AB, BC, CD, DE, EF, FG, GH

7. How many rectangles include only one small rectangle? two small rectangles? Continue until you have counted all the combinations of small rectangles. Make a table to organize your data. **8, 7**
8. What is the total number of rectangles? How can you be sure you counted them all? **36; count until there is 1 rectangle left.**
9. ❷ ***Describe*** the pattern in the table. Did you see this pattern when you used the tracing method? **The number of rectangles decreases by 1 each time.**
10. Use the pattern to find the number of different rectangles in a figure with 5 regions; 15 regions. **15; 120**
11. ***Compare*** the two methods for counting. How are they the same? How are they different? **Answers may vary. Possible answer: Both methods use a pattern to help systematically count the rectangles. In the first, you count overlapping rectangles. In the second, you count combinations.**

■ Now look at the squares in the figure at the left.

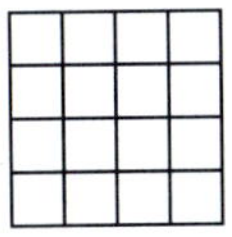

12. ***Explore*** a method of your own or use one of the counting methods above. Which method would you use? Why? **Answers may vary. Possible answer: use different colors.**
13. What is the total number of squares? **30**

Lesson Focus

Materials

Graph paper, math journal

MOTIVATING THE LESSON

Ask students if there is one best method of solving counting problems. Explain that they may find a method that works best for them, as they work through the lesson.

Instruction

❶ Help students observe that there are rectangles that are formed by 2, 3, 4, 5 and 6 smaller rectangles taken together.

❷ It is helpful to solve the problem in parts; that is, to find how many rectangles there are when the smaller rectangles are taken one at a time. (6) Or when they are taken two at a time (5), and so forth. Help students recognize the pattern as a set of triangular numbers. This will help students predict the number of rectangles with 5 and 15 regions.

Closure

Writing in Math Have students make a math journal entry describing the strategy they used to solve the problem and the number pattern they discovered.

Assignment Guide

Basic All
Average All
Enriched All

FOR THE NEXT LESSON

math journal

Teaching Tip

Encourage students to be systematic about reading, writing, and solving problems. Help students understand that being systematic is an important skill to develop for other subject areas.

Lesson Focus

Materials
Math journal

MOTIVATING THE LESSON

Present the following to students: **If there are eight teams in a (single elimination) basketball tournament, how many games must the winning team win to win the tournament?** Mention that one strategy that can be used to solve this problem is to make a diagram. Have students name the eight teams and make a tournament diagram as follows:

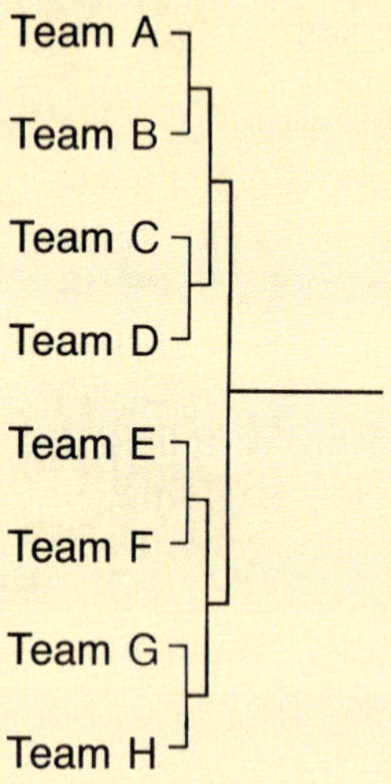

After discussing the diagram, and the possible outcomes, ask students to determine the number of games the winning team would have won to win the tournament. (3) Tell students that today they will solve problems using the strategy: *account for all possibilities* by making an organized list, table or diagram.

OBJECTIVE:
To solve problems by accounting for all possibilities.

4-8 Account for All Possibilities

To account systematically for all possibilities, you can make an organized list, a table, or a diagram.

PROBLEM

Mandy invited Rachel, Sue, Jenny, Pam, Erica, and Latosha for lunch. They decided to take pictures of all seven girls with two in each picture. How many pictures did they take?

SOLUTION

READ Answer these questions to understand the given information.

What do you want to find?	the total number of pictures
How many girls are at the party?	seven
How many girls are in each picture?	two

PLAN One way to account for all the possibilities is to make an organized list of all the pairs of girls.

SOLVE Begin by pairing Mandy with her six friends. Next, pair Latosha with each of the five friends. Since Latosha and Mandy have already been paired, you don't need to count them again.

1

Mandy—Latosha		
Mandy—Pam	Latosha—Pam	
Mandy—Rachel	Latosha—Rachel	Pam—
Mandy—Sue	Latosha—Sue	
Mandy—Jenny	Latosha—Jenny	
Mandy—Erica	Latosha—Erica	

Continue the list to find the total number of pairings. Altogether, there are 21 pairs of girls, so there will be 21 pictures.

LOOK BACK Another way to account systematically for all the possibilities is to use a diagram. Draw line segments to connect all the pairs of girls. If you count the segments as you draw them, you will count 21.

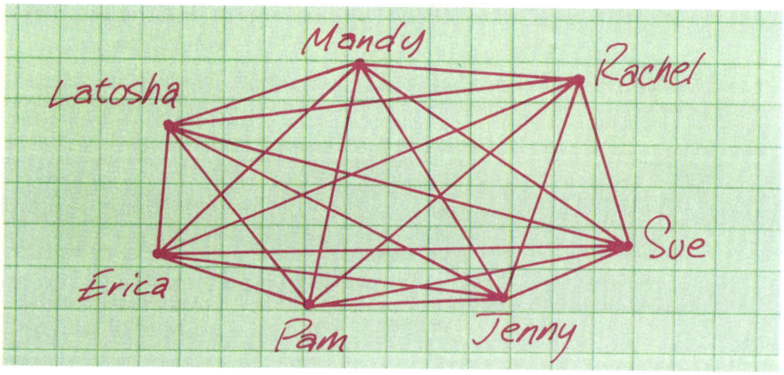

Teaching Tip

The teams for the basketball tournament diagram may be listed vertically on the chalkboard before class, or on a transparency for the overhead projector.

CLASS EXERCISES

Refer to the problem on page 172.

1. Complete the list of paired girls. Count how many pairs are in each group. What pattern do you see? **Each time you start with a new girl, the number decreases by one: 6, 5, 4, . . .**
2. Suppose there were a total of ten girls at Mandy's party. Using the pattern you found, determine how many pictures there would be if there were two girls in each picture. **45 pictures**
3. How many pictures would there be if there were a total of 20 girls at Mandy's party, with two girls in each picture? Use the pattern. **190 pictures**

Solve.

4. André has a job making yogurt sundaes. The yogurt flavors are chocolate and vanilla. The toppings are granola, raisin, cherry, and coconut. How many *different* sundaes consisting of one scoop of yogurt and one topping can André make? **8 sundaes**
5. This year there are seven softball teams in the play-offs. Each team competes against each of the other teams twice. What is the total number of games played? **42 games**

WRITTEN EXERCISES

Use a CALCULATOR where appropriate.

Solve by accounting systematically for all possibilities.

1. Three darts are thrown at the target shown at the right. If each dart lands on the dart board, how many different point totals are possible? **9 (10-1-1 and 4-4-4 are both 12.)**
2. You have four 25-cent stamps and three 30-cent stamps. How many different amounts of postage can you have? **19**
3. Each of the small boxes in the figure at the right is a square. What is the total number of different squares shown in the figure? **17**

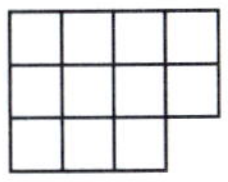

4. You have a collection of coins consisting of one penny, one nickel, one dime, and one quarter. How many different amounts of money can be made using one or more of these coins? **15**

Solve. Use any strategy you wish.

5. Fred has a coordinated wardrobe consisting of three pairs of pants, four shirts, and two sweaters. How many different three-piece outfits can he make? **24**

Instruction

1 It may be helpful for students to visualize the action in the problem. Ask seven volunteers to come to the front of the room and act out the "picture" process. Ask students to record the names of each pair of students in their math journals.

Guided Practice

Class Exercises As students extend the pattern of triangular numbers, they might be able to recognize that they can predict the number of pictures for any number of girls by using a formula. If *n* stands for the number of girls at the party, then $\frac{n(n-1)}{2}$ will yield the number of pictures taken.

A ***common error*** is to double count the pairs. Point out that the order in which elements in a set are arranged does not change the set of elements.

Closure

Writing in Math Have students summarize the solutions to the problem in their math journals. Have them explain the pattern of triangular numbers that enabled them to predict the number of pictures taken for 10 and 20 girls.

Independent Practice

Students may recognize that the solution to question 5 is the product of the number of items: 3(pants) × 4(shirts) × 2(sweaters) = 24 three-piece outfits.

MATH MINUTES

Have students extend the basketball problem to include 16 teams. Ask how many games must the winner of the tournament win. (4)

For question 10 help students see the following: if the square root of the sum of odd numbers is 10, then the sum of the numbers must be 100. Since the sum of the first 10 odd numbers is 100, the 10th number is 19, so, $n = 19$.

Lesson Follow-up

LESSON RESOURCES

TRB Practice D-14
TRB Enrichment D-15

Practice

For use after 4-8 (pp. 172–174)

Account for All Possibilities

Solve by accounting systematically for all possibilities.

1. A baseball team has 4 pitchers and 3 catchers. How many different pitcher-catcher combinations are possible? **12**
2. The baseball team has 2 first basemen, 3 second basemen, and 2 third basemen. How many combinations of the three positions are possible? **12**
3. A quarter is tossed 3 times. In how many different orders can heads and tails be tossed? **8**
4. A quarter is tossed 4 times. In how many different orders can heads and tails be tossed? **16**
5. Curtains are manufactured in 3 different styles and 5 different colors.
   a. How many different style-color combinations are possible? **15**
   b. The curtains are produced in 2 different fabrics. How many different style-color-fabric combinations are possible? **30**
6. A vending machine accepts any combination of nickels, dimes, and quarters equaling 50¢. How many combinations of coins are possible? **10**
7. Joe remembered that the three digits in his locker combination were 3, 5, and 7, but he forgot the order of the numbers. What is the maximum number of combinations he must check in order to open his lock? **6**
8. The first digit of a 3-digit telephone area code is 6 or 7. The second digit is 9. How many codes are possible? **20**

Enrichment

For use after 4-8 (pp. 172–174)

Predicting Coin Tosses

A coin is tossed 2 times. The organized list shows that 4 orders of heads (H) and tails (T) are possible.

1st toss	2nd toss
H	H
H	T
T	H
T	T

1. In how many of the 4 possible orders will the following occur?
   a. 0 heads **1** b. exactly 1 head **2** c. 2 heads **1**

The organized list shows that eight orders of heads and tails are possible if a coin is tossed three times.

1st	2nd	3rd
H	H	H
H	H	T
H	T	H
H	T	T
T	H	H
T	H	T
T	T	H
T	T	T

2. In how many of the eight possible orders will the following occur?
   a. 0 heads **1** b. exactly 1 head **3**
   c. exactly 2 heads **3** d. exactly 3 heads **1**
3. The above results show that there are four possible outcomes when a coin is tossed twice and eight outcomes when it is tossed three times.
   a. Complete: $4 = 2^{\boxed{2}}$ $8 = 2^{\boxed{3}}$
   b. State a rule involving powers of 2 that gives the number of possible outcomes when a coin is tossed n times.
   number of outcomes for n tosses = 2^n

Suppose a coin is tossed four times.

4. How many outcomes are possible? **16**
5. The pattern of numbers shown here is called Pascal's triangle. Explain how the numbers in each row are calculated.
   Each entry is the sum of the two entries above.

1
1 1
1 2 1
1 3 3 1
1 4 6 4 1
1 5 10 10 5 1
1 6 15 20 15 6 1

6. Compare your answers to Exercises 1 and 2 with rows 3 and 4 of Pascal's triangle. What do you notice?
   They are the same.

10. 19; added $1 + 3 + 5 + \ldots$ until the sum was 100 because the square root of 100 is equal to 10.

6. You have mushrooms, onions, green peppers, and olives. How many different pizzas can you make by adding any combinations of the ingredients to a plain pizza? **16 (including a plain pizza)**
7. Without evaluating, what is the final digit of 8^{66}? How did you find the answer? **4; the last digits of the powers of 8 form this pattern: 8, 4, 2, 6**
8. A collector sorted stamps into 2 piles and had 1 stamp left over. When she sorted them into piles of 3, there were 2 left over. For piles of 4, there were 3 left over. What is the least number of stamps the collector could have? **11**
9. The Valley High chorus has 28 members. The band has 52 members. A total of 17 students are in both the band and the chorus. The music teacher wants to distribute the fewest possible tickets to the spring concert. How many tickets must the teacher distribute for each student in band or chorus to get at least one ticket? **63**
10. This is a list of consecutive odd integers from 1 to n: $\{1, 3, 5, \ldots n\}$. The square root of the sum of the numbers is 10. What is the value of n? How did you find the answer?
11. Make a list of the two-digit square numbers. Find the difference between the digits of each number. What do you discover? **They are all prime numbers.**
12. A runner averages 10 km/h. The runner takes a 10 min rest every 5 km. How long will it take to complete a course of 25 km? **190 min**

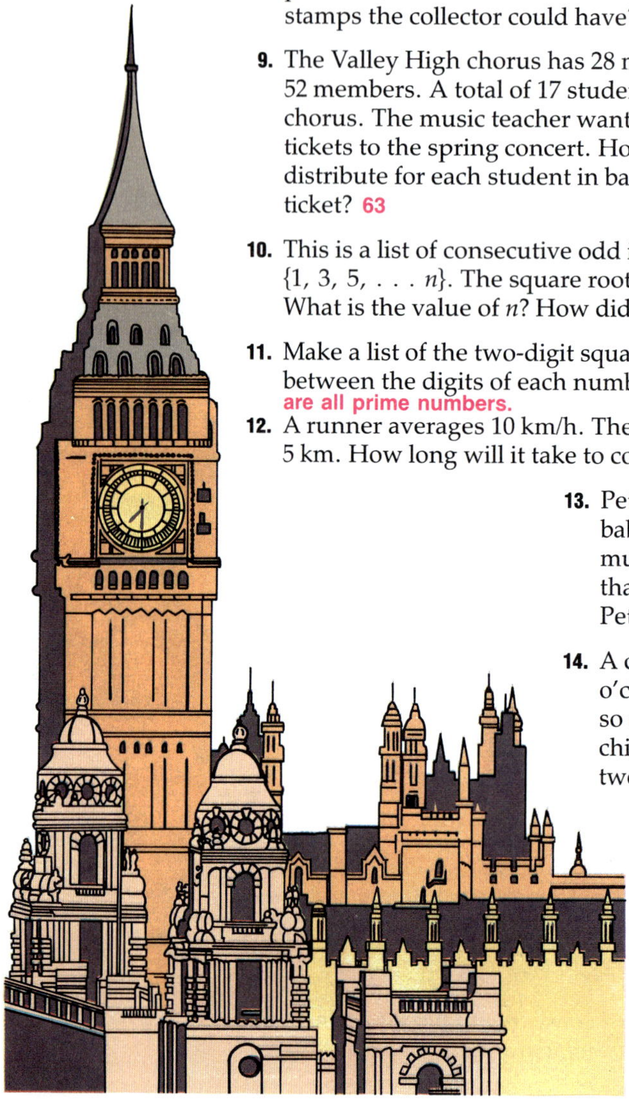

13. Pete and Jack earned $12.50 babysitting. Edward earned half as much as Jack. Pete earned $2.50 less than Jack. How much money does Pete have? **$5.00**
14. A chime clock strikes once at one o'clock, twice at two o'clock, and so on. What is the total number of chimes the clock strikes in a twelve-hour period? **78**

Assignment Guide

Basic All
Average All
Enriched All

FOR THE NEXT LESSON

fraction bars, index cards, overhead projector (optional), math journal

Problem Solving Practice

PROBLEM SOLVING STRATEGIES

Look for a Pattern
Guess and Test
Simplify the Problem
Make an Organized List
Work Backwards
Account for All Possibilities
Make a Table
Write an Equation
Solve by Graphing
Draw a Diagram
Make a Model
Solve Another Way
Simulate the Problem

Solve. The list at the left shows some strategies you can use.

1. A car dealer recommends an oil change every 3,000 mi and a tire rotation every 7,000 mi. When will the oil be changed and the tires rotated at the same time? **every 21,000 mi**

2. Thomas was training for a race for six weeks. Every week he ran one more mile than he ran the week before. Thomas ran a total of 51 miles while in training. How many miles did he run each week? **6 mi, 7 mi, 8 mi, 9 mi, 10 mi, 11 mi**

3. Bonnie had 16 coins in her pocket totaling $1.50. What are two combinations of coins she could have had in her pocket? **Answers will vary. Ex: 14d, 2n; 2q, 6d, 8n**

4. At 5 P.M., the temperature was 65°F. At 5:30 P.M., the temperature was 62°F. At 6 P.M., it was 59°F, at 6:30 P.M., it was 56°F. If the pattern continued, at what time would the temperature go below freezing? **after 10:30 P.M.**

5. How many different pizzas can be made if meatballs, anchovies, green peppers, or olives can be added to a plain pizza? **16 (including a plain pizza)**

6. The florist orders carnations, roses, and tulips. Carnations can be bought only in bunches of sixteen. Roses come in bunches of four, and tulips in bunches of eight. The florist wants the same number of each flower. What is the least number of flowers the florist can order? **48 flowers**

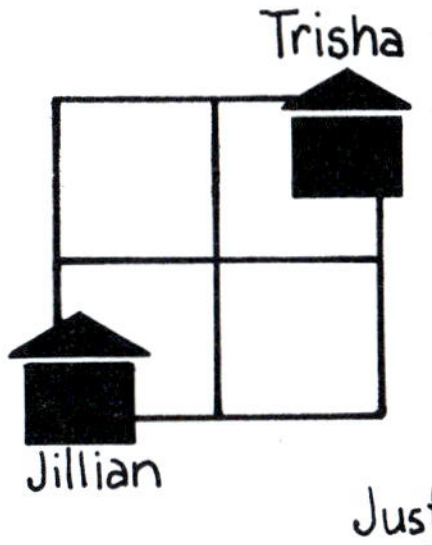

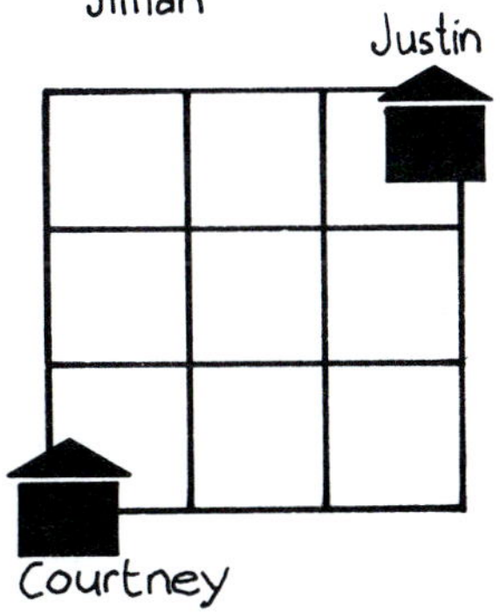

7. Copy the diagrams at the left. **Check students' drawings.**
   a. Using the paths shown, Jillian can walk directly to Trisha's house six different ways. Each route is four blocks long. Draw each route.
   b. There are 20 different routes Courtney can use to walk directly to Justin's house. Each route is six blocks long. How many can you find?

8. In a swim meet, Helen places 7.5 m behind Laney. Laney is 16.5 m ahead of Kay. Kay places 3.75 m behind Julia. How far is Julia behind Laney? **12.75 m**

9. ***DATA FILE 4 (pp. 138–139)***
   a. In what year was the world population five times the population in 1600? **1950**
   b. In what span of years was the world population one sixth of the population in 1975? **1700–1749**

10. Five pears weigh the same as 3 apples and 2 strawberries. An apple weighs the same as 21 strawberries. How many strawberries equal the weight of a pear? **13 strawberries**

Instruction

This page provides a variety of problems that can be used to reinforce and enhance the students' problem solving skills. Encourage students to read each problem carefully. Then have them refer to the list of problem solving strategies to help them decide how to solve the problem.

Point out, however, that not all questions require a strategy for solving, nor are all the strategies in the list used in this lesson.

Study Skills

Good *note-taking* is one technique that helps students develop good review and test-taking skills. The goal of good note-taking is to condense information in an easy-to-review format. Outlining, drawing diagrams, and summarizing information are several methods that can be used.

Outlining has a distinct form. It begins with a title, then a main idea, followed by a series of subtopics. Each level of subtopics is preceded by Roman numerals, capital letters, numbers, and small letters respectively. Outlining aids the students in organizing data in a logical order.

Drawing a diagram provides a pictorial summary for students who think visually. The main idea may be written at the center of the page. Additional data radiates out in any form convenient to each student.

Summarizing information involves writing a short paragraph containing only the main ideas or facts. Then students may be encouraged to develop symbols and phrases that condense the text even more. Notes can be arranged on a page so that the keywords stand out and aid in focusing on a particular fact.

Chapter 4 Review

Match each word with the example that illustrates its meaning.

1. rules of exponents for multiplication **b**
2. base **d**
3. greatest common factor **h**
4. rule of a power raised to a power **f**
5. exponent **a**
6. prime number **c**
7. least common multiple **e**
8. rule of a product raised to a power **g**

a. 3 in 2^3
b. $3^2 \cdot 3^5 = 3^7$
c. 5
d. 2 in 2^3
e. 35 for 5 and 7
f. $(2^3)^2 = 2^6$
g. $(3 \cdot 5)^3 = 3^3 \cdot 5^3$
h. 3 for 12 and 15

Using Exponents 4-1

To evaluate a number that has exponents, remember the base is the number used as a factor and the exponent shows the number of times the base is used as a factor.

$$\text{base} \longrightarrow 3^5 \overset{\swarrow \text{exponent}}{=} 3 \cdot 3 \cdot 3 \cdot 3 \cdot 3 = 243$$

Evaluate

9. 2^3 **8**
10. 5^0 **1**
11. 3^3 **27**
12. 25^1 **25**
13. $(3 + 1)^2$ **16**

Multiplication and Exponents 4-2

To simplify or evaluate multiplication expressions with exponents, use the following rules.

$3^2 \cdot 3^5 = 3^{2+5} = 3^7$ $\quad (3^2)^5 = 3^{2 \cdot 5} = 3^{10}$ $\quad (2 \cdot 3)^3 = 2^3 \cdot 3^3$

14. $a^2 \cdot a^3$ **a^5**
15. $(2a^2)^3$ **$8a^6$**
16. $a(b^3 + b^2)$ **$ab^3 + ab^2$**
17. $(a^2b)^2$ **a^4b^2**
18. $ab^3a^2b^4$ **a^3b^7**

Scientific Notation 4-3

To write a number in scientific notation, express it as a number between 1 and 10 times a power of 10 with exponents.

Write each number in scientific notation.

19. 465,000,000 **4.65×10^8**
20. 13,600,000 **1.36×10^7**
21. 1,280 **1.28×10^3**
22. 5,090,000 **5.09×10^6**

Write each number in standard form.

23. 2.1×10^5 **210,000**
24. 6.13×10^7 **61,300,000**
25. 1.05×10^3 **1,050**
26. 8.35×10^2 **835**

Factors, Multiples, and Divisibility 4-5

To help you to remember the meanings of these words, take a look at the number 15.

3 and 5 are factors of 15.

15 is a multiple of 3: 3, 6, 9, 12, 15. 15 is divisible by 3.
15 is a multiple of 5: 5, 10, 15. 15 is divisible by 5.

True or false. Explain your answer.

27. -378 is divisible by 9. **28.** 800 is a multiple of 5. T; $800 = 5 \cdot 160$ **29.** 12 is a factor of 144. T; $144 = 12^2$

30. 9 is a factor of 93. F **31.** 93 is divisible by 12. F **32.** 95 is a multiple of 19. T; $95 = 19 \cdot 5$

27. T; $-378 \div 9 = -42$

Prime Factorization 4-6

To write a composite number as the product of its prime factors, divide by prime numbers until the quotient is 1, or use a factor tree.

Write the prime factorization using division or a factor tree.

33. 75 $5^2 \cdot 3$ **34.** 420 $2^2 \cdot 3 \cdot 5 \cdot 7$ **35.** 108 $3^3 \cdot 2^2$ **36.** 765 $17 \cdot 5 \cdot 3^2$ **37.** 228 $2^2 \cdot 3 \cdot 19$ **38.** 595 $5 \cdot 7 \cdot 17$

Finding the Greatest Common Factor, the Least Common Multiple 4-7

To find the GCF or LCM, use the prime factorization.

the GCF of 35 and 42	the LCM of 35 and 42
$35 = 5 \cdot 7$	$35 = 5 \cdot 7$
$42 = 2 \cdot 3 \cdot 7$	$42 = 2 \cdot 3 \cdot 7$
GCF = 7	LCM $= 2 \cdot 3 \cdot 5 \cdot 7 = 210$

Find the GCF of each set of numbers.

39. 16, 60 4 **40.** 24, 56 8 **41.** 36, 81, 27 9 **42.** $3x^2y$, $6x^2$ $3x^2$

Find the LCM of each set of numbers.

43. 12, 18 36 **44.** 8, 14 56 **45.** 3, 5, 7 105 **46.** $12x^2y$, $15x^2y^3$ $60x^2y^3$ **47.** $18abc^2$, $22ab^3$ $198ab^3c^2$

Problem Solving 4-8

To account for all possibilities in a word problem, make an organized list, a table, or a diagram.

Solve.

48. There are 10 students competing for the tennis trophy. Each student plays another student once. How many games must be played? 45

The following format is a convenient way to organize math notes. Suggest students use three-ring binders. Notes can be written on a sheet of paper labeled with the chapter number, and ruled or folded into three columns. The columns are labeled TERM, DEFINITION, and EXAMPLE/SKETCH, respectively. With this layout, each item is easily located, defined, and illustrated.

Additional suggestions:

- Encourage students to be consistent in their note-taking style.
- Have students copy notes from the chalkboard to reinforce organized note-taking.
- Emphasize the need for an organized approach.

RESOURCES

TRB D-17–D-23

Scoring Chart

No. Correct	Score
43	100
42	98
41	95
40	92
39	89
38	87
37	85
36	82
35	80
34	78
33	75
32	72
31	70
30	68
29	66
28	64
27	62
26	60
25	57
24	55
23	53
22	50
21	47
20	44
19	42
18	40
17	37
16	35
15	33
14	31
13	28
12	26
11	24
10	22
9	20
8	17
7	15
6	13
5	10
4	8
3	6
2	4
1	2

Chapter 4 Test

Evaluate.

1. 5^3 125 **2.** $2^0 \cdot 2^3$ 8 **3.** $3^2 + 3^3$ 36 **4.** $4^2 \cdot 1^3$ 16 **5.** $1{,}250^1$ 1,250

Evaluate for $a = -2$, $b = 3$.

6. a^2b 12 **7.** $(a \cdot b)^2$ 36 **8.** $b^3 \cdot b^0$ 27 **9.** $(a + b)^5$ 1 **10.** $2(a^2 + b^3)$ 62

Write each number in scientific notation.

11. Every hour 2,500,000 plastic bottles are thrown away in the United States. 2.5×10^6

12. The estimated population of the U.S.S.R. is 287,000,000. 2.87×10^8

Write each number in standard notation.

13. 3.51×10^5 351,000 **14.** 1.9×10^7 19,000,000 **15.** 2.659×10^8 265,900,000 **16.** 4.2×10^3 4,200

List all the factors of each number.

17. 24 1, 2, 3, 4, 6, 8, 12, 24 **18.** 56 1, 2, 4, 7, 8, 14, 28, 56 **19.** 63 1, 3, 7, 9, 21, 63 **20.** 105 1, 3, 5, 7, 15, 21, 35, 105 **21.** 19 1, 19

Tell whether each number is prime or composite.

22. 61 prime **23.** 57 composite **24.** 83 prime **25.** 4,563 composite **26.** 954 composite **27.** 771 composite

Write the prime factorization.

28. 245 $5 \cdot 7^2$ **29.** 378 $2 \cdot 3^3 \cdot 7$ **30.** 242 $11^2 \cdot 2$ **31.** 525 $5^2 \cdot 7 \cdot 3$ **32.** 333 $3^2 \cdot 37$ **33.** 848 $2^4 \cdot 53$

Find the GCF of each set of numbers.

34. 15, 24 3 **35.** 36, 60 12 **36.** 56, 96 8 **37.** $14a^2b^3$, $21ab^2$ $7ab^2$

Find the LCM of each set of numbers.

38. 6, 8 24 **39.** 18, 36 36 **40.** 12, 15 60 **41.** $10x$, $15y$ $30xy$

Solve.

42. Vince has a job making pizzas. There is regular pizza and pan pizza. The choice of toppings is extra cheese, meatballs, or pepperoni. How many different pizzas can he make? 14

43. What is the total number of squares in the figure? 20

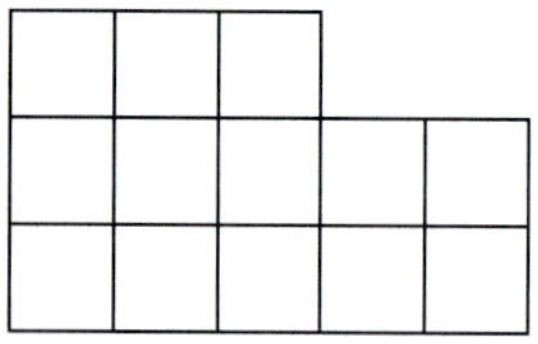

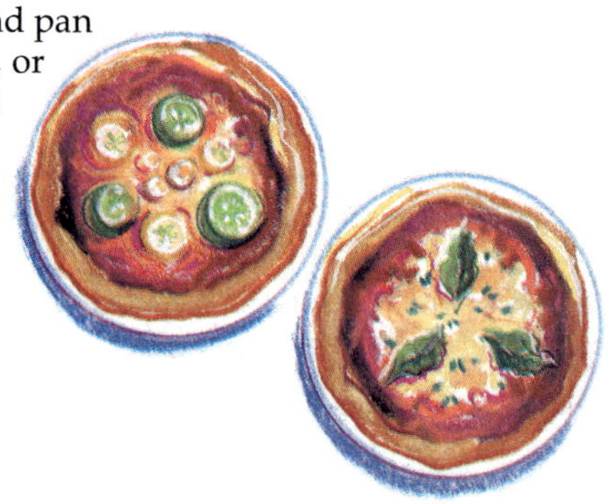

Chapters 1–4 Cumulative Review

Choose the correct answer. Write A, B, C, or D.

1. What relationship does this diagram represent? C

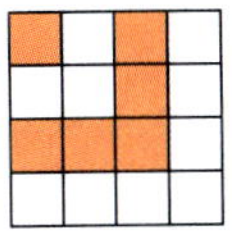

A. $5 - 1 = 7 - 3 = 2^2$
B. $9 + 16 = 5^2$
C. $1 + 3 + 5 + 7 = 4^2$
D. not given

2. Simplify $x^2y \cdot xy^3$. B

A. $2x^2y^3$ **B.** x^3y^4
C. x^2y^3 **D.** not given

3. Find a decimal between -1.5 and -1.3. B

A. 1.4 **B.** -1.4
C. -1.2 **D.** not given

4. Solve $y + 0.5 = 3$. A

A. 2.5 **B.** 1.5
C. 3.5 **D.** not given

5. -4 is a solution of which of the following? C

A. $9x = 36$ **B.** $x - 9 = -5$
C. $x + 9 = 5$ **D.** not given

6. 56,600,000,000 D

A. 5.65×10^8 **B.** 56.5×10^9
C. 5.6×10^{10} **D.** not given

7. Compare $|-1|^3 \;\square\; -(-1)^3$. C

A. < **B.** >
C. = **D.** not given

8. Which number is prime? B

A. 57 **B.** 23
C. 49 **D.** not given

9. Solve. C

A. -7 **B.** 4
C. 1 **D.** not given

10. Which of the following is equal to x^{12}? A

A. $x^6 \cdot x^6$ **B.** $(x^4)^8$
C. $x^6 + x^6$ **D.** not given

11. What is the expression for *the square of the quantity 3 times the absolute value of negative 3?* C

A. $3|(-3)^2|$ **B.** $3|-3|^2$
C. $(3|-3|)^2$ **D.** not given

12. Simplify $3(a + 2b) - 3a$. D

A. $-6a + 6b$ **B.** $3a + 6b$
C. $2b$ **D.** not given

13. Evaluate $(-1)^7 \cdot (-2)^0$. B

A. 2 **B.** -1
C. 1 **D.** not given

14. Which number is divisible by 9? C

A. 1,578 **B.** 5,381
C. 8,622 **D.** not given

15. What is the prime factorization of 90? C

A. $2 \cdot 45$ **B.** $2 \cdot 5 \cdot 9$
C. $2 \cdot 3^2 \cdot 5$ **D.** not given

16. Find the GCF of $4a^3b^2$ and $12ab^3$. A

A. $4ab^2$ **B.** $12a^3b^3$
C. $3a^2b$ **D.** not given

Study Skills

Knowing how to *read a test* is a useful skill for all students. By carefully reading the directions and the problems, students can be taught to interpret words and phrases that can help them understand what is being asked.

Directions tell students how to answer the questions. Have students list direction words such as; *define*, *enumerate*, *name*, *compare*, and *describe*. Then discuss what each word means. Remind students that some directions may require more than one answer as illustrated in this example: "Find the pattern. Name the next number in the sequence."

Qualifying words such as *always*, *never*, *all*, and *sometimes* should be noted since they alter the meaning of questions. Mention to the students that these words are often found in True and False tests.

The following ideas may also help:

- Emphasize the connection between reading and mathematics.
- Remind students to take time to read through each test carefully, underlining important words and phrases.
- Review tests with the class. Discuss questions that students found difficult.

Chapter At a Glance

Chapter 5 introduces students to the concept of fractions as rational numbers. Topics developed in the chapter include using rational numbers in equations and changing rational numbers to decimals.

Reading the Chapter

Chapter 5 contains new terms such as *rational number, mixed number,* and *improper fraction*. Take time to discuss the concepts thoroughly. It may help students to define each term in their own words. Review the terms frequently during the lessons.

Encourage students to organize a strategy for solving equations with rational numbers. Explain that students can review any unclear procedures and then work on computations. Also remind students that understanding the problem is the best preparation for solving it.

Chapter Resources

Practice and Enrichment
TRB E-1 to E-20
Alternate Application TRB E-21

Chapter Test Form A, E-23
Chapter Test Form B, E-25

Computer Test Bank

References

STUDENT BIBLIOGRAPHY

Peter, Rozsa. *Playing with Infinity: Mathematical Exploration and Excursions.* (Dover, New York 1976).

TEACHER BIBLIOGRAPHY

Fisher, Lyle, and William Medigovich. *Problem of the Week* (Dale Seymour Publications, Palo Alto 1981).

SOFTWARE

Blockers and Finders II (Sunburst Communications), for the Apple II family of computers.

FILMS, VIDEOTAPES, AND OTHER MEDIA

The Factory Video (Sunburst Communications, VHS or Beta).

Vocabulary

cross product
equivalent fraction
greatest common factor (GCF)
improper fraction
infinity
least common denominator (LCD)
lowest terms
mixed number
rational number
reciprocal
unit fraction

Materials

fraction models (T*)
pattern blocks (T)
graph paper (T)
math journal
computer and graphing software (optional)

*M—Manipulative Kit
*T—Teacher's Resource Book

Resource Bank

Class Activities

FIELD TRIPS

▼ Jewelry store—to talk about different grades of gold jewelry and the fractional part that is pure gold.

▼ Theater or concert hall—to observe how the shape of the theater and the position of the orchestra pit contribute to quality of sound.

CLASS SPEAKERS

▼ Acoustical physicist or engineer—to discuss how a theater is designed to ensure the sound quality.

▼ Music teacher—to analyze fraction values of notes and rest symbols in simple musical scores.

Bulletin Board

Expand the project on p. 220, question 31, by selecting various types of TV shows. Ask students to collect this data: length of program, commercial time, starting and ending credits time, and actual program time. Have the class display the data on a simulated fraction model (fractions in lowest terms).

Commercials take up the greatest fraction of the time slot for what type of program(s)? (Answers will vary.)

Which of the programs uses the greatest fraction of the time for credits? (Answers will vary.)

Project: Call a local TV station to get any FCC regulations that limit commercial time. Compare your findings with the displayed programs.

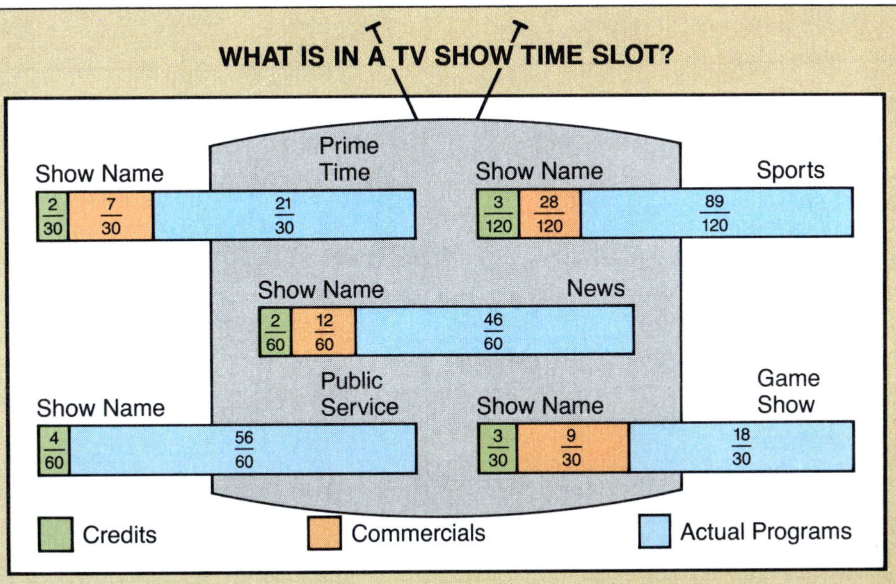

Extra-Credit Problems

BASIC

▼ The king announced that 21 precious stones would be divided among his four children as follows: $\frac{1}{3}$ to the eldest, $\frac{2}{7}$ to each of his twin sons, and $\frac{1}{7}$ to the youngest. Only two were left for the youngest. What went wrong? (The sum is $\frac{22}{21}$, not 1.)

AVERAGE

▼ Write the following fraction in simplest form.

$$1 + \cfrac{1}{2 + \cfrac{1}{1 + \cfrac{1}{2 + \frac{1}{3}}}}$$

$\left(\frac{37}{27} \text{ or } 1\frac{10}{27}\right)$

ENRICHED

▼ What fraction in lowest terms is equivalent to $0.\overline{303}$? to $0.0\overline{45}$? $\left(\frac{101}{333}, \frac{5}{111}\right)$

Chapter Planning Guide

OBJECTIVES	ASSIGNMENT GUIDE			ASSESSMENT	
	Basic	Average	Enriched	Review/ Tests	TRB*
Exploring Fractions To explore fractions using models.	**Manipulative Connection** All	All	All		
5-1 Equivalent Fractions and Lowest Terms To write equivalent fractions and fractions in lowest terms.	1-3, 5, 7-14, 25-32, 36-39, 42-49, MR* All	4-6, 10-24, 27-33, 35, 39-49, MR All	10-24, 30-35, 39-49, MR All	Extra Practice p. 570	Practice E-1 Enrichment E-2
5-2 Fractions and Decimals To write mixed numbers and improper fractions and decimals using models and computation.	1-2, 4-8, 13-16, 21-24, 29-32, 37, 39-40, 43-45, MR All	3-4, 7-10, 15-18, 23-26, 29-37, 40-45, MR All	7-12, 17-20, 25-28, 31-36, 38, 40-45, MR All	Extra Practice p. 570	Practice E-3 Enrichment E-4
5-3 Rational Numbers To study the meaning of rational numbers.	1-8, 11-19, 21-23, 29-34, MR, TY* All	3-6, 9-19, 21-25, 27, 29-34, MR, TY All	11-20, 22-28, 35-36, MR, TY All	Extra Practice p. 570 TY p. 195	Practice E-5 Enrichment E-6 TY E-22
5-4 Comparing and Ordering Rational Numbers To compare and order fractions and decimals.	1-17, 22-23, 25, 35-36, MR All	3-13, 16-19, 23-31, 35-36, MR All	3-13, 18-21, 25-34, MR All	Extra Practice p. 570	Practice E-7 Enrichment E-8
5-5 Adding and Subtracting Rational Numbers To add and subtract fractions and mixed numbers.	1-12, 19-21, 25, 33-34, 39-42, MR, CT* All	3-16, 21-27, 31, 34-36, 39-43, MR, CT All	9-18, 22-26, 28-32, 36-38, 42-43, MR, CT All	Extra Practice p. 570	Practice E-9 Enrichment E-10
5-6 Problem Solving Strategy: Working Backwards To solve problems by working backwards.	**Problem Solving Connection** All	All	All	Problem Solving Pract. p. 225	Practice E-11 Enrichment E-12
5-7 Multiplying and Dividing Rational Numbers To multiply and divide fractions and mixed numbers.	1, 3, 5-14, 17-20, 32-33, 37, MR All	2, 4, 9-16, 20-22, 26-29, 34-35, 37, MR All	10-16, 23-31, 34-36, MR All	Extra Practice p. 570	Practice E-13 Enrichment E-14
Exploring Infinity To use a computer or calculator to study the concept of infinity.	All	All	All		
5-8 Rational Numbers with Exponents To simplify expressions with negative exponents.	1-12, 17-19, 21-26, 29-32, 38-41, MR, CT All	1-14, 17-19, 21, 24-26, 29-34, 38-42, MR, CT All	5-16, 19-20, 27-42, MR, CT All	Extra Practice p. 570	Practice E-15 Enrichment E-16
5-9 Addition and Subtraction Equations To solve equations involving addition and subtraction of rational numbers.	1-11, 20-22, 26-27, 31-33, MR All	1-2, 6-14, 17, 23-32, 34, MR All	9-19, 23-30, 33-34, MR All	Extra Practice p. 570	Practice E-17 Enrichment E-18
5-10 Multiplication Equations To solve equations involving multiplication of rational numbers.	1-8, 13-18, 22, 24, 27-29, 31-39, 44-47, MR, TY All	4-9, 16-19, 22-23, 28-41, 44-49, MR, TY All	8-12, 16-21, 23-26, 30-49, MR, TY All	Extra Practice p. 570 TY p. 221	Practice E-19 Enrichment E-20 TY E-22
5-11 Application: The Stock Market To apply operations with rational numbers to stock market problems.	All	All	All		Alternate Application E-21
Chapter Test	All	All	All	pp. 226-229	Tests E-23–E-29

 *TRB—Teacher's Resource Book MR—Mixed Review TY—Test Yourself CT—Critical Thinking

CONNECTIONS							NCTM CORRELATION
Algebra	Critical Thinking	Problem Solving	Estimation/ Mental Math	Technology	Manipulatives/ Resources	Writing/Reading in Math	
10		4-6	7-10		1-7	1-8	Communication Connections
C*: 6-9, 13-14 W*: 12-24, 30-34, 39-41	C: TD* 1	W: 46-49	W: 17-24	W: 42-45	C: 2-4 W: 3-6, 48	C: 1 W: 1-2	Communication Connections Algebra
W: 37		W: 39-45		C: 13-16 W: 29-36, 44	C: 1-4, TD 2 W: 1-4, 39, 41-45	W: 37, 40	Communication Connections Statistics
C: 15-17 W: 17-28 TY*: 4-5	C: TD 1-3 W: 25-28, 35-36	W: 34			C: 3-6 W: 3-10, 34	W: 29-33, 36	Communication Connections Algebra Math Structure
W: 32-34	C: TD 1-2	W: 35-36	W: 11-13	W: 14-21, 22-27	C: 1-2 W: 1-2, 35		Communication Connections Algebra
C: 9-10 W: 12, 14-15, 18, 31-32, 36-38	C: TD 1-2 W: 32, 41 CT*: 1-7	W: 32, 39, 41-42 CT: 1-7	C: 13-20 W: 19-26, 39-40	W: 27-30	C: 1-4 W: 1-6 CT: 1-7	W: 40 CT: 1-7	Problem Solving Communication Connections Algebra
	W: 7, 9	C: 1-4 W: 1-10	C: 2	W: 1-10	W: 4	W: 1-6, 8, 10	Problem Solving Communication Discrete Math
C: 6-7 W: 9-10, 29-30, 36	C: TD 1-2 W: 36	W: 36	W: 17-25, 37		C: 1-4 W: 1-4		Communication Connections Algebra
8, 9, 11-13		13		1-13	1-13	6, 11-13	Communication Connections Algebra Functions
C: 9-20 W: 9-20, 22-28, 30-31	C: TD 1-3 W: 30-38 CT: 1-5	W: 41-42 CT: 1-4	W: 41		W: 42	W: 29, 41-42 CT: 1-5	Communication Reasoning Connections Algebra
C: 1-11 W: 1-34	C: TD 1	W: 31-34	W: 20-30		C: 1-3 W: 1-2	W: 31-34	Problem Solving Communication Connections Algebra
C: 1-9, TD 1-3 W: 1-30 TY: 1-17	W: 32-43	W: 22-24, 27-31, 45-49	W: 13-15, 36-43, 46-48		W: 31	W: 27-31, 44-49	Problem Solving Communication Algebra Statistics
		C: 12-13 W: 9-13			W: 13	C: 12-13 W: 9-13	Problem Solving Communication

*C—Class Exercises W—Written Exercises CT—Critical Thinking TY—Test Yourself TD—Think and Discuss

Exploring

In this manipulative activity, students use fraction bars to add and subtract with fractions. **See TRB Exploring O-17.**

Exploring Adding and Subtracting Rational Numbers

You can use fraction bars to add and subtract rational numbers.

Make the model at the right. It shows the addition of two fractions.

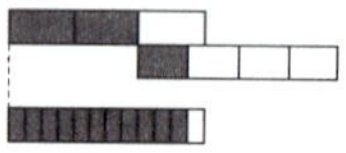

1. What are the addends? **$\frac{2}{3}$ and $\frac{1}{4}$**
2. What is the sum? **$\frac{11}{12}$**

Make the model at the right.

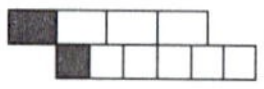

3. What are the addends? **$\frac{1}{4}$ and $\frac{1}{6}$**
4. Is the sum greater than $\frac{1}{4}$? **yes**
5. Is the sum greater than $\frac{1}{2}$? **no**
6. What fraction bar represents the sum? **$\frac{5}{12}$**
7. Why do you think the sum is in twelfths? **Possible answer: 12 is the lowest common multiple of 3 and 4.**

Make the model to find each sum. **Check students' work.**

8. $\frac{5}{12}+\frac{1}{3}=$ **$\frac{9}{12}=\frac{3}{4}$**
9. $\frac{1}{6}+\frac{1}{2}=$ **$\frac{4}{6}=\frac{2}{3}$**
10. $\frac{3}{4}+\frac{1}{12}=$ **$\frac{10}{12}=\frac{5}{6}$**
11. $\frac{1}{4}+\frac{1}{6}=$ **$\frac{10}{12}=\frac{5}{6}$**
12. $\frac{3}{4}+\frac{5}{6}=$ **$\frac{4}{3}=1\frac{1}{3}$**
13. $\frac{2}{3}+\frac{5}{12}=$ **$\frac{13}{12}=1\frac{1}{12}$**

Make the model at the right.

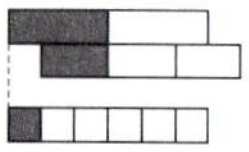

14. What fractions are represented by the two top bars? **$\frac{1}{2}$ and $\frac{1}{3}$**
15. What operation does the model represent? How can you tell? **Subtraction, bottom fraction bar shows an amount less than the first fraction.**

Make the model at the right.

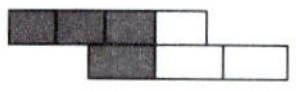

16. What numbers are being subtracted? **$\frac{3}{4}$, $\frac{1}{3}$**
17. What fraction bar shows the difference? **$\frac{5}{12}$**

Make a model to find each difference. **Check students' work.**

18. $\frac{1}{2}-\frac{1}{4}=$ **$\frac{1}{4}$**
19. $\frac{11}{12}-\frac{1}{2}=$ **$\frac{5}{12}$**
20. $\frac{1}{4}-\frac{1}{12}=$ **$\frac{2}{12}=\frac{1}{6}$**
21. $\frac{1}{6}-\frac{1}{12}$ **$\frac{1}{12}$**
22. $\frac{7}{12}-\frac{1}{6}=$ **$\frac{5}{12}$**
23. $\frac{5}{12}-\frac{1}{4}=$ **$\frac{2}{12}=\frac{1}{6}$**

O-17

In this problem solving activity students use a pattern to discover that a series of fractions have a sum that approaches 1. **See TRB Exploring O-18.**

Exploring Patterns with Fractions

To investigate patterns with fractions, you can use a calculator that allows you to input and display fractions, or pencil and paper.

1. Describe the pattern you see in these fractions: $\frac{1}{2}+\frac{1}{4}+\frac{1}{8}$ **Answers may vary. Possible answer: Each fraction is $\frac{1}{2}$ the previous one.**
2. Use the pattern to write the missing fractions. Then find the sum for each group.

	Sum
$\frac{1}{2}+\frac{1}{4}$	**$\frac{3}{4}$**
$\frac{1}{2}+\frac{1}{4}+\frac{1}{8}$	**$\frac{7}{8}$**
$\frac{1}{2}+\frac{1}{4}+\frac{1}{8}+$ **$\frac{1}{16}$**	**$\frac{15}{16}$**
$\frac{1}{2}+\frac{1}{4}+\frac{1}{8}+$ **$\frac{1}{16}$** $+$ **$\frac{1}{32}$**	**$\frac{31}{32}$**

3. What do you notice about the sums as you add each fraction? **Answers may vary. Possible answers: The numerator is one less than the denominator; the sum gets closer and closer to one.**
4. Describe the pattern you see in these fractions: $\frac{1}{3}+\frac{1}{9}+\frac{1}{27}$ **Answers may vary. Possible answer: Each fraction is $\frac{1}{3}$ the previous one.**
5. Now use the pattern you found to write the missing fractions. Then find the sum for each group.

	Sum
$\frac{1}{3}+\frac{1}{9}$	**$\frac{4}{9}$**
$\frac{1}{3}+\frac{1}{9}+\frac{1}{27}$	**$\frac{13}{27}$**
$\frac{1}{3}+\frac{1}{9}+\frac{1}{27}+$ **$\frac{1}{81}$**	**$\frac{40}{81}$**
$\frac{1}{3}+\frac{1}{9}+\frac{1}{27}+$ **$\frac{1}{81}$** $+$ **$\frac{1}{243}$**	**$\frac{121}{243}$**

6. What do you notice about the sums as you add each fraction? **Answers may vary. Possible answer: The sum gets closer and closer to $\frac{1}{2}$.**
7. What do you think will happen if you multiply the fractions instead of adding? Verify your hypothesis by multiplying the fractions in questions 2 and 6. Explain your results. **Answers may vary. Possible answers: The numbers get smaller and smaller. The product is getting closer to zero.**

O-18

In this calculator activity, students use a fraction calculator to evaluate expressions. **See TRB Exploring O-19.**

Exploring Expressions with a Calculator

In order to enter a fraction on many calculators, you first enter the numerator, and then divide by the denominator. For example, $\frac{1}{2}$ is entered 1 [÷] 2. The display reads 0.5.

Some calculators allow you to enter fractions in fraction form. For example, $\frac{1}{2}$ is entered 1 [/] 2 and the display reads 1/2.

Use a calculator. Write the answer.

1. Evaluate the expression $x + y$ for $x=\frac{4}{9}$ and $y=\frac{2}{6}$. **$\frac{14}{18}$ or $0.\overline{7}$**
2. Did you use a fraction calculator? If you answered *yes*, is your answer in lowest terms? **yes, no**

Evaluate each expression by substituting the appropriate values. Let $a=\frac{1}{2}$, $b=\frac{3}{4}$, $c=\frac{3}{8}$, $d=\frac{2}{3}$. If you use a fraction calculator, write your answer in lowest terms.

	Substitution	Calculator Answer	Lowest Terms
3. $a+b$	$\frac{1}{2}+\frac{3}{4}$	$\frac{5}{4}$ or 1.25	$1\frac{1}{4}$
4. $d-c$	$\frac{2}{3}-\frac{3}{8}$	$\frac{7}{24}$ or $0.291\overline{6}$	$\frac{7}{24}$
5. bcd	$\frac{3}{4}\cdot\frac{3}{8}\cdot\frac{2}{3}$	$\frac{18}{96}$ or 0.1875	$\frac{3}{16}$
6. $a \div b$	$\frac{1}{2}\div\frac{3}{4}$	$\frac{4}{6}$ or $0.\overline{6}$	$\frac{2}{3}$

7. Do you think the expressions $a+b-c+d$ and $(a+b)-(c+d)$ will have the same answer? Why or why not? **Answers may vary. Possible answer: No; they are not the same expressions. $a+b-(c+d)=a+b-c-d$.**
8. Evaluate each expression to verify your answer.

Let $a=\frac{1}{2}$, $b=\frac{3}{4}$, $c=\frac{3}{8}$, $d=\frac{2}{3}$.

	Substitution	Calculator Answer	Lowest Terms
$a+b-c+d$	$\frac{1}{2}+\frac{3}{4}-\frac{3}{8}+\frac{2}{3}$	$\frac{37}{24}$ or $1.541\overline{6}$	$\frac{37}{24}$ or $1\frac{13}{24}$
$(a+b)-(c+d)$	$\left(\frac{1}{2}+\frac{3}{4}\right)-\left(\frac{3}{8}+\frac{2}{3}\right)$	$\frac{5}{24}$ or $0.208\overline{3}$	$\frac{5}{24}$

9. Why are the two answers different in question 8? **The expressions are not the same.**

O-19

In this computer activity, students use a spreadsheet program to solve a problem involving rational numbers. **See TRB Exploring O-20.**

Exploring Rational Numbers on Spreadsheets

A computer spreadsheet is useful for solving the kind of problem described below. Read the problem, then answer the questions.

Sam mows a lawn in 20 min. Bob mows the same lawn in 30 min. How long will it take them to do the job together?

1. What fraction of the lawn will Sam have mowed in the given times?
   1 min **$\frac{1}{20}$** 5 min **$\frac{5}{20}$ or $\frac{1}{4}$** 10 min **$\frac{10}{20}$ or $\frac{1}{2}$** x min **$\frac{x}{20}$**
2. What fraction of the lawn will Bob have mowed in the given times?
   1 min **$\frac{1}{30}$** 5 min **$\frac{5}{30}$ or $\frac{1}{6}$** 10 min **$\frac{10}{30}$ or $\frac{1}{3}$** x min **$\frac{x}{30}$**
3. When the entire lawn is mowed, what will be the sum of the part mowed by Sam and the part mowed by Bob? **1**

Set up a spreadsheet. Input the labels shown below.

	A	B	C	D
1	Minutes	Sam's Part	Bob's Part	Total Mowed
2				
3				

Input the following values and formulas in the cells of rows 2 and 3.

Cell	A2	B2	C2	D2
Value/Formula	1	=A2/20	=A2/30	=B2+C2

Cell	A3	B3	C3	D3
Value/Formula	2	=A3/20	=A3/30	=B3+C3

4. What values are recorded in B2? **$\frac{1}{20}$** C2? **$\frac{1}{30}$** D2? **$\frac{5}{60}$ or $\frac{1}{12}$**
5. What values are recorded in B3? **$\frac{2}{20}$ or $\frac{1}{10}$** C3? **$\frac{2}{30}$ or $\frac{1}{15}$** D3? **$\frac{10}{60}$ or $\frac{1}{6}$**

Extend the spreadsheet. Increase the value in the A column by one minute for each row until the value in column D equals 1.

6. Input the values of B, C, and D for each row in your spreadsheet. **Check students' work.**
7. Why do you stop inputting when the column D value is 1? **When the value equals 1, the lawn has been mowed.**
8. How long will it take Sam and Bob to mow the lawn together? **12 min**

O-20

Meeting Individual Needs

GIFTED STUDENTS

Mathematically gifted students often take pride in, and derive satisfaction from, working out mathematical problems and conundrums on their own. While these students benefit from teacher-driven activities and working in cooperative-learning groups, they frequently enjoy opportunities that encourage them toward independent mathematical discovery.

In the following exercise, gifted students, working independently or in small groups with other students, are asked to explore the nature of repeating decimals.

First demonstrate that repeating decimals may be added:

$$\frac{1}{9} + \frac{2}{9} = \frac{3}{9} = \frac{1}{3}, \text{ or } 0.\overline{1} + 0.\overline{2} = 0.\overline{3}$$

Have students use addition of repeating decimals to demonstrate that $1 = 0.\overline{9}$. Possible solutions:

$$\begin{array}{l} \frac{1}{3} = 0.\overline{3} \\ +\frac{2}{3} = 0.\overline{6} \\ \hline \frac{3}{3} = 0.\overline{9}, \text{ or } 1 = 0.\overline{9} \end{array} \qquad \begin{array}{l} \frac{5}{6} = 0.\overline{83} \\ +\frac{1}{6} = 0.\overline{16} \\ \hline \frac{6}{6} = 0.\overline{99}, \text{ or } 1 = 0.\overline{9} \end{array}$$

AT-RISK STUDENTS

For at-risk students, it can often be helpful to spend time reviewing basic concepts and skills. This is particularly important in Chapter 5, where students need to master the concepts governing fractions and decimals before they can work comfortably with rational numbers.

Spend time to ensure that at-risk students have absorbed the first three lessons before moving beyond them. Allow students to review and ask questions. Also give students ample time to use the suggested manipulatives to explore and understand concepts such as equivalent fractions, fractions in lowest terms, improper fractions, and decimal equivalents.

Encourage students to work in pairs to model improper fractions using graph paper. Working in pairs or in small groups, students might also use fraction bars to find and represent fractions equivalent to specific fractions that the teacher has assigned each group.

Even when students have mastered the basic concepts of fractions, they may not be able to use their knowledge to enable them to add and subtract fractions. Have students work with different partners than they had earlier in the chapter, to model questions 1-6 on page 201.

ENGLISH AS A SECOND LANGUAGE

ESL students can gain confidence and enhance both their mathematical and English-language abilities with exercises that help them improve their listening and writing skills as they review mathematical ideas.

To help ESL students as well as other students read and write the improper fractions and mixed numbers, separate the class into pairs. ESL students can be teamed with English-language students who can communicate mathematical ideas well.

Using questions 5-20 on page 190, have one student read a fraction and the other student write and model it. Then the students can switch roles.

Finally, ask students to think of elements of their daily lives that represent part of a whole. Students might see themselves as part of a family, class, or school population, or they might identify one desk as part of a group of desks. Help students name and write the fraction that the part represents.

See TRB Teaching Resources p. 55.

Problem Solving Situation

The question in the "Think About It" box asks whether CDs will ever replace records and tapes. To help students answer this question, have them compare current sales of CDs, tapes, and records with sales of one year ago. Students can also investigate what new types of audio equipment have been developed.

DATA COLLECTION

Data on CDs, tapes, and record sales can be obtained from local record store owners who can provide figures on current sales versus sales from the previous year. If actual figures are not available, the owner may be able to estimate the change in sales. Students can also refer to magazine and newspaper articles.

To find out about new audio technology, students can review stereo magazines and newspaper articles or contact local audio equipment stores.

DATA ANALYSIS

Have students make a graph or chart to explain the data they gathered on sales. The type of graph depends on the extent of the data. Students should compare their charts or graphs to the one on the student page. Based on their information, the students should be able to make predictions about the future sales of CDs, tapes, and records.

THE ORCHESTRA

There are 104 instruments in a typical symphony orchestra. The orchestra is usually arranged in a semicircle with the strings in front and the percussion in back. The orchestra comprises instruments from each instrument family to create a rich, balanced sound.

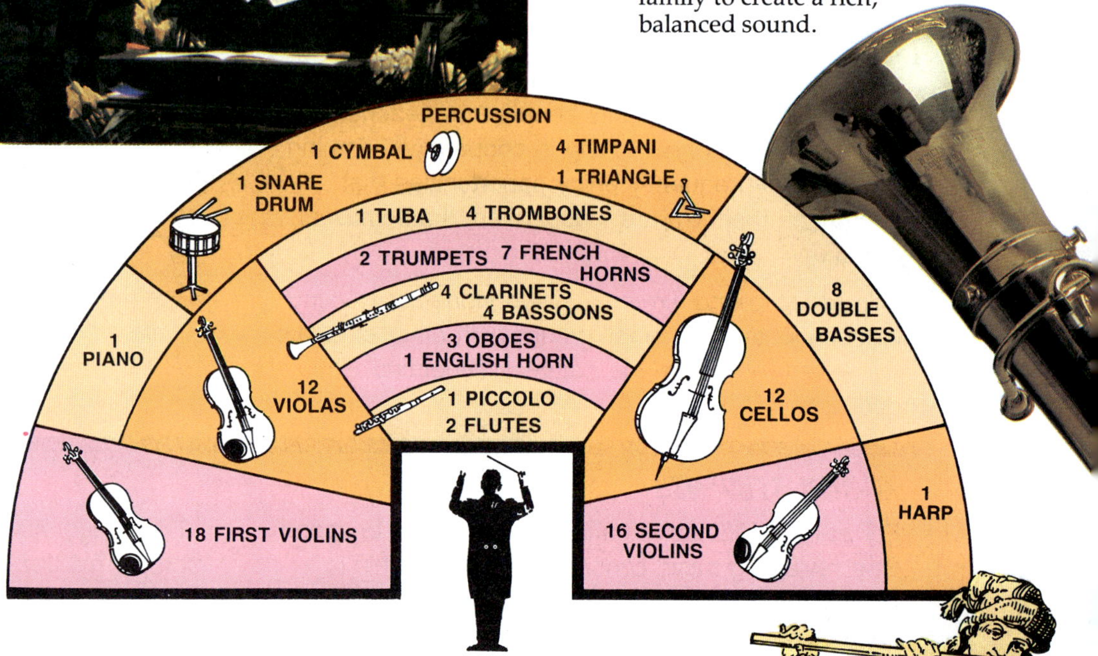

HOT CROSS BUNS

The song HOT CROSS BUNS is written in 4/4 time. In 4/4 time there are four beats to a measure with a quarter note having one beat.

SYMBOL	NOTE	FRACTION	VALUE IN $\frac{4}{4}$ TIME
𝅝	whole	$\frac{1}{1}$	4
𝅗𝅥	half	$\frac{1}{2}$	2
𝅘𝅥	quarter	$\frac{1}{4}$	1
𝅘𝅥𝅮	eighth	$\frac{1}{8}$	$\frac{1}{2}$
𝅘𝅥𝅯	sixteenth	$\frac{1}{16}$	$\frac{1}{4}$

CHAPTER 5

Rational Numbers and Expressions

Think about it...

Look at the triple line graph. Do you think CDs will ever completely replace records and tapes?

See Problem Solving Situation

A portion of the price you pay for a CD is paid to the artist who made the recording. Artists usually receive $\frac{1}{4}$ of the list price of the CD in the form of a royalty.

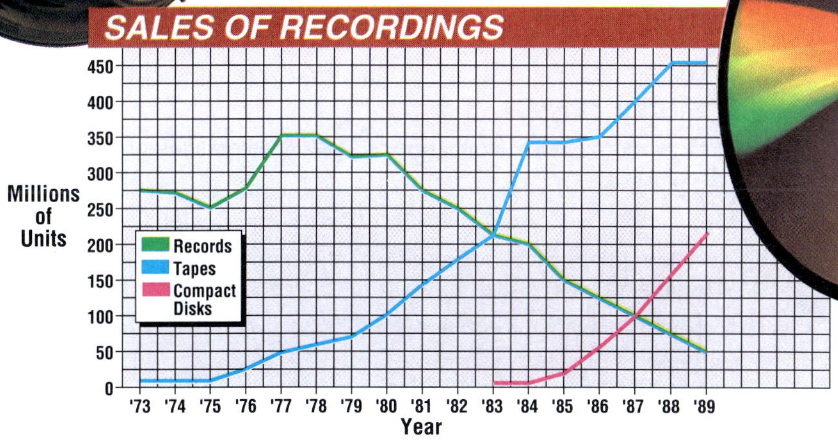

Questions related to this Data File can be found on pages 91, 191, 195, 198, 432, 573, and 577.

Cooperative Learning

This activity can be an interesting one for small groups. The groups can do all the data collection and analysis. As a final project, each group can design an ideal music system.

A very open-ended project like this requires a lot of cooperation. When group members are working, they should be reminded to listen and to consider everyone's opinions and thoughts. To help the groups work together successfully, assign each member a role. Possible roles include Data Collector, Chart Maker, Audio Expert, and Presenter.

Lesson Focus

Materials

Fraction bars, pattern blocks, sets of four strips of paper about 6 in. long, graph paper, math journal

Vocabulary/Symbols

equivalent fractions

MOTIVATING THE LESSON

Ask students to offer their ideas about the uses of fractions. Use their examples to illustrate part of a region and part of a set.

Instruction

1 Have students use fraction bars to model equivalent fractions. If fraction bars are not available, students may make replicas on graph paper. Note that color determines the number of equal parts into which the unit is divided.

2 Some students may have difficulty with this exercise. Having them use the blocks will help them understand why the values represented by certain blocks have changed.

OBJECTIVE: *To explore fractions using models.*

MATERIALS

- Fraction bars
- Pattern blocks
- Graph paper
- Math journal to record work

Exploring Fractions

▼ You can use a variety of models to represent fractions. *Fraction bars* represent fractions as a shaded part of a region.

$\frac{3}{6}$ 

1. ***Explain*** how the numerator (3) and denominator (6) describe the model. **The numerator tells the number shaded. The denominator tells the total number of parts.**

$\frac{2}{4}$ 

▼ Each fraction bar represents the same amount. The fraction bars show *equivalent fractions.*

$\frac{1}{2}$ 

2. a. Find or draw two other fraction bars that show the same fraction. **Check students' work.**

   b. ***Discuss*** Can you find a thirds' bar that shows a fraction equivalent to $\frac{1}{2}$? Why or why not? **No; the fraction bars will not align.**

3. ***Model*** each pair of equivalent fractions. **Check students' work.**

   a. $\frac{2}{3} = \frac{4}{6}$ b. $\frac{4}{12} = \frac{2}{6}$ c. $\frac{6}{6} = \frac{12}{12}$ d. $\frac{8}{6} = \frac{4}{3}$

▼ You can use *pattern blocks* to represent fractions as a part of a whole or part of a set.

4. Suppose one yellow stands for a whole.

   a. Write the fraction for each piece.

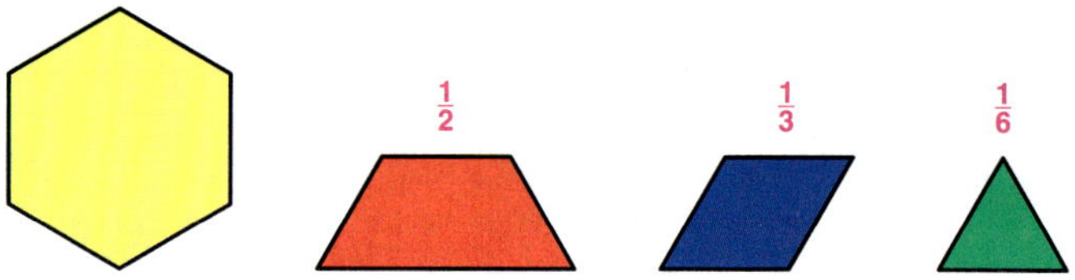

   b. How many greens equal one blue? Write as equivalent fractions. **2 greens = 1 blue $\frac{2}{6} = \frac{1}{3}$**

   c. How many greens equal one red? Write as equivalent fractions. **3 greens = 1 red $\frac{3}{6} = \frac{1}{2}$**

5. Now let one *red* stand for a whole. ***Discuss*** and model.

   a. What pattern block represents $\frac{1}{3}$? What does one blue block represent? **green; $\frac{2}{3}$**

   b. How many greens equal one red? Write as equivalent fractions. **3; $\frac{3}{3} = 1$**

   c. ***Discuss*** Why is one green $\frac{1}{6}$ when the whole is yellow and $\frac{1}{3}$ when the whole is red?

5. c. Yellow is twice the size of red. When yellow is 1, green is $\frac{1}{6}$. When red is doubled to become 1, green is doubled to become $\frac{1}{3}$.

Teaching Tip

Store the pattern blocks in plastic containers by color and make them available for use during all of the fraction lessons.

6. a. Suppose one green represents $\frac{1}{2}$. What piece represents a whole? **blue**

b. Suppose one yellow represents $\frac{1}{2}$. What represents a whole? **2 yellows**

▼ Estimation can help you understand fraction situations.

7. a. Take a sheet of paper and tear off a piece that you think is about $\frac{1}{3}$. Now fold another piece of paper in thirds and compare. How close were you? Compare with another member of your group. **Answers may vary.**

b. Repeat the tearing and comparing for the following fractions.

$\frac{1}{2}$ $\frac{3}{4}$ $\frac{2}{3}$

8. a. A fraction is close to zero when the numerator is small compared to the denominator. 8. b. A fraction is close to $\frac{1}{2}$ when the denominator is about twice the numerator.

▼ You can estimate fractions by comparing the numerator to the denominator.

8. a. These fractions are close to zero. ***Write*** a rule to tell when a fraction is close to zero.

$\frac{1}{14}$ $\frac{3}{17}$ $\frac{2}{25}$ $\frac{7}{125}$

b. These fractions are close to $\frac{1}{2}$. ***Write*** a rule to tell when a fraction is close to $\frac{1}{2}$.

$\frac{3}{8}$ $\frac{6}{14}$ $\frac{11}{23}$ $\frac{51}{100}$

c. These fractions are close to 1. ***Write*** a rule to tell when a fraction is close to 1.

$\frac{99}{100}$ $\frac{3}{4}$ $\frac{45}{50}$ $\frac{79}{91}$

8. c. A fraction is close to 1 when the numerator is close to the denominator.

▼ You can also compare the numerator and the denominator to decide if fractions are greater than, less than, or equal to one.

$\frac{5}{6}$	$\frac{25}{25}$	$\frac{17}{3}$	$\frac{100}{100}$
$\frac{7}{6}$	$\frac{2}{3}$	$\frac{9}{11}$	$\frac{7}{10}$
$\frac{81}{79}$	$\frac{4}{4}$	$\frac{7}{7}$	$\frac{10}{8}$

9. Use the fractions in the box at the right to complete.

a. fractions less than one ▢, ▢, ▢, ▢ **$\frac{5}{6}$, $\frac{2}{3}$, $\frac{9}{11}$, $\frac{7}{10}$**

b. fractions greater than one ▢, ▢, ▢, ▢ **$\frac{17}{3}$, $\frac{7}{6}$, $\frac{81}{79}$, $\frac{10}{8}$**

c. fractions equal to one ▢, ▢, ▢, ▢ **$\frac{25}{25}$, $\frac{100}{100}$, $\frac{4}{4}$, $\frac{7}{7}$**

10. a. Complete to show fractions close to 0. $\frac{▢}{7}$, $\frac{▢}{3}$, $\frac{5}{▢}$, $\frac{17}{▢}$

b. Complete to show fractions close to $\frac{1}{2}$. $\frac{▢}{7}$, $\frac{▢}{9}$, $\frac{13}{▢}$, $\frac{5}{▢}$ **Answers may vary.**

c. Complete to show fractions close to 1. $\frac{▢}{22}$, $\frac{▢}{11}$, $\frac{4}{▢}$, $\frac{20}{▢}$

d. Complete to show fractions greater than 1. $\frac{▢}{2}$, $\frac{▢}{7}$, $\frac{9}{▢}$, $\frac{15}{▢}$

3 Ask students to explain how they could tell whether a fraction is closer to zero, one-half, or one using strips of paper.

Closure

Writing in Math Ask students to summarize how they used models to represent fractions. Have them write a description of equivalent fractions in their math journals.

Assignment Guide

Basic All
Average All
Enriched All

FOR THE NEXT LESSON

fraction bars, index cards, overhead projector (optional), math journal

Lesson Focus

Materials

Fraction bars, index cards, overhead projector (optional), math journal

Vocabulary/Symbols

cross products
equivalent fractions
greatest common factor (GCF)
lowest terms

MOTIVATING THE LESSON

Display three circles (for pizzas) showing quarters, eighths, and twelfths. Ask students what fractions represent half of each pizza. ($\frac{2}{4}$, $\frac{4}{8}$, $\frac{6}{12}$ pizza) Ask students what fractions represent $\frac{1}{4}$ of each pizza. ($\frac{1}{4}$, $\frac{2}{8}$, $\frac{3}{12}$ pizza)

Instruction

1 Ask students to draw a model using a shape other than a circle to show that $\frac{4}{6} = \frac{2}{3}$, or have them use fraction bars.

Address the THINK question by asking students to supply other fractions equivalent to $\frac{4}{6}$ and $\frac{3x}{5y}$. Ask students, in each case, by what factor they are multiplying the numerator and denominator.

OBJECTIVE: *To write equivalent fractions and fractions in lowest terms.*

5-1 Equivalent Fractions and Lowest Terms

▼ Suppose you divide a pizza into eight equal size pieces and eat four of the pieces. You can draw a model and write a fraction to represent the amount eaten.

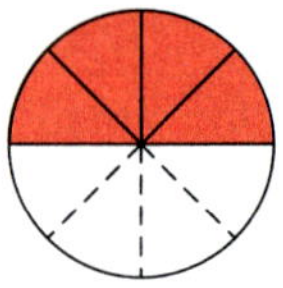

$$\frac{\text{numerator}}{\text{denominator}} \begin{matrix} \rightarrow 4 \leftarrow \text{pieces eaten} \\ \rightarrow 8 \leftarrow \text{pieces in all} \end{matrix}$$

The model shows that the fractions $\frac{4}{8}$ and $\frac{1}{2}$ describe the same part of the pizza. They are *equivalent fractions*.

Equivalent Fractions	You can form equivalent fractions by multiplying or dividing the numerator and denominator by the same nonzero factor.

THINK Why can you multiply the numerator and denominator by $\frac{2}{2}$, $\frac{3}{3}$, or $\frac{n}{n}$? They are all equal to 1. One times any number equals itself.

▼ You can write equivalent fractions using numbers, numbers and variables, or just variables.

Arithmetic	Algebra	
$\frac{1}{2} = \frac{1 \cdot 2}{2 \cdot 2} = \frac{2}{4}$	$\frac{a}{b} = \frac{ac}{bc}$	$(b \neq 0, c \neq 0)$
$\frac{3}{6} = \frac{3 \div 3}{6 \div 3} = \frac{1}{2}$	$\frac{a}{b} = \frac{a \div c}{b \div c}$	$(b \neq 0, c \neq 0)$

Example 1 Write a fraction equivalent to each fraction.

a. $\frac{4}{6}$ **b.** $\frac{3x}{5y}$

Solution **1**

a. We can divide by $1 = \frac{2}{2}$.

$$\frac{4}{6} = \frac{4 \div 2}{6 \div 2} = \frac{2}{3}$$

b. We can multiply by $1 = \frac{2}{2}$.

$$\frac{3x}{5y} = \frac{3x \cdot 2}{5y \cdot 2} = \frac{6x}{10y}$$

THINK Are there other possible solutions? yes, for example $\frac{8}{12}$

Example 2 Replace the variable to form equivalent fractions.

$$\frac{5}{8} = \frac{n}{24}$$

Solution

$$\frac{5 \cdot 3}{8 \cdot 3} = \frac{n}{24}$$ Since 8 · 3 = 24, multiply both numerator and denominator by 3.

$$\frac{5}{8} = \frac{15}{24}$$

So, $n = 15$.

Teaching Tip

Have students make their own sets of paper fraction bars. They can keep the bars together with a paper clip and store them in their notebooks or math journals.

▼ When two fractions are equivalent, their *cross products* are equal. You can use this idea to test for equivalence.

Example 3 **a.** Is $\frac{6}{18} = \frac{7}{21}$? **b.** Is $\frac{3}{15} = \frac{12}{45}$?

$\frac{6}{18} \stackrel{?}{=} \frac{7}{21}$ $\frac{3}{15} \stackrel{?}{=} \frac{12}{45}$

Solution **a.** $6 \cdot 21 \stackrel{?}{=} 18 \cdot 7$
$6 \times 21 = 126$
$18 \times 7 = 126$

The cross products are equal, so the fractions are equivalent.

b. $3 \cdot 45 \stackrel{?}{=} 15 \cdot 12$
$3 \times 45 = 135$
$15 \times 12 = 180$

The cross products are not equal, so the fractions are not equivalent.

(2)

FLASHBACK

$\stackrel{?}{=}$ are the quantities equal?

$\neq$ is not equal to

▼ You can divide both terms of a fraction by the *greatest common factor (GCF)* to write the fraction in *lowest terms.*

FLASHBACK

The *greatest common factor (GCF)* is the greatest factor that two expressions have in common.

Lowest Terms	When a fraction is in lowest terms, the only common factor of the numerator and denominator is 1.

Example 4 **Write $\frac{18}{24}$ in lowest terms.**

Solution $\frac{18}{24} = \frac{18 \div 6}{24 \div 6} = \frac{3}{4}$ Divide by 6, the GCF of 18 and 24.

(3)

▼ You also can write the numerator and denominator as a product of prime factors. Then divide by common factors to write in lowest terms.

Example 5 $\frac{18}{24} = \frac{\cancel{2}^{1} \cdot \cancel{3}^{1} \cdot 3}{\cancel{2}_{1} \cdot 2 \cdot 2 \cdot \cancel{3}_{1}} = \frac{3}{4}$ Divide common factors. Multiply the remaining factors.

(4)

▼ When a fraction includes variables you can divide common variable factors to write the fraction in lowest terms.

Example 6 **Write $\frac{2a^2b}{6ac}$ in lowest terms.**

Solution $\frac{2a^2b}{6ac} = \frac{2 \cdot a \cdot a \cdot b}{2 \cdot 3 \cdot a \cdot c}$ Write as a product of prime factors.

$= \frac{\cancel{2}^{1} \cdot \cancel{a}^{1} \cdot a \cdot b}{\cancel{2}_{1} \cdot 3 \cdot \cancel{a}_{1} \cdot c} = \frac{ab}{3c}$ Divide common factors. Multiply the remaining factors.

(2) Emphasize that finding cross products is usually used as a test for equivalence, rather than for finding equivalent fractions.

(3) Ask students why dividing both terms of a fraction by their GCF gives a fraction in lowest terms. (The new numerator and denominator will only have a common factor of 1.)

(4) Ask students how this method is equivalent to dividing by the GCF of 18 and 24. (Dividing by the common prime factors 2 and 3 is the same as dividing by 6.)

ADDITIONAL EXAMPLES

1. Write an equivalent fraction.
   a. $\frac{3}{7}$
   (Answers may vary; $\frac{6}{14}$)
   b. $\frac{8xy}{10z}$
   (Answers may vary; $\frac{4xy}{5z}$)
2. Replace the variable to form equivalent fractions.
   $\frac{3}{5} = \frac{n}{35}$ $(n = 21)$
3. **a.** $\frac{4}{30} \stackrel{?}{=} \frac{6}{45}$
   (They are equal. $4 \cdot 45 = 180$ and $30 \cdot 6 = 180$)
   b. $\frac{3}{25} \stackrel{?}{=} \frac{9}{65}$
   (They are not equal. $3 \cdot 65 = 195$ and $25 \cdot 9 = 225$)
4. Write $\frac{12}{20}$ in lowest terms. ($\frac{3}{5}$)
5. Write $\frac{14}{56}$ in lowest terms using prime factors. ($14 = 2 \cdot 7$, $56 = 2 \cdot 2 \cdot 2 \cdot 7$; divide by 2 and 7; $\frac{1}{4}$)
6. Write $\frac{6xy^2}{15yz}$ in lowest terms. ($\frac{2xy}{5z}$)

MATH MINUTES

Use sets of index cards to display the fractions below. Ask students to find pairs of equivalent fractions.

$\frac{3}{8}, \frac{a}{b}, \frac{1}{4}, \frac{7}{42}, \frac{2x}{3y}, \frac{4a^2}{10ac}$

$\frac{12}{48}, \frac{9a}{24a}, \frac{8x}{12y}, \frac{2a}{5c}, \frac{4a}{4b}, \frac{1}{6}$

Guided Practice

Class Exercises Have volunteers work the exercises at the chalkboard and explain how they arrived at their answers.

Think and Discuss Encourage students to think of a numerical example that demonstrates the answer to question 2.

A ***common error*** is to divide the terms of a fraction by different numbers when finding lowest terms. For example, for $\frac{12}{20}$, students may incorrectly divide 12 by 4 and 20 by 5. To correct this type of error, have students write the factor by which they are dividing.

$$\frac{12 \div 4}{20 \div 4} = \frac{3}{5}$$

Closure

Writing in Math Ask students to write in their math journals, how to find a fraction equivalent to $\frac{1}{2}$ and how to write $\frac{12}{36}$ in lowest terms.

Independent Practice

For questions 7-16, encourage students to use both multiplication and division to find equivalent fractions.

Have pairs of students complete the Project, question 49, and present their results on one sheet of paper.

THINK AND DISCUSS See below.

1. How are $\frac{1}{2}$ pizza and $\frac{1}{2}$ mi the same? How are they different?

2. Can you write a fraction in lowest terms if you divide by a factor other than the GCF? Explain.

1. They both are $\frac{1}{2}$ of something. The things they measure are very different.
2. No; if you divide by something other than GCF, the numerator and denominator will still have a factor in common.

CLASS EXERCISES

1. During a 30-min radio broadcast there were 7 min of commercials.

a. Write a fraction for the commercial time. $\frac{7}{30}$

b. Write a fraction for the noncommercial time. $\frac{23}{30}$

Check students' work.

Use or draw a model to represent each pair of equivalent fractions.

2. $\frac{2}{4} = \frac{3}{6}$ **3.** $\frac{6}{8} = \frac{9}{12}$ **4.** $\frac{3}{5} = \frac{6}{10}$

Write a fraction equivalent to each fraction. Answers may vary. One example given.

5. $\frac{2}{3}$ $\frac{4}{6}$ **6.** $\frac{5w}{7x}$ $\frac{15w}{21x}$ **7.** $\frac{9}{10x}$ $\frac{9x}{10x^2}$ **8.** $\frac{13c}{26}$ $\frac{c}{2}$ **9.** $\frac{5a}{7a}$ $\frac{5}{7}$

Write in lowest terms.

10. $\frac{3}{9}$ $\frac{1}{3}$ **11.** $\frac{4}{10}$ $\frac{2}{5}$ **12.** $\frac{2}{8}$ $\frac{1}{4}$ **13.** $\frac{2a}{3a}$ $\frac{2}{3}$ **14.** $\frac{4ab^2}{12b}$ $\frac{ab}{3}$

WRITTEN EXERCISES

Write a fraction for each sentence.

1. Three out of thirteen students are in the band. $\frac{3}{13}$

2. A student grew five-eighths inches. $\frac{5}{8}$

3. Copy and complete the chart.

Model	Word Name	Fraction
▲ ▲ ▲ △ △ △	one-half	■ $\frac{1}{2}$
● ● ○ ○ ○ ○ ○ ○	■ one-fourth	$\frac{1}{4}$
■ Check students' work.	seven-eighths	$\frac{7}{8}$
★ ★ ★ ★ ★ ☆ ☆	■ five-sevenths	■ $\frac{5}{7}$

Write a fraction for each shaded region.

4.
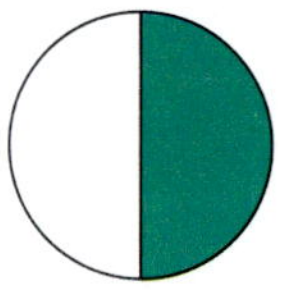
$\frac{1}{2}$

5.

$\frac{6}{10}$ or $\frac{3}{5}$

6.
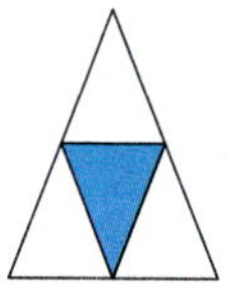
$\frac{1}{4}$

Write three fractions equivalent to each fraction. **Answers may vary. One example given.**

7. $\frac{5}{6}$ $\frac{10}{12}$ 8. $\frac{12}{20}$ $\frac{6}{10}$ 9. $\frac{12}{36}$ $\frac{1}{3}$ 10. $\frac{1}{3}$ $\frac{a}{3a}$ 11. $\frac{2}{5}$ $\frac{4}{10}$

12. $\frac{7b}{14c}$ $\frac{b}{2c}$ 13. $\frac{8k}{9j}$ $\frac{8kx}{9jx}$ 14. $\frac{4t}{100w}$ $\frac{t}{25w}$ 15. $\frac{25x}{75x}$ $\frac{1}{3}$ 16. $\frac{8p}{24p^2}$ $\frac{4p}{12p^2}$

MENTAL MATH Find the value of the variable to form equivalent fractions.

17. $\frac{1}{3} = \frac{n}{6}$ 2 18. $\frac{2}{5} = \frac{10}{y}$ 25 19. $\frac{3}{8} = \frac{a}{16}$ 6 20. $\frac{5}{10} = \frac{1}{b}$ 2

21. $\frac{12}{36} = \frac{x}{3}$ 1 22. $\frac{a}{12} = \frac{1}{4}$ 3 23. $\frac{2}{7} = \frac{w}{14}$ 4 24. $\frac{3}{t} = \frac{1}{3}$ 9

Write in lowest terms.

25. $\frac{5}{25}$ $\frac{1}{5}$ 26. $\frac{7}{14}$ $\frac{1}{2}$ 27. $\frac{6}{9}$ $\frac{2}{3}$ 28. $\frac{11}{22}$ $\frac{1}{2}$ 29. $\frac{25}{75}$ $\frac{1}{3}$

30. $\frac{3a}{6a}$ $\frac{1}{2}$ 31. $\frac{4bc}{12b}$ $\frac{c}{3}$ 32. $\frac{xy}{3y}$ $\frac{x}{3}$ 33. $\frac{5t}{10t^2}$ $\frac{1}{2t}$ 34. $\frac{abc}{5abc}$ $\frac{1}{5}$

35. Use the numbers 3, 5, 6, and 10 to write three pairs of equivalent fractions. $\frac{3}{6} = \frac{5}{10}$ $\frac{3}{5} = \frac{6}{10}$ $\frac{10}{6} = \frac{5}{3}$ $\frac{10}{5} = \frac{6}{3}$

Use prime factors to write each fraction in lowest terms.

36. $\frac{15}{25}$ $\frac{3}{5}$ 37. $\frac{12}{16}$ $\frac{3}{4}$ 38. $\frac{6}{15}$ $\frac{2}{5}$ 39. $\frac{3a^2b}{5b}$ $\frac{3a^2}{5}$ 40. $\frac{6mn}{9m}$ $\frac{2n}{3}$ 41. $\frac{8p^3q}{12p^2}$ $\frac{2pq}{3}$

CALCULATOR Use cross products. Compare using = or ≠.

42. $\frac{18}{30}$ ■ $\frac{15}{25}$ = 43. $\frac{16}{21}$ ■ $\frac{12}{15}$ ≠ 44. $\frac{13}{24}$ ■ $\frac{11}{18}$ ≠ 45. $\frac{18}{32}$ ■ $\frac{27}{48}$ =

Solve.

46. The world production of gold in 1988 was about 58 million troy ounces. The United States produced about 6 million troy ounces. What fraction of the world's production came from the United States? Write your answer in lowest terms. $\frac{3}{29}$

47. For a money-raising project, the senior class baked a huge pizza. The class divided the pizza into 60 pieces. The swim team bought 24 pieces. What fraction of the pizza did the swim team buy? Write your answer in lowest terms. $\frac{2}{5}$

48. A survey of students revealed that the favorite music of half the group was rock. One-third preferred jazz and the rest liked country music. Draw a diagram to represent this data. **Check students' work.**

49. ***PROJECT*** Take a survey of the students in one of your classes. Determine the number of left-handed people. Compare this with the total number of students in the class. Write this information as a fraction.

MIXED REVIEW

Find the factors. **1, 2, 4, 8, 16, 32**

1. 32 2. 27 **1, 3, 9, 27**

Find the prime factors. Then write using exponents.

3. 36 $2^2 \cdot 3^2$ 4. 54 $2 \cdot 3^3$

Estimate.

5. 21.5 · 12.8 **200**

6. 42.8 + 13.6 + 18.23 **70**

7. What are the GCF and the LCM of 18 and 12? **GCF 6 LCM 36**

8. The lowest recorded temperature for Hawaii is 12°F. The lowest recorded temperature for Wisconsin is 66° lower. What is the record for Wisconsin? **−54°F**

Lesson Follow-up

Reteaching Activity Have students use fraction bars to find fractions equivalent to $\frac{1}{2}$, $\frac{1}{3}$, $\frac{2}{3}$, $\frac{1}{4}$, $\frac{3}{4}$, $\frac{1}{6}$, and $\frac{5}{6}$. Then have them tell what factor the terms of each fraction can be multiplied by to make the equivalent fraction they found.

LESSON RESOURCES

TRB Practice E-1
TRB Enrichment E-2

Practice For use after 5-1 (pp. 184–187)

Equivalent Fractions and Lowest Terms

Write a fraction for each sentence.

1. Montana completed 18 passes in 29 attempts. $\frac{18}{29}$
2. One of the 12 light bulbs was defective. $\frac{1}{12}$

Complete the chart.

	Model	Word name	Fraction
3.	○ ○ ○ ○	one-third	$\frac{1}{3}$
4.	□ □ □ □ □	one-half	$\frac{1}{2}$
5.	△ △ △	two-fifths	$\frac{2}{5}$

Write a fraction for each shaded region.

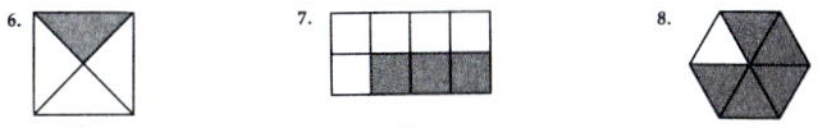

6. $\frac{1}{4}$ 7. $\frac{3}{8}$ 8. $\frac{5}{6}$

Write three fractions equivalent to each fraction. **Answers may vary.**

9. $\frac{1}{4}$ $\frac{2}{8}$ $\frac{3}{12}$ $\frac{4}{16}$ 10. $\frac{2}{3}$ $\frac{4}{6}$ $\frac{6}{9}$ $\frac{8}{12}$
11. $\frac{3}{5}$ $\frac{6}{10}$ $\frac{9}{15}$ $\frac{12}{20}$ 12. $\frac{3}{18}$ $\frac{1}{6}$ $\frac{2}{12}$ $\frac{4}{24}$
13. $\frac{8k}{16k}$ $\frac{1}{2}$ $\frac{2}{4}$ $\frac{4}{8}$ 14. $\frac{3m}{8n}$ $\frac{6m}{16n}$ $\frac{9m}{24n}$ $\frac{12m}{32n}$

MENTAL MATH Rename the variable to form equivalent fractions.

15. $\frac{1}{2} = \frac{6}{n}$ 12 16. $\frac{2}{5} = \frac{m}{20}$ 8 17. $\frac{9}{21} = \frac{3}{x}$ 7
18. $\frac{c}{10} = \frac{12}{40}$ 3 19. $\frac{5}{6} = \frac{15}{x}$ 18 20. $\frac{21}{c} = \frac{7}{8}$ 24

Write in lowest terms.

21. $\frac{10}{15}$ $\frac{2}{3}$ 22. $\frac{18}{36}$ $\frac{1}{2}$ 23. $\frac{27}{36}$ $\frac{3}{4}$ 24. $\frac{12}{15}$ $\frac{4}{5}$
25. $\frac{6xy}{16y}$ $\frac{3x}{8}$ 26. $\frac{24n^2}{28n}$ $\frac{6n}{7}$ 27. $\frac{30hxy}{54kxy}$ $\frac{5h}{9k}$ 28. $\frac{16y^3}{20y^4}$ $\frac{4}{5y}$

Chapter 5 E-1

Enrichment For use after 5-1 (pp. 184–187)

Bear Facts

The world's largest carnivore is a species of bear found in Alaska. The average male is 8 ft long and weighs more than 1,000 pounds.

To find the name of this species, write the missing numerator or denominator to form equivalent fractions. Transfer numbers as indicated by arrows. (The first one has been done for you.) Convert the final number to a letter of the alphabet and write the answer in the answer grid.

1. $\frac{4}{9} = \frac{12}{27}$ → $\frac{9}{27} = \frac{1}{3}$ → $\frac{1}{5} = \frac{11}{55}$ → 11
2. $\frac{12}{32} = \frac{3}{8}$ → $\frac{3}{7} = \frac{21}{49}$ → $\frac{21}{28} = \frac{15}{20}$ → 15
3. $\frac{56}{80} = \frac{7}{10}$ → $\frac{3}{10} = \frac{6}{20}$ → $\frac{6}{8} = \frac{3}{4}$ → 4
4. $\frac{16}{36} = \frac{4}{9}$ → $\frac{4}{7} = \frac{12}{21}$ → $\frac{12}{20} = \frac{9}{15}$ → 9
5. $\frac{35}{56} = \frac{5}{8}$ → $\frac{5}{18} = \frac{10}{36}$ → $\frac{6}{36} = \frac{1}{6}$ → 1
6. $\frac{24}{42} = \frac{8}{14}$ → $\frac{8}{18} = \frac{32}{72}$ → $\frac{32}{44} = \frac{8}{11}$ → 11

Answer: K O D I A K bear

A	B	C	D	E	F	G	H	I	J	K	L	M
1	2	3	4	5	6	7	8	9	10	11	12	13

N	O	P	Q	R	S	T	U	V	W	X	Y	Z
14	15	16	17	18	19	20	21	22	23	24	25	26

E-2 Chapter 5

LESSON QUIZ

Write two fractions equivalent to each given fraction. (Answers may vary. Possible answers given.)

1. $\frac{4}{7}$ $\left(\frac{8}{14}, \frac{12}{21}\right)$
2. $\frac{10}{18}$ $\left(\frac{5}{9}, \frac{20}{36}\right)$
3. $\frac{2x}{3y}$ $\left(\frac{4x}{6y}, \frac{6x}{9y}\right)$

Write in lowest terms.

4. $\frac{24}{32}$ $\left(\frac{3}{4}\right)$
5. $\frac{81}{99}$ $\left(\frac{9}{11}\right)$
6. $\frac{5a}{15a}$ $\left(\frac{1}{3}\right)$

FOR THE NEXT LESSON

fraction bars, pattern blocks, sets of four paper circles cut in half, scissors, math journal

Assignment Guide

Basic 1-3, 5, 7-14, 25-32, 36-39, 42-49, MR All
Average 4-6, 10-24, 27-33, 35, 39-49, MR All
Enriched 10-24, 30-35, 39-49, MR All

Lesson Focus

Materials

Fractions bars, pattern blocks, sets of four paper circles cut in half, scissors, math journal

Vocabulary/Symbols

improper fraction
mixed number

MOTIVATING THE LESSON

Distribute sets of half-circles to small groups of students. Have them use seven half-circles to make circles and ask for the results. (three complete circles and one half circle) Explain to students that they have just used seven half circles to model $3\frac{1}{2}$. Continue the activity with other numbers of halves.

Instruction

1 Ask students to cut their circle pieces so they can model $3\frac{1}{4} = \frac{13}{4}$. If necessary, mention that the circles should be cut into fourths.

2 Ask students how the solution is related to the model in Example 1. (3 is represented by 3 whole bars, then as $\frac{12}{4}$; $\frac{12}{4}$ is added to $\frac{1}{4}$ to show $\frac{13}{4}$.)

OBJECTIVE: *To write mixed numbers, improper fractions, and decimals using models and computation.*

5-2 Fractions and Decimals

▼ If you work more than 40 h/week, your rate of pay may be $1\frac{1}{2}$ times your regular rate. The number $1\frac{1}{2}$ is a *mixed number*.

▼ You can use models to write mixed numbers as *improper fractions*.

Example 1 **Write $3\frac{1}{4}$ as an improper fraction using a model.**

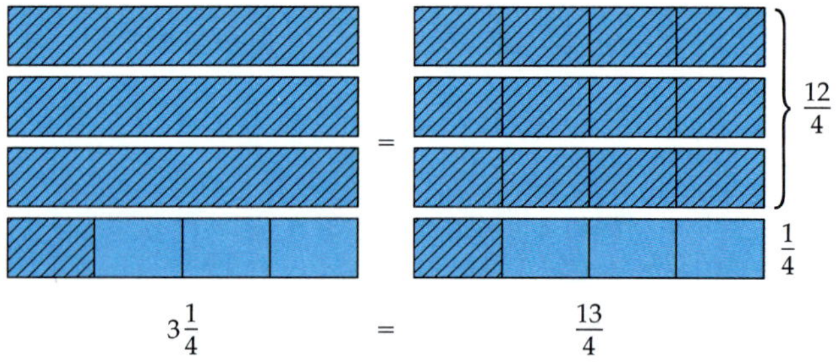

THINK How do you decide which models you will use to replace the whole models?
The denominator of the fractional part tells you which models to use to replace the whole models.

▼ You can also write mixed numbers as improper fractions using equivalent fractions.

Example 2 **Write $3\frac{1}{4}$ as an improper fraction.**

Solution

$$3\frac{1}{4} = \frac{3}{1} + \frac{1}{4}$$
$$= \frac{12}{4} + \frac{1}{4}$$
$$= \frac{13}{4}$$

$\frac{12}{4}$ and $\frac{3}{1}$ are equivalent fractions: $\frac{3}{1} = \frac{3 \cdot 4}{1 \cdot 4} = \frac{12}{4}$

THINK Why is 3 the same as $\frac{3}{1}$?
$\frac{3}{1}$ means 3 divided by 1.
$3 \div 1 = 3$

▼ You can divide to write improper fractions as mixed numbers or decimals.

Example 3 **Write $\frac{7}{2}$ as a mixed number and as a decimal.**

Solution $\frac{7}{2} = 7 \div 2 = 3\frac{1}{2} = 3.5$

▼ You can write decimals as fractions in lowest terms.

Example 4 **Write each decimal as a fraction or mixed number in lowest terms.**

a. 0.12 **b.** 1.625

Solution

a. $0.12 = \frac{12}{100} = \frac{12 \div 4}{100 \div 4} = \frac{3}{25}$

b. $1.625 = 1 + \frac{625}{1{,}000} = 1 + \frac{625 \div 125}{1{,}000 \div 125} = 1\frac{5}{8}$

Teaching Tip

To help students recognize and remember common fraction and decimal equivalents, start a poster on which students can write equivalents. Organize the chart by denominators. When completed, the chart will be a familiar reference to the students.

▼ Sometimes it's necessary to write fractions as decimals before you solve word problems.

Example 5 You are at a delicatessen and ask for $\frac{3}{4}$ lb of potato salad. The scale reads 0.75. Are you getting the amount you requested?

Solution $\frac{3}{4}$ = 3 ⊕ 4 ⊜ 0.75 — Use a calculator. Divide the numerator by the denominator.

Since $\frac{3}{4}$ equals 0.75, you received the right amount.

▼ Some fractions result in decimal patterns that repeat. Three dots at the right of a decimal indicate that digits repeat. You can also write a bar over the digits that repeat.

Example 6 **Use a calculator to write each fraction as a decimal.**

a. $\frac{2}{3}$ **b.** $\frac{15}{11}$

Solution **a.** $\frac{2}{3}$ = 2 ÷ 3 = 0.6666 . . .

$\frac{2}{3} = 0.\overline{6}$ — Write a bar over the 6.

b. $\frac{15}{11}$ = 15 ÷ 11 = 1.3636 . . .

$\frac{15}{11} = 1.\overline{36}$ — Write a bar over the repeating digits 3 and 6.

THINK What is the decimal equivalent for $5\frac{3}{4}$? for $12\frac{3}{4}$? 5.75; 12.75

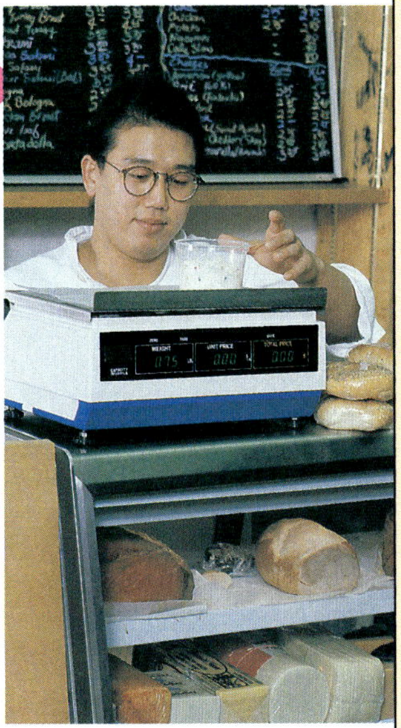

CLASS EXERCISES

Use a model to write as an improper fraction or mixed number.

1. $2\frac{3}{5}$ **2.** $1\frac{5}{8}$ **3.** $\frac{21}{4}$ **4.** $\frac{17}{6}$

Check students' work.

Write as an improper fraction or mixed number.

5. $7\frac{1}{2}$ $\frac{15}{2}$ **6.** $\frac{11}{4}$ $2\frac{3}{4}$ **7.** $\frac{19}{8}$ $2\frac{3}{8}$ **8.** $\frac{21}{5}$ $4\frac{1}{5}$

Write each decimal as a fraction in lowest terms.

9. 0.10 $\frac{1}{10}$ **10.** 0.5 $\frac{1}{2}$ **11.** 2.75 $2\frac{3}{4}$ **12.** 3.25 $3\frac{1}{4}$

CALCULATOR **Write each fraction or mixed number as a decimal.**

13. $\frac{7}{16}$ 0.4375 **14.** $4\frac{7}{16}$ 4.4375 **15.** $\frac{5}{12}$ $0.41\overline{6}$ **16.** $8\frac{5}{12}$ $8.41\overline{6}$

THINK AND DISCUSS See below.

1. Where would you place the bar in the repeating decimal 12.032032 . . . ?
2. Use models to show the relationship between an improper fraction and a mixed number.

1. $12.\overline{032}$
2. Check students' work.

③ It is important for students to become familiar with this procedure, since it will be used for other computations with fractions on a calculator.

ADDITIONAL EXAMPLES

1. Use a model to represent $2\frac{2}{3}$ as an improper fraction. ($\frac{8}{3}$)
2. Write $3\frac{1}{5}$ as an improper fraction. ($\frac{16}{5}$)
3. Write $\frac{11}{4}$ as a mixed number and as a decimal. ($2\frac{3}{4}$, 2.75)
4. Write as a fraction in lowest terms.
   a. 0.375 ($\frac{3}{8}$) **b.** 2.4 ($2\frac{2}{5}$)
5. Write $\frac{4}{5}$ as a decimal (use a calculator). (0.8)
6. Use a calculator to find the decimal.
   a. $\frac{2}{9}$ ($0.\overline{2}$) **b.** $\frac{5}{6}$ ($0.8\overline{3}$)

Guided Practice

Class Exercises To ensure understanding of the concepts, ask students to explain how they arrived at their answers.

Think and Discuss As a follow-up for question 1, ask where the bar should be placed to represent the repeating decimal 1.2303030. . . ($1.2\overline{30}$)

A ***common error*** when writing a fraction as a decimal is to write the numerator and/or denominator with a decimal point. For example, students may write 0.14 for $\frac{1}{4}$. Remind students that the fraction bar means division. Have them read $\frac{1}{4}$ as *one divided by four* and write 1 ÷ 4.

MATH MINUTES

Give each student a fraction or mixed number between zero and two. Ask students to find a decimal equivalent. Have students graph the points that closely correspond to their numbers.

Closure

Ask students to explain how to write a decimal as a fraction and a fraction as a decimal. Then have students draw, in their math journals, a model for $\frac{10}{3} = 3\frac{1}{3}$ and describe how they are related.

Independent Practice

Assign small groups to work on the Project, question 45. Ask groups to represent the data on a circle graph. The sizes of the sectors may be estimated.

Writing in Math Have students record their responses to question 40 in their math journals. Ask if they found any situations in which both a fraction and a decimal can be used.

MIXED REVIEW

Compare. Use <, >, or =.

1. $|5 - (-8)|$ ■ 2^3 >

2. $6 - 15$ ■ $\frac{27}{3}$ <

Write in lowest terms.

3. $\frac{16}{24}$ $\frac{2}{3}$ 4. $\frac{3a}{a^2}$ $\frac{3}{a}$

Write as a decimal.

5. $\frac{3}{5}$ 0.6 6. $1\frac{1}{2}$ 1.5

7. The Lake Meade Reservoir has a capacity of 34,850,000,000 m^3. Write this number in scientific notation. 3.485×10^{10}

WRITTEN EXERCISES

Use each model to answer the questions.

1. 

a. Let each piece represent $\frac{1}{4}$. What improper fraction is shown? $\frac{11}{4}$

b. What mixed number is shown? $2\frac{3}{4}$

2. 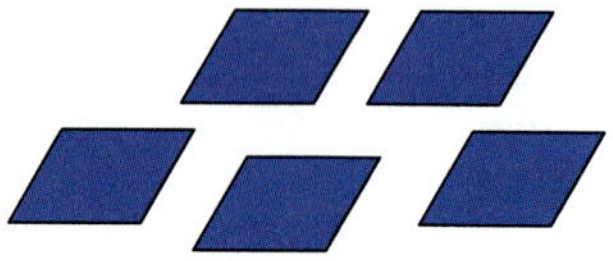

a. Let each piece represent $\frac{1}{3}$. What improper fraction is shown? $\frac{5}{3}$

b. What mixed number is shown? $1\frac{2}{3}$

3. Draw a model to show that $\frac{14}{5} = 2\frac{4}{5}$. Check students' work.

4. Write the length of the segment as an improper fraction and as a mixed number. $\frac{21}{8}$; $2\frac{5}{8}$

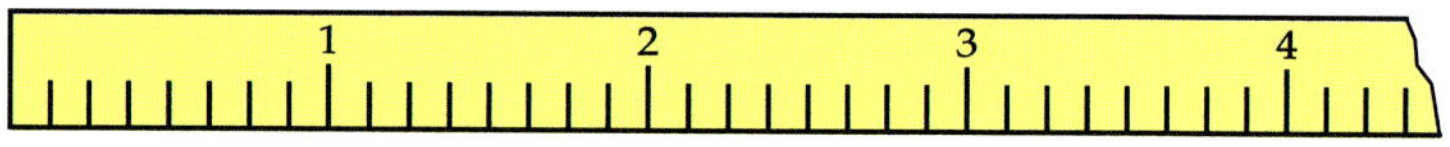

Write each mixed number as an improper fraction.

5. $1\frac{5}{8}$ $\frac{13}{8}$	6. $4\frac{3}{5}$ $\frac{23}{5}$	7. $5\frac{7}{8}$ $\frac{47}{8}$	8. $2\frac{9}{16}$ $\frac{41}{16}$
9. $6\frac{2}{3}$ $\frac{20}{3}$	10. $3\frac{7}{12}$ $\frac{43}{12}$	11. $2\frac{9}{11}$ $\frac{31}{11}$	12. $6\frac{1}{4}$ $\frac{25}{4}$

Write each improper fraction as a mixed number.

13. $\frac{17}{3}$ $5\frac{2}{3}$	14. $\frac{16}{7}$ $2\frac{2}{7}$	15. $\frac{23}{5}$ $4\frac{3}{5}$	16. $\frac{31}{8}$ $3\frac{7}{8}$
17. $\frac{19}{11}$ $1\frac{8}{11}$	18. $\frac{37}{12}$ $3\frac{1}{12}$	19. $\frac{10}{3}$ $3\frac{1}{3}$	20. $\frac{53}{25}$ $2\frac{3}{25}$

Write each decimal as a fraction or mixed number in lowest terms.

21. 0.8 $\frac{4}{5}$	22. 0.17 $\frac{17}{100}$	23. 5.15 $5\frac{3}{20}$	24. 10.01 $10\frac{1}{100}$
25. 2.5 $2\frac{1}{2}$	26. 0.002 $\frac{1}{500}$	27. 6.05 $6\frac{1}{20}$	28. 25.025 $25\frac{1}{40}$

CALCULATOR Write as a decimal.

29. $\frac{7}{25}$ 0.28	30. $\frac{3}{5}$ 0.6	31. $\frac{5}{8}$ 0.625	32. $\frac{9}{20}$ 0.45
33. $\frac{5}{9}$ $0.\overline{5}$	34. $\frac{7}{11}$ $0.\overline{63}$	35. $5\frac{3}{8}$ 5.375	36. $24\frac{7}{15}$ $24.4\overline{6}$

37. Copy and complete the chart. The first row is done for you.

Decimal	Read As	Fraction
0.23	twenty-three hundredths	$\frac{23}{100}$
0.18	eighteen hundredths	$\frac{18}{100}$
5.73	five and seventy-three hundredths	$5\frac{73}{100}$
0.9	nine tenths	$\frac{9}{10}$

38. Write an improper fraction with the greatest possible value using the digits 3, 5, and 6. Write this as a mixed number and as a decimal. Greatest is $\frac{5^6}{3} = 5{,}208\frac{1}{3}$, $5{,}208.\overline{3}$.

39. ***DATA FILE 5 (pp. 180–181)*** Write fractions to represent the number of instruments of each type in a typical symphony orchestra.

a. woodwinds $\frac{15}{104}$ **b.** brasses $\frac{7}{52}$ **c.** percussion $\frac{7}{104}$ **d.** bowed strings $\frac{33}{52}$

40. ***WRITE*** Describe three situations when you would use fractions. Describe three situations when you would use decimals. See right.

40. Answers will vary. Examples are given. Fractions: recipes, sewing, carpentry
Decimals: sales tax, tipping, money

DATA **Use the circle graph at the right.**

41. Did about half the people respond that they see a movie less than once per month? yes

42. Did at least $\frac{1}{4}$ of the people asked say they see a movie two to three times a week? yes

43. Do one out of ten people see a movie once a week? yes

44. ***CALCULATOR***

a. Write each fraction as a decimal. 0.29, 0.56, 0.05, 0.1

b. Do at least 0.5 of the respondents see a movie more than once a week? no

45. ***PROJECT*** Choose five movies or sports teams or a topic of your own. Collect data about the favorites in your class. Write fractions and decimals to describe the choices.

How Often Do You Go to the Movies?

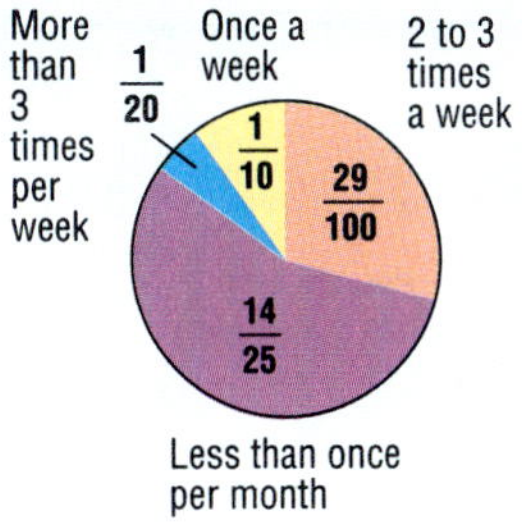

Lesson Follow-up

Reteaching Activity Use pattern blocks to model mixed numbers and improper fractions. Have students use the yellow hexagons as whole units. Ask them to model $\frac{5}{2}$ by placing five red trapezoids on top of the hexagons. Have students observe that $2\frac{1}{2}$ hexagons are covered. Repeat the activity with $\frac{7}{3}$ (blue), and $\frac{15}{6}$ (green).

LESSON RESOURCES

TRB Practice E-3
TRB Enrichment E-4

Practice

For use after 5-2 (pp. 188–191)

Fractions and Decimals

Use each model to answer the questions.

1. △ △ △ △ △ △ △
   a. Let each piece represent $\frac{1}{5}$. What improper fraction is shown? $\frac{7}{5}$
   b. What mixed number is shown? $1\frac{2}{5}$
   c. What decimal is shown? 1.4

2. □ □ □ □ □
   a. Let each piece represent $\frac{1}{2}$. What improper fraction is shown? $\frac{5}{2}$
   b. What mixed number is shown? $2\frac{1}{2}$
   c. What decimal is shown? 2.5

Write each mixed number as an improper fraction.

3. $2\frac{2}{3}$ $\frac{8}{3}$ 4. $4\frac{3}{5}$ $\frac{23}{5}$ 5. $1\frac{6}{7}$ $\frac{13}{7}$ 6. $5\frac{1}{2}$ $\frac{11}{2}$
7. $6\frac{5}{8}$ $\frac{53}{8}$ 8. $2\frac{7}{11}$ $\frac{29}{11}$ 9. $3\frac{1}{9}$ $\frac{28}{9}$ 10. $8\frac{3}{4}$ $\frac{35}{4}$

Write each improper fraction as a mixed number.

11. $\frac{10}{3}$ $3\frac{1}{3}$ 12. $\frac{17}{5}$ $3\frac{2}{5}$ 13. $\frac{29}{4}$ $7\frac{1}{4}$ 14. $\frac{39}{7}$ $5\frac{4}{7}$
15. $\frac{40}{9}$ $4\frac{4}{9}$ 16. $\frac{53}{6}$ $8\frac{5}{6}$ 17. $\frac{81}{10}$ $8\frac{1}{10}$ 18. $\frac{109}{15}$ $7\frac{4}{15}$

Write each decimal as a mixed number or fraction in lowest terms.

19. 0.4 $\frac{2}{5}$ 20. 0.75 $\frac{3}{4}$ 21. 0.16 $\frac{4}{25}$ 22. 1.95 $1\frac{19}{20}$
23. 2.34 $2\frac{17}{50}$ 24. 0.09 $\frac{9}{100}$ 25. 8.8 $8\frac{4}{5}$ 26. 5.008 $5\frac{1}{125}$

CALCULATOR **Write as a decimal.**

27. $\frac{17}{20}$ 0.85 28. $\frac{7}{8}$ 0.875 29. $\frac{9}{16}$ 0.5625 30. $\frac{13}{25}$ 0.52
31. $3\frac{1}{8}$ 3.125 32. $6\frac{9}{32}$ 6.28125 33. $2\frac{87}{125}$ 2.696 34. $4\frac{31}{50}$ 4.62

35. Write an improper fraction with the greatest possible value using each of the digits 5, 7, and 9 once. Write this as a mixed number and as a decimal.
$\frac{97}{5}$; $19\frac{2}{5}$; 19.4

Chapter 5 E-3

Enrichment

For use after 5-2 (pp. 188–191)

Nines in the Denominator

Recall that you can indicate repeating digits in a decimal by writing a bar over the repeating digits.

Example $\frac{1}{3} = 0.33333...$ Shorthand: $\frac{1}{3} = 0.\overline{3}$

Example $\frac{51}{110} = 0.4636363...$ Shorthand: $\frac{51}{110} = 0.4\overline{63}$

Use your calculator to do the following exercises.

Write using a bar.

1. $\frac{17}{18}$ $0.9\overline{4}$ 2. $\frac{3}{11}$ $0.\overline{27}$ 3. $\frac{14}{55}$ $0.2\overline{54}$
4. $\frac{8}{33}$ $0.\overline{24}$ 5. $\frac{68}{153}$ $0.\overline{4}$ 6. $\frac{7{,}067}{990}$ $7.1\overline{38}$

When only nines appear in a denominator, you can predict the decimal equivalent of a fraction.

Write as a decimal.

7. $\frac{4}{9}$ $0.\overline{4}$ 8. $\frac{7}{9}$ $0.\overline{7}$ 9. $\frac{2}{9}$ $0.\overline{2}$ 10. $\frac{5}{9}$ $0.\overline{5}$

11. Predict the decimal equivalent of $\frac{8}{9}$. $0.\overline{8}$ Answers may vary.
12. Find the answer using your calculator. $0.\overline{8}$

Write as a decimal.

13. $\frac{37}{99}$ $0.\overline{37}$ 14. $\frac{82}{99}$ $0.\overline{82}$ 15. $\frac{5}{99}$ $0.\overline{05}$ 16. $\frac{61}{99}$ $0.\overline{61}$

17. Predict the decimal equivalent of $\frac{47}{99}$. $0.\overline{47}$ Answers may vary.
18. Find the answer using your calculator. $0.\overline{47}$

Write as a decimal.

19. $\frac{571}{999}$ $0.\overline{571}$ 20. $\frac{7}{999}$ $0.\overline{007}$ 21. $\frac{365}{999}$ $0.\overline{365}$ 22. $\frac{998}{999}$ $0.\overline{998}$

23. Predict the decimal equivalent of $\frac{135}{999}$. $0.\overline{135}$ Answers may vary.
24. Find the answer using your calculator. $0.\overline{135}$
25. Predict the decimal equivalent of $\frac{8{,}705}{99{,}999}$. $0.\overline{08705}$ Answers may vary.

E-4 Chapter 5

LESSON QUIZ

1. Write $5\frac{3}{7}$ as an improper fraction. ($\frac{38}{7}$)
2. Write $\frac{75}{8}$ as a mixed number. ($9\frac{3}{8}$)
3. Write 2.08 as a fraction in lowest terms. ($2\frac{2}{25}$)
4. Write as a decimal.
   a. $\frac{7}{5}$ (1.4) b. $\frac{9}{2}$ (4.5)
   c. $\frac{1}{4}$ (0.25) d. $\frac{4}{9}$ ($0.\overline{4}$)

Assignment Guide

Basic 1-2, 4-8, 13-16, 21-24, 29-32, 37, 39-40, 43-45, MR All
Average 3-4, 7-10, 15-18, 23-26, 29-37, 40-45, MR All
Enriched 7-12, 17-20, 25-28, 31-36, 38, 40-45, MR All

FOR THE NEXT LESSON

TRB Teaching Resources, rulers, paper, pencil, math journal

Lesson Focus

Materials

TRB Teaching Resources, ruler, paper, pencil, math journal

Vocabulary/Symbols

rational number

MOTIVATING THE LESSON

Use blank number lines or have students draw number lines. Ask student to show these points on the number line: 1, $\frac{1}{2}$, $\frac{3}{4}$, and 2. Have students locate the opposite of 1 (-1) and the opposite of 2 (-2). Ask if students think that $\frac{1}{2}$ and $\frac{3}{4}$ have opposites (yes) and have them suggest locations for $-\frac{1}{2}$ and $-\frac{3}{4}$ on the number line.

Instruction

Mention that a point representing a positive number on the number line has an opposite point. Fractions, whole numbers, and their opposites are all *rational numbers*.

1 Have students approximate the location of any of these numbers on their number lines.

2 For the varying placement of the negative sign, remind students that the fraction bar represents division and that the quotient of a negative number and a positive number is negative.

OBJECTIVE: *To study the meaning of rational numbers.*

5-3 Rational Numbers

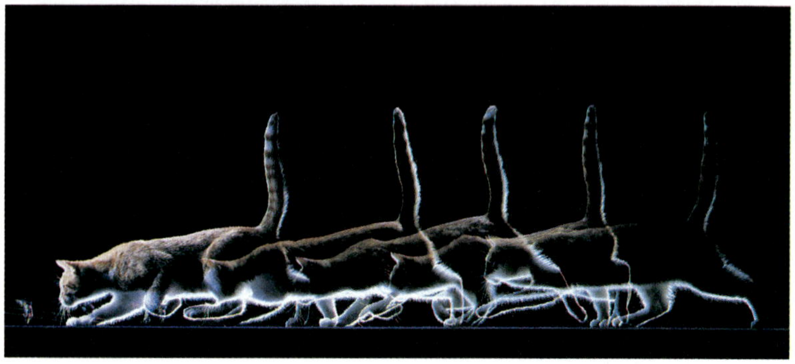

The photograph above was taken with a strobe light that flashed at $\frac{1}{1{,}000}$ of a second intervals. We call the fraction $\frac{1}{1{,}000}$ a *rational number*.

THINK In $\frac{a}{b}$, why can't b equal zero? Division by zero is undefined.

Rational Number	A rational number is a number you write in the form $\frac{a}{b}$, where a is any integer, and b is a nonzero integer.

Fractions are rational numbers. Other numbers are rational if you can express them as fractions.

Proper fractions	$\frac{2}{3}, -\frac{1}{2}$
Improper fractions	$\frac{4}{3}, \frac{10}{8}$
Mixed numbers	$1\frac{5}{6}, 2\frac{3}{11}$
1 **Integers**	$-6 = -\frac{6}{1}, 0 = \frac{0}{1}$
Variables	$\frac{a}{4}, \frac{3a}{4}, \frac{5a}{b}, \frac{2a}{3b}$
Some decimals	$0.5 = \frac{1}{2}, -9.8 = -\frac{98}{10}, 0.\overline{8} = \frac{8}{9}$

THINK What are some equivalent fractions for $\frac{a}{b}$? $\frac{an}{bn}, \frac{a^2}{ab}$

You can express a rational number as a set of equivalent fractions.

Example 1 $\frac{1}{2} = \frac{2}{4} = \frac{3}{6} = \cdots$ Both numerator and denominator are positive.

$\frac{1}{2} = \frac{-1}{-2} = \frac{-2}{-4} = \frac{-3}{-6} = \cdots$ Both numerator and denominator are negative.

You can write negative rational numbers in three ways.

2 ***Example 2*** $-\frac{7}{9} = \frac{-7}{9} = \frac{7}{-9}$

Teaching Tip

Keep a supply of unlabeled number lines available for students to use. Number lines with evenly spaced tic marks will aid students in graphing points correctly.

▼ You can show rational numbers on a number line.

Example 3 **Graph each point on a number line.**

a. 0.5 **b.** $-1\frac{1}{2}$ **c.** $\frac{3}{4}$

Solution

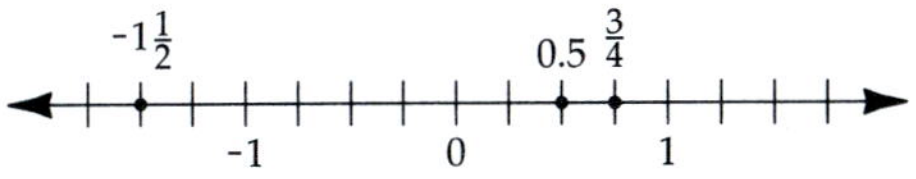

▼ You can find the absolute value of a rational number.

Example 4 The absolute value of $-\frac{5}{6}$ is $\frac{5}{6}$.

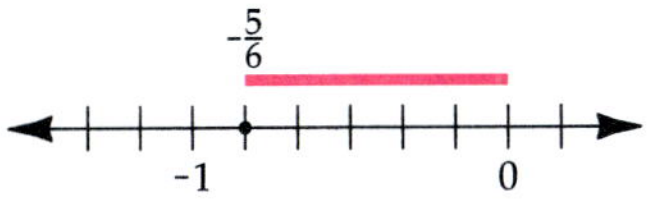

The distance from $-\frac{5}{6}$ to 0 is $\frac{5}{6}$.

Write: $\left|-\frac{5}{6}\right| = \frac{5}{6}$

THINK What is the opposite of $-\frac{a}{b}$? $\frac{a}{b}$

③

▼ The fraction bar is a grouping symbol similar to parentheses.

Example 5 **Evaluate the expression $\frac{x+5}{y}$ for $x = 7$ and $y = -8$.**

Solution

$\frac{x+5}{y} = \frac{7+5}{-8}$ Substitute values.

$= \frac{12}{-8}$ Compute following order of operations.

$= -1\frac{1}{2}$ Write as a mixed number.

④

CLASS EXERCISES

Give a rational number to represent each situation.

1. a loss of one-and-one-half pounds $-1\frac{1}{2}$
2. moving the time ahead a quarter of an hour on a clock $+\frac{1}{4}$

Write a rational number for each point on the number line.

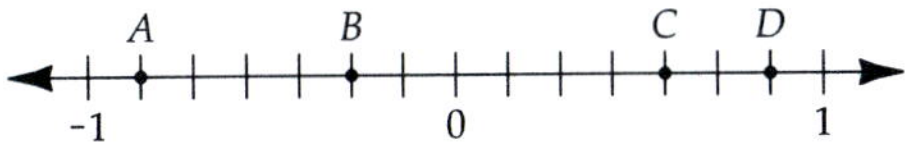

3. A $-\frac{6}{7}$ 4. B $-\frac{2}{7}$ 5. C $\frac{4}{7}$ 6. D $\frac{6}{7}$

Give the opposite and the absolute value.

7. $1\frac{1}{3}$ $-1\frac{1}{3}, 1\frac{1}{3}$
8. $-\frac{3}{5}$ $\frac{3}{5}, \frac{3}{5}$
9. $\frac{5}{-9}$ $\frac{5}{9}, \frac{5}{9}$
10. $-2\frac{2}{3}$ $2\frac{2}{3}, 2\frac{2}{3}$
11. $\frac{1}{4}$ $-\frac{1}{4}, \frac{1}{4}$
12. $\frac{4}{-5}$ $\frac{4}{5}, \frac{4}{5}$
13. $\frac{5}{8}$ $-\frac{5}{8}, \frac{5}{8}$
14. $1\frac{1}{2}$ $-1\frac{1}{2}, 1\frac{1}{2}$

THINK AND DISCUSS

1. Is zero a rational number? Why or why not?
2. Is a set of equivalent fractions equal to the same decimal? yes
3. How many rational numbers are there between 1 and 2? infinite

1. yes $\frac{0}{1}$

MATH MINUTES

Give groups of students an algebraic expression such as $\frac{a^2-5}{b}$. Have students use a spinner to get a value for a and b. Have the entire group agree on an answer for each round.

③ Remind students that the absolute value of a nonzero number is positive because the absolute value is the distance from zero on the number line.

④ Emphasize that it is important to first substitute the values, and then follow the order of operations to evaluate the resulting expression.

ADDITIONAL EXAMPLES

1. Write three fractions equivalent to $-\frac{2}{3}$. ($-\frac{4}{6}$, $-\frac{6}{9}$, $-\frac{8}{12}$)
2. What two other ways can you write $\frac{-5}{6}$? ($\frac{5}{-6}$, $-\frac{5}{6}$)
3. Graph on a number line.
   a. -0.25 **b.** $\frac{-3}{4}$
4. What is the absolute value of $-1\frac{1}{3}$? ($1\frac{1}{3}$)
5. Evaluate $\frac{a^2}{b}$ for $a = -3$ and $b = -2$. ($-4\frac{1}{2}$)

Guided Practice

Class Exercises After working through questions 7-14, ask students what the midpoint of a nonzero rational number and its opposite is on a number line. (zero)

Think and Discuss For question 3, ask a series of questions that establishes the idea that there is always a point between two other points. For example, you might ask students to name a point between $1\frac{9}{10}$ and 2. (Answers will vary; $1\frac{99}{100}$) Then ask students to name a point between $1\frac{99}{100}$ and 2. (Answers will vary; $1\frac{999}{1000}$) Continue until students realize that there are countless, or infinite, numbers of points between 1 and 2.

A ***common error*** is to use $\frac{-2}{-3}$ as a negative quantity. Remind students that $\frac{-2}{-3}$ means $-2 \div -3$ and therefore is equivalent to $\frac{2}{3}$. Give students rational numbers with negative signs in various positions and work with them to identify the positive numbers and the negative numbers.

Closure

Writing in Math Ask small groups of students to think of as many different representations of rational numbers as they can, such as a positive integer, a negative fraction, etc., and to write them in their math journals.

Independent Practice

For questions 25-28, encourage students to substitute both positive and negative values for the variables in order to test the statement. Students may assume that $b \neq 0$ for each question.

Ask students to give an example of when questions 26-28 are false. (Answers may vary. 26: for $a = -1$ and $b = -2$; 27: for $a = -1$ and $b = 2$; 28: for $a = 1$ and $b = 1$)

Have students work in groups to complete questions 35 and 36. If necessary, remind them that natural numbers are sometimes referred to as counting numbers. As a follow-up, ask if all integers, whole numbers, and natural numbers are rational numbers. (yes)

As an extension, refer students to the calculator activity on p. 180E of the Chapter Overview. See also, TRB Exploring O-19.

MIXED REVIEW

Evaluate for x = -3 and y = 2.

1. $x + 3$ 0
2. xy -6
3. $x - y$ -5

Write as a fraction or mixed number in lowest terms.

4. $-\frac{9}{24}$ $-\frac{3}{8}$
5. $\frac{6a^2}{15}$ $\frac{2a^2}{5}$

Write as a decimal.

6. $\frac{4}{5}$ 0.8
7. $-3\frac{1}{4}$ -3.25
8. Consider the pattern $\frac{1}{2}, \frac{2}{4}, \frac{3}{8}, \frac{4}{16}, \ldots$ How does the pattern look when you write the fractions in lowest terms? $\frac{1}{2}, \frac{1}{2}, \frac{3}{8}, \frac{1}{4}$

PROBLEM SOLVING HINT

Try guess and test.

Evaluate for $a = 6$, and $b = -5$. Write in lowest terms.

15. $\frac{a+b}{3}$ $\frac{1}{3}$ 16. $\frac{a-b}{4}$ $2\frac{3}{4}$ 17. $\frac{a+9}{b}$ -3

WRITTEN EXERCISES

Write a rational number to represent each situation.

1. the number of dollars in nine quarters. $\frac{9}{4}$
2. a loss of two dollars and seventy-five cents -2.75

Write a rational number for each point on the number line.

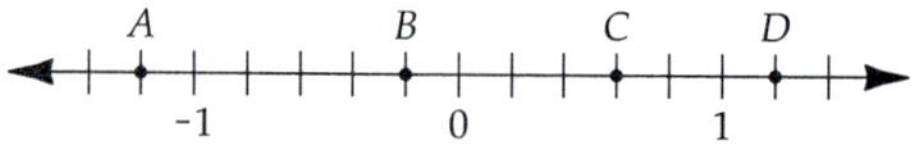

3. A $-1\frac{1}{5}$ 4. B $-\frac{1}{5}$ 5. C $\frac{3}{5}$ 6. D $1\frac{1}{5}$

Graph each point on a number line. Check students' work.

7. $1\frac{1}{2}$ 8. -0.5 9. -3.5 10. $\frac{3}{4}$

Write the opposite and the absolute value.

11. $\frac{-4}{9}$ $\frac{4}{9}, \frac{4}{9}$ 12. -1.73 1.73, 1.73 13. $1\frac{2}{3}$ $-1\frac{2}{3}, 1\frac{2}{3}$ 14. $-2\frac{1}{5}$ $2\frac{1}{5}, 2\frac{1}{5}$

15. Which rational numbers are equal to $-\frac{4}{5}$? a; c
 a. $\frac{4}{-5}$ b. 0.8 c. $-\frac{16}{20}$ d. $\frac{-4}{-5}$

16. Which rational numbers are equal to $\frac{9}{5}$? a; b; d
 a. $\frac{-9}{-5}$ b. $-9 \cdot \frac{1}{-5}$ c. -1.8 d. $\frac{18}{10}$

Evaluate. Write in lowest terms.

17. $\frac{a}{b}$ for $a = 20, b = 25$ $\frac{4}{5}$ 18. $\frac{a^2}{b^2}$ for $a = 4, b = -5$ $\frac{16}{25}$

19. $\frac{a+3}{b}$ for $a = 5, b = -2$ -4 20. $\frac{(a+3)(a-3)}{a^2-9}$ for $a = 5$ 1

Write each rational number in lowest terms.

21. $\frac{4a}{8b}$ $\frac{a}{2b}$ 22. $\frac{2 \cdot 5 \cdot a}{4 \cdot 25 \cdot b}$ $\frac{a}{10b}$ 23. $\frac{4 \cdot a^2}{16 \cdot a \cdot b}$ $\frac{a}{4b}$ 24. $\frac{5 \cdot a^2 \cdot b}{25 \cdot a \cdot b^2}$ $\frac{a}{5b}$

Write *sometimes, always,* or *never* to tell when each is true.

25. $\frac{3a}{3b} = \frac{a}{b}$ always 26. $\frac{a^2}{b} > \frac{a}{b}$ sometimes 27. $\frac{a}{b} = \left|\frac{a}{b}\right|$ sometimes 28. $\frac{a}{b}(-1) > 0$ sometimes

Complete.

29. All ▪ rational numbers are less than zero. (negative)

30. The set of rational numbers is made up of the positive rational numbers, the negative rational numbers, and ▪. (zero)

31. The opposite of a ▪ rational number is positive. (negative)

32. The absolute value of a negative rational number is ▪. (positive)

33. A number is less than zero. Its opposite is a mixed number greater than zero. Is the number a rational number? (yes)

34. ***DATA FILE 5 (pp. 180–181)*** Write fractions that represent these musical notes: whole note, half note, quarter note, eighth note, and sixteenth note. ($\frac{1}{1}, \frac{1}{2}, \frac{1}{4}, \frac{1}{8}, \frac{1}{16}$)

35. Copy and complete the chart. Write *yes* or *no*.

Number	Rational Number	Integer	Whole Number	Natural Number
2	▪ yes	▪ yes	▪ yes	▪ yes
$-\frac{1}{2}$	▪ yes	▪ no	▪ no	▪ no
0.5	▪ yes	▪ no	▪ no	▪ no
$\frac{-12}{-3}$	▪ yes	▪ yes	▪ yes	▪ yes
$0 \cdot \frac{a}{b}$	▪ yes	▪ yes	▪ yes	▪ no

36. ***WRITE*** a sentence about rational numbers. How are rational numbers different from integers? How are they similar?

36. Rational numbers include fractions. Integers do not. Both integers and rational numbers include negative values.

FLASHBACK
Natural numbers are the positive whole numbers.

TEST YOURSELF

Write in lowest terms.

1. $\frac{12}{15}$ ($\frac{4}{5}$) **2.** $\frac{18}{24}$ ($\frac{3}{4}$) **3.** $\frac{6}{20}$ ($\frac{3}{10}$) **4.** $\frac{5a^3b^2}{15ab^2}$ ($\frac{a^2}{3}$) **5.** $\frac{35mn^3}{4m}$ ($\frac{35n^3}{4}$)

Write three fractions equivalent to the given fraction.

6. $\frac{5}{8}$ ($\frac{10}{16}, \frac{15}{24}, \frac{20}{32}$) **7.** $-1\frac{2}{5}$ ($-1\frac{4}{10}, -1\frac{6}{15}, -1\frac{8}{20}$) **8.** $\frac{9}{12}$ ($\frac{3}{4}, \frac{18}{24}, \frac{27}{36}$) **9.** $\frac{2b}{3c}$ ($\frac{4b}{6c}, \frac{8b}{12c}, \frac{6b}{9c}$) **10.** $\frac{16r}{20s}$ ($\frac{4r}{5s}, \frac{8r}{10s}, \frac{32r}{40s}$)

Write the opposite and the absolute value.

11. $\frac{2}{3}$ ($-\frac{2}{3}, \frac{2}{3}$) **12.** $-2\frac{5}{6}$ ($2\frac{5}{6}, 2\frac{5}{6}$) **13.** $1\frac{7}{16}$ ($-1\frac{7}{16}, 1\frac{7}{16}$) **14.** $-2\frac{3}{4}$ ($2\frac{3}{4}, 2\frac{3}{4}$)

Lesson Follow-up

Reteaching Activity Have students graph several positive rational numbers on a number line. Then have them fold the number line at zero and hold the paper to the light to see where the opposites belong on the number line. Have students graph the opposites and observe that the absolute value and the opposite of a negative number is the same positive number.

LESSON RESOURCES

TRB Practice E-5
TRB Enrichment E-6

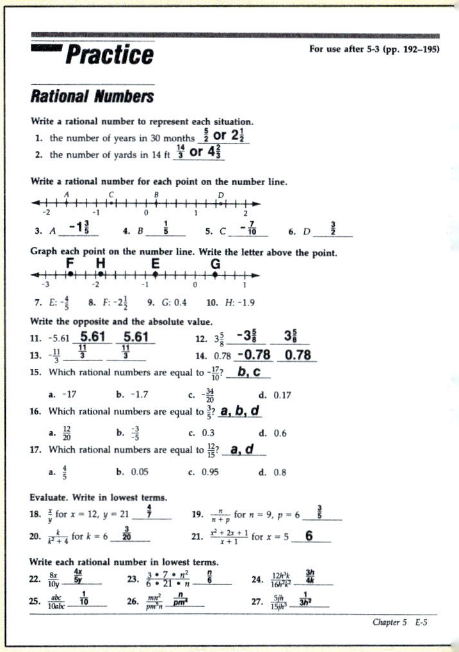

Practice For use after 5-3 (pp. 192–195)

Rational Numbers

Write a rational number to represent each situation.

1. the number of years in 30 months $\frac{5}{2}$ or $2\frac{1}{2}$
2. the number of yards in 14 ft $\frac{14}{3}$ or $4\frac{2}{3}$

Write a rational number for each point on the number line.

3. A $-1\frac{3}{5}$ 4. B $\frac{1}{5}$ 5. C $-\frac{7}{10}$ 6. D $\frac{3}{2}$

Graph each point on the number line. Write the letter above the point.

7. E: $-\frac{4}{5}$ 8. F: $-2\frac{1}{2}$ 9. G: 0.4 10. H: -1.9

Write the opposite and the absolute value.

11. -5.61 5.61 5.61 12. $3\frac{5}{8}$ $-3\frac{5}{8}$ $3\frac{5}{8}$
13. $-\frac{11}{3}$ $\frac{11}{3}$ $\frac{11}{3}$ 14. 0.78 -0.78 0.78

15. Which rational numbers are equal to $-\frac{17}{10}$? b, c
a. -17 b. -1.7 c. $-\frac{34}{20}$ d. 0.17

16. Which rational numbers are equal to $\frac{3}{5}$? a, b, d
a. $\frac{12}{20}$ b. $\frac{-3}{-5}$ c. 0.3 d. 0.6

17. Which rational numbers are equal to $\frac{12}{15}$? a, d
a. $\frac{4}{5}$ b. 0.05 c. 0.95 d. 0.8

Evaluate. Write in lowest terms.

18. $\frac{x}{y}$ for $x = 12$, $y = 21$ $\frac{4}{7}$ 19. $\frac{n}{n+p}$ for $n = 9$, $p = 6$ $\frac{3}{5}$
20. $\frac{k}{k^2+4}$ for $k = 6$ $\frac{3}{20}$ 21. $\frac{x^2+2x+1}{x+1}$ for $x = 5$ 6

Write each rational number in lowest terms.

22. $\frac{8x}{10y}$ $\frac{4x}{5y}$ 23. $\frac{3 \cdot 7 \cdot n^2}{6 \cdot 21 \cdot n}$ $\frac{n}{6}$ 24. $\frac{12h^3k}{16h^2k^2}$ $\frac{3h}{4k}$
25. $\frac{abc}{10abc}$ $\frac{1}{10}$ 26. $\frac{mn^2}{pm^5n}$ $\frac{n}{pm^4}$ 27. $\frac{5jh}{15jh^3}$ $\frac{1}{3h^2}$

Chapter 5 E-5

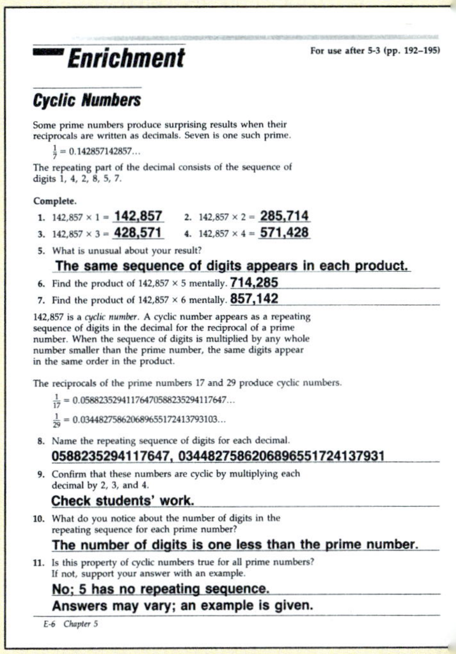

Enrichment For use after 5-3 (pp. 192–195)

Cyclic Numbers

Some prime numbers produce surprising results when their reciprocals are written as decimals. Seven is one such prime.

$\frac{1}{7} = 0.142857142857\ldots$

The repeating part of the decimal consists of the sequence of digits 1, 4, 2, 8, 5, 7.

Complete.

1. 142,857 × 1 = 142,857 2. 142,857 × 2 = 285,714
3. 142,857 × 3 = 428,571 4. 142,857 × 4 = 571,428
5. What is unusual about your result?
The same sequence of digits appears in each product.
6. Find the product of 142,857 × 5 mentally. 714,285
7. Find the product of 142,857 × 6 mentally. 857,142

142,857 is a *cyclic number*. A cyclic number appears as a repeating sequence of digits in the decimal for the reciprocal of a prime number. When the sequence of digits is multiplied by any whole number smaller than the prime number, the same digits appear in the same order in the product.

The reciprocals of the prime numbers 17 and 29 produce cyclic numbers.

$\frac{1}{17} = 0.05882352941176470588235294117647\ldots$

$\frac{1}{29} = 0.0344827586206896551724137931034\ldots$

8. Name the repeating sequence of digits for each decimal.
0588235294117647, 0344827586206896551724137931
9. Confirm that these numbers are cyclic by multiplying each decimal by 2, 3, and 4.
Check students' work.
10. What do you notice about the number of digits in the repeating sequence for each prime number?
The number of digits is one less than the prime number.
11. Is this property of cyclic numbers true for all prime numbers? If not, support your answer with an example.
No; 5 has no repeating sequence.
Answers may vary; an example is given.

E-6 Chapter 5

LESSON QUIZ

Write the opposite and absolute value

1. $-\frac{6}{7}$ $\left(\frac{6}{7}, \frac{6}{7}\right)$
2. 1.8 (-1.8, 1.8)
3. $1\frac{2}{3}$ $\left(1\frac{2}{3}, 1\frac{2}{3}\right)$
4. Which rational numbers are equal to $-\frac{1}{4}$? (b, c)
   a. $\frac{1}{4}$ b. $\frac{1}{-4}$
   c. -0.25 d. -4
5. Evaluate $\frac{2x}{x-y}$ for $x = -3$ and $y = -6$. (-2)

Assignment Guide

Basic 1-8, 11-19, 21-23, 29-34, MR, TY All
Average 3-6, 9-19, 21-25, 27, 29-34, MR, TY All
Enriched 11-20, 22-28, 35-36, MR, TY All

FOR THE NEXT LESSON

fraction bars, math journal

Lesson Focus

Materials

Fraction bars, math journal

Vocabulary/Symbols

Least common denominator (LCD)

MOTIVATING THE LESSON

Write $\frac{3}{7}$ and $\frac{5}{7}$ on the chalkboard. Ask students to write a description in their math journals of how to determine which of the two fractions is greater. As students read their explanations, draw attention to the common denominator.

Instruction

Display $\frac{2}{3}$ and $\frac{3}{4}$ and ask for suggestions of how to compare them. Have students use the fraction bars to demonstrate that $\frac{2}{3} < \frac{3}{4}$. Have them also model $\frac{1}{2} > \frac{5}{12}$, and $\frac{5}{6} > \frac{7}{12}$. Mention that when a model is not available, comparing fractions with different denominators can be done by writing equivalent fractions that have a common denominator.

1 Review the definition of LCM with the students.

2 Encourage the use of calculators to write a fraction as a decimal. This focuses the students' attention on comparing and ordering the numbers.

OBJECTIVE: *To compare and order fractions and decimals.*

5-4 Comparing and Ordering Rational Numbers

▼ The Houston Rockets won 10 out of 12 home games. The Denver Stars won 8 out of 10 games. Which team won a greater fraction of games?

Houston's games → $\frac{10}{12}$

Denver's games → $\frac{8}{10}$

Compare $\frac{10}{12}$ and $\frac{8}{10}$ using fraction strips.

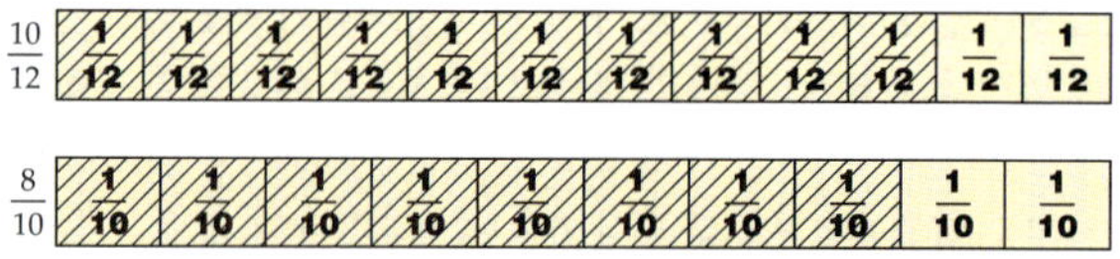

$$\frac{10}{12} > \frac{8}{10}$$

▼ You can also compare fractions by finding their *least common denominator (LCD).*

Least Common Denominator	The least common denominator of two or more fractions is the LCM of the denominators.

Example 1 **Compare $\frac{5}{12}$ and $\frac{4}{9}$.**

Solution

1. Find the LCD. — The LCD of $\frac{5}{12}$ and $\frac{4}{9}$ is 36.
2. Find equivalent fractions. — $\frac{5}{12} = \frac{15}{36}$; $\frac{4}{9} = \frac{16}{36}$
3. Compare. — $\frac{16}{36} > \frac{15}{36}$, so $\frac{4}{9} > \frac{5}{12}$.

1

▼ To compare fractions and decimals, you can first write each in the same form.

Example 2 **Compare 0.8 and $\frac{3}{5}$.**

Solution

1. Use a calculator to write $\frac{3}{5}$ as a decimal.
   3 [÷] 5 [=] 0.6
2. Compare.
   Since $0.8 > 0.6$, $0.8 > \frac{3}{5}$.

2

Teaching Tip

Look at students' math journals regularly. Make comments about the organization of entries, offering suggestions if necessary.

▼ Writing fractions as decimals often makes it easier to order the fractions from least to greatest.

Example 3 **Order $\frac{13}{40}$, $\frac{9}{32}$, $\frac{5}{16}$, $\frac{8}{25}$ from least to greatest.**

Solution Use a calculator to write each fraction as a decimal.

$\frac{13}{40} = 0.325 \quad \frac{9}{32} = 0.28125 \quad \frac{5}{16} = 0.3125 \quad \frac{8}{25} = 0.32$

$0.28125 < 0.3125 < 0.32 < 0.325$ Order the decimals.

$\updownarrow \quad \updownarrow \quad \updownarrow \quad \updownarrow$

$\frac{9}{32} < \frac{5}{16} < \frac{8}{25} < \frac{13}{40}$ Order the fractions.

▼ You can also order fractions on a number line.

Example 4 **Order $\frac{1}{4}$, -0.2, 1.1, and $-\frac{3}{5}$ from least to greatest.**

Solution Use a calculator to write fractions as decimals.

$\frac{1}{4} = 0.25, -\frac{3}{5} = -0.6$

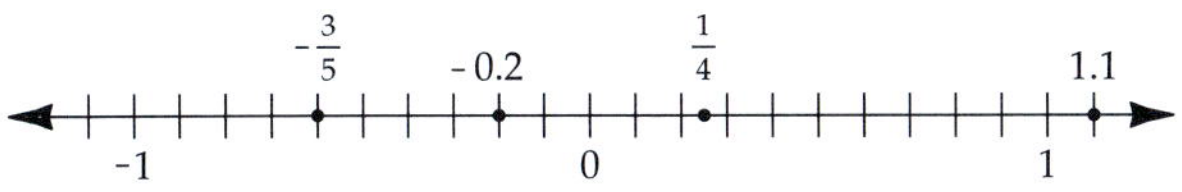

THINK How does writing fractions as decimals help when ordering on the number line?
It is easy to order decimals by the values of the digits in their decimal places.

CLASS EXERCISES

Write the two fractions modeled and compare them.

1. 
$\frac{5}{6} > \frac{3}{4}$

2. 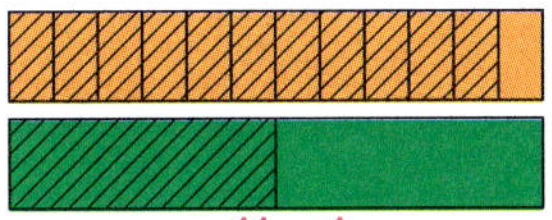
$\frac{11}{12} > \frac{1}{2}$

Compare. Use >, <, or =.

3. $\frac{5}{8}$ ▒ $\frac{3}{4}$ **<**
4. $\frac{7}{15}$ ▒ $\frac{2}{3}$ **<**
5. $\frac{5}{18}$ ▒ $\frac{1}{3}$ **<**
6. $\frac{33}{40}$ ▒ $\frac{5}{8}$ **>**
7. 0.3 ▒ $\frac{-1}{-3}$ **<**
8. 0.22 ▒ $\frac{2}{9}$ **<**
9. 0.63 ▒ $\frac{7}{-11}$ **>**
10. $3\frac{1}{4}$ ▒ 3.2 **>**
11. 4.985 ▒ $4\frac{985}{1,000}$ **=**

Order from least to greatest.

12. $\frac{2}{5}, \frac{7}{20}, \frac{3}{10}$ **$\frac{3}{10}, \frac{7}{20}, \frac{2}{5}$**
13. $\frac{11}{16}, \frac{5}{8}, -\frac{13}{24}$ **$\frac{-13}{24}, \frac{5}{8}, \frac{11}{16}$**
14. $\frac{3}{4}, \frac{7}{10}, \frac{-29}{40}$ **$\frac{-29}{40}, \frac{7}{10}, \frac{3}{4}$**
15. $\frac{6}{11}, \frac{11}{20}, \frac{5}{9}, \frac{14}{-25}$ **$\frac{-14}{25}, \frac{6}{11}, \frac{11}{20}, \frac{5}{9}$**
16. $\frac{7}{33}, \frac{-2}{-9}, \frac{11}{50}, \frac{4}{17}$ **$\frac{7}{33}, \frac{11}{50}, \frac{-2}{-9}, \frac{4}{17}$**
17. $\frac{20}{37}, \frac{6}{11}, \frac{27}{50}, -\frac{51}{90}$ **$\frac{-51}{90}, \frac{27}{50}, \frac{20}{37}, \frac{6}{11}$**

THINK AND DISCUSS **See below.**

1. Suppose you are comparing a fraction and a decimal. Why is it easier to write the fraction as a decimal than to write the decimal as a fraction?
2. Some fractions result in decimals with several places. When comparing, are all of the digits important? Consider $\frac{1}{7}$ ▒ 0.142.

1. You can use a calculator to write the fraction as a decimal.
2. No; you only consider the place value where the values differ. $\frac{1}{7} = 0.\overline{142857}$ compared to $0.\overline{142}$. The ten thousandths' place must be considered.

ADDITIONAL EXAMPLES

1. Compare $\frac{3}{5}$ and $\frac{5}{8}$ by finding the LCD. ($\frac{24}{40} < \frac{25}{40}$; $\frac{3}{5} < \frac{5}{8}$)
2. Compare $\frac{3}{4}$ and 0.7. ($\frac{3}{4} > 0.7$)
3. Order $\frac{4}{5}$, $\frac{-12}{32}$, and $-\frac{1}{3}$ from least to greatest. ($\frac{-12}{32}$, $-\frac{1}{3}$, $\frac{4}{5}$)

Guided Practice

Class Exercises For question 7 suggest that students express both numbers to the hundredths place, which yields 0.30 and 0.33 and then compare the two decimals. Similarly, suggest that students compare 0.220 and $0.2\overline{2}$ in question 8.

Think and Discuss Have students use a calculator to write $\frac{1}{7}$ as $0.\overline{142857}$ for question 2 to show that $\frac{1}{7} > 0.142$.

A ***common error*** is to forget the negative sign when comparing and ordering a list of rational numbers. To avoid and/or correct this error, emphasize this procedure:
1. Find the negative numbers and write them in a group.
2. Find the LCD or write the fractions as decimals to order them.
3. Order the positive numbers.

Closure

Ask student how they would order $\frac{3}{4}$, $\frac{1}{-2}$, 0.12, and -0.8. (-0.8, $-\frac{1}{2}$, 0.12, $\frac{3}{4}$) Encourage them to use a method they are comfortable with. Welcome suggestions, including using mental math.

MATH MINUTES

Assign each member on a team of four students a rational number. At a given signal, team members arrange themselves in order from least to greatest. The first team to order themselves correctly wins.

Independent Practice

For question 32, point out that x is not a whole number.

Lesson Follow-up

Reteaching Activity Give students numbers such as $\frac{3}{4}$, 0.8, $-\frac{1}{8}$, and -0.12 to order from least to greatest, using a calculator. Have them align the decimal points and compare.

($-\frac{1}{8}$, -0.12, $\frac{3}{4}$, 0.8)

LESSON RESOURCES

TRB Practice E-7
TRB Enrichment E-8

Practice

For use after 5-4 (pp. 196–198)

Comparing and Ordering Rational Numbers

Compare. Use >, <, or =.

1. $\frac{2}{3}$ (<) $\frac{7}{9}$ 2. $\frac{3}{5}$ (<) $\frac{7}{10}$ 3. $-\frac{3}{4}$ (>) $-\frac{13}{16}$
4. $\frac{9}{21}$ (=) $\frac{6}{14}$ 5. $-\frac{2}{8}$ (<) $-\frac{7}{32}$ 6. $\frac{7}{9}$ (>) $-\frac{8}{9}$
7. $\frac{5}{8}$ (>) $\frac{7}{12}$ 8. $-\frac{4}{5}$ (>) $-\frac{7}{8}$ 9. $-\frac{4}{18}$ (=) $-\frac{6}{27}$
10. $-\frac{5}{18}$ (<) $-\frac{2}{9}$ 11. $\frac{7}{12}$ (<) $\frac{11}{18}$ 12. $\frac{13}{20}$ (<) $\frac{11}{15}$
13. $-\frac{11}{20}$ (=) $-\frac{22}{40}$ 14. $\frac{6}{5}$ (>) $\frac{1}{5}$ 15. $\frac{11}{28}$ (<) $\frac{14}{19}$

MENTAL MATH Compare. Use >, <, or =.

16. $\frac{8}{17}$ (>) $-\frac{3}{8}$ 17. $-\frac{7}{6}$ (<) $-\frac{6}{7}$ 18. $\frac{-9}{-11}$ (=) $\frac{9}{11}$
19. $\frac{1}{3}$ (>) $-\frac{3}{9}$ 20. $-\frac{12}{6}$ (>) $-\frac{9}{3}$ 21. $\frac{5}{10}$ (<) $\frac{3}{4}$
22. $\frac{7}{11}$ (>) $-\frac{9}{10}$ 23. $-\frac{2}{4}$ (>) $-\frac{5}{8}$ 24. $\frac{5}{12}$ (>) $\frac{3}{10}$

CALCULATOR Compare. Use >, <, or =.

25. $\frac{5}{6}$ (>) $\frac{9}{11}$ 26. $-\frac{11}{17}$ (>) $-\frac{17}{24}$ 27. $\frac{21}{29}$ (<) $\frac{24}{33}$
28. $-\frac{7}{9}$ (<) -0.77 29. $\frac{12}{13}$ (<) 0.9231 30. $\frac{15}{16}$ (<) $\frac{16}{17}$
31. $-\frac{1}{11}$ (<) $-\frac{1}{13}$ 32. $\frac{5}{23}$ (>) $\frac{8}{45}$ 33. $4\frac{11}{19}$ (>) 4.57

CALCULATOR Order from least to greatest.

34. $\frac{3}{8}, \frac{1}{2}, \frac{1}{4}$ **$\frac{1}{4}, \frac{3}{8}, \frac{1}{2}$** 35. $-\frac{11}{15}, -\frac{7}{10}, -\frac{3}{5}$ **$-\frac{11}{15}, -\frac{7}{10}, -\frac{3}{5}$**
36. $\frac{3}{4}, \frac{7}{12}, \frac{11}{16}$ **$\frac{7}{12}, \frac{11}{16}, \frac{3}{4}$** 37. $-\frac{5}{13}, -\frac{7}{15}, -\frac{9}{17}$ **$-\frac{9}{17}, -\frac{7}{15}, -\frac{5}{13}$**
38. $3\frac{11}{15}, 3\frac{19}{24}, 3\frac{13}{18}, 3\frac{17}{21}$ **$3\frac{13}{18}, 3\frac{11}{15}, 3\frac{19}{24}, 3\frac{17}{21}$**

Find a rational number between the given rational numbers. **Answers may vary.**

39. $1\frac{1}{2}$ and 2 **$1\frac{3}{4}$** 40. -3 and -4 **$-3\frac{1}{2}$** 41. $\frac{5}{7}$ and $\frac{13}{14}$ **$\frac{11}{14}$**
42. 0 and $\frac{1}{2}$ **$\frac{1}{4}$** 43. $5\frac{1}{2}$ and 6 **$5\frac{3}{4}$** 44. $\frac{3}{4}$ and $\frac{15}{16}$ **$\frac{13}{16}$**

Chapter 5 E-7

Enrichment

For use after 5-4 (pp. 196–198)

A Rational Maze

Find the correct path from start to finish through the maze. Proceed from one circle to the next only if the second number is greater than the first number.

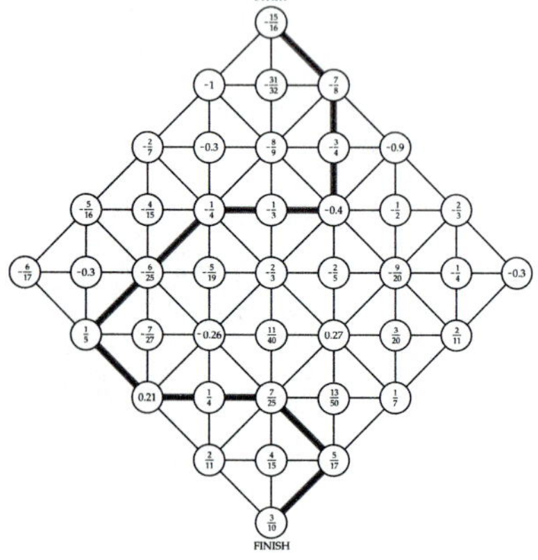

E-8 Chapter 5

MIXED REVIEW

Solve.

1. $0.3x = 2.1$ 7
2. $18.08 = a + 7.5$ 10.58

Find each answer.

3. the opposite of $-7\frac{1}{3}$ $7\frac{1}{3}$
4. $-1\frac{2}{5} + ■ = 0$ $1\frac{2}{5}$
5. $|-2\frac{7}{9}|$ $2\frac{7}{9}$
6. $-4\frac{2}{3} = \frac{■}{3}$ -14
7. Write $-2\frac{3}{8}$ as a decimal. -2.375
8. How many miles will you travel in 3.5 h at a speed of 55 mi/h? 192.5 mi

WRITTEN EXERCISES

Use a model to compare.

1. $\frac{3}{5}$ ■ $\frac{5}{8}$ [<] 2. $\frac{5}{6}$ ■ $\frac{7}{10}$ [>]

Compare. Use >, <, or =.

3. $\frac{13}{18}$ ■ $\frac{7}{9}$ [<] 4. $\frac{11}{12}$ ■ $\frac{5}{6}$ [>] 5. $-\frac{7}{9}$ ■ $-\frac{2}{3}$ [<] 6. $-\frac{5}{6}$ ■ $-\frac{19}{24}$ [<]
7. $\frac{3}{8}$ ■ $\frac{5}{12}$ [<] 8. $\frac{3}{4}$ ■ $\frac{5}{6}$ [<] 9. $-\frac{5}{12}$ ■ $\frac{7}{9}$ [<] 10. $\frac{3}{4}$ ■ $\frac{7}{10}$ [>]

MENTAL MATH **Compare. Use >, <, or =.**

11. $-\frac{3}{19}$ ■ $\frac{1}{200}$ [<] 12. $\frac{-4}{-17}$ ■ $-\frac{5}{2}$ [>] 13. $\frac{(-1)\cdot(-1)}{3}$ ■ $\frac{1}{3}$ [=]

CALCULATOR **Compare. Use >, <, or =.**

14. $\frac{17}{24}$ ■ $\frac{24}{35}$ [>] 15. $-\frac{11}{16}$ ■ $-\frac{19}{28}$ [<] 16. $\frac{15}{22}$ ■ $\frac{23}{34}$ [>] 17. $\frac{11}{25}$ ■ $\frac{17}{30}$ [<]
18. $\frac{3}{8}$ ■ 0.39 [<] 19. $\frac{3}{4}$ ■ 0.752 [<] 20. $-\frac{5}{12}$ ■ -0.34 [<] 21. $2\frac{3}{14}$ ■ 2.22 [<]

CALCULATOR **Order from least to greatest.** 23. $-\frac{5}{12}, -\frac{3}{8}, -\frac{1}{4}$ 24. $-\frac{8}{17}, -\frac{5}{11}, \frac{6}{13}$

22. $\frac{5}{6}, \frac{7}{8}, \frac{19}{24}$ $\frac{19}{24}, \frac{5}{6}, \frac{7}{8}$ 23. $-\frac{5}{12}, -\frac{3}{8}, -\frac{1}{4}$ 24. $-\frac{5}{11}, \frac{6}{13}, -\frac{8}{17}$
25. $-\frac{7}{10}, -\frac{11}{15}, -\frac{7}{12}, -\frac{13}{20}$ $-\frac{11}{15}, -\frac{7}{10}, -\frac{13}{20}, -\frac{7}{12}$ 26. $\frac{17}{20}, \frac{23}{30}, \frac{23}{40}, \frac{13}{24}$ $\frac{13}{24}, \frac{23}{40}, \frac{23}{30}, \frac{17}{20}$ 27. $2\frac{3}{50}, -2\frac{7}{8}, -2\frac{9}{16}, 2\frac{19}{25}$ $-2\frac{7}{8}, -2\frac{9}{16}, 2\frac{3}{50}, 2\frac{19}{25}$

Find a rational number between the given rational numbers.

28. $\frac{1}{2}$ and 1 29. -1 and -2 *30. $2\frac{5}{12}$ and $2\frac{3}{4}$ *31. $-1\frac{3}{4}$ and -2

Answers may vary. 28. $\frac{3}{4}$ 29. $-1\frac{1}{2}$ 30. $2\frac{7}{12}$ 31. $-1\frac{7}{8}$

Compare the rational numbers $\frac{5}{8}$ and $\frac{x}{4}$. Find the values of x that make each statement true.

*32. $\frac{5}{8} = \frac{x}{4}$ $x = 2.5$ *33. $\frac{5}{8} > \frac{x}{4}$ $x < 2.5$ *34. $\frac{5}{8} < \frac{x}{4}$ $x > 2.5$

35. ***DATA FILE 5 (pp. 180–181)*** Write as a fraction in lowest terms the number of tapes sold in 1976 compared with the number of tapes sold in 1986. $\frac{1}{14}$

36. In the high school band there are 15 clarinets. The band has 80 members. Compare the number of clarinets to the number of band members. Write the number as a fraction in lowest terms and as a decimal. $\frac{3}{16}$, 0.1875

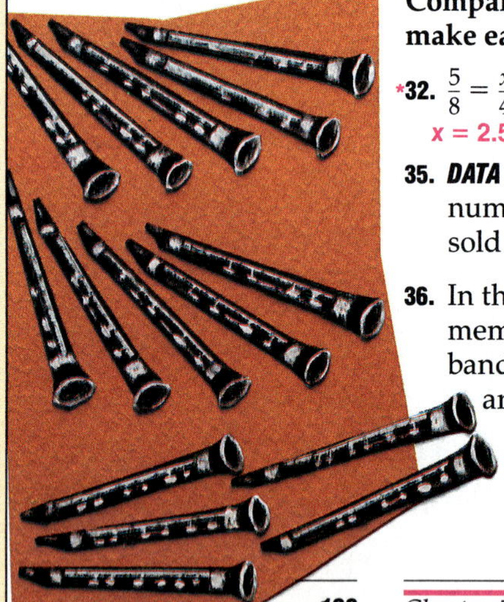

LESSON QUIZ

Compare. Use >, <, or =.

1. $\frac{2}{3}$ ■ $\frac{5}{6}$ (<)
2. $\frac{-1}{8}$ ■ $\frac{1}{-4}$ (>)
3. $\frac{7}{9}$, ■ $-\frac{3}{5}$ (>)
4. 0.4 ■ $\frac{1}{2}$ (<)

Order from least to greatest.

5. $\frac{3}{5}, \frac{-4}{5}, \frac{3}{10}$ $\left(\frac{-4}{5}, \frac{3}{10}, \frac{3}{5}\right)$
6. -0.1, -0.35, $-\frac{1}{4}$ $\left(-0.35, -\frac{1}{4}, -0.1\right)$

FOR THE NEXT LESSON

fraction bars, tangrams, calculator

Assignment Guide

Basic 1-17, 22-23, 25, 35-36, MR All
Average 3-13, 16-19, 23-31, 35-36, MR All
Enriched 3-13, 18-21, 25-34, MR All

OBJECTIVE:
To add and subtract fractions and mixed numbers.

5-5 Adding and Subtracting Rational Numbers

▼ Two thirds of the earth's surface is covered by oceans. Another tenth of the earth's surface is covered by glaciers. What fraction of the earth is covered by oceans and glaciers?

The total amount is the sum of the two fractions.

$\frac{2}{3} + \frac{1}{10} = \frac{20}{30} + \frac{3}{30}$ Find a common denominator.

$= \frac{23}{30}$

▼ You can add rational numbers using a model.

Example 1 $\frac{1}{4} + \frac{1}{6} = ■$

Solution

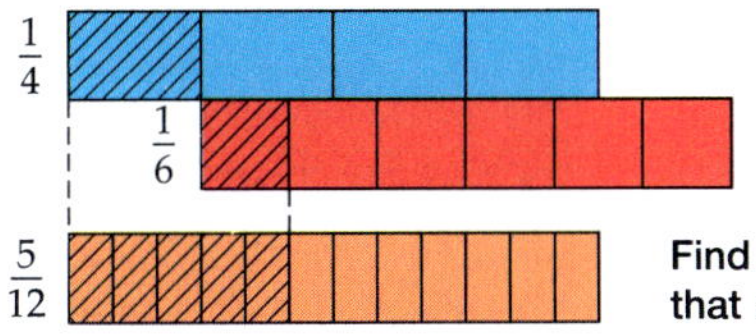

Find the fraction bar that aligns with the sum.

So, $\frac{1}{4} + \frac{1}{6} = \frac{5}{12}$.

THINK Why must the last fraction bar have 12 sections? 12 is the LCM of 4 and 6

▼ You can also subtract rational numbers expressed as fractions.

Example 2 $\frac{1}{12} - \frac{1}{3} = ■$

Solution $\frac{1}{12} - \frac{1 \cdot 4}{3 \cdot 4} = \frac{1}{12} - \frac{4}{12}$ Write equivalent fractions and subtract numerators.

$= -\frac{3}{12}$ Write in lowest terms.

$= -\frac{1}{4}$

THINK Why do you subtract only the numerators? The numerators are the parts to be subtracted. The denominators indicate the number of parts in all.

▼ You can add fractions that contain variables.

Example 3 $\frac{x}{5} + \frac{3}{4} = ■$

Solution $\frac{x \cdot 4}{5 \cdot 4} + \frac{3 \cdot 5}{4 \cdot 5} = ■$ Write equivalent fractions.

$\frac{4x}{20} + \frac{15}{20} = \frac{4x + 15}{20}$ Add numerators.

Lesson Focus

Materials

Fraction bars, tangrams, calculator

MOTIVATING THE LESSON

Ask students to write explanations for why the sum of $\frac{2}{5}$ and $\frac{2}{5}$ is $\frac{4}{5}$, and why $\frac{5}{6} - \frac{4}{6}$ is $\frac{1}{6}$. They may draw models as part of their description.

Instruction

Write $\frac{1}{2} + \frac{1}{3}$ on the chalkboard and ask students to use fraction bars to find the answer. ($\frac{5}{6}$) Help students recognize that the sixths bar must be used in order to interpret the total of $\frac{1}{3}$ put next to $\frac{1}{2}$. Make the connection to writing equivalent fractions that have the LCD, 6. Repeat the activity using $\frac{4}{6} - \frac{1}{2}$. ($\frac{1}{6}$)

1 Have students make the model using their fraction bars. Point out that the twelfths bar can be used to find equivalent fractions for $\frac{1}{4}$ and $\frac{1}{6}$, to show that $\frac{3}{12} + \frac{2}{12} = \frac{5}{12}$.

2 Use a number line to demonstrate subtraction. Remind students that the rule for subtracting integers applies to the numerators, i.e., $1 - 4 = 1 + (-4) = -3$.

Teaching Tip

Encourage students to use visual estimation to check the reasonableness of their answers when using fraction bars.

3 Encourage students to write problems involving mixed numbers in vertical form. Using this format will help them remember to add (or subtract) the integers.

4 Point out that applying rules of integers will also result in a correct answer, but using this method may cause more errors:
$\frac{1}{6} - \frac{4}{6} = -\frac{3}{6}$, $5 - 2 = 3$,
$3 + \left(-\frac{3}{6}\right) = 2\frac{3}{6}$ or $2\frac{1}{2}$.

ADDITIONAL EXAMPLES

1. Use a model to find $\frac{1}{2} - \frac{1}{12}$. $\left(\frac{5}{12}\right)$
2. $\frac{1}{6} - \left(-\frac{2}{3}\right)$ $\left(\frac{5}{6}\right)$
3. $\frac{1}{3}x - \frac{1}{5}$ $\left(\frac{5x - 3}{15}\right)$
4. $1\frac{4}{5} - 3\frac{1}{3}$ $\left(-1\frac{8}{15}\right)$

Guided Practice

Class Exercises For questions 1-4, students may draw models or use fraction bars.

For questions 13-20, have students round to the nearest integer in order to estimate.

Think and Discuss To increase students' awareness of the characteristics of rationals and other sets of numbers ask: **Is the sum or difference of two integers an integer?** (yes) **Is the sum or difference of two whole numbers a whole number?** (No, the difference can be a negative number, which is not a whole number.)

▼ To add mixed numbers, combine the integers and fractions separately.

3

Example 4 $2\frac{3}{4} + 4\frac{5}{12} = \blacksquare$

Solution

$$2\frac{3}{4} = 2\frac{9}{12}$$ Write equivalent fractions.

$$+4\frac{5}{12} = +4\frac{5}{12}$$

$$6\frac{14}{12}$$ Add integers and fractions separately.

$$6 + 1\frac{2}{12}$$ Write improper fractions as mixed numbers and add whole numbers.

So, $2\frac{3}{4} + 4\frac{5}{12} = 7\frac{1}{6}$ Write the fraction in lowest terms.

▼ With subtraction, you may need to rename before subtracting.

FLASHBACK
$5\frac{1}{6} = 4 + 1\frac{1}{6} = 4\frac{7}{6}$

4

Example 5 $5\frac{1}{6} - 2\frac{2}{3} = \blacksquare$

Solution

$$5\frac{1}{6} = 5\frac{1}{6} = 4\frac{7}{6}$$ Write equivalent fractions.

$$-2\frac{2}{3} = -2\frac{4}{6} = -2\frac{4}{6}$$

$$= 2\frac{3}{6}$$

So, $5\frac{1}{6} - 2\frac{2}{3} = 2\frac{1}{2}$ Write the fraction in lowest terms.

2. Yes.
a. $\frac{3}{4} + \frac{1}{2} = \frac{1}{2} + \frac{3}{4}$
$1\frac{1}{4} = 1\frac{1}{4}$
b. $\left(\frac{3}{4} + \frac{1}{2}\right) + \frac{1}{4} = \frac{3}{4} + \left(\frac{1}{2} + \frac{1}{4}\right)$
$\frac{5}{4} + \frac{1}{4} = \frac{3}{4} + \frac{3}{4}$
$\frac{6}{4} = \frac{6}{4}$

THINK AND DISCUSS

1. Is the sum or difference of two rational numbers also a rational number? yes
2. Do the commutative and associative properties also apply to rational numbers? Use examples to support your answer. See above.

CLASS EXERCISES

Use a model to find the sum or difference. Check students' work.

1. $\frac{1}{3} + \frac{1}{6}$ $\frac{1}{2}$
2. $\frac{2}{3} - \frac{1}{6}$ $\frac{1}{2}$
3. $\frac{11}{12} + \frac{5}{6}$ $1\frac{3}{4}$
4. $1\frac{3}{8} - \frac{7}{8}$ $\frac{1}{2}$

Find the sum or difference.

5. $\frac{1}{3} + \frac{3}{4}$ $1\frac{1}{12}$
6. $\frac{4}{5} + 3\frac{7}{10}$ $4\frac{1}{2}$
7. $\frac{3}{4} - \frac{2}{5}$ $\frac{7}{20}$
8. $5\frac{1}{3} - 2\frac{3}{4}$ $2\frac{7}{12}$
9. $\frac{x}{3} + \frac{5}{6}$ $\frac{2x + 5}{6}$
10. $\frac{3}{4}x + \frac{2}{3} + \frac{1}{2}x$ $\frac{15x + 8}{12}$
11. $11\frac{3}{4} - \left(-19\frac{5}{8}\right)$ $31\frac{3}{8}$
12. $5 - 3\frac{3}{4}$ $1\frac{1}{4}$

Estimate the sum or difference.

13. $5\frac{5}{9} + 8\frac{2}{31}$ 14
14. $21.76 - 15\frac{3}{41}$ 7
15. $-\frac{11}{3} + (-3.25)$ -7
16. $15\frac{3}{4} - 38\frac{1}{2}$ -23
17. $2\frac{1}{3} + 7\frac{1}{8}$ 9
18. $-\frac{7}{8} - \left(-\frac{1}{4}\right)$ -1
19. $14.7 + 3\frac{1}{5}$ 18
20. $8\frac{11}{12} + 4\frac{1}{12}$ 13

MATH MINUTES

Draw a large number line on the chalkboard or the floor. Assign groups of three students to model addition or subtraction problems involving rational numbers by moving along the number line.

WRITTEN EXERCISES

Use a model to find the sum or difference.

1. $\frac{3}{8} + \left(-\frac{1}{2}\right)$ **2.** $-\left(\frac{2}{3}\right) - \frac{1}{6}$ **3.** $\frac{3}{4} + \frac{1}{2}$ **4.** $1\frac{1}{2} - \frac{3}{4}$

Check students' work.

Write a number sentence for each model shown.

5.

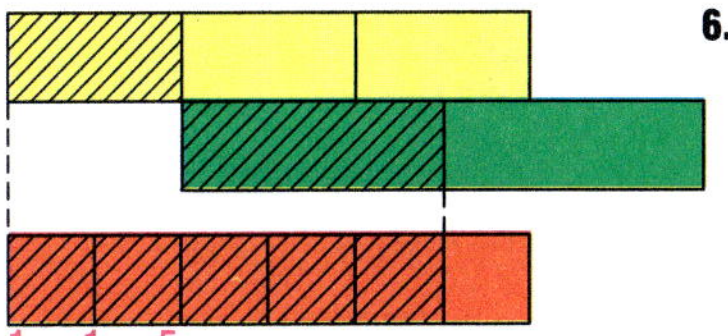

$\frac{1}{3} + \frac{1}{2} = \frac{5}{6}$

6.

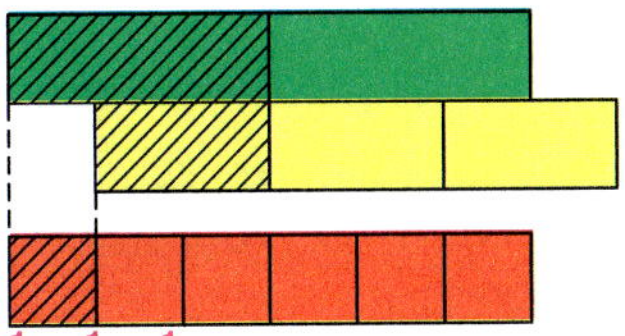

$\frac{1}{2} - \frac{1}{3} = \frac{1}{6}$

Find each sum or difference.

7. $\frac{7}{8} + \frac{5}{12}$ $1\frac{7}{24}$ **8.** $\frac{5}{6} + \frac{-1}{8}$ $\frac{17}{24}$ **9.** $\frac{2}{3} - 1\frac{1}{9}$ $-\frac{4}{9}$ **10.** $4\frac{3}{5} - 2\frac{7}{10}$ $1\frac{9}{10}$

11. $\frac{2}{3} + 1\frac{5}{6}$ $2\frac{1}{2}$ **12.** $\frac{x}{4} + \frac{x}{6}$ $\frac{5x}{12}$ **13.** $14\frac{5}{9} - 5\frac{1}{3}$ $9\frac{2}{9}$ **14.** $\frac{4x}{5} + \left(-\frac{6x}{10}\right)$ $\frac{x}{5}$

15. $\frac{5}{9}y - \frac{1}{6}y$ $\frac{7}{18}y$ **16.** $2\frac{3}{5} + 4\frac{7}{15}$ $7\frac{1}{15}$ **17.** $2\frac{1}{8} - 6\frac{3}{4}$ $-4\frac{5}{8}$ **18.** $\frac{3b}{4} - \frac{5b}{6}$ $\frac{-b}{12}$

Estimate each sum or difference.

▼▼ *SAMPLE* $12\frac{3}{4} - 5\frac{3}{8} \approx 13 - 5 = 8$

19. $28\frac{5}{18} - 12\frac{7}{17}$ 16 **20.** $-145.76 + \left(-76\frac{8}{19}\right)$ -222 **21.** $52.097 - \left(-98\frac{5}{23}\right)$ 150

22. $35.1 - 12\frac{8}{11}$ 22 **23.** $52\frac{25}{48} + 22\frac{7}{16}$ 75 **24.** $42\frac{3}{11} + 57\frac{9}{16}$ 100

MENTAL MATH Find each sum or difference.

25. $5\frac{1}{4} + 19\frac{2}{3} + 4\frac{3}{4}$ $29\frac{2}{3}$ **26.** $7\frac{2}{8} + 4\frac{5}{8} + \left(-6\frac{7}{8}\right)$ 5

CALCULATOR Check whether the following are correct.

27. $25\frac{7}{12} - \left(-18\frac{13}{18}\right) = 44\frac{11}{36}$ yes **28.** $17\frac{11}{24} + \left(-11\frac{17}{30}\right) = 5\frac{107}{120}$ yes

29. $23\frac{5}{8} - 12\frac{3}{5} = 11\frac{1}{3}$ no **30.** $16\frac{7}{12} - \left(-14\frac{3}{8}\right) = 30\frac{23}{24}$ yes

31. Which of the following are equal to $1\frac{1}{2}$? c

a. $4\frac{3}{8} - 3\frac{1}{4}$ **b.** $\frac{-3}{-2} - \frac{1}{2}$ **c.** $2\frac{1}{x} - \frac{3}{x}, x = 4$ **d.** $\frac{x - 1.5}{3.5 - x}, x = \frac{1}{2}$

32. Classify each statement as *sometimes*, *always*, or *never* true.

a. $\left|\frac{4}{5}x\right| \geq \frac{4}{5}x$ always **b.** $\frac{2x}{3} < x$ sometimes **c.** $\frac{1}{x} < x, x \neq 0$ sometimes **d.** $\frac{2}{x} < \frac{3}{x}, x > 0$ always

MIXED REVIEW

Calculate each answer.

1. Write the opposite and the absolute value of $3\frac{1}{5}$.

2. $5\frac{8}{11} + ■ = 0$ $-5\frac{8}{11}$

3. Write the rational number $\frac{-7}{10}$ in two other ways. $-\frac{7}{10}$, $\frac{7}{-10}$

4. $-6\frac{2}{3} = \frac{■}{3}$ -20

5. Compare $-\frac{13}{9}$ and $-\frac{30}{21}$. <

6. When is $\frac{y}{8} > \frac{1}{2}$? when $y > 4$

7. What is the LCM of 15 and 20? 60

8. Sara earns $4.95 an hour and works for 29 hours. Estimate her pay.

1. $-3\frac{1}{5}$, $3\frac{1}{5}$
8. $150

PROBLEM SOLVING HINT

Try guess and test.

A ***common error*** is to forget about the sign(s) of the numbers being added or subtracted. For example, students may get $\frac{1}{6}$ for $\frac{2}{3} - \frac{5}{6}$ instead of $-\frac{1}{6}$. Also, students may get $4\frac{1}{8}$ for $2\frac{7}{8} + (-1\frac{1}{4})$, instead of $1\frac{5}{8}$. Encourage students to look at the operation sign first, and then at the signs of the numbers, to determine what they have to do.

Closure

Write $\frac{1}{4} + (-\frac{7}{10})$, $-3\frac{1}{3} + (-5\frac{1}{6})$, $-2\frac{1}{2} - \frac{3}{4}$ on the chalkboard. Ask students what addition and subtraction rules apply to each example.

Independent Practice

Have students look for sums of one or differences of zero as they do questions 25 and 26.

If students do not have a calculator that computes fractions, encourage them to use their calculators in any way they can to verify the answers in questions 27-30.

Have students explain how they arrived at their answers for question 32. Mention that finding one counter example is enough to categorize the statement as sometimes true. (32b is false when $x \leq 0$ and 32c is false when $-1 \leq x \leq 1$.)

Lesson Follow-up

Reteaching Activity For students who have difficulty with adding and subtracting rational numbers, refer to the manipulative activity on p. 180E of the Chapter Overview. See also TRB Exploring O-17.

LESSON RESOURCES

TRB Practice E-9
TRB Enrichment E-10

Practice For use after 5-5 (pp. 199–202)

Adding and Subtracting Rational Numbers

Find each sum or difference.

1. $\frac{2}{3}+\frac{1}{6}$ **$\frac{5}{6}$**
2. $\frac{5}{8}-\frac{1}{4}$ **$\frac{3}{8}$**
3. $\frac{9}{16}+\frac{3}{4}$ **$1\frac{5}{16}$**
4. $2-\frac{5}{7}$ **$1\frac{2}{7}$**
5. $1\frac{1}{2}-2\frac{4}{5}$ **$-1\frac{3}{10}$**
6. $3\frac{5}{6}+2\frac{3}{4}$ **$6\frac{7}{12}$**
7. $\frac{1}{4}-\frac{1}{3}$ **$-\frac{1}{12}$**
8. $5\frac{7}{8}+3\frac{5}{12}$ **$9\frac{7}{24}$**
9. $2\frac{7}{10}-3\frac{7}{20}$ **$-\frac{13}{20}$**
10. $\frac{x}{3}+\frac{x}{5}$ **$\frac{8x}{15}$**
11. $\frac{2n}{5}+\left(-\frac{n}{6}\right)$ **$\frac{7n}{30}$**
12. $1\frac{2}{3}k+2\frac{1}{4}k$ **$3\frac{11}{12}k$**
13. $8\frac{5}{12}p-9\frac{2}{3}p$ **$-1\frac{1}{4}p$**
14. $2\frac{11}{15}x+3\frac{4}{5}x+4\frac{1}{3}x$ **$10\frac{13}{15}x$**

Estimate each sum or difference. **Estimates may vary.**

15. $13\frac{4}{5}-2\frac{9}{10}$ **11**
16. $18\frac{3}{8}+11\frac{6}{7}$ **30**
17. $-84\frac{3}{11}+103\frac{11}{12}$ **20**
18. $23\frac{6}{13}+32\frac{7}{8}$ **56**
19. $26\frac{9}{10}+72\frac{5}{6}$ **100**
20. $258\frac{17}{20}-119\frac{2}{9}$ **140**

MENTAL MATH Find each sum or difference.

21. $3\frac{3}{8}+2\frac{1}{8}+1\frac{3}{8}$ **$6\frac{7}{8}$**
22. $6\frac{7}{12}-4\frac{5}{12}$ **$2\frac{1}{6}$**
23. $8\frac{3}{16}+2\frac{5}{16}+4\frac{7}{16}$ **$14\frac{15}{16}$**
24. $7\frac{9}{10}-3\frac{3}{10}$ **$4\frac{3}{5}$**

Solve.

25. What fraction of a yard is:
    a. 24 in. **$\frac{2}{3}$ yd** b. 8 in. **$\frac{2}{9}$ yd** c. 1.5 ft **$\frac{1}{2}$ yd**
    d. 14 in. **$\frac{7}{18}$ yd** e. 42 in. **$1\frac{1}{6}$ yd** f. 8 ft **$2\frac{2}{3}$ yd**
    g. 9 in. **$\frac{1}{4}$ yd** h. 11 in. **$\frac{11}{36}$ yd** i. 38 ft **$12\frac{2}{3}$ yd**
26. A picture that is to be framed measures $14\frac{5}{16}$ in. by $9\frac{7}{8}$ in.
    a. How much longer is the picture than it is wide? **$4\frac{7}{16}$ in.**
    b. Find the total length of the four pieces that will be used to construct the frame. **$48\frac{3}{8}$ in.**

Chapter 5 E-9

Enrichment For use after 5-5 (pp. 199–202)

Unit Fractions

A *unit fraction* is a fraction such as $\frac{1}{5}$ which has a numerator of 1.

The ancient Egyptians used only unit fractions. Instead of writing $\frac{3}{4}$, for example, they wrote $\frac{1}{2}$, $\frac{1}{4}$, because $\frac{1}{2}+\frac{1}{4}=\frac{3}{4}$.

Any fraction between 0 and 1 can be written as a sum of two or more unit fractions having unlike denominators. This can be done in more than one way.

Example **Write $\frac{5}{7}$ as a sum of unit fractions.**

Solution

Write an equivalent fraction of your choosing.	$\frac{5}{7}=\frac{20}{28}$
Write the factors of the denominator.	Factors of 28: 1, 2, 4, 7, 14, 28
Write the numerator as a sum of different factors.	$20 = 14 + 4 + 2$
Use the factors to write the required sum.	$\frac{20}{28}=\frac{14+4+2}{28}=\frac{14}{28}+\frac{4}{28}+\frac{2}{28}=\frac{1}{2}+\frac{1}{7}+\frac{1}{14}=\frac{5}{7}$

1. Suppose that $\frac{30}{42}$ had been chosen as the equivalent fraction in Step 1. Use this fraction to write a different sum of unit fractions equal to $\frac{5}{7}$.
   Answers may vary. Possible answer: $\frac{1}{3}+\frac{1}{6}+\frac{1}{7}+\frac{1}{14}$

Write the fraction as a sum of unit fractions in two ways.
Answers may vary. Possible answer given.

2. $\frac{2}{3}$ **$\frac{1}{2}+\frac{1}{6}$** **$\frac{1}{3}+\frac{1}{4}+\frac{1}{12}$**
3. $\frac{1}{2}$ **$\frac{1}{3}+\frac{1}{6}$** **$\frac{1}{4}+\frac{1}{8}+\frac{1}{12}+\frac{1}{24}$**
4. $\frac{7}{8}$ **$\frac{1}{2}+\frac{1}{4}+\frac{1}{8}$** **$\frac{1}{2}+\frac{1}{3}+\frac{1}{24}$**
5. $\frac{15}{36}$ **$\frac{1}{4}+\frac{1}{9}+\frac{1}{18}$** **$\frac{1}{3}+\frac{1}{12}$**
6. $\frac{7}{12}$ **$\frac{1}{3}+\frac{1}{4}$** **$\frac{1}{2}+\frac{1}{12}$**

E-10 Chapter 5

Compare. Use >, <, or =.

33. $\frac{3^2}{5}$ **>** $\frac{17}{20}$
34. $-\frac{5}{14}$ **>** $-\frac{7}{15}$
35. $\frac{3}{8}+\frac{2}{3}$ **>** $\frac{4}{5}$
36. $\frac{a}{3}$ **>** $\frac{a}{4}$, $a > 0$
37. $\frac{a}{5}$ **<** $\frac{a^2}{5}$, $a < 0$
38. $\left|\frac{a}{b}\right|$ **=** $\frac{|a|}{|b|}$, $b \neq 0$

39. Lynn wishes to wallpaper her room. Estimate the perimeter of Lynn's room. The length is $9\frac{3}{8}$ ft and the width is $11\frac{7}{12}$ ft. **42 ft**

40. ***MENTAL MATH*** Make a chart. List the following fractions in three categories: close to 0, close to $\frac{1}{2}$, close to 1.
    $\frac{3}{5}$, $\frac{17}{21}$, $\frac{5}{14}$, $\frac{27}{53}$, $\frac{17}{15}$, $\frac{28}{59}$, $\frac{8}{55}$

0	$\frac{1}{2}$	1
$\frac{8}{55}$	$\frac{3}{5}$, $\frac{27}{53}$, $\frac{28}{59}$, $\frac{5}{14}$	$\frac{17}{21}$, $\frac{17}{15}$

41. What fraction of an hour is fifteen minutes? What fraction is seventy-five minutes? **$\frac{1}{4}$, $\frac{5}{4}$**

PROBLEM SOLVING HINT
Use a Venn diagram.

42. In the lower bass section of the band, 15 students play trombone and 4 play tuba. Two students play both instruments. What fraction of the group play tuba? **$\frac{4}{17}$**

43. What numbers must be in the blanks to form a magic square?

$\frac{1}{2}$	**$\frac{1}{12}$**	$\frac{2}{3}$
$\frac{7}{12}$	$\frac{5}{12}$	**$\frac{1}{4}$**
$\frac{1}{6}$	**$\frac{3}{4}$**	$\frac{1}{3}$

Critical Thinking

EXPLORING VISUALIZATION

An ancient geometric puzzle known as the tangram divides a large square into seven pieces. You can use the pieces to create a variety of geometrical designs. The large triangles are $\frac{1}{4}$ of the tangram. Two of the large triangles equal half of the large square. Written in fractions this would be $\frac{1}{4}+\frac{1}{4}=\frac{1}{2}$.

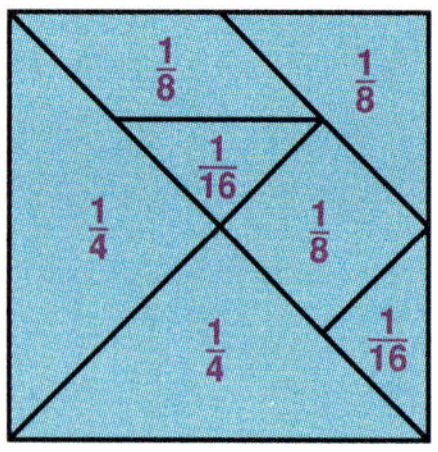

1. Copy the tangram on graph paper. What fraction of the tangram does each piece represent? **See tangram.**

Draw or model the tangram pieces that show each equation.

2. $\frac{1}{4}=\frac{1}{8}+\frac{1}{16}+\frac{1}{16}$
3. $\frac{1}{4}+\frac{1}{4}=\frac{1}{8}+\frac{1}{8}+\frac{1}{8}+\frac{1}{16}+\frac{1}{16}$

Check students' work.

Create a figure to show each expression.

4. $\frac{1}{8}+\frac{1}{16}$
5. $\frac{1}{8}+\frac{1}{16}+\frac{1}{16}$
6. $\frac{1}{4}+\frac{1}{8}+\frac{1}{8}$

Check students' work.

7. Compare your figures with those of other students in the class.

LESSON QUIZ

Add or subtract.

1. $\frac{3}{8}-\frac{1}{12}$ $\left(\frac{7}{24}\right)$
2. $-\frac{3}{5}+\frac{7}{10}$ $\left(\frac{1}{10}\right)$
3. $7\frac{1}{3}-3\frac{5}{6}$ $\left(3\frac{1}{2}\right)$
4. $1\frac{1}{2}+\left(-\frac{7}{8}\right)$ $\left[\frac{5}{8}\right]$

Estimate.

5. $2\frac{3}{20}+\left(-5\frac{5}{6}\right)-4\frac{5}{12}$ [-8]

Assignment Guide

Basic 1-12, 19-21, 25, 33-34, 39-42, MR, CT All
Average 3-16, 21-27, 31, 34-36, 39-43, MR, CT All
Enriched 9-18, 22-26, 28-32, 36-38, 42-43, MR, CT All

FOR THE NEXT LESSON

math journal

PROBLEM SOLVING STRATEGY

OBJECTIVE:
To solve problems by working backwards.

5-6 Working Backwards

With some problems, you have to work backwards from the given information to get an answer.

PROBLEM

1 The Tanaka family is planning a trip to the Grand Canyon. It will take 5 h of driving. In addition, they plan to make three half-hour rest stops. They plan on arriving at 3:30 P.M. What time should they plan to leave?

SOLUTION

READ Answer these questions to understand the given information:

What do you want to find?	The departure time for the trip.
What is the arrival time?	3:30 P.M.
How much time will be spent driving?	5 h
How much time will be spent resting?	$1\frac{1}{2}$ h

PLAN Add up the time needed for the trip.
Work back from the arrival time to find the departure time.

2 **SOLVE** Total time: $5 + 1\frac{1}{2} = 6\frac{1}{2}$
Work backwards from the arrival time:

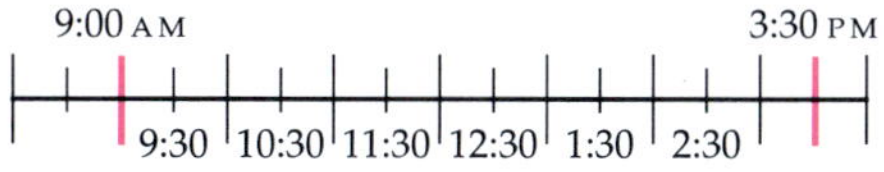

LOOK BACK Did you solve the problem? Count forward from 9:00 A.M. The elapsed time between 9:00 A.M. and 3:30 P.M. is $6\frac{1}{2}$ hours. ✓

CLASS EXERCISES

1. What other strategy could you use to solve this problem? Draw a diagram (of a clock).
2. How could you use estimation to find the approximate departure time? Subtract 7h from 3:30.
3. Suppose it will take $6\frac{1}{2}$ h of driving. At what time should the Tanakas leave? 7:30 A.M.
4. The Tanakas will need $2\frac{1}{2}$ hours to get dressed and have breakfast. They will also need another $1\frac{1}{4}$ h to pack the car. At what time should the Tanakas get up in order to leave at 9:00 A.M.? 5:15 A.M.

Teaching Tip

Blank copies of number lines prepared before the class can be used by students to model many of the Independent Practice questions.

Lesson Focus

Materials
Math journal

MOTIVATING THE LESSON

Present this situation and relate it to the idea of working backwards: Suppose you must leave for school at 7:50 A.M. and you need 45 minutes to get ready for school. **At what time should you wake up?** (7:05 A.M.)

Instruction

1 Ask students what information in the problem can help them decide that working backwards is an appropriate strategy to use. (The end result, or arrival time, is given; the starting (departure) time is what must be found.)

2 As you discuss the solution, focus on the use of a diagram as a way of modeling the problem.

Guided Practice

Class Exercises Ask students to explain how they would solve question 4 and how they would verify their answer.

A ***common error*** is to fail to use an inverse operation as a problem is worked backwards. For example, in Class Exercise 4, students may add the total they find to 9:00 A.M. Recommend modeling the solution and checking that an answer fits the conditions of the problem.

Closure

Writing in Math Ask students to write in their math journals the meaning of working backwards to solve a problem.

Independent Practice

For questions 1-3, remind students to check their answer by substituting it in the problem and reworking the problem to see if they obtain the given end result. Review other problem solving strategies students have used.

Lesson Follow-up

LESSON RESOURCES

TRB Practice E-11
TRB Enrichment E-12

Practice

For use after 5-6 (pp. 203–204)

Working Backwards

Solve each problem by working backwards.

1. Manuel's term paper is due on March 31. He began doing research on March 1. He intends to continue doing research for 3 times as long as he has done already. Then he will spend a week writing the paper and the remaining 3 days typing. What day is it? (Assume he will finish typing on March 30.)
   March 5
2. A disc jockey must allow time for 24 minutes of commercials every hour, along with 4 minutes for news, 3 minutes for weather, and 2 minutes for public-service announcements. If each record lasts an average of 3 minutes, how many records per hour can the DJ play?
   9 records/h
3. Margaret is reading the 713-page novel *War and Peace*. When she has read twice as many pages as she has read already, she will be 119 pages from the end. What page is she on now?
   page 198
4. On Monday the low temperature at the South Pole dropped 9°F from Sunday's low. On Tuesday it fell another 7°, then rose 13° on Wednesday and 17° more on Thursday. Friday it dropped 8° to −50°F. What was Sunday's low temperature?
   -56°F
5. A guide-dog training center took $\frac{1}{2}$ of the puppies at the dog pound. Steve took $\frac{1}{2}$ of those remaining. Tanya took the 5 that were left. How many puppies were there to begin with?
   20 puppies
6. Each problem lists the operations performed on n to produce the given result. Find n.
   a. Multiply by 3, add 4, divide by 5, subtract 6; result, -1.
   $n =$ **7**
   b. Add 2, divide by 3, subtract 4, multiply by 5; result, 35.
   $n =$ **31**
   c. Multiply by 2, add 7, divide by 17; result, 1.
   $n =$ **5**

Chapter 5 E-11

Name ______________ Date ______________

Enrichment

For use after 5-6 (pp. 203–204)

The Nomograph

Addition and subtraction problems can be checked by working backwards.

$$\begin{array}{r}7\\+4\\\hline 11\end{array}\quad\begin{array}{r}11\\-4\\\hline 7\end{array}\qquad\begin{array}{r}15\\-9\\\hline 6\end{array}\quad\begin{array}{r}6\\+9\\\hline 15\end{array}$$

You can use a nomograph to add and subtract fractions and to check solutions to addition and subtraction problems.

Example Add $\frac{5}{16} + \frac{3}{4}$.

Solution Place one end of a straight edge at $\frac{5}{16}$ on scale ①. Place the other end at $\frac{3}{4}$ on scale ③. Read the sum where the straight edge crosses scale ②: $1\frac{1}{16}$

$\frac{5}{16} + \frac{3}{4} = 1\frac{1}{16}$

Add or subtract using the nomograph.

1. $\frac{15}{16} + \frac{1}{2}$ **$1\frac{7}{16}$**
2. $\frac{3}{4} + \frac{7}{8}$ **$1\frac{5}{8}$**
3. $\frac{13}{16} - \frac{5}{8}$ **$\frac{3}{16}$**
4. $1\frac{3}{16} - \frac{3}{4}$ **$\frac{7}{16}$**
5. $\frac{5}{8} - \frac{7}{16}$ **$\frac{3}{16}$**
6. $\frac{1}{2} + \frac{3}{8}$ **$\frac{7}{8}$**
7. Between what two numbers on scale ② can you find:
   a. $\frac{11}{32}$ **$\frac{5}{16}$ and $\frac{3}{8}$**
   b. $1\frac{17}{32}$ **$1\frac{1}{2}$ and $1\frac{9}{16}$**
   c. $1\frac{31}{32}$ **2 and $1\frac{15}{16}$**
   d. $\frac{27}{32}$ **$\frac{7}{8}$ and $\frac{13}{16}$**

Add or subtract using the nomograph.

8. $\frac{3}{4} + \frac{31}{32}$ **$1\frac{23}{32}$**
9. $\frac{15}{32} - \frac{5}{16}$ **$\frac{5}{32}$**
10. $\frac{1}{4} + \frac{7}{32}$ **$\frac{15}{32}$**
11. $\frac{21}{32} - \frac{3}{8}$ **$\frac{9}{32}$**
12. $\frac{11}{16} + \frac{19}{32}$ **$1\frac{9}{32}$**
13. $\frac{5}{32} - \frac{1}{8}$ **$\frac{1}{32}$**

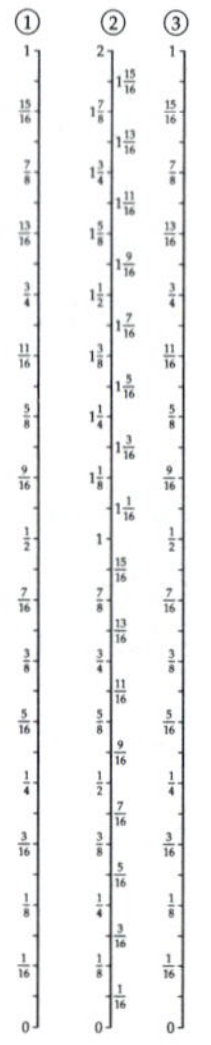

E-12 Chapter 5

WRITTEN EXERCISES

Use a CALCULATOR where appropriate.

Solve each problem by working backwards.

1. Solve this riddle: "I think of a number, add 5, multiply by 3, divide by 4, and subtract 1. The answer is 8." What is the original number? **7**
2. Carla spent $\frac{1}{3}$ of her money at the amusement park. Afterward, she had \$15 left. How much money did she have originally? **\$22.50**
3. A ball is bouncing on the floor. After each bounce, the ball is $\frac{2}{3}$ as high as the previous bounce. On the fifth bounce, the ball is 2 ft off the floor. How high was the ball before the first bounce? **$15\frac{3}{16}$ ft**

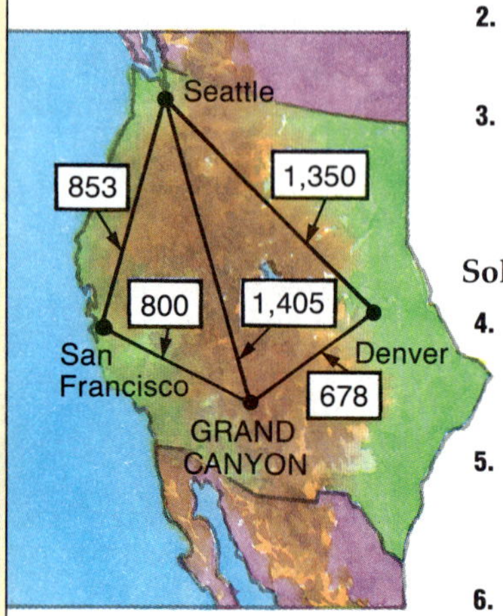

Driving distance in miles

Solve each problem. Use an appropriate strategy.

4. Use the map at the left. It took the Tanakas 7 h to reach the Grand Canyon. If their average speed was 55 mi/h, could Seattle be their home town? **no**
5. Joan is twice as old as her brother Harry. When Joan is twice as old as she is now, she will be six years older than Harry. How old is Joan now? **12 y**
6. Look at the following list of numbers: {1, 1, 1, 2, 3, 5, 6}. How many combinations of numbers will make the following equation true? **4**

   $\frac{■}{■} + \frac{■}{■} = \frac{■}{■}$ **$\frac{1}{2} + \frac{1}{3} = \frac{5}{6}$, $\frac{2}{1} + \frac{3}{1} = \frac{5}{1}$, $\frac{1}{2} + \frac{3}{6} = \frac{1}{1}$, $\frac{1}{3} + \frac{1}{6} = \frac{1}{2}$**

   You cannot use a number on the list more than once for each equation.
7. ***CALCULATOR*** Find the first three quotients. Predict the quotients for the remaining equations.
   a. $1 \div 9 =$ ■ **$0.\overline{1}$**
   b. $2 \div 9 =$ ■ **$0.\overline{2}$**
   c. $3 \div 9 =$ ■ **$0.\overline{3}$**
   d. $4 \div 9 =$ ■ **$0.\overline{4}$**
   e. $5 \div 9 =$ ■ **$0.\overline{5}$**
   f. $6 \div 9 =$ ■ **$0.\overline{6}$**
8. You have two nickels, three dimes, and a quarter. Using at least one of each coin, how many different amounts of money can you make? **6**
9. Look for the number pattern and find the next three numbers.
   $\frac{2}{3}, 1\frac{5}{12}, 2\frac{1}{6}, 2\frac{11}{12}$ **$3\frac{2}{3}, 4\frac{5}{12}, 5\frac{1}{6}$**
10. Alex, Bart, Clarence, and Dan enter a classroom. All four choose one of the four desks at the back. Each day they sit in a different order. How many days can they do this before they must repeat a previous pattern? **24 days**

MATH MINUTES

Ask students to make up a number riddle similar to question 1 of the Written Exercises. Have them exchange riddles with a partner. Each partner takes turns solving his or her riddle.

Assignment Guide

Basic All
Average All
Enriched All

FOR THE NEXT LESSON

math journal, paper, scissors, pattern blocks

Practice

Write in lowest terms.

1. $\frac{3}{15}$ $\frac{1}{5}$
2. $\frac{4}{18}$ $\frac{2}{9}$
3. $\frac{8}{12}$ $\frac{2}{3}$
4. $\frac{20}{45}$ $\frac{4}{9}$
5. $\frac{9}{15}$ $\frac{3}{5}$
6. $\frac{24}{42}$ $\frac{4}{7}$
7. $\frac{15y^2}{35y}$ $\frac{3y}{7}$
8. $\frac{5m}{5m}$ 1
9. $\frac{3a}{9a}$ $\frac{1}{3}$
10. $\frac{6xy}{20x}$ $\frac{3y}{10}$
11. $\frac{12m}{18m^2}$ $\frac{2}{3m}$
12. $\frac{cd}{3c^2d}$ $\frac{1}{3c}$

Write each mixed number as an improper fraction.

13. $2\frac{3}{5}$ $\frac{13}{5}$
14. $1\frac{7}{8}$ $\frac{15}{8}$
15. $4\frac{2}{3}$ $\frac{14}{3}$
16. $5\frac{7}{9}$ $\frac{52}{9}$
17. $8\frac{1}{4}$ $\frac{33}{4}$
18. $12\frac{5}{8}$ $\frac{101}{8}$
19. $20\frac{7}{12}$ $\frac{247}{12}$
20. $15\frac{3}{4}$ $\frac{63}{4}$

Write each improper fraction as a mixed number.

21. $\frac{12}{5}$ $2\frac{2}{5}$
22. $\frac{19}{2}$ $9\frac{1}{2}$
23. $\frac{25}{4}$ $6\frac{1}{4}$
24. $\frac{41}{12}$ $3\frac{5}{12}$
25. $\frac{53}{8}$ $6\frac{5}{8}$
26. $\frac{35}{6}$ $5\frac{5}{6}$
27. $\frac{57}{7}$ $8\frac{1}{7}$
28. $\frac{28}{9}$ $3\frac{1}{9}$

Write the opposite and absolute value of each number. $-24\frac{7}{15}, 24\frac{7}{15}$

29. $2\frac{3}{8}$ $-2\frac{3}{8}, 2\frac{3}{8}$
30. $-5\frac{4}{5}$ $5\frac{4}{5}, 5\frac{4}{5}$
31. $-18\frac{2}{3}$ $18\frac{2}{3}, 18\frac{2}{3}$
32. $24\frac{7}{15}$
33. $-13\frac{5}{9}$ $13\frac{5}{9}, 13\frac{5}{9}$
34. $22\frac{1}{6}$ $-22\frac{1}{6}, 22\frac{1}{6}$
35. $-\frac{11}{15}$ $\frac{11}{15}, \frac{11}{15}$
36. $31\frac{8}{9}$ $-31\frac{8}{9}, 31\frac{8}{9}$

Compare. Use >, <, or =.

37. $\frac{2}{3} < \frac{3}{4}$
38. $-\frac{7}{15} > -\frac{3}{5}$
39. $\frac{7}{8} > \frac{13}{16}$
40. $-\frac{5}{6} > -\frac{7}{8}$
41. $\frac{11}{15} > \frac{7}{10}$
42. $-\frac{5}{9} < \frac{7}{12}$

Find the sum or difference.

43. $\frac{5}{12} + \frac{7}{12}$ 1
44. $\frac{3}{8} - \frac{5}{8}$ $-\frac{1}{4}$
45. $\frac{8}{9} + \frac{2}{9}$ $1\frac{1}{9}$
46. $\frac{3}{5} - \left(-\frac{7}{10}\right)$ $1\frac{3}{10}$
47. $-\frac{5}{8} + \frac{1}{4}$ $-\frac{3}{8}$
48. $\frac{7}{15} - \frac{2}{5}$ $\frac{1}{15}$
49. $-2\frac{1}{2} - 3\frac{3}{4}$ $-6\frac{1}{4}$
50. $7\frac{5}{8} - 2\frac{1}{4}$ $5\frac{3}{8}$
51. $3\frac{5}{6} - 2\frac{3}{8}$ $1\frac{11}{24}$
52. $3\frac{3}{8} - 2\frac{7}{12}$ $\frac{19}{24}$
53. $4\frac{7}{15} + 2\frac{3}{10}$ $6\frac{23}{30}$
54. $-4\frac{2}{3} + 6\frac{5}{8}$ $1\frac{23}{24}$
55. $10 - 3\frac{4}{7}$ $6\frac{3}{7}$
56. $4\frac{3}{8} - 5$ $-\frac{5}{8}$
57. $-5\frac{1}{2} - 5\frac{1}{2}$ -11
58. $-14\frac{5}{8} + 2\frac{3}{5}$ $-12\frac{1}{40}$
59. $18\frac{9}{16} + 11\frac{3}{4}$ $30\frac{5}{16}$
60. $15\frac{3}{8} - 8\frac{3}{4}$ $6\frac{5}{8}$

CAREER

Help Wanted: Acoustical Physicist

Bachelor of Science degree in physics required.

For more information, write to American Institute of Physics, 335 E. 45th St., New York, NY 10017.

Acoustical physicists study sound to design concert halls, theaters, and auditoriums. Acoustical physicists consider the materials used to build and furnish a hall, the shape of the hall, and the placement of people and equipment. The acoustical physicist ensures that sound is distributed evenly so that each person in the theater or auditorium hears equally well.

PROJECT

Find out what kinds of materials absorb sound and what materials efficiently distribute sound.

Career Information

Invite an acoustical physicist (perhaps from a local college or university) to speak to the class. If that is not possible, invite an architect who has had experience designing rooms or buildings that fulfill specific acoustical requirements.

The speaker can discuss topics such as:

- what kinds of courses one needs to take to become an acoustical physicist, and why
- what factors must be considered in deciding how to best design a space according to its acoustical needs
- whether equations are used to determine the kind of design needed for a room or a building
- how the acoustical requirements of a concert hall would be addressed differently from those of a recording studio

Lesson Focus

Materials

Math journal, paper, scissors, pattern blocks

Vocabulary/Symbols

reciprocal

MOTIVATING THE LESSON

Discuss situations that are represented by a fraction of a number, such as $\frac{2}{3}$ of the class are wearing sneakers. Relate these situations to multiplication.

Instruction

1 Make the point that multiplication is usually used to find a fraction of a quantity.

2 Have students draw and cut out the model for $\frac{4}{5}$. To find $\frac{2}{3}$ of $\frac{4}{5}$, have them fold the rectangle into thirds and shade two of the thirds. Mention that even though the unshaded part is not being counted, it is used to tell the fractional value of the small rectangles. ($\frac{1}{15}$) Ask students to describe what this activity shows.

3 Point out that the rules for multiplying positive and negative numbers apply here.

OBJECTIVE: *To multiply and divide fractions and mixed numbers.*

5-7 Multiplying and Dividing Rational Numbers

▼ About $\frac{3}{4}$ of the world's fresh water is found in glaciers. Antarctica has $\frac{9}{10}$ of the world's glaciers. What fraction of the world's fresh water is in Antarctica? To find the solution, multiply $\frac{3}{4}$ by $\frac{9}{10}$.

1

$$\frac{3}{4} \cdot \frac{9}{10} = \frac{3 \cdot 9}{4 \cdot 10}$$
$$= \frac{27}{40}$$

Antarctica has more than half of the world's fresh water.

▼ You can also use a model to multiply rational numbers.

Example 1 $\frac{4}{5} \cdot \frac{2}{3} = \blacksquare$

Solution

2

$\frac{4}{5}$ $\qquad$ $\frac{4}{5} \cdot \frac{2}{3}$

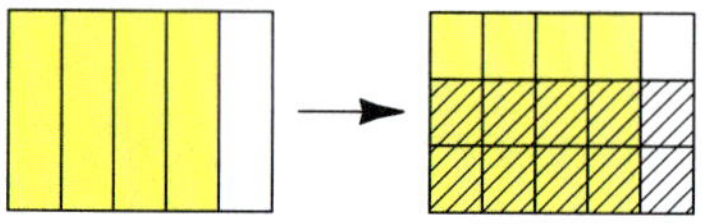

Count the number of rectangles that include both patterns.

So, $\frac{4}{5} \cdot \frac{2}{3} = \frac{8}{15}$.

FLASHBACK

$(+) \cdot (+) = +$
$(+) \cdot (-) = -$
$(-) \cdot (+) = -$
$(-) \cdot (-) = +$

▼ You can multiply fractions that have variables.

Example 2 $-\frac{2}{3} \cdot \frac{x}{7} = \blacksquare$

Solution $-\frac{2}{3} \cdot \frac{x}{7} = -\frac{2 \cdot x}{3 \cdot 7}$ Multiply the numerator and denominator.

$$= -\frac{2x}{21}$$

THINK Why must you multiply both numerators and denominators? **On the model, the product of the numerators gives the number of rectangles. The product of the denominators gives the size of the rectangles.**

Product of Two Rational Numbers	For any two rational numbers $\frac{a}{b}$ and $\frac{c}{d}$, $\frac{a}{b} \cdot \frac{c}{d} = \frac{a \cdot c}{b \cdot d}$ $\quad b \neq 0, d \neq 0$

3

▼ You can multiply mixed numbers.

Example 3 $2\frac{1}{4} \cdot \left(-2\frac{2}{3}\right) = \blacksquare$

Solution $2\frac{1}{4} \cdot \left(-2\frac{2}{3}\right) = \frac{9}{4} \cdot \left(-\frac{8}{3}\right)$ Write mixed numbers as improper fractions.

$$= -\frac{72}{12}$$
$$= -6$$

Teaching Tip

When students are unsure about the meaning of any math terms, encourage them to look them up in the glossary or a dictionary. Remind students to record the definition of these words in their math journals.

▼ You can sometimes simplify fractions before multiplying.

Example 4 $\frac{3}{14} \cdot \frac{2}{3} \cdot \left(-\frac{1}{2}\right) = \blacksquare$

Solution $\frac{\overset{1}{\cancel{3}} \cdot \overset{1}{\cancel{2}} \cdot 1}{14 \cdot \underset{1}{\cancel{3}} \cdot (-\cancel{2})} = -\frac{1}{14}$ Divide common factors from the numerator and denominator.

▼ You can divide two rational numbers.

Example 5 Use pattern blocks to model $\frac{4}{6} \div \frac{1}{3}$.

Solution Two $\frac{1}{3}$ pieces fit in $\frac{4}{6}$.

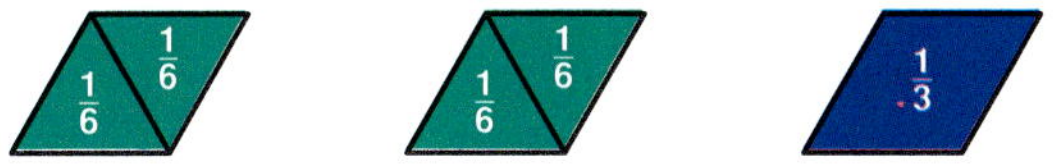

$\frac{4}{6} \div \frac{1}{3} = 2$

▼ You can divide two fractions by multiplying by the *reciprocal* of the second factor.

Example 6 $\frac{2}{5} \div \frac{3}{7} = \blacksquare$

Solution The reciprocal of $\frac{3}{7}$ is $\frac{7}{3}$.

$\frac{2}{5} \cdot \frac{7}{3} = \frac{14}{15}$ Multiply by the reciprocal.

Dividing Two Rational Numbers	For any two rational numbers $\frac{a}{b}$ and $\frac{c}{d}$, $\frac{a}{b} \div \frac{c}{d} = \frac{a}{b} \cdot \frac{d}{c}$ $\quad b \neq 0, c \neq 0, d \neq 0$

FLASHBACK

$\frac{a}{b}$ is the reciprocal of $\frac{b}{a}$.

THINK What is the reciprocal of a whole number? $\frac{1}{\text{number}}$

CLASS EXERCISES

Use a model to find the product or quotient. Check students' work.

1. $\frac{2}{3} \cdot \frac{3}{4}$ $\frac{1}{2}$
2. $\frac{1}{2} \cdot \frac{3}{8}$ $\frac{3}{16}$
3. $\frac{1}{2} \div \frac{1}{3}$ $1\frac{1}{2}$
4. $2\frac{1}{2} \cdot \frac{1}{6}$ $\frac{5}{12}$

Find the product or quotient.

5. $\frac{1}{3} \cdot \frac{1}{2}$ $\frac{1}{6}$
6. $\frac{5a}{7} \cdot -\frac{3a}{5}$ $-\frac{3a^2}{7}$
7. $\frac{5a}{9} \div \frac{4a}{5}$ $\frac{25}{36}$
8. $1\frac{1}{2} \div \frac{3}{8}$ 4
9. $-\frac{5}{8} \div \frac{3}{4}$ $-\frac{5}{6}$
10. $1\frac{1}{5} \cdot \frac{3}{8}$ $\frac{9}{20}$
11. $-4\frac{1}{6} \cdot 1\frac{4}{5}$ $-7\frac{1}{2}$
12. $-3\frac{2}{3} \div \left(-2\frac{4}{9}\right)$ $1\frac{1}{2}$

THINK AND DISCUSS See below.

1. Is the product of two rational numbers a rational number? Is the quotient?
2. If two proper fractions are less than 1, is their product less than 1? yes

1. yes; yes, except zero

MATH MINUTES

Challenge students to evaluate:

$$\frac{2\frac{2}{5} + \frac{1}{2}}{\frac{1}{10}}$$

(29)

ADDITIONAL EXAMPLES

1. Use a model to multiply $\frac{3}{4}$ and $\frac{1}{4}$. $\left(\frac{3}{16}\right)$
2. $-\frac{3}{5} \cdot \left(-\frac{a}{2}\right)$ $\left[\frac{3a}{10}\right]$
3. $-4 \cdot 1\frac{3}{8}$ $\left(-5\frac{1}{2}\right)$
4. Simplify before multiplying. $\frac{2}{3} \cdot \frac{-5}{6} \cdot \frac{4}{-15}$ $\left[\frac{4}{27}\right]$
5. $1\frac{1}{3} \div (-2)$ $\left[-\frac{2}{3}\right]$

Guided Practice

Class Exercises Drawing a model is appropriate for questions 1-4. Pattern blocks may be used for questions 3-4.

Think and Discuss For question 1, ask students when a quotient is undefined. (when zero is the divisor) This is the only instance when a quotient cannot be found from two rational numbers.

For question 2, remind students to test negative fractions before formulating an answer.

A ***common error*** is to forget negative signs when multiplying or dividing. Have students determine the sign of the product or quotient first, record it if negative, and then do the computations.

Closure

Ask the following questions. **Is $2\frac{1}{4} \cdot 1\frac{2}{3}$ greater than 1?** (yes) **than 2?** (yes) **Is $\frac{2}{3} \cdot 4$ greater than 4?** (no) **Is $\frac{2}{3} \cdot \frac{1}{2}$ greater than 1 or less than 1?** (less than 1) **Is $3 \div \frac{1}{2}$ greater than 3?** (yes) **Is $3 \div 1\frac{1}{4}$ greater than 3?** (no)

Independent Practice

For question 37, ask students to write both a multiplication and a division equation for each.

Lesson Follow-up

Reteaching Activity Use pattern blocks to model division. $1 \div \frac{1}{2}$ (2), $1 \div \frac{1}{3}$ (3), $1 \div \frac{1}{6}$ (6), $\frac{1}{2} \div \frac{1}{6}$ (3), and $\frac{1}{3} \div \frac{1}{6}$ (2).

LESSON RESOURCES

TRB Practice E-13
TRB Enrichment E-14

Practice

For use after 5-7 (pp. 206–208)

Multiplying and Dividing Rational Numbers

Find the answer.

1. $\frac{2}{5} \cdot \frac{3}{7}$ **$\frac{6}{35}$**
2. $\frac{1}{2} \div \frac{5}{8}$ **$\frac{4}{5}$**
3. $\frac{5}{9} \cdot \frac{3}{5}$ **$\frac{1}{3}$**
4. $\frac{7}{9} \cdot \frac{6}{13}$ **$\frac{14}{39}$**
5. $-\frac{5}{24} \div \frac{7}{12}$ **$-\frac{5}{14}$**
6. $\frac{3}{8} \div \frac{6}{7}$ **$\frac{7}{16}$**
7. $\frac{5}{6} \cdot \left(-1\frac{3}{10}\right)$ **$-1\frac{1}{12}$**
8. $\frac{15}{19} \div \frac{15}{19}$ **1**
9. $8 \div \frac{4}{5}$ **10**
10. $-4\frac{2}{3}\left(-5\frac{1}{6}\right)$ **$24\frac{1}{9}$**
11. $6\frac{1}{4} \div 2\frac{1}{2}$ **$2\frac{1}{2}$**
12. $2\frac{5}{6}\left(-\frac{2}{5}\right)$ **$-1\frac{2}{15}$**
13. $5\frac{5}{8} \div 1\frac{1}{4}$ **$4\frac{1}{2}$**
14. $4\frac{7}{8} \cdot 6$ **$29\frac{1}{4}$**
15. $2\frac{1}{3} \div \frac{7}{10}$ **$3\frac{1}{3}$**

Estimate the answer by rounding to the nearest integer.

16. $-4\frac{1}{3} \cdot 3\frac{3}{4}$ **-16**
17. $8\frac{7}{9} \div 2\frac{5}{6}$ **3**
18. $-19\frac{13}{16} \div \left(-4\frac{1}{5}\right)$ **5**
19. $8\frac{3}{7} \cdot 8\frac{9}{10}$ **72**
20. $-23\frac{3}{4} \div 15\frac{7}{8}$ **$-1\frac{1}{2}$**
21. $7.73 \cdot 99\frac{2}{3}$ **800**

Compare. Use >, <, or =.

22. $\frac{7}{8} \cdot \frac{4}{5}$ (=) $\frac{7}{8} \div \frac{5}{4}$
23. $\frac{2}{3} \cdot \frac{6}{7}$ (<) $\frac{5}{8} \cdot 1\frac{1}{7}$
24. $-2\frac{1}{3} \cdot 4\frac{1}{8}$ (>) $36 \div \left(-3\frac{3}{5}\right)$
25. $29\frac{1}{4} \div 4\frac{7}{8}$ (<) $4\frac{3}{4} \cdot 1\frac{1}{3}$
26. $1\frac{7}{8} \div 1\frac{7}{8}$ (=) $9 \cdot \frac{1}{9}$
27. $\frac{1}{2} \cdot \frac{2}{3} \cdot \frac{3}{4}$ (>) $\frac{1}{3} \cdot \frac{3}{4} \cdot \frac{4}{5}$

Solve. Write each answer in simplest form.

28. $\frac{1}{2} \cdot \frac{6}{9} \cdot \frac{7}{11} \cdot \frac{1}{7}$ **$\frac{1}{33}$**
29. $\frac{3}{5} \cdot \frac{4}{7} \cdot \frac{8}{9} \cdot \frac{15}{32}$ **$\frac{1}{7}$**
30. $\frac{1}{2} \cdot 1\frac{7}{8} \cdot \frac{1}{5} \cdot \frac{9}{12}$ **$\frac{9}{64}$**
31. $\frac{4}{9} \cdot \frac{3}{7} \cdot \frac{2}{5} \cdot 1\frac{11}{24}$ **$\frac{1}{9}$**
32. $3\frac{1}{4} \cdot 1\frac{3}{5} \cdot 2\frac{5}{8}$ **$13\frac{13}{20}$**
33. $2\frac{1}{3} \cdot 2\frac{5}{8} \cdot 2\frac{2}{7}$ **14**
34. $11 \div 3\frac{1}{7} \div \left(2\frac{3}{4} - \frac{1}{12}\right)$ **$1\frac{5}{16}$**
35. $8 \div 2\frac{3}{4} \div \left(\frac{4}{9} - \frac{4}{5}\right)$ **$-8\frac{2}{11}$**

Chapter 5 E-13

Enrichment

For use after 5-7 (pp. 206–208)

Continued Fractions

The fraction at the right is a continued fraction. A *continued fraction* consists of a series of fractions, each with 1 in the numerator and a sum in the denominator.

$$\cfrac{1}{3 + \cfrac{1}{4 + \cfrac{1}{5}}}$$

To simplify a continued fraction, start at the bottom right and work upwards.

$$\cfrac{1}{3 + \cfrac{1}{4 + \cfrac{1}{5}}} = \cfrac{1}{3 + \cfrac{1}{\frac{21}{5}}} = \cfrac{1}{3 + \frac{5}{21}} = \cfrac{1}{\frac{68}{21}} = \frac{21}{68}$$

Simplify.

1. $\cfrac{1}{1 + \cfrac{1}{1 + \cfrac{1}{2}}}$ **$\frac{3}{5}$**
2. $\cfrac{1}{3 + \cfrac{1}{2 + \cfrac{1}{6}}}$ **$\frac{13}{45}$**
3. $\cfrac{1}{2 + \cfrac{1}{2 + \cfrac{1}{2 + \cfrac{1}{2}}}}$ **$\frac{12}{29}$**
4. $\cfrac{1}{5 + \cfrac{1}{6 + \cfrac{1}{7 + \cfrac{1}{8}}}}$ **$\frac{350}{1{,}807}$**

Example Write $\frac{21}{46}$ as a continued fraction.

Solution Divide the denominator by the numerator. Write the quotient as a whole number plus remainder.

$46 \div 21 = 2$ R 4
$21 \div 4 = 5$ R 1
$4 \div 1 = 4$ R 0

Repeat, following the pattern, until the remainder is zero.

Write the continued fraction using the quotients from the pattern.

$$\frac{21}{46} = \cfrac{1}{2 + \cfrac{1}{5 + \cfrac{1}{4}}}$$

Write as a continued fraction.

5. $\frac{6}{31}$ **$\cfrac{1}{5 + \cfrac{1}{6}}$**
6. $\frac{5}{41}$ **$\cfrac{1}{8 + \cfrac{1}{5}}$**
7. $\frac{7}{31}$ **$\cfrac{1}{4 + \cfrac{1}{2 + \cfrac{1}{3}}}$**
8. $\frac{11}{71}$ **$\cfrac{1}{6 + \cfrac{1}{2 + \cfrac{1}{5}}}$**

E-14 Chapter 5

MIXED REVIEW

Solve.

1. $x - 7.2 = -5.8$ **1.4**
2. $y + (-6.2) = 3.7$ **9.9**
3. $3\frac{1}{3} + \left(-2\frac{5}{6}\right)$ **$\frac{1}{2}$**
4. $\frac{17}{24} \stackrel{?}{=} \frac{13}{18}$ **no**
5. Convert $\frac{13}{40}$ to a decimal. **0.325**
6. What is the GCF of 30 and 24? **6**
7. $\left|-2\frac{1}{2}\right| + \left|2\frac{1}{2}\right|$ **5**
8. Paul's height is $5\frac{11}{12}$ ft, Jim's height is $5\frac{13}{16}$ ft, and Sam's height is $5\frac{7}{8}$ ft. Who is the tallest? **Paul**

FLASHBACK

Order of operations:

1. grouping symbols
2. exponents
3. multiplication and division
4. addition and subtraction

PROBLEM SOLVING HINT

Try guess and test.

WRITTEN EXERCISES

Use a model to find the product or quotient. Check students' work.

1. $\frac{1}{2} \cdot \frac{2}{3}$ **$\frac{1}{3}$**
2. $\frac{5}{6} \cdot \frac{1}{6}$ **$\frac{5}{36}$**
3. $4 \div \frac{1}{2}$ **8**
4. $2\frac{1}{2} \div \frac{1}{2}$ **5**

Find the answer.

5. $\frac{1}{3} \cdot \frac{6}{11}$ **$\frac{2}{11}$**
6. $\frac{8}{15} \div \frac{2}{3}$ **$\frac{4}{5}$**
7. $\frac{3}{8} + \frac{9}{16}$ **$\frac{15}{16}$**
8. $\frac{-9}{10} \div \frac{5}{12} \div \frac{1}{2}$ **$-4\frac{8}{25}$**
9. $\frac{3a}{5} - \frac{7a}{10}$ **$-\frac{a}{10}$**
10. $\frac{5q}{8} \div \frac{3}{5}$ **$\frac{25q}{24}$**
11. $3\frac{1}{3} + 2\frac{1}{2}$ **$5\frac{5}{6}$**
12. $1\frac{3}{8} \div 2\frac{1}{16}$ **$\frac{2}{3}$**
13. $\frac{2}{5} \cdot 2\frac{1}{2}$ **1**
14. $2\frac{1}{3} \cdot \frac{3}{7}$ **1**
15. $\frac{-4}{9}\left(-2\frac{1}{4}\right) - \frac{1}{4}$ **$\frac{3}{4}$**
16. $-1\frac{3}{5} \cdot \left(\frac{-5}{8}\right) + \frac{4}{5}$ **$1\frac{4}{5}$**

Estimate the answer by rounding to the nearest integer.

17. $-12\frac{3}{4} \cdot \left(-3\frac{1}{3}\right)$ **39**
18. $25\frac{1}{10} \div \left(-5\frac{2}{5}\right)$ **-5**
19. $-11\frac{7}{18} + 2\frac{7}{10}$ **-8**
20. $45\frac{3}{8} \cdot \left(-2\frac{6}{7}\right)$ **-135**
21. $-75\frac{1}{12} - \left(-15\frac{1}{10}\right)$ **-60**
22. $-33\frac{1}{16} \cdot \left(-12\frac{5}{7}\right)$
23. $18\frac{2}{5} \div 11\frac{1}{2}$ **$1\frac{1}{2}$**
24. $-42\frac{2}{3} + 65\frac{3}{8}$ **22**
25. $25\frac{3}{8} \cdot 4\frac{5}{7} + 5\frac{1}{3}$

22. 429 25. 130

Compare. Use >, <, or =.

26. $\frac{3}{4} \cdot \frac{4}{5}$ ■ $\frac{3}{4} \div \frac{4}{5}$ **<**
27. $\frac{9}{7} \cdot \left(-\frac{56}{3}\right)$ ■ $-30 + 2\frac{1}{2}$ **>**
28. $\frac{1}{2} \cdot \frac{1}{2} \cdot \frac{1}{2}$ ■ $\frac{1}{2^3}$ **=**
29. $\frac{x}{4} \div 2\frac{1}{2}$ ■ $\frac{2}{3}x$, $x > 0$ **<**
30. $-\frac{2a}{3} \cdot 1\frac{1}{8}$ ■ $\frac{1a}{4} \cdot \left(-1\frac{3}{7}\right)$, $a > 0$ **<**
31. $3\frac{3}{10} \cdot \left(-3\frac{2}{11}\right)$ ■ $-4\frac{2}{3} \cdot 2\frac{1}{4}$ **=**

Solve. Write each answer in simplest form.

32. $\frac{1}{2} \cdot \frac{1}{4} \cdot \left[\frac{1}{3} + \left(-\frac{1}{6}\right)\right]$ **$\frac{1}{48}$**
33. $\frac{4}{5} \div \left(\frac{-4}{9}\right) + \frac{4}{9} \cdot \frac{3}{5}$ **$-1\frac{8}{15}$**
34. $\frac{3}{10} - \frac{4}{5} \cdot \frac{8}{5} \div 2\frac{1}{2}$ **$-\frac{53}{250}$**
35. $\frac{3}{5} + \frac{1}{4} \cdot \left(-\frac{4}{5}\right) \div \frac{2}{15}$ **$-\frac{9}{10}$**

36. Classify the following as *sometimes, always,* or *never* true, assuming $a \neq 0$ and $b \neq 0$.
   a. $\frac{5a}{b} \cdot \frac{4}{25} = \frac{20a}{b}$ **never**
   b. $\frac{5a}{3b} \div \frac{15a}{9b} = 1$ **always**
   c. $\frac{a}{b} \cdot \frac{a}{b} = \left|\frac{a}{b}\right| \cdot \left|\frac{a}{b}\right|$ **always**

37. ***MENTAL MATH*** Solve each problem.
   a. How many quarters are there in \$50? **200**
   b. How many nickels are there in \$25? **500**
   c. How many \$.75 drinks will \$9 buy? **12**

LESSON QUIZ

Find the answer.

1. $\frac{5}{7} \cdot \frac{9}{10}$ $\left(\frac{9}{14}\right)$
2. $2\frac{3}{4} \cdot \left(-5\frac{1}{3}\right)$ $\left[-14\frac{2}{3}\right]$
3. $\frac{-3}{8} \cdot \frac{5}{6} \cdot \frac{4}{35}$ $\left(-\frac{1}{28}\right)$
4. $-1\frac{1}{4} \div \left(-1\frac{7}{8}\right)$ $\left[\frac{2}{3}\right]$
5. Estimate: $-8\frac{4}{9} \cdot 3\frac{7}{8}$ (-32)

Assignment Guide

Basic 1, 3, 5-14, 17-20, 32-33, 37, MR All
Average 2, 4, 9-16, 20-22, 26-29, 34-35, 37, MR All
Enriched 10-16, 23-31, 34-36, MR All

FOR THE NEXT LESSON

math journal, paper and pencil, computer and graphing software or graphing calculator (optional)

OBJECTIVE:
To use a computer or calculator to study the concept of infinity.

MATERIALS

- Paper and pencil
- Computer and graphing software (optional)
- Math journal to record work

Exploring Infinity

■ Using a computer or a graphing calculator, you can create a bar graph of the *unit fractions*, or fractions with a numerator of 1. The first six unit fractions are shown below.

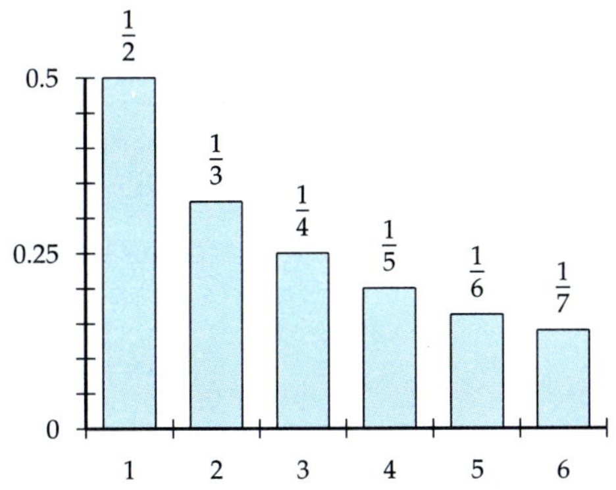

1. Continue the bar graph for the first 10 unit fractions. **Check students' work.**
2. What happens to the size of the bar for each new fraction? **Bars get shorter.**
3. What will the graph look like for 20 unit fractions? **It will look like a curve.**
4. Will this trend continue for 100 unit fractions? for 1,000? **yes, yes**
5. Can a bar have a height of zero? **no**
6. ***Write*** a description of the graph of the unit fractions for an increasing denominator. **The graphs of the unit fractions become smaller.**

■ The graphs of the unit fractions are equivalent to these equations:

$$\text{height of bar } 1 = 1 \cdot \frac{1}{2}$$

$$\text{height of bar } 2 = 1 \cdot \frac{1}{3}$$

$$\text{height of bar } 3 = 1 \cdot \frac{1}{4}$$

$$\text{height of bar } 4 = 1 \cdot \frac{1}{5}$$

$$\text{height of bar } 5 = 1 \cdot \frac{1}{6}$$

$$\text{height of bar } 6 = 1 \cdot \frac{1}{7}$$

7. Why does the height of the bars decrease as the denominator of the unit fraction increases? **As the denominator increases, the value of the fraction decreases.**
8. Write an equation that shows the height for *any* bar. **height of bar $n = 1 \cdot \frac{1}{n+1}$**

Lesson Focus

Materials

Math journal, paper and pencil, computer and graphing software or graphing calculator (optional)

Vocabulary/Symbols

infinity (∞)
unit fraction

MOTIVATING THE LESSON

Display the fractions $\frac{1}{2}, \frac{1}{3}, \frac{1}{4}, \frac{1}{5}, \frac{1}{6}$. Ask students to describe any pattern they find. Ask what happens to the values of the fractions in the pattern. (The values decrease.) Have students give the next three unit fractions. ($\frac{1}{7}, \frac{1}{8}, \frac{1}{9}$)

Instruction

Have students work in small groups and ask them to record their work in their math journals.

To draw their graphs, students can use *Appleworks* for the Apple computer or *Microsoft Works* for the IBM. Other options include: *PFS First Graph* (IBM) and *Master Grapher* (Apple).

Teaching Tip

If graphing software or a graphing calculator are available, provide copies of the graph on page 209 for use with questions 3 and 4. Also provide copies of the procedure needed to create a bar graph from a spreadsheet.

1 Ask what each row of the spreadsheet represents. (A1 is $\frac{1}{2}$, A2 is $\frac{1}{2} \cdot \frac{1}{2}$, A3 is $\frac{1}{2} \cdot \frac{1}{2} \cdot \frac{1}{2}$, and so on.)

2 Ask what happens to the values of expressions 13a and 13c as x approaches infinity. (The values approach infinity.)

Closure

Have groups share what they have found difficult and easy in this lesson. Ask students to explain what is the same about the two bar graphs illustrated in the lesson. (As the denominators of the fractions approach infinity, the heights of the bars approach zero.) Ask what is different about the two graphs. (The rate at which the value of the expression $\frac{1}{2^n}$ decreases is greater than the rate at which $\frac{1}{n}$ decreases.)

As a follow-up to this lesson, you may wish to refer to the problem solving activity on p. 180E of the Chapter Overview. See also TRB Exploring O-18.

■ You can also create a bar graph made up of the product of two unit fractions. For example, look at the spreadsheet shown below.

1

	A
1	= 1/2
2	= A1*A1
3	= A1*A1*A1
4	= A1*A1*A1*A1
5	= A1*A1*A1*A1*A1
6	

A bar graph of the spreadsheet data looks like this.

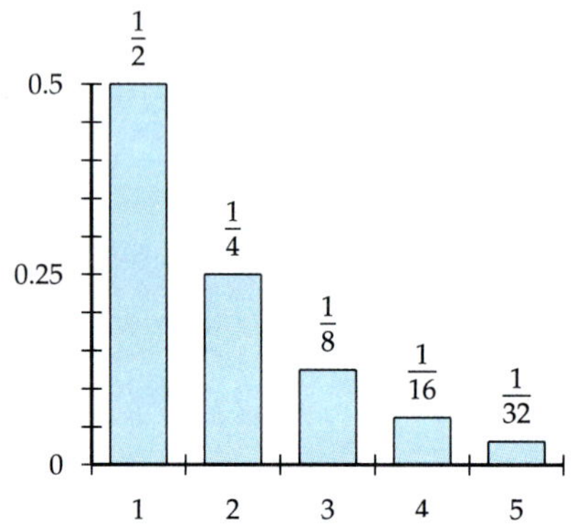

9. How would you continue the spreadsheet data for five more bars? A1*A1*A1*A1*A1*A1, etc.

10. What happens to the bars in the graph? ***Compare*** this to the change in the graph for the unit fractions. The bars become shorter at a faster rate.

11. ***Write*** each of the five fractions on the bar graph using exponents. Then write an equation to show the height of any bar on the graph. $\frac{1}{2^1}, \frac{1}{2^2}, \frac{1}{2^3}, \frac{1}{2^4}, \frac{1}{2^5}$ height of bar $n = \frac{1}{2^n}$

■ For each graph, the height of the bars approaches zero as the denominators become larger, or approach *infinity*. A number is said to approach infinity if the number increases without limit. The symbol for infinity is ∞.

12. Why does the value of a unit fraction approach zero as the denominator approaches infinity? As the denominator becomes greater, the value of the fraction lessens.

2

13. Decide which of the following expressions will approach zero as x approaches infinity. (Use a computer to create a bar graph for $x = 1$ to 5, if necessary.) b, d, e

a. $\frac{1}{3}x$ **b.** $\frac{1}{3x}$ **c.** $\frac{x^2}{2x}$ **d.** $\frac{1}{3^x}$ **e.** $\frac{3}{x}$

Assignment Guide
Basic All
Average All
Enriched All

FOR THE NEXT LESSON
math journal

OBJECTIVE:
To simplify expressions with negative exponents.

5-8 Rational Numbers with Exponents

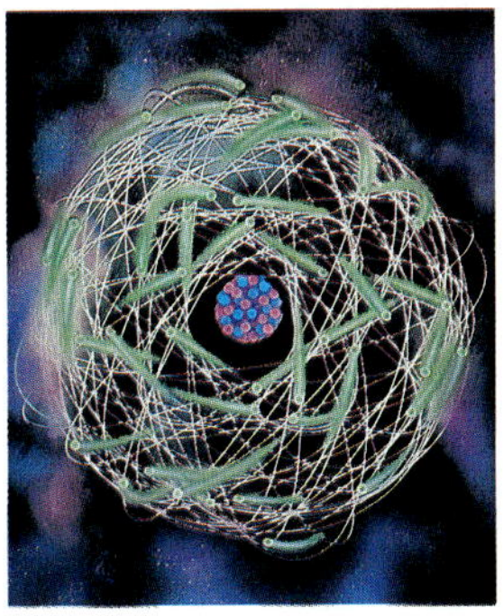

▼ The mass in grams of a proton is 1.67×10^{-24}. To understand numbers like this, you must learn about rational numbers with exponents.

▼ You can divide rational numbers with exponents.

Example 1 **Simplify $7^8 \div 7^3$.**

Solution

$$\frac{7^8}{7^3} = \frac{\overset{1}{\cancel{7}} \cdot \overset{1}{\cancel{7}} \cdot \overset{1}{\cancel{7}} \cdot 7 \cdot 7 \cdot 7 \cdot 7 \cdot 7}{\underset{1}{\cancel{7}} \cdot \underset{1}{\cancel{7}} \cdot \underset{1}{\cancel{7}}}$$ Divide common factors.

$$= \frac{7 \cdot 7 \cdot 7 \cdot 7 \cdot 7}{1} = \frac{7^5}{1}$$

$$= 7^5$$

Notice that $\frac{7^8}{7^3} = 7^{8-3}$, or 7^5. This suggests the following rule.

Rule of Exponents for Division	To divide numbers or variables with the *same* base, subtract exponents.

1

Arithmetic	**Algebra**
$\frac{4^5}{4^2} = 4^{5-2} = 4^3$	$\frac{a^m}{a^n} = a^{m-n} \quad a \neq 0$

Example 2 **Simplify.**

a. $\frac{x^6}{x^3}$ **b.** $\frac{a^4b^2}{a^2b}$

Solution **a.** $x^{6-3} = x^3$ **b.** $\frac{a^4}{a^2} \cdot \frac{b^2}{b} = (a^{4-2})(b^{2-1}) = a^2b$

2 ▼ Sometimes the exponents in the numerator and denominator are equal.

FLASHBACK
Any nonzero number divided by itself is equal to one.

Example 3 **Simplify $\frac{3^4}{3^4}$.**

Solution

1. $3^{4-4} = 3^0$ Use the rules of exponents for division.

2. $\frac{3^4}{3^4} = 1$ Divide common factors.

Since $\frac{3^4}{3^4} = 3^0$ and $\frac{3^4}{3^4} = 1$, you know that $3^0 = 1$. This suggests the following definition of zero as an exponent.

Zero as an Exponent	Any nonzero number with zero as an exponent equals 1. $a^0 = 1$ for all $a \neq 0$.

Lesson Focus

Materials
Math journal

MOTIVATING THE LESSON

Ask students to write a brief explanation, in their math journals, of how they would simplify $5^3 \cdot 5^4$. (5^7) As volunteers read their entries to the class, review that to multiply numbers with the same base, you add the exponents.

Instruction

Ask students to predict how $5^7 \div 5^4$ might be simplified, based on their knowledge of inverse operations. Have them simplify the expression by writing the factors. (5^3) Discuss whether their predictions were on target.

1 Generalize the rule based on students' predictions and observations.

2 To help students understand this concept, expand the expression and simplify it as in Example 1.

Teaching Tip

Create a bulletin board display for each rule showing examples with their expanded factors such as: *To divide numbers with the same base, subtract exponents.* $\frac{2^5}{2^2} = \frac{2 \cdot 2 \cdot 2 \cdot 2 \cdot 2}{2 \cdot 2} = 2^3$

3 Use patterning to obtain the solutions 2^{-1}. Use $\frac{2^2}{2^0}$ (2^2), $\frac{2^2}{2^1}$ (2^1), $\frac{2^2}{2^2}$ (2^0), $\frac{2^2}{2^3}$ (2^{-1}). For each of these examples have students show the expanded factors as well as the use of the rule.

4 For the ***THINK*** question, encourage students to set up an example such as $\frac{0^2}{0^2}$. This would yield 0^0, but 0 cannot be the denominator, so both expressions are undefined.

5 To reinforce the concept of negative exponents, have students simplify 5a by writing the expression using repeated factors. For 5b, help students recognize that they need to apply the rule in order to write the expression with a positive exponent.

ADDITIONAL EXAMPLES

1. Simplify $4^5 \div 4^2$. (4^3)
2. Simplify.
   a. $\frac{z^6}{z^2}$ (z^4)
   b. $\frac{2x^3y^2}{6xy^2}$ $\left(\frac{x^2}{3}\right)$
3. Simplify. $\frac{5^3}{5^3}$ (1)
4. Simplify. $\frac{7^2}{7^6}$ (7^{-4})
5. Write with positive exponents.
   a. $\frac{b^3}{b^9}$ $\left(\frac{1}{b^6}\right)$
   b. a^2b^{-4} $\left(\frac{a^2}{b^4}\right)$

Guided Practice

Class Exercises Work through the exercises orally, asking students to explain how they arrived at certain answers. To point out an advantage of using the rule to simplify question 12, have students note b^{-3} in the denominator, and ask for another way to write the denominator. $\left(\frac{a^6}{b^3}\right)$ Show students that the expression may be written as $a^5b^7 \div \frac{a^6}{b^3}$.

▼ Sometimes the exponent in the denominator is greater than the exponent in the numerator.

Example 4 **Simplify $\frac{2^2}{2^3}$.**

Solution **a.** Use the rule.

$$\frac{2^2}{2^3} = 2^{2-3}$$
$$= 2^{-1}$$

b. Use the meaning of exponents.

$$\frac{2^2}{2^3} = \frac{\overset{1}{\cancel{2}} \cdot \overset{1}{\cancel{2}}}{\underset{1}{\cancel{2}} \cdot \underset{1}{\cancel{2}} \cdot 2}$$
$$= \frac{1}{2}$$

Since $\frac{2^2}{2^3} = 2^{-1}$ and $\frac{2^2}{2^3} = \frac{1}{2}$ you know that $2^{-1} = \frac{1}{2}$. This suggests the following about negative exponents.

Negative Exponents	For any nonzero integers a and n: $a^{-n} = \frac{1}{a^n}$

THINK Why must 0^0 be undefined?
$\frac{0^1}{0^1} = 0^{1-1} = 0^0$
Division by zero is undefined.

Example 5 **Simplify. Write with positive exponents.**

a. $\frac{m^2}{m^5}$ **b.** $4x^{-3}$

Solution **a.** $\frac{m^2}{m^5} = m^{2-5}$
$= m^{-3}$
$= \frac{1}{m^3}$

b. $4x^{-3} = 4 \cdot \frac{1}{x^3}$
$= \frac{4}{x^3}$

THINK AND DISCUSS See below.

1. Can you simplify $\frac{x^3}{y^4}$?
2. How many ways can you write $-8x^{-3}$?
3. Explain why 3^{-2} is not a negative number.

1. No, x and y are different bases.
2. $\frac{-8}{x^3}$, $-\frac{8}{x^3}$, $\frac{8}{-x^3}$
3. The exponent does not affect the sign of the base.

CLASS EXERCISES

Evaluate.

1. 2^{-3} $\frac{1}{8}$ **2.** 5^{-2} $\frac{1}{25}$ **3.** 7^0 1 **4.** -3^0 -1

5. $\frac{2^5}{2^2}$ 8 **6.** $\frac{5^{-3}}{5^{-2}}$ $\frac{1}{5}$ **7.** $8^4 \div 8^2$ 64 **8.** 0^0 undefined

Write with positive or negative exponents. Leave no exponents in the denominator.

9. $\frac{a^3}{a^7}$ a^{-4} **10.** $\frac{m^5}{m^2}$ m^3 **11.** $\frac{b^5}{c^2}$ b^5c^{-2} **12.** $\frac{a^5b^7}{a^6b^{-3}}$ $a^{-1}b^{10}$

Write with positive exponents.

13. a^{-3} $\frac{1}{a^3}$ **14.** $5b^{-7}$ $\frac{5}{b^7}$ **15.** $\frac{6x^2}{x^4}$ $\frac{6}{x^2}$ **16.** $\frac{2y^5}{8y^3}$ $\frac{y^2}{4}$

17. $x^{-3}y^2$ $\frac{y^2}{x^3}$ **18.** 5^{-2} $\frac{1}{5^2}$ **19.** $4a^{-3}$ $\frac{4}{a^3}$ **20.** $15x^2y^{-4}$ $\frac{15x^2}{y^4}$

MATH MINUTES

Have groups order from greatest to least.

$2^6 \div 2^4$, $2^3 \cdot 2^4$, $\frac{2^9}{2}$, $2^{-5} \div 2^0$, $2^{10} \cdot 2^{-9}$
$\left(\frac{2^9}{2}, 2^3 \cdot 2^4, 2^6 \div 2^4, 2^{10} \cdot 2^{-9}, 2^{-5} \div 2^0\right)$

WRITTEN EXERCISES

Evaluate.

1. 6^{-2} **$\frac{1}{36}$**
2. 3^{-1} **$\frac{1}{3}$**
3. $(-2)^0$ **1**
4. $\frac{4^3}{4^5}$ **$\frac{1}{16}$**
5. $2^{-3} \cdot 3$ **$\frac{3}{8}$**
6. -7^0 **-1**
7. $(-5)^{-2}$ **$\frac{1}{25}$**
8. $3^0 \cdot 5^2 \cdot 2^{-4}$ **$\frac{25}{16}$ or $1\frac{9}{16}$**

Write with positive or negative exponents. Leave no exponents in the denominator.

9. $\frac{1}{a^3}$ **a^{-3}**
10. $\frac{b^5}{b^7}$ **b^{-2}**
11. $\frac{5x^2}{10x^{-5}}$ **$\frac{1}{2}x^7$**
12. $\frac{3y^4z}{y^6z^2}$ **$3y^{-2}z^{-1}$**
13. $\frac{x^5y^{-2}}{x^3y^8}$ **x^2y^{-10}**
14. $\frac{y^{-2}z^{-4}}{y^3z^{-2}}$ **$y^{-5}z^{-2}$**
15. $\frac{15b^6c}{3b^2c^{-4}}$ **$5b^4c^5$**
16. $\frac{4xy^{-5}}{20x^7y^{-2}}$ **$\frac{1}{5}x^{-6}y^{-3}$**

Write with positive exponents.

17. $\frac{20m^5}{4m^3}$ **$5m^2$**
18. $\frac{3b^2}{4b^7}$ **$\frac{3}{4b^5}$**
19. $\frac{3x^2y^3}{x^5y}$ **$\frac{3y^2}{x^3}$**
20. $\frac{b^{-3}c^7}{b^5c^{-2}}$ **$\frac{c^9}{b^8}$**

21. Write each of these numbers without an exponent.
   a. -5^2 **-25** b. $(-5)^2$ **25** c. 5^{-2} **$\frac{1}{25}$** d. $(-5)^{-2}$ **$\frac{1}{25}$**

Simplify.

22. $(3a)^2$ **$9a^2$**
23. $(5a)^{-2}$ **$\frac{1}{25a^2}$**
24. $-(2y^4)^0$ **-1**
25. $(a^2b^{-3})^5$ **$a^{10}b^{-15}$ or $\frac{a^{10}}{b^{15}}$**
26. $x^2 \cdot x^{-3}$ **x^{-1} or $\frac{1}{x}$**
27. $\left(\frac{3a}{b}\right)^2 \cdot \left(\frac{a^2}{b^{-3}}\right)$ **$9a^4b$**
28. $(2x^2y^{-3}) \cdot (x^3y^4)$ **$2x^5y$**

29. Study the table at the right.
   a. Describe the pattern in the first column of the table. **See right.**
   b. Describe the pattern in the second column. If this pattern continues, what will be the next three entries?
   c. How are these values related to 3?
   d. Create a similar table for values of n and 2^n. How are the values in the table related to 2?

***True* or *false*? Explain your answer.**

30. $\frac{a^3 \cdot a^4}{a^2} = \frac{a^3 + a^4}{a^2}$ **F**
31. $x^8 \cdot x^2 = x^5 \cdot x^5$ **T**
32. $1^0 = 1^{-1}$ **T**
33. $x^5 \cdot x^3 = x^{15}$ **F**
34. $x^3 \cdot y^4 = (xy)^7$ **F**
35. $5^0 = 7^0$ **T**
36. $(-r^3)^2 > 0$ **T**
37. $-(r^2)^3 < 0$ **T**
38. $1^8 = 1^{23}$ **T**

Write in standard notation.

39. The weight of all the ocean water on Earth is 1.58×10^{18} t. **1,580,000,000,000,000,000**
40. The radius of Earth's orbit is 1.5×10^{11} m. **150,000,000,000 m**

MIXED REVIEW

Find.

1. $\frac{4}{5} \cdot \frac{7}{8}$ **$\frac{7}{10}$**
2. $1\frac{3}{4} \div 4\frac{2}{3}$ **$\frac{3}{8}$**
3. $\frac{5}{8} - 1\frac{3}{4}$ **$-1\frac{1}{8}$**
4. $2\frac{1}{3} + 3\frac{5}{6}$ **$6\frac{1}{6}$**
5. $3^2 - 4$ **5**

Solve each equation.

6. $y + 5 = -7$ **-12**
7. $x - 3 = -8.2$ **-5.2**
8. How many quarters are in $8.50? **34**

29. a. The numbers decrease by 1.
b. The numbers decrease by a power of 3. 1, $\frac{1}{3}$, $\frac{1}{9}$
c. They are values of the powers of 3.
d. They are the values of the powers of 2.

4	81
3	27
2	9
1	3
0	
-1	
-2	

4	16
3	8
2	4
1	2
0	1
-1	$\frac{1}{2}$
-2	$\frac{1}{4}$

Think and Discuss For question 2, encourage students to be creative. Accept all expressions equivalent to $-8x^{-3}$.

For question 3, suggest that students work backwards from 3^{-2} by thinking of a division expression such as $\frac{3^1}{3^3}$.

A ***common error*** is to negate the value of an expression that has a negative exponent. For example, students may incorrectly write -25 for 5^{-2}. Stress that a negative sign with an exponent is a symbol that means a fraction is being represented. Have students always rewrite such an expression as a fraction first, and then evaluate it.

Closure

Writing in Math Have students write the rule of exponents for division and the definition of negative exponents in their math journals. Have them include one example of each.

Independent Practice

Questions 22-28 and 30-38 use all the rules of exponents that have been presented to this point. If necessary, review the rules involving multiplication: $a^m \cdot a^n = a^{m+n}$, and $(a^m)^n = a^{m \cdot n}$.

Critical Thinking Relate powers of 10 to place value, showing that powers of 10 with negative exponents can be used to represent $\frac{1}{10}$ or 0.1, $\frac{1}{100}$ or 0.01, $\frac{1}{1000}$ or 0.001, and so on. Have students work in small groups to complete the questions.

Lesson Follow-up

Reteaching Activity Work with students to simplify expressions. First write all the factors, and then divide out common factors. As the last step, have students write the result so no exponents are in the denominator. As students become proficient, have them use the rules to obtain the same result.

LESSON RESOURCES

TRB Practice E-15
TRB Enrichment E-16

Practice

For use after 5-8 (pp. 211–214)

Rational Numbers with Exponents

Evaluate.

1. 1^{-3} **1**
2. $(-3)^{-3}$ **$-\frac{1}{27}$**
3. 5^{-2} **$\frac{1}{25}$**
4. 9^{-2} **$\frac{1}{81}$**
5. 3^{-2} **$\frac{1}{9}$**
6. $(-5)^{-2}$ **$\frac{1}{25}$**
7. 2^{-5} **$\frac{1}{32}$**
8. 8^0 **1**
9. $(-4)^{-3}$ **$-\frac{1}{64}$**
10. $\frac{3^4}{3^7}$ **$\frac{1}{27}$**
11. 12^{-1} **$\frac{1}{12}$**
12. $(-1)^{-5}$ **-1**
13. $\frac{9^9}{9^4}$ **59,049**
14. $\frac{5^1}{5^5}$ **$\frac{1}{625}$**
15. $\frac{11^8}{11^{13}}$ **$\frac{1}{161,051}$**
16. $\frac{7^0}{7^1}$ **$\frac{1}{7}$**

Write with positive or negative exponents. Leave no exponents in the denominator.

17. $\frac{a^3}{a^7}$ **a^{-4}**
18. $\frac{f^5}{f^6}$ **f^{-1}**
19. $\frac{d^3}{d^{-10}}$ **d^{13}**
20. $\frac{x^7}{x^7}$ **1**
21. $\frac{k^5}{k^9}$ **k^{-4}**
22. $\frac{3y^4}{6y^{-4}}$ **$\frac{1y^8}{2}$**
23. $\frac{9x^8}{12x^5}$ **$\frac{3x^3}{4}$**
24. $\frac{2f^{10}}{f^5}$ **$2f^5$**
25. $\frac{3xy^4}{9xy^{-4}}$ **$\frac{y^8}{3}$**
26. $\frac{4x^2y}{2x^3}$ **$2x^{-1}y$**
27. $\frac{n^{-4}}{n^{-6}}$ **n^2**
28. $\frac{15h^6k^{-3}}{5hk^{-2}}$ **$3h^5k^{-1}$**
29. $\frac{15mn^{-5}}{12m^{-3}n^{-5}}$ **$\frac{5m^4}{4}$**
30. $\frac{32a^3b^{-4}c}{36a^{-7}b^2d^5}$ **$\frac{8a^{10}b^{-6}cd^{-5}}{9}$**

Write with positive exponents.

31. $\frac{24x^6}{6x}$ **$4x^5$**
32. $\frac{7k^3}{k^5}$ **$\frac{7}{k^2}$**
33. $\frac{6m^{-2}n^4}{8m^{-1}n^5}$ **$\frac{3}{4mn}$**
34. $\frac{a^5b^{-4}}{a^{-4}b^5}$ **$\frac{a^9}{b^9}$**
35. $\frac{10c^5d^4}{15cd^4}$ **$\frac{2c^4}{3}$**
36. $\frac{x^0y^{-1}z^2}{x^{-5}y^{-4}z^{-1}}$ **$x^5y^3z^3$**

Write each of these numbers without an exponent.

37. -2^4 **-16**
38. $(-2)^0$ **1**
39. 2^{-4} **$\frac{1}{16}$**
40. $(-2)^{-4}$ **$\frac{1}{16}$**
41. -2^0 **-1**
42. $(-2)^4$ **16**
43. 2^0 **1**
44. $(-2)^{-1}$ **$-\frac{1}{2}$**

Simplify.

45. $(2x)^3$ **$8x^3$**
46. $(-3y^2)^2$ **$9y^4$**
47. $(5ab^{-2})^3$ **$125a^3b^{-6}$**
48. $(12mn)^2$ **$144m^2n^2$**
49. $(-10xy^3)^3$ **$-1,000x^3y^9$**
50. $(9qrs^{-4})^3$ **$729q^3r^3s^{-12}$**
51. $k^{-3} \cdot k^5 \cdot k^{-2}$ **k^0 or 1**
52. $\left(\frac{2x^{-1}}{y}\right)^2\left(\frac{3y^{-2}}{x}\right)^3$ **$\frac{108}{x^5y^8}$**

Chapter 5 E-15

Enrichment

For use after 5-8 (pp. 211–214)

Multi-Magic Squares

The entries in each row, column, and diagonal in the basic magic square add to 15. By adding or subtracting the same number to or from the entries in the basic magic square, you can create a new one. The second magic square, which adds to -6, was created by subtracting 7 from the entries in the basic square.

Basic Magic Square

2	7	6
9	5	1
4	3	8

Subtract 7

-5	0	-1
2	-2	-6
-3	-4	1

Complete the magic square.

1.

37	**42**	**41**
44	**40**	**36**
39	**38**	**43**

Sum = **120**

2.

-17	**-12**	**-13**
-10	**-14**	**-18**
-15	-16	**-11**

Sum = **-42**

3.

-8	**-3**	**-4**
-1	-5	**-9**
-6	**-7**	**-2**

Sum = **-15**

Magic square entries can be used as exponents to create multiplication magic squares—multi-magic squares.

Addition square → Convert to exponents → Complete multi-magic square.

-1	4	3
6	2	-2
1	0	5

2^{-1}	2^4	2^3
2^6	2^2	2^{-2}
2^1	2^0	2^5

$\frac{1}{2}$	16	8
64	4	$\frac{1}{4}$
2	1	32

4. Find the product for each multi-magic square above. **64**

Complete the multi-magic square. In Exercise 5, note that $32 = 2^5$.

5.

$\frac{1}{4}$	**8**	**4**
32	**2**	**$\frac{1}{8}$**
1	$\frac{1}{2}$	**16**

Product = **8**

6.

$\frac{1}{16}$	**2**	**1**
8	**$\frac{1}{2}$**	**$\frac{1}{32}$**
$\frac{1}{4}$	$\frac{1}{8}$	**4**

Product = **$\frac{1}{8}$**

7.

$\frac{1}{8}$	**4**	**2**
16	**1**	**$\frac{1}{16}$**
$\frac{1}{2}$	$\frac{1}{4}$	**8**

Product = **1**

E-16 Chapter 5

Use the article below to answer each question.

> ### Future Glows for Natural Gas
>
> As oil prices continue to rise, American investors look again at an abundant resource, natural gas. In the United States, natural gas now accounts for $\frac{1}{4}$ of daily energy use. Petroleum products account for about $\frac{2}{5}$.
>
> Not only can gas replace oil for home heating, it can also fuel cars and power electric generating plants. Natural Gas burns more cleanly than oil, emitting less carbon dioxide.
>
> When gas replaces coal in electric generating plants, sulphur emissions are immediately cut.

41. How much of the energy of the United States do natural gas and petroleum products provide? $\frac{13}{20}$

42. ***PROJECT*** Research natural gas. Find out how it can be used to fuel automobiles. What are the costs of converting from gasoline to natural gas?

Critical Thinking

EXPLORING PATTERNS IN SCIENTIFIC NOTATION

1. Continue the pattern and describe what happens when the exponents are negative. If n is the exponent, the decimal point moves to the left $|n|$ places.

$$1.2 \times 10^3 = 1,200$$
$$1.2 \times 10^2 = 120$$
$$1.2 \times 10^1 = 12$$
$$1.2 \times 10^0 = \blacksquare \quad 1.2$$
$$1.2 \times 10^{-1} = \blacksquare \quad 0.12$$
$$1.2 \times 10^{-2} = \blacksquare \quad 0.012$$
$$1.2 \times 10^{-3} = \quad 0.0012$$

2. $1.2 \times 10^{-3} = 0.0012$. Extend the pattern above to see if it agrees.

3. Write 3.7×10^{-4} in standard notation. 0.00037

4. 1.67×10^{-24} g is the mass of a proton. How many zeros would follow the decimal point and come before the 1 if this were written in standard notation? 23

5. The mass of the sun is 1.00×10^{30} kg and the mass of an electron is 9.11×10^{-28} g. Why do scientists prefer scientific notation to standard notation for very large and very small numbers? Writing long strings of zeros takes too much space and can be confusing to read.

LESSON QUIZ

1. Evaluate 4^{-2}. $\left(\frac{1}{16}\right)$

Write with positive exponents.

2. $\frac{x^4}{x^7}$ $\left(\frac{1}{x^3}\right)$ 3. $a^2 \cdot a^{-7}$ $\left(\frac{1}{a^5}\right)$

Simplify, leave no exponents in the denominator.

4. $\frac{5^3}{5^{10}}$ (5^{-7})

5. $\frac{a^3b^2c}{ab^3c}$ (a^2b^{-1})

Assignment Guide

Basic 1-12, 17-19, 21-26, 29-32, 38-41, MR, CT All
Average 1-14, 17-19, 21, 24-26, 29-34, 38-42, MR, CT All
Enriched 5-16, 19-20, 27-42, MR, CT All

FOR THE NEXT LESSON

fraction bars

OBJECTIVE:
To solve equations involving addition and subtraction of rational numbers.

5-9 Addition and Subtraction Equations

▼ In four months Jules grew $\frac{2}{3}$ in. How could you represent his new height?

Let x = Jules' height four months ago. You can represent Jules' new height by the following variable expression.

$$x + \frac{2}{3}$$

▼ You can write an equation using a variable expression.

Example 1 **Write an equation to represent the situation.**
At high tide the water level rose $3\frac{1}{2}$ ft to a height of 25 ft.

Solution Let x = the previous height.
Then $x + 3\frac{1}{2} = 25$ is an equation to represent the water level at high tide.

▼ You can solve rational number equations by using or drawing a model.

Example 2 **Solve the equation using a model.**

$$x + \frac{1}{2} = \frac{11}{12}$$

Solution

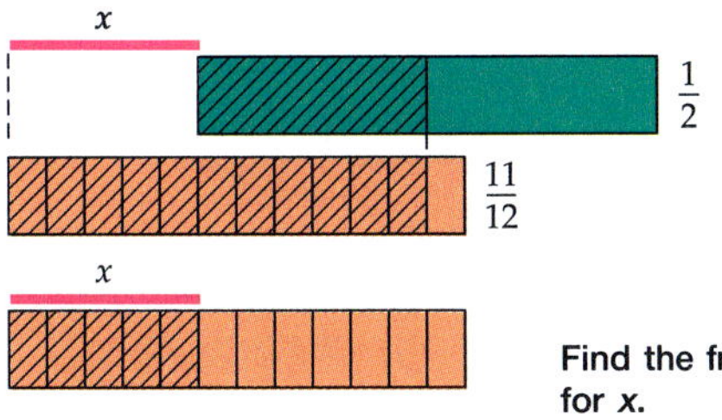

Find the fraction strip for *x*.

So, $x = \frac{5}{12}$.

1

▼ You know how to solve integer equations by isolating the variable. You can use the same procedure for solving equations with rational numbers.

Example 3 **Solve $x + \frac{4}{15} = \frac{7}{10}$.**

Solution

$x + \frac{4}{15} - \frac{4}{15} = \frac{7}{10} - \frac{4}{15}$ Subtract $\frac{4}{15}$ from each side to isolate *x*.

$x = \frac{7}{10} - \frac{4}{15}$

$= \frac{21}{30} - \frac{8}{30}$ Find a common denominator.

$= \frac{13}{30}$

2

NOTES & QUOTES

The ancient Egyptians represented all fractions, except $\frac{2}{3}$, as the sum of *unit fractions.* A unit fraction is a fraction with a numerator of 1, such as $\frac{1}{2}$. The Egyptians would write $\frac{2}{15}$ as $\frac{1}{10} + \frac{1}{30}$. They also used special symbols to denote unit fractions.

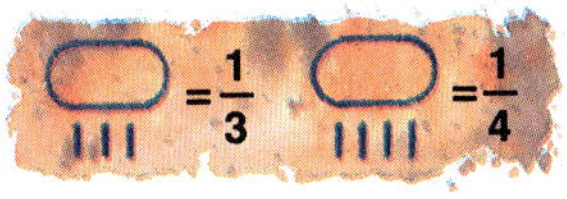

PROJECT Write $\frac{2}{7}$ and $\frac{3}{5}$ using unit fractions.

FLASHBACK

You isolate a variable by getting it alone on one side of the equal sign.

Lesson Focus

Materials
Fraction bars

MOTIVATING THE LESSON

Ask students to write a situation or a problem that represents the equation $n + 1\frac{1}{4} = 2$. (Example: When I added $1\frac{1}{4}$ lb of potato salad to a container, it made 2 lb. How much salad was already in the container?) Ask students to find the value for *n* mentally. ($\frac{3}{4}$)

Instruction

1 Guide students to use fraction bars to model $n + 1\frac{1}{4} = 2$. (Follow the model in the text. Show $1\frac{1}{4}$ above 2 strips, and then show that $\frac{3}{4}$ is needed to fill the space to make 2.) Have students also model Example 2.

2 Make the connection that addition and subtraction equations with fractions are solved the same way as equations with integers and decimals. As you discuss this example, you may wish to review how to find a common denominator.

ADDITIONAL EXAMPLES

1. Write an equation for: After $2\frac{1}{3}$ c of flour were used, there were 2 c left. How many cups were there to start with?
($c - 2\frac{1}{3} = 2$) [$4\frac{1}{3}$]
Use a model to solve.
2. $a + \frac{2}{3} = \frac{5}{6}$ $\left(\frac{1}{6}\right)$
3. $n + \frac{2}{7} = \frac{1}{14}$ $\left(-\frac{3}{14}\right)$
4. $a - \frac{2}{3} = -1\frac{1}{4}$ $\left(-\frac{7}{12}\right)$

Teaching Tip

Encourage students to use manipulative materials whenever they can. Keep materials available for use during class periods.

Guided Practice

Class Exercises Have students work in groups. For questions 4-11, have a group member describe what to do before anyone writes the equation.

Think and Discuss For question 1, guide students to describe the solving process in order to make the comparison.

For question 3, lead students to recognize that changing the fraction to an integer will fit the condition. Multiplying both sides of the equation by 2 will yield $2x + 1 = 20$.

A ***common error*** is to use the negative signs incorrectly while correctly performing required steps to solve an equation. Have students orally work through the solution of an equation. Point out addition or subtraction errors as they occur and have students correct their work as they proceed.

Closure

Ask students to write an equation for this problem: Last December, $5\frac{3}{4}$ in. of snow fell. On December 31, $2\frac{1}{2}$ in. of snow fell. How much snow was there before December 31? ($s + 2\frac{1}{2} = 5\frac{3}{4}$) Have them explain how to solve the equation, and then find the solution. ($s = 3\frac{1}{4}$)

Independent Practice

For questions 17-19 remind students to write the fractions as decimals.

Ask students to explain how they arrived at answers for questions 28-30.

▼ You can also solve equations involving subtraction of rational numbers.

Example 4 **Solve $z - \left(-3\frac{7}{10}\right) = -2\frac{1}{5}$.**

Solution

$$z - \left(-3\frac{7}{10}\right) + \left(-3\frac{7}{10}\right) = -2\frac{1}{5} + \left(-3\frac{7}{10}\right)$$ Add $-3\frac{7}{10}$ to each side.

$$z = -2\frac{1}{5} + \left(-3\frac{7}{10}\right)$$

$$= -2\frac{2}{10} + \left(-3\frac{7}{10}\right)$$ Find a common denominator.

$$= -5\frac{9}{10}$$

THINK AND DISCUSS
See below.

1. How is a rational number equation similar to an equation with integers?
2. How could you write the equation $x + \frac{1}{2} = 1\frac{5}{8}$ as an equation with decimals?
3. How could you write the equation $x + \frac{1}{2} = 10$ as an integer equation?

1. You follow the same procedures to solve both kinds of equations.
2. Change $\frac{1}{2}$ to 0.5 and $1\frac{5}{8}$ to 1.625
3. Multiply both sides of the equation by 2.

CLASS EXERCISES

Use a model to solve each equation.

1. $x + \frac{1}{4} = \frac{2}{3}$ $\frac{5}{12}$
2. $y + \frac{1}{3} = \frac{3}{4}$ $\frac{5}{12}$
3. $z + \left(-\frac{1}{3}\right) = 8$ $8\frac{1}{3}$

Solve each equation.

4. $a + \frac{1}{8} = \frac{5}{6}$ $\frac{17}{24}$
5. $b + \left(-\frac{4}{5}\right) = 6$ $6\frac{4}{5}$
6. $c - \frac{9}{10} = \frac{4}{3}$ $\frac{67}{30}$ or $2\frac{7}{30}$
7. $x + \left(-1\frac{1}{2}\right) = \frac{1}{4}$ $1\frac{3}{4}$
8. $y - 4\frac{7}{8} = -2$ $2\frac{7}{8}$
9. $z + \left(-7\frac{5}{9}\right) = -7\frac{5}{9}$ 0
10. $g - \left(9\frac{2}{3}\right) = -10\frac{4}{5}$ $-1\frac{2}{15}$
11. $h + \left(-12\frac{1}{10}\right) = -12\frac{3}{10}$ $-\frac{1}{5}$

WRITTEN EXERCISES

Write and solve an equation for each model.

1. 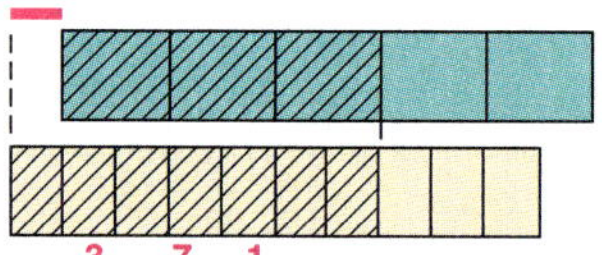

$x + \frac{3}{5} = \frac{7}{10}$; $\frac{1}{10}$

2. 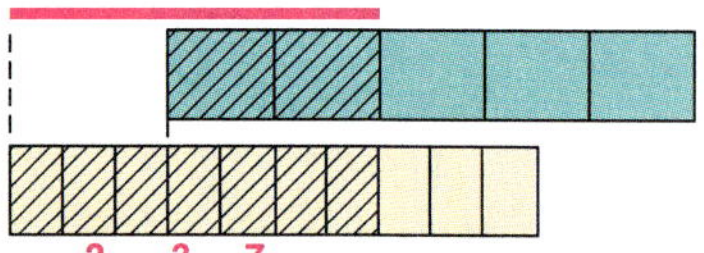

$x - \frac{2}{5} = \frac{3}{10}$; $\frac{7}{10}$

Solve each equation.

3. $m + \frac{3}{4} = \frac{1}{4}$ $-\frac{1}{2}$
4. $p - \frac{2}{3} = 1\frac{1}{3}$ 2
5. $n + \frac{5}{8} = 6$ $5\frac{3}{8}$
6. $a + \frac{5}{8} = \frac{7}{12}$ $-\frac{1}{24}$
7. $b + \left(-\frac{1}{6}\right) = \frac{3}{8}$ $\frac{13}{24}$
8. $c - \frac{3}{16} = -5$ $-4\frac{13}{16}$
9. $d - \left(\frac{-3}{10}\right) = \frac{-7}{8}$ $-1\frac{7}{40}$
10. $e - 6\frac{1}{4} = 3$ $9\frac{1}{4}$
11. $f - \left(-4\frac{5}{12}\right) = 5\frac{3}{8}$ $\frac{23}{24}$

MATH MINUTES

Have small groups determine if the solution is greater than zero or less than zero and explain why.

$x + \left(-\frac{1}{2}\right) = -1\frac{5}{6}$ $(x < 0)$

$x - \frac{3}{4} = 5\frac{5}{8}$ $(x > 0)$

12. $g + 8\frac{1}{6} = 3\frac{4}{9}$ $-4\frac{13}{18}$

13. $h + \left|-2\frac{3}{4}\right| = 5\frac{7}{10}$ $2\frac{19}{20}$

14. $k + 4.5 = 3.3$ -1.2

15. $m + (-0.7) = |-5.4|$ 6.1

16. $n - 7.23 = 10.88$ 18.11

17. $p - 16.5 = -11\frac{1}{2}$ 5

18. $z - 12.3 = 13\frac{1}{5}$ 25.5

19. $q + 6.4 = 12\frac{2}{5}$ 6

MENTAL MATH **Solve each equation.**

20. $a + \frac{3}{5} = \frac{4}{5}$ $\frac{1}{5}$

21. $b - \frac{9}{10} = -\frac{7}{10}$ $\frac{1}{5}$

22. $c + 2\frac{11}{12} = 3\frac{5}{12}$ $\frac{1}{2}$

23. $d + 5\frac{7}{16} = -2\frac{11}{16}$ $-8\frac{1}{8}$

24. $e - \frac{5}{8} = \frac{3}{4}$ $1\frac{3}{8}$

25. $f + \frac{5}{6} = -\frac{7}{12}$ $-1\frac{5}{12}$

26. For which equations does $x = 3\frac{1}{2}$? b

 a. $x + 3\frac{1}{2} = 0$
 b. $x + \left(-\frac{7}{2}\right) = 0$
 c. $x + \left(-\frac{4}{5}\right) = 3\frac{1}{2}$

27. ***WRITE*** Suppose you are solving the equation $x + \frac{1}{2} = \left(-3\frac{4}{5}\right)$. Without solving the equation, how can you tell that x is less than zero? x must be less than zero because the sum will be less than zero.

MENTAL MATH **Which equations have a solution greater than 0?**

28. $x + 4\frac{1}{5} = 5\frac{1}{2}$ yes

29. $x - 5\frac{7}{9} = 6\frac{1}{4}$ yes

30. $x + \left(-5\frac{3}{4}\right) = -5\frac{3}{4}$ no

Write an equation and solve.

31. Billie's tote bag weighed $3\frac{3}{16}$ lb when she left for school. When she returned home it weighed $5\frac{11}{16}$ lb. How much did she add to the weight of her tote bag? Let b represent the number of pounds she added to the weight in her tote bag. $b + 3\frac{3}{16} = 5\frac{11}{16}$, $2\frac{1}{2}$ lb

32. On January 2, 0.26 in. of rain fell. The total rainfall for the year was 3.5 in. How much rain fell on January 1? $r + 0.26 = 3.5$, 3.24 in.

33. In June Jonathan's height was $68\frac{1}{2}$ in. During the school year he had grown $1\frac{5}{8}$ in. What was Jonathan's height the previous September? $h + 1\frac{5}{8} = 68\frac{1}{2}$; $66\frac{7}{8}$ in.

Solve.

34. Some freshmen were trying out for the school track team.
 After Round 1, $\frac{1}{2}$ were eliminated.
 After Round 2, $\frac{1}{3}$ of those remaining were eliminated.
 After Round 3, $\frac{1}{4}$ of those remaining were eliminated.
 After Round 4, $\frac{1}{5}$ of those remaining were eliminated.
 After Round 5, $\frac{1}{6}$ of those remaining were eliminated.
 The 10 who remained became the track team. How many freshmen originally tried out? 60

MIXED REVIEW

Find each answer.

1. *True* or *false*? $\left|-5\frac{3}{5}\right| = \left|5\frac{3}{5}\right|$ T
2. What is the opposite of $-3\frac{8}{13}$? $3\frac{8}{13}$
3. Compare $\frac{13}{36}$ and $\frac{37}{56}$. <
4. Compare $\frac{7}{20}$ and $-\frac{5}{12}$. >
5. $\frac{23}{30} + \left(-\frac{14}{25}\right)$ $\frac{31}{150}$

Simplify.

6. $\frac{a^3}{a^7}$ a^{-4}
7. $\frac{x^3y^7}{x^5y^2}$ $x^{-2}y^5$
8. On Monday it rains $1\frac{1}{2}$ in. and on Tuesday it rains $1\frac{5}{8}$ in. What is the total rainfall for the two days? $3\frac{1}{8}$

PROBLEM SOLVING HINT

Try working backwards.

Lesson Follow-up

Reteaching Activity For students who have difficulty with addition and subtraction equations, refer to the computer activity on p. 180E of the Chapter Overview. See also TRB Exploring O-20.

LESSON RESOURCES

TRB Practice E-17
TRB Enrichment E-18

Practice

For use after 5-9 (pp. 215–217)

Addition and Subtraction Equations

Solve each equation.

1. $x + \frac{5}{8} = \frac{7}{8}$ $\frac{1}{4}$
2. $k + \frac{4}{5} = 1\frac{3}{5}$ $\frac{4}{5}$
3. $4 = \frac{4}{9} + y$ $3\frac{5}{9}$
4. $h + \left(-\frac{5}{8}\right) = -\frac{5}{12}$ $\frac{5}{24}$
5. $n + \frac{2}{3} = \frac{1}{9}$ $-\frac{5}{9}$
6. $e - \frac{11}{16} = -\frac{7}{8}$ $-\frac{3}{16}$
7. $m - \left(-\frac{7}{10}\right) = -1\frac{1}{5}$ $-1\frac{9}{10}$
8. $k - \frac{3}{4} = \frac{2}{5}$ $1\frac{3}{20}$
9. $x - \frac{5}{6} = \frac{1}{10}$ $\frac{14}{15}$
10. $t - \left(-3\frac{1}{6}\right) = 7\frac{2}{3}$ $4\frac{1}{2}$
11. $w - 14\frac{1}{12} = -2\frac{3}{4}$ $11\frac{1}{3}$
12. $v + \left(-4\frac{5}{6}\right) = 2\frac{1}{3}$ $7\frac{1}{6}$
13. $a - 9\frac{1}{6} = -3\frac{19}{24}$ $5\frac{3}{8}$
14. $f + \left|-3\frac{11}{12}\right| = 18$ $14\frac{1}{12}$
15. $z + (-3.4) = |-4.1|$ 7.5
16. $x - \frac{7}{15} = \frac{7}{60}$ $\frac{7}{12}$
17. $h - \left(-6\frac{1}{2}\right) = 14\frac{1}{4}$ $7\frac{3}{4}$
18. $p - 5\frac{3}{8} = -\frac{11}{24}$ $4\frac{11}{12}$

MENTAL MATH **Solve each equation.**

19. $x + \frac{3}{7} = \frac{5}{7}$ $\frac{2}{7}$
20. $k - \frac{8}{9} = -\frac{1}{9}$ $\frac{7}{9}$
21. $a + \frac{1}{9} = \frac{3}{9}$ $\frac{2}{9}$
22. $g - \frac{4}{5} = -\frac{2}{5}$ $\frac{2}{5}$
23. $h + \frac{3}{4} = \frac{7}{8}$ $\frac{1}{8}$
24. $e + 1\frac{13}{16} = 2\frac{5}{16}$ $\frac{1}{2}$
25. $m + \frac{5}{8} = -\frac{3}{16}$ $-\frac{13}{16}$
26. $p - 4\frac{5}{12} = 2\frac{7}{12}$ 7

Write an equation and solve.

27. On Tuesday, the price of a share of XYZ stock fell $5\frac{3}{4}$ from Monday's closing price, to $18\frac{7}{8}$. What was Monday's closing price?
 $n - 5\frac{3}{4} = 18\frac{7}{8}$; $n = 24\frac{5}{8}$
28. Pete's papaya tree grew $3\frac{7}{12}$ ft during the year. If its height at the end of the year was $21\frac{1}{6}$ ft, what was its height at the beginning of the year?
 $h + 3\frac{7}{12} = 21\frac{1}{6}$; $h = 17\frac{7}{12}$ ft
29. Lynn is $1\frac{3}{4}$ ft taller than Jay. If Jay is $4\frac{1}{2}$ ft tall, how tall is Lynn?
 $4\frac{1}{2} + 1\frac{3}{4} = h$; $h = 6\frac{1}{4}$ ft

Chapter 5 E-17

Enrichment

For use after 5-9 (pp. 215–217)

Fractions on a Calculator

Calculators are tailor-made for calculating with decimals. If you wish to use a calculator to solve a problem involving fractions, you must first convert them to decimals.

$\frac{1}{2} + \frac{3}{4} = 0.5$ [+] 0.75 [=] 1.25

If you do not know the decimal equivalent of a fraction, you can find it by dividing the numerator by the denominator.

$\frac{5}{16} = 5$ [÷] 16 [=] 0.3125

Add or subtract using a calculator.

1. $3\frac{3}{8} + 6\frac{7}{16}$ 9.8125
2. $8\frac{9}{10} - 5\frac{3}{4}$ 3.15
3. $19\frac{13}{32} - 9\frac{17}{50}$ 10.06625
4. $6\frac{23}{25} + 14\frac{7}{20}$ 21.27
5. $2\frac{11}{50} + 7\frac{8}{125} + 9\frac{37}{64} + 6\frac{119}{250}$ 25.338125

The disadvantage of the above method is that it does not give answers in fractional form. The following method does.

Example Add using a calculator: $\frac{5}{8} + \frac{23}{32}$

Solution Let $\frac{a}{b}$ = the first fraction. $a = 5, b = 8$

Let $\frac{c}{d}$ = the second fraction. $c = 23, d = 32$

sum $= \frac{ad + bc}{bd}$

$$\text{sum} = \frac{5 \bullet 32 + 8 \bullet 23}{8 \bullet 32} = \frac{160 + 184}{256} = \frac{344}{256} = \frac{43}{32}$$

The difference of two fractions $\frac{a}{b}$ and $\frac{c}{d}$ is $\frac{ad - bc}{bd}$.

Add or subtract using a calculator. Write the answer as a fraction in simplest form.

6. $\frac{8}{9} + \frac{3}{5}$ $\frac{67}{45}$
7. $\frac{5}{9} - \frac{4}{15}$ $\frac{13}{45}$
8. $\frac{3}{4} + \frac{5}{12}$ $\frac{7}{6}$
9. $\frac{8}{21} + \frac{11}{12}$ $\frac{109}{84}$
10. $\frac{17}{54} - \frac{31}{48}$ $-\frac{143}{432}$
11. $\frac{117}{273} - \frac{416}{989}$ $\frac{55}{6,923}$

E-18 Chapter 5

LESSON QUIZ

Solve each equation.

1. $a + \frac{1}{5} = \frac{7}{10}$ $\left(a = \frac{1}{2}\right)$
2. $n + \frac{3}{4} = \frac{1}{8}$ $\left(n = \frac{-5}{8}\right)$
3. $c - \left(\frac{-3}{10}\right) = \frac{-1}{6}$ $\left(c = -\frac{7}{15}\right)$
4. $x + \left(-2\frac{2}{3}\right) = \frac{-4}{9}$ $\left(x = 2\frac{2}{9}\right)$
5. $z - \left(-4\frac{1}{6}\right) = -6\frac{3}{8}$ $\left(z = -10\frac{13}{24}\right)$

Assignment Guide

Basic 1-11, 20-22, 26-27, 31-33, MR All
Average 1-2, 6-14, 17, 23-32, 34, MR All
Enriched 9-19, 23-30, 33-34, MR All

FOR THE NEXT LESSON

math journal

Lesson Focus

Materials

Math journal

MOTIVATING THE LESSON

Have students explain how to solve $3n = -21$. Ask what number they could multiply both sides of the equation by that has the same effect as dividing both sides by 3. ($\frac{1}{3}$) Demonstrate multiplying both sides by $\frac{1}{3}$ to get $n = -7$. Ask how 3 and $\frac{1}{3}$ are related. (They are reciprocals.) Reinforce the point that dividing by a number is the same as multiplying by the reciprocal of a number.

Instruction

1 Encourage students to restate the problem in their own words to set up an equation. Ask students how to solve this equation. Lead students to recognize that dividing both sides by $\frac{3}{4}$ is equivalent to multiplying both sides by $\frac{4}{3}$; it is the next step in the algorithm for dividing by a fraction. Ask what the solution, $x = 4\frac{2}{3}$ represents. (the total weight of the bracelet)

2 Stress that multiplying by the reciprocal is equivalent to dividing both sides by $\frac{7}{8}$. Remind students that the rules for multiplying with negative numbers always apply.

3 As a follow-up to Example 3, ask students to solve $2|x| = 6$. Ask why there is a solution. (The absolute value of x is positive.) Point out that there are two answers, $x = 3$ or $x = -3$.

OBJECTIVE: *To solve equations involving multiplication of rational numbers.*

5-10 Multiplication Equations

▼ An 18-karat gold bracelet is three-fourths pure gold. Suppose a bracelet has $3\frac{1}{2}$ oz of gold. How could you represent the total weight of the bracelet with an equation?

1 Let x = the weight of the bracelet. Then, $\frac{3}{4}\cdot x = 3\frac{1}{2}$ is an equation that represents the situation.

▼ You can solve multiplication equations by isolating the variable and using the multiplication property of equality.

Example 1 Solve $\frac{7}{8}x = -\frac{4}{5}$.

Solution

2

$$\frac{8}{7}\cdot\frac{7}{8}x = \frac{8}{7}\cdot\left(-\frac{4}{5}\right) \quad \text{Multiply both sides by the reciprocal of } \tfrac{7}{8}.$$

$$\frac{\overset{1}{\cancel{8}}\cdot\overset{1}{\cancel{7}}}{\underset{1}{\cancel{7}}\cdot\underset{1}{\cancel{8}}}x = \frac{8\cdot(-4)}{7\cdot 5}$$

$$x = -\frac{32}{35}$$

▼ When equations involve mixed numbers, rewrite them as improper fractions. Then solve the equation.

Example 2 Solve $-\frac{5x}{9} = 3\frac{1}{2}$.

Solution

$$-\frac{5x}{9} = 3\frac{1}{2}$$

$$-\frac{5x}{9} = \frac{7}{2} \quad \text{Write } 3\tfrac{1}{2} \text{ as an improper fraction.}$$

$$\left(-\frac{9}{5}\right)\cdot\left(-\frac{5}{9}\right)x = -\frac{9}{5}\cdot\frac{7}{2} \quad \text{Multiply by the reciprocal.}$$

$$x = -\frac{63}{10}$$

$$= -6\frac{3}{10}$$

FLASHBACK

$-\frac{5x}{9} = -\frac{5}{9}x$

▼ Some equations are false and do not have a solution.

Example 3 Solve $\frac{1}{2}|x| = -\frac{2}{9}$.

Solution

3

$$\frac{1}{2}|x| = -\frac{2}{9}$$

$$\frac{2}{1}\cdot\frac{1}{2}|x| = \frac{2}{1}\cdot\left(-\frac{2}{9}\right) \quad \text{Multiply by the reciprocal.}$$

$$|x| = -\frac{4}{9}$$

Since the absolute value of a number is never negative, $|x| \neq -\frac{4}{9}$. This is a false equation. It has no solution.

Teaching Tip

Encourage students to review the lessons on equations with integers and decimals. They can then relate rational number equations to what is already familiar.

CLASS EXERCISES

Solve each equation.

1. $\frac{1}{5}x = \frac{2}{3}$ $\frac{10}{3}$ or $3\frac{1}{3}$
2. $-\frac{2}{7}x = \frac{3}{8}$ $-\frac{21}{16}$ or $-1\frac{5}{16}$
3. $\frac{7}{8}z = 2\frac{3}{4}$ $3\frac{1}{7}$
4. $-1\frac{6}{7} \cdot x = \frac{9}{10}$ $\frac{-63}{130}$
5. $4\frac{5}{8} \cdot z = 6\frac{2}{5}$ $1\frac{71}{185}$
6. $\frac{2}{3}x = 2$ 3
7. $\frac{2}{3} \cdot |x| = \frac{7}{12}$ $\frac{7}{8}$ or $-\frac{7}{8}$
8. $\frac{5}{9} \cdot |y| = -1\frac{1}{2}$ no solution
9. $1\frac{1}{3} \cdot |m| = 2\frac{2}{3}$ 2, -2

THINK AND DISCUSS See below.

1. How would you solve the equation $\frac{3}{5} = \frac{x}{2}$?
2. How would you solve the equation $\frac{2}{3}x = 3x$? What is the value of x?
3. What is the reciprocal of x? of $\frac{1}{x}$? $\frac{1}{x}$; x

1. Multiply by 2 on each side of the equation.
2. Isolate the variable; $x = 0$.

WRITTEN EXERCISES

Solve each equation.

1. $\frac{2}{7}b = \frac{3}{8}$ $1\frac{5}{16}$
2. $-\frac{5}{7}x = \frac{9}{10}$ $-\frac{63}{50}$ or $-1\frac{13}{50}$
3. $\frac{2}{9}z = 1\frac{4}{5}$ $\frac{81}{10}$ or $8\frac{1}{10}$
4. $6\frac{3}{5} \cdot x = \frac{1}{2}$ $\frac{5}{66}$
5. $\frac{2}{3}x = -8$ -12
6. $1\frac{1}{2} \cdot m = \frac{3}{4}$ $\frac{1}{2}$
7. $\frac{3x}{4} = \frac{3}{8}$ $\frac{1}{2}$
8. $-\frac{7x}{8} = 1$ $-\frac{8}{7}$ or $-1\frac{1}{7}$
9. $\frac{1}{6}x = \frac{2}{3}$ 4
10. $\frac{3}{4}x = -2\frac{1}{3}$ $\frac{-28}{9}$ or $-3\frac{1}{9}$
11. $\frac{1}{2}x = -0.4$ -0.8
12. $\frac{-2}{3}x = 7$ $-\frac{21}{2}$ or $-10\frac{1}{2}$

Without solving, how do you know that $x < 0$?

13. $-\frac{3}{4}x = 6$ neg · neg = pos
14. $\frac{-5}{-7}x = -\frac{3}{5}$ (neg ÷ neg) · neg = neg
15. $\frac{5}{8}x = -1\frac{1}{2}$ pos · neg = neg

For what values of x is each equation true?

16. $\frac{|x|}{5} = \frac{2}{3}$ $\frac{10}{3}, -\frac{10}{3}$
17. $-\frac{3}{5}|x| = -1\frac{2}{3}$ $\frac{25}{9}, -\frac{25}{9}$
18. $\frac{5}{8}|x| = -\frac{4}{5}$ no solution
19. $-2\frac{3}{4} \cdot |x| = 3\frac{1}{7}$ no solution
20. $-1\frac{2}{3} \cdot |x| = \frac{-25}{27}$ $\frac{5}{9}, -\frac{5}{9}$
21. $4\frac{2}{5} \cdot |x| = -2\frac{1}{5}$ no solution

Write an equation and solve each problem.

22. A sheet of plywood is $\frac{3}{4}$-in. thick. How many sheets would make a stack 9 in. high? Let s represent the number of sheets of plywood. $\frac{3}{4}s = 9; s = 12$
23. A fast-growing ivy plant grew $\frac{5}{8}$ in. each day. How many days did it take the plant to grow 12 in? Let d represent the number of days. $\frac{5}{8}d = 12; d = 19.2$
24. How many dimes are in \$12.50? Let d represent the number of dimes. $\frac{1}{10}d = 12.50; d = 125$

MIXED REVIEW See below.

Calculate each answer.

1. $7\frac{1}{3} \cdot \left(-1\frac{3}{11}\right)$ 2. $\frac{-5}{6} \cdot \left(-1\frac{7}{20}\right)$
3. $4\frac{1}{6} \div \frac{3}{8}$ $11\frac{1}{9}$ 4. $2.4 + 3.7$
5. $x + \left(\frac{-5}{12}\right) = \frac{7}{18}$ $\frac{29}{36}$
6. $y - (-1.6) = 3.8$ 2.2
7. Compare $\frac{-7}{16}$ and $\frac{-5}{12}$. <
8. Three-fifths of the freshman class intend to help with a fund raising project. There are 210 students in the class. How many will help? 126

1. $-9\frac{1}{3}$ 2. $1\frac{1}{8}$ 4. 6.1

ADDITIONAL EXAMPLES

1. Solve $-\frac{1}{2}a = -\frac{3}{4}$. ($a = 1\frac{1}{2}$)
2. Solve $2\frac{1}{3}b = -14$. ($b = -6$)
3. Is $\frac{|n|}{-3} = 6$ true or false? (false; $|n| \neq -18$)

Guided Practice

Class Exercises Tell students to inspect questions 7-9 before they attempt to solve them. Ask them to explain how they can find the false equation without solving it. (Question 8 is false because a positive times some number is negative, so the number is negative; $|y|$ cannot be negative.)

For questions 7 and 9, ask why there are two solutions. (A number and its opposite have the same absolute value.)

Think and Discuss In question 2, when students get $2x = 9x$ in the solution, they have two options: subtract $9x$ from both sides to get $-7x = 0$, or $x = 0$; or subtract $2x$ from both sides to get $0 = 7x$, or $x = 0$.

For question 3, ask students what number does not have a reciprocal. (zero; division by zero is undefined.)

MATH MINUTES

Students match each equation with its solution.

$-\frac{1}{2}x = -1\frac{1}{2}$	$x = 3$
$\frac{4}{5}x = -2\frac{2}{5}$	$x = \frac{1}{3}$
$2\frac{1}{3}x = -7$	$x = -3$

($x = 3$, $x = -3$, $x = -3$)

A ***common error*** is to multiply both sides of an equation by the fraction in the equation instead of its reciprocal. For example, students might write:

$$\frac{2}{3}x = \frac{3}{4}$$
$$x = \frac{2}{3} \cdot \frac{3}{4}$$

To remediate this error, ask students to write the number and operation on both sides of the equation, that is,

$$\frac{2}{3}x = \frac{3}{4}$$
$$\frac{3}{2} \cdot \frac{2}{3}x = \frac{3}{2} \cdot \frac{3}{4}$$

Then students may continue to find the solution. ($x = 1\frac{1}{8}$)

Closure

Ask students to describe a way to solve $\frac{3}{4}n = -2$. Welcome all correct responses and ask students how the responses are related.

Independent Practice

Ask students for an equation that fits question 29, noting that the sum, 1, represents all the coins. ($\frac{3}{5} + \frac{1}{3} + p = 1$)

In question 30, help students realize that they need to represent the 6 members who are exactly 15 years old as a fraction. ($\frac{2}{15}$) Then the equation $\frac{2}{15}s = 6$ can be used to find the total number of students.

Solve each equation. Is $a >, <,$ or $=$ to b?

25. $\frac{2}{3}a = \frac{5}{9}, \frac{3}{5}b = \frac{8}{10}$ $a = \frac{5}{6}, b = 1\frac{1}{3}, a < b$

26. $1\frac{1}{2}a = -6\frac{2}{3}, \frac{-3}{8}b = 1\frac{2}{3}$ $a = -4\frac{4}{9}, b = -4\frac{4}{9}, a = b$

Solve.

27. One-fifth of the students at Lincoln High are graduating. This represents 70 students. How many students attend Lincoln High? 350

28. In a tree replanting project, $\frac{2}{3}$ of the trees planted survived the winter. There are 150 trees still living. Find the original number of trees planted. 225

29. Sara has a handful of coins. Three-fifths of the coins are dimes, one-third are nickels, and the rest are pennies. What fraction of the coins are pennies? $\frac{1}{15}$

30. Two-thirds of the science club's members are older than fifteen. One-fifth of the members are younger than 15. Six members are exactly 15 years old. Find the number of students in the science club. 45

31. ***PROJECT*** While watching your favorite half-hour TV program, compute the number of minutes spent on commercials. Then count the number of different products advertised in each set of commercials.

a. What fraction of the 30 min is program time?

b. What fraction of the 30 min is spent on commercials?

c. How many products are advertised?

d. Write the number of products advertised compared to the minutes of commercial time as a fraction.

e. Compare your results with others in the class.

Complete each analogy.

32. one-fifth : terminating decimal : : two-thirds : ■ repeating decimal

33. numerator : denominator : : part : ■ whole

34. multiplication : division : : addition : ■ subtraction

35. sum : product : : addition : ■ multiplication

36. $\frac{1}{10} : 0.10 :: \frac{1}{4} :$ ■ 0.25

37. $0.75 : \frac{3}{4} :: 0.875 :$ ■ $\frac{7}{8}$

38. 11 : prime : : 20 : ■ composite

39. $3^4 : 81 :: 4^3 :$ ■ 64

40. $\frac{2}{4} : \frac{1}{2} :: \frac{3}{6} :$ ■ $\frac{1}{2}$

41. $\frac{15}{8} : \frac{19}{7} :: 1\frac{7}{8} :$ ■ $2\frac{5}{7}$

42. $\frac{1}{3} : 0.\overline{3} :: \frac{2}{3} :$ ■ $0.\overline{6}$

43. $0.\overline{45} : \frac{5}{11} :: 0.\overline{63} :$ ■ $\frac{7}{11}$

Use the article below and the data at the right to answer each question.

Aluminum Recycling on Rise

In 1989, Americans recycled about three-fifths of the aluminum cans produced. They returned 49 billion cans and earned $900 million. The trend toward recycling has steadily increased since 1972 when the number of recycled cans was only $\frac{3}{20}$ of the number produced. Recycling saves money and saves our environment.

Cans Recycled

44. Write each fraction as a decimal. $\frac{3}{20} = 0.15$; $\frac{1}{4} = 0.25$; $\frac{1}{2} = 0.5$; $\frac{3}{5} = 0.6$

45. How much greater is the fraction recycled in 1989 than 1979? $\frac{7}{20}$

46. One year $\frac{2}{5}$ of the cans produced were recycled. Would you predict the year was between 1979 and 1984 or between 1984 and 1989? 1979–1984

47. Would the data suggest that recycling doubled from 1979 to 1984? Why or why not? No; the chart does not tell the number of cans recycled in any year.

48. ***ESTIMATION*** About how much is each recycled can worth? $900 mil ÷ 49 bil ≈ $.02

49. The fraction of recycled cans in 1995 is predicted to be $1\frac{1}{2}$ times the number for 1989. What fraction of the cans produced would you expect to be recycled in 1995? $\frac{9}{10}$

TEST YOURSELF

Solve each equation.

1. $\frac{2}{3} + x = 1\frac{5}{8}$ $\frac{23}{24}$

2. $a - 3\frac{1}{2} = 6\frac{3}{4}$ $10\frac{1}{4}$

3. $m - (-2.5) = 1\frac{5}{8}$ $-\frac{7}{8}$

4. $\frac{2}{3}x = 1\frac{5}{8}$ $2\frac{7}{16}$

5. $-\frac{3}{8}y = \frac{7}{12}$ $-1\frac{5}{9}$

6. $2\frac{2}{3} \cdot p = -2\frac{1}{4}$ $-\frac{27}{32}$

7. $\frac{x}{5} = 3\frac{3}{8}$ $16\frac{7}{8}$

8. $\frac{9}{14} \cdot |x| = -4\frac{1}{2}$ no solution

9. $-\frac{2}{5} \cdot |x| = 3\frac{1}{5}$ no solution

Evaluate.

10. 3^{-2} $\frac{1}{9}$

11. $(-1)^0$ 1

12. $\frac{3^4}{3}$ 27

13. $3^0 \cdot 5^{-1} \cdot 2^3$ $\frac{8}{5}$

Write with positive exponents.

14. $\frac{5n^3}{n^2}$ $5n$

15. $\frac{9a^2}{3a^5}$ $\frac{3}{a^3}$

16. $\frac{8x^3y^2}{xy^5}$ $\frac{8x^2}{y^3}$

17. $\frac{a^{-3}b^2}{a^2b^{-4}}$ $\frac{b^6}{a^5}$

Lesson Follow-up

Reteaching Activity For students who have difficulty with rational number equations, refer to the calculator activity found on p. 180E of the Chapter Overview. See also TRB Exploring O-19.

LESSON RESOURCES

TRB Practice E-19
TRB Enrichment E-20

Practice For use after 5-10 (pp. 218–221)

Multiplication Equations

Solve each equation.

1. $\frac{3}{4}x = \frac{9}{16}$ $\frac{3}{4}$
2. $-\frac{1}{3}p = \frac{1}{4}$ $\frac{-3}{4}$
3. $\frac{4}{7}y = 4$ 7
4. $\frac{-3}{8}k = \frac{1}{2}$ $-1\frac{1}{3}$
5. $\frac{1}{8}h = \frac{1}{10}$ $\frac{4}{5}$
6. $\frac{10}{11}n = \frac{2}{11}$ $\frac{1}{5}$
7. $2\frac{2}{3}e = \frac{3}{4}$ $\frac{9}{32}$
8. $-1\frac{2}{7}m = 6$ $-4\frac{2}{3}$
9. $2\frac{1}{4}f = \frac{6}{5}$ $\frac{8}{15}$
10. $-\frac{1}{4}p = \frac{1}{18}$ $-\frac{2}{9}$
11. $\frac{11}{-12}w = -1$ $1\frac{1}{11}$
12. $\frac{7}{8}c = \frac{7}{6}$ $1\frac{1}{3}$
13. $-3\frac{4}{7}x = 0$ 0
14. $\frac{2}{3}m = 2\frac{2}{9}$ $3\frac{1}{3}$
15. $\frac{1}{5}k = -\frac{1}{3}$ $-1\frac{2}{3}$

For what values of x, if any, is each equation true?

16. $|x| = \frac{5}{8}$ $\frac{5}{8}, \frac{-5}{8}$
17. $|x| = -\frac{2}{3}$ none
18. $\frac{1}{2}|x| = \frac{3}{4}$ $1\frac{1}{2}, -1\frac{1}{2}$
19. $\frac{-4}{5}|x| = -\frac{1}{3}$ $\frac{5}{12}, \frac{-5}{12}$
20. $4\frac{1}{2}|x| = 5\frac{5}{8}$ $1\frac{1}{4}, -1\frac{1}{4}$
21. $-1\frac{2}{3}|x| = 4\frac{1}{6}$ none

Write an equation and solve each problem.

22. There are 3 outs in a baseball inning. Belinsky got 22 outs before he was replaced by a relief pitcher. How many innings did he pitch? Let p represent the number of innings.
$3p = 22$; $p = \frac{22}{3}$ or $7\frac{1}{3}$; $7\frac{1}{3}$ innings

23. Floor boards are $8\frac{3}{8}$ in. wide. How many must be laid side by side to cover a floor 201 in. wide? Let n represent the number .
$8\frac{3}{8}n = 201$; $n = 24$; 24 boards

24. It takes Nancy $1\frac{2}{3}$ min to read 1 page in her social studies book. It took her $22\frac{1}{2}$ min to complete her reading assignment. How long was the assignment? Let m represent the number of pages she read.
$1\frac{2}{3}m = 22\frac{1}{2}$; $m = 13\frac{1}{2}$; $13\frac{1}{2}$ pages

25. It takes Gary three hours to drive to Boston. If the trip is 156 miles, what is Gary's average number of miles per hour? Let x represent the miles per hour.
$x = \frac{156}{3}$; $x = 52$; 52 mi/h

Chapter 5 E-19

Enrichment For use after 5-10 (pp. 218–221)

Musical Notes

In music notation, the duration of a note (the length of time it is intended to last) is indicated by its shape.

𝅝 = whole   𝅗𝅥 = half note   ♩ = quarter note
♪ = eighth note   𝅘𝅥𝅯 = sixteenth note   𝅘𝅥𝅰 = thirty-second note

Note durations are determined relative to one another.

Example The duration of the half note is 1 second. Find the durations of the other notes.

Solution ♩ = $\frac{1}{2}$ 𝅗𝅥 = $\frac{1}{2} \cdot 1 = \frac{1}{2}$ s
𝅝 = 2   𝅗𝅥 = $2 \cdot 1 = 2$ s

A symbol at the beginning of a piece of music indicates the number of one type of note to be played in one minute.

♩ = Play 180 quarter notes in 60 s
Duration of quarter note = $\frac{60}{180} = \frac{1}{3}$ s

Write the fractional duration of each note, in seconds, in the spaces to the right. ♪ = 120

1.  $\frac{1}{2}$ $\frac{1}{2}$ 1 2

 $\frac{1}{4}$ $\frac{1}{4}$ $\frac{1}{8}$ $\frac{1}{8}$ $\frac{1}{8}$ $\frac{1}{8}$ 1 2

2. Suppose that ♩ = 80 appeared at the beginning of the above piece.
a. Find the duration of a quarter note. $\frac{3}{4}$ s
b. Find the duration of a piece of music. 6 s

Placing a dot after a note increases its duration by one-half.

Example Duration of ♩ = $\frac{1}{5}$ s
Solution Duration of 𝅗𝅥. = $\frac{1}{5} + \left(\frac{1}{2} \cdot \frac{1}{5}\right) = \frac{1}{5} + \frac{1}{10} = \frac{3}{10}$ s

3. ♪ = 240. Find the duration of the note in seconds.
♪ $\frac{1}{4}$ s   ♩ $\frac{1}{2}$ s   ♪. $\frac{3}{8}$ s
𝅗𝅥. $1\frac{1}{2}$ s   𝅘𝅥𝅯. $\frac{3}{32}$ s   ♩.. $1\frac{1}{8}$ s

E-20 Chapter 5

LESSON QUIZ

Solve each equation.

1. $\frac{3}{7}x = \frac{1}{2}$ $\left(x = 1\frac{1}{6}\right)$
2. $-\frac{2}{3}a = \frac{4}{9}$ $\left(a = -\frac{2}{3}\right)$
3. $\frac{b}{4} = -\frac{3}{8}$ $\left(b = -1\frac{1}{2}\right)$
4. $2\frac{2}{3}y = -6$ $\left(y = -2\frac{1}{4}\right)$

FOR THE NEXT LESSON

newspaper pages that include the New York Stock Exchange Daily Activity, calculators, math journal

Assignment Guide

Basic 1-8, 13-18, 22, 24, 27-29, 31-39, 44-47, MR, TY All

Average 4-9, 16-19, 22-23, 28-41, 44-49, MR, TY All

Enriched 8-12, 16-21, 23-26, 30-49, MR, TY All

Lesson Focus

Materials

Newspaper pages that include the New York Stock Exchange Daily Activity, calculators, math journal

MOTIVATING THE LESSON

Ask students to explain what they know about the stock markets and how stocks are traded. If possible, arrange for a stock broker or analyst to talk to the class about how to select stocks.

Instruction

1 Review the terms in the table to be sure that students understand them. Have students look at stock activity pages in the newspapers and note that the information displayed is similar to that in the table on page 222.

Guided Practice

Class Exercises For questions 6-11, ask volunteers to find each difference on the chalkboard.

PROBLEM SOLVING APPLICATION

READ · PLAN · SOLVE · LOOK BACK

OBJECTIVE: *To apply operations with rational numbers to stock market problems.*

5-11 The Stock Market

Companies sell shares of stock to raise money. Stock prices appear as mixed numbers such as $3\frac{1}{8}$ for \$3.125. When people buy stock, they usually use a broker who charges a commission for buying or selling stock.

KEY

- **Stock** name of company
- **Div** payment made to investors for each share held
- **High** highest price per share for the day
- **Low** lowest price per share for the day
- **Close** price of a share at the close of the day
- **Chg.** the amount of change from the previous day's closing price

1

Stock	Div	High	Low	Close	Chg.
DQ	1.02	$5\frac{5}{8}$	$5\frac{1}{2}$	$5\frac{1}{2}$	$-\frac{1}{8}$
MCJ	1.26	10	$9\frac{3}{4}$	$9\frac{7}{8}$	$-\frac{1}{8}$
EDL	.90	$11\frac{7}{8}$	$11\frac{5}{8}$	$11\frac{3}{4}$	$+\frac{1}{4}$
JMB	.68	$46\frac{7}{8}$	$45\frac{1}{4}$	$46\frac{5}{8}$	$+1\frac{3}{8}$
BBH	2.76	$53\frac{1}{4}$	$52\frac{3}{4}$	53	$+\frac{1}{4}$

Newspapers publish the results of stock trading daily in lists like the one at the left.

Example 1 Mr. Jitters bought 100 shares of MCJ for $9\frac{3}{4}$ per share plus a broker's fee of \$19.50. He later sold the stock for $11\frac{1}{2}$ per share minus a \$23 broker's fee. Did Mr. Jitters gain or lose money? How much?

Solution

$100\left(9\frac{3}{4}\right) + 19.50 = 100(9.75) + 19.50 = 994.50$ buying price

$100\left(11\frac{1}{2}\right) - 23 = 100(11.50) - 23 = 1,127$ selling price

$1,127 - 994.50 = 132.50$ difference

Mr. Jitters gains \$132.50.

Decision Making ■ **DECISION MAKING** ■ Decision Making ■ Decision Making ■ Decision Making

THE STOCK MARKET

COLLECT DATA Check students' work.

1. Assume you have \$10,000 to invest. You must choose at least four stocks and buy at least 100 shares of each stock. Find out what a PE ratio is and how you might use it to choose a stock. Also consider whether or not the stock pays a dividend. Balance your portfolio by choosing stocks in different industries.
2. Use the financial pages of the newspaper to choose your stocks. Assume a broker's fee of 0.02. Determine the cost of your investment including the broker's fee. Make a table and keep track of your portfolio for three months.

Teaching Tip

Have available books, magazines, and other resources to help students research the stock market and factors that influence the economy. Also encourage students to write to the companies whose stock they are watching and ask for an annual report.

CLASS EXERCISES

Write each share price as a decimal.

1. $9\frac{7}{8}$ 9.875
2. $4\frac{1}{2}$ 4.5
3. $27\frac{3}{8}$ 27.375
4. $198\frac{1}{8}$ 198.125
5. $76\frac{3}{4}$ 76.75

Find each difference. Write the gain or loss in dollars and cents.

6. $22\frac{7}{8}$ to $29\frac{1}{2}$ \$6.625
7. $115\frac{1}{4}$ to $98\frac{3}{4}$ -\$16.50
8. $33\frac{1}{8}$ to $29\frac{7}{8}$ -\$3.25
9. $37\frac{5}{8}$ to $32\frac{1}{2}$ -\$5.125
10. $49\frac{3}{4}$ to $52\frac{1}{8}$ \$2.375
11. $65\frac{1}{2}$ to $67\frac{3}{8}$ \$1.875

Solve.

12. An investor bought 2,500 shares of XYZ Company on Monday. She kept track of the daily changes and on Friday decided to sell. Use the table at the right to answer each question. Each change number is the change from the day before.
    a. What was the price at the end of the week for a share of XYZ stock? $35\frac{1}{8}$
    b. How much money did the investor pay for the 2,500 shares on Monday? \$73,750
    c. How much will the 2,500 shares of stock sell for on Friday? \$87,812.50
    d. What else do you need to know to find out whether the investor made or lost money on the stock? the broker's fee

XYZ Stock

Day	Opening Price	Change
Mon	$29\frac{1}{2}$	
Tues		$-\frac{1}{4}$
Wed		$+4\frac{1}{8}$
Thurs		$+2\frac{1}{2}$
Fri		$-\frac{3}{4}$

13. A share of stock in the ABC Company has an average weekly change of $-\frac{5}{8}$ over 14 weeks. How much less is 100 shares worth at the end of the 14-week period than at the beginning? \$875

Decision Making ■ Decision Making ■ Decision Making ■ Decision Making ■ Decision Making ■ Decision Making

ANALYZE DATA

3. Which of your stocks increased in value? Which decreased?
4. How would factors such as an oil spill or news of a company's new product affect stock prices? What factors influenced the way your stocks changed?

MAKE DECISIONS

5. Stocks are risky investments because their value can increase or decrease. Suppose you invested \$5,000 in stocks and \$5,000 in a savings account. How might your financial position differ at the end of the three months? Do you think it's a good idea to invest all of your money in one stock? Why or why not?

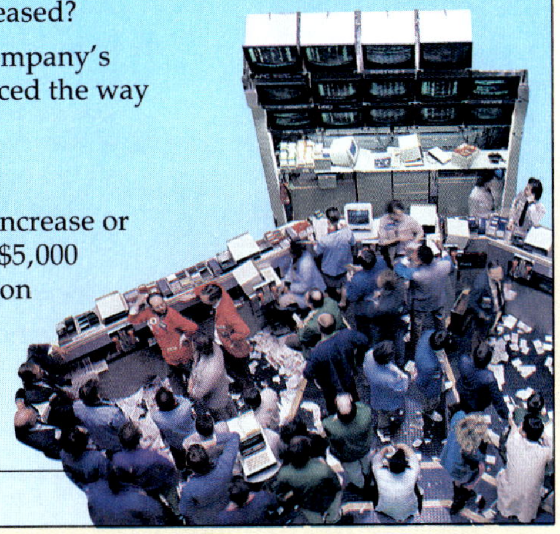

Closure

Writing in Math Ask students to write, in their math journals, the procedure to find the gain or loss for the following: $15\frac{7}{8}$ to $19\frac{3}{4}$ and $12\frac{1}{2}$ to $11\frac{2}{3}$. (gain of $3\frac{7}{8}$, loss of $\frac{5}{6}$)

Independent Practice

Decision Making Ask students to work in small groups to do this activity. For question 1, ask groups to explain the decisions they made and the role various factors, such as economic outlook, had on their decisions.

For question 2, have students update their stock prices weekly. To help them analyze the data, have groups make a line graph for each stock.

MATH MINUTES

Have students write and solve their own problems based on the content of this chapter. Then divide the class into groups and present the problems. Have the groups solve the problems.

For question 9b, assume the stock dividend in the table is an annual dividend. Ask students to find the quarterly dividend on 500 shares of each stock.

Lesson Follow-up

LESSON RESOURCES

TRB Alternate Application E-21

Alternate Application

For use after 5-11 (pp. 223–225)

The Speed of Stars

Because of their great distance, the stars of the night sky seem to be fixed in position. Actually, they are moving at enormous rates of speed.

Astronomers use spectroscopes to measure the speed of stars. Light from a stationary object passing through a spectroscope produces a spectrum crossed by lines in standard positions. The lines of moving objects are shifted away from the standard positions. The greater the shift, the faster the speed.

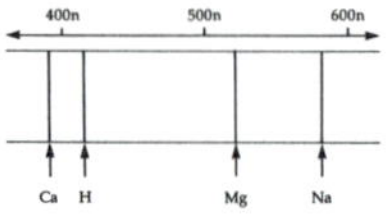

$r = \frac{sc}{0}$ r = rate of speed
s = amount of spectral line shift
c = speed of light = 3×10^8 m/s
p = standard position of line

Line positions are measured in *nanometers* (n).
$1 \text{ n} = 1 \times 10^{-9}$ m

Standard Positions	
Hydrogen (H)	414.6 n
Magnesium (Mg)	520.3 n
Sodium (Na)	582.9 n
Calcium (Ca)	392.4 n

Example The hydrogen line of a star appears at 415.1 in the spectrum. Find the speed of the star.

Solution
1. Find the amount of shift.
   $s = 415.1 - 414.6 = 0.5$ n
2. Apply the formula.
   $r = \frac{sc}{p} = \frac{(0.5 \times 10^{-9})(3 \times 10^8)}{414.6 \times 10^{-9}}$
   $\approx 0.0036 \times 10^8$
   $= 3.6 \times 10^5$ m/s

The star's speed is about 3.6×10^5 m/s.

Solve using your calculator.

1. A sodium line was found in the spectrum of the star Procyon at 582.3 n. Find the speed of Procyon in m/s.
   about 3.1×10^5 m/s
2. Measures of the spectral lines of the star Canopus showed a magnesium line at 519.4 n. Find the speed of Canopus in m/s.
   about 5.2×10^5 m/s
3. The spectrum of the quasar QR-519 showed a calcium line at 245.8. Find the quasar's speed in m/s.
   about 1.12×10^8 m/s

Chapter 5 E-21

WRITTEN EXERCISES

Write each share price as a decimal.

1. $62\frac{5}{8}$ 62.625 **2.** $104\frac{3}{8}$ 104.375 **3.** $77\frac{1}{8}$ 77.125 **4.** $15\frac{3}{4}$ 15.75 **5.** $88\frac{7}{8}$ 88.875

Find each difference. Write the gain or loss in dollars and cents.

6. $17\frac{7}{8}$ to $12\frac{1}{2}$ -\$5.375 **7.** $57\frac{3}{4}$ to $61\frac{3}{8}$ \$3.625 **8.** $96\frac{1}{2}$ to $95\frac{7}{8}$ -\$.625

9. Use the stock table on page 222.
   a. Determine the cost of buying 500 shares of each stock at the closing price. Assume a broker's commission of 0.02.
   b. Determine the dividend earned on 500 shares of each stock.

9. a. DQ $2,805.00
MCJ $5,036.25
EDL $5,992.50
JMB $23,778.75
BBH $27,030.00
b. DQ $510
MCJ $630
EDL $450
JMB $340
BBH $1,380

10. Use the ***DATA*** below to solve.

Stock	*Price* Mon	*Changes* Tues	Wed	Thurs	Fri	a.	b.
VEX	$9\frac{3}{8}$	$+\frac{1}{2}$	$-\frac{1}{8}$	$+2$	$+1\frac{7}{8}$	$+4\frac{1}{4}$	$13\frac{5}{8}$
VYE	$36\frac{1}{2}$	$-1\frac{3}{8}$	$+\frac{3}{4}$	$+\frac{1}{2}$	$-\frac{1}{8}$	$-\frac{1}{4}$	$36\frac{1}{4}$
WITT	$111\frac{1}{4}$	$+\frac{5}{8}$	$+1\frac{1}{8}$	$+\frac{3}{4}$	$-\frac{1}{4}$	$+2\frac{1}{4}$	$113\frac{1}{2}$
WKM	67	$-1\frac{1}{2}$	$-\frac{1}{8}$	$+\frac{1}{4}$	$-1\frac{7}{8}$	$-3\frac{1}{4}$	$63\frac{3}{4}$
X-L	$101\frac{7}{8}$	$-3\frac{1}{4}$	$-1\frac{1}{2}$	$+3\frac{1}{8}$	$+\frac{3}{4}$	$-\frac{7}{8}$	101

   a. What was the net change for each stock for the week?
   b. What was the closing price for each stock on Friday?
   c. Which stock showed the greatest change by Friday? VEX
   d. Which stock had the greatest gain during the week? Explain.
   e. Which stock had the greatest loss during the week? Explain.

10. d. VEX e. WKM

11. Why is the amount of a broker's commission different on the sale of stock than on the purchase when the rate is the same? The commission is figured on different amounts.

12. ***WRITE*** A stockholder in the ABC Company decides to sell all stock in the company based on the steady drop in market price. Is it certain that the investor is making the right move? Describe what things you think might influence the investor's decision. Answers will vary.

13. ***DATA FILE 3 (pp. 96–97)*** Look at the value of the proof sets in 1965, 1968, and 1990. Which proof set seems to be the best investment? Why do you think the value of some sets increased while others decreased? Answers will vary.

Assignment Guide

Basic All
Average All
Enriched All

FOR THE NEXT LESSON

paper, math journal, waste basket

Problem Solving Practice

PROBLEM SOLVING STRATEGIES

Look for a Pattern
Guess and Test
Simplify the Problem
Account for All Possibilities
Make an Organized List
Work Backwards
Make a Table
Write an Equation
Solve by Graphing
Draw a Diagram
Make a Model
Simulate the Problem

Solve. The list at the left shows some possible strategies you can use.

1. In science class there is a jar of bacteria that doubles each day. If the jar is full on the 28th day, on what day is it half full? **27th**

2. Phillip has 4 pairs of pants, 5 shirts, and 2 sweaters. How many different three-piece outfits can he make? **40**

3. What is a four-digit number in which the first digit is half the second, the third digit is the product of the first two, and the last is the sum of the first two? **1,223 or 2,486**

4. You check your coin collection. The total is $16. Surprisingly, you have an equal number of nickels, quarters, and half-dollars. How many coins do you have? **60**

5. In a race Marie was faster than Sophie. Clara beat Lena but lost to Sophie. Who came in last? **Lena**

6. RPM means the number of revolutions a record makes in one minute. How many more revolutions does a 45-rpm record make in 6 min than a $33\frac{1}{3}$-rpm record? **70**

7. Clara had $30. She bought 3 packets of Morning Glory seeds at $1.98 each, 4 packets of Marigold seeds at $2.49 each, tomato plants for $5.95, and fertilizer for $2.89. How much change did she receive? **$5.26**

8. John found the following prices on a list of sports equipment: $3, $2, $6, $4, $9, $8. Although the list seemed a bit odd, John was sure there was a pattern. What are the next three numbers? **12, 16, 15**

9. The magic square shown below has the sum of 15. An anti-magic square uses the numbers from 1–9, but the totals in any direction, including the diagonals, are different. Make an anti-magic square. **Answers may vary. An example is shown.**

Magic Square

4	3	8
9	5	1
2	7	6

Anti-Magic Square

■	■	■
■	■	■
■	■	■

1	3	6
4	5	2
9	8	7

10. ***DATA FILE 9 (pp. 360–361)*** Create a data base of the different types of golf clubs. Include these field names: club, loft, and distance. Arrange the data by distance from least to greatest.

11. Ralph bought $3\frac{1}{2}$ lb of cheese. He used $2\frac{3}{4}$ lb for cheese spread. He used the rest for sandwiches. How much did he use for sandwiches? **$\frac{3}{4}$ lb**

Instruction

This page provides a variety of problems that can be used to reinforce and enhance the students' problem solving skills. Encourage students to read each problem carefully. Then have them refer to the list of problem solving strategies to help them decide how to solve the problem.

Point out, however, that not all questions require a strategy for solving, nor are all the strategies in the list used in this lesson.

ADDITIONAL ANSWERS

10.

Club	loft	distance
wedge	54°, 58°	100
irons 9	47°	110
8	42°	120
7	39°	130
6	35°	140
5	31°	150
4	27°	160
3	23°	170
2	20°	180
1	18°	190
woods 4	19°	190
3	16°	200
2	13°	210
1	10°	220

Study Skills

A personal data file can help students consolidate the facts and concepts they need to know for a test. The data file might be the last few pages in a notebook or binder, or it might be a set of index cards that can be easily transported and studied at quiet moments. The file could include important definitions, formulas, facts, dates, and diagrams.

Computer literate students can keep data in a computer data base.

Some students may wish to study certain subjects by reading aloud and taping the information. Tapes can be played back for instant review. This technique is especially helpful for students who find it easier to learn by listening or whose reading skills are weak.

Here are ways to encourage the creation of data files:

- Provide index cards or other materials.
- Encourage students to use tape recorders.
- Remind students what information should be included in their data file.

Offer students these suggestions:

- Students can decide the kind of data file they will use.
- The file should be simple and easy to access.
- Students can collaborate to create a data file, especially for cooperative study groups.
- The file should be updated throughout the term.

Chapter 5 Review

Complete each statement. Use the vocabulary words given.

VOCABULARY

3 rational number
5 exponents
1 equivalent fractions
2 lowest terms
4 least common denominator

1. You can form ■ by multiplying or dividing the numerator and denominator by the same nonzero factor.
2. When a fraction is in ■, the only common factor of the numerator and denominator is 1.
3. A ■ is a number you write in the form $\frac{a}{b}$ where a is an integer and b is a nonzero integer.
4. The ■ of two or more fractions is the LCM of the denominators.
5. To divide numbers or variables with the same base, subtract ■.

Equivalent Fractions — 5-1

To form equivalent fractions, multiply or divide the numerator and denominator by the same nonzero factor.

Rename the variable to form equivalent fractions.

6. $\frac{3}{4} = \frac{a}{8}$ 6
7. $\frac{2}{3} = \frac{4}{b}$ 6
8. $\frac{3}{x} = \frac{12}{32}$ 8
9. $\frac{y}{5} = \frac{16}{20}$ 4
10. $\frac{5}{6} = \frac{c}{30}$ 25

Fractions and Decimals — 5-2

To write an improper fraction as a mixed number or as a decimal, divide the numerator by the denominator. To write a decimal as a fraction, write the decimal as a fraction with a denominator as a power of ten. Write the fraction in lowest terms.

Write each improper fraction as a mixed number and as a decimal.

11. $\frac{15}{4}$ $3\frac{3}{4}$; 3.75
12. $\frac{3}{2}$ $1\frac{1}{2}$; 1.5
13. $\frac{12}{5}$ $2\frac{2}{5}$; 2.4
14. $\frac{17}{6}$ $2\frac{5}{6}$; $2.8\overline{3}$
15. $\frac{21}{8}$ $2\frac{5}{8}$; 2.625

Write each decimal as a fraction or a mixed number.

16. 0.6 $\frac{3}{5}$
17. 2.375 $2\frac{3}{8}$
18. 5.25 $5\frac{1}{4}$
19. 0.7 $\frac{7}{10}$
20. 0.35 $\frac{7}{20}$

Rational Numbers — 5-3, 5-4

To compare rational numbers, write as fractions with a common denominator and compare the numerators, or write the fractions as decimals and compare.

Compare. Use >, <, or =.

21. $\frac{2}{3}$ ■ $\frac{3}{4}$ <
22. 0.9 ■ $\frac{8}{9}$ >
23. $-\frac{4}{5}$ ■ -0.8 =
24. $\left|\frac{5}{9}\right|$ ■ $\left|\frac{5}{11}\right|$ >

Adding and Subtracting Rational Numbers 5-5

To add or subtract rational numbers, write equivalent fractions with the same denominator, and add or subtract the numerators.

Write each sum or difference.

25. $2\frac{1}{3} + \frac{3}{4}$ $3\frac{1}{12}$

26. $16\frac{4}{5} - 9\frac{2}{3}$ $7\frac{2}{15}$

27. $8\frac{1}{6} + 7\frac{3}{12}$ $15\frac{5}{12}$

28. $11\frac{5}{6} - 5\frac{3}{8}$ $6\frac{11}{24}$

Problem Solving 5-6, 5-11

To solve some problems, you have to work backwards.

To determine the price of 100 shares, multiply the number of shares by the quoted price.

Solve.

29. It will take the Smiths 9 h to drive to Washington. They plan to make five $\frac{1}{2}$-h stops. They plan to arrive at 5:30 P.M. At what time should they plan to leave? 6 A.M.

30. Ms. Nelson bought 500 shares of Plato Publishing at $39\frac{7}{8}$ per share. Find the cost of the 500 shares. $19,937.50

Multiplying and Dividing Rational Numbers 5-7

For any two rational numbers $\frac{a}{b}$ and $\frac{c}{d}$, $\frac{a}{b} \cdot \frac{c}{d} = \frac{a \cdot c}{b \cdot d}$ and $\frac{a}{b} \div \frac{c}{d} = \frac{a}{b} \cdot \frac{d}{c}$.

Find each product or quotient.

31. $\frac{3}{5} \cdot 1\frac{1}{2}$ $\frac{9}{10}$

32. $2\frac{2}{3} \cdot 3\frac{3}{8}$ 9

33. $5\frac{1}{4} \div \frac{7}{8}$ 6

34. $\frac{4}{5} \div 1\frac{3}{5}$ $\frac{1}{2}$

Simplifying Expressions with Exponents 5-8

To divide numbers or variables with the *same* base, subtract exponents. For any nonzero integers a and n: $a^{-n} = \frac{1}{a^n}$.

Write with positive exponents.

35. x^{-5} $\frac{1}{x^5}$

36. $6a^{-1}$ $\frac{6}{a}$

37. $\frac{4m^6}{2m^2}$ $2m^4$

38. $\frac{10b^2}{5b^3}$ $\frac{2}{b}$

39. $\frac{12x^2y^5}{4x^4y^2}$ $\frac{3y^3}{x^2}$

Solving Equations 5-9, 5-10

To solve equations, use the properties of equality.

Solve each equation.

40. $x - \frac{4}{5} = \frac{1}{3}$ $1\frac{2}{15}$

41. $\frac{3}{4}x = 2\frac{1}{2}$ $3\frac{1}{3}$

42. $\frac{2}{5}x = -1\frac{1}{4}$ $-3\frac{1}{8}$

43. $x + 4\frac{2}{3} = 6$ $1\frac{1}{3}$

RESOURCES

TRB E-23–E-29

Scoring Chart	
No. Correct	**Score**
46	100
45	97
44	95
43	93
42	90
41	87
40	85
39	83
38	81
37	79
36	77
35	75
34	73
33	71
32	68
31	66
30	64
29	62
28	60
27	58
26	56
25	54
24	52
23	50
22	47
21	45
20	42
19	39
18	37
17	35
16	32
15	30
14	28
13	26
12	24
11	22
10	20
9	18
8	16
7	14
6	12
5	10
4	8
3	6
2	4
1	2

Chapter 5 Test

Write a rational number for each point on the number line.

E B A C D F

-2 -1 0 1 2

1. *A* $-\frac{1}{6}$ **2.** *B* $-\frac{5}{6}$ **3.** *C* $\frac{1}{2}$ **4.** *D* $1\frac{1}{3}$ **5.** *E* $-1\frac{5}{6}$ **6.** *F* 2

ESTIMATION Tell whether each fraction is close to 0, $\frac{1}{2}$, or 1.

7. $\frac{7}{8}$ 1 **8.** $\frac{7}{12}$ $\frac{1}{2}$ **9.** $\frac{21}{25}$ 1 **10.** $\frac{2}{15}$ 0 **11.** $\frac{16}{31}$ $\frac{1}{2}$ **12.** $\frac{5}{9}$ $\frac{1}{2}$

Write each fraction as a decimal. Write each decimal as a fraction.

13. $\frac{2}{5}$ 0.4 **14.** $\frac{3}{4}$ 0.75 **15.** $\frac{7}{8}$ 0.875 **16.** 0.9 $\frac{9}{10}$ **17.** -0.4 $-\frac{2}{5}$ **18.** 0.75 $\frac{3}{4}$

Compare. Use >, <, or =.

19. $\frac{4}{5}$ ■ $\frac{2}{3}$ > **20.** 0.66 ■ $\frac{2}{3}$ < **21.** $-\frac{7}{10}$ ■ -0.07 < **22.** 0.875 ■ $\frac{7}{8}$ =

Find each answer.

23. $\frac{3}{8}+\frac{5}{6}$ $1\frac{5}{24}$ **24.** $\frac{3}{4}\cdot 2\frac{5}{8}$ $1\frac{31}{32}$ **25.** $3\frac{4}{5}-2\frac{1}{2}$ $1\frac{3}{10}$ **26.** $4\frac{2}{3}\div 1\frac{5}{6}$ $2\frac{6}{11}$

27. $3-\frac{5}{9}$ $2\frac{4}{9}$ **28.** $-1\frac{1}{3}\div\left(-\frac{5}{9}\right)$ $2\frac{2}{5}$ **29.** $-\frac{3}{5}-\left(-1\frac{1}{3}\right)$ $\frac{11}{15}$ **30.** $3\frac{3}{4}\cdot 2\frac{4}{5}$ $10\frac{1}{2}$

31. $4\frac{3}{8}\cdot 2\frac{4}{5}$ $12\frac{1}{4}$ **32.** $4\frac{3}{7}+5\frac{4}{7}$ 10 **33.** $-1\frac{5}{9}\cdot 2\frac{5}{8}$ $-4\frac{1}{12}$ **34.** $-1\frac{7}{8}+\left(-3\frac{5}{6}\right)$ $-5\frac{17}{24}$

Solve each equation.

35. $-\frac{7}{6}+x=\frac{5}{6}$ 2 **36.** $\frac{3}{5}a=9$ 15 **37.** $n+\left(-\frac{7}{8}\right)=\frac{5}{6}$ $1\frac{17}{24}$ **38.** $\frac{2}{3}k=-6$ -9

39. $m-\left(-\frac{3}{4}\right)=1\frac{1}{2}$ $\frac{3}{4}$ **40.** $\frac{3}{8}y=-15$ -40 **41.** $-5b=-3\frac{1}{3}$ $\frac{2}{3}$ **42.** $r-6.5=-9.3$ -2.8

Solve.

43. Suppose you take a number, subtract 8, multiply by 7, add 10, divide by 5, and the answer is 9. What is the original number? 13

44. Josie spent $\frac{3}{4}$ of her money on clothes. She had \$21 left. How much money did she originally have? \$84

45. Write a word problem for the equation $x-1.70=3.50$. Answers will vary.

46. Find the price of 1,000 shares of Universal Tractor stock at $21\frac{5}{8}$ per share. \$21,625

Chapters 1–5 Cumulative Review

Choose the correct answer. Write A, B, C, or D.

1. Evaluate $(-1)^{27}$. C
   - **A.** 1 **B.** 1×10^{27}
   - **C.** -1 **D.** not given

2. Write $\frac{3x^2y}{12xy^2}$ in lowest terms. C
   - **A.** $\frac{1}{4}$ **B.** $\frac{2y}{4xy}$
   - **C.** $\frac{x}{4y}$ **D.** not given

3. Which number is written in scientific notation? B
   - **A.** 0.5×10^6 **B.** 1.5×10^6
   - **C.** 1.5×6^6 **D.** not given

4. Solve $\frac{2}{3}x = 2\frac{2}{9}$. B
   - **A.** $\frac{3}{10}$ **B.** $3\frac{1}{3}$
   - **C.** $2\frac{8}{9}$ **D.** not given

5. Write $\frac{28}{3}$ as a mixed number. D
   - **A.** $3\frac{1}{3}$ **B.** $4\frac{1}{3}$
   - **C.** $7\frac{2}{3}$ **D.** not given

6. (0.0056)(-3.5) C
   - **A.** -0.196 **B.** -0.00196
   - **C.** -0.0196 **D.** not given

7. Identify the property used. B
   $$(x^2y^3)x^0 = x^2y^3$$
   - **A.** associative **B.** identity
   - **C.** commutative **D.** not given

8. The GCF of 25 and 50 is ■. D
   - **A.** 10 **B.** 5
   - **C.** 2 **D.** not given

9. Which number is divisible by 3 and 9? C
   - **A.** 663 **B.** 879
   - **C.** 864 **D.** not given

10. Without computing, state the sign of 2(-4)(-9) + 100. A
    - **A.** positive **B.** negative
    - **C.** zero **D.** not given

11. Linda is 12 years older than Jill. The sum of their ages is 38. Find their ages. D
    - **A.** 12 and 26 **B.** 12 and 38
    - **C.** 5 and 17 **D.** not given

12. Simplify $\frac{5a^3b^{-2}}{15a^2b^3}$. A
    - **A.** $\frac{a}{3b^5}$ **B.** $3ab^{-1}$
    - **C.** $\frac{ab}{3}$ **D.** not given

13. Compare $|2\frac{1}{4}|$ ■ $|-\frac{9}{4}|$. C
    - **A.** $>$ **B.** $<$
    - **C.** $=$ **D.** not given

14. Find the number of times at bat in the formula $a = \frac{h}{n}$ for $a = 0.25$ and $h = 25$. C
    - **A.** 625 **B.** 50
    - **C.** 100 **D.** not given

15. Find the LCD of $\frac{7}{12}$ and $\frac{5}{18}$. D
    - **A.** 35 **B.** 6
    - **C.** 72 **D.** not given

16. $3\frac{1}{2} - 5\frac{3}{8} =$ ■. B
    - **A.** $-2\frac{1}{8}$ **B.** $-1\frac{7}{8}$
    - **C.** $-2\frac{7}{8}$ **D.** not given

Study Skills

Since most standardized tests are written in multiple choice format, knowing how to answer the questions gives the student an advantage. In a multiple choice question, students need to read the choices carefully and consider each possibility in turn. If students are uncertain about the correct answer they can make a guess based on what they know about the problem and on what they observe in the pattern of answer choices. Usually, one choice is obviously incorrect and can be eliminated immediately. Other incorrect choices often reflect common errors and can be eliminated if students are alert to these errors.

Often answer labels provide clues to the correct choice. For example, if a student recognizes that an answer should be expressed in meters and only two of the choices are labeled as such, then the other choices can be eliminated quickly.

Another technique that students can use is to read the question, and before looking at the choices, visualize or estimate the answer. Then, when the students examine the choices, the correct answer may be obvious.

Additional suggestions:

- ▼ Provide time after a test to discuss the questions with the class.
- ▼ Give practice tests to students who tend to "freeze up" in testing situations.
- ▼ Assign one group of students the task of writing a multiple-choice test for another group.

6

RATIOS, PROPORTIONS, AND PERCENT

Chapter At a Glance

Chapter 6 explores the concepts of ratios, proportions, rates, and percents, including applying proportional reasoning to solve problems and to evaluate percents.

Reading the Chapter

Chapter 6 includes the new terms *ratio* and *proportion*. Discuss these terms thoroughly, making sure students know how to read and write ratios and proportions in different forms. For example: 15 to 30, 15:30, and $\frac{15}{30}$ all express the same ratio; 8 is to 12 as 16 is to 24, 8 : 12 : : 16 : 24, and $\frac{8}{12} = \frac{16}{24}$ all express the same proportion. Help students to understand that ratios include two terms and proportions include at least four terms. Explain that when students are reading word problems that involve ratios and/or proportions, it can be helpful to read once for the general idea of the problem, and then, to reread to identify the terms.

You may want to explore a word problem with the students demonstrating how to identify and record the proportional information in the problem. Explain that a proportion is an equation involving at least two ratios. Show students the following form $\frac{?}{?} = \frac{?}{?}$. Tell students that one ratio is always given and have them read carefully, using their investigative skills, to identify and record the given ratio. Then, they can continue their reading to identify the corresponding parts, including the unknown, of the other ratio.

Chapter Resources

Practice and Enrichment TRB F-1 to F-4, F-6 to F-15
Alternate Application TRB F-5

Chapter Test Form A, F-17
Chapter Test Form B, F-19

Computer Test Bank

References

STUDENT BIBLIOGRAPHY

Driscoll, P., et.al. *Ratio, Proportion, and Percent* (Denoyer-Geppert Co., Chicago 1979)

TEACHER BIBLIOGRAPHY

Good, Thomas, et.al. *Effective Mathematics Instruction* (Longman, New York 1989)

SOFTWARE

Creature Cube (Sunburst Communications), for the Apple II family of computers.

FILMS, VIDEOTAPES, AND OTHER MEDIA

Flip-Chip Algebra (Activity Resources Co., Inc.), includes booklet and manipulatives.

Vocabulary

base
percent
percent of change
percentage
rate
ratio
scale drawing

Materials

waste basket
wadded sheet of paper
math journal
graph paper (T*)

*T—Teacher's Resource Book

Resource Bank

Class Activities

FIELD TRIPS

▼ Restaurant—to see how the restaurant plans the amount of food to buy and prepare for one day's business.
▼ Bank—to see how interest is calculated and posted to savings and checking accounts.

CLASS SPEAKERS

▼ Chef—to discuss how recipes are created and then adapted to serve large and small numbers of people.
▼ Bank officer—to discuss saving and borrowing money and how the bank determines the rate of interest.

Bulletin Board

Expand questions 1-6, page 242, by enlarging and duplicating a new-home floor plan, obtained from a local real estate office.

Display the floor plan on the bulletin board. Give students a reduced copy to work with. Display the students' revisions when complete.

How does the architect indicate such things as door openings and windows? (Answers will vary.)

What scale did the architect use? (Answers will vary.)

A client revises the original plan to suit specific needs. Make a list of changes that could be made. (Answers will vary.)

Project: Redesign the floor plan to reflect the client's changes. Be sure to use the same scale the architect used.

REDESIGNING A FLOOR PLAN

Extra-Credit Problems

BASIC

▼ Suppose you and a friend have some money. The ratio of the money you have to your friend's money is 5 : 2. Your friend gets $10 from her parents and you spend $5. Now, you both have the same amount. How much did you each have to start? ($25, $10)

AVERAGE

▼ Two rectangles have a length-to-width ratio of 9 : 5. The perimeter of the second rectangle is 672 in., which is twice the perimeter of the first rectangle. What is the length and width of each rectangle? (216 in., 120 in.; 108 in., 60 in.)

ENRICHED

▼ Study triangle *DEF*. *GH* is parallel to *EF*.

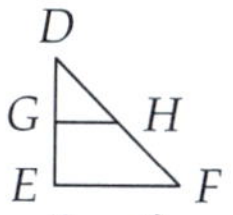

Write proportions to describe the relationships among the line segments. (Some possibilities: *DG* : *GE* : : *GH* : *EF*; *DG* : *DH* : : *GE* : *HF*; *DE* : *DG* : : *DF* : *DH*; *GE* : *DE* : : *HF* : *DF*.)

Chapter Planning Guide

OBJECTIVES	ASSIGNMENT GUIDE			ASSESSMENT	
	Basic	Average	Enriched	Review/ Tests	TRB*
Exploring Ratios Use activities to explore ratios.	All	All	All		
6-1 Ratios, Proportions, and Rates To write ratios and proportions.	1-16, 19-20, 23-27, 32-39, 44-47, 50, MR* All	1-17 odd, 20, 22-23, 24-34 even, 38-48 even, 50-51, MR All	6-8, 12-19, 21-35 odd, 40-46, 49-51, MR All	Extra Practice p. 571	Practice F-1 Enrichment F-2
6-2 Solving Proportions To use proportions to solve problems.	1-6, 13-18, 23, 26-29, 35-36, 40-41, 44-46, 50, MR, CT* All	5-10, 13-20, 23, 26-32, 37-38, 41-42, 46-47, 50, MR, CT All	8-14, 21-26, 29-30, 33-34, 39, 42-43, 48-50, MR, CT All	Extra Practice p. 571	Practice F-3 Enrichment F-4
6-3 Problem Solving Application: Scale Drawing To apply proportional reasoning to problems involving scale plans.	All	All	All		Alternate Application F-5
Exploring Percents Use models to explore percents.	**Manipulative Connection** All	All	All		
6-4 Percent To express ratios and rates as percents.	1-20, 25-26, 29-41, 46-48, 55-60, 66-67, MR All	1-20, 25-26, 29-37, 39-43, 48-51, 55-61, 68-70, MR All	5-24, 27-37, 42-45, 51-58, 61-65, 69-70, MR All	Extra Practice p. 571	Practice F-6 Enrichment F-7
Exploring Number Relationships To explore number relationships.	All	All	All		
6-5 Using Proportions to Find Percent To evaluate percents using proportions.	1-25, 26-38 even, 45-46, 48-50, MR, TY* All	1-35 odd, 36-41, 47-53, MR, TY All	1-41 odd, 42-44, 47-48, 52-53, MR, TY All	Extra Practice p. 571 TY p. 255	Practice F-8 Enrichment F-9 TY F-16
6-6 Percents and Equations To evaluate percents using models and equations.	1-5, 9-14, 21-26, 35-43, 46-47, MR, CT All	1-5, 9-17, 21-29, 35-44, 46-48, MR, CT All	1-8, 12-20, 25-34, 43-50, MR, CT All	Extra Practice p. 571	Practice F-10 Enrichment F-11
6-7 Percent of Change To find percent of increase or decrease.	1-10, 13-14, 17-18, 20, 21-23, 27-29, 37, MR, TY All	1-16 even, 17-20, 22-24, 27-35 odd, 37, MR, TY All	1-19 odd, 20, 25-26, 28-38 even, MR, TY All	Extra Practice p. 571 TY p. 263	Practice F-12 Enrichment F-13 TY F-16
6-8 Problem Solving Strategy: Draw a Diagram To solve problems by drawing a diagram.	**Problem Solving Connection** All	All	All	Problem Solving Practice p. 267	Practice F-14 Enrichment F-15
Chapter Test	All	All	All	pp. 268-271	Tests F-17–F-23

*TRB—Teacher's Resource Book MR—Mixed Review TY—Test Yourself CT—Critical Thinking

CONNECTIONS							NCTM CORRELATION
Algebra	Critical Thinking	Problem Solving	Estimation/ Mental Math	Technology	Manipulatives/ Resources	Writing/Reading in Math	
	5-7	1-7			1-4, 7	6-7	Problem Solving Connections Functions
W*: 47-49	C*: TD* 1-3	C: 4-11 W: 9-15, 23, 44-51		W: 32-35	C: 6-8 W: 19-23, 50	C: 4-11, TD 1-3 W: 9-15, 23, 36-51	Algebra Functions Problem Solving Connections
C: 1-10 W: 1-39	C: TD 1-2 W: 28-31 CT*: 1-8	C: 5-10 W: 13-27, 40, 45-50			W: 24, 27, 45-50 CT: 8	W: 27, 50 CT: 1-7	Algebra Problem Solving Functions Statistics
	W: 4, 5	C: 1-3 W: 1-12		W: 1-12	C: 1-3 W: 1-6	W: 4, 5	Problem Solving Communication Reasoning Connections Geometry
	4, 6		6		1-3, 5	4, 6	Communication Reasoning Functions
	C: TD 1-2 W: 59-64	C: 13 W: 38-45, 65-68	C: 1-8 W: 29-31	W: 25-28	W: 32-37, 65-68	C: 13 W: 55-58, 69	Problem Solving Functions
	1-10	1-10				1-10	Problem Solving Communication Reasoning Connections
C: 1-11 W: 1-25, 35-48 TY*: 3-4, 10-11	C: TD 1-2	C: 9-11 W: 45-53 TY: 1-4, 10-11	W: 17-25, 27-34	W: 35-44	W: 48, 52-53	C: TD 1-2 W: 26, 49-51	Algebra Problem Solving Communication
C: 1-8 W: 1-20	C: TD 1-2 W: 44, 46, 49-50 CT: 1-2	C: 3-8 W: 41-50 CT: 1-2	W: 21-34, 47	W: 35-40	C: 3-6 W: 3-8, 43, 46-50 CT: 2	C: 9, TD 1-2 W: 43	Algebra Functions Statistics Problem Solving
C: 9-10 TY: 1-8	C: TD 1-3	W: 27-38 TY: 7-8	W: 13-16	W: 21-26, 33-38	W: 20, 35-37	C: TD 1-3 W: 20, 37, 38	Algebra Functions Statistics Problem Solving
C: 1-4 W: 1-16		C: 1-4 W: 1-16		W: 1-16	W: 16		Algebra Problem Solving Reasoning

*C—Class Exercises W—Written Exercises CT—Critical Thinking TY—Test Yourself TD—Think and Discuss

Exploring

In this manipulative activity, students explore proportions using a model to represent each ratio. **See TRB Exploring O-21**

Exploring Writing Proportions

Suppose 9 sandwich rolls cost $2. How many rolls can you buy with $6? The models below represent the problem.

1. What ratio does the model on the left represent? **$\frac{9}{2}$**
2. What ratio does the model on the right represent? **$\frac{x}{6}$**
3. Together the two ratios form a proportion. Write the proportion. rolls → / dollars → $\frac{9}{2} = \frac{x}{6}$ ← rolls / ← dollars
4. Suppose you want only 8 rolls. What proportion can you write to find the cost of 8 rolls?
   a. To write a proportion use the ratio you found in problem 1. rolls → / dollars → $\frac{9}{2} = \frac{8}{x}$ ← rolls / ← dollars
   b. Write a second ratio with the same labels and use a variable to represent what you must find.

Write a pair of ratios and a proportion for each.

		First Ratio	Second Ratio	Proportion
5. Suppose 5 jars of honey cost $20. How much will 18 jars cost?	jars → / dollars →	$\frac{5}{20}$	$\frac{18}{x}$	$\frac{5}{20} = \frac{18}{x}$
6. If 12 cans of soup cost $8, how many cans can you buy for $6?	cans → / dollars →	$\frac{12}{8}$	$\frac{x}{6}$	$\frac{12}{8} = \frac{x}{6}$
7. Suppose 9 grapefruits sell for $3. How many can you buy for $21?	grapefruits → / dollars →	$\frac{9}{3}$	$\frac{x}{21}$	$\frac{9}{3} = \frac{x}{21}$
8. If 15 muffins cost $4, how much will 60 muffins cost?	muffins → / dollars →	$\frac{15}{4}$	$\frac{60}{x}$	$\frac{15}{4} = \frac{60}{x}$
9. If 8 pounds of potato salad cost $20, how many pounds can you get for $5?	pounds → / dollars →	$\frac{8}{20}$	$\frac{x}{5}$	$\frac{8}{20} = \frac{x}{5}$
10. If 12 subs cost $48, how many can you buy for $64?	subs → / dollars →	$\frac{12}{48}$	$\frac{x}{64}$	$\frac{12}{48} = \frac{x}{64}$

O-21

Exploring Visual Perception

You can look at a space figure from different points of view.

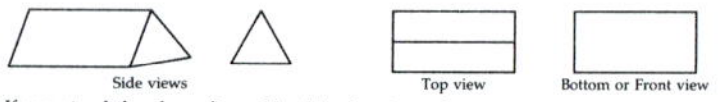

1. If you stand the above figure like this, the views change. Name each view.

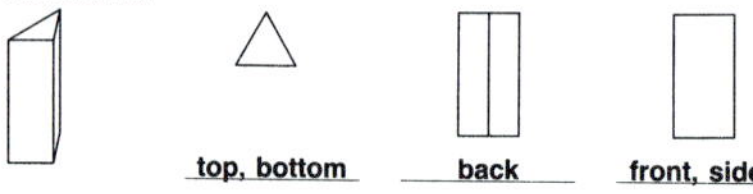

top, bottom **back** **front, sides**

Label the views of each figure *top, bottom, side, front,* or *back.*

2.

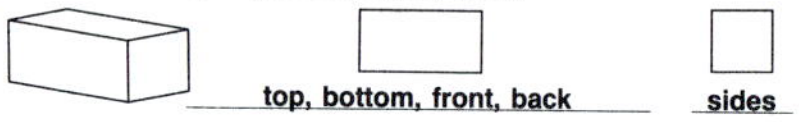

top, bottom, front, back **sides**

3.

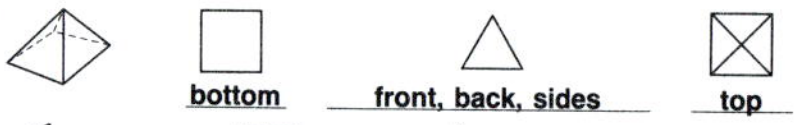

bottom **front, back, sides** **top**

4.

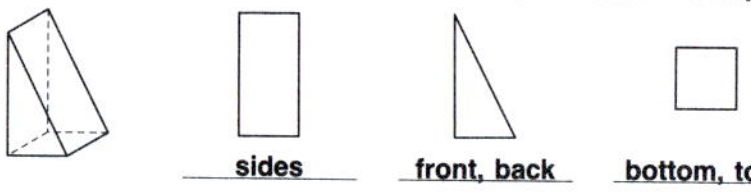

sides **front, back** **bottom, top**

Draw the indicated views of the figure.

5.

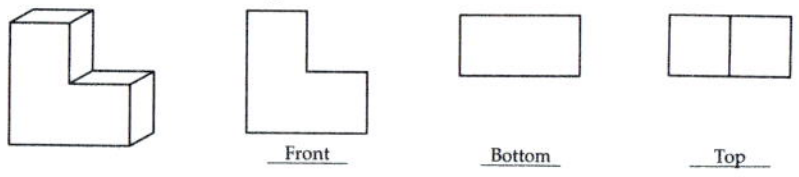

Front Bottom Top

O-22

In this problem solving activity, students explore different views of a space figure. **See TRB Exploring O-22**

In this calculator activity, students explore percents by calculating markdown sale prices. **See TRB Exploring O-23**

Exploring Markdown Using a Calculator

The Shirt Shack is having a 25% markdown sale the first week of the month. This means that the sale price is 25% less than the original price. In the middle of the month Shirt Shack marked down the sale shirts another 15%.

1. Do you think a markdown of 25% followed by another markdown of 15% is the same as a single markdown of 40%? Explain.
   Answers will vary. Possible answer: No, a single markdown of 40% is better because it is taken off the total price.
2. For question 1, which do you think is the better buy and why?
   Answers will vary.
3. What is the sale price of a $25 shirt first marked down 25% and then marked down another 15% (to the nearest cent)?
   $15.94
4. What is the sale price of a $25 shirt marked down 40%?
   $15.00
5. Compare your answer to question 1 with your results for questions 3 and 4. How would you now answer question 1?
   The single markdown is a better buy. By taking 15% off the already reduced price, you are taking 15% off a lesser amount. This is less than a 40% markdown.
6. Successive markdowns of 25% on a $25 shirt and then 15% are equal to a single markdown of what percent? Explain how you would solve the problem using a calculator. Then solve.
   1.00 − [(] 0.75 [×] 0.85 [)] [=] 36.25
7. The Shirt Shack is giving a 50% markdown on sweatshirts. The store, Terrific Tops, has marked the same sweatshirt down three times: 25%, 20%, and 15%. If the shirt originally sold for $35, which store has the better buy? What was the final price at each store?
   The Shirt Shack; final price: $17.50, the final price at Terrific Tops: $17.85.

O-23

Exploring Interest on a Computer Spreadsheet

Interest can be either simple or compound.

Simple Interest

To compute simple interest multiply the principal times the rate times the time.

Compound Interest

To compute compound interest add the interest for each period to the principal before computing the interest for the next period.

The spreadsheet below shows what happens when interest on $200 is compounded annually at 6% for 3 years.

	A	B	C	D	E
1	Period(yr)	Principal	Rate	Interest	Balance
2	1	200	0.06	12	212
3	2	212	0.06	12.72	224.72
4	3	224.72	0.06	13.48	238.20

1. How is the value of cell D2 computed? E2? Write each formula.
   D2 = B2*C2, E2 = B2 + D2
2. What is the relationship between the balance at the end of year 1 and the principal at the beginning of year 2?
   They are the same number.
3. Study the values and formulas in the cells of rows 2 and 3. Input the values and formulas for row 4.

	A	B	C	D	E
2	1	200	0.06	=B2*C2	=B2+D2
3	2	=E2	=C2	=B3*C3	=B3+D3
4	**3**	**E3**	**C2**	**B4*C4**	**B4+D4**
5					

4. What would be the new balance if the interest on $200 was compounded annually at 6% for 4 years?
   252.50
5. How many years would it take for $200 invested at 6% compounded annually to double?
   12
6. What *one* cell needs to be changed to have the spreadsheet compound interest at 12% instead of 6%?
   C2

O-24

In this computer activity, students explore how to compute simple and compound interest using a computer spreadsheet. **See TRB Exploring O-24**

Meeting Individual Needs

GIFTED STUDENTS

Many mathematically gifted students enjoy the challenge of sifting through information to get to the heart of a problem. Such students, intrigued by complexity, also can appreciate working through problems that involve a number of steps.

Chapter 6 offers many opportunities for students to develop and work on multi-step problems involving ratios and rates. To extend the chapter, you can make available to gifted students a variety of county and state maps. Work to familiarize students with the map scales; symbols for state, county, and interstate roads; and the map orientations. Then have students develop multi-step rate-time-distance problems using the maps. Students will need to work with ratio as well as rate-time-distance formulas. Mention that the speed limit on most interstate highways is 55 mi/h, while many other roads and local streets have speed limits as low as 20 mi/h.

AT-RISK STUDENTS

At-risk students are more likely to approach a subject with interest when it is perceived as relevant to their lives. Make a list of things that are popular with the age group of your students and use them in examples when teaching ratios, proportions, and percents. For instance, use a summary of the sale price of videotapes to teach a lesson on percents.

Give the students the following problem to solve: A store sold VCRs at 75% of the regular price of $192.50, or 25% off. Later, the remaining stock was reduced 5% off the sale price. What was the last sale price of the VCRs? What would you guess is the percent off the regular price? What is the actual percent off the regular price? ($137.16, 30%, 29%)

ENGLISH AS A SECOND LANGUAGE

ESL students often find it both reassuring and motivating when a teacher takes the time to recast mathematical ideas and problems in culturally familiar terms. In the section on scale drawings, for example, the ESL students can be asked to use maps of their native countries.

Working in pairs or groups with English-language students can help the ESL students with language skills, such as reading comprehension. Since reading comprehension is important for understanding the lesson on percents and equations, separate the class into pairs. ESL students can be paired with English-language students. Using the triangle diagram on page 256 and "Mental Math" questions 21-34 on page 258, ask the pairs to take turns reading each question aloud. Then ask them to practice rewording the questions to match the form of the diagram. For example, ____ is ____% of ____. This activity can enhance the ESL students' vocabulary and verbal skills, as well as provide an opportunity for social interchange.

See TRB Teaching Resources p. 56.

Problem Solving Situation

Ask students to look at the "Think About It" question. Explain that an index number of 100 means that teens make the decision. Ask students to discuss what clothes they are allowed to buy and what limitations they must adhere to (such as price or style). Also, discuss the role that advertising may have on their choices for purchases.

Lead students to see that advertisers might use the index to gear some advertising to the teens and some to their parents.

Explain that in this activity, students will prepare an ad campaign for a clothing company. They need to decide what styles, designs, and types of clothing teenagers prefer and the kinds of ads that would be particularly appealing to teenagers.

DATA COLLECTION

Ask students to discuss the kind of information needed to design the ad campaign. Then, have students develop at least five multiple choice marketing questions and then use the questions to poll 50 students.

DATA ANALYSIS

Let students categorize and analyze the results of each question asked. Students may wish to separate male and female responses, especially if their ad campaign targets one group only. You may wish to have the students express results as "percents of the total responses" to give additional practice with the content of the chapter.

D A T A   F I L E

CHAPTER 6

Ratios, Proportions, and Percent

TEENAGE RESEARCH UNLIMITED (TRU) of Northbrook, IL, developed the Teenage Buying Control Index (TBC). TRU asked teens whether they or their parents make the buying decision for a list of products. An index number of 100 means the teen makes the decision. An index number of 0 means the parent makes the decision.

TEENAGE BUYING CONTROL INDEX (TBC)

ITEM	MALES	FEMALES
Book	63	63
Poster	78	78
Camera Film	54	53
Health and Beauty Aids	31	42
Food	39	39
Clothing	38	39
Audio/Video	55	48
School Supplies	46	52

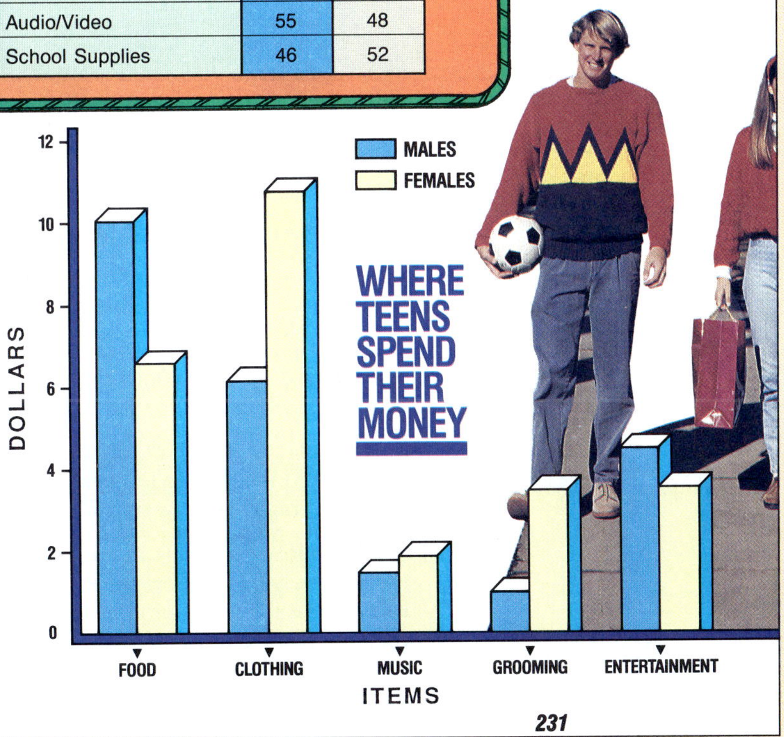

Think about it...

Look at the data from TRU. Why are some items on the TBC rated higher than others? How do you think advertisers might use the index?

See Problem Solving Situation

Cooperative Learning

This activity can be completed by groups of 4-6 students. Each group becomes the *Teen Marketing Co.* The teacher represents the client, *Totally Awesome Clothes*. Based on their analysis of their data, the group will create an ad campaign. For instance, if a group learned that 4 out of 5, or 80%, of the teenagers polled want clothes for school, the group might come up with a slogan such as "For a *totally awesome* day at school, wear our jeans."

Assign roles to members of the group. Roles can include Chartmaker, Account Manager, Statistician, and Checker. Remind the group that they must work together and design one ad campaign. When the ad campaigns are complete, the Account Manager will present the campaign to the class. Encourage students to discuss and evaluate the campaigns and the group process.

Questions related to this Data File can be found on pages 258, 471, 494, and 571.

Lesson Focus

Materials
Paper, math journal, waste basket

Vocabulary/Symbols
ratio

MOTIVATING THE LESSON

Have students use the number of students in the class to write ratios: for example, number of boys to number of girls; number of girls to number of boys; number of girls to total number of students in the class.

Instruction

1 Help students understand that performance can be compared because each player has the same number of attempts. However, if the number of attempts varies, comparing performances is no longer meaningful.

2 As the chart is discussed, mention that the average is the ratio that compares the number of successful baskets to the number of baskets attempted.

3 It should be noted that free throw averages are recorded as percents in record books and sports almanacs.

Closure

Writing in Math To see if students can generalize from this activity, have them research free throw records of leading basketball players and compare the ratios. Have them write their conclusions in their math journals.

Assignment Guide
Basic All
Average All
Enriched All

FOR THE NEXT LESSON

calculator, math journal

Exploring Ratios

OBJECTIVE:
Use activities to explore ratios.

▼ What do batting averages, pass completions, first down statistics, and free-throw averages have in common? They are all *ratios*. Each statistic compares two numbers by division.

MATERIALS
- Waste basket
- Wadded up sheet of paper
- Math journal to record work

1

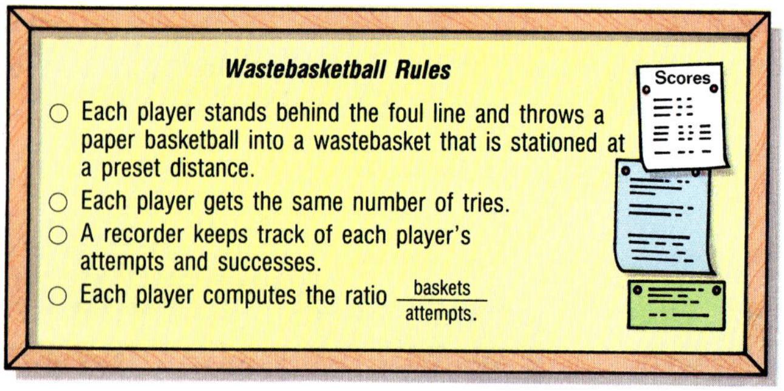

Play a game of wastebasketball and compute your own ratios.

Player	Attempts	Baskets	Free-throw Average (Ratio)

2 Answers will vary.

1. **Compare** with another group. Are the averages close?
2. Change the conditions. Then play more games.
   a. Change the distance to the basket.
   b. Change the number of attempts.
3. Compute new statistics. Which variable has the greatest impact? Is it different for different people? Answers will vary.
4. Compute team statistics. **Compare** with other groups. Are the averages close?
5. **Discuss** what you can conclude from the statistics. If you make 2 baskets out of 2 tries, is your ratio better than someone who gets 7 out of 10? 15 out of 20? The ratio is better but there's no way to tell if you could keep the same ratio with more tries.
6. **Write** Would it be fair to judge a player's ability on one game's free-throw average? Why or why not? No; one game could go very well for a player while another might not.
7. **PROJECT** How does the ratio relate to the number of tries? Look up lifetime free-throw averages for five basketball players. Compare to averages for players in one game and over a season. What impact does the number of attempts have on the average? The greater the number of attempts, the more accurately the average will reflect the player's ability. 3

Teaching Tip

The waste basketball activity can be done effectively by small groups of three or four students. Assign each student a role such as player, score keeper, ball collector, and coach, to ensure that all team members are engaged in the task.

OBJECTIVE: To write ratios and proportions.

6-1 Ratios, Proportions, and Rates

▼ Statistics show that 10 out of 25 people in the United States have brown eyes. The numbers 10 and 25 form a *ratio*. You can write a ratio in three different ways:

$$10 \text{ to } 25; \quad 10:25; \quad \frac{10}{25}$$

You can express a ratio as a fraction in lowest terms.

$$\frac{10}{25} = \frac{2}{5}$$

READ 10 is to 25 as 2 is to 5. You can say that two out of five people have brown eyes.

Ratio	A ratio is a comparison of two quantities by division.	
	Arithmetic	**Algebra**
	1 to 2; 1 : 2; $\frac{1}{2}$	a to b; $a:b$; $\frac{a}{b}$ $\quad b \neq 0$

▼ You can write ratios that compare $\frac{\text{part}}{\text{part}}$, $\frac{\text{part}}{\text{whole}}$, and $\frac{\text{whole}}{\text{part}}$.

Example 1 In a survey of 100 students, 60 reported having after school jobs. Write three ratios for the data.

Solution

$\dfrac{60 \text{ students with jobs}}{40 \text{ students without jobs}}$ $\quad \dfrac{\text{part}}{\text{part}}$

$\dfrac{60 \text{ students with jobs}}{100 \text{ all students surveyed}}$ $\quad \dfrac{\text{part}}{\text{whole}}$

$\dfrac{100 \text{ all students surveyed}}{60 \text{ students with jobs}}$ $\quad \dfrac{\text{whole}}{\text{part}}$

THINK Why don't we write ratios as mixed numbers? **because a ratio is a comparison and it would be difficult to compare with a mixed number.**

▼ When two ratios are equal they form a *proportion.*

Proportion	A proportion is a statement that two ratios are equal. If two ratios are equal, their cross products are equal.	
	Arithmetic	**Algebra**
	6 is to 9 as 8 is to 12	a is to b as c is to d
	6 : 9 : : 8 : 12	$a : b : : c : d$
	$\frac{6}{9} = \frac{8}{12}$	$\frac{a}{b} = \frac{c}{d}$ $\quad b \neq 0, d \neq 0$

Example 2 The directions for making orange juice from concentrate call for 4 cans of water to 1 can of concentrate. So, for 2 cans of concentrate you would need 8 cans of water. The following proportion describes the situation.

$$\frac{1}{4} = \frac{2}{8}$$

Lesson Focus

Materials
Calculator, math journal

Vocabulary/Symbols
proportion
rate
ratio

MOTIVATING THE LESSON

Survey the class by having all students who have brown eyes raise their hands. Then have students write a ratio comparing the number of students with brown eyes to the total number of students in the class.

Instruction

1 As you discuss the concepts of whole and part, point out that whole/part ratios are always greater than 1.

2 Explain that to check if two ratios are in proportion, write the two ratios as equal, then find the cross products. For example: the ratios 12 : 15 and 16 : 20 are in proportion because $12 \cdot 20 = 15 \cdot 16$.

Teaching Tip

Ask students to find examples of the mathematics topic they are studying in magazines and newspapers. For example, while learning about ratios and proportions students might search for examples in baseball statistics, financial reports, and survey results.

3 Remind students that when comparing ratios to determine the unit rate, they must write the two ratios in the same order; that is, the first and third terms must contain the same units, just as the second and fourth terms must contain the same units.

ADDITIONAL EXAMPLES

1. Write three ratios to describe this situation: 14 out of 25 students in the class are baseball fans. (14 : 25; 14 to 25; $\frac{14}{25}$)
2. Express the ratio 12 : 20 in lowest terms. (3 : 5)
Solve.
3. Hank runs one lap of a track in $1\frac{1}{2}$ minutes. How many laps will he run in 1 minute? ($\frac{2}{3}$)
4. A car traveling at a rate of 50 mi/h can stop in 3s. At the same rate of braking, how long will it take a car traveling at 35 mi/h to stop? (2.1s)

Guided Practice

Class Exercises In questions 9-11, check that students use the correct unit rates.

Think and Discuss Students need to realize, in question 1, that the sum of the parts must equal the whole, therefore 3 out of 5 students must *not* have brown eyes.

In question 2, help students see that the cross products result in the equation $ad = bc$.

For question 3, you may wish to have students substitute numbers for the variables, and check the cross products.

THINK What rate describes money earned? mileage? **dollars/hour; miles/gallon**

THINK What proportion would you write to describe the number of minutes it takes to blink once? $\frac{1}{25} = \frac{x}{1}$

▼ You can write *rates* and *unit rates* to describe many situations.

3 **Rate** | A rate is a ratio that compares quantities in different units. A unit rate compares a quantity to one.

Example 3 On average, a person blinks 100 times in 4 min. How many times does a person blink in one minute?

Solution

$\frac{100}{4} = \frac{t}{1} = \frac{\text{number of blinks}}{\text{number of minutes}}$ Write a ratio to describe the the situation.

$4t = 100$ Write cross products.

$\frac{4t}{4} = \frac{100}{4}$ Divide both sides by 4.

$t = 25$

On average, a person blinks 25 times per minute.

Example 4 A car travels 264 mi on 12 gal of gas. Find the unit rate in miles per gallon.

Solution

$\frac{264}{12} = \frac{x}{1}$

264 (÷) 12 (=) 22

$x = 22$ mi/gal

THINK AND DISCUSS **See below.**

1. You know that 2 in 5 people have brown eyes. How many people do *not* have brown eyes? Explain.

2. The ratios $\frac{a}{b}$ and $\frac{c}{d}$ form a proportion. How can you use cross products to express the relationship between a, b, c, and d as an equation without using fractions? **$ad = bc$**

3. A student claims that a ratio would remain unchanged if 1 is added to both the numerator and the denominator of the fraction as in $\frac{a}{b} = \frac{a+1}{b+1}$. Is the student correct? Explain your decision.

1. 3 in 5; $\frac{5}{5}$ is all the people so $\frac{2}{5} + \frac{3}{5} = \frac{5}{5}$
3. Yes if $a = b$; no for all other cases.

CLASS EXERCISES

Write each ratio as a fraction in lowest terms.

1. 9 : 27 **$\frac{1}{3}$** **2.** 10 out of 16 **$\frac{5}{8}$** **3.** 12 is to 8 **$\frac{3}{2}$**

4. 6 people in 50 are over 65 years of age. **$\frac{3}{25}$**

5. 1 person in 18 plays the piano. **$\frac{1}{18}$**

Write three ratios to describe each figure. **Answers will vary. Examples are given.**

6.

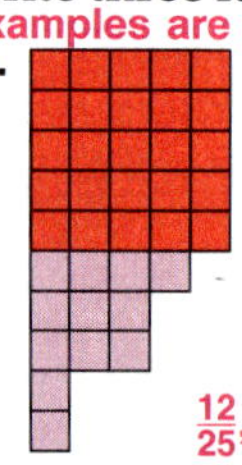

$\frac{12}{25}$, $\frac{25}{37}$, $\frac{12}{37}$

7.

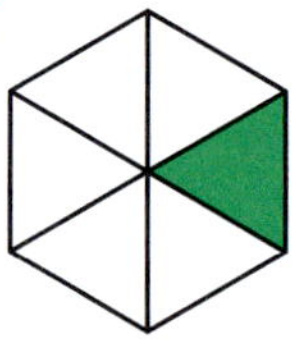

1 to 6
1 to 5
5 to 6

8.

3 to 2
2 to 3
3 to 5

Write each ratio as a unit rate.

9. A bathtub contains 20 gal of water. The tub empties in 4 min. What is the rate of flow per minute? **5 gal/min**

10. A sprinter completes 200 m in 22 s. **$9.\overline{09}$ m/s**

11. A keyboarder types 1,575 words in 25 min. **63 words/min**

MATH MINUTES

Give students the following sets of numbers. Ask them to write as many correct proportions as they can, using cross products to check that the ratios are equal. (Ans. will vary.)
1. 8, 16, 64, 2
2. 1, 39, 3, 13

WRITTEN EXERCISES

Write each ratio as a fraction in lowest terms.

1. 3 : 8 $\frac{3}{8}$
2. 7 is to 9 $\frac{7}{9}$
3. 8 out of 11 $\frac{8}{11}$
4. 14 out of 18 $\frac{7}{9}$
5. 15 : 25 $\frac{3}{5}$
6. 36 is to 48 $\frac{3}{4}$
7. 60 is to 24 $\frac{5}{2}$
8. 16 : 12 $\frac{4}{3}$
9. 25 homes out of 125 have a personal computer. $\frac{1}{5}$
10. 3 out of 12 people live in a rural area. $\frac{1}{4}$
11. 4 people out of 24 attend school. $\frac{1}{6}$
12. 20 homes in 25 have a TV. $\frac{4}{5}$
13. 6 of 24 people live in a household composed of three people. $\frac{1}{4}$
14. 7 of 35 people live in a household composed of four people. $\frac{1}{5}$
15. 6 of 42 people live in a household composed of five or more people. $\frac{1}{7}$

Write three ratios to describe each figure.

16. 
4 to 7 4 to 3 3 to 4

17. 
3 to 2 2 to 3 2 to 5

18. 
2 to 3 1 to 2 2 to 1

Use graph paper to draw a model of each ratio. Check students' work.

19. $\frac{3}{6}$
20. 12 : 36
21. 5 out of 8
22. 9 : 10

23. ***PROJECT*** Survey at least ten people. Find the number of people living in each household. Use your data to write ratios showing the number of people who live in a household composed of three or fewer people, of four people, and of more than four people. Compare your ratios with the statistics in Exercises 13–15.

Compare. Write = or ≠. Then tell which pairs of ratios form a proportion.

24. $\frac{4}{7}$ ■ $\frac{20}{35}$ =; prop.
25. $\frac{3}{2}$ ■ $\frac{16}{10}$ ≠
26. $\frac{3}{4}$ ■ $\frac{12}{15}$ ≠
27. $\frac{8}{3}$ ■ $\frac{56}{21}$ =; prop.
28. $\frac{9}{24}$ ■ $\frac{15}{40}$ = ; prop.
29. $\frac{32}{20}$ ■ $\frac{20}{12}$ ≠
30. $\frac{40}{24}$ ■ $\frac{75}{45}$ = ; prop.
31. $\frac{7}{8}$ ■ $\frac{8}{9}$ ≠

CALCULATOR **Write = or ≠. Then tell which pairs of ratios form a proportion.**

32. $\frac{75}{90}$ ■ $\frac{90}{108}$ = ; prop.
33. $\frac{120}{144}$ ■ $\frac{145}{75}$ ≠
34. $\frac{215}{155}$ ■ $\frac{270}{165}$ ≠
35. $\frac{192}{144}$ ■ $\frac{256}{192}$ = ; prop.

MIXED REVIEW

Find each answer.

1. Which fraction is closer to $\frac{1}{2}$, $\frac{5}{11}$ or $\frac{7}{13}$? $\frac{7}{13}$
2. Compare the fractions $\frac{13}{19}$ and $\frac{19}{28}$. Use >, <, or =. $\frac{13}{19} > \frac{19}{28}$

Solve for x.

3. $\frac{x}{7} = -2$ $x = -14$
4. $2\frac{1}{3}x = -5\frac{1}{4}$ $x = -2\frac{1}{4}$

Are the fractions equivalent?

5. $\frac{4}{12}$ and $\frac{10}{30}$ yes
6. $\frac{7}{8}$ and $\frac{14}{24}$ no

Solve.

7. On a trip, Mark drove 7 mi more than Craig. Craig drove five times as far as Brad. Brad drove 112 mi. How long was the trip? 1,239 mi

A ***common error*** is to change the order of terms in the ratios when writing a proportion. For example, students will write:

$$\frac{\text{miles}}{\text{gallons}} = \frac{\text{gallons}}{\text{miles}}$$

instead of

$$\frac{\text{miles}}{\text{gallons}} = \frac{\text{miles}}{\text{gallons}}$$

It may be helpful if students first set up the proportion using the words for the units. Then they can rewrite the proportion using numbers.

Closure

Writing in Math Ask students to describe the meaning of a proportion in their math journals. Then, using objects in the classroom as examples, have the students write a verbal statement of a proportion, then write the proportion in symbols.

Independent Practice

Have students work in groups to do the Project in question 23. After students compare their responses to questions 13-15, have them compare their responses with other groups.

For questions 32-35, have students write the factors in the cross-products before they enter any values in the calculator.

Lesson Follow-up

Reteaching Activity To help students understand the basic concepts of ratios and proportions, have them model simple examples using manipulatives. (e.g. 3 small cubes: 1 large cube = 9 small cubes: 3 large cubes)

LESSON RESOURCES

TRB Practice F-1
TRB Enrichment F-2

Practice

For use after 6-1 (pp. 233–236)

Ratios, Proportions, and Rates

Write each ratio as a fraction in lowest terms.

1. 7:12 $\frac{7}{12}$
2. 3 is to 6 $\frac{1}{2}$
3. 9 out of 21 $\frac{3}{7}$
4. 10:45 $\frac{2}{9}$
5. 32 out of 40 $\frac{4}{5}$
6. 24 is to 18 $\frac{4}{3}$
7. 36 is to 60 $\frac{3}{5}$
8. 13 out of 14 $\frac{13}{14}$
9. 45:63 $\frac{5}{7}$

Write three ratios each to describe the figures. **Answers may vary.**

10. $\frac{3}{7}, \frac{4}{7}, \frac{3}{4}$
11. $\frac{3}{5}, \frac{2}{5}, \frac{3}{2}$
12. $\frac{2}{3}, \frac{1}{3}, \frac{2}{1}$

Write each ratio as a fraction in lowest terms.

13. 8th-grade boys to 9th-grade boys $\frac{13}{15}$
14. 8th-grade girls to 8th-grade boys $\frac{17}{13}$
15. 8th graders to 9th graders $\frac{15}{13}$
16. boys to girls $\frac{1}{1}$
17. girls to all students $\frac{1}{2}$

	Boys	Girls
8th Grade	26	34
9th Grade	30	22

Compare. Write = or ≠.

18. $\frac{3}{4} = \frac{9}{12}$
19. $\frac{25}{40} = \frac{5}{8}$
20. $\frac{15}{20} \neq \frac{12}{15}$
21. $\frac{8}{12} = \frac{14}{21}$
22. $\frac{13}{15} \neq \frac{4}{5}$
23. $\frac{42}{35} = \frac{36}{30}$
24. $\frac{4}{5} \neq \frac{5}{6}$
25. $\frac{49}{21} = \frac{28}{12}$
26. $\frac{64}{72} = \frac{72}{81}$
27. Of Exercises 18–26, which pairs of ratios form proportions? Write the numbers of the exercises. **18, 19, 21, 23, 25, 26**

CALCULATOR Compare. Write = or ≠.

28. $\frac{138}{161} = \frac{246}{287}$
29. $\frac{136}{221} \neq \frac{159}{265}$
30. $\frac{356}{801} = \frac{292}{657}$

Express as a unit rate.

31. 78 mi on 3 gal **26 mi/gal**
32. \$52.50 in 7 h **\$7.50/h**
33. 416 mi in 8 h **52 mi/h**
34. 9 bull's eyes in 117 throws **13 throws/bull's eye**

Chapter 6 F-1

Enrichment

For use after 6-1 (pp. 233–236)

More on Proportions

Consider the proportion $\frac{3}{5} = \frac{9}{15}$. If the ratio of the numerators is set equal to the ratio of the denominators, another true proportion results: $\frac{3}{9} = \frac{5}{15}$.

Answer these questions to explore whether this is always true.

1. Write three true proportions. **Answers will vary.**
2. Write the proportions that result when you equate the ratio of the numerators to the ratio of the denominators. Then tell whether the proportions are true or false. **Proportions will vary.**
   ______ T or F? **T** ______ T or F? **T**
   ______ T or F? **T**
3. Complete: If $\frac{A}{B} = \frac{C}{D}$, then it appears that $\frac{A}{C} =$ **$\frac{B}{D}$**.

Consider the true proportion $\frac{15}{21} = \frac{20}{28}$.

4. Answer true or false: $\frac{21}{15} = \frac{28}{20}$ **true**
5. Complete: If $\frac{A}{B} = \frac{C}{D}$, then it appears that $\frac{B}{A} =$ **$\frac{D}{C}$**.

Consider the true proportion $\frac{6}{8} = \frac{3}{4}$. Answer true or false.

6. $\frac{6}{4} = \frac{3}{8}$ **false** If $\frac{A}{B} = \frac{C}{D}$, then it appears that $\frac{A}{D} = \frac{B}{C}$. **false**

Consider the true proportion $\frac{3}{8} = \frac{9}{24}$. Answer true or false.

7. $\frac{3+8}{8} = \frac{9+24}{24}$ **true**
8. $\frac{3-8}{8} = \frac{9-24}{24}$ **true**
9. Complete: If $\frac{A}{B} = \frac{C}{D}$, then it appears that $\frac{A+B}{B} =$ **$\frac{C+D}{D}$**, and $\frac{A-B}{B} =$ **$\frac{C-D}{D}$**.

Use the proportions you wrote in Exercise 1 to explore this theorem:

$$\text{If } \frac{A}{B} = \frac{C}{D}, \text{ then } \frac{A+B}{A-B} = \frac{C+D}{C-D}.$$

10. Test the theorem. Write three new proportions. **Answers will vary.**

F-2 Chapter 6

39. 0.18 hits/time at bat
40. 0.57 baskets/throw
44. 5 to 1; 1 to 5; 1 to 6
45. 50 to 2; 2 to 50; 2 to 48

Express each as a unit rate.

36. 20 mi in 5 h **4 mi/h**
37. 42 gal in 7 min **6 gal/min**
38. a fall of 144 ft in 3 s **48 ft/s**
39. 12 hits in 66 times at bat
40. 68 baskets in 119 throws
41. 245 mi in 56 h **4.375 mi/h**
42. 676 mi in 13 h **52 mi/h**
43. 78 hits in 260 times at bat **0.3 hits/time at bat**

Write three ratios to describe each situation. **Answers will vary. Examples are given.**

44. For every five victories, the baseball team lost one game.
45. For every fifty radios sold, two were returned for a refund.
46. A bookstore sells paperbacks, hardbacks, and magazines. For every three paperbacks sold, five hardbacks are also sold. For every ten hardbacks sold, twenty magazines are sold. **3 to 5; 10 to 20; 6 paperbacks to 20 magazines**

Write a ratio to describe each situation. Decide if the rates form a proportion. Solve each problem.

47. In one classroom, 4 of the 24 students are boys. In another classroom, 6 of the 30 students are boys. Is the ratio of boys to total number of students the same in both classes? **$\frac{4}{24} \stackrel{?}{=} \frac{6}{30}$; no**
48. A subcompact car travels 196 mi on 7 gal of gas. A compact travels 336 mi on 12 gal of gas. Is the fuel economy of both cars the same? **$\frac{196}{7} \stackrel{?}{=} \frac{336}{12}$; yes**
49. Two cans of beans cost \$1.69. Five cans cost \$3.98. Is the ratio of cans to cost the same? If not, which is the better buy? Explain. **$\frac{2}{1.69} \stackrel{?}{=} \frac{5}{3.98}$; no; 2 for \$1.69 is \$.85 each; 5 for \$3.98 is \$.80 each.**
50. ***DATA FILE 3 (pp. 96–97)*** Suppose you turn in $\frac{1}{2}$ of a 20-dollar bill. How much will you get? **\$10**
51. Use the article below to answer each question.
   a. Write the ratio of sap to syrup in three different ways. **300 to 7; 300 : 7; $\frac{300}{7}$**
   b. Write the cost of syrup per pint as a ratio. **\$9.00/pt**
   c. Write the cost of syrup per gallon as a ratio. **\$44/gal**

A Sappy Story

Vermont is renowned as a major producer of maple syrup. It is a little known fact, however, that more than 100 farms in Connecticut also produce maple syrup. Sugarers collect sap daily and boil it down to make syrup. In a good year, one small sugarer in Connecticut collects 300 gal of sap from 200 trees. The sap boils down to just 7 gal of syrup. The syrup is sold for \$4.50 per half pint or \$44 per gallon.

LESSON QUIZ

Write each ratio as a fraction in lowest terms.

1. 18 : 24 ($\frac{3}{4}$)
2. 16 is to 12 ($\frac{4}{3}$)

Compare. Write = or ≠.

3. 45 : 54 (≠) 30 : 35
4. 36 to 60 (=) 15 to 25
5. Express 15 miles in 4 hours as a unit rate. ($3\frac{3}{4}$ mi/h)
6. Which rate is faster, 32 ft in 1 s, or 22 mi in 1 h? (22 mi/h)

FOR THE NEXT LESSON

calculator

Assignment Guide

Basic 1-16, 19-20, 23-27, 32-39, 44-47, 50, MR All

Average 1-17 odd, 20, 22-23, 24-34 even, 38-48 even, 50-51, MR All

Enriched 6-8, 12-19, 21-35 odd, 40-46, 49-51, MR All

OBJECTIVE:
To use proportions to solve problems.

6-2 Solving Proportions

▼ An average adult's heart beats 8 times every 6 s. At this rate, how many times does it beat in 120 s? You can write a proportion to describe the situation. Let x = the number of heartbeats in 120 s.

$$\frac{8}{6} = \frac{x}{120} \quad \begin{array}{l}\leftarrow \text{number of heartbeats} \\ \leftarrow \text{number of seconds}\end{array}$$

$6x = 8 \cdot 120$ Write the cross products.
$6x = 960$ Divide both sides by 6.
$x = 160$

The average person's heart beats 160 times in 120 s.

▼ You can use cross products to solve a proportion.

Example 1 A canary's heart beats 130 times in 12 s. At this rate, how many times does it beat in 30 s?

Solution
$\frac{130}{12} = \frac{x}{30}$ Write a proportion.
$130 \cdot 30 = 12x$ Write the cross products.
$3{,}900 \div 12 = 12x \div 12$ Divide both sides by 12.
$325 = x$

A canary's heart beats 325 times in 30 s.

▼ You can use a calculator to help you solve a proportion.

Example 2 **Solve $\frac{x}{3.5} = \frac{35}{8.75}$ for x.**

Solution
$\frac{x}{3.5} = \frac{35}{8.75}$
1 $8.75x = 3.5 \cdot 35$ Write the cross products.
3.5 ⊗ 35 ⊕ 8.75 ⊜ 14 Use a calculator to solve.
$x = 14$

THINK How is writing cross products an application of the multiplication property of equality?
$\frac{a}{b} = \frac{c}{d}$
$(bd)\frac{a}{b} = (bd)\left(\frac{c}{d}\right)$
$ad = bc$

Solving Proportions	To solve a proportion: **1.** Write the cross products. **2.** Solve the equation.

▼ Express quantities in the same units before solving proportions.

Example 3 Ribbon costs \$3 for 15 in. Find the cost of 3 ft of ribbon.

Solution
$\frac{15 \text{ in.}}{\$3} = \frac{3 \text{ ft}}{d}$ Write a proportion.
2 $\frac{15}{3} = \frac{36}{d}$ Write 3 ft as 36 in.
$15d = 108$ Solve the proportion for d.
$d = 7.20$

It costs \$7.20 for 3 ft of ribbon.

Lesson Focus

Materials
Calculator

MOTIVATING THE LESSON

Begin the lesson by measuring the pulse of a student for 6s. Then ask students how you might determine the pulse for 1 min without having to measure it for that length of time. Ask for strategies to solve the problem. Then have the class test their strategies and discuss the results.

Instruction

1 Some students may be able to generalize the solution of a proportion in algebraic terms. For example: $x : a = b : c$ as $x = \frac{ab}{c}$.

2 Remind students, when setting up a proportion to be sure all related quantities are expressed in the same unit.

ADDITIONAL EXAMPLES

1. Hasan can say the alphabet aloud three times in 15 seconds. At this rate how long would it take him to say the alphabet 10 times? (50 seconds)

Solve.

2. $\frac{14}{x} = \frac{20}{170}$ (119)

3. $\frac{1.62}{1.8} = \frac{x}{3}$ (2.7)

Teaching Tip

Show students real-world examples of objects whose sizes are proportional, for example, a photograph and an enlargement, or a floor plan of the classroom.

Guided Practice

Class Exercises As students complete questions 5-8, verify that they express the rates in same units before writing the proportions.

Think and Discuss For question 1, suggest that students prove which proportions are true by substituting these values for the variable:
$a = 4$, $b = 8$, $c = 5$, and $d = 10$.

For question 2, students need to understand that when using a proportion to solve a problem, each ratio in the proportion must compare the same units in the same order.

A ***common error*** occurs when students write proportions and do not compare the same units. Help students see that if they write the proportion in words, it can help them judge if they have set up the proportion in the correct order, and whether or not units should be changed. (For example, for Class Exercise 5 students would write ounces: dollars = ounces : dollars.)

Closure

Have students write a problem that can be solved using a proportion using the following numbers: 16, 10, and 24.

Independent Practice

Have students work in groups to complete question 50 on p. 240. First ask the groups to compare their findings, then have them compare their data with the data in questions 45-49.

THINK AND DISCUSS

See below.

1. If $\frac{a}{b} = \frac{c}{d}$, will $\frac{a}{b} = \frac{d}{c}$? Give examples to justify your answer.
2. Explain why it's important to express quantities in the same units before writing a proportion.

1. no; $\frac{1}{2} = \frac{2}{4}$ but $\frac{1}{2} \neq \frac{4}{2}$
2. Quantities must be in the same units for the comparison to remain in proportion.

MIXED REVIEW

Tell if each equation is true or false.

1. $|-2\frac{1}{4}| + |2\frac{1}{4}| = 0$ F
2. $|-2\frac{1}{4}| - |2\frac{1}{4}| = 0$ T

Write each ratio as a fraction in lowest terms.

3. 30 to 55 $\frac{6}{11}$
4. 125 : 70 $\frac{25}{14}$

Tell whether or not these ratios form a proportion.

5. $\frac{80}{25}$ and $\frac{16}{5}$ yes
6. $\frac{15}{42}$ and $\frac{25}{70}$ yes

Solve.

7. On Saturday, a student bought two cassette tapes for \$8.95 each and a sweater for \$24.95. He received \$20 for mowing a lawn. On Saturday night, he has \$45.12. How much money did he have Saturday morning? \$67.97

CLASS EXERCISES

Solve.

1. $\frac{4}{11} = \frac{x}{16.5}$ $x = 6$
2. $\frac{7}{12} = \frac{17.5}{y}$ $y = 30$
3. $\frac{z}{5.4} = \frac{13}{18}$ $z = 3.9$
4. $\frac{2}{v} = \frac{1}{8}$ $v = 16$

Write a proportion to describe each situation. Then solve.

5. 3 oz of nuts cost \$1.70; 5 oz cost x dollars. $\frac{3}{1.70} = \frac{5}{x}$; $x = \$2.83$
6. A student runs 24 yd in $2\frac{1}{2}$ s; 100 yd in x seconds.
7. 50 calories in 4 oz of orange juice; x calories in 14 oz.
8. A lion's heart beats 12 times in 16 s; x times in 60 s. $\frac{12}{16} = \frac{x}{60}$; $x = 45$

6. $\frac{24}{2.5} = \frac{100}{x}$; $x = 10\frac{5}{12}$ s 7. $\frac{50}{4} = \frac{x}{14}$; $x = 175$

Each pair of figures is in proportion. Find the missing length.

9.
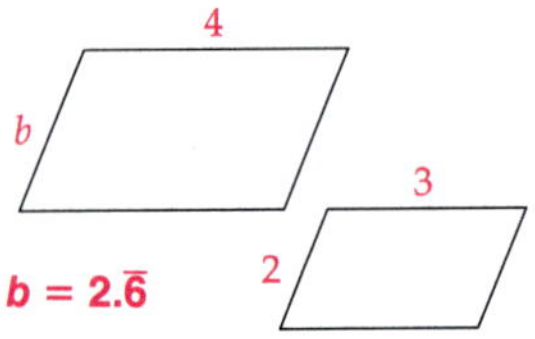

$b = 2.\overline{6}$

10.
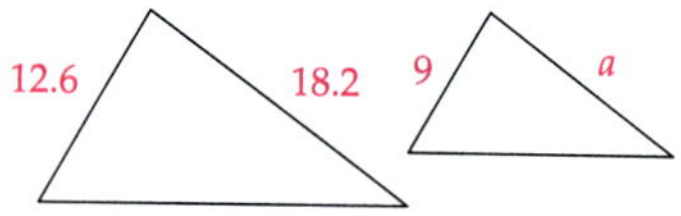

$a = 13$

WRITTEN EXERCISES

Solve.

1. $\frac{4}{15} = \frac{a}{75}$ $a = 20$
2. $\frac{4}{3} = \frac{b}{21}$ $b = 28$
3. $\frac{13}{c} = \frac{39}{60}$ $c = 20$
4. $\frac{3}{6} = \frac{7}{d}$ $d = 14$
5. $\frac{6}{25} = \frac{e}{80}$ $e = 19.2$
6. $\frac{4}{9} = \frac{f}{15}$ $f = 6.\overline{6}$
7. $\frac{3}{8} = \frac{50}{g}$ $g = 133.\overline{3}$
8. $\frac{24}{17} = \frac{108}{h}$ $h = 76.5$
9. $\frac{7}{9} = \frac{j}{22.5}$ $j = 17.5$
10. $\frac{11}{18} = \frac{k}{49.5}$ $k = 30.25$
11. $\frac{6}{13} = \frac{7.8}{m}$ $m = 16.9$
12. $\frac{20}{27} = \frac{1.1}{n}$ $n = 1.485$

Each pair of figures is in proportion. Find the missing length.

13.
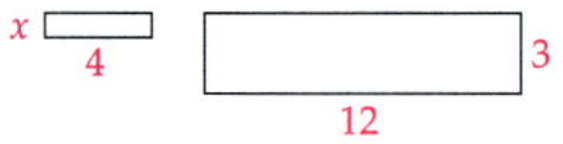

$x = 1$

14.
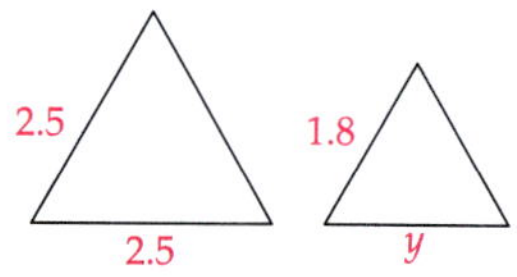

$y = 1.8$

Write a proportion to describe each situation. Then solve.

15. 4 oz of cheese costs \$1.85; $1\frac{1}{2}$ lb costs t dollars. $\frac{4}{1.85} = \frac{24}{t}$; \$11.10
16. A baseball player gets 54 hits in 225 times at bat; x hits in 500 times at bat. $\frac{54}{225} = \frac{x}{500}$; $x = 120$ hits
17. A student runs 5 km in 18 min 36 s; 8 km in v min. $\frac{5}{18.6} = \frac{8}{v}$; $v = 29.76$ min
18. 20 lb of dog food costs \$27.50; 12 lb costs x dollars. $\frac{20}{27.50} = \frac{12}{x}$; $x = \$16.50$

MATH MINUTES

For each proportion find the value for x that will make the proportion true.

1. $4 : x = x : 16$ (8)
2. $x : 3 = 27 : x$ (9)
3. $6 : x = x : 54$ (18)

19. 96 oz costs \$2.25; y pounds costs \$10. $\frac{6}{2.25} = \frac{y}{\$10}$; $y = 26.67$ lb

20. A rectangle measuring 20 cm by 28 cm is reduced to one measuring 9 cm by z cm. $\frac{20}{28} = \frac{9}{z}$; $z = 12.6$ cm

21. Three tea bags are used to make a gallon of iced tea. How many tea bags are needed to make four gallons? $\frac{3}{1} = \frac{x}{4}$; $x = 12$ bags

22. At the Copy Shoppe, 18 copies cost \$1.08. At the same rate, how much will 40 copies cost? $\frac{18}{1.08} = \frac{40}{c}$; $c = \$2.40$

23. The cost of 3 posters is \$9.60. Find the cost of 15 posters. $\frac{3}{9.60} = \frac{15}{p}$; $p = \$48$

24. ***DATA FILE 10 (pp. 404–405)*** What would be the home range, in acres, for a pack of 30 gray wolves? about 2,285,820 acres

25. A microchip inspector found three defects in a batch containing 750 chips. How many defects should the inspector find in a batch of 10,000 chips? $\frac{3}{750} = \frac{x}{10{,}000}$; $x = 40$ defects

26. A truck driver estimated that it would take him 12 h to drive 1,160 km. After 5 h, he had driven 480 km. Is he on target? Explain. Yes; the ratios are approximately equal.

27. ***PROJECT*** How would you describe your class? Use the list of questions at the right or questions of your own. Interview your classmates and tally all the responses. Write ratios to describe the results.

Use the relationship $x \blacktriangleright y = \frac{x + y}{x}$ for exercises 28–35. Evaluate each expression.

▼▼ *SAMPLE* $3 \blacktriangleright 5 = \frac{3 + 5}{3} = \frac{8}{3}$

28. $2 \blacktriangleright 3$ $\frac{5}{2}$ 29. $3 \blacktriangleright 4$ $\frac{7}{3}$ 30. $5 \blacktriangleright -4$ $\frac{1}{5}$ 31. $-7 \blacktriangleright 9$ $-\frac{2}{7}$

Write *yes* or *no* to tell if each is proportional to $\frac{1}{3}$.

32. $-1 \blacktriangleright -3$ no 33. $2 \blacktriangleright 6$ no 34. $-15 \blacktriangleright 10$ yes 35. $5 \blacktriangleright -3$ no

Solve.

36. If $\frac{y}{25} = \frac{6}{30}$, find the ratio of y to 6. $\frac{25}{30}$

37. If $\frac{10}{100} = \frac{r}{30}$, find the ratio of r to 10. $\frac{30}{100}$

38. If $\frac{15}{75} = \frac{30}{f}$, find the ratio of f to 75. $\frac{30}{15}$

39. If $\frac{12}{p} = \frac{1}{3}$, find the ratio of p to 3. $\frac{12}{1}$

40. An artist mixes red and blue paint to make purple paint in the ratio of 2 : 3. This tells you that for every 2 parts of red there are 3 parts of blue. What does each ratio tell you about the mixture of red and blue paint?

 a. 6 : 9 — 6 red : 9 blue
 b. 1 : 1.5 — 1 red : 1.5 blue
 c. 10 : 15 — 10 red : 15 blue
 d. 20 : 30 — 20 red : 30 blue

Lesson Follow-up

Reteaching Activity To help students who are having difficulty writing and solving proportions have them use the following procedure.

First write the unit rates in words, in order.

Second, write the numbers that correspond to each rate.

Third, cross multiply and solve the equation.

For example: 20 lb bird seed for \$3.00. Find the cost of 15 lb.

$$\frac{\text{bird seed}}{\text{cost}} \rightarrow \frac{20}{\$3.00} = \frac{15}{x}$$

$$20x = \$45$$

$$x = \$2.25$$

Refer students to the manipulative activity on p. 230E of the Chapter Overview. See also TRB Exploring O-21.

Critical Thinking You may wish to point out that a number in the intersection of A, B, and C must have the properties of the numbers in all three circles.

LESSON RESOURCES

TRB Practice F-3
TRB Enrichment F-4

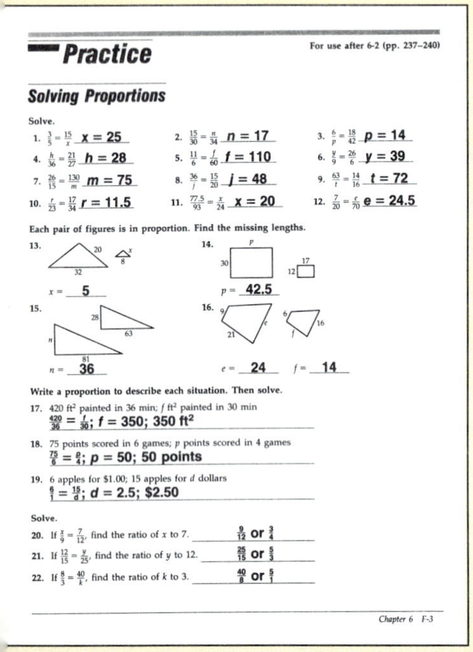

Practice — For use after 6-2 (pp. 237–240)

Solving Proportions

Solve.

1. $\frac{3}{5} = \frac{15}{x}$ **x = 25**
2. $\frac{15}{30} = \frac{n}{34}$ **n = 17**
3. $\frac{6}{p} = \frac{18}{42}$ **p = 14**
4. $\frac{h}{36} = \frac{21}{27}$ **h = 28**
5. $\frac{11}{6} = \frac{f}{60}$ **f = 110**
6. $\frac{y}{9} = \frac{26}{6}$ **y = 39**
7. $\frac{26}{15} = \frac{130}{m}$ **m = 75**
8. $\frac{36}{j} = \frac{15}{20}$ **j = 48**
9. $\frac{63}{t} = \frac{14}{16}$ **t = 72**
10. $\frac{r}{23} = \frac{17}{34}$ **r = 11.5**
11. $\frac{77.5}{93} = \frac{x}{24}$ **x = 20**
12. $\frac{7}{20} = \frac{e}{70}$ **e = 24.5**

Each pair of figures is in proportion. Find the missing lengths.

13. $x =$ **5**
14. $p =$ **42.5**
15. $n =$ **36**
16. $e =$ **24** $f =$ **14**

Write a proportion to describe each situation. Then solve.

17. 420 ft² painted in 36 min; f ft² painted in 30 min
$\frac{420}{36} = \frac{f}{30}$; f = 350; 350 ft²
18. 75 points scored in 6 games; p points scored in 4 games
$\frac{75}{6} = \frac{p}{4}$; p = 50; 50 points
19. 6 apples for \$1.00; 15 apples for d dollars
$\frac{6}{1} = \frac{15}{d}$; d = 2.5; \$2.50

Solve.

20. If $\frac{x}{9} = \frac{7}{12}$, find the ratio of x to 7. **$\frac{9}{12}$ or $\frac{3}{4}$**
21. If $\frac{12}{15} = \frac{y}{25}$, find the ratio of y to 12. **$\frac{25}{15}$ or $\frac{5}{3}$**
22. If $\frac{8}{3} = \frac{40}{k}$, find the ratio of k to 3. **$\frac{40}{8}$ or $\frac{5}{1}$**

Chapter 6 F-3

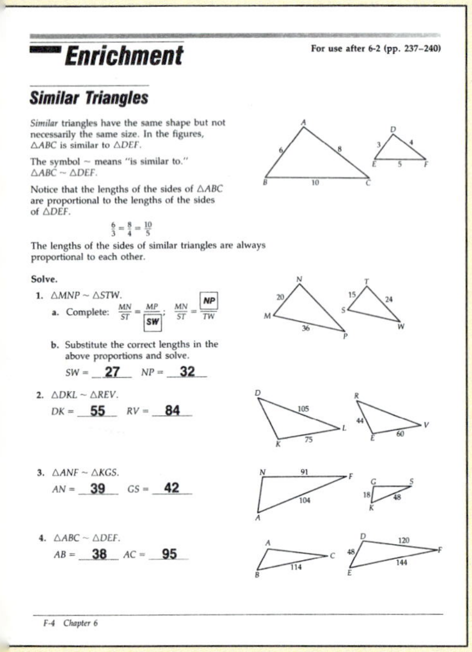

Enrichment — For use after 6-2 (pp. 237–240)

Similar Triangles

Similar triangles have the same shape but not necessarily the same size. In the figures, $\triangle ABC$ is similar to $\triangle DEF$.

The symbol ~ means "is similar to."
$\triangle ABC \sim \triangle DEF$.

Notice that the lengths of the sides of $\triangle ABC$ are proportional to the lengths of the sides of $\triangle DEF$.

$$\frac{6}{3} = \frac{8}{4} = \frac{10}{5}$$

The lengths of the sides of similar triangles are always proportional to each other.

Solve.

1. $\triangle MNP \sim \triangle STW$.
   a. Complete: $\frac{MN}{ST} = \frac{MP}{\boxed{SW}}$; $\frac{MN}{ST} = \frac{\boxed{NP}}{TW}$
   b. Substitute the correct lengths in the above proportions and solve.
   $SW =$ **27** $NP =$ **32**
2. $\triangle DKL \sim \triangle REV$.
   $DK =$ **55** $RV =$ **84**
3. $\triangle ANF \sim \triangle KGS$.
   $AN =$ **39** $GS =$ **42**
4. $\triangle ABC \sim \triangle DEF$.
   $AB =$ **38** $AC =$ **95**

F-4 Chapter 6

Complete.

41. 3 : 15 = 1 : 5 because 15 = 3 · 5 and ■ (15) = ■ (1) · 15
42. 15 : 6 = 5 : 2 because ■ (30) = 15 · 2 and ■ (30) = 6 · ■ (5)
43. 8 : 2 = 16 : 4 because ■ (32) = ■ (8) · 4 and ■ (32) = ■ (16) · 2
44. 6.5 : 2.5 = 19.5 : 7.5 because ■ (48.75) = 6.5 · ■ (7.5) and 2.5 · ■ (19.5)

Human Heartbeats

Age (years)	Heartbeat
Newborn	140/min
1 y	120/min
6 y	100/min
10 y	90/min
12 y	85/min
Adult	80/min

Use the *DATA* at the left to solve each problem.

45. In how many seconds will a newborn's heart beat 35 times? **15 s**
46. In how many seconds will a 12-year-old's heart beat 25 times? **17.65 s**
47. How many times does an adult's heart beat in 270 s? **360 times**
48. How many more times does a newborn's heart beat in 45 s than a 6-year-old child's heart? **30 times**
49. About how many times will an adult's heart beat in one year? **42,048,000 times**
50. ***PROJECT*** Find examples of heartbeat rates during different activities, such as sleeping, running, reading, and so on. Then take your pulse to determine your own heartbeat rate. Compare to see how close your rate is to the average heartbeat rates.

1. The numbers are even. 2. The numbers are prime. 3. The numbers are odd. 4. The even primes are in A and B. 5. The odd primes are in B and C.

Critical Thinking

EXPLORING VENN DIAGRAMS

Study the numbers in each circle. How are they alike? How are they different?

Write a sentence to describe each.

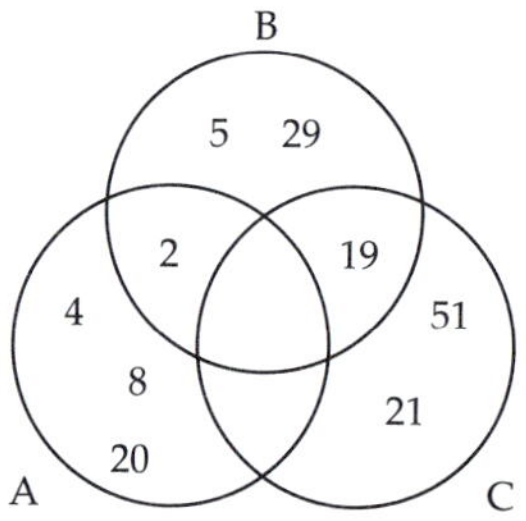

1. the numbers in Circle A
2. the numbers in Circle B
3. the numbers in Circle C
4. Describe the numbers that fit in the intersection of A and B.
5. What numbers fit in the intersection of B and C?
6. Would 11 fit in B? in the intersection of B and C? **yes; yes**
7. Are there other numbers that fit in the intersection of A and B? Why or why not? **No; Two is the only even prime.**
8. Draw your own Venn diagram to classify sets of numbers. Trade with a classmate and describe each other's diagrams.

LESSON QUIZ

Solve.

1. $\frac{y}{1.5} = \frac{4}{15}$ (0.4)
2. $\frac{16.8}{y} = \frac{7.2}{4.2}$ (9.8)
3. Find the ratio of z to 7.5 when: $\frac{12.5}{z} = \frac{2.5}{7.5}$ (5 : 1; z = 37.5)

Complete.

4. 7.5 : 3.5 = 4.5 : 2.1 because ■ = 7.5 × ■ and ■ × 4.5. (15.75; 2.1; 3.5)

Write a proportion and solve.

5. A photograph which is 2 in. by $2\frac{1}{2}$ in. is enlarged to 3 in. by ____ in. ($3\frac{3}{4}$ in.)

FOR THE NEXT LESSON

paper, pencil, centimeter ruler

Assignment Guide

Basic 1-6, 13-18, 23, 26-29, 35-36, 40-41, 44-46, 50, MR, CT All
Average 5-10, 13-20, 23, 26-32, 37-38, 41-42, 46-47, 50, MR, CT All
Enriched 8-14, 21-26, 29-30, 33-34, 39, 42-43, 48-50, MR, CT All

OBJECTIVE: *To apply proportional reasoning to problems involving scale plans.*

PROBLEM SOLVING APPLICATION

6-3 Scale Drawing

Maps and floor plans are examples of *scale drawings.* Sizes in scale drawings are usually smaller than the actual sizes. However, if you sketch something you're observing through a microscope, your measurements will be greater than the actual measurements.

Example 1 On the map below, 1 cm represents about 84 km. What is the air distance between Lubbock and Abilene?

SCALE: 1 cm: 84 km

Solution Measure the distance on the map with a centimeter ruler. The map distance is about 2.9 cm. Write and solve a proportion.

$$\frac{1}{84} = \frac{2.9}{d} \quad \text{Let } d = \text{actual distance.}$$

$$1 \cdot d = 84 \cdot 2.9 \quad \text{Use cross products.}$$

$$d = 243.6$$

The air distance is about 244 km.

CLASS EXERCISES

Use the map above. Round to the nearest 5 km.

1. What is the air distance from Midland to Dallas? 520 km
2. Which city on the map is about 285 km from Ft. Worth? Austin
3. A plane flies from San Angelo to Houston and then on to Galveston. About how far does it travel in all? 620 km

Lesson Focus

Materials
Paper, pencil, centimeter ruler

Vocabulary/Symbols
scale drawing

MOTIVATING THE LESSON

Begin the lesson by providing a number of maps for students to examine. Discuss the scales that are contained on each of the maps. Point out that a map scale is a ratio that compares the distance on the map to the actual distance.

Instruction

You may wish to mention that, when traveling by car, the distances between cities may be different than when traveling by airplane, because ground travel depends on the network of roads. Although air routes are also networked, they are usually more direct than ground routes. Comment also that using a scale to find distance, will result in an approximate answer, since all measurements are approximations.

Class Exercises You may wish to have students work in pairs to complete these exercises.

A ***common error*** that occurs when working with scale drawings is to set up the proportion incorrectly. Since two measurements are involved for each ratio, the students may not recognize when they have made a mistake. Have students write a reminder like those given below for each problem they work.

$$\frac{\text{map distance}}{\text{actual distance}} \text{ or } \frac{\text{drawing length}}{\text{actual length}}$$

Teaching Tip

Have students explore science resource materials to find examples of scale drawings that are enlargements. Some possible examples might be single-celled animals, plant cells, or blood cells.

Closure

Ask students to explain why a map scale is a ratio. Then have them suggest some common map scales.

Independent Practice

As students complete questions 1-4, you might suggest that they copy the diagram and make a scale model for the bed.

Decision Making As an alternative to choosing a room in their home, some students may wish to design a room and make scale models of furniture.

For students who have difficulty visualizing a three-dimensional space in two dimensions, refer them to the problem solving activity on p. 230E of the Chapter Overview. See also TRB Exploring O-22.

BEDROOM AND BATH AREA

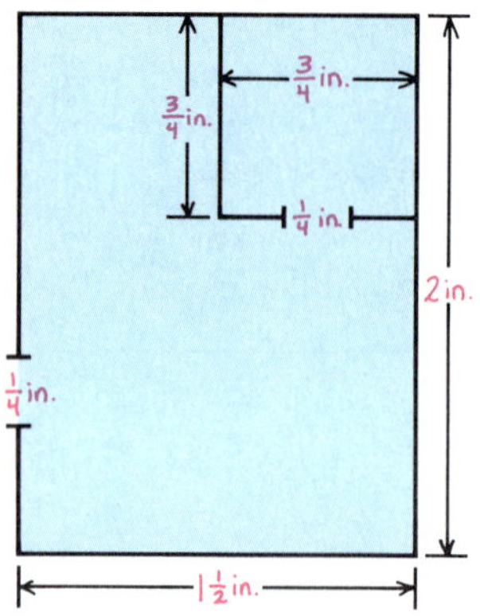

WRITTEN EXERCISES

Use a CALCULATOR where appropriate.

The length of the bedroom and bath area in the diagram at the left is 2 in. The actual length is 20 ft. Use the diagram to solve.

1. What is the scale for the diagram? **1 in. : 10 ft**
2. How many feet wide are the doors leading into the bedroom and into the bath? **$2\frac{1}{2}$ ft**
3. How wide is the widest part of the bedroom? **20 ft**
4. Could a bed 6 ft long and 3 ft wide fit into the narrow section of the bedroom? Does it matter along which wall the bed is placed? Explain. **Yes; no; Each wall is $7\frac{1}{2}$ ft long.**
5. Suppose you want to make a scale drawing of a rectangular dance floor that is 90 ft long and 75 ft wide. Can you fit the drawing on a piece of paper that measures $8\frac{1}{2}$ in. by 11 in. if your scale is 1 in. : 9 ft? Explain. **Yes; The scale of the dance floor is 10 : $8.\overline{3}$.**
6. In a scale drawing, Marco plans to use $\frac{1}{2}$ in. to represent 1 ft. The room he wants to show in the drawing is 16 ft 3 in. long. To the nearest half-inch, what will be the length in the scale drawing? **8 in.**

Decision Making ■ **DECISION MAKING** ■ ***Decision Making*** ■ ***Decision Making*** ■ ***Decision Making***

SCALE DRAWING

■ ***COLLECT DATA*** **Check students' work.**

1. Choose a furnished room in your home. Measure its length and width to the nearest inch. Measure the width of each door and window. Then measure the distance that each door, window, and unmovable piece of furniture is from the nearest corner.
2. Measure the movable furniture so that you can tell how much floor space it occupies.

■ ***ANALYZE DATA***

3. Make a scale drawing of the room, showing windows, doors, and any unmovable furniture. Then make and cut out a scale model for each piece of movable furniture. Position the cutouts on the scale drawing to show the location of each piece of furniture.

Each piece of a model railroad built on the HO scale is $\frac{1}{87}$ the size of an actual railroad part. The N scale, where models are $\frac{1}{160}$ the size of the real thing, is another popular scale.

7. Which models are smaller, HO or N scale models? N

8. Each car on a full-size passenger train is 80 ft long.
   a. What is the length, in inches, of a model passenger car, using the HO scale? using the N scale? HO; about 11 in. N; 6 in.
   b. What is the length of a model train with eight cars, using the HO scale? using the N scale? HO, about 88 in.; N, about 48 in.

9. A diesel electric locomotive is 60 ft long. How long is the model locomotive using the N scale? 4.5 in.

10. A boxcar on a freight train is 40 ft long. A model boxcar is $\frac{1}{4}$ ft long. In what scale was the model built, HO or N? N

11. You are building a table for your HO-scale model train set. Your railroad includes a passenger train 12 cars long. Each car is 1.2 ft long. How long must your table be for the cars to fit end-to-end? 14.4 ft

12. Toy trains are larger than most model trains. Using the O scale, a toy train is $\frac{1}{48}$ the size of an actual train. A toy locomotive is 1.04 ft long. How long is the real locomotive? 49.92 ft

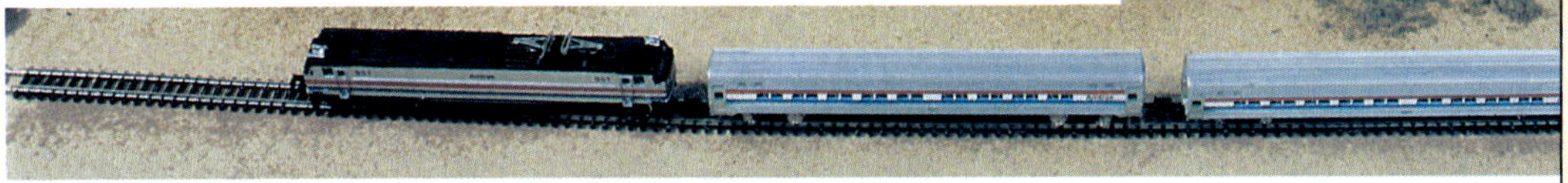

Decision Making ■ Decision Making ■ Decision Making ■ Decision Making ■ Decision Making ■ Decision Making ■

4. Look at the furniture in the room. Analyze the placement of each piece to make the most efficient use of light, traffic patterns, and so on.

5. Move your cutouts to try different furniture arrangements. What floor plan do you like the best? Why?

■ **MAKE DECISIONS**

6. Decide on two pieces of furniture that might be good additions to the room. Find out their dimensions as accurately as you can.

7. Make flat, to-scale cutouts for the new furniture. Work with the cutouts to decide on one or two good ways to arrange all the items in the room. You may remove two other pieces of furniture from the room if necessary. State some of the points you considered in making your decision.

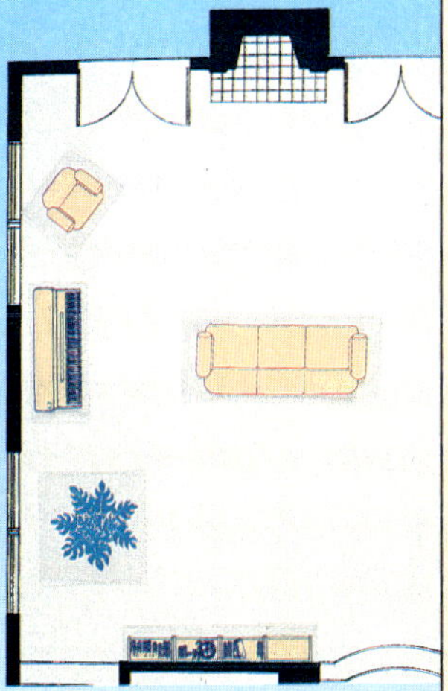

Lesson Follow-up

LESSON RESOURCES

TRB Alternate Application F-5

Alternate Application

For use after 6-3 (pp. 241–243)

Sampling

A sample is a small group chosen to represent a large group. Public opinion surveys of small samples of the population, for example, are assumed to represent the entire population. Using proportions, the results from the small group can be applied to the large one.

A toothpaste manufacturer conducted a survey of consumers to find their preferences in toothpaste.

Toothpaste Survey	
Brand	Number Preferring
Delightful	138
New Shine	23
Peppy	184
Other	115

1. How many people were surveyed? 460

Write each ratio in lowest terms:

2. number preferring Delightful to total $\frac{3}{10}$
3. number preferring New Shine to total $\frac{1}{20}$
4. number preferring Peppy to total $\frac{2}{5}$
5. number preferring Other to total $\frac{1}{4}$

The number of toothpaste users in one state is approximately 8,464,000.

6. Write and solve a proportion to find the number of people in that state who might be expected to prefer Peppy brand.
   $\frac{x}{8{,}464{,}000} = \frac{2}{5}$; $x = 3{,}385{,}600$
7. The manufacturer estimated that 100,500 people in one city might be expected to prefer Delightful. Find the estimated number of toothpaste users in that city.
   335,000

Results of a poll of residents of Alphaville are shown.

Do you favor widening Smith Road to 6 lanes?	
Yes	56
No	70
Not sure	42

8. Based on the poll, how many of the village's 4,320 registered voters might be expected to:
   a. favor widening Smith Road? 1,440
   b. oppose widening the road? 1,800
9. On the South Side, 405 people oppose widening the road. Estimate the number of South Side residents, based on the poll results.
   about 972

Chapter 6 F-5

MATH MINUTES

Have students work in groups and provide each group with a road map. Ask students to measure the distance between various cities, using a centimeter ruler. Have them find the air distance between cities using proportions.

Assignment Guide

Basic All
Average All
Enriched All

FOR THE NEXT LESSON

graph paper, math journal, overhead projector

Lesson Focus

Materials

graph paper, math journal, overhead projector

Vocabulary/Symbols

percent

MOTIVATING THE LESSON

Display a 10 by 10 grid using the overhead projector. Shade in 35 squares. Ask students to write a ratio that compares the shaded squares to the total number of squares. ($\frac{35}{100}$) Comment that the ratio $\frac{35}{100}$ can also be written as 35% because a percent is a ratio that compares a number to 100.

Instruction

1 Help students to understand that percents are ratios that compare a part to a whole amount. Show students that every percent can be expressed as a fraction in which the numerator represents the part and the denominator represents the whole. Stress that percents written as fractions can be expressed in lowest terms.

2 ***Writing in Math*** Students need to realize that any ratio that compares a quantity less than 1 to 100 will indicate a percent less than 1%.

Closure

Have students write the meaning of percent in their math journals. Then have them give examples of ratios that can be expressed as percents.

Assignment Guide

Basic All
Average All
Enriched All

FOR THE NEXT LESSON

math journal

OBJECTIVE:
Use models to explore percents.

MATERIALS

- Graph paper
- Math journal to record work

Exploring Percents

▼ A *percent* is a ratio that compares a number to 100. You can use a decimal model to show percents.

1. Each grid has 100 squares. Write a ratio and a percent to describe the shaded part.

   a. $\frac{25}{100}$ 25%
   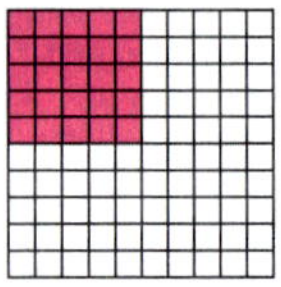
   b. $\frac{19}{100}$ 19%
   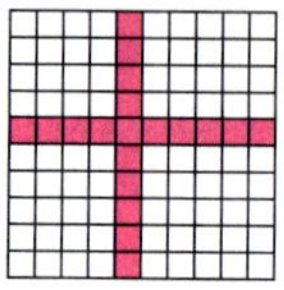
   c. $\frac{1}{100}$ 1%
   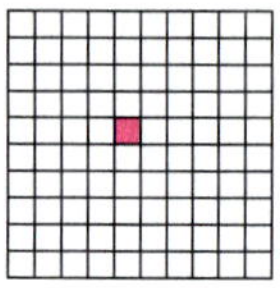

2. Use graph paper to model each percent. **Check students' work.**
   a. 20% **b.** 60% **c.** 73% **d.** 5%

3. Compare with others in your group. Do all your models look the same? Do they all show the same percents? **Answers will vary.**

4. Write a ratio and a percent to describe the shaded part of each figure.

   a.
   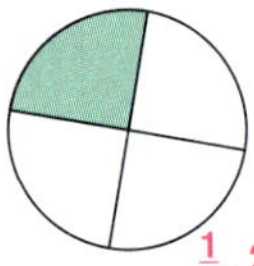
   $\frac{1}{4}$, 25%
   b.
   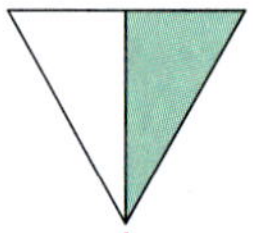
   $\frac{1}{2}$, 50%
   c.
   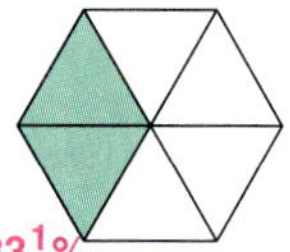
   $\frac{2}{6}$, $33\frac{1}{3}$%
   d.
   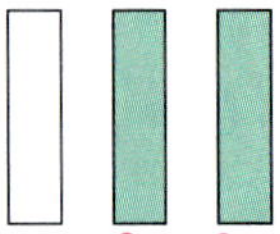
   $\frac{2}{3}$, $66\frac{2}{3}$%

   1 **e.** ***Discuss*** how ratios and percents are like fractions. **they both compare quantities**

▼ You can model percents greater than 100 and less than 1.

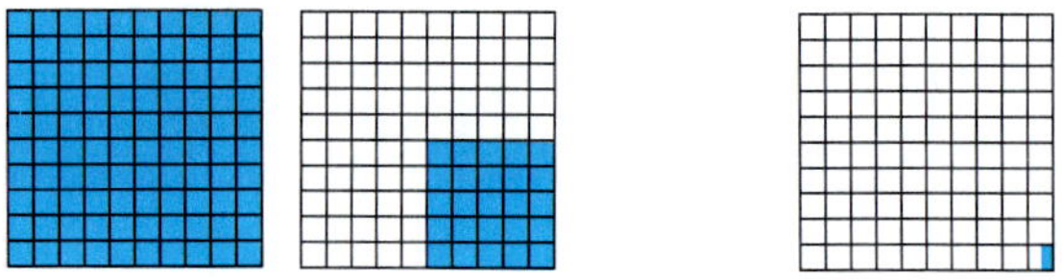

100% + 25% = 125% $\frac{1}{2}$ of 1% = 0.5%

5. Use graph paper to model each percent. **Check students' work.**
   a. 150% **b.** 170% **c.** 0.25% **d.** 0.75%

▼ You can use number sense to estimate percents. You know that $\frac{1}{100}$ is 1% so $\frac{1}{200}$ is less than 1%.

2 6. Which fractions are less than 1%? ***Write*** a rule to tell when a fraction equals a percent less than 1%. **a, c**

   a. $\frac{2}{400}$ **b.** $\frac{300}{500}$ **c.** $\frac{1}{1,000}$ **d.** $\frac{500}{1,000}$

   When the ratio of numerator to denominator is less than $\frac{1}{100}$, the percent is less than 1%.

Teaching Tip

To review the concept of percent, you may wish to have students look through newspapers and magazines to find examples of percents used in everyday situations.

OBJECTIVE:
To express ratios and rates as percents.

6-4 Percent

In a recent survey, 25 people out of 100 said they buy a newspaper every day. When you compare a number to 100, you are finding a *percent*.

Percent (%) means *per hundred*. You can express 25 out of 100 as 25% (25 percent). You can say that 25% of the people buy a daily newspaper.

Percent	A percent is a ratio that compares a number to 100.

You can express a ratio as a percent.

Example 1 **Express each ratio as a percent.**

a. 5 out of 100 **b.** 4 out of 5

1 ***Solution*** **a.** $\frac{5}{100} = 5\%$ **b.** $\frac{4}{5} = \frac{x}{100}$ Write and solve a proportion.

$= \frac{80}{100}$

$= 80\%$

You can write proportions to solve percent problems.

Example 2 In a survey, 57 out of 90 ninth graders said they received an allowance. What percent is this?

Solution

$\frac{57}{90} = \frac{n}{100}$ Write a proportion.

$90n = 5{,}700$ Write cross products.

$\frac{90n}{90} = \frac{5{,}700}{90}$ Divide both sides by 90.

$n = 63.\overline{3}$

$\approx 63\%$ Write a percent.

You can write a decimal as a fraction with a denominator of 100 before writing as a percent.

Example 3 **Write each decimal as a percent.**

a. 0.62 **b.** 0.03 **c.** 0.005 **d.** 1.25

2 ***Solution***

a. $0.62 = \frac{62}{100} = 62\%$

b. $0.03 = \frac{3}{100} = 3\%$

c. $0.005 = \frac{0.5}{100} = 0.5\%$

d. $1.25 = \frac{125}{100} = 125\%$

Teaching Tip

Prepare (or have a group of students prepare) a poster of common decimal/fraction/percent equivalents to display in the classroom. Suggest that the students memorize the equivalencies because they will be using them often in solving percent problems.

Lesson Focus

Materials
Math journal

Vocabulary/Symbols
percent (%)
per hundred

MOTIVATING THE LESSON

Recall the experiments completed in the previous lesson and have students explain how they would show a 110% and 0.5% on a model. Have the students look at the historical note (on page 246) on the derivation of the word *percent* to reinforce the idea that percent is the ratio that compares a number to 100.

Instruction

1 Mention to students that, for Example 1b, the proportion was solved by finding an equivalent fraction with a denominator of 100. Ask students by what factor 4 and 5 were multiplied to get 80 and 100. (20)

2 Have the students study the decimals for a, b, c, and d. Ask students which decimal represents a fraction that is greater than 1. (1.25) Point out that this also means that the percent equivalent will also be greater than 100%. Then ask which decimal is closest to zero. (0.005) Point out that 0.005 represents a percent that is less than 1%.

3 Ask the students why the number representing the numerator is entered into the calculator first when changing a fraction to a percent. (because the numerator is divided by the denominator to get a decimal equivalent)

For Example 4c, have a volunteer explain why $1\frac{2}{3}$ is entered as 5 ÷ 3. (because $1\frac{2}{3}$ was written as an improper fraction to be able to divide on the calculator)

ADDITIONAL EXAMPLES

1. Express the ratio as a percent. 2 out of 10 (20%)
2. Write a proportion to find the percent. 28 teachers out of 52 (about 54%)
3. Use the rule to write each decimal as a percent.
   - a. 1.76 (176%)
   - b. 0.075 (7.5%)
   - c. 0.0001 (0.01%)

Guided Practice

Class Exercises For question 13, ask how the problem could be solved mentally. (Think: $\frac{4}{16} = \frac{1}{4}$, $\frac{1}{4}$ = 25%; if 25% live in a rural area then 100% − 25%, or 75%, do not live in a rural area.)

Think and Discuss Emphasize to students that 0.1 and 0.1% are not the same amount. The percent symbol means that in the fractional equivalent, the denominator is 100. With decimals, the denominator of the fractional equivalent depends on the decimal place.

NOTES & QUOTES

During the time of Columbus, the words *per cento,* derived from the Latin form *per centum,* were used to indicate *per hundred.* Later, the words were abbreviated and the term slowly changed its form to the symbol we use today.

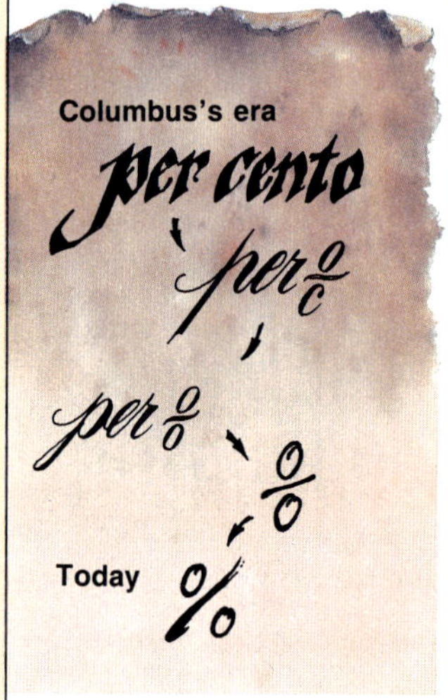

THINK AND DISCUSS

1. Why is it possible to write a decimal as a percent by moving the decimal two places to the right? Justify your answer and give examples.
2. Which do you think is easiest to compare: ratios, decimals, or percents? Explain. Answers will vary.

1. Percent means per 100 and a two-place decimal represents hundredths. 0.67 = 67% 0.582 = 58.2%

▼ You can use the following rule to write decimals as percents.

Decimals to Percents	To write decimals as percents, move the decimal point two places to the right and write the percent sign.

▼ You can use a calculator to write fractions as decimals before writing as percents.

3 ***Example 4*** **Write as a percent.**

a. $\frac{3}{8}$ b. $\frac{5}{400}$ c. $1\frac{2}{3}$ d. $2\frac{1}{4}$

Solution

a. 3 ÷ 8 = 0.375 = 37.5%

b. 5 ÷ 400 = 0.0125 = 1.25%

c. 5 ÷ 3 = 1.6 . . . ≈ 167%

d. 9 ÷ 4 = 2.25 = 225%

CLASS EXERCISES

Write each fraction as a percent. Round to the nearest tenth.

1. $\frac{23}{100}$ 23%
2. $\frac{1}{4}$ 25%
3. $\frac{11}{20}$ 55%
4. $\frac{1}{6}$ 16.7%

Write each ratio as a percent. Round to the nearest tenth.

5. 28 : 40 70%
6. 60:150 40%
7. 75 : 39 192.3%
8. 20 : 36 55.6%

Write each decimal as a percent.

9. 1.68 168%
10. 0.36 36%
11. 0.70 70%
12. 0.002 0.2%
13. Write a percent to describe each situation.
    - a. 4 out of 16 people live in a rural area. 25%
    - b. What percent do *not* live in a rural area? 75%

WRITTEN EXERCISES

Write each ratio as a percent. Round to the nearest tenth.

1. 15 : 20 75%
2. 6 : 30 20%
3. 30 : 48 62.5%
4. 22 : 80 27.5%
5. 28 : 48 58.3%
6. 32 : 56 57.1%
7. 84 : 60 140%
8. 72 : 54 133.3%

Write each decimal as a percent. Round to the nearest tenth.

9. 0.33 33%
10. 0.35 35%
11. 0.06 6%
12. 0.0075 0.8%

MATH MINUTES

On flag day, students wore red, white, or blue shirts. What percent does each amount represent?

Color	Number of Students
red	15 (50%)
white	10 ($33\frac{1}{3}$%)
blue	5 (about $16\frac{2}{3}$%)

13. 0.045 4.5% **14.** 0.375 37.5% **15.** 1.88 188% **16.** 2.59 259%

Write each fraction as a percent. Round to the nearest tenth.

17. $\frac{79}{100}$ 79% **18.** $\frac{29}{50}$ 58% **19.** $\frac{3}{10}$ 30% **20.** $\frac{7}{20}$ 35%

21. $\frac{17}{25}$ 68% **22.** $\frac{3}{50}$ 6% **23.** $\frac{111}{100}$ 111% **24.** $\frac{27}{80}$ 33.8%

CALCULATOR Write each fraction as a percent. Round to the nearest tenth.

25. $\frac{2}{9}$ 22.2% **26.** $\frac{5}{6}$ 83.3% **27.** $\frac{7}{16}$ 43.8% **28.** $\frac{18}{11}$ 163.6%

ESTIMATION About what percent of each flag is red? Accept reasonable estimates.

29. 80%

Tennessee

30. 35%

North Carolina

31. 25%

Arizona

Copy and complete the table.

	Fraction	Decimal	Percent
32.	$\frac{4}{5}$	0.80	80%
33.	$\frac{1}{10}$	0.10	10%
34.	$\frac{1}{2}$	0.5	50%
35.	$\frac{3}{4}$	0.75	75%
36.	$\frac{67}{100}$	0.67	67
37.	$\frac{1}{4}$	0.25	25

Solve. Round to the nearest tenth.

38. Jamie scored 31 correct on a 45-item test. The passing grade is 70%. Did Jamie pass? 68.9%; no

39. A scale drawing needs to be enlarged by a factor of 1.12. Express this as a percent. 112%

40. A map was drawn to a scale of 0.725% of the original. Express this as a decimal. 0.00725

41. A student committee has 15 members. Nine voted in favor of a smoking ban. What percent is this? 60%

MIXED REVIEW

Solve for x.

1. $0.85x = 39.95$ $x = 47$

2. $4.8x = -0.84$ $x = -0.175$

3. $\frac{25}{48} = \frac{x}{132}$ $x = 68.75$

4. $\frac{52}{x} = \frac{14}{490}$ $x = 1{,}820$

Write a fraction and a percent for each ratio.

5. 42 : 168 $\frac{1}{4}$ = 25%

6. 176 : 224 $\frac{11}{14}$ = 78.6%

Solve.

7. The average of three test scores is 85. One test score is 90; another is 72. What is the third? 93

A ***common error*** when writing a ratio as a percent is to set up the ratio incorrectly. For example, 31 out of 44, may be written as $\frac{44}{31}$ and the percent found to be about 142%. Have the students first estimate to determine if the result will be greater or less than 100%. Students should reason that 31 out of 44 is about $\frac{30}{40}$ or $\frac{3}{4}$; $\frac{3}{4}$ is 75%, so the answer should be about 75%.

Closure

Writing in Math Have students explain, in their math journals, when they would use a proportion to write a decimal as a percent, and when they would use the rule.

Independent Practice

For questions 17-24, suggest that students estimate before finding the actual percent.

Before students answer questions 25-28, have volunteers tell which number is entered into the calculator first.

Lesson Follow-up

Reteaching Activity Provide graph paper. Have students outline 10 unit × 10 unit squares and make models of ratios such as 2 out of 100, 2 out of 10, 35 out of 100, 7 out of 10. Have students write the decimal/fraction/percent equivalent for each ratio.

LESSON RESOURCES

TRB Practice F-6
TRB Enrichment F-7

Practice For use after 6-4 (pp. 245–248)

Percent

Write each ratio as a percent.

1. 9:12 75%
2. 16:20 80%
3. 17:34 50%
4. 49:70 70%
5. 19:76 25%
6. 81:90 90%
7. 21:35 60%
8. 21:56 37.5%
9. 39:60 65%
10. 27:75 36%
11. 42:48 87.5%
12. 65:52 125%

Write each decimal as a percent.

13. 0.16 16%
14. 0.72 72%
15. 0.08 8%
16. 0.99 99%
17. 0.4 40%
18. 1 100%
19. 0.036 3.6%
20. 0.777 77.7%
21. 3.04 304%
22. 0.002 0.2%
23. 0.0004 0.04%
24. 5.009 500.9%

Write each fraction as a percent.

25. $\frac{7}{10}$ 70%
26. $\frac{3}{5}$ 60%
27. $\frac{11}{20}$ 55%
28. $\frac{17}{25}$ 68%
29. $\frac{1}{5}$ 20%
30. $\frac{39}{100}$ 39%
31. $\frac{1}{20}$ 5%
32. $\frac{13}{50}$ 26%
33. $\frac{24}{25}$ 96%
34. $\frac{31}{40}$ 77.5%
35. $\frac{111}{200}$ 55.5%
36. $\frac{403}{1,000}$ 40.3%

CALCULATOR Write each fraction as a percent. Round to the nearest tenth of a percent.

37. $\frac{4}{7}$ 57.1%
38. $\frac{57}{99}$ 57.6%
39. $\frac{40}{13}$ 307.7%

Compare. Use <, >, or =.

40. $\frac{1}{3}$ [>] 33%
41. $\frac{3}{100}$ [=] 3%
42. 55% [<] 55
43. 0.7 [>] 7%
44. 80% [=] $\frac{4}{5}$

F-6 Chapter 6

Enrichment For use after 6-4 (pp. 245–248)

Fish Story

What do you call a sketch of a fish's fins?

To find out, write the answers to Exercises 1–13 in the empty circles. Then use a ruler to connect each circle with a circle on the right containing an equivalent ratio, fraction, or decimal. (Connect the dots on the circles.) Write the letter of the appropriate circle above the number of the exercise in the answer grid at the bottom.

Write as a percent.

Exercise	Answer	Circle	Letter
1. 2:5	40%	0.35	A
2. $\frac{7}{20}$	35%	0.2	G
3. 9:30	30%	0.8	C
4. 0.08	8%	$\frac{3}{8}$	N
5. 0.25	25%	$\frac{2}{25}$	A
6. 7:35	20%	1:50	D
7. 9:27	$33\frac{1}{3}$%	0.4	E
8. $\frac{4}{5}$	80%	$\frac{1}{3}$	R
9. 0.375	37.5%	$\frac{3}{10}$	A
10. 0.02	2%	6:3	W
11. 3:100	3%	0.03	S
12. 2	200%	0.025	L
13. 10:400	2.5%	$\frac{1}{4}$	I

Answer: A S C A L E D R A W I N G
2 11 8 4 13 1 10 7 3 12 5 9 6

Chapter 6 F-7

NOTES & QUOTES

The proportion of gold to other metals is marked in karats. Pure gold is marked 24 K. This is read as 24 karats. Gold is often mixed with other metals to make it more durable.

Karats	Gold	Other	Total
24 K	24	0	24
18 K	18	6	24
14 K	14	10	24
10 K	10	14	24

42. In a ball-throwing contest, Player A scores 30 hits out of 35 shots. Player B scores 0.85 of her throws and Player C's rate is 85.5%. Which player has the best record? **Player A**

43. At a local high school, 560 out of 1,060 students voted to support a community fund drive. What percent of the students did not support the fund drive? **47.2%**

44. A crowd filled the 8,000 seats at a football stadium. There were 1,500 children and 5,600 men present. Write a ratio and a percent that describes how many seats were filled by:

a. men $\frac{7}{10}$ = 70% **b.** children $\frac{3}{16}$ = 18.8% **c.** women $\frac{9}{80}$ = 11.3%

45. A student is a member of a 4-person relay team. What percent of the distance will she run in a race? **25%**

Compare. Use >, <, or =.

46. $\frac{7}{12}$ ■ 60% **<** **47.** 0.0325 ■ 32.5% **<** **48.** $\frac{5}{8}$ ■ 0.625 **=**

49. 0.05% ■ 50% **<** **50.** $\frac{7}{8}$ ■ 68% **>** **51.** 15 : 30 ■ 85% **<**

52. $\frac{3}{2}$ ■ 1.5 **=** **53.** 140 : 130 ■ 104% **>** **54.** 0.1756 ■ 176% **<**

WRITE Does each sentence makes sense? Explain why or why not.

55. A student ran 150% farther today than yesterday. **yes**

56. Since 15% of the students play tennis, 85% of the students do not play tennis. **Yes, 15% + 85% = 100%**

57. A student got 200% of the items correct on a test. **No; Getting 100% is a perfect score.**

58. A student missed 12 items on a test and got an A. **Yes; The grade depends on the ratio of missed to total.**

Complete each analogy.

59. dime : dollar : : 10 : ■ **100**

60. foot : yard : : ■ : 3 **1**

61. 50 : 20 : : 25 : ■ **10**

62. 10 : decade : : 100 : ■ **century**

63. 1 : 7 : : day : ■ **week**

64. 1 : 365 : : ■ : year **day**

Use the *DATA* at the left. Write a ratio of pure gold to total karat weight for jewelry that has each marking.

65. 10 K $\frac{5}{12}$ **66.** 14 K $\frac{7}{12}$ **67.** 18 K $\frac{3}{4}$ **68.** 24 K $\frac{1}{1}$

69. ***WRITE*** a sentence that tells what percent of gold each of the markings indicate. **Check students' journals. 10 K = 41.7% gold; 14 K = 58.3% gold; 18 K = 75% gold; 24 K = 100% gold**

70. **a.** Which label indicates that less than 50% of the item consists of gold? **10 K**

b. Which label indicates that an item is 100% gold? **24 K**

LESSON QUIZ

Write each ratio as a percent.

1. 10 : 15 ($66\frac{2}{3}$%)
2. 92 : 64 (143.75%)

Write each decimal as a percent.

3. 0.175 (17.5%)
4. 3.01 (301%)

Compare. Use >, <, or =.

5. $\frac{12}{15}$ ■ 75% (>)
6. 0.135 ■ 110% (<)
7. $\frac{24}{6}$ ■ 2.42 (>)
8. 0.01% ■ $\frac{3}{4}$% (<)

FOR THE NEXT LESSON

math journal

Assignment Guide

Basic 1-20, 25-26, 29-41, 46-48, 55-60, 66-67, MR All
Average 1-20, 25-26, 29-37, 39-43, 48-51, 55-61, 68-70, MR All
Enriched 5-24, 27-37, 42-45, 51-58, 61-65, 69-70, MR All

Practice

Write as a fraction in lowest terms.

1. 18 is to 45 $\frac{2}{5}$
2. 17 : 24 $\frac{17}{24}$
3. 24 out of 60 $\frac{2}{5}$
4. 92 : 38 $\frac{46}{19}$
5. 21 out of 49 $\frac{3}{7}$
6. 14 is to 52 $\frac{7}{26}$

Compare. Use = or ≠. Then tell if the pair of ratios form a proportion.

7. $\frac{11}{9}$ ■ $\frac{55}{54}$ ≠
8. $\frac{5}{15}$ ■ $\frac{3}{9}$ = ; prop.
9. $\frac{6}{9}$ ■ $\frac{24}{36}$ = ; prop.
10. $\frac{52}{4}$ ■ $\frac{26}{2}$ = ; prop.
11. $\frac{4}{5}$ ■ $\frac{20}{25}$ = ; prop.
12. $\frac{12}{4}$ ■ $\frac{9}{3}$ = ; prop.
13. $\frac{7}{28}$ ■ $\frac{4}{16}$ = ; prop.
14. $\frac{16}{26}$ ■ $\frac{8}{12}$ ≠

Solve.

15. $\frac{a}{7} = \frac{1}{3.5}$ $a = 2$
16. $\frac{6}{7.2} = \frac{5}{b}$ $b = 6$
17. $\frac{3}{c} = \frac{2}{8}$ $c = 12$
18. $\frac{7}{5} = \frac{d}{45}$ $d = 63$
19. $\frac{22}{18} = \frac{2.75}{e}$ $e = 2.25$
20. $\frac{3a}{39} = \frac{3}{6}$ $a = 6.5$
21. $\frac{8}{5g} = \frac{16}{20}$ $g = 2$
22. $\frac{2.4}{3} = \frac{12}{h}$ $h = 15$

Write as a percent. Round to the nearest tenth.

23. $\frac{9}{5}$ 180%
24. $\frac{1}{3}$ 33.3%
25. $\frac{3}{8}$ 37.5%
26. $\frac{125}{100}$ 125%
27. $\frac{46}{60}$ 76.7%
28. $\frac{7.5}{10}$ 75%
29. $\frac{11}{20}$ 55%
30. $\frac{5}{6}$ 83.3%

Write as a percent. Round to the nearest tenth.

31. 2 : 30 6.7%
32. 15 : 5 300%
33. 34 : 40 85%
34. 2 : 3 66.7%
35. 5 : 8 62.5%
36. 1 : 25 4%
37. 4 : 32 12.5%
38. 7 : 35 20%

Write as a decimal.

39. $22\frac{1}{2}\%$ 0.225
40. 165% 1.65
41. 73.6% 0.736
42. 9% 0.09

Express as a fraction in lowest terms, as a decimal, and as a percent. Round to the nearest tenth of a percent.

43. 15 : 45 $\frac{1}{3}$, $0.\overline{3}$, 33.3%
44. 6 : 10 $\frac{3}{5}$, 0.6, 60%
45. 3 : 9 $\frac{1}{3}$, $0.\overline{3}$, 33.3%
46. 13 : 10 $\frac{13}{10}$, 1.3, 130%
47. 48 : 72 $\frac{2}{3}$, $0.\overline{6}$, 66.7%
48. 19 : 95 $\frac{1}{5}$, 0.2, 20%
49. 32 : 54 $\frac{16}{27}$, $0.\overline{592}$, 59.3%
50. 16 : 28 $\frac{4}{7}$, $0.\overline{571428}$, 57.1%

Compare the following. Use >, <, or =.

51. $\frac{24}{36}$ ■ 65% >
52. $\frac{11}{12}$ ■ 90% >
53. $0.45\frac{1}{3}$ ■ $45\frac{1}{3}\%$ =
54. 0.52 ■ $\frac{13}{25}$ =
55. 19.6% ■ 0.195 >
56. 5 : 18 ■ 28% <
57. 0.75 ■ $\frac{3}{4}$ =
58. 0.25 ■ $\frac{1}{5}$ >
59. 38% ■ 0.038 >

CAREER

Help Wanted: Chef

High school diploma required. Vocational training or 2–4 years of college desirable.

For more information, write to American Culinary Federation, P.O. Box 3466, St. Augustine, FL 32084.

Preparing food for others is a delicate yet rigorous undertaking. A chef must know how to combine ingredients to create the desired taste. Will the meal serve 10 or 100? Chefs must understand proportions to adjust a recipe. They also need to use ratios if they need to substitute ingredients. Chefs even use percents to help determine what types and quantities of food to prepare.

PROJECT

Find a recipe for your favorite food. Rewrite the recipe so that it will serve your entire class.

Career Information

Arrange a tour for students in a large restaurant or food service company. Request that the tour be conducted by a chef who will later speak to students about his or her occupation.

Suggest the following as topics for discussion:

- the duties or responsibilities of a chef, including menu planning, ordering food supplies, food preparation, and kitchen management
- the preparation necessary for a career as a chef, including education and/or on-the-job training, and the skills and abilities that are job prerequisites
- the working conditions, such as the hours and physical demand of the worksite, that are common to the profession
- how technology and changes in American culture have affected the profession
- current job opportunities and projections for the future

Lesson Focus

Materials

Math journal

MOTIVATING THE LESSON

Have students write these ratios in lowest terms:

a. 6 out of 10 homes in the U.S. have cable television ($\frac{3}{5}$)

b. 10 oranges for \$2.40. ($\frac{1}{\$0.24}$)

c. 280 out of 700 students are freshmen ($\frac{2}{5}$)

Instruction

1 As you discuss the problem with the students, point out that the answer will be expressed as a fraction, not as a whole number.

2 Help students to realize that the number of boys must be a number that is compatible with $\frac{2}{3}$.

3 Comment that twice the number of boys who have dates with ninth-grade girls will be the total number of ninth-grade girls.

OBJECTIVE:
To explore number relationships.

Exploring Number Relationships

1 ■ At a homecoming dance, two-thirds of the ninth-grade boys have dates with a ninth-grade girl. Half of the ninth-grade girls have dates with a ninth-grade boy. What part of the entire ninth-grade class are boys who do *not* have dates with ninth-grade girls?

You can organize and analyze the data to make it easier to decide on a strategy to solve the problem.

1. ***Analyze*** the data and answer these questions.
   a. Do you know the number of ninth-grade students that are in the school? no
   b. Do you know whether there are more boys or girls in the ninth grade at the school? ***Explain*** how you know and use a model to justify your answer. There are more girls since it takes $\frac{2}{3}$ boys to equal $\frac{1}{2}$ girls.
2. List some strategies you think might help you solve the problem. Answers will vary. Example: guess and test.

■ Sometimes substituting a number in a problem can help you understand the relationships in a problem.

3. Is this technique like any of the strategies you listed? If so, describe the ones that are similar. Answers will vary.
4. a. Choose a possible number of ninth-grade boys and substitute this number into the problem. ***Compare*** with another group.
   2 b. Could there be a total of exactly ten boys in the ninth grade? exactly twelve boys? ***Explain.*** No; $\frac{2}{3}$ of 10 is not a whole number. Yes; $\frac{2}{3}$ of 12 is 8.

5. Answers will vary. Answers given are based on 30 ninth-grade boys.

5. ***Analyze*** the data after you substitute a number into the problem.
   a. How many ninth-grade boys have dates with ninth-grade girls? 20
   3 b. How many ninth-grade girls have dates with ninth-grade boys? ***Explain*** how you found the number of girls when you know the number of boys. Use a model if necessary. 20; The number of boys with dates is the same as the number of girls.
   c. What is the total number of ninth-grade girls at the school? 40
   d. What is the total number of ninth-grade students? 70
   e. How many ninth-grade boys do *not* have a date with a ninth-grade girl? ***Describe*** how you found this answer. ***Compare*** with another group. 10; Ans. will vary. One-third the number of ninth-grade boys
   f. What fractional part of the entire ninth grade is boys who do not have a date with a ninth-grade girl? $\frac{1}{7}$
   g. ***Write*** a ratio in three different forms to describe your answer. 1 to 7, 1 : 7, $\frac{1}{7}$

Teaching Tip

As students solve problems, provide an opportunity for class discussion. Encourage students to explain each problem in their own words and to suggest some strategies they might use to solve the problem.

6. Choose a *different* possible number of ninth-grade boys and try substituting this number into the problem.
   a. Complete each step in Exercise 5 using your new chosen number. Answers will vary.
   b. ***Compare*** the results you get from Exercises 5 and 6. What do you notice about the ratios? They are equal.
7. Refer to the problem. ***Write*** a ratio that describes each of the following.
   a. the part of the entire ninth-grade class that are girls who do *not* have a date with a ninth-grade boy $\frac{2}{7}$
   b. the part of the entire ninth-grade class that are boys who do have dates with a ninth-grade girl $\frac{2}{7}$
   c. the part of the entire class that are girls who do have a date with a ninth-grade boy $\frac{2}{7}$
   d. a ratio of boys to girls $\frac{3}{4}$

■ You can often use more than one strategy to solve a problem.

8. Try using some of the other strategies you listed to solve the same problem. Check students' work.
9. ***Describe*** how you might use a diagram or draw a picture to help you solve the problem. Check students' work.
   a. What kind of picture would be appropriate?
   b. How can your picture display the ratio of boys to girls?
10. ***Summarize*** your findings. Check students' work.
   a. ***Explain*** how substituting numbers can sometimes help to clarify the relationships within the problem.
   b. List the strategies that were effective in solving the problem. ***Discuss*** which ones were helpful and which ones were not. ***Explain.***
   c. ***Write*** a similar problem. Trade with a partner and solve each other's problem.

Closure

Writing in Math Have students write in their math journals the strategies they used to solve the problem. Then have them compare their strategies with others in the class.

Assignment Guide
Basic All
Average All
Enriched All

FOR THE NEXT LESSON
calculator

Lesson Focus

Materials

Calculator

Vocabulary/Symbols

base
percentage
rate

MOTIVATING THE LESSON

Begin the lesson by formulating a percent statement about the class. For example, suppose 14 out of 25 students in the class are involved in extracurricular activities. Students can write the ratio as an equivalent ratio with a denominator of 100: $\frac{14}{25} = \frac{56}{100}$. Have students determine what percent $\frac{56}{100}$ equals. (56%)

Instruction

1 Stress that the formula $\frac{p}{b} = \frac{n}{100}$ can be used to solve for any one of the three variables when the other two are given.

OBJECTIVE: *To evaluate percents using proportions.*

6-5 Using Proportions to Find Percent

▼ In a recent year, there were 8,763 commercial radio stations in the United States. Of these, 295 stations had a rock format. To find the percent of stations with a rock format, you can find the ratio of some number to 100.

You can write a proportion to solve the problem.

$\frac{295}{8,763} = \frac{n}{100}$	Write a proportion.
$8,763n = 29,500$	Write cross products.
$\frac{8,763n}{8,763} = \frac{29,500}{8,763}$	Divide both sides by 8,763.
$n = 3.366$	Solve for n.
$= 3\%$	Round to the nearest whole number. Write as a percent.

Proportions and Percents To find the ratio of a number to 100, use the formula. $\frac{\text{part}}{\text{whole}} = \frac{n}{100}$

▼ You can use proportions to solve other percent problems.

Example 1 About 5% of the 8,763 radio stations had a golden oldie format. How many stations played golden oldies?

Solution

$\frac{p}{8,763} = \frac{5}{100}$	Write a proportion.
$8,763\left(\frac{p}{8,763}\right) = 8,763\left(\frac{5}{100}\right)$	Multiply both sides by 8,763.
$p = \frac{43,815}{100}$	
$= 438.15 \approx 438$	

1

About 438 radio stations played golden oldies.

Example 2 Memphis, TN, had 12 AM radio stations in one year. These stations accounted for 60% of the AM and FM stations in Memphis. How many AM and FM stations did Memphis have at this time?

Solution

$\frac{12}{b} = \frac{60}{100}$	Write a proportion.
$60b = 1,200$	Write the cross products.
$\frac{60b}{60} = \frac{1,200}{60}$	Divide both sides by 60.
$b = 20$	Solve for b.

Memphis had 20 AM and FM radio stations.

Teaching Tip

It may be helpful if students remember this proportion:

$\frac{\text{part}}{\text{whole}} = \frac{\text{percent}}{100}$

▼ To compare ratios and percents, write both in the same form.

Example 3 At Pineapples, all books and posters are marked down 30%. At Avocados, the same items are marked $\frac{1}{3}$ off. Which store offers the greater markdown?

Solution To compare the sale prices, write $\frac{1}{3}$ as a percent.

$$\frac{1}{3} = \frac{x}{100}$$

$$x = 100\left(\frac{1}{3}\right)$$

$$x = 33\frac{1}{3}\%$$

Since $33\frac{1}{3}\% > 30\%$, Avocados offers the greater markdown.

CLASS EXERCISES

1. $\frac{n}{100} = \frac{30}{40}$; 75% 2. $\frac{40}{20} = \frac{x}{100}$; 200% 3. $\frac{x}{20} = \frac{80}{100}$; 16 4. $\frac{x}{50} = \frac{300}{100}$; 150 5. $\frac{x}{75} = \frac{33\frac{1}{3}}{100}$; x = 25 6. $\frac{x}{60} = \frac{40}{100}$; 24

Write and solve a proportion. Round to the nearest tenth.

1. What percent of 40 is 30?
2. What percent of 20 is 40?
3. Find 80% of 20.
4. Find 300% of 50.
5. What is $33\frac{1}{3}\%$ of 75?
6. What is 40% of 60?
7. 25% of f is 8. What is f? $\frac{8}{f} = \frac{25}{100}$; 32
8. 250% of t is 50. What is t? $\frac{50}{t} = \frac{250}{100}$; 20

Solve.

9. A bank account balance of $400 earns $24 interest in one year. What is the rate of interest? 6%
10. A bicycle cost $200 last year. The same bike costs $250 this year. What percent of last year's cost is this year's cost? 125%
11. A student pole vaulted 5 ft yesterday. Today she vaulted 20% higher. How high was her vault today? 6 ft

THINK AND DISCUSS See below.

1. Is a% of b the same as b% of a? Explain.
2. When might it be easier to solve a percent problem by writing it as a fraction with a denominator other than 100?

1. yes; $\frac{a}{100}(b) = \frac{b}{100}(a)$

$\frac{ab}{100} = \frac{ab}{100}$

2. when the percent has a fractional part such as $33\frac{1}{3}\%$

WRITTEN EXERCISES

Write a proportion to solve. Round to the nearest tenth.

1. What percent of 25 is 13? 52%
2. Find 18% of 150. 27
3. 116% of a is 125. What is a? 107.8
4. Find 116% of 75. 87
5. What percent is 40 of 120? 33.3%
6. 49% of b is 31.85. What is b? 65
7. Find 60% of 15. 9
8. $12\frac{1}{2}\%$ of n is 6. What is n? 48

ADDITIONAL EXAMPLES

1. What is 15% of 60? (9)
2. What percent of 40 is 25? ($62\frac{1}{2}\%$)
3. 48 is 80% of what number? (60)

Guided Practice

Class Exercises For questions 1-11, have students identify the base, rate, and percentage. The unknown can be replaced by the corresponding variable.

Think and Discuss For question 1, have students substitute values for a and b.

In question 2, for percents less than 1% sometimes it is more convenient to express the fraction as a power of ten. For example, $\frac{1}{2}\%$ can be expressed as $\frac{5}{1,000}$.

MATH MINUTES

Have students find a newspaper ad for a new car. Have them identify the list price of the car, and how much money they would need for a 15% down payment. Then have them calculate the amount of the balance.

A ***common error*** occurs when students substitute values for variables in the formula incorrectly. Suggest to students that before they substitute values, they should identify which numbers in a problem are the part, the whole, and the percent.

Closure

Writing in Math Check students' mastery of the lesson by having them write as many statements as they can about percents using the numbers 50, 100, and 200.

Independent Practice

For questions 41 and 42, it may be helpful for students to think: 80 increased by 65% is 165% of 80; and 36 reduced by $16\frac{2}{3}\%$ is $83\frac{1}{3}\%$ of 36.

MIXED REVIEW

Find each answer.

1. $x^2 + 3$ for $x = 7$ 52
2. $a + b^2$ for $a = 3$ and $b = 6$ 39

Write each as a percent.

3. $2\frac{7}{12}$ $258\frac{1}{3}\%$ 4. 4.523 452.3%
5. 0.08 8% 6. 45.6 4,560%
7. The ski club has 50 members. There are 19 students going on a ski trip. What percent of the club members are going on the trip? 38%

9. Find $58\frac{1}{3}\%$ of 54. 31.5
10. What percent of 250 is 75? 30%
11. What percent of 20 is 7? 35%
12. What percent of 80 is 130? 162.5%
13. Find 125% of 16. 20
14. Find 92% of 625. 575
15. What percent of 40 is 70? 175%
16. 35% of x is 52.5. What is x? 150

Compare. Use >, <, or =.

17. $\frac{14}{25}$ ■ 56% =
18. 0.3125 ■ 32.5% <
19. $\frac{1}{9}$ ■ 0.111 >
20. 0.4205 ■ 42.5% <
21. 30 : 36 ■ 85% <
22. $\frac{8}{7}$ ■ 114% >
23. $\frac{13}{8}$ ■ 1.625 =
24. 125 : 120 ■ 104% >
25. $\frac{7}{8}$ ■ 75% >

26. ***WRITE*** a paragraph explaining why a proportion is frequently set up with at least one ratio having a denominator of 100. Check students' journals.

MENTAL MATH **Solve.**

27. What percent of 50 is 10? 20%
28. What percent of 10 is 15? 150%
29. What percent is 6 of 24? 25%
30. What percent is 18 of 9? 200%
31. Find 120% of 20. 24
32. Find 9% of 300. 27
33. 200% of p is 24. What is p? 12
34. $\frac{1}{4}\%$ of n is 3.75. What is n? 1,500

CALCULATOR **Write a proportion and solve. Round to the nearest tenth.**

35. What percent of 92 is 17? 18.5%
36. What percent of 68 is 89? 130.9%
37. Find 93% of 47.89. 44.5
38. Find 53% of 76,550. 40,571.5
39. Find 138% of 61. 84.2
40. Find 189% of 82. 155
41. Find 80 increased by 65%. 132
42. Find 36 reduced by $16\frac{2}{3}\%$. 30
43. 43% of q is 18.06. What is q? 42
44. 2.5% of z is 912.5. What is z? 36,500

Solve.

45. The interest on an account of $1,500 is $120 for one year. What is the rate of interest? 8%

46. The population of a city was 28,000 in 1950 and 70,000 in 1990. What percent is the 1990 population of the 1950 population? 250%

47. A salesperson gets an 8% commission on sales after the first $5,000 per month. What commission would the salesperson receive on sales of $20,000? $1,200

48. ***DATA FILE 4 (pp. 138–139)*** About what percent of the world's population speaks Chinese? what percent speaks English? 14%; 8%

WRITE a sentence telling whether or not the statement makes sense.

49. Sam got a 25% markdown on the price of a new car. Pat got \$2,500 off. Pat got a better deal. **See below.**

50. On a recent math test, 15 students passed in one class and 22 passed in another. The second class did better on the test. **See below.**

51. The human body is about 60% water by weight. That means that a person who weighs 100 lb is about 60 lb water. **It makes sense.**

To Insure Proper Service

Some say TIPS means *To Insure Proper Service.* No matter what your translation, you need to know the rules. On a long awaited cruise vacation, you just want to have fun. You think you've accounted for everything. Everything, that is, except for the small print that reads GRATUITIES NOT INCLUDED. Don't despair. Follow these simple guidelines for tipping and you'll be in for a smooth sailing vacation!

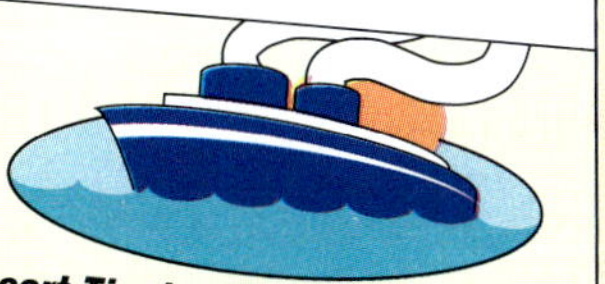

Resort Tipping Guidelines

Restaurants	15%-20% of the check before taxes
Caddies	\$4-\$5 per bag or 15% of the greens fees
Hairdresser	15%
Laundry	15%-20%
Taxis	15% of the fare (\$.25 minimum)
Bellhops and porters	\$.50-\$1 per bag

Use the chart at the right to solve.

52. What would be the tip for a \$25 haircut? **\$3.75**

53. You tipped your server \$4.50. About what was the cost of your meal? **15% tip; \$30.00 20% tip; \$22.50**

49. You can't tell who got the better deal without knowing the original price.
50. You can't tell who did better unless you know how many students are in each class.

TEST YOURSELF

Write each ratio as a fraction in lowest terms.

1. 7 out of 21 students have a TV in their bedroom. $\frac{1}{3}$

2. 5 out of 100 students plan to be doctors. $\frac{1}{20}$

Write a proportion to describe each situation. Then solve.

3. One pint of paint costs \$2.89. Find the cost of a quart of paint. **\$5.78**

4. Three light bulbs cost \$5.25. Find the cost of one light bulb. **\$1.75**

Write each as a percent.

5. $\frac{3}{4}$ **75%** **6.** 0.89 **89%** **7.** $\frac{7}{8}$ **87.5%** **8.** 0.03 **3%** **9.** 0.007 **0.7%**

Write and solve a proportion.

10. Find 18% of 30. **5.4** **11.** 25% of n is 25. Find n. $n = 100$

Lesson Follow-up

Reteaching Activity Have students write a proportion before solving each problem.

1. Find 20% of 48. $(\frac{p}{48} = \frac{20}{100})$

2. 50 is what percent of 75? $(\frac{50}{75} = \frac{n}{100})$

3. 57 is 30% of what number? $(\frac{57}{b} = \frac{30}{100})$

LESSON RESOURCES

TRB Practice F-8
TRB Enrichment F-9

Practice

For use after 6-5 (pp. 252–255)

Using Proportions to Find Percent

Write and solve a proportion for each.

1. What percent of 40 is 12? **30%**
2. What percent of 75 is 60? **80%**
3. What percent of 48 is 18? **37.5%**
4. What percent is 17 of 25? **68%**
5. What percent is 54 of 60? **90%**
6. What percent of 68 is 51? **75%**
7. What percent is 39 of 50? **78%**
8. What percent of 52 is 65? **125%**
9. Find 80% of 25. **20**
10. Find 75% of 76. **57**
11. Find 150% of 74. **111**
12. Find $27\frac{1}{2}$% of 96. **26.4**
13. Find 44% of 375. **165**
14. Find 24% of 120. **28.8**
15. Find 65% of 180. **117**
16. Find 260% of 30. **78**
17. 40% of x is 28. What is x? **70**
18. 75% of p is 12. What is p? **16**
19. 9% of k is 27. What is k? **300**
20. $62\frac{1}{2}$% of t is 35. What is t? **56**
21. 38% of n is 33.44. What is n? **88**
22. 120% of y is 42. What is y? **35**
23. 300% of m is 600. What is m? **200**
24. 1.5% of h is 12. What is h? **800**

Solve.

25. The Eagles won 70% of the 40 games that they played. How many games did they win? **28**
26. Thirty-five of 40 students surveyed said that they favored recycling. What percent of those surveyed favored recycling? **87.5%**
27. Candidate Carson received 2,310 votes, 55% of the total. How many total votes were cast? **4,200**

F-8 Chapter 6

Enrichment

For use after 6-5 (pp. 252–255)

Opinion Survey

A survey of recreation preferences was conducted. The table lists the results by gender and age categories.

Respondent Summary	
Male	160
Female	240
Under 20	80
20–39	152
40–59	140
60 and over	28

1. How many people were polled? **400**
2. What percent of those polled were: female? **60%** aged 40–59? **35%** aged 60 and over **7%**
3. Which group listed in the table contains 38% of those polled? **20–39**
4. The number in the 40–59 age bracket represents what percent of the number in the 60 and over bracket? **500%**
5. Seventy-five percent of the males polled said that they participate in a sport. How many males participate in a sport? **120**
6. Thirty females said that tennis is their favorite sport.
   a. What percent of all females prefer tennis to other sports? **12.5%**
   b. In a more extensive poll, 768 of the females questioned (or 12%) said that they prefer tennis. How many females were questioned in the larger survey? **6,400**
7. Twenty-five percent of those over 40 who were questioned named bowling as their favorite sport. Twelve were in the 60 and over category. How many were in the 40–59 category? **30**
8. Twenty-eight of those under 20 named basketball as their favorite sport.
   a. What percent in this category prefer basketball? **35%**
   b. Of those in this category who do not prefer basketball, 26.9% (to the nearest tenth of a percent) prefer baseball. How many named baseball as their favorite sport? **14**

Chapter 6 F-9

LESSON QUIZ

Write and solve a proportion for each.

1. What percent of 120 is 80? ($66\frac{2}{3}$%)
2. What percent of 50 is 80? (160%)
3. Find $83\frac{1}{3}$% of 60. (50)
4. Find 125% of 200 (250)
5. 50 is 200% of what number? (25)

Assignment Guide

Basic 1-25, 26-38 even, 45-46, 48-50, MR, TY All
Average 1-35 odd, 36-41, 47-53, MR, TY All
Enriched 1-41 odd, 42-44, 47-48, 52-53, MR, TY All

FOR THE NEXT LESSON

calculator, math journal

Lesson Focus

Materials

Calculator, math journal

MOTIVATING THE LESSON

As a brief review have students write the decimal equivalents for the following percents:
35% (0.35) 2% (0.02) 18% (0.18)
$61\frac{1}{2}$% (0.615) 134% (1.34)

Instruction

1 Ask students how they know which operation to use to find the value of x. Help them see that when the variable appears in the part of the diagram that reads "■%" or "of ■" they must use the inverse of multiplication, or division.

2 Point out that the word "of" in "18% of 90,400,000" indicates multiplication.

ADDITIONAL EXAMPLES

Draw a diagram and write an equation. Solve.

1. Find 20% of 45.
Answer: $x = 9$

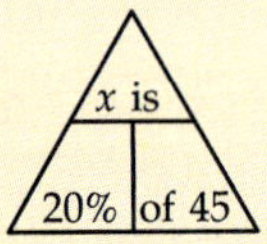

2. 21 is 30% of what number?
Answer: $x = 70$

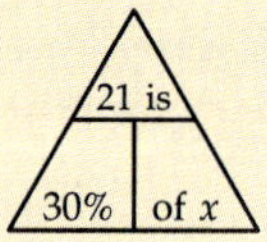

3. What percent of 30 is 18?
Answer: $x = 60\%$

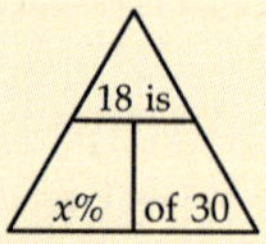

OBJECTIVE: *To evaluate percents using models and equations.*

6-6 Percents and Equations

▼ You can use a triangle diagram to relate the parts of a percent problem.

You know that 12 is 50% of 24. You can read this from the diagram as $\frac{12}{50\%} = 24$. You can also see the following relationships from the diagram.

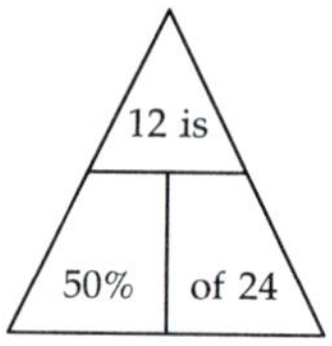

50% of 24 is 12 → $50\% \cdot 24 = 12$

50% equals 12 divided by 24 → $50\% = \frac{12}{24}$

▼ You can draw a diagram to model the parts of any percent problem. The diagram always has the same form. To use the diagram, decide which part of the problem fits in each section. Then write and solve the appropriate equation.

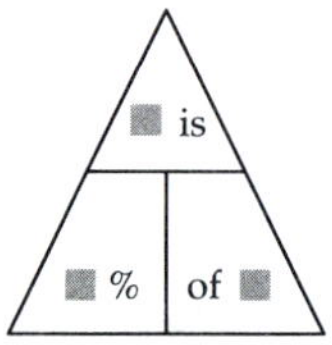

Example 1 In a recent survey, about 28% of the households surveyed were watching the top-rated TV program. Researchers used the information to estimate that about 25,312,000 households were watching the program. This assumes how many households in the total viewing population?

Solution Let x = the number of households

1 Draw a diagram and use it to write an equation.

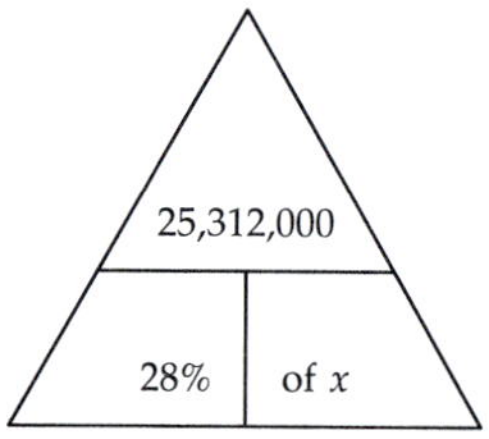

$\frac{25,312,000}{28\%} = x$

25,312,000 ÷ 0.28 = 90,400,000

There are about 90,400,000 households in the viewing population.

Example 2 About 18% of the 90,400,000 households in the viewing population watched the thirteenth-place program. About how many households watched this program?

Solution Let x = the number of households that watched the program.

2 Draw a diagram and use it to write an equation.

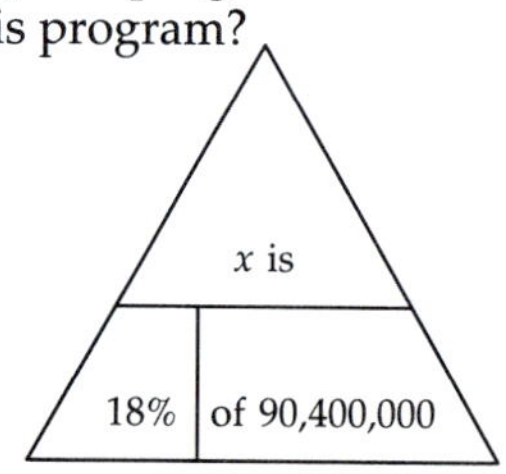

18% of $90,400,000 = x$

0.18 × 90,400,000 ÷ 16,272,000

About 16,272,000 households watched the thirteenth-place program.

Teaching Tip

Have posters available of triangle diagrams with the variable in a different part on each. When students need reinforcement, use the posters as flash cards, asking students to name the operation.

Example 3 About 21,114,600 households in the viewing population watched the eighth-place program. What percent of the 90,400,000 households watched this program?

Solution Let x = the percent viewing the program. Draw a diagram and use it to write an equation.

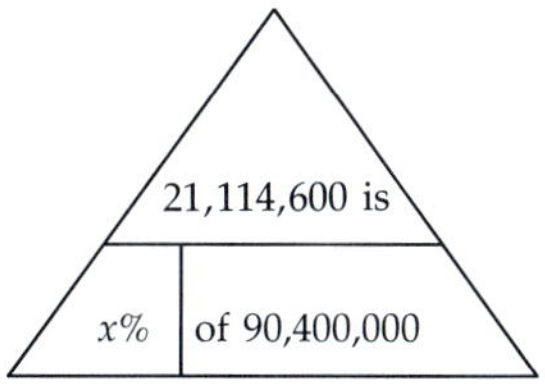

$$x = \frac{21{,}114{,}600}{90{,}400{,}000} \approx 0.234$$

About 23.4% of the households saw the program.

CLASS EXERCISES

Write and solve an equation for each triangle diagram.

1.

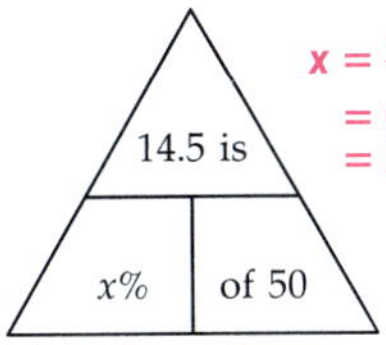

$x = \frac{14.5}{50} = 0.29 = 29\%$

2.

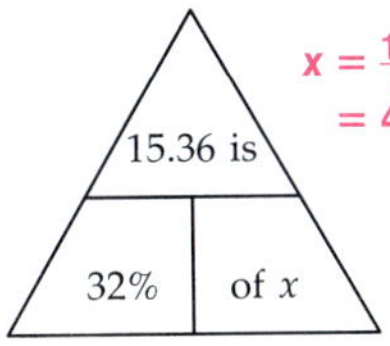

$x = \frac{15.36}{0.32} = 48$

Draw a triangle diagram and use it to solve. Check students' diagrams.

3. What percent of 25 is 40? 160%

4. Find 30% of 30. 9

5. What is $66\frac{2}{3}\%$ of 63? 42

6. Find 150% of 90. 135

7. Sneakers are on sale for 60% of their regular price. They are selling for \$36. What is their regular price? \$60

8. A banker loaned \$650 last year. The balance is \$400 this year. What percent of the original loan is still outstanding? 61.5%

9. ***WRITE*** In your own words, describe how to use a triangle diagram to solve percent problems. Check students' journals.

WRITTEN EXERCISES

Write and solve an equation for each triangle diagram.

1.

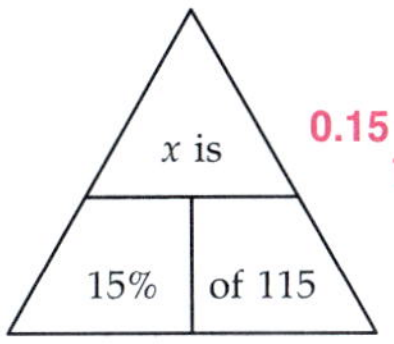

$15\% = \frac{x}{115}$

$0.15 \cdot 115 = x$

$17.25 = x$

2.

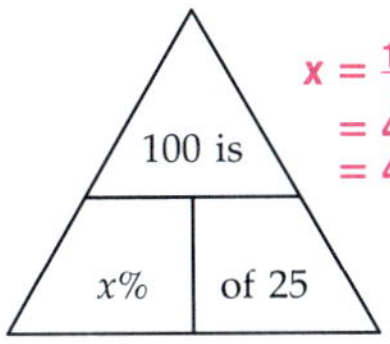

$x = \frac{100}{25} = 4.00 = 400\%$

THINK AND DISCUSS See below.

1. When might it be easier to use the proportion method than the equation method to solve a percent problem?

2. Describe a shortcut for finding the sale price of an item marked 20% off.

1. When finding what percent one number is of another.
2. Think of the sale price as 100% − 20% = 80% of the original price. Multiply 0.80 by the original price.

Guided Practice

Class Exercises For questions 1-8, have students label each diagram with the correct operation before writing the equations.

Think and Discuss Help students understand that the proportion and equation methods for solving percent problems are equivalent. Encourage students to discuss why they prefer one method rather than the other.

A ***common error*** occurs when students diagram a percent problem incorrectly. Show them that the diagram reads like an answer. For example have students read the question "What percent of 25 is 40?" Then ask the students to reword the question as an answer including a variable (e.g., forty is *x* percent of 25). Students can then see how the problem fits into the diagram.

Closure

Writing in Math Have students describe in their math journals, a procedure for determining the proper operation to solve percent problems. Ask them to support their procedures with examples.

MATH MINUTES

Have students estimate the percent of a number:

	1%	20%	50%
20	(0.2)	(4)	(10)
120	(1.2)	(24)	(60)
32	(0.32)	(6.4)	(16)
250	(2.5)	(50)	(125)

Independent Practice

For questions 3-8 remind students that the diagram always has the same form. Review the diagram if necessary.

As students complete questions 21-34, have them explain their thinking.

For calculator questions 35-40 students should understand that using the [%] key eliminates the need for pressing the [=] key, and that the percent is displayed in decimal form.

As a follow-up to the Written Exercises, have students work the computer activity on p. 230E of the Chapter Overview. See also TRB Exploring O-24.

MIXED REVIEW

Simplify.

1. $|43| + |a|$ for $a = -9$ 52
2. $p^2 + p^3$ for $p = 5$ 150

Use proportions to find each percent.

3. What percent of 360 is 45? 12.5%
4. Find 35% of 60. 21

Subtract.

5. 0.85 − 0.23 0.62
6. 1.25 − 0.351 0.899
7. A book was marked at 85% of list price. It sold for \$12.75. What was the list price? \$15.00

Draw a triangle diagram and use it to solve.

3. What percent of 20 is 11? 55%
4. Find 56% of 75. 42
5. 135% of t is 63. What is t? $46\frac{2}{3}$
6. What percent of 25 is 17? 68%
7. Find 500% of 12. 60
8. 85% of z is 106,250. What is z? 125,000

Solve. Round to the nearest tenth.

9. What percent of 4 is 9? 225%
10. $33\frac{1}{3}\%$ of s is $7\frac{1}{2}$. What is s? 22.5
11. Find $26\frac{2}{3}\%$ of 81. 21.6
12. $12\frac{1}{4}\%$ of b is 9.1875. What is b? 75
13. Find 5.5% of 44. 2.4
14. $16\frac{2}{3}\%$ of m is 6. What is m? 36
15. What percent of 150 is 96? 64%
16. What percent of 45 is 24? 53.3%
17. Find 15% of 150. 22.5
18. What percent of 8 is 20? 250%
19. Find 225% of 36. 81
20. 35% of d is 105,000. What is d? 300,000

MENTAL MATH **Solve.**

21. What percent of 60 is 30? 50%
22. Find 20% of 20. 4
23. 100% of t is 100. What is t? 100
24. What percent of 3 is 30? 1,000%
25. Find $33\frac{1}{3}\%$ of 66. 22
26. 200% of g is 24. What is g? 12
27. What percent of 55 is 11? 20%
28. Find 5% of 10. 0.5
29. 18% of n is 18. What is n? 100
30. What percent of 100 is 35? 35%
31. Find 10% of 125. 12.5
32. 25% of z is 200. What is z? 800
33. What percent of 9 is 3? $33\frac{1}{3}\%$
34. Find 20% of 80. 16

CALCULATOR **Solve. Round to the nearest tenth.**

35. What percent of 117 is 54? 46.2%
36. 18% of a is 15. What is a? 83.3
37. Find 16% of 83. 13.3
38. What percent of 59 is 176? 298.3%
39. 23% of a is 23. What is a? 100
40. Find 86% of 29. 24.9

Solve.

41. The rate of interest on a savings account is $5\frac{1}{4}\%$. What is the simple interest for a year on \$750? \$39.38
42. An item on sale for \$7.50 is 70% of the original price. What was the original price? \$10.71
43. ***DATA FILE 6 (pp. 230–231)*** Is it accurate to say that females spend about 50% more per week on clothes than males do? no
44. Identical sweaters are on sale in two different stores. In the first store, the sweater is on sale at 70% of the list price of \$25. In the second store, the sweater is on sale at 60% of the list price of \$30. Which sweater is the better buy? 70% of 25 = \$17.50

45. A tie and a wallet were originally the same price. The tie is now 75% of its original price; the wallet is now 70% of its original price. The wallet has a sale price $17.50. What was the original price of each item? What is the sale price of the tie? $25; $18.75

DATA **Use the table at the right.**

46. True or false? Write *T* or *F*.

   a. The number of households with VCRs increased more than 100% from 1978 to 1980. T

   b. The number of households with VCRs in 1978 is less than 1% of the number of households with VCRs in 1986. T

47. ***ESTIMATION*** Suppose the average cost of a VCR was $1,200 in 1978. Estimate the value of the sales of VCRs in 1978. $240,000,000

48. Use the data to make a bar graph. Check students' graphs.

49. The number of households with VCRs in 1981 was about 57% of the households with VCRs in 1982. How many households had VCRs in 1981? 1,442,100

50. Which of the following could you *not* conclude from the data? b.

   a. VCRs are very popular in the United States.

   b. More VCRs are produced in the United States than in any other country.

   c. The number of movies available on videotape probably increased from 1978 to 1986.

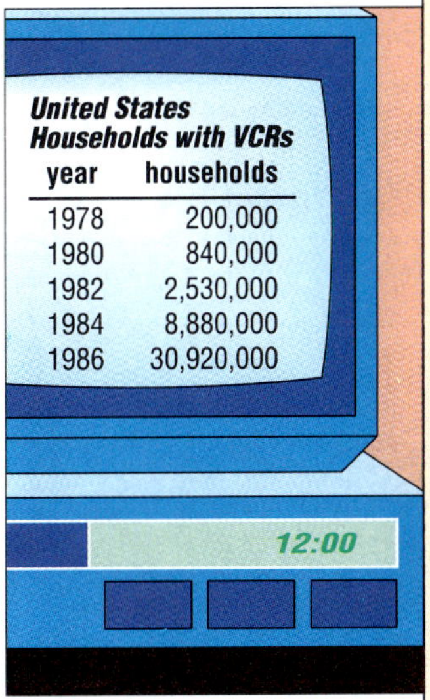

United States Households with VCRs

year	households
1978	200,000
1980	840,000
1982	2,530,000
1984	8,880,000
1986	30,920,000

Critical Thinking

EXPLORING VISUAL PERCEPTION

1. Study the unfolded pattern. Then decide which, if any, of the three-dimensional figures would result if you fold the pattern. a. and c.

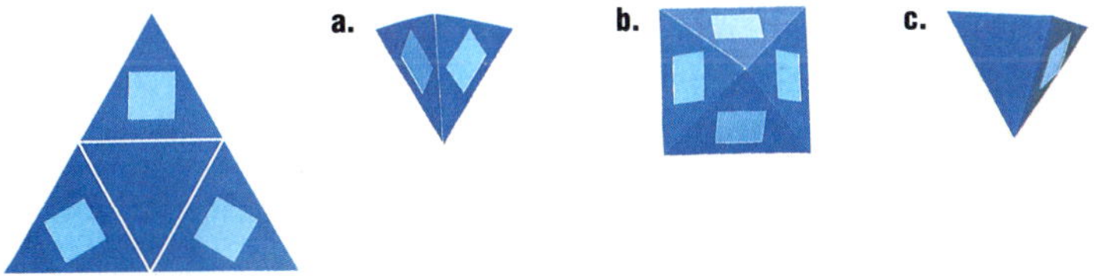

2. Use graph paper to make a pattern for a visual perception problem like the one above. Trade with a classmate and try to solve the problem. Check students' work.

Lesson Follow-up

Reteaching Activity If students have difficulty placing the parts of a percent problem in a diagram, have them label the sections of the diagram as *part*, *percent*, and *whole*.

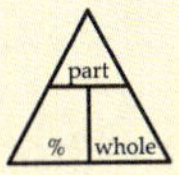

Then have students practice identifying and diagramming the parts of several problems.

LESSON RESOURCES

TRB Practice F-10
TRB Enrichment F-11

Practice

For use after 6-6 (pp. 256–259)

Percents and Equations

Write and solve an equation to find each. Use a triangle diagram if necessary.

1. What percent of 25 is 17? 68%
2. What percent of 40 is 26? 65%
3. What percent is 10 of 8? 125%
4. What percent of 32 is 28? 87.5%
5. What percent is 63 of 84? 75%
6. What percent is 84 of 60? 140%
7. What percent is 3 of 600? 0.5%
8. What percent is 22 of 4? 550%
9. Find 45% of 60. 27
10. Find 37.5% of 104. 39
11. Find 325% of 52. 169
12. Find 68% of 150. 102
13. Find $66\frac{2}{3}$% of 87. 58
14. Find 0.4% of 25. 0.1
15. Find 1% of 3,620. 36.2
16. Find 180% of 65. 117
17. $62\frac{1}{2}$% of x is 5. What is x? 8
18. 300% of k is 42. What is k? 14
19. $33\frac{1}{3}$% of p is 19. What is p? 57
20. 70% of c is 49. What is c? 70
21. 15% of n is 1,050. What is n? 7,000
22. 38% of y is 494. What is y? 1,300
23. $183\frac{1}{3}$% of m is 209. What is m? 114
24. 0.4% of h is 276. What is h? 69,000

Solve.

25. How much interest will be earned in one year on a deposit of $1,400 invested at $6\frac{1}{2}$% simple interest?
   $91
26. A camera regularly priced at $295 was placed on sale at $236. What percent of the regular price was the sale price?
   80%
27. 936 students, 65% of the entire student body, attended the football game. Find the size of the student body.
   1,440

F-10 Chapter 6

Enrichment

For use after 6-6 (pp. 256–259)

Estimating Percents

You can use compatible numbers to estimate percents mentally.

Example Of the 88 members of the school orchestra, 68.2% practice at least 1 hour a day. Mentally estimate the numbers of players who practice at least 1 hour.

Solution
1. Choose a simple fraction close to the given percent. — $\frac{2}{3} = 66\frac{2}{3}\%$
2. Choose an appropriate base close to the given base. — Ninety is close to 88 and is divisible by 3.
3. Solve mentally. — Think: $\frac{2}{3} \times 90 = 60$

About 60 members practice at least 1 hour per day.

1. To use this method, you must know some basic percent-fraction equivalents. Complete the table of equivalents.

Percent-Fraction Equivalents													
10%	$12\frac{1}{2}$%	20%	25%	30%	$33\frac{1}{3}$%	40%	50%	60%	70%	75%	80%	$83\frac{1}{3}$%	90%
$\frac{1}{10}$	$\frac{1}{8}$	$\frac{1}{5}$	$\frac{1}{4}$	$\frac{3}{10}$	$\frac{1}{3}$	$\frac{2}{5}$	$\frac{1}{2}$	$\frac{3}{5}$	$\frac{7}{10}$	$\frac{3}{4}$	$\frac{4}{5}$	$\frac{5}{6}$	$\frac{9}{10}$

Estimate each percent mentally. **Answers may vary.**

2. 41.7% of 26 10
3. 73.9% of 43.1 33
4. $82\frac{1}{2}$% of 37 30
5. 59% of 198 120
6. 34% of 61 20
7. 48.7% of 498.7 250
8. 91% of 68.7 63
9. 11.7% of 7,977 800
10. The sizes of the four sections of the school orchestra are given. Estimate how many of the 88 members are in each.
   string, 48.9% 44 woodwind, 17% 18
   brass, 23.9% 22 percussion, 10.2% 9
11. Of the 26 eighth graders in the orchestra, 15 play stringed instruments. Estimate the percent who play strings.
   60%
12. 588 people attended the orchestra's second concert, 73.2% of the number that attended the first concert. Estimate the number that attended the first concert.
   800

Chapter 6 F-11

LESSON QUIZ

Name the operation. Solve.
1. What percent is 20 of 25? (division; 80%)
2. What is 20% of 25? (multiplication; 5)
3. What percent of 35 is 7? (division; 20%)
4. 25 is 20% of what number? (division; 125)

Assignment Guide

Basic 1-5, 9-14, 21-26, 35-43, 46-47, MR, CT All
Average 1-5, 9-17, 21-29, 35-44, 46-48, MR, CT All
Enriched 1-8, 12-20, 25-34, 43-50, MR, CT All

FOR THE NEXT LESSON

calculator

Lesson Focus

Materials

Calculator

MOTIVATING THE LESSON

To begin the lesson, present students with the following problem:
A compact disc that once sold for \$16 is now selling for \$12. What was the percent of decrease in the price? As the problem is discussed, ask students what the amount of decrease was. (\$4) Then ask what ratio compares the amount of decrease to the original price. ($\frac{4}{16}$) Then ask for the percent of 16 that is 4. (25%) A similar example can be used to illustrate the percent of increase.

Instruction

1 The ratio that expresses *amount of change/original amount* should be recognized by students as the rate or percent of change.

ADDITIONAL EXAMPLES

Find the percent of change.
1. 48 increased to 60 (25%)
2. 48 decreased to 42 ($12\frac{1}{2}$%, or 12.5%)
3. 54 decreased to 18 ($66\frac{2}{3}$%)
4. 60 increased to 360 (500%)

OBJECTIVE:
To find percent of increase or decrease.

6-7 Percent of Change

▼ Over a ten-year period, the average tuition at a private four-year college increased from \$2,476 to \$7,693. You can write and solve an equation to find the *percent of change.*

1. Subtract to find the amount of change. $7{,}693 - 2{,}476 = 5{,}217$

2. Write an equation.
percent of change = $\frac{\text{amount of change}}{\text{original amount}}$ $n = \frac{5{,}217}{2{,}476}$

3. Solve the equation. Round to the nearest hundredth. $5{,}217 \div 2{,}476 = 2.107\ldots$ $n \approx 2.11$

4. Write as a percent. 211%

Tuition rose by about 211% over the ten-year period.

Percent of Change	Use the following equation to find percent of change. percent of change = $\frac{\text{amount of change}}{\text{original amount}}$

1

▼ You can use the same approach to find the percent of decrease.

Example 1 High school enrollment fell from 13.3 million students to 11.4 million in 20 y. Find the percent of change.

Solution

$13.3 - 11.4 = 1.9$	Find the amount of change.
$d = \frac{1.9}{13.3}$	Write an equation.
$1.9 \div 13.3 = 0.1428\ldots$	Solve the equation.
$d \approx 0.14$	Round to the nearest hundredth.
14%	Write as a percent.

High school enrollment fell about 14%.

▼ Always use the same units when working with measures.

FLASHBACK
16 oz = 1 lb

Example 2 An animal weighed 5 lb at birth. A week later it weighed 5 lb 6 oz. What was the percent of change?

Solution

$86 - 80 = 6$	Write measures in the same units. Find the amount of change.
$c = \frac{6}{80}$	Write an equation.
$6 \div 80 = 0.075$	Solve the equation.
$c = 0.075$	
7.5%	Write as a percent.

The animal had a 7.5% gain in weight.

Teaching Tip

To help students appreciate the usefulness of finding the percent of change, display newspaper clippings advertising percent-off sales. Write an appropriate equation to demonstrate the sale price of an item listed in the advertisement.

▼ The percent of change may be part of a multi-step problem.

Example 3 An investment grew from \$100 to \$110 in the past year because of interest earned. The investment will earn interest at the same rate for the next year. What will be the value of the investment at the end of next year?

Solution
1. Find the percent of increase: $\frac{10}{100} = 10\%$.
2. Find the interest for the next year: 10% of \$110 = \$11.
3. Find the value of the investment: \$110 + \$11 = \$121.

At the end of next year, the investment will be worth \$121.

THINK Why is the denominator always the original amount in percent of change problems? The original amount is what you are comparing to.

CLASS EXERCISES

Find the percent of increase.

1. 30 is increased to 39. 30%
2. 45 is increased to 144. 220%

Find each percent of decrease.

3. 60 is decreased to 48. 20%
4. 96 is decreased to 78. 18.75%

Find each percent of change.

5. 72 to 99 37.5%
6. 64 to 40 37.5%
7. 144 to 300 $108\frac{1}{3}\%$
8. 400 to 120 70%

Find the percent of change for the first set of numbers. Use the same percent of change to complete the second set of numbers.

9. 50 to 60 : : 60 to ▒. 72
10. 250 to 200 : : 200 to ▒. 160

THINK AND DISCUSS See below.

1. 100 is increased by 10%. Then the result is decreased by 10%. Is the final result 100? Explain.
2. 100 is decreased by 10%. Then the result is increased by 10%. Is the final result 100? Explain.
3. Are the values of the answers to Questions 1 and 2 the same? Explain.

1. No, because the original amount changes. The original amount is first 100, then 110.
2. No; the original amount is first 100, then 90.
3. Yes; both are 99.

WRITTEN EXERCISES

Find each percent of increase. Round to the nearest tenth.

1. 50 is increased to 66. 32%
2. 80 is increased to 95. 18.8%
3. 32 is increased to 76. 137.5%
4. 45 is increased to 105. 133.3%

Find each percent of decrease. Round to the nearest tenth.

5. 90 is decreased to 75. 16.7%
6. 64 is decreased to 24. 62.5%
7. 120 is decreased to 95. 20.8%
8. 280 is decreased to 126. 55%

Find each percent of change. Round to the nearest tenth.

9. 38 to 95 150%
10. 111 to 74 33.3%
11. 27 to 72 166.7%
12. 180 to 54 70%

Guided Practice

Class Exercises Students can use estimation to check the reasonableness of their answers for questions 1-8.

Think and Discuss For questions 1 and 2, to help students understand that they are finding the percent of change for two different numbers, have them write an equation for each part of the question. Ask students what numbers are used for the original amount in each equation. (100, 110)

A ***common error*** occurs when students incorrectly write a ratio comparing the amount of change to the resulting amount. In example 3, students may write the incorrect ratio $\frac{10}{110}$. Suggest that students write the formula: percent of change = amount of change/original number, and then substitute the appropriate numbers.

Closure

Writing in Math Have students write the formula for percent of change in their own words. Then have them write two examples, one to find the percent of increase and one to find the percent of decrease.

MATH MINUTES

Present the following problem to students: **If a rectangle has a length of 12 in. and a width of 8 in., what is the percent of increase if the area of the rectangle is increased to 120 in.^{2}?** (25%)

Independent Practice

For question 20, it may be helpful for students to write the ratio "difference/1980 price" to remind themselves that the lower price per item does not necessarily correspond to the original amount.

As students use the calculator in questions 21-26, remind them to find the amount of change first and then divide the difference by the original amount.

20. calc. 69.23% D
cam. 30% D
radio 65% D
game 241.38% I
sneakers 79.31% I
candy 100% I
movie 125% I
tape 11.25% I
watch 25% I
pen 34.48% I

Item	*Price ($)* 1980	1990
calculator	13	4
110 camera	40	28
radio with headphones	100	35
video game system	29	99
sneakers	29	52
candy bar	0.25	0.50
movie ticket	2	4.50
cassette tape	8.98	9.99
basic watch	16	20
ballpoint pen	0.29	0.39

MIXED REVIEW

Evaluate.

1. $|-3| + |x|$ for $x = -5$ 8
2. $[(3 + 12)(8 \div 2)]^2$ 3,600

Compare. Use >, <, or =.

3. 42 : 54 < 56 : 64.
4. $\frac{3}{11}$ > $27\frac{1}{4}\%$
5. Find 45% of 45. 20.25
6. There are about 20,000 human-made objects in orbit around Earth. All but 300 are junk. What percent are junk? 98.5%

MENTAL MATH **Find each percent of change. Label your answer as increase or decrease.**

13. 25 to 30 20% increase **14.** 40 to 45 12.5% increase **15.** 50 to 45 10% decrease **16.** 40 to 30 25% decrease

Find the percent of change from the first number to the second. Use the same rate of change to find the next number in the pattern. Round each answer to the same number of decimal places.

17. 38, 57, ■ 86 **18.** 70.6, 105.9, ■ 158.9 **19.** 103.6, 77.7, ■ 58.3

DATA **Use the table at the left.**

20. a. Find the percent of change for each item. Round each answer to the nearest hundredth. Label your answer as increase or decrease.

b. ***WRITE*** Which items increased in price? Which items decreased in price? Why do you think some items increased in price and some decreased? Check students' work.

CALCULATOR **Find each percent of change. Label your answer as increase or decrease. Round to the nearest tenth.**

21. 87 to 108 24.1% I **22.** 59 to 127 115.3% I **23.** 77 to 13 83.1% D

24. 132.8 to 93.3 29.7% D **25.** 131.75 to 40.45 69.3% D **26.** 18 to 47.69 164.9% I

Solve. Round each answer to the nearest tenth.

27. A stock increased in value from $130 to $166. What was the percent of change? 27.7%

28. The value of artwork appreciated from $295 to $495. What was the percent of change? 67.8%

29. Investments decreased in value from $1,750 to $1,232. What was the percent of change? 29.6%

30. A stock traded at the following prices each day for a week: Monday, $846; Tuesday, $819; Wednesday, $838; Thursday, $864; Friday, $850. Tues. -3.2%; Wed. +2.3%; Thurs. +3.1%; Fri. -1.6%.

a. Find the percent of change for each day. See above.

b. On which day was the percent of change the greatest? Tues.

c. On which day was the percent of change the least? Fri.

d. What was the percent of change from the value on Monday to the value on Friday? 0.5%

31. A worker received a raise from $22,000 to $25,000. What is the percent of change? +13.6%

32. A family bought a house for $78,000 in 1985. In 1991, the family sold the house for $88,900. Find the percent of change. 14%

33. A gallon of gas cost $.32 in 1972. In 1991, a gallon of gas cost $1.58. What is the percent of change? +393.8%

34. The population of Growtown increased from 10,000 to 13,000 in one year, while the population of Slowtown decreased from 30,000 to 24,000 in one year. Growtown +30%; Slowtown -20%

a. Find the respective rates of increase or decrease.

b. If both towns maintain the same rate of change, when will the population of Growtown exceed that of Slowtown? in two years

35. ***DATA FILE 3 (pp. 96–97)*** Find the percent of change for the value of each proof set from 1965 to 1968. 1936: 1,175%; 1937: 2,087.5%; 1940: 2,300%; 1955: 445.5%; 1961: 566.7%

36. ***DATA*** Use the table on page 259. Find the percent of change for each time period.

a. from 1978 to 1980 320%

b. from 1980 to 1982 201.2%

c. from 1982 to 1984 251%

d. from 1984 to 1986 248.2%

37. ***DATA*** Use the table at the right to solve.

a. Find the percent of change for each occupation from 1986 to 2000. Round to the nearest whole percent.

b. Is the percent of change the only factor that you would consider when making a career decision? no

c. ***WRITE*** What impact might these projections have on a person's career planning? Check students' journals.

38. ***PROJECT*** Research the expected employment demands in the profession of your choice. Find out what training and education is necessary to qualify for this position.

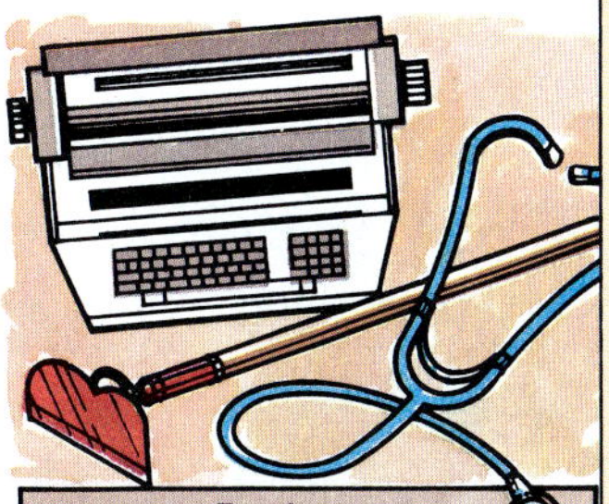

Employment (in thousands)

Career	1986	2000
Nurse	1,406	2,018
Cashier	2,165	2,740
Medical asst.	132	251
Farmer	1,182	850
Typist	178	128

37.
Nurse	+44%
cashier	+27%
med. asst.	+90%
farmer	-28%
typist	-28%

TEST YOURSELF

Write and solve an equation.

1. Find 33% of 120. 39.6
2. Find 125% of 42. 52.5
3. What percent of 50 is 10? 20%
4. 15% of q is 9.75. What is q? 65
5. What percent of 12 is 8? $66\frac{2}{3}\%$
6. 80% of w is 120. What is w? 150

Solve. Round to the nearest tenth of a percent.

7. A pair of sneakers was reduced from $125 to $85. What was the percent of change? 32%
8. Tickets to regular season games are $36. Tickets to playoff games are $50. What is the percent of change? 38.9%

Lesson Follow-up

Reteaching Activity For students who have difficulty finding percent increase and decrease, refer to the calculator activity on p. 230E of the Chapter Overview. See also TRB Exploring O-23.

LESSON RESOURCES

TRB Practice F-12
TRB Enrichment F-13

Practice

For use after 6-7 (pp. 260–263)

Percent of Change

Find each percent of increase.

1. 40 is increased to 45. 12.5%
2. 33 is increased to 55. $66\frac{2}{3}\%$
3. 15 is increased to 34. $126\frac{2}{3}\%$
4. 11 is increased to 88. 700%
5. 72 is increased to 117. $62\frac{1}{2}\%$
6. 28 is increased to 49. 75%
7. 35 is increased to 49. 40%
8. 48 is increased to 132. 175%

Find each percent of decrease.

9. 60 is decreased to 15. 75%
10. 56 is decreased to 35. $37\frac{1}{2}\%$
11. 140 is decreased to 77. 45%
12. 96 is decreased to 64. $33\frac{1}{3}\%$
13. 99 is decreased to 69. $30.\overline{30}\%$
14. 50 is decreased to 44. 12%
15. 83 is decreased to 0. 100%
16. 475 is decreased to 152. 68%

Find each percent of change. Label your answer as increase (I) or decrease (D).

17. 24 to 21 $12\frac{1}{2}\%$; D
18. 64 to 80 25%; I
19. 100 to 113 13%; I
20. 50 to 41 18%; D
21. 63 to 105 $66\frac{2}{3}\%$; I
22. 42 to 168 300%; I
23. 80 to 24 70%; D
24. 200 to 158 21%; D

MENTAL MATH **Find each percent of change. Label your answer as increase (I) or decrease (D).**

25. 20 to 24 20%; I
26. 44 to 22 50%; D
27. 16 to 12 25%; D
28. 10 to 100 900%; I
29. 20 to 40 100%; I
30. 10 to 50 400%; I
31. 12 to 16 $33\frac{1}{3}\%$; I
32. 80 to 100 25%; I

CALCULATOR **Find each percent of change. Round your answer to the nearest hundredth. Label your answer as increase (I) or decrease (D).**

33. 56 to 71 26.79%; I
34. 127 to 84 33.86%; D
35. 63.7 to 42.9 32.65%; D
36. 119 to 306.4 157.48%; I
37. 69 to 117 69.57%; I
38. 19 to 9 52.63%; D
39. 95 to 145 52.63%; I
40. 88 to 26 70.45%; D

F-12 Chapter 6

Enrichment

For use after 6-7 (pp. 260–263)

Consumer Price Index and Inflation

Each month, the United States Bureau of Labor Statistics publishes cost-of-living figures called the Consumer Price Index (CPI). The CPI measures how the prices of goods and services change over time. Current prices are given relative to 1967 prices, which are arbitrarily set at 100.

Example The CPI for dairy products in 1986 was 360.9. Find the 1986 cost of a quart of milk which cost $.23 in 1967.

Solution $\text{Cost} = \frac{\text{CPI}}{100} \times 1967 \text{ price}$

$= \frac{360.9}{100} \times 0.23$

≈ 0.83

The cost of a quart of milk in 1986 was about $0.83.

Find the cost in the given year to the nearest cent. The 1967 price and the year's CPI are given in parentheses.

1. 1986, hospital room ($86; 763.0) $656.18
2. 1983, loaf of bread ($0.44; 292.5) $1.29
3. 1987, apartment rent ($160; 368.2) $589.12
4. 1980, winter coat ($56; 155.1) $86.86

Inflation measures annual increase in price. You can find the rate of inflation by calculating the percent of increase in the CPI.

Example The CPI for transportation rose from 249.7 in 1980 to 304.2 in 1986. Find the rate of inflation for the period.

Solution $304.2 - 249.7 = 54.5$
$54.5 \div 249.7 = 0.218\ldots$

The rate of inflation was about 21.8%.

Find the inflation rate to the nearest tenth. Beginning and ending CPIs are in parentheses.

5. food, 1970–1980 (114.9; 254.6) $121.6%
6. medical care, 1986–1987 (443.9; 459.1) 3.4%
7. entertainment, 1982–1985 (235.8; 265) 12.4%
8. gasoline, 1982–1983 (389.3; 376.3) -3.3%
9. all items, 1967–1987 (100; 339.3) 239.3%

Chapter 6 F-13

LESSON QUIZ

Find the percent of change.

1. 45 to 54 (20%)
2. 40 to 25 ($37\frac{1}{2}\%$)
3. 20 to 50 (150%)
4. 50 to 10 (80%)
5. Find the percent of change for the first pair of numbers. Use the same rate of change to complete the second pair of numbers. 48 to 60, 60 to ■ (75)

Assignment Guide

Basic 1-10, 13-14, 17-18, 20, 21-23, 27-29, 37, MR, TY All

Average 1-16 even, 17-20, 22-24, 27-35 odd, 37, MR, TY All

Enriched 1-19 odd, 20, 25-26, 28-38 even, MR, TY All

FOR THE NEXT LESSON

magazines or science books, math journal

Lesson Focus

Materials

Magazines or science books, math journal

MOTIVATING THE LESSON

Begin the lesson by reviewing and discussing the problem solving strategies learned so far: *Look for a pattern, guess and test, simplify the problem, account for all possibilities,* and *working backwards.* Then introduce the idea of using a diagram to solve a problem and mention that they will use this strategy in this lesson.

Instruction

1 Reproduce the diagram from the lesson on the chalkboard.

Explain that the diagram will help students understand all the mathematical relationships. Ask a student for the distance from Boise to Seattle. (700 mi) Ask students what two facts are unknown about the trip from Salt Lake City to Seattle. (the distance in miles from Salt Lake City to Boise and the percent that the bus trip represents)

Ask students to discuss why the diagram is shown with intervals of 10%. (We know that the total distance is 100% of the trip, and the diagram needs to represent the whole trip. It makes sense to show 10% intervals because 10% is an easy number to work with.)

2 Have the students tell why the divisor 7 is used to find how many miles each interval represents. (7 is the number of intervals between the point for Boise and the point for Seattle)

3 Ask a volunteer to show that the proportion is true by finding the cross products.

PROBLEM SOLVING STRATEGY

OBJECTIVE: *To solve problems by drawing a diagram.*

6-8 Draw a Diagram

You can solve many types of problems using diagrams.

PROBLEM

A sales representative travels by train from Salt Lake City, Utah, to Boise, Idaho. This distance is about 30% of her entire trip. She then flies 400 mi to Spokane, Washington. She completes the remaining 300 mi of her trip to Seattle, Washington, by bus. What percent of her trip did she cover by bus?

SOLUTION

Answer these questions to understand the given information.

READ What facts do you have?
- Salt Lake City to Boise → 30% by train
- Boise to Spokane → 400 mi by air
- Spokane to Seattle → 300 mi by bus

What do you want to find? What percent of the trip (the part covered by bus) does 300 mi represent?

PLAN Make a diagram to organize the information.

1

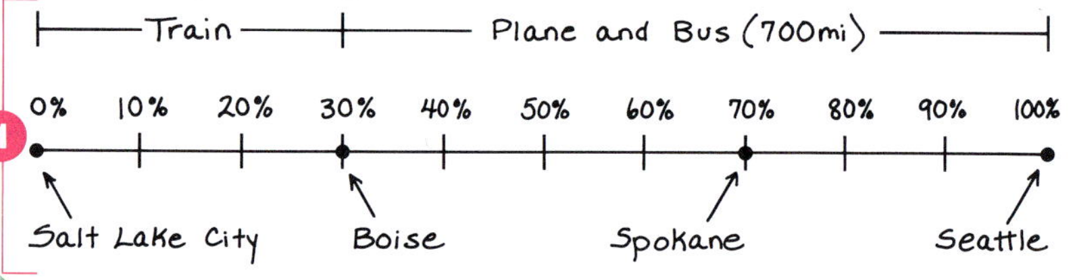

SOLVE

How far did she travel by air and by bus?	700 mi
What percent represents the total she traveled by air and bus?	70%
2 The distance between two marks on your diagram represents what percent of the total distance?	10%
How many miles does the distance between two marks represent?	100 mi (700 ÷ 7)
How many miles did she travel in all?	1,000 mi (100 · 10)
What percent of the trip did she cover by bus?	$\frac{300}{1,000} = 30\%$

LOOK BACK Does your answer make sense? Look at the problem another way to check.

Think of a proportion.

70% : 700 as 30% : 300. So, 300 mi by bus is 30%.

3

Teaching Tip

Have available almanacs, books of world records, atlases, and other sources of interesting data for students to use to generate problems.

CLASS EXERCISES

1. Refer to the problem in the example on page 264.
   a. What percent of the trip was covered by plane? 40%
   b. How many miles were covered by train? 300 mi
2. A student has 80 miles to walk. She walks 50% of the distance the first day and 25% of the remaining distance the second day. How far does she have to walk to finish her trip? 30 mi
3. A student put points P and Q on a line to the right of point X. Three times the distance from point X to point P is four times the distance from point X to point Q. What is the ratio of $PQ : XQ$? 1 : 3
4. There are 25 students in an algebra class. Ten students are members of the math club. Twelve students are members of the chess club. Five students are members of both clubs. How many of the students in the algebra class are members of neither club? 8 students

WRITTEN EXERCISES

Use a CALCULATOR where appropriate.

Solve using a diagram.

1. Container A has twice the capacity of container B. Container A is full of sand and container B is empty. Suppose $\frac{1}{8}$ of the sand in container A is poured into container B. What fractional part of container B will contain sand? $\frac{1}{4}$
2. Six students are in the finals of the chess tournament. Each student plays one game against every other student. How many games will be played? 15 games
3. Points P, Q, R, and S appear in that order on a line. The ratio of $PQ : QR$ is $3 : 4$, and the ratio of $QR : RS$ is $2 : 5$. The length of $PQ = 6$ in. Find the length of PS. 34 in.
4. A wooden cube that measures 5 cm along each edge is painted blue. The painted cube is then cut into centimeter cubes. How many of the small cubes are painted on three faces? 8 cubes
5. A homeowner is building a fence around a rectangular lot that measures 48 ft by 60 ft. He digs a post hole at each corner and one every 4 ft in between corners. How many post holes must he dig? 54 post holes

MATH MINUTES

Have students write a problem such as: In a photo, a giraffe's neck is three times as long as its head. The length of its legs and body together is equal to the length of its neck. The giraffe's head measures 1 cm. How tall is the giraffe?

Guided Practice

Class Exercises For question 3 suggest that students pick a distance such as 3 units (intervals) to represent the distance from X to Q. They need to understand, however, that when choosing the distance for XQ that $4(XQ) = 3(XP)$.

A ***common error*** is to draw a diagram that is not proportional to the information in the problem. For example, students might quickly read question 3 of the Class Exercises and draw the figure as:

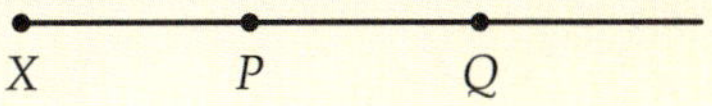

Suggest that students read problems more than once, simplify any wording that may seem complicated, and make any necessary revisions to expressions and diagrams.

Closure

Ask students for what types of problems the strategy, *draw a diagram,* might be especially useful. (Answers will vary; those containing geometric measurement, distance relationships, whole to part comparisons, and so forth.) Ask students to record their answers in their math journals.

Independent Practice

Ask students what other strategy they could use to solve problem 2. (Answers may vary; account for all possibilities.)

For question 4, some students may be able to visualize the cube, but others may need help drawing an isometric model. Ask how many cubes are in each row. (5)

You may wish to have students work in pairs to find the answers to questions 6-15. Ask students to record the strategy they used to solve each problem. Encourage the class to develop their own strategies as they work through the exercises.

Lesson Follow-up

LESSON RESOURCES

Practice F-14
Enrichment F-15

Practice

For use after 6-8 (pp. 264–266)

Draw a Diagram

Solve using a diagram.

1. A hot-air balloon drifted 13 mi south, 8 mi east, 17 mi north, 15 mi west, and 4 mi south. How far was the balloon from its starting point?
   7 mi
2. Four students are standing in line. Art is behind Doug. Brenda is ahead of Connie. Doug is behind Connie. Find the order of the four.
   Brenda, Connie, Doug, Art
3. Tom weighs 20% more than Fred, 10% less than Pete, and 20% less than Mike. Fred weighs 120 lb. Find the weights of Tom, Pete, and Mike.
   Tom: 144 lb; Pete: 160 lb; Mike: 180 lb
4. The diagram shows that 2 diagonals can be drawn from a given vertex (corner) in a 5-sided figure.

   a. How many diagonals can be drawn from a given vertex of a 6-sided figure; a 7-sided figure?
   3; 4
   b. Predict the number of diagonals that can be drawn from a given vertex of a 147-sided figure.
   144
5. Two cars started at the same point and traveled in opposite directions. One went 50 mi/h, the other 60 mi/h. After 1 h, the faster car turned around and started chasing the slower. Two hours later, how far apart were the cars and which one was in the lead?
   90 mi; the slower car
6. The Hawaiian volcano Mauna Kea is 33,476 ft high, with 19,680 ft of its height below the ocean. The top of a lighthouse standing at sea level is 13,711 ft below the top of Mauna Kea. A shark is swimming 19,277 ft above the base of the volcano. Find the difference in elevation from the shark to the top of the lighthouse.
   488 ft

F-14 *Chapter 6*

Enrichment

For use after 6-8 (pp. 264–266)

Where's the Fire?

A taxicab picked up a passenger at City Hall at 6 P.M. and drove to the airport. The cab arrived at 6:45, waited 15 min, then returned to City Hall at the same rate.

The first two parts of the trip are drawn on the graph. Finish the graph of the cab trip and of these trips:

1. A jogger left his house at the 18-mile marker on Airport Road at 5:45 P.M. and jogged to the 8-mile marker, arriving at 7:15 P.M.
2. A fire engine left the 2-mile marker at 6:45 P.M. and sped to a fire at the 14-mile marker in 15 min. It remained for an hour, then returned to the station.
3. A dog left the 2-mile marker at 5:45 P.M. and trotted to the fire. It arrived just as the fire engine was leaving.

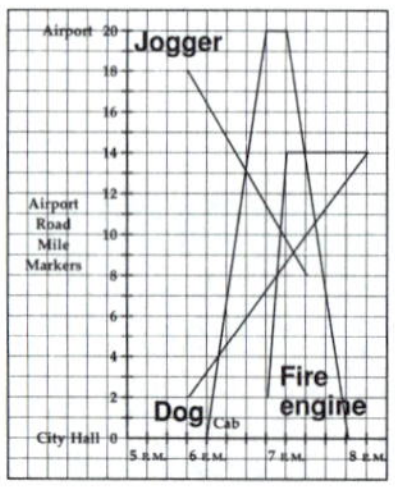

Use the graph to solve.

4. At what time did the jogger pass the cab? **6:30 p.m.**
5. Where were the jogger and the cab when they passed? **13-mi marker**
6. Where did the jogger meet the fire engine? **near the 10-mi marker**
7. Where did the dog meet the jogger?
   9-mi marker
8. Where did the dog meet the fire engine the first time?
   8-mi marker
9. Where did the dog meet the cab on the way to the airport?
   4-mi marker
10. List the order in which the following passed the $8\frac{1}{2}$-mile marker: cab, dog, fire engine, jogger
    cab, fire engine, dog, jogger

Chapter 6 F-15

Solve. Use any strategy you wish.

6. After George poured 48 gal of water into an empty tank, the tank was 75% full. How many gallons does the tank hold? **64 gal**
7. Maureen cut a 20-cm wire into exactly three pieces. The first piece is 3 cm shorter than the second piece. The third piece is 4 cm shorter than the second piece. Find the length of the shortest piece. **5 cm**
8. At a local high school, 60% of the students are girls. Of these girls, 75% own a cassette player. What percent of all the students are girls who do *not* own cassette players? **15%**

9. Twelve of Ms. Brown's students tried out for both the gymnastics and the baseball teams. Half of the students made the baseball team. One-third made the gymnastics team. One-fourth made neither team. How many students made both teams? **1**
10. Belinda takes the bus $\frac{2}{5}$ of the distance from home to school. She then walks 2 blocks to her friend's house and rides with her friend the remaining 7 blocks to school. What fractional part of the trip did Belinda ride with her friend? $\frac{7}{15}$
11. A student has 5 mi to walk. The student walks half the distance the first hour and a fourth of the remaining distance the next hour. How far does the student have to walk to finish the trip? $1\frac{7}{8}$ **mi**
12. On Saturday, a student pays off a loan of $12.50. He then earns $20 mowing a lawn. At the mall, he buys two pairs of socks for $3.98 each and a pair of shoes for $45.79. He has $24.32 left. How much money did he have to start with? **$70.57**
13. A student was standing in the middle of a line. Twenty-three students were ahead of him. How many students were in the line? **47 students**
14. A board is cut in half. Then each piece is cut in half again. Then each of these pieces is cut in half.
    a. How many cuts are made? **7 cuts**
    b. How many pieces are there? **8 pieces**
15. A chain fence encloses a square pen. The 8 posts on each side of the pen are spaced 6 ft apart and are 6 in. in diameter. There is a post at each corner of the pen.
    a. How many posts are there? **28**
    b. What are the dimensions of the pen? **46 ft × 46 ft**
16. ***DATA FILE 9 (pp. 360–361)*** Suppose you are teeing off on the 8th hole at St. Andrew's golf club. Would you be more likely to use a 1 wood or a 3 wood? **3**

Assignment Guide

Basic All
Average All
Enriched All

FOR THE NEXT LESSON

algebra tiles or blue and white squares of paper and red paper rectangles, overhead projector, math journal

Problem Solving Practice

PROBLEM SOLVING STRATEGIES

Look for a Pattern
Guess and Test
Simplify the Problem
Account for All Possibilities
Make an Organized List
Work Backwards
Make a Table
Write an Equation
Solve by Graphing
Draw a Diagram
Make a Model
Simulate the Problem

Solve. The list at the left shows some strategies you may use.

1. What is the maximum possible number of digits in the product of two positive integers each having two digits? 4
2. To make an orange dye, 3 parts of red dye are mixed with 2 parts of yellow dye. To make a purple dye, 2 parts of blue dye are mixed with 1 part of red dye. Suppose equal amounts of orange and purple are mixed. What fractional part of the mixture is red dye? $\frac{7}{15}$
3. When n is divided by 7, the remainder is 4. What is the remainder when $2n$ is divided by 7? 1
4. The ratio of Clark's weight to Kim's weight is 3 : 2. The ratio of Kim's weight to Janine's weight is 1 : 2. Compare Clark's weight with Janine's weight. 3 : 4
5. Working together, two painters earned $1,000 for painting a new house. The first painter worked for seven days. The second painter worked for three days. What is the first painter's fair share of the $1,000? $700
6. Refer to the two circle graphs at the left.
   a. What percent of Earth's water supply is fresh water? 2.8%
   b. What percent of Earth's fresh water is found in rivers, lakes, and streams? 0.03%
   c. What percent of Earth's total water supply is found in rivers, lakes, and streams? How did you find the answer? 0.00084%
   d. What is the ratio of salt water to fresh water? the ratio of salt water to the water from icecaps and glaciers? $\frac{97.2}{2.8}$, $\frac{97.2}{2.2}$
7. ***DATA FILE 4 (pp. 138–139)***
   a. In 1989, the world population was about 5 billion. Approximately what percent of the population spoke Japanese? 2.3%
   b. Is this more than the percent of the population that spoke both English and Spanish? no
8. ***DATA FILE 10 (pp. 404–405)***
   a. What is the total area (in mi^2) of the earth covered by oceans? Of the total area, approximately what percent is the Pacific Ocean? the Atlantic Ocean? 129,096,200 mi^2; 49.4%; 24.6%
   b. The total surface area of Earth is 1.96951×10^8 mi^2. Approximately what percent of Earth's surface area do the oceans cover? $\approx$ 66%

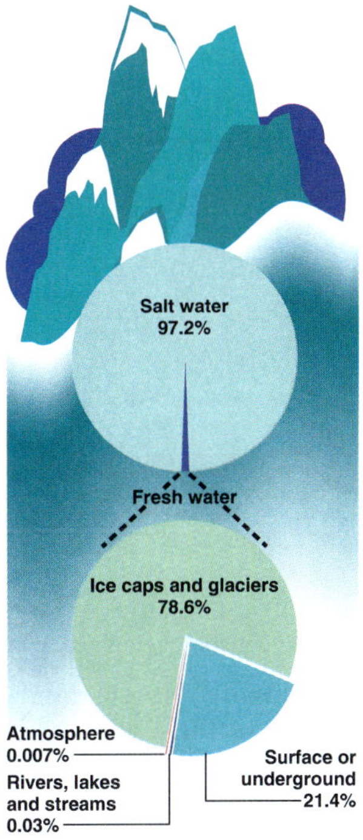

Instruction

This page provides a variety of problems that can be used to reinforce and enhance the students' problem solving skills. Encourage students to read each problem carefully. Then have them refer to the list of problem solving strategies to help them decide how to solve the problem.

Point out, however, that not all questions require a strategy for solving, nor are all the strategies in the list used in this lesson.

Study Skills

Students may find that *cooperative studying* is an effective and motivating way to review material. Groups of two or three students can either meet before class to prepare for the lesson, or after class to review the lesson and revise notes. Using class notes, outlines, or texts, students can discuss what they've learned, share insights and ideas, or pretest themselves and check their answers.

Each group member can be assigned a specific role, such as Researcher, Answer Checker, Note Taker, or Explainer. The group's tasks are then structured so that each student can make a valuable contribution.

Here are some ideas for promoting cooperative study:

- ▼ Teach cooperative skills in the classroom.
- ▼ Provide study guides before tests to help groups focus their study tasks.
- ▼ Provide time for groups to meet.
- ▼ Arrange for groups to use the classroom or school library, if possible.

Offer students the following suggestions:

- ▼ Make a list of rules that will help the group interact successfully.
- ▼ Accept responsibility for helping everyone in the group succeed.
- ▼ Discuss with parents how the students will get to the group's meeting place.

Chapter 6 Review

Write true or false. If false, change the underlined word to make the statement true.

1. A proportion is a comparison of two quantities by division. false; ratio
2. If two ratios are equal, their cross products are equal. true
3. $\frac{\text{miles}}{\text{hour}}$ is an example of a ratio. true
4. A rate is a ratio that compares a number to 100. false; percent
5. The percent of change is the amount of change divided by the original amount. true

Ratios, Proportions, and Rates — 6-1

To compare two quantities, write a ratio. To determine if two ratios form a proportion, multiply the cross products. If the cross products are equal, the ratios form a proportion.

To compare quantities measured in different units, use a rate.

Compare. Write = or ≠. Then tell whether or not the ratios form a proportion.

6. $\frac{3}{8}$ ■ $\frac{9}{24}$ =; proportion
7. $\frac{5}{7}$ ■ $\frac{20}{35}$ ≠
8. $\frac{3}{4}$ ■ $\frac{15}{20}$ =; proportion
9. $\frac{2}{3}$ ■ $\frac{18}{17}$ ≠
10. $\frac{5}{6}$ ■ $\frac{100}{130}$ ≠

Write each ratio as a unit rate.

11. 150 mi in 3 h 50 mi/h
12. 115 mi on 5 gal 23 mi/gal
13. 270 words in 3 min 90 words/min
14. $9.45 for 5 lb $1.89/lb

Solving Proportions — 6-2

To solve a proportion, write the cross products, then solve.

Solve.

15. $\frac{5}{6} = \frac{n}{42}$ $n = 35$
16. $\frac{3}{2} = \frac{18}{x}$ $x = 12$
17. $\frac{15}{a} = \frac{30}{98}$ $a = 49$
18. $\frac{y}{9} = \frac{30}{90}$ $y = 3$
19. $\frac{21}{25} = \frac{m}{150}$ $m = 126$

Scale Drawing — 6-3

To use a scale drawing, measure the distance on the drawing. Then use the scale given to write and solve a proportion.

Use the scale 1 cm : 75 km.

20. The distance on the map from Centerville to Toptown is 2.5 cm. What is the actual distance? 187.5 km
21. The actual distance from Summit to Crown Heights is 37.5 km. What is the map distance? 0.5 cm

Writing Percents 6-4

To write decimals as percents, move the decimal point two places to the right and write a percent sign. To write a fraction as a percent, first write the fraction as a decimal. Then write the decimal as a percent.

Write each decimal as a percent.

22. 0.05 5% **23.** 0.98 98% **24.** 1.45 145%

Write each fraction as a percent.

25. $\frac{3}{4}$ 75% **26.** $\frac{5}{8}$ 62.5% **27.** $\frac{3}{25}$ 12%

Using Proportions to Find Percent 6-5

To find the ratio of a number to 100, use the following formula.

$$\frac{\text{part}}{\text{whole}} = \frac{n}{100}$$

Write and solve a proportion.

28. Find $12\frac{1}{2}\%$ of 48. 6

29. 20% of x is 30. What is x? $x = 150$

30. What percent is 90 of 270? $33\frac{1}{3}\%$

Using Equations to Find Percent 6-6

Use a triangle diagram to relate the parts of a percent problem.

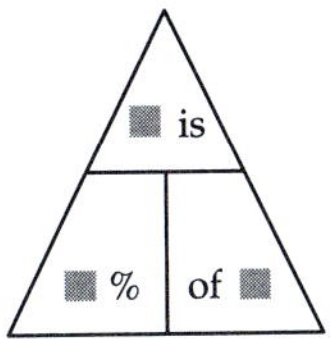

Write and solve an equation. Round to the nearest tenth.

31. 35% of a is 70. What is a? $a = 200$

32. Find 68% of 300. 204

33. What percent is 9 of 180? 5%

34. What percent of 55 is 10? 18.2%

35. Find 3% of 89. 2.7

36. 125% of y is 100. What is y? $y = 80$

Finding Percent of Change 6-7

$$\text{percent of change} = \frac{\text{amount of change}}{\text{original amount}}$$

Find the percent of change.

37. 18 to 24 $33\frac{1}{3}\%$ **38.** 120 to 90 25% **39.** 148 to 37 75% **40.** 285 to 342 20% **41.** 1,000 to 250 75%

Problem Solving 6-8

To solve a problem, draw a diagram to represent the information.

Solve.

42. Alicia rides 25% of a 100-mi trip the first day. She rides $33\frac{1}{3}\%$ of the remaining distance the second day. What percent of the original distance does she have to ride to finish her trip? 50%

43. ***DATA*** Refer to the circle graphs on page 267. What percent of Earth's fresh water supply is on the surface or underground? 21.4%

RESOURCES

TRB F-17–F-23

Scoring Chart	
No. Correct	**Score**
32	100
31	96
30	93
29	90
28	87
27	84
26	81
25	78
24	75
23	71
22	68
21	65
20	62
19	59
18	56
17	53
16	50
15	47
14	43
13	40
12	37
11	34
10	31
9	27
8	24
7	21
6	18
5	15
4	12
3	9
2	6
1	3

Chapter 6 Test

Write each ratio as a unit rate.

1. A car travels 84 mi on 3 gal of gas. **28 mi/gal**
2. A car travels 220 mi in 4 h. **55 mi/h**

Compare. Write = or ≠.

3. $\frac{7}{8}$ ■ $\frac{42}{40}$ ≠
4. $\frac{3}{5}$ ■ $\frac{45}{75}$ =
5. $\frac{12}{18}$ ■ $\frac{18}{12}$ ≠
6. $\frac{5}{9}$ ■ $\frac{25}{81}$ ≠

Write a proportion to describe each situation. Then solve.

7. Three cans of dog food sell for 99¢. Find the cost of 15 cans. **$4.95**
8. A photo that measures 5 in. by 7 in. is enlarged to 7.5 in. by b. **b = 10.5 in.**
9. A student reads 45 pages in 2 h; x pages in 3 h. **$67\frac{1}{2}$ pages**

The length of the kitchen in the diagram at the right is $1\frac{1}{4}$ in. The actual length is 20 ft. Use the diagram to solve.

10. What is the scale of the diagram? **1 in. : 16 ft**
11. What is the actual size of the dining area? **12 ft by 12 ft**
12. How wide is the kitchen at its narrowest part? **4 ft**

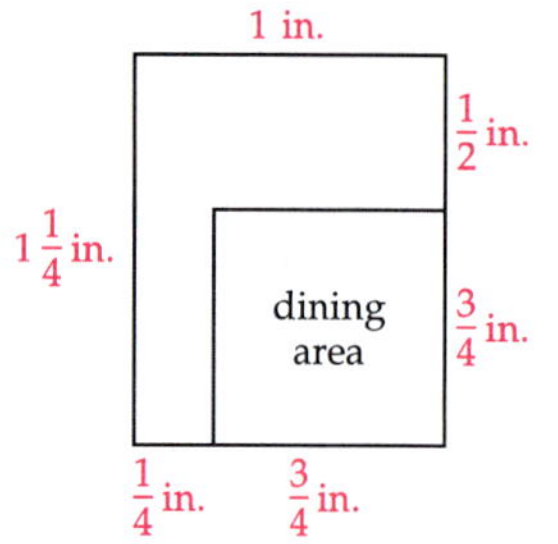

Write each decimal as a percent.

13. 0.37 **37%**
14. 0.005 **0.5%**
15. 1.02 **102%**

Write each fraction as a percent.

16. $\frac{5}{8}$ **62.5%**
17. $\frac{2}{3}$ **$66\frac{2}{3}$%**
18. $\frac{7}{9}$ **$77.\overline{7}$%**

19. ***WRITE*** a paragraph describing how to use a triangle diagram to solve a percent problem.

Solve.

20. What percent of 400 is 20? **5%**
21. Find $33\frac{1}{3}$% of 12. **4**
22. 20% of c is 24. What is c? **120**
23. What percent of 3 is 15? **500%**
24. Find 125% of 50. **62.5**
25. 60% of y is 75. What is y? **125**

Find each percent of change. Round to the nearest tenth.

26. 60 to 36 **40%**
27. 18 to 24 **33.3%**
28. 15 to 25 **66.7%**
29. 85 to 50 **41.2%**
30. 88 to 300 **240.9%**

Solve.

31. Suppose you have posts of length 6 in., 9 in., and 11 in. How can you use the posts to measure a 4-in. length?

32. Two oranges weigh the same as an apple and a grape. An apple weighs the same as 11 grapes. How many grapes equal the weight of an orange? **6**

Chapters 1–6 Cumulative Review

Choose the correct answer. Write A, B, C, or D.

1. Which number is divisible by 3 and 5? C

A. 725 **B.** 726
C. 720 **D.** not given

2. Write $\frac{7}{8}$ as a percent. A

A. $87\frac{1}{2}\%$ **B.** 70%
C. 78% **D.** not given

3. Which is *not* a ratio that describes C

□□□□ △△△?

A. 4:3 **B.** 4:7
C. 4 + 3 **D.** not given

4. Simplify $\frac{5a^3b^4}{10a^2b^3}$. B

A. $\frac{a}{2b^{-1}}$ **B.** $\frac{1}{2}ab$
C. $\frac{1}{2}a^{5b-1}$ **D.** not given

5. Evaluate $\frac{2x^3 - 0.01}{x^2}$ for $x = -0.2$. D

A. 0.65 **B.** -0.41
C. 0.065 **D.** not given

6. The figures are in proportion. Find the missing length. A

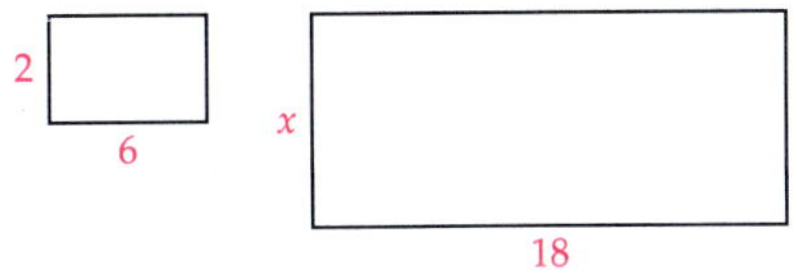

A. 6 **B.** 10
C. 12 **D.** not given

7. Solve $1.5 = 0.2x - 0.5$. D

A. 1 **B.** 5
C. -10 **D.** not given

8. Compare $\frac{\sqrt{25}}{2^3}$ ■ $\frac{\sqrt{9}}{2^2}$. B

A. > **B.** <
C. = **D.** not given

9. Find the LCM of $6x^3y^2$ and $8x^2y^5$. B

A. $2x^2y^2$ **B.** $24x^3y^5$
C. $48x^5y^7$ **D.** not given

10. Evaluate $3 + 4 \cdot 5 - 5 + 6 \div 3$. A

A. 20 **B.** 12
C. 8 **D.** not given

11. The sum of two numbers is 9. The product is 20. What are the numbers? C

A. 3 and 6 **B.** 2 and 10
C. 4 and 5 **D.** not given

12. Identify the model. C

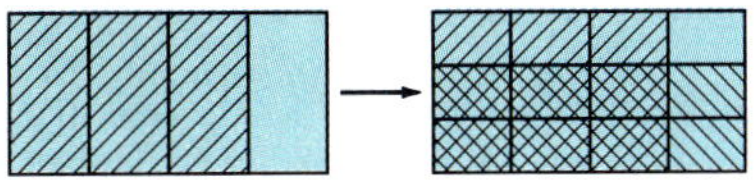

A. $\frac{3}{4} + \frac{2}{3}$ **B.** $\frac{3}{4} - \frac{6}{12}$
C. $\frac{3}{4} \cdot \frac{2}{3}$ **D.** not given

Study Skills

Objective test formats include *true-false* questions. This format usually requires specific answers.

Help students see that they may be able to make an informed guess about an answer. By analyzing the statement and by using what they already know about the topic, students may be able to eliminate one of the two possible answers. Consider the statement "An octagon has *eight* sides." Students may remember a similar word, *octopus*, a sea creature with *eight* legs, and correctly mark the statement true. Point out, that this kind of guessing may be to their advantage unless they are penalized for wrong answers, as with some standardized tests.

Students should look for qualifying words such as *rarely*, *usually*, *always*, and *none* in the statement. To be true, a statement must be true 100% of the time. If the students can think of any exception, the statement is false. Since many statements have exceptions, students should be cautious when answering statements that contain the qualifiers *always* and *never*.

Additional suggestions:

- Discuss techniques students may have developed to answer true-false questions.
- Provide practice tests, then analyze any clues that suggest the answers.
- Have groups of students make up and exchange some true-false tests and then discuss the results.

Chapter At a Glance

In Chapter 7, students learn to solve two-step equations. They also learn to identify and solve inequalities.

Reading the Chapter

Although Chapter 7 does not contain many new vocabulary words or terms, it does require students to translate words into symbols and interpret symbolic sentences. It may help to review with students the idea that math is a language in its own right and that it requires some patience to go back and forth between words and the symbolic language of math. Remind students that equations are mathematical sentences that may take time for students to understand. Encourage students to use algebra tiles and to draw number lines to model equations. When students are asked to solve word problems, ask them to be sure to translate each part of the problem into the appropriate symbols before trying to find a solution.

Chapter Resources

Practice and Enrichment TRB G-1 to G-14
Alternate Application TRB G-15
Chapter Test Form A, G-17
Chapter Test Form B, G-19
Computer Test Bank

References

STUDENT BIBLIOGRAPHY

Laycock, Mary, and Margaret Smart. *Solid Sense of Mathematics* (Activity Resource Co., Inc., Hayward, Calif. 1981)

TEACHER BIBLIOGRAPHY

Litwiller, Bonnie, and David Duncan. *Activities for the Maintenance of Computational Skills and the Discovery of Patterns* (NCTM, Reston, Va. 1980).

SOFTWARE

Safari Search (Sunburst Communications), for the Apple II and IBM family of computers.

FILMS, VIDEOTAPES, AND OTHER MEDIA

Survival Study Skills (Sunburst), activity cards for developing students' study skills.

Vocabulary

addition and subtraction properties of inequality
inequalities
periodic table

Materials

algebra tiles (M*)
graph paper (T*)
ruler

*M—Manipulative Kit
*T—Teacher's Resource Bank

Resource Bank

Class Activities

FIELD TRIPS

▼ Advertising agency—to learn what factors determine strategies for an ad campaign.

▼ Television station—to find out how television shows are broadcast.

CLASS SPEAKERS

▼ Marketing director—to discuss what kind of data is needed to make marketing decisions.

▼ Programming director—to talk about how decisions are made regarding what T.V. programs are aired and in what time slots.

Bulletin Board

Prepare a chart that shows the amount of calcium, phosphorus, and sodium in a selected group of common foods. Ask students to use the data in the table to write an equation for and then solve the problems below. After students have solved the problems, invite them to use the chart to write and solve problems of their own.

Leigh's snack contained 50 mg of calcium. She ate 2 servings of oatmeal raisin cookies. What else did she eat? ($2 \cdot 21 + x = 50$; a pear)

Eli's lunch had 731 mg of sodium. He bought pizza and two different fruits with equal amounts of sodium. What fruits did he buy? ($729 + 2x = 731$; apple and orange)

Project: Have students work in groups to research the sodium content of other foods and plan a low-sodium menu. (Check students' work.)

AMOUNTS OF THREE MINERALS in Selected Foods*

Food	Calcium (milligrams)	Phosphorus (milligrams)	Sodium (milligrams)
Apple	7	10	1
Carrot	37	36	47
Cottage cheese	94	152	229
Hot dog	5	102	—
Milk	33	14	16
Nectarine	4	24	6
Oatmeal raisin cookies	21	102	162
Oranges	41	20	1
Pear	8	11	2
Pizza	17	92	729
Potato salad	19	63	480
Roll, plain	74	85	506
Tomato soup	6	14	396

*(Serving size - 3.5 oz.)

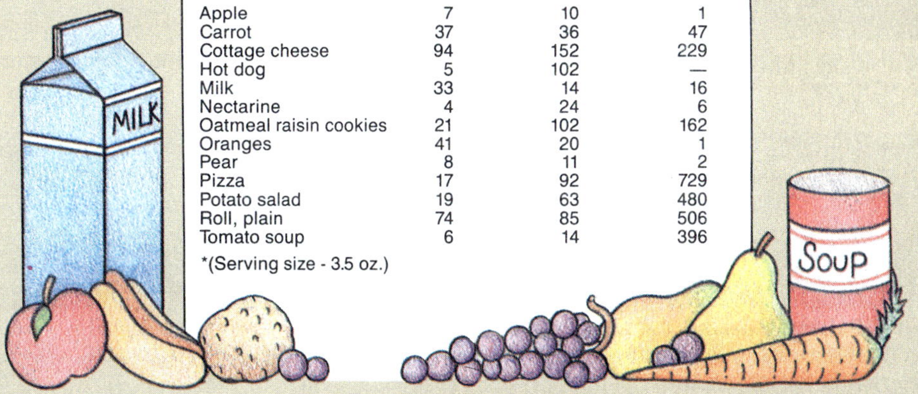

Extra-Credit Problems

BASIC

▼ A jar has 62 plastic, metal, and vinyl paper clips. There are twice as many plastic clips as metal clips. The number of plastic and vinyl clips are consecutive even integers. How many of each type of clip are in the jar? (m: 12, p: 24, v: 26)

AVERAGE

▼ The sum of four numbers is 228. The third number is 6 times the first number. The second, third, and fourth numbers are in increments of 20. What are the four numbers? Hint: The four numbers share a common digit. (12, 52, 72, 92)

ENRICHED

▼ Meg sold half as many hats as Joe. Lou sold 8 more than Joe. The sum of the number of hats sold by Lou and Bré, divided by 6, equals the number of hats Meg sold. How many hats did they sell if the total is 198? (Meg 22, Joe 44, Lou 52, Bré 80)

Chapter Planning Guide

OBJECTIVES	ASSIGNMENT GUIDE			ASSESSMENT	
	Basic	Average	Enriched	Review/ Tests	TRB*
Exploring Two-step Equations Use models to explore and solve two-step equations.	**Manipulative Connection** All	All	All		
7-1 Two-step Equations To solve simple two-step equations.	1-17 odd, 22-24, 28, 30, 32-37, 39, 41-43, 47, MR* All	2-16 even, 23-25, 30-37, 40-44, 47-49, MR All	15-21, 25-27, 29-35, 37-40, 44-49, MR All	Extra Practice p. 572	Practice G-1 Enrichment G-2
7-2 Simplifying and Solving Equations To solve two-step equations by combining like terms and using the distributive property on one side.	1-10, 21-24, 28, 30-33, 38, 40, 42, MR, TY* All	11-17, 21-25, 28-31, 34-35, 38-42, MR, TY All	15-23, 26-31, 36-42, MR, TY All	Extra Practice p. 572 TY p. 282	Practice G-3 Enrichment G-4 TY G-16
7-3 Problem Solving Strategy: Writing Equations To solve problems by writing equations.	**Problem Solving Connection** All	All	All	Problem Solving Practice p. 283	Practice G-5 Enrichment G-6
7-4 Equations with Variables on Both Sides To solve equations with an unknown on both sides.	1, 3-4, 7-23 odd, 25-26, 29, 31-32, 38, MR All	1, 4-5, 8-24 even, 26-27, 29, 33-34, 37-38, MR All	2, 5-6, 16-24, 27-28, 30, 35-38, MR All	Extra Practice p. 572	Practice G-7 Enrichment G-8
Exploring Inequalities To use models to explore inequalities.	**Manipulative Connection** All	All	All		
7-5 Inequalities and Their Graphs To write and graph simple inequalities with one variable.	1-3, 6-7, 10-11, 16-22, 31-34, 37, 39-40, 43-44, 49-51, 55-57, MR All	2-4, 7-8, 12-13, 16-18, 23-26, 33-37, 40-41, 44-45, 47, 51-52, 55-57, MR All	3-5, 8-9, 14-18, 27-30, 33-42, 45-48, 53-57, MR All	Extra Practice p. 572	Practice G-9 Enrichment G-10
7-6 Solving One-step Inequalities To solve one-step inequalities using properties for inequalities.	1-5, 8-10, 12-15, 24-28, MR, TY All	1-3, 5-6, 9-11, 16-19, 24-28, MR, TY All	2-3, 5-7, 10-11, 19-28, MR, TY All	Extra Practice p. 572 TY p. 289	Practice G-11 Enrichment G-12 TY G-16
7-7 Solving Two-step Inequalities To solve simple two-step inequalities.	1-2, 5-29 odd, 31-34, 37-41, MR, CT* All	2-3, 6-30 even, 31-32, 34-35, 37-41, MR, CT All	3-4, 18-32, 35-41, MR, CT All	Extra Practice p. 572	Practice G-13 Enrichment G-14
7-8 Problem Solving Application: Buying a Car To apply mathematical skills when purchasing a car.	All	All	All	Extra Practice p. 572	Alternate Application G-15
Chapter Test	All	All	All	pp. 308-311	Tests G-17–G-23

 *TRB—Teacher's Resource Book MR—Mixed Review TY—Test Yourself CT—Critical Thinking

			CONNECTIONS				NCTM CORRELATION
Algebra	Critical Thinking	Problem Solving	Estimation/ Mental Math	Technology	Manipulatives/ Resources	Writing/Reading in Math	
All	All	2			All	1, 3b, 4b, 6b	Communication Reasoning Algebra
C*: All W*: All	C: 1-6 W: 34-35	C: 5-6 W: 34-35, 49	W: 22-27	W: 28-31	W: 49	W: 32-33, 39-46	Problem Solving Reasoning Communication Algebra
C: All W: All TY*: All	C: 1-4, 11-13, TD* All W: 30-41 TY: 6	C: 11-13 W: 32-39, 42 TY: 6	W: 21-23	W: 24-29	W: 42	W: 30-31, 34-39 TY: 6	Problem Solving Algebra Reasoning
C: All W: All	CT*: All	C: All W: All		W: 1-19	W: 17	C: All W: All	Algebra Problem Solving Reasoning
C: All W: All	C: 1-4, TD All W: 25-28	W: 31-38	W: 3-6	W: 29-30	W: 37-38	W: 31-37, 38b	Functions Structure Algebra
	All				All	6	Reasoning Communication Connections
C: 1-3, 7-12 W: 10-56	C: 4-10, TD: All W: 1-9, 16-18, 57c	C: 11-12 W: 31-37, 50-57		W: 6-9, 57a-c	C: 7-10 W: 16-30, 37-39, 56-57	C: 1-3, 11-12 W: 10-15, 31-36, 50-56	Algebra Connections Communication
C: All W: 1-24 TY: All	C: TD All, 1-12 W: 1-11	W: 24-28		W: 26-27	W: 24-28	W: 1-3, 24 TY: 11-13	Algebra Reasoning Connections
C: All W: 1-40	C: 1-4, TD: All W: 1-4, 31-32 CT: All	C: 14 W: 33-41	W: 37-38		W: 41	C: 14 W: 33-36 CT: All	Algebra Reasoning Problem Solving
		C: 1-4 W: 1-12	W: 5-8	W: 1-12	W: 9-10		Problem Solving Algebra Functions

*C—Class Exercises W—Written Exercises CT—Critical Thinking TY—Test Yourself TD—Think and Discuss

Exploring

In this manipulative activity, students use algebra tiles to model solutions to equations with variables on both sides of the equal sign. **See TRB Exploring O-25.**

Exploring Equations with a Variable on Both Sides

You can use algebra tiles to model equations that have a variable on both sides.

1. Follow the steps to model the solution for $x + 3 + 3x = 5x$. Write the resulting equation.

Steps	Model	Equation
a. Model the equation.		$x + 3 + 3x = 5x$
b. Combine like terms.		$4x + 3 = 5x$
c. Subtract 4x from both sides.		$3 = x$

2. What did you do to isolate the variable? **4x was subtracted from both sides.**
3. What is the solution? **x = 3**
4. Model and solve $2a + 4 + a = 4a + 2$. Write the equation that results from each step of your solution. **$3a + 4 = 4a + 2$, $4 = a + 2$, $2 = a$**
5. You needed two steps to isolate the variable after you combined like terms. What were they? **Subtract 3a from both sides; subtract 2 from both sides.**
6. Write the equation for this model. **$3x - 2 + 2x = x + 4 + 3x$**
7. Model the solution and write the equation for each step you model. **$5x - 2 = 4x + 4$, $x - 2 = 4$, $x = 6$**
8. Model and solve $4 + 2b - 3 + 3b = b + 7 + 2b$. **b = 3**

O-25

In this critical thinking activity, students consider pairs of expressions to develop conclusions about the values in the expressions. **See TRB Exploring O-26.**

Exploring Quantitative Comparisons

To compare expressions such as $a + 4$ and $a + 3$ you need to account for all possible values the expressions may contain.

	Trial	Result
Example Evaluate the expression for $a = 0$.	$0 + 4 > 0 + 3$	$a + 4 > a + 3$
1. What happens when $a > 0$? Try $a = 2$.	**$2 + 4 > 2 + 3$**	**$a + 4 > a + 3$**
2. Try another positive value.	**Answers may vary.**	
3. What happens when $a < 0$? Try $a = -6$.	**$-6 + 4 > -6 + 3$**	**$a + 4 > a + 3$**
4. Try another negative value.	**Answers may vary.**	

5. What can you conclude? **a + 4 is always greater than a + 3.**

Compare 3n and 2n.

	Trial	Result
6. Evaluate 3n and 2n for $n = 0$.	**$3 \cdot 0 = 2 \cdot 0$**	**$3n = 2n$**
7. What happens when $n > 0$? Choose a value and evaluate the expressions. **Answers may vary.**	**$3 \cdot 1 > 2 \cdot 1$**	**$3n > 2n$**
8. Choose a value for n less than zero and evaluate the expressions. **Answers may vary.**	**$3(-1) < 2(-1)$**	**$3n < 2n$**

9. Do you think that choosing a positive or a negative fraction for n changes your results? Show two examples. **no, $3(\frac{1}{2}) > 2(\frac{1}{2})$, $3(-\frac{1}{2}) < 2(-\frac{1}{2})$**
10. What can you conclude about comparing 3n and 2n? **3n can be greater than, equal to, or less than 2n, depending on the value of n.**

Compare $3x + 2 - x$ and $x + 5 + x - 3$.

11. Simplify the expressions. **$2x + 2$, $2x + 2$**
12. What can you conclude? **The expressions are equal.**

Compare $\frac{2}{x}$ and $\frac{3}{x}$.

13. What happens when $x = 0$? **Both expressions are undefined.**

	Trial	Result
14. Evaluate for $x > 0$. Choose a value. **Answers may vary.**	**$\frac{2}{3} < \frac{3}{3}$**	**$\frac{2}{x} < \frac{3}{x}$**
15. Evaluate for $x < 0$. Choose a value. **Answers may vary.**	**$\frac{2}{-1} > \frac{3}{-1}$**	**$\frac{2}{x} > \frac{3}{x}$**

16. Write a conclusion about the comparison. **When x is positive, $\frac{2}{x} < \frac{3}{x}$; when x is negative, $\frac{2}{x} > \frac{3}{x}$**

O-26

In this problem solving activity, students analyze the meaning of quantitative statements. **See TRB Exploring O-27.**

Exploring Writing Inequalities

Many situations can be represented by writing inequalities.

Situation:
Jack is allowed to spend at most \$10 on a gift for his friend.

1. What does the phrase *at most \$10* mean? **He can spend up to and including \$10 on the gift.**
2. Does *at most \$10* include the amount \$10? **yes**
3. Let g be the cost of the gift. How does $g \le 10$ represent this situation? **The cost g can be \$10 or less than \$10.**

Answer the questions about each situation.

Situation: Andrea said she has a minimum of 25 CDs.

4. What does the phrase *minimum of 25* mean? **The least number of CDs she has is 25; she has 25 or more CDs.**
5. Does a *minimum of 25* include 25? **yes**
6. Let a be the number of CDs. Write an inequality for the situation. **$a \ge 25$**

Situation: Seven less than twice Kim's age is not greater than 35.

7. What does the phrase *is not greater than 35* mean? **a quantity that is 35 or less than 35**
8. Does the phrase include 35? **yes**
9. What inequality symbol will you use? **≤**
10. Let a be Kim's age. Write an inequality for the situation. **$2a - 7 \le 35$**

Match the verbal sentence with the inequality.

11. The value of x is less than 3. **c** — a. $x > 3$
12. The value of x is at most 3. **d** — b. $x \ge 3$
13. The value of x is not less than 3. **b** — c. $x < 3$
14. The value of x is greater than 3. **a** — d. $x \le 3$

O-27

In this computer activity, students use a spreadsheet to evaluate expressions and find solutions to equations. **See TRB Exploring O-28.**

Exploring Equations on a Computer Spreadsheet

You can use a spreadsheet to evaluate an expression and to find solutions to equations.

Set up the computer spreadsheet, inputting the labels shown.

	A	B	C
1	x	2x − 5	3(x − 1)
2			

Input the following values and formulas in the cells of row 2.

Cell	A2	B2	C2
Value/Formula	-2	=2*A2 - 5	=3*(A2 - 1)

1. What value is stored in each cell? B2 **-9** C2 **-9**

Now input these values and formulas in the cells of row 3.

Cell	A3	B3	C3
Value/Formula	-1	=2*A3 - 5	=3*(A3 - 1)

2. What value is stored in each cell? B3 **-7** C3 **-6**

Use the values at the right for column A in rows 4 through 6. Find the values of the expressions $2x - 5$ and $3(x - 1)$ for $x = 0, 1, 2$. Then analyze the spreadsheet to answer the following questions.

Row	A
4	0
5	1
6	2

3. When $x = 1$ what is the value of $2x - 5$? $3(x - 1)$? **-3; 0**
4. How can you use the information in the spreadsheet to solve the equation $2x - 5 = -1$? **Find where 2x − 5 has the value -1. The x value substituted is the solution, x = 2.**
5. Write two other equations involving the expression $2x - 5$ that can be solved using the information in the spreadsheet. **Answers will vary.**
6. Explain how you know that 0 is the solution to $3(x - 1) = -3$. **The expression 3(x − 1) has the value -3 when x has the value 0.**
7. What can you say about the solution to $2x - 5 = -4$? **The solution is between 0 and 1.**

O-28

Meeting Individual Needs

GIFTED STUDENTS

Mathematically gifted students may find the section of this chapter that deals with inequalities particularly stimulating, as it introduces them to mathematical relationships that are less determinate than equations. By working with inequalities in mixed ability groups, gifted students have the opportunity to share their knowledge at the same time they're being stretched by demanding material.

After separating the class into small groups of 4 to 6 students, with one gifted student per group, explain that, in the following activity, groups will brainstorm to find solutions for inequalities. Review the fact that an equation such as $|x - 2| = 3$ is satisfied by 5 or -1, because they are both 3 units from 2 on the number line.

If $|x - 2| > 3$ means that the solutions are all numbers on the number line that are more than 3 units from 2, the solution is written $x > 5$ or $x < -1$. You can write on the chalkboard that the graph of the solution would look like this:

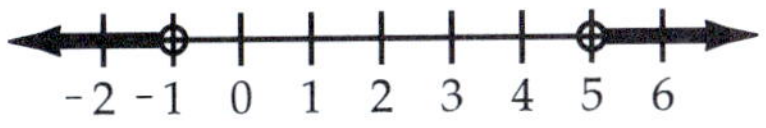

Then ask the groups to solve these inequalities and graph the solutions on a number line:

$|x - 5| \leq 2$

[$3 \leq x \leq 7$; 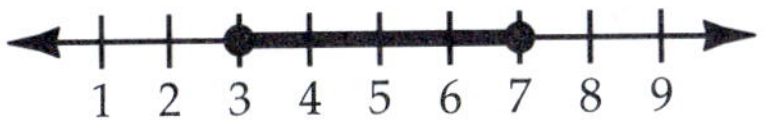]

$|x + 3| > 1.5$

[$x < -4.5$ or $x > -1.5$; 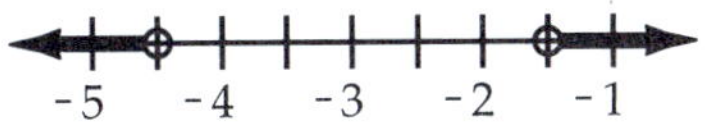]

AT-RISK STUDENTS

At-risk students are more likely to be motivated to solve mathematical problems when they are presented in an entertaining context. Part of what can make a context appealing for such students is its practicality or relevance, as well as its appeal to the students' value systems. This is especially true of the concept of inequalities, which can be very difficult for at-risk students.

Have the students work with a partner. Assign a topic of interest to each pair such as rock concerts, videotapes, or a sporting event. Ask each pair to create a word problem involving their topic. Refer the students to question 35 on page 282 as a sample. Then have the pairs write an equation for their problem and solve.

ENGLISH AS A SECOND LANGUAGE

Chapter 7 requires that students use the many properties they have learned. ESL students might find it particularly helpful if terms, such as the *subtraction property of equality* and *the distributive property*, are reviewed in detail before the start of the lesson. Use models wherever possible to illustrate a concept.

As the equations in the chapter become more complex, it is important to remember that ESL students can be mathematically gifted, just as native language students can be. To help determine if an ESL student's difficulties are language-based or math-related, pair each ESL student with a native language student. Have partners work on questions 3-10 on page 285. Monitor each pair closely to see how much the ESL student participates in the language-related skill of deciphering the problem and in the math-related computational skill. ESL students, who understand the math but are having trouble with the language, can subsequently be paired with English-speaking students with good communication skills. Both students will benefit from each other's talents.

See TRB Teaching Resources p. 57.

Problem Solving Situation

The questions in the "Think About It" box ask about the elements that might make one airport busier than another. To help students formulate ideas, encourage them to look at the data and then make reasonable guesses. For example students might consider the location of the airport: Is it close to large cities? Is it near railways and waterways? Is it near major highways?

DATA COLLECTION

Data can be obtained from general encyclopedias or geography textbooks, depending on the nature of factors students propose. Students might also write or telephone the public relations office of the airport and make a request for an information packet. Students can also query the public relations person about what factors contribute to airport traffic.

DATA ANALYSIS

Have students prepare a report on their findings. Ask them to structure the report by listing the factors that were researched, explaining each factor, displaying and evaluating the research, and providing concluding remarks. The report would be enhanced by any visual device that clarifies ideas, including maps.

WHEN STANDARD TIME was first established in 1884, there were 24 time zones—one for every 15° of longitude. The prime meridian at 0° longitude is used to determine noon. Time for other zones depends on the number of zones east or west of the prime meridian.

The line opposite 0° at 180° (called the international date line) is used to determine midnight. To make standard time practical for daily use, many nations, states, and cities changed the time zones to include local boundaries.

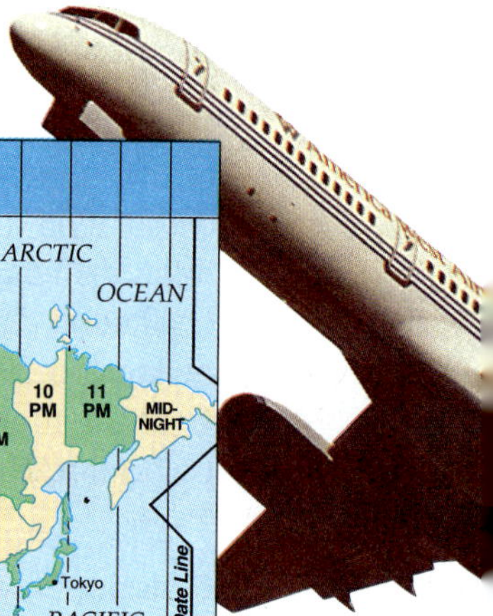

World Time Zones

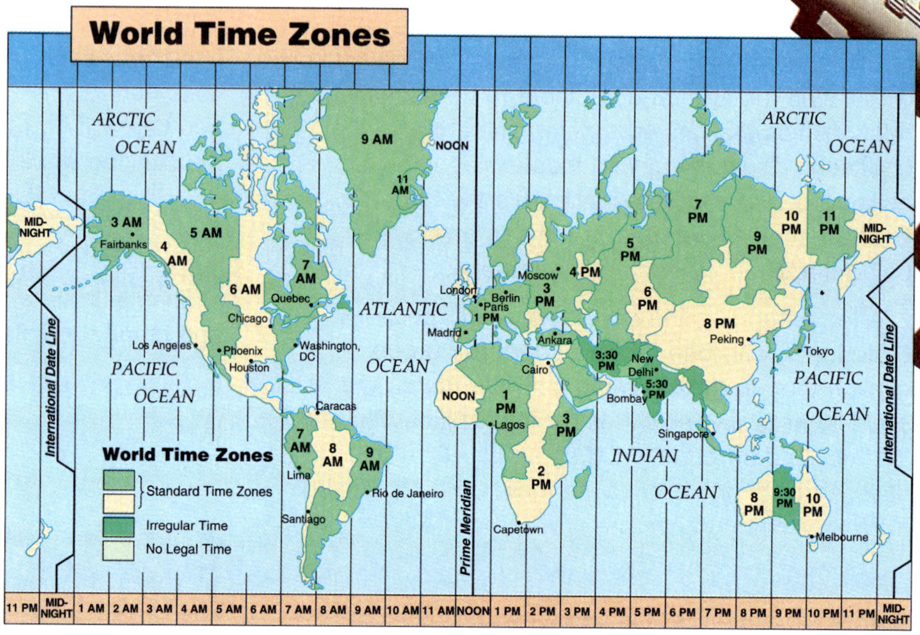

Federal regulations ban smoking on all domestic flights of less than 6 h.

Passengers 11 years of age, or younger, and 62 years of age, or older, are eligible for discounts on most major airlines. Children under two fly for free.

Air Travel Information

272

CHAPTER 7

Equations and Inequalities

UNITED STATES BUSIEST AIRPORTS

AIRPORT	PASSENGER BOARDINGS (Millions of Passengers) 1987	PASSENGER BOARDINGS (Millions of Passengers) 2000	RANK 1987	RANK 2000
Chicago O'Hare	27.5	43.6	1	1
Atlanta	23.9	34.6	2	3
Los Angeles	21.2	27.1	3	4
Dallas/Ft. Worth	20.8	23.6	4	5
Denver	16.1	35.7	5	2
New York/Kennedy	14.4	17.6	6	12
San Francisco	14.0	20.3	7	7
Newark	12.2	23.4	8	6
New York/La Guardia	11.6	16.0	9	15
Miami	11.6	18.9	10	9

Think about it...

Look at the data about passenger boardings. What factors might make one airport busier than another? What might happen to increase or decrease the passenger boardings from year to year?

See Problem Solving Situation

Cooperative Learning

This activity lends itself to cooperative brainstorming sessions. Remind the groups that when they brainstorm, they need to encourage each other's participation without judging the value of the idea. If judgments come too early in brainstorming, it may cause inhibitions and prevent usable and productive ideas from emerging.

When brainstorming sessions are completed, groups can turn their efforts toward researching and preparing their reports. Students may need help in assembling the questions they want to research and may also need guidance on where the research can be done. If the students plan to call a public relations office of an airport, they may want some coaching on how to conduct themselves on the telephone, so their inquiries will be effective in getting a response.

It may be desirable to assign the role of Interviewer to a student who does not seem uncomfortable when speaking to adults or to groups of his or her peers. Other roles that can be assigned are Writer, Illustrator, and Researcher.

Questions related to this Data File can be found on pages 295, 299, 413, 422, 500, 572, and 576.

Lesson Focus

Materials

Algebra tiles or blue and white squares of paper and red paper rectangles, overhead projector, math journal

MOTIVATING THE LESSON

Review modeling and solving one-step equations. Have students use algebra tiles to model and solve equations such as $3x = -6$ and $x + 2 = 5$. Ask students to summarize the process for solving a one-step equation. (Perform the same operation on both sides of the equation to isolate the variable on one side.) Write the equation $3x - 4 = 20$ on the chalkboard. Ask students what operations are involved in the equation. (multiplication and subtraction) Next, ask students how they think they would find the value of x. (Undo both operations.)

Instruction

Divide the class into groups of 3-4 students and distribute algebra tiles to each group.

(1) Remind groups to use the blue tiles for negative numbers and white tiles for positive numbers.

(2) Have each group analyze and discuss how to use tiles to solve a two-step equation and agree on the procedure they used. Have one member of each group describe the procedure they should use. The class can discuss variations on the procedures.

Closure

Writing in Math Ask students to write in their math journals how they use tiles to solve a two-step equation.

OBJECTIVE: *Use models to explore and solve two-step equations.*

MATERIALS

- Algebra tiles
- Math journal to record work

Exploring Two-step Equations

▼ You can use algebra tiles to model a two-step equation.

a. $2x + 3 = 5$

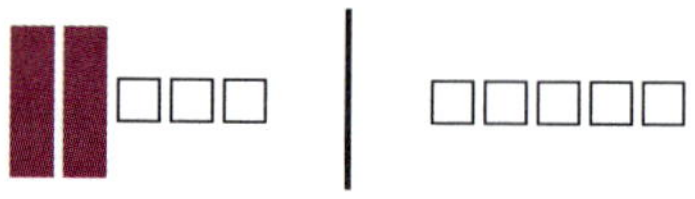

b. $2x - 4 = 6$

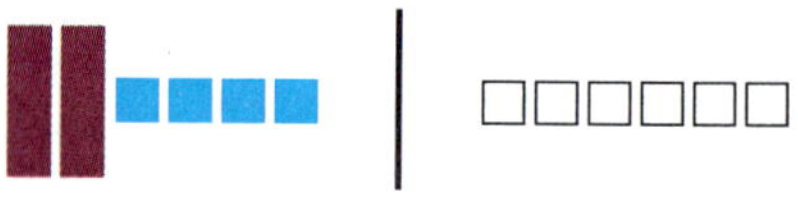

1. ***Describe*** what each tile represents in each equation. ***Analyze*** how these equations and models are different from the ones you have seen before.
2. ***Model*** each equation. ***Compare*** your models with those of another group. **Check students' work.** (1)
   a. $3n + 4 = 7$ **b.** $4x - 2 = 6$ **c.** $5r - 2 = 8$

▼ You can also use algebra tile models to help you solve a two-step equation.

3. ***Analyze*** the model for $2x + 3 = 5$. To solve the equation, you must isolate the variable.
   a. ***Discuss*** what isolating a variable means.
   b. ***Describe*** the first step you would take in isolating the variable expression $2x$. What operation did you perform?
   c. ***Discuss*** what you need to do to solve the resulting equation for x. What is the value of x? **Divide the white tiles evenly among the variables; 1**
4. ***Analyze*** the model for $2x - 4 = 6$.
   a. ***Discuss*** why it is helpful to think of this equation as $2x + (-4) = 6$. **so that you can model subtraction**
   b. ***Describe*** the two steps needed to solve the equation. **add 4; divide by 2**
   c. What is the solution of $2x - 4 = 6$? **5**
5. Use models to solve each equation. **Check students' work.** (2)
   a. $3r + 4 = 10$ **b.** $2b - 7 = 3$ **c.** $4n - 6 = -2$
6. ***Analyze*** your solutions for Exercise 5. **a. Sub. 4. b. Add 7. c. Add 6.**
   a. What was the first thing you did to solve each equation?
   b. ***Summarize*** Write a rule describing the steps you used to solve a two-step equation. **First, undo addition or subtraction. Second, undo multiplication.**

1. The dark rectangles represent variables, white squares represent positive numbers, and blue squares represent negative numbers. The variable has a coefficient.
3. a. to get the variable alone on one side of the equation
3. b. Remove three white tiles from each side; subtraction

Teaching Tip

Allow students to use the overhead projector to share models and equations with the class.

Assignment Guide

Basic All
Average All
Enriched All

FOR THE NEXT LESSON

calculator, math journal

OBJECTIVE:
To solve simple two-step equations.

7-1 Two-step Equations

A student pays $.86 for three school stickers and a school emblem. An emblem costs $.29. What is the price for each sticker? If we represent the price of one sticker with variable p, we can write the following equation.

$$3p + 29 = 86$$

FLASHBACK
You can use inverse operations to solve an equation.

You can solve this equation by using the subtraction and division properties of equality.

1

$$3p + 29 = 86$$
$$3p + 29 - 29 = 86 - 29 \quad \text{Subtract 29 from both sides.}$$
$$3p = 57$$
$$\frac{3p}{3} = \frac{57}{3} \quad \text{Divide both sides by 3.}$$
$$p = 19$$

Check $3p + 29 = 86$

3 ⊠ 19 ⊞ 29 ⊟ 86 ✓ — Replace p with 19.

Each sticker costs $.19.

You can also solve two-step equations using the addition and multiplication properties of equality.

Example 1 **Write an equation to describe the situation. Solve.**

Suppose you divide a number by 4 and then subtract 5 to get 8. What is the original number?

Solution Let n = the original number.
Then the equation $\frac{n}{4} - 5 = 8$ describes the situation.

2

$$\frac{n}{4} - 5 = 8$$
$$\frac{n}{4} - 5 + 5 = 8 + 5 \quad \text{Add 5 to both sides.}$$
$$\frac{n}{4} = 13$$
$$4\left(\frac{n}{4}\right) = 13(4) \quad \text{Multiply both sides by 4.}$$
$$n = 52$$

Check $\frac{n}{4} - 5 = 8$

52 ÷ 4 − 5 = 8 ✓ — Replace n with 52.

Solving a Simple Two-step Equation	To solve a simple two-step equation: **1.** Undo addition or subtraction. **2.** Undo multiplication or division.

Lesson Focus

Materials
Calculator, math journal

MOTIVATING THE LESSON

Ask students what steps would solve these equations: $-5n = 15$; $\frac{x}{3} = 14$. Then ask how this equation is different: $-5n + 5 = 15$. (the equation contains $+5$) Explain to the students that they will learn to solve equations like this one.

Instruction

1 Ask students to describe the type of equation that results when 29 is subtracted from both sides. (one-step equation)

2 Help students decide what operations should be done first, and suggest that they work backwards. The last step in the situation is "subtract 5," so the first step is to undo that operation.

Teaching Tip

As the content of the material becomes more difficult, review information, such as the properties of equality, which were taught earlier in the year.

EE: 1B, 1C, 2F, 3A, 3B, 3F, 3G, 3H; TAAS: 2, 6, 7, 8, 9, 12

3 Ask students what the $20 covers. (A third of the canoe rental plus paddle.)

ADDITIONAL EXAMPLES

1. Write an equation to describe the situation. Solve. Suppose you multiply a number by -3 and then add 7 to get -2. What was the original number? ($-3n + 7 = -2$; 3)
2. Choose the correct equation and solve. At a T-shirt sale, Sharla bought three shirts and used a coupon to save $4. Her bill, before sales tax, was $20. What was the marked price of each T-shirt?
   a. $3m - 4 = 20$
   b. $3m + 4 = 20$
   c. $\frac{n}{3} + 4 = 20$
   d. $\frac{n}{3} - 4 = 20$
   (a, $8)

Guided Practice

Class Exercises Some students may need help with question 11. Have them rewrite the equation so the solution is on the right. Then ask students if $x = -9$ and $-9 = x$ are the same. (yes)

Think and Discuss If students need help with question 1, have them substitute values for *a*, *b*, and *c*.

In question 2, be sure students understand that they are adding -2 and multiplying by $\frac{1}{5}$, the reciprocal of 5.

In question 3, students should see that both values are positive, so *n* must be positive.

Example 2 **Choose the correct equation. Solve.**

3 Three friends rent a canoe. Each person also rents a paddle for $4. Each person pays a total of $20. What is the cost of renting the canoe without the paddles?

a. $3n - 4 = 20$ b. $3n + 12 = 20$

c. $\frac{n}{3} + 4 = 20$ d. $\frac{n}{3} - 4 = 20$

Solution Let n = the total cost of the canoe rental. Then the correct equation is *c*.

$\frac{n}{3} + 4 = 20$

$\frac{n}{3} + 4 - 4 = 20 - 4$ Subtract 4 from each side.

$\frac{n}{3} = 16$

$3\left(\frac{n}{3}\right) = 16(3)$ Multiply both sides by 3.

$n = 48$ The check is left for you.

The cost of renting a canoe without the paddles is $48.

1. Add *b* to each side.
2. Add -2 to each side; mult. each side by $\frac{1}{5}$.
3. No; the equation becomes $18n = 403$. A positive number times a positive number equals a positive number.

THINK AND DISCUSS See above.

1. What is the first step in solving the equation $ax - b = c$ for x?
2. How can you solve the equation $5x + 2 = -28$ using addition and multiplication?
3. Is the solution of the equation $18n - 1 = 402$ a negative number? Explain.

CLASS EXERCISES

State the first step in solving each equation.

1. $\frac{a}{6} + 9 = 13$ Subtract 9.
2. $4b - 6 = 2$ Add 6.
3. $2c + 1 = 5$ Subtract 1.
4. $\frac{d}{3} - 10 = 5$ Add 10.

Choose the correct equation. Solve.

a. $3x + 1 = 14$ b. $x - 1 = 14$

c. $\frac{x}{3} + 1 = 14$ d. $\frac{x}{3} - 1 = 14$

5. Sarita, Clara, and Joe baked muffins, which they shared equally. Clara ate one on the way home. She had 14 muffins left. How many muffins did Sarita, Clara, and Joe bake? d; 45
6. Kyle sent away for three tapes. The cost of the order, with $1 for shipping, was $14. What was the price of each tape? a; $4.33

Solve each equation.

7. $9x - 15 = 39$ 6
8. $\frac{y}{7} - 6 = 8$ 98
9. $\frac{2}{3}n + \frac{3}{8} = \frac{15}{16}$ $\frac{27}{32}$
10. $4 - \frac{z}{3} = 13$ -27
11. $-35 = 4x + 1$ -9
12. $2.4r - 5.6 = 11.2$ 7

MATH MINUTES

Match the equivalent equations.

1. $2x + 7 = 9$ a. $4x = 29$
2. $3x - 5 = 7$ b. $\frac{x}{3} = 15$
3. $4x - 9 = 20$ c. $2x = 2$
4. $\frac{x}{3} - 5 = 10$ d. $3x = 12$

(1-*c*, 2-*d*, 3-*a*, 4-*b*)

13. $6n - 5 = 55$ 10 **14.** $\frac{x}{3} + 2 = 14$ 36 **15.** $16d - 28 = 174$ $12\frac{5}{8}$

WRITTEN EXERCISES

Solve each equation.

1. $3n + 5 = 23$ 6 **2.** $18 = 4t + 2$ 4 **3.** $-86 + 4k = 102$ 47

4. $30 = 18 + 2b$ 6 **5.** $5 + \frac{k}{9} = -31$ -324 **6.** $2 + \frac{m}{3} = 0$ -6

7. $12d - 6 = 138$ 12 **8.** $4x - 2 = 28.4$ 7.6 **9.** $15 = 6 + \frac{m}{6}$ 54

10. $-19 = 4 + 3x$ $-7\frac{2}{3}$ **11.** $4 - \frac{k}{5} = 18$ -70 **12.** $15 = -11c + 4$ -1

13. $12 - 11s = 45$ -3 **14.** $10 = 3 + \frac{d}{2}$ 14 **15.** $\frac{-3}{4}n + \frac{1}{4} = 1\frac{3}{4}$ -2

16. $\frac{x}{-6} + 7 = 0$ 42 **17.** $0 = 91 + 13t$ -7 **18.** $5p - 0.48 = 0.12$ 0.12

19. $\frac{2n}{5} - 23 = 11$ 85 **20.** $2.1 + 3b = 1.8$ -0.1 **21.** $\frac{t}{8} - \frac{3}{4} = \frac{1}{2}$ 10

MENTAL MATH Solve each equation.

22. $2n + 3 = 15$ 6 **23.** $\frac{y}{5} - 2 = 10$ 60 **24.** $4x - 1 = 27$ 7

25. $\frac{m}{10} + 3 = 6$ 30 **26.** $2x - 7 = 11$ 9 **27.** $3a - 2 = 13$ 5

CALCULATOR Solve each equation.

28. $31.5 - 4.2x = -65.1$ 23 **29.** $238.7 + 1.8k = 3.02 \cdot 10^5$ $167{,}645.1\overline{6}$

30. $0 = -5.67x + 0.25$ 0.0440917 **31.** $47.2 = \frac{4r}{5} + 81.9$ -43.375

Write a situation for each equation. Check students' work.

32. $3g + 4 = 16$ **33.** $\frac{r}{4} + 0.35 = 5.15$

Choose the correct equation. Solve.

34. A student bought some pencils for $.39 each and a pad of drawing paper for $1.19. The total cost for the supplies was $3.92. How many pencils did the student buy? c; 7

a. $39x + 1.19 = 3.92$ **b.** $1.19x + 39y = 3.92$
c. $0.39x + 1.19 = 3.92$ **d.** $0.039x + 1.19 = 3.92$

35. A student is saving $15 each week from earnings and already has $150. In how many weeks will the balance be $210? c; 4

a. $150 + 210 = 15n$ **b.** $150 - 15n = 210$
c. $150 + 15n = 210$ **d.** $15n - 150 = 210$

MIXED REVIEW

Solve each equation.

1. $12n + 60 = 300$ 20
2. $25n - 30 = 70$ 4

Find each answer.

3. Solve $\frac{12}{x} = \frac{18}{27}$. 18
4. What is 15% of $42? $6.30
5. 24 is 25% of what number? 96

Simplify each expression.

6. $4m + 5 - m$ $3m + 5$
7. $13 - 5x + 2$ $15 - 5x$
8. How much warmer is it when the temperature is 90° than when the temperature is -6°? 96°

A ***common error*** in solving an equation such as $\frac{n}{4} - 5 = 8$ is to subtract 5 from the right side of the equation. To avoid this error have students write the operations to use on both sides of the equation. It may help to remind students that each side of the equation must be balanced.

Closure

Writing in Math Ask students to write in their math journals the steps they use to solve questions 13 and 14 of the Class Exercises.

Independent Practice

For questions 28-31, remind students that they have to write out each step in order, to know which numbers to enter in the calculator.

Encourage students to share the situations they create in questions 32-33 with the class.

If students have difficulty with questions 47-48, suggest that they replace the variables with numbers to help them solve the equations.

Lesson Follow-up

Reteaching Activity To help students who are having difficulty with the order of the operations when solving two-step equations, have them write sentences describing what happens during each step. Emphasize that they want to undo the operations.

LESSON RESOURCES

TRB Practice G-1
TRB Enrichment G-2

Practice

For use after 7-1 (pp. 275–278)

Two-step Equations

Solve each equation.

1. $4x - 17 = 31$ **x = 12**
2. $15 = 2m + 3$ **m = 6**
3. $\frac{k}{3} + 3 = 8$ **k = 15**
4. $7 = 3 + \frac{h}{6}$ **h = 24**
5. $9n + 18 = 81$ **n = 7**
6. $5 = \frac{y}{3} - 9$ **y = 42**
7. $14 = 5k - 31$ **k = 9**
8. $\frac{t}{9} - 7 = -5$ **t = 18**
9. $\frac{v}{8} - 9 = -13$ **v = -32**
10. $25 - 13f = -14$ **f = 3**
11. $18p - 45 = 0$ **$p = 2\frac{1}{2}$**
12. $\frac{2}{3}y - 6 = 2$ **y = 12**
13. $40 - 5n = -2$ **n = 8.4**
14. $\frac{7}{8}h - \frac{5}{8} = 2$ **h = 3**

MENTAL MATH **Solve each equation.**

15. $3p + 5 = 14$ **p = 3**
16. $\frac{k}{2} - 5 = 1$ **k = 12**
17. $\frac{m}{7} - 3 = 0$ **m = 21**
18. $10v - 6 = 24$ **v = 3**

CALCULATOR **Solve each equation.**

19. $9w - 16.3 = 5.3$ **w = 2.4**
20. $88.1 - 2.3f = 72.46$ **f = 6.8**
21. $-15.3 = -7.5k + 55.2$ **k = 9.4**
22. $26e + 891 = -71$ **e = -37**

Choose the correct equation. Solve.

23. Tehira has read 110 pages of a 290-page book. She reads 20 pages each day. How many days will it take to finish?
   a. $20 + 110p = 290$ b. $20p + 290 = 110$
   c. $110 + 20p = 290$ d. $290 = 110 - 20p$
   c; p = 9; 9 d

Write an equation to describe the situation. Solve.

24. A waitress earned $73 for 6 hours of work. The total included $46 in tips. What was her hourly wage?
   6h + 46 = 73; h = 4.5; $4.50/h
25. A car rented for $29 per day plus $0.08 per mile. Julia paid $46.12 for a one-day rental. How far did she drive?
   29 + 0.08m = 46.12; m = 214; 214 mi

Enrichment

For use after 7-1 (pp. 275–278)

Solving Number Problems

You can use two-step equations to solve problems involving odd and even numbers and multiples of whole numbers.

The sum of three consecutive odd numbers is 99. Answer these questions to learn a method for finding the numbers.

1. Let n = the first number. Write an expression for each of the other two numbers in terms of n.
   $n + 2$, $n + 4$
2. Write and solve an equation to find the three numbers.
   $n + (n + 2) + (n + 4) = 99$; $3n + 6 = 99$; $n = 31$; 31, 33, 35

Find the numbers described.

3. three consecutive odd numbers with a sum of 261 **85, 87, 89**
4. two consecutive even numbers with a sum of 894 **446, 448**
5. an odd number which when multiplied by 4 and added to the next odd number produces a sum of 177
   35, 37

Use the same principle to solve problems involving multiples of whole numbers.

6. m is a multiple of 6. Write the next two multiples of 6 in terms of m.
   $m + 6$, $m + 12$

Solve.

7. The sum of three consecutive multiples of 5 is 90. Find the numbers.
   25, 30, 35
8. The sum of four consecutive multiples of 7 is 378. Find the numbers.
   84, 91, 98, 105
9. The sum of three times a multiple of 4 and the next two multiples of 4 is 272. Find the numbers.
   52, 56, 60

Write a two-step equation with each number as its solution. Check students' work.

36. −2 **37.** 7 **38.** 0.3

Write an equation to describe the situation. Do not solve.

39. Seven less than three times a number equals 19. Find the number. $3n - 7 = 19$

40. Linda had $235 in her savings account. She withdrew the same amount each week for 15 weeks. Her balance was then $55. How much money did Linda withdraw each week? $235 - 15d = 55$

Write an equation to describe the situation. Solve.

41. Two is twelve times a number less four. Find the number. $2 = 12n - 4; \frac{1}{2}$

42. Phillip wants to buy a bicycle for $189. He has $24 and plans to save $15 each week. In how many weeks will he be able to buy the bicycle? $15d + 24 = 189; 11$

43. Phyllis wants to buy a camera for $78. She has already saved $36. She plans to save $8 each week. In how many weeks will she be able to buy the camera? $8d + 36 = 78; 5.25$

44. A taxi ride costs $.40 for each quarter-mile and $.85 for each additional passenger. Yolanda and Cara paid $5.25 altogether. How far did they travel? $0.4m + 0.85 = 5.25$; 2.75 miles

45. Thirty is three times a number less nine. Find the number. $30 = 3n - 9; 13$

46. Greg bought four greeting cards, all at the same price, and a package of wrapping paper for $1.79. He spent a total of $5.19. How much was each greeting card? $4c + 1.79 = 5.19$; $.85

Solve each equation for x.

* **47.** $ax + b = c$ $x = \frac{c - b}{a}$ * **48.** $\frac{x}{a} - b = c,\ a \neq 0$ $x = (c + b)a$

49. ***DATA FILE 3 (pp. 96–97)*** Nicholas turned in five halves of one-dollar bills that had accidentally been torn, along with $6 worth of pennies. How much money will he receive? $8.50

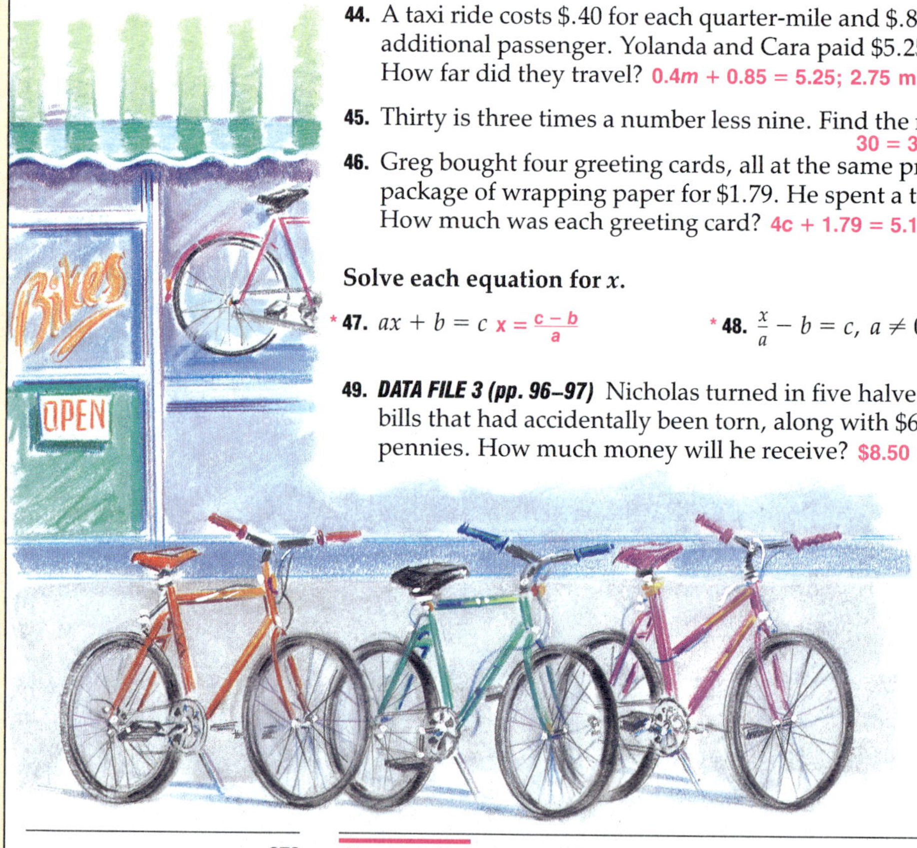

LESSON QUIZ

Solve.

1. $4 - 3x = 16$ (−4)

2. $\frac{x}{6} - 1 = 3$ (24)

Write an equation to find the number. Solve.

3. 15 is 3 less than half a number. ($15 = \frac{n}{2} - 3$; 36)

4. 13 is 5 more than twice a number. ($13 = 2g + 5$; 4)

Assignment Guide

Basic 1-17 odd, 22-24, 28, 30, 32-37, 39, 41-43, 47, MR All

Average 2-16 even, 23-25, 30-37, 40-44, 47-49, MR All

Enriched 15-21, 25-27, 29-35, 37-40, 44-49, MR All

FOR THE NEXT LESSON

calculator, index cards or worksheets, math journal

OBJECTIVE: *To solve two-step equations by combining like terms and using the distributive property on one side.*

7-2 Simplifying and Solving Equations

A bowler scores 20 points more in her second game than in her first game. Her total for both games is 310. What is her score in the first game? Let s represent the score in the first game and $s + 20$ represent the score in the second game. Then the following equation describes the situation.

$$s + s + 20 = 310$$

You can solve this equation by combining like terms.

$s + s + 20 = 310$	
$2s + 20 = 310$	Combine like terms.
$2s + 20 - 20 = 310 - 20$	Subtract 20 from both sides.
$2s = 290$	
$\frac{2s}{2} = \frac{290}{2}$	Divide both sides by 2.
$s = 145$	
Check $s + s + 20 = 310$	
145 (+) 145 (+) 20 (=) 310 ✓	Replace s with 145.

Sometimes you need to use the distributive property before combining like terms.

Example 1 $2(x + 7) - 4x = 8$

Solution

$2x + 14 - 4x = 8$	Distributive property
$-2x + 14 = 8$	Combine like terms.
$-2x + 14 - 14 = 8 - 14$	Subtract 14 from each side.
$-2x = -6$	
$\frac{-2x}{-2} = \frac{-6}{-2}$	Divide both sides by -2.
$x = 3$	
Check $2(x + 7) - 4x = 8$	Replace x with 3.
$2(3 + 7) - 4(3) = 8$	
$2(10) - 12 = 8$	
$20 - 12 = 8$	
$8 = 8$ ✓	

Solving a Multi-step Equation

To solve a multi-step equation:
1. Remove parentheses using the distributive property.
2. Combine like terms.
3. Undo addition or subtraction.
4. Undo multiplication or division.

THINK What if we let s represent the score in the second game. How would the score of the first game be expressed? What would be the new equation?
$s - 20$; $s + s - 20 = 310$

Lesson Focus

Materials
Calculator, index cards or worksheets, math journal

Vocabulary/Symbols
multi-step equation

MOTIVATING THE LESSON
As a warm-up for solving multi-step equations, have students simplify expressions such as the following:
1. $x + 3x + 1 - 2x$ $(2x + 1)$
2. $5 - x + 3x - 7$ $(2x - 2)$
3. $4(3x + 1)$ $(12x + 4)$
4. $3(2x - 5) - 7x$ $(-x - 15)$
5. $2 - 5(1 - 2x) + x - 3$ $(11x - 6)$

Instruction

1 Ask students to solve the new equation $s + s - 20 = 310$ and to describe the result. ($s = 165$; the answer gives the bowler's score in the second game; it does not answer the question stated in the problem.) Point out that students must subtract 20 from 165 to get the bowler's score in the first game.

2 Help students to understand why like terms were combined. (to obtain a simpler two-step equation)

3 Point out that once the distributive property has been applied, the equation resembles the initial one in the preceding example.

Teaching Tip
Have available index cards or worksheet with pairs of equivalent equations. Students can refer to these equivalent equations when solving problems.

4 Remind students that percent means part of a hundred, so 20% is $\frac{20}{100}$ or $\frac{1}{5}$ in simplest form.

ADDITIONAL EXAMPLES

Solve and check.

1. $3x + 9 - 5x = -23$ ($x = 16$)
2. $3(b + 3) = 6$ ($b = -1$)
3. $5(x - 6) + 2x + 1 = 13$ ($x = 6$)
4. Write an equation. Solve. Find two consecutive integers whose sum is 863. ($x + x + 1 = 863$; 431 and 432)
5. A discount store advertises that it sells every item at 30% less than the list price. If the discount store sells a portable TV for \$98, what is its list price? ($p - 0.3p = 98$; \$140)

Guided Practice

Class Exercises To simplify the left side of the equation in question 4, students will have to find the LCD of the fractions. Have students rewrite the equation as $\frac{1}{3}x - \frac{1}{6}x = -7$ and then as $(\frac{1}{3} - \frac{1}{6})x = -7$. This leads readily to the simplified form of $\frac{1}{6}x = -7$.

Think and Discuss Have volunteers explain to the class how they arrived at their answers.

THINK How would you write the equation in Example 2 using a fraction instead of a decimal for 20%? $p - \frac{1}{5}p = 48$ **4**

THINK AND DISCUSS See below.

1. How can you solve the equation $25x + 2 = -73$ using addition and multiplication?
2. Is the solution of the equation $14(2x - 1) = 56$ a negative number? Explain.

1. add -2 to each side; mult. each side by $\frac{1}{25}$
2. No; the equation becomes $28x = 70$. A positive number times a positive number equals a positive number.

▼ You can use an equation to solve word problems.

Example 2 The sale price of a sweater is \$48. The price is 20% less than the original price. What was the original price?

Solution Let p = the original price.
Then $p - 0.2p = 48$ describes the situation.

$p - 0.2p = 48$	
$0.8p = 48$	Combine like terms.
$\frac{0.8p}{0.8} = \frac{48}{0.8}$	Divide both sides by 0.8.
$p = 60$	
Check $p - 0.2p = 48$	
60 ⊟ 0.2 ⊠ 60 ⊜ 48 ✓	Replace p with 60.

CLASS EXERCISES

Simplify the left side of each equation. Do not solve.

1. $4x + 5x = 45$ $9x$
2. $-2(x - 7) = 8$ $-2x + 14$
3. $5 + x - 2x = 8$ $5 - x$
4. $\frac{x}{3} - \frac{x}{6} = -7$ $\frac{x}{6}$

Solve and check each equation.

5. $9x - 2x = -42$ -6
6. $4x + 1 - x = 19$ 6
7. $15 = x - 7x$ $-2\frac{1}{2}$
8. $3(n - 2) = 36$ 14
9. $1.2x + 2.6x = 4.56$ 1.2
10. $\frac{x}{2} - \frac{x}{4} = -\frac{1}{8}$ $-\frac{1}{2}$

Choose the equation that describes the situation. Solve.

11. This year's soybean crop of 224,000 t represents an increase of 40% over last year's crop. How many tons of soybeans were produced last year? c; 160,000
    - a. $n = 224{,}000 - 0.4n$
    - b. $n = 224{,}000 + 0.4n$
    - c. $n + 0.4n = 224{,}000$
    - d. $n - 0.4n = 224{,}000$
12. Sally paid \$22.40 for a sweatshirt. It had been discounted by 30%. What was the original price of the sweatshirt? c; \$32
    - a. $x + 0.30x = 22.40$
    - b. $x + 30x = 22.40$
    - c. $x - 0.30x = 22.40$
    - d. $x = 22.40(0.30x)$
13. Joe bought 2 cartons of milk on Monday and 3 cartons on Tuesday. He spent \$1.75. How much was each carton of milk? d; \$.35
    - a. $2c + 2c = 1.75$
    - b. $2c + 3c = 0.175$
    - c. $5c = 17.5$
    - d. $1.75 = 2c + 3c$

MATH MINUTES

What was done to the first equation to get the second equation?
$5(n - 2) = 20$, $n - 2 = 4$ (divide both sides by 5).
$-2(x - 7) = 12$, $-2x + 14 = 12$ (use the distributive property).

WRITTEN EXERCISES

Solve and check each equation.

1. $5x - x = 11$ $2\frac{3}{4}$
2. $-4(y - 1) = 28$ -6
3. $7 = 2(y + 6)$ $-2\frac{1}{2}$
4. $4(y + 2) = 2$ $-1\frac{1}{2}$
5. $9 = \frac{1}{3}(h - 4)$ 31
6. $0.9t + 2.3t = -6.4$ -2
7. $16 = 2(y - 1) - 6$ 12
8. $6 = a + a + 4$ 1
9. $n + 2 - 3n = -8$ 5
10. $9 - b + 8b = 23$ 2
11. $9(2m + 5) = -14$ $-3\frac{5}{18}$
12. $2x + 4 + 3x = -26$ -6
13. $36 = y - 5y - 12$ -12
14. $\frac{1}{5}(x + 2) = 2$ 8
15. $8 - 3(x - 4) = 6$ $4\frac{2}{3}$
16. $21 = 9 - 2(4a + 2)$ -2
17. $7(2k - 1) + 4 = 7$ $\frac{5}{7}$
18. $15 = -8(c - 1) + 9$ $\frac{1}{4}$
19. $-0.5x + 4 + 2x = 9$ $3\frac{1}{3}$
20. $\frac{2}{3}n + \frac{3}{8}n = \frac{15}{16}$ $\frac{9}{10}$

MENTAL MATH **Find each value.**

21. If $\frac{1}{6}(4x + 5) = 0.5$, find the value of $4x + 5$. 3
22. If $3(x + 5) = 18$, find the value of $3x$. 3
23. If $21 = 2(4 + x) + 7$, find the value of $2x$. 6

CALCULATOR **Solve each equation.** 24. $-3.\overline{814}$

24. $25.8x + 17.3 - 4.2x = -65.1$
25. $238.7 + 1.8(k - 0.2) = 371.9$ 74.2
26. $0 = 2.4x + 9.8x - 0.25$ 0.0204918
27. $-13.5 = 0.8x + 6.7 - 1.3x$ 40.4
28. $12.3 + 18.6x - 3.5(2.8x + 8.5) = 7(2.55 + 2.5)$ 6
29. $2.34(2.2x - 4.66) + 2.352x + 2.4044 = 0.25 + 1.25$ $1.\overline{3}$

Write a situation for each equation. Check students' work.

30. $x + 2x + 3x = 18$
31. $5(x - 1) = 10$

Choose the equation which describes the situation. Solve.

32. Julio bought some pencils for \$.39 each and the same number of erasers for \$.19 each. The total for his supplies was \$4.06. How many pencils did Julio buy? b; 7
    - **a.** $39x + 19 = 4.06$
    - **b.** $0.39x + 0.19x = 4.06$
    - **c.** $0.39 + 0.19x = 4.06$
    - **d.** $0.39x + (0.39)(0.19) = 4.06$
33. Use the table at the right. What was the original price of a pair of running shoes? a; \$70
    - **a.** $0.3p = 21$
    - **b.** $30p = 21$
    - **c.** $\frac{p}{30} = 21$
    - **d.** $\frac{p}{0.3} = 21$

NOTES & QUOTES

The procedure for solving an equation is known as an *algorithm.* The term is based on the name of a Persian mathematician, al-Khowârizmî (c. 780–850).

SALE
ATHLETIC SHOES
30% OFF

Shoe Type	Savings
Basketball	\$15
Track	\$ 9
Running	\$21

A **common error** occurs when students use the distributive property incorrectly. (For example, to solve $5 - 3(1 - x) = 23$, students may write $5 - 3 - 3x = 23$, forgetting that $-3x$ is being subtracted, which results in $3x$. Write these equations on the chalkboard and ask students to find the false equation. (c)
a. $5 - 3(1 - x) = 5 - 3 + 3x$
b. $6 - 2(5 - y) = 6 - 10 + 2y$
c. $7 - 5(2 - c) = 7 - 10 - 5c$
d. $8 - 6(x - 3) = 8 - 6x + 18$

Closure

Writing in Math Ask students to write in their math journals a detailed solution, describing each step, for the equation $5x - (1 - x) = -27$.

Independent Practice

For questions 21-23, point out that it is not necessary to solve for x to find the required values. Ask students what the quickest way is to find the value in question 21. (Multiply both sides of the equation by 6 to get $4x + 5 = 3.0$)

Encourage students' creativity for questions 30-31 by allowing them to share their "stories" with the entire class.

Encourage students to explore equations on a computer spreadsheet. See p. 272E of the Chapter Overview. See also TRB Exploring O-28.

Lesson Follow-up

Reteaching Activity If students have difficulty solving multi-step equations, have them practice using the distributive property, combining like terms, and solving equations of the form $ax + b = c$ separately, before putting them all together.

LESSON RESOURCES

TRB Practice G-3
TRB Enrichment G-4

Practice

For use after 7-2 (pp. 279–282)

Simplifying and Solving Equations

Solve and check each equation.

1. $\frac{p}{3} - 7 = -2$ **$p = 15$**
2. $2(n - 7) + 3 = 9$ **$n = 10$**
3. $0 = 5(k + 9)$ **$k = -9$**
4. $4h + 7h - 16 = 6$ **$h = 2$**
5. $3(2n - 7) = 9$ **$n = 5$**
6. $-27 = 8x - 5x$ **$x = -9$**
7. $4p + 5 - 7p = -1$ **$p = 2$**
8. $7 - y + 5y = 9$ **$y = \frac{1}{2}$**
9. $8e + 3(5 - e) = 10$ **$e = -1$**
10. $h + 3h + 4h = 100$ **$h = 12\frac{1}{2}$**
11. $1.2m + 7.5m + 2.1 = 63$ **$m = 7$**
12. $14 = \frac{2}{3}(9y - 15)$ **$y = 4$**
13. $-37 = 3x + 11 - 7x$ **$x = 12$**
14. $9 - 3(n - 5) = 30$ **$n = -2$**
15. $\frac{1}{6}(y + 42) - 15 = -3$ **$y = 30$**
16. $0.7n - 1.5 + 7.3n = 14.5$ **$n = 2$**

MENTAL MATH Find each value.

17. If $2(x + 7) = 24$, find the value of $2x$. **$2x = 10$**
18. If $y + 4y + 8y = 19$, find the value of $13y$. **$13y = 19$**
19. If $\frac{1}{3}(5k + 2) = 7$, find the value of $5k + 2$. **$5k + 2 = 21$**
20. If $7n + 3n = 46$, find the value of n. **$n = 4.6$**

CALCULATOR Solve each equation.

21. $16.3k + 19.2 + 7.5k = -64.1$ **$k = -3.5$**
22. $93.96 = 4.7p + 8.7p - 2.6p$ **$p = 8.7$**
23. $2.3(x + 1.4) = -9.66$ **$x = -5.6$**
24. $\frac{1}{8.3}(x - 17.7) + 19.6 = 27.8$ **$x = 85.76$**

Write an equation to describe the situation. Solve.

25. Jolene bought three blouses at one price and 2 blouses priced \$3 below the others. The total cost was \$91.50. Find the prices of the blouses.
$3x + 2(x - 3) = 91.50$; \$19.50 and \$16.50
26. Jack's overtime wage is \$3 per hour more than his regular hourly wage. He worked for 5 hours at his regular wage and 4 hours at the overtime wage. He earned \$66. Find his regular wage.
$5h + 4(h + 3) = 66$; \$6/h

Chapter 7 G-3

Enrichment

For use after 7-2 (pp. 279–282)

Simplifying Equations with Fractions

You can use a shortcut to solve equations containing fractions.

Example Solve: $\frac{2}{3}x - \frac{7}{6} = \frac{7}{2}$

First, use the normal method to solve. Complete:

1. $\frac{2}{3}x - \frac{7}{6} + \boxed{\frac{7}{6}} = \frac{7}{2} + \boxed{\frac{7}{6}}$
2. $\frac{2}{3}x = \boxed{\frac{14}{3}}$
3. $\frac{2}{3}x \div \boxed{\frac{2}{3}} = \boxed{\frac{14}{3}} \div \boxed{\frac{2}{3}}$
4. $x = \boxed{7}$

To use the shortcut, first find the LCM of all the denominators appearing in the equation.

5. LCM of 3, 6, and 2 = **6**
6. Multiply both sides of the equation by the LCM.
$\boxed{6}\left(\frac{2}{3}x - \frac{7}{6}\right) = \boxed{6}\frac{7}{2}$
7. Apply the distributive property.
$\boxed{6}\frac{2}{3}x - \boxed{6}\frac{7}{6} = \boxed{21}$
8. Multiply. $\boxed{4x} - \boxed{7} = \boxed{21}$
9. Solve. $x = \boxed{7}$

Solve using the shortcut.

10. $\frac{3}{5}k - \frac{7}{10} = \frac{1}{2}$ **$k = 2$**
11. $\frac{1}{3}m + \frac{1}{6} = \frac{4}{9}$ **$m = \frac{5}{6}$**
12. $\frac{5}{8}y - \frac{1}{2} = \frac{3}{4}$ **$y = 2$**
13. $\frac{5}{7}n - \frac{3}{14} = \frac{1}{2}$ **$n = 1$**
14. $\frac{3}{8}h + \frac{7}{12} = \frac{3}{4}$ **$h = \frac{4}{9}$**
15. $\frac{13}{30}e - \frac{7}{10} = \frac{3}{5}$ **$e = 3$**

G-4 Chapter 7

MIXED REVIEW

Find the LCM of each set of numbers.

1. 6, 8 **24**
2. 2, 5, 6 **30**

Solve and check.

3. $3n + 8 = -44$ **$-17\frac{1}{3}$**
4. $16 = \frac{n}{4}$ **64**

Write an algebraic expression.

5. Ten less than one-fourth of a number **$\frac{1}{4}n - 10$**
6. 12 times a number decreased by 5 times the same number **$12n - 5n$**
7. A student pays \$3.75 for lunch and \$6.92 for a snack after school. He has \$16.48 left. How much money did he start with? **\$27.15**

Write an equation to describe the situation. Solve.

34. I'm thinking of a number. If I subtract 7 and multiply the result by 3, I get 18. What is the number? **$3(n - 7) = 18$; 13**
35. Team A defeated Team B by 13 points. The total number of points scored by both teams was 171. How many points were scored by Team B? **$p + p + 13 = 171$; 79**
36. Find two consecutive integers whose sum is -39. **$x + x + 1 = -39$; -19 and -20**
37. Bill paid \$53 for a sweater. It was 20% off the original price. What was the original price of the sweater? **$p - 0.2p = 53$; \$66.25**
38. Jo Ann paid \$16.25 for an alarm clock. The original price was \$25. What was the percent discount? **$25 - 25d = 16.25$; 35%**
39. I'm thinking of a number. If the number is decreased by 8 and the result is tripled, I get 36. What is the number? **$3(n - 8) = 36$; 20**

Each of these solutions contains a common error. Find each error. State the correct solution.

40. $7x - 1 = 15$
$7x = 14$
$x = 2$
Step 2 should be $7x = 16$; $x = 2\frac{2}{7}$.

41. $3(x - 1) - 5 = 14$
$3x - 1 - 5 = 14$
$3x - 6 = 14$
$3x = 20$
$x = \frac{20}{3}$
Step 2 should be $3x - 3 - 5 = 14$; $x = 7\frac{1}{3}$.

42. ***DATA FILE 3 (pp. 96–97)*** An employee at the United States Mint worked 3.2 h on Monday, 6.8 h on Tuesday, and 6 h on Wednesday. How many sheets of one-dollar bills were produced in this time? How many one-dollar bills were produced? **128,000; 4,096,000**

TEST YOURSELF

Solve each equation.

1. $4a - 7 = -15$ **-2**
2. $52 = \frac{b}{3} - 2$ **162**
3. $8.35 + 0.12s = 9.07$ **6**
4. $3x + 3x + 9 = -9$ **-3**
5. $-8 = -2(y + 5)$ **-1**

Write an equation to describe the situation. Solve.

6. Carmen had \$185 in her account. She withdrew the same amount each week for 11 weeks. She then deposited \$40. At that time, her balance was \$93. How much money did Carmen withdraw each week? **$185 - 11x + 40 = 93$; \$12**

LESSON QUIZ

Solve.

1. $-13(3x - 11) + 2 = 28$
2. $6 - 4\left(2 - \frac{1}{2}x\right) + x = 22$

Write an equation. Solve.

3. I'm thinking of a number. If I subtract 7 and then multiply the result by 3, I get 18. What was my original number?

QUIZ ANSWERS

1. $(x = 3)$
2. $(x = 8)$
3. $[3(n - 7) = 18; 13]$

FOR THE NEXT LESSON

math journal, index cards, five- and twelve-hour clock faces

Assignment Guide

Basic 1-10, 21-24, 28, 30-33, 38, 40, 42, MR, TY All
Average 11-17, 21-25, 28-31, 34-35, 38-42, MR, TY All
Enriched 15-23, 26-31, 36-42, MR, TY All

Problem Solving Practice

PROBLEM SOLVING STRATEGIES

Look for a Pattern
Guess and Test
Simplify the Problem
Make an Organized List
Work Backwards
Account for All Possibilities
Make a Table
Write an Equation
Solve by Graphing
Draw a Diagram
Make a Model
Solve Another Way
Simulate the Problem

Solve. Use an appropriate strategy or a combination of strategies.

1. A student is building a square pen 21 ft long on each side. He puts one post at each corner and one post every 3 ft in between. How many posts will he use? **28**
2. Find two whole numbers whose sum is 125 and whose difference is 23. **51 and 74**
3. There are six basketball teams in a tournament. Each team will compete twice against each of the other teams. What is the total number of games to be played? **30**
4. A student counts 18 legs on the chairs and three-legged stools at the Science Fair exhibit booth. How many chairs are there? how many stools? **3 chairs; 2 stools**
5. The perimeter of a rectangle is 340 cm. The length is 20 more than twice the width of the rectangle. Find the length. **120 cm**
6. A supermarket clerk stacks cans of beans for a display. The clerk puts 10 cans on the bottom row, 9 cans on top of them, and so on. How many cans does the clerk use if the top layer has one can? **55**
7. A student bought some cassette tapes for $7.95 each and a cassette player for $45.98. The total is $77.78. How many cassette tapes did the student buy? **4**
8. A ninth-grade student rides her bike every other day for exercise. Her friend rides his bike every third day. They both rode their bikes on January 2. How many other days in January will they both ride their bikes on the same day? **4**
9. In a competition, one runner runs 280 m farther than another. Together they run 2,870 m. How many meters did each run? **1,295 m and 1,575 m**
10. A student bought a skirt for $21.49 and two blouses for $18.98 each. She then had $23.15. How much money did she have before she made the purchases? **$82.60**
11. The sum of the numbers on two facing pages is 149. Their product is 5,550. What are the page numbers? **74 and 75**
12. There are 156 students at band practice. There are twice as many females as males at the practice session. How many males are there? **52**
13. A painter can paint a square wall that measures 100 ft on a side in 1 h. How long will it take to paint a square wall that measures 50 ft on a side? **15 min**

Instruction

This page provides a variety of problems that can be used to reinforce and enhance the students' problem solving skills. Encourage students to read each problem carefully. Then have them refer to the list of problem solving strategies to help them decide how to solve the problem.

Point out, however, that not all questions require a strategy for solving, nor are all the strategies in the list used in this lesson.

Lesson Focus

Materials

Math journal, index cards, five- and twelve-hour clock faces

MOTIVATING THE LESSON

Review the problem solving strategies students have been using to solve problems. (Such as, guess and test, draw a diagram, find a pattern, write an equation, work backwards, and so on.) Then present this problem and have students decide which strategy would be most appropriate to solve it.

Jessica is paid her regular hourly rate for her 40-hour work week and *time and a half* (1.5 times the normal hourly rate) for overtime. In a week when she worked 44 hours, her pay was \$239.20. Find her regular hourly rate of pay. (Most students will probably decide on the strategy *write an equation*.) If you want students to solve the problem, Jessicas' regular hourly rate is \$5.20/h, which can be obtained from an equation such as $40d + 4(1.5d) = 239.20$

Instruction

1 Be sure students understand what the variable represents.

2 Point out that sometimes the value found for the variable does not answer the question. Ask students to check that their solution answers the question.

PROBLEM SOLVING STRATEGY

OBJECTIVE:
To solve problems by writing equations.

7-3 Writing Equations

You can solve many types of problems using equations.

PROBLEM

The Standard Oil Building has eight times as many stories as the first skyscraper. The Sears Tower has 110 stories, which is 20 more than the total number of stories of the other two skyscrapers. How many stories does the first skyscraper have?

SOLUTION

READ Answer these questions to understand the given information:

What do you want to find?	how many stories are in the first skyscraper
How many stories are in the Standard Oil Building?	eight times as many as are in the first skyscraper
How many are in the Sears Tower?	110 stories
How many more stories are in the Sears Tower than in the other two skyscrapers?	20 more stories

PLAN Write an equation.

1 Let n = the number of stories in the first skyscraper.
Then $8n$ = the number of stories in the Standard Oil Building.
The equation $n + 8n + 20 = 110$ describes the situation.

SOLVE Solve the equation.

2

$$n + 8n + 20 = 110$$
$$9n + 20 = 110 \quad \text{Combine like terms.}$$
$$9n + 20 - 20 = 110 - 20 \quad \text{Subtract 20 from each side.}$$
$$9n = 90$$
$$\frac{9n}{9} = \frac{90}{9} \quad \text{Divide both sides by 9.}$$
$$n = 10$$

The first skyscraper has 10 stories.

LOOK BACK How many stories are in the Standard Oil Building? 80 stories

CLASS EXERCISES

Write an equation for each sentence. Do not solve.

1. Two-thirds of a number is decreased by 7 to obtain 13. $\frac{2}{3}n - 7 = 13$
2. 45 less twice a number is equal to 15. $45 - 2n = 15$

Teaching Tip

Students can use this checklist to make sure they have answered a word problem:

- Read the problem.
- Write the equation.
- Simplify the equation.
- Solve the equation.
- Answer the original question.

WRITTEN EXERCISES

Use a CALCULATOR where appropriate.

Write an equation. Do not solve.

1. The difference between 8 times a number and $\frac{1}{2}$ of the number is 16. Find the number. $8n - \frac{1}{2}n = 16$
2. When half of n is added to three times n, the sum is thirty-five. Find n. $\frac{1}{2}n + 3n = 35$

Write an equation. Solve.

3. A pair of boots cost \$5 more than a pair of shoes. The total cost for both is \$114.90. Find the price of the boots. $n + (n - 5) = 114.90$, where n = price of boots; \$59.95
4. A coin bank contains \$2.80 in dimes and quarters. The bank contains the same number of each coin. How many of each coin does the bank contain? *Hint:* Let n = the number of dimes. Since each dime has a value of \$.10, the value of n dimes is \$.10n. Since the number of quarters is also n, the value of n quarters is \$.25n. $0.10n + 0.25n = 2.80$; 8
5. A pencil and an eraser together cost \$.95. The pencil costs \$.45 more than the eraser. Find the cost of the eraser. $c + c + 0.45 = 0.95$; \$.25
6. A piggy bank contains the same number of pennies, dimes, and quarters for a total of \$13.32. How many of each kind of coin is in the bank? $0.01c + 0.10c + 0.25c = 13.32$; 37
7. A cheese pizza costs \$8.75. Each additional topping costs \$1.25. If a pizza costs \$12.50, how many toppings are on the pizza? $8.75 + 1.25t = 12.50$; 3
8. A wire of uniform thickness and composition weighs 48 lb. The wire is cut into two pieces. The piece that is 120 yd long weighs 32 lb. Find the original length of the wire. $\frac{32}{120} = \frac{48}{x}$; 180 yd
9. The booster club sold 75% of the tickets printed for the raffle. They did not sell 175 tickets. How many tickets did they have printed? $x - 0.75x = 175$; 700
10. A car rental agency charges \$27.95 a day plus \$.14/mi. Pat's bill for 3 days was \$154.83. How many miles did she drive? $3(27.95) + 0.14x = 154.83$; 507 m
11. Jan has \$240 in the bank to pay for her tuba lessons. A lesson costs \$15. How many lessons can she afford with her savings? $15x = 240$; 16
12. The perimeter of a rectangle is 64 cm. The length is 4 cm less than twice the width. Find the length and width. $w + 2w - 4 + w + 2w - 4 = 64$; 12 cm and 20 cm
13. A number s is multiplied by $\frac{2}{3}$. Then $\frac{2}{5}$ is subtracted from the product. The result is $\frac{11}{45}$. What number is s? $\frac{2}{3}s - \frac{2}{5} = \frac{11}{45}$; $\frac{29}{30}$
14. Water flows over the Niagara Falls at a rate of 1.5 million gal/s. How many gallons flow over the falls in 1 h 15 s? 5,422.5 million gal $[(60 \cdot 60) + 15]1.5 = x$

Guided Practice

Class Exercises Discuss reading word problems carefully. Point out that changing a single word could make a difference in the result. To illustrate this, ask students to insert the word *than* between *less* and *twice* in question 2 and to write the equation for the new sentence. (45 less than twice a number is equal to 15; $2n - 45 = 15$)

A ***common error*** occurs when students solve the equation and fail to answer the question posed in the problem. Remind students to check that the solution to the equation also answers the question.

Closure

Writing in Math Have students describe in their math journals the essential steps for using the strategy writing equations.

Independent Practice

For question 8, remind students that a proportion is an equation.

To write the equation in question 12, students need to use the perimeter formula.

To reinforce that more than one problem solving strategy can be used to solve a problem, have students suggest alternative strategies for questions 13–19. (For example, guess and test for questions 7 and 13.)

MATH MINUTES

Divide the class into small groups. Give each group sets of word problem and equation cards and have students take turns matching the word problem cards with the appropriate equation cards.

Critical Thinking Point out that adding or subtracting groups of 12 on a 12-hour clock has no effect on the time. Therefore, adding 16 hours to a displayed time of 10 o'clock is the same as 16 − 1(12) = 4 hours or 16 − 2(12) = -8 hours because you arrive at 2 o'clock whether you count forward 4 hours, or back 8 hours, from 10 o'clock.

Lesson Follow-up

LESSON RESOURCES

TRB Practice G-5
TRB Enrichment G-6

Practice

For use after 7-3 (pp. 284–286)

Writing Equations

Write an equation. Do not solve.

1. The sum of half of a number and 8 less than the number is 25.
   $\frac{1}{2}n + (n - 8) = 25$
2. The product of 6 and 3 more than k is 48.
   $6(k + 3) = 48$

Write an equation. Solve.

3. Bill purchased 4 pens for $3.32, including $.16 sales tax. Find the cost of 1 pen.
   $4p + 0.16 = 3.32$; $p = 0.79$; $.79
4. Arnold had $1.70 in dimes and quarters. He had 3 more dimes than quarters. How many of each coin did he have?
   $25q + 10(q + 3) = 170$; $q = 4$; 4 quarters, 7 dimes
5. A baby weighed 3.2 kg at birth. She gained 0.17 kg per week. How old was she when she weighed 5.75 kg?
   $3.2 + 0.17w = 5.75$; $w = 15$; 15 weeks
6. In the parking lot at a truck stop there were 6 more cars than 18-wheel trucks. There were 134 wheels in the parking lot. How many cars and trucks were there?
   $18t + 4(t + 6) = 134$; $t = 5$; 5 trucks, 11 cars
7. A bottle and a cap together cost $1.10. The bottle cost $1 more than the cap. How much did each cost?
   $c + (c + 1) = 1.1$; $c = 0.05$; cap = $.05, bottle = $1.05
8. Orlando worked for $6/h one week and $7/h the next week. He worked 5 more hours the second week than the first and earned $347 for the 2 weeks of work. How many hours did he work each week?
   $6h + 7(h + 5) = 347$; $h = 24$; 24h and 29h
9. A triangle has two sides equal in length and a third side 5 in. longer than half the length of each of the other two sides. If the perimeter of the triangle is 50 in., how long is each side?
   $s + s + (\frac{1}{2}s + 5) = 50$; $s = 18$; 18 in., 18 in., 14 in.

Chapter 7 G-5

Enrichment

For use after 7-3 (pp. 284–286)

Work Problems

You can use equations to solve work problems.

Marie can clean a garage in 3 h. Megan can clean it in 4 h. How long would it take them to clean the garage if they worked together?

The key to solving work problems is to look at what happens in 1 h.

	Amount Marie does in 1 h	+	Amount Megan does in 1 h	+	Amount they do together in 1 h
	$\frac{1}{3}$	+	$\frac{1}{4}$	=	$\frac{1}{h}$
Add the fractions.			$\frac{7}{12}$	=	$\frac{1}{h}$
Cross multiply.			$7h$	=	12
Solve.			h	=	$\frac{12}{7}$ or $1\frac{5}{7}$

Cleaning the garage will take them $1\frac{5}{7}$ h if they work together.

Write an equation. Solve.

1. Marjorie can wash the car in 2 h. Brad can wash it in 4 h. How long will the job take if they work together?
   $\frac{1}{2} + \frac{1}{4} = \frac{1}{h}$; $h = 1\frac{1}{3}$; $1\frac{1}{3}$ h
2. Joe can wax the gym in 3 h. Steve can wax it in 2 h. How long will the job take if they work together?
   $\frac{1}{3} + \frac{1}{2} = \frac{1}{h}$; $h = 1\frac{1}{5}$; $1\frac{1}{5}$ h
3. Jessie can paint a barn in 8 h. Jerome needs 10 h to do the same job. How long will it take them working together?
   $\frac{1}{8} + \frac{1}{10} = \frac{1}{h}$; $h = 4\frac{4}{9}$; $4\frac{4}{9}$ h
4. Lucy can mow the golf course in 6 h. Stan needs 8 h to do the same job, and Yolanda needs 12 h. How long will it take them to mow the course if they work together?
   $\frac{1}{6} + \frac{1}{8} + \frac{1}{12} = \frac{1}{h}$; $h = 2\frac{2}{3}$; $2\frac{2}{3}$ h
5. A swimming pool has 3 drains. One can drain the pool in 5 h, one can drain it in 8 h, and one can drain it in 10 h. How long would it take to drain the pool if all three were open at once?
   $\frac{1}{5} + \frac{1}{8} + \frac{1}{10} = \frac{1}{h}$; $h = 2\frac{6}{17}$; $2\frac{6}{17}$ h

G-6 Chapter 7

Solve.

15. Miss Zawtocki teaches a total of 40 students. The ratio of female students to male students is 1 to 4. How many female students does she teach? **8**
16. The weight of an object on the moon is about $\frac{1}{6}$ of its weight on Earth. If an astronaut weighs 134 lb on Earth, how much would she weigh on the moon to the nearest tenth of a pound? **22.3 lb**
17. ***DATA FILE 1 (pp. 2–3)*** What is the danger of frostbite if the air temperature is 10°F and the wind speed is 25 mi/h? **increased danger**
18. Approximately 27 million acres of rain forest are cut and burned each year. If this rate continues, how many acres will be destroyed in the next decade? **270 million acres**
19. Mr. Macintosh has 42 rows of apple trees in his orchard. Each row contains 24 trees. Mr. Macintosh is expecting to harvest 18,144 bu of apples this year. On the average, how many bushels of apples does each tree produce? **18 bu**

Critical Thinking

EXPLORING CLOCK ARITHMETIC

Visualize a 12-hour clock face. Start at 10:00. Count forward 16 h. What time will it be?

To find the time using clock arithmetic, follow these steps.

1. To find the *clock time,* add the starting time to the number. 10 + 16 = 26
2. Use the following formula to determine what the time will be.
   clock time − (hours on clock face)(groups of 12 in clock time)
   26 − (12)(2) = 2, so the time will be 2 o'clock.

Solve.

1. It is 3:00. What time will it be in 7 h? **10:00**
2. It is 6:00. What time will it be in 17 h? **11:00**
3. It is 4:00. How many hours ago was it 9 o'clock? **7 h**
4. **a.** Suppose a clock shows five hours on its face. How could you adapt the formula so that you could use clock arithmetic? Write the formula. **clock time − (5)(groups of 5 in clock time)**
   b. It is 4 o'clock. What time will it be in 17 h? **1:00**
   c. It is 1 o'clock. What time was it 29 hours ago? **2:00**

Assignment Guide

Basic All
Average All
Enriched All

FOR THE NEXT LESSON

overhead projector, algebra tiles or blue and white squares of paper and red paper rectangles, math journal, calculator

OBJECTIVE:
To solve equations with an unknown on both sides.

7-4 Equations with Variables on Both Sides

▼ The Ricardos have three children whose ages are consecutive even integers. The sum of the children's ages is equal to four times the youngest child's age. What are the ages of the Ricardos' children?

Let a = the youngest child's age.
Then $a + 2$ = the middle child's age.
Then $a + 4$ = the oldest child's age.

We can write the following equation to describe this situation.

$$a + a + 2 + a + 4 = 4a$$

You can solve this equation by using a model.

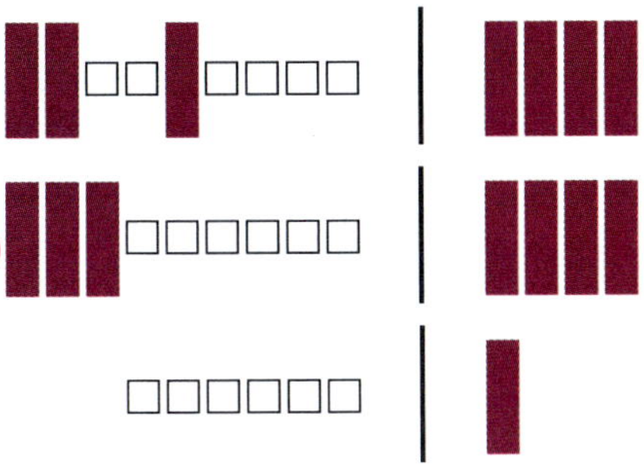

Model the equation.

Combine like terms.

Subtract $3a$ from each side of the equation.

The value of a is 6. So the ages of the Ricardo children are 6, 8, and 10.

▼ You can also solve equations with variables on both sides by using algebraic properties.

Example 1 **Solve $3x + 7 = 5x - 1$.**

Solution

$3x + 7 = 5x - 1$

$3x + 7 - 3x = 5x - 1 - 3x$ — Subtract $3x$ from each side.

$7 = 2x - 1$

$7 + 1 = 2x - 1 + 1$ — Add 1 to each side.

$8 = 2x$

$\frac{8}{2} = \frac{2x}{2}$ — Divide each side by 2.

$4 = x$

Check $3x + 7 = 5x - 1$

$3(4) + 7 = 5(4) - 1$ — Replace x with 4.

$12 + 7 = 20 - 1$

$19 = 19$ ✓

Lesson Focus

Materials

Overhead projector, algebra tiles or blue and white squares of paper and red paper rectangles, math journal, calculator

MOTIVATING THE LESSON

Write these equations on the chalkboard.

$5x = -15$
$5x + 3 = -12$
$2x + 3x + 3 = -12$
$2x + 3(x + 1) = -12$

Ask students to copy the equations, one at a time, in their math journals and to describe the first step for solving each equation. Ask what happens after the first step is applied. (A simpler equation is obtained.) Then mention that today students will solve equations such as $m + m + 3 + m - 2 = 5m$ and $x + 4(x - 2) = x + 15$. Ask students to tell what they notice about these equations. (The variable appears on both sides of the equation.)

Instruction

1 Use the overhead projector and the tiles to model the equation and its solution. Use a piece of paper with a vertical bar to represent the equal sign. At the same time, have groups of students use tiles to model the equation and the solution process at their desks.

2 It may be useful to have groups of students use the tiles to model the action of each step of the solution process.

Teaching Tip

Have the equations for *Motivating the Lesson* written on the chalkboard in preparation for the lesson.

3 Ask students how this equation differs from the one in Example 1. (More than one *term* on each side of the equation contains the variable.)

4 ***Writing in Math*** Have students write in their math journals, the equations for Examples 1-3 along with the descriptions of the first steps in solving them.

5 Encourage students to refer to this summary when solving equations. Help students to see that, depending upon the equation to be solved, any one of the four steps can be used as the first step in solving it.

ADDITIONAL EXAMPLES

Solve and check.

1. $3x + 4 = 4x - 1$ $(x = 5)$
2. $7x - 2 = -3x + 18$ $(x = 2)$
3. $11m - 13 - 5m = 8m + 3$ $(m = -8)$
4. $3(x + 1) + 1 = 4x - 1$ $(x = 5)$

Guided Practice

Class Exercises For students who have difficulty in using the distributive property with fractions in question 10, suggest that they begin by multiplying both sides of the equation by 4 to eliminate the fractions in the first step as shown:

$\frac{1}{4}(d + 2) = \frac{3}{4}d - 6$
$4[\frac{1}{4}(d + 2)] = 4(\frac{3}{4}d - 6)$
$d + 2 = 3d - 24.$

Think and Discuss For question 1, ask students to describe two ways to get the x terms on the left side of the equation. (Subtract x from both sides or add $-x$ to both sides of the equation.)

▼ When more than one term on a side of an equation contains a variable, you need to combine like terms before you add or subtract.

Example 2 Solve $5x - 2 - 3x = 2x + 7 - x$.

Solution

3

$5x - 2 - 3x = 2x + 7 - x$	
$2x - 2 = x + 7$	Combine like terms.
$2x - 2 - x = x + 7 - x$	Subtract x from each side.
$x - 2 = 7$	
$x - 2 + 2 = 7 + 2$	Add 2 to each side.
$x = 9$	

Check

$5x - 2 - 3x = 2x + 7 - x$	
$5(9) - 2 - 3(9) = 2(9) + 7 - 9$	Replace x with 9.
$45 - 2 - 27 = 18 + 7 - 9$	
$16 = 16$ ✓	

▼ You may need to use the distributive property to remove parentheses before combining like terms.

Example 3 Solve $n + 2(n + 2) = n + 16$.

Solution

4

$n + 2(n + 2) = n + 16$	
$n + 2n + 4 = n + 16$	distributive property
$3n + 4 = n + 16$	Combine like terms.
$3n + 4 - n = n + 16 - n$	Subtract n from each side.
$2n + 4 = 16$	
$2n + 4 - 4 = 16 - 4$	Subtract 4 from each side.
$2n = 12$	
$\frac{2n}{2} = \frac{12}{2}$	Divide both sides by 2.
$n = 6$	The check is left for you.

THINK How is adding and subtracting variables similar to adding and subtracting numbers from both sides of an equation? **You combine like terms by adding or subtracting their coefficients. This is like combining numbers.**

5

Solving Equations with Variables on Both Sides

To solve equations with variables on both sides:
1. Use the distributive property to remove parentheses.
2. Combine like terms.
3. Use the addition and/or subtraction properties so that variables are on one side and constants on the other.
4. Use the multiplication or division property.

1. Subtract x from each side.
2. First, solve for the variable term. Second, solve for the variable.

THINK AND DISCUSS **See above.**

1. To solve the equation $6x = x - 10$, what should you do first?
2. How are the steps for solving equations related to the strategy, *Simplify the Problem?*

CLASS EXERCISES

State the first step in solving each equation.

1. $4x = 9x + 50$ **Subtract 9x.**
2. $2x - 9 = 27$ **Add 9.**
3. $4 - x + 6x = 10 + x - 1$ **Combine like terms on each side.**
4. $3(2x - 0.3) = 15 - (x + 2)$ **Distribute 3 and -1.**

MATH MINUTES

What is the first step?
1. $3x - 1 + 9x = 6x + 7 + 8x$
2. $3(2 - 3x) - 4(x - 1) = 2 - 5x$

(1. Combine like terms.
2. Remove parentheses, use the distributive property.)

Write an equation for each diagram. Solve.

5.

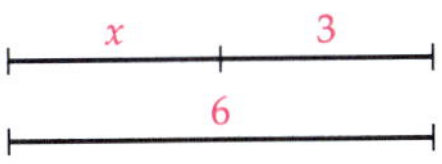

$x + 3 = 6$; 3

6.

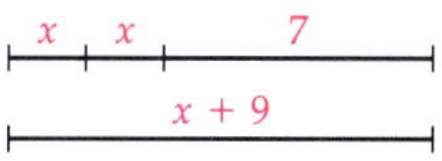

$2x + 7 = x + 9$; 2

Solve and check each equation.

7. $n + 12 = 5n$ 3

8. $-2r + 7 = r - 8$ 5

9. $9 - (x - 4) = 3(x - 1)$ 4

10. $\frac{1}{4}(d + 2) = \frac{3}{4}d - 6$ 13

WRITTEN EXERCISES

Write an equation for each diagram. Solve.

1.

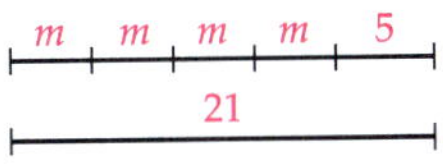

$4m + 5 = 21$; 4

2.

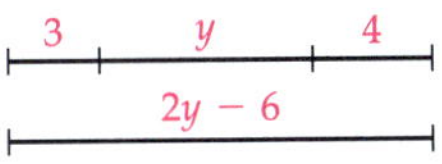

$y + 7 = 2y - 6$; 13

MENTAL MATH Solve.

3. $x + x + x = x + 6$ 3

4. $3x + 20 = 8x$ 4

5. $x + 7 = 2x + 6$ 1

6. $2(x + 4) = 3x$ 8

Solve each equation.

7. $5x + 8 = 7x$ 4

8. $3a = a + 22$ 11

9. $8.6 + 2.1x = -0.05x$ -4

10. $m + m + 18 = 4m$ 9

11. $4w + 8 = 6w - 4$ 6

12. $7r = 2(r - 10)$ -4

13. $6(h + 3) = -2(h + 31)$ -10

14. $-2(y + 6) = y + 3 + 2y$ -3

15. $9 - (2y - 3) = y$ 4

16. $4 - 7t = 2(t - 7)$ 2

17. $2g + 6 = -g - 8$ $-4\frac{2}{3}$

18. $5(n - 3) = 2n - 6$ 3

19. $7b = b + 16 + 2b$ 4

20. $0.3k + 1.4 = 4.2 - 0.1k$ 7

21. $4(8 - k) = 2k + 16$ $2\frac{2}{3}$

22. $m - 16 = 3m + 18 + 2m$ $-8\frac{1}{2}$

23. $\frac{1}{5}(x + 8) = \frac{4}{5}x - \frac{1}{5}$ 3

24. $\frac{1}{2}(2h + 4) = \frac{1}{3}(h - 4)$ -5

Is the given number a solution of the equation?

25. $b + 6 = 3(b - 4)$, $b = 9$ yes

26. $10 - 6m = 2(m - 3)$, $m = -3$ no

27. $-f + 3f = f + 27$, $f = 27$ yes

28. $a - 8 - 2 = \frac{1}{2}(a - 2)$, $a = 18$ yes

CALCULATOR Solve.

29. $3.6a - 6.2 - 0.1a = 1.5 + 0.2(a + 6)$ 2.69

30. $4 - 0.6a = 3(1.5a + 0.9) + 7.93$ -1.3

MIXED REVIEW

Write an algebraic expression. See below.

1. twice the result of increasing a number by 4

2. \$14 more than the cost of a cassette $c + 14$

Write an equation. Solve. See below.

3. Six less than three times a number is twelve.

4. Fifty-six is eight more than six times a number.

Replace each ■ with <, >, or =.

5. $\frac{12}{-4}$ ■ $-2 - 1$ =

6. 3 ■ $\sqrt{9}$ =

7. A student pays \$168 for a coat and jacket. The jacket costs \$18 less than the coat. What is the price of the coat? \$93

1. $2(n + 4)$
3. $3n - 6 = 12$; 6
4. $56 = 6x + 8$; 8

A ***common error*** occurs when students try to combine like terms from both sides of an equation, such as, $m - 16 + 2m = 3m + 18 + 4m$, instead of simplifying each side separately. Have students cover the equal sign and one side of the equation. Stress that they must simplify each side of the equation separately. They can solve the equation when there are no more than two terms, a variable term and a constant term, on each side of the equal sign.

Closure

Present the equation $\frac{1}{4}(d + 2) = \frac{1}{3}d - 8$. Ask students how they might apply the alternative method of eliminating the fractions, as discussed in Class Exercise 10, to this equation. (Multiply both sides of the equation by 12.)

Independent Practice

Ask students how they could eliminate the decimals in the equations in question 9 or question 20, before they apply the steps for solving equations with the variable on both sides. (Multiply both sides of the equation by the appropriate power of ten.)

Lesson Follow-up

Reteaching Activity For students who have difficulty with this lesson refer to the manipulative activity found on page 272E of the Chapter Overview. See also TRB Exploring O-25.

LESSON RESOURCES

TRB Practice G-7
TRB Enrichment G-8

Write an equation to describe the situation. Solve.

31. Find three consecutive integers whose sum is 165. $x + x + 1 + x + 2 = 165$; 54, 55, and 56

32. Find three consecutive odd integers whose sum is 87. $x + x + 2 + x + 4 = 87$; 27, 29, and 31

33. One more than one-half of a number is one less than two-thirds of the number. What is the number? $\frac{1}{2}n + 1 = \frac{2}{3}n - 1$; 12

34. If a number is subtracted from 18, the result is four less than the number. What is the number? $18 - n = n - 4$; 11

35. Twice a number less eight is 16 less than three times the number. What is the number? $2n - 8 = 3n - 16$; 8

36. Find four consecutive integers whose sum is negative two. $x + x + 1 + x + 2 + x + 3 = -2$; -2, -1, 0, and 1

37. ***DATA FILE 9 (pp. 360–361)*** Twice the measure of the loft of a 6 iron minus x is equal to the measure of the loft of a 9 iron. $2(35) - x = 47$; $x = 23$

Use the article below.

38. a. Use Diophantine symbols. Write the numerals 56, 129, and 683. *nf; ski; xqc*

b. ***WRITE*** Will a letter in any number written in Diophantine symbolism ever come before a letter of later rank in the alphabet? Explain.

c. Determine how many years Diophantus lived. *Hint:* Let his age a equal the expression in the article below. Simplify the expression and solve for a. 84y

38. b. No; higher numbers correspond with letters near the end of the alphabet.

Diophantine Symbolism Using Roman Signs

1	2	3	4	…	9
a	b	c	d	…	i
10	20	30	…	90	
j	k	l	…	r	
100	200	300	…		
s	t	u	…		

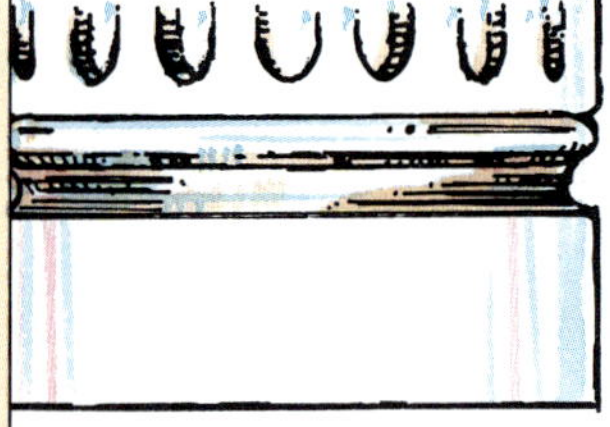

The Father of Algebra

Diophantus was a Greek mathematician who lived in the third century. He is called the father of algebra because he was the first mathematician to use symbols. For example, Diophantus expressed numerals as shown in the table at the left. The number 234 would have been written as *tld*.

We know how long Diophantus lived from a description of his life in terms of an algebraic riddle. The riddle states, "Diophantus' youth lasted one sixth of his life. He grew a beard after one twelfth more. After one seventh more of his life he married. Five years later he had a son. The son lived exactly one-half as long as his father and Diophantus died four years after his son. All of this adds up to the years Diophantus lived."

Adding the parts of Diophantus' life results in the following expression:

$$\frac{1}{6}a + \frac{1}{12}a + \frac{1}{7}a + 5 + \frac{1}{2}a + 4$$

Practice

For use after 7-4 (pp. 287–290)

Equations with Variables on Both Sides

MENTAL MATH Solve.

1. $n + n + n + n = n + n + n + 10$ **n = 10**
2. $5x + 7 = 6x$ **x = 7**
3. $k + 12 = 3k$ **k = 6**
4. $8m = 5m + 12$ **m = 4**
5. $3p - 9 = 4p$ **p = -9**

Solve each equation.

6. $3k + 16 = 5k$ **k = 8**
7. $5e = 3e + 36$ **e = 18**
8. $n + 4n - 22 = 7n$ **n = -11**
9. $2(x - 7) = 3x$ **x = -14**
10. $8h - 10h = 3h + 25$ **h = -5**
11. $7n + 6n - 5 = 4n + 4$ **n = 1**
12. $11(p - 3) = 5(p + 3)$ **p = 8**
13. $9(m + 2) = -6(m + 7)$ **m = -4**
14. $y + 2(y - 5) = 2y + 2$ **y = 12**
15. $-9x + 7 = 3x + 19$ **x = -1**
16. $0.2n + 13 = 1.3n - 14.5$ **n = 25**
17. $-6(4 - t) = 12t$ **t = -4**
18. $\frac{1}{2}(e - 6) = \frac{1}{4}(e + 6)$ **e = 18**
19. $5m + 9 = 3(m - 5) + 7$ **$m = -\frac{17}{2}$**
20. $\frac{2}{3}p + 12 = \frac{3}{5}p + 10$ **p = -30**
21. $x + 7x + 15x = 29x + 18$ **x = -3**
22. $3(x + 7) + 2(x - 5) = x - 5$ **x = -4**
23. $5g = 6g$ **g = 0**

Is the given number a solution of the equation?

24. $k + 9 = 6(k - 11)$, $k = 15$ **yes**
25. $2(x + 7) = 5(x - 7)$, $x = 3$ **no**
26. $8(7 - p) - 8 = -16(p - 2)$, $p = -2$ **yes**

CALCULATOR Solve.

27. $3.2c + 9.4 - 0.8c = 3(c - 2.4) + 7.6$ **c = 15**
28. $6.9(3.5x - 2.7) = 2.3(2.1x + 17.1)$ **x = 3**

Write an equation. Solve.

29. The difference when 7 less than a number is subtracted from twice the number is 12. What is the number?
$2n - (n - 7) = 12$; $n = 5$
30. Four less than three times a number is three more than two times the number. What is the number?
$3n - 4 = 2n + 3$; $n = 7$

Chapter 7 G-7

Enrichment

For use after 7-4 (pp. 287–290)

Equations with Two Variables

An equation containing one variable has only one solution.

Equation: $3x + 6 = 21$
Solution: $x = 5$

An equation containing two variables may have many solutions. The table lists four solutions of the equation $2x + y = 30$. For example, $x = 1$ and $y = 28$ is a solution of the equation because $2(1) + (28) = 30$.

x	y
1	28
2	26
3	24
4	22

1. Find three more solutions of the equation $2x + y = 30$. **Answers may vary.**
$x =$ **5**, $y =$ **20**, $x =$ **6**, $y =$ **18**, $x =$ **7**, $y =$ **16**

Complete each table for the given equation and values of x.

2. $x - y = 12$

x	y
15	3
14	2
10	-2
8	-4

3. $x + y = 4$

x	y
-1	5
2	2
5	-1
6	-2

4. $3x + y = 19$

x	y
3	10
4	7
5	4
7	-2

5. $5x - y = 11$

x	y
4	9
3	4
0	-11
-1	-16

6. $2x - 3y = 16$

x	y
17	6
14	4
11	2
5	-2

7. $3x + 2y = 13$

x	y
1	5
3	2
5	-1
7	-4

Find an equation with the solutions given in the table. **Equations may vary.**

8.

x	y
8	3
6	5
4	7
2	9
11	0

Equation: **$x + y = 11$**

9.

x	y
20	6
16	2
14	0
9	-5
7	-7

Equation: **$x - y = 14$**

10.

x	y
1	24
2	22
3	20
4	18
5	16

Equation: **$2x + y = 26$**

G-8 Chapter 7

LESSON QUIZ

Solve and check.

1. $5x + 5 = 7x + 13$ $(x = -4)$

2. $3x - 1 + 9x = 6x + 7 + 8x$ $(x = -4)$

3. $3(2 - 3x) - 4(x - 1) = 2 - 5x$ $(x = 1)$

4. Find three consecutive even integers if the third is three times the first.
(2, 4, and 6)

Assignment Guide

Basic 1, 3-4, 7-23 odd, 25-26, 29, 31-32, 38, MR All

Average 1, 4-5, 8-24 even, 26-27, 29, 33-34, 37-38, MR All

Enriched 2, 5-6, 16-24, 27-28, 30, 35-38, MR All

FOR THE NEXT LESSON

paper, pencil, ruler, math journal

OBJECTIVE:
Use models to explore inequalities.

MATERIALS

- Paper
- Ruler
- Math journal to record work

Exploring Inequalities

▼ Study the number line models to see what happens to each inequality sign for $-2 < 4$ when you multiply each side of the inequality by a positive or negative number.

a. Add 3, a positive number.

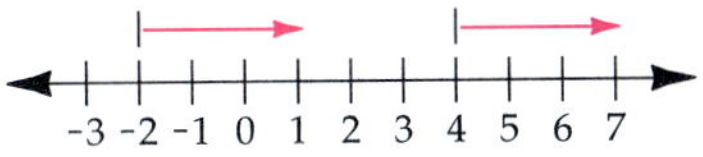

$-2 < 4$
$-2 + 3 \,\blacksquare\, 4 + 3$
$1 < 7$

b. Add -3, a negative number.

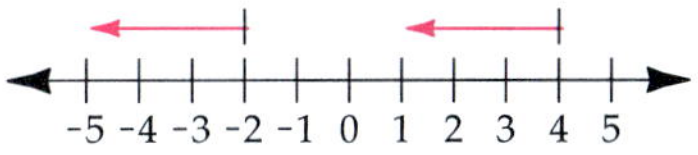

$-2 < 4$
$-2 + (-3) \,\blacksquare\, 4 + (-3)$
$-5 < 1$

c. Multiply each side by 2, a positive number.

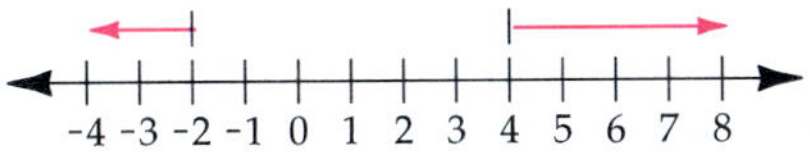

$-2 < 4$
$2 \cdot (-2) \,\blacksquare\, 2 \cdot 4$
$-4 < 8$

d. Multiply each side by -1, a negative number.

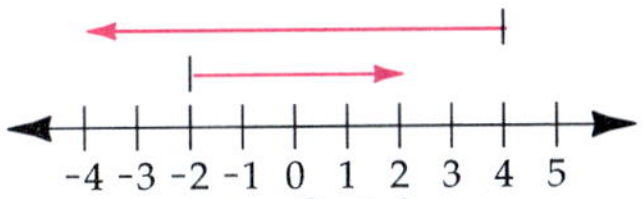

$-2 < 4$
$-1 \cdot (-2) \,\blacksquare\, -1 \cdot 4$
$2 > -4$

1. ***Describe*** what you notice about the direction of the arrows in the inequalities modeled above.

1. When a pos. or neg. number was added to both sides, the signs stayed the same. When both sides were mult. by a pos. number, the signs stayed the same. When both sides were mult. by a neg. number, the signs changed directions.

(1) 2. Try adding other positive and negative numbers to the inequality $-2 < 4$. ***Model*** on the number line. ***Explain*** what happens to the inequality sign. The sign stays the same.

3. ***Explore*** what happens to the inequality sign if you subtract the same number from each side of an inequality. The sign stays the same.

4. Try multiplying each side of $-2 < 4$ by other positive and negative numbers and model on the number line. ***Explain*** what happens to the inequality sign. The signs stay the same when mult. by a pos. number. They change direction when mult. by a neg. number.

5. ***Explore*** what happens if you divide each side of the inequality $-2 > 4$ by 2 and by -2. The sign stays the same when div. by 2. It changes direction when div. by -2.

(2) 6. ***Summarize*** Write a rule that tells what happens to an inequality sign if you perform the following operations on each side of the inequality. Check students' work

   a. Add or subtract either a positive number or a negative number. stays the same
   b. Multiply or divide by a positive number. stays the same
   c. Multiply or divide by a negative number. changes direction

(3) 7. How are the rules for inequalities similar to the rules for equations? How are they different?

7. You can add., sub., mult., and divide values on each side of an inequality or equation. When mult. or dividing both sides of an inequality by a neg. number, the direction of the sign changes.

Lesson Focus

Materials
Paper, pencil, ruler, math journal

MOTIVATING THE LESSON

Write several pairs of integers on the chalkboard, two positive numbers, one positive and one negative, and two negative numbers. For each pair, ask students which value is greater. Then write each pair as an inequality on the chalkboard and read each inequality aloud to the class.

Instruction

Writing in Math Have students work in groups and write their answers in their math journals.

(1) Encourage groups to explore several examples before they generalize.

(2) Remind groups that everyone must agree on the rules they write. Ask one member of each group to read the group's rules to the class and discuss the rules.

(3) Discuss the similarities and differences with the class.

Closure

Have students review the rules they have written in their math journals and make any necessary corrections. Encourage students to use symbols as well as words.

Assignment Guide
Basic All
Average All
Enriched All

FOR THE NEXT LESSON

calculator, math journal, computer (optional)

Teaching Tip

Have number lines with integers clearly marked displayed around the room so that students can verify their answers and reinforce number theory concepts.

Lesson Focus

Materials

Calculators, math journal, computer (optional)

Vocabulary/Symbols

inequalities

MOTIVATING THE LESSON

Review with students the meaning of the symbols < and >, reminding students that the wider part points to the greater number. Ask students to make each statement true.

1. -5 ■ -1 (<)
2. $\sqrt{4}$ ■ 2 (=)
3. If 3 is to the right of -5 on a number line, then 3 ■ -5. (>)

Have students illustrate their answers on a number line.

Instruction

1 Remind students that every point on a real number line corresponds to a decimal, either rational (terminating or repeating) or irrational (non-terminating and non-repeating).

2 If students are having difficulty, ask them how many numbers satisfy the inequality $y < 3$. (infinitely many)

3 Ask students to read the inequality $x \neq 4$ and to explain what it means. (x is not, or does not, equal 4; x can be any real number except 4.) Have students describe how they would graph $x \neq 4$. (Place an open dot at 4; show all numbers to the left of 4 and all numbers to the right of 4.)

OBJECTIVE: *To write and graph simple inequalities with one variable.*

7-5 Inequalities and Their Graphs

No one under 17 is admitted to theaters showing R-rated films unless they are accompanied by an adult. If you let a represent age, you can write the *inequalities* $a < 17$ and $a > 0$ to describe this situation.

Inequality	An inequality is a statement that two expressions are not equal.

You can graph an inequality on a number line.

Example 1 **Graph each inequality.**

a. $y < 3$ **b.** $x > -1$

Solution **a.**

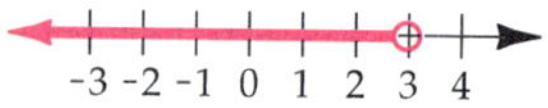

1

The value for y is any real number less than 3. We place an open dot above 3 to show that 3 is not a solution.

b.

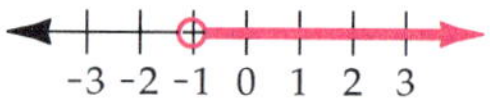

The value for x is any real number greater than -1. We place an open dot above -1 to show that -1 is not a solution.

THINK Why is graphing an inequality an easier way to show the solutions than listing them? There are too many values to list.

2

FLASHBACK

≤ is less than or equal to

≥ is greater than or equal to

3

You can also graph inequalities that include an equal sign.

Example 2 **Graph each inequality.**

a. $a \leq -2$ **b.** $g \geq -6$

Solution **a.**

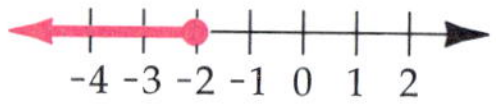

The value for a is -2 or any real number less than -2. We place a solid dot above -2 since -2 is a solution.

b.

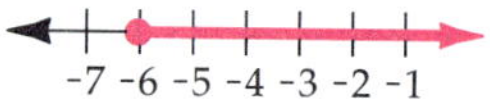

The value for g is -6 or any real number greater than -6. We place a solid dot above -6 since -6 is a solution.

Example 3 **State whether the endpoint of the graph of each inequality would be a solid dot or an open dot.**

a. $y + 3 > 12$ **b.** $3a \geq -21$ **c.** $6x \leq 18$

Solution **a.** open **b.** solid **c.** solid

Teaching Tip

Have the overhead projector ready for students to share their graphs or demonstrate graphing procedures with the class.

▼ You can write an inequality for a graph.

Example 4 **Write an inequality for each graph.**

a. 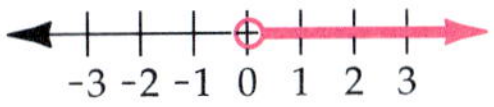

b. 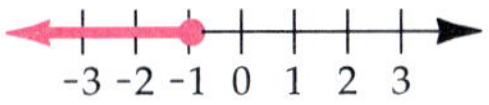

Solution a. $x > 0$ b. $x \leq -1$

▼ You can write an inequality for a sentence or a sentence for an inequality.

Example 5 a. Write an inequality for p is greater than -2.

b. Write a sentence for the inequality $b \leq 8$.

Solution a. $p > -2$ b. b is less than or equal to 8.

▼ You can write an inequality to describe a situation.

Example 6 **Write an inequality to describe the situation.**

There are more than 15 girls in the class.
Let g = the number of girls in the class.

Solution $g > 15$

4

CLASS EXERCISES

Read each inequality aloud.

1. $t > -16$
2. $57 \leq n$
3. $y < 28$

Tell whether each inequality is true or false.

4. $-2 + 7 > 6$ false
5. $0.03 > 0.1$ false
6. $-2^4 < (-2)^4$ true

Match each inequality with the appropriate graph.

7. $x \geq -4$ b
8. $x \leq -4$ c
9. $x > 0.4$ d
10. $t \geq -0.4$ a

a. 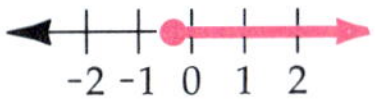

b. 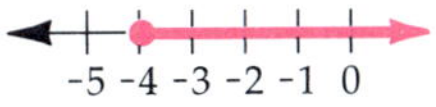

c. 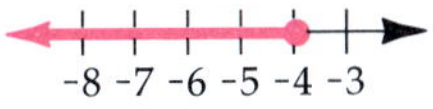

d. 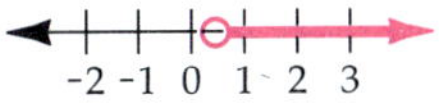

Write an inequality for each word phrase.

11. x is less than 5. $x < 5$
12. y is more than -3. $y > -3$

THINK AND DISCUSS

1. Is the inequality $a > b$ equivalent to $b < a$? yes
2. How is an inequality with one variable different from an equation?

2. The value of the variable in an inequality will have many values.

MATH MINUTES

Write an inequality.
1. No more than 30 students watched the game after school. ($s \leq 30$)
2. Less than 50 students saw the play. ($s < 50$)
3. More than 100 tickets were sold. ($t > 100$)

4 Ask students to describe the replacement set for g and the graph of $g > 15$. (Since g is the number of girls, g must be a non-zero whole number. The graph would be a series of solid dots, 16, 17, 18, and so on.)

ADDITIONAL EXAMPLES

Graph each inequality.

1. $x < -3$ 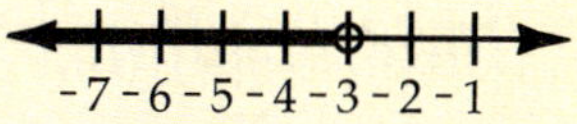
2. $p \geq -1.5$ 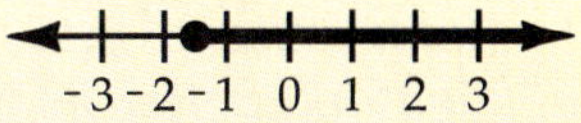

Write an inequality for each graph.

3. 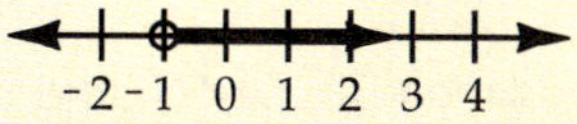 ($x > -1$)

Write an inequality for each sentence.

4. n is negative ($n < 0$)
5. Ms. Perez lives more than three miles from the office. Let d = the distance from Ms. Perez' home to the office. ($d > 3$)

Guided Practice

Class Exercises For question 9, help students develop test taking skills by asking them why they can eliminate three of the choices immediately. (The graph must show an open dot and only one of the choices has an open dot.)

Think and Discuss Encourage students to relate what they have learned about equations to inequalities.

A ***common error*** occurs with the use of the symbols ≥ and ≤. Students think that $4 \geq 4$ is not a true statement because they do not recognize that the inequality *includes* the equal sign. To remediate this error, ask students to write the inequalities that satisfy these conditions: a number equal to or greater than 5 ($x \geq 5$), a number less than or equal to 7 ($n \leq 7$).

Closure

Writing in Math Ask students to write a verbal description and to draw a graph for each of the following inequalities in their math journals:
$a \leq -5$, $a \geq -5$, $a \neq -5$,
$a < -5$, $a > -5$

Independent Practice

Some students may need to be reminded of the definition of the absolute value of a number for questions 2, 9, and 13.

For questions 50, 52-53, and 55, ask students to state the phrase, or phrases in each sentence, that is the situation to be described. For example, in question 50, the phrase is *fewer than 45 people*.

Students can do question 57 with or without a computer.

Lesson Follow-up

Reteaching Activity For students who continue to have difficulty with writing and interpreting inequalities, refer to the exploring activity found on page 272E of the Chapter Overview. See also TRB Exploring O-27.

MIXED REVIEW

Solve.

1. $2x - 5 = 29$ $x = 17$

2. $x - 3(5 - x) = 2 + 7x$ $x = -5.\overline{6}$ or $-5\frac{2}{3}$

Write as an expression.

3. 4 less than twice a number $2n - 4$

4. 14 less a number $14 - x$

Simplify.

5. $\sqrt{16} - |-4|$ 0

6. $n + (n + 2) + (n + 4)$ $3n + 6$

Solve.

7. A recipe calls for $\frac{2}{3}$ c of milk. Jane needs to triple the recipe. How much milk will she need? 2 cups

WRITTEN EXERCISES

Tell whether each inequality is true or false.

1. $-5 + 2 < -2$ true
2. $|-8 \times 2| \geq 10$ true
3. $\sqrt{16} \leq 4$ true
4. $3(-5) + 1 > -2(-6) - 18$ false
5. $(0.5)(2 + 8) < -2^2 + 2$ false

CALCULATOR **True or false.**

6. $-8.46 - 4.51 < 5(1.13) - 1.76$ true
7. $3(-0.04 + 7.12) > -2.87(-2.35)$ true
8. $-0.8(8.3 + 6.8) \geq 4.3(-3)$ true
9. $|-14.3 + 4.9| < 4.47(2.1)$ false

Write each inequality as a word sentence. Check students' work.

10. $2.5 > m$
11. $6.2 \leq j$
12. $5 \leq k$
13. $|x| \geq 2$
14. $8r < 29$
15. $72 \geq g$

State an inequality for each graph.

16. 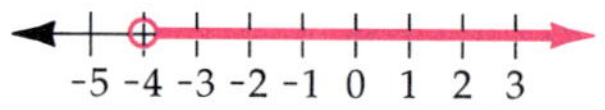

$x > -4$

17. 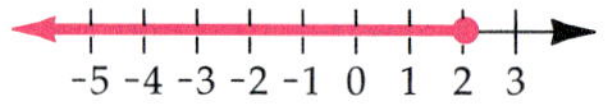

$x \leq 2$

18. 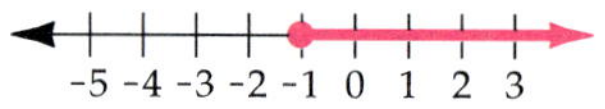

$x \geq -1$

Graph each inequality on a number line. Check students' work.

19. $x < 7$
20. $y > 2$
21. $a \geq -2$
22. $j \leq 0$
23. $x > -1$
24. $a < 2$
25. $m > -5$
26. $b \geq 6$
27. $p \leq 4$
28. $x > 1$
29. $j \geq -1$
30. $c < 1$

Write an inequality for each sentence.

31. Three is less than ten. $3 < 10$
32. The total t is greater than seven. $t > 7$
33. A number p is positive. $p > 0$
34. A number c is at least a dozen. $c \geq 12$
35. The price p is not more than \$30. $p \leq 30$
36. The number of seats s in the auditorium is no more than 3,500. $s \leq 3{,}500$
37. ***DATA FILE 4 (pp. 138–139)*** Three times x is less than the population of Monaco per square kilometer. $3x < 15{,}000$

State two inequalities for each graph.

38. 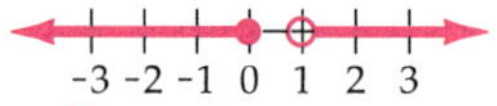

$x \leq 0$; $x > 1$

39. 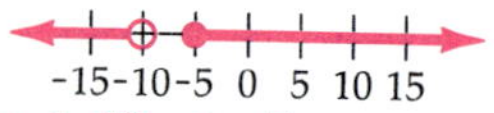

$x < -10$; $x \geq -5$

Replace ■ with <, >, or = to make the statement $3 + x$ ■ $7 + x$ true.

40. if $x = 4$ < **41.** if $x = 0$ < **42.** if $x = -5$ < **43.** if $x < 0$ <

Replace ■ with <, >, or = to make the statement $3x$ ■ $7x$ true.

44. if $x = 7$ < **45.** if $x = 0$ = **46.** if $x = -2$ > **47.** if $x < -1$ >

Replace ■ with < or > to make each statement true.

48. If $a < b$, then b ■ a. > **49.** If $x > y$, and $y > z$, then x ■ z. >

Write an inequality to describe each situation.

50. Fewer than 45 people attended the show. Let n equal the number of people who attended the show. $n < 45$

51. A student has $5 but does not have enough money to purchase three pairs of socks. Let m equal the cost of three pairs of socks. $m > 5$

52. At least 127 students attended the rock concert. Let s equal the number of students. $s \geq 127$

53. A student picked at least 15 bushels of apples. Let b equal the number of bushels picked. $b \geq 15$

54. A student pays for three movie tickets with a twenty-dollar bill. The change is less than $1. Let t equal the cost of a movie ticket. $19 < 3t < 20$

55. No more than 50 students walked in the walkathon. Let s equal the number of students. $s \leq 50$

56. ***DATA FILE 7 (pp. 272–273)*** Let h equal the flight time in hours.

a. On flights of how many hours is smoking not permitted? $h < 6$

b. On flights of how many hours is smoking permitted? $h \geq 6$

COMPUTER You can use a computer to compare data. The table at the right shows yearly salaries for six people. Each person's tax is based on salary. Let s equal salary.

Name	Salary	Tax
Smith, J.	$25,700	$5,140
Chien, H.	32,500	6,500
Garcia, R.	22,000	4,400
O'Malley, M.	19,500	2,925
Jones, K.	38,500	7,700
Strauss, J.	17,000	2,550

Tax Rates
If $s < 20{,}000$, 15%
If $s > 20{,}000$, 20%

57. a. Run the following BASIC program. Use the data in the table for S. Check students' work.

```
10 PRINT "SALARY";
20 INPUT S
30 IF S < 20000 THEN PRINT "TAX = ";S*.15
40 IF S > 20000 THEN PRINT "TAX = ";S*.2
```

b. ***CALCULATOR*** Did the program calculate the correct tax for each employee? Yes, if the program was input correctly.

c. The IF command allows the program to decide which tax rate to use. What role do the inequalities play in the program? They determine how the IF command decides what tax rate to use.

LESSON RESOURCES

TRB Practice G-9
TRB Enrichment G-10

FOR THE NEXT LESSON

calculator, math journal, computer, and spreadsheet and graphics software (optional)

Practice

For use after 7-5 (pp. 292–295)

Inequalities and Their Graphs

Tell whether each inequality is true or false. Write T or F.

1. $-8 > 5 - 13$ F 2. $0 \leq -4 + 6$ T 3. $|9 - 17| \geq 8$ T
4. $2(7 - 9) > 3(-6) + 15$ F 5. $\frac{2}{3}(10 - 25) \leq (-2)^3$ T

CALCULATOR Write T or F.

6. $|7.65 - 11.03| \leq 3.5(5.74 - 4.83)$ F
7. $-4.65(-2.91) > 3.86(5.71 - 2.36)$ T
8. $(2.7 - 5.3)(6.4 - 11.9) \leq 0.4|14.91 - 50.66|$ T

Write each inequality as a word sentence.

9. $17 < k$
17 is less than k or k is greater than 17.
10. $x \leq 2.9$
x is less than or equal to 2.9.
11. $|p| \geq 13$
The absolute value of p is greater than or equal to 13.

State an inequality for each graph.

12. $x \geq -3$ 13. $x < 2$
14. $x \leq -7$ 15. $x > -11$

Graph each inequality on the number line.

16. $m = 5$ 17. $k > 0$
18. $y < -2$ 19. $c \geq -4$

Write an inequality for each word phrase.

20. A number n is negative. $n < 0$
21. The sum of x and y is less than or equal to -2. $x + y \leq -2$
22. A number h is not less than 7. $h \geq 7$

Chapter 7 G-9

Enrichment

For use after 7-5 (pp. 292–295)

Conjunctions and Disjunctions

Two statements connected by the word "and" form a *conjunction*. Two statements connected by the word "or" form a *disjunction*. You can use inequality symbols to write conjunctions and disjunctions.

Conjunction: x is greater than 5 *and* x is less than 8.

Symbols: $5 < x < 8$

Disjunction: p is less than -3 *or* p is greater than or equal to 4.

Symbols: $-3 > p$ or $p \geq 4$

Use symbols to write each statement.

1. The number n is greater than 7 and it is less than 10. $7 < n < 10$
2. k is less than -2 or k is greater than 0. $k < -2$ or $k > 0$
3. y is greater than or equal to -6 and less than or equal to 5. $-6 \leq y \leq 5$
4. h is less than or equal to -10 or greater than -3. $h \leq -10$ or $h > -3$

List the integers that satisfy each statement.

5. $1 < m \leq 6$ 2, 3, 4, 5, 6 6. $-8 \leq f < -2$. -8, -7, -6, -5, -4, -3
7. $4 \geq c \geq 0$ 0, 1, 2, 3, 4 8. $13 < e < 15$. 14
9. $p > 6$ and $p \leq 9$ 7, 8, 9 10. $h > -3$ and $h < -4$. none

You can graph conjunctions and disjunctions on a number line

Statement	Graph
$x > 5$ and $x > 3$ (Only numbers greater than 5 satisfy both.)	
$p < -2$ or $p \geq 1$	

Graph each statement.

11. $x > 7$ or $x \leq 4$ 12. $x > -2$ and $x \leq 3$
13. $x \leq 3$ or $x > 5$ 14. $x < 9$ and $x > 6$

G-10 Chapter 7

LESSON QUIZ

Graph each inequality.

1. $p < 2.5$

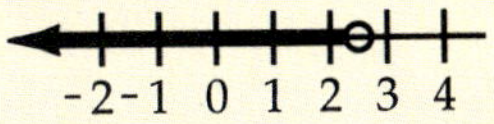

Write an inequality for each graph.

2.

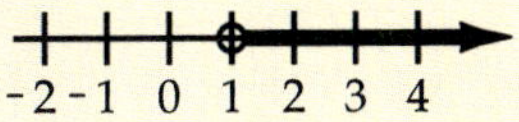

$(x > 1)$

3.

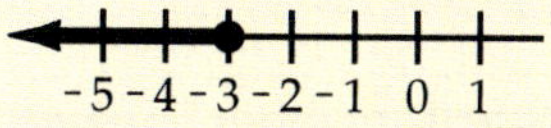

$(x \leq -3)$

Write an inequality for each sentence.

4. p is positive. $(p > 0)$

5. r is less than zero. $(r < 0)$

6. Mario's height h is greater than 156 cm. $(h > 156)$

Assignment Guide

Basic 1-3, 6-7, 10-11, 16-22, 31-34, 37, 39-40, 43-44, 49-51, 55-57, MR All

Average 2-4, 7-8, 12-13, 16-18, 23-26, 33-37, 40-41, 44-45, 47, 51-52, 55-57, MR All

Enriched 3-5, 8-9, 14-18, 27-30, 33-42, 45-48, 53-57, MR All

Lesson Focus

Materials

Calculator, math journal, computer, and spreadsheet and graphics software (optional)

Vocabulary/Symbols

Addition Properties for Inequalities
Division Properties for Inequalities
Multiplication Properties for Inequalities
Subtraction Properties for Inequalities

MOTIVATING THE LESSON

Write the following on the chalkboard and ask students to replace the box with an inequality symbol to make the sentence a true inequality.

1. $-3 < -1$ so $-3 + 5$ ■ $-1 + 5$ (<)
2. $5 > -2$ so $(-3)(5)$ ■ $(-3)(-2)$ [<]
3. $-11 > -15$ so $-11 - 3$ ■ $-15 - 3$ (>)
4. $12 < 24$ so $12 \div -4$ ■ $24 \div -4$ (>)

Ask students to summarize the results of performing the operations on the inequalities. Have students write the summary in their math journals.

OBJECTIVE: *To solve one-step inequalities using properties for inequalities.*

NOTES & QUOTES

Dimitri Ivanovich Mendeleev (1834–1907) was a Russian chemist. He classified the elements by their similarities. Mendeleev left gaps in his periodic table for elements that would be discovered in later years.

7-6 Solving One-step Inequalities

▼ In 1869, Dimitri Mendeleev arranged the 63 known elements into the first *periodic table*. By 1984, advances in science had gradually increased the size of the table to 109 elements. How many elements could have been added to Mendeleev's table by 1982?

We can let n equal the number of elements added to the periodic table. Then the following inequality represents the number of elements that could have been added to the table by 1982.

$$63 + n < 109$$
$$-63 + 63 + n < 109 - 63$$
$$n < 46$$

Fewer than 46 elements could have been added to the table by 1982.

▼ You can solve inequalities by using the *addition properties for inequalities.*

Example 1 Solve the inequality $n - 15 < 73$.

1 **Solution**
$$n - 15 < 73$$
$$n - 15 + 15 < 73 + 15 \quad \text{Add 15 to each side.}$$
$$n < 88$$

Addition Properties for Inequalities	**Arithmetic** **1.** $7 > 3$, so $7 + 4 > 3 + 4$. **2.** $2 < 5$, so $2 + 6 < 5 + 6$. **Algebra** **1.** If $a > b$, then $a + c > b + c$. **2.** If $a < b$, then $a + c < b + c$.

▼ You can solve inequalities by using the *subtraction properties for inequalities.*

2 **Example 2** Solve the inequality $-26 > y + 15$.

Solution
$$-26 > y + 15$$
$$-26 - 15 > y + 15 - 15 \quad \text{Subtract 15 from each side.}$$
$$-41 > y$$

Subtraction Properties for Inequalities	**Arithmetic** **1.** $12 > 4$, so $12 - 3 > 4 - 3$. **2.** $8 < 9$, so $8 - 2 < 9 - 2$. **Algebra** **1.** If $a > b$, then $a - c > b - c$. **2.** If $a < b$, then $a - c < b - c$.

Teaching Tip

Have worksheets with number lines on them available for the students to model inequalities.

You can solve inequalities which involve multiplication using the *multiplication properties for inequalities.*

Multiplication Properties for Inequalities	Arithmetic
	1. $3 < 4$, so $3(5) < 4(5)$.
	2. $7 > 2$, so $7(4) > 2(4)$.
	3. $6 < 7$, so $6(-2) > 7(-2)$.
	4. $7 > 5$, so $7(-3) < 5(-3)$.
	Algebra
	1. If c is positive and $a < b$, then $ac < bc$.
	2. If c is positive and $a > b$, then $ac > bc$.
	3. If c is negative and $a < b$, then $ac > bc$.
	4. If c is negative and $a > b$, then $ac < bc$.

Example 3 Solve the inequality $\frac{x}{6} > -4$.

Solution

$\frac{x}{6} > -4$

$6\left(\frac{x}{6}\right) > -4(6)$ Multiply each side by 6.

$x > -24$

THINK How are the rules for solving inequalities similar to those for solving equations? How are they different? **First, undo add. or sub. Second, undo mult. or division. The direction of the signs may change in an inequality.**

3

You can solve inequalities which involve division using the *division properties for inequalities.*

Division Properties for Inequalities	Arithmetic
	1. $3 < 6$, so $3 \div 3 < 6 \div 3$.
	2. $6 > 2$, so $6 \div 2 > 2 \div 2$.
	3. $6 < 12$, so $6 \div (-3) > 12 \div (-3)$.
	4. $16 > 8$, so $16 \div (-4) < 8 \div (-4)$.
	Algebra
	1. If c is positive and $a < b$, then $\frac{a}{c} < \frac{b}{c}$.
	2. If c is positive and $a > b$, then $\frac{a}{c} > \frac{b}{c}$.
	3. If c is negative and $a < b$, then $\frac{a}{c} > \frac{b}{c}$.
	4. If c is negative and $a > b$, then $\frac{a}{c} < \frac{b}{c}$.

Example 4 Solve the inequality $-5x < 20$.

Solution

$-5x < 20$

$\frac{-5x}{-5} > \frac{20}{-5}$ Divide each side by -5. Change the direction of the inequality symbol.

$x > -4$

You need to change the direction of the inequality sign only when multiplying or dividing by a negative number.

Instruction

1 Remind students that *solving an inequality* means finding all permissible values of the variable that make the inequality true.

2 After you present the formal properties for inequalities, have students refer to their math journals and compare the formal definitions with the summary they wrote for Motivating the Lesson. Students should refine their definitions.

3 Help students understand that the only difference is that the inequality sign must be reversed when both sides of an inequality are multiplied or divided by the same negative number.

ADDITIONAL EXAMPLES

Solve each inequality.

1. $x + 13 < 21$ ($x < 8$)
2. $3x \geq -9$ ($x \geq -3$)
3. $x - 5 > -1$ ($x > 4$)
4. $\frac{1}{4}y \leq \frac{1}{2}$ ($y \leq 2$)
5. $-5w > 15$ ($w < -3$)

A ***common error*** occurs when students reverse the inequality symbol whenever they see a negative sign. For example, students will solve $6x > -12$ as $x < -2$. To help students avoid this error, encourage them to substitute values into the inequality. They will realize that $6(-3) > -12$ is a false statement.

MATH MINUTES

Working in pairs one student states a one-step inequality and an operation. (e.g., $3x < 15$; multiply by -2) The other student performs the operation and states the result. ($-6x > -30$)

Guided Practice

Class Exercises For questions 1-6, ask students to give an example for each operation.

Think and Discuss In question 1, help students realize that multiplying both sides by zero results in the equation 0 = 0, which is an identity. Students should know that you can't divide by zero.

In question 2, students need to realize that they check an inequality the same way they check an equation.

Closure

Writing in Math Have students write in their math journals why they must reverse the inequality symbol, whenever they multiply or divide both sides of an inequality by the same negative number. Then have them compare their explanations with those of other students.

Independent Practice

Some students may be confused by question 5 because the variable appears on the right. Although the procedure for solving inequalities with the variable on the right side is the same as when the variable is on the left side, students may find it easier to rewrite the inequality.

Have students work in pairs to answer questions 25-28. Partners should compare their data for question 28 with other pairs of students.

For question 27, students can use *Appleworks* for an Apple computer or *Microsoft Works* for an IBM. They can also refer to the computer lessons in Chapters 1 and 2.

THINK AND DISCUSS

1. The rules for multiplying and dividing both sides of an inequality do not mention zero. Discuss these cases.
2. Is -6 a solution of the inequality $x \div 3 > 2$? Explain.

1. When both sides are mult. by zero, the value of each side is zero. Division by zero is undefined.
2. No; A neg. number is less than a pos. number.

MIXED REVIEW

Find the GCF of each set of numbers.

1. 10, 35 **5**
2. 4, 7, 8 **1**

Graph each inequality.

3. $x < 2$ **Check students' work.**
4. $x \geq -5$

Solve each equation.

5. $3x - 5 = 4 - 2x$ **$1\frac{4}{5}$**
6. $3(1 - x) = 6x + 11$ **$-\frac{8}{9}$**
7. Find two whole numbers whose sum is 164 and whose difference is 12. **76 and 88**

CLASS EXERCISES

State whether the inequality symbol remains the same or is reversed when you do the following to each side of an inequality.

1. add -5 **same**
2. multiply by -7 **reversed**
3. divide by 3 **same**
4. subtract 12 **same**
5. divide by -1 **reversed**
6. add 8 **same**

What was done to both sides of the first inequality to obtain the second?

7. $x - 5 \geq 6; x \geq 11$ **Add 5.**
8. $8 > -4x; -2 < x$ **Divide by -4.**

Tell whether each number is a solution of $3 - x > 1$.

9. 3 **no**
10. 2 **no**
11. 0 **yes**
12. -4 **yes**

Solve each inequality.

13. $\frac{x}{-6} > 3$ **$x < -18$**
14. $-3.2x < 14.4$ **$x > -4.5$**
15. $x - 8.4 > -2.7$ **$x > 5.7$**

WRITTEN EXERCISES

State whether the inequality symbol remains the same or is reversed when you do the following to each side of an inequality.

1. A negative number is subtracted from both sides. **same**
2. Both sides are multiplied by a positive number. **same**
3. Both sides are divided by a negative number. **reversed**

What was done to both sides of the first inequality to obtain the second?

4. $x + 8 \leq 11; x \leq 3$ **Sub. 8.**
5. $9 > -3x; -3 < x$ **Divide by -3.**
6. $4x \geq 48; x \geq 12$ **Divide by 4.**
7. $\frac{1}{3}x \leq 18; x \leq 54$ **Mult. by 3.**

Tell whether each number is a solution of $5 - 2x \leq 1$.

8. 3 **yes**
9. 2.5 **yes**
10. 0 **no**
11. -4 **no**

Solve each inequality.

12. $x + 6 \geq 7$ **$x \geq 1$**
13. $3 \leq x - 5$ **$8 \leq x$**
14. $\frac{x}{3} \geq -5$ **$x \geq -15$**
15. $-3x < 0$ **$x > 0$**
16. $-4x \leq -16$ **$x \geq 4$**
17. $x - 9 > -5$ **$x > 4$**
18. $\frac{1}{2}x \geq -3$ **$x \geq -6$**
19. $x - 7 < -15$ **$x < -8$**
20. $9x \leq 27$ **$x \leq 3$**
21. $5 + x > -7$ **$x > -12$**
22. $\frac{b}{3} \leq -31$ **$b \leq -93$**
23. $-3 \geq \frac{g}{-7}$ **$21 \leq g$**

24. **DATA FILE 7 (pp. 272–273)** Write each as an inequality.

a. Use two inequalities to describe the passengers eligible for discounted airline fares. Let p = the age of passengers eligible for discounted fares. $p \leq 11$; $p \geq 62$

b. Two times the number of passengers boarding in Miami in 1987 is less than the number boarding in Chicago/O'Hare in 1987. 2(11,600,000) < 27,500,000

DATA The circle graph at the right shows the source of energy for every 100 kilowatts (kw) of energy produced in the United States.

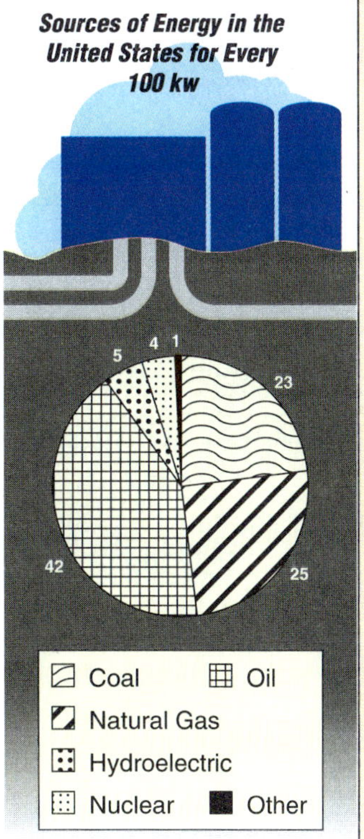

25. Let x and y be two sources of energy from the graph. Determine the source related to x and y based on the inequalities: $x < y$; $x + y > 10$; $xy > 200$; $y - x \geq 35$. hydroelectric and oil

26. What percent of the energy is obtained from oil and natural gas? 67%

27. **COMPUTER** How many more kilowatts of nuclear energy would be required to increase nuclear energy's share to at least 25%? Assume that the other sources of energy have the same number of kilowatts shown on the circle graph. *Hint:* use a spreadsheet and a graphics program. 28

28. **PROJECT** What sources of energy are not included in the graph? Check students' work.

TEST YOURSELF

Solve each equation.

1. $12w = 7w + 25$ 5
2. $8 - 3t = t - 2t - 6$ 7
3. $(y + 6)3 + 2y = 2y + 6$ -4
4. $12 - 2x = -(x - 4)$ 8

Solve each inequality.

5. $y - 3 > -7$ $y > -4$
6. $y + 4 < 8$ $y < 4$
7. $\frac{s}{-6} > -7$ $s < 42$
8. $7h > \frac{1}{3}$ $h > \frac{1}{21}$
9. $64 \leq -8k$ $-8 \geq k$
10. $y - (-5) > -7$ $y > -12$

Write an inequality to describe the situation. Solve.

11. Seven less than a number is greater than negative two. Find the number. $n - 7 > -2$; $n > 5$

12. When this number is divided by negative four, the result is at least 30. Find the number. $\frac{n}{-4} \geq 30$; $n \leq -120$

13. Negative eight is greater than or equal to a number divided by negative three. Find the number. $-8 \geq \frac{n}{-3}$; $n \geq 24$

Lesson Follow-up

Reteaching Activity To help students understand properties, provide them with several inequalities, like $-3 < 6$. Have students perform each operation using the values 3 and -3.

LESSON RESOURCES

TRB Practice G-11
TRB Enrichment G-12

Practice For use after 7-6 (pp. 296–299)

Solving One-step Inequalities

Answer *true* or *false*.

1. When a negative number is added to both sides of an inequality, the inequality is reversed. false
2. When both sides of an inequality are multiplied by a negative number, the inequality is reversed. true

What was done to both sides of the first inequality to obtain the second?

3. $p - 7 > 14$; $p > 21$ added 7
4. $4 \leq -4y$; $-1 \geq y$ divided by -4
5. $\frac{1}{8}n < -2$; $n < -16$ multiplied by 8

Tell whether each number is a solution of $7 \geq 3k - 2$.

6. 4 no
7. -2 yes
8. 0 yes
9. 3 yes

Solve each inequality for the variable x.

10. $7 + x \geq 9$ $x \geq 2$
11. $-5x < 10$ $x > -2$
12. $\frac{x}{4} > 1$ $x > 4$
13. $-5 \leq x - 6$ $x \geq 1$
14. $-8 < -8x$ $x < 1$
15. $\frac{1}{3}x > -2$ $x > -6$
16. $0 \geq x + 12$ $x \leq -12$
17. $x - 15 \leq -8$ $x \leq 7$
18. $\frac{2}{3}x < -6$ $x < -9$
19. $-2.6 < \frac{x}{-10}$ $x < 26$
20. $13 + x \geq 13$ $x \geq 0$
21. $-x \leq 2$ $x \geq -2$

Write an inequality for each word phrase. Solve.

22. Six less than n is less than -4. Find n. $n - 6 < -4$; $n < 2$
23. The product of k and -5 is no more than 30. Find k. $-5k \leq 30$; $k \geq -6$
24. Half of p is at least -7. Find p. $\frac{1}{2}p \geq -7$; $p \geq -14$
25. Six more than 3 times n is greater than -7. Find n. $3n + 6 > -7$; $n > -4\frac{1}{3}$
26. The product of k and 9 is no more than 15. Find k. $9k \leq 15$; $k \leq 1\frac{2}{3}$
27. One-third of p is at least -17. Find p. $\frac{1}{3}p \geq -17$; $p \geq -51$
28. The opposite of g is at least -5. Find g. $-g \geq -5$; $g \leq 5$

Chapter 7 G-11

Enrichment For use after 7-6 (pp. 296–299)

Discovering Heptagons

What is a heptagon?

To find out, solve each inequality in the diagram below. Use a ruler to connect each inequality with its solution. When you have finished, you will have drawn a heptagon in the center of the square. Shade the heptagon and tell what it is in the space below the drawing.

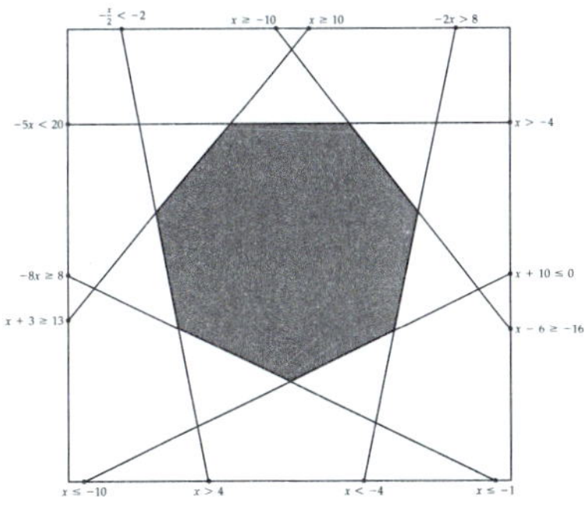

A heptagon is a 7-sided polygon (or 7-sided figure).

G-12 Chapter 7

LESSON QUIZ

State whether the inequality symbol remains the same or is reversed when you do the following to each side of an inequality.

1. Multiply by $\frac{-2}{3}$. (reversed)
2. Add -6. (same)
3. Divide by -3. (reversed)
4. Subtract 9. (same)

Tell whether each number is a solution of $3 - 4x > 7$.

5. -2 (yes)
6. 0 (no)

Solve each inequality.

7. $\frac{-x}{5} \leq 0$ ($x \geq 0$)
8. $x - 15 > 10$ ($x > 25$)
9. $x + 5 < 12$ ($x < 7$)
10. $3x \geq -12$ ($x \geq -4$)

Assignment Guide

Basic 1-5, 8-10, 12-15, 24-28, MR, TY All
Average 1-3, 5-6, 9-11, 16-19, 24-28, MR, TY All
Enriched 2-3, 5-7, 10-11, 19-28, MR, TY All

FOR THE NEXT LESSON

math journal

Lesson Focus

Materials
Math journal

MOTIVATING THE LESSON

Have students solve the following equations.

1. $7x - 8 = 27$ ($x = 5$)
2. $9x + 7 - 3x = 49$ ($x = 7$)

Ask students to discuss the procedures they used to solve the equations. Mention that in this lesson more than one inequality property will be used to solve inequalities like $3x - 2 - 5x \geq 18$.

Instruction

1 To help students see the similarities, have them solve $250 + 75h = 1{,}000$. ($h = 10$)

2 Ask students why the symbol $\leq$ is used in the inequality. (*No more than* is another way to say less than or equal to.)

3 Ask students to explain whether or not they could find the two *largest* consecutive integers whose sum exceeds 55. (no, because there are infinitely many pairs)

ADDITIONAL EXAMPLES

Solve each inequality.

1. $2y - 3 \geq 7$ ($y \geq 5$)
2. $5 - \frac{1}{2}x > 9$ ($x < -8$)
3. $1 + \frac{1}{3}z \leq -4$ ($z \leq -15$)

Write an inequality. Solve.

4. Sue put 4 apples (all the same weight) and a 3-lb box of rice in one pan of a balance scale and a 5-lb weight in the other pan; the apples and rice were heavier. What was the weight of one apple? ($4a + 3 > 5$; greater than $\frac{1}{2}$ lb)
5. Find the two smallest consecutive even integers whose sum is at least 46. ($x + x + 2 \geq 46$; 22 and 24)

OBJECTIVE:
To solve simple two-step inequalities.

7-7 Solving Two-step Inequalities

▼ A research company budgeted \$1,000 for an on-line computer information service. The service charges a \$250 monthly service fee plus \$75 for each hour of use. How many hours can the research company use the service and stay within budget?

Let h = the number of hours.
Then $250 + 75h \leq 1{,}000$ describes the situation.

The steps for solving a two-step inequality are similar to the steps for solving a two-step equation.

1

$$250 + 75h \leq 1{,}000$$
$$250 + 75h - 250 \leq 1{,}000 - 250 \quad \text{Subtract 250 from each side.}$$
$$75h \leq 750$$
$$\frac{75h}{75} \leq \frac{750}{75} \quad \text{Divide each side by 75.}$$
$$h \leq 10$$

The company can use the computer service for, at most, 10 h.

FLASHBACK
When multiplying by a negative number, the inequality symbol is reversed.

Example 1 **Solve each inequality.**

a. $6 - x > 3$

b. $\frac{1}{2}y - 3 \leq -5$

Solution

a.
$$6 - x - 6 > 3 - 6$$
$$-x > -3$$
$$(-1)(-x) < (-1)(-3)$$
$$x < 3$$

b.
$$\frac{1}{2}y - 3 + 3 \leq -5 + 3$$
$$\frac{1}{2}y \leq -2$$
$$(2)\left(\frac{1}{2}y\right) \leq (2)(-2)$$
$$y \leq -4$$

▼ You can use inequalities to describe situations.

Example 2 Divide a number by -5. Then add 4 to the quotient. The result is no more than 7. Find the number.

Solution Let n = the original number.
Then $\frac{n}{-5} + 4 \leq 7$ describes the situation.

2

$$\frac{n}{-5} + 4 \leq 7$$
$$\frac{n}{-5} + 4 - 4 \leq 7 - 4 \quad \text{Subtract 4 from each side.}$$
$$\frac{n}{-5} \leq 3$$
$$-5 \cdot \frac{n}{-5} \geq -5 \cdot 3 \quad \text{Multiply each side by -5.}$$
$$n \geq -15$$

The number is greater than or equal to -15.

Teaching Tip

Divide the class into small groups to work through the examples in the text and the exercises. This procedure is especially helpful for the exercises involving word problems.

▼ Sometimes you will use an inequality to solve a problem with only one correct answer.

Example 3 **Find the two smallest consecutive integers whose sum is greater than 55.**

Solution Let i = the lesser of the two integers.
Then $i + 1$ = the next consecutive integer.
The inequality $i + i + 1 > 55$ describes the situation.

$i + i + 1 > 55$

$2i + 1 > 55$ Combine like terms.

$2i + 1 - 1 > 55 - 1$ Subtract 1 from each side.

$2i > 54$

$\frac{2i}{2} > \frac{54}{2}$ Divide each side by 2.

$i > 27$

The two consecutive integers are 28 and 29.

3

CLASS EXERCISES

What was done to each side of the first inequality to obtain the second?

1. $2 + x \leq 9; x \leq 7$ Sub. 2.
2. $16 + x > -5; x > -21$ Sub. 16.
3. $-3x > 24; x < -8$ Div. -3.
4. $8x - 3 \leq 19; 8x \leq 22$ Add 3.

Solve each inequality for x.

5. $\frac{2x}{3} + 1 > 3$ $x > 3$
6. $4x - 1 \leq 3$ $x \leq 1$
7. $6x + 2 \geq 0$ $x \geq -\frac{1}{3}$
8. $\frac{1}{2}x - 3 < 1$ $x < 8$
9. $20 - 3x > 2$ $x < 6$
10. $3x - 1 < 11$ $x < 4$
11. $5 - (x - 1) \geq 9$ $x \leq -3$
12. $-4(2x + 7) < -12$ $x > -2$
13. $3 + x > -6$ $x > -9$

Write an inequality to describe the situation. Solve.

14. Find the two greatest consecutive odd integers whose sum is less than -11. $2x + 2 < -11$; -7 and -5

THINK AND DISCUSS See below.

1. Why is addition or subtraction usually performed before multiplication or division when solving an inequality?
2. Are there any situations where it might be easier to multiply or divide before adding and subtracting? Explain.
3. Find two positive and two negative solutions of the inequality $|x + 3| < 5$.

1. Inequalities follow the rules for order of operations.
2. Yes; $3(x + 5) > 15$, you can divide each side by 3.
3. Accept reasonable answers.

WRITTEN EXERCISES

What was done to each side of the first inequality to obtain the second?

1. $8 - x \leq 11; -x \leq 3$ Sub 8.
2. $6 + 3x > 5; 3x > -1$ Sub 6.
3. $4x + 7 > 0; 4x > -7$ Sub 7.
4. $\frac{2x}{5} < 2; x < 5$ Mult. $\frac{5}{2}$.

Guided Practice

Class Exercises Students who have difficulty solving inequalities can replace the inequality sign with an equal sign. Once they have solved the equation, they can replace with the appropriate inequality sign. (They should remember the effects of negative numbers.)

Think and Discuss If students have difficulty with question 1, ask them what the goal is in solving an inequality. (to isolate the variable on one side of the inequality)

For question 2, ask students in which Class Exercise it might be easier to multiply or divide before adding or subtracting. (question 12)

For question 3, be sure that students' answers are between -8 and 2.

A ***common error*** is that students forget to reverse the inequality symbol when they multiply or divide both sides of an inequality by a negative number. Use the inequality $-3 < 6$ to show that if the inequality symbol *is not* reversed when both sides of the given inequality are multiplied by -4 or divided by -3, the resulting statements, $12 < -24$ and $1 < -2$, are false.

Closure

Writing in Math Ask students to write a summary of the procedure for solving two-step inequalities in their math journals.

MATH MINUTES

Have students work in pairs. Each student makes up three two-step inequalities for his or her partner to solve. The partners check each other's work.

Independent Practice

Point out that to solve some of the inequalities in questions 5-30, more than two steps are necessary. Ask students to identify those questions and to state what they would do first to solve the inequalities. (13, 16, 22, 24, 25, 29; combine like terms on the left side of the inequality.)

For questions 31-36, stress that students should check their solutions in the conditions stated in the problems. Point out that if their solutions do not match the conditions stated in questions 31-32, they chose the incorrect inequality and should reread the question and try a different inequality.

Critical Thinking For quantitative comparisons involving variables, suggest that students substitute a number for the variable in each pair of expressions to see whether the two expressions *can* be equal, *must* be equal, *cannot* be equal, and so on.

As a follow-up, refer students to the critical thinking activity found on page 272E of the Chapter Overview. See also TRB Exploring O-26.

MIXED REVIEW

Write an equation. Solve. See below.

1. Two times a number is 5 more than the number.

2. Five less than double the number is 121. $2n - 5 = 121$; 63

Solve and check.

3. $-2x \geq 6$ $x \leq -3$

4. $\frac{1}{2}x \leq 16$ $x \leq 32$

Solve for x if y is 2.

5. $x - y = 16$ $x = 18$

6. $3x + y = 17$ $x = 5$

7. The length of a rectangle is 5 m longer than the width. The perimeter is 40 m. What are the length and width?

1. $2n = n + 5$; 5
7. $7\frac{1}{2}$ m, $12\frac{1}{2}$ m

Solve each inequality for variable x.

5. $\frac{x}{3} + 2 > 3$ $x > 3$ **6.** $9x + 3 \geq 21$ $x \geq 2$ **7.** $\frac{1}{2}x - 4 \geq -1$ $x \geq 6$

8. $\frac{1}{2}x - \frac{1}{4} < -\frac{3}{4}$ $x < -1$ **9.** $-3x + 15 < 0$ $x > 5$ **10.** $2x - 9 > -7$ $x > 1$

11. $5x \leq 15 - 8$ $x \leq 1\frac{2}{5}$ **12.** $11 - 3x > 5$ $x < 2$ **13.** $6x - 10 - x > 14$ $x > 4\frac{4}{5}$

14. $8 + 2x \leq 10$ $x \leq 1$ **15.** $-5x + 3 \geq 28$ $x \leq -5$ **16.** $2x + x + 5 > 18$ $x > 4\frac{1}{3}$

17. $1 + 8x > 25$ $x > 3$ **18.** $4 + 7x \geq 32$ $x \geq 4$ **19.** $\frac{1}{9}x + 13 \geq 5$ $x \geq -72$

20. $2x + \frac{1}{2} > \frac{3}{2}$ $x > \frac{1}{2}$ **21.** $-21 - 3x < 0$ $x > -7$ **22.** $10x - 8 - x < 19$ $x < 3$

23. $19 - 8x > -5$ $x < 3$ **24.** $7 + 2x - x \geq 9$ $x \geq 2$ **25.** $18x - 5x - 4 > 22$ $x > 2$

26. $6 - 4x > 14$ $x < -2$ **27.** $\frac{1}{3}x + 11 < 31$ $x < 60$ **28.** $\frac{1}{6}x - 2 < 4$ $x < 36$

29. $2x + 2x + 2 < 18$ $x < 4$ **30.** $11x + 13 > -86$ $x > -9$

Choose the inequality which describes the situation. Solve.

31. You divide a number by -3. Then you subtract 1 from the quotient. The result is at most 5. Find the number. (Let x = the number.) b; $x \geq -18$

a. $\frac{-3}{x} - 1 < 5$ **b.** $\frac{x}{-3} - 1 \leq 5$

c. $\frac{x}{3} - 1 \leq -5$ **d.** $\frac{x}{-3} - 1 > 5$

32. Laura has \$16 in her savings account. She earns \$4.50 per hour babysitting. Laura wants to purchase a sweater for \$55. What is the least number of hours Laura must babysit in order to buy the sweater? Let x = the number of hours Laura must babysit. c; 9

a. $16x + 4.50 \geq 55$ **b.** $4.50x \geq 55 + 16$

c. $4.50x + 16 \geq 55$ **d.** $55 \geq 16 + 4.50x$

Write an inequality to describe the situation. Solve.

33. Five less than twice a number is at least 13. Find the number. $2n - 5 \geq 13$; $n \geq 9$

34. An artist withdrew \$14 from a bank in each of the last three weeks and still has more than \$65. How much did he start with? $x - 3(14) > 65$; $x > 107$

35. Students in a math class need an average test score of at least 90 points to earn an A. A student's test scores are 88, 91, and 85. What could the student score on the next test to have an A average? $\frac{x + 88 + 91 + 85}{4} \geq 90$; $x \geq 96$

36. A salesperson earns a salary of \$600 per month, plus a commission of 2% of sales. How much must the salesperson sell to have a monthly income of at least \$1,700? $600 + 0.02x \geq 1{,}700$; $x \geq 55{,}000$

MENTAL MATH

37. Find the least whole number solution of $6x - 19 > -7$. 3

38. Find the greatest integer solution of $-5x + 7 > 22$. -4

Write a problem that could be solved using the inequality.

39. $n + n + 2 < 15$

40. $50h + 40h > 360$
Check students' work.

41. ***DATA FILE 10 (pp. 404–405)***

a. What is the least number of acres of rain forests cut down in an eight-hour working day? 24,000 acres

b. How many eight-hour days would it take to destroy an area of rain forest equivalent to the home range of a grizzly bear? 39.1

Critical Thinking

EXPLORING QUANTITATIVE COMPARISONS

▼ A *quantitative comparison* is a type of question that appears on some standardized tests. This type of question requires that you evaluate two quantities, compare them, and choose the correct lettered response. Common answer choices are listed below.

(A) The number in Column A is greater.

(B) The number in Column B is greater.

(C) The two numbers are equal.

(D) The relationship cannot be determined based on the available information.

Compare each pair of expressions. The first one is done for you.

	Column A	*Column B*	*Response Choice*
1.	$x + 2$	$x + 5$	B, no matter what replacement you choose for x, $x + 5 > x + 2$.
2.	$2x$	$5x$	D
3.	$\lvert -5 - 2\rvert$	$\lvert 5\rvert - \lvert 2\rvert$	A
4.	$3x$	$\frac{x}{3}$	D
5.	$5(x - 3)$	$5x - 15$	C
6.	$\frac{x-2}{2-x}; x \neq 2$	$4 - \sqrt{16}$	B

Lesson Follow-up

Reteaching Activity Have students solve an inequality and its corresponding equation, for example, $-3x + 5 < -7$ and $-3x + 5 = -7$, side by side. The solutions are parallel and the inequality symbol is reversed as a result of dividing both sides by -3.

LESSON RESOURCES

TRB Practice G-13
TRB Enrichment G-14

Practice

For use after 7-7 (pp. 300–303)

Solving Two-step Inequalities

What was done to each side of the first inequality to obtain the second?

1. $4 - 7x \le 2$; $-7x \le -2$ subtracted 4
2. $\frac{x}{-3} > 1$; $x < -3$ multiplied by -3
3. $5x - 6 \ge 2$; $5x \ge 8$ added 6
4. $-6x < 12$; $x > -2$ divided by -6

Solve for x.

5. $2x - 5 > 1$ $x > 3$
6. $5x + 2 \le 17$ $x \le 3$
7. $3 < \frac{1}{2}x + 1$ $x > 4$
8. $7x + 2x \ge 21 - 3$ $x \ge 2$
9. $-8x + 18 > -22$ $x < 5$
10. $9 - x > 10$ $x < -1$
11. $\frac{1}{5}x + 6 > -3$ $x > -45$
12. $19 + 8 \le 6 + 7x$ $x \ge 3$
13. $\frac{x}{-4} > 0$ $x < 0$
14. $9x - 7 \le 38$ $x \le 5$
15. $3x + 11x \le -42$ $x \le -3$
16. $-12 < -12x$ $x < 1$
17. $\frac{1}{7}x + 5 > 0$ $x > -35$
18. $50 < 8 - 6x$ $x < -7$
19. $3(x - 9) > -18$ $x > 3$
20. $30 \ge -6(5 - x)$ $x \le 10$

Write an inequality that describes each situation. Then solve.

21. Nine more than half the number n is no more than -8. Find n.
$\frac{1}{2}n + 9 \le -8$; $n \le -34$

22. Judith drove h hours at a rate of 55 mi/hr. She did not reach her goal of driving 385 miles for the day. How long did she drive?
$55h < 385$; $h < 7$; less than 7h

23. The average age of the four Berger children is at least 15. The ages of three of the children are 19, 14, and 12. What is the age of the fourth child?
$\frac{1}{4}(19 + 14 + 12 + n) \ge 15$; $n \ge 15$; at least 15

24. Five less than three times the number k is neither less than nor equal to 19. Find k.
$3k - 5 > 19$; $k > 8$

Chapter 7 G-13

Enrichment

For use after 7-7 (pp. 300–303)

The Nation of Inequalia

The map provides information about the area, population, and amount of wooded terrain in the 7 states of the nation of Inequalia.

The Nation of Inequalia
Area, Population and Wooded Terrain

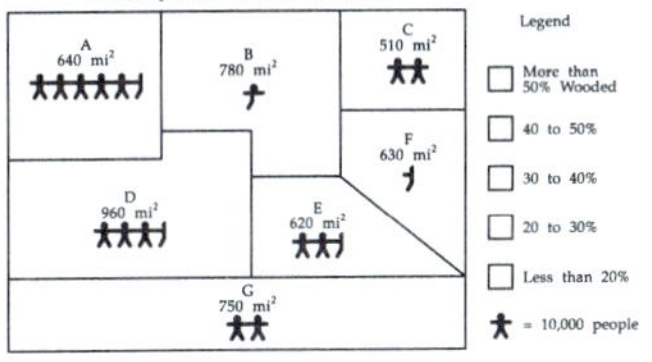

Write >, <, or =. P = population, W = amount of wooded terrain, S = area. States are lettered A-G.

1. $P(B)$ [>] $P(F)$
2. $W(E)$ [<] $W(G)$
3. $S(G)$ [<] $S(B)$
4. $W(A)$ [<] $W(E)$
5. $P(G)$ [=] $P(C)$
6. $S(F)$ [>] $S(E)$

Solve.

7. Name the states that have populations of at most 20,000. B, C, F, G
8. Which states are no more than 40% wooded? A, C, D, E
9. Which states might have less than 100 mi² (square miles) of wooded terrain?
A, D
10. Which states have 300 or more mi² of wooded terrain? B, F, G

Answer *true* or *false*.

11. At least 390 mi² of B are wooded. true
12. G could have more wooded terrain than F. true
13. D could have more wooded terrain than E. true

G-14 Chapter 7

LESSON QUIZ

Solve each inequality.

1. $13 - 7x < -1$ $(x > 2)$
2. $3x - 1 \le -16$ $(x \le -5)$
3. $-2(x - 4) \ge 10$ $(x \le -1)$
4. $\frac{3}{4}x + 2 > 2$ $(x > 0)$

Write an inequality. Solve.

5. Lee is thinking of a number. When she subtracts the number from 5 and doubles the result, the answer is greater than 14. What is the number?
[$2(5 - n) > 14$; any number less than -2, that is, -3, -4, -5, -6, . . .]

FOR THE NEXT LESSON

classified sections of various newspapers, math journal

Assignment Guide

Basic 1-2, 5-29 odd, 31-34, 37-41, MR, CT All

Average 2-3, 6-30 even, 31-32, 34-35, 37-41, MR, CT All

Enriched 3-4, 18-32, 35-41, MR, CT All

Career Information

Plan to invite a director of marketing or a marketing analyst from a nearby company to talk to the class. Ask students to compile a list of questions to ask the speaker about the skills that are needed for a career in the field of marketing. Ask the speaker to cover math-related topics such as:

- the training necessary to qualify for a position in a marketing department
- the role of computers and new technology in marketing analysis and sales predictions
- the skills needed to predict, calculate, and coordinate data for buying, pricing, and moving materials or finished goods
- the relationship between sales and marketing and how the two areas of a company interrelate

CAREER

Help Wanted: Director of Marketing

Bachelor's degree in marketing or business.

For more information, write to Sales and Marketing Executives, 446 Statler Office Tower, Cleveland, OH 44115.

The director of marketing relies on equations and inequalities to predict market conditions. The director uses predictions to determine what actions the company will take to remain competitive. The director may consider how much the cost of materials can increase before the company must raise the price of its goods or services. The director may also use predictions to determine the effect of price changes on the quantity of goods sold.

PROJECT

Find out how statistical and quantitative data are useful in making marketing decisions.

Practice

Solve and check each two-step equation.

1. $6x + 4 = 40$ 6
2. $7x - 10 = 25$ 5
3. $57 - 11x = 13$ 4
4. $6 + 5x = 66$ 12
5. $\frac{2}{9}x - 9 = 45$ 243
6. $7 - \frac{3}{5}x = 13$ -10
7. $19 + 2x = 57$ 19
8. $7x - 17 = -3$ 2
9. $-21 = -6t - 3$ 3

Solve and check each multi-step equation.

10. $8x - 20 + 2x = 60$ 8
11. $42 = -6(5 - x)$ 12
12. $\frac{2}{7}(x + 5) = 8$ 23
13. $\frac{2}{3}x - \frac{1}{12}x = 4$ $6\frac{6}{7}$
14. $5x - 2x + 11 = 59$ 16
15. $19 = x + x - 7$ 13
16. $3(3x - 8) = 21$ 5
17. $1.5x - 0.7x = 8$ 10
18. $18 - 3x = 4(x + 8)$ -2
19. $6.5 - (0.1x + 2) = -1.6x$ -3
20. $2(6 + x) = 17 - 3x$ 1
21. $x + 9 = 2x - 43 + x$ 26
22. $\frac{3}{4}(x - 4) = \frac{1}{4}x + 23$ 52
23. $\frac{1}{2}(6x - 6) = \frac{1}{4}(x + 54)$ 6
24. $8x - 7 = 2x - 1$ 1
25. $7x + 5 = 8x - 3$ 8

Graph each inequality. Check students' work.

26. $x \leq -3$
27. $x < 6$
28. $x > 0$
29. $x \geq 5$
30. $x > -4$
31. $x \leq 10$

Solve each inequality for x.

32. $x + 6 > 2$ $x > -4$
33. $-2x < 8$ $x > -4$
34. $x - 7 \leq 6$ $x \leq 13$
35. $\frac{1}{3}x > -6$ $x > -18$
36. $x - \frac{1}{5} < \frac{4}{5}$ $x < 1$
37. $6x \leq 18$ $x \leq 3$
38. $10 > 5x$ $2 > x$
39. $8x \geq 32$ $x \geq 4$
40. $x - 2 > 5$ $x > 7$
41. $6 + 3x > 12$ $x > 2$
42. $18 - 5x < -2$ $x > 4$
43. $x + x - 7 \leq 21$ $x \leq 14$
44. $\frac{1}{2}x - 9 < 27$ $x < 72$
45. $\frac{4}{5}x + 4 \geq 16$ $x \geq 15$
46. $8x + 54 < 53$ $x < -\frac{1}{8}$

Write an equation to describe the situation. Solve.

47. Martha is saving \$22 each week from her paycheck. She already has \$47 in her account. In how many weeks will her balance be \$201? $22x + 47 = 201$; 7
48. Fifty-three is four times a number minus nineteen. Find the number. $53 = 4n - 19$; 18
49. Find four consecutive integers whose sum is -490. $x + x + 1 + x + 2 + x + 3 = -490$; -121, -122, -123 and -124
50. Four times the sum of a number and five is equal to 100. Find the number. $4(n + 5) = 100$; 20

OBJECTIVE: *To apply mathematical skills when purchasing a car.*

7-8 Buying a Car

Buying a car is one of the first major purchases an individual makes. There are several things to consider when making such an expensive purchase.

A car dealership will give you a *base price* on a car. This is the price of the car without any additional features. Additional features, or *options*, have an additional charge. The table at the left lists the cost of several options that dealers frequently offer.

Option	Price
Air Conditioning	$744
AM/FM Radio-cassette	$155
Automatic Transmission	$732
Metallic Color	$91
Sunroof	$549

Example 1 Mr. Perry wants to buy a car with automatic transmission, an AM/FM radio-cassette, and a sunroof. What is the total cost of these options?

Solution 732 + 155 + 549 = 1,436 Add the cost of the options.

The total cost of these options is $1,436.

1 You will automatically pay a *destination* and *delivery charge* on a new car. This charge pays for the cost of shipping the car from the manufacturer and for preparing the automobile for sale. Most states require you to pay a *sales tax* on a new car.

THINK What must a dealership do to a car to prepare it for sale? wash and wax the car; attach the dealership nameplate

Example 2 James is purchasing a car for $8,667. The sales tax is 6%. How much sales tax will James pay?

2 **Solution** 6% = 0.06 Write 6% as a decimal.

8,667(0.06) = 520.02 Multiply the cost of the car by 0.06.

James will pay $520.02 in sales tax.

Many people take out a loan to help pay for a car. They then make monthly payments. To determine the monthly payment, multiply the number of thousands of dollars being borrowed by a factor which can be found in an *amortization table* like the one shown below.

3

Amortization Table (per $1,000)

number of months	interest rate 10%	11%	12%
12	87.92	88.39	88.85
24	46.15	46.61	47.08
36	32.27	32.74	33.22
48	25.37	25.85	26.34

Example 3 What is the factor for a 10% loan over 36 mo?

Solution Find the number of months. Read across to the 10% column. The factor is 32.27.

Teaching Tip

List problem solving steps in the classroom for students' reference. These steps should include Planning, Collecting Data, Solving the Problem, and Looking Back and Ahead.

Lesson Focus

Materials

Classified section of various newspapers, math journal

Vocabulary/Symbols

amortization chart
base price
destination and delivery charge
options
sales tax

MOTIVATING THE LESSON

Distribute automobile advertisements to small groups of students. Have the groups review the ads. Have students discuss why a manufacturer might offer 0% interest on loans, or a rebate on the final cost, or a 50,000 mi (or 3-year) warranty on a car. Ask students what factors might affect the choice one makes in buying a car.

Instruction

1 Delivery charges include cleaning the car, checking the car's mechanical and electronic parts, and doing any other special work to the car. Ask students what other costs there are when buying a car. (Possible answers include insurance, license, and registration.)

2 Review that sales tax is a tax charged by states on goods and services purchased within the state. The sales tax is used to raise money for the state, and the rate varies from state to state. Ask students what the sales tax is in their state.

3 Explain that an amortization chart shows the repayment of the loan (principal) and the interest charged.

Guided Practice

Class Exercises For question 4, ask students what would be the monthly loan payment and total payment on a $10,000 loan. ($263.40; $12,643.20) Then ask them to find the monthly and total loan payments if the interest rate is 11% over 36 mo. ($327.40; $11,786.40) Ask why they might choose one loan payment over another.

Closure

Writing in Math Ask students to describe in their math journals, what each number in the amortization chart represents and what the monthly payment found by using the amortization chart includes.

Independent Practice

Decision Making Have students work in small groups. For questions 1 and 3, students may use newspaper ads to collect the data. Students may want to compare costs in different parts of the country. Have groups display their charts in the classroom and compare the cost of different types of cars.

In question 5, ask students why a large dealership might have a lower base price than a small dealership. (Possible answers: volume discount, more efficient dealer delivery process)

Expand question 11 to help students realize the total cost of owning a car. Have students find insurance cost, maintenance costs, and gasoline costs.

Example 4 Lynn wants to borrow $12,000 from her bank. The bank is offering an 11% interest rate. Lynn plans to take out a 36-mo loan. What will be her monthly payment?

Solution

1. Determine the number of thousands of dollars that are being borrowed on $12,000. — 12,000 is the same as 12 thousands.
2. Find the factor on an 11% interest rate over 36 mo using the amortization table. — 32.74
3. Multiply the number of thousands of dollars being borrowed by the factor. — $12 \cdot 32.74 = 392.88$

Lynn's monthly payment will be $392.88.

CLASS EXERCISES

Refer to the options list on page 305.

1. What would a sunroof and an AM/FM radio-cassette cost? **$704**
2. How much sales tax must you pay on a $10,000 car at 6%? **$600**
3. How many thousands are in $14,500? **14.5**
4. What is the factor on a loan at 12% interest over 48 mo? **26.34**

Decision Making **DECISION MAKING** Decision Making Decision Making Decision Making

BUYING A CAR

COLLECT DATA **Check students' work.**

1. Choose a car you would like to own. Interview salespeople at three dealerships. Organize your data in a chart.

Dealership	base price	options	price of options	dest/del charge

2. Find out the sales tax in your state.
3. Contact three banks in your area and find out what interest rates they offer on new car loans.

ANALYZE DATA

The base price of a car includes the dealership's profit.

4. Did the base prices on the car you chose vary?

MATH MINUTES

Have students use the amortization table on p. 305 to determine a payment schedule for a college loan.

WRITTEN EXERCISES

Use a CALCULATOR where appropriate.

Determine the sales tax on each amount at a rate of 6%.

1. \$8,675 **\$520.50**
2. \$7,988 **\$479.28**
3. \$16,654 **\$999.24**
4. \$12,560 **\$753.60**

Determine how many thousands each amount represents.

5. \$9,000 **9**
6. \$8,500 **8.5**
7. \$11,750 **11.75**
8. \$14,250 **14.25**

Determine each factor using the amortization table on page 305.

9. a loan at 12% over 24 mo **47.08**
10. a loan at 11% over 48 mo **25.85**
11. Gerald wants to buy a car with a base price of \$9,518. The car is metallic black. Gerald is having a sunroof installed. The destination and delivery charge is \$342.
    a. Gerald has enough money saved to pay for the sales tax. How much sales tax will Gerald pay at a rate of 6%? **\$630**
    b. Gerald took out a loan at 10% over 48 mo for the remaining cost of the car. What will he pay per month? **\$266.39**
12. Alicia bought a car with a base price of \$14,670. She had an AM/FM radio-cassette installed. The destination and delivery charge was \$397. Alicia paid 5% sales tax. She withdrew \$4,483.10 from her savings and took out a 12% loan over 36 mo for the rest. What does Alicia pay per month? **\$382.03**

Decision Making ■ Decision Making ■ Decision Making ■ Decision Making ■ Decision Making ■ Decision Making ■

5. Why might one dealership offer a lower base price than another?
6. Did the prices on the options you chose vary from place to place?
7. Is there a relationship between the base price that a dealership offered and the price that they charged for the options?
8. Why might loan rates differ at different banks?

■ **MAKE DECISIONS**

Suppose you are buying the car of your choice. You may add any options. You have enough money saved for sales tax.

9. What is the price of the car, including the options you chose, the destination and delivery charge, and sales tax?
10. Determine what your monthly payment would be if you took out a loan for the full amount needed at 11.5% over 48 mo.
11. How much did you pay the bank for their loan service?

Lesson Follow-up

LESSON RESOURCES

TRB Alternate Application G-15

Assignment Guide

Basic All
Average All
Enriched All

FOR THE NEXT LESSON

geoboards, rubber bands, math journal

Alternate Application

For use after 7-8 (pp. 305–307)

Meeting the Mean

The *mean* of a set of values, also called the average, is the sum of the values divided by the number of values.

Example Find the mean of these ages: 8, 27, 37

Solution
1. Add. $8 + 27 + 37 = 72$
2. Divide by the number of values. $\frac{72}{3} = 24$

The mean age is 24.

Find the mean.

1. Pulse rates (beats/min): 75, 81, 68, 67, 66, 70, 84 **73 beats/min**
2. Building heights (ft): 714, 485, 619, 809, 333 **592 ft**
3. Ocean depths (m): 3,930; 3,566; 3,831; 1,035 **3,090.5 m**

If you know the mean and all but one of the values, you can use an inequality to find the missing value.

Example The mean of 5 temperatures is at least 87°F. Four of the temperatures are 81°F, 82°F, 88°F, and 96°F. Find the fifth temperature.

Solution
1. Let n = the missing value.
2. Use the variable to write an inequality. $\frac{81 + 82 + 88 + 96 + n}{5} \geq 87$
3. Solve. $\frac{347 + n}{5} \geq 87$
   $347 + n \geq 435$
   $n \geq 88$

The temperature must be greater than or equal to 88°F.

Find the missing value.

4. Test scores: 78, 83, 76, 75, 80, 84, 82, 71, 80
   Mean must be at least 80 to earn a B. **score ≥ 91**
5. Monthly savings deposits: \$275, \$295, \$270, \$350, \$340
   Mean must be at least \$325 to meet vacation goal. **deposit ≥ \$420**
6. Weight gain (kg): 2.7, 3.1, 3.5, 2.1, 3.5, 2.1, 2.2, 2.6, 2.7, 3.4
   Mean must be no more than 2.8 kg to meet goal. **gain ≤ 2.9 kg**

Chapter 7 G-15

Study Skills

Knowing where to *look for data* is an essential part of the studying process. Encourage students to seek information beyond that given in their textbooks. Often, another point of view will help to enhance appreciation of a topic or provide the key to understanding a concept.

Challenge students to discover the kinds of data available in the school and/or public library. Discussions should include information available in the book stacks, on the library computers, and in the reference and media areas.

Then suggest a topic and have students "brainstorm" other potential sources. For example, to learn more about computers, students may suggest contacting a local computer society, computer retailers, universities, and computer book publishers.

Encourage students to continue to use the sources they have discovered throughout the school year.

Other sources of information include (depending on data):

- Government Printing Office
- Census Bureau
- utility companies
- League of Women Voters
- chambers of commerce
- consumer and business organizations

Chapter 7 Review

Complete each statement. Use the vocabulary words given.

VOCABULARY
negative
combine
operations
distributive
equation
not equal

1. A two-step equation involves two ■. operations
2. To solve $3(x + 2) - 2x$, first remove the parentheses using the ■ property. distributive
3. To solve $2x - 3 + 4x = x + 2 - 3x$, first ■ like terms. combine
4. An inequality is a statement that two expressions are ■. not equal
5. If c is ■ and $a > b$, then $ac < bc$. negative
6. The steps for solving a two-step inequality are similar to the steps for solving a two-step ■. equation

Two-step Equations — 7-1

To solve a two-step equation,
1. undo addition and subtraction;
2. undo multiplication and division.

Solve each equation.

7. 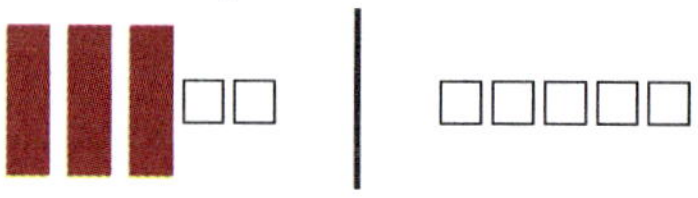 $x = 1$

8.  $x = 2$

9. $5a + 3 = 28$ $a = 5$
10. $\frac{x}{4} - 9 = -6$ $x = 12$
11. $7n - 2 = 19$ $n = 3$
12. $\frac{5b}{6} + 7 = 22$ $b = 18$

Simplifying and Solving Equations — 7-2

To solve a multi-step equation,
1. remove parentheses using the distributive property;
2. combine like terms;
3. undo addition and subtraction;
4. undo multiplication and division.

Solve each equation.

13. $3(x - 5) + 2x = 20$ $x = 7$
14. $2x - 8 - 3x = 2$ $x = -10$
15. $(2x - 1)2 + 5x = 1$ $x = \frac{1}{3}$
16. $9 - 5x + 1 = 15$ $x = -1$
17. $5(x + 2) - 3x = 38$ $x = 14$
18. $8x + 5 - 5x = 3$ $x = -\frac{2}{3}$

Problem Solving Strategy: Writing Equations — 7-3

To write an equation for a word problem, you need to recognize words that imply the variable(s), the operation(s), and the equality.

Solve.

19. The difference between 3 times n and 2 times the quantity $n + 5$ is 3. Find n. $3n - 2(n + 5) = 3$; $n = 13$
20. A sweater cost $12 more than a shirt. The total cost of two sweaters and three shirts is $144. Find the cost of the shirt. $2(x + 12) + 3x = 144$; $x = 24$

Equations with Variables on Both Sides **7-4**

To solve equations with variables on each side,

1. use the distributive property to remove parentheses;
2. combine like terms;
3. use the addition and subtraction property so that all variables are on one side and constants on the other;
4. use the multiplication and division property.

Solve each equation.

21. 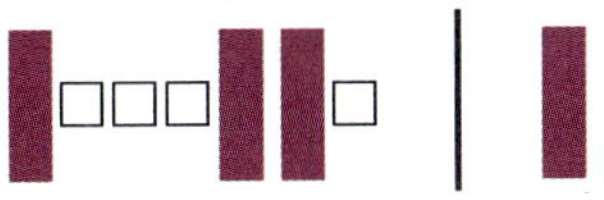 x = -2

22. $3x + 5 = 7x - 11$ x = 4

23. $n + 3(n - 2) = 5n + 4$ n = -10

Inequalities and Their Graphs **7-5**

To graph an inequality, use an open dot for $>$ and $<$, use a closed dot for $\geq$ and $\leq$. If the symbol is $>$ or $\geq$, graph all points to the right of the boundary. If the symbol is $<$ or $\leq$, graph all points to the left of the boundary.

Match each inequality with the appropriate graph.

a. 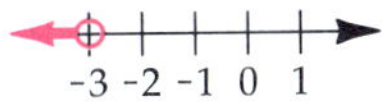

b. 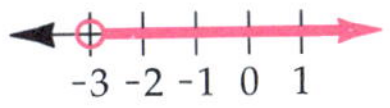

c. 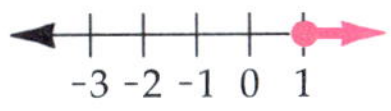

d. 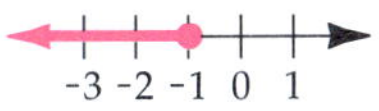

24. $x \geq 1$ c **25.** $x < -3$ a **26.** $x > -3$ b **27.** $x \leq -1$ d

Solving One-step Inequalities **7-6**

To solve an addition or subtraction inequality, add or subtract the same value from both sides of the inequality.

To solve a multiplication or division inequality, multiply or divide both sides of the inequality by the same nonzero value. If the value is positive, keep the inequality symbol. If the value is negative, reverse the inequality symbol.

Solve each inequality.

28. $x - 3 < -8$ $x < -5$ **29.** $x + 2 \geq -1$ $x \geq -3$ **30.** $-3x > 12$ $x < -4$ **31.** $\frac{x}{4} \leq -2$ $x \leq -8$

Solving Two-step Inequalities **7-7**

To solve a simple two-step inequality,

1. undo addition and subtraction;
2. undo multiplication and division.

Solve each inequality.

32. $5x - 7 \leq 18$ $x \leq 5$ **33.** $2y + 4 > 12$ $y > 4$ **34.** $\frac{b}{-3} + 5 > 8$ $b < -9$ **35.** $\frac{2a}{3} - 1 < 7$ $a < 12$

309

RESOURCES

TRB G-17–G-20

Scoring Chart	
No. Correct	**Score**
35	100
34	97
33	94
32	91
31	88
30	85
29	82
28	79
27	76
26	73
25	70
24	67
23	64
22	61
21	58
20	55
19	53
18	50
17	47
16	44
15	41
14	38
13	35
12	32
11	29
10	26
9	23
8	20
7	17
6	14
5	11
4	8
3	5
2	3
1	1

Chapter 7 Test

Solve each equation.

1. 

2. 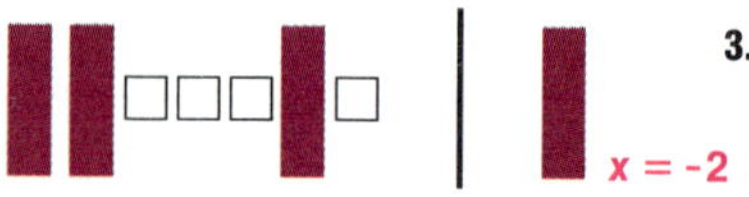

3. 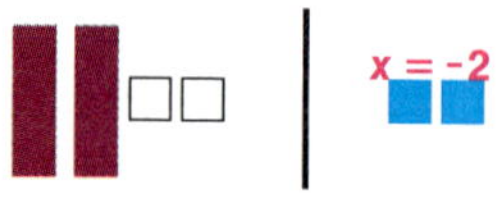

4. $5x + 9 = -6$ **x = -3**

5. $0.5 + 2n = 3$ **n = 1.25**

6. $\frac{3}{4}y - 5 = 7$ **y = 16**

7. $7p - 3 = 18$ **p = 3**

8. $-3(b - 6) = 27$ **b = -3**

9. $1.5(c + 2) = 6$ **c = 2**

10. $7z + 8 - 2z = 23$ **z = 3**

11. $5x - 9 = 3x$ **$x = 4\frac{1}{2}$**

12. $2(6 - 2x) = 5x - 6$ **x = 2**

Write an equation to describe each situation. Solve.

13. Kendra bought a scarf for \$35.75. The original price was \$55. What was the percent discount? **55 – 55x = 35.75; 35%**

14. The perimeter of a rectangle is 132 cm. The length is 3 cm more than twice the width. Find the length and width.
w + (2w + 3) + w + (2w + 3) = 132
length = 45 cm
width = 21 cm

Match each inequality with the appropriate graph.

15. $x \geq 2$ **c**

16. $x > -2$ **b**

17. $x > 2$ **d**

18. $x \leq -2$ **a**

a. 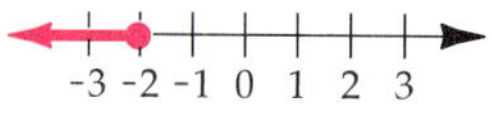

b. 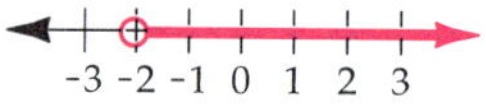

c. 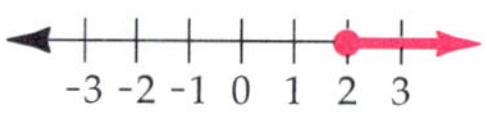

d. 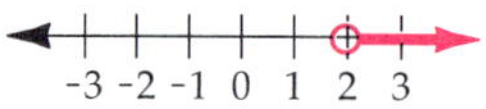

Solve each inequality.

19. $5 \leq x + 1$ **$x \geq 4$**

20. $3a > 4$ **$a > 1\frac{1}{3}$**

21. $y - 6 < 9$ **$y < 15$**

22. $-2n \leq 10$ **$n \geq -5$**

23. $\frac{b}{3} \geq \frac{1}{3}$ **$b \geq 1$**

24. $\frac{p}{-2} < -5$ **$p > 10$**

25. $-2x + 14 < 6$ **$x > 4$**

26. $9y - 8 > -17$ **$y > -1$**

27. $\frac{1}{5}c - 1 \geq 2$ **$c \geq 15$**

28. $-9 + 6r \leq -33$ **$r \leq -4$**

29. $-7m + 6 < 48$ **$m > -6$**

30. $9k + 5 > -67$ **$k > -8$**

Write an inequality for each sentence. Graph each inequality.

31. The total, t, is greater than 5. **$t > 5$** 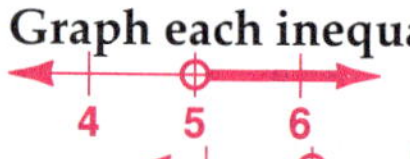

32. The perimeter, p, is less than 64. **$p < 64$** 

33. The number of passengers, p, on the bus is less than or equal to 45. **$p \leq 45$** 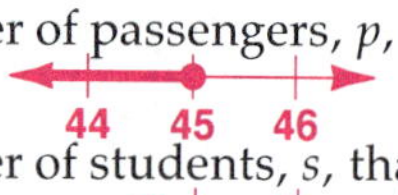

34. The number of students, s, that ran in the road race was not less than 55. **$s \geq 55$** 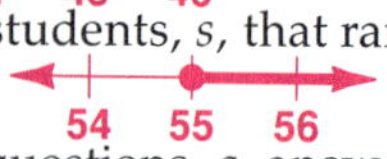

35. The number of questions, q, answered correctly is less than or equal to 35. **$q \leq 35$** 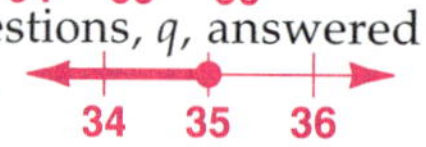

Chapters 1–7 Cumulative Review

Choose the correct answer. Write A, B, C, or D.

1. Write an equation for the model. B

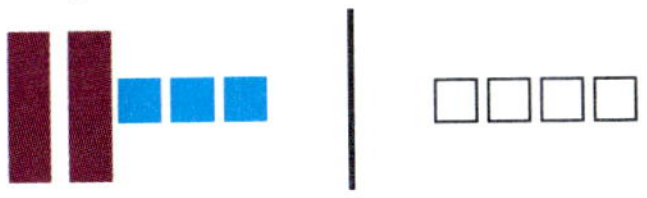

A. $2x + 3 = 4$ **B.** $2x - 3 = 4$
C. $2x = 3 - 4$ **D.** not given

2. What is the prime factorization of 108? C

A. $2 \cdot 3$ **B.** $2^3 \cdot 3^2$
C. $2^2 \cdot 3^3$ **D.** not given

3. Solve $\frac{10}{15} = \frac{x}{3}$. A

A. 2 **B.** $\frac{1}{2}$
C. 5 **D.** not given

4. Write $\frac{3a^3b}{12ac^2}$ in lowest terms. C

A. $\frac{3ab}{4ac}$ **B.** $\frac{ab}{4c}$
C. $\frac{a^2b}{4c^2}$ **D.** not given

5. Solve $3x - 1 + 2x = x + 11$. A

A. 3 **B.** 2
C. 0 **D.** not given

6. Write 0.0050 as a percent. C

A. 50% **B.** 5%
C. 0.5% **D.** not given

7. Find the next number. C
0.5, 1.2, 2, 2.9, 3.9, . . .

A. 4.9 **B.** 4
C. 5 **D.** not given

8. Count the number of squares. B

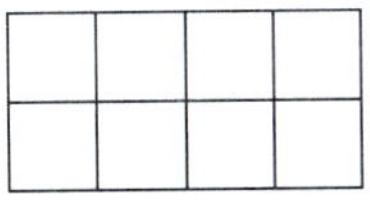

A. 8 **B.** 11
C. 10 **D.** not given

9. Simplify $x^2(x - 24)$. D

A. $x^3 - x^2y$ **B.** $x^3 - 2x^2y$
C. $x^3 - 2y^2$ **D.** not given

10. Find a number between $-1\frac{1}{2}$ and -2. C

A. $-\frac{3}{4}$ **B.** $-1\frac{1}{4}$
C. $-1\frac{3}{4}$ **D.** not given

11. What percent is 35 of 105? C

A. 3% **B.** 33%
C. $33\frac{1}{3}$% **D.** not given

12. If c is negative and $a < b$, then A

A. $ac > bc$ **B.** $ac < bc$
C. $ac \leq bc$ **D.** not given

13. Simplify $(a^2b^3)^2(ab^2)$. B

A. a^5b^7 **B.** a^5b^8
C. a^3b^5 **D.** not given

14. There are 12 students that belong to the Math Club and 18 that belong to the Science Club. There are 5 students that belong to both clubs. How many notices need to be printed for a joint meeting? D

A. 30 **B.** 13
C. 7 **D.** not given

Study Skills

Some students may find *sentence-completion tests* particularly difficult to handle. Help students realize that they can develop strategies for giving correct responses.

Suggest that students look for the most obvious clues first. The number of words in the answer may be indicated by the number of blanks used; the extent of an answer may be shown by the length of the blank. Caution students, however, that this may not always be reliable. Modifiers and verbs can also be clues. The word "a" before the blank suggests that the answer begins with a consonant, whereas, "an" suggests a vowel. Some verbs, such as "is", suggest a singular noun or one answer, while others like "are", may suggest a plural noun or more than one answer.

Have students note the words or labels immediately following the blanks. A word such as "meters" suggests that the answer is a measure, and that more than one meter is involved.

You may wish to:

- Use a variety of test formats.
- After each test, discuss the techniques used by students to answer the questions.

Remind students to:

- Write the answers they know first.
- Use time wisely in order to analyze the remaining questions.
- Leave time to check all answers.

Chapter At a Glance

Chapter 8 introduces the students to graphing in the coordinate plane. The use of graphs to model and solve equations is explored.

Reading the Chapter

A problem is often easier to solve when the reader can visualize the relationship between all the parts of the problem. The coordinate plane introduced in this chapter is a useful tool for visualizing the parts of an equation. When students explore using graphs to solve problems, remind them to look for negative numbers in an equation, then to take care to move in the appropriate direction on each axis on the graph.

Review important terminology, such as *ordered pair*, *slope*, *intercept*, and *quadrant*, prior to each lesson.

Chapter Resources

Practice and Enrichment
TRB H-1 to H-16
Alternate Application TRB H-17

Chapter Test Form A, H-19
Chapter Test Form B, H-21

Computer Test Bank

References

STUDENT BIBLIOGRAPHY

Spencer, Donald D. *Exploring the World of Computers* (Camelot Publishing, Ormond Beach, Fla. 1982).

TEACHER BIBLIOGRAPHY

Alberti, Del and Mary Laycock. *The Correlations of Activity Centered Science and Mathematics* (Activity Resources Co., Hayward, Calif. 1975).

SOFTWARE

Green Globs and Graphing Equations (Sunburst Communications), for Apple II and IBM computers.

FILMS, VIDEOTAPES, AND OTHER MEDIA

The Whatsit Corporation (Sunburst Communications), computer disk, worksheets, and overhead transparencies. A simulated company is started and run by students.

Vocabulary

- coordinate plane
- intercept
- linear equations
- nautical mile
- ordered pair
- origin
- quadrants
- simulate
- slope
- solution
- system of linear equations
- systems of linear inequalities
- trajectory
- *x*-axis
- *y*-axis

Materials

- geoboard (M*)
- math journal
- graph paper (T*)
- computer and graphing software
- graphing calculator
- measuring tape or ruler
- rubber ball
- stopwatch

*M—Manipulative Kit
*T—Teacher's Resource Book

Resource Bank

Class Activities

FIELD TRIPS

▼ Engineering firm—to see how slope and coordinates are indicated on the blueprints of a road project.
▼ Town or county planning board—to see how coordinate mapping is used to regulate land use.

CLASS SPEAKERS

▼ Civil engineer—to discuss how slope affects the design of roads and river embankments.
▼ Surveyor—to discuss how coordinates are used to develop topographic maps.

Bulletin Board

The four color map theory, proven by Appel and Hoken in 1976, shows that every map, on a flat surface or a sphere, can be colored using no more than four colors. The theory assumes that every map contains at least one of the four configurations below. Display an uncolored map of the United States and each of the four configurations.

Every "normal map" includes at least one of these four configurations. How many can you find in the U.S. map? (all 4 are present.)

A "normal map" cannot include an island or a state which is completely enclosed by another. Are there any areas on the U.S. map not considered "normal"? (Michigan, in two parts)

Project: Color the map using as few colors as possible. States that share a common border cannot be the same color, but states that meet at only a single point can be colored the same.

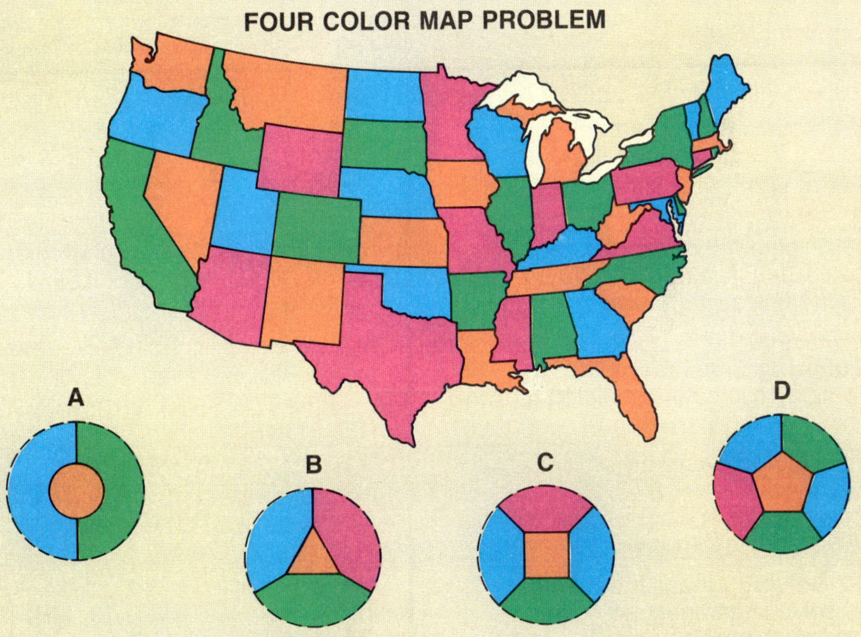

Extra-Credit Problems

BASIC

▼ Draw the figures described below. Then name four coordinates of each.
1. Circle with a radius of 7 units.
2. Right triangle with base of 6.
(Answers will vary.)

AVERAGE

▼ The sum of two numbers is 63. The first number divided by $\frac{1}{2}$ the second equals 5. Write the system of linear equations. Solve by graphing. Hint: both numbers are divisible by 9. (45, 18)

ENRICHED

▼ Graph a line with a slope of 1 and a y-intercept of (0, 4). Name 3 sets of coordinates on the line. Write an inequality for the area to the left of the line. ($y = x + 4$; $y > x + 4$; answers will vary.)

Chapter Planning Guide

OBJECTIVES	ASSIGNMENT GUIDE			ASSESSMENT	
	Basic	Average	Enriched	Review/ Tests	TRB*
Exploring Verbal Communication To explore communication problems.	**Manipulative Connection** All	All	All		
8-1 The Coordinate Plane To locate and graph a point given the coordinates and to identify the coordinates of a given point.	1-17 odd, 18-21, 28-32, 34, MR*, CT* All	2-20 even, 28-34, MR, CT All	17-34, MR, CT All	Extra Practice p. 573	Practice H-1 Enrichment H-2
8-2 Solving Equations To solve linear equations in two variables.	1-13, 16-17, 22-24, 28-29, MR, TY* All	1-25 odd, 28-29, MR, TY All	6-26 even, 27-29, MR, TY All	Extra Practice p. 573 TY p. 323	Practice H-3 Enrichment H-4 TY H-18
8-3 Graphing Linear Equations To graph a linear equation and find *x*- and *y*-intercepts.	1-7, 12-15, 20-25, 32, MR All	1-9, 12-15, 20-26, 32, MR All	8-23, 26-32, MR All	Extra Practice p. 573	Practice H-5 Enrichment H-6
Exploring Slope To use coordinate graphs to explore linear equations, intercepts, and slope.	All	All	All		
8-4 Slope and *y*-intercept To find the slope and *y*-intercept for the graph of a linear equation.	1-21 odd, 25-27, 31a-b, 32-33, MR All	2-5, 8-24 even, 28-31c, 33-36, MR All	2-18 even, 22-24, 29-31, 34-38, MR All	Extra Practice p. 573	Practice H-7 Enrichment H-8
8-5 Problem Solving Strategy: Solve by Graphing To solve problems using a graph.	**Problem Solving Connection** All	All	All	Problem Solving Practice p. 337	Practice H-9 Enrichment H-10
Exploring Simulations To study computer-simulated motion.	All	All	All		
8-6 Solving Systems of Linear Equations To solve two equations in two variables by graphing.	1-17, 22, 25, MR, TY All	1-15 odd, 16-20, 23-25, MR, TY All	7-21 odd, 22-26, MR, TY All	Extra Practice p. 573 TY p. 343	Practice H-11 Enrichment H-12 TY H-18
8-7 Solving Linear Inequalities To solve linear inequalities in two variables.	1-5, 8-10, 18-19, MR, CT All	3-7, 11-14, 17, 19, MR, CT All	2-10 even, 13-17, 19, MR, CT All	Extra Practice p. 573	Practice H-13 Enrichment H-14
8-8 Graphing Linear Inequalities To graph a linear inequality in two variables and to explore simple systems of inequalities.	1-21 odd, 22-27, 38-39, 43-46, 50, MR All	2-24, even, 36-42, 44-48, 50, MR All	11-14, 28-35, 40-46 even, 48-50, MR All	Extra Practice p. 573	Practice H-15 Enrichment H-16
8-9 Problem Solving Application: Direct and Indirect Variation To solve problems involving direct and indirect variation.	All	All	All		Alternate Application H-17
Chapter Test	All	All	All	pp. 356-359	Tests H-19–H-25

*TRB—Teacher's Resource Book MR—Mixed Review CT—Critical Thinking TY—Test Yourself

CONNECTIONS							NCTM CORRELATION
Algebra	**Critical Thinking**	**Problem Solving**	**Estimation/ Mental Math**	**Technology**	**Manipulatives/ Resources**	**Writing/Reading in Math**	
	8a				1-8	1-8	Communications
C*: 1-15 W*: 1-33 CT*: 1-3	CT: 1-4 C: 15, TD* 1-4 W: 33		C: 9-10 W*: 18-22		W: 9-16, 29-32, 33, 34 CT: 2-3	C: TD 2-4 CT: 1, 4	Communications Algebra Connections
C: 1-12 W: 1-27, 29 TY*: 1-11	C: TD 1-3	W: 29	W: 1-6	W: 7-9		W: 28, 29	Communication Problem Solving Algebra
C: 1-11 W: 1-32	C: TD 1-4	W: 25-32	W: 20-24		W: 8-19, 25-32	C: TD 1-4	Problem Solving Algebra
1, 2, 3, 4, 5, 6, 8, 9, 10	2, 3, 4, 5, 7, 9				8	2, 3, 4, 5, 7, 9, 10	Communication Algebra
C: 4-9 W: 7-12, 19-24, 31-38	C: TD 1-4 W: 31, 37, 38	W: 37-38	C: 1-3 W: 1-3		C: 10 W: 4-6, 13-18, 25-31, 37		Problem Solving Algebra
	C: 3 W: 2d	C: 1-3 W: 1-15	W: 1, 11b		W: 1, 2, 14, 15	W: 2d, 14c	Problem Solving Communications Connections
3, 5	1, 2, 4-6		7a	3-5	1-2, 4, 5, 7b	1, 2, 4-7	Communications Connections Reasoning
C: 1-12 W: 1-24 TY: 1-12	C: TD 1-3 W: 22-24	C: 11-12 W: 16-21, 25-26			C: 8-12 W: 4-21, 25	C: TD 1-3	Problem Solving Connections Algebra Communications
C: 1-10, 11c W: 1- 16, 18c, 19	C: TD 1-3 W: 19 CT: 1-2	C: 11 W: 17-18		W: 5-8		C: TD 1-3 W: 19	Problem Solving Algebra
C: 1-17 W: 1-48	C: TD 2 W: 44-45	C: 13-17 W: 37-40, 44-48			C: 10-17 W: 22-48, 50	W: 49, 50	Problem Solving Communication Algebra Connections
C: 1-3 W: 1-9		W: 5-9					Communication Algebra Problem Solving Connections

*C—Class Exercises W—Written Exercises CT—Critical Thinking TY—Test Yourself TD—Think and Discuss

Exploring

In this manipulative activity, students explore slope by manipulating the *y*-intercept on a graph. **See TRB Exploring O-29**.

Exploring More About Slope

Lay the edge of a ruler on the line. Rotate it in a clockwise direction keeping the *y*-intercept (0, 1), until it is parallel to the *x*-axis.

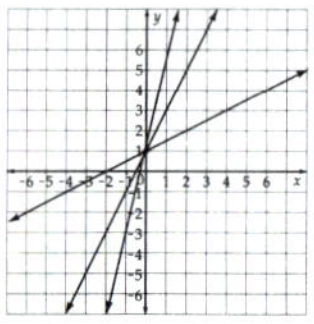

1. What is the slope of the original line? **2**
2. Did the line become steeper or flatter? What happened to the slope?
   Flatter; it decreased.
3. Draw a line on the graph. Find its slope to demonstrate your answer in question 2.
   Answers may vary; a line with slope $\frac{1}{2}$.
4. The slope of the line horizontal to the *x*-axis is 0. Why?
   The change in *y* is zero, so the rate is $\frac{0}{\text{horizontal change}}$.

Lay a ruler on the original line in the graph above. Turn it in a counter-clockwise direction (not past the *y*-axis) keeping the *y*-intercept (0, 1).

5. What happened to the steepness of the line? What happened to the slope?
   It became steeper; it increased.
6. Draw a line on the graph and find its slope to demonstrate your answer to question 5.
   Answers may vary; a line with slope 4.
7. Rotate the ruler to a position where you think the slope is 100; 1,000. Describe what happens.
   It comes very close to the *y*-axis; it is almost vertical.
8. A vertical line has no slope because its ratio is undefined. Why is the ratio undefined?
   The change in x is 0, so the ratio is $\frac{\text{vertical change}}{0}$.

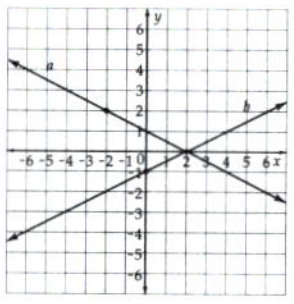

Use the graph at the right.

9. What are the slopes of the lines? **a: $-\frac{1}{2}$; b: $\frac{1}{2}$**
10. Describe a line with a negative slope.
    The line goes down from left to right.

O-29

In this problem solving activity, students use graphing to solve a problem. **See TRB Exploring O-30**.

Exploring Graphing as a Strategy

Kevin and Mike are bicycling. Kevin begins his trip at 8:00 A.M., traveling at 15 mi/h. At 8:12 A.M. Mike starts out to follow Kevin. Mike's speed is 19 mi/h. If they travel exactly the same route, when will Mike pass Kevin?

1. How far has Kevin traveled by the time Mike begins?
   3 mi
2. At 8:20 A.M. how long will Kevin have been bicycling and how far will he have gone? What about Mike?
   Kevin: 20 min, 5 mi
   Mike: 8 min, about 2.5 mi
3. Guess how long it will take Mike to catch up to Kevin.
   Answers will vary.

The solution can be found by graphing the two situations on one graph. Complete the chart to find the coordinates.

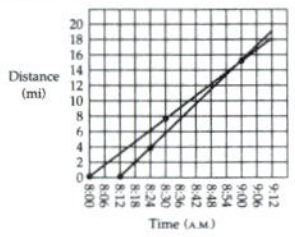

4.

Kevin's Bike Trip			
Time	8:00	8:30	9:00
Distance (mi)	**0**	**7.5**	**15**

5. What three coordinates can you plot on the graph? Plot the points and connect them.
   (8:00, 0), (8:30, 7.5), (9:00, 15)
6. At 8:12 A.M. what is Mike's distance?
   0
7. When Mike has traveled 12 min, what time will it be? What distance will he have covered?
   8:24; 3.8 mi
8. What two points can you plot for Mike?
   (8:12, 0), (8:24, 3.8)
9. How can you tell when Mike overtakes Kevin?
   The lines intersect.
10. Give the approximate time and distance traveled when they meet.
    Answers may vary. Exact time and distance are 8:57, 14.25 mi.

O-30

In this calculator activity, students explore the graphing of functions using a graphing calculator. **See TRB Exploring O-31**.

Exploring the Graphing of Relationships

A T-shirt company has regular expenses of \$5,000 per week and a silkscreening expense of \$1.25 per shirt. The company receives \$500 a week from investments and charges \$5 per shirt wholesale. Find the number of shirts that must be sold to break even.

The break-even point for a business is that point where the expenses equal the revenues. When revenues are greater, the business operates at a profit. When the expenses are greater, the business operates at a loss.

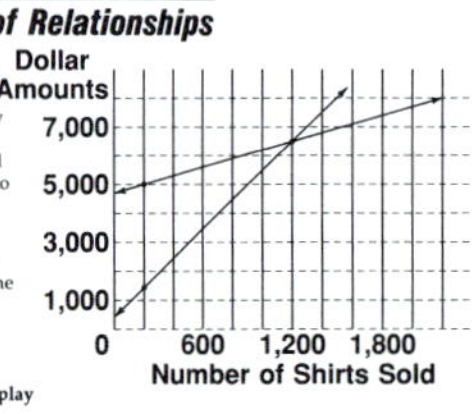

You can use a graphics calculator to display graphs, or you can draw the graphs on graph paper or use the computer.

1. Set up an equation that defines the expenses.
   Let y = expenses and x = the number of shirts made. **$y = 5{,}000 + 1.25x$**
2. Set up an equation that defines the revenues.
   Let y = revenues and x = the number of shirts sold. **$y = 500 + 5x$**

Graph the expense line and the revenue line on the same set of axes. Adjust the scale and the *x* and *y* axis to fit the data.

3. How can you determine from the graph where the break-even point occurs?
   at the point of intersection of the graphs

Use the graph to answer the following questions.

4. How many T-shirts must the company sell to break even? **1,200 shirts**
5. What are the expenses and revenues at the break-even point? **\$6,500**
6. If the company sells 1,300 T-shirts, approximately how much profit should the company make?
   \$375 since Revenue − Expenses = Profit
7. Suppose the weekly expenses increase to \$7,500 and the silkscreening costs double. If the investment returns remain the same and the company raises the wholesale price per shirt to \$7.50, how many shirts would the company have to sell to break even?
   1,400 shirts

O-31

In this computer activity, students use a spreadsheet to solve nonroutine problems. **See TRB Exploring O-32**.

Exploring Problem Solving Using a Spreadsheet

You can find the solution to coin problems using a spreadsheet.

Example Joe has \$2.05 in nickels and dimes. He has four more dimes than nickels. How many nickels and dimes does Joe have?

Set up the computer spreadsheet as shown.

	A	B	C	D	E
1	No. of Nickels	No. of Dimes	$ Amt: Nickels	$ Amt: Dimes	Total Amt
2					

1. If Joe has one nickel, then how many dimes does he have?
   5 dimes
2. What formula would define the value of B2 in terms of A2?
   =A2+4
3. If Joe has one nickel, what amount of money does he have in nickels? **\$.05**
4. What formula do you input into:
   a. C2 to determine the amount of money Joe has in nickels? **=0.05*A2**
   b. D2 to determine the amount of money Joe has in dimes? **=0.10*B2**
   c. E2 to determine the total amount of money Joe has? **=C2+D2**
5. How do you know that Joe has more than one nickel?
   1 nickel means 4 dimes; this gives a total of only 55¢

Increase the number of nickels by 1 and input the new value into A3. Input the appropriate formulas for cells B3 through E3.

6. What values do you have for B3? **6** C3 **0.10** D3 **0.60** E3? **0.70**
7. Continue increasing the number of nickels. Find the solution.
   How many nickels and dimes does Joe have? **11 nickels, 15 dimes**

Solve using a spreadsheet.

8. A collection of nickels and quarters contains 25 coins. If there are 3 more nickels than quarters, what is the total value? **\$3.45**
   How many nickels and quarters are there? **11 quarters, 14 nickels**
9. What formulas were used for the cells in row 2?
   1, =A2+3, =0.25*A2, =0.05*B2, =C2+D2

O-32

Meeting Individual Needs

GIFTED STUDENTS

The gifted student needs to master basic skills as much as any other student; however, the training should be provided at advanced levels. Many practical problems in business, science, and industry are solved using a technique known as linear programming, which involves graphing several inequalities on the same coordinate axes. Ask students to write four inequalities from the information below and then graph the solution of the system of inequalities.

A pet store owner plans to buy some hamsters and rabbits. She wants to purchase at least 16 animals. Each hamster costs $10 and each rabbit costs $20. She decides to spend at least $200. Remember, the number of hamsters and rabbits must be at least one.

Students should determine the coordinates of the corner points on the graph, because they are part of the solution set, and are where the boundary lines intersect. Ask students: If it costs 13 cents a day to feed each hamster and 30 cents a day to feed each rabbit, how many of each animal should be purchased if daily feeding costs are to be kept to a minimum? Have students use the coordinates of the corner points to calculate the answer. (12 hamsters, 4 rabbits)

AT-RISK STUDENTS

At-risk students often need to be reminded of the many ways mathematics affects their daily lives and their futures. Ask students to think of events they are anticipating in the next few years. For example, some students may anticipate buying a car.

One consideration in buying a car is the operating costs. Ask students to compare the operating costs for a truck and a car to decide which is the better vehicle to operate.

Have students use the following information:

- Expected Average Annual Mileage is 14,000 miles.
- The car gets 32 miles per gallon and uses premium gasoline that costs $1.63 per gallon.
- The truck averages 25% fewer miles per gallon than the car, but runs on regular gasoline that costs $1.52 per gallon. (car; annual car cost is $713.13, annual truck cost is $886.67)

ENGLISH AS A SECOND LANGUAGE

ESL students will be challenged in this chapter to verbalize concepts and understand the explanations of other students. Help ESL students work successfully with the chapter's demands by modeling ideas whenever possible and encouraging students to do the same.

When playing the game "What's My Shape?" in the lesson "Exploring Verbal Communication," you can match an ESL student with an English-language student. You might want to relax the game's rules to allow the ESL student to hear the directions in several different ways. This game can be adapted for other areas of the chapter.

For the lesson "Exploring Slope," it might be helpful to have ESL students verbalize their explanations first, and then, as a separate assignment, practice writing the skills in their math journals. You can record their oral explanations for the students to use as a reference when writing. **See TRB Teaching Resources p. 58.**

Problem Solving Situation

For the Think About It question, students should recognize that the center of population has continued to move westward. While it is possible for the center of population to move east, the westward trend makes it unlikely.

Data File 8 can be extended by explaining to students that cartography—the art and science of map making—relies heavily on mathematical skills, such as proportions, coordinate graphing, and algebraic methods of representing and analyzing geometric curves. To help students better understand the issues involved in making maps, have them investigate the history and uses of maps. Students can report on either an important figure in map making history (Ptolemy, Martin Waldseemuller, Juan de la Rosa, Gerhardus Mercator) or a type of map (nautical or aeronautical charts, geographical maps, etc.).

DATA COLLECTION

Information on maps and their history can be found in books on the subject. If a nearby store specializes in new and antique maps, students might arrange an interview with the owner. Students might try calling the National Geographic Society's public information office.

DATA ANALYSIS

Students assigned to explore a historical figure can report on the individual's contribution to the history of map making. Students assigned to report on a type of map can explore the way a cartographer collects the data and prepares the map. Reports can include examples of maps, pictures of cartographers, and other visual aids.

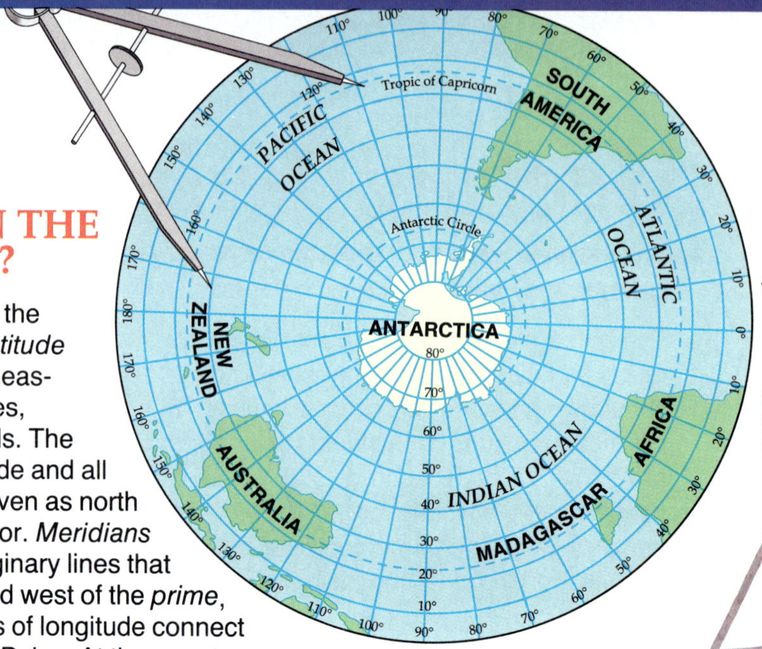

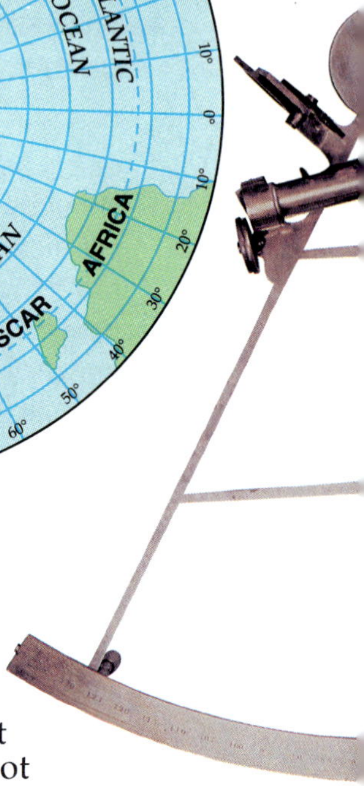

WHERE IN THE WORLD?

We locate points on the globe by means of *latitude* and *longitude.* We measure latitude in degrees, minutes, and seconds. The equator is at 0° latitude and all other latitudes are given as north or south of the equator. *Meridians* of longitude are imaginary lines that locate points east and west of the *prime*, or 0°, meridian. Lines of longitude connect the North and South Poles. At the equator, 1° of longitude equals 69.15 mi. At the Arctic Circle, 1° of longitude equals 27.65 mi.

Getting around

You can stand in four states in the United States at the same time if you're on the spot where Arizona, New Mexico, Utah, and Colorado meet.

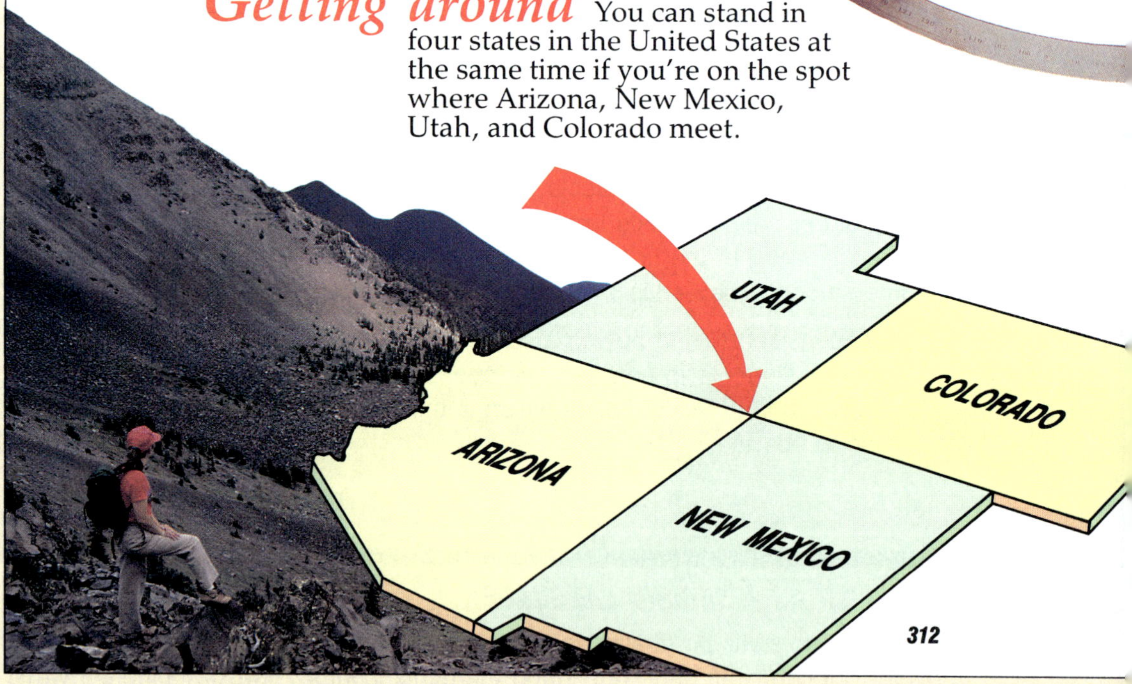

CHAPTER 8

Graphing in the Coordinate Plane

UNITED STATES CENSUS DATA

YEAR	POPULATION
1790	3,929,214
1800	5,308,483
1810	7,239,881
1820	9,638,453
1830	12,866,020
1840	17,069,453
1850	23,191,876
1860	31,443,321
1870	39,818,449
1880	50,155,783
1890	62,947,714
1900	75,994,575
1910	91,972,266
1920	105,710,620
1930	122,775,046
1940	131,669,275
1950	150,697,361
1960	179,323,175
1970	203,302,031
1980	226,545,805
1990	249,632,692

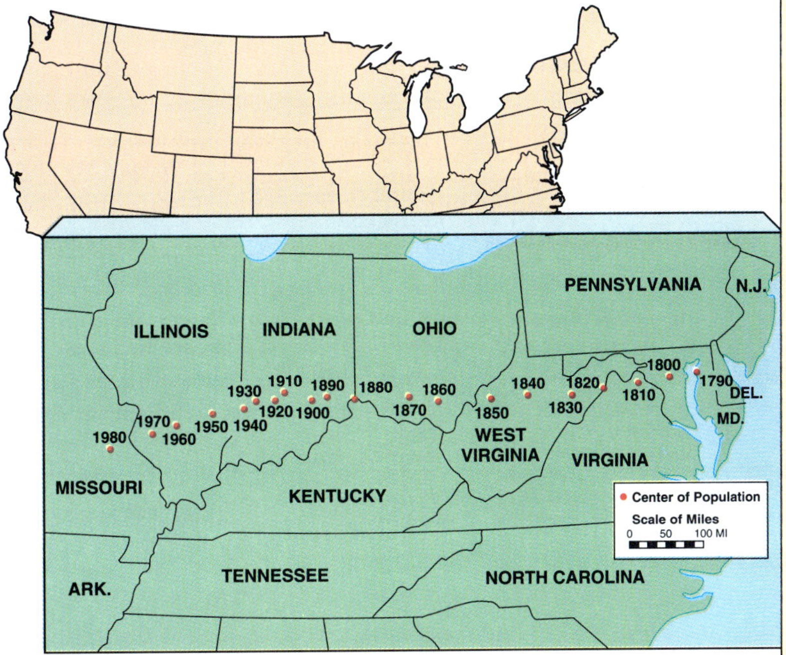

THE CENTER OF POPULATION is the point in the United States at which the population is evenly balanced. This means that if the country were flat, and everyone had the same weight, the country would exactly balance at the center of population.

Think about it...

Look at the Center of Population map. Why do you think the center of population moves? Do you think the center of population will ever stop moving west?

See Problem Solving Situation

313

Cooperative Learning

The research project can be enjoyable and challenging for groups of 4 to 6 students. All group members can do the research and agree upon what information to include in the report. Remind students that because a lot of data may be collected, it will be important for the group to decide on a set of criteria for evaluating what material to include.

Possible roles for group members include the following positions: Information Collector, Writer, Interviewer, and Presenter.

Questions related to this data file can be found on pages 20, 343, 456, and 553.

Lesson Focus

Materials

Geoboards, rubber bands, math journal

MOTIVATING THE LESSON

Divide the class into pairs and distribute geoboards and rubber bands to each student. Allow students to experiment with materials for a few minutes, encouraging them to create various geometric shapes. Explain to the students that in today's lesson they will be exploring aspects of oral communication.

Instruction

1 To be sure that students understand the rules, demonstrate a number of correct and incorrect shapes on the geoboard.

2 After each student has had an opportunity to play both roles and the pairs have discussed how to give clear instructions, have the class discuss ways to describe shapes on the geoboard. The discussion should include the basics of a rectangular coordinate system, for example, "the second peg in the fourth row."

3 Encourage students to give directions using north, south, east, and west and relate these compass points to coordinate graphs and maps.

OBJECTIVE:
To explore communication problems.

MATERIALS

- Geoboard
- Math journal to record work

Exploring Verbal Communication

▼ Oral communication is an important tool. It involves an accurate portrayal of a situation by the speaker and careful analysis by the listener.

▼ You can improve your communication skills by choosing a partner and playing this game on a geoboard.

What's My Shape?

Number of Players: 2

Objective: Each player will end up with the same shape in the same location on their geoboard.

Rules:

1. Players sit back-to-back so that neither player can see the other's geoboard.
2. Player 1 creates a shape on the geoboard using a rubber band. The rubber band may not intersect itself and it may not double back. Examples are shown below.

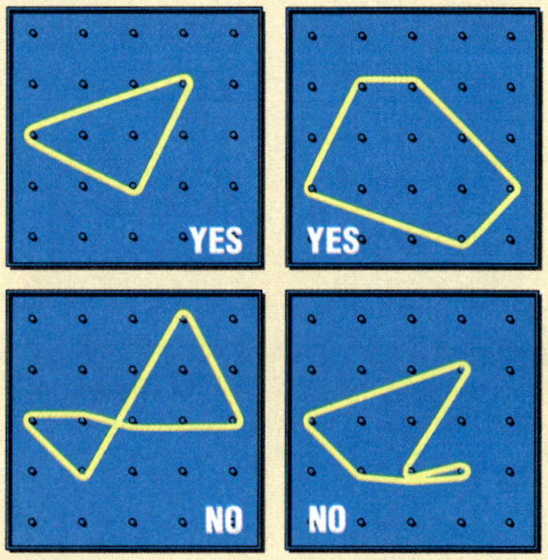

3. Player 1 describes the shape to player 2.
4. Player 2 listens carefully to the description and tries to duplicate the shape in the correct location on his or her geoboard.
5. Player 2 may ask player 1 to repeat the instructions, but may not ask any other questions.
6. After player 2 has completed the shape, both players turn around and compare the shapes on their geoboards.

Teaching Tip

If geoboards are unavailable, students can complete the lesson using dot paper. Students should observe the same rules with dot paper as those used with geoboards.

Check students' work for Questions 1–8.

1. ***Write*** a description of the method player 1 used to communicate the location of the points on the geoboard. Include answers to the following questions in your description.
   a. Did player 1 give instructions to move right or left?
   b. Did player 1 give instructions to move up or down?
   c. Did player 1 refer to columns and rows?
2. a. ***Describe*** some of the problems player 1 had in describing the shape and location.
   b. ***Explain*** how player 1 could improve his or her oral communication skills.
3. ***Describe*** some of the problems player 2 had in interpreting player 1's instructions.
4. Together, write a set of clear instructions so that the shape and location can be duplicated without any problem.
5. Reverse roles and repeat the activity. Answer Questions 1–4. **2**

▼ Good communication skills are necessary when you are giving directions to someone.

6. a. ***Explain*** how playing this game is similar to giving directions to get from your school to your home.
   b. ***Explain*** how it is different.
   c. ***Write*** a clear set of instructions telling someone how to get from your school to your home. **3**
   d. ***DATA FILE 1 (pp. 2–3) Write*** a clear set of instructions telling someone how to use the wind chill chart to determine the degree of danger of frostbite if the air temperature is -15°F and the wind speed is 25 mi/h.

▼ You can try other games to improve listening skills.

7. One student reads a short news article to the listener. The listener writes a short paragraph describing what he or she heard.
   a. ***Compare*** what is written with the original article.
   b. ***Explain*** how the listener can improve his or her listening skills.
8. Form a group of ten students. One student whispers the contents of a short newspaper story to another. That student whispers to another what he or she heard. Repeat the process until the last student has heard the story. That student then relates the story he or she heard to the group.
   a. ***Compare*** the accuracy of the last story with the original.
   b. ***Describe*** how the group can improve its communication skills.

Closure

Writing in Math Ask students to write in their math journals the most effective method they found for locating points on the geoboard. (Answers may vary. Possible answer: identifying the pegs in rows and columns.) Then ask the students to describe what information is important when providing directions. (Possible answer: identifying a starting point, using compass directions, identifying landmarks.)

Assignment Guide

Basic All
Average All
Enriched All

FOR THE NEXT LESSON

graph paper, state road map with town index, overhead projector, transparency of a coordinate plane, math journal

Lesson Focus

Materials

Graph paper, state road map with town index, overhead projector, transparency of a coordinate plane, math journal

Vocabulary/Symbols

coordinate plane
ordered pair
origin
x-axis
y-axis

MOTIVATING THE LESSON

Display the state map to students. Discuss the map key and the coordinate grid of letters and numbers. Have students name a city in the state and locate the coordinates in the index. Then have a student use the grid coordinates to locate the city on the map. Repeat the activity, making sure students understand the procedure for using the grid.

Instruction

1 Present the coordinate plane on the overhead projector. Distribute graph paper to students and have them duplicate the coordinate plane. Be sure they number the graph correctly and indicate the *x*- and *y*-axes.

2 Discuss the quadrants with students and the meaning of an ordered pair. Mention that in an ordered pair the *x*-coordinate is always named first.

3 Help students realize that to locate a point, they begin at the origin and move right on the *x*-axis if *x* is positive, or left on the *x*-axis if *x* is negative. From that point, they move up on the *y*-axis if *y* is positive, or down on the *y*-axis if *y* is negative.

OBJECTIVE: *To locate and graph a point given the coordinates and to identify the coordinates of a given point.*

8-1 The Coordinate Plane

▼ Mapmakers use letters and numbers to designate regions on maps. According to the index, Southeastern Avenue in Indianapolis, Indiana, is in Region B2.

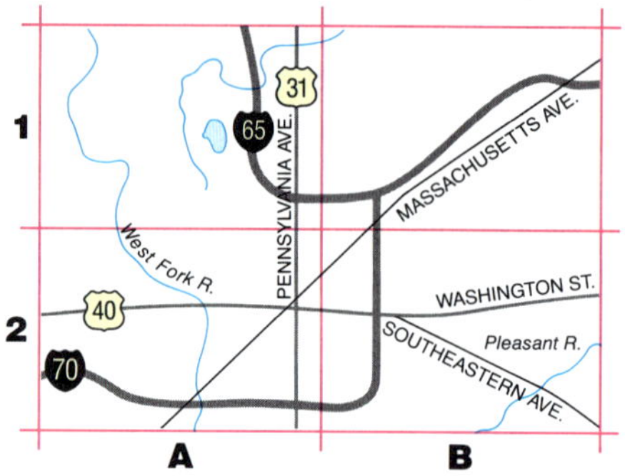

▼ Mathematicians represent a region using a *coordinate plane*.

Coordinate Plane	A coordinate plane is the plane which results when two perpendicular number lines intersect at their zero points. The number lines form a grid on the plane.

1 ▼ We call the horizontal number line the *x-axis*, with the positive direction to the right. We call the vertical number line the *y-axis*, with the positive direction upward. The *x*- and *y*-axes intersect at the *origin*. The *x*- and *y*-axes divide the coordinate plane into four *quadrants*.

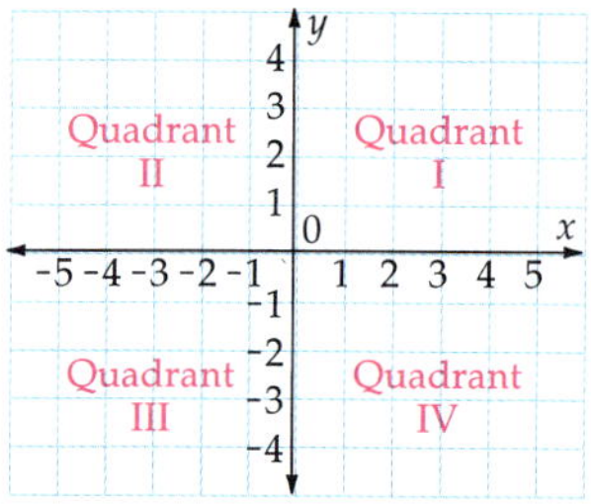

▼ We use *ordered pairs* to locate points in the coordinate plane.

2

Ordered Pair	An ordered pair is a pair of numbers (*x*,*y*) assigned to a point on a coordinate plane.

In the ordered pair (x,y), the value that corresponds with x is called the x-coordinate. The value that corresponds with y is called the y-coordinate. You can locate a point on the coordinate plane when given an ordered pair.

FLASHBACK
The absolute value of a number is its distance from zero on a number line.

Locating a Point on the Coordinate Plane	To locate $P(x,y)$ on the coordinate plane: **1.** Begin at origin. **2.** Locate x on the x-axis. **3.** Move up or down the absolute value of y units.

3

Teaching Tip

Allow students to work on transparencies and with the overhead projector. This lesson on the coordinate grid provides opportunities to locate ordered pairs on a transparency grid.

▼ Use the coordinate plane below for the examples.

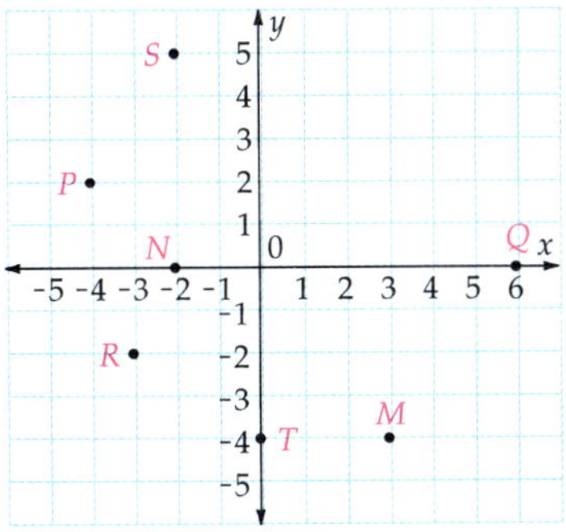

Example 1 **Locate each point on the coordinate plane. State the letter of the point.**

a. (-2,5) **b.** (0,-4) **c.** (-3,-2) **d.** (6,0)

Solution
a. Start at the origin. Move 2 units left and up 5. *S*
b. Start at the origin. Move zero units left or right and 4 units down. *T*
c. Start at the origin. Move 3 units left and 2 units down. *R*
d. Start at the origin. Move 6 units right and zero units up or down. *Q*

▼ Identifying the quadrant in which a point falls is similar to identifying a region on a map.

Example 2 In which quadrant or on which axis does each point fall?

a. (2,6) **b.** (-4,0) **c.** (0,0)
d. (3,-1) **e.** (-5,-2) **f.** (0,5)

Solution
a. quadrant I **b.** *x*-axis **c.** both axes
d. quadrant IV **e.** quadrant III **f.** *y*-axis

▼ You can use an ordered pair to locate a point on the coordinate plane.

Example 3 **Name the coordinates of each point.**

a. point *M* **b.** point *P* **c.** point *N*

Solution
a. Point *M* is 3 units to the right (positive) of the origin and 4 down (negative). The ordered pair is (3,-4).
b. Point *P* is 4 units to the left (negative) of the origin and 2 units up (positive). The ordered pair is (-4,2).
c. Point *N* is 2 units to the left (negative) of the origin and zero units up or down. The ordered pair is (-2, 0).

NOTES & QUOTES

In the seventeenth century, René Descartes (1596–1650) developed coordinate graphing.

Legend has it that Descartes was watching a fly on a tiled ceiling when it occurred to him to describe the fly's position in relation to its distance from the intersection of the tiles.

4 Before completing Example 2, have students describe the *x* and *y* coordinates for points in each quadrant. [I (+, +); II (-, +); III (-, -); and IV (+, -)]

ADDITIONAL EXAMPLES

1. Locate and label the following points on the coordinate plane.
A (-3, 6) B (5, 2)
C (0, 5) D (4, -1)
E (-4, -1) F (-3, 0)
[Check students' graphs.]

2. On which axis or in which quadrant is each point from question 1 located?
(A II; B I; C *y*-axis; D IV; E III; F *x*-axis)

MATH MINUTES

Each student writes 10 ordered pairs involving all four quadrants. Working in pairs, one student names an ordered pair and the other names the quadrant. Taking turns, students identify all 20 ordered pairs.
(Check students' work.)

Guided Practice

Class Exercises For question 15, points P, Q, and *R* are not the points on the coordinate grid used for questions 1-8.

Think and Discuss For question 2, remind students that a negative *x* coordinate is to the left of the origin, and a negative *y* coordinate is below the origin.

Have students substitute values for *a* and *b* for question 3, and describe the quadrant for each ordered pair.

A ***common error*** occurs when students reverse the order in an ordered pair and plot the points beginning on the *y*-axis. Review with students that an ordered pair always indicates the *x*-axis first, then the *y*-axis. Remind students that when plotting points, begin at the origin and move along the *x*-axis first.

Closure

Writing in Math Ask students to describe in their math journals how to locate a point in the coordinate plane. Then have students compare locating a city on a map and locating a point in a coordinate plane.

Independent Practice

Suggest that students graph each point in questions 23-28 to check their answers.

Critical Thinking Have students work in small groups. For question 3, have students start with the original points for each part of the question. Transformations occur as follows: reflection in the *y*-axis, reflection in the *x*-axis, rotation of 180°, and enlargement of the original rectangle.

THINK AND DISCUSS

1. What ordered pair names the origin? **(0,0)**
2. Describe how to locate point *Q*(4,-3).
3. Do (*a*,*b*) and (*b*,*a*) describe the same point? Assume that *a* and *b* are not the same number. Explain.
4. Complete: *P*(*a*,*b*) is in quadrant II. The value of *a* must be ■. The value of *b* must be ■.

2. Start at the origin. Move 4 units to the right and 3 units down.
3. No; (*a*,*b*) means *a* units left or right of the origin on the x-axis and *b* units up or down. (*b*,*a*) means *b* units left or right of the origin on the x-axis and *a* units up or down.
4. negative; positive

CLASS EXERCISES

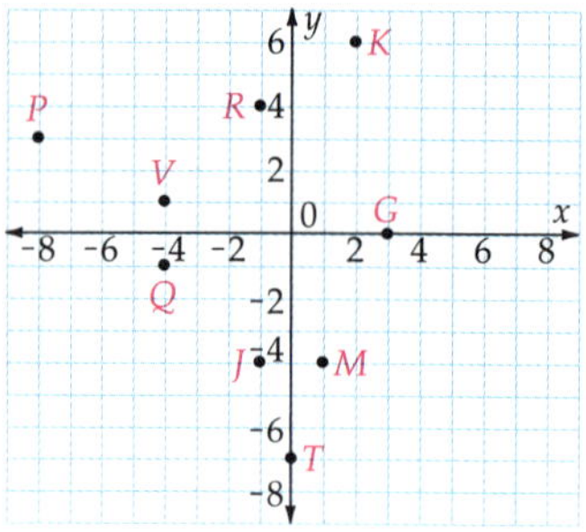

State the letter of the point named by each ordered pair.

1. (-1,-4) **J**
2. (3,0) **G**
3. (-1,4) **R**
4. (-4,-1) **Q**

State the coordinates of each point.

5. *T* **(0,-7)**
6. *V* **(-4,1)**
7. *M* **(1,-4)**
8. *K* **(2,6)**

MENTAL MATH State the coordinates of each point described.

9. the point which is 5 units to the left of the *y*-axis and down 2 units from the *x*-axis **(-5,-2)**
10. the point on the *y*-axis that is 4 units below the *x*-axis **(0,-4)**

In which quadrant or on which axis does each point fall?

11. (-2.6,3.4) **II**
12. $\left(5\frac{1}{4},2\frac{1}{2}\right)$ **I**
13. $\left(0,-4\frac{2}{3}\right)$ **y-axis**
14. (1.36,19.41) **I**

15. *P*(-1,3), *Q*(4,3), and *R*(4,-2) are three vertices of a square. Find the coordinates of the fourth vertex. **(-1,-2)**

WRITTEN EXERCISES

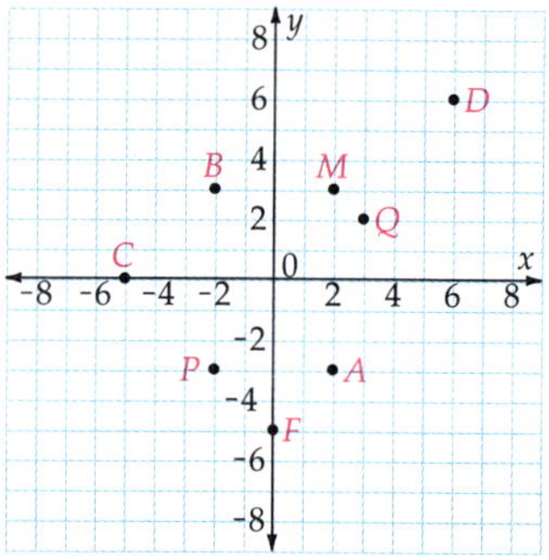

State the letter of the point named by each ordered pair.

1. (3,2) **Q**
2. (0,-5) **F**
3. (2,3) **M**
4. (-2,-3) **P**

State the coordinates of each point.

5. *A* **(2,-3)**
6. *B* **(-2,3)**
7. *C* **(-5,0)**
8. *D* **(6,6)**

Draw a coordinate plane. Graph each point. **Check students' work.**

9. *F*(-3,2)
10. *G*(-1.5,0)
11. *H*(1,7)
12. *J*(-3.5,-4)
13. *K*(5,-6)
14. *L*(0,0)
15. $M\left(\frac{1}{2},3\right)$
16. *N*(7,0)

17. State the coordinates of four points in the coordinate plane that are 3 units from the origin. **(3,0)(0,3)(-3,0)(0,-3)**

ADDITIONAL ANSWERS

4. When you multiply the *x*-coordinates by -1, the figure moves one unit to the right (a slide). When you multiply the *y*-coordinates by -1, the figure is reflected over the *x*-axis. When you multiply each coordinate by -1, the resulting figure is a reflection over the *x*-axis and a slide one unit to the right. When you multiply each coordinate by 2, the resulting figure moves up one unit and the length and width are doubled.

MENTAL MATH State in which quadrant or on which axis $P(x,y)$ lies if the following conditions are true.

18. x is negative and y is zero. x-axis

19. x is positive and y is negative. IV

20. x is positive and y is positive. I

21. x is negative and y is positive. II

22. x is negative and y is negative. III

In which quadrant or on which axis does each point fall?

23. (13,25) I

24. (−17.654,−0.02) III

25. $(0,|-2|)$ y-axis

26. (x,y) if $x > 0$, $y < 0$ IV

27. (x,y) if $x < 0$, $y < 0$ III

28. (x,y) if $x = 0$, $y > 0$ y-axis

Check students' work.

Graph and connect each point in the order given. Connect the last point to the first. Name the figure.

29. (−1,2), (1,5), (7,5), (5,2) parallelogram

30. (2,2), (2,−1), (−5,−1), (−5,2) rectangle

31. (−4,1), (1,1), (−3,−1) triangle

32. (2,−4), (7,−1), (4,4), (−1,1) square

33. $P(0,5)$, $Q(5,0)$, and $R(-5,0)$ are three vertices of a square. Find the coordinates of the fourth vertex. (0,−5)

34. ***PROJECT*** Draw a dot-to-dot picture on a coordinate grid. Write the coordinates of each point in order. Exchange with a classmate and draw each other's picture. Check students' work.

MIXED REVIEW

Solve for y.

1. $4.5 + 7y = 18.5$ 2

2. $5a - 10 = 30$ 8

Solve for x.

3. $-3x + 8 > 29$ $x < -7$

4. $\frac{2}{3}x - 19 \leq -13$ $x \leq 9$

Solve for y if x = 3.

5. $7x + y = 24$ 3

6. $21x + 2y = -7$ −35

Solve.

7. The perimeter of a rectangle is 62 in. Its length is twice its width, less 5 in. Find the actual length and width.
12 in., 19 in.

Critical Thinking

EXPLORING GRAPHING

1. ***Describe*** what you think will happen to a figure if you perform an operation on one or both coordinates of each point. Check students' work.

2. Graph the points (−2,1), (−2,3), (1,3), (1,1). Connect them in the order given. Connect the last point to the first. Check students' work.

3. Find the new coordinates and graph.

a. Multiply each x-coordinate by −1. (2,1)(2,3)(−1,3)(−1,1)

b. Multiply each y-coordinate by −1. (−2,−1)(−2,−3)(1,−3)(1,−1)

c. Multiply each coordinate by −1. (2,−1)(2,−3)(−1,−3)(−1,−1)

d. Multiply each coordinate by 2. (−4,2)(−4,6)(2,6)(2,2)

4. ***Compare*** the figures in Exercise 3. ***Write*** a short paragraph describing your results. See Additional Answers.

Lesson Follow-up

Reteaching Activity Provide students with a coordinate plane. Ask them to identify the signs in each quadrant of the plane [I (+, +); II (−, +); III (−, −); IV (+, −)]. Then ask them to locate the following points: (6, −1), (4, 0), (−3, −5), (2, −2), (0, 0), (3, 5). [Check students' graphs.]

LESSON RESOURCES

TRB Practice H-1
TRB Enrichment H-2

Practice

For use after 8-1 (pp. 316–319)

The Coordinate Plane

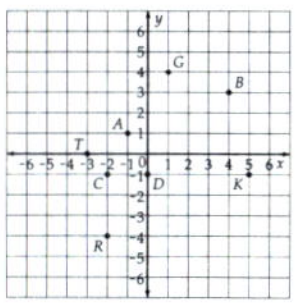

Write the letter of the point named by each ordered pair.

1. (1, 4) G
2. (−3, 0) T
3. (5, −1) K
4. (−2, −4) R

Write the coordinates of each point.

5. A (−1, 1)
6. B (4, 3)
7. C (−2, −1)
8. D (0, −1)

Graph each point. Write the letter beside the point.

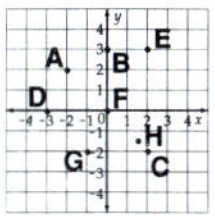

9. A(−2, 2)
10. B(0, 3)
11. C(2, −2)
12. D(−3, 0)
13. E(2, 3)
14. F(0, 0)
15. G(−1, −2)
16. $H\left(1\frac{1}{2}, -1\frac{1}{2}\right)$

Complete using positive or negative. Write p or n.

17. In Quadrant II, x is n and y is p.
18. In Quadrant III, x is n and y is n.
19. Point P is below the x-axis and right of the y-axis. The x-coordinate of P is ■ and the y-coordinate is ■.
p; n

In which quadrant or on what axis does each point fall?

20. (3, −5) IV
21. $(|-7|, |-23|)$ I
22. (0, 35.7) y-axis
23. (−83, 0) x-axis
24. $P(-2, 3)$, $Q(5, 3)$, and $R(-2, -6)$ are three vertices of a rectangle. Find the coordinates of the fourth vertex.
(5, −6)
25. A circle with center at the origin passes through the points $A(-7, 0)$, $B(7, 0)$, and $C(0, x)$. Find the two possible values of x.
7, −7
26. The point $G(x, y)$ is directly above $E(-2, 7)$ and directly to the right of $F(-4, 12)$. Find x and y.
$x = -2$, $y = 12$

Chapter 8 H-1

Enrichment

For use after 8-1 (pp. 316–319)

Latitude and Longitude

Geographers divide the earth into a coordinate grid using *latitude* and *longitude* lines. Latitude lines are parallel to the equator and run from 90°N (the North Pole) to 90°S (the South Pole). Longitude lines are measured east and west of the *prime meridian*, the 0° longitude line which runs through Greenwich, England. Point *A* in the figure has coordinates (15°S, 45°W).

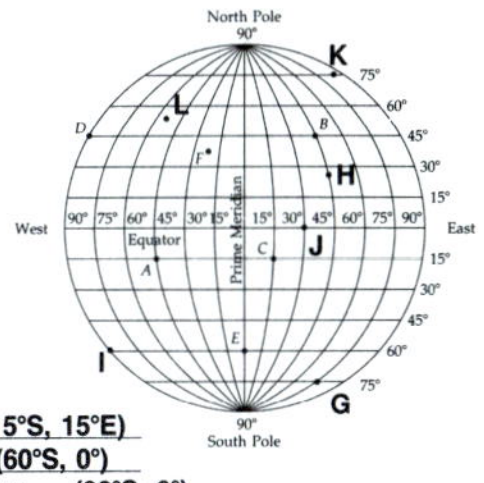

Write the coordinates, giving the north-south coordinate first.

1. B (45°N, 45°E)
2. C (15°S, 15°E)
3. D (45°N, 90°W)
4. E (60°S, 0°)
5. F $(37\frac{1}{2}°N, 22\frac{1}{2}°W$
6. South Pole (90°S, 0°)

Graph each point on the coordinate grid above. Write the letter beside the point.

7. G(75°S, 75°E)
8. H(25°N, 45°E)
9. I(60°S, 90°W)
10. J(0°, 30°E)
11. $K\left(75°N, 82\frac{1}{2}°E\right)$
12. L(50°N, 55°W)
13. How many degrees of latitude separate Halifax, Nova Scotia (45°N, 65°W), and Cordoba, Argentina (32°S, 65°W)?
77°
14. How many degrees of longitude separate Baku, U.S.S.R. (41°N, 50°E), and New Haven, CT (41°N, 73°W)?
123°
15. One degree of longitude at the equator equals 69.2 mi. How far is it from Quito, Ecuador (0°, 79°W), to Kampala, Uganda $(0°, 32\frac{1}{2}°E)$?
7,715.8 mi
16. Belem, Brazil (0°, 48°W), is located due west of Libreville, Gabon. The distance between the cities is 3,944.4 mi. Give the latitude and longitude of Libreville.
(0°, 9°E)

H-2 Chapter 8

LESSON QUIZ

Draw a coordinate plane. Graph each point.

1. (3, −3) **2.** (0, −15)
3. (−4, 5) **4.** (2, −5)
5. (−5, 0) **6.** (−2, −5)
[Check students' graphs.]

In which quadrant or on which axis does each point fall?

7. (−2, −7) [III]
8. (0, 3) [*y*-axis]
9. (−3, 4) [II]
10. (−4, 0) [*x*-axis]

Assignment Guide

Basic 1-17 odd, 18-21, 28-32, 34, MR, CT All
Average 2-20 even, 28-34, MR, CT All
Enriched 17-34, MR, CT All

FOR THE NEXT LESSON

calculator, math journal, number cubes

Lesson Focus

Materials
Calculator, math journal, number cubes

Vocabulary/Symbols
solution

MOTIVATING THE LESSON

Discuss with students that at times they may need two variables to describe a problem situation.

Present the following problem to students: If decals sell for 10 cents each and stickers sell for 5 cents each, how many decals and stickers can be purchased for 60 cents?

Discuss how an equation could be written for the problem by letting x = the number of decals and y = the number of stickers. $(10x + 5y = 60)$ Mention that to solve this equation, students may substitute a value for x and a value for y, such that 10 times the value of x plus 5 times the value of y is 60.

Have students substitute values for the variables to find solutions to the equation. $(x = 4, y = 4; x = 3, y = 6$ are some possible solutions.) Point out that the solutions to this equation are ordered pairs.

Instruction

1. When $x = 0$, the term containing x is equal to zero; therefore, y and a constant remain.

2. An equation with two variables has an infinite number of solutions. Here students are asked to find only three solutions.

OBJECTIVE: *To solve linear equations in two variables.*

8-2 Solving Equations

▼ The normal low temperature at the base of Mt. Rushmore in July is 21°C. The temperature drops an average of 1°C for every 100 m of vertical ascent. You can write an equation using two variables to describe this situation. Let y = temperature and let x = meters above the foot of the mountain.

$$y = 21 - 0.01x$$

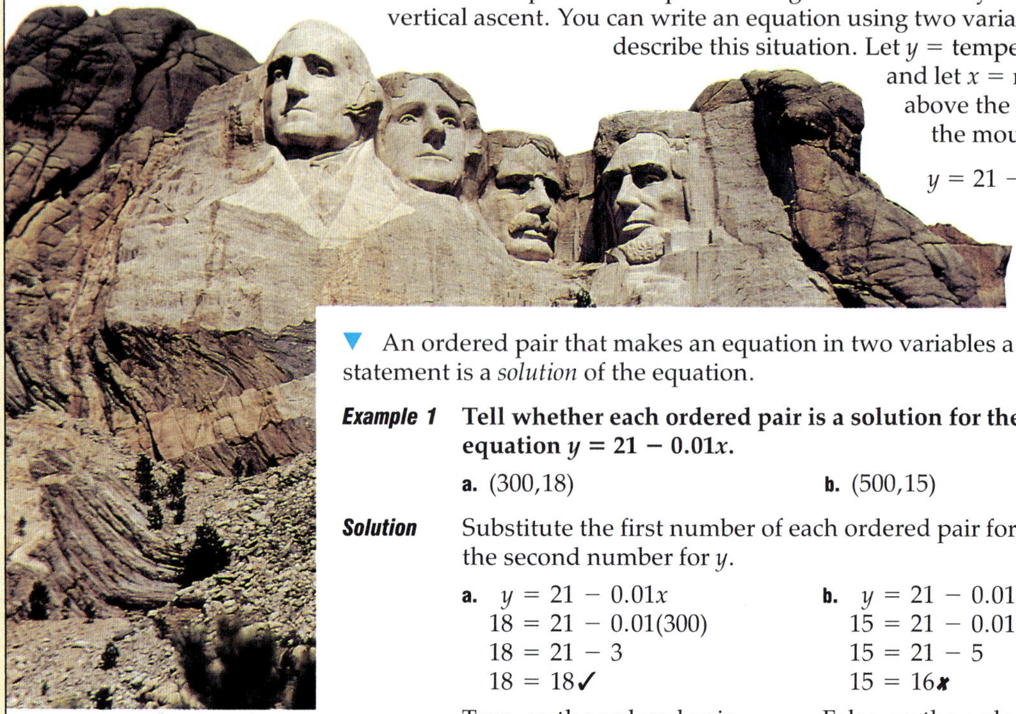

What is the average temperature 300 m above the base of Mt. Rushmore in July? 18°C

THINK Why is it sometimes helpful to choose 0 as a value for x? You can determine where the graph crosses the y-axis. (1)

▼ An ordered pair that makes an equation in two variables a true statement is a *solution* of the equation.

Example 1 **Tell whether each ordered pair is a solution for the equation $y = 21 - 0.01x$.**

a. (300,18) **b.** (500,15)

Solution Substitute the first number of each ordered pair for x and the second number for y.

a.
$y = 21 - 0.01x$
$18 = 21 - 0.01(300)$
$18 = 21 - 3$
$18 = 18$ ✓

True, so the ordered pair (300,18) is a solution.

b.
$y = 21 - 0.01x$
$15 = 21 - 0.01(500)$
$15 = 21 - 5$
$15 = 16$ ✗

False, so the ordered pair (500,15) is not a solution.

▼ You can find solutions for an equation with two variables by using a table to organize your data.

(2)

Example 2 **Find three solutions to the equation $y = 3x + 6$.**

Solution

1. Choose three values for x. Try -3, 0, and 4.
2. Use a table to organize your data.
3. Substitute x values into the equation to find y values.

x	y	(x,y)
-3	■	(■,■)
0	■	(■,■)
4	■	(■,■)

$y = 3x + 6$
$y = 3(-3) + 6$
$y = -9 + 6$
$y = -3$

$y = 3x + 6$
$y = 3(0) + 6$
$y = 0 + 6$
$y = 6$

$y = 3x + 6$
$y = 3(4) + 6$
$y = 12 + 6$
$y = 18$

The ordered pairs (-3,-3), (0,6), and (4,18) are solutions.

Teaching Tip

Encourage students to use estimation when testing if a coordinate is a solution to an equation. Estimation is useful for checking the reasonableness of an answer.

▼ You may need to solve an equation for y in order to find solutions of the equation.

Example 3 **Solve the equation $3x + 5y = 30$ for y in terms of x.**

Solution

$$3x + 5y = 30$$
$$3x + 5y - 3x = 30 - 3x \quad \text{Subtract } 3x \text{ from each side.}$$
$$5y = 30 - 3x$$
$$\frac{5y}{5} = \frac{30 - 3x}{5} \quad \text{Divide each side by 5.}$$
$$y = \frac{30}{5} - \frac{3x}{5}$$
$$y = 6 - \frac{3}{5}x$$
$$y = -\frac{3}{5}x + 6$$

FLASHBACK

Solving an equation for a specific variable means to isolate that variable on one side of the equation.

Example 4 **Find three solutions of the equation $4x - \frac{2}{3}y = 6$.**

Solution 1. Solve for y in terms of x.

$$4x - \frac{2}{3}y = 6$$
$$3\left(4x - \frac{2}{3}y\right) = 6(3) \quad \text{Multiply each side by 3.}$$
$$12x - 2y = 18$$
$$12x - 2y - 12x = 18 - 12x \quad \text{Subtract } 12x \text{ from each side.}$$
$$-2y = 18 - 12x$$
$$\frac{-2y}{-2} = \frac{18 - 12x}{-2} \quad \text{Divide each side by -2.}$$
$$y = \frac{18}{-2} + \frac{-12x}{-2}$$
$$= -9 + 6x$$
$$= 6x - 9$$

2. Find three solutions of the equation $y = 6x - 9$.
   a. Choose three values for x. Try -3, 0, and 2.
   b. Use a table to organize your data.

x	y	(x,y)
-3	■	(■,■)
0	■	(■,■)
2	■	(■,■)

   c. Substitute x values into the equation to find y values.

$y = 6x - 9$	$y = 6x - 9$	$y = 6x - 9$
$y = 6(-3) - 9$	$y = 6(0) - 9$	$y = 6(2) - 9$
$y = -18 - 9$	$y = 0 - 9$	$y = 12 - 9$
$y = -27$	$y = -9$	$y = 3$

The ordered pairs (-3,-27), (0,-9), and (2,3) are solutions of the equation $y = 6x - 9$.

ADDITIONAL EXAMPLES

1. Tell whether each ordered pair is a solution of the equation $2x + 3y = 12$.
   a. (9, -2) [yes] b. (2, 3) [no]
2. Find three solutions of the equation $2x + y = 5$. [possible answers: (2, 1); (1, 3); (0, 5)]
3. Solve for y in terms of x. $6x - 3y = 18$ ($y = 2x - 6$)
4. Find three solutions of the equation $4x + \frac{2}{5}y - 10 = 0$ [possible answers: (2, 5); (5, -25); (0, 25)]

MATH MINUTES

Have students toss a number cube twice. The first number is the x-coordinate; the second is the y-coordinate. Have students determine if the ordered pair is a solution to these equations: $y = x + 4$; $y = 2x + 1$; $y = 3x - 2$; $2x + y = 3$.

Guided Practice

Class Exercises Calculators may be useful for many of these exercises.

For questions 10-12, list at least 6 different solutions for each on the chalkboard.

Think and Discuss In question 2, to clear an equation of decimals, multiply each term on both sides of the equation by the appropriate power of 10. For the given equation, multiply by 100: $25x + 75y = 675$.

For question 3, substitute the coordinates of the given point for x and y in the equation to obtain a simple equation involving k.

a. $-3 + k \cdot 3 = 6 \qquad k = 3$

b. $4 - k \cdot 2 = 12 \qquad k = -4$

A ***common error*** occurs when students incorrectly substitute values for x in an equation, such as $y = -x - 1$, to find ordered pairs. To help students find ordered pairs correctly, suggest that they construct a table, for example:

$y = -x - 1$

x	-x	−1	y
2	-2	−1	-3
1	-1	−1	-2
0	-0	−1	-1

Closure

Have students solve this equation for y : $y - 2x = -4$. Then have them make a table of solutions (x, y) for $x = -2$, 0, and 1.

Independent Practice

Examples of the uses of coordinate systems in real world situations for question 28 might include locations in a parking lot, seats in a theater, latitude and longitude on a globe, radar, etc.

THINK AND DISCUSS

1. How is a table helpful in determining ordered pair solutions? See below.

2. How can you clear the decimals in $0.25x + 0.75y = 6.75$?

3. Find the value of k that makes the given point a solution of the equation:

a. $x + ky = 6$; (-3,3)

b. $x - ky = 12$; (4,2)

3. Check students' work.
4. negative; negative
5. Answers may vary.

MIXED REVIEW

Solve.

1. $5x + 9 = 6x + 25$ -16

2. $12 - \frac{3}{4}y = 34 + 2y$ -8

3. Graph (-1,3), (2,0), and (1,1) in the coordinate plane. See above.

4. In the third quadrant, the x-coordinate is always ▒ and the y-coordinate is always ▒.

5. Write three ordered pairs whose y-coordinate is twice the x-coordinate.

6. Name the coordinates of the point on the y-axis 8 units above the x-axis. (0,8)

7. A student gets out of bed and spends 48 min getting ready for school, 25 min walking to school, and 55 min in her first class. The class ends at 9:25 A.M. At what time did the student get up? 7:17 A.M.

CLASS EXERCISES

Tell whether each ordered pair is a solution of $4x - 3y = 6$.

1. (5,7) no **2.** (3,2) yes **3.** (0.5,-1.3) no

Find the value of y that corresponds to each value of x for the equation $6x + 2y = 12$.

4. $x = -3$ 15 **5.** $x = 2.5$ -1.5 **6.** $x = 0$ 6

Solve for y in terms of x.

7. $-5x + \frac{y}{3} = 9$ $y = 15x + 27$ **8.** $-4x - 0.5y = -2$ $y = -8x + 4$ **9.** $\frac{1}{4}x + \frac{1}{4}y = \frac{1}{2}y$ $y = x$

Find four solutions for each equation. Answers may vary.

10. $x - y = 9$ **11.** $3x + y = 12$ **12.** $3x + y = 24$

1. It helps to organize the x- and y-coordinates. When you read across you have the ordered pair. 2. Multiply both sides by 100. 3. a. $k = 3$ b. $k = -4$

WRITTEN EXERCISES

MENTAL MATH **Is each ordered pair a solution of $x + 2y = 57$?**

1. (5,0) no **2.** (-1,28) no **3.** (57,0) yes

MENTAL MATH **Is each ordered pair a solution of $x + 2y = 5$?**

4. (-1,2) no **5.** (-3,4) yes **6.** (2,-3) no

CALCULATOR **Is each ordered pair a solution of $x + 2y = 5$?**

7. (1.2,1.9) yes **8.** (-4.5,4.5) no **9.** (13.2,4.1) no

Find the value of y that corresponds to each value of x.

10. $3x - 4y = 24$ if $x = -2$ -7.5

11. $5 - y = \frac{1}{2}x$ if $x = 12$ -1

12. $2x + y = 5$ if $x = 2$ 1

13. $x + 8y + 6 = 0$ if $x = -22$ 2

14. $2.9x + 2y = 5$ if $x = 0$ 2.5

15. $0.25x + 0.5y = 6.75$ if $x = 4$ 11.5

Solve for y in terms of x.

16. $x = 11 - 3y$ $y = \frac{x - 11}{-3}$

17. $x + y - 5 = 4x$ $y = 3x + 5$

18. $5x + 3y = 2x - 10$ $y = -x - \frac{10}{3}$

19. $2(x - y) = -x + 10$

20. $x - y - 8 = 6(2x + 4)$ $y = -11x - 32$

21. $7y - x - 1 = y - (2x + 1)$ $y = -\frac{1}{6}x$

19. $y = \frac{3}{2}x - 5$

Solve for y in terms of x. Find four solutions of each equation. Answers may vary.

22. $4x + \frac{1}{2}y = 3$ $y = -8x + 6$

23. $x + 2y = -5$ $y = -\frac{1}{2}x - \frac{5}{2}$

24. $3x - y = -1$ $y = 3x + 1$

25. $2x - 3y = 12$ $y = \frac{2}{3}x - 4$

26. $x + y = 32$ $y = -x + 32$

27. $x + 4y = 16$ $y = -\frac{1}{4}x + 4$

28. ***PROJECT*** Many situations rely on coordinate systems. As you press B5 in a vending machine, the letter refers to a row of items and the number to a position in the row. Find other examples in which a coordinate system is used to locate positions. Write the results in your math journal. **Check students' work.**

Use the article below to answer each question.

Mountains Under the Sea

A mountain range exists in the Pacific Ocean, far, far beneath the surface of the water. In 1960, Jacque Piccard and Donald Walsh made a record dive in these mountains at the site of the Marianas Trench. Piccard and Walsh descended to 10,916 m. The depth of the dive is remarkable because of the tremendous amount of pressure that exists at these depths. The pressure of the air at sea level is 1 kg/cm^2, but it increases 0.1 kg/cm^2 for every meter an object descends in salt water.

29. The equation $P = 1 + 0.1x$ gives the pressure in kg/cm^2 at x m below sea level.

a. Find four possible solutions for this equation. **Answers may vary.**

b. Find the pressure at the record-breaking depth achieved by Piccard and Walsh. **1,092.6 kg/cm^2**

c. ***DATA FILE 10 (pp. 404–405)*** Find the pressure at the greatest known depth of the Atlantic Ocean. **2,838.4 kg/cm^2**

TEST YOURSELF

State the letter of the point named by each ordered pair.

1. (-3,3) **F**
2. (0,-2) **G**
3. (-2,1) **H**
4. (-3,0) **E**

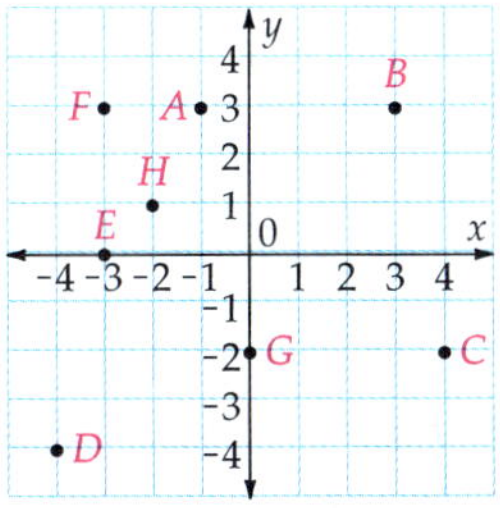

State the coordinates of each point.

5. A **(-1,3)**
6. B **(3,3)**
7. C **(4,-2)**
8. D **(-4,-4)**

Answers may vary.

Solve for y in terms of x. Find three solutions of each equation.

9. $3x + 2y = 4$ **$y = -\frac{3}{2}x + 2$** **10.** $\frac{1}{2}x + y = -3$ **$y = -\frac{1}{2}x - 3$** **11.** $x + 3y = 7$ **$y = -\frac{1}{3}x + \frac{7}{3}$**

Lesson Follow-up

Reteaching Activity Have students solve a set of equations for y in terms of x. Then have students use a table to find ordered pairs which are solutions to the equations.

LESSON RESOURCES

TRB Practice H-3
TRB Enrichment H-4

Practice

For use after 8-2 (pp. 320–323)

Solving Equations

MENTAL MATH Is each ordered pair a solution of $3x - 2y = 12$? Write *yes* or *no*.

1. (0, 4) **no**
2. (6, 3) **yes**
3. (4, 0) **yes**

Is each ordered pair a solution of $3x - 2y = 12$?

4. (-3, 2) **no**
5. (10, 9) **yes**
6. (-16, -30) **yes**
7. (8.4, 6.6) **yes**
8. (-12.6, -24.9) **yes**
9. (-3.5, -11.25) **yes**

Find the value of y that corresponds to each value of x.

10. $4x + 2y = 2$ if $x = -1$ $y =$ **3**
11. $\frac{2}{3}x - y = 11$ if $x = 9$ $y =$ **-5**
12. $3x + y - 12 = 0$ if $x = 3.5$ $y =$ **1.5**
13. $-2y + 21 = 7x$ if $x = 5$ $y =$ **-7**
14. $-\frac{1}{2}x + 5y = 47$ if $x = -14$ $y =$ **8**
15. $1.8x + 2.6y = 11.6$ if $x = 18$ $y =$ **-8**

Solve for y in terms of x.

16. $3y = 15x - 12$ $y =$ **$5x - 4$**
17. $5x + 10 = 10y$ $y =$ **$\frac{1}{2}x + 1$**
18. $3y - 21 = 12x$ $y =$ **$4x + 7$**
19. $5y + 3 = 2y - 3x + 5$ $y =$ **$-x + \frac{2}{3}$**
20. $-2(x + 3y) = 18$ $y =$ **$-\frac{1}{3}x - 3$**
21. $5(x + y) = 20 + 3x$ $y =$ **$-\frac{2}{5}x + 4$**

Solve for y in terms of x. Find three solutions of each equation. **Solutions will vary.**

22. $x + \frac{1}{3}y = 2$ $y =$ **$-3x + 6$** **(0, 6)**, **(2, 0)**, **(1, 3)**
23. $8x + 5 = 2y + 1$ $y =$ **$4x + 2$** **(0, 2)**, **(1, 6)**, **(-1, -2)**
24. $6y - 3x + 18 = 0$ $y =$ **$\frac{1}{2}x - 3$** **(0, -3)**, **(6, 0)**, **(2, -2)**
25. $7\frac{1}{2}x + 1\frac{1}{2}y = 15$ $y =$ **$-5x + 10$** **(0, 10)**, **(1, 5)**, **(-1, 15)**

Chapter 8 H-3

Enrichment

For use after 8-2 (pp. 320–323)

What's for Dessert?

What did the mathematician do when the waiter informed her that the restaurant had no more peaches for dessert?

To find the answer, decide which of the two given ordered pairs is a solution of the equation. Write the letter of the correct pair above the number of the exercise in the answer grid at the bottom of the page.

1. $4x - y = 6$ $E(2, 2)$ $A(6, 3)$
2. $2x - 3 = 3y$ $I(2, 1)$ $E(-3, -3)$
3. $5x - 4y = -4$ $M(1, 0)$ $E(0, 1)$
4. $2(x - y) = -4$ $R(3, 5)$ $A(1, -2)$
5. $x = 2y - 10$ $N(3, -4)$ $D(-4, 3)$
6. $13 - x - y = 8$ $S(12, -7)$ $T(-2, 8)$
7. $3x + 2y = 5y - 3$ $H(-3, -4)$ $R(-4, -3)$
8. $7y = 3x + 1$ $A(-5, -2)$ $C(4, 2)$
9. $\frac{1}{2}x = \frac{1}{3}y + 1$ $E(4, 6)$ $D(8, 9)$
10. $x + 5y - 13 = 2x$ $R(-3, 2)$ $S(-4, 1)$
11. $4(3y - 2x) = 4$ $O(-2, -1)$ $L(-1, -2)$
12. $2.4x + 3.5y = -1.8$ $B(-4, -8)$ $P(8, -6)$

Answer: She **O R D E R E D** (11 4 9 1 7 3 5) **P E A R S** (12 2 8 10 6).

H-4 Chapter 8

LESSON QUIZ

1. Tell whether each ordered pair is a solution of the equation $3x - 4y = 12$.
a. (1, -2) [no] **b.** (2, -1.5) [yes]

2. Find three solutions of the equation $2x + 4y = 8$. [possible answer: (2, 1); (0, 2); (-2, 3)]

3. Solve for y in terms of x.
$3x + 2y + 1 = 4$
($y = 1.5 - 1.5x$)

4. Find three solutions of $6x - \frac{3}{2}y - 12 = 0$ [possible answer: (2, 0); (3, 4); (1, -4)]

Assignment Guide

Basic 1-13, 16-17, 22-24, 28-29, MR, TY All
Average 1-25 odd, 28-29, MR, TY All
Enriched 6-26 even, 27-29, MR, TY All

FOR THE NEXT LESSON

graph paper, ruler, math journal

Lesson Focus

Materials

Graph paper, ruler, math journal

Vocabulary/Symbols

linear equation
standard form
x-intercept
y-intercept

MOTIVATING THE LESSON

Review with students the definition of a coordinate plane. Then discuss how in the previous lesson, students solved equations in two variables for y in terms of x and found the solutions to be ordered pairs. Mention that there are an infinite number of solutions for an equation such as $2x + y = 5$. In this lesson students will show some of the solutions by graphing them on the coordinate plane.

Instruction

1 Have students locate another point on the line, for example, (3, -3), and check that the ordered pair is a solution. Students should realize that every point on the line is a solution of the equation.

2 Every point on the x-axis has a y-coordinate of zero. Therefore, letting $y = 0$ in the equation helps to find the x-coordinate at the point where the graph crosses the x-axis. Similarly, every point on the y-axis has an x-coordinate of zero. So we substitute 0 for x to find the y-coordinate at the point where the graph crosses the y-axis.

OBJECTIVE: *To graph a linear equation and find x and y-intercepts.*

8-3 Graphing Linear Equations

▼ You can graph an equation to show the solutions.

Example 1 **Graph the equation $2x + y = 3$.**

Solution **1.** Solve for y in terms of x.

$$\begin{aligned} 2x + y &= 3 \\ 2x + y - 2x &= 3 - 2x \\ y &= 3 - 2x \\ y &= -2x + 3 \end{aligned}$$

2. Find three solutions to the equation.

a. Choose three values. Try -1, 0, and 2.

b. Use a table to organize your data.

x	y	(x,y)
-1	■	(■,■)
0	■	(■,■)
2	■	(■,■)

THINK How many points must you have to draw a line? 2

c. Substitute each x value into the equation to find y.

$y = -2x + 3$	$y = -2x + 3$	$y = -2x + 3$
$y = -2(-1) + 3$	$y = -2(0) + 3$	$y = -2(2) + 3$
$y = 2 + 3$	$y = 0 + 3$	$y = -4 + 3$
$y = 5$	$y = 3$	$y = -1$

Three solutions are (-1,5), (0,3), and (2,-1).

1

3. Graph the points. Draw a line connecting them.

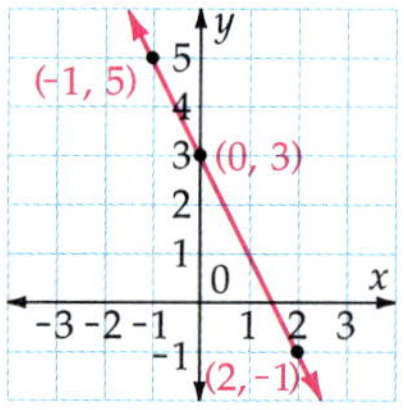

Each ordered pair on the graph is a solution of the linear equation $2x + y = 3$.

▼ We call the equations you have been solving *linear equations.*

Linear Equation	A linear equation is an equation for which the graph is a line. The standard form of a linear equation is $Ax + By = C$, where A, B, and C are real numbers and A and B are not both equal to zero.

Example 2 **Write each linear equation in standard form.**

a. $x = 12 - 2y$ **b.** $x + y - 2 = 0$

Solution **a.**

$$\begin{aligned} x &= 12 - 2y \\ x + 2y &= 12 - 2y + 2y \\ x + 2y &= 12 \end{aligned}$$

b.

$$\begin{aligned} x + y - 2 &= 0 \\ x + y - 2 + 2 &= 0 + 2 \\ x + y &= 2 \end{aligned}$$

Teaching Tip

Encourage students to extend the graph of a linear equation so that the x- and y-intercepts are included.

Example 3 Is the point (6,1) on the graph of $x - 2y = 4$?

Solution

$$\begin{aligned} x - 2y &= 4 \quad \text{Substitute 6 for } x \text{ and 1 for } y.\\ 6 - 2(1) &= 4\\ 6 - 2 &= 4\\ 4 &= 4 ✓ \end{aligned}$$

True, so (6,1) is on the graph of $x - 2y = 4$.

▼ You can use the x- and y-intercepts when graphing an equation.

x-intercept	The x-intercept is the x-coordinate of a point where a graph crosses the x-axis.
y-intercept	The y-intercept is the y-coordinate of a point where a graph crosses the y-axis.

Example 4 **Find the x-intercept and the y-intercept for the equation $2x - 3y = 12$. Use the intercepts to sketch the graph.**

Solution To find the x-intercept, substitute 0 for y.

$$\begin{aligned} 2x - 3y &= 12\\ 2x - 3(0) &= 12\\ 2x - 0 &= 12\\ 2x &= 12\\ \frac{2x}{2} &= \frac{12}{2}\\ x &= 6 \end{aligned}$$

The x-intercept is 6.

To find the y-intercept, substitute 0 for x.

$$\begin{aligned} 2x - 3y &= 12\\ 2(0) - 3y &= 12\\ 0 - 3y &= 12\\ -3y &= 12\\ \frac{-3y}{-3} &= \frac{12}{-3}\\ y &= -4 \end{aligned}$$

The y-intercept is -4.

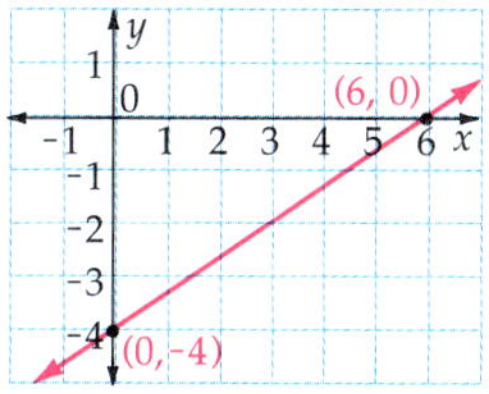

▼ You can graph an equation which contains only one variable.

Example 5 **Graph each equation on the same coordinate plane.**

a. $4x = 12$ **b.** $3y = -6$

Solution **a.** When you solve this equation for x, you get $x = 3$. Any ordered pair having an x-coordinate equal to 3 will be a solution.

b. When you solve this equation for y, you get $y = -2$. Any ordered pair having a y-coordinate equal to -2 will be a solution.

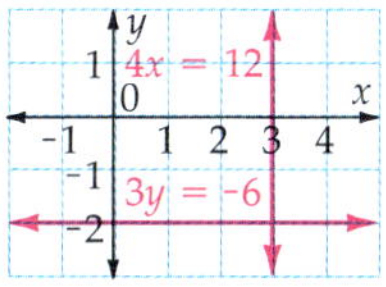

THINK What are two solutions of the equation $4x = 12$? of the equation $3y = -6$? **Answers may vary.**

3 The equation $4x = 12$ is equivalent to the equation $x = 3$. We want all the points in the coordinate plane that have $x = 3$; the y-coordinate could be any number. These points lie on a vertical line, three units to the right of the y-axis.

The equation $3y = -6$ is equivalent to $y = -2$. The graph is a horizontal line, two units below the x-axis.

ADDITIONAL EXAMPLES

1. Graph $2x + \frac{1}{2}y = -3$. (See below.)

2. Write each in standard form.

a. $7x - 2 = 3y + 8$ $(7x - 3y = 10)$

b. $2(x - 3) - 3(y + 1) - 12 = 0$ $(2x - 3y = 21)$

3. Is the point (-1, 3) on the graph of $4x + y = -1$? (yes)

4. Find the x-intercept and the y-intercept for the equation $x - 2y = 4$. (x-intercept is 4, y-intercept is -2) Use the intercepts to sketch the graph.

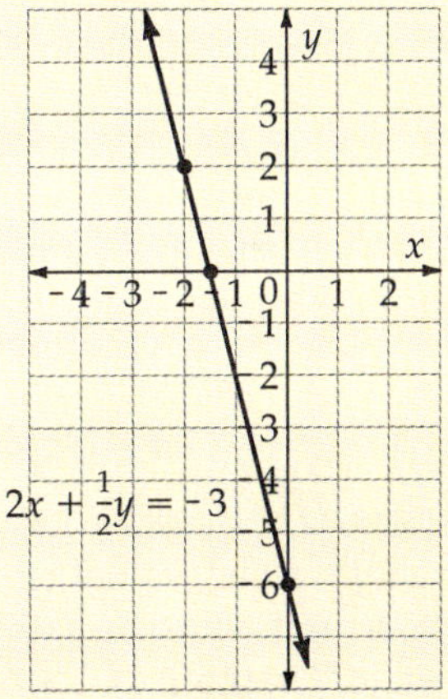

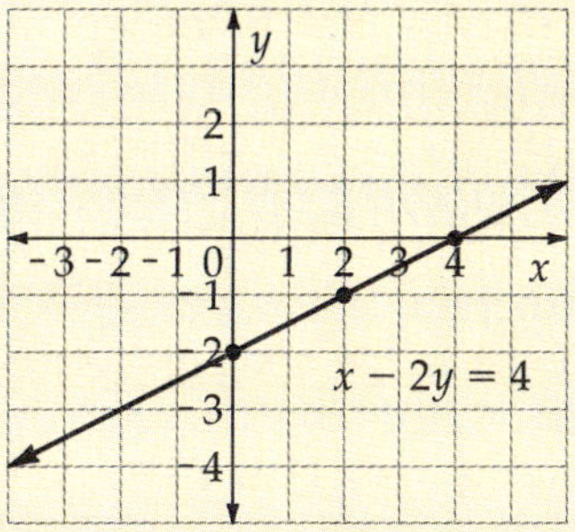

MATH MINUTES

Graph a line on the chalkboard. Have students determine the equation for the line. Have them analyze the x- and y-intercepts. Students can also use the *guess and test* strategy. (Check students' work.)

Guided Practice

Class Exercises For question 2, help students understand there are two ways to tell if each point is on the graph: (1) check if the point *appears* to be on the graph; (2) check whether or not the ordered pair is a solution of the equation.

Think and Discuss For question 1a, any line equation in the form $x = k$, where k is a constant, will be an equation which is a vertical line parallel to the y-axis. For question 1b, any equation in the form $y = k$, will be an equation of a horizontal line parallel to the x-axis.

In question 2, (2, 0), (-2, 0) and (0, -4) are the intercepts. Have students plot several other points. The graph is a curve called a parabola.

A ***common error*** occurs when students incorrectly graph the line for $x = k$ parallel to the x-axis and $y = k$ parallel to the y-axis. Have students find at least two solutions of the equation to avoid this mistake.

Closure

Writing in Math Have students write in their math journals the procedures for graphing a linear equation. Have them compare and discuss their procedures.

Independent Practice

To extend question 28, refer students to the computer activity on p. 312E of the Chapter Overview. See also TRB Exploring O-32.

For questions 30 and 32, discuss the possible replacements for the variables in the equations before students find the solutions.

THINK AND DISCUSS

1. Describe and give an example of an equation of a line with
   a. no y-intercept.
   b. no x-intercept.
2. Is the graph of $y = x^2 - 4$ a line? Use intercepts to help you decide. Explain.
3. Is the graph of the equation $xy = 12$ a line? Use guess and test to find four ordered pairs that are solutions. Graph the points.
4. Explain how the x- and y-intercepts can be helpful when graphing.

1. An equation with no y-intercept is parallel to the y-axis. An example is $x = 4$. An equation with no x-intercept is parallel to the x-axis. An example is $y = 3$.
2. No; The graph of $y = x^2 - 4$ has two x-intercepts. Therefore, it is not a line.
3. No; Check students' work.
4. They give you two points to determine a line.

CLASS EXERCISES

The figure at the right shows the graph of the equation $y = 3x - 2$.

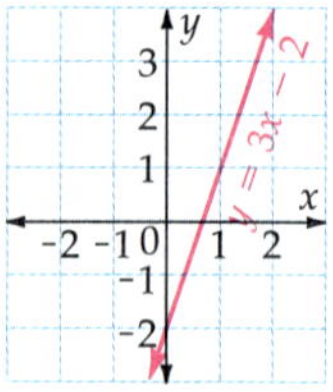

1. Write the equation in standard form. $-3x + y = -2$
2. Tell whether each point is on the graph.
   a. (0,−2) yes **b.** (1,0) no
   c. $\left(\frac{4}{3},2\right)$ yes **d.** (−0.5,−3.5) yes
3. What are the x-intercept and the y-intercept? $\frac{2}{3}$ and -2

Find the x-intercept and y-intercept for each equation.

4. $4x - 3y = 12$ 3; −4
5. $2x + \frac{1}{2}y = -3$ $-\frac{3}{2}$; −6
6. $\frac{1}{3}x = 2$ 6; none
7. $1.5y + 6 = 0$ none; −4

Solve for y in terms of x. Graph each equation. Check students' work.

8. $8x + 4y = 16$ $y = -2x + 4$
9. $9.3x + 3.1y = 15.5$ $y = -3x + 5$
10. $\frac{1}{4}x - 3y = 6$ $y = \frac{1}{12}x - 2$
11. $6x + \frac{2}{3}y = 8$ $y = -9x + 12$

WRITTEN EXERCISES

The figure at the right shows the graph of the equation $3 + y = -\frac{1}{2}x$.

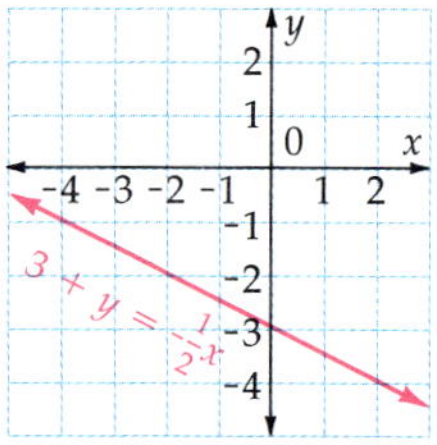

1. Write the equation in standard form. $\frac{1}{2}x + y = -3$
2. Tell whether each point is on the graph.
   a. (1,2.5) no **b.** (−2,2) no
   c. (−4,−1) yes **d.** (−2,−2) yes
3. What are the x-intercept and the y-intercept? −6 and −3

Find the x-intercept and y-intercept for each equation.

4. $3x + 5y = 15$ 5; 3
5. $0.75x - 0.25y + 1 = 0$ $-1\frac{1}{3}$; 4
6. $6x + 8 = 0$ $-\frac{4}{3}$; none
7. $y = \frac{2}{3}x$ 0; 0

QUIZ ANSWERS

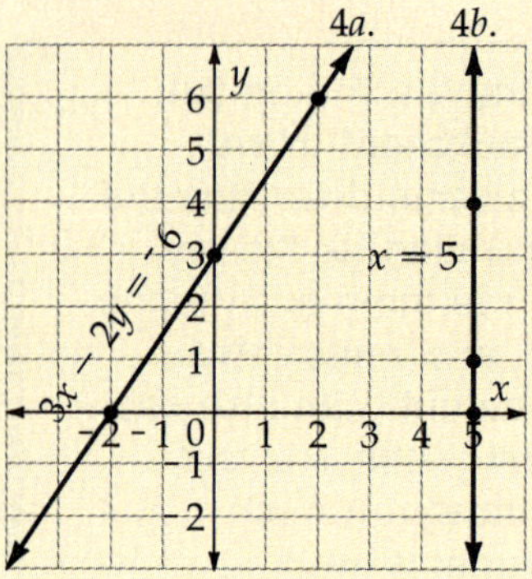

Solve for y in terms of x. Graph each equation. Check students' work.

8. $2x - \frac{1}{3}y = 1$ $y = 6x - 3$

9. $10 - 2(x + 2y) = -3y - (x - 8)$ $y = -x + 2$

10. $y:x = 3:2$ $y = \frac{3}{2}x$

11. $y - 3(0.25x + 1) = -8 - \frac{3}{4}x$ $y = -5$

Graph each equation on a separate set of coordinate axes. Check students' work.

12. $3x - 5y = 15$

13. $3x + \frac{1}{2}y = -3$

14. $\frac{1}{3}x + 1 = -2$

15. $y = 0$

16. $|x| = 2$

17. $|y| = 3$

18. $|x| + |y| = 3$

19. $|x + y| = 3$

MENTAL MATH **Write an equation for each line described.**

20. the line parallel to the y-axis and 2 units to the left of the y-axis $x = -2$

21. the line 2 units to the right of the line $x = -3$ $x = -1$

22. the line 7 units below the line $y = 2$ $y = -5$

23. the line parallel to the x-axis and 6 units below the x-axis $y = -6$

24. the line perpendicular to the x-axis and passing through the point (-3,0) $x = -3$

Write an equation using two variables for each situation. Graph the equation on a coordinate plane. Use the graph to find one solution of the word problem. Check students' work. Answers may vary. An example is given.

25. Find two numbers whose difference is 3. $x - y = 3$; 3 and 0

26. Find two numbers such that one is three times the other. $3x = y$; 4 and 12

27. Louis buys six pieces of fruit. Some are apples and some are oranges. Let x equal the number of oranges and y equal the number of apples. $x + y = 6$; 2 oranges and 4 apples

28. A collection of nickels and dimes is valued at \$1. Let x equal the number of nickels and y equal the number of dimes. $0.05x + 0.10y = 1$; 2 nickels and 9 dimes

29. Ben cut a ribbon so that one piece was twice as long as the other. Let x equal the shorter piece and y equal the longer piece. $2x = y$; 4 and 8

30. Gina's sister earned \$6 at her lemonade stand. She sold small cups of lemonade for \$.25 and large cups for \$.40. Let x equal the number of small cups sold and y equal the number of large cups sold. $0.25x + 0.40y = 6$; 16 small cups and 5 large cups

31. A 12-m fence encloses a rectangular garden. Let x equal the width of the garden and y equal the length of the garden. $2x + 2y = 12$; Width is 1 and length is 5.

32. ***DATA FILE 3 (pp. 96–97)*** A collection of 1936 and 1961 proof sets was valued at \$20,264 in 1990. Let x equal the number of 1936 proof sets and y equal the number of 1961 proof sets. $5{,}050x + 16y = 20{,}264$; 4 sets of each

MIXED REVIEW

Find the GCF for each set of numbers.

1. 12, 20, 44 4

2. 36, 28, 84 4

Solve for y in terms of x.

3. $2x + 5y = 11$

4. $-\frac{3}{2}x - y - 6 = 0$

Plot each set of points on a coordinate grid.

5. (-3,1), (-6,2), (3,-1)

6. (-7,-3), (-4,0), (1,5)

Check students' work.

Solve.

7. Find two consecutive numbers whose product is 1,190. 34 and 35

3. $y = \frac{-2x + 11}{5}$

4. $y = \frac{-3}{2}x - 6$

THINK In Exercise 27, can the values of x and y be negative numbers? fractions? Why or why not? No; no; You cannot buy a negative amount or a fraction of a piece of fruit.

FLASHBACK

perimeter of a rectangle = $2l + 2w$

LESSON QUIZ

1. Write each equation in standard form.

a. $y = 3x - 1$ $(-3x + y = -1)$

b. $y = \frac{1}{4}x + 2$ $(-x + 4y = 8)$

2. Find the x-intercept and the y-intercept for $2x + 3y = 6$. (x-intercept is 3, y-intercept is 2)

3. Is the point (-1, -4) on the graph of $-2x + y = -2$? (yes)

4. Graph each equation.

a. $3x - 2y = -6$ **b.** $x = 5$

(See answers on p. 326.)

Assignment Guide

Basic 1-7, 12-15, 20-25, 32, MR All

Average 1-9, 12-15, 20-26, 32, MR All

Enriched 8-23, 26-32, MR All

FOR THE NEXT LESSON

graph paper, ruler, math journal

Lesson Follow-up

Reteaching Activity Have students write the equation $y = x + 2$ in standard form. Then have students find the x-intercept and y-intercept. (-2 and 2) Discuss the procedures for making a table to find the solutions of the equation. Then have students graph the equation.

LESSON RESOURCES

TRB Practice H-5

TRB Enrichment H-6

Practice For use after 8-3 (pp. 324–327)

Graphing Linear Equations

Find the x-intercept and y-intercept for each equation.

1. $3x + 3y = 9$ (3, 0); (0, 3)
2. $-4y = 16x - 8$ $(\frac{1}{2}, 0)$; (0, 2)
3. $\frac{3}{4}x = y$ (0, 0); (0, 0)
4. $-2(x - y) = 10$ (-5, 0); (0, 5)

Graph.

5. $x + y = 3$

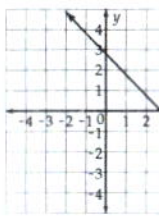

6. $y = 2x - 1$

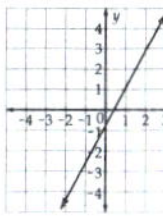

7. $x = -4$

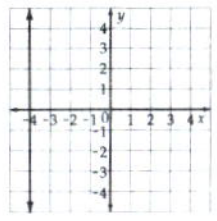

8. $y = 1.5$

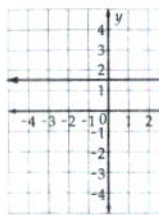

9. $3y = 2(x + 3)$

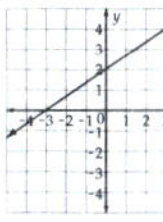

10. $x + y = 8(x + 1) - 4 + x$

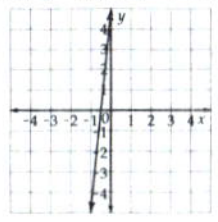

MENTAL MATH **Write an equation for each line described.**

11. the line 5 units above the x-axis $y = 5$
12. the line 3 units to the left of the line $x = -4$ $x = -7$
13. the line 6 units above the line $y = -8$ $y = -2$
14. the y-axis $x = 0$

Write an equation using two variables. Graph.

15. Find two numbers whose sum is 2. Equation: $x + y = 2$

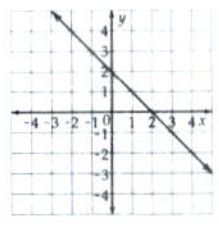

Chapter 8 H-5

Enrichment For use after 8-3 (pp. 324–327)

Quadratic Equations

The graphs of all equations are not lines. An important class of nonlinear equations, called *quadratic equations*, contains variables that are squared.

Follow these steps to draw the graph of the quadratic equation $x^2 + y^2 = 25$.

1. Complete the table to find ordered pairs that are solutions of $x^2 + y^2 = 25$. Notice that for each given value of x or y there are two possible values of the other variable.

x	0	0	3	3	4	4	-3	-3	-4	-4	5	-5
y	5	-5	4	-4	3	-3	4	-4	3	-3	0	0

CALCULATOR **Find y-coordinates of points with these x-coordinates on the graph of $x^2 + y^2 = 25$. Round to the nearest tenth.**

2. (1, 4.9) (-1, 4.9) (2, 4.6) (-2, 4.6)
(1, -4.9) (-1, -4.9) (2, -4.6) (-2, -4.6)

3. Graph the points you found in Exercises 1–2.

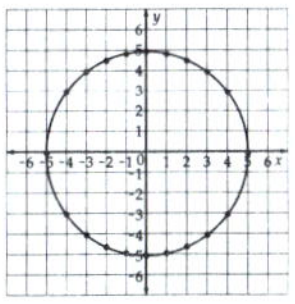

4. What kind of figure does the graph of $x^2 + y^2 = 25$ resemble? a circle

5. Connect the points to show the complete graph.

H-6 Chapter 8

Lesson Focus

Materials

Graph paper, ruler, math journal

Vocabulary/Symbols

slope

MOTIVATING THE LESSON

Review the procedures for graphing an equation. Then discuss how, by carefully observing a given graph, an equation for the graph can be written. Tell students that today's lesson will involve writing an equation from information obtained from a graph.

Instruction

1 Help students understand that positive and negative signs are used to show the change in direction from one point to the next point.

2 It may be beneficial to review that $\frac{-1}{3} = \frac{1}{-3} = -\frac{1}{3}$.

OBJECTIVE:
To use coordinate graphs to explore linear equations, intercepts, and slope.

MATERIALS

- Graph paper
- Math journal to record work

Exploring Slope

▼ The graph below shows the linear equation $-2x + y = 4$.

Copy and complete a table like the one at the left.

Standard form	Solved for y	y-intercept	Ratio

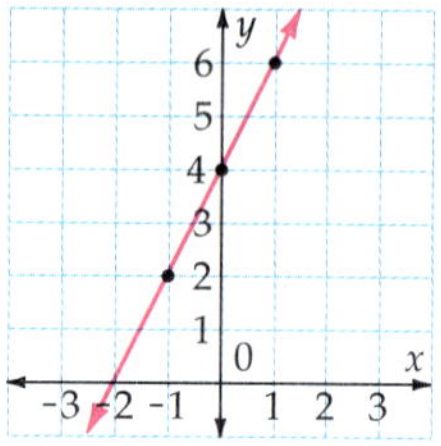

1. a. In what form is the equation written? Write the equation in the table. Standard form
   b. Solve the equation for y in terms of x. Write the equation in the table. $y = 2x + 4$
2. a. ***Describe*** what a y-intercept is. where the line crosses the y-axis
   b. Determine from the graph what the y-intercept of the equation $y = 2x + 4$ is. Write the y-intercept in the table. 4
3. a. ***Write*** the coordinates of three points found on the line. (-1,2) (0,4) and (1,6)
   b. ***Explain*** how to get from the first point to the second. Use only two sentences. Give vertical directions first using the words *up* and *down*. Then give horizontal directions, using *left* and *right*. What is the ratio in fraction form of the vertical change to the horizontal change? Go up 2. (+2) Go right 1. (+1) The ratio is $\frac{2}{1}$.

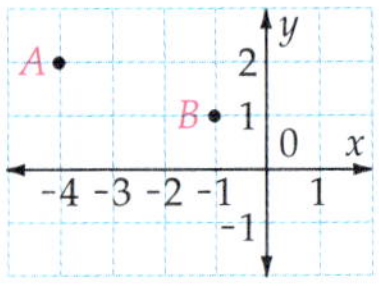

Example

To get from B to A:	To get from A to B:
1. Go up 1. (+1)	1. Go down 1. (-1)
2. Go left 3. (-3)	2. Go right 3. (+3)
3. The ratio is $\frac{1}{-3}$ or $-\frac{1}{3}$.	3. The ratio is $\frac{-1}{3}$ or $-\frac{1}{3}$.

   c. ***Explain*** how to get from the second point to the third. What is the ratio of the vertical change to the horizontal change?
   d. ***Explain*** how to get from the third point to the first. What is the ratio of the vertical change to the horizontal change?
   e. ***Discuss*** how the ratios in (b), (c), and (d) are alike. How are they different? The ratios are all 2 when simplified. They are different when in fraction form.
   f. ***Write*** the ratio of the vertical change to the horizontal change in simplest form. Write the result in your table. 2
4. Repeat Steps 1–3 for each equation shown. Check students' work.

3\. c. Go up 2. (+2) Go right 1. (+1) The ratio is $\frac{2}{1}$.
d. Go down 4. (-4) Go left 2. (-2) The ratio is $\frac{-4}{-2}$.

4\. $y = 2x - 2$; slope = 2
$y = -x - 1$; slope = -1
$y = \frac{3}{2}x - 1$; slope = $\frac{3}{2}$

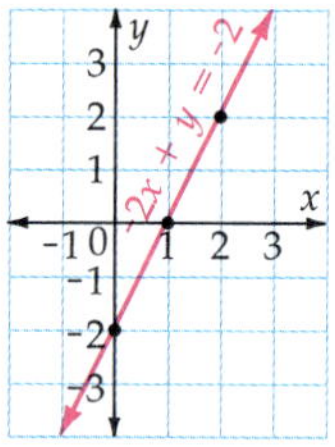

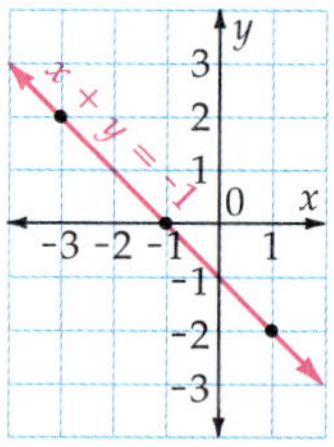

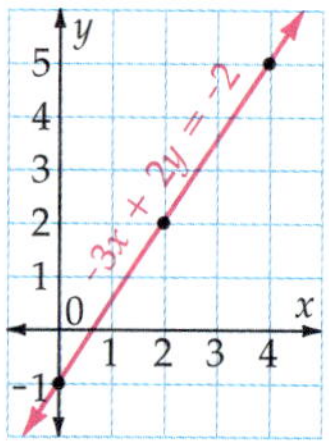

Teaching Tip

It may be helpful to present an overhead transparency of the graph shown on the top of page 328. You can have students come to the overhead projector to determine, by counting, the change in the x- and y-coordinates when finding the slope.

▼ We call the ratio of the vertical change to the horizontal change the *slope* of the line.

5. a. ***Analyze*** the table and discuss the relationship between each equation when it is solved for y in terms of x and its y-intercept. **The y-intercept is the constant term in the equation.**

b. ***Analyze*** the table and discuss the relationship between each equation when it is solved for y in terms of x and its slope. **The slope is the coefficient of the x term.**

c. Is the y-intercept the same for the equation $y = 3x - 4$ and $y = 3x + (-4)$? **yes**

3

6. Name the y-intercept and slope of each linear equation.

a. $y = 12x - 4$ **−4; $\frac{12}{1}$ or 12**

b. $y = 2x + 14$ **14; $\frac{2}{1}$ or 2**

▼ Knowing the y-intercept and the slope are useful when sketching the graph of a line.

7. a. ***Discuss*** why solving an equation for y is helpful when graphing the equation. **You can easily determine the slope and y-intercept of the equation.**

8. Sketch each equation by using the y-intercept and slope.

a. $-4x + y = 10$

b. $-5x + y = -10$

Check students' work.

4

▼ You can write the equation of a line by studying its graph.

9. a. What is the y-intercept? **−3**

b. What is the slope? **−2**

c. ***Write*** an equation for the line graphed at the right. **$y = -2x - 3$**

d. ***Discuss*** how you could check that the equation is correct. **Determine points that are solutions of the equation and see if they are on the line.**

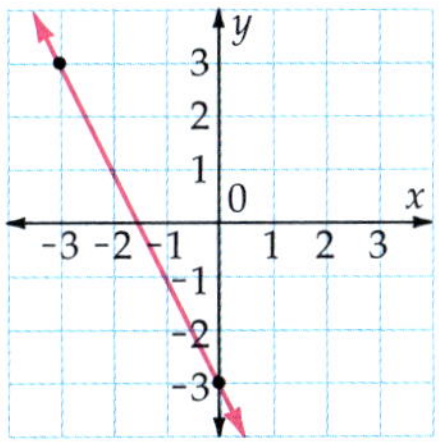

▼ You can write the equation of a line when you know the y-intercept and the slope.

10. ***Write*** the equation of each line described.

a. The y-intercept is 7 and the slope is 8. **$y = 8x + 7$**

b. The y-intercept is −1 and the slope is 6. **$y = 6x - 1$**

3 When an equation is written in the form $y = mx + b$ help students understand that m = slope and b = the y-intercept.

4 The first point can be determined by locating the y-intercept. The second point can be determined by writing the slope as a ratio and moving from the point containing the y-intercept in the directions indicated by the ratio.

Closure

Writing in Math Have students write in their math journals a description of the procedures used to determine the slope of a line and to write an equation for a graph when the slope and y-intercept are given.

As an extension, refer students to the manipulative activity on p. 312E of the Chapter Overview. See also TRB Exploring O-29.

Assignment Guide

Basic All
Average All
Enriched All

FOR THE NEXT LESSON

graph paper, math journal

Lesson Focus

Materials
Graph paper, math journal

Vocabulary/Symbols
slope
slope-intercept form

MOTIVATING THE LESSON

On the chalkboard, draw two pictures to represent hills as shown below.

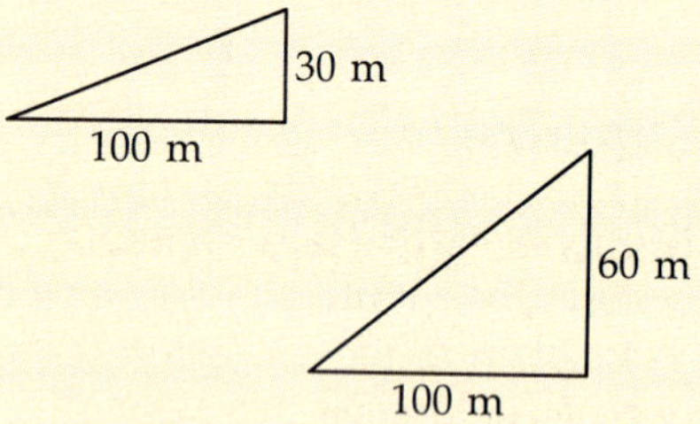

Ask students which hill would require more effort to climb. (the one on the right) Then ask why it would be more difficult to climb. (because it is steeper) Comment that the steepness of a hill is called its *slope* and that the slope is the mathematical ratio of the rise over the run of the hill.

Instruction

1 Explain that the sign in the numerator is positive if the move is up, negative if the move is down; the sign in the denominator is positive if the move is to the right, negative if the move is to the left.

2 Beginning at Q, the slope is $\frac{-3}{2}$, which is equal to $-\frac{3}{2}$. When the line runs down moving from left to right, the slope is negative.

3 Beginning at P, the slope is $\frac{-2}{-4}$, which is equal to $\frac{1}{2}$. When the line runs up moving from left to right, the slope is positive.

OBJECTIVE:
Find the slope and y-intercept for the graph of a linear equation.

8-4 Slope and y-intercept

▼ The ski hill pictured at the left has a vertical rise of 4 m for every 20 m of horizontal run. You can find the steepness or *slope* of the ski hill by using the ratio of the vertical change to the horizontal change. The ratio for the ski hill is $\frac{4}{20}$ or $\frac{1}{5}$.

Slope	The slope of a line is the ratio of the vertical change in y to the corresponding horizontal change in x.

▼ You can also find the slope of a straight line in the coordinate plane by counting the units of vertical change and the units of horizontal change from one point to another.

Example 1 **Find the slope of the line shown on each graph.**

a. 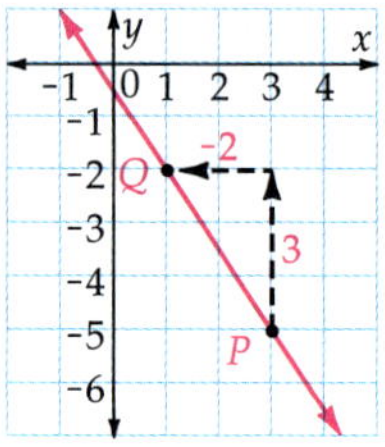

b. 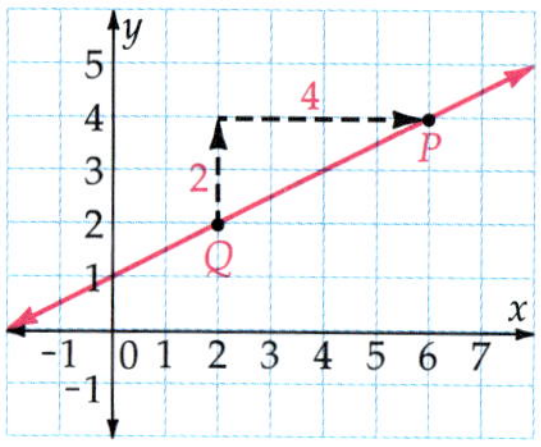

THINK What is the slope if you begin at point Q? $\frac{-3}{2}$ or $-\frac{3}{2}$ **(2)**

THINK What is the slope if you begin at point P? $\frac{-2}{-4}$ or $\frac{1}{2}$ **(3)**

(1) Solution **a.** Choose any two points on the line. Try $P(3,-5)$ and $Q(1,-2)$. If you begin at P, and move three units up and two units to the left, you reach Q.

$$\text{slope} = \frac{\text{vertical change}}{\text{horizontal change}} = \frac{3}{-2} \text{ or } -\frac{3}{2}$$

b. Choose any two points on the line. Try $P(6,4)$ and $Q(2,2)$. If you begin at Q, and move 2 units up and 4 units to the right, you reach P.

$$\text{slope} = \frac{\text{vertical change}}{\text{horizontal change}} = \frac{2}{4} \text{ or } \frac{1}{2}$$

▼ You can use a formula to find the slope if you know the coordinates of any two points on the line.

Slope Formula	Use the following formula to calculate slope. $\text{slope} = \frac{\text{difference in } y\text{-coordinates}}{\text{difference in } x\text{-coordinates}}$

(4) Example 2 **Find the slope of a line that contains the points $R(-2,1)$ and $S(4,3)$.**

Solution
$$\text{slope} = \frac{\text{difference in } y\text{-coordinates}}{\text{difference in } x\text{-coordinates}} = \frac{3-1}{4-(-2)} = \frac{2}{6} \text{ or } \frac{1}{3}$$

Teaching Tip

Write a formula for finding the slope of a line using the Greek letter Δ (delta) on the chalkboard: $m = \frac{\Delta y}{\Delta x}$.
Explain that this formula describes the difference of the y values divided by the difference of the x values.

Example 3 **Determine the slope of the line containing the given points. Graph the line on a coordinate plane.**

a. (2,4) and (-3,4) **b.** (3,2) and (3,-1)

Solution **a.** slope $= \frac{4-4}{-3-2} = \frac{0}{-5} = 0$

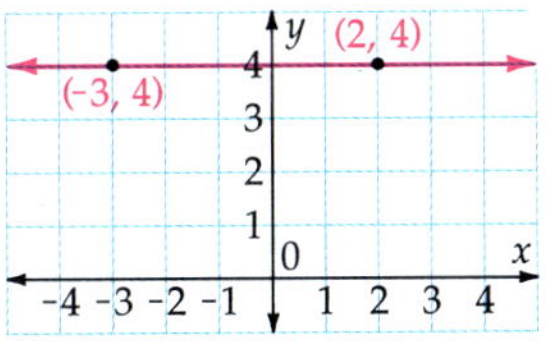

The slope of a horizontal line is zero.

b. slope $= \frac{-1-2}{3-3} = \frac{-3}{0}$

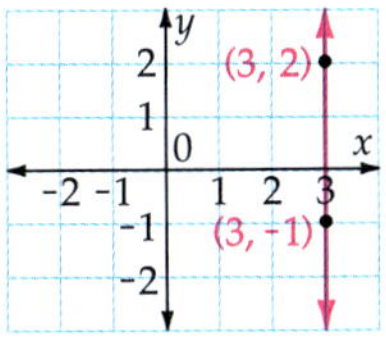

Because division by zero is undefined, a vertical line has no slope. **5**

▼ You can draw the graph of a line if you know the slope of the line and a point on the line. **6**

Example 4 **Graph the line with slope $\frac{3}{4}$ and y-intercept -4.**

Solution Locate the point (0,-4) and label it P. Move 3 units up and 4 units to the right. You are now at (4,-1), or Q. Draw the line containing P and Q.

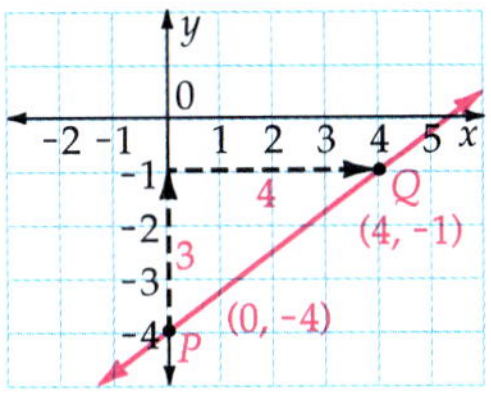

> **FLASHBACK**
> The y-intercept is the y-coordinate of a point where a graph crosses the y-axis.

▼ When you solve an equation for y it is in *slope-intercept* form.

Slope-intercept Form	A linear equation in the form $y = mx + b$ is in slope-intercept form. The slope is m and the y-intercept is b.

Example 5 **Sketch the graph of the equation $y = 3x + 1$.**

Solution The slope is 3 or $\frac{3}{1}$. The y-intercept is 1. Locate the point (0,1) and label it P. Move 3 units up and 1 unit to the right. You are now at (1,4), or Q. Draw the line containing P and Q.

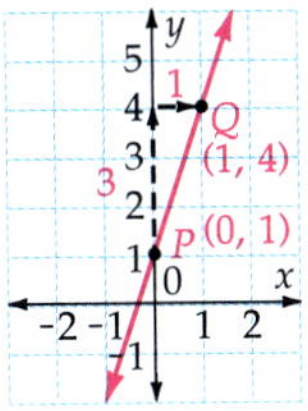

4 Help students understand that for the two points $R(x_1, y_1)$ and $S(x_2, y_2)$, the slope can be found in either of the following two ways, but the differences must be found in the same order.

$$\text{slope} = \frac{y_2 - y_1}{x_2 - x_1} = \frac{y_1 - y_2}{x_1 - x_2}$$

5 It can also be said that the slope of a vertical line is undefined.

6 Explain that if the given slope were negative, for example, $-\frac{3}{4}$, the negative sign can be attached to either the numerator or denominator; so $-\frac{3}{4} = \frac{-3}{4} = \frac{3}{-4}$. To find the point, move down 3, and 4 to the right.

ADDITIONAL EXAMPLES

1. Plot the points (1, 2) and (5, -4) and find the slope of the line containing them. (slope $= \frac{-3}{2}$)

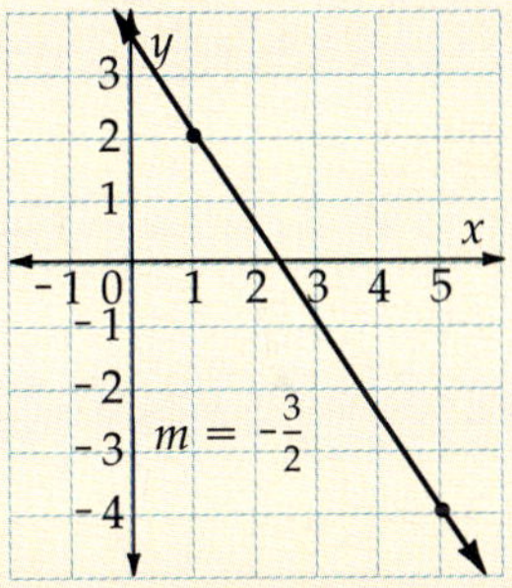

2. Find the slope of a line that contains the points $R(-3, 1)$ and $S(2, -2)$. (slope $= \frac{-3}{5}$)

3. Graph the line with a slope of 3 containing the point (1, 3). (see below)

4. Sketch the graph of the equation $y = \frac{3}{2}x + 2$. (see below)

Answers

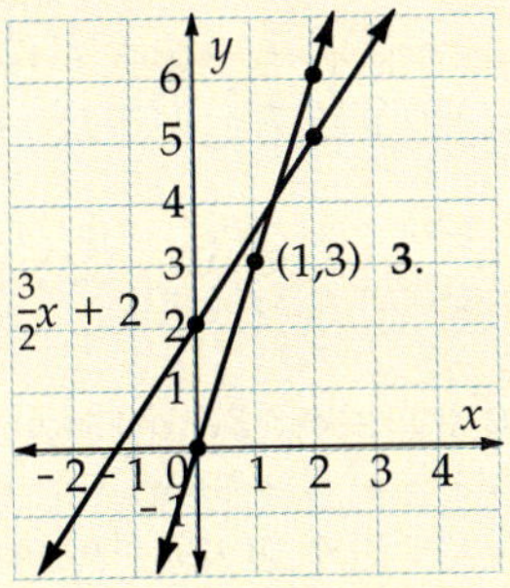

MATH MINUTES

Have students work in pairs. Have one student use the formula: slope $= \frac{y_2 - y_1}{x_2 - x_1}$ and the other student use slope $= \frac{y_1 - y_2}{x_1 - x_2}$. Find the slope.

1. (5, 3) (-1, 1) [$m = \frac{1}{3}$]
2. (1, 0) (-2, 3) [$m = -1$]
3. (4, 1) (3, -2) [$m = \frac{3}{1}$ or 3]

Guided Practice

Class Exercises After students find the slope of each line in questions 4-6, have them plot the points and check the slope.

For question 10, start at the point (1, -3) and use the fact that the slope of -2 can be written as $\frac{-2}{1}$ or $\frac{2}{-1}$.

Think and Discuss For question 1, help students realize that steepness is determined by the absolute value of the slope. A line with a slope of -2 is steeper than a line with a slope of $\frac{1}{2}$. For question 3, a line having a y-intercept of zero passes through the origin (0, 0), which is on both axes.

A ***common error*** occurs when students forget to find the difference in the y-coordinates and x-coordinates in the same order when determining the slope of a line. In Example 2, students might find $\frac{3-1}{-2-4} = \frac{2}{-6} = -\frac{1}{3}$. To help avoid this problem, have students label the points (x_1, y_1) (x_2, y_2), plot the two points, and sketch the change from x_2 to x_1 and from y_2 to y_1.

Closure

Writing in Math Have students write in their math journals the definition of slope of a line. Then have them describe the direction of a line with a positive slope and the direction of a line with a negative slope.

Independent Practice

Have students graph the points and lines for questions 7-12.

Extend question 37 by having students find the rates for 5 mi and 10 mi and graph the results.

THINK AND DISCUSS

True or false?

1. The greater the absolute value of m, the greater the steepness of the line. true
2. A line with a positive slope runs downward from left to right. false
3. If a line has a y-intercept of zero, it does not cross the x-axis. false
4. Given $y = mx + b$, the slope of the line is m and the y-intercept is b. true

CLASS EXERCISES

Find the steepness and y-intercept of each line.

1.

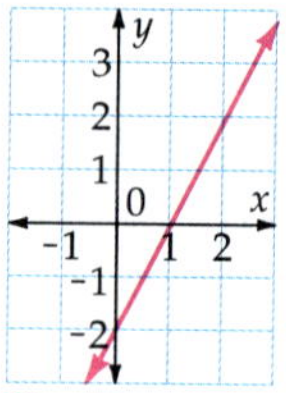

2; -2

2.

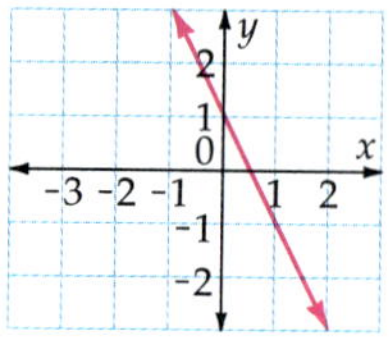

-2; 1

3.

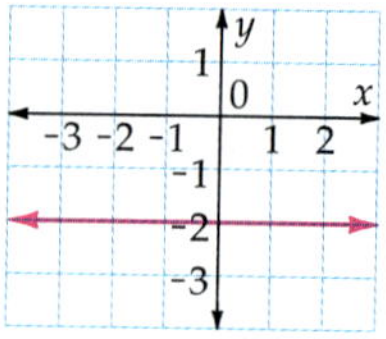

0; -2

Find the slope of the line containing the given points.

4. (4,5) (6,13) 4
5. (-3,5) (-4,-1) 6
6. (-10,-6) (-13,-10) $\frac{4}{3}$

Write each linear equation in slope-intercept form. Name the slope and y-intercept.

7. $3x + y = 3$ $y = -3x + 3$; -3; 3
8. $2y + 4x = 12$ $y = -2x + 6$; -2; 6
9. $-10 + y = -2x$ $y = -2x + 10$; -2; 10
10. Graph the line with slope -2 and containing the point (1,-3). Check students' work.

MIXED REVIEW

Find the LCM.

1. 4, 10, 16 80
2. 7, 12, 18 252

Tell whether or not each equation is linear. Explain how you know.

3. $Ax + By + C = 0$ yes
4. $y = -2$ yes
5. $xy = -12$ no
6. $x + y = -6$ yes
7. $3(x - 2y) + y = 7 - x$ yes
8. $x = y^2$ no

Solve.

9. The price of gold per oz increases from \$350 to \$420. Find the percent of increase. 20%

WRITTEN EXERCISES

MENTAL MATH Find the steepness and y-intercept of each line.

1.

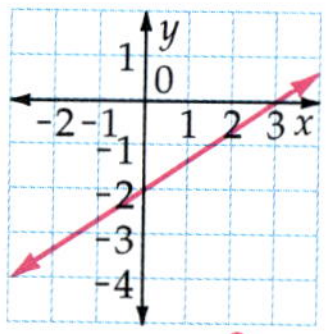

$\frac{2}{3}$; -2

2.

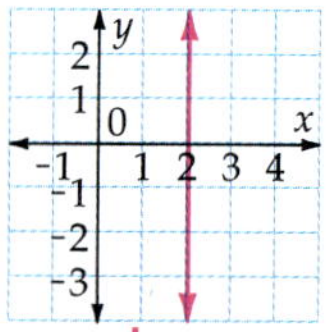

no slope; no y-intercept

3.

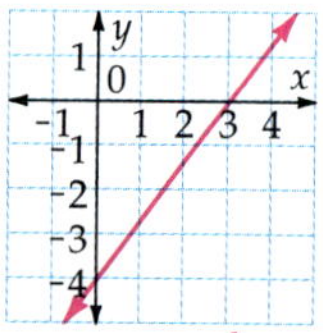

$\frac{4}{3}$; -4

Plot each pair of points. Draw the line containing them. Then find its slope and y-intercept. Check students' work.

4. (-1,3), (1,-1) -2; 1
5. (-6,2), (0,4) $\frac{1}{3}$; 4
6. (0,3), (5,0) $-\frac{3}{5}$; 3

Find the slope of the line containing the given points.

7. (2,1) (3,1) 0
8. (-2,5) (-2,-1) no slope
9. (3,5) (6,15) $\frac{10}{3}$
10. (-5,-2) (1,4) 1
11. (1,-2) (2,-4) -2
12. (-1,15) (3,5) $\frac{-5}{2}$

Graph each line described. Check students' work.

13. the line having slope -3 and containing the point (-6,5)
14. the line having slope 5 and containing the point (-4,-5)
15. the line having slope $\frac{2}{3}$ and containing the point (0,-4)

QUIZ ANSWERS

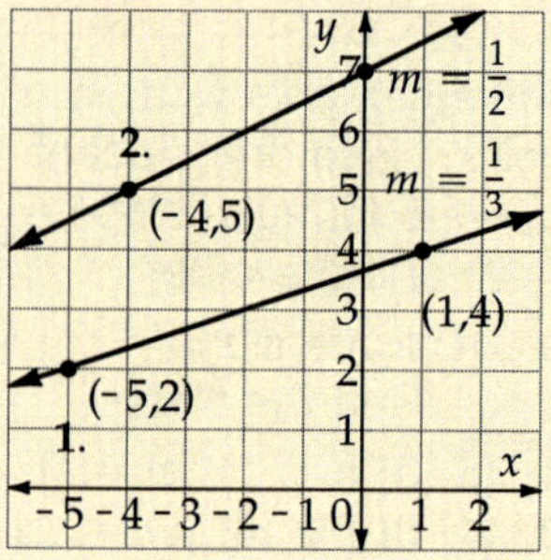

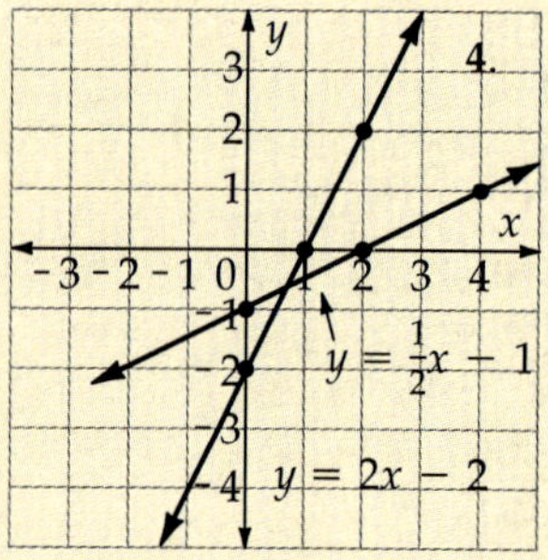

Graph each line described. Check students' work.

16. the line having slope 0 and containing the point (5,3)

17. the line having no slope and containing the point (4,-2)

18. the line having slope -3 and containing the origin

Write each linear equation in slope-intercept form. Name the slope and y-intercept.

19. $-2x + y = 1$ $y = 2x + 1$; 2; 1

20. $6x + y = \frac{1}{6}$ $y = -6x + \frac{1}{6}$; -6; $\frac{1}{6}$

21. $3 + y = -2x$ $y = -2x - 3$; -2; -3

22. $-2x + 3y = 18$ $y = \frac{2}{3}x + 6$; $\frac{2}{3}$; 6

23. $-3x + 4y - 1 = 0$ $y = \frac{3}{4}x + \frac{1}{4}$; $\frac{3}{4}$; $\frac{1}{4}$

24. $4y = 2x$ $y = \frac{1}{2}x$; $\frac{1}{2}$; 0

Sketch the graph of each equation. Check students' work.

25. $y = 2x + 4$

26. $y = \frac{1}{2}x$

27. $y = -5x - 3$

28. $y = -x + 1$

29. $y = -\frac{2}{5}x - 2$

30. $y = x - \frac{3}{4}$

31. a. Graph each pair of lines on one coordinate grid. Check students' work.

Pair 1	Pair 2
$y = 2x - 5$	$y = -3x + 1$
$y = 2x + 3$	$y = -3x - 2$

b. Describe the lines. The lines are parallel.

c. Compare the slopes of each pair of equations. They are the same.

d. Draw a conclusion. When two lines have the same slope and different y-intercepts, they are parallel.

Write an equation in standard form for each line described. ***Hint:*** **First write the equation in slope-intercept form.**

32. having slope 3 and y-intercept -4 $-3x + 1y = -4$

33. having slope 0 and y-intercept 6 $0x + 1y = 6$

34. having slope -2 and y-intercept -1. $2x + 1y = -1$

35. having no slope and containing the point (-3,0). $1x + 0y = -3$

36. having slope -8 and y-intercept $\frac{1}{2}$ $8x + 1y = \frac{1}{2}$

37. ***PROJECT*** Call a local taxi company to find out the rates for trips of distances up to 1.5 mi. Make a table and draw the graph. Find the slope and the y-intercept. Compare with other members of your group. Check students' work.

38. The table at the right shows a student's savings between the sixth and tenth weeks. The student saved at the same rate during this time.

Student's Savings

Weeks (x)	Savings (y)
6	$5
10	$11

a. Find the rate of savings per week. $1.50 per week.

b. Find the slope of the line passing through points (6,5) and (10,11). How does the slope compare with the savings? The slope and the rate of savings are the same.

Lesson Follow-up

Reteaching Activity For students having difficulty with the concept in this lesson, assign *Exploring More About Slope*, TRB 0-29. See Chapter Overview p. 312E.

LESSON RESOURCES

TRB Practice H-7
TRB Enrichment H-8

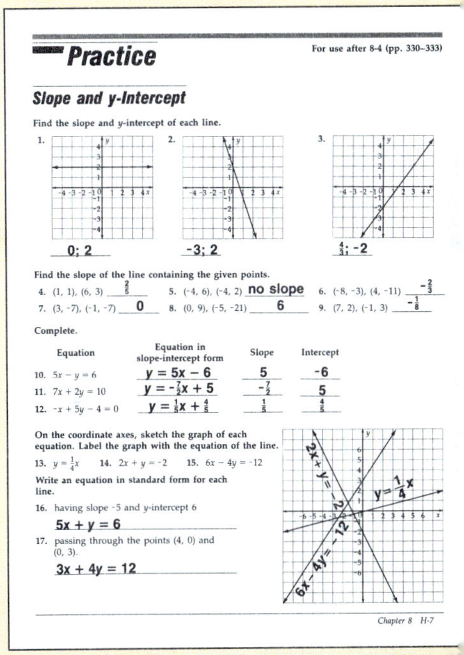

Practice For use after 8-4 (pp. 330–333)

Slope and y-Intercept

Find the slope and y-intercept of each line.

1. 0; 2
2. -3; 2
3. $\frac{4}{3}$; -2

Find the slope of the line containing the given points.

4. (1, 1), (6, 3) $\frac{2}{5}$
5. (-4, 6), (-4, 2) no slope
6. (-8, -3), (4, -11) $-\frac{2}{3}$
7. (3, -7), (-1, -7) 0
8. (0, 9), (-5, -21) 6
9. (7, 2), (-1, 3) $-\frac{1}{8}$

Complete.

	Equation	Equation in slope-intercept form	Slope	Intercept
10.	$5x - y = 6$	$y = 5x - 6$	5	-6
11.	$7x + 2y = 10$	$y = -\frac{7}{2}x + 5$	$-\frac{7}{2}$	5
12.	$-x + 5y - 4 = 0$	$y = \frac{1}{5}x + \frac{4}{5}$	$\frac{1}{5}$	$\frac{4}{5}$

On the coordinate axes, sketch the graph of each equation. Label the graph with the equation of the line.

13. $y = \frac{1}{4}x$ 14. $2x + y = -2$ 15. $6x - 4y = -12$

Write an equation in standard form for each line.

16. having slope -5 and y-intercept 6
$5x + y = 6$

17. passing through the points (4, 0) and (0, 3).
$3x + 4y = 12$

Chapter 8 H-7

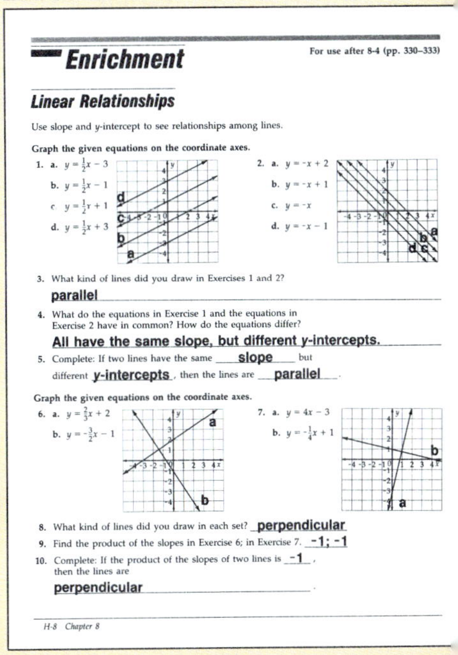

Enrichment For use after 8-4 (pp. 330–333)

Linear Relationships

Use slope and y-intercept to see relationships among lines.

Graph the given equations on the coordinate axes.

1. a. $y = \frac{1}{2}x - 3$
b. $y = \frac{1}{2}x - 1$
c. $y = \frac{1}{2}x + 1$
d. $y = \frac{1}{2}x + 3$

2. a. $y = -x + 2$
b. $y = -x + 1$
c. $y = -x$
d. $y = -x - 1$

3. What kind of lines did you draw in Exercises 1 and 2?
parallel

4. What do the equations in Exercise 1 and the equations in Exercise 2 have in common? How do the equations differ?
All have the same slope, but different y-intercepts.

5. Complete: If two lines have the same slope but different y-intercepts, then the lines are parallel.

Graph the given equations on the coordinate axes.

6. a. $y = \frac{2}{3}x + 2$
b. $y = -\frac{3}{2}x - 1$

7. a. $y = 4x - 3$
b. $y = -\frac{1}{4}x + 1$

8. What kind of lines did you draw in each set? perpendicular

9. Find the product of the slopes in Exercise 6; in Exercise 7. -1; -1

10. Complete: If the product of the slopes of two lines is -1, then the lines are
perpendicular.

H-8 Chapter 8

LESSON QUIZ

(See p. 332 for answers to 1, 2, and 4.)

1. Plot the points (-5, 2) and (1, 4) and find the slope of the line containing them. ($m = \frac{1}{3}$)

2. Graph the line having a slope of $\frac{1}{2}$ containing the point (-4, 5).

3. Write the equation in slope intercept form. Name the slope and the y-intercept. $3x + y = 4$ ($y = -3x + 4$, $m = -3$, y intercept = 4)

4. Sketch the graph of each equation.
$y = 2x - 2$ $y = \frac{1}{2}x - 1$

Assignment Guide

Basic 1-21 odd, 25-27, 31a-b, 32-33, 37, MR All

Average 2-5, 8-24 even, 28-31c, 33-36, MR All

Enriched 2-18 even, 22-24, 29-31, 34-38, MR All

FOR THE NEXT LESSON

graph paper, math journal

Lesson Focus

Materials

Graph paper, math journal

MOTIVATING THE LESSON

Present students with the following problem: Last year the cost of a personal compact disc player was $200. This year the cost of the same player is $225. If the rate of increase is the same, what will be the cost of the player next year?

Comment that the cost of the player increased at a rate of $25 per year, and at the same rate the cost of the player next year will be $225 + $25 or $250. Tell students that problems such as these can be solved by using the strategy of making a graph.

Instruction

1 Review other possible strategies, including *make a table* and *write an equation*.

2 Help students realize that the graph cannot extend beyond quadrant I, because neither the age of the car nor the value of the car can be less than zero.

3 The problem can be solved by using the strategy of writing an equation:
$y = 12{,}000 - 1{,}500x$,
where x is the age of the car and y is the dollar value of the car. In three years the car will be 5 years old; therefore,
$y = 12{,}000 - 1{,}500(5) = 4{,}500$
Solve the problem by using a table:

x	y
0	12,000
1	10,500
2	9,000
3	7,500
4	6,000
5	4,500

PROBLEM SOLVING STRATEGY

OBJECTIVE:
To solve problems using a graph.

8-5 Solve by Graphing

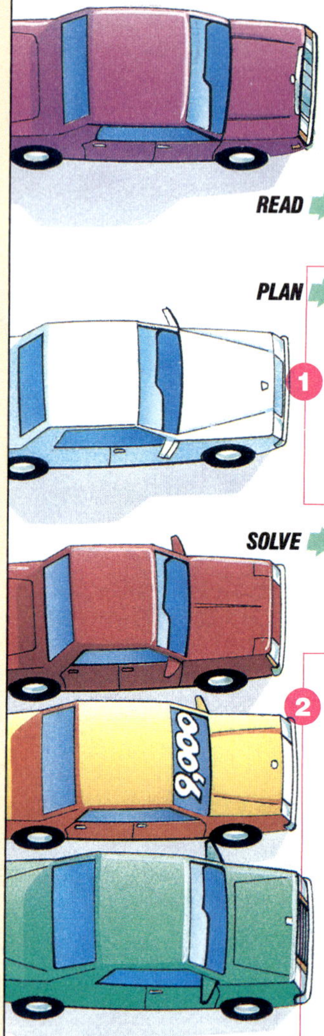

■ You can solve many types of problems using graphs.

PROBLEM

Two years ago the value of a new car was $12,000. Its current value is $9,000. Predict the value of the car three years from now if it continues to depreciate at the same rate.

SOLUTION

READ What do you want to find? — the value of the car in three years

PLAN Decide on a strategy. — Make a graph. Let x = the age of the car and y = the dollar value.

1 What was the value of x when the car was new? the value of y? — $x = 0$, $y = 12{,}000$

What is the current value of x when y is 9,000? — $x = 2$

SOLVE What will be the value of x in three years? — $x = 2 + 3 = 5$

Use the information to write two ordered pairs. — (0, 12,000) and (2, 9,000)

2 Plot the points. Connect them. Extend the line so that you can find other values.

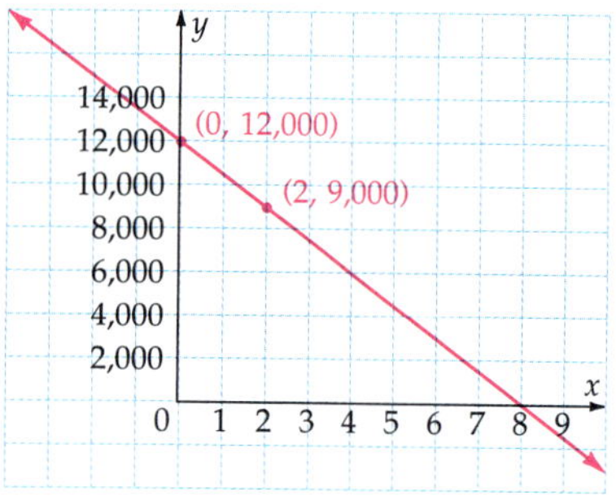

Find the value of y when $x = 5$ — $y = 4{,}500$

The value of the car in three years will be $4,500.

LOOK BACK Check your answer by solving another way.

3 In two years the car depreciates $3,000, or $1,500 per year. If it drops at the same rate for three more years, it would be worth 9,000 − 3(1,500) or $4,500.

Teaching Tip

Have newspapers and magazines available for students to review and select data that can be used to create problems that can be solved by making a graph.

CLASS EXERCISES

Refer to the problem on page 334.

1. What will the car be worth five years from now? $1,500
2. When the car is worth $6,000, how old will the car be? 4 years old
3. How old must the car be before its value has depreciated to $0? Explain why this is probably not an accurate prediction. 8 years; The car will still have some value after 8 years.

WRITTEN EXERCISES

Use a CALCULATOR where appropriate.

Solve by using the strategy of graphing.

1. The relationship between the Fahrenheit and Celsius temperature scales is linear. The freezing point of water is 32°F or 0°C. The boiling point of water is 212°F or 100°C.
   a. Make a graph showing this information. Check students' work.
   b. What is the approximate Fahrenheit temperature that is equivalent to 10°C? 50°F
   c. What is the approximate Celsius temperature that is equivalent to 70°F? 20°C
2. The Jackson family bought a house in 1968 for $32,000. In 1988 they sold it for $192,000. Assume that the increase in value was constant over the 20-year period.
   a. Make a graph showing this information. Check students' work.
   b. What was the value of the house in 1980? $128,000
   c. Assume the value of the house continues to increase at the same rate. Predict the value of the house in the year 2000. $288,000
   d. Is it reasonable to assume that the increase in value over time is constant? Explain. No; The economy changes from year to year.

Solve using any strategy.

3. The temperature at 6:00 A.M. was 48°F. At 9:00 A.M. it was 60°F. The temperature climbed at a constant rate from 6:00 A.M. to 11:00 A.M. What was the temperature at 10:00 A.M.? 64°F
4. An airplane flying at an altitude of 30,000 ft begins its descent at the rate of 1,500 ft/min. Assume the plane continues to descend at the same rate. In how many minutes will the plane be on the ground? 20 min
5. If 4 is subtracted from three times a number, the result is two more than the number. Find the number. 3

Guided Practice

Class Exercises For question 1, extend the graph to seven years.

A ***common error*** occurs when students extend graphs beyond logical limits. For example, in the graph on page 334, students might extend the graph beyond quadrant I, resulting in negative years or a negative value for the car. Remind students that they must judge the reasonableness of their answers.

Closure

Writing in Math Ask students to describe in their math journals situations when graphs and equations of lines are used to represent the relationship between real world quantities. Some possibilities include time and distance, temperature and time, and quantity and cost.

Independent Practice

Ask students to identify the strategy they used to solve questions 3-15. For question 5, students may use the strategy *write an equation.*

As an extension, refer students to the problem solving activity on p. 312E of the Overview. See also TRB Exploring O-30.

MATH MINUTES

List several sets of points, for example (-2, -1), (-1, 1), (1, 2), (0, 3). For each set, have groups graph the points, find the slope between various pairs of points, and determine which point is not on the line. [(1, 2)]

For questions 7, 9, and 10, students may choose the strategy *write and solve a proportion.* The strategy *make a table* is useful for question 8. Students may *write an equation* for questions 11 and 12, and use the strategy *guess and test* for question 13.

Lesson Follow-up

LESSON RESOURCES

TRB Practice H-9
TRB Enrichment H-10

Practice — For use after 8-5 (pp. 334–336)

Solve by Graphing

Solve by using the strategy of graphing.

1. The water pressure at a depth of 100 ft in the ocean is 45-lb/ft². At a depth of 500 ft the pressure is 225 lb/ft².
   a. Graph this information.
   b. What is the approximate pressure at a depth of 800 ft? **360 lb/ft²**
   c. What is the approximate depth where the pressure is 180 lb/ft²? **400 ft**

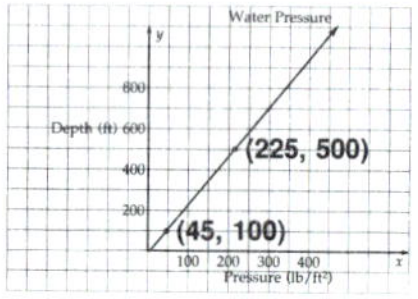

2. Tanya opened a savings account in 1984. She had $4,000 in the account by 1988 and $5,500 by 1991.
   a. Graph this information. Assume the value of the account increases constantly with time.
   b. What was the approximate amount in the account in 1986? **$3,000**
   c. About how much will be in the account in 1995? **$7,500**

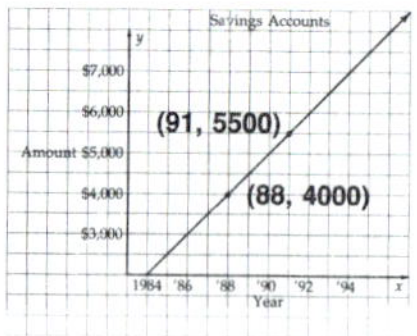

3. The driver of a car slowed to a stop at a constant rate. After 5 s the car was traveling 30 mi/h. After 8 s the car was traveling 12 mi/h.
   a. Graph this information.
   b. About how fast was the car moving when the driver first applied the brakes? **60 mi/h**
   c. About how long did it take the car to stop? **10 s**

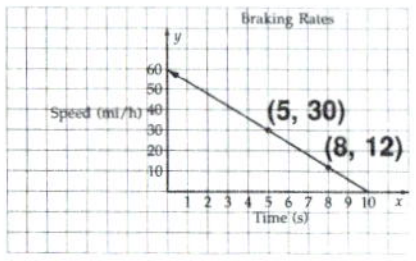

Chapter 8 H-9

Enrichment — For use after 8-5 (pp. 334–336)

Find the Equation from the Graph

A giraffe was 7 ft tall at the age of 4 and $11\frac{1}{2}$ ft tall at the age of 7.

1. Make a graph showing this information. Assume the rate of growth was constant.
2. Choose two points on the line. Find the slope using the formula slope = $\frac{\text{difference in } y}{\text{difference in } x}$. **$\frac{3}{2}$**
3. Find the y-intercept by finding where the graph crosses the vertical axis. **(0, 1)**

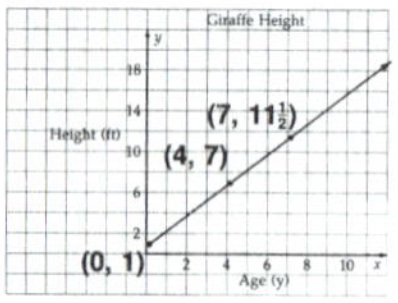

4. Write the equation of the line in slope-intercept form. **$y = \frac{3}{2}x + 1$**
5. Use the equation to find the following information.
   a. the giraffe's height at the age of 5 **$8\frac{1}{2}$ ft**
   b. the age at which the giraffe was 16 ft tall **10 y**

A hippopotamus weighed 700 lb at the age of 1 and 1,900 lb at the age of 3.

6. Make a graph showing this information. Assume the rate of growth was constant.
7. Find the slope of the line. **600**
8. Find the y-intercept. **(0, 100)**
9. Write the equation of the line. **$y = 600x + 100$**
10. Use the equation to find the following information.
    a. the hippo's weight at the age of 8 **4,900 lb**
    b. the age at which the hippo weighed 7,900 lb **13**

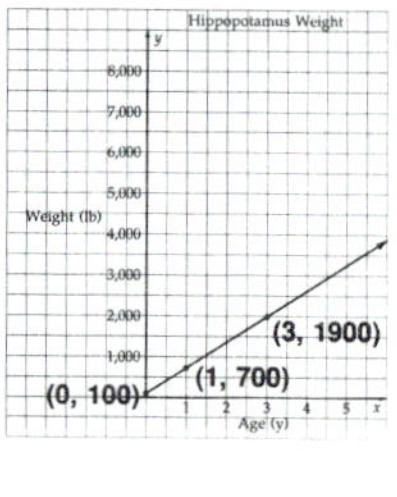

H-10 Chapter 8

6. A supermarket charges \$1.19 for a 12-oz jar of salsa and \$1.89 for a 20-oz jar. The manufacturer has just come out with a 16-oz size. What would you suggest the supermarket should charge for the new size? Justify your answer. **\$1.55**
7. The sales tax on an item costing \$17.50 is \$1.23. Find the sales tax on an item that costs \$30. Round your answer to the nearest cent. **\$2.11**
8. A shoe store employs high-school students as part-time salespeople. The starting pay is \$3.85/h. Every six months a worker is eligible for a raise of \$.35 an hour, if the work has been satisfactory. What can a good salesperson expect to be earning per hour two years after starting at the store? **\$5.25**
9. A slope of $\frac{1}{10}$ is suitable for a ramp to allow wheelchair access to a building. How far from a doorway will a ramp extend if the doorway is $3\frac{1}{2}$ ft above the ground? **35 ft**
10. A delivery van travels 240.8 mi using 10.6 gal of gas. The tank holds 13.6 gal. How many miles can the van expect to travel on a full tank of gas? Round your answer to the nearest mile. **309 mi**
11. A store is holding its annual 30% off sale.
    a. What will you save on an item if the regular price is \$89.95? Round your answer to the nearest dollar. **\$27**
    b. A student saved \$50 on a dress she bought at the sale. Estimate its original price. **\$150**
12. The high school hired a rock band for a concert. The school guaranteed the band a fee of \$1,500, plus \$4.50 for each ticket sold. There are 1,132 seats in the auditorium. What is the greatest possible amount of money the band can earn for its single concert performance? the least amount? **\$6,594; \$1,500**
13. A student has \$5.90 in dimes and quarters. There are 32 coins altogether. How many of each coin does the student have? **18 quarters and 14 dimes**
14. ***PROJECT*** Look up record times for the 100-m dash in the Olympics for the years 1980 and 1984. **Check students' work.**
    a. Assume that the record time continues to decrease at the same rate. Graph the data from the first two years. Use the horizontal axis to represent the year. Use the vertical axis to represent the time.
    b. Extend the graph to predict the record time for the 1988 Olympic games. Compare your result with the actual Olympic record in 1988.
    c. ***WRITE*** a paragraph about your findings in your journal.
15. ***DATA FILE 9 (pp. 360–361)*** Ron played eighteen holes of golf at St. Andrew's golf club for five consecutive days. About how many miles did Ron walk on the golf course? **20 mi**

Assignment Guide

Basic All
Average All
Enriched All

FOR THE NEXT LESSON

computer and graphing software or graphing calculator, measuring tape, ball, stopwatch, math journal

Problem Solving Practice

PROBLEM SOLVING STRATEGIES

Look for a Pattern
Guess and Test
Simplify the Problem
Make an Organized List
Work Backwards
Account for All Possibilities
Make a Table
Write an Equation
Solve by Graphing
Draw a Diagram
Make a Model
Solve Another Way
Simulate the Problem

Solve. Use an appropriate strategy or a combination of strategies.

1. In a collection of dimes and quarters, there are seven more quarters than there are dimes. How many dimes and quarters are there if the collection is worth $3.50? **12 quarters and 5 dimes**
2. Anna has $75 in the bank. She saves $5 the first week, $10 the second week, and $15 the third week. At this rate, how much money will she have in the bank at the end of 12 weeks? **$465**
3. A landscaper wants to plant a bush every 1.5 ft around the edge of a circular garden. It is 72 ft around the edge of the garden. How many plants should the landscaper purchase? **48**
4. A student withdraws $15 from his bank account. He then buys lunch for $4.75, a ticket to the movies for $7.50, and a snack after the movie for $3.45. He has $14.23 left. How much money did he have before he withdrew the $15? **$14.93**
5. The sum of three consecutive integers is 111. What are the integers? **36, 37, 38**
6. A plumber charges $35 for a service call, plus $60 per hour for her time.
   a. Find the cost of a two-hour service call. **$155**
   b. How much time was spent on a call if the bill was $125? **$1\frac{1}{2}$ h**
7. The height of a toddler is 36 in. The mother's height is double the toddler's height, less 7 in. How tall is the mother? **65 in. or 5 ft 5 in.**
8. You can buy 12 pencils for $.80. How much will you pay for 27 pencils? **$1.80**
9. One square has sides four times as long as the sides of a second square. The combined area of the squares is 272 ft^2. Find their dimensions. **4 ft × 4 ft and 16 ft × 16 ft**
10. The sum of the squares of two consecutive integers is 145. Find the integers. **8 and 9 or -8 and -9**
11. Mr. Harrow earns $22,500 a year. He gets paid weekly. He pays 6.3% of his salary to social security. How much money is taken out of each pay check for social security? **$27.26**
12. There are four candidates running for president of the student council. Three candidates are running for vice-president. How many different ways can the two offices be filled? **12**
13. A grocery store sells apples at $1.92 a dozen. Oranges are $2.16 a dozen. What is the cost of 4 apples and 3 oranges? **$1.18**

Instruction

This page provides a variety of problems that can be used to reinforce and enhance the students' problem solving skills. Encourage students to read each problem carefully. Then have them refer to the list of problem solving strategies to help them decide how to solve the problem.

Point out, however, that not all questions require a strategy for solving, nor are all the strategies in the list used in this lesson.

Lesson Focus

Materials

Computer and graphing software or graphing calculator, measuring tape, ball, stop watch, math journal

Vocabulary/Symbols

simulate

MOTIVATING THE LESSON

Ask students why scientists and engineers might use computer simulations. (Possible answer: to test situations that would be difficult to do in real life)

Then explain that in this lesson the class will be exploring the effect of speed on an object in motion. Review the situation in question 1, and ask students what effect they think increasing the speed of the ball will have on the distance the ball travels. (Answers may vary. The distance increases.)

Instruction

1 Have students work in small groups to model the experiment for each of the conditions and record their results in their math journals. Compare the results in a class discussion.

2 Students may find it difficult to believe that the time it takes the ball to hit the floor remains constant regardless of the horizontal speed of the ball. Have students continue this part of the experiment until they feel comfortable with the result. Ask students to judge the reasonableness of their answers.

OBJECTIVE:
To study computer-simulated motion.

MATERIALS

- Computer and graphing software or a graphing calculator
- Measuring tape or ruler
- Rubber ball or tennis ball
- Stopwatch
- Math journal to record work

DATAPOINT

Keep track of your data in a table.

d	D	h	t
⋮	⋮	⋮	⋮

Exploring Simulations

Scientists and engineers use computers to *simulate*, or re-create the motion of an object such as rolling a ball off a table.

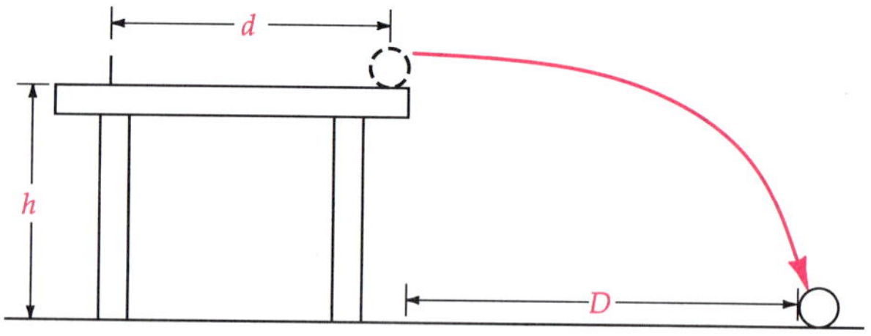

1 **1. Model** the situation shown above. Measure the distances d and h. Roll a ball off a table along distance d. Use a stopwatch to measure t, the time it takes the ball to hit the ground after it rolls off the table. Measure D, the distance from the table to where the ball hits the ground.

a. Continue the experiment, but each time increase the speed of the ball. What happens to the time it takes the ball to hit the ground? What happens to D? **The time stays the same. D increases.**

b. When does $D = 0$? Is there a limit to the value of D? **when you drop the ball; no**

2 **2.** Increase the value of h by using a higher table. Repeat the experiment.

a. How does increasing h affect t? How does increasing h affect D? **t increases; D increases**

b. How does increasing the speed of the ball affect D? **increases D**

You can use a computer or graphing calculator to graph the *trajectory*, or path, of the ball. The equation for this trajectory is $y = -16x^2 + 10$. The figure below shows the part of the graph that simulates the ball's trajectory.

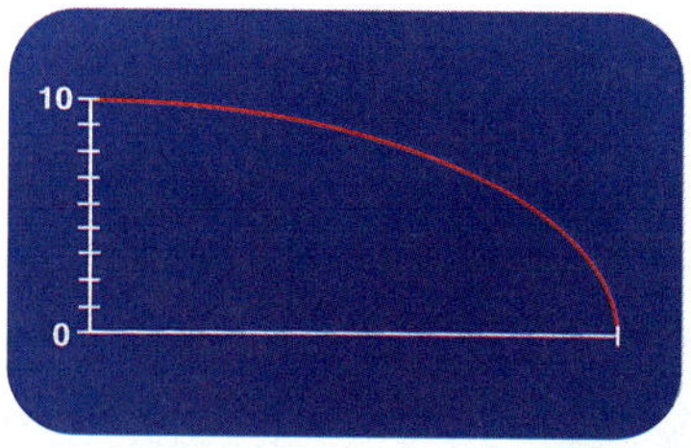

3. In the equation $y = -16x^2 + 10$, the value 10 indicates the height of the table. Write equations for table heights of 15 ft, 20 ft, and 35 ft. **$y = -16x^2 + 15$; $y = -16x^2 + 20$; $y = -16x^2 + 35$**

4. Graph the equations from Exercise 3. What do you notice about the trajectory as h increases? **Check students' work. The distance, D, increases as h increases.**

Teaching Tip

Computer graphing software can be helpful in providing an exciting visual presentation for the data students have collected in their experiments. Have manuals for the software available for students to use.

■ The equation $y = -16x^2 + 10$ assumes a speed of 1 ft/s. The equation for *any* speed is $y = -\frac{16}{v^2}x^2 + h$, where h is the height of the table and v is the speed of the ball.

5. a. ***Write*** equations for each speed and height.

Speed (ft/s)	1	2	3	4	4	4	6	7	8
Height (ft)	10	10	10	10	15	20	20	20	20

b. Graph each equation. **Check students' work.**

c. ***Analyze*** what happens to D as the speed of the ball increases but the height remains the same. Is this what you would expect to happen? **increases; yes**

d. What happens to D as the height of the table increases but the speed of the ball remains the same? Is this what you would expect to happen? **increases; yes**

e. What happens to D as both the table height and the speed of the ball increase? Is this what you would expect to happen? **increases; yes**

6. ***Explore*** What happens to D for very large values of v? What happens to the ball's trajectory for large values of v? **increases; It approaches the horizontal.**

DATAPOINT

You can use a spreadsheet and graphics package to graph equations. Use the spreadsheet format below to create a line graph for as many points as you want.

	A	B
1	0	= -16*A1^2 +10
2	0.01	= -16*A2^2 +10
3	0.02	= -16*A3^2 +10
4	0.03	= -16*A4^2 +10
5	0.04	= -16*A5^2 +10
6	0.05	= -16*A6^2 +10
7	0.06	= -16*A7^2 +10
8	0.07	= -16*A8^2 +10
9	0.08	= -16*A9^2 +10
10	0.09	= -16*A10^2 +10
11	0.1	= -16*A11^2 +10
12		

■ A computer or graphing calculator simulation can help you predict the outcome of an experiment. The three trajectories below are for a ball moving with speeds 1 ft/s, 2 ft/s, and 3 ft/s off a 4-ft high table.

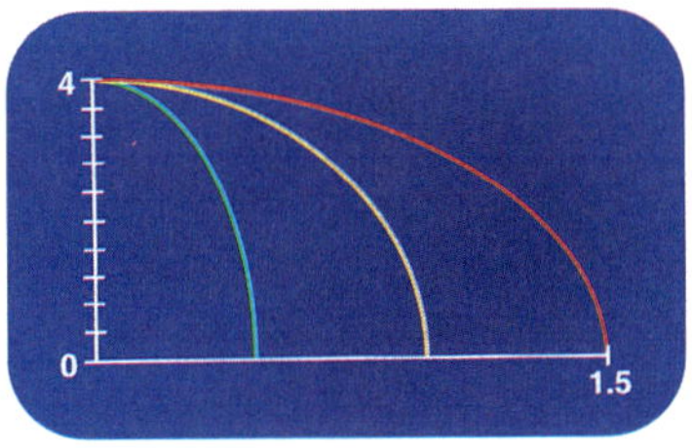

7. a. ***Estimation*** What are the values for D in each case? **0.5, 1, 1.5**

b. ***Model*** each situation in the graph. To get the right speed, measure d and use a stopwatch. Time the motion of the ball to 1 s, 2 s, or 3 s, depending on the speed. **Check students' work.**

c. How do your results compare to those of the computer simulation? How do they differ? **Results should be similar.**

d. ***Discuss*** the advantages to using a computer to simulate an experiment. **You can run more trials in a given period of time.**

DATAPOINT

A satellite in orbit around Earth is actually falling. It never hits the ground because it is moving forward fast enough to keep it from dropping.

3 Students who use a computer or graphing calculator should set the range to show only the first quadrant of the coordinate grid; this way the simulated trajectories of the balls will look more realistic.

Suggest software packages to use:

- *Master Grapher*
- *PFS First Graph*
- *Appleworks*
- *Lotus 1-2-3*
- *Excel*

4 Emphasize to students that computer simulations allow for exploration of various scenarios. Encourage students to explore trajectories other than those they have graphed.

Closure

Writing in Math Have students compare in their math journals the results of a computer simulation with the results of their own experiment. Have them consider what might account for differences.

ADDITIONAL ANSWERS

5. a. $y = -16x^2 + 10$
$y = -4x^2 + 10$
$y = -\frac{16}{9}x^2 + 10$
$y = -x^2 + 10$
$y = -x^2 + 15$
$y = -x^2 + 20$
$y = -\frac{4}{9}x^2 + 20$
$y = -\frac{16}{49}x^2 + 20$
$y = -\frac{16}{64}x^2 + 20$

Assignment Guide

Basic All
Average All
Enriched All

FOR THE NEXT LESSON

graph paper, math journal

Lesson Focus

Materials

Graph paper, math journal

Vocabulary/Symbols

solution
system of linear equations

MOTIVATING THE LESSON

Present the following problem to students: Find two numbers whose sum is 10 and whose difference is 4. Discuss how information in the problem can be expressed as two equations using the same variables: $x + y = 10$; and $x - y = 4$. Point out to students that two equations of this type together are called a system of linear equations, and that the solution of the equations is an ordered pair of numbers that makes both equations true.

Instruction

1 Ask whether the following is a system of linear equations, $xy = 12$ and $x + y = 12$. Students may realize that this is not a system of linear equations since $xy = 12$ is not a linear equation.

2 Ask students if the equations were presented in slope-intercept form, would they be able to determine, without graphing, if there is no solution. (Yes, because the equations will have the same slope but different y-intercepts. The graphs will be parallel lines.)

OBJECTIVE: *To solve two equations in two variables by graphing.*

8-6 Solving Systems of Linear Equations

▼ Two linear equations using the same variables form a *system of linear equations*.

System of Linear Equations	A system of linear equations is two or more linear equations using the same variables.

Example 1 **State whether each pair of linear equations is a system of linear equations.**

a. $2x + y = 8$
$x - 3y = -9$

b. $x + 8y = 12$
$6a + b = -3$

1

Solution **a.** Yes, this is a system of linear equations. Both equations contain variables x and y.

b. No, this is not a system of linear equations. The equations do not contain the same variables.

▼ A *solution* of a system of linear equations makes all of the equations in the system true.

Solution	A solution of a system of linear equations is any ordered pair of numbers that satisfies all equations in the system.

Example 2 **Tell whether the ordered pair (2,5) is a solution for the system of linear equations $2x + y = 9$ and $4x - y = 3$.**

Solution Substitute the ordered pair into each equation.

$2x + y = 9$ | $4x - y = 3$
$2(2) + 5 = 9$ | $4(2) - 5 = 3$
$4 + 5 = 9$ | $8 - 5 = 3$
$9 = 9$ ✓ | $3 = 3$ ✓

Since (2,5) makes both equations true, it is a solution of the system of linear equations.

FLASHBACK
Every point on the graph of an equation represents an ordered pair of numbers that is a solution of the equation.

▼ You can solve a system of linear equations by graphing.

Example 3 **Solve the system $2x + 3y = 6$ and $3x - y = -2$ by graphing.**

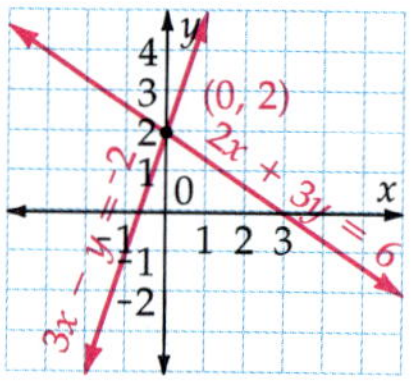

Solution Graph each equation on the same set of axes.
The lines intersect at (0,2).
The point (0,2) is the only solution of the system.

Teaching Tip

You may wish to display this definition.
A system of two equations with one common solution is a **system of consistent equations;** one with no common solution is a **system of inconsistent equations.**

▼ Some systems of linear equations have no solution.

Example 4 **Solve the system of linear equations $x + y = 1$ and $x + y = 4$ by graphing.**

Solution Graph each equation on the same set of axes. The lines are parallel and do not intersect. There is no solution to this system. (2)

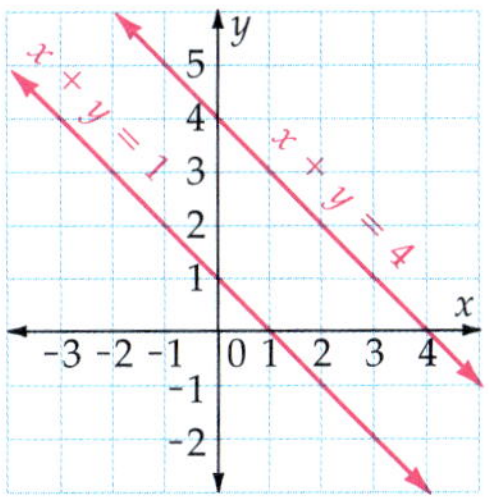

▼ Some systems of equations have infinitely many solutions.

Example 5 **Solve the system of linear equations $x - 2y = 4$ and $2x - 4y = 8$ by graphing.**

Solution Graph each equation on the same set of axes. The graph of each equation is the same line. Therefore, every point on the line satisfies both equations. There are infinitely many solutions to this system. (3)

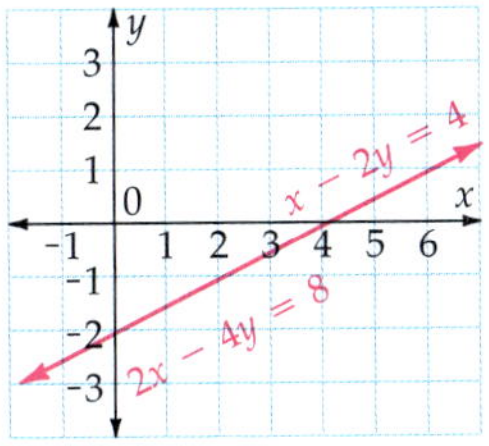

▼ Systems of equations can help you solve word problems.

Example 6 **Find two numbers whose sum is 6 and difference is 4.**

Solution Let x represent the first number and y represent the second number. Write a system of linear equations.

$x + y = 6$ ←**Sum**
$x - y = 4$ ←**Difference**

Graph each equation on the same set of axes. The lines intersect at (5,1). The numbers are 5 and 1. (4)

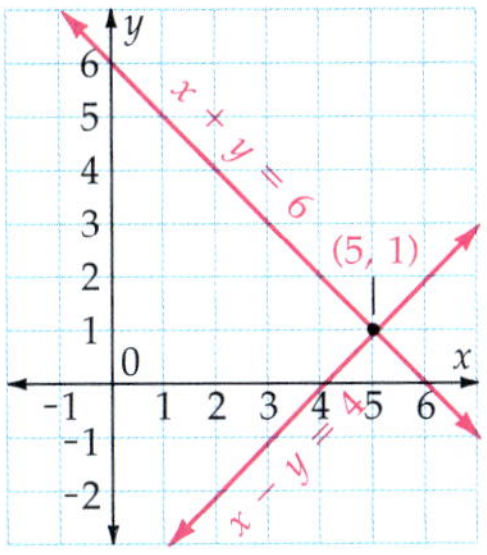

Check $x + y = 6$ $\quad$ $x - y = 4$
$5 + 1 = 6$ $\quad$ $5 - 1 = 4$
$6 = 6$ ✓ $\quad$ $4 = 4$ ✓

CLASS EXERCISES

Find the solution for each system.

1. $y = x + 1$ and $y = 3x - 7$ **(4,5)**
2. $x + y = -3$ and $y = 3x - 7$ **(1,-4)**

Write the coordinates of each point.

3. a point that satisfies $y = x + 1$ but not $y = 3x - 7$ **Answers may vary.**
4. a point that satisfies $y = 3x - 7$ but not $x + y = -3$ **Answers may vary.**

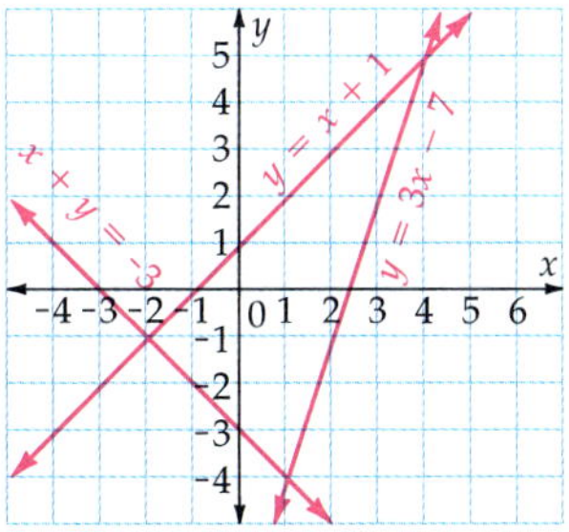

MATH MINUTES

Have students graph the following equations on the same graph.

$x - y = 1$
$x + 2y = 10$
$2x - 3y = -1$

Then ask, is it possible to have more than two equations in a system of linear equations? (yes)

(3) Students should see that $x - 2y = 4$, and $2x - 4y = 8$ are equivalent equations.

(4) Help students to recognize that if y represents the first number and x represents the second number, the solution would be the same, but the graph of the second number would be different. The point of intersection would be (1, 5).

ADDITIONAL EXAMPLES

1. State whether each pair of equations is a system of linear equations.
  a. $y = 2x$ $\quad$ **b.** $12a - 5b = 6$
  $y = 3x - 3$ $\quad$ $3x + 5y = 15$
  (yes) $\quad$ (no)

2. Tell whether (3, -2) is a solution of the system:
  $2x - 3y = 12$
  $x - y = 1$ (no)

3. Solve by graphing.
  a. $x - y = 4$ $\quad$ **b.** $x = 2 - 2y$
  $x + 3y = 12$ $\quad$ $\frac{1}{2}x + y = 4$
(see graphs below)

4. Find two numbers whose sum is -1 and whose difference is 11. (5, -6)

3a.

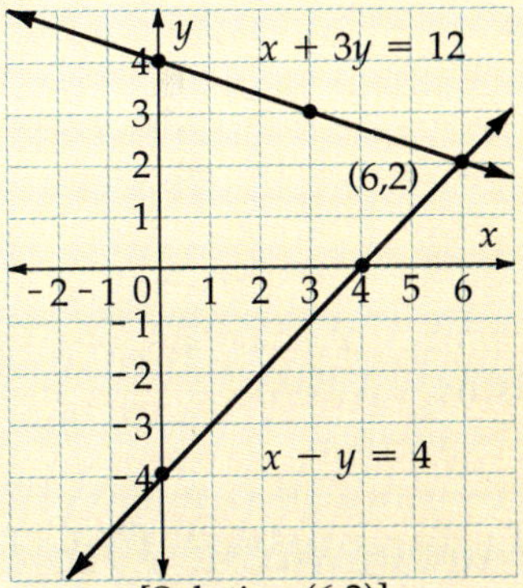

[Solution (6,2)]

3b.

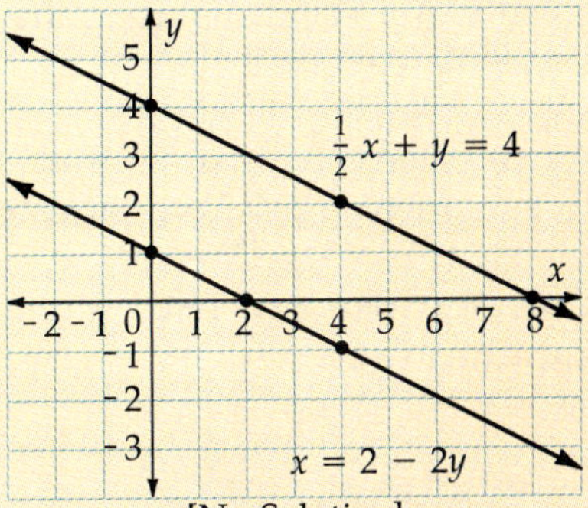

[No Solution]

Guided Practice

Class Exercises For questions 8-12, have students check their solutions by substitution.

Think and Discuss For question 1, if the solution is not a common point on the graph, it would be very difficult to read from the graph.

A ***common error*** occurs when students incorrectly graph the equations and accept the point of intersection as the solution of the system. Encourage students to check their solutions by substituting the ordered pair in each equation.

Closure

Writing in Math Have students describe in their math journals the graph for the following situations when solving a system of equations: exactly one solution; an infinite number of solutions; and no solution. Then have them explain how the slope-intercept forms of a system of equations will help to determine if there are an infinite number of solutions, or if there is no solution.

Independent Practice

For questions 4-15, you may wish to have students graph some of the equations using the slope-intercept method.

As an extension, refer students to the graphing calculator activity on p. 312E of the Chapter Overview. See also TRB Exploring O-31.

THINK AND DISCUSS See below.

1. All the examples had integer solutions. When do you think graphing might not be useful for solving a system of equations?
2. A system of two linear equations in two variables has no solution. Describe the graphs of the equations.
3. Two distinct ordered pairs are solutions of a system of two linear equations. Describe the graphs of the equations.

1. when the solution(s) are not integers
2. The lines are parallel.
3. The lines are the same.

MIXED REVIEW

Simplify.

1. $\frac{15x^3}{5x}$ $3x^2$
2. $\frac{20x^2y^6}{8xy^3}$ $\frac{5xy^3}{2}$

Find the slope and y-intercept.

3. $x = y$ 1; 0
4. $2x + 3y = 12$ $-\frac{2}{3}$; 4

Solve each inequality.

5. $2x + 3 < 9$ $x < 3$
6. $3 - (2x + 1) \le 8$ $x \ge -3$

Solve.

7. A coat regularly sells for \$125. It is on sale for 40% off. What is the sale price? \$75

Tell whether the ordered pair is a solution of the system.

5. $3x - 2y = 8$, $x = -3y$; (3,-1) no
6. $x - 2y = 0$, $2x + y = 4$; (2,1) no
7. $x + 2y = 3$, $y = 2x - 1$; (1,1) yes

Solve each system by graphing. Check your solutions. Check students' work.

8. $2x - y = 4$, $y = -2x$ (1,-2)
9. $3y - 2x = 6$, $y = x + 1$ (3,4)
10. $2x + 3y = 6$, $y + 2 = 0$ (6,-2)

Write a system of linear equations. Solve by graphing.

11. The sum of two numbers is 19. Their difference is 5. $x + y = 19$; $x - y = 5$; 7 and 12
12. The sum of two numbers is 10. Twice the larger decreased by three times the smaller is 5. Find the two numbers. $x + y = 10$; $2x - 3y = 5$; 3 and 7

WRITTEN EXERCISES

Tell whether the ordered pair is a solution of the system.

1. $y = x + 2$, $x - 3y = 4$; (-5,-3) yes
2. $y = x - 1$, $x = 4y$; $\left(\frac{4}{3}, \frac{1}{3}\right)$ yes
3. $x + y = 2$, $-x + y = -4$; (3,1) no

Solve each system by graphing. Check your solutions.

4. $y = x + 5$, $2x + y = 8$ (1,6)
5. $x = y - 4$, $x + y = 6$ (1,5)
6. $2x - y = 2$, $2y = 4x - 4$ infinite
7. $x + 4y = 6$, $x - 2 = 0$ (2,1)
8. $x + y = 3$, $2x = 10 - 2y$ no solution
9. $x = y$, $x + y = 4$ (2,2)
10. $y = x - 8$, $y = 3x$ (-4,-12)
11. $y = x - 2$, $x + 3y = 6$ (3,1)
12. $2x + 3y = 6$, $2x + y + 2 = 0$ (-3,4)
13. $x + y = 1$, $y = -x - 1$ no solution
14. $y = 2x - 4$, $4x - 2y = 8$ infinite
15. $y = 2x - 2$, $x = 3$ (3,4)

Write a system of linear equations. Solve by graphing. Check students' work.

16. The sum of two numbers is 120. Their difference is 20. Find the numbers. $x + y = 120$; $x - y = 20$; 70 and 50
17. The sum of two numbers is 55. Their difference is 15. Find the numbers. $x + y = 55$; $x - y = 15$; 20 and 35
18. The difference of two numbers is 5. The greater number decreased by twice the lesser number is 9. Find the numbers. $x - y = 5$; $x - 2y = 9$; -4 and 1
19. A 144-m rope is cut into 2 pieces. One piece is three times as long as the other. How long is each piece of rope? $x + y = 144$; $x = 3y$; 108 m and 36 m

QUIZ ANSWERS

3.
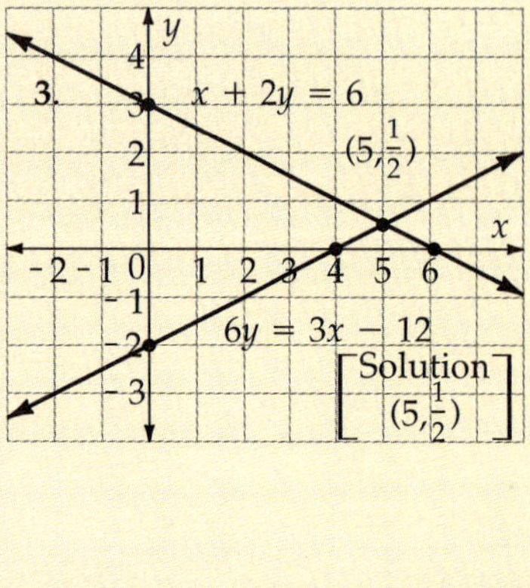

4.
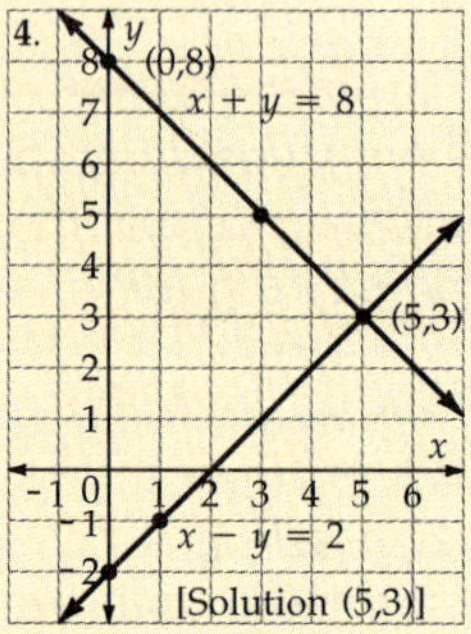

20. Cliff is 6 years older than Claire. In two years, Cliff will be twice Claire's age. Find their present ages. $x + 6 = y$; $y + 2 = 2(x + 2)$; Claire is 4 and Cliff is 10

21. Questions on a 16-item test are worth either 5 points or 10 points each. There are a total of 100 points on the test. How many items of each point value appear on the test? $x + y = 16$; $5x + 10y = 100$; 12 five-point questions and 4 ten-point questions

Describe the nature of each system of linear equations without graphing. Note that they are all written in the form $y = mx + b$.

22. $y = x + 2$
$y = -x + 2$
intersect at one point

23. $y = -x + 2$
$y = -x + 5$
parallel

24. $y = x - 1$
$3y = 3x - 3$
the same line

25. ***DATA FILE 8 (pp. 312–313)*** In what year was the United States population approximately one fourth of the population in 1910? 1850

26. The unit of distance a boat uses to navigate is called a *nautical mile.* One nautical mile is 6,076.1 ft. Saona Island and Mona Island are located in the Caribbean Sea. They are 45 mi apart from each other.

a. To the nearest tenth, how many nautical miles apart are Saona Island and Mona Island? *Hint:* 5,280 ft is 1 mi. 39.1

b. A boat measures its speed in knots. One knot is equal to one naut mi/h. If your boat is traveling at a rate of 12 naut mi/h, how long will it take, to the nearest hour, to go from Saona Island to Mona Island? 3 h

TEST YOURSELF

Find the x-intercept and the y-intercept for each equation.

1. $2x + 5y = 20$ 10; 4
2. $3y = 6 + 9x$ $-\frac{2}{3}$; 2
3. $5x + 15 - y = 5$ -2; 10

Find the slope of the line containing the given points.

4. (7,-3) (7,4) no slope
5. (-12,6) (4,-2) $-\frac{1}{2}$
6. (-5,-3) (6,-1) $\frac{2}{11}$

Write each equation in slope-intercept form. Name the slope and y-intercept. Sketch the graph of the equation.

7. $4x + 2y = 14$ $y = -2x + 7$; -2; 7
8. $3y = -2$ $y = -\frac{2}{3}$; 0; $-\frac{2}{3}$
9. $-4 + 6y = -2x$ $y = -\frac{1}{3}x + \frac{2}{3}$; $-\frac{1}{3}$; $\frac{2}{3}$

Solve each system of linear equations by graphing.

10. $2x + y = 4$
$y = \frac{-1}{2}x + 7$
(-2,8)

11. $2x - 3y = 12$
$4x - 24 = 6y$
infinite

12. $2x - 3y = 12$
$x - 3 = 0$
(3,-2)

LESSON QUIZ

1. Tell whether (6, 4) is a solution of the system.
$4x - 3y = 12$
$x - y = 12$ (no)

2. Tell whether (4, 8) is a solution of the system.
$x + y = 12$
$x - y = 4$ (no)

3. Solve by graphing. Check your solutions.
$x + 2y = 6$ and $6y = 3x - 12$

4. Solve by graphing. The sum of two numbers is 8. Their difference is 2. Find the numbers. (5, 3) (see graphs to the left)

FOR THE NEXT LESSON

calculator, math journal

Assignment Guide

Basic 1-17, 22, 25, MR, TY All

Average 1-15 odd, 16-20, 23-25, MR, TY All

Enriched 7-21 odd, 22-26, MR, TY All

Lesson Follow-up

Reteaching Activity Have students graph the following system on the same coordinate graph, $3x - 2y = 6$ and $x - y = 2$. Review that each point on the line is an ordered pair which is a solution of the equation. The point of intersection is a pair of coordinates which is a solution of both equations. [(2, 0)] Have students check the solution.

LESSON RESOURCES

TRB Practice H-11

TRB Enrichment H-12

Practice

For use after 8-6 (pp. 340–343)

Solving Systems of Linear Equations

Tell whether the ordered pair is a solution of the system. Write *yes* or *no*.

1. $y = 6x + 12$
$2x - y = 4$
(-4, -12) yes

2. $y = -3x$
$x = 4y + \frac{1}{2}$
$\left(-\frac{1}{2}, \frac{3}{2}\right)$ no

3. $x + 2y = 2$
$2x + 5y = 2$
(6, -2) yes

Solve each system by graphing. Check your solutions.

4. $x + y = 3$
$x - y = -1$
Solution: (1, 2)

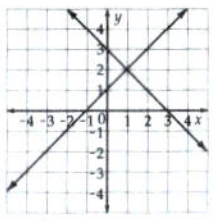

5. $2x + y = 1$
$x - 2y = 3$
Solution: (1, -1)

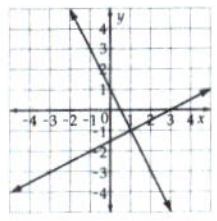

6. $y + 2 = 0$
$2x + y = 0$
Solution: (1, -2)

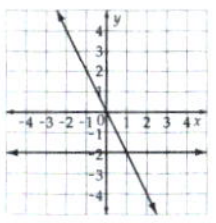

7. $3x + 2y = -6$
$x + 3y = -2$
Solution: (-2, 0)

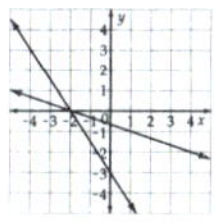

Write a system of linear equations. Solve by graphing.

8. The sum of two numbers is 3. Their difference is 1. Find the numbers.

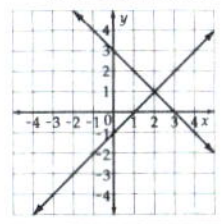

$x + y = 3$
$x - y = 1$
(2, 1)

Chapter 8 H-11

Enrichment

For use after 8-6 (pp. 340–343)

Step Graphs

Some graphs are not continuous, unbroken lines. Instead they jump from one segment to the next. They are called *step graphs.*

The step graph shows the cost of mailing a first-class letter in the country of Stampia. Circles indicate that the point is not on the graph. Solid dots are on the graph.

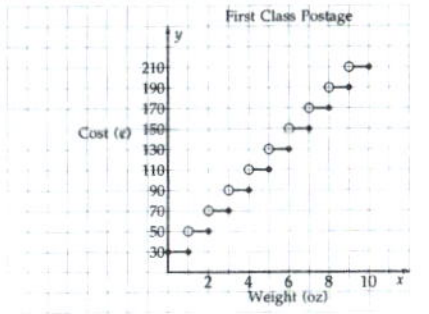

Find the postage for letters of each weight.

1. 4 oz 90¢
2. 7 oz $1.50
3. 10 oz $2.10
4. $8\frac{3}{16}$ oz $1.90
5. a little more than 6 oz $1.50
6. a little less than 10 oz $2.10
7. Write a range of weights that you can mail for $1.30.
5 lbs < weight ≤ 6 lbs
8. Speedy Taxi charges $1.40 for a ride of one half-mile or less and $.25 for each additional quarter-mile. Write values for the y-axis. Draw the step graph for each distance on the x-axis.

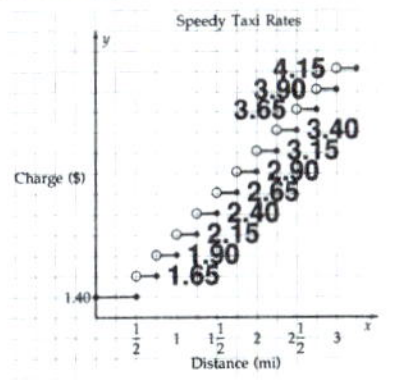

9. Use the graph to find the charge for each distance.
a. $2\frac{1}{8}$ mi $3.15
b. between $\frac{3}{4}$ and 1 mi $1.90
c. 2.7 mi $3.65
d. $1\frac{13}{32}$ mi $2.40
e. Write a range of distances you can ride for $4.15.
3 mi < distance ≤ $3\frac{1}{4}$ mi

H-12 Chapter 8

Lesson Focus

Materials

Calculator, math journal

Vocabulary/Symbols

linear inequality

MOTIVATING THE LESSON

Present the following problem to students: **Tacos cost \$1.25 and a carton of juice costs \$.60. Julie wants to buy at least one of each item for lunch. How many of each can she buy for less than \$4.00?** Ask students how they might solve this problem. (Answers will vary.) Discuss with students how the number of tacos and the number of cartons of juice can be represented by the variables x and y. Since Julie wants to spend less than \$4.00, an inequality can be written to express the information in the problem. ($1.25x + .60y < 4.00$) Point out that the solution to the inequality is an ordered pair of positive integers that makes the inequality true.

Instruction

1 Ask students if the ordered pair (-5, -2) from part c, although a true solution to the inequality, is a reasonable answer to the problem. (no) Remind students that they must judge the reasonableness of their answers.

2 Help students realize that there are an infinite number of solutions to this inequality.

OBJECTIVE: *To solve linear inequalities in two variables.*

8-7 Solving Linear Inequalities

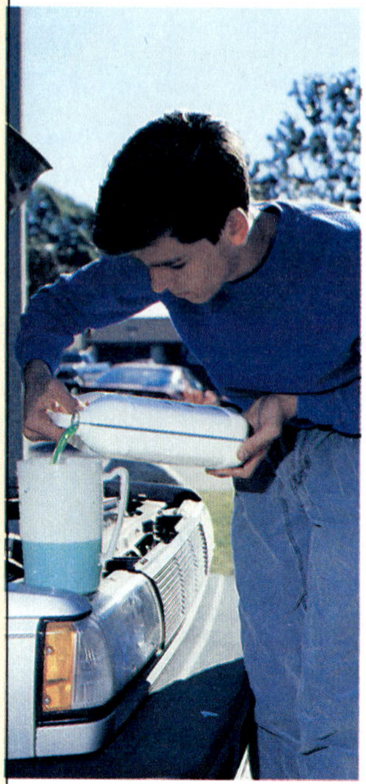

▼ The radiator of a car requires a 40% solution of antifreeze to protect it to -24°F. You can write an equation in two variables to describe the amount of antifreeze needed for your car. Let x represent the capacity of your radiator and y represent the amount of antifreeze solution needed.

$$y = 0.4x$$

You may put more antifreeze in the radiator than is needed to be sure your car is protected. You can describe the situation with an inequality in two variables.

$$y \geq 0.4x$$

▼ An ordered pair that makes an inequality in two variables a true statement is a solution of the inequality.

Example 1 **Tell whether each ordered pair is a solution of the inequality $y \geq 0.4x$.**

a. (4,5) **b.** (5,1) **c.** (-5,-2)

Solution

a. $y \geq 0.4x$
$5 \geq 0.4(4)$
$5 \geq 1.6$ ✓
True, so (4,5) is a solution.

b. $y \geq 0.4x$
$1 \geq 0.4(5)$
$1 \geq 2$ ✗
False, so (5,1) is not a solution.

c. $y \geq 0.4x$
$-2 \geq 0.4(-5)$
$-2 \geq -2$ ✓
True, so (-5,-2) is a solution.

1

▼ You can find solutions for a linear inequality in two variables by solving for y in terms of x.

2

Example 2 **Find three solutions of the inequality $x - y > 1$.**

Solution Solve for y in terms of x.

$x - y > 1$
$x - y - x > 1 - x$ Subtract x from each side.
$-y > 1 - x$
$-1(-y) < -1(1 - x)$ Multiply each side by -1.
$y < -1 + x$
$y < x - 1$ Substitute a value, say 0, for x.
$y < 0 - 1$
$y < -1$

When x is 0, y can be any number less than -1. So, (0,-2), (0,-3), and (0,-4) are three solutions of the inequality. There may be infinitely many solutions of an inequality.

FLASHBACK

When both sides of an inequality are multiplied by a negative number, the inequality sign is reversed.

Teaching Tip

To help students see the limits of a problem involving an inequality have them model the problem. Modeling can help students visualize the problems.

▼ You can solve inequalities to find solutions for real world problems.

Example 3 William has \$36. He wants to buy some tropical fish. Red Oscars are \$6 each. Blue Acaras are \$12 each. How many of each can William buy if he wants to buy at least one of each?

Solution

1. Write an inequality to describe the situation.

Let x equal the number of Red Oscars.
So, $6x$ equals the amount of money for Red Oscars.
Let y equal the number of Blue Acaras.
So, $12y$ equals the amount of money for Blue Acaras.
The inequality $6x + 12y \leq 36$ describes William's situation.

2. Solve the inequality for y.

$$6x + 12y \leq 36$$
$$6x + 12y - 6x \leq 36 - 6x \quad \text{Subtract } 6x \text{ from each side.}$$
$$12y \leq 36 - 6x$$
$$\frac{12y}{12} \leq \frac{36 - 6x}{12} \quad \text{Divide each side by 12.}$$
$$y \leq \frac{36}{12} - \frac{6x}{12}$$
$$\leq 3 - \frac{1}{2}x$$

3. Substitute values for x. Use a chart to organize the data.

x	y	(x,y)
0	3	(0,3)
1	2.5	(1,2.5)
2	2	(2,2)
3	1.5	(3,1.5)
4	1	(4,1)
5	0.5	(5,0.5)
6	0	(6,0)
7	-0.5	(7,-0.5)

4. Analyze the chart to make reasonable conclusions.

The ordered pairs (2,2) and (4,1) are solutions. William can buy 2 Red Oscars and 2 Blue Acaras or 4 Red Oscars and 1 Blue Acara.

The ordered pairs (0,3) and (6,0) are *not* solutions. William wants to buy *at least* one of each type.

The ordered pairs (1,2.5), (3,1.5), and (5,0.5) are *not* solutions. William cannot buy a fraction of a fish.

The ordered pair (7,-0.5) is *not* a solution. William can only buy positive numbers of items.

3

3 Mention that to analyze the data in the chart, students will need to reread the problem, then check that each ordered pair makes sense within the context of the problem.

ADDITIONAL EXAMPLES

1. Tell whether each ordered pair is a solution of the inequality $y - 2x \geq 2$.
a. (-1, 0) [yes] **b.** (2, -4) [no]
c. (0, 2) [yes] **d.** (1, 4) [yes]

2. Find three solutions of the inequality $2x + 3y \leq 12$. [(2, 1), (1, -2), (0, 4)]

3. Oranges cost \$.35 each and apples cost \$.30 each. How many of each fruit can be bought for less than \$3.00? (Answers will vary. Let x = number of oranges, y = number of apples; therefore, $0.35x + 0.3y < 3$.)

MATH MINUTES

Have students look through a store catalog and identify two items they would like to purchase. Have them write and solve an inequality for various budgeted amounts.

Guided Practice

Class Exercises Remind students that the solutions for question 11d must be reasonable. The ordered pair (11, 33.4) may be a true solution, but is not reasonable.

Think and Discuss For question 2, review the concept of absolute value. Have students try several ordered pairs to help confirm that there is no solution. Students can simplify the problem by letting $z = x + y$: $|z| < 0$ may be easier for them to understand.

A ***common error*** occurs when students forget to reverse the inequality symbol when multiplying or dividing by a negative number, as they had to do in Example 2. To help students avoid this error, have them substitute the solutions in the inequality to see that the symbol must be reversed when multiplying or dividing by a negative number.

Closure

Writing in Math Have students make up a problem that can be solved by writing and solving a linear inequality. Have students solve the problem, and then exchange problems and check the solutions.

Independent Practice

For questions 9-16, have students substitute the ordered pairs they found in the original inequalities.

For question 19, have students discuss their answers in small groups before they record their answers in their math journals.

Critical Thinking Ask students to explain how they know what type of slope each graph has.

THINK AND DISCUSS

1. List three ordered pairs that satisfy the inequality $y > 2x$. Answers may vary.
2. Describe the solution set of the inequality $|x + y| < 0$.
3. Describe the solution set of the inequality $|x| + |y| \le 0$. (0,0)

2. no solution

CLASS EXERCISES

Tell whether each ordered pair is a solution of the inequality.

1. $|x + y| > 3$; (0,-5) yes
2. $2x + y \le -1$; (2,3) no
3. $3x - 5y \ge 36$; (7,-4) yes
4. $|x| - |y| > 0$; (-1,-11) no

Solve each inequality for y in terms of x. Write three ordered pairs that are solutions of the inequality. Answers may vary.

5. $3x - 2y > -12$
6. $4x + 5y \ge 15$
7. $x \le -3y$
8. $x - 2y < -1$ $y > \frac{1}{2}x + \frac{1}{2}$
9. $-6x + 8y > 48$ $y > \frac{3}{4}x + 6$
10. $x - y < 1$ $y > x - 1$
11. Students are selling tickets for a play. Student tickets cost \$3. General admission tickets cost \$5. How many of each kind must the students sell to raise at least \$200 for costumes?
    a. Suppose only student tickets are sold. How many must the students sell to purchase costumes? 67
    b. Suppose only general admission tickets are sold. How many must the students sell to purchase costumes? 40
    c. Let x equal the number of student tickets sold and y equal the number of general admission tickets sold. Write an inequality that describes the situation. $3x + 5y \ge 200$
    d. Find three solutions. Assume at least one of each type of ticket must be sold. Answers may vary.

5. $y < \frac{3}{2}x + 6$ 6. $y \ge -\frac{4}{5}x + 3$ 7. $y \le -\frac{1}{3}x$

WRITTEN EXERCISES

Tell whether each ordered pair is a solution of the inequality.

1. $|x + y| > 3$; (-2,1) no
2. $2x + 3y \le 12$; (-2,5) yes
3. $x - 5y < 0$; (7,-3) no
4. $|x| + |y| > |x + y|$; (-4,10) yes

CALCULATOR **Tell whether each ordered pair is a solution of the inequality.**

5. $|x| - 6.8y < -27.09$; (-3.75,4.5) no
6. $2x - 9.4y \ge 3.7x$; (8.96,-1.73) yes
7. $3.2x + y < 0.35$; (-1.25,4.5) no
8. $13.85x + 7.94y > 0$; (3.91,-6.72) yes

Solve each inequality for y in terms of x. Write three ordered pairs that are solutions of the inequality. Answers may vary.

9. $5x - 2y < 10$ $y > \frac{5}{2}x - 5$
10. $x + 2y + 13 \ge 5x - y - 6$ $y \ge \frac{4}{3}x - \frac{19}{3}$

11. $(x - 2y) > x + y + 1$ **12.** $6 - (3x - y) \leq 12$

13. $|x| - y \geq 4$ **14.** $11 - 2\left(x - \frac{1}{2}y\right) - 3y < 0$

15. $3 - y \geq x - 2$ $y \leq -x + 5$ **16.** $-x > 3y - 5(2y - 3)$ $y > \frac{1}{7}x + \frac{15}{7}$

Solve. 11. $y < \frac{-1}{3}$ 12. $y \leq 3x + 6$ 13. $y \leq |x| - 4$ 14. $y > -x + \frac{11}{2}$

17. Ray bought five identical pencils and two identical pens. He spent not more than $2. What could be the cost of each item? **Answers may vary.**

18. A collection of nickels and dimes is worth less than $1.70. Determine how many coins of each type are in the collection.

a. Suppose the collection contains only nickels. How many nickels could there be? **Less than 34**

b. Suppose the collection contains only dimes. How many dimes could there be? **Less than 17**

c. Let x equal the number of nickels in the collection and y equal the number of dimes in the collection. Write an inequality that describes the problem situation. $0.05x + 0.10y < 1.70$

d. Find ten possible solutions. Assume there is at least one nickel and one dime in the collection. **Answers will vary.**

19. Write a journal entry explaining how you know how many solutions each inequality has. **Check students' work.**

a. $|x| + |y| \geq 0$ **infinite** **b.** $|x - y| < 0$ **no solution**

Critical Thinking

EXPLORING GRAPHING

▼ Classify each graph.

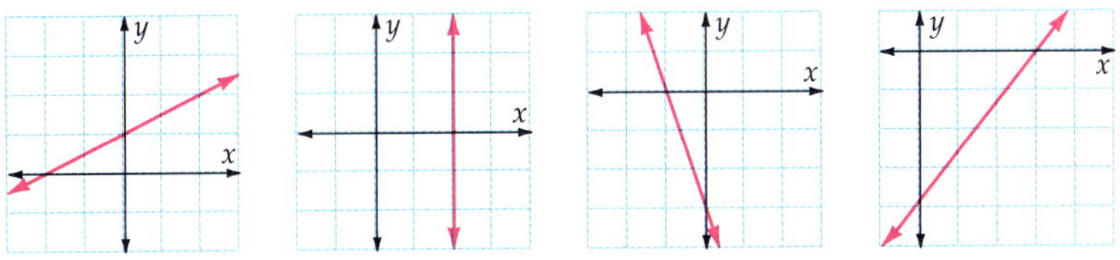

A B C D

1. a. Which graphs have a positive slope? **A and D**

b. Which graphs have a negative slope? **C**

2. a. Which graphs have a positive y-intercept? **A**

b. Which graphs have a negative x-intercept? **A and C**

MIXED REVIEW

Write each fraction as a decimal and as a percent.

1. $\frac{3}{40}$ 0.075; 7.5% **2.** $\frac{2}{9}$ $0.\overline{2}$; $22.\overline{2}\%$

Solve by graphing.

3. $4x - y = 3$ (1,1)
$x + 2y = 3$

4. $y = x - 4$ (6,2)
$x + 3y = 12$

Find each answer.

5. Graph the equation: $y = -x + 3$. **Check students' work.**

6. Is the point (0,0) above or below the graph of the equation $y = -x + 1$? **below**

Solve.

7. The interest on an account of $2,500 for one year is $300. What is the rate of interest? **12%**

Lesson Follow-up

Reteaching Activity Review the steps for solving linear inequalities. Ask students to write each step needed to solve each equation and inequality:

$2x + 5y = 10$ $x - 3y = 6$
$2x + 5y > 10$ $x - 3y \geq 6$

Have students find three ordered pairs for each inequality and substitute the values in the original inequality.

LESSON RESOURCES

TRB Practice H-13
TRB Enrichment H-14

Practice

For use after 8-7 (pp. 344–347)

Solving Linear Inequalities

Tell whether the ordered pair is a solution of the inequality. Write *yes* or *no*.

1. $-x + y \leq 15$; (20, -5) **yes**
2. $x - 7y > 0$; (1, 4) **no**
3. $x + 3y \leq 9$; (3, 2) **yes**
4. $-5x - 4y > 6$; (-3, 2) **yes**
5. $|x - y| > 7$; (-4, 5) **yes**
6. $|x| - |y| > |x + y|$; (-15, 8) **no**
7. $3x \geq 4y$; (-4, -3) **yes**
8. $2(5 - 3x) > -7y$; (5, 3) **yes**

CALCULATOR Tell whether the ordered pair is a solution of the inequality. Write yes or no.

9. $3.2x - 4.67y > 0$; (1.04, -2.2) **yes**
10. $9.47x - 3.68y > 0$; (2.54, 7.16) **no**
11. $4.8x \leq 48 - 3.4y$; (-7.2, 3.6) **yes**
12. $|5.1x + 6.8y| \geq 50$; (-4.7, -3.9) **yes**
13. $4.9(8.2x - 12.6y) < 120$; (0.7, -1.5) **no**

Solve each inequality for y in terms of x. Write three ordered pairs that are solutions of the inequality. **Ordered pairs may vary.**

14. $12x - 3y < 21$ $y > 4x - 7$ ____, ____, ____
15. $x + 10 < 5y$ $y > \frac{1}{5}x + 2$ ____, ____, ____
16. $-2y + 8x \geq 22$ $y \leq 4x - 11$ ____, ____, ____
17. $24 + 2(x - 3y) < 0$ $y > \frac{1}{3}x + 4$ ____, ____, ____
18. $\frac{3}{4}y > 9x + 12$ $y > 12x + 16$ ____, ____, ____
19. $-6x + 3(7 - y) < -3$ $y > -2x + 8$ ____, ____, ____

Joe spent less than $2.80 buying some $.25 stamps and some $.20 stamps. Let x = the number of $.25 stamps he bought. Let y = the number of $.20 stamps he bought.

20. Write an equality describing Joe's purchase. $0.25x + 0.20y < 2.80$
21. Solve the inequality for y in terms of x. $y < -\frac{5}{4}x + 14$
22. Find y for the given values of x.
   a. $x = 8$ $y < 4$
   b. $x = 4$ $y < 9$

Chapter 8 H-13

Enrichment

For use after 8-7 (pp. 344–347)

The Greatest Integer Relationship

A pair of brackets ([ ]) indicates the *greatest integer relationship*.

$y = [x]$ means

"y is the greatest integer that is less than or equal to x."

Examples If $x = 3.7$, $y = 3$; If $x = -2$, $y = -2$; If $x = -6.2$, $y = -7$

1. $y = [x]$. Complete the table of ordered pairs.

x	3.5	0	$-3\frac{1}{2}$	$2\frac{3}{5}$	-4	-0.1	2	$-2\frac{3}{19}$
y	3	0	-4	2	-4	-1	2	-3

2. The graph of the greatest integer relationship is a step graph. Complete the graph at right, using solid dots to indicate points that are on the graph and circles to indicate points that are not on the graph.

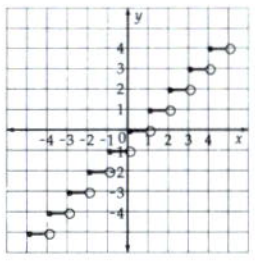

3. Suppose $y = [x + 1]$. Complete the table of ordered pairs.

x	3.5	0	$-3\frac{1}{2}$	$2\frac{3}{5}$	-4	-0.1	2	$-2\frac{6}{19}$
y	4	1	-3	3	-3	0	3	-2

4. Draw the graph of $y = [x + 1]$.

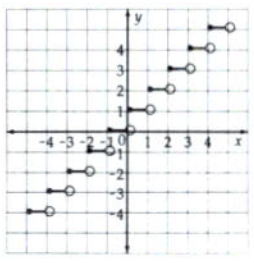

H-14 Chapter 8

LESSON QUIZ

1. Tell whether each ordered pair is a solution of the inequality $y - 2x > 0$.
(2, -4) [no] (2, 3) [no]
(-2, 0) [yes] (5, -1) [no]

2. Muffins cost $.80 each and a carton of milk costs $.40. How many of each can be bought spending no more than $3.00? (Answers will vary. $y < 7.5 - 2x$)

3. Find three solutions of $2x - y > 8$. [possible answer: (5, 0), (3, -3), (4, -3)]

FOR THE NEXT LESSON

graph paper, ruler, math journal

Assignment Guide

Basic 1-5, 8-10, 18-19, MR, CT All
Average 3-7, 11-14, 17, 19, MR, CT All
Enriched 2-10 even, 13-17, 19, MR, CT All

Lesson Focus

Materials

Graph paper, ruler, math journal

Vocabulary/Symbols

system of linear inequalities

MOTIVATING THE LESSON

Discuss with students situations when linear inequalities can be used to solve a problem. (any situation that requires a range of numbers as the solution) Point out that the solution of an inequality can be represented on a graph.

Instruction

1 Mention that at times there are a finite number of solutions for an inequality. However, the examples shown here have an infinite number of solutions.

2 Since the inequality indicates that $2x - 3y$ is greater than 6, the ordered pair (5, 1), which falls below the line, is a solution. The ordered pair (0, -2) is not a solution, because it falls on the dotted line. Students can also check the solution by substituting the ordered pairs in the original inequality.
$2(5) - 3(1) > 6$ is true,
$2(0) - 3(-2) > 6$ is false.

OBJECTIVE: *To graph a linear inequality in two variables and to explore simple systems of inequalities.*

8-8 Graphing Linear Inequalities

▼ You can use the graph of the equation $x + y = 5$ to solve the inequality $x + y \leq 5$.

The line shows the solutions for $x + y = 5$.

The shaded region of the graph shows the solutions for $x + y < 5$.

The line and the shaded region show the solutions for $x + y \leq 5$.

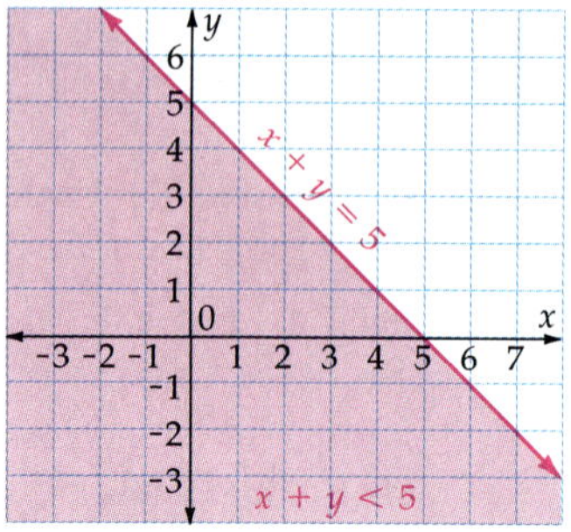

THINK Is it possible to list all the solutions for an inequality? Explain. **No; There are an infinite number of solutions to an inequality.** (1)

▼ Sometimes the line may include points that are not solutions to the inequality.

Example 1 **Graph the inequality $2x - 3y > 6$.**

Solution

1. Substitute = for >.

$$2x - 3y > 6$$
$$2x - 3y = 6$$

2. Solve the equation for y.

$$2x - 3y - 2x = 6 - 2x$$
$$-3y = 6 - 2x$$
$$\frac{-3y}{-3} = \frac{6 - 2x}{-3}$$
$$y = \frac{6}{-3} - \frac{2x}{-3}$$
$$y = -2 + \frac{2}{3}x$$
$$y = \frac{2}{3}x - 2$$

3. Graph the equation. The inequality does not include *is equal to*, so use a dotted line to show that the line itself is not part of the solution set.

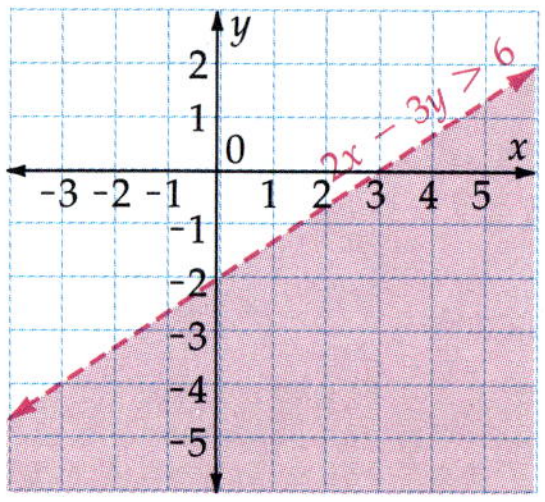

THINK Is the ordered pair (5,1) a solution of the inequality $2x - 3y > 6$? is (0,-2)? **yes; no** (2)

Check Choose any point above or below the line. Determine if the point is a solution. Try (0,0).

$$2x - 3y > 6$$
$$2(0) - 3(0) > 6$$
$$0 - 0 > 6$$
$$0 > 6 \text{ False}$$

Since (0,0) is not a solution, shade the area below the line. Any point that is shaded is a solution of the inequality.

Teaching Tip

Have an assortment of pencils of different colors available for students to use to shade the graphs of inequalities and systems of inequalities.

▼ When you graph linear inequalities in two variables, the line is part of the solution if the inequality symbol is $\leq$ or $\geq$. The line is not part of the solution if the inequality symbol is $<$ or $>$.

Example 2 **Write the equation of the line you would graph for each inequality. Tell whether the graph of the equation would be drawn as a solid or a dotted line.**

a. $y > 4x + 3$ **b.** $3x - 7y \leq -21$ **c.** $3x \geq 2y + 1$

Solution **a.** Graph the equation $y = 4x + 3$. The graph is a dotted line.

b. Graph the equation $3x - 7y = -21$. The graph is a solid line.

c. Graph the equation $3x = 2y + 1$. The graph is a solid line.

▼ You can show the solution of a *system of linear inequalities* by graphing both inequalities on the same set of coordinate axes.

System of Linear Inequalities	A system of linear inequalities is two or more linear inequalities using the same variables.

Example 3 **Solve the system of linear inequalities $y > x$ and $y \leq 2$.**

Solution **1.** Graph $y > x$. Shade the region above the dotted line.

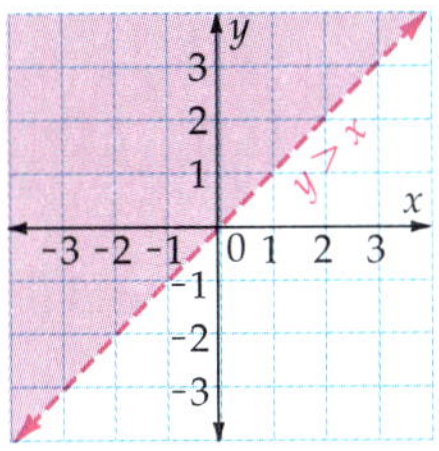

2. Graph $y \leq 2$ on the same set of axes. Shade below the solid line.

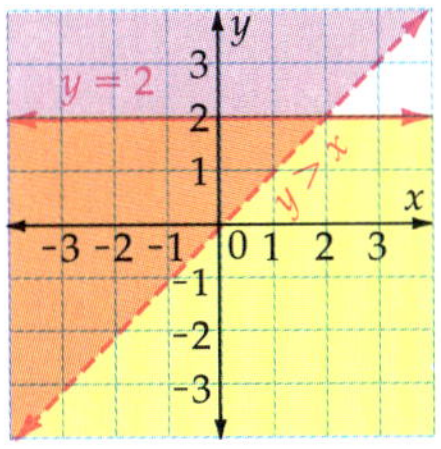

The graph of the solution is the part shaded in both colors.

1. $x + 2y = 5$; dotted 2. $3x - y = 1$; solid 3. $x = 6$; dotted 4. $y + 2x = -1$; solid 5. $x - 3y = 9$; solid 6. $y = 7$; dotted

CLASS EXERCISES

Write the equation of the line you would graph for each inequality. Tell whether the graph would be a solid or a dotted line.

1. $x + 2y > 5$ **2.** $3x - y \leq 1$ **3.** $x < 6$

4. $y + 2x \leq -1$ **5.** $x - 3y \geq 9$ **6.** $y > 7$

THINK AND DISCUSS

1. Is (-2,-1) a solution of the inequality $y > x$? yes

2. If (p,q) is a solution of $y > x$, what must be true about $\frac{p}{q}$? it is less than one

3. State whether (0,0) is a solution.

a. $4x - 11y \leq 15$ yes

b. $-x + y > -1.45$ yes

c. $x \geq -3$ yes

ADDITIONAL EXAMPLES

1. Write the equation of the line you would draw for each inequality. Tell whether the graph would be a solid or a dotted line.

a. $x - y \geq 1$ ($y = x - 1$ solid) **b.** $2x - 3y < 12$ ($y = \frac{2}{3}x - 4$ dotted)

2. Graph the inequality. $3x - y \geq 4$ (See below.)

3. Graph the solution of the system of linear inequalities. $y \geq 2x$ $x < 3$ (See below.)

2.

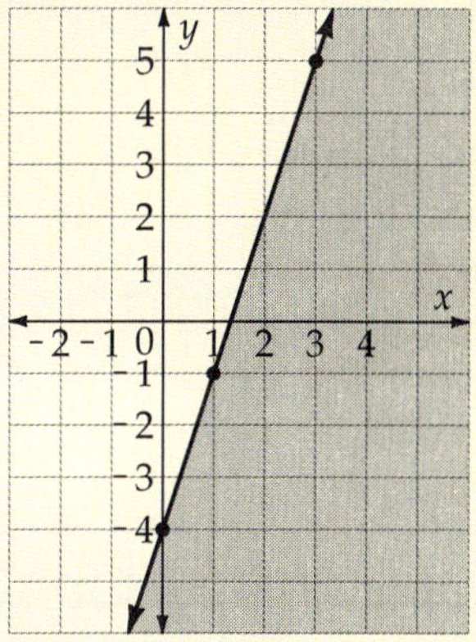

3.

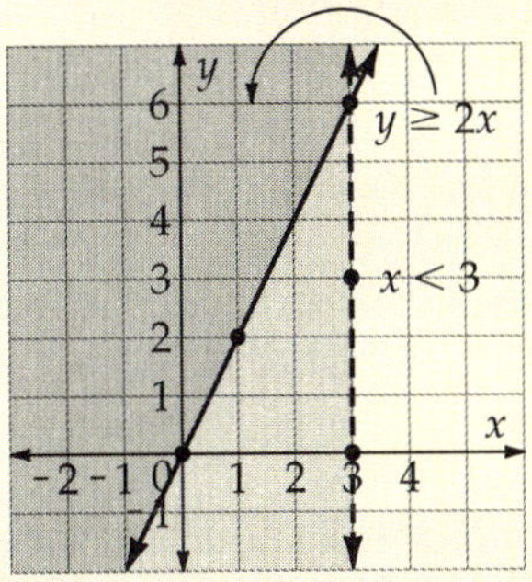

Guided Practice

Class Exercises Be sure students understand when the graph of an inequality is a solid line and when the graph is a dashed line.

Think and Discuss Have students answer questions 1-3 and then have them graph the inequalities to confirm their answers.

MATH MINUTES

List several systems of linear equations on the chalkboard. Have students work in pairs to solve and graph each system of linear equations. Then ask students to compare their graphs.

A ***common error*** occurs when students draw a solid line instead of a dotted line (or vice versa) in graphing an inequality. Remind students that a solid line is drawn when the inequality is $\leq$ or $\geq$ because the line includes possible solutions. A dotted line is drawn when the inequality is $<$ or $>$ because the line does not include the possible solutions.

Closure

Writing in Math Have students write in their math journals the procedure for graphing $3x - 2y > 12$. Have students compare their procedures and make any corrections.

Independent Practice

Remind students that the solutions of the inequalities in questions 39 and 48 is a set of whole numbers.

The solution for question 40 must be a set of positive numbers.

Ask volunteers to share what they wrote for question 49 with the class.

Point out, for question 50, that linear programming is not a form of computer programming, although a computer may be used as a problem solving tool. Mention that it is a problem solving research technique involving many variables, and the best solution, among many solutions, must be chosen.

MIXED REVIEW

Determine whether each system has one solution, no solution, or infinitely many solutions.

1. $4x - 2y = 6$
$2x - y = 3$ infinite

2. $4x - 2y = 6$
$2x - y = 1$ no solution

3. $4x - 2y = 6$
$2x + y = 1$ one

Write three ordered pairs that are solutions of each inequality. Answers may vary.

4. $4x - y \leq 6$

5. $5 - 2x - y > x$

Solve.

6. A student has \$120 to spend on clothes. Shirts cost \$15 and sweaters cost \$24.

a. What is the maximum number of shirts she can buy? 8

b. What is the maximum number of sweaters she can buy? 5

c. Assume she wants to buy at least one shirt and one sweater. How many of each can she buy?

6. c.

Shirts	Sweaters
1	4
1	3
1	2
1	1
2	3
2	2
2	1
3	3
3	2
3	1
4	2
4	1
5	1
6	1

Tell whether the region containing the origin (0,0) would be shaded in the graph of each inequality.

7. $x + 2y > 5$ no **8.** $3x - y \leq 1$ yes **9.** $x < 6$ yes

Graph each inequality. Check students' work.

10. $x + 2y > 5$ **11.** $3x - y \leq 1$ **12.** $x < 6$

Check students' work.

Write an inequality for each word sentence. Graph the inequality.

13. The sum of two numbers is greater than 3. $x + y > 3$

14. A number is greater than or equal to three times another. $x \geq 3y$

15. The y-coordinate of a point is less than twice the x-coordinate. $y < 2x$

Check students' work.

Write two inequalities. Graph the system of inequalities.

16. Find two numbers such that one is greater than four times the other and their sum is greater than 14. $x > 4y$ or $y > 4x$; $x + y > 14$

17. Find two numbers such that the sum is greater than 3 and the difference is less than 5. $x + y > 3$; $x - y < 5$ or $y - x < 5$

1. $2x + y = 3$; solid 2. $\frac{1}{2}x - y = 4$; dotted 3. $y = -2$; solid
4. $-3x = 6$; dotted 5. $x - 4y = 1$; dotted 6. $3x + 2y = 4$; dotted
7. $5x - 3y = 2$; dotted 8. $2x - y = 7$; solid 9. $x = 9$; solid

WRITTEN EXERCISES

Write the equation of the line you would graph for each inequality. Tell whether the graph would be a solid or a dotted line. See above

1. $2x + y \geq 3$ **2.** $\frac{1}{2}x - y < 4$ **3.** $y \geq -2$

4. $-3x < 6$ **5.** $x - 4y > 1$ **6.** $3x + 2y > 4$

7. $5x - 3y < 2$ **8.** $2x - y \geq 7$ **9.** $x \geq 9$

10. $x - 2y \leq 4$ $x - 2y = 4$; solid **11.** $3x + y < 2$ $3x + y = 2$; dotted **12.** $x + 4y \leq 5$ $x + 4y = 5$; solid

Tell whether the region containing the origin would be shaded in the graph of each inequality.

13. $2x + y \geq 6$ no **14.** $\frac{1}{2}x - 2y < 4$ yes **15.** $y \geq -4$ yes

16. $-3x < 18$ yes **17.** $x > y$ no **18.** $2x - 2y \leq 1$ yes

19. $y > -2x - 1$ yes **20.** $x + 2y > 4$ no **21.** $y > 3x + 1$ no

Graph each inequality. Check students' work.

22. $2x - y \geq -4$ **23.** $\frac{3}{4}x + 4y < -8$ **24.** $y \geq 2$

25. $-2x < 6$ **26.** $-1 \leq x - 3$ **27.** $10 - y \leq 1$

28. $y + 4 < 3x$ **29.** $y > x + 4$ **30.** $2x + 3y < 9$

31. $x + 6 < 3y$ **32.** $x - y \leq 4$ **33.** $2x - y \geq -2$

34. $x < 6$ **35.** $y > 0$ **36.** $3x > 4y$

QUIZ ANSWERS

2.

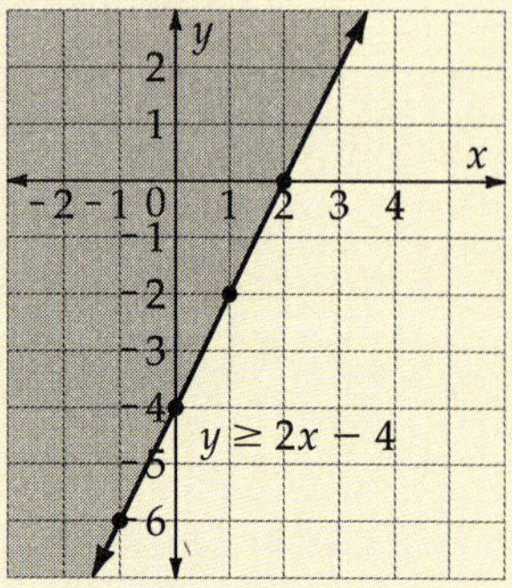

3.

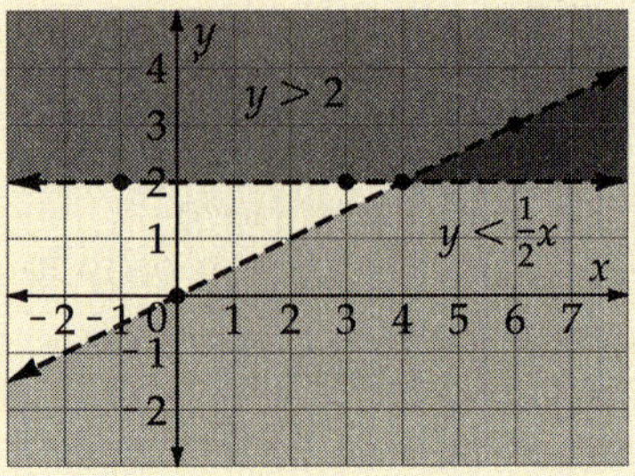

Write an inequality for each word sentence. Graph the inequality. Check students' work.

37. Find two numbers whose difference is greater than 3. $x - y > 3$ or $y - x > 3$

38. Find two numbers where one is at least three times the other. $x \geq 3y$ or $y \geq 3x$

39. Timothy has a collection of nickels and dimes valued at less than \$1. Let x equal the number of nickels and y equal the number of dimes. $0.05x + 0.10y < 1$

40. A gift wrapper cut a ribbon so that one piece was more than twice as long as the other. Let x equal the shorter piece and y equal the longer piece. $y > 2x$

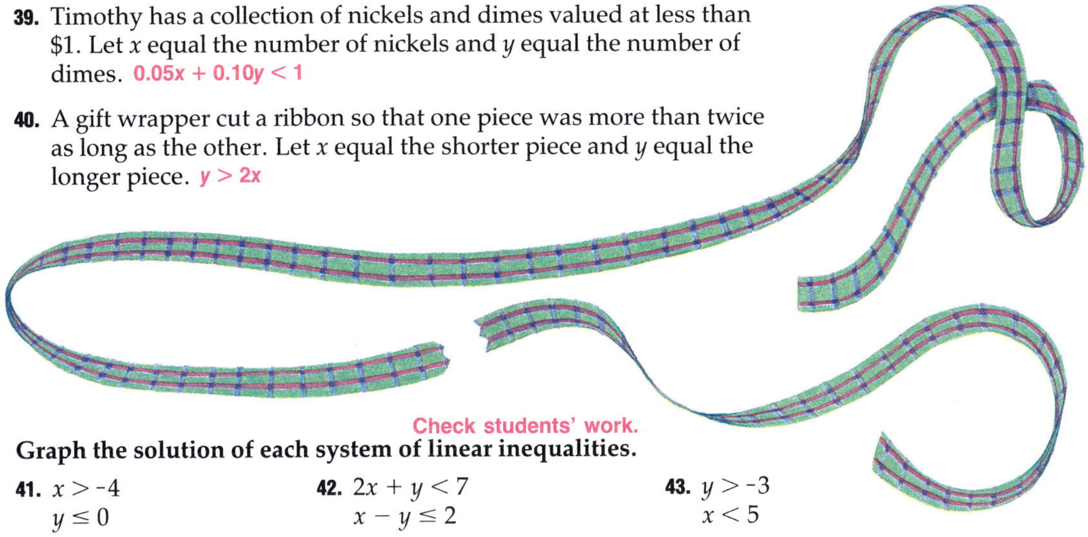

Graph the solution of each system of linear inequalities. Check students' work.

41. $x > -4$, $y \leq 0$

42. $2x + y < 7$, $x - y \leq 2$

43. $y > -3$, $x < 5$

Use the system of linear inequalities $2x - y \geq 5$ and $\frac{1}{3}x + y < 2$ to find each point described. Answers may vary.

44. a point which satisfies the first inequality but not the second

45. a point which satisfies the second inequality but not the first

Write two inequalities. Graph the system of inequalities. Check students' work.

46. Find two numbers where one is more than three times the other and their sum is greater than 12. $x > 3y$ or $y > 3x$; $x + y > 12$

47. Find two numbers such that their sum is not more than 10 and the larger is greater than twice the smaller. Let y equal the larger number. $x + y \leq 10$; $y > 2x$

48. A student has a collection of more than 18 dimes and quarters. It is valued at more than \$2.80. Let x equal the number of dimes and y equal the number of quarters. $x + y > 18$; $0.10x + 0.25y > 2.80$

49. ***WRITE*** two different situations that you could represent by the inequality $y \geq x + 3$. Check students' work.

50. ***PROJECT*** *Linear programming* is an interesting topic related to inequalities that has numerous real world applications. Research this topic and give some examples of situations where linear programming might be useful. Check students' work.

Lesson Focus

Reteaching Activity Provide students with the steps to solve and graph an inequality. Have students solve the inequality $4x + 2y > 10$, showing their work next to each step you have indicated. ($y > -2x + 5$)

LESSON RESOURCES

TRB Practice H-15
TRB Enrichment H-16

Practice

For use after 8-8 (pp. 348–351)

Graphing Linear Inequalities

Write the equation of the line you would draw for each inequality. Tell whether the graph of the equation would be drawn as a solid or a dotted line. Write *solid* or *dotted*.

1. $4x - 5y \geq 3$ **$4x - 5y = 3$; solid**
2. $y < \frac{2}{3}x + 2$ **$y = \frac{2}{3}x + 2$; dotted**
3. $2x > 3 - 2y$ **$2x = 3 - 2y$; dotted**
4. $y \leq -5$ **$y = -5$; solid**
5. $-x + 2y > -3$ **$-x + 2y = -3$; dotted**
6. $8 \leq 5y - 3x$ **$8 = 5y - 3x$; solid**

Tell whether the region containing the origin would be shaded in the graph of each inequality. Write yes or no.

7. $4x - 5y \geq 3$ **no**
8. $y < \frac{2}{3}x + 2$ **yes**
9. $2x > 3 - 2y$ **no**
10. $y \leq -5$ **no**
11. $-x + 2y > -3$ **yes**
12. $8 \leq 5y - 3x$ **no**

Graph each inequality.

13. $y < x$

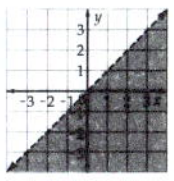

14. $x + y < 2$

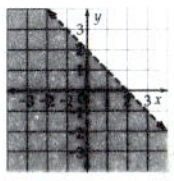

15. $y \leq x - 1$

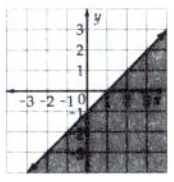

16. $x > -2$

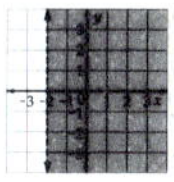

17. $-3x \geq 6 - 2y$

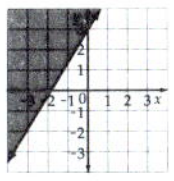

18. $x + 2y \geq 4$

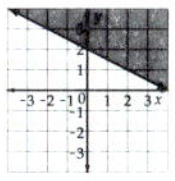

Chapter 8 H-15

Enrichment

For use after 8-8 (pp. 348–351)

Inequalities

London's Great Fire of 1666 changed a professor of mathematics and physics into a world-renowned architect. This learned man oversaw the rebuilding of much of London, including St. Paul's Cathedral. The name of this mathematician/architect is **Sir Christopher Wren**."

To find the answer, decide which of the two inequalities has the given ordered pair as a solution. Write the letter of the inequality above the number of the exercise in the answer grid at the bottom of the page.

1. (1, 6)	R: $y < 2x + 5$	A: $y > 2x + 5$
2. (-3, -11)	E: $y \geq 3x - 2$	N: $y > 3x - 2$
3. (-10, 0)	C: $y \leq \frac{1}{2}x + 5$	I: $y < \frac{1}{2}x + 5$
4. (3, -20)	W: $y > 8 - 12x$	O: $y < 8 - 12x$
5. (-1, 0)	M: $5 + 4x < y$	S: $5 + 4x > y$
6. (-8, -7)	I: $x \leq y$	G: $x \geq y$
7. (6, -5)	O: $x > -3$	C: $x < -3$
8. (-2, 2)	S: $y - 4 > x$	T: $y - 4 \geq x$
9. (-3, 3)	R: $x + y > y$	H: $x + y > x$
10. $\left(-\frac{4}{5}, \frac{1}{5}\right)$	H: $-x > 0$	P: $-x < 0$
11. (2, 1.2)	E: $x < 3y - 2$	R: $x > 3y - 2$
12. (4.5, 3.4)	R: $3x < 4y$	T: $3x > 4y$
13. (-5, 3)	M: $2x + 3y > 0$	I: $2x + 3y < 0$
14. (2.7, 4.8)	S: $-3x + 5y < 16$	R: $-3x + 5y > 16$
15. (0.8, 0.2)	W: $8 - 8x > 9y$	E: $8 - 8x < 9y$
16. (-3.9, -6.5)	S: $7x - 4y < -2$	R: $7x - 4y > -2$
17. (5, 8)	P: $x \leq 5$	A: $x > 5$
18. (-3, -10)	N: $y \leq x + 7$	P: $y \geq x - 5$

S	I	R		C	H	R	I	S	T	O	P	H	E	R		W	R	E	N
14	6	12		3	9	1	13	5	8	7	17	10	2	16		4	11	15	18

H-16 Chapter 8

LESSON QUIZ

1. Write the equation of the line you would draw for each inequality:

a. $3x - y \leq 6$ ($y = 3x - 6$)

b. $4x - 2y < 12$ ($y = 2x - 6$)

2. Graph the inequality $2x - y \leq 4$ (See p. 350.)

3. Graph the solution of the system of inequalities: $y < \frac{1}{2}x$ and $y > 2$ (See p. 350.)

Assignment Guide

Basic 1-21 odd, 22-27, 38-39, 43-46, 50, MR All

Average 2-24 even, 36-42, 44-48, 50, MR All

Enriched 11-14, 28-35, 40-46 even, 48-50, MR All

FOR THE NEXT LESSON

math journal

Career Information

Arrange for a civil engineer to visit the class. Civil engineers work for a variety of firms, including major construction companies, building or highway design companies, the Army Corp of Engineers, Environmental Protection Agency (EPA), or a local environmental agency. Ask students to prepare questions about how the speaker uses mathematics.

Ask the speaker to discuss the following topics:

- education and training required, especially courses in mathematics
- job responsibilities and how the job differs in various industries
- computer hardware and software used in designing, scheduling, and fulfilling other responsibilities
- licensing requirements for the professional engineer designation
- career options and future job possibilities for civil engineers

CAREER

Help Wanted: Civil Engineer

Bachelor's degree in engineering required. Some jobs available to college graduates with math and science degrees.

For more information, write to American Society of Civil Engineers, 345 E. 47th St. New York, NY 10017.

Civil engineers plan, construct, and maintain structures such as airports and bridges. Engineers must consider the slope of a river embankment or road to insure proper drainage after a heavy rainfall. Slope is also a major consideration when designing roads over mountainous terrain.

PROJECT

Locate a contour map for a nearby area. Write a paragraph describing the terrain of the area.

Practice

In which quadrant or on what axis does each point fall?

1. (4,5) I **2.** (-3,0) x-axis **3.** (11,-7) IV **4.** (-6,-2) III

State the coordinates of each point described. Begin at the origin.

5. the point which is 4 units to the left of the y-axis and down 3 units from the x-axis (-4,-3)

6. the point which is 3 units to the right of the y-axis and up 7 units from the x-axis. (3,7)

Solve for y in terms of x. Find three solutions of each equation. Answers may vary.

7. $3x + y = -10$ $y = -3x - 10$ **8.** $2x + 4y = 8$ $y = -\frac{1}{2}x + 2$ **9.** $4x + 2y - 8 = 0$ $y = -2x + 4$

Find the x-intercept and y-intercept for each equation.

10. $3x - 4y = 12$ 4,-3 **11.** $y = 6x$ 0,0 **12.** $\frac{4}{5}x - y = 20$ 25,-20

Graph each equation on a separate set of coordinate axes. Check students' work.

13. $x = 8$ **14.** $-4x + y = 16$ **15.** $2y + 6 = -4x$

16. $y = -\frac{1}{2}$ **17.** $2y = -6$ **18.** $|y| = 5$

Find the slope of the line containing the given points.

19. (1,-1)(-1,1) -1 **20.** (2,-5)(2,4) no slope **21.** (2,-9)(12,3) $\frac{6}{5}$

Graph each line described. Check students' work.

22. the line having slope -5 and containing the point (-3,2) $y = -5x - 13$

23. the line having no slope and containing the point (6,-1) $x = 6$

Write each linear equation in slope-intercept form. Name the slope and y-intercept. Sketch the graph of each equation. Check students' work.

24. $6x + y = 24$ $y = -6x + 24$; -6; 24 **25.** $x - 4y = 12$ $y = \frac{1}{4}x - 3$; $\frac{1}{4}$; -3 **26.** $-x - y = -1$ $y = -x + 1$; -1; 1

Solve each system by graphing. Check your solutions. Check students' work.

27. $y = x + 2$, $3x + 4y = 22$ (2,4) **28.** $x = y - 8$, $x - y = 2$ no solution **29.** $2x + 3y = 8$, $y = 2x$ (1,2)

Solve each inequality for y in terms of x. Write three ordered pairs that are solutions of each. Answers may vary.

30. $7x - 2y < 20$ $y > \frac{7}{2}x - 10$ **31.** $4x + 5y > -6$ $y > -\frac{4}{5}x - \frac{6}{5}$ **32.** $12x + 2y - 8 > 0$ $y > -6x + 4$

Graph each inequality. Check students' work.

33. $y > -x + 3$ **34.** $x - y \leq -6$ **35.** $2x + y \geq 1$

PROBLEM SOLVING APPLICATION

8-9 Direct and Indirect Variation

OBJECTIVE:
To solve problems involving direct and indirect variation.

FLASHBACK
Volume is the amount of space an object occupies.

- The temperature and volume of a gas vary *directly*.

Direct Variation	Direct variation means that as one factor increases the other factor also increases. We represent direct variation by an equation in the form $y = kx$, where k is not zero. k is the constant of variation.

- *Charles's law* states the relationship between the temperature in degrees Kelvin and volume of a gas.

Charles's Law	The volume of a fixed amount of gas varies directly with the temperature of the gas.

NOTES & QUOTES
Jacques Charles (1746–1823) was a French scientist. He discovered the relationship between the temperature of a gas and its volume in the late 1700s.

Example 1 A gas has a volume of 250 mL at 300°K. The temperature of the gas decreases to 240°K. What is its volume?

Solution Let x = temperature
Let y = volume

$y = kx$ Find k, the constant of variation.
$250 = k(300)$
$\frac{250}{300} = k$
$\frac{5}{6} = k$

$y = \frac{5}{6}x$ Rewrite the formula.
$= \frac{5}{6}(240)$ Substitute 240 for x.
$= 200$

The volume of the gas is 200 mL.

1

- The volume and pressure of a gas vary *indirectly*.

Indirect Variation	In indirect variation, one factor increases as the other factor decreases. The equation $xy = k$ represents an indirect variation. k is the constant of variation.

- *Boyle's law* states the relationship between the volume and pressure of a gas.

Boyle's Law	The volume of a fixed amount of gas varies indirectly with the pressure of the gas.

NOTES & QUOTES
Robert Boyle (1627–1691) was an Irish scientist. In 1662, he reported the relationship between the pressure and volume of a gas. This is now known as Boyle's law.

Teaching Tip

It may be helpful to have a science teacher (or high school physics teacher) prepare a simple demonstration of Charles's and Boyle's laws. If this is not possible, prepare a poster with simple drawings to illustrate key points in the lesson.

Lesson Focus

Materials
Math journal

Vocabulary/Symbols
Boyle's Law
Charles's Law
constant of variation
direct variation
indirect variation

MOTIVATING THE LESSON

Explain the meaning of *gas* and then give real-world examples of direct and indirect variation. (*direct:* inflating a balloon, the volume increases as the pressure inside the balloon increases; *indirect*: the faster a car travels, the less time it takes to get to a destination.)

Instruction

A coordinate plane can be used to illustrate Examples 1 and 2.

For Example 1, graph temperature along the x-axis and volume along the y-axis. After students find the constant of variation k, have them find the volume for 293°K. Then have them find the volume for two more temperatures such as 283°K and 343°K. After plotting the points on the graph, ask students to predict the volumes at other temperatures.

Repeat the procedure for Example 2. Have students note the differences on the graph.

1 Another method for solving the problem in Example 1 is to use a proportion. To help students understand, have them consider a balloon at temperature T_1 and volume V_1. If the temperature is changed to T_2, then according to Charles's law, the volume is V_2. So,

$$\frac{T_1}{T_2} = \frac{V_1}{V_2}$$

Then for Example 1,

$$\frac{300}{240} = \frac{250}{V_2} \quad (V_2 = 200)$$

2 A proportion may be used for Boyle's law also. Use the balloon example again. Have the students consider a balloon with pressure P_1 and volume V_1. If the pressure is changed to P_2, then according to Boyle's law,

$$\frac{P_1}{P_2} = \frac{V_2}{V_1}$$

To solve the problem in Example 2 write,

$$\frac{5}{10} = \frac{V_2}{60}$$
$$V_2 = 30$$

Guided Practice

Class Exercises Ask students to work in small groups. Have them guess, by looking at the patterns in the tables, if the data varies directly or indirectly. Then ask students to write equations to check their guesses.

Closure

Writing in Math Have students describe real world examples of Charles's law and Boyle's law in their math journals. Also ask students to write each law in their own words.

Independent Practice

For question 8, mention that *atmosphere* is a unit of measure: 1 atmosphere is equal to the pressure of air at sea level; therefore at 20 atmospheres the pressure is 20 times greater than the pressure at sea level.

Example 2 The volume of a gas is 60 ft^3 under 5 lb of pressure. What is the gas's volume under 10 lb of pressure?

Solution **2**

Let x = pressure
Let y = volume

$xy = k$ — Find k, the constant of variation.

$5(60) = k$

$300 = k$

$xy = 300$ — Rewrite the formula.

$10y = 300$ — Substitute 10 for y.

$\frac{10y}{10} = \frac{300}{10}$

$y = 30$

The volume of the gas is 30 ft^3.

CLASS EXERCISES

State whether the data varies directly or indirectly. Write an equation to describe each variation. State the constant of variation.

1.

x	y
2	4
3	6
4	8

direct; $y = 2x$; 2

2.

x	y
6	3
2	9
1	18

indirect; $xy = 18$; 18

3.

x	y
40	120
60	180
80	240

direct; $y = 3x$; 3

Decision Making **DECISION MAKING** Decision Making Decision Making Decision Mak

DIRECT AND INDIRECT VARIATION

A product is packaged under pressure by inserting a gas into the container.

COLLECT DATA

1. Find five products that are packaged under pressure. Check students' work.
2. Read each warning label. Write a paragraph telling how the labels are similar to each other. Check students' work.
3. What does packaging a product under pressure allow the contents to do? to spray

ANALYZE DATA

Support each answer using either Charles's law or Boyle's law.

4. What do you think will happen to a container that is stored above the recommended temperature? By Charles's law, the volume will increase and possibly cause the container to explode.
5. When pressure is applied to the container, what will happen to the contents of the container? By applying pressure, the contents are able to be released. Therefore, by Boyle's law, the volume decreases.

MATH MINUTES

Have students think of a way to diagram the relationships in each law. One possibility is shown.

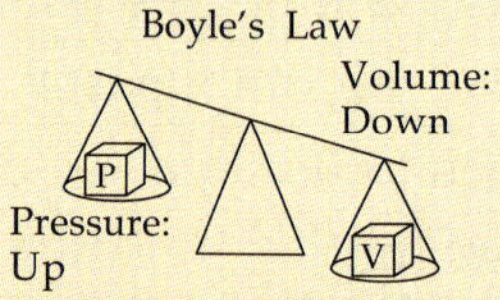

WRITTEN EXERCISES

Use a CALCULATOR where appropriate.

State whether each equation is a direct or indirect variation. For each variation, state the constant of variation.

1. $y = 30x$ direct; 30
2. $xy = 58$ indirect; 58
3. $8.5x = y$ direct; 8.5
4. $x = \frac{10}{y}$ indirect; 10

Assume that y varies directly as x.

5. An object weighs 6 times more on Earth than it does on the moon. Ian weighs 165 lb on Earth. What would he weigh on the moon? 27.5 lb
6. A gas has a volume of 150 mL at 320°K. The gas's temperature is increased to 360°K. What is its volume to the nearest tenth? 168.8 mL

Assume that y varies indirectly as x.

7. A piano string 40 in. long vibrates at a frequency of 520 cycles/s. Find the frequency of the string if it were shortened to 18 in. Round to the nearest whole unit. 1,156 cycles/s
8. Pressure acting on 12 m^3 of a gas is 20 atmospheres. The pressure is reduced until the volume is 15 m^3. What is the new pressure acting on the gas? 16 atmospheres
9. Amy drove for 4 h at a rate of 40 mi/h. To the nearest hour, how long would it have taken Amy if she drove at 55 mi/h? 3 h

Decision Making ■ Decision Making ■ Decision Making ■ Decision Making ■ Decision Making ■ Decision Making ■

6. Why do container labels suggest that you use pressurized products in well-ventilated areas? A well-ventilated area has more volume into which the gas can escape.
7. Why do airlines recommend that you not bring contents that are under pressure on board a plane? See below.
8. Do you think that there is a minimum temperature that pressurized products must be stored under? Explain. No.; As the temperature decreases the volume decreases.

■ MAKE DECISIONS

■ Some products that are packaged under pressure contain gases that are harmful to the environment.

9. What environmental concerns arise from the use of chlorofluorocarbons in a container stored under pressure? They may damage the ozone layer.
10. What are alternative packaging methods that manufacturers could use? Check students' work.
11. What can we do as individuals to promote these alternatives? Check students' work.

7. As a plane increases its altitude, the pressure decreases. Therefore, by Boyle's law, the volume of the gas increases and may cause the container to explode.

Decision Making Have students work in small groups. Explain that they can use what they know about Charles's law and Boyle's law to answer the questions.

Lesson Follow-up

LESSON RESOURCES

TRB Alternate Application H-17

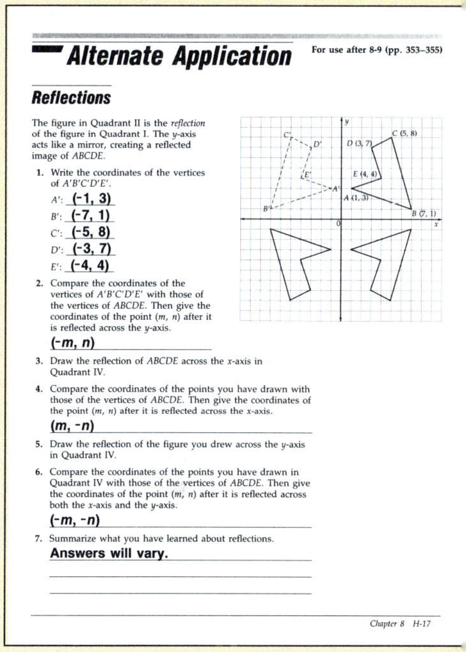

Alternate Application For use after 8-9 (pp. 353–355)

Reflections

The figure in Quadrant II is the *reflection* of the figure in Quadrant I. The y-axis acts like a mirror, creating a reflected image of $ABCDE$.

1. Write the coordinates of the vertices of $A'B'C'D'E'$.
   A': (-1, 3)
   B': (-7, 1)
   C': (-5, 8)
   D': (-3, 7)
   E': (-4, 4)
2. Compare the coordinates of the vertices of $A'B'C'D'E'$ with those of the vertices of $ABCDE$. Then give the coordinates of the point (m, n) after it is reflected across the y-axis.
   (-m, n)
3. Draw the reflection of $ABCDE$ across the x-axis in Quadrant IV.
4. Compare the coordinates of the points you have drawn with those of the vertices of $ABCDE$. Then give the coordinates of the point (m, n) after it is reflected across the x-axis.
   (m, -n)
5. Draw the reflection of the figure you drew across the y-axis in Quadrant IV.
6. Compare the coordinates of the points you have drawn in Quadrant IV with those of the vertices of $ABCDE$. Then give the coordinates of the point (m, n) after it is reflected across both the x-axis and the y-axis.
   (-m, -n)
7. Summarize what you have learned about reflections.
   Answers will vary.

Chapter 8 H-17

Assignment Guide

Basic All
Average All
Enriched All

FOR THE NEXT LESSON

paper, pencil, dot or graph paper, magazines or newspapers, math journal

Study Skills

Memorization is one of the ways to transfer information into memory. It is a time-honored method, but one that is often misapplied by students who believe that it is the only way to learn (and the only way to study for a test), instead of seeing it as a useful tool for absorbing only certain kinds of information. Such students create unnecessary anxiety for themselves as they seek to memorize, for example, the entire contents of a chapter. Memorization in math is generally most useful for information such as formulas, definitions, correct spelling, and other rules. Memorization is not particularly helpful for building deeper cognitive understanding of material.

There are a number of ways to help students improve their ability to memorize. The most basic of these involve attentive listening in class, taking notes, and involvement in class discussions.

Here are some other memorization techniques:

- Do memory work in short intervals of no more than 15 minutes each.
- Write down or re-read what has to be memorized.
- Record what needs to be memorized on note cards or an audio cassette. Read the cards or listen to the tape often.

Chapter 8 Review

Write *true* or *false*. If false, change the underlined word(s) to make a true statement.

1. A coordinate pair is a pair of numbers (x,y) assigned to a point on a coordinate plane. **false; an ordered pair**
2. A linear equation is an equation for which the graph is a line. **true**
3. The x-intercept is the x-coordinate of the point at which the graph of a linear equation intersects the y-axis. **false; x-axis**
4. A linear equation in the form $y = mx + b$ is in slope-intercept form. The slope is b, and the y-intercept is m. **false; *m*; *b***
5. When you graph a linear inequality in two variables, the line is part of the solution if the inequality symbol is $<$ or $>$. **false; is not part of**

The Coordinate Plane — 8-1

To locate $P(x,y)$ on the coordinate plane,

1. Begin at the origin.
2. Move x units along the x-axis.
3. Move $|y|$ units up or down.

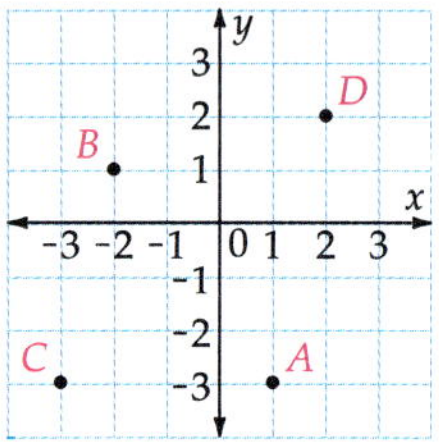

State the coordinates of each point.

6. A **(1,-3)**
7. B **(-2,1)**
8. C **(-3,-3)**
9. D **(2,2)**

Solving Equations — 8-2

To solve an equation in two variables, choose a value for x. Then substitute the x value into the equation to find the y value. Write the solution as an ordered pair.

Write the solution for the given value of x.

10. $2x + 5y = 12; x = 1$ **(1,2)**
11. $3x = -\frac{1}{2}y + 5; x = 2$ **(2,-2)**
12. $3(x - 1) = 2y; x = 3$ **(3,3)**

Graphing Linear Equations — 8-3

To graph a linear equation,

1. Solve for y in terms of x.
2. Find three solutions to the equation.
3. Plot the points and draw a straight line.

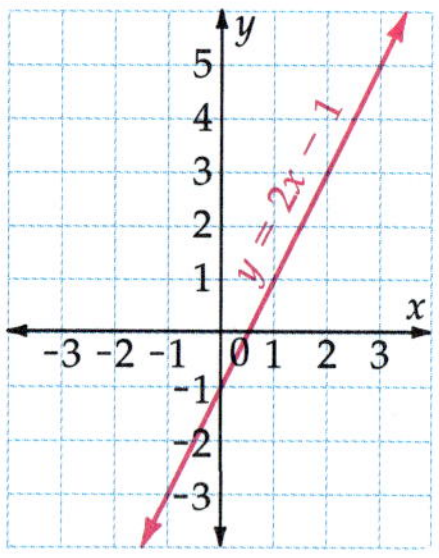

Determine whether each point is on the graph.

13. (1,1) **yes**
14. (2,2) **no**
15. (-1,-3) **yes**
16. (3,5) **yes**

Slope and y-intercept 8-4

To find the slope of a line when you know the coordinates of any two points on the line, use the slope formula.

$$\text{slope} = \frac{\text{difference in } y\text{-coordinates}}{\text{difference in } x\text{-coordinates}}$$

An equation in the form $y = mx + b$ is in slope-intercept form. The slope is m and the y-intercept is b.

Find the slope of the line containing the given points.

17. (1,3)(2,5) 2 **18.** (-1,0)(1,3) $\frac{3}{2}$ **19.** (2,-2)(3,-3) -1 **20.** (-6,7)(4,7) 0

Write each linear equation in slope-intercept form. Name the slope and y-intercept.

21. $y - 2x = 3$
$y = 2x + 3$; 2; 3

22. $\frac{1}{2}x + y = -2$
$y = -\frac{1}{2}x - 2$; $-\frac{1}{2}$; -2

23. $2x + 2y = 10$
$y = -x + 5$; -1; 5

24. $5 - 2y = 10x$
$y = -5x + \frac{5}{2}$; -5; $\frac{5}{2}$

Problem Solving 8-5

To solve a word problem using a graph, write the given information as ordered pairs, plot the points, and draw a line. Find the missing information by reading the graph.

25. A car uses 10 L of gasoline to travel 90 km. How much gasoline will the car use to travel 198 km? 22 L

Solving Systems of Linear Equations 8-6

To solve a system of linear equations, graph both equations on the same coordinate plane. A solution is any ordered pair that satisfies all equations in the system.

Tell whether the ordered pair is a solution of the system.

26. $y = 2x - 1$
$3y + 2x = 13$
(2,3) yes

27. $3x - 2y = 10$
$x + y = 0$
(-2,-8) no

28. $\frac{1}{3}x + 2y = 1$
$\frac{3}{4}x + \frac{1}{4}y = -2$
(-3,1) yes

Solving and Graphing Linear Inequalities 8-7, 8-8

A solution of a linear inequality is an ordered pair that makes the inequality true.

To graph a linear inequality, graph the related equation. Determine if the line should be solid or dotted and whether the solutions are above or below the line. Shade in the appropriate region.

Graph each inequality. Give three ordered pair solutions. Check students' work.

29. $y \geq x + 2$ **30.** $y < x - 3$ **31.** $y > x + 3$ **32.** $y < x + 3$

RESOURCES

TRB H-19–H-22

Scoring Chart	
No. Correct	**Score**
23	100
22	96
21	91
20	87
19	83
18	78
17	74
16	70
15	65
14	61
13	57
12	52
11	48
10	43
9	39
8	35
7	30
6	26
5	22
4	17
3	13
2	8
1	4

Chapter 8 Test

In which quadrant or on which axis does each point fall?

1. (-5,7) II

2. (0,-4) y-axis

3. (-8,-6) III

Write an equation using two variables. Find three solutions. Accept reasonable answers.

4. Mrs. Jones bought eight cans of juice. Some were orange juice and some were apple juice. Let x = the number of cans of orange juice and let y = the number of cans of apple juice. $x + y = 8$

5. A collection of nickels and dimes is valued at $2.50. Let x = the number of nickels and y = the number of dimes. $0.05x + 0.10y = 2.50$

Find the slope of the line containing the given points.

6. (5,1)(8,-2) -1

7. (6,3)(-2,4) $-\frac{1}{8}$

8. (-4,3)(6,-5) $-\frac{4}{5}$

Write each linear equation in slope-intercept form. Name the slope and y-intercept.

9. $x + \frac{1}{2}y = 4$
$y = -2x + 8$; -2; 8

10. $6x - 3y = 6$
$y = 2x - 2$; 2; -2

11. $3x = 4y + 1$
$y = \frac{3}{4}x - \frac{1}{4}$; $\frac{3}{4}$; $-\frac{1}{4}$

Graph each equation on a separate set of coordinate axes. Check students' work.

12. $3x + y = 4$

13. $2x - y = 1$

14. $y = 3x + 1$

Write a system of linear equations. Find the solution.

15. A piece of ribbon 30 in. long is cut into 2 pieces. One piece is 5 times as long as the other. How long is each piece of ribbon? $x + y = 30$; $x = 5y$; 5 in. and 25 in.

16. The sum of two numbers is 35. When the greater number is decreased by 3 times the lesser number, the result is 15. $x + y = 35$; $x - 3y = 15$; 5 and 30

Solve each system by graphing. Check your solutions. Check students' work.

17. $x + y = 3$, $x - y = 1$ (2,1)

18. $x + 3y = 6$, $x + 3y = 9$ no solution

19. $y = 2x + 3$, $3y - 6x = 9$ infinite solutions

Write an inequality. Find three solutions. Accept reasonable answers.

20. Find two numbers whose sum is less than 5. $x + y < 5$

Graph each inequality or system. Check students' work.

21. $x + y > 1$

22. $x - y < 2$

23. $y + 2 < x$, $y - 3x > 2$

Chapters 1–8 Cumulative Review

Choose the correct answer. Write A, B, C, or D.

1. In which quadrant does $\left(-1\frac{1}{2},2\right)$ fall? B

A. I **B.** II
C. III **D.** not given

2. Find a solution of $2x + y = 3$. C

A. (-1,2) **B.** (-1,1)
C. (2,-1) **D.** not given

3. State the inequality of which (3,1) is a solution. A

A. $2x - y \geq 5$ **B.** $2x - y < 5$
C. $2x - y > 5$ **D.** not given

4. Find the missing digit so that the resulting number is divisible by 9. C

5,76□,239

A. 1 **B.** 9
C. 4 **D.** not given

5. Evaluate $\frac{3}{7} \div 1\frac{2}{5}$. D

A. $\frac{3}{5}$ **B.** $3\frac{4}{5}$
C. $\frac{5}{7}$ **D.** not given

6. Compare $\frac{5}{8}$ □ $\frac{3}{5}$. A

A. $>$ **B.** $<$
C. $=$ **D.** not given

7. 37.5% of a is 36. What is a? A

A. 96 **B.** 48
C. 64 **D.** not given

8. Write the inequality. B

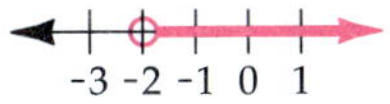

A. $x < -2$ **B.** $x > -2$
C. $x \geq -2$ **D.** not given

9. Solve $-2(x - 1) \leq -6$. D

A. $x \leq 3$ **B.** $x \leq -3$
C. $x \geq 3$ **D.** not given

10. Write 132% as a decimal. C

A. 0.132 **B.** 13.2
C. 1.32 **D.** not given

11. Find the slope of the line containing the points (3,2) and (1,-2). B

A. -2 **B.** 2
C. 1 **D.** not given

12. Write an inequality to describe the situation. A movie costs \$5. A drink costs \$1. Aaron did not spend more than \$7. B

A. $5x + 1y = 7$ **B.** $5x + 1y \leq 7$
C. $5x + 1y < 7$ **D.** not given

13. Write 0.28 as a fraction in lowest terms. B

A. $\frac{28}{100}$ **B.** $\frac{7}{25}$
C. $\frac{2}{25}$ **D.** not given

14. Write 9.05×10^8 in standard notation. A

A. 905,000,000 **B.** 90,500,000
C. 90,500,000,000 **D.** not given

15. Simplify $\frac{8a^2b^{-3}c}{24a^3bc^{-2}}$. A

A. $\frac{c^3}{3ab^4}$ **B.** $3a^5b^{-2}c$
C. $\frac{b^{-2}}{3ac}$ **D.** not given

16. Find the y-intercept of the equation $x + 2y = -4$. B

A. -4 **B.** -2
C. $-\frac{1}{2}$ **D.** not given

Study Skills

Test-taking is often a time so charged with anxiety that a student will inadvertently give an incorrect answer, even though he or she knows how to approach solving a particular problem. In the rush to answer questions quickly, students might not be thinking about *accuracy* and might make mistakes such as writing down incorrect numbers, failing to organize numbers in the appropriate columns, or providing an answer in incorrect units.

These common errors can be avoided if students understand that by not paying attention and hurrying to find solutions, they run the risk that the test will not reflect their true level of competence. Students should be encouraged to spend enough time with a problem to check that the computation and answer have been recorded accurately.

Some techniques for ensuring accuracy include:

- labeling units as the problem is worked as well as in the answer
- drawing and labeling diagrams
- making sure that digits are recorded in correct columns before computation. Hint: Have students turn lined scrap paper on its side and use the columns to align the digits.

Chapter At a Glance

Chapter 9 introduces measurement of geometric figures. Teaching emphasis is on discovering and exploring the properties of geometric figures.

Reading the Chapter

Students will be introduced to a variety of geometric terms. Review their derivation. Explain how understanding the words helps students visualize geometric figures. For example, *polygon* comes from the Greek words *poly*, meaning "many," and *gon*, meaning "angled." So a polygon is a *many-angled figure*.

Chapter Resources

Practice and Enrichment TRB I-1 to I-18
Alternate Application TRB I-19
Chapter Test Form A, I-21
Chapter Test Form B, I-23
Computer Test Bank

References

STUDENT BIBLIOGRAPHY

Baldwin, Margaret and Gary Pack. *Computer Graphics* (Franklin Watts, New York 1984).

TEACHER BIBLIOGRAPHY

Greenes, Carole, John Gregory, and Dale Seymour. *Successful Problem Solving Techniques* (Dale Seymour Publications, Palo Alto, Calif. 1977).

SOFTWARE

LOGO (Terrapin, Inc.), for Apple II and Macintosh computers.

FILMS, VIDEOTAPES, AND OTHER MEDIA

The Geometric Supposer (Sunburst Communications), computer disks and lab pack for Apple II and IBM computers.

Vocabulary

acute angle
acute triangle
adjacent angles
angle
central angle
chord
circle
circumference
complementary angles
concave polygons
congruent
congruent polygons
convex polygons
corresponding
diagonal
diameter
equiangular triangle
equilateral triangle
isosceles triangle
line
line symmetry
obtuse angle
obtuse triangle
parallel lines
parallelogram
parallel planes
perimeter
pi (π)
plane
point
polygon
pure tessellation
ray
radius (radii)
rectangle
regular polygon
rhombus
right angle
right triangle
scalene triangle
segment
semi-regular tessellation
side
similar
skew lines
square
straight angle
supplementary angles
the measure of (*m*)
trapezoid
vertex
vertical angles

Materials

geoboard (M*)
dot paper (T*)
colored pencils
math journal
graph paper (T)
LOGO software
computer
string
measuring stick or tape
large grid paper (T)
coin or marker

M*—Manipulative Kit
T*—Teacher's Resource Book

Resource Bank

Class Activities

FIELD TRIPS

▼ Public building—to locate and identify geometric figures used in the design and decor of the building.
▼ Supermarket—to identify and categorize the different types of packaging that are used.

CLASS SPEAKERS

▼ Package designer—to discuss how packages are designed and manufactured.
▼ Optician—to discuss how lenses are measured and made.

Bulletin Board

Have students draw the triangular design using rulers and protractors. Display the designs.

Draw $\triangle PQR$. $\overline{PQ}$ and $\overline{QR}$ are 4 in. long. $m\angle PQR = 15°$. Then draw the reflection of $\triangle PQR$ around $\overline{QR}$ to form $\triangle RQS$. What is the measure of $\angle RQS$? (15°)

Now draw $\triangle MQR$ over $\triangle PQR$. $\overline{MQ}$ and $\overline{QR}$ are 4 in., $m\angle MQR = 30°$. Repeat this process six times, increasing the measure of $\angle Q$ by 15° each time.

Project: Repeat the process. This time increase the length of the new side of the triangle by $\frac{1}{2}$ in. each time. (Design should be circular)

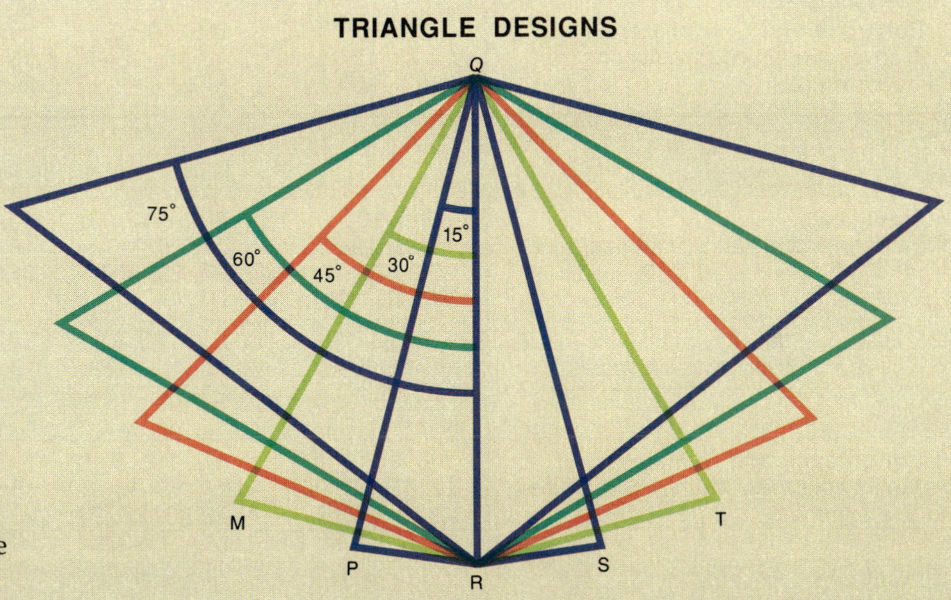

Extra-Credit Problems

BASIC

▼ What is the perimeter of the four attached hexagons, if each side is 3 in.? (54 in.)

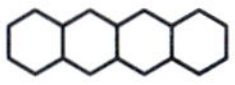

▼ What is the perimeter of 12 attached hexagons? Explain your answer. (150 in., $12(n) + 6$ where n = number of hexagons)

AVERAGE

▼ Find the perimeter of the rectangle. ($P = 48$ in.)

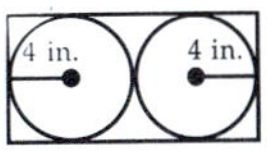

▼ Explain how you found the answer. (Answers will vary.)

ENRICHED

▼ The radius of $\odot B = 3$ in. Find the circumference of $\odot B$, C, and A. Find the perimeter of $\triangle ADE$. Use $\pi \approx 3.14$. ($\odot B$, 18.84 in., $\odot C$, 18.84 in., $\odot A$, 37.68 in., $\triangle ADE$, 18 in.)

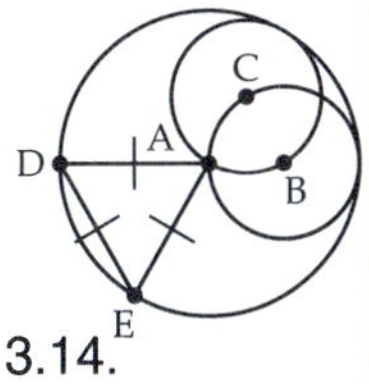

Chapter Planning Guide

OBJECTIVES	ASSIGNMENT GUIDE			ASSESSMENT	
	Basic	Average	Enriched	Review/ Tests	TRB*
9-1 Introduction to Geometry To understand geometric concepts, symbols, and vocabulary.	1-4, 9, 11-12, 15, MR* All	3-6, 9, 12-13, 15, MR All	5-8, 10, 13-15, MR All	Extra Practice p. 574	Practice I-1 Enrichment I-2
9-2 Angles To investigate angle measures and relationships.	1-4, 9, 11-12, 15, 17-18, 21-24, 29-32, MR All	3-6, 9, 12-13, 15, 18-19, 23-26, 31-34, MR All	5-8, 10, 13-14, 16, 19-20, 25-28, 33-36, MR All	Extra Practice p. 574	Practice I-3 Enrichment I-4
Exploring Segments and Angles To explore segments and angles using geoboards.	**Manipulative Connection** All	All	All		
9-3 Polygons and Quadrilaterals To discover properties of polygons and quadrilaterals.	1-4, 9-10, 12-13, 15-16, 20a-b, MR All	3-6, 10-13, 16-18, 20b-c, MR All	5-8, 10-11, 13-14, 18-19, 20b-d, MR All	Extra Practice p. 574	Practice I-5 Enrichment I-6
Exploring with LOGO To explore segments, angles, and polygons using LOGO.	All	All	All		
9-4 Triangles To discover properties of triangles.	1-3, 7-10, 15-16, 19-20, 22-25, 30, MR, CT* All	3-5, 9-12, 16-17, 20-21, 24-27, 30, MR, CT All	4-6, 11-14, 17-18, 20-21, 26-31, MR, CT All	Extra Practice p. 574	Practice I-7 Enrichment I-8
9-5 Circles To discover the properties of circles.	1-3, 5-8, 12-14, 18-19, 21, 23-24, MR All	2-4, 7-9, 13-15, 18-19, 21, 23-24, MR All	3-5, 9-11, 15-17, 19-20, 22-24, MR All	Extra Practice p. 574	Practice I-9 Enrichment I-10
9-6 Congruence and Symmetry To identify congruent figures and figures with line symmetry.	1-2, 4-5, 7, 10-12, 16, 18, 20, 22-23, 25, 27, MR, TY* All	1-2, 5-6, 8, 12-14, 16, 18, 20, 22-23, 25, 27, MR, TY All	1, 3, 6, 8-9, 13-15, 17, 19, 21, 23-24, 26, 27, MR, TY All	Extra Practice p. 574 TY p. 383	Practice I-11 Enrichment I-12 TY I-20
9-7 Similar Figures To determine correspondences in similar figures and to find missing measures in similar figures.	1, 3, 5-6, 9, 11-12, 14-15, MR, CT All	1, 3, 6-7, 10-12, 14-15, MR, CT All	2, 4, 7-8, 10-11, 13-16, MR, CT All	Extra Practice p. 574	Practice I-13 Enrichment I-14
Exploring Pi To explore pi as a ratio.	All	All	All		
9-8 Perimeter and Circumference To find the perimeter or circumference of a figure.	1-6, 8-9, 12-17, 20-21, 24, 26, 28, 31-32, 34, MR, TY All	2-6, 10-14, 17-18, 21-26, 28-29, 31-34, MR, TY All	3-4, 6-7, 10-11, 14-19, 22-23, 26-27, 29-34, MR, TY All	Extra Practice p. 574 TY p. 393	Practice I-15 Enrichment I-16 TY I-20
9-9 Problem Solving Strategy: Draw a Diagram. To solve problems by drawing a diagram.	**Problem Solving Connection** All	All	All	Problem Solving Practice p. 396	Practice I-17 Enrichment I-18
9-10 Problem Solving Application: Tessellations To understand and create tessellations.	All	All	All		Alternate Application I-19
Chapter Test	All	All	All	pp. 400-403	Tests I-21–I-27

 *TRB—Teacher's Resource Book MR—Mixed Review TY—Test Yourself CT—Critical Thinking

	CONNECTIONS						NCTM CORRELATION
Algebra	Critical Thinking	Problem Solving	Estimation/ Mental Math	Technology	Manipulatives/ Resources	Writing/Reading in Math	
W*: 9-10	C*: TD* 1-4 W: 4-6, 11-14	W: 4-6			C: 4-7 W: 4, 15	C: 8, TD 1-4 W: 7, 15	Algebra Communication Problem Solving Reasoning Geometry
W: 11-16 34-36	C: 3, TD 1-3 W: 7	W: 15-16	W: 21-33		W: 8, 17-20	C: TD 1-3 W: 7	Algebra Geometry Problem Solving Communication
	1-6	1-6			1-6	1-6	Geometry Problem Solving Reasoning
	C: TD 1-3 W: 1-8	W: 15-20			W: 15-18, 20		Geometry Problem Solving Reasoning
	1-3			1-3	1-3		Geometry Connections
W: 28-29	C: TD 1-3 W: 7-14 CT*: 1-5	C: 4-7 W: 15-18		W: 22-29	C: 4-7 W: 15-18	C: TD 1-3 CT: 1-5	Algebra Geometry Problem Solving Reasoning
W: 6-17	C: TD 1-2 W: 1-5, 18, 20	W: 18-24	W: 6-11	W: 12-17, 21-23	W: 19, 21, 22, 24	C: TD 1-2	Algebra Geometry Problem Solving Connections
W: 16-17	C: 1-2, 5 TD 1-3 W: 1-3 TY: 1-4	W: 4-9, 16, 17, 26	W: 22-24		C: 3-4 W: 18-21, 25-27	C: 5, TD 1-3 W: 27	Algebra Geometry Problem Solving Reasoning
C: 3-4 W: 5-8	C: 1-2, TD 1-3 W: 9-10 CT: 1-2	W: 5-8, 12-16			C: 1 W: 1-2, 14-16	C: TD 1-3 W: 11, 16 CT: 2	Geometry Problem Solving Communication Reasoning
		1-3		1c	1-3		Problem Solving Connections
C: 2-3 W: 1-30 TY: 1-3	C: TD 1-2	W: 1-11, 24-34 TY: 1-3	C: 1-3	W: 16-19	C: 1-3 W: 31-34	C: TD 1-2	Algebra Problem Solving Reasoning Connections
W: 2, 4, 6	W: 2-3, 5-6	C: 1-2 W: 1-6	C: 1				Problem Solving Functions Geometry Communication
	C: 1-3 W: 1-3, 6-7	W: 4			C: 1-3 W: 1-3, 5-9	W: 9	Problem Solving Functions Geometry

*C—Class Exercises W—Written Exercises CT—Critical Thinking TY—Test Yourself TD—Think and Discuss

Exploring

In this manipulative activity, students use straws to construct and classify triangles. **See TRB Exploring O-33.**

Exploring Triangles

Triangles can be classified according to their sides. A *scalene triangle* has no two sides that are equal. An *isosceles triangle* has two equal sides. An *equilateral triangle* has three equal sides.

Cut drinking straws in 1-inch lengths. Each length represents one unit of length. Use the pieces of straw to construct triangles.

Build a triangle using 3 units for each side of the triangle.

1. a. Draw a picture of the triangle.
   b. What kind of triangle did you form? **equilateral**

Build a triangle with sides of 3 units, 3 units, and 1 unit.

2. a. Draw a picture of the triangle.
   b. What kind of triangle did you form? **isosceles**

Build a scalene or isosceles triangle.

3. a. Draw a picture of the triangle.
   b. What dimensions did you use? **Answers will vary.**
   c. Can you build a triangle with dimensions 1, 1, 5? Why or why not? **No; one side is too long.**

Use the straws. Determine if triangles can be made with the dimensions below. If yes, write the kind of triangle.

	Longest Side	Shortest Side	Other Side	Triangle?	Kind
4.	5	2	5	**yes**	**isosceles**
5.	5	3	4	**yes**	**scalene**
6.	6	1	5	**no**	**—**
7.	4	4	4	**yes**	**equilateral**
8.	7	2	3	**no**	**—**

9. What relationship must there be among the measures of the sides to form a triangle? **The sum of the shortest and other side must be greater than the longest side.**
10. Find five more sets of dimensions that will form a triangle. **Answers will vary. 6, 8, 10; 3, 3, 3; 4, 5, 7; 4, 4, 3; 8, 12, 13**

O-33

Exploring Congruence

You know that congruent figures have the same size and shape. Congruent figures coincide if they are placed one upon the other.

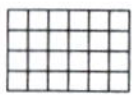

Draw a 6 x 4 rectangular region on graph paper to answer the questions.

Think about cutting the 6 x 4 region into two congruent figures.

1. Draw one of the two congruent figures. **Answers will vary.**
2. How do you know what size each figure should be?
   6 x 4 = 24 squares   24 ÷ 2 = 12 squares in each figure.

Now think about cutting the region into three congruent figures.

3. Draw one of the three congruent figures.
4. How do you know what size each figure should be?
   6 x 4 = 24 squares   24 ÷ 3 = 8 squares in each figure

Using the same region, try to form four congruent figures.

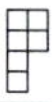

5. Draw one of the four congruent figures. **Answers will vary.**
6. What size will each figure be? **6 squares**
7. Think about other ways to draw four congruent figures. Draw one way. **Answers will vary.**

Now try using the same region to form six congruent figures.

8. Draw one of the six congruent figures. **Answers will vary.**
9. What size will each figure be? **4 squares**
10. Think about other ways to draw six congruent figures. Draw one way. **Answers will vary.**

Use what you have learned to answer these questions.

11. Is it possible for two figures to have the same area, but not have the same shape? If so, show an example.
    Answers will vary.

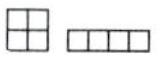

12. Choose a square region of a different size. Find as many ways as you can to cut the region into congruent figures.
    Region size: **Answers will vary.**
    Possible congruent figures: **Answers will vary.**

O-34

In this problem solving activity, students use models to explore the congruence of regions. **See TRB Exploring O-34.**

In this computer activity, students use a spreadsheet to explore regular polygons. Using a pattern, they find the sum of the interior angles and the measure of each angle. **See TRB Exploring O-35.**

Exploring Regular Polygons

The sum of the measures of the interior angles of any triangle is 180°. Use this to find the sum of the measures of the angles for any regular polygon.

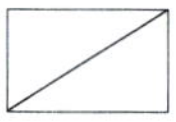

1. Draw a diagonal in the quadrilateral to form two triangles. How can you use the triangles to find the sum of the measures of the angles of the quadrilateral?
   Multiply 180° times the number of triangles in the figure.
2. Find the sum of the measures of the angles of the pentagon. **180° × 3 = 540°**

Count the triangles and the sides in each figure.

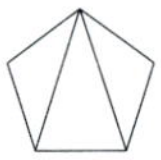

3. Write the relationship between the number of triangles and the number of sides.
   There are two fewer triangles than sides.

Use a spreadsheet to record the data and show the relationships.

	A	B	C	D	E
1	Polygon	Number	Number of	Sum of	Interior
2		of Sides	Triangles	Angles	Angle
3	Quadril.	4	2	360	90
4	Pentagon	5	3	540	108
5	Hexagon	6	4	720	120
6	Heptagon	7	5	900	128.571429
7	Octagon	8	6	1080	135
8	Nonagon	9	7	1260	140
9	Decagon	10	8	1440	144

4. Write a formula to find the value of cell C3 and D3.
   =B3-2; =C3*180
5. For a regular polygon, how can you find the measure of one interior angle, given the sum of all the angles? **Divide the sum of the interior angles by the number of sides or angles.**
6. Write a formula to find the value of cell E3. **D3/B3**
7. Complete the spreadsheet for a quadrilateral through a decagon.
8. Find the sum of the angle measures and the measure of an interior angle for a regular polygon with n sides.
   sum = $(n - 2)180$; interior angle = $[(n - 2)180] \div n$

O-35

Exploring Polygons with LOGO

The two procedures below are written in LOGO. One will produce an equilateral triangle and one will not. Study both procedures. Answer the questions.

LOGO Commands	
FD forward	BK back
RT right turn	LT left turn
REPEAT	

Procedure 1

```
TO TRIANGLE
REPEAT 3 [FD 20 RT 60]
END
```

Procedure 2

```
TO TRIANGLE
REPEAT 3 [FD 20 RT 120]
END
```

1. How do the procedures differ from one another?
   In procedure 1 the right turn is 60°; in procedure 2 the right turn is 120°.
2. Which will result in an equilateral triangle? **Answers will vary.**

Use a computer or graph paper to test both procedures. If you use a graph, substitute the command "FD 5" for "FD 20."

3. Did procedure 1 produce an equilateral triangle? Why?
   No; the turtle did not turn far enough to form a closed figure, that is, the equilateral triangle.
   Did procedure 2 produce an equilateral triangle? Why?
   Yes; the 120° turns the turtle made formed a closed figure, an equilateral triangle.
5. What were the total number of degrees needed to return the turtle to the starting point?
   360°
6. In the command, REPEAT 3 [FD 20 RT 120], what is the relationship between the number of sides the figure has and the number of degrees the turtle is instructed to turn?
   The number of degrees equals 360° divided by the number of sides.

Write a procedure that will produce each regular polygon. Use a computer or graph paper to test your answers.

7. Square **TO SQUARE**
   REPEAT 4[FD 20 RT 90]
   END
8. Octagon **TO OCTAGON**
   REPEAT 8[FD 20 RT 45]
   END
9. Nonagon **TO NONAGON**
   REPEAT 9[FD 20 RT 40]
   END
10. Decagon **TO DECAGON**
    REPEAT 10[FD 20 RT 36]
    END

O-36

In this computer activity, students use LOGO to explore the construction of polygons with emphasis on determining the measure of the interior angles. **See TRB Exploring O-36.**

Meeting Individual Needs

GIFTED STUDENTS

Mathematically gifted students may be intrigued to discover for themselves how the mathematical terms for triangular and pentagonal numbers are generated.

Display the arrays below and explain that dots may be arranged in arrays shaped like regular polygons. Comment that the largest triangular array shown contains all the smaller triangular arrays.

Explain that students can see from these figures that 1, 3, 6, 10, and 15 are all triangular numbers.

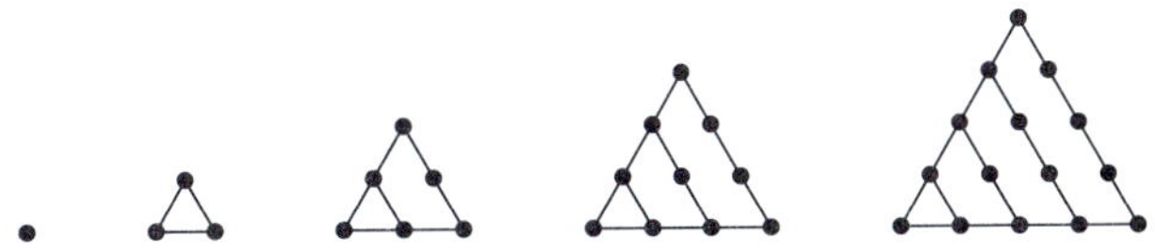

Ask students to add dots to the last array to reveal the next triangular number. (21)

Next display a 5-dot pentagonal array. Have students use what they learned about triangular arrays to draw the next pentagonal array. (12 dots) Ask students to predict and draw the next two pentagonal arrays. (22, 35) Have students determine the number of dots on each side if there are 70 dots in the array. (7)

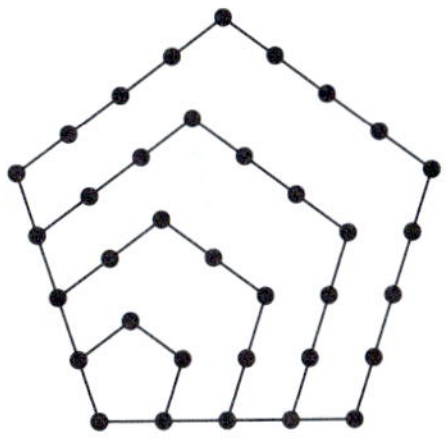

AT-RISK STUDENTS

Manipulatives and visual aids can be helpful to at-risk students, who often appreciate when mathematical terms can be made concrete. Symmetry is particularly effective for manipulative and visual exploration.

Review reflectional symmetry and rotational symmetry with students. Remind students that an object with reflectional symmetry has at least one line of symmetry and an object with rotational symmetry has a point around which the object is rotated. Some objects have both types of symmetry.

Display the figures below and identify the type of symmetry each possesses. (reflectional symmetry: 1, 2, 4, rotational symmetry: 2, 3, 4 both: 2, 4)

Then ask students to find two objects that have rotational symmetry, two that have reflectional symmetry, and two that have both types of symmetry. Have students identify the line or lines of symmetry or the point around which the objects can be rotated.

1.
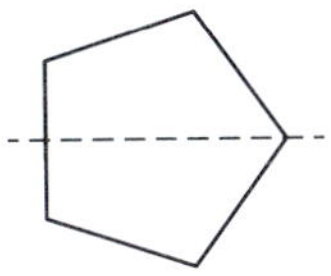

2.

3.

4.
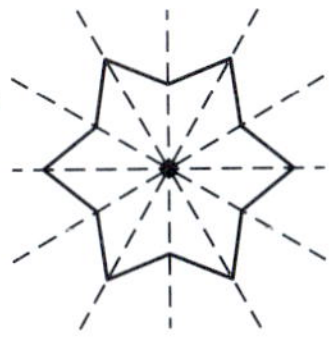

ENGLISH AS A SECOND LANGUAGE

Provide activities that allow the ESL student to demonstrate proficiency in math without relying on verbal skills. This activity emphasizes listening, comprehension, and visual skills. Give the ESL student a page of shapes and a list of terms such as *polygon*, *similar triangle*, and *acute angle*. Have the student match the correct name with the shape. Then ask the student to draw the shapes as you name them. As the student shows more proficiency, increase the amount of information to be included in each drawing. For instance, ask the student to draw a polygon labeled *ABCD*.

See TRB Teaching Resources p. 59.

Problem Solving Situation

The "Think About It" question asks which hole is most difficult and which is the least difficult. Difficulty is indicated by par, which is determined by sand traps, trees, and other obstacles. Par is the number of strokes needed to get the ball from the tee, or starting point, into the hole. It is determined by the distance the ball must cover and the difficulty in getting the ball into the hole. Golfers want to get the ball into the hole at par or less.

As a follow-up activity, have students choose a golf course and decide which golf clubs to use on each hole.

DATA COLLECTION

Students can select a local golf course or the course used for one of these tournaments: The Masters; U.S. Open; British Open; or PGA Championship. Students are to find the par and distance from the tee to the hole for each hole of their chosen golf course. They can find this information in newspapers, books, and magazines, or they can call the information bureau of the golf course. Maps or designs of the golf course will enable students to see the location of obstacles.

DATA ANALYSIS

Ask students to make a chart, similar to the one on page 361, for the course they chose. Using the information on page 360 and the chart they have made, students will decide which club or combination of clubs to use to score par on each hole.

There are two different types of golf clubs—woods and irons. Woods have a bit more weight in the head and thus hit the ball farther. Irons are used for accuracy.

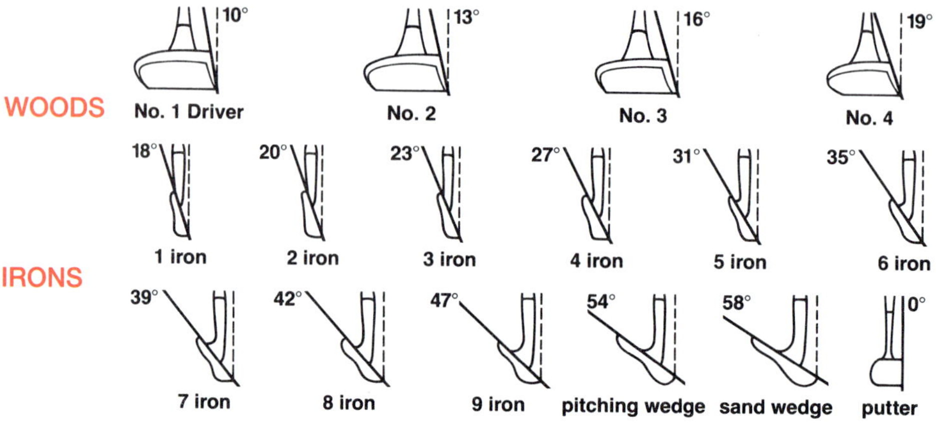

The *loft* of a club is the angle at which a club is set from vertical. The height and distance a ball travels are a function of the loft.

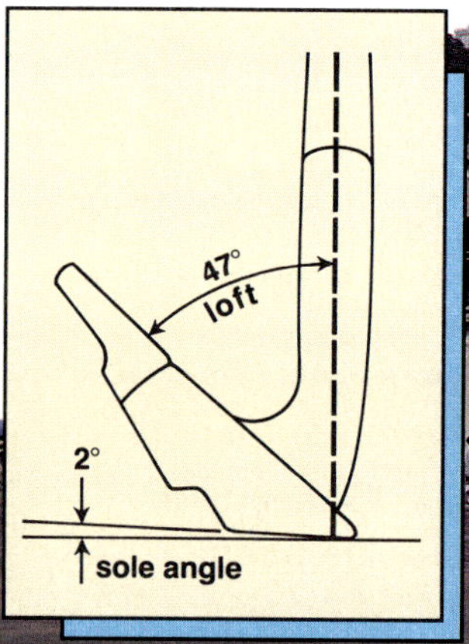

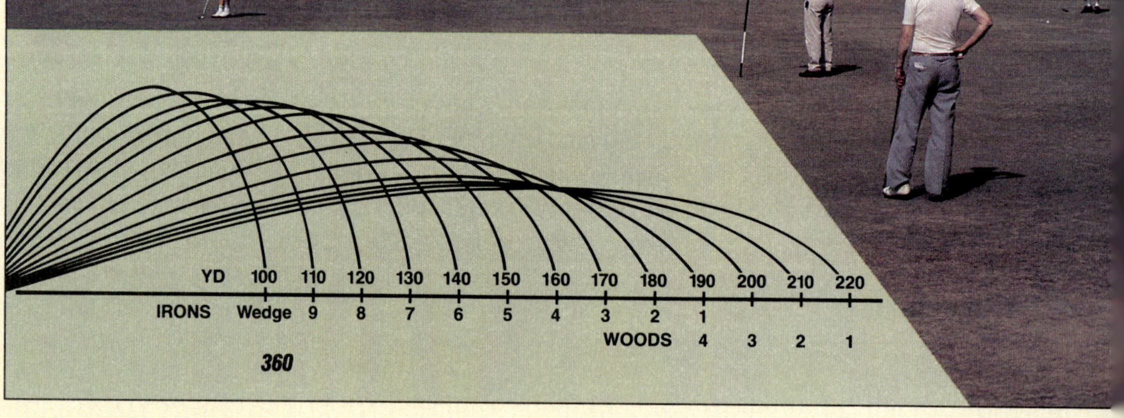

360

CHAPTER 9

Algebra in Geometry and Measurement

Think about it...

Look at the data for St. Andrews golf club. Which hole do you think is the most difficult? the least difficult? Give reasons for your answers.

See Problem Solving Situation

THE ROYAL AND ANCIENT GOLF CLUB
St. Andrews, Scotland

St. Andrews golf club is the world's oldest golf course. The course has 140 sand traps and 7 greens that are each shared by 2 holes.

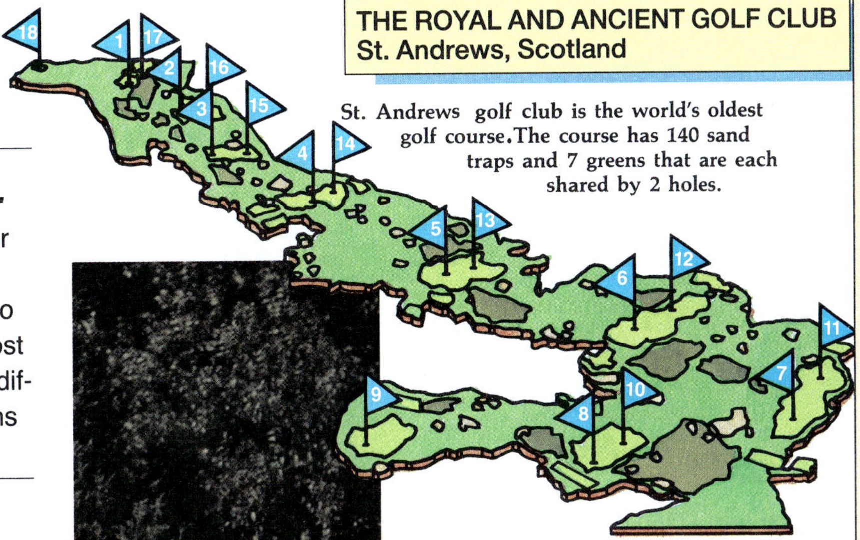

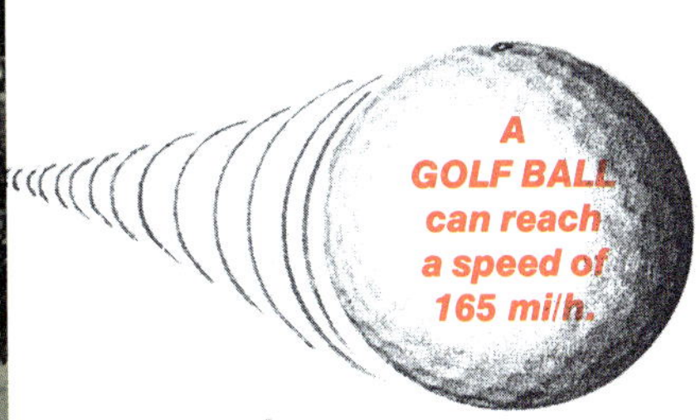

Par for the Royal and Ancient Golf Club, St. Andrews, Scotland

Hole	1	2	3	4	5	6	7	8	9	10	11	12	13	14	15	16	17	18
Par	4	4	4	4	5	4	4	3	4	4	3	4	4	5	4	4	4	4
Yards	370	411	371	463	564	416	372	178	356	342	172	316	425	567	413	382	461	354

Cooperative Learning

This activity can be done in cooperative learning groups of 3 or 4 students.

The groups will decide which golf clubs to use. Remind students that every member of the group needs to be actively involved in the decision making and to understand how the choices were made. Possible roles include Checker, Summarizer, Reporter, Illustrator, and Recorder.

Each group should present its choices to the class. To extend this activity, provide the class with golf clubs to put some of its decisions into practice.

Questions related to this Data File can be found on pages 25, 29, 47, 107, 225, 266, 290, 336, 574, and 578.

Lesson Focus

Materials

Paper, pencil, dot or graph paper, magazines or newspapers, math journal

Vocabulary/Symbols

line
parallel lines
parallel planes
plane
point
ray
segment
skew lines

MOTIVATING THE LESSON

Write the terms *point*, *line*, *plane*, *segment*, and *ray* on the chalkboard. Then ask students to identify examples of each term in the classroom. Possibilities might include: a sheet of paper (plane), a flashlight beam (ray), the end of a push pin (point), the edge of a desk (line segment), marks on a ruler (segment). Have students describe how the properties of each object match the term.

Instruction

1 Ask students to analyze the relationships among a line, a segment, and a ray by telling how they are alike and different.

2 Mention that there are more than three answers. Ask students to work in groups to list different ways to name each figure. Then groups can compare their results.

3 Work through the solution to Example 2 orally. Then ask students to demonstrate solutions different from the one in the text and to verify that their solutions meet the given criteria.

OBJECTIVE: *To understand geometric concepts, symbols, and vocabulary.*

9-1 Introduction to Geometry

▼ Geometric shapes are evident in many man-made and natural structures. The hexagonal design of the snowflake or honeycomb and the spiral design of the snail are but two examples of geometry in nature.

▼ The table illustrates some basic geometric figures.

1

Figure	Properties	Example	Symbol	Read as
Point	• represents a position in space	• *A*	*A*	*point A*
Line	• continues without end in opposite directions	←•—•→ with points *A*, *B*	$\overleftrightarrow{AB}$	*line AB*
Plane	• is a flat surface with no thickness that continues without end in all directions	parallelogram *ABCD* labeled *M*	*ABCD* *M*	*plane ABCD* or *plane M*
Segment	• is part of a line with two endpoints	•—• with endpoints *A*, *B*	$\overline{AB}$	*segment AB*
Ray	• is part of a line with only one endpoint • continues without end in one direction	•—•→ with points *A*, *B*	$\overrightarrow{AB}$	*ray AB*

▼ You can name a line by any two points on the line.

Example 1 **Use the figure at the right.**

a. Give three ways, other than line *AB*, to name the line.
b. Name three different segments.
c. Name three different rays.
d. Are there points other than *A*, *B*, *C*, and *X* on the line?

Solution **2**

a. $\overleftrightarrow{AC}$, $\overleftrightarrow{BA}$, and $\overleftrightarrow{BX}$ are three possible ways to name the line.
b. $\overline{AC}$, $\overline{BC}$, and $\overline{BX}$ are three different segments.
c. $\overrightarrow{AX}$, $\overrightarrow{BX}$, and $\overrightarrow{CA}$ are three possible rays.
d. Yes. There are an infinite number of points on the line.

NOTES & QUOTES

One need not know the profoundest mysteries of geometry to be able to discern its usefulness.
–Robert Boyle (1627–1691)

Teaching Tip

Several days before you plan to teach lessons on geometric figures, ask the class to bring in objects from home that can be used to represent the figures. Prepare a list of objects and post it in the classroom or ask volunteers to choose items to bring in.

▼ When lines, rays, and segments intersect, the intersection is a point. *Parallel lines* (∥ lines) are lines that are always the same distance apart and never intersect.

Parallel Lines	Two lines are parallel if they lie in the same plane and do not intersect.
Parallel Planes	Two planes are parallel if they do not intersect.
Skew Lines	Skew lines are lines that do not lie in the same plane and do not intersect.

▼ Unless you are given other instructions, you will be judging the geometric properties of figures in this book by appearance.

Example 2 **Use graph paper to draw and label a figure containing three lines. Make two of the lines parallel.**

Solution

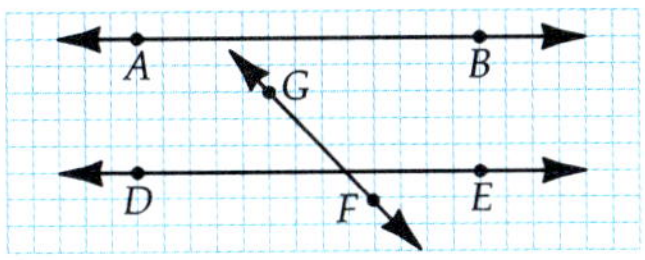

$\overleftrightarrow{AB} \parallel \overleftrightarrow{DE}$

1. point—where floor tiles intersect; line—edge of desk; skew lines—the line formed by the intersection of two walls and the line formed by the intersection of wall and floor; plane—ceiling, floor, wall

CLASS EXERCISES

Answers may vary. One example is given.

Use the figure at the right.

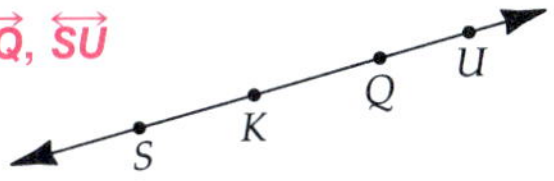

1. Name the line in three ways. $\overleftrightarrow{SK}$, $\overleftrightarrow{KQ}$, $\overleftrightarrow{SU}$
2. Name three different segments. $\overline{SK}$, $\overline{KQ}$, $\overline{SQ}$
3. Name three rays. $\overrightarrow{KS}$, $\overrightarrow{QU}$, $\overrightarrow{SU}$

Use dot paper or graph paper to draw and label a figure to fit each description. Check students' drawings.

4. intersecting lines $\overleftrightarrow{AB}$ and $\overleftrightarrow{CD}$
5. parallel rays $\overrightarrow{MN}$ and $\overrightarrow{OP}$
6. point Q on $\overline{MN}$ and $\overline{XY}$
7. $\overleftrightarrow{CD}$ containing point X

Check students' work.

8. ***WRITE*** a description of the streets in your neighborhood. Use the street names and the terms *parallel* and *intersecting* as appropriate. Could streets in a town form skew lines? Why or why not? No; streets are not in different planes.

Railroad tracks are an example of parallel lines. Can you think of an example of parallel planes? ceiling and floor

3

2. Parallel lines on a sphere would be concentric circles. Parallels of latitude would be an example.
3. 3 points determine a plane
4. no, could be skew

THINK AND DISCUSS See above.

1. Look around your classroom. Give an example of a point, a line, skew lines, and a plane.
2. Describe how parallel lines would look on a sphere.
3. Why would a three-legged table stand firmly on the ground while a four-legged table might wobble?
4. If lines are not parallel, must they intersect?

ADDITIONAL EXAMPLES

1. Draw a line. Label points X, Y, and Z, with Z between X and Y. If $XZ = 3$ and $XY = 10$, find ZY. (7)
2. Name the figure that has two endpoints. (segment)

Guided Practice

Class Exercises For question 5, ask students if the endpoints of both rays must be on the same end. Have students draw figures to support their ideas. (Endpoints may be either on the opposite or the same end.)

Think and Discuss For question 4, if students answer *no*, have them give an explanation or draw a figure to support their answers. (Students should mention or draw skew lines.) Then, have students explain under what conditions the answer might be *yes*. (if both lines are in the same plane)

A ***common error*** is to use the wrong symbols for line, ray, and line segment. Point out to students that the arrows in the symbol indicate direction; so the two arrows on the symbol for line represent the figure *extending in both directions* without end; the one arrow on the symbol for ray represents the figure extending in only one direction without end; there are no arrows on the symbol for segment, which indicates that the figure does not extend beyond the endpoints.

MATH MINUTES

Students work together to write true and false statements, using the words *sometimes*, *never*, or *always*, about the geometric figures in this lesson. For example: A line *always* has two endpoints. (false)

Closure

Writing in Math Have students write in their math journals the properties of each basic geometric figure and their symbols as presented in this lesson. Ask students to include, when possible, examples from the real world for each figure.

Independent Practice

For question 6 encourage students to draw diagrams.

Lesson Follow-up

Reteaching Activity Have students make a chart similar to the one on page 362 to include parallel lines, parallel planes, and skew lines.

LESSON RESOURCES

TRB Practice I-1
TRB Enrichment I-2

Practice — For use after 9-1 (pp. 362–364)

Introduction to Geometry

Complete. Use *point, line, plane, segment,* or *ray.*

1. A **line** continues without end in opposite directions.
2. A **segment** is part of a line and has two endpoints.
3. A **point** represents a position in space.
4. A **plane** is a flat surface with no thickness.
5. A **ray** is part of a line and has one endpoint.

True or *false?*

6. Parallel lines lie in the same plane. **true**
7. Skew lines may be parallel. **false**
8. $\overleftrightarrow{AB}$ and $\overleftrightarrow{CB}$ name the same line. **true**
9. $\overrightarrow{AB}$ and $\overrightarrow{BA}$ name the same ray. **false**
10. $\overrightarrow{AB}$ and $\overrightarrow{AC}$ name the same ray. **true**

11. Name all possible segments in the figure. **$\overline{AP}$, $\overline{PC}$, $\overline{AC}$, $\overline{NP}$, $\overline{PB}$, $\overline{NB}$**
12. Name a line containing N and B. **$\overleftrightarrow{NB}$**
13. How do you know that $\overleftrightarrow{AC}$ and $\overleftrightarrow{NB}$ are not parallel? **They intersect at *P*.**

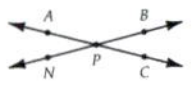

Complete, using the three-dimensional figure. Use parallel lines, skew lines, parallel planes, or intersecting planes.

14. $\overleftrightarrow{AD}$ and $\overleftrightarrow{EH}$ are **parallel lines**.
15. $ABCD$ and $AEFB$ are **intersecting planes**.
16. $\overleftrightarrow{EF}$ and $\overleftrightarrow{AD}$ are **skew lines**.

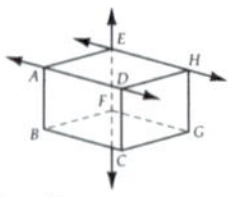

Find the length of the indicated segments.

17. 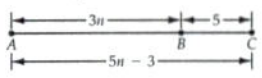

$AB =$ **12 units** $AC =$ **17 units**

18. 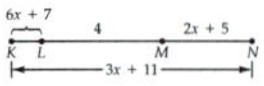

$KL =$ **1 unit** $MN =$ **3 units** $KN =$ **8 units**

Chapter 9 I-1

Enrichment — For use after 9-1 (pp. 362–364)

Lines and Curves

You can use straight lines to produce the illusion of curves. To see how, connect each pair of identical letters, using a ruler.

1. 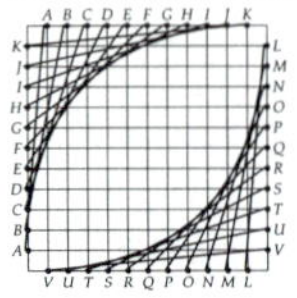

2. 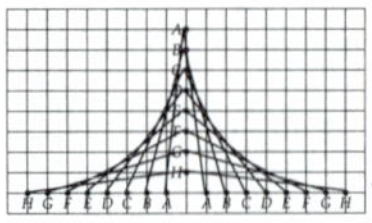

3. 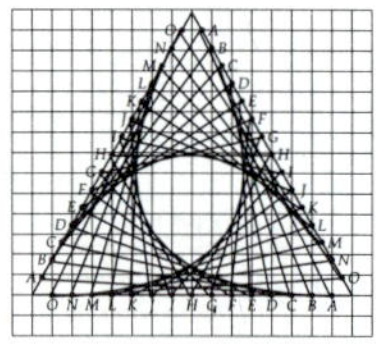

4. Create your own design using the method shown above.
Check students' drawings.

I-2 Chapter 9

MIXED REVIEW

1. Graph the points $A(5,0)$, $B(6,3)$, $C(1,1)$. Connect A to B and B to C.
2. $a = \frac{b}{2} + c - 2$. Find a if $b = 11$ and $c = 3$. **$6\frac{1}{2}$**
3. What percent of 73 is 25? **34%**
4. A 12-oz can of juice costs 69¢. What would you expect to pay for a 16-oz can? **92¢**

Solve.

5. $2x + 8 = 12$ **$x = 2$**
6. $3 + \frac{a}{8} = 11$ **$a = 64$**
7. How many three-digit numbers, greater than 500, can you form using the digits 2, 6, and 8 exactly once each? **4**

WRITTEN EXERCISES

1. Name all possible segments in the figure below using the points labeled. **$\overline{BZ}$, $\overline{BT}$, $\overline{BM}$, $\overline{MT}$, $\overline{ZM}$, $\overline{ZT}$**

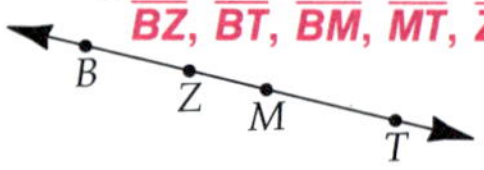

2. Name four different rays in the figure below.

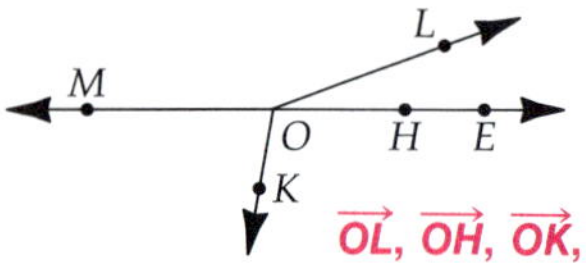

$\overrightarrow{OL}$, $\overrightarrow{OH}$, $\overrightarrow{OK}$, $\overrightarrow{OM}$

3. Points A, B, C, and D are collinear (all on a line) and in the same order. Name $\overrightarrow{DA}$ three different ways. **$\overrightarrow{DC}$, $\overrightarrow{DB}$, $\overrightarrow{DA}$**
4. Draw a figure in which $\overleftrightarrow{ZR}$ contains P and $\overline{PQ}$ contains R.
   a. Is only one figure possible? If *no*, show an alternate figure. **no**
   b. Is R between Q and P? **yes**
   c. Must $\overrightarrow{ZR}$ contain P? **no**
   d. Could $\overrightarrow{ZR}$ contain P? **yes**

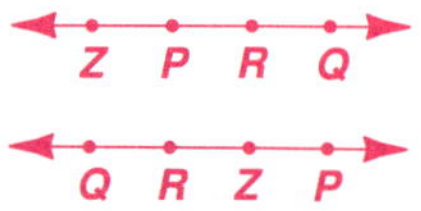

5. How many lines can you draw that contain a given point Q? that contain two given points X and Y? **an infinite number; one**
6. a. Suppose a town installs a mailbox at point M. How many straight roads can the town build leading to M? **an infinite number**
   b. Suppose a town installs mailboxes at points M and A. How many straight roads might the town build that pass by both mailboxes? **one**
7. ***WRITE*** Remember to use complete sentences.
   a. Describe the intersection of two planes. **The intersection of two planes is a line.**
   b. Give a physical example of the intersection of two planes. **the intersection of a floor and a wall.**
8. Planes M and N do not intersect. What can you conclude? **Planes *M* and *N* are parallel.**

Write an equation and then find the length of each segment.

9. 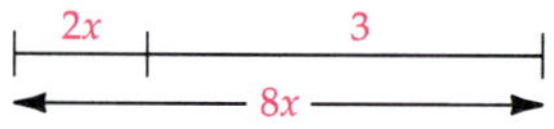

$2x + 3 = 8x$; 4

10. 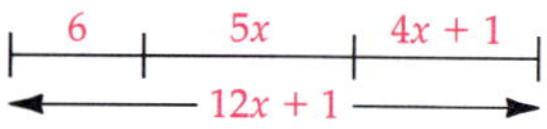

$6 + 5x + 4x + 1 = 12x + 1$; 25

True or false?

11. Skew lines never intersect. **true**
12. A ray has two endpoints. **false**
13. A line has no endpoints. **true**
14. A segment has two endpoints. **true**
15. ***PROJECT*** Collect at least five pictures from magazines and catalogs that show geometry in the real world. ***WRITE*** a sentence for each picture to explain the connection. **Check students' work.**

LESSON QUIZ

Name each geometric figure.

1. 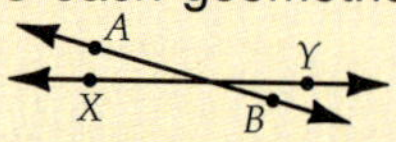

(line AB intersects line XY)
2. jump rope pulled tightly (segment)
3. 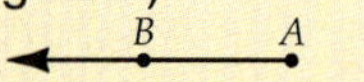

(Ray AB)
4. floor and ceiling of a room (parallel planes)
5. 

(segment MN)

Assignment Guide

Basic 1-4, 9, 11-12, 15, MR All
Average 3-6, 9, 12-13, 15, MR All
Enriched 5-8, 10, 13-15, MR All

FOR THE NEXT LESSON

paper, pencil, ruler or straight edge, protractor, street map, math journal

OBJECTIVE: *To investigate angle measures and relationships.*

9-2 Angles

▼ When lines or parts of lines intersect, they form angles. Different angles contribute to designs on quilts, tile floors, stained glass windows, and other forms of art and architecture.

▼ You can name an angle by the vertex and points on the sides, by a number, or by the vertex alone. When using three letters, the middle letter always names the vertex.

Angle	Two rays with a common endpoint form an angle.

Example 1 **Name the angle shown in four different ways.**

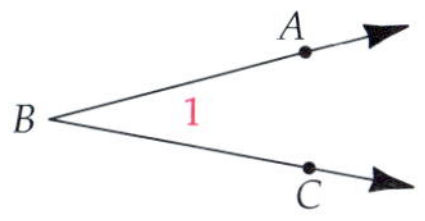

Solution $\angle ABC$, $\angle 1$, $\angle B$, $\angle CBA$

> **FLASHBACK**
> The symbol ⊾ indicates a right angle. Perpendicular lines form right angles.

▼ We classify angles by their measure in degrees (°). The notation $m\angle ABC$ means *the measure of angle ABC.*

1

acute	*obtuse*	*right*	*straight*
less than 90°	between 90° and 180°	equals 90°	equals 180°
$m\angle 2 < 90°$	$90° < m\angle 3 < 180°$	$m\angle PQR = 90°$	$m\angle AOR = 180°$

▼ We use a protractor to measure and draw angles.

2

Example 2 **Use a protractor to measure $\angle XYZ$.**

1. Place the center point of the protractor on Y, the vertex of the angle.
2. Position the protractor so that $\overrightarrow{YZ}$ passes through zero on the protractor scale. Estimate to decide which scale to read. Is $\angle XYZ$ acute or obtuse? **acute**
3. Read the angle measure at the point where $\overrightarrow{YX}$ passes through the protractor scale.

Solution $m\angle XYZ = 29°$

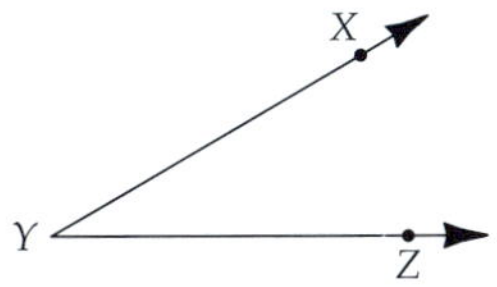

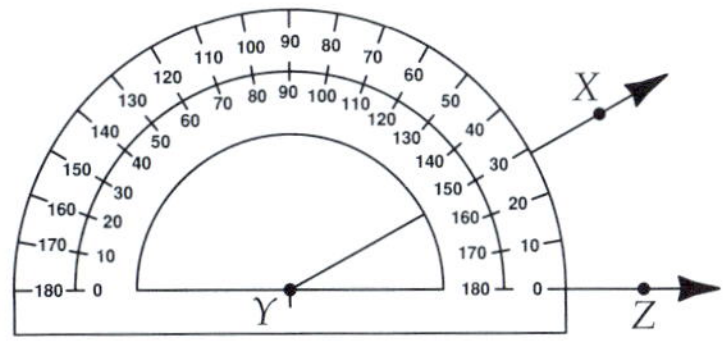

Lesson Focus

Materials

Paper, pencil, ruler or straight edge, protractor, street map, math journal

Vocabulary/Symbols

acute angle
adjacent angles
angle ∠
complementary angles
obtuse angle
right angle
straight angle
supplementary angles
the measure of (m)
vertical angles

MOTIVATING THE LESSON

Draw acute, obtuse, right, and straight angles on the chalkboard and have volunteers name the angles. If practical and available, have students identify examples of the angles in the classroom. Then, help students develop definitions for the angles and compare their definitions to those given in the text.

Instruction

1 Ask students how the four angles are alike and different. (All have two rays with a common endpoint. The measure of the angles are different.) Demonstrate that the length of either ray does not affect the measure of the angle.

2 Demonstrate how to measure an angle using the overhead projector. Mention that $\angle XYZ$ may be called $\angle Y$.

Teaching Tip

Have available angle makers prepared from strips of index cards and paper fasteners. Students who have trouble measuring angles with a protractor can gain additional practice by placing the angle on *top* of the protractor. Students can then read the correct measure.

3 Have students draw and measure angles with a protractor and exchange their drawings. Ask students to check the measures of the angles that were drawn.

4 Discuss with students that supplementary angles and complementary angles are not always adjacent angles. (For example, two 45° angles in a triangle are complementary but not adjacent).

ADDITIONAL EXAMPLES

1. Classify each angle.
- **a.** 165° (obtuse)
- **b.** 90° (right)
- **c.** 23° (acute)

2. Draw a pair of supplementary angles whose difference is 20°. (100° and 80°; check students' drawings.)

3. What is the total number of degrees of the measures of two angles that are complementary? (90°)

Guided Practice

Class Exercises For question 1, have students give the two names orally. (∠*ZYX*, ∠*XYZ*)

For questions 2 and 5 have students describe each step used to find the solutions.

Think and Discuss For question 3, have students draw vertical angles, measure them, and use the definition to verify that vertical angles have the same angle measure and are not adjacent.

A ***common error*** occurs when students do not use the correct scale on the protractor. Remind students to estimate the size of an angle before measuring to decide which scale to use and to judge the reasonableness of an answer.

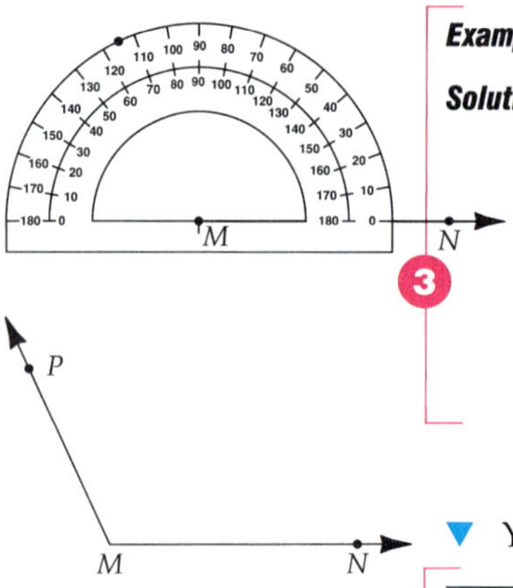

Example 3 **Use a protractor to draw an angle with measure 115°.**

Solution
1. Draw and label a ray.
2. Place the center point of the protractor on the endpoint of the ray. Line up the ray so that it passes through zero.
3. Mark a point at 115° on the protractor scale.
4. Remove the protractor. Draw a ray connecting the endpoint of the ray and the point marked at 115°.
5. $m\angle PMN = 115°$

▼ You can add or subtract the measures of *adjacent angles*.

Adjacent Angles — Two angles that have the same vertex and have a common side but no interior points in common form adjacent angles.

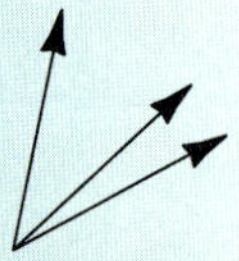

Supplementary Angles — Two angles are supplementary angles if the sum of their measures is 180°.

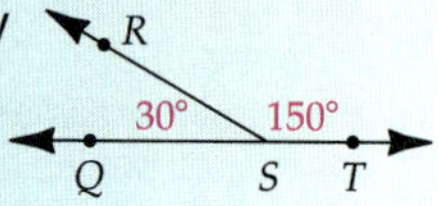

Complementary Angles — Two angles are complementary angles if the sum of their measures is 90°.

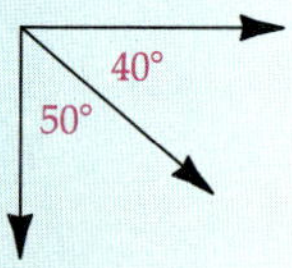

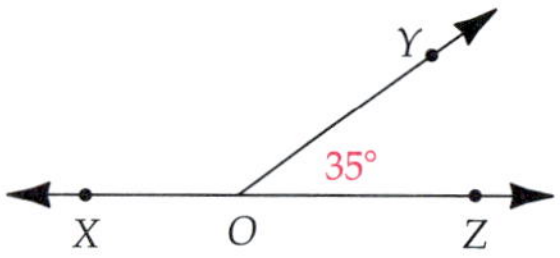

Example 4 $m\angle XOZ = 180°$. **Find** $m\angle YOX$.

Solution

$35° + m\angle YOX = 180°$ — ∠*XOY* and ∠*YOZ* are supplementary.

$35° + m\angle YOX - 35° = 180° - 35°$ — Subtract 35° from each side.

$m\angle YOX = 145°$

▼ When segments, rays, or lines intersect, they form vertical angles.

Vertical Angles — Two intersecting lines form two pairs of vertical angles. The measures of vertical angles are equal.

Example 5 **Find the value of** x **and** y.

Solution $x = 28$ $y = 152$

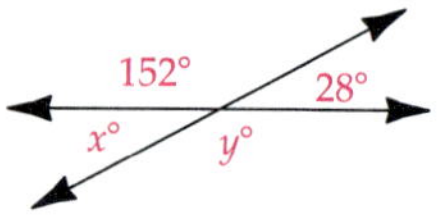

MATH MINUTES

Provide students with copies of the street map of their area. Have students work in groups to list street intersections that form angles, and identify the angles the intersections form. (Check students' work.)

Example 6 Find the measure of each numbered angle.

Solution
$m\angle 1 = 90° - 26° = 64°$
$m\angle 2 = 180° - 90° = 90°$
$m\angle 3 = 26°$
$m\angle 4 = m\angle 1 = 64°$
$m\angle 5 = m\angle 2 = 90°$

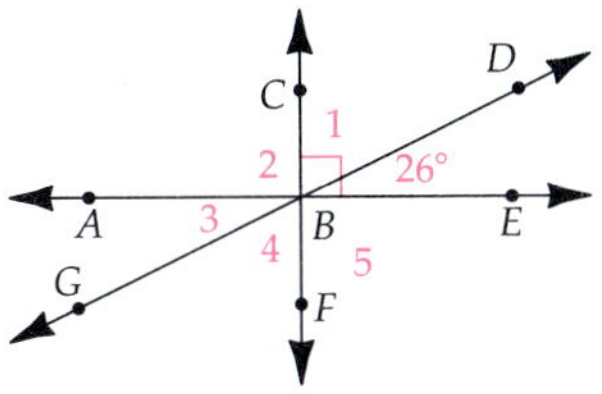

THINK AND DISCUSS See below.

1. Describe a real world example of vertical angles.
2. How can you measure an angle if one ray is not on the base line of the protractor?
3. Can a pair of vertical angles be adjacent? Why or why not?

CLASS EXERCISES

Refer to the figure at the right.

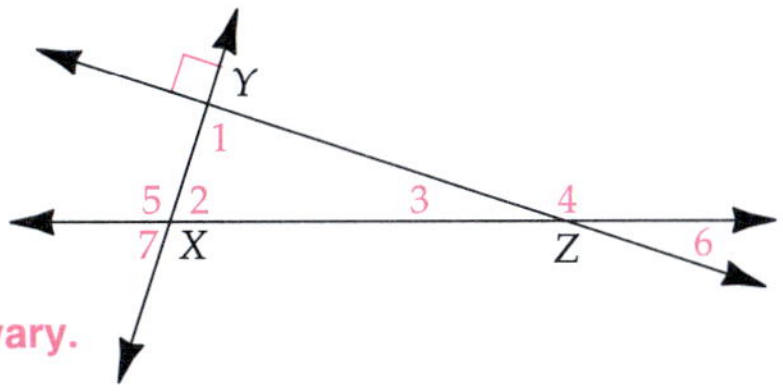

1. Give another name for $\angle 1$. $\angle XYZ$
2. Find $m\angle 5 + m\angle 2$. 180°
3. Name a pair of supplementary angles. Answers will vary.
4. Name a pair of adjacent angles. Answers will vary.
5. Is $\angle Y$ acute, obtuse, or right? right
6. State the relationship between $\overleftrightarrow{XZ}$ and $\overleftrightarrow{ZY}$.
7. Name two pairs of vertical angles.
8. $m\angle 3 = t°$ and $m\angle 4 = 9t°$. Find the value of t. $t = 18$

6. $\overleftrightarrow{XZ}$ and $\overleftrightarrow{ZY}$ intersect to form vertical angles.
7. Answers may vary. One example is given. $\angle 7$ and $\angle 2$, $\angle 3$ and $\angle 6$

1. street intersection
2. Take the two readings and subtract.
3. No; They don't share a common side.

WRITTEN EXERCISES

Classify each angle as acute, obtuse, right, or straight.

1.  obtuse
2. 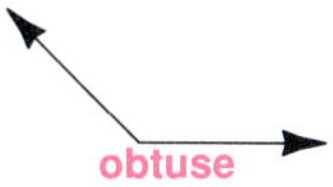 obtuse
3.  straight
4.  right
5. 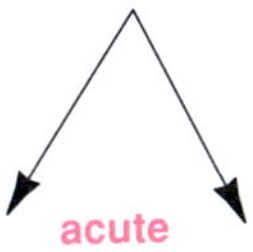 acute
6. 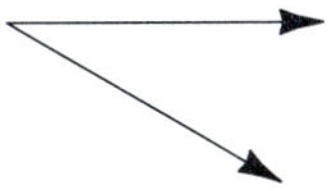 acute
7. Under what conditions are two angles supplementary? if the sum of their measures is 180°
8. Use a drawing to illustrate vertical angles. Check students' work.

Name two pairs of vertical angles in each figure. Answers may vary. One example is given.

9. 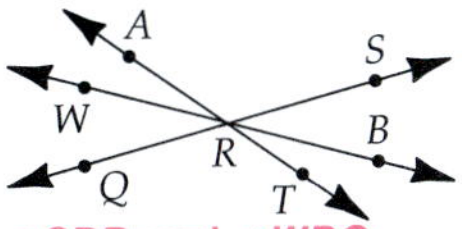

$\angle SRB$ and $\angle WRQ$
$\angle WRA$ and $\angle TRB$

10. 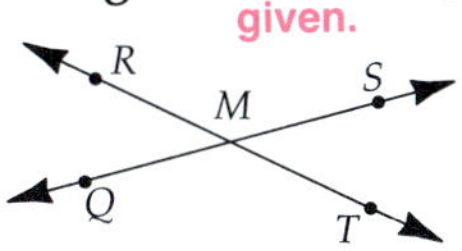

$\angle QMR$ and $\angle SMT$
$\angle QMT$ and $\angle SMR$

Describe the angles in the photograph. Answers will vary.

Closure

Writing in Math Have students write a description in their math journals for each of the terms presented in this lesson. Ask students to include examples from the real world to illustrate each term.

Independent Practice

For question 9, there are six correct answers. Have students work with partners to name all six pairs of vertical angles.

For questions 11-14 have students write equations and explain how they can be used to find the solutions.

Lesson Follow-up

Reteaching Activity Provide these problems for students to solve.

a. Explain how the properties of supplementary angles can be used to find the unknown measure.

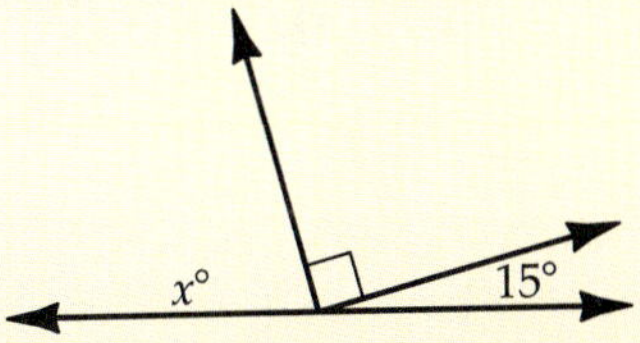

(The sum must be 180°, so $180° - 15° = 165°, 165° - 90° = 75°$)

b. Explain how the properties of supplementary and vertical angles can be used to find the unknown measures.

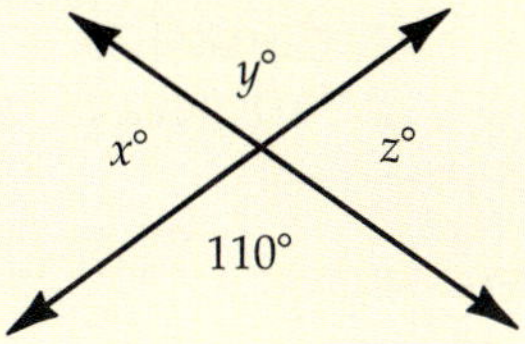

(Vertical angles have equal measures; so $y = 110$. Straight angles are supplementary angles; so $180° - 110° = 70°$, and $x = 70$ and $z = 70$)

LESSON RESOURCES

TRB Practice I-3
TRB Enrichment I-4

Find the measure of each indicated angle.

11. $\angle ACD$ and $\angle ACB$ 45° 135°

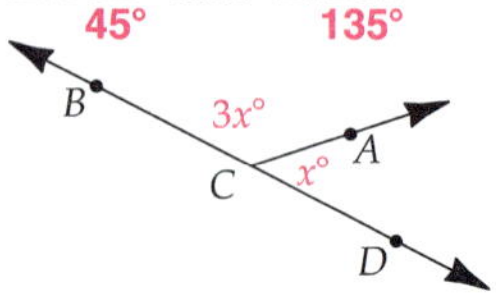

12. $\angle QRT$ and $\angle QRU$ 62° 152°

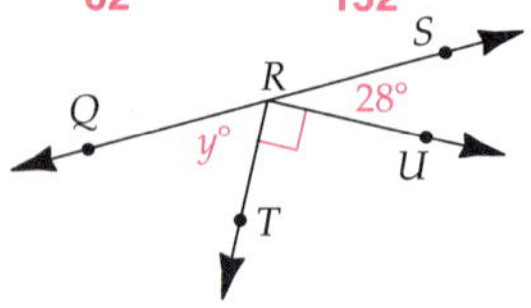

13. $\angle TMX$ 70°

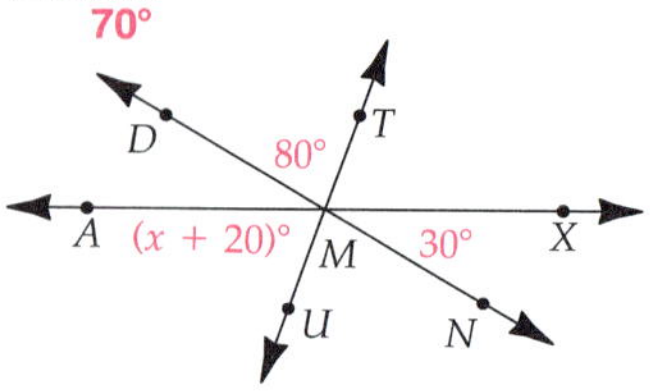

14. $\angle AMT$ and $\angle TMH$ 42° 48°

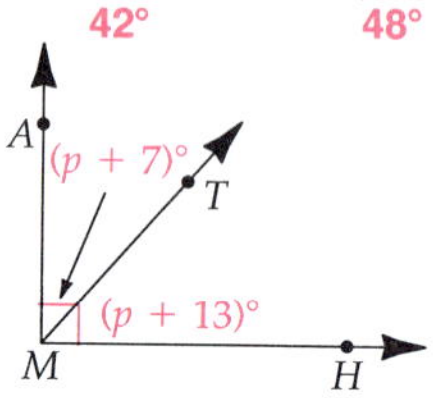

PROBLEM SOLVING HINT

Try guess and test.

15. Find an angle with a measure that is twice as great as a supplement. 120°

16. Find a pair of supplementary angles such that the difference of their measures is 56°. 62° and 118°

Use a protractor to measure each angle.

17.

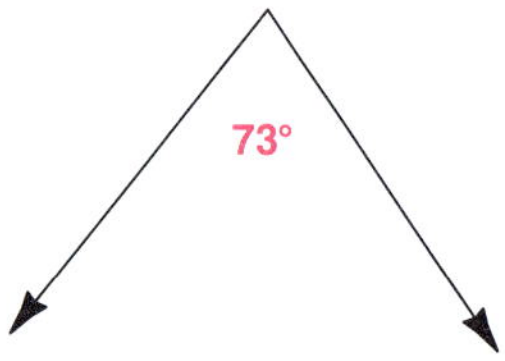

18.

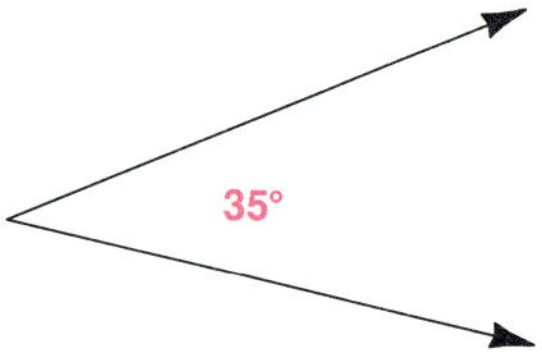

19.

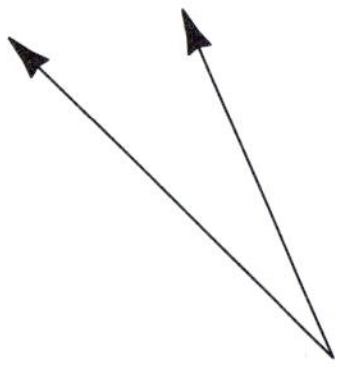

20.

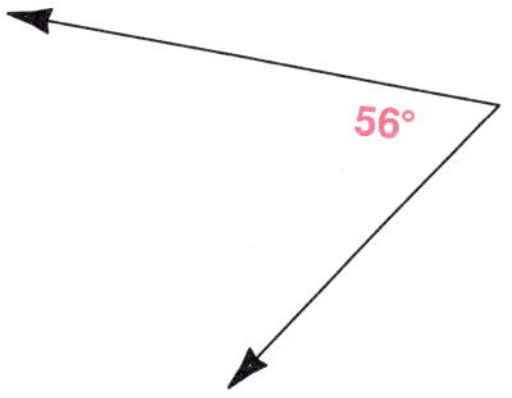

MIXED REVIEW

Solve for y in terms of x.

1. $x = 4y - 8$ $y = \frac{1}{4}x + 2$

2. $x - y + 12 = 2x$ $y = -x + 12$

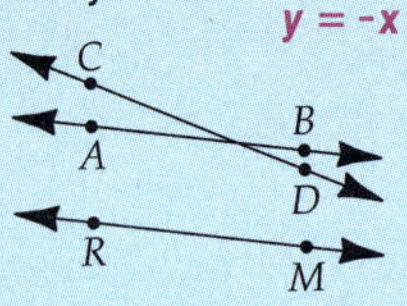

3. Name one pair of parallel segments. $\overline{AB}$ and $\overline{RM}$

4. $\overline{CD}$ and $\overline{AB}$ are called ▒ segments. intersecting

5. The Jones' monthly mortgage payment is \$1,237. There is a 2% penalty if the payment is late. How much is the penalty? \$24.74

Tell whether each angle is acute, right, obtuse, or straight.

21. 65° acute **22.** 45° acute **23.** 90° right **24.** 180° straight

25. 125° obtuse **26.** 27° acute **27.** 21° acute **28.** 108° obtuse

Find the measure of a complement and a supplement of each angle, if possible.

29. 90° none; 90° **30.** 18° 72°, 162° **31.** 115° none; 65° **32.** 89° 1°, 91°

33. 43° 47°, 137° **34.** $x°$ 90° − x°, 180° − x° **35.** $(y - 20)°$ (110 − y)°, (200 − y)° **36.** $(3a)°$ 90° − (3a)°, 180° − (3a)°

Practice

For use after 9-2 (pp. 365–368)

Angles

Classify each angle as acute, obtuse, right, or straight.

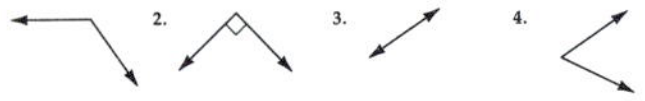

1. obtuse 2. right 3. straight 4. acute

Name the angle that forms a vertical angle with the angle named.

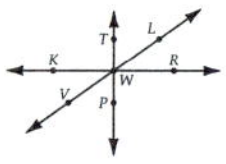

5. $\angle RWP$ $\angle KWT$ 6. $\angle KWV$ $\angle LWR$
7. $\angle TWL$ $\angle VWP$

Find the measure of each indicated angle.

8. $m\angle ABD$ 43°
$m\angle DBC$ 47°

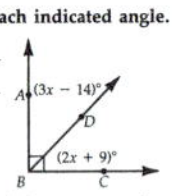

9. $m\angle RNP$ 107°
$m\angle QNP$ 73°

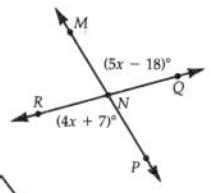

Use a protractor to find the measure of each angle.

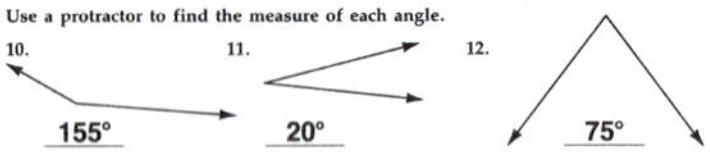

10. 155° 11. 20° 12. 75°

Complete.

	Measure of Angle	Acute, Right, Obtuse, or Straight	Measure of Supplement	Measure of Complement (if possible)
13.	21°	acute	159°	69°
14.	119°	obtuse	61°	none
15.	178°	obtuse	2°	none
16.	57.6°	acute	122.4°	32.4°

17. Find a pair of complementary angles such that the difference of their measures is 12°. 51°, 39°

Chapter 9 I-3

Enrichment

For use after 9-2 (pp. 365–368)

Clock Angles

Complete.

The minute hand of a clock rotates through:

1. 90 degrees in 15 minutes
2. 6 degrees in 1 minute

The hour hand of a clock rotates through:

3. 30 degrees in 1 hour
4. $\frac{1}{2}$ degree in 1 minute

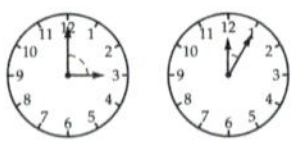

Find how long it takes the minute hand to rotate through these angles.

5. 18° 3 min 6. 78° 13 min 7. 135° $22\frac{1}{2}$ min

Find how long it takes the hour hand to rotate through these angles.

8. 24° 48 min 9. 63° 126 min 10. $155\frac{1}{2}°$ 311 min

Find the measure of the angle formed by the hour and minute hands.

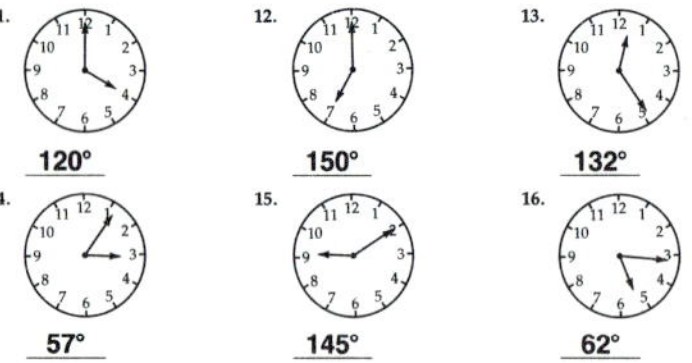

11. 120° 12. 150° 13. 132°

14. 57° 15. 145° 16. 62°

I-4 Chapter 9

LESSON QUIZ

Classify the angles in each example.

1. a scissors fully opened (obtuse and acute)

2. a corner of a sheet of notebook paper (right)

3. a referee's arms signaling a touchdown (acute)

4. a person pointing east and west at the same time (straight)

Assignment Guide

Basic 1-4, 9, 11-12, 15, 17-18, 21-24, 29-32, MR All

Average 3-6, 9, 12-13, 15, 18-19, 23-26, 31-34, MR All

Enriched 5-8, 10, 13-14, 16, 19-20, 25-28, 33-36, MR All

FOR THE NEXT LESSON

geoboards and colored rubber bands, or dot paper and colored pencils, math journal

OBJECTIVE:
To explore segments and angles using geoboards.

Exploring Segments and Angles

▼ You can use a geoboard to explore many geometric concepts. As you complete the activities that follow, think about the ideas of segments and angles. The geoboard at the right shows a segment *one unit in length.*

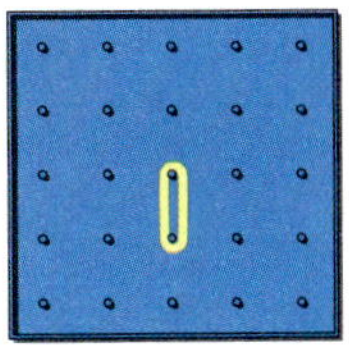

1. Use your geoboard or dot paper to represent a segment of the given length. Check students' work.
   a. 2 units **b.** 4 units **c.** 5 units
2. What is the longest segment you can show? Explain how you know you have the longest possible segment. Diagonal of the geoboard; By rotating the band about the endpoints, you can see that this is the longest.

▼ In geometry, the word *congruent* means *same size and same shape.* Check students' work.

3. ***Model*** the following on your geoboard or draw on dot paper.
   a. two or more congruent segments
   b. two intersecting segments that are not congruent
   Discuss the ways your segments might intersect.
   c. two parallel segments; three parallel segments
   Discuss the ways your segments might be parallel.
   d. two perpendicular segments
   Discuss the ways your segments might be perpendicular.
4. Show each figure below on your geoboard or on dot paper. ***Classify*** each angle as acute, obtuse, right, or straight. Check students' work.

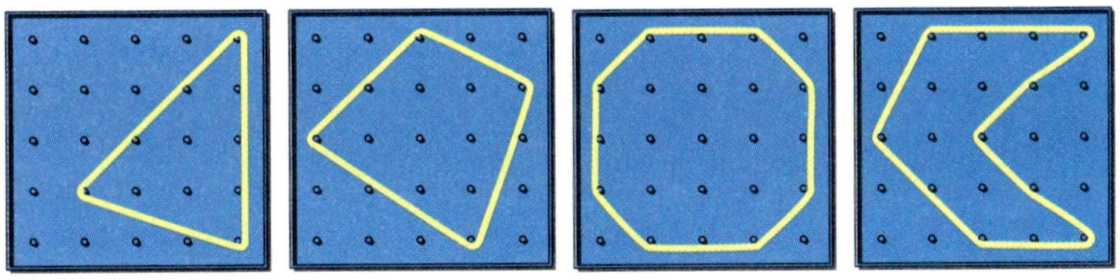

▼ Shapes you make with one rubber band in which edges do not cross or touch are called *polygons.* A place where the band turns a corner (or is hooked on a peg) is called a *vertex.* The segment joining two vertices is called a *side.*

5. Make several polygons on your geoboard or dot paper. Make each one a different color. State the number of vertices and sides for each polygon you made. Check students' work.
6. What are some limitations you experience when modeling figures on a geoboard? Answer may vary. One example is given. You cannot make circles.

MATERIALS

- Geoboards and colored rubber bands or
- Dot paper and colored pencils
- Math journal to record work

Assignment Guide

Basic All
Average All
Enriched All

FOR THE NEXT LESSON

paper, pencil, newspapers, magazines, shape tracers, tangram, math journal

Teaching Tip

Use lessons in which students work with manipulative materials to encourage experimentation. Ask, *What would happen if . . . ?* and *How could you find out . . . ?* Have additional manipulatives available to encourage students to expand their understanding.

Lesson Focus

Materials

Geoboards and colored rubber bands, or dot paper and colored pencils, math journal

Vocabulary/Symbols

congruent
polygon
side
vertex

MOTIVATING THE LESSON

Ask students what they know about segments, polygons, and angles. Then distribute the geoboards and allow students to construct figures of their own choice on the geoboards.

Instruction

Have students record their answers in their math journals.

1 Ask students for suggestions on how to establish segment length when the rubber band is not parallel with one of the sides of the geoboard.

2 Encourage students to explore alternatives by asking, *Is there another way?* or *What other ways can you use?*

3 Students can use the geoboard to make the figures and use dot paper to record and label them.

Closure

Writing in Math Have students write the meaning of these terms in their math journals: *congruent, parallel, perpendicular, side, vertex,* and *angle: acute, obtuse, right, straight.*

Lesson Focus

Materials

Paper, pencil, newspapers, magazines, shape tracers, tangram, math journal

Vocabulary/Symbols

concave polygons
convex polygons
diagonal
parallelogram
polygon
rectangle
regular polygon
rhombus
square
trapezoid

MOTIVATING THE LESSON

Have students find logos containing geometric figures in newspaper and magazine ads. Discuss how the figures are used in the logo designs. Note students' previous knowledge of terms and properties involving quadrilaterals and other polygons.

Instruction

1 Review the definition of *collinear* with students and ask them to write the definition in their math jounals. Then, ask students to draw a polygon and write, in their own words, why the figure is a polygon.

2 Have students work in groups to draw polygons and their diagonals to determine if the polygons are convex or concave. Then, groups exchange their examples and check each others' work.

3 If available, have students use shape tracers to draw two or three regular polygons different from those in the text. Ask students to verify that the figures they drew are regular polygons.

OBJECTIVE: *To discover properties of polygons and quadrilaterals.*

9-3 Polygons and Quadrilaterals

▼ Many of the shapes you see in the world around you and in art are examples of *polygons*.

Describe the polygons you see in the picture of the pyramidal entrance to the Louvre Museum.

Polygon	A polygon is a closed plane figure such that no two segments with a common endpoint are collinear and segments intersect only at the endpoints.

1

Example 1 **State whether or not each figure is a polygon.**

a. **b.** **c.**

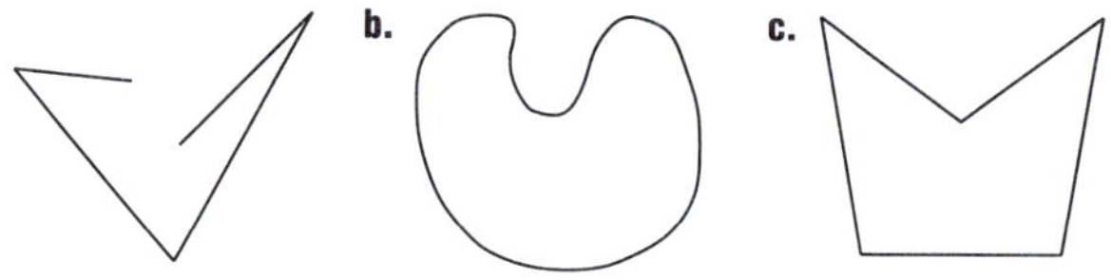

Solution **a.** no (not closed) **b.** no (not segments) **c.** yes

▼ Polygons may be *convex* or *concave*.

Convex and Concave Polygons	A polygon is convex if all points on the diagonals are inside the polygon. Otherwise, the polygon is concave.

FLASHBACK
A *diagonal* is a segment that joins two nonconsecutive vertices.

2

Example 2 **Determine whether each polygon is convex or concave.**

a. **b.** **c.**

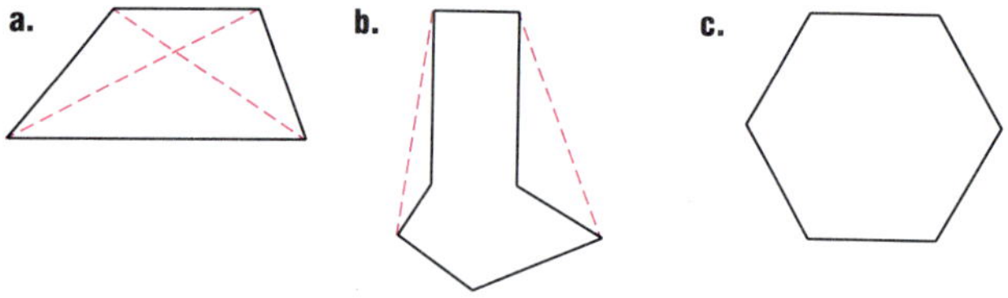

Solution **a.** convex **b.** concave **c.** convex

Teaching Tip

Lessons in geometry benefit from hands-on activities, which students respond to with a high level of attention. Novel materials (such as tangrams, plastic geometry templates, geoboards) motivate interest and help to reinforce concepts.

▼ A polygon is *regular* if the measures of all sides and all angles are equal.

Example 3 **Decide if the polygon is regular.**

a. **b.** **c.**

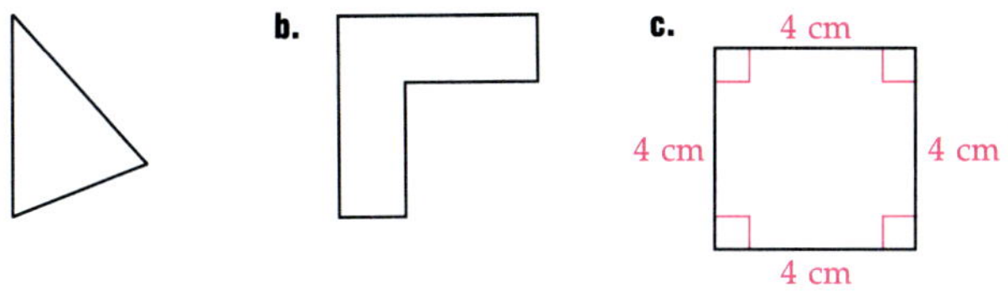

Solution **a.** no **b.** no **c.** yes

3

▼ Some quadrilaterals have special names and properties.

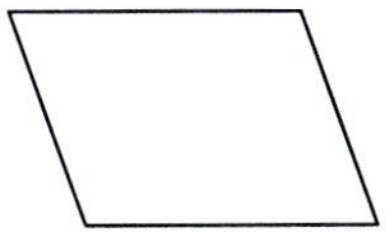

parallelogram
two pairs of opposite parallel sides

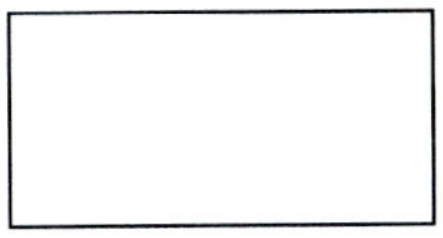

rectangle
parallelogram with four right angles

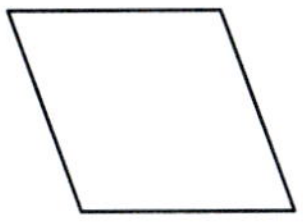

rhombus
parallelogram with all sides equal

square
a parallelogram that is both a rectangle and a rhombus

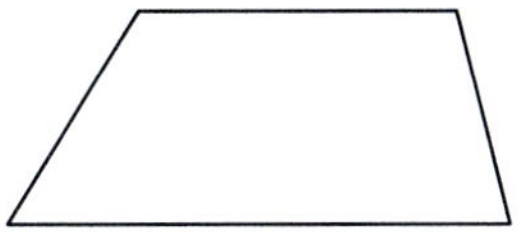

trapezoid
exactly one pair of parallel sides

CLASS EXERCISES

State whether or not the figure is a polygon.

1.

no

2.

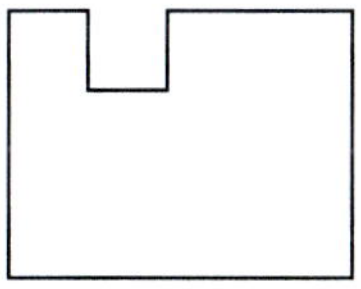

yes

3.

no

Give three correct names for each figure. Choose from polygon, quadrilateral, parallelogram, rectangle, square, or trapezoid.

4.

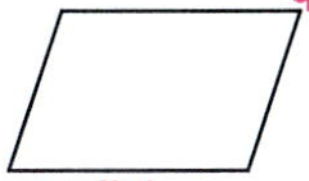

parallelogram polygon quadrilateral

5.

quadrilateral parallelogram polygon square rectangle

6.

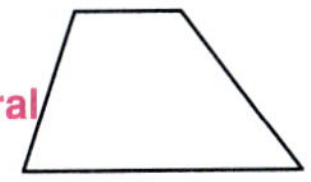

quadrilateral trapezoid polygon

FLASHBACK

We name polygons by their number of sides.

Number of Sides	Polygon
3	triangle
4	quadrilateral
5	pentagon
6	hexagon
7	heptagon
8	octagon
9	nonagon
10	decagon
12	dodecagon
n	n-gon

THINK Is a trapezoid a parallelogram? No, a trapezoid has only one pair of parallel sides.

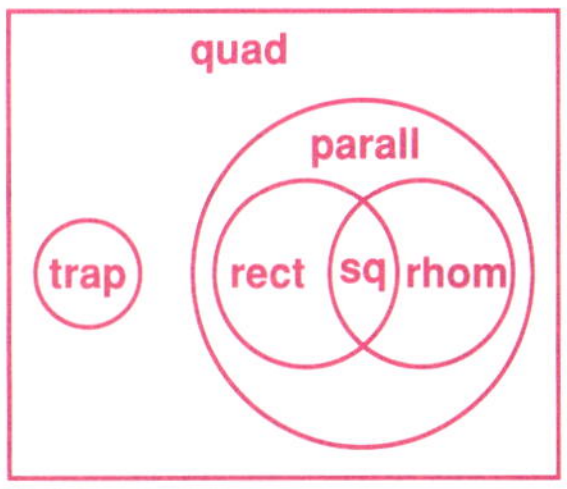

THINK AND DISCUSS

1. Use a Venn diagram to show the relationship between quadrilaterals, squares, parallelograms, trapezoids, rhombuses, and rectangles. See above.

2. Can a regular polygon be concave? no

3. Which quadrilateral is a regular polygon? square

ADDITIONAL EXAMPLES

1. Draw five different quadrilaterals. Name them. (rectangle, square, parallelogram, rhombus, trapezoid)

2. Draw a rhombus with a right angle. Name the figure. (square)

3. Draw three figures that are not polygons. State the reasons why each is not a polygon. (Check students' work.)

Guided Practice

Class Exercises For questions 1-3, have students explain why each figure is or is not a polygon.

Think and Discuss For question 1, work through the Venn diagram on the chalkboard. Then, have students duplicate the process in their math journals.

A ***common error*** is to confuse the names and properties of polygons. It may be helpful to have students enlarge the Venn diagram (from *Think and Discuss*) on posterboard. Then, small groups of students can discuss the diagram and give examples of each figure.

Closure

Writing in Math Have students write in their math journals the properties of each figure presented in this lesson. Ask students to illustrate each description.

Independent Practice

To answer questions 1-8 on page 372, have students refer to the Venn diagram in their math journals.

MATH MINUTES

Have students work in small groups. Ask one student to describe a specific polygon, such as "a quadrilateral with only one pair of parallel sides." (trapezoid) The other students each draw the figure, name it, and compare their answers.

For question 19, have students repeat the activity for three-, four-, and five-sided figures.

Lesson Follow-up

Reteaching Activity Assign *Exploring Regular Polygons*, found on p. 360E of the Chapter Overview. See also TRB Exploring O-35.

LESSON RESOURCES

TRB Practice I-5
TRB Enrichment I-6

Practice

For use after 9-3 (pp. 370–372)

Polygons and Quadrilaterals

Is the figure a polygon? Write *yes* or *no*.

1. no
2. yes
3. no

Tell whether the statement is *true* or *false*.

4. A quadrilateral is a polygon with four sides. **true**
5. All quadrilaterals are parallelograms. **false**
6. A rhombus is an equilateral parallelogram. **true**
7. A square is a rectangle. **true**
8. No trapezoid is a parallelogram. **true**
9. All squares are convex polygons. **true**

Give two other correct names for each figure. Answers may vary.

10. trapezoid *PMKR* **PRKM, MKRP**
11. parallelogram *RTGH* **RHGT, TGHR**

Write whether the polygon is concave or convex. Then classify the polygon by the number of sides.

12. concave quadrilateral
13. convex pentagon
14. concave octagon

Draw a figure to fit each description. Figures may vary.

15. convex hexagon
16. triangle
17. concave quadrilateral

Chapter 9 I-5

Enrichment

For use after 9-3 (pp. 370–372)

Networks

The figure shows a *network* of connected segments called *paths*. A point where paths intersect is called a *vertex*. Vertex *F* is *even* because an even number of paths (4) meet there. Vertex *C* is *odd* because an odd number of paths (3) meet there.

Tell whether the vertex in the above figure is even or odd.

1. *A* **even**
2. *B* **even**
3. *D* **odd**
4. *E* **even**

Some networks can be traced without lifting your pencil or retracing any path. The figure on the left is traceable but the one on the right is not. (Check this for yourself.)

Write the number of even or odd vertices in each figure. Then tell if the figure is traceable. Write *yes* or *no*.

5. even 4, odd 0, traceable yes
6. even 2, odd 0, traceable yes
7. even 0, odd 4, traceable no
8. even 2, odd 2, traceable yes
9. even 2, odd 2, traceable yes
10. even 1, odd 4, traceable no

11. Study your results in Exercises 5–10. Then state a rule that tells whether or not a network is traceable depending on the number of even and odd vertices that it has.
A network is traceable if and only if it has no odd vertices or exactly 2 odd vertices.

I-6 Chapter 9

WRITTEN EXERCISES

True or false? Write *T* or *F*.

1. Every rhombus is a regular quadrilateral. **F**
2. All quadrilaterals are parallelograms. **F**
3. Some trapezoids are squares. **F**
4. All squares are rectangles. **T**
5. All rectangles are squares. **F**
6. Some rectangles are rhombuses. **T**
7. Some parallelograms are squares. **T**
8. All parallelograms are quadrilaterals. **T**

Give two other correct names for each figure.

9. square *ABCD*

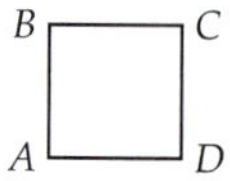

**rectangle *ABCD*
polygon *ABCD***

10. parallelogram *BASE*

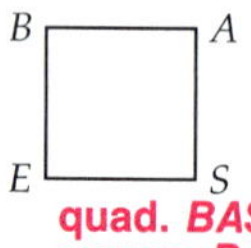

**quad. *BASE*
square *BASE***

11. trapezoid *QRTS*

**quad. *QRTS*
polygon *QRTS***

Determine whether the polygon is concave or convex. Then classify the polygon by the number of sides.

12. 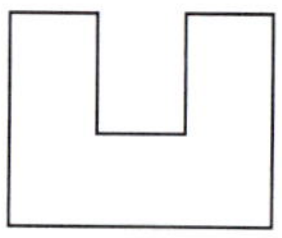

concave, octagon

13. 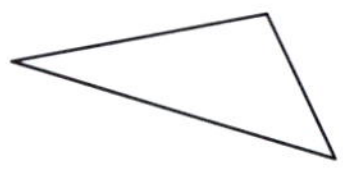

convex, triangle

14. 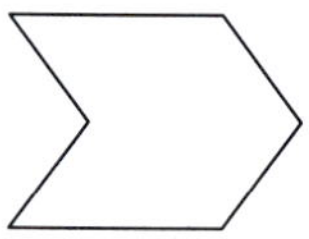

concave, hexagon

Draw a figure to fit each description. **Check students' work.**

15. a convex quadrilateral
16. a concave hexagon
17. a pentagon
18. a rhombus

19. Look at convex hexagon *ABCDEF*. How many diagonals can you draw from vertex *A*? How many triangles are formed? **3, 4**

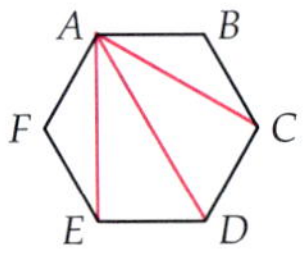

20. ***PROJECT*** Use or make a tangram. **Check students' work.**
 a. Classify the pieces in as many ways as possible.
 b. Use any number of pieces to form a square, a hexagon, a pentagon, a trapezoid, and a parallelogram.
 c. Record the number of pieces used to make each figure.
 d. Compare with a classmate. Did you both use the same pieces to make the figures? Is there more than one way to make each figure?

MIXED REVIEW

Simplify.

1. x^2 for $x = 2.5$ **6.25**
2. $2y^3$ for $y = 0.8$ **1.024**
3. Find a supplement to a 120° angle. **60°**
4. Find a complement to a 43° angle. **47°**
5. Name the vertex in angle *ABC*. ***B***

Solve.

6. $3a - 4 > 7$ $a > \frac{11}{3}$
7. $4y = 17 + 3y$ $y = 17$
8. Joe bought 7 more tickets than Ellen. Together they bought 19 tickets. How many tickets did each buy? **Joe 13; Ellen 6**

LESSON QUIZ

(Check students' work for questions 1-3.)

1. Draw a convex polygon.
2. Draw a concave polygon.
3. Draw a convex octagon.
4. Give three other names for a rectangle. (parallelogram, quadrilateral, polygon)
5. How many sides are found in a *n*-gon? (*n*)

Assignment Guide

Basic 1-4, 9-10, 12-13, 15-16, 20a-b, MR All
Average 3-6, 10-13, 16-18, 20b-c, MR All
Enriched 5-8, 10-11, 13-14, 18-19, 20b-d, MR All

FOR THE NEXT LESSON

graph paper, LOGO software and computer (optional), math journal

OBJECTIVE:
To explore segments, angles, and polygons using LOGO.

MATERIALS

- graph paper
- LOGO software and a computer (optional)
- Math journal to record work

LOGO COMMANDS

FD	forward
BK	back
RT	right turn
LT	left turn
PU	pen up
PD	pen down

Exploring with LOGO

■ You can use LOGO to draw segments, angles, and polygons.

1. a. Sketch what you think the following commands will produce. Then, if possible, try them on a computer. **Check students' work.**

FD 25
RT 45
FD 25
RT 45

b. Classify the angle from part(a). **obtuse**

c. What command determines the measure of the angle? What numbers would give an acute angle? an obtuse angle? **RT 45; 91 to 179; 1 to 89**

d. Suppose you change the second and fourth lines from RT to LT. Does this change your angle measure? **no**

e. Write a procedure to make an animal from segments and angles. Use the commands at the left. **Check students' work.**

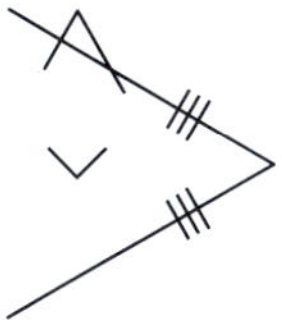

2. a. Sketch what you think the following commands will produce. Then, if possible, try them on a computer.

FD 65 RT 90 FD 65 RT 90
FD 65 RT 90 FD 65 RT 90

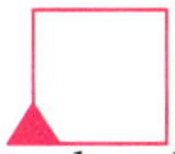

b. What would you change to make a larger square? **FD 65**

c. What would you change to draw a rectangle that is not a square? **The second and fourth measures in FD.**

d. What would you change to draw a rhombus? **the angle measures**

■ You can use the REPEAT command to shorten your procedures.

3. Type in the following POLYGON procedure.

TO POLYGON :LENGTH :SIDES
REPEAT :SIDES [FD :LENGTH RT 360 / :SIDES]
END

a. Choose a value for LENGTH. Replace SIDES with 3 and then with 4. What types of polygons did you get? **triangle; quadrilateral**

b. What would SIDES be for a hexagon? **6**

c. Try greater and greater values for SIDES. ***Describe*** what happens to your polygon. **It gets closer and closer to a circle.**

Lesson Focus

Materials

Graph paper, LOGO software and computer (optional), math journal

MOTIVATING THE LESSON

Ask students to describe how they would tell someone unfamiliar with geometric figures to draw a triangle, an angle of 60°, and a square. Encourage students to think of each step involved. Have volunteers draw the figures on the chalkboard.

Instruction

Have students record their work in their math journals.

1 Remind students that the distance 25 is 25 unit lengths on the computer screen. When working on graph paper, students should define their unit length. Also, remind students to press the RETURN key after they enter each command.

2 Encourage students to revise their procedure as they make their animal.

3 For *b-d*, encourage students to experiment with the length of the sides and the measures of the angles.

4 In the POLYGON procedure, LENGTH is the length of each side of the polygon, and SIDES represents the number of sides.

For example, to create a 5-sided figure with 25 as the length of each side, type POLYGON 25 5 and press RETURN.

Closure

Writing in Math Have students summarize, in their math journals, the effect that changing the side lengths, angle measures, and number of sides had on their geometric figures.

ssignment Guide
asic All
verage All
nriched All

OR THE NEXT LESSON
onstruction paper, pencil, cissors, protractor, math urnal

Teaching Tip

When students are dealing with technology that is new or unfamiliar to them, review the technical terms and commands. Allow students to practice with terms and commands before they attempt the questions.

Lesson Focus

Materials

Construction paper, pencil, scissors, protractor, math journal

Vocabulary/Symbols

acute triangle
equiangular triangle
equilateral triangle
isosceles triangle
obtuse triangle
right triangle
scalene triangle
counterexample

MOTIVATING THE LESSON

Write the following statement on the chalkboard:

The sum of the measures of the interior angles is 180° for (no, some, all) triangles.

Have students choose the word they think makes the statement true. Then, discuss how they could determine if their choice is correct. Ask students to draw and cut out the angles of triangles made from construction paper and compare their results.

Instruction

1 Mention that triangles are sometimes classified by sides *and* angles. One example is an obtuse isosceles triangle.

2 If the experiment is done as a class exercise, suggest to students that they construct their triangles to include a variety of sizes and types. Remind students that they are testing the statement for ALL triangles.

ADDITIONAL EXAMPLES

1. Is it possible to have a triangle with two right angles? (No, the sum of the measure of the interior angles of a triangle is 180°.)

OBJECTIVE: *To discover properties of triangles.*

9-4 Triangles

▼ The framework of a geodesic dome is made up of triangles. Engineers use triangles to lend stability to bridges and other structures.

1 ▼ We classify triangles by their angles or sides.

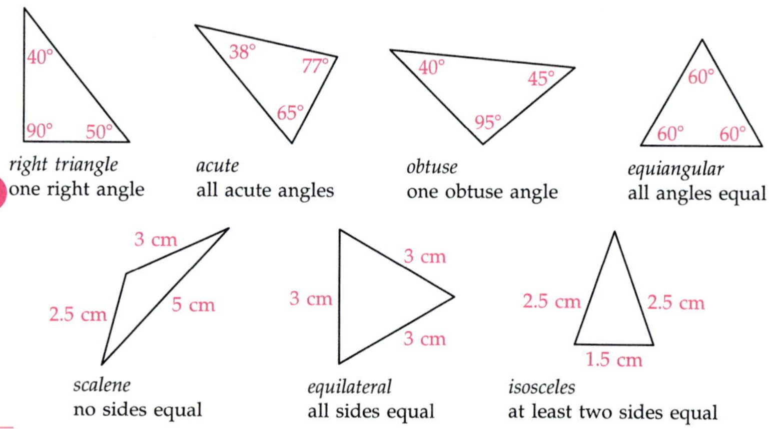

THINK How could you use a protractor to find the sum of the measures of the interior angles of a triangle? **Measure each angle. Add the three measures.**

2 ▼ You can do an experiment to show that the sum of the measures of the interior angles of a triangle is 180°.

a. Draw and label any triangle ABC. Tear off any two angles. (B and C for example)

b. Position B and C next to A, bringing vertices together. Notice that the angles form a straight angle.

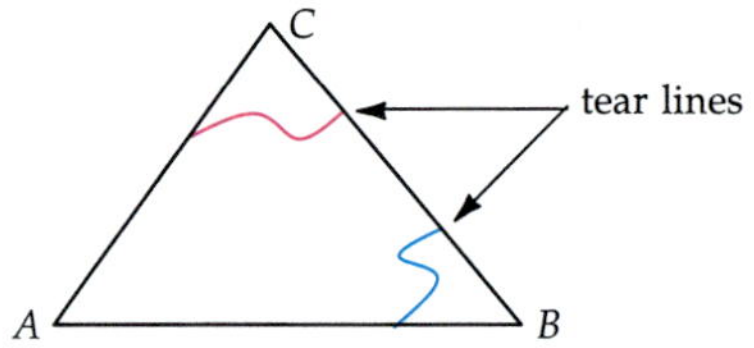

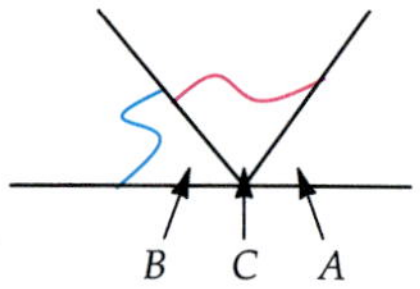

$m\angle A + m\angle B + m\angle C = 180°$

Teaching Tip

Asking the right kinds of questions can help students develop the higher order thinking skills needed to solve complex mathematical problems. Questions such as *How do you know that?* or *Why does that work?* encourage students to explain.

Example Find $m\angle 1$.

Solution

$$m\angle 1 + 47^\circ + 63^\circ = 180^\circ$$
$$m\angle 1 + 110^\circ = 180^\circ$$
$$m\angle 1 = 180^\circ - 110^\circ$$
$$= 70^\circ$$

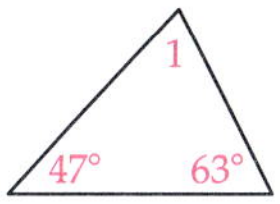

CLASS EXERCISES

Classify each triangle as acute, obtuse, or right.

1. 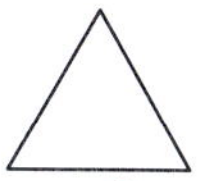 acute
2. 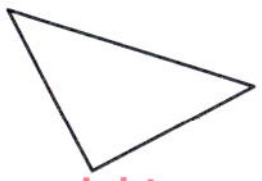 right
3. 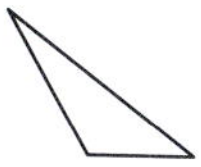 obtuse

Draw a figure to fit each description. Check students' work.

4. right triangle ABC
5. an obtuse scalene triangle
6. isosceles triangle TRI
7. a regular triangle

WRITTEN EXERCISES

Classify each triangle by its angles or its sides.

1. 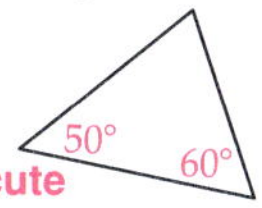

acute

2. 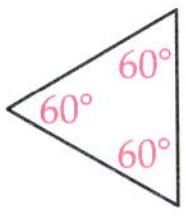

equiangular

3. 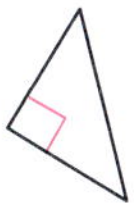
right

4. 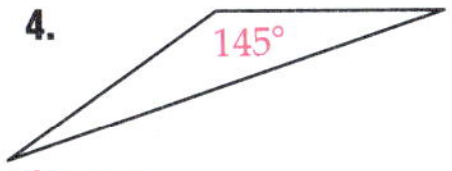

obtuse

5. 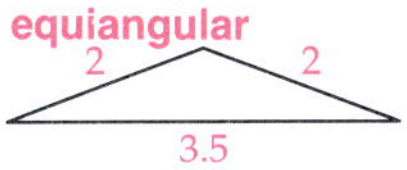

isosceles

6. 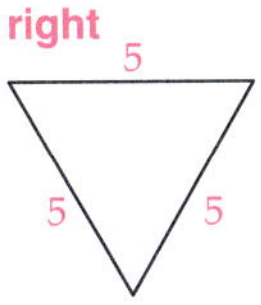

equilateral or equiangular

True or false? Write T or F.

7. No equilateral triangles are scalene. T
8. All right triangles are isosceles. F
9. Some right triangles are isosceles. T
10. Some right triangles are obtuse. F
11. All triangles have three angles. T
12. Some triangles are regular. T
13. Some parallelograms are triangles. F
14. All triangles are equilateral. F

THINK AND DISCUSS

1. Does a triangle have diagonals? Explain.
2. Are all equilateral triangles isosceles? Are all isosceles triangles equilateral? Explain.
3. What is a common name for a regular triangle?

1. No; there are no two nonconsecutive vertices in a triangle.
2. Yes; they have at least two congruent sides. No; isosceles triangles need only two congruent sides.
3. Equilateral or equiangular

MATH MINUTES

Students use paper triangles to find the true statements.
1. All equilateral triangles are acute. (T)
2. No equilateral triangle is right. (T)
3. Some equilateral triangles are obtuse. (F)

2. Must all angles in an obtuse triangle be obtuse? Explain. (No, only one angle can be obtuse, since the sum of the measures of the interior angles of a triangle is 180°. If one angle is greater than 90° then the sum of the other two angles must be less than 90°.)

Guided Practice

Class Exercises Have students explain their answers for questions 1-3.

Think and Discuss Extend question 2 by asking if all equilateral triangles are equiangular (yes) and if all equiangular triangles are equilateral. (yes)

A ***common error*** occurs when students classify triangles only by their angle measures or by their sides. For example, a right isosceles triangle may be classified as a right triangle only. Suggest that students first identify the angle measures, then identify the sides, and finally name the triangle.

Closure

Writing in Math Have students use their math journals to describe each classification of triangle and the relationships between equiangular and equilateral triangles.

Independent Practice

For questions 7-14, have students explain their answers in writing.

As an extension, refer to the manipulative activity on p. 360E of the Chapter Overview. See also TRB Exploring O-33.

Critical Thinking For questions 2-4, have students explain why one counterexample is sufficient. (One counterexample is proof that the statement is false.)

Lesson Follow-up

Reteaching Activity Assign *Exploring Polygons with LOGO*, found on p. 360E of the Chapter Overview. See also TRB Exploring O-36.

LESSON RESOURCES

TRB Practice I-7
TRB Enrichment I-8

Practice — For use after 9-4 (pp. 374–376)

Triangles

Classify each triangle by its angles or its sides.

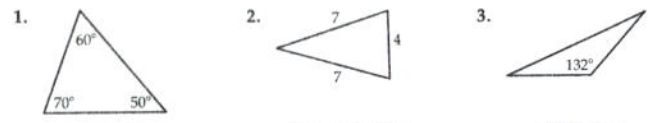

1. acute 2. isosceles 3. obtuse

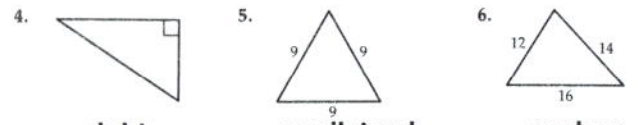

4. right 5. equilateral 6. scalene

Tell whether the statement is *true* or *false*.

7. An equilateral triangle is isosceles. true
8. A triangle can have more than one right angle. false
9. A scalene triangle may be isosceles. false
10. An equilateral triangle cannot be obtuse. true
11. A right triangle may be an acute triangle. false
12. A triangle cannot have more than one obtuse angle. true

Find the measure of the indicated angle.

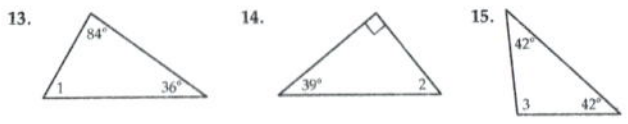

$m\angle 1 =$ 60° $m\angle 2 =$ 51° $m\angle 3 =$ 96°

CALCULATOR **Find the measure of the third angle of a triangle that has two angles with the given measures.**

16. 117°, 58° 5° 17. 9°, 23° 148° 18. 155.5°, 23° 1.5°

19. The measures of the angles of a triangle are $(3x + 15)°$, $(4x - 21)°$, and $(2x + 24)°$. Find the measure of each angle. 69°, 51°, 60°

Chapter 9 I-7

Enrichment — For use after 9-4 (pp. 374–376)

Angles of a Polygon

The sum of the measures of the angles of a triangle is 180°. You can find the sum of the measures of the angles of a polygon by dividing it into triangles and then multiplying the number of triangles by 180°.

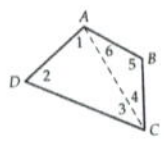

1. Diagonal $\overline{AC}$ divides quadrilateral $ABCD$ into triangles.
   a. Find $m\angle 1 + m\angle 2 + m\angle 3$. 180°
   b. Find $m\angle 4 + m\angle 5 + m\angle 6$. 180°
   c. Find $m\angle A + m\angle B + m\angle C + m\angle D$. 360°

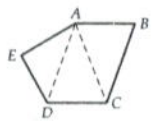

2. Diagonals $\overline{AC}$ and $\overline{AD}$ divide pentagon $ABCDE$ into triangles.
   a. How many triangles are there? 3
   b. Describe how you can find the sum of the measures of the angles of the pentagon. Multiply 3 by 180°.
   c. What is the sum? 540°
3. Complete.

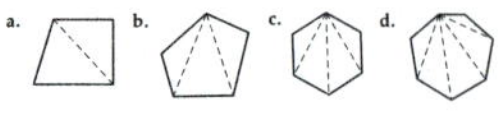

	a.	b.	c.	d.
Number of sides	4	5	6	7
Number of triangles	2	3	4	5
Sum of measures of angles	360°	540°	720°	900°

4. Write a formula for the sum of the measures of the angles of a polygon with n sides. $(n - 2)180°$
5. Find the sum of the measures of the interior angles of a polygon with the given number of sides.
   a. 8 1,080° b. 13 1,980° c. 29 4,860° d. 185 32,940°
6. The sum of the measures of the angles of a polygon is given. Find the number of sides.
   a. 1,620 11 b. 2,520 16 c. 3,960 24

I-8 Chapter 9

MIXED REVIEW

Insert parentheses to make each equation true.

1. $6^2 + (18 + 9) \cdot -2 = -18$
2. $(150 + 17) \cdot 10 = 1{,}670$

Draw a figure to fit each description. Check students' work.

3. a regular triangle
4. a pentagon

Write as a decimal.

5. 23.5% 0.235
6. 5.9% 0.059
7. A mail-order catalog offers computer disks at 100 for $121.50 or 50 for $62.50. Which is the lower price per disk? 100 for $121.50

Draw a figure to fit each description. Check students' work.

15. a convex parallelogram
16. a trapezoid
17. an obtuse scalene triangle
18. a triangle with one obtuse angle

Find the measure of each numbered angle.

19. 90° 20. 80° 21. 50°

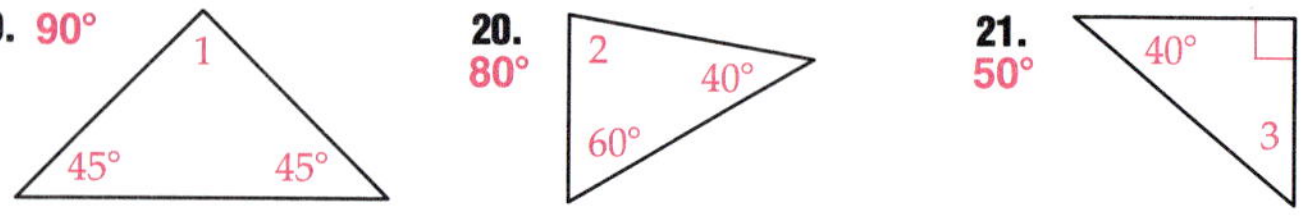

CALCULATOR **Find the measure of the third angle of a triangle that has two angles with the given measures.**

22. 110°, 35° 35°
23. 25°, 65° 90°
24. 45°, 45° 90°
25. 30°, 60° 90°
26. 126°, 20° 34°
27. 42°, 78° 60°
28. $(2y)°$, 60° $120° - (2y)°$
29. $(x + 4)°$, 96° $(80 - x)°$

Classify each triangle as acute, right, or obtuse.

30. $\triangle MNO$ with $m\angle M = 35°$, $m\angle N = 70°$, and $m\angle O = 75°$ acute
31. $\triangle ABC$ with $m\angle A = 104°$, $m\angle B = 46°$, and $m\angle C = 30°$ obtuse

Critical Thinking

EXPLORING INDUCTIVE REASONING

▼ You use *inductive reasoning* to draw conclusions based on specific examples.

1. Does the expression $n^2 + n + 5$ always produce a prime number when n is a positive integer?
   a. Try several values of n.
   b. A *counterexample* is an example that proves your conclusion is false. What value of n is a counterexample? 4

n	$n^2 + n + 5$
1	7
2	11
3	17
4	25

Use inductive reasoning to tell whether the statement is true or false. If false, give a counterexample.

2. All squares are rectangles. T
3. All rectangles are squares. F
4. All equilateral triangles are equiangular. T
5. ***DISCUSS*** How do you know when you have tried enough examples to make an inductive conclusion?

5. You can make an inductive conclusion after trying numerous examples but you cannot prove the conclusion true. If you have proved the statement to be false, you are done once you find a counterexample.

LESSON QUIZ

1. Find the measure of the third angle for each triangle.
   (a) 23° and 53° (104°)
   (b) 125° and 50° (5°)
   (c) 60° and 60° (60°)
2. Classify each triangle in question 1. (obtuse triangle, obtuse triangle, equiangular triangle)

Assignment Guide

Basic 1-3, 7-10, 15-16, 19-20, 22-25, 30, MR, CT All
Average 3-5, 9-12, 16-17, 20-21, 24-27, 30, MR, CT All
Enriched 4-6, 11-14, 17-18, 20-21, 26-31, MR, CT All

FOR THE NEXT LESSON

paper, pencil, compass, protractor, pocket mirror, construction paper, math journal

OBJECTIVE:
To discover the properties of circles.

9-5 Circles

▼ Compact disks are shaped like circles. The circular shape allows the disk to store thousands of bits of information on a spiral track. If unraveled, the hair-thin track could stretch for several miles.

Circle	A circle (⊙) is the set of all points the same distance from a given point called the center.

▼ The table shows some parts of circles and their properties.

1

Term	*Description*
chord	a segment that has endpoints on the circle
diameter	a chord that passes through the center of the circle; The diameter (d) is the length of such a chord.
radius	a segment that has endpoints at the center of the circle and on the circle; The radius (r) is the length of such a segment.
central angle	an angle with vertex at the center of the circle

▼ We usually name a circle by its center.

Example 1 **Name all the radii, diameters, chords, and central angles shown in ⊙O.**

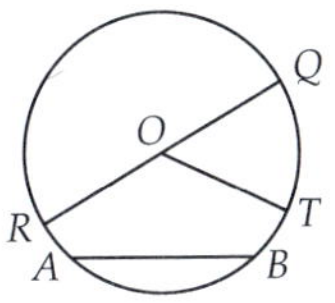

2

Solution

radii	$\overline{OT}(\overline{TO})$; $\overline{OQ}(\overline{QO})$; and $\overline{OR}(\overline{RO})$
diameter	$\overline{RQ}(\overline{QR})$
chord	$\overline{AB}(\overline{BA})$; $\overline{RQ}(\overline{QR})$
central angles	$\angle TOQ(\angle QOT)$; $\angle ROT(\angle TOR)$

▼ In any circle, the diameter d is twice the radius r.

FLASHBACK
We denote the length of $\overline{AB}$ as AB.

Example 2 In a given circle, $\overline{YZ}$ and $\overline{XW}$ are diameters.

a. If $OZ = 10$ cm, find YZ.

b. If $WX = 25$ cm, find OX.

3

Solution Draw a figure to show the given information.

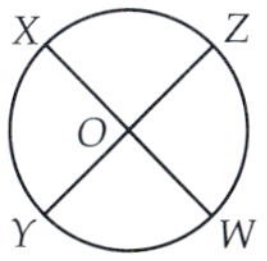

a. $YZ = 2(OZ)$
$= 2(10) = 20$ cm

b. $OX = \frac{WX}{2}$
$= \frac{25}{2} = 12.5$ cm

Lesson Focus

Materials
Paper, pencil, compass, protractor, pocket mirror, construction paper, math journal

Vocabulary/Symbols
central angle
chord
circle
diameter d
radius (radii) r

MOTIVATING THE LESSON
Draw a circle on the chalkboard. Show a diameter, radius, and chord without naming the parts. Ask students to copy the figure in their math journals and to name and label the parts they know.

Instruction

1 Have students use the table to check their labels for the circle in their math journals.

2 Have students work through the solutions orally. Then, ask if all diameters are chords (yes) and if all chords are diameters. (no)

3 Ask students why it is necessary in Example 2a to multiply by 2 and in Example 2b to divide by 2. (The diameter YZ is twice the radius OZ; the radius OX is one-half the diameter WX.)

ADDITIONAL EXAMPLES
1. Draw a circle.
   a. Draw and label two diameters.
   b. Label one chord that is not a diameter.
   c. Draw and label one radius that is part of a diameter.
   d. Name one central angle. (Check students' work.)
2. If a diameter is 15 cm, how long is the radius? (7.5 cm)

Teaching Tip
Evaluate your teaching style by occasionally asking students for comments. Ask them what you did that was most helpful and have students give suggestions for future lessons that might help them learn a concept. (Suggestions might include using models, and so forth.)

Guided Practice

Class Exercises Help students understand that *radii* is the plural form of *radius*. For question 4, ask students what percent equals $\frac{1}{5}$. (20%)

Think and Discuss Question 1 offers an opportunity to reinforce the concept of counterexample presented in the previous lesson.

A ***common error*** occurs when students use *diameter* for *radius* and *radius* for *diameter*. Ask students which is longer, the radius or the diameter. Ask students for ways of remembering which is the radius and which is the diameter. Have them suggest memory aids.

Closure

Explore with students the use of the circle in curriculum areas other than math. For example, discuss what a circle might symbolize in art and why it is a common shape in design.

Independent Practice

Have students rewrite the statement in question 1 to make it true.

Before students begin questions 21-22, review the first two steps for making a circle graph. (Write the data in decimal form, and multiply each decimal by 360.) Ask students what the product represents. (central angle)

For question 24, suggest that students limit their data to seven items or less.

Makeup of Gold Jewelry

copper	22%
silver	25%
gold	53%

Metals in Gold Jewelry

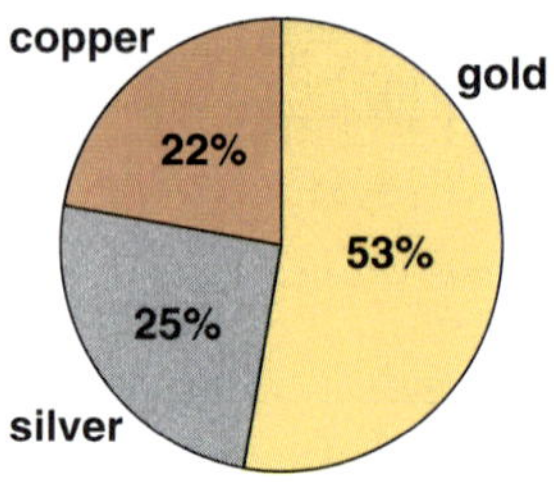

▼ There are 360° in any circle. You can use this fact to construct circle graphs.

Example 3 **Construct a circle graph for the data at the left.**

Solution

1. Write the data in decimal form. $22\% = 0.22$, $25\% = 0.25$, $53\% = 0.53$
2. Multiply each decimal by 360 to determine the measure of each central angle. Round to the nearest degree. Make sure the total is 360. 360 ⊗ 0.22 ⊜ 79.2 ≈ 79; 360 ⊗ 0.25 ⊜ 90; 360 ⊗ 0.53 ⊜ 190.8 ≈ 191
3. Draw a circle with a compass and any radius. Draw each angle with a protractor.
4. Label each section. Give your graph a title.

chords: $\overline{QR}$, $\overline{BP}$, $\overline{ML}$ diameters: $\overline{ML}$, $\overline{BP}$
radii: $\overline{OM}$, $\overline{OB}$, $\overline{OL}$, $\overline{OP}$ cent. angles: ∠MOP, ∠LOP, ∠LOB, ∠BOM

THINK AND DISCUSS

1. Is every chord a diameter? Explain. See below.
2. How many chords, radii, and diameters does a circle have?

1. No; only the chord that passes through the center is a diameter.
2. an infinite number

CLASS EXERCISES

1. Name all chords, radii, diameters, and central angles in the circle at the right.

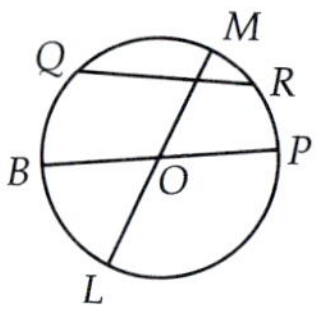

2. If $PO = 15$ in., find PB. 30 in.
3. If $LM = 37.5$ in., find OM. 18.75 in.
4. $\angle MOP$ is $\frac{1}{5}$ of circle O. Find $m\angle MOP$. 72°

NOTES & QUOTES

The ancient Sumerians believed that Earth revolved around the sun. The Sumerians assumed that Earth traveled in a circular path that took 360 days (one year) to complete. They used this relationship to define the degree as the distance traveled in one day, or $\frac{1}{360}$ of the circle.

WRITTEN EXERCISES

True or false? Write *T* or *F*.

1. If a chord of a circle is 20 cm long, the radius is 10 cm. F
2. The longest chord in any circle is the diameter. T
3. All radii of a given circle are the same length. T
4. All chords of a circle are the same length. F
5. If two circles have the same radius, they have the same diameter. T

MENTAL MATH **For each length, find the radius or the diameter.**

6. $r = 50$ cm, $d =$ ■ 100 cm
7. $r = 42.5$ in., $d =$ ■ 85 in.
8. $d = 100$ cm, $r =$ ■ 50 cm
9. $d = 70$ cm, $r =$ ■ 35 cm
10. $r = 36$ ft, $d =$ ■ 72 ft
11. $d = 500x$ ft, $r =$ ■ 250x ft

MATH MINUTES

Have students use compasses to draw overlapping circles forming a variety of shapes. Encourage students to color their designs and to create a bulletin board display of their work.

CALCULATOR For each length, find the radius or the diameter.

12. $r = 32.2$ mm
$d =$ ▒ 64.4 mm

13. $r = 15.75$ in.
$d =$ ▒ 31.5 in.

14. $d = 17.09$ cm
$r =$ ▒ 8.545 cm

15. $d = 0.58$ cm
$r =$ ▒ 0.29 cm

16. $r = 7{,}832$ ft
$d =$ ▒ 15,664 ft

17. $d = 90{,}089$ mi
$r =$ ▒ 45,044.5 mi

18. Circle O has a diameter of 12 in. Point A is 6 in. from point O.

a. Is point A inside, outside, or on the circle? on the circle

b. Is $\overline{OA}$ a radius of $\odot O$? yes

c. Point B is on $\odot O$. Is $\overline{AB}$ a chord? yes

d. What must be true of $\overline{OB}$ for $\overline{AB}$ to be a diameter? *B, O, and A must be collinear.*

19. Draw $\odot O$, with $r = 3$ cm. Label two points as C and D inside the circle.

a. Is $\overline{CD}$ a chord of $\odot O$? no

b. Can $\overline{CD}$ be a diameter of $\odot O$? why or why not? no; C and D are not on the circle.

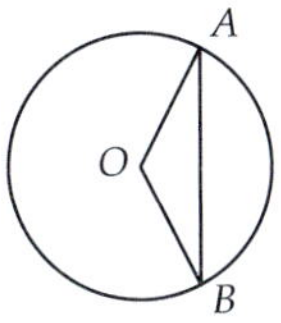

20. In $\odot O$ at the right, $\overline{OA}$ and $\overline{OB}$ are radii. What kind of triangle is $\triangle AOB$? How do you know?
Isosceles; since $\overline{OA}$ and $\overline{OB}$ are radii, they are equal.

CALCULATOR Use the data to construct a circle graph.
Check students' graphs.

21. Recommended Diet Components

Carbohydrates	48%
Refined sugars	10%
Saturated fats	10%
Monosaturated fats	10%
Polyunsaturated fats	10%
Protein	12%

22. Teens' Career Choices

Health care	16%
Trade	8%
Teacher	7%
Performer	6%
Sports	6%
Business/law	6%
Other	20%
Undecided	31%

23. **CALCULATOR** In a survey about favorite colors, 28% responded that blue was their favorite color. Another 56% chose red. All other respondents chose yellow. Draw a circle graph to display the results of the survey. Check students' work.

24. **PROJECT** Choose a topic from the list below, or one of your own. Collect data in your class and construct a circle graph to display the results. Check students' work.

a. career choice
b. favorite sport
c. favorite subject
d. own TV, camera, phone, computer

PROBLEM SOLVING HINT

Drawing a diagram may help.

MIXED REVIEW

Solve.

1. $\frac{6}{8} = \frac{n}{12}$ 9 **2.** $\frac{a}{24} = \frac{30}{48}$ 15

Evaluate.

3. $y^2 + 5$ for $y = 12$ 149

4. $5a^4 - a$ for $a = 10$ 49,990

5. Find $m\angle 1$ and $m\angle 2$.
$m\angle 1 = 100°$; $m\angle 2 = 60°$

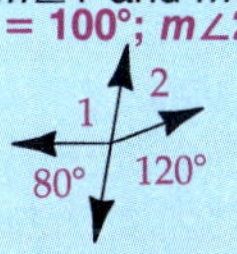

Use the formula $d = rt$.

6. A satellite orbiting Earth travels at a rate of 28,000 km/h. How far will it travel in 24 h? 672,000 km

Lesson Follow-up

Reteaching Activity Have students orally work through each step in constructing a circle graph. Then have them make a circle graph with four sections labeled 10%, 20%, 30%, and x respectively. Ask students what percent the x section represents and how many degrees it measures. (40%, 144°)

LESSON RESOURCES

TRB Practice I-9
TRB Enrichment I-10

Practice

For use after 9-5 (pp. 377–379)

Circles

True or false?

1. All chords of a circle pass through the center. false
2. A diameter of a circle has length twice that of a radius. true
3. A diameter is a chord. true
4. All radii of a circle have the same length. true
5. The center of a circle is part of the circle. false

MENTAL MATH For each length, find the radius or the diameter.

6. $d = 60$ m, $r =$ 30 m
7. $r = 12.5$ in., $d =$ 25 in.
8. $r = 35$ yd, $d =$ 70 yd
9. $d = 2{,}400$ mm, $r =$ 1,200 mm
10. $r = 90$ ft, $d =$ 180 ft
11. $d = 500$ m, $r =$ 250 m

CALCULATOR For each length, find the radius or the diameter.

12. $r = 38.9$ cm, $d =$ 77.8 cm
13. $d = 0.77$ m, $r =$ 0.385 m
14. $r = 587.9$ m, $d =$ 1,175.8 m
15. $r = 64{,}395$ in., $d =$ 128,790 in.
16. $d = 1{,}365$ ft, $r =$ 682.5 ft
17. $d = 147{,}681$ km, $r =$ 73,840.5 km

Name the indicated part of circle O.

18. $\overline{MK}$ chord
19. $\overline{OP}$ radius
20. $\overline{MP}$ diameter
21. $\angle KOP$ central angle
22. What kind of triangle is $\triangle OMK$? isosceles
23. Use the data to construct a circle graph.

Voter Preference for Senator

Peterson	40%
Washington	30%
Gomex	15%
Thomson	10%
Miller	5%

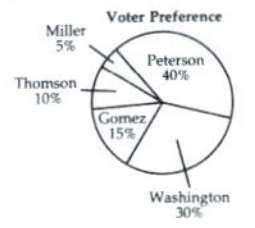

Chapter 9 I-9

Enrichment

For use after 9-5 (pp. 377–379)

Where's the Quake?

Earthquakes produce two types of shock waves. Primary (P) waves flow through the earth at 4.3 mi/s. Secondary (S) waves flow through the earth at 2.4 mi/s. Following an earthquake, the faster P waves arrive first at distant seismic stations, followed later by S waves.

A seismic station was located 103.2 mi from an earthquake.

1. How long did it take the P waves to reach the station? 24 s
2. How long did it take the S waves to arrive? 43 s
3. How much time elapsed between the arrival of the P waves and the S waves? 19 s

You can find the distance from an earthquake to a seismic station using the formula $D = 5.43t$, where D is the distance in miles, and t is the number of seconds that elapse between the arrival of P and S waves.

Confirm that the formula predicts your answer to Exercise 3.

Shock waves from an earthquake arrived at three seismic stations. Find the distance of the quake from each station. Round to the nearest mile.

Station	P-S wave time difference (s)
San Francisco	221
Houston	82.9
Salt Lake City	193.3

4. San Francisco 1,200 mi
5. Houston 450 mi
6. Salt Lake City 1,050 mi
7. Use the distances to draw circles with the given cities as center. Mark the location of the quake with the letter E.

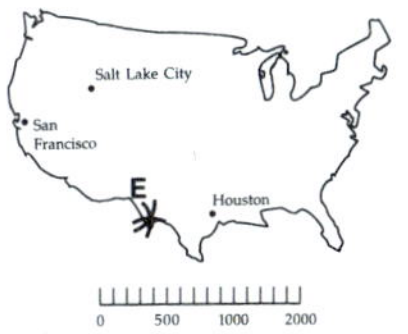

I-10 Chapter 9

LESSON QUIZ

1. If chord AB is not equal to diameter XY, which is longer? (XY)
2. Suppose chord AB is equal to diameter XY. Which is longer, chord AB or radius CD? (AB)
3. Find 60% of 360°. (216°)
4. After conducting a survey, the data are changed to percents. What is the total percent? (100%)
5. The data from question 4 are converted to degrees to form a circle graph. What is the total number of degrees? (360°)

Assignment Guide

Basic 1-3, 5-8, 12-14, 18-19, 21, 23-24, MR All
Average 2-4, 7-9, 13-15, 18-19, 21, 23-24, MR All
Enriched 3-5, 9-11, 15-17, 19-20, 22-24, MR All

FOR THE NEXT LESSON

ruler, scissors, pocket mirrors, tracing paper, construction paper, math journal

Lesson Focus

Materials

Ruler, scissors, pocket mirrors, tracing paper, construction paper, math journal

Vocabulary/Symbols

congruent polygons
corresponding
corresponds to
is congruent to
line symmetry

MOTIVATING THE LESSON

Put the symbol ≅ on the chalkboard. Ask students if they know what the symbol represents. (is congruent to) If some students know, have them explain to others what *is congruent to* means. (Two figures have the same size and shape.) If students do not know, ask them to guess. Point out that part of the symbol looks like an equal sign. Write the students' guesses on the chalkboard, then have the class read the first half of page 380 to determine which guesses are close to the definition.

Instruction

1 Note that two congruent triangles have six pairs of corresponding parts. (3 pairs of angles and 3 pairs of sides) Have students look at the congruent figures *SLOW* and *TIME* at the top of the page. Ask how many pairs of corresponding angles there are. (4) Have students name the pairs of corresponding sides. (*WS*↔*ET*, *SL*↔*TI*, *LO*↔*IM*, *OW*↔*ME*)

2 Have students locate lines of symmetry for objects in the classroom; for example, the binding on a book is a line of symmetry for the book.

OBJECTIVE: *To identify congruent figures and figures with line symmetry.*

9-6 Congruence and Symmetry

▼ Two figures with the same size and shape are *congruent*. We use the symbol ≅ to mean *is congruent to*. When two figures are congruent their *corresponding* parts are equal. The symbol ↔ means *corresponds to*.

▼ Think of sliding one figure on top of the other. You can see the following correspondences.

$S \leftrightarrow T$	$\angle S \leftrightarrow \angle T$
$L \leftrightarrow I$	$\angle L \leftrightarrow \angle I$
$O \leftrightarrow M$	$\angle O \leftrightarrow \angle M$
$W \leftrightarrow E$	$\angle W \leftrightarrow \angle E$

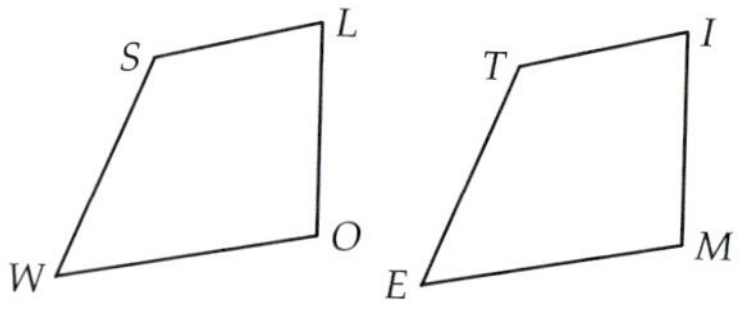

SLOW ≅ *TIME*

Describe figures in the kaleidoscope picture that appear to be congruent. **Answers will vary.**

▼ The definition of congruent figures follows.

Congruent Polygons	Two polygons are congruent if there is a correspondence between their vertices such that the corresponding sides and corresponding angles are congruent.

Example 1 **Do the figures appear to be congruent? Explain.**

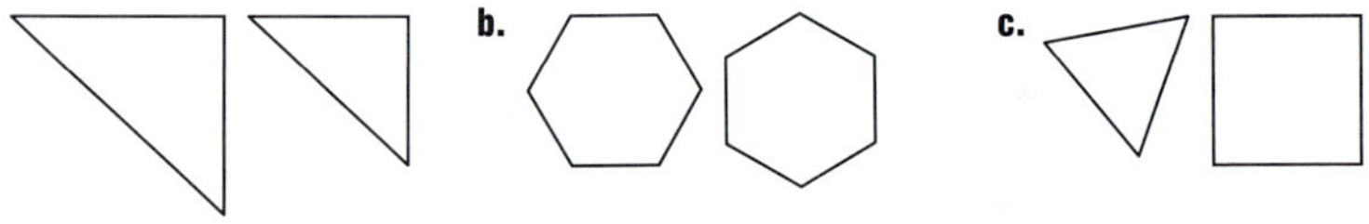

Solution
a. no; not the same size
b. yes; same size and same shape
c. no; not the same shape

FLASHBACK
We use hatch marks and arcs on figures to indicate congruent parts.

1

Example 2 **$\triangle ABC \cong \triangle XYZ$. Write congruence statements for corresponding parts.**

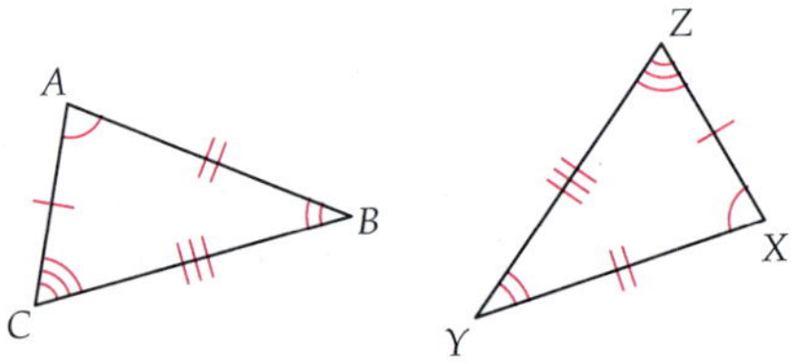

Solution

$\overline{AB} \cong \overline{XY}$	$\angle A \cong \angle X$
$\overline{BC} \cong \overline{YZ}$	$\angle B \cong \angle Y$
$\overline{CA} \cong \overline{ZX}$	$\angle C \cong \angle Z$

Teaching Tip

Reading in mathematics requires a slower pace than other types of reading. Allow time for students to absorb the content on a page. Before students begin reading, mention ideas to look for, connections to make, and questions students can ask themselves.

▼ If you can draw a line through a figure so that one side is a reflection of the other, the figure is said to have *line symmetry*. Some figures have more than one line of symmetry. Some figures have no lines of symmetry.

Example 3 **Is each dotted line a line of symmetry?**

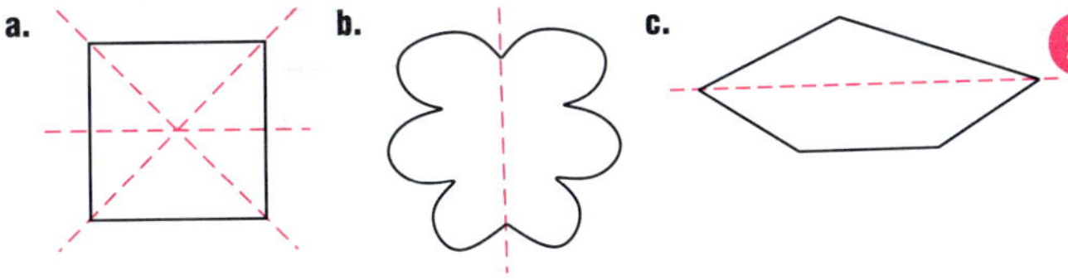

Solution **a.** yes **b.** yes **c.** no

2

CLASS EXERCISES

Write congruence statements for corresponding parts. ∠DAC ≅ ∠BAC, ∠B ≅ ∠D, ∠BCA ≅ ∠DCA

1.

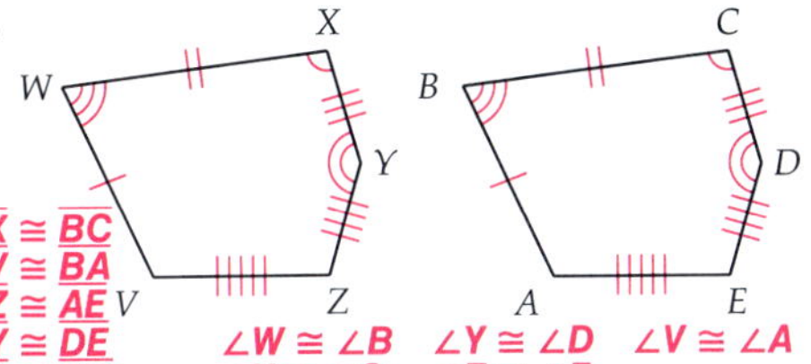

$\overline{WX} \cong \overline{BC}$, $\overline{WV} \cong \overline{BA}$, $\overline{VZ} \cong \overline{AE}$, $\overline{ZY} \cong \overline{DE}$, $\overline{YX} \cong \overline{DC}$; ∠W ≅ ∠B, ∠X ≅ ∠C, ∠Y ≅ ∠D, ∠Z ≅ ∠E, ∠V ≅ ∠A

2.

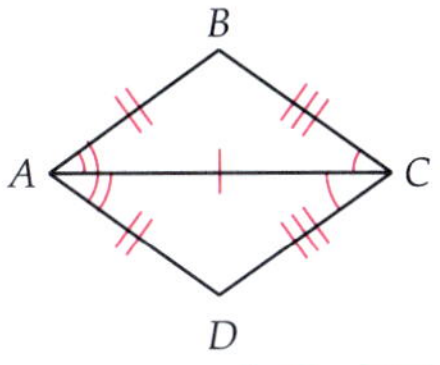

$\overline{BC} \cong \overline{CD}$, $\overline{AB} \cong \overline{AD}$, $\overline{AC} \cong \overline{AC}$

Trace each figure and draw all lines of symmetry.

3.

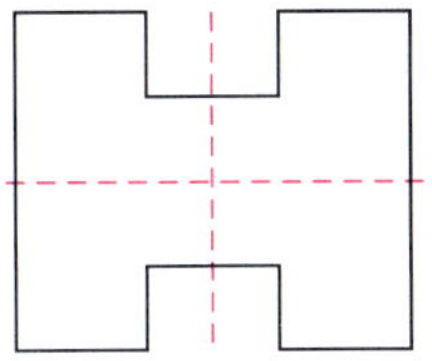

4.

5. ***WRITE*** Why would it be important for machines to produce congruent parts? **so all products are uniform in size; so all parts will fit together properly**

WRITTEN EXERCISES

1. Which figure appears not to be congruent to the other three? **B**

THINK AND DISCUSS

1. Do congruent figures have to be polygons? Can segments, angles, circles, curves, or shapes from nature be congruent? **no; yes**

2. Are all right angles congruent? all acute angles? **yes; no**

3. How many lines of symmetry does a circle have? **an infinite number**

MATH MINUTES

Ask students to draw half of a figure that includes a dotted line of symmetry on a piece of tracing paper, then trade papers with their partners and complete the partners' drawings.
(Check students' work.)

ADDITIONAL EXAMPLES

1. Name some examples of figures from nature that appear to be congruent. (Accept reasonable responses such as the left hand ≅ the right hand; the cells of a honeycomb are all congruent.)

2. How many pairs of corresponding parts exist in a pair of congruent pentagons? (10)

Guided Practice

Class Exercises For question 1, ask students how many corresponding parts will be listed. Repeat for question 2.

Think and Discuss Ask students how a counterexample will help them answer question 2.

A ***common error*** occurs when students incorrectly match corresponding parts of congruent figures. Students might match top to top or left side to left side rather than mentally rotating or flipping the figure. Provide figures for students to rotate or flip. With enough concrete practice, students can develop the ability to rotate and flip the figures mentally.

Closure

Writing in Math Ask students to describe the terms *congruent* and *symmetry* in their math journals. Then, have them explain the process they can use to decide if two figures are congruent.

Independent Practice

Before students answer questions 4-9, have students trace the triangles and indicate corresponding sides and angles.

For questions 10-15, suggest that students draw the figures.

For questions 20-21, it may be useful for students to place a mirror on the dotted line of symmetry to view how the completed figure should look.

As an extension, refer to the problem solving activity on p. 360E of the Chapter Overview. See also TRB Exploring O-34.

Lesson Follow-up

Reteaching Activity Give students figures cut from construction paper: an equilateral triangle, a right isosceles triangle, a square, a regular pentagon. Have students trace one figure on a sheet of paper. Then encourage them to fold the figure in various ways to identify all the possible lines of symmetry. Have students record the lines of symmetry as dotted lines on the traced figure. Repeat the process with the other shapes.

MIXED REVIEW

1. What is 33% of 98? 32.34
2. Make a factor tree for 130. See below.

Find the diameter.

3. $r = 13.5$ mm 27 mm
4. $r = \frac{7}{8}$ in. $1\frac{3}{4}$ in.

Solve each proportion.

5. $\frac{4}{12} = \frac{n}{15}$ 5
6. $\frac{a}{21} = \frac{4}{28}$ 3
7. An angle whose measure is between 0 and 90 is ■. acute
8. How many rectangles are in the figure below? 25

2. 130 → 10, 13; 10 → 2, 5

Write congruence statements for corresponding parts.

2. ∠OTH ≅ ∠OSX; ∠THO ≅ ∠SXO

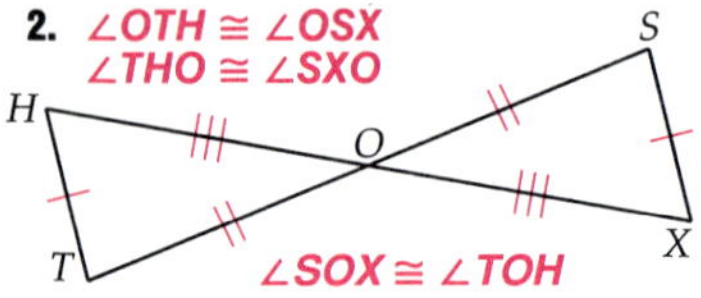

∠SOX ≅ ∠TOH

$\overline{HT} \cong \overline{SX}$ $\overline{TO} \cong \overline{SO}$ $\overline{HO} \cong \overline{XO}$

3.

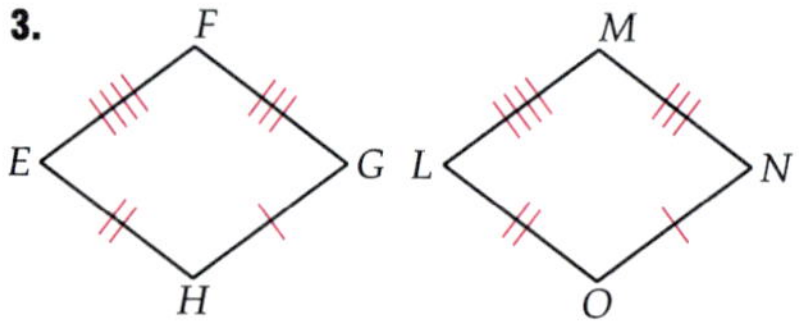

$\overline{EF} \cong \overline{LM}$; $\overline{FG} \cong \overline{MN}$; $\overline{HG} \cong \overline{ON}$; $\overline{EH} \cong \overline{LO}$

3. ∠F ≅ ∠M, ∠G ≅ ∠N, ∠H ≅ ∠O, ∠E ≅ ∠L

△XYZ ≅ △ABC. Write each angle measure or segment length.

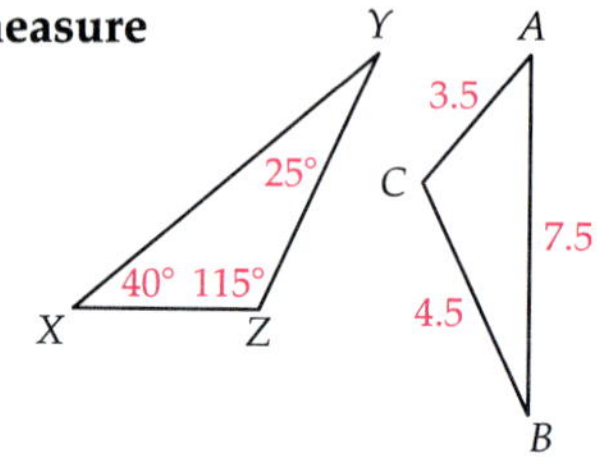

4. ∠C 115°
5. ∠A 40°
6. ∠B 25°
7. XZ 3.5
8. XY 7.5
9. YZ 4.5

Suppose RSTU ≅ WXYZ. Which congruence statements must be true?

10. ∠R ≅ ∠W T
11. $\overline{ST} \cong \overline{XY}$ T
12. ∠R ≅ ∠Z F
13. $\overline{UT} \cong \overline{ZY}$ T
14. $\overline{RS} \cong \overline{XY}$ F
15. ∠T ≅ ∠W F

Find the measure of each numbered angle.

16. △ABC ≅ △XYZ

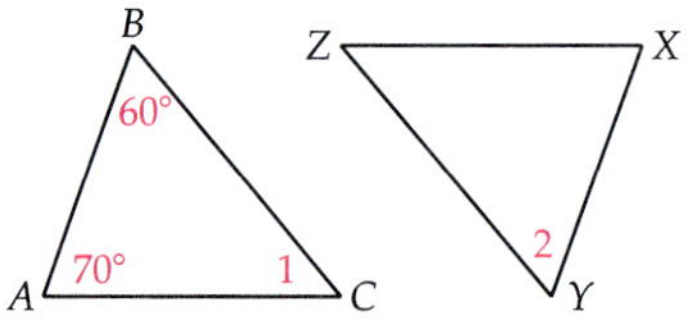

m∠1 = 50° m∠2 = 60°

17. △PQR ≅ △MNO

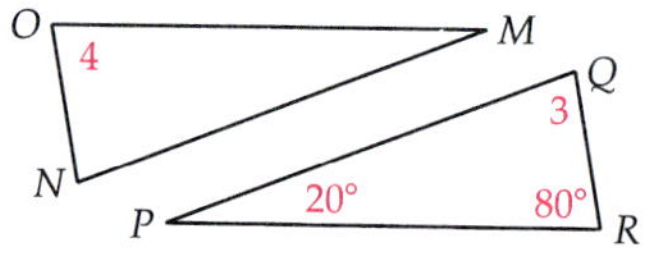

m∠4 = 80° m∠3 = 80°

Trace each pair of symmetric figures. Fold to find the line of symmetry. Check students' drawings.

18.

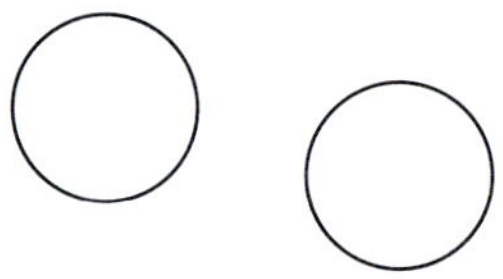

19.

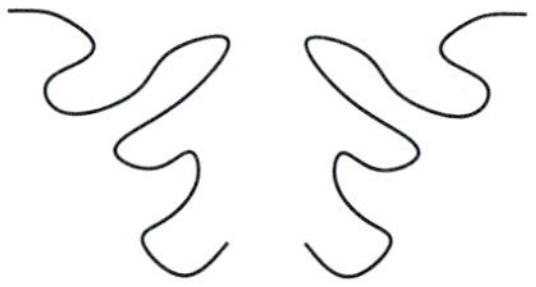

The dotted line is a line of symmetry for each figure. Trace the figure and complete the drawing. Check students' drawings.

20.

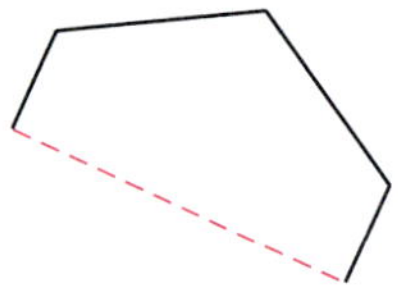

21.

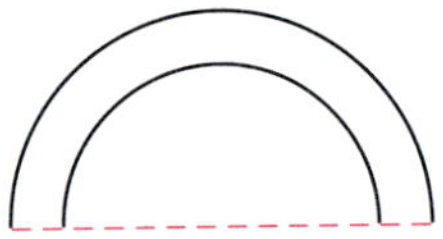

Which figures have line symmetry?

22.

yes

23.

yes

24.

no

25. Fold a plain sheet of paper in half lengthwise. Write your name along the fold. Turn the paper over and trace your name through the paper. Is the fold a line of symmetry? Are the two names congruent? **Check students' work.**

26. Which letters of the alphabet have only one line of symmetry when written in capital block form? Sketch your responses with the symmetry lines. **A, B, C, D, E, K, M, T, V, W, Y**

27. ***PROJECT*** Carpenters and artisans use a tool called a level to insure that a surface is flat. You can make a level that uses the properties of congruent triangles. **Check students' work.**

Cut three congruent rectangles out of oak tag. Join the rectangles at the short edges to form a triangle. Mark the midpoint of the triangle's base. Attach a string weighted with a button to the vertex at the top of the triangle. When you stand the level on a flat surface, $\triangle FLT \cong \triangle ALT$. Rest the triangle on a non-level surface. Do you still see congruent triangles? Explain.
No; the weight will swing downwards.

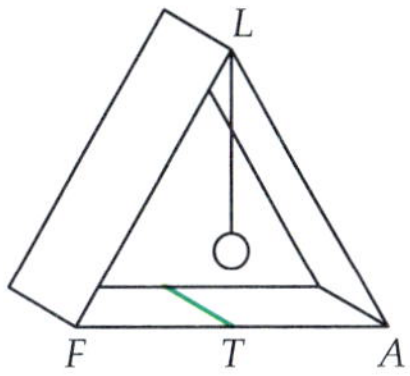

TEST YOURSELF

1. Name three lines, three rays, three segments, and three angles in the figure at the right. **Answers may vary.**
$\overleftrightarrow{WY}$, $\overleftrightarrow{VU}$, $\overleftrightarrow{ZX}$; $\overrightarrow{MZ}$, $\overrightarrow{MX}$, $\overrightarrow{MY}$; $\overline{WM}$, $\overline{UM}$, $\overline{ZX}$; $\angle XMY$, $\angle VMX$, $\angle ZMU$

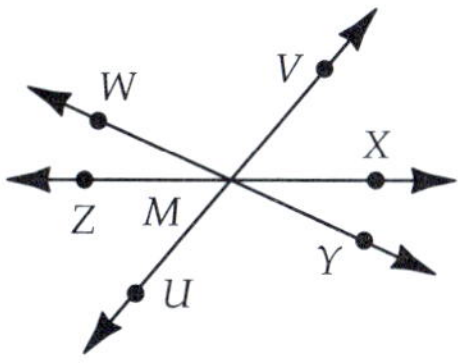

Use the figure at the right.

2. Classify *ABCD* as a square, a rectangle, or a parallelogram. **parallelogram**

3. Classify $\triangle ABD$ as *acute*, *obtuse*, or *right*. **acute**

4. Write a congruence statement for corresponding parts in $\triangle ABD$ and $\triangle CDB$. **$\overline{AB} \cong \overline{CD}$ $\overline{BD} \cong \overline{BD}$ $\angle ABD \cong CDB$ $\overline{BC} \cong \overline{AD}$ $\angle BAD \cong \angle DCB$ $\angle CBD \cong \angle ADB$**

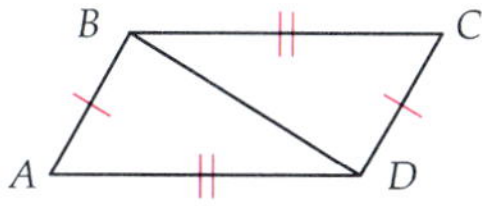

For each given length, find the radius or diameter.

5. $r = 26$ ft, $d =$ ▒ **52 ft**

6. $d = 60$ in., $r =$ ▒ **30 in.**

7. $r = 15.5$ m, $d =$ ▒ **31 m**

8. $d = 22.2$ cm, $r =$ ▒ **11.1 cm**

LESSON RESOURCES

TRB Practice I-11
TRB Enrichment I-12

FOR THE NEXT LESSON

paper, pencil, ruler, construction paper, scissors, math journal

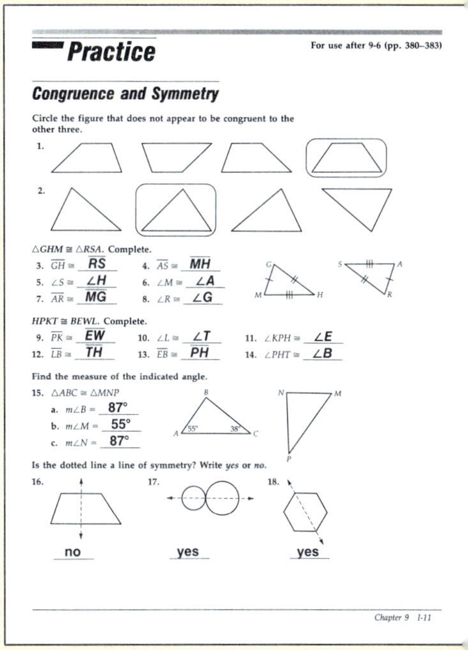

Practice — For use after 9-6 (pp. 380–383)

Congruence and Symmetry

Circle the figure that does not appear to be congruent to the other three.

1.

2.

$\triangle GHM \cong \triangle RSA$. Complete.

3. $\overline{GH} \cong$ **$\overline{RS}$** 4. $\overline{AS} \cong$ **$\overline{MH}$**
5. $\angle S \cong$ **$\angle H$** 6. $\angle M \cong$ **$\angle A$**
7. $\overline{AR} \cong$ **$\overline{MG}$** 8. $\angle R \cong$ **$\angle G$**

HPKT ≅ *BEWL*. Complete.

9. $\overline{PK} \cong$ **$\overline{EW}$** 10. $\angle L \cong$ **$\angle T$** 11. $\angle KPH \cong$ **$\angle E$**
12. $\overline{LB} \cong$ **$\overline{TH}$** 13. $\overline{EB} \cong$ **$\overline{PH}$** 14. $\angle PHT \cong$ **$\angle B$**

Find the measure of the indicated angle.

15. $\triangle ABC \cong \triangle MNP$
a. $m\angle B =$ **87°**
b. $m\angle M =$ **55°**
c. $m\angle N =$ **87°**

Is the dotted line a line of symmetry? Write *yes* or *no*.

16. **no** 17. **yes** 18. **yes**

Chapter 9 I-11

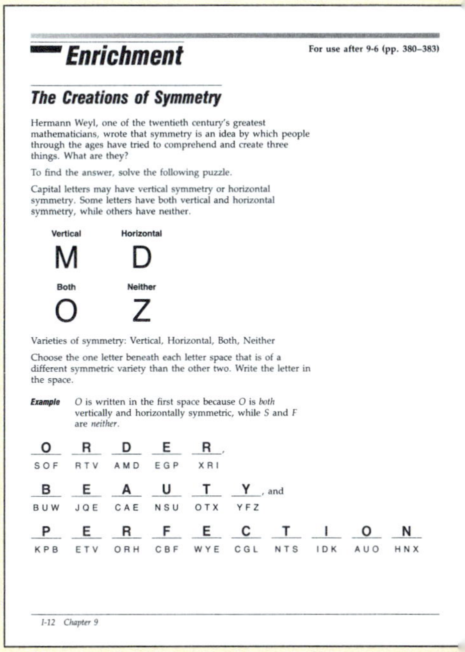

Enrichment — For use after 9-6 (pp. 380–383)

The Creations of Symmetry

Hermann Weyl, one of the twentieth century's greatest mathematicians, wrote that symmetry is an idea by which people through the ages have tried to comprehend and create three things. What are they?

To find the answer, solve the following puzzle.

Capital letters may have vertical symmetry or horizontal symmetry. Some letters have both vertical and horizontal symmetry, while others have neither.

Vertical	Horizontal
M	D
Both	**Neither**
O	Z

Varieties of symmetry: Vertical, Horizontal, Both, Neither

Choose the one letter beneath each letter space that is of a different symmetric variety than the other two. Write the letter in the space.

Example *O* is written in the first space because *O* is *both* vertically and horizontally symmetric, while *S* and *F* are *neither*.

O	R	D	E	R,					
SOF	RTV	AMD	EGP	XRI					
B	E	A	U	T	Y, and				
BUW	JQE	CAE	NSU	OTX	YFZ				
P	E	R	F	E	C	T	I	O	N
KPB	ETV	ORH	CBF	WYE	CGL	NTS	IDK	AUO	HNX

I-12 Chapter 9

LESSON QUIZ

1. Are all circles congruent? Explain. (No; circles can be different sizes even though they all have a degree measure of 360°.)

2. In two equilateral triangles, one pair of corresponding sides is congruent. Are the triangles congruent? (yes)

3. In two equiangular triangles, all the pairs of corresponding angles are congruent. Are the triangles congruent? Explain. (Maybe; the sides of the two triangles could be different lengths.)

4. Show the lines of symmetry, if any, for the capital letter A. (one line of symmetry)

Assignment Guide

Basic 1-2, 4-5, 7, 10-12, 16, 18, 20, 22-23, 25, 27, MR, TY All

Average 1-2, 5-6, 8, 12-14, 16, 18, 20, 22-23, 25, 27, MR, TY All

Enriched 1, 3, 6, 8-9, 13-15, 17, 19, 21, 23-24, 26, 27, MR, TY All

Lesson Focus

Materials

paper, pencil, ruler, construction paper, scissors, math journal

Vocabulary/Symbols

similar ~

MOTIVATING THE LESSON

Draw a large figure and a smaller, similar stick figure on the chalkboard. Ask the students to compare the smaller figure to the larger figure. On the chalkboard write the phrases students use to compare the figures. Suggest to students that the difference between the two figures is size.

Instruction

1 Remind students that a proportion is an equation. Cross multiplication is used to find the unknown (or variable) in a proportion.

2 Provide students with the measures of two angles of a triangle, such as 34° and 68°. Ask each student to draw and cut out a triangle with those angle measures. Have groups compare their triangles and identify which are similar and which are congruent.

3 After working through example 4, ask students to make a drawing in their math journals of similar triangles that look like the sketch. Have students label their drawings and then write the proportion for the base and height of the triangles.

OBJECTIVE: *To determine correspondences in similar figures and to find missing measures in similar figures.*

9-7 Similar Figures

▼ Designers often make scale models of new products. The models are the same shape but not the same size as the final product. We say the models and the real product are *similar*.

FLASHBACK
A proportion is an equation stating that two ratios are equal.

Similar Figures	Two figures are similar (~) if corresponding angles are congruent and corresponding sides are in proportion.

Example 1 **Do the pairs of figures appear to be similar?**

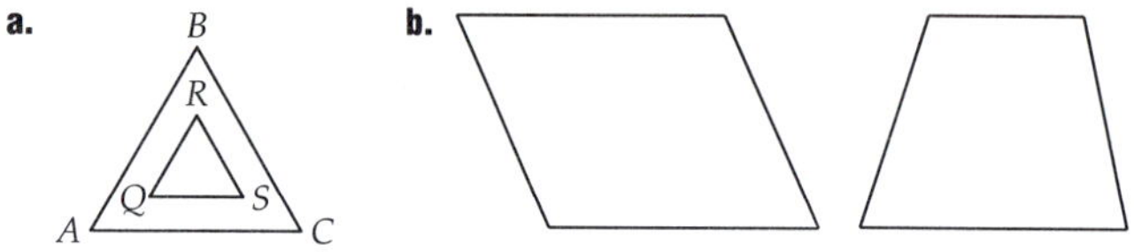

Solution **a.** yes **b.** no; not the same shape

▼ In similar figures, the lengths of the sides may change but the measures of the angles remain the same. You can use proportions to find an unknown length when two figures are similar.

Example 2 ***ABCDE* ~ *VWXYZ*. Find *AB*.**

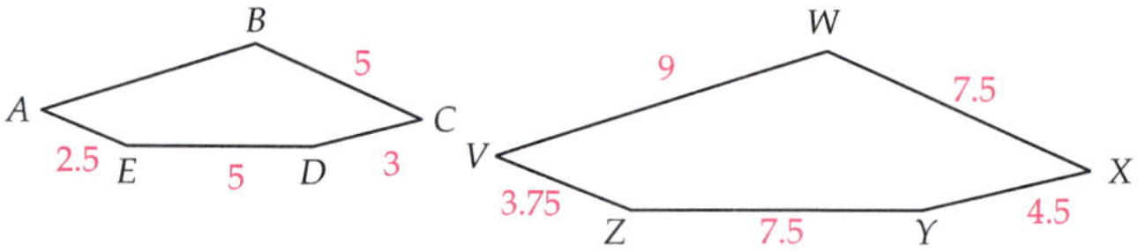

Solution **1** Since the figures are similar, corresponding sides are in proportion.

$\frac{AB}{VW} = \frac{BC}{WX}$ Write a proportion.

$\frac{AB}{9} = \frac{5}{7.5}$ Substitute values.

$AB = \frac{5}{7.5} \cdot 9$ Solve.

$= 5$ ÷ 7.5 × 9 = 6

$AB = 6$

Teaching Tip

Order and neatness are important elements for written work in math. As you walk around the classroom, look for students' works that illustrate an organizational technique. Have the students explain to the class how they decided to use a particular technique.

▼ When you know the measures of the angles, you can determine whether or not two triangles are similar.

Similar Triangles	Two triangles are similar if two angles of one are congruent to two angles of another.

Example 3 **Determine whether or not the triangles are similar. If *yes*, write a similarity statement.**

Solution $m\angle Y = 180° - (110° + 30°)$ since $m\angle B = m\angle Y$

$= 40°$ and $m\angle C = m\angle Z$

$\triangle ABC \sim \triangle XYZ$

Two angles of one triangle are congruent to two angles of the other. The triangles are similar.

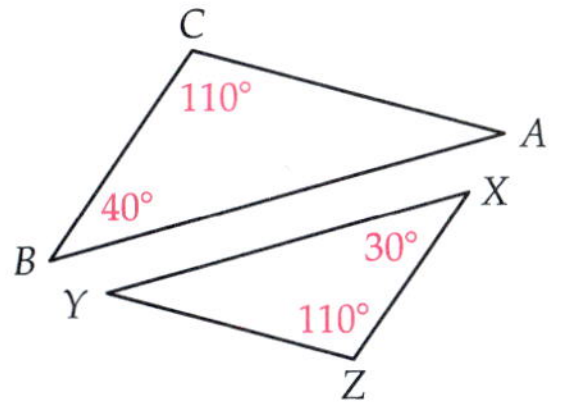

2

▼ You can use similar triangles to measure distances that you would be unable to measure directly.

Example 4 A 6-ft tall man, standing near a tree, casts a shadow 14 ft long. The tree casts a shadow 40 ft long. Use similar triangles to find the height of the tree to the nearest tenth of a foot.

Solution Draw a sketch to display the information in the problem.

$m\angle N = m\angle L$ Right angles are congruent.

$m\angle A = m\angle A$ An angle is congruent to itself.

$\triangle TAL \sim \triangle MAN$

$\frac{AL}{AN} = \frac{TL}{MN}$

$\frac{40}{14} = \frac{TL}{6}$

$40 \cdot 6 = 14 \cdot TL$ Cross multiply to solve.

240 ⊕ 14 ⊜ 17.142 . . .

$TL \approx 17.1$ ft Round to the nearest tenth.

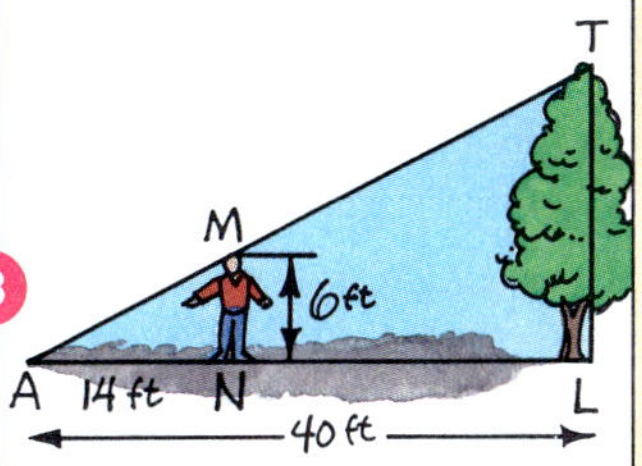

3

1. yes; they're the same shape.
2. yes. No; measure of acute angles and sides may vary.
3. yes; they're the same shape.

CLASS EXERCISES

1. Draw three different equilateral triangles. Are all your triangles similar? Check students' work; yes.
2. In similar figures, what ratios are always equal? corresponding parts

HGFE ~ *MNOP*.

3. Find *EF*. 2.4
4. Find *MN*. 2.5

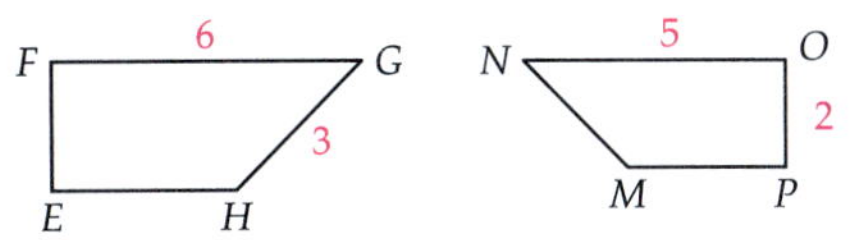

THINK AND DISCUSS See above.

1. Are all circles similar? Explain.
2. Are all right angles similar? All right triangles? Explain.
3. Are congruent figures always similar? Explain.

ADDITIONAL EXAMPLES

1. Draw a stop sign. Explain what makes your drawing similar to actual stop signs. (The angles are congruent and the sides are proportional.)

2. $\triangle ABC$ has angles 30°, 70°, and *x*. $\triangle XYZ$ has angles 80°, 30°, and *y*. Are the triangles similar? (yes)

3. Suppose the shadow in Example 4 is 12 ft. What is the height of the tree? (20 ft)

Guided Practice

Class Exercises Have students work through these questions orally. For questions 3-4, ask students to describe the procedure they would use to find the unknown side.

Think and Discuss Extend question 3 by asking if its converse is true. **Are similar figures always congruent?** (no) Then ask, **If two polygons are similar, which parts are congruent and which parts are proportional?** (corresponding angles; corresponding sides)

A ***common error*** occurs when students set up a proportion for similarity incorrectly. Students choose measures that do not correspond and set them equal to each other. For example, if $\triangle ABC \sim \triangle XYZ$, students may identify the incorrect proportion, $\frac{AB}{XY} = \frac{ZY}{CB}$, instead of $\frac{AB}{XY} = \frac{CB}{ZY}$.

MATH MINUTES

Have students write proportions about the similar figures in this design.

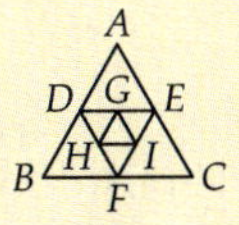

(Possible answers: $\frac{DE}{BC} = \frac{FE}{AB}$, $\frac{HI}{BC} = \frac{GI}{AC}$)

Closure

Have students work together in small groups. Ask them to write as many true statements as they can about polygons that are similar. (Possibilities include: corresponding angles of similar polygons are congruent; corresponding sides of similar polygons are porportional; use proportions to find unknown lengths of similar polygons.)

Independent Practice

For questions 9-10, ask students to explain *yes* or *no* answers before continuing.

For questions 12-15, encourage students to draw simple diagrams and label all the known parts before attempting to write the necessary proportions.

Critical Thinking Have students work in small groups to compare their solutions to question 1. If they disagree, encourage them to share their reasons for thinking as they do. Ask students to come to an agreement about the answer.

Lesson Follow-up

Reteaching Activity Give students construction paper models of similar triangles and have them identify the corresponding sides. Then, write a ratio and have the students supply a second ratio to complete a proportion.

WRITTEN EXERCISES

Use graph paper to sketch a figure similar to, but not congruent to, each figure. Check students' work.

1.

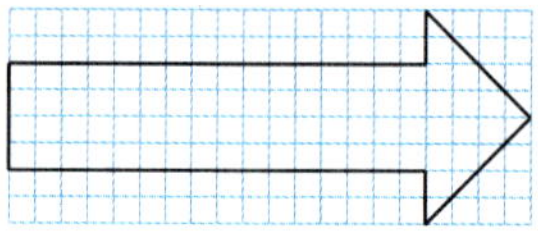

2.

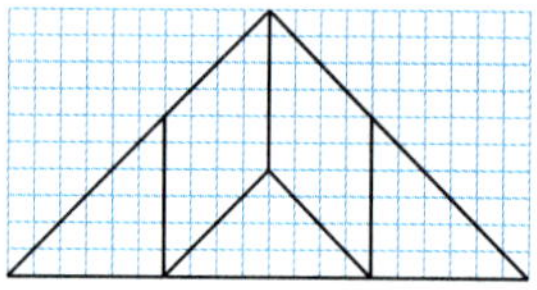

3. $MORE \sim TRYS$. Name four equal ratios. $\frac{MO}{RT} = \frac{OR}{RY} = \frac{ER}{YS} = \frac{ME}{TS}$

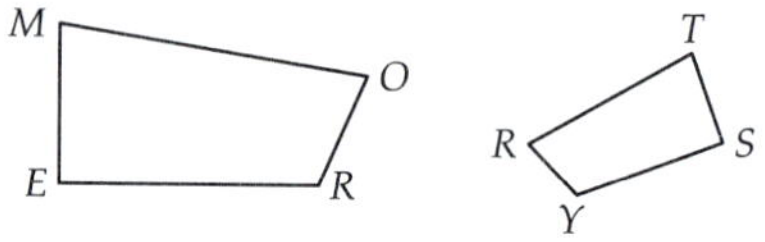

4. $\triangle HOT \sim \triangle PIE$. Name three equal ratios. $\frac{HO}{PI} = \frac{HT}{PE} = \frac{OT}{IE}$

Assume each pair of triangles is similar. Find the missing length.

5.

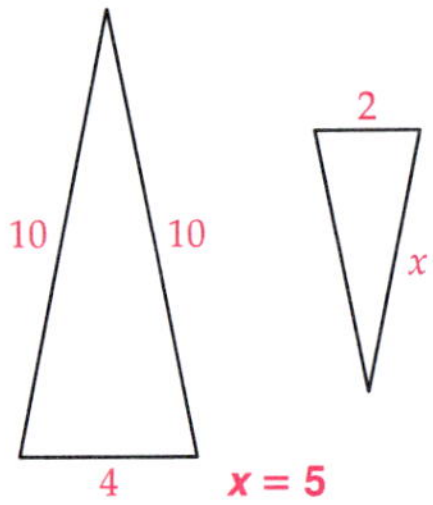

$x = 5$

6.

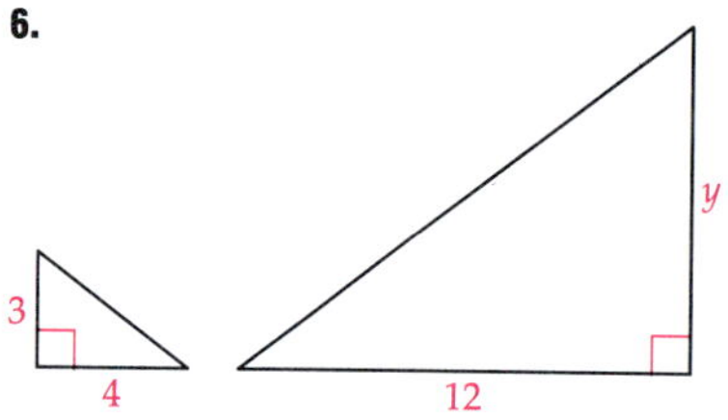

$y = 9$

7.

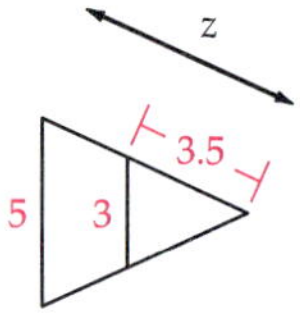

$z = 5.8\overline{3}$

8.

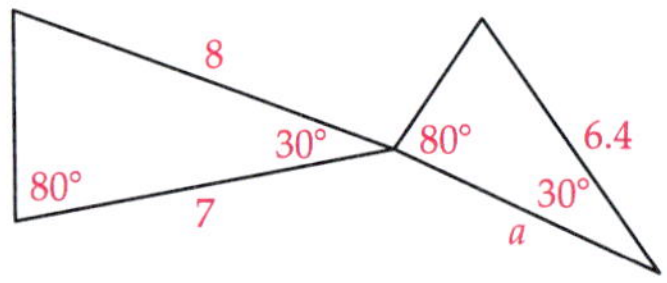

$a = 5.6$

Is each pair of triangles similar? If *yes,* identify three pairs of corresponding sides.

9. yes

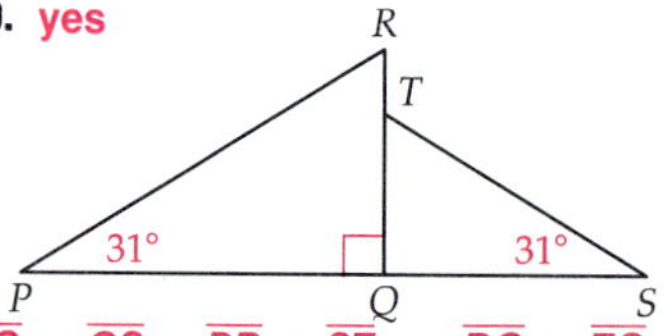

$\overline{PQ} \leftrightarrow \overline{QS}$ $\overline{PR} \leftrightarrow \overline{ST}$ $\overline{RQ} \leftrightarrow \overline{TQ}$

10. yes

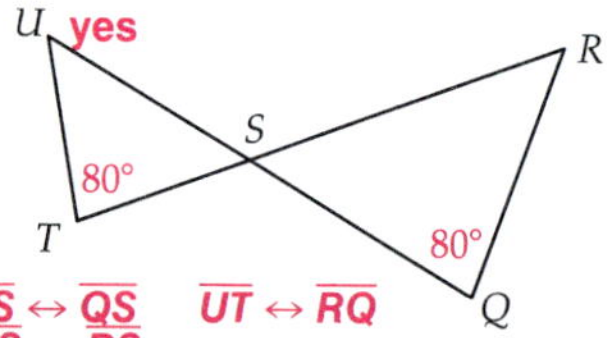

$\overline{TS} \leftrightarrow \overline{QS}$ $\overline{UT} \leftrightarrow \overline{RQ}$
$\overline{US} \leftrightarrow \overline{RS}$

11. Suppose you wish to make a mailbox which is a model of your home. ***WRITE*** a paragraph which describes your plan. Include as many details as possible. Check students' work.

MIXED REVIEW

Write each fraction as a decimal.

1. $\frac{3}{8}$ 0.375 **2.** $\frac{5}{12}$ $0.41\overline{6}$

3. $\triangle ABC \cong \triangle XYZ$. Write congruence statements for corresponding parts. See above.

3. $\overline{AB} \cong \overline{XY}$ $\angle A \cong \angle X$
$\overline{BC} \cong \overline{YZ}$ $\angle B \cong \angle Y$
$\overline{AC} \cong \overline{XZ}$ $\angle C \cong \angle Z$

4. $\triangle BRC \sim \triangle PNT$. Find RC. 9

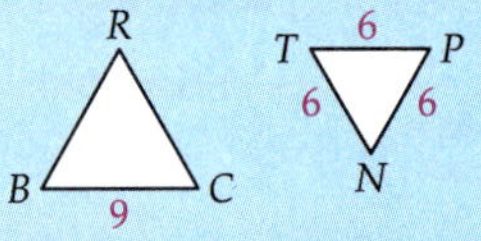

Simplify.

5. 6^2 36 **6.** t^2 for $t = 1.2$ 1.44

7. On the math team, the ratio of girls to boys is 3:2. There are 15 girls. How many boys are on the team? 10

Solve.

12. A tree casts a shadow 10 ft long. A 5-ft tall person casts a shadow 3 ft long. How tall is the tree? **$16\frac{2}{3}$ ft**

13. A photographic slide is 35 mm wide and 22 mm high. The projected image is 85 cm wide. How high is the image? **53.4 cm**

14. ***PROJECT*** Use similar triangles to measure the height of your school, flagpole, or a tall tree. **Check students' work.**

15. ***DATA FILE 3 (pp. 96–97)*** Find the dimensions of a dollar bill. What are the possible dimensions for a sheet of bills?

> ***PROBLEM SOLVING HINT***
> Draw a diagram.

Check Your Proportions

What do paintings by Leonardo DaVinci, buildings of ancient Greece, and your physical proportions have in common? They all reflect the *Golden ratio.* The golden ratio (calculated as approximately 1.61) is the ratio of the length to the width of a golden rectangle, which is thought to be the most pleasing to the eye. Examples of the golden ratio are found in art, architecture, nature, and in the human skeleton. An American researcher found that the ratio of a person's height to the distance from the ground to his waist approaches the golden ratio. How do you measure up?

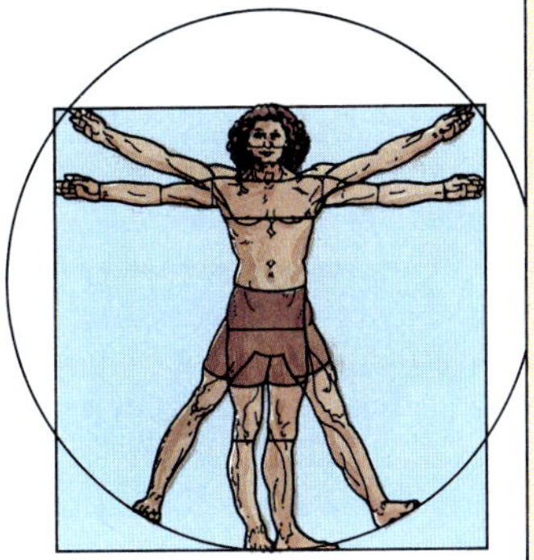

16. Use the article above. Measure, then compute each ratio. Do your ratios approach the golden ratio? **Check students' work.**

a. $\frac{\text{head to ground}}{\text{waist to ground}}$ **b.** $\frac{\text{shoulder to finger tips}}{\text{elbow to finger tips}}$

15. Answers will vary. 24.5 in. × 21 in.

Critical Thinking

EXPLORING DEDUCTIVE REASONING

▼ When you reach conclusions after reasoning from accepted information, you are using *deductive reasoning*.

1. Each card has either a triangle, circle, or square on the back. Use the given information to decide the figure on each card.

a. The figure on card 1 is not a quadrilateral.

b. The figure on the even-numbered card is not a polygon.

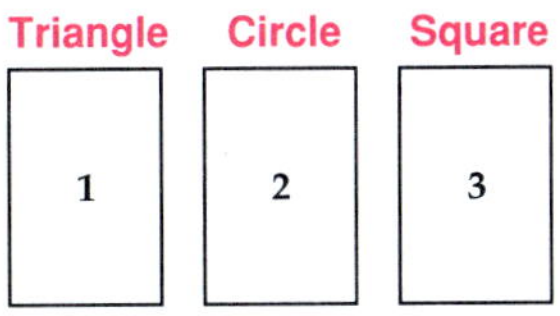

2. ***WRITE*** a problem of your own that requires deductive reasoning. Give it to a classmate to solve. **Check students' work.**

LESSON RESOURCES

TRB Practice I-13
TRB Enrichment I-14

FOR THE NEXT LESSON

string and measuring stick or measuring tape, circular objects (cans, pails, jars), large grid, penny, calculator, math journal

Practice

For use after 9-7 (pp. 384–387)

Similar Figures

Sketch a figure similar to, but not congruent to, each figure. **Figures may vary.**

1\.
2\.

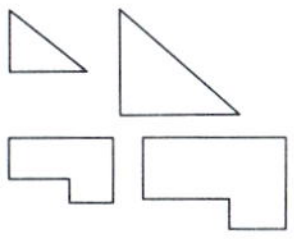

KLSN ~ *PAGT*. Complete each proportion.

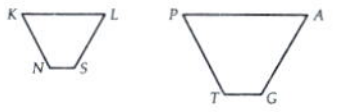

3\. $\frac{LS}{AG} = \frac{?}{PT}$ **KN** 4. $\frac{LK}{?} = \frac{SL}{GA}$ **AP**

5\. $\frac{TG}{NS} = \frac{PA}{?}$ **KL** 6. $\frac{?}{NS} = \frac{PT}{KN}$ **TG**

Assume each pair of triangles is similar. Find *x* and *y*.

7\. $x =$ **28 units**
$y =$ **12 units**

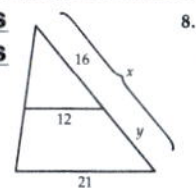

8\. $x =$ **28 units**
$y =$ **48 units**

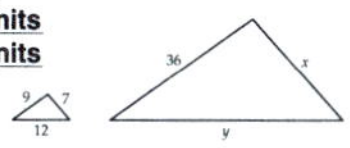

Solve.

9\. A rectangular painting is 56 in. long and 36 in. wide. A photo of the painting is 4.5 in. wide. How long is the photo?
7 in.

10\. At the same time a lighthouse casts a 99-ft shadow, a 24-ft telephone pole casts a 54-ft shadow. How tall is the lighthouse?
44 ft

Chapter 9 I-13

Enrichment

For use after 9-7 (pp. 384–387)

More on the Golden Ratio

The symbol ϕ is used to denote the golden ratio. To six decimal places, $\phi = 1.618034$. To discover some of the fascinating properties of the golden ratio, answer these questions.

1\. a. Use your calculator to find ϕ^2. $\phi^2 =$ **2.618034**

b. What is unusual about your result?
The decimal parts of ϕ and ϕ^2 are the same.

2\. The first seven terms of the Fibonacci sequence are 1, 1, 2, 3, 5, 8, 13.

a. Describe how each new term in the sequence is found.
Add the previous two terms.

b. Find the next five terms of the Fibonacci sequence.
21, 34, 55, 89, 144

3\. Use your calculator to complete the table.

$\phi = 1.618034$

$\phi^2 =$ **2.618034** $= \phi +$ **1**
$\phi^3 =$ **4.236068** $= 2\phi +$ **1**
$\phi^4 =$ **6.854102** $= 3\phi +$ **2**
$\phi^5 =$ **11.090170** $= 5\phi +$ **3**
$\phi^6 =$ **17.944272** $= 8\phi +$ **5**

4\. Describe the patterns in the last column above.
The coefficients of ϕ and the addends are Fibonacci numbers.

5\. The following ratios consist of a Fibonacci number divided by the previous Fibonacci number.

$\frac{1}{1} = 1$ $\frac{2}{1} = 2$ $\frac{3}{2} = 1.5$ $\frac{5}{3} = 1.\overline{6}$ $\frac{8}{5} = 1.6$

a. Continue the pattern, writing the next five ratios. Write the complete decimal given by your calculator.
1.625 **1.6153846** **1.6190476** **1.617647** **1.6181818**

b. To what number do the ratios appear to be getting closer and closer?
ϕ

I-14 Chapter 9

LESSON QUIZ

(Check students' work for questions 1-3.)

1. Draw two similar triangles. Label each angle.

2. Write a proportion that is true about your triangles.

3. Draw two similar polygons other than triangles. Label each angle. Write a proportion that is true about your polygons.

4. Solve. $\frac{4}{31} = \frac{5}{x}$ Round to the nearest tenth. (38.8)

Assignment Guide

Basic 1, 3, 5-6, 9, 11-12, 14-15, MR, CT All

Average 1, 3, 6-7, 10-12, 14-15, MR, CT All

Enriched 2, 4, 7-8, 10-11, 13-16, MR, CT All

Career Information

Arrange for the class to visit an optician's lab or store and observe how lenses are measured and made.

Ask the optician to discuss topics such as:

- ▼ What equipment is used to make the measurements for the lens?
- ▼ What education, specific courses, as well as licensing, is required to become an optician?
- ▼ What different information is needed to make eyeglass lenses and contact lenses?
- ▼ How do opticians check the fit of frames and the fit of contact lenses?
- ▼ How is mathematics used in the daily work of an optician?

CAREER

Help Wanted: Optician

High school diploma and vocational training required. Courses in algebra, geometry, and mechanical drawing are helpful.

For more information, write to Opticians Association of America, 10341 Democracy Lane, PO Box 10110, Fairfax, VA 22030.

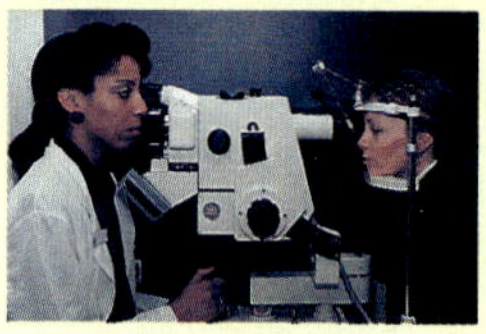

An optician fills the prescription after an optometrist prescribes lenses. An optician measures either the distance between the centers of the pupils or the corneas. Opticians assist in frame selection and then either make the lenses or instruct a lab in making them. Opticians use special instruments and microscopes to check the lenses and adjust the fit of the frames.

PROJECT

Find out how technology helps opticians make lenses with greater speed and accuracy.

Practice

Find the measure of each indicated angle.

1. $\angle PNM$ 55°

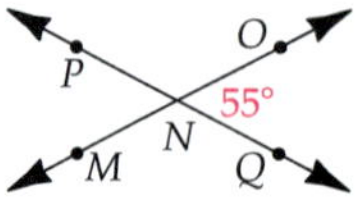

2. $\angle CBD$ 27°

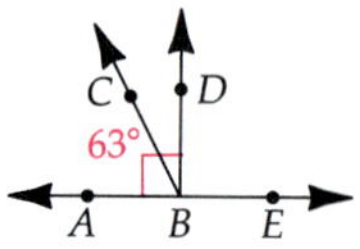

3. $\angle UEN$ 75°

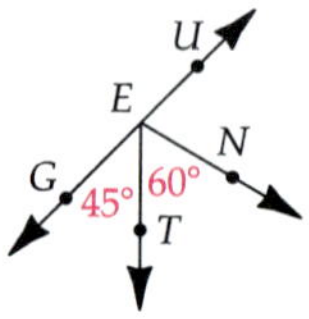

Find the measure of a complement and a supplement of each angle, if possible.

4. 12° 78°, 168°
5. 57° 33°, 123°
6. 101° none; 79°
7. 43° 47°, 137°
8. 128° none; 52°
9. 16° 74°, 164°
10. 92° none; 88°
11. 179° none; 1°

Find the measure of the third angle of a triangle that has two angles with the given measures.

12. 55°, 65° 60°
13. 45°, 45° 90°
14. 60°, 70° 50°
15. 80°, 28° 72°
16. 58°, 65° 57°
17. 110°, 25° 45°
18. 78°, 19° 83°
19. 103°, 33° 44°

For each length, find the radius or the diameter.

20. $r = 17$ in. $d =$ ■ 34 in.
21. $r = 56.8$ m $d =$ ■ 113.6 m
22. $d = 85.5$ cm $r =$ ■ 42.75 cm
23. $d = 0.38$ km $r =$ ■ 0.19 km
24. $r = 28.6$ mi $d =$ ■ 57.2 mi
25. $r = 67{,}385$ mi $d =$ ■ 134,770 mi

Name all chords, diameters, radii, and central angles shown in each circle. See Additional Answers.

26.

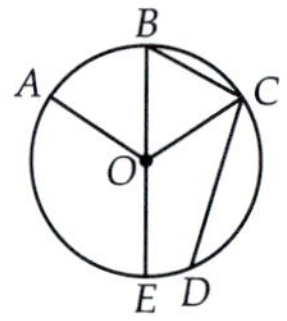

27.

28.

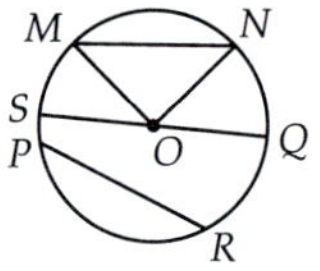

Write congruence statements for the corresponding parts of each pair of polygons.

29.

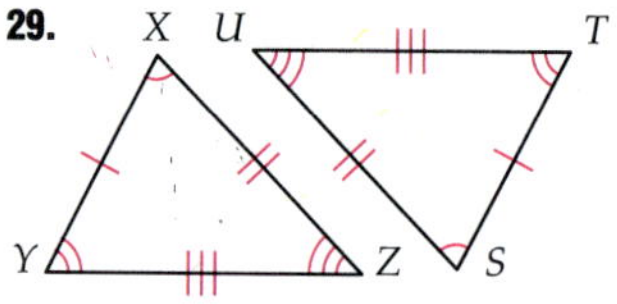

$\overline{YX} \cong \overline{TS}$, $\overline{XZ} \cong \overline{SU}$, $\overline{ZY} \cong \overline{UT}$, $\angle Y \cong \angle T$, $\angle X \cong \angle S$, $\angle Z \cong \angle U$

30.

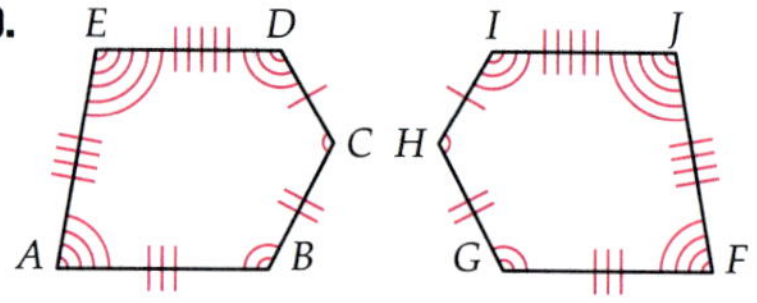

$\overline{DC} \cong \overline{IH}$, $\overline{CB} \cong \overline{HG}$, $\overline{BA} \cong \overline{GF}$, $\overline{AE} \cong \overline{FJ}$, $\overline{ED} \cong \overline{IJ}$, $\angle A \cong \angle F$, $\angle B \cong \angle G$, $\angle C \cong \angle H$, $\angle D \cong \angle I$, $\angle E \cong \angle J$

ADDITIONAL ANSWERS

26. diameter: $\overline{BE}$
central angles: $\angle AOB$, $\angle AOC$, $\angle AOE$, $\angle BOC$, $\angle COE$
radii: $\overline{OA}$, $\overline{OB}$, $\overline{OC}$, $\overline{OE}$
chords: $\overline{BC}$, $\overline{CD}$, $\overline{BE}$

27. radii: $\overline{OH}$, $\overline{OJ}$, $\overline{OK}$, $\overline{OG}$
chords: $\overline{LK}$, $\overline{JI}$, $\overline{GJ}$, $\overline{HK}$
diameters: $\overline{GJ}$, $\overline{KH}$
central angles: $\angle GOH$, $\angle HOJ$, $\angle JOK$, $\angle KOG$

28. central angles: $\angle SOM$, $\angle SON$, $\angle MON$, $\angle NOQ$, $\angle MOQ$
radii: $\overline{OS}$, $\overline{OM}$, $\overline{ON}$, $\overline{OQ}$
chords: $\overline{MN}$, $\overline{PR}$, $\overline{SQ}$
diameter: $\overline{SQ}$

OBJECTIVE:
To explore pi as a ratio.

MATERIALS

- String and a measuring stick or a measuring tape
- Several circular objects (cans, records, CDs, etc.)
- Large grid paper
- Coin or other marker
- Calculator
- Math journal to record work

Exploring Pi

▼ Pi (π) is a fascinating ratio. You may use it in many calculations with circles. The activities that follow will help you discover pi.

1 **1.** Choose at least five circular objects. Complete steps (a–c) and record your data in a chart.

a. Measure the circumference of each object.

b. Measure the diameter of each object.

Object	Circumference (C)	Diameter (d)	$\frac{C}{d}$

c. Use a calculator to find the ratio $\frac{C}{d}$ for each object. What is the approximate value for the ratio?

▼ Here is another experiment that approximates pi.

2 **2.** Use a penny and a large grid like the one below. Toss the coin so that it lands on the grid. Count only tosses that land completely on the grid. **Check students' work.**

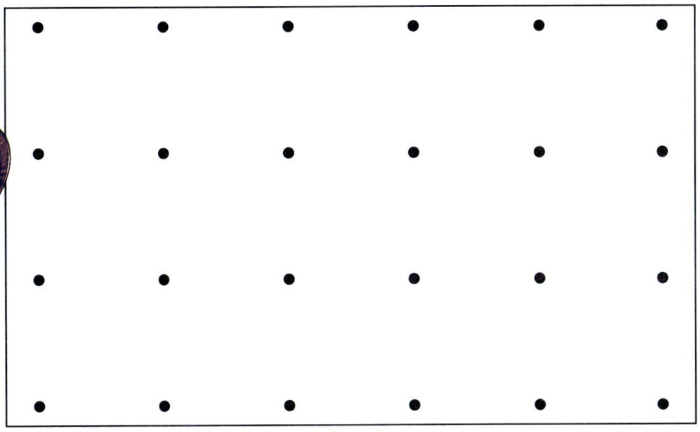

Tosses						
1–10	*h*	*m*	*m*	*h*	*h*	*h*
11–20						

a. Make a 10 × 10 chart to record your tosses. Count a hit when the coin touches or covers a dot. Make an *h* in a box on your chart. When the coin misses, mark an *m* on your chart. Continue filling in your grid for 100 tosses.

b. Use your data from (a) to complete a chart like the one at the left.

	h	*t*	$\frac{4h}{t}$
20 tosses		20	
40 tosses		40	
60 tosses			
80 tosses			
100 tosses			

c. Combine your data with data from other groups. Use your data to complete the formula $\frac{4h}{t} = ■$. **3.1415**

3. Is your result close to 3.1416? Do you think more tosses will result in a closer approximation of π? **yes; yes**

Lesson Focus

Materials

String and measuring stick or measuring tape, circular objects (cans, pails, jars), large grid, penny, calculator, math journal

Vocabulary/Symbols

pi (π)

MOTIVATING THE LESSON

Mention that the Greek letter π is used to denote the ratio $\frac{C}{d}$ (circumference to diameter). Comment that the class will conduct experiments in which the ratio is expressed, and determine if this ratio is constant for all circles.

Instruction

1 Have the students work in small groups. Set up areas in the room with objects to be measured. Groups move from area to area (at your signal) to complete the measuring. Then groups analyze the results.

2 Groups work at their desks. This experiment is a derivation of one of the historical attempts to generate pi. A penny must be used to validate the experiment. If a larger coin is used, the grid should be larger than the one shown on page 389.

Closure

Have students record the definition of pi in their math journals. Ask students to record 3.1416, 3.14, and $\frac{22}{7}$ as commonly used approximate values of π.

Assignment Guide

Basic All
Average All
Enriched All

FOR THE NEXT LESSON

Paper, pencil, graph paper, ruler, math journal

Teaching Tip

When conducting cooperative experiments, display a flowchart or outline of the tasks to be completed (and the order in which they are to be done), so students will understand what is expected of them. The flowchart helps to keep the groups focused.

Lesson Focus

Materials

Paper, pencil, graph paper, ruler, math journal

Vocabulary/Symbols

circumference
perimeter

MOTIVATING THE LESSON

Have students work in small groups to brainstorm a list of situations in which one needs to find the perimeter. In addition to rectangular regions or objects, suggest that circular and irregularly-shaped perimeters should also be included in their lists. Possible situations include: fences and borders, picture frames, molding for windows, doors, ceilings and floors, path around a lake.

Instruction

1 Ask students why 1 roll 7 ft long is not a reasonable answer. (Less than a full roll cannot be purchased and one roll is too little.) Reinforce that the answer must be reasonable and fit the situation.

2 Have different students work through the solutions for 2a, 2b, and 2c orally. Encourage them to explain the reasons for each step in the process. Note that the answers are written in terms of pi. When written this way, the answers are more precise than when a value is substituted for pi.

OBJECTIVE: *To find the perimeter or circumference of a figure.*

9-8 Perimeter and Circumference

▼ A seamstress buying lace for the bottom of a dress needs to know the *perimeter* of the hem.

Perimeter	Perimeter is the distance around a figure.

▼ You may need to find the perimeter of a common shape.

Example 1 A homeowner wants to buy a wallpaper border for a rectangular room. The room is 9 ft 2 in. long and 12 ft 3 in. wide. There is a 3-ft wide doorway. Borders are sold in 11-yd lengths. How many rolls of border does the homeowner need?

1

Solution

1. Find the perimeter. $P = 2l + 2w$
$P = 2(9 \text{ ft } 2 \text{ in.}) + 2(12 \text{ ft } 3 \text{ in.})$
$= 42 \text{ ft } 10 \text{ in.}$

2. Subtract the width of the doorway. $42 \text{ ft } 10 \text{ in.} - 3 \text{ ft} = 39 \text{ ft } 10 \text{ in.}$

3. Round to the nearest foot. $39 \text{ ft } 10 \text{ in.} \approx 40 \text{ ft}$

4. Compare. $11 \text{ yd} = 33 \text{ ft}$

Since one roll is not enough, the homeowner must purchase two rolls of border paper.

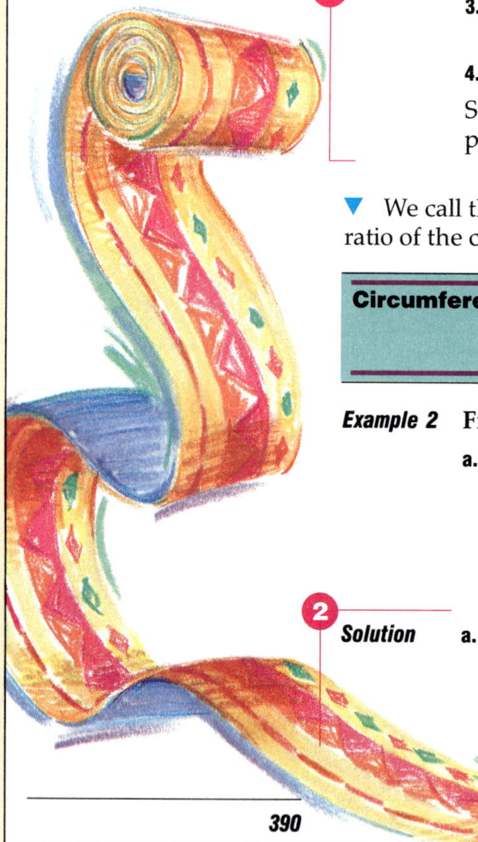

▼ We call the distance around a circular figure *circumference*. The ratio of the circumference to the diameter is pi (π).

Circumference	The circumference (C) is the distance around a circle. Use the formula $C = \pi d$ to compute circumference.

Example 2 **Find each circumference.**

a.

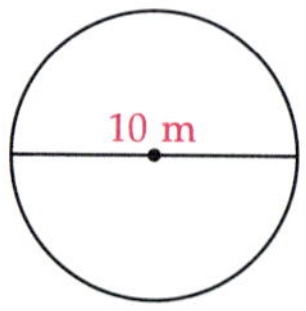

b.

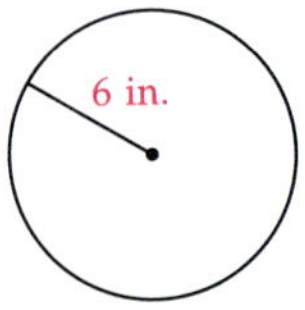

c.

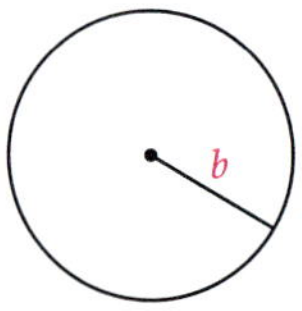

2

Solution

a. $C = \pi d$
$= 10\pi \text{ m}$

b. $C = \pi d$
$= \pi(2r)$
$= \pi(2 \cdot 6)$
$= 12\pi \text{ in.}$

c. $C = \pi d$
$= \pi(2r)$
$= \pi(2b)$
$= 2b\pi \text{ units}$

Teaching Tip

Encourage students to draw diagrams to help clarify questions. Diagrams can be especially helpful in answering questions and finding solutions involving geometric concepts.

▼ If you know the circumference of a circle, you can find the radius and diameter.

Example 3 **The circumference of a circle is 43 π ft. Find the diameter and the radius.**

Solution

$C = \pi d$

$43\pi = \pi d$

$\frac{43\pi}{\pi} = \frac{\pi d}{\pi}$ Divide both sides by π.

$43 = d$

$d = 2r$

$43 = 2r$ Substitute 43 for *d*.

43 ÷ 2 = *r* Divide both sides by 2.

$21.5 = r$

▼ We frequently use 3.14 or $\frac{22}{7}$ as approximations for π. Unless you are asked to approximate your answer, or given a value for π, leave answers in terms of π.

Example 4 **Find the circumference of a circle with the given radius or diameter. Round measures to the nearest tenth.**

a. $d = 12$ mi (Use 3.14 for π.)

b. $r \approx 7$ ft (Use $\frac{22}{7}$ for π.)

Solution

a. $C = \pi d$
$\approx (3.14)(12)$
$= 37.68$
≈ 37.7 mi

b. $C = \pi\, 2r$
$\approx \frac{22}{7}(2)(7)$
$= \frac{22}{7}(14)$
$= 2(22)$
$= 44$ ft

CLASS EXERCISES

1. Choose five objects or areas in the classroom. Estimate and then measure the perimeter of each object. Choose an appropriate unit for measuring each item. Record your results in a table like the one below. Check students' work.

Item	Unit	Estimate	Measure	Difference	% Error

NOTES & QUOTES

Using only paper and pencil, William Shanks (1812–1882) spent 20 years calculating π to 707 decimal places. In 1945 it was discovered that Mr. Shanks had made an error in the 528th place. Today, supercomputers can calculate the value of π to millions of places in seconds.

THINK AND DISCUSS See below.

1. Describe a method for finding the perimeter of some irregularly shaped objects such as a leaf or an oil spill.

2. Suppose you are asked to find the perimeter of your math book. Would you rather measure in inches or feet? Explain your choice.

1. Measure with string and then measure the string. 2. Inches; you'd have to use fractions if you measure in feet.

3 Ask students what formula to use (a) if they know the diameter of a circle and want to find the circumference; (b) if they know the radius and want to find the circumference; (c) if they know the circumference and want to find the diameter. [(a) $C = \pi d$; (b) $C = \pi 2r$; (c) $C = \pi d$]

ADDITIONAL EXAMPLES

1. Find the perimeter of a regular octagon with a side of 7.5 in. (60 in.)
2. Choose a circular object such as a clock. Measure the circumference. Calculate the diameter. Check your answer by measuring. (Check students' work.)
3. Find the precise circumference of a circle with a radius of 14 feet. (28π ft)
4. Find an approximate answer for question 3. ($2 \cdot 14 \cdot 3.14 = 87.92$ ft)

Guided Practice

Class Exercises Encourage students to choose objects that have substantially different shapes and perimeters. One object should be circular; another might be an irregular shape and so forth.

Think and Discuss To extend question 2, offer other measuring situations for which students can determine an appropriate unit with which to measure. Possibilities include: fencing in a schoolyard, (feet, yards, or meters); installing molding around the ceiling of a dollhouse livingroom (inches or centimeters); shore line of Lake Erie. (kilometers)

MATH MINUTES

Divide the class into teams. Have each team come up with a method for estimating the perimeter of the classroom. Teams then compare methods and estimates. Students should decide on the best estimation method.

A ***common error*** occurs when students use radius with $C = \pi d$ and diameter with $C = 2\pi r$. In Example 4b, on page 391, students might substitute 7 in the formula $C = \pi d$ instead of $C = \pi 2r$. Encourage students to draw a simple diagram and label the given measures. Tell them, then, to write the formula and to refer to their diagrams as they substitute values for the terms in the formula.

Closure

Writing in Math Ask students to summarize in their math journals how the terms *circumference* and *perimeter* are alike and different. (They are both measures of the distance around a figure. Perimeter is the distance around the sides of a polygon. Circumference is the distance around a circle.) Then ask members of the class to explain the steps involved in finding the perimeters of various polygons, such as a square, a rectangle, a trapezoid, an irregular pentagon, a regular hexagon, and the circumference of a circle.

Independent Practice

For questions 1-4, suggest that students draw a diagram and label each side. Remind students to include the unit of measure as part of their answers.

For questions 5-6, suggest that students copy the figures and label all the sides, then find the total of the measures.

In question 31, ask students to find several solutions.

Estimate the circumference of each circle in centimeters. Then measure each diameter and compute the circumference. Use 3.14 for π. Answers may vary; accept any reasonable estimate.

2.

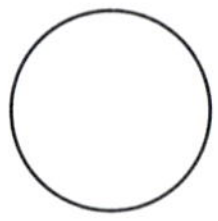

4.71 cm

3.

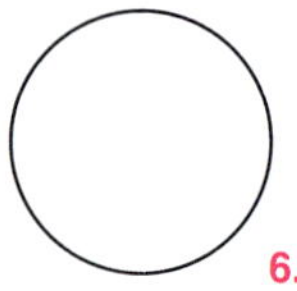

6.28 cm

WRITTEN EXERCISES

Find the perimeter of each figure.

1. a square with side 9 ft 36 ft
2. an equilateral triangle with side 5 yd 15 yd
3. a trapezoid with bases 3 ft and 3 yd, and sides 5 ft 7 yd 1 ft
4. a regular hexagon with side 3.7 in. 22.2 in.

Find the perimeter of each polygon.

5.

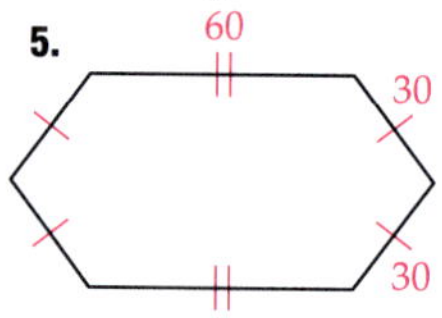

240

6.

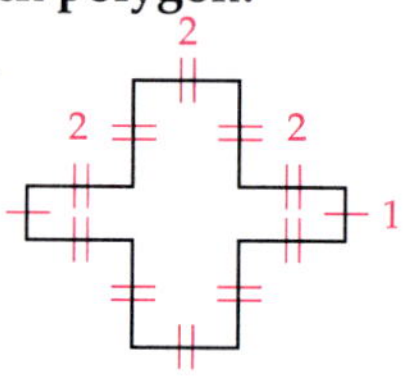

22

7.

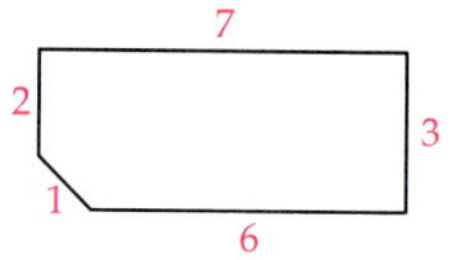

19

Find the amount of tape needed to wrap each figure. Use 3.14 for π.

8.

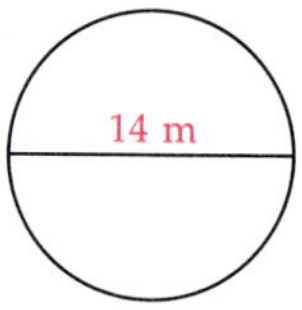

43.96 m

9.

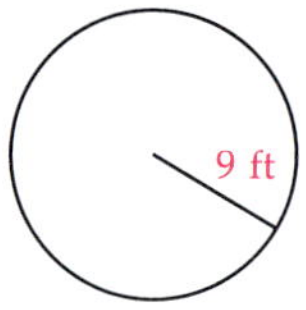

56.52 ft

10.

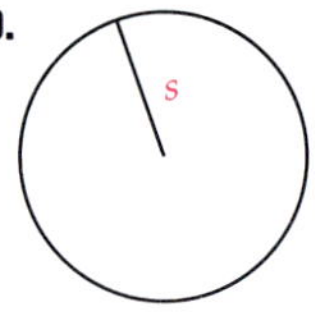

6.28 s

11.

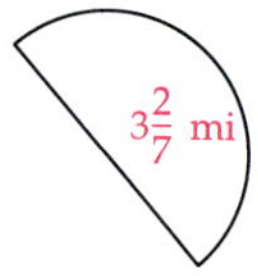

8.4 mi

Find the diameter and the radius of a circle with the given circumference.

12. $C = 15\pi$ m 15 m, 7.5 m
13. $C = 5a\pi$ yd 5a yd, 2.5a yd
14. $C \approx 19$ in. ≈6.1 in., ≈3 in.
15. $C \approx 2.5x$ units ≈ 0.8x units, ≈ 0.4x units

CALCULATOR **Find the circumference to the nearest tenth of a unit. Use 3.14 for π.**

16. $r = 5$ 31.4
17. $r = 3.5$ 22.0
18. $d = 17.8$ 55.9
19. $d = 0.625$ 2.0

Find the circumference. Use $\frac{22}{7}$ for π.

20. $r = \frac{1}{4}$ $1\frac{4}{7}$
21. $r = \frac{3}{11}$ $1\frac{5}{7}$
22. $d = 49$ 154
23. $d = 181$ $568\frac{6}{7}$

MIXED REVIEW

1. Multiply $\frac{1}{2}(5)(8)$. 20
2. What percent of 135 is 125? 93%
3. $|1 - 3.5| = \blacksquare$ 2.5
4. $ABCD \sim HIJK$. Write ratios relating corresponding sides.

Find each answer.

5. $5.5n^2$ for $n = 2.2$ 26.62
6. $a^2 + b^2$ for $a = 5$ and $b = 12$ 169
7. How many integers between 100 and 1,000 use only the digits 3, 4, or 5? 27

4. $\frac{AB}{HI} = \frac{BC}{IJ} = \frac{CD}{JK} = \frac{DA}{KH}$

Find the unknown length.

24. $P = 75$
7, 14

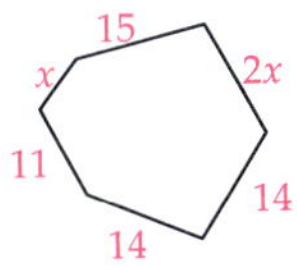

25. $P = 91$
27

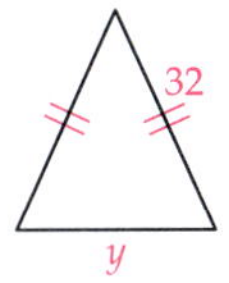

26. $P = 10$
$1\frac{1}{4}$, $3\frac{3}{4}$

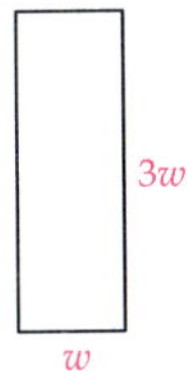

27. $P = 37$
6

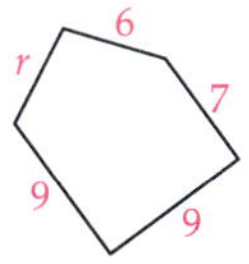

Write an expression for the perimeter of each polygon.

28.

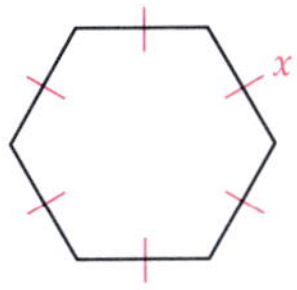

$6x = P$

29.

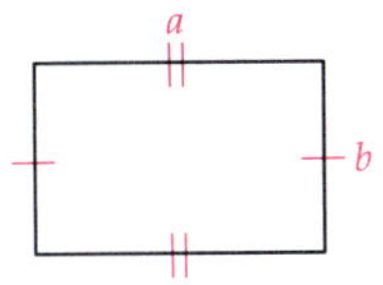

$2a + 2b = P$

30.

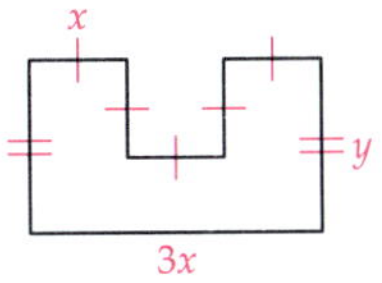

$3x + 5x + 2y = P$

Use graph paper to draw a polygon that fits each description. Check students' work.

31. a rectangle with a perimeter of 12 units

32. a square with a perimeter of 60 units

33. a hexagon with a perimeter of 8 units

34. ***DATA FILE 3 (pp. 96–97)*** Find two possible perimeters for a sheet of dollar bills. Answers will vary. 91 in.

TEST YOURSELF

1. Find BC. $\triangle ABC \sim \triangle DEF$.
6

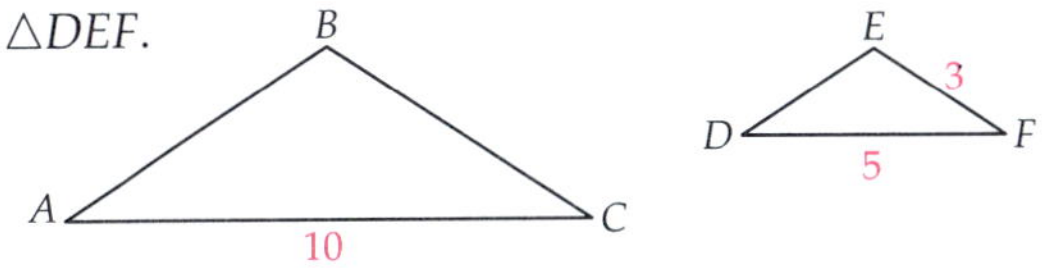

Find the perimeter or circumference. Use 3.14 for π.

2.
33.4

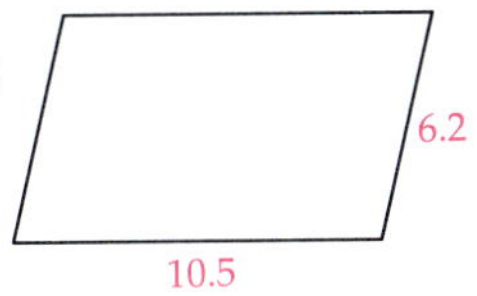

3.
94.2

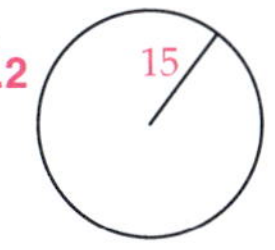

NOTES & QUOTES

In about 200 B.C., using only his knowledge of geometry, Eratosthenes estimated Earth's circumference as 46,250 km. In 1700, the French mathematician, Jean Picard used a telescope to refine the estimate to 40,040 km. Today's technology uses the orbits of satellites to measure Earth's circumference to an accuracy of 0.5 m. Modern instruments show that Earth is pear-shaped with the diameter from North Pole to South Pole about 43 km less than at the equator.

Lesson Follow-up

Reteaching Activity Provide problems that contain one piece of information, such as the radius of the circle, and space to write the formula, and each step of the solution.

LESSON RESOURCES

TRB Practice I-15
TRB Enrichment I-16

FOR THE NEXT LESSON

tube or circle, tape measure, math journal

Practice

For use after 9-8 (pp. 390–393)

Perimeter and Circumference

Find the perimeter of each figure.

1. a rectangle with length 17 in. and width 9 in. **52 in.**
2. an isosceles triangle with base 11.3 cm and sides 14.2 cm **39.7 cm**
3. a scalene triangle with sides 2 ft 4 in., 3 ft 5 in., and 4 ft 6 in. **10 ft 3 in.**

Find the perimeter of each polygon.

4.

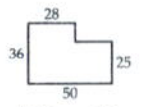

172 units

5.

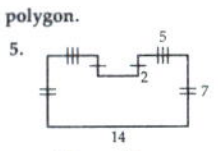

46 units

6.

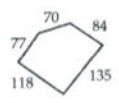

484 units

Find the circumference of each figure. Use 3.14 for π.

7.

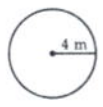

25.12 m

8.

188.4 cm

9.

15.072 m

10.

263.76 cm

Find the diameter and radius of a circle with the given circumference.

11. $C = 8\pi$ in. $d =$ **8 in.** $r =$ **4 in.**
12. $C = 56.2\pi$ cm $d =$ **56.2 cm** $r =$ **28.1 cm**

CALCULATOR Find the circumference to the nearest tenth of a unit. Use 3.14 for π.

13. $d = 25.8$ $C =$ **81.0 units**
14. $r = 9.1$ $C =$ **57.1 units**
15. $r = 0.07$ $C =$ **0.4 unit**

Find the circumference. Use $\frac{22}{7}$ for π.

16. $d = 14$ $C =$ **44 units**
17. $d = 1\frac{3}{11}$ $C =$ **4 units**
18. $r = \frac{7}{8}$ $C =$ **$5\frac{1}{2}$ units**

19. A rectangle with length 17.3 cm has perimeter 46.8 cm. Find the width.
6.1 cm

Chapter 9 I-15

Enrichment

For use after 9-8 (pp. 390–393)

The Distance and Size of the Sun

To measure the distance to the sun, two astronomers at positions A and B photographed the sun when the sky was dark during a solar eclipse. Later measurements of the photos showed that, due to the different camera locations, the sun appeared in slightly different positions against the background stars. The angle of difference was 0.00244°.

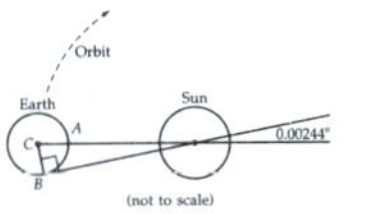

(not to scale)

1. Why is $m\angle CSB$ also 0.00244?
Vertical angles are congruent.

The radius of the earth, CB is 3,965 mi. The astronomers reasoned that for every 0.00244° the earth revolves in its orbit about the sun, it travels the distance of one radius, or 3,965 miles.

2. How many 0.00244° angles are there in a full 360° orbit about the sun? Round to the nearest thousand.
148,000

3. Multiply your answer by the earth's radius to find the circumference of the earth's orbit. Round to the nearest million.
587,000,000 mi

4. Use $C = 2\pi r$ to find r, the distance from the earth to the sun. Use $\pi = 3.14$ and round to the nearest million.
93,000,000 mi

To find the radius of the sun, the astronomers stood at A and measured the angle between the center of the sun and the edge. The angle was 0.25°.

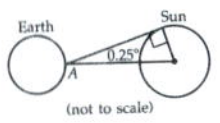

(not to scale)

5. If the sun orbited the earth (it doesn't!), its orbit would equal the earth's orbit around the sun. How many 0.25° angles would it move through during one orbit?
1,440

6. Use your answers to Exercises 3 and 5 to find the radius of the sun. Round to the nearest thousand.
408,000 mi

I-16 Chapter 9

LESSON QUIZ

1. Find the perimeter of a square with a side of 13 in. (52 in.)

2. Find the perimeter of a regular hexagon used in a quilt. A side is 5 in. (30 in.)

3. Find the circumference of a circle with a radius of 45 cm. (282.6 cm)

4. Find the approximate diameter of a circle in which the circumference is about 47 ft. (Accept reasonable estimates; 14.97 ft)

Assignment Guide

Basic 1-6, 8-9, 12-17, 20-21, 24, 26, 28, 31-32, 34, MR, TY All

Average 2-6, 10-14, 17-18, 21-26, 28-29, 31-34, MR, TY All

Enriched 3-4, 6-7, 10-11, 14-19, 22-23, 26-27, 29-34, MR, TY All

Lesson Focus

Materials

Tube or circle, tape measure, math journal

MOTIVATING THE LESSON

Review the different problem solving strategies that have been taught this year. Ask students to describe problems for which drawing a picture helps in finding the solution. (Possible answers: area, perimeter, and volume problems.)

Instruction

1 Ask students to mark a starting point on the tube and a corresponding point on the floor. Have students roll the tube one complete rotation and mark the end point. Students should measure the distance the tube rolled. Then have students determine the diameter of the tube and the circumference. The circumference and the distance the tube rolled should be approximately the same.

Guided Practice

Class Exercises Ask students to solve question 2 if the radius of the tire is 17 in. (same answer)

A ***common error*** occurs when students draw an incorrect picture and then solve the problem based on the picture. Encourage students to "read" their picture and confirm that it represents the problem.

Closure

Writing in Math Have students write a problem that can be solved by using the strategy *draw a picture*. Have students exchange and solve each other's problems.

PROBLEM SOLVING STRATEGY

OBJECTIVE: *To solve problems by drawing a diagram.*

9-9 Draw a Diagram

For many geometric problems it helps to draw a diagram to represent the given information.

PROBLEM

The wheels of a bicycle have a diameter of 60 cm. How many revolutions do the wheels make during a 4,000 m race?

SOLUTION

READ What information do you have? — The wheels have a diameter of 60 cm. The race is 4,000 m long.

What do you want to find? — How many revolutions (turns) do the wheels make in a race?

PLAN Decide on a strategy. — Draw a picture to represent the information.

1

What is a complete turn? — the circumference of the wheel

What units will you use? — Find the circumference in centimeters. Then write the circumference in meters.

SOLVE Use the formula $C = \pi d$. Use 3.14 for π.

$$C = \pi d \approx 3.14(60) = 188.4 \text{ cm} = 1.884 \text{ m}$$

Divide the total distance by the distance of one revolution. Round to the nearest whole number.

$$4{,}000 \div 1.884 = 2{,}123.142\ldots \approx 2{,}123$$

LOOK BACK Estimate to check your answer. — Each revolution was 1.8 m or about 2 m. The race was 4,000 m, so each wheel turned about 2,000 times.

The answer of 2,123 turns is reasonable.

Teaching Tip

Make a poster that lists all the problem solving strategies taught in the textbook. Add new strategies as they are introduced. Include a sample problem for each strategy.

CLASS EXERCISES

Refer to the problem on page 394. Use 3.14 for π.

1. Suppose the race was 6,000 m long. Estimate the number of revolutions for the tires. 3,185

2. Suppose a race car has tires with diameter of 34 in.
   a. How far would the tires turn in one complete revolution? 106.76 in.
   b. How many revolutions would the tires make in a 500 mi race? about 300,000

WRITTEN EXERCISES

Use a CALCULATOR where appropriate.

Solve using a picture. Use 3.14 for π. Round to the nearest whole number.

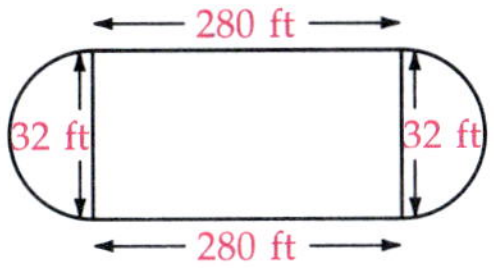

1. How many laps must you run on the track pictured at the right to cover 1 mi? *Hint:* 5,280 ft = 1 mi ≈ 8

2. Sam is building a circular table for eight people. Each person needs 2 ft of table edge. What is the diameter of the table? 5 ft

3. Two intersecting circles can form two or three regions. Three intersecting circles can form from three to seven regions. What is the greatest number of regions you can form with four intersecting circles? 13

Solve using any strategy.

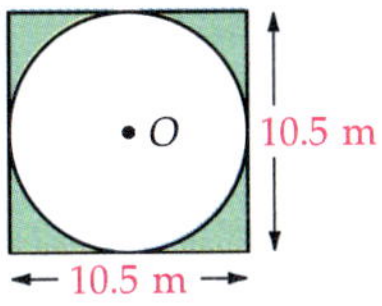

4. Find the circumference of circle O. Use 3.14 for π. 32.97 m

5. Suppose you want to make a scale model of Earth, the sun, and the moon. Use the ***DATA*** at the right to determine the dimensions of the model. Assume that Earth will have a diameter of 3 in.
   a. What will be the diameter of the sun and the moon? sun = 324 in. moon = 0.825 in.
   b. In the model, how far from Earth must each be? sun = 34,875 in. moon = 90 in.

	d (mi)	*distance from Earth (mi)*
Earth	8,000	0
Sun	864,000	93,000,000
Moon	2,200	240,000

6. The figures shown are all composed of equal size squares. Figure A has a perimeter of 36. Find the perimeter of each figure.

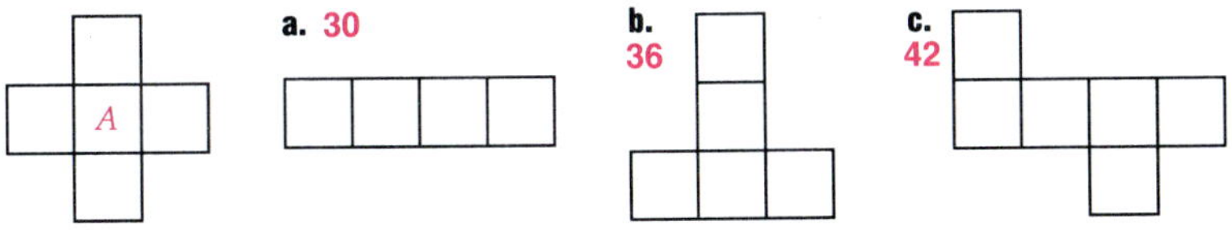

a. 30 b. 36 c. 42

Independent Practice

For questions 1-3, have students include their drawings with their solutions.

For question 5, help students realize that the ratio of the scale model is 3 : 8,000 and the ratio must remain consistent.

Lesson Follow-up

LESSON RESOURCES

TRB Practice I-17
TRB Enrichment I-18

Practice For use after 9-9 (pp. 394–395)

Draw a Diagram

Solve using a diagram. Use 3.14 for π.

1. An old-fashioned "high-wheel" bicycle had a small wheel with diameter 8 in. in back and a large wheel with diameter 56 in. in front. A circular bicycle track had diameter 112 ft.
   a. Find the circumference in inches.
   back wheel 25.12 in. front wheel 175.84 in. track 4,220.16 in.
   b. Find the number of complete wheel revolutions in one circuit of the track.
   back wheel 168 revolutions front wheel 24 revolutions
   c. A rider made 100 circuits of the track. How many more revolutions did the back wheel make than the front?
   14,400

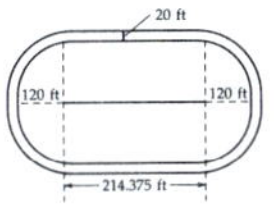

2. Two runners ran once around the track, one on the very inside, one on the very outside. How much farther did the outside runner run than the inside runner did? (Hint: Use 120 ft as a radius.)
   125.6 ft

3. A circular lighthouse has radius 10 m. A circular fence around the outside of the lighthouse has circumference 12.56 m greater than that of the lighthouse.
   a. Find the circumference of the lighthouse. 62.8 m
   b. Find the circumference of the fence. 75.36 m
   c. Calculate the radius of the fence. 12 m
   d. How far is the fence from the lighthouse? 2 m

4. To make a pencil holder, Maria tied pieces of colored string around the outside of a cylindrical can 4 in. in diameter and 6 in. high. The string was $\frac{1}{12}$ in. in diameter. Maria used 1 in. of string to tie each knot. How much string did she use if she covered the entire outside of the can with circles of string?
   976.32 in.

5. Four circular pies fit exactly into a square box measuring 18 in. on a side. Find the circumference of one of the pies.
   28.26 in.

Chapter 9 I-17

Enrichment For use after 9-9 (pp. 394–395)

Ten-Speed Bicycles

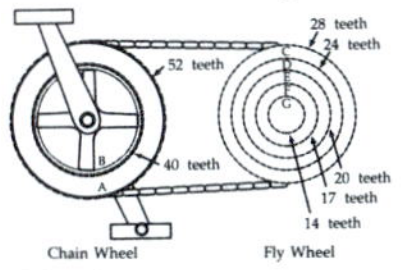

The *gear ratio* of a 10-speed bicycle tells you how many times the wheels revolve each time the pedals revolve once. You can find the gear ratio by dividing the number of teeth engaged on the chain wheel by the number engaged on the flywheel.

Find the gear ratio when the given gears are engaged. Round to the nearest hundredth.

1. B and E 2
2. A and C 1.86
3. B and F 2.35
4. What is the highest gear ratio? 3.71 the lowest? 1.43

A rider's *cadence* is the number of pedal revolutions per minute.

5. A rider's cadence when the B and D gears are engaged is 32. Find the number of revolutions of the wheels in 5 min. Round to the nearest whole number.
   267 revolutions
6. The diameter of a 10-speed bicycle is 29 in., including tires. What is the circumference? Use 3.14 for π. Round to the nearest whole number.
   91 in.

A rider's cadence is 25. The A and F gears are engaged. **7–10. Answers may vary due to rounding.**

7. How far does the bicycle travel in 1 min? Round to the nearest whole number.
   6,959 in.
8. How far does the bicycle travel in 1 h? 417,540 in.
9. Find the rider's speed in mi/h (1 mi = 63,360 in.). Round to the nearest tenth.
   6.6 mi/h
10. A rider's cadence is 80. The B and E gears are engaged. Find the rider's speed in mi/h.
   13.8 mi/h

I-18 Chapter 9

MATH MINUTES

Ask students to work in pairs to draw and describe each classification of triangle: right, acute, obtuse, equiangular, scalene, equilateral, and isosceles. (Check students' work.)

Assignment Guide

Basic All
Average All
Enriched All

FOR THE NEXT LESSON

graph paper, scissors, math journal

Instruction

This page provides a variety of problems that can be used to reinforce and enhance the students' problem solving skills. Encourage students to read each problem carefully. Then have them refer to the list of problem solving strategies to help them decide how to solve the problem.

Point out, however, that not all questions require a strategy for solving, nor are all the strategies in the list used in this lesson.

Problem Solving Practice

READ PLAN SOLVE LOOK BACK

PROBLEM SOLVING STRATEGIES

Look for a Pattern
Guess and Test
Simplify the Problem
Account for All Possibilities
Make an Organized List
Work Backwards
Make a Table
Write an Equation
Solve by Graphing
Draw a Diagram
Make a Model
Simulate the Problem

Exchange Rates

	Equivalent In U.S. Dollars
Yen	0.007432
Pound	1.9075
Franc	0.1935

Solve. The list at the left shows some strategies you might use.

1. The scores in a game were -7, 8, 7, 15, -6, 12, and -6. Was the final score positive or negative? **positive**
2. State sales tax on new car purchases is 5%. Find the sales tax on a car costing \$22,489. **\$1124.45**
3. The Blazers lost 35% of the 20 games they played this season. How many games did they win? **13 games**
4. You open a book and the product of the page numbers is 16,002. To what pages have you opened? **126, 127**
5. Two buses leave a terminal at the same time and travel in opposite directions. One bus travels at 55 mi/h. The other travels at 48 mi/h. How far apart will the buses be in 3 h? **309 mi**
6. Sabrina can run 5 km in 1 h. In the same time, Lucy can run a 3-km distance. How much of a head start does Lucy need for both girls to finish an 8-km course at the same time? **64 min**
7. A stamp collector has 53 rare stamps. This is 12 less than 5 times the number he had a year ago. How many stamps did the collector have a year ago? **13**
8. Suppose you want to find the thickness of one sheet of paper. Describe a method that uses a ruler. **Measure a stack of paper. Divide your answer by the number of sheets.**
9. ***DATA*** Use the data at the left. How many yen would you get for \$100? how many pounds? how many francs? **13,455.33 yen; 52.42 pounds; 516.8 francs**
10. At a track meet, finishers received 5 points for blue ribbons and 3 points for red ribbons. How many and what color ribbons were won for the following points?

    a. 12 **4 red** **b.** 14 **1 blue 3 red** **c.** 18 **3 blue; 1 red or 6 red** **d.** 15 **3 blue or 5 red**
11. A real estate agent earns a 6% commission on the sale of a house. Find the commission on a \$289,000 house. **\$17,340.00**
12. ***DATA FILE 10 (pp. 404–405)*** Find the circumference of the circular home range for a snowshoe hare. **180π m or 565.2 m**
13. Two friends rented a canoe for 10 days. One friend used the canoe for 6 days. The other friend used the canoe for 4 days. How much of the \$150 rental fee should each friend pay? **\$90; \$60**
14. José bought two posters for \$9. The posters were on sale at one for full price and the second at half price. How much was the full price poster? **\$6**

OBJECTIVE:
To understand and create tessellations.

NOTES & QUOTES

"All my works are games. Serious games."
M. C. Escher

Maurice Cornelius Escher (1898–1972), a Dutch artist, used tessellations in many of his sketches. He created at least 150 different tessellations without ever having any formal mathematics training.

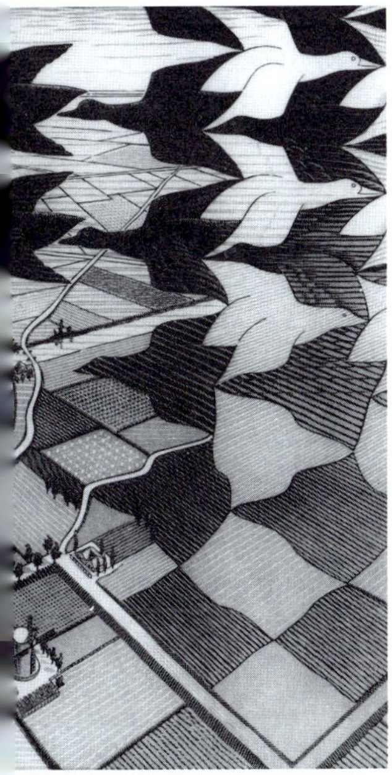

PROBLEM SOLVING APPLICATION

9-10 Tessellations

The *tessellation* at the right is made from congruent pentagons. You could see this tessellation in the street tiling of portions of Cairo, Egypt. When a tessellation uses only one shape, we call it a *pure tessellation*.

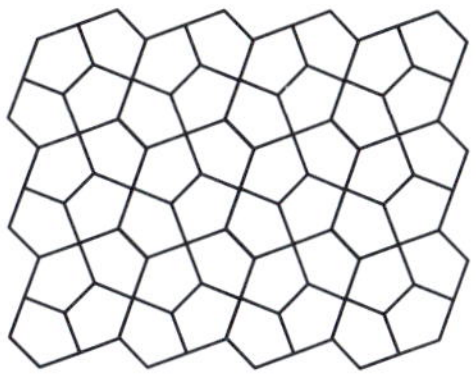

Tessellation	A tessellation is a design that covers a plane with no gaps and no overlaps.

You can determine if a figure forms a pure tessellation by using graph paper to represent a plane.

Example 1 **Determine whether the figure at the right forms a pure tessellation.**

Solution

1

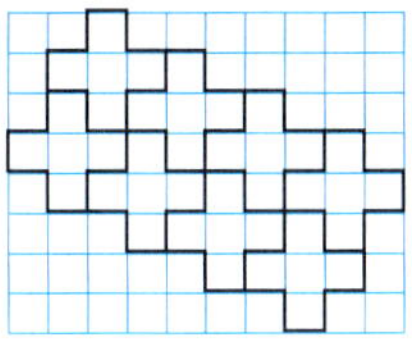

Yes, the figure forms a pure tessellation.

Tiled floors often use a pattern in which two figures tessellate. The patterns are *semiregular tessellations*. You can use a pattern to determine whether two figures form a semiregular tessellation.

Semiregular Tessellation	A semiregular tessellation is a design that covers a plane using more than one shape.

Example 2 **Determine whether the figures at the right form a semiregular tessellation.**

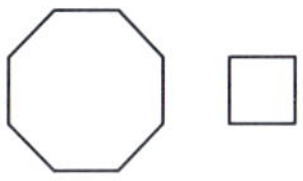

Solution

2

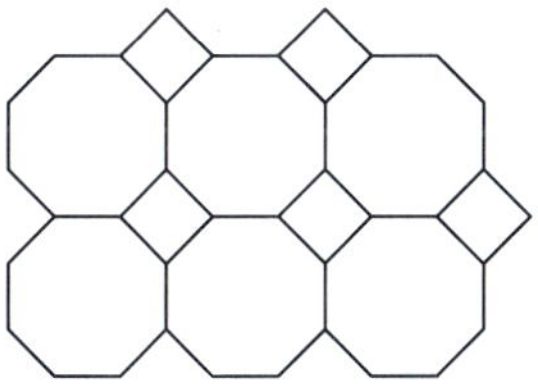

Yes, the figures form a semiregular tessellation.

Lesson Focus

Materials
Graph paper, scissors, math journal

Vocabulary/Symbols
pure tessellation
semi-regular tessellation

MOTIVATING THE LESSON

Have students look at patterns that repeat. Provide examples of tessellations to students and have them describe all that they observe about the patterns. Then explain that these are special patterns that will be investigated in this lesson.

Instruction

1 Have students draw the figure on graph paper to confirm that no other figure is needed to fill the plane and that there are no gaps in the design.

2 Mention that the shapes in a semi-regular tessellation must be regular polygons with sides of the same length. In addition, all the vertices are identical.

Teaching Tip

Have posters, art books, or slides of M.C. Escher's works available for students to review and evaluate.

(3) Students may choose a shape that is different from the one in the text. As students attempt to create a pure tessellation, mention that the sum of the measures of the angles, wherever points meet, must be 360°.

Guided Practice

Class Exercises For question 3, encourage students to share their designs with their classmates.

Closure

Writing in Math Have students describe and draw a pure and semi-regular tessellation in their math journals.

Independent Practice

For question 4, have students graph six triangles around point *P*.

Have students share their designs for question 9.

Decision Making Have students work in small groups.

For question 1, suggest students look in nature books or magazines.

Suggest that students use floor tile flyers for question 3 if they are unable to visit a store.

For question 4b, have students create a pure tessellation using hexagons. They can then determine what polygon is formed.

THINK What are four shapes that will tessellate? **Answers may vary.**

■ You can create a tessellating figure by beginning with a shape that is known to tessellate.

Example 3 **Create a tessellating figure using a square.**

Solution (3)

1. Begin with a square.

2. Cut a shape out of one side. Slide the shape to the opposite side. Tape the shape in place.

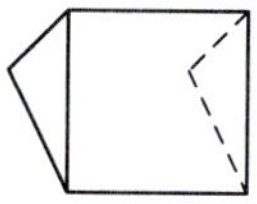

3. Use the new shape as a pattern. Trace the shape in different places to form a tessellation.

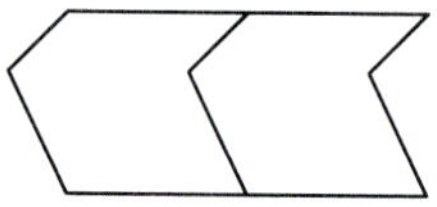

THINK The Greeks proved that only three *regular* polygons will tessellate a plane. Which regular polygons tessellate? **triangle, square, hexagon**

CLASS EXERCISES

Use graph paper to determine whether each figure forms a pure tessellation.

1. yes

2. no

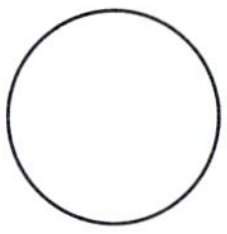

3. yes

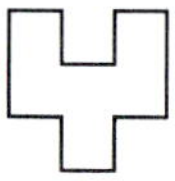

Decision Making ■ **DECISION MAKING** ■ Decision Making ■ Decision Making ■ Decision Making

TESSELLATIONS

Tessellations are frequently found in nature. Architects and designers often use nature as a model for their work.

■ ***COLLECT DATA*** **Check students' work.**

1. List three places in nature where tessellations occur.
2. List three places where you see man-made tessellations in your home and community.
3. Visit a store that sells floor tiles. In what shapes do the tiles come? Write down the cost of each tile.

■ ***ANALYZE DATA***

4. a. Why do you think bees use a hexagon shape for their honeycombs instead of an octagon? **Hexagons will tessellate. Therefore, they make the most use of the area of the honeycomb.**

MATH MINUTES

Have students use shape tracers to draw several different polygons (at least five copies of each figure). Students cut out the shapes to make a semi-regular tessellation. (Check students' work.)

WRITTEN EXERCISES

Use a CALCULATOR where appropriate.

Use a pattern to determine whether the figures form a pure tessellation.

1. yes 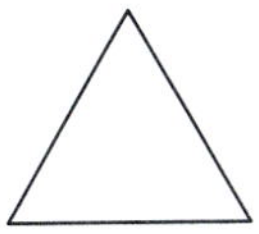
2. no 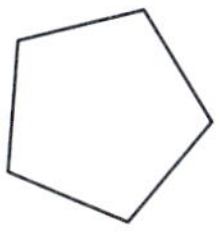
3. no 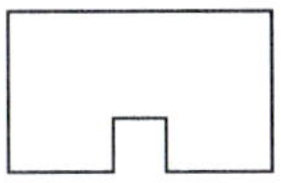
4. A triangle tessellates six times around point *P*. What must be the size of each angle around point *P*? 60°
5. Create a pure tessellation on graph paper. Check students' work.

Use a pattern to determine whether the figures form a semiregular tessellation.

6. no 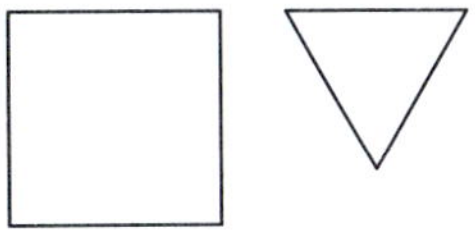
7. yes 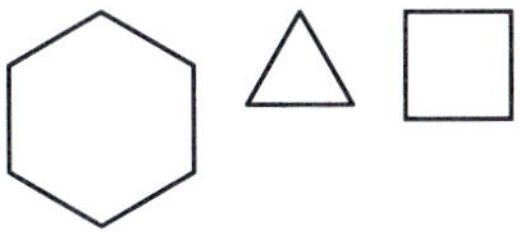
8. Create a semiregular tessellation on graph paper. Check students' work.
9. ***PROJECT*** A tessellation of equilateral triangles is shown at the right. You can form another tessellation when you connect the center of each triangle across the common sides of the tessellating triangles. The new tessellation is called a *dual*. Every tessellation of regular polygons has a dual. Create a dual using any regular polygon. Describe your results. Check students' work.

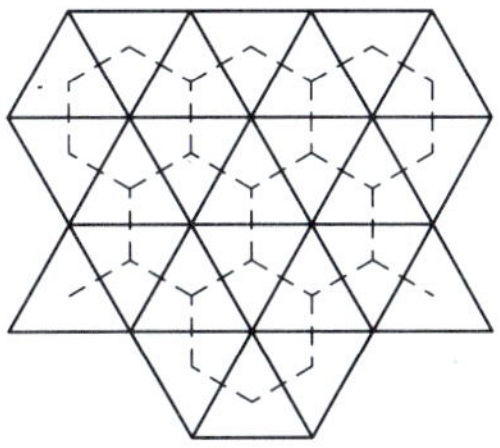

THINK Some patterns will tessellate around a sphere rather than a plane. What shapes tessellate on a soccer ball? pentagons and hexagons

Decision Making ■ Decision Making ■ Decision Making ■ Decision Making ■ Decision Making ■ Decision Making ■

b. Each hexagon in a honeycomb can be broken down into smaller regular polygons. What kind of regular polygons are they? triangles

5. What is the sum of the measures of all the angles that come together at a point in a pure tessellation? in a semiregular tessellation? 360; 360
6. Why do you think architects frequently use a triangular truss to build bridges? Triangles will tessellate across a given area. They give a stable flat surface onto which a road can be built.

■ ***MAKE DECISIONS***

7. A customer wants you to design a tiling pattern for a rectangular foyer. The foyer is 15 ft by 15 ft. Create two designs on graph paper. Make the first a pure tessellation and the second a semiregular tessellation. For each design, determine how many of each shape tile the customer will need. Then find the total cost of each design. Check students' work.

Lesson Follow-up

LESSON RESOURCES

TRB Alternate Application I-19

Alternate Application

For use after 9-10 (pp. 397–399)

How Old Are You?

Mercury, the planet nearest the sun, orbits the sun at an average distance of 3.6×10^7 mi. Its average speed in orbit is 1.07×10^5 mi/h.

Solve. Use 3.14 for π. Give answers in scientific notation rounded to three decimal places unless otherwise instructed.

1. Find the circumference of Mercury's orbit. 2.261×10^8 mi
2. Find the number of hours it takes Mercury to complete one orbit of the sun. 2.113×10^3 h
3. Convert the above answer to days. Round to the nearest day and express your answer in standard form. 88 da

A "year" on a planet is the length of time it takes the planet to travel once around the sun. One Earth year is about 365 days.

4. About how many Mercury years are there in one Earth year? Round to the nearest year. 4 y
5. Find your age in Mercury years. Student's age × 4

Pluto, the planet farthest from the sun, orbits the sun at an average distance of 3.65×10^9 mi. Its average speed in orbit is 1.05×10^4 mi/h.

Solve.

6. What is the circumference of Pluto's orbit? 2.292×10^{10} mi
7. What is the length of a year on Pluto in days, rounded to the nearest day? 90,952 da
8. What is the number of Earth years in one Pluto year, rounded to the nearest year? 249 y
9. What is your age in Pluto years? Student's age ÷ 249

Chapter 9 I-19

Assignment Guide

Basic All
Average All
Enriched All

FOR THE NEXT LESSON

geoboards and rubber bands or dot paper, metric ruler, pencil, scissors, math journal

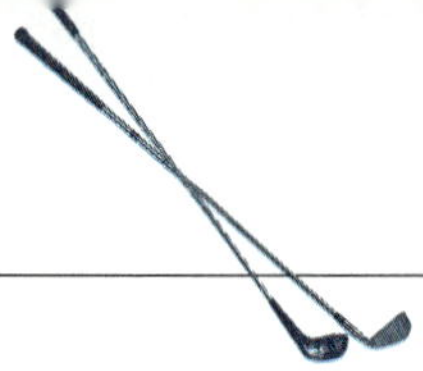

Study Skills

Since a *textbook* is designed as a teaching tool, it is important that students take the time to examine it and take advantage of every feature.

The table of contents is an outline of all the information the book contains. Students can view it as a study outline, since they will cover most of this material during the school year.

Often there are pages at the beginning of the book that explain the author's purpose in writing the book and give suggestions on how to use it. These pages are helpful when writing a report.

Questions that may appear at the end of the chapter provide a handy self-testing opportunity. The questions highlight the important ideas taught in the chapter.

Maps, charts, tables, and graphs reinforce the ideas in the text visually, presenting data in a concise, easy-to-understand format. Using these aids can save valuable study time.

A glossary is an invaluable study aid. It highlights the important words in the book and their definitions.

Chapter 9 Review

1. Classify the following words. Use the categories basic geometric figures, angles, triangles, quadrilaterals, polygons, or circles. Some words belong in more than one category. **See additional answers.**

acute	diameter	obtuse	ray	square
adjacent	equiangular	parallelogram	rectangle	straight
chord	equilateral	pentagon	rhombus	supplementary
circumference	hexagon	plane	right	triangle
complementary	isosceles	point	scalene	trapezoid
convex	line	radius	segment	vertical

Introduction to Geometry — 9-1

The basic geometric figures include point, line ($\overleftrightarrow{AB}$), segment ($\overline{AB}$), ray ($\overrightarrow{AB}$), and plane.

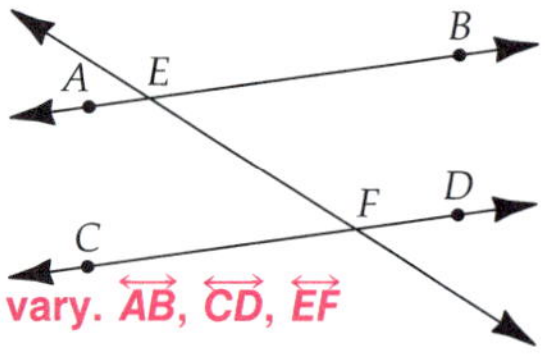

Refer to the figure at the right.

2. Name two parallel lines. **$\overleftrightarrow{AB} \parallel \overleftrightarrow{CD}$**
3. Name three lines. **Ans. may vary. $\overleftrightarrow{AB}$, $\overleftrightarrow{CD}$, $\overleftrightarrow{EF}$**
4. Name five different segments. **Ans. may vary. $\overline{AE}$; $\overline{EB}$; $\overline{EF}$; $\overline{FD}$; $\overline{CF}$**
5. Name eight different rays. **Ans. may vary. $\overrightarrow{EF}$, $\overrightarrow{CD}$, $\overrightarrow{BA}$, $\overrightarrow{AB}$, $\overrightarrow{EB}$, $\overrightarrow{FD}$, $\overrightarrow{FC}$, $\overrightarrow{EA}$, $\overrightarrow{FE}$**

Angles — 9-2

We classify angles by their measures. Acute angles have measures greater than 0° and less than 90°. Obtuse angles have measures greater than 90° and less than 180°. Right angles measure 90°.

Two angles are supplementary if the sum of their measures is 180. Two angles are complementary if the sum of their measures is 90.

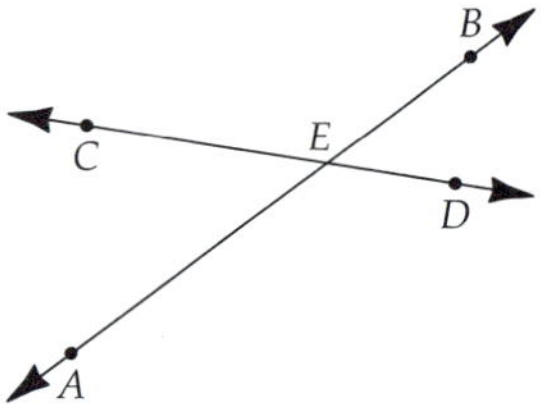

Refer to the figure at the right.

6. $\angle AEC$ is an ■ angle. **acute**
7. $\angle AEC$ and $\angle BEC$ are ■ angles. **supplementary**
8. $\angle AEC$ ■ $\angle BED$. **$\cong$**
9. If $m\angle CEB = 120°$, $m\angle AED =$ ■. **120°**
10. If $m\angle BED = 55°$, $m\angle AED =$ ■. **125°**

Polygons and Quadrilaterals — 9-3

Some quadrilaterals have special names and special properties.

***True* or *false*? Write *T* or *F*.**

11. A trapezoid is a parallelogram. **F**
12. All squares are rhombuses. **T**
13. A rhombus is a rectangle. **F**
14. Some rhombuses are rectangles. **T**
15. All rectangles are parallelograms. **T**
16. A rhombus can be a square. **T**

ADDITIONAL ANSWERS

basic figures	angles	triangles	quadrilaterals	polygons	circles
line	acute	acute	parallelogram	convex	chord
plane	adjacent	equiangular	rectangle	hexagon	circumference
point	complementary	equilateral	rhombus	parallelogram	diameter
ray	obtuse	isosceles	square	pentagon	radius
segment	right	obtuse	trapezoid	rectangle	
	straight	right		rhombus	
	supplementary	scalene		square	
	vertical			triangle	
				trapezoid	

Triangles 9-4

We classify triangles by angles and by sides. The sum of the measures of the interior angles of a triangle is 180°.

Classify each triangle by its angles.

17. 30°, 60°, and 90° **right** **18.** 40°, 37°, and 103° **obtuse** **19.** 55°, 55°, and 70° **acute**

20. A triangle has two angles with measures of 50° and 70°. What is the measure of the third angle? **60°**

Circles 9-5

In any circle, the diameter *d* is twice the radius *r*.

For each length, find the radius or the diameter.

21. $r = 25$ cm, $d =$ ▒ **50 cm** **22.** $d = 40$ cm, $r =$ ▒ **20 cm** **23.** $r = 3\frac{1}{2}$ in., $d =$ ▒ **7 in.**

24. $r = 4.2$ cm, $d =$ ▒ **8.4 cm** **25.** $d = 30$ ft, $r =$ ▒ **15 ft** **26.** $d = 9$ ft, $r =$ ▒ **4.5 ft**

Congruent Figures 9-6

Two figures are congruent if their corresponding angles and corresponding sides are congruent.

27. $ABCD \cong WXYZ$. Write congruence statements for corresponding parts. **$\overline{AB} \cong \overline{WX}$; $\overline{BC} \cong \overline{XY}$; $\overline{CD} \cong \overline{YZ}$; $\overline{AD} \cong \overline{WZ}$; $\angle A \cong \angle W$; $\angle B \cong \angle X$; $\angle C \cong \angle Y$; $\angle D \cong \angle Z$**

Similar Figures 9-7

Two figures are similar if the corresponding angles are congruent and the corresponding sides are in proportion.

Assume each pair of triangles is similar. Find the missing length.

28. **x = 9**

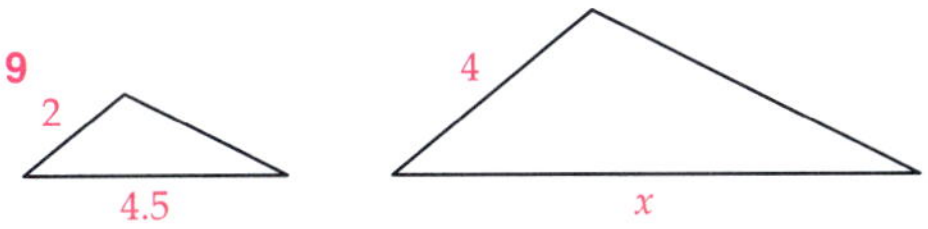

29.

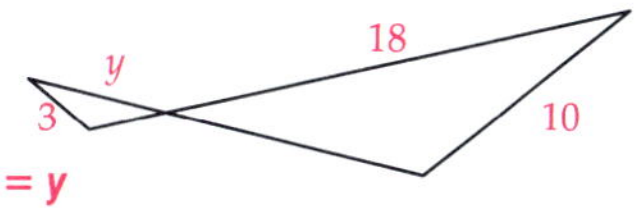

5.4 = y

Perimeter and Circumference 9-8

To find the perimeter of a polygon, add the lengths of the sides.
To find the circumference of a circle, use the formula $C = \pi d$.

Find the perimeter or the circumference. Use 3.14 for π.

30. rectangle with sides 24 in. and 18 in. **84 in.** **31.** circle with $d = 150$ cm **471 cm**

The appendix section of a book contains additional information, tables, or references that can also be helpful. Students should be aware that authors often provide additional sources for data.

Finally, the index is a quick and ready source for any topic in the book. Subtopics enable students to locate an item in a particular context.

Here are some suggestions:

- Provide time for students to examine the textbook.
- Refer students to these features when appropriate.
- Challenge students to find other features in the book that save time.

RESOURCES

TRB I-21–I-23

Scoring Chart	
No. Correct	**Score**
25	100
24	96
23	92
22	88
21	84
20	80
19	76
18	72
17	68
16	64
15	60
14	56
13	52
12	48
11	44
10	40
9	36
8	32
7	28
6	24
5	20
4	16
3	12
2	8
1	4

Chapter 9 Test

Use the figure at the right. Answers may vary.

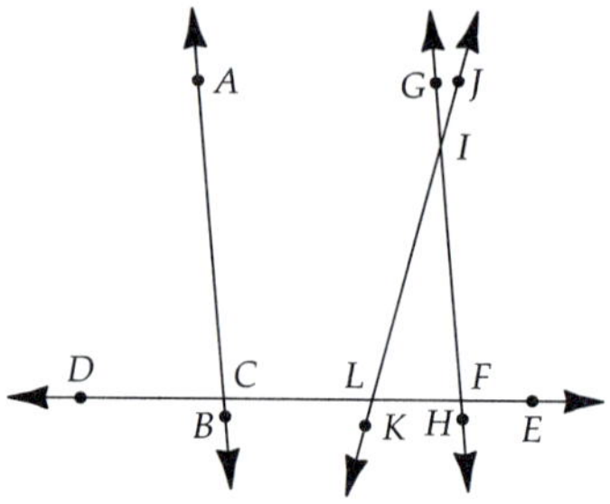

1. Name $\overleftrightarrow{AB}$ in three ways. $\overleftrightarrow{AC}$; $\overleftrightarrow{CB}$; $\overleftrightarrow{BA}$
2. Name three different segments on $\overleftrightarrow{CD}$. $\overline{CD}$; $\overline{CL}$; $\overline{LE}$
3. Name three rays on $\overleftrightarrow{GH}$. $\overrightarrow{GH}$; $\overrightarrow{IG}$; $\overrightarrow{IH}$

Complete each sentence.

4. $\angle ACL$ is an ■ angle. obtuse
5. $\angle ACD$ and $\angle LCB$ are ■ angles. vertical
6. $m\angle GIL$ ■ $m\angle JIF$. =
7. $\overleftrightarrow{AB}$ is ■ to $\overleftrightarrow{GH}$. parallel
8. $\angle JLF$ is an ■ angle. acute
9. $m\angle GFL = 85°$, $m\angle GFE =$ ■. 95°
10. ***WRITE*** Describe the intersection of two lines. The intersection of two lines is a point.

Draw a figure to fit each description. See students' figures.

11. a regular quadrilateral
12. a hexagon
13. an acute triangle

Find the measure of each numbered angle. Then classify each triangle by its angles and its sides.

14. 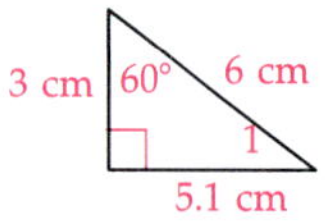

$\angle 1 = 30°$; right scalene

15. 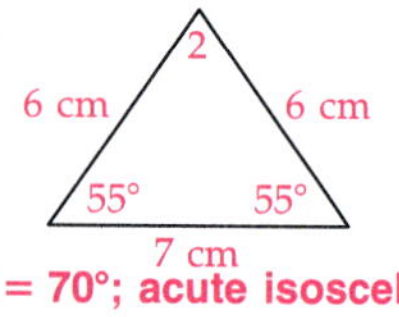

$\angle 2 = 70°$; acute isosceles

16. 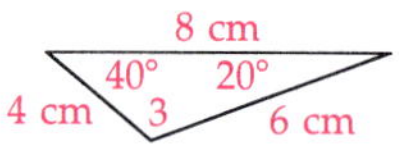

$\angle 3 = 120°$; obtuse scalene

Find the radius or diameter for each given length.

17. $r = 12$ in. $d =$ ■ 24 in.
18. $d = 15$ cm $r =$ ■ 7.5 cm
19. $r = 30$ mm $d =$ ■ 60 mm
20. Which figure appears *not* to be congruent to the other three? c

a. 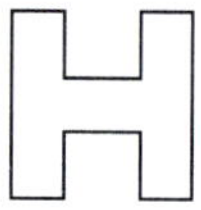

b. 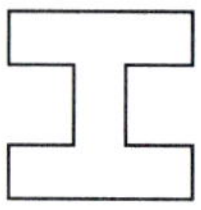

c. 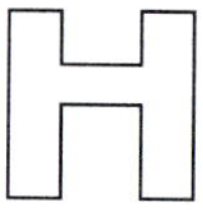

d. 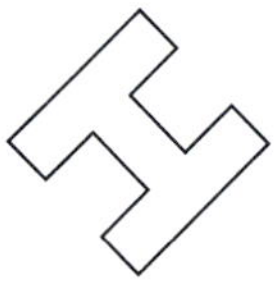

Each pair of triangles is similar. Find the missing length.

21. 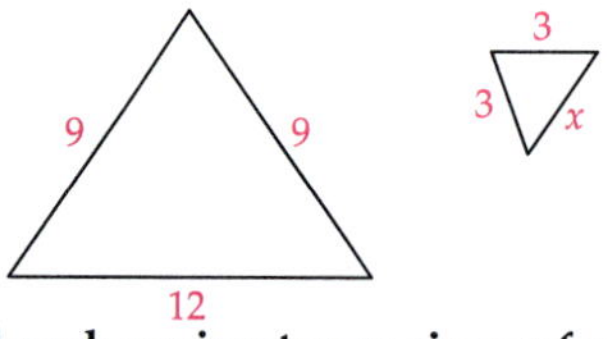

x = 4

22. 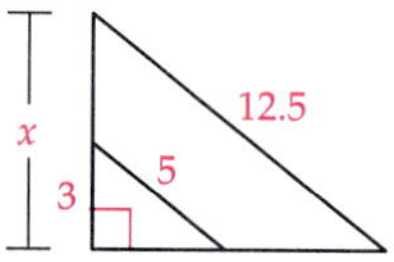

x = 7.5

Find each perimeter or circumference. Use 3.14 for π.

23. a rectangle with length 12 in. and width 8 in. 40 in.
24. a circle with radius 0.5 cm 3.14 cm
25. an isosceles triangle with side 60 mm and base 30 mm 150 mm

Chapters 1–9 Cumulative Review

Choose the correct answer. Write A, B, C, or D.

1. Order $\frac{1}{2}, \frac{5}{6}, \frac{2}{3}, \frac{3}{8}$ from least to greatest. C

A. $\frac{1}{2}, \frac{2}{3}, \frac{3}{8}, \frac{5}{6}$ **B.** $\frac{3}{8}, \frac{2}{3}, \frac{1}{2}, \frac{5}{6}$

C. $\frac{3}{8}, \frac{1}{2}, \frac{2}{3}, \frac{5}{6}$ **D.** not given

2. Write $\frac{5}{8}$ as a decimal. B

A. 0.58 **B.** 0.625

C. 0.875 **D.** not given

3. Solve. C

A. 3 **B.** -4

C. -2 **D.** not given

4. $d = 4.5$. What is the radius? B

A. 9 **B.** 2.25

C. 3.14 **D.** not given

5. $|-0.5| - |1.5| = $ ■ A

A. -1 **B.** 2

C. -2 **D.** not given

6. $b(2b + 3) - b^2 = $ ■ B

A. $5b - b^2$ **B.** $2b^2 + 3b - b^2$

C. $3b^2 + 3b$ **D.** not given

7. Describe the solution to the system. A

$2y - x = 3; 3y = \frac{3}{2}x + 9$

A. no solution **B.** one solution

C. infinite **D.** not given

8. Classify the triangle. A

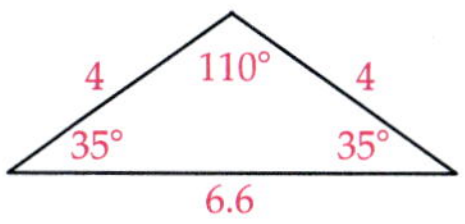

A. obtuse isosceles

B. right scalene

C. acute isosceles

D. not given

9. Write $1\frac{2}{3}$ as a percent. C

A. $16\frac{2}{3}\%$ **B.** 123%

C. $166\frac{2}{3}\%$ **D.** not given

10. 3,256,134 is divisible by ■. C

A. 8 **B.** 2, 3

C. 2, 3, 6 **D.** not given

11. Name the figure. B

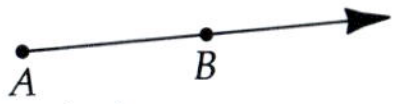

A. $\overleftrightarrow{AB}$ **B.** $\overrightarrow{AB}$

C. $\overline{AB}$ **D.** not given

12. $3x - 2y = 8$. What is the slope? B

A. 3 **B.** $\frac{3}{2}$

C. -2 **D.** not given

13. $ABCD \sim EFGH$. Find x. B

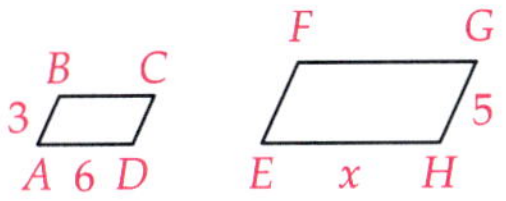

A. 8 **B.** 10

C. 4 **D.** not given

14. What percent of 80 is 20? D

A. 20% **B.** 80%

C. 60% **D.** not given

15. The 6 members of a chess club play each other once. How many games are played? C

A. 6 **B.** 12

C. 15 **D.** not given

16. A plane climbs to 30,000 ft then descends 5,000 ft. The plane then climbs another 7,000 ft. What is the plane's altitude? B

A. 6,000 ft **B.** 32,000 ft

C. 28,000 ft **D.** not given

Study Skills

Essay questions are considered subjective questions because they often ask the students for an opinion, a judgement, or a conclusion. Several skills are tested: the ability to understand what is being asked, to recall pertinent information, to organize an appropriate answer, and to write the answer in a clear and concise manner.

The test should be read carefully several times. The first reading will help students determine which questions can be answered easily. For the second reading, students can make formulas to use in the margins. Finally, in the third reading, students can analyze each question and underline words that indicate what is being asked.

Suggest that students plan their answers by recalling all they know about the topic, and then outlining their answers briefly in the margin. A simple way to begin the answer is to rephrase the question. For example, "What do you think the result will be? I think the result will be . . . ." Point out that ideas and opinions need to be supported with facts and examples.

Remind students, answers that are neat, brief, and to the point save students' time, and the teacher's time as well. Answers should be checked to see that all parts of the question have been answered.

Here are some ideas:

- Ask students to list words, such as *define, contrast, enumerate, describe,* and *compare.* Then discuss the kind of answer that is appropriate for each word.
- Have students practice answering essay-type questions.

10 AREA AND VOLUME FORMULAS

Chapter At a Glance

In Chapter 10 students learn to find the area and the volume of many figures by using formulas. Students also learn to recognize the relationship among different figures.

Reading the Chapter

This chapter introduces many new geometric terms and concepts to the students. Since students need to build on their understanding of each lesson as it is presented, it is important that they commit these new definitions and concepts to memory. Encourage students to read slowly and carefully, and to draw models when they encounter written descriptions of two- and three-dimensional figures. Review new terms regularly, and urge students to use their math journals to write the definitions of new terms and to draw models of the figures or concepts being defined.

Chapter Resources

Practice and Enrichment TRB J-1 to J-14, J-16 to J-19
Alternate Application TRB J-15

Chapter Test Form A, J-21
Chapter Test Form B, J-23

Computer Test Bank

References

STUDENT BIBLIOGRAPHY

Horn, Delton T. *Smart Apples: 31 Artificial Intelligence Experiments with the Apple II, II+, IIe, IIc and IIgs* (Tab, Blue Ridge Summit, Pa. 1987).

TEACHER BIBLIOGRAPHY

Greenes, C., G. Immerzeel, et al. *Techniques of Problem Solving (TOPS)* (Dale Seymour Publications, Palo Alto, Calif. 1981).

SOFTWARE

Elastic Lines: The Electronic Geoboard (Sunburst Communications), for the Apple II family of computers.

FILMS, VIDEOTAPES, AND OTHER MEDIA

Fractal Fantasy (Computer Art Resource, Mill Valley, Calif.), a 35-min sample of computer-generated graphics. Color, VHS.

Vocabulary

altitude
area
base
base edge
cone
cylinder
inclined plane
lateral area
lateral edge
lateral face
net
polyhedron
prism
pyramid
slant height
sphere
surface area
vertex
volume

Materials

geoboard (M*)
colored rubber bands
dot paper (T*)
colored pencils
math journal
graph paper (T)

*M—Manipulative Kit
*T—Teacher's Resource Book

Resource Bank

Class Activities

FIELD TRIPS

▼ Natural gas storage facility—to learn how natural gas is stored and its volume measured.

▼ Landmarks commission office—to explore restrictions on building and renovating in landmark areas.

CLASS SPEAKERS

▼ Operations research analyst—to discuss how mathematical models are used to represent solutions to the problems of organizations.

▼ Landfill manager—to learn how shrinking landfill sites are being utilized.

Bulletin Board

Have students use the figures below to make scale models of the total area of the Amazonian rain forest and the estimates of the portion of the rainforest that has been destroyed.

Have students determine the percentages that represent the estimated portions of the rain forest that have been destroyed.

Ask students to research and display statistics, such as the ones shown, on the Amazonian rain forest and the effects of its destruction.

Project: Have students research other large scale changes in land use and their environmental effects. (Check students' work.)

Total Area of the Amazonian Rain forest = 2,700,000 sq mi

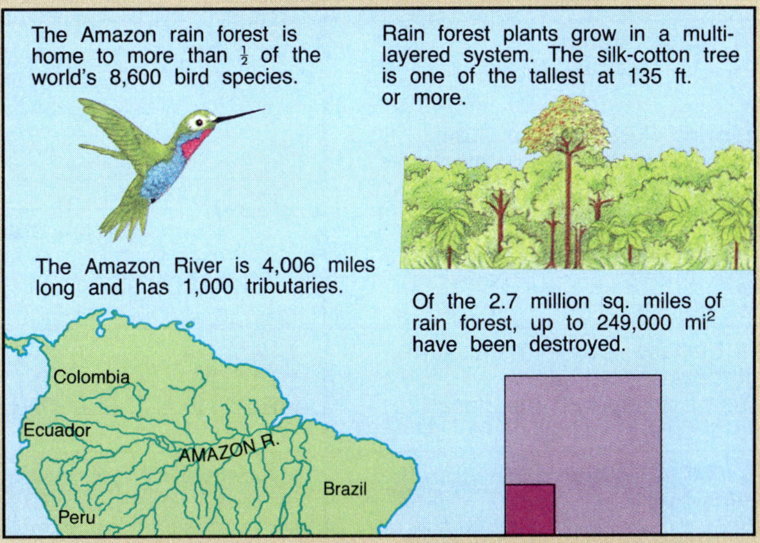

Extra-Credit Problems

BASIC

▼ Penny is planting 3 triangular beds in a rectangular garden. She needs 4 seedlings for every square foot of garden. How many seedlings does she need for the shaded section? (120)

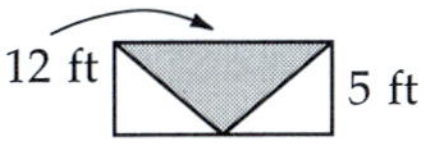

AVERAGE

▼ Kurt built an L-shaped deck. A light at Point A lights part of the square-shaped patio. Find the area of the unlit part of the patio. ($A \approx 13.76$ ft^2)

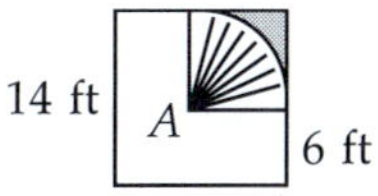

ENRICHED

▼ Vic is tarring a 160 ft^2 roof. He needs $\frac{1}{2}$ gal for every square foot. If 231 in.^{3} equals 1 gal, does a container with a radius of 12 in. and a height of 42 in. hold enough tar to complete the job? (yes; about 82.2 gal)

Chapter Planning Guide

OBJECTIVES	ASSIGNMENT GUIDE			ASSESSMENT	
	Basic	Average	Enriched	Review/ Tests	TRB*
10-1 Area of Rectangles and Parallelograms To find the area of rectangles and parallelograms.	1-10, 13-14, 16-19, 25, 31-35, MR* All	1-6, 10-15, 18-20, 24, 28-35, MR All	1-6, 10-15, 20-31, 33-35, MR All	Extra Practice p. 575	Practice J-1 Enrichment J-2
10-2 Area of Triangles and Trapezoids To find areas of triangles and trapezoids.	2-12 even, 13-14, 16-19, 21-26, MR All	1-11 odd, 13-14, 16-26, MR All	1-3, 7-9, 14-26, MR All	Extra Practice p. 575	Practice J-3 Enrichment J-4
10-3 Area of Circles To find the areas of circles.	1-5, 10-17, 20, 22, 25-26, 31, 35-36, MR, TY* All	4-8, 10-18, 20-23, 26-28, 32-36, MR, TY All	6-15, 17-21, 23-24, 28-30, 32-37, MR, TY All	Extra Practice p. 575 TY p. 417	Practice J-5 Enrichment J-6 TY J-20
Exploring Pick's Theorem To develop Pick's theorem using geoboards.	**Manipulative Connection** All	All	All		
10-4 Space Figures To identify common polyhedra.	1-6, 8-24, MR, CT* All	1-5, 7, 10-21, 25-26, 28, MR, CT All	6-19, 24-28, MR, CT All	Extra Practice p. 575	Practice J-7 Enrichment J-8
Exploring Patterns in a Cube To explore spatial visualization and patterns in a cube.	All	All	All		
10-5 Problem Solving Strategy: Make a Model To solve problems by making a model.	**Problem Solving Connection** All	All	All	Problem Solving Practice p. 445	Practice J-9 Enrichment J-10
10-6 Surface Area—Prisms and Cylinders To use a model to find surface area of prisms and cylinders.	1, 3-5, 7-8, 10-11, 13, MR All	2-4, 6-7, 9-11, 13, MR All	6-13, MR All	Extra Practice p. 575	Practice J-11 Enrichment J-12
10-7 Surface Area—Pyramids, Cones, Spheres To use a model to find surface area of pyramids, cones, and spheres.	1-2, 4-5, 7, 9, 13, 15-16, MR All	2-3, 5-6, 8-9, 12, 14-16, MR All	7-16, MR All	Extra Practice p. 575	Practice J-13 Enrichment J-14
10-8 Problem Solving Application: Parade Floats To use area formulas to plan parade floats.	All	All	All		Alternate Application J-15
10-9 Volume—Prisms and Cylinders To find the volume of prisms and cylinders.	3-6, 10-11, 13-14, 17, 20-22, MR, TY All	1-3, 5-8, 14-16, 20-22, MR, TY All	1, 3, 8-9, 12, 14-22, MR, TY All	Extra Practice p. 575 TY p. 440	Practice J-16 Enrichment J-17 TY J-20
10-10 Volume—Pyramids, Cones, and Spheres To find the volume of pyramids, cones, and spheres.	3-7, 9-10, 12, 15-17, MR, CT All	1-2, 5-10, 12-13, 15-16, MR, CT All	1-3, 7-8, 10-11, 13-18, MR, CT All	Extra Practice p. 575	Practice J-18 Enrichment J-19
Chapter Test	All	All	All	pp. 446-449	Tests J-21–J-27

*TRB—Teacher's Resource Book. MR—Mixed Review TY—Test Yourself CT—Critical Thinking.

			CONNECTIONS				
Algebra	Critical Thinking	Problem Solving	Estimation/ Mental Math	Technology	Manipulatives/ Resources	Writing/Reading in Math	NCTM CORRELATION
C*: 8 W*: 15, 22-23	W: 24	W: 33-35	W: 24	W: 7-12, 16-23	W: 32, 35	W: 33-35	Communication Connections Geometry
C: 9 W: 17	W: 21-26	W: 20	W: 13-15	W: 1-12		W: 21-26	Communication Connections Geometry
C: 9 W: 7, 9, 30, 37 TY* 1-5	C: TD* 2 W: 21, 37	W: 16-21, 31-37	C: 1-6 W: 1-9	C: 8 W: 20-29, 36-37	W: 20, 37	W: 16-19	Communication Connections Problem Solving
5	6-9	6-10	3		1-2, 4-5	3, 6c, 8	Problem Solving Communication Reasoning Connections
	W: 8-13, CT*: 1-4	CT: 1-4			C: 5-7 W: 4-5, 26-28	W: 8-19, 24-25	Communication Connections Geometry
	4-8	4-8			1-3	1-9	Problem Solving Communication Connections
W: 3-4, 6, 8-9		C: 1-3 W: 1-11		W: 1-11	C: 1-3	C: 1-3 W: 1-11	Problem Solving Communication Reasoning Connections
C: 6 W: 4-8, 10-12	C: TD 2	W: 9-13		W: 4-8, 10-12	C: 2-5 W: 1-3, 9, 13	W: 9-13	Problem Solving Communication Connections Geometry
C: 4-8 W: 7-14	C: TD 1-2	C: 6-7 W: 11-16	C: 7	W: 7-16	W: 15	C: 6-7 W: 11-16	Problem Solving Connections Geometry
		C: 1 W: 1-2		W: 1-2	W: 1-2	C: 1 W: 1-2	Problem Solving Communication Connections
C: 4 W: 3, 6, 12	C: TD 3-4	W: 13-17, 21-22	C: 1-2	W: 7-17	W: 20-22	W: 13-17, 21-22	Problem Solving Connections Geometry
C: 11-12 W: 1-3, 9-11	C: 6-10 CT: 1-5	C: 11-13 W: 7, 12-18 CT: 1-5		W: 1-8, 12		W: 7-8, 12-18	Problem Solving Reasoning Connections Geometry

*C—Class Exercises W—Written Exercises CT—Critical Thinking TY—Test Yourself TD—Think and Discuss

Exploring

In this manipulative activity, students use dot paper to explore the relationship between the area of a rectangle and the area of a triangle. **See TRB Exploring O-37.**

Exploring Area

You can use dot paper or a geoboard to model the relationship between the area of a triangle and the area of a rectangle.

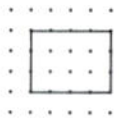

Draw a 3 x 4 rectangle

1. How many square units are inside the figure? **12**
2. What is the area of the rectangle? **12 square units**

Draw a right triangle with one leg 3 units long and the other leg 4 units long.

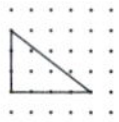

3. Does the figure contain only whole unit squares? **no**
4. Can you determine the area easily by counting the squares? **no**
5. How is this triangle related to the rectangle in question 1? **It is half of the rectangle.**
6. How can you use this relationship to find the area of a right triangle? **Find $\frac{1}{2}$ the area of the 3 x 4 rectangle.**

Look at the triangle at the right.

7. What kind of triangle do you see? **isosceles**
8. Estimate the area of the triangle. **Answers will vary.**

One way to find the area of the triangle is to draw a rectangle around the triangle. Draw a rectangle with the base and height equal to that of the triangle.

9. What is the area of the rectangle? **4 square units**
10. What is the area of each of the right triangles adjacent to the original triangle? **1 square unit each**
11. How is the area of the rectangle related to the area of the three triangles contained in it? **It is equal to the sum of the three areas.**
12. If you know the areas of the two right triangles and the rectangle, how can you find the area of the original triangle? **Subtract the areas of the two right triangles from the area of the rectangle.**
13. What is the area of the original triangle? **2 square units (4 − 2 = 2)**

Use this method to find the area of the following figures.

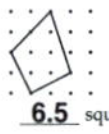

6.5 square units

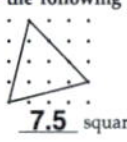

7.5 square units

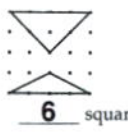

6 square units

O-37

In this problem solving activity, students explore the relationships among the number of faces, edges, and vertices in a polyhedron. **See TRB Exploring O-38.**

Exploring Euler's Formula

A polyhedron is a 3-dimensional figure in which all faces are polygons. All polyhedrons contain faces (F), edges (E), and vertices (V).

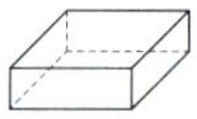

1. This figure is a rectangular prism. How many faces does the figure contain? **6**
2. What polygon does the prism have for its faces? **rectangles**
3. How many vertices does the rectangular prism contain? How many edges? **8; 12**
4. What relationship can you observe among the number of faces, edges, and vertices? **Answers will vary. The sum of the number of faces and number of vertices is two more than the number of edges.**
5. This figure is called a tetrahedron. Write the number of faces, vertices, and edges it contains. F **4** V **4** E **6**

6. Is the number of edges greater than the number of vertices? **yes**
7. Is the number of edges greater than the number of faces? **yes**
8. What relationship can you observe among the number of faces, edges, and vertices? **Answers will vary. The sum of the number of faces and number of vertices is two more than the number of edges.**

For each polyhedron count and record the faces, edges, and vertices.

	Diagram	Name	Number of Vertices	Number of Faces	Number of Edges
9.		Cube	8	6	12
10.		Square Pyramid	5	5	8
11.		Pentagonal Prism	10	7	15

12. Look at the results for each polyhedron. What relationship do you observe between the number of edges and the sum of the number of vertices and the number of faces? **In each case the sum of the number of vertices and the number of faces are two more than the number of edges.**
13. Express the relationship using mathematical symbols. **F + V = E + 2**

O-38

In this calculator activity, students use a square piece of cardboard to make a container with the greatest possible volume. **See TRB Exploring O-39.**

Exploring Volume

An open container can be made by cutting corners of equal size from a square piece of cardboard.

If the cardboard is 20 in. on a side, what size piece should be cut from the corners to make the container with the greatest possible volume?

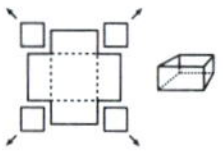

Use a calculator to find the answer.

1. Imagine that a 1-in. square is cut from each corner. Now write the dimensions of each figure. What is the volume of the resulting container? **324 in.³**
2. If you cut a 2-in. square from each corner of the cardboard, do you think the volume will be greater or less than the volume you just calculated? **Answers will vary.**
   Write the dimensions and volume of the container. **2 in. × 16 in. × 16 in., 512 in.³**
3. If you cut a 3-in. square from each corner of the cardboard, do you think the volume will be greater or less than the volume you just calculated? **Answers will vary.**
   Write the dimensions and volume of the container. **3 in. × 14 in. × 14 in., 588 in.³**
4. If you cut a 4-in. square from each corner of the cardboard, do you think the volume will be greater or less than the volume you just calculated? **Answers will vary.**
   Write the dimensions and volume of the container. **4 in. × 12 in. × 12 in., 576 in.³**
5. What do you notice about the last volume you calculated? **The volume is less.**
6. Explain how you would determine the size of the corner squares to be removed to get a container with the greatest volume. **Calculate the vol. for dimensions between 3 in. and 4 in. until the greatest vol. is found.**
   Now find the dimensions and volume to the nearest tenth. **3.3 in. × 13.4 in. × 13.4 in., 592.5 in.³**

O-39

In this computer activity, students use a spreadsheet to explore the relationships among the length and width of a rectangle, its perimeter, and its area. **See TRB Exploring O-40.**

Exploring Perimeter and Area on a Spreadsheet

A rectangular-shaped region has a perimeter of 24 feet. What should the length and width be to enclose the greatest area?

Use graph paper to illustrate the problem.

1. Draw a rectangle that could represent the region. What are the dimensions? **Answers will vary. Possible answers are given.**
   length **1 ft** width **11 ft** perimeter **24 ft** area **11 ft²**
2. Draw another rectangle that could represent the region.
   length **2 ft** width **10 ft** perimeter **24 ft** area **20 ft²**
3. Draw a third rectangle.
   length **3 ft** width **9 ft** perimeter **24 ft** area **27 ft²**
4. Look at the length and width of each rectangle. What can you say about the sum of the length and the width? **Answers may vary. They have a sum of 12.**
5. Use what you observed to write the dimensions for six different rectangles. Give only whole number dimensions. **1, 11; 2, 10; 3, 9; 4, 8; 5, 7; 6, 6**

Use a computer spreadsheet to record and interpret the data.

	A	B	C	D
1	LENGTH	WIDTH	PERIMETER	AREA
2	1	11	24	11
3	2	10	24	20
4				

6. What formulas will compute the perimeter for cell C2? **=2*(A2+B2)**
   C3 **=2*(A3+B3)**
7. What formulas will compute the area for cell D2? **=A2*B2**
   D3 **=A3*B3**

Input the values you found in question 5. Then find the perimeter and area for all the possible rectangles.

8. What are the dimensions of the rectangle with the greatest area? **6 ft × 6 ft**
9. Find the rectangle with the greatest area that has a perimeter of 28 ft. What is the length and width? **7 ft, 7 ft**
10. For maximum area, what is the relationship between the length and width of a rectangle for any given perimeter? **They must be equal.**

O-40

Meeting Individual Needs

GIFTED STUDENTS

Students who are mathematically gifted enjoy challenges that allow them to go beyond the material presented in a lesson. Present the following problems to your gifted students when they have mastered the material in Chapter 10. After they have solved the problems, invite the students to explain the problems and solutions to the entire class.

1. A square prism has a height of 20 in. If you divide the base into 4 triangles by drawing the diagonals, and then draw a perpendicular line from where the diagonals intersect, to the side, the length of the perpendicular line is 5 in. Find the length of a side, the area of the base, and the volume of the prism. (10 in.; 100 in.^{2}; 2,000 in.^{3})

2. The height of a regular hexagonal prism is 8 m and the length of the sides of the base is 12 m. The perpendicular distance from the center of the hexagon to the side is 10.392 m. Find the area of the base and the volume of the prism. (374.112 m^2; 2,992.896 m^3)

AT-RISK STUDENTS

Relevance to everyday life and practicality are two characteristics of math lessons that engage the interest of at-risk students. The following example may be used to illustrate the kind of "everyday" problem that may be solved using the concepts and formulas taught in Chapter 10. Encourage students to use their personal experiences to create and solve similar problems.

You are planning to build an L-shaped house. An excavator will be hired to clear the land and dig and prepare a hole 8 ft deep for the foundation. The plans for the foundation are shown below. If excavation costs \$2.60/ft^3, what will be the cost of excavation? (\$29,286.40)

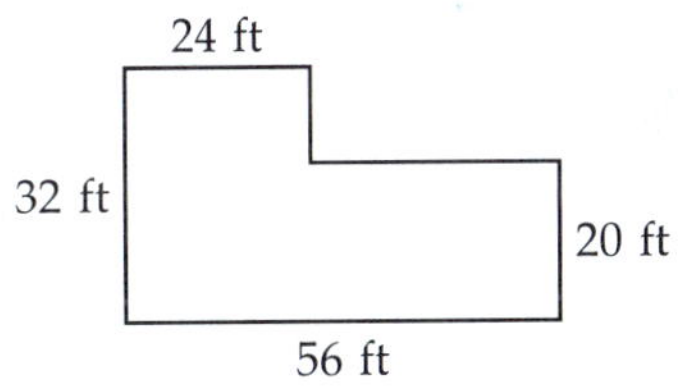

ENGLISH AS A SECOND LANGUAGE

ESL students might appreciate a review of new terms and concepts before and during each lesson. Students might also find it helpful if the teacher can provide a display of the geometric shapes discussed. The display can reinforce concepts and enhance language skills for ESL students and provide a visual reference for the rest of the class.

In addition to reviewing frequently the formula for area and the terms "base" and "height," you can have the class work in groups to find the area of classroom objects such as a globe, a desktop, or a chalkboard.

ESL students can be paired with English-language students to work on Lesson 10-5. Encourage the pairs to model each situation, if necessary. When deciding which pairs of students might work well together, consider matching an ESL student who has good math skills with an English-language student who is a good communicator but needs help with math. Each student can then benefit from the other's strengths.

See TRB Teaching Resources p. 60.

Problem Solving Situation

The "Think About It" question asks whether there is a relationship between the size of an animal and its home range. To help students answer this question, ask them to compare what they know about the size of an animal with the size of the home ranges listed. For instance, bears are large, and their home range is also large. Ask students to pick five animals not listed on the chart. Have them explore and present data for three factors, other than size, that shape the relationships between the animals and the size of their home ranges.

DATA COLLECTION

Information about animals can be gathered from encyclopedias and natural history magazines and journals. Students might also talk with biology teachers, or interview zoologists from nearby zoos or universities.

DATA ANALYSIS

Have students make charts to explain the data they have found and the relationships they discovered. Invite students to discuss what might happen to an animal if the size of its habitat is altered through a natural disaster such as fire, or through the development of housing and other buildings. Students might want to include pictures of the animals they have chosen.

Tropical rain forests cover only about 2% of the earth's surface and yet are home to half the world's wild plant and animal life. Though many of the rain forest's plants are vital to medical research, the rain forests are being cut down at the rate of 50 acres per minute.

Corcovado, Costa Rica Rain Forest	
area	163 mi^2
average rainfall	220 in./y
species of trees	500
species of mammals	140
species of insects	6,000
species of butterflies	123

EARTH'S OCEANS

OCEAN	AREA (mi^2)	GREATEST WIDTH (mi)	GREATEST KNOWN DEPTH (ft)	AVERAGE DEPTH (ft)
Arctic	5,105,700	2,630	17,880	4,360
Indian	28,350,500	6,200	25,344	12,780
Atlantic	31,820,000	4,150	28,374	14,000
Pacific	63,820.000	11,000	36,198	14,050

404

CHAPTER 10

Area and Volume Formulas

WILDLIFE HOME RANGES

An animal's home range is the amount of space the animal needs to fulfill its requirements for food, breeding, and so forth. Home range can be expressed in acres or by the radius of a circular area.

Home Ranges for Mammals in Oregon and Washington

MAMMAL	HOME RANGE (Acres)	CIRCULAR HOME RANGE (Radius in Meters)
Black bear	4,382.4	2,370
Cougar	123,753	12,600
Gray wolf (pack of five)	380,970	22,100
Grizzly bear	938,730	34,600
Mule deer	1,045.8	1,160
Northern flying squirrel	0.082	32
Otter	7,494.9	3,100
Porcupine	86.9	333
Red fox	154.38	718
Snowshoe hare	6.35	90
Striped skunk	251.49	567

Think about it...

Look at the home range data.
Do you think there is a relationship between the size of an animal and the size of its home range?

See Problem Solving Situation

Cooperative Learning

This activity can be interesting and entertaining for students working in cooperative groups. Each group member can be responsible for checking a different information source. Possible group roles might include Chart Maker, Presenter, Artist, and Data Collector. You might want to extend the activity by having each group select one of the animals they've chosen and prepare a full report on the animal, its habitat, and the prospects for the animal's future.

Questions related to this Data File can be found on pages 78, 103, 151, 170, 239, 267, 303, 323, 396, 416, 445, 481, 553, and 575.

Lesson Focus

Materials

Geoboards and rubber bands or dot paper, metric ruler, pencil, scissors, math journal

Vocabulary/Symbols

altitude
area

MOTIVATING THE LESSON

Have students use a geoboard or dot paper to make the top left rectangle shown in the margin of page 406. Discuss the concept of area and how to count square units. Then have students make the remaining figures and describe how they estimated the area of each.

Instruction

1 Use Example 1 to lead to the formula for the area of a rectangle. Point out that the number of squares is equal to the product of the number of rows and the number in each row.

2 Point out that base and height can be interchanged when working with rectangles. Have students also note that when no unit names are given in a figure, the length is referred to as units and the area is square units. Note the abbreviations used for other units of area.

3 Have students draw the figure on dot paper, cut it out, and then cut a piece to form a rectangle with the same base and height.

OBJECTIVE: *To find the area of rectangles and parallelograms.*

10-1 Area of Rectangles and Parallelograms

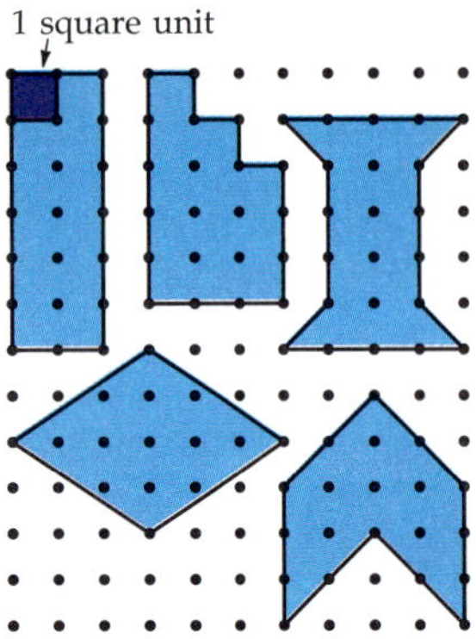

▼ Geoboards and dot paper are helpful models for understanding area. Each figure at the left takes up 12 square units of area.

Area	Area is the amount of surface inside a region.

▼ You can find the area of a rectangle on a geoboard by counting the square units within the figure.

Example 1 **Find the area of each rectangle.**

a. 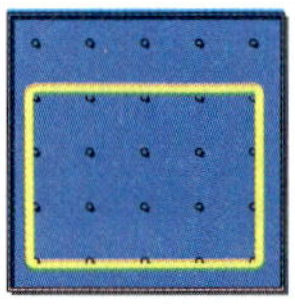

b. 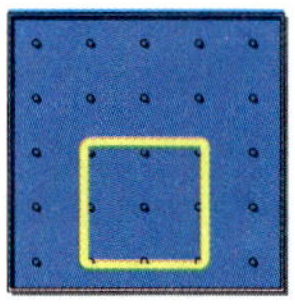

Solution a. The area is 12 square units. b. The area is 4 square units.

▼ You can find the area of a rectangle by multiplying the base length (b) by the height (h). Either side may be the base or the height.

Area of a Rectangle	The area (A) of a rectangle equals the product of its base length (b) and its height (h). $A = bh$

Example 2 **Find the area of each rectangle.**

a. 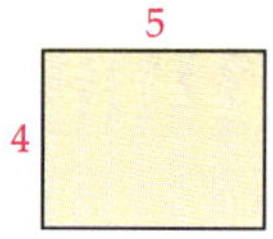

b. 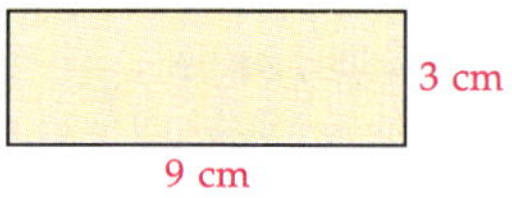

Solution

a. $A = bh$
$= 4 \cdot 5$
$= 20$
The area is 20 square units.

b. $A = bh$
$= 9 \cdot 3$
$= 27$
The area is 27 cm^2.

THINK How many rectangles can you form that have an area of 48 ft^2? **an infinite number**

Example 3 **Use the area formula to find the missing information.**

a. b is x and h is $5x$. Find A.

b. A is 21 cm^2 and h is 7 cm. Find b.

Solution

a. $A = bh$
$= x \cdot 5x$
$= 5x^2$
The area is $5x^2$ square units.

b. $A = bh$
$21 = b \cdot 7$
$b = 3$
The base is 3 cm.

Teaching Tip

Keep a supply of dot paper or grid paper available for students to draw models of geometric figures.

▼ A rectangle is a special kind of parallelogram. You can make a parallelogram into a rectangle by rearranging the pieces.

Example 4 **Use a geoboard or dot paper. Show that a parallelogram and a rectangle with height 3 and base length 7 have the same area.**

Solution

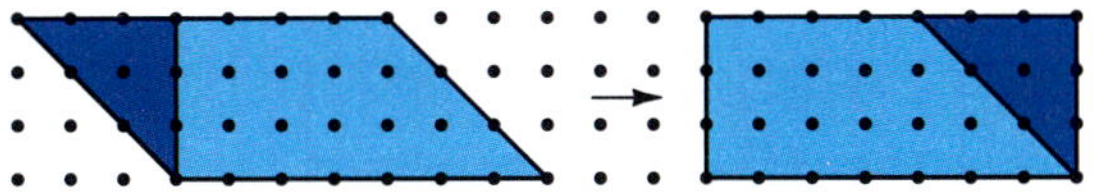

Copy the triangle formed on the left. Place the same shape on the right.

$A = bh$ — Find the area of the rectangle.

$= 7 \cdot 3$

$= 21$ square units

The parallelogram and rectangle each have an area of 21 square units.

3

▼ Example 4 illustrates that you find the area of a parallelogram the same way you find the area of a rectangle.

Area of a Parallelogram	The area of a parallelogram equals the product of its base length (b) and its height (h). $A = bh$

▼ The height of a parallelogram is the length of an *altitude*.

Altitude	An altitude is a segment from one vertex perpendicular to the line containing the opposite side, called the base.

Example 5 **State the height and the measure of the base. Find the area.**

a.

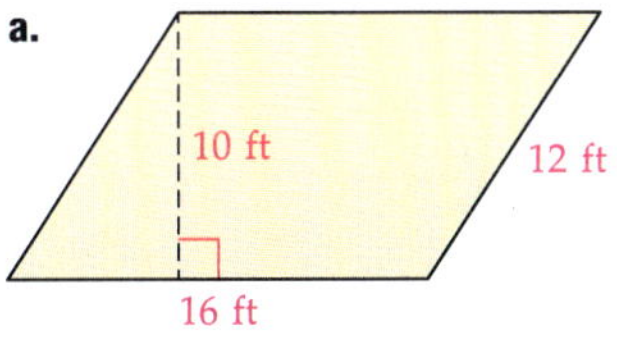

b.

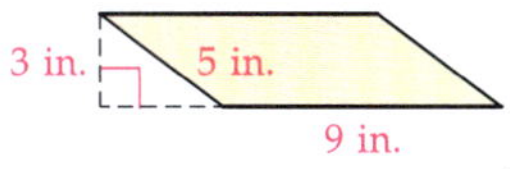

Solution

a. $h = 10$ ft,
$b = 16$ ft
$A = bh$
$= 16 \cdot 10$
$= 160\ ft^2$

b. $h = 3$ in.
$b = 9$ in.
$A = bh$
$= 9 \cdot 3$
$= 27\ in.^2$

THINK When is the height of a parallelogram equal to the length of each of the sides? when the parallelogram is a square

ADDITIONAL EXAMPLES

Find the area.

1.

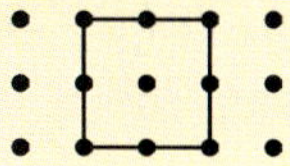

(4 square units)

2.

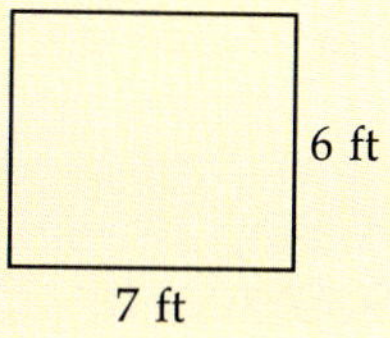

($42\ ft^2$)

3. a. Find A if the base is $2a$ and the height is $3a$. ($A = 6a^2$)

b. Find h if $A = 24\ in.^2$ and $b = 8$ in. ($h = 3$ in.)

4. Find the area.

a.

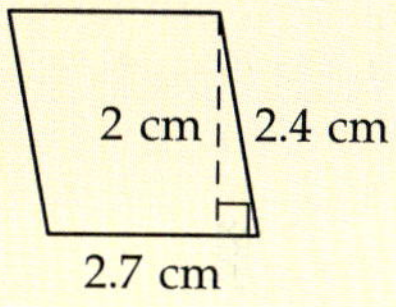

($5.4\ cm^2$)

b.

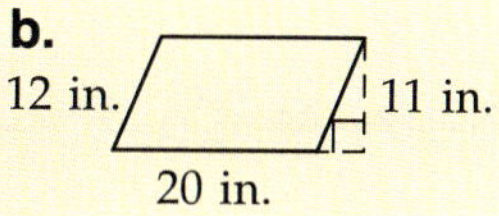

($220\ in.^2$)

MATH MINUTES

Have students draw all rectangles that have an area of 36 square units, using whole numbers for sides.(1×36, 2×18, etc.) Then have them identify the greatest perimeter (1×36) and the smallest perimeter (6×6).

Guided Practice

Class Exercises Select students to measure the room for question 10.

Think and Discuss Have students draw a square millimeter and a square centimeter in their math journals before they answer question 2. Have a student draw a square meter on the chalkboard. Repeat the question, using square inch, square foot, and square yard as the units of measure.

A ***common error*** is to use the side of a parallelogram instead of the height to find area. Emphasize that the height must be perpendicular to the base. Have students look at the parallelograms in this lesson and point out the bases and the heights.

Closure

Writing in Math Ask students to describe how to find the area of a parallelogram, in their math journals. Then ask them to think of practical situations in which area is used, such as wall surfaces to be painted or papered.

Independent Practice

Question 24 may be demonstrated by making a rectangle with paper strips and paper fasteners. Students can move the sides of the figure to make the angle smaller and observe what happens to the enclosed area as the height approaches zero.

THINK AND DISCUSS

1. We use a unit square as the basic unit for measuring area. Why is this a better choice than a unit circle? See below.
2. Choose mm^2, cm^2, or m^2 as the most appropriate unit of area measure for each.
   a. textbook cover cm^2
   b. thumbprint mm^2
   c. classroom floor m^2
   d. pan pizza cm^2
   e. football field m^2
   f. postage stamp mm^2

CARPET SALE $9.99 per square yard, including pad. Buy today and we will install tomorrow.

4. $h = 18$ m
   $b = 24$ m
   $A = 432\ m^2$
5. $h = 12$ cm
   $b = 10$ cm
   $A = 120\ cm^2$
6. $h = 10$ in.
   $b = 8$ in.
   $A = 80\ in.^2$

CLASS EXERCISES

Find the area of each figure.

1. 

$56\ ft^2$

2. 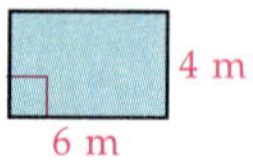

$24\ m^2$

3. 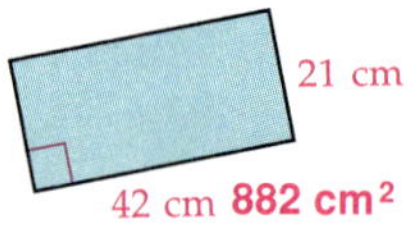

$882\ cm^2$

State the height and the measure of the base. Find the area. See left.

4. 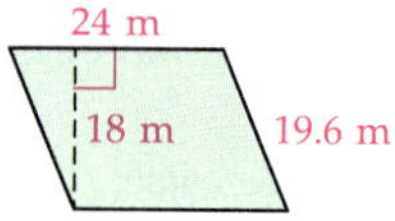

5. 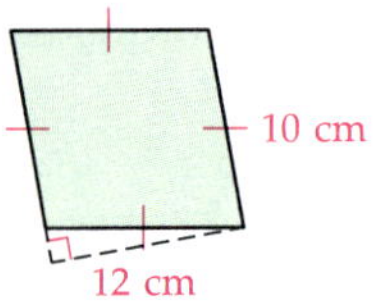

6. 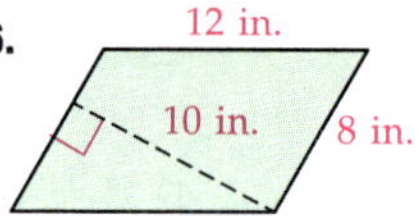

Complete.

7. parallelogram,
   Area = $8\ m^2$
   $b = 3$ m, $h = ■$ $2\frac{2}{3}$ m
8. rectangle,
   Area = $75x^2$
   $h = 15x$, $b = ■$ $5x$
9. parallelogram,
   Area = $525\ cm^2$
   $b = 25$ cm, $h = ■$ 21 cm
10. Use the advertisement at the left. What would it cost for both carpet and pad for your classroom? Answers will depend on classroom size.

1. Squares fit next to each other to cover a figure without gaps.

WRITTEN EXERCISES

Answers will vary.

Name something you would measure with the given unit.

1. 1 square centimeter
2. 1 square foot
3. 1 square meter
4. 1 square millimeter
5. 1 square inch
6. 1 square mile

CALCULATOR **Find the area of each figure.**

7. 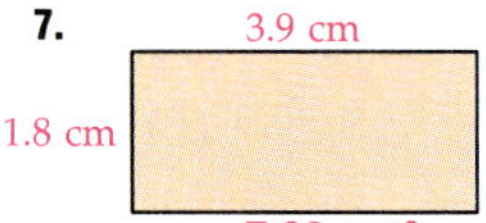

$7.02\ cm^2$

8. 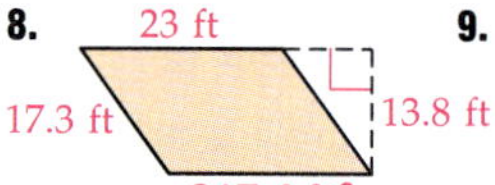

$317.4\ ft^2$

9. 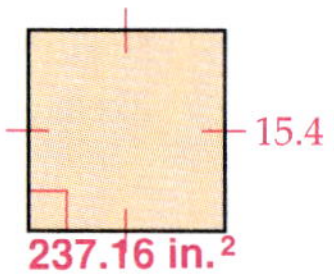

$237.16\ in.^2$

10. 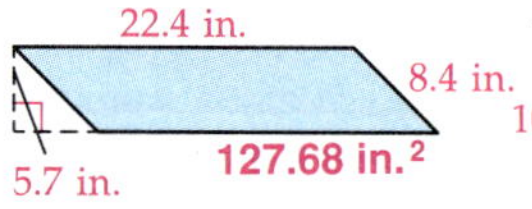

$127.68\ in.^2$

11. 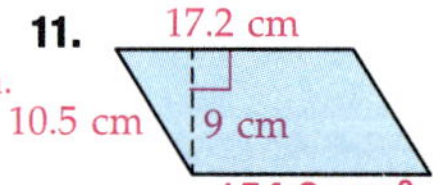

$154.8\ cm^2$

12. 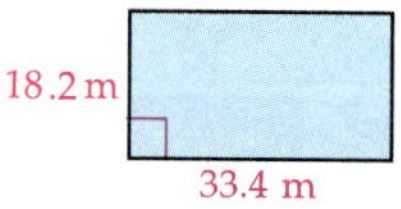

$607.88\ m^2$

Complete.

13. rectangle,
    $A = 22\ cm^2$
    $b = 4$ cm, $h = ■$ 5.5 cm
14. parallelogram,
    $A = 24.8\ ft^2$
    $h = 16$ ft, $b = ■$ 1.55 ft
15. parallelogram,
    $A = 120x^2$
    $b = 8x$, $h = ■$ $15x$

CALCULATOR Find the area of each parallelogram with the given base and height. Round to the nearest hundredth.

16. $b = 8.2$ cm, $h = 11.4$ cm **93.48 cm²**
17. $b = 15$ ft, $h = 36$ ft **540 ft²**
18. $b = 7.4$ m, $h = 0.008$ m **0.06 m²**
19. $b = 5.879$ km, $h = 10$ km **58.79 km²**
20. $b = 29.9$ in., $h = 32.67$ in. **976.83 in.²**
21. $b = 55$ cm, $h = 10.5$ cm **577.5 cm²**
22. $b = 3x + 7$, $h = 3x$ **$9x^2 + 21x$ square units**
23. $b = 6x$, $h = 7x$ **$42x^2$ square units**

24. Find the area of each parallelogram.

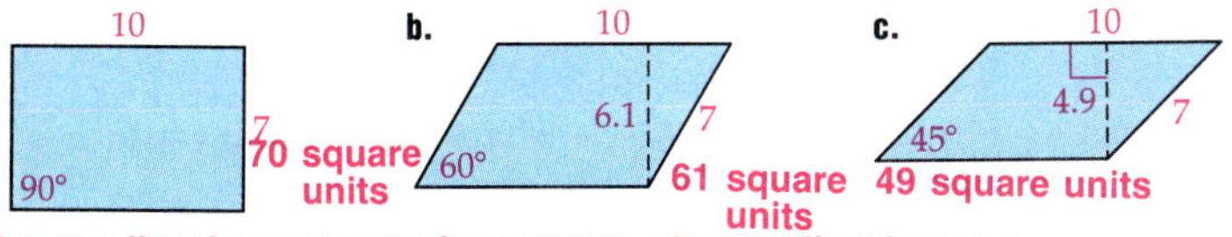

d. Does the angle measure affect the area measure? **Yes, the smaller the acute angle measure, the smaller the area.**

e. Two parallelograms have sides of lengths 10 and 7. The parallelograms have altitudes to the same base. One height is 5 and the other is 4.99. Without calculating, which parallelogram has the greater area? **the parallelogram with height 5**

25. The area of a square is 64 ft². What is its perimeter? **32 ft**

26. The area of a square is $144x^2$ m². What is its perimeter? **48x m**

27. The perimeter of a square is 28 cm. What is its area? **49 cm²**

28. The area of a parallelogram is 36 ft². Its height is 4 ft. Find its perimeter. **26 ft**

29. The base length of a rectangle is 5 times as great as its height. The area is 320 ft². What is the perimeter of the rectangle? **96 ft**

30. The area of a rectangle is 56 cm². Name three different dimensions that the rectangle could have. **Answers may vary. Three possible answers: 1 × 56, 2 × 28, 4 × 14.**

31. How many square feet are in a square yard? Draw a diagram to illustrate your answer. **9 ft²** **Check students' diagrams.**

32. PROJECT Cut the label from any cylindrical can. Make sure the cut is perpendicular to the ends. Tape the label to your paper and find the area to the nearest square centimeter. **Check students' work.**

Use the article at the right.

33. You wish to erect a political sign that is 6 ft high. What is the greatest width it can be? **$5\frac{1}{3}$ ft.**

34. Draw and label the dimensions for three different political signs that use the maximum possible area. **Answers may vary.**

35. PROJECT Research the sign laws in your community. Design a political campaign sign for a candidate of your choice. **Check students' work.**

MIXED REVIEW

1. DATA FILE 1 (pp. 2–3) Write the freezing point of salt in scientific notation. **1.474×10^3**

2. Simplify $2^2 \cdot 3^{-4}$. **$\frac{4}{81}$**

3. 3^4 ■ 4^3. **>**

4. Between what two integers is the square root of 41? **6 and 7**

5. Find the circumference of a circle with $r = 8$. **50.24**

6. What is the sum of the measures of the angles of a triangle? **180°**

7. A 3-ft tall child casts an 8-ft long shadow. How high is the tree next to the child if the tree's shadow is 60 ft? **22.5 ft**

Sign Regs for Politicos

The city council issued the following regulations for political signs:

1. Only one sign may be posted on a single parcel of land.
2. Signs may not exceed 32 ft².
3. Detached signs cannot be more than 6 ft high.
4. All signs must be removed within 15 days after the election.

LESSON QUIZ

Find the area.

1.

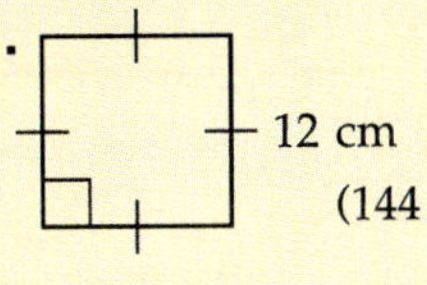

(144 cm²)

2.

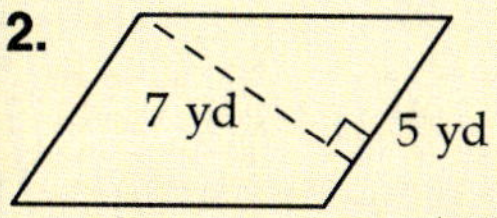

(35 yd²)

3. Find the area of a rectangle with $b = 4.1$ and $h = 5$. ($A = 20.5$ square units)

4. The area of a parallelogram is 84 ft² and the height is 6 ft. What is the base? (14 ft)

Assignment Guide

Basic 1-10, 13-14, 16-19, 25, 31-35, MR All

Average 1-6, 10-15, 18-20, 24, 28-35, MR All

Enriched 1-6, 10-15, 20-31, 33-35, MR All

FOR THE NEXT LESSON

dot paper, ruler, scissors, math journal

Lesson Follow-up

Reteaching Activity For students having difficulty with the areas of triangles and rectangles, refer to the exploring activity found on page 404E of the Chapter Overview. See also TRB Exploring O-37.

LESSON RESOURCES

TRB Practice J-1
TRB Enrichment J-2

Practice For use after 10-1 (pp. 406–409)

Area of Rectangles and Parallelograms

CALCULATOR Find the area of each figure.

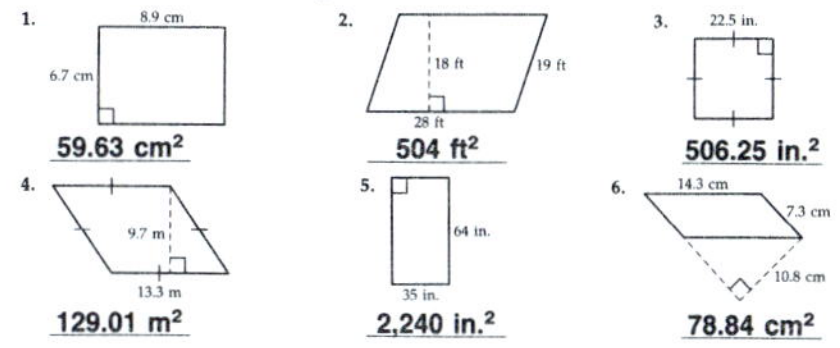

Find the missing measures.

7. A parallelogram
Area = 133 in.²
b = 19 in.
h = **7 in.**

8. A rectangle
Area = 76.56 cm²
h = 8.8 cm
b = **8.7 cm**

9. A parallelogram
Area = 96 km²
h = 12 km
b = **8 km**

Find the area for the given rectangles or parallelograms. Use a calculator where necessary. Round to the nearest hundredth.

10. $l = 5.09$ m, $w = 4.3$ m **21.89 m²**
11. $l = 35.5$ mi, $w = 8.25$ mi **292.88 mi²**
12. $l = 14x$, $w = 9$ **126x units²**
13. $l = x + 5$, $w = 24$ **24x + 120 units²**
14. $b = 84$ in., $h = 67$ in. **5,628 in.²**
15. $b = 0.094$ cm, $h = 0.07$ cm **0.01 cm²**
16. $b = x + 4x$, $h = 9x - 6x$ **15x² units²**
17. $b = 2x - 7$, $h = 8$ **16x − 56 units²**
18. The area of a rectangle is 156 yd². The length is 13 yd. What is its perimeter? **50 yd**
19. The perimeter of a square is 72 in. What is its area? **324 in.²**
20. Surveyors sometimes measure in links and rods. There are 25 links in a rod. How many square links are in a square rod? **625 square links**

Chapter 10 J-1

Enrichment For use after 10-1 (pp. 406–409)

Squaring a Rectangle

To solve the four-square puzzle at the right, you must reason logically about the way squares fit together.

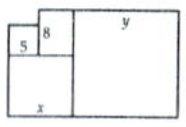

1. Use the given measurements to find x. **13**
2. Find y. **21**
3. How can you find the area of the four-square puzzle? **Add the areas of the squares together.**

Now apply this method to the figure below, which consists of ten squares fitted together to form a rectangle. The length of the sides of two of the squares are given. Write the dimensions of each of the remaining squares in the given square. Write the dimensions and area of the rectangle in the space below the figure.

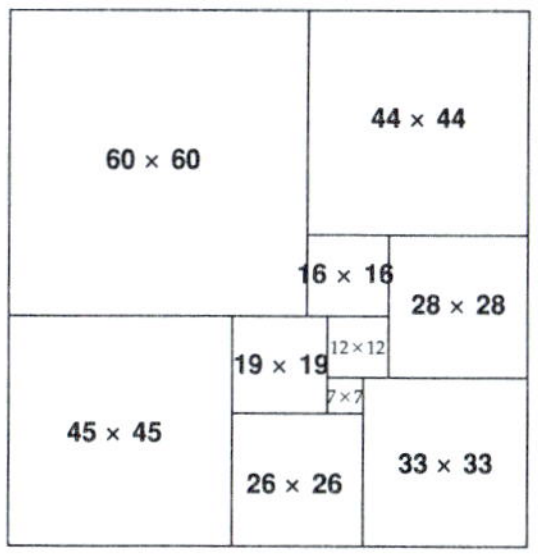

Dimensions of rectangle: **104 × 105** Area: **10,920 units²**

J-2 Chapter 10

Lesson Focus

Materials

Dot paper, ruler, scissors, math journal

MOTIVATING THE LESSON

Ask students to draw any size parallelogram on dot paper. Have them draw one diagonal and observe the triangles formed. Ask what is special about the triangles. (They are congruent.) Have students cut out the triangles to demonstrate congruence.

Instruction

Have students use the triangles from above to form the original parallelogram and find its area. Ask what the area of each triangle must be. (half the area of the parallelogram) Ask students to use the formula for area of a parallelogram ($A = bh$) to write a formula for the area of a triangle. ($A = \frac{1}{2}bh$)

1 All the forms of $\frac{1}{2}bh$ are equal because multiplication is commutative and associative.

2 Have students identify base and height in each figure. Remind them that base and height must be perpendicular.

3 If students have difficulty visualizing the height of triangle 2, extend the drawing to the right to illustrate the height perpendicular to the 8-cm base.

4 Explain how subscripts are used in the formulas to refer to the parts of the figure.

5 To help students visualize the figures, ask them to sketch the figures.

OBJECTIVE: *To find areas of triangles and trapezoids.*

10-2 Area of Triangles and Trapezoids

▼ You can form two congruent triangles by drawing a diagonal in any parallelogram.

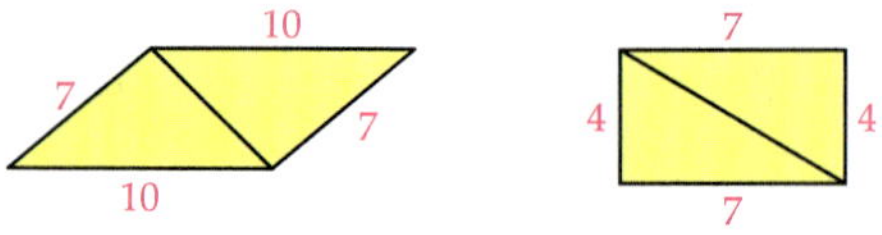

The figures suggest that the area of a triangle is half the area of a parallelogram.

THINK Why are the triangles formed by the diagonal of a parallelogram congruent? The corresponding angles and sides are equal.

Example 1 **Find the area of the parallelogram. Then find the area of each triangle formed by the diagonal.**

Solution $A = bh$
$= 9 \cdot 4$
$= 36$

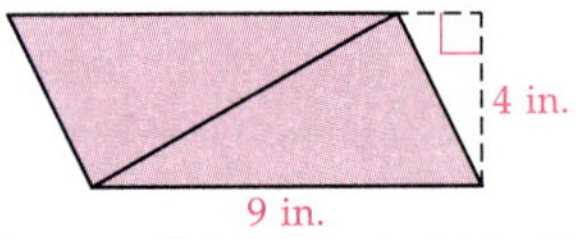

The area of the parallelogram is 36 in.^{2}.
The area of each triangle is $\frac{1}{2} \cdot 36$ or 18 in.^{2}.

THINK Are $\frac{1}{2}bh$, $\frac{bh}{2}$, $b \cdot \frac{h}{2}$, and $h \cdot \frac{b}{2}$ equal? yes

1

Area of a Triangle	The area of a triangle equals half the product of the base length (b) and the height (h). $A = \frac{1}{2}bh$

▼ The height (h) of a triangle is the measure of an altitude. You can draw an altitude from each vertex of a triangle. An altitude may be inside, outside, or on the triangle.

Example 2 **State the height and the measure of the base. Find the area.**

2

a.

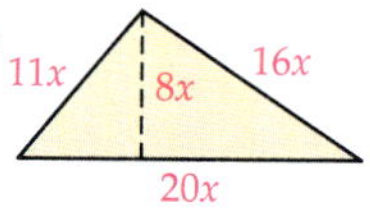

b.

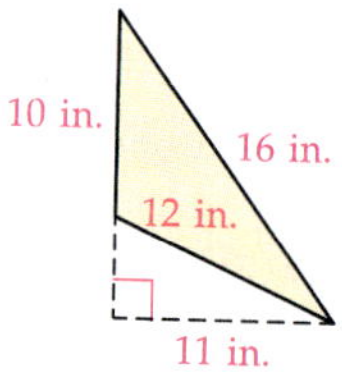

Solution

a. $h = 8x$; $b = 20x$
$A = \frac{1}{2}bh$
$= \frac{1}{2} \cdot 20x \cdot 8x$
$= 80x^2$ square units

b. $h = 11$ in.; $b = 10$ in.
$A = \frac{1}{2}bh$
$= \frac{1}{2} \cdot 10 \cdot 11$
$= 55$ in.^{2}

Teaching Tip

Have students keep a page in their math journals for area and volume formulas and a simple example for each formula. Students can then look up a formula easily when they need to.

▼ A diagonal through a trapezoid forms two triangles having different areas.

Example 3 **Find the area of each triangle and of the trapezoid.**

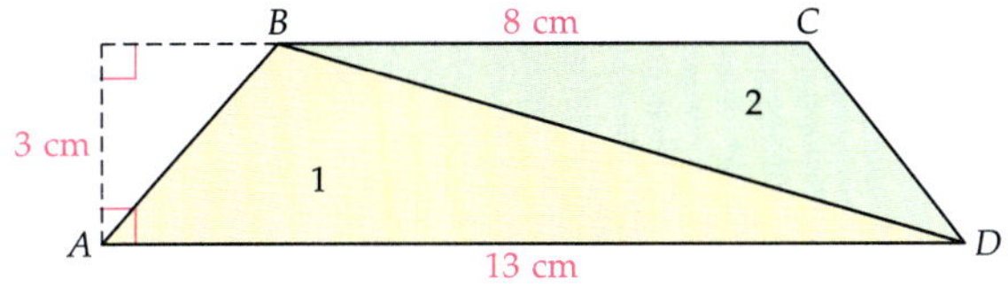

Solution

Area of $\triangle ABD$

$A = \frac{1}{2}bh$

$= \frac{1}{2} \cdot 13 \cdot 3$

$= 19.5 \text{ cm}^2$

Area of $\triangle BCD$

$A = \frac{1}{2}bh$

$= \frac{1}{2} \cdot 8 \cdot 3$

$= 12 \text{ cm}^2$

The area of the trapezoid is $19.5 + 12 = 31.5 \text{ cm}^2$.

THINK Why is the height of both triangles the same? For each triangle, 3 cm is the length from the vertex to an extension of the opposite side.

3

4

▼ The area of a trapezoid is the sum of the areas of the two triangles formed by a diagonal. In a trapezoid, we label the bases (the parallel sides) as b_1 and b_2.

$$A = \frac{1}{2}b_1h$$

$$+ \; A = \frac{1}{2}b_2h$$

Area of trapezoid $= \frac{1}{2}b_1h + \frac{1}{2}b_2h$

$$= \frac{1}{2}h(b_1 + b_2)$$

Area of a Trapezoid	The area of a trapezoid equals half the product of the height (h) and the sum of the bases (b_1 and b_2). $A = \frac{1}{2}h(b_1 + b_2)$

Example 4 **Use the formulas to find the missing information.**

a. triangle, $h = 9$ mm, $A = 67.5 \text{ mm}^2$ Find b.

b. trapezoid, $b_1 = 12$ cm, $b_2 = 16$ cm, $A = 112 \text{ cm}^2$ Find h.

Solution

a. $A = \frac{1}{2}bh$

$67.5 = \frac{1}{2}b \cdot 9$

$67.5 = 4.5b$

$b = 15$ mm

b. $A = \frac{1}{2}h(b_1 + b_2)$

$112 = \frac{1}{2}h(12 + 16)$

$112 = \frac{1}{2}h \cdot 28$

$112 = 14h$

$h = 8$ cm

5

ADDITIONAL EXAMPLES

1. Find the areas of the parallelogram and the two triangles. (27 cm^2, 13.5 cm^2)

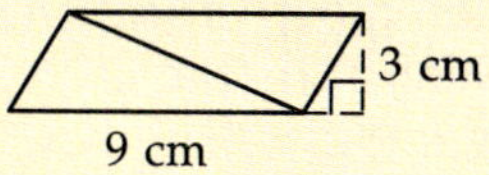

2. Find the area of the triangles.
   a. $h = 3x$, $b = 4x$ ($A = 6x^2$)
   b. $h = 5$ ft, $b = 5$ ft ($A = 12.5 \text{ ft}^2$)
3. Find the area of each triangle and of the trapezoid. (1 : 20mm^2, 2 : 16mm^2, trapezoid: 36mm^2)

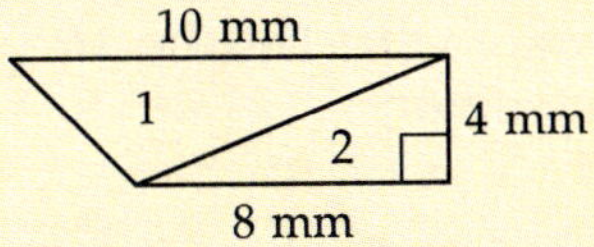

4. **a.** Triangle: $h = 10$ in., $A = 11 \text{ in.}^2$, find b. ($b = 2.2$ in.)
   b. Trapezoid: $b_1 = 3$ ft, $b_2 = 4$ ft, $A = 42 \text{ ft}^2$, find h. ($h = 12$ ft)
5. Triangle: $b = 10$ m, $h = 240$ cm, find A. ($A = 12 \text{ m}^2$ or $120{,}000 \text{ cm}^2$)

MATH MINUTES

Cut out a triangle and a trapezoid from construction paper. Ask students to estimate the areas in square centimeters or in square inches. Then have them measure the figures and calculate the areas. (Check students' work.)

Guided Practice

Class Exercises Have students work in small groups. Explain that each group is responsible for making sure that all members understand how to find the answers.

Think and Discuss Have students copy the figure in question 2 on dot paper. One method is to subtract the outside areas from the total area of the grid. Another method is to divide the figure into triangles and/or rectangles and find the area of each section.

A ***common error*** occurs when students omit the parentheses in the trapezoid formula. Have them copy the formula and substitute the measurements directly below the variables each time they work an area problem.

Closure

Writing in Math Ask students to write in their math journals how the area of a parallelogram and the area of a triangle are related. Then, have them describe how to find the area of a trapezoid.

Independent Practice

For questions 21-26, have students explain how they formed their conclusions. For question 21, mention that in an equilateral triangle there is an altitude from each side, but the altitudes are all the same height. However, an isosceles triangle has two distinct heights.

As a follow-up, refer to the computer activity found on page 404E of the Chapter Overview. See also TRB Exploring O-40.

THINK AND DISCUSS See below.

1. Are $\left(\frac{1}{2} \cdot 3\right) \cdot 8$ and $3 \cdot \left(\frac{1}{2} \cdot 8\right)$ equal? How can this help in finding the area of a triangle?
2. Find the area pictured. Is there more than one method?

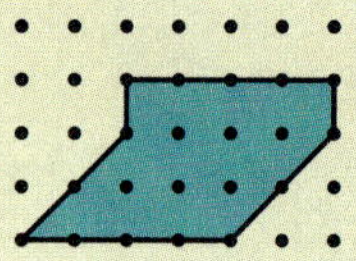

3. Why can the sides be the altitude and base in a right triangle?

1. Yes; The order when multiplying can be changed.
2. 12 square units; method 1. count squares, method 2. use formulas
3. The sides of a right triangle are perpendicular.

▼ When you find the area of a figure, all units must be the same.

Example 5 **Find the area of a trapezoid with $b_1 = 5$ mm, $b_2 = 1$ cm, and $h = 4$ mm.**

Solution $b_2 = 1 \text{ cm} = 10 \text{ mm}$ Write 1 cm as 10 mm so the bases are in the same units.

$$A = \frac{1}{2}h(b_1 + b_2)$$
$$= \frac{1}{2} \cdot 4(5 + 10)$$
$$= 30 \text{ mm}^2$$

CLASS EXERCISES

Find the area of a triangle with the given base and height.

1. $b = 12$ cm, $h = 7$ cm **42 cm²**
2. $h = 3$ in., $b = 1$ ft **18 in.²**
3. $b = 2.5$ mm, $h = 5$ mm **6.25 mm²**

4. 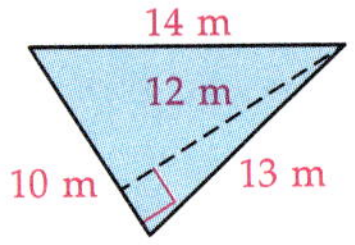

60 m²

5. 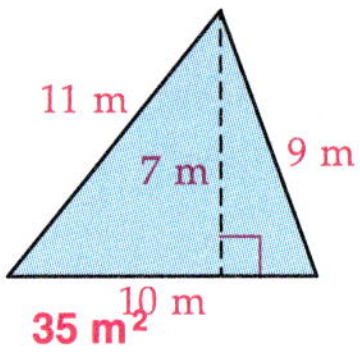

35 m²

6. 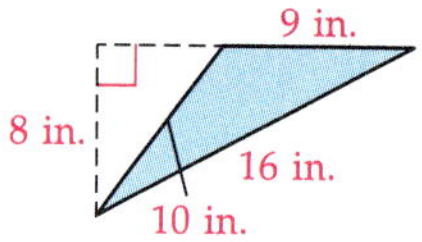

36 in.²

Find the area of a trapezoid with the given base lengths and height.

7. $b_1 = 6$ cm, $b_2 = 10$ cm, $h = 5$ cm **40 cm²**
8. $b_1 = 6$, $b_2 = 2$, $h = 5$ **20**
9. $h = 3$, $b_1 = x$, $b_2 = x + 2$ **3x + 3**

10. 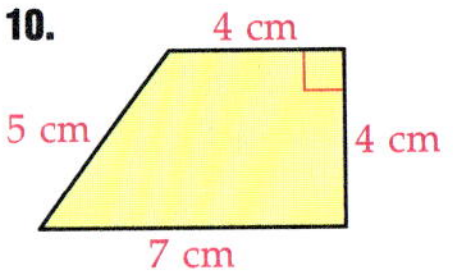

22 cm²

11. 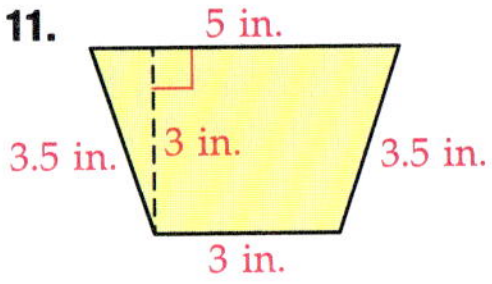

12 in.²

12. 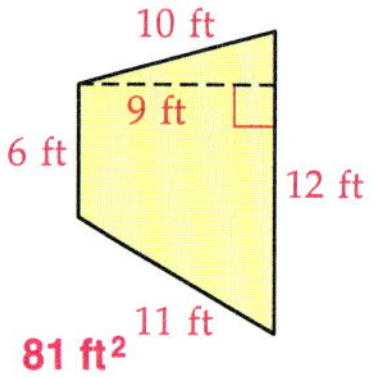

81 ft²

WRITTEN EXERCISES

CALCULATOR **Find the area of a triangle with the given base and height.**

1. $b = 8.2$ m, $h = 9.6$ m **39.36 m²**
2. $h = 35$ mm, $b = 20$ mm **350 mm²**
3. $b = 49$ in., $h = 18$ in. **441 in.²**

4.

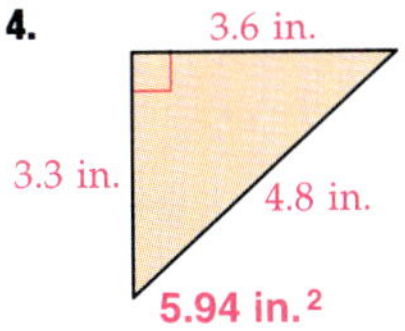

5.94 in.^{2}

5.

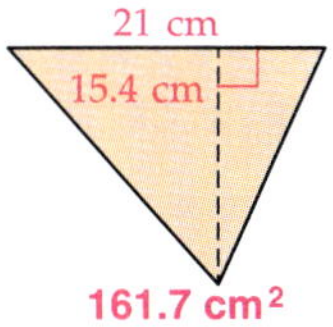

161.7 cm^2

6.

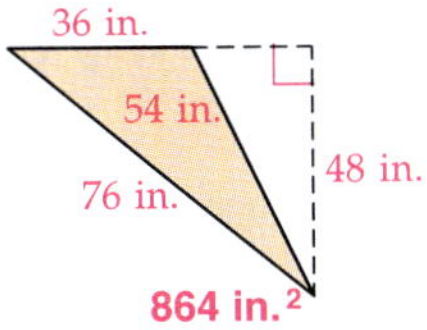

864 in.^{2}

CALCULATOR **Find the area of a trapezoid with the given base lengths and height.**

7. $b_1 = 40$ cm, $b_2 = 35$ cm, $h = 20$ cm **750 cm^2**

8. $b_1 = 54$ m, $b_2 = 80$ m, $h = 15$ m **1,005 m^2**

9. $h = 62.4$ ft, $b_1 = 30.5$ ft, $b_2 = 14.2$ ft **1,394.64 ft^2**

10.

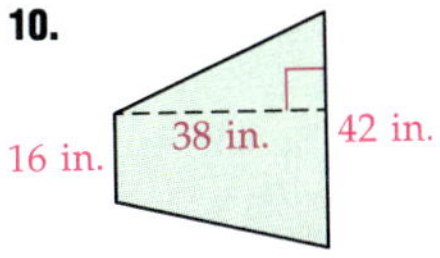

1,102 in.^{2}

11.

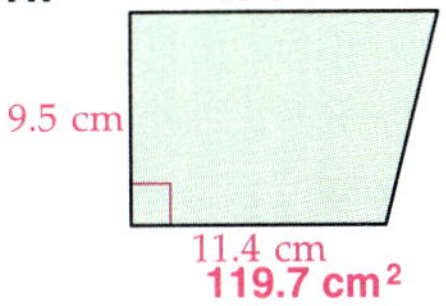

119.7 cm^2

12.

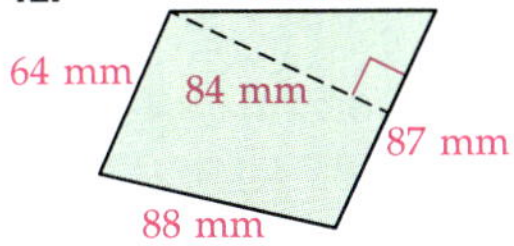

6,342 mm^2

Measure each figure to the nearest tenth of a centimeter and find the area. **Answers may vary slightly.**

13.

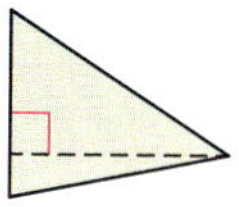

1.4 cm^2

14.

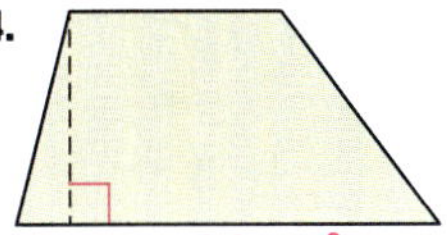

4.3 cm^2

15.

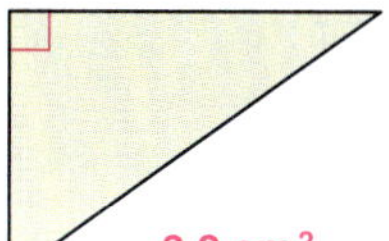

3.2 cm^2

Find the missing values.

Figure	***h***	***b_1***	***b_2***	***A***
16. triangle	0.1 m	4 cm	—	■ **20 cm^2**
17. square	$9x$	—	—	■ **$81x^2$ square units**
18. trapezoid	■ **3 ft**	8 ft	2 ft	15 ft^2
19. parallelogram	■ **1.5 ft**	10 ft	10 ft	15 ft^2

20. Find the area of the yellow square tile at the right. **50 square units**

True or false? Write T or F.

21. A triangle has three altitudes. **T**

22. A square is always a parallelogram. **T**

23. A trapezoid and a parallelogram can never have the same area. **F**

24. A square crossed by two diagonals will form four congruent triangles. **T**

25. An altitude is always perpendicular to a side of a triangle. **T**

26. Two rectangles with the same perimeter always have the same area. **F**

MIXED REVIEW

Find the area.

1. rectangle, $b = 15$ cm, $h = 12$ cm **180 cm^2**

2. parallelogram, $b = 38.2$ mm, $h = 20$ mm **764 mm^2**

3. ***DATA FILE 7 (pp. 272–273)*** Find the percent change, to the nearest tenth, of passenger boardings in Denver from 1987 to 2000. **121.7%**

4. Write 3,120,000 × 4,600 using scientific notation.

5. Find the circumference of a circle with radius $3x$. **18.84x**

Solve.

6. $-8x - 12 = 28$ **-5**

7. $\frac{2}{3}x = \frac{5}{8}$ **$\frac{15}{16}$**

8. A rectangular yard is 20 ft by 40 ft. Your lawn mower will mow a 2-ft wide path. What is the least number of paths you must make to mow the lawn? **10**

4. 1.4352×10^{10}

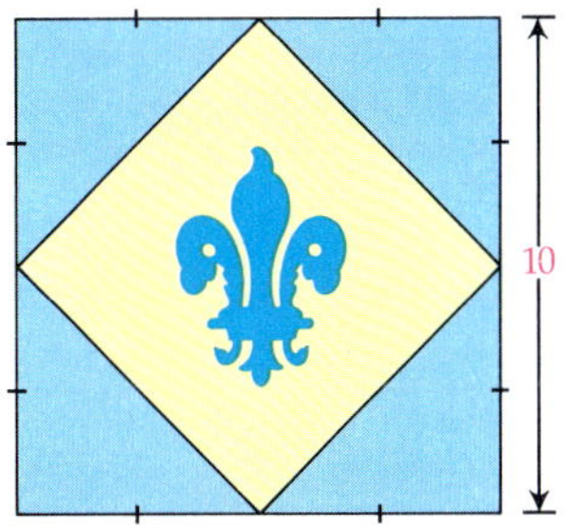

Lesson Follow-up

Reteaching Activity Have students draw triangles and trapezoids on dot paper or grid paper. Have them estimate the areas by counting, and then find the bases and heights and use the formulas to find the areas. Have them compare the actual areas to their estimates.

LESSON RESOURCES

TRB Practice J-3
TRB Enrichment J-4

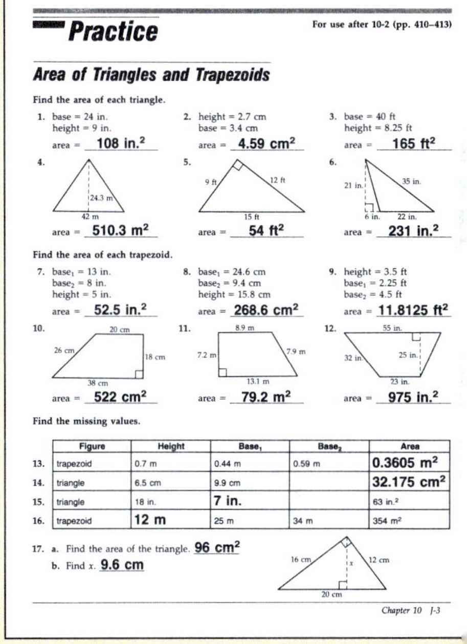

Practice For use after 10-2 (pp. 410–413)

Area of Triangles and Trapezoids

Find the area of each triangle.

1. base = 24 in., height = 9 in. area = **108 in.^{2}**
2. height = 2.7 cm, base = 3.4 cm area = **4.59 cm^2**
3. base = 40 ft, height = 8.25 ft area = **165 ft^2**
4. (24.3 m, 42 m) area = **510.3 m^2**
5. (9 ft, 12 ft, 15 ft) area = **54 ft^2**
6. (21 in., 35 in., 6 in., 22 in.) area = **231 in.^{2}**

Find the area of each trapezoid.

7. base$_1$ = 13 in., base$_2$ = 8 in., height = 5 in. area = **52.5 in.^{2}**
8. base$_1$ = 24.6 cm, base$_2$ = 9.4 cm, height = 15.8 cm area = **268.6 cm^2**
9. height = 3.5 ft, base$_1$ = 2.25 ft, base$_2$ = 4.5 ft area = **11.8125 ft^2**
10. (20 cm, 26 cm, 18 cm, 38 cm) area = **522 cm^2**
11. (8.9 m, 7.2 m, 7.9 m, 13.1 m) area = **79.2 m^2**
12. (55 in., 32 in., 25 in., 23 in.) area = **975 in.^{2}**

Find the missing values.

	Figure	Height	Base$_1$	Base$_2$	Area
13.	trapezoid	0.7 m	0.44 m	0.59 m	**0.3605 m^2**
14.	triangle	6.5 cm	9.9 cm		**32.175 cm^2**
15.	triangle	18 in.	**7 in.**		63 in.^{2}
16.	trapezoid	**12 m**	25 m	34 m	354 m^2

17. a. Find the area of the triangle. **96 cm^2**
b. Find x. **9.6 cm**
(16 cm, 12 cm, 20 cm)

Chapter 10 J-3

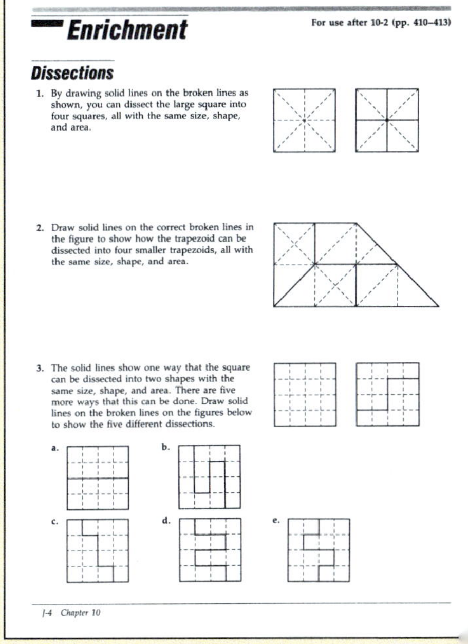

Enrichment For use after 10-2 (pp. 410–413)

Dissections

1. By drawing solid lines on the broken lines as shown, you can dissect the large square into four squares, all with the same size, shape, and area.

2. Draw solid lines on the correct broken lines in the figure to show how the trapezoid can be dissected into four smaller trapezoids, all with the same size, shape, and area.

3. The solid lines show one way that the square can be dissected into two shapes with the same size, shape, and area. There are five more ways that this can be done. Draw solid lines on the broken lines on the figures below to show the five different dissections.

a. b. c. d. e.

J-4 Chapter 10

LESSON QUIZ

Find the area.

1.

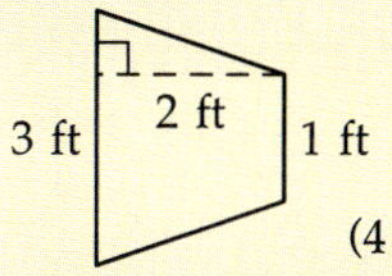

(4 ft^2)

2.

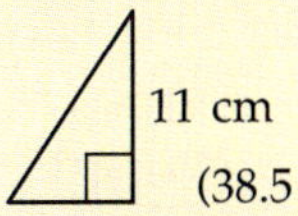

(38.5 cm^2)

3. Find the base of a triangle that has an area of 15 cm^2 and a 4-cm height. (7.5 cm)

4. Find the area of this trapezoid: $b_1 = 12$ in., $b_2 = 15$ in., $h = 5$ in. (67.5 in.^{2})

Assignment Guide

Basic 2-12 even, 13-14, 16-19, 21-26, MR All

Average 1-11 odd, 13-14, 16-26, MR All

Enriched 1-3, 7-9, 14-26, MR All

FOR THE NEXT LESSON

graph paper, compass, ruler, pencil, scissors, string

Lesson Focus

Materials

Graph paper, compass, ruler, pencil, scissors, string

MOTIVATING THE LESSON

Have students draw two circles of different sizes with a compass on graph paper. Ask them to count square units to estimate the areas of the circles. Discuss the ways students counted the squares and any problems they found.

Instruction

1 Have students make the model shown by drawing a circle on a piece of paper, then folding and cutting the paper to get congruent sectors. Have students point out where the circumference and the radius of the original circle are on the new figure.

2 After discussing the formula, ask students to go back to the circles they made on graph paper and calculate the areas. Have them compare the areas to the estimates they made by counting.

3 Remind students that the exact area is expressed in terms of pi and the approximate area is calculated using 3.14 for pi. Point out the *approximately equal* symbol in the solution.

4 Review how to find the square root of a perfect square. Also review how to use the square root key on a calculator.

OBJECTIVE: *To find the areas of circles.*

10-3 Area of Circles

▼ Notice what happens when you cut a circle into several equal size pieces and then rearrange the pieces.

1

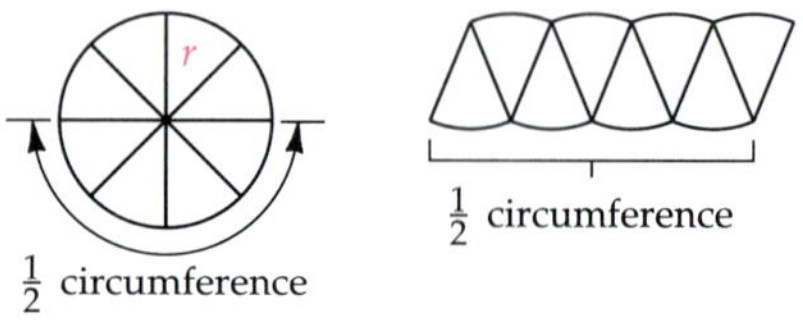

FLASHBACK
$A = bh$ (parallelogram)
$C = \pi d = 2\pi r$

The new shape closely resembles a parallelogram where the height, h, is about the same as the radius of the circle. The base length, b, is about half the circumference (C) of the circle. You can use the formula for the area of a parallelogram to find the formula for the area of a circle.

$b = \frac{1}{2}C$ — Substitute the formula for circumference.
$= \frac{1}{2}2\pi r$
$= \pi r$
$A = bh$ — Substitute πr for b and r for h.
$= \pi r \cdot r$
$= \pi r^2$ — Simplify.

2

Area of a Circle	The area of a circle equals the product of π and the square of the radius (r). $A = \pi r^2$

Example 1 **Find the area of a circle with diameter 12 cm. Give both an exact and an approximate answer. Use 3.14 for π.**

3

Solution
$d = 12, r = 6$ — The radius is half the diameter.
$A = \pi r^2$
$= \pi(6)^2$
$= 36\pi\ \text{cm}^2$, exact answer
$\approx 36 \cdot 3.14$ — Substitute 3.14 for π.
$= 113.04\ \text{cm}^2$

THINK Why does substituting 3.14 for π give you an approximate answer? The exact value of π is not 3.14.

▼ If you know the area of a circle, you can find the radius.

Example 2 **Find the radius of a circle with area 452.16 cm². Use 3.14 for π.**

4

Solution
$A = \pi r^2$
$452.16 = 3.14 \cdot r^2$ — Divide each side by 3.14.
$144 = r^2$ — Find the square root of 144.
$r = 12$ cm

Teaching Tip

Have students work with models in small groups. They will be able to share their thoughts and locate and correct errors more easily.

CLASS EXERCISES

Find the area of each circle. Give an exact answer and an approximate answer. Use 3.14 for π.

1. $r = 3$ ft 9π ft^2 28.26 ft^2
2. $d = 10$ m 25π m^2 78.5 m^2
3. $r = 20$ cm 400π cm^2 1,256 cm^2

4. 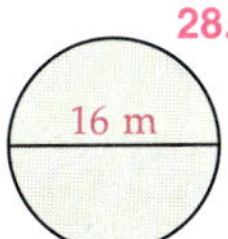

64π m^2 200.96 m^2

5. 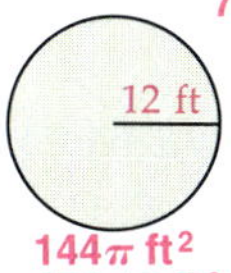

144π ft^2 452.16 ft^2

6. 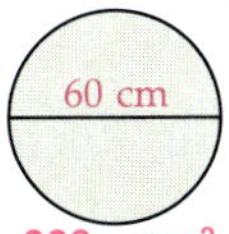

900π cm^2 2826 cm^2

Find the radius of a circle with the given area. Use 3.14 for π.

7. 49π cm^2 7 cm
8. 254.34 yd^2 9 yd
9. $225x^2\pi$ square units 15x units

THINK AND DISCUSS See below.

1. Name two real world situations where area of a circle is important.
2. Suppose you didn't know the formula for the area of a circle. How could you approximate the number of square units in a circle?

1. Answers may vary.
2. Answers may vary. One possible answer is "Form a square around a circle using the diameter. Estimate how much to subtract from the corners."

WRITTEN EXERCISES

2. $\frac{1}{4}\pi$ m^2, 0.79 m^2 4. 36π in.^{2}, 113.04 in.^{2} 5. 2.56π ft^2, 8.04 ft^2 6. 17.64π mm^2, 55.39 mm^2

Find the area of each circle. Give an exact answer and an approximate answer. Use 3.14 for π. Round to nearest hundredth.

1. $r = 11$ mi 121π mi^2, 379.94 mi^2
2. $r = \frac{1}{2}$ m See above.
3. $d = 1.2$ in. 0.36π in.^{2}, 1.13 in.^{2}
4. $d = 12$ in.
5. $d = 3.2$ ft
6. $d = 8.4$ mm
7. $r = 5x$ $25\pi x^2$ square units, $78.5x^2$ square units
8. $d = 6\sqrt{2}$ 18π square units, 56.52 square units
9. $d = 4.2x$ $4.41\pi x^2$ square units, $13.85x^2$ square units

Match the item listed with the most reasonable area.

	Item		*Area*
10.	dinner plate f	a.	3.14 cm^2
11.	quarter e	b.	113.04 in.^{2}
12.	circle on basketball floor d	c.	28.26 cm^2
13.	jar lid c	d.	36π ft^2
14.	shirt button a	e.	452.16 mm^2
15.	12-in. pizza b	f.	78.5 in.^{2}

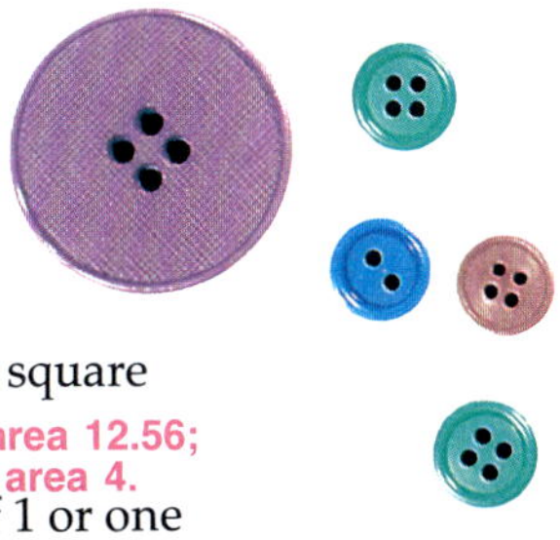

16. Which has a greater area, a circle with a radius of 2 or a square with side of length 2? Explain your answer. Circle has area 12.56; square has area 4.
17. Which has a greater area, four circles with the radius of 1 or one circle with the radius of 4? Explain your answer. 1 circle of radius 4 is larger. Area is 50.24; area of 4 circles is 12.56.
18. Write a formula for the area of a circle that uses the diameter instead of the radius. $\left(\frac{d}{2}\right)^2\pi$
19. What is the area of the largest circle that will fit in a square with area 64 cm^2? 50.24 cm^2

ADDITIONAL EXAMPLES

1. Give both an exact and an approximate area of a circle with a diameter of 15 cm. (56.25π cm^2, 176.625 cm^2)
2. Find the radius of a circle with area 28.26 ft^2. (3 ft)

Guided Practice

Class Exercises Have students work in small groups to complete the questions. Then ask groups to explain their answers.

Think and Discuss After all suggestions are discussed for question 1, have students describe one real-world situation in their math journals.

Welcome varied and creative suggestions for question 2.

MATH MINUTES

Distribute 12-in. strings and graph paper to groups of students. Ask them to find the shape that encloses the greatest area. (Circle: $A = 11.46$ in.^{2})

A ***common error*** is to use the diameter, instead of the radius, when finding the area of a circle. For example, students may interpret r^2 as $2r$. Have students write the formula each time they use it. Point out that r^2 in the formula should remind them to inspect the given measurement to see whether it is the diameter or the radius.

Closure

Write the formula for area of a circle on the chalkboard. Ask students to think of a way that will help them remember this formula. (For example, area is measured in square units; the area formula has a square in it.)

Independent Practice

For question 32, suggest that students trace the figure and draw a horizontal diameter. This will enable them to see two smaller semicircles which can be interchanged, resulting in the shaded area being half of the large circle.

Help students recognize that for question 33, the area of the outer ring is found by subtracting the area of the middle circle from the area of the largest circle.

For question 37, students can use *Appleworks*, *Microsoft Works*, or *Excel* for their spreadsheets. An alternate way to answer parts a and b is to construct bar graphs from the data.

r (units)	*A* (square units)
1	3.1
3	28.3
5	78.5
7	153.9

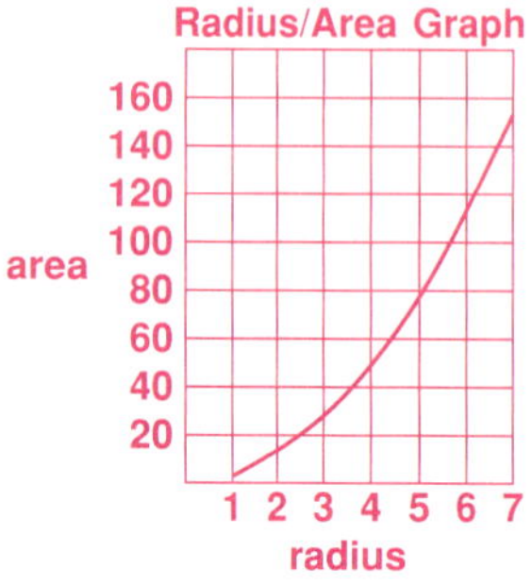

20. ***DATA FILE 10 (pp. 404–405)*** Find the approximate area of the circular home range for each animal. Use 3.14 for π.
a. cougar 498,506,400 m^2 **b.** red fox 1,618,700 m^2 **c.** Northern flying squirrel 3,200 m^2

21. Copy and complete the table at the left. Round to the nearest tenth. Use 3.14 for π.
a. Graph the results. Use the x-axis for radius and the y-axis for area. See graph at left.
b. Predict the radius of a circle with area 125 square units. approximately 6 units

Find the radius of a circle with the given area. Use 3.14 for π.

22. 81π cm^2 9 cm **23.** 803.84 in.^{2} 16 in. **24.** 7.065 mm^2 1.5 mm

Find the circumference and area of each circle. Use 3.14 for π. Round to the nearest tenth.

25. $r = 5.2$ cm 32.7 cm 84.9 cm^2 **26.** $d = 7.8$ in. 24.5 in. 47.8 in.^{2} **27.** $r = 18.6$ m 116.8 m 1,086.3 m^2

28. $r = \frac{3}{4}$ cm 4.7 cm, 1.8 cm^2 **29.** $r = 8.7$ m 54.6 m, 237.7 m^2 **30.** $d = 15xy$ 47.1xy, 176.6x^2y^2 sq. units

Find the area of each shaded portion. Use 3.14 for π.

31.

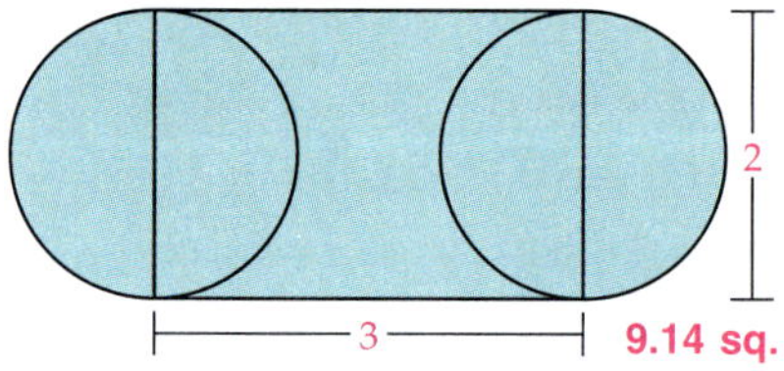

9.14 sq. units

32.

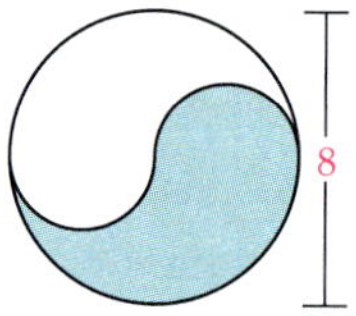

25.12 sq. units

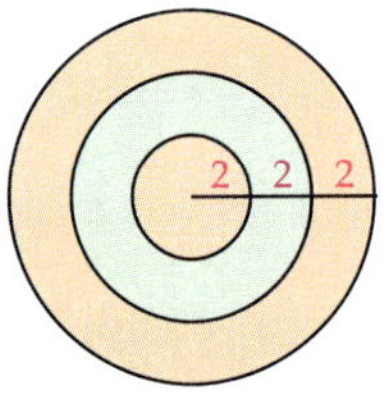

33. Find the area of the outer ring of the figure at the left. 20π sq. units or 62.8 sq. units

34. Manufacturers of tin cans stamp the lids from rectangular sheets of tin.

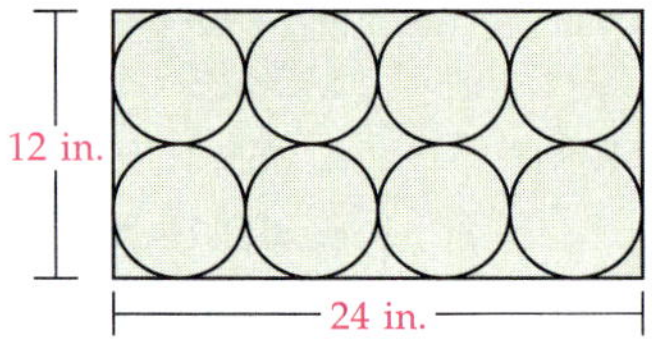

a. What is the radius of each lid? 3 in.
b. What is the total area used by the lids? 226.08 in.^{2}
c. How much of the sheet of tin is wasted? 61.92 in.^{2}

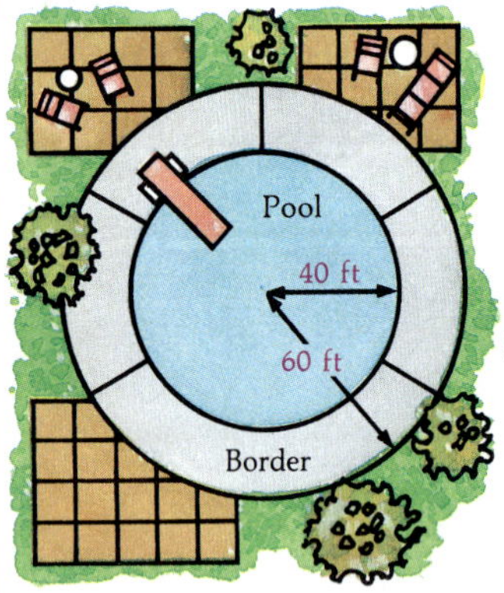

35. You wish to carpet the border of the pool at the left.
a. How many square feet do you need for the border? 6,280 ft^2
b. Carpets are sold by the square yard. How many square feet are in a square yard? 9 ft^2
c. How many square yards of carpet should you purchase? 698 yd^2

36. **a.** What is the area of each pizza pictured below?

b. What is the price per square inch?

c. Is the largest pizza always the best buy?

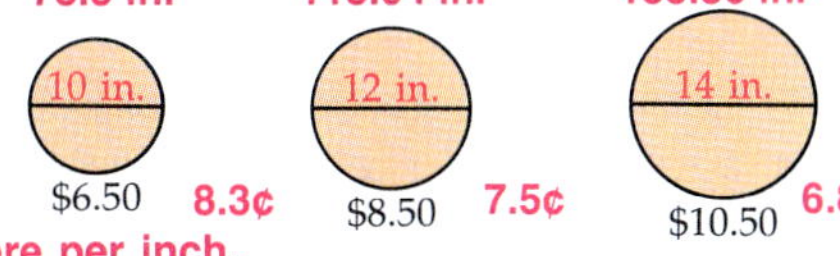

No, in some cases it can cost more per inch.

37. ***COMPUTER*** Use a spreadsheet to compare the radius, circumference, and area of different circles. Use the format shown below.

	A	B	C
1	Radius	Circumference	Area
2	1	=2*A2*3.14	=3.14*A2^2
3	2	=2*A3*3.14	=3.14*A3^2
4	3	=2*A4*3.14	=3.14*A4^2
5			

Complete cells A2 to A10 for radii from 1 to 9.

a. What happens to the circumference of a circle if the radius is doubled? if the radius is tripled? The circumference is doubled, tripled.

b. If you double the radius of a circle, do you double the area? No, the area is multiplied by four.

MIXED REVIEW

1. How many degrees are in each angle of an equilateral triangle? 60°

2. $\sqrt{121} = \blacksquare$ 11

3. $2^3 = \blacksquare$ 8

Find the area.

4. square, $s = 12$ 144

5. parallelogram, $b = 6$, $h = 10$ 60

6. trapezoid, $b_1 = 8$, $b_2 = 12$, $h = 7$ 70

7. triangle, $b = 9$, $h = 15$ 67.5

8. The front of Sam's house looks like a rectangle with a triangle above the rectangle. The rectangle is 20 ft by 10 ft and the triangle is 20 ft by 8 ft. Find the area of the front of the house. 280 ft²

TEST YOURSELF

State the formula for the area of each figure.

1. circle πr^2 **2.** triangle $\frac{1}{2}bh$ **3.** rectangle bh

4. trapezoid $\frac{1}{2}h(b_1 + b_2)$ **5.** parallelogram bh

Find the area of each figure. Use 3.14 for π.

6.

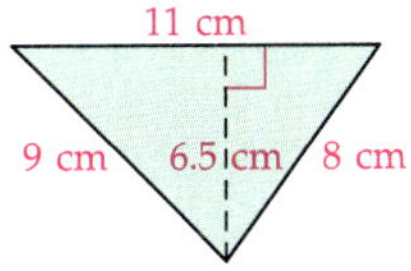

35.75 cm²

7.

286 ft²

8.

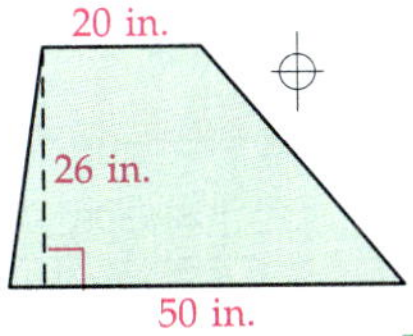

910 in.²

9.

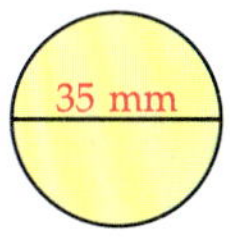

961.625 mm²

10.

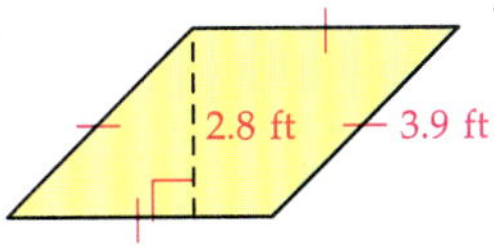

10.92 ft²

11.

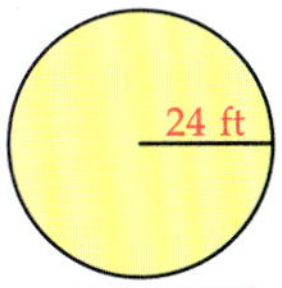

1,808.64 ft²

Lesson Follow-up

Reteaching Activity Have students draw circles on graph paper, estimate their areas by counting square units, and then calculate the areas by using the formula.

LESSON RESOURCES

TRB Practice J-5
TRB Enrichment J-6

Practice

For use after 10-3 (pp. 414–417)

Area of Circles

Find the area of each circle. Give an exact answer and an approximate answer. Use 3.14 for π.

1. $r = 7$ m
$A =$ **49π m²**
$A \approx$ **153.86 m²**

2. $d = 18$ cm
$A =$ **81π cm²**
$A \approx$ **254.34 cm²**

3. $d = 42$ m
$A =$ **441π m²**
$A \approx$ **1384.74 m²**

4. $r = 35$ km
$A =$ **$1{,}225\pi$ km²**
$A \approx$ **3,846.5 km²**

5. $d = 22$ cm
$A =$ **121π cm²**
$A \approx$ **379.94 cm²**

6. $r = 25$ ft
$A =$ **625π ft²**
$A \approx$ **1962.5 ft²**

7.

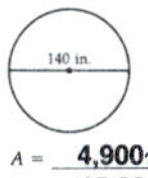

$A =$ **$4{,}900\pi$ in.²**
$A \approx$ **15,386 in.²**

8.

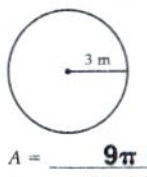

$A =$ **9π m²**
$A \approx$ **28.26 m²**

9.

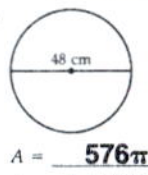

$A =$ **576π cm²**
$A \approx$ **1,808.64 cm²**

10. $d = 8x$
$A =$ **$16x^2\pi$ units²**
$A \approx$ **$50.24x^2$ units²**

11. $r = 10x$
$A =$ **$100\pi x^2$ units²**
$A \approx$ **$314x^2$ units²**

12. $r = 2y$
$A =$ **$4y^2\pi$ units²**
$A \approx$ **$12.56y^2$ units²**

Find the radius of each circle.

13. $A = 144\pi$ ft² $r =$ **12 ft**

14. $A = 676\pi$ cm² $r =$ **26 cm**

15. $A = 3{,}600\pi$ km² $r =$ **60 km**

Find the circumference and area of each circle. Give an exact answer.

16. $r = 13$ in.
$C =$ **26π in.**
$A =$ **169π in.²**

17. $d = 46$ ft
$C =$ **46π ft**
$A =$ **529π ft²**

18. $r = 5kx$
$C =$ **$10kx\pi$ units**
$A =$ **$25k^2x^2\pi$ units²**

19. A circle has a circumference of 78π yd. What is the area of the circle? **$1{,}521\pi$ yd²**

20. Circle 1 has a radius of 12 in. Circle 2 has a radius of 24 in. Find the ratio of the area of the smaller circle to the area of the larger. **$\frac{1}{4}$**

Chapter 10 J-5

Enrichment

For use after 10-3 (pp. 414–417)

Ratios of Circumference and Area

Complete. Use 3.14 form.

	Circle 1			Circle 2			Ratios		
	radius (r_1)	circumference (C_1)	area (A_1)	radius (r_2)	circumference (C_2)	area (A_2)	$\frac{r_2}{r_1}$	$\frac{C_2}{C_1}$	$\frac{A_2}{A_1}$
1.	2	12.56	12.56	4	25.12	50.24	2	2	4
2.	3	18.84	28.26	6	37.68	113.04	2	2	4
3.	4	25.12	50.24	8	50.24	200.96	2	2	4
4.	5	31.4	78.5	10	62.8	314	2	2	4
5.	2	12.56	12.56	6	37.68	113.04	3	3	9
6.	3	18.84	28.26	9	56.52	254.34	3	3	9
7.	4	25.12	50.24	12	75.36	452.16	3	3	9
8.	5	31.4	78.5	15	94.2	706.5	3	3	9
9.	2	12.56	12.56	10	62.8	314	5	5	25
10.	3	18.84	28.26	15	94.2	706.5	5	5	25
11.	4	25.12	50.24	20	125.6	1256	5	5	25
12.	5	31.4	78.5	25	157	1962.5	5	5	25

Study the patterns in the last three columns.

13. Write a statement comparing the ratio of the radii of two circles to the ratio of their circumferences.
The ratios will be the same.

14. Write a statement comparing the ratio of the radii of two circles to the ratio of their areas.
When the ratio of the radii of two circles is squared, the result is the ratio of their areas.

15. The radius of one circle is 12 times the radius of a smaller circle.
a. Predict the ratio of the circumferences in the circles. **12:1**
b. Predict the ratio of the area of the circles. **144:1**

16. The ratio of the areas of two circles is 49. How many times longer is the radius of the larger circle than the smaller circle? **7**

J-6 Chapter 10

LESSON QUIZ

Find the area of each circle. Use 3.14 for pi.

1. $r = 8$ in. (200.96 in.²)

2. $d = 2.2$ cm (3.7994 cm²)

3. Find the radius of a circle with area 314 mm². (10mm)

4. Find the diameter of a circle with area 225π ft.² (30 ft.)

Assignment Guide

Basic 1-5, 10-17, 20, 22, 25-26, 31, 35-36, MR, TY All

Average 4-8, 10-18, 20-23, 26-28, 32-36, MR, TY All

Enriched 6-15, 17-21, 23-24, 28-30, 32-37, MR, TY All

FOR THE NEXT LESSON

geoboards and colored rubber bands or dot paper and colored pencils, math journal, overhead projector (optional)

Lesson Focus

Materials

Geoboards and colored rubber bands or dot paper and colored pencils, math journal, overhead projector (optional)

MOTIVATING THE LESSON

Ask students to make a figure of their choice on the geoboard or dot paper. Have them count the units to estimate the area. Ask if anyone drew a figure whose area was difficult to count. Tell students there is a formula they can use to find areas of figures on geoboards. They will develop the formula in this lesson.

Instruction

Have students work in pairs or small groups. Model the first set of shapes on the overhead projector and discuss counting the boundary dots, interior dots, and area.

1 Tell students they are not looking for a formula here. Any relationship can be expressed as an approximation. For example, the number of interior dots seems to be about half of the boundary dots, and the area seems to be close to the number of boundary dots.

2 Encourage students to guess and test each rule to find the correct answer.

OBJECTIVE: *To develop Pick's theorem using geoboards.*

MATERIALS

- Geoboards and colored rubber bands or
- Dot paper and colored pencils
- Math journal to record work

$1\frac{1}{2}$ $2\frac{1}{2}$ $3\frac{1}{2}$ $4\frac{1}{2}$

Exploring Pick's Theorem

▼ You can use a geoboard to develop a formula or rule about area. Each square on the geoboard represents 1 square unit of area. Use dot paper to copy the shapes shown.

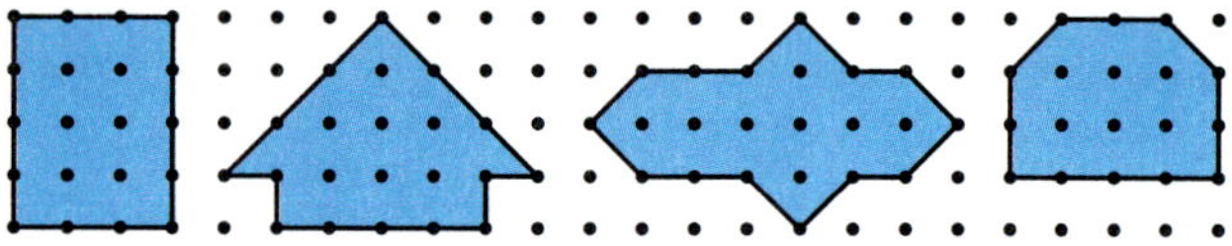

A B C D

1. Find the area of each shape. 12, 13, 14, 11
2. Copy and complete the chart for each shape. Interior dots are those inside the shape. Boundary dots are those on the side or at corners of the shape.

	Boundary Dots	Interior Dots	Area
A	14	6	12
B	14	7	13
C	14	8	14
D	12	6	11

1

3. ***Describe*** the relationship between the number of interior dots, the number of boundary dots, and the area. Answers may vary.
4. ***Model*** several figures with five boundary dots and up to four interior dots. Find the area of each figure. ***Discuss*** how the dots are related to the area. As the number of interior dots increases, the area increases 1 unit for each dot.
5. Copy the shapes below on a geoboard or dot paper.

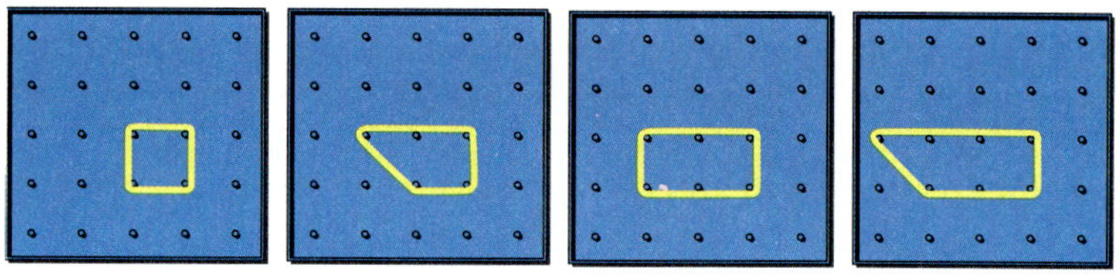

a. Copy and complete the chart.

Boundary Dots	Interior Dots	Area
4	0	1
5	0	$1\frac{1}{2}$
6	0	2
7	0	$2\frac{1}{2}$

2

b. Which rule relates boundary dots (B) to area when the number of interior dots is zero? III

I. $\frac{B}{2} + 1$ **II.** $\frac{B}{2}$ **III.** $\frac{B}{2} - 1$

Teaching Tip

It may be useful to use a geoboard for the overhead projector for this lesson. Students can discuss their ideas as they analyze the patterns on the screen.

6. Copy the shapes below on a geoboard or dot paper.

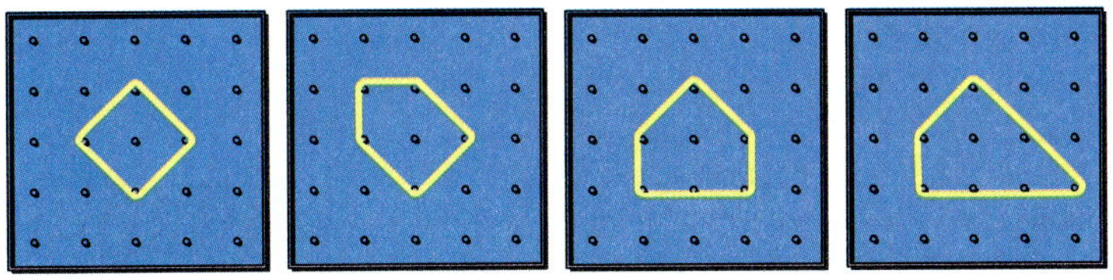

Boundary Dots	Interior Dots	Area
4	1	2
5	1	$2\frac{1}{2}$
6	1	3
7	1	$3\frac{1}{2}$

a. Complete a chart for one interior dot.

b. ***Predict*** the area of a shape with 12 boundary dots and one interior dot. 6

c. ***Write*** a rule that relates boundary dots to area when there is one interior dot. $\frac{B}{2}$

7. Copy the shapes below on a geoboard or dot paper.

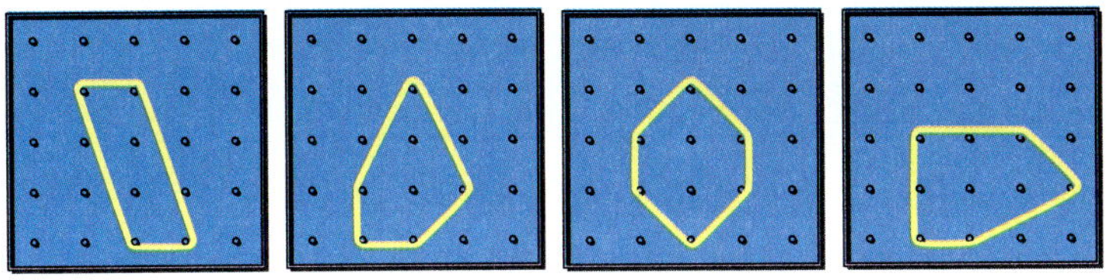

Boundary Dots	Interior Dots	Area
4	2	3
5	2	$3\frac{1}{2}$
6	2	4
7	2	$4\frac{1}{2}$

a. Complete a chart for two interior dots.

b. ***Predict*** the area of a shape with 12 boundary dots and two interior dots. 7

c. ***Write*** a rule that relates boundary dots to area when there are two interior dots. $\frac{B}{2} + 1$

8. ***Write*** a rule for finding area for each number of dots.

a. three interior dots **b.** four interior dots **c.** n interior dots

9. Test your rule. Find the area of each figure using area formulas. Then use your rule. 8. a. $\frac{B}{2} + 2$ b. $\frac{B}{2} + 3$ c. $\frac{B}{2} + n - 1$

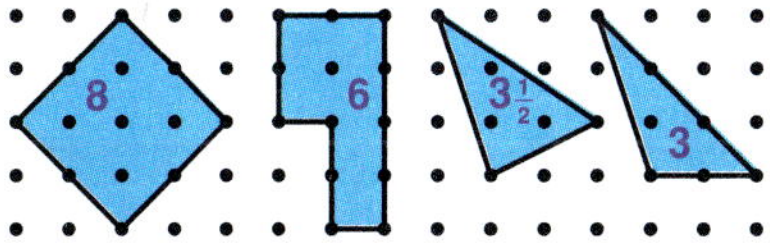

10. Use Pick's theorem to find the area of each shaded region.

3 Monitor groups as they answer these questions. If they are not close to getting Pick's Formula, $(\frac{B}{2} + I - 1)$, help them analyze their data from previous questions and help them see the intended relationships. For example, use the formula for no interior dots, $\frac{B}{2} - 1$, and ask students to look at the effect the number of interior dots has on each area.

Closure

Writing in Math Ask students to briefly describe in their math journals what they did in this lesson. Ask them to also write Pick's Formula and explain how it works.

Assignment Guide

Basic All
Average All
Enriched All

FOR THE NEXT LESSON

an assortment of objects that are pyramids, prisms, cones, cylinders, and spheres, graph paper, ruler, pencil, string, math journal

Lesson Focus

Materials

An assortment of objects that are prisms, pyramids, cones, cylinders, and spheres, graph paper, ruler, pencil, string, math journal

Vocabulary/Symbols

base
base edge
cone
cylinder
lateral edge
lateral face
net
polyhedron
prism
pyramid
sphere
vertex

MOTIVATING THE LESSON

Display a box, a ball, a piece of chalk, or any other objects that are space figures mentioned in this lesson. Ask groups of students to make a list in their math journals of the characteristics of each object. Then discuss the groups' lists.

Instruction

As you discuss each category of space figures, ask students to name the space figures in the collection of objects they wrote about. Have them identify faces, vertices, bases (if any), and edges.

1 Ask students to think of the name of another prism that is not shown. (Examples: rectangular prism, octagonal prism) Have them rely on the statement that tells how to name a prism. Also ask if a cube is a prism. (yes)

2 Ask students to name another pyramid that is not shown. (Examples: pentagonal pyramid, octagonal pyramid)

OBJECTIVE: *To identify common polyhedra.*

10-4 Space Figures

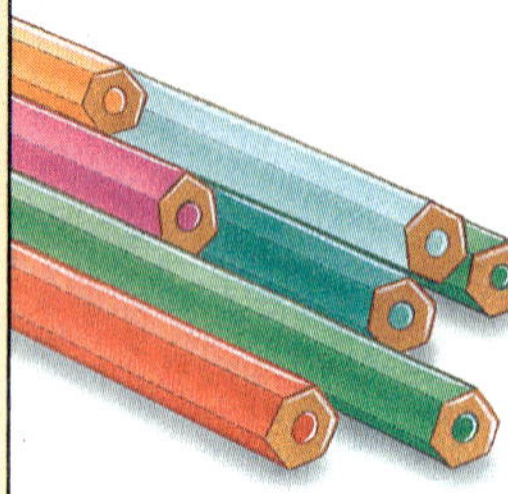

▼ An interior designer must be aware of the interplay of shapes in a room. Three-dimensional or space figures interact to produce different effects.

▼ *Prisms* are a type of *polyhedron*. An unsharpened six-sided pencil is in the shape of a prism. The diagram shows some of the terms we use to describe the parts of a prism.

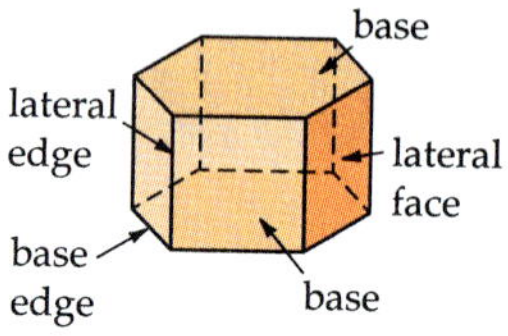

Polyhedron	A polyhedron is a space figure in which all faces are polygons.
Prism	A prism is a polyhedron with two parallel bases that are congruent polygons and sides that are parallelograms.

We name a prism by the shape of its bases.

Example 1

1

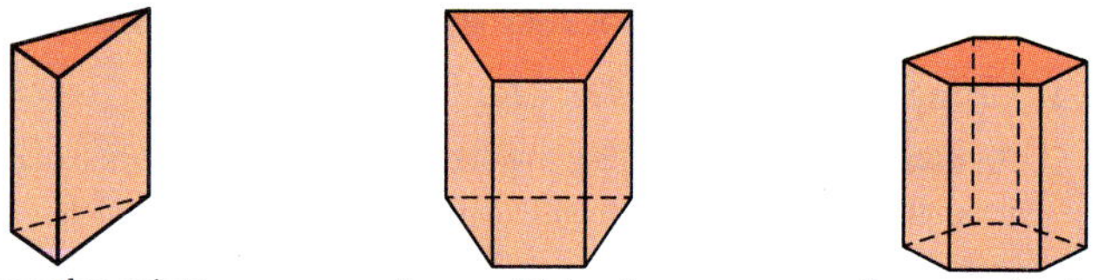

triangular prism — trapezoidal prism — hexagonal prism

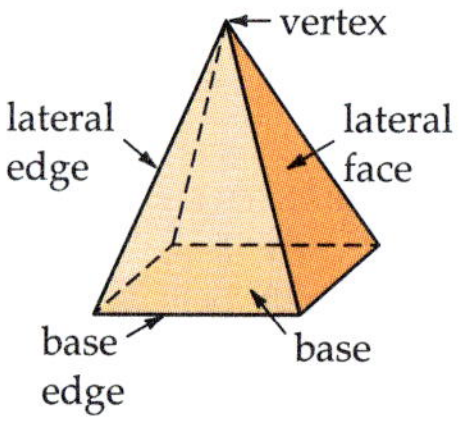

▼ *Pyramids* are a type of polyhedron. The ancient Egyptians built pyramids for tombs. You can also find a pyramid on a dollar bill. The diagram at the left shows the parts of a pyramid.

Pyramid	A pyramid is a polyhedron with triangular sides that meet at a vertex. The base of a pyramid is a polygon.

We name a pyramid by the shape of its base.

THINK Why do we use dashed lines when drawing space figures? **to show depth**

Example 2

2

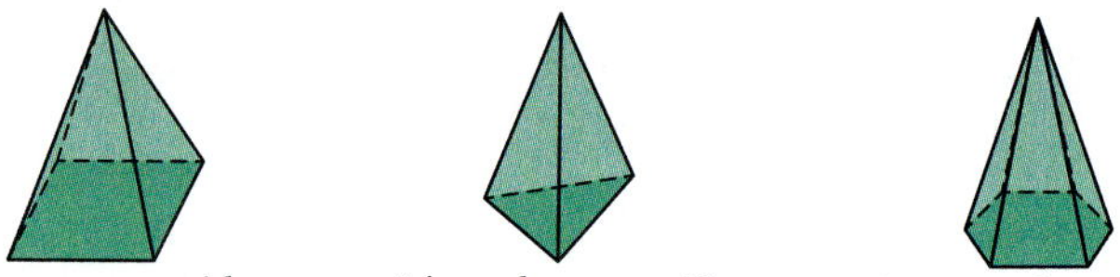

square pyramid — triangular pyramid — hexagonal pyramid

Teaching Tip

Allow students to use models of figures as often as possible. Handling and inspecting models of space figures allows students to see the actual shapes of the faces, as well as to become familiar with other characteristics.

▼ Cylinders, cones, and spheres are space figures that contain circles.

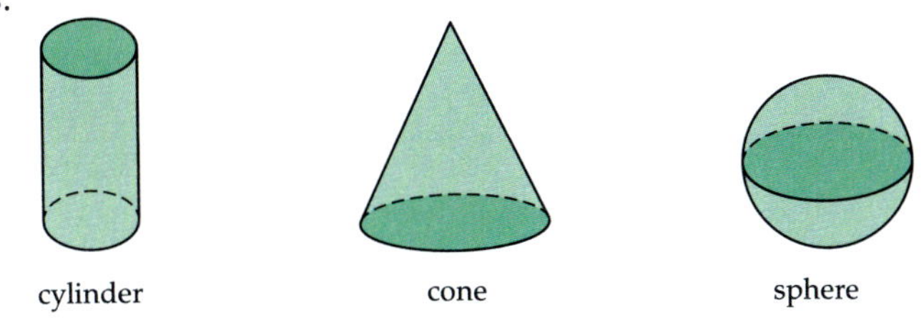

Cylinder	A cylinder is a space figure with two circular, parallel, and congruent bases.
Cone	A cone is a space figure with one circular base and one vertex.
Sphere	A sphere is the set of all points in space that are the same distance from a given point called the center.

(3) **THINK** How is the definition of a sphere similar to the definition of a circle? **Both are a set of all points the same distance from a point. A circle is in a plane. A sphere is in space.**

▼ A *net* is a pattern you can fold into a space figure.

Example 3 **Name the space figure you can form from each net.**

a.

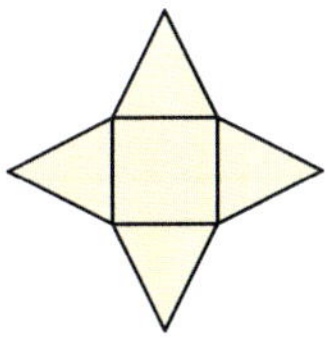

b.

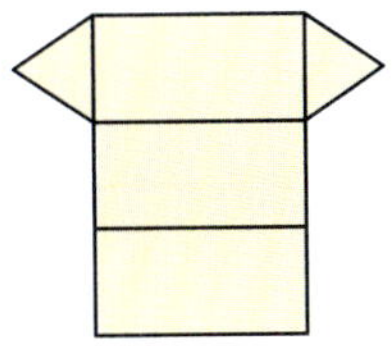

Solution **a.** square pyramid **b.** triangular prism

(4) **1. There are six nets.**

CLASS EXERCISES

1. Give another name for a cube. **a regular rectangular prism**
2. What polyhedron is a cone most like? **pyramid**
3. What polyhedron is a cylinder most like? **prism**
4. How many faces of a regular hexagonal pyramid are congruent? What is the name of the shape of a face? **6; triangle**
5. Draw a net for a pentagonal pyramid. **Check students' drawings.**
6. Draw a net for a hexagonal prism.
7. Draw a net for a cylinder.

THINK AND DISCUSS

1. How many nets can you draw for a cube? Draw them. **See above.**
2. How do the bases of a cylinder look when they are drawn? Draw a cylinder and a cone. **oval** **Check students' drawings.**
3. What is a mathematical name for a brick? **rectangular prism**

MATH MINUTES

Display three or four different space figures. Ask students to draw the top view and a side view of each. For example, the top view of a triangular prism is a triangle, and the side view is a rectangle.
(Check students' work.)

(3) Demonstrate the definition of a circle. Hold one end of a string at the center and use chalk at the other end to draw a circle. Demonstrate the definition of a sphere by holding the end of a string in front of you and moving the other end in all directions that would be points on the sphere.

(4) Display a cut-out net for each figure and ask volunteers to fold them to see what space figures result.

ADDITIONAL EXAMPLES

1. Name the space figure you can form from the net. (cube or square prism)

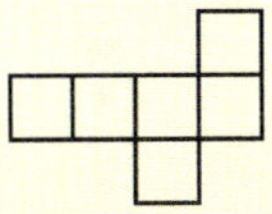

Guided Practice

Class Exercises Have students work in small groups to complete the exercises. Suggest that students attempt to sketch the pyramid in question 4 in order to answer the question.
Have students use graph paper to draw the nets in questions 5-7.

Think and Discuss For question 1, point out that two nets with the same configuration but different orientation, count as one net. For example, these nets are the same:

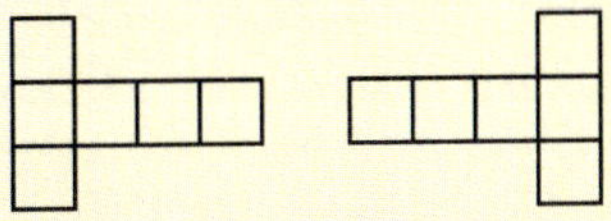

For question 2, discuss that shapes are drawn differently to give the figure perspective.
Likewise, some faces on a rectangular prism appear to be parallelograms rather than rectangles.

A ***common error*** is to confuse the terms prism and pyramid. Have students make a side-by-side list of characteristics that will help them remember the figures correctly. For example, students may write:

prism	*pyramid*
2 bases	1 base
"flat" top	"pointed" top
rectangular faces	triangular faces

Closure

Writing in Math Review the characteristics of prisms and pyramids. Then ask what distinguishes a cylinder, a cone, and a sphere from prisms and pyramids. (They are not polyhedra.)

Independent Practice

Provide as many space figures as possible for students to use for questions 14-19.

In question 25, mention that polyhedra is the plural of polyhedron.

As a follow-up, refer to the problem solving activity found on page 404E of the Chapter Overview. See also, TRB Exploring O-38.

MIXED REVIEW

1. Find the area of a circle with diameter 6*a* units. See below.
2. Find the area of a parallelogram with base 2*x* and height 4. 8x square units
3. Find the area of a circle with radius 5 in. 25π in.^{2}
4. Find the area of a triangle with $b = 7$ cm and $h = 10$ cm. 35 cm^2

Solve.

5. $2x + 7 = -19$ -13
6. $-3x + 8 = 29$ -7
7. $\frac{5}{8}x - \frac{1}{2} = \frac{3}{8}$ $1\frac{2}{5}$
8. ***DATA FILE 7 (pp. 272–273)*** Suppose it is 11:00 A.M. in Rio de Janeiro, Brazil. What time is it in Fairbanks, Alaska? 5:00 A.M.
9. Two angles of a triangle measure 74° and 86°. Find the measure of the third angle and classify the triangle by its angles. 20° acute

1. $9a^2\pi$ square units

WRITTEN EXERCISES

Name each polyhedron.

1. 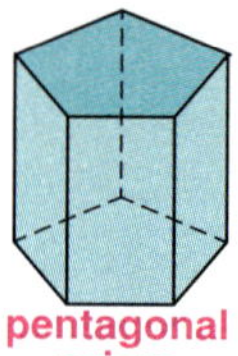
pentagonal prism

2. 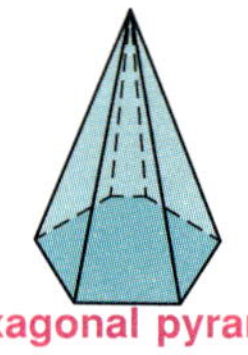
hexagonal pyramid

3. 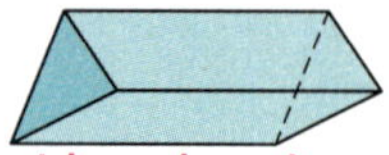
triangular prism

Name the space figure you can form from each net.

4. 
rectangular prism

5. 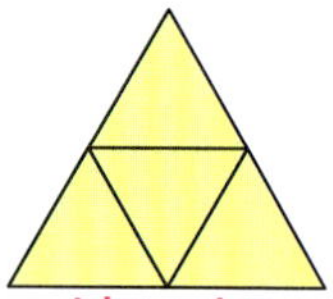
triangular pyramid

Draw a net for each space figure named.

6. hexagonal pyramid
Check students' work.
7. octagonal prism

True or false? Write T or F.

8. A cone has two bases. F
9. A square pyramid has four triangles. T
10. The lateral faces of a prism are parallelograms. T
11. A cylinder is an example of a polyhedron. F
12. The lateral faces of a regular pyramid are isosceles triangles. T
13. A pyramid can have a circular base. F

Complete each statement.

14. A triangular prism has ▒ triangles. 2
15. An octagonal prism has 8 ▒ and 2 ▒. rectangles, octagons
16. A hexagonal prism has ▒ lateral edges. 6
17. A square pyramid has ▒ base edges. 4
18. A cone has ▒ vertex. 1
19. A ▒ has 5 parallelograms for faces and 2 ▒ for bases. pentagonal prism, pentagons

Write the mathematical name for each object.

20. soup can
cylinder
21. shoe box
rectangular prism
22. tepee
cone
23. basketball
sphere
24. ***WRITE*** Examine a soccer ball. Describe it using mathematical names. The soccer ball is a sphere. On the surface of the ball are regular pentagons and hexagons.

25. ***PROJECT*** A tetrahedron, octahedron, hexahedron, dodecahedron, and icosahedron are regular polyhedra.

a. Find out how many faces each regular polyhedron has and what polygons make the faces.

b. These shapes are often called Platonic solids. Research what each solid represented for Plato. tetrahedron—fire, octahedron—air, icosahedron—water, cube—earth, dodecahedron—universe

25. a. tetrahedron, 4 triangles; octahedron, 8 triangles; hexahedron, 6 squares; dodecahedron, 12 pentagons; icosahedron, 20 triangles

Copy each of the figures below on graph paper and write the mathematical name for each. Check students' work.

26.

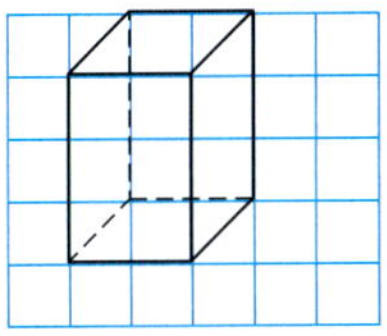

rectangular prism

27.

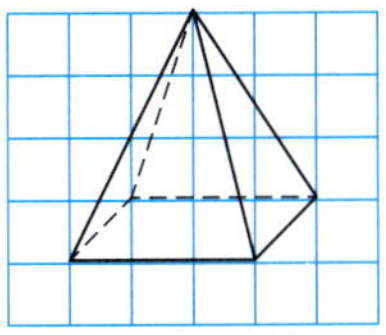

square pyramid

28. ***PROJECT*** Draw the nets for four polyhedra on heavy paper or cardboard. Fold to form the polyhedron and label with the mathematical name.

Critical Thinking

EXPLORING VISUALIZATION

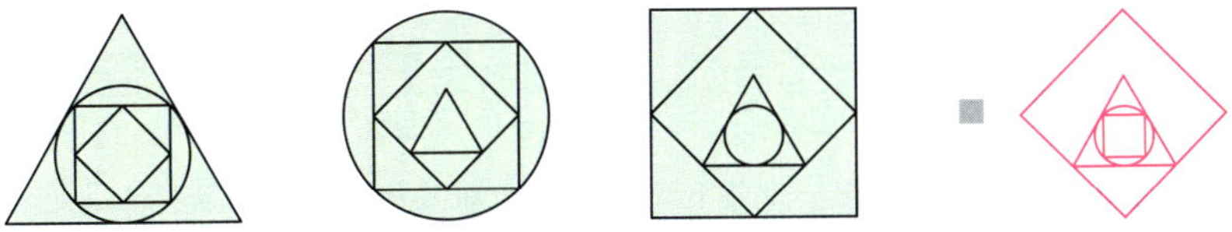

A *B* *C* *D*

1. Guess the pattern for *D*. Answers may vary.
2. Name each shape in pattern *A* in order from greatest area to least. Do this for pattern *B* and pattern *C*. See right.
3. Predict the pattern for *D* and draw it. See above.
4. Select the pattern that best completes the series below. C

A: triangle, circle, square, square
B: circle, square, square, triangle
C: square, square, triangle, circle
D: square, triangle, circle, square

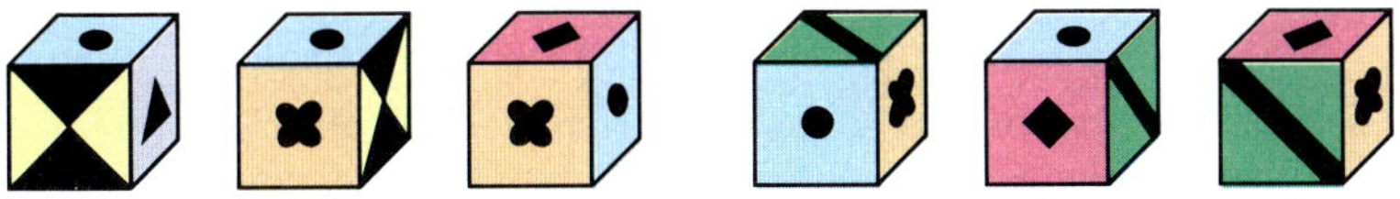

A *B* *C*

Lesson Follow-up

Reteaching Activity Give students three or four nets for various space figures. Have them count the number of faces and bases, and then write the shapes of the faces and bases. Students are to match this information with the definition of each figure to determine the name of the figure.

LESSON RESOURCES

TRB Practice J-7
TRB Enrichment J-8

Practice For use after 10-4 (pp. 420–423)

Space Figures

Name each space figure.

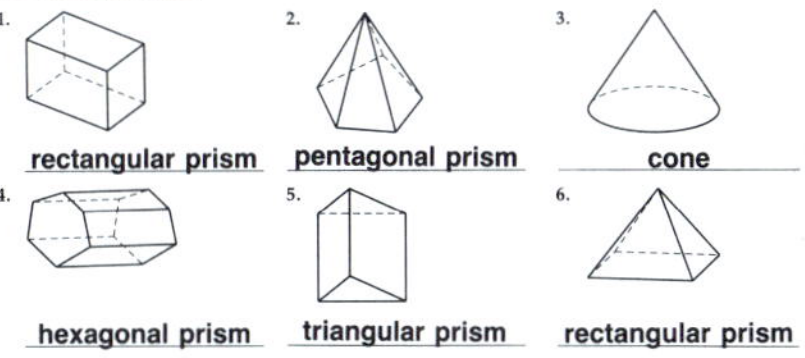

1. rectangular prism
2. pentagonal prism
3. cone
4. hexagonal prism
5. triangular prism
6. rectangular prism

Complete by naming the missing polygon.

7. Each side of a triangular prism is a **parallelogram**
8. The base of a heptagonal pyramid is a **heptagon**
9. Each side of a rectangle-based pyramid is a **triangle**
10. Each side of a trapezoidal prism is a **parallelogram**
11. Each base of a hexagonal prism is a **hexagon**

Name the space figure that can be formed from each net.

12. rectangular prism
13. hexagonal prism
14. pentagonal prism
15. Name the regular polyhedron that results if the bases of two square-based pyramids are placed together. **octahedron**
16. Name the regular polyhedron that is a pyramid. **tetrahedron**
17. Name the regular polyhedron that is a prism. **hexahedron**

Chapter 10 J-7

Enrichment For use after 10-4 (pp. 420–423)

Euler's Formula

The sides of a polyhedron are called *faces*. A segment common to two faces is an *edge*. A point where edges intersect is a *vertex*.

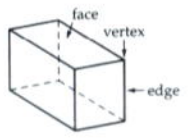

In the eighteenth century, Leonard Euler, a Swiss mathematician, discovered a remarkable relationship among the faces, edges, and vertices of any convex polyhedron. To discover the relationship, complete the chart for the given polyhedra.

	Figure	Number of faces (F)	Number of vertices (V)	Number of edges (E)
1.		5	5	8
2.	hexahedron	6	8	12
3.		7	10	15
4.	triangular pyramid	4	4	6
5.		9	9	16

6. Calculate $F + V - E$ for each polyhedron. What do you find? **In each case, the result is 2.**

Euler proved that your result is true for any convex polyhedron.

7. A convex polyhedron has 18 edges and 12 vertices. How many faces does the polyhedron have? **8**
8. A convex polyhedron has 10 faces and 16 vertices. How many edges does the polyhedron have? **24**

J-8 Chapter 10

LESSON QUIZ

1. A textbook resembles what polyhedron? (rectangular prism)
2. How many bases does a square pyramid have? (one)
3. If the bases of a hexagonal prism are regular polygons, what is true about the lateral faces? (They are congruent parallelograms or rectangles).
4. Name a solid figure that has a circular base. (cone or cylinder)

FOR THE NEXT LESSON

graph paper, math journal, centimeter or unit cubes.

Assignment Guide

Basic 1-6, 8-24, MR, CT All
Average 1-5, 7, 10-21, 25-26, 28, MR, CT All
Enriched 6-19, 24-28, MR, CT All

Lesson Focus

Materials

Graph paper, math journal, centimeter or unit cubes

MOTIVATING THE LESSON

Review the characteristics of a cube and have students tell how many faces (6), edges (12), and vertices (8) a cube has.

Instruction

Have students work in small groups. Encourage them to use unit cubes to model larger cubes with edges two units long and three units long in questions 2-3.

(1) Ask if any cubes would have four, five, or six sides painted. (no) Point out that the sum of the numbers in the four right columns must be the same as the total number of cubes.

(2) Ask why this number is always eight. (The only cubes with three sides painted are the corners.)

(3) If some groups finish early, ask them to make a chart that depicts the new results.

Closure

Have groups discuss and compare their results. If any group was able to determine a rule or a formula for a column in the chart in question 3, ask them to share it with the class.

Assignment Guide

Basic All
Average All
Enriched All

FOR THE NEXT LESSON

wood, stiff cardboard for a ramp, two cylinders with different diameters and the same height, string, weights, math journal

OBJECTIVE:
To explore spatial visualization and patterns in a cube.

MATERIALS

- Graph paper
- Math journal to record work

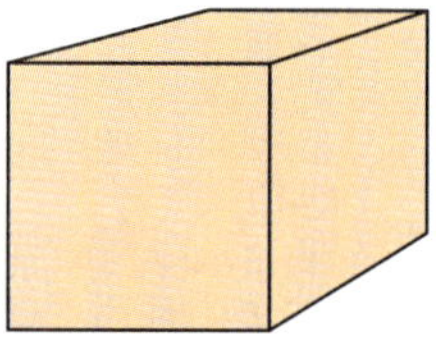

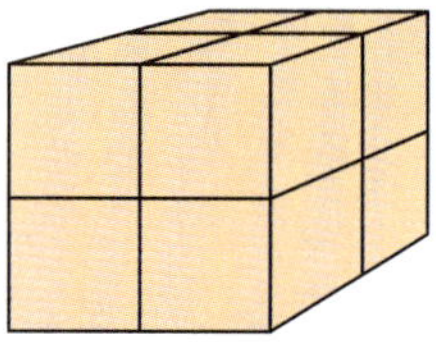

4. 0 column—cubic numbers; 1 column—6 times square numbers; 2 column—multiples of 12; 3 column—all the same.

Exploring Patterns in a Cube

You can explore the characteristics of three-dimensional objects by thinking about how they appear and by sketching them.

1. Consider a cube.
   a. How many faces does it have? 6
   b. You tie a string to the cube and dip it into paint. How many faces have you painted? 6
2. Suppose you cut the painted cube in half vertically and horizontally to make smaller cubes. Each edge is half the length of the original cube.
   a. How many smaller cubes do you have? 8
   b. How many cube faces have painted surfaces? 24
   c. How many faces of each of the smaller cubes would be painted? 3
3. Complete the chart and extend it for a cube cut into smaller cubes of equal size.

(1)

Number of Segments on Each Edge	Total Number of Cubes	Total Expressed as a Number Cubed	Number of Cubes with the Given Number of Sides Painted 0	1	2	3
2	8	2^3	0	0	0	8
3	27	3^3	1	6	12	8
4	64	4^3	8	24	24	8
5	125	5^3	27	54	36	8
6	216	6^3	64	96	48	8

4. ***Discuss*** the patterns you find in each column of the chart.
5. ***Predict*** the total number of cubes if there are 12 smaller cubes on each edge. 1,728 cubes
6. (2) There are 100 cubes on each edge. How many will be painted on three sides? 8
7. ***Predict*** the number that have two sides painted if there are 10 cubes on each edge. How many will have no paint at all? 96, 8^3 or 512
8. (3) ***Analyze*** Suppose you could paint only the sides of the cube that you can see without moving it. How many sides of the original cube would be painted? How will this change all of your results? More sides will have no paint; only 4 cubes will have 3 sides painted.
9. ***Write*** Was it necessary to draw a sketch of each cube? Explain when you were able to anticipate the results of the next row. Answers may vary.

Teaching Tip

Allow students to use manipulative materials as much as possible in this lesson. Constructing the cubes will help them with the spatial visualization needed to answer the questions.

PROBLEM SOLVING STRATEGY

OBJECTIVE:
To solve problems by making a model.

10-5 Make a Model

Sometimes solving a problem is made simpler by a model.

PROBLEM

A delivery person is unloading drums of oil along a ramp. In order to prevent the drums from being damaged, the ramp is not too steep. The more slowly a moving drum rolls, the less likely it is that it will be damaged. What kind of drum is better to use, one with a large or small radius?

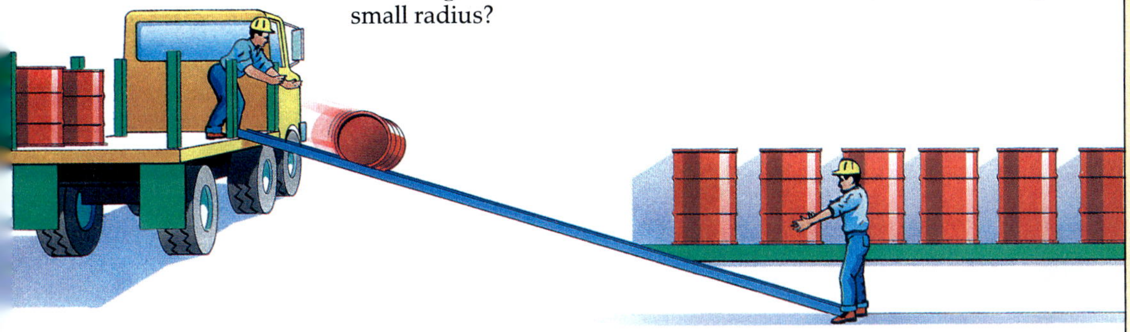

SOLUTION

READ What do you want to find? — which size of cylindrically shaped drum travels more slowly down an inclined plane

PLAN Use a wooden plank as an inclined plane.
(1) Take some cylinders such as an oatmeal box or the cardboard center from paper towels. Make sure that the cylinders have different radii but the same height. Roll the object down the inclined plane. Determine which cylinder takes longer to roll down the plane.

SOLVE What do you notice? — Cylinders with a greater radius take longer to travel down the plane.

LOOK BACK What can you conclude? — Cylindrical drums with a greater radius should be used because they travel down a ramp more slowly.

CLASS EXERCISES

1. The more area contacting the surface of the ramp, the more slowly the cylinder will roll.

Answer each question.

1. How might the area touching the ramp affect the speed at which a cylinder rolls down the ramp?

Lesson Focus

Materials
Wood, stiff cardboard for a ramp, two cylinders with different diameters and the same height, string, weights, math journal

Vocabulary/Symbols
inclined plane

MOTIVATING THE LESSON

Display two cylinders that have different bases but are the same length. Ask students to predict which one will roll down a ramp more slowly.

Instruction

Explain that performing the action in a problem using other materials is called modeling or simulating the problem.

(1) Use two cylinders made of similar materials to avoid having other variables, such as weight, affect the result of the experiment. Perform several trials using different pairs of cylinders.

Guided Practice

Class Exercises For question 1, increasing the diameter of a cylinder, but not its height, increases the area of the curved surface that rolls along the ramp. So, the increase in area might also cause it to roll more slowly.

A ***common error*** is to interpret the details of a problem incorrectly. Encourage students to reread the problem and list the details along with a quick sketch.

Closure

Ask students to describe the advantages of using a model to solve a problem, and discuss where problems might occur.

Teaching Tip

Display the problem you are working with on the overhead projector throughout the discussion, so that students may readily refer to the original problem.

EE: 1A, 1B, 1C, 1D, 1E; TAAS: 11, 13

Independent Practice

Have students make a model for question 1 by tying a weight to a length of string. They can observe the results by shortening or lengthening the string.

Lesson Follow-up

LESSON RESOURCES

TRB Practice J-9
TRB Enrichment J-10

Practice For use after 10-5 (pp. 425–426)

Make a Model

1. Eighty phone numbers are chosen at random from a telephone book.
   a. Guess approximately how many numbers will end in 3, 4, or 5.
   Answers may vary.
   b. Conduct an experiment, choosing 80 numbers from a phone book. How many end in 3, 4, or 5?
   Theoretically, 24, but answers may vary.
2. A narrow strip of paper is twisted once, then joined at the ends with glue or tape. The strip is then cut lengthwise along the dotted line shown.

   a. Guess the results.
   Answers may vary.
   b. Make and cut a model as directed. What are the results?
   A single loop results.
3. The midpoint of a segment is the point that divides the segment into two segments of equal length. A quadrilateral with unequal sides is drawn. The midpoints of the four sides are found and connected in order.

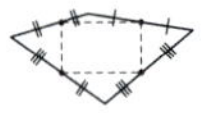

   a. Guess what kind of quadrilateral is formed.
   Answers may vary.
   b. Draw four quadrilaterals with unequal sides and connect the midpoints of adjacent sides. What kind of quadrilaterals appear to have been formed?
   parallelograms
4. A penny with Lincoln's head upright is rolled along the edge of another penny as shown in the figure.

   a. At the end, do you think Lincoln will be right-side-up or upside-down?
   Answers may vary.
   b. Conduct an experiment to find out. What are your results?
   Lincoln is upright.

Chapter 10 J-9

Enrichment For use after 10-5 (pp. 425–426)

Models of Polyhedra

You can use the patterns on this page to construct models of three of the five regular polyhedra.

1. Name the polyhedron that can be built from each pattern.
   a. tetrahedron
   b. octahedron
   c. dodecahedron
2. Name the polyhedra that cannot be built from the patterns.
   hexahedron, icosahedron
3. Cut out the patterns. Fold on the dotted lines and fasten the edges with tape to construct the models.

a.

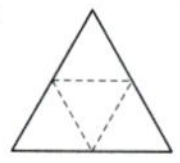

b.

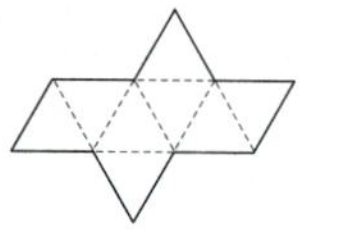

c.

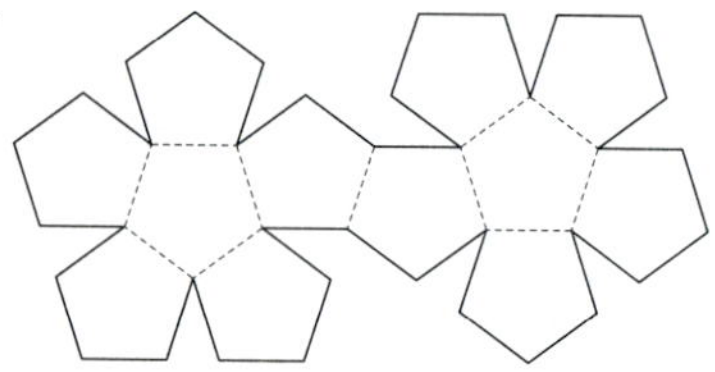

J-10 Chapter 10

2. How can you find the area of the part of the cylinder that touches the ramp? Use circumference of top as one dimension and height as other.
3. Use cylinders with the same radii but with different lengths. What affect does length have on the rate a cylinder travels down a ramp? The longer the cylinder is, the slower it is.

WRITTEN EXERCISES

Use a CALCULATOR where appropriate.

1. Use 2 strings of different lengths with equal weights on the end. Test the time of swings.

Explain how a model could be made for each situation.

1. You wish to find out the effect the length of a pendulum has on the amount of time the pendulum will swing.
2. You must decide which bridge design will be the strongest for a new bridge across Niagara Falls. Make models of designs and test for strength.

Use any strategy to solve each problem.

3. The junior class sold tickets for a pancake breakfast. One hundred twenty people came to the breakfast. This amount accounted for 60% of the tickets sold. How many tickets were sold? 200
4. The length of a rectangle is twice the width. The perimeter of the rectangle is 42 cm. What are the length and width? 7, 14
5. ***DATA FILE 2 (pp. 52–53)*** A treasure hunter must determine the depth of a sunken ship. She sends a sonar wave towards the location of the ship. It takes 9.2 s for the sound to return. How far beneath the ocean is the sunken ship? 7,042.6 m
6. The difference between two numbers is 18. The sum of the two numbers is 34. What are the two numbers? 26, 8
7. Eight teams are in a soccer tournament. When a team wins, it goes on to play another team. A team that loses is out of the tournament. How many games must be played in this tournament? 7
8. You fill a container $\frac{3}{4}$ full of water. The amount of water now in the container is 6 quarts. How much can the container hold? 8 qt
9. A number n is multiplied by $\frac{5}{8}$. Then the result is subtracted from $\frac{2}{3}$. The answer is $\frac{7}{12}$. What is n? $\frac{2}{15}$
10. Sara rented a car for two days. The rate was \$22.50 per day and \$.32 per mile. Sara traveled 150 mi. How much was she charged? \$93
11. Troy is digging post holes for his ranch. He has a triangular plot that is 100 yd by 300 yd by 250 yd. He digs a hole every 10 yd. How many holes must he dig? 65

MATH MINUTES

Have students make these paper models:
1. Make 5 folds that result in 6 sections. (accordion folds)
2. Make 5 folds that result in 32 sections. (Double the paper on each fold.)

Assignment Guide

Basic All
Average All
Enriched All

FOR THE NEXT LESSON

rectangular prism models, sheets of paper, math journal

426

OBJECTIVE: *To use a model to find surface area of prisms and cylinders.*

10-6 Surface Area—Prisms and Cylinders

▼ The packages for most food items are prisms and cylinders. The cost of the package is part of the price of the item. Manufacturers consider the surface area of a package when calculating the price.

Surface Area	Surface area (*SA*) is the sum of the areas of the base(s) and the side(s).

▼ You can use a net to help you find the surface area.

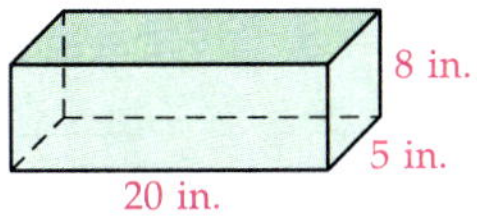

1

Example 1 **Find the surface area of the rectangular prism at the left using a net.**

Solution

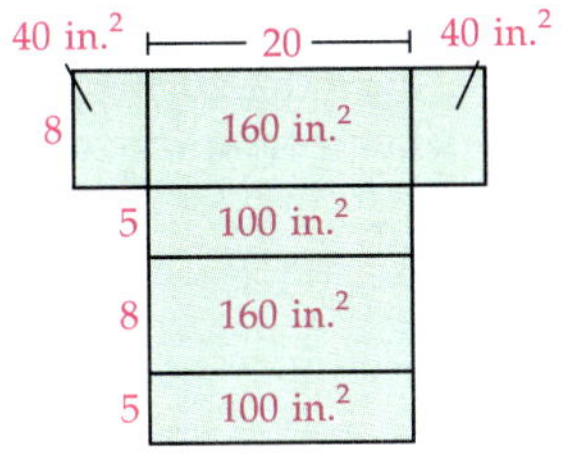

Draw and label a net.

Write the area of each rectangle on the net.

$40 + 40 + 160 + 100 + 160 + 100 = 600$ Add the areas.

The surface area is 600 in².

▼ You can use formulas to find the surface area.

Lateral Area of a Prism	The lateral area (*LA*) of a prism is the product of the perimeter of the base (*P*) and the height of the prism (*h*). $A = Ph$

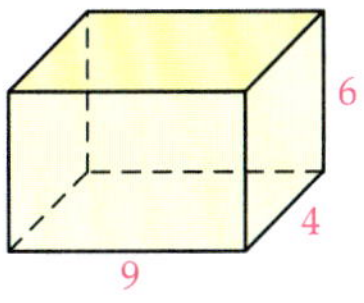

2

Example 2 **Find the surface area of the rectangular prism at the left using formulas.**

Solution

1. State each dimension. $b = 4, h = 6$; height of prism $= 9$
2. Find the perimeter. $P = 2(b + h) = 2(4 + 6) = 20$
3. Find the lateral area. $LA = Ph = 20 \cdot 9 = 180$
4. Find the base area. $A = bh = 4 \cdot 6 = 24$
5. Find the sum of the two base areas and the lateral area. $SA = 2 \cdot 24 + 180 = 228$

The surface area is 228 square units.

Lesson Focus

Materials

Rectangular prism models, sheets of paper, math journal

Vocabulary/Symbols

lateral area
surface area

MOTIVATING THE LESSON

Display an object that is a rectangular prism. Ask what comprises the surface of the prism. (the two bases and four lateral faces) Review the term lateral face by having students point out the lateral faces.

Instruction

Discuss the meaning of surface area of a prism and ask students to think of situations in which finding surface area is necessary. (Possible answer: to determine how much paint is needed for a room) Ask students to suggest how the surface area of the displayed object (prism) might be found. (Find the area of each face and then add the areas.)

1 Discuss that drawing a net is one way to account for all faces of a prism.

2 Mention that sometimes it is more convenient to use a formula to find the surface area. Draw the net of the figure in the text. Show students that the formula for lateral area comes from using the distributive property. The perimeter $(6 + 4 + 6 + 4)$ multiplied by the height, 9, is equivalent to finding the area of each face separately.

Teaching Tip

Start a poster that exhibits area formulas for polygons and surface area formulas for space figures. Have volunteers fill in formulas and draw an example of each. The poster will help students remember the formulas as needed.

3 Have students roll a sheet of paper to form a cylinder. They will see by unrolling the sheet that the length of one side is the circumference of the base.

ADDITIONAL EXAMPLES

1. Use a net to find the surface area. ($342\ cm^2$)

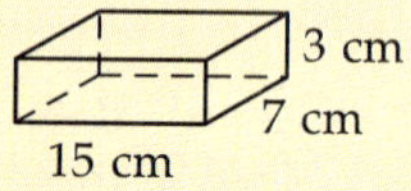

2. Use the formula to find the surface area of a prism whose base is 2 cm by 5 cm and whose height is 4 cm. ($76\ cm^2$)

3. Find the surface area of a cylinder with radius 3 in. and height 5 in. ($150.72\ in^2$)

Guided Practice

Class Exercises Have students work in small groups to ensure that they ask questions and compare their responses.

Think and Discuss Ask students to list two situations in their math journals for question 1.

For question 2, have volunteers use a model to show the two heights.

A ***common error*** is to omit a face or one of the bases when finding the surface area of a figure. After students have found the surface area of a figure, have them write what faces have been accounted for, next to their calculations.

Closure

Ask students to describe how the formula for surface area applies to rectangular prisms and cylinders. Then have them stand the two figures on a base and note similarities. (The figures have two bases; the lateral area involves the surfaces related to the height.)

NOTES & QUOTES

By the year 2,000 B.C., the Babylonians and Egyptians had fairly accurate calculations for π.

Babylonian value	$3\frac{1}{8}$
Egyptian value	$4\left(\frac{8}{9}\right)^2$

What is the decimal equivalent for each calculation of π? **3.125, 3.1604938 . . .**

▼ You can cut a label from a can of soup to see that it is a rectangle. The height of the rectangle is the height of the can. The base length of the rectangle is the circumference of the can. The area of the rectangle is the *lateral area* of the can.

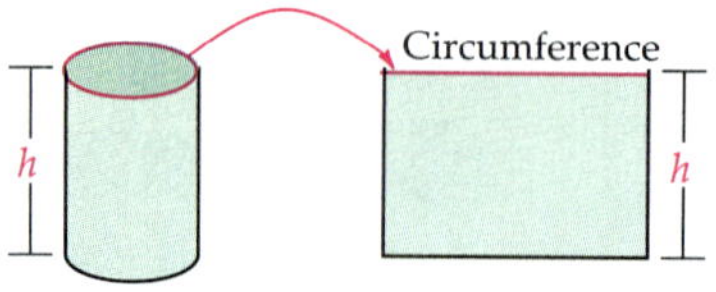

3

Lateral Area of a Cylinder The lateral area (*LA*) of a cylinder is the product of the circumference of the base (*C*) and the height of the cylinder (*h*).

$$A = Ch$$

Example 3 **Find the surface area of the oatmeal box at the left to the nearest square centimeter. Use 3.14 for π.**

Solution Draw and label a net.

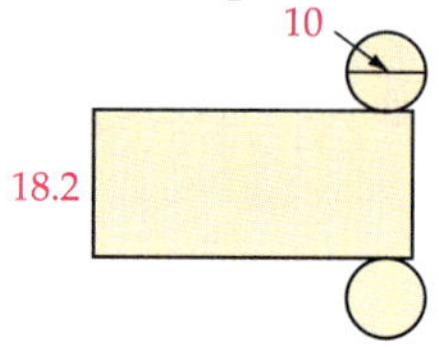

Area of bases	$= 2(\pi r^2) = 2(\pi \cdot 5^2) \approx 50(3.14)$	$= 157$
LA	$= Ch = \pi \cdot 10 \cdot 18.2 \approx 182(3.14)$	$= 571.48$
Total		$= 728.48$

The surface area of the oatmeal box is about $728\ cm^2$.

THINK AND DISCUSS

1. Name a real world situation where you need to know surface area.

2. Explain the difference between the height of a base and the height of a prism.

1. Answers may vary.
2. Height of base is only for the base surface. Height of prism is for the lateral surfaces.

CLASS EXERCISES

1. Find the surface area of a cube that is 10 ft on each edge. **600 ft^2**
2. Draw and label a net for a hexagonal prism with base edge 7 and height 13. **Check students' work.**
3. Draw and label a net and find the surface area for a cylinder with radius 8 and height 12. Use 3.14 for π. **1,004.8 sq. units**
4. Draw and label a net and find the surface area of a rectangular prism. The base is 3 in. × 5 in. and the height is 11 in. **206 $in.^2$**
5. Draw and label a net and find the surface area of a cylindrical water tank with radius 20 ft and height 30 ft. Use 3.14 for π. **6,280 ft^2**
6. Use formulas to find the surface area of a square prism with base edge 7 m and height 15 m. **518 m^2**

MATH MINUTES

Ask students to predict whether the surface area of a cylinder doubles if its height is doubled. (no) Then have them sketch a model, provide measurements, and test their predictions. Have students discuss their results and predictions.

WRITTEN EXERCISES

Find the surface area of the space figure shown in each net.

1. 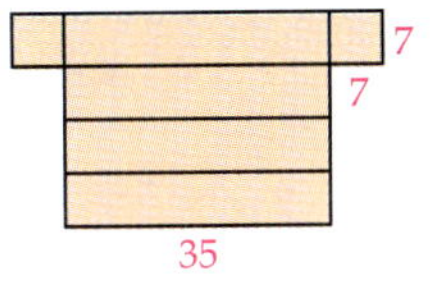

1,078 sq. units

2. 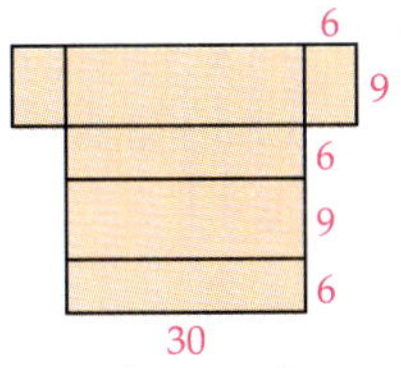

1,008 sq. units

3. 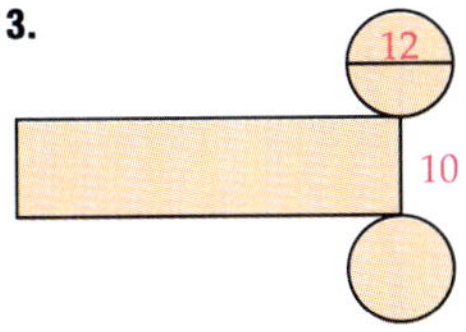

602.88 sq. units

CALCULATOR Find the surface area of each space figure.

4. 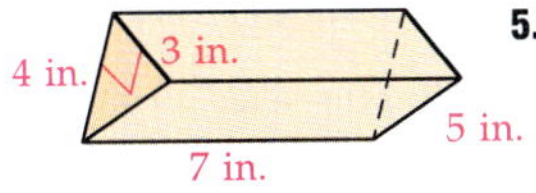

96 in.^2

5. 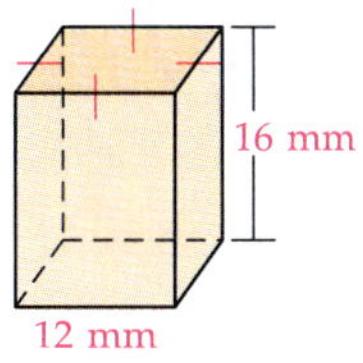

$1{,}056\text{ mm}^2$

6. 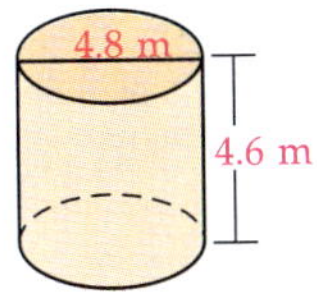

105.504 m^2

Find the surface area.

7. a triangular prism with all base edges 9 cm, base height 7.8 cm, and height of the prism 15 cm 475.2 cm^2
8. a cylinder with radius 8 cm and height 10 cm; Use 3.14 for π. 904.32 cm^2
9. Use cubes of base 1, 2, and 3 units to answer the following questions.
   a. Find the surface area of each cube. 6 sq. units, 24 sq. units, 54 sq. units
   b. If the length of the cube is doubled, the surface area is ■. quadrupled If the length is tripled, the surface area is ■. nine times larger
10. The neighborhood swimming pool needs to be resurfaced. The pool is 40 ft by 60 ft. The depth of the pool is 6 ft.
    a. How many sides need to be resurfaced? how many bases? 4, 1
    b. What is the total number of square feet to be resurfaced? $3{,}600\text{ ft}^2$
    c. The materials for resurfacing the pool cost $\$1.75/\text{ft}^2$. What is the cost of resurfacing the pool? $6,300
11. A cylindrical storage tank needs painting. The radius of the tank is 18 ft and its height is 30 ft. The paint covers $350\text{ ft}^2/\text{gal}$. How many gallons of paint are needed? Use 3.14 for π. 16 gal
12. Which cylinder at the right will be the more expensive to paint? Explain your answer. *A* has area 282.6 ft^2; *B* has area 150.72 ft^2. *A* takes more paint.

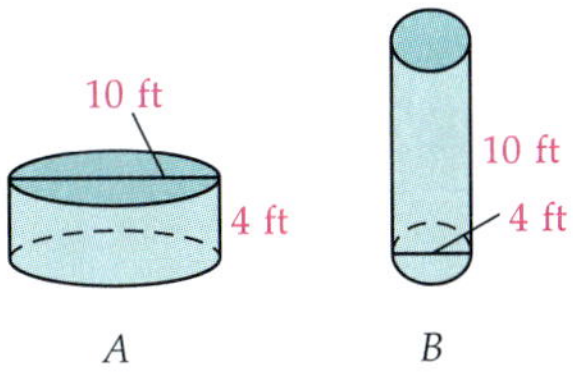

13. **PROJECT** Draw and label a net and find the number of square feet in the surface area of a room in your home. Answers may vary.

MIXED REVIEW

1. $(x)(2x)(3x)$ $6x^3$
2. $a(4a + 7)$ $4a^2 + 7a$
3. Solve for x. $\frac{3x}{4} = \frac{5}{6}$ $1\frac{1}{9}$
4. $\sqrt[2]{144}$ 12
5. 16^2 256

Find the area.

6. circle, $r = 12$ 144π or 452.16
7. parallelogram, $b = 8$ and $h = 14$ 112
8. ***DATA FILE 2 (pp. 52–53)*** Express the sound intensity of a rocket launch in bels. 1,500 bels
9. The wheels of a racing bike are about 70 cm in diameter. What is the circumference of a wheel? 219.8 cm

Independent Practice

For exercise 10, have students use an open box as a model to see which sides should be resurfaced.

Lesson Follow-up

Reteaching Activity Provide selected figures and their nets for students to measure and find surface areas.

LESSON RESOURCES

TRB Practice J-11
TRB Enrichment J-12

Practice

For use after 10-6 (pp. 427–429)

Surface Area—Prisms and Cylinders

Name the polyhedron and find the surface area of each net.

1. 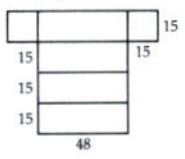

rectangular prism
3,330 sq. units

2. 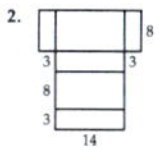

rectangular prism
356 sq. units

3. 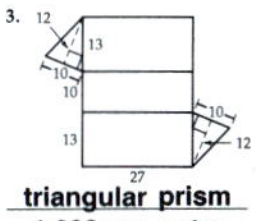

triangular prism
1,092 sq. units

Draw and label a net and find the surface area. Use 3.14 for π.

4. 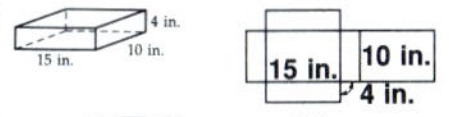

surface area = 500 in.^2

5. 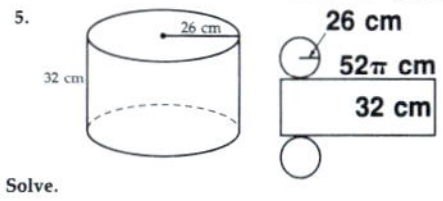

surface area = $947{,}024\text{ cm}^2$

Solve.

6. A room is 18 ft long, 14 ft wide, and 8 ft high.
   a. Find the cost of painting the four walls with two coats of paint costing \$9.50 per gallon. Each gallon covers 256 ft^2. $38
   b. Find the cost of carpeting the floor with carpet costing $\$3/\text{ft}^2$. $756
   c. Find the cost of covering the ceiling with acoustic tile costing $\$7.50/\text{ft}^2$. $1,890
   d. Find the total cost of renovating the room. $2,684

Chapter 10 J-11

Enrichment

For use after 10-6 (pp. 427–429)

Packing Pineapples

Sam Edwards is a packer for Peterson's Pineapples. Sam's job is to package pineapples for the lowest price possible.

Pineapples are packed in cubical packages measuring 1 ft on a side.

Sam's present assignment is to gift-wrap eight pineapples in wrapping paper costing $\$.032/\text{ft}^2$.

1. Suppose he packages the pineapples in eight separate packages.
   a. Find the total surface area of the packages. 48 ft^2
   b. Find the cost to wrap the packages. $1.54

2. Suppose he wraps four packages each containing two pineapples.
   a. Find the total surface area of the packages. 40 ft^2
   b. Find the cost to wrap the packages. $1.28

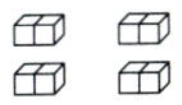

3. a. Find the surface area and cost if Sam wraps a single package in the following configurations.

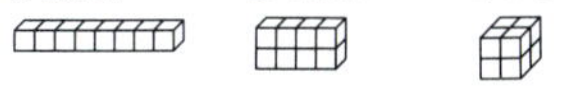

	a. 8 × 1 × 1	b. 4 × 2 × 1	c. 2 × 2 × 2
surface area	34 ft^2	28 ft^2	24 ft^2
cost	$1.09	$.90	$.77

   b. How much would he save by sending the pineapples in a 2 × 2 × 2 package instead of an 8 × 1 × 1 package? $.32

4. Find the surface area and cost of wrapping 27 pineapples in the following configurations. Paper costs $\$.027/\text{ft}^2$.

	a. three 3 × 3 × 1 packages	b. one 3 × 3 × 2 and one 3 × 3 × 1 package	c. one 3 × 3 × 3 package
surface area	90 ft^2	72 ft^2	54 ft^2
cost	$2.43	$1.94	$1.46

5. Find the lowest possible cost of wrapping 64 apples in paper costing $\$.035/\text{ft}^2$. $3.36 (one 4 × 4 × 4 package)

J-12 Chapter 10

LESSON QUIZ

Find the surface area.

1. 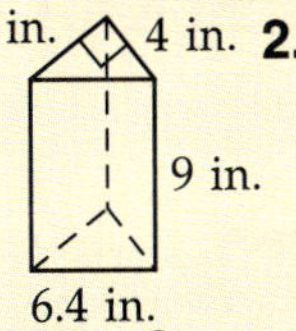

(158.6 in^2)

2. 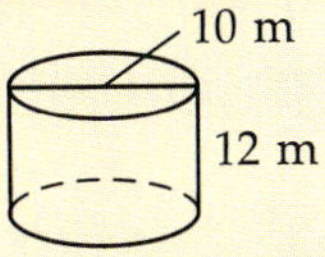

(533.8 m^2)

3. A rectangular prism measures 4 ft by 3 ft by 7 ft. What is the surface area? (122 ft^2)

Assignment Guide

Basic 1, 3-5, 7-8, 10-11, 13, MR All
Average 2-4, 6-7, 9-11, 13, MR All
Enriched 6-13, MR All

FOR THE NEXT LESSON

ruler, pencil, scissors, tape, TRB Teaching Resources, compass, one orange and knife (optional), math journal

Lesson Focus

Materials

Ruler, pencil, scissors, tape, TRB Teaching Resources, compass, one orange and knife (optional), math journal

Vocabulary/Symbols

slant height

MOTIVATING THE LESSON

Give each student (or pair of students) a copy of the net for a square pyramid (TRB Teaching Resources). Ask them to identify the figure the net forms and to list some characteristics. (Possible answer: the lateral faces are four congruent triangles.)

Instruction

Ask students how the surface area of the net might be found. Then have them measure to the nearest tenth of a centimeter and find the total surface area. When they finish, have them cut out the net and tape it together to make the pyramid. Point out that the height of each triangle becomes the slant height of the pyramid, and the area of the four faces is the lateral area.

1 Have students use a compass to draw a sector of a circle that is smaller than a semicircle. Students then cut out the sector and hold it together to form a cone. Have them look at the circular base and then open the cone to demonstrate that the curved edge is the circumference of the base. Relate the formula for lateral area to students' models.

2 Remind students that the surface area is an approximation because 3.14 is an approximation for pi.

OBJECTIVE: *To use a model to find surface area of pyramids, cones, and spheres.*

10-7 Surface Area—Pyramids, Cones, Spheres

▼ To find the surface area of pyramids and cones, you must use the height of a face, called the *slant height* (l), to find the area of the lateral faces.

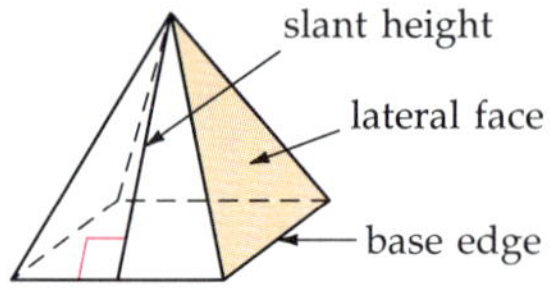

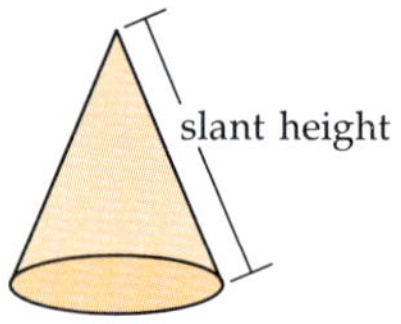

▼ You can draw a net to help find the surface area of a pyramid.

Example 1 A pyramid has a base edge of 720 ft. The slant height is 584 ft. Find the surface area.

Solution

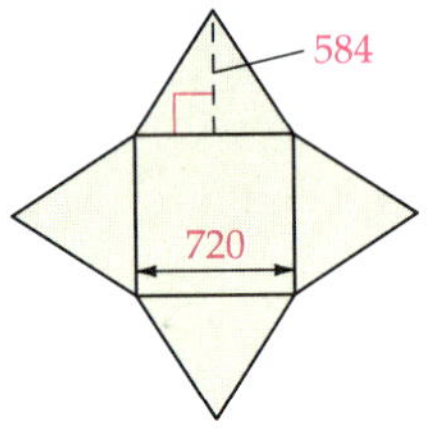

Draw and label a net.

$$\text{Area of lateral faces} = 4\left(\tfrac{1}{2}bl\right)$$
$$= 4\left(\tfrac{1}{2} \cdot 720 \cdot 584\right) = 840{,}960 \text{ ft}^2$$
$$\text{Area of base} = (bh) = (720 \cdot 720) = 518{,}400 \text{ ft}^2$$
$$\text{Total} = 1{,}359{,}360 \text{ ft}^2$$

The surface area is 1,359,360 ft^2.

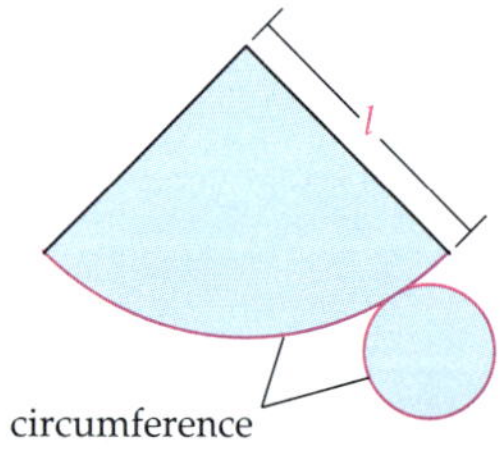

▼ The curved surface of a cone is its *lateral area.* A cone that is cut and flattened may remind you of a triangle with the height of l and the base equal to the circumference of the circular base. If you substitute C for b and l for h in the triangle area formula, the result is the formula for the lateral area of a cone.

$$\text{Triangle Area} = \tfrac{1}{2}bh$$

$$\text{Lateral Area} = \tfrac{1}{2}Cl$$

Lateral Area of a Cone	The lateral area (LA) of a cone equals half the product of the circumference (C) and slant height (l). $LA = \frac{1}{2}Cl$

1

Teaching Tip

Have students sketch geometric figures and nets on graph paper. Using graph paper will help students draw the figures correctly and label heights and other dimensions more easily.

Example 2 **Find the surface area of the cone. Use 3.14 for π.**

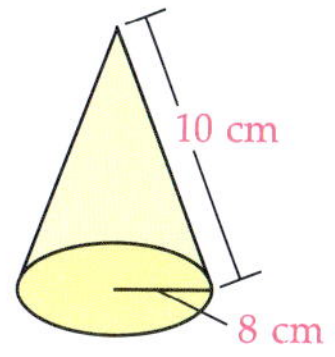

Solution Draw and label a net.

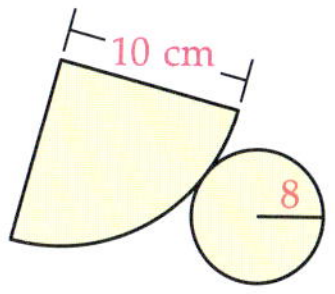

Lateral area	$= \frac{1}{2}Cl$	$= \frac{1}{2} \cdot 16\pi \cdot 10$	$\approx 80(3.14)$	$= 251.2$	
Base area	$= \pi r^2$	$= \pi \cdot 8^2$	$\approx 64(3.14)$	$= 200.96$	
Total				$= 452.16 \text{ cm}^2$	

The total surface area of the cone is about 452.16 cm^2.

FLASHBACK
$C = \pi d$

▼ Many sports, including basketball, tennis, soccer, and golf rely on spheres.

Surface Area of a Sphere The surface area of a sphere equals the product of 4π and the square of the radius (r).

$$A = 4\pi r^2$$

Example 3 **Calculate the surface area of a basketball. Use 3.14 for π.**

Solution

$$\begin{aligned} A &= 4\pi r^2 \\ &= 4\pi(12)^2 \\ &\approx 576(3.14) \\ &= 1{,}808.64 \end{aligned}$$

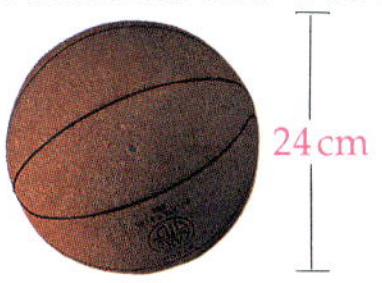

The surface area of a basketball is about $1{,}808.64 \text{ cm}^2$.

CLASS EXERCISES

Check students' nets.

Draw and label a net and find the surface area. Use 3.14 for π.

1.

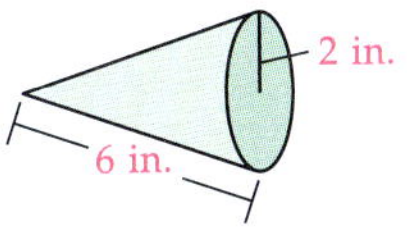

50.24 in.^{2}

2.

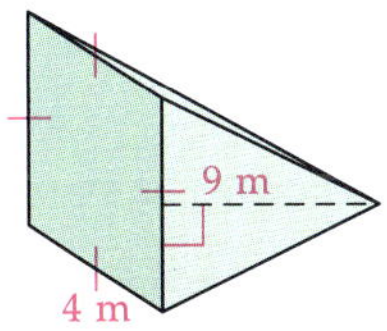

88 m^2

3.

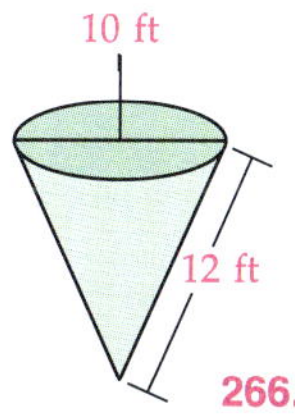

266.90 ft^2

4. Find the surface area of a cone with radius 3 and slant height 8. Use 3.14 for π. 103.62 sq. units

5. Find the surface area of a square pyramid with base 5 and slant height 8. 105 sq. units

THINK AND DISCUSS

1. Name two uses of cones. Answers may vary.

2. How is the slant height of a pyramid different from the height of a prism?

3. Could you find the lateral area of a pyramid in a way that is similar to finding the lateral area of a cone? Explain. $\frac{1}{2}Pl$

2. Slant height is height of a face. Height of prism is through the center of the figure.

3 Find the surface area of an orange. Have a volunteer find the radius of the orange. Then score the peel, cut it into eight triangles, give groups of students one triangle to measure and find the area. Show that the sum of the areas of the triangles is close to the surface area calculated by using the formula.

ADDITIONAL EXAMPLES

Find the surface area.

1. A square pyramid with a base edge of 5 cm and slant height of 4 cm (65 cm^2)

2. A cone with a radius of 3 ft and a slant height of 7 ft (94.2 ft^2)

3. Marble with a diameter of 12 mm (452.16 mm^2)

Guided Practice

Class Exercises Have students work in small groups and sketch the geometric figures to help them visualize each problem.

Think and Discuss Have students make nets to help them answer questions 1-2.

A ***common error*** is to omit the area of the base when finding the surface area of a pyramid or a cone. In Example 2, students find the lateral area of the cone and give as the answer 251.2 cm^2. When finding the surface area of any figure, have students ask these questions and write the answers as notes: Does the figure have any bases? How many? Are there lateral faces? What shape and how many of each are there?

Closure

Writing in Math Have students describe in their math journals how to find the surface area of a pyramid and a cone, and compare the processes to those of finding the surface area of prisms and cylinders.

MATH MINUTES

Have students draw rectangular and square pyramids or their nets and describe how finding their surface areas differ. (four congruent lateral faces versus two pairs of congruent lateral faces)

Independent Practice

For questions 4-14, suggest that students sketch the figure or the net and label it.

Lesson Follow-up

Reteaching Activity Provide selected figures and their nets for students to measure and find surface areas.

LESSON RESOURCES

TRB Practice J-13
TRB Enrichment J-14

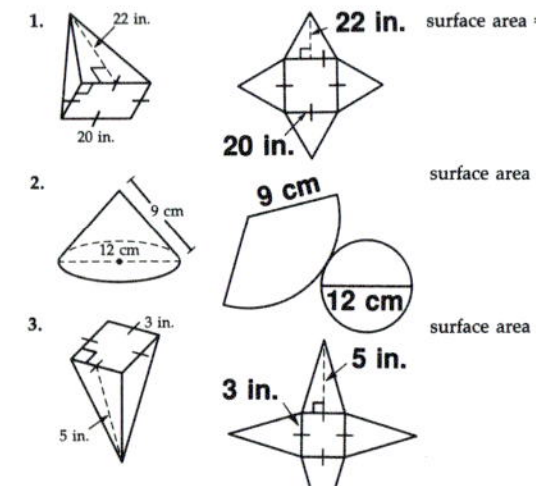

Practice For use after 10-7 (pp. 430–432)

Surface Area—Pyramids, Cones, and Spheres

Draw and label a net and find the surface area. Use 3.14 for π.

1. (22 in., 20 in.) surface area = 1,280 in.^{2}

2. (9 cm, 12 cm) surface area = 282.6 cm^2

3. (3 in., 5 in.) surface area = 39 in.^{2}

Find the surface area. Give an exact answer.

4. A cone $d = 36$ in., $l = 50$ in. 1,224π in.^{2}
5. A square-based pyramid $s = 8$ in., $l = 12$ in. 256π in.^{2}
6. A cone $r = 64$ cm, $l = 90$ cm 9,856π cm^2
7. A sphere with diameter 44 cm 1,936π cm^2
8. A hemisphere with radius 5 in. 50π in.^{2}
9. A sphere with radius 27 in. 2,916π in.^{2}
10. A hemisphere with diameter 70 cm 2,450π cm^2
11. A cone and a square-based pyramid have slant heights of 6 in. The diameter for the cone and the base edge of the pyramid are both 8 in.
    a. Which space figure has the greater surface area? pyramid
    b. By how much does the surface area of the greater space figure exceed that of the smaller? Use 3.14 for π. 34.4 in.^{2}

Chapter 10 J-13

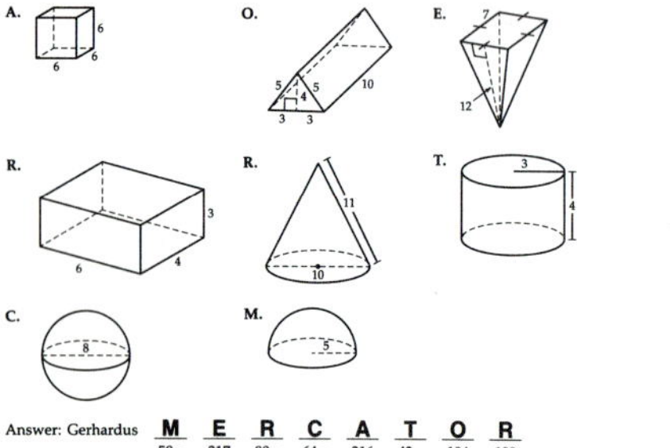

Enrichment For use after 10-7 (pp. 430–432)

Flattening the Earth

Unlike other space figures, a sphere cannot be drawn as a net. (Think of how an orange peel splits if you try to flatten it.) This poses a dilemma for mapmakers trying to draw the spherical Earth on flat paper. No matter how they draw the map, it will suffer from distortion.

In 1569, a Flemish geographer drew a "flat" Earth map with its distortions confined to the polar regions, where they were of little consequence. His map is still widely used today. To learn his name, find each surface area. Write the letter of the exercise above the correct surface area in the grid at the bottom of the page.

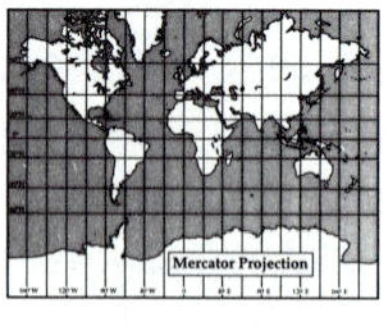

A. (6, 6, 6) O. (5, 5, 4, 3, 3, 10) E. (7, 12)

R. (6, 4, 3) R. (11, 10) T. (3, 4)

C. (8) M. (5)

Answer: Gerhardus

M	E	R	C	A	T	O	R
50π	217	80π	64π	216	42π	184	108

J-14 Chapter 10

8. Both figures have a hexagon for a base. The sides of the prism are rectangles. The sides of the pyramid are triangles.

MIXED REVIEW

Solve.

1. $2x - 5 = 39$ 22
2. $18 - 3x = 42$ -8
3. ***DATA FILE 5 (pp. 180–181)*** The list price on a CD is $11.97. How much will the artist receive in royalties, to the nearest cent, on each CD sold? $2.99
4. $\frac{x}{12} = \frac{4}{15}$ $3\frac{1}{5}$

Find the perimeter.

5. a regular hexagon with side 13.2 cm 79.2 cm
6. an equilateral triangle with side 8.6 in. 25.8 in.
7. Use scientific notation to find $(3.2 \times 10^4)(3.2 \times 10^4)$.
8. Explain how a hexagonal pyramid and a hexagonal prism are alike and how they are different. See above.

7. 1.024×10^9

6. Which has the greater surface area, a cylinder with radius 2 and height 2 or a sphere with radius 2? The surface areas are the same.
7. Which has the greater surface area, a square prism with base edge 4 and height 5 or a square pyramid with base edge 4 and slant height 5? The prism has greater surface area.
8. Find the surface area of a tennis ball with diameter of 2.5 in. 19.625 in.^{2}

WRITTEN EXERCISES

Draw and label a net and find the surface area. Use 3.14 for π.

1. 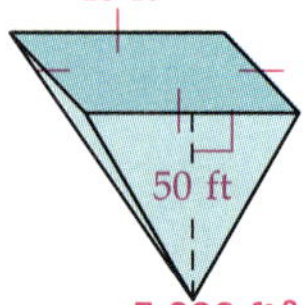

5,600 ft^2

2. 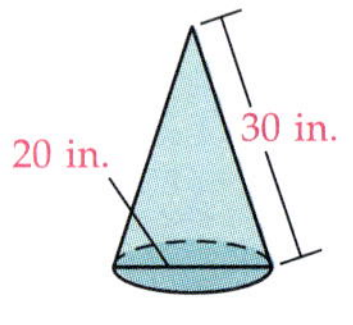

1,256 in.^{2}

3. 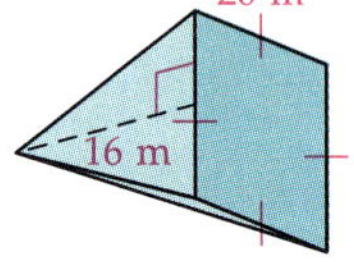

1,040 m^2

4. square pyramid, $s = 35$ in. $l = 42$ in. 4,165 in.^{2}
5. cone, $r = 14$ m $l = 25$ m 1,714.44 m^2
6. cone, $d = 22$ cm $l = 34$ cm 1,554.3 cm^2

CALCULATOR **Find the surface area. Use 3.14 for π.**

7. a sphere with radius 15 in. 2,826 in.^{2}
8. a sphere with diameter 18 m 1,017.36 m^2
9. a hemisphere (half of a sphere) with radius 27 m 4,578.12 m^2
10. a hemisphere with diameter 42 cm 2,769.48 cm^2
11. Find the surface areas of a sphere, $r = 2$, and a sphere, $r = 5$. How do the ratios of the surface areas compare with the ratios of the radii? radii $\frac{2}{5}$; areas $\frac{4}{25}$. Ratio of areas is the square of the ratio of the radii.
12. Which is greater, the surface area of a cone with radius 5 and slant height 10 or the surface area of a sphere with radius 5? The sphere has twice as much area as the cone.
13. A spherical satellite is 3 m in diameter. What is the approximate area of the material covering its surface? 28.26 m^2
14. Find the surface area of a triangular pyramid that has a base congruent to the lateral faces. The base edge is 24 cm and the slant height is 16 cm. 768 cm^2

Earth 197,289,301 mi^2; moon 14,649,984 mi^2

15. ***PROJECT*** Research the radius of both Earth and the moon. Calculate the surface area of each.
16. Spaceship Earth at Epcot Center in Florida is a 180-ft geosphere. Estimate the approximate surface area by assuming it is a sphere with diameter 180 ft. 101,736 ft^2

LESSON QUIZ

Find the surface area.

1. A cone with a radius of 3 in. and a slant height of 5 in. (75.36 in.^{2})
2. A sphere with a diameter of 40 cm (5,024 cm^2)
3. About how many square yards of material are needed to make the sides and bottom of a tent that is a square pyramid with a base 2 yd and slant height 2 yd? (12 yd^2)

Assignment Guide

Basic 1-2, 4-5, 7, 9, 13, 15-16, MR All
Average 2-3, 5-6, 8-9, 12, 14-16, MR All
Enriched 7-16, MR All

FOR THE NEXT LESSON

calculator

PROBLEM SOLVING APPLICATION

OBJECTIVE:
To use area formulas to plan parade floats.

10-8 Parade Floats

Thousands of dollars are spent each year to construct the spectacular floats that appear in the Rose Bowl Parade. Many months of planning and building go into each one of these elaborate floats.

Example Sunnydale Farms is going to have a float in a parade. On the float they plan to have a large sun symbol like the one at the left. The interior circle will have a diameter of 4 ft. Each triangle will have a height of 4 ft. The base will be 1.25 ft. What will be the total area of the sun symbol?

Solution Use the formulas for the area of a circle and the area of a triangle to find the total area of the sun symbol. Use 3.14 for π.

1

$A = \pi r^2$ Find the area of a circle.
$\approx 3.14 \cdot 2^2$
$= 3.14 \cdot 4$
$= 12.56 \text{ ft}^2$

$A = \frac{1}{2}bh$ Find the area of a triangle.
$= \frac{1}{2} \cdot 1.25 \cdot 4$
$= 2.5 \text{ ft}^2$

$2.5 \cdot 10 = 25 \text{ ft}^2$ Find the area of 10 triangles.
$12.56 + 25 = 37.56 \text{ ft}^2$ Find the total area.

The total area of the sun symbol is 37.56 ft^2.

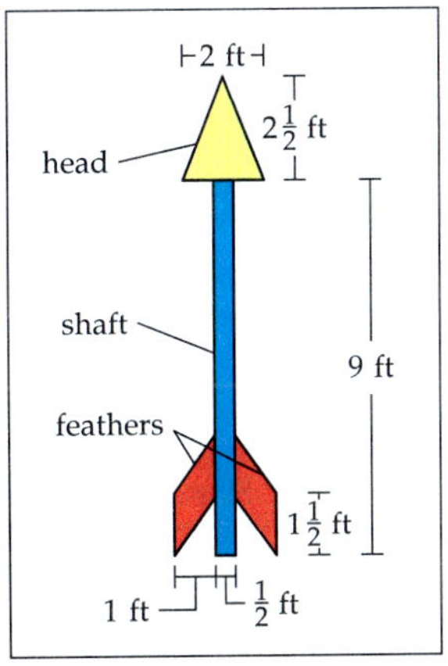

CLASS EXERCISE

Use the *DATA* in the figure at the left to solve.

1. A float will have a child holding a large arrow. All parts of the arrow will be made from heavy cardboard.
   a. What is the area of the head of the arrow? 2.5 ft^2
   b. What is the area of the shaft of the arrow? 4.5 ft^2
   c. What is the area of each section of feathers in the arrow? 1.5 ft^2
   d. What is the total area of the arrow? 10 ft^2
   e. The cardboard costs $\$.27/\text{ft}^2$. What is the cost to build the arrow? $2.70
   f. Covering the arrow with flowers will cost $\$3.60/\text{ft}^2$. What is the cost of the flowers? What is the total cost? $36; $38.70

Teaching Tip

Ask volunteers to use graph paper to model problems about area. Use the models to provide a visual representation of the problems as well as to check answers.

Lesson Focus

Materials
Calculator

MOTIVATING THE LESSON

Begin by explaining that the Rose Bowl parade takes place in Pasadena, California. It is part of the festivities whose highlight is the annual New Year's Day Rose Bowl football game. The game is played by the two leading college football teams.

During the week before the parade it is possible to visit the site where the floats are being prepared.

Instruction

1 Ask a volunteer to record the facts on the chalkboard as the class reads through the problem. Have the student sketch the circle and the triangle, and label them appropriately. Have students note that the number of triangles is not given in the problem, but can be determined by counting the number in the illustration.

Guided Practice

Class Exercises Suggest that students change fractions in the diagram to decimals and use a calculator to find the answers. As the area of each part of the arrow is found, it can be added to memory.

For question 1e, students can recall the total area from memory and multiply by 0.27 to find the cost to build the arrow.

For question 1f, students can recall the total area and multiply by 3.60 to find the cost of the flowers.

Closure

List on the chalkboard the area formulas for each of the figures used in the lesson. Have students discuss how each formula reflects the general idea that length times width results in the area of a figure. If necessary, recall that the parts of a circle can be rearranged to form a parallelogram.

Independent Practice

You may wish to challenge students to predict, before they calculate, whether or not the heart will be covered with roses on both sides. Have pairs of students work together to estimate the answer.

Decision Making Have students form groups. Suggest roles such as Designer, Survey Taker, Mathematician, Cost Analyst. You may wish to have groups compare quotes for the cost of building materials and discuss which suppliers provide the best buy.

WRITTEN EXERCISES

Use a CALCULATOR where appropriate.

Use the *DATA* below to solve.

1. A float will have a large heart made of plywood. To make the heart, builders will take a circle with a diameter of 5 ft and cut it in half. They will then attach both pieces to one side of an equilateral triangle. The triangle has height 8.6 ft and side 10 ft. The builders will cover the heart with red roses. To control expenses, they do not want the area covered to exceed 275 ft^2.

a. Make a sketch of the way the heart will look when it has been constructed as described. Include measurements in your sketch. **Check students' work.**

b. How many square feet of plywood will have to be covered with roses if both sides are covered? Will this be within the limit of 275 ft^2? **125.25 ft^2; yes**

c. The cost of covering the float with red roses is \$10/ft^2. What is the cost for covering one side of the heart with flowers? both sides? **\$626.25; \$1,252.50**

Decision Making* DECISION MAKING *Decision Making* *Decision Making* *Decision Maki

PARADE FLOATS

COLLECT DATA

1. You are on the committee to design a float for your school's homecoming parade. You want an imaginative design that is simple to build. You also want to use real flowers.

a. Decide what kind of float platform you will use. Find the dimensions of the platform.

b. Sketch several figures that you could build for the float.

c. Find the cost of a square foot of building materials.

d. Survey local florists. Find the cost of various flowers. Determine which flowers will stay fresh the longest. Record your results in a chart like the one below.

Types of Flowers	Number of Blossoms to Cover 1 ft^2	Life Expectancy of Flowers
Roses		
Daisies		
Carnations		

MATH MINUTES

Have students form groups and write problems that members of other groups can solve mentally. Example: Estimate the area of the largest circle that can fit inside an 18 ft by 14 ft rectangle. (about 150 ft^2)

Use the ***DATA*** below to solve.

2. Every year near the end of November, Chicago has a parade. A department store plans to build a large gift box for the middle of its float. The gift box will measure 6 ft by 9 ft by 5 ft. The platform of the float will be the bottom of the box. This will save the cost of materials for one side of the box.

a. Make a sketch of the gift box that the store plans to build. Include measurements. **Check students' work.**

b. To save as much building material as possible, which side of the box should be the bottom? Give its length, width, and area. **9 ft × 6 ft; 54 ft^2**

c. What will be the total area of the sides that need to be decorated? **204 ft^2**

d. Building materials cost \$.55/ft^2. What is the cost of building the box? **\$112.20**

e. The cost of decorations is \$2.25/ft^2. What is the cost of decorating the gift box? **\$459.00**

f. What is the total cost of the gift box? **\$571.20**

Decision Making ■ Decision Making ■ Decision Making

ANALYZE DATA

2. Calculate the areas of the shapes you plan to build. Find the cost of the building materials.

3. Calculate the areas of the shapes you plan to cover with each type of flower.

4. Calculate the cost of flowers for each part of the float you are designing.

5. Find the total cost of the design.

MAKE DECISIONS

6. What is the cost of decorating your float? What changes can you make to lower the cost?

7. Suppose you have an unlimited budget. What design would you use for your float? What would be the total cost?

Lesson Follow-up

LESSON RESOURCES

TRB Alternate Application J-15

Assignment Guide

Basic All
Average All
Enriched All

FOR THE NEXT LESSON

centimeter or unit cubes, graph paper, ruler, empty cans with labels, math journal

Alternate Application

For use after 10-8 (pp. 433–435)

Minimizing Packaging Costs

A designer has been hired to design a new container for a laundry soap manufacturer. The container must be a rectangular prism with volume 216 in^3.

The designer begins by considering four possible shapes.

1. To minimize packaging costs, the designer will use the container with the least possible surface area. Find the height and surface area of each container.

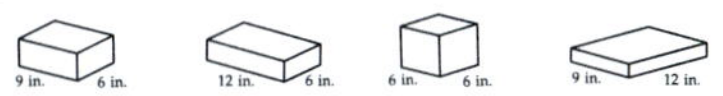

a. **4 in.** **228 in.^{2}** b. **3 in.** **252 in.^{2}** c. **6 in.** **216 in.^{2}** d. **2 in.** **300 in.^{2}**

2. Which container should the designer choose? **c**
3. Each of these containers has volume 1,000 cm^3. Find the surface area of each.

a. **730 cm^2** b. **780 cm^2** c. **600 cm^2** d. **850 cm^2**

4. Answer true or false: If two containers have the same volume, then they have the same surface area.
false
5. What shape should a manufacturer choose in order to minimize the surface area of a rectangular-prism-shaped container for a given volume?
cube
6. For each volume, give the dimensions of the container with the minimum surface area.
a 8 ft^3 **2 ft × 2 ft × 2 ft**
b. 125 in.^{3} **5 in. × 5 in. × 5 in.**
c. 512 cm^3 **8 cm × 8 cm × 8 cm**

Chapter 10 J-15

Career Information

Arrange for an operations research analyst to visit the class. Have students prepare questions to ask about the speaker's background, experience, and use of mathematics.

Ask the speaker to discuss the following topics:

- educational background required for the position of operations research analyst
- explanation of the duties and responsibilities of his or her position
- the type of computer hardware and software programs used to assist in making changes in the organization
- how the responsibilities and knowledge required differs among industries, for example, service industries versus manufacturing industries

Practice

CAREER

Help Wanted: Operations Research Analyst

Bachelor's degree in business, mathematics, or computer science required.

For more information, write to Operations Research Society of America, Mount Royal and Guilford Ave., Baltimore, MD 21202.

Operations research analysts work to improve the efficiency of organizations. Responsibilities include inventory control, distribution systems, and adjusting personnel schedules. Operations research analysts construct mathematical models to represent situations. Then they do research and computer analysis in order to recommend policies that will benefit the company.

PROJECT

Call a hospital. Find out how an operations research analyst handles scheduling patient admissions.

Find each area. Round to the nearest tenth.

1. trapezoid: $b_1 = 2.5$ m; $b_2 = 3.4$ m; $h = 9.9$ m **29.2 m²**
2. triangle: $b = 5$ in.; $h = 7.2$ in. **18 in.²**
3. rectangle: $b = 3\frac{1}{2}$ ft; $h = 8\frac{3}{4}$ ft **30.6 ft² or $30\frac{5}{8}$ ft²**
4. square: $s = 87\frac{7}{8}$ cm **7,722.0 cm²**
5. parallelogram: $b = 4.8$ yd; $h = 9.2$ yd **44.2 yd²**
6. circle: $d = 6.5$ cm **33.2 cm²**
7. trapezoid: $b_1 = 3.8$ m; $b_2 = 5.2$ m; $h = 2,800$ cm **126 m² or 1,260,000 cm²**
8. circle: $r = 3.25$ cm **33.2 cm²**

9. 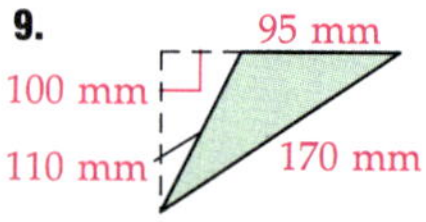

4,750 mm²

10. 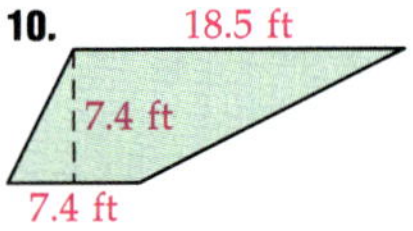

95.8 ft²

11. 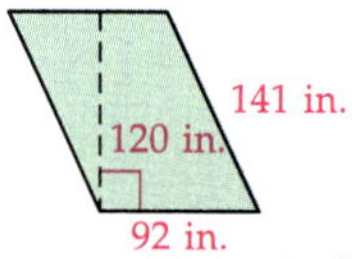

11,040 in.²

Name the space figure that each net can form.

12. 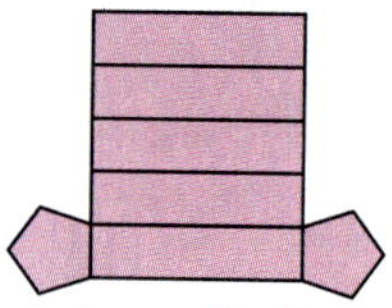
pentagonal prism

13. 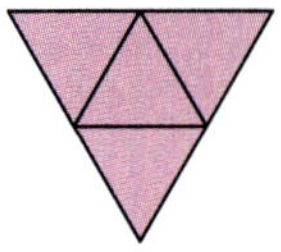
triangular pyramid

14. 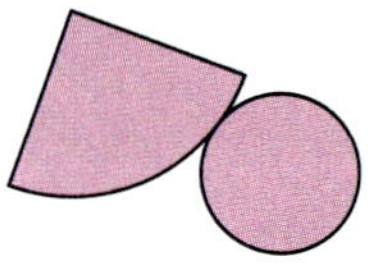
cone

Find each surface area. Use 3.14 for π.

15. 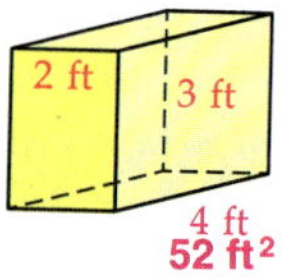

52 ft²

16. 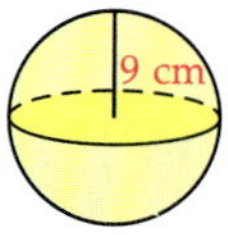

1,017.36 cm²

17. 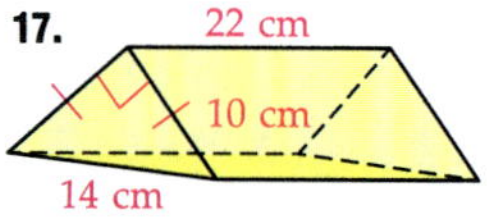

848 cm²

18. 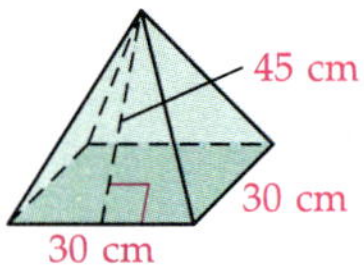

3,600 cm²

19. 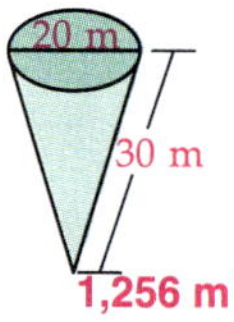

1,256 m²

20. 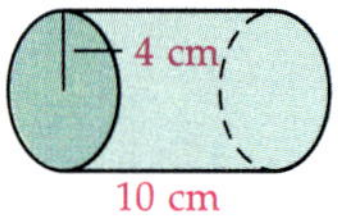

351.68 cm²

21. 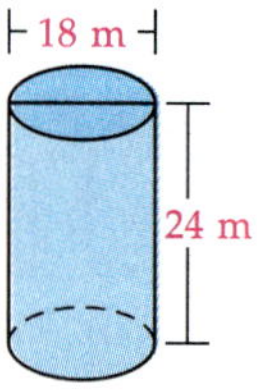

1,865.16 m²

22. 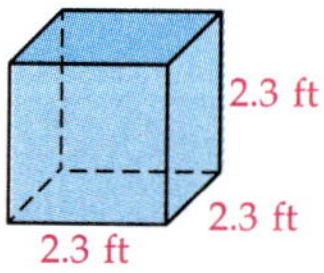

31.74 ft²

23. 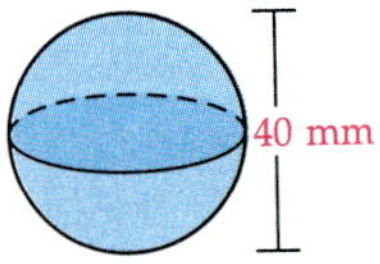

5,024 mm²

OBJECTIVE:
To find the volume of prisms and cylinders.

10-9 Volume—Prisms and Cylinders

1

A gallon and a liter are liquid measures, but they also measure volume. A gallon of milk occupies 231 in.^{3} of space. A liter of milk occupies 1,000 cm^3 of space.

Volume	Volume is the measure of the space inside a space figure. We measure volume in cubic units.

THINK Why do scientists prefer to use metric measure? **It is easier to change from one measure to another in the metric system because you are using multiples of ten.**

2

The figure below shows a rectangular prism. The base is covered by 12 cubes. The height is 2, allowing for 2 layers of 12 cubes. The volume of the prism is 24 cubic units.

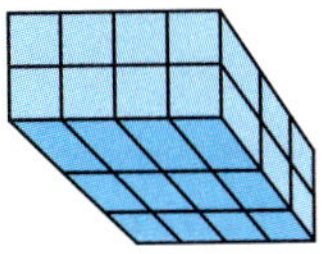

This example suggests the following formula.

Volume of a Prism or Cylinder	The volume (V) of a prism or a cylinder is base area (B) times the height (h). $V = Bh$

THINK To find the volume of a rectangular prism you can multiply length × width × height. How is this the same as $V = Bh$? **Length × width is the same as the area of the base.**

In formulas for volume, B represents the base area, while b is the length of the base edge.

Example 1 **Find the volume of the triangular prism.**

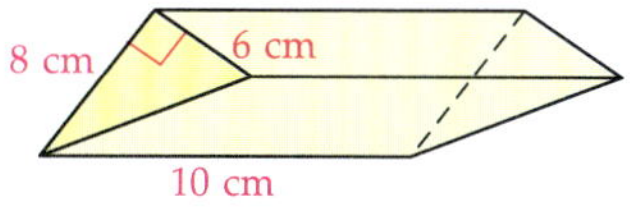

3

Solution

$B = \frac{1}{2}bh$ Find the area of the base.

$= \frac{1}{2}(8 \times 6)$ Substitute 8 for b and 6 for h. Simplify.

$= 24$

$V = Bh$ Find the volume.

$= (24)10$ Substitute 24 for B and 10 for h.

$= 240$ Simplify.

The volume is 240 cm^3.

THINK Why is $h = 6$ used to find base area and $h = 10$ used to find volume? **Because 6 is the height of the base and 10 is the height of the prism.**

Lesson Focus

Materials

Centimeter or unit cubes, graph paper, ruler, empty cans with labels, math journal

Vocabulary/Symbols

volume

MOTIVATING THE LESSON

Discuss the concept of volume and ask students to give real-world examples of where knowing or calculating volume is necessary. (Examples: space inside a refrigerator is given in cubic feet; the dimensions of a pool are used to calculate volume, which is converted into gallons.)

Instruction

1 Tell students that by equating units of capacity with units of volume, they can find the dimensions of containers needed to hold certain amounts of liquids or dry goods.

2 Have students model the prism using unit cubes. Help them to recognize that 12, or 3 × 4, are needed for one layer. For two layers the volume is 24 cubic units. Ask what the volume would be if there were 5 layers of cubes. (60 cubic units) Have students discuss the formula with respect to their models.

3 Ask students what finding the area of the triangular base is equivalent to when modeling. (making one layer of cubes) Note that the triangular base is a right triangle.

Address the THINK question by having students point out that 6 (or 8) is the height of the triangle and 10 is the height of the prism.

Teaching Tip

Have students make models of a cubic inch, a cubic foot, a cubic millimeter, a cubic centimeter, and, if space is available, a cubic yard and a cubic meter. Keep them on display to reinforce the actual sizes of the units.

EE: 1A, 1C, 1D, 4A, 4B, 4E; TAAS: 4, 11

(4) Help students realize that finding the volume of the cylinder is similar to finding the volume of prisms. You find the area of the base, or B, and multiply it by the height. Have students use rulers and cans to model this example. Measure the diameter and the height to calculate the volume in cubic centimeters, and compare it to the number of milliliters on the label. Remind students that 1 ml = 1 cm^3 of water. If a measuring cup and water are available, fill one of the cans with water and measure the number of milliliters. This number should be close to the calculated volume.

ADDITIONAL EXAMPLES

1. Find the volume of a triangular prism with a height of 5 in. and a triangular base with a height of 4 in. and a base of 3 in. (30 $in.^3$)

2. Find the volume of a cylinder with a height of 12 ft and a base with a diameter of 5 ft. (235.5 ft^3)

Guided Practice

Class Exercises Ask students to explain how to find the volume for questions 3-6.

Urge students to sketch and label cubes to help them visualize the units in questions 7-14.

Think and Discuss As a follow-up to question 2, ask students to find the mass of one liter of water. (1,000 g or 1 kg)

For question 3, have students use a model to demonstrate that the prism can be oriented so three different sides can be bases. Also, note that the commutative and associative properties of multiplication can be applied.

▼ Finding the volume of a cylinder is similar to finding the volume of a prism.

Example 2 Find the volume of a juice can in centimeters. Use 3.14 for π.

Solution (4)

$B = \pi r^2$ — Find the area of the base.
$= \pi \cdot 3^2$
$= 9\pi$

$V = Bh$ — Use the formula for volume.
$= 9\pi \cdot 12.5$
$= 112.5\pi$
$\approx 3.14 \times 112.5$
$= 353.25 \text{ cm}^3$

The volume is about 353.25 cm^3.

THINK AND DISCUSS

1. Why is volume usually labeled in gallons or liters rather than cubic inches or cubic centimeters? custom

2. In metric units, volume and mass are easily changed from one unit to another. For water, 1 ml = ■ cm^3 = ■ g. 1, 1

3. When finding the volume of a rectangular prism with dimensions 5 × 8 × 9, does it matter which two dimensions represent the base of the figure? no

4. Name at least one prism and one cylinder where you need to know the volume. Answers may vary.

CLASS EXERCISES

MENTAL MATH **Calculate the volume of each rectangular prism.**

1.
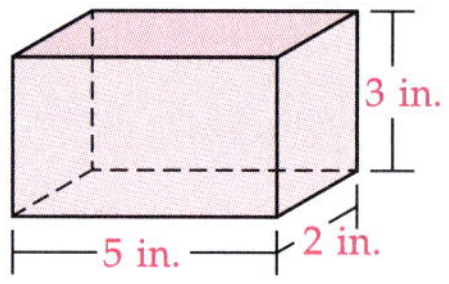

30 $in.^3$

2.
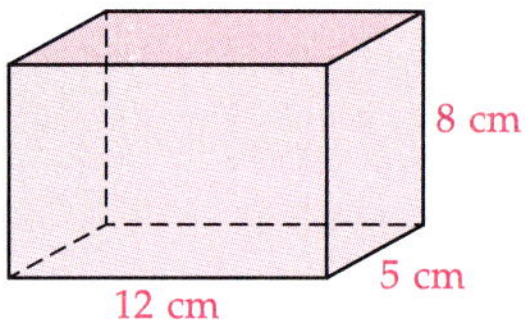

480 cm^3

Find the volume of each prism.

3.
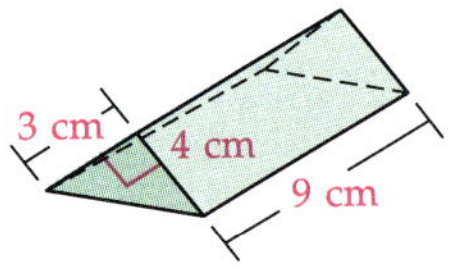

54 cm^3

4.
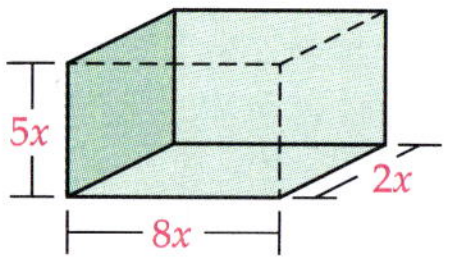

$80x^3$ cu. units

Find the volume of each cylinder. Use 3.14 for π.

5.
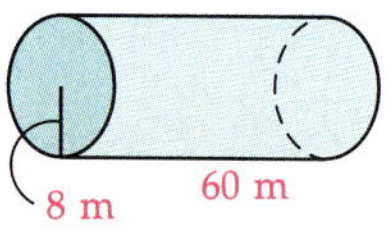

12,057.6 m^3

6.
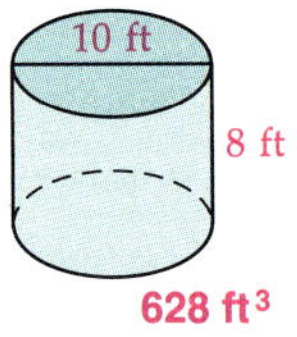

628 ft^3

Complete.

7. 1 ft^2 = ■ $in.^2$ 144
8. 1 yd^2 = ■ ft^2 9
9. 1 cm^2 = ■ mm^2 100
10. 1 m^2 = ■ cm^2 10,000
11. 1 ft^3 = ■ $in.^3$ 1,728
12. 1 yd^3 = ■ ft^3 27
13. 1 cm^3 = ■ mm^3 1,000
14. 1 m^3 = ■ cm^3 1,000,000

MATH MINUTES

Have groups of students predict the volume of a stack of 10 coins (whatever coins are available). Then have them measure, using the unit of their choice, and calculate the volume. (Check students' work.)

WRITTEN EXERCISES

Find the volume of a prism with the given dimensions.

1. square base: 7 in. by 7 in. height: 13 in. **637 in.^{3}**
2. rectangular base: 9 ft by 5 ft height: 36 in. **135 ft^3**
3. cube: sides: $3a$ **$27a^3$**

4. 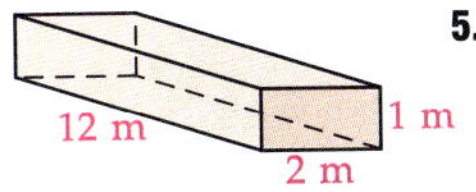

24 m^3

5. 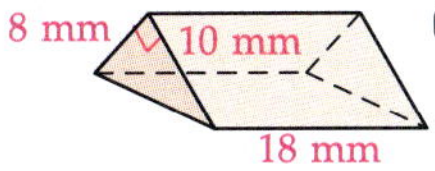

720 mm^3

6. 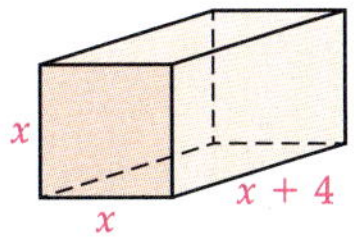

$x^3 + 4x^2$ cu. units

CALCULATOR **Find the volume of a cylinder with the given dimensions to the nearest hundredth. Use 3.14 for π.**

7. radius: 12 ft height: 15 ft **6,782.4 ft^3**
8. diameter: 3.8 m height: 18 m **204.04 m^3**
9. radius: 7.6 cm height: 32 cm **5,803.72 cm^3**

10. 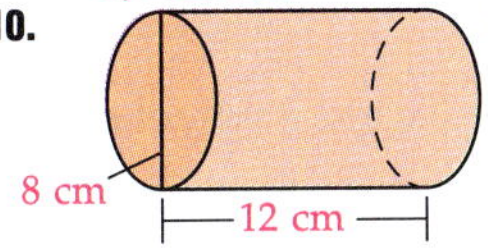

602.88 cm^3

11. 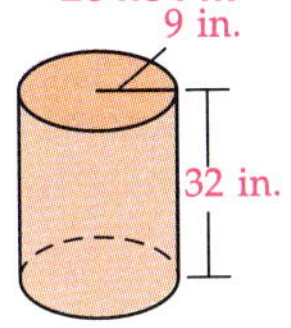

8,138.88 in.^{3}

12. 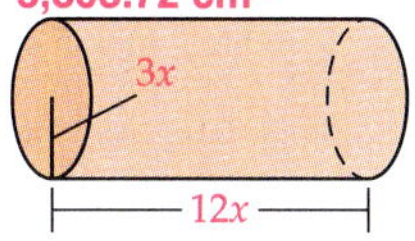

$339.12x^3$ cu. units

13. Wood for wood-burning stoves is sold by the cord. A cord is 8 ft × 4 ft × 4 ft. How many cubic feet is a cord? **128 ft^3**
14. When purchasing concrete, the price quoted is *per yard*, which means per cubic yard. Concrete costs $52 a yard. How much would it cost to pour a 14 ft × 16 ft × 6 in. slab for a patio? **$215.70**
15. Cylinder A has radius 1 and height 3. Cylinder B has radius 1 and height 6. What is the ratio of the volumes of the two cylinders? How does doubling the height affect the volume? **$\frac{1}{2}$; doubles the volume**
16. Cylinder A has radius 1 and height 3. Cylinder B has radius 2 and height 3. What is the ratio of the volumes of the two cylinders? How does doubling the radius affect the volume? **$\frac{1}{4}$; quadruples the volume**
17. How much juice can a drinking straw hold? A straw has a diameter of 6 mm and a length of 208 mm. **5,878.08 mm^3**
18. Write a formula for the volume of a cube that has side x. **x^3**
19. Write a formula for the volume of a cylinder with radius x and height x. **πx^3**
20. ***PROJECT*** Pop half a cup of popcorn.
    a. How many cups of popped corn do you have?
    b. What is the ratio of popped to raw popcorn?
    Answers may vary.

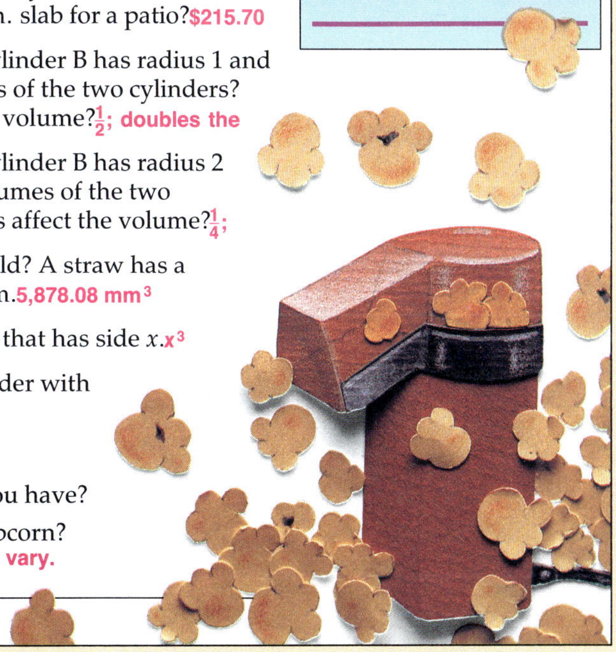

MIXED REVIEW

1. Write an expression for the supplement of a $3x°$ angle. **$(180 - 3x)°$**
2. What is the ratio of a dime to a quarter, in simplest terms? **$\frac{2}{5}$**
3. Find the circumference of a circle with r = 15 cm. **94.2 cm**
4. Find $\sqrt[3]{27}$. **3**
5. Use scientific notation to write the area of a square that is 5.6×10^6 units on a side. **3.136×10^{13} sq. units**

Find the surface area.

6. a triangular prism with dimensions base edges = 8 in., h = 6.9 in., and height of prism 14 in. **391.2 in.^{2}**
7. a cone with r = 6 ft and l = 8 ft. **263.76 ft^2**
8. Juan has $3.80 in coins in his pocket. He has 6 quarters and 12 dimes. The rest are nickels. How many nickels does he have? **22**

A ***common error*** occurs when students use a lateral face instead of the base when attempting to find the volume of a triangular prism. When a triangular prism is drawn with a rectangular face in the base position, such as in Example 1, pg 437, have students either sketch the figure so it "stands" on a triangular base or turn their books to change the orientation of the figure.

Closure

Writing in Math Ask students to describe similarities in finding the volumes of rectangular prisms, triangular prisms, and cylinders. Have them write the descriptions and examples in their math journals.

Independent Practice

Suggest that students sketch and label figures for questions 13-17.

For question 17, have students express the volume in cubic millimeters and in milliliters.

To extend question 20, have students use the ratio to predict how much raw popcorn, when it is popped, would be needed to fill the classroom. (Answers will vary.)

As a follow-up, refer to the calculator activity found on page 404E of the Chapter Overview. See also, TRB Exploring O-39.

Lesson Follow-up

Reteaching Activity Have students model rectangular prisms using unit cubes. Have them count cubes and then use the formula to find the volumes.

LESSON RESOURCES

TRB Practice J-16
TRB Enrichment J-17

Practice

For use after 10-9 (pp. 437–440)

Volume of Prisms and Cylinders

Find the volume of each prism.

1. rectangular base: 8 in. by 6 in. height: 7 in. **336 in.^{3}**
2. square base: 3.5 ft on a side height: 6 ft **73.5 ft^3**
3. cube: sides: $13y$ **2,197y^3 units^3**
4. 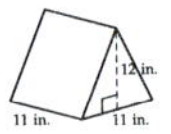

726 in.^{3}
5. 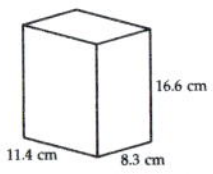

1,570.692 cm^3
6. 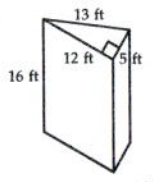

480 ft^3

Find the volume of each cylinder.

7. radius: 14 in. height: 18 in. **3,528π in.^{3}**
8. radius: 5.2 cm height: 11.2 cm **302.848π cm^3**
9. diameter: 5 ft height: 9 ft **56.25π ft^3**
10. 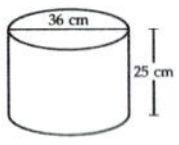

8,100π cm^3
11. 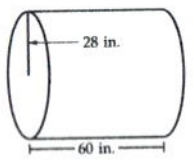

47,040π in.^{3}
12. 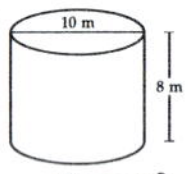

200π m^3
13. The height of a prism with volume 621 in.^{3} is 23 in. Find the area of the base. **27 in.^{2}**
14. Find the length of each side of a square-base prism if the height is 11 cm and the volume is 539 cm^3. **7 cm**
15. A cylinder has height 14 cm and volume 1,134π cm^3.
   a. Find the area of the base. **81π cm^2**
   b. Find the radius. **9 cm**

J-16 Chapter 10

Enrichment

For use after 10-9 (pp. 437–440)

Paper Towel Thickness

Each sheet in a roll of paper towels measures 11.5 in. by 9 in. There are 120 sheets in the roll.

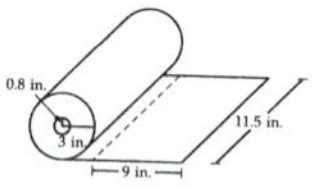

Answer these questions to find the thickness of each sheet. Use 3.14 for π.

1. The radius of the entire paper towel cylinder is 3 in. The radius of the hollow cylinder in the middle is 0.8 in. Describe how you can find the volume of paper on the roll. **Subtract the volume of the hollow cylinder from the volume of the entire cylinder.**
2. Use the method you described to find the volume of paper on the roll. Round to the nearest hundredth. **301.88 in.^{3}**

Each sheet of paper on the roll is a very thin rectangular prism. Let h = height of the prism.

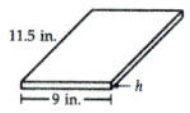

3. Find the volume of each sheet of paper in terms of h. **103.5h**
4. Find the volume of paper on the roll in terms of h. **12,420h**

Your answers to Exercises 2 and 4 give you two different expressions for the volume of paper on the roll.

5. Write an equation using the two expressions. **12,420h = 301.88**
6. Solve the equation to find h, the thickness of one sheet of paper. Round to the nearest thousandth. **0.024 in.**

Chapter 10 J-17

Sailing Across the Prairie

American pioneers traveled west of the Mississippi in prairie schooners, or covered wagons. For two thousand miles and several months, the wagon was home for the pioneer family. A prairie schooner was about 4 ft wide and 10 ft long. Most wagons were about 8 ft high. In this small space, pioneer families carried all of the supplies needed to survive the trip and to start new lives.

Use the article at the left to answer each question.

21. What is the volume of a rectangular prism the dimensions of a prairie schooner? **320 ft^3**
22. A wagon wheel is about 4 ft in diameter.
   a. What is its circumference? Use 3.14 for π. **12.56 ft**
   b. A mile is 5,280 ft. About how many turns would a wheel make in a mile? in 2,000 mi? **420 turns in a mile; 840,000 in 2,000 miles**

TEST YOURSELF

Find the surface area of each figure.

1. 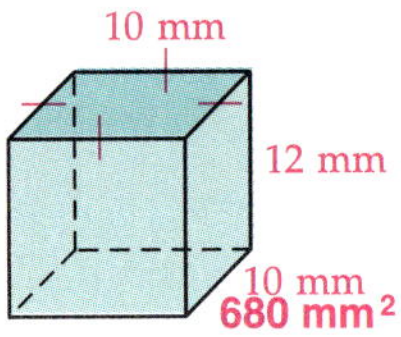

680 mm^2

2. 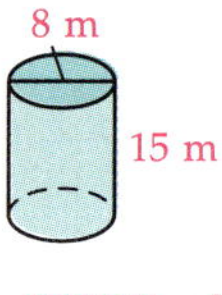

477.28 m^2

3. 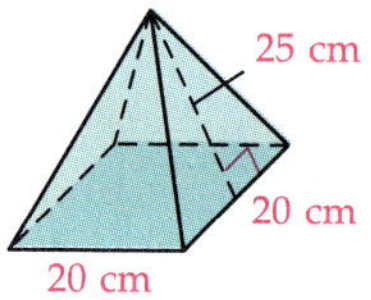

1,400 cm^2

Find the volume of each figure.

4. 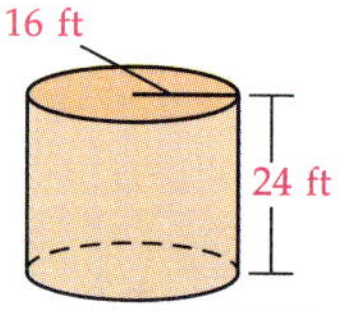

19,292.16 ft^3

5. 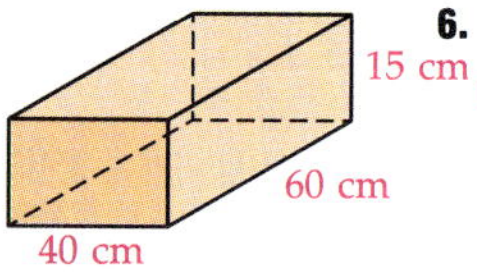

36,000 cm^3

6. 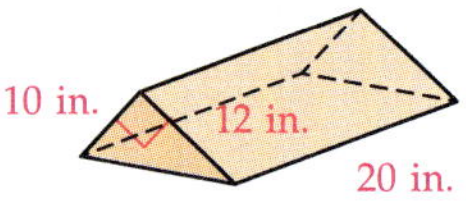

1,200 in.^{3}

7. 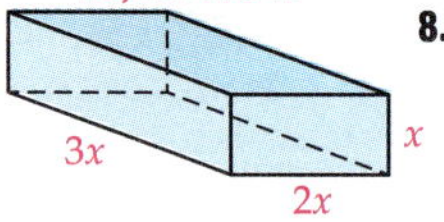

6x^3 cu. units

8. 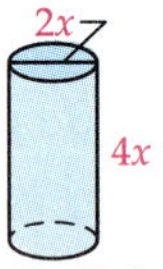

12.56x^3 cu. units

9. 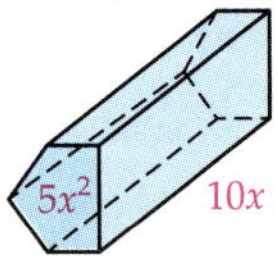

50x^3 cu. units

LESSON QUIZ

Find the volume of each figure. Use 3.14 for π.

1. A square prism with a base edge of 12 cm and a height of 14 cm. (2,016 cm^3)
2. 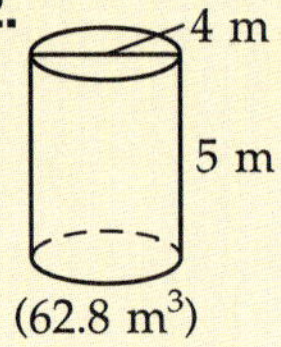

(62.8 m^3)
3. 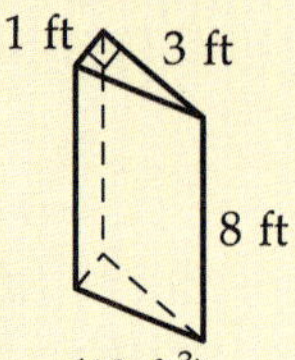

(12 ft^3)
4. A rectangular prism has a volume of 270 m^3, a base that measures 5 m by 6 m. What is the height? (9 m)

FOR THE NEXT LESSON

paper, compass, scissors, tape, rice or similar dry material, math journal

Assignment Guide

Basic 3-6, 10-11, 13-14, 17, 20-22, MR, TY All
Average 1-3, 5-8, 14-16, 20-22, MR, TY All
Enrichment 1, 3, 8-9, 12, 14-22, MR, TY All

OBJECTIVE: *To find the volume of pyramids, cones, and spheres.*

10-10 Volume—Pyramids, Cones, and Spheres

1

▼ You can fill three cones with water and pour the contents into a cylinder with the same height and radius. The cylinder will be completely filled.

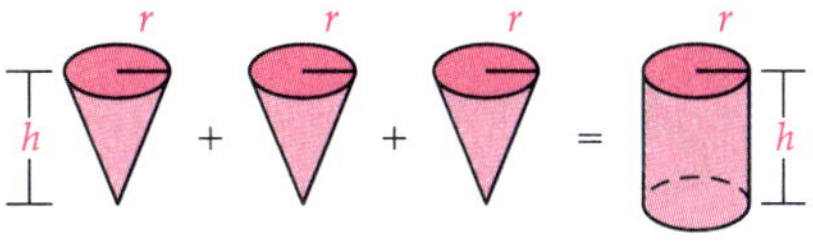

The volume of the cone is $\frac{1}{3}$ the volume of the cylinder. The same relationship is true of a pyramid and a prism with the same base and height.

> **FLASHBACK**
> The formula for the volume of a prism and a cylinder is Bh.

Volume of a Cone and a Pyramid	The formula for the volume of a cone and pyramid is $\frac{1}{3}$ the base area (B) times the height (h). $V = \frac{1}{3}Bh$

▼ The height of a regular pyramid or cone is the measure of the altitude from the vertex to the center of the base.

Example 1 **Find the volume of each figure.**

a.

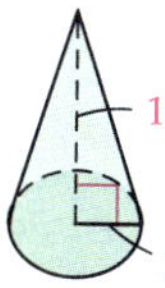

b.

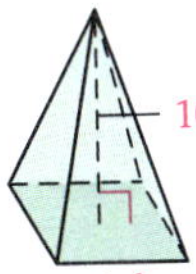

Solution

2

a.
$$B = \pi r^2 = \pi \cdot 3^2 = 9\pi$$
$$V = \frac{1}{3}Bh = \frac{1}{3} \cdot 9\pi \cdot 10 = 30\pi \approx 30 \cdot 3.14 = 94.2$$

b.
$$B = bh = 6 \cdot 6 = 36$$
$$V = \frac{1}{3}Bh = \frac{1}{3} \cdot 36 \cdot 10 = 120$$

The volume of the cone is about 94.2 ft^3.
The volume of the square pyramid is 120 ft^3.

> **THINK** Why is the volume of the cone less than the volume of the pyramid? The base is smaller. The circle will fit inside the square.

3

▼ Below is the formula for the volume of a sphere.

Volume of a Sphere	The volume (V) of a sphere with radius r is $V = \frac{4}{3}\pi r^3$.

Lesson Focus

Materials

Paper, compass, scissors, tape, rice or similar dry material, math journal

MOTIVATING THE LESSON

Display models of a cone and square pyramid. Ask students to describe similarities between the figures. Then ask how they might find the volume for the figures. Lead them to suggest that the procedures for finding the volumes might be similar.

Instruction

1 Have students make models to demonstrate this relationship. Draw a sector using a compass, cut it out, and tape it together to form a cone. Then trace the circumference of the base and make a cylinder with the same circumference. Cut the cylinder so it is the same height as the cone and tape the base to one end. Fill the cone with rice and empty it into the cylinder. It should take about three cones to fill the cylinder, leaving room for error in measuring, cutting, and taping.

2 Ask students to explain the only difference between finding the volume of a cone and the volume of a pyramid. (the way the area of the base is found)

3 Without calculating, ask students to predict which figure would have the greater volume, if the diameter of the cone was 3 ft instead of 6 ft. (Pyramid: a circle with diameter equal to three would fit "inside" the square base of the pyramid.)

Teaching Tip

Have regular pyramids or cones made out of paper available for the students to use to measure the altitude. Students can tie a string around a pencil and drop the string from the base of the figure through the vertex to measure the altitude.

ADDITIONAL EXAMPLES

1. Find the volume of a cone with a height of 4 m and a diameter of 2.2 m. (5.07 m^3, to the nearest hundredth)
2. Find the volume of a square pyramid with a base edge of 2.4 cm and a height of 6 cm. (11.52 cm^3)
3. Find the volume of a sphere that has a diameter of 14 ft. (1,436.03 ft^3, to the nearest hundredth)

Guided Practice

Class Exercises Review that the exact volume of a cone or a sphere is expressed in terms of π, and the approximate volume results from using 3.14 for π.

Suggest that students sketch the figures in questions 11-12. Point out that they should keep their answers in terms of π. This will make it easier to determine the ratios.

Think and Discuss Ask what properties allow the formula in question 1 to be written in various ways. (commutative and associative properties of multiplication)

For question 3, have students make a chart to show what happens to the volume when the radius is tripled and quadrupled. (The volume is 27, or 3^3, times the original volume; the volume is 64, or 4^3, times the original volume.)

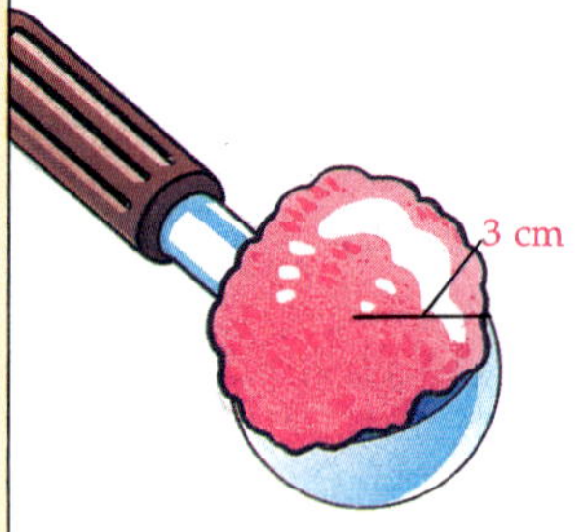

Example 2 Find the volume of a spherical scoop of ice cream with radius 3 cm.

Solution
$$V = \frac{4}{3}\pi r^3$$
$$\approx \frac{4}{3} \cdot 3.14 \cdot 3^3 \quad \text{Use 3.14 for } \pi.$$
$$= \frac{4}{3} \cdot 3.14 \cdot 27$$
$$= 113.04 \text{ cm}^3$$
The volume is about 113.04 cm^3.

THINK AND DISCUSS

1. Does $\frac{1}{3}Bh = \frac{1}{3}hB = \frac{Bh}{3}$? How can this help in finding volume of a pyramid or a cone? See below.

2. Compare the formula for the surface area of a sphere to the formula for the volume of a sphere. How are they alike? How are they different?

3. In a sphere, if the radius is doubled, how is the volume changed?

1. yes; The order of numbers can be changed.
2. Both formulas multiply by 4π. If you know the surface area, you could find the volume by multiplying by $\frac{r}{3}$.
3. increased 8 times

CLASS EXERCISES

Find the volume of each figure. Use 3.14 for π.

1.
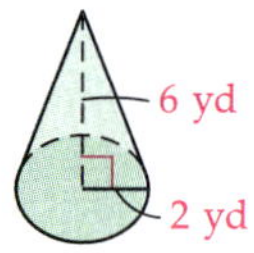

25.12 yd^3

2.
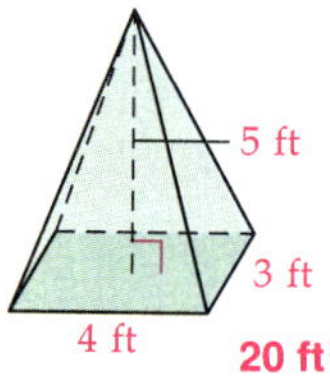

20 ft^3

3.
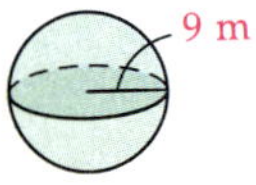

3,052.08 m^3

4. A cone has a radius 1 and height 1. What is the exact volume of the cone? $\frac{\pi}{3}$

5. The volume of a pyramid is 25 m^3. What is the volume of a prism with the same base and same height? 75 m^3

Complete each analogy.

6. square : cube : : circle : ▒ sphere

7. pyramid : cone : : polygonal base : ▒ circular base

8. volume : cubic units : : area : ▒ square units

9. cylinder : prism : : cone : ▒ pyramid

10. perimeter : area : volume : : cm : ▒ : ▒ cm^2 cm^3

11. Cone A has $h = 5$ and $r = 3$. Cone B has $h = 5$ and $r = 6$. What is the ratio of the volumes? How does doubling the radius affect the volume? $\frac{1}{4}$; quadruples the volume

12. Cone A has $h = 5$ and $r = 3$. Cone B has $h = 10$ and $r = 3$. What is the ratio of the volumes? How does doubling the height affect the volume? $\frac{1}{2}$; doubles the volume

13. How many cones of radius 1 and height 1 equal the volume of a sphere with radius 1? 4

MATH MINUTES

Ask students to find the volume of this figure. (446.4 in^3)

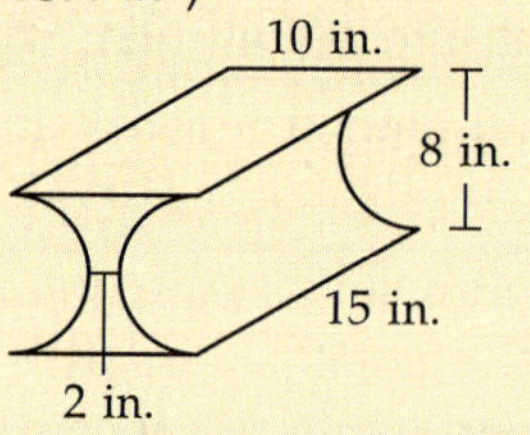

WRITTEN EXERCISES

CALCULATOR **Find the volume. Use 3.14 for π.**

1. sphere, $r = 6$ cm 904.32 cm^3
2. square-based pyramid, 720 m^3 $s = 12$ m, $h = 15$ m
3. cone, $r = 9$ ft, $h = 10$ ft 847.8 ft^3

4. 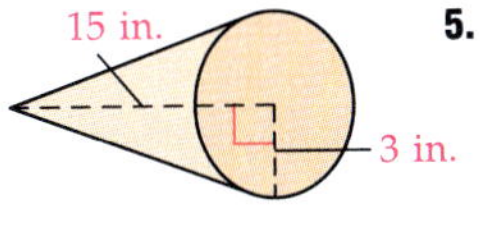

141.3 in.^{3}

5. 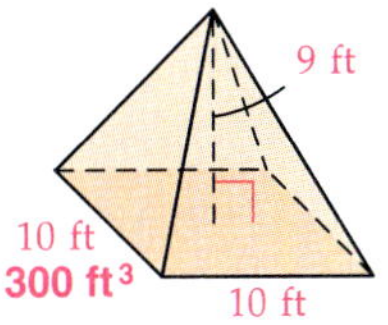

300 ft^3

6. 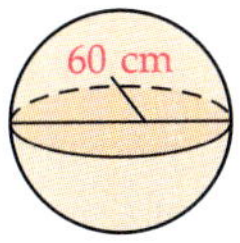

113,040 cm^3

7. Theater A sells popcorn in prism-shaped boxes. Theater B sells popcorn in pyramid-shaped boxes. The base and height of both popcorn boxes are the same. Which box holds more popcorn? How much more? The box from theater A holds 3 times more.
8. Find the volume of a square-based pyramid with base edge 9 in. and height 1 ft. 324 in.^{3}

Find the radius of a sphere for each volume.

9. $\frac{256}{3}\pi$ m^3 4 m
10. $\frac{4}{3}\pi$ cm^3 1 cm
11. $\frac{500}{3}\pi$ ft^3 5 ft

12. **CALCULATOR** The diameter of Earth is about 7,926.6 mi.
    a. Find the surface area. approximately 197,290,000 mi^2
    b. Find the volume of Earth. You may need to use scientific notation. approximately 2.606×10^{11} mi^3
    c. Find the surface area and volume of Earth in the encyclopedia. Compare your answers. Area: 196,950,000 mi^2, volume: 2.69×10^{11} mi^3
13. Tennis balls with a diameter of 2.5 in. are sold in cans of three. The can is a cylinder. What is the volume of the space in the can not occupied by tennis balls? Assume the balls touch the can on the sides, top, and bottom. $\approx$ 12.27 in.^{3}
14. *You place a steel ball with diameter 4 cm, in a water-filled cylinder that is 5 cm in diameter and 10 cm high. How much water will spill? $\approx$ 33 cm^3
15. The diameter of the world's largest ball of string is 12 ft 9 in. Francis A. Johnson of Darwin, Minnesota, collected the ball of string between 1950 and 1978.
    a. What is the circumference of the ball of string? 480.42 in. or 40.035 ft
    b. What is the surface area? 73,504.3 in.^{2} or 510.45 ft^2
    c. What is the volume? $\approx$ 1,874,000 in.^{3} or $\approx$1,085 ft^3

MIXED REVIEW

1. Find the volume of a compact disc with diameter 14 cm and thickness 4 mm. 61,544 mm^3
2. Find the circumference of a circle that has $d = 13.8$ in. 43.332 in.
3. Find the surface area of a cylinder with $r = 5$ in. and $h = 12$ in. 533.8 in.^{2}
4. $(3 \times 10^2)(4 \times 10^2) = \blacksquare$ 1.2×10^5

Solve.

5. $-4.2y + 3.4 = 7.6$ −1
6. $2.8x - 5.6 = 16.8$ 8
7. $3x - \frac{3}{8} = 1\frac{5}{12}$ $\frac{43}{72}$
8. A box is 25.5 cm by 17 cm by 5 cm. What volume of dishwasher detergent can it hold? 2,167.5 cm^3

A ***common error*** is to forget to multiply by one-third when finding the volume of a pyramid or a cone. Require students to write the volume formula each time they need to use it, substitute the values, and then multiply.

Closure

Writing in Math Have students write the formulas in this lesson in their math journals along with a simple worked example that they can use for reference. Ask what is common in all the volume formulas they have learned so far (except the sphere). (You need to find the area of the base of the figure and multiply it by the height of the figure.) Ask which of the figures require more than finding Bh to find the volume. (pyramids and cones, multiply Bh by $\frac{1}{3}$)

Independent Practice

For questions 9-11, suggest that students write an equation and/or use trial and error to find the answers.

For question 14, help students to recognize that the ball will displace its volume in water. If a graduated cylinder or measuring cup and water are available, demonstrate this concept with any irregularly shaped object that sinks.

Critical Thinking Have students work in small groups to form a hypothesis and then test the hypothesis.

Lesson Follow-up

Reteaching Activity Provide a labeled drawing of a cone and a pyramid. Have students write the volume formula as $\frac{1}{3} \times$ *area of the base* $\times$ *height* for each and list the dimensions they will be using to find the volume.

LESSON RESOURCES

TRB Practice J-18
TRB Enrichment J-19

Practice — For use after 10-10 (pp. 441-444)

Volume of Pyramids, Cones, and Spheres

Find the volume. Use 3.14 for π.

1. square-based pyramid $s = 9$ in. $h = 12$ in. **324 in.^{3}**
2. cone $r = 8$ cm $h = 15$ cm **1,004.8 cm^3**
3. sphere $r = 6$ in. **904.32 in.^{3}**
4. cone $B = 93$ ft^2 $h = 7$ ft **217 ft^3**
5. sphere $r = \frac{3}{4}$ in. **1.76625 in.^{3}**
6. pyramid $B = 774$ cm^2 $h = 42$ cm **10,836 cm^3**
7. 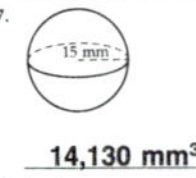

14,130 mm^3
8. 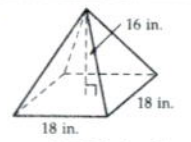

1,728 in.^{3}
9. 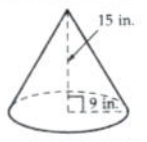

1,271.7 in.^{3}
10. An ice-cream cone has radius 3 cm and height 12 cm. Ice cream fills the cone and forms a hemisphere at the opening. Find the total volume of ice cream. Use 3.14 for π. **169.56 cm^3**
11. A cone with base area 372 in.^{2} has volume 2,852 in.^{3}. Find the height of the cone. **23 in.**
12. Find the radius of a sphere that has volume $\frac{32}{3}\pi$ in.^{3}. **2 in.**
13. The base of a pyramid is a square with sides 12 in. long. The pyramid is 16 in. high. A cone has a radius of 6 in. and a height of 16 in.
    a. Which figure has the greater volume? **the pyramid**
    b. The figure with greater volume is filled with sand and some of this sand is poured into the smaller figure. When the smaller figure is full, how much sand remains in the larger figure? Use 3.14 for π. **165.12 in.^{3}**

J-18 Chapter 10

Enrichment — For use after 10-10 (pp. 441–444)

Density

The *density* of an object is its mass per unit of volume:

$$\text{density} = \frac{\text{mass}}{\text{volume}}$$

Example **A cube of zinc measuring 4 cm on a side has a mass of 454 g. Find the density.**

Solution volume $= 4 \times 4 \times 4 = 64$ cm^3
density $= \frac{454}{64} = 7.1$ g/cm^3 (nearest tenth)

Find the density. Round to the nearest tenth. Use 3.14 for π.

	Material	Shape	Dimensions (cm)	Mass (g)	Density
1.	copper	rectangular prism	$7.1 \times 4 \times 2.9$	716	**8.7 g/cm^3**
2.	tungsten	triangular prism	$B = 42$ $h = 24$	19,454	**19.3 g/cm^3**
3.	iron	sphere	$r = 9$	24,111	**7.9 g/cm^3**

Find the mass. Round to the nearest tenth. Use 3.14 for π.

4. a rectangular prism of silver measuring 4 cm $\times$ 2.5 cm $\times$ 1.8 cm **189 g**
5. a lead cylinder with radius 6 cm and height 9.1 cm. **11,623.9 g**
6. a chromium cone with radius 5 cm and height 8.1 cm **1,504.8 g**

Table of Densities	
Material	Density (g/cm^3)
Carbon	2.3
Chromium	7.1
Cobalt	8.9
Gold	19.3
Lead	11.3
Magnesium	1.7
Osmium	22.4
Silver	10.5
Titanium	4.5

Identify the material described.

7. a rectangular prism measuring 6 cm $\times$ 5 cm $\times$ 3.7 cm, with a mass of 189 g **magnesium**
8. a pyramid with a square base 2.5 cm on a side, height 4.2 cm, with a mass of 196 g **osmium**
9. a cube measuring 1.2 cm on a side, with a mass of 33.4 g **gold**

Chapter 10 J-19

16. ***WRITE*** Which is the better buy on oranges? Explain how you made your selection.

4/\$1.00 3/\$1.00

17. A pyramid is 460 ft high. It is 760 ft on each side of its square base. What is the volume of the pyramid? **88,565,333$\frac{1}{3}$ ft^3**

18. You want to fill the top half of an hourglass $\frac{2}{3}$ full of salt. The height of the hourglass is 12 cm and the radius is 3 cm. Find the volume of salt needed. **37.68 cm^3**

16. Larger oranges are a better buy.
3 oranges with radius 3.5 in. have volume 538.5 in.^{3}
4 oranges with radius 3 in. have volume 452.16 in.^{3}

Critical Thinking

EXPLORING A HYPOTHESIS

A hypothesis is an educated guess. To find if a hypothesis is true, you must test it.

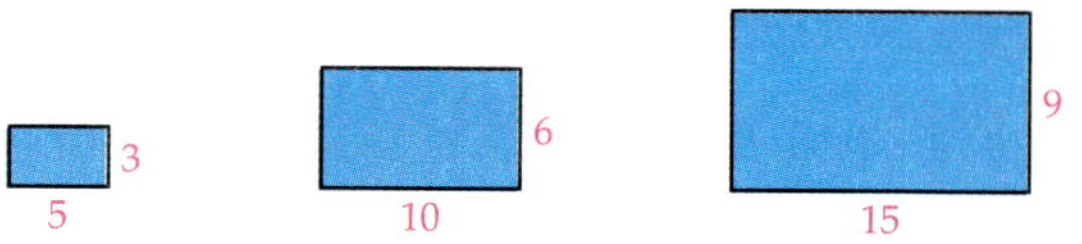

A B C

1. Write a hypothesis about the ratio of the perimeters and areas of two similar figures. **Answers may vary.**

	A	B	C
b	5	10	15
h	3	6	9
p	16	32	48
A	15	60	135

2. Make a table with entries for base, height, perimeter, and area and find this information for similar figures A, B, and C.
3. Use the information in your chart for figures B and C to answer the following questions. Write all fractions in lowest terms.
   a. What is the ratio of the heights? **$\frac{2}{3}$**
   b. What is the ratio of the bases? **$\frac{2}{3}$**
   c. What is the ratio of the perimeters? **$\frac{2}{3}$**
   d. What is the ratio of the areas? **$\frac{4}{9}$**
4. Do you need to revise your hypothesis? Do so, if necessary. Test it on figures A and C. **A possible response: "Square of ratio of heights equals ratio of areas."**
5. Predict the ratio of the perimeters and areas of two triangles if the ratio of their heights is $\frac{5}{8}$. **$\frac{5}{8}$, $\frac{25}{64}$**

LESSON QUIZ

Find the volume. Use 3.14 for pi.

1. A sphere with a radius of 5 cm. (523.33 cm^3, to the nearest hundredth)

2. square pyramid $s = 9$ in., $h = 8$ in. (216 in.^{3})

3. cone $r = 10$ ft, $h = 9$ ft. (942 ft^3)

4. A kickball with a diameter of 8 in. is in a box that is a cube. The ball touches the box on each face. How much empty space is inside the box? (244.05 in.^{3})

Assignment Guide

Basic 3-7, 9-10, 12, 15-17, MR, CT All
Average 1-2, 5-10, 12-13, 15-16, MR, CT All
Enriched 1-3, 7-8, 10-11, 13-18, MR, CT All

FOR THE NEXT LESSON

algebra tiles or graph paper, calculator, math journal, paper, pencil

Problem Solving Practice

PROBLEM SOLVING STRATEGIES

Look for a Pattern
Guess and Test
Simplify the Problem
Account for All Possibilities
Make an Organized List
Work Backwards
Make a Table
Write an Equation
Solve by Graphing
Draw a Diagram
Make a Model
Simulate the Problem

Solve. Use an appropriate strategy or combination of strategies.

1. A rectangular piece of tin measures 26 in. × 20 in. A square measuring 2 in. × 2 in. is cut out of each corner and the sides are folded to form a box. What is the volume of the box? **704 in.^{3}**
2. Susan and Karen are in a marathon race. Susan's average speed is 5 mi/h and Karen's is 8 mi/h. If they start at 10 A.M., when will they be $4\frac{1}{2}$ mi apart? **11:30 A.M.**
3. Frank Goodshot usually hits his mark when throwing free throws. However, in the last few games he has made only 15 shots out of 24. How many consecutive free throws must Frank make to raise his record to 80%? **21**
4. A 12-m by 15-m rectangular garden has a walk 1 m wide around it. What is the area of the walk? **58 m^2**
5. Three boys, Jack, Jim, and Jay, planned to share a bag of apples equally. Jack found the bag on the doorstep first and took his share. Jim came along later and took what he thought was his share. Later yet, Jay arrived, taking what he thought was his share. He left 8 apples. How many apples were there to start with? **27**
6. Five economically minded girls decided to share wardrobes. Each bought a different three-piece outfit, a skirt, a blouse, and a vest. Then they traded pieces around. How many different outfits can they make? **125**
7. Juan and his younger brother Kimo picked apples. The average of what they picked was 10 bushels. Juan picked three times as many apples as Kimo. How many bushels did each pick? **Juan picked 15 bushels. Kimo picked 5 bushels.**
8. ***DATA (p. 128)*** Use the formula for cricket chirps and temperature. How many chirps does a cricket make per minute when the temperature is 56°? **76**
9. Six girls ran a 100-yd race. Fran beat Clara by 8 yd. Clara finished 12 yd behind Teresa. Marie finished 16 yd behind Fran but 2 yd ahead of Cindy. Vivian finished exactly between the first and last runner. In what order did the girls finish the race? **Teresa, Fran, Vivian, Clara, Marie, Cindy**
10. ***DATA FILE 10 (pp. 404–405)*** About how many Arctic Oceans could fit into the Pacific Ocean? Consider the volume of water in each, based on the average depth. **40**
11. A rectangle is 8 ft longer than it is wide. The area of the rectangle is 240 ft^2. What are the dimensions? **12 ft × 20 ft**

Instruction

This page provides a variety of problems that can be used to reinforce and enhance the students' problem solving skills. Encourage students to read each problem carefully. Then have them refer to the list of problem solving strategies to help them decide how to solve the problem.

Point out, however, that not all questions require a strategy for solving, nor are all the strategies in the list used in this lesson.

Study Skills

Learning is the process by which information is transferred from short-term memory to *long-term memory*. Short-term memory is the brain's clearinghouse; it stores information that is only needed for a little while (for example, today's weather) and then discards it to make way for other information. Only when the brain is signaled that a piece of information is important will that information be encoded into long-term memory.

Many students believe that if they pay attention in class and take notes, they have done all they can to facilitate the learning process. Of course, paying attention and note-taking are both keys to getting information into long-term memory; however, there are many other equally important techniques for enhancing that process. All of these techniques encourage the students to interact with the material in ways that help students to fully absorb the meaning of what is being taught. Only when information is internalized in this way has the learning process been accomplished.

Chapter 10 Review

Complete each statement. Use the vocabulary words given.

VOCABULARY
area
cone
volume
cylinder
sphere
polygons
prism
faces
slant height
vertex
triangles
center

1. The ■ is the amount of surface inside a region. **area**
2. A polyhedron is a space figure that has ■ for all faces. **polygons**
3. The sides of a pyramid are ■ that meet at a common point. **triangles**
4. The sides of a ■ are parallelograms. **prism**
5. A ■ has two circular bases that are parallel and congruent. **cylinder**
6. A ■ has one circular base and one ■. **cone, vertex**
7. A ■ is the set of all points in space that are the same distance from a given point called the ■. **sphere, center**
8. Surface area is the sum of the area of the ■. **faces**
9. The ■ ■ is the height of a face. **slant height**
10. ■ is the measure of the space inside a space figure. **volume**

Finding Areas of Polygons — 10-1, 10-2

To find the area of a polygon, use the appropriate formula.

parallelogram	triangle	trapezoid
$A = bh$	$A = \frac{1}{2}bh$	$A = \frac{1}{2}(b_1 + b_2)h$

Find the area of each figure.

11. 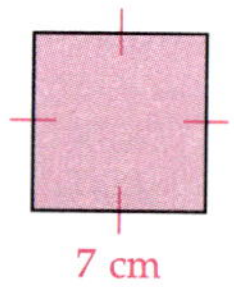

49 cm²

12. 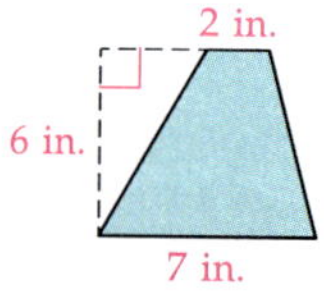

27 in.²

13. 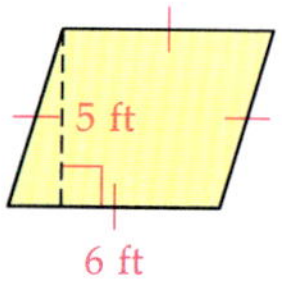

30 ft²

14. 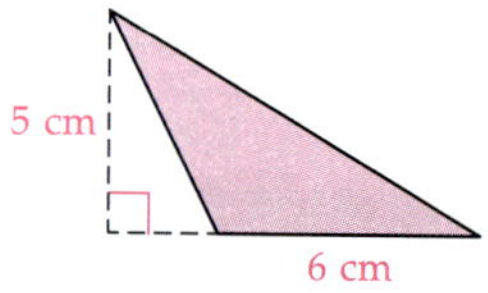

15 cm²

Finding the Area of a Circle — 10-3

To find the area of a circle, multiply π by the square of the radius. $A = \pi r^2$.

Find the area of each figure. Use 3.14 for π.

15. 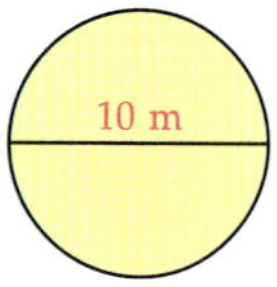

78.5 m²

16. 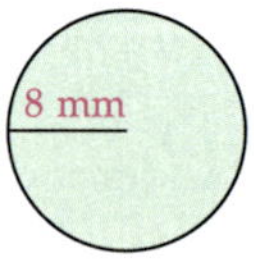

200.96 mm²

17. 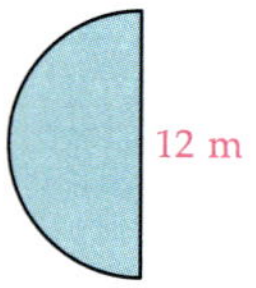

56.52 m²

18. 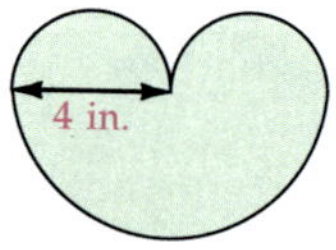

37.68 in.²

Space Figures 10-4

We name pyramids and prisms by the shape of their bases.

Name the space figure that each net can form.

19. 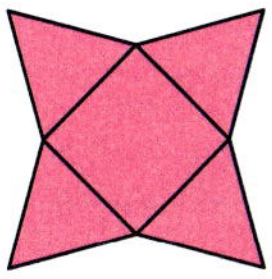

square pyramid

20. 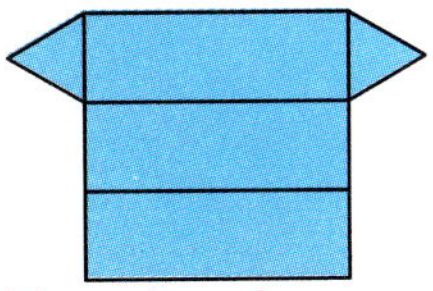

triangular prism

21. 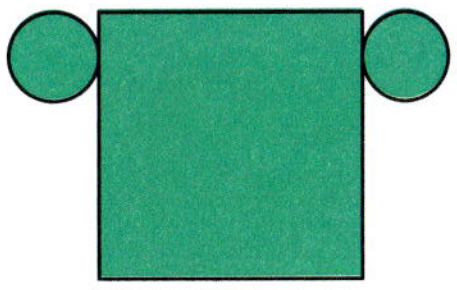

cylinder

Problem Solving 10-5

To solve some problems, make a model.

22. What are the dimensions of a rectangular sheet of wrapping paper that will cover a 6 in. × 6 in. × 6 in. gift box with no more than a $\frac{1}{2}$ in. overlap? 13 in. × $24\frac{1}{2}$ in.

Finding Surface Area 10-6, 10-7

To find the surface area of a prism or a cylinder, add the areas of the base(s) and the side(s).

To find the surface area of a cone or a pyramid, use the slant height. The surface area is the sum of the lateral area and the base area.

The surface area of a sphere is $A = 4\pi r^2$.

Find the surface area. Use 3.14 for π.

23. 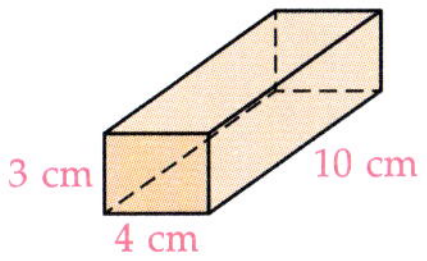

164 cm^2

24. 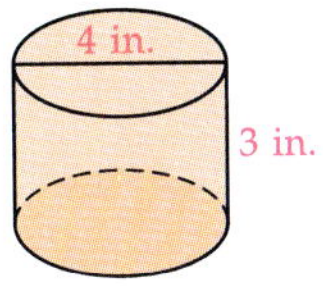

62.8 $in.^2$

25. 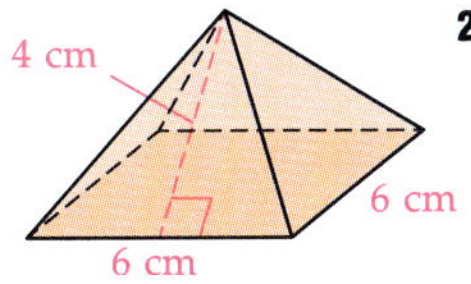

84 cm^2

26. 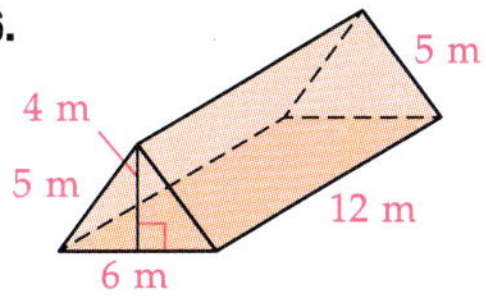

216 m^2

Finding Volume 10-9, 10-10

To find the volume of a space figure, use the appropriate formula.

prisms and cylinders	pyramids and cones	spheres
$V = Bh$	$V = \frac{1}{3}Bh$	$V = \frac{4}{3}\pi r^3$

Find the volume. Use 3.14 for π.

27. 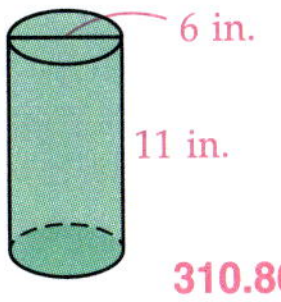

310.86 $in.^3$

28. 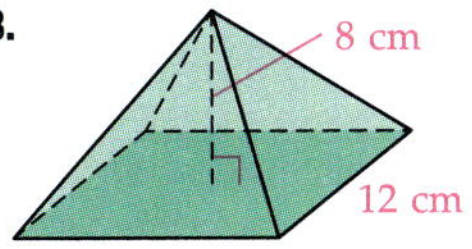

384 cm^3

29. 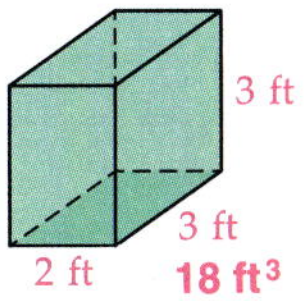

18 ft^3

30. 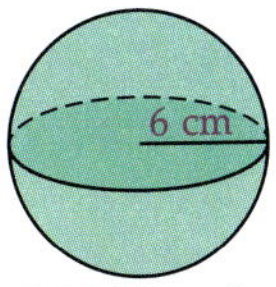

904.32 cm^3

Here are some of the techniques for fostering long-term memory:

- Students can try to make connections between the ideas being taught and other ideas they've learned in other disciplines.
- Students can use manipulatives, drawings, or other visual aids that let them "see" the information.
- Students can restate information in another way.
- Students can take notes and ask questions about anything they don't understand.

RESOURCES

TRB J-21–J-24

Scoring Chart	
No. Correct	**Score**
20	100
19	95
18	90
17	85
16	80
15	75
14	70
13	65
12	60
11	55
10	50
9	45
8	40
7	35
6	30
5	25
4	20
3	15
2	10
1	5

Chapter 10 Test

Find the area of each figure. Use 3.14 for π.

1.
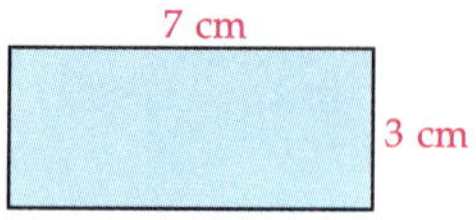

21 cm^2

2.
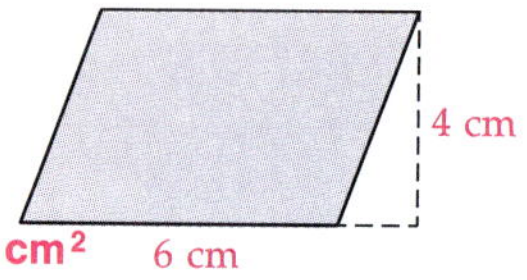

24 cm^2

3.
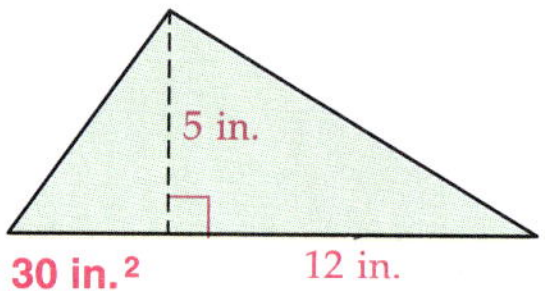

30 in.^{2}

4.
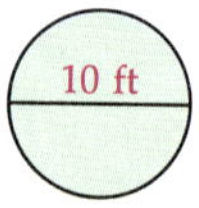

78.5 ft^2

5.
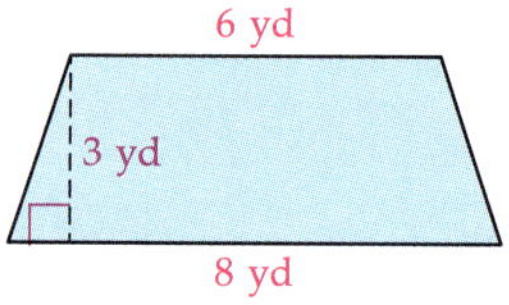

21 yd^2

6.
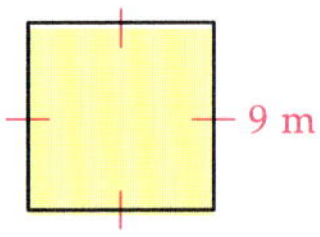

81 m^2

Find the missing measures.

7. a square
$A = 121$ m^2
$b = ■$ 11 m

8. a triangle
$A = 28$ m^2
$b = 7$ m
$h = ■$ 8 m

9. a circle
$A = 64\pi$ cm^2
$r = ■$ 8 cm

10. a parallelogram
$A = 48$ in.^{2}
$b = 16$ in.
$h = ■$ 3 in.

Find the surface area of each figure. Use 3.14 for π.

11.
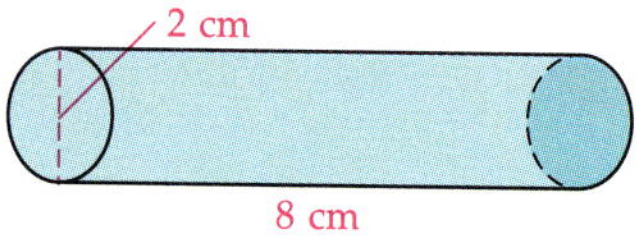

56.52 cm^2

12.
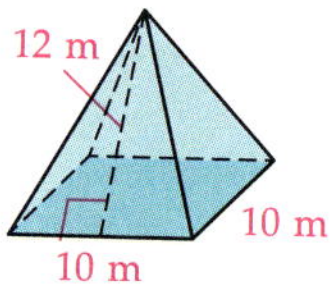

340 m^2

13.
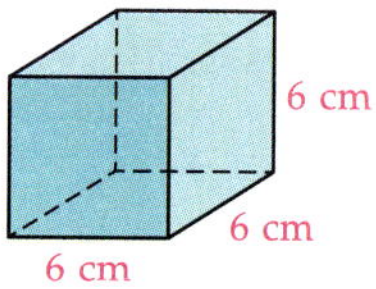

216 cm^2

Find the volume of each figure. Use 3.14 for π.

14.
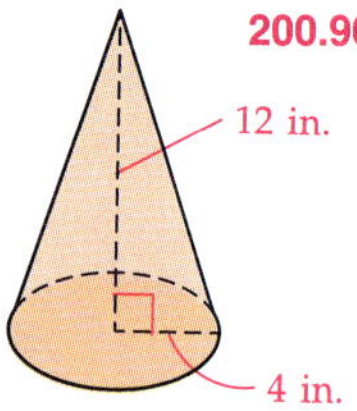

200.96 in.^{3}

15.
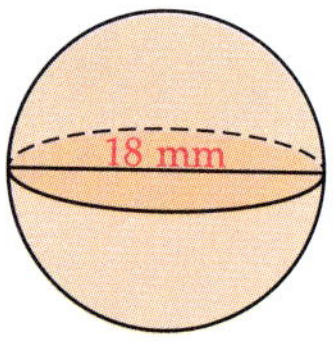

3,052.08 mm^3

16.
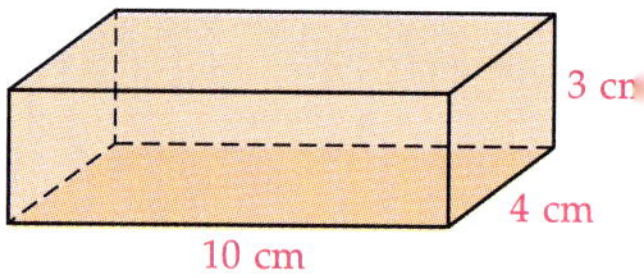

120 cm^3

Solve.

17. The height of a rectangle is doubled. How does this affect the area? **The area is doubled.**

18. What is the surface area of a sphere with radius 5 ft? **314 ft^2**

19. How much greater is the volume of a cone with radius 6 ft and height 10 ft than the volume of a square pyramid with base edge 6 ft and height 10 ft? Use 3.14 for π. **256.80 ft^3**

20. ***WRITE*** How is the formula for the volume of a prism or cylinder the same as the formula for volume of a pyramid or cone? How are they different? **The formulas for the volume of a prism and a pyramid are alike except that for a pyramid you multiply by $\frac{1}{3}$. The same is true for a cylinder and a cone.**

Chapters 1–10 Cumulative Review

Choose the correct answer. Write A, B, C, or D.

1. What is the volume? B

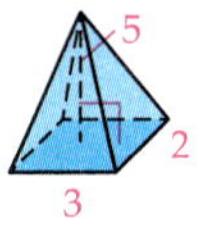

A. 30 **B.** 10
C. 15 **D.** not given

2. $\triangle ABC \sim \triangle DBE$. Find x. A

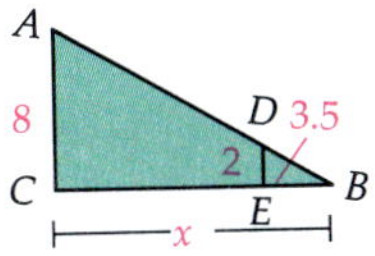

A. 14 **B.** 5
C. 12.5 **D.** not given

3. $\triangle ABC \cong \triangle DEF$. Which is not true? C

A. $\angle A \cong \angle D$ **B.** $\overline{AC} \cong \overline{FD}$
C. $\angle B \cong \angle F$ **D.** not given

4. Simplify $\frac{3x^5y^3z^{-2}}{24x^6y^2z}$. A

A. $\frac{y}{8xz^3}$ **B.** $8xyz^{-3}$
C. $\frac{y^5}{8xz^3}$ **D.** not given

5. What is the prime factorization of 84? D

A. $2^3 \cdot 7$ **B.** $2 \cdot 3^2 \cdot 7$
C. $2 \cdot 3 \cdot 7$ **D.** not given

6. A \$250 coat was on sale for \$200. What was the percent decrease? C

A. 25% **B.** 80%
C. 20% **D.** not given

7. Marcia has 13 dimes and quarters worth \$2.80. How many of each does she have? C

A. 10 *d*, 3 *q* **B.** 6 *d*, 7 *q*
C. 3 *d*, 10 *q* **D.** not given

8. Find the area of the shaded region. A

A. 32 **B.** 64
C. 16 **D.** not given

9. Name the space figure the net will form. B

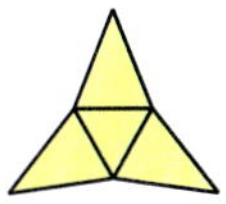

A. prism **B.** pyramid
C. cylinder **D.** not given

10. Solve $5 - 3x > 17$. C

A. $x < 4$ **B.** $x > -4$
C. $x < -4$ **D.** not given

11. What percent of 23 is 4.6? B

A. 80% **B.** 20%
C. 46% **D.** not given

12. Find the LCM of 18 and 24. B

A. 6 **B.** 72
C. 42 **D.** not given

13. What is the equation for a line having slope -2 and y-intercept 3? A

A. $2x + y = 3$ **B.** $y - 2x = 3$
C. $3x + y = -2$ **D.** not given

14. Which point lies on the line $2y - 3x = 14$? A

A. (-4,1) **B.** (1,-4)
C. (0,-7) **D.** not given

Study Skills

When students are asked to *problem solve*, particularly in a test situation, it is important that they be able to approach the problem solving process in an organized way. This helps to ensure that students will work as efficiently and accurately as possible. An overall approach to problem solving consists of four basic steps. First, students should read the problem to gather the facts and decide exactly what the problem is asking them to do. Next, students need to form a plan for solving the problem. Then students need to solve the problem and check their computations.

Ask students to discuss some of the techniques they have used for understanding and planning how to solve problems. List these approaches on the chalkboard, along with the following suggestions:

- look for a pattern
- guess and test
- simplify the problem
- account for all possibilities
- work backwards
- make a table
- write an equation
- solve by graphing
- draw a diagram
- make a model
- simulate the problem

11 Chapter At a Glance

In Chapter 11 students learn to work with square roots, right triangles, the Pythagorean theorem, and trigonometric ratios.

Reading the Chapter

The concepts and methods presented in this chapter give students an opportunity to develop the skills to visualize and draw geometric figures. Urge students to spend time analyzing statements or directions so that what they visualize and draw will be accurate.

Mention that these skills are needed to solve problems involving geometric ideas as well as other concepts developed in the book. If a student is having difficulty drawing a figure, pair him or her with an artistically talented student.

Chapter Resources

Practice and Enrichment TRB K-1 to K-12
Alternate Application TRB K-13

Chapter Test Form A, K-15
Chapter Test Form B, K-17

Computer Test Bank

References

STUDENT BIBLIOGRAPHY

Dods, Stuart C. *A Collection of Points: Mathematics Problem Solving with a Computer* (Trilium Press, Monroe, New York 1988).

TEACHER BIBLIOGRAPHY

Chazan, Daniel and Richard Houde. *How to Use Conjecturing and Microcomputers to Teach Geometry* (NCTM, Reston, Va.).

SOFTWARE

Plane View (Sunburst Communications), for the Apple II family of computers.

MacDraw (Claris Corporation), for the Macintosh family of computers.

FILMS, VIDEOTAPES, AND OTHER MEDIA

The Geometric Supposer Video (Sunburst Communications), VHS or Beta.

Vocabulary

converse of the Pythagorean theorem
cosine (cos)
irrational number
isosceles right triangle
perfect square
principal square root
Pythagorean theorem
Pythagorean triple
real numbers
simulate
sine (sin)
square root
Table of Trigonometric Ratios
tangent (tan)
trigonometric ratio
trigonometry

Materials

algebra tiles (M*)
graph paper (T*)
calculator
math journal

*M—Manipulative Kit
*T—Teacher's Resource Book

Resource Bank

Class Activities

FIELD TRIPS

▼ Local college or university—to allow students to observe how drafting and engineering students create technical illustrations.

▼ Local museums—to observe the relationship of similar triangles in a stained glass window.

CLASS SPEAKERS

▼ Landscape architect—to discuss what factors are taken into consideration in landscaping a public park.

▼ Stained glass artist—to discuss how triangles are used in a stained glass arrangement.

Bulletin Board

Have students bring in and display pictures of pyramids. After students complete Lesson 11-4, have them label the diagram with additional data from the results of question 17 and then answer the following questions.

How do the height and slant height of the Great Pyramid differ? (The height is measured from the center of the square base to the top, 481 ft. The slant height is the altitude of the triangular face, 611.45 ft.)

Use trigonometric ratios to find the angle measures of a triangular face. (about 58°, 64° or 63°, 58°)

What can you say about the triangles formed by the altitude of a face? (congruent)

Project: Ask students to research and compare the dimensions and the purposes of pyramids built by different cultures.

PYRAMIDS

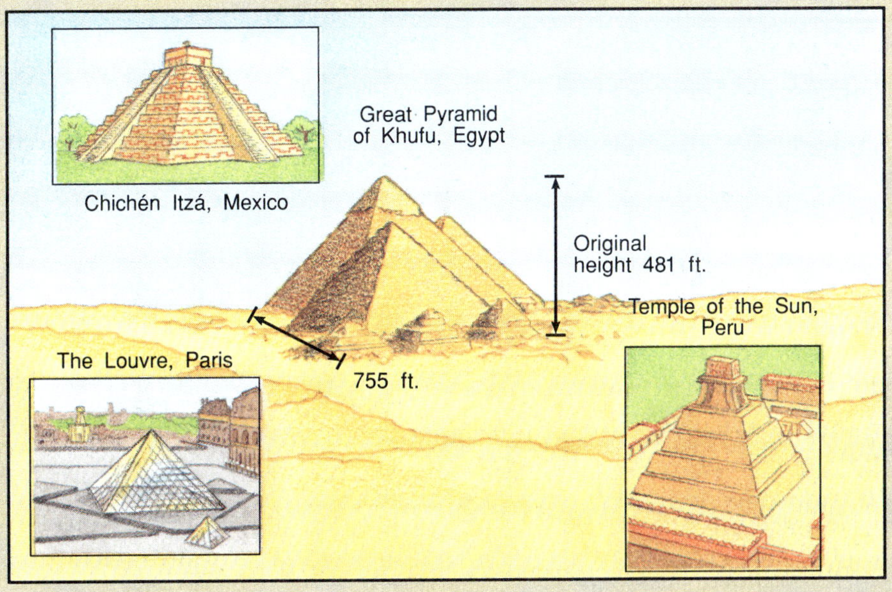

Extra-Credit Problems

BASIC

▼ Triangles *LMN* and *XYZ* are similar. What is the ratio of the area of the triangles if the ratio of side *MN* to *YZ* is 1 : 4? if the ratio of the sides is 1 : 5? (1 : 16, 1 : 25)

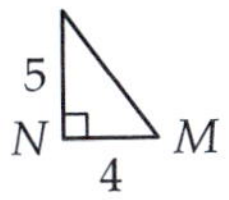

AVERAGE

▼ Find the area of an equilateral triangle with sides of 16 cm. Hint: cut the equilateral triangle in half. ($64\sqrt{3}$ cm^2.)

ENRICHED

▼ Triangles *ABC* and *QRS* are similar. What is the ratio of the areas if $\frac{BC}{RS} = \frac{1}{3}$? if $\frac{BC}{RS} = \frac{1}{5}$? (1 : 9, 1 : 25)

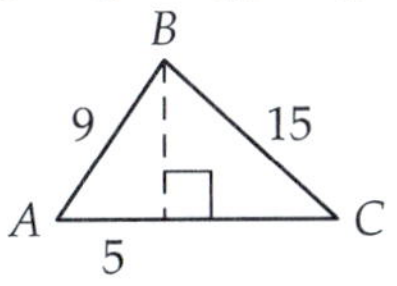

Chapter Planning Guide

OBJECTIVES	ASSIGNMENT GUIDE			ASSESSMENT	
	Basic	Average	Enriched	Review/ Tests	TRB*
Exploring Square Roots To explore square roots using several methods.	**Manipulative Connection** All	All	All		
11-1 Finding Square Roots Finding square roots of numbers and expressions.	1-8 odd, 9-12, 17-18, 21-24, 33-36, 41-44, 49-51, MR*, CT* All	2-16 even, 18-19, 25-28, 35-38, 43-46, 50-53, MR, CT All	4-8, 13-16, 19-20, 29-32, 37-40, 45-48, 53-56, MR, CT All	Extra Practice p. 576	Practice K-1 Enrichment K-2
11-2 Problem Solving Strategy: Simulating a Problem To solve a problem by simulation.	**Problem Solving Connection** All	All	All	Problem Solving Practice p. 481	Practice K-3 Enrichment K-4
Exploring Right Triangles To explore properties of right triangles.	All	All	All		
11-3 Pythagorean Theorem To use the rule of Pythagoras to verify right triangles and to find missing values of right triangles.	1-2, 5-6, 9, 11-12, 14-16, 20, 23-26, 31-36, MR All	2-3, 6-7, 10-12, 15-17, 21, 26-28, 33-38, MR All	3-4, 7-8, 10, 12-13, 18-22, 28-30, 35-40, MR All	Extra Practice p. 576	Practice K-5 Enrichment K-6
11-4 Similar Right Triangles To use parts of similar triangles in problem solving.	1-2, 5, 8-9, 11-12, 14, 17, MR, TY* All	2-3, 6, 8-9, 11-12, 14-15, 17, MR, TY All	3-4, 7, 9-10, 13-17, MR, TY All	Extra Practice p. 576 TY p. 468	Practice K-7 Enrichment K-8 TY K-14
11-5 Special Right Triangles To discover and use the properties of 45°-45°-90° and 30°-60°-90° right triangles.	1-2, 5-6, 9, 12, 15-18, 20, MR, CT All	2-3, 6-7, 10, 13, 17-20, MR, CT All	3-4, 7-8, 11, 14, 17-20, MR, CT All	Extra Practice p. 576	Practice K-9 Enrichment K-10
11-6 Trigonometric Ratios To define and use sine, cosine, and tangent of acute angles in right triangles.	1-3, 7-11, 13-14, 16-17, 22-24, MR, TY All	3-5, 7-11, 13-14, 18-19, 23-24, MR, TY All	4-10, 12, 14-15, 20-21, 24-25, MR, TY All	Extra Practice p. 576 TY p. 476	Practice K-11 Enrichment K-12 TY K-14
11-7 Problem Solving Application: Precision Drawing To apply right triangle concepts and trigonometric ratios to precision drawings.	All	All	All		Alternate Application K-13
Chapter Test	All	All	All	pp. 482-485	Tests K-15–K-21

*TRB—Teacher's Resource Book MR—Mixed Review CT—Critical Thinking TY—Test Yourself

			CONNECTIONS				NCTM CORRELATION
Algebra	**Critical Thinking**	**Problem Solving**	**Estimation/ Mental Math**	**Technology**	**Manipulatives/ Resources**	**Writing/Reading in Math**	
			13-15	15, 17-18	1, 4, 7, 8, 11	8f	Communication Reasoning Functions Connections
C*: 17 W*: 13, 16, 20, 24, 26-28, 30, 32	C: TD* 1-4 W: 1-8 CT*: 1-4	CT: 1-4	W: 9-20, 33-40	C: 10-21 W: 21-32, 41-48 CT: 4	C: 10-17 W: 49-56 CT: 4	C: 1-9	Problem Solving Functions
		C: 1-2 W: 1-8		W: 1-8	W: 8	W: 1-8	Problem Solving
4, 6		1-7			3-5	4d, 6c, 7	Communication Connections Synthetic Geometry Algebraic Geometry
C: 3-5 W: 9-22	C: TD 1-3	C: 7 W: 31-40	W: 1-4, 11-13	W: 5-8, 34		C: 7 W: 31-40	Problem Solving Algebraic Geometry
C: 1-4, 8 W: 1-17, TY*: 1-3	C: TD 1-3	W: 14-17 TY: 4-5	W: 1-4	W: 11-13	C: 5-8 W: 17	W: 14-17	Problem Solving Algebraic Geometry Algebra
C: 10-12 W: 1-8	CT: 1-5	W: 13-20	W: 12	W: 18	W: 1-8, 20	W: 13-20	Problem Solving Algebraic Geometry
C: 1-12 W: 1-9, 13-21 TY: 4-6	C: TD 1-2	W: 22-25	W: 10-12		C: 1-12 W: 1-25 TY: 1-6	W: 22-25	Problem Solving Algebraic Geometry Trigonometry
C: 4-6		C: 1-6 W: 1-11		W: 1-11 W: 1-11	C: 1-6		Communication Algebraic Geometry Trigonometry

*C—Class Exercises W—Written Exercises CT—Critical Thinking TY—Test Yourself TD—Think and Discuss

Exploring

In this manipulative activity, students explore the Pythagorean theorem by using tangrams. **See TRB Exploring O-41.**

Exploring the Pythagorean Theorem

You can use a tangram to illustrate the Pythagorean theorem. (Refer to page 202 of the student text for a tangram.) Trace the medium-sized triangle from the tangram. The legs of the triangle should be parallel to the edges of the paper. Label the legs *a* and *b* and the hypotenuse *c*.

1. What other two pieces from the tangram can be put together to make a square on the hypotenuse of the triangle? Trace the square on the hypotenuse. **2 large triangles**
2. Which tangram pieces can be put together to make a square on leg *a*? Trace the square on leg *a*. **Answers may vary. Possible answer: a medium triangle and two small triangles**
3. a. What kind of triangle is the medium-sized triangle from the tangram? **an isosceles right triangle**
   b. What is true about legs *a* and *b* on the triangle? **They are congruent.**
   c. What is true about the squares constructed on these legs? **They are congruent.**
   d. Which tangram pieces can be put together to make a square on leg *b*? Trace the square on leg *b*. **a medium triangle and two small triangles**
4. Find the number of small triangles that fit inside the squares constructed on the hypotenuse and on each of the legs of the triangle. **8 small triangles on the hypotenuse, 4 small triangles on each leg**
5. Look at the three numbers that result. What relationship do you see? **The sum of the two lesser numbers equals the third number.**
6. Now measure and find the area of the squares adjacent to the original triangle. What relationship do you see? **The area of the square of the hypotenuse equals the sum of the areas of the square of the legs.**
7. Express the relationship as an equation using the terms *c*, *b*, and *a*. $c^2 = a^2 + b^2$

O-41

In this problem solving activity, students apply the Pythagorean theorem to line segments. **See TRB Exploring O-42.**

Exploring Applications of the Pythagorean Theorem

Many different problems can be solved using the Pythagorean theorem. Here is one kind of problem.

Use graph paper. Draw a coordinate graph for each.

Plot the points (3,4) and (3,7) on the grid. Connect the points with a line segment.

1. What type of line segment is formed? **vertical line segment**
2. What is the length of the line segment? **3 units**
3. How did you determine the length of the segment? **Answers may vary.**

Plot the points (4,1) and (-4,1) on the grid. Connect the points with a line segment.

4. What type of line segment is formed? **horizontal line segment**
5. What is the length of the line segment? **8 units**
6. How did you determine the length of the segment? **Answers may vary.**

Plot the points (1,4) and (4,1) on the grid. Connect the points with a line segment.

7. Describe how this line segment differs from the other line segments? **slanting line segment**
8. Is is possible to find the length in the same way you did before? **no**
9. What do you estimate the length of the line segment to be? **Answers will vary.**

Find the length of the line segment. Construct a right triangle using the line segment as the hypotenuse of a triangle.

10. What are the coordinates of the vertex of the right angle if (1,4) and (4,1) are the coordinates of the hypotenuse? **(1,1)**
11. What are the lengths of the legs of the right triangle formed? **(3 units, 4 units)**
12. You know the length of the legs. Describe how to find the hypotenuse. **Use Pythagorean theorem.**
13. What is the length of the hypotenuse? **5 units**
14. Do you think that you can use this method to find the length of any slanting line in the coordinating plane? **yes**
15. What is the length of the line segment that connects points (-4,5) and (8,0)? **13 units**

O-42

In this calculator activity, students explore perfect squares by using a calculator. **See TRB Exploring O-43.**

Exploring Perfect Squares

One way to find a perfect square number is to multiply a counting number by itself. Use a calculator to discover another rule.

1. Find the product of 8 and 9.
   a. Is it a perfect square? Why or why not? **No; the square root of the product is not an integer.**
   b. If you add 1 to the product, will you get a perfect square? **no**
2. Try to find a pair of consecutive whole numbers whose product is a perfect square. Describe what you found. **none found**
3. Can you find a pair of consecutive whole numbers whose product plus 1 is a perfect square? **no**

Try three consecutive whole numbers.

4. Find the product of 8 × 9 × 10. Is it a perfect square? **no**
5. Add 1 to the product, is it a perfect square? **no**
6. Try to find three consecutive whole numbers whose product is a perfect square. Describe what you found. **none found**
7. Can you find three consecutive whole numbers whose product plus 1 is a perfect square? **yes; possible answer: 2 • 3 • 4 + 1**

Now try four consecutive whole numbers.

8. Find the product of 8 × 9 × 10 × 11. Is it a perfect square? **no**
9. Add 1 to the product. Is it a perfect square? **yes; $\sqrt{7920} = 89$**
10. Try to find four consecutive whole numbers whose product is a perfect square. Describe what you found. **none found**
11. Test other sets of four consecutive whole numbers plus 1. Describe what you found. **They are always perfect squares.**
12. Write a rule to find a perfect square using four consecutive numbers. **To find a perfect square, add 1 to the product of four consecutive whole numbers.**

O-43

In this computer activity, students use a spreadsheet to discover Pythagorean triples. **See TRB Exploring O-44.**

Exploring Pythagorean Triples with a Spreadsheet

A Pythagorean triple is a set of three positive integers that can be expressed as $a^2 + b^2 = c^2$.

1. What is the third number in this Pythagorean triple? 5, 12, **13**
2. Find a set of Pythagorean triples that has 8 as one of the three numbers. **Possible answer: 6, 8, 10 or 8, 15, 17**

Finding Pythagorean triples by trial and error is possible but not easily done. Here is one way to find them. **Answers will vary.**

3. a. Select two positive integers. Record their value. ______
   b. What is the value of twice their product? ______
   c. What is the value of the sum of their squares? ______
   d. What is the value of the difference of their squares? (Subtract the lesser from the greater.) ______
   e. Could the numbers you obtained for questions b through d represent the measures of the sides of a right triangle? **yes**

To find Pythagorean triples you can use a spreadsheet and the formulas suggested in question 3.

Set up a spreadsheet as shown. Let a and b represent the numbers selected (a > b).

	A	B	C	D	E
1	a	b	2ab	a^2 – b^2	a^2 + b^2
2	2	1	4	3	5
3	3	2			
4	4	2			

4. What formulas should be assigned to cell C2? **=2*A2*B3** D2? **=A2^2−B2^2** E2? **=A2^2+B2^2**
5. Why must a > b? **so the difference of squares will be positive.**
6. Which formula expresses the value of the square of the hypotenuse? $a^2 + b^2$

Add column F to the spreadsheet to compute the sum of the legs squared. Add column G to compute the square of the hypotenuse.

7. Verify that the formulas in columns C through E produce Pythagorean triples. What must be the relationship between the values in column F and column G for each row? **They must be equal.**

O-44

Meeting Individual Needs

GIFTED STUDENTS

Mathematically gifted students often enjoy the creative experience of exploring new ways to solve a problem. Students can use the Pythagorean theorem to help them do this exercise. Give the students a copy of $\triangle ABC$. Have them draw an equilateral triangle on each side of $\triangle ABC$. Then ask the students to show that the sum of the area of the two smaller equilateral triangles equals the area of the larger equilateral triangle. ($2.25\sqrt{3}$, $4\sqrt{3}$, $6.25\sqrt{3}$) Repeat the exercise using $\triangle ABC$, but this time label side BC 9 and side CA 12. ($20.25\sqrt{3}$, $36\sqrt{3}$, $56.25\sqrt{3}$)

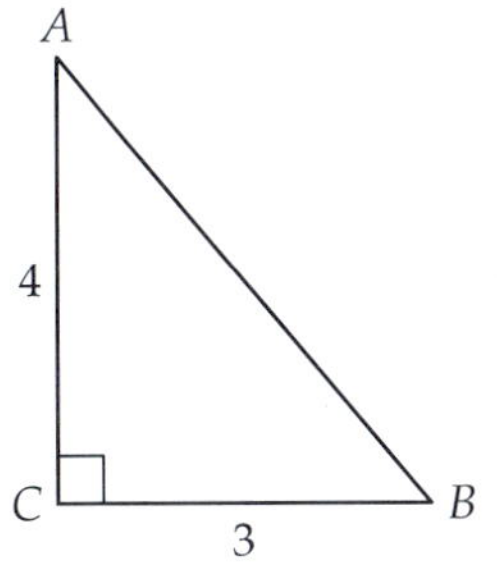

AT-RISK STUDENTS

At-risk students often benefit from organizational aids to help them remember rules, formulas, and mathematical procedures. Since there are many procedures to remember in this chapter, it would be helpful to provide a study guide for these students to use as a reference. Give students a copy of the table at the right. Have small groups peruse the chapter to find the information needed to complete the chart. When students have completed their charts, suggest that they keep their charts for reference.

Title	Formula	When to Use
Pythagorean theorem	$a^2 + b^2 = c^2$	To find side or hypotenuse of right triangle
sine		
cosine		
tangent		
Converse of the Pythagorean theorem		
Similar triangles		
45°-45°-90° triangle		
30°-60°-90° triangle		

ENGLISH AS A SECOND LANGUAGE

Building English vocabulary is probably the most important skill ESL students can develop. Vocabulary can be strengthened through several methods; one effective method is modeling. Modeling can be used to reinforce a concept. For example, students can model a squared number. Provide unit cubes and have students arrange the cubes in a rectangular array to show 4^2. The result should be a 4×4 array. Help students realize that the model shows that $4^2 = 16$ and $\sqrt{16} = 4$. Students can also draw and label the various triangles discussed in the chapter.

See TRB Teaching Resources p. 61.

Problem Solving Situation

The "Think About It" question asks students to consider whether the size and shape of a parachute affect the rate at which the parachute descends. Students may correctly guess that the size and shape do affect the speed of the parachute. In order to find out how size and shape affect speed, have students research different parachute designs. Have them relate differences in size, shape, materials, and other design features to the purposes, e.g., material or recreational use.

Once they are familiar with the basic designs, students should build scale models of different sizes and shapes.

DATA COLLECTION

By attaching a weighted object to each parachute, students can measure the speed of the parachute by dropping it from a height. Remind students that the speed (s) is found by dividing the height (h) from which the parachute was dropped by the time (t) it takes the parachute to hit the ground.

DATA ANALYSIS

Have students chart the data they obtained on the various design features of parachutes and their effects on the rate of descent. If enough information is available on a particular feature, have students display it in graph form. Students should conclude that a parachute with a larger (surface) area will descend at a slower speed.

PARACHUTING TERMS

TERM	DEFINITION
HOLDING	facing canopy into the wind to minimize ground speed
FREE FALL	a jump in which the parachute is activated manually
TARGET	the landing area
TERMINAL VELOCITY	the greatest speed at which a body falls through the air; about 176 ft/s which a jumper reaches after 12 s of free fall
WIND DRIFT INDICATOR	determines the strength and direction of the wind; usually a windsock that is 19.5 ft long with a diameter that varies from 36.25 in. to 12 in.
RUNNING	directing the canopy downwind to maximize ground speed

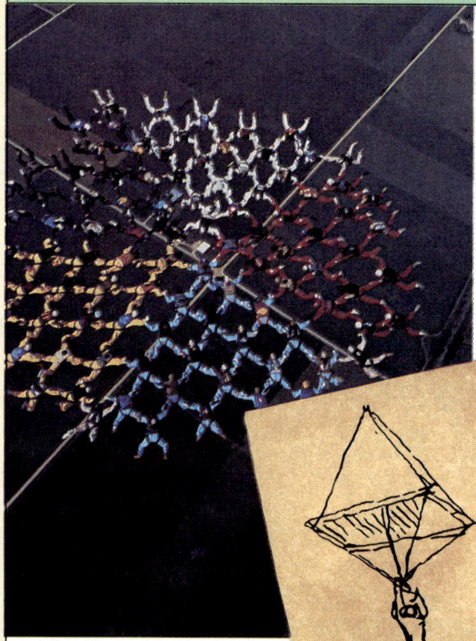

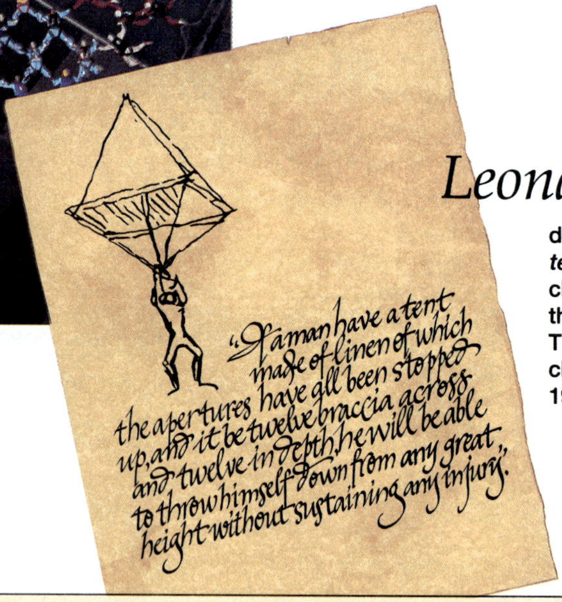

Leonardo daVinci designed a parachute, called a *tent roof* in 1495. Today, parachuting is a popular sport in the United States and Europe. The first sport parachute championships were held in 1951.

CHAPTER

11 Right Triangles in Algebra

IF A PARACHUTIST descends at a rate of 10 ft/s, and travels 2,460 ft straight down, it takes 4.1 min to reach the ground.

You can use right triangles to determine the angle the parachutist makes with the ground, the distance traveled in the jump, and the time it takes to reach the ground.

Up, up, and away

You can use the tangent ratio and the table of trigonometric ratios on page 580 to find the angle the jumper makes with the ground.

$$\tan x = \frac{2{,}460}{5{,}280} \approx .46591$$

$$x \approx 25°$$

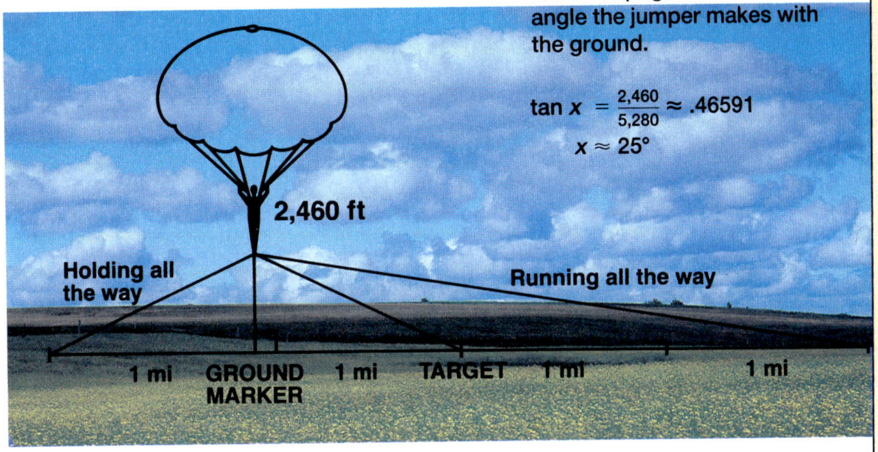

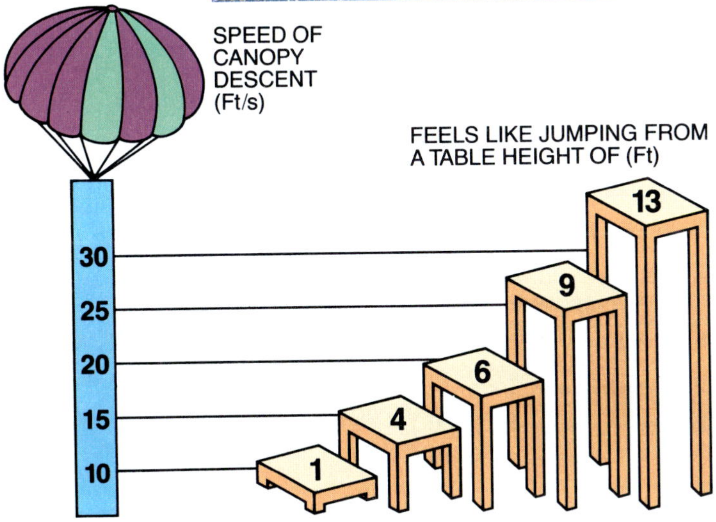

Think about it...

Do you think the size and shape of the parachute canopy affect the rate at which a parachute descends? Describe an experiment to test your hypothesis.

See Problem Solving Situation

Questions related to this Data File can be found on pages 472 and 476.

Cooperative Learning

This activity will be more rewarding and more manageable for students if they work together in small groups. Groups may divide their research into specific tasks or target areas and assign one area to each group member. Consider assigning specific roles to individuals in the group. For example, a Coordinator could be given the responsiblity of monitoring the progress of individual group members. A Data Manager could review every member's research findings and eliminate duplication of results, and a Presenter could be selected to present the group's results to the entire class. Roles such as these can help to ensure group interdependence and individual accountability.

Lesson Focus

Materials

Algebra tiles or graph paper, calculator, math journal, paper, pencil

Vocabulary/Symbols

square root

MOTIVATING THE LESSON

To begin the lesson, ask students to share what they know about a square. (For example, a square is a rectanglular figure with four congruent sides.) Then ask students if they can make a square with 30 algebra tiles. (no)

Instruction

Have students work in small groups. Distribute algebra tiles or graph paper and calculators to each group. Ask students to record their answers in their math journals.

1 Have students model several squares and write the corresponding equations. Help students understand that they can use an exponent to express the total number of tiles in a square.

2 Emphasize that although $5^2 = 25$ and $(-5)^2 = 25$, the notation $\sqrt{25}$ indicates the positive square root. Remind students that if negative numbers are permitted, there will be a negative sign in front of the square root sign, for example, $-\sqrt{25} = -5$.

3 Have groups build models for each question before they decide on the answer.

OBJECTIVE: *To explore square roots using several methods.*

MATERIALS

- Algebra tiles or graph paper
- Standard calculator
- Math journal to record work

Exploring Square Roots

1

▼ Suppose you build a square having length of 5 tiles.

1. ***Model*** the square using algebra tiles. **Check students' work.**
2. How many tiles make up the length of each side of the square? **5**
3. How many tiles do you need to build the square? **25**
4. **a.** ***Write*** an equation using an exponent to describe the relationship between the number of tiles on each side of the square and the total number in the square. **$5^2 = 25$**
   b. ***Model*** other squares using algebra tiles. Write an equation for each square as you did in (a). **Check students' work.**
5. What part of the equation corresponds to the number of tiles on each side of the square? **the base of the exponent**
6. ***Discuss*** how you can determine the number of tiles in a square, given the number of tiles on each side. **Square the number of tiles on each side.**

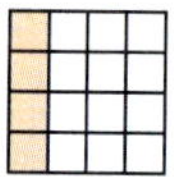

$4^2 = 16$

4 is a square root of 16

▼ The opposite of squaring a number is finding its *square root*. We refer to the number of tiles on each side of the square as the square root of the total number of tiles in the square.

7. **a.** ***Model*** a square with 9 tiles. **Check students' work.**
   b. How many tiles are on each side? **3**
   c. What is the square root of 9? **3**
8. ***Model*** the square represented by each equation using graph paper. Shade in the square root. State the square root of the product. **Check students' models.**
   a. $6^2 = 36$ **6** **b.** $49 = 7^2$ **7** **c.** $1^2 = 1$ **1** **d.** $3.5^2 = 12.25$ **3.5**
   e. ***Describe*** the position in each equation of the square root of the product. **the base of the exponent**
   f. ***Write*** a sentence relating your answers in (5) and (8e).

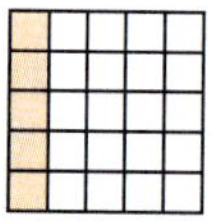

5 is a square root of 25

$\sqrt{25} = 5$

▼ We say *five is a square root of twenty-five* and write $\sqrt{25} = 5$. The notation $\sqrt{25}$ means to find a positive number that when squared is equal to 25. We call the symbol $\sqrt{\ }$ the square root sign.

2

9. **a.** Is there a number other than positive five, that when squared gives you twenty-five? **-5**
   b. ***Explain*** why this number is not a correct value of $\sqrt{25}$. **The $\sqrt{\ }$ symbol denotes a positive value.**
10. Find each square root.
    a. $\sqrt{81}$ **9** **b.** $\sqrt{100}$ **10** **c.** $\sqrt{144}$ **12** **d.** $\sqrt{20.25}$ **4.5**

8. f. The base of the exponent represents both the number of tiles on each side of the square and the square root of the number of tiles in the square.

Teaching Tip

To encourage students to cooperate when working in groups, assign specific roles. Possible roles might be Model Builder, Calculator Operator, and Mathematician. However, remind students that everyone is reponsible for the group's answers.

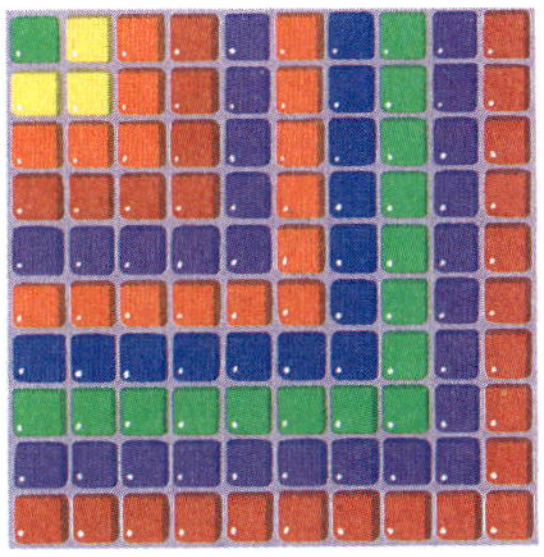

▼ Suppose you are given 21 tiles with which to build a square.
A square cannot be built with exactly 21 tiles.

11. a. ***Model*** the square. Describe your results.

b. Does 21 have a whole number square root? Explain. No; A square could not be built with exactly 21 tiles.

12. Complete each statement.

a. If I could eliminate ■ tiles, I would have a ■ × ■ square. 5; 4; 4

b. If I could add ■ more tiles, I would have a ■ × ■ square. 4; 5; 5

13. a. Between what two positive integers does the $\sqrt{21}$ lie? 4 and 5

b. To what integer is $\sqrt{21}$ closer? Base your answer on the number of squares you need to add or subtract. 5

14. Each square root lies between what two integers? Circle the integer to which it is closer.

a. $\sqrt{5}$ (2), 3 **b.** $\sqrt{76}$ 8, (9) **c.** $\sqrt{147}$ (12), 13

3

▼ You can use a standard calculator to estimate a square root.

Estimate the value of $\sqrt{76}$ to the nearest tenth. You already know that $\sqrt{76}$ lies between the integers 8 and 9.

Try 8.3	8.3	⊗	8.3	⊜	68.89	**too low**
Try 8.6	8.6	⊗	8.6	⊜	73.96	**too low**
Try 8.7	8.7	⊗	8.7	⊜	75.69	**very close**

So, to the nearest tenth, $\sqrt{76}$ is approximately 8.7.

15. Use a calculator to find an approximate value, to the nearest tenth, for each square root. Do not use the square root key.

a. $\sqrt{75}$ 8.7 **b.** $\sqrt{29}$ 5.4 **c.** $\sqrt{94}$ 9.7 **d.** $\sqrt{186}$ 13.6

16. a. ***Describe*** a method for finding an approximate value of a square root to the nearest hundredth.

b. Find the value of $\sqrt{131}$ to the nearest hundredth. 11.45

a. Trap the value of the square root between two integers. Try different values that are approximated to the nearest hundredth.

▼ You can also use a square root key on a calculator.

17. ***Explore*** how to find the square root of a number using a square root key. Start with a square root you already know, such as $\sqrt{9}$.

18. Use a calculator to find the value of each square root.

a. $\sqrt{25}$ 5 **b.** $\sqrt{19}$ 4.3588989 **c.** $\sqrt{0}$ 0 **d.** $\sqrt{-36}$ none

An error message appears.

19. a. ***Describe*** the result when you took the square root of −36.

b. Is it possible to take the square root of a negative number? No; When a negative number is squared, the result is a positive value.

Closure

Have students work in pairs. One partner names a number. The other partner gives the whole number square root, if it exists, or the two whole numbers between which the square root lies. Then the students reverse roles and repeat the process.

Writing in Math Ask students to describe in their math journals the procedure they used to find the square root.

Assignment Guide

Basic All
Average All
Enriched All

FOR THE NEXT LESSON

paper, pencil, calculator, math journal

Lesson Focus

Materials

Paper, pencil, calculator, math journal

Vocabulary/Symbols

irrational number
perfect square
principal square root
real numbers
square root

MOTIVATING THE LESSON

Have students review the square numbers (perfect squares) less than 1,000 by asking them to complete the pattern.

$$\begin{array}{cccc} 1^2, & 2^2, & 3^2, \ldots, & 31^2 \\ \downarrow & \downarrow & \downarrow & \downarrow \\ 1, & 4, & 9, \ldots, & 961 \end{array}$$

Instruction

1 Ask students to explain the difference between finding the *square root* and finding the *principal square root*. (To find the square root means to find both square roots, one positive and one negative, of the number. To find the principal square root means to find the positive square root.)

OBJECTIVE: *Finding square roots of numbers and expressions.*

FLASHBACK

A square is a quadrilateral with four congruent sides and four right angles.

11-1 Finding Square Roots

▼ Each of the quadrilaterals in the puzzle is a square, except the surrounding quadrilateral. Here are two possible ways to find the length of a side of the shaded square if its area is 81.

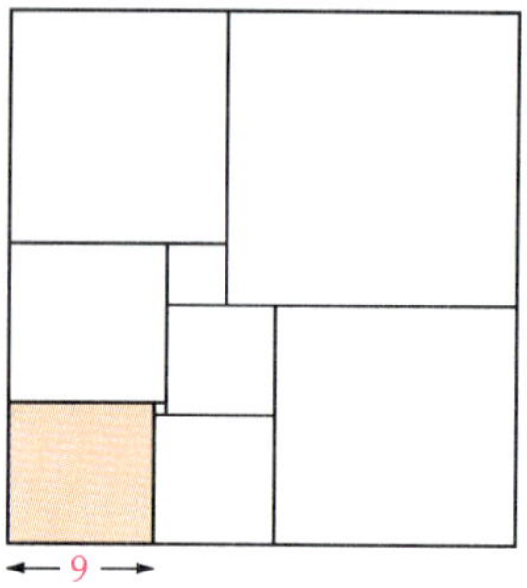

1. Think, *What number squared gives me 81?* Use guess and test with several numbers until you arrive at the correct answer, which is 9.
2. Think, *What is the square root of 81?* The result is also 9.

▼ The inverse of squaring a number is finding its *square root.*

Square Root	The square root of a number, n, is a if $a^2 = n$.

Example 1 **Find the square root of 64.**

Solution **a.** $(-8)^2 = 64$, so -8 is a square root.

b. $8^2 = 64$, so 8 is also a square root.

▼ We use the symbol $\sqrt{\ }$ to denote the positive square root or *principal square root* of a number.

1 **Principal Square Root**	The principal square root of a number is its positive square root. The principal square root is denoted by the symbol $\sqrt{\ }$.

Example 2 **Find the principal square root of each expression.**

a. 10,000 **b.** $121x^2$

Solution **a.** $100^2 = 10{,}000$, so the principal square root is 100.

b. $(11x)^2 = 121x^2$, so the principal square root is $11x$.

Example 3 **Evaluate.**

a. $\sqrt{196}$ **b.** $\sqrt{225p^6}$

Solution **a.** $14^2 = 196$, so 14 is the value of the expression.

b. $(15p^3)^2 = 225p^6$, so $15p^3$ is the value of the expression.

Teaching Tip

The contents of this chapter have a rich history. Have students research such topics as: pi or π, square numbers, the square root symbol, $\sqrt{\ }$, rational and irrational numbers, and the Pythagoreans.

▼ Rational numbers such as 144, 64, and 1 are all perfect squares. They have whole number square roots of 12, 8, and 1, respectively. Rational numbers such as 27 and 65 are not perfect squares. They do not have whole number square roots. We call the square roots of these numbers *irrational numbers*. The set of *real numbers* is made up of the rational numbers and the irrational numbers.

Irrational Numbers	Irrational numbers are numbers which we cannot express as either terminating or repeating decimals.

Example 4 **Determine whether each number is rational or irrational. Use a calculator, if necessary.**

a. 15 **b.** 0.1212 . . . **c.** π **d.** $\sqrt{1.21}$

Solution
a. rational; 15 terminates
b. rational; the decimal has a repeating pattern
c. π 3.1415927 . . . ; irrational; the decimal does not terminate or repeat
d. 1.21 [√] 1.1; rational; 1.1 terminates

▼ You can use a table of square roots to find an approximate value for irrational square roots.

Example 5 **Find the approximate value of $\sqrt{29}$. Use the table on p. 579.**

Solution $\sqrt{29} \approx 5.385$

1. yes; The solutions for $x^2 = 16$ are +4 and -4. The solution for $x = \sqrt{16}$ is only +4 because $\sqrt{\ }$ denotes a positive value.
2. No; One is positive and one is negative.
3. $3^2 + (-4)^2 = 5^2$
4. the number under the square root sign

CLASS EXERCISES

Complete.

1. The name for the mathematical symbol $\sqrt{\ }$ is ■. the square root symbol
2. *The square root of forty-nine* can be written using mathematical symbols as ■. $\sqrt{49}$
3. A ■ number is implied by the symbol $\sqrt{\ }$. positive
4. The set of real numbers is made up of ■ and ■ numbers. rational; irrational
5. A repeating decimal is a(n) ■ number. rational
6. If a square has an area of 49 in.², each side has length of ■. 7 in.
7. The two square roots of 100 are ■ and ■. The principal square root is ■. -10; 10; 10
8. $\sqrt{92}$ is between the integers ■ and ■. 9; 10
9. The decimal 0.121231234 . . . is a(n) ■ number. irrational

NOTES & QUOTES

We use the term *radical* to refer to the square root sign. The term comes from the Latin word *radix*, which means root.

THINK Why does a table of square roots only show approximate values? Irrational numbers do not terminate, so they are rounded to the nearest thousandth.

THINK AND DISCUSS See above.

1. Are the solutions for $x^2 = 16$ and $x = \sqrt{16}$ different? Why?
2. Do the expressions $\sqrt{25}$ and $-\sqrt{25}$ have the same value? Explain.
3. Rewrite the equation $3^2 + -4^2 = 5^2$ to make it true.
4. What number results when you square a square root?

2 Emphasize that a number is a perfect square only if it has a whole number square root.

ADDITIONAL EXAMPLES

1. Find the square roots of 25. (-5, 5)
2. Find the principal square root of 144. (12)
3. Evaluate: $\sqrt{36x^{12}}$ $(6x^6)$
4. Determine whether $\sqrt{5}$ is rational or irrational. (irrational)
5. Find the approximate square root using the square root table. $\sqrt{38}$ (6.164)

Guided Practice

Class Exercises For question 6, be sure that students understand that they are to find the principal square root.

If students have difficulty with question 14, on p. 456, ask them what fraction multiplied by itself will give $\frac{4}{9}$.

Think and Discuss For question 2, help students understand that $-\sqrt{25}$ means "the opposite of the principle square root of 25," which is -5.

MATH MINUTES

Which terms are irrational numbers?

$-\sqrt{35}$, $\sqrt{35}$, 3.555 . . ., 3.5050050005 . . ., $\sqrt{\frac{2}{3}}$, 3.14 (All are irrational except 3.555 . . . and 3.14.)

A ***common error*** occurs when students confuse principal square root with square root. Help them to differentiate between the two by having them complete statements such as: The two square roots of 169 are ■ and ■. The principal square root is ■. (13, -13; 13)

Closure

Writing in Math Have students write a brief summary of the lesson in their math journals. Ask them to differentiate between square root and principal square root and rational and irrational numbers in their summaries.

Independent Practice

Discuss students' answers to questions 1-8. To be sure they understand the concepts, ask students to explain their answers.

For questions 9-16, recall the difference between squaring and finding the square root.

As an extension to this lesson, assign the calculator activity that explores perfect squares on p. 450E of the Chapter Overview. See also TRB Exploring O-43.

MIXED REVIEW

1. Find the surface area of a cube whose side is 8.2 cm. $403.44\ cm^2$
2. Find the volume of a cylinder with $r = 0.03$ m and $h = 0.9$ m. $0.00254\ m^3$

Solve and check.

3. $8n - 2 \le 5$ $n \le \frac{7}{8}$
4. $\frac{1}{3}x + 5 \ge \frac{1}{2}$ $x \ge -13\frac{1}{2}$
5. Find the slope of the line $-4x - 7y = 10$. $-\frac{4}{7}$
6. Solve the proportion $\frac{x}{7} = \frac{7}{3}$. $x = 16\frac{1}{3}$
7. ***DATA FILE 8 (pp. 312–313)*** What is the percent increase in the United States population from 1790 to 1890? 1,502%
8. Laura bought two sweaters for \$36. The sweaters were on sale at one for full price and the second at half price. How much was the half-price sweater? \$12

Find each square root. Use the square root table on p. 579 or a calculator, if necessary. Round answers to the nearest thousandth.

10. $\sqrt{100}$ 10 **11.** $\sqrt{1}$ 1 **12.** $\sqrt{13}$ 3.606 **13.** $\sqrt{50}$ 7.071

14. $\sqrt{\frac{4}{9}}$ $\frac{2}{3}$ **15.** $\sqrt{12.25}$ 3.5 **16.** $\sqrt{15^2}$ 15 **17.** $\sqrt{m^2}$ m

State whether each number is rational or irrational. Use a calculator, if necessary.

18. $\sqrt{0}$ rational **19.** $\sqrt{87}$ irrational **20.** $-\sqrt{16}$ rational **21.** 4.1010010001 . . . irrational

WRITTEN EXERCISES

True or false.

1. All real numbers are integers. false
2. All integers are real numbers. true
3. The principal square roots of 36 are -6 and 6. false
4. $\sqrt{36} + \sqrt{64} = 14$ true
5. $\sqrt{36 + 64} = \sqrt{100} = 10$ true
6. $\sqrt{79}$ is between 8 and 9. true
7. If a square has an area of 225 cm^2, the side is 15 cm. true
8. An approximation for $\sqrt{3}$ is 1.723. false

MENTAL MATH **Square each term.**

9. 16 256 **10.** -8 64 **11.** $\frac{2}{3}$ $\frac{4}{9}$ **12.** 11 121

13. $5x^3$ $25x^6$ **14.** $\sqrt{2}$ 2 **15.** $\sqrt{9^2}$ 81 **16.** $\sqrt{2x}$ 2x

MENTAL MATH **Evaluate.**

17. $\sqrt{10{,}000}$ 100 **18.** $\sqrt{169}$ 13 **19.** $\sqrt{59 + 5}$ 8 **20.** $\sqrt{p^2}$ p

Find each square root. If necessary, use your calculator. Round decimal answers to the nearest thousandth.

21. $\sqrt{49}$ 7 **22.** $\sqrt{81}$ 9 **23.** $\sqrt{\frac{16}{25}}$ $\frac{4}{5}$ **24.** $\sqrt{(r + 3)^2}$ $r + 3$

25. $\sqrt{49.49}$ 7.035 **26.** $\sqrt{196x^6}$ $14x^3$ **27.** $\sqrt{8y^5 \cdot 2y^5}$ $4y^5$ **28.** $\sqrt{3 \cdot 12x^{16}}$

29. $\sqrt{256}$ 16 **30.** $\sqrt{7y^4}$ $2.646y^2$ **31.** $\sqrt{100 \div 5}$ 4.472 **32.** $\sqrt{a^4b^6c^8}$ $a^2b^3c^4$

28. $6x^8$

Between what two integers does each square root lie? Circle the integer to which the square root is closer.

33. $\sqrt{51}$ (7), 8 **34.** $\sqrt{93}$ 9, (10) **35.** $\sqrt{5}$ (2), 3 **36.** $-\sqrt{22}$ (-5), -4

37. $-\sqrt{5}$ -3, (-2) **38.** $\sqrt{132}$ (11), 12 **39.** $\sqrt{7 + 11}$ (4), 5 **40.** $\sqrt{99}$ (10), 9

State whether each term is rational or irrational. Use a calculator, if necessary. 41. rational 42. rational 43. irrational 44. rational

41. $\sqrt{625}$ **42.** $-\sqrt{36}$ **43.** $\sqrt{32}$ **44.** $\sqrt{0}$

45. 198.94762 rational **46.** $\sqrt{53}$ irrational **47.** 4.33333 . . . rational **48.** $\sqrt{5+11}$ rational

Use the square root table on p. 579 to approximate each value.

49. $\sqrt{50}$ 7.071 **50.** $\sqrt{77}$ 8.775 **51.** $\sqrt{99}$ 9.950 **52.** $\sqrt{2}$ 1.414

53. $\sqrt{43}$ 6.557 **54.** $\sqrt{17}$ 4.123 **55.** $\sqrt{54}$ 7.348 **56.** $\sqrt{87}$ 9.327

Critical Thinking

EXPLORING PATTERNS

Sir Isaac Newton devised a method for finding a square root. It is often referred to as the *divide and average method.*

Problem Approximate the value of $\sqrt{47}$.

Solution Use the *divide and average method:*

1. Trap between two integers. $6 < \sqrt{47} < 7$
2. Identify the closer integer. $\sqrt{47}$ is closer to 7
3. Estimate to the nearest tenth. 6.8
4. Divide the number that you are taking the square root of by the estimate. Carry to whatever place accuracy is desired in the answer. $47 \div 6.8 \approx 6.91$
5. Find the mean average of the quotient and the divisor. $(6.8 + 6.91) \div 2 \approx 6.86$
6. Use the mean as the new divisor. $47 \div 6.86 \approx 6.85$
7. If necessary, repeat Steps 4, 5, and 6 until the divisor is very close to the quotient.

You can use a BASIC program to simulate Newton's method.

```
10 PRINT "Square root of what number";
20 INPUT A
30 PRINT "How many averages";
40 INPUT B
50 PRINT "What is your first estimate";
60 INPUT C
70 FOR X = 1 to B
80 PRINT (A/C + C)/2
90 NEXT X
```

1. Continue this pattern for $\sqrt{47}$. Explain your results. How can you get the answer? You keep getting 6.85 and 6.86. Go halfway between the numbers to get 6.855.
2. Use Newton's method to find the approximate value of $\sqrt{19}$. 4.359
3. Why do you think people no longer use Newton's method? Calculators and computers can calculate square roots instantly.
4. ***PROJECT*** Use the computer program shown at the right to find the square roots of various numbers. Check students' work.

Critical Thinking Point out that this alternative method can be used if tables and/or calculators are not available.

Lesson Follow-up

Reteaching Acitivity To help students to estimate square roots, have them:

a. Pick a number.
b. Guess the square root.
c. Check the guess by using a calculator.

LESSON RESOURCES

TRB Practice K-1
TRB Enrichment K-2

Practice

For use after 11-1 (pp. 454–457)

Finding Square Roots

True or false?

1. Negative numbers are neither rational nor irrational. false
2. $p^2 = c$. Therefore, $c = \sqrt{p}$. false
3. One square root of $\frac{16}{25}$ is $-\frac{4}{5}$. true
4. $\sqrt{1}$ is an irrational number. false
5. $\sqrt{29}$ is between 5 and 6. true
6. A number that can be expressed as a repeating decimal is rational. true
7. All real numbers are rational. false
8. Every positive number has two square roots. true

MENTAL MATH Square each term.

9. 15 225 10. $\frac{1}{2}$ $\frac{1}{4}$ or 0.25 11. 0.2 0.04 12. -9 81
13. y^3 y^6 14. $\sqrt{7}$ 7 15. $3m$ $9m^2$ 16. $\sqrt{a+b}$ $a+b$

MENTAL MATH Simplify.

17. $\sqrt{900}$ 30 18. $\sqrt{x^{16}}$ x^8 19. $\sqrt{144}$ 12 20. $\sqrt{9+16}$ 5

Find each square root. If necessary, use your calculator. Round decimal answers to the nearest thousandth.

21. $\sqrt{169}$ 13 22. $-\sqrt{100}$ -10 23. $\sqrt{0.16}$ 0.4
24. $\sqrt{1369}$ 37 25. $\sqrt{\frac{121}{144}}$ $\frac{11}{12}$ or $0.91\overline{6}$ 26. $\sqrt{83.7}$ 9.149
27. $\sqrt{225x^8}$ $15x^4$ 28. $\sqrt{(x+7)^{12}}$ $(x+7)^6$ 29. $\sqrt{18p^3 \cdot 2p}$ $6p^2$
30. $\sqrt{81x^{10}y^{14}}$ $9x^5y^7$ 31. $\sqrt{18 \cdot 8}$ 12 32. $\sqrt{343 \div 7}$ 7

Each square root is between what two integers? Circle the integer to which it is closer.

33. $\sqrt{18}$ (4), 5 34. $\sqrt{60}$ 7, (8) 35. $-\sqrt{8}$ (-3), -2
36. $\sqrt{90}$ (9), 10 37. $\sqrt{29+8}$ (6), 7 38. $-\sqrt{21}$ -4, (-5)

Is the term rational or irrational? Write *R* or *I*.

39. $\sqrt{289}$ R 40. 5.7777... R 41. $\sqrt{41}$ I

Chapter 11 K-1

Enrichment

For use after 11-1 (pp. 454–457)

Patterns of Roots

Since $5^3 = 5 \cdot 5 \cdot 5 = 125$, 5 is the *cube root* of 125. This is written $5 = \sqrt[3]{125}$.

Find the cube root.

1. $\sqrt[3]{8}$ 2 2. $\sqrt[3]{1}$ 1 3. $\sqrt[3]{27}$ 3 4. $\sqrt[3]{64x^{12}}$ $4x^4$

Complete the table.

n	1	2	3	4	5	6	7	8	9	10
n^3	1	8	27	64	125	216	343	512	729	1,000

5. Write the ones' digits of the numbers in the n^3 row. Tell what is interesting about these digits.
1, 8, 7, 4, 5, 6, 3, 2, 9, 0; each of the digits 0–9 occurs once.
6. Use your calculator to find these cubes.
a. 17^3 4,913 b. 27^3 19,683 c. 57^3 185,193
d. 19^3 6,859 e. 49^3 117,649 f. 79^3 493,039
7. Describe a method for predicting the ones' digit of the cube root of a perfect cube simply by looking at the perfect cube. Answers may vary. Find the cube in the table with the same ones' digit as the given cube. The two roots will have the same ones' digit.

MENTAL MATH Write the ones' digit of the cube root.

8. $\sqrt[3]{681{,}472}$ 8 9. $\sqrt[3]{140{,}608}$ 2 10. $\sqrt[3]{12{,}167}$ 3

$2^5 = 2 \cdot 2 \cdot 2 \cdot 2 \cdot 2 = 32$. 2 is the *fifth root* of 32. $2 = \sqrt[5]{32}$.

Complete the table.

n	1	2	3	4	5	6	7	8	9	10
n^5	1	32	243	1,024	3,125	7,776	16,807	32,768	59,049	100,000

11. Describe how to predict the ones' digit of the fifth root of a perfect fifth power.
The fifth power and its root have the same ones' digit.

MENTAL MATH Write the ones' digit of the fifth root.

12. $\sqrt[5]{2{,}476{,}099}$ 9 13. $\sqrt[5]{229{,}345{,}007}$ 7 14. $\sqrt[5]{33{,}554{,}432}$ 2

K-2 *Chapter 11*

LESSON QUIZ

Find each square root.

1. $-\sqrt{49}$ (-7)
2. $\sqrt{49}$ (7)
3. $\sqrt{\frac{16}{81}}$ $\left(\frac{4}{9}\right)$

Each square root is between what two integers?

4. $\sqrt{75}$ (8 and 9)
5. $-\sqrt{125}$ (-11 and -12)

State whether the number is rational or irrational.

6. -55 (rational)
7. $-\sqrt{55}$ (irrational)
8. $\sqrt{55}$ (irrational)

FOR THE NEXT LESSON

blocks, straws or other small objects, math journal

Assignment Guide

Basic 1-8 odd, 9-12, 17-18, 21-24, 33-36, 41-44, 49-51, MR, CT All

Average 2-16 even, 18-19, 25-28, 35-38, 43-46, 50-53, MR, CT All

Enriched 4-8, 13-16, 19-20, 29-32, 37-40, 45-48, 53-56, MR, CT All

Lesson Focus

Materials
Blocks, straws or other small objects, math journal

Vocabulary/Symbols
simulate

MOTIVATING THE LESSON

Ask students to tell what strategy can be used to solve this problem: Paula, Tom, Vivian, Mario, Elaine, and Harold, in that order, are seated (evenly spaced) around a circular table. Who is seated directly across from Vivian? (Students may draw a picture or act it out.)

Instruction

1 If there are at least 25 students in your class, have them act out the solution. Otherwise, have students use objects to demonstrate each step.

Guided Practice

Class Exercises Students can benefit by working in small groups.

A ***common error*** occurs when students choose an inappropriate strategy to solve a problem. *Drawing a picture* may suggest a solution that is not readily found in *making a table.* Have the class compile a chart of problem-solving strategies and discuss when each is useful.

Closure

Writing in Math Have students describe why simulation was an appropriate strategy for the problem. Then have them list the strategy in their math journals and summarize how it aided in finding the solution.

OBJECTIVE: *To solve a problem by simulation.*

PROBLEM SOLVING STRATEGY

11-2 Simulating a Problem

Sometimes when solving a problem, it is helpful to act out or *simulate* the problem.

PROBLEM

A class of 25 seated students counted off by ones beginning with the number one. Each student who counted a multiple of four stood up. Then the students who were still seated counted off by ones again. Each student who counted a multiple of four stood up. This process was repeated one more time. How many students were standing after the third counting? **1**

SOLUTION

READ What do you want to find? The number of students standing after the third counting

PLAN Start out with 25 students. Act out each step of the problem.

SOLVE Have 25 students sit in their seats.

First Counting
1. All students count off by ones.
2. All students who counted a multiple of four stand up.

Second Counting
1. All students seated count off by ones.
2. All students who counted a multiple of four stand up.

Third Counting
1. All students seated count off by ones.
2. All students who counted a multiple of four stand up.

LOOK BACK There will be 13 students standing after the third counting.

Teaching Tip

Computers are especially useful for simulating problems. Consider having students use computers in small groups to solve problems similar to the one in this lesson.

CLASS EXERCISES

Refer to the problem on page 458.

1. How many students will be standing after the fifth counting? 18
2. After how many countings will three students be sitting? 9

WRITTEN EXERCISES

Use a CALCULATOR where appropriate.

Solve by simulating the problem.

1. Suppose you purchase a rare coin for $15, sell it for $23, and then buy it back for $31. How much money did you make or lose in selling and repurchasing this coin? Lost $8.
2. Jim is hosting a dinner party. Jim greets the first guest and they shake hands. The second guest arrives and shakes hands with Jim and the first guest. The third guest arrives and shakes hands with Jim, the first guest, and the second guest. This pattern continues. How many handshakes have taken place after the ninth guest arrives? 45

Solve. Use any strategy.

3. Sandy and Toby are 18 mi apart. They begin to walk toward one another. Sandy walks at 3 mi/h and Toby walks at 2 mi/h. After 1 h of walking Toby decides to rest for 1 h. In how many hours will they meet? How far will Sandy have walked? Toby? 4 h; 12 mi; 6 mi
4. Jon's bicycle license is a three digit number. The product of the digits is 140. The sum of the digits is 16. The numbers appear in descending order. What is Jon's license number? 754
5. Chris was thinking of a number. He added 4, multiplied the sum by -5, then subtracted 12. He then doubled the result to get -34. Of what number was Chris thinking? -3
6. There are 180 children standing around a parachute (assume this is a circle). The children are spaced evenly and numbered consecutively from 1 to 180. Ann is number 7. Tara is standing directly opposite Ann. What is Tara's number? 97
7. A hot air balloon is 2,200 ft in the air. It is scheduled to land at 3:22 P.M. The balloon descends at a rate of 110 ft/min. At what time should the descent begin? 3:02 P.M.
8. ***DATA FILE 12 (pp. 486–487)*** Milwaukee, WI, has an altitude of 635 ft. Boston, MA, has an altitude of 21 ft. About how much farther will an average home run travel in Milwaukee than in Boston? ≈4.5 ft

Independent Practice

Have students share their solutions for questions 3-8 to reinforce the idea that there is usually more than one way to solve a problem.

Lesson Follow-up

LESSON RESOURCES

TRB Practice K-3
TRB Enrichment K-4

Practice — For use after 11-2 (pp. 458–459)

Simulating a Problem

Solve by simulating the problem.

1. Twenty people seated in a circle counted to seven, beginning with the number one. The seventh person dropped out and those remaining counted to seven again. If every seventh person dropped out, what was the number of the last person remaining in the circle? Use the number circle to simulate the problem.

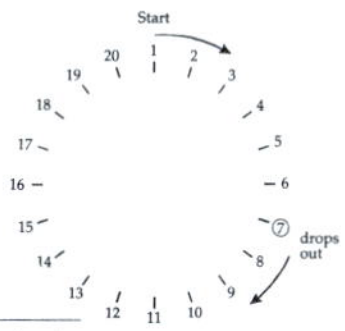

3

2. The two digits of Luther's age are the same. The tens' and ones' digits of his house number, which is between 400 and 500, show his age. His house number is divisible by his daughter's age, 13.
   a. Find the first number between 400 and 500 which is divisible by 13. 403
   b. By adding 13 successively, write the remaining numbers between 400 and 500 that are divisible by 13. 416, 429, 442, 455, 468, 481, 494
   c. Which of the numbers that you wrote is Luther's house number? 455
3. The Rockets played their first volleyball game on Friday, October 18, and played a game every Friday thereafter.
   a. What was the date of their ninth game? December 13
   b. What was the number of the game they played on February 7? 17
4. Five coins are placed side by side as shown. A move consists of sliding two adjacent coins to an open spot without changing the order of the two coins. (The move "2-3 right" is illustrated.) Find three successive moves that will leave the coins in this order: 3-1-5-2-4

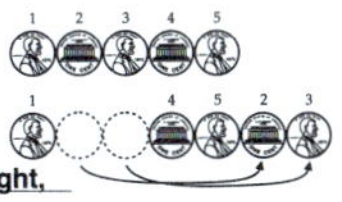

Answers may vary. 1-2 right, 4-5 right, 2-4 right

Chapter 11 K-3

Enrichment — For use after 11-2 (pp. 458–459)

Calculator Simulations

By programming your calculator to generate successive multiples, you can simulate number searches. First, make sure you can generate multiples on your calculator.

Example Program a calculator to generate multiples of 17.

	Standard	Scientific
	17 (+) 17 (=) (=) (=) (=)	17 (+) [K] 17 (=) (=) (=) (=)
Result:	34, 51, 68, 85	34, 51, 68, 85

Each time you press (=), the next multiple of 17 is displayed.

Example A number divisible by 179 contains three identical digits. Find the smallest such number.

Solution Program your calculator to generate multiples of 179. Press (=) until you find the number. Answer: 1,611

Solve using calculator simulations.

1. A number that is divisible by 679 contains four consecutive digits that appear in order from right to left. Find the smallest such number. 5,432
2. A number that is divisible by 129 contains three 2s. Find the smallest such number. 2,322
3. A number that is divisible by 251 contains three 7s. Find the smallest such number. 6,777
4. A multiple of 733 contains consecutive digits that appear in order. Find the smallest such multiple. 23,456

By using (×) instead of (+), you can program your calculator to generate powers of numbers. Answers may vary.

5. Find a power of 6 containing three 7s in a row. 7,776
6. Find a power of 8 containing three 7s in a row. 16,777,216
7. Find a power of 11 containing three 8s in a row. 214,358,881

K-4 Chapter 11

MATH MINUTES

Have students work in small groups. Each group makes up a word problem and then asks other groups to suggest strategies for analyzing it, and then solving it.

Assignment Guide

Basic All
Average All
Enriched All

FOR THE NEXT LESSON

graph paper, math journal, paper, pencil, scissors

Lesson Focus

Materials

Graph paper, math journal, paper, pencil, scissors

MOTIVATING THE LESSON

Draw several right triangles on the chalkboard. Ask students to identify the legs, hypotenuse, and right angle for each triangle.

Instruction

1 Have groups brainstorm ideas to find the length of the hypotenuse. If necessary, tell them that they can cut the graph paper. Have the groups conclude that they can measure the hypotenuse against a strip of graph paper.

2 Explain that because of its angle, the square on the hypotenuse cannot be drawn directly on the graph paper. A 5-by-5 square of graph paper can be cut out and attached to the hypotenuse to show an area of 25 square units.

3 Have the class discuss the relationships they observed. Then explain that the relationships are expressed as the Pythagorean theorem.

Closure

Ask students to draw three right triangles and follow the instructions in questions 6-7 for each triangle. Then ask students if their results are consistent with the Pythagorean theorem.

Assignment Guide

Basic All
Average All
Enriched All

FOR THE NEXT LESSON

overhead projector, geoboard or dot paper, graph paper, pencil, ruler, calculator, math journal

OBJECTIVE:
To explore properties of right triangles.

MATERIALS

- Graph paper
- Math journal to record work

4. c. Draw the square separately on a piece of graph paper. Cut it out. Mount the square on the hypotenuse.

Exploring Right Triangles

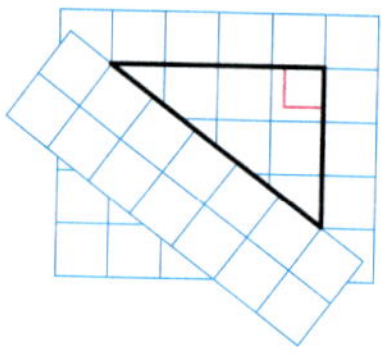

■ Look at the right triangle shown. We call the side opposite the right angle the *hypotenuse.* We call the other two sides *legs.*

1. How many units long is each leg? 3 units and 4 units
2. a. How many units long is the hypotenuse? 5 units
   b. ***Explain*** how the length of the hypotenuse was determined using graph paper. Cut off a strip of graph paper and use it as a ruler.

(1)

■ A square is drawn on each side of the right triangle.

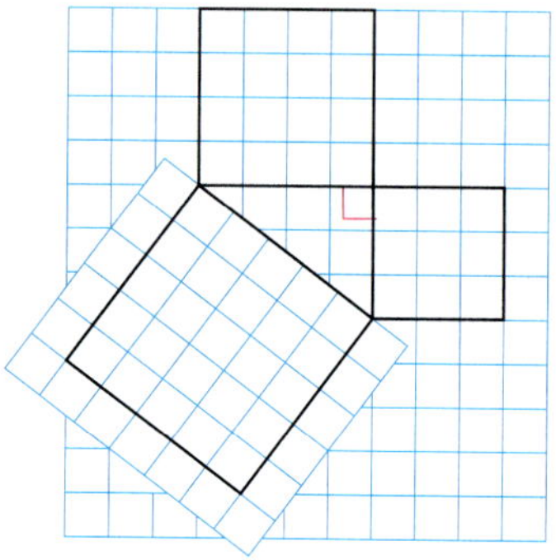

Egyptian rope stretchers used right triangles to relocate property lines after the annual flooding of the Nile river.

3. What is the length of a side on each square? ***Describe*** what it means to square a side of a triangle. 3, 4, and 5; to draw a square on each side of a triangle having the same dimensions as the side
4. a. What are the areas of the squares on each leg? 9 and 16
   b. What is the area of the square on the hypotenuse? 25
   (2) c. ***Explain*** how to draw the square on the hypotenuse.
   (3) d. ***Discuss*** the relationship between your findings in (a) and (b). When you add the areas of the squares on the legs you get the area of the square on the hypotenuse.

■ A right triangle has side lengths 6, 8, and 10.

5. ***Model*** the right triangle and square each side. Check students' work.
6. a. What are the areas of the squares on each leg? 36 and 64
   b. What is the area of the square on the hypotenuse? 100
   c. ***Discuss*** the relationship between your findings in (a) and (b). When you add the areas of the squares on the legs you get the area of the square on the hypotenuse.

■ The sides of a right triangle share a special relationship.

7. ***Describe*** the relationship between the sides of any right triangle. If the legs are a and b and the hypotenuse is c, then $a^2 + b^2 = c^2$.

Teaching Tip

Periodically, review the dynamics of group interaction with the class. Discuss ways to improve how groups work together and to enhance the successful interaction among group members.

OBJECTIVE: *To use the rule of Pythagoras to verify right triangles and to find missing values of right triangles.*

11-3 Pythagorean Theorem

Pythagoras is most famous for discovering the relationship between the lengths of the sides of a right triangle. This relationship is known as the *Pythagorean theorem.*

Pythagorean Theorem	In any right triangle with legs a and b, and hypotenuse c, $a^2 + b^2 = c^2$.

1

NOTES & QUOTES

Pythagoras was a Greek mathematician born about 500 B.C. He founded a school to promote the study of philosophy, mathematics, and natural science.

You can model the Pythagorean theorem using graph paper. The sum of the squares on each side equals the number of squares on the hypotenuse.

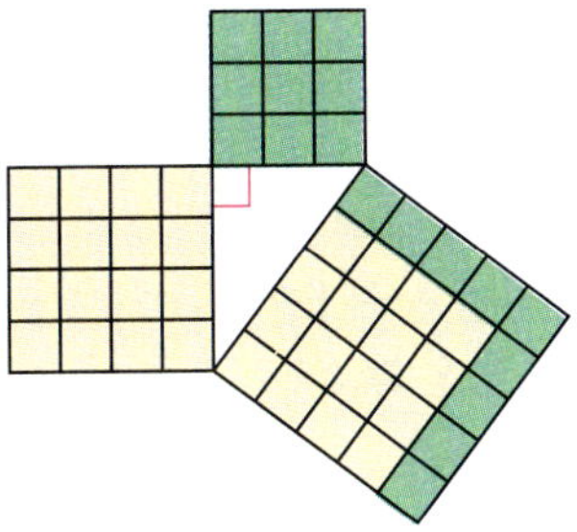

You can use the Pythagorean theorem to find unknown lengths of a triangle.

Example 1 **Find c, the length of the hypotenuse.**

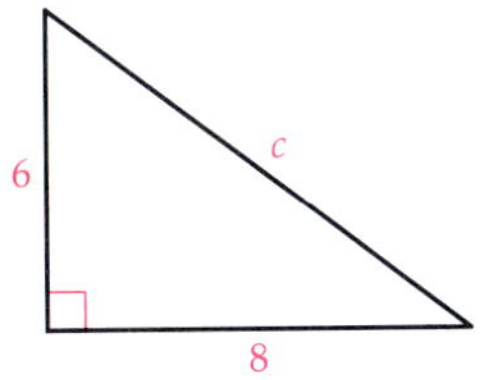

2

Solution

$a^2 + b^2 = c^2$ Write the theorem.

$6^2 + 8^2 = c^2$ Substitute values for the variables.

$36 + 64 = c^2$

$100 = c^2$

$\sqrt{100} = \sqrt{c^2}$ Find the square root of each side.

$10 = c$

The hypotenuse, c, has length 10.

Check

$a^2 + b^2 = c^2$ Substitute values.

$6^2 + 8^2 = 10^2$

$36 + 64 = 100$

$100 = 100$ ✓

Lesson Focus

Materials

Overhead projector, geoboard or dot paper, graph paper, pencil, ruler, calculator, math journal

Vocabulary/Symbols

converse of the Pythagorean theorem
Pythagorean theorem
Pythagorean triple

MOTIVATING THE LESSON

Help students recall the relationship among the sides of a right triangle by using a geoboard or dot paper with an overhead projector. Show a right triangle with the squares drawn on the sides.

Instruction

1 Point out that $a^2 + b^2$ represents the sum of the areas of the squares on the legs and c^2 represents the area of the square on the hypotenuse.

2 Have students verify the theorem by actually counting the number of unit squares in each square.

Teaching Tip

Geoboards, dot paper, or graph paper are especially useful for representing right triangles. Students can use these aids to verify answers obtained using the Pythagorean theorem and vice versa.

③ Point out that since both the theorem and its converse are true, the theorem can be used to find unknown measures of a triangle, as in Example 1, and also to determine if a triangle is a right triangle, as in Example 2.

④ Emphasize the importance of organizing the solution step by step.

ADDITIONAL EXAMPLES

1. The lengths of the legs of a right triangle are 7 and 12 units. Find the length of the hypotenuse to the nearest tenth. (13.9 units)
2. Do side lengths of 9, 12, and 5 units form a right triangle? (no)
3. Find the area of a triangle where leg b measures 3 units and the hypotenuse measures 17 units. (25.1 square units)

▼ The converse of the Pythagorean theorem is also true.

③ **Converse of Pythagorean Theorem**	If $a^2 + b^2 = c^2$, then the triangle with sides a, b, and c is a right triangle.

THINK When given three side lengths of a right triangle, how do you know which one is the length of the hypotenuse? The hypotenuse is the longest side of a right triangle.

Example 2 **Determine if the triangle with the given side lengths is a right triangle.**

a. 12, 16, 20 **b.** 7, 8, 9

Solution

a. $12^2 + 16^2 = 20^2$
$144 + 256 = 400$
$400 = 400$ ✓
This is a right triangle.

b. $7^2 + 8^2 = 9^2$
$49 + 64 = 81$
$113 = 81$ ✗
This is not a right triangle.

We say that a set of integers such as 12, 16, 20 is a *Pythagorean triple.*

④ ▼ You can use the Pythagorean theorem to help solve a problem.

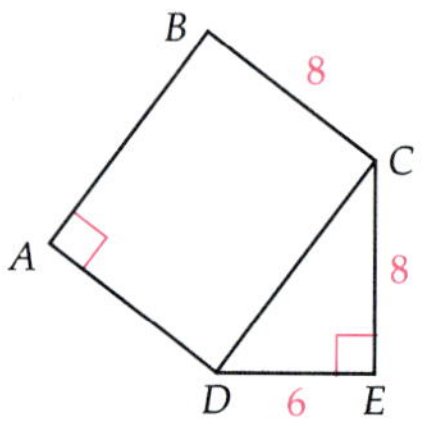

Example 3 **Find the area of *ABCED*.**

Solution

1. Find CD using the Pythagorean theorem.
$$6^2 + 8^2 = CD^2$$
$$36 + 64 = CD^2$$
$$100 = CD^2$$
$$\sqrt{100} = \sqrt{CD^2}$$
$$10 = CD$$

2. Find the area of the triangle.
$$A = \frac{1}{2}bh$$
$$24 = \frac{1}{2}(6)(8)$$

3. Find the area of the rectangle.
$$A = lw$$
$$80 = (10)(8)$$

4. Find the total area. Add areas together.
$$24 + 80 = 104$$

The area of *ABCED* is 104 square units.

THINK AND DISCUSS

1. How could the Pythagorean theorem be used to find the altitude of an equilateral triangle with side length 12?
2. Is it possible for all three numbers in a Pythagorean triple to be even? odd? yes; no
3. A cube has a side length of 5. Find the length of the diagonal on one of the faces to the nearest hundredth. 7.07

1. The hypotenuse is 12. A leg is 6. Solve for the other leg, the altitude, using the equation $6^2 + x^2 = 12^2$.

CLASS EXERCISES

Name the legs and the hypotenuse.

1. $\overline{AB}$ and $\overline{BC}$; $\overline{AC}$

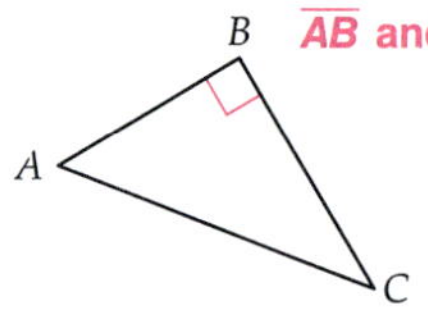

2. $\overline{WX}$ and $\overline{XY}$; $\overline{WY}$

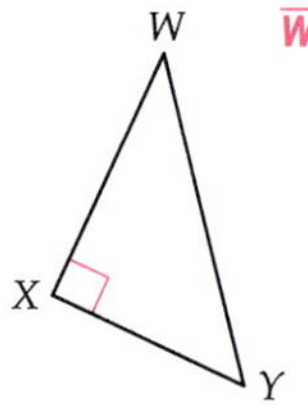

MATH MINUTES

Which group *is not* a Pythagorean triple?

1. a. 3, 6, $\sqrt{45}$
b. 1, 2, 4
c. 5, 12, 13 (b)
2. a. 6, 5, 11
b. 4, 8, $\sqrt{80}$
c. 20, 21, 29 (a)

Write an equation. Solve for x.

3.

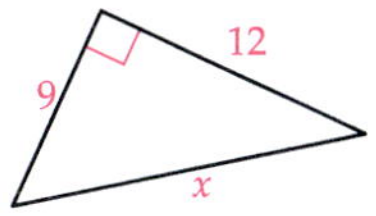

4.

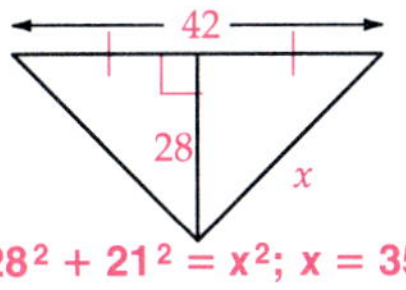

5.

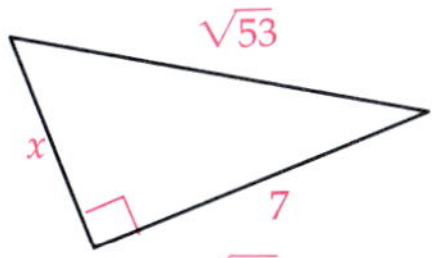

$9^2 + 12^2 = x^2$; $x = 15$ $28^2 + 21^2 = x^2$; $x = 35$ $7^2 + x^2 = (\sqrt{53})^2$; $x = 2$

6. Is $1, 1, \sqrt{2}$ a Pythagorean triple? yes

7. Two hikers started their trip from base camp by walking 15 m due east. They then turned due north, walking 17 m to a large pond. How far is the pond from base camp to the nearest tenth of a meter? 22.7 m

WRITTEN EXERCISES

MENTAL MATH **Simplify.**

1. $\sqrt{5^2}$ 5 **2.** $\sqrt{144}$ 12 **3.** $\left(\frac{3}{5}\right)^2$ $\frac{9}{25}$ **4.** $(2\sqrt{3})^2$ 12

CALCULATOR **Find each value to the nearest thousandth.**

5. $\sqrt{63}$ 7.937 **6.** $\sqrt{12}$ 3.464 **7.** $\sqrt{32}$ 5.657 **8.** $\sqrt{95}$ 9.747

Name the legs and the hypotenuse.

9.

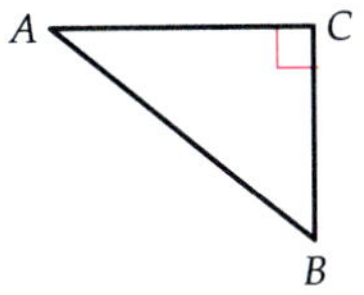

$\overline{AC}$ and $\overline{CB}$; $\overline{AB}$

10.

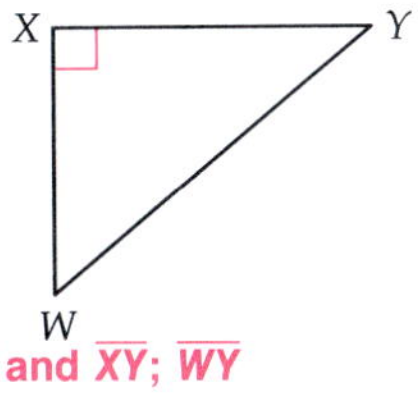

$\overline{WX}$ and $\overline{XY}$; $\overline{WY}$

State which value is the length of the hypotenuse.

11. 10, 6, 8 10 **12.** $\sqrt{9}, \sqrt{25}, \sqrt{16}$ $\sqrt{25}$ **13.** $5, \sqrt{56}, 9$ 9

Write an equation. Solve for x. Round to the nearest hundredth.

14.

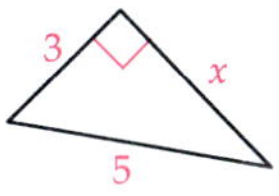

15.

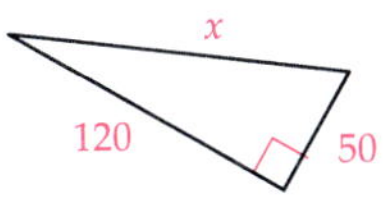

16.

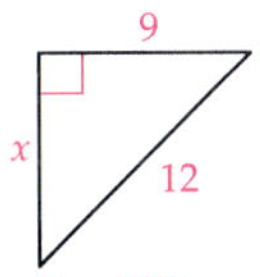

$3^2 + x^2 = 5^2$; $x = 4$ $50^2 + 120^2 = x^2$; $x = 130$ $9^2 + x^2 = 12^2$; $x = 7.94$

17.

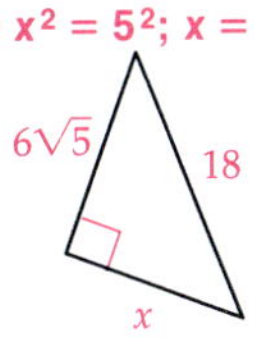

18.

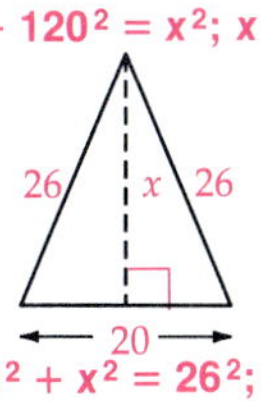

19.

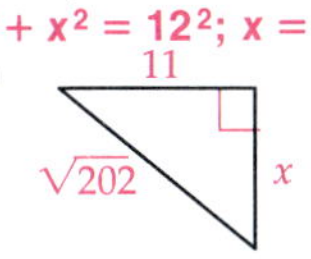

$x^2 + (6\sqrt{5})^2 = 18^2$; $x = 12$ $10^2 + x^2 = 26^2$; $x = 24$ $11^2 + x^2 = \sqrt{202}^2$; $x = 9$

MIXED REVIEW

Find the principal square root.

1. $\sqrt{625a^2b^4}$ $25ab^2$

2. $\sqrt{x^2y^2}$ xy

3. Is 8.2020 . . . a rational number? yes

Find a complement and a supplement.

4. 57° **5.** 32°

6. How many lines of symmetry does a square have? 4

7. How many square units are found in a square with side length 16? 256

8. ***DATA FILE 1 (pp. 2–3)*** What is the average temperature of the water in a pond below the icy surface? 36°F

4. 33°; 123°
5. 58°; 148°

Guided Practice

Class Exercises Suggest that students draw a diagram for question 7.

Think and Discuss For question 1, have students draw a diagram.

For question 2, help students recall properties of odd and even numbers, such as odd + even = odd and odd × odd = odd.

As you discuss question 3, students may need to be reminded that each face of a cube is a square.

A ***common error*** is to reverse the positions of a leg and the hypotenuse in the theorem. Stress that the hypotenuse is the longest side and is usually represented by *c*.

Closure

Writing in Math Have students write both the Pythagorean theorem and its converse in their math journals. Then have them discuss the uses for the theorem and list them as well.

Independent Practice

Students could also use a table for questions 5-8.

For questions 14-19, remind students that the hypotenuse is always opposite the right angle. Suggest that students write the theorem and then substitute the values into the theorem, as shown in Example 1.

As an extension, assign the problem solving activity on p. 450E of the Chapter Overview. See also TRB Exploring O-42.

To answer question 31, suggest that students begin with $1^2 + 1^2 = a^2$.

Lesson Follow-up

Reteaching Activity Assign *Exploring the Pythagorean Theorem*, TRB O-41. See page 450E of the Chapter Overview.

LESSON RESOURCES

TRB Practice K-5
TRB Enrichment K-6

Practice

For use after 11-3 (pp. 461–464)

Pythagorean Theorem

CALCULATOR Find each value to the nearest thousandth.

1. $\sqrt{48}$ **6.928** 2. $\sqrt{7}$ **2.646** 3. $\sqrt{85}$ **9.220**

Write an equation. Solve for x. Calculate each length to the nearest hundredth if the length is not a whole number.

4. $6^2 + 8^2 = x^2$; $x =$ 10
5. $x^2 + 24^2 = 26^2$; $x =$ 10
6. $x^2 + 7^2 = 9^2$; $x =$ 5.66
7. $x^2 + 15^2 = 17^2$; $x =$ 8
8. $x^2 + 40^2 = 41^2$; $x =$ 9
9. $x^2 + 5^2 = (\sqrt{146})^2$; $x =$ 11
10. $4^2 + 6^2 = x^2$; $x =$ 7.21
11. $12^2 + 9^2 = x^2$; $x =$ 15
12. $1^2 + 1^2 = x^2$; $x =$ 1.41

Determine which are Pythagorean triples. Write yes or no.

13. 20, 21, 29 **yes** 14. 7, 11, 12 **no** 15. 10, $2\sqrt{11}$, 12 **yes**
16. 28, 45, 53 **yes** 17. m, n, $\sqrt{m^2 + n^2}$ **yes** 18. 10, 15, 20 **no**
19. A rectangular park measures 300 ft by 400 ft. A sidewalk runs diagonally from one corner to the opposite corner. Find the length of the sidewalk. **500 ft**

Chapter 11 K-5

Enrichment

For use after 11-3 (pp. 461–464)

Finding Pythagorean Triples

There are several methods for finding Pythagorean triples.

I. Calculate multiples of known Pythagorean triples.

$7 \times (3) = 21$ $7 \times (4) = 28$ $7 \times (5) = 35$

Therefore, 21, 28, 35 is a Pythagorean triple: $21^2 + 28^2 = 35^2$

Calculate multiples of the 3, 4, 5 triple that include the given numbers.

1. 18, **24**, **30** 2. **27**, 36, **45** 3. **150**, **200**, 250

Calculate multiples of the 5, 12, 13 triple that include the given numbers.

4. 20, **48**, **52** 5. **10**, 24, **26** 6. **45**, **108**, 117

II. Substitute whole number values of n in the expressions $2n + 1$, $2n^2 + 2n$, and $2n^2 + 2n + 1$.

Using $n = 7$: $2n + 1 = 2(7) + 1 = 15$
$2n^2 + 2n = 2(7)^2 + 2(7) = 112$
$2n^2 + 2n + 1 = 2(7)^2 + 2(7) + 1 = 113$

Therefore, 15, 112, 113 is a Pythagorean triple.

Find a Pythagorean triple using the given value of n.

7. $n = 1$ **3**, **4**, **5** 8. $n = 2$ **5**, **12**, **13**
9. $n = 5$ **11**, **60**, **61** 10. $n = 4$ **9**, **40**, **41**

III. Let $a = m^2 - n^2$, $b = 2mn$, and $c = m^2 + n^2$. Choose values of m and n, so that $|m| > |n|$, one factor is odd and one even, and m and n do not share a common factor.

Using $m = 8$ and $n = 3$: $a = m^2 - n^2 = 8^2 - 3^2 = 55$
$b = 2mn = 2 \cdot 8 \cdot 3 = 48$
$c = m^2 + n^2 = 8^2 + 3^2 = 73$

Therefore, 55, 48, 73 is a Pythagorean triple.

Use this method to calculate two Pythagorean triples. Write the values of m and n that you choose. **Answers will vary. Possible answers:**

11. **$m = 9$; $n = 2$; 77, 36, 85**
12. **$m = 6$; $n = 5$; 11, 60, 61**

K-6 Chapter 11

Write an equation. Solve for x. Round to the nearest hundredth.

20.

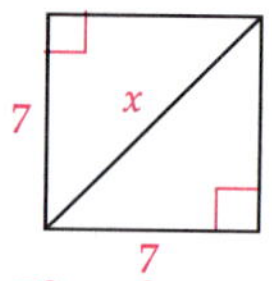

$7^2 + 7^2 = x^2$; $x = 9.90$

***21.**

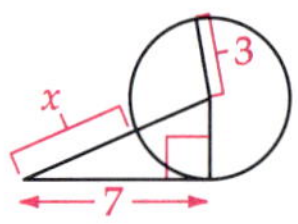

$3^2 + 7^2 = h^2$; $x = 4.62$

22.

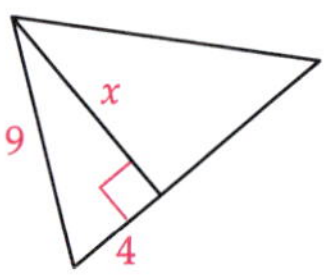

$4^2 + x^2 = 9^2$; $x = 8.06$

Determine whether each is a Pythagorean triple.

23. 7, 24, 25 **yes** **24.** 5, 12, 13 **yes** **25.** 1, 2, $2\sqrt{3}$ **no** **26.** 1, 0.24, 0.26 **no**

27. 4, 5, 6 **no** **28.** $1\frac{3}{5}$, $1\frac{1}{5}$, 2 **no** **29.** $3p$, $4p$, $5p$ **yes for $p > 0$** **30.** $\sqrt{5}$, $\sqrt{3}$, $\sqrt{2}$ **no**

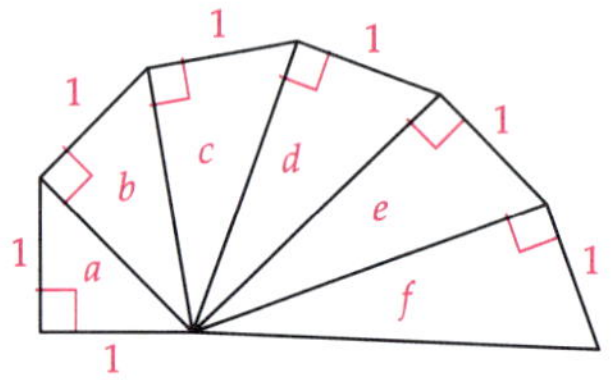

31. Find the lengths of the sides, a, b, c, d, e, f. Your answer may contain radical signs. **$\sqrt{2}$, $\sqrt{3}$, 2, $\sqrt{5}$, $\sqrt{6}$, $\sqrt{7}$**

32. A carpenter was building a square deck. He measured the sides and the diagonal of the deck and got measures of 8 ft, 9 ft, and 12 ft. Did the carpenter fulfill his contract? **no**

33. The recommended dimensions for a basketball court are 94 ft long and 50 ft wide. To the nearest foot, how long is the diagonal of the basketball court? **106 ft**

34. **CALCULATOR** A desk top is 75 cm wide and 130 cm long. How long is the diagonal of the desk top to the nearest hundredth? **150.08 cm**

35. Each side of a square is 15 in. long. Find the length of the diagonal to the nearest hundredth. **21.21 in.**

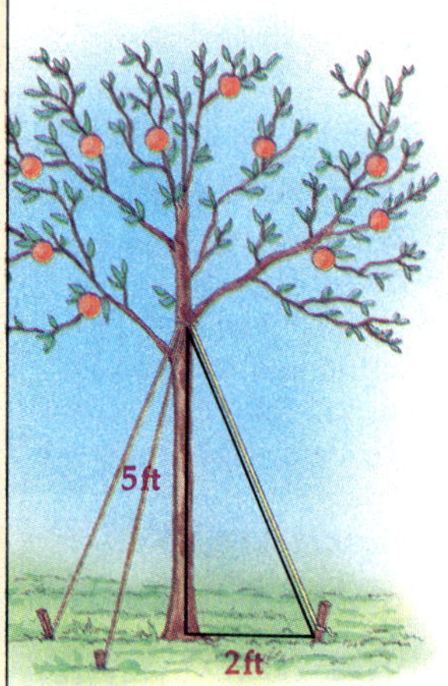

36. You are working for a landscaping company and you must plant and stake a tree. The stakes must be 2 ft from the base of the tree and the wires must extend 5 ft up the trunk. How long will the wires be to the nearest tenth? **5.4 ft**

37. Each side of a cube is 10 ft long. Find the length of the diagonal on one of the faces to the nearest tenth. **14.1 ft**

38. A ladder is 6 ft away from the base of a building. The ladder is 18 ft long. How many feet above the ground is the top of the ladder to the nearest hundredth? **16.97 ft**

39. The base of a pyramid is square. Each side of the base is 8 in. The slant height of the pyramid is 22 in.
   a. Find the height of the pyramid to the nearest hundredth. **21.63 in.**
   b. Find the volume of the pyramid to the nearest hundredth. **461.44 in.^{3}**

40. A cone has a diameter of 10 in. and a slant height of 16 in.
   a. Find the height of the cone to the nearest thousandth. **15.199 in.**
   b. Find the volume of the cone to the nearest thousandth. Use 3.14 for π. **397.707 in.^{3}**

LESSON QUIZ

1. Name one Pythagorean triple. (Possible answers: 3, 4, 5; 20, 21, 29)
2. Find the length, to the nearest unit, of one leg of a right triangle, if the other leg is 11 units and the length of the hypotenuse is 32 units. (30 units)
3. Find the length of an altitude of an equilateral triangle if the length of a side is 6 units. ($\sqrt{27}$ or 5.2)

FOR THE NEXT LESSON

calculator, math journal

Assignment Guide

Basic 1-2, 5-6, 9, 11-12, 14-16, 20, 23-26, 31-36, MR All

Average 2-3, 6-7, 10-12, 15-17, 21, 26-28, 33-38, MR All

Enriched 3-4, 7-8, 10, 12-13, 18-22, 28-30, 35-40, MR All

OBJECTIVE:
To use parts of similar triangles in problem solving.

11-4 Similar Right Triangles

▼ Jack plans to cut down a tree. Since he does not want the tree to land on his house, pool, or shed, he must fell the tree towards the fence. To make sure the tree does not damage the fence, which is 49 ft away, Jack must know the height of the tree. Jack is 6 ft tall and casts a shadow 17 ft long. The tree casts a shadow 102 ft long. Using these measures and his knowledge of similar triangles, Jack finds the height of the tree.

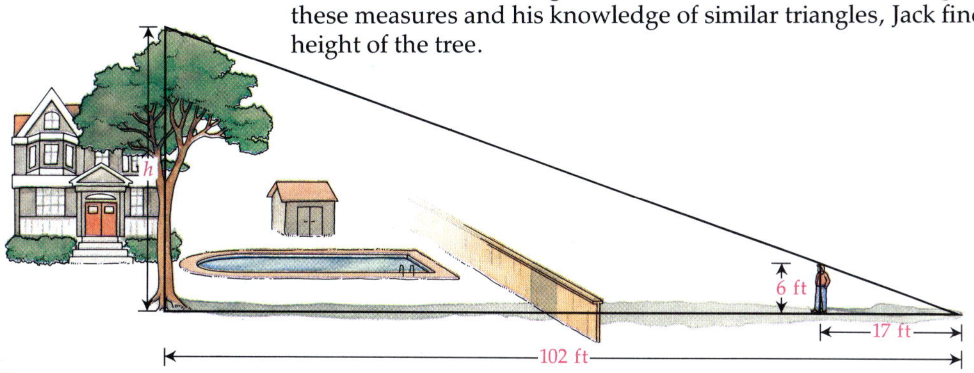

THINK What would be the proportion if Jack's height were 5 ft 9 in? $\frac{h}{102} = \frac{5.75}{17}$

You can use similar triangles to set up a proportion.

$$\frac{\text{height of tree } (h)}{\text{length of tree shadow}} = \frac{\text{Jack's height}}{\text{length of Jack's shadow}}$$

Solve the proportion for h.

$$\frac{h}{102} = \frac{6}{17} \quad \text{Substitute values.}$$
$$17h = 6(102) \quad \text{Cross multiply.}$$
$$17h = 612$$
$$\frac{17h}{17} = \frac{612}{17} \quad \text{Divide each side by 17.}$$
$$h = 36$$

The tree is 36 ft tall. Since $36 < 49$, the fence will not be damaged.

▼ There is often more than one way to solve a problem.

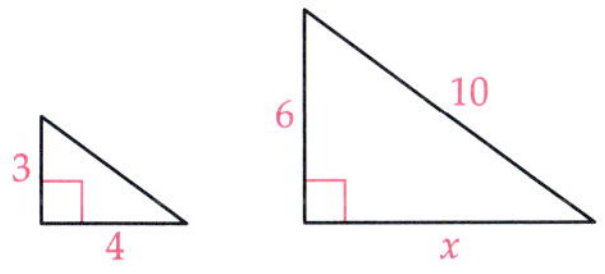

Example 1 **Find x using two different methods. The right triangles are similar.**

1

Solution **Method 1** Use the Pythagorean theorem.

$$a^2 + b^2 = c^2 \quad \text{Write the theorem.}$$
$$6^2 + x^2 = 10^2 \quad \text{Substitute values.}$$
$$36 + x^2 = 100$$
$$36 + x^2 - 36 = 100 - 36$$
$$x^2 = 64$$
$$\sqrt{x^2} = \sqrt{64} \quad \text{Take each square root.}$$
$$x = 8$$

Lesson Focus

Materials
Calculator, math journal

MOTIVATING THE LESSON

Help students recall what is meant by similar triangles (all corresponding angles are of equal measure and the corresponding sides are in proportion). Discuss also the minimum conditions necessary for two triangles to be similar.

Instruction

1 Discuss the practicality of both methods for finding lengths of sides of right triangles. Ask students why the Pythagorean theorem was not used in the tree problem. (The measures of two sides of the triangle are unknown.)

Teaching Tip

Have a poster ready for class that shows a pair of similar triangles with a list of the corresponding angles and the proportional sides.

EE: 1A, 1B, 1C, 3A, 3B, 3E, 3G, 3H, 3I, 4A, 4C, 4D, 4E; TAAS: 1, 2, 3, 4, 5, 6, 7, 8, 9, 10, 11, 12

2 Ask students to explain whether or not CB could have been found by using similar triangles. (no, because the proportion would have contained two unknowns)

ADDITIONAL EXAMPLES

The triangles are similar.

1. Find x. (26)

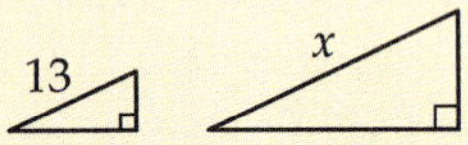

2. Find AE. (20)

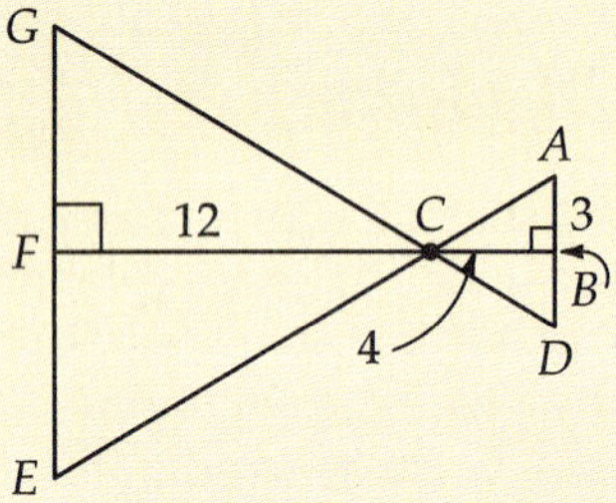

Guided Practice

Think and Discuss For question 1, ask students how the three angle measures of a triangle are related. (Their sum is always 180°.) Then ask what the sum is of the angle measures of two acute angles in a right triangle. (90°)

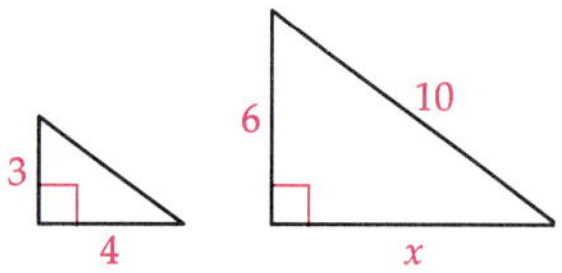

Method 2 Use similar triangles.

$\frac{3}{4} = \frac{6}{x}$ Write a proportion.

$24 = 3x$ Cross multiply.

$\frac{24}{3} = \frac{3x}{3}$ Divide each side by 3.

$8 = x$

The value of x is 8 no matter what method you use.

▼ You may need to use both methods to solve a problem.

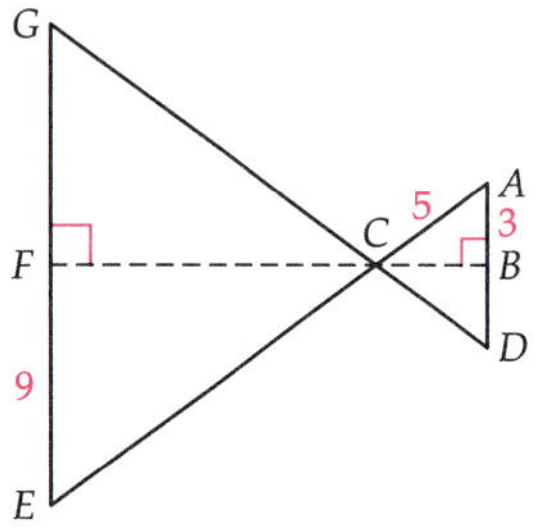

Example 2 **Find FB. $\overline{FC}$ and $\overline{CB}$ are altitudes of the triangles.**

Solution

2

1. Find CB using the Pythagorean theorem.

$a^2 + b^2 = c^2$ Write the theorem.

$3^2 + CB^2 = 5^2$ Substitute values.

$9 + CB^2 = 25$

$9 + CB^2 - 9 = 25 - 9$ Subtract 9 from each side.

$CB^2 = 16$

$\sqrt{CB^2} = \sqrt{16}$ Find each square root.

$CB = 4$

2. Find FC using similar triangles.

$\frac{9}{3} = \frac{FC}{4}$ Write a proportion.

$4 \cdot \frac{9}{3} = \frac{FC}{4} \cdot 4$ Multiply each side by 4.

$\frac{36}{3} = FC$

$12 = FC$

3. $FB = CB + FC$

$= 4 + 12$

$= 16$

THINK AND DISCUSS

1. How are the acute angles of a right triangle related?
2. If one acute angle of a right triangle has measure $x°$, how would you express the measure of the other acute angle? $(90 - x)°$
3. In a triangle, if two pairs of corresponding angles are congruent, are three pairs of corresponding angles congruent? Explain.

1. They are complementary.
3. Yes; when you add the measures of the two angles in each triangle and subtract the sum from 180, the result will be the same in each triangle.

CLASS EXERCISES

Solve for x.

1. $\frac{x}{5} = \frac{12}{30}$ 2
2. $\frac{2.5}{x} = \frac{7}{6}$ $2\frac{1}{7}$
3. $\frac{9}{22} = \frac{x}{154}$ 63
4. $\frac{a}{b} = \frac{x}{y}$ $\frac{ay}{b}$

Complete each proportion. $\triangle ABC \sim \triangle DEC$.

5. $\frac{BC}{EC} = \frac{AC}{■}$ DC
6. $\frac{■}{CB} = \frac{DC}{AC}$ EC
7. $\frac{EC}{BC} = \frac{■}{AB}$ DE

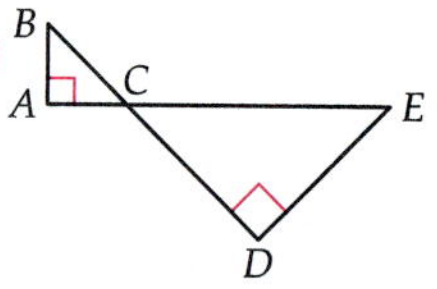

MATH MINUTES

Find all triangles that are similar to triangle ABC.

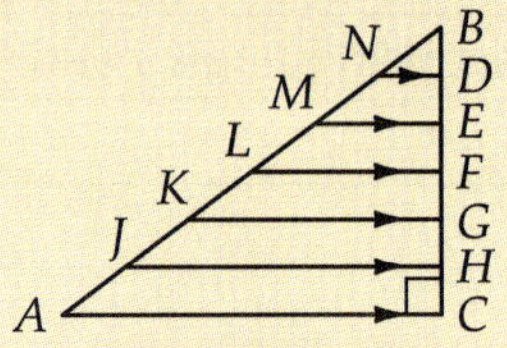

(Δ JBH, Δ KBG, Δ LBF, Δ MBE, Δ NBD)

8. a. Find x using the Pythagorean theorem. 12

b. Find y using the Pythagorean theorem. 9

c. $\triangle ABC \sim$ ■ $\sim$ ■ $\triangle CBD$; $\triangle ACD$

d. $\triangle DBC \sim$ ■ $\sim$ ■ $\triangle CBA$; $\triangle DCA$

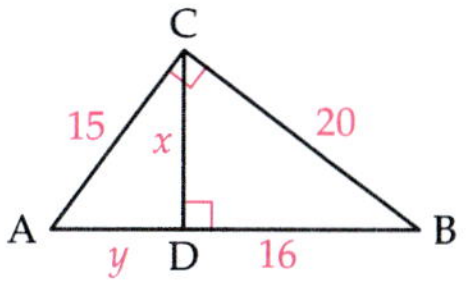

WRITTEN EXERCISES

MENTAL MATH Solve for variable x.

1. $\frac{4}{9} = \frac{12}{x}$ 27 **2.** $\frac{x}{9} = \frac{7}{6}$ $10\frac{1}{2}$ **3.** $\frac{1.5}{4} = \frac{x}{8}$ 3 **4.** $\frac{y}{x} = \frac{a}{b}$ $\frac{yb}{a}$

Solve for x using the Pythagorean theorem. Use the square root table on p. 579 to find the length to the nearest thousandth.

5.

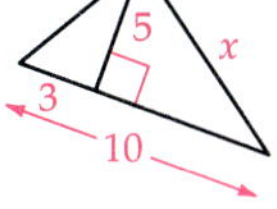

$x = 8.602$

6.

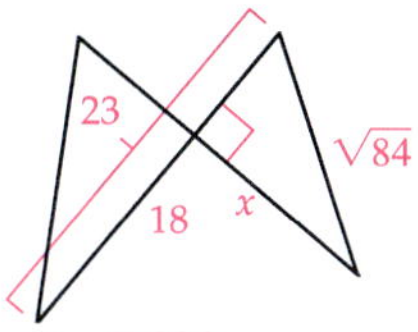

$x = 7.681$

7.

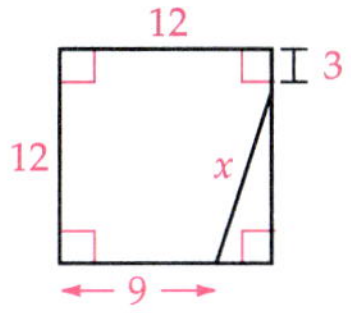

$x = 9.487$

Solve for x using similar triangles. Each pair of triangles is similar.

8.

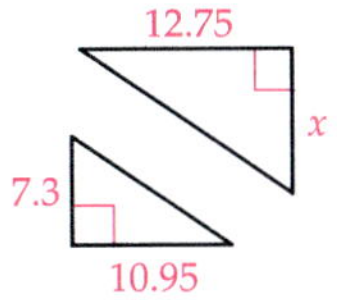

$x = 8.5$

9.

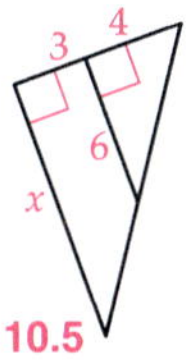

$x = 10.5$

10.

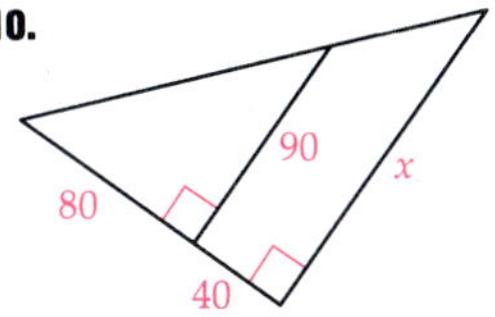

$x = 135$

CALCULATOR Solve for x and y. Round to the nearest hundredth.

11.

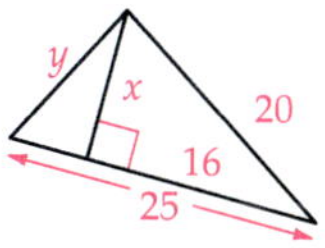

$x = 12$
$y = 15$

12.

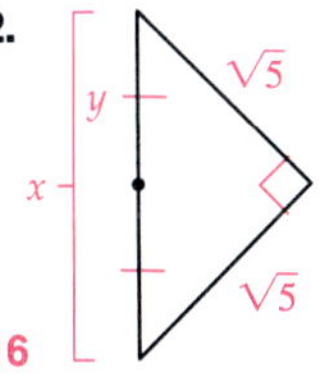

$x = 3.16$
$y = 1.58$

13.

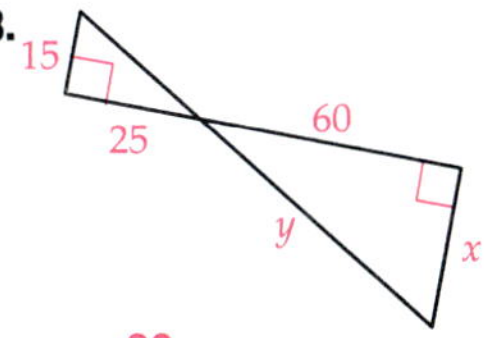

$x = 36$
$y = 69.97$

Solve.

14. A ladder is leaning against a building in such a way that the ladder touches the top of an 8-ft fence. The bottom of the ladder is 4 ft from the base of the fence. The fence is 8 ft from the building. How high up the building is the top of the ladder? 24 ft

MIXED REVIEW

Determine if each is a Pythagorean triple.

1. 16, 30, 34 yes

2. 16, 12, 15.5 no

Write in scientific notation.

3. 8,960,000,000,000 8.96×10^{12}

Find the GCF.

4. 18,48 6 **5.** 76,38 38

Write an equation and solve.

6. Four divided by twice x equals one-eighth of 12.

7. Half the product of 6 and x is four greater than x.

8. Pat missed 2% of the 50 balls pitched. How many did he miss?

6. $\frac{4}{2x} = \frac{1}{8} \cdot 12$; $x = 1\frac{1}{3}$
7. $\frac{6x}{2} = x + 4$; $x = 2$
8. $0.02 \times 50 = x$; 1

PROBLEM SOLVING HINT

Draw a diagram to help visualize the problem.

Class Exercises For question 8, have students work in groups. Then have the groups discuss their answers with the class.

A ***common error*** occurs when students confuse the correspondence when working with similar triangles. Have students practice identifying the pairs of corresponding angles and the corresponding sides *opposite* the corresponding angles.

Closure

Writing in Math Ask students to copy this diagram of similar triangles into their math journals and to outline the steps they would use to find x and y.

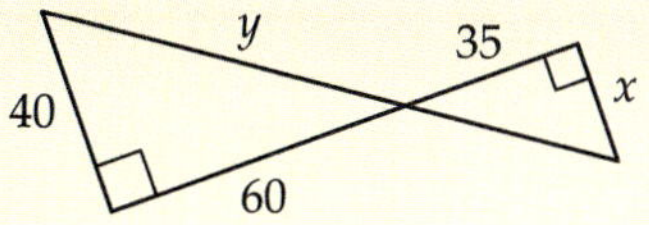

Independent Practice

Remind students to choose the appropriate method for finding the unknown measures unless stated otherwise. Have students work with a partner for question 17c.

Lesson Follow-up

Reteaching Activity Select a tall object that can be measured indirectly using shadows and similar triangles. Have students draw a diagram and find the height of the object.

LESSON RESOURCES

TRB Practice K-7
TRB Enrichment K-8

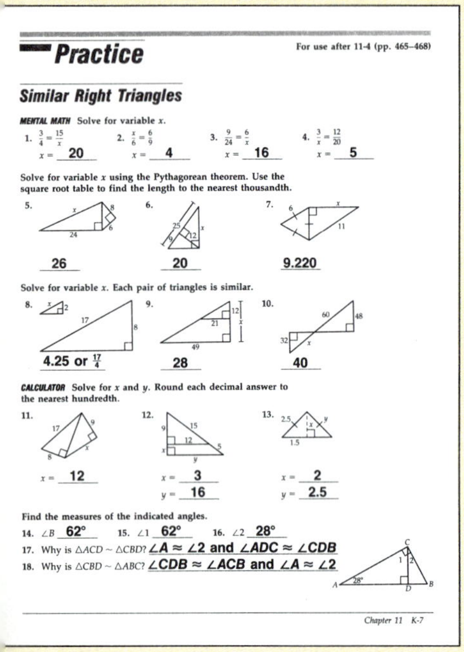

Practice For use after 11-4 (pp. 465–468)

Similar Right Triangles

MENTAL MATH Solve for variable x.

1. $\frac{3}{4} = \frac{15}{x}$ $x =$ 20
2. $\frac{x}{6} = \frac{6}{9}$ $x =$ 4
3. $\frac{9}{24} = \frac{6}{x}$ $x =$ 16
4. $\frac{3}{x} = \frac{12}{20}$ $x =$ 5

Solve for variable x using the Pythagorean theorem. Use the square root table to find the length to the nearest thousandth.

5. 26
6. 20
7. 9.220

Solve for variable x. Each pair of triangles is similar.

8. 4.25 or $\frac{17}{4}$
9. 28
10. 40

CALCULATOR Solve for x and y. Round each decimal answer to the nearest hundredth.

11. $x =$ 12
12. $x =$ 3, $y =$ 16
13. $x =$ 2, $y =$ 2.5

Find the measures of the indicated angles.

14. $\angle B$ 62°
15. $\angle 1$ 62°
16. $\angle 2$ 28°
17. Why is $\triangle ACD \sim \triangle CBD$? $\angle A \approx \angle 2$ and $\angle ADC \approx \angle CDB$
18. Why is $\triangle CBD \sim \triangle ABC$? $\angle CDB \approx \angle ACB$ and $\angle A \approx \angle 2$

Chapter 11 K-7

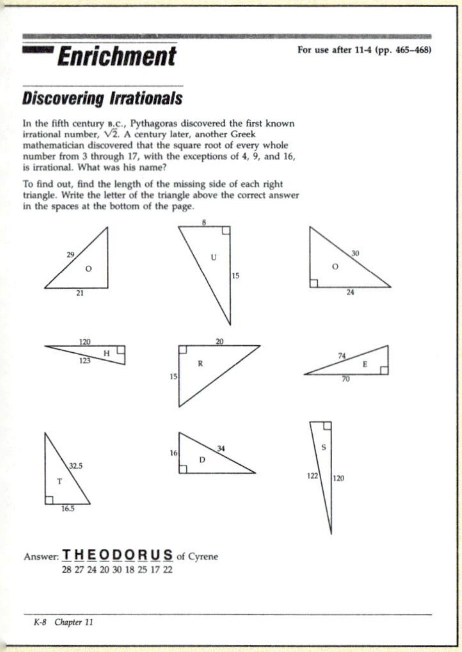

Enrichment For use after 11-4 (pp. 465–468)

Discovering Irrationals

In the fifth century B.C., Pythagoras discovered the first known irrational number, $\sqrt{2}$. A century later, another Greek mathematician discovered that the square root of every whole number from 3 through 17, with the exceptions of 4, 9, and 16, is irrational. What was his name?

To find out, find the length of the missing side of each right triangle. Write the letter of the triangle above the correct answer in the spaces at the bottom of the page.

Answer: THEODORUS of Cyrene
28 27 24 20 30 18 25 17 22

K-8 Chapter 11

15. A radio tower is 25 ft high and casts a shadow of 40 ft. At the same time, a taller tower casts a shadow of 70 ft. What is the height of the taller radio tower? 43.75 ft

16. Ian is 160 cm tall. His image on the film of a camera is 1.6 cm. Suppose the film is 2 cm from the camera lens. How far is Ian from the camera? 200 cm

The Great Pyramid

The Great Pyramid of Khufu near Gîza was built about 2700 B.C. The pyramid has a square base. The Egyptians were able to form right angles by using an instrument called a gromma. The base length of each of the pyramid's four sides is 755 ft. Its height originally reached 481 ft but has grown smaller over the years from erosion. The pyramid is built from 2.3 million blocks of sandstone. Each block weighs about 2.5 t.

Use the article at the left to answer each question.

17. a. Determine the slant height of the faces of the pyramid to the nearest hundredth. 611.45 ft
    b. Find the area of the base of the pyramid. 570,025 ft^2
    c. ***PROJECT*** Research the gromma. Tell how Egyptians used this instrument to construct right angles. Check students' work.

TEST YOURSELF

Solve for variable x. Round to the nearest tenth.

1. 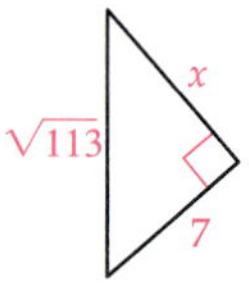

$x = 8$

2. 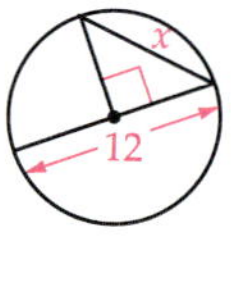

$x = 8.5$

3. $x = 14.7$
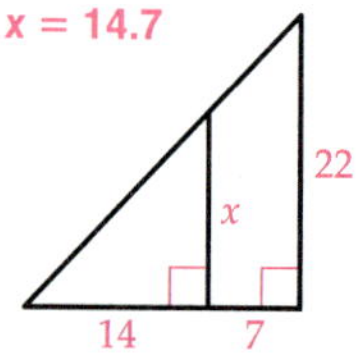

4. A rectangular box has a square base with area 16 cm^2. The height of the box is 7 cm. Find the length of the diagonal of a rectangular side. Use the square root table on p. 579. 8.062 cm

5. A 42-ft tree casts a shadow of 63 ft. How long is the shadow cast by a 5-ft tall girl? 7.5 ft

LESSON QUIZ

Find DB and FB.
($DB = 8$, $FB = 28$)

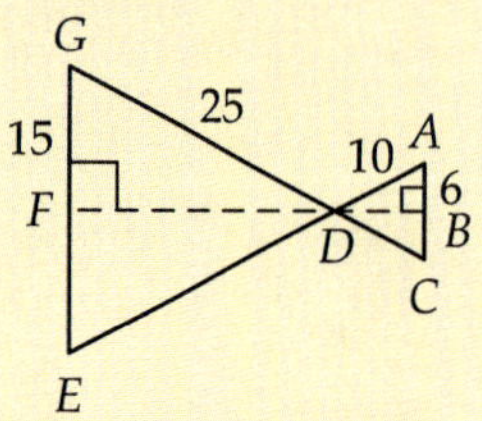

Assignment Guide

Basic 1-2, 5, 8-9, 11-12, 14, 17, MR, TY All
Average 2-3, 6, 8-9, 11-12, 14-15, 17, MR, TY All
Enriched 3-4, 7, 9-10, 13-17, MR, TY All

FOR THE NEXT LESSON

protractor, ruler, calculator, graph paper, math journal

OBJECTIVE: *To discover and use the properties of 45°-45°-90° and 30°-60°-90° right triangles.*

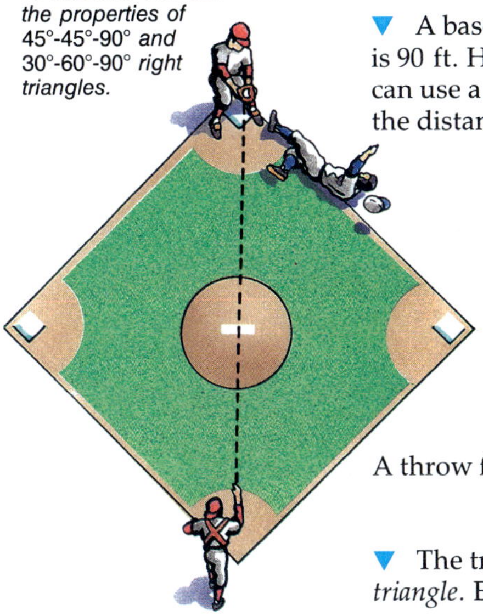

11-5 Special Right Triangles

▼ A baseball diamond is a square. The distance between the bases is 90 ft. How long is a throw from home plate to second base? You can use a special right triangle and the Pythagorean theorem to find the distance.

$a^2 + b^2 = c^2$	Write the theorem.
$90^2 + 90^2 = c^2$	Substitute values.
$8{,}100 + 8{,}100 = c^2$	
$16{,}200 = c^2$	
$\sqrt{16{,}200} = \sqrt{c^2}$	Take each square root.
$\sqrt{16{,}200} = c$	Simplify the right side.
$\sqrt{8{,}100 \cdot 2} = c$	
$\sqrt{8{,}100} \cdot \sqrt{2} = c$	
$90\sqrt{2} = c$	

A throw from home plate to second base is $90\sqrt{2}$ ft long.

▼ The triangle formed by the diagonal of a square is an *isosceles right triangle.* Each base angle has a measure of 45°.

45°-45°-90° Triangle

In a 45°-45°-90° right triangle, the lengths of the sides have the following relationships.

$$\text{hypotenuse} = \sqrt{2} \cdot \text{leg}$$

$$\text{leg} = \frac{\text{hypotenuse} \cdot \sqrt{2}}{2}$$

▼ You can use the properties of a 45°-45°-90° triangle to determine missing lengths.

Example 1 **Find x. The answer may contain a square root sign.**

a.

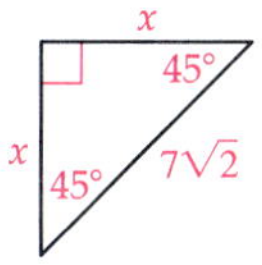

b.

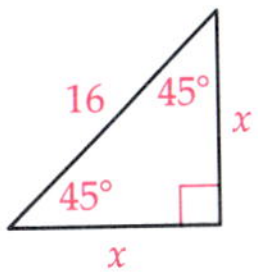

THINK Why is $\frac{x\sqrt{2}}{\sqrt{2}}$ equal to x? $\frac{\sqrt{2}}{\sqrt{2}} = 1$ and $x \cdot 1 = x$

Solution 1

a. $\text{hypotenuse} = \sqrt{2} \cdot \text{leg}$

$7\sqrt{2} = x\sqrt{2}$

$\frac{7\sqrt{2}}{\sqrt{2}} = \frac{x\sqrt{2}}{\sqrt{2}}$ Divide each side by $\sqrt{2}$.

$7 = x$

b. $\text{leg} = \frac{\text{hypotenuse} \cdot \sqrt{2}}{2}$

$x = \frac{1}{2}(16)(\sqrt{2})$

$x = 8\sqrt{2}$

Lesson Focus

Materials

Protractor, ruler, calculator, math journal, graph paper

Vocabulary/Symbols

isosceles right triangle

MOTIVATING THE LESSON

Have students draw these triangles and ask them to find x. ($\sqrt{2}$, $\sqrt{3}$)

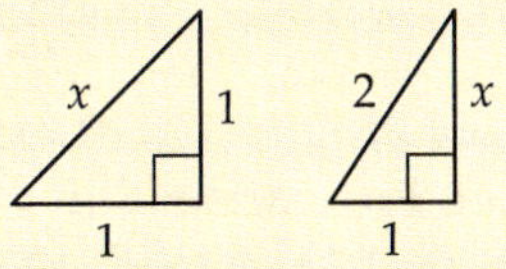

Then ask the students to measure the acute angles in each triangle. (45°, 45°; 30°, 60°) Mention that these are special right triangles and that in this lesson students will learn some shortcuts for finding lengths of sides for such triangles.

Instruction

1 Have students verify the results by using the converse of the Pythagorean theorem. [$7^2 + 7^2 = (7\sqrt{2})^2$, $49 + 49 = 98$, $98 = 98$; $(\frac{16}{\sqrt{2}})^2 + (\frac{16}{\sqrt{2}})^2 = 16^2$, $\frac{256}{2} + \frac{256}{2} = 256$, $128 + 128 = 256$, $256 = 256$]

Teaching Tip

Prepare more than one way to present or demonstrate the concept to be taught. If students cannot grasp the initial presentation, you can then rely on an alternate approach.

ADDITIONAL EXAMPLES

1. Find the length of the hypotenuse of an isosceles right triangle if each leg is 33 cm. long. ($33\sqrt{2}$ cm)

2. Find the length of the hypotenuse of a 30°-60°-90° triangle if the longer leg is $12\sqrt{3}$ in. long. (24 in.)

Guided Practice

Class Exercises Have students answer questions 1-12 orally.

Think and Discuss For question 1 point out to students that the lengths of a 45°-45°-90° triangle form the ratio $1:1:\sqrt{2}$.

A ***common error*** occurs when students forget which situation calls for $\sqrt{2}$ and which calls for $\sqrt{3}$. Have students draw $1:1:\sqrt{2}$ and $1:\sqrt{3}:2$ right triangles in their math journals. Then students can use them for reference.

Closure

Writing in Math Have students select a real-life example to test the 45°-45°-90° relationship. Have them describe their choice, the measurements, and their findings in their math journals.

THINK In a 30°-60°-90° right triangle, how do you know which leg is longer? the leg opposite the 60° angle

▼ The 30°-60°-90° right triangle is another special right triangle.

30°-60°-90° Triangle	In a 30°-60°-90° triangle, the lengths of the sides have the following relationships. hypotenuse = 2(shorter leg) longer leg = shorter leg($\sqrt{3}$)

You can use these relationships to determine missing lengths.

Example 2 **Find x and y. The answer may have a square root sign.**

a.

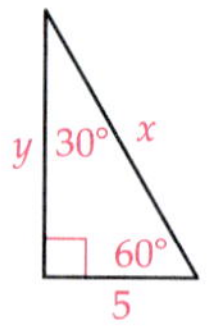

b.

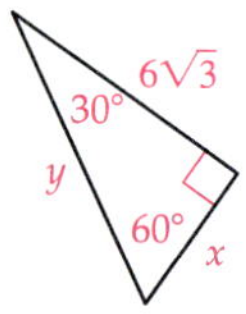

Solution

a. hypotenuse = 2(shorter leg) — Solve for x.

$x = 2(5)$

$x = 10$

longer leg = shorter leg($\sqrt{3}$) — Solve for y.

y = shorter leg($\sqrt{3}$)

$y = 5\sqrt{3}$

b. longer leg = shorter leg($\sqrt{3}$) — Solve for x.

$6\sqrt{3} = (x)(\sqrt{3})$

$$\frac{6\sqrt{3}}{\sqrt{3}} = \frac{\sqrt{3}x}{\sqrt{3}}$$

$6 = x$

hypotenuse = 2(shorter leg) — Solve for y.

$y = 2(6)$

$y = 12$

THINK AND DISCUSS

1. In a 45°-45°-90° triangle, by what do you multiply each leg to get the length of the hypotenuse? $\sqrt{2}$

2. In a 30°-60°-90° triangle, what operation do you perform on the hypotenuse to get the length of the side opposite the 30° angle? Divide by 2.

3. In a 30°-60°-90° triangle, what operation do you perform on the side opposite the 60° angle to get the length of the side opposite the 30° angle? Divide by $\sqrt{3}$.

CLASS EXERCISES

Use the triangle at the right.

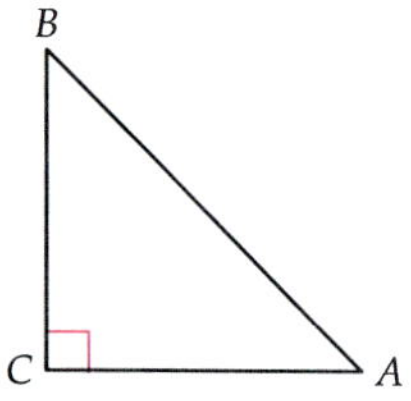

1. Which side is opposite $\angle A$? $\overline{BC}$
2. Which side is opposite $\angle B$? $\overline{CA}$
3. Which side is opposite $\angle C$? $\overline{BA}$
4. Which side is the hypotenuse? $\overline{BA}$
5. Which sides are the legs? $\overline{BC}$ and $\overline{CA}$
6. Which side of the triangle is the longest? $\overline{BA}$

MATH MINUTES

Have students create riddles about the special right triangles for other students to solve. For example: My hypotenuse is 4 and my acute angles are 30° and 60°. What are the measures of my legs? (2, $2\sqrt{3}$)

Tell whether a triangle with sides of the given lengths is 45°-45°-90°, 30°-60°-90°, or neither.

7. 6, 8, 10 neither
8. 5, 5, $5\sqrt{2}$ 45°-45°-90°
9. 15, $7.5\sqrt{3}$, 7.5 30°-60°-90°

Find each value. Answers may contain square root signs.

10.

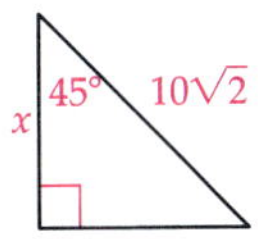

x = 10

11.

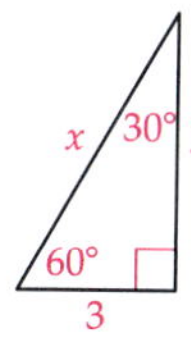

x = 6
$y = 3\sqrt{3}$

12.

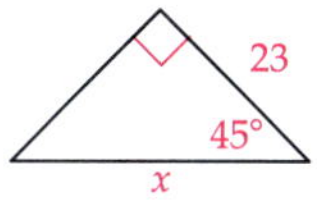

$x = 23\sqrt{2}$

WRITTEN EXERCISES

The length of one side of the triangle is given. Find the missing measures. Answers may contain square root signs.

	a	b	c	
1.	9	■ 9	■	$9\sqrt{2}$
2.	■ 5.4	5.4	■	$5.4\sqrt{2}$
3.	■ 4	■ 4	$4\sqrt{2}$	
4.	■ $\frac{3}{\sqrt{2}}$	■ $\frac{3}{\sqrt{2}}$	3	

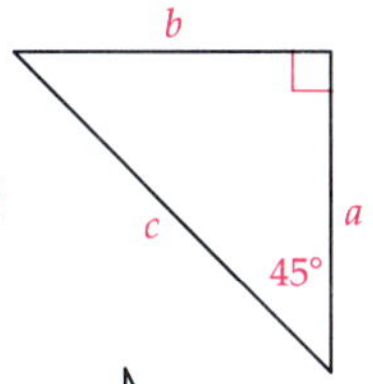

	d	e	f
5.	■ $2\sqrt{3}$	2	■ 4
6.	■ $5\sqrt{3}$	■ 5	10
7.	■ $7\sqrt{3}$	7	■ 14
8.	$8\sqrt{3}$	■ 8	■ 16

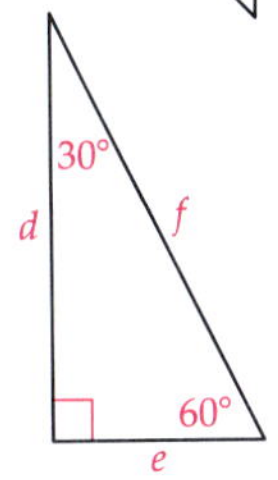

Find each value. Answers may contain square root signs.

9.

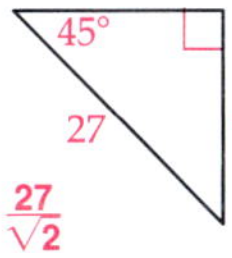

$x = \frac{27}{\sqrt{2}}$

10.

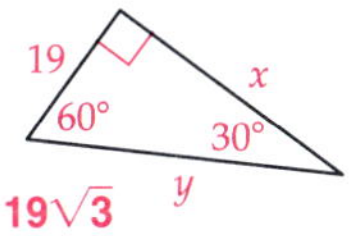

$x = 19\sqrt{3}$
y = 38

11.

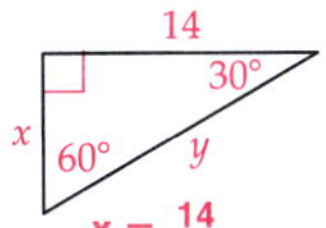

$x = \frac{14}{\sqrt{3}}$
$y = \frac{28}{\sqrt{3}}$

12. ***MENTAL MATH*** A square has a diagonal $6\sqrt{2}$ in. long.
  a. Find the length of a side of the square. 6 in.
  b. Find the perimeter of the square. 24 in.

13. Each face on a cube has a diagonal 12 in. long.
  a. Find the length of a side of the cube to the nearest tenth. 8.5 in.
  b. Find the surface area of the cube to the nearest hundredth. 433.5 in.^{2}

MIXED REVIEW

Solve and check.

1. $\frac{x}{4} - 2 = 7 + x$ -12

2. $-3.5x - 4x + 85 = x$ 10

Solve the proportion.

3. $\frac{9.5}{x} = \frac{1.5}{0.75}$ x = 4.75

4. Simplify $-(4x^3)^2$. $-16x^6$

5. -12 − (-18) 6

6. A polygon with five sides is a ■. pentagon

7. ***DATA FILE 6 (pp. 230–231)*** What percent of teens study for 2 or more hours each night? 27%

8. All books are on sale for 40% off. Judy paid \$19.99 for the new best seller. What was the original price? \$33.32

Independent Practice

Mention that square root signs may be left in denominators, since the skill for rationalizing denominators has not been taught yet. Teaching this skill is optional at this time.

For questions 12-16, 18-20, suggest that students begin by drawing a diagram.

Three-dimensional objects would be helpful for questions 13-14.

As an extension, assign the computer activity on p. 450E of the Chapter Overview. See also TRB Exploring O-44.

Critical Thinking If students have difficulty in getting started, ask them what the significance of the first clue is. (Lisa does not clean the kitchen; so *no* can be written in the first cell.) Students can then proceed in this manner and by the process of elimination will be able to tell who cleans each room.

Lesson Follow-up

Reteaching Activity Have students draw 30°-60°-90° and 45°-45°-90° triangles. Have them measure the lengths of the sides, label the diagrams, and record the data in the chart. Then have students test the values using the relationships described in the lesson.

LESSON RESOURCES

TRB Practice K-9
TRB Enrichment K-10

Practice

For use after 11-5 (pp. 469–472)

Special Right Triangles

Use the triangle to answer each question.

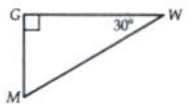

1. Which side is opposite ∠W? $\overline{GM}$
2. Which side is opposite ∠G? $\overline{MW}$
3. Which side is opposite ∠M? $\overline{GW}$
4. Which sides are legs? $\overline{GM}$, $\overline{GW}$
5. Which side is the hypotenuse? $\overline{MW}$

The length of one side of the triangle is given. Find the missing measures. Answers may contain square root signs.

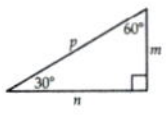

	m	n	p
6.	14	$14\sqrt{3}$	28
7.	18	$18\sqrt{3}$	36
8.	9	$9\sqrt{3}$	18
9.	5	$5\sqrt{3}$	10

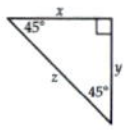

	x	y	z
10.	11	11	$11\sqrt{2}$
11.	8.7	8.7	$8.7\sqrt{2}$
12.	7	7	$7\sqrt{2}$
13.	17	17	$17\sqrt{2}$

In the figure, $BD = 6\sqrt{2}$. Find each value.

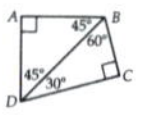

14. AB 6
15. AD 6
16. BC $3\sqrt{2}$
17. CD $3\sqrt{6}$

18. ***MENTAL MATH*** One leg of a 45°-45°-90° right triangle measures 14 cm. Find the perimeter. $28 + 14\sqrt{2}$ cm

19. The figure is a cube with diagonal $\overline{CE}$.

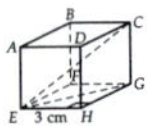

a. Find CG. 3 cm
b. Find GE. $3\sqrt{2}$ cm
c. Use the Pythagorean theorem in △CGE to find CE to the nearest tenth. 5.2 cm

Enrichment

For use after 11-5 (pp. 469–472)

Solving Space Figures

You can use your knowledge of special right triangles to find information about space figures.

The base of the pyramid is a square with sides 10 units long. The slant height is EF and the height of the pyramid is EG. m∠BEF = m∠CEF = 30°.

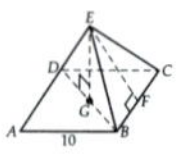

Find the following information about the pyramid. Answers may contain square root signs.

1. m∠EBF 60°
2. m∠ECF 60°
3. BF 5 units
4. FC 5 units
5. EB 10 units
6. slant height EF $5\sqrt{3}$ units
7. DB $10\sqrt{2}$ units
8. GB $(GB = \frac{1}{2} \cdot DB)$ $5\sqrt{2}$ units
9. pyramid height EG (Use the Pythagorean theorem in △EGB.) $\sqrt{50}$ or $5\sqrt{2}$ units
10. What type of special right triangle is △EGB? 45°-45°-90°
11. m∠GEB 45°
12. m∠GBE 45°

In Exercises 13–16, use $\sqrt{3} = 1.73$ and $\sqrt{2} = 1.41$.

13. area of square ABCD 110 sq. units
14. area of △EBC 43.25 sq. units
15. surface area of pyramid 273 sq. units
16. volume of pyramid 235 cu. units

The radius of the cone's base is 12. m∠OPQ = 30°. Find the following information about the cone. Use $\pi = 3.14$ and $\sqrt{3} = 1.73$. Give answers to the nearest hundredth.

17. height OP 20.76 units
18. slant height PQ 24 units
19. area of base 452.16 sq. units
20. lateral area 904.32 sq. units
21. surface area 1,356.48 sq. units
22. volume 3,128.95 cu. units

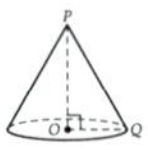

14. The base of a cone has a diameter of 8. The cone slants on a 60° angle and has a slant height of 8. Find the height of the cone to the nearest thousandth. Use the square root table. 6.928

15. △ABC is equilateral. Each side has a length of 52 in. Find the altitude. The answer may contain a square root sign. $26\sqrt{3}$ in.

16. The length of the hypotenuse of a 30°-60°-90° triangle is 20*a*. What is the length of the shorter leg? What is the length of the longer leg? Your answers may contain square root signs. 10*a*; $10a\sqrt{3}$

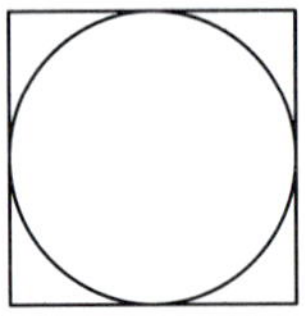

17. Use the diagram at the left. The diameter of the circle is 16 cm. How much larger is the area of the square than the area of the circle? Use 3.14 for π. 55.04 cm^2

18. ***CALCULATOR*** A 15-ft ladder is leaning against a wall at a 45° angle. To the nearest tenth, how high above the ground is the ladder touching the wall? 10.6 ft

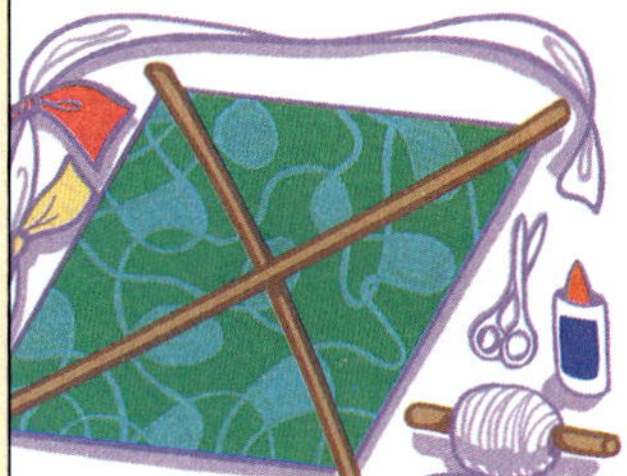

19. Janet is building a kite in the shape of a square. Each side of the square is 18 in. To the nearest inch, how much wood does Janet need to make the diagonals? 51 in.

20. ***DATA FILE 11 (pp. 450–451)*** A skydiver jumps out of a plane at 2,750 ft. He wants to hit a marker 1.25 mi away. He falls at a rate of 10 ft/s. To the nearest tenth of a minute, how long will it take the jumper to hit the marker? 11.9 min

Critical Thinking

EXPLORING LOGIC

▼ To solve a logic problem, organize the information in a chart like the one at the left. Each of Mrs. Stephan's five children cleans one room each week. The rooms cleaned are the living room, family room, kitchen, bathroom, and a bedroom. Who cleans each room?

	Lisa				
Kitchen					

Lisa—bedroom
Nicole—bathroom
Robert—Living room
Eric—family room
Jo Ann—Kitchen

Copy and complete the chart to solve.

1. Lisa does not have to scrub the kitchen floor.
2. Nicole never makes her bed.
3. Robert and Eric always vacuum under the couches.
4. The family room joins the kitchen. Eric and Jo Ann enjoy talking to each other while they clean.
5. Only the living room and family room contain couches.

LESSON QUIZ

1. If the hypotenuse of a right triangle is 10 and one leg is 5, what is the length of the other leg? ($5\sqrt{3}$)
2. Draw an equilateral triangle with sides 5 cm. long. Draw one of the altitudes and find its length to the nearest hundredth of a centimeter. (4.33 cm)

Assignment Guide

Basic 1-2, 5-6, 9, 12, 15-18, 20, MR, CT All
Average 2-3, 6-7, 10, 13, 17-20, MR, CT All
Enriched 3-4, 7-8, 11, 14, 17-20, MR, CT All

FOR THE NEXT LESSON

overhead projector, calculator with trigonometric ratio keys or tables of trigonometric ratios, protractor, ruler, math journal

OBJECTIVE: *To define and use sine, cosine, and tangent of acute angles in right triangles.*

FLASHBACK

A ratio is a comparison of two quantities by division.

11-6 Trigonometric Ratios

▼ *Trigonometry* means triangle measurement. A *trigonometric ratio* is a ratio of the measures of two sides of a right triangle.

Example 1 **Find each ratio.**

a. $\frac{CA}{CB}$ **b.** $\frac{CA}{AB}$ **c.** $\frac{CB}{AB}$

Solution **a.** $\frac{3}{4}$ **b.** $\frac{3}{5}$ **c.** $\frac{4}{5}$

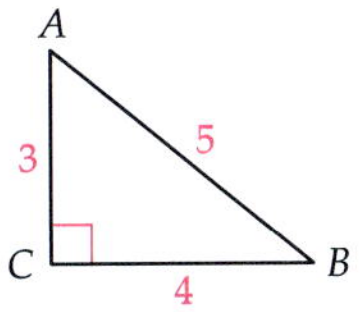

▼ Trigonometric ratios have special names.

1

Trigonometric Ratios	
	tangent of $\angle A = \frac{\text{length of side opposite } \angle A}{\text{length of side adjacent to } \angle A}$
	sine of $\angle A = \frac{\text{length of side opposite } \angle A}{\text{hypotenuse}}$
	cosine of $\angle A = \frac{\text{length of side adjacent to } \angle A}{\text{hypotenuse}}$

You can write the terms tangent, sine, and cosine using the abbreviations *tan, sin,* and *cos.* A shorter version of the trigonometric ratios follows.

$$\tan = \frac{\text{opposite}}{\text{adjacent}} \qquad \sin = \frac{\text{opposite}}{\text{hypotenuse}} \qquad \cos = \frac{\text{adjacent}}{\text{hypotenuse}}$$

Example 2 **Find each trigonometric ratio.**

a. $\tan Y$ **b.** $\cos X$

Solution **a.** $\tan Y = \frac{\text{opposite}}{\text{adjacent}} = \frac{12}{5}$

b. $\cos X = \frac{\text{adjacent}}{\text{hypotenuse}} = \frac{12}{13}$

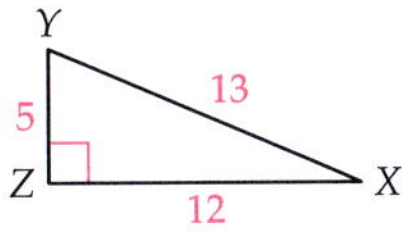

THINK Why are the values found in the Table of Trigonometric Ratios only approximate values? Some of the ratios are irrational numbers.

▼ If you know the measure of an acute angle of a right triangle, you can use a table of trigonometric ratios to approximate values of the tangent, sine, and cosine of the angle.

Example 3 **Find the tangent, sine, and cosine of 42° using the table on p. 580.**

Solution $\tan 42° \approx 0.9004$ $\sin 42° \approx 0.6691$ $\cos 42° \approx 0.7431$

Angle	Sin	Cos	Tan
51°	.7771	.6293	1.2349
52°	.7880	.6157	1.2799
53°	.7986	.6018	1.3270
54°	.8090	.5878	1.3764
55°	.8192	.5736	1.4281

▼ You can use a table of trigonometric ratios to approximate the measure of an angle when you know the sine, cosine, or tangent.

Example 4 **Sin A = .7986. Find $m\angle A$.**

Solution Use the table. Find .7986 in the sin column.
$m\angle A = 53°$.

Lesson Focus

Materials

Overhead projector, calculator with trigonometric ratio keys or tables of trigonometric ratios, protractor, ruler, math journal

Vocabulary/Symbols

cosine (cos)
sine (sin)
Table of Trigonometric Ratios
tangent (tan)
trigonometric ratio
trigonometry

MOTIVATING THE LESSON

Review writing a ratio as a decimal by having students find a decimal to the nearest ten-thousandth for each of the following:

1. $\frac{5}{12}$ (0.4167)
2. $\frac{8}{15}$ (0.5333)
3. $\frac{13}{12}$ (1.0833)
4. $\frac{15}{16}$ (0.9375)

Instruction

1 Using the overhead projector, draw a right triangle *ABC*, with right angle at *C*. Label the hypotenuse, then label the legs as opposite $\angle A$ and adjacent to $\angle A$.

Teaching Tip

Have scientific calculators with trigonometric ratio keys available for students to use in class.

2 Ask students what other trigonometric ratio could have been used to find z. ($\sin 52° = \frac{z}{33}$ or $\cos 38° = \frac{z}{33}$)

ADDITIONAL EXAMPLES

1. Find the sine, cosine, and tangent of 53°. (0.7986, 0.6018, 1.3270)

2. Find x, y, and z. Round to the nearest thousandth. ($x = 66$, $y = 26.842$, $z = 60.291$)

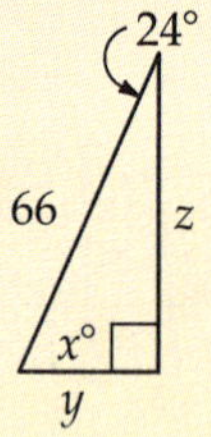

Guided Practice

Class Exercises If students are using calculators, have them compare table values with calculator values. Ask why the values are not exactly the same. (The calculator does not round values to the nearest ten-thousandth as the table does.)

Think and Discuss For question 2, ask students between which two sine values 0.638 lies. (sine 39° and sine 40°, or 0.6293 and 0.6428) Help students conclude that an angle can be measured in degrees, minutes, and seconds.

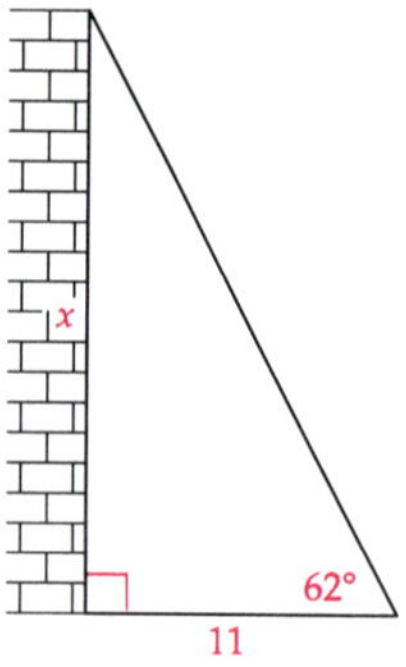

Example 5 A carpenter needs a ladder to do repair work on a building. The base of the ladder is 11 ft away from the building and forms a 62° angle with the ground. To the nearest foot, how high up on the building must the ladder rest?

Solution Draw a diagram.

$\tan 62° = \frac{x}{11}$ — Use the tangent ratio.

$1.8807 \approx \frac{x}{11}$

$11(1.8807) = \left(\frac{x}{11}\right)(11)$ — Multiply each side by 11.

$20.6877 = x$

$21 \approx x$ — Round to the nearest whole number.

The ladder must rest about 21 ft up on the building.

▼ You may need to use more than one trigonometric ratio to find measures in a triangle.

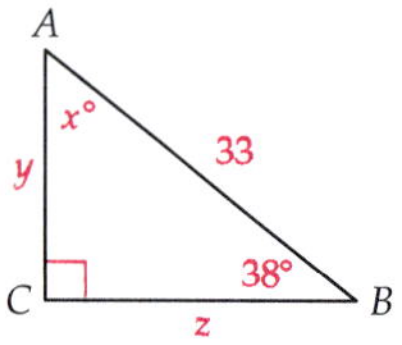

Example 6 **Use the triangle at the left. Find x, y, and z in $\triangle ABC$. Round to the nearest hundredth.**

Solution **1.** Find x.

$x = 90 - 38$
$= 52$ — $\angle A$ and $\angle B$ are complementary.

2. Find y.

$\sin 38° = \frac{y}{33}$ — $\sin = \frac{\text{opposite}}{\text{hypotenuse}}$

$0.6157 \approx \frac{y}{33}$ — Substitute the value from the table.

$33(0.6157) \approx \left(\frac{y}{33}\right)(33)$ — Multiply each side by 33.

$20.3181 \approx y$ — Round.

$20.32 \approx y$

Angle	Sin	Cos	Tan
36°	.5878	.8090	.7265
37°	.6018	.7986	.7536
38°	.6157	.7880	.7813
39°	.6293	.7771	.8098
40°	.6428	.7660	.8391

2 **3.** Find z.

$\tan 38° = \frac{20.3181}{z}$ — $\tan = \frac{\text{opposite}}{\text{adjacent}}$

$0.7813 \approx \frac{20.3181}{z}$ — Substitute the value from the table.

$z(0.7813) = \left(\frac{20.3181}{z}\right)(z)$ — Multiply each side by z.

$0.7813z = 20.3181$

$\frac{0.7813z}{0.7813} = \frac{20.3181}{0.7813}$ — Divide each side by 0.7813.

$z = 26.005504$

≈ 26.01 — Round.

MATH MINUTES

Display the triangle below. Students write the ratios for the tan, sin, and cos of $\angle A$ and $\angle B$.

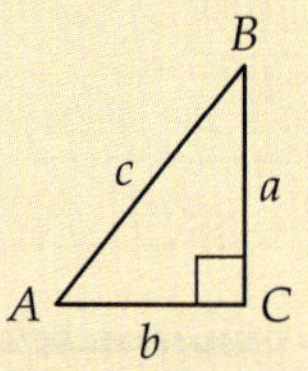

CLASS EXERCISES

Use the table of trigonometric ratios on p. 580 to find each value.

1. sin 86° 0.9976 **2.** cos 16° 0.9613 **3.** tan 53° 1.3270

Find $m\angle A$.

4. $\sin A = 0.8480$ 58° **5.** $\tan A = 3.7321$ 75° **6.** $\cos A = 0.5$ 60°

State a trigonometric equation using the given values.

7.

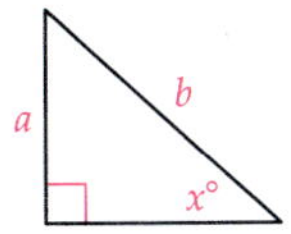

8.

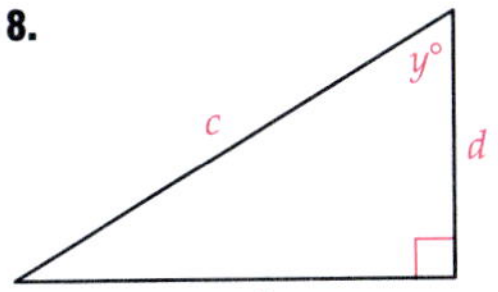

9.

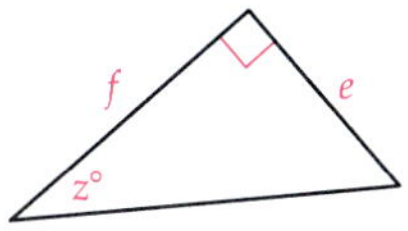

$\sin x° = \frac{a}{b}$ $\cos y° = \frac{d}{c}$ $\tan z° = \frac{e}{f}$

Find x, y, and z for each triangle. Round to the nearest hundredth. Use the table of trigonometric ratios on p. 580.

10.

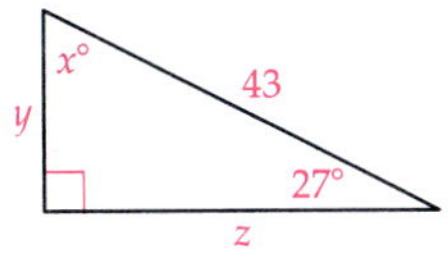

$x = 63$
$y = 19.52$
$z = 38.31$

11.

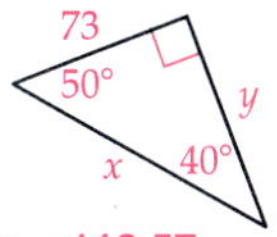

$x = 113.57$
$y = 87$

12.

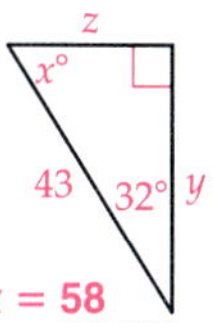

$x = 58$
$y = 36.46$
$z = 22.79$

WRITTEN EXERCISES

Use the table of trigonometric ratios on p. 580 to find each value.

1. sin 10° 0.1736 **2.** cos 80° 0.1736 **3.** tan 15° 0.2679

4. tan 55° 1.4281 **5.** sin 78° 0.9781 **6.** cos 60° 0.5000

Find $m\angle B$. Use the table of trigonometric ratios on p. 580.

7. $\sin B = 0.9945$ 84° **8.** $\cos B = 0.7660$ 40° **9.** $\tan B = 1.9626$ 63°

MENTAL MATH **Use 45°-45°-90° and 30°-60°-90° right triangles to find each value. Your answer may contain a square root sign.**

10. tan 45° 1 **11.** tan 60° $\sqrt{3}$ **12.** sin 30° $\frac{1}{2}$

State a trigonometric equation using the given values.

13.

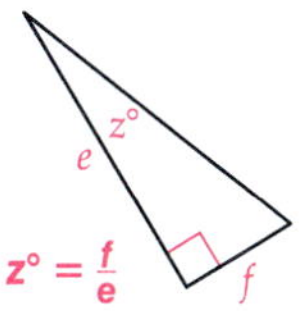

$\tan z° = \frac{f}{e}$

14.

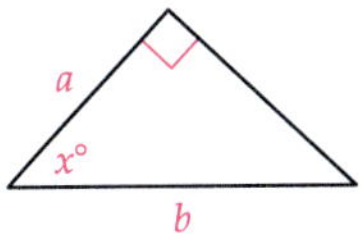

$\cos x° = \frac{a}{b}$

15.

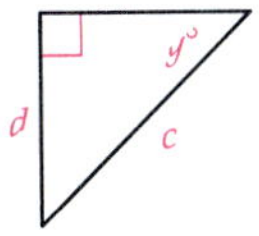

$\sin y° = \frac{d}{c}$

THINK AND DISCUSS See below.

1. Why does the trigonometric table only go to 89°?

2. No angle has sine 0.638. If you compute that ratio, how do you decide which angle to use? Do angles have measure in divisions smaller than degrees?

MIXED REVIEW

A square has a diagonal $3.4\sqrt{2}$ ft long.

1. Find the length of a side of the square. 3.4 ft

2. Find the area of the square. 11.56 ft^2

3. The angles of a triangle are 30°, 60°, 90°. The hypotenuse is 16 in. long. Find the lengths of the other two sides. See below.

4. Simplify $(-x)^2 - (x)^2$. 0

5. A triangle with all angles less than 90° is called a(n) ■ triangle. acute

6. Find the radius of a circle with area $25\pi m^2$. 5 m

7. Find three consecutive integers with a sum of -42. -13, -14, -15

1. An acute angle of a right triangle cannot be 90°.
2. Determine which angle has a sine closest to 0.638. Yes; degrees are divided into minutes.
3. 8 in., $8\sqrt{3}$ in.

A ***common error*** occurs when students confuse the trigonometric ratios. You can suggest this memory device:

$\sin = \dfrac{\text{opposite}}{\text{hypotenuse}}$ **soh**

$\cos = \dfrac{\text{adjacent}}{\text{hypotenuse}}$ **cah**

$\tan = \dfrac{\text{opposite}}{\text{adjacent}}$ **toa**

Closure

Writing in Math Have students draw this triangle in their math journals and then describe how they would find x and y. Have students discuss their methods with the class.

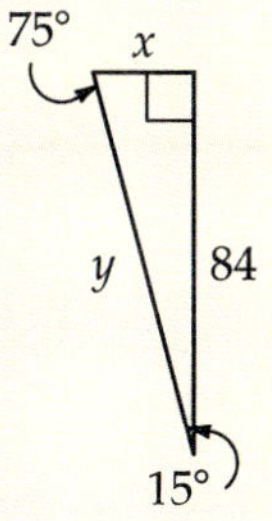

Independent Practice

Students may use a calculator instead of the tables for questions 1-9.

For questions 10-12, encourage students to draw a sketch to help them visualize the relationships of the sides.

Lesson Follow-up

Reteaching Activity Have students draw several right triangles, measure the sides and angles, and label the triangles. Ask students to find the values of the trigonometric ratios, using the definitions, for each acute angle. Finally, have them compare the values with the values in the trigonometric table.

LESSON RESOURCES

TRB Practice K-11
TRB Enrichment K-12

Practice

For use after 11-6 (pp. 473–476)

Trigonometric Ratios

Use the table of trigonometric ratios to find each value.

1. cos 20° **0.9397** 2. tan 64° **2.0503** 3. sin 41° **0.6561**
4. tan 8° **0.1405** 5. sin 88° **0.9994** 6. cos 53° **0.6018**

Find $m\angle A$.

7. $\tan A = 5.1446$ **79°** 8. $\sin A = 0.0872$ **5°** 9. $\cos A = 0.3746$ **68°**

MENTAL MATH Write each ratio using square root signs. Use your knowledge of 45°-45°-90° and 30°-60°-90° right triangles.

10. tan 30° $\frac{1}{\sqrt{3}} = \frac{\sqrt{3}}{3}$ 11. cos 45° $\frac{1}{\sqrt{2}} = \frac{\sqrt{2}}{2}$ 12. sin 60° $\frac{\sqrt{3}}{2}$

Find the trigonometric ratios.

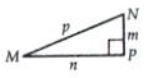

13. sin N $\frac{n}{p}$ 14. tan M $\frac{m}{n}$
15. cos N $\frac{m}{p}$

Write the trigonometric ratio to the nearest hundredth.

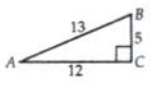

16. sin B **0.92** 17. cos B **0.38**
18. tan A **0.42**

Find x, y, and z for each triangle. Write each decimal answer to the nearest hundredth.

19. $x =$ **39°**, $y =$ **18.65**, $z =$ **15.10**

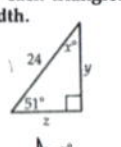

20. $x =$ **54°**, $y =$ **6.54**, $z =$ **11.12**

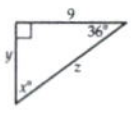

21. $x =$ **22°**, $y =$ **68°**, $z =$ **269.63**

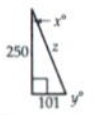

22. $x =$ **55°**, $y =$ **20.92**, $z =$ **17.14**

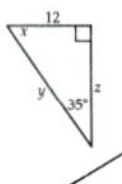

23. A surveyor standing 2,250 ft from the base of the World Trade Center in New York City measured a 31° angle to the topmost point. To the nearest ft, how tall is the World Trade Center?
1,352 ft

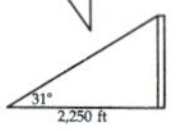

Chapter 11 K-11

Enrichment

For use after 11-6 (pp. 473–476)

Trigonometric Patterns

1. Use the table of trigonometric ratios to find each value.

sin 20° **0.3420**	sin 32° **0.5299**	sin 71° **0.9455**	sin 83° **0.9925**
cos 70° **0.3420**	cos 58° **0.5299**	cos 19° **0.9455**	cos 7° **0.9925**

2. How are the two angles in each of the above boxes related?
They are complementary angles; their sum is 90°.
3. Predict the answers without using the trigonometric table.
sin 50° = cos **40°** cos 75° = sin **15°** cos 27° = sin **63°**

CALCULATOR Complete. Use the table of trigonometric values. Round answers to the nearest ten-thousandth.

	∠A	sin A	cos A	(sin A)²	(cos A)²	(sin A)² + (cos A)²
4.	5°	0.0872	0.9962	0.0076	0.9924	1
5.	27°	0.4540	0.8910	0.2061	0.7939	1
6.	41°	0.6561	0.7547	0.4304	0.5696	1

7. What do you notice about the results in the last column? **They all equal 1.**
8. Predict the answers without using the trigonometric table.
$(\sin 16°)^2 + (\cos 16°)^2$ **1** $(\sin 57°)^2 + (\cos 57°)^2$ **1**

CALCULATOR Complete. Use the table of trigonometric values. Round answers to the nearest ten-thousandth.

	∠A	sin A	cos A	$\frac{\sin A}{\cos A}$	tan A
9.	13°	0.2250	0.9744	0.2309	0.2309
10.	34°	0.5592	0.8290	0.6745	0.6745
11.	51°	0.7771	0.6293	1.2349	1.2349

12. Describe the results in the last two columns. **They are equal.**
13. Write a statement relating the tangent of an angle A to its sine and cosine. $\tan A = \frac{\sin A}{\cos A}$

K-12 Chapter 11

Find x, y, and z for each triangle. Round to the nearest hundredth. Use the table of trigonometric ratios on p. 580.

16.

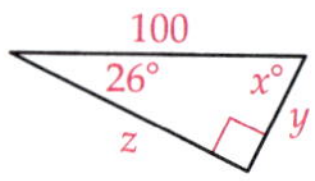

$x = 64$
$y = 43.84$
$z = 89.88$

***17.**

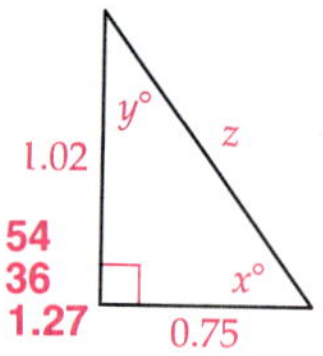

$x \approx 54$
$y \approx 36$
$z = 1.27$

18.

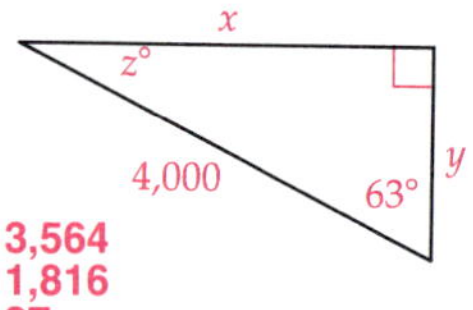

$x = 3{,}564$
$y = 1{,}816$
$z = 27$

19.

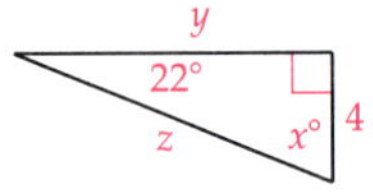

$x = 68$
$y = 9.90$
$z = 10.68$

20. $x = 6.93$

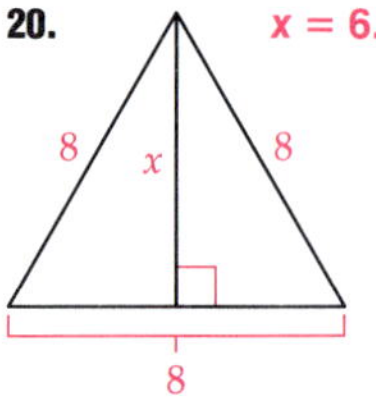

21. $x \approx 28$
$y \approx 62$

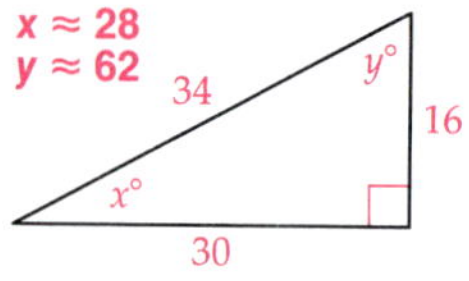

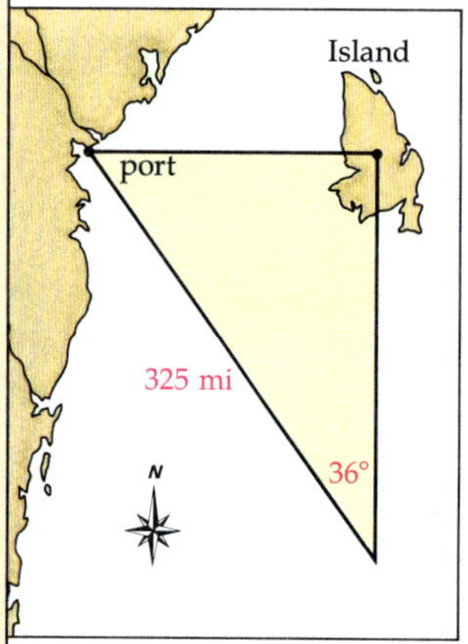

22. A navigator brings you a map charting a ship's course. She wants to know how far it is, to the nearest mile, from the island to the mainland port. Use the diagram at the left. **191 m**

23. A Boeing 747 climbs continuously at a 30° angle to a height of 35,000 ft. To the nearest tenth of a mile, how far has the plane traveled to reach that elevation? *Hint:* 5,280 ft = 1 mi. **13.3 mi**

24. ***DATA FILE 11 (pp. 450–451)*** A skydiver jumps from 22,500 ft. He hits a marker 10,224 ft away. To the nearest degree, what angle does the skydiver make with the ground? **66°**

25. A man 6 ft tall paces 75 ft from the base of a tree. He uses a protractor to approximate the angle from his eye to the top of the tree. He finds that this angle is about 25°. Find the height of the tree to the nearest foot. **41 ft**

TEST YOURSELF

Tell whether a triangle with sides of the given lengths is 45°-45°-90°, 30°-60°-90°, or neither.

1. 10, 5, $5\sqrt{3}$ **30°-60°-90°** **2.** 8, $8\sqrt{2}$, 8 **45°-45°-90°** **3.** 6, 8, 10 **neither**

Find x, y, and z for each triangle. Round to the nearest hundredth.

4.

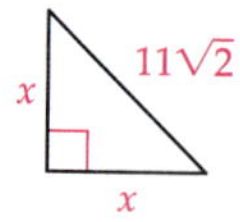

$x = 11$

5.

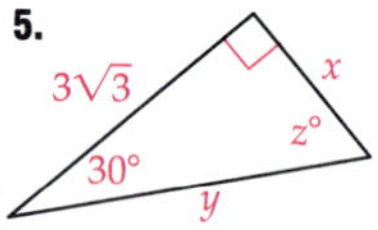

$x = 3$
$y = 6$
$z = 60$

6.

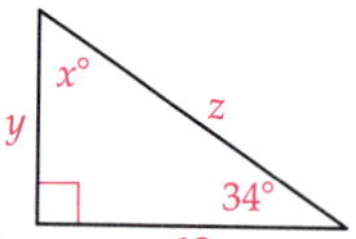

$x = 56$
$y = 8.09$
$z = 14.48$

LESSON QUIZ

1. If the hypotenuse of a right triangle is 10 units and the legs are 5 and $5\sqrt{3}$ units, what are the sin, cos, and tan of the smallest angle? The answers may contain square root signs. ($\frac{5}{10}$ or $\frac{1}{2}$, $\frac{5\sqrt{3}}{10}$ or $\frac{1}{2}\sqrt{3}$, $\frac{5}{5\sqrt{3}}$ or $\frac{1}{\sqrt{3}}$)

2. Find the cos of 38°. (0.788)

3. Find x, y and z. Round to the nearest hundredth. (z: 62°, x: 53.17, y: 113.26)

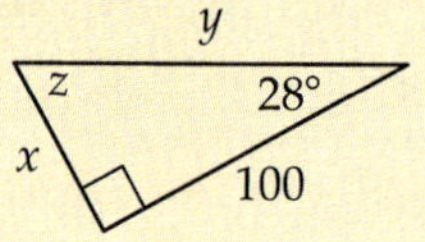

Assignment Guide

Basic 1-3, 7-11, 13-14, 16-17, 22-24, MR, TY All
Average 3-5, 7-11, 13-14, 18-19, 23-24, MR, TY All
Enriched 4-10, 12, 14-15, 20-21, 24-25, MR, TY All

FOR THE NEXT LESSON

compass, ruler, protractor, projector, math journal

OBJECTIVE: *To apply right triangle concepts and trigonometric ratios to precision drawings.*

PROBLEM SOLVING APPLICATION

11-7 Precision Drawing

Architects and designers often require accurate plans and drawings. Professionals frequently use computers with graphic capability to help them with their work. Even so, there are many occasions when precision drawing must be done by hand.

THINK What are some professions that require precision drawing? architect; technical artist

You can draw an angle accurately using a compass and a ruler.

Example 1 **Draw a right angle using a ruler and a compass.**

Solution A triangle whose sides measure 3 units, 4 units, and 5 units is a right triangle. Draw a right triangle with these dimensions.

1. Draw $\overline{AB}$, 3 cm long.

2. Use a compass with a radius of 4 cm. With center at B, draw an arc above B.

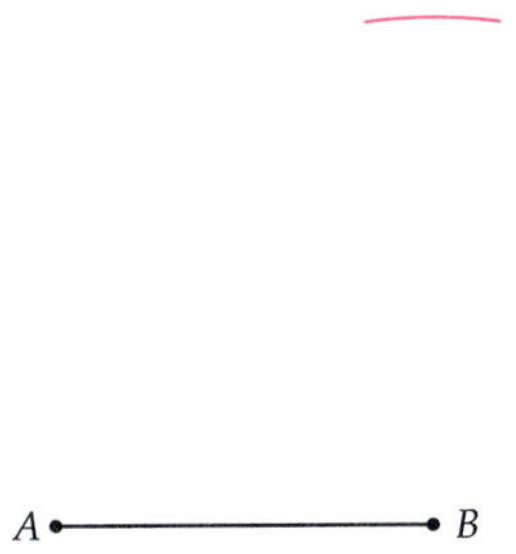

3. Use a compass with a radius of 5 cm. With center at A, draw an arc that crosses the arc above B. Label the point of intersection as C.

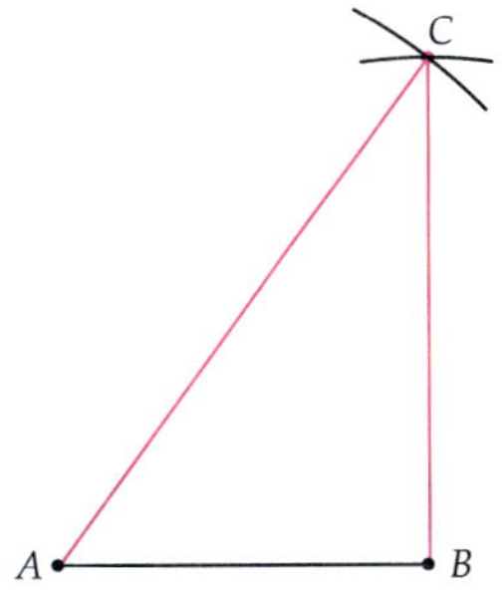

Connect the three points to form $\triangle ABC$. By the converse of the Pythagorean theorem, $\triangle ABC$ is a right triangle. Therefore, $\angle ABC$ is a right angle.

Lesson Focus

Materials

Compass, ruler, protractor, projector, math journal

MOTIVATING THE LESSON

Ask students to describe situations in which they might need precise drawings to complete tasks at home. Examples might include building a work bench and drawing a pattern for a costume. Have students speculate on the consequences of making an inaccurate drawing.

Instruction

1 Use the chalkboard or an overhead projector to demonstrate the solution to Example 1. Since some students may have difficulty using the compass, comment aloud on each of your actions as you use the compass. Have students substitute the given values for the variables in the Pythagorean theorem to prove that ΔABC is a right triangle.

Teaching Tip

Suggest that students use a ruler or straightedge to match an angle to its tangent in reading across the table of trigonometric ratios. In this way, students will avoid substituting erroneous values for the tangent.

2 Demonstrate each step of the solution to Example 2 on the chalkboard or overhead projector. Remind students that it is important to label each part of the triangle as it is drawn. When the triangle is completed, use the protractor to verify that $\angle A$ is 23°.

Guided Practice

Class Exercises Allow students to use calculators to solve for x in questions 4-6.

Closure

Writing in Math Have students use their math journals to describe each step involved in drawing a right triangle when the side measurements are given and in drawing a right triangle that has an angle with a given measure.

Independent Practice

For questions 7-10, have students use the concept of supplementary angles to draw the obtuse angles by first drawing the acute angle and then extending the leg of the right triangle that is adjacent to the acute angle.

Decision Making This feature allows students to consider the practical issues involved in making precise drawings. For question 6, students may need assistance to understand that measurements with ratios that are greater may result in smaller units being used, and, therefore, a more accurate drawing will result.

■ You can draw any angle by finding the tangent of the angle and by predetermining the length of the adjacent side.

Example 2 **Draw a 23° angle. Use a protractor, a ruler, and the table of trigonometric ratios on p. 580.**

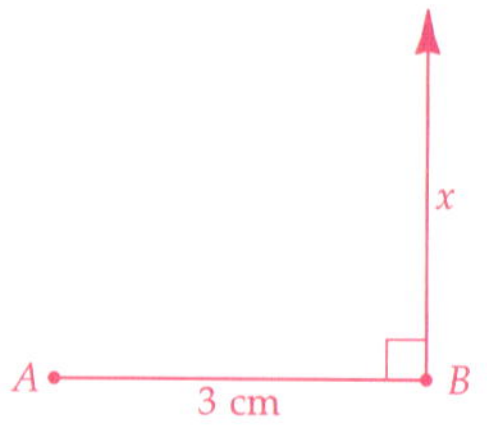

Solution

1. Choose a length, say 3 cm, for the side adjacent to the 23° angle. Draw $\overline{AB}$ with length 3 cm. Use a protractor to draw $\angle B$ with measure 90°. Let x represent the unknown length.
2. Solve for x to determine the length of the side opposite the 23° angle.

2

$\tan = \frac{\text{opposite}}{\text{adjacent}}$	Use the tangent ratio.
$\tan 23° = \frac{x}{3}$	Substitute values.
$0.4245 = \frac{x}{3}$	Find tan 23° in the table.
$3(0.4245) = \frac{x}{3}(3)$	Multiply each side by 3.
$1.2735 = x$	Round to the nearest tenth.
$x \approx 1.3$	

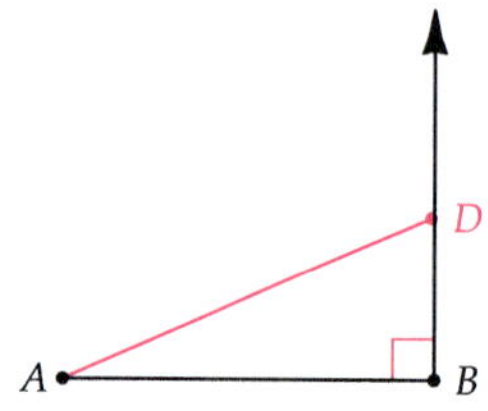

3. Mark a point 1.3 cm from B. Label the point as D. Then draw $\overline{AD}$.

$\angle A \approx 23°$.

■■■■■■ Decision Making ■ **DECISION MAKING** ■ Decision Making ■ Decision Making ■ Decision Making

PRECISION DRAWING

■ **COLLECT DATA** Check students' work.

1. Visit a store that sells drafting supplies. Take along the compass, ruler, and protractor that you use in school.
   a. Do the inch and centimeter marks on your ruler perfectly match those on the best quality ruler found in the store?
   b. Is your protractor the same size as the best one in the store?

■ **ANALYZE DATA**

2. Can the quality of the instruments you use affect the accuracy of the drawings you make? Explain. **Yes; The higher quality instruments have more accurate markings.**
3. Can the width of the markings on a ruler or compass affect the accuracy of a drawing? Explain. **Yes; The thinner the markings on a ruler, the more accurate the drawing will be.**
4. Can the fineness of your pencil point affect the accuracy of a drawing? Explain. **Yes; The sharper the pencil point, the more accurate the drawing will be.**

MATH MINUTES

Display triangle ABC. Have students use the Pythagorean theorem and the table of square roots to find the length of x. ($x = 4.899$)

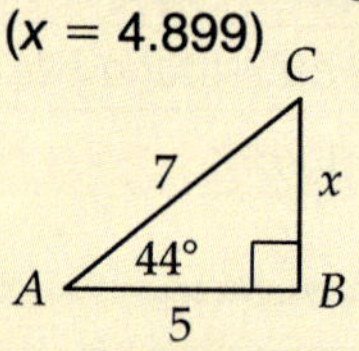

CLASS EXERCISES

Find the tangent of each angle. Use the table of trigonometric ratios on p. 580.

1. 87° 19.0811
2. 15° 0.2679
3. 72° 3.0777

Solve for x. Use the table of trigonometric ratios on p. 580.

4. $\tan 47° = \frac{x}{4}$ $x = 4.2896$
5. $\tan 63° = \frac{x}{7}$ $x = 13.7382$
6. $\frac{x}{5} = \tan 36°$ $x = 3.6325$

WRITTEN EXERCISES

Use a CALCULATOR where appropriate.

Check students' work.

Use a compass, a ruler, and the table of trigonometric ratios to draw an angle with the given measure. Check using a protractor.

1. 20°
2. 25°
3. 85°
4. 50°
5. 76°
6. 35°
7. 145°
8. 176°
9. 138°
10. 121°

11. Use a compass and a ruler to draw a right triangle with sides of measure 5 units, 12 units, and 13 units.
    a. Label the triangle as shown. Use the table of trigonometric ratios on p. 580 to find $m\angle A$ and $m\angle B$. $m\angle A \approx 23°$ $m\angle B \approx 67°$
    b. Check your results using a protractor.

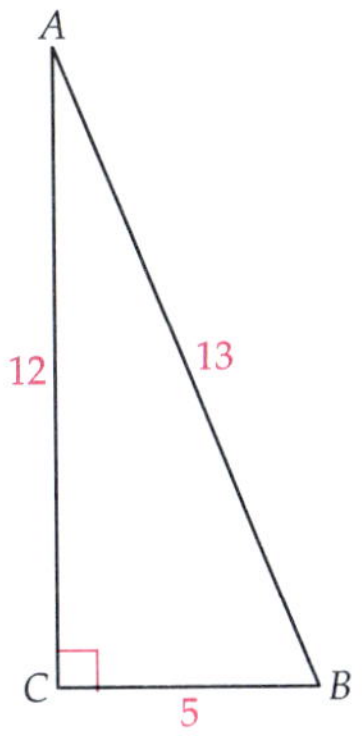

Decision Making ■ Decision Making ■ Decision Making ■ Decision Making ■ Decision Making ■ Decision Making ■

5. Would inaccuracies in angles and lengths in a drawing be more easily detected in small or in large drawings? Explain. Large drawings; inaccuracies are more obvious when drawings are larger.

■ ***MAKE DECISIONS***

6. Do you think it would be more accurate to draw a right triangle using a compass and ruler with the measurements in the ratio 3 : 4 : 5, 6 : 8 : 10, or 9 : 12 : 15? Explain. 9 : 12 : 15; The compass is easier to maneuver at larger settings.
7. Suppose you are using a compass, ruler, and trigonometric ratios to draw angles. Would a table that expresses the ratios to hundredths or one that rounds ratios to the nearest tenth help you make a more accurate drawing? Explain. Ratios to the nearest hundredth are more accurate because this is a more precise measurement.

Lesson Follow-up

LESSON RESOURCES

TRB Alternate Application K-13

Assignment Guide

Basic All
Average All
Enriched All

FOR THE NEXT LESSON

pencil, paper, math journal, newspapers

Alternate Application

For use after 11-7 (pp. 477–479)

Polar Coordinates

In Chapter 8, you graphed points using rectangular coordinates. You can also graph points using *polar coordinates*. Point *A* has polar coordinates (3, 60°). The first number is the point's *distance* from origin. The second number is the *angle* formed by a line segment drawn to the point and the positive horizontal axis.

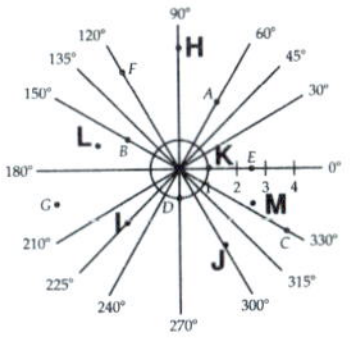

Write the polar coordinates of the point.

1. *B* (2, 150)
2. *C* (4, 330°)
3. *D* (1, 270°)
4. *E* $(2\frac{1}{2}, 0°)$
5. *F* $(3\frac{1}{2}, 120°)$
6. *G* (4, 195°)

Graph these points. Write the letter beside the point.

7. *H*(4, 90°)
8. *I*(2, 225°)
9. *J*(3, 300°)
10. *K*(1, 0°)
11. *L*(3, 165°)
12. $M(2\frac{1}{2}, 345°)$

Point *P* has polar coordinates (r, $x°$) and rectangular coordinates (a, b).

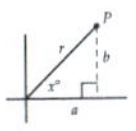

$\frac{a}{r} = \cos x$ $\frac{b}{r} = \sin x$

$a = r \cos x$ $b = r \sin x$

You can use these equations to convert polar to rectangular coordinates.

Example A point has polar coordinates (5, 30°). Write the rectangular coordinates.

Solution

$a = r \cos x$
$= 5 \cos 30°$
$= 5(0.8660)$
$a = 4.33$

$b = r \sin x$
$= 5 \sin 30°$
$= 5(0.5)$
$b = 2.5$

The rectangular coordinates are (4.33, 2.5).

Write the rectangular coordinates of the point with these polar coordinates. Round to the nearest hundredth.

13. (6, 45°) (4.24, 4.24)
14. (10, 62°) (4.69, 8.83)
15. (2, 73°) (0.58, 1.91)
16. (8.1, 24°) (7.40, 3.29)
17. (4.2, 81°) (0.66, 4.15)
18. (16, 8°) (15.84, 2.23)

Chapter 11 K-13

Career Information

Arrange for a landscape architect to talk to the class. Encourage students to ask questions about the skills necessary for a career as a landscape architect.

Ask the speaker to cover topics such as:

- the training necessary to qualify as a landscape architect, including course work and licensing requirements
- how a landscape architect uses right triangles and trigonometric ratios on the job
- computer software that is used to assist the landscape architect in making decisions and planning designs
- how to prepare and interpret landscape maps
- other landscaping issues, such as environmental impact and resource management

CAREER

Help Wanted: Landscape Architect

Bachelor's degree in landscape architecture required. Master's degree preferable. For more information, write to American Society of Landscape Architects, 1733 Connecticut Ave., NW, Washington, DC 20009.

Landscape architects design residential areas, industrial and public parks, shopping centers, and college campuses. Landscape architects design sites to be attractive as well as functional. Before sketching a design for a site, landscape architects consider natural elements, effects of existing structures, the purpose of the project, and funding.

PROJECT

Visit a local park. What factors had to be considered in its design?

Practice

Find each square root.

1. $\sqrt{169}$ **13**
2. $\sqrt{4x \cdot 4x}$ **4x**
3. $\sqrt{\frac{81}{4}}$ **$\frac{9}{2}$**
4. $\sqrt{a^2b^2}$ **ab**
5. $\sqrt{\frac{x^2}{y^2}}$ **$\frac{x}{y}$**
6. $\sqrt{625x^{10}y^8}$ **$25x^5y^4$**
7. $\sqrt{400x^6}$ **$20x^3$**
8. $\sqrt{529c^2}$ **23c**
9. $\sqrt{121c^4}$ **$11c^2$**
10. $\sqrt{2{,}500}$ **50**

Approximate each square root to the nearest integer.

11. $\sqrt{2}$ **1**
12. $-\sqrt{5}$ **-2**
13. $\sqrt{10}$ **3**
14. $\sqrt{47}$ **7**
15. $\sqrt{96}$ **10**

State whether each number is rational or irrational.

16. $\sqrt{11}$ **irrational**
17. $-\sqrt{9}$ **rational**
18. 4.1472 . . . **irrational**
19. $\sqrt{0.04}$ **rational**
20. $\sqrt{81}$ **rational**

Determine whether each is a Pythagorean triple.

21. 10, 11, 12 **no**
22. 2, 3, 4 **no**
23. 1.2, 0.5, 1.3 **no**
24. 30, 40, 50 **yes**
25. $\sqrt{16}, \sqrt{25}, \sqrt{36}$ **no**
26. 5, 12, 13 **yes**
27. 17, 8, 15 **yes**
28. 1, 3, 5 **no**

Identify each triangle as 45°-45°-90°, 30°-60°-90°, or neither.

29. 5.2, $5.2\sqrt{2}$, 5.2 **45°-45°-90°**
30. 8, 4, $4\sqrt{3}$ **30°-60°-90°**
31. 15, 20, 25 **neither**

Use the Pythagorean theorem to find x.

32. 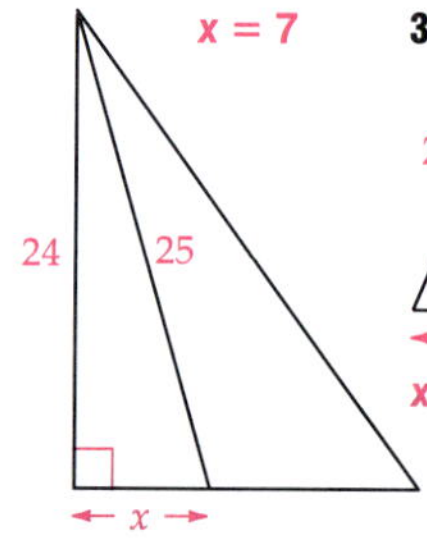

x = 7

33. 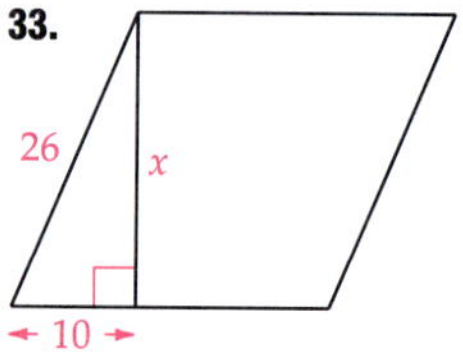

x = 24

34. 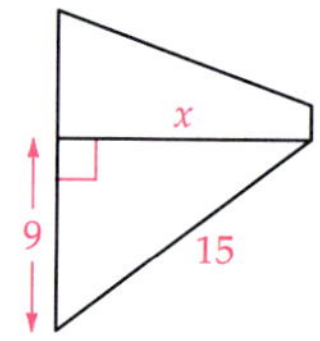

x = 12

Use similar triangles to find y.

35. 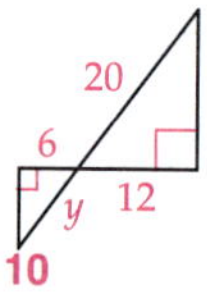

y = 10

36. 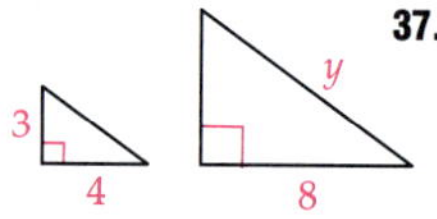

y = 10

37. 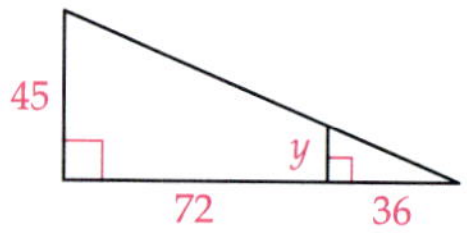

y = 15

Find each missing length. Answers may contain square root signs.

38. 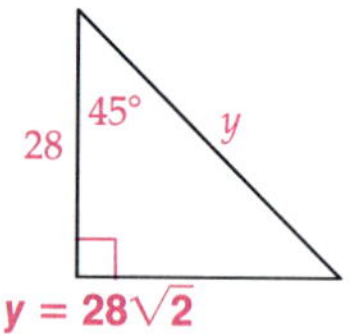

$y = 28\sqrt{2}$

39. 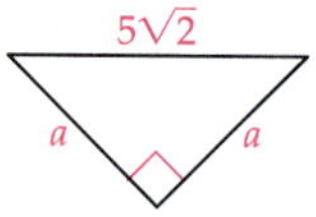

a = 5

40. 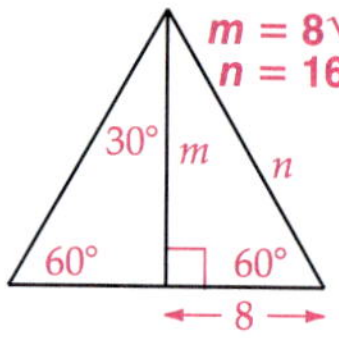

$m = 8\sqrt{3}$
n = 16

Problem Solving Practice

PROBLEM SOLVING STRATEGIES

Look for a Pattern
Guess and Test
Simplify the Problem
Make an Organized List
Work Backwards
Account for All Possibilities
Make a Table
Write an Equation
Solve by Graphing
Draw a Diagram
Make a Model
Solve Another Way
Simulate the Problem

Solve. Use an appropriate strategy or a combination of strategies.

1. A car manufacturer offers exterior colors of white, blue, red, black, and silver. The manufacturer offers the same colors, plus tweed, for the interior. The manufacturer offers pinstripes, but only for cars with black or silver exteriors. How many different styles are there to choose from? **42**

2. *The Battle of Gettysburg* is the largest painting in the world. The painting is 410 ft. long and 70 ft. wide. A tennis court is 78 ft. by 27 ft. How many tennis courts can fit inside the world's largest painting? **See Additional Answers.**

3. In $\triangle XYZ$, $m\angle X$ is 29° less than $m\angle Y$, and $m\angle Z$ is 52° less than $m\angle Y$. Find the measure of each angle. **58°, 87°, and 35°**

4. Mt. Vesuvius is a volcano in Italy. Its crater is 7,920 ft in diameter. What is the area of the crater? **49, 240, 224 ft²**

5. ***CALCULATOR*** The length of the Golden Gate Bridge in San Francisco, California, is 4,200 ft. How long, to the nearest hundredth of a minute, would it take a car driving at a continuous rate of 35 mi/h to cross the bridge? *Hint:* 5,280 ft = 1 mi. **1.36 min**

6. ***DATA FILE 10 (pp. 404–405)*** How much more circular area, to the nearest square meter, does a snowshoe hare require than a Northern flying squirrel requires? Use 3.14 for π. **22,219 m²**

7. First-class letters cost \$.29 for the first ounce and \$.23 for each additional ounce. Warren spent \$2.82 to send a first-class letter. How many ounces did his letter weigh? **12**

8. I am a number less than 100. I am the product of two prime numbers. If you reverse my digits I am prime. The sum of my digits is a one-digit prime. One of my digits is a square number. What number am I? **34 or 14**

9. Each face of cube A has a diagonal of $10\sqrt{2}$ cm. Each face of cube B has a diagonal of $7\sqrt{2}$ cm. How much more liquid will cube A hold than cube B? **657 cm³**

10. Madison Square Garden in New York City is built in the shape of a circle. Its diameter is 404 ft. The stadium accommodates 20,234 spectators. To the nearest hundredth square foot, how much area does this allow for each person? Use 3.14 for π. **6.33 ft²**

11. Three integers have a sum of 48. The greater integer is four more than the middle integer. The least integer is ten less than the middle integer. Find the integers. **8, 18, and 22**

Instruction

This page provides a variety of problems that can be used to reinforce and enhance the students' problem solving skills. Encourage students to read each problem carefully. Then have them refer to the list of problem solving strategies to help them decide how to solve the problem.

Point out, however, that not all questions require a strategy for solving, nor are all the strategies in the list used in this lesson.

ADDITIONAL ANSWERS

2. 10 (see diagram)

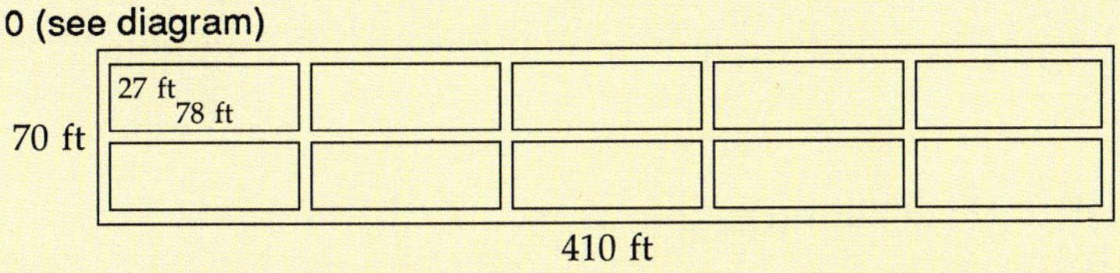

Study Skills

Successful students are often those who know when and how to ask for assistance. Asking for help not only serves the students, but enables the teacher to assess where students are in the learning process.

Discuss with the class the value of asking for assistance. Explain that when a student asks a question, the answer benefits the class as a whole, as well as the teacher who can determine if additional instruction is necessary. Often questions can help enrich the lesson with unique insights, ideas, and solutions.

Prepare guidelines with the class that indicate when and how students may ask for help. The suggestions may include time after class or before a test for students to confer with the teacher, a discussion period after a lesson, or study periods where cooperative-study teams can provide additional help.

Guidelines may include the requirement that students attempt to solve the problem first, and then ask for help. Agree on a protocol for asking and answering questions, as well as a procedure for seeking information or help outside the classroom.

Chapter 11 Review

Match each word with the example that illustrates its meaning.

1. square root symbol g	**a.** hypotenuse = $\sqrt{2}$(length of leg)
2. principal square root d	**b.** $a^2 + b^2 = c^2$
3. irrational number f	**c.** $\frac{\text{opposite}}{\text{adjacent}}$
4. Pythagorean theorem b	**d.** $\sqrt{4} = +2$
5. in a 45°-45°-90° triangle a	**e.** $\frac{\text{opposite}}{\text{hypotenuse}}$
6. in a 30°-60°-90° triangle h	**f.** 2.6457513 . . .
7. tangent A c	**g.** $\sqrt{}$
8. cosine A i	**h.** hypotenuse = 2(shorter side)
9. sine A e	**i.** $\frac{\text{adjacent}}{\text{hypotenuse}}$

Finding Square Roots — 11-1

To find the square root, think of the squares. You can also use a calculator or a table of squares.

Find each square root. Use the square root table on p. 579.

10. $\sqrt{196}$ 14 **11.** $\sqrt{64p^2}$ 8p **12.** $\sqrt{3}$ 1.732 **13.** $\sqrt{(x-2)^2}$ (x − 2) **14.** $\sqrt{625}$ 25 **15.** $\sqrt{31 + 5}$ 6

Problem Solving — 11-2

To solve a problem, sometimes it is helpful to act it out.

16. A class of 31 students counted off by 1s beginning with the number one. Each student who counted a multiple of 5 stood up. This process was repeated. Those who were standing were skipped. After the fourth counting, how many students were standing? 18

Pythagorean Theorem — 11-3

To use the Pythagorean theorem to find unknown parts of a right triangle, substitute the known values in the formula $a^2 + b^2 = c^2$, and solve for the unknown.

Write an equation. Solve for x.

17. 5

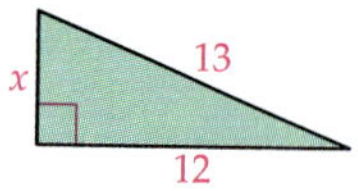

18. 15

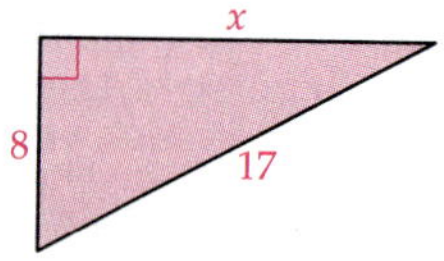

19. 25

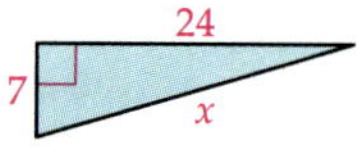

20. 6.708

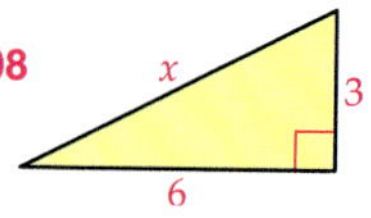

Similar Right Triangles 11-4

To find values in similar right triangles, you can use the Pythagorean theorem or write a proportion. Sometimes you will use both.

Solve for x. Each pair of triangles is similar.

21. 39

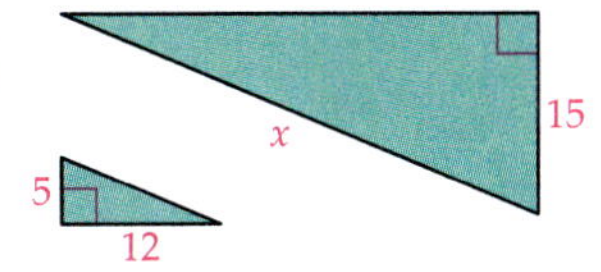

22. 9

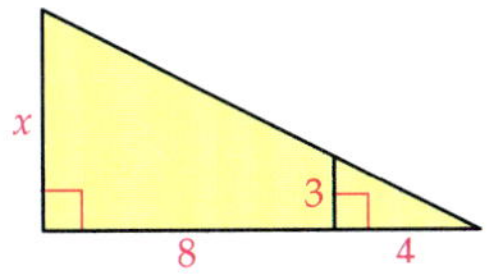

23. 25

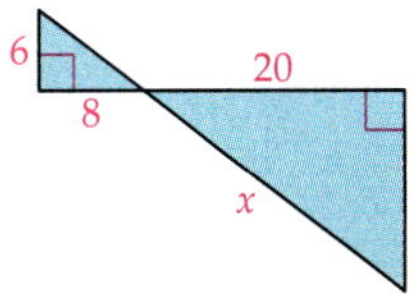

Special Right Triangles 11-5

To find missing lengths in a 45°-45°-90° triangle, use the relationships

$\text{hypotenuse} = \sqrt{2} \cdot \text{leg}$ $\qquad$ $\text{leg} = \frac{\text{hypotenuse} \cdot \sqrt{2}}{2}$.

To find missing lengths in a 30°-60°-90° triangle, use the relationships

$\text{hypotenuse} = 2(\text{shorter leg})$ $\qquad$ $\text{longer leg} = \text{shorter leg}(\sqrt{3})$.

Find each value. Answers may contain square root signs.

24. 32

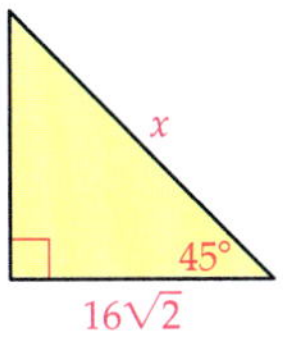

25. 18

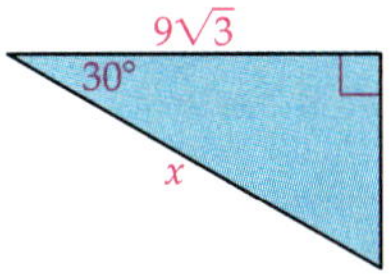

26. $8\sqrt{3}$

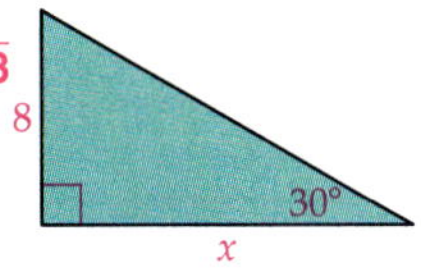

27. 20

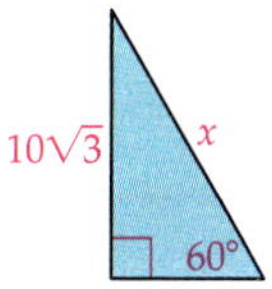

Trigonometric Ratios 11-6

To find measures in a right triangle, use a trigonometric ratio.

$\tan A = \frac{\text{opposite}}{\text{adjacent}}$ $\qquad$ $\cos A = \frac{\text{adjacent}}{\text{hypotenuse}}$ $\qquad$ $\sin A = \frac{\text{opposite}}{\text{hypotenuse}}$

Find each missing length or angle measure. Round to the nearest thousandth. Use the table of trigonometric ratios on p. 580.

28.

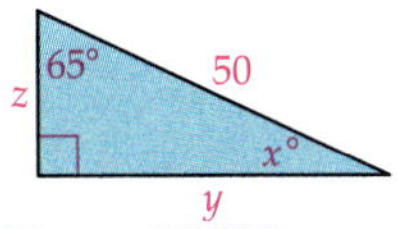

x = 25; y = 45.315; z = 21.13

29. x = 70; y = 26.312; z = 9.576

30.

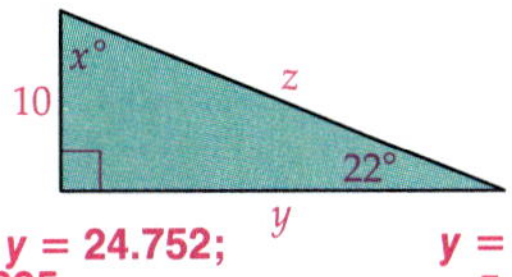

x = 68; y = 24.752; z = 26.695

31.

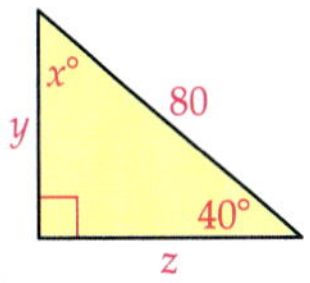

x = 50; y = 51.424; z = 61.28

32. A ladder is 8 ft away from a building and forms a 54° angle with the ground. What is the length of the ladder? 13.610 ft

33. A train in the mountains rises 8 ft for every 200 ft it moves along the track. Find the angle of elevation of the tracks. 2°

Here are some additional ideas:

- Be aware that students may be uncomfortable about asking for help.
- Appropriate help may require referral to school resources, such as the librarian, nurse, or counselor.
- Conclude lessons by having students explain how they solved problems and tell what difficulties they may have encountered.

RESOURCES

TRB K-15–K-18

Scoring Chart	
No. Correct	**Score**
26	100
25	96
24	92
23	88
22	84
21	80
20	76
19	72
18	68
17	64
16	60
15	57
14	54
13	50
12	46
11	42
10	38
9	34
8	30
7	26
6	22
5	18
4	14
3	10
2	6
1	2

Chapter 11 Test

Find each square root.

1. $\sqrt{64}$ 8 **2.** $\sqrt{169}$ 13 **3.** $\sqrt{\frac{36}{49}}$ $\frac{6}{7}$ **4.** $\sqrt{(x+5)^2}$ $x + 5$

Solve for x. Calculate each length to the nearest hundredth.

5. 25

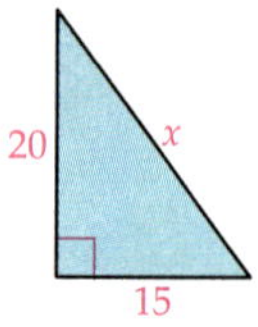

6. 6.93

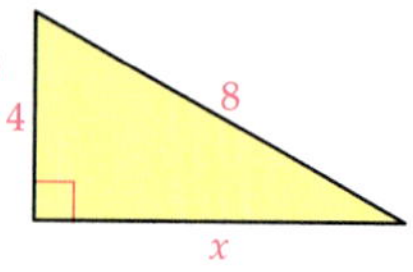

7. 14.42

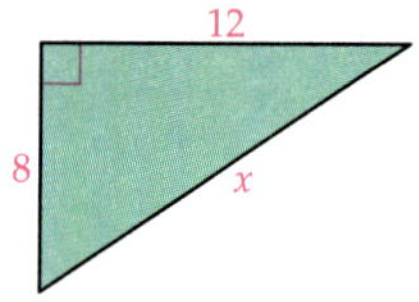

8. 25.10

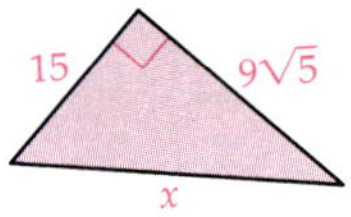

Determine if the lengths can be the sides of a right triangle.

9. 9, 12, 14 no **10.** 1.1, 6.0, 6.1 yes **11.** $\frac{3}{5}, \frac{4}{5}, 1$ yes **12.** 8.1, 15.2, 18.6 no

Solve for x. Answers may contain square root signs.

13. $4\sqrt{2}$

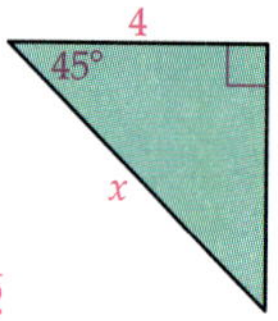

14. 15

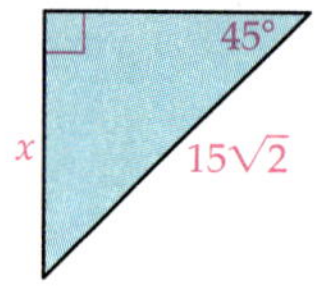

15. $4\sqrt{3}$

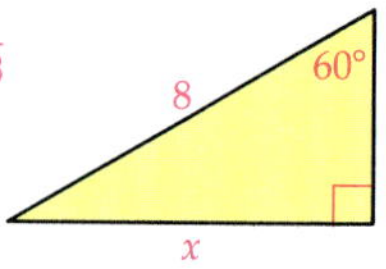

16. 10

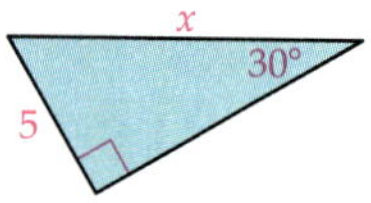

Use the table of trigonometric ratios on p. 580 to find each value.

17. cos 25° 0.9063 **18.** tan 40° 0.8391 **19.** sin 73° 0.9563 **20.** tan 88° 28.6363

Find each missing length or angle measure. Round angles to the nearest whole number and lengths to the nearest thousandth. Use the table of trigonometric ratios on p. 580.

21.

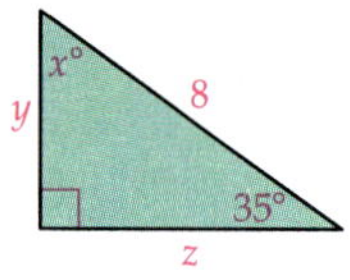

$x = 55$; $y = 4.589$; $z = 6.554$

22.

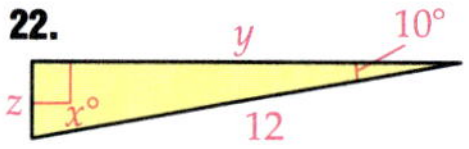

$x = 80$; $y = 11.818$; $z = 2.083$

23.

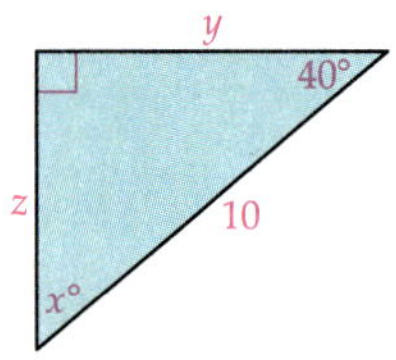

$x = 50$; $y = 7.660$; $z = 6.428$

24.

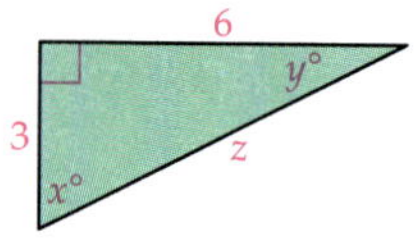

$x \approx 63$; $y \approx 27$; $z = 6.708$

Solve.

25. A square is 6 in. on each side. What is the measure of the diagonal of the square? $6\sqrt{2}$ in. or 8.485 in.

26. The captain of a ship sights a lighthouse. The angle of elevation is 12°. The captain knows that the lighthouse is 24 m above sea level. What is the distance from the ship to the lighthouse? 112.9 m

Chapters 1–11 Cumulative Review

Choose the correct answer. Write A, B, C, or D.

1. Solve $\frac{2}{3}x - 5 = 3x + 9 - \frac{1}{3}x$. A

A. -7 **B.** 7
C. $3\frac{1}{2}$ **D.** not given

2. Solve. B

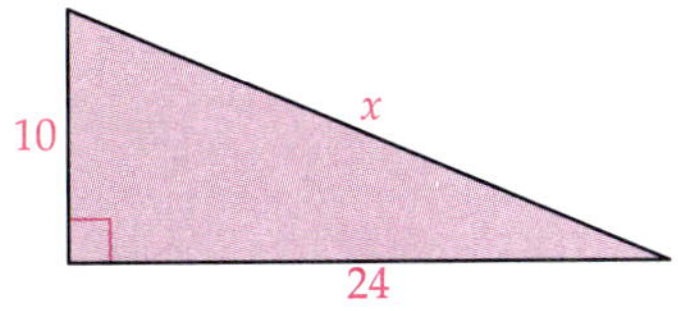

A. 17 **B.** 26
C. 28 **D.** not given

3. Find the y-intercept.
$2x + 3y = 12$ D

A. 2 **B.** 12
C. 3 **D.** not given

4. Solve. B

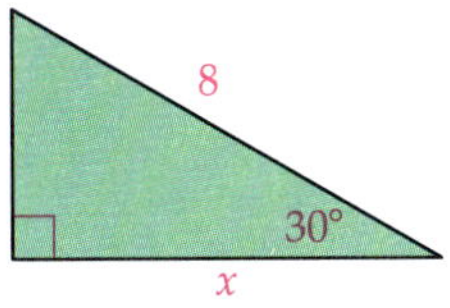

A. 4 **B.** $4\sqrt{3}$
C. $8\sqrt{3}$ **D.** not given

5. $\angle ABC$ and $\angle CBD$ are complementary angles. $m\angle ABC = 73°$. Find $m\angle CBD$. B

A. 107° **B.** 17°
C. 27° **D.** not given

6. Simplify $\sqrt{32x^3 \cdot 2x^5}$. C

A. $64x^8$ **B.** $16x^2\sqrt{2}$
C. $8x^4$ **D.** not given

7. Solve $\frac{a}{16} = \frac{15}{48}$. A

A. 5 **B.** 10
C. 24 **D.** not given

8. Find the area. C

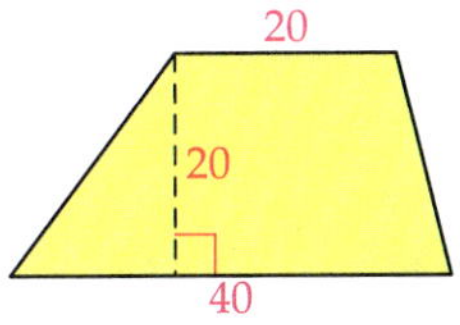

A. 400 sq. units **B.** 300 sq. units
C. 600 sq. units **D.** not given

9. The area of a triangle is 20 cm². The height is 5 cm. What is the base? B

A. 4 cm **B.** 8 cm
C. 10 cm **D.** not given

10. Solve. A

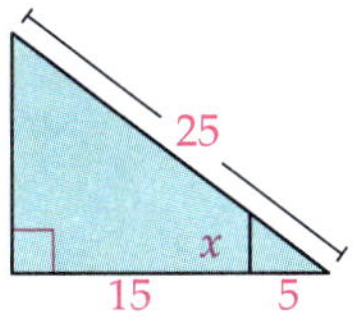

A. 3.75 **B.** 5
C. 8.5 **D.** not given

11. Name the inequality for the model. A

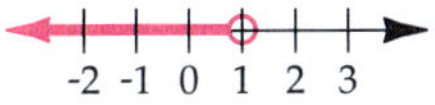

A. $x < 1$ **B.** $x \geq 1$
C. $x \leq 1$ **D.** not given

12. Determine the slope of the line containing (-2,3) and (5,-1). B

A. $\frac{5}{6}$ **B.** $-\frac{4}{7}$
C. 2 **D.** not given

Study Skills

A common objective test format is the *matching* format. Generally, test-taking skills can be applied successfully to test formats that supply the answers. Fortunately, since most of the matching choices will be answers for other questions, the use of distractors are limited.

Some matching tests have an even number of questions and choices, others offer more choices than are needed. When there are more choices than are needed, it makes sense to begin with the shorter column to save time. Students should answer the questions they know first. As each choice is used, it should be crossed out.

When all the questions with obvious choices have been answered, the remaining questions should be carefully read again. Students should look for clues in context and in grammar, and try to recall all the data they learned about the topic. If a student has marked an answer, but begins to doubt his or her choice, recommend that he or she leave the answer as is, because the first answer that comes to mind is often the correct one.

Have students recall the following ideas:

- Maintain a positive attitude.
- Use test-taking strategies to help relieve test anxiety.
- Do physical exercise to help reduce pre-test stress and to increase alertness.

12 STATISTICS AND PROBABILITY

Chapter At a Glance

Chapter 12 introduces students to concepts of statistics and probability. Students learn to evaluate different types of statistical data and to calculate probability.

Reading the Chapter

In this chapter, students are introduced to a variety of statistical concepts and terms. Some of these terms, such as *mode*, *median*, *mean*, *dependent* and *independent events*, are introduced in the same lesson. Suggest students visualize the words and their meanings as they read each term.

As students read and solve the problems in this chapter, ask them to think of examples in their daily lives that require similar problem solving skills. Help students develop the ability to understand statistical data in their lives.

Chapter Resources

Practice and Enrichment TRB L-1 to L-16
Alternate Application TRB L-17
Chapter Test Form A, L-19
Chapter Test Form B, L-21
Computer Test Bank

References

STUDENT BIBLIOGRAPHY

Riedel, Manfred G. *Odds and Chances for Kids* (Prentice Hall, Englewood Cliffs, N.J. 1979).

TEACHER BIBLIOGRAPHY

Teaching Statistics and Probability: 1981 Yearbook (NCTM, Reston, Va. 1981).

SOFTWARE

Digits By Chance (Sunburst Communications), for the Apple II family of computers.

FILMS, VIDEOTAPES, AND OTHER MEDIA

Against All Odds (Annenberg), a set of seven 30-minute tapes, VHS or Beta.

Vocabulary

average
bimodal
box and whisker plot
compound event
counting principle
dependent events
event
factorial
frequency distribution
histogram
independent events
leaves
line plot
mean
measure of central tendency
median
mode
Monte Carlo simulation
probability
quartile
random event
range
sample
sample space
stem
stem and leaf plot

Materials

number cubes
spinners (M*)
counters
coins
math journal

*M—Manipulative Kit

Resource Bank

Class Activities

FIELD TRIPS

▼ Insurance company—to see how statistical computer programs are used.
▼ Market research company—to learn how statistical data are used to identify target markets.

CLASS SPEAKERS

▼ Actuary—to discuss how statistics are used to develop actuarial tables.
▼ Pollster—to explain how poll results are representative of the population.

Bulletin Board

Have students work in groups to design a board game. Then have groups exchange games and play the games. Display several games and rules.

Was the game that your group designed fair? (Answers will vary.)

Was the game you played fair? (Answers will vary.)

How important is skill or experience at playing the game to winning the game your group designed? (Answers will vary.)

Which games would you want to play and why? (Answers will vary.)

Project: Research some of the theories of probability to determine how games of chance contributed to the development of the theories of probability.

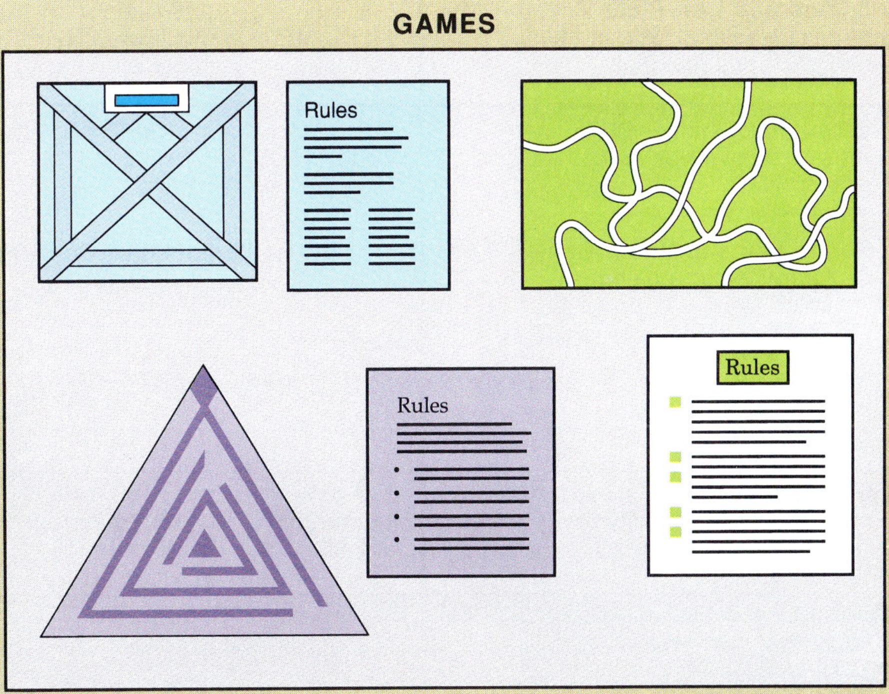

Extra-Credit Problems

BASIC

▼ How many games must be held if all 12 participants play every other participant once? How did you solve this problem? (66; counting principle)

AVERAGE

▼ Zip codes are made up of 9 digits. How many combinations are possible, if all nine digits may not be the same and the sixth digit cannot be zero? (899,999,991)

ENRICHED

▼ The volleyball team plays 84 practice games. Six players are required for each game and no game has the same combination of players. How many members are on the team? (9)

Chapter Planning Guide

OBJECTIVES	ASSIGNMENT GUIDE			ASSESSMENT	
	Basic	Average	Enriched	Review/ Tests	TRB*
12-1 Mean, Median, and Mode To calculate the measures of central tendency and select the most appropriate measure for a given situation.	1-2, 5-6, 9, 12, 15-17, 22-24, 25a-b, d, MR* All	2-3, 6-7, 10, 13, 17-19, 22-25, MR All	3-4, 7-8, 11, 14, 19-25, MR All	Extra Practice p. 577	Practice L-1 Enrichment L-2
12-2 Line Plots and Frequency Tables To arrange data into line plots and frequency tables.	1-3, 7-8, 11-12, 15, 17-22, MR All	2-4, 8-9, 12-13, 15, 17-22, MR All	4-6, 9-10, 13-14, 16-22, MR All	Extra Practice p. 577	Practice L-3 Enrichment L-4
Exploring Misuse of Statistics To draw valid conclusions from statistical data.	All	All	All		
12-3 Stem and Leaf Plots To present statistical data in the form of stem and leaf plots.	1-3, 7-8, 11-15, MR, CT* All	2-4, 8-9, 11-15, MR, CT All	4-6, 9-15, MR, CT All	Extra Practice p. 577	Practice L-5 Enrichment L-6
12-4 Box and Whisker Plots To present data in box and whisker plots.	1-2, 5-6, 10-11, 13-18, MR, TY* All	2-3, 6-7, 10-11, 13-18, MR, TY All	3-4, 8-10, 12-18, MR, TY All	Extra Practice p. 577 TY p. 505	Practice L-7 Enrichment L-8 TY L-18
Exploring Fair and Unfair Games To explore whether a game is fair or unfair.	**Manipulative Connection** All	All	All		
12-5 Counting Principle To calculate the number of outcomes generated by a given event.	1-4, 9-10, 13-14, 17-18, 22-24, 27-31, MR, CT All	3-6, 10-11, 14-15, 18-21, 24-31, MR, CT All	5-8, 11-12, 15-16, 19-20, 22, 25-31, MR, CT All	Extra Practice p. 577	Practice L-9 Enrichment L-10
12-6 Probability To calculate the probability of occurrence of a given event.	1, 3-4, 7-11, 18-22, MR All	1, 4-5, 9-13, 18-22, MR All	2, 5-6, 13-22, MR All	Extra Practice p. 577	Practice L-11 Enrichment L-12
Exploring Pascal's Triangle To explore probabilities using Pascal's triangle.	All	All	All		
12-7 Problem Solving Strategy: Simulate the Problem To solve problems using simulation.	**Problem Solving Connection** All	All	All	Problem Solving Practice p. 526	Practice L-13 Enrichment L-14
12-8 Independent and Dependent Events To calculate probabilities of independent and dependent events.	1-2, 5-6, 9-10, 13-17, MR, TY All	2-3, 6-7, 10-11, 13-17, MR, TY All	3-4, 7-8, 11-17, MR, TY All	Extra Practice p. 577 TY p. 525	Practice L-15 Enrichment L-16 TY L-18
12-9 Problem Solving Application: Making Predictions To apply capture/recapture methods to wildlife management.	All	All	All	Extra Practice p. 577	Alternate Application L-17
Chapter Test	All	All	All	pp. 530-533	Tests L-19–L-25

 *TRB—Teacher's Resource Book MR—Mixed Review CT—Critical Thinking TY—Test Yourself

CONNECTIONS							
Algebra	**Critical Thinking**	**Problem Solving**	**Estimation/ Mental Math**	**Technology**	**Manipulatives/ Resources**	**Writing/Reading in Math**	**NCTM CORRELATION**
	C*: 6-9, TD* All W*: 15-21	C: 1-5 W: 1-4, 12-14, 23-25	W: 5-8	W: 9-11	W: 23-24	C: 6-9 W: 15-25	Reasoning Statistics Problem Solving
	C: 7-10, TD 1-2 W: 17-21	W: 15-22			W: 17, 19-22	W: 19	Reasoning Statistics Problem Solving
	All	All				All	Reasoning Communication Statistics
	C: 1-5, TD All W: 11-15 CT*: 1-3	C: All W: All			C: 6-11 W: 1-12	C: 1-5 W: 11-15 CT: 1-3	Statistics Problem Solving Reasoning
	C: 5, TD All	W: 10-20			W: 13, 15-20	C: 3-5 W: 5-20	Statistics Reasoning Communication
	All	All			2, 4b, 9	All	Reasoning Problem Solving Probability
	C: 5-8, TD All W: 17-20, 28-31 CT: All	C: 9-10 W: 21-31	W: 13-16	W: 9-12	W: 29-31 CT: All	C: 9-10 W: 28	Probability Reasoning Problem Solving
	C: TD All W: 11d, 12c, 13c, 15f, 18b, 22	C: All W: All				C: All W: All	Probability Problem Solving Reasoning
	All	All				All	Probability Problem Solving Reasoning
		C: All W: All	W: 10	W: 1-17	C: 1-2 W: 1	C: 1-2 W: 1-17	Problem Solving Reasoning Communication Functions
	C: 1-3, TD All W: 1-4, 9-17	C: All W: All			W: 13-17	C: All W: 1-4, 9-17	Problem Solving Probability Reasoning
C: 1 W: All		C: All W: All		W: 1-5	W: 5		Functions Problem Solving Connections

*C—Class Exercises TD—Think and Discuss W—Written Exercises CT—Critical Thinking

Exploring

In this manipulative activity, students use cubes to build towers and explore mean, median, and mode. **See TRB Exploring O-45.**

Exploring Mean, Median, and Mode

Use unit cubes to model a set of towers with heights 7,7,3,6,7.

1. If the towers are arranged from least height to greatest height, what is the height of the first tower?
   What is the height of the tower in the middle? **3; 7**
2. In this set, is there a height that occurs more often than another? If yes, which height?
   yes; 7

Rearrange the unit cubes in the towers to obtain five towers of equal height.

3. What is the new height of each tower? **6**
4. What did you need to do to make the towers equal heights?
   remove cubes from higher towers,
   add cubes to lower towers
5. Do the heights of the new towers represent the mean, median, or mode of the height of the first set of towers?
   mean
6. If you find the mean of the heights of the new towers arithmetically, what value do you get?
   6
7. How would you describe the tower with the height 3, to the mean of the set of new towers?
   The tower with the height 3 is below the mean, or below average.

Add three unit cubes to each tower in a set, with heights 7,7,3,6,7.

8. What are the heights of these towers? **10,10,6,9,10**
9. How do you think the mean is affected? **The mean is increased by 3.**
10. What is the mean of this set of towers? **9**
11. The height of a set of five towers has a mean of 5 and a mode of 2. What are the heights of the five towers?
    possible answers: 2,2,6,7,8; 2,2,2,9,10
12. The height of a set of five towers has a mean of 5, a mode of 2 and a median of 4. What are the heights of the five towers?
    possible answers: 2,2,4,7,10; 2,2,4,8,9

O-45

In this problem solving activity, students use "real world" data to graph and analyze a scattergram. **See TRB Exploring O-46.**

Exploring Scattergrams

A class was trying to decide if there is a reasonable formula that relates height and weight. The students decided to collect data and analyze it. They randomly polled 12 individuals to record their heights and weights.

Person	Height (in.)	Weight (lb)
1	65	140
2	68	148
3	50	80
4	71	170
5	54	88
6	60	128

Person	Height (in.)	Weight (lb)
7	72	163
8	66	140
9	69	155
10	62	130
11	65	136
12	56	100

Plot the data on the graph at the right.

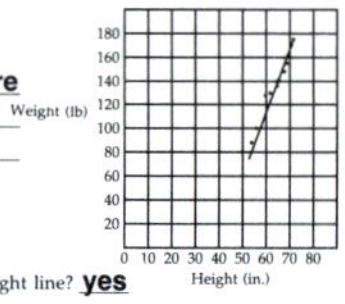

1. Describe the graph that results. **The points are scattered.**
2. Does the data appear to fit an equation? **no**

This coordinate graph is called a *scattergram*. Most information from the "real world" cannot be represented by an equation. Often, however, the data suggests a pattern.

3. Do the plotted points seem to cluster around a straight line? **yes**

Although no line passes through all the data points, you can find a line that describes the pattern suggested by the data. Draw a line that seems closest to all the points.

If the line has a positive slope, the data is positively related. If the line has a negative slope, the data is negatively related. If the data is scattered about with no apparent clustering, the data does not suggest a relationship.

4. How would you classify the relationship between the height and weight? **positive relationship**
5. Using the line you drew, approximate the weight of a 5 ft person. **about 120 lbs**
   Approximate the height of a 180 lb person. **76 in.**
6. If you graphed the following relationships, would you expect to get a positive, negative or no correlation?
   a. age of a car and resale value **negative**
   b. food preference and math grade **no**
   c. times at bat and hits **positive**

O-46

In this calculator activity, students use the calculator to find the product of factorials and are introduced to the factorial key. **See TRB Exploring O-47.**

Exploring Factorials

A number followed by an exclamation point is called a *factorial*. The symbol is mathematical shorthand for expressing the product of all the whole numbers from the given number to 1.

Example: 5 × 4 × 3 × 2 × 1 is expressed as 5!

To find the product, you can use a calculator to multiply or use the factorial key [n!].

1. Multiply 5 × 4 × 3 × 2 × 1. **120**
2. Find 5[n!]. **120**
3. Why do you think a factorial key is important on a calculator? **Answers may vary. Possible answers: It is convenient; less chance of mistakes during calculation.**

Find the product of each factorial.

4. 6! **720**
5. 9! **362,880**
6. 10! **3,628,880**
7. 13! **6.2270208 E9**
8. Suppose you find 8! on your calculator. How could you use the result to find 9! without using the factorial key again? **Multiply 8! by 9.**
9. Can you add, subtract, multiply, and divide factorials? Choose factorials to test your answer. Demonstrate each operation.
   Answers may vary. Possible answers:
   yes; 8! + 2! = 40,322; 8! ÷ 2! = 20,160;
   8! − 2! = 40,318; 8! × 2! = 80,640

Sometimes factorials can be used to solve counting problems. Solve these problems. If you cannot use a factorial, explain why.

10. In how many ways could you arrange:
    a. The digits in the number 987,654? **6! or 720 ways**
    b. The letters in the word MULTIPLY? **8! or 40,320 ways**
    c. The letters of your first name? **Answers will vary.**
11. Each box at the right can hold one digit of a 5-digit number. Write in each box the number of possible digits that can be used in that position if you use digits 0 through 9 without repeating any digits. Remember: 0 can not be in the greatest place value position.
    [9] [9] [8] [7] [6]
12. In question 13 can you use a factorial to find out how many different 5-digit numbers can be made using digits 0 through 9 if the digits do not repeat? Why or why not?
    No; numbers are not an ordered arrangement to 1.

O-47

In this computer activity, students use the computer to explore the probability of a coin toss. **See TRB Exploring O-48.**

Exploring a Coin Toss Simulation

The following computer program simulates the toss of a coin. The program instructs the computer to pick random numbers between 0 and 0.999999999. It assigns "heads" to those random numbers with values less than 0.5 and "tails" to those greater than or equal to 5. You can control the number of tosses by inputting the number when you are prompted by the computer.

Enter the program carefully into the computer.

```
10  PRINT "THIS PROGRAM SIMULATES A COIN TOSS"
20  INPUT "HOW MANY TIMES DO YOU WANT THE COIN TOSSED "; N  ← [Input the Number]
30  A=0:B=0
40  FOR C = 1 TO N
50  T = RND (1)
60  IF T < .5 THEN PRINT "HEADS":A=A+1
70  IF T >= .5 THEN PRINT "TAILS":B=B+1
80  NEXT C
90  PRINT "THE TOTAL NUMBER OF HEADS IS ";A
100 PRINT "THE TOTAL NUMBER OF TAILS IS ";B
110 END
```

For exercises 1 through 11, answers will vary.

1. The probability of getting heads when tossing a coin is 1/2. If a coin is tossed 10 times, how many times can you expect heads to result?
   ______
2. Toss a coin 10 times by hand. What results did you get. heads ____ tails ____
3. Instruct the computer to toss the coin 10 times. Write the results. heads ____ tails ____
4. How do the results compare to what you expected? ______
5. Instruct the computer to toss the coin 100 times. Write the results. heads ____ tails ____
6. Are the results closer to what you expected? **They should be.**
7. Do you think the results of the tosses will be closer to the expected probability if the number of tosses is increased? **yes**
8. Test your answer. Run the program for 1,000 tosses. Write the results. heads ____ tails ____
9. Run the program for 1,000 tosses five times. Find the average number of heads and tails tossed. heads ____ tails ____
10. How did the average compare to the expected results? **They should be closer.**
11. Do you think tossing one coin 1,000 times is the same as tossing 1,000 coins once?
    yes

O-48

Meeting Individual Needs

GIFTED STUDENTS

Gifted students often enjoy the opportunity to extend the content of the chapter. Working in small groups, gifted students can share their insight with others and apply the chapter content to other situations.

Ask students to list all the possible sums of two number cubes, numbered 1 to 6. (2, 3, 4, 5, 6, 7, 8, 9, 10, 11, 12) Then ask groups to determine probability of each sum. (2:$\frac{1}{36}$, 3:$\frac{1}{18}$, 4:$\frac{1}{12}$, 5:$\frac{1}{9}$, 6:$\frac{5}{36}$, 7:$\frac{1}{6}$, 8:$\frac{5}{36}$, 9:$\frac{1}{9}$, 10:$\frac{1}{12}$, 11:$\frac{1}{18}$, 12:$\frac{1}{36}$)

Have students roll the number cubes 50 times to test their predictions.

Then ask groups to determine the probability for each situation.

- The sum is an even number. ($\frac{1}{2}$)
- The sum is a multiple of 3. ($\frac{1}{3}$)
- If the number on the first cube is 6, the probability that the number on the second cube will also be a 6. ($\frac{1}{6}$)

AT-RISK STUDENTS

Some students may feel that the study of statistics is irrelevant to their daily lives; this can be especially true of at-risk students. If students understand how their lives are affected and influenced by the interpretation of statistical data, it can be more interesting and meaningful to them.

Expand the project described in question 5 on page 497. Ask students to identify all the statistical data they see or hear over 2 or 3 days. For each type of statistical data they identify, they should describe how it was used, what they think it meant, and whether it could be misleading. In small groups, have students categorize the types of statistical data they found, such as graphs, percents, or ratios.

Next, have the groups chart or graph their categories of statistical data. Ask groups if they think the types of statistical data they identified are representative of how statistical data are used in our society today. (probably not, sample too small)

ENGLISH AS A SECOND LANGUAGE

ESL students may need additional review of the concepts and phrases used in this chapter. Students are introduced to many terms that may be new to them. Understanding the terms is essential for success in the chapter. In addition to the terms, students are introduced to phrasing that is unique to statistics and probability, for example, "The probability of choosing a blue sweater at random." ESL students may find these phrases difficult to understand. To help clarify new ideas, paraphrase the terminology and provide a visual demonstration of the concept. When presenting several terms together, such as *mean*, *mode*, and *median*, have students work in pairs to find each measure with the same data. This will help reinforce the meanings.

To help students become comfortable with the phrases used in probability, have them work in groups to solve the questions on pages 514-515. Each group should read the question aloud and decide what the sample space is and what the question is asking before they solve it.

See TRB Teaching Resources p. 62.

Problem Solving Situation

The "Think About It" question asks students to speculate on the reason(s) that Olympic records for the 400-m freestyle swimming event have improved steadily since 1924. Ask students to investigate the possible causes for changes that have occurred in sports records. Students' speculations on possible causes can be a starting point for their inquiries. For example, students may hypothesize that a more scientific approach to nutrition and training is the reason for the improved times. Ask them to research their hypotheses and report results to the class. Have students research Olympic sports records to determine if a similar trend exists in other swimming events and other sports.

DATA COLLECTION

Encyclopedia, world and sports almanacs, and other reference books available at the public library contain sports statistics. Non-fiction books on the Olympics and its history may provide explanations for changes in sports records. The U.S. Olympic Committee may make information available to those who call or write.

DATA ANALYSIS

Have students display their data in charts and/or graphs. To display a correlation between a change in sports records and another factor, such as increased training time, have students make a double bar or line graph, if the data are sufficient. Ask students to forecast whether the factor they chose is likely to increase or continue its influence on changes in sports records.

DATA FILE

BASEBALL STADIUMS

TEAM	STADIUM	SEATING CAPACITY
ATLANTA BRAVES	Atlanta–Fulton County Stadium	52,003
BALTIMORE ORIOLES	Memorial Stadium	54,017
BOSTON RED SOX	Fenway Park	34,182
CALIFORNIA ANGELS	Anaheim Stadium	64,593
CHICAGO CUBS	Wrigley Field	39,600
CHICAGO WHITE SOX	Comiskey Park	44,087
CINCINNATI REDS	Riverfront Stadium	52,392
CLEVELAND INDIANS	Cleveland Stadium	74,483
DETROIT TIGERS	Tiger Stadium	52,416
HOUSTON ASTROS	Astrodome	45,000
KANSAS CITY ROYALS	Royals Stadium	40,625
LOS ANGELES DODGERS	Dodger Stadium	56,000
MILWAUKEE BREWERS	Milwaukee County Stadium	53,192
MINNESOTA TWINS	Hubert H. Humphrey Metrodome	55,883
MONTREAL EXPOS	Olympic Stadium	59,149
NEW YORK METS	Shea Stadium	55,300
NEW YORK YANKEES	Yankee Stadium	57,545
OAKLAND A's	Oakland Coliseum	49,219
PHILADELPHIA PHILLIES	Veterans Stadium	64,538
PITTSBURGH PIRATES	Three Rivers Stadium	58,727
SAN DIEGO PADRES	Jack Murphy Stadium	58,433
SAN FRANCISCO GIANTS	Candlestick Park	58,000
SEATTLE MARINERS	Kingdome	58,150
ST. LOUIS CARDINALS	Busch Stadium	54,224
TEXAS RANGERS	Arlington Stadium	43,508
TORONTO BLUE JAYS	Skydome	53,000

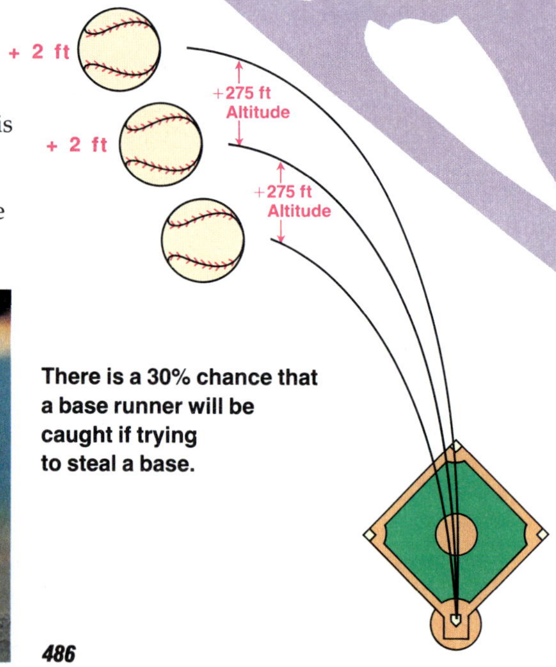

At higher altitudes where the air is thinner, a baseball faces less resistance. A 275-ft rise in altitude adds about 2 ft to the distance an average home run will travel.

There is a 30% chance that a base runner will be caught if trying to steal a base.

486

CHAPTER

12 Statistics and Probability

YEAR	1924	1928	1932	1936	1948	1952	1956	1960	1964	1968	1972	1976	1980	1984	1988
MALE	5:04.2	5:01.6	4:48.4	4:44.5	4:41.0	4:30.7	4:27.3	4:18.3	4:12.2	4:09.0	4:00.27	3:51.93	3:51.31	3:51.23	3:46.25
FEMALE	6:02.2	5:42.8	5:28.5	5:26.4	5:17.8	5:12.1	4:54.6	4:50.6	4:43.3	4:31.8	4:19.44	4:09.89	4:08.76	4:07.10	4:03.85

OLYMPIC RECORD TIMES FOR 400-m FREESTYLE SWIMMING (min.)

Think about it...

Look at the data for the Olympic records in the 400-m freestyle swimming competition. Why do you think the times for this Olympic event have gone down since 1924? Do you think the times will continue to go down?

See Problem Solving Situation

USA

1988 WINTER OLYMPICS

Brian Boitano's scores for Technical Merit in Men's Figure Skating:

5.9
5.9
5.9
5.9
5.9
5.8
5.8
5.8
5.8

Cooperative Learning

Have students work in pairs or groups of three to complete this activity. Consider students' personalities and individual strengths and weaknesses when assigning students to partnerships or groups. For example, pairing a creative student who does not have good organizational skills with a student whose work habits are more consistently well organized may work to the advantage of both students. Accordingly, assign a role to each student that elicits his or her strengths. Possible roles include Data Researcher and Data Organizer. Ask students to share in the presentation of their results to the whole class and to evaluate the success of their partnership or group.

Questions related to this Data File can be found on pages 459, 491, 495, 504, and 560.

Lesson Focus

Materials

Pencil, paper, math journal, newspapers

Vocabulary/Symbols

average
bimodal
mean
measures of central tendency
median
mode
range

MOTIVATING THE LESSON

Review with students the definition of average. Have students share how they use the term average in common everyday situations. (Students will probably give examples of averages that have to be found by mathematical computation: e.g., average test score, bowling average, batting average.) Mention that this lesson will discuss some averages that do not involve mathematical computations.

Instruction

1 Ask students to describe how the data are arranged. (Wages of the same amount are grouped together and the wages are arranged from least to greatest.)

2 Help students to reach the conclusion that grouping wages of the same amount together makes the number that occurs most often obvious.

3 Students will probably recognize the mean as the average with which they are most familiar.

4 Discuss with students that the range of a set of data indicates the *spread* of the data and is not the average.

OBJECTIVE: *To calculate the measures of central tendency and select the most appropriate measure for a given situation.*

12-1 Mean, Median, and Mode

▼ A company's employees stated that average weekly wages were $200. Owners argued that wages averaged $300. The personnel department insisted that wages were $250. Each group is correct.

In statistics we use three *measures of central tendency* to describe data characteristics: *mean, median,* and *mode.* The company's weekly salary data are shown below.

1

Employee Wages (in dollars)	200, 200, 200, 200, 200, 200, 200, 200, 200 250, 250, 250, 250, 250, 250 350, 350 500, 500 1,000

▼ To find the mode, group all wages of the same amount together. Wages of $200 appear most often. The mode is 200. The employees used the mode as the average.

THINK Why does it help to group wages of the same amount together? **It helps in finding the number that appears most often.**

2

Mode The mode is the data item that occurs most often.

3 ▼ The owners divided the total wages by the total number of employees, 6,000 ÷ 20 = 300, to get the mean.

▼ To find the median, order the wages for all the employees. Find the middle number. If there is an even number of data items, add the two middle numbers. Then divide by 2 to find the midpoint. The median is 250. Personnel used the median as the average.

THINK The best average in a situation is the one that best reflects what is most typical. Which average reflects a typical salary? Why? **Answers may vary. One possible answer: "The median reflects a typical salary because it is the middle value."**

Median The median is the middle value in a set of data.

▼ Some data have more than one mode. Some data have no mode.

Example 1 **Find the mode for each set of data.**

a. 2 5 3 3 4 5 6 5 3 4 2 7

b. 6 3 2 5 6 4 4 2 3 5

Solution **a.** 7 6 5 5 5 4 4 3 3 3 2 2 Group like numbers.

Bimodal data have two modes. The modes of these data are 5 and 3.

b. 6 6 5 5 4 4 3 3 2 2 Group like numbers.

There is no mode. All numbers are listed the same number of times.

Teaching Tip

Have newspapers, almanacs, and other sources of data available to help students gain a better understanding of how the three measures of central tendency are used in everyday situations.

▼ You can use the measures of central tendency to describe various situations.

Example 2 **Is the mean, median, or mode the best average for each situation? Explain.**

a. the favorite rock group of the Freshman class

b. the time it takes each student to get to school

c. the cost of houses in your community

Solution **a.** Mode; The mode is useful in determining the most frequently chosen category. Mode is the preferred statistic when the data are not numerical.

b. Mean; The mean is useful in situations that do not use unusually large or small numbers that distort the results.

c. Median; This measure tells that half the data items are above and half below the average. The median is used when extreme measures distort the mean.

▼ *Range* is a measure of data dispersion.

Range	The range of a set of data is the difference between the greatest and least values in the set.

Example 3 **Find the range in the following set of data.**

car prices: $8,750; $24,560; $16,230; $26,990; $12,400

Solution $26,990 − $8,750 = $18,240   Subtract the least value from the greatest.

4
1. Not at all. Removing one number from each end will not affect the middle.
2. It will affect the mode only if the mode is one of the two numbers eliminated.
3. The greatest and least numbers were part of the original average. Dropping them is likely to change the average.

CLASS EXERCISES

Most answers are rounded to the nearest tenth.

Find the mean, median, mode, and range of each of the following.

1. 1 1 2 2 3 3 4 4 4 4 5 5 6 7 8 3.9; 4; 4; 7

2. 10 13 15 15 16 16 14.2; 15; 15 and 16; 6

3. 50 50 50 50 60 70 70 70 90 90 100 100 100 100 75; 70; 50 and 100; 50

4. 3 3 3 3 3; 3; 3; 0

5. 59 63 48 50 85 61; 59; no mode; 37

Answers may vary. Samples given below.

What is the best average for each situation? Explain.

6. the average height of the students in the class
Median. Extremes in height will affect the mean.

7. the average scores of candidates on a scholarship examination
Mean. No extremes are likely to distort the average.

THINK AND DISCUSS
See above.

The greatest number and the least number are eliminated from a set of data.

1. How will this affect the median? Explain.

2. How could this affect the mode? Explain.

3. How could this affect the mean? Explain.

ADDITIONAL EXAMPLES

Find the mean, median, mode, and range for each set of data.

1. 27 28 29 29 29 30 30 30 31 31 31 31 32 32 33 ($30\frac{1}{5}$ or 30.2, 30, 31, 6)

2. 8 5 4 8 6 9 5 8 5 6 4 9 2 5 8 4 9 3 7 10 ($6\frac{1}{4}$ or 6.25, 6, 8 and 5, 8)

3. 90 80 75 80 75 90 90 75 80 ($81\frac{2}{3}$ or $81.\overline{6}$, 80, none, 15)

Guided Practice

Class Exercises If students need help with question 6, have them supply sample data.

Think and Discuss Have students explore a variety of scenarios to illustrate the differing effects on the mode and the median.

A ***common error*** occurs when students proceed to find the median simply by identifying the middle term in a list of data as presented. Emphasize that students must check to see if the data are arranged from least to greatest and if not, they must put the data in this order before identifying the middle term. Have them practice arranging lists of data, such as 90 65 85 70 90 60 75, in numerical order.

Closure

Ask students to write in their own words a definition for each of the three measures of central tendency: mean, median, and mode. Have them write the definitions in their math journals.

MATH MINUTES

Have students find statements using the term *average* in newspaper articles. Then have them determine which measure of central tendency is being used. (Check students' work.)

Independent Practice

The data in question 14 are representative of data that might be collected from your students. Collecting similar data from your students will enhance their identification and understanding of the topic.

Writing in Math Have students write their paragraphs for questions 22 and 25d in their math journals. Then have students share their answers for question 22 with the class.

Have students work in small groups on the project in question 23. Encourage the groups to suggest other data collection projects. Then have each group work on a different project and present the findings to the entire class.

Lesson Follow-up

Reteaching Activity For students who have difficulty with mean, median, and mode, refer to the manipulative activity on p. 486E of the Chapter Overview. See also TRB Exploring O-45.

MIXED REVIEW

Find each answer.

1. Find the area of a circle with a diameter of 6.5 in. 33.2 in.^{2}
2. Find the surface area of a rectangular prism that is 3 ft × 6 ft × 18 ft. 360 ft^2

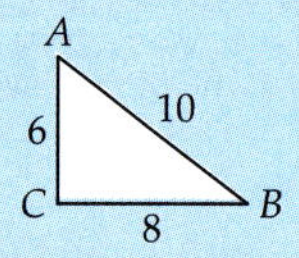

3. Find sin *A*. $\frac{4}{5}$
4. Find tan *B*. $\frac{3}{4}$

Use the table for Questions 5 and 6.

Rainfall (in.)	1	1	3	3	2
Day	1	2	3	4	5

5. On which day or days did the most rain fall?
6. On which day or days did the least rain fall?
7. The cost of taking a taxi is \$1 plus \$.11 per $\frac{1}{10}$ mi. What is the cost of a trip that is $5\frac{3}{5}$ mi? \$7.16

5. 3rd and 4th
6. 1st and 2nd

8. the average scores of candidates on a driver's education test
9. the average earnings of 14-year-old students

Sample answers: 8. Mode. This will tell you what score was most often made. 9. Mode. This will tell you what salary is most often earned.

WRITTEN EXERCISES

Find the mean, median, mode, and range for each set of data.

1. golf scores 5 5 5 6 3 3 4 7
   4.8; 5; 5; 4
2. diving scores 9.7 9.8 9.2 9.9 8.9 8.7 8.8
   9.3; 9.2; no mode; 1.2
3. miles per gallon 17.8 22.5 27.0 23.5 18.9 16.7 24.8 19.0 23.0
   21.5; 22.5; no mode; 10.3
4. allowance 3.50 5.50 2.00 5.00 2.75 3.00 4.00 4.50 3.00 3.50
   3.68; 3.5; 3 and 3.5; 3.5

MENTAL MATH **Find the mean, median, mode, and range for each set of data.**

5. 0 0 1 1 2 2 2 3 3 4 4 2; 2; 2; 4
6. 1 1 2 2 3 3 3 4 5 6 3; 3; 3; 5
7. 2 2 2 3 4 5 6 6 7 8 4.5; 4.5; 2; 6
8. 1 3 3 4 6 6 6 8 8 5; 6; 6; 7

CALCULATOR **Find the mean, median, mode, and range for each set of data.**

9. 76 84 88 90 78 80 84 88 92 80 86 84
   84.2; 84; 84; 16
10. 135 170 165 170 185 165 170 175 160 150 145
   162.7; 165; 170; 50
11. 98 97 101 104 105 102 103 100 101 99
   101; 101; 101; 8

Solve.

12. A student timed the length of telephone calls made over one weekend.

    Minutes of calls 10 2 11 4 20 12 16 9 14 2
    16 13 35 5 18 4
    11.9; 11.5; 2, 4, and 16

    a. Find the mean, median, and mode for the length of the calls.
    b. Which average best reflects the typical phone call? Explain.
    c. Find the range of the length of the calls. 33

    12. b. The median. It is not influenced by the extreme of 35 minutes.

13. A basketball player made the following points per game.
    10 5 8 15 7 9 3 30 3
    a. Find the mean, the median, and the mode. 10; 8; 3
    b. Which average best reflects a typical game? Explain. The median. It is not influenced by the extreme of 30 points.

14. In a classroom experiment each student estimated when one minute had elapsed. The results of their estimates, in seconds, are listed below.

57 59 56 54 61 60 63 55 59 51 65 58 69 62 63
57 54 64 58 55 64 61 63 60

a. Find the mean, the median, and the mode. 59.5; 59.5; 63

b. Which average best reflects a typical estimate? Explain. Mean. There are no large or small numbers to distort the mean.

Would you use the mean, the median, or the mode for each situation? Explain. Answers may vary.

15. the heights of the members of the basketball team

16. the distance members of your class live from the school

17. the rainfall for the month of August in your community

18. the number of times each class member went to a mall last week

19. the cost of new cars

20. the amount of TV each student watches each week

21. the amount of time spent doing homework by the members of your class

22. ***WRITE*** a paragraph describing three situations in which the mode, the mean, and the median would each be the most appropriate measure of central tendency. Answers may vary.

23. ***PROJECT*** Record the age of each member of your class in years and months. Find the mean, the median, and the mode. Which measure is the most representative? Check students' work.

24. ***DATA FILE 12 (pp. 486–487)*** Use the score for Brian Boitano to calculate each of the following.

a. Find the mean, the median, and the mode of his scores. 5.86; 5.9; 5.9

b. Which average best reflects a typical score? Explain. Mean. It accounts for all tries.

25. A student had the following scores on exams in her history class: 83, 76, 92, 76, 93.

a. The teacher allows the student to decide which measure of central tendency to use as an average. Which measure do you recommend? Explain. The mean, because it is the highest average.

b. There is one more exam. What score must the student make to raise her average to 85 if using median as the average? 87

c. What score must the student make to raise the average to 85 if using the mean as the average? 90

d. ***WRITE*** a paragraph explaining which measure of central tendency you consider the most representative of the student's grades. What measure would you choose for your grades? Answers may vary.

LESSON QUIZ

Find the mean, median, mode, and range for each set of data.

1. 6 11 14 12 8 9 11 13 15 (11, 11, 11, 9)
2. 102 101 104 101 104 105 101 105 105 102 (103, 103, 101 and 105, 4)

Would you use the mean, the median, or the mode for each situation? Explain. (Possible answer shown.)

3. the number of automobiles in each home (mode)
4. the distance each student can make a paper airplane fly (median)

Assignment Guide

Basic 1-2, 5-6, 9, 12, 15-17, 22-24, 25a-b, d, MR All

Average 2-3, 6-7, 10, 13, 17-19, 22-25, MR All

Enriched 3-4, 7-8, 11, 14, 19-25, MR All

LESSON RESOURCES

TRB Practice L-1
TRB Enrichment L-2

FOR THE NEXT LESSON

pencil, paper, graph paper (optional), math journal, spinners

Practice

For use after 12-1 (pp. 448–491)

Mean, Median, and Mode

Complete for the given sets of data.

Data	Mean	Median	Mode	Range
1. 8, 15, 9, 7, 4, 5, 9, 11	8.5	8.5	9	11
2. 70, 61, 28, 40, 60, 72, 25, 31, 64, 63	51.4	60.5	none	47
3. 4.9, 5.7, 6.0, 5.3, 4.8, 4.9, 5.3, 4.7, 4.9, 5.6, 5.1	5.2	5.1	4.9	1.3
4. 271, 221, 234, 240, 271, 234, 213, 253, 196	237	234	234, 271	75
5. 0, 2, 3, 3, 3, 4, 4, 5	3	3	3	5
6. $\frac{5}{8}, \frac{1}{2}, \frac{1}{4}, \frac{1}{8}$	$\frac{3}{8}$	$\frac{3}{8}$	none	$\frac{1}{2}$
7. 1,216; 4,891; 2,098; 3,662; 5,748	3,523	3,662	none	4,532
8. 1, 2, 3, 4, 5, 6, 7, 8, 9, 10	5.5	5.5	none	9
9. 5, 10, 15, 20, 25, 30, 35, 40	22.5	22.5	none	35
10. 85, 73, 93, 74, 71, 101, 71, 90, 98	84	85	71	30

11. There were 8 judges at a gymnastics competition. Kathleen received these scores for her performance on the uneven parallel bars:

8.9, 8.7, 8.9, 9.2, 8.8, 8.2, 8.9, 8.8

a. Find these statistics: mean 8.8 median 8.85 mode 8.9

b. Which statistic reflects a typical score? Explain. Answers may vary. All are reasonable; the mean is slightly lowered by the score of 8.2.

c. Why do you think that the highest and lowest judge's scores are disregarded in tallying the total score in a gymnastics competition? This eliminates scores that are not representative of the majority.

Chapter 12 L-1

Enrichment

For use after 12-1 (pp. 488–491)

More on Mean, Median, Mode, and Range

Solve.

1. Leon must have a mean score of 90 on his math quizzes to earn an A. So far, he has received grades of 88, 85, 91, 92, 94, 81, 86, and 98. What grade must he earn on his final quiz? 95
2. The median of the following set of data is 6: 1, 8, 7, 1, 6, 8, 3, 6. What number from the set of data can you subtract 1 from so that the median of the resulting set will be $5\frac{1}{2}$? 6
3. There are twelve girls and eight boys in a math class. The girls' mean score on the final exam was 83.5. The boys' mean score was 81.5. What was the mean for the entire class? Hint: First find the total points scored by all members of the class. 82.7
4. The mode of the following set of data is 23: 19, 5, 23, x, 17. What is the mean? 17.4
5. The median of the following set of data is 26: 17, y, 49, 13. Find y. 35
6. The total weight of all the students in a class is 3,159 lb. The mean weight of the students is 117 lb. How many students are there in the class? 27 students
7. Ramon, Jake, and Pearl are all less than 50 years old. The mode of their ages is 37. The range is 15. What is their mean age? What is their median age? 32; 37
8. There are five one-digit numbers in a set of data. The mean of the numbers is 3. A mode is 1. The median is 4. What are the numbers? 1, 1, 4, 4, 5

L-2 Chapter 12

Lesson Focus

Materials

Pencil, paper, graph paper (optional), math journal, spinners

Vocabulary/Symbols

frequency distribution
line plot

MOTIVATING THE LESSON

Write the names of the months of the year in a row at the bottom of the chalkboard. Call students, a row at a time, to the board and have each student in turn write an x above the month of his or her birthday. Mention that students have just organized the data on a line plot. Then ask students which measures of central tendency they could find directly from the line plot. (median and mode)

Instruction

1 To ensure that students understand the line plot, ask them how many students were carrying four books (4) and how they could find the median for the set of data. (Find the column that contains the middle number, which is the 13th x.)

2 Advise students to check off each item in the set of data as they tally it to make sure that all items are recorded.

3 Help students to understand that the algorithm is a short form for adding the 15 individual items in the frequency distribution. Have them verify that the sum of the items is 29.

OBJECTIVE: *To arrange data into line plots and frequency tables.*

12-2 Line Plots and Frequency Tables

In a survey, 25 students were stopped in the hallway and asked how many books they were carrying. Their responses were: 2, 0, 4, 1, 2, 3, 1, 0, 1, 6, 4, 1, 0, 2, 5, 1, 4, 3, 1, 6, 2, 5, 4, 3, 1.

The *line plot* shows the data. An × represents one student.

The greatest number of ×s are above the number 1. The mode is 1.

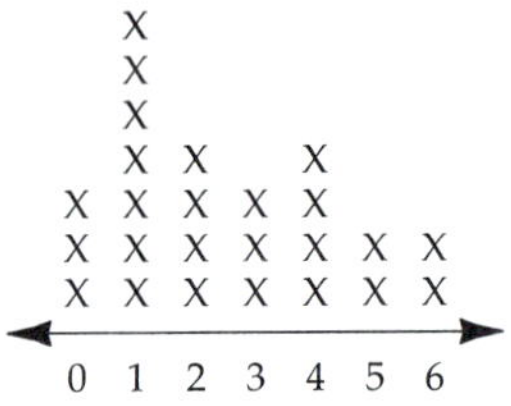

THINK What does more than one x in a column represent? More than one student carries the number of books specified on the number line.

1

Line Plot	A line plot shows data on a number line. You place an x for each response above the category of the response.

NOTES & QUOTES

Statistical thinking will one day be as necessary for efficient citizenship as the ability to read and write.
–H. G. Wells (1866–1946)

A *frequency distribution* is another way to organize data. Instead of the ×s of the line plot, you write the frequency of a particular response to correspond to the type of response.

The table shows the frequency distribution for the book survey.

Number of books	n	0	1	2	3	4	5	6
Number of students	f	3	7	4	3	4	2	2

You can see that three students carried no books, seven carried 1 book, and so on. The greatest frequency is seven, so 1 is the mode.

Frequency Distribution	A frequency distribution is a listing of data that pairs each data item with the number of times it occurs.

You can construct a frequency table from a tally.

Example 1 **Arrange the data in a frequency table.**

7 2 5 4 1 6 5 2 5 1
3 6 2 3 4 5 2 6 3 4

Solution 2

1. Determine the values for n. Then prepare a tally chart.
2. Count the tally marks for each value. Record that number as the frequency.

n	1	2	3	4	5	6	7
Tally	//	////	///	///	////	///	/
f	2	4	3	3	4	3	1

Teaching Tip

Allow students to use calculators to find the mean of large numbers, and/or when many items are involved, to save time and to ensure greater accuracy.

▼ Sometimes you need to find the mean, median, and mode from a frequency distribution.

Example 2 **Find the mean, median, and mode.**

n	0	1	2	3	4
f	2	3	5	4	1

Solution **a.** To find the mean, multiply the number (n) times the frequency (f) for all n. Add the products. Divide by the total of the f.

$$\frac{2(0) + 3(1) + 5(2) + 4(3) + 1(4)}{2 + 3 + 5 + 4 + 1} = \frac{29}{15} = 1.9\overline{3}$$

The mean is $1.9\overline{3}$.

b. To find the median, list each number (n) f times. Then find the middle number.

0 0 1 1 1 2 2 2 2 2 3 3 3 3 4

There are 15 items. The eighth one is in the middle. The median is 2.

c. The mode is 2, the number with the highest frequency.

3

CLASS EXERCISES

Check students' plots.

Draw a line plot for each set of data. Find the mean, median, and mode.

1. 7 11 10 10 8 11 9 7 9 8 11 11 $9.\overline{3}$; 9.5 and 11

2. 5 0 2 1 4 3 4 0 2 5 4 3 2 0 4 2.6; 3; 4

Find the mean, median, and mode for each.

3.

x	f
0	3
1	2
2	2
3	1
4	2

4.

x	f
16	2
17	5
18	4
19	2
20	1

5.

x	f
2	1
4	3
6	2
8	3
10	1
12	3

6.

x	f
1	1
2	2
3	2
4	2
5	2
6	1

3. 1.7; 1.5; 0 4. 17.6; 17.5; 17 5. 7.4; 8; 4, 8, and 12 6. 3.5; 3.5; 2, 3, 4 and 5

Make a frequency table for each set of data. Determine the mode of each distribution, if one exists. Check students' plots.

7. 25 29 28 28 30 25 26 28 27 29 26 30 28

8. 10 30 20 30 50 10 40 30 50 40 30 50 30

9. 1 4 0 3 0 1 3 2 2 4 no mode

10. 6 2 8 7 9 3 5 4 8 2 4 6 4 1 4

THINK AND DISCUSS See below.

1. Would you expect a set of exam scores to have more than one mode? Explain.

2. Describe the line plot of a distribution with no mode.

1. It is reasonable to expect several scores to be made an equal number of times.
2. An equal number of Xs will be in each category.

ADDITIONAL EXAMPLES

1. a. Draw a line plot of the data.

14 11 8 8 13 11 9 10
12 9 12 10 11 9 12 13
11 9 8 14

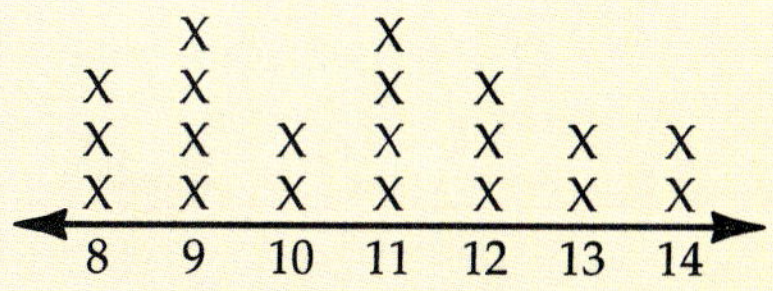

b. Arrange the data from 1a. in a frequency table.

n	8	9	10	11	12	13	14
Tally	|||	||||	||	||||	|||	||	||
f	3	4	2	4	3	2	2

2. Find the mean, median, and mode.

n	5	6	7	8	9	10
f	1	3	8	7	5	1

(7.6, 8, 7)

Guided Practice

Class Exercises For questions 1-2, ask students which measure of central tendency is the easiest to find on a line plot. (The mode, if there is one.)

Think and Discuss If students have difficulty with question 2, ask them when a set of data would have no mode. (when each item occurred the same number of times)

MATH MINUTES

Have students work in groups of five or six. Each student in turn spins a spinner twice. Students arrange the numbers in a frequency distribution and find the mean, median, and mode of the set of data. (Check students' work.)

A ***common error*** occurs when students forget to multiply the number (n) by the frequency (f) when they calculate the mean from a frequency distribution. Have students include another row labeled $n \cdot f$ in their frequency distribution and write each subtotal in that row.

Closure

Ask students to describe how to obtain the median and the mode for a set of data directly from a line plot and how to calculate the mean using grouped data in a frequency distribution. Have them write the descriptions in their math journals.

Independent Practice

Have students share and compare their findings for the project in question 19d.

Encourage students to work in small groups for the project in question 20 and to design and present a project using similar data.

Lesson Follow-up

Reteaching Activity Have students who have difficulty calculating the mean from the frequency table rewrite the data to show each item as many times as it occurs. Then have them find the mean and compare the result with the result obtained by following the steps of the algorithm.

MIXED REVIEW

Find each answer.

1. Solve the equation $2x + 16 = 5$. -5.5

2. Write $\frac{5}{8}$ as a percent and as a decimal. 62.5%, 0.625

Use the data: 12, 14, 16, 16, 18, 18.

3. Find the mean, the median, and the mode. 15.7; 16; 16 and 18

Use the data: 8, 15, 22, 9, 11, 16, 20.

4. Find the mean, the median, and the mode. 14.4; 15; no mode

Find each square root.

5. $\sqrt{25}$ 5 **6.** $-\sqrt{\frac{49}{64}}$ $-\frac{7}{8}$

7. *DATA FILE 6 (pp. 230–231)* One hundred and sixty-six teenage girls will be asked if they do the food shopping for their families. Make a prediction as to how many of the girls will respond *yes*. 156

1. 4; 4 2. 4; 4 and 5
3. 3.5; 1 and 6
4. 3.5; no mode

15.a.

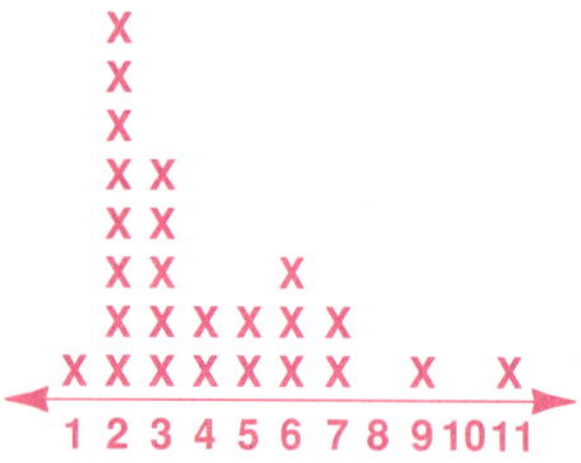

WRITTEN EXERCISES

Draw a line plot for each frequency distribution. Find the median and the mode. Check students' work for plot. See left.

1.

x	1	2	3	4	5	6
f	2	5	7	8	4	3

2.

x	1	2	3	4	5	6
f	1	3	5	8	8	5

3.

x	1	2	3	4	5	6
f	7	5	3	2	6	7

4.

x	1	2	3	4	5	6
f	5	5	5	5	5	5

5.

x	1	2	3	4	5	6
f	1	8	6	7	5	3

3.5; 2

6.

x	1	2	3	4	5	6
f	1	2	9	8	9	1

4; 3 and 5

Make a frequency table for each set of data. Find the mean, the median, and the mode. Check students' work for table.

7. baseball scores — 9 6 6 7 10 6 8 6 8 7 7.3, 7, 6

8. the ages of club members — 14 16 14 16 14 16 13 12 15 16 12 12 15 14 15 15 14.3; 14.5; 14, 15, and 16

9. the heights of test plants — 25 25 20 25 20 25 30 25 31 26 28 30 25.8; 25; 25

10. a class set of test scores — 100 90 70 60 95 65 85 70 70 75 80 85 75 70 100 90 80; 77.5; 70

Construct the frequency table from each line plot. Find the median and the mode. Check students' work for table.

11.

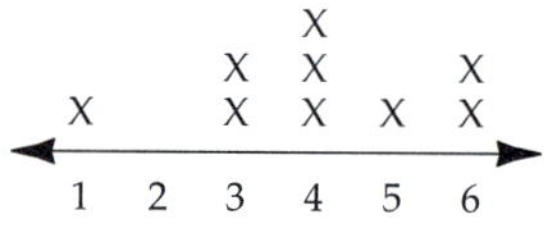

4; 4

12.

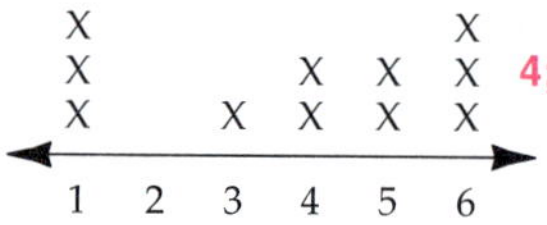

4; 1 and 6

13.

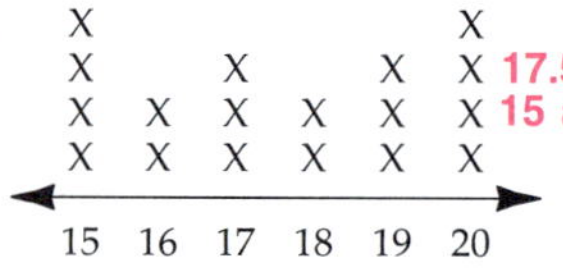

17.5; 15 and 20

14.

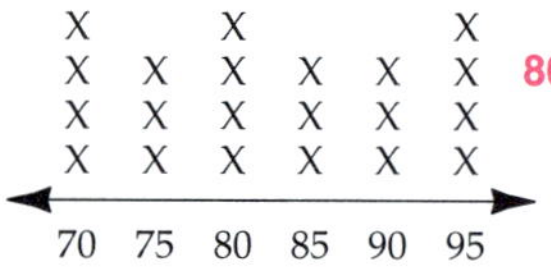

80; 70, 80 and 95

15. The number of letters in each of the first twenty-five words in a given passage of text are shown below.

2 7 6 3 1 2 7 6 3 2 2 6 9
5 3 2 2 4 5 4 3 2 3 2 11

a. Draw the line plot for the data. at left

b. Make a frequency table. Check students' work.

c. Find the mean, the median, and the mode for the data. 4.1; 3; 2

QUIZ ANSWERS

(1.)

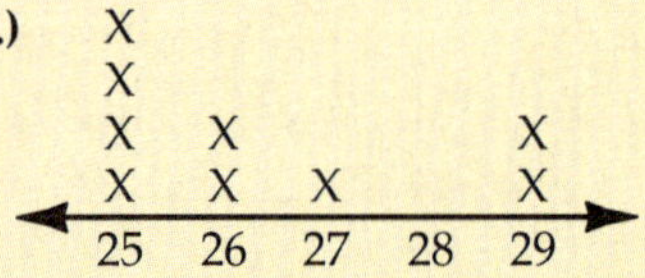

(2.)

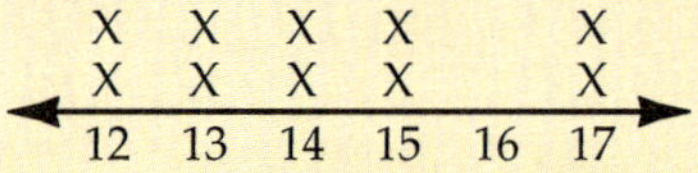

16. The following figures represent the weekly earnings in dollars of the employees at Yanktown Industrial Enterprise.

160 160 160 200 200 200 200 200
200 240 240 240 360 360 360 520

a. Draw a line plot. Check students' work.

b. Draw a frequency table. Check students' work.

c. Find the mean, median, mode, and range. 250; 200; 200; 360

17. Use the ***DATA*** at the right.

a. Draw a frequency table. Check students' work.

b. Find the mean, median, and mode. 2.7; 2; 0

c. Which measure of central tendency best reflects the average number of gold medals won? Explain your choice. Possible answer: Mode. This indicates the number of medals most often won.

18. Babe Ruth's home runs from 1920 to 1934 are shown below.

54 59 35 41 46 25 47 60 54
46 49 46 41 34 22

a. Draw a frequency table. Check students' work.

b. Find the mean, median, and mode. 43.9; 46; 46

c. Which average best reflects a typical year? Explain your choice. Possible answer: Mean. No extreme numbers distort the mean.

19. ***PROJECT*** Select a 50-word passage from a reading of your choice.

a. Tally the number of times each letter of the alphabet is used. Check students' work.

b. Draw a frequency table.

c. Find the mean, the median, and the mode of the distribution.

d. *E* is the most frequently used letter in English. Is this true of your data?

20. ***PROJECT*** Survey car usage. Count the number of people in each vehicle that passes a given point for a half-hour each weekday. Repeat at the same time on a weekend day. Construct two frequency tables. Are the results the same during the weekend as they are during the week? How could you use this information to determine the need for a traffic light?

21. Use the ***DATA*** at the right.

a. What age has the highest frequency for each group? male, 40–44; female, 30–34

b. ***ANALYZE*** How would you find the median age for a large set of data? Find the median for men and women. See below.

22. ***DATA FILE 12 (pp. 486–487)*** Round the seating capacity for each baseball stadium to the nearest thousand. Use the rounded data to draw a frequency table and to find the median and the mode of the data. Check students' work. 54,000; 58,000

21. b. 1. Total number entered. 2. Divide by 2 to find middle number. 3. Add number entered one group at a time, starting at the top, until middle number is reached or surpassed. 4. Find corresponding age. Men: 35–39 women: 35–39

Distribution of Gold Medals 1988 Winter Games

Country	*Medals*
Soviet Union	11
East Germany	9
Switzerland	5
Austria	3
West Germany	2
Finland	4
Netherlands	3
Sweden	4
United States	2
Italy	2
Norway	0
Canada	0
Yugoslavia	0
Czechoslovakia	0
France	1
Japan	0
Liechtenstein	0

1990 Boston Marathon

Age	Number Entered	
	Male	Female
Under 20	19	3
20–24	251	96
25–29	828	330
30–34	1,318	396
35–39	1,533	371
40–44	1,681	297
45–49	1,026	147
50–54	605	76
55–59	266	16
60 & over	140	11

LESSON RESOURCES

TRB Practice L-3
TRB Enrichment L-4

FOR THE NEXT LESSON

Newspapers, magazines with statistical data, math journal

Practice

For use after 12-2 (pp. 492–495)

Line Plots and Frequency Tables

Draw a line plot for each frequency distribution. Find the mean, the median, and the mode.

1.

x	1	2	3	4	5	6
f	2	0	4	1	2	4

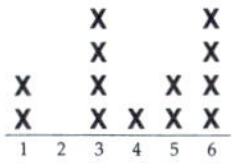

mean 4
median 4
mode 3, 6

2.

x	1	2	3	4	5	6
f	4	4	0	0	3	2

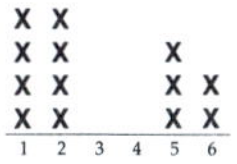

mean 3
median 2
mode 1, 2

Make a frequency table for each set of data. Find the mean, the median, and the mode.

3. 5 1 4 6 2 6 4 5
1 3 2 6 4 5 4 6

n	1	2	3	4	5	6
f	2	2	1	4	3	4

mean 4
median 4
mode 4, 6

4. 4 3 1 2 1 3
3 6 1 3 2 1

n	1	2	3	4	6
f	4	2	4	1	1

mean 2.5
median 2.5
mode 1, 3

Construct the frequency table from each of the following line plots. Find the mean, the median, and the mode.

5.

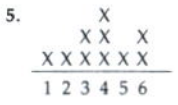

n	1	2	3	4	5	6
f	1	1	2	3	1	2

mean 3.8
median 4
mode 4

Chapter 12 L-3

Enrichment

For use after 12-2 (pp. 492–495)

From Line Plot to Graph

Use the data in the rainfall table to construct line plots for each of the four cities.

Average Monthly Rainfall (in.)

City	J	F	M	A	M	J	J	A	S	O	N	D
Albuquerque, NM	0	0	1	0	1	1	1	2	1	1	0	1
Charleston, SC	3	3	4	3	4	7	7	7	5	3	2	3
Eureka, CA	7	5	5	3	2	1	0	0	1	3	6	6
Wilmington, DE	3	3	4	3	3	4	4	4	4	3	3	4

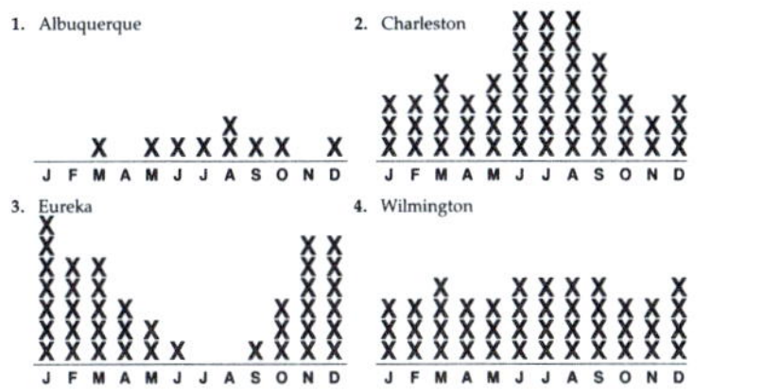

5. A line plot can be converted to a line graph by adding a vertical scale, connecting the tops of the rows of *x*s, and then eliminating the *x*s. Draw four line graphs on the axes, one for each city's rainfall. Use the type of line shown.

Albuquerque ——— Charleston ------
Eureka ·········· Wilmington -·-·-·-

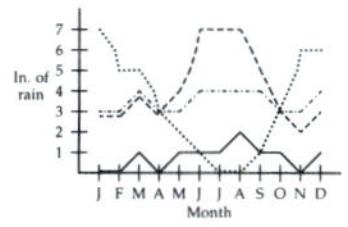

6. How does a line graph differ from a line plot?
Several sets of data can be shown at one time.

7. How does a line graph show that two cities have the same amount of rainfall in the same month?
Two lines contain the same point.

L-4 Chapter 12

LESSON QUIZ

Draw a line plot for each set of data. Arrange the data in a frequency table. Find the mode for each set of data.

1. 25 29 26 26 25 27 25 29 25 (mode: 25)

2. 14 17 12 13 15 17 13 15 12 14 (no mode)

Find the mean, median, and mode for each frequency distribution.

3.

x	0	1	2	3	4
f	1	0	2	4	5

(3, 3, 4)

4.

x	10	20	30	40	50	60
f	3	6	5	3	6	1

(32.5, 30, 20 and 50)

Assignment Guide

Basic 1-3, 7-8, 11-12, 15, 17-22, MR All

Average 2-4, 8-9, 12-13, 15, 17-22, MR All

Enriched 4-6, 9-10, 13-14, 16-22, MR All

Lesson Focus

Materials

Newspapers, magazines with statistical data, math journal

Vocabulary/Symbols

sample

MOTIVATING THE LESSON

Have students share advertising claims they may have read or heard on TV or radio which they think might illustrate a misuse of statistics. (e.g., Four out of five people use *Smooth* eye drops.) Mention that the purpose of this lesson is to examine ways of uncovering such misuses.

Instruction

1 Discuss which audiences advertisers are trying to reach when they sponsor a particular TV show. Assist students by asking questions, such as, **Why are the daytime shows called "Soaps"?** and **Who do advertisers target on Saturday mornings?** Ask students to observe which advertisers sponsor the prime-time evening shows and discuss why they might do this.

2 If students have already conducted surveys of this type, have them review the results. If not, you might plan to have them conduct this survey to get real answers to the questions posed.

Exploring Misuse of Statistics

OBJECTIVE:
To draw valid conclusions from statistical data.

▼ Statistics is a powerful tool when used to influence the way a person perceives a situation. You might cite statistics to influence a friend's opinion. In a similar manner, companies often use statistics to present their best image. Advertisers also use statistics to influence you to select their product.

▼ When you use statistics, you must analyze the problem before choosing the data that meet your needs.

1. A toy manufacturer wants to sell more toys. The manufacturer decides to advertise on the most frequently watched television program. Using data collected by a rating company, the manufacturer determines that an adult comedy program is the most popular.
   1 a. ***Analyze*** Did the manufacturer find the group of people that might want to own the toy (the target market)? no
   b. ***Discuss*** What factors need to be considered when deciding where to advertise? age of target market
   c. ***Decide*** Who is in the target market? What types of shows would reach more people in the target market? depends on toy; probably children's shows

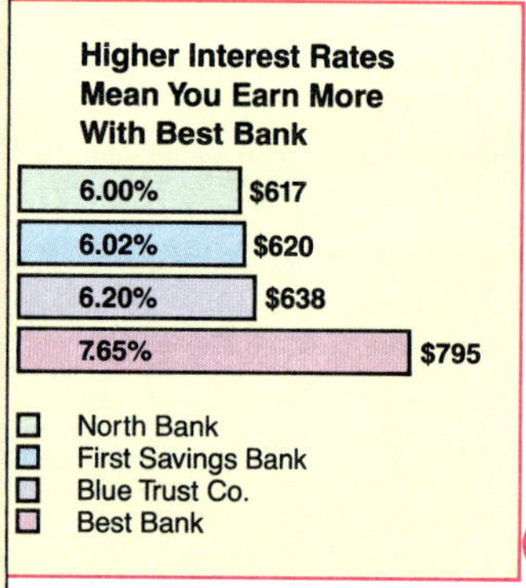

Explain how the graph is misleading.

▼ The most accurate way to collect data is to get everyone's opinion. This is usually impractical. A *sample* of the target group can often provide sufficient information from which to make decisions.

2. Suppose you want to find the number of boys in your school who watch football on Sunday. Check students' work.
   a. ***Describe*** the group about which you want information.
   b. Suppose your sample includes all the boys in your class. Does your sample represent the boys in the school? ***Explain.***
   c. ***Discuss*** What might be a better sample? Do you think the size of the sample has any effect on the outcome? Would ten boys be enough? ***Explain.***
   2 d. ***Discuss*** Where would you take the survey? When would be a good time to take the survey?
   e. ***Discuss*** methods of sampling. Would an interview survey be as accurate as a written survey? How would you word your survey?
   f. Would the time of year in which you took the survey make a difference? the part of the country? ***Explain.***
   g. What other factors could influence your results?

Teaching Tip

Newspapers and magazines contain many examples of graphs that mislead, whether intentionally or not. Have newspapers and magazines available for students to use.

▼ Sometimes people present accurate data in a misleading manner to encourage the conclusions they support.

3. ***Compare*** and analyze the data in the graphs below.

Wages Skyrocket

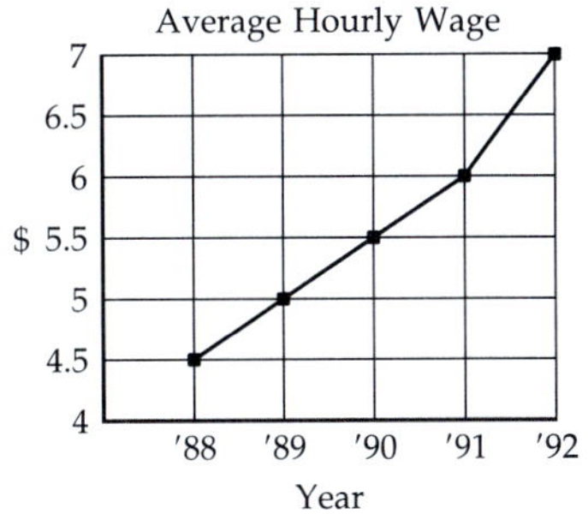

Wages Show Modest Gain

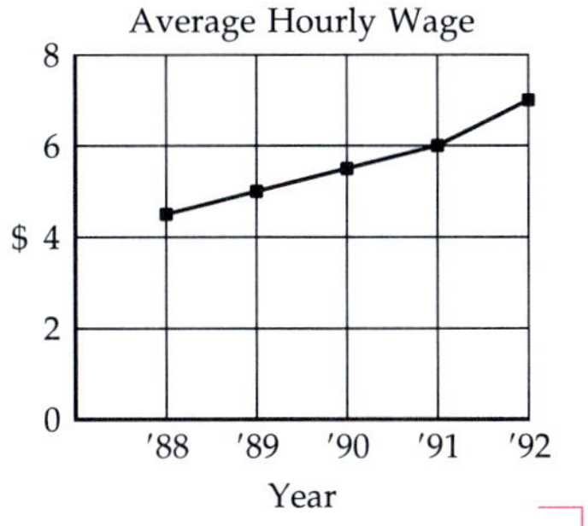

a. ***Discuss*** What is the difference in the way the vertical scales are used?

b. Do you think the information was presented accurately by each graph? ***Explain.*** Answers may vary.

c. ***Write*** a short paragraph telling which graph you think was prepared by the personnel department of the company and which was prepared by the workers. Support your viewpoint. Answers may vary.

See right. Possible answers given.

4. An ad stated that 3 out of 4 dentists recommend *Smile* toothpaste.

a. ***Decide*** Do you need to know how many dentists were included in the sample?

b. ***Decide*** How could the city where the survey was taken affect the validity of the sample?

c. ***Discuss*** whether the dentists also recommend other brands. ***Explain*** why this might affect your opinion of the data.

5. ***PROJECT*** Find examples in a newspaper or a magazine where statistical data is used. Check students' work.

a. ***Explain*** Does the collection of data appear to be accurate?

b. Is the information presented in a distorted manner?

c. ***Discuss*** How could you present the data more fairly?

6. ***PROJECT*** Write a survey question. Check students' work.

a. ***Decide*** What is your target market?

b. Can you survey the entire target market or is it more practical to take a sample? Take the survey.

c. ***Analyze*** How can you accurately present the survey results? Make a presentation with a graph.

3. a. The vertical scale on second graph goes from 0 to 8, diminishing the effect of change.
4. a. Yes. Five dentists would be too small a sample.
4. b. If the dentists surveyed lived in the town where *Smile* is made, they may not be interested in other brands.
4. c. This is important. *Smile* advertisers may not be giving us all the information.

3 Share examples of graphs from newspapers or magazines so students will see how changing the scale and omitting data changes the impression created by the graph.

4 These activities allow students to share their ideas and findings with the class. When engaged in these activities, students apply language skills, as well as problem solving skills.

Closure

To see whether students can generalize from the activities of this lesson, have them work in small groups to explain how to create a valid sample, plan the collection of data using a sample, and explain the meaning of "pseudo-statistical" statements like "3 out of 4 dentists recommend."

Assignment Guide
Basic All
Average All
Enriched All

FOR THE NEXT LESSON
paper and pencil, $\frac{1}{4}$-inch or centimeter graph paper, math journal, overhead projector

Lesson Focus

Materials

Paper and pencil, $\frac{1}{4}$-inch or centimeter graph paper, math journal, overhead projector

Vocabulary/Symbols

histogram
leaves
stem and leaf plot
stem

MOTIVATING THE LESSON

Have students review and discuss the methods they have used to organize statistical data. (e.g., line plots, frequency distribution tables) Note that two other methods for organizing such data are discussed in this lesson.

Instruction

1 Ask students what they can conclude about the fat content of the cheese pizza just by looking at the plot. (The pizza contains less fat than most of the other foods on the list.)

2 Point out that it is necessary to identify what the stem represents and what the leaves represent because the same numbers would be used for the stem and leaf for 103 as are used for 10.3.

3 Have students discuss why the back-to-back stem and leaf plot is a more effective way of comparing the men's and women's times than the table is. (The data are arranged in numerical order, and identical scores are grouped together.)

OBJECTIVE: *To present statistical data in the form of stem and leaf plots.*

12-3 Stem and Leaf Plots

▼ The table at the left gives the fat content in one serving of various foods. How does the fat content for a cheese pizza compare with other items in the list?

Fat Content in One Serving

Food	Fat Content in Grams
Deluxe, large hamburger	33
Plain, small hamburger	10
Cheeseburger	17
Roast beef sandwich	15
Beef and cheese sandwich	22
Fish sandwich	26
Fried chicken pieces	21
Deluxe pizza	26
Cheese pizza	12
Taco	26
Apple turnover	24
French fries	12

It is difficult to compare the fat content from the table. You can reorganize the data by constructing a *stem and leaf plot.*

Choose the value for the *stem* by finding the least value and the greatest value for the fat content totals. The least value is 10. The greatest value is 33. Both numbers have two digits. The tens' digits will become the stem. Therefore, the stems for this data will be the digits 1 through 3.

Write the digits from 1 to 3 in a column. Draw a line to their right.

1	
2	
3	

The *leaves* are the ones' digits associated with the tens' values. For a cheeseburger with the fat content of 17 g, the stem is 1 and the leaf is 7. For an apple turnover with fat content of 24 g, the stem is 2 and the leaf is 4. Record each food's fat content to obtain the plot below.

1	0 7 5 2 2
2	2 6 1 6 6 4
3	3

Arrange the leaves on each stem in order from least to greatest.

1	0 2 2 5 7
2	1 2 4 6 6 6
3	3

1 The mode is the greatest number of repeated leaves. For this data the mode is 26. Since the data are in order, the median is the midpoint of the twelve items. The median is the average of the sixth item, 21, and the seventh item, 22. The median is 21.5.

Now it is easier to evaluate the data. A cheese pizza, with fat content of 12 g for a serving, is below the median and the mode for fat content of the foods listed.

▼ In summary, to make a stem and leaf plot:

1. Choose a stem.
2. Write all values between and including the least stem and the greatest stem in order in a column. Draw a line to their right.
3. Write the leaf values to the right of the stem values.
4. Rewrite the plot, putting the values in order.

Teaching Tip

To encourage students to organize their work neatly, have $\frac{1}{4}$-inch or centimeter graph paper available for students' use in making stem and leaf plots.

▼ The stem may be more than one digit. It can be any number that will provide a useful way to organize the data.

Example 1 Use the table at the right to construct a stem and leaf plot to compare the Olympic times for the 80-m hurdles from 1932–1968. Then find the mode.

Solution The stem will be the whole number of seconds. The leaves will be the tenths of a second.

10	3 5 7 8 9
11	2 7 7

The mode is 11.7 s.

▼ A back-to-back stem and leaf plot records two sets of data. The side-by-side display makes the data easier to compare.

Example 2 Draw a back-to-back stem and leaf plot for the times in the 100-m dash in the Olympic Games from 1928–1968. Find each median and mode.

Solution Use seconds for the stem and tenths of seconds for the leaves.

Men's Time (tenths of seconds)	Stem (seconds)	Women's Time (tenths of seconds)
9	9	
8 5 4 3 3 3 2 0	10	
	11	0 0 4 5 5 5 9 9
	12	2

Men's Scores		Women's Scores
10.3 s	Median	11.5 s
10.3 s	Mode	11.5 s

Women's 80-m Hurdles

Year	Time (in seconds)
1932	11.7
1936	11.7
1948	11.2
1952	10.9
1956	10.7
1960	10.8
1964	10.5
1968	10.3

There were no Olympics in 1940 and 1944.

100-m Dash Time (in seconds)

Year	Men	Women
1928	10.8	12.2
1932	10.3	11.9
1936	10.3	11.5
1948	10.3	11.9
1952	10.4	11.5
1956	10.5	11.5
1960	10.2	11.0
1964	10.0	11.4
1968	9.9	11.0

There were no Olympics in 1940 and 1944.

CLASS EXERCISES

Use the stem and leaf plot below to answer each question.

6	1 1 3 5 5
7	0 2 2 4
8	4 5 8 9
9	3 6 7 9 9 9

1. What numbers make up the stem? 6, 7, 8, 9
2. What numbers make up the leaves for the first stem? 1, 3, 5
3. What is the mode? 99
4. What is the median? 84
5. ***WRITE*** Describe a situation for which the data in the stem and leaf plot might apply. Possible answer: a set of exam grades.

ADDITIONAL EXAMPLES

Make a stem and leaf plot for each set of data. Then find the median and the mode.

1. 117 122 124 128 131 134 137 139 140 141 145

2. 1.8 3.1 2.0 2.5 2.6 2.1 2.3 1.9 3.0 3.2 2.5 2.7 2.6 2.5 2.8

3. Make a back-to-back stem and leaf plot for the data in Sets A and B. Then find the median and the mode for each set.
Set A: 52 53 47 51 52 49 50 54 52 51 53;
Set B: 53 56 60 58 54 57 56 55 57 61 56

Answers:

1.

11	7
12	2 4 8
13	1 4 7 9
14	0 1 5

134; none

2.

1	8 9
2	0 1 3 5 5 5 6 6 7 8
3	0 1 2

2.5; 2.5

3.

Set A (ones)	Stem (tens)	Set B (ones)
9 7	4	
4 3 3 2 2 2 1 1 0	5	3 4 5 6 6 6 7 7 8
	6	0 1

Set A 52; 52;
Set B 56; 56

Guided Practice

Class Exercises Encourage students to be creative in answering question 5. Have them share their answers with the class.

MATH MINUTES

Find the median and mode for each set.

Set A		Set B
9 8	1	
3 2 2 1	2	
7 6 6 6	3	2 3 8
	4	0 5 5 6 7 9
	5	4 5

(A: 22.5 36; B: 45 45)

Think and Discuss For question 1, help students recall that the mean must be calculated, and therefore, it would not be especially useful to make a stem and leaf plot just to find the mean.

A ***common error*** occurs when students do not organize the leaves in numerical order in a stem and leaf plot. Have students make two stem and leaf plots for a set of data, one presenting the data as originally organized, the other presenting the data in numerical order.

Closure

Writing in Math Have students explain in their math journals why a stem and leaf plot is a useful way to organize statistical data in order to find its median and mode.

Independent Practice

Remind students to indicate what the stems and leaves represent when the leaves do not represent ones.

For question 11a, have students explain their answer. (These animals have a life span of less than 10 years.)

Critical Thinking Use the overhead projector to illustrate that if the stem and leaf plot is drawn on graph paper and rectangles are drawn around the leaves, the drawing can be turned 90° counterclockwise to obtain the histogram. Point out that in a histogram, the individual scores are no longer available and the median and the mode cannot be determined.

THINK AND DISCUSS

1. Is the stem and leaf plot a useful method of organizing data if you only want to find the mean? Explain. See below.

2. A set of data contains numbers in the 20s, 30s, and 50s only. Is it necessary to put a 4 on the stem in a stem and leaf plot? Explain. See below.

3. In a stem and leaf plot, can a leaf have no stem? Explain.

1. No. The chart makes adding the numbers inconvenient.
2. Yes. The stem will show a gap in the data.
3. No. Every leaf must have a stem. To do a plot for numbers like 9, 8, 10, 12, 15, 20, the stem for 9 and 8 would be zero.

Life Span for Animals

Animal	*Years*
Black bear	32
Condor	52
Cow	15
Dog	12
Giraffe	10
Gorilla	20
House cat	12
Human	71
Kangaroo	7
Lion	13
Opossum	1
Pig	10
Zebra	15

Make a stem and leaf plot from each set of data. Then find the median and the mode. Check students' work.

6. 15, 22, 25, 10, 36, 15, 28, 35, 18 22; 15

7. 105, 115, 95, 97, 86, 89, 102, 107, 114, 113, 113, 113 106; 113

8. 786, 789, 791, 777, 771, 781, 796, 797, 800, 801, 781 789; 781

Make a back-to-back stem and leaf plot from each of the sets of data. Then find the median and the mode for each set of data.

9. Set A: 25, 23, 33, 36, 42, 44 — Set B: 19, 16, 23, 34, 26

10. Set C: 236, 237, 241, 250, 242 — Set D: 262, 251, 248, 243, 257

11. Set E: 9.1, 8.2, 7.3, 6.4, 7.3, 8.5 7.75; 7.3 — Set F: 7.6, 9.2, 8.2, 8.3, 9.7, 7.6 8.25; 7.6

9. Set A: 34.5; no mode Set B: 23; no mode 10. Set C: 241; no mode Set D: 251; no mode

WRITTEN EXERCISES

7. Set A: 62.5; 63 Set B: 57; no mode 8. Set C: 147; 142 Set D: 148; no mode 9. Set E: 5.85; no mode Set F: 6.45; no mode

Make a stem and leaf plot for each set of data. Then find the median and the mode. Check students' work.

1. 785, 785, 776, 772, 792, 788, 761, 768, 768, 750 774; 768 and 785

2. 4.5, 4.3, 0.8, 3.5, 2.6, 1.4, 0.2, 0.8, 4.3, 6.0 3.05; 0.8 and 4.3

3. 89, 70, 102, 82, 74, 74, 78, 105, 108, 107, 75 82; 74

4. 47, 41, 60, 75, 85, 53, 57, 76, 79, 81, 84, 86 75.5; no mode

5. 11.5, 11.8, 10.6, 10.4, 9.5, 12.2, 11.8, 11.8, 11.8, 10.6 11.65; 11.8

6. 225, 220, 221, 222, 231, 231, 219, 219, 215, 229, 236 222; 219 and 231

Make a back-to-back stem and leaf plot for each set of data. Then find the median and the mode. Check students' work.

7. Set A: 63, 62, 63, 52, 58, 63 — Set B: 45, 48, 53, 57, 61, 58, 65

8. Set C: 156, 158, 142, 142, 147 — Set D: 141, 145, 148, 156, 157

9. Set E: 5.2, 5.8, 6.7, 6.3, 5.9, 4.1 — Set F: 7.1, 6.4, 6.5, 6.8, 5.1, 5.2

10. Set G: 206, 205, 210, 215, 222 210; no mode — Set H: 218, 219, 228, 208, 209 218; no mode

11. Use the ***DATA*** at the left to solve.

  a. What number will make up the stem for the kangaroo and the opossum? 0

  b. Construct a stem and leaf plot. Check students' work.

  c. Find the mode. 10, 12, and 15

  d. Find the median. 13

12. ***DATA FILE 7 (pp. 272–273)*** Draw a stem and leaf plot for the passenger boardings in 2000. Find the median. 23.5 million passengers

Use the stem and leaf plot below to answer each question. The plot is for the length of time, in minutes, two classes spent on homework.

Class A		Class B
7 4 3	6	1 1 3 5 5
9 9 8 5 4 4	7	0 2 2 4
5 2 1 0	8	4 5 8 9
7 6 6 4 2	9	3 6 7 9 9 9

13. What numbers make up the stem? 6, 7, 8, 9

14. What is the lowest time for each set of data? 63 min, 61 min

15. What is the median and the mode for each set of data? Class A 79.5, 74, 79 and 96, Class B 84; 99

Critical Thinking

EXPLORING HISTOGRAMS

▼ A *histogram* uses rectangles to show frequency data. You can make a histogram from a stem and leaf plot. The horizontal axis corresponds to the stem. The vertical axis reflects the number of items in the leaf (the frequency).

4	8 9
5	2 5 6 7
6	3
7	4 5

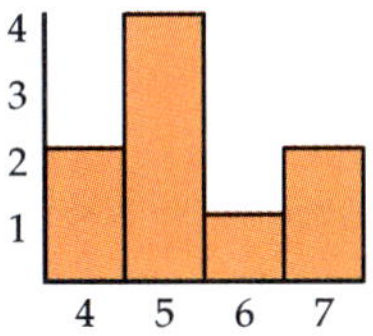

▼ You can divide the information on the horizontal scale of the histogram into smaller groupings.

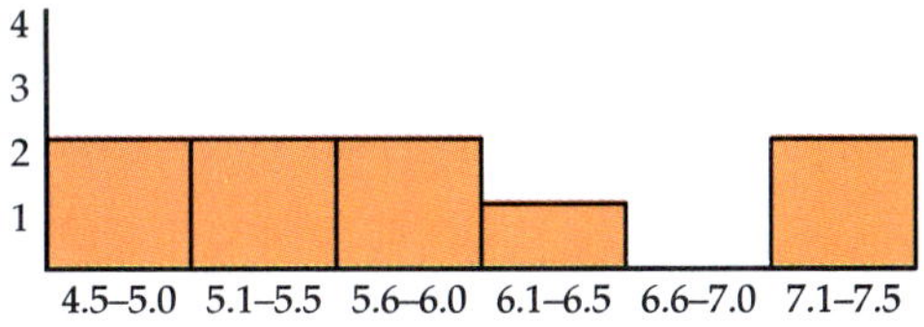

1. How does a change in the horizontal scale of a histogram affect the vertical scale? A change in the groupings of the horizontal scale changes the heights on the vertical scale.
2. How can the change in scales affect how people understand the information on the histogram? The shape of the graph is changed, giving a different impact.
3. Draw two histograms with different horizontal scales for the data at the right. Which scale more accurately reflects the information? Explain. Check students' work.

MIXED REVIEW

Solve.

1. $2x^2 = 72$ 6, -6

2. Find the surface area of a cube with sides that measure 2.5 in. 37.5 in.^{2}

Use the frequency table to find each answer.

x	1	2	3	4
f	6	7	4	2

3. Find the median. 2

4. Find the mode. 2

Find $\frac{1}{2}$, $\frac{1}{4}$, and $\frac{3}{4}$ of each number.

5. 36 18, 9, 27 **6.** 92 46, 23, 69

7. *DATA FILE 13 (pp. 534–535)* Find the mean, median, and mode of the global carbon emissions. Your answer will be in millions of tons. 852.375; 803.5; 774

Record High Temperatures in Western States (°F)

State	Temperature
Alaska	100
Arizona	127
California	134
Colorado	118
Hawaii	100
Idaho	118
Montana	117
Nevada	122
Oregon	119
Utah	116
Washington	118
Wyoming	114

Lesson Follow-up

Reteaching Activity Display these data: 653 658 660 662 654 661 657 660 651 663 675. Have students make a stem and leaf plot by following these steps: (1) identify the least and greatest values; (2) identify the stems; (3) record the leaves on each stem; (4) sequence each leaf from least to greatest.

LESSON RESOURCES

TRB Practice L-5
TRB Enrichment L-6

Practice For use after 12-3 (pp. 498–501)

Stem and Leaf Plots

Make a stem and leaf plot for each set of data, then find the median and the mode. Check students' work.

1. 50, 39, 40, 48, 38, 42, 53, 42, 38, 48, 47, 43, 51, 48, 43
   median 43 mode 48
2. 6.2, 7.0, 6.3, 6.6, 7.2, 7.3, 6.7, 7.2, 7.0, 6.1, 6.7, 6.4, 6.9, 7.2, 6.2, 6.8, 7.1
   median 6.8 mode 7.2
3. 14, 19, 23, 30, 26, 10, 7, 16, 28, 17, 24, 1, 28, 19, 10, 16, 19, 5, 25
   median 19 mode 19
4. 91, 106, 86, 93, 100, 84, 108, 95, 99, 98, 83, 96, 109, 84, 105, 102, 87, 94
   median 95.5 mode 84

Make a back-to-back stem and leaf plot from each of the sets of data. Then find the median and the mode for each set of data.

5. Set A: 25, 28, 31, 20, 26, 22, 21, 25
   Set B: 31, 28, 19, 20, 22, 30, 18, 25, 27, 19
   A: median 25 mode 25
   B: median 23.5 mode 19

A		B
	1	8 9 9
8 6 5 5 2 1 0	2	0 2 5 7 8
1	3	0 1

6. Set A: 175, 186, 169, 180, 178, 183, 176, 184, 179
   Set B: 191, 175, 178, 187, 180, 178, 186, 182, 179
   A: median 179 mode none
   B: median 180 mode 178

A		B
9	16	
9 8 6 5	17	5 8 8 9
6 4 3 0	18	0 2 6 7
	19	1

Chapter 12 L-5

Enrichment For use after 12-3 (pp. 498–501)

What's Holding Everything Up?

What prevents a stem-and-leaf tree from falling in a tornado?

To find the answer, make a stem and leaf plot from the given data. Write leaves in the blanks beginning on the left. When the plot is finished, look at each stem in turn. Find the letter in "QUESTION MARK" that appears above the final leaf. Write the letter above the stem in the answer grid.

Example

	Q	U	E	S	T	I	O
3	1	1	3	5	6		

Write T above 3 in the answer grid.

Ages of Contestants in Trivia Competition

76	44	27	40	85	69	14	9	73	97	71	25	12
104	53	16	46	35	20	95	54	42	12	49	78	31
78	43	31	8	86	68	44	97	18	33	76	92	14
95	40	21	75	19	6	71	36	12	26	71	53	28
57	93	22	45	71	96	61	74	17	49	68	15	44

	Q	U	E	S	T	I	O	N	M	A	R	K
0	6	8	9									
1	2	2	2	4	4	5	6	7	8	9		
2	0	1	2	5	6	7	8					
3	1	1	3	5	6							
4	0	0	2	3	4	4	4	5	6	9	9	
5	3	3	4	7								
6	1	8	8	9								
7	1	1	1	1	3	4	5	6	6	8	8	
8	5	6										
9	2	3	5	5	6	7	7					
10	4											

Answer: SQUARE ROOTS (6 10 8 1 4 0 7 2 9 3 5)

L-6 Chapter 12

LESSON QUIZ

Make a stem and leaf plot from each set of data. Then find the median and the mode.

1. 9 8 10 13 12 10 9 14 10 13 7 (10; 10)

2. 3.5 6.5 6.8 5.3 3.8 4.2 6.0 5.9 6.2 4.7 4.9 6.3 5.6 5.9 (5.75; 5.9)

3. Make a back-to-back stem and leaf plot from the sets of data. Then find the median and the mode for each set of data.
Set A: 27 23 28 32 35 39 41 37 44 36 40 45 (36.5; none)
Set B: 28 33 40 22 40 31 23 37 38 29 25 42 (32; 40)

Assignment Guide

Basic 1-3, 7-8, 11-15, MR, CT All
Average 2-4, 8-9, 11-15, MR, CT All
Enriched 4-6, 9-15, MR, CT All

FOR THE NEXT LESSON

graph paper (optional), math journal

Lesson Focus

Materials

Graph paper (optional), math journal

Vocabulary/Symbols

box and whisker plot
quartile

MOTIVATING THE LESSON

Write this set of scores in a column from greatest to least on the chalkboard.

100 90 85 80 74 73 72 72
71 70 68 67 60 60 55 50

Ask students to draw a rectangle around the middle 50% of the scores (67-74) and to give the range of those scores (7) and the range of the entire set of scores. (50)

Instruction

1 Ask students whether or not the box would be affected if the upper 25% of the scores were in the low 80s and if the lower 25% of the scores went as low as zero. (No; in each case the box represents the middle 50%, which is not affected by extremes.)

OBJECTIVE: *To present data in box and whisker plots.*

12-4 Box and Whisker Plots

SCORE
HOME 2
VISITORS 1

Boston Bruins Scorers' points for year

Bourque	61
Brickley	35
Burridge	61
Carpenter	40
Carter	22
Crowder	33
Galley	29
Hawgood	40
Janney	62
Johnston	21
Joyce	49
Linseman	72
Neely	75
Sweeney	28
Thelven	21
Wesley	54

▼ The chart at the left shows the top sixteen scorers for the Boston Bruins during a recent hockey season. You can make a *box and whisker plot* to show the data. A box and whisker plot is especially useful in showing the distribution of data in each *quartile,* that is, in each 25% of the data.

1. Arrange the data in order from least to greatest. Then find the median. For these data, the median is 40.
2. Separate the data into four groups. Find the medians of the lower and upper halves. The median of the lower half is 28.5, and the median of the upper half is 61. These values are the first and third quartiles.
3. Draw a number line to display the data. Mark the number line with the quartile values.
4. Draw a box that extends from the first to the third quartiles. Mark the median with a vertical line across the box. Then draw whiskers from the box to the highest and lowest scores.

75
72
62
61 — Third Quartile
61
54
49
Median→ 40 — Second Quartile
40
35
33
29 — First Quartile
28
22
21
21

1

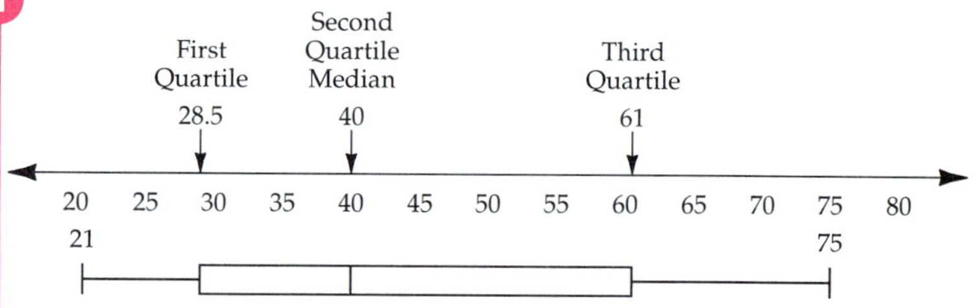

THINK Why aren't the quartiles marked at $\frac{1}{4}$ of the distance on the number line? **The box and whisker plot divides the data into 4 equal groups rather than the number line into 4 equal parts.**

▼ Some trends are easy to recognize in a box and whisker plot.

Example 1 **Describe the data in the plot below.**

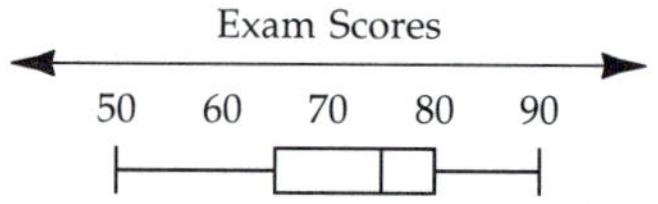

Solution The highest score is 90 and the lowest is 50. Of the scores, 25% are greater than 80 and 25% are less than 65. The plot has a small box indicating that half of the scores are clustered around the median, which is 75.

Teaching Tip

To gain insight into students' understanding of organizing statistical data in a box and whisker plot, have them think out loud as they determine the quartiles and construct the plots.

▼ You can use a box and whisker plot to compare two sets of data, such as scores for teams or individual players.

Example 2 Use the box and whisker plots to compare the leading scores for the Los Angeles Kings and the New York Islanders. What conclusions can you draw?

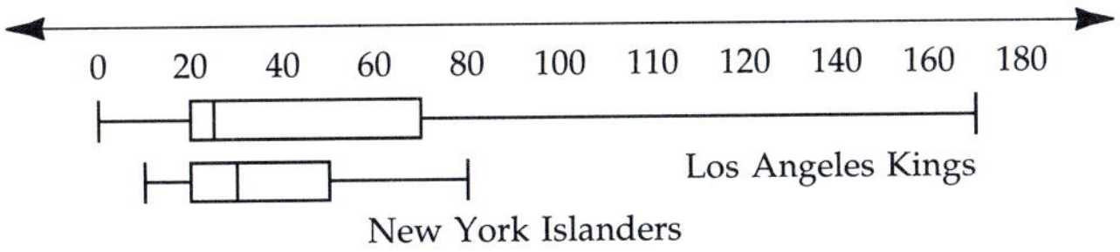

Solution The first and second groups for both teams look fairly similar. The New York Islanders have a slightly higher median score. The top half is very different. The Kings have 25% of their scores between 70 and 170. The top 25% for the Islanders is between 50 and 80. The top group shows that the Kings outscore the Islanders by a large margin.

CLASS EXERCISES

Make a box and whisker plot for each set of data. Check students' work.

1. 16, 18, 59, 75, 29, 34, 25, 49, 27, 16, 21, 58, 71, 19, 31, 50
2. 138, 149, 200, 101, 128, 196, 186, 150, 129, 176, 192, 190, 107, 175, 171, 163

Use the box and whisker plot below to answer each question

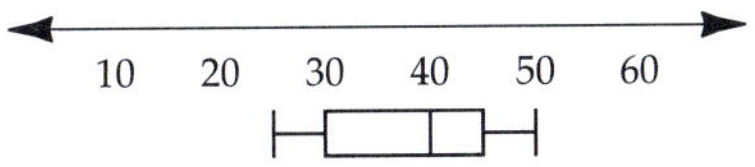

3. What is the median? 40
4. What percent of the numbers are contained in the box? 50%
5. Are the data evenly distributed? Explain. No. The median is not in the middle of the box and the box from 30 to 40 is twice as large as the other box.

Make a box and whisker plot for each set of data. Use a single number line. See additional answers.

6. 1st set: 12, 16, 62, 48, 16, 59, 43, 39
   2nd set: 34, 92, 73, 71, 59, 68, 49, 84
7. 1st set: 36, 9, 4, 3, 12, 29, 50, 16, 25, 21
   2nd set: 18, 22, 7, 4, 11, 16, 40, 18, 33, 9

THINK AND DISCUSS See below.

1. Explain how you can find the quartiles for a set of data.
2. Describe a set of data that has a long box and short whiskers.
3. How is a box and whisker plot like a stem and leaf plot? How is it different?
4. Can you tell what the mean, median, and mode of a set of data would be by looking at a box and whisker plot? Explain.

1. Divide the data into two parts. The medians of each of these parts are the quartiles.
2. Example: Exam scores 70–90, 50%; 65–70, 25%; 90–95, 25%.
3. Both plots show where data clusters. B-W gives visual picture of each quarter's spread. S-L shows concentration along stem.
4. only median

ADDITIONAL EXAMPLES

1. Describe the data in the plot below.

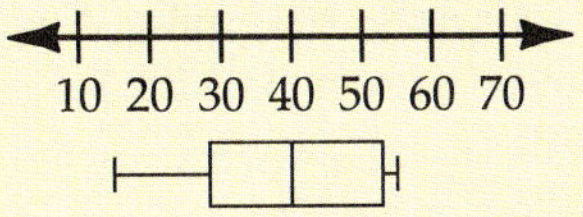

(The highest score is 55 and the lowest is 15. Of the scores, 25% are greater than 50 and 25% are less than 30. Half the scores are clustered around the median, 40.)

2. Compare the two sets of data.

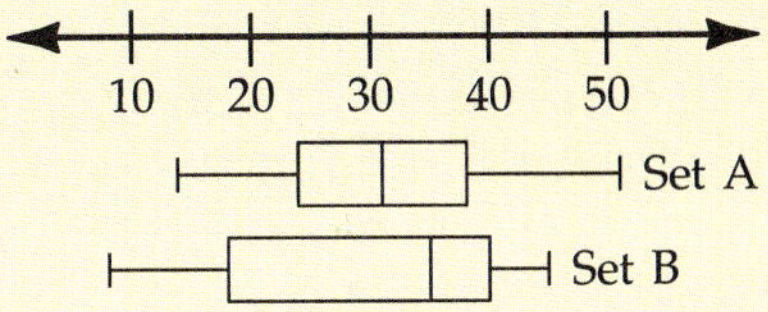

(Answers will vary. Possible answers: Set B has the higher median. The upper 25% for Set A is between 38 and 50 and for Set B between 40 and 45. The range for the lower 25% of each set is the same, 10.)

Guided Practice

Class Exercises For question 5, have students discuss situations where the data might be evenly distributed throughout.

Think and Discuss For question 3, help students recall the methods for organizing statistical data and state which methods identify individual data items. (line plots, frequency tables, stem and leaf plots)

MATH MINUTES

Use this set of data to complete each sentence:
55 46 43 42 41 40 39 35 35 31 29 27 26 24 13 10

1. The median is _______. (35)
2. 26.5 is the _______. (first or lower quartile)
3. The third quartile is _______. (41.5)

ADDITIONAL ANSWERS

6.

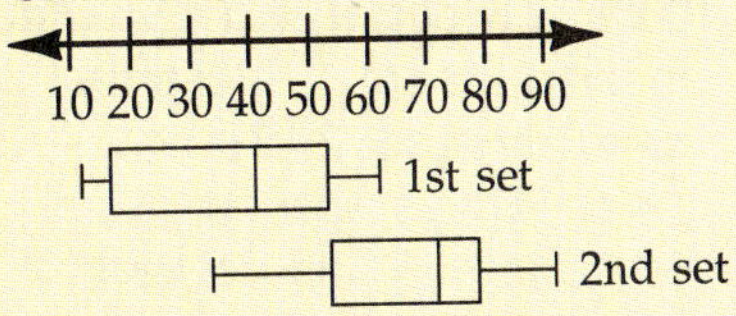

7.

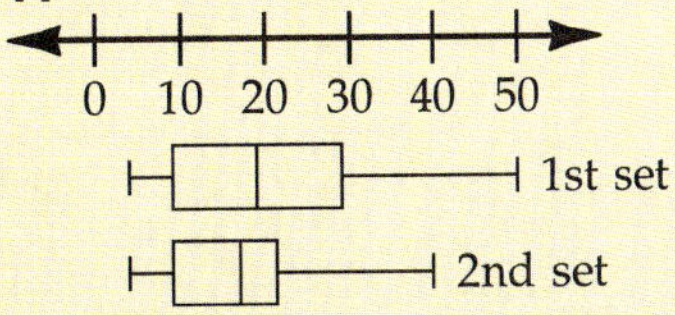

A ***common error*** occurs when students do not correctly identify the quartile splits. Emphasize that a quartile occurs at the median of each half of the set of data. Provide practice by having students separate sets of data, such as 1 5 6 8 10 13 15 17 18 20 25 29, into two equal subsets and then find the median of each subset.

Closure

Writing in Math Have students write a description in their math journals of how a box and whisker plot provides useful information about data.

Independent Practice

For question 12, ask students to include an analysis of the data by describing what the median and the quartiles are and how the upper and lower 25% of the data are distributed.

MIXED REVIEW

1. $29\frac{1}{2}\%$ of 400 is what number? 118
2. What percent of 48 is 18? 37.5%

Use the set of test scores: 92, 84, 76, 68, 90, 67, 82, 71, 79, 85, 79.

3. Find the median and range. 79; 25
4. Find the mean. 79.4
5. Find the mode. 79
6. Make a stem and leaf plot. Check students' work.

Solve.

7. Find $5 \cdot 4 \cdot 3 \cdot 2 \cdot 1$ 120
8. ***DATA FILE 12 (pp. 486–487)*** Draw a back-to-back stem and leaf plot for the Olympic records for 400-m freestyle swimming. Round the time to the nearest second. Check students' work.

Maximum Speed of Animals for a Quarter Mile (mi/h)

Cheetah	70
Lion	50
Quarter horse	47.5
Coyote	43
Hyena	40
Rabbit	35
Giraffe	32
Grizzly bear	30
Cat (domestic)	30
Man	27.89
Elephant	25
Squirrel	12

WRITTEN EXERCISES

Make a box and whisker plot for each set of data.

1. 4, 6, 9, 4, 5, 12, 16, 21, 38, 5, 2, 27 Check students' work.
2. 4, 5, 3, 2, 1, 6, 7, 8, 4, 5, 5, 6, 2, 1, 9, 3
3. 38, 49, 16, 21, 48, 36, 29, 52, 31, 25, 49, 36
4. 47, 19, 98, 16, 25, 38, 55, 49, 86, 79, 15, 91, 57

Use the box and whisker plot to answer each question.

Prices of 20 Walk-around Stereos

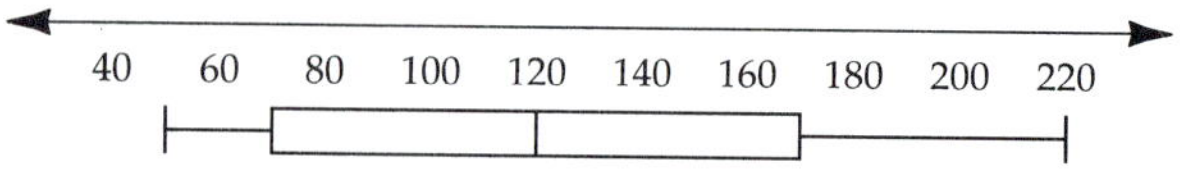

5. What are the highest and lowest prices for stereos? $220, $50
6. What is a median price for a stereo? $120
7. What percent of prices are greater than $70? 75%
8. What percent of stereos cost more than $170? 25%
9. Are the data evenly distributed? no

WRITE **Make a box and whisker plot for each set of data. Use a single number line. Then write a comparison.** See additional answers.

10. 1st Set: 3, 7, 9, 12, 2, 1, 6, 5, 4, 3, 7, 10, 13, 8, 1, 9
    2nd Set: 9, 8, 1, 7, 6, 3, 7, 9, 8, 6, 4, 7, 8, 9, 10, 10
11. 1st Set: 34, 25, 19, 38, 49, 16, 38, 49, 56, 24, 42, 36
    2nd Set: 38, 49, 52, 39, 50, 55, 46, 45, 40, 39, 51, 42
12. Use the data at the left to make a box and whisker plot for the maximum speed of animals. Check students' work.
13. ***PROJECT*** Measure the heights of the students in your class. Make a box and whisker plot of the data, by gender, to compare the heights of male and female students. Check students' work.
14. ***WRITE*** Make a box and whisker plot for each set of data on a single number line. Compare. Then write your conclusions. Check students' work.

National Hockey League High Scorers

Buffalo Sabres	18, 88, 57, 44, 1, 18, 70, 28, 6, 60, 16, 41, 18, 52, 49, 52, 20, 43, 44, 17
St. Louis Blues	7, 67, 13, 70, 84, 34, 29, 26, 45, 17, 52, 55, 23, 24, 25, 26, 19, 43, 42, 20

15. ***DATA FILE 13 (pp. 534–535)*** Make a box and whisker plot for the global carbon emissions per person. Check students' work.

Set 1: the lower 25% is between 1 and 3, the top 25% between 9 and 13. The median for set 1 is slightly lower than set 2. Set 2: lower 25% is between 1 and 6, the top 25% is between 9 and 10.

ADDITIONAL ANSWERS

10.

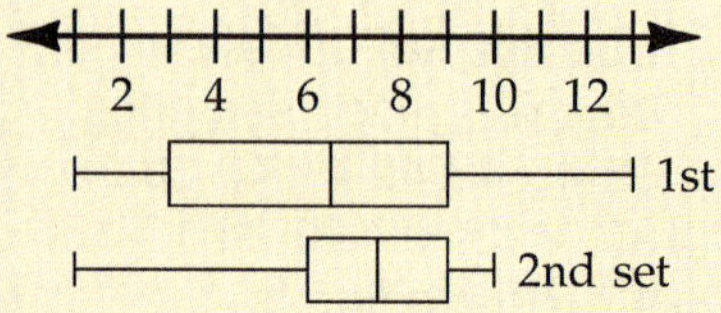

11.

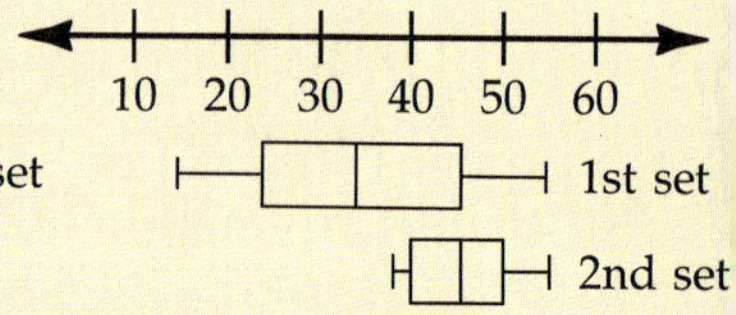

16. ***PROJECT*** Find scores for your two favorite teams for the past season. Make a box and whisker plot for each set of scores on a single number line. Compare. Then write your conclusions about the data. **Check students' work.**

Use the article below and the chart at the right.

Earthquake!

A scientific journal recently reported that earthquakes can occur even in areas considered stable, far away from the edges of the earth's rigid plates. Stable regions make up about two-thirds of the continental crust.

The journal reported that the most reliable gauge of an earthquake's size is the moment-magnitude scale (M). The moment-magnitude scale is based directly on the physical process in the center of an earthquake.

Earthquakes in Stable Areas

Place	M
New Madrid, 1812	8.3
New Madrid, 1811	8.2
New Madrid, 1812	8.1
Kutch, 1819	7.8
Baffin Bay, 1933	7.7
Taiwan Straits, 1604	7.7
South Carolina, 1886	7.6
Nanai, 1918	7.4
Grand Banks, 1929	7.4
Basel, 1356	7.4

17. What is the mean, median, and mode of the magnitude of the earthquake? Which measure is the most representative? **7.76; 7.7; 7.4; mean**
18. Draw a line plot for the information in the chart. **Check students' work.**
19. Make a frequency table of the information in the chart. **Check students' work.**
20. Explain why a box and whisker plot is not helpful in understanding the information in the chart. **The information is too closely packed for a b-w plot to be helpful.**

TEST YOURSELF

Use the data below for Exercises 1–3.

Average Monthly Temperatures(°F) for Washington, DC

44, 45, 55, 65, 75, 83, 86, 84, 78, 67, 56, 45

1. Find the mean, median, mode, and range. **65.3; 66; 45; 42**
2. Draw a stem and leaf plot. **Check students' work.**
3. Draw a box and whisker plot. **Check students' work.**

Use the data below for each exercise.

3, 2, 5, 7, 2, 4, 3, 1, 2, 5, 3, 4

4. Draw a line plot. **Check students' work.**
5. Make a frequency table. **Check students' work.**

Lesson Follow-up

Reteaching Activity For students having difficulty evaluating data, refer to the problem solving activity on p. 486E of the Chapter Overview. See also, TRB Exploring O-46.

LESSON RESOURCES

TRB Practice L-7
TRB Enrichment L-8

Practice

For use after 12-4 (pp. 502–505)

Box and Whisker Plots

Make a box and whisker plot for each set of data.

1. 16, 20, 30, 15, 23, 11, 15, 21, 30, 29, 13, 16
2. 9, 12, 10, 3, 2, 3, 9, 11, 5, 1, 10, 4, 7, 12, 3, 10
3. 70, 77, 67, 65, 79, 82, 70, 68, 75, 73, 69, 66, 70, 73, 89, 72

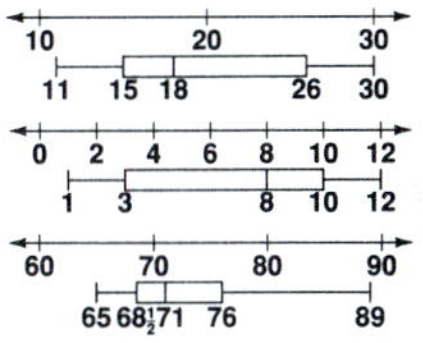

Use the box and whisker plot to answer each question.

Weekly Mileage Totals, 24 Runners

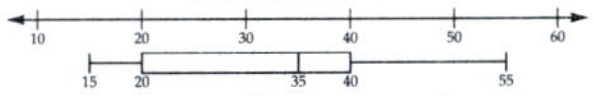

4. What is the highest weekly total? **55** the lowest? **15**
5. What is the median weekly total? **35**
6. What percent of runners run less than 40 miles a week? **75%**
7. How many runners run less than 20 miles a week? **6**

WRITE Make a box and whisker plot for each set of data. Use a single number line. Then write a comparison.

8. 1st set: 7, 12, 25, 3, 1, 29, 30, 7, 15, 2, 5, 10, 29, 1, 10, 30, 18, 8, 7, 29
   2nd set: 37, 17, 14, 43, 27, 19, 32, 1, 8, 48, 26, 16, 28, 6, 25, 18

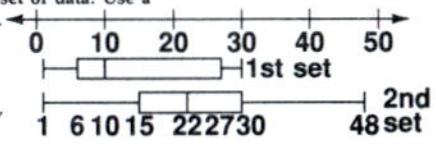

The data in the second set is more spread out, yet 50% of the data is between 15 and 30. In the first set, the data is about evenly spaced for the first, second and fourth quartile. The third quartile is 3 times larger than any other.

Chapter 12 L-7

Enrichment

For use after 12-4 (pp. 502–505)

Plot the Answer

To answer each of the following questions, make a box and whisker plot for the given set of data. Then write the first quartile, the median, and the third quartile in the answer boxes.

Example

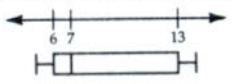

6 7 1 . 3

Check students' box and whisker plots.

1. What was the year of the birth of Pythagoras, the discoverer of the Pythagorean Theorem?
   Data: 6, 10, 2, 7, 6, 12, 8, 6, 4, 13, 2, 6
   5 6 9 B.C.
2. The Caspian Sea is the largest lake in the world. What is its area?
   Data: 38, 7, 50, 17, 11, 39, 25, 55, 8, 27, 42, 5, 37, 57, 46, 23
   1 4 3 , 2 4 4 mi^2
3. The Pacific Ocean, the world's largest, is also the deepest. What is the average depth of the Pacific Ocean?
   Data: 26, 53, 32, 0, 33, 1, 13, 1, 36, 34, 41, 0
   1 2 , 9 3 5 ft
4. The greatest recorded snowfall in the United States in a single year fell at Mount Rainier, Washington, in 1971–1972. What was the total amount of snow that fell?
   Data: 58, 9, 14, 33, 60, 12, 15, 37 59, 13, 60, 42, 40, 7, 12, 4
   1 , 2 2 4 . 5 0 in.

L-8 Chapter 12

LESSON QUIZ

Make a box and whisker plot for each set of data.

1. 1 4 5 5 6 8 10 11 12 12 13 15 (lower quartile 5, median 9, upper quartile 12)

2. Make a box and whisker plot for each set of data. Use a single number line. Then write a comparison.

1st Set: 27 23 28 32 35 39 41 37 44 36 40 45
2nd Set: 28 33 40 22 40 31 23 37 38 29 42 25
(*1st Set:* quartiles 30 and 40.5, median 36.5; *2nd Set:* quartiles 26.5 and 39, median 32; both sets are similar and ranges are about the same.)

Assignment Guide

Basic 1-2, 5-6, 10-11, 13-18, MR, TY All
Average 2-3, 6-7, 10-11, 13-18, MR, TY All
Enriched 3-4, 8-10, 12-18, MR, TY All

FOR THE NEXT LESSON

number cubes, spinners, counters, coins, math journal

Lesson Focus

Materials

Number cubes, spinners, counters, coins, math journal

MOTIVATING THE LESSON

Ask students to describe a *fair* and an *unfair* game. (In a fair game, every player of equal skill has an equal chance of winning; in an unfair game, some players will win more often, regardless of their ability.)

Draw these spinners on the chalkboard:

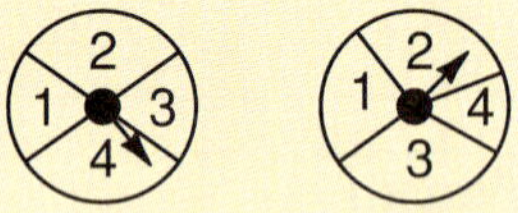

Ask students which spinner they would choose if they can win a game only when the number 3 is spun, and explain why. (the one on the right, because 3 occupies a larger area on that spinner than it does on the left spinner) Then ask students if they think the spinner on the right is a fair spinner. (no)

Instruction

Have students work in pairs. Give each pair two number cubes, two counters, a spinner, and three coins. Have students analyze the rules for each game, and before they play the game, guess whether the game will be fair or unfair and why. Students should record their predictions in their math journals.

(1) To help students identify all the possible outcomes, encourage them to be systematic. One approach is to list all the possible combinations, starting with 1 × 1, 1 × 2, etc.

OBJECTIVE:
To explore whether a game is fair or unfair.

MATERIALS

- Number cubes
- Spinners
- Counters
- Coins
- Math journal to record work

Exploring Fair and Unfair Games

▼ Playing a game is fun if the game is fair. It helps to analyze the rules of a game to see if each player has an equally likely chance of winning. Read the rules of the game below.

The Good Times and the Bad Times

Players: Player A and Player B

Materials: Two number cubes

Rules:
- Players take turns tossing the number cubes and then finding the product of the numbers on the two cubes.
- If the product of the numbers is even, Player A scores a point. If the product of the numbers is odd, Player B scores a point.
- The player with the most points at the end of 20 rounds is the winner. A round consists of each player tossing the number cubes once.

1. ***Discuss*** Is one player more likely to win the game than the other player? ***Explain***. **Answers may vary.**
2. Choose a partner and play the game four times. Record your results.
3. ***Discuss*** Which player won more games? Based on your results, do you think each player has an equal chance of winning?
4. Make a list or draw a diagram to find all possible outcomes. (1)
   a. How many possible outcomes are there? **36**
   b. In how many ways can a player toss a product that is even? a product that is odd? **27; 9**
5. ***Analyze*** Is the game fair or unfair? ***Explain***. **Unfair. The chances of A winning are greater than for B to win.**
6. ***Discuss*** How could you change the game to make it fair? Use lists or diagrams to support your conclusions. **A possible answer: Instead of products, find the sum of the numbers.**
7. (2) ***Discuss*** How do other factors, such as the age and experience of the players, contribute to making a contest fair or unfair? Is a game of basketball fair as long as all players play by the same rules? Is a spelling bee always fair? **Answers may vary.**

Teaching Tip

Have available the rules for several well known games. (e.g., tic-tac-toe, checkers, chess) Ask students to analyze the rules to decide if some players benefit more than others.

8. ***Analyze*** each game. Make a conjecture about whether the game is fair or unfair. Explain your reasoning. Then play each game.

Match or Not

Players: Player A and Player B

Materials: Two counters. One counter with both sides labeled **X** and the other with one side labeled **X** and the other side labeled **Y**.

Rules:
- Each player takes turns tossing the counters.
- If both counters match, Player A gets one point. If the counters do not match, Player B gets the point.
- The first player to get 20 points wins.

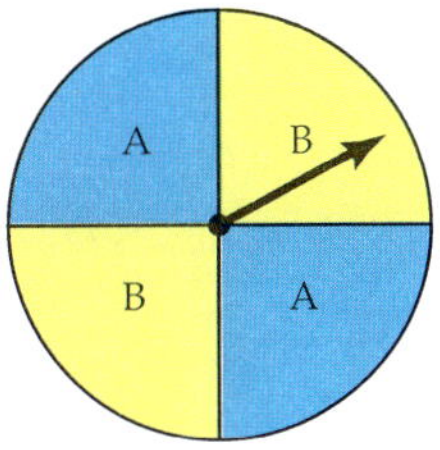

Spin Around

Players: Player A and Player B

Materials: Spinner

Rules:
- Each player spins the spinner twice.
- Player A scores if the spinner lands on the same letter both times. Player B scores if the spinner lands on different letters each time.
- The player with the most points after each player has had 50 spins is the winner.

A Lucky Trio of Coins

Players: Player A and Player B

Materials: Three coins

Rules:
- Each player takes turns tossing three coins.
- If all three coins show tails or all three coins show heads, Player A gets one point. If not, Player B gets the point.
- The player with the most points after 20 tosses is the winner.

9. ***Summarize*** Which of the three games is the most fair? the most unfair? ***Explain***. Support your conclusions with lists or diagrams.

3 Match or Not and Spin Around are both fair. Each player has the same chance to win. A Lucky Trio is unfair. There are only 2 out of 8 ways to get a match with all three coins.

10. ***PROJECT*** Make up a fair game and an unfair game. Use number cubes, coins, or spinners. Trade with a partner. ***Analyze*** the rules of each game. Decide which is fair and which is unfair. Play the games to test your conjectures.

2 Discuss questions 6-7 with the entire class.

3 Have pairs of students discuss their conclusions with other pairs. Students should be able to justify their conclusions.

Closure

Writing in Math Have students compare, in their math journals, their predictions and conclusions about the games. Ask them to identify factors that mislead them. Then ask students to describe what is needed for a fair game. (Possible answers: equal chance, equal opportunities to play)

Assignment Guide
Basic All
Average All
Enriched All

FOR THE NEXT LESSON
paper, pencil, calculator, math journal, scissors, tape

Lesson Focus

Materials

Paper, pencil, calculator, math journal, scissors, tape

Vocabulary/Symbols

counting principle
factorial (!)

MOTIVATING THE LESSON

Ask students to list all the three-digit numbers that can be formed using each of the digits 1, 2, and 3, once in each number. (123, 132, 213, 231, 312, 321) Mention that in this lesson they will learn a more efficient method than listing for determining the number of different choices from a large number of possibilities.

Instruction

1 To ensure that students understand how to make a tree diagram, have them make a diagram with the branches spreading down, to show the available course selections.

2 Have students make an organized list of all the possible arrangements. Then ask for volunteers to act out different ways for filling the positions.

3 Ask students to write the answer to Example 1 in factorial notation. (4!)

4 Ask students why it is necessary to multiply the numerator and denominator by 4, 3, 2, and 1. (To write the numerator as 9!, it was necessary to multiply it by 4, 3, 2, and 1; then the denominator had to be multiplied by the same sequence of numbers to have a fraction equivalent to the original product of the factors.)

OBJECTIVE: *To calculate the number of outcomes generated by a given event.*

12-5 Counting Principle

1

THINK How could you make a tree diagram with the branches spreading down rather than across?

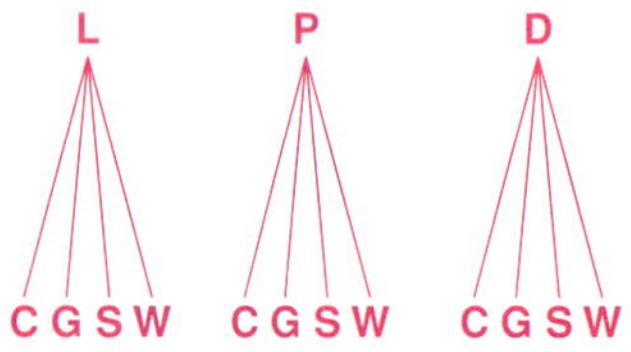

▼ Suppose you can fulfill your English requirement by taking Literature (L), Poetry (P), or Drama (D) first semester and Composition (C), Speech (S), Grammar (G), or Creative Writing (W) second semester. How many different ways can you choose?

You can draw a tree diagram to display the possible choices.

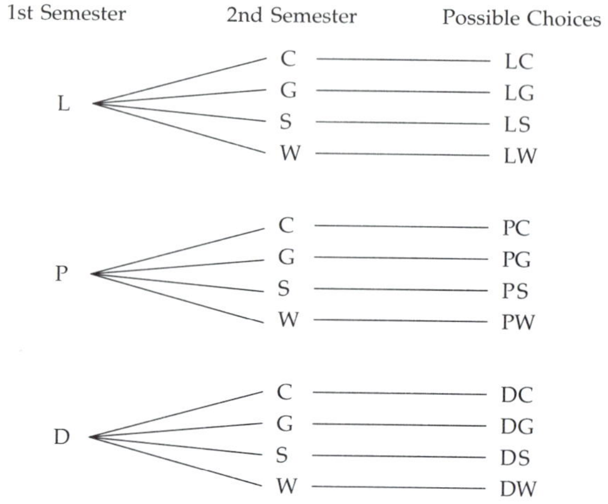

▼ You can also use the *counting principle.*

Number of Choices First Semester		Number of Choices Second Semester		Possible Choices
3	×	4	=	12

THINK Could you use the counting principle for more than two events? Explain. **Yes. If there are 3 events a, b, and c, the number of outcomes would be a · b · c.**

Counting Principle The number of outcomes for an event with two or more stages equals the product of the number of outcomes at each stage.

▼ You can use the counting principle when you must make choices in an ordered arrangement.

2

Example 1 Four Olympic gold medalists will pose together for a promotional photograph. How many different ways can they stand side-by-side in the photo?

Solution

1st Position		2nd Position		3rd Position		4th Position
4 choices		3 choices		2 choices		1 choice
4	×	3	×	2	×	1 = 24

Teaching Tip

To evaluate an expression written in factorial notation, students can use a scientific calculator with a factorial key. Students can also use a standard calculator, although this method requires a repeated use of the multiplication key.

▼ *Factorials* are a mathematical shorthand for situations such as finding the product of numbers in an ordered arrangement.

Example 2 **Find the value of 6 factorial.**

Solution $6! = 6 \times 5 \times 4 \times 3 \times 2 \times 1 = 720$

Factorial	A factorial is the product of all whole numbers from n to 1. We write this as $n!$

3

THINK Will 10^n be greater than or less than $n!$? For $n \le 24$, $10^n > n!$ For $n > 24$, $10^n < n!$ $10^{25} = 1 \times 10^{25}$ $25! \approx 1.55 \times 10^{25}$

▼ You can use factorial expressions for products of all whole numbers from n to m, where m is a whole number less than n. Algebraically, this is $n(n-1)(n-2)(n-3)\ .\ .\ .\ (m)$.

Example 3 **Express $9 \cdot 8 \cdot 7 \cdot 6 \cdot 5$ as a factorial.**

Solution

$\frac{9 \times 8 \times 7 \times 6 \times 5}{1}$ Express as a fraction.

$\frac{9 \times 8 \times 7 \times 6 \times 5 \times 4 \times 3 \times 2 \times 1}{4 \times 3 \times 2 \times 1} = \frac{9!}{4!}$ Multiply numerator and denominator by $4 \cdot 3 \cdot 2 \cdot 1$.

$9 \cdot 8 \cdot 7 \cdot 6 \cdot 5 = \frac{9!}{4!}$

4

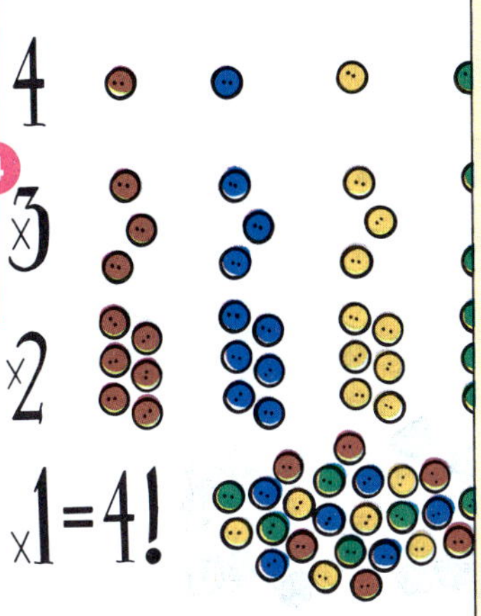

Example 4 **Evaluate $\frac{10!}{6!}$.**

Solution

$\frac{10!}{6!} = \frac{10 \times 9 \times 8 \times 7 \times 6 \times 5 \times 4 \times 3 \times 2 \times 1}{6 \times 5 \times 4 \times 3 \times 2 \times 1}$ Divide common terms.

$\frac{10!}{6!} = \frac{10 \times 9 \times 8 \times 7}{1}$ Simplify.

$\frac{10!}{6!} = 5{,}040$

CLASS EXERCISES

Find the value of each factorial.

1. $4!$ 24
2. $8!$ 40,320
3. $\frac{11!}{8!}$ 990
4. $\frac{22!}{17!}$ 3,160,080

Write each expression as a factorial.

5. $5 \cdot 4 \cdot 3 \cdot 2 \cdot 1$ $5!$
6. $12 \cdot 11 \cdot 10 \cdot 9 \cdot 8 \cdot 7$ $\frac{12!}{6!}$
7. $15 \cdot 14 \cdot 13 \cdot 12$ $\frac{15!}{11!}$
8. $8 \cdot 7 \cdot 6$ $\frac{8!}{5!}$

Solve. Make a tree diagram to check your answer.

9. There are 6 roads leading from Seymour to Clarksville and 3 roads leading from Clarksville to Belleview. How many possible routes are there from Seymour to Belleview through Clarksville? 18

THINK AND DISCUSS See below.

1. When is it impractical to make a tree diagram?
2. How is using factorials a modification of the counting principle?
3. Does $6! - 2! = 4!$? no

1. when there is a large number of choices
2. Factorials require that the number of choices decrease by 1 for each successive choice. The counting principle is not limited in this way.

ADDITIONAL EXAMPLES

1. Five math books are to be placed on a shelf. How many ways can the books be arranged? (120 ways)
2. Evaluate. 7! (5,040)
3. Evaluate. $\frac{8!}{3!}$ (6,720)
4. Express $10 \cdot 9 \cdot 8$ as a factorial. $\left(\frac{10!}{7!}\right)$

Guided Practice

Class Exercises After students have solved questions 9 and 10, ask them to contrast this use of the counting principle with the ordered arrangement in Example 1.

Think and Discuss Have students explain how they arrived at their conclusion for question 3.

A ***common error*** occurs when students incorrectly express a sequence of numbers that does not end in 1 as a factorial. For example, students may express the sequence in Example 3 as 9!. Remind students that the last factor in the numerator must be 1 and that once the sequence of factors for the numerator has been determined, the same sequence of factors must be used in the denominator to evaluate the sequence as a factorial. To help students avoid this error, have them find the value of 8! and then determine the sequence of factors needed to write an expression like $8 \cdot 7 \cdot 6$ as a factorial.

Closure

Have students write the meanings of these expressions in their math journals: $5!$, $\frac{5!}{5}$, $\frac{5!}{3!}$. Then have them write a problem that can be solved with a tree diagram and include the solution.

MATH MINUTES

Have students fill in the boxes with the appropriate number. Not all numbers are in factorial notation.

$\frac{9!}{\boxed{8!}} = \frac{4! \cdot 5!}{\boxed{320}} = 9$

$3! \cdot 5! \cdot \boxed{1} = 6! \cdot \frac{8!}{\boxed{8!}} = 720$

Independent Practice

For questions 9-12, review the use of the factorial key. If students do not use the factorial key, ask them to record their key strokes.

Have students show their work for questions 17-20.

Writing in Math Have students write their answers to question 28 in their math journals.

For questions 29-31, encourage students to create situations analogous to those provided in the data and to develop related problems.

As an extension, refer to the calculator activity on p. 486E of the Chapter Overview. See also, TRB Exploring O-47.

Critical Thinking Have students work in small groups. Have groups solve question 1 and then have them make a cube and test their conclusion. The groups can then solve question 2, using what they learned in question 1.

10. A student has 4 blouses and 5 skirts. How many different blouse skirt combinations can she wear? 20

MIXED REVIEW

Write using scientific notation.

1. 100,000,000,000 1×10^{11}

2. 0.0001 1×10^{-4}

Use the data to solve the problem: 11, 15, 18, 19, 22, 27, 30, 8, 45.

3. Find the median. 19

4. Draw a box and whiskers plot for the data. Check students' work.

Simplify.

5. $\frac{9}{27}$ $\frac{1}{3}$ **6.** $\frac{40}{88}$ $\frac{5}{11}$

Solve.

7. A storekeeper marks a \$38 sweater 25% off. What is the new price? \$28.50

WRITTEN EXERCISES

Find the value of each factorial.

1. 5! 120 **2.** 7! 5,040 **3.** 10! 3,628,800 **4.** 2! 2

5. $\frac{5!}{2!}$ 60 **6.** $\frac{6!}{3!}$ 120 **7.** $\frac{10!}{5!}$ 30,240 **8.** $\frac{13!}{7!}$ 1,235,520

CALCULATOR **Find the value of each factorial.**

9. 12! 479,001,600 **10.** $8! \times 3!$ 241,920 **11.** $\frac{15!}{10!}$ 360,360 **12.** $9! \cdot 9$ 3,265,920

MENTAL MATH **Find the value of each factorial.**

13. $\frac{6!}{5!}$ 6 **14.** $\frac{100!}{99!}$ 100 **15.** $\frac{10!}{8!}$ 90 **16.** 3! 6

Tell whether each is true or false. If it is false, explain why.

17. $5! \times 2! = 5! + 5!$ T **18.** $5! \times 2! = 10!$ F

19. $\frac{12!}{4!} = 3!$ F **20.** $\frac{5!}{4!} = 5$ T

Solve using a tree diagram.

21. There are 3 ways of performing Task A. There are 4 ways of performing Task B. How many ways are there of performing Task A and then Task B? 12

22. There are 8 roads leading from Marsh to Taft and 5 roads leading from Taft to Polk. How many possible routes are there to take from Marsh to Polk through Taft? 40

Solve using the counting principle.

23. A student has 5 pairs of pants, 8 shirts, and 2 ties. How many different outfits does he have to choose from? Each pant/shirt/tie combination is considered a different outfit. 80

24. You can buy a pizza with thin crust or thick crust. You have a choice of six toppings. How many different combinations can you make if you choose one type of crust and one type of topping for your pizza? 12

25. An automobile manufacturer makes 4 different car styles. Each style comes in 11 different colors. Each car can have 5 different interior styles and automatic or standard transmission. Jamal wishes to order one of each kind of car for his car lot. How many cars must Jamal order? 440

26. You wish to have your picture taken with 5 friends. In how many ways can you line up for the photograph? 120

27. There are seven people eligible for three different positions on the student council. Their names have been placed in a paper bag.

a. How many people are eligible for the first position? 7

b. How many people are eligible for the second position once the first has been selected? 6

c. How many people are eligible for the third position once the first two have been selected? 5

d. In how many ways can three people out of seven be selected for three positions on the student council? 210

28. ***WRITE*** a paragraph defining the counting principle. Give an example of a situation that uses the principle to solve a problem. Check students' work.

Use the *DATA* at the right to determine how many different costumes each character could wear in the class play.

29. A hobo must wear a hat, a jacket, and carry a suitcase. 36

30. A scarecrow must wear a hat, a scarf, and a jacket. 24

31. The mystery person must wear a hat, a scarf, and carry a suitcase. 18

Costume Props

hat	*scarf*
straw	knit
baseball	kerchief
derby	
Jacket	***Suitcase***
denim	satchel
plaid	briefcase
striped	duffel bag
leather	

Critical Thinking

EXPLORING VISUAL THINKING

1. Look at the three views of a number cube shown below. What number is opposite the number 6? 2

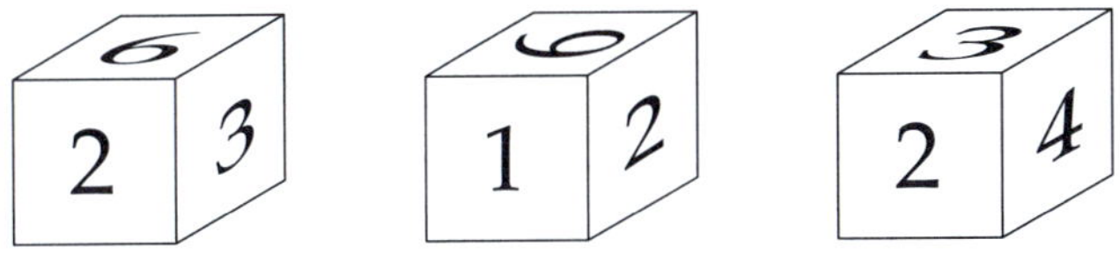

2. Look at the cube below. Side A is red, side B is blue, side C is green, and side D is yellow. What color is opposite side A? red

PROBLEM SOLVING HINT

1. Draw a net.

2. A number may be repeated.

Lesson Follow-up

Reteaching Activity Review ratios of factorials. Then have students act out situations, such as choosing three people from a group of five to stand side-by-side. Then have students find the value of the factorial.

LESSON RESOURCES

TRB Practice L-9
TRB Enrichment L-10

Practice

For use after 12-5 (pp. 508–511)

Counting Principle

Find the value of each factorial.

1. 3! 6 2. 6! 720 3. 4! 24 4. $\frac{7!}{5!}$ 42

5. $\frac{8!}{7!}$ 8 6. $\frac{6!}{2!}$ 360 7. $\frac{3!}{4!}$ $\frac{1}{4}$ 8. $\frac{12!}{10!}$ 132

CALCULATOR **Find the value of each factorial.**

9. 11! 39,916,800 10. 4! × 7! 120,960

11. $\frac{13!}{5!}$ 51,891,840 12. $\frac{5!8!}{6!}$ 6,720

MENTAL MATH **Find the value of each factorial.**

13. $\frac{5!}{3!}$ 20 14. $\frac{18!}{17!}$ 18 15. $\frac{11!}{9!}$ 110

Solve each problem using a tree diagram.

17. A ballot offered 3 choices, for president (*A*, *B*, *C*) and 2 choices for vice president (*M*, *N*). How many choices for a combination of the two offices did it offer? List them.
6; *AM, AN, BM, BN, CM, CN*

18. The Cougar baseball team has 4 pitchers (*P*1, *P*2, *P*3, *P*4) and 2 catchers (C1, C2). How many pitcher-catcher combinations are possible? List them.
8; *P1C1, P1C2, P2C1, P2C2, P3C1, P3C2, P4C1, P4C2*

Solve each problem using the counting principle.

19. There are 5 roads from Allen to Baker, 7 roads from Baker to Carlson, and 4 roads from Carlson to Dodge. How many different routes from Allen to Dodge by way of Baker and Carlson are possible?
140 routes

20. Drapery is sold in 4 different fabrics. Each fabric comes in 13 different patterns. Each pattern is offered in 9 different colors. How many fabric-pattern-color combinations are there?
468 combinations

Chapter 12 L-9

Enrichment

For use after 12-5 (pp. 508–511)

Counting Trails

The diagram shows trails leading from Camp Alpha (*A*), where you are camped, to the river. You may walk only in a southerly direction.

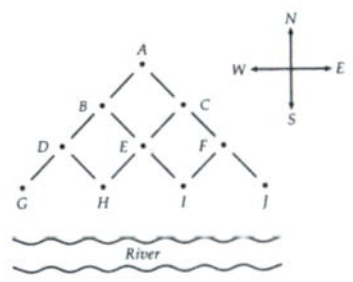

How many routes can you choose to get from *A* to:

1. *B*? 1 2. *C*? 1 3. *D*? 1

4. *F*? 1 5. *G*? 1 6. *J*? 1

7. Find the number of routes you can choose to get from *A* to *E*. Number (*A*-*E*) 2

8. Find the number of choices you have for walking from *A* to *H* and from *A* to *I*. Number (*A*-*H*) 3
Number (*A*-*I*) 3

Use the above method to find the number of trails from Camp Alpha to the Rainbow Desert. Write the number of choices to each point in the open circles.

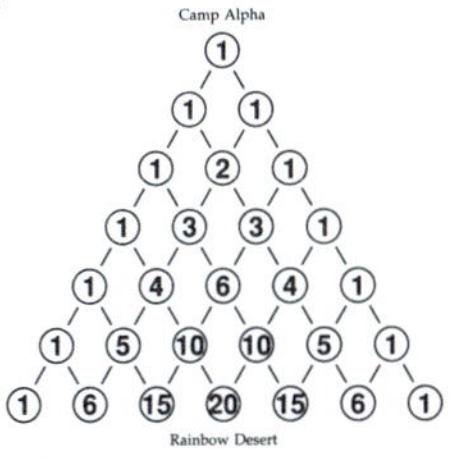

You may recognize the triangle of numbers you have written as Pascal's triangle, an array with wide applications in mathematics.

L-10 Chapter 12

LESSON QUIZ

1. If you can choose from 4 activities for the morning and 5 for the afternoon, how many combinations of activities can you choose for the day? (20)

2. Evaluate. 9! (362,880)

3. Evaluate. $\frac{11!}{8!}$ (990)

4. Express 12 · 11 · 10 · 9 · 8 as a factorial. ($\frac{12!}{7!}$)

FOR THE NEXT LESSON

paper, pencil, math journal, spinners, number cubes, coins

Assignment Guide

Basic 1-4, 9-10, 13-14, 17-18, 22-24, 27-31, MR, CT All

Average 3-6, 10-11, 14-15, 18-21, 24-31, MR, CT All

Enriched 5-8, 11-12, 15-16, 19-20, 22, 25-31, MR, CT All

Lesson Focus

Materials

Paper, pencil, math journal, spinners, number cubes, coins

Vocabulary/Symbols

event
probability
random event
sample space

MOTIVATING THE LESSON

Begin the lesson by discussing actual insurance rates for different types of cars. Discuss why premiums differ. (for example, the chance, or probability, of that type of car being involved in an accident)

Instruction

1 Have students explain what is meant by equally likely outcomes. (outcomes that have equal chances of occurring)

2 Introduce the concept $P(E) + P(\text{not } E) = 1$ by having students find sums of probabilities such as in tossing a coin, and getting heads, $P(\text{heads}) + P(\text{not heads})$, or $P(\text{tails})$, $(\frac{1}{2} + \frac{1}{2} = 1)$ and in tossing a number cube and getting a 3, $P(3) + P(\text{not } 3)$. $(\frac{1}{6} + \frac{5}{6} = 1)$

OBJECTIVE: *To calculate the probability of occurrence of a given event.*

12-6 Probability

Car owners pay insurance premiums to protect against loss or damage. If there is an accident, the insurance company pays damages. Usually the premium is less than the value of the car.

To set the fee, mathematicians use *probability* theory to review past claims payouts and estimate future payouts. Then they use the expected payout to determine insurance rates.

Probability	Probability is the likelihood that a certain *event*, or set of outcomes, will occur. $P(E) = \frac{\text{number of favorable outcomes}}{\text{total number of possible outcomes}}$
Sample Space	The set of possible outcomes is the sample space.

Example 1 When tossing a number cube, there are six possible outcomes, 1, 2, 3, 4, 5, and 6.

THINK How many favorable outcomes are there for the event of tossing an even number? 3

Probability relies on events occurring *randomly*. All outcomes must be equally likely to occur for the calculations to be valid.

Example 2 What is the probability that a letter chosen at random from the word **MISSISSIPPI** is the letter I?

1

Solution

1. Find the size of the sample space. — Number of letters → 11
2. Find the number of favorable outcomes. — Number of I's → 4
3. Find the probability. — $P(I) = \frac{\text{Number of I's}}{\text{Number of letters}} = \frac{4}{11}$

The probability of choosing the letter I is $\frac{4}{11}$.

You can find the probability for more than one favorable outcome.

Example 3 You draw a card at random from a hat containing cards numbered 1–6. Find the probability that the card is even.

Solution

1. Find the size of the sample space. — Cards → 6
2. Find the favorable outcomes for the event. — Even cards, (2, 4, 6) → 3.
3. Find the probability. — $P(\text{even}) = \frac{\text{Even cards}}{\text{All cards}} = \frac{3}{6} = \frac{1}{2}$.

The probability of drawing an even card is $\frac{1}{2}$.

NOTES & QUOTES

In the small number of things we are able to know with any certainty, the principal means of ascertaining truth are based on probabilities.
—Pierre Simon de Laplace (1749–1827)

Teaching Tip

To help students better understand the concepts of random events and probabilities, have concrete objects such as spinners, number cubes, and coins available for students to use to compare results with the theoretical probabilities.

▼ You can determine the probability of an event *not* occurring.

Example 4 There are 3 blue marbles, 2 yellow marbles, and 4 red marbles in a bag. What is the probability that a marble chosen at random is *not* a red marble?

Solution

1. Find the size of the sample space. — Number of marbles → 9
2. Find the favorable outcomes. — Number of not red marbles → 5
3. Find the probability. — $P(\text{not red}) = \frac{\text{not red}}{\text{all marbles}} = \frac{5}{9}$

The probability of *not* drawing a red marble is $\frac{5}{9}$.

Getting a red marble or not getting a red marble covers all the possibilities. Their sum must be 1. You can subtract one of them from 1 to find the other.

2

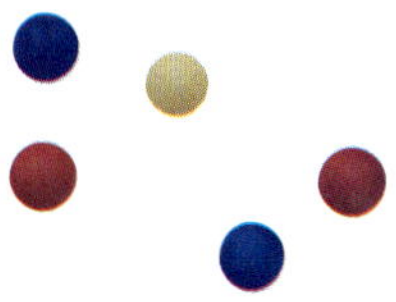

THINK The probability of getting a red marble is $\frac{4}{9}$. Why is $P(\text{not red}) = 1 - P(\text{red})$?

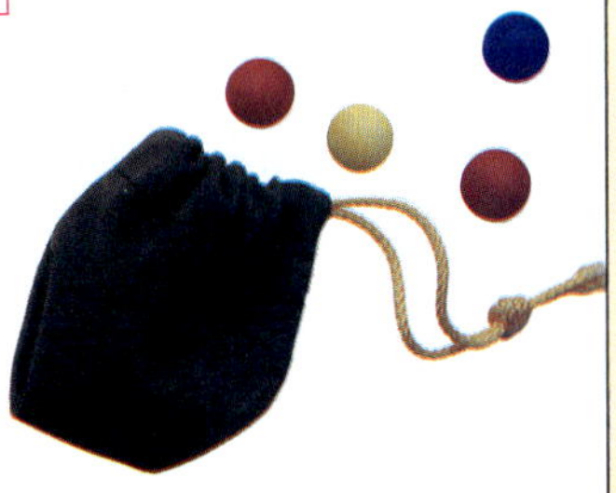

CLASS EXERCISES

Use the word ARKANSAS to answer each question.

1. You wish to know the probability of selecting the letter A.
   a. What is the number of possible outcomes in the sample space for selecting the letter A? 8
   b. What is the number of favorable outcomes for the event of selecting the letter A? 3
   c. What is the probability of selecting the letter A? $\frac{3}{8}$
   d. What is the probability of *not* selecting the letter A? $\frac{5}{8}$
2. What is the probability of selecting a vowel? $\frac{3}{8}$
3. What is the probability of selecting the letter C? 0

Mon., Tues., Wed., Thur., Fri., Sat., Sun.

Find each sample space.

4. choosing a Monday from all the days of the week
5. choosing the letter X from all the letters in the alphabet a, b, c . . . z

Find each probability.

6. What is the probability that a digit selected at random from the number 364,892 is a multiple of 3? $\frac{3}{6} = \frac{1}{2}$
7. A math class has 10 boys and 15 girls. What is the probability that a student chosen at random is a girl? $\frac{15}{25} = \frac{3}{5}$
8. In a class of 24 students, 8 are saving to buy a camera. What is the probability that a randomly selected student is *not* saving to buy a camera? $\frac{16}{24} = \frac{2}{3}$
9. Find the probability that a student chosen at random from your math class has blue eyes. Answers may vary.

1. 1; Answers may vary.
2. 0; Answers may vary.

THINK AND DISCUSS

1. What is the probability of an event that is certain to occur? Give an example of such an event.
2. What is the probability of an event that is impossible? Give an example of such an event.
3. Can a probability be greater than 1? Explain. no
4. Can a probability be less than 0? Explain. no

MATH MINUTES

Present these possible outcomes: 1, 2, d, f, 3, 3, 4, a, a, a, f, 8, 9, 0 10, m, c. Working in pairs, one student names an event, the other gives the probability of that event. Students then reverse roles.
(Check students' work.)

ADDITIONAL EXAMPLES

1. Find the probability that a letter selected at random from the word DICTIONARY is an I. ($\frac{2}{10}$ or $\frac{1}{5}$)
2. Ten cards are numbered 1, 2, 3, . . . , 10. What is the probability that a card chosen at random contains a prime number? ($\frac{4}{10}$ or $\frac{2}{5}$)
3. In a bag of beans, 3 are black, 5 are brown, and 6 are red. What is the probability that a bean chosen at random is *not* red? ($\frac{8}{14}$ or $\frac{4}{7}$)

Guided Practice

Class Exercises Have students work through the exercises orally to assess their understanding of probability.

Think and Discuss Foreshadow the next lesson by asking students to find the probability of obtaining an even number or a prime number by tossing a fair number cube. ($\frac{5}{6}$) Then ask students to explain why $P(\text{even or prime}) \neq P(\text{even}) + P(\text{prime})$. [There are six numbers on the cube. The number 1 is neither prime nor even, and the number 2 is both even and prime. So, $P(\text{even: 2, 4, and 6}) = \frac{1}{2}$, $P(\text{prime: 2, 3, and 5}) = \frac{1}{2}$; which equals 1 or $\frac{6}{6}$.]

A ***common error*** occurs when students do not remember that $P(E) + P(\text{not } E)$ must equal 1. Have them make two lists, one titled *favorable outcomes*, the other *unfavorable outcomes*, to show that $P(E) + P(\text{not } E) = 1$.

Closure

Writing in Math Have students write a description of the meaning of probability in their math journals.

Independent Practice

For question 15, have students poll their classmates and then compare the results of their polls to the poll results in the text.

As a variation for question 18, have students experiment with a foreign coin to see what effect, if any, it has on the outcome of the experiment.

Writing in Math Have students share their explanations for question 22 with the class.

MIXED REVIEW

Give each number.

1. opposite of -9 **9**
2. $|-5|$ **5**

Find the factorial.

3. 6! **720** 4. $\frac{7!}{3!}$ **840**

Use the data for the questions below: 8, 9, 5, 12, 4, 6, 4, 10.

5. Find the mean. **7.25**
6. Draw a box and whisker plot. **Check students' work.**

Write an equation. Then solve.

7. Students paid $850 for tickets to the Spring Fling. Each ticket cost $5. How many tickets did the students purchase? **$5x = 850$, $x = 170$**

WRITTEN EXERCISES

Find the probability of each event.

1. that a digit selected at random from the number 164,743 is a multiple of 2 $\frac{3}{6} = \frac{1}{2}$
2. that a randomly chosen month has 30 days $\frac{4}{12} = \frac{1}{3}$

Find each probability when a letter is chosen at random from the word MATHEMATICS.

3. choosing a consonant $\frac{7}{11}$
4. choosing the letter M $\frac{2}{11}$
5. choosing a letter that occurs more than once $\frac{6}{11}$
6. choosing the letter K. **0**

Find each probability.

7. What is the probability that a state selected at random from a list of the 50 United States begins with the letter M? $\frac{4}{25}$
8. A lab class has 8 boys and 10 girls. What is the probability that a student chosen at random is a boy? $\frac{8}{18} = \frac{4}{9}$
9. Students in a class were asked to name their preferred type of motor vehicle. Eight students preferred a pickup, 12 preferred a sports car, and eight preferred a convertible. What is the probability that a randomly selected student preferred a sports car? $\frac{12}{28} = \frac{3}{7}$
10. What is the probability that any letter of the alphabet is not included in the sentence *"The quick brown fox jumps over the lazy dog"*? **0**
11. There are 14 boys and 12 girls in a math class. There are 60 boys and 65 girls in the freshman class. In the school there are 375 boys and 360 girls. Find the probability that a female student is chosen at random.
    a. from the math class $\frac{12}{26} = \frac{6}{13}$
    b. from the freshman class $\frac{65}{125} = \frac{13}{25}$
    c. from the school $\frac{360}{735} = \frac{24}{49}$
    d. Are the answers for (a), (b), and (c) different? Why or why not? **Yes. The sample space is different for each question.**
12. Find the probability of each event when a day of the week is chosen at random.
    a. a day of the week has six letters $\frac{3}{7}$
    b. the day of the week has more than six letters $\frac{4}{7}$
    c. What is the sum of the answers to parts (a) and (b)? Explain. **1 All of the days have six or more letters.**

13. Find the probability of each event when a prime number less than 100 is chosen.

a. It has one digit. $\frac{4}{25}$ b. It has more than one digit. $\frac{21}{25}$

c. Why is the sum of parts (a) and (b) equal to 1? **Any prime number has 1 or more digits.**

14. The figure at the right illustrates a pattern of floor tiling. You drop a coin onto this portion of the floor tile. Find the probability for each event.

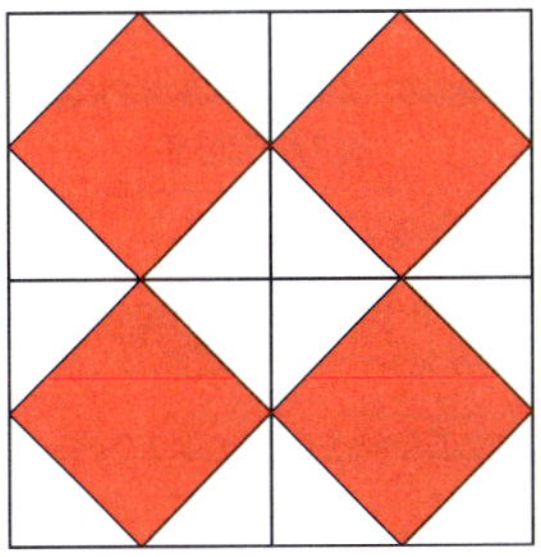

a. The coin lands on a red tile. $\frac{1}{2}$ b. The coin lands on a white tile. $\frac{1}{2}$

15. A group of students was polled to find out their favorite sports. Four of the students chose skiing, eight chose baseball, ten chose basketball, four chose football, and two chose track. Find the probability of a randomly chosen student preferring each event.

a. skiing $\frac{1}{7}$ b. baseball $\frac{2}{7}$ c. basketball $\frac{5}{14}$

d. football $\frac{1}{7}$ e. track $\frac{1}{14}$

f. What is the sum of the five probabilities? Explain. **1**

16. A set of flash cards is numbered from 1 to 36. A card is chosen at random. Find the probability for each event.

a. an even number $\frac{1}{2}$ b. a multiple of 3 $\frac{1}{3}$

c. a multiple of both 2 and 3 $\frac{1}{6}$ d. a multiple of 2 or 3 $\frac{2}{3}$

e. a prime number $\frac{11}{36}$ f. a square number $\frac{1}{6}$

17. A number is chosen at random. Find the probability of the last digit of its square being each number listed below.

a. 1 $\frac{1}{5}$ b. 4 $\frac{1}{5}$ c. 5 $\frac{1}{10}$ d. 6 $\frac{1}{5}$ e. 9 $\frac{1}{5}$

f. Why is the sum of these answers not equal to 1? **The event of having the digit 0 as the last digit is not included.**

18. **PROJECT** What is the probability that a coin will land with the head face up? Record your answer. **Check students' answers.**

a. Toss a coin 100 times. Record the number of times the coin lands face up. Find the probability of the coin landing face up.

b. Do the results of tossing the coin agree with the probability you expected? Why do you think there might be a difference?

Sweaters	
Color	*Style*
Blue	Cardigan
Pink	Pullover
Red	Vest
Brown	
Black	

Use the ***DATA*** at the right to solve.

19. What is the size of the sample space if each style comes in each color? **15**

20. What is the probability of choosing a blue sweater at random? $\frac{1}{5}$

21. What is the probability of choosing a sweater vest at random? $\frac{1}{3}$

22. **WRITE** Suppose you have a bag containing an equal number of nickels, dimes, and quarters. Suppose you reach into the bag and choose a coin. Is it equally likely that you will pick a dime or a quarter? Explain. **No. When you reach into the bag you can tell what kind of coin you are getting by its size.**

Lesson Follow-up

Reteaching Activity Students who have difficulty with complementary events, $P(E)$ and $P(\text{not } E)$, should begin by listing all favorable events and all unfavorable events separately.

LESSON RESOURCES

TRB Practice L-11
TRB Enrichment L-12

Practice

For use after 12-6 (pp. 512–515)

Probability

Find each probability.

1. A letter is chosen at random from the word PROBABILITY. Find the probability of each event.

a. selecting a *B* $\frac{2}{11}$ b. selecting a *P* $\frac{1}{11}$

c. selecting an *A* or an *I* $\frac{3}{11}$ d. not selecting a *P* $\frac{10}{11}$

2. A child is chosen at random from the Erb and Smith families. Find the probability that the child is:

	Erb family	Smith family
Girls	2	5
Boys	4	3

a. a boy $\frac{1}{2}$ b. an Erb $\frac{3}{7}$

c. an Erb girl $\frac{1}{7}$ d. a child **1**

e. not a Jones boy **1** f. a Jones **0**

3. A spinner numbered from 1 to 20 is spun randomly. Find the probability that the spinner lands on:

a. 17 $\frac{1}{20}$ b. an odd number $\frac{1}{2}$

c. a number divisible by 5 $\frac{1}{5}$ d. 26 **0**

e. a number with a 1 in it $\frac{11}{20}$ f. a prime number $\frac{2}{5}$

g. a square number $\frac{1}{5}$ h. a number **1**

i. a number that is not less than 17 $\frac{1}{5}$ j. a number divisible by 3 or 4 $\frac{1}{2}$

4. A drawer contains 6 red socks, 4 blue socks, and 14 white socks. A sock is pulled from the drawer at random. Find the probability that the sock is:

a. red $\frac{1}{4}$ b. blue $\frac{1}{6}$

c. red or white $\frac{5}{6}$ d. red, white, or blue **1**

e. not red $\frac{3}{4}$ f. green **0**

5. A box contains 7 red, 14 yellow, 21 green, 42 blue, and 84 purple marbles. A marble is drawn at random from the box. Find the probability that the marble is:

a. red $\frac{1}{24}$ b. yellow $\frac{1}{12}$

c. green or blue $\frac{3}{8}$ d. purple, yellow, or red $\frac{5}{8}$

Chapter 12 L-11

Enrichment

For use after 12-6 (pp. 512–515)

The Probability of π

You and a partner can conduct a probability experiment to approximate the value of π. First, recall this definition: Two numbers are *relatively prime* if their only common factor is 1.

Each partner should choose a 2-digit number at random from a phone book or the page numbers of a book. Record the numbers as Trial 1. Conduct as many trials as possible. The more pairs you choose, the closer will be your approximation to π. **Check students' tables.**

Trial	First number	Second number	Trial	First number	Second number
1			21		
2			22		
3			23		
4			24		
5			25		
6			26		
7			27		
8			28		
9			29		
10			30		
11			31		
12			32		
13			33		
14			34		
15			35		
16			36		
17			37		
18			38		
19			39		
20			40		

Now find the following. **Answers will vary.**

1. number of trials (*N*) ______ 2. number of relatively prime pairs (*R*) ______
3. To approximate π, find $\sqrt{\frac{6N}{R}}$ using your calculator. ______
4. To the nearest thousandth, how far off is your approximation? Use π = 3.142.

______

L-12 Chapter 12

LESSON QUIZ

1. A set of 12 cards is labeled with the names of the months. What is the probability that a card chosen at random contains the name of a month having more than 6 letters? ($\frac{1}{2}$)

2. What is the probability that a letter chosen at random from the word WINTER is a vowel? ($\frac{2}{6}$ or $\frac{1}{3}$)

3. What is the probability that the name of a state selected at random from the 50 states of the United States, does not begin with a consonant? ($\frac{12}{50}$ or $\frac{6}{25}$)

Assignment Guide

Basic 1, 3-4, 7-11, 18-22, MR All
Average 1, 4-5, 9-13, 18-22, MR All
Enriched 2, 5-6, 13-22, MR All

FOR THE NEXT LESSON

calculator, coins, math journal

Lesson Focus

Materials
Calculator, coins, math journal

Vocabulary/Symbols
Pascal's triangle

MOTIVATING THE LESSON

Have students toss a penny and a nickel. Ask volunteers for the outcomes. Write each different possible outcome as an ordered pair on the chalkboard. [(H,H), (H, T), (T, H), (T, T)] Mention that in today's lesson, students will be doing a coin toss experiment to discover a numerical pattern.

Instruction

Have students work in small groups. Distribute coins (pennies, nickels, dimes, and quarters) to each group. Ask students to write the responses to the questions in their math journals.

1 To ensure that students understand the notation, ask them which *ordered triple* shows that the penny and dime turned up heads and the nickel turned up tails. (H, T, H)

2 Point out the symmetry of each row (numbers that are equidistant from the endpoints of the row are the same). Ask students if they can find a pattern for the numbers, other than the numeral 1 appearing at the beginning and end of each row. (Each number is the sum of the two numbers that are closest to it in the row directly above it.)

OBJECTIVE:
To explore probabilities using Pascal's triangle.

MATERIALS

- Coins
- Math journal to record work

Exploring Pascal's Triangle

■ In probability problems where the underlying experiment has two equally likely outcomes, you can often find solutions using Pascal's triangle.

Problem If three fair coins are tossed, what is the probability of obtaining exactly two heads?

Plan You need to list all possible outcomes and count them to find the denominator. You need to identify the successful outcomes and count them to find the numerator. Then put these numbers into a fraction to write the probability.

$$P = \frac{\text{number of successful outcomes}}{\text{total number of outcomes}}$$

■ Suppose you have a penny, a nickel, and a dime. How can you list all possible outcomes?

1

1. List the set of all possible outcomes as ordered triples (a, b, c) where a is the outcome of the penny, b the nickel, and c the dime.

   (H,H,H), (H,H,T), (H,T,H), (H,T,T), (T,H,H), (T,H,T), (T,T,H), (T,T,T)

   There are eight possible outcomes.

2. Make a tree diagram in which you consider the outcomes one coin at a time.

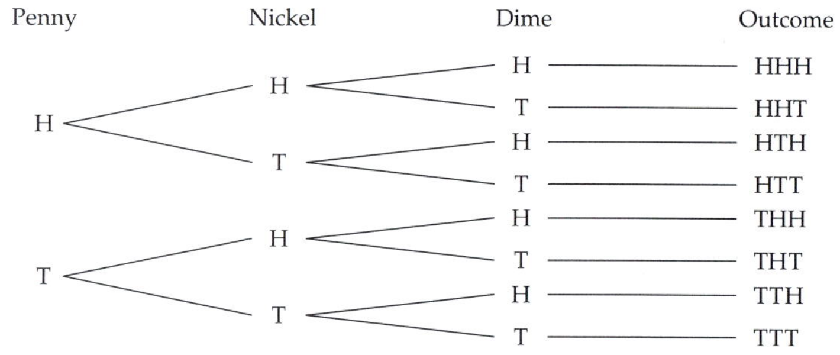

3. Identify the successful outcomes: HHT, HTH, THH.
4. Write the probability: $P(\text{two heads}) = \frac{3}{8}$.
5. Compare Row 3 in Pascal's triangle with the results of the coin tossing problem.

2

Row 0					1				
Row 1				1		1			
Row 2			1		2		1		
Row 3		1		3		3		1	
Row 4	1		4		6		4		1

Teaching Tip

The overhead projector is especially useful for illustrating the symmetry and construction of Pascal's triangle.

6. Find the sum of the numbers in Row 3. 8
7. How does the sum compare with the total number of outcomes in the coin-tossing problem? same
8. Look back at the list of outcomes for tossing three coins. In how many outcomes were there exactly zero heads? Notice that the first number in Row 3 is 1.
9. Look at the list of outcomes for tossing three coins. In how many outcomes were there exactly 1 head? Notice that the second number in Row 3 is 3. We know that there were 3 outcomes with exactly 2 heads. The third number in Row 3 is 3.
10. In how many outcomes were there exactly 3 heads? The fourth number in Row 3 is 1. Check students' answers.

■ The numbers in Row 3 of Pascal's triangle tell how many of the possible outcomes in tossing three coins represent exactly 0, 1, 2, or 3 heads.

Use Row 4 of Pascal's triangle to answer each question. Check your results by listing the possible outcomes of the experiment.

11. How many possible outcomes are there for the experiment consisting of tossing four different coins? 16
12. In how many ways can you obtain exactly 0, 1, 2, 3, or 4 heads? 1, 4, 6, 4, 1
13. What is the probability of obtaining exactly 2 heads when tossing four coins? $\frac{3}{8}$
14. What is the probability of obtaining exactly 3 heads when tossing four coins? $\frac{1}{4}$
15. What is the probability of obtaining exactly 3 tails when tossing four coins? $\frac{1}{4}$
16. In a family of four children, what is the probability that all four children are girls? *Hint:* There are two equally likely outcomes at the birth of each child, so the situation has the same mathematical structure as flipping a coin. $\frac{1}{16}$
17. In a family of five children, what is the probability that exactly two of the children are boys? $\frac{5}{16}$
18. ***Write*** a similar problem. Trade with a classmate and solve each other's problem. Check students' answers.

PROBLEM SOLVING HINT
Extend Pascal's triangle to Row 5.

Closure

Ask students to give two different probability problems—involving situations not used in the lesson—that can be solved by using Pascal's triangle. Have students also give two problems that cannot be solved using Pascal's triangle. (e.g., a true/false quiz; a spinner with four sections, red, yellow, blue, white)

Assignment Guide

Basic All
Average All
Enriched All

FOR THE NEXT LESSON

coins, index cards, math journal, spinner

Lesson Focus

Materials

Coins, index cards, math journal, spinner

Vocabulary

Monte Carlo simulation

MOTIVATING THE LESSON

Begin the lesson by discussing the idea of simulation. Ask a volunteer to look up the definition in a dictionary. Help students conclude that simulating a problem is "imitating" the situation, using a model and the same mathematical relationships. Have students note that a computer is a valuable tool for simulating problems.

Instruction

1 Have students restate the steps in their own words to ensure that they understand the terminology of statistics.

2 Stress that one trial consists of drawing the *five* pairs of cards. A successful outcome requires that at least two of the five pairs match.

PROBLEM SOLVING STRATEGY

READ PLAN SOLVE LOOK BACK

12-7 Simulate the Problem

OBJECTIVE: *To solve problems using simulation.*

You can solve many probability problems using a *Monte Carlo simulation*. Monte Carlo methods make it possible to model a situation for which a test is impractical and probability formulas are cumbersome. Use the steps below to develop a model.

1

1. Assign success to one outcome and failure to another.
2. Choose a random device, such as number cubes or a spinner.
3. Determine the probability of the outcome of one trial.
4. Decide the definition of a trial.
5. Perform a sufficient number of trials.
6. Compute the simulated probability $\frac{\text{number of successful trials}}{\text{number of trials}}$.

PROBLEM

You have been given a quiz written in ancient Sanskrit. You are expected to match five words with their definitions. What is the probability that you will get 1 out of 5 answers correct?

SOLUTION

READ	What do you want to find?	the probability that you will get 1 out of 5 answers correct
PLAN	Decide on a strategy.	Use a Monte Carlo simulation.
	What random device can you use to simulate the problem?	Use a set of cards with the letters a, b, c, d, and e to represent the words. Use a second set of cards with A, B, C, D, and E to represent the definitions.
	What is a *successful* outcome?	obtaining two or more correct answers by matching upper and lowercase letters
	What is the probability of a match on each draw?	Each probability is $\frac{1}{5}$.
	What will represent one *trial?*	drawing five pairs of cards, one card from each pile
SOLVE (2)	How many trials are sufficient?	at least 100 trials
	Tally the results.	One student completed 19 successes out of 100 trials.
	Give the expected probability.	$\frac{19}{100} = \frac{\text{number of successes}}{\text{number of trials}}$
LOOK BACK	Did you solve the problem?	The probability is estimated to be $\frac{19}{100}$.

Teaching Tip

Allow students to make up and solve their own simulation problems. Students are highly motivated to solve probability problems concerning situations about which they are most curious.

CLASS EXERCISES

1. Perform the experiment. **Check students' work.**
   a. Compare your results with those in the example.
   b. Why do you think the results of your experiment might differ from those of the student in the example?
2. Repeat the experiment to find the probability of getting two correct answers for each of these matching quizzes in Sanskrit.
   a. a 4-word quiz b. a 6-word quiz
   Check students' work.

1. तुल्य	A. half
2. गण्	B. first
3. अर्ध	C. equal
4. आदिम	D. count

WRITTEN EXERCISES

Use a CALCULATOR where appropriate.

Solve by simulating the problem.

1. Suppose you take a true-false test. You don't know the answers to any of the ten questions. What is the probability you will get 7 out of 10 correct? **Check students' work.**
   a. Model the situation using a coin. Let heads represent a true statement and tails represent a false statement.
   b. Let another student prepare an *answer key* by tossing the coin ten times and recording each *correct answer*.
   c. A trial occurs when you toss the coin ten times to represent the ten questions on the quiz. A successful trial occurs when you get seven or more answers that match the *answer key*.
   d. You need to try about 100 trials.
   e. Write the probability: $\frac{\text{number of successful trials}}{\text{number of trials}}$. **Check students' work.**
2. Suppose you take a ten-question multiple-choice test. Each question has four choices. You don't know any of the answers.
   a. What is the probability you will get 7 out of 10 correct? Model the situation using a spinner with four equal sections.
   b. Compare your answer with Exercise 1. Explain why it is different.
   c. What is the probability you will get 4 out of 10 correct?

Solve using any strategy.

3. What is the probability that exactly three children in a family of five children will be boys? $\frac{5}{16}$
4. What is the probability that exactly two children in a family of six children will be girls? $\frac{15}{64}$

MATH MINUTES

Have groups prepare cards for Written Exercise 2, to test the probability that they found and to construct a frequency table with the results. (Check students' work.)

Guided Practice

Class Exercises Have students form groups to perform the experiment. Then have them compare each group's results.

For question 2, you may wish to challenge those students with a knowledge of BASIC computer language to write a program that will simulate the problem.

A ***common error*** is to repeat the trials too few times. For example, it would have been inappropriate to repeat the trials only ten times in the original problem. Repeating trials 100 times is usually appropriate for solutions of exercises.

Closure

Writing in Math Ask students to summarize the solution process in the lesson. Then have them write the definition for Monte Carlo simulation and list the solution steps in their math journals.

Independent Practice

Have the class continue working in groups.

For question 1b, ask students to list the numbers 1 through 10 on a piece of paper. Have them record true (for heads), or false (for tails), next to each of the numbers as the coin is tossed. This list will represent the "answer key," against which each toss in the simulation will be matched.

As an extension, refer to the computer activity on p. 486E of the Chapter Overview. See also, TRB Exploring O-48.

Lesson Follow-up

LESSON RESOURCES

TRB Practice L-13
TRB Enrichment L-14

Practice

For use after 12-7 (pp. 518–520)

Simulate the Problem

1. Suppose that before the 1999 World Series you attempt to predict the winners of the seven games. What is the probability that you will get 4 or more out of 7 correct?

1999 World Series	
Game	Winner
1	Red Sox
2	Cubs
3	Red Sox
4	Cubs
5	Cubs
6	Red Sox
7	Red Sox

Model the situation using a coin. Let heads represent Red Sox and tails represent Cubs. A trial occurs when you toss the coin seven times to represent the seven games in the series. A successful trial occurs when you get four or more outcomes that match the true outcomes listed in the table.

Work with a partner. Carry out 50 trials. Write the probability after the given number of trials.

a. 10 ______ **Answers will vary.**
b. 20 ______
c. 30 ______
d. 40 ______
e. 50 ______

2. An irresponsible TV weatherperson forecasts the weather by throwing a number cube and consulting the weather key shown here. The weather during one 5-day stretch is given in the table. What is the probability that the forecaster was right at least 3 days out of 5?

Weather Key
1—clear and warm
2—clear and cool
3—cloudy and cool
4—intermittent showers
5—continual rain
6—snow

Use a number cube to simulate the forecaster's predictions. A successful trial occurs when you roll the correct weather three or more times out of five.

Mon	Tue	Wed	Thu	Fri
continual rain	continual rain	clear and cool	cloudy and cool	snow

Work with a partner. Carry out 50 trials. Write the probability after the given number of trials.

a. 10 ______ **Answers will vary.**
b. 20 ______
c. 30 ______
d. 40 ______
e. 50 ______

Chapter 12 L-13

Enrichment

For use after 12-7 (pp. 518–520)

Simulations with Random Numbers

Photos of the ten starters in the NBA All-Star game are given out at random to ticket buyers, one photo per ticket. How many tickets would you have to buy to obtain all ten photos?

You can simulate the purchase of tickets using random numbers generated by computer. Let each of the digits 0 through 9 represent a different photo. A trial occurs when you scan the digits left to right, counting digits until one of each has appeared.

76245	40901	97017	70762
88449	01912	07436	50813
36607	11458	30440	52639
32378	68038	89351	37005
59960	70408	29812	83126
78071	17456	96104	18327
70009	23233	65438	59685
88598	47821	00265	82525
43989	90770	22965	44247
76210	22467	83275	32286
36782	00268	97121	57676
36270	77786	89578	21059
58581	95331	78629	73344
07648	70164	34994	67662
49475	50558	34698	71800
54728	81972	58975	30761
79178	33692	57352	72862
21252	12746	37554	97775
45557	96345	33271	53464
46717	72364	86954	55580
05822	46760	44294	07672
01055	79044	19308	83623
49044	95495	81256	53214
30883	89660	96142	18354
24290	01551	80092	82118

1. Begin the first trial at the upper left of the table. According to the first trial, how many tickets must you buy in order to obtain all ten photos?
34 tickets

2. Conduct as many trials as possible. Begin each trial where the last one left off. Record the results in the table.

Trial number	Number of Tickets Purchased	Trial number	Number of Tickets Purchased
1	34	8	29
2	26	9	46
3	28	10	26
4	21	11	19
5	23	12	45
6	19	13	51
7	33	14	33

3. Find the mean number of ticket purchases you must make in order to obtain all ten photos. Round to the nearest whole number.
31

4. If you used a different table of random digits, would you get the same answer?
The average of many trials using different tables would result in approximately the same answer.

5. How can you get a more accurate answer?
by conducting more trials

L-14 Chapter 12

5. Thirteen of 25 students are going on a field trip. Six students are traveling in a van. What is the probability that a student chosen at random is *not* traveling in a van? $\frac{19}{25}$

6. What is the probability that a student will draw a card at random showing A or B from these cards? $\frac{3}{11}$

P	R	O	B	A	B	I	L	I	T	Y

7. A student is going to choose a date from the month of January at random. What is the probability that the student will choose January 1? $\frac{1}{31}$

8. Suppose you toss three coins. What is the probability that all the coins will land heads up? $\frac{1}{8}$

9. A student has a mean of 92 for two quizzes. What grade does the student need on the next quiz to have a mean of 94? **98**

10. A student tells you that she receives an average salary of \$3.50/h. Her manager tells you he pays an average of \$4/h. The chain they work for says the pay averages \$4.75/h. Explain how they could all be telling the truth. **The student, the manager, and the chain are using different kinds of averages.**

11. A store advertises a jacket for \$72. During a sale the store reduces the jacket price by 25%. After the sale, the store raises the jacket price by 25%. What is the price of the jacket after the sale? **\$67.50**

12. The sum of five consecutive numbers is 380. What are the numbers? **74, 75, 76, 77, 78**

13. Double a number minus half the number is five minus the number. What is the number? **2**

14. A student uses 24 yd of fencing to make a rectangular pen. The pen is 6 yd longer than it is wide. What are the dimensions of the pen? **3 yd × 9 yd**

15. The circumference of the peg below is 3 in. Will the peg go through the hole? Explain. **Yes. The circumference of the hole is 3.14 in.**

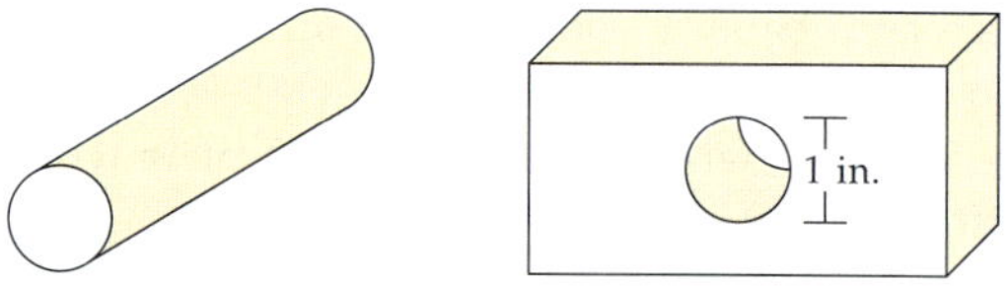

16. The average American generates 3.5 lb of waste per day. Of all waste, about 6.5% is plastic. About how much plastic waste does the average American generate each day? each year? **0.2275 lb; 83.0375 lb.**

17. In a recent year, the number of major trash composting projects in the United States rose from 42 to 75. Find the percent of change. **78.6%**

Assignment Guide

Basic All
Average All
Enriched All

FOR THE NEXT LESSON

paper, pencil, calculator, math journal

Practice

Find the mean, median, mode, and range for each set of data.

1. 6, 7, 8, 7, 6, 15, 24, 36, 28, 26, 18, 11 16; 13; 6 and 7; 30

2. 1, 2, 5, 6, 7, 3, 8, 9, 1, 1, 2, 9, 8, 7, 6 5; 6; 1; 8

3. 2, 5, 6, 5, 4, 2, 1, 4, 4, 7, 2, 4, 3, 7 4; 4; 4; 6

Draw a line plot and make a frequency table for each set of data. Find the mean, the median, and the mode. Check students' work for plot and table.

4. 11, 12, 13, 15, 11, 12, 14, 10, 28 14; 12; 11 and 12

5. 6, 5, 6, 5, 6, 2, 8, 0, 0, 0, 1, 6 3.75; 5; 6

6. 20, 30, 40, 50, 20, 30, 20, 50, 60, 10, 20 31.8; 30; 20

Make a back-to-back stem and leaf plot for each set of data. Find the median and the mode. Check students' work for plot.

7. Set A: 23, 24, 25, 26, 23, 22 23.5, 23 Set B: 19, 22, 25, 26, 17, 23, 21 22, no mode

8. Set C: 45, 46, 50, 51, 48, 49 48.5, no mode Set D: 52, 53, 52, 54, 50, 51, 52 52, 52

Make a box and whisker plot for each set of data. Use a single number line. Then compare. Check students' work for plot.

9. Set A: 2, 4, 6, 1, 2, 3, 4, 12, 15, 10, 10, 8
Set B: 8, 9, 10, 7, 10, 8, 8, 7, 10, 9, 11, 11
The data for Set A is spread out. Set B data is consistent, staying close to the median, which is 9.

10. Set C: 30, 40, 50, 20, 30, 70, 80, 100, 60, 50, 50, 60
Set D: 15, 25, 75, 75, 15, 35, 45, 65, 75, 25, 25, 45 The plots are very similar in shape. Set C is higher on the scale for every quartile than Set D.

Write each expression as a factorial.

11. $4 \cdot 3 \cdot 2 \cdot 1$ 4!

12. $9 \cdot 8 \cdot 7 \cdot 6$ $\frac{9!}{5!}$

13. $25 \cdot 24 \cdot 23$ $\frac{25!}{22!}$

14. $100 \cdot 99 \cdot 98 \cdot 97$ $\frac{100!}{96!}$

Evaluate.

15. $\frac{10!}{7!}$ 720

16. $\frac{7!}{4!}$ 210

17. $\frac{50!}{48!}$ 2,450

Use the counting principle to solve.

18. A student has 3 clean shirts, 4 clean pairs of socks, and 2 clean pairs of pants. How many different outfits can he wear? 24

Find the probability of each event when a letter is chosen at random from the word BEEKEEPER.

19. selecting the letter K $\frac{1}{9}$

20. selecting the letter E $\frac{5}{9}$

21. selecting a consonant $\frac{4}{9}$

22. selecting a vowel $\frac{5}{9}$

23. selecting a letter of the English alphabet 1

24. selecting a number 0

CAREER

Help Wanted: Actuary

Bachelor's degree in math plus advanced training in statistics required. For more information, write to American Academy of Actuaries, 1720 I St. NW, 7th Floor, Washington, DC 20006.

Actuaries calculate insurance risks and premiums. Actuaries compute the probability of events and determine the premiums to charge for insuring these risks. They also evaluate the risk of damage or loss of property and the resulting payouts. Most actuaries work for insurance companies.

PROJECT

Design and conduct a survey of your classmates' weekly expenses. Analyze the data and determine the amount of money your classmates will require on a weekly basis to cover their expenses.

Career Information

Invite an actuary from a local insurance company to speak to the class about how their mathematical skills are used in the insurance industry.

Ask the speaker to address topics such as:

- what kind of insurance the actuary is involved with (e.g., auto, health, homeowners, life)
- what kind of information actuaries use and how it is gathered
- how the information is used to calculate insurance risks and premiums
- what other industries employ actuaries
- the educational background required for a career as an actuary, including advanced training and mathematical expertise

Lesson Focus

Materials

Paper, pencil, calculator, math journal

Vocabulary/Symbols

compound event
dependent events
independent events

MOTIVATING THE LESSON

Ask students if they can think of any probability situations that might be a combination of two single events. (e.g., getting "heads" on a penny and "tails" on a nickel; getting a 3, then a 5 when tossing a number cube; randomly choosing a red scarf, then a blue scarf from a drawer containing 6 blue and 4 red scarves)

Instruction

1 To help students see the two alternative methods of doing the same thing, have them draw a tree diagram to show that (Walk, Car) occurs once out of 6 possible outcomes.

2 Mention that $P(B, \text{given } A)$ indicates that B is dependent on A and that event A has occurred.

3 To help students visualize the situation, have them act it out. Then ask students to draw a tree diagram to represent the possible outcomes and to verify the probability.

OBJECTIVE: *To calculate probabilities of independent and dependent events.*

12-8 Independent and Dependent Events

▼ Some events are a combination of two or more single events. A trip from New York to Seattle via Chicago is actually a combination of two trips: one from New York to Chicago and one from Chicago to Seattle. We call this a *compound event*.

Compound Event	A compound event is a combination of two or more events.

▼ A compound event can be a combination of *independent events.*

Independent Events	Independent events are events in which the outcome of one has no effect on the outcome of the other. For two independent events A and B, $P(\text{A and B}) = P(\text{A}) \cdot P(\text{B})$.

FLASHBACK
You can use a tree diagram to find a sample space.

Example 1 A student can walk, take the bus, or ride to school with a friend. After school, he can walk or ride with a friend to the library. The student is equally likely to make any of these choices. What is the probability that he will walk to school and then ride with a friend to the library after school?

Solution *Method 1.* Find the probability of two independent events using the counting principle.

Ways to School		Ways to Library		Possible Outcomes
3	×	2	=	6

Favorable outcome, write the probability as:

$\frac{\text{favorable outcome}}{\text{possible outcomes}} = \frac{1}{6}.$

1 *Method 2.* Find the probability of two independent events using the formula.

$P(\text{Walk}) = \frac{1}{3}$ $\quad P(\text{Car}) = \frac{1}{2}$

$P(\text{Walk, Car}) = \frac{1}{3} \times \frac{1}{2}$

$= \frac{1}{6}$

The probability of walking to school and then riding to the library with a friend after school is $\frac{1}{6}$.

Teaching Tip

To enhance students' understanding of the probabilities of independent and dependent events, have them act out the problems, using objects where appropriate.

▼ A compound event can be a combination of *dependent events*.

Dependent Events	Dependent events are events in which the outcome of one depends on the outcome of the other. For two dependent events A and B, where B is dependent on A, $P(A \text{ and } B) = P(A) \cdot P(B, \text{given } A)$.

Example 2 Three girls and two boys are running for class president and vice president. Slips of paper with the five names are put into a bag. The first name drawn from the bag will be president. The second will be vice president. What is the probability that both officers will be girls?

Solution $P(\text{Girl}) = \frac{3}{5}$ First draw.

$P(\text{Girl}) = \frac{2}{4} = \frac{1}{2}$ Second draw. Assume a girl has been chosen on the first draw. There are 4 names left, so 4 is the number of possible outcomes.

$\frac{3}{5} \times \frac{1}{2} = \frac{3}{10}$ Multiply the probabilities.

The probability that both officers will be girls is $\frac{3}{10}$.

THINK How are dependent events different from independent events? Independent events do not influence the outcome of successive events. For dependent events, the outcome of an event does influence the outcome of successive events.

2

3

1. The result of multiplying the probabilities is the same as counting favorable outcomes and dividing by the possible outcomes. 2. The sample space of the second event is decreased as a result of the outcome of the first event.

CLASS EXERCISES

Are the events independent or dependent? Explain.

1. You select a card. Without putting the card back, you select a second card. dependent
2. You select a card. After putting it back, you select a second card. independent
3. You roll a number cube. You roll it again. independent
4. Cards numbered 5, 5, 3, 7, and 4 are placed face down on a table.
   a. What is the probability that you select a 5 at random? $\frac{2}{5}$
   b. You do not replace the 5. What is the probability that your next selection is a 4? $\frac{1}{4}$
   c. You put the cards back on the table before making a second selection. What is the probability of selecting a 5, then a 4? $\frac{2}{25}$
5. A student has 5 blue socks and 4 orange socks. Find the probability that she will randomly select these items.
   a. a blue sock, then an orange sock $\frac{5}{18}$ b. two blue socks $\frac{5}{18}$
   c. an orange sock, then a blue sock $\frac{5}{18}$ d. two orange socks $\frac{1}{6}$

THINK AND DISCUSS See above.

1. Explain why you can multiply probabilities for two independent or two dependent events.
2. When computing the probability of dependent events, how does the probability of the second event show that the events are dependent?

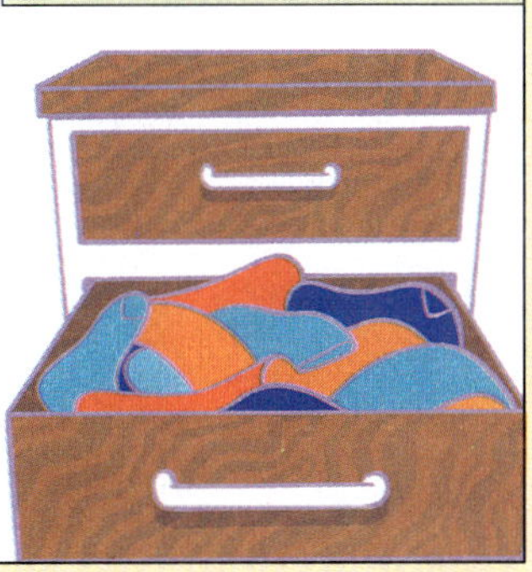

MATH MINUTES

Ask groups of four to write a problem involving the probability of a compound event. Then have students decide whether the events are independent or dependent and solve the problem.
(Check students' work.)

ADDITIONAL EXAMPLES

1. From a choice of 2 bran, 1 corn, and 2 oatmeal muffins, and, a choice of orange, apple, and grape juice, what is the probability that you will randomly choose a bran muffin and orange juice? ($\frac{2}{15}$)
2. The names of 3 girls and 2 boys are placed in a bag. You draw a name but you do not put it back before you draw another name. What is the probability that both names will be boys' names? ($\frac{1}{10}$)

Guided Practice

Class Exercises Have students give their explanations orally to determine whether they can distinguish between independent and dependent events.

Think and Discuss For question 2, help students conclude that the denominator of the second probability ratio (before written in simplest form) shows that there is one less possible outcome after the first event has occurred.

A ***common error*** occurs when students use an incorrect ratio for the probability of the second of two dependent events. Have students list the possible outcomes that remain, after the first event's probability has been decided.

Closure

Writing in Math Have students write a paragraph explaining the distinction between dependent and independent events in their math journals.

Independent Practice

For the project suggested in question 17, have students brainstorm questions to include in the opinion poll.

6. A caterer is serving sandwiches from a tray containing 3 chicken, 2 cheese, and 2 peanut butter sandwiches. What is the probability that she will randomly serve the following sandwiches?
   a. 2 chicken sandwiches $\frac{1}{7}$
   b. a cheese sandwich, then a peanut butter sandwich $\frac{2}{21}$
   c. a chicken sandwich, a cheese sandwich, then another chicken sandwich $\frac{2}{35}$

WRITTEN EXERCISES

Are the events independent or dependent? Explain.

1. Two number cubes are thrown. The result on one is 3 and on the other is 5. independent
2. It has rained for the last three Saturdays. Rain is forecast for next Saturday. independent
3. You exercise daily. You make the tennis team. dependent
4. You choose a white marble from a bag and do not put it back. You choose again and get another white marble. dependent

For each exercise, assume that events A and B are independent. Then find *P*(A and B).

5. $P(A) = \frac{1}{2}$, and $P(B) = \frac{1}{5}$. $\frac{1}{10}$
6. $P(A) = \frac{1}{3}$, and $P(B) = \frac{2}{7}$. $\frac{2}{21}$
7. $P(A) = \frac{2}{9}$, and $P(B) = \frac{3}{8}$. $\frac{1}{12}$
8. $P(A) = \frac{7}{16}$, and $P(B) = \frac{8}{35}$. $\frac{1}{10}$

Tell whether each problem involves dependent or independent events. Then find the probability.

9. A refrigerator contains 12 orange drinks, 4 grape drinks, and 25 apple drinks. Ann is first in line. Mark is second. What is the probability that Ann gets an apple drink and Mark a grape drink if they choose their drinks at random? dependent $\frac{5}{82}$
10. Suppose you roll a number cube twice. What is the probability that you will roll each of the following pairs of numbers?
    a. 6, then 5 $\frac{1}{36}$
    b. 6, then 6 $\frac{1}{36}$
    c. 4, then 3, then 1 independent $\frac{1}{216}$
11. A student's wallet contains three one-dollar bills, two five-dollar bills, and three ten-dollar bills. The student selects two bills from her wallet at random. What is the probability that the student selects these bills? dependent
    a. a one-dollar bill, then a ten-dollar bill $\frac{9}{56}$
    b. a ten-dollar bill, then a five-dollar bill $\frac{3}{28}$

MIXED REVIEW

Find each answer.

1. $\frac{1}{2}$ of 390 195
2. $\frac{3}{4}$ of 176 132

Find the probability for each outcome.

3. a coin landing face up $\frac{1}{2}$
4. rolling a 6 on a number cube $\frac{1}{6}$

Write an expression.

5. 48 times a number $48n$
6. 29 times a number less 5 $29n - 5$

Solve.

7. A student's age is three times that of her brother. In 15 years, her brother's age will be $\frac{2}{3}$ her age. How old is each now? 5, 15

12. Suppose you spin a spinner like the one at the right. What is the probability that you get each outcome? independent

a. C, then A $\frac{1}{12}$ b. B, then C, then A $\frac{1}{36}$

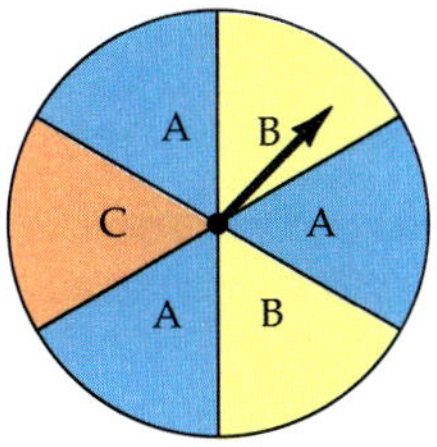

Use the ***DATA*** at the right.

13. How many students were in the survey? Would a sample be an appropriate way to measure preference in this case? Why or why not? 24; No; The class is small enough to ask each person for his/her preference.

14. What is the probability that a student selected at random from the class prefers pizza? $\frac{3}{8}$

15. What is the probability that a boy prefers pizza? $\frac{2}{5}$

16. What is the probability that a girl does not prefer pizza? $\frac{9}{14}$

Votes for Class Party

	Barbecue	Pizza	Total
Boys	6	4	10
Girls	9	5	14

17. ***PROJECT*** Choose a topic from the list below or one of your own and conduct an opinion poll. Survey at least 30 people. Check students' work.
- quality of cafeteria food
- curfew times for 9th graders
- study habits

a. Record the frequency of each response in a table.

b. Find the probability for each response (outcome).

c. ***WRITE*** three problems based on your poll. Trade with a friend. Solve each other's problem.

TEST YOURSELF

Make a stem and leaf plot for the data. Find the mean, median, and mode. Check students' work.

1. 33, 35, 32, 28, 28, 21, 27, 35, 39, 40, 22, 24 30.3; 30; 28 and 35

Solve.

2. From Compt there are 4 ways to get to Murch. From Murch there are 5 ways to get to Toll. How many ways are there from Compt to Toll through Murch? 20

3. You roll a number cube. What is the probability of rolling a 4 and then rolling another 4? $\frac{1}{36}$

4. You have 12 socks in a drawer and choose 2 at random. Five socks are blue, three brown, and four black. What is the probability that you choose a black sock and then choose a brown one? $\frac{1}{11}$

Lesson Follow-up

Reteaching Activity Encourage those students who have difficulty in finding the probabilities of compound events to check their answers by drawing tree diagrams.

LESSON RESOURCES

TRB Practice L-15
TRB Enrichment L-16

Name ____________ Date ____________

Practice

For use after 12-8 (pp. 522–525)

Independent and Dependent Events

Determine whether each of the pairs of events is independent or dependent. Write *I* or *D* and explain.

1. A guest at a party takes a sandwich from a tray. A second guest then takes a sandwich.
D; the second guest's choice is limited by the first's.

2. Sam flips a coin and gets heads. He flips again and gets tails.
I; the second outcome is not affected by the first.

For each exercise, assume that events *A and B* are independent. Then find *P*(*A* and *B*).

3. $P(A) = \frac{2}{3}$, and $P(B) = \frac{3}{4}$ $\frac{1}{2}$
4. $P(A) = \frac{1}{2}$, and $P(B) = \frac{8}{9}$ $\frac{4}{9}$
5. $P(A) = \frac{5}{14}$, and $P(B) = \frac{7}{15}$ $\frac{1}{6}$
6. $P(A) = 1$, and $P(B) = 0$ 0

Determine whether each problem involves dependent or independent events. Then find the probability.

7. A shelf holds 3 novels, 2 biographies and 1 history book. Two students in turn choose a book at random. What is the probability that the students chose each of the following? Dependent
a. both novels $\frac{1}{5}$
b. both biographies $\frac{1}{15}$
c. a history, then a novel $\frac{1}{10}$
d. a novel, then a history $\frac{1}{10}$

8. Meg flipped a penny the given number of times. What is the probability the results were as follows?
Independent
a. 2; two heads $\frac{1}{4}$
b. 2; two tails $\frac{1}{4}$
c. a tail, then a head $\frac{1}{4}$
d. 5; five tails $\frac{1}{32}$

9. Two puppies are chosen at random from a box at the mall. What is the probability of these outcomes?
Dependent

Free Puppies for Adoption!
5 black retrievers
3 brown hounds
4 black setters

a. both black $\frac{6}{11}$
b. both brown $\frac{1}{22}$
c. a setter, then a hound $\frac{1}{11}$
d. a retriever, then a setter $\frac{5}{33}$
e. both setters $\frac{1}{11}$

Chapter 12 L-15

Enrichment

For use after 12-8 (pp. 522–525)

Odds

Odds are a way of comparing favorable and unfavorable outcomes.

$$\text{Odds in favor} = \frac{\text{number of favorable outcomes}}{\text{number of unfavorable outcomes}}$$

$$\text{Odds against} = \frac{\text{number of unfavorable outcomes}}{\text{number of favorable outcomes}}$$

Example A number cube is tossed. What are the odds in favor of rolling a number divisible by 3? What are the odds against this outcome?

Solution There are two favorable outcomes, 3 and 6.
There are four unfavorable outcomes, 1, 2, 4 and 5.
Odds in favor of the outcome $= \frac{2}{4} = \frac{1}{2}$.
Odds against the outcome $= \frac{4}{2} = \frac{2}{1}$.

Odds are often given in this form:
The odds in favor are 1 to 2. The odds against are 2 to 1.

Find the odds. Give the answer in the form "*a* to *b*."

1. A spinner is spun. Find the odds:

a. in favor of a 5 1 to 7
b. in favor of an even number 1 to 1
c. against a number divisible by 3 3 to 1
d. against a prime number 1 to 1

2. A record is chosen at random from a sale bin. Find the odds:

Record Sale!
8 jazz
6 classical
2 rock

a. in favor of classical 3 to 5
b. in favor of rock 1 to 7
c. against jazz 1 to 1

3. Two spinners are spun together. Find the odds:

a. in favor of a sum of 7 1 to 7
b. against a sum of 6 13 to 3

L-16 Chapter 12

LESSON QUIZ

The lunch wagon has 6 turkey, 4 ham, and 2 chicken sandwiches; 5 bottles of grape juice, 3 bottles of fruit punch, and 2 bottles of tomato juice.

1. What is the probability of randomly choosing a ham sandwich and fruit punch? ($\frac{1}{10}$)

2. What is the probability of randomly choosing a turkey sandwich and then a chicken sandwich? ($\frac{1}{11}$)

Assignment Guide

Basic 1-2, 5-6, 9-10, 13-17, MR, TY All
Average 2-3, 6-7, 10-11, 13-17, MR, TY All
Enriched 3-4, 7-8, 11-17, MR, TY All

FOR THE NEXT LESSON

math journal

Instruction

This page provides a variety of problems that can be used to reinforce and enhance the students' problem solving skills. Encourage students to read each problem carefully. Then have them refer to the list of problem solving strategies to help them decide how to solve the problem.

Point out, however, that not all questions require a strategy for solving, nor are all the strategies in the list used in this lesson.

Problem Solving Practice

READ PLAN SOLVE LOOK BACK

PROBLEM SOLVING STRATEGIES

Look for a Pattern
Guess and Test
Simplify the Problem
Account for All Possibilities
Make an Organized List
Work Backwards
Make a Table
Write an Equation
Solve by Graphing
Draw a Diagram
Make a Model
Simulate the Problem

Solve. Use an appropriate strategy or combination of strategies.

1. A slope of $\frac{1}{10}$ is suitable for a ramp to allow wheelchair access to a building. How far from a doorway will a ramp extend if the doorway is $10\frac{1}{2}$ ft above the ground? **105 ft**
2. The sales tax on an item costing \$23.50 is \$1.88. What is the sales tax on an item costing \$47? **\$3.76**
3. A student has a coordinated wardrobe consisting of four blouses, two sweaters, and three skirts. How many three-piece outfits can she make? **24**
4. In a collection of dimes and nickels, there are 9 more nickels than dimes. The collection is worth \$1.65. How many nickels and dimes are there? **17 nickels, 8 dimes**
5. A classroom is 15 ft high. Its floor has an area of 642 ft^2. Each student needs 300 ft^3 of air. How many students can be assigned to the room? **32 students**
6. The number of calls to a weather information number for fifteen consecutive days was 79, 75, 84, 103, 129, 95, 89, 114, 128, 112, 115, 105, 120, 127, and 101.
   a. Find the median number of calls. **105**
   b. Draw a stem and leaf plot. **Check students' work.**
   c. What must the number of calls be for the next day to change the median to 107? **109**
7. The sum of two numbers is 11. Twice the greater subtracted from 3 times the lesser equals 3. What are the numbers? **5 and 6**
8. A couple pays a babysitter \$4.75/h plus \$5 taxi fare and expenses. What is the greatest whole number of hours the babysitter can work and still receive less than \$25? **4 h**
9. The mean score of four bowlers is 140. Three scores are the same. The third is twice that of the fourth. Find the scores. **160, 160, 160, 80 or 112, 112, 224, 112**
10. Two salespeople drove a distance of 600 mi in one day. The first salesperson drove at least 40 mi more than twice the distance driven by the second. What is the greatest distance that the second salesperson might have driven? **$186\frac{2}{3}$ mi**
11. A band director wished to arrange the band members in pairs for a marching pattern. He found that he was one person short. He tried to arrange by fives and sevens and was still one person short. What is the least number of people in the marching band? **69**

PROBLEM SOLVING APPLICATION

READ PLAN SOLVE LOOK BACK

OBJECTIVE: *To apply capture/recapture methods to wildlife management.*

12-9 Making Predictions

Many species of wildlife in the United States have become extinct, including the North Carolina parakeet, the passenger pigeon, and the California grizzly bear. The increased number of wildlife in danger of becoming extinct is a concern for environmentalists.

Capture/recapture is a method for estimating a total population from a sample. Researchers capture, tag, and set free animals of a certain type. At a later time, animals in the same species are recaptured. You can use the following formula to estimate the number of animals in the total population.

$$\frac{\text{tagged animals}}{\text{total population}} = \frac{\text{tagged animals recaptured}}{\text{total animals recaptured}}$$

Example National park rangers wish to know the number of coyotes in a certain section of the Yellowstone National Park. Park rangers capture, tag, and set free 24 coyotes. Two weeks later, rangers capture 38 coyotes. Eight of the coyotes have tags. Estimate the number of coyotes in the section of the park surveyed.

(1) **Solution** $\frac{\text{tagged animals}}{\text{total population}} = \frac{\text{tagged animals recaptured}}{\text{total animals recaptured}}$

$\frac{24}{n} = \frac{8}{38}$ Substitute values in the formula.

$8n = 912$ Write cross products.

$n = 114$

The estimated number of coyotes is 114.

CLASS EXERCISES

Solve.

1. A naturalist and his assistants capture, tag, and set free 32 spotted deer. A week later the scientists capture 45 deer. Twelve have tags.
   a. What is the ratio of tagged animals originally captured to the number of animals in the total population (n)? $\frac{32}{n}$
   b. What is the ratio of tagged animals recaptured to the total number of animals recaptured? $\frac{12}{45}$
   c. Estimate the total population of spotted deer. 120

Lesson Focus

Materials
Math journal

MOTIVATING THE LESSON

Explain that the lesson involves observing a *sample*, or part of a group, to make a prediction about the group as a whole. You may wish to discuss situations where sampling is used. (market research, population surveys, polls)

Instruction

(1) Comment that the solution represents an estimate of the total population. Mention that the estimate is the best guess that the environmentalists can make.

Guided Practice

Class Exercises Remind students that a proportion represents equivalent ratios.

Teaching Tip

Some students may think one consideration is sufficient to make a decision. Be sure that the students examine all possibilities before they make a decision.

Closure

Writing in Math Have students restate, in more general terms, the formula used in the example. Then have them record the formula in their math journals for later use.
For example:

$$\frac{\text{sample}}{\text{total}} = \frac{\text{part of sample}}{\text{part of total}}$$

Independent Practice

For questions 1 and 4, be sure that the students realize that there is no way to establish the total population.

Decision Making Have a student read the paragraph to the class. Ask, what does the store manager want to know? (how many of each item, size, and color to buy) Discuss with students the ramifications of overstocking merchandise.

Have students work in groups of 4-6 for questions 1-4. Limit color choices to the school colors.

Before students answer question 8, ask if the groups are satisfied with their questionnaires, and the results of their survey. Ask what changes they would make, if any.

WRITTEN EXERCISES

Use a CALCULATOR where appropriate.

Solve.

1. In a study of catfish in Beaver Lake, workers for the state extension service caught, tagged, and set free 124 catfish. A few weeks later the workers caught 140 catfish. Thirty-five had tags. Estimate the number of catfish in the lake. **496**
2. There are an unknown number of marbles in a bag. You take 10 marbles out and mark them. You put the marked marbles back into the bag and mix the contents well. You take out 25 marbles. Five of them are marked. Estimate the number of marbles in the bag. **50**
3. You are in a large city. You wish to estimate the number of yellow cabs. You count 75 yellow cabs and you keep track of their license numbers. The next day you count 84 cabs, 20 of which are repeats. Estimate the number of yellow cabs. **315**
4. The ecology class is helping the local conservation society to determine the number of raccoons in a nearby forest. In early October, the students and society members captured, tagged, and set free 68 raccoons. Three weeks later, 84 raccoons were captured, and 16 had tags. Estimate the number of raccoons in the forest. **357**

Decision Making* DECISION MAKING *Decision Making Decision Making Decision Maki

MAKING PREDICTIONS

COLLECT DATA

The school bookstore plans to stock sweatshirts, hats, and jackets. It is important not to overstock. The store manager asks you to determine the number, color, and size of each item to order.

1. Write a survey questionnaire to find out student interest. Design the questionnaire so that you can estimate the sales by color and size of each item.
2. Who will you survey? Will a random survey suit your purposes or will selecting survey groups from each grade level be a more accurate method of determining your market?
3. Where will you conduct the survey? Will verbal responses be as helpful as written responses?
4. Conduct the survey.

MATH MINUTES

Choose several students and count those wearing blue socks. Have the class use the sample to predict the number of students in class wearing blue socks. Compare predictions to the actual number. Discuss the results.

Use the *DATA* below and at the right to solve.

5. The ecology class designed a T-shirt to sell to students as a fund-raiser. The profits are to go to the local conservation society. To determine the number of shirts to order, the class conducted a marketing survey. The results are on the chart at the right. The formula the class used to determine the expected sales is

$$\frac{\text{expected sales}}{\text{total in target group}} = \frac{\text{number of yes responses}}{\text{number of students surveyed}}.$$

One hundred fifty students were surveyed. The target group was the entire school population of 2,000 students.

a. What are the expected sales for the small T-shirt? **240**

b. What are the expected sales for the medium T-shirt? **400**

c. What are the expected sales for the large T-shirt? **600**

d. The class did not want to order more T-shirts than they could actually sell. The group decided to order only 80% of the expected sales found in the survey. How many T-shirts in each size did the ecology class order? **192; 320; 480**

e. The amount of money the club makes on each shirt is $1.25. How much money will the ecology club give to the conservation society if the club sells all of the T-shirts ordered? **$1,240**

Yes Responses to Marketing Survey for T-shirts

small	18
medium	30
large	45

Decision Making ■ Decision Making ■ Decision Making ■ Decision Making ■ Decision Making ■ Decision Making

■ ANALYZE DATA

5. Calculate the number of expected sales for each item by color and size.

6. In collecting data, was interest expressed in items not on your list? Do you need to expand your choices and do another survey?

■ MAKE DECISIONS

7. Decide if you will order exactly the number of items you have found to be your expected sales. Should you order more? less? Explain your decision.

8. Contact a supplier to find the wholesale cost of each item. Determine what price you will set for each item.

9. Present your results to the school bookstore or to some other group for a fund-raising project.

Lesson Follow-up

LESSON RESOURCES

TRB Alternate Application L-17

Alternate Application For use after 12-9 (pp. 527–529)

Permutations

A *permutation* is an ordered arrangement of objects. There are six permutations of the letters *A*, *B*, and *C* to form two-letter words:

AB AC BA BC CA CB

Three-digit numbers are to be formed using only the digits 1, 2, 4, 7, and 9. No digit may be repeated. Answer these questions to find how many such permutations are possible.

1st digit ↓ 2nd digit ↓ 3rd digit ↓ ___ ___ ___

1. How many choices of digits do you have:
   a. for the first space? **5**
   b. for the second space? (Since digits cannot be repeated, you may not use the digit you used above.) **4**
   c. for the third space? **3**
2. Multiply your three answers to find the number of possible permutations of the digits. **60**

Find the number of permutations.

3. Art, Becky, Carl, and Denise are lined up to buy tickets. How many different permutations of the four are possible? **24**
4. Suppose Ed was also in line. How many permutations would there be? **120**

Numbers are to be formed using the digits 1, 2, 3, 4, 5, and 6. No digit may be repeated.

5. How many two-digit numbers can be formed? **30**
6. How many three-digit numbers can be formed? **120**
7. How many four-digit numbers can be formed? **360**
8. A combination lock has 36 numbers on it. How many different 3-number combinations are possible if no number may be repeated? **42,840**

Chapter 12 L-17

Assignment Guide

Basic All
Average All
Enriched All

FOR THE NEXT LESSON

algebra tiles or paper squares in two sizes and paper rectangles, math journal

Study Skills

Homework provides students with an opportunity to improve their study skills and habits. It allows students a chance to develop their ability to manage time, materials, and the information being studied on their own. Although students often balk at homework's intrusion on their free time, they might be more inclined to see its virtues if they can be enlisted to create an overall plan, addressing the goals and amount of homework they will be asked to do.

Lead students in a discussion about homework, emphasizing its positive aspects. Encourage them to share their thoughts about homework—its benefits and drawbacks. Work with students to draft a class statement about what homework should accomplish.

The class statement might include some of the following ideas:

- Homework should include time for reviewing the day's material, and previewing the next day's lesson.
- Homework should enhance student's ability to work independently.
- The teacher can occasionally provide homework outside of the textbook to offer students opportunities for more creative exploration of the material being taught.

Chapter 12 Review

Write *true* or *false*. If false, change the underlined word(s) to make the statement true.

1. The median is the number that occurs most often in a set of data. false, mode
2. A box and whisker plot is useful in showing the distribution of data by quartiles. true
3. A line plot shows data on a number line. true
4. Dependent events are events in which the outcome of one has no effect on the outcome of the other. false, independent
5. A multiple, written as $n!$, is the product of all whole numbers from n to 1. false, factorial

Finding Mean, Median, and Mode — 12-1

In a set of data,

The *mode* is the number that occurs most often.

The *mean* is the sum of n numbers divided by n.

The *median* is the middle value.

To find the *range*, subtract the least value from the greatest value.

Find the mean, median, mode, and range of each of the following.

6. 5, 7, 9, 6, 7, 8, 6, 8, 9, 7, 8, 9, 8 7.46; 8; 8; 4
7. 128, 111, 102, 107, 115, 125, 98, 135, 119 115.56; 115; no mode; 37

Making Line Plots and Frequency Tables — 12-2

To make a line plot, use a number line.

To make a frequency distribution, make a table.

Data: 1, 0, 3, 3, 2, 1, 4, 4, 5, 7, 2, 3

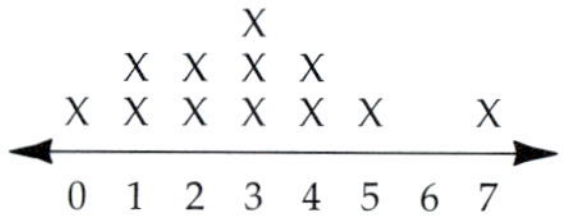

Line plot

n	0	1	2	3	4	5	6	7
f	1	2	2	3	2	1	0	1

Frequency distribution

Draw a line plot and a frequency distribution for the following.

8. 8, 4, 5, 1, 8, 4, 7, 9, 10, 5, 0, 5, 3, 4, 2

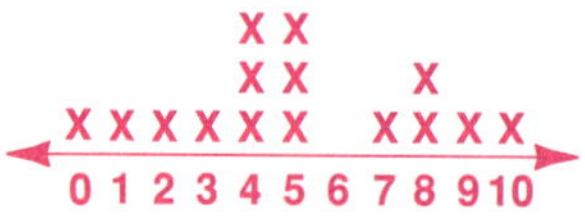

n	0	1	2	3	4	5	6	7	8	9	10
f	1	1	1	1	3	3	0	1	2	1	1

Making a Stem and Leaf Plot and a Box and Whisker Plot 12-3, 12-4

To make a stem and leaf plot, arrange the data along a stem using any reasonable choice of place value.

To make a box and whisker plot, separate the data into four groups by finding the median and the medians of the upper and lower sections of data. Draw a box to extend from the first to the third quartiles. Draw whiskers from the box to the highest and lowest scores.

Data: 70, 65, 72, 83, 85, 78, 85, 78, 82, 74, 68, 76

6	5 8
7	0 2 4 6 8 8
8	2 3 5 5

Stem and Leaf Plot

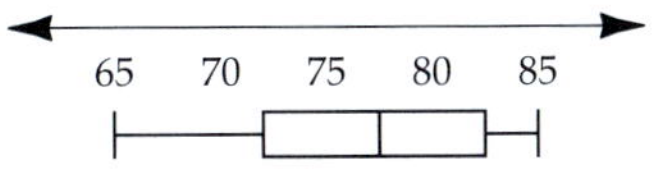

Box and Whisker Plot

Make a stem and leaf plot and a box and whisker plot for the following. At right.

5	0 5 5
6	0 0 5 5
7	0 5 5 5 5
8	0 0 5 5 5
9	5

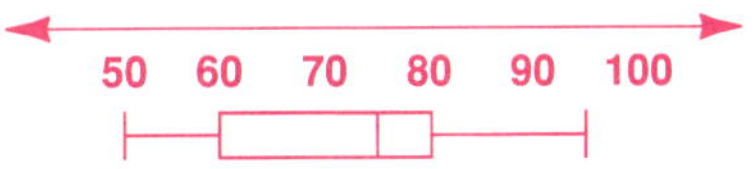

9. 75, 70, 80, 85, 85, 55, 60, 60, 65, 85, 75, 95, 50, 55, 75, 80, 65, 75

Using the Counting Principle 12-6

Use the counting principle to find the number of outcomes for two or more events. Find the product of the outcomes for each event.

Use factorials when the number of outcomes must be in an ordered arrangement.

five factorial → $5! = 5 \cdot 4 \cdot 3 \cdot 2 \cdot 1$

Find the value of each factorial.

10. $3!$ 6 **11.** $7!$ 5,040 **12.** $\frac{4!}{2!}$ 12 **13.** $\frac{8!}{4!}$ 1,680

Solve.

14. Jan has 5 pairs of pants and 7 shirts. How many different pant/shirt combinations are there for him to wear? 35

Using Probability and Finding the Probability of Independent and Dependent Events 12-7, 12-8

To determine probability, divide the number of favorable outcomes by the total number of possible outcomes.

To find the probability of independent and dependent events, multiply the probability of each event. For dependent events, the probability of the second event is affected by the first event.

Find the probability of each event if a card is drawn at random.

H	A	P	P	Y

15. selecting an H $\frac{1}{5}$ **16.** selecting a P $\frac{2}{5}$ **17.** selecting an A $\frac{1}{5}$ **18.** not selecting an H $\frac{4}{5}$

19. selecting a Y, replacing it, and then selecting a P $\frac{2}{25}$

20. selecting a Y, not replacing it, and then selecting a P $\frac{1}{10}$

RESOURCES

TRB L-19–L-22

Scoring Chart

No. Correct	Score
20	100
19	95
18	90
17	85
16	80
15	75
14	70
13	65
12	60
11	55
10	50
9	45
8	40
7	35
6	30
5	25
4	20
3	15
2	10
1	5

Chapter 12 Test

Find the mean, median, mode, and range of the following.

1. 15, 18, 23, 22, 19, 15, 17, 22, 29, 20 20; 19.5; 15 and 22; 14

2. 42, 40, 39, 45, 41, 43 41.67; 41.5; no mode; 6

Draw a line plot for the following frequency distributions. Check students' work.

3.

x	1	2	3	4	5	6
f	0	4	2	1	3	2

4.

x	12	13	14	15	16	17
f	4	7	3	1	2	5

Arrange the set of data into a frequency table.

n	92	95	97	98	100	101	105	108
f	3	2	1	2	2	2	2	1

5. the weight of school children 98, 101, 105, 95, 108, 92, 95, 100, 101, 98, 97, 105, 92, 92, 100

Make a stem and leaf plot for the following. Check students' work.

6. 17, 25, 32, 18, 22, 31, 27, 16, 19, 22, 35, 28, 25, 24

7. Make a back-to-back stem and leaf plot for the data below. Check students' work.

200-Meter Dash Time (in seconds)	Year	1960	1964	1968	1972	1976	1980	1984	1988
	Men	20.5	20.3	19.83	20.00	20.23	20.19	19.80	19.75
	Women	24.0	23.0	22.5	22.4	22.37	22.03	21.81	21.34

Make a box and whisker plot for the following. Check students' work.

8. 12, 8, 5, 9, 7, 12, 6, 8, 7, 9, 10, 12

9. 58, 63, 45, 82, 55, 79, 59, 77, 54, 83, 58

Find the value of each factorial.

10. 4! 24

11. 5! 120

12. $\frac{9!}{3!}$ 60,480

13. $\frac{7!}{2!}$ 2,520

Solve.

14. At the school picnic the children had a choice of a hot dog or a hamburger. They had a choice of one of three toppings. How many choices did they have in all? 6

15. You and three friends want to have your picture made together. If you line up shoulder to shoulder, in how many ways can the picture be made? 24

You pick one marble at random from a bag with 4 red marbles, 3 green marbles, and 5 blue marbles. Find each probability.

16. selecting a red marble $\frac{1}{3}$

17. selecting a yellow marble 0

18. selecting a green marble $\frac{1}{4}$

19. selecting a red marble, replacing it, and then selecting another red marble $\frac{1}{9}$

20. selecting a red marble, not replacing it, and then selecting another red marble $\frac{1}{11}$

Chapters 1–12 Cumulative Review

Choose the correct answer. Write A, B, C, or D.

1. Name the opposite of $|-3 + (-2)^3|$. B

A. 11 **B.** -11
C. 5 **D.** not given

2. The square root of a number cubed is 8. What is the number? C

A. 8 **B.** 2
C. 4 **D.** not given

3. Find a decimal between $(-0.1)^2$ and 0.05. A

A. 0.03 **B.** 0.2
C. 0.3 **D.** not given

4. Simplify $\frac{x^3y^4}{(x^2y^3)^2}$. C

A. xy^2 **B.** $\frac{x}{y^2}$
C. $x^{-1}y^{-2}$ **D.** not given

5. Find the median. B

1	6
2	5 7
3	4 4 4 6 6
4	1 2 8 9
5	4

A. 36.6 **B.** 36
C. 34 **D.** not given

6. Find the mean. A

250, 280, 240, 230, 270,
240, 270, 240, 230, 250

A. 250 **B.** 245
C. 240 **D.** not given

7. Evaluate $\frac{a^5b^3c}{a^6b^2}$ for $a = 2$, $b = -3$, and $c = -4$. B

A. 12 **B.** 6
C. -6 **D.** not given

8. Find the slope for $2x - 3y = 15$. D

A. 2 **B.** -2
C. -3 **D.** not given

9. In $\triangle ABC$, $\angle A = 55°$, $\angle C = 15°$. Name the triangle by angles. B

A. acute **B.** obtuse
C. right **D.** not given

10. 25% of r is 200. What is r? C

A. 25 **B.** 50
C. 800 **D.** not given

11. $3\frac{1}{5} \cdot 1\frac{1}{4} \div 2\frac{2}{3}$ A

A. $1\frac{1}{2}$ **B.** 4
C. $4\frac{4}{5}$ **D.** not given

12. Find the area of $\triangle CDE$. A

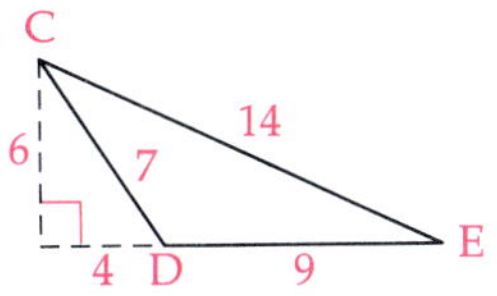

A. 27 **B.** 36
C. 12 **D.** not given

13. You have 10 red cards, 5 yellow cards, and 3 green cards. What is the probability of picking a yellow card? A

A. $\frac{5}{18}$ **B.** $\frac{3}{18}$
C. $\frac{5}{10}$ **D.** not given

14. Write an inequality for *the number t is at least 35.* C

A. $t > 35$ **B.** $t < 35$
C. $t \geq 35$ **D.** not given

Study Skills

One of the rewarding aspects of mathematics is that it is possible for a student to know whether or not an answer is correct. Taking the time to *check answers* can help students discover and correct mistakes. This can enhance not only their test scores but their understanding of the material.

Students can ask themselves questions such as: Did I ignore any important data? Did I label elements of the problem correctly? Can I substitute numbers in an algebraic answer and get the correct answer? Is the answer in the correct units?

Other suggestions for checking answers include:

- Re-read the question to understand what is being asked.
- Use inverse operations.
- Determine whether or not the answer is reasonable.
- Estimate.
- If an answer is incorrect, try the problem again without referring to the answer. This technique helps eliminate repeating a mistake.

POLYNOMIALS

13 Chapter At a Glance

In this chapter students learn to identify and simplify polynomials. Students explore addition, subtraction and multiplication of polynomials using models.

Reading the Chapter

In this chapter students distinguish between monomials, binomials, trinomials, and polynomials. Review the definitions of the prefixes *mono-*, *bi-*, *tri-*, and *poly-*, pointing out that the prefix of each word is a clue to its meaning.

Remind students of the importance of reading problems carefully. Encourage them to look for like terms in both mathematical expressions and word problems. It may help some students to identify and write down the "given" in a problem before finding a solution. This often helps to focus on a strategy for solving the problem.

Chapter Resources

Practice and Enrichment TRB M-1 to M-8, M-10 to M-11
Alternate Application TRB M-9
Chapter Test Form A, M-13
Chapter Test Form B, M-15
Computer Test Bank

References

STUDENT BIBLIOGRAPHY

Wells, David. *The Penguin Dictionary of Curious and Interesting Numbers.* (Viking Penguin, New York 1986).

Dupré, Jean-Paul. *Barron's Junior Fact Finder.* (Barron's Educational Series, New York 1989).

TEACHER BIBLIOGRAPHY

Howden, Hilde. *Algebra Tiles for the Overhead Projector.* (Cuisenaire Co. of America, Inc., New Rochelle, New York 1985).

SOFTWARE

Tobbs Learns Algebra (Sunburst Communications), for the Apple II family of computers.

FILMS, VIDEOTAPES, AND OTHER MEDIA

Opening Doors: An Introduction to Peer Coaching. (Association for Supervision and Curriculum Development, Alexandria, Va.).

Vocabulary

binomial
FOIL method
monomial
matrix
polynomial
trinomial
Cramer's rule
determinant

Materials

algebra tiles (*M)
colored paper
math journal

*M—Manipulative Kit

Resource Bank

Class Activities

FIELD TRIPS

▼ Pharmaceutical manufacturing plant—to learn how chemical formulas are used in the manufacturing process.

▼ Advertising agency—to discuss how demographic information influences advertising decisions.

CLASS SPEAKERS

▼ Finance Manager—to discuss how the financial health of a company is measured.

▼ Public health statistician—to explore how health statistics are gathered and analyzed.

Bulletin Board

Have groups of students research and display data on a variety of topics related to ecological issues and environmental hazards. Some possible ideas are shown below.

Have students write critical thinking questions for their topic and exchange their questions with another group. Possible examples:

1. What activities can you suggest to decrease water consumption?
2. What factors may contribute to the endangerment of the animal species shown?

Project: Have students undertake a long-term project, e.g., a recycling project, that aims at reducing an adverse environmental situation.

OUR ENVIRONMENT - Facts to Consider

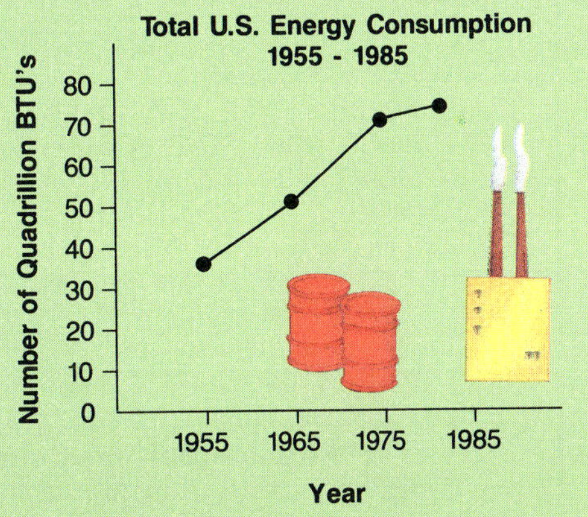

The Amount of Water Used
1. To shower — 25-50 gal
2. To brush teeth with the water Running — 2 gal
3. To wash dishes by hand — 20 gal
4. To run dishwasher — 20 gal

Each and Every Day
- The U.S. produces 1.5 billion pounds of hazardous waste.
- American public school students throw out $3 million worth of lunch food.

Selected List of North American Endangered Species
Brown or grizzly bear, Southern sea otter, Utah prairie dog, California Condor, American Crocodile

Extra-Credit Problems

BASIC

▼ A rectangle is made up of right triangles A and B. The height of $\triangle A$ is x and the base is $x + 8$. The area of $\triangle B$ is $y + 30$. Write an equation. Solve for y if $x = 12$. [$x(x + 8) = 2(y + 30)$; $y = 90$]

AVERAGE

▼ Window 1 is a rectangle 12 in. wider than it is high. Window 2 is also a rectangle with area 340 in.^{2} greater than window 1. If window 2 is 3 in. wider and 4 in. higher than window 1, find the dimensions of both. (40 in. by 52 in.; 44 in. by 55 in.)

ENRICHED

▼ A circular clock has a border around it that is 2.5 cm wide. If the area of the border is 176.625 cm^2, what is the diameter of the clock? (20 cm)

Chapter Planning Guide

OBJECTIVES	ASSIGNMENT GUIDE			ASSESSMENT	
	Basic	Average	Enriched	Review/ Tests	TRB*
Exploring Polynomials To explore polynomials using models.	**Manipulative Connection** All	All	All		
13-1 Polynomials To define, identify, and simplify polynomials.	1-2, 5-7, 11-13, 19-21, 25-26, 29-32, 36-38, MR*, CT* All	2-3, 6-8, 12-14, 20-22, 26-29, 31-33, 37-39, MR, CT All	3-4, 8-10, 14-18, 22-24, 27-29, 33-35, 39-41, MR, CT All	Extra Practice p. 578	Practice M-1 Enrichment M-2
13-2 Adding and Subtracting Polynomials To simplify polynomials by combining like terms.	1, 3-4, 7-8, 11, 13-14, 19-24, 30, 32-33, 36-37, MR All	1, 4-5, 8-9, 11, 15-16, 23-28, 32-34, 37, 39, MR All	2, 5-6, 9-10, 12, 17-18, 25-32, 37-38, 40, MR All	Extra Practice p. 578	Practice M-3 Enrichment M-4 TY M-12
13-3 Multiplying a Polynomial by a Monomial To multiply a polynomial by a monomial	1-3, 7-9, 11-20, 31-34, 38-39, MR All	2-4, 7-10, 11-29 odd, 32-35, 39-41, MR All	4-7, 10, 12-30 even, 34-37, 40-41, MR All	Extra Practice p. 578	Practice M-5 Enrichment M-6
Exploring Mind Reading Tricks To explore algebraic expressions.	All	All	All		
13-4 Multiplying Binomials To find the product of two binomials.	1, 4-6, 10-12, 16-24, 34-36, 40-41, 44, 46-47, 49-50, 53, MR, TY* All	2, 5-7, 11-13, 19-27, 35-37, 41-42, 45-47, 49-51, 53-54, MR, TY All	3, 7-9, 13-15, 25-33, 37-39, 42-43, 45-46, 48-54, MR, TY All	Extra Practice p. 578 TY p. 553	Practice M-7 Enrichment M-8 TY M-12
13-5 Problem Solving Application: Using Matrices To solve systems of equations using matrices.	All	All	All		Alternate Application M-9
13-6 Problem Solving Strategy: Using Multiple Strategies To solve problems using one or more strategies.	**Problem Solving Connection** All	All	All	Problem Solving Practice p. 561	Practice M-10 Enrichment M-11
Chapter Test	All	All	All	pp. 562-565	Tests M-13–M-19

 *TRB—Teacher's Resource Book MR—Mixed Review CT—Critical Thinking TY—Test Yourself

CONNECTIONS							NCTM CORRELATION
Algebra	Critical Thinking	Problem Solving	Estimation/ Mental Math	Technology	Manipulatives/ Resources	Writing/Reading in Math	
All	All				1, 3-4, 6, 8-10	2, 4 7, 9	Algebra Reasoning Communication
C*: All W*: All	C: 1-8, 14-16, TD*, CT*: All		W: 30-35	W: 36-41	C: 9-13 W: 19-28	W: 29	Algebra Reasoning Communication
C: All W: All	C: TD All W: 31	C: 11 W: 32-39			C: 1-2, 7-8 W: 1-2, 11-12	C: 11 W: 31-39	Algebra Reasoning Communication
C: All W: All	C: 13-16, TD All W: 7-10, 38-41	W: 31-37	W: 7	W: 8-10	C: 1-6 W: 1-6	W: 7 38-41	Algebra Geometry Problem Solving
All	All				All	2	Algebra Reasoning Communication
C: All W: All TY*: All	C: TD All W: 40-43 TY: 1-14, 19	W: 44-54	W: 34-39		C: 1-4 W: 1-3, 53-54	W: 40-48, 53-54, TY: 19, 20	Algebra Communication Geometry
C: All W: All	C: 1-4 W: 4-6	W: 7-11		W: All		W: 10-11	Algebra Reasoning Problem Solving Communication
	C: All W: All	C: All W: All		W: All		C: All W: All	Problem Solving Reasoning Communication

*C—Class Exercises W—Written Exercises TD—Think and Discuss CT—Critical Thinking TY—Test Yourself

Exploring

In this manipulative activity, students explore the square of a sum by repeatedly folding a square sheet of paper. **See TRB Exploring O-49.**

Exploring the Square of a Sum

This activity shows how an algebraic expression can be illustrated geometrically.

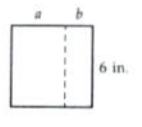

Start with a square sheet of paper measuring 6 in. on a side. Fold $\frac{1}{3}$ of the square to form a vertical crease parallel to the edge.

1. What is the length of *a*? of *b*? **4 in.; 2 in.**
2. Write an equation that represents the side measure as the sum of the two lengths.
   4 + 2 = 6 or *a* + *b* = 6

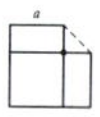

Fold the corner of length *b* onto the vertical crease to locate a point. Mark the point. Fold a horizontal crease through the point parallel to the edge of the square.

3. What is the shape of the region that has *b* as a side? What are the dimensions of this figure?
   square; 2 in. × 2 in.
4. What is the shape of the region that has *a* and *b* as sides? What are the dimensions of the figure?
   rectangle; 2 in. × 4 in.

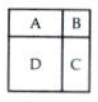

Your square should look like the figure at the right. Label the dimensions and identify the figures formed for each region.

Region	Dimensions	Type of Figure Formed
A	**2 × 4**	**rectangle**
B	**2 × 2**	**square**
C	**4 × 2**	**rectangle**
D	**4 × 4**	**square**

5. If the areas of these regions are added together, what must they equal?
   the area of the original square; $(a + b)^2$ or 36
6. What is true about the area of regions A and C? **They are the same.**
7. Complete.
   area of original square = area of **A** + area of **B** + area of **C** + area of **D**
   area of original square = area of **B** + 2 × area of **A or C** + area of **D**
8. Substitute *a* for the longer dimension and *b* for the shorter dimension.
   What is the area of region A? **ab** B? **b^2** C? **ab** D? **a^2** original square? **$(a + b)^2$**
9. Complete the expression: $(a + b)^2$ = **a^2** + 2 **ab** + **b^2**

O-49

Exploring Patterns

Consider the rectangular array of numbers.

1	2	3	4	5	6	7
8	9	10	11	12	13	14
15	16	17	18	19	20	21
22	23	24	25	26	27	28

1. Draw a box around the numbers 8, 9, 10, 15, 16, 17, 22, 23, and 24.
   a. Find the product of the numbers in the opposite corners of the box.
   8 × 24 **192** 10 × 22 **220**
   b. What is the difference between the products? **28**
2. Draw a box around the numbers 10, 11, 12, 17, 18, 19, 24, 25, and 26.
   a. Find the product of the numbers in opposite corners of the box. **10 × 26 = 260** **24 × 12 = 288**
   b. What is the difference between the products? **28**
3. Suppose you draw a box around any other nine numbers in the array. Do you think the difference between the products of numbers in opposite corners will also be 28?
   Answers may vary.
   Possible answer: yes.

You can multiply polynomials to test the idea in question 3. Refer to the nine numbers in question 1. The middle number of the array is 16.

4. Let $n = 16$. What expression can you write with *n* to represent 23?
   $n + 7$
5. Write an expression in terms of *n* to represent each number.
   8 **$n - 8$** 9 **$n - 7$** 10 **$n - 6$** 15 **$n - 1$**
   17 **$n + 1$** 22 **$n + 6$** 24 **$n + 8$**
6. Find the product of the expression that represents the opposite corners.
   $(n - 8)$(**$n + 8$**) = **$n^2 - 64$**
   $(n - 6)$(**$n + 6$**) = **$n^2 - 36$**
7. Did you find that the difference between the products in opposite corners of an array of nine numbers is always 28?
   yes

O-50

In this problem solving activity, students explore patterns in an array of numbers. **See TRB Exploring O-50.**

In this calculator activity, students explore the pattern in the difference between successive squares. **See TRB Exploring O-51.**

Exploring Successive Squares

Mathematicians find patterns in numbers and use these patterns to make predictions.

Use the x^2 key on the calculator to simplify numbers. Look for patterns to help you predict the answers.

1. Find 20^2. **400** Find 21^2. **441**
   What is the difference between the two square numbers? **41**
   How is the difference related to 20? **twice 20 + 1**
2. Find 40^2. **1,600** Find 41^2. **1,681**
   What is the difference between the two square numbers? **81**
   How is the difference related to 40? **twice 40 + 1**
3. What do you think will be the difference between 80^2 and 81^2? **Answers will vary.**
   Find 80^2. **6,400** Find 81^2. **6,561**
   What is the difference between the two square numbers?
   161
   How is the difference related to 80? **twice 80 + 1**
4. What do you think will be the difference between 25^2 and 26^2? **Answers will vary.**
   Find 25^2. **625** Find 26^2. **676**
   How is the difference related to 25? **twice 25 + 1**
5. If the first number is *x*, write the expression for its square.
   x^2
   Write the expression for the square of the next consecutive number.
   $(x + 1)^2$
   Find the difference between the two consecutive square numbers.
   $(x + 1)^2 - x^2 = x^2 + 2x + 1 - x^2 = 2x + 1$
6. How would you describe in words what the difference is between consecutive square numbers?
   The difference between two consecutive squares is twice the square root of the lesser square plus one.
7. What do you think the difference will be between 3.4^2 and 4.4^2?
   2(3.4) + 1 = 7.8

O-51

Exploring the Meaning of a Polynomial

The following problem can help you understand polynomials.

Jason is 16 years old. Each year, beginning with his 13th birthday, his mother has given him $100. He invests the money at an annual yield of 6%.

1. How much money did Jason get on his 13th birthday? **$100**

If 6% is the interest rate, we can express the interest as 100(0.06). We can express the total amount in the account at the end of the first year as 100 + 100(0.06). Another way is 106% of $100 or 100(1.06).

2. What does Jason add to the account on his 14th birthday? **$100**
   Money in the account after Jason's 14th birthday: 100(1.06) + **100**
3. During the year the amount in the account earns 6% interest. Use the distributive property to write the addition expression another way so that it represents the total amount in the account.
   Money in the account after earning 6% interest: 1.06[100(1.06) + 100] =
   $100(1.06)^2 + 100(1.06)$
4. What does Jason add to the account on his 15th birthday? **$100**
   Money in the account after Jason's 15th birthday: $100(1.06)^2 + 100(1.06)$ + **100**
5. During the year the amount is invested at 6%. Use the distributive property to write the addition expression another way so that it represents the total amount in the account now.
   Money in the account after earning 6% interest: $1.06[100(1.06)^2 + 100(1.06) + 100]$ =
   $100(1.06)^3 + 100(1.06)^2 + 100(1.06)$

A spreadsheet can be used to organize and evaluate the data.

	A	B	C	D
1	amount before	birthday	amount	total
2	birthday	year	added	
3	0	13	100	100
4	106	14	100	206

6. Write a formula for each of the following cells. A4 = **D3*1.06** D4 = **A4+C4**
   A5 = **D4*1.06** D5 = **A5+C5** A6 = **D5*1.06** D6 = **A6+C6**
7. How much money is in his account after his 16th birthday? **$437.46**
8. Explain how the polynomial, $100y^4 + 100y^3 + 100y^2 + 100y$ represents the amount of money in Jason's account just before his 17th birthday.
   Since *y* represents 1.06, the polynomial represents $437.46 plus another year of interest for a total of $463.71.

O-52

In this computer activity, students explore the meaning of a polynomial by writing and evaluating equations using a spreadsheet. **See TRB Exploring O-52.**

Meeting Individual Needs

GIFTED STUDENTS

Gifted students often enjoy the opportunity to apply what they have learned in one area of mathematics to other areas. Have students work in pairs to solve each problem.

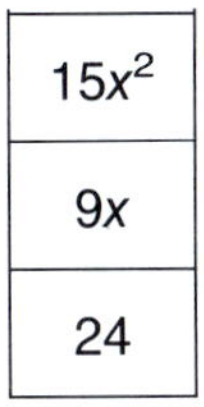

The area of this rectangle is $15x^2 + 9x + 24$. What are the possible lengths and widths? $[3(5x^2 + 3x + 8)]$

If $x = 4$ in., what is the area of the rectangle? (300 in.^{2})

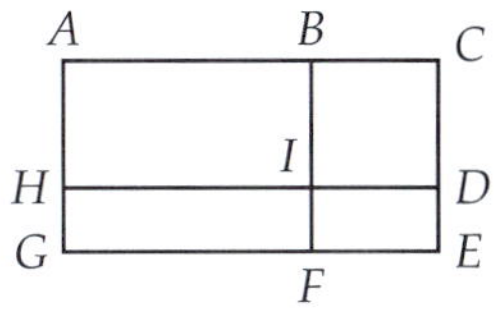

Use the rectangle above to solve each problem. Let $CD = 2x$, $DE = x$, $FE = y$, and $GF = 2y$.

1. Find the perimeter of rectangle *ACDH*. $(6y + 4x)$

2. Find the area of rectangle *ACEG*. $(9xy)$

AT-RISK STUDENTS

At-risk students may not feel that they can apply what they have learned about polynomials to their daily lives. To reinforce the skills needed to multiply binomials, have students practice with actual numbers.

Ask students to work in groups to find the product 7×7.99 mentally. Encourage the groups to think of 7.99 as a binomial $(8 - 0.01)$ and then solve: $7(8 - 0.01) = 56 - 0.07 = 55.93$.

Next, ask the groups to use two binomials to find the product of two actual numbers: (78)(82).

Help groups identify these binomials and then solve: $78 = 80 - 2$ and $82 = 80 + 2$; therefore,

$$\begin{aligned}(78)(82) &= (80 - 2)(80 + 2)\\ &= 6{,}400 + 160 - 160 - 4\\ &= 6{,}400 - 4\\ &= 6{,}396.\end{aligned}$$

Ask groups to use these methods to solve these problems: **1.** (57)(63) [3,591]; **2.** (49)(51) [2,499]; **3.** (9)(10.98) [98.82]; **4.** (6)(9.92) [59.52]

ENGLISH AS A SECOND LANGUAGE

In this chapter, students are introduced to the terms *monomial, binomial, trinomial,* and *polynomial*. For the ESL student, these similar sounding words may present an extra challenge. Before introducing these words, review the prefixes: *mono-* (one), *bi-* (two), *tri-* (three), and *poly-* (many).

As each word is introduced, write it on the chalkboard with several examples. As the word is used during the lesson, refer to the chalkboard. Encourage ESL students to use the words during the class.

To reinforce an understanding of each type of expression, have students work in pairs. The first student states an expression, for example, binomial. The second student then writes that type of expression, and the partners check the expression together. Have students repeat this several times, exchanging roles. This exercise can be expanded by having the first student ask for an expression with a monomial and binomial.

See TRB Teaching Resources p. 63.

Problem Solving Situation

The chart on global carbon emissions indicates that North America has a significantly higher level of carbon emissions than Latin America, the U.S.S.R., and Western Europe. (Possible answers: cold winters, high level of industrialization, and dense population.) Have students research the reason(s) for the higher levels. Students should focus on the types of fuels used throughout each country.

DATA COLLECTION

Information on energy consumption is available in periodicals, almanacs, encyclopedia, and other reference books available at the public library. In addition, students could write to the U.S. Energy Information Administration for detailed statistics.

DATA ANALYSIS

Have students compare statistics on the consumption of different kinds of fuel or energy sources within a particular country or region. If data are available, have students compare statistics on the differences in consumption of a single fuel by two or more countries or regions. Ask students to display their data in tables or graphs, and to report orally on the effects of energy consumption on the environment.

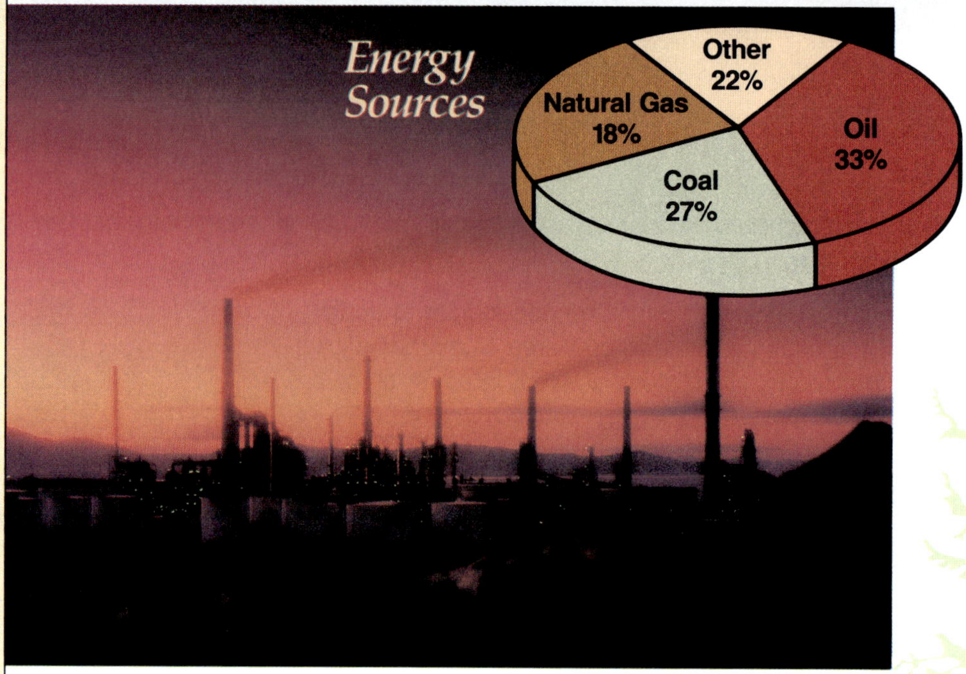

COAL, OIL, AND NATURAL GAS are *fossil fuels*. They are formed from plants and animals that lived millions of years ago. *Carbon avoidance* is the comparative cost of a coal-fired electrical plant. The fuel and operating costs of a coal plant are about $.02/kW·h. Pollution costs are about $.015/kW·h. Thus, anything greater than $.035/kW·h is a carbon avoidance cost.

Fossil Fuel Alternative	Carbon Reduction	Carbon Avoidance Cost (per ton)
Improving Energy Efficiency	100 %	$0–19
Wind Power	100 %	$107
Geothermal Energy	99 %	$123
Wood Power	100 %	$141
Steam-injected Gas Turbine	61 %	$109–200
Solar with Gas	7.9 %	$216
Nuclear	86 %	$535

CHAPTER
13 Polynomials

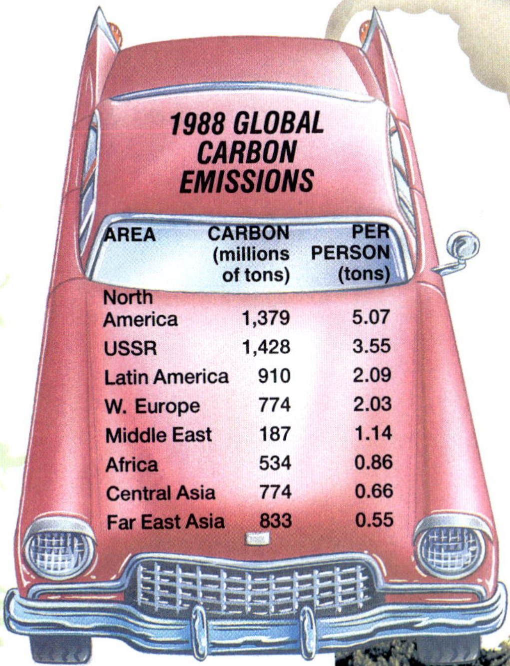

1988 GLOBAL CARBON EMISSIONS

AREA	CARBON (millions of tons)	PER PERSON (tons)
North America	1,379	5.07
USSR	1,428	3.55
Latin America	910	2.09
W. Europe	774	2.03
Middle East	187	1.14
Africa	534	0.86
Central Asia	774	0.66
Far East Asia	833	0.55

GLOBAL WARMING ■ Burning fossil fuels created 5.66 billion t of carbon waste in 1988, more than one ton for every person on the planet. Burning 1 t of carbon releases 3.7 t of carbon dioxide (CO_2). Carbon dioxide is a chief contributor to global warming. Some scientists believe that Earth's average temperature could rise 3°F to 9°F by the year 2050. This could result in the oceans rising, in coastal flooding, and the loss of farm land due to too much salt water.

CARBON DATA

Think about it...

Look at the Global Carbon Emissions data. Why do you think North America has greater carbon emissions per person than any other part of the world?

See Problem Solving Situation

Questions related to this data file can be found on pages 501, 504, and 561.

Cooperative Learning

Since there are several directions students can take to research this problem, have students work together in groups of four to six. This will allow them to research several approaches to the problem without overwhelming the resources of any one student. In their initial meeting, have students divide their research into several manageable parts. Each student could also assume an overall management role; e.g., Task Analyst, Project Manager, Data Coordinator, Data Analyst, and Data Display Specialist. Review roles with students to ensure that the work load is divided equitably: interdependence is integral to the group's organizational plan.

Lesson Focus

Materials

Algebra tiles or paper squares in two sizes and paper rectangles, math journal

MOTIVATING THE LESSON

Discuss with students how algebraic expressions, such as $3x + 2$, are modeled using algebra tiles. (Use three rectangular tiles that are each 1 unit by x units and two unit squares.) Then ask students if they can use the same tiles to model $3x^2 + 2$ and, if not, to explain why not. (No; $3x$ and $3x^2$ are not like terms, so another tile is needed to represent the square of a variable.)

Instruction

Have students work in small groups. Distribute algebra tiles or paper squares and rectangles to each group. Ask students to record their answers in their math journals. To ensure the cooperation of each student, assign roles such as Model Builder, Summary Giver, Recorder, or Mathematician, but emphasize that every student is responsible for the group's answers.

1 Help students see the relationship between the tiles for x^2, x, and 1 and the base-ten models for 100, 10, and 1. Then ask students to state an expression for the total area shown. ($x^2 + x + 1$)

2 Ask students what properties they are applying when they manipulate the models to physically combine like terms. (commutative and associative properties)

OBJECTIVE: *To explore polynomials using models.*

Exploring Polynomials

MATERIALS

- Algebra tiles or colored paper to represent integers, variables, and a variable squared
- Math journal to record work

▼ You know you can use algebra tiles to represent algebraic expressions such as $4x$, $x + 3$, and $2x + 1$. You can also use algebra tiles to represent expressions such as $x^2 + 2$ and $2x^2 + 3x$.

1. ***Compare*** the models.

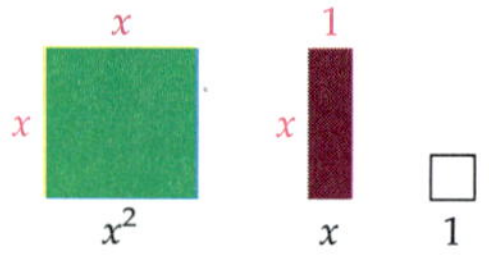

a. ***Describe*** the area of each model. The area of the large square is x^2. The area of the rectangle is x. The area of the small square is 1.

b. ***Discuss*** how the area describes the value of each model. The area is the amount of region covered by the model.

2. ***Write*** the expression represented by each model.

a. 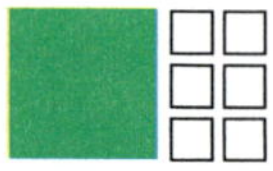 $x^2 + 6$

b. 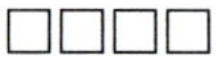 4

c.  $2x^2 + 3x$

d.  $x^2 + 2x + 1$

3. ***Model*** each expression. Check students' work.

a. $3x^2 + 2$ **b.** $2x + 3$ **c.** $4x^2 + 3x$

d. $x^2 + x + 4$ **e.** $3x^2 + x + 2$ **f.** $5x^2 + 4x$

▼ You can use models to help you combine like terms.

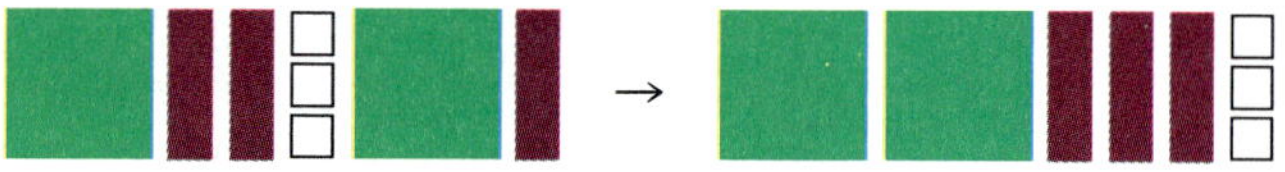

4. Use a model to represent each expression. Then use the model to combine like terms. Write the resulting expression. Check students' models.

a. $2x^2 + 5 + 3x + x^2$ $3x^2 + 3x + 5$

b. $4x + 3x^2 + 5 + x^2$ $4x^2 + 4x + 5$

c. $7 + 3x^2 + 4x + 3 + 2x$ $3x^2 + 6x + 10$

d. $5x + x^2 + 2x + 6$ $x^2 + 7x + 6$

e. $3x^2 + 2 + 2x + x^2 + 5$ $4x^2 + 2x + 7$

f. $12 + 3x^2 + 5x + 2 + x^2 + 4x$ $4x^2 + 9x + 14$

g. $7 + 2x^2 + 4x + 2 + 2x^2 + 3x$ $4x^2 + 7x + 9$

Teaching Tip

To gain insight into students' understanding and to provide practice in communicating mathematical procedures and concepts, have students explain what they are doing as they manipulate tiles to find a sum or difference.

5. ***Describe*** how you can use what you know about pairing like terms to find $(2x^2 + 3x + 4) + (x^2 + x + 3)$. **Combine the x^2 terms, the x terms, and the constants. The result is $3x^2 + 4x + 7$.**

6. Use models to find each sum. **Check students' work.** (3)
   a. $(3x^2 + 2x - 7) + (2x^2 + 4x + 2)$ **$5x^2 + 6x - 5$**
   b. $(2x^2 + 3x) + (5x^2 + 8x + 12)$ **$7x^2 + 11x + 12$**
   c. $(x^2 + 2) + (3x^2 + x)$ **$4x^2 + x + 2$**
   d. $(x^2 + 3x + 2) + (2x^2 + x + 3)$ **$3x^2 + 4x + 5$**

7. ***Summarize*** Write a rule for finding the sum of two algebraic expressions. **Combine like terms from each algebraic expression.** (4)

▼ You can use algebra tiles to help you find the difference of two algebraic expressions.

8. ***Explore*** what happens when you subtract $x^2 + x + 3$ from $3x^2 + 2x + 5$.
   a. Model $3x^2 + 2x + 5$. **Check students' work.**

   b. Remove tiles that represent $x^2 + x + 3$.
   c. The remaining tiles represent the difference which is $2x^2 + x + 2$.

9. ***Write*** the subtraction expression for each model.
   a.

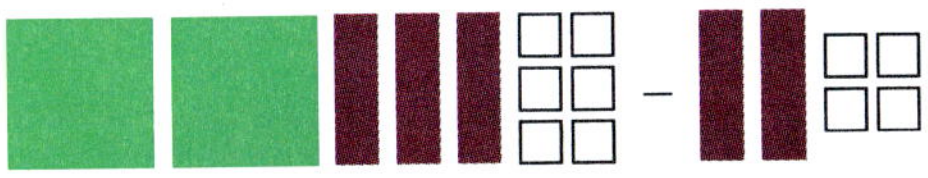

$2x^2 + x + 2$

   b.

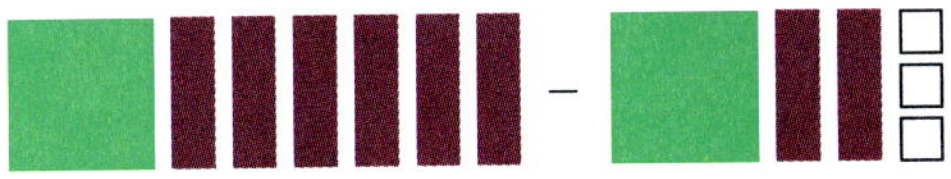

$4x - 3$

10. ***Model*** each difference. Write the resulting expression. **Check students' work.** (5)
   a. $(4x^2 + 3x + 5) - (2x^2 + x + 3)$ **$2x^2 + 2x + 2$**
   b. $(5x^2 + 2x + 1) - (4x^2 + 2x + 1)$ **x^2**
   c. $(3x^2 + 6x + 8) - (x^2 + 2x + 6)$ **$2x^2 + 4x + 2$**
   d. $(6x^2 + 4x + 7) - (6x^2 + 3x + 7)$ **x**
   e. $(4x^2 + 2x + 7) - (x^2 + 2x + 3)$ **$3x^2 + 4$**
   f. $(x^2 + 3x + 2) - (x^2 + 2x + 1)$ **$x + 1$**
   g. $(5x^2 + 5x + 5) - (4x^2 + 4x + 4)$ **$x^2 + x + 1$**

(3) Students may suggest placing the tiles for the second polynomial under the first polynomial. Allow them to do so if it helps them to keep track of the like terms.

(4) Have students use their rule to find sums such as $(x^2 + 5x + 3) + (3x^2 + 2x + 1)$ and then verify the result using tiles. $(4x^2 + 7x + 4)$

(5) Ask students if they can model the difference $(5x^2 + 4x + 6) - (3x^2 + 2x + 9)$ and, if not, to explain why not. (Yes; they must add zero pairs to show 6 − 9.)

Closure

Have the groups share their summaries for modeling the sum or difference of two algebraic expressions. Then ask students to make whatever adjustments are necessary to the notes they recorded in their math journals.

Assignment Guide

Basic All
Average All
Enriched All

FOR THE NEXT LESSON

algebra tiles or paper squares in two sizes and paper rectangles, math journal, index cards, calculator, number cubes

Lesson Focus

Materials

Algebra tiles or paper squares in two sizes and paper rectangles, math journal, index cards, calculator, number cubes

Vocabulary/Symbols

binomial
monomial
polynomial
trinomial

MOTIVATING THE LESSON

Write algebraic expressions such as the following on the chalkboard and ask students to simplify them by combining like terms.

1. $3x^2 + 2x - x^2$
$(2x^2 + 2x)$

2. $5x^2 - 3x^2 + 7$
$(2x^2 + 7)$

3. $8 - x + 3x - 5 + x^2$
$(3 + 2x + x^2)$

Instruction

1 Have students give more examples of monomials so that they recognize which kinds of expressions are monomials and which are not.

2 Students might notice that a polynomial is simply an algebraic expression. Stress that although polynomials are algebraic expressions, not all algebraic expressions are polynomials, as illustrated in Example 2c: $\frac{7}{a^2} - 2b^2$.

OBJECTIVE: *To define, identify, and simplify polynomials.*

13-1 Polynomials

▼ Mathematicians use algebraic expressions to represent real world situations. The expression $4.8s^2$ represents the distance in meters that an object falls in s seconds. The expression $4.8s^2$ is a *monomial*.

Monomial	A monomial is a real number, a variable, or the product of a real number and one or more variables.

A monomial cannot contain any operation other than multiplication and cannot have a variable as an exponent.

Example 1 **Tell which expressions are monomials.**

a. $3x^2y$ **b.** 8 **c.** $8 + a$ **d.** $\frac{a}{7y}$

Solution

a. monomial; It is the product of a real number (3) and one or more variables (x, x, y).

b. monomial; It is a real number.

c. not a monomial; It is the sum of a real number and a variable.

d. not a monomial; The denominator cannot contain a variable.

▼ Some algebraic expressions that include monomials are *polynomials*.

Polynomial	A polynomial is a monomial or a sum or difference of monomials.

Example 2 **Tell which expressions are polynomials.**

a. $x + 3y$ **b.** $\frac{7}{8}a^2 - 2b^2$ **c.** $\frac{7}{a^2} - 2b^2$

Solution

a. polynomial; It is the sum of two monomials.

b. polynomial; It is the difference of two monomials.

c. not a polynomial; The expression $\frac{7}{a^2}$ is *not* a monomial because the denominator contains a variable.

FLASHBACK
Terms are the parts of an expression that are separated by addition and subtraction symbols.

▼ We call the monomials that make up a polynomial its *terms*.

Example 3 **How many terms does each polynomial have?**

a. $3x^2y$ **b.** $2x - 4y$ **c.** $4a^2 + 2ab - 5b^2$

Solution **a.** one **b.** two **c.** three

Teaching Tip

Have each student write a monomial on a card. Then collect the cards so that they can be used for the Reteaching Activity.

▼ Some polynomials have special names that identify the number of terms in the polynomial.

Polynomial	Terms	Examples
monomial	one	$0.08x$, mn
binomial	two	$a - 3b$, $x^2y + 4$
trinomial	three	$2a - 4 + 6b$, $xy^3 - 0.2xy + 1.55y^4$

Example 4 **Identify each expression as a monomial, binomial, or trinomial.**

a. $3x - 2y$ **b.** $8x^3yz$ **c.** $7x^5y + 8y + 18$ **d.** $12 - n$

Solution **a.** binomial **b.** monomial **c.** trinomial **d.** binomial

▼ You can write a polynomial for a model. 3

Example 5 **Write the polynomial represented by the model.**

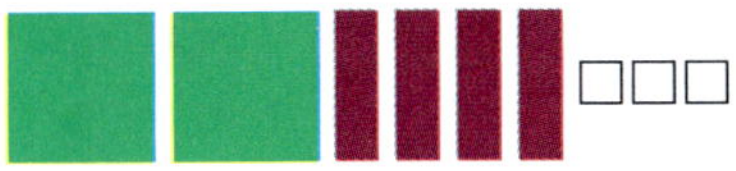

Solution $2x^2 + 4x + 3$

▼ You can model a given polynomial.

Example 6 **Show a model for $3x^2 + x + 2$.**

Solution

▼ You can evaluate any polynomial when given values of the variables. 4

Example 7 **Evaluate each polynomial for $m = 8$ and $p = -3$.**

a. $3m - 2p$ **b.** $m^2 + 3m - 6$ **c.** $3p^2 - 4m + 15$

Solution

a. $3m - 2p = 3(8) - 2(-3)$
$= 24 - (-6)$
$= 24 + 6$
$= 30$

b. $m^2 + 3m - 6 = 8^2 + 3(8) - 6$
$= 64 + 24 - 6$
$= 82$

c. $3p^2 - 4m + 15 = 3(-3)^2 - 4(8) + 15$
$= 3(9) - 4(8) + 15$
$= 27 - 32 + 15$
$= 10$

If you can jump 3 ft on Earth, you can jump 10 ft on Mercury. Let j = height to which you can jump on Earth. What polynomial would express the height to which you could jump on Mercury? $\frac{10j}{3}$

3 Ask students to describe how the model relates to the polynomial it represents. (The large squares correspond to the x^2 term; the rectangles correspond to the x term; and the small squares correspond to the constant term.)

4 Emphasize that the procedure for evaluating polynomials is identical to the procedure for evaluating any other algebraic expression.

ADDITIONAL EXAMPLES

Which of the following are polynomials?

1. $3x^3 - 2xy + z$ (polynomial)

2. $7 - \frac{8}{n}$ (not a polynomial)

3. How many terms does the polynomial $4xy + y$ have? Identify it as a monomial, binomial, or trinomial. (two terms; binomial)

4. Show a model for the polynomial $4x^2 + 5x + 3$.

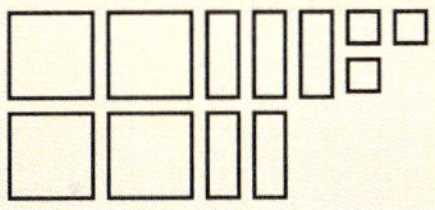

5. Evaluate $n^2 - 2n - 5$ for $n = -2$. (3)

MATH MINUTES

Working in pairs, one student writes a polynomial in one variable, such as $2a^2 - a + 5$. The other student tosses a number cube and evaluates the polynomial for the number that lands face up.

Guided Practice

Class Exercises To check for understanding, have students discuss their answers.

Think and Discuss For question 1, ask students to name the prefixes. (*mono-*, *bi-*, *tri-*, *poly-*)

For question 2, in order to classify the expression as a polynomial, some students may need to be reminded that each term in an algebraic expression must be a monomial.

A ***common error*** occurs when students attempt to model a monomial such as $3x^2$ as

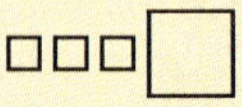

Point out that this model means $3 + x^2$ and that just as 3 is modeled as $1 + 1 + 1$, $3x^2$ is modeled as $x^2 + x^2 + x^2$.

Closure

Writing in Math Ask students to select three things they learned in this lesson and to write a brief description of each in their math journals.

Independent Practice

For questions 19-24 have the students work in groups using algebra tiles to model the polynomials.

As a follow-up, refer to the computer activity on p. 534E of the Chapter Overview. See also TRB Exploring O-52.

Critical Thinking Students can find the answers by using intuitive reasoning or by solving the inequalities, then by listing the integers that satisfy each inequality and looking for the solutions that are common to both inequalities.

THINK AND DISCUSS

1. Find the meaning of the prefixes used for the expressions: monomial, binomial, trinomial, and polynomial. Do the prefixes reflect the expressions they represent? What would you call a polynomial with four terms? **See below.**

2. Is $y^2 + 3y + \frac{7}{y}$ a trinomial? Explain.

1. mono- means one, bi- means two, tri- means three, poly- means many. Yes, the prefixes do reflect the expressions they represent. quadrinomial
2. No; it is not a trinomial because $\frac{7}{y}$ is not a monomial.

11. binomial
12. monomial
13. monomial
14. trinomial
15. monomial
16. binomial
17. binomial
18. trinomial

CLASS EXERCISES

Which of the following are monomials?

1. $2 + x$ **no** **2.** $18ab^2$ **yes** **3.** $\frac{4}{b}$ **no** **4.** 1 **yes**

Identify each expression as a monomial, binomial, or trinomial.

5. $3xy + 4y^3$ **binomial** **6.** $0.8x$ **monomial** **7.** $1.7y^2 + 2.4y - 9$ **trinomial** **8.** 658 **monomial**

Write the polynomial represented by each model.

9. 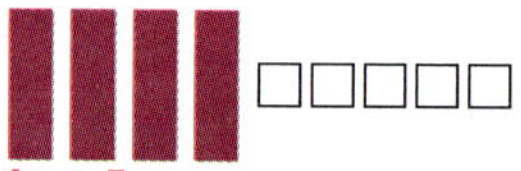 **4x + 5**

10.  **$2x^2 + 2x + 3$**

Model each polynomial. **Check students' work.**

11. $3x^2 + 2x + 4$ **12.** $5x^2 + 3^2$ **13.** $2a^2 + a + 7$

Evaluate each polynomial for $a = 2$ and $b = 4$.

14. $5a + 7b$ **38** **15.** $2a^2 - b + 4$ **8** **16.** $ab^2 + 5$ **37**

WRITTEN EXERCISES

Which of the following are monomials?

1. $2x$ **yes** **2.** $-0.3y + 9.35$ **no** **3.** $\frac{a}{3}$ **yes** **4.** 8 **yes**

How many terms are in each expression?

5. $-x^2 + 3x$ **2** **6.** 5 **1** **7.** $27 + x - 4xy$ **3**

8. $16 - 3xy + c + x^2$ **4** **9.** x **1** **10.** $x^2yz - 1$ **2**

Identify each expression as a monomial, binomial, or trinomial.

11. $3x^2 + 2x$ **12.** 21 **13.** $7p^2$ **14.** $1 + 4x - xy$

15. $5x$ **16.** $56 - x$ **17.** $4.5 + 3.7a$ **18.** $x^2 + 7x + 4$

Model each polynomial. **Check students' work.**

19. $2x^2 + x + 4$ **20.** $x^2 + 3x + 1$ **21.** $4x^2 + 2x$

22. $3x^2 + 5$ **23.** $2x + 6$ **24.** $4x^2 + 3x$

Write the polynomial represented by each model.

25. 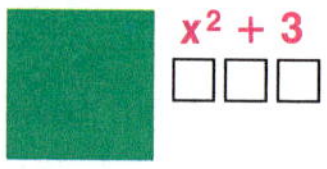 **$x^2 + 3$**

26. 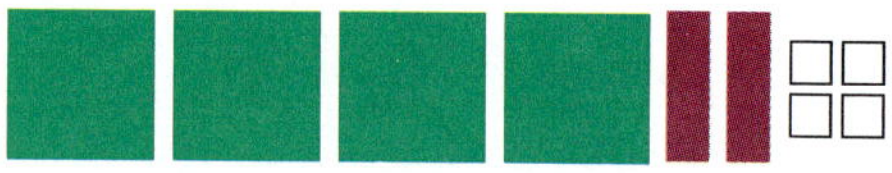 **$4x^2 + 2x + 4$**

27. 
$x^2 + 2x + 3$

28. 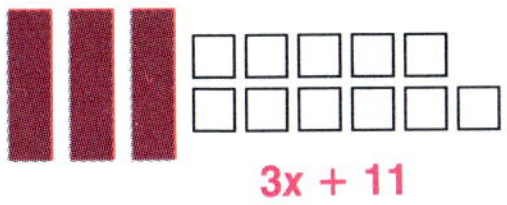
$3x + 11$

29. ***WRITE*** at least five words that begin with mono-, bi-, tri-, or poly-. Give the meaning of each word. **Answers will vary.**

MENTAL MATH Evaluate each polynomial for $a = 1$, $b = 2$, and $c = -1$.

30. $4a + b$ **6**
31. $a^2 + 2a + 3$ **6**
32. $b^2 + 6$ **10**
33. $3c + b$ **-1**
34. $c^2 + c - 1$ **-1**
35. $a^2 + c$ **0**

CALCULATOR Evaluate each polynomial for $d = 12$, $e = -11$, and $m = 15$.

36. $2d^2 + d$ **300**
37. $d^2 + 3d + 7$ **187**
38. $3d + 4e$ **-8**
39. $m^2 - 2m$ **195**
40. $e^2 + 4e - 6$ **71**
41. $8d + 7m$ **201**

Critical Thinking

EXPLORING CONCLUSIONS

For each Given, tell whether or not the Conclusions are possible.

1. Given: $x - 7 < 0$
   $x + 3 > 0$

   Conclusions:
   a. $x = 0$ **yes**
   b. $x = -1$ **yes**
   c. $x < 7$ and $x > -3$ **yes**
   d. $x > 7$ **no**
   e. $x < -7$ **no**

2. Given: $3y < 100$
   $\frac{y}{2} > 5$

   Conclusions:
   a. $y = 5$ **no**
   b. $y = -5$ **no**
   c. $y < 0$ **no**
   d. $y \geq 0$ **no**
   e. $y > 10$ and $y < 33\frac{1}{3}$ **yes**

3. Given: $a^2 > 90$
   $a + 3 < 20$

   Conclusions:
   a. $a < 20$ **no**
   b. $a > -10$ **no**
   c. $a = 7$ **no**
   d. $a = 12$ **yes**
   e. $a < -10$ **yes**

4. Given: $n^2 + 6 > 30$
   $n + 6 > 5$

   Conclusions:
   a. $n > 0$ **no**
   b. $n < 0$ **no**
   c. $n = 0$ **no**
   d. $n > 5$ **yes**
   e. $n < 5$ **no**

MIXED REVIEW

Simplify.

1. x^2x^3 **x^5**
2. $(4m^2)^3$ **$64m^6$**

Are the events dependent or independent?

3. $P(A) = \frac{3}{7}$ $P(B) = \frac{1}{3}$ $P(A \text{ and } B) = \frac{1}{7}$ **independent**
4. $P(A) = \frac{5}{6}$ $P(B) = \frac{2}{5}$ $P(A \text{ and } B) = \frac{5}{24}$ **dependent**

Simplify.

5. $2x^2 + 3x^2 + x^2$ **$6x^2$**
6. $3(x + 3) - 2x - x$ **9**

Solve.

7. Four class officers and their advisor want to stand side-by-side for a photograph. How many poses are possible? **120 poses**

Lesson Follow-up

Reteaching Activity Display cards with one monomial written on each. Ask students how to construct a binomial using the cards. (connect two cards with a plus or minus sign) Then ask the students how many binomials can be constructed from two cards and to identify those binomials.
(three; $a + b$, $a - b$, and $b - a$)

LESSON RESOURCES

TRB Practice M-1
TRB Enrichment M-2

Practice

For use after 13-1 (pp. 538–541)

Polynomials

Tell whether the expression is a monomial. Write *yes* or *no*.

1. n **yes**
2. $7 - 3m - m^2$ **no**
3. $\frac{1}{x}$ **no**
4. $7.27k^9$ **yes**
5. 1,000 **yes**
6. $\frac{1}{9}x^2y^4z$ **yes**

Identify each expression as a monomial, binomial, or trinomial. Write *m*, *b*, or *t*.

7. $36abc$ **m**
8. $10 - h^3$ **b**
9. $95xy + y$ **b**
10. $a^2 + b^2 + cd$ **t**
11. $3k$ **m**
12. $-12e + 12f^2$ **b**

Model each polynomial.

13. $x^2 + 2x + 2$

14. $2x^2 + x + 1$

Write the polynomial for each model.

15. 
$2x^2 + 2x + 3$

16. 
$3x^2 + 2$

17. 
$4x + 6$

18. 
$4x^2 + 3x$

MENTAL MATH Evaluate each polynomial for $x = -1$, $y = 3$, and $z = 2$.

19. $x^2 + z$ **3**
20. $3y + x$ **8**
21. $2z + y$ **7**
22. $x + y + z$ **4**
23. $x^2 + y^2$ **10**
24. $z - x - y$ **0**

CALCULATOR Evaluate each polynomial for $m = 21$, $n = -9$, and $p = 28$.

25. $3m - 2p$ **7**
26. $2n^2 - 5m$ **57**
27. $m^2 - n^2$ **360**
28. $n^2 + 5n - 6$ **30**
29. $5p^2 - 5p$ **3,780**
30. $7m + 6p$ **315**

Chapter 13 M-1

Enrichment

For use after 13-1 (pp. 538–541)

Using Polynomials

Solve using the given polynomials.

1. Find the surface area of a cylinder with radius 8 cm and height 14 cm. Use $\pi = 3.14$.
   1,105.28 cm²
   ($S = 2\pi r^2 + 2\pi rh$; S = surface area, r = radius, h = height)

2. Find the number of diagonals that can be drawn in a polygon with 24 sides.
   252 diagonals
   ($N = \frac{1}{2}n^2 - \frac{3}{2}n$; N = number of diagonals, n = number of sides)

3. a. Find the Fahrenheit temperature when the Celsius temperature is 345°C.
   653°F
   b. Find the Celsius temperature when the Fahrenheit temperature is 401°F.
   205°C
   ($F = 1.8C + 32$; F = Fahrenheit temperature, C = Celsius temperature)

4. a. A rectangle has length 27.1 cm and width 19.8 cm. Find the perimeter.
   93.8 cm
   b. A rectangle has perimeter 347 yd. The length is 118 yd. Find the width.
   55.5 yd
   ($p = 2l + 2w$; p = perimeter, l = length, w = width)

5. A rock thrown from the top of a cliff at an initial velocity of 3 m/s takes 6.2 s to reach the bottom. To the nearest meter, how tall is the cliff?
   170 m
   ($d = 4.9t^2 - vt$; d = distance fallen, t = time falling, v = initial velocity)

6. a. 9, 21, 33, 45, . . . is an *arithmetic sequence*. The first term is 9. The difference between terms is 12. There are 118 terms in the sequence. Find the last term.
   1,413
   b. There are 275 terms in the arithmetic sequence 981, 945, 909, 873, . . . Find the last term.
   -8,883
   ($l = a + nd - d$; l = last term, a = first term, n = number of terms, d = difference between terms)

M-2 Chapter 13

LESSON QUIZ

Identify as a monomial, binomial, or trinomial.

1. $3xy - 2z$ (binomial)
2. $12x^3yz$ (monomial)
3. Model $4x^2 + 3x + 6$.

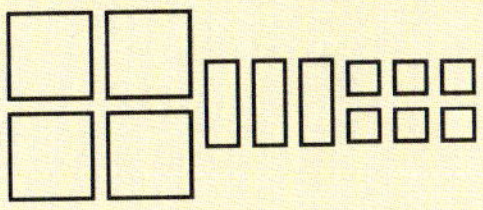

4. Evaluate $2y - 4x^2 + 1$ for $x = -2$ and $y = 3$. (-9)

Assignment Guide

Basic 1-2, 5-7, 11-13, 19-21, 25-26, 29-32, 36-38, MR, CT All

Average 2-3, 6-8, 12-14, 20-22, 26-29, 31-33, 37-39, MR, CT All

Enriched 3-4, 8-10, 14-18, 22-24, 27-29, 33-35, 39-41, MR, CT All

FOR THE NEXT LESSON

algebra tiles or paper squares in two sizes and paper rectangles, math journal, paper strips, scissors, vertically ruled paper

Lesson Focus

Materials

Algebra tiles or paper squares in two sizes and paper rectangles, math journal, paper strips, scissors, vertically ruled paper

MOTIVATING THE LESSON

Write the expression $3x^2 + 2x + 8 + 5x^2 + 2$ on the chalkboard and ask students whether it is in simplest form. (no) Then have students write the expression in simplest form. $(8x^2 + 2x + 10)$

Instruction

1 Have students copy the drawing and verify that its area can be represented by $4x^2 - 4 + x^2$.

2 Students will probably note that the models are used in the same way that they were used to combine like terms in a single expression.

3 Students may find the column format easier to use because it is easier to identify the sign associated with each term. Encourage students to use whichever format is easier for them.

13-2 Adding and Subtracting Polynomials

OBJECTIVE: *To simplify polynomials by combining like terms.*

THINK How would you use the associative and commutative properties to group like terms? **Use the commutative property to rearrange the terms and the associative property to regroup the terms so that like terms can be combined.**

1 ▼ The figure at the left represents a lawn. You can express the area of the lawn by the polynomial $4x^2 - 4 + x^2$ or simplify it as $5x^2 - 4$.

▼ You can model addition of polynomials.

Example 1 **Use models to find $(2x^2 + 3x + 7) + (x^2 + x + 3)$.**

2 **Solution**

$2x^2 + 3x + 7$

$x^2 + x + 3$

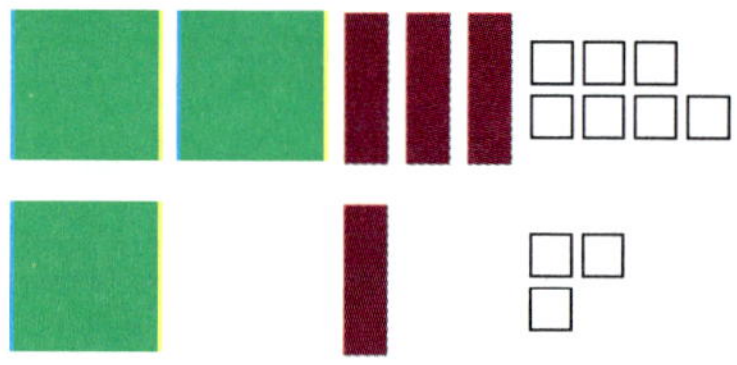

The sum is $3x^2 + 4x + 10$.

▼ You can add polynomials using number properties to combine like terms.

Example 2 **Find $(5y^2 + 3y + 9) + (2y^2 + 5y - 7)$.**

Solution

$(5y^2 + 2y^2) + (3y + 5y) + (9 - 7)$ — Group like terms.

$(5 + 2)y^2 + (3 + 5)y + (9 - 7)$ — Use the distributive property.

$7y^2 + 8y + 2$

▼ You can add polynomials in a column by aligning like terms and then combining them.

Example 3 **Find the sum of $2z^2 + 5xz - x^2$ and $4z^2 - 3xz + x^2$.**

3 **Solution**

$2z^2 + 5xz - x^2$ — Align like terms.

$+\ 4z^2 - 3xz + x^2$ — Add the terms in each column.

$6z^2 + 2xz + 0 = 6z^2 + 2xz$

▼ You can model subtraction of polynomials.

Example 4 **Use models to find $(4x^2 + 5x) - (2x^2 + 4x)$.**

Solution

$4x^2 + 5x$

Remove tiles for the second polynomial.
Count the remaining tiles.

The difference is $2x^2 + x$.

Teaching Tip

The overhead projector is especially useful for modeling concepts with small objects. Allow students to use the overhead projector to share their models with the class.

▼ You can subtract polynomials by adding the opposite of each term in the second polynomial.

Example 5 **Find $(5x - 9 + y) - (2x + 4 - 3y)$.**

Solution

$5x - 9 + y + (-2x) + (-4) + 3y$	Add the opposite of the second polynomial.
$5x + (-2x) - 9 + (-4) + y + 3y$	Group like terms.
$(5 - 2)x - 9 - 4 + (1 + 3)y$	Use the distributive property.
$3x - 13 + 4y$	

THINK Why is subtracting a polynomial the same as adding its opposite? 4 Adding a negative is the same as subtracting.

1. Models, vertical addition, and grouping like terms. Preferences will vary.
2. Place like terms one under the other and add the opposite of the bottom terms.

▼ You can use these methods to add and subtract polynomials when solving equations.

Example 6 **a.** Write a polynomial for the perimeter of the polygon. Simplify.

b. If $P = 26$ and $b = 3$, find a.

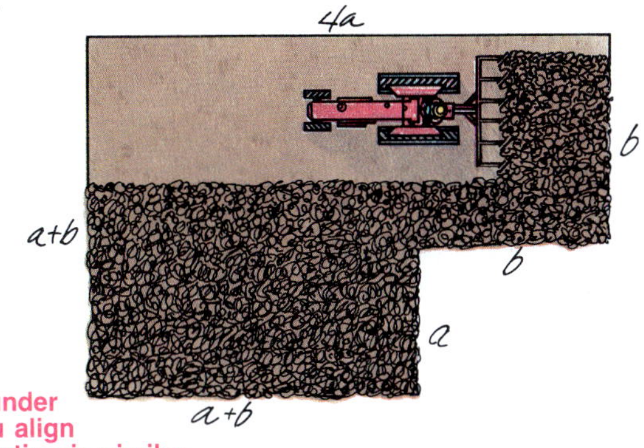

Solution **a.** $4a + b + b + a + a + b + a + b$

$7a + 4b$ Combine like terms.

b.

$7a + 4b = 26$	Write an equation.
$7a + 4(3) = 26$	Substitute values.
$7a + 12 = 26$	
$7a = 14$	
$a = 2$	

4. When adding whole numbers you align hundreds under hundreds, tens under tens, ones under ones, etc. You align terms of a polynomial according to like terms. Subtraction is similar.

CLASS EXERCISES

Use a model to find each sum. Check students' work.

1. $(x^2 + 3x + 1) + (x^2 + x + 6)$ $2x^2 + 4x + 7$
2. $(x^2 + 5x + 2) + (3x^2 + x + 1)$ $4x^2 + 6x + 3$

Find each sum.

3. $(3x - 2y) + (5x + 4y)$ $8x + 2y$
4. $(x^2 + 3x - 7) + (x^2 - 6x - 9)$ $2x^2 - 3x - 16$
5. $\begin{array}{r} 5a + 7b \\ + -3a + 2b \\ \hline \end{array}$ $2a + 9b$
6. $\begin{array}{r} x^4 + 3x^3 - x^2 + x - 2 \\ + \quad 7x^3 + x^2 - 5x - 9 \\ \hline \end{array}$ $x^4 + 10x^3 - 4x - 11$

Use a model to find each difference. Check students' work.

7. $(2x^2 + 3x) - (x^2 + 2x)$ $x^2 + x$
8. $(x^2 + 3x + 5) - (x^2 + x + 2)$ $2x + 3$

Find each difference by adding opposites.

9. $(8j - 3k + 6m) - (-2j + 3m)$ $10j - 3k + 3m$
10. $(-11a^2 + 2a - 1) - (7a^2 + 4a - 1)$ $-18a^2 - 2a$
11. $(9x^2 - 4y + 5z) - (-4x^2 - 15z)$ $13x^2 - 4y + 20z$

THINK AND DISCUSS

1. Describe the three methods you could use to add polynomials. Which do you prefer? Explain. See above.
2. Explain how you could use a column format to subtract polynomials. See above.
3. What is true about the sum of a polynomial and its opposite? The sum is 0.
4. Compare and contrast the methods for adding whole numbers with those for adding polynomials. Then repeat for the subtraction methods. See left.

4 Ask students to give the opposite of 7 and of $7x$. $(-7, -7x)$ Then ask them how they would find the opposite of a polynomial such as $3x^2 - 7x + 7$. (Find the opposite of each term in the polynomial.)

ADDITIONAL EXAMPLES

Use models to find each sum or difference.

1. $(4x^2 + 3x + 5) + (2x^2 + x + 1)$ $[6x^2 + 4x + 6]$
2. $(2y^2 - 4y - 3) - (5y + 9)$ $[2y^2 - 9y - 12]$

Find each sum or difference.

3. $(3x + 2y) + (x - 4y)$ $[4x - 2y]$
4. $(5x^2 - x) - (3x^2 + 4x - 7)$ $[2x^2 - 5x + 7]$

Guided Practice

Class Exercises For questions 3-11, encourage students who are having difficulty to use models, and to choose the format, column or horizontal, they prefer.

Think and Discuss For question 2, help students to see that either of the following ways can be used: write the opposite of the second polynomial under the first and then add; or write the two polynomials in column format and subtract term by term.

A ***common error*** occurs when students disregard the negative signs associated with the terms of a polynomial when adding or subtracting. Have students write the polynomials on a strip of paper and cut the terms apart, keeping the negative signs with the appropriate terms. Then align the strips with like terms and find the sum or difference.

MATH MINUTES

Have students solve this problem. The sum of two polynomials is $3x^2 - x - 7$. One polynomial is $x^2 + 2x + 5$. What is the other? $(2x^2 - 3x - 12)$ Have students write and solve similar problems. (Check students' work.)

Closure

Writing in Math Ask students to describe in their math journals the processes for adding and subtracting polynomials.

Independent Practice

For questions 3-10 and 13-31, students may use either column or horizontal format.

Writing in Math For question 32, have students write the paragraph in their math journals. Then have them share their ideas with the class.

For questions 33-35, have students name multiples of 2, consecutive even numbers, and consecutive multiples of 5 to help them understand the algebraic representations.

For questions 36-38, remind students that perimeter is the distance around a figure and that it can be found by adding the lengths of the sides.

MIXED REVIEW

Find the area of each polygon.

1. 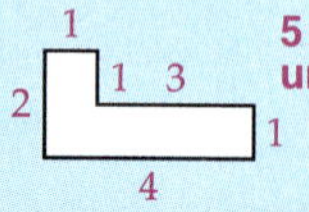

5 square units

2. 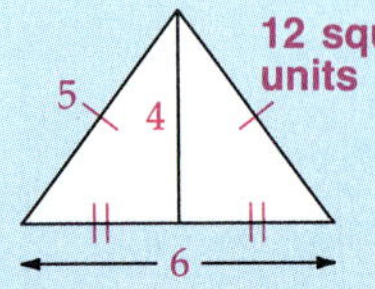

12 square units

State the number of terms in each polynomial.

3. $8yz + 1$ 2

4. $5 + 4xy - z$ 3

Simplify.

5. $6x + 3(x - 2) + 8x$ $17x - 6$

6. $2(12x + 3 - 10y)$ $24x + 6 - 20y$

7. A student participates in a walk for charity. His friends pledged a total of $3.20 for each mile he walked. The student earned $22.80 for the charity. How many miles did he walk? 7.125 miles

19. $m - 9$
20. $2j^2 + 3j$
21. $4ab - 10$
22. $15d^2q + 2dq^2$
23. $w^2 + 7w - 6$
24. $8t^2$

12. Three numbers are consecutive multiples of 4.

a. Write an expression for their sum. Use polynomials for each term. Then simplify the expression. $4x + (4x + 4) + (4x + 8)$; $12x + 1$

b. Find the numbers if the sum is 108. 32, 36, 40

WRITTEN EXERCISES

Use a model to find each sum. Check students' work.

1. $(x^2 + 3x - 2) + (3x^2 + 2x + 4)$ $4x^2 + 5x + 2$

2. $(x^2 + 2x + 1) + (x^2 + 3x + 4)$ $2x^2 + 5x + 5$

Find each sum by combining like terms.

3. $3x + 2 + (-4x + 3)$ $-x + 5$

4. $5x^2 + 3x + 7 + (7x - 2)$ $5x^2 + 10x + 5$

5. $-4x^2 + 2x - 1 + (x^2 - x + 8)$ $-3x^2 + x + 7$

6. $7x^3 + 4x^2 + 3x - 1 + (8x^3 - 10x + 18)$ $15x^3 + 4x^2 - 7x + 17$

7.
$$\begin{array}{r} x^2 + 4x - 2 \\ +\ 8x^2 - 3x + 7 \\ \hline 9x^2 + x + 5 \end{array}$$

8.
$$\begin{array}{r} xy + 5x - 2y + 4 \\ +\ 2xy - 3x - 3y - 8 \\ \hline 3xy + 2x - 5y - 4 \end{array}$$

9.
$$\begin{array}{r} x^3 + 5x^2 + 3x - 2 \\ +\ x^3 \qquad\quad - 2x + 6 \\ \hline 2x^3 + 5x^2 + x + 4 \end{array}$$

10.
$$\begin{array}{r} 4x^2 - 5xy \qquad\quad + 7 \\ +\ 8x^2 + 3xy - 3y - 4 \\ \hline 12x^2 - 2xy - 3y + 3 \end{array}$$

Use a model to find each difference.

11. $(5x + 9) - (2x + 1)$ $3x + 8$

12. $(3x^2 + x + 7) - (2x^2 + x + 2)$ $x^2 + 5$

Subtract each by adding the opposite of the second polynomial.

13. $(6y - 8) - (2y + 7)$ $4y - 15$

14. $(x^2 - 3x - 9) - (5x - 4)$ $x^2 - 8x - 5$

15. $(mn^2 + 4m - n^2) - (-3mn^2 + 2m + n^2)$ $4mn^2 + 2m - 2n^2$

16. $(6a^2b + 5ab^2 - 8) - (2a^2b - 3ab^2 + 1)$ $4a^2b + 8ab^2 - 9$

17. $(4a^2 + 3ab + b) - (2a^2 - 2ab - b)$ $2a^2 + 5ab + 2b$

18. $(7p^2q^2 + 5pq - 8) - (4pq - 5)$ $7p^2q^2 + pq - 3$

Add or subtract. See left.

19. $(3m - 8) - (2m + 1)$

20. $(8j^2 + 2j) - (6j^2 - j)$

21. $(ab - 4) + (3ab - 6)$

22. $(13d^2q - 3dq^2) + (2d^2q + 5dq^2)$

23. $(w^2 + 5w) + (2w - 6)$

24. $(11t^2 + 2) - (3t^2 + 2)$

25. $(x^2 - 5x - 9) + (-4x^2 - 3x + 17)$ $-3x^2 - 8x + 8$

26. $(ab + b - 4a) + (-2ab + 6b + 2a)$ $-ab + 7b - 2a$

27. $(y - 3x + 1) - (5x - 9 + y)$ $-8x + 10$

28. $(-3x^4y^3 - 5xy + 2) + (x^4y^3 + x^2 + xy + 1)$ $-2x^4y^3 + x^2 - 4xy + 3$

29. $(9a^7 - 7a^4 + a^2 - 8) + (8a^7 + 15a^4 + 12)$ $17a^7 + 8a^4 + a^2 + 4$

30. $(m^3n - 3m^2n^2 + 8mn - 6) - (m^3n - 3m^2n^2 + 8mn - 6)$ 0

31. $(4a^3b^2 - 9a^2b + 2ab + 11) + (7a^3b^2 - 6a^2b - 4ab + 12)$ $11a^3b^2 - 15a^2b - 2ab + 23$

32. ***WRITE*** a paragraph explaining how you can use the commutative, associative, and distributive properties in the addition and subtraction of polynomials. Check students' work.

Write an expression for each phrase. Use the same variable in each term of the expression.

33. the sum of three consecutive multiples of 2 $2a + (2a + 2) + (2a + 4)$
    a. Simplify the expression. $6a + 6$
    b. Find the numbers if their sum is 36. 10, 12, 14

34. the sum of four consecutive even numbers $x + (x + 2) + (x + 4) + (x + 6)$
    a. Simplify the expression. $4x + 12$
    b. Find the numbers if their sum is 84. 18, 20, 22, 24

35. the sum of five consecutive multiples of 5 $5a + (5a + 5) + (5a + 10) + (5a + 15) + (5a + 20)$
    a. Simplify the expression. $25a + 50$
    b. Find the numbers if their sum is 375. 65, 70, 75, 80, 85

Write the perimeter of each figure as a polynomial. Simplify.

36.

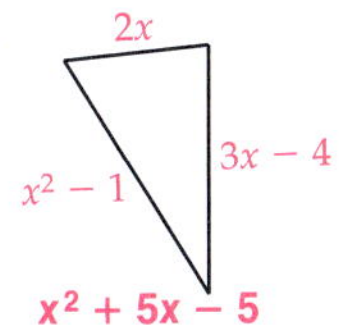

$x^2 + 5x - 5$

37.

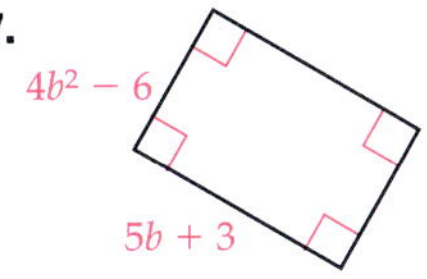

$8b^2 + 10b - 6$

38. $2a^2 - 1$

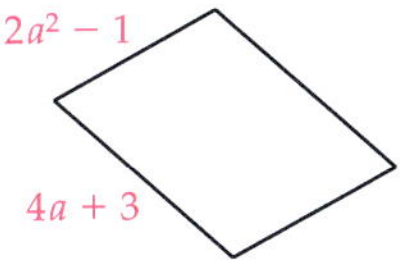

$4a^2 + 8a + 4$

39. The sum of the interior angles of a convex polygon with n sides is $180(n - 2)$. Find the sum of the interior angles for each polygon.
    a. square 360
    b. pentagon 540
    c. hexagon 720
    d. decagon 1,440

40. Use the article at the right.
    a. Write a polynomial to represent the amount of fuel the space shuttle burns in m minutes. $8{,}000\,m$
    b. Write the speed of the space shuttle in miles per hour. Round to the nearest whole mile. *Hint:* 5,280 ft = 1 mi. 2,523 mi/h
    c. The air distance from New York to Paris is 3,624 mi. Suppose you could fly from New York to Paris on the space shuttle. How long would the trip take? Round to the nearest tenth of an hour. What is the percent of change from Lindbergh's flight? Round to the nearest percent. 1.4 h; -96% change

Exploring Takes Energy

Charles Lindbergh's historic flight from New York to Paris in 1927 took 33.5 h. Lindbergh's plane, the *Spirit of St. Louis,* had no front window. An extra tank of gas carrying 450 gal of gas weighing 4,000 lb took the window's place. Today's space shuttle uses about 8,000 lb of fuel per minute at liftoff alone. In flight, the space shuttle flies at a speed of 3,700 ft/s at an altitude 100 mi above Earth.

Lesson Follow-up

Reteaching Activity Have students use vertically ruled paper to practice adding and subtracting polynomials by writing only one term in each column. For example to simplify $(x^2 - 8x + 9) - (-2x^2 - 3x - 5)$, students would proceed as follows:

x^2	$-8x$	$+9$
$-(-2x^2)$	$-(-3x)$	$-(-5)$
$3x^2$	$-5x$	$+14$

LESSON RESOURCES

TRB Practice M-3
TRB Enrichment M-4

Practice

For use after 13-2 (pp. 542–545)

Adding and Subtracting Polynomials

Use a model to find each sum. Check students' models.

1. $(2x^2 + 5x + 4) + (x^2 - 3x - 3)$ $3x^2 + 2x + 1$
2. $(-x^2 + 3x - 1) + (3x^2 - x + 2)$ $2x^2 + 2x + 1$

Find each sum by combining like terms.

3. $(5x + 3) + (2x - 7)$ $7x - 4$
4. $2x^2 + 4 + (3x^2 - 4x - 5)$ $5x^2 - 4x - 1$
5. $(-2x^2 + 4x - 5) + (8x + 5x^2 + 6)$ $3x^2 + 12x + 1$
6. $4x^2 + 1 + (3x - 5x^2 - 2)$ $-x^2 + 3x - 1$
7. $6x^2 + 5x - 5$; $+\ x^2 - 8x + 3$ — $7x^2 - 3x - 2$
8. $2x^3 - 5x^2 \quad -5$; $+3x^3 + 7x^2 + 9x$ — $5x^3 + 2x^2 + 9x - 5$
9. $-4x^2y^2 + 3xy + x^2 - 4y^2$; $+\ x^2y^2 - 6xy - x^2 - 5y^2$ — $-3x^2y^2 - 3xy \quad - 9y^2$
10. $12x^2y + 9xy^2 - 2x^2 + 5y^2$; $+\ x^2y - 8xy^2 + 2x^2 - y^2$ — $13x^2y + xy^2 \quad + 4y^2$

Use a model to find each difference.

11. $(3x - 2) - (4x + 3)$ $-x - 5$
12. $(2x^2 - 4x + 1) - (x^2 - 2x + 1)$ $x^2 - 2x$

Subtract each pair of polynomials by adding the opposite of the second polynomial.

13. $(10m - 4) - (3m - 5)$ $7m + 1$
14. $(k^2 - 2k + 5) - (k^2 + 5k + 3)$ $-7k + 2$
15. $(2x^2 + 7x - 4) - (x^2 - 4)$ $x^2 + 7x$
16. $(3x^2y^2 + 2xy + 5y) - (-2x^2y^2 - 4x + 5y)$ $5x^2y^2 + 2xy + 4x$
17. $(7x^3 - 5x^2 - 3x + 8) - (10x^3 - 4x^2 + 5x + 9)$ $-3x^3 - x^2 - 8x - 1$
18. $(x^2 + 2y + 5) - (4x + 4y)$ $x^2 - 4x - 2y + 5$
19. $(-4a^2b + 7ab^2 - 9a - 6b + 13) - (-6a^2b + 8a + 10b - 18)$ $2a^2b + 7ab^2 - 17a - 16b + 31$

Chapter 13 M-3

Enrichment

For use after 13-2 (pp. 542–545)

A Greek Translation

The word *polynomial* comes from the Greek words *poly* and *nominus*. What does *polynomial* mean?

To find out, solve each problem by adding or subtracting polynomials. Write the answer above the number of the exercise in the blanks at the bottom of the page.

1. John is $e^3 + 4e^2 - 3$ years old. Darlene is $4e^2 - e + e^3 - 3$ years old. How much older is John than Darlene?
2. The radius of a circle is $\frac{1}{2}a$. Find the diameter.
3. The length of a rectangle is $ab + b^2c + \frac{1}{4}a$. The width is $-b^2c + \frac{1}{4}a - ab$. Find the perimeter.
4. 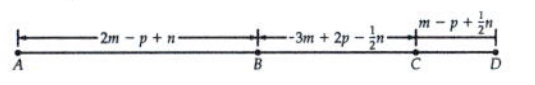

   Find AD.
5. $m\angle GSJ = 4w - 2m + 5h$. Find $m\angle GSH$.

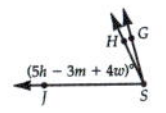

6. Write the exponent when the following expression is simplified: $x^{2k^2+3e+8m} \bullet x^{5e-4k^2-4m} \bullet x^{2k^2-3m-8e}$
7. Arsenio was born in the year $x^2 - 7y - 3r$. How old was he in the year $-3r - 6y + x^2$?
8. Garden City is located at mile post $5e + 7s - 7t + 6h$ on the county highway. Roseville is at a higher numbered post, $8s - 7t + 6h + 5e$. How far apart are the towns?
9. A triangle has sides of length $-\frac{3}{2}a - b + \frac{3}{2}n$, $3a - b$, and $-\frac{3}{2}a + 2b - \frac{1}{2}n$. Find the perimeter.

Answer: having m (6) a (2) n (4) y (7)   n (9) a (3) m (5) e (1) s (8)

M-4 *Chapter 13*

LESSON QUIZ

Find each sum or difference.

1. $(2x^2 + 3x - 3) + (3x^2 - 8x - 1)$ $[5x^2 - 5x - 4]$
2. $(ab + 3b - a) - (ab - 4b)$ $[7b - a]$
3. $(4xy^2 - y + 8x) - (-xy^2 + 2y)$ $[5xy^2 - 3y + 8x]$
4. The sum of three consecutive even numbers is 66. Find the numbers. (20, 22, 24)

Assignment Guide

Basic 1, 3-4, 7-8, 11, 13-14, 19-24, 30, 32-33, 36-37, MR All

Average 1, 4-5, 8-9, 11, 15-16, 23-28, 32-34, 37, 39, MR All

Enriched 2, 5-6, 9-10, 12, 17-18, 25-32, 37-38, 40, MR All

FOR THE NEXT LESSON

algebra tiles or paper squares, math journal, paper strips, scissors

Lesson Focus

Materials

Algebra tiles or paper squares, math journal, paper strips, scissors

MOTIVATING THE LESSON

Ask students how they would complete this multiplication; $3(a + 3)$. Then ask them what they think the result would be if they completed this multiplication $a(a + 3)$. (Use the distributive property: $3 \cdot a + 3 \cdot 3 = 3a + 9$; $a^2 + 3a$.)

Instruction

1 Have students begin with a rectangle x units by x units and draw four rectangles each 1 unit by x units along one side to model the product. Then have students repeat the model a second time below the first model. Ask students to model other similar products to be sure they understand how to use and interpret the model.

2 Students should notice that multiplying a binomial or a polynomial of more than two terms by a monomial is a simple application of the distributive property.

ADDITIONAL EXAMPLES

Simplify.

1. $x(x + 7)$ $[x^2 + 7x]$
2. $2m(m - 4)$ $[2m^2 - 8m]$
3. $-3y(4 + 7y - 3x - 5z)$ $[-12y - 21y^2 + 9xy + 15yz]$
4. $6a^2(4ab - 5a + 11)$ $[24a^3b - 30a^3 + 66a^2]$

OBJECTIVE: *To multiply a polynomial by a monomial.*

13-3 Multiplying a Polynomial by a Monomial

▼ You can write a polynomial to describe the area of the rectangle at the left. You can use a model to illustrate both the area and a way to simplify the resulting expression.

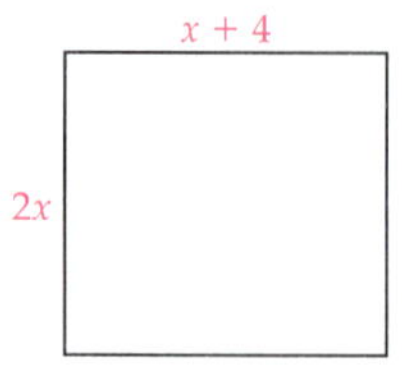

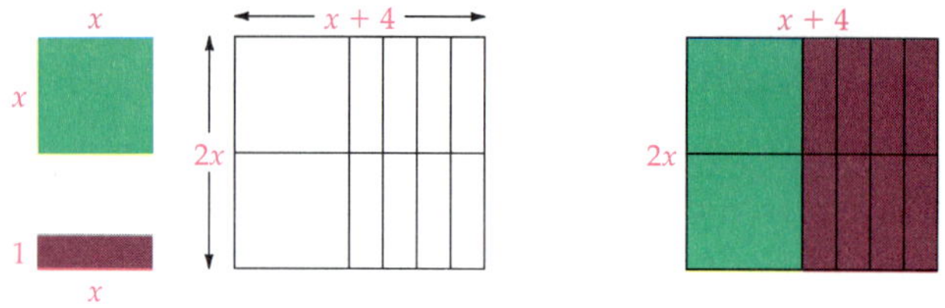

$$2x(x + 4) = 2x^2 + 8x$$

▼ You can use the distributive property to find the product of a monomial and a binomial.

Example 1 Simplify $3x(x - 4)$.

Solution

$3x(x) + 3x(-4)$	Use the distributive property.
$3x^2 - 12x$	Simplify.

▼ You can also use the distributive property to find the product of a monomial and a polynomial with more than two terms.

Example 2 Simplify $-2x(z^2 - 3x + y - 7)$.

Solution

$-2x(z^2) - 2x(-3x) - 2x(y) - 2x(-7)$
$-2xz^2 + 6x^2 - 2xy + 14x$

▼ You can use the rules of exponents to simplify the product of a monomial and a polynomial.

Example 3 Simplify $3x^2(8x^2 - 5xy + 2y^3)$.

Solution

$3x^2(8x^2) + 3x^2(-5xy) + 3x^2(2y^3)$
$3(8)x^2x^2 + 3(-5)x^2xy + 3(2)x^2y^3$
$24x^4 - 15x^3y + 6x^2y^3$

FLASHBACK

When multiplying powers with the same base, add exponents.

The expression $-2(8t^2 - 70t)$, where t equals time in seconds, represents the height in feet at which a fireworks burst will explode. At what height would a burst explode after 7s? 196 ft

546

Teaching Tip

To help develop an understanding of the procedure for multiplying a binomial by a monomial, provide algebra tiles or paper squares for students to manipulate while they describe what happens in each step.

CLASS EXERCISES

Use a model to find each product.

1. $2x(x + 4)$ $2x^2 + 8x$
2. $x(2x + 3)$ $2x^2 + 3x$
3. $3x(x + 1)$ $3x^2 + 3x$
4. $x(x + 5)$ $x^2 + 5x$
5. $2x(x + 3)$ $2x^2 + 6x$
6. $2x(3x + 1)$ $6x^2 + 2x$

Use the distributive property to find each product.

7. $3x(x + 5)$ $3x^2 + 15x$
8. $-4xy(2x - 3y)$ $-8x^2y + 12xy^2$
9. $5x(-3x^2 + 2x)$ $-15x^3 + 10x^2$
10. $4x(7x^6 - 3x^5 + 2x^2 + 1)$
11. $xy(x^2 + 2xy + y^2)$
12. $3x^2y(2x^2 - xy + y^2)$

10. $28x^7 - 12x^6 + 8x^3 + 4x$ 11. $x^3y + 2x^2y^2 + xy^3$ 12. $6x^4y - 3x^3y^2 + 3x^2y^3$

Use the distributive property to find each product. Then evaluate the polynomial for $x = 5$ and $y = 7$.

13. $x^2(y + 7)$ $x^2y + 7x^2$; 350
14. $-2x(3x + xy - y^2)$ $-6x^2 - 2x^2y + 2xy^2$; -10
15. $2x^2(y + 3)$ $2x^2y + 6x^2$; 500
16. $2y(x + y)$ $2xy + 2y^2$; 168

Write an expression for the area of each shaded region.

17.
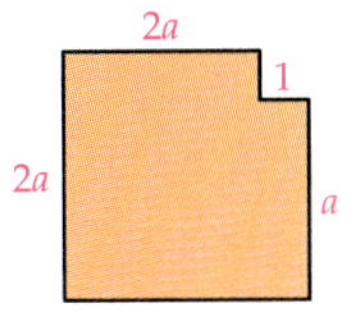

$4a^2 + a$

18.
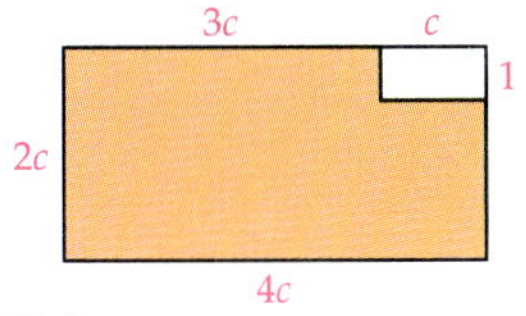

$8c^2 - c$

THINK AND DISCUSS See below.

1. Explain how you could use the column form to find the product of a monomial and a polynomial.
2. How many terms are in the product of $3x$ and $2x + 5y + 7$? Explain.
3. Explain how to find the product of a polynomial with five terms and a monomial.

1. You could use the column form to find the product in the same way you would multiply a 2- or 3-digit number by a 1-digit number.

Ex:
$$\begin{array}{r} x + 4 \\ 2x \\ \hline 2x^2 + 8x \end{array}$$

2. The number of terms in the product is equal to the number of terms in the polynomial since you are multiplying by a monomial.
3. Multiply each term in the polynomial by the monomial.

WRITTEN EXERCISES

Use a model to find each product. Check students' work.

1. $2x(x + 6)$ $2x^2 + 12x$
2. $x(2x + 6)$ $2x^2 + 6x$
3. $2x(3x - 1)$ $6x^2 - 2x$
4. $3x(2x + 4)$ $6x^2 + 12x$
5. $x(5x + 3)$ $5x^2 + 3x$
6. $2x(4x + 7)$ $8x^2 + 14x$

7. ***MENTAL MATH*** Complete each exercise using two different methods. First substitute the value for the variable, then multiply. Next, multiply the polynomial first, then substitute the value for the variable. Evaluate each of the following for $x = 1$, $y = -1$, and $z = 2$.

   a. $3x(-6y + z)$ 24 b. $y^2(x - y)$ 2 c. $5z(3x + y^2)$ 40 d. $2x^2(z^2 - y^2)$ 6

   e. ***WRITE*** a paragraph telling which method of solving the expressions above you found to be the easiest. Explain. Check students' work.

CALCULATOR Find each product. Then evaluate the expression for $x = 12$ and $y = -15$.

8. $2x^2(x^2 + y^2)$ $2x^4 + 2x^2y^2$; 106,272
9. $-4xy(xy - x^2)$ $-4x^2y^2 + 4x^3y$; $-233{,}280$
10. $x^2(5x + y^2 - y)$ $5x^3 + x^2y^2 - x^2y$; 43,200

MATH MINUTES

Underline the correct answer.

1. $a(2a - 5) = 2a^2 - 5$ or $\underline{2a^2 - 5a}$
2. $-4xy(3x - 2y) =$ $\underline{-12x^2y + 8xy^2}$ or $-12x^2y - 8xy^2$
3. $-3xy(5x - 3y - 7) =$ $-15x^2y - 3y - 7$ or $\underline{-15x^2y + 9xy^2 + 21xy}$

Guided Practice

Class Exercises As a check for questions 13-16, have students evaluate the expression before and after they perform the multiplication.

For questions 17-18, suggest that students copy each figure and divide it into parts to help them find the total area. Students should discover that it is easier to find the total area, then subtract the part that is cut out in question 18.

Think and Discuss Once students have proposed a method for question 1, have them try it on questions 7-12 of the Class Exercises to see if they get the same results as they did using the distributive property. (They should.)

A ***common error*** occurs when students forget to multiply the monomial by each term in the polynomial. Have students anticipate the number of terms in the product of $4x(3y + xy + 6)$. [3] Then, as they find the product, remind them of the number of terms they will find. (3; $12xy + 4x^2y + 24x$)

Closure

Writing in Math Ask students to describe in their math journals how they would find the product of a monomial and a binomial. Ask students to include in the description a decision on the number of terms in the answer.

Independent Practice

For question 7e, tally students' choices to see if there was a majority for either method.

Lesson Follow-up

Reteaching Activity For students having difficulty with polynomials, refer to the manipulative activity on p. 534E of the Chapter Overview. See also TRB Exploring O-49.

LESSON RESOURCES

TRB Practice M-5
TRB Enrichment M-6

Practice

For use after 13-3 (pp. 546–548)

Multiplying a Polynomial by a Monomial

Use a model to find each product.

1. $x(4x + 2)$ $4x^2 + 2x$
2. $x(3x - 2)$ $3x^2 - 2x$
3. $2x(x - 2)$ $2x^2 - 4x$
4. $3x(x + 5)$ $3x^2 + 15x$
5. $2x(3x + 1)$ $6x^2 + 2x$
6. $3x(3x - 1)$ $9x^2 - 3x$

CALCULATOR Find each product. Then evaluate the expression for $x = -9$ and $y = 14$.

7. $x^2(x + 2y)$ $x^3 + 2x^2y$; 1539
8. $-xy(2x - y)$ $-2x^2y + xy^2$; -4,032
9. $y(4x + y - 2x^2)$ $4xy + y^2 - 2x^2y$; -2,576
10. $3y(5y - 2x + 4xy)$ $15y^2 - 6xy + 12xy^2$; -17,472

Use the distributive property to simplify.

11. $4x(3x - 5)$ $12x^2 - 20x$
12. $-8x(x - 7)$ $-8x^2 + 56x$
13. $6y(x^2 + 2y)$ $6x^2y + 12y^2$
14. $7xy^2(y - 2x + x^2)$ $7xy^3 - 14x^2y^2 + 7x^3y^2$
15. $3xy(2xy + 5)$ $6x^2y^2 + 15xy$
16. $-9xyz(-2xy + 3yz - 4xz)$ $18x^2y^2z - 27xy^2z^2 + 36x^2yz^2$
17. $-\frac{1}{2}m(4m^2 - 2mn)$ $-2m^3 + m^2n$
18. $12ab(-\frac{1}{2}b + \frac{1}{4}a^3)$ $-6ab^2 + 3a^4b$
19. $-15a^2(a - b + 3c)$ $-15a^3 + 15a^2b - 45a^2c$
20. $-3x^2a^2(2a^3 + ab - x)$ $-6x^2a^5 - 3x^2a^3b + 3x^3a^2$
21. $3x(2x - 4y) + 4y(3x - y)$ $6x^2 - 4y^2$
22. $x(5x + 2y + 4xy) - 3y(2x - 4x^2 + 3y)$ $5x^2 - 4xy + 16x^2y - 9y^2$

Write an expression for the area of each figure. Simplify.

23. 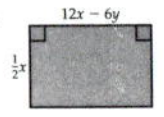

$6x^2 - 3xy$

24. 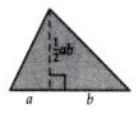

$\frac{1}{4}a^2b + \frac{1}{4}ab^2$

25. 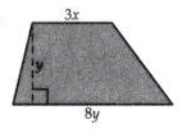

$\frac{3}{2}xy + 4y^2$

Enrichment

For use after 13-3 (pp. 546–548)

Factoring Polynomials

Factoring is writing a number or expression as a product. For example, one way to factor the number 42 is to write it as 2 • 3 • 7. You can factor polynomials by using the distributive property in reverse.

To factor $3x + 3y$: Find the common factor. The common factor is 3; therefore, $3x + 3y = 3(x + y)$.

To factor $a^3 + ab^2$: Find the common factor. The common factor is a; therefore, $a^3 - ab^2 = a(a^2 - b^2)$

Complete to show how the given expression can be factored.

1. $5x + 5y = 5($ x + y $)$
2. $-3m - 3n = -3($ m + n $)$
3. $7a + 7b =$ 7 $(a + b)$
4. $x^2 + xy =$ x $(x + y)$
5. $4x^3 + 4x^2y =$ $4x^2$ $(x + y)$
6. $6ab + 12b =$ 6b $(a + 2)$
7. $5m + 10m^2 = 5m($ 1 + 2m $)$
8. $-21x^3 + 28xy = -7x($ $3x^2$ − 4y $)$

Factor the given expression.

9. $8x + 8y$ $8(x + y)$
10. $13a - 13b$ $13(a - b)$
11. $2x^3 + 2x^2$ $2x^2(x + 1)$
12. $11a + 11b + 11c$ $11(a + b + c)$
13. $-2m - 2n - 2p$ $-2(m + n + p)$
14. $2x + 4y - 6z$ $2(x + 2y - 3z)$
15. $5h + 15h^3$ $5h(1 + 3h^2)$
16. $6p - 24pt$ $6p(1 - 4t)$
17. $-9x^2 - 9y^2$ $-9(x^2 + y^2)$
18. $20m + 25n - 35k$ $5(4m + 5n - 7k)$
19. $x^3y^2 + x^2y^3 + x^4y$ $x^2y(xy + y^2 + x^2)$
20. $-12ab^2c + 18a^2bc^2 - 30ab^3c^3$ $-6abc(2b - 3ac + 5b^2c^2)$
21. $90w^3x + 144w^2$ $18w^2(5wx + 8)$
22. $161mn - 133$ $7(23mn - 19)$
23. $51x - 68$ $17(3x - 4)$
24. $95a + 171$ $19(5a + 9)$

MIXED REVIEW

Find each answer.

1. Find the median. -8.3, 7.6, -2.1, 9.5, 8.4, -2.6 **2.75**

2. Find the area of a rectangle that measures 3.2 m by 8.6 m. **27.52 m²**

Find each sum or difference.

3. $(2x + 8) + (3x^2 + 5x - 2)$ $3x^2 + 7x + 6$

4. $(-7x^2 - 8x + 4) - (2x^2 - 3x - 9)$ $-9x^2 - 5x + 13$

Find each product.

5. $x \cdot x \cdot x^2$ x^4

6. $x^3 \cdot x^6 \cdot x$ x^{10}

Solve.

7. A college student received a bank statement. The balance was $200. It showed deposits of $400, interest of $1, and checks totaling $650. What was the beginning balance? **$449**

Use the distributive property to find each product.

11. $2x(4x - 1)$ $8x^2 - 2x$
12. $x(x^2 + 3x)$ $x^3 + 3x^2$
13. $x^2(2x - 5)$ $2x^3 - 5x^2$
14. $x^2(x^2 - 9)$ $x^4 - 9x^2$
15. $y^2(y + 2y^2 - 3)$
16. $-3(2x^2 - 3x - 1)$
17. $5xy(x + 5 - y)$
18. $5a^2bc^2(abc - a^2b^2c^2 + 6a^2bc)$
19. $-3xy(2x^2y + xy + y^2 - 3)$
20. $4z(2z^6 - 3z^5 - 12z^2 + 8)$
21. $8xyz(12x^2y^2 + 3x^3z^5)$
22. $-2x^2b^2(-4xb^3 + 3x^3b^2)$
23. $12x^2(x + y^2 + z)$
24. $-4x^2y(25x^5y^2z + 8x^3y)$
25. $\frac{1}{2}y(x + xy^2 + 5)$
26. $3y\left(x^2 - xy - \frac{1}{3}x\right)$
27. $7x^2(2x^2 + y^2 - xy)$
28. $4x^2(x^2 + xy - y^2)$
29. $3x(x^2 + 5) + 2x(x - 3)$ $3x^3 + 2x^2 + 9x$
30. $2y(-y + 4) + 3(y - 5)$ $-2y^2 + 11y - 15$

Solve. Use only one variable for each expression.

31. Assume e is an even integer. Write an expression to represent the product of e and the next consecutive even integer. Simplify. $e(e + 2)$; $e^2 + 2e$

32. Assume m is a multiple of 6. Write an expression to represent the product of m and the next consecutive integer that is a multiple of 6. Simplify. $m(m + 6)$; $m^2 + 6m$

33. The width of a rectangle is $\frac{1}{2}$ the length plus 7. Write an expression to represent its area. Then simplify the expression.

34. The length of a rectangle is 5 less than 4 times its width. Write an expression to represent its area. Simplify. $(4w - 5)w$; $4w^2 - 5w$

35. The base length of a triangle is $8x$. The triangle's height is twice that plus 5. Write an expression for the area of the triangle. Simplify. $\frac{1}{2}(8x)(16x + 5)$; $64x^2 + 20x$

36. The height of an isosceles triangle is $\frac{1}{3}$ its base less 3. Write an expression to represent its area. Simplify. $\frac{1}{2}b(\frac{1}{3}b - 3)$; $\frac{1}{6}b^2 - \frac{3}{2}b$

37. Express the number 792 in expanded form. Let $x = 10$. Write a polynomial to represent the number. Then write a polynomial in x to represent 40. Find the product of the two polynomials. Simplify it and substitute 10 for x. Is your answer equal to 40(792)?

Write an expression for the area of the shaded region.

38. 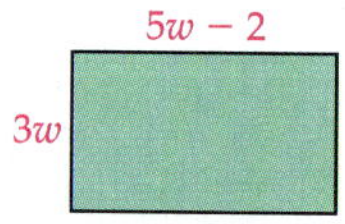

$15w^2 - 6w$

39. 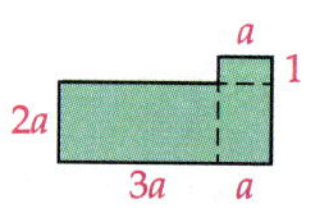

$8a^2 + a$

40. 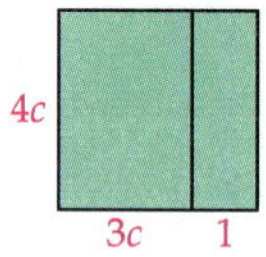

$12c^2 + 4c$

41. 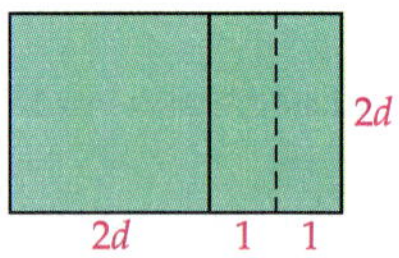

$4d^2 + 4d$

15. $y^3 + 2y^4 - 3y^2$
16. $-6x^2 + 9x + 3$
17. $5x^2y + 25xy - 5xy^2$
18. $5a^3b^2c^3 - 5a^4b^3c^4 + 30a^4b^2c^3$
19. $-6x^3y^2 - 3x^2y^2 - 3xy^3 + 9xy$
20. $8z^7 - 12z^6 - 48z^3 + 32z$
21. $96x^3y^3z + 24x^4yz^6$
22. $8x^3b^5 - 6x^5b^4$
23. $12x^3 + 12x^2y^2 + 12x^2z$
24. $-100x^7y^3z - 32x^5y^2$
25. $\frac{1}{2}xy + \frac{1}{2}xy^3 + \frac{5}{2}y$
26. $3x^2y - 3xy^2 - xy$
27. $14x^4 + 7x^2y^2 - 7x^3y$
28. $4x^4 + 4x^3y - 4x^2y^2$
33. $l(\frac{1}{2}l + 7)$; $\frac{1}{2}l^2 + 7l$
37. $7 \times 10^2 + 9 \times 10 + 2$; $7x^2 + 9x + 2$; $4x$; $(7x^2 + 9x + 2)4x = 28x^3 + 36x^2 + 8x$; $28(10)^3 + 36(10)^2 + 8(10) = 31{,}680 = 40(792)$; yes

LESSON QUIZ

Find each product.

1. $3x(2x - 8)$
2. $a^2(3a^2 - 4a - 2)$
3. $-5xy(2x^2 + 4xy - y^2)$
4. Find the product $x(3x - 5y)$ then evaluate it for $x = 3$ and $y = -2$.
5. The base of a triangle is 4 more than twice its height. Write an expression for its area, then simplify it.

QUIZ ANSWERS

1. $(6x^2 - 24x)$
2. $(3a^4 - 4a^3 - 2a^2)$
3. $(-10x^3y - 20x^2y^2 + 5xy^3)$
4. $(3x^2 - 5xy; 57)$
5. $[\frac{1}{2}h\ (2h + 4); h^2 + 2h]$

FOR THE NEXT LESSON

algebra tiles or paper squares and rectangles, math journal

Assignment Guide

Basic 1-3, 7-9, 11-20, 31-34, 38-39, MR All
Average 2-4, 7-10, 11-29 odd, 32-35, 39-41, MR All
Enriched 4-7, 10, 12-30 even, 34-37, 40-41, MR All

OBJECTIVE:
To explore algebraic expressions.

MATERIALS

- Algebra tiles or colored paper
- Math journal to record work

Exploring Mind-reading Tricks

■ You have probably heard of mind-reading tricks. Most such tricks are simple applications of algebraic expressions.

1. Think of a number. Add 4. Multiply by 2. Subtract 6. Divide by 2. Subtract your original number. What is your final result?

a. ***Model*** each step of the process with algebra tiles. The number you think of is the variable. Use the rectangle to represent this number. Use positive tiles to represent the units.

①

Start with your number.

Add 4.

Multiply the result by 2.

Subtract 6.

Divide by 2.

Subtract your number.
The result is always 1.

b. Use a variable to replace the tiles.

Start with your number.	n
Add 4.	$n + 4$
Multiply the result by 2.	$2(n + 4) = 2n + 8$
Subtract 6.	$2n + 8 - 6 = 2n + 2$
Divide by 2.	$\frac{2n + 2}{2} = n + 1$
Subtract your number. The result is always 1.	$n + 1 - n = 1$

2. ***Explore*** Use algebra tiles and then variables to find the result of another trick. **Check students' work.**

② Think of a number. Triple it. Add 14. Subtract 5. Divide by 3. Subtract your original number. What do you get?

a. ***Write*** an equation to summarize your exploration of the trick.

b. Make up a mind-reading trick of your own and try it on a classmate. Then explain how the trick works.

Lesson Focus

Materials

Algebra tiles or paper squares and rectangles, math journal

MOTIVATING THE LESSON

Begin the lesson by asking students to follow directions such as: think of a number; multiply by five; divide by five; state the result and the reason for the result. (the original number; multiplying by five and dividing by five are inverse operations)

Instruction

Have students work in small groups. Distribute algebra tiles or paper squares and rectangles to each group. Ask students to write their answers in their math journals.

① Ask students how the multiplication was accomplished using the tiles (by adding tiles, or repeating the addition shown) and how the division was accomplished. (by removing tiles)

② Help students to see that the reason for the result is that the constant term, 14 − 5, is 9.

Closure

Have students share their mind reading tricks with the entire class.

Assignment Guide

Basic All
Average All
Enriched All

FOR THE NEXT LESSON

algebra tiles or paper squares, math journal

Teaching Tip

Encourage students to analyze the mind reading tricks created by other students to help them develop higher-order thinking skills, which are necessary for everyday problem solving.

Lesson Focus

Materials

Algebra tiles or paper squares, math journal

Vocabulary/Symbols

FOIL method

MOTIVATING THE LESSON

Draw a square like the one below on the chalkboard.

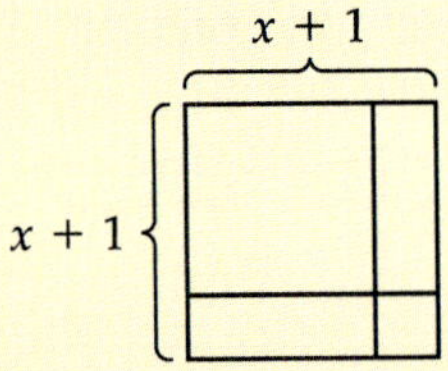

Ask students to determine the area of the square. $(x^2 + 2x + 1)$ Then ask them to state a product indicating the area of the square. $[(x + 1)(x + 1)$ or $(x + 1)^2]$

Instruction

1 Point out that students can use squares and strips to model the area or they can simply mark off the rectangle as shown. Discuss how to find the product from the factors. (Multiply each term in the second binomial by each term in the first binomial.)

2 If students have difficulty understanding this use of the distributive property, have them first represent the binomial $(x - 3)$ with a frame and write ■$(x + 4)$ = ■x + ■4. Then have them replace each frame with $(x - 3)$.

3 Mention that the FOIL method is just a simplified way of writing the distributive property. Have students observe that *First* and *Outside* result from the product x times $(x + 6)$; *Inside* and *Last* result from the product 2 times $(x + 6)$.

OBJECTIVE:
To find the product of two binomials.

NOTES & QUOTES

Isaac Newton (1642–1727) invented the binomial theorem, a formula for finding the product of the expression $(x + y)^2$. By the age of 23, Newton also had made discoveries about the nature of light, had invented calculus, and had established the theory of universal gravitation. When hailed for his achievements, Newton modestly replied, "If I have seen a little farther than others, it is because I have stood on the shoulders of giants."

DISCUSS the meaning of Newton's famous quote.

13-4 Multiplying Binomials

▼ Suppose a is an even integer. You can represent the product of the next two consecutive odd integers, $a + 1$ and $a + 3$, by the expression $(a + 1)(a + 3)$.

1 You can use a model to help find the product.

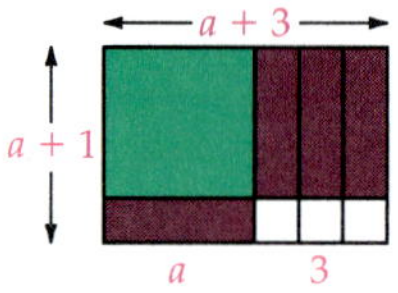

$$(a + 1)(a + 3) = a^2 + 4a + 3$$

▼ If you think of one binomial as a single expression, you can use the distributive property twice to multiply two binomials.

Example 1 **Multiply $(x - 3)(x + 4)$.**

2

Solution

$(x - 3)x + (x - 3)4$	Multiply the first binomial by each term in the second.
$x(x) - 3(x) + x(4) - 3(4)$	Multiply each binomial and monomial.
$x^2 - 3x + 4x - 12$	Combine like terms.
$x^2 + x - 12$	Simplify.

▼ The FOIL method is another way to find the product of two binomials.

1. Multiply the First terms in each binomial.
2. Multiply the Outside terms in each binomial.
3. Multiply the Inside terms in each binomial.
4. Multiply the Last terms in each binomial.
5. Add the products.

Example 2 **Multiply $(x + 2)(x + 6)$.**

3

Solution

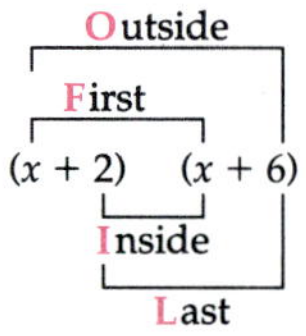

First	Outside	Inside	Last
$x(x)$ +	$x(6)$ +	$2(x)$ +	$2(6)$

$x^2 + 6x + 2x + 12$	Combine like terms.
$x^2 + 8x + 12$	Simplify.

Teaching Tip

Make a poster illustrating the memory device *FOIL* and display it to help students remember how to multiply two binomials.

▼ When you use FOIL to multiply the sum and difference of the same terms, or to square a binomial, the product has interesting characteristics.

Example 3 **Multiply $(a + 5)(a - 5)$.**

Solution

$$(a + 5)(a - 5) = a^2 - 5a + 5a - 25$$
$$= a^2 - 25$$

Example 4 **Multiply.**

a. $(a + 7)^2$ **b.** $(a - 3)^2$

Solution

a. $(a + 7)(a + 7) = a^2 + 14a + 49$

b. $(a - 3)(a - 3) = a^2 - 6a + 9$

> ***FLASHBACK***
> The square of a number is the product of the number and itself.

The examples lead to the following rules for finding special products.

Product of $(a + b)(a - b)$	To find the product of the sum and difference of two terms, square the first term and subtract the square of the second term. $(a + b)(a - b) = a^2 - b^2$
Squaring Binomials	To square a binomial, square the first term then add or subtract twice the product of the two terms and add the square of the second term. $(a + b)^2 = a^2 + 2ab + b^2$ $(a - b)^2 = a^2 - 2ab + b^2$

4

CLASS EXERCISES

Find each product using models. Check students' work.

1. $(x + 2)(x + 1)$ $x^2 + 3x + 2$
2. $(x + 1)(x + 4)$ $x^2 + 5x + 4$
3. $(x + 2)(x + 2)$ $x^2 + 4x + 4$
4. $(x + 2)(x + 3)$ $x^2 + 5x + 6$

Find each product using the distributive property.

5. $(x + 3)(x - 2)$ $x^2 + x - 6$
6. $(x + 7)(x + 9)$ $x^2 + 16x + 63$
7. $(y - 2)(y - 4)$ $y^2 - 6y + 8$

Find each product using the FOIL method.

8. $(a - 9)(a + 1)$ $a^2 - 8a - 9$
9. $(x + 3)(x - 4)$ $x^2 - x - 12$
10. $(x - 7)(x + 5)$ $x^2 - 2x - 35$

Find each product.

11. $(x + 5)^2$ $x^2 + 10x + 25$
12. $(x - 2)(x + 2)$ $x^2 - 4$
13. $(x - 5)^2$ $x^2 - 10x + 25$

1. Each term of one factor is multiplied by (distributed) each term of the other factor.
2. The product of two binomials does not always have three terms. It could have two terms if it were in the form $(a + b)(a - b)$. It would have four terms if it were in the form $(x + y)(x + 1)$. No, it couldn't have more than four terms.

THINK AND DISCUSS See above.

1. How is using FOIL similar to using the distributive property? Which method do you prefer?
2. Does the product of two binomials always have three terms? Could it have two? four? more than four? Explain.

MATH MINUTES

Correct each equation.

1. $(a - 5)(a + 2) = a^2 + 3a - 10$ $[a^2 - 3a - 10]$
2. $(a - 5)^2 = a^2 - 10a - 25$ $[a^2 - 10a + 25]$
3. $(a + b)(c + d) = ac + abcd + bd$ $[ac + ad + bc + bd]$

4 Have students explore the products $(a + b)^2$ and $(a + b)(a - b)$ after discussing this example. See if students can use the patterns to give several such products mentally.

ADDITIONAL EXAMPLES

Find each product.

1. $(x + 6)(x + 2)$ $[x^2 + 8x + 12]$
2. $(2a + 3)(a - 6)$ $[2a^2 - 9a - 18]$
3. $(4x - 2)(3x - 8)$ $[12x^2 - 38x + 16]$
4. $(x - 7)^2$ $[x^2 - 14x + 49]$
5. $(x + 6)(x - 6)$ $[x^2 - 36]$

Guided Practice

Class Exercises Discuss students' answers to assess their understanding of the process for multiplying two binomials. If students have difficulty using one method, encourage them to try another.

Think and Discuss To help students answer question 2, have them try to find a variety of binomial pairs whose products have two, three, or four terms.

A ***common error*** occurs when students forget the middle terms and simply multiply the first and last terms of the binomials. For example, students may give the product of $(x - 3)(x + 4)$ as $x^2 - 12$. Assign values for the variables and have students evaluate the indicated products. Then have them check their work by using the same values to evaluate their answers.

Closure

Writing in Math Ask students to describe in their own words each of the methods used in this lesson to multiply two binomials and to write this description in their math journals.

Independent Practice

For questions 4-15, encourage students to check their answers by finding the products using a different method (for example, check the use of the distributive property with FOIL and vice versa).

If students are having difficulty writing the expressions for questions 44-48, have them work first with actual numbers to determine how the numbers are related. From this knowledge, they should then be able to write the algebraic expressions.

As a follow up, refer to the calculator activity on p. 534E of the Chapter Overview. See also TRB Exploring O-51.

20. $y^2 - 20y + 100$
21. $4a^2 - 25$
22. $a^2 - a - 56$
23. $4a^2 + 4a + 1$
24. $a^2 - 225$
25. $x^2 + 15x - 756$
26. $x^2 - 13x + 40$
27. $3a^2 - 2a - 8$
28. $x^2 + 9x - 22$
29. $x^2 - 3.5x - 6.24$
30. $3x^3 + 6x + x^2y + 2y$

MIXED REVIEW

Find each answer.

1. 12.5% of 16 2
2. 15% of 90 13.5

Find each product.

3. $2x(3x + 1)$ $6x^2 + 2x$
4. $-5x(x^2 - 2x)$ $-5x^3 + 10x^2$

Find the greatest common factor.

5. 60, 100 20
6. 12, 78 6

Solve.

7. A landscape architect wanted to plant trees around 3 sides of a square yard. The trees were to be planted 6 ft apart. The architect ordered 37 trees. What are the dimensions of the yard? 72 ft by 72 ft

WRITTEN EXERCISES

Find each product using models. Check students' work.

1. $(x + 1)(x + 6)$ $x^2 + 7x + 6$
2. $(x + 2)(x + 4)$ $x^2 + 6x + 8$
3. $(a + 1)(a + 7)$ $a^2 + 8a + 7$

Find each product using the distributive property.

4. $(a - 1)(a + 6)$ $a^2 + 5a - 6$
5. $(a + 3)(a - 2)$ $a^2 + a - 6$
6. $(y - 16)(y + 20)$ $y^2 + 4y - 320$
7. $(a + 3)(a + 8)$ $a^2 + 11a + 24$
8. $(x + 4)(2x + 1)$ $2x^2 + 9x + 4$
9. $(x + a)(y + b)$ $xy + xb + ay + ab$

Find each product using FOIL. 10. $y^2 + 10y + 16$

10. $(y + 2)(y + 8)$
11. $(x + 1)(x + 12)$ $x^2 + 13x + 12$
12. $(x - 8)(x - 3)$ $x^2 - 11x + 24$
13. $(3 + x)(5 - x)$ $15 + 2x - x^2$
14. $(3x + 1)(2x - 4)$ $6x^2 - 10x - 4$
15. $(2a + b)(4c - 2d)$ $8ac - 4ad + 4bc - 2bd$

Find each product. 16. $y^2 + 10y + 25$ 17. $x^2 - 9$ 18. $a^2 - 64$ 19. $b^2 - 14b + 49$

16. $(y + 5)^2$
17. $(x - 3)(x + 3)$
18. $(a - 8)(a + 8)$
19. $(b - 7)^2$
20. $(y - 10)^2$
21. $(2a + 5)(2a - 5)$
22. $(a + 7)(a - 8)$
23. $(2a + 1)^2$
24. $(a - 15)(a + 15)$
25. $(x - 21)(x + 36)$
26. $(x - 5)(x - 8)$
27. $(3a + 4)(a - 2)$
28. $(x - 2)(x + 11)$
29. $(x + 1.3)(x - 4.8)$
30. $(3x + y)(x^2 + 2)$
31. $(b - 1)^3$ $b^3 - 3b^2 + 3b - 1$
32. $(x + 1)(x^2 + 2x - 3)$ $x^3 + 3x^2 - x - 3$
33. $-4(2a + 1)(a - 3)$ $-8a^2 + 20a + 12$

MENTAL MATH **Find each product mentally.**

34. $(x - 1)^2$ $x^2 - 2x + 1$
35. $(x + 1)^2$ $x^2 + 2x + 1$
36. $(x + 3)^2$ $x^2 + 6x + 9$
37. $(x - y)(x + y)$ $x^2 - y^2$
38. $(x + 1)(x - 1)$ $x^2 - 1$
39. $(x + 2)(x + 2)$ $x^2 + 4x + 4$

Write an expression for the area of each rectangle. Simplify the expression.

40. $2x^2 + 7x + 3$

$x + 3$ by $2x + 1$

41. $10x^2 + 26x + 12$

$2x + 4$ by $5x + 3$

42. $4x - 5$ by $3x + 1$

$12x^2 - 11x - 5$

43. $10x - 2$ by $2x + 1$

$20x^2 + 6x - 2$

Solve. Write an expression to represent each problem. Use one variable in each expression. Then simplify each expression.

44. Assume e is an even integer. Find the product of the next two consecutive even integers. $(e + 2)(e + 4) = e^2 + 6e + 8$

45. Assume w is an integer that is a multiple of 3. Find the product of the previous two consecutive integers. $(w - 1)(w - 2) = w^2 - 3w + 2$

46. The base of a parallelogram is $w + 5$ centimeters. The height is 2 cm less. Find the area of the parallelogram. $(w + 5)(w + 3) = w^2 + 8w + 15$

47. The side of a square is $(t - 6)$ meters. Find the area. $(t - 6)^2 = t^2 - 12t + 36$

48. Two right triangles are joined to form a rectangle. The base of the triangle is m units. The height of the triangle is $m + 2$ units. Find the area of the rectangle. $m(m + 2) = m^2 + 2m$

Solve each equation.

49. $3(x + 5) - 2(x + 6) = 5$ $x = 2$

50. $4(a - 3) + 5(a + 7) = 14$ $a = -1$

51. $-(w + 7) - 6(w + 6) = -43$ $w = 0$

52. $5(b - 6) + 4(6 - b) = -8$ $b = -2$

53. ***DATA FILE 8 (pp. 312–313)*** By about how many miles does the center of population move each year? about 50 mi

54. ***DATA FILE 10 (pp. 404–405)*** The area of Rhode Island is about 1,055 mi^2. About how long does it take to destroy an area of rain forests equal to the area of Rhode Island? about 9.4 days

FLASHBACK

27,878,400 ft^2 = 1 mi^2
1 acre = 43,560 ft^2

TEST YOURSELF

Is each a monomial, binomial, trinomial, or not a polynomial?

1. $4x$
2. $5y + 6$
3. $\frac{7}{w}$
4. $m^2 + 3m - 7$
5. $186.5p$
6. $3g + g^2$
7. $14v7v$
8. $\frac{a}{3}$

1. monomial
2. binomial
3. not a polynomial
4. trinomial
5. monomial
6. binomial
7. monomial
8. monomial

Evaluate each polynomial for $x = 7$, $y = 10$, and $z = -5$.

9. $x^2 + 5x - 3$ 81
10. $y^2 + z$ 95
11. $z^2 + 5z - x$ -7
12. $x^2 - y^2$ -51
13. $y^2 - z + x$ 112
14. $z^2 + x^2 - y^2$ -26

Add or subtract.

15. $(7a + 3) + (4a - 3)$ $11a$
16. $(p + 7) + (p^2 + 5)$ $p^2 + p + 12$
17. $(3ab^2 + 9) + (2a + b)$ $3ab^2 + 2a + b + 9$
18. $(rs - 6) - (r^2s^2 + 10)$ $rs - r^2s^2 - 16$

19. Write an expression for the area of square that has sides of length $a^2 + b$. Simplify the expression. $(a^2 + b)(a^2 + b) = a^4 + 2a^2b + b^2$

20. ***WRITE*** a sentence explaining how to use the FOIL method to find the product of two polynomials. Check students' work.

Lesson Follow-up

Reteaching Activity For students having difficulty with multiplication of binomials, refer to problem solving activity on p. 534E of the Chapter Overview. See also TRB Exploring O-50.

LESSON RESOURCES

TRB Practice M-7
TRB Enrichment M-8

Practice

For use after 13-4 (pp. 550–553)

Multiplying Binomials

Find each product using models.

1. $(x + 2)(x + 3)$ $x^2 + 5x + 6$
2. $(x + 5)(x + 1)$ $x^2 + 6x + 5$

Find each product using the distributive property.

3. $(x + 4)(x + 5)$ $x^2 + 9x + 20$
4. $(x + 7)(x + 2)$ $x^2 + 9x + 14$
5. $(x + 1)(x - 6)$ $x^2 - 5x - 6$
6. $(x + 8)(x - 3)$ $x^2 + 5x - 24$
7. $(2x + 5)(x + 3)$ $2x^2 + 11x + 15$
8. $(x - 4)(x - 6)$ $x^2 - 10x + 24$

Find each product using FOIL.

9. $(y - 7)(y - 6)$ $y^2 - 13y + 42$
10. $(x - 9)(x - 5)$ $x^2 - 14x + 45$
11. $(x - 10)(x + 3)$ $x^2 - 7x - 30$
12. $(2x + 3)(3x + 2)$ $6x^2 + 13x + 6$
13. $(4x - 1)(2x + 7)$ $8x^2 + 26x - 7$
14. $(x + a)(x + b)$ $x^2 + ax + bx + ab$

Find each product

15. $(y - 9)^2$ $y^2 - 18y + 81$
16. $(x - 4)(x + 4)$ $x^2 - 16$
17. $(3m - n)(m + n)$ $3m^2 + 2mn - n^2$
18. $(a - 14)(a + 8)$ $a^2 - 6a - 112$
19. $(k - 6)(k + 6)$ $k^2 - 36$
20. $(p + 5)^2$ $p^2 + 10p + 25$
21. $(2x - 7)(2x + 7)$ $4x^2 - 49$
22. $(m - 15)(m - 20)$ $m^2 - 35m + 300$
23. $(3k + 4)^2$ $9k^2 + 24k + 16$
24. $(x - 20)(x + 20)$ $x^2 - 400$
25. $(5n + 4)(4n - 5)$ $20n^2 - 9n - 20$
26. $(10x - 1)^2$ $100x^2 - 20x + 1$

MENTAL MATH Find each product mentally.

27. $(x + 2)(x - 2)$ $x^2 - 4$
28. $(x - 3)^2$ $x^2 - 6x + 9$
29. $(a + b)(a - b)$ $a^2 - b^2$
30. $(x + 1)^2$ $x^2 + 2x + 1$
31. $(a - b)(a - b)$ $a^2 - 2ab + b^2$
32. $(x + 4)(x - 4)$ $x^2 - 16$
33. A rectangle has length $4x + 3$ and height $3x - 7$. Find the area of the rectangle. $12x^2 - 19x - 21$ units^2
34. Mark worked for $6d - 9$ days and earned $8d + 25$ dollars per day. How much did he earn altogether? $48d^2 + 78d - 225$ dollars

Chapter 13 M-7

Enrichment

For use after 13-4 (pp. 550–553)

Sum and Product Factoring

When two binomials are multiplied together, an interesting pattern occurs in the product. To understand the pattern, study these examples.

$(x + 3)(x + 5) = x^2 + 8x + 15$ — 3 and 5; 15 is the product of 3 and 5; 8 is the sum of 3 and 5

$(x + 7)(x - 2) = x^2 + 5x - 14$ — 7 and -2; 14 is the product of 7 and -2; 5 is the sum of 7 and -2

You can use this pattern to factor many trinomials.

Example Factor $x^2 + 9x + 20$.

Solution *THINK* What two numbers have a sum of 9 and a product of 20? $x^2 + 9x + 20$ (9: sum; 20: product)

ANSWER: 4 and 5

Therefore, $x^2 + 9x + 20 = (x + 4)(x + 5)$.

1. a. Write two numbers with a sum of 8 and a product of 12. 6, 2
   b. Use the numbers to complete: $x^2 + 8x + 12 = (x + 6)(x + 2)$
2. a. Write two numbers with a sum of -7 and a product of 10. -2, -5
   b. Use the numbers to complete: $x^2 - 7x + 10 = (x - 2)(x - 5)$
3. a. Write two numbers with a sum of -3 and a product of -40. 5, -8
   b. Use the numbers to complete: $x^2 - 3x - 40 = (x - 8)(x + 5)$

Factor each trinomial.

4. $x^2 + 5x + 6$ $(x + 2)(x + 3)$
5. $x^2 + 10x + 21$ $(x + 3)(x + 7)$
6. $x^2 + 13x + 36$ $(x + 4)(x + 9)$
7. $x^2 + 13x + 42$ $(x + 6)(x + 7)$
8. $x^2 - 12x + 20$ $(x - 10)(x - 2)$
9. $x^2 - 4x + 3$ $(x - 1)(x - 3)$
10. $x^2 - 8x + 15$ $(x - 3)(x - 5)$
11. $x^2 - 17x + 72$ $(x - 9)(x - 8)$
12. $x^2 - x - 30$ $(x - 6)(x + 5)$
13. $x^2 + 4x - 21$ $(x + 7)(x - 3)$
14. $x^2 - 4x - 5$ $(x - 5)(x + 1)$
15. $x^2 + x - 56$ $(x + 8)(x - 7)$
16. $x^2 + 13x - 30$ $(x + 15)(x - 2)$
17. $x^2 - 16x + 55$ $(x - 5)(x - 11)$

M-8 Chapter 13

LESSON QUIZ

Find each product.

1. $(m + 3)(m + 5)$ $[m^2 + 8m + 15]$
2. $(2a + 3)(a - 9)$ $[2a^2 - 15a - 27]$
3. $(3c - 8)(2c - 3)$ $[6c^2 - 25c + 24]$
4. $(x - 4)(x + 4)$ $[x^2 - 16]$
5. $(m - 6)^2$ $[m^2 - 12m + 36]$

Assignment Guide

Basic 1, 4-6, 10-12, 16-24, 34-36, 40-41, 44, 46-47, 49-50, 53, MR, TY All

Average 2, 5-7, 11-13, 19-27, 35-37, 41-42, 45-47, 49-51, 53-54, MR, TY All

Enriched 3, 7-9, 13-15, 25-33, 37-39, 42-43, 45-46, 48-54, MR, TY All

FOR THE NEXT LESSON

math journal

Career Information

Arrange a tour for students to a corporation. Request that a representative from the Finance department be available to answer students' questions.

Suggest the following topics for discussion:

- The factors the company uses to set prices for its products.
- What factors determine how much it would cost to run a particular department within the company.
- What formulas are used to determine if the company is profitable.
- How fluctuations in loan interest influence future expansion.
- How the company phases out unprofitable products.
- How employee pensions are funded.
- What financial considerations the company uses to replace old or worn-out equipment.

CAREER

Help Wanted: Finance Manager

A Bachelor's degree in accounting or finance, or a business degree with emphasis on accounting or finance is required.

For more information, write: American Financial Services Association, Fourth Floor, 1101 14th St. NW, Washington, DC, 20005.

Financial managers plan and analyze operations in a business. They evaluate a company's performance and determine the company's monetary needs. Financial managers use formulas to compile data to make decisions.

PROJECT

Analyze the cost of running a car wash. Decide what materials you will need and what they will cost. Use the information to decide on a price to charge per car.

Practice

Tell how many terms are in each of the following expressions. Then identify each as a monomial, binomial or trinomial.

1. $3x^2 + x$ 2; binomial
2. xy 1; monomial
3. $5x + 3$ 2; binomial
4. $7x^2 + x + 4$ 3; trinomial
5. 72 1; monomial
6. $2xy + x + y$ 3; trinomial

Add or subtract.

7. $(2r^2 - 5) + (4r^2 + 2)$ $6r^2 - 3$
8. $(5ab - 2) - (ab + 6)$ $4ab - 8$
9. $(-8d + 2) - (-4d + 3)$ $-4d - 1$
10. $(6c^2 - 3a) + (-5c^2 - 2a)$ $c^2 - 5a$
11. $(-7x^2y - 2xy + 7) - (5x^2 + 4xy - 6)$ $-7x^2y - 6xy - 5x^2 + 13$
12. $(3mn^2 - mn + 6) + (4mn^2 + mn - 2)$ $7mn^2 + 4$
13. $(5rt^2 + 4r^2t - 2t) - (3rt^2 - 2r^2t - 4t)$ $2rt^2 + 6r^2t + 2t$
14. $(2jk^2 + 5jk + j - 4) + (jk^2 + 3jk - j - 1)$ $3jk^2 + 8jk - 5$

Use the distributive property to find each product.

15. $3x(7y + 2)$ $21xy + 6x$
16. $-6xy(4x^2 - 2y^2)$ $-24x^3y + 12xy^3$
17. $5xy(4x + 3y)$ $20x^2y + 15xy^2$
18. $7ab^2(a - b + 4)$ $7a^2b^2 - 7ab^3 + 28ab^2$
19. $2a^2b^2(5a^2 - 5b^2)$ $10a^4b^2 - 10a^2b^4$
20. $6x^2y(7x^2 + 3y^2 + 2xy - 5)$
21. $4cd(2c + 3d - 6)$ $8c^2d + 12cd^2 - 24cd$
22. $3x^2yz(4x + 5y + 6z - 2)$ $12x^3yz + 15x^2y^2z + 18x^2yz^2 - 6x^2yz$

20. $42x^4y + 18x^2y^3 + 12x^3y^2 - 30x^2y$

Find each product.

See Additional Answers.

23. $(x + 5)^2$
24. $(x - 7)(2x + 9)$
25. $(x - 1.5)(x - 3)$
26. $(2x + 3)^2$
27. $(x + 5)(x - 5)$ $x^2 - 25$
28. $(2x - 4)(2x + 4)$
29. $(x - 5)^2$
30. $(4x - 10)(x - 8)$
31. $(x^2 - 2)(x - 5)$
32. $(x + 1)(x + 1)$ $x^2 + 2x + 1$
33. $(6x + 2)(2x + 6)$
34. $(x^2 - x)(x^2 - 8)$
35. $(-5x + 2)^2$ $25x^2 - 20x + 4$
36. $9(x + 4)(x - 2)$
37. $(10x^2 - 2)(3x^2 + x)$
38. $9(x + 4)^2$ $9x^2 + 72x + 144$
39. $3(x + 6)(x^2 - 4x)$ $3x^3 + 6x^2 - 72x$
40. $\left(\frac{1}{2}x - 67\right)\left(\frac{1}{3}x + 12\right)$ $\frac{1}{6}x^2 - \frac{49}{3}x - 804$

Evaluate each polynomial for $x = 3$, $y = -2$, and $z = 1$.

41. $x^2 + 4$ 13
42. $3x^2 - 2y + z$ 32
43. $4x^2y^2 - 4x + 8$ 140
44. $7xy - 3z$ -45
45. $-3z^2y + 2xy$ -6
46. $-x^2 + 2y - z + 8$ -6
47. $30x^5 - 10z^{10}$ 7,280
48. $-22xyz^3$ 132
49. $2x^2 + y - 50$ -34
50. $25x^3 + 5y^2 + 10z - 1$ 704
51. $z^5 + x^2 - y$ 12
52. $5x^2 - 2y^2$ 37
53. $y^2 + xyz$ -2
54. $3x + z^2$ 10
55. $y^2 - z + x^2$ 12
56. $(x + y)^2$ 1

ADDITIONAL ANSWERS

23. $x^2 + 10x + 25$
24. $2x^2 - 5x - 63$
25. $x^2 - 4.5x + 4.5$
26. $4x^2 + 12x + 9$
27. $x^2 - 25$
28. $4x^2 - 16$
29. $x^2 - 10x + 25$
30. $4x^2 - 42x + 80$
31. $x^3 - 5x^2 - 2x + 10$
33. $12x^2 + 40x + 12$
34. $x^4 - x^3 - 8x^2 + 8x$
36. $9x^2 + 18x - 72$
37. $30x^4 + 10x^3 - 6x^2 - 2x$

PROBLEM SOLVING APPLICATION

OBJECTIVE: *To solve systems of equations using matrices.*

13-5 Using Matrices

You can use a *matrix* to solve a system of linear equations.

Matrix	A matrix is an array of numbers written in brackets.

The matrix of a system of two linear equations in two variables is square and is made up of the coefficients from each equation.

Example 1 **Determine the matrix of coefficients for the system of linear equations.**

$3x + 5y = 13$
$2x + 3y = 8$

Solution $\begin{bmatrix} 3 & 5 \\ 2 & 3 \end{bmatrix}$ Write the coefficients for x in one column and the coefficients for y in the other column.

A *determinant* is real number value associated with a square matrix.

Determinant	We define the determinant (D) of a square matrix as $D = \begin{vmatrix} a_1 & b_1 \\ a_2 & b_2 \end{vmatrix} = a_1b_2 - a_2b_1$.

Example 2 **Find the determinant of the matrix below.**

$\begin{bmatrix} 8 & 3 \\ 4 & 5 \end{bmatrix}$

Solution $D = \begin{vmatrix} 8 & 3 \\ 4 & 5 \end{vmatrix} = 8(5) - 4(3) = 28$

You can use *Cramer's rule* and determinants to find the solution of a system of equations.

Cramer's Rule	Cramer's rule uses the following determinants. $D = \begin{vmatrix} a_1 & b_1 \\ a_2 & b_2 \end{vmatrix}$ $D_x = \begin{vmatrix} c_1 & b_1 \\ c_2 & b_2 \end{vmatrix}$ $D_y = \begin{vmatrix} a_1 & c_1 \\ a_2 & c_2 \end{vmatrix}$ The x value of the solution is $\frac{D_x}{D}$ and the y value is $\frac{D_y}{D}$. The letter c in each matrix represents the constant term in each equation.

THINK How is evaluating a determinant similar to cross multiplication? **You multiply the first and fourth terms and also the second and third terms.**

THINK How do you know whether a matrix or a determinant is being shown? **A matrix is enclosed with brackets. A determinant is enclosed by two bars.**

Lesson Focus

Materials
Math journal

Vocabulary/Symbols
Cramer's Rule
determinant
matrix

MOTIVATING THE LESSON

Write the equations from the text on the chalkboard and box the coefficients and the x and y variables.

$\boxed{3x + 5y} = 13$
$\boxed{2x + 3y} = 8$

Ask students what they notice about these equations. (Possible answers: the coefficients for x and y are lined up; there is a coefficient for each x and y variable.)

Tell students that today they will learn about matrices and how to use them to solve systems of linear equations.

Instruction

1. Erase the x's and y's from the equation you wrote on the chalkboard to create the matrix in the text. Explain that the first column contains the x coefficients and the second column contains the values for the y coefficients.

2. Write these two linear equations to help students remember where to get the values:

$$a_1x + b_1y = c_1$$
$$a_2x + b_2y = c_2$$

3. Keep the equation you wrote for item 2 on the chalkboard. Using the equations from the beginning of the lesson, ask students to assign the values for a_1 (3), b_1 (5), c_1 (13), a_2 (2), b_2 (3), c_2 (8). Students can then use these values with Cramer's Rule.

Teaching Tip

Encourage students to share their thought processes for problem solving with the class. This provides an opportunity to improve students' self-esteem, demonstrate valid alternatives, and to correct any erroneous assumptions or conclusions.

4 Mention to students that the coefficient of y in the first equation is 1. Be sure that students understand what each value represents.

Guided Practice

Class Exercises Mention to students that for question 2, the coefficient of x for the second equation is -1.

For questions 2-3, be sure students use the correct values for x and y.

Closure

Writing in Math Have students describe Cramer's rule in their math journals. Then have students write the steps needed to solve a system of linear equations using Cramer's rule.

Independent Practice

For questions 10 and 11, have students identify their equations and the value for a_1, a_2, b_1, b_2, c_1, and c_2 before they solve each problem.

Decision Making For question 1b, have students find the daily rate for one person in the hotel room and two people in the hotel room.

For question 2, remind students that the franc is used in France and the peso is used in Mexico.

For question 3 have students determine whether their answers would be different for departures from the East Coast or from the West Coast.

For question 6, suggest that the students let x equal the hotel cost and y equal the airfare cost.

Example 3 **Solve the system of equations using Cramer's rule.**

$3x + y = 5$
$2x + 3y = 8$

Solution **4**

1. Solve for D. $D = \begin{vmatrix} 3 & 1 \\ 2 & 3 \end{vmatrix} = 3(3) - 2(1) = 7$

2. Solve for D_x. $D_x = \begin{vmatrix} 5 & 1 \\ 8 & 3 \end{vmatrix} = 5(3) - 8(1) = 7$

3. Solve for D_y. $D_y = \begin{vmatrix} 3 & 5 \\ 2 & 8 \end{vmatrix} = 3(8) - 2(5) = 14$

4. Solve for x and y.

$x = \frac{D_x}{D} = \frac{7}{7} = 1 \qquad y = \frac{D_y}{D} = \frac{14}{7} = 2$

The solution of the system of equations is (1,2).

THINK How could you check to make sure that (1,2) is the solution? **Determine if (1,2) is a solution of each equation in the system.**

CLASS EXERCISES

Determine the matrix of coefficients for each system of equations. Evaluate the determinant of each matrix. **See side column.**

1. $2x + 6y = 22$
$4x + 3y = 17$

2. $x + 2y = 3$
$3y - x = 2$

3. $-4y + 6x = 16$
$-2x + 3y = -2$

4. What will be the value of the determinant if all elements in one column are zero? **zero**

1. $\begin{bmatrix} 2 & 6 \\ 4 & 3 \end{bmatrix}$; -18
2. $\begin{bmatrix} 1 & 2 \\ -1 & 3 \end{bmatrix}$; 5
3. $\begin{bmatrix} 6 & -4 \\ -2 & 3 \end{bmatrix}$; 10

Decision Making ■ DECISION MAKING ■ Decision Making ■ Decision Making ■ Decision Making

USING MATRICES

■ You can solve real life problems using matrices.

■ ***COLLECT DATA*** **Check students' work.**

1. a. Visit a travel agent and find the prices of a round-trip coach ticket to Paris and to Mexico City. Record the data in a chart like the one below.

City	Travel Cost	Hotel Cost
Paris		
Mexico City		

b. Your travel agent will be able to recommend a hotel in the area. Record the price of the hotel in your chart.

2. Look in the newspaper and find the exchange rate from dollars to francs and from dollars to pesos.

MATH MINUTES

Identify each matrix if $D = -52.5$, $D_x = 42$, and $D_y = 80$

$\begin{bmatrix} 5 & 12 \\ -15 & -20 \end{bmatrix}$ (D_y) $\quad \begin{bmatrix} 12 & 3 \\ -20 & -1.5 \end{bmatrix}$ (D_x)

$(D) \begin{bmatrix} 5 & 3 \\ 15 & -1.5 \end{bmatrix}$

WRITTEN EXERCISES

Use a CALCULATOR where appropriate.

Write the matrix of coefficients for each system of linear equations.

1. $x + 2y = 16$
$2x + 3y = 26$ $\begin{bmatrix} 1 & 2 \\ 2 & 3 \end{bmatrix}$

2. $y = 2x$
$-x + y = 1$ $\begin{bmatrix} -2 & 1 \\ -1 & 1 \end{bmatrix}$

3. $y = -3x + 17$
$2y + 6 = 2x$ $\begin{bmatrix} 3 & 1 \\ -2 & 2 \end{bmatrix}$

Evaluate each determinant.

4. $\begin{vmatrix} 3 & 4 \\ 2 & 9 \end{vmatrix}$ 19

5. $\begin{vmatrix} 8 & 4 \\ -6 & -2 \end{vmatrix}$ 8

6. $\begin{vmatrix} -7 & -2 \\ 0 & 11 \end{vmatrix}$ -77

Solve each system of linear equations using Cramer's rule.

7. $2x + 4y = 12$
$3x + 5y = 14$
(-2,4)

8. $-4x + 7y = 1$
$25 = 2x + 5y$
(5,3)

9. $x + 2y = 3$
$3y - x = 2$
(1,1)

Write a system of linear equations for each situation. Solve using Cramer's rule.

10. At the grocery store, one box of laundry detergent and two bottles of fabric softener cost \$7.75. Two boxes of laundry detergent and one bottle of fabric softener cost \$8.75. Find the cost of each product. A box of laundry detergent is \$3.25 and a container of fabric softener is \$2.25.

11. A manufacturer sells packages of pens and pencils. A package of 3 pens and 5 pencils costs \$1.65. A package of 5 pens and 10 pencils costs \$3.00. Find the price of one pen and two pencils. Each pen is \$.30 and each pencil is \$.15. One pen and two pencils cost \$.60.

Decision Making ■ Decision Making ■ Decision Making ■ Decision Making ■

■ **ANALYZE DATA** Check students' work.

3. Which city is less expensive to visit? Why might this be true?

4. How much would a meal cost in United States dollars if it costs \$16 in Mexico City?

5. Write a system of linear equations using the collected data.

■ **MAKE DECISIONS**

6. A travel agency offers a package deal to Freeport, Bahamas. The four day-three night package is \$590 per person. The seven day-six night package is \$815 per person. Each price includes the cost of airfare and hotel.

a. Write a system of linear equations for the situation. Let a = airfare, b = cost of hotel per night. $a + 3b = 590$, $a + 6b = 815$

b. Solve the system using Cramer's rule. $a = 365$, $b = 75$

c. Find the cost of each airfare per person. \$365

d. Find the cost of the hotel per night. \$75

Lesson Focus

LESSON RESOURCES

TRB Alternate Application M-9

Assignment Guide

Basic All
Average All
Enriched All

FOR THE NEXT LESSON

math journal, geoboard, dot paper, or graph paper

Alternate Application

For use after 13-5 (pp. 555–557)

Simplifying Formulas

You can use what you have learned about polynomials when working with formulas.

1. Complete to show how the polynomial $2l + 2w$ can be written as a product:
$p = 2l + 2w = 2(\underline{l} + \underline{w})$
rectangle — perimeter $p = 2l + 2w$

2. A rectangle has length 13.7 and width 6.3.
a. Find the perimeter using the formula $p = 2l + 2w$. 40
b. Find the perimeter using the new formula you write in Exercise 1.
40

3. Which formula is easier to use? Explain your answer.
Answers may vary; the second method involves fewer operations.

4. Complete to show how the polynomial $2\pi r^2 + 2\pi rh$ can be written as a product:
$S = 2\pi r^2 + 2\pi rh = 2\pi r(\underline{r} + \underline{h})$
cylinder — surface area $S = 2\pi r^2 + 2\pi rh$

5. A cylinder has radius 3 and height 5.
a. Find the surface area using the formula $S = 2\pi r^2 + 2\pi rh$. Use $\pi = 3.14$.
150.72 units²
b. Find the surface area using the new formula you wrote in Exercise 4.
150.72 units²

6. Which formula is easier to use? Explain your answer.
Answers may vary; the second method involves fewer operations.

7. The formula for the amount (A) of money after a principle (p) is invested for t years at an interest rate of r is $A = p + prt$. Give another way to write the formula:
$A = p + prt = \underline{p}(\underline{1} + \underline{rt})$

Chapter 13 M-9

Lesson Focus

Materials

Math journal, geoboard, dot paper, or graph paper

MOTIVATING THE LESSON

Have students solve this problem: **What are the dimensions of a rectangle if its perimeter is 12 ft?** Students should see that there are several possible answers. (1 ft by 5 ft, 2 ft by 4 ft, 3 ft by 3 ft, 4 ft by 2 ft, 5 ft by 1 ft) Then ask students to share the strategies they used to solve the problem. (possible answers: *write an equation, guess and test,* and *make a table)*

Instruction

1 Ask students how the values for *b* and *h* were chosen. (so that their sum would be 52; i.e., one half of the perimeter)

2 Ask students to give the dimensions of the rectangle with the greatest area if the perimeter is 24 ft. (6 ft by 6 ft)

PROBLEM SOLVING STRATEGY

OBJECTIVE: *To solve problems using one or more strategies.*

13-6 Using Multiple Strategies

Sometimes you may need to use more than one strategy to solve a problem.

PROBLEM

A rancher bought 104 ft of fencing to make a rectangular corral. The rancher wants the corral to have the greatest possible area. What dimensions should the rancher use?

SOLUTION

READ What do you want to find? — the base and height of a rectangle with $P = 104$ ft and the greatest possible area

What do you know? — $A = bh$

$P = 2b + 2h$
$= 104$ ft

PLAN Decide on strategies.

1. Draw a diagram to represent the information.
2. Use guess and test.
3. Make a table.

1

SOLVE Draw a diagram.

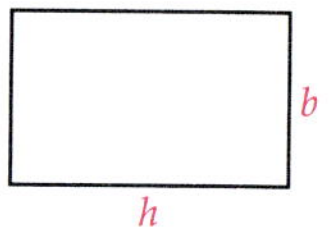

Choose some possible dimensions. Make a table to organize your data.

Look for a pattern. Guess some higher numbers and see how long the area continues to increase.

b	h	P(ft)	A(ft²)
10	42	104	420
11	41	104	451
12	40	104	480
20	32	104	640
22	30	104	660
24	28	104	672
26	26	104	676
28	24	104	672

The corral with the greatest area is a square that is 26 ft on a side.

2

LOOK BACK Describe the pattern. — As the dimensions of the rectangle get closer to those of a square, the area approaches the greatest possible area.

Teaching Tip

Have students work in small groups to help them gain an understanding of how and when strategies may be combined and used to solve problems.

CLASS EXERCISES

Solve.

1. A gardener wants to fence in the greatest possible area using 200 ft of fencing. What should be the base length and height of the garden? 50 ft by 50 ft
2. ***CALCULATOR*** A circle and a square both have an area of 144 square units. Use 3.14 for π.
   a. What is the circumference of the circle? Round to the nearest tenth. 42.5 units
   b. What is the perimeter of the square? 48 units
   c. Which figure is the most economical if purchasing fencing materials to surround the figure? circle
   d. Fencing material is $6.80/unit. How much is saved by choosing the most economical figure? $37.40

WRITTEN EXERCISES

Use a CALCULATOR where appropriate.

Solve.

1. A student playing a computer chess game gets 5 points every time he wins the game. The computer gets 3 points every time it wins the game. They play 128 games and end with a tie score. How many games did the computer win? 80 games
2. Two people on bicycles leave home at 10 A.M. and ride towards each other. Their homes are 56 mi apart. The first cyclist pedals at 16 mi/h. The second pedals at 12 mi/h. At what time will they meet? 12:00 P.M.
3. A painter places an 8.5-ft ladder against a wall. The bottom of the ladder is 4 ft from the base of the wall. How high up on the wall does the ladder reach? 7.5 ft
4. There are 27 white cubes assembled to form a large cube. The outside surface of the large cube is then painted red. The large cube is then separated into a set of smaller cubes. How many of the small cubes will have exactly two red faces? 12 cubes
5. A student weighs his hamsters two at a time. Together, Sandy and White Ears weigh 209 g. White Ears and Sport weigh 223 g together. Sandy and Sport weigh 216 g together. How much does each hamster weigh? Sandy = 101 g; White Ears = 108 g; Sport = 115 g
6. A grocer is arranging cans in a pyramid. She uses 9 cans on each side of the base of the pyramid. How many cans will be in the pyramid? 285 cans

Guided Practice

Class Exercises For question 2, have students also use $\frac{22}{7}$ as the value for π.

A ***common error*** occurs when students choose values for both variables without regard to their relationship. Have students write the perimeter equation, $2a + 2b = 104$, substitute a value for a or b, then solve the equation to find the value of the other variable.

Closure

Writing in Math Have students summarize in their math journals how the combined strategies *guess and test, make a table,* and *look for a pattern* were used to solve the problem.

Independent Practice

To reinforce the idea that there is usually more than one way to arrive at a solution, encourage students to share the strategies they used to solve each problem.

MATH MINUTES

Students work in small groups to make up a problem that can be solved by using two or more strategies. Groups then exchange problems, analyze them, and state the strategies that might be used to solve them. (Check students' work.)

If students have difficulty with question 11, have them label the vertex and one point on each ray so that they can identify the different angles.

Lesson Follow-up

LESSON RESOURCES

TRB Practice M-10
TRB Enrichment M-11

7. A room has an area of 1,025 ft^2 and a 10-ft ceiling. Occupancy guidelines recommend at least 200 ft^3 per person. What should be the maximum number of people allowed in the room? **51 people**

8. A man who won the lottery gave his daughter half of the money. He gave his brother half as much as he gave his daughter and kept $3.8 million for himself. How much did the man win? **$15.2 million**

9. A student has $8 to spend on a phone call to a friend. The cost of a call is $.34 for the first minute and $.24 for each additional minute. How long can she talk to her friend? **32.9 min**

10. A student decided to purchase a new telephone. He could choose from 8 different models, 2 different cord lengths, and 4 different colors. How many possible choices did he have if he can choose only one of each model, cord length, and color. **64 choices**

11. How many different angles can you find in the figure? **10 angles**

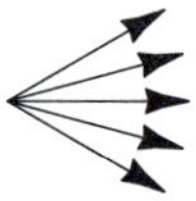

12. A clerk starts working at a beginning salary of $10,400 with an annual increase of $400. The clerk hires an assistant at a starting salary of $9,600 per year with an annual increase of $600. At this rate, in how many years will the assistant be earning more money than the clerk? **5 years**

13. A lot measures 50 ft by 100 ft. The house on the lot measures 25 ft by 50 ft. What is the area of the lawn? **3,750 ft^2**

14. A bus left Freetown at noon traveling 40 mi/h. A car left Freetown at 1:30 P.M. traveling 60 mi/h.
    a. At what time did the car catch up with the bus? **4:30 P.M.**
    b. How many miles from Freetown were the car and the bus when they met? **180 mi**

15. A student spends $\frac{1}{3}$ of her money on a movie and $\frac{1}{6}$ of the remaining amount on a snack after the movie. She now has $16. How much money did she originally have? **$28.80**

16. A boy jogs in the park every other day. His sister jogs every third day. They both jogged together on April 2. How many more days in April can they jog together if they maintain this schedule? **4**

17. In how many different ways can you give change from a $100 bill for a $78 purchase if the customer will accept no more than seven singles? **6 ways**

18. ***DATA FILE 12 (pp. 486–487)*** A base runner tried to steal a base 73 times during the baseball season. Approximately how many times was the base runner out? **≈22**

Assignment Guide

Basic All
Average All
Enriched All

Practice

For use after 13-6 (pp. 558–560)

Using Multiple Strategies

Solve.

1. The product of two whole numbers is 36. What is the greatest possible sum that the numbers can have?
   37 (1 and 36)
2. The sum of two numbers is 14. What is the greatest possible product that the numbers can have?
   49 (7 and 7)
3. A rectangle has length $(x-3)^2$ and width 4. The perimeter of the rectangle is 40. Find the length.
   16
4. A rectangle has length $k+6$ and width $k-6$. The area of the rectangle is 64. Find the length and the width.
   16 and 4
5. A rectangular prism has length $x+2$, width $x+1$, height 4, and volume 24. Find the length and the width.
   3 and 2
6. A piece of cardboard measures 12 ft by 12 ft. Corners are to be cut from it as shown by the broken lines, and the sides folded up to make a box with an open top. What size corners should be cut from the cardboard to make a box with the greatest possible volume?
   2 ft × 2 ft

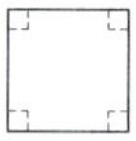

7. What is the maximum number of small boxes that can fit inside the large box?
   64 boxes

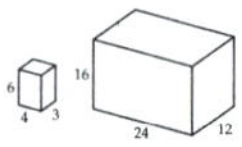

8. The perimeter of a right triangle is 24 in. Find the dimensions of the triangle if the sides are all whole-number lengths.
   6 in., 8 in., 10 in.

M-10 Chapter 13

Enrichment

For use after 13-6 (pp. 558–560)

Graphing the Maximum

In the text lesson you used multiple strategies to find the greatest possible area that can be enclosed by a given amount of fencing. One way to solve such problems is to draw a graph.

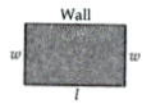

Marnie has 16 m of fencing with which to enclose three sides of a rectangular garden adjacent to a wall. The wall will form the fourth side of the garden. Marnie wants the garden to have the greatest possible area. Answer these questions to find a graphical solution of the problem.

1. Complete the table. Remember that 16 m of fencing will be used to enclose two widths and one length of the garden. (That is, $2w + l = 16$.)

width	$\frac{1}{2}$	1	2	3	4	5	6	7	$7\frac{1}{2}$
length	15	14	12	10	8	6	4	2	1
area	$7\frac{1}{2}$	14	24	30	32	30	24	14	$7\frac{1}{2}$

2. Use values of width and area from the table as ordered pairs to draw a graph. Connect the points in the way that seems most reasonable to you.

The graph you have drawn is called a *parabola*. It clearly illustrates that as the width increases, the area increases too, but only to a certain point. After that the area begins to decrease.

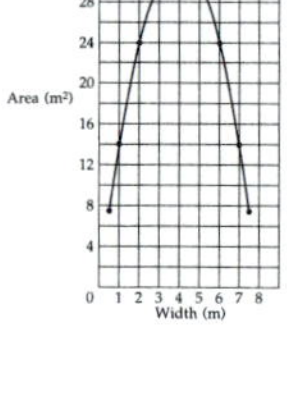

3. What value of the width gives the greatest possible area?
   4 m

Chapter 13 M-11

Problem Solving Practice

PROBLEM SOLVING STRATEGIES

Draw a Diagram
Make a Table
Look for a Pattern
Guess and Test
Write an Equation
Simplify the Problem
Work Backwards
Account for All Possibilities
Make an Organized List
Solve by Graphing
Make a Model
Simulate the Problem

Solve. Use an appropriate strategy or combination of strategies.

1. There are 7 roads from Mayville to Scottsburg and 4 roads from Scottsburg to Dunlap. How many possible routes can you take from Mayville to Dunlap if you go through Scottsburg? **28 possible routes**
2. A singles tennis court is 75% as wide as a doubles tennis court. The singles court is 27 ft wide. How wide is the doubles court? **36 ft**
3. A student's average grade for six math quizzes is 85. He received these grades on five math quizzes: 82, 88, 94, 72, 88. What grade did he receive on the other quiz? **86**
4. There are 48 students in the band. Of these students, 24 have blonde hair, and 18 have blue eyes. There are 16 students who do *not* have blonde hair or blue eyes.
   a. How many students have both blonde hair and blue eyes? **10 students**
   b. How many students do *not* have blond hair? **24 students**
   c. How many students do *not* have blue eyes? **30 students**
5. An astronaut's spacesuit weighs 10.02 lb on the moon. This is 16.7% of its weight on Earth. How much does the spacesuit weigh on Earth? **60 lb**
6. A student weighs 10 lb more than his sister. Together they weigh 260 lb. How much does each weigh? **135 lb, 125 lb**
7. The area of rectangle A exceeds the area of square B by 24 yd^2. Find the dimensions of each figure. **A is 3 yd by 11 yd**

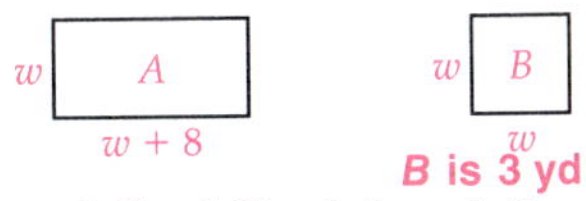

B is 3 yd by 3 yd

8. A student has 4 one-dollar bills, 2 five-dollar bills, and 5 ten-dollar bills. What is the probability that he will select the following bills at random?
   a. a one-dollar bill, then a ten-dollar bill $\frac{2}{11}$
   b. a ten-dollar bill, then a five-dollar bill $\frac{1}{11}$
   c. two ten-dollar bills $\frac{2}{11}$
9. One hot chocolate for each of the 29 students on a field trip would cost \$21.75. The bill for hot chocolate was \$27.75. How many students had two hot chocolates? **8 students**
10. ***DATA FILE 13 (pp. 534–535)*** About how much carbon is released per year by the 17 million people of Texas? Express your answer in pounds using scientific notation. $\approx 1.7238 \times 10^{11}$

Instruction

This page provides a variety of problems that can be used to reinforce and enhance the students' problem solving skills. Encourage students to read each problem carefully. Then have them refer to the list of problem solving strategies to help them decide how to solve the problem.

Point out, however, that neither all questions require a strategy for solving nor are all the strategies in the list used in this lesson.

Study Skills

Teaching is a partnership between students and teachers. In a good *student-teacher partnership*, the student offers a willingness to learn, to respect the teacher, and to make a diligent effort for all tasks; the teacher offers the understanding that effort is almost as important as being correct, and displays the willingness to answer students' questions patiently and fully. Additionally, the teacher works hard to present the material in as creative and stimulating a manner as possible.

Encourage students to discuss what they expect from their relationship with a teacher. If students seem reluctant to speak openly, reassure them that the candor will benefit both you and them. You might have the class draft a teacher-student contract establishing reasonable expectations on both sides.

Here are a few items the contract might include:

- ▼ Teachers can expect students to be respectful and polite. They can also expect students to challenge them if students feel an explanation is inadequate or incorrect.
- ▼ Teachers can expect students to be cooperative, punctual, attentive, and willing to work hard.
- ▼ Students can expect teachers to take time to answer questions and deal with individual problems.

Chapter 13 Review

Write an explanation for each of the following.

1. Explain the term monomial and write two examples. **A monomial is a real number, variable, or the product of a real number and one or more variables. Examples: $6x^2$; $3x$.**
2. Explain the term polynomial and write two examples. **A polynomial is a monomial or a sum or difference of monomials. Examples: $5x - 2$; $12x^2 - 6x - 5$.**
3. Explain how to use the distributive property to find $x(2x - 1)$. **Multiply x times each term within the parentheses. $x(2x - 1) = x \cdot 2x - x \cdot 1 = 2x^2 - x$**
4. Explain how you use FOIL to find the product of $(x + 3)(2x + 1)$. **Multiply the first terms, $x \cdot 2x$, the outer terms, $x \cdot 1$, the inner terms, $3 \cdot 2x$, and the last terms, $3 \cdot 1$. Then add the products. $2x^2 + x + 6x + 3 = 2x^2 + 7x + 3$**
5. Explain how to find the product of $(a + b)(a - b)$. **Square the first term and subtract the square of the second term. $(a + b)(a - b) = a^2 - b^2$**
6. Explain the phrase *square a binomial* and write an example. **To square a binomial like $(a + b)^2$ or $(a - b)^2$, square the first term. Add or subtract twice the product of the two terms and add the square of the last term. $(a + b)^2 = a^2 + 2ab + b^2$; $(a - b)^2 = a^2 - 2ab + b^2$**

Polynomials — 13-1

To name a polynomial, count the number of terms.

Write the polynomial for each model. Then write the name of each polynomial.

7. 
$2x^2 + 3x$; binomial

8. 
$x^2 + 2x + 4$; trinomial

9. 
$3x^2$; monomial

Evaluate each polynomial for $a = 1$ and $b = -2$.

10. $3b^2 - ab + 1$ **15**
11. $a^2b + 3b - a$ **-9**
12. $b^3 + ab^2 + 3b$ **-10**

Adding and Subtracting Polynomials — 13-2

To add or subtract polynomials, use models, number properties, or align like terms in columns.

To add polynomials, combine like terms.

To subtract polynomials, add the opposite of each term in the second polynomial.

Use a model to find each sum or difference.

13. $(5x^2 + 3x) + (2x^2 + x)$ **$7x^2 + 4x$**
14. $(3x^2 + 2x + 4) - (x^2 - x + 3)$ **$2x^2 + 3x + 1$**

Add or subtract.

15. $(3x + 2y) + (5x + 3y)$ **$8x + 5y$**
16. $(5m - 3n) - (2m - n)$ **$3m - 2n$**
17. $(4a^2 + 6b) + (2a^2 - 3b)$ **$6a^2 + 3b$**
18. $(-r^2 + 2s) - (2r^2 - 3s)$ **$-3r^2 + 5s$**
19. $(3c^2 + 5c - 2) - (-c^2 + 3c - 1)$ **$4c^2 + 2c - 1$**
20. $(a^2b + b^2 - a) + (a^2b - 2b^2 + 3a)$ **$2a^2b - b^2 + 2a$**

Multiplying a Polynomial by a Monomial — 13-3

To multiply a polynomial by a monomial, you can use models, or you can use the distributive property. Use the rules of exponents to simplify the product.

Use a model to find each product. Check students' work.

21. $2x(x + 5)$ $2x^2 + 10x$

22. $3x(2x + 1)$ $6x^2 + 3x$

Use the distributive property to find each product.

23. $x(x - 5)$ $x^2 - 5x$

24. $3x(x + 2)$ $3x^2 + 6x$

25. $2x(x - 2)$ $2x^2 - 4x$

26. $4x(x^2 + 3x)$ $4x^3 + 12x^2$

27. $-5x(-2x^2 + 2x - 3)$ $10x^3 - 10x^2 + 15x$

28. $-2xy(x^2 - xy + y^2)$ $-2x^3y + 2x^2y^2 - 2xy^3$

Multiplying a Polynomial by a Binomial — 13-4

To multiply a polynomial by a binomial, you can use a model, you can use the distributive property twice, or you can use the FOIL method. FOIL means: multiply the First terms, Outside terms, Inside terms, and Last terms. Then add the products.

Write a multiplication expression for the model. Simplify.

29.

$(x + 2)(x + 2) = x^2 + 4x + 4$

30.

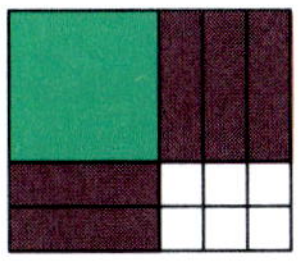

$(x + 2)(x + 3) = x^2 + 5x + 6$

Find each product.

31. $(x - 1)^2$ $x^2 - 2x + 1$

32. $(a + 3)^2$ $a^2 + 6a + 9$

33. $(y - 2)(y + 2)$ $y^2 - 4$

34. $(b + 3)(b - 2)$ $b^2 + b - 6$

35. $(m + 5)(m - 3)$ $m^2 + 2m - 15$

36. $(2n - 4)(n + 5)$ $2n^2 + 6n - 20$

Problem Solving — 13-5, 13-6

To solve a problem, use an appropriate strategy or combination of strategies.

37. The number of bacteria doubles each minute. A bottle is completely filled after 5 minutes. After how many minutes was the bottle half full? 4 min

38. Fourteen boxes contain 152 classic and rock CDs. There are 10 classic CDs in a box and 12 rock CDs in a box. How many boxes contain each kind of CD? 8 classic; 6 rock

39. The number 72 can be divided into three numbers in the ratio 1 to 2 to 3. Find these three numbers. 12, 24, 36

40. A gardener plans to use 196 ft of fencing to enclose a garden. What is the largest possible area of the garden? 2,401 ft^2

RESOURCES

TRB M-13–M-16

Scoring Chart	
No. Correct	**Score**
33	100
32	97
31	94
30	90
29	87
28	84
27	81
26	78
25	75
24	72
23	69
22	66
21	63
20	60
19	57
18	55
17	53
16	50
15	46
14	42
13	38
12	35
11	32
10	29
9	26
8	23
7	20
6	17
5	14
4	11
3	8
2	5
1	2

Chapter 13 Test

Use a model to represent each polynomial. Check students' work.

1. $3x + 2x^2 + 3$

2. $3x^2 + 2x + 1$

Write the polynomial represented by each model.

3. 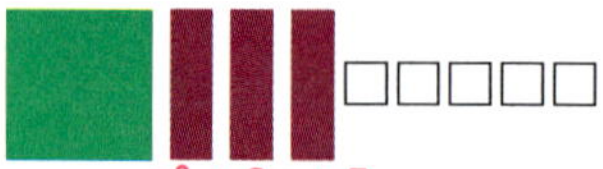

$x^2 + 3x + 5$

4. 

$2x^2 + x + 3$

Identify each expression as a monomial, binomial, or trinomial.

5. $7x - 5$ binomial

6. $-3x^2 + 3x + 1$ trinomial

7. $x^2 + 9$ binomial

8. x^2 monomial

Evaluate each polynomial for $a = 2$ and $b = -3$.

9. $7a + 3b$ 5

10. $a^2 - b^2 + 1$ -4

11. $(b - 1)^2$ 16

12. $a^2 + 2b - 5$ -7

Find each sum.

13.
$$\begin{array}{r} 2x^3 - 3x^2 + x - 1 \\ + \quad x^3 + 2x^2 - 3x + 2 \\ \hline 3x^3 - x^2 - 2x + 1 \end{array}$$

14.
$$\begin{array}{r} 5x^3 + 2x^2 - 3x + 5 \\ + \; 2x^3 - 5x^2 + 6x - 9 \\ \hline 7x^3 - 3x^2 + 3x - 4 \end{array}$$

Add or subtract.

15. $(5x^2 - 2x) - (3x^2 + x)$ $2x^2 - 3x$

16. $(3x^2 + 5x - 3) + (x^2 - 2x + 1)$ $4x^2 + 3x - 2$

17. $(7x^2 + 5x + 3) - (4x^2 + 7x)$ $3x^2 - 2x + 3$

18. $(9x^2 - 4x - 8) + (3x^2 + 6x + 3)$ $12x^2 + 2x - 5$

19. $(-x^2 + 3x + 4) + (2x^2 - 5x + 1)$ $x^2 - 2x + 5$

20. $(2x^2 - 3x + 4) - (x^2 - 2x - 1)$ $x^2 - x + 5$

Find each product. 24. $2x^3 + 6x^2 - 4x$ 25. $3x^3 + 6x^2y - 3x^2$

21. $(x + 3)^2$ $x^2 + 6x + 9$

22. $(x - 2)^2$ $x^2 - 4x + 4$

23. $x(3x^2 - 2x + 5)$ $3x^3 - 2x^2 + 5x$

24. $2x(x^2 + 3x - 2)$

25. $3x^2(x + 2y - 1)$

26. $xy(x^2 - 2y + y^2)$ $x^3y - 2xy^2 + xy^3$

27. $(x - 2)(x + 3)$ $x^2 + x - 6$

28. $(x + 2)(x - 5)$ $x^2 - 3x - 10$

29. $(2x - 3)(x + 4)$ $2x^2 + 5x - 12$

Solve.

30. The width of a rectangle is $\frac{1}{3}$ the length plus 5. The perimeter is 34. Find the length and the width. 9, 8

31. A pair of jeans are on sale for 35% off the original price. The sale price is \$19.50. What is the original price? \$30.00

32. The sum of three consecutive multiples of 4 is 60. What is the product of the three multiples? 7,680

33. Two students are chosen at random to do a project together. There are 15 girls and 10 boys in the class. What is the probability that both students chosen for the project are boys? $\frac{3}{20}$

Chapters 1–13 Cumulative Review

Choose the correct answer. Write A, B, C, or D.

1. Add $(3x^2 + 2x - 5) + (2x^2 - 3x + 1)$. B

A. $5x^2 + 5x + 6$ **B.** $5x^2 - x - 4$
C. $x^2 + 5x - 6$ **D.** not given

2. Find the volume of a cone, $r = 4$, $h = 12$. C

A. 16π **B.** 192π
C. 64π **D.** not given

3. Multiply $(x - 2)(x - 3)$. A

A. $x^2 - 5x + 6$ **B.** $x^2 - 5x - 6$
C. $x^2 - 5x - 5$ **D.** not given

4. Find the mode of the following data: 31, 29, 31, 23, 35, 34, 19, 35, 23, 29, 35. C

A. 23 **B.** 29
C. 35 **D.** not given

5. Find a solution of $x - 2y = 3$ and $3x + y = 2$. A

A. (1,-1) **B.** (-1,1)
C. (3,2) **D.** not given

6. Find the circumference of a circle with $r = 3$. B

A. 3π **B.** 6π
C. 9π **D.** not given

7. Write the fraction for $37\frac{1}{2}\%$. C

A. $\frac{1}{2}$ **B.** $\frac{75}{2}$
C. $\frac{3}{8}$ **D.** not given

8. A set of (d) dimes and (n) nickels is worth \$3.20. There are 52 coins. How many of each coin are there? B

A. 12 n, 40 d **B.** 40 n, 12 d
C. 26 n, 26 d **D.** not given

9. Solve $-3x + 1 < 25$. B

A. $x < -8$ **B.** $x > -8$
C. $x > 8$ **D.** not given

10. 32% of b = 10,000. Find b. B

A. 3,200 **B.** 31,250
C. 3,125 **D.** not given

11. Multiply $(a - b)^2$. A

A. $a^2 - 2ab + b^2$ **B.** $a^2 + 2ab + b^2$
C. $a^2 + 2ab - b^2$ **D.** not given

12. Find the distance Joe traveled if he traveled 50 mi/h for $2\frac{1}{2}$ h. D

A. 150 mi **B.** 100 mi
C. 250 mi **D.** not given

13. What is the probability of tossing heads twice with a fair coin? C

A. $\frac{1}{2}$ **B.** 1
C. $\frac{1}{4}$ **D.** not given

14. Find the next number. 0, -1, $\sqrt{1}$, -2, $\sqrt{4}$, -3, . . . B

A. $\sqrt{5}$ **B.** $\sqrt{9}$
B. $\sqrt{7}$ **D.** not given

15. Write 568,000,000 in scientific notation. C

A. 5.68×10^6 **B.** 56.8×10^7
C. 5.68×10^8 **D.** not given

16. Write the expression for the model. B

A. $2(x^2 + 2x) + 3$ **B.** $2x^2 + 2x + 3$
C. $x^2 + 2x + 3$ **D.** not given

Study Skills

Students can be encouraged to view tests as an *opportunity to learn*. This attitude can be fostered if a teacher is sympathetic to the fact that not all students perform well on tests, and believes that an overall classroom grade should also account for a student's hard work, eagerness to learn, class participation, and general grasp of the material. If a student feels confident that tests are not the sole determinant of his or her grade, the student is more willing to view a test as a chance to evaluate skills, knowledge, and test-taking techniques.

Encourage students to take time to review their tests. Remind them that tests are a valuable guide to determining what material is most important to learn and whether the student needs to better understand the material.

Here are some questions students might want to ask themselves when reviewing tests:

- Were errors made in reading or understanding?
- Was pre-test preparation adequate?
- Are there particular areas of weakness the test identifies?
- Were mistakes made due to not enough care in organizing and writing the answers?
- Are there areas of special strengths shown on the test?

Chapter 1 Extra Practice

Write an integer.

1. opposite of 4 -4
2. opposite of -8 8
3. $|10|$ 10
4. $|-6|$ 6

Compare. Use <, >, or =.

5. 5 ▒ -1 >
6. -7 ▒ -3 <
7. -2 ▒ -2 =
8. -9 ▒ 0 <

Find each answer.

9. $-11 + 1$ -10
10. $-4 - (-9)$ 5
11. $-3(-3)$ 9
12. $44 \div (-4)$ -11
13. $28 \cdot (-4)$ -112
14. $-64 \div (-8)$ 8
15. $98 - (-12)$ 110
16. $-52 + (-11)$ -63
17. $-2 + (-7) + 15$ 6
18. $-2 \cdot 10 \cdot (-4)(-1)$ -80
19. $120 \div (-4) \div 10 \div (-1)$ 3
20. $9 \cdot 7 \div (4 - 1)$ 21
21. $5 + 7 \cdot 3 + 1$ 27
22. $3(12 + 6) - 9 \cdot 6$ 0
23. $|12 - (-5)|3$ 51
24. $1 + (-4) \cdot (-7) + 10 - 2$ 37
25. $30 - 7 \cdot 2(22 - 12) \div 5$ 2

Write an expression for each word phrase.

26. a number increased by eight $x + 8$
27. the product of negative seven and x $-7x$
28. twelve less than the absolute value of negative three $|-3| - 12$
29. the opposite of the quantity seven less than y $-(y - 7)$
30. three times the sum of negative eight and twelve $3(-8 + 12)$
31. the absolute value of the difference of negative three and ten $|-3 - 10|$
32. the quotient of twenty-five and negative five, minus two $25 \div (-5) - 2$
33. ten times the quantity seventeen minus negative eleven $10[17 - (-11)]$

Evaluate each expression for the given values of the variables.

34. $4a + 7$, for $a = -2$ -1
35. $8m + 13 + 6n$, for $m = 5$, $n = 3$ 71
36. $7|x - y| + y$, for $x = 3$, $y = 11$ 67
37. $-14 - 2(a - b)$, for $a = 6$, $b = 4$ -18

Solve.

38. A group of children line up in a row. The first child takes 1 step forward. The second child takes 2 steps forward. The third child takes 5 steps, the fourth takes 14, and so on. Following this pattern, how many steps will the next two children take? 41 and 122

39. Concert tickets cost \$20 and concert T-shirts cost \$12.
    a. Write an expression for the cost of x tickets and y shirts. $20x + 12y$
    b. Find the cost of 4 tickets and 6 shirts. \$152
    c. How many tickets can you buy for \$60? how many T-shirts? 3; 5

40. ***DATA FILE 1 (pp. 2–3)*** The air temperature is 5°F. What is the maximum wind speed, in miles per hour, before there is an increased danger of frostbite? 10 mi/h

41. After being kicked forward 20 m, a ball is pushed backward 7 m. Describe the position of the ball using an integer. +13

Chapter 2 Extra Practice

Replace each variable with the given value. State whether the equation is true or false.

1. $17 - a = 7, a = 8$ false
2. $4 + m = 2m + 8, m = -4$ true
3. $8 - 2q = 3q + 1, q = 0$ false
4. $-x + 3 = 2x, x = 1$ true
5. $4x + y = 20, x = -4, y = 5$ false
6. $3a = 2b - 9c, a = -3, b = 0, c = 1$ true

Which of the numbers -1, 0, 1, 3, 5 is a solution?

7. $0 = 121y$ 0
8. $-8 = 2m - 18$ 5
9. $4n = 3n - (-3)$ 3
10. $x(3 - 8) = 5(-x)$ -1, 0, 1, 3, 5

Evaluate.

11. $26 + 8 + 10$ 44
12. $33 \cdot 2 \cdot 5 + 7$ 337
13. $5(8 - 3)$ 25
14. $6[8 - 2 - (-7)]$ 78
15. $5[(-8) + 17 - (2)3]$ 15
16. $510 \div (-5) \cdot 2$ -204
17. $5(8 \div 8) - 33 \cdot 2$ -61
18. $18 + 7 \cdot 4 - 3 \div (-1)$ 49
19. $4(3 + 7) \div 2$ 20

Simplify each expression.

20. $-3(2c)$ -6c
21. $41 - 2(m + 1) - m$ -3m + 39
22. $3(a + b + 2c)$ 3a + 3b + 6c
23. $3q + 2(q + 1)$ 5q + 2
24. $(8 \div 4)r - 3r$ -1r
25. $8 + (3s + 2)(-2)$ -6s + 4

Solve each equation.

26. $x - 33 = 0$ x = 33
27. $j + (-22) = 4$ j = 26
28. $14 = z - 9$ z = 23
29. $z(3 - 1) = 40$ z = 20
30. $5q = 26 - 1$ q = 5
31. $2|n| = 10$ n = 5 or n = -5

Write an equation for each word sentence.

32. Twice the sum of a number and one is twenty-two. $2(n + 1) = 22$
33. Negative three divided by negative one is three. $-3 \div (-1) = 3$
34. The sum of three and four times a number x is equal to $6x$. $3 + 4x = 6x$
35. The product of two numbers is the absolute value of negative twenty-four. $xy = |-24|$

Solve.

36. Terry's age is half Bobby's age. The sum of their ages is 36. How old are Terry and Bobby? 12 and 24
37. ***DATA FILE 2 (pp. 52–53)*** A lightning flash is seen 7 s before the thunder is heard. How far away is the lightning? 2,422 m
38. There are 10 children on a school bus. At the next stop, 5 get on and 1 gets off. Then x get on and none get off. This leaves a total of $3x$ children on the bus. Solve for x. How many children are on the bus? 21
39. There are twice as many golf balls as ping-pong balls. If 9 golf balls are taken away, there will be the same amount of golf balls and ping-pong balls. How many golf balls and ping-pong balls are there altogether? 27

Chapter 3 Extra Practice

Order from greatest to least.

1. 2.012, 2.12, 2.011 **2.12, 2.012, 2.011**

2. -0.03, -0.33, -3 **-0.03, -0.33, -3**

3. 0.00004, 1.00009, 0.000045 **1.00009, 0.000045, 0.00004**

4. 278, 2.78, 27.8 **278, 27.8, 2.78**

Round each decimal to the indicated place.

5. 0.38, nearest tenth **0.4**

6. -3.089, nearest hundredth **-3.09**

7. 238.079, nearest whole number **238**

Estimate using the technique which seems best. **Accept reasonable answers.**

8. -42.039 ÷ 10.99 **-4**

9. \$3.74 + \$12.12 + \$3.00 **\$19**

10. -97.3 + (-33.9776) + (-28.0549) **-160**

11. 83.6 + 7.98 + (-2.09) **90**

12. 397 ÷ 1.9 **200**

13. \$238.01 + \$449.99 + \$302.55 **\$1,000**

Use estimation to place the decimal point in each answer.

14. 4.003 · 0.64 = 2‸56192

15. 4.32 ÷ 0.02 = 216‸

16. 12.38 + 8.02 + 10.99 = 31‸39

17. -28 ÷ 0.24 = -116‸667

18. 23.05 · 0.07 = 1‸6135

19. 525.1324 ÷ 5.2 = 100‸987

Evaluate each expression for $a = 3.02$ and $b = -12.3$.

20. $a + b$ **-9.28**

21. $a - b$ **15.32**

22. $2a - b$ **18.34**

23. $\frac{a-b}{2}$ **7.66**

24. $\frac{2(a+b)}{4}$ **-4.64**

25. $\frac{-a-b}{5 \div (-1)}$ **-1.856**

26. $5b - 2a$ **-67.54**

27. $|a| + |b|$ **15.32**

Solve each equation using any method.

28. $a + 0.03 = 3.75$ **a = 3.72**

29. $48.9 + y = 50$ **y = 1.1**

30. $q - 4.099 = 2.33$ **q = 6.429**

31. $m - (-1.2) = 3.09$ **m = 1.89**

32. $5m = -95$ **m = -19**

33. $0.5 = \frac{x}{33}$ **x = 16.5**

34. $y \div 0.3 = -15$ **y = -4.5**

35. $98.53 = 0.9853n$ **n = 100**

36. $3.3x = 13.2$ **x = 4**

Write an equation for each problem.

37. Three hundredths times a number is equal to the opposite of three and one hundredth. **0.03x = -3.01**

38. x is equal to four hundred ninety-five and seventy-three hundredths. **x = 495.73**

39. The absolute value of negative nine tenths is equal to two times a number. **|-0.9| = 2x**

40. Five divided by two and five hundredths is equal to q. **5 ÷ 2.05 = q**

Solve.

41. A student is assigned pages 38–130. The student reads a page every 20 s. How long will it take the student to finish the assignment? **31 min**

42. A painter can paint one window in 17 min. There are 7 windows to paint. How long will it take? Round to the nearest hour. **2 h**

43. A pound of peanuts costs \$2.87. How many whole pounds can be bought with \$20? **6**

44. ***DATA FILE 3 (pp. 96–97)*** A collector spent \$7.24 on proof sets in 1936. How much profit would the collector make if he sold them all in 1965? **\$1,592.76**

Extra Practice

Evaluate.

1. 4^5 1,024
2. $(-2)^4$ 16
3. -2^4 -16
4. -12^0 -1
5. $5x^2 \cdot 3x$ $15x^3$
6. $-(2mn)^3$ $-8m^3n^3$
7. $(4y^4)^3$ $64y^{12}$
8. $(x^2)(x^3)(-x^1)$ $-x^6$
9. $[5 + (-7)]^4$ 16
10. $-2(-3 + 4)^5$ -2
11. y^3 for $y = 2$ 8
12. x^4 for $x = -1$ 1
13. $-q^5$ for $q = 1$ -1
14. $(2y)^2$ for $y = -2$ 16
15. $-(a^0bc)$ for $a = 2, b = 3, c = 4$ -12
16. $(r + 2s + t)^2$ for $r = 3, s = -1, t = 5$ 36
17. $x^2 + 3x + y^2$ for $x = 7, y = -8$ 134

Find the value of a.

18. $4^a = 64$ 3
19. $a^5 = 1$ 1
20. $11^a = 1$ 0
21. $6^a = 36$ 2
22. $a^4 = 625$ 5
23. $3^3 = a$ 27

Write each answer in scientific and standard notation.

24. $(4 \times 10^3)(3 \times 10^2)$ 1.2×10^6; 1,200,000
25. $1.32 \times 40{,}000$ 5.28×10^4; 52,800
26. $(1.4 \times 10^5)2.32$ 3.248×10^5; 324,800
27. $7(2.6 \times 10^6)$ 1.82×10^7; 18,200,000

State which of the numbers 2, 3, 4, 5, or 9 are divisors.

28. 25 5
29. 72 2, 3, 4, 9
30. 18 2, 3, 9
31. 135 3, 5, 9
32. 24,270 2, 3, 5

Find the GCF of each set of numbers.

33. 58, 34 2
34. 21, 63 21
35. 18, 36, 38 2
36. $35x^3y^2, 70x^6y^4$ $35x^3y^2$
37. $27a^2b^2, 9ab^2$ $9ab^2$

Find the LCM of each set of numbers.

38. 60, 12 60
39. 55, 100 1,100
40. 16, 20, 36 720
41. $12a^5b, 6ab$ $12a^5b$
42. $x^4y^3, 200x^2y^4$ $200x^4y^4$

Solve.

43. ***DATA FILE 4 (pp. 138–139)*** How many babies are born in 10 s? How many people die in 10 s? Round to the nearest integer. 45; 15

44. A professional basketball player runs 8 mi in an average game. About how many miles does a player run during games in a season of 82 games? 656 mi

45. A popcorn popper pops 2 kernels/s for 10 s. After 10 s it pops 4 kernels/s for 20 s. After 20 s it pops 6 kernels/s for 30 s, and so on. At the end of 2 min and 30 s, how many kernels have popped? 1,100

46. There are 5 children who want to play a game. Only 2 can play the game at a time. In how many different ways can the children be paired? 10

47. Find one pair of numbers that satisfies both conditions.
    a. their product is 80
    b. their LCM is 20 20 and 4

48. Find one pair of numbers that satisfies both conditions.
    a. their sum is 40
    b. their GCF is 5 5 and 35

Chapter 5 Extra Practice

Write in lowest terms.

1. $\frac{7}{14}$ $\frac{1}{2}$
2. $\frac{xyz}{2xz}$ $\frac{y}{2}$
3. $\frac{4mn}{20mn}$ $\frac{1}{5}$
4. $\frac{25x}{75y}$ $\frac{x}{3y}$
5. $\frac{3xy}{9xy}$ $\frac{1}{3}$
6. $\frac{12abc}{144ab}$ $\frac{c}{12}$

Write each decimal as a fraction or mixed number in lowest terms.

7. 0.2 $\frac{1}{5}$
8. 4.1 $4\frac{1}{10}$
9. 20.08 $20\frac{2}{25}$
10. 0.17 $\frac{17}{100}$
11. 0.005 $\frac{1}{200}$
12. 1.125 $1\frac{1}{8}$

Write as a decimal.

13. $\frac{4}{10}$ 0.4
14. $\frac{11}{20}$ 0.55
15. $\frac{3}{5}$ 0.6
16. $\frac{5}{8}$ 0.625
17. $\frac{3}{4}$ 0.75
18. $\frac{1}{3}$ $0.\overline{3}$

Order from least to greatest.

19. $\frac{8}{4}, \frac{1}{2}, -\frac{3}{4}$ $-\frac{3}{4}, \frac{1}{2}, \frac{8}{4}$
20. $-\frac{15}{20}, -\frac{3}{10}, -\frac{4}{5}$ $-\frac{4}{5}, -\frac{15}{20}, -\frac{3}{10}$
21. $\frac{2^2}{5}, 2\frac{3}{10}, \frac{9}{10}$ $\frac{2^2}{5}, \frac{9}{10}, 2\frac{3}{10}$
22. $\frac{a}{3}, \frac{a}{6}, \frac{3a}{2}$ for $a > 0$ $\frac{a}{6}, \frac{a}{3}, \frac{3a}{2}$
23. $\frac{xy}{15}, \frac{-xy}{5}, \frac{3xy}{10}$ for $xy > 0$ $\frac{-xy}{5}, \frac{xy}{15}, \frac{3xy}{10}$

Find each answer. Write in lowest terms.

24. $\frac{5}{6} + \frac{3}{8}$ $1\frac{5}{24}$
25. $\frac{2}{3} - 1\frac{1}{2}$ $-\frac{5}{6}$
26. $\frac{x}{4} - \frac{x}{10}$ $\frac{3x}{20}$
27. $\frac{8}{9}x + \left(\frac{-3}{6}x\right)$ $\frac{7}{18}x$
28. $-5 + \left(-\frac{5}{10}\right)$ $-5\frac{1}{2}$
29. $8 - \frac{2}{4}$ $7\frac{1}{2}$
30. $\frac{1}{4} \cdot \frac{5}{9}$ $\frac{5}{36}$
31. $\frac{8}{11} \div \frac{7}{9}$ $\frac{72}{77}$
32. $2\frac{3}{4} \cdot \frac{3}{7}$ $1\frac{5}{28}$
33. $\frac{3x}{5} \cdot \frac{7x}{9}$ $\frac{7x^2}{15}$
34. $5\frac{4}{7} \div \left(\frac{-3}{14}\right)$ -26
35. $\frac{4}{5} + 1\frac{2}{15}$ $1\frac{14}{15}$
36. $\frac{7}{8}q - \frac{5}{5}q$ $-\frac{1}{8}q$
37. $\left(\frac{3}{8}\right)\left(-2\frac{3}{4}\right) - \frac{1}{2}$ $-1\frac{17}{32}$
38. $\frac{9}{11} \div \frac{3}{4} \div \frac{2}{7}$ $3\frac{9}{11}$
39. $\frac{-1}{3} \div (-3)$ $\frac{1}{9}$

Simplify.

40. $\frac{15y^7}{33y^3}$ $\frac{5y^4}{11}$
41. $\frac{a^3b^2c^0}{abc}$ $\frac{a^2b}{c}$
42. $\frac{9m^7n^2}{3m^3n}$ $3m^4n$
43. $\frac{-y^3z^5}{y^4z^5}$ $\frac{-1}{y}$
44. $\frac{16q^2m^3}{12q^4m}$ $\frac{4m^2}{3q^2}$

Solve each equation. Write in lowest terms.

45. $h + 3\frac{1}{3} = 4\frac{7}{9}$ $1\frac{4}{9}$
46. $x - |-3.1| = 8\frac{1}{4}$ $11\frac{7}{20}$
47. $c + \left(\frac{-3}{8}\right) = -\frac{7}{10}$ $\frac{-13}{40}$
48. $d + \frac{3}{7} = -2\frac{3}{14}$ $-2\frac{9}{14}$
49. $\frac{3}{10}c = \frac{3}{4}$ $2\frac{1}{2}$
50. $-\frac{7}{6}z = -4$ $3\frac{3}{7}$
51. $\frac{-9}{10}q = 4$ $-4\frac{4}{9}$
52. $\frac{-5}{9}y = -\frac{3}{6}$ $\frac{9}{10}$

Solve.

53. ***DATA FILE 4 (pp. 138–139)*** How many millions of people speak Chinese, Spanish, German, or Arabic? 1,205
54. You have $20.50 in quarters, dimes, nickels, and pennies. You have an equal number of each coin. How many of each coin do you have? 50
55. Four children are told to line up and hold hands as they cross the street. How many different ways can they line up? 24
56. A ball is pushed down a flight of 50 stairs. The ball rolls down four stairs per second. How long will it take the ball to roll down $\frac{4}{5}$ of the stairs? 10 s

Chapter 6 Extra Practice

Write each ratio as a fraction in lowest terms.

1. 2:6 $\frac{1}{3}$ **2.** 9:3 $\frac{3}{1}$ **3.** 7:28 $\frac{1}{4}$ **4.** 25:45 $\frac{5}{9}$

5. 3 people out of 27 are wearing hats. $\frac{1}{9}$

6. 18 dogs out of 81 have fleas. $\frac{2}{9}$

7. 6 out of the 30 TVs are on sale. $\frac{1}{5}$

8. 7 out of the 31 athletes made the team. $\frac{7}{31}$

Write a proportion to describe each situation. Then solve.

9. An athlete swims 20 laps in 30 min; x laps in 2 hours. $\frac{20}{30} = \frac{x}{120}$; $x = 80$

10. 38 lb of soil cost \$3.20; 95 lb cost y dollars. $\frac{38}{3.20} = \frac{95}{y}$; $y = 8$

11. 20 stamps cost 95¢; z stamps cost 19¢. $\frac{20}{95} = \frac{z}{19}$; $z = 4$

12. A car can travel 282 mi on 14 gal of gas; m miles on 42 gal of gas. $\frac{282}{14} = \frac{m}{42}$; $m = 846$

Write a ratio and percent for each.

13. 17 questions right out of 20 $\frac{17}{20}$; 85%

14. 12 questions right out of 18 $\frac{12}{18}$; $66\frac{2}{3}$%

15. 40 questions wrong out of 120 $\frac{40}{120}$; $33\frac{1}{3}$%

16. 2 questions wrong out of 20 $\frac{2}{20}$; 10%

Solve.

17. What percent of 80 is 36? 45%

18. Find 39% of 66. 25.74

19. 33% of q is 109. What is q? $330.\overline{30}$

20. What percent of 33 is 99? 300%

21. Find 23% of 28. 6.44

22. 75% of k is 15. What is k? 20

Find each percent of change. Round to the nearest tenth.

23. from 18 to 12 -33.3%

24. from 88 to 125 +42%

25. from 15.5 to 25.5 +64.5%

26. from 100 to 88 -12%

27. from 2.4 to 8.6 +258.3%

28. from 11 to 17 +54.5%

29. from 92 to 98 +6.5%

30. from 34 to 49 +44.1%

Solve.

31. An athlete must swim 18 laps. The athlete swims $\frac{1}{2}$ of the laps and then swims $\frac{1}{3}$ of those remaining. How many laps are left? 6

32. A florist cuts a wire into three pieces. The first piece is 20% of the second. The third piece is 110% of the second. The wire is 23 cm long. Find the length of each piece. 2 cm, 10 cm, 11 cm

33. There are 10 girls and 15 boys. What percent of the group is boys? 60%

34. ***DATA FILE 6 (pp. 230–231)*** What percent of teens study an average of more than 2 h a day? 19%

35. In a town with 12,000 residents, 2 out of 10 voted in the last election. How many residents voted? 2,400

36. In a sample of 2,500 ballpoint pens, 15 were found to be defective. How many pens would you expect to be defective in a shipment of 10,000 pens? 60

37. ***DATA*** Use the data on page 130. Find the percent of change in the price of a loaf of bread and a half gallon of milk from 1890 to 1980. bread: +4,533% milk: +679%

38. A salesperson earns \$800 per week plus 5% commission on sales over \$10,000. How much would the salesperson earn in a week when sales were \$25,000? \$1,550

Chapter 7 Extra Practice

Solve each equation. Check.

1. $3x + (-12) = x$ $x = 6$
2. $2m + -3m - 8 = 26$ $m = -34$
3. $4(-y + 2) = -1$ $y = 2\frac{1}{4}$
4. $-0.2 - q = 9q$ $q = -0.02$
5. $\frac{3}{4}x - \frac{1}{2}x = 5$ $x = 20$
6. $32 = 4(a - 2) + 10$ $a = 7\frac{1}{2}$
7. $4(b - 2.1) = b + 0.6$ $b = 3$
8. $\frac{1}{4}(x - 8) = \frac{3}{4}x$ $x = -4$
9. $48 = \frac{1}{2}(8x - 14) + 15$ $x = 10$

Graph each inequality on a number line. Write each inequality as a word sentence. Check students' work.

10. $y < 9$
11. $-3 > q$
12. $0.009 \geq p$
13. $a \geq -2$
14. $c > 99$
15. $c < -99$
16. $n < 7.3$
17. $h \geq -1.2$
18. $-11 \leq b$

Solve each inequality.

19. $y + 3 \geq 9$ $y \geq 6$
20. $y - 3 > -7$ $y > -4$
21. $\frac{y}{4} \leq -9$ $y \leq -36$
22. $3q > 0$ $q > 0$
23. $-9x \leq -5$ $x \geq \frac{5}{9}$
24. $\frac{x}{7} \geq -3$ $x \geq -21$
25. $19 - 3x > -2$ $x < 7$
26. $-314 \leq x + 1$ $x \geq -315$
27. $\frac{1}{2}(x - 6) \leq 22$ $x \leq 50$
28. $-5(a - 3) \leq 45$ $a \geq -6$
29. $-\frac{x}{3} + 3 \geq -27$ $x \leq 90$
30. $\frac{2}{3}(4a + 12) \geq \frac{1}{3}(6a - 10)$ $a \geq -17$

Write an inequality to describe the situation. Solve.

31. When a is divided by 8, the result is at most 13. Find a. $\frac{a}{8} \leq 13$; $a \leq 104$
32. Four less than n is greater than negative six. Find n. $n - 4 > -6$; $n > -2$
33. Five less than seven times p is at least twenty-three. Find p. $7p - 5 \geq 23$; $p \geq 4$
34. Three times q plus negative twenty-two is less than q. Find q. $3q + (-22) < q$; $q < 11$

Solve.

35. Each month Gil saves \$22.27. How long does it take him to save \$267.24? 12 mo
36. Four times a number is fifty-two minus forty. Find the number. 3
37. A student buys three movie tickets for \$6.60 each. The student pays with \$20. What is the change? \$.20
38. You open a book. The product of the two page numbers is 9,702. What are the page numbers? 98 and 99
39. Nita has test scores of 92, 84, and 87. She needs a 90 average to get an A. What is the lowest score she can get on her next test and still have an A average? 97
40. ***DATA FILE 7 (pp. 272–273)*** Estimate which is greater using the figures from 1987: the total of passengers boarding at Chicago/O'Hare, Atlanta, and Miami or the total boarding in Denver, Los Angeles, and San Francisco. passengers boarding at Chicago/O'Hare, Atlanta, and Miami

Chapter 8 Extra Practice

In which quadrant or on which axis does each point fall?

1. (-3,18) II **2.** (0,44) y-axis **3.** (22,3) I **4.** (-18,-5) III **5.** (-0.33,-5) III

Solve for y in terms of x. Find four solutions of each equation. Answers will vary.

6. $5y - 10x = 15$ $y = 2x + 3$

7. $4 - y = \frac{1}{2}x$ $y = -\frac{1}{2}x + 4$

8. $2x + 2y = -4$ $y = -x - 2$

9. $\frac{1}{2}y - x = 12$ $y = 2x + 24$

10. $-4x - 0.5y = 4$ $y = -8x - 8$

11. $\frac{1}{3}y - x = 1$ $y = 3x + 3$

Graph each equation. Name the slope, x-intercept, and y-intercept. Check students' work.

12. $y = -8$ 0; none; -8

13. $y = 2x - 5$ 2; $2\frac{1}{2}$; -5

14. $x = y$ 1; 0; 0

15. $x + y = 5$ -1; 5; 5

16. $-y + 3x = \frac{1}{2}$ 3; $\frac{1}{6}$; $-\frac{1}{2}$

17. $\frac{1}{4}x + y = -3$ $-\frac{1}{4}$; -12; -3

Solve each system by graphing. Check your solutions.

18. $y = x + 4$
$y = x - 2$ no solution

19. $2x + y = 3$
$-2y = 14 - x$ (4,-5)

20. $y = -3$
$2x + 3y = 6$
$(7\frac{1}{2},-3)$

Solve each inequality for y in terms of x. Write three ordered pairs that are solutions of the inequality. Answers will vary.

21. $3x - y < 5$ $y > 3x - 5$

22. $-x + 4y + 12 \geq 0$ $y \geq \frac{1}{4}x - 3$

23. $x - 5 + 3y < 44 + 2y + 3x$ $y < 2x + 49$

24. $-x > 2y - 5(x + y)$ $y > -\frac{4}{3}x$

Graph each inequality. Check students' work.

25. $y > 7$

26. $x + 3y < 2$

27. $4x + 4y > 10$

28. $x - \frac{1}{3}y \leq \frac{1}{3}$

29. $\frac{3}{4}x - \frac{1}{4}y \leq \frac{1}{4}$

30. $3x + 6y > 15$

Solve.

31. At a party 20 guests consume 5 bags of popcorn and 10 gal of juice. At this rate, how much popcorn and juice will 28 guests consume? 7 bags of popcorn and 14 gal of juice

32. A line has a slope of $\frac{2}{5}$ and passes through the point (-10,7). Find the equation of the line. State the quadrants which the line will pass through. $y = \frac{2}{5}x + 11$; quadrants I, II, and III

33. Find two consecutive numbers such that the greater number times 3 is 9 less than 5 times the lesser number. 6 and 7

34. ***DATA FILE 5 (pp. 180–181)*** Write a ratio in lowest terms to compare the number of violas to the number of second violins. $\frac{3}{4}$

35. A train that is 500 m long is traveling at a speed of 125 km/h. How long will it take the train to entirely pass through a tunnel that is 2 km long? 1 min 12 s

36. Shoes are $10 off. A customer buys three pairs of shoes for $110. What percent did the customer save on the total purchase? Round to the nearest whole number. 21%

Tell whether each angle is acute, right, obtuse, or straight.

1. 2° acute **2.** 75° acute **3.** 99° obtuse **4.** 90° right **5.** 180° straight **6.** 77° acute

Find the measure of a complement and a supplement of each angle, if possible.

7. 90° none; 90 **8.** 1° 89; 179 **9.** 45° 45; 135 **10.** 33° 57; 147 **11.** 140° not possible; 40° **12.** 101° not possible; 79

Draw a figure to fit each description. Check students' work.

13. a concave octagon **14.** an obtuse triangle **15.** a quadrilateral **16.** a convex rhombus

Find the measure of the third angle of a triangle that has two angles with the given measures.

17. 30°, 60° 90 **18.** $x°$, $x°$ 180 – 2x **19.** 40°, 60° 80 **20.** 60°, 60° 60 **21.** 39°, 57° 84 **22.** $(x + 2)°$, $(x - 4)°$ 180 – (2x – 2)

Find each radius or diameter.

23. $r = 29.3$ cm $d = 58.6$ cm **24.** $d = 40$ yd $r = 20$ yd **25.** $d = 78$ in. $r = 39$ in. **26.** $r = 10$ km $d = 20$ km **27.** $r = 228.5$ m $d = 457$ m

***ABCDE* ≅ *MNOPQ*. Tell whether each statement is true or false.**

28. $\angle A \cong \angle P$ false **29.** $\overline{AB} \cong \overline{MN}$ true **30.** $\angle D \cong \angle P$ true **31.** $\overline{CE} \cong \overline{OP}$ false **32.** $\angle B \cong \angle N$ true **33.** $\angle A \cong \angle M$ true

Find the circumference. Use 3.14 for π.

34.

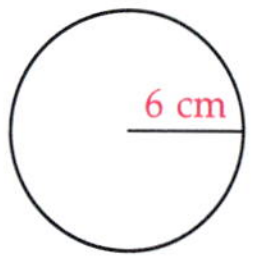

37.68 cm

35.

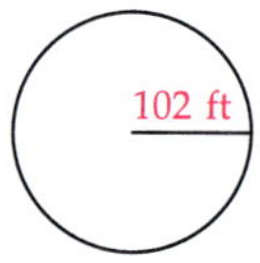

640.56 ft

36.

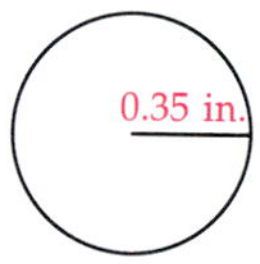

2.198 in.

Find the perimeter of each figure.

37. a regular hexagon with side 2.8 in. 16.8 in.

38. a parallelogram with sides 4 ft and 7 ft 22 ft

39. an equilateral triangle with side 9 m 27 m

40. a figure with sides 4 cm, 8 cm, 12 cm, and 12 cm 36 cm

Solve.

41. Sol and Julia run a race. Sol runs 7.3 mi/h. Julia runs 6.8 mi/h. After 2 h, how far apart will they be? 1 mi

42. A coat is on sale for 75% off the regular price. If the sale price is $60, what was the original price? $240

43. Carlos wants to trim a circular carpet in his house. The carpet has a diameter of 20 yd. How much trim does he need? Use 3.14 for π. 62.8 yd

44. ***DATA FILE 9 (pp. 360–361)*** At St. Andrew's golf club a golfer takes a break after the 13th hole. If the golfer has a score of 57, what must the golfer average on the remaining holes to receive a score of par? 3

Chapter 10 Extra Practice

Find the area of each parallelogram with the given base and height. Round to the nearest tenth.

1. $b = 6.3$ cm, $h = 2.9$ cm 18.3 cm²

2. $b = 13$ ft, $h = 19$ ft 247 ft²

3. $b = 0.8$ m, $h = 1.1$ m 0.9 m²

4. $b = 55$ in., $h = 22$ in. 1,210 in.²

5. $b = 17x$, $h = 33x$ $561x^2$

6. $b = x$, $h = 2x$ $2x^2$

Find the area of each circle with the given diameter or radius. Round to the nearest tenth. Use 3.14 for π.

7. $r = 8.1$ cm 206 cm²

8. $d = 0.39$ in. 0.1 in.²

9. $r = 13.2$ m 547.1 m²

10. $r = 0.75$ in. 1.8 in.²

11. $r = 5.6$ m 98.5 m²

12. $d = 16z$ $201z^2$

13. $r = 14.5$ cm 660.2 cm²

14. $d = 5.5x$ $23.7x^2$

Find each surface area. Use 3.14 for π.

15. a rectangular prism with base edges 7 in. and 9 in., and height 5 in. 286 in.²

16. a cylinder with radius 7 cm and height 12 cm 835.24 cm²

17. a square pyramid with edge 59 m and slant height 66 m 11,269 m²

18. a cone with radius 9 cm and slant height 15 cm 678.24 cm²

19. a sphere with radius 12 in. 1,808.64 in.²

20. a hemisphere with radius 5 cm 157 cm²

Find each volume. Use 3.14 for π.

21. a triangular prism with base 4 ft, height 7 ft, and a prism height of 10 ft 140 ft³

22. a rectangular prism with length 3 in., width 8 in., and height 9 in. 216 in.³

23. a cylinder with $r = 11$ cm and $h = 6$ cm 2,279.64 cm³

24. a cone with radius 9 m and height 13 m 1,102.14 m³

25. a square pyramid with side 5 cm and height 12 cm 100 cm³

26. a sphere with radius 10 in. 4,186.67 in.³

Solve.

27. A dress is on sale for 20% off the original price of \$89. Another dress not on sale costs \$61. Which dress costs less? the \$61 dress

28. Ten players line up to shoot baskets. Each player makes at least 60% of her shots. Altogether, there were 300 attempts. What's the smallest number of baskets made? 180

29. A juice can has radius 1.1 in. and height 4.2 in. What is the volume of the can? Round to the nearest unit. Use 3.14 for π. 16 in.³

30. ***DATA FILE 10 (pp. 404–405)*** What is the area in square meters of the black bear's home range? Use 3.14 for π. 17,637,066 m²

Chapter 11 Extra Practice

Find each square root. If necessary, use your calculator. Round decimal answers to the nearest thousandth.

1. $\sqrt{51}$ 7.141 **2.** $\sqrt{36}$ 6 **3.** $\sqrt{144}$ 12 **4.** $\sqrt{\frac{16}{36}}$ $\frac{2}{3}$

5. $\sqrt{49x^2}$ 7x **6.** $\sqrt{101.101}$ 10.055 **7.** $\sqrt{(a+b)^2}$ a + b **8.** $\sqrt{a^4b^{12}c^8}$ $a^2b^6c^4$

Determine whether each is a Pythagorean triple.

9. 3, 4, 5 yes **10.** 9, 7, 12 no **11.** $2\sqrt{12}$, 5, $\sqrt{37}$ no

12. 6, 7, 8 no **13.** 5, 12, 13 yes **14.** $3x^2$, $4x^2$, $5x^2$ yes

Find the missing length.

15.

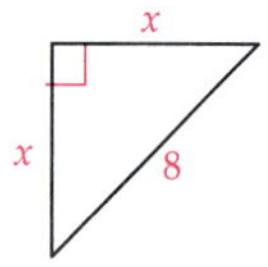

$x = 4\sqrt{2}$

16.

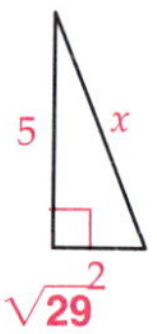

$x = \sqrt{29}$

17.

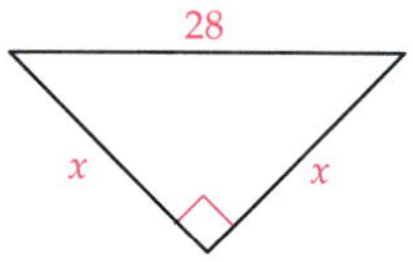

$x = 14\sqrt{2}$

18.

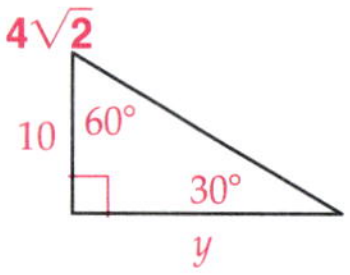

$y = 10\sqrt{3}$

19.

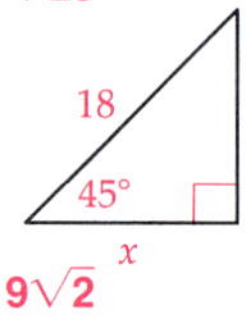

$x = 9\sqrt{2}$

20.

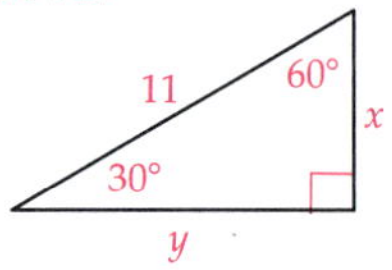

$x = 5.5$
$y = 5.5\sqrt{3}$

State the trigonometric ratio using the given values.

21. tan $k°$

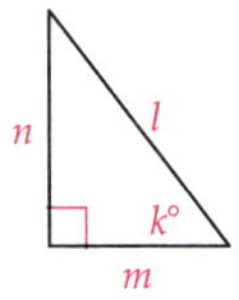

$\frac{n}{m}$

22. sin $x°$

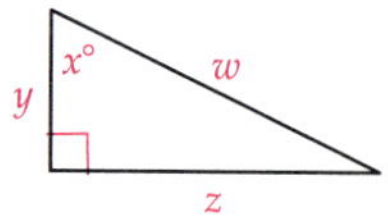

$\frac{z}{w}$

23. cos $a°$

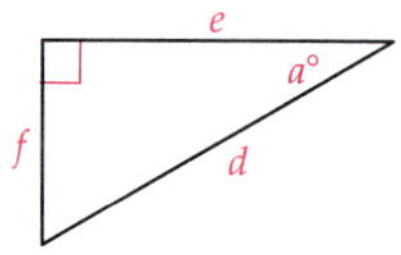

$\frac{e}{d}$

Solve.

24. The diameter of a circular ring is 18 yd. What is the circumference of the ring? What is the area? Use 3.14 for π. 56.52 yd; 254.34 yd^2

25. At takeoff, an airplane forms a 44° angle with the runway. How many miles will the airplane fly before reaching an altitude of 28,000 ft? Round to the nearest mile. 8 mi

26. It takes Jonathan 1 min 20 s to walk around the track. Twelve laps equal 1 mile. How long will it take Jonathan to walk 1 mile? 3 A.M.

27. ***DATA FILE 7 (pp. 272–273)*** It is 12 A.M. in Fairbanks, Alaska. What time is it in Chicago, Illinois? 3 A.M.

Chapter 12 Extra Practice

Find the mean, median, mode, and range of each of the following. Round to the nearest tenth.

1. 13, 12, 15, 13, 9, 6, 5 10.4; 12; 13; 10

2. 28, 1, 5, 6, 9, 1, 3, 10, 1 7.1; 5; 1; 27

3. 100, 100, 100, 100, 100, 0, 50 78.6; 100; 100; 100

4. 9; 9; 8; 3

x	*f*
8	5
9	2
10	3
11	1

5. Make a line plot for the table in Exercise 4.

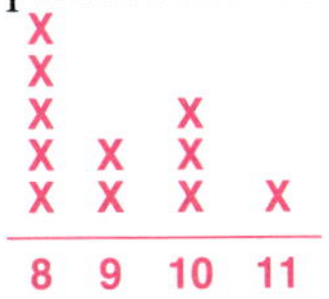

Make a stem and leaf plot for each set of data. Then find the median and the mode for each set of data. Check students' work for plots.

6. 30, 15, 19, 35, 20, 20 20; 20

7. 10, 16, 22, 13, 25, 13, 13, 10 13; 13

8. 11, 55, 30, 32, 55, 12, 13, 55 31; 55

9. 33, 99, 82, 66, 72, 66 69; 66

Make a box and whisker plot for each set of data. Check students' work.

10. 40, 43, 48, 48, 50, 66, 60, 61, 70, 69, 45, 46

11. 40, 11, 30, 12, 28, 17, 29, 19

Find the value of each factorial.

12. 8! 40,320

13. $\frac{5!}{4!}$ 5

14. 5! 120

15. 6! − 3! 714

Suppose you have a deck of 52 cards. Find the probability of selecting each of the following. Cards will always be replaced. An ace is not considered to be a face card.

16. a red card $\frac{1}{2}$

17. a heart $\frac{1}{4}$

18. a king $\frac{1}{13}$

19. a face card $\frac{3}{13}$

20. a face card first and then an ace $\frac{3}{169}$

21. the jack of hearts first and a king second $\frac{1}{676}$

Solve.

22. ***DATA FILE 5 (pp. 180–181)*** If you choose one musician at random from a symphony orchestra, what is the probability that he or she will play the cello? $\frac{3}{26}$

23. A ball was thrown up a hill 22 ft. It then rolled back 28 ft. Represent this as an integer. −6

24. Find the probability of rolling one number cube and getting a multiple of 2 first and a 3 second. $\frac{1}{12}$

25. Find the volume of a sphere with radius 7.5 cm. 1,766.25 cm^3

26. A right triangle has side lengths 5, $5\sqrt{3}$, and 10. What are the angle measures? 30, 60, 90

27. The hypotenuse of a 45°-45°-90° triangle has length 12. What is the length of each leg? $6\sqrt{2}$

28. The sum of three consecutive even integers is 30. Their mean is 10. Find the integers. 8, 10, and 12

29. The sum of the squares of two consecutive positive integers is 85. Find the integers. 6 and 7

577

Chapter 13 Extra Practice

Simplify.

1. $(x^2 + 3x + 4) + (2x^2 + x + 1)$ $3x^2 + 4x + 5$
2. $(3x^2 + x + 5) + (x^2 + 2x + 1)$ $4x^2 + 3x + 6$
3. $(mn^2 + 3n - 8) - (3mn^2 - m + n - 5)$ $-2mn^2 + m + 2n - 3$
4. $18p^2q^2 + 19 - 18p$ $18p^2q^2 - 18p + 19$
5. $(a^3n^2 - 3a^2n + 5a - 9) - (4a^3n^2 + a^2n - a - 9)$ $-3a^3n^2 - 4a^2n + 6a$
6. $(-3w^4y^5 - 18w^3y^4 + 2w^2y^3 - 3wy^2) + (3w^4y^5 + 18w^3y^4 - 2w^2y^3 + 3wy^2)$ 0

Simplify.

7. $18x(4x^2 - 2x + 9)$ $72x^3 - 36x^2 + 162x$
8. $-m^4(m^2 - 14m + 5)$ $-m^6 + 14m^5 - 5m^4$
9. $\frac{1}{2}a(5a^5 - a^3 - 33)$ $\frac{5}{2}a^6 - \frac{1}{2}a^4 - \frac{33}{2}a$
10. $4abc(a^4b + 10abc - 8)$
11. $(x + 1)(x + 3)$ $x^2 + 4x + 3$
12. $(x - 2)(x + 5)$ $x^2 + 3x - 10$
13. $(x - 9)(x + 9)$ $x^2 - 81$
14. $(x - 5)^2$ $x^2 - 10x + 25$
15. $(x + 2)(2x^2 + x + 1)$ $2x^3 + 5x^2 + 3x + 2$
16. $-5(x + 9)(x - 1)$ $-5x^2 - 40x + 45$

10. $4a^5b^2c + 40a^2b^2c^2 - 32abc$

Write an expression using one variable to represent each product. Then simplify.

17. If m is an even integer, find the product of the next three consecutive even integers. $(m + 2)(m + 4)(m + 6)$; $m^3 + 12m^2 + 44m + 48$
18. If w is a multiple of 5, find the product of the next two consecutive multiples of 5. $(w + 5)(w + 10)$; $w^2 + 15w + 50$
19. If z is a multiple of 4, find the product of the previous two consecutive multiples of 4. $(z - 4)(z - 8)$; $z^2 - 12z + 32$
20. If m is an integer, find the product of the next three consecutive integers. $(m + 1)(m + 2)(m + 3)$; $m^3 + 6m^2 + 11m + 6$

Solve.

21. A ball has a diameter of 12 cm. Find the volume. Use 3.14 for π. 904.32 cm^3
22. A car traveled 192 mi in 3 h. What was the car's average speed? 64 mi/h
23. Ron has a choice of 3 sweaters, 2 pairs of pants, and 4 pairs of shoes. How many different outfits can Ron choose? 24
24. ***DATA FILE 9 (pp. 360–361)*** What is the average distance that a ball will go if hit with a club having a loft of 31°? 150 yd
25. Three consecutive integers have a sum of 291. What is the sum of the largest and smallest of these integers? 194
26. Janet is four years older than Frank. The sum of their ages is 76. How old was Janet eight years ago? 32
27. A sock drawer contains 4 red socks, 3 green socks, 8 blue socks, and 10 black socks. What is the probability that a blue sock is chosen first and another blue sock second? $\frac{7}{75}$
28. An average honeybee hive produces 350 oz of honey. How many honeybee hives would be needed to produce 2,275 oz of honey? 6.5

578

Table 1: Squares and Square Roots

N	N^2	$\sqrt{N}$	N	N^2	$\sqrt{N}$
1	1	1	51	2,601	7.141
2	4	1.414	52	2,704	7.211
3	9	1.732	53	2,809	7.280
4	16	2	54	2,916	7.348
5	25	2.236	55	3,025	7.416
6	36	2.449	56	3,136	7.483
7	49	2.646	57	3,249	7.550
8	64	2.828	58	3,364	7.616
9	81	3	59	3,481	7.681
10	100	3.162	60	3,600	7.746
11	121	3.317	61	3,721	7.810
12	144	3.464	62	3,844	7.874
13	169	3.606	63	3,969	7.937
14	196	3.742	64	4,096	8
15	225	3.873	65	4,225	8.062
16	256	4	66	4,356	8.124
17	289	4.123	67	4,489	8.185
18	324	4.243	68	4,624	8.246
19	361	4.359	69	4,761	8.307
20	400	4.472	70	4,900	8.367
21	441	4.583	71	5,041	8.426
22	484	4.690	72	5,184	8.485
23	529	4.796	73	5,329	8.544
24	576	4.899	74	5,476	8.602
25	625	5	75	5,625	8.660
26	676	5.099	76	5,776	8.718
27	729	5.196	77	5,929	8.775
28	784	5.292	78	6,084	8.832
29	841	5.385	79	6,241	8.888
30	900	5.477	80	6,400	8.944
31	961	5.568	81	6,561	9
32	1,024	5.657	82	6,724	9.055
33	1,089	5.745	83	6,889	9.110
34	1,156	5.831	84	7,056	9.165
35	1,225	5.916	85	7,225	9.220
36	1,296	6	86	7,396	9.274
37	1,369	6.083	87	7,569	9.327
38	1,444	6.164	88	7,744	9.381
39	1,521	6.245	89	7,921	9.434
40	1,600	6.325	90	8,100	9.487
41	1,681	6.403	91	8,281	9.539
42	1,764	6.481	92	8,464	9.592
43	1,849	6.557	93	8,649	9.644
44	1,936	6.633	94	8,836	9.695
45	2,025	6.708	95	9,025	9.747
46	2,116	6.782	96	9,216	9.798
47	2,209	6.856	97	9,409	9.849
48	2,304	6.928	98	9,604	9.899
49	2,401	7	99	9,801	9.950
50	2,500	7.071	100	10,000	10

579

Table 2: Table of Trigonometric Ratios

Angle	Sine	Cosine	Tangent	Angle	Sine	Cosine	Tangent
1°	0.0175	0.9998	0.0175	46°	0.7193	0.6947	1.0355
2°	0.0349	0.9994	0.0349	47°	0.7314	0.6820	1.0724
3°	0.0523	0.9986	0.0524	48°	0.7431	0.6691	1.1106
4°	0.0698	0.9976	0.0699	49°	0.7547	0.6561	1.1504
5°	0.0872	0.9962	0.0875	50°	0.7660	0.6428	1.1918
6°	0.1045	0.9945	0.1051	51°	0.7771	0.6293	1.2349
7°	0.1219	0.9925	0.1228	52°	0.7880	0.6157	1.2799
8°	0.1392	0.9903	0.1405	53°	0.7986	0.6018	1.3270
9°	0.1564	0.9877	0.1584	54°	0.8090	0.5878	1.3764
10°	0.1736	0.9848	0.1763	55°	0.8192	0.5736	1.4281
11°	0.1908	0.9816	0.1944	56°	0.8290	0.5592	1.4826
12°	0.2079	0.9781	0.2126	57°	0.8387	0.5446	1.5399
13°	0.2250	0.9744	0.2309	58°	0.8480	0.5299	1.6003
14°	0.2419	0.9703	0.2493	59°	0.8572	0.5150	1.6643
15°	0.2588	0.9659	0.2679	60°	0.8660	0.5000	1.7321
16°	0.2756	0.9613	0.2867	61°	0.8746	0.4848	1.8040
17°	0.2924	0.9563	0.3057	62°	0.8829	0.4695	1.8807
18°	0.3090	0.9511	0.3249	63°	0.8910	0.4540	1.9626
19°	0.3256	0.9455	0.3443	64°	0.8988	0.4384	2.0503
20°	0.3420	0.9397	0.3640	65°	0.9063	0.4226	2.1445
21°	0.3584	0.9336	0.3839	66°	0.9135	0.4067	2.2460
22°	0.3746	0.9272	0.4040	67°	0.9205	0.3907	2.3559
23°	0.3907	0.9205	0.4245	68°	0.9272	0.3746	2.4751
24°	0.4067	0.9135	0.4452	69°	0.9336	0.3584	2.6051
25°	0.4226	0.9063	0.4663	70°	0.9397	0.3420	2.7475
26°	0.4384	0.8988	0.4877	71°	0.9455	0.3256	2.9042
27°	0.4540	0.8910	0.5095	72°	0.9511	0.3090	3.0777
28°	0.4695	0.8829	0.5317	73°	0.9563	0.2924	3.2709
29°	0.4848	0.8746	0.5543	74°	0.9613	0.2756	3.4874
30°	0.5000	0.8660	0.5774	75°	0.9659	0.2588	3.7321
31°	0.5150	0.8572	0.6009	76°	0.9703	0.2419	4.0108
32°	0.5299	0.8480	0.6249	77°	0.9744	0.2250	4.3315
33°	0.5446	0.8387	0.6494	78°	0.9781	0.2079	4.7046
34°	0.5592	0.8290	0.6745	79°	0.9816	0.1908	5.1446
35°	0.5736	0.8192	0.7002	80°	0.9848	0.1736	5.6713
36°	0.5878	0.8090	0.7265	81°	0.9877	0.1564	6.3138
37°	0.6018	0.7986	0.7536	82°	0.9903	0.1392	7.1154
38°	0.6157	0.7880	0.7813	83°	0.9925	0.1219	8.1443
39°	0.6293	0.7771	0.8098	84°	0.9945	0.1045	9.5144
40°	0.6428	0.7660	0.8391	85°	0.9962	0.0872	11.4301
41°	0.6561	0.7547	0.8693	86°	0.9976	0.0698	14.3007
42°	0.6691	0.7431	0.9004	87°	0.9986	0.0523	19.0811
43°	0.6820	0.7314	0.9325	88°	0.9994	0.0349	28.6363
44°	0.6947	0.7193	0.9657	89°	0.9998	0.0175	57.2900
45°	0.7071	0.7071	1.0000				

Glossary

A

absolute value (p. 5) The absolute value of an integer is its distance from zero on a number line.

acute angle (p. 365) An acute angle has measure less than 90°.

acute triangle (p. 374) An acute triangle is one in which all angles have measure less than 90°.

adding two integers with different signs (p. 10) To add two integers with different signs, find the *difference* of the absolute values of the addends. The sum has the sign of the integer with the greater absolute value.

adding two integers with the same sign (p. 10) To add two integers with the same sign, add the absolute values of the integers. The sum has the same sign as the addends.

addition properties for inequalities (p. 296)

1. If $a > b$, then $a + c > b + c$.
2. If $a < b$, then $a + c < b + c$.

addition property of equality (p. 73) You can add the same value to both sides of an equation.

If $a = b$, then $a + c = b + c$.

additive identity (p. 59) The additive identity is zero. $a + 0 = a$

adjacent angles (p. 366) Two angles that have the same vertex and have a common side but no interior points in common form adjacent angles.

altitude (p. 407) An altitude is a segment from one vertex of a polygon perpendicular to the line containing the opposite side, called the base.

angle (p. 365) Two rays with a common endpoint form an angle.

area (p. 406) Area is the amount of surface inside a region. We measure area in square units.

area of a circle (p. 414) The area of a circle equals the product of π and the square of the radius (r). $A = \pi r^2$

area of a parallelogram (p. 407) The area of a parallelogram equals the product of its base length (b) and its height (h). $A = bh$

area of a rectangle (p. 406) The area of a rectangle equals the product of its base length (b) and its height (h). $A = bh$

area of a trapezoid (p. 411) The area of a trapezoid equals half the product of the height (h) and the sum of the bases (b_1 and b_2). $A = \frac{1}{2}h(b_1 + b_2)$

area of a triangle (p. 410) The area of a triangle equals half the product of the base length (b) and the height (h). $A = \frac{1}{2} bh$

associative property of addition (p. 58) You can change the grouping and then add without changing the sum. $(a + b) + c = a + (b + c)$

associative property of multiplication (p. 59) You can change the grouping and then multiply without changing the product. $(ab)c = a(bc)$

B

base (p. 141) A base is a number used as a factor. For a^n, a is the base.

box and whisker plot (p. 502) A box and whisker plot is an organization of data. It is especially useful to show the distribution of each 25% of data.

C

central angle (p. 377) A central angle is an angle with the vertex at the center of a circle.

chord (p. 377) A chord is a segment with endpoints on a circle.

circle (p. 377) A circle (⊙) is the set of all points the same distance from a given point called the center.

circumference (p. 390) The circumference (C) is the distance around a circle. Use the formula $C = \pi d$ to compute circumference.

common factor (p. 167) The factors that are the same for a given set of numbers are the common factors. A common factor of 12 and 18 is 6.

common multiples (p. 168) The multiples that are the same for a given set of whole numbers are the common multiples. A common multiple of 6 and 8 is 24.

commutative property of addition (p. 58) You can add in any order without changing the sum. $a + b = b + a$

commutative property of multiplication (p. 58) You can multiply in any order without changing the product. $a \cdot b = b \cdot a$

compatible numbers (p. 105) Compatible numbers are two numbers that are easy to compute mentally.

complementary angles (p. 366) Two angles are complementary angles if the sum of their measures is 90°.

composite number (p. 163) A composite number is a whole number greater than one with more than two factors.

compound event (p. 522) A compound event is a combination of two or more events.

cone (p. 421) A cone is a space figure with one circular base and one vertex.

congruent polygons (p. 380) Two polygons are congruent if there is a correspondence between their vertices such that the corresponding sides and corresponding angles are congruent.

converse of Pythagorean theorem (p. 462) If $a^2 + b^2 = c^2$, then the triangle with sides a, b, and c is a right triangle.

convex and concave polygons (p. 370) A polygon is convex if all points on the diagonals are inside the polygon. Otherwise, the polygon is concave.

coordinate plane (p. 316) A coordinate plane is the plane which results when two perpendicular number lines intersect at their zero points. The number lines form a grid on the plane.

cosine ratio (p. 473) In a right triangle, the cosine of $\angle A = \frac{\text{length of side adjacent to } \angle A}{\text{hypotenuse}}$.

counting principle (p. 508) The counting principle states that the number of outcomes for an event with two or more stages equals the product of the number of outcomes at each stage.

cross products (p. 185) Finding cross products is a method of checking equivalence of fractions or ratios.

$\frac{a}{b} = \frac{c}{d}$, if $a \cdot d = b \cdot c$

cylinder (p. 421) A cylinder is a space figure with two circular, parallel, and congruent bases.

D

data base (p. 122) A data base is a collection of information.

dependent events (p. 523) Dependent events are events in which the outcome of one depends on the outcome of the other.

For two dependent events A and B, where B is dependent on A,

$P(\text{A and B}) = P(\text{A}) \cdot P(\text{B, given A})$.

diagonal (p. 370) A diagonal is a segment that joins two nonconsecutive vertices of a polygon.

diameter (p. 377) A diameter is a chord that passes through the center of a circle. The diameter (d) is the length of such a segment.

direct variation (p. 353) Direct variation means that as one factor increases the other factor also increases. We represent direct variation by an equation in the form $y = kx$, where k is not zero. k is the constant of variation.

distributive property of multiplication over addition (p. 62) You can distribute a factor to each term inside a set of parentheses.

$a(b + c) = ab + ac \qquad (b + c)a = ba + ca$

distributive property of multiplication over subtraction (p. 62) You can distribute a factor to each term inside a set of parentheses.

$a(b - c) = ab - ac \qquad (b - c)a = ba - ca$

dividing integers (p. 26) To divide two integers, find the quotient of the absolute values of the integers. Then use these rules.

1. The quotient of two integers with the same sign is positive.
   $(+) \div (+) = + \qquad (-) \div (-) = +$
2. The quotient of two integers with different signs is negative.
   $(+) \div (-) = - \qquad (-) \div (+) = -$

dividing two rational numbers (p. 207) For any two rational numbers $\frac{a}{b}$ and $\frac{c}{d}$,

$\frac{a}{b} \div \frac{c}{d} = \frac{a}{b} \cdot \frac{d}{c} \qquad b \neq 0, c \neq 0, d \neq 0.$

divisible (p. 159) A number is divisible by a second number if the second number divides the first with no remainder.

divisible by 3 (p. 160) A number is divisible by 3 if the sum of its digits is divisible by 3.

divisible by 9 (p. 160) A number is divisible by 9 if the sum of its digits is divisible by 9.

division properties for inequalities (p. 297)

1. If c is positive and $a < b$, then $\frac{a}{c} < \frac{b}{c}$.
2. If c is positive and $a > b$, then $\frac{a}{c} > \frac{b}{c}$.
3. If c is negative and $a < b$, then $\frac{a}{c} > \frac{b}{c}$.
4. If c is negative and $a > b$, then $\frac{a}{c} < \frac{b}{c}$.

division property of equality (p. 76) You can divide both sides of an equation by the same nonzero value.

If $a = b$, then $a \div c = b \div c, \frac{a}{c} = \frac{b}{c}, c \neq 0$.

E

equation (p. 54) An equation is a mathematical sentence with an equal sign.

equiangular triangle (p. 374) An equiangular triangle is a triangle in which all angles have equal measure.

equilateral triangle (p. 374) An equilateral triangle is a triangle in which all sides have equal measure.

equivalent fractions (p. 184) You can form equivalent fractions by multiplying or dividing the numerator and denominator by the same nonzero factor.

evaluate an expression (p. 44) To evaluate an expression, replace each variable with a number. Then compute, following order of operations.

exponent (p. 141) An exponent shows the number of times a base is used as a factor. For a^n, n is the exponent.

F

factor (p. 159) One number is a factor of another if it divides that number with no remainder.

factorial (p. 509) A factorial is the product of all whole numbers from n to 1. We write this as $n!$

FOIL method (p. 550) Use the FOIL method to find the product of two binomials.

1. Multiply the **F**irst terms in each binomial.
2. Multiply the **O**utside terms in each binomial.
3. Multiply the **I**nside terms in each binomial.
4. Multiply the **L**ast terms in each binomial.
5. Add the products.

formula (p. 128) A formula is an equation that shows the relationship between two or more variables.

frequency distribution (p. 492) A frequency distribution is a listing of data that pairs each data item with the number of times it occurs.

front-end estimation (p. 105) To use front-end estimation:

1. Add the front-end digits.
2. Adjust by estimating the sum of the remaining digits.
3. Add the two values.

G

greatest common factor (GCF) (p. 167) The greatest common factor of a set of numbers is the greatest number that is a factor of the given numbers.

grouping symbols (p. 40) Grouping symbols include parentheses, ( ), brackets, [ ], absolute value symbols, and a division bar. These are used to group expressions.

H

height (p. 407) The height of a figure is the length of its altitude.

hypotenuse (p. 460) The hypotenuse is the side of a right triangle opposite the right angle.

I

improper fraction (p. 188) A fraction that has a numerator equal to or greater than the denominator is an improper fraction.

independent events (p. 522) Independent events are events in which the outcome of one has no affect on the outcome of the other.

For two independent events A and B,

$P(\text{A and B}) = P(\text{A}) \cdot P(\text{B})$.

indirect variation (p. 353) In indirect variation, one factor increases as the other factor decreases. The equation $xy = k$ represents an indirect variation. k is the constant of variation.

inequality (p. 292) An inequality is a statement that two expressions are not equal.

integers (p. 4) The whole numbers and their opposites form the set of integers.

. . . , −4, −3, −2, −1, 0, 1, 2, 3, 4, . . .
negative zero positive

inverse operations (p. 72) Inverse operations are operations that undo each other. Addition and subtraction are inverse operations. Multiplication and division are inverse operations.

irrational numbers (p. 455) Irrational numbers are numbers which we cannot express as either terminating or repeating decimals.

isosceles triangle (p. 374) An isosceles triangle has at least two sides with equal measure.

L

lateral area of a cone (p. 430) The lateral area (LA) of a cone equals half the product of the circumference (C) and slant height (l).
$LA = \frac{1}{2}Cl$

lateral area of a cylinder (p. 428) The lateral area (LA) of a cylinder is the product of the circumference of the base (C) and the height of the cylinder (h).
$A = Ch$

lateral area of a prism (p. 427) The lateral area (LA) of a prism is the product of the perimeter of the base (P) and the height of the prism (h).
$A = Ph$

least common denominator (LCD) (p. 196) The least common denominator of two or more fractions is the LCM of the denominators.

least common multiple (LCM) (p. 168) The least common multiple is the least number that is a common multiple of two or more given numbers.

legs of a right triangle (p. 460) The legs of a right triangle are the two sides that form the right angle.

like terms (p. 66) Like terms have the same variable(s).

line (p. 362) A line continues without end in opposite directions. We denote line AB by $\overleftrightarrow{AB}$.

line plot (p. 492) A line plot shows data on a number line. You place an × for each response above the category of the response.

line symmetry (p. 381) A figure is said to have line symmetry if a line can be drawn through the figure so that one side is a mirror image of the other.

linear equation (p. 324) A linear equation is an equation for which the graph is a line. The standard form of a linear equation is $Ax + By = C$, where A, B, and C are real numbers and A and B are not both equal to zero.

locating a point on the coordinate plane (p. 316) To locate $P(x, y)$ on the coordinate plane:

1. Begin at origin.
2. Locate x on the x-axis.
3. Move up or down the absolute value of y units.

lowest terms (p. 185) When a fraction is in lowest terms, the only common factor of the numerator and denominator is 1.

M

mean (p. 27) The mean is the sum of a set of numbers divided by the number of items in the set.

measures of central tendency (p. 488) The measures of central tendency are statistics used to describe data characteristics. These measures are mean, median, and mode.

median (p. 488) The median is the middle value in a set of data.

mixed number (p. 188) A mixed number is a number that includes an integer and a fraction.

mode (p. 488) The mode is the data item that occurs most often.

monomial (p. 538) A monomial is a real number, a variable, or the product of a real number and one or more variables.

multiple (p. 159) A multiple of a number is the product of that number and any other whole number.

multiplication properties for inequalities (p. 297)

1. If c is positive and $a < b$, then $ac < bc$.
2. If c is positive and $a > b$, then $ac > bc$.
3. If c is negative and $a < b$, then $ac > bc$.
4. If c is negative and $a > b$, then $ac < bc$.

multiplication property of equality (p. 77) You can multiply both sides of an equation by the same value. If $a = b$, then $ac = bc$.

multiplicative identity (p. 59) The multiplicative identity is one. $a \cdot 1 = a$

multiplying integers (p. 22) To multiply two integers, find the product of the absolute values of the integers. Then use these rules.

1. The product of two integers with the same sign is positive.
   $(+)(+) = +$ $(-)(-) = +$
2. The product of two integers with different signs is negative.
   $(+)(-) = -$ $(-)(+) = -$

N

negative exponents (p. 212) For any nonzero integers a and n: $a^{-n} = \frac{1}{a^n}$.

numerical coefficient (p. 66) A numerical coefficient is a number that is multiplied by a variable.

numerical expression (p. 34) A numerical expression names a number. A numerical expression does not contain variables.

O

obtuse angle (p. 365) An obtuse angle is an angle with measure between 90° and 180°.

obtuse triangle (p. 374) An obtuse triangle has one obtuse angle.

open equation (p. 54) An open equation is an equation that contains one or more variables.

opposites (p. 4) Opposites are two integers the same distance from zero on a number line, but in opposite directions.

order of operations (p. 40)

1. Do all operations within grouping symbols.
2. Evaluate powers.
3. Multiply and divide from left to right.
4. Add and subtract from left to right.

ordered pair (p. 316) An ordered pair is a pair of numbers (x,y) assigned to a point on a coordinate plane.

origin (p. 316) The origin is the intersection point of the x- and y-axes in a coordinate plane. The coordinates of the origin are (0,0).

P

parallel lines (p. 363) Two lines are parallel if they lie in the same plane and do not intersect.

parallel planes (p. 363) Two planes are parallel if they do not intersect.

parallelogram (p. 371) A parallelogram is a quadrilateral with two pairs of opposite parallel sides.

percent (p. 245) A percent is a ratio that compares a number to 100.

percent equation (p. 256) Use a triangle to solve percent problems.

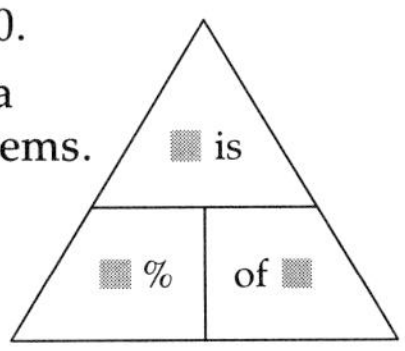

percent of change (p. 260) Use the following formula to find percent of change.

$$\text{percent of change} = \frac{\text{amount of change}}{\text{original amount}}$$

perimeter (p. 390) Perimeter is the distance around a figure.

pi (π) (p. 389) Pi is the ratio of the circumference of a circle to its diameter.

plane (p. 362) A plane is a flat surface with no thickness that continues without end in all directions.

point (p. 362) A point represents a position in space.

polygon (p. 370) A polygon is a closed plane figure such that no two segments with a common endpoint are collinear and segments intersect only at the endpoints.

polyhedron (p. 420) A polyhedron is a space figure in which all faces are polygons.

polynomial (p. 538) A polynomial is a monomial or a sum or difference of monomials.

prime factorization (p. 164) Prime factorization is an expression showing a composite number as a product of its prime factors.

prime number (p. 163) A prime number is a whole number greater than 1 with exactly two factors, 1 and the number itself.

principal square root (p. 454) The principal square root of a number is its positive square root. The principal square root is denoted by the symbol $\sqrt{\ }$.

prism (p. 420) A prism is a polyhedron with two parallel bases that are congruent polygons and sides that are parallelograms.

probability (p. 512) Probability is the likelihood that a certain event, or set of outcomes, will occur.

$$P(\text{E}) = \frac{\text{number of favorable outcomes}}{\text{total number of possible outcomes}}$$

product of $(a + b)(a - b)$ (p. 551) To find the product of the sum and difference of two terms, square the first term and subtract the square of the second term.

$$(a + b)(a - b) = a^2 - b^2$$

product of two rational numbers (p. 206) For any two rational numbers $\frac{a}{b}$ and $\frac{c}{d}$,

$$\frac{a}{b} \cdot \frac{c}{d} = \frac{a \cdot c}{b \cdot d},\ b \neq 0,\ d \neq 0.$$

585

proportion (p. 233) A proportion is a statement that two ratios are equal. If two ratios are equal, their cross products are equal.

a is to *b* as *c* is to *d*

$a : b : : c : d, \frac{a}{b} = \frac{c}{d}, b \neq 0, d \neq 0$

proportions and percents (p. 252) To find the ratio of a number to 100, use the following formula.

$\frac{\text{part}}{\text{whole}} = \frac{n}{100}$

pyramid (p. 420) A pyramid is a polyhedron with triangular sides that meet at a vertex. The base of a pyramid is a polygon.

Pythagorean theorem (p. 461) In any right triangle with legs *a* and *b*, and hypotenuse *c*,

$a^2 + b^2 = c^2$.

Q

quadrant (p. 316) A quadrant is one of four sections into which the *x*- and *y*-axes divide the coordinate plane.

quadrilateral (p. 371) A quadrilateral is a polygon with four sides.

R

radius (p. 377) A radius is a segment that has endpoints at the center of a circle and on the circle. The radius (*r*) is the length of such a segment.

range (p. 489) The range of a set of data is the difference between the greatest and least values in the set.

rate (p. 234) A rate is a ratio that compares quantities in different units. A unit rate compares a quantity to one.

ratio (p. 233) A ratio is a comparison of two quantities by division.

a to b; $a : b$; $\frac{a}{b}$; $b \neq 0$

rational number (p. 192) A rational number is a number you write in the form $\frac{a}{b}$, where *a* is any integer, and *b* is a nonzero integer.

ray (p. 362) A ray is part of a line with only one endpoint that continues without end in one direction. We denote ray *AB* as $\overrightarrow{AB}$.

rectangle (p. 371) A rectangle is a parallelogram with four right angles.

regular polygon (p. 371) A polygon is regular if the measures of all sides and all angles are equal.

rhombus (p. 371) A rhombus is a parallelogram with all sides equal.

right angle (p. 365) A right angle is an angle that measures 90°.

right triangle (p. 374) A right triangle is a triangle with one right angle.

rule of a power raised to a power (p. 145) To raise a power to a power, multiply the exponents.

$(a^m)^n = a^{m \cdot n}$

rule of a product raised to a power (p. 146) To raise a product to a power, raise each factor to the power and then use the rule of exponents for multiplication. $(ab)^m = a^m b^m$

rule of exponents for division (p. 211) To divide numbers or variables with the *same* base, subtract exponents.

$\frac{a^m}{a^n} = a^{m-n}, a \neq 0$

rule of exponents for multiplication (p. 145) To multiply numbers or variables with the *same* base, add exponents. $a^m \cdot a^n = a^{m+n}$

S

sample space (p. 512) The set of possible outcomes is the sample space.

scalene triangle (p. 374) A scalene triangle is a triangle that has no sides equal.

scientific notation (p. 149) A number is in scientific notation when it is written as the product of a number greater than or equal to 1 and less than 10, and a power of 10.

segment (p. 362) A segment is part of a line with two endpoints. We denote segment *AB* by $\overline{AB}$.

similar figures (p. 384) Two figures are similar (~) if corresponding angles are congruent and corresponding sides are in proportion.

similar triangles (p. 385) Two triangles are similar if two angles of one are congruent to two angles of another.

simplify an expression (p. 67) To simplify an expression, replace it with an equivalent expression that contains no like terms or parentheses.

sine ratio (p. 473) In a right triangle, the

sine of $\angle A = \frac{\text{length of side opposite } \angle A}{\text{hypotenuse}}$.

skew lines (p. 363) Skew lines are lines that do not lie in the same plane and do not intersect.

slant height (p. 430) A slant height of a cone or a pyramid is the height of a face.

slope (p. 330) The slope of a line is the ratio of the vertical change in y to the corresponding horizontal change in x. Use the following formula to calculate slope.

$$\text{slope} = \frac{\text{difference in } y \text{ coordinates}}{\text{difference in } x \text{ coordinates}}$$

slope-intercept form (p. 331) A linear equation in the form $y = mx + b$ is in slope-intercept form. The slope is m and the y-intercept is b.

solution (p. 55) A solution is a number that replaces a variable to make an open equation true.

solution of a system of linear equations (p. 340) A solution of a system of linear equations is any ordered pair of numbers that satisfies all equations in the system.

solving a multi-step equation (p. 279) To solve a multi-step equation:

1. Remove parentheses using the distributive property.
2. Combine like terms.
3. Undo addition or subtraction.
4. Undo multiplication or division.

solving a simple two-step equation (p. 275) To solve a simple two-step equation:

1. Undo addition or subtraction.
2. Undo multiplication or division.

solving proportions (p. 237) To solve a proportion:

1. Write the cross products.
2. Solve the equation.

sphere (p. 421) A sphere is the set of all points in space that are the same distance from a given point called the center.

square (p. 371) A square is a parallelogram that is both a rectangle and a rhombus.

square root (p. 454) The square root of a number, n, is a if $a^2 = n$.

squaring binomials (p. 551) To square a binomial, square the first term. Then add or subtract twice the product of the two terms and add the square of the second term.

$(a + b)^2 = a^2 + 2ab + b^2$

$(a - b)^2 = a^2 - 2ab + b^2$

stem and leaf plot (p. 498) A stem and leaf plot is an organization of data that groups data into categories based on place values.

subtracting integers (p. 15) To subtract an integer, add its opposite.

subtraction properties for inequalities (p. 296)

1. If $a > b$, then $a - c > b - c$.
2. If $a < b$, then $a - c < b - c$.

subtraction property of equality (p. 72) You can subtract the same value from both sides of an equation.

If $a = b$, then $a - c = b - c$.

supplementary angles (p. 366) Two angles are supplementary angles if the sum of their measures is 180°.

surface area (p. 427) Surface area (SA) is the sum of the areas of the base(s) and the side(s). Surface area is measured in square units.

surface area of a sphere (p. 431) The surface area of a sphere equals the product of 4π and the square of the radius (r).

$A = 4\pi r^2$

system of linear equations (p. 340) A system of linear equations is two or more linear equations using the same variables.

system of linear inequalities (p. 349) A system of linear inequalities is two or more linear inequalities using the same variables.

T

tangent ratio (p. 473) In a right triangle, the tangent of $\angle A = \frac{\text{length of side opposite } \angle A}{\text{length of side adjacent to } \angle A}$.

term (p. 66) A term is a part of an expression. Terms are separated by addition and subtraction symbols.

tessellation (p. 397) A tessellation is a design that covers a plane with no gaps and no overlaps.

trapezoid (p. 371) A trapezoid is a quadrilateral with exactly one pair of parallel sides.

triangle (p. 371) A triangle is a polygon that has three sides.

triangle, 30°-60°-90° (p. 470) In a 30°-60°-90° triangle, the lengths of the sides have the following relationships.

hypotenuse = 2(shorter leg)

longer leg = shorter leg $(\sqrt{3})$

shorter leg = $\frac{\text{longer leg}}{\sqrt{3}}$

triangle, 45°-45°-90° (p. 469) In a 45°-45°-90° right triangle, the lengths of the sides have the following relationships.

hypotenuse = $\sqrt{2} \cdot$ leg

leg = $\frac{\text{hypotenuse} \cdot \sqrt{2}}{2}$

trigonometric ratio (p. 473) A trigonometric ratio is a ratio of the measures of two sides of a right triangle.

V

variable (p. 34) A variable is a symbol (usually a letter) that stands for a number.

variable expression (p. 34) A variable expression is an expression that contains at least one variable.

vertex (p. 365) The vertex of an angle is the common endpoint of the two rays forming the angle.

vertical angles (p. 366) Two intersecting lines form two pairs of vertical angles. The measures of vertical angles are equal.

volume (p. 437) Volume is the measure of the space inside a space figure. We measure volume in cubic units.

volume of a cone and a pyramid (p. 440) The formula for the volume of the cone and pyramid is base area (B) times one-third the height (h).

$V = \frac{1}{3}Bh$

volume of a cylinder or a prism (p. 437) The volume (V) of a prism or a cylinder is base area (B) times the height (h).

$V = Bh$

volume of a sphere (p. 441) The volume (V) of a sphere with radius r is $V = \frac{4}{3}\pi r^3$.

X

***x*-axis** (p. 316) The x-axis is the horizontal number line on a coordinate plane.

***x*-intercept** (p. 325) The x-intercept is the x-coordinate of a point where a graph crosses the x-axis.

Y

***y*-axis** (p. 316) The y-axis is the vertical number line on a coordinate plane.

***y*-intercept** (p. 325) The y-intercept is the y-coordinate of a point where a graph crosses the y-axis.

Z

zero as an exponent (p. 211) Any nonzero number with zero as an exponent equals 1.

$a^0 = 1$ for all $a \neq 0$.

Selected Answers

CHAPTER 1

Integers and Expressions

1-1 pages 6–7
Written Exercises **1.** 110 **3.** -300 **5.** -8 **13.** -6 **15.** 3 **17.** 4 **19.** -9 **21.** -12,500; -15,617; 0 **23.** -2 **25.** -8 **27.** 8 **33.** > **35.** < **37.** < **39.** = **41.** zero **43.** negative **53. a.** -27°; **b.** -20°; -71°
Mixed Review **1.** 907 **2.** 814 **3.** 1,088 **4.** 42 **5.** 7,872 **6.** 13

1-2 pages 12–13
Written Exercises **1.** -20 + 18 = -2; still owe $2 **3.** -10 + (-2) + 8 + (-5) + (-13) + 1 = -21; temp. is 21° below. **5.** -9 **7.** 4 **9.** 9 **11.** -8 **13.** -40 **15.** -847 **17.** -5 **19.** 0 **21.** 3 **23.** -13 **25.** 15 **27.** -13 **29.** 4 **31.** -6 **33.** 9 **35.** -6 < 2 **37.** -2 + (−7) = −9 **39.** 3 + (−8) = −5 **41.** < **43.** < ; < **45.** 10-yd loss
Mixed Review **1.** -8 **2.** 12 **3.** 10 **4.** 16 **5.** < **6.** > **7.** 12 **8.** -13
Critical Thinking **1.** -7, 525, -47 **2.** -19, -50, -198 **3.** -7, -78, -47 **4.** all neg. **5.** -47, -19, -7; -5, -25, -40

1-3 pages 15–17
Written Exercises **1.** 3 − (-2) = 5 **3.** 3 − 5 = -2 **5.** 3,000 − 600 **7.** -5 **9.** -16 **11.** -60 **13.** 150 **15.** 66 **17.** -196 **19.** 178 **21.** 913 **23.** -31 **25.** 25 **27.** -175 **29.** -422 **31.** 15 **33.** 191 **35.** -101 **37.** 56 **39.** 356 **41.** 38 **43.** 16; 20; 24; 28; 28 **45.** -7, -6, -1, -3, -8; -15 **47. a.** The temperature decreases.; **b.** 24°; **c.** decrease; **d.** 21,000 m **49.** -15 **57.** 180 **59.** -70 **61.** 2,400 **63.** -3,600
Mixed Review **1.** -29 **2.** 65 **3.** 7 **4.** -6 **5.** -6, -7, -8, or -9 **6.** -6 **7.** -925 ft

1-4 pages 20–21
Written Exercises **1.** 78 students **3.** 4, 3, 1, 16, 15, 1; 9, 8, 1, 25, 24, 1; **a.** 11 · 11; 1 **b.** subtract 1; 2,208; **c.** add 1; 4,225 **5.** 1810 and 1820; 467,174 people **7. a.** $59; $21; **b.** 10
Critical Thinking **1.** 107 **2.** 234 **3.** 37

1-5 pages 24–25
Written Exercises **1.** -60 **3.** -30 **5.** 21 **7.** 20 **9.** A **11.** B **13.** 4,661 **15.** 46,354 **17.** 15 **19.** -96 **21.** -220 **23.** 12,288 **25.** -200 **27.** -56 **29.** -7 **31.** 0 **33.** -26 **35.** -81 **37.** -19 **39.** 8(-5) = -40 **41.** 6(−9) = −54 **43.** > **45.** = **47. a.** -$36; **b.** $40 **49.** -3 and -4 **51.** 5 and -1 **53. a.** 5,000 ft at 40°; **b.** no
Mixed Review **1.** < **2.** > **3.** < **4.** -11 **5.** 3,003 **6.** |-20| **7.** 6 **8.** 180°

1-6 pages 27–29
Written Exercises **1.** -9(10) = -90 **3.** 8(7) = 56 **5.** -7 **7.** -5 **9.** -9 **11.** 126 **13.** 56 **15.** 4 **17.** 19 **19.** -1 **21.** -15 **23.** -384 **25.** -59 **27.** -225 **29.** -80 **31.** -35 **33.** 64 **35.** 3,375 **37.** -9 **39.** -42(3); -126 **41.** -25 − 200; -225 **43.** $3 **45.** 0 **47.** < **49.** < **55.** -15 **57.** 2 ft/s
Mixed Review **1.** -45 **2.** 24 **3.** -48 **4.** 30 **5.** 13, 18, 23 **6.** 25, 36, 49
Test Yourself **1.** > **2.** < **3.** = **4.** 8 **5.** -85 **6.** -12 **7.** -8 **8.** 20 **9.** -45 **10.** 8 **11.** 3 **12.** 44

Practice page 30
1. 55 **3.** -23 **5.** -19 **7.** 133 **9.** -334 **11.** 141 **13.** -100 **15.** 63 **17.** -89 **19.** -62 **21.** -238 **23.** -224 **25.** -705 **27.** 63 **29.** -84 **31.** -162 **33.** 375 **35.** -896 **37.** 288 **39.** -496,000 **41.** -9 **43.** -6 **45.** -5 **47.** 37 **49.** -37 **51.** 153 **53.** 8 **55.** 142 **57.** -20 **59.** -3 **61.** 0 **63.** 55 **65.** 10 **67.** -1

1-7 page 33
Written Exercises **1.** 90 C **3.** 230 C **5.** 120 C **7.** -580 C **9.** 660 C; about 5 h **11.** 900 C; 18,000 C

1-8 pages 36–37
Written Exercises **1.** $3x - 3$ **3.** $3z + (-2)$ **9.** 23(-9) **11.** -6 − 8 **13.** $19 + m$ **15.** $12x$ **17.** $n \div (-1)$ **19.** $g \cdot 4r$ **21.** $10a$ **23.** $t + 200$ **33. a.** 7 · 1; **b.** 7 · 4; **c.** $7w$ **35. a.** 15 − 3; **b.** $15 - p$; **c.** 15 + 10; **d.** $15 + f$ **37.** $d - 20$ **41. a.** (2 · 25) + (4 · 12); **b.** $25j + 12t$ **43.** b **45.** d
Mixed Review **1.** -5 **2.** 20 **3.** 0 **4.** -23 **5.** 100 **6.** 8 **7.** $2.95

1-9 pages 41–43
Written Exercises **1.** addition **3.** subtraction inside absolute value symbols **5.** -1 **7.** 3 **9.** 4 **11.** -20 **13.** -30 **15.** -13 **17.** 243 **19.** 4 **21.** -394 **23.** -8 **25.** -2[(7 + 8) ÷ 5 + 5] = -16 **27.** > **29.** (7 + 4) · 6 = 66 **31.** 3 · (8 − 2 + 5 − 12) = -3 **33.** no **35.** yes **37.** 25 h **39.** Alice 103; Ray 118 **41.** A possible answer is [-6 + (-8)](6 + 4) − 2 **45.** 5 + (4)(9); 41 **47.** 17 − (25 ÷ 5); 12 **49.** 130 + (116 − 8); 238 **55.** 74 + 5*9 + -7 = 112 **57.** 70 + 8*-9 = -2 **59.** 2,087*37 − 1,951 = 75,268 **61.** Yes, computers follow order of operations.
Mixed Review **1.** $6n$ **2.** $x - 6$ **3.** $a + |-7|$ **4.** < **5.** < **6.** < **7.** 36

1-10 pages 45–46
Written Exercises **1.** -24 **3.** -5 **5.** -12 **7.** 11 **9.** 21 **11.** 7 **13.** 20 **15.** -35 **17.** 21 **19.** 1 **21.** 425 **23.** 18 **25.** 117 **27.** A possible answer is 2. **29.** 8, -8 **31.** -6 **33.** 0 **35.** 3 **37. a.** $265m$; **b.** 1,590; **c.** 381,600 **39. a.** $14m$; **b.** 350 C **41.** $400 **43.** 6, 12, -12, 36, 4 **45.** 55 **47. a.** 13; **b.** 24

Mixed Review 1. -54 2. -48 3. -5 4. -24 5. the sum of 6 and a number 6. twice the quantity of a number minus 2 7. the opposite of 12 times a number 8. 83

Problem Solving Practice page 47
1. a. Each month the interest increases by a penny more than the previous month.; b. May $1.04, $105.10; June $1.05, $106.15; July $1.06, $107.21; August $1.07, $108.28 3. thermosphere, mesosphere, stratosphere, troposphere 5. -39°F

Chapter 1 Review pages 48–49
1. grouping symbols; order of operations 2. opposite 3. integers 4. absolute value 5. variable expression; variable 6. mean 7. > 8. > 9. = 10. < 11. < 12. > 13. -7 14. 12 15. -9 16. 14 17. -12 18. -16 19. -27 20. -15 21. -1 22. -42 23. -5 24. 72 25. 7 26. -3 27. -165 28. $x - 25$ 29. $3rn$ 30. $y + 2$ 31. 36 32. 33 33. 25 34. 24 35. 3 36. 4 37. 19 38. 6 39. 16 40. 14 41. -450 42. 16 43. $112 44. She will gain weight.

Chapter 1 Cumulative Review page 51
1. A 3. C 5. D 7. B 9. B 11. B 13. A 15. D

CHAPTER 2

Solving Equations

2-1 pages 56–57
Written Exercises 1. yes 3. yes 5. open 7. false 9. false 11. true 13. true 15. true 17. false 19. true 21. yes 23. yes 25. yes 27. no 29. yes 31. 0 33. -2 35. 4 37. > 39. < 41. > 43. 30; 48 45. 15; 28 47. $0 \cdot (-7) = -7$; false 49. $15 + n = 50$; open 51. $(-7) + 12 = -5$; false 53. $3 \cdot 32 = 96$; true 57. $13b = f$
Mixed Review 1. -1 2. 10 3. -3 4. 13 5. 13, 21, 34 or 12, 17, 23 6. 15, 8, 3 7. 12h

2-2 pages 60–61
Written Exercises 1. c 3. d 5. f 7. d 9. c 11. $z \cdot 25$ 13. $5(a + b)$ 15. $3 \cdot (25 \cdot 4)$ 17. $4(ab)$ 19. 47,000 21. 2,800 23. 100 25. 3; 7 27. 730; 270 29. 58 31. 60 33. 60,000 35. a. No, it would change the order of operations. b. Yes, it does not matter in what order you multiply 6, 5, and -4.
Mixed Review 1. true 2. open 3. no 4. yes 5. no 6. yes 7. $3 \cdot (4 - 7) + 6 = -3$ 8. 4 and 9

2-3 pages 64–65
Written Exercises 1. $4(w + 7)$ 3. $7(x + y + 5)$ 5. -4 7. c 9. $-w$ 11. -9; $-9(x + y)$ 13. e; $(b - c - d)e$ 15. -1; $-1(a + b)$ 17. w, y 19. -2, -4 21. 5, -2 23. 1,120 25. 5,075 27. -12 29. -30 31. 7 33. 285 35. 16 37. 3,098 mi
Mixed Review 1. C 2. AI 3. A 4. MI 5. < 6. = 7. < 8. 605

Critical Thinking 1. a. 3; b. 0; c. 1 2. a. =; b. yes 3. yes 4. yes; zero 5. It is an operation which is associative and commutative. It has an identity of zero.

2-4 pages 68–69
Written Exercises 1. 2; 5, 8; $5a$, $8a$; none 3. 2; 2; none; -7 5. 2; -7; none; 3 7. 4; 6, 4, 1; $6ab$, $4ba$, ab; 8 9. $9a$ 11. $5b$ 13. A 15. D 17. A 19. $5a$ 21. 6; 4; 10 23. $1m$ 25. $6a$ 27. $54k + 5$ 29. $5g + 15$ 31. $8x - 32w$ 33. $-83x$ 35. $44a - 59b + 19c$ 37. $-12y$; 60 39. $3y + 2z - 16$; -17 41. $6x + 7x + 14$; $13x + 14$ 43. $3x + 2x + 189$; $5x + 189$
Mixed Review 1. 7 2. 3 3. 5 4. 3 5. -4 6. 86 7. 46
Test Yourself 1. yes 2. no 3. no 4. C 5. A 6. MI 7. C 8. C 9. D 10. $50b$ 11. 173 12. $18y$ 13. $9a$ 14. -2 15. $16w - 6$

2-5 pages 74–75
Written Exercises 1. $x = 2$ 3. $w = -2$ 5. $-2 = x - 3$; 1 7. -5 9. -54 11. 626 13. 82 15. -13 17. 10,221 19. 41 21. 39 23. -11 25. -300 27. 0 29. 308 31. 1,364,615 33. 77,098 35. 500 37. 65 39. -20 41. 1,598 43. 25 45. $b - a$ 47. $d + 5 = 17$; $d = 12$ 49. $20 + d = 150$; $d = 130$ 51. $5{,}200 = 2{,}680 + h$; $h = 2{,}520$ 53. $700 = 119 + p$; $p = 581$ million
Mixed Review 1. 3 2. 3 3. $4x + 4$ 4. $2q + 6$ 5. 3 6. 12 7. 512 min

2-6 pages 77–79
Written Exercises 1. $g = 4$ 3. $h = -3$ 5. yes 7. no 9. $8 = 4x$; 2 11. -15 13. 6 15. -4 17. 5 19. 52 21. -16 23. 300 25. -24 27. 42 29. -15,000 31. 382,300 33. -42,336 35. -1,586 37. -768 39. -8 41. 504 43. 15 45. 31 47. 11 49. 0 51. 875,000 53. $12d$ 55. -6, 6 57. $\frac{b}{a}$ 59. $b + a$
Mixed Review 1. -11 2. 17 3. -19 4. $a - 3 + b$ 5. $7(9 - w)$ 6. $|-8 + q|$ 7. 60 s or 1 min
Critical Thinking 1. they get thinner 2. where all of the lines meet 3. All of the dot is on the missing piece. 4. 2 5. 3

2-7 pages 81–82
Written Exercises 1. c 3. $n - 24 = -9$ 5. $15c = 30$ 9. $x + 46 + 54 = 150$; 50 mm
Mixed Review 1. 515 2. 540 3. Divide each side by 5. 4. Subtract 5 from each side. 5. Add 5 to each side. 6. Multiply each side by 5. 7. 1,028 ft
Test Yourself 1. 20 2. -68 3. -34 4. 32 5. -23 6. -13 7. $n + 12 = 20$; 8 8. $(c - 6) - 3 = -10$; -1

2-8 page 86
Written Exercises 1. 12, 84 and 8, 56 3. 11, 12 5. -2, -3, -4 7. 12:40 P.M. 9. -10 11. 30 ft, 40 ft, 50 ft

590

2-9 pages 88–89
Written Exercises **1.** $2,860,000,000 **3. a.** about 2 billion lb; **b.** about 16 billion kW · h **5.** $325,000

Practice page 90
1. open **3.** true **5.** open **7.** no **9.** yes **11.** no **13.** A **15.** C **17.** MI **19.** -476 **21.** 216 **23.** -22 **25.** 26 **27.** -16 **29.** $-15k$ **31.** $5g + 8$ **33.** $4x + 4y - 4z$ **35.** $-3w + 4$ **37.** 27 **39.** 12 **41.** 9 **43.** -39 **45.** 3,072 **47.** $n + 5 = -123$; -128

Problem Solving Practice page 91
1. a. 4; **b.** 8; **c.** yes; 16; 2 **3. a.** total wages for each employee; **b.** = SUM(B2:B4); =SUM(D2:D4); place in B5 and D5, respectively **5. a.** $260; **b.** $1\frac{1}{2}$ h

Chapter 2 Review pages 92–93
1. f **2.** a **3.** e **4.** i **5.** k **6.** j **7.** b **8.** c **9.** d **10.** g **11.** h **12.** 547 **13.** 80 **14.** 700 **15.** 6,500 **16.** 115 **17.** 192 **18.** 80 **19.** 0 **20.** $8x + 5y$ **21.** $9a + 16$ **22.** $x = -1$ **23.** $x = 3$ **24.** $a = 35$ **25.** $x = 11$ **26.** $y = 480$ **27.** $n = 20$ **28.** $x = -1$ **29.** $x = 2$ **30.** $m = -72$ **31.** $b = 12$ **32.** $c = 288$ **33.** $k = 18$ **34.** $2x + 28 = 54$ **35.** $x - 17 = 12$ **38.** 5 pkgs of 15 plates; 4 pkgs of 20 plates **39.** 80 billion lb

Chapters 1–2 Cumulative Review page 95
1. B **2.** C **3.** C **4.** D **5.** A **6.** A **7.** A **8.** C **9.** C **10.** B **11.** B **12.** D **13.** C **14.** C **15.** B **16.** B

CHAPTER 3

Decimals and Equations

3-1 pages 102–103
Written Exercises **5.** 512.73 km/h **11.** > **13.** = **15.** > **17.** > **19.** 4.5, 4.05, 4.049 **21.** 3.003, 0.3002, 0.30, 0.030008, 0.03 **29.** 0.8 **31.** 0.36 **33.** 365,987 **37. a.** lesser; **b.** greater; **c.** when the digit to the right is greater than the digit to the left **39.** If the absolute value of a negative number is less than the absolute value of another negative number, the first number is greater than the second number.
Mixed Review **1.** -5 **2.** 3 **3.** -9 **4.** -45 **5.** 3 and 7 **6.** $d - 40 = 182$; $222

3-2 pages 106–107
Written Exercises **1.** 0.25 **3.** 60 **5.** 40 **7.** -18,000 **9.** 300 **11.** 8 **13.** 27 **15.** 3 **17.** 4 **19.** 0.02 **21.** $2,000 **23.** 0.23 **25.** 200 **27.** -20 **29.** 22.4256 **31.** 213.76 **33.** 109.571 **35.** 67.3436 **37.** yes **39.** food **41.** $2 **43.** $3,200 **45.** about 3,500 yd
Mixed Review **1.** -0.88 **2.** 4.13 **3.** 24 **4.** -2 **5.** -108 **6.** 11 **7.** 36

3-3 pages 109–110
Written Exercises **1.** 11.5 h **3.** 14.375 h **5.** -7.8 **7.** 15.1 **9.** 6.14 **11.** -20.62 **13.** -25.48 **15.** -14.965 **17.** -3.07 **19.** -25 **21.** -20.44 **23.** 3.001 **25.** 7.69 **27.** $-40.416w$ **29.** $-0.8m$ **31.** $7.6x$ **33.** $17.247a - 24.228$ **35.** $-1.8102x$ **37.** -90 **39.** 3 **41. a.** $12.3a - 1.08$; **b.** 102.2523 **43.** $21.67
Mixed Review **1.** $200 **2.** 20,000 **3.** -5 **4.** 0.1 **5.** 4 **6.** 35 **7.** $.22
Critical Thinking **1.** 0.125 ÷ 0.625 = 0.2; 0.2 ÷ 0.125 = 1.6; 1.6 ÷ 0.2 = 8 **2.** 2.5 and 2; 2 ÷ 2.5 = 0.8; 0.8 ÷ 2 = 0.4; 0.4 ÷ 0.8 = 0.5; 0.5 ÷ 0.4 = 1.25; 1.25 ÷ 0.5 = 2.5; 2.5 ÷ 1.25 = 2; 2 ÷ 2.5 = 0.8; By following the pattern, you arrive at the numbers you began with; no. **3.** 2 and -0.2; -0.2 ÷ 2 = -0.1; -0.1 ÷ (-0.2) = 0.5; 0.5 ÷ (-0.1) = -5; -5 ÷ 0.5 = -10; -10 ÷ (-5) = 2; 2 ÷ (-10) = -0.2; -0.2 ÷ 2 = -0.1; By following the pattern, you arrive at the numbers you began with; no. **4.** By following this pattern, you arrive at the numbers you began with. It does not matter whether you use decimals, whole numbers, positive numbers, or negative numbers.

3-4 pages 112–114
Written Exercises **1.** -13.391 **3.** 1.63 **5.** -0.3698 **7.** -0.0233 **9.** 12.58 **11.** 1.032 **13.** 13.31 **15.** -23.12 **17.** 4.66 **19.** 2 **21.** 60,392.0034 **23.** 7.131 **25.** 3.9 **27.** 0 **29.** -5.4 **31.** 2.9 **33.** 0.4 **35.** -59.9 **37. a.** apple, grapefruit, orange, grape; **b.** blue **39.** $b - c = a$ **41.** $x = 0.08$ and $y = -0.05$ **43.** $n - 0.058 = 0.58$; 0.638 **45.** $1.0099 = n + 2$; $n = -0.9901$ **47.** $3 + n = 8.16$; $n = 5.16$
Mixed Review **1.** 7.629 **2.** -1.6642 **3.** -2.46 **4.** -4.754 **5.** $-39.56x$ **6.** $-0.28y$ **7.** $1.82
Test Yourself **1.** < **2.** = **3.** > **4.** -9.7 **5.** 4.3 **6.** 18.0 **7.** 0.3 **8.** 24 **9.** 250 **10.** -13.34 **11.** 3.8 **12.** -1.2 **13.** $x - 0.09$ **14.** $-12.56y$ **15.** -6.71 **16.** 0.111

3-5 pages 116–118
Written Exercises **1.** -2.44 **3.** 0.044 **5.** 1.2 **7.** 42.6 **9.** 0.374 **11.** -14.85 **13.** -5.4 **15.** 86.7 **17.** 3.0772 **19.** -708 **21.** 0.048308 **23.** 194.0 **25.** -42.1 **27.** 9.50 **29.** 236.03 **31.** 36.9 **33.** -0.2 **35.** -6 **37.** $x = zy$ **39.** $x = yz$ **41.** $0.004n = 0.88$; 220 **43.** $\frac{n}{-2.35} = 400.9$; -942.115 **47. a.** $2,537.50; **b.** 205; **c.** yes
Mixed Review **1.** 1.446 **2.** 84.72 **3.** 0 **4.** 0.111 **5.** -1,056 **6.** -360 **7.** 14
Critical Thinking **1.** The outside shape becomes shaded inside; the shaded shape inside becomes the outside shape. **2.** c and d

Practice page 119
1. sixty-seven hundredths

591

3. six hundred thirty-seven and four ten-thousandths **5.** 215.74 **7.** 42.07 **9.** > **11.** < **13.** > **15.** = **17.** = **19.** 20 **21.** 1 **23.** \$56 **25.** 6.62 **27.** -94 **29.** -41.706 **31.** -1 **33.** $x = 7.75$ **35.** $a = 0$ **37.** $z = 30$ **39.** $t = -1.2$ **41.** $y = 42.71$ **43.** $j = -4.036$ **45.** $x = -1.94$ **47.** $r = -18.57$ **49.** $m = np$ **51.** $m = \frac{n}{p}$ **53.** $m = \frac{p}{n}$

Problem Solving Practice page 124

1. Christine, Lisa, Nicole, JoAnn **3.** 8.2 mi **5.** 276.8 m **7.** on; off **9.** 1.2 h

3-6 pages 126–128

Written Exercises **1.** 66.8 g **3.** water: ethyl alcohol, gasoline; mercury: copper, gasoline, ethyl alcohol, rubber, and iron **5.** 110 g **7.** 0.917 g/cm^3 **9.** 1,050 g

3-7 pages 129–131

Written Exercises **1.** 87°F **3.** 52°F **5.** 465.85 mi **7.** 1,704 cm **9.** 6,700 mi; high **11.** 0.347 **13.** 0.366 **15.** 0.345 **17.** Cobb, Hornsby, Jackson, Browning, Delahanty, Keeler **19.** \$536.25 **21.** \$683.98 **23.** -128.2°F **25.** \$.43 **27.** **a.** 7.875; 64.05; **b.** \$.32 and \$2.56; **c.** \$116.48 **31.** $V = IR$; $R = \frac{V}{I}$

Mixed Review **1.** -30.67 **2.** -2.236 **3.** 34 **4.** -33.8 **5.** 4.03 **6.** \$2,600

Test Yourself **1.** 2.68 **2.** -0.186875 **3.** 20.24372 **4.** 77.26952 **5.** 40 **6.** 3.03 **7.** $a = \frac{c}{b}$ **8.** $a = b - c$ **9.** $a = cb$ **10.** **a.** 168; **b.** 11; **c.** 9.7 h

3-8 page 133

Written Exercises **1.** 87 **3.** 23,471.4 **5.** 23 **7.** 8 airplanes; 16 spaceships **9.** 717

Chapter 3 Review pages 134–135

1. F, hundredths **2.** F, estimating **3.** T **4.** T **5.** F, variables **6.** < **7.** = **8.** > **9.** < **10.** < **11.** 0.25 **12.** 0.5 **13.** 0.25 **14.** 53 **15.** 0.75 **16.** \$24 **17.** 0.2 **18.** 12,905 **19.** \$79 **20.** -1.1 **21.** 0.4 **22.** 4.1 **23.** 11 **24.** $4.5a + 3.1$ **25.** $\frac{x}{2}$ **26.** $8a + 13ab$ **27.** -5 **28.** 0.55 **29.** 7.35 **30.** 11 **31.** 10 **32.** 7.3 **33.** 3 **34.** 3.8 **35.** 192.5 mi **36.** 36 ft **37.** 9.125 h **38.** 1,400 children **39.** 10.5 g/cm^3

Chapters 1–3 Cumulative Review page 137

1. B **2.** C **3.** A **4.** B **5.** C **6.** D **7.** A **8.** C **9.** A **10.** A **11.** A **12.** B **13.** C **14.** B

CHAPTER 4

Number Theory

4-1 pages 142–144

Written Exercises **1.** 8^3 **3.** $2r^4s^2$ **5.** x^2y^2z **7.** n^{30} **9.** a^a **11.** $(a + 1)^3$ **13.** **a.** 1; **b.** 1,000,000 **15.** **a.** -16; **b.** 1 **17.** 36 **19.** -4 **21.** 1 **23.** 7 **25.** 9 **27.** > **29.** = **31.** < **33.** > **35.** d **37.** (-1) raised to an even power is positive; 1 **39.** The power tells you the number of zeros in the product. $10^6 = 1{,}000{,}000$; $10^{10} = 10{,}000{,}000{,}000$ **41.** **a.** 0, 1, 0; 4, 4, 1; 8, 16, 16; 12, 64, 81; 16, 256, 256 **b.** $n = 2, 4$; $n = 3$; $n = 0, 1, n > 4$ **43.** $n = 0$ **45.** $x =$ any pos. value **47.** $x = 3$ **49.** **a.** It is half as tall.; yes; **b.** Each bar is twice as tall as the previous bar.

Mixed Review **1.** 19.5 **2.** -19 **3.** 123.75 mi **4.** 280 km **5.** 285.75 mi **6.** $4x - 2y$ **7.** $-3w - 7$ **8.** 83

Critical Thinking **2.** 8; 27; 64 **3.** 1; 8; 27; 64 **4.** 1,000

4-2 pages 146–148

Written Exercises **1.** -8 **3.** 8 **5.** x^9 **7.** $10x^9$ **9.** 72 **11.** $-x^8$ **13.** $16a^6$ **15.** $1{,}296y^{12}$ **17.** $-27y^{12}$ **19.** -8 **21.** 531,441 **23.** 9 **25.** 17 **27.** -216 **29.** F; add exponents; $x^5 \cdot x^3 = x^8$ **31.** T; 1 = 1 **33.** T; $(r^2)^3$ is positive, so the opposite is less than 0. **35.** no **37.** no **39.** no **41.** yes **43.** < **45.** < **47.** < **49.** 3^{50}; $2^{75} = (2^3)^{25}$ and $3^{50} = (3^2)^{25}$ **51.** $x = 2$ **53.** **a.** 1,024; **b.** 10^3; **c.** 2^3; **d.** $2^{10} \cdot 2^{10}$; **e.** $2^{20} \cdot 2^3$

Mixed Review **1.** 4 **2.** 360 **3.** $-5a^3b^2$ **4.** $-14c^2d^3$ **5.** -12 **6.** 6 **7.** -1 **8.** 8

Critical Thinking **1.** the two numbers diagonally above it in the preceding row **2.** 1, 6, 15, 20, 15, 6, 1 **3.** 8 **4.** 1,048,576

4-3 pages 150–152

Written Exercises **1.** Ex: 6.25×10^8; 62.5×10^7; 625×10^6 **3.** 6×10^{13} **5.** 6.382×10^7 **7.** 100,000 **9.** 7,654 **11.** 600.32 **13.** 4,060 **15.** 1.5×10^5; 150,000 **17.** 9.9×10^{12}; 9,900,000,000,000 **19.** 4.7×10^6; 4,700,000 **21.** 3 **23.** 8.45 **25.** 0.000845 **27.** 2.49×10^{21} **29.** 7.89×10^{20} **31.** 3.22×10^{30} **33.** 7×10^8 **37.** 5.79×10^7, 1.082×10^8, 1.496×10^8, 2.279×10^8, 7.783×10^8, 1.427×10^9, 2.869×10^9, 4.497×10^9, 5.9×10^9 **39.** 4.06×10^{13} km **43.** 7.6×10^8

Mixed Review **1.** multiplicative identity **2.** commutative prop. for addition **3.** additive identity **4.** distributive property **5.** $3m^7$ **6.** n^6 **7.** $25y^3$ **8.** 2^5 or 32

Test Yourself **1.** a^2b^3 **2.** $4x^3y$ **3.** 64 **4.** 144 **5.** -243 **6.** 1 **7.** -625 **8.** 512 **9.** 1×10^4 **10.** 1×10^5 **11.** 1×10^7 **12.** 7.5×10^4 **13.** 8.54×10^5 **14.** 1.645123×10^6

4-4 pages 154–155

Written Exercises **1.** 1.3125×10^{10} **3.** **a.** 2.8×10^8; **b.** about 1,393 da **5.** 7.3×10^8 **7.** 7.498×10^6

Practice page 158

1. $3^2 \cdot 5^3$ **3.** $(-3a)^5$ **5.** 512 **7.** -243 **9.** 64 **11.** 16 **13.** 40,353,607 **15.** 256 **17.** -147 **19.** 6,561 **21.** 8,388,608 **23.** b^3 **25.** k^{13} **27.** c^{11} **29.** y^6 **31.** $243r^{15}$ **33.** $405m^{10}$ **35.** 104 **37.** 200 **39.** -81 **41.** -16,384 **43.** 240,000,000 **45.** 10,000,000 **47.** 98,367.5 **49.** 3.392×10^6 km **51.** 5.88×10^{21} **53.** 1×10^{19} m

592

4-5 pages 161–162
Written Exercises 1. yes 3. yes 5. yes 7. no 9. yes 11. yes 13. 1, 2, 3, 5, 6, 10, 15, 30 15. 1, 5, 11, 55 17. 1, 29 19. 0, 12, 24, 36, 48 21. 0, 25, 50, 75, 100 23. 2, 3, 5, 9 25. 2, 3 27. 5 29. 3
31. a. 78, no, no, 96, yes, yes; b. If the last two digits are divisible by 4, the number is divisible by 4. 33. 7
35. odd 37. yes
Mixed Review 1. $n - 5 = -24$ 2. $3n - 10 = 57$ 3. -23
4. 36 5. -19 6. 3.48×10^6 7. 2.5×10^2 8. 95°F
Test Yourself 1. 9,604 2. 12,300
3. 28,560,000,000,000,000 4. 9.65×10^6 5. 5.48×10^2
6. 3×10^6 7. 1, 3, 9, 27 8. 1, 3, 5, 9, 15, 45
9. 1, 2, 3, 4, 5, 6, 10, 12, 15, 20, 30, 60 10. 3, 5, 9
11. 2, 3, 5 12. 3, 9 13. 4^4 14. 17^3 15. z^4 16. 20
17. 50 18. 29 19. -9

4-6 pages 165–166
Written Exercises 1. composite 3. prime
5. composite 7. $5^2 \cdot 17$ 9. $2 \cdot 3 \cdot 31$ 11. $3^2 \cdot 5^2 \cdot 7$
13. 1,056 15. 9,274,720 17. $3^3 \cdot 23$ 19. $11 \cdot 23$
21. even 23. 6, 14, 21, 42 25. 841; 960
27. a. China; b. Brazil
29. a. 3, 5; 5, 7; 11, 13; 17, 19; 29, 31; 41, 43; 59, 61; b. They are all odd.; c. yes
Mixed Review 1. 1, 2, 4, 8 2. 0, 8, 16, 24
3. 1, 2, 3, 4, 6, 9, 12, 18, 36 4. -9 5. 45 6. 4
7. -57 to -59°F
Critical Thinking 1. They are all located in columns 2, 4, and 6.
2. They are all located in columns 3 and 6.
3. They are in diagonal lines going from right to left, starting with 5, 30, 60, and 90. 4. 7
5. The multiples of 11 are multiples of 2, 3, 5, or 7.
6. They are all primes. All other numbers are crossed out because they are multiples of 2, 3, 5, and 7.
7. 2, 3, 5, 7, 11, 13, 17, 19, 23, 29, 31, 37, 41, 43, 47, 53, 59, 61, 67, 71, 73, 79, 83, 89, 97, 101, 103, 107, 109, 113, 127, 131, 137, 139, 149, 151, 157, 163, 167, 173, 179, 181, 191, 193, 197, 199; Stop at 13 on the sieve.

4-7 pages 169–170
Written Exercises 1. 7 3. 13 5. x^2y 7. $30a$ 9. 3
11. 60 13. 1,260 15. $24a^3b^2$ 17. 180 19. 4; 96
21. 5; 37,800 23. If a is the LCM of 8 and x, then a is divisible by both 8 and x. Since $8 = 2^3$, a is divisible by 2^3. 25. 2 times per minute 27. 2 tables that seated 5 people; 7 tables that seated 8 people 29. 21 ft. 31. 59
Mixed Review 1. $2^2 \cdot 3$ 2. $3 \cdot 41$ 3. 1, 2, 3, 4, 6, 12
4. 0, 12, 24, 36 5. 4 6. -11.4 7. -16 8. 8, 3

4-8 pages 173–174
Written Exercises 1. 9 3. 17 5. 24 7. 4; the last digits of the powers of 8 form this pattern: 8, 4, 2, 6
9. 63 tickets 11. They are all prime numbers. 13. $5

Problem Solving Practice page 175
1. every 21,000 mi 5. 16 (including a plain pizza)
9. a. 1950 b. 1700–1749

Chapter 4 Review pages 176–177
1. b 2. d 3. h 4. f 5. a 6. c 7. e 8. g 9. 8
10. 1 11. 27 12. 25 13. 16 14. a^5 15. $8a^6$
16. $ab^3 + ab^2$ 17. a^4b^2 18. a^3b^7 19. 4.65×10^8
20. 1.36×10^7 21. 1.28×10^3 22. 5.09×10^6
23. 210,000 24. 61,300,000 25. 1,050 26. 835
27. T 28. T 29. T 30. F 31. F 32. T
33. $3 \cdot 5^2$ 34. $2^2 \cdot 3 \cdot 5 \cdot 7$ 35. $2^2 \cdot 3^3$ 36. $3^2 \cdot 5 \cdot 17$
37. $2^2 \cdot 3 \cdot 19$ 38. $5 \cdot 7 \cdot 17$ 39. 4 40. 8 41. 9
42. $3x^2$ 43. 36 44. 56 45. 105 46. $60x^2y^3$
47. $198ab^3c^2$ 48. 45

Chapters 1–4 Cumulative Review page 179
1. C 2. B 3. B 4. A 5. C 6. D 7. C 8. B 9. C
10. A 11. C 12. D 13. B 14. C 15. C 16. A

CHAPTER 5

Rational Numbers and Expressions

5-1 pages 186–187
Written Exercises 1. $\frac{3}{13}$
3. $\frac{1}{2}$; one-fourth; five-sevenths; $\frac{5}{7}$ 5. $\frac{6}{10}$ or $\frac{3}{5}$ 17. 2
19. 6 21. 1 23. 4 25. $\frac{1}{5}$ 27. $\frac{2}{3}$ 29. $\frac{1}{3}$ 31. $\frac{c}{3}$ 33. $\frac{1}{2t}$
37. $\frac{3}{4}$ 39. $\frac{3a^2}{5}$ 41. $\frac{2pq}{3}$ 43. ≠ 45. = 47. $\frac{2}{5}$
Mixed Review 1. 1, 2, 4, 8, 16, 32 2. 1, 3, 9, 27
3. 2, 3; $2^2 \cdot 3^2$ 4. 2, 3; $2 \cdot 3^3$ 5. 200 6. 70
7. GCF 6; LCM 36 8. -54°F

5-2 pages 190–191
Written Exercises 1. a. $\frac{11}{4}$; b. $2\frac{3}{4}$ 5. $\frac{13}{8}$ 7. $\frac{47}{8}$ 9. $\frac{20}{3}$
11. $\frac{31}{11}$ 13. $5\frac{2}{3}$ 15. $4\frac{3}{5}$ 17. $1\frac{8}{11}$ 19. $3\frac{1}{3}$ 21. $\frac{4}{5}$ 23. $5\frac{3}{20}$
25. $2\frac{1}{2}$ 27. $6\frac{1}{20}$ 29. 0.28 31. 0.625
33. $0.\overline{5}$ 35. 5.375
37. 0.18, $\frac{18}{100}$; 5.73, five and seventy-three hundredths; 0.9, $\frac{9}{10}$ 39. a. $\frac{15}{104}$; b. $\frac{7}{52}$; c. $\frac{7}{104}$; d. $\frac{33}{52}$ 41. yes
43. yes
Mixed Review 1. > 2. < 3. $\frac{2}{3}$ 4. $\frac{3}{a}$ 5. 0.6 6. 1.5
7. 3.485×10^{10}

5-3 pages 194–195
Written Exercises 1. $\frac{9}{4}$ 3. $-1\frac{1}{5}$ 5. $\frac{3}{5}$ 11. $\frac{4}{9}$, $\frac{4}{9}$
13. $-1\frac{2}{3}$, $1\frac{2}{3}$ 15. a; c 17. $\frac{4}{5}$ 19. -4 21. $\frac{a}{2b}$ 23. $\frac{a}{4b}$
25. always 27. sometimes 29. negative 31. negative
33. yes 35. yes, yes, yes, yes; yes, no, no, no; yes, no, no, no; yes, yes, yes, yes; yes, yes, yes, no
Mixed Review 1. 0 2. -6 3. -5 4. $-\frac{3}{8}$ 5. $\frac{2a^2}{5}$ 6. 0.8
7. -3.25 8. $\frac{1}{2}$, $\frac{1}{2}$, $\frac{3}{8}$, $\frac{1}{4}$
Test Yourself 1. $\frac{4}{5}$ 2. $\frac{3}{4}$ 3. $\frac{3}{10}$ 4. $\frac{a^2}{3}$ 5. $\frac{35n^3}{4}$ 11. $-\frac{2}{3}$, $\frac{2}{3}$
12. $2\frac{5}{6}$, $2\frac{5}{6}$ 13. $-1\frac{7}{16}$, $1\frac{7}{16}$ 14. $2\frac{3}{4}$, $2\frac{3}{4}$

5-4 page 198
Written Exercises 1. < 3. < 5. < 7. < 9. <
11. < 13. = 15. < 17. < 19. < 21. <

23. $-\frac{5}{12}, -\frac{3}{8}, -\frac{1}{4}$ **25.** $-\frac{11}{15}, -\frac{7}{10}, -\frac{13}{20}, -\frac{7}{12}$
27. $-2\frac{7}{8}, -2\frac{9}{16}, 2\frac{3}{50}, 2\frac{19}{25}$ **33.** $x < 2.5$ **35.** $\frac{1}{14}$
Mixed Review **1.** 7 **2.** 10.58 **3.** $7\frac{1}{3}$ **4.** $1\frac{2}{5}$ **5.** $2\frac{7}{9}$ **6.** -14 **7.** -2.375 **8.** 192.5 mi

5-5 pages 201–202

Written Exercises **5.** $\frac{1}{3} + \frac{1}{2} = \frac{5}{6}$ **7.** $1\frac{7}{24}$ **9.** $-\frac{4}{9}$ **11.** $2\frac{1}{2}$ **13.** $9\frac{2}{9}$ **15.** $\frac{7}{18}y$ **17.** $-4\frac{5}{8}$ **19.** 16 **21.** 150 **23.** 75 **25.** $29\frac{2}{3}$ **27.** yes **29.** no **31.** C **33.** $>$ **35.** $>$ **37.** $<$ **39.** 42 ft **41.** $\frac{1}{4}, \frac{5}{4}$ **43.** $\frac{1}{12}, \frac{7}{12}, \frac{1}{4}, \frac{3}{4}$
Mixed Review **1.** $-3\frac{1}{5}, 3\frac{1}{5}$ **2.** $-5\frac{8}{11}$ **3.** $-\frac{7}{10}, \frac{7}{-10}$ **4.** -20 **5.** $<$ **6.** when $y > 4$ **7.** 60 **8.** $150

5-6 page 204

Written Exercises **1.** 7 **3.** $15\frac{3}{16}$ ft **5.** 12 y **7.** **a.** $0.\overline{1}$; **b.** $0.\overline{2}$; **c.** $0.\overline{3}$; **d.** $0.\overline{4}$; **e.** $0.\overline{5}$; **f.** $0.\overline{6}$ **9.** $3\frac{2}{3}, 4\frac{5}{12}, 5\frac{1}{6}$

Practice page 205

1. $\frac{1}{5}$ **3.** $\frac{2}{3}$ **5.** $\frac{3}{5}$ **7.** $\frac{3y}{7}$ **9.** $\frac{1}{3}$ **11.** $\frac{2}{3m}$ **13.** $\frac{13}{5}$ **15.** $\frac{14}{3}$ **17.** $\frac{33}{4}$ **19.** $\frac{247}{12}$ **21.** $2\frac{2}{5}$ **23.** $6\frac{1}{4}$ **25.** $6\frac{5}{8}$ **27.** $8\frac{1}{7}$ **29.** $-2\frac{3}{8}, 2\frac{3}{8}$ **31.** $18\frac{2}{3}, 18\frac{2}{3}$ **33.** $13\frac{5}{9}, 13\frac{5}{9}$ **35.** $\frac{11}{15}, \frac{11}{15}$ **37.** $<$ **39.** $>$ **41.** $>$ **43.** 1 **45.** $1\frac{1}{9}$ **47.** $-\frac{3}{8}$ **49.** $-6\frac{1}{4}$ **51.** $1\frac{11}{24}$ **53.** $6\frac{23}{30}$ **55.** $6\frac{3}{7}$ **57.** -11 **59.** $30\frac{5}{16}$

5-7 page 208

Written Exercises **1.** $\frac{1}{3}$ **3.** 8 **5.** $\frac{2}{11}$ **7.** $\frac{15}{16}$ **9.** $-\frac{a}{10}$ **11.** $5\frac{5}{6}$ **13.** 1 **15.** $\frac{3}{4}$ **17.** 39 **19.** -8 **21.** -60 **23.** $1\frac{1}{2}$ **25.** 130 **27.** $>$ **29.** $<$ **31.** $=$ **33.** $-1\frac{8}{15}$ **35.** $-\frac{9}{10}$ **37.** **a.** 200; **b.** 500; **c.** 12
Mixed Review **1.** 1.4 **2.** 9.9 **3.** $\frac{1}{2}$ **4.** $\neq$ **5.** 0.325 **6.** 6 **7.** 5 **8.** Paul

5-8 pages 213–214

Written Exercises **1.** $\frac{1}{36}$ **3.** 1 **5.** $\frac{3}{8}$ **7.** $\frac{1}{25}$ **9.** a^{-3} **11.** $\frac{1}{2}x^7$ **13.** x^2y^{-10} **15.** $5b^4c^5$ **17.** $5m^2$ **19.** $\frac{3y^2}{x^3}$ **21.** **a.** -25 **b.** 25 **c.** $\frac{1}{25}$ **d.** $\frac{1}{25}$ **23.** $\frac{1}{25a^2}$ **25.** $a^{10}b^{-15}$ or $\frac{a^{10}}{b^{15}}$ **27.** $9a^4b$ **29.** **a.** The numbers decrease by 1.; **b.** The numbers decrease by a power of 3. 1, $\frac{1}{3}, \frac{1}{9}$; **c.** They are values of the powers of 3.; **d.** They are the values of the powers of 2. **31.** T **33.** F **35.** T **37.** T **39.** 1,580,000,000,000,000,000 **41.** $\frac{13}{20}$
Mixed Review **1.** $\frac{7}{10}$ **2.** $\frac{3}{8}$ **3.** $-1\frac{1}{8}$ **4.** $6\frac{1}{6}$ **5.** 5 **6.** -12 **7.** -5.2 **8.** 34
Critical Thinking **1.** If n is the exponent, the decimal point moves to the left $|n|$ places. **2.** 1.2, 0.12, 0.012, $1.2 \times 10^{-3} = 0.0012$ **3.** 0.00037 **4.** 23 **5.** Writing long strings of zeros takes too much space and can be confusing to read.

5-9 pages 216–217

Written Exercises **1.** $x + \frac{3}{5} = \frac{7}{10}; \frac{1}{10}$ **3.** $-\frac{1}{2}$ **5.** $5\frac{3}{8}$ **7.** $\frac{13}{24}$ **9.** $-1\frac{7}{40}$ **11.** $\frac{23}{24}$ **13.** $2\frac{19}{20}$ **15.** 6.1 **17.** 5 **19.** 6 **21.** $\frac{1}{5}$ **23.** $-8\frac{1}{8}$ **25.** $-1\frac{5}{12}$ **27.** x must be less than zero because the sum will be less than zero. **29.** yes **31.** $b + 3\frac{3}{16} = 5\frac{11}{16}, 2\frac{1}{2}$ lb **33.** $h + 1\frac{5}{8} = 68\frac{1}{2}, 66\frac{7}{8}$ in.
Mixed Review **1.** T **2.** $3\frac{8}{13}$ **3.** $<$ **4.** $>$ **5.** $\frac{31}{150}$ **6.** a^{-4} **7.** $x^{-2}y^5$ **8.** $3\frac{1}{8}$

5-10 pages 219–221

Written Exercises **1.** $1\frac{5}{16}$ **3.** $\frac{81}{10}$ or $8\frac{1}{10}$ **5.** -12 **7.** $\frac{1}{2}$ **9.** 4 **11.** -0.8 **13.** neg · neg = pos **15.** pos · neg = neg **17.** $\frac{25}{9}, -\frac{25}{9}$ **19.** no solution **21.** no solution **23.** $\frac{5}{8}d = 12; d = 19.2$ **25.** $a = \frac{5}{6}, b = 1\frac{1}{3}, a < b$ **27.** 350 **29.** $\frac{1}{15}$ **33.** whole **35.** multiplication **37.** $\frac{7}{8}$ **39.** 64 **41.** $2\frac{5}{7}$ **43.** $\frac{7}{11}$ **45.** $\frac{7}{20}$ **47.** No; the chart does not tell the number of cans recycled in any year. **49.** $\frac{9}{10}$
Mixed Review **1.** $-9\frac{1}{3}$ **2.** $1\frac{1}{8}$ **3.** $11\frac{1}{9}$ **4.** 6.1 **5.** $\frac{29}{36}$ **6.** 2.2 **7.** $<$ **8.** 126
Test Yourself **1.** $\frac{23}{24}$ **2.** $10\frac{1}{4}$ **3.** $-\frac{7}{8}$ **4.** $2\frac{7}{16}$ **5.** $-1\frac{5}{9}$ **6.** $-\frac{27}{32}$ **7.** $16\frac{7}{8}$ **8.** no solution **9.** no solution **10.** $\frac{1}{9}$ **11.** 1 **12.** 27 **13.** $\frac{8}{5}$ **14.** $5n$ **15.** $\frac{3}{a^3}$ **16.** $\frac{8x^2}{y^3}$ **17.** $\frac{b^6}{a^5}$

5-11 page 224

Written Exercises **1.** 62.625 **3.** 77.125 **5.** 88.875 **7.** $3.625 **9.** **a.** DQ $2,805.00, MCJ $5,036.25, EDL $5,992.50, JMB $23,778.75, BBH $27,030.00; **b.** DQ $510, MCJ $630, EDL $450, JMB $340, BBH $1,380
11. The commission is figured on different amounts.

Problem Solving Practice page 225

1. 27th **3.** 1,223 or 2,486 **5.** Lena **7.** $5.26 **11.** $\frac{3}{4}$ lb

Chapter 5 Review pages 226–227

1. equivalent fractions **2.** lowest terms **3.** rational number **4.** least common denominator **5.** exponents **6.** 6 **7.** 6 **8.** 8 **9.** 4 **10.** 25 **11.** $3\frac{3}{4}$; 3.75 **12.** $1\frac{1}{2}$; 1.5 **13.** $2\frac{2}{5}$; 2.4 **14.** $2\frac{5}{6}$; $2.8\overline{3}$ **15.** $2\frac{5}{8}$; 2.625 **16.** $\frac{3}{5}$ **17.** $2\frac{3}{8}$ **18.** $5\frac{1}{4}$ **19.** $\frac{7}{10}$ **20.** $\frac{7}{20}$ **21.** $<$ **22.** $>$ **23.** $=$ **24.** $>$ **25.** $3\frac{1}{12}$ **26.** $7\frac{2}{15}$ **27.** $15\frac{5}{12}$ **28.** $6\frac{11}{24}$ **29.** 6 A.M. **30.** $19,937.50 **31.** $\frac{9}{10}$ **32.** 9 **33.** 6 **34.** $\frac{1}{2}$ **35.** $\frac{1}{x^5}$ **36.** $\frac{6}{a}$ **37.** $2m^4$ **38.** $\frac{2}{b}$ **39.** $\frac{3y^3}{x^2}$ **40.** $1\frac{2}{15}$ **41.** $3\frac{1}{3}$ **42.** $-3\frac{1}{8}$ **43.** $1\frac{1}{3}$

Chapters 1–5 Cumulative Review page 229

1. C **2.** C **3.** B **4.** B **5.** D **6.** C **7.** B **8.** D **9.** C **10.** A **11.** D **12.** A **13.** C **14.** C **15.** D **16.** B

CHAPTER 6

Ratios, Proportions, and Percent

6-1 pages 235–236

Written Exercises **1.** $\frac{3}{8}$ **3.** $\frac{8}{11}$ **5.** $\frac{3}{5}$ **7.** $\frac{5}{2}$ **9.** $\frac{1}{5}$ **11.** $\frac{1}{6}$ **13.** $\frac{1}{4}$ **15.** $\frac{1}{7}$ **17.** 3 to 2; 2 to 3; 2 to 5 **25.** ≠ **27.** =; prop. **29.** ≠ **31.** ≠ **33.** ≠ **35.** =; prop. **37.** 6 gal/min **39.** 0.18 hits/time at bat **41.** 4.375 mi/h **43.** 0.3 hits/time at bat **45.** 50 to 2; 2 to 50; 2 to 48 **47.** $\frac{4}{24} \stackrel{?}{=} \frac{6}{30}$; no **49.** $\frac{2}{1.69} \stackrel{?}{=} \frac{5}{3.98}$; no, 2 for \$1.69 is \$.85 each; 5 for \$3.98 is \$.80 each. **51.** **a.** 300 to 70; 300 : 70; $\frac{300}{70}$; **b.** \$9.00/pt **c.** \$22/gal

Mixed Review **1.** $\frac{7}{13}$ **2.** $\frac{13}{19} > \frac{19}{28}$ **3.** $x = -14$ **4.** $x = -2\frac{1}{4}$ **5.** yes **6.** no **7.** 1,239 mi

6-2 pages 238–240

Written Exercises **1.** $a = 20$ **3.** $c = 20$ **5.** $e = 19.2$ **7.** $g = 133.\overline{3}$ **9.** $j = 17.5$ **11.** $m = 16.9$ **13.** $x = 1$ **15.** $\frac{4}{1.85} = \frac{24}{t}$; \$11.10 **17.** $\frac{5}{18.6} = \frac{8}{v}$; $v = 29.76$ min **19.** $\frac{6}{2.25} = \frac{y}{\$10}$; $y = 26.67$ lb **21.** $\frac{3}{1} = \frac{x}{4}$; $x = 12$ bags **23.** $\frac{3}{\$9.60} = \frac{15}{p}$; $p = \$48$ **25.** $\frac{3}{750} = \frac{x}{10{,}000}$; $x = 40$ defects **29.** $\frac{7}{3}$ **31.** $-\frac{2}{7}$ **33.** no **35.** no **37.** $\frac{30}{100}$ **39.** $\frac{12}{1}$ **41.** 15; 1 **43.** 32; 8; 32; 16 **45.** 15 s **47.** 360 times **49.** 42,048,000 times

Mixed Review **1.** F **2.** T **3.** $\frac{6}{11}$ **4.** $\frac{25}{14}$ **5.** yes **6.** yes **7.** \$67.97

Critical Thinking **1.** even numbers **2.** prime numbers **3.** odd numbers **4.** the even primes **5.** the odd primes **6.** yes; yes **7.** No; Two is the only even prime.

6-3 pages 242–243

Written Exercises **1.** 1 in. : 10 ft **3.** 20 ft **5.** Yes; the scale of the dance floor is 10 : $8.\overline{3}$. **7.** N **9.** 4.5 in. **11.** 14.4 ft

6-4 pages 246–248

Written Exercises **1.** 75% **3.** 62.5% **5.** 58.3% **7.** 140% **9.** 33% **11.** 6% **13.** 4.5% **15.** 188% **17.** 79% **19.** 30% **21.** 68% **23.** 111% **25.** 22.2% **27.** 43.8% **29.** 80% **31.** 25% **33.** $\frac{1}{10}$; 10% **35.** 0.75; 75% **37.** $\frac{1}{4}$; 0.25 **39.** 112% **41.** 60% **43.** 47.2% **45.** 25% **47.** < **49.** < **51.** < **53.** > **55.** yes **57.** No; 100% is a perfect grade. **59.** 100 **61.** 10 **63.** week **65.** $\frac{5}{12}$ **67.** $\frac{3}{4}$

Mixed Review **1.** $x = 47$ **2.** $x = -0.175$ **3.** $x = 68.75$ **4.** $x = 1{,}820$ **5.** $\frac{1}{4} = 25\%$ **6.** $\frac{11}{14} = 78.6\%$ **7.** 93

Practice page 249

1. $\frac{2}{5}$ **3.** $\frac{2}{5}$ **5.** $\frac{3}{7}$ **7.** ≠ **9.** =; prop. **11.** =; prop. **13.** =; prop. **15.** $a = 2$ **17.** $c = 12$ **19.** $e = 2.25$ **21.** $g = 2$ **23.** 180% **25.** 37.5% **27.** 76.7% **29.** 55% **31.** 6.7% **33.** 85% **35.** 62.5% **37.** 12.5% **39.** 0.225 **41.** 0.736 **43.** $\frac{1}{3}$; $0.\overline{3}$; 33.3% **45.** $\frac{1}{3}$, $0.\overline{3}$, 33.3% **47.** $\frac{2}{3}$, $0.\overline{6}$, 66.7% **49.** $\frac{16}{27}$, $0.\overline{592}$, 59.3% **51.** > **53.** = **55.** > **57.** = **59.** >

6-5 pages 253–255

Written Exercises **1.** 52% **3.** 107.8 **5.** 33.3% **7.** 9 **9.** 31.5 **11.** 35% **13.** 20 **15.** 175% **17.** = **19.** > **21.** < **23.** = **25.** > **27.** 20% **29.** 25% **31.** 24 **33.** 12 **35.** 18.5% **37.** 44.5 **39.** 84.2 **41.** 132 **43.** 42 **45.** 8% **47.** \$1,200 **49.** You can't tell who got the better deal without knowing the original price. **51.** It makes sense. **53.** 15% tip: \$30.00; 20% tip: \$22.50

Mixed Review **1.** 52 **2.** 39 **3.** $258\frac{1}{3}\%$ **4.** 452.3% **5.** 8% **6.** 4,560% **7.** 38%

Test Yourself **1.** $\frac{1}{3}$ **2.** $\frac{1}{20}$ **3.** \$5.78 **4.** \$1.75 **5.** 75% **6.** 89% **7.** 87.5% **8.** 3% **9.** 0.7% **10.** 5.4 **11.** $n = 100$

6-6 pages 257–259

Written Exercises **1.** $15\% = \frac{x}{115}$; $0.15 \cdot 115 = x$; $17.25 = x$ **3.** 55% **5.** $46\frac{2}{3}$ **7.** 60 **9.** 225% **11.** 21.6 **13.** 2.4 **15.** 64% **17.** 22.5 **19.** 81 **21.** 50% **23.** 100 **25.** 22 **27.** 20% **29.** 100 **31.** 12.5 **33.** $33\frac{1}{3}\%$ **35.** 46.2% **37.** 13.3 **39.** 100 **41.** \$39.38 **43.** no **45.** \$25; \$18.75 **47.** \$240,000,000 **49.** 1,442,100

Mixed Review **1.** 52 **2.** 150 **3.** 12.5% **4.** 21 **5.** 0.62 **6.** 0.899 **7.** \$15.00

Critical Thinking **1.** a and c

6-7 pages 261–263

Written Exercises **1.** 32% **3.** 137.5% **5.** 16.7% **7.** 20.8% **9.** 150% **11.** 166.7% **13.** 20% increase **15.** 10% decrease **17.** 86 **19.** 58.3 **21.** 24.1%; I **23.** 83.1%; D **25.** 69.3%; D **27.** 27.7% **29.** 29.6% **31.** +13.6% **33.** +393.8% **35.** 1936: 1,175%; 1937: 2,087.5%; 1940: 2,300%; 1955: 445.5%; 1961: 566.7% **37.** **a.** +44%, +27%, +90%, −28%, −28% **b.** no

Mixed Review **1.** 8 **2.** 3,600 **3.** < **4.** > **5.** 20.25 **6.** 98.5%

Test Yourself **1.** 39.6 **2.** 52.5 **3.** 20% **4.** 65 **5.** $66\frac{2}{3}\%$ **6.** 150 **7.** 32% **8.** 38.9%

6-8 pages 265–266

Written Exercises **1.** $\frac{1}{4}$ **3.** 34 in. **5.** 54 post holes **7.** 5 cm **9.** 1 **11.** $1\frac{7}{8}$ mi **13.** 47 students **15.** **a.** 28; **b.** 46 ft × 46 ft

Problem Solving Practice page 267

1. 4 **3.** 1 **5.** \$700 **7.** **a.** 2.3% **b.** no

Chapter 6 Review pages 268–269

1. false; ratio **2.** true **3.** true **4.** false; percent **5.** true **6.** =; prop. **7.** ≠ **8.** =; prop. **9.** ≠ **10.** ≠ **11.** 50 mi/h **12.** 23 mi/gal **13.** 90 words/min **14.** \$1.89/lb **15.** $n = 35$ **16.** $x = 12$ **17.** $a = 49$ **18.** $y = 3$ **19.** $m = 126$ **20.** 187.5 km **21.** 0.5 cm **22.** 5% **23.** 98% **24.** 145% **25.** 75% **26.** 62.5%

27. 12% **28.** 6 **29.** $x = 150$ **30.** $33\frac{1}{3}\%$ **31.** $a = 200$ **32.** 204 **33.** 5% **34.** 18.2% **35.** 2.7 **36.** $y = 80$ **37.** $33\frac{1}{3}\%$ **38.** 25% **39.** 75% **40.** 20% **41.** 75% **42.** 50% **43.** 21.4%

Chapters 1–6 Cumulative Review page 271
1. C **2.** A **3.** C **4.** B **5.** D **6.** A **7.** D **8.** B **9.** B **10.** A **11.** C **12.** C

CHAPTER 7

Equations and Inequalities

7-1 pages 277–278
Written Exercises **1.** 6 **3.** 47 **5.** -324 **7.** 12 **9.** 54 **11.** -70 **13.** -3 **15.** -2 **17.** -7 **19.** 85 **21.** 10 **23.** 60 **25.** 30 **27.** 5 **29.** $167{,}645.1\overline{6}$ **31.** -43.375 **35.** C; 4 **39.** $3n - 7 = 19$ **41.** $2 = 12n - 4; \frac{1}{2}$ **43.** $8d + 36 = 78$; 5.25 **45.** $30 = 3n - 9$; 13 **47.** $x = \frac{c - b}{a}$ **49.** $8.50
Mixed Review **1.** 20 **2.** 4 **3.** 18 **4.** $6.30 **5.** 96 **6.** $3m + 5$ **7.** $15 - 5x$ **8.** 96°

7-2 pages 281–282
Written Exercises **1.** $2\frac{3}{4}$ **3.** $-2\frac{1}{2}$ **5.** 31 **7.** 12 **9.** 5 **11.** $-3\frac{5}{18}$ **13.** -12 **15.** $4\frac{2}{3}$ **17.** $\frac{5}{7}$ **19.** $3\frac{1}{3}$ **21.** 3 **23.** 6 **25.** 74.2 **27.** 40.4 **29.** $1.\overline{3}$ **33.** a; $70 **35.** $p + p + 13 = 171$; 79 **37.** $p - 0.2p = 53$; $66.25 **39.** $3(n - 8) = 36$; 20 **41.** Step 2 should be $3x - 3 - 5 = 14$; $x = 7\frac{1}{3}$.
Mixed Review **1.** 24 **2.** 30 **3.** $-17\frac{1}{3}$ **4.** 64 **5.** $\frac{1}{4}n - 10$ **6.** $12n - 5n$ **7.** $27.15
Test Yourself **1.** -2 **2.** 162 **3.** 6 **4.** -3 **5.** -1 **6.** $185 - 11x + 40 = 93$; $12

Problem Solving Practice page 283
1. 28 **3.** 30 **5.** 120 cm **7.** 4 **9.** 1,295 m and 1,575 m **11.** 74 and 75 **13.** 15 min

7-3 pages 285–286
Written Exercises **1.** $8n - \frac{1}{2}n = 16$ **3.** $n + (n - 5) = 114.90$, where n = price of boots; $59.95 **5.** $c + c + 0.45 = 0.95$; $.25 **7.** $8.75 + 1.25t = 12.50$; 3 **9.** $x - 0.75x = 175$; 700 **11.** $15x = 240$; 16 **13.** $\frac{2}{3}s - \frac{2}{5} = \frac{11}{45}; \frac{29}{30}$ **15.** 8 **17.** increased danger **19.** 18 bu
Critical Thinking **1.** 10:00 **2.** 11:00 **3.** 7 h **4.** **a.** clocktime − (5)(groups of 5 in clocktime); **b.** 1:00; **c.** 2:00

7-4 pages 289–290
Written Exercises **1.** $4m + 5 = 21$; 4 **3.** 3 **5.** 1 **7.** 4 **9.** -4 **11.** 6 **13.** -10 **15.** 4 **17.** $-4\frac{2}{3}$ **19.** 4 **21.** $2\frac{2}{3}$ **23.** 3 **25.** yes **27.** yes **29.** $2.\overline{69}$ **31.** $x + x + 1 + x + 2 = 165$; 54, 55, and 56 **33.** $\frac{1}{2}n + 1 = \frac{2}{3}n - 1$; 12 **35.** $2n - 8 = 3n - 16$; 8 **37.** $2(35) - x = 47$; $x = 23$
Mixed Review **1.** $2(n + 4)$ **2.** $c + 14$ **3.** $3n - 6 = 12$; 6 **4.** $56 = 6x + 8$; 8 **5.** = **6.** = **7.** $93

7-5 pages 294–295
Written Exercises **1.** true **3.** true **5.** false **7.** true **9.** false **17.** $x \le 2$ **31.** $3 < 10$ **33.** $p > 0$ **35.** $p \le 30$ **37.** $3x < 15{,}000$ **39.** $x < -10$; $x \ge -5$ **41.** $<$ **43.** $<$ **45.** = **47.** $>$ **49.** $>$ **51.** $m > 5$ **53.** $b \ge 15$ **55.** $s \le 50$ **57.** **a.** $5,140; $6,500; $4,400; $2,925; $7,700; $2,550; **b.** Yes, if the program was input correctly.; **c.** They determine how the IF command decides what tax rate to use.
Mixed Review **1.** $x = 17$ **2.** $x = -5.\overline{6}$ or $-5\frac{2}{3}$ **3.** $2n - 4$ **4.** $14 - x$ **5.** 0 **6.** $3n + 6$ **7.** 2 cups

7-6 pages 298–299
Written Exercises **1.** same **3.** reversed **5.** Divide by -3. **7.** Mult. by 3. **9.** yes **11.** no **13.** $8 \le x$ **15.** $x > 0$ **17.** $x > 4$ **19.** $x < -8$ **21.** $x > -12$ **23.** $21 \le g$ **25.** hydroelectric and oil **27.** 28
Mixed Review **1.** 5 **2.** 1 **5.** $1\frac{4}{5}$ **6.** $-\frac{8}{9}$ **7.** 76 and 88
Test Yourself **1.** 5 **2.** 7 **3.** -4 **4.** 8 **5.** $y > -4$ **6.** $y < 4$ **7.** $s < 42$ **8.** $h > \frac{1}{21}$ **9.** $-8 \ge k$ **10.** $y > -12$ **11.** $n - 7 > -2$; $n > 5$ **12.** $\frac{n}{-4} \ge 30$; $n \le -120$ **13.** $-8 \ge \frac{n}{-3}$; $n \ge 24$

7-7 pages 301–303
Written Exercises **1.** Sub. 8. **3.** Sub. 7. **5.** $x > 3$ **7.** $x \ge 6$ **9.** $x > 5$ **11.** $x \le 1\frac{2}{5}$ **13.** $x > 4\frac{4}{5}$ **15.** $x \le -5$ **17.** $x > 3$ **19.** $x \ge -72$ **21.** $x > -7$ **23.** $x < 3$ **25.** $x > 2$ **27.** $x < 60$ **29.** $x < 4$ **31.** b; $n \ge -18$ **33.** $2n - 5 \ge 13$; $n \ge 9$ **35.** $\frac{x + 88 + 91 + 85}{4} \ge 90$; $x \ge 96$ **37.** 3 **41.** **a.** 24,000 acres; **b.** 39.1
Mixed Review **1.** $2n = n + 5$; 5 **2.** $2n - 5 = 121$; 63 **3.** $x \le -3$ **4.** $x \le 32$ **5.** $x = 18$ **6.** $x = 5$ **7.** $7\frac{1}{2}$ m, $12\frac{1}{2}$ m
Critical Thinking **2.** D **3.** A **4.** D **5.** C **6.** B

Practice page 304
1. 6 **3.** 4 **5.** 243 **7.** 19 **9.** 3 **11.** 12 **13.** $6\frac{6}{7}$ **15.** 13 **17.** 10 **19.** -3 **21.** 26 **23.** 6 **25.** 8 **33.** $x > -4$ **35.** $x > -18$ **37.** $x \le 3$ **39.** $x \ge 4$ **41.** $x > 2$ **43.** $x \le 14$ **45.** $x \ge 15$ **47.** $22x + 47 = 201$; 7 **49.** $x + x + 1 + x + 2 + x + 3 = -490$; -121, -122, -123, and -124

7-8 page 307
1. $520.50 **3.** $999.24 **5.** 9 **7.** 11.75 **9.** 47.08 **11.** **a.** $630; **b.** $266.39

Chapter 7 Review pages 308–309
1. operations **2.** distributive **3.** combine **4.** not equal **5.** negative **6.** equation **7.** $x = 1$ **8.** $x = 2$ **9.** $a = 5$ **10.** $x = 12$ **11.** $n = 3$ **12.** $b = 18$ **13.** $x = 7$ **14.** $x = -10$ **15.** $x = \frac{1}{3}$ **16.** $x = -1$ **17.** $x = 14$ **18.** $x = -\frac{2}{3}$ **19.** $3n - 2(n + 5) = 3$; $n = 13$ **20.** $2(x + 12) + 3x = 144$; $x = 24$ **21.** $x = -2$ **22.** $x = 4$ **23.** $n = -10$ **24.** c **25.** a **26.** b **27.** d **28.** $x < -5$

29. $x \geq -3$ **30.** $x < -4$ **31.** $x \leq -8$ **32.** $x \leq 5$
33. $y > 4$ **34.** $b < -9$ **35.** $a < 12$

Chapters 1–7 Cumulative Review page 311
1. B **2.** C **3.** A **4.** C **5.** A **6.** C **7.** C **8.** B
9. D **10.** C **11.** C **12.** A **13.** B **14.** D

CHAPTER 8

Graphing in the Coordinate Plane

8-1 pages 318–319
Written Exercises **1.** Q **3.** M **5.** (2,-3) **7.** (-5,0)
17. (3,0)(0,3)(-3,0)(0,-3) **19.** IV **21.** II **23.** I
25. y-axis **27.** III **29.** parallelogram **31.** triangle
33. (0,-5)
Mixed Review **1.** 2 **2.** 8 **3.** $x < -7$ **4.** $x \leq 9$ **5.** 3
6. -35 **7.** 12 in., 19 in.
Critical Thinking **3.** **a.** (2,1)(2,3)(-1,3)(-1,1); **b.** (-2,-1)(-2,-3)(1,-3)(1,-1); **c.** (2,-1)(2,-3)(-1,-3)(-1,-1); **d.** (-4,2)(-4,6)(2,6)(2,2) **4.** When the x-coordinate is multiplied by -1, the figure slides to the right 1 unit. When the y-coordinate is multiplied by -1, the figure is reflected over the x-axis. When each coordinate is multiplied by -1, the figure slides to the right 1 unit and is reflected over the x-axis. When each coordinate is multiplied by 2, the figure moves up 1 unit and each side is twice the length of the original figure.

8-2 pages 322–323
Written Exercises **1.** no **3.** yes **5.** yes **7.** yes **9.** no
11. -1 **13.** 2 **15.** 11.5 **17.** $y = 3x + 5$ **19.** $y = \frac{3}{2}x - 5$
21. $y = -\frac{1}{6}x$ **23.** $y = -\frac{1}{2}x - \frac{5}{2}$ **25.** $y = \frac{2}{3}x - 4$
27. $y = -\frac{1}{4}x + 4$ **29.** **b.** 1,092.6 kg/cm^2 **c.** 2,838.4 kg/cm^2
Mixed Review **1.** -16 **2.** -8 **4.** neg., neg. **6.** (0,8)
7. 7:17 A.M.
Test Yourself **1.** F **2.** G **3.** H **4.** E **5.** (-1,3)
6. (3,3) **7.** (4,-2) **8.** (-4,-4) **9.** $y = -\frac{3}{2}x + 2$
10. $y = -\frac{1}{2}x - 3$ **11.** $y = -\frac{1}{3}x + \frac{7}{3}$

8-3 pages 326–327
Written Exercises **1.** $\frac{1}{2}x + y = -3$ **3.** -6 and -3 **5.** $-1\frac{1}{3}$; 4
7. 0; 0 **9.** $y = -x + 2$ **11.** $y = -5$ **21.** $x = -1$ **23.** $y = -6$ **25.** $x - y = 3$; 3 and 0 **27.** $x + y = 6$; 2 oranges and 4 apples **29.** $2x = y$; 4 and 8 **31.** $2x + 2y = 12$; width is 1 and length is 5.
Mixed Review **1.** 4 **2.** 4 **3.** $y = -\frac{2}{5}x + \frac{11}{5}$
4. $y = -\frac{3}{2}x - 6$ **7.** 34 and 35

8-4 pages 332–333
Written Exercises **1.** $\frac{2}{3}$; -2 **3.** $\frac{4}{3}$; -4 **5.** $\frac{1}{3}$; 4 **7.** 0
9. $\frac{10}{3}$ **11.** -2 **19.** $y = 2x + 1$; 2; 1
21. $y = -2x - 3$; -2; -3 **23.** $y = \frac{3}{4}x + \frac{1}{4}$; $\frac{3}{4}$; $\frac{1}{4}$
31. **b.** The lines are parallel. **c.** They are the same. **d.** When two lines have the same slope and different y-intercepts, they are parallel.
33. $0x + 1y = 6$ **35.** $1x + 0y = -3$

Mixed Review **1.** 80 **2.** 252 **3.** yes **4.** yes **5.** no
6. yes **7.** yes **8.** no **9.** 20%

8-5 pages 335–336
Written Exercises **1.** **b.** 50°F; **c.** 20°C **3.** 64°F **5.** 3
7. \$2.11 **9.** 35 ft **11.** **a.** \$27; **b.** \$150
13. 18 quarters and 14 dimes **15.** 20 mi

Problem Solving Practice page 337
1. 12 quarters and 5 dimes **3.** 48 **5.** 36, 37, 38
7. 65 in. or 5 ft 5 in. **9.** 4 ft × 4 ft and 16 ft × 16 ft
11. \$27.26 **13.** \$1.18

8-6 pages 342–343
Written Exercises **1.** yes **3.** no **5.** (1,5) **7.** (2,1)
9. (2,2) **11.** (3,1) **13.** no solution **15.** (3,4)
17. $x + y = 55$; $x - y = 15$; 20 and 35
19. $x + y = 144$; $x = 3y$; 108 m and 36 m
21. $x + y = 16$; $5x + 10y = 100$; 12 five-point questions and 4 ten-point questions **23.** parallel **25.** 1850
Mixed Review **1.** $3x^2$ **2.** $\frac{5xy^3}{2}$ **3.** 1; 0 **4.** $-\frac{2}{3}$; 4
5. $x < 3$ **6.** $x \geq -3$ **7.** \$75
Test Yourself **1.** 10; 4 **2.** $-\frac{2}{3}$; 2 **3.** -2; 10 **4.** no slope
5. $-\frac{1}{2}$ **6.** $\frac{2}{11}$ **7.** $y = -2x + 7$; -2; 7 **8.** $y = -\frac{2}{3}$; 0; $-\frac{2}{3}$
9. $y = -\frac{1}{3}x + \frac{2}{3}$; $-\frac{1}{3}$; $\frac{2}{3}$ **10.** (-2,8) **11.** infinite
12. (3,-2)

8-7 pages 346–347
Written Exercises **1.** no **3.** no **5.** no **7.** no
9. $y > \frac{5}{2}x - 5$ **11.** $y < -\frac{1}{3}$ **13.** $y \leq |x| - 4$
15. $y \leq -x + 5$ **19.** **a.** infinite; **b.** no solution
Mixed Review **1.** 0.075; 7.5% **2.** $0.\overline{2}$; $22.\overline{2}$% **3.** (1,1)
4. (6,2) **6.** below **7.** 12%
Critical Thinking **1.** A and D **2.** C **3.** A **4.** A and C

8-8 pages 350–351
Written Exercises **1.** $2x + y = 3$; solid **3.** $y = -2$; solid
5. $x - 4y = 1$; dotted **7.** $5x - 3y = 2$; dotted
9. $x = 9$; solid **11.** $3x + y = 2$; dotted **13.** no
15. yes **17.** no **19.** yes **21.** no **37.** $x - y > 3$
39. $0.05x + 0.10y < 1$ **47.** $x + y \leq 10$; $y > 2x$
Mixed Review **1.** infinite **2.** no solution **3.** one
6. **a.** 8; **b.** 5; **c.** (shirts,sweaters); (1,4)(1,3)(1,2)(1,1)(2,3)(2,2)(2,1)(3,3)(3,2)(3,1)(4,2)(4,1)(5,1)(6,1)

Practice page 352
1. I **3.** IV **5.** (-4,-3) **7.** $y = -3x - 10$ **9.** $y = -2x + 4$
11. 0; 0 **19.** -1 **21.** $\frac{6}{5}$ **23.** $x = 6$ **25.** $y = \frac{1}{4}x - 3$; $\frac{1}{4}$; -3
27. (2,4) **29.** (1,2) **31.** $y > -\frac{4}{5}x - \frac{6}{5}$

8-9 pages 353–355
Written Exercises **1.** direct; 30 **3.** direct; 8.5 **5.** 27.5 lb
7. 1,156 cycles/s **9.** 3 h

Chapter 8 Review pages 356–357
1. false; an ordered pair **2.** true **3.** false; x-axis
4. false; m; b **5.** false; is not part of **6.** (1,-3)
7. (-2,1) **8.** (-3,-3) **9.** (2,2) **10.** (1,2) **11.** (2,-2)

12. (3,3) **13.** yes **14.** no **15.** yes **16.** yes **17.** 2 **18.** $\frac{3}{2}$ **19.** -1 **20.** 0 **21.** $y = 2x + 3$; 2; 3 **22.** $y = -\frac{1}{2}x - 2$; $-\frac{1}{2}$; -2 **23.** $y = -x + 5$; -1; 5 **24.** $y = -5x + \frac{5}{2}$; -5; $\frac{5}{2}$ **25.** 22 liters **26.** yes **27.** no **28.** yes

Chapters 1–8 Cumulative Review page 359

1. B **2.** C **3.** A **4.** C **5.** D **6.** A **7.** A **8.** B **9.** D **10.** C **11.** B **12.** B **13.** B **14.** A **15.** A **16.** B

CHAPTER 9

Algebra in Geometry and Measurement

9-1 page 364

Written Exercises **1.** $\overline{BZ}$, $\overline{BT}$, $\overline{BM}$, $\overline{MT}$, $\overline{ZM}$, $\overline{ZT}$ **3.** $\overrightarrow{DC}$, $\overrightarrow{DB}$, $\overrightarrow{DA}$ **5.** an infinite number; one **7. a.** The intersection of two planes is a line.; **b.** the intersection of a floor and wall **9.** $2x + 3 = 8x$; 4 **11.** true **13.** true

Mixed Review **2.** $6\frac{1}{2}$ **3.** 34% **4.** 92¢ **5.** $x = 2$ **6.** a = 64 **7.** 4

9-2 pages 367–368

Written Exercises **1.** obtuse **3.** straight **5.** acute **7.** if the sum of their measures is 180° **9.** $\angle SRB$ and $\angle WRQ$; $\angle WRA$ and $\angle TRB$ **11.** 45°; 135° **13.** 70° **15.** 120° **17.** 73° **19.** 22° **21.** acute **23.** right **25.** obtuse **27.** acute **29.** none; 90° **31.** none; 65° **33.** 47°; 137° **35.** $(110 - y)°$; $(200 - y)°$

Mixed Review **1.** $y = \frac{1}{4}x + 2$ **2.** $y = -x + 12$ **3.** $\overline{AB}$ and $\overline{RM}$ **4.** intersecting **5.** $24.74

9-3 page 372

Written Exercises **1.** F **3.** F **5.** F **7.** T **9.** rectangle *ABCD*; polygon *ABCD* **11.** quad. *QRTS*; polygon *QRTS* **13.** convex; triangle **19.** 3, 4

Mixed Review **1.** 6.25 **2.** 1.024 **3.** 60° **4.** 47° **5.** B **6.** $a > \frac{11}{3}$ **7.** $y = 17$ **8.** Joe 13; Ellen 6

9-4 pages 375–376

Written Exercises **1.** acute **3.** right **5.** isosceles **7.** T **9.** T **11.** T **13.** F **19.** 90° **21.** 50° **23.** 90° **25.** 90° **27.** 60° **29.** $(80 - x)°$ **31.** obtuse

Mixed Review **1.** $6^2 + (18 + 9) \cdot -2 = -18$ **2.** $(150 + 17) \cdot 10 = 1{,}670$ **5.** 0.235 **6.** 0.059 **7.** 100 for $121.50

Critical Thinking **1. b.** 4 **2.** T **3.** F **4.** T **5.** You can make an inductive conclusion after trying numerous examples, but you cannot prove the conclusion true. If you have proved the statement to be false, you are done once you find a counterexample.

9-5 pages 378–379

Written Exercises **1.** F **3.** T **5.** T **7.** 85 in. **9.** 35 cm **11.** 250*x* ft **13.** 31.5 in. **15.** 0.29 cm **17.** 45,044.5 mi **19. a.** no; **b.** no; C and D are not on the circle.

Mixed Review **1.** 9 **2.** 15 **3.** 149 **4.** 49,990 **5.** $m\angle 1 = 100°$; $m\angle 2 = 60°$ **6.** 672,000 km

9-6 pages 381–382

Written Exercises **1.** B **3.** $\overline{EF} \cong \overline{LM}$; $\overline{FG} \cong \overline{MN}$; $\overline{HG} \cong \overline{ON}$; $\overline{EH} \cong \overline{LO}$; $\angle F \cong \angle M$; $\angle G \cong \angle N$; $\angle H \cong \angle O$; $\angle E \cong \angle L$; $EFGH \cong LMNO$ **5.** 40° **7.** 3.5 **9.** 4.5 **11.** T **13.** T **15.** F **17.** $m\angle 3 = 80°$; $m\angle 4 = 80°$ **23.** yes

Mixed Review **1.** 32.34 **3.** 27 mm **4.** $1\frac{3}{4}$ in. **5.** 5 **6.** 3 **7.** acute **8.** 25

Test Yourself
1. Answers may vary. $\overleftrightarrow{WY}$, $\overleftrightarrow{VU}$, $\overleftrightarrow{ZX}$; $\overrightarrow{MZ}$, $\overrightarrow{MX}$, $\overrightarrow{MY}$; $\overline{WM}$, $\overline{UM}$, $\overline{ZX}$; $\angle XMY$, $\angle VMX$, $\angle ZMU$ **2.** parallelogram **3.** acute **4.** $\overline{AB} \cong \overline{CD}$; $\overline{BD} \cong \overline{BD}$; $\overline{BC} \cong \overline{AD}$; $\angle BAD \cong \angle DCB$; $\angle ABD \cong \angle CDB$; $\angle CBD \cong \angle ADB$ **5.** 52 ft **6.** 30 in. **7.** 31 m **8.** 11.1 cm

9-7 pages 386–387

Written Exercises **3.** $\frac{MO}{RT} = \frac{OR}{RY} = \frac{ER}{YS} = \frac{ME}{TS}$ **5.** $x = 5$ **7.** $z = 5.8\overline{3}$ **9.** yes; $\overline{PQ} \leftrightarrow \overline{QS}$; $\overline{PR} \leftrightarrow \overline{ST}$; $\overline{RQ} \leftrightarrow \overline{TQ}$ **13.** 53.4 cm **15.** A possible answer is 24.5 in. × 21 in.

Mixed Review **1.** 0.375 **2.** $0.41\overline{6}$ **3.** $\overline{AB} \cong \overline{XY}$; $\overline{BC} \cong \overline{YZ}$; $\overline{AC} \cong \overline{XZ}$; $\angle A \cong \angle X$; $\angle B \cong \angle Y$; $\angle C \cong \angle Z$ **4.** 9 **5.** 36 **6.** 1.44 **7.** 10

Critical Thinking card 1 is a triangle; card 2 is a circle; card 3 is a square

Practice page 388

1. 55° **3.** 75° **5.** 33°, 123° **7.** 47°, 137° **9.** 74°, 164° **11.** none; 1° **13.** 90° **15.** 72° **17.** 45° **19.** 44° **21.** 113.6 m **23.** 0.19 km **25.** 134,770 mi **27.** chords: $\overline{LK}$, $\overline{JI}$; $\overline{GJ}$, $\overline{KH}$ diameters: $\overline{GJ}$, $\overline{KH}$; radii: $\overline{OH}$, $\overline{OJ}$, $\overline{OK}$, $\overline{OG}$; central angles: $\angle GOH$, $\angle HOJ$, $\angle JOK$, $\angle KOG$ **29.** $\overline{YX} \cong \overline{TS}$, $\overline{XZ} \cong \overline{SU}$, $\overline{ZY} \cong \overline{UT}$, $\angle Y \cong \angle T$, $\angle X \cong \angle S$, $\angle Z \cong \angle U$

9-8 pages 392–393

Written Exercises **1.** 36 ft **3.** 7 yd 1 ft **5.** 240 **7.** 19 **9.** 56.52 ft **11.** 8.4 mi **13.** 5*a* yd, 2.5*a* yd **15.** ≈0.8*x* units, ≈0.4*x* units **17.** 22.0 **19.** 2.0 **21.** $1\frac{5}{7}$ **23.** $568\frac{6}{7}$ **25.** 27 **27.** 6 **29.** $2a + 2b = P$

Mixed Review **1.** 20 **2.** 93% **3.** 2.5 **4.** $\frac{AB}{HI} = \frac{BC}{IJ} = \frac{CD}{JK} = \frac{DA}{KH}$ **5.** 26.62 **6.** 169 **7.** 27

Test Yourself **1.** 6 **2.** 33.4 **3.** 94.2

9-9 page 395

Written Exercises **1.** about 8 **3.** 13 **5. a.** sun = 324 in.; moon = 0.825 in. **b.** sun = 34,875 in.; moon = 90 in.

Problem Solving Practice page 396
1. positive **3.** 13 games **5.** 309 mi
7. 13 **9.** 13,455.33 yen; 52.42 pounds; 516.8 francs
11. $17,340.00 **13.** $90; $60

9-10 pages 399–400
Written Exercises **1.** yes **3.** no **7.** yes

Chapter 9 Review page 400
2. $\overleftrightarrow{AB} \parallel \overleftrightarrow{CD}$ **3.** Answers may vary. $\overleftrightarrow{AB}$, $\overleftrightarrow{CD}$, $\overleftrightarrow{EF}$
4. Answers may vary. $\overline{AE}$; $\overline{EB}$; $\overline{EF}$; $\overline{FD}$; $\overline{CF}$ **5.** Answers may vary. $\overrightarrow{EF}$, $\overrightarrow{CD}$, $\overrightarrow{BA}$, $\overrightarrow{AB}$, $\overrightarrow{EB}$, $\overrightarrow{FD}$, $\overrightarrow{FC}$, $\overrightarrow{EA}$, $\overrightarrow{FE}$
6. acute **7.** supplementary **8.** $\cong$ **9.** 120° **10.** 125°
11. F **12.** T **13.** F **14.** T **15.** T **16.** T **17.** right
18. obtuse **19.** acute **20.** 60° **21.** 50 cm **22.** 20 cm
23. 7 in. **24.** 8.4 cm **25.** 15 ft **26.** 4.5 ft **27.** $\overline{AB} \cong \overline{WX}$; $\overline{BC} \cong \overline{XY}$; $\overline{CD} \cong \overline{YZ}$; $\overline{AD} \cong \overline{WZ}$; $\angle A \cong \angle W$; $\angle B \cong \angle X$; $\angle C \cong \angle Y$; $\angle D \cong \angle Z$ **28.** $x = 9$ **29.** $y = 5.4$
30. 84 in. **31.** 471 cm

Chapters 1–9 Cumulative Review page 403
1. C **2.** B **3.** C **4.** B **5.** A **6.** B **7.** A **8.** A **9.** C
10. C **11.** B **12.** B **13.** B **14.** D **15.** C **16.** B

CHAPTER 10

Area and Volume Formulas

10-1 pages 408–409
Written Exercises **7.** 7.02 cm^2 **9.** 237.16 $in.^2$
11. 154.8 cm^2 **13.** 5.5 cm **15.** $15x$ **17.** 540 ft^2
19. 58.79 km^2 **21.** 577.5 cm^2 **23.** $42x^2$ sq. units
25. 32 ft **27.** 49 cm^2 **29.** 96 ft **31.** 9 ft^2 **33.** $5\frac{1}{3}$ ft
Mixed Review **1.** 1.474×10^3 **2.** $\frac{4}{81}$ **3.** > **4.** 6 and 7
5. 50.24 **6.** 180° **7.** 22.5 ft

10-2 pages 412–413
Written Exercises **1.** 39.36 m^2 **3.** 441 $in.^2$
5. 161.7 cm^2 **7.** 750 cm^2 **9.** 1,394.64 ft^2
11. 119.7 cm^2 **13.** 1.4 cm^2 **15.** 3.2 cm^2
17. $81x^2$ sq. units **19.** 1.5 ft **21.** T **23.** F **25.** T
Mixed Review **1.** 180 cm^2 **2.** 764 mm^2 **3.** 121.7%
4. 1.4352×10^{10} **5.** $18.84x$ **6.** -5 **7.** $\frac{15}{16}$ **8.** 10

10-3 pages 415–416
Written Exercises **1.** 121π mi^2; 379.94 mi^2
3. 0.36π $in.^2$; 1.13 $in.^2$ **5.** 2.56π ft^2; 8.04 ft^2
7. $25\pi x^2$ sq. units; $78.5x^2$ sq. units **9.** $4.41\pi x^2$ sq. units; $13.85x^2$ sq. units **11.** e **13.** c **15.** b **17.** 1 circle of radius 4 is larger. Area is 50.24. Area of 4 circles is 12.56. **19.** 50.24 cm^2 **21.** **b.** approximately 6 units
23. 16 in. **25.** 32.7 cm and 84.9 cm^2 **27.** 116.8 m and 1,086.3 m^2 **29.** 54.6 m and 237.7 m^2 **31.** 9.14 sq. units
33. 20π sq. units or 62.8 sq. units **35.** **a.** 6,280 ft^2; **b.** 9 ft^2; **c.** 698 yd^2 **37.** **a.** The circumference is doubled, tripled.; **b.** No. The area is multiplied by 4.

Mixed Review **1.** 60° **2.** 11 **3.** 8 **4.** 144 **5.** 60 **6.** 70
7. 67.5 **8.** 280 ft^2
Test Yourself **1.** πr^2 **2.** $\frac{1}{2}bh$ **3.** bh **4.** $\frac{1}{2}h(b_1 + b_2)$
5. bh **6.** 35.75 cm^2 **7.** 286 ft^2 **8.** 910 $in.^2$
9. 961.625 mm^2 **10.** 10.92 ft^2 **11.** 1,808.64 ft^2

10-4 pages 422–423
Written Exercises **1.** pentagonal prism
3. triangular prism **5.** triangular pyramid **9.** T
11. F **13.** F **15.** rectangles, octagons **17.** 4
19. pentagonal prism, pentagons **21.** rectangular prism
23. sphere **27.** square pyramid
Mixed Review **1.** $9a^2\pi$ square units **2.** $8x$ square units
3. 25π $in.^2$ **4.** 35 cm^2 **5.** -13 **6.** -7 **7.** $1\frac{2}{5}$ **8.** 5:00 A.M.
9. 20°; acute
Critical Thinking **2.** Pattern A: triangle, circle, square, square; Pattern B: circle, square, square, triangle; Pattern C: square, square, triangle, circle.
3. Pattern D: square, triangle, circle, square. **4.** C

10-5 page 426
Written Exercises **1.** Use 2 strings of different lengths with equal weights on the end. Test the time of swings.
3. 200 **5.** 7,042.6 m **7.** 7 **9.** $\frac{2}{15}$ **11.** 65

10-6 page 429
Written Exercises **1.** 1,078 sq. units **3.** 602.88 sq. units
5. 1,056 mm^2 **7.** 475.2 cm^2 **9.** **a.** 6 sq. units, 24 sq. units, 54 sq. units; **b.** quadrupled; nine times larger
11. 16 gal
Mixed Review **1.** $6x^3$ **2.** $4a^2 + 7a$ **3.** $1\frac{1}{9}$ **4.** 12 **5.** 256
6. 144π or 452.16 **7.** 112 **8.** 1,500 bels **9.** 219.8 cm

10-7 page 432
Written Exercises **1.** 5,600 ft^2 **3.** 1,040 m^2
5. 1,714.44 m^2 **7.** 2,826 $in.^2$ **9.** 4,578.12 m^2
11. radii $\frac{2}{5}$; areas $\frac{4}{25}$; Ratio of areas is the square of the ratio of the radii. **13.** 28.26 m^2
Mixed Review **1.** 22 **2.** -8 **3.** $2.99 **4.** $3\frac{1}{5}$ **5.** 79.2 cm
6. 25.8 in. **7.** 1.024×10^9 **8.** Both figures have a hexagon for a base. The sides of the prism are rectangles. The sides of the pyramid are triangles.

10-8 pages 434–435
Written Exercises **1.** **b.** 125.25 ft^2; yes; **c.** $626.25; $1,252.50

Practice page 436
1. 29.2 m^2 **3.** 30.6 ft^2 **5.** 44.2 yd^2 **7.** 126 m^2 or 1,260,000 cm^2 **9.** 4,750 mm^2 **11.** 11,040 $in.^2$
13. triangular pyramid **15.** 52 ft^2 **17.** 848 cm^2
19. 1,256 m^2 **21.** 1,865.16 m^2 **23.** 5,024 mm^2

10-9 pages 439–440
Written Exercises **1.** 637 $in.^3$ **3.** $27a^3$ **5.** 720 mm^3
7. 6,782.4 ft^3 **9.** 5,803.72 cm^3 **11.** 8,138.88 $in.^3$
13. 128 ft^3 **15.** $\frac{1}{2}$; doubles the volume
17. 5,878.08 mm^3 **19.** πx^3 **21.** 320 ft^3

Mixed Review 1. $(180 - 3x)°$ 2. $\frac{2}{5}$ 3. 94.2 cm 4. 3 5. 3.136×10^{13} square units 6. 391.2 $in.^2$ 7. 263.76 ft^2 8. 22
Test Yourself 1. 680 mm^2 2. 477.28 m^2 3. 1,400 cm^2 4. 19,292.16 ft^3 5. 36,000 cm^3 6. 1,200 $in.^3$ 7. $6x^3$ cu. units 8. $12.56x^3$ cu. units 9. $50x^3$ cu. units

10-10 pages 443–444

Written Exercises 1. 904.32 cm^3 3. 847.8 ft^3 5. 300 ft^3 7. The box from theater A holds 3 times more. 9. 4 m 11. 5 ft 13. ≈12.27 $in.^3$ 15. a. 480.42 in. or 40.035 ft; b. 73,504.3 $in.^2$ or 510.45 ft^2; c. ≈1,874,000 $in.^3$ or ≈1,085 ft^3 17. $88{,}565{,}333\frac{1}{3}$ ft^3
Mixed Review 1. 61,544 mm^3 2. 43.332 in. 3. 533.8 $in.^2$ 4. 1.2×10^5 5. -1 6. 8 7. $\frac{43}{72}$ 8. 2,167.5 cm^3
Critical Thinking 3. a. $\frac{2}{3}$; b. $\frac{2}{3}$; c. $\frac{2}{3}$; d. $\frac{4}{9}$ 5. $\frac{5}{8}$, $\frac{25}{64}$

Problem Solving Practice page 445

1. 704 $in.^3$ 3. 21 5. 27
7. Juan picked 15 bushels. Kimo picked 5 bushels.
9. Teresa, Fran, Vivian, Clara, Marie, Cindy
11. 12 ft × 20 ft

Chapter 10 Review pages 446–447

1. area 2. polygons 3. triangles 4. prism 5. cylinder 6. cone, vertex 7. sphere, center 8. faces 9. slant height 10. volume 11. 49 cm^2 12. 27 $in.^2$ 13. 30 ft^2 14. 15 cm^2 15. 78.5 m^2 16. 200.96 mm^2 17. 56.52 m^2 18. 37.68 $in.^2$ 19. square pyramid 20. triangular prism 21. cylinder 22. 13 in. × $24\frac{1}{2}$ in. 23. 164 cm^2 24. 62.8 $in.^2$ 25. 84 cm^2 26. 216 m^2 27. 310.86 $in.^3$ 28. 384 cm^3 29. 18 ft^3 30. 904.32 cm^3

Chapters 1–10 Cumulative Review page 449

1. B 2. A 3. C 4. A 5. D 6. C 7. C 8. A 9. B 10. C 11. B 12. B 13. A 14. A

CHAPTER 11

Right Triangles in Algebra

11-1 pages 456–457

Written Exercises 1. false 3. false 5. true 7. true 9. 256 11. $\frac{4}{9}$ 13. $25x^6$ 15. 81 17. 100 19. 8 21. 7 23. $\frac{4}{5}$ 25. 7.035 27. $4y^5$ 29. 16 31. 4.472 33. (7), 8 35. (2), 3 37. (-2), -3 39. (4), 5 41. rational 43. irrational 45. rational 47. rational 49. 7.071 51. 9.950 53. 6.557 55. 7.348
Mixed Review 1. 403.44 cm^2 2. 0.00254 m^3 3. $n \le \frac{7}{8}$ 4. $x \ge -13\frac{1}{2}$ 5. $-\frac{4}{7}$ 6. $x = 16\frac{1}{3}$ 7. 1,502% 8. \$12
Critical Thinking 1. You keep getting 6.85 and 6.86. Go halfway between the numbers to get 6.855. 2. 4.359 3. Calculators and computers can calculate square roots instantly.

11-2 page 459

Written Exercises 1. lost \$8 3. 4 h; 12 mi; 6 mi 5. -3 7. 3:02 P.M.

11-3 pages 463–464

Written Exercises 1. 5 3. $\frac{9}{25}$ 5. 7.937 7. 5.657 9. $\overline{AC}$ and $\overline{CB}$; $\overline{AB}$ 11. 10 13. 9 15. $50^2 + 120^2 = x^2$; $x = 130$ 17. $x^2 + (6\sqrt{5})^2 = 18^2$; $x = 12$ 19. $11^2 + x^2 = \sqrt{202}^2$; $x = 9$ 21. $3^2 + 7^2 = h^2$; $x = 4.62$ 23. yes 25. no 27. no 29. yes for $p > 0$ 31. $\sqrt{2}$, $\sqrt{3}$, 2, $\sqrt{5}$, $\sqrt{6}$, $\sqrt{7}$ 33. 106 ft 35. 21.21 in. 37. 14.1 ft 39. a. 21.63 in.; b. 461.44 $in.^3$
Mixed Review 1. $25ab^2$ 2. xy 3. yes 4. 33°; 123° 5. 58°; 148° 6. 4 7. 256 8. 36°F

11-4 pages 467–468

Written Exercises 1. 27 3. 3 5. $x = 8.602$ 7. $x = 9.487$ 9. $x = 10.5$ 11. $x = 12$; $y = 15$ 13. $x = 36$; $y = 69.97$ 15. 43.75 ft 17. a. 611.45 ft; b. 570,025 ft^2
Mixed Review 1. yes 2. no 3. 8.96×10^{12} 4. 6 5. 38 6. $\frac{4}{2x} = \frac{1}{8} \cdot 12$; $x = 1\frac{1}{3}$ 7. $\frac{6x}{2} = x + 4$; $x = 2$ 8. $0.02 \times 50 = x$; 1
Test Yourself 1. $x = 8$ 2. $x = 8.5$ 3. $x = 14.7$ 4. 8.062 cm 5. 7.5 ft

11-5 pages 471–472

Written Exercises 1. 9; $9\sqrt{2}$ 3. 4; 4 5. $2\sqrt{3}$; 4 7. $7\sqrt{3}$; 14 9. $x = \frac{27}{\sqrt{2}}$ 11. $x = \frac{14}{\sqrt{3}}$; $y = \frac{28}{\sqrt{3}}$ 13. a. 8.5 in.; b. 433.5 in^2 15. $26\sqrt{3}$ in. 17. 55.04 cm^2 19. 51 in.
Mixed Review 1. -12 2. 10 3. $x = 4.75$ 4. $-16x^6$ 5. 6 6. pentagon 7. 27% 8. \$33.32
Critical Thinking Lisa–bedroom; Nicole–bathroom; Robert–living room; Eric–family room; JoAnn–kitchen

11-6 pages 475–476

Written Exercises 1. 0.1736 3. 0.2679 5. 0.9781 7. 84° 9. 63° 11. $\sqrt{3}$ 13. $\tan z° = \frac{f}{e}$ 15. $\sin y° = \frac{d}{c}$ 17. $x \approx 54$; $y \approx 36$; $z = 1.27$ 19. $x = 68$; $y = 9.90$; $z = 10.68$ 21. $x \approx 28$; $y \approx 62$ 23. 13.3 mi 25. 41 ft
Mixed Review 1. 3.4 ft 2. 11.56 ft^2 3. 8 in.; $8\sqrt{3}$ in. 4. 0 5. acute 6. 5 m 7. -13, -14, -15
Test Yourself 1. 30°-60°-90° 2. 45°-45°-90° 3. neither 4. $x = 11$ 5. $x = 3$; $y = 6$; $z = 60$ 6. $x = 56$; $y = 8.09$; $z = 14.48$

11-7 page 479

Written Exercises 11. a. $m\angle A \approx 23°$; $m\angle B \approx 67°$

Practice page 480

1. 13 3. $\frac{9}{2}$ 5. $\frac{x}{y}$ 7. $20x^3$ 9. $11c^2$ 11. 1 13. 3 15. 10 17. rational 19. rational 21. no 23. no 25. no 27. yes 29. 45°-45°-90° 31. neither 33. $x = 24$ 35. $y = 10$ 37. $y = 15$ 39. $a = 5$

Problem Solving Practice page 481

1. 42 3. 35°, 58°, and 87° 5. 1.36 min 7. 12 9. 657 cm^3 11. 8, 18, and 22

Chapter 11 Review pages 482–483

1. g **2.** d **3.** f **4.** b **5.** a **6.** h **7.** c **8.** i **9.** e **10.** 14 **11.** $8p$ **12.** 1.732 **13.** $x - 2$ **14.** 25 **15.** 6 **16.** 18 **17.** 5 **18.** 15 **19.** 25 **20.** 6.708 **21.** 39 **22.** 9 **23.** 25 **24.** 32 **25.** 18 **26.** $8\sqrt{3}$ **27.** 20 **28.** $x = 25$; $y = 45.315$; $z = 21.13$ **29.** $x = 70$; $y = 26.312$; $z = 9.576$ **30.** $x = 68$; $y = 24.752$; $z = 26.695$ **31.** $x = 50$; $y = 51.424$; $z = 61.28$ **32.** 13.610 ft **33.** 2°

Chapters 1–11 Cumulative Review page 485

1. A **2.** B **3.** D **4.** B **5.** B **6.** C **7.** A **8.** C **9.** B **10.** A **11.** A **12.** B

CHAPTER 12

Statistics and Probability

12-1 pages 490–491

Written Exercises **1.** 4.8, 5, 5, 4 **3.** 21.5, 22.5, no mode, 10.3 **5.** 2, 2, 2, 4 **7.** 4.5, 4.5, 2, 6 **9.** 84.2, 84, 84, 16 **11.** 101, 101, 101, 8 **13.** **a.** 10; 8; 3 **b.** The median. It is not influenced by the extreme of 30 points. **25.** **a.** The mean, because it is the highest average.; **b.** 87 **c.** 90

Mixed Review **1.** 33.2 in.^{2} **2.** 360 ft^2 **3.** $\frac{4}{5}$ **4.** $\frac{3}{4}$ **5.** 3rd and 4th **6.** 1st and 2nd **7.** $7.16

12-2 pages 494–495

Written Exercises **1.** 4; 4 **3.** 3.5; 1 and 6 **5.** 3.5, 2 **7.** 7.3, 7, 6 **9.** 25.8, 25, 25 **11.** 4, 4 **13.** 17.5, 15 and 20 **15.** **c.** 4.1, 3, 2 **17.** **b.** 2.7, 2, 0 **21.** **a.** men 40–44, women 30–34; **b.** Total number entered. Divide by 2 to find middle number. Add numbers entered one group at a time. Start at the top until the middle number is reached or surpassed. Find corresponding age. Men: 35–39. Women 35–39.

Mixed Review **1.** -5.5 **2.** 62.5%, 0.625 **3.** 15.7, 16, 16 and 18 **4.** 14.4, 15, no mode **5.** 5 **6.** $-\frac{7}{8}$ **7.** 156

12-3 pages 500–501

Written Exercises **1.** 774, 768 and 785 **3.** 82, 74 **5.** 11.65, 11.8 **7.** Set A: 62.5, 63; Set B: 57, no mode **9.** Set E: 5.85, no mode; Set F: 6.45, no mode **11.** **a.** 0; **c.** 10, 12, and 15; **d.** 13 **13.** 6, 7, 8, 9 **15.** Class A: 79.5, 74, 79 and 96; Class B: 84, 99

Mixed Review **1.** 6,-6 **2.** 37.5 in.^{2} **3.** 2 **4.** 2 **5.** 18, 9, 27 **6.** 46, 23, 69 **7.** 852.375; 803.5; 774

Critical Thinking **1.** A change in the groupings of the horizontal scale changes heights on the vertical scale. **2.** The shape of the graph is changed, giving a different impact.

12-4 pages 504–505

Written Exercises **5.** $220, $50 **7.** 75% **9.** no **17.** 7.76, 7.7, 7.4, mean

Mixed Review **1.** 118 **2.** 37.5% **3.** 79, 25 **4.** 79.4 **5.** 79 **7.** 120

Test Yourself **1.** 65.3, 66, 45, 42

12-5 pages 510–511

Written Exercises **1.** 120 **3.** 3,628,800 **5.** 60 **7.** 30,240 **9.** 479,001,600 **11.** 360,360 **13.** 6 **15.** 90 **17.** T **19.** F **21.** 12 **23.** 80 **25.** 440 **27.** **a.** 7; **b.** 6; **c.** 5; **d.** 210 **29.** 36 **31.** 18

Mixed Review **1.** 1×10^{11} **2.** 1×10^{-4} **3.** 19 **5.** $\frac{1}{3}$ **6.** $\frac{5}{11}$ **7.** $28.50

Critical Thinking **1.** 2 **2.** red

12-6 pages 514–515

Written Exercises **1.** $\frac{1}{2}$ **3.** $\frac{7}{11}$ **5.** $\frac{6}{11}$ **7.** $\frac{4}{25}$ **9.** $\frac{3}{7}$ **11.** **a.** $\frac{6}{13}$; **b.** $\frac{13}{25}$; **c.** $\frac{24}{49}$; **d.** Yes. The sample space is different for each question. **13.** **a.** $\frac{4}{25}$; **b.** $\frac{21}{25}$; **c.** Any prime number has 1 or more digits. **15.** **a.** $\frac{1}{7}$; **b.** $\frac{2}{7}$; **c.** $\frac{5}{14}$; **d.** $\frac{1}{7}$; **e.** $\frac{1}{14}$; **f.** 1 **17.** **a.** $\frac{1}{5}$; **b.** $\frac{1}{5}$; **c.** $\frac{1}{10}$; **d.** $\frac{1}{5}$; **e.** $\frac{1}{5}$; **f.** The event of having the digit 0 as the last digit is not included. **19.** 15 **21.** $\frac{1}{3}$

Mixed Review **1.** 9 **2.** 5 **3.** 720 **4.** 840 **5.** 7.25 **7.** $5x = 850$; 170 tickets

12-7 pages 519–520

Written Exercises **3.** $\frac{5}{16}$ **5.** $\frac{19}{25}$ **7.** $\frac{1}{31}$ **9.** 98 **11.** $67.50 **13.** 2 **15.** Yes. The circumference of the hole is 3.14 in. **17.** 78.6%

Practice page 521

1. 16, 13, 6 and 7, 30 **3.** 4, 4, 4, 6 **5.** 3.75, 5, 6 **7.** Set A: 23.5, 23 Set B: 22, no mode **9.** The data for Set A is spread out. Set B data is consistent, staying close to the median, which is 9. **11.** 4! **13.** $\frac{25!}{22!}$ **15.** 720 **17.** 2,450 **19.** $\frac{1}{9}$ **21.** $\frac{4}{9}$ **23.** 1

12-8 pages 524–525

Written Exercises **1.** independent **3.** dependent **5.** $\frac{1}{10}$ **7.** $\frac{1}{12}$ **9.** Dependent $\frac{5}{82}$ **11.** dependent **a.** $\frac{9}{56}$; **b.** $\frac{3}{28}$ **13.** 24; No. The class is small enough to ask each person for his/her preference. **15.** $\frac{2}{5}$

Mixed Review **1.** 195 **2.** 132 **3.** $\frac{1}{2}$ **4.** $\frac{1}{6}$ **5.** $48n$ **6.** $29n - 5$ **7.** 5 y, 15 y

Test Yourself **1.** 30.3, 30, 28 and 35 **2.** 20 **3.** $\frac{1}{36}$ **4.** $\frac{1}{11}$

Problem Solving Practice page 526

1. 105 ft **3.** 24 **5.** 32 students **7.** 5 and 6 **9.** 160, 160, 160, 80 or 112, 112, 224, 112 **11.** 69

12-9 pages 528–529

Written Exercises **1.** 496 **3.** 315 **5.** **a.** 240; **b.** 400; **c.** 600; **d.** 192, 320, 480; **e.** $1,240

Chapter 12 Review pages 530–531

1. false, mode **2.** true **3.** true **4.** false, independent **5.** false, factorial **6.** 7.46, 8, 8, 4 **7.** 115.56, 115, no mode, 37 **10.** 6 **11.** 5,040

601

12. 12 **13.** 1,680 **14.** 35 **15.** $\frac{1}{5}$ **16.** $\frac{2}{5}$ **17.** $\frac{1}{5}$ **18.** $\frac{4}{5}$ **19.** $\frac{2}{25}$ **20.** $\frac{1}{10}$

Chapters 1–12 Cumulative Review page 533
1. B **2.** C **3.** A **4.** C **5.** B **6.** A **7.** B **8.** D **9.** B **10.** C **11.** A **12.** A **13.** A **14.** C

CHAPTER 13

Polynomials

13-1 pages 540–541
Written Exercises **1.** monomial **3.** monomial **5.** 2 **7.** 3 **9.** 1 **11.** binomial **13.** monomial **15.** monomial **17.** binomial **25.** $x^2 + 3$ **27.** $x^2 + 2x + 3$ **31.** 6 **33.** -1 **35.** 0 **37.** 187 **39.** 195 **41.** 201
Mixed Review **1.** x^5 **2.** $64m^6$ **3.** independent **4.** dependent **5.** $6x^2$ **6.** 9 **7.** 120 poses
Critical Thinking **1. a.** yes; **b.** yes; **c.** yes; **d.** no; **e.** no **2. a.** no; **b.** no; **c.** no; **d.** no; **e.** yes **3. a.** no; **b.** no; **c.** no; **d.** yes; **e.** yes **4. a.** no; **b.** no; **c.** no; **d.** yes; **e.** no

13-2 pages 544–545
Written Exercises **1.** $4x^2 + 5x + 2$ **3.** $-x + 5$ **5.** $-3x^2 + x + 7$ **7.** $9x^2 + x + 5$ **9.** $2x^3 + 5x^2 + x + 4$ **11.** $3x + 8$ **13.** $4y - 15$ **15.** $4mn^2 + 2m - 2n^2$ **17.** $2a^2 + 5ab + 2b$ **19.** $m - 9$ **21.** $4ab - 10$ **23.** $w^2 + 7w - 6$ **25.** $-3x^2 - 8x + 8$ **27.** $-8x + 10$ **29.** $17a^7 + 8a^4 + a^2 + 4$ **31.** $11a^3b^2 - 15a^2b - 2ab + 23$ **33.** $2a + (2a + 2) + (2a + 4)$; **a.** $6a + 6$; **b.** 10, 12, 14 **35.** $5a + (5a + 5) + (5a + 10) + (5a + 15) + (5a + 20)$; **a.** $25a + 50$; **b.** 65, 70, 75, 80, 85 **37.** $8b^2 + 10b - 6$ **39. a.** 360; **b.** 540; **c.** 720; **d.** 1,440
Mixed Review **1.** 5 sq. units **2.** 12 sq. units **3.** 2 **4.** 3 **5.** $17x - 6$ **6.** $24x + 6 - 20y$ **7.** 7.125 mi

13-3 pages 547–548
Written Exercises **1.** $2x^2 + 12x$ **3.** $6x^2 - 2x$ **5.** $5x^2 + 3x$ **7. a.** 24; **b.** 2; **c.** 40; **d.** 6 **9.** $-4x^2y^2 + 4x^3y$; -233,280 **11.** $8x^2 - 2x$ **13.** $2x^3 - 5x^2$ **15.** $2y^4 + y^3 - 3y^2$ **17.** $5x^2y + 25xy - 5xy^2$ **19.** $-6x^3y^2 - 3x^2y^2 - 3xy^3 + 9xy$ **21.** $96x^3y^3z + 24x^4yz^6$ **23.** $12x^3 + 12x^2y^2 + 12x^2z$ **25.** $\frac{1}{2}xy + \frac{1}{2}xy^3 + \frac{5}{2}y$ **27.** $14x^4 + 7x^2y^2 - 7x^3y$ **29.** $3x^3 + 2x^2 + 9x$ **31.** $e(e + 2)$; $e^2 + 2e$ **33.** $l\left(\frac{1}{2}l + 7\right)$; $\frac{1}{2}l^2 + 7l$ **35.** $\frac{1}{2}(8x)(16x + 5)$; $64x^2 + 20x$ **37.** $7 \times 10^2 + 9 \times 10 + 2$; $7x^2 + 9x + 2$; $4x$; $(7x^2 + 9x + 2)4x = 28x^3 + 36x^2 + 8x$; $28(10)^3 + 36(10)^2 + 8(10) = 31{,}680 = 40(792)$; yes **39.** $8a^2 + a$ **41.** $4d^2 + 4d$
Mixed Review **1.** 2.75 **2.** 27.52 m² **3.** $3x^2 + 7x + 6$ **4.** $-9x^2 - 5x + 13$ **5.** x^4 **6.** x^{10} **7.** $449

13-4 pages 552–553
Written Exercises **1.** $x^2 + 7x + 6$ **3.** $a^2 + 8a + 7$ **5.** $a^2 + a - 6$ **7.** $a^2 + 11a + 24$ **9.** $xy + xb + ay + ab$ **11.** $x^2 + 13x + 12$ **13.** $15 + 2x - x^2$ **15.** $8ac - 4ad + 4bc - 2bd$ **17.** $x^2 - 9$ **19.** $b^2 - 14b + 49$ **21.** $4a^2 - 25$ **23.** $4a^2 + 4a + 1$ **25.** $x^2 + 15x - 756$ **27.** $3a^2 - 2a - 8$ **29.** $x^2 - 3.5x - 6.24$ **31.** $b^3 - 3b^2 + 3b - 1$ **33.** $-8a^2 + 20a + 12$ **35.** $x^2 + 2x + 1$ **37.** $x^2 - y^2$ **39.** $x^2 + 4x + 4$ **41.** $10x^2 + 26x + 12$ **43.** $20x^2 + 6x - 2$ **45.** $(w - 1)(w - 2) = w^2 - 3w + 2$ **47.** $(t - 6)^2 = t^2 - 12t + 36$ **49.** $x = 2$ **51.** $w = 0$ **53.** about 50 mi
Mixed Review **1.** 2 **2.** 13.5 **3.** $6x^2 + 2x$ **4.** $-5x^3 + 10x^2$ **5.** 20 **6.** 6 **7.** 72 ft by 72 ft
Test Yourself **1.** monomial **2.** binomial **3.** not a polynomial **4.** trinomial **5.** monomial **6.** binomial **7.** monomial **8.** monomial **9.** 81 **10.** 95 **11.** -7 **12.** -51 **13.** 112 **14.** -26 **15.** $11a$ **16.** $p^2 + p + 12$ **17.** $3ab^2 + 2a + b + 9$ **18.** $-r^2s^2 + rs - 16$ **19.** $(a^2 + b)(a^2 + b) = a^4 + 2a^2b + b^2$

Practice page 554
1. 2; binomial **3.** 2; binomial **5.** 1; monomial **7.** $6r^2 - 3$ **9.** $-4d - 1$ **11.** $-7x^2y - 6xy - 5x^2 + 13$ **13.** $2rt^2 + 6r^2t + 2t$ **15.** $21xy + 6x$ **17.** $20x^2y + 15xy^2$ **19.** $10a^4b^2 - 10a^2b^4$ **21.** $8c^2d + 12cd^2 - 24cd$ **23.** $x^2 + 10x + 25$ **25.** $x^2 - 4.5x + 4.5$ **27.** $x^2 - 25$ **29.** $x^2 - 10x + 25$ **31.** $x^3 - 5x^2 - 2x + 10$ **33.** $12x^2 + 40x + 12$ **35.** $25x^2 - 20x + 4$ **37.** $30x^4 + 10x^3 - 6x^2 - 2x$ **39.** $3x^3 + 6x^2 - 72x$ **41.** 13 **43.** 140 **45.** -6 **47.** 7,280 **49.** -34 **51.** 12 **53.** -2 **55.** 12

13-5 page 557
Written Exercises **1.** $\begin{bmatrix} 1 & 2 \\ 2 & 3 \end{bmatrix}$ **3.** $\begin{bmatrix} 3 & 1 \\ -2 & 2 \end{bmatrix}$ **5.** 8 **7.** (-2,4) **9.** (1,1) **11.** One pen and two pencils cost $.60.

13-6 pages 559–560
Written Exercises **1.** 80 games **3.** 7.5 ft **5.** Sandy = 101 g; White Ears = 108 g; Sport = 115 g **7.** 51 people **9.** 32.9 min **11.** 10 angles **13.** 3,750 ft² **15.** $28.80 **17.** 6 ways

Problem Solving Practice page 561
1. 28 possible routes **3.** 86 **5.** 60 lb **7.** A is 3 yd by 11 yd; B is 3 yd by 3 yd **9.** 8 students

Chapter 13 Review pages 562–563
1. A monomial is a real number, variable, or the product of a real number and one or more variables. Examples: $6x^2$; $3x$. **2.** A polynomial is a monomial or a sum or difference of monomials. Examples: $5x - 2$; $12x^2 - 6x - 5$. **3.** Multiply x by each term within the parentheses. $x(2x - 1) = x \cdot 2x - x \cdot 1 = 2x^2 - x$ **4.** Multiply the first terms, $x \cdot 2x$, the outer terms, $x \cdot 1$, the inner terms, $3 \cdot 2x$, and the last terms, $3 \cdot 1$. Then add the products. $2x^2 + x + 6x + 3 = 2x^2 + 7x + 3$ **5.** Square the first term and subtract the square of the

second term. $(a + b)(a - b) = a^2 - b^2$ **6.** To square a binomial like $(a + b)^2$ or $(a - b)^2$, square the first term. Add or subtract twice the product of the two terms and add the square of the last term.
$(a + b)^2 = a^2 + 2ab + b^2$; $(a - b)^2 = a^2 - 2ab + b^2$
7. $2x^2 + 3x$; binomial **8.** $x^2 + 2x + 4$; trinomial
9. $3x^2$; monomial **10.** 15 **11.** -9 **12.** -10
13. $7x^2 + 4x$ **14.** $2x^2 + 3x + 1$ **15.** $8x + 5y$
16. $3m - 2n$ **17.** $6a^2 + 3b$ **18.** $-3r^2 + 5s$
19. $4c^2 + 2c - 1$ **20.** $2a^2b - b^2 + 2a$ **21.** $2x^2 + 10x$
22. $6x^2 + 3x$ **23.** $x^2 - 5x$ **24.** $3x^2 + 6x$ **25.** $2x^2 - 4x$
26. $4x^3 + 12x^2$ **27.** $10x^3 - 10x^2 + 15x$
28. $-2x^3y + 2x^2y^2 - 2xy^3$
29. $(x + 2)(x + 2) = x^2 + 4x + 4$
30. $(x + 2)(x + 3) = x^2 + 5x + 6$ **31.** $x^2 - 2x + 1$
32. $a^2 + 6a + 9$ **33.** $y^2 - 4$ **34.** $b^2 + b - 6$
35. $m^2 + 2m - 15$ **36.** $2n^2 + 6n - 20$ **37.** 4 min
38. 8 classic; 6 rock **39.** 12, 24, 36 **40.** 2,401 ft^2

Chapters 1–13 Cumulative Review page 565
1. B **2.** C **3.** A **4.** C **5.** A **6.** B **7.** C **8.** B **9.** B **10.** B **11.** A **12.** D **13.** C **14.** B **15.** C **16.** B

CHAPTER 1

Extra Practice

page 566
1. -4 **3.** 10 **5.** > **7.** = **9.** -10 **11.** 9 **13.** -112 **15.** 110 **17.** 6 **19.** 3 **21.** 27 **23.** 51 **25.** 2 **27.** $-7x$ **29.** $-(y - 7)$ **31.** $|-3 - 10|$ **33.** $10[17 - (-11)]$ **35.** 71 **37.** -18 **39. a.** $20x + 12y$; **b.** $152; **c.** 3; 5 **41.** +13

CHAPTER 2

Extra Practice

page 567
1. false **3.** false **5.** false **7.** 0 **9.** 3 **11.** 44 **13.** 25 **15.** 15 **17.** -61 **19.** 20 **21.** $-3m + 39$ **23.** $5q + 2$ **25.** $-6s + 4$ **27.** $j = 26$ **29.** $z = 20$ **31.** $n = 5$ or $n = -5$ **33.** $-3 \div (-1) = 3$ **35.** $xy = |-24|$ **37.** 2,422 m **39.** 27

CHAPTER 3

Extra Practice

page 568
1. 2.12, 2.012, 2.011 **3.** 1.00009, 0.000045, 0.00004 **5.** 0.4 **7.** 238 **9.** $19 **11.** 90 **13.** $1,000 **15.** 216 **17.** -116.667 **19.** 100.987 **21.** 15.32 **23.** 7.66 **25.** -1.856 **27.** 15.32 **29.** $y = 1.1$ **31.** $m = 1.89$ **33.** $x = 16.5$ **35.** $n = 100$ **37.** $0.03x = -3.01$ **39.** $|-0.9| = 2x$ **41.** 31 min **43.** 6

CHAPTER 4

Extra Practice

page 569
1. 1,024 **3.** -16 **5.** $15x^3$ **7.** $64y^{12}$ **9.** 16 **11.** 8 **13.** -1 **15.** -12 **17.** 134 **19.** 1 **21.** 2 **23.** 27 **25.** 5.28×10^4; 52,800 **27.** 1.82×10^7; 18,200,000 **29.** 2,3,4,9 **31.** 3,5,9 **33.** 2 **35.** 2 **37.** $9ab^2$ **39.** 1,100 **41.** $12a^5b$ **43.** 45; 15 **45.** 1,100 **47.** 20 and 4

CHAPTER 5

Extra Practice

page 570
1. $\frac{1}{2}$ **3.** $\frac{1}{5}$ **5.** $\frac{1}{3}$ **7.** $\frac{1}{5}$ **9.** $20\frac{2}{25}$ **11.** $\frac{1}{200}$ **13.** 0.4 **15.** 0.6 **17.** 0.75 **19.** $-\frac{3}{4}, \frac{1}{2}, \frac{8}{4}$ **21.** $\frac{2^2}{5}, \frac{9}{10}, 2\frac{3}{10}$ **23.** $\frac{-xy}{5}, \frac{xy}{15}, \frac{3xy}{10}$ **25.** $-\frac{5}{6}$ **27.** $\frac{7}{18}x$ **29.** $7\frac{1}{2}$ **31.** $\frac{72}{77}$ **33.** $\frac{7x^2}{15}$ **35.** $1\frac{14}{15}$ **37.** $-1\frac{17}{32}$ **39.** $\frac{1}{9}$ **41.** $\frac{a^2b}{c}$ **43.** $\frac{-1}{y}$ **45.** $1\frac{4}{9}$ **47.** $\frac{-13}{40}$ **49.** $2\frac{1}{2}$ **51.** $-4\frac{4}{9}$ **53.** 1,205 million people **55.** 24

CHAPTER 6

Extra Practice

page 571
1. $\frac{1}{3}$ **3.** $\frac{1}{4}$ **5.** $\frac{1}{9}$ **7.** $\frac{1}{5}$ **9.** $\frac{20}{30} = \frac{x}{120}$; $x = 80$ **11.** $\frac{20}{95} = \frac{z}{19}$; $z = 4$ **13.** $\frac{17}{20}$; 85% **15.** $\frac{40}{120}$; $33\frac{1}{3}$% **17.** 45% **19.** $330.\overline{30}$ **21.** 6.44 **23.** -33.3% **25.** +64.5% **27.** +258.3% **29.** +6.5% **31.** 6 **33.** 60% **35.** 2,400 **37.** bread: +4,533%; milk: +679%

CHAPTER 7

Extra Practice

page 572
1. $x = 6$ **3.** $y = 2\frac{1}{4}$ **5.** $x = 20$ **7.** $b = 3$ **9.** $x = 10$ **19.** $y \geq 6$ **21.** $y \leq -36$ **23.** $x \geq \frac{5}{9}$ **25.** $x < 7$ **27.** $x \leq 50$ **29.** $x \leq 90$ **31.** $\frac{a}{8} \leq 13$; $a \leq 104$ **33.** $7p - 5 \geq 23$; $p \geq 4$ **35.** 12 months **37.** $.20 **39.** 97

CHAPTER 8

Extra Practice

page 573
1. II **3.** I **5.** III **7.** $y = -\frac{1}{2}x + 4$ **9.** $y = 2x + 24$

603

11. $y = 3x + 3$ 13. 2; $2\frac{1}{2}$; -5 15. -1; 5; 5
17. $-\frac{1}{4}$; -12; -3 19. (4,-5) 21. $y > 3x - 5$
23. $y < 2x + 49$
31. 7 bags of popcorn and 14 gal of juice
33. 6 and 7 35. 1 min 12 s

CHAPTER 9

Extra Practice

page 574
1. acute 3. obtuse 5. straight 7. not possible; 90
9. 45; 135 11. not possible; 40 17. 90 19. 80 21. 84
23. $d = 58.6$ cm 25. $r = 39$ in. 27. $d = 457$ m
29. true 31. false 33. true 35. 640.56 ft
37. 16.8 in. 39. 27 m 41. 1 mi 43. 62.8 yd

CHAPTER 10

Extra Practice

page 575
1. 18.3 cm^2 3. 0.9 m^2 5. $561x^2$ 7. 206 cm^2
9. 547.1 m^2 11. 98.5 m^2 13. 660.2 cm^2 15. 286 $in.^2$
17. 11,269 m^2 19. 1,808.64 $in.^2$ 21. 140 ft^3
23. 2,279.64 cm^3 25. 100 cm^3 27. the $61 dress
29. 16 $in.^3$

CHAPTER 11

Extra Practice

page 576
1. 7.141 3. 12 5. $7x$ 7. $a + b$ 9. yes 11. no
13. yes 15. $x = 4\sqrt{2}$ 17. $x = 14\sqrt{2}$ 19. $x = 9\sqrt{2}$
21. $\frac{n}{m}$ 23. $\frac{e}{d}$ 25. 8 mi 27. 3 A.M.

CHAPTER 12

Extra Practice

page 577
1. 10.4; 12; 13; 10 3. 78.6; 100; 100; 100 7. 13; 13
9. 69; 66 13. 5 15. 714 17. $\frac{1}{4}$ 19. $\frac{3}{13}$ 21. $\frac{1}{676}$
23. -6 25. 1,766.25 cm^3 27. $6\sqrt{2}$ 29. 6 and 7

CHAPTER 13

Extra Practice

page 578
1. $3x^2 + 4x + 5$ 3. $-2mn^2 + m + 2n - 3$
5. $-3a^3n^2 - 4a^2n + 6a$ 7. $72x^3 - 36x^2 + 162x$
9. $\frac{5}{2}a^6 - \frac{1}{2}a^4 - \frac{33}{2}a$ 11. $x^2 + 4x + 3$ 13. $x^2 - 81$
15. $2x^3 + 5x^2 + 3x + 2$
17. $(m + 2)(m + 4)(m + 6)$; $m^3 + 12m^2 + 44m + 48$
19. $(z - 4)(z - 8)$; $z^2 - 12z + 32$ 21. 904.32 cm^3 23. 24
25. 194 27. $\frac{7}{75}$

Index

A

Absolute value
- integers, 5, 141
- rational number, 192

Addition
- associative property, 58–59, 108
- commutative property, 58–59, 108
- equations, 72–75, 111–114, 215–217
- exponents, 145
- integers, 8, 10–13
- polynomials, 542–545
- properties of inequalities, 296
- property of equality, 73, 111
- rational numbers, 199–202

Additive identity, 59
Adjacent angles, 366
Algebraic factorization, 168
Algebraic properties, in solving equations, 287–288

Altitude
- parallelogram, 407
- triangle, 410

Angles, 365–369
- acute, 365
- adjacent, 366
- central, 377
- complementary, 366
- exploring, 369
- interior, 374
- measuring, 365
- obtuse, 365
- right, 365
- straight, 365
- supplementary, 366
- vertical, 366

Applications, *See also* Careers; Computer; Data; Data File; Exploring; Functions; Geometry; Problem Solving Applications; Problem Solving Practice; Projects; Write

Area, 62
- circles, 414–417
- parallelograms, 406–409, 414
- rectangles, 406–409
- trapezoids, 410–413
- triangles, 410–413
- *See also* Surface area

Associative property
- of addition, 58, 108

Associative property (cont.)
- of multiplication, 59

Average, 27, 488–491. *See also* Mean

B

Base
- and exponent, 141
- cylinders, 428
- parallelograms, 407
- prisms, 420
- rectangles, 406
- trapezoids, 411
- triangles, 410

Bimodal data, 488
Binomials, 539
- squaring, 551

Box and whisker plots, 502–505
Boyle's law, 353

C

Calculator, 7, 12, 16, 24, 28, 42, 45, 46, 56, 75, 78, 109, 113, 117, 130, 131, 143, 147, 151, 157, 165, 170, 187, 189, 190, 191, 198, 201, 204, 247, 254, 258, 262, 277, 281, 289, 294, 295, 322, 346, 379, 392, 408, 409, 413, 429, 432, 443, 463, 464, 467, 472, 481, 490, 510, 541, 547
Calorie, 31

Careers
- acoustical physicist, 205
- actuary, 521
- airline pilot, 30
- chef, 249
- civil engineer, 352
- demographer, 158
- director of marketing, 304
- finance manager, 554
- landscape architect, 480
- medical record technician, 119
- operations research analyst, 436
- optician, 388
- urban and regional planner, 90

Cells, spreadsheet, 38–39
Center, 377
- spheres, 421

Central angle, 377
Central tendency, measures of, 488, 489
Change, percent of, 260–263
Chapter reviews. *See* Reviews, chapter
Chapter tests. *See* Tests, chapter
Charles's law, 353
Chord, 377
Circles, 377–379
- area, 414–417
- parts of, 377

Circumference, 390–393, 414, 428, 430

Classification
- exploring, 13
- graphs, 347
- integers, 13
- prime numbers, 166
- triangles, 374

Clock arithmetic, 286
Coefficient, 66, 80
Common multiples, 168

Commutative property
- of addition, 58–59, 108
- of multiplication, 58–59

Compatible numbers, 105
Complementary angles, 366
Composite numbers, 163
Compound events, 522, 523
Computer, 38, 43, 91, 118, 144, 147, 209, 295, 299, 338, 373, 417, 457
- data bases, 122–123
- exploring, 38–39, 83–84, 122–123
- graphing, 83–84, 209–210
- simulations, 338–339
- spreadsheets, 38–39, 83–84

Concave polygon, 370
Conclusions, exploring, 541
Cones, 421
- surface area, 430–431
- volume, 441–444

Congruence, 380–383, 385
Congruent figures, 380–383
Constant, 66
Convex polygon, 370
Coordinate plane, 316–319
Corresponding parts, 380, 384
Cosine, 473
Counterexample, 376
Counting principle, 508–511
Cramers' rule, 555
Critical thinking, exploring
- a hypothesis, 444
- classification, 13, 166
- clock arithmetic, 286
- conclusions, 541
- decision making, 32–33, 88–89, 126–127, 154–155, 222–223, 242–243, 306–307, 354–355, 398–399, 434–435, 478–479, 528–529, 556–557
- deductive reasoning, 387

Critical thinking (cont.)
graphing, 319, 347
histograms, 501
inductive reasoning, 376
logic, 65, 472
logical families, 118
number patterns, 144
Pascal's triangle, 148
patterns, 457
patterns in division, 110
patterns in scientific notation, 214
quantitative comparisons, 303
Venn diagrams, 21, 240
visual thinking, 79, 511
visual perception, 259
visualization, 202, 423
Cross products
fractions, 185
proportions, 237
Cube, exploring patterns in a, 424
Cumulative reviews. *See* Reviews, cumulative
Cylinder, 421, 428
surface area, 428–429
volume, 437–440, 441

D

Data, 7, 12, 17, 20, 25, 29, 37, 43, 46, 47, 50, 57, 65, 75, 78, 82, 87, 91, 103, 107, 124, 126, 136, 151, 153, 154, 165, 170, 175, 191, 195, 198, 224, 225, 236, 239, 240, 248, 254, 258, 259, 262, 263, 266, 267, 278, 282, 286, 290, 294, 295, 299, 303, 327, 343, 387, 393, 396, 416, 445, 459, 472, 476, 481, 491, 495, 500, 511, 515, 525, 553, 560, 561
bimodal, 488
box and whisker plots, 502–505
in tables, 120–121
line plots, 492–495
range, 489
stem and leaf plots, 498–501
Data Bases, exploring, 122–123
Data file
Chap. 1, 2–3
Chap. 2, 52–53
Chap. 3, 96–97
Chap. 4, 138–139
Chap. 5, 180–181
Chap. 6, 230–231
Chap. 7, 272–273
Chap. 8, 312–313
Chap. 9, 360–361
Chap. 10, 404–405
Data file (cont.)
Chap. 11, 450–451
Chap. 12, 486–487
Chap. 13, 534–535
Datapoint, 38, 39, 83, 84, 122, 123, 338, 339
Decimal point, 99
Decimal square, 98
Decimals, 98–110
and fractions, 188–191, 196–197
and percents, 244, 245–248
estimating, 104–107
exploring, 98–99
in expressions, 108–110
models of, 98–99, 101, 244
repeating, 189
rounding, 103, 104
Decision making, 32, 88, 126, 154, 222, 242, 306, 354, 398, 434, 478, 528, 556
Deductive reasoning, exploring, 387
Denominator, 182–183, 185
Dependent events, 522–523
Determinant, 555
Diagonal, 370
parallelograms, 410
trapezoids, 411
Diagrams
drawing, 264–266, 394–395, 477–479
tree, 508, 522
triangle, 256–257
Diameter, 377, 390–391
Direct variation, 353
Distributive property
and solving equations, 279, 288
of multiplication over addition, 62, 108
of multiplication over subtraction, 62
multiplying binomials, 550
multiplying polynomials, 546
Divide and average method, 457
Divisibility, 159–162
Division
equations, 115–118
exploring, 110
integers, 26–29
patterns in, 110
properties for inequalities, 297
property of equality, 76–77
rational numbers, 207–208, 211
with exponents, 211
Drawing
a diagram, 264–266, 394–395, 477–479
scale, 241–243

E

Ecology, 87–89
decision making, 88–89
Equations, 54
addition, 72–75, 111–114, 215–217
addition property, 73, 111
division, 76–79, 115–118
division property, 76
equivalent, 72
exploring, 70–71, 274
false, 54, 218
graphing, 324–327
linear, 324–327, 340–343
models of, 70–71, 215, 274, 287
multiplication, 76–79, 115–118, 218–221
multiplication property, 77, 115
open, 54
percents and, 256–259
simplifying and solving, 275–278, 279–282
subtraction, 72–75, 111–114, 215–217
subtraction property, 72–73, 111
two-step, 274–278
variables and, 54–57
with variables on both sides, 287–290
Equivalent
decimals, 99
fractions, 182, 184–187, 188, 192
Estimation, 17, 88, 89, 104–107, 110, 130, 147, 183, 200, 201, 221, 247, 259, 339, 453
decimals, 104–107
fractions, 183
percents, 244, 247
Events, 512
compound, 522, 523
independent and dependent, 522–525
random, 512
Exploring
a hypothesis, 444
classification, 13, 166
clock arithmetic, 286
conclusions, 541
counting problems, 171
data bases, 122–123
data in tables, 120–121
decimals, 98
deductive reasoning, 387
equations, 70–71
factors and multiples, 156–157
fair and unfair games, 506–507
fractions, 182–183

Exploring (cont.)
graphing, 83–84, 319, 347
histograms, 501
inductive reasoning, 376
inequalities, 291
infinity, 209–210
integers, 8–9
logic, 472
logical families, 118
logical thinking, 65
mind-reading tricks, 549
misuse of statistics, 496–497
number patterns, 18, 144
number relationships, 250–251
Pascal's triangle, 148, 516–517
patterns, 457
patterns in a cube, 424
patterns in division, 110
patterns in scientific notation, 214
percents, 244
pi (π), 389
Pick's theorem, 418–419
polynomials, 536–537
quantitative comparisons, 303
ratios, 232
right triangles, 460
segments and angles, 369
simulations, 338–339
slope, 328–329
spreadsheets, 38–39
square numbers, 140
square roots, 452–453
two-step equations, 274
Venn diagrams, 21, 240
verbal communication, 314–315
visual perception, 259
visual thinking, 79, 511
visualization, 202, 423
with LOGO, 373
Exponents, 141–155, 546
negative, 212
rational numbers with, 211–214
rules, 145–148
scientific notation, 149–152
zero as, 211
Expressions
evaluating, 44–46
numerical, 34
polynomials as, 536
variable, 34, 66–69
with decimals, 108–110
Extra practice
Chap. 1, 566
Chap. 2, 567
Chap. 3, 568
Chap. 4, 569
Chap. 5, 570
Extra practice (cont.)
Chap. 6, 571
Chap. 7, 572
Chap. 8, 573
Chap. 9, 574
Chap. 10, 575
Chap. 11, 576
Chap. 12, 577
Chap. 13, 578

F

Factorials, 509
Factors, 156–157, 159–162
exploring, 156–157
exponents, 141
greatest common factor (GCF), 167–170
prime, 163–166, 185
False equations, 54, 218
Fitness and health, decision making, 32–33
FOIL method, multiplying binomials, 550
Formulas, 128–131
batting averages, 129
circumference, 390
distance, 128
surface area, 427
Fraction bars, 182
Fractions, 182–191
and decimals, 188–191, 196–197, 245
and percents, 245
comparing, 196–198
denominator, 183, 185
equivalent, 182, 184–187, 188, 192
estimation, 183
exploring, 182–183
improper, 188, 218
lowest terms, 185
models of, 182, 184
numerator, 182, 183, 185
ordering, 197
ratio expressed as, 233
rational numbers as, 192
reciprocals, 207
with variables, 184, 185, 199, 206
Frequency distribution, 492–495
Frequency tables, 492–495
Front-end estimation, decimals, 105
Functions
applications, 42–44, 46, 57, 87–89, 140, 209–210, 353–355, 360
computer graphing, 83
direct variation, 353
evaluating, 42–44, 56–57
Functions (cont.)
graphing, 324–327, 331, 340–343
indirect variation, 353
linear, 324–327, 331, 340–343
trigonometric, 473–476

G

games, fair and unfair, 506–507
exploring problem solving, 506–507
Geometry and measurement, 360–399, 404–445, 460–479
applications, 53, 61, 62, 63, 64, 66, 80, 104, 106, 110, 125–127, 153–155, 159, 163, 206, 234, 236, 238, 241–243, 244, 247, 259, 312, 353–357, 360–361, 394–395, 397–399, 404, 405, 425–426, 433–435, 440, 451, 454, 461, 465, 477–479, 520, 534–535, 536–537, 539–541, 542, 543, 545, 546, 547, 548, 550, 552, 558. *See also* individual topics
GCF. *See* Greatest common factor (GCF)
Glossary, 581–588
Graphing
computer, 83–84, 209–210
exploring, 83–84, 319, 347
inequalities, 291, 292
linear equations, 324–327, 331, 340-343
problem solving by, 334–336
system of linear equations, 340–343
Graphs
bar, 83–84, 209–210
circle, 84, 191, 378
classifying, 347
double bar, 84
line, 83
Greatest common factor (GCF), 167–170, 185

H

Height
cylinders, 428
prisms, 427,
rectangles, 406
triangles, 410. *See also* Altitude; Slant height
Hexagonal prism, 420
Hexagonal pyramid, 420
Histograms, exploring, 501
Hundredths, 98

Hypotenuse, 460, 461, 469, 470
Hypothesis, exploring a, 444

I

Identities
additive, 59
multiplicative, 59
Improper fractions, 188, 218
Independent events, 522–525
Indirect variation, 353
Inductive reasoning, exploring, 376
Inequalities, 291, 303
and addition and subtraction properties for, 296
and multiplication and division properties for, 297
exploring, 291
graphing, 291, 292–295, 348–351
linear, 344–351
one-step, 296–299
solving, 296–299, 300–303, 344–347
two-step, 300–303
Infinity, exploring, 209–210
Integers, 4–33
absolute value, 5
addition, 8–9, 10–13
and number line, 4–5
classification, 13
comparing, 5
consecutive, 86
division, 26–29
exploring, 8–9
models of, 8
multiplying, 22–25
opposite, 4–5, 8–9
positive and negative, 4, 8
set of, 4
subtraction, 14–17
Intersection
perpendicular lines, 316
points of, 363
segments, rays, or lines, 367
Venn diagram, 21
Inverse operations, 26, 72
Irrational numbers, 455
Isosceles right triangle, 469

L

Lateral area, 427. *See also* Surface area
LCD, 196. *See* Least common denominator (LCD)
LCM, 168–170. *See* Least common multiple
Least common denominator (LCD), 196
Least common multiple (LCM), 168–170
Legs, triangle, 460, 461, 470
Light year, 152
Like terms, 66, 80
combining, 279, 288
Line plots, 492–495
Line symmetry, 381
Linear equations
graphing, 324–327, 340–343
slope-intercept form, 331
solving systems of, 340–343
system of, 340
Linear inequalities
graphing, 348–351
solving, 344–347, 349
system of, 349
Linear programming, 351
Lines
coordinate plane, 316
graphing, 331
graphs of linear equation, 324
parallel, 363
perpendicular, 316
properties, 362
skew, 363
slope, 329, 330
symbol, 362
Logic, exploring, 472
Logical thinking, exploring, 65
LOGO, exploring with, 373
Lowest terms, 185

M

Maps, 241, 316
Matrices, 555
Mean, 27, 488–491, 493
Median, 488–491, 493
Mental math, 12, 16, 24, 28, 42, 45, 56, 60, 61, 64, 75, 77, 78, 109, 113, 117, 143, 187, 198, 201, 202, 208, 217, 254, 258, 262, 277, 281, 289, 303, 318, 319, 322, 327, 332, 378, 438, 456, 463, 467, 471, 475, 490, 510, 541, 547, 552
Metric system, 125–127
Mind-reading tricks, exploring, 549
Mixed number, 188, 200, 206, 218
Mode, 488–491, 492, 493
Models
decimals, 98, 99, 101, 244
equations, 70–71, 215, 274, 287
fractions, 182, 184
integers, 8
Models (cont.)
number sentences, 8
percents, 244
polynomials, 539, 542
problem solving with, 425–426
Pythagorean theorem, 461
rational numbers, 199, 206, 215
simulations, 338, 339
variable expressions, 34
zero, 8
Monomials, 538
multiplying by polynomials, 546–548
Monte Carlo simulations, 518
Multiples, 156–157, 159–162
common, 168
exploring, 156–157
factors and, 156–157, 159–162
least common (LCM), 168–170, 196
Multiplication
associative property, 58–59, 108
binomials, 550–553
commutative property, 58–59
equations, 115–118, 218–221
exponents, 145
FOIL method, 550
in monomials, 538
integers, 22–25
polynomials, 546–548
properties for inequalities, 297
property of equality, 77
rational numbers, 206–208
sum and difference of two terms, 551
Multiplicative identity, 59

N

Naming
circles, 377
polygons, 371
Natural numbers, 195
Nautical mile, 343
Net, 421, 427, 430
Notes & quotes, 10, 16, 20, 28, 35, 56, 67, 85, 102, 142, 161, 165, 166, 196, 215, 246, 248, 281, 296, 317, 353, 362, 378, 391, 393, 397, 428, 455, 461, 492, 512, 550
Number line
comparing integers, 5
fractions, 197
inequalities, 291, 292
integers on, 4, 10
line plots, 492–495
opposite integers, 4
rational numbers, 193

Number patterns, 18
exploring, 144
Number relationships, exploring, 250–251
Number sentences, models, 8
Numbers
compatible, 105
mixed, 188, 200, 206, 218
natural, 195
negative, 4, 8, 10, 300
prime and composite, 163
rational, 192–202
relatively prime, 167
square, 140
Numerator, 183, 185
Numerical expressions, 34–47, 54
evaluating, 40–46

O

One-step inequalities, 296–299
Open equation, 54, 73
Operations
inverse, 26, 72
order of, 40–43, 56, 208
properties of, 58–61
Opposite, 4
integers, 4–5, 8, 9, 74
solving equations, 112
square/square roots, 452
Order of operations, 40–43, 56, 208
Ordered pairs, 316, 317, 320, 340
Origin, 316
Ounces, 260

P

Parallel lines, 363
Parallel planes, 363
Parallelograms
area, 407–409, 414
diagonal, 410
Pascal's triangle, 148
exploring, 516–517
Pattern blocks, 182
Patterns, 19–21
exploring, 424, 457
in division, 110
in scientific notation, 214
number, 18, 144
repeating decimal, 189
Percent(s), 244–248
and decimals, 244, 245–246
and equations, 256–259
and fractions, 245
as ratio, 244, 245, 253
estimating, 244, 247
Percent(s) (cont.)
exploring, 244
models, 244
of change, 260–263
proportions and, 245, 252–255
Perfect number, 161
Perimeter, 327, 390–393
Periodic table, 296
Pi (π), 389, 390, 391, 414, 431
exploring, 389
Pick's theorem, exploring, 418–419
Planes
parallel, 363
properties, 362
symbol, 362. *See also* Coordinate plane
Point
center of circle, 377
intersections as, 363
properties, 362
symbol, 362
Polygons, 370–372
convex and concave, 370
naming, 371
regular, 371
Polyhedrons, 420
Polynomials, 536–541
addition and subtraction, 542–545
exploring, 536–537
multiplication by monomials, 546–548
Pound, 260
Powers, 141, 145, 146, 546,
of ten, 149
Prime factorization, 163–166, 167
Prime numbers, 163
classification, 166
Principal square root, 454
Prisms, 420
base, 420
lateral area, 427
surface area, 427
types, 420
volume, 437–440, 441
Probability, 512–515
dependent events, 523
exploring, Pascal's triangle, 516–517
independent events, 522
Monte Carlo simulation, 518–519
Problem solving applications
buying a car, 305–307
direct and indirect variation, 353–355
ecology, 87–89
fitness and health, 31–33
making predictions, 527–529
parade floats, 433–435
Problem solving applications (cont.)
precision drawing, 477–479
scale drawing, 241–243
simulations, 518–519
stock market, 222–224
tessellations, 397–399
using matrices, 555–557
using the metric system, 125–127
water resources, 153–155
Problem solving practice, 47, 91, 124, 175, 225, 267, 283, 337, 396, 445, 481, 526, 561
Problem solving strategies
account for all possibilities, 172–174
draw a diagram, 264–266, 394–395
guess and test, 85–86
look for a pattern, 19–21
make a model, 425–426
multiple strategies, 558
simplify the problem, 132
simulating a problem, 458–459, 518–520
solve by graphing, 334–337
working backwards, 203–204
writing equations, 284–286
Product, 22, 34, 146
Projects, 6, 7, 12, 29, 30, 39, 75, 82, 84, 89, 90, 113, 119, 123, 131, 144, 151, 152, 155, 158, 187, 191, 205, 214, 220, 232, 235, 239, 240, 249, 263, 299, 304, 319, 323, 333, 336, 351, 364, 383, 387, 388, 409, 423, 429, 436, 439, 480, 497, 504, 505, 507, 515, 521, 525, 554
Properties
addition and subtraction for inequalities, 296
addition and subtraction of equality, 72–75, 111
algebraic, 287
associative, 58–59, 108
commutative, 58–59, 108
distributive, 62–65, 108, 279, 288, 546, 550
identity, 59
line, 362
multiplication and division for inequalities, 297
multiplication and division of equality, 76–79, 115
plane, 362
point, 362
ray, 362
segment, 362
Proportions, 233–236, 384
and percents, 245, 252–255
same units, 237

Proportions (cont.)
solving, 237–240
Protractor, 365
Pyramids, 420
surface area, 430–432
types, 420
volume, 441–444
Pythagorean theorem, 461–464, 469
converse, 462
model, 461

Q

Quadrants, 316, 317
Quadrilaterals, 371. *See also* Parallelograms; Rectangles; Squares
Quantitative comparisons, exploring, 303
Quartile, 502
Quotient, 26

R

Radius, 377, 391, 414, 431, 441
Random events, 512
Range, 489
Rates, 234–236
unit, 234
Rational numbers, 192–202
absolute value, 193
addition and subtraction, 199–202
equations, 215
irrational numbers distinguished, 455
multiplication and division, 206–208, 211
negative, 192
on number lines, 193
with exponents, 211–214
Ratios, 232–236, 473
as a fraction, 233
exploring, 232
percents as, 244, 245, 253
proportions, 233
rates, 234
slope as, 330
trigonometric, 473–476
Rays, 362
properties, 362
symbol, 362
Reasoning
deductive, 387
inductive, 376
Reciprocal, 207
Rectangles
area, 406–409
Rectangles (cont.)
base, 406
height, 406
Region, coordinate plane and, 316
Regular polygon, 371
Relatively prime numbers, 167
Reviews
Chapter, 48–49, 92–93, 134–135, 176–177, 226–227, 268–269, 308–309, 356–357, 400–401, 446–447, 482–483, 530–531, 562–563
Cumulative, 51, 95, 137, 179, 229, 271, 311, 359, 403, 449, 485, 533, 565
Right triangles
exploring, 460
45°-45°-90°, 469
isosceles, 469
Pythagorean theorem, 461–464, 469
similar, 465–468
30°-60°-90°, 470
Rounding decimals, 103, 104

S

Sample, 496
Sample space, 512, 522,
Scale drawing, 241–243
maps, 241
Scientific notation, 149–152
exploring patterns in, 214
standard notation compared, 149
Segments
exploring, 369
properties, 362
symbol, 362
Set, of integers, 4
Sieve, 166
Similar, 384
figures, 384–387
right triangles, 465–468
triangles, 385
Simplifying
equations, 279–282
problems, 132–133
variable expressions, 66–69
Simulations
exploring, 338–339
in problem solving, 458–459, 518–520
Monte Carlo, 518
Sine, 473
Slant height, 430
Slope, 328–333
and y-intercept, 330–333
exploring, 328–329
formula, 330
Solution
of an equation, 55, 73
of an equation in two variables, 320
of a system of linear equations, 340
Space figures, 420–423
Spheres, 421
surface area, 431
volume, 441–444
Spreadsheets, 83–84
cells, 38–39
exploring, 38–39
Square numbers
binomials, 551
exploring, 140
opposite of square roots, 452
Square pyramids, 420
Square root(s)
exploring, 452–453
finding, 454–457
table, 579
Squares, 371
diagonal of, 370. *See also* Quadrilaterals; Square numbers
Statistics, 486–527. *See also* individual topics
Stem and leaf plots, 498–501
back-to-back, 499
Stock market, 222–224
Subtraction
equations, 72–75, 111–114, 215-217
integers, 14–17
polynomials, 542–545
properties for inequalities, 296
property of equality, 72–73, 111
rational numbers, 199–202
Supplementary angles, 366
Surface area, 427
cones, 430–432
cylinders, 428–429
prisms, 427–429
pyramids, 430–432
spheres, 431
Symbols
approximate equality, 104
congruence, 380
division, 26
equality, 5, 55, 185
factorials, 509
grouping, 40, 58, 141
inequality, 5, 99, 292, 300, 344
line, 362
multiplication, 22
plane, 362
point, 362
ray, 362
repeating decimal, 189

Symbols (cont.)
- right angle, 365
- segment, 362
- similar, 384
- square root, 452, 454

Symmetry, line, 381

T

Tables
- data in, 120–121
- exploring, 120–121
- frequency, 492–495
- periodic, 296
- square root, 579
- trigonometric, 580
- using to solve equations, 320

Tangent, 473
Tangram, 202, 372
Tenths, 98
Terms, 66
- combining like, 67–69, 279, 288, 542
- monomials as, 538
- product of sum and difference, 551

Tessellations, 397
Tests, chapter, 50, 94, 136, 178, 228, 270, 310, 358, 402, 448, 484, 532, 564
Think about it, 3, 53, 97, 139, 181, 231, 273, 313, 361, 405, 451, 487, 535
Thousandths, 100
Ton, 87
Trajectory, 338
Trapezoidal prisms, 420
Trapezoids
- area, 411–413
- bases, 411
- diagonal, 411

Tree diagram, 508, 522
Triangle diagrams, 256–258
Triangles, 374–376
- area, 410–413
- base, 410
- classifying, 374
- congruent, 380
- exploring, 460
- right, 460, 461–472
- similar, 385
- sum of measures of interior angles, 374

Triangular prisms, 420
Triangular pyramids, 420
Trigonometric ratios, 473–476
- cosine, 473
- sine, 473
- tangent, 473

Trigonometric table, 580
Trinomial, 539
Two-step equations, 274–278
- exploring, 274
- simplifying and solving, 279–282
- writing, 284–286

Two-step inequalities, 300–303

U

Unit fractions, 209, 215
Unit rates, 234
Units
- cubic, 437
- using same, 237, 260, 412

V

Variable expressions, 34–37, 54, 80, 215
- models, 34
- simplifying, 66–71

Variables, 34, 80, 108
- and equations, 54–57
- and expressions, 34–39
- and fractions, 184, 185, 199, 206
- in system of linear equations, 340
- isolating in equations, 215, 321
- on both sides of equations, 287–290

Venn diagrams
- exploring, 21, 240

Verbal communication, exploring, 314–315
Vertex, 365, 366
- cones, 421
- pyramids, 420

Vertical angles, 366
Visual perception, exploring, 259
Visual thinking, exploring, 79, 511
Visualization, 202
- exploring, 423

Volume
- cones or pyramids, 441–444
- cubic units, 437
- prisms or cylinders, 437–440
- spheres, 441

W

Water resources, 153–155
- decision making, 154–155

Word problems, using equations to solve, 284–286
Write, 7, 8, 13, 15, 18, 25, 43, 46, 57, 61, 65, 70, 82, 99, 103, 107, 110, 114, 117, 140, 143, 144, 152, 157, 191, 195, 209, 217, 224, 232, 244, 248, 251, 254, 255, 257, 262, 263, 290, 315, 328, 329, 336, 351, 363, 364, 386, 387, 419, 422, 424, 444, 452, 491, 497, 499, 504, 511, 515, 517, 525, 536, 537, 541, 545, 547 549, 553

X

***x*-axis**, 316
***x*-coordinate**, 316
***x*-intercept**, 325

Y

***y*-axis**, 316
***y*-coordinate**, 316
***y*-intercept**, 325, 331

Z

Zero, 9, 73
- as exponent, 211
- division, 26
- in decimals, 100
- model, 8

Acknowledgments

PHOTO CREDITS

KEY TO PHOTO SOURCE ABBREVIATIONS
Bruce Coleman, Inc.= BC; Freelance Photographers Guild = FPG; Ken Karp = KK; Russ Lappa = RL; Larry Lawfer = LL; Picture Cube = PC; PhotoEdit = PE; Photo Researchers, Inc. = PR; Tom Stack & Associates = TSA; Stock Market = SM; Tony Stone Worldwide = TSW; Woodfin Camp & Associates = WC.

KEY TO PHOTO POSITION ON TEXT PAGE
T=Top; **M**=Middle; **B**=Bottom; **L**=Left; **R**=Right.

Back Cover: Top, James H. Carmichael/BC; Center, PR; Bottom, Hank Morgan/PR.

Front Matter: i T, Photo by Mark Richards; **i B,** Hank Morgan/PR; **vi TR,** Patrick Aventurier/Gamma-Liaison; **vi BL,** Rolf Sorensen/TSW; **vii TMR,** Nancy Sheehan; **vii TR,** Chris Hackett/The Image Bank; **viii TL,** Tony Freeman/PE; **ix TL,** David Ball/SM; **ix BR,** TSW; **x MR,** J. J. Raynal/PR; **xi TM,** Brownie Harris/SM; **xi BL,** European Space Agency/PR.

CHAPTER ONE 2, Annie Griffiths/BC; **2-3,** PR; **3,** Rolf Sorensen/TSW; **4,** Keith Lanpher; **9,** Richard Haynes; **13,** David Madison/BC; **17,** David Austen/TSW; **23,** Greg Vaughn/TSA; **30,** Thomas Braise/SM; **33,** FPG; **40,** Bob Daemmrich/TSW; **42,** Photo by Mark Richards; **47,** Jack Finch/PR.

CHAPTER TWO 52 M, Armando Jenik/The Image Bank; **52 B,** Carl Roessler/Animals Animals; **52 MR,** Carl Roessler/FPG; **52-53,** John L. Pontier/Animals Animals; **53,**FPG; **61 T,** Clyde H. Smith/FPG; **61 (inset),** Lee Foster/FPG; **67,** The Granger Collection; **69,** Peter Menzel; **71, LL**; **76,** Michael Melford/The Image Bank; **80,** LL/PC; **89,** Arnold John Kaplan/PC; **90,** Brownie Harris/SM.

CHAPTER THREE 96, Patrick Aventurier/Gamma-Liaison; **96-97,** Nancy Sheehan; **97 T,** B, Nancy Sheehan; **98, LL; 99,** Meral Dabcovich; **100,** Jack Dermid/BC; **102,** The Granger Collection; **111,** FPG; **115,** Steve Ogden/TS; **119,** Pete Saloutos/SM; **120,** Peter Steiner/SM; **124,** TSW; **127 T,** Bob and Clara Calhoun/BC; **127 B,** Meral Dabcovich; **129,** UPI/Bettmann Newsphotos; **133,** Doug Armand/TSW.

CHAPTER FOUR 139 T, Peter Miller/The Image Bank; **139 BR,** Chris Hackett/The Image Bank; **139 (inset),** Nancy Sheehan; **141,** Robert Knauft/PR; **144,** M. Richards/PE; **150,** FPG; **152,** NASA; **155,** KK; **156,** Michal Heron/WC; **167, 168,** KK.

CHAPTER FIVE 180, Milton Feinberg/PC; **180-181,** Dave Schaefer/PC; **181, LL; 181 T,** Richard Anders/FPG; **181 B,** Phil Degginger/BC; **183,** Richard Haynes; **187, LL; 189,** KK; **192,** Kim Taylor/BC; **195,** Sarah Putnam/PC; **196,** Mackson/FPG; **198,** Photo by Mark Richards; **199,** Franz Lazi/FPG; **205,** Michael Keller/FPG; **206,** TSW; **211,** Chris Bjornberg/PR; **213,** Photo by Mark Richards; **214,** Thomas Kitchin/TS; **218,** Meral Dabcovich; **223,** Alan Klehr/TSW; **228,** Ken Karp.

CHAPTER SIX 230, Tony Freeman/PE; **231,** David Young-Wolff/PE; **232,** KK; **236,** Kindra Clineff PC; **237,** Zur Veroffentlichung/FPG; **240,** Robert Huntzinger/SM; **243 T,** LL; **243 B,** Nancy Sheehan; **245,** Michal Heron/WC; **248,** Brian Seed/TSW; **249,** Steve Liss/Gamma-Liaison; **250-251,** Richard Hutchings/ InfoEdit; **252,** Martin Rogers/TSW; **259,** LL; **260,** Richard Laird/FPG; **265,** Nancy Sheehan; **266,** Joe Baraban/SM.

CHAPTER SEVEN 272, John Blaustein/WC; **272-273,** Joe Tower/SM; **273,** Chris Sorenson/SM; **274,** Richard Haynes; **276,** J. H. Robinson/PR; **279,** Tom Campbell/FPG; **282,** T. J. Florian/Rainbow; **284,** Scott Deitrich/TSW; **292,** Gary Buss/FPG; **295,** Frank Siteman/PC; **296,** Novosti/Science Photo Library/PR; **300,** Jon Feingersh/TSA; **303,** Alan Carey/PR; **304,** Dick Luria/FPG; **307,** Charles West/SM.

CHAPTER EIGHT 312, Tom Bean/SM; **312-313,** The Granger Collection; **313,** Jeffry Myers/FPG; **314 (all),** RL; **315,** Richard Haynes; **318,** Keith Olson/TSW; **320,** TSW; **323,** Charles Seaborn/WC; **329,** Cathlyn Melloan/TSW; **333,** Wesley Bocxe/PR; **335,** Peter Gridley/FPG; **336,** David Young-Wolff/PE; **337,** David Young-Wolff/PE; **338, 339 L,** Lily Yamamoto; **339 R,** European Space Agency/Science Photo Library/PR; **344,** Tony Freeman/PE; **346,** All photos by Mark Richards; **352,** Richard Hutchings/PR; **354,** J. Mejuto/FPG; **355,** KK.

CHAPTER NINE 361, Henley and Savage/TSW; **360-361,** Nancy Sheehan; **360,** David Ball/SM; **362,** D. Wilder/TSA; **367,** S.L. Craig/BC; **369 (all),** RL; **370,** John Lamb/TSW; **374,** David Ball/PC; **377 T,** Stanley Rowin/PC; **377 B,** Dr. Jeremy Burgess/Science Photo Library/PR; **380,** Tony Freeman/PE; **383 L,** Ray Coleman/PR; **383 M,** Brian Parker/TSA; **383 R,** Michael Keller/SM; **384,** Dan McCoy/Rainbow; **388,** Paulette Brunner/TSA; **389,** FPG; **394,** Vandystadt/PR; **395,** Michael Dunn/SM; **397,** Art Resource; **399,** Aga Khan Program Archives, M.I.T., Photo by George J. Kostaras, 1983.

CHAPTER TEN 404, FPG; **404-405,** FPG; **405,** TSW; **406,** RL; **411,** Tom Tracy/FPG; **414,** David Ball/PC; **415,** Meral Dabcovich; **417,** Photo by Mark Richards; **418, 419 (all),** RL; **421,** Richard Haynes; **430,** Hugh Sitton/TSW; **431 L,** Frank Cezus/FPG; **431 R,** Bob Brudd/TSW; **432,** Alan Smith/TSW; **434,** R. B. Sanchez/SM; **435 T,** Joe Sohn/SM; **436,** Dan McCoy/Rainbow; **439,** Photo by Mark Richards; **440,** The Granger Collection; **441,** Tom Tracy/FPG; **443,** RL; **445,** José Carrillo/TSW.

CHAPTER ELEVEN 450, Tom Sanders/SM; **450-451,** J. J. Raynal/PR; **451,** Richard Burda/FPG; **455,** RL; **457,** Bill Sanderson/Science Photo Library/PR; **458,** John Terence Turner/FPG; **459,** Henley and Savage/TSW; **461,** Photoworld/FPG; **468,** K&G Photo/FPG; **477,** David Jeffrey/The Image Bank; **479,** Stock Imagery; **480,** Bill Losh/FPG; **482,** J. J. Raynal/PR.

CHAPTER TWELVE 486 T, Tim Davis/Duomo; **486 B, 487,** David Madison/Duomo; **491,** Morris Lane/SM; **492,** David Conklin/PE; **503,** Mitchell Layton/Duomo; **504,** David Weintraub/PR; **507 T, B,** Richard Haynes; **508,** David Madison/Duomo; **513,** KK; **514,** Bob Peterson/FPG; **521,** Tom Tracy/FPG; **527,** Bonnie L. Lange/Stock Imagery; **528,** Alan Carey/PR; **529,** KK.

CHAPTER THIRTEEN 534, FPG; **535,** Stock Imagery; **537,** Richard Haynes; **545 T,** The Bettmann Archive; **545 B,** Robert P. Morrison/FPG; **546 T, B,** John Gillmoure/SM; **550,** The Granger Collection; **554,** Ed Lettau/FPG; **560,** KK; **561,** NASA/SB/FPG.

ILLUSTRATION CREDITS

Technical art by York Graphic Services, Inc., Synergy 2000 Series.

Bob Barner: 42, 86, 148, 198, 213, 281, 346, 417, 439

Eliot Bergman: 12, 25, 31, 53, 84, 103, 113, 128, 138, 139, 154, 181, 191, 221, 255, 259, 267, 299, 378, 396

Boston Graphics, Inc.: 17, 18, 27, 79, 150, 230, 231, 242, 264, 316, 360, 413, 423, 451, 465, 476, 486, 496, 500, 534, 545

John W. Cataldo: 450

Donald Doyle: 246

Function Thru Form, Inc., Guilbert Gates and Kathleen Katims: 241, 272, 312, 313, 404

Andrea Grassi: Decision Making logo, 458: figure icons, 313: Statue of Liberty icon

Mark Herman: 44, 110, 173, 217, 275, 319, 334, 363, 425, 433, 442, 469, 489, 557

Fran Jarvis: 507: spinner, 460: grid designs, 487: 511: 534

Barbara Maslen: 21, 32, 63, 104, 117, 151, 184, 219, 278, 343, 390, 506, 515, 524

Eve Melnechuk: Problem Solving logo

Terry Presnall: Calculator logo, computer logo, 2, 6, 10, 11, 19, 20, 29, 34, 52, 66, 88, 96, 101, 107, 111, 130, 131, 145, 153, 163, 172, 180, 204, 215, 230, 231, 232, 233, 239, 246, 247, 263, 273, 290, 312, 317, 361, 383, 385, 387, 416, 460, 487, 495, 499, 502, 519, 522, 534, 542, 543

Susan Spellman: 121, 287, 345, 517, 539, 559

Gary Torrisi: 2, 14, 46, 58, 72, 156, 157, 158, 210, 225, 330, 364, 404, 405, 420, 428, 438, 464, 497, 510, 535, 552

C. A. Trachok: 43, 75, 82, 100, 108, 123, 170, 174, 286, 351, 487, 511, 520

Cameron Wasson: 37, 54, 114, 203, 220, 235, 250, 251, 264, 285, 347, 426, 444, 453, 472, 509, 523, 538, 557

Any photo or illustration acknowledgment inadvertently omitted will be amended upon notification.

Index

A

Absolute value
  integers, 5, 141
  rational number, 192
Addition
  associative property, 58-59, 108
  commutative property, 58-59, 108
  equations, 72-75, 111-114, 215-217
  exponents, 145
  integers, 8-9, 10-13
  polynomials, 542-545
  properties of inequalities, 296
  property of equality, 73, 111
  rational numbers, 199-202
Additive identity, 59
Adjacent angles, 366
Algebraic factorization, 168
Algebraic properties, in solving equations, 287-288
Altitude
  parallelogram, 407
  triangle, 410
Angles, 365-369
  acute, 365
  adjacent, 366
  central, 377
  complementary, 366
  exploring, 369
  interior, 374
  measuring, 365
  obtuse, 365
  right, 365
  straight, 365
  supplementary, 366
  vertical, 366
Applications, *See also* Careers; Computer; Data; Data File; Exploring Functions; Geometry; Problem Solving Applications; Problem Solving Practice; Projects; Write
Area, 62
  circles, 414-417
  parallelograms, 406-409, 414
  rectangles, 406-409
  trapezoids, 410-413
  triangles, 410-413
  See also Surface area
Associative property
  of addition, 58, 108
  of multiplication, 59
At-risk students, 2F, 52F, 96F, 138F, 180F, 230F, 272F, 312F, 360F, 404F, 450F, 486F, 534F
Average, 27, 488-491. *See also* Mean

B

Base
  and exponent, 141
  cylinders, 428
  parallelograms, 407
  prisms, 420
  rectangles, 406
  trapezoids, 411
  triangles, 410
Bibliography. *See* Student bibliography; Teacher bibliography
Bimodal data, 488
Binomials, 539
  squaring, 551
Box and whisker plots, 502-505
Boyle's Law, 353
Bulletin board, 2B, 52B, 96B, 138B, 180B, 230B, 272B, 312B, 360B, 404B, 450B, 486B, 534B

C

Calculator, 7, 12, 16, 24, 28, 42, 45, 46, 56, 75, 78, 109, 113, 117, 130, 131, 143, 147, 151, 157, 165, 170, 187, 189, 190, 191, 198, 201, 204, 235, 247, 254, 258, 262, 277, 281, 289, 294, 295, 322, 346, 376, 379, 392, 408, 409, 412, 413, 429, 432, 439, 443, 463, 464, 467, 472, 481, 490, 510, 541, 547, 559
Calculator activities, 52E, 96E, 138E, 180E, 230E, 312E, 404E, 450E, 486E, 534E
Calorie, 31
Careers
  acoustical physicist, 205
  actuary, 521
  airline pilot, 30
  chef, 249
  civil engineer, 352
  demographer, 158
  director of marketing, 304
  finance manager, 554
  landscape architect, 480
  medical record technician, 119
  operations research analyst, 436
  optician, 388
  urban and regional planner, 90
Cells, spreadsheet, 38-39
Center, 377
  spheres, 421
Central angle, 377
Central tendency, measures of, 488, 489
Change, percent of, 260-263
Chapter resources, 2A, 52A, 96A, 138A, 180A, 230A, 272A, 312A, 360A, 404A, 450A, 486A, 534A
Chapter reviews. *See* Reviews, chapter
Chapter tests. *See* Tests, chapter
Charles's Law, 353
Chord, 377
Circles, 377-379
  area, 414-417
  parts of, 377
Circumference, 390-393, 414, 428, 430
Class activities. *See* Class speakers; Field trips
Classification
  exploring, 13
  graphs, 347
  integers, 13
  prime numbers, 166
  triangles, 374
Class speakers, 2B, 52B, 96B, 138B, 180B, 230B, 272B, 312B, 360B, 404B, 450B, 486B, 534B
Clock arithmetic, 286
Coefficient, 66, 80
Common multiples, 168
Commutative property
  of addition, 58-59, 108
  of multiplication, 58-59
Compatible numbers, 105
Complementary angles, 366
Composite numbers, 163
Compound events, 522, 523
Computer, 38, 43, 91, 118, 144, 147, 209, 295, 299, 338, 373, 417, 457
  data bases, 122-123
  exploring, 38-39, 83-84, 122-123
  graphing, 83-84, 209-210
  simulations, 338-339
  spreadsheets, 38-39, 83-84, 210
Computer activities, 2E, 52E, 96E, 180E, 230E, 272E, 312E, 360E, 450E, 486E, 534E
Concave polygon, 370
Conclusions, exploring, 541
Cones, 421
  surface area, 430-431
  volume, 441-444
Congruence, 380-383, 385
Congruent figures, 380-383
Constant, 66
Convex polygon, 370
Cooperative learning, 3, 53, 97, 139, 181, 231, 273, 313, 361, 405, 451, 487, 535
Coordinate plane, 316-319
Corresponding parts, 380, 384
Cosine, 473
Counterexample, 376
Counting principle, 508-511
Cramers' rule, 555
Critical thinking
  a hypothesis, 444
  classification, 13, 166
  clock arithmetic, 286
  conclusions, 541
  decision making, 32-33, 88-89, 126-127, 154-155, 222-223,

Critical Thinking (contd.)
242-243, 306-307, 354-355, 398-399, 434-435, 478-479, 528-529, 556-557
deductive reasoning, 387
graphing, 319, 347
histograms, 501
inductive reasoning, 376
logic, 472
logical families, 118
logical thinking, 65
number patterns, 144
Pascal's triangle, 148
patterns, 457
patterns in division, 110
patterns in scientific notation, 214
quantitative comparisons, 303
Venn diagrams, 21, 240
visual perception, 259
visual thinking, 79, 511
visualization, 202, 423
Cross products
fractions, 185
proportions, 237
Cube, exploring patterns in a, 424
Cumulative reviews. *See* Reviews, cumulative
Cylinder, 421, 428
surface area, 428-429
volume, 437-440, 441

D

Data, 7, 12, 17, 20, 25, 29, 37, 43, 46, 47, 50, 57, 65, 75, 78, 82, 87, 88, 91, 103, 107, 124, 126, 136, 147, 151, 153, 154, 165, 170, 175, 191, 195, 198, 224, 225, 236, 239, 240, 248, 254, 258, 259, 262, 263, 266, 267, 269, 278, 282, 286, 290, 294, 295, 299, 303, 323, 327, 336, 343, 387, 393, 395, 396, 416, 445, 459, 472, 476, 481, 491, 494, 495, 500, 501, 504, 511, 515, 525, 529, 553, 560, 561
bimodal, 488
box and whisker plots, 502-505
in tables, 120-121
line plots, 492-495
range, 489
stem and leaf plots, 498-501
Data analysis. See Data file
Data Bases, exploring, 122-123
Data collection. See Data file
Data file
Chap. 1, 2-3
Chap. 2, 52-53
Chap. 3, 96-97
Chap. 4, 138-139
Chap. 5, 180-181
Chap. 6, 230-231
Chap. 7, 272-273
Chap. 8, 312-313
Chap. 9, 360-361
Chap. 10, 404-405
Chap. 11, 450-451
Chap. 12, 486-487
Chap. 13, 534-535
***see also* Cooperative learning; Problem solving situation**
Datapoint, 38, 39, 83, 84, 122, 123, 338, 339
Decimal point, 99
Decimal square, 98
Decimals, 98-110
and fractions, 188-191, 196-197
and percents, 244, 245-248
estimating, 104-107
exploring, 98-99
in expressions, 108-110
models of, 98-99, 101, 244
repeating, 189
rounding, 103, 104
Decision making, 32, 88, 126, 154, 222, 242, 306, 354, 398, 434, 478, 528, 556
Deductive reasoning, exploring, 387
Denominator, 182-183, 185
Dependent events, 522-523
Determinant, 555
Diagonal, 370
parallelograms, 410
trapezoids, 411
Diagrams
drawing, 264-266, 394-395, 477-479
tree, 508, 522
triangle, 256-257
Diameter, 377, 390-391
Direct variation, 353
Distributive property
and solving equations, 279, 288
of multiplication over addition, 62, 108
of multiplication over subtraction, 62
multiplying binomials, 550
multiplying polynomials, 546
Divide and average method, 457
Divisibility, 159-162
Division
equations, 115-118
exploring, 110
integers, 26-29
patterns in, 110
properties for inequalities, 297
property of equality, 76-77
rational numbers, 207-208, 211
with exponents, 211
Drawing
a diagram, 264-266, 394-395, 477-479
scale, 241-243

E

Ecology, 87-89
decision making, 88-89
English as a second language, 2F, 52F, 96F, 138F, 180F, 230F, 272F, 312F, 360F, 404F, 450F, 486F, 534F
Equations, 54
addition, 72-75, 111-114, 215-217
addition property, 73-74, 111
division, 76-79, 115-118
division property, 76-77
equivalent, 72
exploring, 70-71, 274
false, 54, 218
graphing, 324-327
linear, 324-327, 340-343
models of, 70-71, 215, 274, 287
multiplication, 76-79, 115-118, 218-221
multiplication property, 77, 115
open, 54
percents and, 256-259
simplifying and solving, 275-278, 279-282
subtraction, 72-75, 111-114, 215-217
subtraction property, 72-73, 111
two-step, 274-278
variables and, 54-57
with variables on both sides, 287-290
Equivalent
decimals, 99
fractions, 182, 184-187, 188, 192
Estimation, 17, 88, 89, 104-107, 110, 130, 147, 183, 200, 201, 221, 247, 259, 339, 453
decimals, 104-107
fractions, 183
percents, 244, 247
Events, 512
compound, 522, 523
independent and dependent, 522-525
random, 512
Exploring
a hypothesis, 444
classification, 13, 166
clock arithmetic, 286
computer data bases, 122-123
conclusions, 541
counting problems, 171
data in tables, 120-121
decimals, 98
deductive reasoning, 387
equations, 70-71
factors and multiples, 156-157
fair and unfair games, 506-507
fractions, 182-183
graphing, 83-84, 319, 347
histograms, 501
inductive reasoning, 376
inequalities, 291
infinity, 209-210
integers, 8-9
logic, 472

logical families, 118
logical thinking, 65
mind-reading tricks, 549
misuse of statistics, 496-497
number patterns, 18, 144
number relationships, 250-251
Pascal's triangle, 148, 516-517
patterns, 457
patterns in a cube, 424
patterns in division, 110
patterns in scientific notation, 214
percents, 244
pi (π), 389
Pick's theorem, 418-419
polynomials, 536-537
quantitative comparisons, 303
ratios, 232
right triangles, 460
segments and angles, 369
simulations, 338-339
slope, 328-329
spreadsheets, 38-39
square numbers, 140
square roots, 452-453
TRB see Calculator activities; Computer activities; Manipulative activities; Problem solving activities
two-step equations, 274
Venn diagrams, 21, 240
verbal communication, 314-315
visual perception, 259
visual thinking, 79, 511
visualization, 202, 423
with LOGO, 373
Exponents, 141-155, 546
negative, 212
rational numbers with, 211-214
rules, 145-148
scientific notation, 149-152
zero as, 211
Expressions
evaluating, 44-46
numerical, 34
polynomials as, 536
variable, 34, 66-69
with decimals, 108-110
Extra-credit problems, 2B, 52B, 96B, 138B, 180B, 230B, 272B, 312B, 360B, 404B, 450B, 486B, 534B
Extra practice
Chap. 1, 566
Chap. 2, 567
Chap. 3, 568
Chap. 4, 569
Chap. 5, 570
Chap. 6, 571
Chap. 7, 572
Chap. 8, 573
Chap. 9, 574
Chap. 10, 575
Chap. 11, 576
Chap. 12, 577
Chap. 13, 578

F

Factorials, 509
Factors, 156-157, 159-162
exploring, 156-157
exponents, 141
greatest common factor (GCF), 167-170
prime, 163-166, 185
False equations, 54, 218
Field trips, 2B, 52B, 96B, 138B, 180B, 230B, 272B, 312B, 360B, 404B, 450B, 486B, 534B
Films, videotapes and other media, 2A, 52A, 96A, 138A, 180A, 230A, 272A, 312A, 360A, 404A, 450A, 486A, 534A
Fitness and health, decision making, 32-33
FOIL method, multiplying binomials, 550
Formulas, 128-131
batting averages, 129
circumference, 390
distance, 128
surface area, 427
Fraction bars, 182
Fractions, 182-191
and decimals, 188-191, 196-197, 245
and percents, 245
comparing, 196-198
denominator, 182-183, 185
equivalent, 182, 184-187, 188, 192
estimation, 183
exploring, 182-183
improper, 188, 218
lowest terms, 185
models of, 182, 184
numerator, 182-183, 185
ordering, 197
ratio expressed as, 233
rational numbers as, 199
reciprocals, 207
with variables, 184, 185, 199, 206
Frequency distribution, 492-495
Frequency tables, 492-495
Front-end estimation, decimals, 105
Functions
applications, 42-44, 46, 57, 87-89, 140, 209-210, 353-355, 360
computer graphing, 83
direct variation, 353
evaluating, 42-44, 56-57
graphing, 324-327, 331, 340-343
indirect variation, 353
linear, 324-327, 331, 340-343
trigonometric, 473-476

G

Games, fair and unfair, 506-507
exploring problem solving, 506-507
Geometry and measurement, 360-399, 404-445, 460-479
applications, 53, 61, 62, 63, 64, 66, 80, 104, 106, 110, 125-127, 153-155, 159, 163, 206, 234, 236, 238, 241-243, 244, 247, 259, 312, 353-357, 360-361, 394-395, 397-399, 404, 405, 425-426, 433-435, 440, 451, 454, 461, 465, 477-479, 520, 534-535, 536-537, 539-541, 542, 543, 545, 546, 547, 548, 550, 552, 558. *See also* individual topics
GCF. *See* Greatest common factor (GCF)
Gifted students, 2F, 52F, 96F, 138F, 180F, 230F, 272F, 312F, 360F, 404F, 450F, 486F, 534F
Glossary, 581-588
Graphing
computer, 83-84, 209-210
exploring, 83-84, 319
inequalities, 291, 292
linear equations, 324-327, 331, 340-343
problem solving by, 334-336
system of linear equations, 340-343
Graphs
bar, 83-84, 209-210
circle, 84, 378
classifying, 347
double bar, 84
line, 83
Greatest common factor (GCF), 167-170, 185

H

Height
cylinders, 428
prisms, 427
rectangles, 406
triangles, 410
See also Altitude; Slant height
Hexagonal prism, 420
Hexagonal pyramid, 420
Histograms, exploring, 501
Hundredths, 98
Hypotenuse, 460, 461, 469, 470
Hypothesis, exploring a, 444

I

Identities
additive, 59
multiplicative, 59
Improper fractions, 188, 218
Independent events, 522-525
Indirect variation, 353
Inductive reasoning, exploring, 376

Inequalities, 291-303
  and addition and subtraction properties for, 296
  and multiplication and division properties for, 297
  exploring, 291
  graphing, 291, 292-295, 348-351
  linear, 344-351
  one-step, 296-299
  solving, 296-299, 300-303, 344-347
  two-step, 300-303
Infinity, exploring, 209-210
Integers, 4-33
  absolute value, 5
  addition, 8-9, 10-13
  and number line, 4-5
  classification, 13
  comparing, 5
  consecutive, 86
  division, 26-29
  exploring, 8-9
  models of, 8
  multiplying, 22-25
  opposite, 4-5, 8-9
  positive and negative, 4, 8
  set of, 4
  subtraction, 14-17
Intersection
  perpendicular lines, 316
  points of, 363
  segments, rays, or lines, 367
  Venn diagram, 21
Inverse operations, 26, 72
Irrational numbers, 455
Isosceles right triangle, 469

L

Lateral area. 427 *See also* Surface area
LCD, 196. *See* Least common denominator (LCD)
LCM. 168-170 *See* Least common multiple
Least common denominator (LCD), 196
Least common multiple (LCM), 168-170
Legs, triangle, 460, 461, 470
Light year, 152
Like terms, 66, 80
  combining, 279, 288
Linear equations
  graphing, 324-327, 340-343
  slope-intercept form, 331
  solving systems of, 340-343
  system of, 340
Linear inequalities
  graphing, 348-351
  solving, 344-347, 349
  system of, 349
Linear programming, 351
Line plots, 492-495
Lines
  coordinate plane, 316
  graphing, 331
  graphs of linear equation, 324
  parallel, 363
  perpendicular, 316
  properties, 362
  skew, 363
  slope, 329, 330
  symbol, 362
Line symmetry, 381
Logic, exploring, 472
Logical thinking, exploring, 65
LOGO, exploring with, 373
Lowest terms, 185

M

Manipulative activities, 2E, 52E, 96E, 138E, 180E, 230E, 272E, 312E, 360E, 404E, 450E, 486E, 534E
Maps, 241, 316
Materials, 2A, 52A, 96A, 138A, 180A, 230A, 272A, 312A, 360A, 404A, 486A, 534A
Matrices, 555
Mean, 27, 488-491, 493
Media. *See* Films, videotapes, and other media
Median, 488-491, 493
Mental math, 12, 16, 24, 28, 42, 45, 56, 60, 61, 64, 75, 77, 78, 109, 113, 117, 143, 187, 198, 201, 202, 208, 217, 254, 258, 262, 277, 281, 289, 303, 318, 319, 322, 327, 332, 378, 438, 456, 463, 467, 471, 475, 490, 510, 541, 547, 552
Metric system, 125-127
Mind-reading tricks, exploring, 549
Mixed number, 188, 200, 206, 218
Mode, 488-491, 492, 493
Models
  decimals, 98, 99, 101, 244
  equations, 70-71, 215, 274, 287
  fractions, 182, 184
  integers, 8-9
  number sentences, 8-9
  percents, 244
  polynomials, 539, 542
  problem solving with, 425-426
  Pythagorean theorem, 461
  rational numbers, 199, 206, 215
  simulations, 338, 339
  variable expressions, 34
  zero, 8
Monomials, 538
  multiplying by polynomials, 546-548
Monte Carlo simulations, 518
Multiples, 156-157, 159-162
  common, 168
  exploring, 156-157
  factors and, 156-157, 159-162
  least common (LCM), 168-170, 196
Multiplication
  associative property, 58-59, 108
  binomials, 550-553
  commutative property, 58-59
  equations, 115-118, 218-221
  exponents, 145
  FOIL method, 550
  in monomials, 538
  integers, 22-25
  polynomials, 546-548
  properties for inequalities, 297
  property of equality, 77
  rational numbers, 206-208
  sum and difference of two terms, 551
Multiplicative identity, 59

N

Naming
  circles, 377
  polygons, 371
Natural numbers, 195
Nautical mile, 343
Net, 421, 427, 430
Notes & quotes, 10, 16, 20, 28, 35, 56, 67, 85, 102, 142, 161, 165, 166, 196, 215, 246, 248, 281, 296, 317, 353, 362, 378, 391, 397, 428, 455, 461, 492, 512, 550
Number line
  comparing integers, 5
  fractions, 197
  inequalities, 291, 292
  integers on, 4, 10
  line plots, 492-495
  opposite integers, 4
  rational numbers, 193
Number patterns, 18
  exploring, 144
Number relationships, exploring, 250-251
Numbers
  compatible, 105
  mixed, 188, 200, 206, 218
  natural, 195
  negative, 300
  prime and composite, 163
  rational, 192-202
  relatively prime, 167
  square, 140
Number sentences, models, 8
Numerator, 183, 185
Numerical expressions, 34-47, 54
  evaluating, 40-47

O

One-step inequalities, 296-299
Open equation, 54, 73

Operations
inverse, 26, 72
order of, 40-43, 56, 208
properties of, 58-61
Opposite, 4
integers, 4-5, 8, 9, 74
solving equations, 112
square/square roots, 452
Order of operations, 40-43, 56, 208
Ordered pairs, 316, 317, 320, 340
Origin, 316
Ounces, 260

P

Parallel lines, 363
Parallel planes, 363
Parallelograms
area, 407-409, 414
diagonal, 410
Pascal's triangle, 148
exploring, 516-517
Pattern blocks, 182
Patterns, 19-22
exploring, 424, 457
in division, 110
in scientific notation, 214
looking for 19-21
number, 18, 144
repeating decimal, 189
Percent(s), 244-248
and decimals, 244, 245-246
and equations, 256-259
and fractions, 245
as ratio, 244, 245, 253
estimating, 244, 247
exploring, 244
models, 244
of change, 260-263
proportions and, 245, 252-255
Perfect number, 161
Perimeter, 327, 390-393
Periodic table, 296
Pi (π), 389, 390, 391, 414, 431
exploring, 389
Pick's theorem, exploring, 418-419
Planes
parallel, 363
properties, 362
symbol, 362. *See also* Coordinate plane
Point
center of circle, 377
intersections as, 363
properties, 362
symbol, 362
Polygons, 370-372
convex and concave, 370
naming, 371
regular, 371
Polyhedrons, 420
Polynomials, 538-541
addition and subtraction, 542-545
exploring, 536-537
multiplication by monomials, 546-548
Pound, 260
Powers, 141, 145, 146, 546,
of ten, 149
Prime factorization, 163-166, 167
Prime numbers, 163
classification, 166
Principal square root, 454
Prisms, 420
base, 420
lateral area, 427
surface area, 427
types, 420
volume, 437-440, 441
Probability, 512-515
dependent events, 523
exploring Pascal's Triangle, 516-517
independent events, 522
Monte Carlo simulation, 518-520
Problem solving
exploring, fair and unfair games, 506-507
with patterns, 19-22
Problem solving activities, 2E, 52E, 96E, 138E, 180E, 230E, 272E, 312E, 360E, 404E, 450E, 486E, 534E
Problem solving applications,
buying a car, 305-307
direct and indirect variation, 353-355
ecology, 87-89
fitness and health, 31-33
making predictions, 527-529
parade floats, 433-435
precision drawing, 477-479
scale drawing, 241-243
simulations, 518-519
stock market, 222-224
tessellations, 397-399
using matrices, 555-557
using the metric system, 125-127
water resources, 153-155
Problem solving practice, 47, 91, 124, 175, 225, 267, 283, 337, 396, 445, 481, 526, 561
Problem solving situation, 2, 52, 96, 138, 180, 230, 272, 312, 360, 404, 450, 486, 534
Problem solving strategies,
account for all possibilities, 172-174
draw a diagram, 264-266, 394-395
guess and test, 85-86
look for a pattern, 19-21
make a model, 425-426
multiple strategies, 558
simplify the problem, 132
simulate a problem, 458-459, 518-520
solve by graphing, 334-337
working backwards, 203-204
writing equations, 284-286
Product, 22, 34, 146, 159
Projects, 6, 7, 12, 29, 30, 39, 75, 82, 84, 89, 90, 113, 119, 123, 131, 144, 151, 152, 155, 158, 187, 191, 205, 214, 215, 220, 232, 235, 239, 240, 249, 263, 299, 304, 319, 323, 333, 336, 351, 352, 364, 372, 379, 383, 387, 388, 399, 409, 423, 429, 432, 436, 439, 457, 468, 480, 491, 495, 497, 504, 505, 507, 515, 521, 525, 554
Properties
addition and subtraction of equality, 72-75, 111
addition and subtraction for inequalities, 296-299
algebraic, 287
associative, 58-59, 108
commutative, 58-59, 108
distributive, 62-65, 108, 279, 288, 546, 550
identity, 59
line, 362
multiplication and division of equality, 76-79, 115
multiplication and division for inequalities, 297
plane, 362
point, 362
ray, 362
segment, 362
Proportions, 233-236, 384
and percents, 245, 252-255
same units, 237
solving, 237-240
Protractor, 365
Pyramids, 420
surface area, 430-432
types, 420
volume, 441-444
Pythagorean theorem, 461-464, 469
converse, 462
model, 461

Q

Quadrants, 316, 317
Quadrilaterals, 371. *See also* Parallelograms; Rectangles; Squares
Quantitative comparisons, exploring, 303
Quartile, 502
Quotient, 26

R

Radius, 377
Random events, 512
Range, 489
Rates, 234-236
  unit, 234
Rational numbers, 192-202
  absolute value, 193
  addition and subtraction, 199-202
  equations, 215
  irrational numbers distinguished, 455
  multiplication and division, 206-208, 211
  negative, 192
  on number lines, 193
  with exponents, 211-214
Ratios, 232-236, 473
  as a fraction, 233
  exploring, 232
  percents as, 244, 245, 253
  proportions, 233
  rates, 234
  slope as, 330
  trigonometric, 473-476
Rays, 362
  properties, 362
  symbol, 362
Reasoning
  deductive, 387
  inductive, 376
Reciprocal, 207
Rectangles
  area, 406-409
  base, 406
  height, 406
Region, coordinate plane and, 316
Regular polygon, 371
Relatively prime numbers, 167
Reviews
  Chapter, 48-49, 92-93, 134-135, 176-177, 226-227, 268-269, 308-309, 356-357, 400-401, 446-447, 482-483, 530-531, 562-563
  Cumulative, 51, 95, 137, 179, 229, 271, 311, 359, 403, 449, 485, 533, 565
Right triangles
  exploring, 460
  45°-45°-90°, 469
  isosceles, 469
  Pythagorean theorem, 461-464, 469
  similar, 465-468
  30°-60°-90°, 470
Rounding decimals, 103, 104

S

Sample, 496
Sample space, 512, 522
Scale drawing, 241-243
  maps, 241
Scientific notation, 149-152
  exploring patterns in, 214
  standard notation compared, 149
Segments
  exploring, 369
  properties, 362
  symbol, 362
Set, of integers, 4
Sieve, 166
Similar, 384
  figures, 384-387
  right triangles, 465-468
  triangles, 385
Simplifying
  equations, 279-282
  problems, 132-133
  variable expressions, 66-69
Simulations
  exploring, 338-339
  in problem solving, 458-459, 518-520
  Monte Carlo, 518
Sine, 473
Slant height, 430
Slope, 328-333
  and *y*-intercept, 330-333
  exploring, 328-329
  formula, 330
Software, 2A, 52A, 96A, 138A, 180A, 230A, 272A, 312A, 360A, 404A, 450A 486A, 534A
Solution
  of an equation, 55, 73
  of an equation in two variables, 320
  of a system of linear equations, 340
Space figures, 420-423
Spheres, 421
  surface area, 431
  volume, 441-444
Spreadsheets, 83-84
  cells, 38-39
  exploring, 38-39
Square numbers
  binomials, 551
  exploring, 140
  opposite of square roots, 452
Square pyramids, 420
Square root(s)
  exploring, 452-453
  finding, 454-457
  table, 579
Squares, 371
  diagonal of, 370
  See also Quadrilaterals; Square numbers
Statistics, 486-527. *See also* individual topics
Stem and leaf plots, 498-501
  back-to-back, 499
Stock market, 222-224
Student bibliography, 2A, 52A, 96A, 138A, 180A, 230A, 272A, 312A, 360A, 404A, 450A, 486A, 534A
Study skills, 48, 51, 92, 95, 134, 137, 176, 179, 226, 229, 268, 271, 308, 311, 356, 359, 400, 403, 446, 449, 482, 485, 530, 533, 562, 565
Subtraction
  equations, 72-75, 111-114, 215-217
  integers, 14-17
  polynomials, 542-545
  properties for inequalities, 296
  property of equality, 72-73, 111
  rational numbers, 199-202
Supplementary angles, 366
Surface area, 427
  cones, 430-431
  cylinders, 428-429
  prisms, 427-429
  pyramids, 430-432
  spheres, 431
Symbols
  approximate equality, 104
  congruence, 380
  division, 26
  equality, 5, 55, 185
  factorials, 509
  grouping, 40, 58, 141
  inequality, 5, 99, 185, 292, 300, 344
  line, 362
  multiplication, 22
  plane, 362
  point, 362
  ray, 362
  repeating decimal, 189
  right angle, 365
  segment, 362
  similar, 384
  square root, 452, 454
Symmetry, line, 381

T

Tables
  data in, 120-121
  exploring, 120-121
  frequency, 492-495
  periodic, 296
  square root, 579
  trigonometric, 580
  using to solve equations, 320
Tangent, 473

Tangram, 202, 372
Teacher bibliography, 2A, 52A, 96A, 138A, 180A, 230A, 272A, 312A, 360A, 404A, 450A, 486A, 534A
Tenths, 98
Terms, 66
combining like, 67-69, 279, 288, 542
monomials as, 538
product of sum and difference, 551
Tessellations, 397
Tests, chapter, 50, 94, 136, 178, 228, 270, 310, 358, 402, 448, 484, 532, 564
Think about it, 3, 53, 97, 139, 181, 231, 273, 313, 361, 405, 451, 487, 535
Thousandths, 100
Ton, 87
Trajectory, 338
Trapezoidal prisms, 420
Trapezoids
area, 411-413
bases, 411
diagonal, 411
Tree diagram, 508, 522
Triangle diagrams, 256-258
Triangles, 374-376
area, 410-413
base, 410
classifying, 374
congruent, 380
exploring, 460
right, 460, 461-472
similar, 385
sum of measures of interior angles, 374
Triangular prisms, 420
Triangular pyramids, 420
Trigonometric ratios, 473-476
cosine, 473
sine, 473
tangent, 473
Trigonometric table, 580
Trinomial, 539
Two-step equations, 274-278
exploring, 274
simplifying and solving, 279-282
writing, 284-286
Two-step inequalities, 300-303

U

Unit fractions, 209, 215
Unit rates, 234
Units
cubic, 437
using same, 237, 260, 412

V

Variable expressions, 34-37, 54, 80, 215
models, 34
simplifying, 66-71
Variables, 34, 80, 108
and equations, 54-57
and expressions, 34-39
and fractions, 184, 185, 199, 206
in system of linear equations, 340
isolating in equations, 215, 321
on both sides of equations, 287-290
Venn diagrams, 202
exploring, 21, 240
Verbal communication, exploring, 314-315
Vertex, 365, 366
cones, 421
pyramids, 420
Vertical angles, 366
Videotapes. *See* Films, videotapes and other media
Visual perception, exploring, 259
Visual thinking, exploring, 79, 511
Visualization, 202
exploring, 423
Vocabulary, 2A, 52A, 96A, 138A, 180A, 230A, 272A, 312A, 360A, 404A, 450A, 486A, 534A
Volume
cones or pyramids, 441-444
cubic units, 437
prisms or cylinders, 437-440
spheres, 441

W

Water resources, 153-155
decision making, 154-155
Word problems, using equations to solve, 280
Write, 7, 8, 13, 15, 18, 25, 43, 46, 57, 61, 65, 70, 82, 99, 103, 107, 110, 114, 117, 140, 143, 144, 152, 157, 161, 162, 191, 195, 209, 210, 217, 224, 232, 244, 248, 250, 251, 254, 255, 257, 262, 263, 290, 315, 328, 329, 336, 339, 351, 363, 364, 386, 387, 419, 423, 424, 444, 452, 491, 497, 499, 504, 511, 515, 517, 525, 536, 537, 541, 545, 547, 549, 553

X

x-axis, 316
x-coordinate, 316
x-intercept, 325

Y

y-axis, 316
y-coordinate, 316
y-intercept, 325, 331

Z

Zero, 9, 73
as exponent, 211
division, 26
in decimals, 100
model, 8